PROJECTS

The following projects employ various technologies and provide the basis for individual study or for laboratory assignments.

C. HENRY EDWARDS
The University of Georgia, Athens

DAVID E. PENNEY
The University of Georgia, Athens

Prentice Hall
Upper Saddle River, NJ 07458

Library of Congress Cataloging-in-Publication Data
Edwards, C. H. (Charles Henry)
Calculus with analytic geometry/C. H. Edwards, Jr., David E. Penney.--5th ed.
p. cm.
Includes bibliographical references and index.
ISBN (invalid) 0-13-736331-1 (hardcover)
1. Calculus. 2. Geometry, Analytic. I. Penney, David E. II. Title.
QA303.E223 1998
515'. 15--dc21 97-22635
CIP

Acquisitions Editor: George Lobell
Editorial Assistant: Gale Epps
Assistant Editor: Audra Walsh
Editorial Director: Tim Bozik
Editor-in-Chief: Jerome Grant
Assistant Vice President of Production and Manufacturing: David W. Riccardi
Editorial/Production Supervision: Jack Casteel
Senior Managing Editor: Linda Mihatov Behrens
Executive Managing Editor: Kathleen Schiaparelli
Manufacturing Buyer: Alan Fischer
Manufacturing Manager: Trudy Pisciotti
Director of Marketing: John Tweeddale
Marketing Manager: Melody Marcus
Marketing Assistants: Diana Penha, Jennifer Pan
Creative Director: Paula Maylahn
Art Manager: Gus Vibal
Art Director: Maureen Eide
Cover and Interior Design/Layout: Lorraine Castellano
Copy Editor: Joyce Grandy
Cover Image: Tomio Ohachi/PPS

Printed in the United States of America
10 9 8 7 6 5 4

Reprinted with corrections July, 1999

ISBN 0-13-736331-1

Prentice-Hall International (UK) Limited, London
Prentice-Hall of Australia Pty. Limited Sydney.
Prentice-Hall Canada Inc., Toronto
Prentice-Hall Hispanoamericana, S.A., Mexico
Prentice-Hall of India Private Limited, New Delhi
Prentice-Hall of Japan, Inc. Tokyo
Prentice-Hall Asia Pte, Ltd., Singapore
Editora Prentice-Hall do Brasil, Ltda., Rio de Janiero

CONTENTS

CHAPTER 10 POLAR COORDINATES AND PLANE CURVES 567

CHAPTER 11 INFINITE SERIES 623

CHAPTER 12 VECTORS, CURVES, AND SURFACES IN SPACE 707

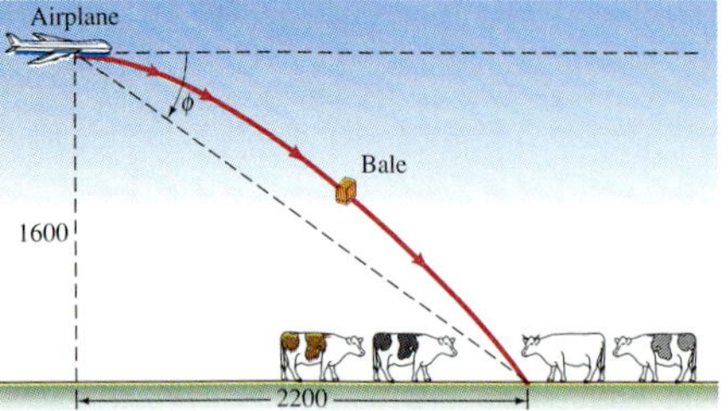

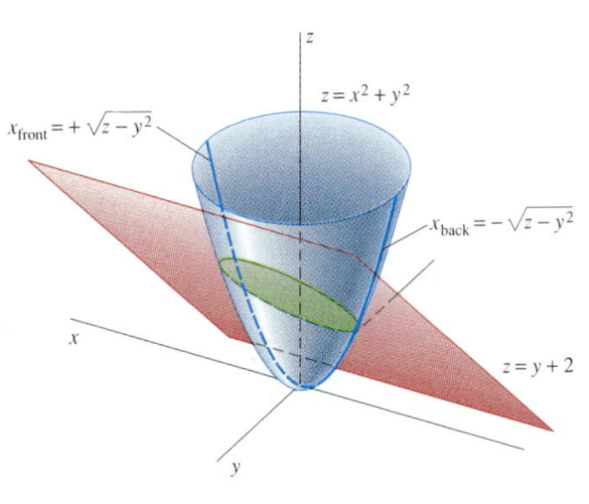

z
$z = x^2 + y^2$
$x_{\text{front}} = +\sqrt{z - y^2}$
$x_{\text{back}} = -\sqrt{z - y^2}$
$z = y + 2$
x
y

PHOTO CREDITS **p.1** Giraudon **p. 53** Hulton/Corbis-Bettmann; David Exton/ National Museum of Science & Industry Science Museum, London; U. S. Navy **p. 99** Culver Pictures **p. 161** Richard Menga/ Fundamental Photographs **p. 199** Museum fur Geschichte der Stadt Leipzig; David E. Penney/From "A Source Book in Mathematics, 1200-1800" by D. J. Struik, Harvard University Press, 1969 **p.267** Culver Pictures **p. 279** Richard Menga/Fundamental Photographs **p. 361** Culver Pictures **p. 407** Focus on Sports **p. 410** Louis Villota/Stock Market **p. 419** Courtesy of Ross Karlin; David E. Penney **p. 460** Jean - Mark Barey, Agence Vandystadt/ Photo Researchers **p. 474** The Granger Collection **p. 475** Granger Collection **p. 511** New York Public Library at Lincoln Center; David E. Penney **p. 567** Stock Montage, Inc./Historical Pictures Collection; Stephen Gerard/Photo Researchers; Stephen Gerard/Photo Researchers; **p. 623** University of Cambridge **p. 707** Library of Congress; Gene Bayers/Focus on Sports **p. 744** Robert Garvey/Black Star **p. 795** New York Public Library at Lincoln Center **p. 885** Bibliothèque Nationale **p. 885** Tony Tomsic/Time Inc. Magazines, Sports Illustrated **p. 913** Bettmann Archives/Juloan Baum **p. 959** Julian Baum/Corbis Bettmann; Science Photo/Photo Researchers.

ABOUT THE AUTHORS

C. Henry Edwards, University of Georgia, received his Ph.D. from the University of Tennesee in 1960. He then taught at the University of Wisconsin for three years and spent a year at the Institute for Advanced Study (Princeton) as an Alfred P. Sloan Research Fellow. Professor Edwards has just completed his thirty-third year of teaching at Georgia (including teaching calculus almost every year) and has received numerous University-wide teaching awards (including his recent selection as the single 1997 recipient of the state-wide Georgia Regents award for research university faculty teaching excellence). His scholarly career has ranged from research and dissertation direction in topology to the history of mathematics to computing and technology in mathematics (his focus in recent years). In addition to his calculus, advanced calculus, linear algebra, and differential equations textbooks, he is also well-known to calculus instructors as author of *The Historical Development of the Calculus* (Springer-Verlag, 1979). He has served as a principal investigator on three recent NSF-supported projects: (1) A project to introduce technology throughout the mathematics curricula in two northeast Georgia public school systems (including *Maple* for beginning algebra students), (2) A *Calculus-with-Mathematica* pilot program at the University of Georgia, (3) A *Matlab*-based computer lab project for upper division numerical analysis and applied mathematics students. Currently he is leading the development of a technology-intensive web-based freshman mathematics course for non-science majors.

David E. Penney, University of Georgia, completed his Ph.D. at Tulane University in 1965 while teaching at the University of New Orleans. Earlier he had worked in experimental biophysics at Tulane University and the Veteran's Administration Hospital in New Orleans. He began teaching calculus in 1957 and has taught the course almost every term since then. He joined the mathematics department at the University of Georgia in 1966 and has since received numerous university-wide teaching awards as well as directing several doctoral dissertations and undergraduate research projects. He is the author of research papers in number theory and topology and is author or co-author of books on calculus, differential equations, linear algebra, and liberal arts mathematics.

PREFACE

The role and practice of mathematics in the world at large is now undergoing a revolution that is driven largely by computational technology. Calculators and computer systems provide students and teachers with mathematical power that no previous generation could have imagined. We read even in daily newspapers of stunning mathematical events like the proof of Fermat's last theorem, finally completed since the fourth edition of this text appeared. Surely *today* is the most exciting time in all history to be mathematically alive! So in preparing this new edition of ***Calculus with Analytic Geometry,*** we wanted first of all to bring a sense of this excitement to the students who will use it.

We also realize that the calculus course is a principal gateway to technical and professional careers for a still increasing number of students in an ever widening range of curricula. Wherever we look—in business and government, in science and technology—almost every aspect of professional work in the world involves mathematics. We therefore have re-thought once again the goal of providing calculus students the solid foundation for their subsequent work that they deserve to get from their calculus textbook.

The text for this edition has been reworked from start to finish. Discussions and explanations have been rewritten throughout in language that (we hope) today's students will find lively and accessible. Seldom-covered topics have been trimmed to accommodate a leaner calculus course. Historical and biographical notes have been added to show students the human face of calculus. Graphics calculator and computer lab projects (with *Derive, Maple,* and *Mathematica* options) for key sections throughout the text have been added. Indeed, a new spirit and flavor reflecting the prevalent interest in graphics calculators and computer systems will be discernible throughout this edition. Consistent with the graphical emphasis of the current calculus reform movement, several hundred new computer-generated figures have been added. Many of these additional figures serve to illustrate a more deliberative and exploratory approach to problem-solving. Our own teaching experience suggests that the use of contemporary technology can make calculus more concrete and accessible to many students.

FIFTH EDITION FEATURES

In preparing this edition, we have benefitted from many valuable comments and suggestions from users of the first four editions. This revision was so pervasive that the individual changes are too numerous to be detailed in a preface, but the following paragraphs summarize those that may be of widest interest.

Additional Problems This revision incorporates the most substantial additional of new problems since the first edition was published in 1982. Over 1250 of the fifth edition's approximately 6700 problems are new for this edition. Almost all of these new problems lie in the intermediate range of difficulty, neither highly theoretical nor computationally routine. Many of them have a new technology flavor, suggesting (if not requiring) the use of technology ranging from a graphing calculator to a computer algebra system.

New Examples and Computational Details Throughout we have rewritten discussions and explanations in language that today's students will find more lively and accessible. The extent of this revision in text content is illustrated by the fact that approximately 20% of the fifth edition's over 700 in-text examples are new. Moreover, we have inserted an additional line or two of computational detail in many of the worked-out examples to make them easier for student readers to follow. The purpose of these computational changes is to make the computations themselves less of a barrier to conceptual understanding.

Project Material Each chapter now contains several supplementary projects—a total of more than 50, many of them new for this edition. Each project typically employs some aspect of modern computational technology to illustrate the principal ideas of the preceding section, and typically contains additional problems intended for solution with the use of a graphics calculator or computer. Figures and data illustrate the use of graphics calculators and computer systems such as *Derive, Maple,* and *Mathematica.* This project material is suitable for use in a computer or calculator lab conducted in association with a standard calculus course, perhaps meeting weekly. It can also be used as a basis for graphics calculator or computer assignments that students will complete outside of class, or for individual study.

Computer Graphics An increased emphasis on graphical visualization along with numeric and symbolic understanding is provided by the computer-generated artwork, about 25% of which is new for this edition. Over 550 MATLAB-generated figures (half of them new for this edition) illustrate the kind of figures that students using graphics calculators can produce for themselves. Many of these are included with new graphical problem material. *Mathematica*-generated color graphics are included to highlight all sections involving three-dimensional material.

Historical Material Historical and biographical chapter openings offer students a sense of the development of our subject by real, live human beings. Both authors are fond of the history of mathematics and believe that it can favorably influence both our teaching and students' learning of mathematics. For this reason numerous historical comments appear in the text itself.

Introductory Chapters Chapters 1 and 2 have been streamlined for a leaner and quicker start on calculus. Chapter 1 concentrates on functions and graphs. It includes two sections cataloging the elementary functions of calculus and provides a foundation for an early emphasis on transcendental functions. Chapter 1 concludes with a section addressing the question "What *is* calculus?" Chapter 2, on limits, begins with a section on tangent lines to motivate the official introduction of limits in Section 2.2. Trigonometric limits are treated throughout Chapter 2 in order to encourage a richer and more visual introduction to the limit concept.

Differentiation Chapters The sequence of topics in Chapters 3 and 4 varies a bit from the most traditional order. We attempt to build student confidence by introducing topics more nearly in order of increasing difficulty. The chain rule appears quite early (in Section 3.3) and we cover the basic techniques for differentiating algebraic functions before discussing maxima and minima in Sections 3.5 and 3.6. The appearance of inverse functions is delayed until Chapter 7. Section 3.7 treats the derivatives of all six trigonometric functions. Implicit differentiation and related rates are combined in a single section (Section 3.8). The mean value theorem and its applications are deferred to Chapter 4. Sections 4.4 on the first derivative test and 4.6 on higher derivatives and concavity have been simplified and streamlined. A great deal of new graphic material has been added in the curve-sketching sections that conclude Chapter 4.

Integration Chapters New and simpler examples have been inserted throughout Chapters 5 and 6. Antiderivatives (formerly at the end of Chapter 4) now begin Chapter 5. Section 5.4 (Riemann sums) has been simplified greatly, with upper and lower sums eliminated and endpoint and midpoint sums emphasized instead. Many instructors now believe that the first applications of integration ought not be confined to the standard area and volume computations; Section 6.5 is an optional section that introduces separable differential equations. To eliminate redundancy, the material on centroids and the theorems of Pappus is delayed to Chapter 14 (Multiple Integrals), where it can be treated in a more natural context.

Early Transcendentals Functions Options An "early transcendental functions" version of this book is also available. In the present version, the flexible organization of Chapter 7 offers a variety of options to those instructors who favor an earlier treatment of transcendental functions. Section 7.1 begins with the "high school" approach to exponential functions, followed by the idea of a logarithm as "the power to which the base a must be raised to get the number x." On this basis, Section 7.1 carries out a low-key review of the laws of exponents and of logarithms, and investigates informally the differentiation of exponential and logarithmic functions. This section on the elementary differential calculus of exponentials and logarithms can be covered any time after Section 3.3 (on the chain rule). If this is done, then Section 7.2—based on the definition of the logarithm as an integral—can be covered any time after the integral has been defined in Chapter 5 (along with as much of the remainder of Chapter 7 as the instructor desires). The remaining transcendental functions—inverse trigonometric and hyperbolic—are now treated in Chapter 8, which includes also indeterminate forms and l'Hôpital's rule (much earlier than in the third edition).

Thus the text offers a variety of ways to accommodate a course syllabus that includes exponential functions early in differential calculus, and/or logarithmic functions early in integral calculus.

Streamlining Techniques of Integration Chapter 9 is organized to accommodate those instructors who feel that methods of formal integration now require less emphasis, in view of modern techniques for both numerical and symbolic integration. Integration by parts (Section 9.3) now precedes trigonometric integrals (Section 9.4). The method of partial fractions appears in Section 9.5, and trigonometric substitutions and integrals involving quadratic polynomials follow in Sections 9.6 and 9.7. Improper integrals appear in Section 9.8, and the more specialized rationalizing substitutions have been relegated to the Chapter 9 Miscellaneous Problems. This rearrangement of Chapter 9 makes it more convenient to stop wherever the instructor desires.

Vectors The major reorganization for the fifth edition is a response to numerous user suggestions to combine the treatments of two-dimensional vectors and three-dimensional vectors, which appeared in separate chapters of the fourth edition. In this reorganization we have also amalgamated the treatments of polar curves and parametric curves, which also appeared in separate chapters in the fourth edition. As a consequence, the contents of three chapters in the fourth edition have been efficiently combined in two chapters of this revision—Chapter 10 on Polar Coordinates and Plane Curves, and Chapter 12 on Vectors, Curves, and Surfaces in Space.

Infinite Series After the usual introduction to convergence of infinite sequences and series in Sections 11.2 and 11.3, a combined treatment of Taylor polynomials and Taylor series appears in Section 11.4. This makes it possible for the instructor to experiment with a much briefer treatment of infinite series, but still offer exposure to the Taylor series that are so important for applications.

Differential Equations Many calculus instructors now believe that differential equations should be seen as early and as often as possible. The very simplest differential equations (of the form $dy/dx = f(x)$) appear in a subsection at the end of Section 5.2 (Antiderivatives). Section 6.5 illustrates applications of integration to the solution of separable differential equations. Section 9.5 includes applications of the method of partial fractions to population problems and the logistic equation. In such ways we have distributed enough of the spirit and flavor of differential equations throughout the text that it seemed expeditious to eliminate the (former) final chapter devoted solely to differential equations. But those who so desire can arrange with the publisher to obtain for supplemental use appropriate sections of Edwards and Penney, *Differential Equations: Computing and Modeling* (Englewood Cliffs, N.J.: Prentice Hall, 1996).

Linear Algebra Notation and Terminology An innovation for the fifth edition is the inclusion (for optional coverage) of matrix terminology and notation in the multivariable portion of the text—for example, in the treatment of quadric surfaces in Chapter 12 and of directional derivatives and the multivariable chain rule in Chapter 13. These subsections will enhance the understanding of multivariable concepts for those students who are familiar with matrix notation at the level of the definition of the product of two matrices.

MAINTAINING TRADITIONAL STRENGTHS

While many new features have been added, five related objectives remained in constant view: **concreteness**, **readability**, **motivation**, **applicability**, and **accuracy**.

- ▼ *CONCRETENESS* The power of calculus is impressive in its precise answers to realistic questions and problems. In the necessary conceptual development of the subject, we keep in sight the central question: How does one actually *compute* it? We place special emphasis on concrete examples, applications, and problems that serve both to highlight the development of the theory and to demonstrate the remarkable versatility of calculus in the investigation of important scientific questions.

- ▼ *READABILITY* Difficulties in learning mathematics often are complicated by language difficulties. Our writing style stems from the belief that crisp exposition, both intuitive and precise, makes mathematics more accessible—and hence more readily learned—with no loss of rigor. We hope our language is clear and attractive to students and that they can and actually will read it, thereby enabling the instructor to concentrate class time on the less routine aspects of teaching calculus.

- ▼ *MOTIVATION* Our exposition is centered around examples of the use of calculus to solve real problems of interest to real people. In selecting such problems for examples and exercises, we took the view that stimulating interest and motivating effective study go hand in hand. We attempt to make it clear to students how the knowledge gained with each new concept or technique will be worth the effort expended. In theoretical discussions, especially, we try to provide an intuitive picture of the goal before we set off in pursuit of it.

▼ *APPLICATIONS* Its diverse applications are what attract many students to calculus, and realistic applications provide valuable motivation and reinforcement for all students. This book is well-known for the broad range of applications that we include, but it is neither necessary nor desirable that the course cover all of the applications in the book. Each section or subsection that may be omitted without loss of continuity is marked with an asterisk. This provides flexibility for each instructor to determine his or her own flavor and emphasis.

▼ *ACCURACY* Our coverage of calculus is complete (although we hope it is somewhat less than encyclopedic). Still more than its predecessors, this edition was subjected to a comprehensive reviewing process to help ensure accuracy. For example, essentially every problem answer appearing in the Answers section at the back of the book in this edition has been verified using *Mathematica.* With regard to the selection and sequence of mathematical topics, our approach is traditional. But close examination of the treatment of standard topics may betray our own participation in the current movement to revitalize the teaching of calculus. We continue to favor an intuitive approach that emphasizes both conceptual understanding and care in the formulation of definitions and key concepts of calculus. Some proofs that may be omitted at the discretion of the instructor are placed at the ends of sections and others are deferred to the book's appendices. In this way we leave ample room for variation in seeking the proper balance between rigor and intuition.

SUPPLEMENTARY MATERIAL

A variety of electronic and printed supplements are provided by the publisher, including a WWW site that consitutes an on-line calculator/computer guide for calculus. This web site at `www.prenhall.com/edwards` is designed to assist calculus students as they work on the book's projects using graphing calculators and computer algebra systems such as *Derive, Maple, Mathematica,* and MATLAB. The authors will maintain and expand this site to provide calculus students with new and evolving supplementary materials on a continuing basis, and to explore the use of emerging technology for new channels of communication and more active learning experiences.

Answers to most of the odd-numbered problems appear in the back of the book. Solutions to most problems (other than those odd-numbered ones for which an answer alone is sufficient) are available in the *Instructor's Solutions Manual.* A subset of that manual, containing solutions to problems numbered 1, 4, 7, 10, . . . is available as a *Student Solutions Manual.* A collection of some 1700 additional problems suitable for use as test questions, the *Calculus Test Item File,* is available (in both electronic and hard-copy form) for use by instructors. Finally, an *Instructor's Edition* including section-by-section teaching outlines and suggestions is available to those who are using this book to teach calculus.

ACKNOWLEDGMENTS

All experienced textbook authors know the value of critical reviewing during the preparation and revision of a manuscript. In our work on this edition of the book we have benefitted greatly from the advice of the following exceptionally able reviewers:

- André Adler, Illinois Institute of Technology
- John R. Akeroyd, University of Arkansas-Fayetteville
- Marcia Birken, Rochester Institute of Technology
- Stephen Bricher, Linfield College
- Robert D. Davis, University of Nevada at Reno
- Jeff Dodd, Jacksonville State University
- Larry Dornhoff, University of Illinois at Urbana-Champaign
- Stephen Dragosh, Michigan State University
- Michael Gilpin, Michigan Tech
- Heini Halberstam, University of Illinois at Urbana-Champaign
- Thomas Hern, Bowling Green State University
- Lisa Lorentzen, Norwegian University of Science & Technology
- Daniel McCallum, University of Arkansas at Little Rock
- Pallasena Narayanaswami, Memorial University of Newfoundland
- Roger Pinkham, Stevens Institute of Technology
- Irwin Pressman, Carleton University
- Zhong-Jim Ruan, University of Illinois at Urbana-Champaign

Many of the best improvements that have been made must be credited to colleagues and users of the first four editions throughout the United States, Canada, Europe, and South America. We are grateful to all those, especially students, who have written to us, and hope that they will continue to do so. We thank Mary and Nancy Toscano, who checked the accuracy of every example and odd-numbered answer. We also believe that the quality of the finished book itself is adequate testimony to the skill, diligence, and talent of an exceptional staff at Prentice Hall; we owe special thanks to George Lobell, our mathematics editor; Jack Casteel, production editor; Tony Palermino, developmental editor; Lorraine Costellano, designer; and Network Graphics, illustrator. Finally, we again are unable to thank Alice Fitzgerald Edwards and Carol Wilson Penney for their continued assistance, encouragement, support, and patience.

C.H.E.
hedwards@math.uga.edu
Athens, Georgia, U.S.A.

D.E.P.
dpenney@math.uga.edu
Athens, Georgia, U.S.A.

CHAPTER 1

FUNCTIONS AND GRAPHS

René Descartes (1596–1650)

The seventeenth-century French scholar René Descartes is perhaps better remembered today as a philosopher than as a mathematician. But most of us are familiar with the "Cartesian plane" in which the location of a point P is specified by its coordinates (x, y).

As a schoolboy Descartes was often permitted to sleep late because of allegedly poor health. He claimed that he always thought most clearly about philosophy, science, and mathematics while lying comfortably in bed on cold mornings. After graduating from college, where he studied law (apparently with little enthusiasm), Descartes traveled with various armies for a number of years, but more as a gentleman soldier than as a professional military man.

After finally settling down (in Holland), Descartes published, in 1637, his famous philosophical treatise *Discourse on the Method* (of Reasoning Well and Seeking Truth in the Sciences). One of three appendices to this work sets forth his new "analytic" approach to geometry. His principal idea (published almost simultaneously by his countryman Pierre de Fermat) was the correspondence between an *equation* and its *graph,* generally a curve in the plane. The equation could be used to study the curve and vice versa.

Suppose that we want to solve the equation $f(x) = 0$. Its solutions are the intersection points of the graph of $y = f(x)$ with the x-axis, so an accurate picture of the curve shows the number and approximate locations of the solutions of the equation. For instance, the graph

$$y = x^3 - 3x^2 + 1$$

has three x-intercepts, showing that the equation

$$x^3 - 3x^2 + 1 = 0$$

has three real solutions—one between -1 and 0, one between 0 and 1, and one between 2 and 3. A modern graphics calculator or computer graphing program can approximate these solutions more accurately by magnifying the regions in which they are located. For instance, the magnified center region shows that the corresponding solution is $x \approx 0.65$.

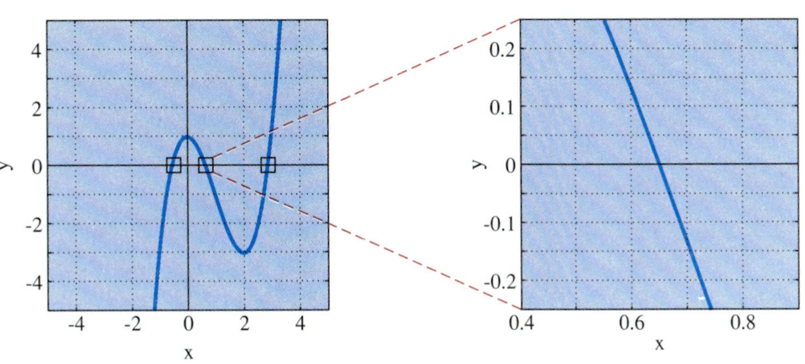

The graph $y = x^3 - 3x^2 + 1$

1.1 FUNCTIONS AND MATHEMATICAL MODELING

Calculus is one of the supreme accomplishments of the human intellect. This mathematical discipline stems largely from the seventeenth-century investigations of Isaac Newton (1642–1727) and Gottfried Wilhelm Leibniz (1646–1716). Yet some of its ideas date back to the time of Archimedes (287–212 B.C.) and originated in cultures as diverse as those of Greece, Egypt, Babylonia, India, China, and Japan. Many of the scientific discoveries that have shaped our civilization during the past three centuries would have been impossible without the use of calculus.

The principal objective of calculus is the analysis of problems of change (of motion, for example) and of content (the computation or area and volume, for instance). These problems are fundamental because we live in a world of ceaseless change, filled with bodies in motion and phenomena of ebb and flow. Consequently, calculus remains a vibrant subject, and today this body of computational technique continues to serve as the principal quantitative language of science and technology.

Functions

Most applications of calculus involve the use of real numbers or *variables* to describe changing quantities. The key to the mathematical analysis of a geometric or scientific situation is typically the recognition of relationships among the variables that describe the situation. Such a relationship may be a formula that expresses one variable as a *function* of another. For example:

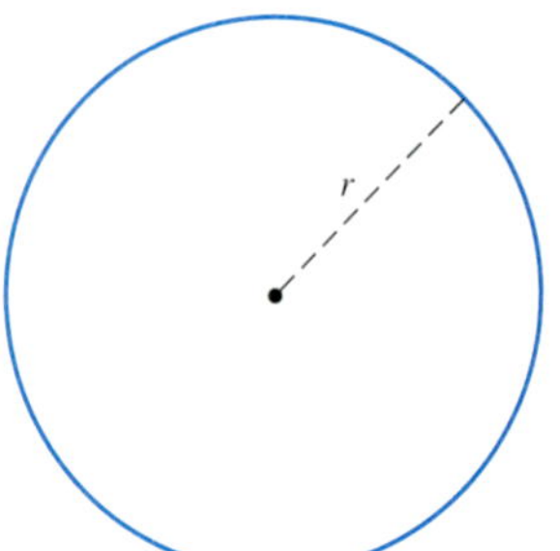

Fig. 1.1.1 Circle: area $A = \pi r^2$, circumference $C = 2\pi r$.

- The area A of a circle of radius r is given by $A = \pi r^2$ (Fig. 1.1.1). The volume V and surface area S of a sphere of radius r are given by

$$V = \tfrac{4}{3}\pi r^3 \quad \text{and} \quad S = 4\pi r^2,$$

  respectively (Fig. 1.1.2).
- After t seconds (s) a body that has been dropped from rest has fallen a distance

$$s = \tfrac{1}{2}gt^2$$

  feet (ft) and has speed $v = gt$ feet per second (ft/s), where $g \approx 32$ ft/s^2 is gravitational acceleration.
- The volume V (in liters, L) of 3 grams (g) of carbon dioxide at 27°C is given in terms of its pressure p in atmospheres (atm) by $V = 1.68/p$.

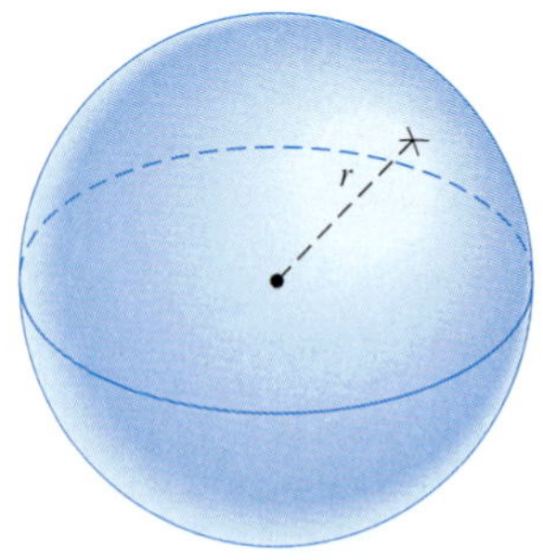

Fig. 1.1.2 Sphere: volume $V = \frac{4}{3}\pi r^3$, surface area $S = 4\pi r^2$.

Definition ***Function***

A real-valued **function** f defined on a set D of real numbers is a rule that assigns to each number x in D exactly one real number, denoted by $f(x)$.

The set D of all numbers for which $f(x)$ is defined is called the **domain** (or **domain of definition**) of the function f. The number $f(x)$, read "f of x," is called the **value** of f at the number (or point) x. The set of all values $y = f(x)$ is called the **range** of f. That is, the range of f is the set

$$\{y : y = f(x) \quad \text{for some } x \text{ in } D\}.$$

In this section we will be concerned more with the domain of a function than with its range.

EXAMPLE 1 The squaring function defined by

$$f(x) = x^2$$

assigns to each real number x its square x^2. Because every real number *can* be squared, the domain of f is the set $\boldsymbol{R}$ of all real numbers. But only nonnegative numbers are squares. Moreover, if $a \geqq 0$, then $a = (\sqrt{a})^2 = f(\sqrt{a})$, so a is a square. Hence the range of the squaring function f is the set $\{y : y \geqq 0\}$ of all nonnegative real numbers. ■

Functions can be described in various ways. A *symbolic* description of the function f is provided by a formula that specifies how to compute the number $f(x)$ in terms of the number x. Thus the symbol $f(\ )$ may be regarded as an operation that is to be performed whenever a number or expression is inserted between the parentheses.

EXAMPLE 2 The formula

$$f(x) = x^2 + x - 3 \tag{1}$$

defines a function f whose domain is the entire real line $\boldsymbol{R}$. Some typical values of f are $f(-2) = -1$, $f(0) = -3$, and $f(3) = 9$. Other values of the function f include

$$\begin{aligned} f(4) &= 4^2 + 4 - 3 = 17, \\ f(c) &= c^2 + c - 3, \\ f(2 + h) &= (2 + h)^2 + (2 + h) - 3 \\ &= (4 + 4h + h^2) + (2 + h) - 3 = h^2 + 5h + 3, \quad \text{and} \\ f(-t^2) &= (-t^2)^2 + (-t^2) - 3 = t^4 - t^2 - 3. \end{aligned}$$

■

When we describe the function f by writing a formula $y = f(x)$, we call x the **independent variable** and y the **dependent variable** because the value of y depends—through f—on the choice of x. As the independent variable x changes, or varies, then so does the dependent variable y. The way that y varies is determined by the rule of the function f. For example, if f is the function of Eq. (1), then $y = -1$ when $x = -2$, $y = -3$ when $x = 0$, and $y = 9$ when $x = 3$.

You may find it useful to visualize the dependence of the value $y = f(x)$ on x by thinking of the function f as a kind of machine that accepts as input a number x and then produces as output the number $f(x)$, perhaps displayed or printed (Fig. 1.1.3).

Fig. 1.1.3 A "function machine."

One such machine is the square root key of a simple pocket calculator. When a nonnegative number x is entered and this key is pressed, the calculator displays (an approximation to) the number $\sqrt{x}$. Note that the domain of this *square root function* $f(x) = \sqrt{x}$ is the set $[0, +\infty)$ of all nonnegative real numbers, because no negative number has a real square root. The range of f is also the set of all nonnegative real numbers, because the symbol $\sqrt{x}$ always denotes the *nonnegative* square root of x. The calculator illustrates its knowledge of the domain by displaying an error message if we ask it to calculate the square root of a negative number (unless it's one of the more sophisticated calculators, such as the TI-85 or HP-48S, that handle complex numbers).

EXAMPLE 3 Not every function has a rule expressible as a simple one-part formula such as $f(x) = \sqrt{x}$. For instance, if we write

$$h(x) = \begin{cases} x^2 & \text{if } x \geqq 0, \\ \sqrt{-x} & \text{if } x < 0, \end{cases}$$

then we have defined a perfectly good function with domain $\boldsymbol{R}$. Some of its values are $h(-4) = 2$, $h(0) = 0$, and $h(2) = 4$. By contrast, the function g in Example 4 is defined initially by means of a verbal description rather than by means of formulas. ■

EXAMPLE 4 For each real number x, let $g(x)$ denote the greatest integer that is less than or equal to x. For instance, $g(2.5) = 2$, $g(0) = 0$, $g(-3.5) = -4$, and $g(\pi) = 3$. If n is an integer, then $g(x) = n$ for every number x such that $n \leqq x < n + 1$. This function g is called the **greatest integer function** and is often denoted by

$$g(x) = \lfloor x \rfloor.$$

Thus $\lfloor 2.5 \rfloor = 2$, $\lfloor -3.5 \rfloor = -4$, and $\lfloor \pi \rfloor = 3$. Note that although $\lfloor x \rfloor$ is defined for all x, the range of the greatest integer function is not all of $\boldsymbol{R}$, but the set $\boldsymbol{Z}$ of all *integers*. ■

It is not necessary that the name of a function be a single letter such as f or g. For instance, think of the trigonometric functions $\sin(x)$ and $\cos(x)$ with the names sin and cos.

EXAMPLE 5 Another descriptive name for the greatest integer function of Example 4 is

$$\text{FLOOR}(x) = \lfloor x \rfloor. \tag{2}$$

(We think of the integer n as the "floor" beneath the real numbers lying between n and $n + 1$.) Similarly, we may use $\text{ROUND}(x)$ to name the familiar function that "rounds off" the real number x to the nearest integer n, except that $\text{ROUND}(x) = n + 1$ if $x = n + \frac{1}{2}$ (so we "round upward" in case of ambiguity). Round off enough different numbers to convince yourself that

$$\text{ROUND}(x) = \text{FLOOR}(x + \tfrac{1}{2}) \tag{3}$$

for all x.

Closely related to the FLOOR and ROUND functions is the "ceiling function" used by the U.S. Postal Service; $\text{CEILING}(x)$ denotes the least integer that is not less than the number x. In 1997 the postage rate for a first-class letter was 32¢ for the first ounce and 23¢ for each additional ounce or fraction thereof. For a letter weighing $w > 0$ ounces, the number of "additional ounces" involved is $\text{CEILING}(w) - 1$. Therefore the postage $s(w)$ due on this letter is given by

$$s(w) = 32 + 23 \cdot [\text{CEILING}(w) - 1] = 9 + 23 \cdot \text{CEILING}(w).$$ ■

Domains and Intervals

The function f and the value or expression $f(x)$ are different in the same sense that a machine and its output are not the same. Nevertheless, it is common to use an expression like "the function $f(x) = x^2$" to define a function merely by writing its formula. In this situation the domain of the function is not specified. Then, by convention, **the domain of the function f is the set of all real numbers x for which the expression $f(x)$ makes sense and produces a real number y.** For instance, the domain of the function $h(x) = 1/x$ is the set of all nonzero real numbers (because $1/x$ is defined precisely when $x \neq 0$).

Domains of functions frequently are described in terms of *intervals* of real numbers (Fig. 1.1.4). (Interval notation is reviewed in Appendix A.) Recall that a **closed interval** $[a, b]$ contains both its endpoints $x = a$ and $x = b$, whereas the **open interval**

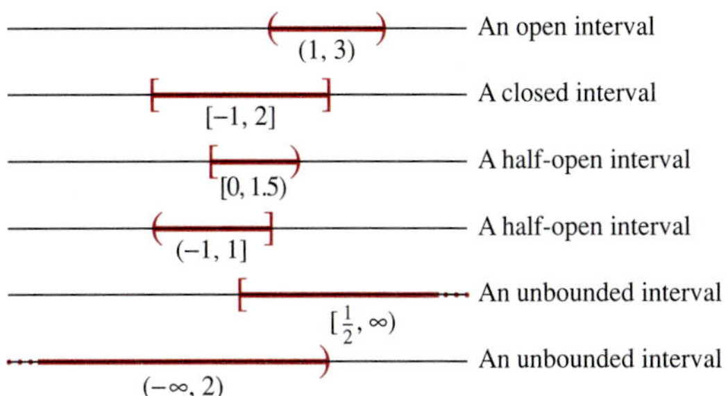

Fig. 1.1.4 Some examples of intervals of real numbers.

(a, b) contains neither endpoint. Each of the **half-open intervals** $[a, b)$ and $(a, b]$ contains exactly one of its two endpoints. The **unbounded interval** $[a, \infty)$ contains its endpoint $x = a$, whereas $(-\infty, a)$ does not. The previously mentioned domain of $h(x) = 1/x$ is the *union* of the unbounded intervals $(-\infty, 0)$ and $(0, \infty)$.

EXAMPLE 6 Find the domain of the function $g(x) = \dfrac{1}{2x + 4}$.

Solution Division by zero is not allowed, so the value $g(x)$ is defined precisely when $2x + 4 \neq 0$. This is true when $2x \neq -4$, and thus when $x \neq -2$. Hence the domain of g is the set $\{x : x \neq 2\}$, which is the union of the two unbounded open intervals $(-\infty, -2)$ and $(-2, \infty)$, shown in Fig. 1.1.5. ■

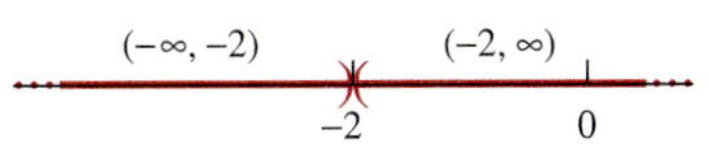

Fig. 1.1.5 The domain of $g(x) = 1/(2x + 4)$ is the union of two unbounded open intervals.

EXAMPLE 7 Find the domain of $h(x) = \dfrac{1}{\sqrt{2x + 4}}$.

Solution Now it is necessary not only that the quantity $2x + 4$ be nonzero, but also that it be positive, in order that the square root $\sqrt{2x + 4}$ be defined. But $2x + 4 > 0$ when $2x > -4$, and thus when $x > -2$. Hence the domain of h is the single unbounded open interval $(-2, \infty)$. ■

Mathematical Modeling

The investigation of an applied problem often hinges on defining a function that captures the essence of a geometrical or physical situation. Examples 8 and 9 illustrate this process.

EXAMPLE 8 A rectangular box with a square base has volume 125. Express its total surface area A as a function of the edge length x of its base.

Solution The first step is to draw a sketch and to label the relevant dimensions. Figure 1.1.6 shows a rectangular box with square base of edge length x and with height y. We are given that the volume of the box is

$$V = x^2 y = 125. \tag{4}$$

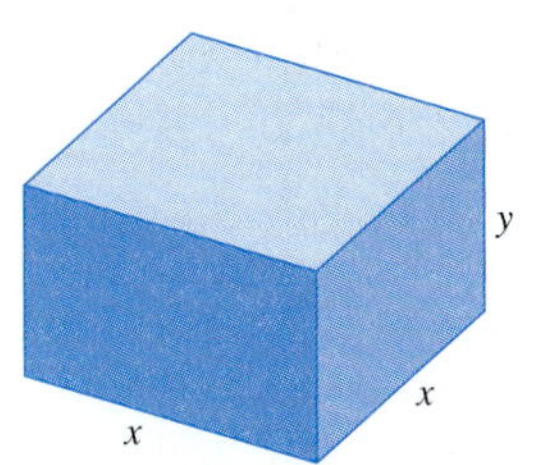

Fig. 1.1.6 The box of Example 8.

Both the top and bottom of the box have area x^2, and each of its four vertical sides has area xy, so its total surface area is

$$A = 2x^2 + 4xy. \tag{5}$$

But this is a formula for A in terms of the *two* variables x and y rather than a function of the *single* variable x. To eliminate y and thereby obtain A in terms of x alone, we solve Eq. (4) for $y = 125/x^2$ and then substitute this result in Eq. (5) to obtain

$$A = 2x^2 + 4x \cdot \frac{125}{x^2} = 2x^2 + \frac{500}{x}.$$

Thus the surface area is given as a function of the edge length x by

$$A(x) = 2x^2 + \frac{500}{x}, \qquad 0 < x < +\infty. \tag{6}$$

It is necessary to specify the domain, because negative values of x make sense in the *formula* in (6) but do not belong in the domain of the *function* A. Because every $x > 0$ determines such a box, the domain does, in fact, include all positive real numbers. ■

COMMENT In Example 8 our goal was to express the dependent variable A as a *function* of the independent variable x. Initially, the geometric situation provided us instead with

1. The *formula* in Eq. (5) expressing A in terms of both x and the additional variable y, and
2. The *relation* in Eq. (4) between x and y, which we used to eliminate y and thereby express A as a function of x alone.

We will see that this is a common pattern in many different applied problems, such as the one that follows.

THE ANIMAL PEN PROBLEM You must build a rectangular holding pen for animals. To save material, you will use an existing wall as one of its four sides. The fence for the other three sides costs \$5/ft, and you must spend \$1/ft to paint the portion of the wall that forms the fourth side of the pen. If you have a total of \$180 to spend, what dimensions will maximize the area of the pen you can build?

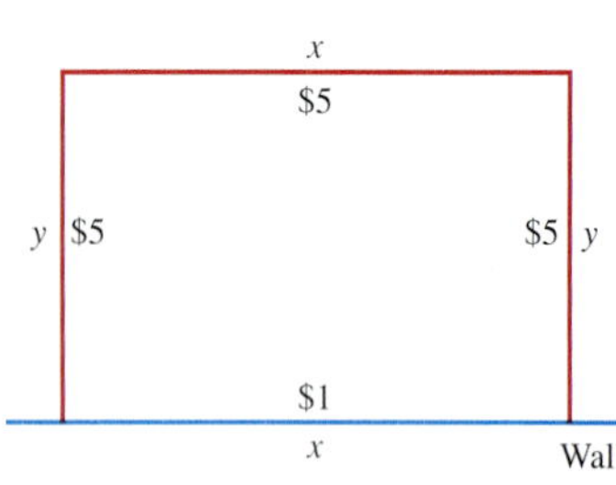

Fig. 1.1.7 The animal pen.

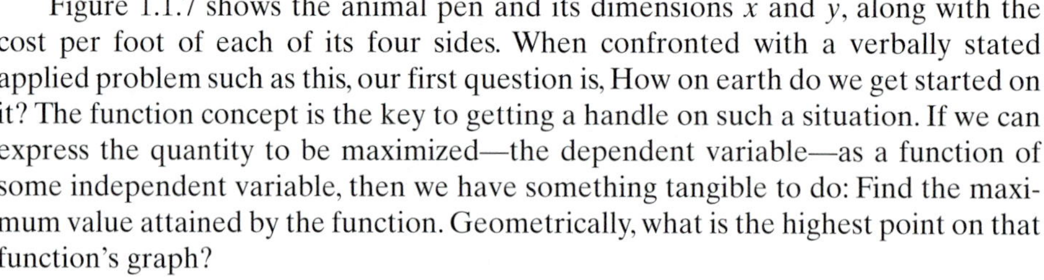

Figure 1.1.7 shows the animal pen and its dimensions x and y, along with the cost per foot of each of its four sides. When confronted with a verbally stated applied problem such as this, our first question is, How on earth do we get started on it? The function concept is the key to getting a handle on such a situation. If we can express the quantity to be maximized—the dependent variable—as a function of some independent variable, then we have something tangible to do: Find the maximum value attained by the function. Geometrically, what is the highest point on that function's graph?

EXAMPLE 9 In connection with the animal pen problem, express the area A of the pen as a function of the length x of its wall side.

Solution The area A of the rectangular pen of length x and width y is

$$A = xy. \tag{7}$$

By multiplying the length of each side in Fig. 1.1.7 by its cost per foot and then adding the results, we find that the total cost C of the pen is

$$C = x + 5y + 5x + 5y = 6x + 10y.$$

So

$$6x + 10y = 180, \tag{8}$$

because we are given $C = 180$. Choosing x to be the independent variable, we use the relation in Eq. (8) to eliminate the additional variable y from the area formula in Eq. (7). We solve Eq. (8) for y and substitute the result

$$y = \tfrac{1}{10}(180 - 6x) = \tfrac{3}{5}(30 - x) \tag{9}$$

in Eq. (7). Thus we obtain the desired function

$$A(x) = \tfrac{3}{5}(30x - x^2)$$

that expresses the area A as a function of the length x.

In addition to this formula for the function A, we must also specify its domain. Only if $x > 0$ will actual rectangles be produced, but we find it convenient to include the value $x = 0$ as well. This value of x corresponds to a "degenerate rectangle" of base length zero and height

$$y = \tfrac{3}{5} \cdot 30 = 18,$$

a consequence of Eq. (9). For similar reasons, we have the restriction $y \geqq 0$. Because

$$y = \tfrac{3}{5}(30 - x),$$

it follows that $x \leqq 30$. Thus the complete definition of the area function is

$$A(x) = \tfrac{3}{5}(30x - x^2), \quad 0 \leqq x \leqq 30. \tag{10}$$

■

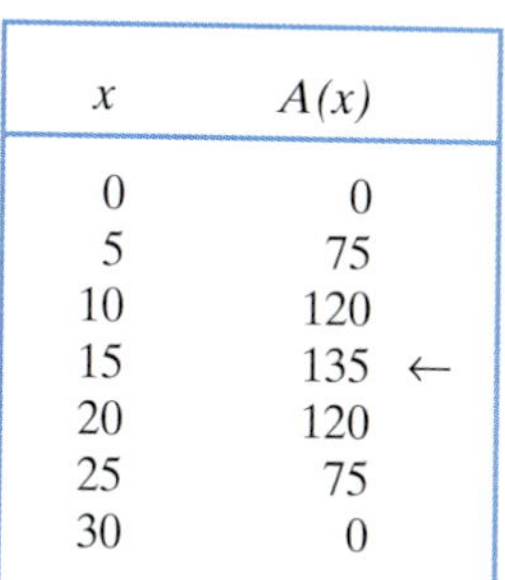

x	$A(x)$
0	0
5	75
10	120
15	135 ←
20	120
25	75
30	0

Fig. 1.1.8 Area $A(x)$ of a pen with side of length x.

COMMENT The domain of a function is a necessary part of its definition, and for each function we must specify the domain of values of the independent variable. In applications, we use the values of the independent variable that are relevant to the problem at hand.

Example 9 illustrates an important part of the solution of a typical applied problem—the formulation of a **mathematical model** of the physical situation under study. The area function $A(x)$ defined in (10) provides a mathematical model of the animal pen problem. The shape of the optimal animal pen can be determined by finding the maximum value attained by the function A on its domain of definition.

x	$A(x)$
10	120
11	125.4
12	129.6
13	132.6
14	134.4
15	135 ←
16	134.4
17	132.6
18	129.6
19	125.4
20	120

Fig. 1.1.9 Further indication that $x = 15$ yields maximal area $A = 135$.

Numerical Investigation

Armed with the result of Example 9, we might attack the animal pen problem by calculating a table of values of the area function $A(x)$ in Eq. (10). Such a table is shown in Fig. 1.1.8. The data in this table suggest strongly that the maximum area is $A = 135$ ft^2, attained with side length $x = 15$ ft, in which case Eq. (9) yields $y = 9$ ft. This conjecture appears to be corroborated by the more refined data shown in Fig. 1.1.9.

Thus it seems that the animal pen with maximal area (costing \$180) is $x = 15$ ft long and $y = 9$ ft wide. The tables in Figs. 1.1.8 and 1.1.9 show only *integral* values of x, however, and it is quite possible that the length x of the pen of maximal area is *not* an integer. Consequently, numerical tables alone do not settle the matter. A new mathematical idea is needed to *prove* that $A(15) = 135$ is the maximum value of

$$A(x) = \tfrac{3}{5}(30x - x^2), \quad 0 \leqq x \leqq 30$$

for *all* x in its domain. We attack this problem again in Section 1.2.

Tabulation of Functions

Many scientific and graphics calculators allow the user to program a given function for repeated evaluation and thereby to compute painlessly tables like those in Figs. 1.1.8 and 1.1.9. For instance, Figs. 1.1.10 and 1.1.11 show displays of a calculator prepared to calculate values of the dependent variable

$$\texttt{Y1} = A(x) = (3/5)(30x - x^2),$$

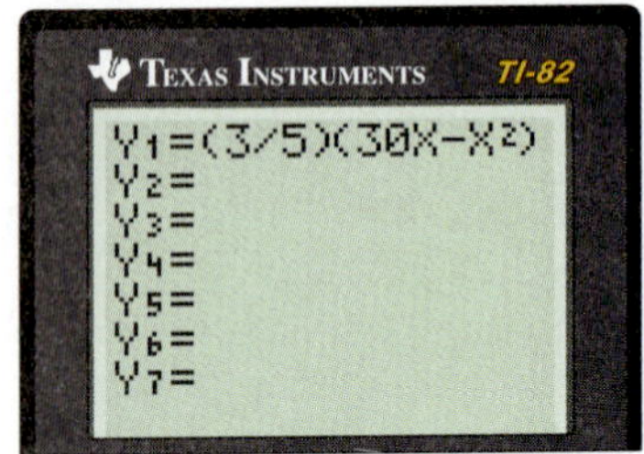

Fig. 1.1.10 A calculator programmed to evaluate $A(x) = (3/5)(30x - x^2)$.

and Fig. 1.1.12 shows the calculator's resulting version of the table in Fig. 1.1.9.

The use of a calculator or computer to tabulate values of a function is a simple technique with surprisingly many applications. Here we illustrate a method of solving an equation of the form $f(x) = 0$ by *repeated tabulation* of values $f(x)$ of the function f.

To give a specific example, suppose that we ask what value of x in Eq. (10) yields an animal pen of area $A = 100$. Then we need to solve the equation

$$A(x) = \tfrac{3}{5}(30x - x^2) = 100,$$

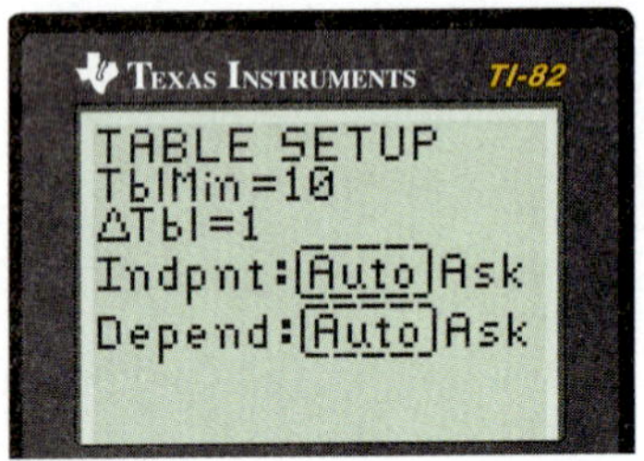

Fig. 1.1.11 The table setup.

which is equivalent to the equation

$$f(x) = \tfrac{3}{5}(30x - x^2) - 100 = 0. \tag{11}$$

This is a quadratic equation that could be solved using the quadratic formula of basic algebra, but we want to take a more direct, numerical approach. The reason is that the numerical approach is applicable even when no simple formula (such as the quadratic formula) is available.

The data in Fig. 1.1.8 suggest that one value of x for which $A(x) = 100$ lies somewhere between $x = 5$ and $x = 10$ and that a second such value lies between $x = 20$ and $x = 25$. Indeed, substitution in Eq. (11) yields

$$f(5) = -25 < 0 \quad \text{and} \quad f(10) = 20 > 0.$$

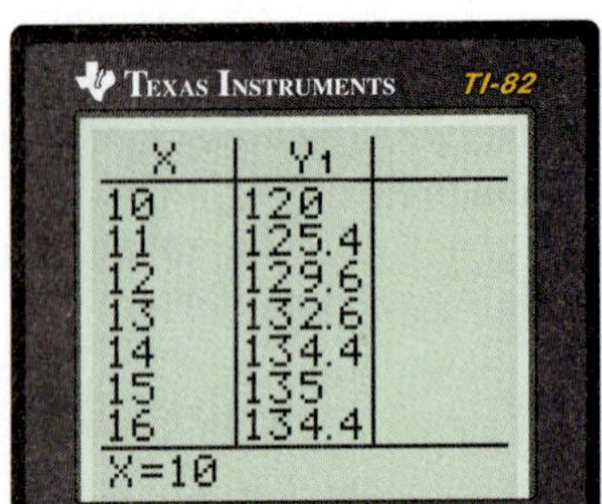

Fig. 1.1.12 The resulting table.

The fact that $f(x)$ is *negative* at one endpoint of the interval $[5, 10]$ but *positive* at the other endpoint suggests that $f(x)$ is *zero* somewhere between $x = 5$ and $x = 10$.

To see *where,* we tabulate values of $f(x)$ on $[5, 10]$. In the table of Fig. 1.1.13 we see that $f(7) < 0$ and $f(8) > 0$, so we focus next on the interval $[7, 8]$. Tabulation of $f(x)$ on $[7, 8]$ gives the table of Fig. 1.1.14, where we see that $f(7.3) < 0$ and $f(7.4) > 0$.

We therefore tabulate $f(x)$ once more, this time on the interval $[7.3, 7.4]$. In Fig. 1.1.15 we see that

$$f(7.36) \approx -0.02 \quad \text{and} \quad f(7.37) \approx 0.07.$$

Because $f(7.36)$ is considerably closer to zero than is $f(7.37)$, we conclude that the desired solution of Eq. (11) is given approximately by $x \approx 7.36$, accurate to two decimal places. If greater accuracy were needed, we could continue to tabulate $f(x)$ on smaller and smaller intervals.

If we were to begin with the interval $[20, 25]$ and proceed similarly, we would find the second value $x \approx 22.64$ such that $f(x) = 0$. (You should do this for practice.)

Finally, let's calculate the corresponding values of the width y of the animal pen such that $A = xy = 100$:

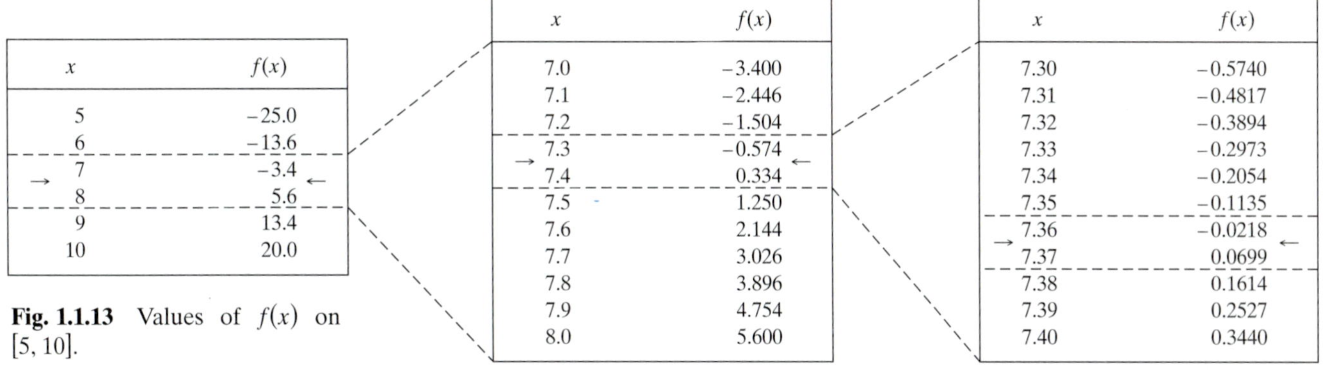

x	$f(x)$
5	−25.0
6	−13.6
→ 7	−3.4 ←
8	5.6
9	13.4
10	20.0

Fig. 1.1.13 Values of $f(x)$ on $[5, 10]$.

x	$f(x)$
7.0	−3.400
7.1	−2.446
7.2	−1.504
→ 7.3	−0.574 ←
7.4	0.334
7.5	1.250
7.6	2.144
7.7	3.026
7.8	3.896
7.9	4.754
8.0	5.600

Fig. 1.1.14 Values of $f(x)$ on $[7, 8]$.

x	$f(x)$
7.30	−0.5740
7.31	−0.4817
7.32	−0.3894
7.33	−0.2973
7.34	−0.2054
7.35	−0.1135
→ 7.36	−0.0218 ←
7.37	0.0699
7.38	0.1614
7.39	0.2527
7.40	0.3440

Fig. 1.1.15 Values of $f(x)$ on $[7.3, 7.4]$.

- If $x \approx 7.36$, then $y \approx 13.59$.
- If $x \approx 22.64$, then $y \approx 4.42$.

Thus, under the cost constraint of the animal pen problem, we can construct either a 7.36-ft by 13.59-ft or a 22.64-ft by 4.42-ft rectangle, both of area 100 ft^2.

The layout of Figs. 1.1.13 through 1.1.15 suggests the idea of repeated tabulation as successive numerical magnification. This method of repeated tabulation can be applied to a wide range of equations of the form $f(x) = 0$. If the interval $[a, b]$ contains a solution and the endpoint values $f(a)$ and $f(b)$ differ in sign, then we can approximate this solution by tabulating values on successively smaller subintervals. Problems 57 through 66 and the projects at the end of this section are applications of this concrete numerical method for the approximate solution of equations.

1.1 PROBLEMS

In Problems 1 through 4, find and simplify each of the following values: (a) $f(-a)$; (b) $f(a^{-1})$; (c) $f(\sqrt{a})$; (d) $f(a^2)$.

1. $f(x) = \dfrac{1}{x}$

2. $f(x) = x^2 + 5$

3. $f(x) = \dfrac{1}{x^2 + 5}$

4. $f(x) = \sqrt{1 + x^2 + x^4}$

In Problems 5 through 10, find all values of a such that $g(a) = 5$.

5. $g(x) = 3x + 4$

6. $g(x) = \dfrac{1}{2x - 1}$

7. $g(x) = \sqrt{x^2 + 16}$

8. $g(x) = x^3 - 3$

9. $g(x) = \sqrt[3]{x + 25}$

10. $g(x) = 2x^2 - x + 4$

In Problems 11 through 16, compute and then simplify the quantity $f(a + h) - f(a)$.

11. $f(x) = 3x - 2$

12. $f(x) = 1 - 2x$

13. $f(x) = x^2$

14. $f(x) = x^2 + 2x$

15. $f(x) = \dfrac{1}{x}$

16. $f(x) = \dfrac{2}{x + 1}$

In Problems 17 through 20, find the range of values of the given function.

17. $f(x) = \begin{cases} \dfrac{x}{|x|} & \text{if } x \neq 0, \\ 0 & \text{if } x = 0 \end{cases}$

18. $f(x) = [\![3x]\!]$ (Recall that $[\![x]\!]$ is the largest integer not exceeding x.)

19. $f(x) = (-1)^{[\![x]\!]}$

20. $f(x)$ is the first-class postage (in cents) for a letter mailed in the United States and weighing x ounces, $0 < x < 12$. In 1997 the postage rate for such a letter was 32¢ for the first ounce plus 23¢ for each additional ounce or fraction thereof.

In Problems 21 through 35, find the largest domain (of real numbers) on which the given formula determines a (real-valued) function.

21. $f(x) = 10 - x^2$

22. $f(x) = x^3 + 5$

23. $f(t) = \sqrt{t^2}$

24. $g(t) = (\sqrt{t})^2$

25. $f(x) = \sqrt{3x - 5}$

26. $g(t) = \sqrt[3]{t + 4}$

27. $f(t) = \sqrt{1 - 2t}$

28. $g(x) = \dfrac{1}{(x + 2)^2}$

29. $f(x) = \dfrac{2}{3 - x}$

30. $g(t) = \sqrt{\dfrac{2}{3 - t}}$

31. $f(x) = \sqrt{x^2 + 9}$

32. $h(z) = \dfrac{1}{\sqrt{4 - z^2}}$

33. $f(x) = \sqrt{4 - \sqrt{x}}$

34. $f(x) = \sqrt{\dfrac{x + 1}{x - 1}}$

35. $g(t) = \dfrac{t}{|t|}$

36. Express the area A of a square as a function of its perimeter P.

37. Express the circumference C of a circle as a function of its area A.

38. Express the volume V of a sphere as a function of its surface area S.

39. Given: 0°C is the same as 32°F, and a temperature change of 1°C is the same as a change of 1.8°F. Express the Celsius temperature C as a function of the Fahrenheit temperature F.

40. Show that if a rectangle has base x and perimeter 100 (Fig. 1.1.16), then its area A is given by the function

$$A(x) = x(50 - x), \qquad 0 \leqq x \leqq 50.$$

41. A rectangle with base of length x is inscribed in a circle of radius 2 (Fig 1.1.17). Express the area A of the rectangle as a function of x.

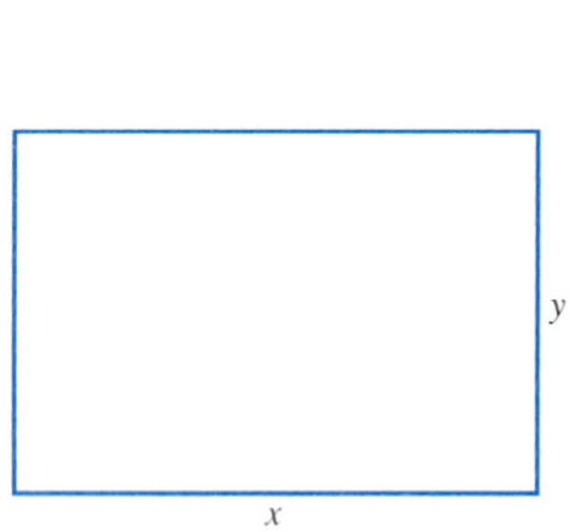

Fig. 1.1.16 $A = xy$ (Problem 40).

Fig. 1.1.17 $A = xy$ (Problem 41).

42. An oil field containing 20 wells has been producing 4000 barrels of oil daily. For each new well that is drilled, the daily production of each well decreases by 5 barrels per day. Write the total daily production of the oil field as a function of the number x of new wells drilled.

43. Suppose that a rectangular box has volume 324 cm^3 and a square base of length x centimeters. The material for the base of the box costs 2¢/cm^2, and the material for its top and four sides costs 1¢/cm^2. Express the total cost of the box as a function of x. See Fig. 1.1.18.

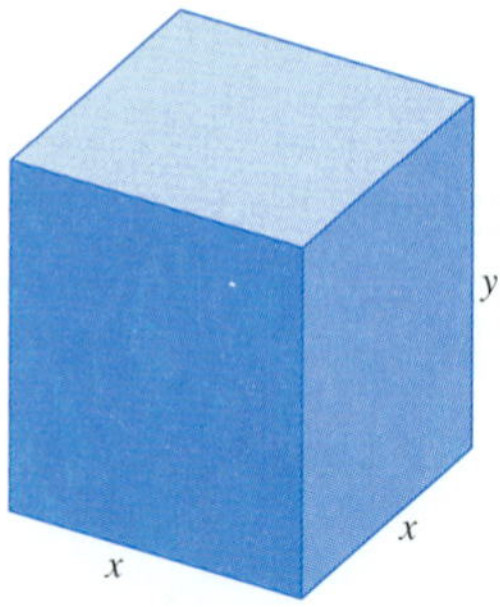

Fig. 1.1.18 $V = x^2y$ (Problem 43).

44. A rectangle of fixed perimeter 36 is rotated around one of its sides S to generate a right circular cylinder. Express the volume V of this cylinder as a function of the length x of the side S. See Fig. 1.1.19.

45. A right circular cylinder has volume 1000 in.^3, and the radius of its base is r inches. Express the total surface area A of the cylinder as a function of r. See Fig. 1.1.20.

46. A rectangular box has total surface area 600 cm^2 and a square base with edge length x centimeters. Express the volume V of the box as a function of x.

47. An open-topped box is to be made from a square piece of cardboard of edge length 50 in. First, four small squares, each of edge length x inches, are cut from the corners of the cardboard (Fig. 1.1.21). Then the four resulting flaps are turned up—folded along the dotted lines—to form the four sides of the box, which will thus have a square base and a depth of x inches (Fig. 1.1.22). Express its volume V as a function of x.

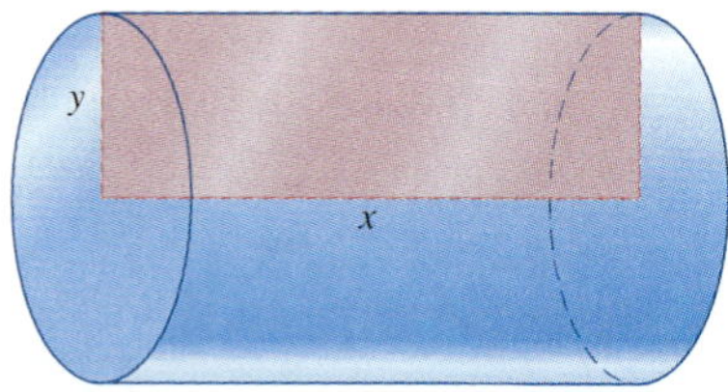

Fig. 1.1.19 $V = \pi xy^2$ (Problem 44).

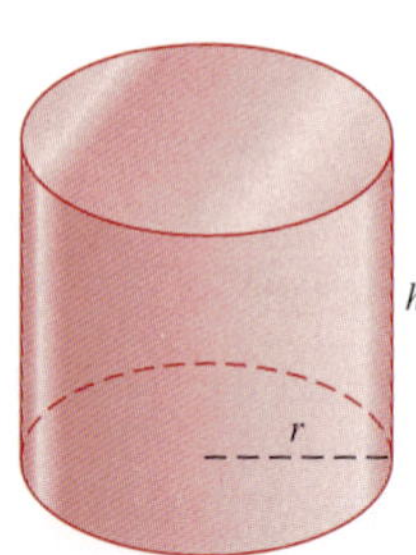

Fig. 1.1.20 $V = \pi r^2h$ (Problem 45).

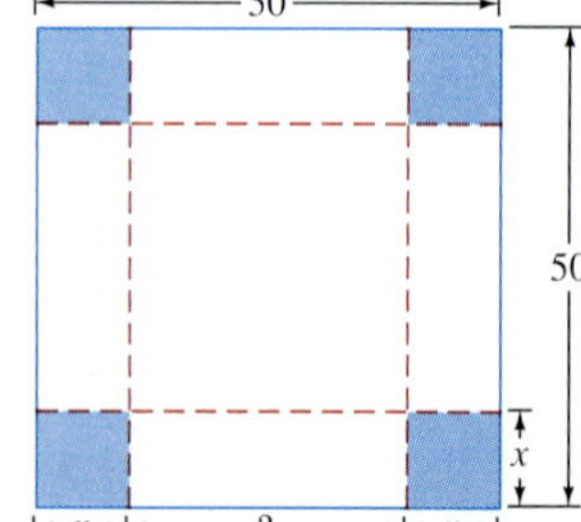

Fig. 1.1.21 Fold the edges up to make a box (Problem 47).

Fig. 1.1.22 The box of Problem 47.

48. Continue Problem 40 by numerically investigating the area of a rectangle of perimeter 100. What dimensions (length and width) would appear to maximize the area of such a rectangle?

49. Determine numerically the number of new oil wells that should be drilled to maximize the total daily production of the oil field of Problem 42.

50. Investigate numerically the total surface area A of the rectangular box of Example 8. Assuming that both $x \geqq 1$ and $y \geqq 1$, what dimensions x and y would appear to minimize A?

Problems 51 through 56 deal with the functions CEILING, FLOOR, and ROUND of Example 5.

51. Show that $\text{CEILING}(x) = -\text{FLOOR}(-x)$ for all x.

52. Suppose that k is a constant. What is the range of the function $g(x) = \text{ROUND}(kx)$?

53. What is the range of the function $g(x) = \frac{1}{10}\text{ROUND}(10x)$?

54. Recalling that $\pi \approx 3.14159$, note that $\frac{1}{100}\text{ROUND}(100\pi) = 3.14$. Hence define (in terms of ROUND) a function $\text{ROUND2}(x)$ that gives the value of x rounded accurate to two decimal places.

55. Define a function $\text{ROUND4}(x)$ that gives the value of x rounded accurate to four decimal places, so that $\text{ROUND4}(\pi) = 3.1416$.

56. Define a function $\text{CHOP4}(x)$ that "chops off" (or discards) all decimal places of x beyond the fourth one, so that $\text{CHOP4}(\pi) = 3.1415$.

In Problems 57 through 66, a quadratic equation $ax^2 + bx + c = 0$ and an interval $[p, q]$ containing one of its solutions are given. Use the method of repeated tabulation to approximate this solution with two digits correct or correctly rounded to the right of the decimal. Check that your result agrees with one of the two solutions given by the quadratic formula,

$$x = \frac{-b \pm \sqrt{b^2 - 4ac}}{2a}.$$

57. $x^2 - 3x + 1 = 0, \quad [0, 1]$

58. $x^2 - 3x + 1 = 0, \quad [2, 3]$

59. $x^2 + 2x - 4 = 0, \quad [1, 2]$

60. $x^2 + 2x - 4 = 0, \quad [-4, -3]$

61. $2x^2 - 7x + 4 = 0, \quad [0, 1]$

62. $2x^2 - 7x + 4 = 0, \quad [2, 3]$

63. $x^2 - 11x + 25 = 0, \quad [3, 4]$

64. $x^2 - 11x + 25 = 0, \quad [7, 8]$

65. $3x^2 + 23x - 45 = 0, \quad [1, 2]$

66. $3x^2 + 23x - 45 = 0, \quad [-10, -9]$

1.1 PROJECT: A SQUARE WADING POOL

Suppose that you have a square piece of tin with its edge measuring a feet, where a denotes the largest digit in your student I.D. number. You plan to make a wading pool in the manner suggested by Figs. 1.1.21 and 1.1.22—by cutting a square of edge x from each of the four corners of the piece of tin, then turning up each of the four resulting flaps to form an open-topped square box that will hold water after you solder its four vertical edges. You want to determine x so that the wading pool so constructed will have the largest possible volume V. Begin by expressing the box's volume $V = f(x)$ as a function of its height x, then use the method of repeated tabulation to find the maximum value $V_{\max}$ (rounded off accurate to two decimal places) attained by the function f on the interval $[0, a/2]$. (Why is this the appropriate domain of f?)

Some computer systems have "one-liners" for tabulation of functions. The table in Fig. 1.1.23 lists commands for several common systems that can be used to tabulate

BASIC	`For x=a to b STEP h : Print x,f(x) :NEXT`
Derive	`[VECTOR(x,x,a,b,h), VECTOR(f(x),x,a,b,h)]``
Maple	`for x from a by h to b do print (x,f(x)) od`
Mathematica	`Table[ {x,f[x]}, {x,a,b,h} ] // TableForm`
MATLAB	`x=a:h:b; y=f(x); [x;y]'`

Fig. 1.1.23 Commands for tabulating the function $f(x)$.

values of the function $f(x)$ on the interval $[a, b]$ with subintervals of length h. (You are to insert the formula for your function $f(x)$ and the desired numerical values of a, b, and h.)

For a second investigation, suppose you decide instead that you want your pool to have exactly *half* the maximum possible volume V_{max}. Note first that a tabulation of $f(x)$ on the interval $[0, a/2]$ indicates that this will hold for *two* different values of x. Find both of them (rounded off accurate to two decimal places).

Write the results of your investigations in the form of a carefully organized report consisting of complete sentences (plus pertinent equations and data tables) explaining your results in detail and telling precisely what you did to solve your problems.

1.2 GRAPHS OF EQUATIONS AND FUNCTIONS

Graphs and equations of *straight lines* in the *xy*-coordinate plane are reviewed in Appendix B. Recall the **slope-intercept equation**

$$y = mx + b \tag{1}$$

of the straight line with **slope** $m = \tan\phi$, **angle of inclination** ϕ, and y-**intercept** b (Fig. 1.2.1). The "rise over run" definition

$$m = \frac{\text{rise}}{\text{run}} = \frac{\Delta y}{\Delta x} = \frac{y_2 - y_1}{x_2 - x_1} \tag{2}$$

of the slope (Fig. 1.2.2) leads to the **point-slope equation**

$$y - y_0 = m(x - x_0) \tag{3}$$

of the straight line with slope m that passes through the point (x_0, y_0)—see Fig. 1.2.3. In either case a point (x, y) in the xy-plane lies on the line if and only if its coordinates x and y satisfy the indicated equation.

If $\Delta y = 0$ in Eq. (2), then $m = 0$ and the line is *horizontal*. If $\Delta x = 0$, then the line is *vertical* and (because we cannot divide by zero) the slope of the line is not defined. Thus:

- **Horizontal lines have slope zero.**
- **Vertical lines have no defined slope at all.**

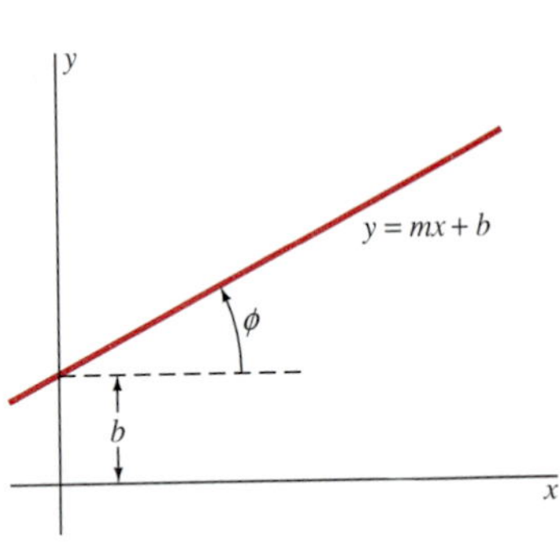

Fig. 1.2.1 A line with y-intercept b and inclination angle ϕ.

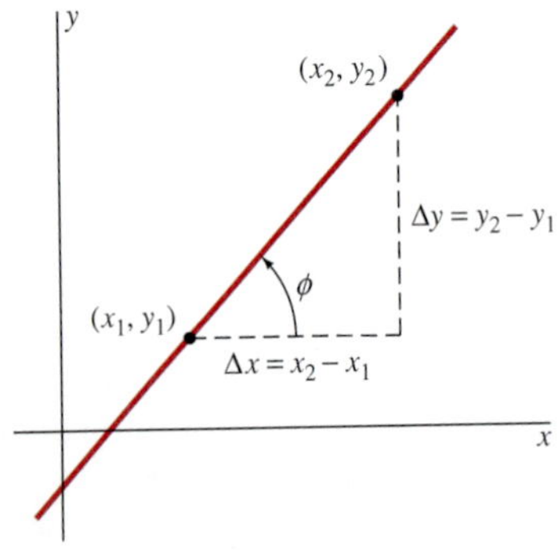

Fig. 1.2.2 Slope $m = \tan\phi = \dfrac{\Delta y}{\Delta x}$.

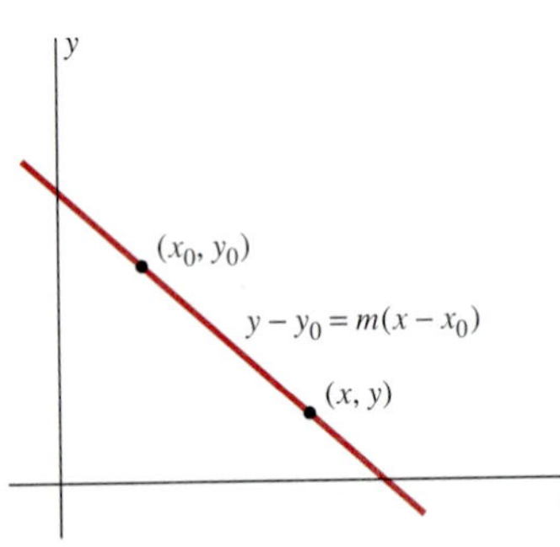

Fig. 1.2.3 The line through (x_0, y_0) with slope m.

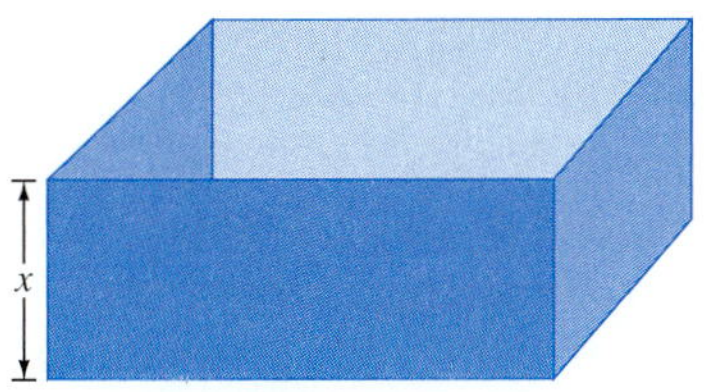

Fig. 1.1.22 The box of Problem 47.

48. Continue Problem 40 by numerically investigating the area of a rectangle of perimeter 100. What dimensions (length and width) would appear to maximize the area of such a rectangle?

49. Determine numerically the number of new oil wells that should be drilled to maximize the total daily production of the oil field of Problem 42.

50. Investigate numerically the total surface area A of the rectangular box of Example 8. Assuming that both $x \geqq 1$ and $y \geqq 1$, what dimensions x and y would appear to minimize A?

Problems 51 through 56 deal with the functions CEILING, FLOOR, and ROUND of Example 5.

51. Show that $\text{CEILING}(x) = -\text{FLOOR}(-x)$ for all x.

52. Suppose that k is a constant. What is the range of the function $g(x) = \text{ROUND}(kx)$?

53. What is the range of the function $g(x) = \frac{1}{10}\text{ROUND}(10x)$?

54. Recalling that $\pi \approx 3.14159$, note that $\frac{1}{100}\text{ROUND}(100\pi) = 3.14$. Hence define (in terms of ROUND) a function $\text{ROUND2}(x)$ that gives the value of x rounded accurate to two decimal places.

55. Define a function $\text{ROUND4}(x)$ that gives the value of x rounded accurate to four decimal places, so that $\text{ROUND4}(\pi) = 3.1416$.

56. Define a function $\text{CHOP4}(x)$ that "chops off" (or discards) all decimal places of x beyond the fourth one, so that $\text{CHOP4}(\pi) = 3.1415$.

In Problems 57 through 66, a quadratic equation $ax^2 + bx + c = 0$ and an interval $[p, q]$ containing one of its solutions are given. Use the method of repeated tabulation to approximate this solution with two digits correct or correctly rounded to the right of the decimal. Check that your result agrees with one of the two solutions given by the quadratic formula,

$$x = \frac{-b \pm \sqrt{b^2 - 4ac}}{2a}.$$

57. $x^2 - 3x + 1 = 0, \quad [0, 1]$

58. $x^2 - 3x + 1 = 0, \quad [2, 3]$

59. $x^2 + 2x - 4 = 0, \quad [1, 2]$

60. $x^2 + 2x - 4 = 0, \quad [-4, -3]$

61. $2x^2 - 7x + 4 = 0, \quad [0, 1]$

62. $2x^2 - 7x + 4 = 0, \quad [2, 3]$

63. $x^2 - 11x + 25 = 0, \quad [3, 4]$

64. $x^2 - 11x + 25 = 0, \quad [7, 8]$

65. $3x^2 + 23x - 45 = 0, \quad [1, 2]$

66. $3x^2 + 23x - 45 = 0, \quad [-10, -9]$

1.1 PROJECT: A SQUARE WADING POOL

Suppose that you have a square piece of tin with its edge measuring a feet, where a denotes the largest digit in your student I.D. number. You plan to make a wading pool in the manner suggested by Figs. 1.1.21 and 1.1.22—by cutting a square of edge x from each of the four corners of the piece of tin, then turning up each of the four resulting flaps to form an open-topped square box that will hold water after you solder its four vertical edges. You want to determine x so that the wading pool so constructed will have the largest possible volume V. Begin by expressing the box's volume $V = f(x)$ as a function of its height x, then use the method of repeated tabulation to find the maximum value V_{max} (rounded off accurate to two decimal places) attained by the function f on the interval $[0, a/2]$. (Why is this the appropriate domain of f?)

Some computer systems have "one-liners" for tabulation of functions. The table in Fig. 1.1.23 lists commands for several common systems that can be used to tabulate

BASIC	`For x=a to b STEP h : Print x,f(x) :NEXT`
Derive	`[VECTOR(x,x,a,b,h), VECTOR(f(x),x,a,b,h)]``
Maple	`for x from a by h to b do print (x,f(x)) od`
Mathematica	`Table[ {x,f[x]}, {x,a,b,h} ] // TableForm`
MATLAB	`x=a:h:b; y=f(x); [x;y]'`

Fig. 1.1.23 Commands for tabulating the function $f(x)$.

values of the function $f(x)$ on the interval $[a, b]$ with subintervals of length h. (You are to insert the formula for your function $f(x)$ and the desired numerical values of a, b, and h.)

For a second investigation, suppose you decide instead that you want your pool to have exactly *half* the maximum possible volume V_{max}. Note first that a tabulation of $f(x)$ on the interval $[0, a/2]$ indicates that this will hold for *two* different values of x. Find both of them (rounded off accurate to two decimal places).

Write the results of your investigations in the form of a carefully organized report consisting of complete sentences (plus pertinent equations and data tables) explaining your results in detail and telling precisely what you did to solve your problems.

1.2 GRAPHS OF EQUATIONS AND FUNCTIONS

Graphs and equations of *straight lines* in the *xy*-coordinate plane are reviewed in Appendix B. Recall the **slope-intercept equation**

$$y = mx + b \tag{1}$$

of the straight line with **slope** $m = \tan\phi$, **angle of inclination** ϕ, and y-**intercept** b (Fig. 1.2.1). The "rise over run" definition

$$m = \frac{\text{rise}}{\text{run}} = \frac{\Delta y}{\Delta x} = \frac{y_2 - y_1}{x_2 - x_1} \tag{2}$$

of the slope (Fig. 1.2.2) leads to the **point-slope equation**

$$y - y_0 = m(x - x_0) \tag{3}$$

of the straight line with slope m that passes through the point (x_0, y_0)—see Fig. 1.2.3. In either case a point (x, y) in the xy-plane lies on the line if and only if its coordinates x and y satisfy the indicated equation.

If $\Delta y = 0$ in Eq. (2), then $m = 0$ and the line is *horizontal*. If $\Delta x = 0$, then the line is *vertical* and (because we cannot divide by zero) the slope of the line is not defined. Thus:

- **Horizontal lines have slope zero.**
- **Vertical lines have no defined slope at all.**

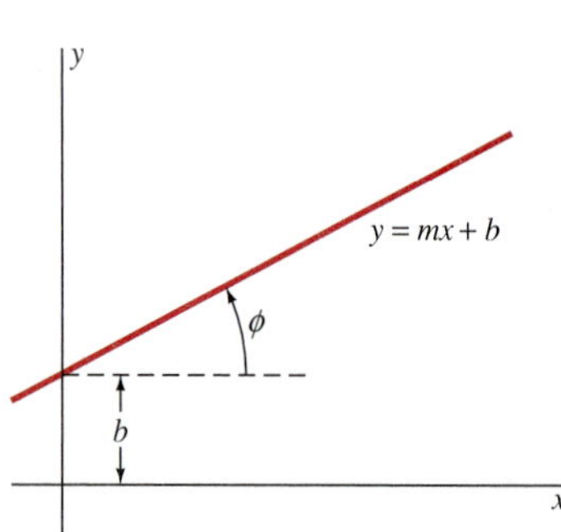

Fig. 1.2.1 A line with y-intercept b and inclination angle ϕ.

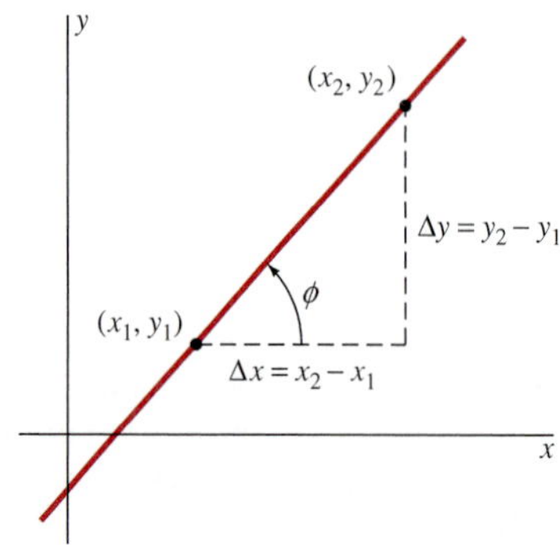

Fig. 1.2.2 Slope $m = \tan\phi = \dfrac{\Delta y}{\Delta x}$.

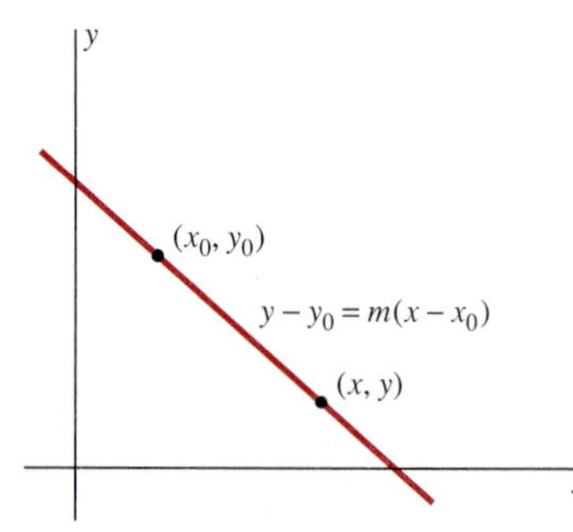

Fig. 1.2.3 The line through (x_0, y_0) with slope m.

EXAMPLE 1 Write an equation of the line L that passes through the point $P(3, 5)$ and is parallel to the line having equation $y = 2x - 4$.

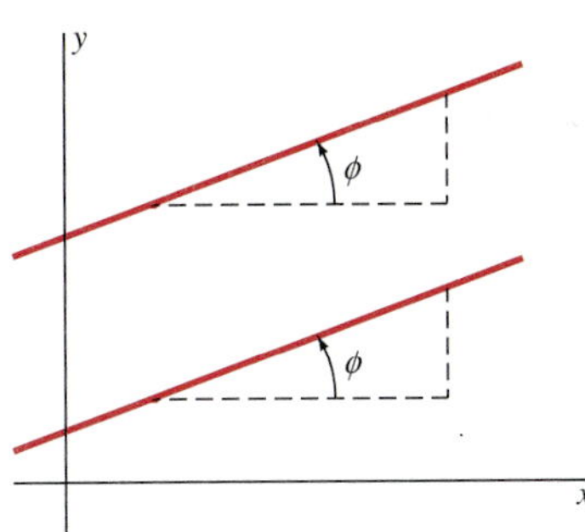

Fig. 1.2.4 Parallel lines have the same slope $m = \tan\phi$.

Solution The two parallel lines have the same angle of inclination ϕ (Fig. 1.2.4) and therefore have the same slope m. Comparing the given equation $y = 2x - 4$ with the slope-intercept equation in (1), we see that $m = 2$. The point-slope equation therefore gives

$$y - 5 = 2(x - 3)$$

—alternatively, $y = 2x - 1$, for an equation of the line L.

Both Eqs. (1) and (3) can be put into the form of the general linear equation

$$Ax + By = C. \tag{4}$$

Conversely, if $B \neq 0$, then we can divide the terms in Eq. (4) by B and solve for y, thereby obtaining the point-slope equation of a straight line. If $A = 0$, then the resulting equation has the form $y = H$, the equation of a *horizontal line* with slope zero. If $B = 0$ but $A \neq 0$, then Eq. (4) can be solved for $x = K$, the equation of a *vertical line* (having no slope at all). In summary, we see that if the coefficients A and B are not both zero, then Eq. (4) is the equation of some straight line in the plane. ■

Graphs of More General Equations

A straight line is a simple example of the graph of an equation. By contrast, a computer-graphing program produced the exotic curve shown in Fig. 1.2.5 when asked to picture the set of all points (x, y) satisfying the equation

$$x^2 + y^2 = (x^2 + y^2 - 2x)^2.$$

Both a straight line and this complicated curve are examples of *graphs* of equations.

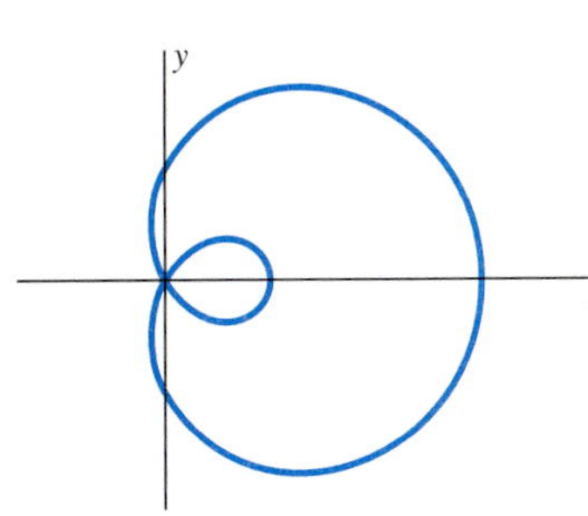

Fig. 1.2.5 The graph of the equation $x^2 + y^2 = (x^2 + y^2 - 2x)^2$.

> **Definition** *Graph of an Equation*
>
> The **graph** of an equation in two variables x and y is the set of all points (x, y) in the plane that satisfy the equation.

For example, the distance formula of Fig. 1.2.6 tells us that the graph of the equation

$$x^2 + y^2 = r^2 \tag{5}$$

is the circle of radius r centered at the origin $(0, 0)$. More generally, the graph of the equation

$$(x - h)^2 + (y - k)^2 = r^2 \tag{6}$$

is the circle of radius r with center (h, k). This also follows from the distance formula, because the distance between the points (x, y) and (h, k) in Fig. 1.2.7 is r.

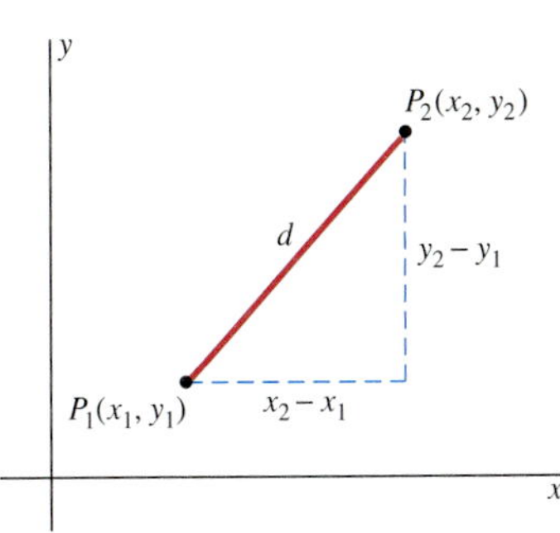

Fig. 1.2.6 The Pythagorean theorem implies the **distance formula** $d = \sqrt{(x_2 - x_1)^2 + (y_2 - y_1)^2}$.

EXAMPLE 2 The equation of the circle with center $(3, 4)$ and radius 10 is

$$(x - 3)^2 + (y - 4)^2 = 100,$$

which may also be written in the form

$$x^2 + y^2 - 6x - 8y - 75 = 0.$$

■

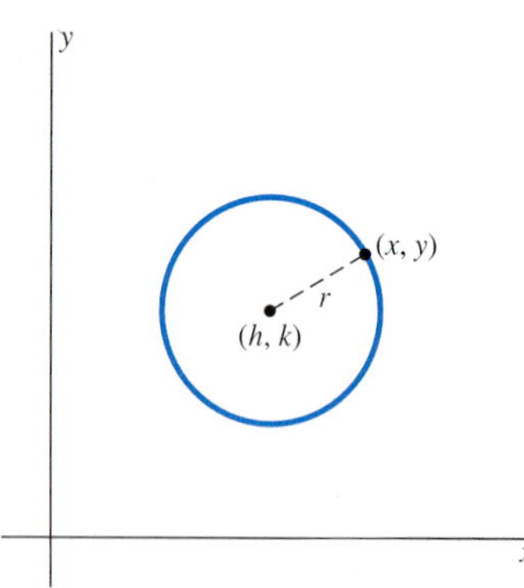

Fig. 1.2.7 A translated circle

Translates of Graphs

Suppose that the xy-plane is shifted rigidly (or *translated*) by moving each point h units to the right and k units upward. (A negative value of h or k corresponds to a leftward or downward movement.) That is, each point (x, y) of the plane is moved to the point $(x + h, y + k)$; see Fig. 1.2.8. Then the circle with radius r and center $(0, 0)$ is translated to the circle with radius r and center (h, k). Thus the general circle described by Eq. (6) is a *translate* of the origin-centered circle. Note that the equation of the translated circle is obtained from the original equation by replacing x with $x - h$ and y with $y - k$. This observation illustrates a general principle that describes equations of translated (or "shifted") graphs.

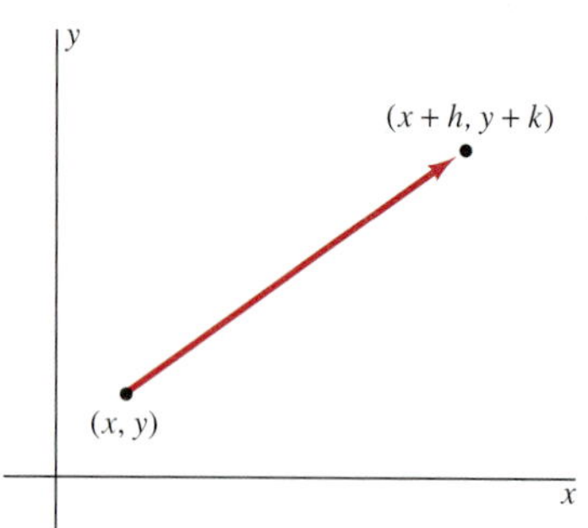

Fig. 1.2.8 Translating a point

Translation Principle

When the graph of an equation is translated h units to the right and k units upward, the equation of the translated curve is obtained from the original equation by replacement of x with $x - h$ and of y with $y - k$.

Observe that we can write the equation of a translated circle in Eq. (6) in the general form

$$x^2 + y^2 + ax + by = c. \tag{7}$$

What, then, can we do when we encounter an equation already of the form in Eq. (7)? We first recognize that it is an equation of a circle. Next, we can discover its center and radius by the technique of *completing the square*. To do so, we note that

$$x^2 + ax = \left(x + \frac{a}{2}\right)^2 - \frac{a^2}{4},$$

which shows that $x^2 + ax$ can be made into a perfect square by adding to it the square of *half* the coefficient of x.

EXAMPLE 3 Find the center and radius of the circle that has the equation

$$x^2 + y^2 - 4x + 6y = 12.$$

Solution We complete the square separately for both variables x and y. This gives

$$(x^2 - 4x + 4) + (y^2 + 6y + 9) = 12 + 4 + 9;$$
$$(x - 2)^2 + (y + 3)^2 = 25.$$

Hence the circle—shown in Fig. 1.2.9—has center $(2, -3)$ and radius 5. Solution of the last equation for y gives

$$y = -3 \pm \sqrt{25 - (x - 2)^2}.$$

Thus the whole circle consists of the graphs of the *two* equations

$$y = -3 + \sqrt{25 - (x - 2)^2}$$

and

$$y = -3 - \sqrt{25 - (x - 2)^2}$$

that describe its upper and lower semicircles. ■

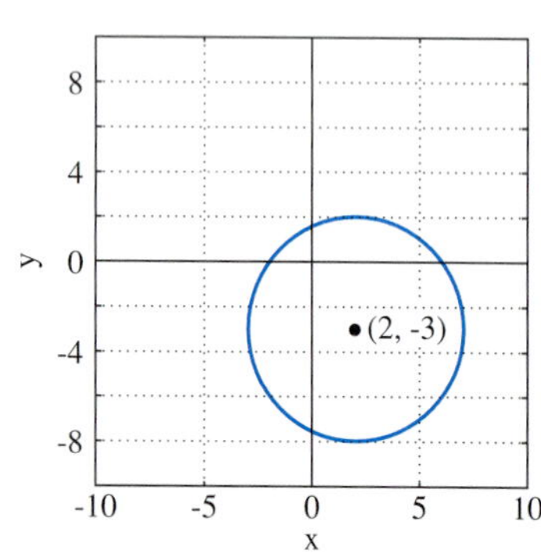

Fig. 1.2.9 The circle of Example 3

Graphs of Functions

The graph of a function is a special case of the graph of an equation.

> **Definition** ***Graph of a Function***
> The **graph** of the function f is the graph of the equation $y = f(x)$.

Thus the graph of the function f is the set of all points in the plane that have the form $(x, f(x))$, where x is in the domain of f (see Fig. 1.2.10). Because the second coordinate of such a point is uniquely determined by its first coordinate, we obtain the following useful principle.

> **The Vertical Line Test**
> Each vertical line through a point in the domain of a function meets its graph in exactly one point.

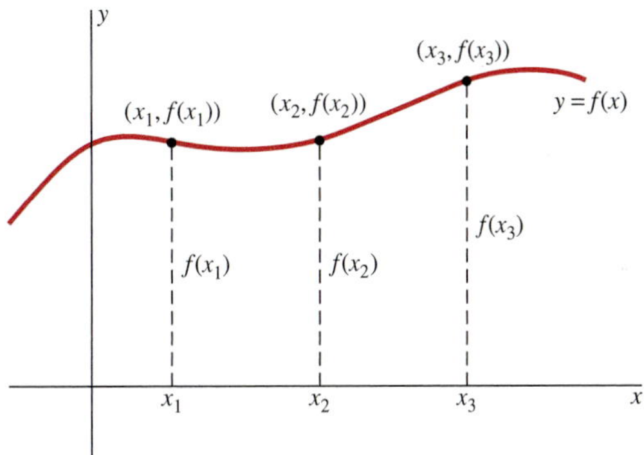

Fig. 1.2.10 The graph of the function f

Thus no vertical line can intersect the graph of a function in more than one point. For instance, it follows that the curve in Fig. 1.2.5 cannot be the graph of a *function,* although it *is* the graph of an equation. Similarly, a circle cannot be the graph of a function.

EXAMPLE 4 Construct the graph of the absolute value function $f(x) = |x|$.

Solution Recall that

$$|x| = \begin{cases} x & \text{if } x \geqq 0, \\ -x & \text{if } x < 0. \end{cases}$$

So the graph of $y = |x|$ consists of the right half of the line $y = x$ together with the left half of the line $y = -x$, as shown in Fig. 1.2.11. ■

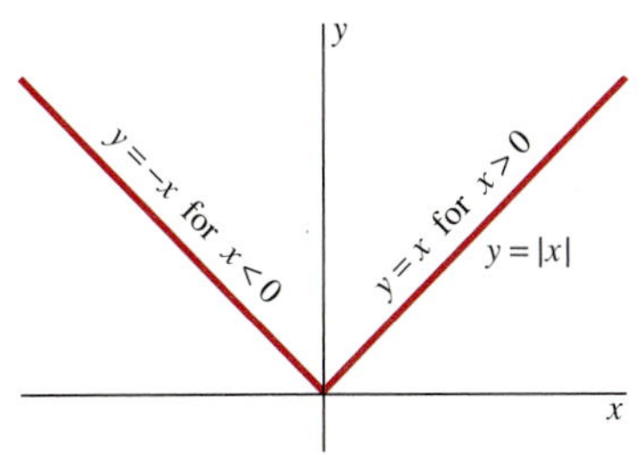

Fig. 1.2.11 The graph of the absolute value function $y = |x|$ of Example 4

EXAMPLE 5 Sketch the graph of the reciprocal function.

$$f(x) = \frac{1}{x}.$$

Solution Let's examine four natural cases.

1. When x is positive and numerically large, $f(x)$ is small and positive.
2. When x is positive and near zero, $f(x)$ is large and positive.
3. When x is negative and numerically small (negative and close to zero), $f(x)$ is large and negative.
4. When x is large and negative (x is negative but $|x|$ is large), $f(x)$ is small and negative (negative and close to zero).

To get started with the graph, we can plot a few points, for example, $(1, 1)$, $(-1, -1)$, $(10, 0.1)$, $(0.1, 10)$, $(-10, -0.1)$, and $(-0.1, -10)$. The result of the information displayed here suggests that the actual graph is much like the one shown in Fig. 1.2.12. ■

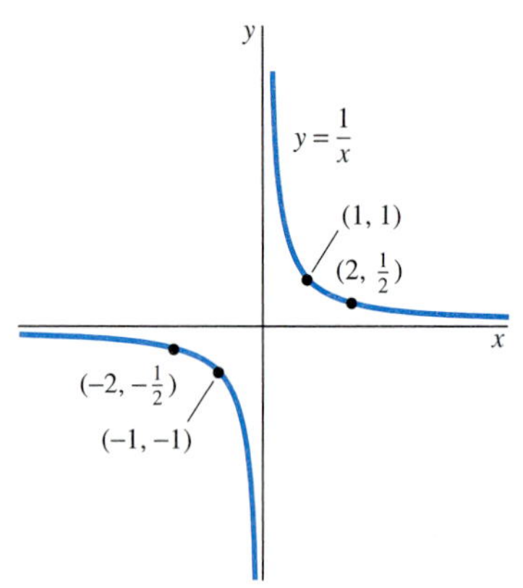

Fig. 1.2.12 The graph of the reciprocal function $y = 1/x$ of Example 5

Figure 1.2.12 exhibits a "gap," or "discontinuity," in the graph of $y = 1/x$ at $x = 0$. Indeed, the gap is called an *infinite discontinuity* because y increases without bound as x approaches zero from the right, whereas y decreases without bound as x approaches zero from the left. This phenomenon generally is signaled by the presence of denominators that are zero at certain values of x, as in the case of the functions

$$f(x) = \frac{1}{1 - x} \quad \text{and} \quad f(x) = \frac{1}{x^2},$$

which we ask you to graph in the problems.

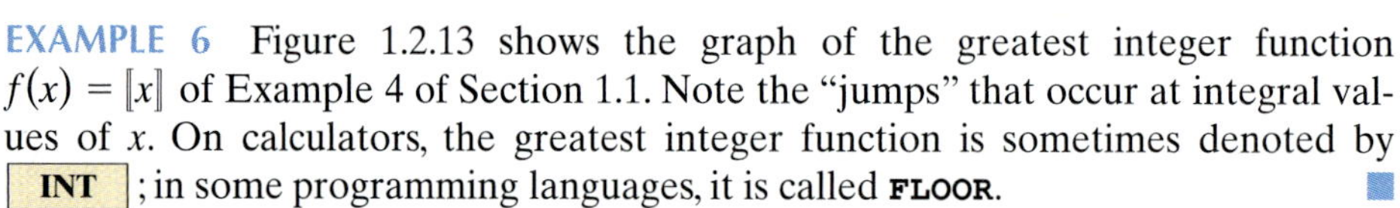
EXAMPLE 6 Figure 1.2.13 shows the graph of the greatest integer function $f(x) = [\![x]\!]$ of Example 4 of Section 1.1. Note the "jumps" that occur at integral values of x. On calculators, the greatest integer function is sometimes denoted by **INT**; in some programming languages, it is called **FLOOR**. ■

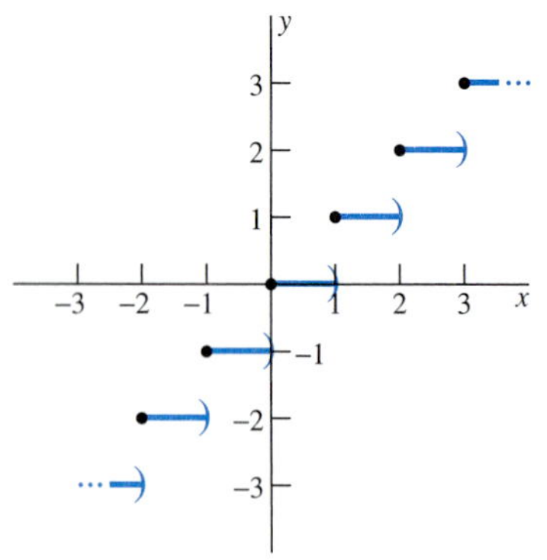

Fig. 1.2.13 The graph of the greatest integer function $f(x) = [\![x]\!]$ of Example 6

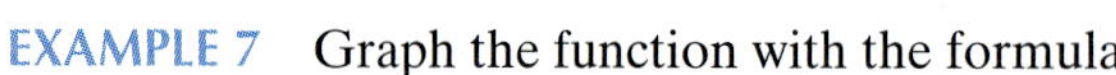
EXAMPLE 7 Graph the function with the formula

$$f(x) = x - [\![x]\!] - \tfrac{1}{2}.$$

Solution Recall that $[\![x]\!] = n$, where n is the greatest integer not exceeding x; $n \leqq x < n + 1$. Hence if n is an integer, then

$$f(n) = n - n - \tfrac{1}{2}.$$

This implies that the point $(n, -\frac{1}{2})$ lies on the graph of f for each integer n. Next, if $n \leqq x < n + 1$ (where, again, n is an integer), then

$$f(x) = x - n - \tfrac{1}{2}.$$

Because $y = x - n - \frac{1}{2}$ has as its graph a straight line of slope 1, it follows that the graph of f takes the form shown in Fig. 1.2.14. This *sawtooth function* is another example of a discontinuous function. The values of x where the value of $f(x)$ makes a jump are called **points of discontinuity** of the function f. Thus the points of discontinuity of the sawtooth function are the integers. As x approaches the integer n from the left, the value of $f(x)$ approaches $+\frac{1}{2}$, but $f(x)$ abruptly jumps to the value $-\frac{1}{2}$ when $x = n$. A precise definition of continuity and discontinuity for functions appears in Section 2.4. Figure 1.2.15 shows a graphics calculator prepared to graph the sawtooth function. ■

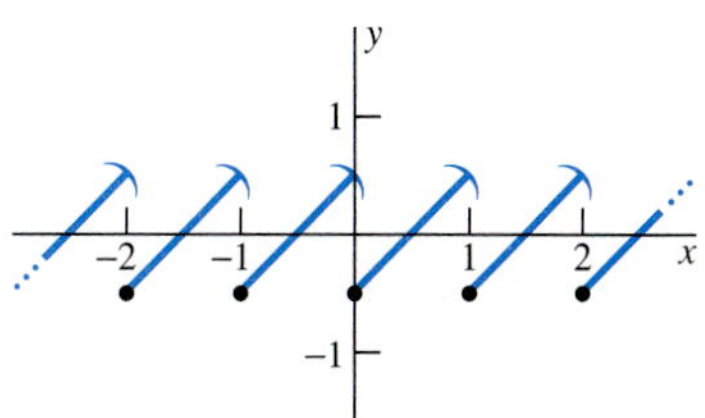

Fig. 1.2.14 The graph of the sawtooth function $f(x) = x - [\![x]\!] - \frac{1}{2}$ of Example 7

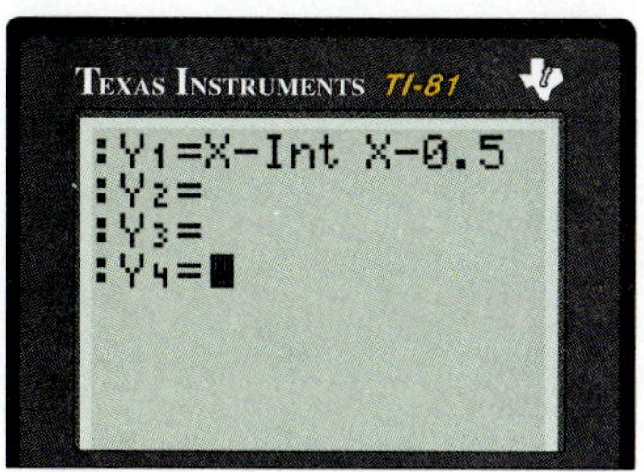

Fig. 1.2.15 A graphics calculator prepared to graph the sawtooth function of Example 7

Parabolas

The graph of a *quadratic* function of the form

$$f(x) = ax^2 + bx + c \quad (a \neq 0) \tag{8}$$

is a *parabola* whose shape resembles that of the particular parabola in Example 8.

EXAMPLE 8 Construct the graph of the parabola $y = x^2$.

Solution We plot some points in a short table of values.

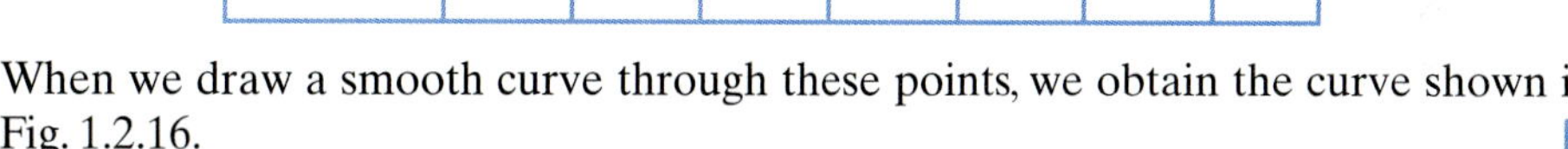

x	-3	-2	-1	0	1	2	3
$y = x^2$	9	4	1	0	1	4	9

When we draw a smooth curve through these points, we obtain the curve shown in Fig. 1.2.16. ■

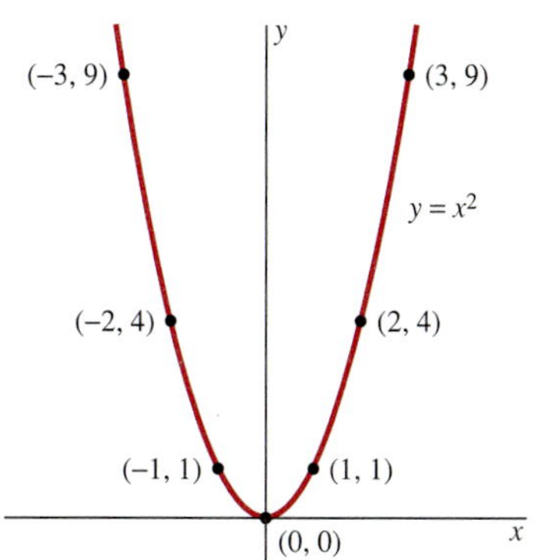

Fig. 1.2.16 The graph of the parabola $y = x^2$ of Example 8

The parabola $y = -x^2$ would look similar to the one in Fig. 1.2.16 but would open downward instead of upward. More generally, the graph of the equation

$$y = ax^2 \tag{9}$$

is a parabola with its *vertex* at the origin, provided that $a \neq 0$. This parabola opens upward if $a > 0$ and downward if $a < 0$. [For the time being, we may regard the vertex of a parabola as the point at which it "changes direction." The vertex of a parabola of the form $y = ax^2$ $(a \neq 0)$ is always at the origin. A precise definition of the *vertex* of a parabola appears in Chapter 10.]

EXAMPLE 9 Construct the graphs of the functions $f(x) = \sqrt{x}$ and $g(x) = -\sqrt{x}$.

Solution After plotting and connecting points as in Example 8, we obtain the parabola $y^2 = x$ shown in Fig. 1.2.17. This parabola opens to the right. The upper half is the graph of $f(x) = \sqrt{x}$; the lower half is the graph of $g(x) = -\sqrt{x}$. Thus the union of the graphs of these two functions is the graph of the *single* equation $y^2 = x$. (Compare this with the circle of Example 3.) More generally, the graph of the equation

$$x = by^2 \tag{10}$$

is a parabola with its vertex at the origin, provided that $b \neq 0$. This parabola opens to the right if $b > 0$ (as in Fig. 1.2.17), but to the left if $b < 0$. ■

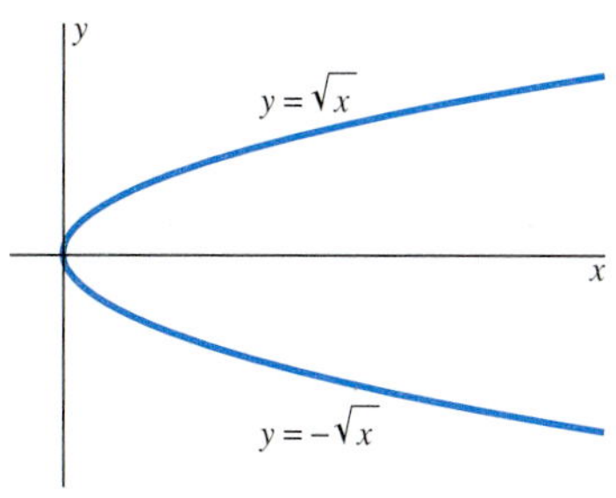

Fig. 1.2.17 The graph of the parabola $x = y^2$ of Example 9

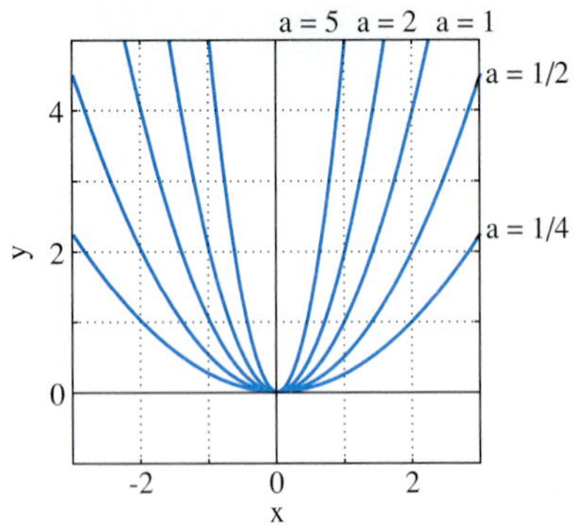

Fig. 1.2.18 Parabolas with different widths

The *size* of the coefficient a in Eq. (9) [or of b in Eq. (10)] determines the "width" of the parabola; its *sign* determines the direction in which the parabola opens. Specifically, the larger $a > 0$ is, the steeper the curve rises and hence the narrower the parabola is (see Fig. 1.2.18).

The parabola in Fig. 1.2.19 has the shape of the "standard parabola" of Example 8, but its vertex is located at the point (h, k). In the indicated uv-coordinate system, the equation of this parabola is $v = u^2$, in analogy with Eq. (9) with $a = 1$. But the uv-coordinates and xy-coordinates are related as follows:

$$u = x - h, \quad v = y - k.$$

Hence the xy-coordinate equation of this parabola is

$$y - k = (x - h)^2. \tag{11}$$

Thus when the parabola $y = x^2$ is translated h units to the right and k units upward, the equation in (11) of the translated parabola is obtained by replacing x with $x - h$ and y with $y - k$. This is another instance of the *translation principle* that we observed in connection with circles.

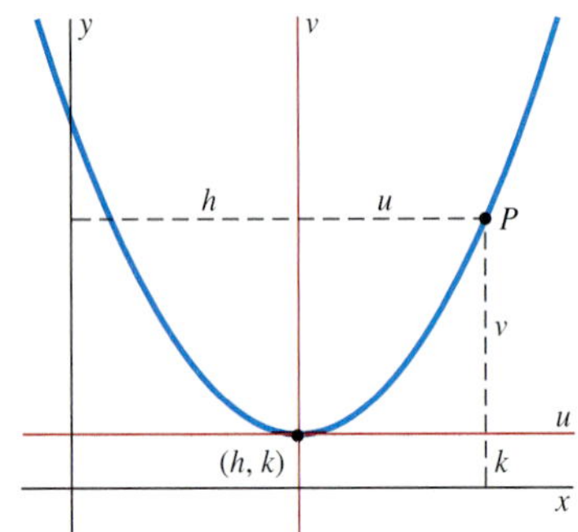

Fig. 1.2.19 A translated parabola

More generally, the graph of any equation of the form

$$y = ax^2 + bx + c \quad (a \neq 0) \tag{12}$$

can be recognized as a translated parabola by first completing the square in x to obtain an equation of the form

$$y - k = a(x - h)^2. \tag{13}$$

The graph of this equation is a parabola with its vertex at (h, k).

EXAMPLE 10 Determine the shape of the graph of the equation

$$y = 2x^2 - 4x - 1. \tag{14}$$

Solution If we complete the square in x, Eq. (14) takes the form

$$y = 2(x^2 - 2x + 1) - 3;$$
$$y + 3 = 2(x - 1)^2.$$

Hence the graph of Eq. (14) is the parabola shown in Fig. 1.2.20. It opens upward and its vertex is at $(1, -3)$. ■

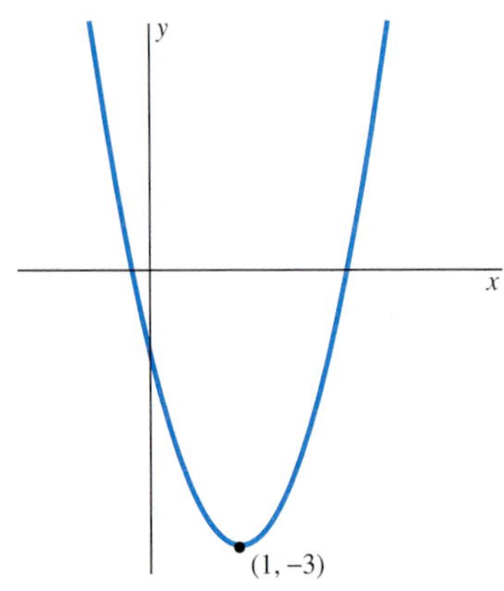

Fig. 1.2.20 The parabola $y = 2x^2 - 4x - 1$ of Example 10

Applications of Quadratic Functions

In Section 1.1 we saw that a certain type of applied problem may call for us to find the maximum or minimum attained by a certain function f. If the function f is quadratic as in Eq. (8), then the graph of $y = f(x)$ is a parabola. In this case the maximum (or minimum) value of $f(x)$ corresponds to the highest (or lowest) point of the parabola. We can therefore find this maximum (or minimum) value graphically—at least approximately—by zooming in on the vertex of the parabola.

For instance, recall the animal pen problem of Section 1.1. In Example 9 there we saw that the area A of the pen (see Fig. 1.2.21) is given as a function of its base length x by

$$A(x) = \tfrac{3}{5}(30x - x^2), \quad 0 \leqq x \leqq 30. \tag{15}$$

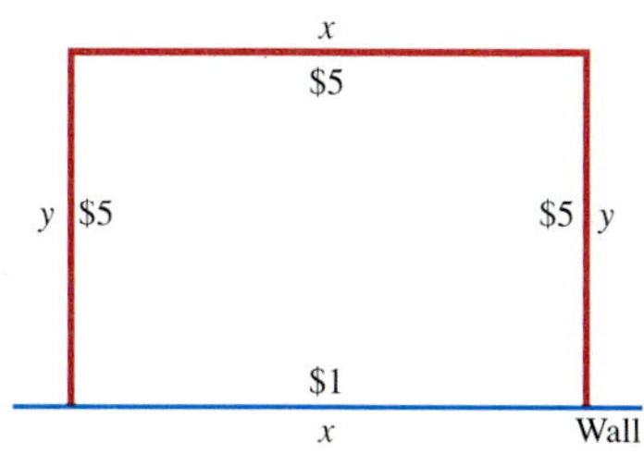

Fig. 1.2.21 The animal pen

Figure 1.2.22 shows the graph $y = A(x)$, and Figs. 1.2.23, 1.2.24, and 1.2.25 show successive magnifications of the region near the high point (vertex) of the parabola. The dashed rectangle in each figure is the viewing window for the next. Figure 1.2.25 makes it *seem* that the maximum area of the pen is $A(15) = 135$. It is clear from the figure that the maximum value of $A(x)$ is within 0.001 of $A = 135$.

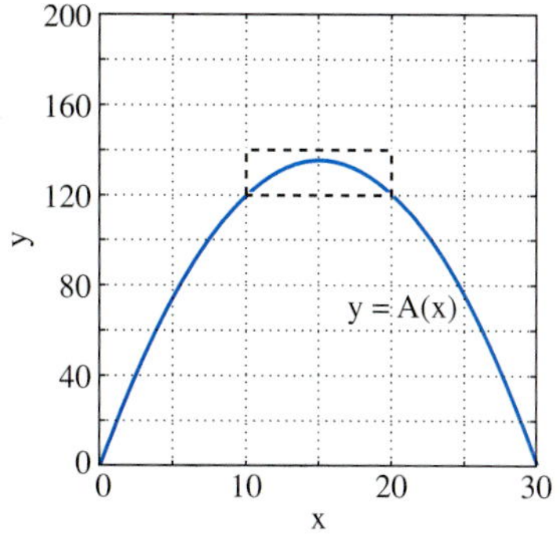

Fig. 1.2.22 The graph $y = A(x)$.

Fig. 1.2.23 The first zoom.

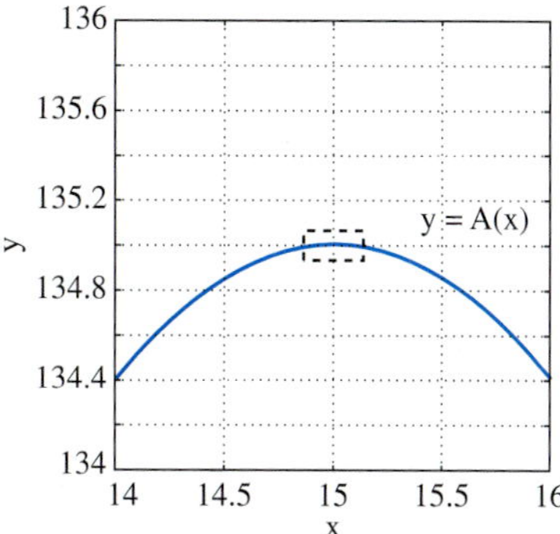

Fig. 1.2.24 The second zoom.

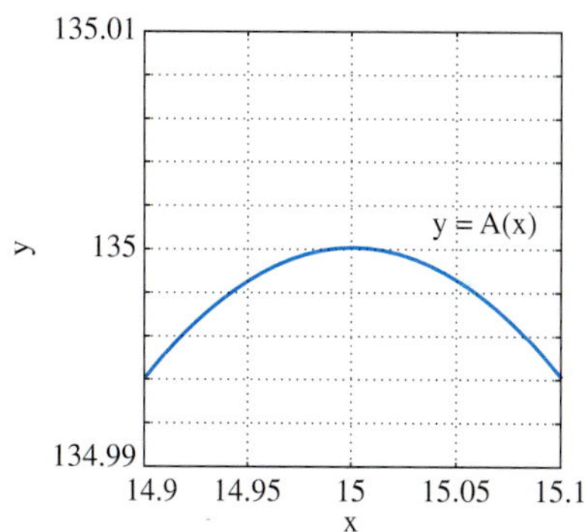

Fig. 1.2.25 The third zoom.

We can verify by completing the square as in Example 10 that the maximum value is *precisely* $A(15) = 135$:

$$A = -\tfrac{3}{5}(x^2 - 30x) = -\tfrac{3}{5}(x^2 - 30x + 225 - 225)$$
$$= -\tfrac{3}{5}(x^2 - 30x + 225) + 135;$$

that is,

$$A - 135 = -\tfrac{3}{5}(x - 15)^2. \tag{16}$$

It follows from Eq. (16) that the graph of Eq. (15) is the parabola shown in Fig. 1.2.26, which opens downward from its vertex (15, 135). This *proves* that the maximum value of $A(x)$ on the interval $[0, 30]$ is the value $A(15) = 135$, as both our numerical investigations in Section 1.1 and our graphical investigations here suggest. And when we glance at Eq. (16) in the form

$$A(x) = 135 - \tfrac{3}{5}(x - 15)^2,$$

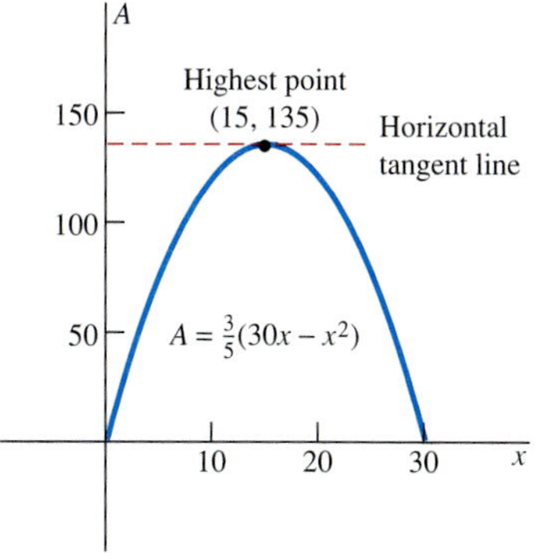

Fig. 1.2.26 The graph of $A(x) = \frac{3}{5}(30x - x^2)$ for $0 \leqq x \leqq 30$

it's clear and unarguable that the maximum possible value of $135 - \frac{3}{5}u^2$ is 135 when $u = x - 15 = 0$—that is, when $x = 15$.

The technique of completing the square is quite limited: It can be used to find maximum or minimum values only of *quadratic* functions. One of the goals in calculus is to develop a more general technique that can be applied to a far wider variety of functions.

The basis of this more general technique lies in the following observation. Visual inspection of the graph of

$$A(x) = \tfrac{3}{5}(30x - x^2)$$

in Fig. 1.2.26 suggests that the line tangent to the curve at its highest point is horizontal. If we *knew* that the tangent line to a graph at its highest point must be horizontal, then our problem would reduce to showing that $(15, 135)$ is the only point of the graph of $y = A(x)$ at which the tangent line is horizontal.

But what do we mean by the *tangent line* to an arbitrary curve? We pursue this question in Section 2.1. The answer will open the door to the possibility of finding maximum and minimum values of a wide variety of functions.

Graphic, Numeric, and Symbolic Viewpoints

An equation $y = f(x)$ provides a *symbolic* description of the function f. A table of values of f (like those in Section 1.1) is a *numeric* representation of the function, whereas this section deals largely with *graphic* representations of functions. Interesting applications often involve looking at the same function from at least two of these three viewpoints.

EXAMPLE 11 Suppose that a car begins (at time $t = 0$ h) in Athens, Georgia (position $x = 0$ mi) and travels to Atlanta (position $x = 60$ mi) with a constant speed of 60 mi/h. The car stays in Atlanta for exactly 1 h, then returns to Athens, again with a constant speed of 60 mi/h. Describe the car's "position function" both graphically and symbolically.

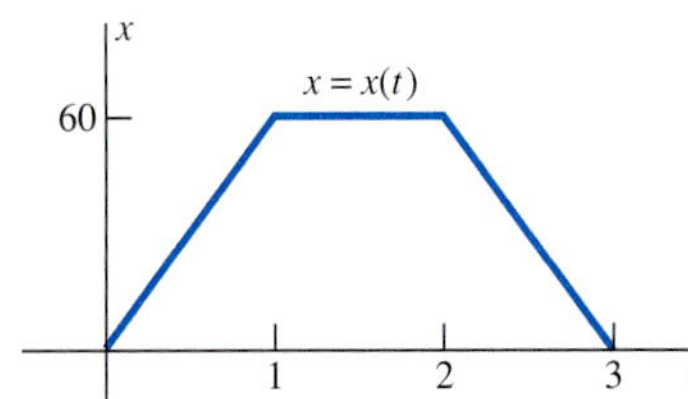

Fig. 1.2.27

Solution It's fairly clear that $x = 60t$ during the 1-h trip from Athens to Atlanta; for instance, after $t = \frac{1}{2}$ h the car has traveled halfway, so $x = 30 = \frac{1}{2} \cdot 60$. During the next hour, $1 \leqq t \leqq 2$, the car's position is constant, $x \equiv 60$. And perhaps you can see that during the return trip of the third hour, $2 \leqq t \leqq 3$, the car's position is given by

$$x = 60 - 60(t - 2) = 180 - 60t$$

(so that $x(2) = 60$ and $x(3) = 0$). Thus the position function $x(t)$ is defined symbolically by

$$x(t) = \begin{cases} 60t & \text{if } 0 \leqq t \leqq 1, \\ 60 & \text{if } 1 \leqq t \leqq 2, \\ 180 - 60t & \text{if } 2 \leqq t \leqq 3. \end{cases}$$

The domain of this function is the t-interval $[0, 3]$, and its graph is shown in Fig. 1.2.27, where we denote both the function and the dependent variable by the same symbol x (an abuse of notation that's not uncommon in applications). ■

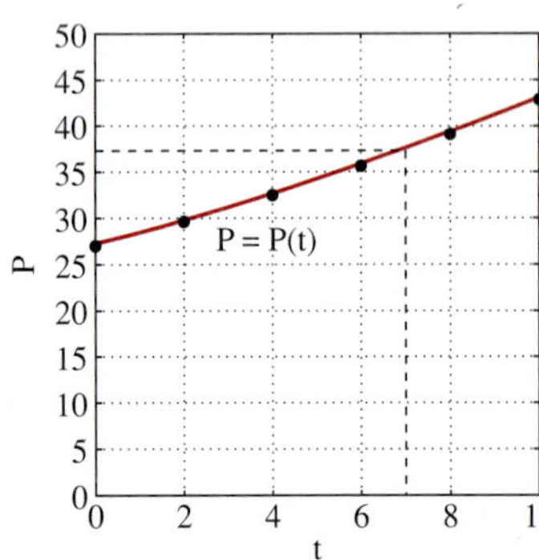

Fig. 1.2.28 The population function of Example 12

EXAMPLE 12 During the decade of the 1980s the population P (in thousands) of a small but rapidly growing city was recorded in the following table.

Year	1980	1982	1984	1986	1988	1990
t	0	2	4	6	8	10
P	27.00	29.61	32.48	35.62	39.07	42.85

Estimate the population of this city in the year 1987.

Solution Figure 1.2.28 shows a graph of the population function $P(t)$ obtained by connecting the six given data points $(t, P(t))$ with a smooth curve. A careful measurement of the height of the point on this curve at which $t = 7$ yields the approximate population $P(7) \approx 37.4$ (thousand) of the city in 1987. ■

1.2 PROBLEMS

In Problems 1 through 10, write an equation of the line L described and sketch its graph.

1. L passes through the origin and the point $(2, 3)$.

2. L is vertical and has x-intercept 7.

3. L is horizontal and passes through $(3, -5)$.

4. L has x-intercept 2 and y-intercept -3.

5. L passes through $(2, -3)$ and $(5, 3)$.

6. L passes through $(-1, -4)$ and has slope $\frac{1}{2}$.

7. L passes through $(4, 2)$ and has angle of inclination 135°.

8. L has slope 6 and y-intercept 7.

9. L passes through $(1, 5)$ and is parallel to the line with equation $2x + y = 10$.

10. L passes through $(-2, 4)$ and is perpendicular to the line with equation $x + 2y = 17$.

Sketch the translated circles in Problems 11 through 16. Indicate the center and radius of each.

11. $x^2 + y^2 = 4x$

12. $x^2 + y^2 + 6y = 0$

13. $x^2 + y^2 + 2x + 2y = 2$

14. $x^2 + y^2 + 10x - 20y + 100 = 0$

15. $2x^2 + 2y^2 + 2x - 2y = 1$

16. $9x^2 + 9y^2 - 6x - 12y = 11$

Sketch the translated parabolas in Problems 17 through 22. Indicate the vertex of each.

17. $y = x^2 - 6x + 9$

18. $y = 16 - x^2$

19. $y = x^2 + 2x + 4$

20. $2y = x^2 - 4x + 8$

21. $y = 5x^2 + 20x + 23$

22. $y = x - x^2$

The graph of the equation $(x - h)^2 + (y - k)^2 = C$ is a circle if $C > 0$, is the single point (h, k) if $C = 0$, and contains no points if $C < 0$. (Why?) Identify the graphs of the equations in Problems 23 through 26. If the graph is a circle, give its center and radius.

23. $x^2 + y^2 - 6x + 8y = 0$

24. $x^2 + y^2 - 2x + 2y + 2 = 0$

25. $x^2 + y^2 + 2x + 6y + 20 = 0$

26. $2x^2 + 2y^2 - 2x + 6y + 5 = 0$

Sketch the graphs of the functions in Problems 27 through 50. Take into account the domain of definition of each function, and plot points as necessary.

27. $f(x) = 2 - 5x,\ \ -1 \leqq x \leqq 1$

28. $f(x) = 2 - 5x,\ \ 0 \leqq x < 2$

29. $f(x) = 10 - x^2$

30. $f(x) = 1 + 2x^2$

31. $f(x) = x^3$

32. $f(x) = x^4$

33. $f(x) = \sqrt{4 - x^2}$

34. $f(x) = -\sqrt{9 - x^2}$

35. $f(x) = \sqrt{x^2 - 9}$

36. $f(x) = \dfrac{1}{1 - x}$

37. $f(x) = \dfrac{1}{x + 2}$

38. $f(x) = \dfrac{1}{x^2}$

39. $f(x) = \dfrac{1}{(x - 1)^2}$

40. $f(x) = \dfrac{|x|}{x}$

41. $f(x) = \dfrac{1}{2x + 3}$

42. $f(x) = \dfrac{1}{(2x + 3)^2}$

43. $f(x) = \sqrt{1 - x}$

44. $f(x) = \dfrac{1}{\sqrt{1 - x}}$

45. $f(x) = \dfrac{1}{\sqrt{2x + 3}}$

46. $f(x) = |2x - 2|$

47. $f(x) = |x| + x$

48. $f(x) = |x - 3|$

49. $f(x) = |2x + 5|$

50. $f(x) = \begin{cases} |x| & \text{if } x < 0, \\ x^2 & \text{if } x \geqq 0 \end{cases}$

Graph the functions given in Problems 51 through 56. Indicate any points of discontinuity.

51. $f(x) = \begin{cases} 0 & \text{if } x < 0, \\ 1 & \text{if } x \geqq 0 \end{cases}$

52. $f(x) = \begin{cases} 1 & \text{if } x \text{ is an integer}, \\ 0 & \text{otherwise} \end{cases}$

53. $f(x) = [\![2x]\!]$

54. $f(x) = \dfrac{x - 1}{|x - 1|}$

55. $f(x) = [\![x]\!] - x$ **56.** $f(x) = [\![x]\!] + [\![-x]\!] + 1$

In Problems 57 through 64, use a graphics calculator or computer to find (by zooming) the highest or lowest (as appropriate) point P on the given parabola. Determine the coordinates of P with two digits to the right of the decimal correct or correctly rounded. Then verify your result by completing the square to find the actual vertex of the parabola.

57. $y = 2x^2 - 6x + 7$ **58.** $y = 2x^2 - 10x + 11$

59. $y = 4x^2 - 18x + 22$ **60.** $y = 5x^2 - 32x + 49$

61. $y = -32 + 36x - 8x^2$ **62.** $y = -53 - 34x - 5x^2$

63. $y = 3 - 8x - 3x^2$ **64.** $y = -28 + 34x - 9x^2$

In Problems 65 through 68, use the method of completing the square to graph the appropriate function and thereby determine the maximum or minimum value requested.

65. If a ball is thrown straight upward with initial velocity 96 ft/s, then its height t second later is $y = 96t - 16t^2$ (ft). Determine the maximum height that the ball attains.

66. Find the maximum possible area of the rectangle described in Problem 40 of Section 1.1.

67. Find the maximum possible value of the product of two positive numbers whose sum is 50.

68. In Problem 42 of Section 1.1, you were asked to express the daily production of a specific oil field as a function $P = f(x)$ of the number x of new oil wells drilled. Construct the graph of f, and use it to find the value of x that maximizes P.

In Problems 69 through 72 write a symbolic description of the function whose graph is pictured. You may use the greatest integer function of Example 4.

69. Figure 1.2.29 **70.** Figure 1.2.30

71. Figure 1.2.31 **72.** Figure 1.2.32

Each of Problems 73 through 76 describes a trip you made along a straight road connecting two cities 120 miles apart. Sketch the graph of the distance x from your starting point (in miles) as a function of the time t elapsed (in hours). Also describe the function $x(t)$ symbolically.

73. You traveled for one hour at 45 mi/h, then realized you were going to be late, and therefore traveled at 75 mi/h for the next hour.

74. You traveled for one hour at 60 mi/h, stopped for a half-hour while a herd of bison crossed the road, then drove on toward your destination for the next hour at 60 mi/h.

75. You traveled for one hour at 60 mi/h, were suddenly engulfed in a dense fog, and drove back home at 30 mi/h.

76. You traveled for a half hour at 60 mi/h, suddenly remembered you had left your wallet at home, drove back at 60 mi/h to get it, and finally drove for two hours at 60 mi/h toward your original destination.

77. Suppose that the cost C of printing a pamphlet of at most 100 pages is a linear function of the number p of pages it contains. It costs \$1.70 to print a pamphlet with 34 pages, whereas a pamphlet with 79 pages costs \$3.05. (a) Express C as a function of p. Use this function to find the cost of printing a pamphlet with 50 pages. (b) Sketch the straight line graph of the function $C(p)$. Tell what the slope and the C-intercept of this line mean—perhaps in terms of the

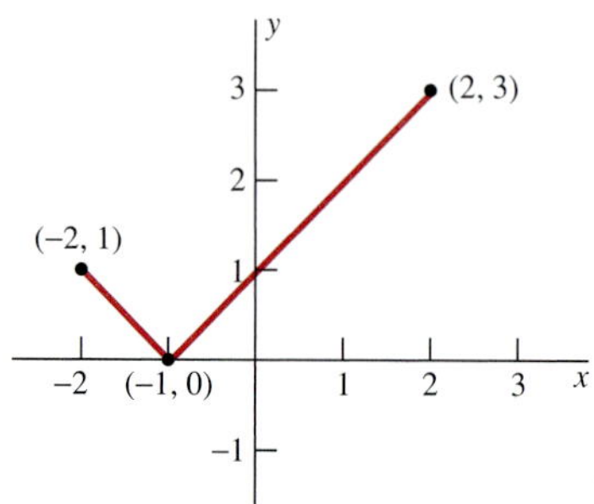

Fig. 1.2.29 Problem 69

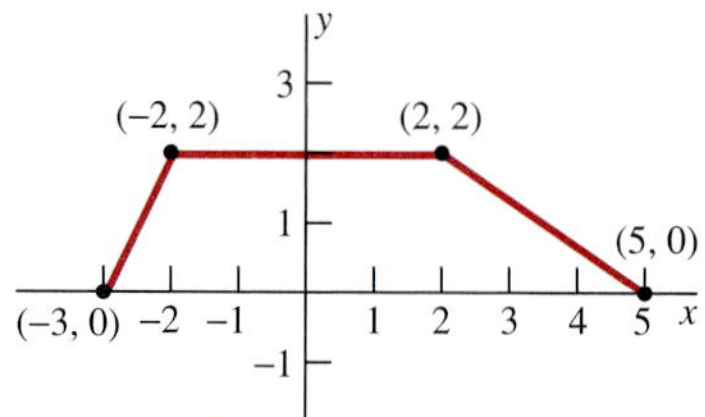

Fig. 1.2.30 Problem 70

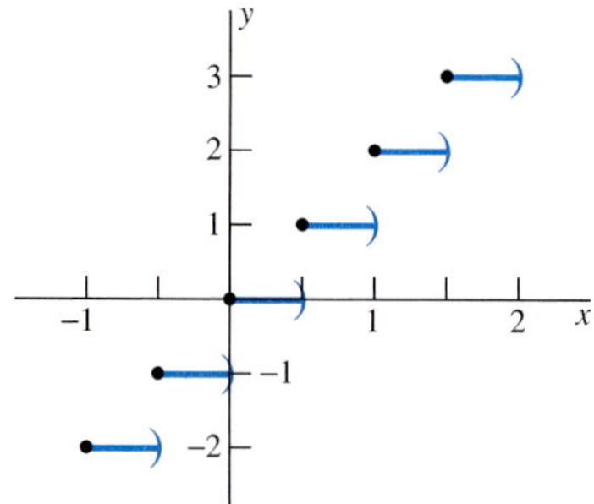

Fig. 1.2.31 Problem 71

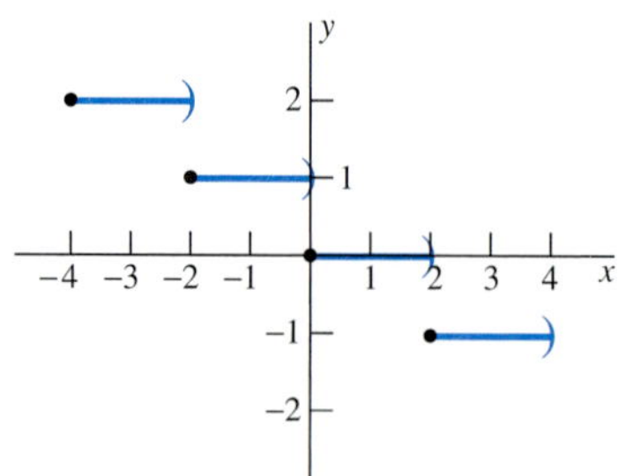

Fig. 1.2.32 Problem 72

"fixed cost" to set up the press for printing and the "marginal cost" of each additional page printed.

78. Suppose that the cost C of renting a car for a day is a linear function of the number x of miles you drive that day. On day 1 you drove 207 miles and the cost was \$99.45. On day 2 you drove 149 miles and the cost was \$79.15. (a) Express C as a function of x. Use this function to find the cost for day 3 if you drove 175 miles. (b) Sketch the straight line graph of the function $C(x)$. Tell what the slope and the C-intercept of this line mean—perhaps in terms of fixed and marginal costs as in Problem 77.

79. For a Federal Express letter weighing at most one pound sent to a certain destination, the charge C is \$8.00 for the first 8 ounces plus 80¢ for each additional ounce or fraction thereof. Sketch the graph of this function C of the total number x of ounces, and describe it symbolically in terms of the greatest integer function of Example 4 in Section 1.1.

80. In a certain city, the charge C for a taxi trip of at most 20 miles is \$3.00 for the first 2 miles (or fraction thereof), plus 50¢ for each half-mile (or part thereof) up to a total of 10 miles, plus 50¢ for each mile (or part thereof) over 10 miles. Sketch the graph of this function C of the number x of miles, and describe it symbolically in terms of the greatest integer function of Example 4 in Section 1.1.

81. The volume V (in liters) of a sample of 3 g of carbon dioxide at 27°C was measured as a function of its pressure p (in atmospheres) with the results shown in the following table.

p	0.25	1.00	2.50	4.00	6.00
V	6.72	1.68	0.67	0.42	0.27

Sketch the graph of the function $V(p)$ and use the graph to estimate the volumes of the gas sample at pressures of 0.5 and 5 atmospheres.

82. The average temperature T (in °F) on a day t days after July 15 in Athens, Georgia, was measured at two-month intervals, with the results shown in the following table.

t	July 15	Sept. 15	Nov. 15	Jan. 15	Mar. 15	May 15
T	79.1	70.2	52.3	43.4	52.2	70.1

Sketch the graph of the function $T(t)$, and use the graph to estimate the average temperature on October 15 and on April 15.

1.2 PROJECT: A BROKEN TREE

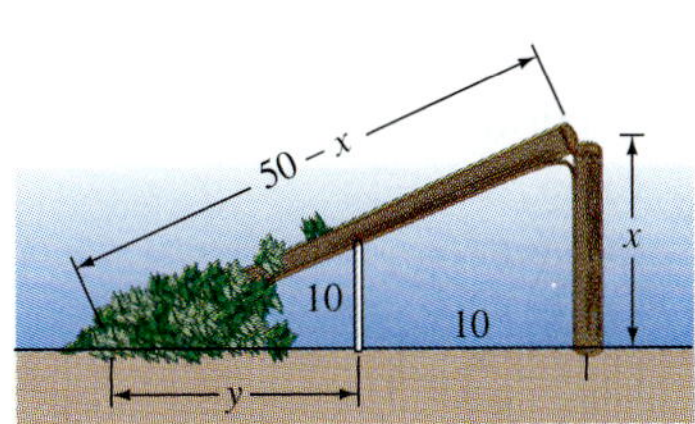

Fig. 1.2.33 The broken tree

A 50-ft tree stands 10 ft from a 10-ft fence. The tree is suddenly "broken" at a height of x feet (Fig. 1.2.33). You are to determine the height x such that the tree falls so that its trunk barely touches the top of the fence when the tip of the tree strikes the ground on the other side of the fence.

Use similar triangles and the Pythagorean theorem to show that the distances x and y indicated in Fig. 1.2.33 satisfy the two equations

$$y = \frac{100}{x - 10}, \tag{1}$$

$$(y + 10)^2 = -100(x - 25). \tag{2}$$

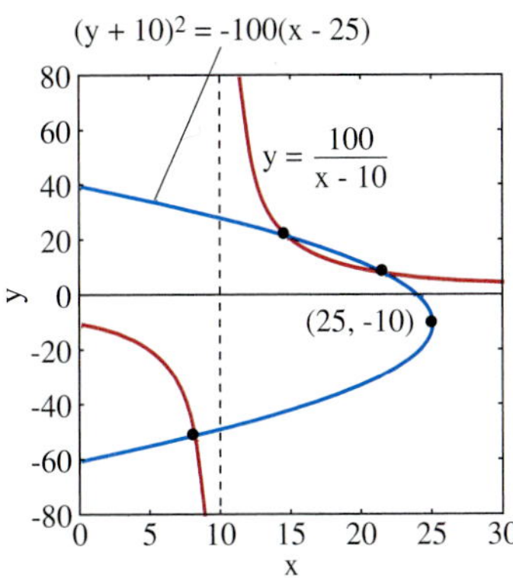

Fig. 1.2.34 The hyperbola and parabola in the broken tree investigation

The graph of Eq. (1) is a translated "rectangular hyperbola" that resembles the graph $y = 1/x$ (Fig. 1.2.12) and the graph of Eq. (2) is a parabola with vertex $(25, -10)$ that opens to the left, as indicated in Fig. 1.2.34.

Determine graphically the coordinates (accurate to two rounded decimal places) of the three points of the intersection of the hyperbola and parabola that appear in Fig. 1.2.34. Which of these intersections corresponds to an actual broken tree as in Fig. 1.2.33? Is there more than one answer?

Note With a typical graphics calculator you can carry out successive magnifications of the region near an intersection point simply by pressing the **ZOOM** key repeatedly. Figure 1.2.35 lists common computer system commands to graph $y = f(x)$ in the "viewing window" $a \leqq x \leqq b, c \leqq y \leqq d$.

For your own personal investigation, begin instead with a tree of height $h = 50 + 10p$ feet, where p denotes the last nonzero digit in your student I.D. number. Write up the results of your investigation in the form of a carefully organized report

Derive	`Author f(x) , then use Plot and Zoom commands`
Maple	`plot( f(x) , x=a..b, y=c..d)`
Mathematica	`Plot[ f[x], {x,a,b} , PlotRange ->{c,d} ]`
MATLAB	`x = a : h : b; plot(x,f(x)), axis([a b c d])`

Fig. 1.2.35 Commands for graphing the function $f(x)$.

consisting of complete sentences (including pertinent equations and figures) explaining your results in detail and telling precisely what you did to solve your problem. Don't just "solve the equations"! Answer the original question completely as to the possible height or heights at which the tree can be broken to fit the physical picture.

1.3 A BRIEF CATALOG OF FUNCTIONS, PART 1

In this section and the next we briefly survey a variety of functions that are used in applications of calculus to describe changing phenomena in the world around us. Our viewpoint here is largely graphical. The objective is for you to attain a general understanding of major differences between different types of functions. In later chapters we use calculus to investigate further the graphs presented here.

Power Functions

A function of the form $f(x) = x^k$ (where k is a constant) is called a **power function**. If $k = 0$, then we have the constant function $f(x) \equiv 1$. The appearance of the graph of a power function with exponent $k = n$, a positive integer, depends on whether n is even or odd.

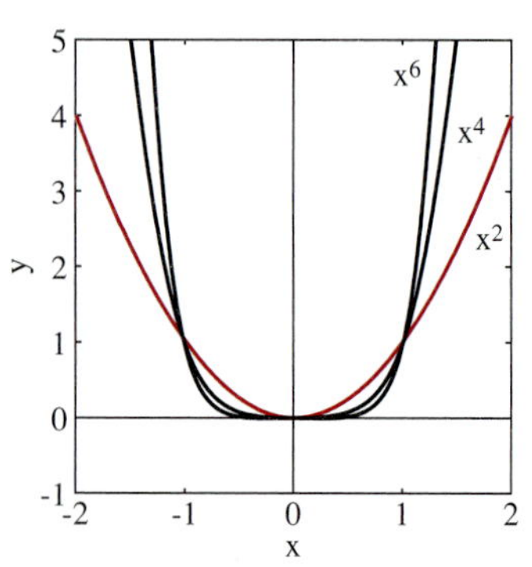

Fig. 1.3.1 Graphs of power functions of even degree (Example 1)

EXAMPLE 1 The graphs of the *even*-degree power functions x^2, x^4, x^6, ... "cup upward," as indicated in Fig. 1.3.1. If $n > 2$ is an *even* integer, then the graph $y = x^n$ resembles the parabola $y = x^2$ but is flatter near the origin and steeper when $|x| > 1$.

The graphs of the *odd*-degree power functions x^1, x^3, x^5, ... go "from southwest to northeast," as indicated in Fig. 1.3.2. If $n > 1$ is an *odd* integer, then the graph $y = x^n$ resembles that of $y = x^3$ but again is flatter near the origin and steeper when $|x| > 1$. ■

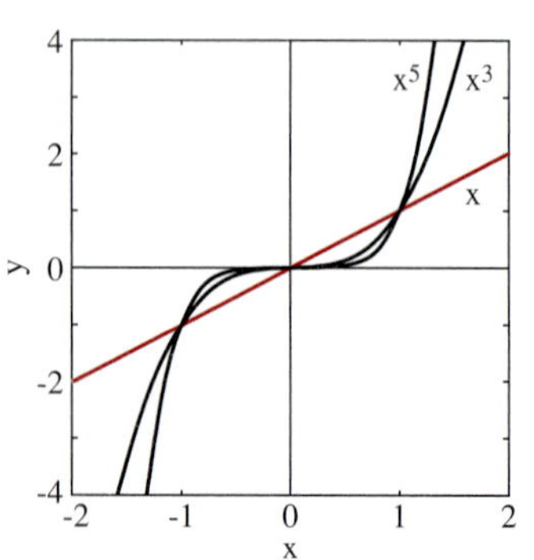

Fig. 1.3.2 Graphs of power functions of odd degree (Example 1)

Note that all the power function graphs in Figs. 1.3.1 and 1.3.2 pass through the origin, through the point $(1, 1)$, and either through $(-1, 1)$ or $(-1, -1)$, depending on whether n is even or odd. In either case, x^n increases numerically (either positively or negatively) as x does. Would you agree that, when $|x|$ is large, the notation

$$x^n \to +\infty \quad \text{as} \quad x \to +\infty, \qquad x^n \to \begin{cases} +\infty & \text{as } x \to -\infty \quad \text{if } n \text{ is even,} \\ -\infty & \text{as } x \to -\infty \quad \text{if } n \text{ is odd} \end{cases}$$

(with the arrow signifying "goes to") provides a convenient and suggestive description of the general features of the graphs in Fig. 1.3.1 and 1.3.2?

The graph $y = x^k$ may have a quite different appearance if the exponent k is not a positive integer. If k is a negative integer—say, $k = -m$ where m is a positive integer—then

$$f(x) = x^k = x^{-m} = \frac{1}{x^m},$$

so in this case the power function is the reciprocal of a function like those in Example 1. Figures 1.3.3 and 1.3.4 show the graphs of

$$y = x^{-1} = \frac{1}{x} \quad \text{and} \quad y = x^{-2} = \frac{1}{x^2},$$

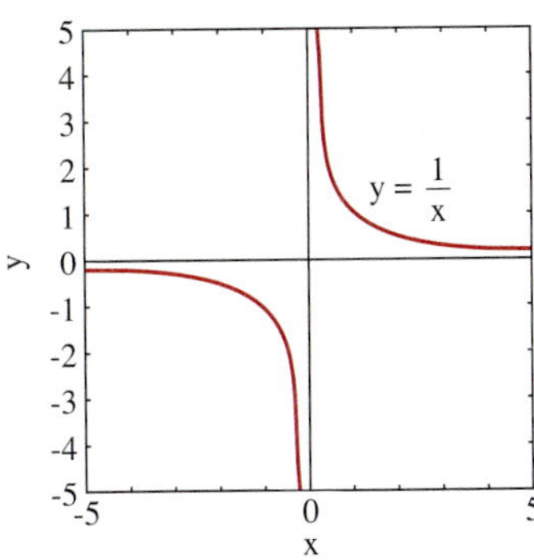

Fig. 1.3.3 $y = \dfrac{1}{x}$

respectively. Observe that $x = 0$ is not in the domain of such a function. Moreover, the reciprocal of a number close to zero is very large in magnitude, which explains the behavior of these graphs near zero: In both graphs, $|y|$ is very large—so the point (x, y) is either very high or very low—when x is close to zero.

The situation is still more complicated if the exponent k is not an integer. We do not (at the present time) attempt to define the expression x^k if k is irrational—that is, not a quotient of integers. But if $k = m/n$ is rational, with the integers m and $n > 0$ having no common factor larger than 1, then we can write

$$x^k = x^{m/n} = \sqrt[n]{x^m},$$

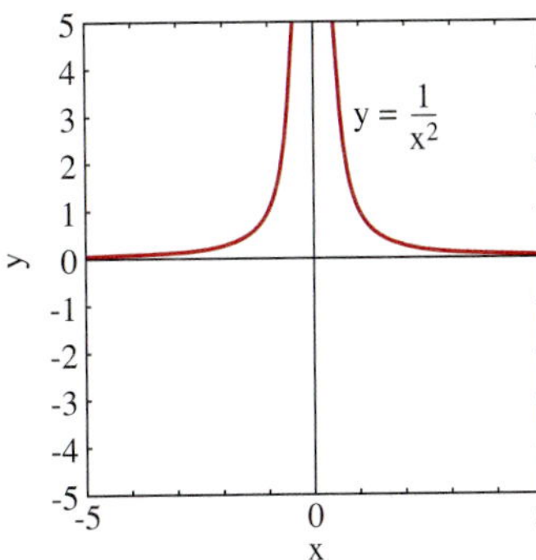

Fig. 1.3.4 $y = \dfrac{1}{x^2}$

and thereby interpret $f(x) = x^k$ as a "root function." If n is odd then $\sqrt[n]{x^m}$ is defined for all real x if m is positive and for all nonzero x if m is negative. But if n is even and m is odd, then the root $\sqrt[n]{x^m}$ is not defined for x negative.

The typical behavior of such root functions is illustrated by the graphs $y = x^{1/2} = \sqrt{x}$ and $y = x^{1/3} = \sqrt[3]{x}$ shown in Figs. 1.3.5 and 1.3.6. The square root $\sqrt{x}$ is defined only for $x \geqq 0$. The cube root $\sqrt[3]{x}$ is defined for all x, but observe that its graph appears to be tangent to the y-axis at the origin.

Combinations of Functions

Many varied and complicated functions can be assembled out of simple "building-block functions." Here we discuss some ways of combining functions to obtain new ones.

Suppose that f and g are functions and that c is a fixed real number. The **(scalar) multiple** cf, the **sum** $f + g$, the **difference** $f - g$, the **product** $f \cdot g$, and the **quotient** f/g are the new functions with these formulas:

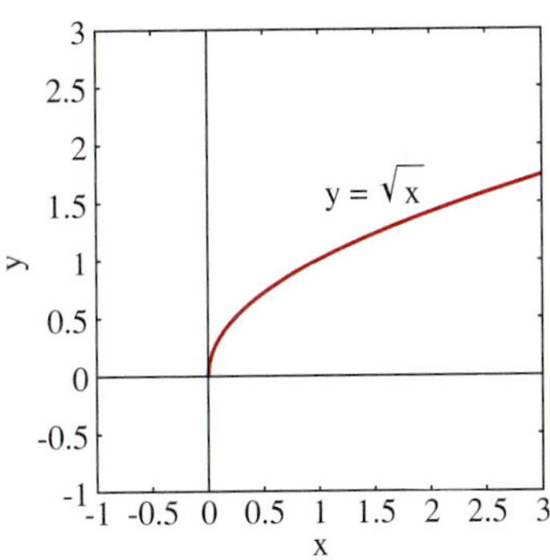

Fig. 1.3.5 $y = x^{1/2}$

$$(cf)(x) = c \cdot f(x), \tag{1}$$

$$(f + g)(x) = f(x) + g(x), \tag{2}$$

$$(f - g)(x) = f(x) - g(x), \tag{3}$$

$$(f \cdot g)(x) = f(x) \cdot g(x), \quad \text{and} \tag{4}$$

$$\left(\frac{f}{g}\right)(x) = \frac{f(x)}{g(x)}. \tag{5}$$

The combinations in Eqs. (2) through (4) are defined for every number x that lies both in the domain of f and in the domain of g. In Eq. (5) we must also require that $g(x) \neq 0$.

EXAMPLE 2 Let $f(x) = x^2 + 1$ and $g(x) = x - 1$. Then

$$(3f)(x) = 3(x^2 + 1),$$

$$(f + g)(x) = (x^2 + 1) + (x - 1) = x^2 + x,$$

$$(f - g)(x) = (x^2 + 1) - (x - 1) = x^2 - x + 2,$$

$$(f \cdot g)(x) = (x^2 + 1)(x - 1) = x^3 - x^2 + x - 1, \quad \text{and}$$

$$\left(\frac{f}{g}\right)(x) = \frac{x^2 + 1}{x - 1} \qquad (x \neq 1).$$ ■

Fig. 1.3.6 $y = x^{1/3}$

EXAMPLE 3 If $f(x) = \sqrt{x - 1}$ for $x \leqq 1$ and $g(x) = \sqrt{1 + x}$ for $x \geqq -1$, then the sum and product of f and g are defined where *both* f and g are defined. Thus the domain of both

$$f(x) + g(x) = \sqrt{1 - x} + \sqrt{1 + x}$$

and

$$f(x) \cdot g(x) = \sqrt{1 - x}\ \sqrt{1 + x} = \sqrt{1 - x^2}$$

is the closed interval $[-1, 1]$. But the domain of the quotient

$$\frac{f(x)}{g(x)} = \frac{\sqrt{1 - x}}{\sqrt{1 + x}} = \sqrt{\frac{1 - x}{1 + x}}$$

is the half-open interval $(-1, 1]$, because $g(-1) = 0$. ■

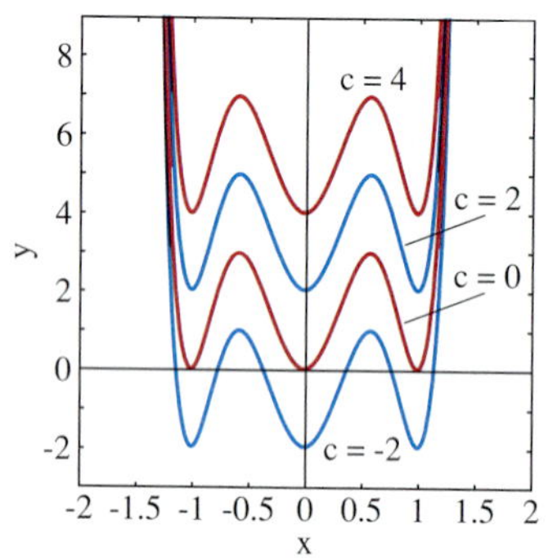

Fig. 1.3.7 $y = 20x^2(x^2 - 1)^2 + c$ for $c = -2,\ 0,\ 2,\ 4$

The results of algebraic operations can sometimes be visualized with the aid of geometric interpretations of the operations. Figures 1.3.7 through 1.3.10 show the results of various operations involving the function $f(x) = 20x^2(x^2 - 1)^2$. Adding a constant simply shifts the graph vertically, as in Fig. 1.3.7, which shows $y = f(x) + c$ for $c = -2, 0, 2$, and 4. Multiplication by a positive constant c expands (if $c > 1$) or contracts (if $0 < c < 1$) the graph in the vertical direction, as in Fig. 1.3.8, which shows $y = cf(x)$ for $c = 1, 2$, and 3. Figure 1.3.9 shows $y = f(x)$ and the parabola $y = 2x^2$, whereas Fig. 1.3.10 shows the graph $y = 2x^2 + f(x)$, obtained by adding the ordinates of the two curves.

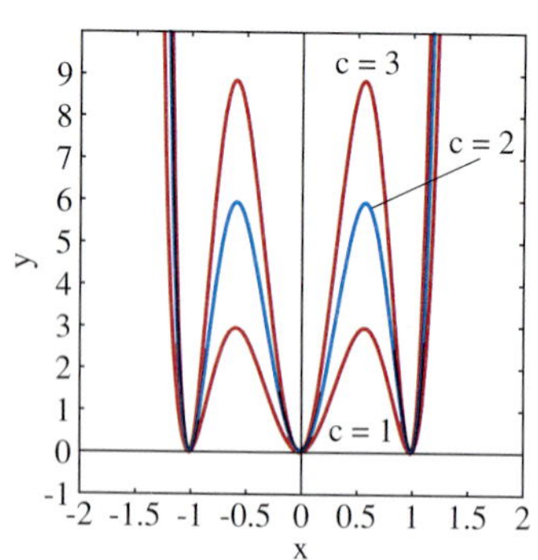

Fig. 1.3.8 $y = c \cdot 20x^2(x^2 - 1)^2$ for $c = 1, 2, 3$

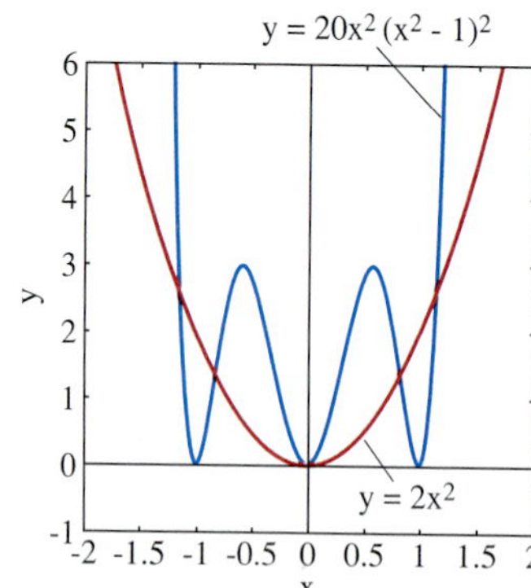

Fig. 1.3.9 $y = 2x^2$ and $y = 20x^2(x^2 - 1)^2$

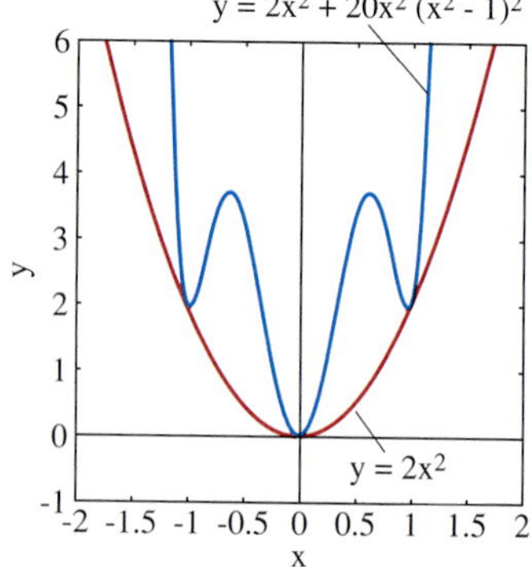

Fig. 1.3.10 $y = 2x^2$ and $y = 2x^2 + 20x^2(x^2 - 1)^2$

Polynomials

A **polynomial** of **degree** n is a function of the form

$$p(x) = a_n x^n + a_{n-1}x^{n-1} + \cdots + a_2 x^2 + a_1 x + a_0, \tag{6}$$

where the coefficients $a_0, a_1, \ldots, a_n$ are fixed real numbers and $a_n \neq 0$. Thus an nth-degree polynomial is a sum of constant multiples of the **power functions**

$$1, \quad x, \quad x^2, \ldots, x^{n-1}, \quad x^n.$$

A first-degree polynomial is simply a *linear function* $a_1x + a_0$ whose graph is a straight line. A second-degree polynomial is a *quadratic function* whose graph $y = a_2x^2 + a_1x + a_0$ is a parabola (see Section 1.2).

Recall that a **zero** of the function f is a solution of the equation

$$f(x) = 0.$$

Is it obvious to you that *the zeros of $f(x)$ are precisely the x-intercepts of the graph*

$$y = f(x)?$$

Indeed, one major reason that we are interested in the graph of a function is to see the number and approximate locations of its zeros.

A key to understanding graphs of higher-degree polynomials is the *fundamental theorem of algebra*. It states that every nth-degree polynomial has n zeros (possibly complex, possibly repeated). It follows that an nth-degree polynomial has *no more than n* distinct real zeros.

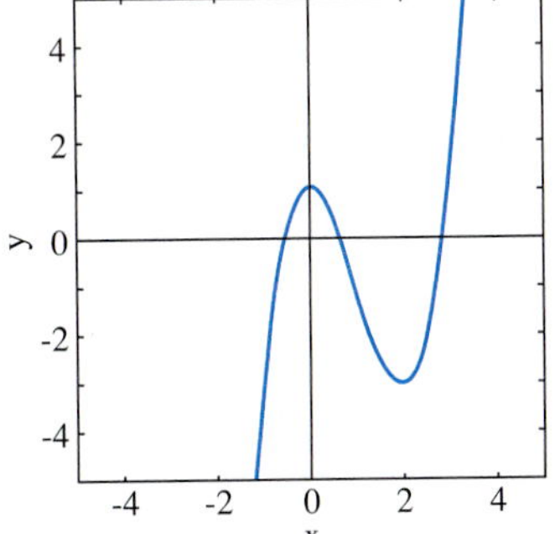

Fig. 1.3.11 $f(x) = x^3 - 3x^2 + 1$ has three real zeros (Example 4).

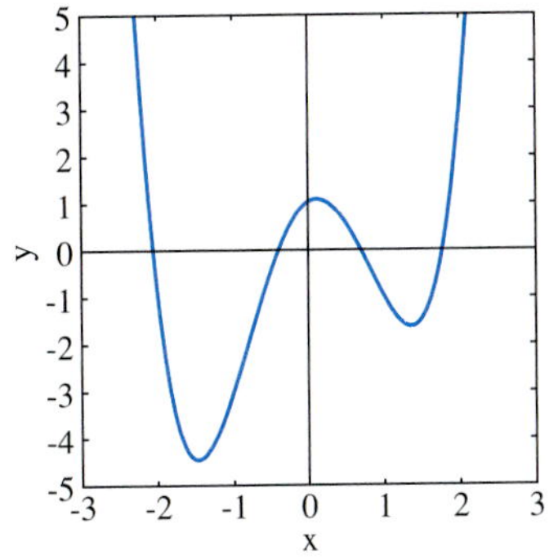

Fig. 1.3.12 $f(x) = x^4 - 4x^2 + x + 1$ has four real zeros (Example 4).

EXAMPLE 4 Figures 1.3.11 and 1.3.12 exhibit polynomials that have the maximum number of real zeros allowed by the fundamental theorem of algebra. But the power function graphs in Fig. 1.3.1 and 1.3.2 show that a high-degree polynomial may have only a single real zero. And the quadratic function

$$f(x) = x^2 + 4x + 13 = (x + 2)^2 + 9$$

has no real zeros at all. (Why not?) Figure 1.3.7 includes graphs of sixth-degree polynomials having six, three, or no zeros. Indeed, an nth-degree polynomial can have any number of zeros from 0 to n if n is even (from 1 to n if n is odd). ■

A polynomial behaves "near infinity"—that is, outside an interval on the x-axis containing its real zeros—in much the same way as a power function of the same degree. If $p(x)$ is a polynomial of *odd* degree, then $y = p(x)$ goes in opposite (vertical) directions as x goes to $-\infty$ and to $+\infty$ (like the cubic polynomial graph in Fig. 1.3.11). But if $p(x)$ is a polynomial of *even* degree, then $y = p(x)$ goes in the same (vertical) direction as x goes to $-\infty$ and to $+\infty$ (like the fourth-degree polynomial graph in Fig. 1.3.12).

Between the extremes to the left and right, where $|x|$ is large, an nth-degree polynomial has at most $n - 1$ "bends"—like the two bends of the third-degree polynomial graph in Fig. 1.3.11 and the three bends of the fourth-degree polynomial graph in Fig. 1.3.12. In Chapter 4 we will use calculus to see why this is so (and to make precise the notion of a "bend" in a curve).

Calculator/Computer Graphing

A typical calculator or computer graphing utility shows (on its graphics screen or monitor) only that portion of a graph $y = f(x)$ that lies within a selected rectangular **viewing window** of the form

$$\{(x, y) : a \leqq x \leqq b \quad \text{and} \quad c \leqq y \leqq d\}.$$

The parts of the graph that lie outside this viewing window remain unseen (Fig. 1.3.13). Using a calculator, we can enter the maximum and minimum x- and y-values explicitly in a form such as

```
Xmin = a    Ymin = c
Xmax = b    Ymax = d
```

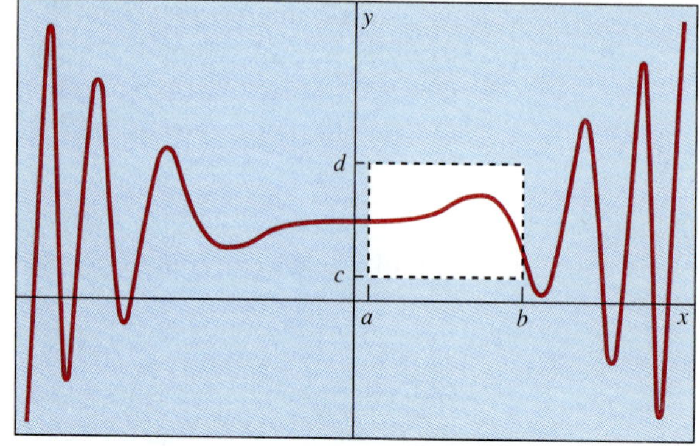

Fig. 1.3.13 The viewing window $a \leqq x \leqq b, \quad c \leqq x \leqq d$

Frequently the user must specify the x-**range** $[a, b]$ and the y-**range** $[c, d]$ carefully so that the viewing window will show the desired portion of the graph. The calculator or computer's "default window" may provide only a starting point.

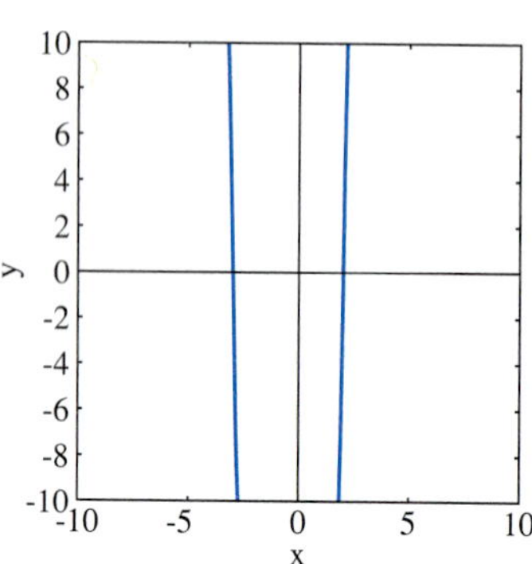

Fig. 1.3.14 $y = x^3 + 12x^2 + 5x - 66$ with viewing window $-10 \leqq x \leqq 10, -10 \leqq y \leqq 10$

EXAMPLE 5 Construct a graph that exhibits the principal features of the cubic polynomial

$$y = x^3 + 12x^2 + 5x - 66. \tag{7}$$

Solution We anticipate a graph that looks somewhat like the cubic graph in Fig. 1.3.11, one that goes "from southwest to northeast," perhaps with a couple of bends in between. But when we enter Eq. (7) in a typical graphing calculator with default viewing window $-10 \leqq x \leqq 10$, $-10 \leqq y \leqq 10$, we get the result shown in Fig. 1.3.14. Evidently our viewing window is not large enough to show the expected behavior.

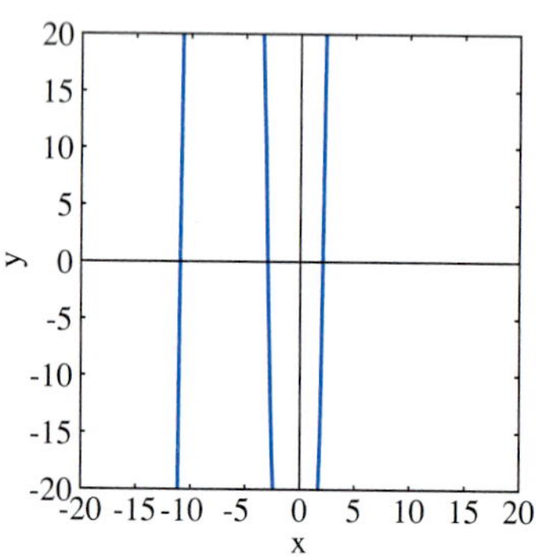

Fig. 1.3.15 $y = x^3 + 12x^2 + 5x - 66$ with viewing window $-20 \leqq x \leqq 20, -20 \leqq y \leqq 20$

Doubling each dimension of the viewing window, we get the result in Fig. 1.3.15. Now we see the three zeros that a cubic polynomial can have, as well as some possibility of two bends, but it appears that magnification in the y-direction is indicated. Perhaps we need a y-range measuring in the hundreds rather than the tens. With the viewing window $-20 \leqq x \leqq 20$, $-200 \leqq y \leqq 200$, we finally get the satisfying graph shown in Fig. 1.3.16.

Once we have zoomed out to see the "big picture," we can zoom in on points of interest. For instance, Fig. 1.3.16 indicates "zoom boxes" locating the three zeros of the polynomial in (7). Apparently these zeros are located at or near the points $x = -11$, $x = -3$, and $x = 2$. Each can be approximated graphically as closely as you please (subject to the limitations of your computer) by the method of successive magnifications. (See whether you can convince yourself that these three zeros are *exactly* the indicated integers. How could you verify that this actually is true?) ■

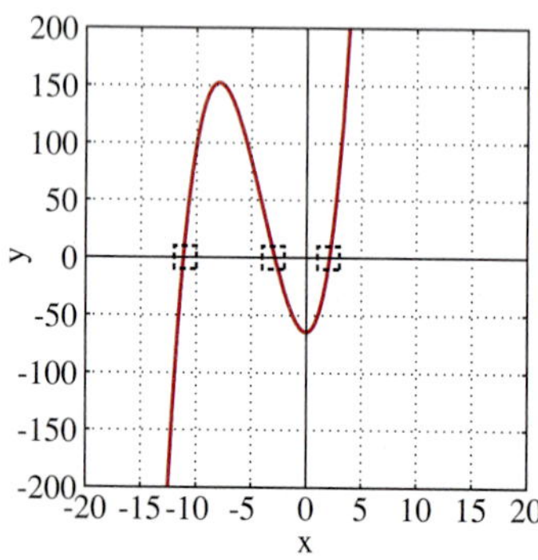

Fig. 1.3.16 $y = x^3 + 12x^2 + 5x - 66$ with viewing window $-20 \leqq x \leqq 20, \; -200 \leqq y \leqq 200$

EXAMPLE 6 Investigate the graph of the quartic (fourth-degree) polynomial

$$f(x) = (x^2 - 1)(x - 10)(x - 10.1) = x^4 - 20.1x^3 + 100x^2 + 20.1x - 101. \tag{8}$$

Solution Here we know the zeros $x = -1, 1, 10$, and 10.1 in advance, so it makes sense to choose an x-range that includes all four. Noting that $f(0) = -101$, we suspect that a y-range measuring in the hundreds is indicated. Thus with the viewing window $-5 \leqq x \leqq 15$, $-1000 \leqq y \leqq 1000$, we get the attractive graph in Fig. 1.3.17. Observe that with its three bends it resembles the quartic graph in Fig. 1.3.12.

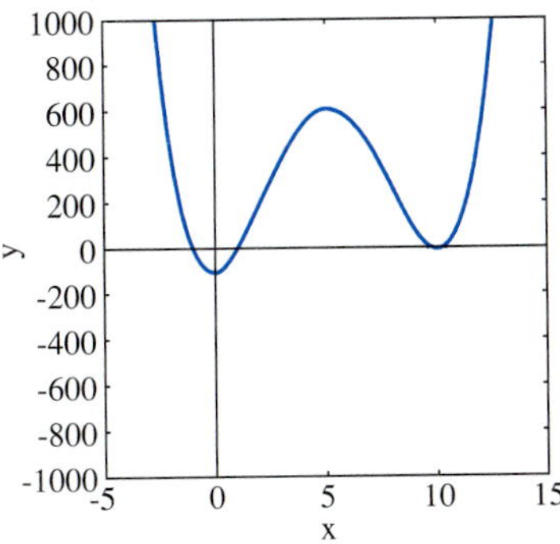

Fig. 1.3.17 $y = (x^2 - 1)(x - 10)(x - 10.1)$ with viewing window $-5 \leqq x \leqq 15$, $-1000 \leqq y \leqq 1000$

Fig. 1.3.18 $y = (x^2 - 1)(x - 10)(x - 10.1)$ with viewing window $9.5 \leqq x \leqq 10.5$, $-1 \leqq y \leqq 1$

But now the behavior of the graph near the point $x = 10$ is unclear. Does it dip beneath the x-axis? We select the viewing window $9.5 \leqq x \leqq 10.5$, $-1 \leqq y \leqq 1$ to magnify this area and get the result shown in Fig. 1.3.18. This case appears to require different plots on different scales to show all the behavior of the graph. ■

Our graphs in Examples 5 and 6 exhibit the maximum possible number of zeros and bends for the polynomials in Eqs. (7) and (8), so we are fairly confident that our investigations reveal the main qualitative features of the graphs of these polynomials. But only with the calculus techniques of Chapter 4 can we be certain of the structure of a graph. For instance, a polynomial graph can exhibit fewer than the maximum possible number of bends, but at this stage we cannot be certain that more bends are not hidden somewhere, perhaps visible only on a scale different from that of the viewing window we have selected.

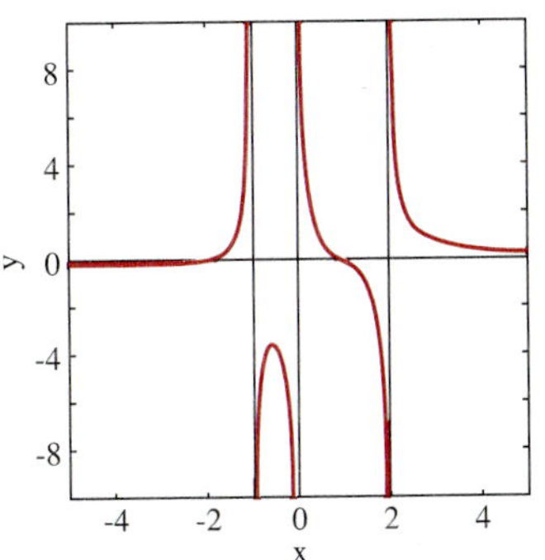

Fig. 1.3.19 The graph of the rational function in Eq. (10) (Example 7)

Rational Functions

Just as a rational number is a quotient of two integers, a **rational function** is a quotient

$$f(x) = \frac{p(x)}{q(x)} \tag{9}$$

of two polynomials $p(x)$ and $q(x)$. Graphs of rational functions and polynomials have several features in common. For instance, a rational function has only a finite number of zeros, because $f(x)$ in Eq. (9) can be zero only when the numerator polynomial $p(x)$ is zero. Similarly, the graph of a rational function has only a finite number of bends.

But the denominator polynomial $q(x)$ in Eq. (9) may have a zero at a point $x = a$ where the numerator is nonzero. In this case the value of $f(x)$ will be very large in magnitude when x is close to a. This observation implies that the graph of a rational function may have a feature that no polynomial graph can have—an *asymptote.*

Fig. 1.3.20 The graph of the rational function in Eq. (11) (Example 8)

EXAMPLE 7 Figure 1.3.19 shows the graph of the rational function

$$f(x) = \frac{(x + 2)(x - 1)}{x(x + 1)(x - 2)}. \tag{10}$$

Note the x-intercepts $x = -2$ and $x = 1$, corresponding to the zeros of the numerator $(x + 2)(x - 1)$. The vertical lines $x = -1$, $x = 0$, and $x = 2$ shown in the graph correspond to the zeros of the denominator $x(x+1)(x-2)$. These vertical lines are *asymptotes* of the graph of f. ■

EXAMPLE 8 Figure 1.3.20 shows the graph of the rational function

$$f(x) = \frac{x(x+2)(x-1)}{(x+1)(x-2)}. \tag{11}$$

The x-intercepts $x = -2$, $x = 0$, and $x = 1$ correspond to the zeros of the numerator, whereas the asymptotes $x = -1$ and $x = 2$ correspond to the zeros of the denominator. ■

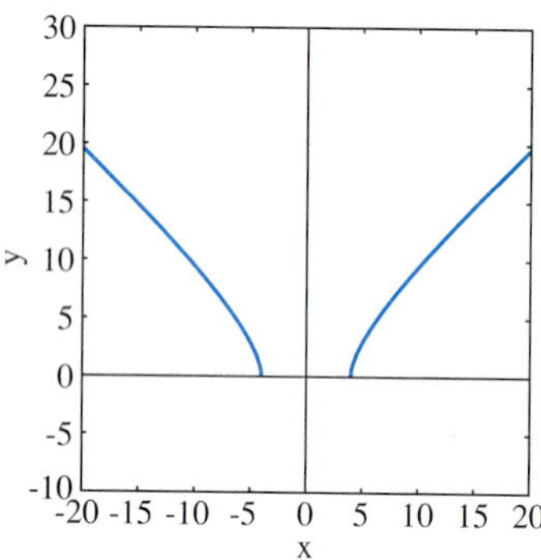

Fig. 1.3.21 $y = \sqrt[4]{16 - x^4}$ on $[-2, 2]$

It should be clear that—by counting x-intercepts and asymptotes—you could match the rational functions in Eqs. (10) and (11) with their graphs in Figs. 1.3.19 and 1.3.20 without knowing in advance which was which.

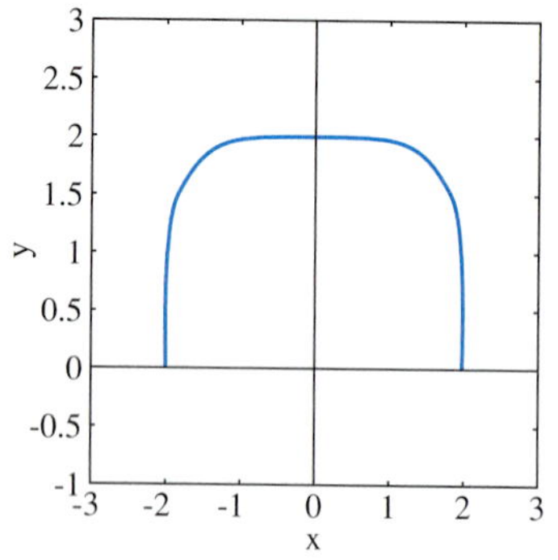

Algebraic Functions

An **algebraic function** is one whose formula can be constructed beginning with power functions and using the algebraic operations of addition, subtraction, multiplication by a real number, multiplication, division, and taking roots. Thus polynomials and rational functions are algebraic functions. But whereas every polynomial is defined everywhere on the real line, and every rational function is defined except at the (finitely many) real zeros of its denominator (which correspond to vertical asymptotes), the domain of definition of an algebraic function may be quite limited. For instance, Figs. 1.3.21 and 1.3.22 show the graphs of the algebraic functions

$$f(x) = \sqrt[4]{16 - x^4} \quad \text{and} \quad g(x) = \sqrt{x^2 - 16}$$

on the bounded and unbounded intervals (respectively) where they are defined.

The graph of every polynomial or rational function looks "smooth" at every point where it is defined, but the graph of an algebraic function may exhibit "corners" or sharp "cusps" where it does not look smooth. For instance, look at the graphs in Figs. 1.3.23 and 1.3.24 of the algebraic functions

$$f(x) = \sqrt{x^2} = |x| \quad \text{and} \quad g(x) = \sqrt[3]{x^2(x-2)^2}.$$

Fig. 1.3.22 $y = \sqrt{x^2 - 16}$ on $(-\infty, -4) \cup (4, \infty)$

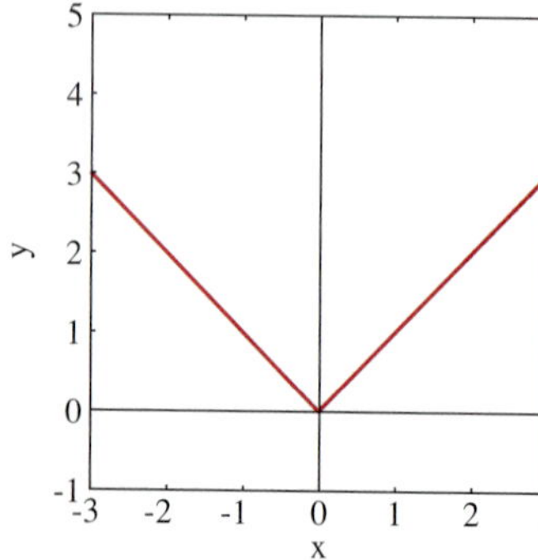

Fig. 1.3.23 $y = |x|$ with a "corner" at the origin

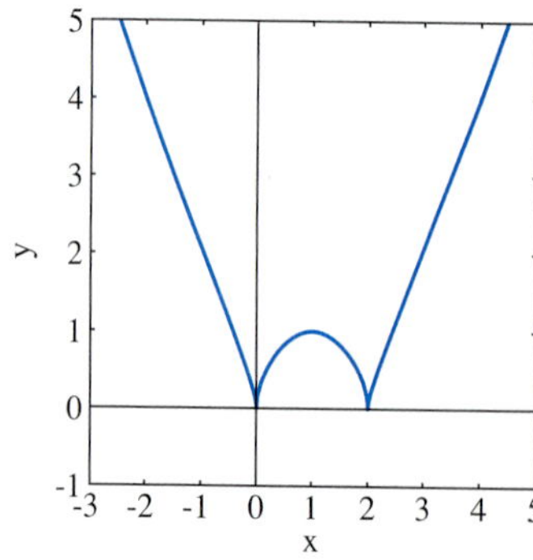

Fig. 1.3.24 $g(x) = \sqrt[3]{x^2(x-2)^2}$ with "cusps" at $(0, 0)$ and $(2, 0)$

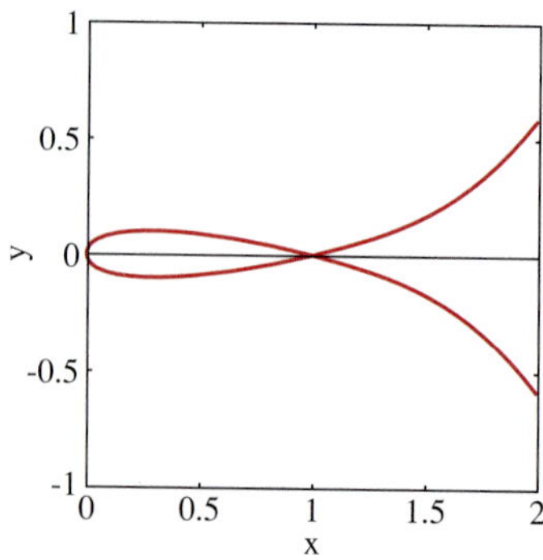

Fig. 1.3.25 $y = \pm(0.2969\sqrt{x} - 0.126\,x - 0.3516\,x^2 + 0.2843\,x^3 - 0.10151\,x^4)$

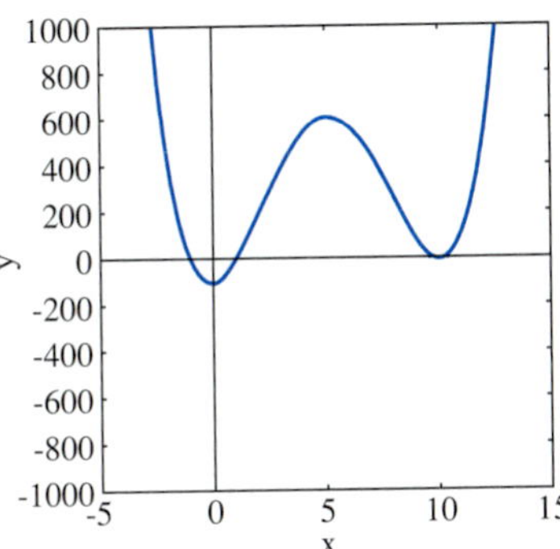

Fig. 1.3.17 $y = (x^2 - 1)(x - 10)(x - 10.1)$ with viewing window $-5 \leqq x \leqq 15$, $-1000 \leqq y \leqq 1000$

Fig. 1.3.18 $y = (x^2 - 1)(x - 10)(x - 10.1)$ with viewing window $9.5 \leqq x \leqq 10.5$, $-1 \leqq y \leqq 1$

But now the behavior of the graph near the point $x = 10$ is unclear. Does it dip beneath the x-axis? We select the viewing window $9.5 \leqq x \leqq 10.5$, $-1 \leqq y \leqq 1$ to magnify this area and get the result shown in Fig. 1.3.18. This case appears to require different plots on different scales to show all the behavior of the graph. ■

Our graphs in Examples 5 and 6 exhibit the maximum possible number of zeros and bends for the polynomials in Eqs. (7) and (8), so we are fairly confident that our investigations reveal the main qualitative features of the graphs of these polynomials. But only with the calculus techniques of Chapter 4 can we be certain of the structure of a graph. For instance, a polynomial graph can exhibit fewer than the maximum possible number of bends, but at this stage we cannot be certain that more bends are not hidden somewhere, perhaps visible only on a scale different from that of the viewing window we have selected.

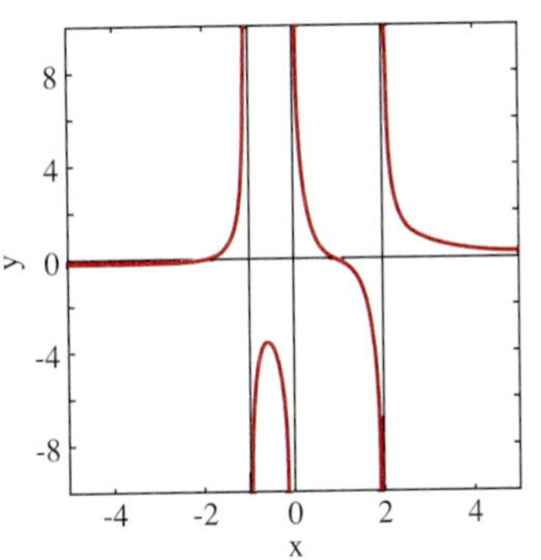

Fig. 1.3.19 The graph of the rational function in Eq. (10) (Example 7)

Rational Functions

Just as a rational number is a quotient of two integers, a **rational function** is a quotient

$$f(x) = \frac{p(x)}{q(x)} \tag{9}$$

of two polynomials $p(x)$ and $q(x)$. Graphs of rational functions and polynomials have several features in common. For instance, a rational function has only a finite number of zeros, because $f(x)$ in Eq. (9) can be zero only when the numerator polynomial $p(x)$ is zero. Similarly, the graph of a rational function has only a finite number of bends.

But the denominator polynomial $q(x)$ in Eq. (9) may have a zero at a point $x = a$ where the numerator is nonzero. In this case the value of $f(x)$ will be very large in magnitude when x is close to a. This observation implies that the graph of a rational function may have a feature that no polynomial graph can have—an *asymptote.*

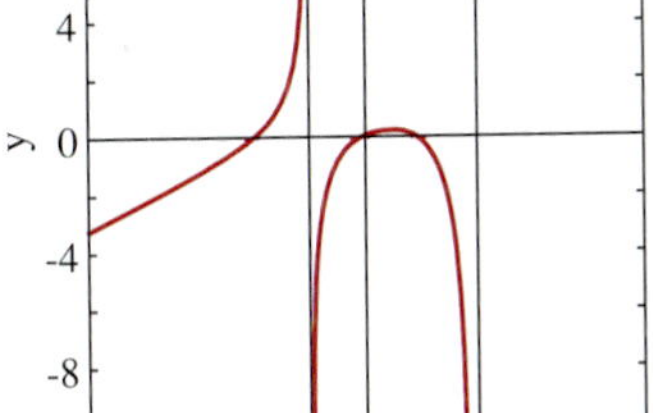

Fig. 1.3.20 The graph of the rational function in Eq. (11) (Example 8)

EXAMPLE 7 Figure 1.3.19 shows the graph of the rational function

$$f(x) = \frac{(x + 2)(x - 1)}{x(x + 1)(x - 2)}. \tag{10}$$

Note the x-intercepts $x = -2$ and $x = 1$, corresponding to the zeros of the numerator $(x + 2)(x - 1)$. The vertical lines $x = -1$, $x = 0$, and $x = 2$ shown in the graph

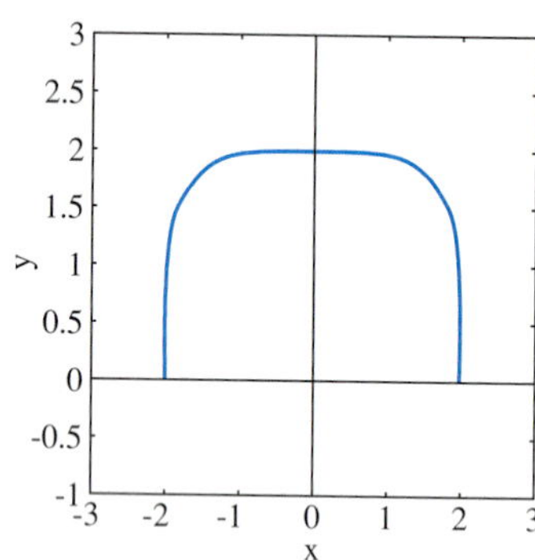

Fig. 1.3.21 $y = \sqrt[4]{16 - x^4}$ on $[-2, 2]$

correspond to the zeros of the denominator $x(x + 1)(x - 2)$. These vertical lines are *asymptotes* of the graph of f.

EXAMPLE 8 Figure 1.3.20 shows the graph of the rational function

$$f(x) = \frac{x(x + 2)(x - 1)}{(x + 1)(x - 2)}. \tag{11}$$

The x-intercepts $x = -2$, $x = 0$, and $x = 1$ correspond to the zeros of the numerator, whereas the asymptotes $x = -1$ and $x = 2$ correspond to the zeros of the denominator.

It should be clear that—by counting x-intercepts and asymptotes—you could match the rational functions in Eqs. (10) and (11) with their graphs in Figs. 1.3.19 and 1.3.20 without knowing in advance which was which.

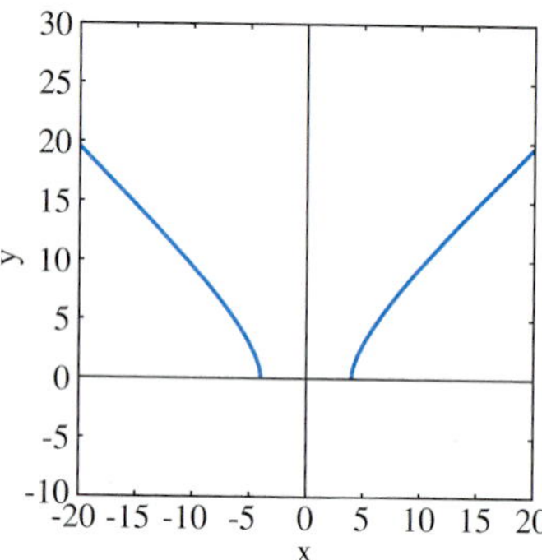

Fig. 1.3.22 $y = \sqrt{x^2 - 16}$ on $(-\infty, -4) \cup (4, \infty)$

Algebraic Functions

An **algebraic function** is one whose formula can be constructed beginning with power functions and using the algebraic operations of addition, subtraction, multiplication by a real number, multiplication, division, and taking roots. Thus polynomials and rational functions are algebraic functions. But whereas every polynomial is defined everywhere on the real line, and every rational function is defined except at the (finitely many) real zeros of its denominator (which correspond to vertical asymptotes), the domain of definition of an algebraic function may be quite limited. For instance, Figs. 1.3.21 and 1.3.22 show the graphs of the algebraic functions

$$f(x) = \sqrt[4]{16 - x^4} \quad \text{and} \quad g(x) = \sqrt{x^2 - 16}$$

on the bounded and unbounded intervals (respectively) where they are defined.

The graph of every polynomial or rational function looks "smooth" at every point where it is defined, but the graph of an algebraic function may exhibit "corners" or sharp "cusps" where it does not look smooth. For instance, look at the graphs in Figs. 1.3.23 and 1.3.24 of the algebraic functions

$$f(x) = \sqrt{x^2} = |x| \quad \text{and} \quad g(x) = \sqrt[3]{x^2(x - 2)^2}.$$

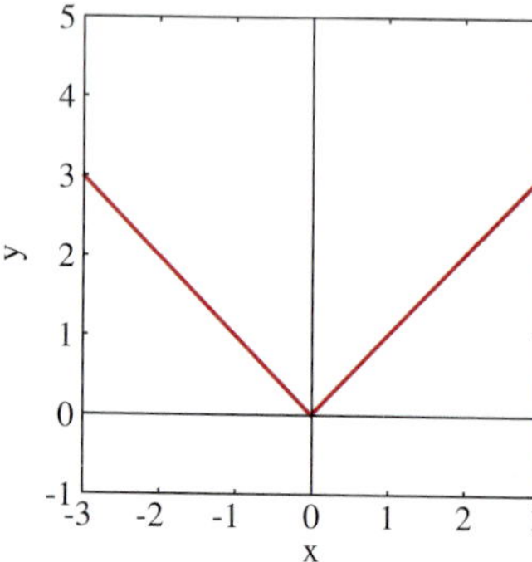

Fig. 1.3.23 $y = |x|$ with a "corner" at the origin

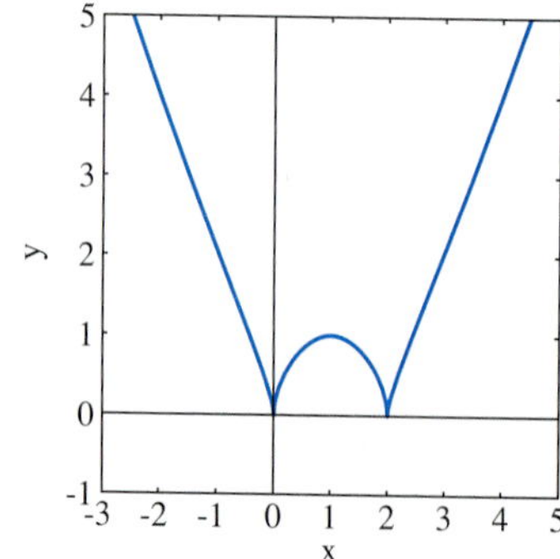

Fig. 1.3.24 $g(x) = \sqrt[3]{x^2(x - 2)^2}$ with "cusps" at $(0, 0)$ and $(2, 0)$

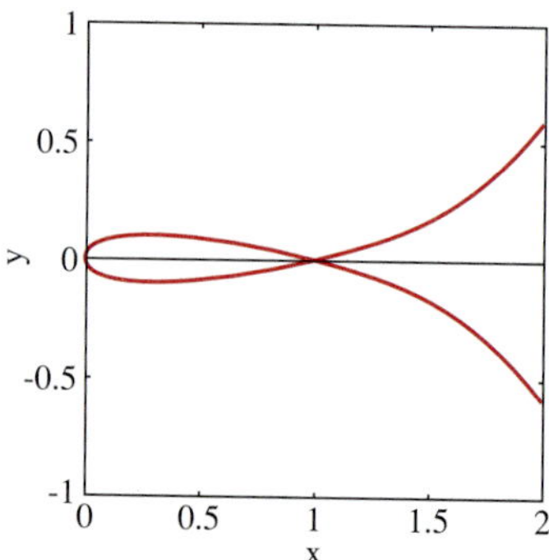

Fig. 1.3.25 $y = \pm(0.2969\sqrt{x} - 0.126\,x - 0.3516\,x^2 + 0.2843\,x^3 - 0.10151\,x^4)$

In Chapter 3 we will use concepts of calculus to say precisely what is meant by a smooth graph.

Figure 1.3.25 shows the graphs of the two algebraic functions defined by

$$y = \pm(0.2969\sqrt{x} - 0.126x - 0.3516x^2 + 0.2843x^3 - 0.10151x^4). \qquad \textbf{(12)}$$

The loop describes the cross-sectional profile shape of the NASA 0012 airfoil as designed by aeronautical engineers.

1.3 PROBLEMS

In Problems 1 through 6, find $f + g$, $f \cdot g$, and f/g, and give the domain of definition of each of these new functions.

1. $f(x) = x + 1,\ g(x) = x^2 + 2x - 3$

2. $f(x) = \dfrac{1}{x-1},\ g(x) = \dfrac{1}{2x+1}$

3. $f(x) = \sqrt{x},\ g(x) = \sqrt{x-2}$

4. $f(x) = \sqrt{x+1},\ g(x) = \sqrt{5-x}$

5. $f(x) = \sqrt{x^2+1},\ g(x) = \dfrac{1}{\sqrt{4-x^2}}$

6. $f(x) = \dfrac{x-1}{x-2},\ g(x) = \dfrac{x+1}{x+2}$

In Problems 7 through 12, match the given polynomial with its graph among those shown in Figs. 1.3.26 through 1.3.31. Do not use a graphing calculator or a computer. Instead, consider the degree of the polynomial, its indicated number of zeros, and its behavior for $|x|$ large.

7. $f(x) = x^3 - 3x + 1$

8. $f(x) = 1 + 4x - x^3$

9. $f(x) = x^4 - 5x^3 + 13x + 1$

10. $f(x) = 2x^5 - 10x^3 + 6x - 1$

11. $f(x) = 16 + 2x^2 - x^4$

12. $f(x) = x^5 + x$

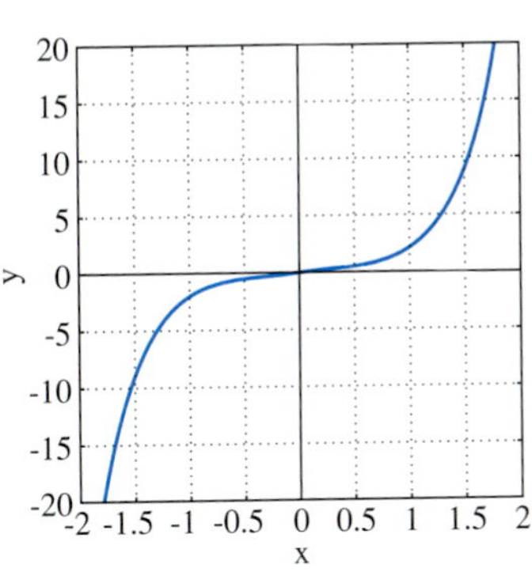

Fig. 1.3.26

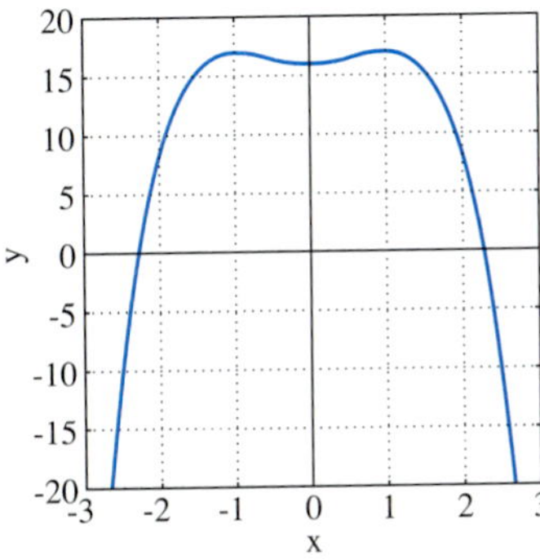

Fig. 1.3.27

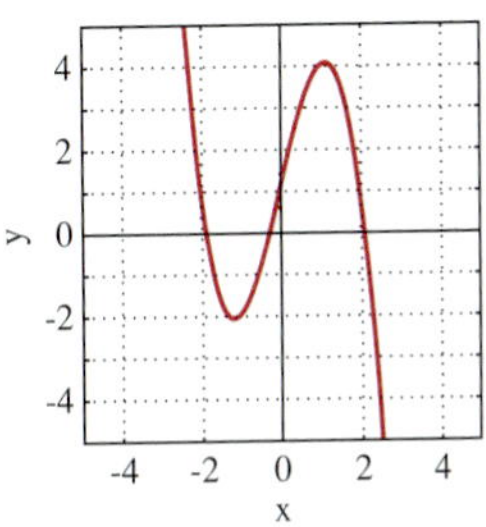

Fig. 1.3.28

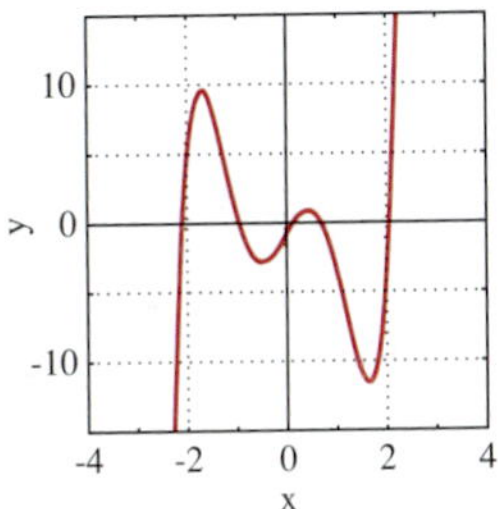

Fig. 1.3.29

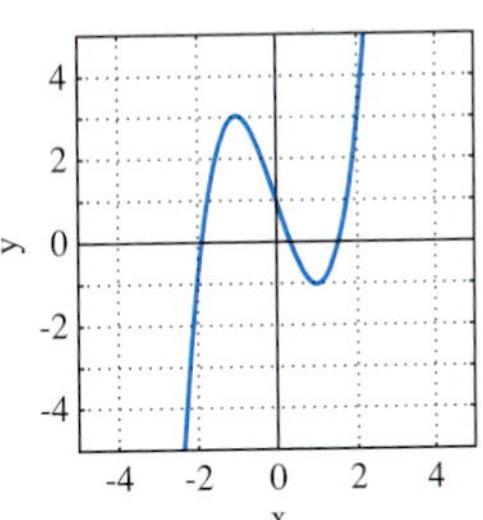

Fig. 1.3.30

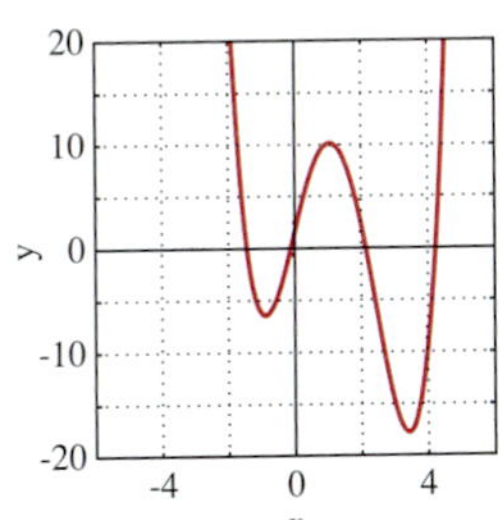

Fig. 1.3.31

In Problems 13 through 16, use the vertical asymptotes of the given rational function (rather than a graphing calculator or computer) to match it with its graph among those shown in Figs. 1.3.32 through 1.3.35.

13. $f(x) = \dfrac{1}{(x+1)(x-2)}$

14. $f(x) = \dfrac{x}{x^2-9}$

15. $f(x) = \dfrac{3}{x^2+1}$

16. $f(x) = \dfrac{x^2+1}{x^3-1}$

In Problems 17 through 20, use primarily the domain of definition of the given algebraic function (rather than a graphing

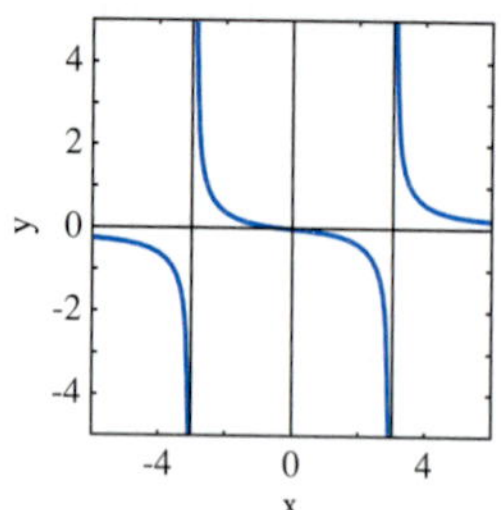

Fig. 1.3.32

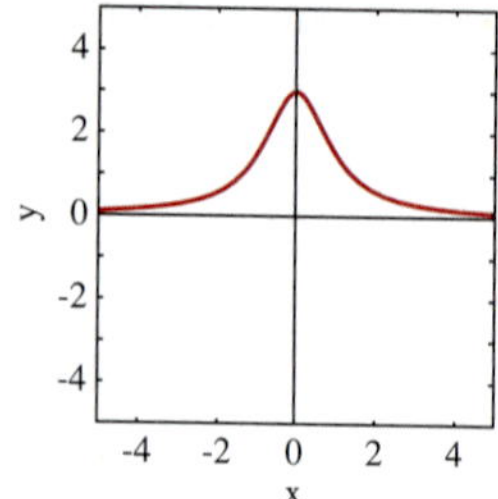

Fig. 1.3.33

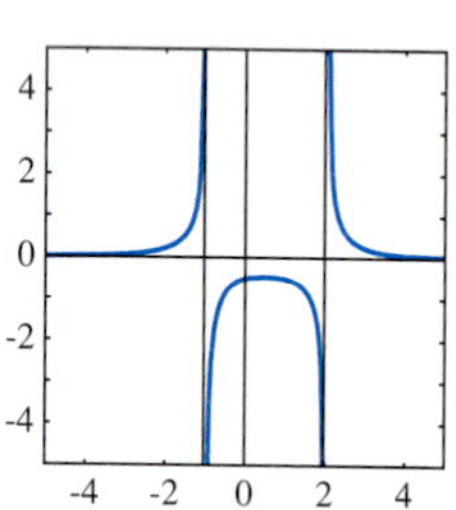

Fig. 1.3.34

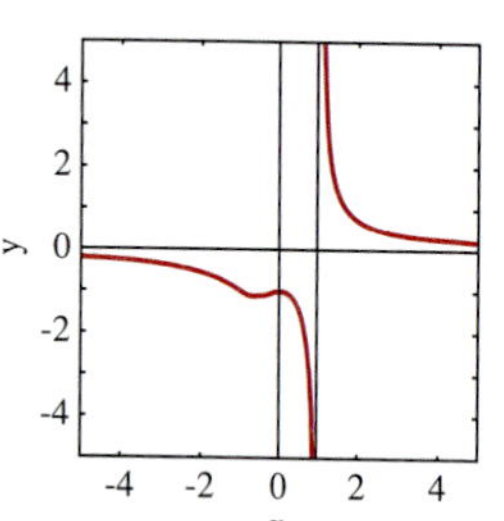

Fig. 1.3.35

calculator or computer) to match it with its graph among those in Figs. 1.3.36 through 1.3.39.

17. $f(x) = x\sqrt{x+2}$

18. $f(x) = \sqrt{2x - x^2}$

19. $f(x) = \sqrt{x^2 - 2x}$

20. $f(x) = 2\sqrt[3]{x^2 - 2x}$

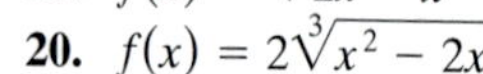

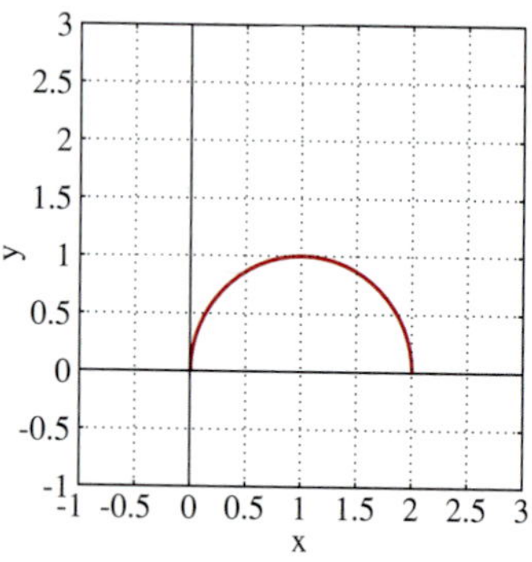

Fig. 1.3.36

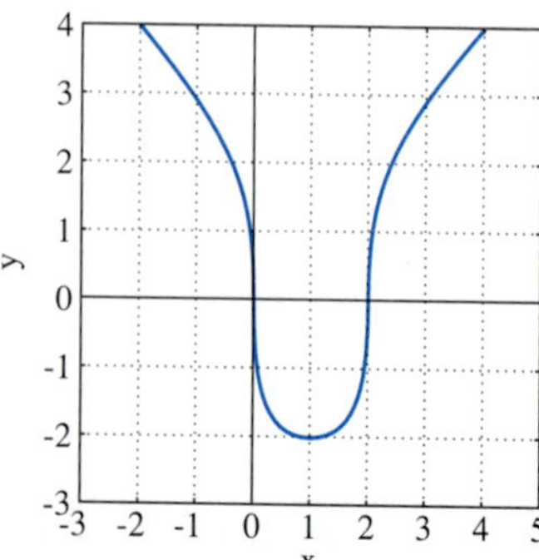

Fig. 1.3.37

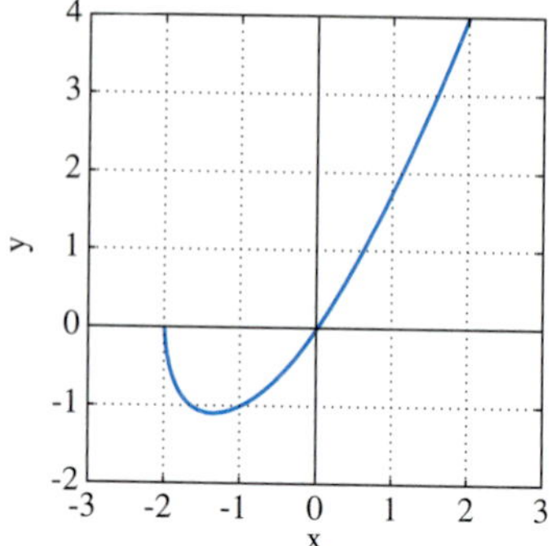

Fig. 1.3.38

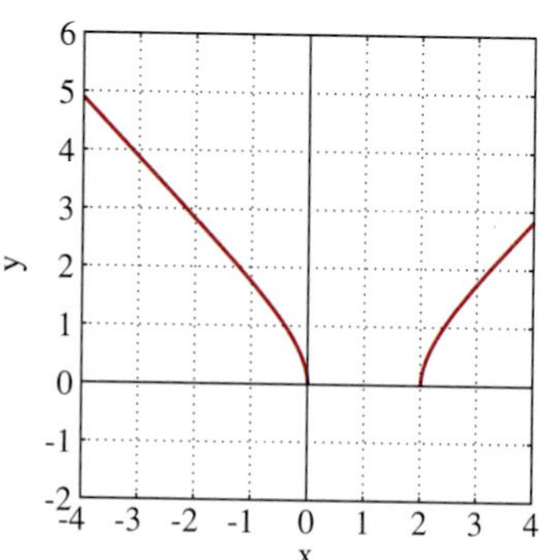

Fig. 1.3.39

In Problems 21 through 30 use a graphing calculator or computer to determine one or more appropriate viewing windows to exhibit the principal features of the graph $y = f(x)$. In particular, determine thereby the number of real solutions of the equation $f(x) = 0$ and the approximate location (to the nearest integer) of each of these solutions.

21. $f(x) = x^3 - 3x + 1$

22. $f(x) = x^3 - 3x + 2$

23. $f(x) = x^3 - 3x + 3$

24. $f(x) = 2x^4 - 6x^3 + 10x - 5$

25. $f(x) = 2x^4 - 6x^3 + 10x - 6$

26. $f(x) = 2x^4 - 6x^3 + 10x - 7$

27. $f(x) = x^3 - 50x - 100$

28. $f(x) = x^4 + 20x^3 - 50x - 30$

29. $f(x) = x^5 + 5x^4 - 100x^3 - 200x^2 + 2500x - 3500$

30. $f(x) = x^6 - 250x^4 + 2500x^2 - 2500$

In Problems 31 through 37, determine how the graph $y = f(x)$ changes when the value of c is changed. With a graphing calculator or computer you should be able to plot graphs with different values of c on the same screen.

31. $f(x) = x^3 - 3x + c, \quad -5 \leqq c \leqq 5$

32. $f(x) = x^3 + cx, \quad -5 \leqq c \leqq 5$

33. $f(x) = x^3 + cx^2, \quad -5 \leqq c \leqq 5$

34. $f(x) = x^4 + cx^2, \quad -5 \leqq c \leqq 5$

35. $f(x) = x^5 + cx^3 + x, \quad -5 \leqq c \leqq 5$

36. $f(x) = \dfrac{1}{1 + cx^2}, \quad 1 \leqq c \leqq 10$

37. $f(x) = \sqrt{\dfrac{x^2}{c^2 - x^2}}, \quad 1 \leqq c \leqq 10,\ x \text{ in } (-c, c)$

38. Use the graphical method of repeated magnifications to find both the length and the maximum width of the airfoil shown in Fig. 1.3.25. Determine each accurate to three decimal places.

1.3 PROJECT: A LEANING LADDER

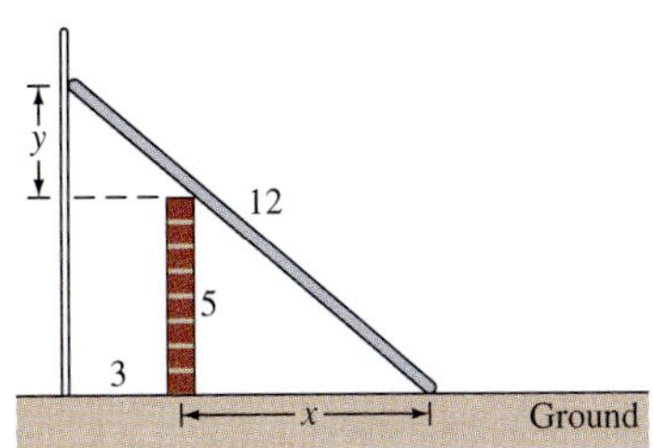

Fig. 1.3.40 The leaning ladder

A 12-ft ladder leans across a 5-ft fence and touches a high wall located 3 ft behind the fence (Fig. 1.3.40). You are to find the distance x from the foot of this ladder to the bottom of the fence.

Use similar triangles and the Pythagorean theorem to show that the distances x and y in Fig. 1.3.40 satisfy the two equations

$$xy = 15, \tag{1}$$

$$(x + 3)^2 + (y + 5)^2 = 144. \tag{2}$$

Eliminate y by substituting $y = 15/x$ from Eq. (1) into Eq. (2). Then simplify the resulting equation to obtain the quartic equation

$$x^4 + 6x^3 - 110x^2 + 150x + 225 = 0. \tag{3}$$

Use the graphical method of successive magnifications to determine each of the real solutions (accurate to two rounded decimal places) of Eq. (3). Which of these solutions corresponds to a physically possible position of the leaning ladder as shown in Fig. 1.3.40? Is there more than one answer?

For your own personal investigation, let p and q denote the last two nonzero digits in your student I.D. number (with p the larger) and begin instead with a ladder of length $L = 3p$ feet. Suppose that the fence is p feet high and stands q feet from the high wall. Write the results of your investigation in the form of a carefully organized report consisting of complete sentences (and including pertinent equations and figures) explaining your results in detail. Tell precisely what you did to solve your problem. Don't merely "solve the equation"! Answer completely the original question as to the possible placement (or placements) of the foot of the ladder to fit the physical picture.

1.4 A BRIEF CATALOG OF FUNCTIONS, PART 2

Continuing the survey of elementary functions begun in Section 1.3, we now review briefly the most familiar nonalgebraic functions that are studied in calculus. They include the *trigonometric functions* that are used to model periodic phenomena—phenomena of ebb and flow—involving quantities that oscillate with the passage of time—and the *exponential* and *logarithmic functions* that are used to model phenomena of growth and decay—involving quantities that either increase steadily or decrease steadily as time passes. We also introduce *composition* of functions, a new way (in addition to the algebraic operations of Section 1.3) of combining familiar functions to form new ones.

Trigonometric Functions

A review of trigonometry is included in Appendix C. In elementary trigonometry a trigonometric function such as $\sin A$, $\cos A$, or $\tan A$ ordinarily is first defined as a function of an *angle A* in a right triangle. But here a trigonometric function of a *real number x* corresponds to that function of an angle measuring x radians. Thus

$$\sin\frac{\pi}{6} = \frac{1}{2}, \quad \cos\frac{\pi}{6} = \frac{\sqrt{3}}{2}, \quad \text{and} \quad \tan\frac{\pi}{6} = \frac{\sin\frac{\pi}{6}}{\cos\frac{\pi}{6}} = \frac{1}{\sqrt{3}}$$

because $\pi/6$ is the radian measure of an angle of 30°. Recall that

$$\pi \text{ radians} = 180 \text{ degrees}, \tag{1}$$

so

$$1 \text{ rad} = \frac{180}{\pi} \text{ deg} \quad \text{and} \quad 1 \text{ deg} = \frac{\pi}{180} \text{ rad}.$$

Figures 1.4.1 and 1.4.2 show the graphs $y = \sin x$ and $y = \cos x$ of the sine and cosine functions, respectively. The value of each oscillates between $+1$ and -1, exhibiting the characteristic *periodicity* of the trigonometric functions:

$$\sin(x + 2\pi) = \sin x \quad \text{and} \quad \cos(x + 2\pi) = \cos x \tag{2}$$

for all x.

If we translate the graph $y = \cos x$ by $\pi/2$ units to the right, we get the graph $y = \sin x$. This observation corresponds to the familiar relation

$$\cos\left(x - \frac{\pi}{2}\right) = \cos\left(\frac{\pi}{2} - x\right) = \sin x. \tag{3}$$

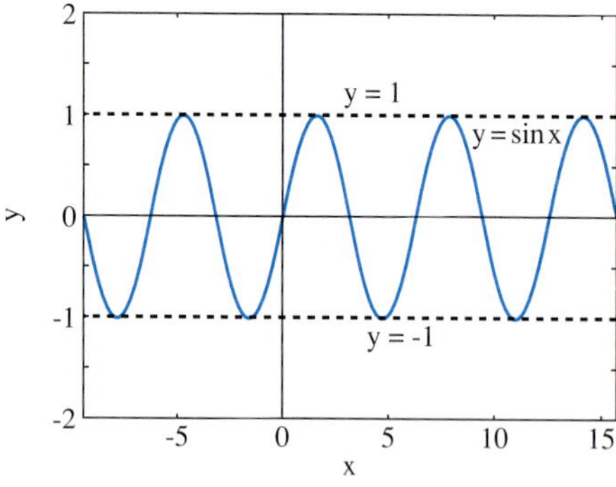

Fig. 1.4.1 $y = \sin x$

Fig. 1.4.2 $y = \cos x$

EXAMPLE 1 Figure 1.4.3 shows the translated sine curve obtained by translating the origin to the point $(1, 2)$. Its equation is obtained upon replacing x and y in $y = \sin x$ with $x - 1$ and $y - 2$, respectively:

$$y - 2 = \sin(x - 1); \quad \text{that is,}$$

$$y = 2 + \sin(x - 1).$$

The world around us is full of quantities that oscillate like the trigonometric functions. Think of the alternation of day and night, the endless repetition of the seasons, the monthly cycle of the moon, the rise and fall of the tides, the beat of your heart.

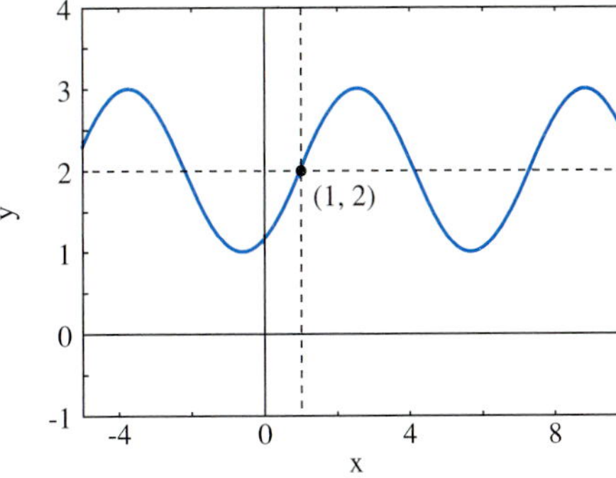

Fig. 1.4.3 The translated sine curve $y - 2 = \sin(x - 1)$

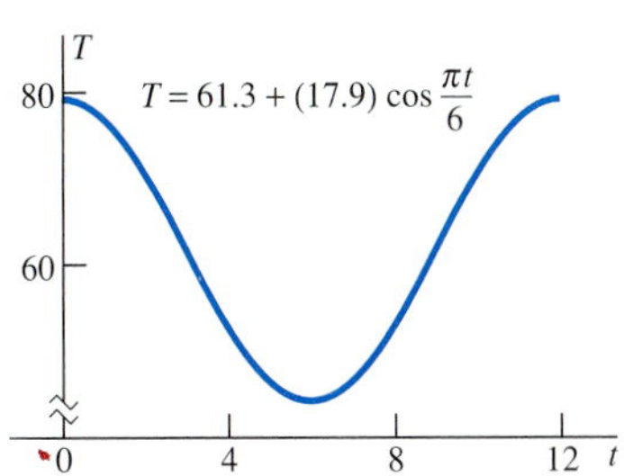

Fig. 1.4.4 Average daily temperature in Athens, Georgia, t months after July 15, (Example 2)

EXAMPLE 2 Figure 1.4.4 shows the cosine-like behavior of temperatures in Athens, Georgia. The average temperature T (in $^\circ$F) during a 24-h day t months after July 15 is given approximately by

$$T = T(t) = 61.3 + 17.9 \cos\frac{\pi t}{6}. \tag{4}$$

For instance, on a typical October 15 (three months after July 15) the average temperature is

$$T(3) = 61.3 + 17.9 \cos\frac{3\pi}{6} = 61.3 \quad (^\circ\text{F})$$

because $\cos(3\pi/6) = \cos(\pi/2) = 0$. Thus the "midpoint" of fall weather in Athens—when the average daily temperature is midway between summer's high and winter's low—occurs about three weeks after the official beginning of fall (on or about September 22). Note also that

$$T(t + 12) = 61.3 + 17.9 \cos\left(\frac{\pi t}{6} + 2\pi\right) = 61.3 + 17.9 \cos\left(\frac{\pi t}{6}\right) = T(t)$$

(why?), in agreement with the yearly 12-month cycle of average weather. ■

The periodicity and oscillatory behavior of the trigonometric functions make them quite unlike polynomial functions. Because

$$\sin n\pi = 0 \quad \text{and} \quad \cos\left(\frac{2n + 1}{2}\pi\right) = 0 \tag{5}$$

for $n = 0, 1, 2, 3, \ldots$, we see that the simple trigonometric equations

$$\sin x = 0 \quad \text{and} \quad \cos x = 0 \tag{6}$$

have *infinitely many solutions.* In contrast, a polynomial equation can have only a finite number of solutions.

Figure 1.4.5 shows the graph of $y = \tan x$. The x-intercepts correspond to the zeros of the numerator $\sin x$ in the relation

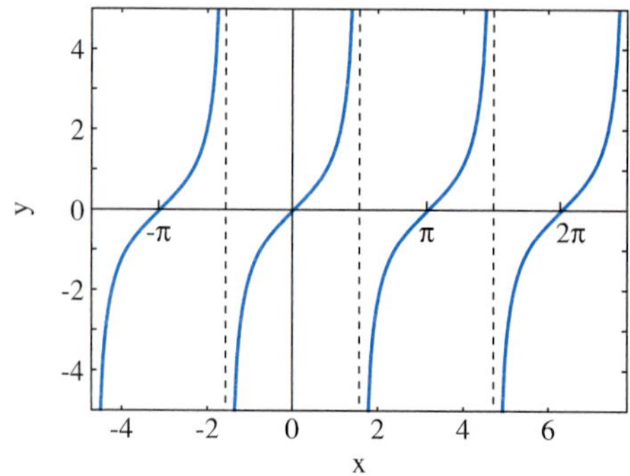

Fig. 1.4.5 $y = \tan x$

$$\tan x = \frac{\sin x}{\cos x}, \tag{7}$$

whereas the asymptotes correspond to the zeros of the denominator $\cos x$. Observe the "infinite gaps" in the graph $y = \tan x$ at these odd integral multiples of $\pi/2$. We call these gaps *discontinuities,* phenomena we discuss further in Chapter 2.

Composition of Functions

Many varied and complex functions can be "put together" by using quite simple "building-block" functions. In addition to adding, subtracting, multiplying, or dividing two given functions, we can also combine functions by letting one function act on the output of the other.

Definition *Composition of Functions*

The **composition** of the two functions f and g is the function $h = f \circ g$ defined by

$$h(x) = f(g(x)) \tag{8}$$

for all x in the domain of g such that $u = g(x)$ is in the domain of f. (The right-hand side in Eq. (8) is read "f of g of x.")

Thus the output $u = g(x)$ of the function g is used as the input to the function f (Fig. 1.4.6). We sometimes refer to g as the *inner function* and to f as the *outer function* in Eq. (8).

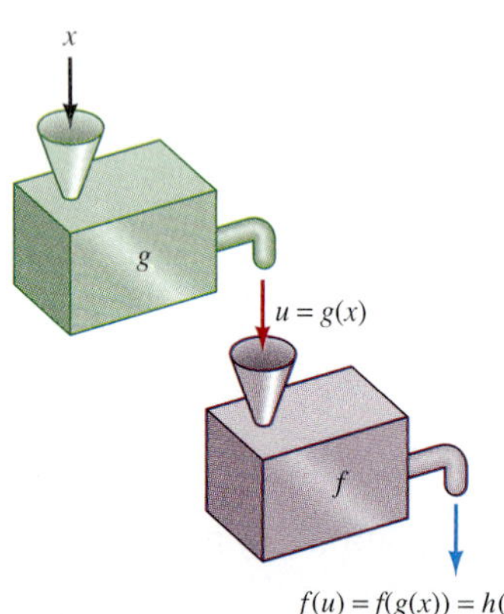

Fig. 1.4.6 The composition of f and g

EXAMPLE 3 If $f(x) = \sqrt{x}$ and $g(x) = 1 - x^2$, then

$$f(g(x)) = \sqrt{1 - x^2} \quad \text{for } |x| \leqq 1,$$

whereas

$$g(f(x)) = 1 - (\sqrt{x})^2 = 1 - x \quad \text{for } x \geqq 0.$$

■

The $f(g(x))$ notation for compositions is most commonly used in ordinary computations, but the $f \circ g$ notation emphasizes that the composition may be regarded as a new kind of combination of the functions f and g. But Example 3 shows that $f \circ g$ is quite unlike the product fg of the two functions f and g, for

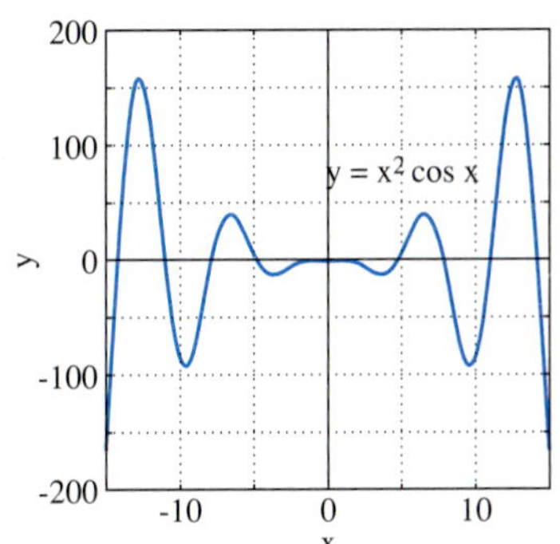

Fig. 1.4.7 $y = x^2 \cos x$ (Example 4)

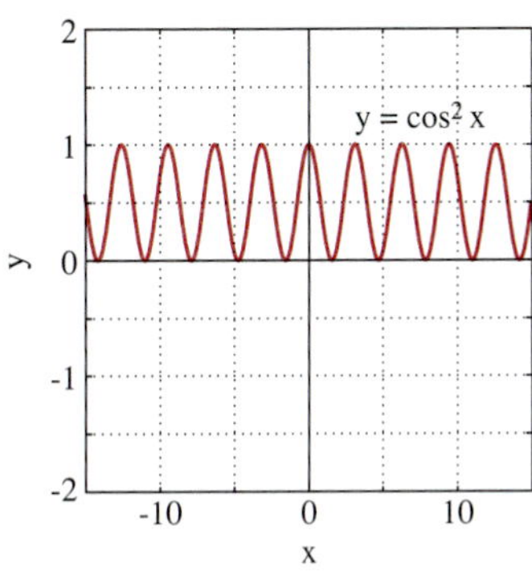

Fig. 1.4.8 $y = \cos^2 x$ (Example 4)

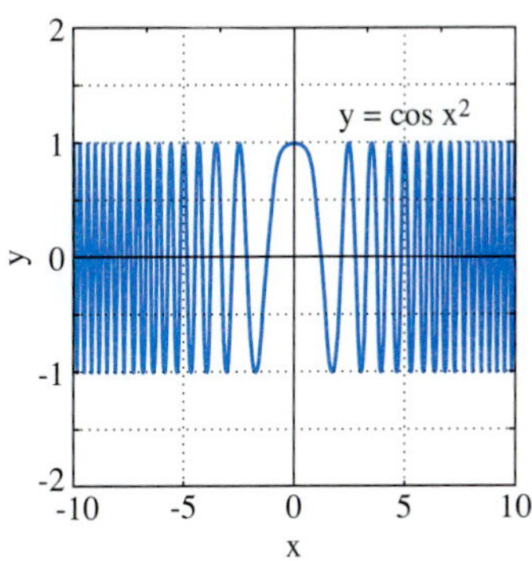

Fig. 1.4.9 $y = \cos x^2$ (Example 4)

$$f \circ g \neq g \circ f,$$

whereas $fg = gf$ [because $f(x) \cdot g(x) = g(x) \cdot f(x)$ whenever $f(x)$ and $g(x)$ are defined]. So remember that composition is quite different in character from ordinary multiplication of functions.

EXAMPLE 4 If

$$f(x) = x^2 \quad \text{and} \quad g(x) = \cos x,$$

then the functions

$$\begin{aligned} f(x)g(x) &= x^2 \cos x, \\ f(g(x)) &= \cos^2 x = (\cos x)^2, \quad \text{and} \\ g(f(x)) &= \cos x^2 = \cos(x^2) \end{aligned}$$

are defined for all x. Figures 1.4.7 through 1.4.9 illustrate vividly how different these three functions are. ■

EXAMPLE 5 Given the function $h(x) = (x^2 + 4)^{3/2}$, find two functions f and g such that $h(x) = f(g(x))$.

Solution It is technically correct—but useless—simply to let $g(x) = x$ and $f(u) = (u^2 + 4)^{3/2}$. We seek a nontrivial answer here. To calculate $(x^2 + 4)^{3/2}$, we must first calculate $x^2 + 4$. So we choose $g(x) = x^2 + 4$ as the inner function. The last step is to raise $u = g(x)$ to the power $\frac{3}{2}$, so we take $f(u) = u^{3/2}$ as the outer function. Thus if

$$f(x) = x^{3/2} \quad \text{and} \quad g(x) = x^2 + 4,$$

then $f(g(x)) = f(x^2 + 4) = (x^2 + 4)^{3/2} = h(x)$. ■

Exponential Functions

An exponential function is a function of the form

$$f(x) = a^x, \tag{9}$$

where the **base** a is a fixed positive real number—a constant. Note the difference between an exponential function and a power function. In the power function x^n, the *variable* x is raised to a *constant* power; in the exponential function a^x, a *constant* is raised to a *variable* power.

Many computers and programmable calculators use the notation **a ∧ x** to denote the exponential a^x (a few use **a ↑ x**). If $a > 1$, then the graph $y = a^x$ looks much like those in Fig. 1.4.10, which shows $y = 2^x$ and $y = 10^x$. The graph of an exponential function with base a, $a > 1$, increases steadily from left to right. Therefore, such a graph is nothing like the graph of a polynomial or trigonometric function. The larger the base a, the more rapid the rate at which the curve $y = a^x$ rises (for $x > 0$). Thus $y = 10^x$ climbs more steeply than $y = 2^x$.

EXAMPLE 6 Every exponential function (with base $a > 1$) increases *very rapidly* when x is large. The following table comparing values of x^2 with 2^x exhibits vividly the rapid rate of increase of the exponential function 2^x, even compared with the power function x^2, which increases at a more restrained rate as x increases.

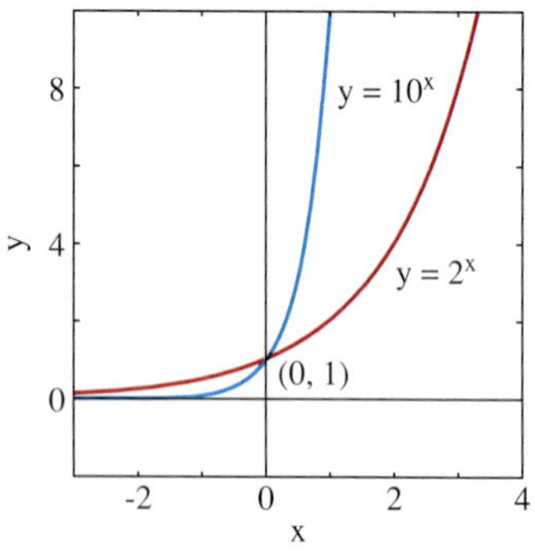

Fig. 1.4.10 Increasing exponential functions $y = 2^x$ and $y = 10^x$

x	x^2	2^x
10	100	1024
20	400	1048576
30	900	1073741824
40	1600	1099511627776
50	2500	1125899906842624
60	3600	1152921504606846976
70	4900	1180591620717411303424
80	6400	1208925819614629174706176
90	8100	1237940039285380274899124224
100	10000	1267650600228229401496703205376

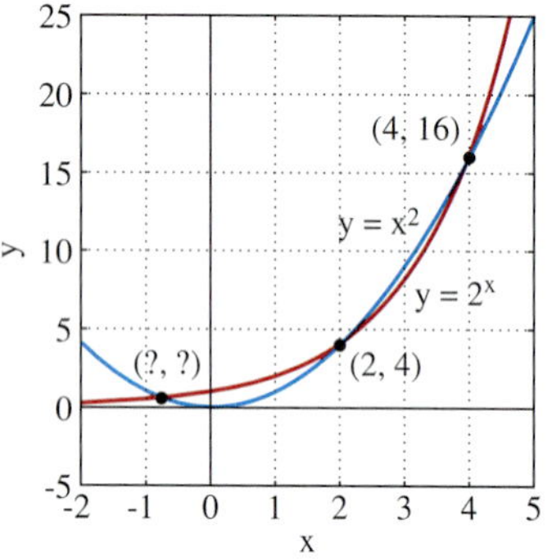

Fig. 1.4.11 $y = x^2$ and $y = 2^x$

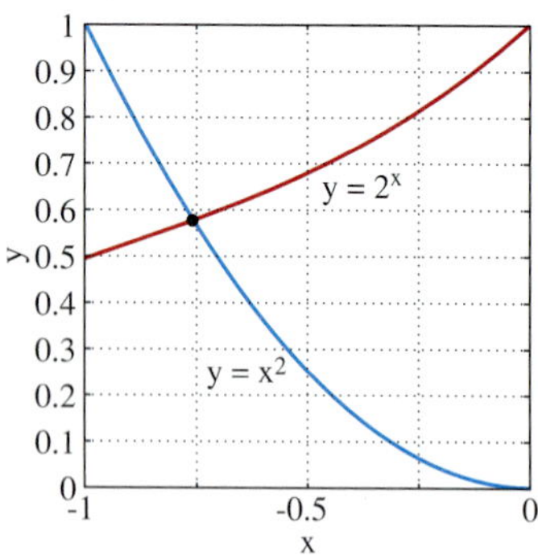

Fig. 1.4.12 The negative solution of $x^2 = 2^x$

The comparison between x^2 and 2^x for smaller values of x is interesting in a different way. The graphs of $y = x^2$ and $y = 2^x$ in Fig. 1.4.11 indicate that the equation $x^2 = 2^x$ has three solutions between $x = -2$ and $x = 5$. Is it clear to you that $x = 2$ and $x = 4$ are *exact* solutions? The "zoom" shown in Fig. 1.4.12 indicates that the negative solution is a bit less than -0.75. Perhaps you can zoom once more and find the value of this negative solution accurate to at least two decimal places. ■

If we replace x in Eq. (9) with $-x$, we get the function a^{-x}. Its graph $y = a^{-x}$ *decreases* from left to right if $a > 1$. Figure 1.4.13 shows the graphs $y = 3^{-x}$ and $y = 7^{-x}$.

Whereas trigonometric functions are used to describe periodic phenomena of ebb and flow, exponential functions are used to describe natural processes that always increase or always decrease.

EXAMPLE 7 Let $P(t)$ denote the number of rodents after t months in a certain prolific population that doubles every month. If there are $P(0) = 10$ rodents initially, then there are

- $P(1) = 10 \cdot 2^1 = 20$ rodents after 1 month,
- $P(2) = 10 \cdot 2^2 = 40$ rodents after 2 months,
- $P(3) = 10 \cdot 2^3 = 80$ rodents after 3 months,

and so forth. Thus the rodent population after t months is given by the exponential function

$$P(t) = 10 \cdot 2^t \tag{10}$$

if t is a nonnegative integer. Under appropriate conditions, Eq. (10) gives an accurate approximation to the rodent population even when t is not an integer. For instance, this formula predicts that after $t = 4\frac{1}{2}$ months, there will be

$$P(4.5) = 10 \cdot 2^{4.5} \approx 226.27 \approx 226 \quad \text{rodents.}$$ ■

EXAMPLE 8 Suppose that you invest \$5000 in a money-market account that pays 8% interest compounded annually. This means that the amount in the account is multiplied by 1.08 at the end of each year. Let $A(t)$ denote the amount in your account at the end of t years. Then:

- $A(1) = 5000 \cdot 1.08^1$ (\$5400.00) after 1 yr,
- $A(2) = 5000 \cdot 1.08^2$ (\$5832.00) after 2 yr,
- $A(3) = 5000 \cdot 1.08^3$ (\$6298.56) after 3 yr,

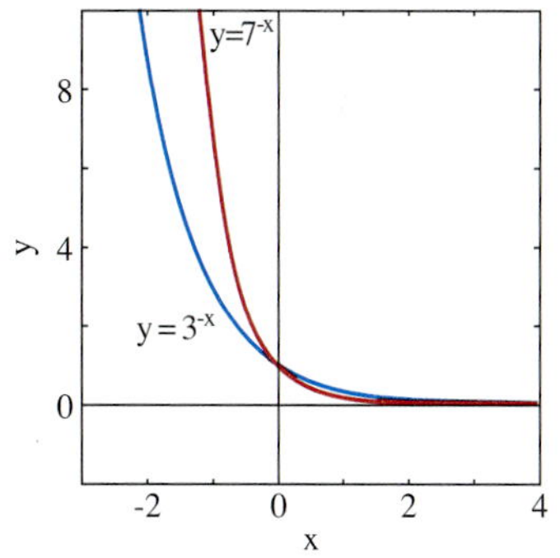

Fig. 1.4.13 Decreasing exponential functions $y = 3^{-x}$ and $y = 7^{-x}$

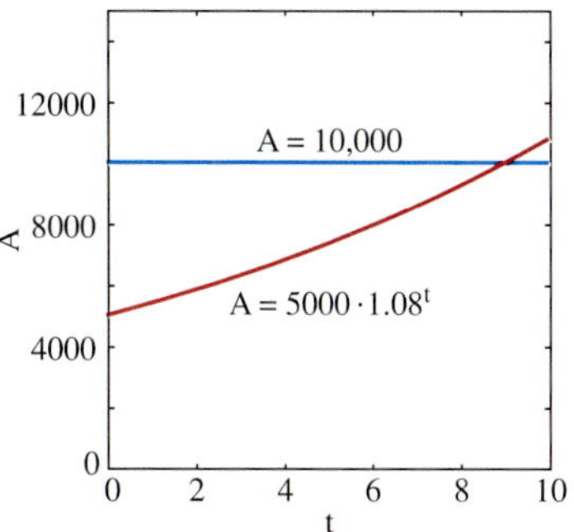

Fig. 1.4.14 The graph for Example 8

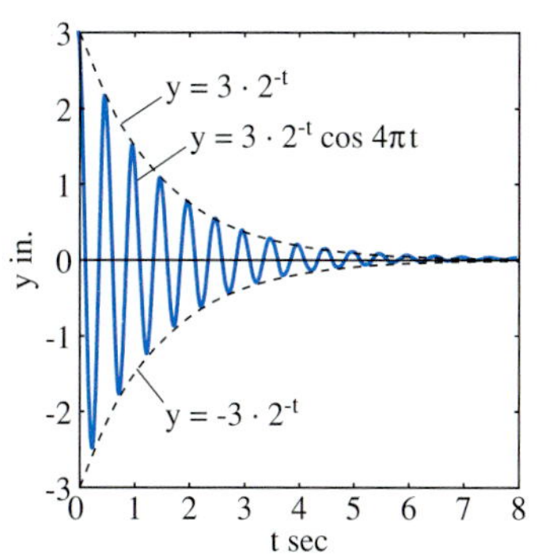

Fig. 1.4.15 $y(t) = 3{\cdot}2^{-t}\cos 4\pi t$ (Example 9)

and so on. Thus after t years (t a nonnegative integer), the amount in your account is given by the exponential function

$$A(t) = 5000 \cdot 1.08^t. \tag{11}$$

In the case of a slightly lower interest rate (about 7.696%) compounded *continuously* (rather than annually), this formula also holds when t is not an integer. Figure 1.4.14 shows the graph $A(t) = 5000 \cdot 1.08^t$ as well as the horizontal line $A = 10000$. From this graph we see, for instance, that the amount in the account has doubled (to \$10000) after approximately $t = 9$ yr. We could approximate the "doubling time" t more accurately by magnifying the graph near the intersection of the horizontal line and the rising curve. ■

Example 9 exhibits a function that combines the steady decrease of an exponential function with negative exponent with the oscillation of a trigonometric function.

EXAMPLE 9 The function

$$y(t) = 3 \cdot 2^{-t} \cos 4\pi t \tag{12}$$

might describe the amplitude y, in inches, of the up-and-down vibrations of a car with very poor shock absorbers t seconds after it hits a deep pothole. Can you see that Eq. (12) describes an initial ($t = 0$) amplitude of 3 inches that halves every second, while two complete up-and-down oscillations occur every second? (The factor $3 \cdot 2^{-t}$ is the decreasing amplitude of the vibrations, while the function $\cos 4\pi t$ has period $\frac{1}{2}$ s.) Figure 1.4.15 shows the graph of $y(t)$. The curve described in Eq. (12) oscillates between the two curves $y(t) = \pm 3 \cdot 2^{-t}$. It appears that the car's vibrations subside and are negligible after 7 or 8 seconds. ■

Logarithmic Functions

In analogy with the inverse trigonometric functions that you saw in trigonometry, logarithms are "inverse" to exponential functions. The **base a logarithm** of the positive number x is the power to which a must be raised to get x. That is,

$$y = \log_a x \quad \text{if} \quad a^y = x. \tag{13}$$

The **LOG** key on most calculators gives the base 10 (*common*) *logarithm* $\log_{10} x$. The **LN** key gives the *natural logarithm*

$$\ln x = \log_e x,$$

where e is a special irrational number:

$$e = 2.71828182845904523536\ldots.$$

You'll see the significance of this strange-looking base in Chapter 7.

Figure 1.4.16 shows the graphs $y = \ln x$ and $y = \log_{10} x$. Both graphs pass through the point $(1, 0)$ and rise steadily (though slowly) from left to right. Because exponential functions never take on zero or negative values, neither zero nor any negative number is in the domain of any logarithm function.

The facts that $\log_{10} 100{,}000 = 5$ and $\log_{10} 1{,}000{,}000 = 6$ indicate that the function $\log x = \log_{10} x$ increases quite slowly as x increases. Whereas Example 6 illustrates the fact that an exponential function a^x (with $a > 1$) increases more rapidly than any power function as $x \to \infty$, Example 10 illustrates the fact that a logarithmic function increases *more slowly* than any power function.

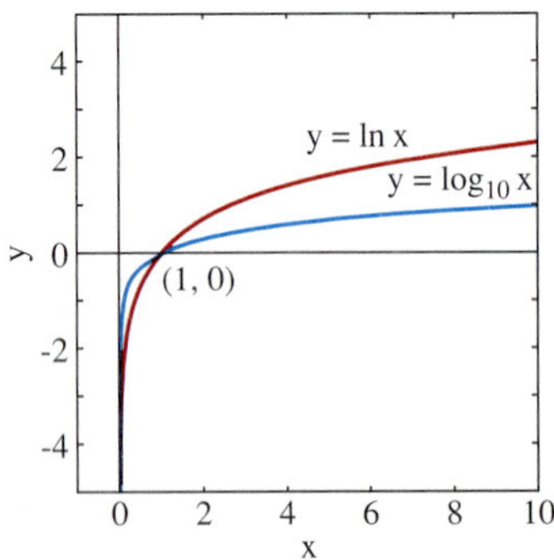

Fig. 1.4.16 The common and natural logarithm functions

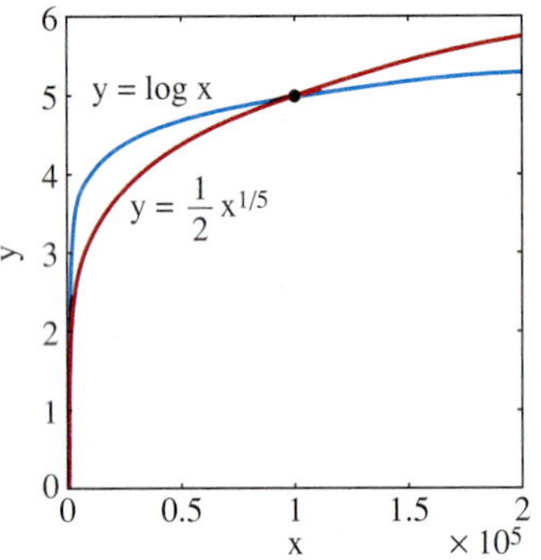

Fig. 1.4.17 $\frac{1}{2}x^{1/5}$ passes $\log x$

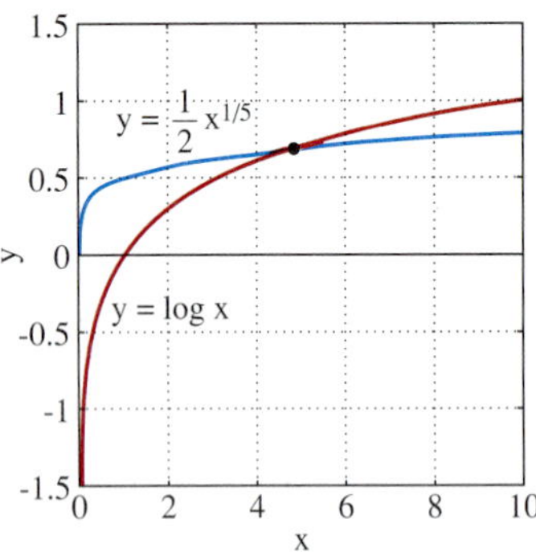

Fig. 1.4.18 $\log x$ passes $\frac{1}{2}x^{1/5}$

EXAMPLE 10 In the following table we compare the rate of growth of the power function $f(x) = \frac{1}{2}x^{1/5}$ with that of the logarithm function $g(x) = \log x$.

x	$f(x) = \frac{1}{2}x^{1/5}$	$g(x) = \log x$
20000	3.62390	4.30103
40000	4.16277	4.60206
60000	4.51440	4.77815
80000	4.78176	4.90309
100000	5	5
120000	5.18569	5.07918
140000	5.34805	5.14613
160000	5.49280	5.20412
180000	5.62373	5.25527
200000	5.74349	5.30103

It appears here and in Fig. 1.4.17 that $\log x$ is smaller than $\frac{1}{2}x^{1/5}$ when $x > 100{,}000$. Figure 1.4.18 shows that $\log x$ initially is smaller than $\frac{1}{2}x^{1/5}$ but "catches up and passes" $\frac{1}{2}x^{1/5}$ somewhere around (although a bit less than) $x = 5$. Then $\frac{1}{2}x^{1/5}$ in turn catches up and passes $\log x$ at $x = 100{,}000$. When $x = 10^{50}$, $\frac{1}{2}x^{1/5} = 5{,}000{,}000{,}000$, but the value of $\log x$ is only 50. ■

Transcendental Equations

The trigonometric, exponential, and logarithmic functions are called *transcendental* functions. As we saw in Eq. (6), an equation that includes transcendental functions can have infinitely many solutions. But it also may have only a finite number of solutions; which of these two possibilities is the case can be difficult to determine. One approach is to write the given equation in the form

$$f(x) = g(x), \tag{14}$$

where both the functions f and g are readily graphed. Then the solutions of Eq. (14) correspond to the intersections of the two graphs $y = f(x)$ and $y = g(x)$.

EXAMPLE 11 The single point of intersection of the graphs $y = x$ and $y = \cos x$, shown in Fig. 1.4.19, indicates that the equation

$$x = \cos x$$

has only a single solution. Moreover, from the graph you can glean the additional information that the solution lies in the interval $(0, 1)$. ■

EXAMPLE 12 The graphs of $y = 1 - x$ and $y = 3\cos x$ are shown in Fig. 1.4.20. In contrast with Example 11, there are three points of intersection of the graphs. This makes it clear that the equation

$$1 - x = 3\cos x$$

has one negative solution and two positive solutions. They could be approximated by (separately) zooming in on the three intersection points. ■

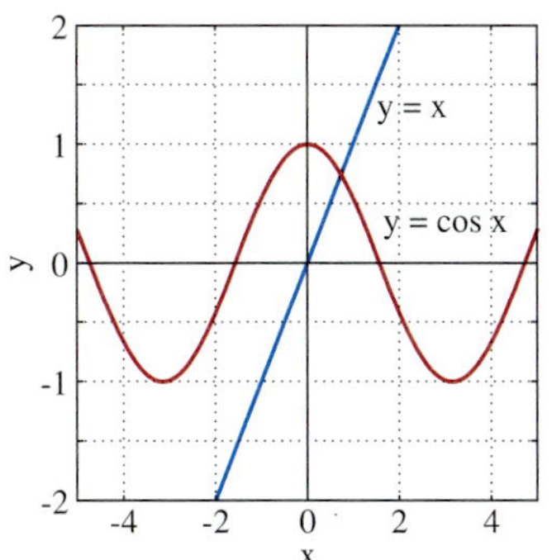

Fig. 1.4.19 Solving the equation $x = \cos x$ of Example 11

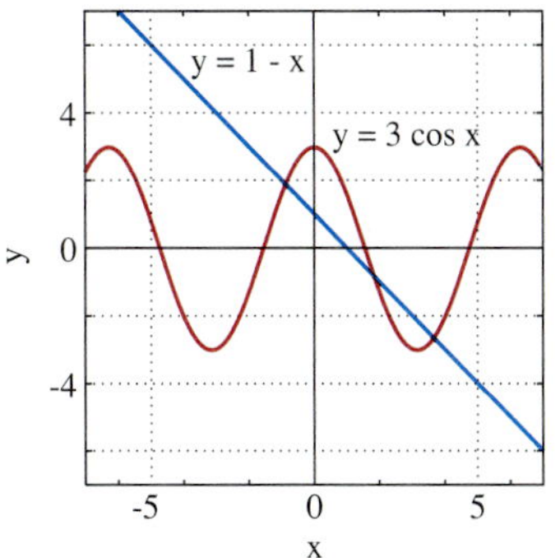

Fig. 1.4.20 Solving the equation $1 - x = 3\cos x$ of Example 12

Can You Believe What You See on Your Calculator/Computer Screen?

The next examples show that the short answer is "Not always." One reason is that a typical graphing calculator or simple computer program plots only a finite number of equally spaced points on the curve $y = f(x), a \leqq x \leqq b$, joining the selected points with straight line segments. If the plotted points are sufficiently close, then the resulting graph may look to the unaided eye like a smooth curve, but it may miss some essential features that would be revealed if more points were plotted.

EXAMPLE 13 A 1-ampere alternating current with frequency 60 Hz (Hertz; cycles per second) is described by the function

$$I(t) = \sin 120\pi t. \tag{15}$$

The absolute value $|I(t)|$ gives the magnitude (in amperes) of the current at time t, which flows in one direction when $I > 0$, and in the opposite direction when $I < 0$. A simple computer program was used to plot the alleged graphs of $I(t)$ shown in Figs. 1.4.21 through 1.4.23. The graph in Fig. 1.4.21 is plotted on the interval $-1 \leqq t \leqq 1$, where we should see 120 complete oscillations because the period of $I(t)$ in Eq. (15) is $1/60$ s. But instead the figure shows exactly one oscillation, so something has gone badly wrong. The graph in Fig. 1.4.22 is plotted on the interval $-\frac{1}{2} \leqq t \leqq \frac{1}{2}$, and whatever it is has gone from merely wrong to outright bizarre. Finally, in Fig. 1.4.23 the graph is plotted on the interval $-\frac{1}{30} \leqq t \leqq \frac{1}{30}$ of length $\frac{4}{60}$, so we should see exactly four complete oscillations. And indeed we do, so we've finally got a correct graph of the current function in Eq. (15). ■

Here's an explanation of what went wrong at first in Example 13. The computer was programmed to plot values at exactly 120 equally spaced points of the interval desired. So in Fig. 1.4.21 we're plotting only 1 point per cycle—not nearly enough to capture the actual shape of the curve—and only 2 points per cycle in Fig. 1.4.22. But in Fig. 1.4.23 we're plotting 30 points per cycle, which gives an accurate representation of the actual graph.

The incorrect graph in Fig. 1.4.21—which seems to portray an oscillation with the incorrect period of 2 s, instead of $\frac{1}{60}$ s—is an example of the phenomenon of *aliasing*. Another example of aliasing, occasionally seen in old movie westerns, is the wagon wheel that appears to rotate slowly in the wrong direction.

Figure 1.4.22 consists largely of line segments joining consecutive points that are far apart. In Fig. 1.4.24, points 1, 3, 5, 7, ..., 117, 119 in the interval $[-0.5, 0.5]$ are

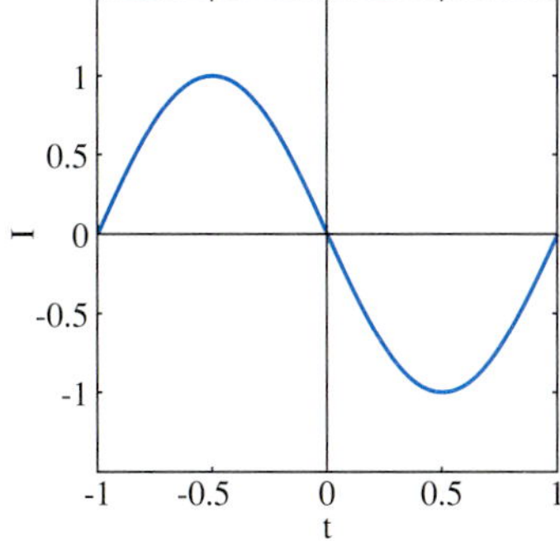

Fig. 1.4.21 On the interval $[-1, 1]$ it's wrong.

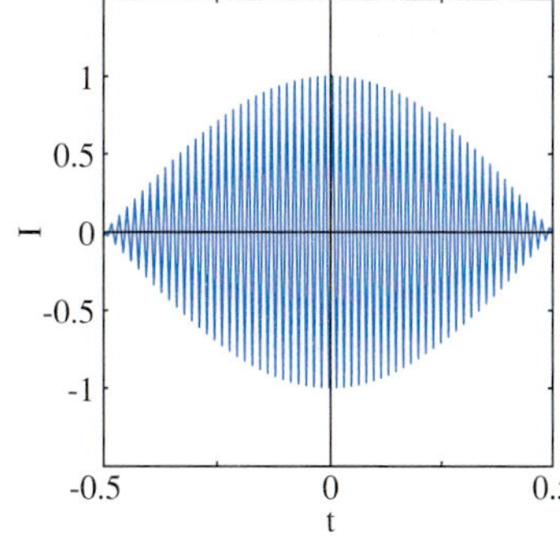

Fig. 1.4.22 On the interval $[-1/2, 1/2]$ it's bizarre.

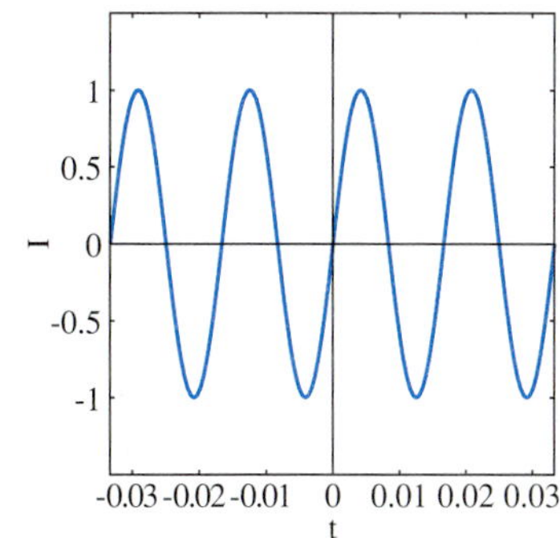

Fig. 1.4.23 On the interval $[-1/30, 1/30]$ it's correct!

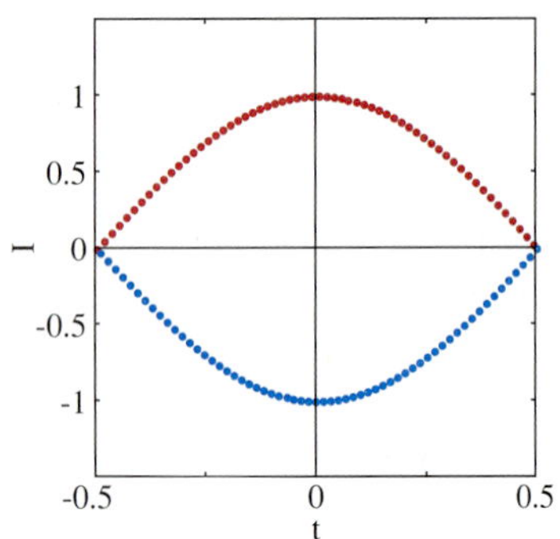

Fig. 1.4.24 Individual plotted points that are joined by line segments in Fig. 1.4.22

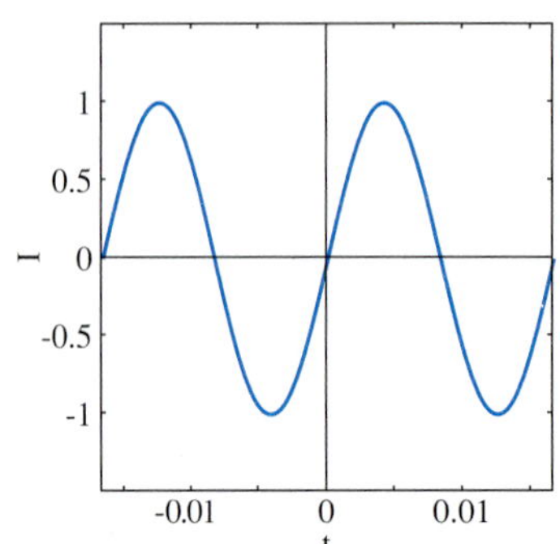

Fig. 1.4.25 $I(t) = \sin 120\pi t + 0.01 \sin 12000\pi t$ on the interval $-1/60 \leqq t \leqq 1/60$

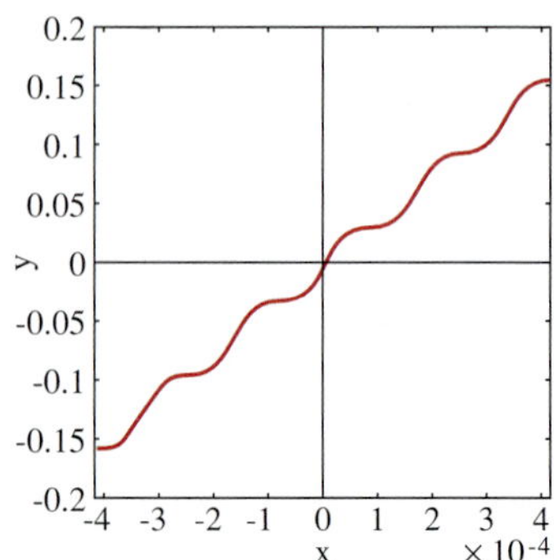

Fig. 1.4.26 $I(t) = \sin 120\pi t + 0.01 \sin 12000\pi t$ on the interval $-1/2400 \leqq t \leqq 1/2400$

plotted in red, while points 2, 4, 6, ..., 118, 120 are plotted in blue. Now you can see what happened when the computer plotted line segments joining point 1 to point 2, point 2 to point 3, and so forth.

One moral of Example 13 is that it pays to know what you're looking for in a graph. If the graph looks markedly different in windows of different sizes, this is a clue that something's wrong.

Whereas in Example 13 we got anomalous results by plotting the graph in windows of different sizes, the next example illustrates a situation where we *must* plot graphs on different scales in order to see the whole picture.

EXAMPLE 14 Now suppose that a high-frequency (6000 Hz) current of 0.01 ampere is added to the current in Eq. (15), so the resulting current is described by

$$I(t) = \sin 120\pi t + (0.01)\sin 12000\pi t. \tag{16}$$

When we plot Eq. (16) on the interval $-\frac{1}{60} \leqq t \leqq \frac{1}{60}$, we get the graph shown in Fig. 1.4.25. It looks like two cycles of the original current in (15), although the plot is perhaps a bit "fuzzy." To see the effect of the added second term in Eq. (16), we must plot the graph on a much magnified scale, as in Fig. 1.4.26. The "fuzz" in Fig. 1.4.25 has now been magnified to show clearly the high-frequency oscillations with period $\frac{1}{6000}$ s. ■

1.4 PROBLEMS

In Problems 1 through 10, match the given function with its graph among those shown in Figs. 1.4.27 through 1.4.36. Try to do this without turning on your graphing calculator or computer.

1. $f(x) = 2^x - 1$

2. $f(x) = 2 - 3^{-x}$

3. $f(x) = 1 + \cos x$

4. $f(x) = 2 - 2\sin x$

5. $f(x) = 1 + 2\cos x$

6. $f(x) = 2 - \sin x$

7. $f(x) = \dfrac{x}{2^x}$

8. $f(x) = \dfrac{\log x}{x}$

9. $f(x) = \dfrac{1 + \cos 6x}{1 + x^2}$

10. $f(x) = 2^{-x}\sin 10x$

In Problems 11 through 20, find $f(g(x))$ and $g(f(x))$.

11. $f(x) = 1 - x^2, \quad g(x) = 2x + 3$

12. $f(x) = -17, \quad g(x) = |x|$

13. $f(x) = \sqrt{x^2 - 3}, \quad g(x) = x^2 + 3$

14. $f(x) = x^2 + 1, \quad g(x) = \dfrac{1}{x^2 + 1}$

15. $f(x) = x^3 - 4, \quad g(x) = \sqrt[3]{x + 4}$

16. $f(x) = \sqrt{x}, \quad g(x) = \cos x$

17. $f(x) = \sin x, \quad g(x) = x^3$

18. $f(x) = \sin x, \quad g(x) = \cos x$

19. $f(x) = 1 + x^2, \quad g(x) = \tan x$

20. $f(x) = 1 - x^2, \quad g(x) = \sin x$

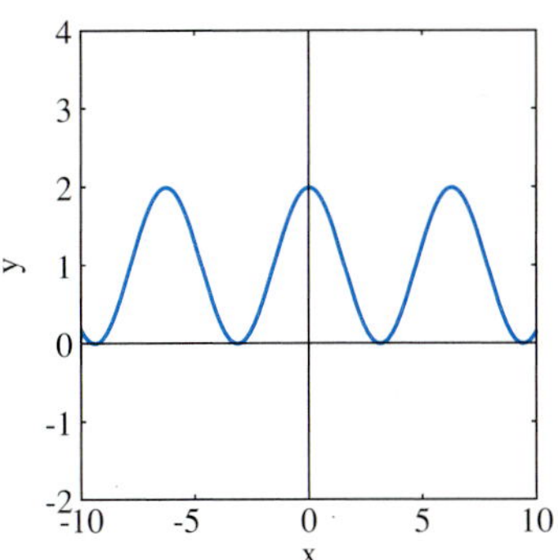

Fig. 1.4.27

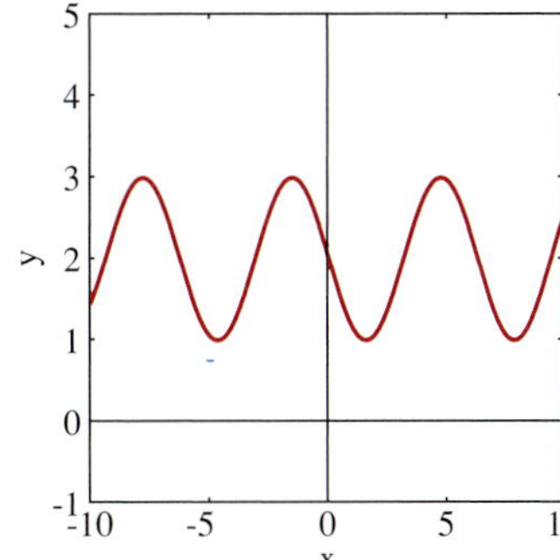

Fig. 1.4.28

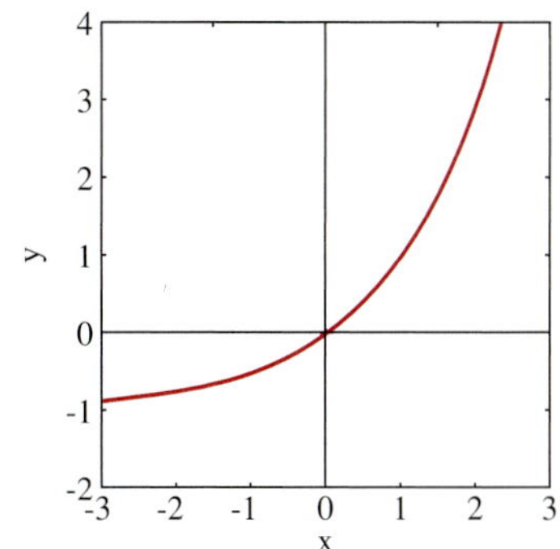

Fig. 1.4.29

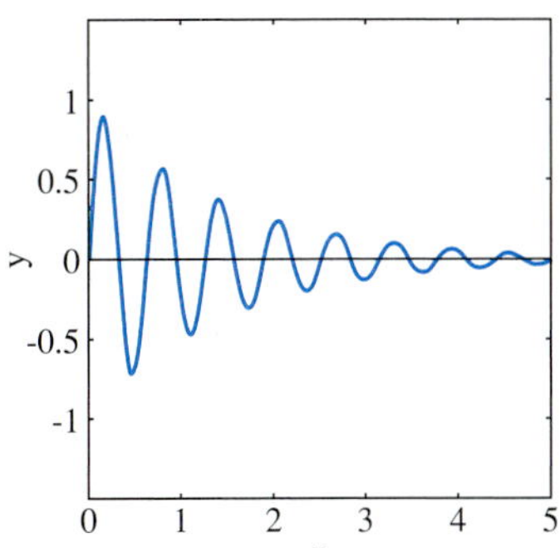

Fig. 1.4.30

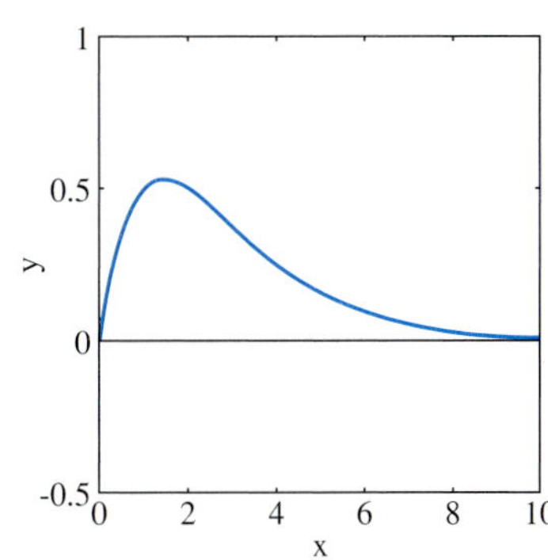

Fig. 1.4.31

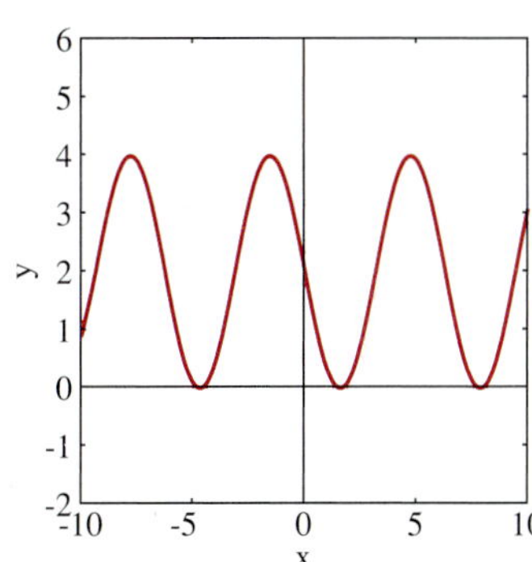

Fig. 1.4.32

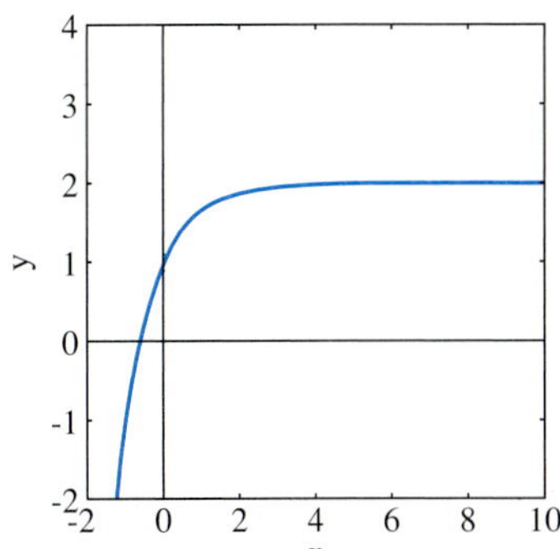

Fig. 1.4.33

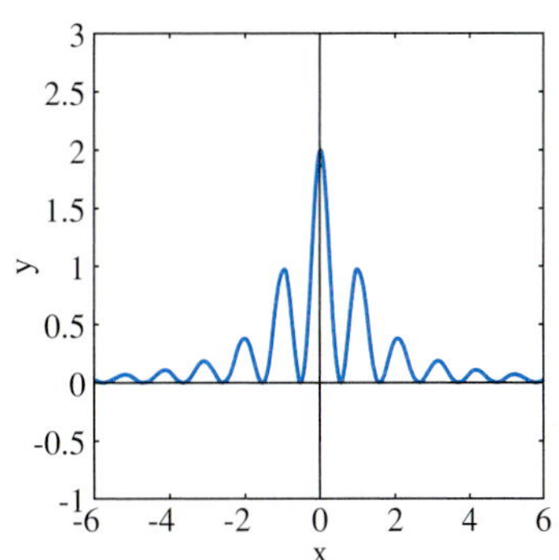

Fig. 1.4.34

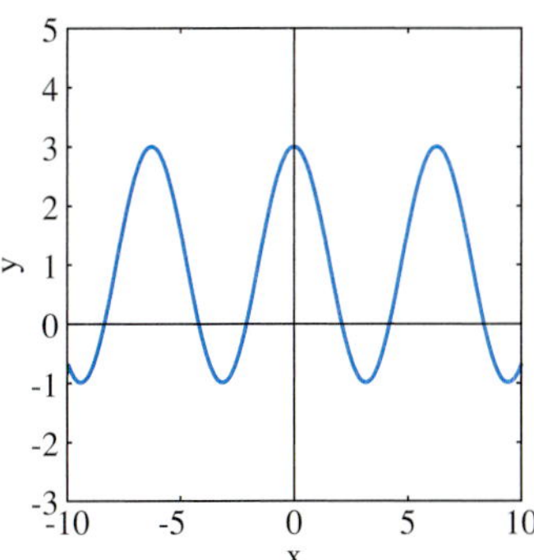

Fig. 1.4.35

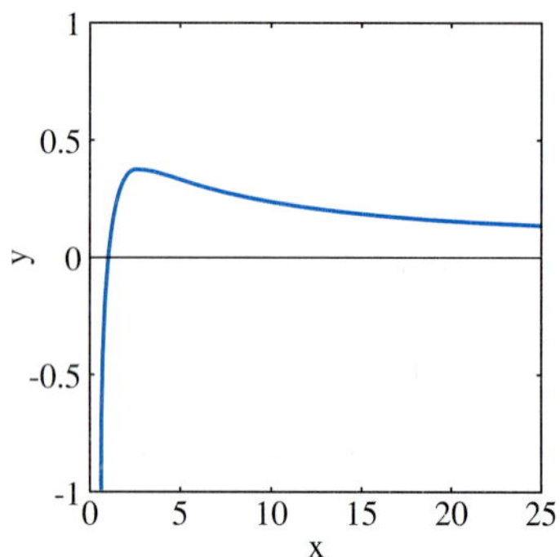

Fig. 1.4.36

In Problems 21 through 30, find a function of the form $f(x) = x^k$ (you must specify k) and a function g such that $f(g(x)) = h(x)$.

21. $h(x) = (2 + 3x)^2$

22. $h(x) = (4 - x)^3$

23. $h(x) = \sqrt{2x - x^2}$

24. $h(x) = (1 + x^4)^{17}$

25. $h(x) = (5 - x^2)^{3/2}$

26. $h(x) = \sqrt[3]{(4x - 6)^4}$

27. $h(x) = \dfrac{1}{x + 1}$

28. $h(x) = \dfrac{1}{1 + x^2}$

29. $h(x) = \dfrac{1}{\sqrt{x + 10}}$

30. $h(x) = \dfrac{1}{(1 + x + x^2)^3}$

Use a graphing calculator or computer to determine by inspection of graphs the number of real solutions of the equations in Problems 31 through 40.

31. $x = 2^{-x}$

32. $x + 1 = 3\cos x$

33. $x - 1 = 3\cos x$

34. $x = 5\cos x$

35. $x = 7\cos x$

36. $2\log_{10} x = \cos x \quad (x > 0)$

37. $\log_{10} x = \cos x \quad (x > 0)$

38. $x^2 = 10\cos x$

39. $x^2 = 100\sin x$

40. $x = 5\cos x + 10\log_{10} x \quad (x > 0)$

41. Consider the population of Example 7 in this section, which starts with 10 rodents and doubles every month. Determine graphically (that is, by zooming) how long it will take this population to grow to 100 rodents. (Assume that each month is 30 days long, and obtain an answer correct to the nearest day.)

42. Consider the money-market account of Example 8, which pays 8% annually. Determine graphically how long it will take the initial investment of \$5000 to triple.

43. In 1980 the population P of Mexico was 67.4 million and was growing at the rate of 2.6% per year. If the population continues to grow at this rate, then t years after 1980 it will be $P(t) = 67.4 \cdot (1.026)^t$ (millions). Determine graphically how long it then will take the population of Mexico to double.

44. Suppose that the amount A of ozone in the atmosphere decreases at the rate of 0.25% per year, so that after t years the amount remaining is $A(t) = A_0(0.9975)^t$, where A_0 denotes the initial amount. Determine graphically how long it will take for only half the original amount of ozone to be left. Does the numerical value of A_0 affect this answer?

45. The nuclear accident at Chernobyl left the surrounding region contaminated with strontium-90, which initially was emitting radiation at approximately 12 times the level safe for human habitation. When an atom of strontium-90 emits radiation, it decays to a nonradioactive isotope. In this way, about 2.5% of the strontium-90 disappears each year. Then the amount of radiation left after t years will be $A(t) = 12 \cdot (0.975)^t$ (measured in "safe units" of radiation). Determine graphically how long (after the original accident) it will be until the region measures only 1 safe unit and it is therefore safe for humans to return.

46. Refer to Example 6 of this section; determine graphically the value (accurate to three decimal places) of the negative solution of the equation $x^2 = 2^x$.

47. Refer to Example 10 of this section; determine graphically the value (accurate to three decimal places) of the solution near $x = 5$ of the equation $\log_{10} x = \frac{1}{2}x^{1/5}$.

48. The equation $x^{10} = 3^x$ has three real solutions. Graphically approximate each of them accurate to two decimal places.

1.4 PROJECT: A SPHERICAL ASTEROID

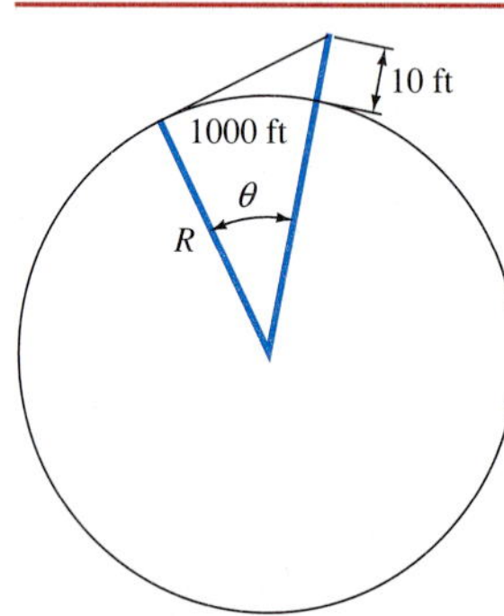

Fig. 1.4.37 The asteroid problem

You land your space ship on a spherical asteroid between Earth and Mars. Your copilot walks 1000 feet away along the smooth surface carrying a 10-ft rod and thereby vanishes over the horizon. When she places one end of the rod on the ground and holds it straight up and down, you—lying flat on the ground—can just barely see the tip of the rod. What is the radius R of this asteroid? First find R accurate to the nearest foot and then convert to miles.

Suggestion: Show first that the radius R and the angle θ (in radians) indicated in Fig. 1.4.37 satisfy the equations

$$R\theta = 1000 \quad \text{and} \quad (R + 10)\cos\theta = R.$$

To get started on a graphical solution, solve each of these equations for R in terms of θ.

For your own personal investigation, let p be the largest and q the next largest digits in your student I.D. number. Suppose instead that your copilot's rod is p feet long and that she walks $100q$ feet away before all but the tip of the rod vanishes from your view. Write the results of your investigation in the form of a carefully organized report consisting of complete sentences (together with pertinent equations and figures) and explaining your results in detail. Tell precisely what you did to solve your problem.

1.5 PREVIEW: WHAT IS CALCULUS?

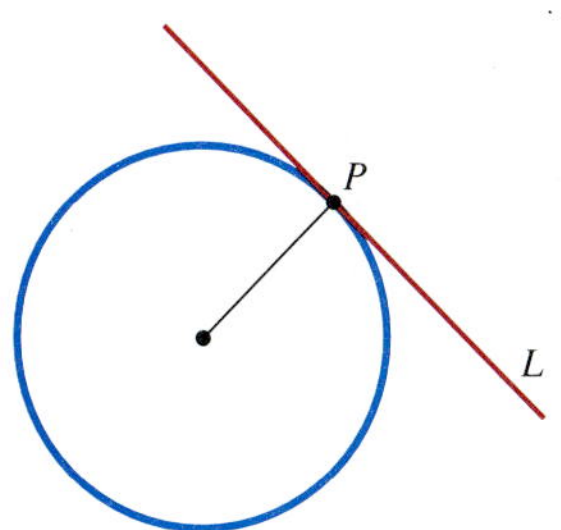

Fig. 1.5.1 The tangent L touches the circle at the point P.

Surely this question is on your mind as you begin a study of calculus that may extend over two or three terms. Following our review of functions and graphs in Sections 1.1 through 1.4, we can preview here at least the next several chapters, where the central concepts of calculus are developed.

The Two Fundamental Problems

The body of computational technique that constitutes "the calculus" revolves around two fundamental geometric problems that people have been investigating for more than 2000 years. Each problem involves the graph $y = f(x)$ of a given function.

The first fundamental problem is this: What do we mean by the *line tangent* to the curve $y = f(x)$ at a given point? The word *tangent* stems from the Latin word *tangens,* for "touching." Thus a line tangent to a curve is one that "just touches" the curve. Lines tangent to circles (Fig. 1.5.1) are well known from elementary geometry. Figure 1.5.2 shows the line tangent to the parabola $y = x^2$ at the point $(1, 1)$. We will see in Section 2.1 that this particular tangent line has slope 2, so its point-slope equation is

$$y - 1 = 2 \cdot (x - 1); \quad \text{that is,} \quad y = 2x - 1.$$

Our first problem is how to find tangent lines in more general cases.

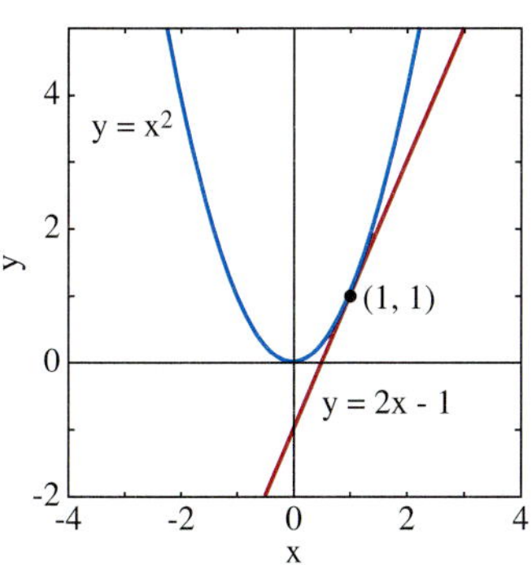

Fig. 1.5.2 The line tangent to the parabola $y = x^2$ at the point $(1, 1)$

> **The Tangent Problem**
>
> Given a point $P(x, f(x))$ on the curve $y = f(x)$, how do we calculate the slope of the tangent line at P (Fig. 1.5.3)?

We begin to explore the answer to this question in Chapter 2. If we denote by $m(x)$ the slope of the tangent line at $P(x, f(x))$, then m is a *new function.* It might informally be called a *slope predictor* for the curve $y = f(x)$. In calculus this slope predictor function is called the **derivative** of the function f. In Chapter 3 we learn to calculate derivatives of a variety of functions, and in both Chapter 3 and Chapter 4 we see numerous applications of derivatives in solving real-world problems. This gives an introduction to the part of calculus called *differential calculus*.

The tangent problem is a geometric problem—a purely mathematical question. But its answer (in the form of derivatives) is the key to the solution of diverse applied problems in many scientific and technical areas. Examples 1 and 2 may suggest to you the *connections* that are the key to the pivotal role of calculus in science and technology.

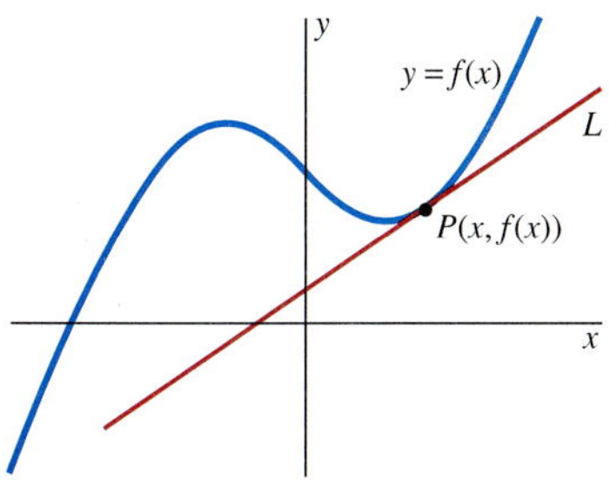

Fig. 1.5.3 What is the slope of the line L tangent to the graph $y = f(x)$ at the point $P(x, f(x))$?

EXAMPLE 1 Suppose that you're driving a car along a long, straight road (Fig. 1.5.4). If $f(t)$ denotes the *distance* (in miles) the car has traveled at time t (in hours), then the slope of the line tangent to the curve $y = f(t)$ at the point $(t, f(t))$ (Fig. 1.5.5) is the *velocity* (in miles per hour) of the car at time t. ■

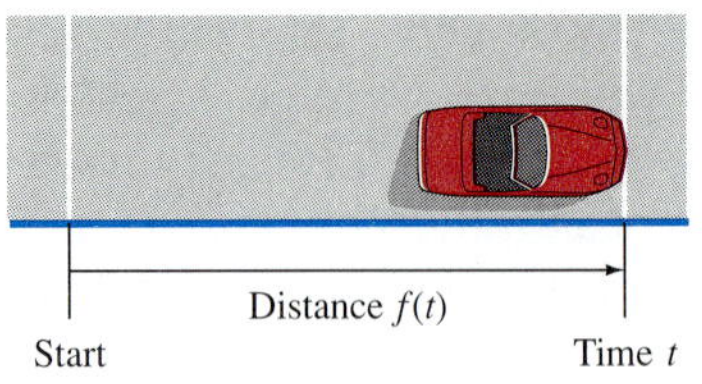

Fig. 1.5.4 A car on a straight road (Example 1)

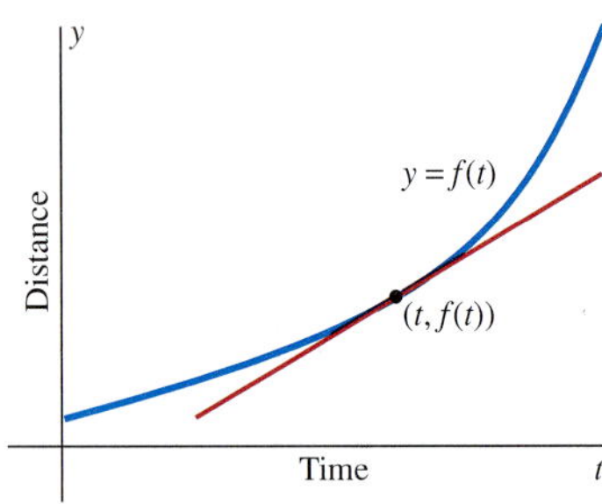

Fig. 1.5.5 The slope of the tangent line at the point $(t, f(t))$ is the velocity at the time t (Example 1).

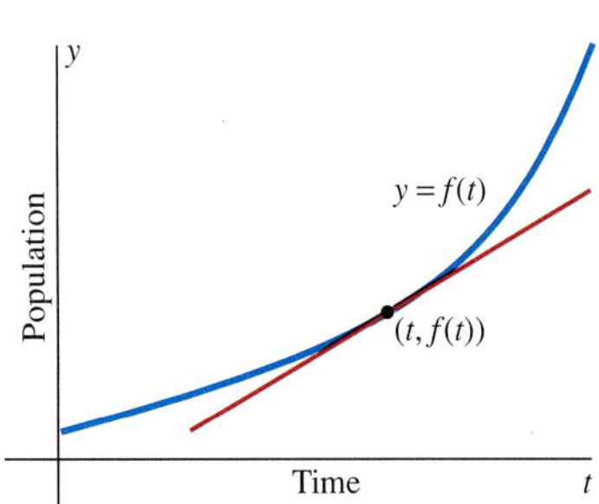

Fig. 1.5.6 The rate of growth of $f(t)$ at the time t is the slope of the tangent line at the point $(t, f(t))$ (Example 2).

EXAMPLE 2 Suppose that $f(t)$ denotes the number of people in the United States who have a certain serious disease at time t (measured in days from the beginning of the year). Then the slope of the line tangent to the curve $y = f(t)$ at the point $(t, f(t))$ (Fig. 1.5.6) is the *rate of growth* (the number of persons newly affected per day) of the diseased population at time t. ■

Note The truth of the statements made in these two examples is *not* obvious. To understand such things is one reason you study calculus! We return to the concepts of velocity and rate of change at the beginning of Chapter 3.

Here we will be content with the observation that the slopes of the tangent lines in Examples 1 and 2 at least have the correct *units*. If in the time–distance plane of Example 1 we measure time t (on the horizontal axis) in seconds and distance y (on the vertical axis) in feet (or meters), then the slope (ratio of rise to run) of a straight line has the dimensions of feet (or meters) per second—the proper units for velocity (Fig. 1.5.7). Similarly, if in the ty-plane of Example 2 time t is measured in months and y is measured in persons, then the slope of a straight line has the proper units of persons per month for measuring the rate of growth of the afflicted population (Fig. 1.5.8).

The second fundamental problem of calculus is the problem of *area*. Given the graph $y = f(x)$, what is the area between the graph of f and the x-axis over the interval $[a, b]$?

The Area Problem

If $f(x) \geqq 0$ for x in the interval $[a, b]$, how do we calculate the area A of the plane region that lies between the curve $y = f(x)$ and the x-axis over the interval $[a, b]$ (Fig. 1.5.9)?

We begin to explore the answer to this second question in Chapter 5. In calculus the area A is called an *integral* of the function f. Chapters 5 and 6 are devoted to the calculation and application of integrals. This gives an introduction to the other part of calculus, which is called *integral calculus*.

Like the tangent problem, the area problem is a purely mathematical geometric question, but its answer (in the form of integrals) has extensive ramifications of practical importance. Examples 3 and 4 have an obvious kinship with Examples 1 and 2.

EXAMPLE 3 If $f(t)$ denotes the *velocity* of a car at time t, then the area under the curve $y = f(t)$ over the time interval $[a, b]$ is equal to the *distance* traveled by the car between time $t = a$ and time $t = b$ (Fig. 1.5.10). ■

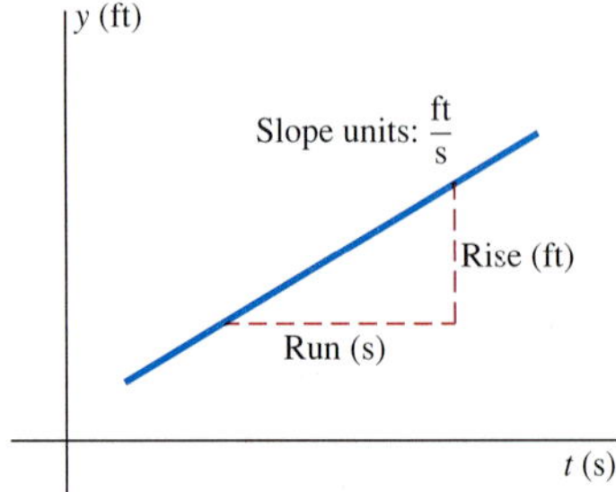

Fig. 1.5.7 Here slope has the dimensions of velocity (ft/s).

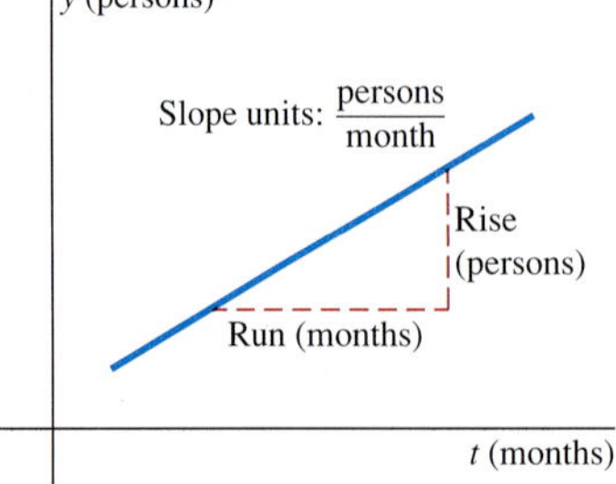

Fig. 1.5.8 Here slope has the dimensions of rate of change of population.

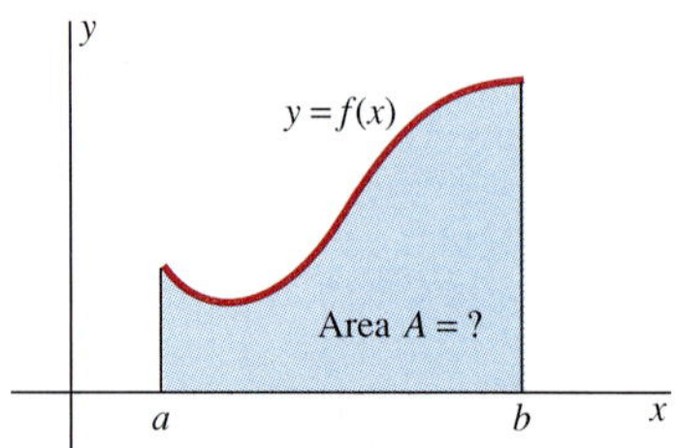

Fig. 1.5.9 The area problem

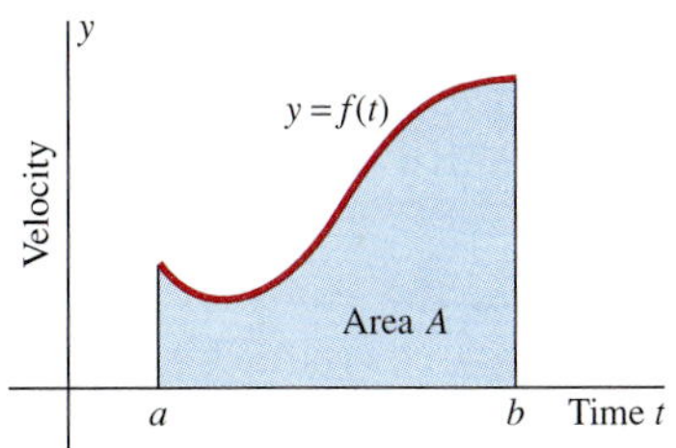

Fig. 1.5.10 The area A under the velocity curve is equal to the distance traveled during the time interval $a \leqq t \leqq b$ (Example 3).

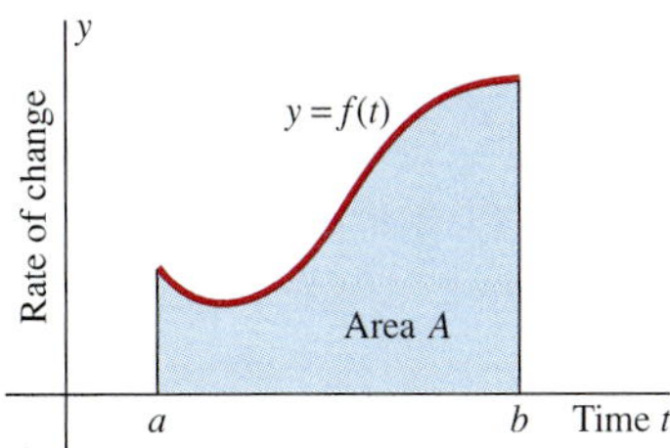

Fig. 1.5.11 The area A under the rate-of-change curve is equal to the net change in the population from time $t = a$ to $t = b$ (Example 4).

EXAMPLE 4 If $f(t)$ denotes the *rate of growth* of a diseased population at time t, then the area under the curve $y = f(t)$ over the time interval $[a, b]$ is equal to the net change in the *size* of this population between time $t = a$ and time $t = b$ (Fig. 1.5.11). ■

When we discuss integrals in Chapter 5, you will learn why the statements in Examples 3 and 4 are true.

The Fundamental Relationship

Examples 1 and 3 are two sides of a certain coin: There is an "inverse relationship" between the *distance* traveled and the *velocity* of a moving car. Examples 2 and 4 exhibit a similar relationship between the *size* of a population and its *rate of change*.

Both the distance–velocity relationship and the size–rate-of-change relationship illustrated by Examples 1 through 4 are consequences of a deep and fundamental relationship between the tangent problem and the area problem. This more general relationship is described by the *fundamental theorem of calculus,* which we discuss in Section 5.6. It was discovered in 1666 by Isaac Newton at the age of 23 while he was still a student at Cambridge University. A few years later it was discovered independently by Gottfried Wilhelm Leibniz, who was then a German diplomat in Paris who studied mathematics privately. Although the tangent problem and the area problem had, even then, been around for almost 2000 years, and much progress on separate solutions had been made by predecessors of Newton and Leibniz, their joint discovery of the fundamental relationship between the area and tangent problems made them famous as "the inventors of the calculus."

Applications of Calculus

So calculus centers on the computation *and application* of derivatives and integrals—that is, of tangent line slopes and areas under graphs. Throughout this textbook, you will see concrete applications of calculus to different areas of science and technology. The following list of a baker's dozen such applications gives just a brief indication of the extraordinary range and real-world power of calculus.

- Suppose that you make and sell tents. How can you make the biggest tent from a given amount of cloth and thereby maximize your profit? (Section 3.6)
- You throw into a lake a cork ball that has one-fourth the density of water. How deep will it sink in the water? (Section 3.9)
- A driver involved in an accident claims he was going only 25 mi/h. Can you determine from his skid marks the actual speed of his car at the time of the accident? (Section 5.2)
- The great pyramid of Khufu at Gizeh, Egypt, was built well over 4000 years ago. No personnel records from the construction remain, but nevertheless we can calculate the approximate number of laborers involved. (Section 6.6)
- If the earth's population continues to grow at its present rate, when will there be "standing room only"? (Section 7.5)
- The factories polluting Lake Erie are forced to cease dumping wastes into the lake immediately. How long will it take for natural processes to restore the lake to an acceptable level of purity? (Section 7.6)
- In 1845 the Belgian demographer Verhulst used calculus to predict accurately the course of U.S. population growth (to within 1%) well into the twentieth century, long after his death. How? (Section 9.5)

- Suppose that you win the Florida lottery and decide to use part of your winnings to purchase a "perpetual annuity" that will pay you and your heirs (and theirs, ad infinitum) $10,000 per year. What is a fair price for an insurance company to charge you for such an annuity? (Section 9.8)
- What explains the fact that a well-positioned reporter can eavesdrop on a quiet conversation between two diplomats 50 feet away in the Whispering Gallery of the U.S. Senate, even if this conversation is inaudible to others in the same room? (Section 10.7)
- Suppose that Paul and Mary alternately toss a fair six-sided die in turn until one wins the pot by getting the first "six." How advantageous is it to be the one who tosses first? (Section 11.3)
- How can a submarine traveling in darkness beneath the polar icecap keep accurate track of its position without being in radio contact with the rest of the world? (Section 12.4)
- Suppose that your club is designing an unpowered race car for the annual downhill derby. You have a choice of solid wheels, bicycle wheels with thin spokes, or even solid spherical wheels (like giant ball bearings). Can you determine (without time-consuming experimentation) which will make the race car go the fastest? (Section 14.5)
- Some bullets have flattened tips. Is it possible that an artillery shell with a flat-tipped "nose cone" may experience less air resistance—and therefore travel farther—than a shell with a smoothly rounded tip? (Section 15.5)

CHAPTER 1 REVIEW: *Definitions and Concepts*

Use this list as a guide to key topics that you may need to review or discuss.

1. The definition of a function
2. The domain and range of a function
3. Dependent and independent variables
4. Open and closed interval notation
5. The idea of a mathematical model
6. Numerical investigation of values of functions by repeated tabulation
7. Slope-intercept and point-slope equations of straight lines
8. The graph of an equation
9. The graph of a function
10. The translation principle for graphs of equations
11. The vertical line test for graphs of functions
12. Equations and translates of circles
13. Parabolas and graphs of quadratic functions
14. Contrasting graphic, numeric, and symbolic views of functions
15. Graphical solution of equations by the method of repeated magnification
16. Algebraic combinations of functions
17. The maximum possible number of zeros of a polynomial
18. The behavior of the polynomial $p(x)$ for $|x|$ large
19. Location of the vertical asymptotes of a rational function
20. Qualitative comparison of graphs of power functions, polynomials, rational functions, and algebraic functions
21. Periods of trigonometric functions
22. Rates of growth of exponential and logarithmic functions
23. Qualitative comparison of the graphs of trigonometric, exponential, and logarithmic functions
24. Composition of functions
25. Determination of a viewing window that exhibits the principal features of the graph of a function
26. Possible pitfalls of calculator and computer graphing

CHAPTER 1 *Miscellaneous Problems*

In Problems 1 through 10, find the domain of definition of the function with the given formula.

1. $f(x) = \sqrt{x - 4}$

2. $f(x) = \dfrac{1}{2 - x}$

3. $f(x) = \dfrac{1}{x^2 - 9}$

4. $f(x) = \dfrac{x}{x^2 + 1}$

5. $f(x) = (1 + \sqrt{x})^3$

6. $f(x) = \dfrac{x + 1}{x^2 - 2x}$

7. $f(x) = \sqrt{2 - 3x}$

8. $f(x) = \dfrac{1}{\sqrt{9 - x^2}}$

9. $f(x) = (x - 2)(4 - x)$

10. $f(x) = \sqrt{(x - 2)(4 - x)}$

11. In accord with Boyle's law, the pressure p (lb/in.^2) and volume V (in.^3) of a certain gas satisfy the condition $pV = 800$. What is the range of possible values of p, given $100 \leqq V \leqq 200$?

12. The relationship between the Fahrenheit temperature F and the Celsius temperature C is given by

$$F = 32 + \tfrac{9}{5}C.$$

If the temperature on a given day ranges from a low of 70°F to a high of 90°F, what is the range of temperature in degrees Celsius?

13. An electric circuit contains a battery that supplies E volts in series with a resistance of R ohms (Fig. 1.MP.1). Then the current of I amperes that flows in the circuit satisfies Ohm's law, $E = IR$. If $E = 100$ and $25 < R < 50$, what is the range of possible values of I?

14. The period T (in seconds) of a simple pendulum of length L (in feet) is given by $T = 2\pi\sqrt{L/32}$. If $3 < L < 4$, what is the range of possible values of T?

15. Express the volume V of a cube as a function of its total surface area S.

16. The height of a certain right circular cylinder is equal to its radius. Express its total surface area A (including both ends) as a function of its volume V.

17. Express the area A of an equilateral triangle as a function of its perimeter P.

18. A piece of wire 100 in. long is cut into two pieces of lengths x and $100 - x$. The first piece is bent into the shape of a square, the second into the shape of a circle. Express as a function of x the sum A of the areas of the square and circle.

In Problems 19 through 24, write an equation of the straight line L described.

19. L passes through $(-3, 5)$ and $(1, 13)$.

20. L passes through $(4, -1)$ and has slope -3.

21. L has slope $\frac{1}{2}$ and y-intercept -5.

22. L passes through $(2, -3)$ and is parallel to the line with equation $3x - 2y = 4$.

23. L passes through $(-3, 7)$ and is perpendicular to the line with equation $y - 2x = 10$. (Appendix B reviews slopes of perpendicular lines.)

24. L is the perpendicular bisector of the segment joining $(1, -5)$ and $(3, -1)$.

In Problems 25 through 34, match the given function with its graph among those shown in Figs. 1.MP.2 through 1.MP.11. Try to do so without turning on your graphing calculator or computer. Instead, rely on your knowledge of the general characteristics of polynomial, rational, algebraic, trigonometric, exponential, and logarithmic functions.

25. $f(x) = 2 - 2x - x^2$

26. $f(x) = x^3 - 4x^2 + 5$

27. $f(x) = x^4 - 4x^3 + 5$

28. $f(x) = \dfrac{5}{x^2 - x - 6}$

29. $f(x) = \dfrac{5}{x^2 - x + 6}$

30. $f(x) = \sqrt{8 + 2x - x^2}$

31. $f(x) = 2^{-x} - 1$

32. $f(x) = \log_{10}(x + 1)$

33. $f(x) = 1 + 3\sin x$

34. $f(x) = x + 3\sin x$

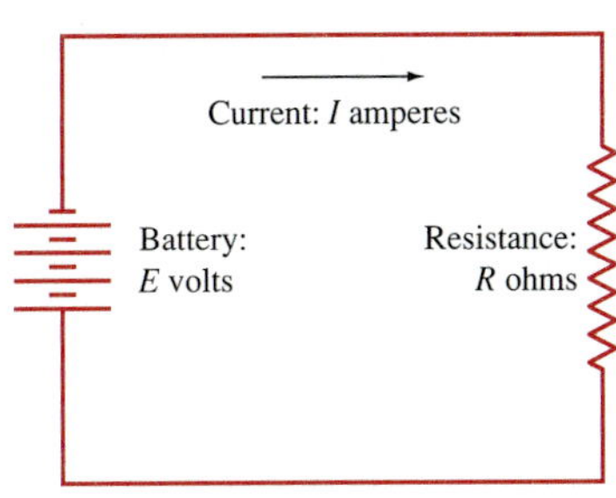

Fig. 1.MP.1 The simple electric circuit of Problem 13

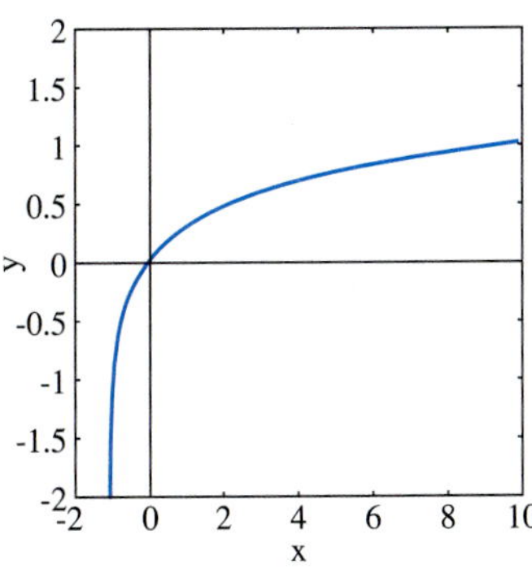

Fig. 1.MP.2

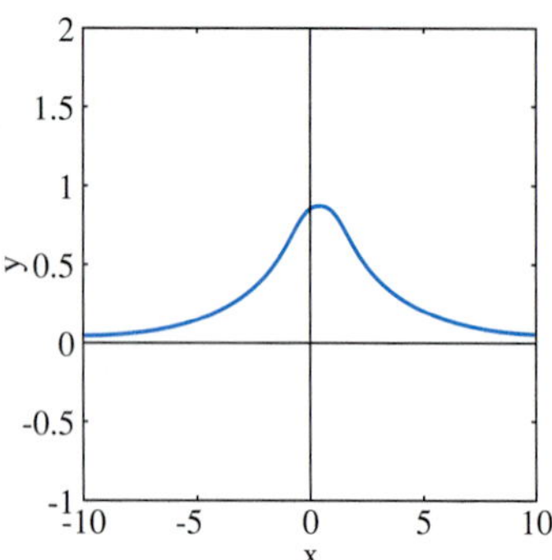

Fig. 1.MP.3

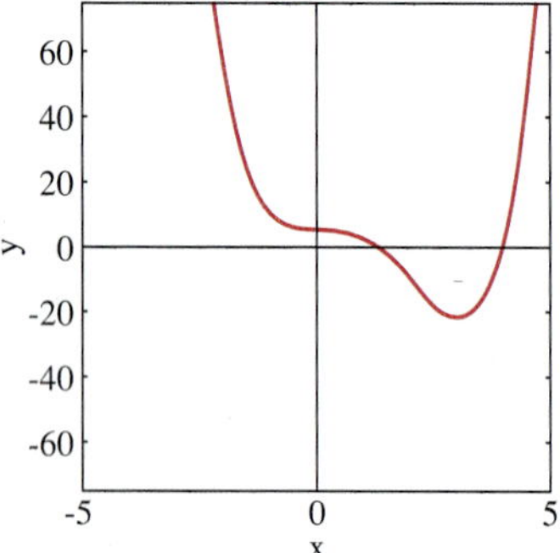

Fig. 1.MP.4

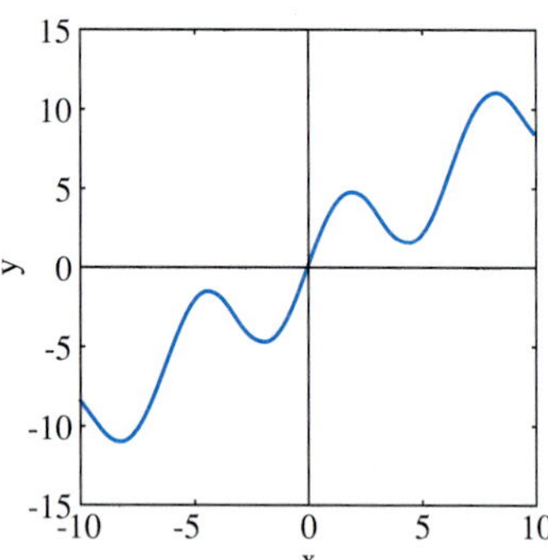

Fig. 1.MP.5

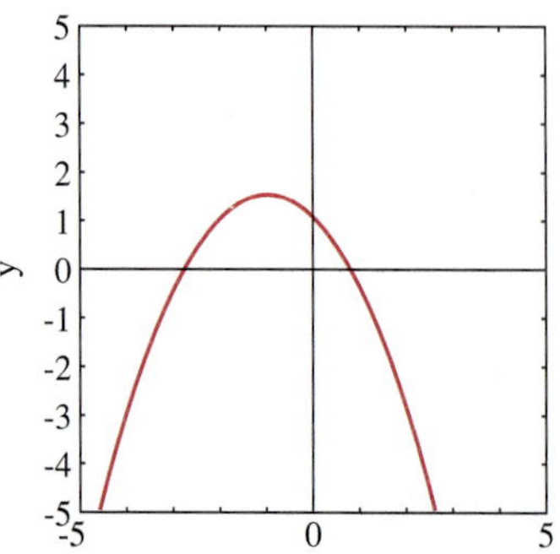

Fig. 1.MP.6

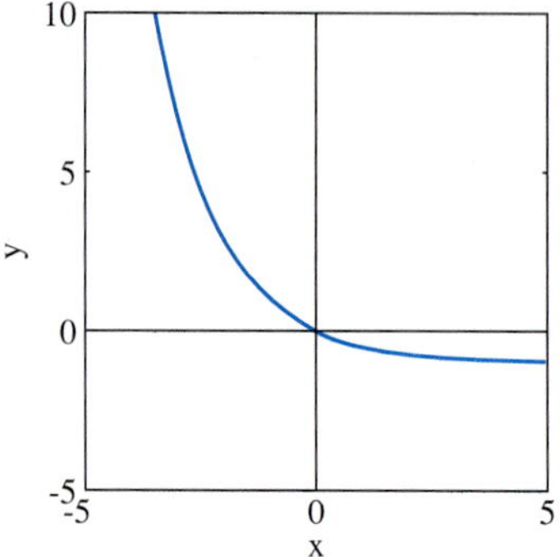

Fig. 1.MP.7

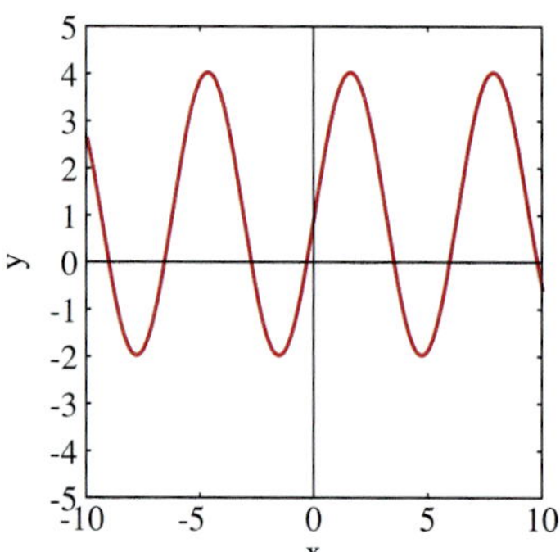

Fig. 1.MP.8

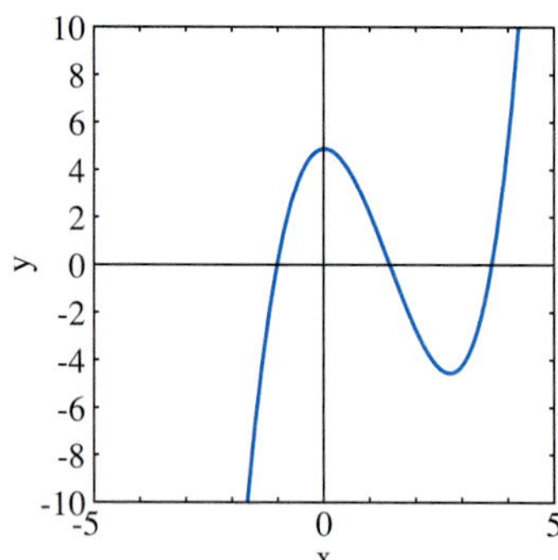

Fig. 1.MP.9

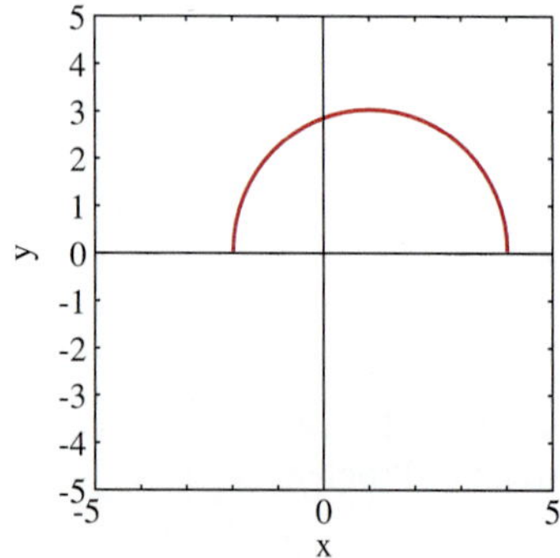

Fig. 1.MP.10

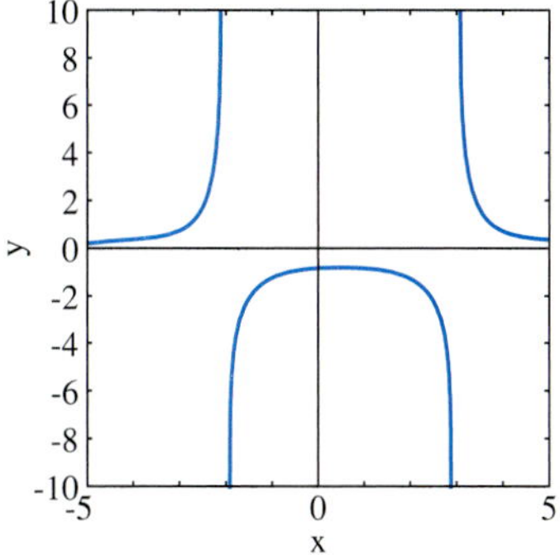

Fig. 1.MP.11

Sketch the graphs of the equations and functions given in Problems 35 through 44.

35. $2x - 5y = 7$

36. $|x - y| = 1$

37. $x^2 + y^2 = 2x$

38. $x^2 + y^2 = 4y - 6x + 3$

39. $y = 2x^2 - 4x - 1$

40. $y = 4x - x^2$

41. $f(x) = \dfrac{1}{x + 5}$

42. $f(x) = \dfrac{1}{4 - x^2}$

43. $f(x) = |x - 3|$

44. $f(x) = |x - 3| + |x + 2|$

45. Apply the triangle inequality (of Appendix A) twice to show that

$$|a + b + c| \leqq |a| + |b| + |c|$$

for arbitrary real numbers a, b, and c.

46. Write $a = (a - b) + b$ to deduce from the triangle inequality that

$$|a| - |b| \leqq |a - b|$$

for arbitrary real numbers a and b.

47. Solve the inequality $x^2 - x - 6 > 0$. [*Suggestion:* Conclude from the factorization

$$x^2 - x - 6 = (x - 3)(x + 2)$$

that the quantities $x - 3$ and $x + 2$ are either both positive or both negative. Consider the two cases separately to conclude that the solution set is $(-\infty, -2) \cup (3, +\infty)$.]

Use the method of Problem 47 to solve the inequalities in Problems 48 through 50.

48. $x^2 - 3x + 2 < 0$

49. $x^2 - 2x - 8 > 0$

50. $2x \geqq 15 - x^2$

The remaining problems require the use of an appropriate calculator or computer. In Problems 51 through 56, use either the method of repeated tabulation or the method of successive zooms (or both) to find the two roots (with three digits to the right of the decimal correct or correctly rounded) of the given algebraic equation. You may check your work with the aid of the quadratic formula and an ordinary calculator.

51. $x^2 - 5x - 7 = 0$

52. $3x^2 - 10x - 11 = 0$

53. $4x^2 - 14x + 11 = 0$

54. $5x^2 + 24x - 35 = 0$

55. $8x^2 + 33x - 36 = 0$

56. $9x^2 + 74x - 156 = 0$

In Problems 57 through 62, apply either the method of repeated tabulation or the method of successive zooms (or both) to find the lowest point on the given parabola. You may check your work by completing the square.

57. $y = x^2 - 5x + 7$

58. $y = 3x^2 - 10x + 11$

59. $y = 4x^2 - 14x + 11$

60. $y = 5x^2 + 24x + 35$

61. $y = 8x^2 + 33x + 35$

62. $y = 9x^2 + 74x + 156$

63. Figure 1.MP.12 shows a 10-cm by 7-cm portrait with a border of width x on the top and bottom and of width $2x$ on either side. The area of the border is itself 20 cm^2. Use either repeated tabulation or successive zooms to find x.

64. A mail-order catalog lists a 60-in. by 35-in. tablecloth that shrinks 7% in area when first washed. The catalog description also implies that *both* the length and width will decrease by the same amount x. Use numerical (tabulation) or graphical (zoom) methods to find x.

Determine graphically the number of real solutions of each equation in Problems 65 through 70.

65. $x^3 - 7x + 3 = 0$

66. $x^4 - 3x^2 + 4x - 5 = 0$

67. $\sin x = x^3 - 3x + 1$

68. $\cos x = x^4 - x$

69. $\cos x = \log_{10} x$

70. $10^{-x} = \log_{10} x$

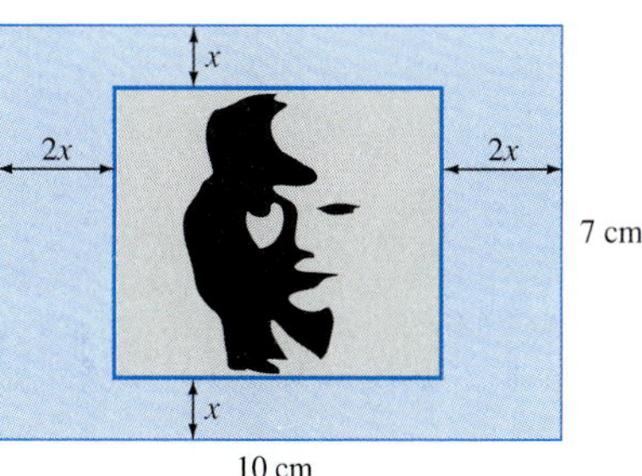

Fig. 1.MP.12 The bordered portrait of Problem 63

CHAPTER 2

PRELUDE TO CALCULUS

Ada Byron (1815–1852)

The modern computer programming language Ada is named in honor of Ada Byron, daughter of the English poet Lord Byron. Her interest in science and mathematics led her around 1840 to study the Difference Engine, a gear-based mechanical calculator that the mathematician Charles Babbage had built to compute tables of values of functions. By then he was designing his much more advanced Analytic Engine, an elaborate computing machine that would have been far ahead of its time had it been completed. In 1843 Ada Byron wrote a series of brief essays explaining the planned operation of the Analytical Engine and its underlying mathematical principles. She included a prototype "computer program" to illustrate how its calculations were to be "programmed" in advance, using a deck of punched cards to specify its instructions.

The Difference Engine

Calculus has been called "the calculating engine par excellence." But in our own time the study and applications of calculus have been reshaped by electronic computers. Throughout this book we illustrate concepts of calculus by means of graphic, numeric, and symbolic results generated by computers. In Chapter 2 we exploit computational technology systematically in the investigation of limits.

Almost exactly a century after the death of Ada Byron, the first modern computer compiler—for translation of human language programs into machine language instructions—was developed by Grace Murray Hopper. As a mathematician and U.S. Navy officer, Hopper had worked with the very first modern electronic computers developed during and immediately after World War II. In 1967 she was recalled to active duty to lead efforts to standardize the computer language COBOL for the Navy. In 1985 at the age of 79, she became Rear Admiral Grace Hopper. In 1986 she was retired—as the Navy's oldest commissioned officer on active duty—in a ceremony held aboard the U.S.S. *Constitution*, the Navy's oldest commissioned warship.

Grace Murray Hopper (1906–1992)

2.1 TANGENT LINES AND SLOPE PREDICTORS

In Sections 1.2 and 1.5 we saw that certain applied problems raise the question of what is meant by the *tangent line* at a specified point of a general curve $y = f(x)$. In this section we see that this "tangent line problem" leads to the limit concept, which we pursue further in Section 2.2.

In elementary geometry the line tangent to a circle at a point P is defined as the straight line through P that is perpendicular to the radius OP to that point (Fig. 2.1.1). A general graph $y = f(x)$ has no radius for us to use, but the line tangent to the graph at the point P should be the straight line through P that has—in some sense—the same direction at P as the curve itself. Because a line's "direction" is determined by its slope, our plan for defining a line tangent to a curve amounts to finding an appropriate "slope–prediction formula" that will give the proper slope of the tangent line. Example 1 illustrates this approach in the case of one of the simplest of all non-straight curves, the parabola with equation $y = x^2$.

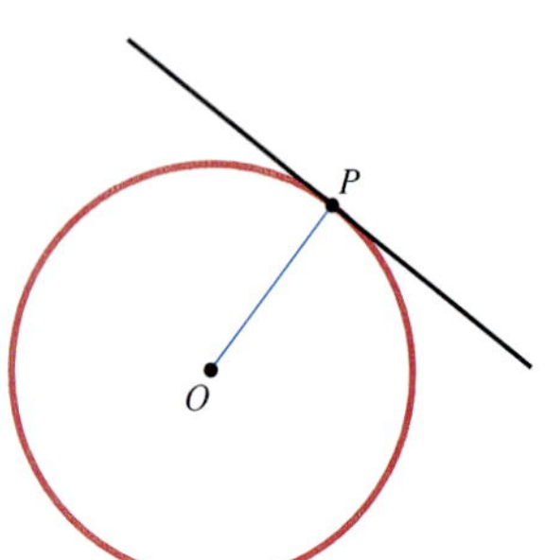

Fig. 2.1.1 The line tangent to the circle at the point P is perpendicular to the radius OP.

EXAMPLE 1 Determine the slope of the line L tangent to the parabola $y = x^2$ at the point $P(a, a^2)$.

Solution Figure 2.1.2 shows the parabola $y = x^2$ and a typical point $P(a, a^2)$ on it. The figure also shows a visual guess of the direction of the desired tangent line L at P. We must find the slope of L.

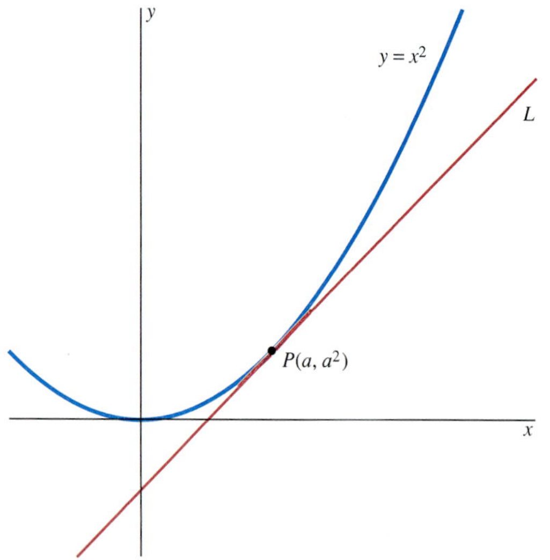

Fig. 2.1.2 The tangent line at P should have the same direction as the curve does at P (Example 1).

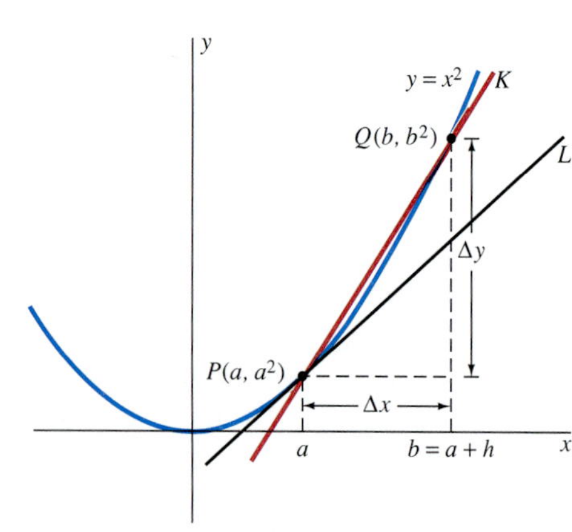

Fig. 2.1.3 The secant line K passes through the two points P and Q, which we can use to determine its slope (Example 1).

We cannot immediately calculate the slope of L, because we know the coordinates of only one point $P(a, a^2)$ of L. Hence we begin with another line whose slope we can compute. Figure 2.1.3 shows the **secant line** K that passes through the point P and the nearby point $Q(b, b^2)$ of the parabola $y = x^2$. Let us write

$$h = \Delta x = b - a$$

for the difference of the x-coordinates of P and Q. (The notation Δx is as old as calculus itself, and it means now what it did 300 years ago: an **increment,** or *change,* in the value of x.) Then the coordinates of Q are given by the formulas

$$b = a + h \quad \text{and} \quad b^2 = (a + h)^2.$$

Hence the difference in the y-coordinates of P and Q is

$$\Delta y = b^2 - a^2 = (a + h)^2 - a^2.$$

Because P and Q are two different points, we can use the definition of slope to calculate the slope m_{PQ} of the secant line K through P and Q. If you change the value of $h = \Delta x$, you change the line K and thereby change its slope. Therefore, m_{PQ} depends on h:

$$\begin{aligned} m_{PQ} &= \frac{\Delta y}{\Delta x} = \frac{(a + h)^2 - a^2}{(a + h) - a} \\ &= \frac{(a^2 + 2ah + h^2) - a^2}{h} = \frac{2ah + h^2}{h} = \frac{h(2a + h)}{h}. \end{aligned} \tag{1}$$

Because h is nonzero, we may cancel it in the final fraction. Thus we find that the slope of the secant line K is given by

$$m_{PQ} = 2a + h. \tag{2}$$

Now imagine what happens as you move the point Q closer and closer to the point P. (This situation corresponds to h approaching zero.) The line K still passes through P and Q, but it pivots around the fixed point P. As h approaches zero, the secant line K comes closer to coinciding with the tangent line L. This phenomenon is suggested in Fig. 2.1.4, which shows the secant line K approaching the tangent line L.

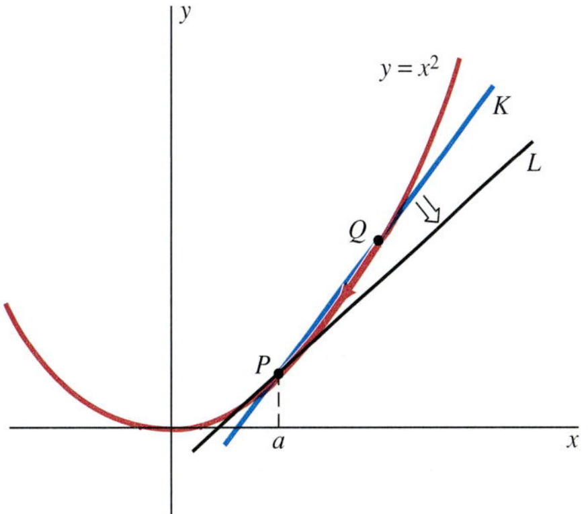

Fig. 2.1.4 As $h \to 0$, Q approaches P, and K moves into coincidence with the tangent line L (Example 1).

The tangent line L must, by definition, lie in the *limiting position* of the secant line K. To see precisely what this means, examine what happens to the slope of K as K pivots into coincidence with L:

As h approaches zero,
Q approaches P, and so
K approaches L; meanwhile,
the slope of K approaches the slope of L.

Hence our question is this: As the number h approaches zero, what value does the slope $m_{PQ} = 2a + h$ approach? We can state this question of the "limiting value" of $2a + h$ by writing

$$\lim_{h \to 0} (2a + h) = ? \qquad \textbf{(3)}$$

Here, "lim" is an abbreviation for the word "limit," and "$h \to 0$" is an abbreviation for the phrase "h approaches zero." Thus (3) asks, "What is the limit of $2a + h$ as h approaches zero?"

For any specific value of a we can investigate this question numerically by calculating values of $2a + h$ with values of h that become closer and closer to zero—such as the values $h = 0.1$, $h = -0.01$, $h = 0.001$, $h = -0.0001, \ldots$, or the values $h = 0.5$, $h = 0.1$, $h = 0.05$, $h = 0.01, \ldots$. For instance, the tables of values in Figs. 2.1.5 and 2.1.6 indicate that with $a = 1$ and $a = -2$ we should conclude that

$$\lim_{h \to 0} (2 + h) = 2 \quad \text{and} \quad \lim_{h \to 0} (-4 + h) = -4.$$

h	$2 + h$
0.1	2.1
0.01	2.01
0.001	2.001
0.0001	2.0002
↓	↓
0	2

Fig. 2.1.5 As $h \to 0$ (first column), $2 + h$ approaches 2 (second column).

h	$-4 + h$
0.5	−3.5
0.1	−3.9
0.05	−3.95
0.01	−3.99
0.005	−3.995
0.001	−3.999
↓	↓
0	−4

Fig. 2.1.6 As $h \to 0$ (first column), $-4 + h$ approaches -4 (second column).

h	$2a + h$
0.01	$2a$+0.01
0.001	$2a$+0.001
⋮	⋮
↓	↓
0	$2a$

Fig. 2.1.7 As $h \to 0$ (first column), $2a + h$ approaches $2a$ (second column) (Example 1).

More generally, it seems clear from the table in Fig. 2.1.7 that

$$\lim_{h \to 0} m_{PQ} = \lim_{h \to 0} (2a + h) = 2a. \qquad \textbf{(4)}$$

This, finally, answers our original question: The slope $m = m(a)$ of the line tangent to the parabola $y = x^2$ at the point (a, a^2) is given by

$$m = 2a. \qquad \textbf{(5)}$$

■

The formula in Eq. (5) is a "slope predictor" for (lines tangent to) the parabola $y = x^2$. Once we know the slope of the line tangent to the curve at a given point of the curve, we can then use the point-slope formula to write an equation of this tangent line.

EXAMPLE 2 With $a = 1$, the slope predictor in Eq. (5) gives $m = 2$ for the slope of the line tangent to $y = x^2$ at the point $(1, 1)$. Hence an equation of this line is

$$y - 1 = 2(x - 1); \quad \text{that is,} \quad y = 2x - 1.$$

With $a = -3$, Eq. (5) gives $m = -6$ as the slope of the line tangent at $(-3, 9)$, so an equation of the line tangent to the curve at this point is

$$y - 9 = -6(x + 3); \quad \text{that is,} \quad y = -6x - 9.$$

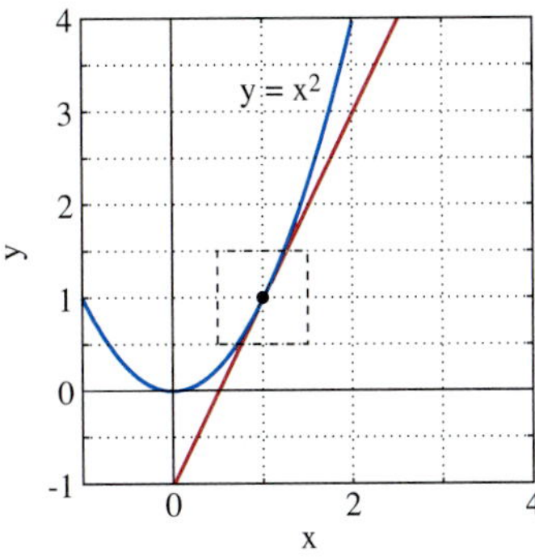

Fig. 2.1.8 The parabola $y = x^2$ and its tangent line at $P(1, 1)$

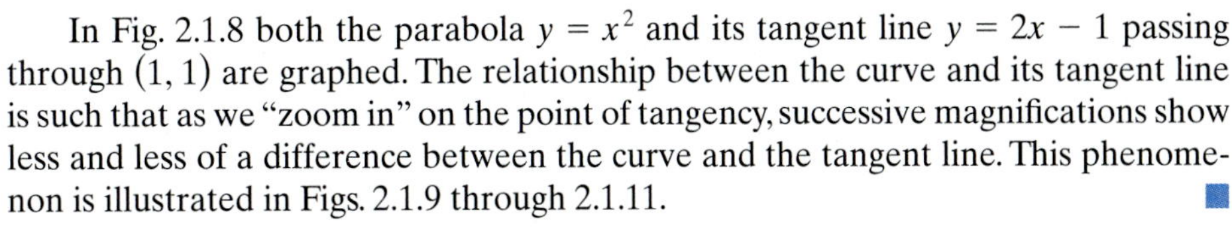

In Fig. 2.1.8 both the parabola $y = x^2$ and its tangent line $y = 2x - 1$ passing through $(1, 1)$ are graphed. The relationship between the curve and its tangent line is such that as we "zoom in" on the point of tangency, successive magnifications show less and less of a difference between the curve and the tangent line. This phenomenon is illustrated in Figs. 2.1.9 through 2.1.11. ■

REMARK In Example 1 we proceeded as though the concept of a tangent line to a curve were self-evident. The actual meaning of the slope-predictor result $m = 2a$ in Eq. (5) is this: Whatever is meant by the line tangent to the parabola $y = x^2$ at the point $P(a, a^2)$, it can only be the unique straight line through P with slope $m = 2a$. Thus we must *define* the line tangent to $y = x^2$ at P to be the line whose point-slope equation is $y - a^2 = 2a(x - a)$. Pictures like those in Figs. 2.1.8 through 2.1.11 certainly support our conviction that this definition is the correct one.

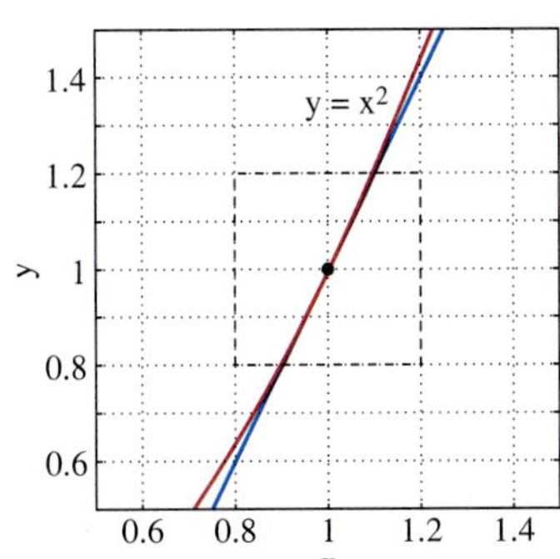

Fig. 2.1.9 First magnification

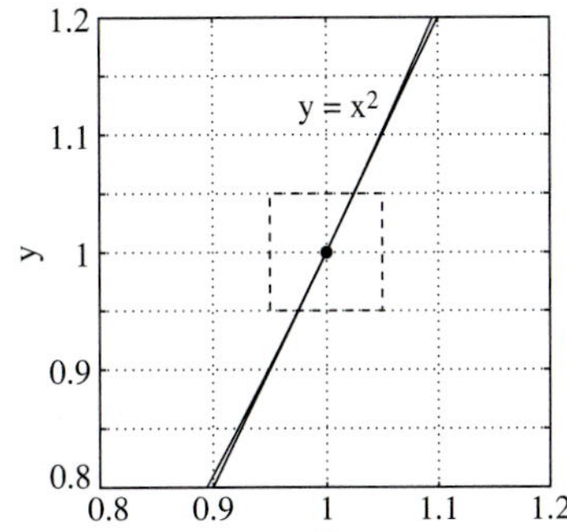

Fig. 2.1.10 Second magnification

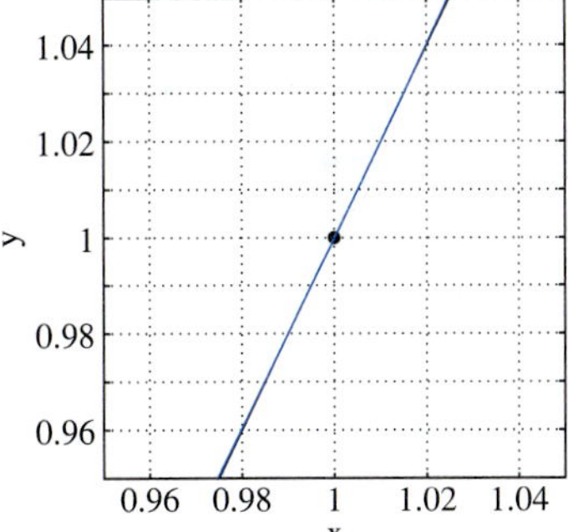

Fig. 2.1.11 Can you see the difference?

More General Slope Predictors

The general case of the line tangent to a curve $y = f(x)$ is scarcely more complicated than the special case $y = x^2$ of Example 1. Given the function f, suppose that we want to find the slope of the line L tangent to $y = f(x)$ at the point $P(a, f(a))$. As indicated in Fig. 2.1.12, let K be the secant line passing through the point P and the nearby point $Q(a + h, f(a + h))$ on the graph. The slope of this secant line is the **difference quotient**

$$m_{PQ} = \frac{\Delta y}{\Delta x} = \frac{f(a + h) - f(a)}{h} \qquad (\text{with } h \neq 0). \tag{6}$$

We now force Q to approach the fixed point P along the curve $y = f(x)$ by making h approach zero. We ask whether m_{PQ} approaches some limiting value m as $h \to 0$. If so, we write

$$m = \lim_{h \to 0} \frac{f(a + h) - f(a)}{h}$$

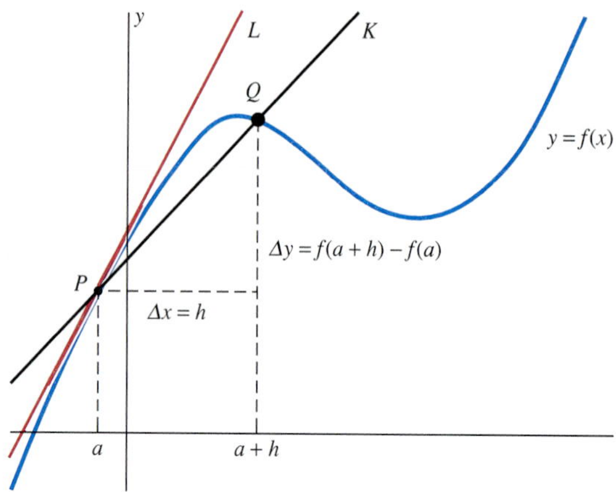

Fig. 2.1.12 As $h \to 0$, $Q \to P$, and the slope of K approaches the slope of the tangent line L.

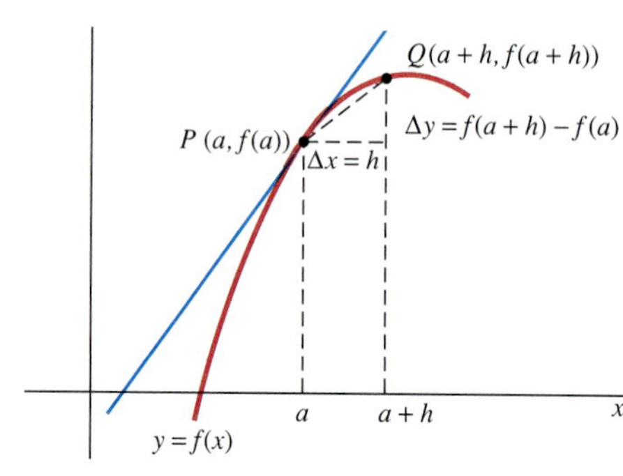

Fig. 2.1.13 The slope of the tangent line at $(a, f(a))$ is $m(a) = \lim_{h\to 0} \dfrac{f(a+h) - f(a)}{h}$.

and conclude that this number m is the slope of the line tangent to the graph $y = f(x)$ at the point $(a, f(a))$. Actually, this slope depends on a and we can indicate this by writing

$$m(a) = \lim_{h\to 0} \frac{f(a+h) - f(a)}{h}. \tag{7}$$

If we can express the limiting value on the right explicitly in terms of a, then Eq. (7) yields a *slope predictor* for lines tangent to the curve $y = f(x)$. In this case the line tangent to the curve at the point $P(a, f(a))$ is *defined* to be the straight line through P that has slope $m(a)$. This tangent line is indicated in Fig. 2.1.13.

In Chapter 3 we will recognize the fact that the slope $m(a)$ is somehow "derived" from the function f by calling this number the **derivative** of the function f at the point a. Indeed, much of Chapter 3 will be devoted to methods of calculating derivatives of various familiar functions. Most of these methods are based on the limit techniques of Sections 2.2 and 2.3, but the case of quadratic functions is sufficiently simple for inclusion here. Recall from Section 1.2 that the graph of any quadratic function is a parabola that opens either upward or downward.

Theorem ***Parabolas and Tangent Lines***

Consider the parabola $y = f(x)$ where

$$f(x) = px^2 + qx + r \tag{8}$$

(with $p \neq 0$). Then the line tangent to this parabola at the point $P(a, f(a))$ has slope

$$m(a) = 2pa + q. \tag{9}$$

PROOF The slope of the secant line given in (6) may be simplified as follows:

$$m_{PQ} = \frac{f(a+h) - f(a)}{h}$$

$$= \frac{[p(a+h)^2 + q(a+h) + r] - [pa^2 + qa + r]}{h}$$

$$= \frac{[p(a^2 + 2ah + h^2) + q(a+h) + r] - [pa^2 + qa + r]}{h}$$

$$= \frac{2pah + ph^2 + qh}{h},$$

and therefore

$$m_{PQ} = 2pa + q + ph.$$

The numbers $p, q,$ and a are fixed, so as $h \to 0$ the product ph approaches zero, much as in our computations in Example 1. Thus

$$m(a) = \lim_{h\to 0} m_{PQ} = \lim_{h\to 0} (2pa + q + ph) = 2pa + q,$$

as claimed in Eq. (9). ■

Remark Thus the formula $m(a) = 2pa + q$ provides a ready slope predictor for lines tangent to the parabola

$$y = px^2 + qx + r. \tag{10}$$

Given the numbers $p, q,$ and a, we need only substitute in the slope-predictor formula to obtain the slope $m(a)$ of the line tangent to the parabola at the point whose x-coordinate is a. We needn't repeat the computational steps that were carried out in the derivation of the slope-predictor formula.

EXAMPLE 3 Find an equation of the line that is tangent to the parabola $y = 2x^2 - 3x + 5$ at the point where $x = -1$.

Solution Here we have $p = 2, q = -3, r = 5$. Thus with $a = -1$, we find that $f(x) = px^2 + qx + r$ yields $f(-1) = 2\cdot(-1)^2 - 3\cdot(-1) + 5 = 10$. The slope predictor in (9) then gives

$$m(-1) = 2\cdot(2)\cdot(-1) + (-3) = -7$$

for the slope of the tangent line at the point $(-1, 10)$ of the parabola. The point-slope equation of the tangent line is therefore

$$y - 10 = (-7)(x + 1); \qquad \text{that is,} \quad y = -7x + 3. \quad ■$$

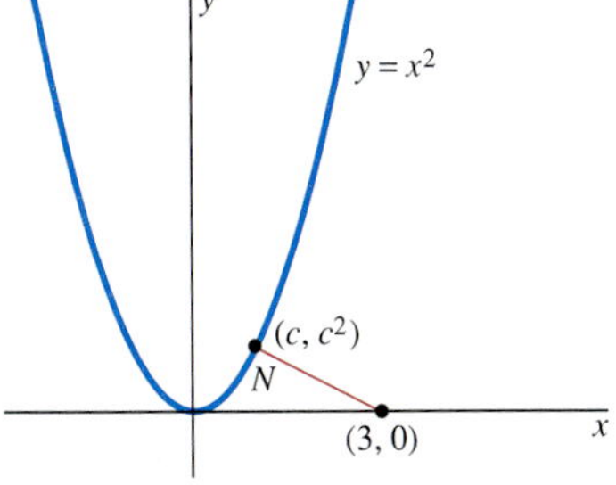

Fig. 2.1.14 The normal line N from the point $(3, 0)$ to the point (c, c^2) on the parabola $y = x^2$

Normal Lines

How would you find the point $P(c, c^2)$ that lies on the parabola $y = x^2$ and is closest to the point $(3, 0)$? Intuitively, the line segment N with endpoints $(3, 0)$ and P should be perpendicular, or *normal,* to the parabola's tangent line at P (Fig. 2.1.14). But if the slope of the tangent line is m, then—by Theorem 2 in Appendix B—the slope of the normal line is

$$m_N = -\frac{1}{m}. \tag{11}$$

(Theorem 2 tells us that if two perpendicular lines have nonzero slopes m_1 and m_2, then $m_1 m_2 = -1$.) More precisely, the **normal line** at a point P of a curve where the

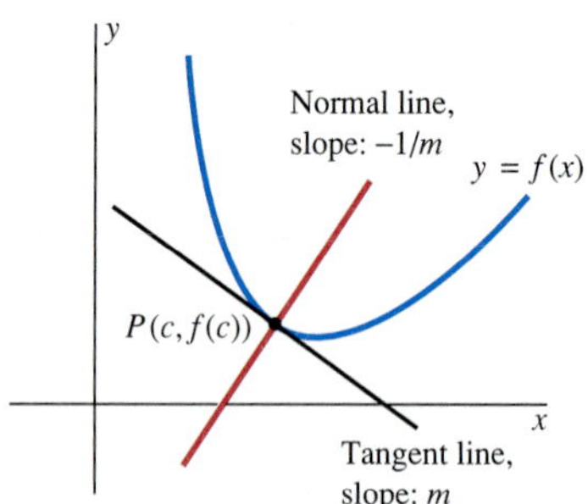

Fig. 2.1.15 The tangent line and normal line through the point P on a curve

tangent line has slope m is *defined* to be the line through P with slope $m_N = -1/m$ (Fig. 2.1.15). Consequently, the parabolic slope predictor in (9) enables us to write equations of lines normal to parabolas as easily as equations of tangent lines.

EXAMPLE 4 In Example 3 we found that the line tangent to the parabola $y = 2x^2 - 3x + 5$ at the point $(-1, 10)$ has slope -7. Therefore the slope of the line normal to that parabola at P is $m_N = -1/(-7) = \frac{1}{7}$. So the point-slope equation of the normal line is

$$y - 10 = \frac{1}{7}(x + 1); \quad \text{that is,} \quad y = \frac{1}{7}x + \frac{71}{7}.$$

The Animal Pen Problem Completed

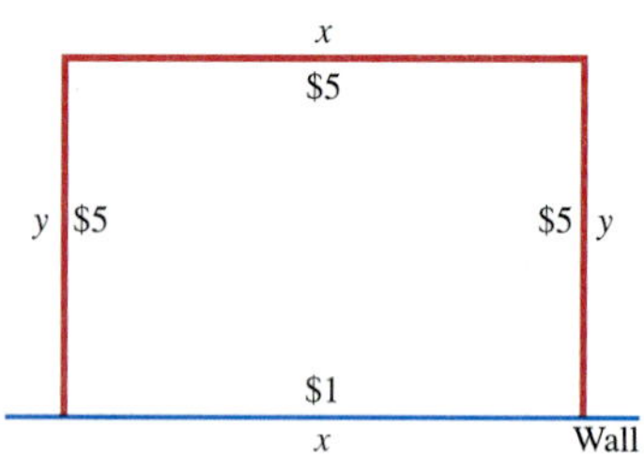

Fig. 2.1.16 The animal pen

Now we can apply our newfound knowledge of slope-predictor formulas to wrap up our continuing discussion of the animal pen problem of Section 1.1. In Example 9 there we found that the area A of the pen (see Fig. 2.1.16) is given as a function of its base length x by

$$A(x) = \tfrac{3}{5}(30x - x^2) = -\tfrac{3}{5}x^2 + 18x \tag{12}$$

for $0 \leqq x \leqq 30$. Therefore our problem is to find the maximum value of $A(x)$ for x in the closed interval $[0, 30]$.

Let us accept as intuitively obvious—we will see a proof in Chapter 3—the fact that the maximum value of $A(x)$ occurs at the high point where the line tangent to the parabola $y = A(x)$ is *horizontal,* as indicated in Fig. 2.1.17. But the function $A(x)$ in Eq. (12) is quadratic with $p = -\frac{3}{5}$ and $q = 18$ (comparing (12) with (8)). Therefore the slope predictor in (9) implies that the slope of the tangent line at an arbitrary point $(x, A(x))$ of the parabola is given by

$$m = m(x) = 2px + q = -\tfrac{6}{5}x + 18.$$

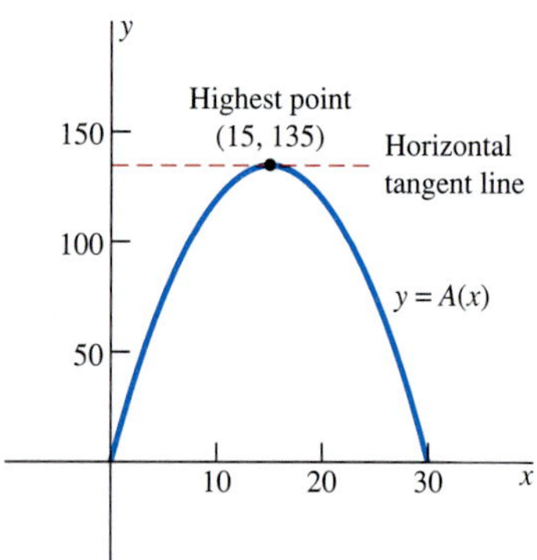

Fig. 2.1.17 The graph of $y = A(x)$, $0 \leqq x \leqq 30$

We ask when $m = 0$ and find that this happens when

$$-\tfrac{6}{5}x + 18 = 0,$$

and thus when $x = 15$. In agreement with the result found by algebraic methods in Section 1.2, we find that the maximum possible area of the pen is

$$A(15) = \tfrac{3}{5}(30 \cdot 10 - 15^2) = 135 \quad (\text{ft}^2).$$

Numerical Investigation of Slopes

Suppose that you are given the function f and a specific numerical value of a. You can then use a calculator to investigate the value

$$m = \lim_{h \to 0} \frac{f(a + h) - f(a)}{h} \tag{13}$$

of the slope of the line tangent to the curve $y = f(x)$ at the point $(a, f(a))$. Simply calculate the values of the difference quotient

$$\frac{f(a + h) - f(a)}{h} \tag{14}$$

with successively smaller nonzero values of h to see whether a limiting numerical value is apparent.

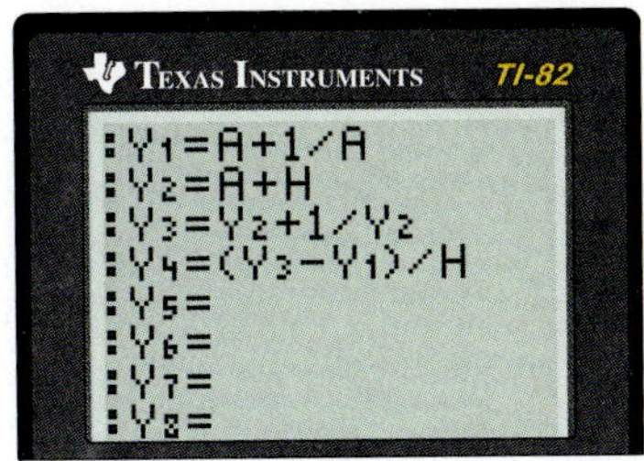

Fig. 2.1.18 A calculator prepared to calculate $\dfrac{f(a+h)-f(a)}{h}$ with $f(x) = x + \dfrac{1}{x}$

EXAMPLE 5 Find by numerical investigation (an approximation to) the line tangent to the graph of

$$f(x) = x + \frac{1}{x} \tag{15}$$

at the point $(2, \frac{5}{2})$.

Solution Figure 2.1.18 shows a TI calculator prepared to calculate the difference quotient in (14) with the function f in Eq. (15). As indicated in Fig. 2.1.19, successive values of this quotient can then be calculated by brief "one-liners." Figure 2.1.20 shows an HP calculator prepared to define the same quotient; then evaluation of the expression `'M(2,0.0001)'` yields the approximate value $m \approx 0.75001$. In this way we get the table shown in Fig. 2.1.21, which suggests that the slope of the line tangent to the graph of $f(x)$ at the point $(2, \frac{5}{2})$ is $m = \frac{3}{4}$. If so, then the tangent line at this point has the point-slope equation

$$y - \tfrac{5}{2} = \tfrac{3}{4}(x - 2); \quad \text{that is,} \quad y = \tfrac{3}{4}x + 1.$$

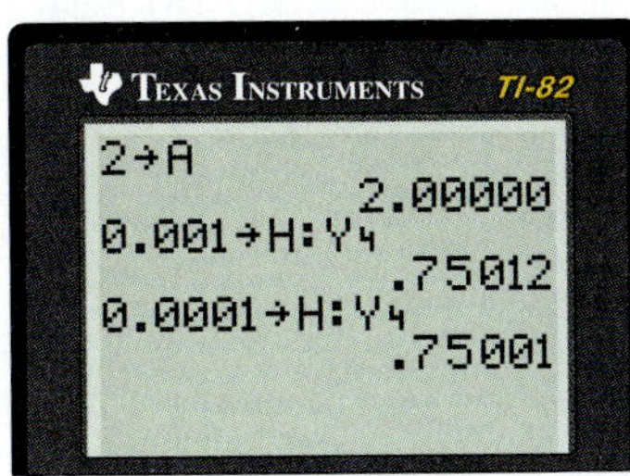

Fig. 2.1.19 Approximating $\displaystyle\lim_{h\to 0} \frac{f(a+h)-f(a)}{h}$

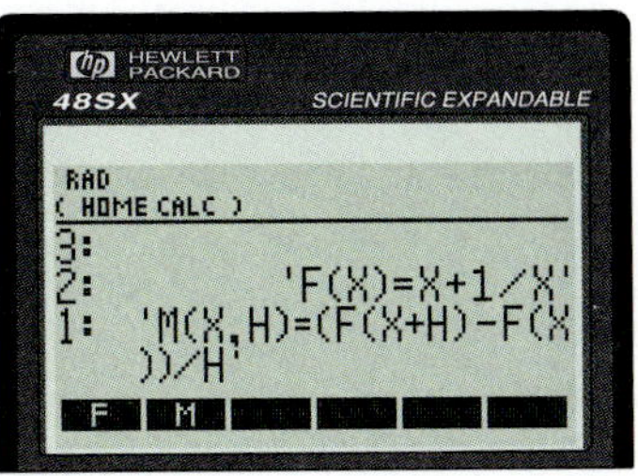

Fig. 2.1.20 A calculator prepared to compute $\dfrac{f(x+h)-f(x)}{h}$

h	$\dfrac{f(2+h)-f(2)}{h}$
0.1	0.76190
0.01	0.75124
0.001	0.75012
0.0001	0.75001
0.00001	0.75000
↓	↓
0	$\dfrac{3}{4}$

Fig. 2.1.21 Numerical investigation of the limit in (13) with $f(x) = x + \dfrac{1}{x}$, $a = 2$

Our numerical investigation does not constitute a rigorous proof that this actually is the desired tangent line, but Figs. 2.1.22 and 2.1.23 showing the computer-generated graphs

$$y = x + \frac{1}{x} \quad \text{and} \quad y = \frac{3}{4}x + 1$$

are strong evidence that we've got it right. (Do you agree?) ■

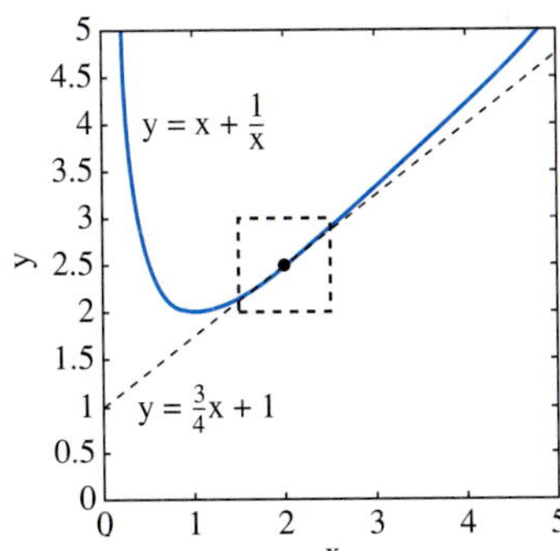

Fig. 2.1.22 The curve and its tangent line (Example 5)

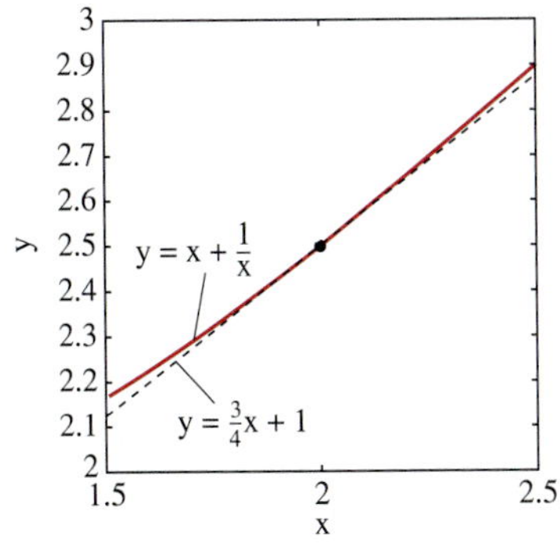

Fig. 2.1.23 The curve and its tangent line magnified near $(2, \frac{5}{2})$

2.1 PROBLEMS

In Problems 1 through 14, first apply the slope-predictor formula in (9) for quadratic functions to write the slope $m(a)$ of the line tangent to $y = f(x)$ at the point where $x = a$. Then write an equation of the line tangent to the graph of f at the point $(2, f(2))$.

1. $f(x) \equiv 5$
2. $f(x) = x$
3. $f(x) = x^2$
4. $f(x) = 1 - 2x^2$
5. $f(x) = 4x - 5$
6. $f(x) = 7 - 3x$
7. $f(x) = 2x^2 - 3x + 4$
8. $f(x) = 5 - 3x - x^2$
9. $f(x) = 2x(x + 3)$
10. $f(x) = 3x(5 - x)$
11. $f(x) = 2x - \left(\dfrac{x}{10}\right)^2$
12. $f(x) = 4 - (3x + 2)^2$
13. $f(x) = (2x + 1)^2 - 4x$
14. $f(x) = (2x + 3)^2 - (2x - 3)^2$

In Problems 15 through 24, find all points of the curve $y = f(x)$ at which the tangent line is horizontal.

15. $y = 10 - x^2$
16. $y = 10x - x^2$
17. $y = x^2 - 2x + 1$
18. $y = x^2 + x - 2$
19. $y = x - \left(\dfrac{x}{10}\right)^2$
20. $y = x(100 - x)$
21. $y = (x + 3)(x - 5)$
22. $y = (x - 5)^2$
23. $y = 70x - x^2$
24. $y = 100\left(1 - \dfrac{x}{10}\right)^2$

In Problems 25 through 35, use the slope-predictor formula for quadratic functions as necessary. In Problems 25 through 27 write equations for both the line tangent to, and the line normal to, the curve $y = f(x)$ at the given point P.

25. $y = x^2$; $P(-2, 4)$

26. $y = 5 - x - 2x^2$; $P(-1, 4)$

27. $y = 2x^2 + 3x - 5$; $P(2, 9)$

28. Prove that the line tangent to the parabola $y = x^2$ at the point (x_0, y_0) intersects the x-axis at the point $(x_0/2, 0)$. See Fig. 2.1.24.

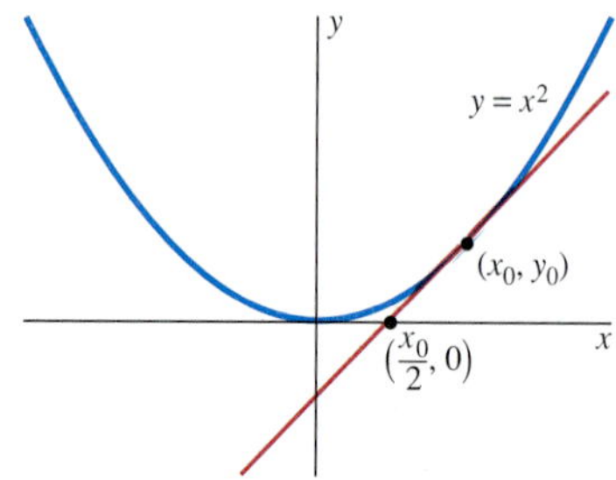

Fig. 2.1.24 The parabola and tangent line of Problem 28

29. If a ball is thrown straight upward with initial velocity 96 ft/s, then its height t seconds later is $y(t) = 96t - 16t^2$ feet. Determine the maximum height the ball attains by finding the point on the parabola $y = 96t - 16t^2$ where the tangent line is horizontal.

30. According to Problem 40 of Section 1.1, the area of a rectangle with base of length x and perimeter 100 is $A(x) = x(50 - x)$. Find the maximum possible area of this rectangle by finding the point on the parabola $A = x(50 - x)$ at which the tangent line is horizontal.

31. Find the maximum possible value of the product of two positive numbers whose sum is 50.

32. Suppose that a projectile is fired at an angle of 45° from the horizontal. Its initial position is the origin in the xy-plane, and its initial velocity is $100\sqrt{2}$ ft/s (Fig. 2.1.25). Then its trajectory will be the part of the parabola $y = x - (x/25)^2$ for which $y \geqq 0$. (a) How far does the projectile travel (horizontally) before it hits the ground? (b) What is the maximum height above the ground that the projectile attains?

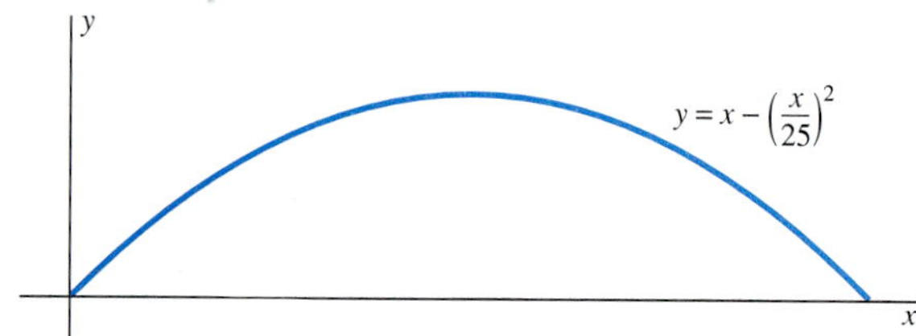

Fig. 2.1.25 The trajectory of the projectile of Problem 32

33. One of the two lines that pass through the point $(3, 0)$ and are tangent to the parabola $y = x^2$ is the x-axis. Find an equation for the *other* line. (*Suggestion*: First find the value of the number a shown in Fig. 2.1.26.)

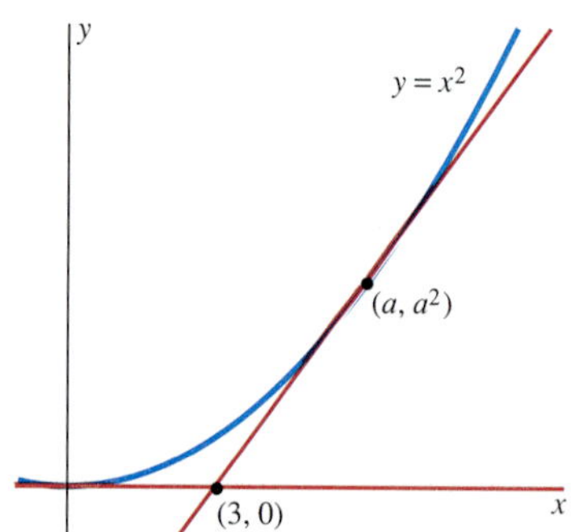

Fig. 2.1.26 Two lines tangent to the parabola of Problem 33

34. Write equations for the two straight lines that pass through the point $(2, 5)$ and are tangent to the parabola $y = 4x - x^2$. (*Suggestion:* Draw a figure like Fig. 2.1.26.)

35. Between Examples 3 and 4 we raised—but did not answer—the question of how to locate the point on the graph of $y = x^2$ closest to the point $(3, 0)$. It's now time for you to find that point. (*Suggestion:* Draw a figure like Fig. 2.1.26. The cubic equation you should obtain has one solution apparent by inspection.)

In Problems 36 through 42 use a calculator to investigate (as in the Numerical Investigation of Slopes at the end of this section) the slope of the line tangent to $y = f(x)$ at the point $(a, f(a))$. Then check your result by graphing both the curve $y = f(x)$ and the alleged tangent line.

36. $f(x) = x^2$; $a = -1$

37. $f(x) = x^3$; $a = 2$

38. $f(x) = x^3$; $a = -1$

39. $f(x) = \sqrt{x}$; $a = 1$

40. $f(x) = \sqrt{x}$; $a = 4$

41. $f(x) = \dfrac{1}{x}$; $a = 1$

42. $f(x) = \dfrac{1}{x}$; $a = -\dfrac{1}{2}$

2.1 PROJECT: NUMERICAL SLOPE INVESTIGATIONS

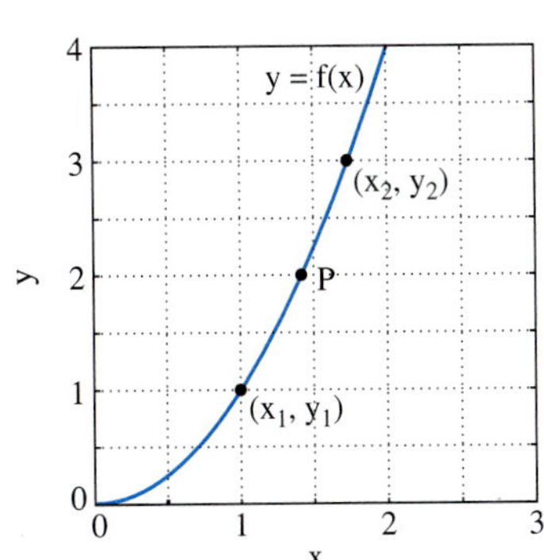

Fig. 2.1.27 Points on either side of P

The following problems are motivated by Figs. 2.1.8 through 2.1.11, which illustrate the fact that a curve and its tangent line are virtually indistinguishable at a sufficiently high level of magnification. The problems require a graphic utility (calculator or computer) with a *trace facility* that enables you to move the cross hairs (cursor) and to read the xy-coordinates of selected points on a graph. For example, with some graphics calculators, you can simply press the **TRACE** key and then move the cursor along the graph. With some computer graphing programs, the xy-coordinates of the current position of the cursor are automatically displayed at the bottom of the monitor screen.

Each problem lists a curve $y = f(x)$ and a point $P(a, f(a))$ at which the slope $m(a)$ of its tangent line is to be approximated graphically. Zoom in on P (magnifying at least six times) until the graph $y = f(x)$ in the viewing window appears to be a straight line. At each zoom, record the coordinates (x_1, y_1) and (x_2, y_2) of two points located on either side of P (as in Fig. 2.1.27). Then let

$$m_k = \frac{\Delta y}{\Delta x} = \frac{y_2 - y_1}{x_2 - x_1}$$

denote the approximate slope that results at the kth zoom. Is it clear which limiting value the approximate slopes $m_1, m_2, m_3, \dots,$ are approaching? [In each case $m(a)$ should be an integer or an integral multiple of $\frac{1}{2}$, $\frac{1}{4}$, or $\frac{1}{8}$.]

1. $f(x) = x^2$, $P = P(-2, 4)$; $m(-2) = ?$

2. $f(x) = \sqrt{x}$, $P = P(1, 1)$; $m(1) = ?$

3. $f(x) = \dfrac{1}{x}$, $P = P(2, \frac{1}{2})$; $m(2) = ?$

4. $f(x) = \dfrac{12}{x^2}$, $P = P(-4, \frac{3}{4})$; $m(-4) = ?$

5. $f(x) = \sqrt{x^2 - 9}$, $P = P(5, 4)$; $m(5) = ?$

2.2 THE LIMIT CONCEPT

In Section 2.1 we defined the slope m of the line tangent to the graph $y = f(x)$ at the point $P(a, f(a))$ to be

$$m = \lim_{h \to 0} \frac{f(a + h) - f(a)}{h}. \qquad \textbf{(1)}$$

The graph that motivated this definition is repeated in Fig. 2.2.1, with $a + h$ relabeled as x (so that $h = x - a$). We see that x approaches a as h approaches zero, so Eq. (1) can be written in the form

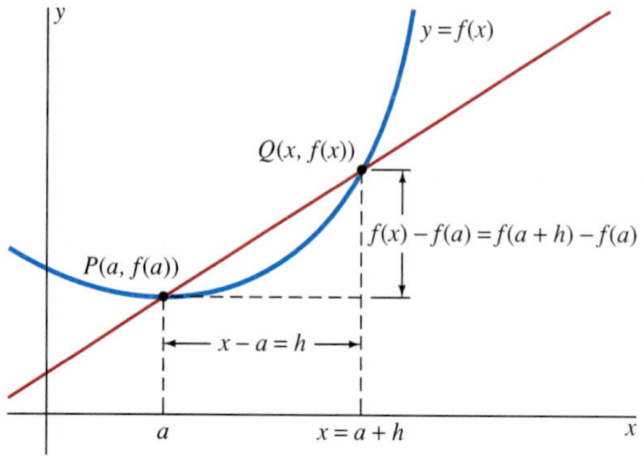

Fig. 2.2.1 The slope m at $P(a, f(a))$ can be defined in this way: $m = \lim_{x \to a} \dfrac{f(x) - f(a)}{x - a}$.

$$m = \lim_{x \to a} \frac{f(x) - f(a)}{x - a}. \tag{2}$$

Thus the computation of m amounts to the determination of the limit, as x approaches a, of the function

$$g(x) = \frac{f(x) - f(a)}{x - a}. \tag{3}$$

Developing general methods for calculating such limits requires that we investigate more fully the meaning of the statement

$$\lim_{x \to a} f(x) = L. \tag{4}$$

This is read "the limit of $f(x)$ as x approaches a is L." We sometimes write Eq. (4) in the concise form

$$f(x) \to L \quad \text{as} \quad x \to a.$$

The function f need not be defined at the point $x = a$ for us to discuss the limit of f at a. The actual value of $f(a)$—if any—actually is immaterial. It suffices for $f(x)$ to be defined for all points *other than* a in some open interval containing a. This is exactly the situation for the function in Eq. (3), which is defined *except* at a (where the denominator is zero). The following statement presents the meaning of Eq. (4) in intuitive language.

Idea of the Limit

We say that the number L is the *limit* of $f(x)$ as x approaches a provided that we can make the number $f(x)$ as close to L as we please merely by choosing x sufficiently near, though not equal to, the number a.

What this means, roughly, is that $f(x)$ tends to get closer and closer to L as x gets closer and closer to a. Once we decide how close to L we want $f(x)$ to be, it is necessary that $f(x)$ be that close to L for *all* x sufficiently close to (but not equal to) a.

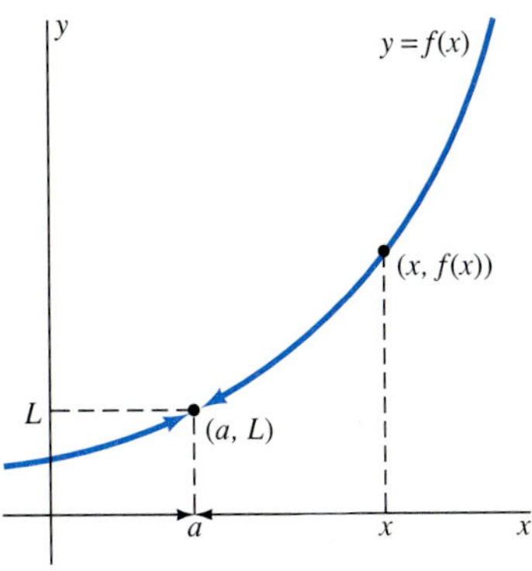

Fig. 2.2.2 Graphical interpretation of the limit concept

Figure 2.2.2 shows a graphical interpretation of the limit concept. As x approaches a (from either side), the point $(x, f(x))$ on the graph $y = f(x)$ must approach the point (a, L).

In this section we explore the idea of the limit, mainly through the investigation of specific examples. A precise statement of the definition of the limit appears at the end of this section.

EXAMPLE 1 Investigate the value of $\lim_{x \to 3} \dfrac{x - 1}{x + 2}$.

Investigation This is an investigation (rather than a solution), because numerical calculations may strongly suggest the value of a limit but cannot establish its value with certainty. The table in Fig. 2.2.3 gives values of

$$f(x) = \frac{x - 1}{x + 2},$$

correct rounded to six decimal places, for values of x that approach 3 (but are not equal to 3). The two columns of the table show values of x that approach 3 both from the left and from the right.

x	$\dfrac{x-1}{x+2}$	x	$\dfrac{x-1}{x+2}$
2	0.250000	4	0.500000
2.9	0.387755	3.1	0.411765
2.99	0.398798	3.01	0.401198
2.999	0.399880	3.001	0.400120
2.9999	0.399988	3.0001	0.400012
↓	↓	↓	↓
3	0.4	3	0.4

Fig. 2.2.3 Investigating the limit in Example 1

Now examine the table—read down the columns for x, because *down* is the table's direction for "approaches"—to see what happens to the corresponding values of $f(x)$. The data clearly suggest that

$$\lim_{x \to 3} \frac{x - 1}{x + 2} = 0.4. \qquad ■$$

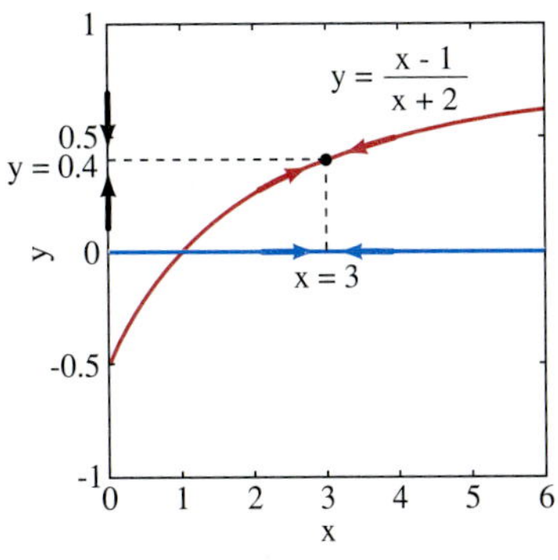

Fig. 2.2.4 The limit in Example 1

Remark 1 The graph of $f(x) = (x - 1)/(x + 2)$ in Fig. 2.2.4 reinforces our guess that $f(x)$ is near 0.4 when x is near 3. For still more reinforcement you can use a graphing calculator or computer to zoom in on the point where $x = 3$.

Remark 2 *Note that we did not simply substitute the value $x = 3$ into the function $f(x) = (x - 1)/(x + 2)$ to obtain the apparent value $\frac{2}{5} = 0.4$ of the limit.* Although such substitution would produce the correct answer in this particular case, in many limits it produces either an incorrect answer or no answer at all. (See Examples 2 and 3 and Problems 19 through 36 and 47 through 56.)

EXAMPLE 2 Investigate the value of $\displaystyle\lim_{x\to 2}\frac{x^2-4}{x^2+x-6}$.

Investigation The numerical data shown in Fig. 2.2.5 certainly suggest that

$$\lim_{x\to 2}\frac{x^2-4}{x^2+x-6}=0.8.$$ ■

x	$\dfrac{x^2-4}{x^2+x-6}$	x	$\dfrac{x^2-4}{x^2+x-6}$
1	0.750000	3	0.833333
1.5	0.777778	2.5	0.818182
1.9	0.795918	2.1	0.803922
1.99	0.799599	2.01	0.800399
1.999	0.799960	2.001	0.800040
1.9999	0.799996	2.0001	0.800004
↓	↓	↓	↓
2	0.8	2	0.8

Fig. 2.2.5 Investigating the limit in Example 2

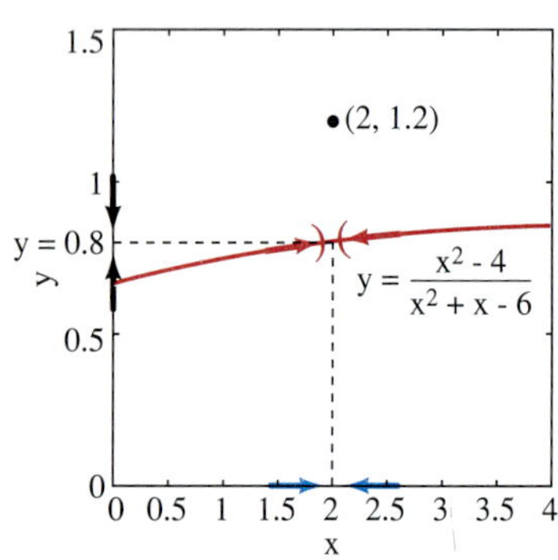

Fig. 2.2.6 The limit in Example 2

Remark The function

$$f(x)=\frac{x^2-4}{x^2+x-6}$$

is not defined at $x = 2$, so there is no question of simply substituting 2 for x. But if we let

$$g(x)=\begin{cases}\dfrac{x^2-4}{x^2+x-6} & \text{if } x\neq 2,\\ 1.2 & \text{if } x=2,\end{cases}$$

then $g(x)$ *is* defined at $x = 2$ (and agrees with $f(x)$ elsewhere). Is it clear to you that f and g must have the same limit at $x = 2$? Figure 2.2.6 shows the graph $y = g(x)$, including the isolated point $(2, 1.2)$ on its graph.

EXAMPLE 3 Investigate the value of $\displaystyle\lim_{t\to 0}\frac{\sqrt{t+25}-5}{t}$.

Investigation Here we cannot make a guess by substituting $t = 0$, because the fraction is meaningless when $t = 0$. But the numerical data shown in Fig. 2.2.7 indicate that

$$\lim_{t\to 0}\frac{\sqrt{t+25}-5}{t}=0.1.$$ ■

t	$\dfrac{\sqrt{t+25}-5}{t}$
1.0	0.099020
0.5	0.099505
0.1	0.099900
0.05	0.099950
0.01	0.099990
0.005	0.099995
↓	↓
0	0.1

Fig. 2.2.7 Investigating the limit in Example 3

Remark Can you see that, upon dividing each number in the second column of Fig. 2.2.7 by 10000, one might well suspect that

$$\lim_{t\to 0}\frac{\sqrt{t+25}-5}{10000t}=0? \quad \text{(Wrong!)}$$

In fact, the value of this limit (as we will see in Example 11) is exactly $10^{-5} = 0.00001$, not zero. This fact constitutes a warning that numerical investigations of limits are not conclusive.

The numerical investigation in Example 3 is incomplete because the table in Fig. 2.2.7 shows values of the function $g(t)$ on only one side of the point $t = 0$. But in order that $\lim_{x\to a} f(x) = L$, it is necessary for $f(x)$ to approach L *both* as x approaches a from the left *and* as x approaches a from the right. If $f(x)$ approaches different values as x approaches a from different sides, then $\lim_{x\to a} f(x)$ does not exist. In Section 2.3 we discuss *one-sided* limits in more detail.

EXAMPLE 4 Investigate $\lim_{x\to 0} f(x)$, given

$$f(x) = \frac{x}{|x|} = \begin{cases} 1 & \text{if } x > 0, \\ -1 & \text{if } x < 0. \end{cases}$$

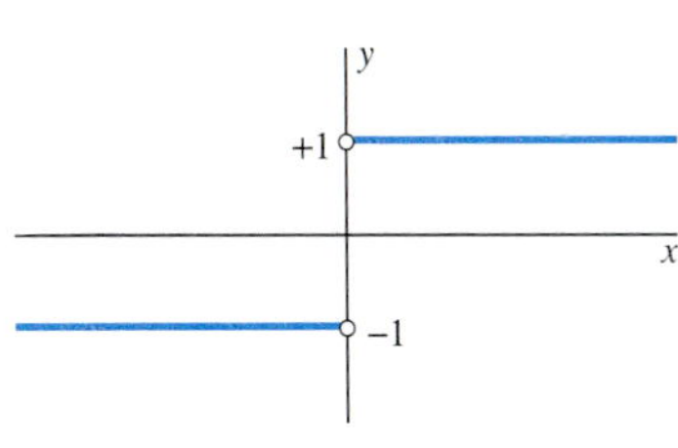

Fig. 2.2.8 The graph of $f(x) = \dfrac{x}{|x|}$ (Example 4)

Solution From the graph of f shown in Fig. 2.2.8, it is apparent that $f(x) \to 1$ as $x \to 0$ from the right and that $f(x) \to -1$ as $x \to 0$ from the left. In particular, there are positive values of x as close to zero as we please such that $f(x) = 1$ and negative values of x equally close to zero such that $f(x) = -1$. Hence we cannot make $f(x)$ as close as we please to any *single* value of L *merely* by choosing x sufficiently close to zero. Therefore,

$$\lim_{x\to 0} \frac{x}{|x|} \quad \text{does not exist.}$$

■

In Example 5 the value obtained by substituting $x = a$ in $F(x)$ to find $\lim_{x\to a} F(x)$ is incorrect.

EXAMPLE 5 Evaluate $\lim_{x\to 0} F(x)$ where

$$F(x) = \begin{cases} 1 & \text{if } x \neq 0, \\ 0 & \text{if } x = 0. \end{cases}$$

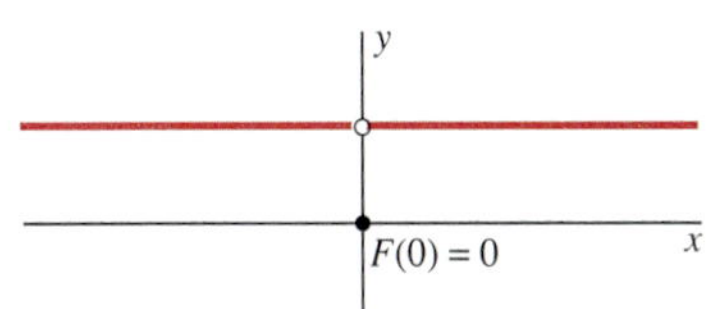

Fig. 2.2.9 The graph of the function F of Example 5

The graph of F is shown in Fig. 2.2.9.

Solution The fact that $F(x) = 1$ for *every* value of $x \neq 0$ in any neighborhood of zero implies that

$$\lim_{x\to 0} F(x) = 1.$$

But note that the value of the limit at $x = 0$ is *not* equal to the functional value $F(0) = 0$ there. ■

The Limit Laws

Numerical investigations such as those in Examples 1 through 3 provide us with an intuitive feeling for limits and typically suggest the correct value of a limit. But most limit computations are based neither on merely suggestive (and imprecise) numerical estimates nor on direct (but difficult) applications of the definition of limit. Instead, such computations are performed most easily and naturally with the aid of the *limit laws* that we give next. These "laws" actually are *theorems,* whose proofs (based on the precise definition of the limit) are included in Appendix D.

Constant Law

If $f(x) \equiv C$, where C is a constant [so $f(x)$ is a **constant function**], then

$$\lim_{x \to a} f(x) = \lim_{x \to a} C = C. \tag{5}$$

Sum Law

If both of the limits

$$\lim_{x \to a} f(x) = L \quad \text{and} \quad \lim_{x \to a} g(x) = M$$

exist, then

$$\lim_{x \to a} \left[f(x) \pm g(x)\right] = \left[\lim_{x \to a} f(x)\right] \pm \left[\lim_{x \to a} g(x)\right] = L \pm M. \tag{6}$$

(The limit of a sum is the sum of the limits; the limit of a difference is the difference of the limits.)

Product Law

If both of the limits

$$\lim_{x \to a} f(x) = L \quad \text{and} \quad \lim_{x \to a} g(x) = M$$

exist, then

$$\lim_{x \to a} \left[f(x)g(x)\right] = \left[\lim_{x \to a} f(x)\right]\left[\lim_{x \to a} g(x)\right] = LM. \tag{7}$$

(The limit of a product is the product of the limits.)

Quotient Law

If both of the limits

$$\lim_{x \to a} f(x) = L \quad \text{and} \quad \lim_{x \to a} g(x) = M$$

exist *and* if $M \neq 0$, then

$$\lim_{x \to a} \frac{f(x)}{g(x)} = \frac{\lim_{x \to a} f(x)}{\lim_{x \to a} g(x)} = \frac{L}{M}. \tag{8}$$

(The limit of a quotient is the quotient of the limits, provided that the limit of the denominator is not zero.)

Root Law

If n is a positive integer and if $a > 0$ for even values of n, then

$$\lim_{x \to a} \sqrt[n]{x} = \sqrt[n]{a}. \tag{9}$$

The case $n = 1$ of the root law is obvious:

$$\lim_{x\to a} x = a. \tag{10}$$

Examples 6 and 7 show how the limit laws can be used to evaluate limits of polynomials and rational functions.

EXAMPLE 6

$$\begin{aligned}\lim_{x\to 3}(x^2+2x+4) &= \left(\lim_{x\to 3}x^2\right)+\left(\lim_{x\to 3}2x\right)+\left(\lim_{x\to 3}4\right)\\ &= \left(\lim_{x\to 3}x\right)^2+2\left(\lim_{x\to 3}x\right)+\left(\lim_{x\to 3}4\right)=3^2+2\cdot 3+4=19.\end{aligned}$$ ■

EXAMPLE 7

$$\begin{aligned}\lim_{x\to 3}\frac{2x+5}{x^2+2x+4} &= \frac{\lim_{x\to 3}(2x+5)}{\lim_{x\to 3}(x^2+2x+4)}\\ &= \frac{2\cdot 3+5}{3^2+2\cdot 3+4}=\frac{11}{19}.\end{aligned}$$ ■

Note In Examples 6 and 7, we systematically applied the limit laws until we could simply substitute 3 for $\lim_{x\to 3} x$ at the final step. To determine the limit of a quotient of polynomials, we must verify before this final step that the limit of the denominator is not zero. If the denominator limit is zero, then the limit *may* fail to exist.

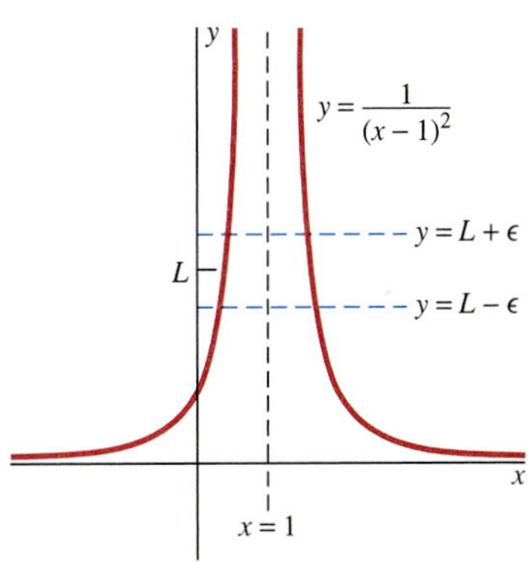

Fig. 2.2.10 The graph of $y = \dfrac{1}{(x-1)^2}$ (Example 8)

EXAMPLE 8 Investigate $\displaystyle\lim_{x\to 1}\frac{1}{(x-1)^2}$.

Solution Because $\lim_{x\to 1}(x-1)^2 = 0$, we cannot apply the quotient law. Moreover, we can make $1/(x-1)^2$ arbitrarily large by choosing x sufficiently close to 1. Hence $1/(x-1)^2$ cannot approach any (finite) number L as x approaches 1. Therefore the limit in this example does not exist. You can see the geometric reason if you examine the graph of $y = 1/(x-1)^2$ in Fig. 2.2.10. As x approaches 1, the corresponding point on the curve must leave the indicated strip between the two horizontal lines $y = L - \epsilon$ and $y = L + \epsilon$ that bracket the proposed limit L and hence cannot approach the point $(1, L)$. ■

EXAMPLE 9 Investigate $\displaystyle\lim_{x\to 2}\frac{x^2-4}{x^2+x-6}$.

Solution We cannot immediately apply the quotient law (as we did in Example 7) because the denominator approaches zero as x approaches 2. If the numerator were approaching some number other than zero as $x \to 2$, then the limit would fail to exist (as in Example 8). But here the numerator *does* approach zero, so there is a possibility that a factor of the numerator can be canceled with a factor of the denominator, thus removing the zero-denominator problem. Indeed,

$$\begin{aligned}\lim_{x\to 2}\frac{x^2-4}{x^2+x-6} &= \lim_{x\to 2}\frac{(x-2)(x+2)}{(x-2)(x+3)}\\ &= \lim_{x\to 2}\frac{x+2}{x+3}=\frac{4}{5}.\end{aligned}$$

We can cancel the factor $x - 2$ because it is nonzero: $x \neq 2$ when we evaluate the limit as x approaches 2. Moreover, this verifies the numerical limit of 0.8 that we found in Example 3. ■

Substitution of Limits

It is tempting to write

$$\lim_{x \to -4} \sqrt{x^2 + 9} = \sqrt{\lim_{x \to -4} (x^2 + 9)}$$
$$= \sqrt{(-4)^2 + 9} = \sqrt{25} = 5. \quad \textbf{(11)}$$

But can we simply "move the limit inside the radical" in Eq. (11)? To analyze this question, let us write

$$f(x) = \sqrt{x} \quad \text{and} \quad g(x) = x^2 + 9.$$

Then the function that appears in Eq. (11) is the composite function

$$f(g(x)) = \sqrt{g(x)} = \sqrt{x^2 + 9}.$$

(The left-hand expression in this equation is read "f of g of x.") Hence our question is whether

$$\lim_{x \to a} f(g(x)) = f\left(\lim_{x \to a} g(x)\right).$$

The next limit law answers this question in the affirmative, provided that the "outside function" f meets a certain condition; if so, then the limit of the composite function $f(g(x))$ as $x \to a$ may be found by substituting into the function f the limit of $g(x)$ as $x \to a$.)

Substitution Law *Limits of Compositions*

Suppose that

$$\lim_{x \to a} g(x) = L \quad \text{and that} \quad \lim_{x \to L} f(x) = f(L).$$

Then

$$\lim_{x \to a} f(g(x)) = f\left(\lim_{x \to a} g(x)\right) = f(L). \quad \textbf{(12)}$$

Thus the condition under which Eq. (12) holds is that the limit of the *outer* function f not only exists at $x = L$, but also is equal to the "expected" value of f—namely, $f(L)$. In particular, because

$$\lim_{x \to -4} (x^2 + 9) = 25 \quad \text{and} \quad \lim_{x \to 25} \sqrt{x} = \sqrt{25} = 5,$$

this condition is satisfied in Eq. (11). Hence the computations shown there are valid.

In this section we use only the following special case of the substitution law. With $f(x) = x^{1/n}$, where n is a positive integer, Eq. (12) takes the form

$$\lim_{x \to a} \sqrt[n]{g(x)} = \sqrt[n]{\lim_{x \to a} g(x)}, \quad \textbf{(13)}$$

under the assumption that the limit of $g(x)$ exists as $x \to a$ (and is positive if n is even). With $g(x) = x^m$, where m is a positive integer, Eq. (13) in turn yields

$$\lim_{x \to a} x^{m/n} = a^{m/n}, \tag{14}$$

with the condition that $a > 0$ if n is even. Equations (13) and (14) may be regarded as generalized root laws. Example 10 illustrates the use of these special cases of the substitution law.

EXAMPLE 10

$$\begin{aligned}
\lim_{x \to 4} \left(3x^{3/2} + 20\sqrt{x}\right)^{1/3} &= \left(\lim_{x \to 4} \left(3x^{3/2} + 20\sqrt{x}\right)\right)^{1/3} && \text{(using Eq. (13))} \\
&= \left(\lim_{x \to 4} 3x^{3/2} + \lim_{x \to 4} 20\sqrt{x}\right)^{1/3} && \text{(using the sum law)} \\
&= \left(3 \cdot 4^{3/2} + 20\sqrt{4}\right)^{1/3} && \text{(using Eq. (14))} \\
&= (24 + 40)^{1/3} = \sqrt[3]{64} = 4.
\end{aligned}$$

■

Slope-Predictor Functions

Our discussion of limits began with the slope

$$m(a) = \lim_{h \to 0} \frac{f(a + h) - f(a)}{h} \tag{15}$$

of the line tangent to the graph $y = f(x)$ at the point $(a, f(a))$. The lines tangent to $y = f(x)$ at different points have different slopes. Thus if we replace a with x in Eq. (15), we get a *new function* defined by

$$m(x) = \lim_{h \to 0} \frac{f(x + h) - f(x)}{h}. \tag{16}$$

This function m may be regarded as a "slope predictor" for lines tangent to the graph $y = f(x)$. It is a new function *derived* from the original function $f(x)$, and in Chapter 3 we will call it the *derivative* of f.

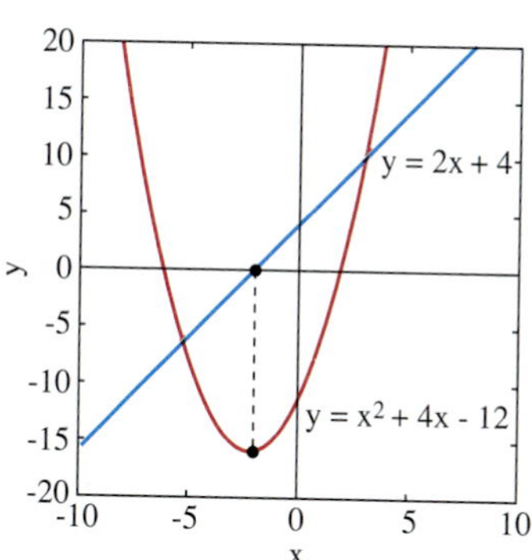

Fig. 2.2.11 The parabola $y = x^2 + 4x - 12$ and its slope predictor $m(x) = 2x + 4$

EXAMPLE 11 In Section 2.1 we saw that the line tangent to the graph $y = px^2 + qx + r$ at the point where $x = a$ has slope $m(a) = 2pa + q$. Hence, replacing a with x, we see that the slope-prediction function for the quadratic function

$$f(x) = px^2 + qx + r \tag{17}$$

is the linear function

$$m(x) = 2px + q. \tag{18}$$

Figure 2.2.11 illustrates the case $p = 1, q = 4, r = -12$. It is worth noting that the x-intercept where $m(x) = 0$ corresponds to the point of the parabola $y = f(x)$ where the tangent line is horizontal. ■

The slope-predictor definition in Eq. (16) calls for us to carry out the following four steps.

1. Write the definition of $m(x)$.
2. Substitute into this definition the formula of the given function f.
3. Make algebraic substitutions until step 4 can be carried out.
4. Determine the value of the limit as $h \to 0$.

Note that x may be thought of as a *constant* throughout this computation—it is h that is the variable in this four-step process.

EXAMPLE 12 Find the slope-predictor function for the function

$$f(x) = x + \frac{1}{x}$$

that was investigated numerically in Example 5 of Section 2.1.

Solution The first two steps in the preceding list yield

$$m(x) = \lim_{h \to 0} \frac{f(x+h) - f(x)}{h} = \lim_{h \to 0} \frac{\left(x + h + \dfrac{1}{x+h}\right) - \left(x + \dfrac{1}{x}\right)}{h}.$$

We cancel the two copies of x in the numerator and proceed to simplify algebraically, first finding a common denominator in the numerator:

$$m(x) = \lim_{h \to 0} \frac{h + \dfrac{1}{x+h} - \dfrac{1}{x}}{h}$$

$$= \lim_{h \to 0} \frac{h(x+h)x + x - (x+h)}{h(x+h)x}$$

$$= \lim_{h \to 0} \frac{h(x+h)x - h}{h(x+h)x}.$$

Now we can divide numerator and denominator by h (because $h \neq 0$) and finally apply the sum, product, and quotient laws to take the limit as $h \to 0$:

$$m(x) = \lim_{h \to 0} \frac{h(x+h)x - h}{h(x+h)x}$$

$$= \lim_{h \to 0} \frac{(x+h)x - 1}{(x+h)x} = \frac{x^2 - 1}{x^2} = 1 - \frac{1}{x^2}.$$

For instance, the slope of the line tangent to

$$y = x + \frac{1}{x}$$

at the point $(2, \frac{5}{2})$ is $m(2) = \frac{3}{4}$ (thus verifying the result of Example 5 in Section 2.1). ■

Example 13 illustrates an algebraic device often used in "preparing" functions before taking limits. This device can be applied when roots are present and resembles the simple computation

$$\frac{1}{\sqrt{5}-\sqrt{2}} = \frac{1}{\sqrt{5}-\sqrt{2}} \cdot \frac{\sqrt{5}+\sqrt{2}}{\sqrt{5}+\sqrt{2}}$$

$$= \frac{\sqrt{5}+\sqrt{2}}{5-2} = \frac{\sqrt{5}+\sqrt{2}}{3}.$$

EXAMPLE 13 Find the slope-predictor function for the function $f(x) = \sqrt{x}$.

Solution

$$m(x) = \lim_{h\to 0} \frac{\sqrt{x+h}-\sqrt{x}}{h}. \tag{19}$$

To prepare the fraction for the evaluation of the limit, we first multiply the numerator and denominator by the *conjugate* $\sqrt{x+h}+\sqrt{x}$ of the numerator:

$$m(x) = \lim_{h\to 0} \frac{\sqrt{x+h}-\sqrt{x}}{h} \cdot \frac{\sqrt{x+h}+\sqrt{x}}{\sqrt{x+h}+\sqrt{x}}$$

$$= \lim_{h\to 0} \frac{(x+h)-x}{h(\sqrt{x+h}+\sqrt{x})}$$

$$= \lim_{h\to 0} \frac{1}{\sqrt{x+h}+\sqrt{x}}.$$

Thus

$$m(x) = \frac{1}{2\sqrt{x}}. \tag{20}$$

(In the final step we used the sum, quotient, and root laws—we did not simply substitute 0 for h.) ■

Note that if we equate the right-hand sides of Eqs. (19) and (20) and take $x = 25$, then we get the limit in Example 4:

$$\lim_{h\to 0} \frac{\sqrt{25+h}-5}{h} = \frac{1}{10}.$$

(The t in Example 4 has been replaced with h.) And if we divide both sides by 10000 we find that

$$\lim_{h\to 0} \frac{\sqrt{25+h}-5}{10000h} = \frac{1}{100000} = 0.00001,$$

as claimed in the remark following Example 3.

The Definition of the Limit

There are few people for whom a full understanding of the limit concept comes quickly or easily. Indeed, the precise meaning of the statement "$F(x)$ approaches L as x approaches a" was debated vigorously (and sometimes acrimoniously) for hundreds of years—until late in the nineteenth century. Then the German mathematician Karl Weierstrass (1815–1897) finally formulated the rigorous definition of the limit that is accepted today.

Definition ***Definition of the Limit***

The number L is the **limit** of $F(x)$ as x approaches a provided that, given any number $\epsilon > 0$, there exists a number $\delta > 0$ such that

$$|F(x) - L| < \epsilon$$

for all x such that

$$0 < |x - a| < \delta.$$

Figure 2.2.12 illustrates this definition. The points on the graph of $y = F(x)$ that satisfy the inequality $|F(x) - L| < \epsilon$ are those points that lie between the two horizontal lines $y = L - \epsilon$ and $y = L + \epsilon$. The points on this graph that satisfy the inequality $|x - a| < \delta$ are those points that lie between the two vertical lines $x = a - \delta$ and $x = a + \delta$. Consequently, the definition implies that $\lim_{x\to a} F(x) = L$ if and only if the following is true:

> Suppose that the two horizontal lines $y = L - \epsilon$ and $y = L + \epsilon$ (with $\epsilon > 0$) are given. Then it is possible to choose two vertical lines $x = a - \delta$ and $x = a + \delta$ (with $\delta > 0$) with the following property: Every point on the graph of $y = F(x)$ (with $x \neq a$) that lies between the two vertical lines must also lie between the two horizontal lines.

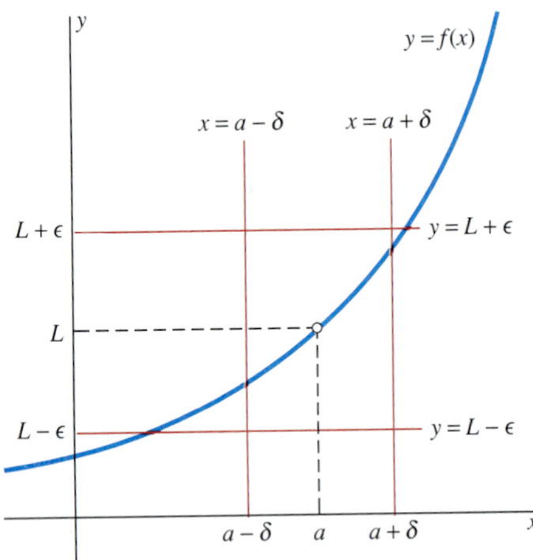

Fig. 2.2.12 Geometric illustration of the limit definition

Figure 2.2.12 suggests that the closer together the two horizontal lines, the closer together the two vertical lines will need to be. This is what we mean by "making $F(x)$ closer to L by making x closer to a."

2.2 PROBLEMS

Apply the limit laws of this section to evaluate the limits in Problems 1 through 18. Justify each step by citing the appropriate limit law.

1. $\lim_{x\to 3} (3x^2 + 7x - 12)$

2. $\lim_{x\to -2} (x^3 - 3x^2 + 5)$

3. $\lim_{x\to 1} (x^2 - 1)(x^7 + 7x - 4)$

4. $\lim_{x\to -2} (x^3 - 3x + 3)(x^2 + 2x + 5)$

5. $\lim_{x\to 1} \dfrac{x+1}{x^2+x+1}$

6. $\lim_{t\to -2} \dfrac{t+2}{t^2+4}$

7. $\lim_{x\to 3} \dfrac{(x^2+1)^3}{(x^3-25)^3}$

8. $\lim_{z\to -1} \dfrac{(3z^2+2z+1)^{10}}{(z^3+5)^5}$

9. $\lim_{x\to 1} \sqrt{4x+5}$

10. $\lim_{y\to 4} \sqrt{27-\sqrt{y}}$

11. $\lim_{x\to 3} (x^2-1)^{3/2}$

12. $\lim_{t\to -4} \sqrt{\dfrac{t+8}{25-t^2}}$

13. $\lim_{z\to 8} \dfrac{z^{2/3}}{z-\sqrt{2z}}$

14. $\lim_{t\to 2} \sqrt[3]{3t^3+4t-5}$

15. $\lim_{w\to 0} \sqrt{(w-2)^4}$

16. $\lim_{t\to -4} \sqrt[3]{(t+1)^6}$

17. $\lim_{x\to -2} \sqrt[3]{\dfrac{x+2}{(x-2)^2}}$

18. $\lim_{y\to 5} \left(\dfrac{2y^2+2y+4}{6y-3}\right)^{1/3}$

In Problems 19 through 28, note first that the numerator and denominator have a common algebraic factor (as in Example 9). Use this fact to help evaluate the given limit.

19. $\lim_{x\to -1} \dfrac{x+1}{x^2-x-2}$

20. $\lim_{t\to 3} \dfrac{t^2-9}{t-3}$

21. $\lim_{x\to 1} \dfrac{x^2+x-2}{x^2-4x+3}$

22. $\lim_{y\to -1/2} \dfrac{4y^2-1}{4y^2+8y+3}$

23. $\lim_{t\to -3} \dfrac{t^2+6t+9}{t^2-9}$

24. $\lim_{x\to 2} \dfrac{x^2-4}{3x^2-2x-8}$

25. $\lim_{z\to -2} \dfrac{(z+2)^2}{z^4-16}$

26. $\lim_{t\to 3} \dfrac{t^3-9t}{t^2-9}$

27. $\lim_{x\to 1} \dfrac{x^3-1}{x^4-1}$

28. $\lim_{y\to -3} \dfrac{y^3+27}{y^2-9}$

In Problems 29 through 36, evaluate those limits that exist.

29. $\lim_{x\to 3} \dfrac{\frac{1}{x}-\frac{1}{3}}{x-3}$

30. $\lim_{t\to 0} \dfrac{\frac{1}{2+t}-\frac{1}{2}}{t}$

31. $\lim_{x\to 4} \dfrac{x-4}{\sqrt{x}-2}$

32. $\lim_{x\to 9} \dfrac{3-\sqrt{x}}{9-x}$

33. $\lim_{t\to 0} \dfrac{\sqrt{t+4}-2}{t}$

34. $\lim_{h\to 0} \dfrac{1}{h}\left(\dfrac{1}{\sqrt{9+h}}-\dfrac{1}{3}\right)$

35. $\lim_{x\to 4} \dfrac{x^2-16}{2-\sqrt{x}}$

36. $\lim_{x\to 0} \dfrac{\sqrt{1+x}-\sqrt{1-x}}{x}$

In Problems 37 through 46, use the four-step process illustrated in Examples 12 and 13 to find a slope-predictor function for the given function $f(x)$. Then write an equation for the line tangent to the curve $y = f(x)$ at the point where $x = 2$.

37. $f(x) = x^3$

38. $f(x) = \dfrac{1}{x}$

39. $f(x) = \dfrac{1}{x^2}$

40. $f(x) = \dfrac{1}{x+1}$

41. $f(x) = \dfrac{2}{x-1}$

42. $f(x) = \dfrac{x}{x-1}$

43. $f(x) = \dfrac{1}{\sqrt{x+2}}$

44. $f(x) = x^2 + \dfrac{3}{x}$

45. $f(x) = \sqrt{2x+5}$

46. $f(x) = \dfrac{x^2}{x+1}$

In Problems 47 through 56, the actual value of the given limit $\lim_{x\to a} f(x)$ is a rational number that is a ratio of two single-digit integers. Guess this limit on the basis of a numerical investigation in which you calculate $f(x)$ for $x = a \pm 0.1, x = a \pm 0.05, x = a \pm 0.01, x = a \pm 0.005$, and so on. Use other similar values of x near a as you wish.

47. $\lim_{x\to 0} \dfrac{(1+x)^2-1}{x}$

48. $\lim_{x\to 1} \dfrac{x^4-1}{x-1}$

49. $\lim_{x\to 0} \dfrac{\sqrt{x+9}-3}{x}$

50. $\lim_{x\to 4} \dfrac{x^{3/2}-8}{x-4}$

51. $\lim_{x\to 0} \dfrac{1}{x}\left[\dfrac{2}{(2+x)^3}-\dfrac{1}{4}\right]$

52. $\lim_{x\to 0} \dfrac{(3+x)^{-1}-(3-x)^{-1}}{x}$

53. $\lim_{x\to 0} \dfrac{\sin x}{x}$

54. $\lim_{x\to 0} \dfrac{1-\cos x}{x^2}$

55. $\lim_{x\to 0} \dfrac{x-\sin x}{x^3}$

56. $\lim_{x\to 0} \left(1+\dfrac{1}{|x|}\right)^x$

57. In contrast with the rational-valued limits in Problems 47 through 56, the value of the limit

$$\lim_{x\to 0} (1+x)^{1/x}$$

is the famous irrational number e (of Chapter 7), whose three-place decimal approximation is $e \approx 2.718$. Numerically investigate this limit to approximate e accurate to five decimal places. Corroborate this value graphically by zooming in on the y-intercept of the curve $y = (1+x)^{1/x}$.

58. Verify graphically the limit

$$\lim_{x\to 0} \frac{\sin x}{x}$$

of Problem 53 by zooming in on the y-intercept of the curve $y = (\sin x)/x$.

59. Investigate the limit

$$\lim_{x\to 0} \frac{x-\tan x}{x^3}$$

both numerically and graphically. Determine its value accurate to four decimal places.

60. The value of

$$\lim_{x\to 0} \frac{\sin 2x}{\tan 5x}$$

is the ratio of two single-digit integers. Determine this value both numerically and graphically.

61. Calculate the value of

$$f(x) = \sin \frac{\pi}{x}$$

for $x = \frac{1}{2}, \frac{1}{4}, \frac{1}{8}, \frac{1}{16}, \ldots$. What do you now conjecture to be the value of

$$\lim_{x \to 0} \sin \frac{\pi}{x}?$$

Next calculate $f(x)$ for $x = \frac{2}{3}, \frac{2}{9}, \frac{2}{27}, \frac{2}{81}, \ldots$. Now what do you conclude?

62. To investigate the limit of $f(x) = \sin x + 10^{-5} \cos x$ as $x \to 0$, set your graphing calculator or computer to display exactly four digits to the right of the decimal point. After calculating $f(x)$ with $x = 0.1,\ 0.001,\ 0.00001,\ 0.0000001, \ldots$ what do you conclude? (Your answer may depend on how your particular calculator works.) Now zoom in on the y-intercept of $y = f(x)$ sufficiently to show that the value of the limit is nonzero. What is it?

63. Investigate numerically or graphically (or both) the value of the limit

$$\lim_{x \to 0} \left(\log_{10} \frac{1}{|x|} \right)^{-1/32}.$$

The actual value of this limit is zero, so you'll see that your calculator or computer cannot always be believed.

2.2 PROJECT: SLOPES AND LOGARITHMS

Show first that if

$$f(x) = a^x \tag{1}$$

(where a is a positive constant), then the general slope-predictor limit in Eq. (1) of this section gives

$$m_a = \lim_{x \to 0} \frac{a^x - 1}{x} \tag{2}$$

for the slope of the line tangent to the graph $y = a^x$ at its y-intercept $(0, 1)$. (We write m_a to indicate the dependence of this limit on the value of the constant a.)

To investigate this limit using a TI calculator with $a = 10$, for instance, you can enter **1** $\to$**x** and then repeatedly enter the commands

```
x/4 → x : (10^x-1)/x
```

The analogous *Mathematica* and *Maple* commands for repeated execution (beginning with **x** = **1**) would be

```
x = x/4;    (10^x-1)/x // N
```

and

```
x := x/4:    evalf((10^x-1)/x);
```

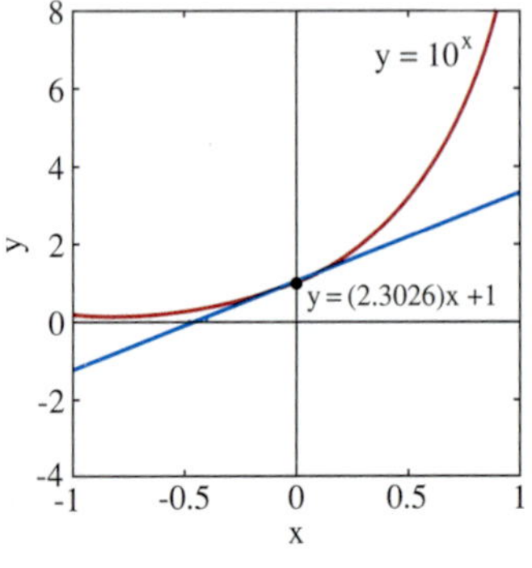

Fig. 2.2.13 $y = 10^x$ and its tangent line at (0,1)

respectively. You should find that $m_{10} \approx 2.3026$ accurate to four decimal places. Figure 2.2.13 shows the graph $y = 10^x$ and the tangent line with slope m_{10}.

Now let p and q be the first two digits of your student I.D. number that are greater than 1. Numerically determine (as suggested above) the values of the slopes m_p, m_q, and m_{pq}. Are your results consistent with the conjecture that

$$m_p + m_q = m_{pq}? \tag{3}$$

Do you see the analogy with the law of logarithms

$$\log p + \log q = \log pq? \tag{4}$$

At present the connection between Eqs. (3) and (4) is surely an enigma rather than an explanation. The mystery will be explained in Chapter 7, when we study "natural" (base e) exponential functions and logarithms. (For a clue, use the LN key on your calculator to find $\ln p$, $\ln q$, and $\ln 10$. Compare these values with the slopes m_p, m_q, and m_{10}.)

2.3 MORE ABOUT LIMITS

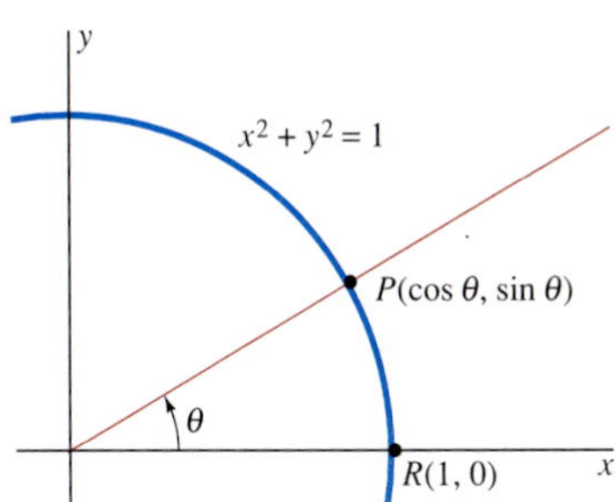

Fig. 2.3.1 An angle θ

To investigate limits of trigonometric functions, we begin with Fig. 2.3.1, which shows an angle θ with its vertex at the origin, its initial side along the positive x-axis, and its terminal side intersecting the unit circle at the point P. By the definition of the sine and cosine functions, the coordinates of P are $P(\cos\theta, \sin\theta)$. From geometry we see that, as $\theta \to 0$, the point $P(\cos\theta, \sin\theta)$ approaches the point $R(1, 0)$. Hence $\cos\theta \to 1$ and $\sin\theta \to 0$ as $\theta \to 0$ through positive values. A similar picture gives the same result for negative values of θ, so we see that

$$\lim_{\theta\to 0} \cos\theta = 1 \quad \text{and} \quad \lim_{\theta\to 0} \sin\theta = 0. \tag{1}$$

Equation (1) says simply that the *limits* of the functions $\cos\theta$ and $\sin\theta$ as $\theta \to 0$ are equal to their respective values at $\theta = 0$: $\cos 0 = 1$ and $\sin 0 = 0$.

The limit of the quotient $(\sin\theta)/\theta$ as $\theta \to 0$ plays a special role in the calculus of trigonometric functions. For instance, it is needed to find slopes of lines tangent to trigonometric graphs such as $y = \cos x$ and $y = \sin x$.

Note that the value of the quotient $(\sin\theta)/\theta$ is not defined when $\theta = 0$. (Why not?) But a calculator set in *radian mode* provides us with the numerical evidence shown in Fig. 2.3.2. This table strongly suggests that the limit of $(\sin\theta)/\theta$ is 1 as $\theta \to 0$. This conclusion is supported by the graph of $y = (\sin x)/x$ shown in Fig. 2.3.3, where it appears that the point (x, y) on the curve is near $(0, 1)$ when x is near zero. At the end of this section we provide a geometric proof of the following result.

θ	$\dfrac{\sin\theta}{\theta}$
± 1.0	0.84147
± 0.5	0.95885
± 0.1	0.99833
± 0.05	0.99958
± 0.01	0.99998
± 0.005	1.00000
± 0.001	1.00000
⋮	⋮
↓	↓
0	1

Fig. 2.3.2 The numerical data suggest that $\lim_{\theta\to 0} \dfrac{\sin\theta}{\theta} = 1$.

Theorem 1 ***The Basic Trigonometric Limit***

$$\lim_{x\to 0} \frac{\sin x}{x} = 1. \tag{2}$$

As in Examples 1 and 2, many other trigonometric limits can be reduced to the one in Theorem 1.

EXAMPLE 1 Show that

$$\lim_{x\to 0} \frac{1 - \cos x}{x} = 0. \tag{3}$$

Solution We multiply the numerator and denominator in Eq. (3) by the *conjugate* $1 + \cos x$ of the numerator $1 - \cos x$. Then we apply the identity $1 - \cos^2 x = \sin^2 x$. This gives

$$\lim_{x\to 0} \frac{1 - \cos x}{x} = \lim_{x\to 0} \frac{1 - \cos x}{x} \cdot \frac{1 + \cos x}{1 + \cos x} = \lim_{x\to 0} \frac{\sin^2 x}{x(1 + \cos x)}$$

$$= \left(\lim_{x\to 0} \frac{\sin x}{x}\right)\left(\lim_{x\to 0} \frac{\sin x}{1 + \cos x}\right) = 1 \cdot \frac{0}{1 + 1} = 0.$$

In the last step we used *all* the limits in Eqs. (1) and (2). ■

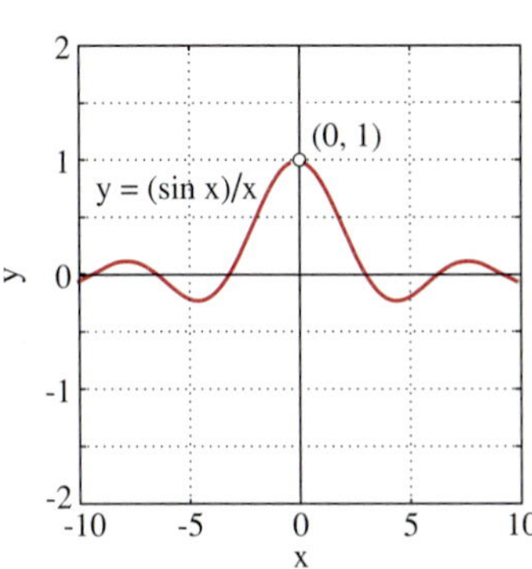

Fig. 2.3.3 $y = \dfrac{\sin x}{x}$ for $x \neq 0$

x	$\sin\frac{\pi}{x}$
1	0
0.5	0
0.1	0
0.05	0
0.01	0
0.005	0
0.001	0

Fig. 2.3.4 Do you think that $\lim_{x\to 0} \sin\frac{\pi}{x} = 0$ (Example 3)?

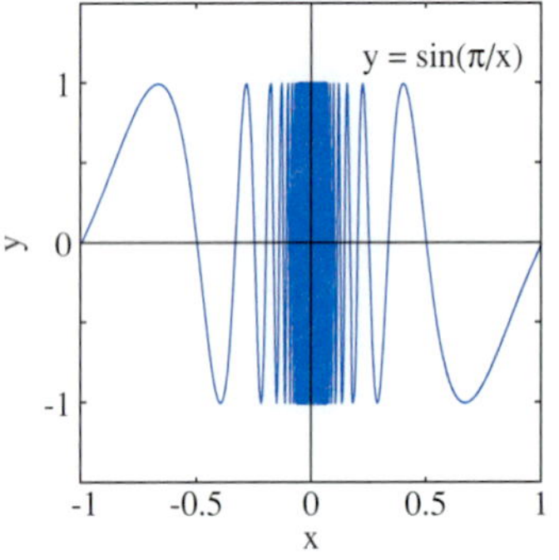

Fig. 2.3.5 The graph of $y = \sin\frac{\pi}{x}$ shows infinite oscillation as $x \to 0$ (Example 3).

x	$\sin\frac{\pi}{x}$
$\frac{2}{9}$	+1
$\frac{2}{11}$	−1
$\frac{2}{101}$	+1
$\frac{2}{103}$	−1
$\frac{2}{1001}$	+1
$\frac{2}{1003}$	−1

Fig. 2.3.6 Verify the entries in the second column (Example 3).

EXAMPLE 2 Evaluate $\displaystyle\lim_{x\to 0}\frac{\tan 3x}{x}$.

Solution

$$\lim_{x\to 0}\frac{\tan 3x}{x} = 3\left(\lim_{x\to 0}\frac{\tan 3x}{3x}\right) = 3\left(\lim_{\theta\to 0}\frac{\tan\theta}{\theta}\right) \qquad (\theta = 3x)$$

$$= 3\left(\lim_{\theta\to 0}\frac{\sin\theta}{\theta\cos\theta}\right) \qquad \left(\text{because } \tan\theta = \frac{\sin\theta}{\cos\theta}\right)$$

$$= 3\left(\lim_{\theta\to 0}\frac{\sin\theta}{\theta}\right)\left(\lim_{\theta\to 0}\frac{1}{\cos\theta}\right) \qquad \text{(by the product law of limits)}$$

$$= 3\cdot 1\cdot\frac{1}{1} = 3.$$

We used the fact that $\tan\theta = (\sin\theta)/(\cos\theta)$ as well as some of the limits in Eqs. (1) and (2). ■

Example 3 constitutes a *warning:* The results of numerical investigation can be misleading unless they are interpreted with care.

EXAMPLE 3 The numerical data shown in the table of Fig. 2.3.4 suggest that the limit

$$\lim_{x\to 0}\sin\frac{\pi}{x} \tag{4}$$

has the value zero. But it appears in the graph of $y = \sin(\pi/x)$ (for $x \neq 0$), shown in Fig. 2.3.5, that the value of $\sin(\pi/x)$ oscillates infinitely often between $+1$ and -1 as $x \to 0$. Indeed, this fact follows from the periodicity of the sine function, because π/x increases without bound as $x \to 0$. Hence $\sin(\pi/x)$ cannot approach zero (or any other number) as $x \to 0$. Therefore the limit in (4) *does not exist.*

We can explain the potentially misleading results tabulated in Fig. 2.3.4 as follows: Each value of x shown there just happens to be of the form $1/n$, the reciprocal of an integer. Therefore

$$\sin\frac{\pi}{x} = \sin\frac{\pi}{1/n} = \sin n\pi = 0$$

for every nonzero integer n. But with a different selection of "trial values" of x, we might have obtained the results shown in Fig. 2.3.6, which immediately suggest the nonexistence of the limit in (4). ■

The Squeeze Law of Limits

A final property of limits that we will need is the *squeeze law* (also known as the "sandwich theorem"). It is related to the fact that taking limits preserves inequalities among functions.

Figure 2.3.7 illustrates how and why the squeeze law works and how it got its name. The idea is that $g(x)$ is trapped between $f(x)$ and $h(x)$ near a; both $f(x)$ and $h(x)$ approach the same limit L, so $g(x)$ must approach L as well. A formal proof of the squeeze law can be found in Appendix D.

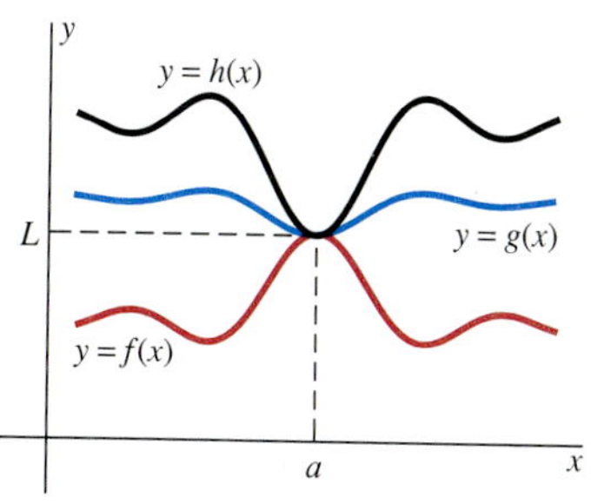

Fig. 2.3.7 How the squeeze law works

Squeeze Law

Suppose that $f(x) \leqq g(x) \leqq h(x)$ for all $x \neq a$ in some neighborhood of a and also that

$$\lim_{x \to a} f(x) = L = \lim_{x \to a} h(x).$$

Then

$$\lim_{x \to a} g(x) = L$$

as well.

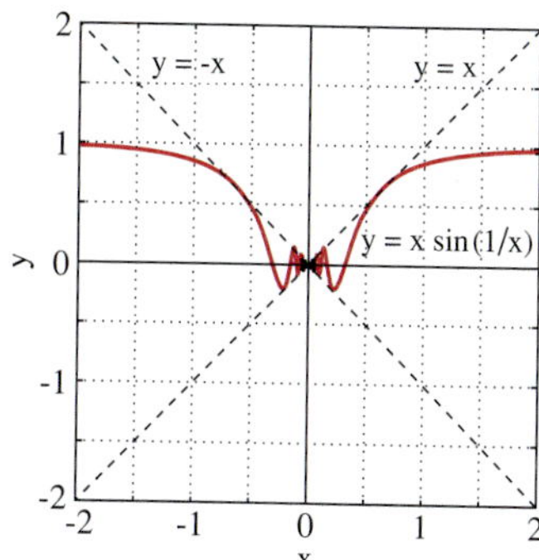

Fig. 2.3.8 The graph of $g(x) = x \sin \frac{1}{x}$ for $x \neq 0$ (Example 4)

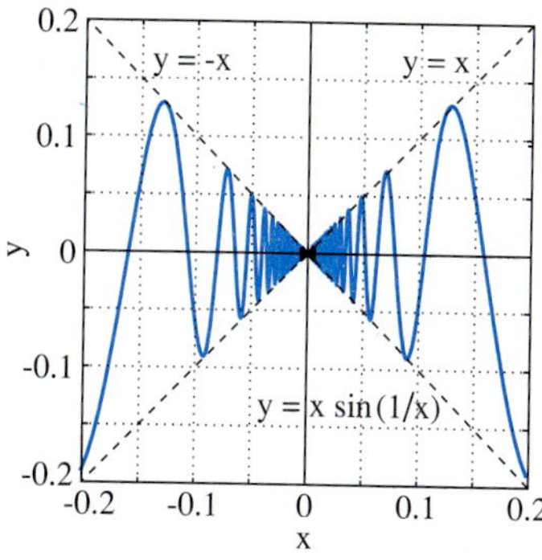

Fig. 2.3.9 The graph magnified near the origin (Example 4)

EXAMPLE 4 Figures 2.3.8 and 2.3.9 show two views of the graph of the function g defined for $x \neq 0$ by

$$g(x) = x \sin \frac{1}{x}.$$

As in Example 3, $\sin(1/x)$ oscillates infinitely often between $+1$ and -1 as $x \to 0$. Therefore $g(x)$ oscillates infinitely often between the values $+x$ and $-x$. Because $|\sin(1/x)| \leqq 1$ for all $x \neq 0$,

$$-|x| \leqq x \sin \frac{1}{x} \leqq +|x|$$

for all $x \neq 0$. Moreover, $\pm |x| \to 0$ as $x \to 0$, so with $f(x) = -|x|$ and $h(x) = +|x|$, it follows from the squeeze law of limits that

$$\lim_{x \to 0} x \sin \frac{1}{x} = 0. \qquad \textbf{(5)}$$

■

QUESTION Why *doesn't* the limit in Eq. (5) follow from the product law of limits with $f(x) = x$ and $g(x) = \sin(1/x)$?

One-Sided Limits

In Example 4 of Section 2.2 we examined the function

$$f(x) = \frac{x}{|x|} = \begin{cases} 1 & \text{if } x > 0, \\ -1 & \text{if } x < 0. \end{cases}$$

The graph of $y = f(x)$ is shown in Fig. 2.3.10. We argued that the limit of $f(x)$ as $x \to 0$ does not exist because $f(x)$ approaches $+1$ as x approaches zero from the right, whereas $f(x) \to -1$ as x approaches zero from the left. A natural way of describing this situation is to say that at $x = 0$ the *right-hand limit* of $f(x)$ is $+1$ and the *left-hand limit* of $f(x)$ is -1.

Here we define and investigate such one-sided limits. Their definitions will be stated initially in the informal language we used in Section 2.2 to describe the "idea of the limit." To define the right-hand limit of $f(x)$ at $x = a$, we must assume that f is defined on an open interval immediately to the right of a. To define the left-hand limit, we must assume that f is defined on an open interval immediately to the left of a.

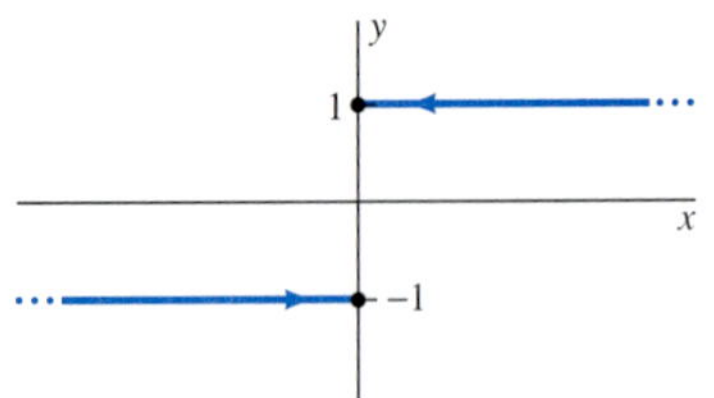

Fig. 2.3.10 The graph of $f(x) = \dfrac{x}{|x|}$ again

The Right-Hand Limit of a Function

Suppose that f is defined on the interval (a, c) immediately to the right of a. Then we say that the number L is the **right-hand limit** of $f(x)$ as x approaches a, and we write

$$\lim_{x \to a^+} f(x) = L, \tag{6}$$

provided that we can make the number $f(x)$ as close to L as we please merely by choosing the point x in (a, c) sufficiently close to a.

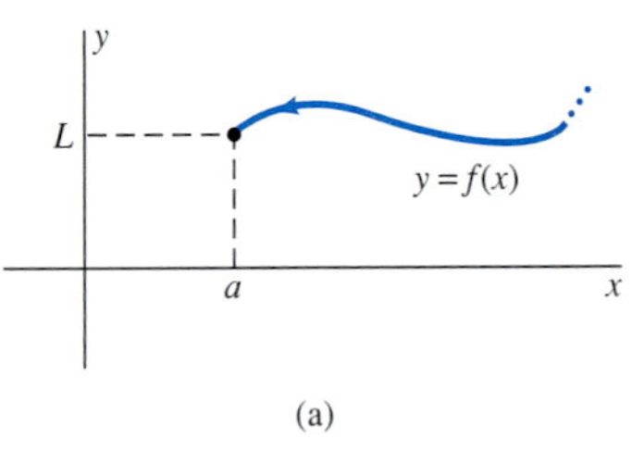

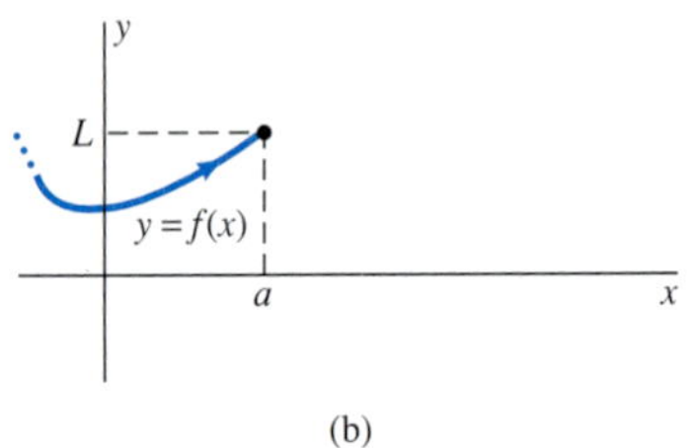

Fig. 2.3.11 (a) The right-hand limit of $f(x)$ is L. (b) The left-hand limit of $f(x)$ is L.

We may describe the right-hand limit in Eq. (6) by saying that $f(x) \to L$ as $x \to a^+$, or as x approaches a from the right. (The symbol a^+ denotes the right-hand, or positive, side of a.) More precisely, Eq. (6) means that, given $\epsilon > 0$, there exists $\delta > 0$ such that

$$|f(x) - L| < \epsilon \tag{7}$$

for all x such that

$$a < x < a + \delta. \tag{8}$$

See Fig. 2.3.11 for a geometric interpretation of one-sided limits.

The Left-Hand Limit of a Function

Suppose that f is defined on the interval (c, a) immediately to the left of a. Then we say that the number L is the **left-hand limit** of $f(x)$ as x approaches a, and we write

$$\lim_{x \to a^-} f(x) = L, \tag{9}$$

provided that we can make the number $f(x)$ as close to L as we please merely by choosing the point x in (c, a) sufficiently close to a.

A consequence of these definitions is that the value of $f(a)$ itself is not relevant to the existence or value of the one-sided limits, just as it is not relevant to the existence or value of the (regular) two-sided limit.

We may describe the left-hand limit in Eq. (9) by saying that $f(x) \to L$ as $x \to a^-$, or as x approaches a from the left. (The symbol a^- denotes the left-hand, or negative, side of a.) We get a precise definition of the left-hand limit in Eq. (9) merely by changing the open interval in (8) to the interval $a - \delta < x < a$.

Our preliminary discussion of the function $f(x) = x/|x|$ amounts to saying that this function's one-sided limits at $x = 0$ are

$$\lim_{x \to 0^+} \frac{x}{|x|} = 1 \quad \text{and} \quad \lim_{x \to 0^-} \frac{x}{|x|} = -1.$$

We argued further in Example 4 of Section 2.2 that, because these two limits are not equal, the two-sided limit of $f(x)$ as $x \to 0$ does not exist. The following theorem is intuitively obvious and can be proved by using the precise definitions of all the limits involved.

Theorem 2 ***One-Sided Limits and Two-Sided Limits***

Suppose that the function f is defined for $x \neq a$ in some open interval containing the point a. Then the two-sided limit

$$\lim_{x \to a} f(x)$$

exists and is equal to the number L if and only if both one-sided limits

$$\lim_{x \to a^+} f(x) \quad \text{and} \quad \lim_{x \to a^-} f(x)$$

exist and are equal to L.

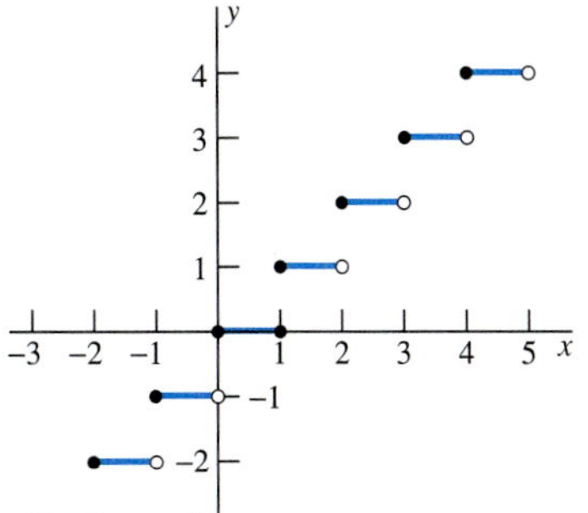

Fig. 2.3.12 The graph of the greatest integer function $f(x) = [\![x]\!]$ (Example 5).

Theorem 2 is particularly useful in showing that certain (two-sided) limits do *not* exist, frequently by showing that the left-hand and right-hand limits are not equal to each other.

EXAMPLE 5 The graph of the greatest integer function $f(x) = [\![x]\!]$ is shown in Fig. 2.3.12. It should be apparent that if a is not an integer, then

$$\lim_{x \to a^+} [\![x]\!] = \lim_{x \to a^-} [\![x]\!] = \lim_{x \to a} [\![x]\!] = [\![a]\!].$$

But if $a = n$, an integer, then

$$\lim_{x \to n^-} [\![x]\!] = n - 1 \quad \text{and} \quad \lim_{x \to n^+} [\![x]\!] = n.$$

Because these left-hand and right-hand limits are not equal, it follows from Theorem 2 that the limit of $f(x) = [\![x]\!]$ does not exist as x approaches an integer n. ■

EXAMPLE 6 According to the root law in Section 2.2,

$$\lim_{x \to a} \sqrt{x} = \sqrt{a} \quad \text{if } a > 0.$$

But the limit of $f(x) = \sqrt{x}$ as $x \to 0^-$ is not defined because the square root of a negative number is undefined. Hence f is undefined on every open interval containing zero. What we can say in the case $a = 0$ is that

$$\lim_{x \to 0^+} \sqrt{x} = 0,$$

and that the left-hand limit

$$\lim_{x \to 0^-} \sqrt{x}$$

does not exist. ■

To each of the limit laws in Section 2.2 there correspond two *one-sided limit laws*—a right-hand version and a left-hand version. You may apply these one-sided limit laws in the same way you apply the two-sided limit laws in the evaluation of limits.

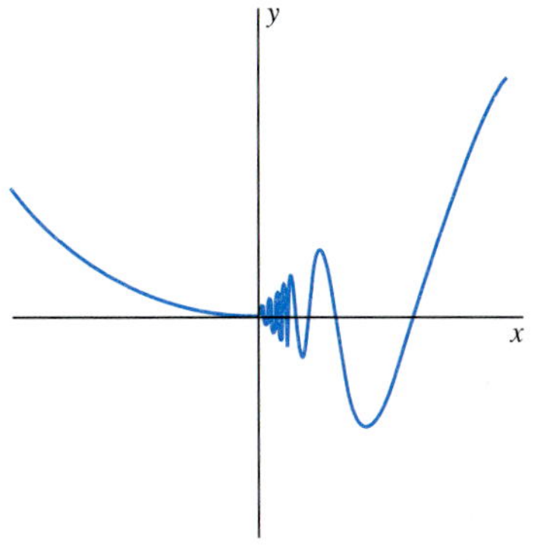

Fig. 2.3.13 $y = f(x)$ (Example 7)

EXAMPLE 7 Figure 2.3.13 shows the graph of the function f defined by

$$f(x) = \begin{cases} x^2 & \text{if } x \leqq 0, \\ x \sin \dfrac{1}{x} & \text{if } x > 0. \end{cases}$$

Clearly

$$\lim_{x\to 0^-} f(x) = 0 \quad \text{and} \quad \lim_{x\to 0^+} f(x) = 0$$

by a one-sided version of the squeeze law (as in Example 4). It therefore follows from Theorem 2 that

$$\lim_{x\to 0} f(x) = 0.$$

EXAMPLE 8 Upon applying the appropriate one-sided limit laws, we find that

$$\lim_{x\to 3^-}\left(\frac{x^2}{x^2+1}+\sqrt{9-x^2}\right) = \frac{\lim_{x\to 3^-} x^2}{\lim_{x\to 3^-}(x^2+1)} + \sqrt{\lim_{x\to 3^-}(9-x^2)}$$

$$= \frac{9}{9+1} + \sqrt{0} = \frac{9}{10}.$$

Note that the two-sided limit at 3 is not defined, because $\sqrt{9-x^2}$ is not defined when $x > 3$.

Existence of Tangent Lines

Recall that the slope of the line tangent to the graph $y = f(x)$ at the point $P(a, f(a))$ is defined to be

$$m = \lim_{x\to a}\frac{f(x)-f(a)}{x-a} \qquad \textbf{(10)}$$

provided that this (two-sided) limit exists. In this case an equation of the line tangent to the graph $y = f(x)$ at $P(a, f(a))$ is

$$y - f(a) = m(x-a).$$

If the limit in (10) does not exist, then we say that the curve $y = f(x)$ does not have a tangent line at the point P. The following example gives perhaps the simplest example of a function whose graph has a tangent line everywhere except at a single isolated point.

EXAMPLE 9 Show that the graph $y = |x|$ has no tangent line at the origin.

Solution Figure 2.3.14 shows the graph of the function $f(x) = |x|$. The sharp corner at the point $(0, 0)$ makes it intuitively clear that there can be no tangent line there—surely no single straight line can be a good approximation to the shape of the graph at the origin. To verify this intuitive observation, note that when $a = 0$ we have

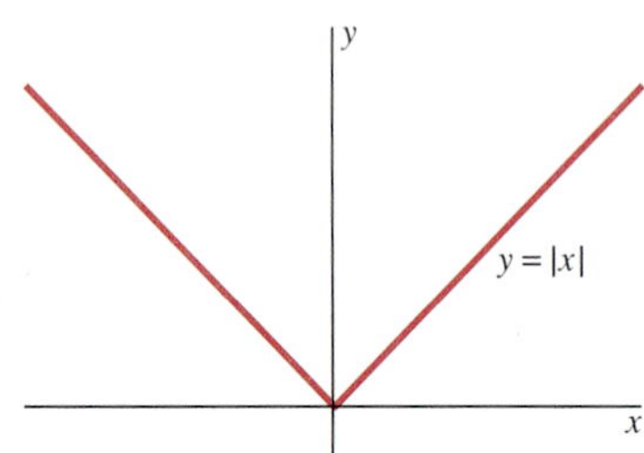

Fig. 2.3.14 The graph of $f(x) = |x|$ has a corner point at (0,0).

$$\frac{f(a+h)-f(a)}{h} = \frac{|h|}{h} = \begin{cases} -1 & \text{if } h < 0, \\ +1 & \text{if } h > 0. \end{cases}$$

Hence the left-hand limit of the quotient is -1, whereas the right-hand limit is $+1$. Therefore the two-sided limit in (10) does not exist, so the graph $y = |x|$ has no tangent line at the origin, where $a = 0$.

QUESTION Does Fig. 2.3.14 make it clear to you that for $f(x) = |x|$ and $a \neq 0$, the value of the "slope limit" in (10) is given by

$$m = \begin{cases} -1 & \text{if } a < 0, \\ +1 & \text{if } a > 0? \end{cases}$$

It follows (as is apparent from Fig. 2.3.14) that the line $y = x$ is tangent to the graph $y = |x|$ at any point of the graph to the right of the origin, and that the line $y = -x$ is the tangent line at any point of the graph to the left of the origin.

Infinite Limits

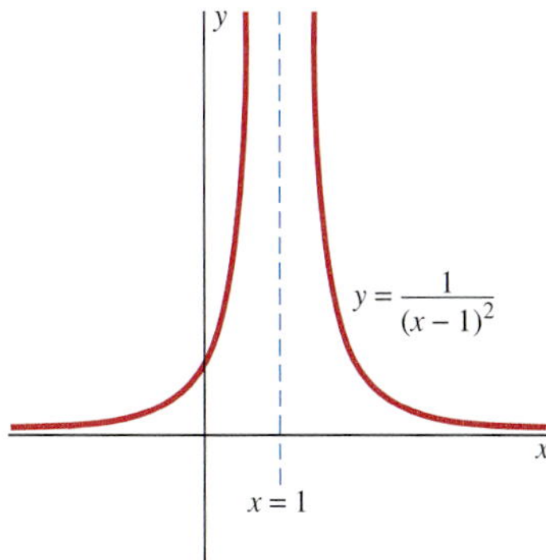

Fig. 2.3.15 The graph of the function $f(x) = \dfrac{1}{(x-1)^2}$

In Example 8 of Section 2.2, we investigated the function $f(x) = 1/(x-1)^2$; the graph of f is shown in Fig. 2.3.15. The value of $f(x)$ *increases without bound* (that is, eventually exceeds any preassigned number) as x approaches 1 either from the right or from the left. This situation can be described by writing

$$\lim_{x \to 1^-} \frac{1}{(x-1)^2} = +\infty = \lim_{x \to 1^+} \frac{1}{(x-1)^2}, \tag{11}$$

and we say that each of these one-sided limits is equal to "plus infinity."

CAUTION The expression

$$\lim_{x \to 1^+} \frac{1}{(x-1)^2} = +\infty \tag{12}$$

does not mean that there exists an "infinite real number" denoted by $+\infty$—there does not! Neither does it mean that the limit on the left-hand side in Eq. (12) exists—it does not! Instead, Eq. (12) is just a convenient way of saying *why* the right-hand limit in Eq. (12) does not exist: because the quantity $1/(x-1)^2$ increases without bound as $x \to 1^+$.

With similar provisos we may write

$$\lim_{x \to 1} \frac{1}{(x-1)^2} = +\infty \tag{13}$$

despite the fact that the (two-sided) limit in Eq. (13) does not exist. The expression in Eq. (13) is merely a convenient way of saying that the limit in Eq. (13) does not exist because $1/(x-1)^2$ increases without bound as $x \to 1$ from either side.

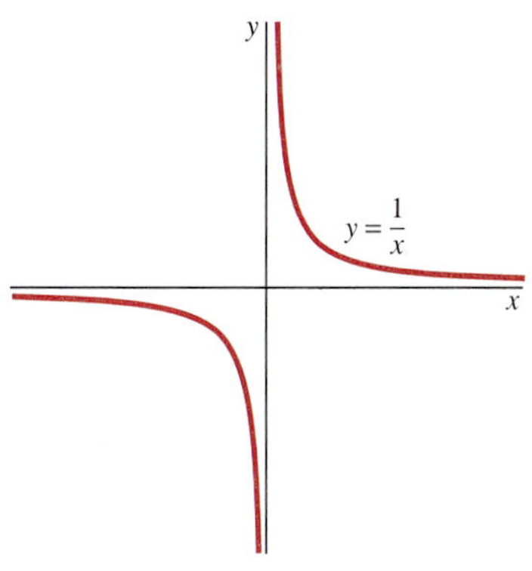

Fig. 2.3.16 The graph of the function $f(x) = \dfrac{1}{x}$

Now consider the function $f(x) = 1/x$; its graph is shown in Fig. 2.3.16. This function increases without bound as x approaches zero from the right but decreases without bound—it becomes less than any preassigned negative number—as x approaches zero from the left. We therefore write

$$\lim_{x \to 0^-} \frac{1}{x} = -\infty \quad \text{and} \quad \lim_{x \to 0^+} \frac{1}{x} = +\infty. \tag{14}$$

There is no shorthand for the two-sided limit in this case. We may say only that

$$\lim_{x \to 0} \frac{1}{x} \quad \text{does not exist.}$$

EXAMPLE 10 Investigate the behavior of the function

$$f(x) = \frac{2x+1}{x-1}$$

near the point $x = 1$, where the limit of $f(x)$ does not exist.

Solution First we look at the behavior of $f(x)$ just to the right of the number 1. If x is greater than 1 but close to 1, then $2x + 1$ is close to 3 and $x - 1$ is a small *positive* number. In this case the quotient $(2x + 1)/(x - 1)$ is a large positive number, and the closer x is to 1, the larger this positive quotient will be. For such x, $f(x)$ increases without bound as x approaches 1 from the right. That is,

$$\lim_{x \to 1^+} \frac{2x + 1}{x - 1} = +\infty, \tag{15}$$

as the data in Fig. 2.3.17 suggest.

x	$\dfrac{2x+1}{x-1}$	x	$\dfrac{2x+1}{x-1}$
1.1	32	0.9	−28
1.01	302	0.99	−298
1.001	3002	0.999	−2998
1.0001	30002	0.9999	−29998
⋮	⋮	⋮	⋮
↓	↓	↓	↓
1	$+\infty$	1	$-\infty$

Fig. 2.3.17 The behavior of $f(x) = \dfrac{2x + 1}{x - 1}$ for x near 1 (Example 10)

If instead x is less than 1 but still close to 1, then $2x + 1$ is still close to 3, but now $x - 1$ is a *negative* number close to zero. In this case the quotient $(2x + 1)/(x - 1)$ is a (numerically) large negative number and decreases without bound as $x \to 1^-$. Hence we conclude that

$$\lim_{x \to 1^-} \frac{2x + 1}{x - 1} = -\infty. \tag{16}$$

The results in Eqs. (15) and (16) provide a concise description of the behavior of $f(x) = (2x + 1)/(x - 1)$ near the point $x = 1$. Finally, to remain consistent with Theorem 2 on one-sided and two-sided limits, we say in this case that

$$\lim_{x \to 1} \frac{2x + 1}{x - 1} \quad \text{does not exist.}$$ ■

The Basic Trigonometric Limit

We conclude this section with a geometric proof that

$$\lim_{\theta \to 0} \frac{\sin \theta}{\theta} = 1. \tag{17}$$

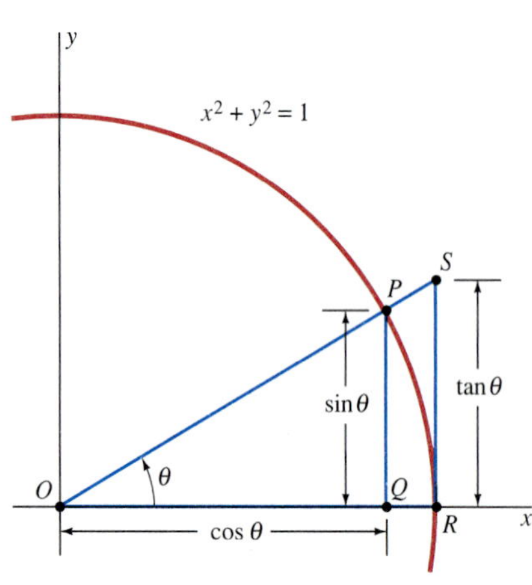

Fig. 2.3.18 Aid to the proof of the basic trigonometric limit

PROOF Figure 2.3.18 shows the angle θ, the triangles OPQ and ORS, and the circular sector OPR that contains the triangle OPQ and is contained in the triangle ORS. Hence

$$\text{area}(\triangle OPQ) < \text{area}(\text{sector } OPR) < \text{area}(\triangle ORS).$$

In terms of θ, this means that

$$\frac{1}{2}\sin\theta\cos\theta < \frac{1}{2}\theta < \frac{1}{2}\tan\theta = \frac{\sin\theta}{2\cos\theta}.$$

Here we use the standard formula for the area of a triangle to obtain the area of $\triangle OPQ$ and $\triangle ORS$. We also use the fact that the area of a circular sector in a circle of radius r is $A = \frac{1}{2}r^2\theta$ if the sector is subtended by a central angle of θ radians; here, $r = 1$. If $0 < \theta < \pi/2$, then we can divide each member of the last inequality by $\frac{1}{2}\sin\theta$ to obtain

$$\cos\theta < \frac{\theta}{\sin\theta} < \frac{1}{\cos\theta}.$$

We take reciprocals, which reverses the inequalities:

$$\cos\theta < \frac{\sin\theta}{\theta} < \frac{1}{\cos\theta}.$$

Now we apply the squeeze law of limits with

$$f(\theta) = \cos\theta, \qquad g(\theta) = \frac{\sin\theta}{\theta}, \qquad \text{and} \qquad h(\theta) = \frac{1}{\cos\theta}.$$

Because it is clear from Eq. (1) (at the beginning of this section) that both $f(\theta)$ and $h(\theta)$ approach 1 as $\theta \to 0^+$, so does $g(\theta) = (\sin\theta)/\theta$. This geometric argument shows that $(\sin\theta)/\theta \to 1$ for *positive* values of θ that approach zero. But the same result follows for negative values of θ, because $\sin(-\theta) = -\sin\theta$. So we have proved Eq. (17). ■

2.3 PROBLEMS

Find the trigonometric limits in Problems 1 through 24. If you have a graphics calculator or a computer with graphing facility, verify that graphical evidence supports your answer.

1. $\lim_{\theta\to 0} \dfrac{\theta^2}{\sin\theta}$

2. $\lim_{\theta\to 0} \dfrac{\sin^2\theta}{\theta^2}$

3. $\lim_{\theta\to 0} \dfrac{1-\cos\theta}{\theta^2}$

4. $\lim_{\theta\to 0} \dfrac{\tan\theta}{\theta}$

5. $\lim_{t\to 0} \dfrac{2t}{(\sin t) - t}$

6. $\lim_{\theta\to 0} \dfrac{\sin(2\theta^2)}{\theta^2}$

7. $\lim_{x\to 0} \dfrac{\sin 5x}{x}$

8. $\lim_{z\to 0} \dfrac{\sin 2z}{z\cos 3z}$

9. $\lim_{x\to 0} \dfrac{\sin x}{\sqrt{x}}$

10. $\lim_{x\to 0} \dfrac{1-\cos 2x}{x}$

11. $\lim_{x\to 0} \dfrac{1}{x}\sin\dfrac{x}{3}$

12. $\lim_{\theta\to 0} \dfrac{(\sin 3\theta)^2}{\theta^2\cos\theta}$

13. $\lim_{x\to 0} \dfrac{1-\cos x}{\sin x}$

14. $\lim_{x\to 0} \dfrac{\tan 3x}{\tan 5x}$

15. $\lim_{x\to 0} x \sec x \csc x$

16. $\lim_{\theta\to 0} \dfrac{\sin 2\theta}{\theta}$

17. $\lim_{\theta\to 0} \dfrac{1-\cos\theta}{\theta\sin\theta}$

18. $\lim_{\theta\to 0} \dfrac{\sin^2\theta}{\theta}$

19. $\lim_{z\to 0} \dfrac{\tan z}{\sin 2z}$

20. $\lim_{x\to 0} \dfrac{\tan 2x}{3x}$

21. $\lim_{x\to 0} x\cot 3x$

22. $\lim_{x\to 0} \dfrac{x-\tan x}{\sin x}$

23. $\lim_{t\to 0} \dfrac{1}{t^2}\sin^2\left(\dfrac{t}{2}\right)$

24. $\lim_{x\to 0} \dfrac{\sin 2x}{\sin 5x}$

Use the squeeze law of limits to find the limits in Problems 25 through 28. Also illustrate each of these limits by graphing the functions f, g, and h (in the notation of the squeeze law) on the same screen.

25. $\lim_{x\to 0} x^2\cos 10x$

26. $\lim_{x\to 0} x^2\sin\dfrac{1}{x^2}$

27. $\lim_{x\to 0} x^2\cos\dfrac{1}{\sqrt[3]{x}}$

28. $\lim_{x\to 0} \sqrt[3]{x}\sin\dfrac{1}{x}$

Use one-sided limit laws to find the limits in Problems 29 through 48 or to determine that they do not exist.

29. $\lim_{x\to 0^+} (3-\sqrt{x})$

30. $\lim_{x\to 0^+} (4+3x^{3/2})$

31. $\lim_{x\to 1^-} \sqrt{x-1}$

32. $\lim_{x\to 4^-} \sqrt{4-x}$

33. $\lim_{x\to 2^+} \sqrt{x^2-4}$

34. $\lim_{x\to 3^+} \sqrt{9-x^2}$

35. $\lim_{x\to 5^-} \sqrt{x(5-x)}$

36. $\lim_{x\to 2^-} \sqrt{4-x^2}$

37. $\lim_{x\to 4^+} \sqrt{\dfrac{4x}{x-4}}$

38. $\lim_{x\to -3^+} \sqrt{6-x-x^2}$

39. $\lim_{x\to 5^-} \dfrac{x-5}{|x-5|}$

40. $\lim_{x\to -4^+} \dfrac{16-x^2}{\sqrt{16-x^2}}$

41. $\lim_{x\to 3^+} \dfrac{\sqrt{x^2-6x+9}}{x-3}$

42. $\lim_{x\to 2^+} \dfrac{x-2}{x^2-5x+6}$

43. $\lim_{x\to 2^+} \dfrac{2-x}{|x-2|}$

44. $\lim_{x\to 7^-} \dfrac{7-x}{|x-7|}$

45. $\lim_{x\to 1^+} \dfrac{1-x^2}{1-x}$

46. $\lim_{x\to 0^-} \dfrac{x}{x-|x|}$

47. $\lim_{x\to 5^+} \dfrac{\sqrt{(5-x)^2}}{5-x}$

48. $\lim_{x\to -4^-} \dfrac{4+x}{\sqrt{(4+x)^2}}$

For each of the functions in Problems 49 through 58, there is exactly one point a where both the right-hand and left-hand limits of $f(x)$ fail to exist. Describe (as in Example 10) the behavior of $f(x)$ for x near a.

49. $f(x) = \dfrac{1}{x-1}$

50. $f(x) = \dfrac{2}{3-x}$

51. $f(x) = \dfrac{x-1}{x+1}$

52. $f(x) = \dfrac{2x-5}{5-x}$

53. $f(x) = \dfrac{1-x^2}{x+2}$

54. $f(x) = \dfrac{1}{(x-5)^2}$

55. $f(x) = \dfrac{|1-x|}{(1-x)^2}$

56. $f(x) = \dfrac{x+1}{x^2+6x+9}$

57. $f(x) = \dfrac{x-2}{4-x^2}$

58. $f(x) = \dfrac{x-1}{x^2-3x+2}$

In Problems 59 and 60, find the left-hand and right-hand limits of $f(x)$ at $a = 2$. Does the two-sided limit of f exist there? Sketch the graph of $y = f(x)$.

59. $f(x) = \dfrac{x^2 - 4}{|x - 2|}$

60. $f(x) = \dfrac{x^4 - 8x + 16}{|x - 2|}$

In Problems 61 through 68, do the following: (a) Sketch the graph of the given function f. (b) For each integer n, evaluate the one-sided limits

$$\lim_{x \to n^-} f(x) \quad \text{and} \quad \lim_{x \to n^+} f(x)$$

in terms of n. (c) Determine those values of a for which $\lim_{x \to a} f(x)$ exists. Recall that $[\![x]\!]$ denotes the greatest integer that does not exceed x.

61. $f(x) = \begin{cases} 2 & \text{if } x \text{ is not an integer,} \\ 2 + (-1)^x & \text{if } x \text{ is an integer.} \end{cases}$

62. $f(x) = \begin{cases} x & \text{if } x \text{ is not an integer,} \\ 0 & \text{if } x \text{ is an integer.} \end{cases}$

63. $f(x) = [\![10x]\!]$

64. $f(x) = (-1)^{[\![x]\!]}$

65. $f(x) = x - [\![x]\!] - \frac{1}{2}$

66. $f(x) = \left[\!\!\left[\dfrac{x}{2}\right]\!\!\right]$

67. $f(x) = [\![x]\!] + [\![-x]\!]$

68. $f(x) = \begin{cases} \dfrac{[\![x]\!]}{x} & \text{if } x \neq 0, \\ 0 & \text{if } x = 0. \end{cases}$

69. If $g(x) = \frac{1}{10}[\![10x]\!]$, the value of x to one decimal place rounded *down*, sketch the graph of g and determine the values of a such that $\lim_{x \to a} g(x)$ exists.

70. The sign function $\operatorname{sgn}(x)$ is defined as follows:

$$\operatorname{sgn}(x) = \begin{cases} \dfrac{x}{|x|} & \text{if } x \neq 0, \\ 0 & \text{if } x = 0. \end{cases}$$

Use the sign function to define two functions f and g whose limits as $x \to 0$ do not exist, but such that (a) $\lim_{x \to 0} [f(x) + g(x)]$ does exist; (b) $\lim_{x \to 0} f(x) \cdot g(x)$ does exist.

71. Let

$$f(x) = \begin{cases} x^2 & \text{if } x \text{ is rational,} \\ 0 & \text{if } x \text{ is irrational.} \end{cases}$$

Use the squeeze law of limits to show that $\lim_{x \to 0} f(x) = f(0) = 0$.

72. Sketch the graph of the function

$$f(x) = \frac{1}{1 + 2^{1/x}}$$

for $x \neq 0$. Then determine whether $\lim_{x \to 0} f(x)$ exists.

In Problems 73 and 74, first examine the value of $f(x)$ on intervals of the form

$$\frac{1}{n+1} < x < \frac{1}{n},$$

where n is an interger. Then determine whether $\lim_{x \to 0} f(x)$ exists. If your graphing calculator or computer has a greatest integer (or "floor") function, graph f to corroborate your answer.

73. $f(x) = x \cdot \left[\!\!\left[\dfrac{1}{x}\right]\!\!\right]$

74. $f(x) = x^2 \cdot \left[\!\!\left[\dfrac{1}{x}\right]\!\!\right]$

2.4 THE CONCEPT OF CONTINUITY

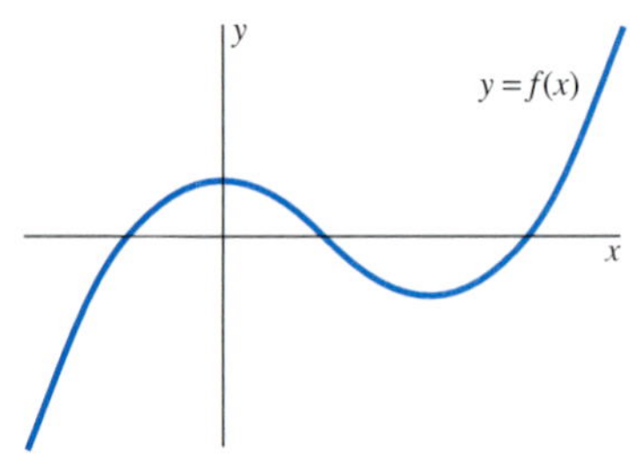

Fig. 2.4.1 A continuous graph.

Anyone can see a drastic difference between the graphs in Figs. 2.4.1 and 2.4.2. Figure 2.4.1 is intended to suggest that the graph $y = f(x)$ can be traced with a continuous motion—without any jumps—of the pen from left to right. But in Fig. 2.4.2 the pen must make a sudden jump at $x = a$.

The concept of continuity isolates the property that the function f of Fig. 2.4.1 possesses but that the function g of Fig. 2.4.2 lacks. We first define *continuity* of a function at a single point.

Definition *Continuity at a Point*

Suppose that the function f is defined in a neighborhood of a. We say that f is **continuous at** a provided that $\lim_{x \to a} f(x)$ exists and moreover that the value of this limit is $f(a)$. In other words, f is continuous at a provided that

$$\lim_{x \to a} f(x) = f(a). \qquad \textbf{(1)}$$

Briefly, continuity of f at a means this:

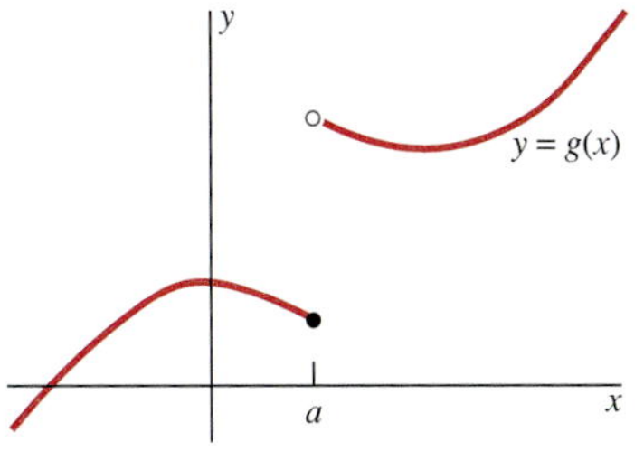

Fig. 2.4.2 A graph that is not continuous

The limit of f at a is equal to the value of f at a.

Another way to put it is this: The limit of f at a is the "expected" value—the value that you would assign if you knew the values of $f(x)$ for $x \neq a$ in an open interval containing a and you knew f to be "predictable." Alternatively, continuity of f at a means this: When x is close to a, $f(x)$ is close to $f(a)$.

Analysis of the definition of continuity shows us that to be continuous at the point a, the function f must satisfy the following three conditions:

1. The function f must be defined at a [so that $f(a)$ exists].
2. The limit of $f(x)$ as x approaches a must exist.
3. The numbers in conditions 1 and 2 must be equal:

$$\lim_{x \to a} f(x) = f(a).$$

If any one of these conditions is not satisfied, then f is not continuous at a. Examples 1 through 3 illustrate these three possibilities for *discontinuity* at a point. If the function f is *not* continuous at a, then we say that it is **discontinuous** there or that a is a **discontinuity** of f. Intuitively, a discontinuity of f is a point where the graph of f has a "gap," or "jump," of some sort.

EXAMPLE 1 Figure 2.4.3 shows the graph of the function f defined by

$$f(x) = \frac{1}{x-2} \quad \text{for } x \neq 2.$$

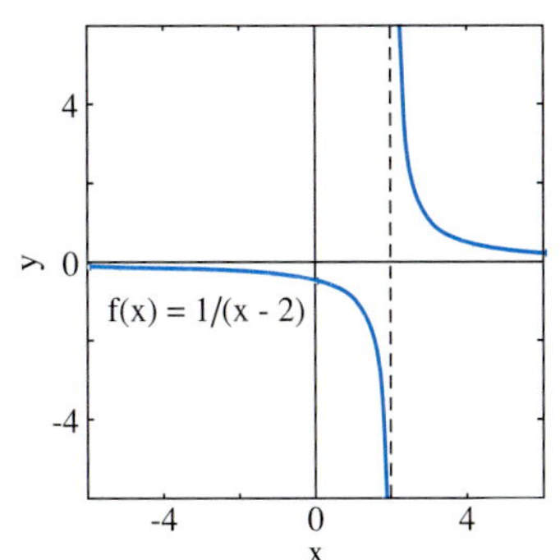

Fig. 2.4.3 The function $f(x) = 1/(x-2)$ has an infinite discontinuity at $x = 2$ (Example 1).

Because f is not defined at the point $x = 2$, it is not continuous there. Moreover, f has what might be called an *infinite discontinuity* at $x = 2$. ■

EXAMPLE 2 Figure 2.4.4 shows the graph of the function g defined by

$$g(x) = \operatorname{sgn}(x) = \begin{cases} +1 & \text{if } x \geqq 0, \\ -1 & \text{if } x < 0. \end{cases}$$

Its left-hand and right-hand limits at $x = 0$ are unequal, so $g(x)$ has no limit as $x \to 0$. Consequently, the function g is not continuous at $x = 0$; it has what might be called a *finite jump discontinuity* there. ■

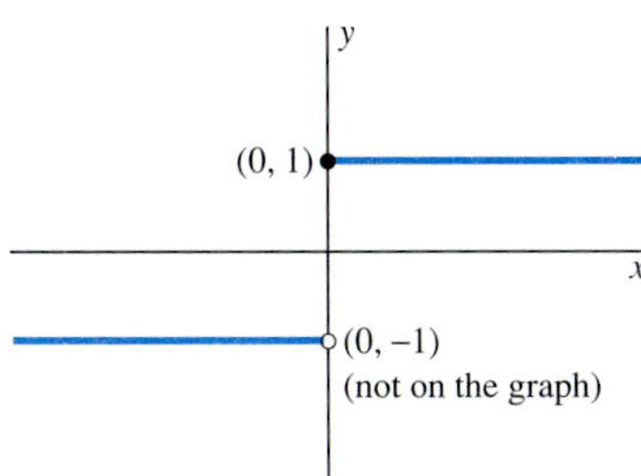

Fig. 2.2.4 The function g has a finite jump discontinuity at $x = 0$ (Example 2).

EXAMPLE 3 Figure 2.4.5 shows the graph of the function h defined by

$$h(x) = \begin{cases} \dfrac{\sin x}{x} & \text{if } x \neq 0, \\ 0 & \text{if } x = 0. \end{cases}$$

Because we saw in Section 2.3 that

$$\lim_{x \to 0} h(x) = \lim_{x \to 0} \frac{\sin x}{x} = 1,$$

whereas $h(0) = 0$, we see that the limit and the value of h at $x = 0$ are not equal. Thus the function h is not continuous there. As x moves from negative values through $x = 0$ to positive values, the value of $h(x)$ jumps from "near 1" to zero and back again. ■

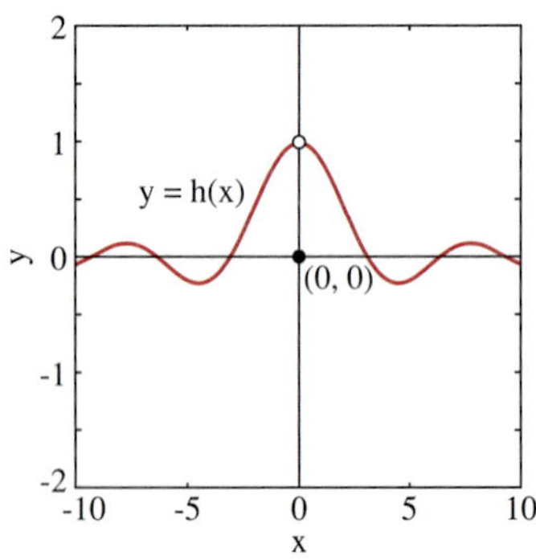

Fig. 2.4.5 The point $(0, 0)$ is on the graph; the point $(0, 1)$ is not (Example 3).

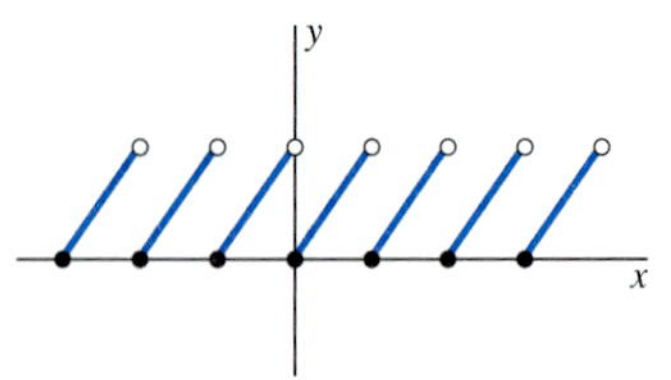

Fig. 2.4.6 The "sawtooth function" of Example 4

EXAMPLE 4 Figure 2.4.6 shows the graph of the function f defined by

$$f(x) = x - [\![x]\!].$$

As before, $[\![x]\!]$ denotes the largest integer no greater than x. If $x = n$, an integer, then $[\![n]\!] = n$, so $f(n) = 0$. On the open interval $(n, n + 1)$, the graph of f is linear and has slope 1. It should be clear that f is

- Continuous at x is x if *not* an integer;
- Discontinuous at each integer point on the x-axis.

Combinations of Continuous Functions

Frequently we are most interested in functions that *are* continuous. Suppose that the function f is defined on an open interval or a union of open intervals. Then we say simply that f is **continuous** if it is continuous at each point of its domain of definition.

It follows readily from the limit laws in Section 2.2 that *any sum or product of continuous functions is continuous.* That is, if the functions f and g are continuous at $x = a$, then so are $f + g$ and $f \cdot g$. For instance, if f and g are continuous at $x = a$, then

$$\lim_{x \to a} [f(x) + g(x)] = \left(\lim_{x \to a} f(x) \right) + \left(\lim_{x \to a} g(x) \right) = f(a) + g(a).$$

EXAMPLE 5 Because $f(x) = x$ and constant-valued functions are clearly continuous everywhere, it follows that the cubic polynomial function

$$f(x) = x^3 - 3x^2 + 1 = x \cdot x \cdot x + (-3) \cdot x \cdot x + 1$$

is continuous everywhere.

More generally, it follows in a similar way that *every* **polynomial function**

$$p(x) = b_n x^n + b_{n-1} x^{n-1} + \cdots + b_1 x + b_0$$

is continuous at each point of the real line. In short, every polynomial is continuous everywhere.

If $p(x)$ and $q(x)$ are polynomials, then the quotient law for limits and the continuity of polynomials imply that

$$\lim_{x \to a} \frac{p(x)}{q(x)} = \frac{\lim_{x \to a} p(x)}{\lim_{x \to a} q(x)} = \frac{p(a)}{q(a)}$$

provided that $q(a) \neq 0$. Thus every **rational function**

$$f(x) = \frac{p(x)}{q(x)} \tag{2}$$

is continuous wherever it is defined—that is, wherever the denominator polynomial is nonzero. More generally, the quotient of any two continuous functions is continuous at every point where the denominator is nonzero.

At a point $x = a$ where the denominator in Eq. (2) is zero, $q(a) = 0$, there are two possibilities:

- If $p(a) \neq 0$, then f has an infinite discontinuity (as in Figs. 2.4.3 and 2.4.7) at $x = a$.
- Otherwise, f *may* have a *removable discontinuity* at $x = a$.

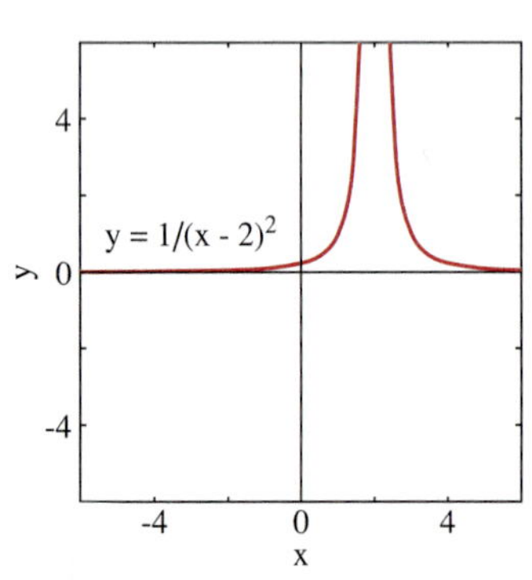

Fig. 2.4.7 The function $f(x) = 1/(x - 2)^2$ has an infinite discontinuity at $x = 2$.

The point $x = a$ where the function f is discontinuous is called a **removable discontinuity** of f provided that there exists a function F such that

- $F(x) = f(x)$ for all $x \neq a$ in the domain of f, and
- This new function F is continuous at $x = a$.

Thus, by adjoining the single point $(a, F(a))$ to the graph of f, we "remove" the discontinuity, obtaining the graph of the function F that is continuous at $x = a$.

EXAMPLE 6 Suppose that

$$f(x) = \frac{x - 2}{x^2 - 3x + 2}. \tag{3}$$

We factor the denominator: $x^2 - 3x + 2 = (x - 1)(x - 2)$. This shows that f is not defined at $x = 1$ and at $x = 2$. Thus the rational function defined in Eq. (3) is continuous except at these two points. Because cancellation gives

$$f(x) = \frac{x - 2}{x^2 - 3x + 2} = \frac{x - 2}{(x - 1)(x - 2)} = \frac{1}{x - 1}$$

except at the single point $x = 2$, the new function

$$F(x) = \frac{1}{x - 1} \tag{4}$$

agrees with $f(x)$ if $x \neq 2$ but is continuous at $x = 2$ also, where $F(2) = 1$. Thus f has a removable discontinuity at $x = 2$; the discontinuity at $x = 1$ is not removable. See Fig. 2.4.8. ■

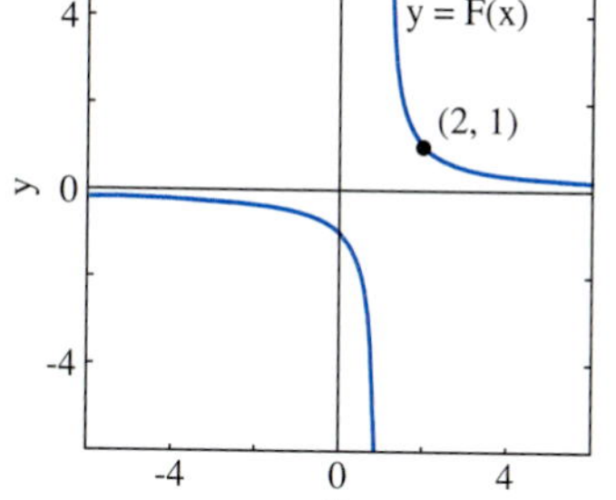

Fig. 2.4.8 In Example 6, the graph $y = F(x)$ consists of the graph $y = f(x)$ with the single point $(2, 1)$ adjointed.

Continuity of Trigonometric Functions

At the beginning of Section 2.3 we noted that

$$\lim_{x \to 0} \cos x = 1 \quad \text{and} \quad \lim_{x \to 0} \sin x = 0. \tag{5}$$

Because $\cos 0 = 1$ and $\sin 0 = 0$, the sine and cosine functions are continuous at $x = 0$ by definition. But this fact implies that they are continuous everywhere.

Theorem 1 ***Continuity of Sine and Cosine***

The functions $f(x) = \sin x$ and $g(x) = \cos x$ are continuous functions of x on the whole real line.

Proof We give the proof only for $\sin x$; the proof for $\cos x$ is similar (see Problem 67). We want to show that $\lim_{h \to 0} \sin x = \sin a$ for every real number a. If we write $x = a + h$, so that $h = x - a$, then $h \to 0$ as $x \to a$. Thus we need only show that

$$\lim_{h \to 0} \sin(a + h) = \sin a.$$

But the addition formula for the sine function yields

$$\lim_{h \to 0} \sin(a + h) = \lim_{h \to 0} (\sin a \cos h + \cos a \sin h)$$

$$= (\sin a)\left(\lim_{h\to 0}\cos h\right) + (\cos a)\left(\lim_{h\to 0}\sin h\right) = \sin a$$

as desired; we used the limits in Eq. (5) in the last step. ■

It now follows that the function

$$\tan x = \frac{\sin x}{\cos x}$$

is continuous except where $\cos x = 0$—that is, except when x is an *odd* integral multiple of $\pi/2$. As illustrated in Fig. 2.4.9, $\tan x$ has an infinite discontinuity at each such point.

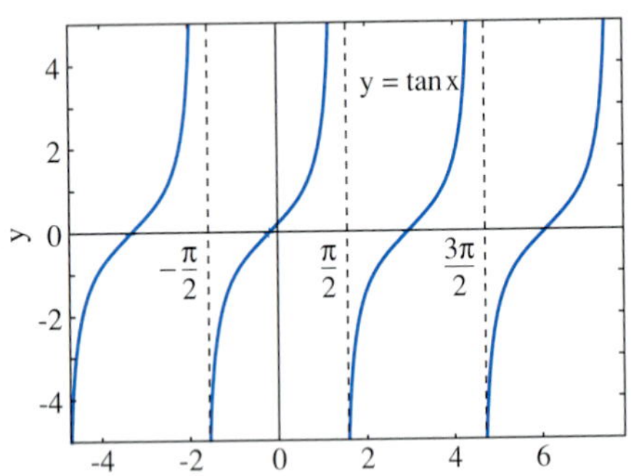

Fig. 2.4.9 The function $\tan x$ has infinite discontinuities at $x = \pm\pi/2, \pm 3\pi/2, \ldots$.

Composition of Continuous Functions

Recall from Section 1.4 that the **composition** of the two functions f and g is the function $h = f \circ g$ defined by

$$h(x) = f(g(x)) \qquad \textbf{(6)}$$

for all x in the domain of g such that $u = g(x)$ is in the domain of f. Theorem 2 implies that functions built by forming compositions of *continuous* functions are themselves continuous.

Theorem 2 ***Continuity of Compositions***

The composition of two continuous functions is continuous. More precisely, if g is continuous at a and f is continuous at $g(a)$, then $f \circ g$ is continuous at a.

PROOF The continuity of g at a means that $g(x) \to g(a)$ as $x \to a$, and the continuity of f at $g(a)$ implies that $f(g(x)) \to f(g(a))$ as $g(x) \to g(a)$. Hence the substitution law for limits (Section 2.2) yields

$$\lim_{x\to a} f(g(x)) = f\left(\lim_{x\to a} g(x)\right) = f(g(a)),$$

as desired. ■

Recall from the root law in Section 2.2 that

$$\lim_{x\to a} \sqrt[n]{x} = \sqrt[n]{a}$$

under the conditions that n is an integer and that $a > 0$ if n is even. Thus the nth-root function $f(x) = \sqrt[n]{x}$ is continuous everywhere if n is odd; f is continuous for $x > 0$ if n is even.

We may combine this result with Theorem 2. Then we see that a root of a continuous function is continuous wherever it is defined. That is, the composition

$$g(x) = [f(x)]^{1/n}$$

is continuous at a if f is, assuming that $f(a) > 0$ if n is even (so that $\sqrt[n]{f(a)}$ is defined).

EXAMPLE 7 Show that the function

$$f(x) = \left(\frac{x - 7}{x^2 + 2x + 2}\right)^{2/3}$$

is continuous on the whole real line.

SOLUTION Note first that the denominator

$$x^2 + 2x + 2 = (x + 1)^2 + 1$$

is never zero. Hence the rational function

$$r(x) = \frac{x - 7}{x^2 + 2x + 2}$$

is defined and continuous everywhere. It then follows from Theorem 2 and the continuity of the cube root function that

$$f(x) = \left([r(x)]^2\right)^{1/3}$$

is continuous everywhere. Hence, for example,

$$\lim_{x \to -1} \left(\frac{x - 7}{x^2 + 2x + 2}\right)^{2/3} = f(-1) = (-8)^{2/3} = 4.$$

Continuous Functions on Closed Intervals

An applied problem typically involves a function whose domain is a *closed interval.* For example, in the animal pen problem of Section 1.1, we found that the area A of the rectangular pen in Fig. 2.4.10 was expressed as a function of its base length x by

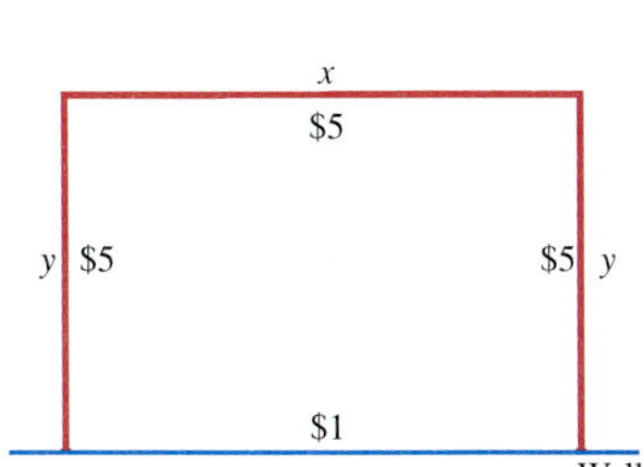

Fig. 2.4.10 The animal pen

$$A = f(x) = \tfrac{3}{5}\,x(30 - x).$$

Although this formula for f is meaningful for all x, only values in the closed interval $[0, 30]$ correspond to actual rectangles, so only such values are pertinent to the animal pen problem.

The function f defined on the closed interval $[a, b]$ is said to be **continuous on** $[a, b]$ provided that it is continuous at each point of the open interval (a, b) *and* that

$$\lim_{x \to a^+} f(x) = f(a) \quad \text{and} \quad \lim_{x \to b^-} f(x) = f(b).$$

The last two conditions mean that at each endpoint the value of the function is equal to its limit from within the interval. For instance, every polynomial is continuous on every closed interval. The square root function $f(x) = \sqrt{x}$ is continuous on the closed interval $[0, 1]$ even though f is not defined for $x < 0$.

Continuous functions defined on closed intervals have very special properties. For example, every such function has the *intermediate value property* of Theorem 3. (A proof of this theorem is given in Appendix E.) We suggested earlier that continuity of a function is related to the possibility of tracing its graph without lifting the pen from the paper. Theorem 3, the intermediate value theorem, expresses this fact with precision.

Theorem 3 ***Intermediate Value Property***

Suppose that the function f is continuous on the closed interval $[a, b]$. Then $f(x)$ assumes every intermediate value between $f(a)$ and $f(b)$. That is, if K is any number between $f(a)$ and $f(b)$, then there exists at least one number c in (a, b) such that $f(c) = K$.

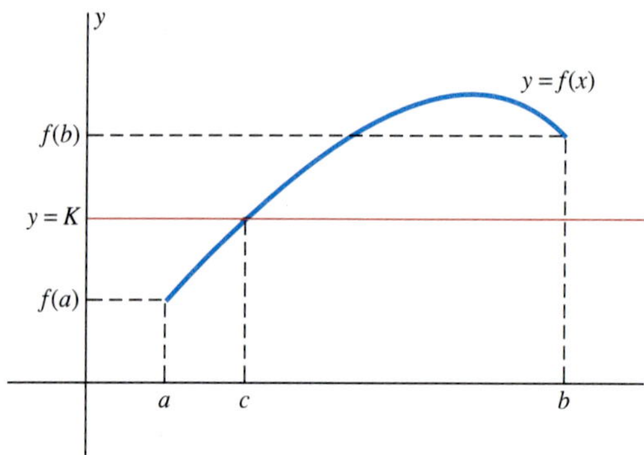

Fig. 2.4.11 The continuous function f attains the intermediate value K at $x = c$.

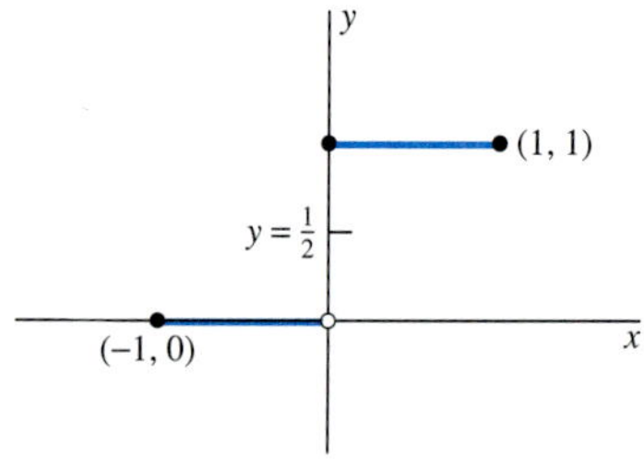

Fig. 2.4.12 This discontinuous function does not have the intermediate value property (Example 8).

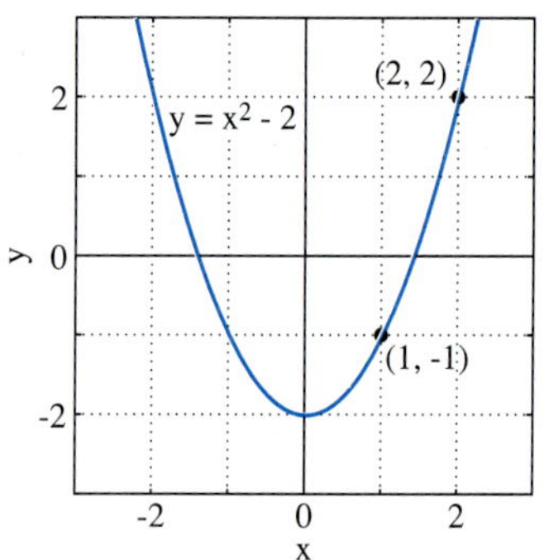

Fig. 2.4.13 The graph of $f(x) = x^2 - 2$ (Example 9).

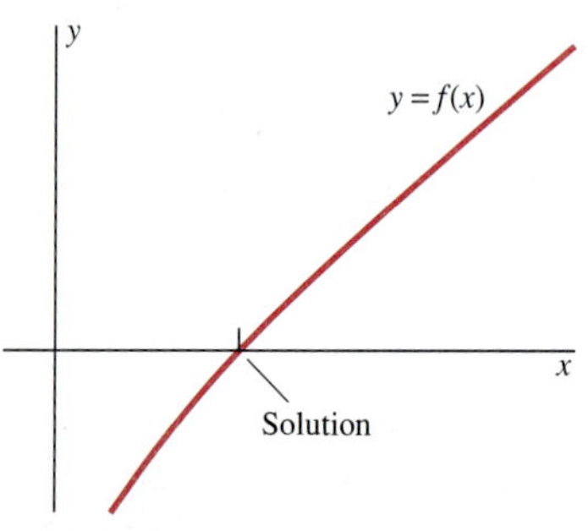

Fig. 2.4.14 The solution of the equation $f(x) = 0$

Figure 2.4.11 shows the graph of a typical continuous function f whose domain is the closed interval $[a, b]$. The number K is located on the y-axis, somewhere between $f(a)$ and $f(b)$. In the figure $f(a) < f(b)$, but this is not important. The horizontal line through K must cross the graph of f somewhere, and the x-coordinate of the point where graph and line meet yields the value of c. The number c is the one whose existence is guaranteed by the intermediate value property of the continuous function f.

Thus the intermediate value property implies that each horizontal line meeting the y-axis between $f(a)$ and $f(b)$ must cross the graph of the continuous function f somewhere. This is a way of saying that the graph has no gaps or jumps, suggesting that the idea of being able to trace such a graph without lifting the pen from the paper is accurate.

EXAMPLE 8 The discontinuous function defined on $[-1, 1]$ as

$$f(x) = \begin{cases} 0 & \text{if } x < 0, \\ 1 & \text{if } x \geqq 0 \end{cases}$$

does *not* attain the intermediate value $\frac{1}{2}$. See Fig. 2.4.12. ■

Existence of Solutions of Equations

An important application of the intermediate value property is the verification of the existence of solutions of equations written in the form

$$f(x) = 0. \tag{7}$$

EXAMPLE 9 You could attempt to approximate the number $\sqrt{2}$ graphically by zooming in on the intersection of the parabola $y = x^2 - 2$ with the positive x-axis (Fig. 2.4.13). The x-coordinate of the intersection yields the positive solution of the equation

$$f(x) = x^2 - 2 = 0. \tag{8}$$

Perhaps it makes no sense to zoom in on this point unless we know that it's "really there." But we can see from Eq. (8) that

$$f(1) = -1 < 0, \quad \text{whereas} \quad f(2) = 2 > 0.$$

We note that the function f is continuous on $[1, 2]$ (it is continuous everywhere) and that $K = 0$ is an intermediate value of f on the interval $[1, 2]$. Therefore, it follows from Theorem 3 that $f(c) = 0$ for some number c in $(1, 2)$—that is, that

$$c^2 = 2.$$

This number c is the desired square root of 2. Thus it is the intermediate value property of continuous functions that guarantees the existence of the number $\sqrt{2}$: There *is* a real number whose square is 2. ■

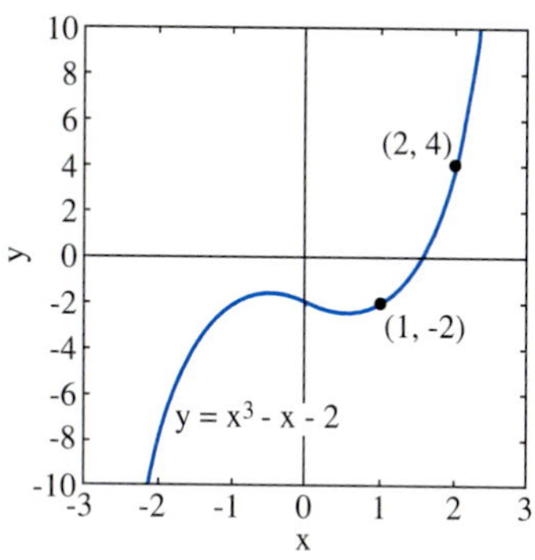

Fig. 2.4.15 The equation $x^3 - x - 2 = 0$ of Example 10 appears to have a solution somewhere between $x = 1$ and $x = 2$.

As indicated in Fig. 2.4.14, the solutions of Eq. (7) are simply the points where the graph $y = f(x)$ crosses the x-axis. Suppose that f is continuous and that we can find a closed interval $[a, b]$ (such as the interval $[1, 2]$ of Example 9) such that the value of f is positive at one endpoint of $[a, b]$ and negative at the other. That is, suppose that $f(x)$ *changes sign* on the closed interval $[a, b]$. Then the intermediate value property ensures that $f(x) = 0$ at some point of $[a, b]$.

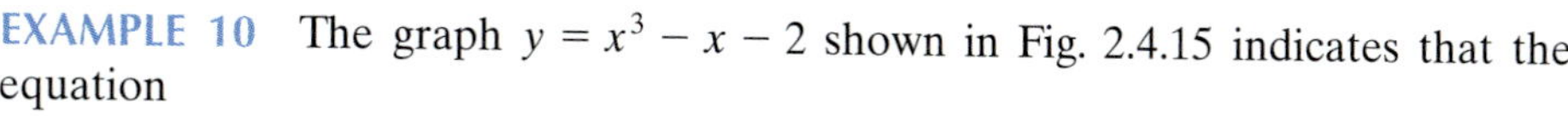

EXAMPLE 10 The graph $y = x^3 - x - 2$ shown in Fig. 2.4.15 indicates that the equation

$$f(x) = x^3 - x - 2 = 0$$

has a solution somewhere between $x = 1$ and $x = 2$. Apply the intermediate value theorem to show that this actually is so.

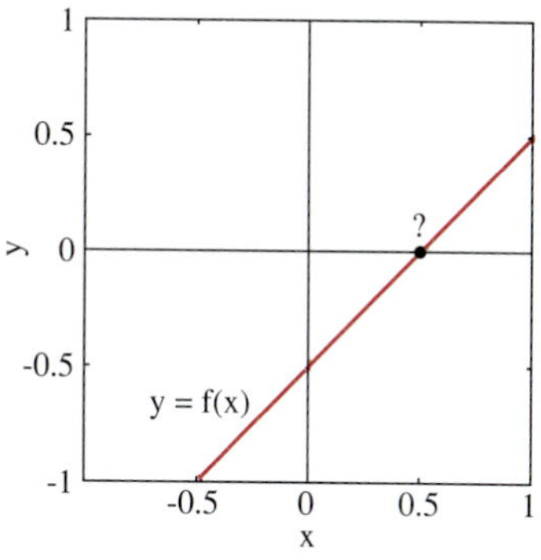

Fig. 2.4.16 The graph $y = f(x)$ of Example 11 appears to have x-intercept $x = 0.5$.

Solution The function $f(x)$ is continuous on $[1, 2]$ because it is a polynomial and, therefore, is continuous everywhere. Because $f(1) = -2$ and $f(2) = +4$, the intermediate value theorem implies that every number between -2 and $+4$ is a value of $f(x)$ on $[1, 2]$. In particular,

$$-2 = f(1) < 0 < f(2) = +4,$$

so the intermediate value property of f implies that f attains the value 0 at some number c between $x = 1$ and $x = 2$. That is,

$$f(c) = c^3 - c - 2 = 0,$$

so $x = c$ is a solution in $(1, 2)$ of the equation $x^3 - x - 2 = 0$. ■

The following example shows that not every suspected root of an equation $f(x) = 0$ that seems to be visible on a computer-plotted figure is actually there. Indeed, a graphing calculator or computer ordinarily is programmed to plot close but isolated points on the desired graph $y = f(x)$ and then join these points with line segments so short that the result looks like a smooth curve. In effect, the computer is *assuming* that the function f is continuous, whether or not it actually is continuous.

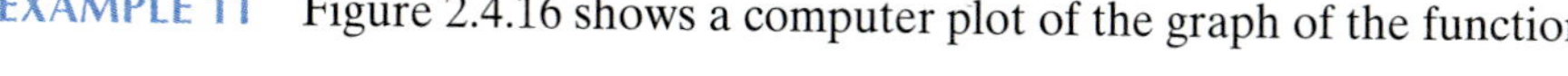

EXAMPLE 11 Figure 2.4.16 shows a computer plot of the graph of the function

$$f(x) = \frac{10 \cdot [\![1000x]\!] - 4995}{10000}.$$

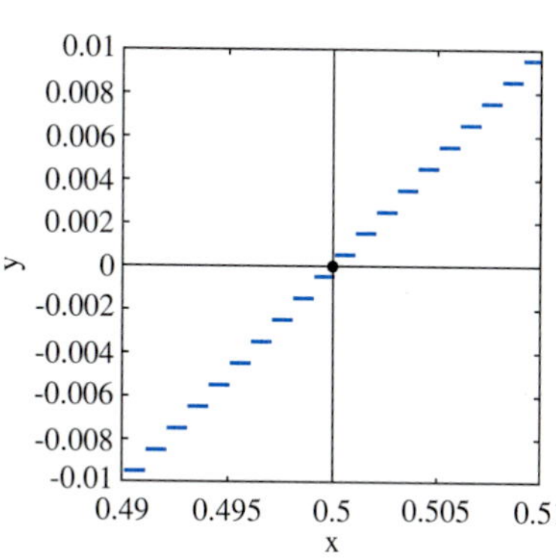

Fig. 2.4.17 The graph in Example 11 jumps across the x-axis—there is no x-intercept.

The graph $y = f(x)$ appears indistinguishable from the line $y = x - \frac{1}{2}$, and in particular it appears that the equation $f(x) = 0$ has the solution $x = \frac{1}{2}$. But when we zoom in near this alleged solution we see the graph shown in Fig. 2.4.17. Now we see that the function f is discontinuous and its graph actually "jumps" across the x-axis without intersecting it. Thus the equation $f(x) = 0$ has no solution at all. ■

2.4 PROBLEMS

In Problems 1 through 8, apply the limit laws and the theorems of this section to show that the given function is continuous for all x.

1. $f(x) = 2x^5 - 7x^2 + 13$

2. $f(x) = 7x^3 - (2x + 1)^5$

3. $g(x) = \dfrac{2x - 1}{4x^2 + 1}$

4. $g(x) = \dfrac{x^3}{x^2 + 2x + 5}$

5. $h(x) = \sqrt{x^2 + 4x + 5}$

6. $h(x) = \sqrt[3]{1 - 5x}$

7. $f(x) = \dfrac{1 - \sin x}{1 + \cos^2 x}$

8. $g(x) = \sqrt[4]{1 - \sin^2 x}$

In Problems 9 through 14, apply the limit laws and the theorems of this section to show that the given function is continuous on the indicated interval.

9. $f(x) = \dfrac{1}{x + 1}, \quad x > -1$

10. $f(x) = \dfrac{x - 1}{x^2 - 4}, \quad -2 < x < 2$

11. $g(t) = \sqrt{9 - 4t^2}, \quad -\frac{3}{2} \leqq t \leqq \frac{3}{2}$

12. $h(z) = \sqrt{(z - 1)(3 - z)}, \quad 1 \leqq z \leqq 3$

13. $f(x) = \dfrac{x}{\cos x}, \quad -\frac{1}{2}\pi < x < \frac{1}{2}\pi$

14. $g(t) = \sqrt{1 - 2\sin t}, \quad -\frac{1}{6}\pi < t < \frac{1}{6}\pi$

In Problems 15 through 36, determine where the function with the given formula is continuous. Recall that when the domain of a function is not specified, it is the set of all real numbers for which the formula of the function is meaningful.

15. $f(x) = 2x + \sqrt[3]{x}$

16. $g(x) = x^2 + \dfrac{1}{x}$

17. $f(x) = \dfrac{1}{x + 3}$

18. $f(t) = \dfrac{5}{5 - t}$

19. $f(x) = \dfrac{1}{x^2 + 1}$

20. $g(z) = \dfrac{1}{z^2 - 1}$

21. $f(x) = \dfrac{x - 5}{|x - 5|}$

22. $h(x) = \dfrac{x^2 + x + 1}{x^2 + 1}$

23. $f(x) = \dfrac{x^2 + 4}{x - 2}$

24. $f(t) = \sqrt[4]{4 + t^4}$

25. $f(x) = \sqrt[3]{\dfrac{x + 1}{x - 1}}$

26. $F(u) = \sqrt[3]{3 - u^3}$

27. $f(x) = \dfrac{3}{x^2 - x}$

28. $f(z) = \sqrt{9 - z^2}$

29. $f(x) = \dfrac{x}{\sqrt{4 - x^2}}$

30. $f(x) = \sqrt{\dfrac{1 - x^2}{4 - x^2}}$

31. $f(x) = \dfrac{\sin x}{x^2}$

32. $g(\theta) = \dfrac{\theta}{\cos\theta}$

33. $f(x) = \dfrac{1}{\sin 2x}$

34. $f(x) = \sqrt{\sin x}$

35. $f(x) = \sin|x|$

36. $G(u) = \dfrac{1}{\sqrt{1 + \cos u}}$

In Problems 37 through 48, find the points where the given function is not defined and is therefore not continuous. For each such point a, tell whether this discontinuity is removable.

37. $f(x) = \dfrac{x}{(x + 3)^3}$

38. $f(t) = \dfrac{t}{t^2 - 1}$

39. $f(x) = \dfrac{x - 2}{x^2 - 4}$

40. $G(u) = \dfrac{u + 1}{u^2 - u - 6}$

41. $f(x) = \dfrac{1}{1 - |x|}$

42. $h(x) = \dfrac{|x - 1|}{(x - 1)^3}$

43. $f(x) = \dfrac{x - 17}{|x - 17|}$

44. $g(x) = \dfrac{x^2 + 5x + 6}{x + 2}$

45. $f(x) = \begin{cases} -x & \text{if } x < 0, \\ x^2 & \text{if } x > 0 \end{cases}$

46. $f(x) = \begin{cases} x + 1 & \text{if } x < 1, \\ 3 - x & \text{if } x > 1 \end{cases}$

47. $f(x) = \begin{cases} 1 + x^2 & \text{if } x < 0, \\ \dfrac{\sin x}{x} & \text{if } x > 0 \end{cases}$

48. $f(x) = \begin{cases} \dfrac{1 - \cos x}{x} & \text{if } x < 0, \\ x^2 & \text{if } x > 0 \end{cases}$

In Problems 49 through 52, find a value of the constant c so that the function $f(x)$ is continuous for all x.

49. $f(x) = \begin{cases} x + c & \text{if } x < 0, \\ 4 - x^2 & \text{if } x \geqq 0 \end{cases}$

50. $f(x) = \begin{cases} 2x + c & \text{if } x \leqq 3, \\ 2c - x & \text{if } x > 3 \end{cases}$

51. $f(x) = \begin{cases} c^2 - x^2 & \text{if } x < 0, \\ 2(x - c)^2 & \text{if } x \geqq 0 \end{cases}$

52. $f(x) = \begin{cases} c^3 - x^3 & \text{if } x \leqq \pi, \\ c\sin x & \text{if } x > \pi \end{cases}$

In Problems 53 through 58, apply the intermediate value property of continuous functions to show that the given equation has a solution in the given interval.

53. $x^2 - 5 = 0$ on $[2, 3]$

54. $x^3 + x + 1 = 0$ on $[-1, 0]$

55. $x^3 - 3x^2 + 1 = 0$ on $[0, 1]$

56. $x^3 = 5$ on $[1, 2]$

57. $x^4 + 2x - 1 = 0$ on $[0, 1]$

58. $x^5 - 5x^3 + 3 = 0$ on $[-3, -2]$

In Problems 59 and 60, show that the given equation has three distinct roots by calculating the values of the left-hand side at $x = -3, -2, -1, 0, 1, 2,$ and 3 and then applying the intermediate value property of continuous functions on appropriate closed intervals.

59. $x^3 - 4x + 1 = 0$

60. $x^3 - 3x^2 + 1 = 0$

61. Suppose that you accept a job now (time $t = 0$) at an annual salary of \$25000 and are promised a 6% raise at the end of each year of employment. Explain why your salary in thousands of dollars after t years is given by the formula

$$S(t) = 25 \cdot (1.06)^{[\![t]\!]}.$$

Graph this function for the first five years and comment on its continuity.

62. Suppose that you accept the same job as in Problem 61, but now you are promised a 1.5% raise at the end of each quarter (three months). (a) Write a formula for your salary (in thousands of dollars) after t years. (b) Graph this new salary function and comment on its continuity. (c) Which is the better deal, the promised salary of Problem 61 or the one of this problem?

63. Suppose that f and g are two functions both continuous on the interval $[a, b]$, and such that $f(a) = g(b) = p$ and $f(b) = g(a) = q$ where $p \neq q$. Sketch typical graphs of two such functions. Then apply the intermediate value theorem to the function $h(x) = f(x) - g(x)$ to show that $f(c) = g(c)$ at some point c of (a, b).

64. Suppose that today you leave your home in Estes Park, Colorado, at 1 P.M. and drive to Grand Lake, arriving at 2 P.M. Tomorrow you leave your destination in Grand Lake at 1 P.M. and retrace the same route, arriving home at 2 P.M. Use Problem 63 as a suggestion to show that at some instant between 1 and 2 P.M. you are at precisely the same point on the road both days. What must you assume about the functions describing your location as a function of time each day?

65. Apply the intermediate value property of continuous functions to show that every positive number a has a square root. That is, given $a > 0$, prove that there exists a number r such that $r^2 = a$.

66. Apply the intermediate value property to prove that every real number has a cube root.

67. Show that the cosine function is continuous on the set of all real numbers. (*Suggestion:* Alter the proof of Theorem 1 of the continuity of the sine function.)

68. Determine where the function $f(x) = x + [\![x]\!]$ is continuous.

69. Suppose that $f(x) = 0$ if x is a rational number, whereas $f(x) = 1$ if x is irrational. Prove that f is discontinuous at every real number.

70. Suppose that $f(x) = 0$ if x is a rational number, whereas $f(x) = x^2$ if x is irrational. Prove that f is continuous only at the single point $x = 0$.

71. Figure 2.4.18 suggests that the equation $x = \cos x$ has a solution in the interval $(0, \pi/2)$. Use the intermediate value theorem to show that this is true. Then use your calculator to approximate this solution accurate to two decimal places.

72. Figure 2.4.19 suggests that the equation $x = -5 \cos x$ has at least three distinct solutions. Use the intermediate value theorem to show that this is true. Then use your calculator to approximate each of these solutions accurate to two decimal places.

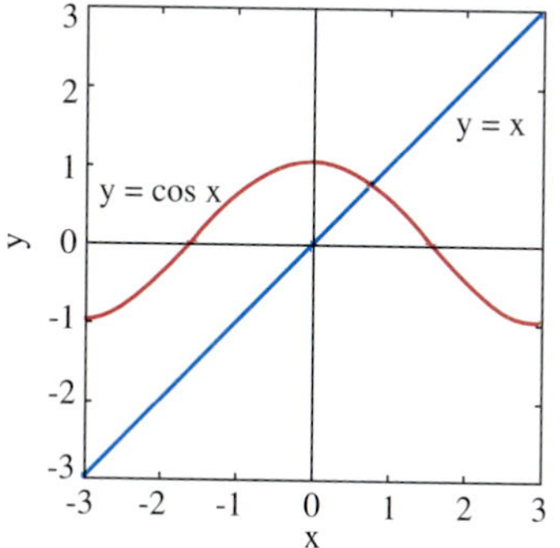

Fig. 2.4.18 The graphs $y = x$ and $y = \cos x$ (Problem 71)

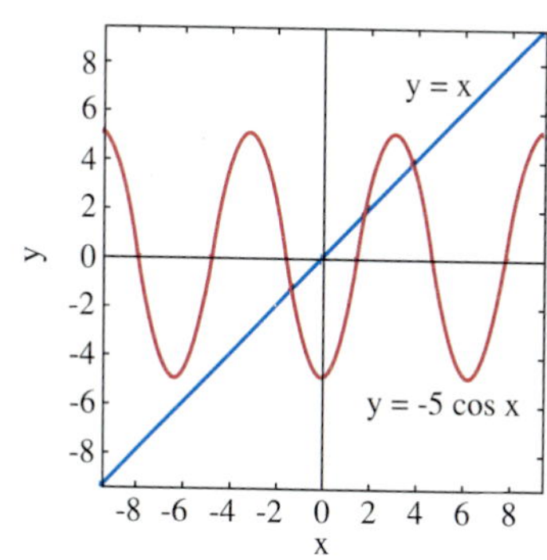

Fig. 2.4.19 The graphs $y = x$ and $y = -5 \cos x$ (Problem 72)

2.4 PROJECT: THE BROKEN TREE AGAIN

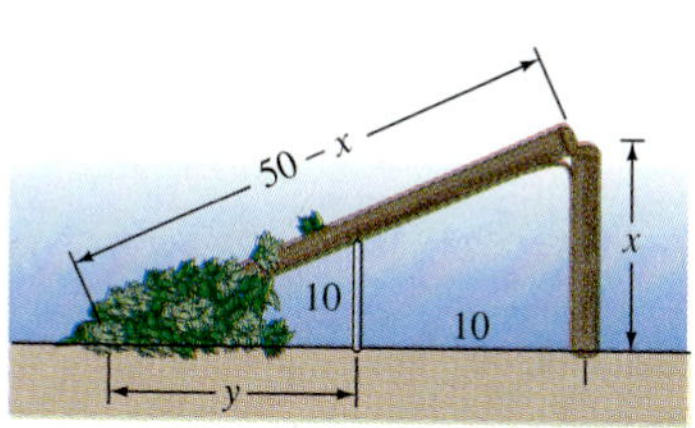

Fig. 2.4.20 The return of the broken tree.

A tree 50 ft high stands 10 ft from a 10-ft fence. Then the tree is "broken" at a height of x feet (Fig. 2.4.20). In Project 1.2 you were asked to show that the indicated lengths x and y satisfy the two equations

$$y = \frac{100}{x - 10}, \tag{1}$$

$$(y + 10)^2 = -100(x - 25). \tag{2}$$

Now eliminate y to obtain the single equation

$$f(x) = x^3 - 44x^2 + 600x - 2500 = 0. \tag{3}$$

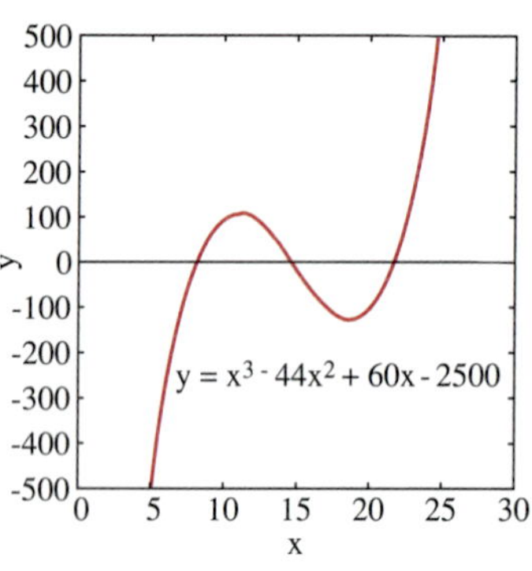

Fig. 2.4.21 The graph of the broken-tree equation

The graph $y = f(x)$ shown in Fig. 2.4.21 indicates that Eq. (3) has three real solutions. Apply the intermediate value theorem to prove that this is so. Then zoom in on these three solutions to determine each accurate to two decimal places. Which of these solutions corresponds to an actual broken tree as in Fig. 2.4.20?

For your own personal investigation, begin instead with a tree of height $h = 50 + 10p$ (feet), where p denotes the last nonzero digit in your student I.D. number. Write the results of your investigation in the form of a carefully organized report consisting of complete sentences (together with pertinent equations and figures) and explaining your results in detail. Tell precisely what you did to solve your problem; don't just "solve the equation." Answer completely the original question as to the possible height (or heights) at which the actual tree can be broken.

CHAPTER 2 REVIEW: *Definitions, Concepts, Results*

Use the following list as a guide to topics that you may need to review.

1. Slope predictors for tangent lines
2. Lines tangent to parabolas
3. Numerical investigation of slopes of lines tangent to more general curves
4. The idea of the limit of $f(x)$ as x approaches a
5. Numerical and graphical investigations of limits
6. The limit laws: constant, addition, product, quotient, root, substitutions, and squeeze
7. Slope-predictor functions
8. The four-step process for finding slope-predictor functions
9. The precise definition of the limit
10. The basic trigonometric limit
11. Evaluation of limits of trigonometric functions
12. Right-hand and left-hand limits
13. The relation between one-sided and two-sided limits
14. Existence of tangent lines
15. Infinite limits
16. Continuity of a function at a point
17. Combinations of continuous functions
18. Continuity of polynomials and rational functions
19. Removable discontinuities
20. Continuity of trigonometric functions
21. Compositions of continuous functions
22. The intermediate value property of continuous functions
23. Existence of solutions of equations

CHAPTER 2 *Miscellaneous Problems*

Apply the limit laws to evaluate the limits in Problems 1 through 40 or to show that the indicated limit does not exist, as appropriate.

1. $\lim_{x\to 0}(x^2 - 3x + 4)$

2. $\lim_{x\to -1}(3 - x + x^3)$

3. $\lim_{x\to 2}(4 - x^2)^{10}$

4. $\lim_{x\to 1}(x^2 + x - 1)^{17}$

5. $\lim_{x\to 2}\dfrac{1 + x^2}{1 - x^2}$

6. $\lim_{x\to 3}\dfrac{2x}{x^2 - x - 3}$

7. $\lim_{x\to 1}\dfrac{x^2 - 1}{1 - x}$

8. $\lim_{x\to -2}\dfrac{x + 2}{x^2 + x - 2}$

9. $\lim_{t\to -3}\dfrac{t^2 + 6t + 9}{9 - t^2}$

10. $\lim_{x\to 0}\dfrac{4x - x^3}{3x + x^2}$

11. $\lim_{x\to 3}(x^2 - 1)^{2/3}$

12. $\lim_{x\to 2}\sqrt{\dfrac{2x^2 + 1}{2x}}$

13. $\lim_{x\to 3}\left(\dfrac{5x + 1}{x^2 - 8}\right)^{3/4}$

14. $\lim_{x\to 1}\dfrac{x^4 - 1}{x^2 + 2x - 3}$

15. $\lim_{x\to 7}\dfrac{\sqrt{x + 2} - 3}{x - 7}$

16. $\lim_{x\to 1^+}(x - \sqrt{x^2 - 1})$

17. $\lim_{x\to -4}\dfrac{\dfrac{1}{\sqrt{13 + x}} - \dfrac{1}{3}}{x + 4}$

18. $\lim_{x\to 1^+}\dfrac{1 - x}{|1 - x|}$

19. $\lim_{x\to 2^+}\dfrac{2 - x}{\sqrt{4 - 4x + x^2}}$

20. $\lim_{x\to -2^-}\dfrac{x + 2}{|x + 2|}$

21. $\lim_{x\to 4^+}\dfrac{x - 4}{|x - 4|}$

22. $\lim_{x\to 3^-}\sqrt{x^2 - 9}$

23. $\lim_{x\to 2^+}\sqrt{4 - x^2}$

24. $\lim_{x\to -3}\dfrac{x}{(x + 3)^2}$

25. $\lim_{x\to 2}\dfrac{x+2}{(x-2)^2}$

26. $\lim_{x\to 1^-}\dfrac{x}{x-1}$

27. $\lim_{x\to 3^+}\dfrac{x}{x-3}$

28. $\lim_{x\to 1^-}\dfrac{x-2}{x^2-3x+2}$

29. $\lim_{x\to 1^-}\dfrac{x+1}{(x-1)^3}$

30. $\lim_{x\to 5^+}\dfrac{25-x^2}{x^2-10x+25}$

31. $\lim_{x\to 0}\dfrac{\sin 3x}{x}$

32. $\lim_{x\to 0}\dfrac{\tan 5x}{x}$

33. $\lim_{x\to 0}\dfrac{\sin 3x}{\sin 2x}$

34. $\lim_{x\to 0}\dfrac{\tan 2x}{\tan 3x}$

35. $\lim_{x\to 0^+}\dfrac{x}{\sin\sqrt{x}}$

36. $\lim_{x\to 0}\dfrac{1-\cos 3x}{2x}$

37. $\lim_{x\to 0}\dfrac{1-\cos 3x}{2x^2}$

38. $\lim_{x\to 0} x^3\cot x\csc x$

39. $\lim_{x\to 0}\dfrac{\sec 2x\tan 2x}{x}$

40. $\lim_{x\to 0} x^2\cot^2 3x$

In Problems 41 through 46, apply your knowledge of lines tangent to parabolas (Section 2.1) to write a slope-predictor formula for the given curve $y = f(x)$. Then write an equation for the line tangent to $y = f(x)$ at the point $(1, f(1))$.

41. $f(x) = 3 + 2x^2$

42. $f(x) = x - 5x^2$

43. $f(x) = 3x^2 + 4x - 5$

44. $f(x) = 1 - 2x - 3x^2$

45. $f(x) = (x-1)(2x-1)$

46. $f(x) = \dfrac{x}{3} - \left(\dfrac{x}{4}\right)^2$

In Problems 47 through 53, use the "four-step process" of Section 2.2 to find a slope-predictor formula for the graph $y = f(x)$.

47. $f(x) = 2x^2 + 3x$

48. $f(x) = x - x^3$

49. $f(x) = \dfrac{1}{3-x}$

50. $f(x) = \dfrac{1}{2x+1}$

51. $f(x) = x - \dfrac{1}{x}$

52. $f(x) = \dfrac{x}{x+1}$

53. $f(x) = \dfrac{x+1}{x-1}$

54. Find a slope-predictor formula for the graph

$$f(x) = 3x - x^2 + |2x + 3|$$

at the points where a tangent line exists. Find the point (or points) where no tangent line exists. Sketch the graph of f.

55. Write equations of the two lines through $(3, 4)$ that are tangent to the parabola $y = x^2$. (*Suggestion:* Let (a, a^2) denote either point of tangency; first solve for a.)

56. Write an equation for the circle with center $(2, 3)$ that is tangent to the line with equation $x + y + 3 = 0$.

In Problems 57 through 60, explain why each function is continuous wherever it is defined by the given formula. For each point a where f is not defined by the formula, tell whether a value can be assigned to $f(a)$ in such a way to make f continuous at a.

57. $f(x) = \dfrac{1-x}{1-x^2}$

58. $f(x) = \dfrac{1-x}{(2-x)^2}$

59. $f(x) = \dfrac{x^2+x-2}{x^2+2x-3}$

60. $f(x) = \dfrac{|x^2-1|}{x^2-1}$

61. Apply the intermediate value property of continuous functions to prove that the equation $x^5 + x = 1$ has a solution.

62. Apply the intermediate value property of continuous functions to prove that the equation $x^3 - 4x^2 + 1 = 0$ has three different solutions.

63. Show that there is a number x between 0 and $\pi/2$ such that $x = \cos x$.

64. Show that there is a number x between $\pi/2$ and π such that $\tan x = -x$. (*Suggestion:* First sketch the graphs of $y = \tan x$ and $y = -x$.)

65. Find how many straight lines through the point $(12, \frac{15}{2})$ are normal to the graph of $y = x^2$ and find the slope of each. (*Suggestion:* The cubic equation you should obtain has one root evident by inspection.)

66. A circle of radius r is dropped into the parabola $y = x^2$. If r is too large, the circle will not fall all the way to the bottom; if r is sufficiently small, the circle will touch the parabola at its vertex $(0, 0)$ (see Fig. 2.4.22). Find the largest value of r so that the circle will touch the vertex of the parabola.

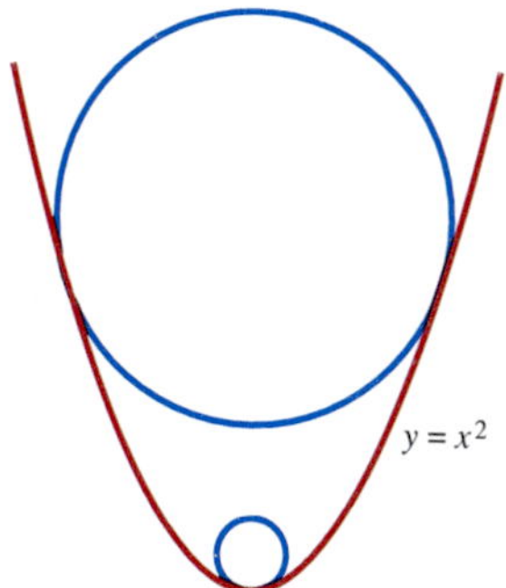

Fig. 2.4.22 If the circle is too large, it cannot touch the bottom of the parabola (Problem 66).

CHAPTER 3

THE DERIVATIVE

Isaac Newton (1642–1727)

Isaac Newton was born in a rural English farming village on Christmas Day in 1642, three months after his father's death. When the boy was three, his mother remarried and left him with his grandmother. Nothing known about his childhood and early schooling hinted that his life and work would constitute a turning point in the history of humanity.

But due to the influence of an uncle who suspected hidden potential in young Isaac, Newton was able to enter Cambridge University in 1661. During the years 1665 and 1666, when Cambridge closed because of the bubonic plague then sweeping Europe, he returned to his country home and there laid the foundations for the three towering achievements of his scientific career—the invention of the calculus, the discovery of the spectrum of colors in light, and the theory of gravitation. Of these two years he later wrote, "In those days I was in the prime of my age of invention and minded mathematics and philosophy more than at any time since." Indeed, his thirties were devoted more to smoky chemical (and even alchemical) experiments than to serious mathematical investigations.

In his forties, while a mathematics professor at Cambridge, Newton wrote the *Principia Mathematica* (1687), perhaps the single most influential scientific treatise ever published. In it he applied the concepts of the calculus to explore the workings of the universe, including the motions of the earth, moon, and planets about the sun. A student is said to have remarked, "There goes the man that wrote a book that neither he nor anyone else understands." But it established for Newton such fame that upon his death in 1727 he was buried alongside his country's greats in Westminster Abbey with such pomp that the French philosopher Voltaire remarked, "I have seen a professor of mathematics . . . buried like a king who had done good to his subjects."

Shortly after his Cambridge graduation in 1665, Newton discovered a new method for solving an equation of the form $f(x) = 0$. Unlike special methods such as the quadratic formula that apply only to equations of special form, *Newton's method* can be used to approximate numerical solutions of virtually any equation. In Section 3.9 we present an iterative formulation of Newton's method that is especially adaptable to calculators and computers. There we describe how the combination of Newton's method with modern computer graphics has led to the generation of striking fractal images associated with the science of *chaos*. The pictures here result from the application of a complex-number version of Newton's method to the simple equation $x^3 + 1 = 0$.

3.1 THE DERIVATIVE AND RATES OF CHANGE

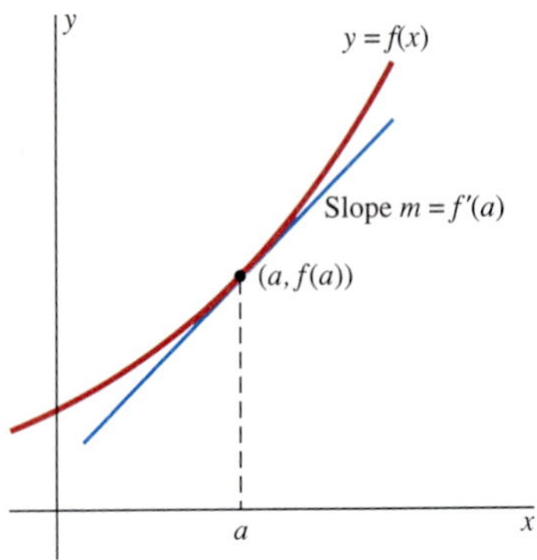

Fig. 3.1.1 The geometric motivation for the definition of the derivative

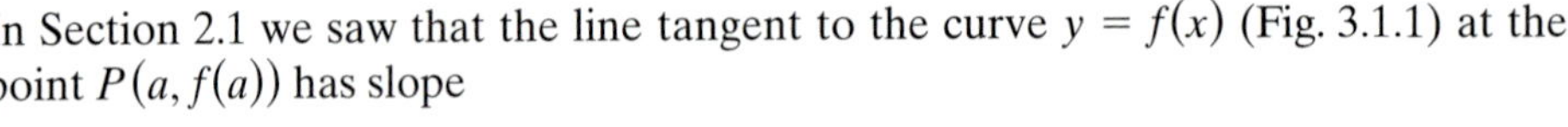

In Section 2.1 we saw that the line tangent to the curve $y = f(x)$ (Fig. 3.1.1) at the point $P(a, f(a))$ has slope

$$m = m(a) = \lim_{h \to 0} \frac{f(a + h) - f(a)}{h} \tag{1}$$

provided that this limit exists. As in the slope-prediction formulas of Section 2.2, we get a new function f'—the *derivative* of the original function f—when we replace the constant a in (1) with the independent variable x.

Definition *The Derivative*

The **derivative** of the function f is the function f' defined by

$$f'(x) = \lim_{h \to 0} \frac{f(x + h) - f(x)}{h} \tag{2}$$

for all x for which this limit exists.

It is important to understand that, when the limit in (2) is evaluated, we hold x fixed while h approaches zero. When we are specifically interested in the value $f'(a)$ of the derivative f' at the number $x = a$, we sometimes rewrite Eq. (2) in the form

$$f'(a) = \lim_{h \to 0} \frac{f(a + h) - f(a)}{h} = \lim_{x \to a} \frac{f(x) - f(a)}{x - a}. \tag{3}$$

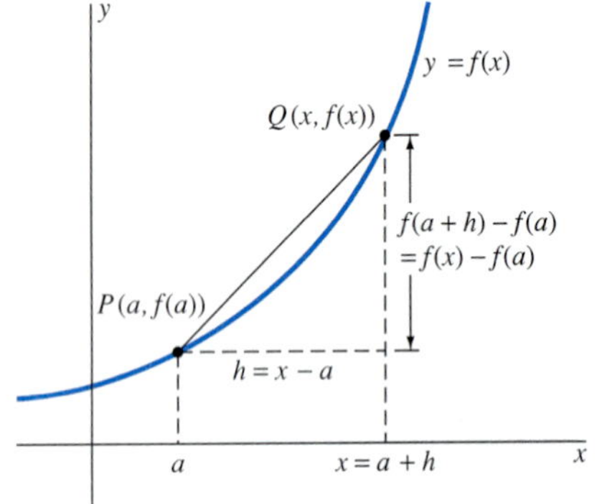

Fig. 3.1.2 The notation in Eq. (3)

The second limit in Eq. (3) is obtained from the first by writing $x = a + h$, $h = x - a$, and by noting that $x \to a$ as $h \to 0$ (Fig. 3.1.2). The statement that these equivalent limits exist can be abbreviated as "$f'(a)$ exists." In this case we say that the function f is **differentiable** at $x = a$. The process of finding the derivative f' is called **differentiation** of f.

However it is found, the derivative f' is a slope predictor for lines tangent to the graph $y = f(x)$ of the original function f (Fig. 3.1.1).

The Derivative as Slope Predictor

The slope m of the line tangent to the graph $y = f(x)$ at the point $(a, f(a))$ where $x = a$ is

$$m = f'(a). \tag{4}$$

Application of the point-slope formula gives

$$y - f(a) = f'(a) \cdot (x - a) \tag{5}$$

as an equation of this tangent line.

Differentiating a given function f by direct evaluation of the limit in Eq. (3) requires four steps:

1. Write the definition in Eq. (2) of the derivative.

2. Substitute the expressions $f(x + h)$ and $f(x)$ as determined by the particular function f.
3. Simplify the result by algebraic methods until it is possible to ...
4. ... Apply appropriate limit laws to finally evaluate the limit.

In Section 2.2 we used this same "four-step process" to calculate several slope-predictor functions—that is, derivatives. The limit calculations of Example 12 and 13 there, where we found the derivatives of the functions

$$f(x) = x + \frac{1}{x} \quad \text{and} \quad f(x) = \sqrt{x},$$

illustrate algebraic simplification techniques that frequently are useful in the evaluation of derivatives directly from the definition in Eq. (2).

EXAMPLE 1 First apply the definition of the derivative directly to differentiate the function

$$f(x) = \frac{x}{x + 3}.$$

Then find the line tangent to the graph of f at the origin, where $f(0) = 0$.

Solution The preceding steps 1 and 2 give

$$f'(x) = \lim_{h \to 0} \frac{f(x + h) - f(x)}{h} = \lim_{h \to 0} \frac{\dfrac{x + h}{(x + h) + 3} - \dfrac{x}{x + 3}}{h}.$$

Then an algebraic simplification suggested by the common-denominator calculation

$$\frac{\dfrac{a}{b} - \dfrac{c}{d}}{h} = \frac{\dfrac{ad - bc}{bd}}{h} = \frac{ad - bc}{hbd}$$

yields

$$\begin{aligned} f'(x) &= \lim_{h \to 0} \frac{(x + h)(x + 3) - x(x + h + 3)}{h(x + h + 3)(x + 3)} \\ &= \lim_{h \to 0} \frac{3h}{h(x + h + 3)(x + 3)} = \lim_{h \to 0} \frac{3}{(x + h + 3)(x + 3)} \\ &= \frac{3}{\left(\lim\limits_{h \to 0} (x + h + 3)\right)\left(\lim\limits_{h \to 0} (x + 3)\right)}. \end{aligned}$$

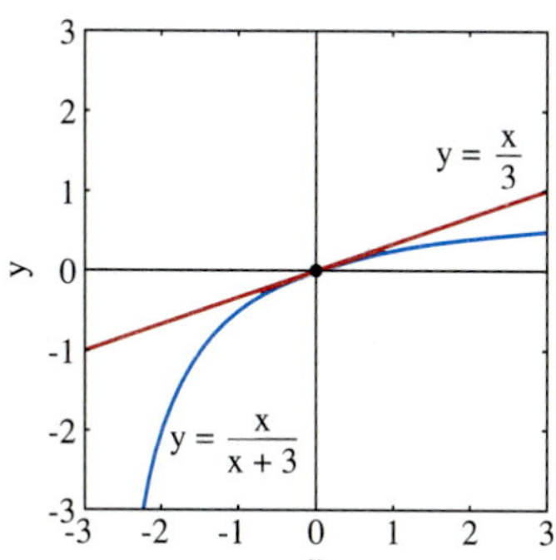

Fig. 3.1.3 The tangent line $y = \frac{1}{3}x$ to the curve $y = x/(x + 3)$ at the origin

We therefore find finally that

$$f'(x) = \frac{3}{(x + 3)(x + 3)} = \frac{3}{(x + 3)^2}.$$

Substitution of $a = 0$, $f(0) = 0$, and $f'(0) = \frac{1}{3}$ in Eq. (5) gives the equation $y = \frac{1}{3}x$ of the line tangent to the graph $y = x/(x + 3)$ at the origin $(0, 0)$ (Fig. 3.1.3). ■

Even when the function f is rather simple, this four-step process for computing f' directly from the definition of the derivative can be time consuming. Also, step 3 may require considerable ingenuity. Moreover, it would be very repetitious to continue relying on this process. To avoid tedium, we want a fast, easy, short method for computing $f'(x)$.

That new method is one focus of this chapter: the development of systematic methods ("rules") for differentiating those functions that occur most frequently. Such functions include polynomials, rational functions, the trigonometric functions $\sin x$ and $\cos x$, and combinations of such functions. Once we establish these general differentiation rules, we can apply them formally, almost mechanically, to compute derivatives. Only rarely should we need to return to the definition of the derivative.

In Example 11 of Section 2.2 we saw that a slope-predictor function for lines tangent to the graph $y = f(x)$ of the quadratic function $f(x) = ax^2 + bx + c$ is the linear function $y = 2ax + b$. Restated in the language of derivatives, this is an example of a "differentiation rule."

Differentiation of Quadratic Functions

The derivative of the quadratic function

$$f(x) = ax^2 + bx + c \tag{6}$$

is the linear function

$$f'(x) = 2ax + b. \tag{7}$$

It may be instructive to derive the differentiation formula in (7) directly from the definition of the derivative:

$$\begin{aligned} f'(x) &= \lim_{h\to 0} \frac{f(x+h) - f(x)}{h} \\ &= \lim_{h\to 0} \frac{[a(x+h)^2 + b(x+h) + c] - [ax^2 + bx + c]}{h} \\ &= \lim_{h\to 0} \frac{(ax^2 + 2ahx + ah^2 + bx + bh + c) - (ax^2 + bx + c)}{h} \\ &= \lim_{h\to 0} \frac{2ahx + ah^2 + bh}{h} \\ &= \lim_{h\to 0} (2ax + ah + b). \end{aligned}$$

Therefore

$$f'(x) = 2ax + b.$$

Once we know this rule, we need never again apply the definition of the derivative to differentiate a quadratic function.

EXAMPLE 2 If $f(x) = 3x^2 - 4x + 5$, we can apply Eq. (7) to write the following immediately, without having to go through the four-step process:

$$f'(x) = 2\cdot 3x + (-4) = 6x - 4.$$

Similarly, if $g(t) = 2t - 5t^2$, then

$$g'(t) = 2 + 2\cdot(-5t) = 2 - 10t.$$

It makes no difference what the name for the function is or whether we write x or t for the independent variable. This flexibility is valuable—in general, it is such adaptability that makes mathematics applicable to virtually every other branch of human knowledge. In any case, you should learn every differentiation rule in a form independent of the notation used to state it.

We develop additional differentiation rules in Sections 3.2 through 3.4. First, however, we must introduce new notation and a new interpretation of the derivative.

Differential Notation

An important alternative notation for the derivative originates from the early custom of writing Δx in place of h (because $h = \Delta x$ is an increment in x) and

$$\Delta y = f(x + \Delta x) - f(x)$$

for the resulting change (or increment) in y. The slope of the secant line K of Fig. 3.1.4 is then

$$m_{\text{sec}} = \frac{\Delta y}{\Delta x} = \frac{f(x + h) - f(x)}{h},$$

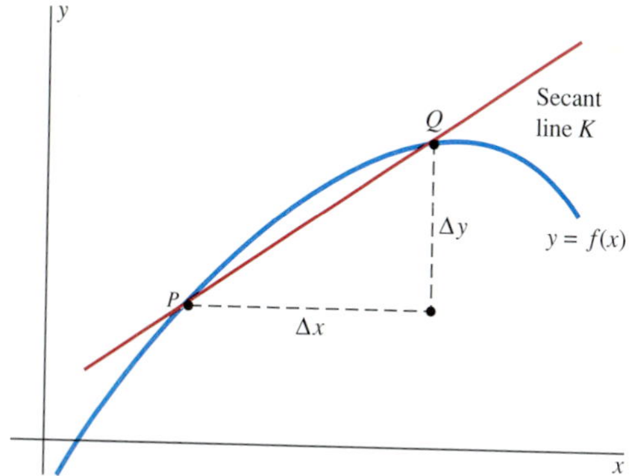

Fig. 3.1.4 Origin of the dy/dx notation

and the slope of the tangent line is

$$m = \frac{dy}{dx} = \lim_{\Delta x \to 0} \frac{\Delta y}{\Delta x}. \tag{8}$$

Hence, if $y = f(x)$, we often write

$$\frac{dy}{dx} = f'(x). \tag{9}$$

(The so-called *differentials* dy and dx are discussed carefully in Chapter 4.) The symbols $f'(x)$ and dy/dx for the derivative of the function $y = f(x)$ are used almost interchangeably in mathematics and its applications, so you need to be familiar with both versions of the notation. You also need to know that dy/dx is a single symbol representing the derivative; it is *not* the quotient of two separate quantities dy and dx.

EXAMPLE 2 (CONTINUED) If $y = ax^2 + bx + c$, then the derivative in Eq. (7) in differential notation takes the form

$$\frac{dy}{dx} = 2ax + b.$$

Consequently,

$$\text{if} \quad y = 3x^2 - 4x + 5, \quad \text{then} \quad \frac{dy}{dx} = 6x - 4;$$

$$\text{if} \quad z = 2t - 5t^2, \quad \text{then} \quad \frac{dz}{dt} = 2 - 10t.$$

The letter d in the notation dy/dx stands for the word *differential*. Whether we write dy/dx or dz/dt, the dependent variable appears "upstairs," and the independent variable "downstairs."

Rates of Change

The derivative of a function serves as a slope predictor for straight lines tangent to the graph of that function. Now we introduce the equally important interpretation of the derivative of a function as the rate of change of that function with respect to the independent variable.

We begin with the *instantaneous rate of change* of a function whose independent variable is time t. Suppose that Q is a quantity that varies with time t, and write $Q = f(t)$ for the value of Q at time t. For example, Q might be

- The size of a population (such as kangaroos, people, or bacteria);
- The number of dollars in a bank account;
- The volume of a balloon being inflated;
- The amount of water in a reservoir with variable inflow and outflow;
- The amount of a chemical product produced in a reaction; or
- The distance traveled t hours after the beginning of a journey.

The change in Q from time t to time $t + \Delta t$ is the **increment**

$$\Delta Q = f(t + \Delta t) - f(t).$$

The **average rate of change** of Q (per unit of time) is, by definition, the ratio of the change ΔQ in Q to the change Δt in t. Thus it is the quotient

$$\frac{\Delta Q}{\Delta t} = \frac{f(t + \Delta t) - f(t)}{\Delta t} \tag{10}$$

illustrated in Fig. 3.1.5.

We define the **instantaneous rate of change** of Q (per unit of time) to be the limit of this average rate as $\Delta t \to 0$. That is, the instantaneous rate of change of Q is

$$\lim_{\Delta t \to 0} \frac{\Delta Q}{\Delta t} = \lim_{\Delta t \to 0} \frac{f(t + \Delta t) - f(t)}{\Delta t}. \tag{11}$$

But the right-hand limit in Eq. (11) is simply the derivative $f'(t)$. So we see that the instantaneous rate of change of $Q = f(t)$ is the derivative

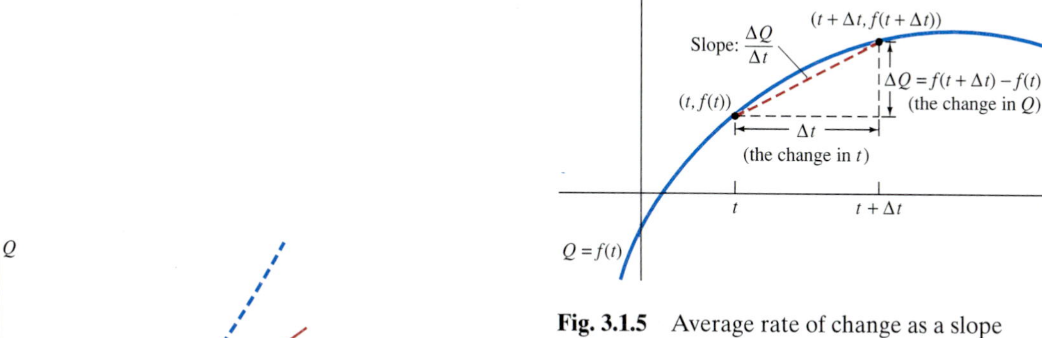

Fig. 3.1.5 Average rate of change as a slope

$$\frac{dQ}{dt} = f'(t). \tag{12}$$

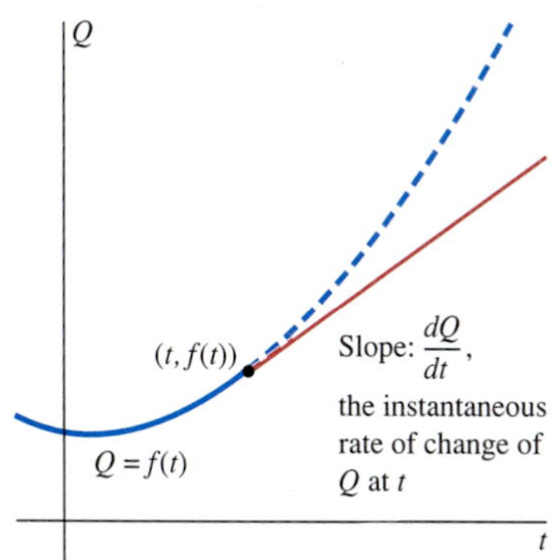

Fig. 3.1.6 The relation between the tangent line at $(t, f(t))$ and the instantaneous rate of change of f at t.

To interpret intuitively the concept of instantaneous rate of change, think of the point $P(t, f(t))$ moving along the graph of the function $Q = f(t)$. As Q changes with time t, the point P moves along the curve. But if suddenly, at the instant t, the point P begins to follow a straight line path, then the new path of P would appear as in Fig. 3.1.6. The dashed curve in the figure corresponds to the "originally planned" behavior of Q (before P decided to fly off along the straight line path). But the straight line path of P (of constant slope) corresponds to the quantity Q "changing at a constant rate." Because the straight line is tangent to the graph $Q = f(t)$, we can interpret dQ/dt as the instantaneous rate of change of the quantity Q at the instant t:

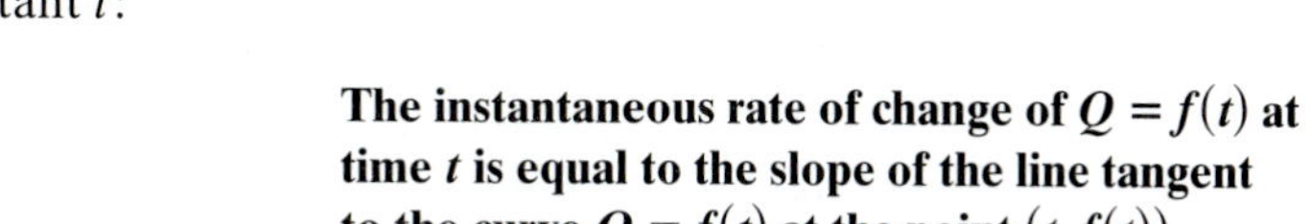

The instantaneous rate of change of $Q = f(t)$ at time t is equal to the slope of the line tangent to the curve $Q = f(t)$ at the point $(t, f(t))$.

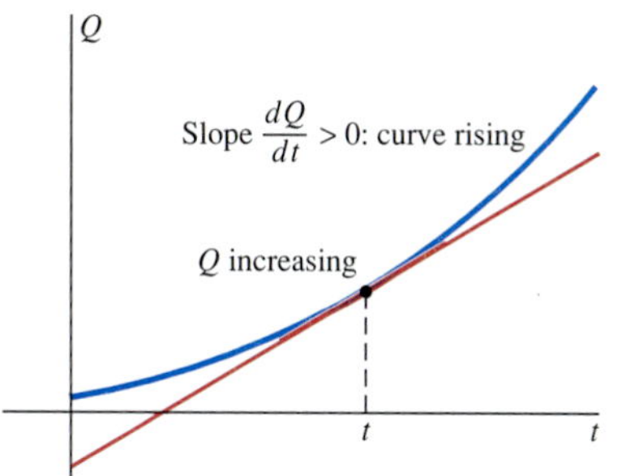

Fig. 3.1.7 Quantity increasing—derivative positive

We can draw additional important conclusions. Because a positive slope corresponds to a rising tangent line and a negative slope corresponds to a falling tangent line (as in Figs. 3.1.7 and 3.1.8), we say that

$$Q \text{ is } \textit{increasing} \text{ at time } t \text{ if } \frac{dQ}{dt} > 0;$$

$$Q \text{ is } \textit{decreasing} \text{ at time } t \text{ if } \frac{dQ}{dt} < 0. \tag{13}$$

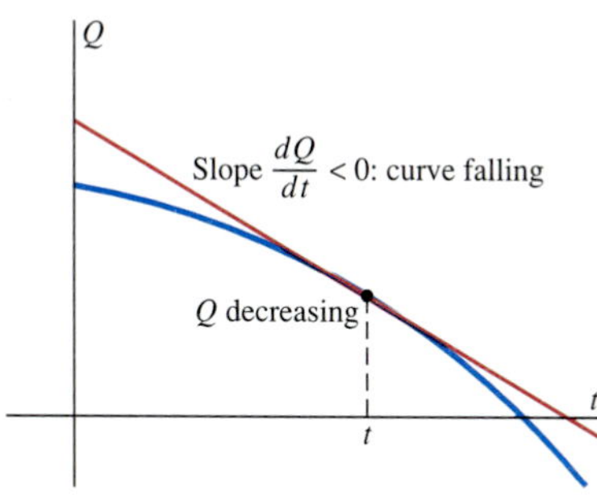

Fig. 3.1.8 Quantity decreasing—derivative negative

Note The meaning of the phrase "$Q = f(t)$ is increasing *over* (or *during*) *the time interval from* $t = a$ *to* $t = b$" should be intuitively clear. The expressions in (13) give us a way to make precise what we mean by "$Q = f(t)$ is increasing *at time* t"—that is, at the instant t.

EXAMPLE 3 The cylindrical tank in Fig. 3.1.9 has a vertical axis and is initially filled with 600 gal of water. This tank takes 60 min to empty after a drain in its bottom is opened. Suppose that the drain is opened at time $t = 0$. Suppose also that the volume V of water remaining in the tank after t minutes is

$$V(t) = \frac{1}{6}(60 - t)^2 = 600 - 20t + \frac{1}{6}t^2$$

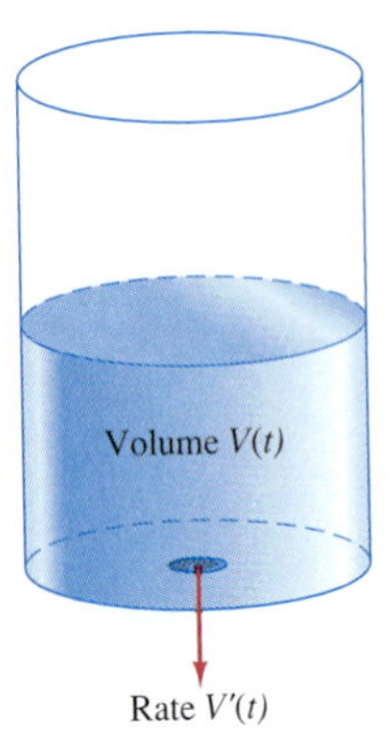

Fig. 3.1.9 The draining tank of Example 3

gallons. Find the instantaneous rate at which water is flowing out of the tank at time $t = 15$ (min) *and* at time $t = 45$ (min). Also find the average rate at which water flows out of the tank during the half hour from $t = 15$ to $t = 45$.

Solution The instantaneous rate of change of the volume $V(t)$ of water in the tank is given by the derivative

$$\frac{dV}{dt} = -20 + \frac{1}{3}t.$$

At the instants $t = 15$ and $t = 45$ we obtain

$$V'(15) = -20 + \frac{1}{3}\cdot 15 = -15$$

and

$$V'(45) = -20 + \frac{1}{3}\cdot 45 = -5.$$

The units here are gallons per minute (gal/min). The fact that $V'(15)$ and $V'(45)$ are negative is consistent with the observation that V is a decreasing function of t (as t increases, V decreases). One way to indicate this is to say that after 15 min, the water is flowing *out* of the tank at 15 gal/min; after 45 min, the water is flowing *out* at 5 gal/min. The instantaneous rate of change of V at $t = 15$ is -15 gal/min, and the instantaneous rate of change of V at $t = 45$ is -5 gal/min. We could have predicted the units, because $\Delta V/\Delta t$ is a ratio of gallons to minutes, and therefore its limit $V'(t) = dV/dt$ must be expressed in the same units.

During the time interval of length $\Delta t = 30$ min from time $t = 15$ to time $t = 45$, the *average* rate of change of the volume $V(t)$ is

$$\begin{aligned}\frac{\Delta V}{\Delta t} &= \frac{V(45) - V(15)}{45 - 15}\\ &= \frac{\frac{1}{6}(60 - 45)^2 - \frac{1}{6}(60 - 15)^2}{45 - 15} = \frac{-300}{30}.\end{aligned}$$

Each numerator in the last equation is measured in gallons—this is especially apparent when you examine the second numerator—and each denominator is measured in minutes. Hence the ratio in the last fraction is a ratio of gallons to minutes, so the average rate of change of the volume V of water *in* the tank is -10 gal/min. Thus the average rate of flow of water *out* of the tank during this half-hour interval is 10 gal/min. ■

Our examples of functions up to this point have been restricted to those with formulas or verbal descriptions. Scientists and engineers often work with tables of values obtained from observations or experiments. Example 4 shows how the instantaneous rate of change of such a tabulated function can be estimated.

EXAMPLE 4 The table in Fig. 3.1.10 gives the U.S. population P (in millions) in the nineteenth century at 10-yr intervals. Estimate the instantaneous rate of population growth in 1850.

Solution We take $t = 0$ (yr) in 1800, so $t = 50$ corresponds to the year 1850. In Fig. 3.1.11 we have plotted the given data and then added a freehand sketch of a smooth curve that fits these data.

t	Year	U.S. population (millions)
0	1800	5.3
10	1810	7.2
20	1820	9.6
30	1830	12.9
40	1840	17.1
50	1850	23.2
60	1860	31.4
70	1870	38.6
80	1880	50.2
90	1890	62.9
100	1900	76.0

Fig. 3.1.10 Data for Example 4

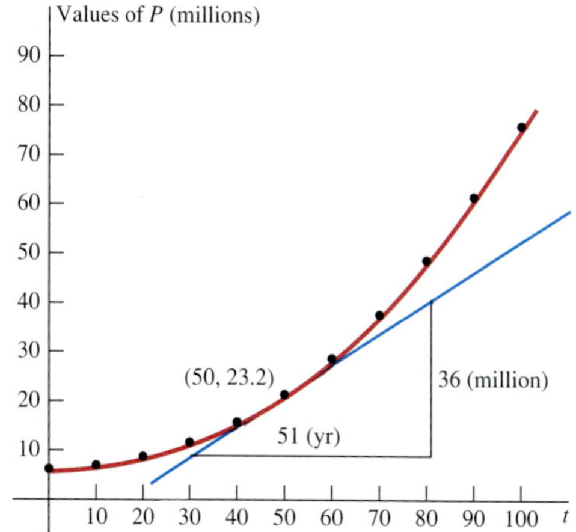

Fig. 3.1.11 A smooth curve that fits the data of Fig. 3.1.10 well (Example 4)

However we obtain it, a curve that fits the data should be a good approximation to the true graph of the unknown function $P = f(t)$. The instantaneous rate of change dP/dt in 1850 is the slope of the tangent line at the point $(50, 23.2)$. We draw the tangent line as accurately as we can by visual inspection and then measure the base and height of the triangle in Fig. 3.1.11. In this way we approximate the slope of the tangent at $t = 50$ as

$$\frac{dP}{dt} \approx \frac{36}{51} \approx 0.71$$

millions of people per year (in 1850). Although there was no national census in 1851, we would expect the U.S. population then to have been approximately $23.2 + 0.7 = 23.9$ million. ■

Velocity and Acceleration

Suppose that a particle moves along a horizontal straight line, with its location x at time t given by its **position function** $x = f(t)$. Thus we make the line of motion a coordinate axis with an origin and a positive direction; $f(t)$ is merely the x-coordinate of the moving particle at time t (Fig. 3.1.12).

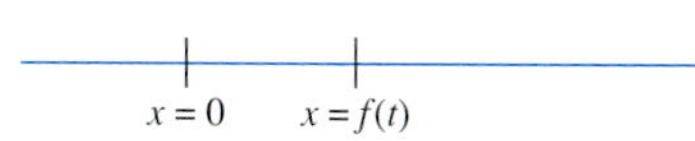

Fig. 3.1.12 The particle in motion is at the point $x = f(t)$ at time t.

Think of the time interval from t to $t + \Delta t$. The particle moves from position $f(t)$ to position $f(t + \Delta t)$ during this interval. Its displacement is then the increment

$$\Delta x = f(t + \Delta t) - f(t).$$

We calculate the *average velocity* of the particle during this time interval exactly as we would calculate average speed on a long motor trip: We divide the distance by the time to obtain an average speed in miles per hour. In this case we divide the displacement of the particle by the elapsed time to obtain the **average velocity**

$$\bar{v} = \frac{\Delta x}{\Delta t} = \frac{f(t + \Delta t) - f(t)}{\Delta t}. \tag{14}$$

(The overbar is a standard symbol that usually connotes an average of some sort.) We define the **instantaneous velocity** v of the particle at the time t to be the limit of the average velocity $\bar{v}$ as $\Delta t \to 0$. That is,

$$v = \lim_{\Delta t \to 0} \frac{\Delta x}{\Delta t} = \lim_{\Delta t \to 0} \frac{f(t + \Delta t) - f(t)}{\Delta t}. \tag{15}$$

We recognize the limit on the right in Eq. (15)—it is the definition of the derivative of f at time t. Therefore the velocity of the moving particle at time t is simply

$$v = \frac{dx}{dt} = f'(t). \tag{16}$$

Thus *velocity is instantaneous rate of change of position.* The velocity of a moving particle may be positive or negative, depending on whether the particle is moving in the positive or negative direction along the line of motion. We define the **speed** of the particle to be the *absolute value* $|v|$ of the velocity.

EXAMPLE 5 Figure 3.1.13 shows a car moving along the (horizontal) x-axis. Suppose that its position (in feet) at time t (in seconds) is given by

$$x(t) = 5t^2 + 100.$$

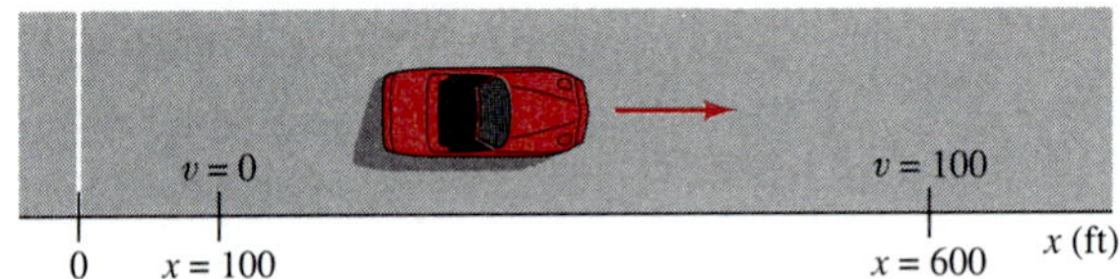

Fig. 3.1.13 The car of Example 5

Then its velocity at time t is

$$v(t) = x'(t) = 10t.$$

Because $x(0) = 100$ and $v(0) = 0$, the car starts at time $t = 0$ from rest—$v(0) = 0$—at the point $x = 100$. Substituting $t = 10$, we see that $x(10) = 600$ and $v(10) = 100$, so after 10 s the car has traveled 500 ft (from its starting point $x = 100$), and its speed then is 100 ft/s.

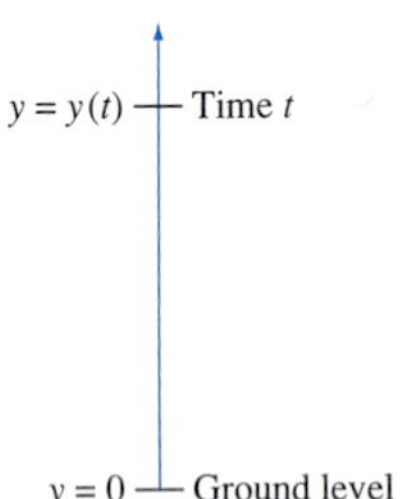

Fig. 3.1.14 Vertical motion with position function $y(t)$

Vertical Motion

In the case of vertical motion—such as that of a ball thrown straight upward—it is common to denote the position function by $y(t)$ rather than by $x(t)$. Typically, $y(t)$ denotes the height above the ground at time t, as in Fig. 3.1.14. But velocity is still the derivative of position:

$$v(t) = \frac{dy}{dt}.$$

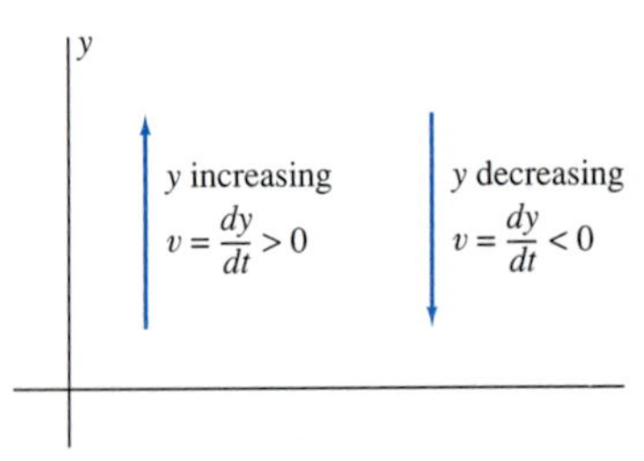

Fig. 3.1.15 Upward motion and downward motion

Upward motion with y increasing corresponds to *positive velocity,* $v > 0$ (Fig. 3.1.15). *Downward motion* with y decreasing corresponds to *negative velocity,* $v < 0$.

The case of vertical motion under the influence of constant gravity is of special interest. If a particle is projected straight upward from an initial height y_0 (ft) above the ground at time $t = 0$ (s) and with initial velocity v_0 (ft/s) *and* if air resistance is negligible, then its height y (in feet above the ground) at time t is given by a formula known from physics,

$$y(t) = -\frac{1}{2}gt^2 + v_0 t + y_0. \tag{17}$$

Here g denotes the *acceleration* due to the force of gravity. Near the surface of the earth, g is nearly constant, so we assume that it is exactly constant, and at the surface of the earth $g \approx 32\ \text{ft/s}^2$, or $g \approx 9.8\ \text{m/s}^2$.

If we differentiate y with respect to time t, we obtain the velocity of the particle at time t:

$$v(t) = \frac{dy}{dt} = -gt + v_0. \tag{18}$$

The **acceleration** of the particle is defined to be the instantaneous time rate of change (derivative) of its velocity:

$$a = \frac{dv}{dt} = -g. \tag{19}$$

Your intuition should tell you that a body projected upward in this way will reach its maximum height at the instant that its velocity becomes zero—when $v(t) = 0$. (We shall see in Section 3.5 why this is true.)

EXAMPLE 6 Find the maximum height attained by a ball thrown straight upward from the ground with initial velocity $v_0 = +96$ ft/s. Also find the velocity with which it hits the ground upon its return.

Solution To begin the solution of a motion problem such as this, we sketch a diagram like that in Fig. 3.1.16, indicating both the given data and the data that are unknown at the time instants in question. Here we focus on the time $t = 0$ when the ball leaves the ground ($y = 0$), the unknown time when it reaches its maximum height with velocity $v = 0$, and the unknown time when it returns to the ground.

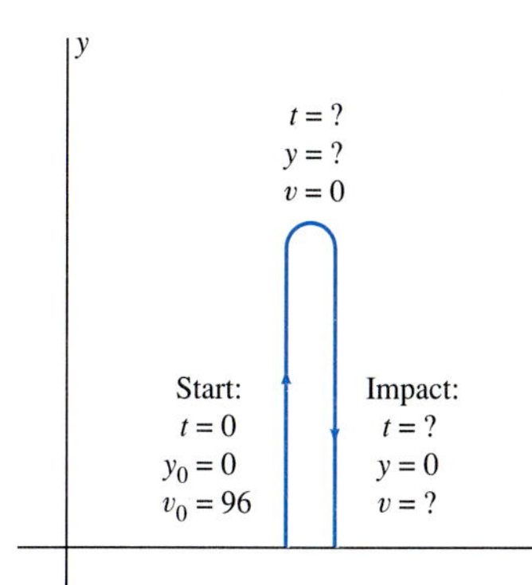

Fig. 3.1.16 Data for the ball of Example 6

We begin with Eq. (18), with $v_0 = 96$ and $g = 32$. We find the velocity of the ball at time t to be

$$v(t) = -32t + 96$$

while it remains aloft. The ball attains its maximum height when $v(t) = 0$; that is, when

$$-32t + 96 = 0.$$

This occurs when $t = 3$ (s). Upon substituting this value of t into the altitude function in Eq. (17) and taking $y_0 = 0$, we find that the maximum height of the ball is

$$y_{\max} = y(3) = (-16)(3^2) + (96)(3) = 144 \quad \text{(ft)}.$$

The ball returns to the ground when $y(t) = 0$. The equation

$$y(t) = -16t^2 + 96t = -16t(t - 6) = 0$$

has the two solutions $t = 0$ and $t = 6$. Thus the ball returns to the ground at time $t = 6$. The velocity with which it strikes the ground is

$$v(6) = (-32)(6) + 96 = -96 \quad \text{(ft/s)}.$$

Other Rates of Change

The derivative of any function—not merely a function of time—may be interpreted as its instantaneous rate of change with respect to the independent variable. If $y = f(x)$, then the **average rate of change** of y (per unit change in x) on the interval $[x, x + \Delta x]$ is the quotient

$$\frac{\Delta y}{\Delta x} = \frac{f(x + \Delta x) - f(x)}{\Delta x}.$$

The **instantaneous rate of change** of y **with respect to** x is the limit, as $\Delta x \to 0$, of the average rate of change. Thus the instantaneous rate of change of y with respect to x is

$$\lim_{\Delta x \to 0} \frac{\Delta y}{\Delta x} = \frac{dy}{dx} = f'(x). \quad \textbf{(20)}$$

Example 7 illustrates the fact that a dependent variable may sometimes be expressed as two different functions of two different independent variables. The derivatives of these functions are then rates of change of the dependent variable with respect to the two different independent variables.

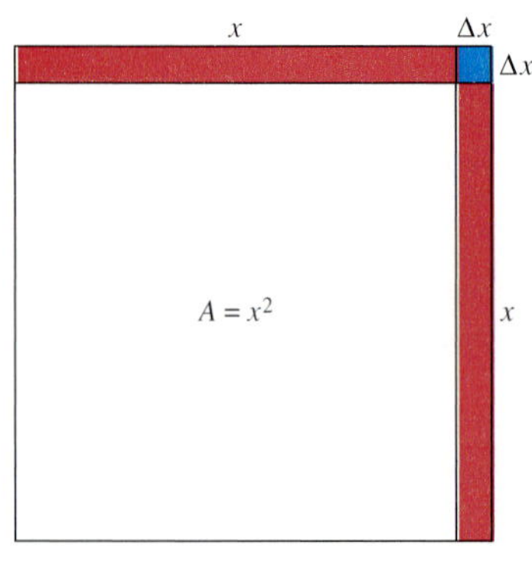

Fig. 3.1.17 The square of Example 7:

$$A + \Delta A = (x + \Delta x)^2;$$

$$\Delta A = 2x\Delta x + (\Delta x)^2;$$

$$\frac{\Delta A}{\Delta x} = 2x + \Delta x;$$

$$\frac{dA}{dx} = 2x.$$

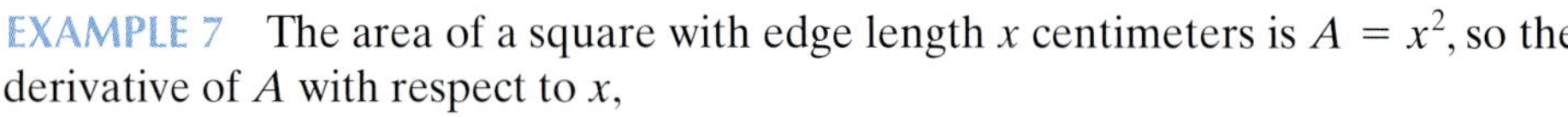

EXAMPLE 7 The area of a square with edge length x centimeters is $A = x^2$, so the derivative of A with respect to x,

$$\frac{dA}{dx} = 2x, \quad \textbf{(21)}$$

is the rate of change of its area A with respect to x. (See the computations in Fig. 3.1.17.) The units of dA/dx are square centimeters *per centimeter.* Now suppose that the edge length of the square is increasing with time: $x = 5t$, with time t in seconds. Then the area of the square at time t is

$$A = (5t)^2 = 25t^2.$$

The derivative of A with respect to t is

$$\frac{dA}{dt} = 2 \cdot 25t = 50t; \quad \textbf{(22)}$$

this is the rate of change of A with respect to time t, with units of square centimeters *per second.* For instance, when $t = 10$ (so $x = 50$), the values of the two derivatives of A in Eqs. (21) and (22) are

$$\left.\frac{dA}{dx}\right|_{x=50} = 2 \cdot 50 = 100 \quad (\text{cm}^2/\text{cm})$$

and

$$\left.\frac{dA}{dt}\right|_{t=10} = 50 \cdot 10 = 500 \quad (\text{cm}^2/\text{s}).$$

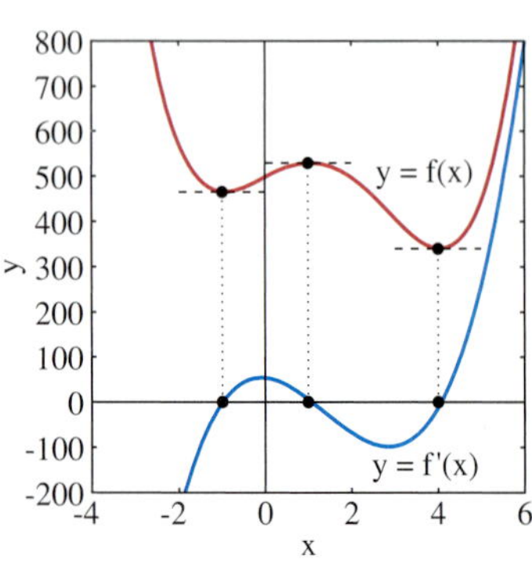

Fig. 3.1.18 Correspondence between the function graph $y = f(x)$ and the derivative graph $y = f'(x)$

The notation dA/dt for the derivative suffers from the minor inconvenience of not providing a "place" to substitute a particular value of t, such as $t = 10$. The last lines of Example 7 illustrate one way around this difficulty.

Just as we can speak of whether the quantity $Q(t)$ is increasing or decreasing at time $t = a$—according as $Q'(a) > 0$ or $Q'(a) < 0$—we can ask whether the function $y = f(x)$ is an increasing or decreasing function of x. Thinking of rising tangent lines with positive slopes, and falling tangent lines with negative slopes, we say in analogy with (13) that

y is *increasing* at the point $x = a$ if $f'(a) > 0$;

y is *decreasing* at the point $x = a$ if $f'(a) < 0$. **(23)**

EXAMPLE 8 Figure 3.1.18 shows the graphs $y = f(x)$ of a function and $y = f'(x)$ of its derivative. Observe that

- $y = f(x)$ has a horizontal tangent line at points where $f'(x) = 0$;
- $f(x)$ is increasing on open intervals where $f'(x) > 0$; and
- $f(x)$ is decreasing on open intervals where $f'(x) < 0$. ■

3.1 PROBLEMS

In Problems 1 through 10, find the indicated derivative by using the differentiation rule in Eqs. (6) and (7):

$$\text{If} \quad f(x) = ax^2 + bx + c, \quad \text{then} \quad f'(x) = 2ax + b.$$

1. $f(x) = 4x - 5$; find $f'(x)$.
2. $g(t) = 100 - 16t^2$; find $g'(t)$.
3. $h(z) = z(25 - z)$; find $h'(z)$.
4. $f(x) = 16 - 49x$; find $f'(x)$.
5. $y = 2x^2 + 3x - 17$; find dy/dx.
6. $x = 16t - 100t^2$; find dx/dt.
7. $z = 5u^2 - 3u$; find dz/du.
8. $v = 5y(100 - y)$; find dv/dy.
9. $x = -5y^2 + 17y + 300$; find dx/dy.
10. $u = 7t^2 + 13t$; find du/dt.

In Problems 11 through 20, apply the definition of the derivative (as in Example 1) to find $f'(x)$.

11. $f(x) = 2x - 1$
12. $f(x) = 2 - 3x$
13. $f(x) = x^2 + 5$
14. $f(x) = 3 - 2x^2$
15. $f(x) = \dfrac{1}{2x + 1}$
16. $f(x) = \dfrac{1}{3 - x}$
17. $f(x) = \sqrt{2x + 1}$
18. $f(x) = \dfrac{1}{\sqrt{x + 1}}$
19. $f(x) = \dfrac{x}{1 - 2x}$
20. $f(x) = \dfrac{x + 1}{x - 1}$

Problems 21 through 25 give the position function $x = f(t)$ of a particle moving in a horizontal straight line. Find its location x when its velocity v is zero.

21. $x = 100 - 16t^2$
22. $x = -16t^2 + 160t + 25$
23. $x = -16t^2 + 80t - 1$
24. $x = 100t^2 + 50$
25. $x = 100 - 20t - 5t^2$

Problems 26 through 29 give the height $y(t)$ (in feet at time t seconds) of a ball thrown vertically upward. Find the maximum height that the ball attains.

26. $y = -16t^2 + 160t$
27. $y = -16t^2 + 64t$
28. $y = -16t^2 + 128t + 25$
29. $y = -16t^2 + 96t + 50$

In Problems 30 through 35 (Figs. 3.1.19 through 3.1.24), match the given graph of the function f with that of its derivative, which appears among those in Fig. 3.1.25, parts (a) through (f).

30. Figure 3.1.19
31. Figure 3.1.20
32. Figure 3.1.21
33. Figure 3.1.22
34. Figure 3.1.23
35. Figure 3.1.24

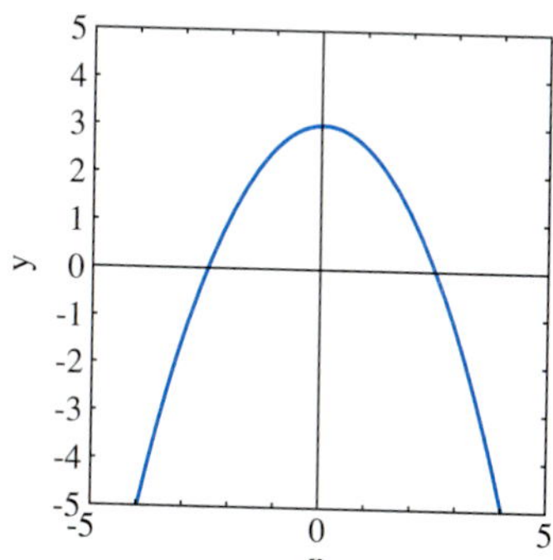

Fig. 3.1.19

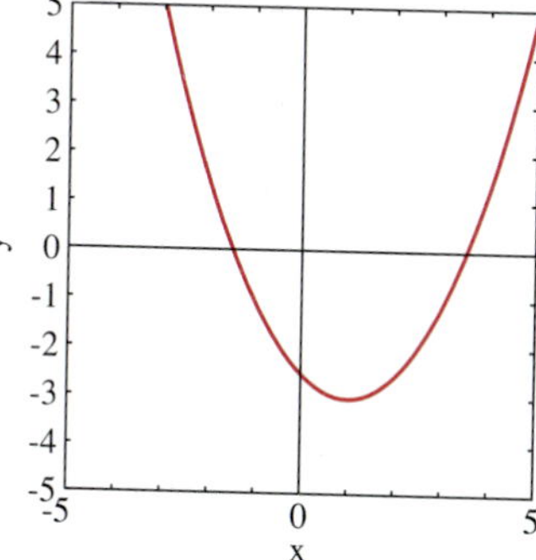

Fig. 3.1.20

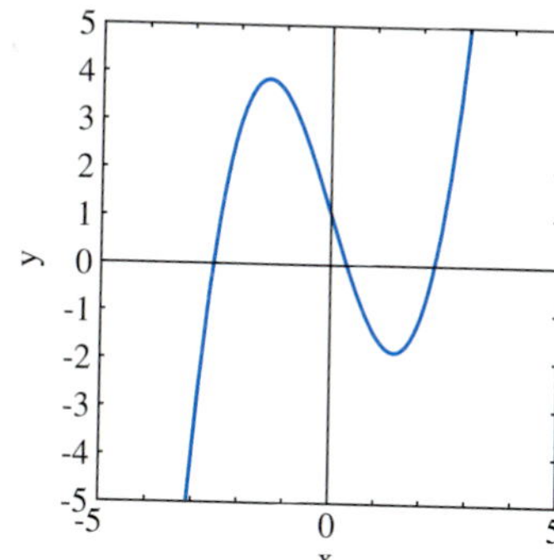

Fig. 3.1.21

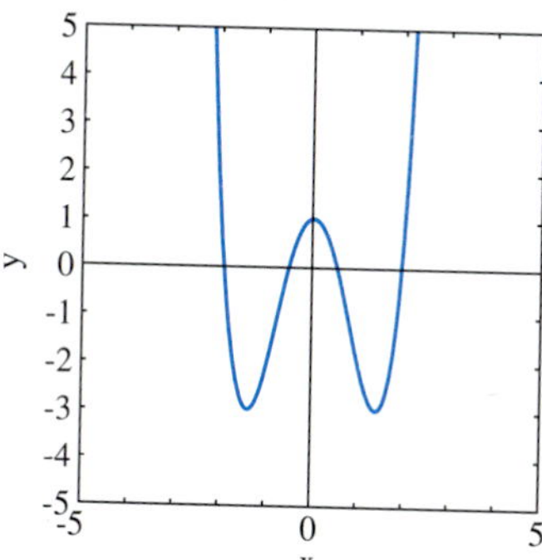

Fig. 3.1.22

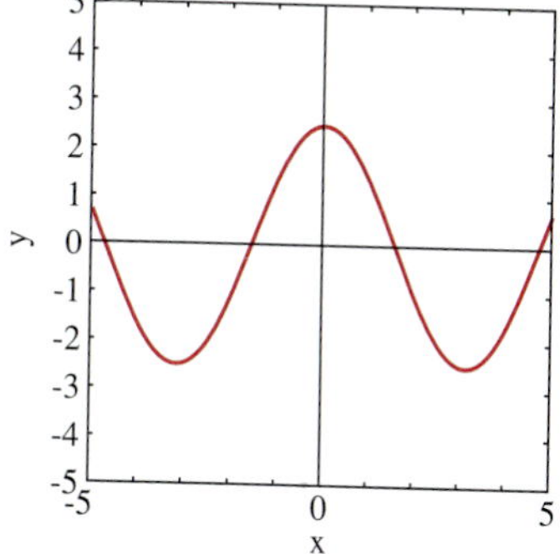

Fig. 3.1.23

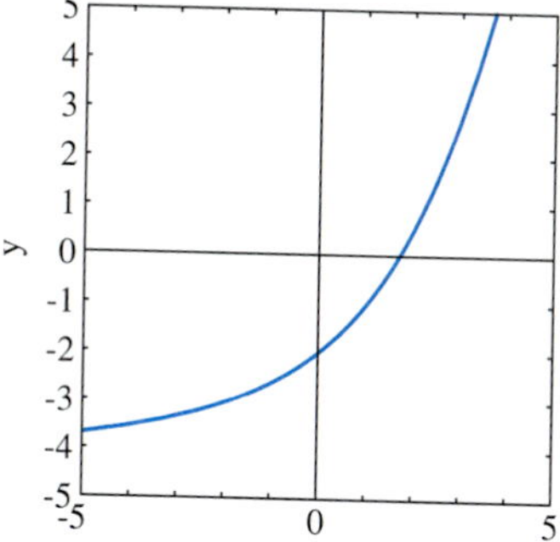

Fig. 3.1.24

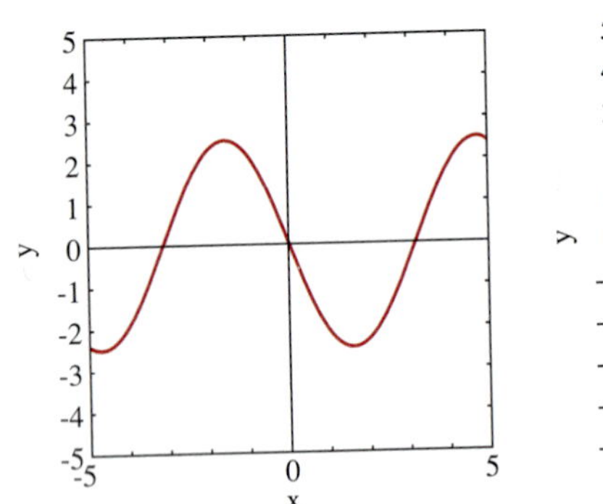
Fig. 3.1.25(a)

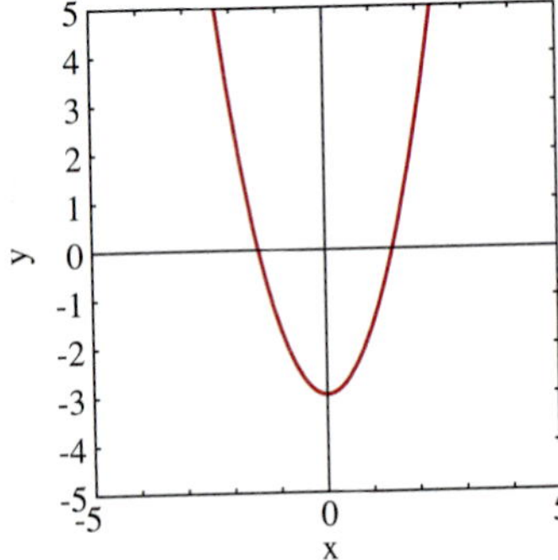
Fig. 3.1.25(b)

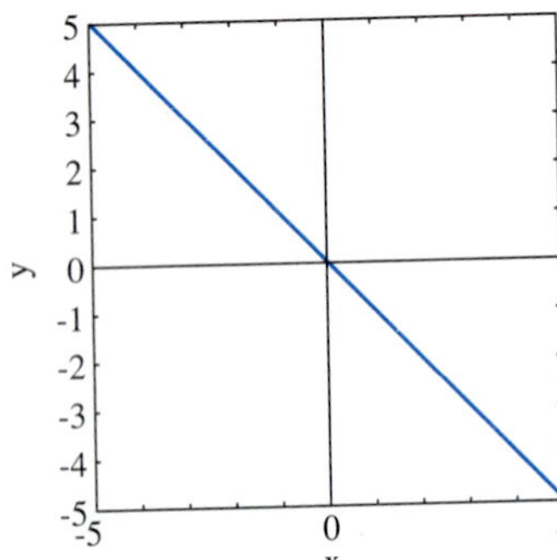
Fig. 3.1.25(c)

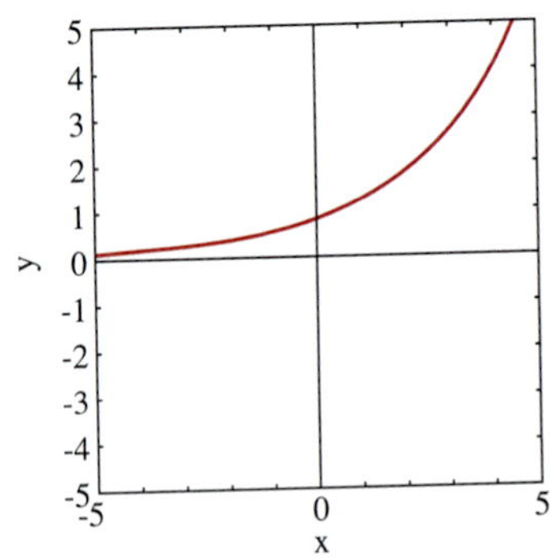
Fig. 3.1.25(d)

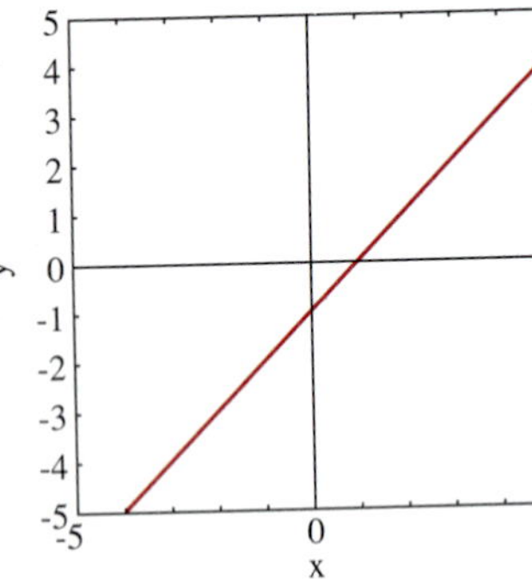
Fig. 3.1.25(e)

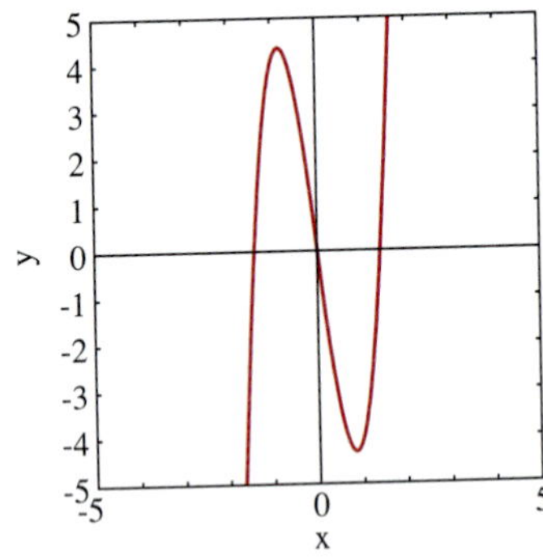
Fig. 3.1.25(f)

36. The Celsius temperature C is given in terms of the Fahrenheit temperature F by $C = \frac{5}{9}(F - 32)$. Find the rate of change of C with respect to F and the rate of change of F with respect to C.

37. Find the rate of change of the area A of a circle with respect to its circumference C.

38. A stone dropped into a pond at time $t = 0$ s causes a circular ripple that travels out from the point of impact at 5 m/s. At what rate (in square meters per second) is the area within the circle increasing when $t = 10$?

39. A car is traveling at 100 ft/s when the driver suddenly applies the brakes ($x = 0, t = 0$). The position function of the skidding car is $x(t) = 100t - 5t^2$. How far and for how long does the car skid before it comes to a stop?

40. A water bucket containing 10 gal of water develops a leak at time $t = 0$, and the volume V of water in the bucket t seconds later is given by

$$V(t) = 10\left(1 - \frac{t}{100}\right)^2$$

until the bucket is empty at time $t = 100$. (a) At what rate is water leaking from the bucket after exactly 1 min has passed? (b) When is the instantaneous rate of change of V equal to the average rate of change of V from $t = 0$ to $t = 100$?

41. A population of chipmunks moves into a new region at time $t = 0$. At time t (in months) the population numbers

$$P(t) = 100[1 + (0.3)t + (0.04)t^2].$$

(a) How long does it take for this population to double its initial size $P(0)$? (b) What is the rate of growth of the population when $P = 200$?

42. The following data describe the growth of the population P (in thousands) of Gotham City during a 10-year period. Use the graphical method of Example 4 to estimate its rate of growth in 1989.

YEAR	1984	1986	1988	1990	1992	1994
P	265	293	324	358	395	427

43. The following data give the distance x in feet traveled by an accelerating car (that starts from rest at time $t = 0$) in the first t seconds. Use the graphical method of Example 4 to estimate its speed (in miles per hour) when $t = 20$ and again when $t = 40$.

t	0	10	20	30	40	50	60
x	0	224	810	1655	2686	3850	5109

In Problems 44 through 49, use the fact (proved in Section 3.2) that the derivative of $y = ax^3 + bx^2 + cx + d$ is $dy/dx = 3ax^2 + 2bx + c$.

44. Prove that the rate of change of the volume V of a cube with respect to its edge length x is equal to half the surface area S of the cube (Fig. 3.1.26).

45. Show that the rate of change of the volume V of a sphere with respect to its radius r is equal to its surface area S (Fig. 3.1.27).

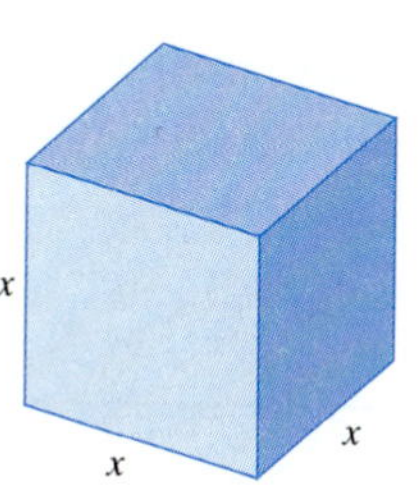

Fig. 3.1.26 The cube of Problem 44: volume $V = x^3$, surface area $S = 6x^2$.

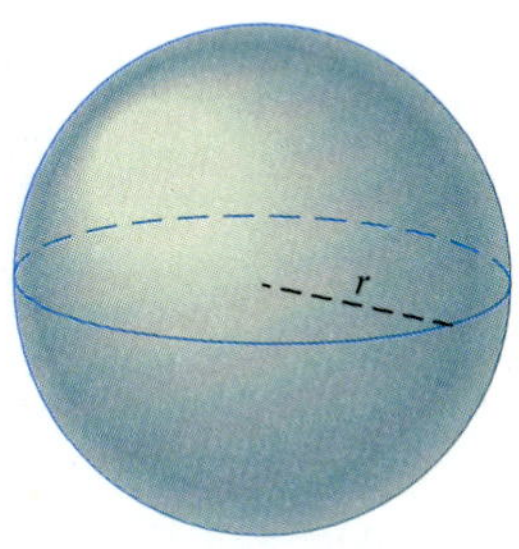

Fig. 3.1.27 The sphere of Problem 45: volume $V = \frac{4}{3}\pi r^3$ surface area $S = 4\pi r^2$.

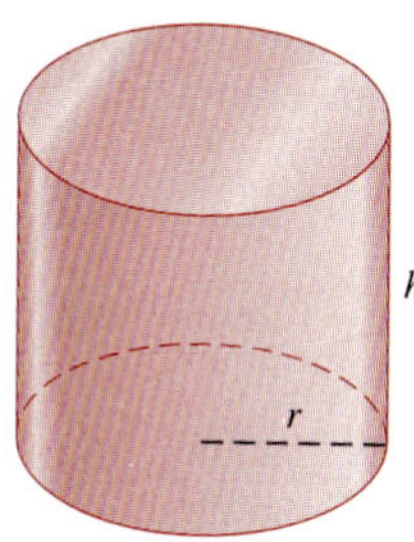

Fig. 3.1.28 The cylinder of Problem 46: volume $V = \pi r^2 h$, surface area $S = 2\pi r^2 + 2\pi rh$

46. The height h of a certain cylinder whose height changes is always twice its radius r. Show that the rate of change of its volume V with respect to r is equal to its total surface area S (Fig. 3.1.28).

47. A spherical balloon with an initial radius r of 5 in. begins to leak at time $t = 0$, and its radius t seconds later is $r = (60 - t)/12$ in. At what rate (in cubic inches per second) is air leaking from the balloon when $t = 30$?

48. The volume V (in liters) of 3 g of CO_2 at 27°C is given in terms of its pressure p (in atmospheres) by the formula $V = 1.68/p$. What is the rate of change of V with respect to p when $p = 2$ (atm)? [*Suggestion:* Use the fact that the derivative of $f(x) = c/x$ is $f'(x) = -c/x^2$ if c is a constant; you can establish this by using the definition of the derivative.]

49. As a snowball with an initial radius of 12 cm melts, its radius decreases at a constant rate. It begins to melt when $t = 0$ (h) and takes 12 h to disappear. (a) What is its rate of change of volume when $t = 6$? (b) What is its average rate of change of volume from $t = 3$ to $t = 9$?

50. A ball thrown vertically upward at time $t = 0$ (s) with initial velocity 96 ft/s and with initial height 112 ft has height function $y(t) = -16t^2 + 96t + 112$. (a) What is the maximum height attained by the ball? (b) When and with what impact speed does the ball hit the ground?

51. A spaceship approaching touchdown on the planet Gzyx has height y (meters) at time t (seconds) given by $y = 100 - 100t + 25t^2$. When and with what speed does it hit the ground?

52. The population (in thousands) of the city Metropolis is given by

$$P(t) = 100[1 + (0.04)t + (0.003)t^2],$$

with t in years and with $t = 0$ corresponding to 1980. (a) What was the rate of change of P in 1986? (b) What was the average rate of change of P from 1983 to 1988?

3.1 PROJECT: A CITY'S POPULATION GROWTH

This project involves a graphical analysis of the population growth of a small city during the decade of the 1990s.

For your own small city, choose a positive integer k with $k \leqq 9$ (perhaps the last nonzero digit of your student I.D. number). Then suppose that the population P of the city t years after 1990 is given (in thousands) by

$$P(t) = 10 + t - (0.1)t^2 + (0.001)(k + 5)t^3.$$

Investigate the following questions.

1. Does the graph of $P(t)$ indicate that the population is increasing throughout the 1990s? Explain your answer.
2. Does the graph of the derivative $P'(t)$ confirm that $P(t)$ is increasing through the 1990s? What property of this graph is pertinent to the question?
3. What points on the graph of $P'(t)$ correspond to the time (or times) at which the instantaneous rate of change of P is equal to its average rate of change between the years 1990 and 2000? Apply the method of successive magnification to find each such time accurate to two decimal places.

4. What points on the graph of the derivative $P'(t)$ correspond to the time (or times) at which the population $P(t)$ is increasing the slowest? the fastest? Apply the method of successive magnification to find each such time accurate to two decimal places.

3.2 BASIC DIFFERENTIATION RULES

Here we begin our development of formal rules for finding the derivative f' of the function f:

$$f'(x) = \lim_{h \to 0} \frac{f(x+h) - f(x)}{h}. \tag{1}$$

Some alternative notation for derivatives will be helpful.

When we interpreted the derivative in Section 3.1 as a rate of change, we found it useful to employ the dependent-independent variable notation

$$y = y(x), \qquad \Delta x = h, \qquad \Delta y = f(x + \Delta x) - f(x). \tag{2}$$

This led to the "differential notation"

$$\frac{dy}{dx} = \lim_{\Delta x \to 0} \frac{\Delta y}{\Delta x} = \lim_{\Delta x \to 0} \frac{f(x + \Delta x) - f(x)}{\Delta x} \tag{3}$$

for the derivative. When you use this notation, remember that the symbol dy/dx is simply another notation for the derivative $f'(x)$; it is *not* the quotient of two separate entities dy and dx.

A third notation is sometimes used for the derivative $f'(x)$; it is $D_x f(x)$. Here, think of D_x as a "machine" that operates on the function f to produce its derivative $D_x f$ with respect to x (Fig. 3.2.1). Thus we can write the derivative of $y = f(x) = x^3$ in any of three ways:

$$f'(x) = \frac{dy}{dx} = D_x x^3 = 3x^2.$$

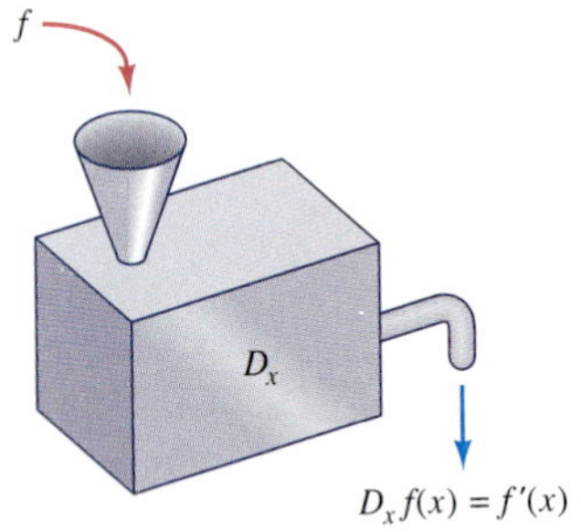

These three notations for the derivative—the function notation $f'(x)$, the differential notation dy/dx, and the operator notation $D_x f(x)$—are used almost interchangeably in mathematical and scientific writing, so you need to be familiar with each.

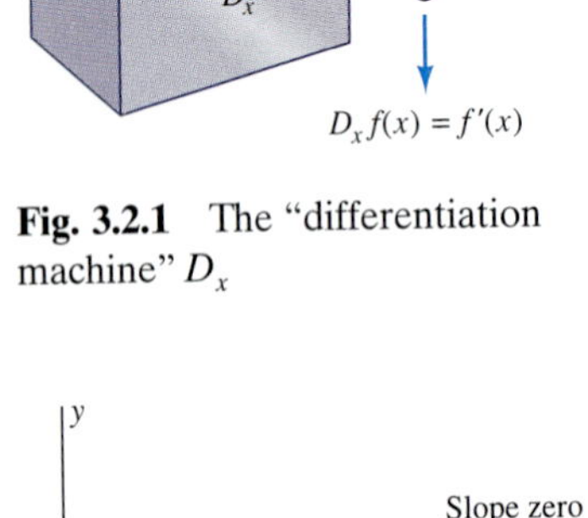

Fig. 3.2.1 The "differentiation machine" D_x

The Derivative of a Constant

Our first differentiation rule says that *the derivative of a constant function is identically zero*. Geometry makes this obvious, because the graph of a constant function is a horizontal straight line that is its own tangent line, with slope zero at every point (Fig. 3.2.2).

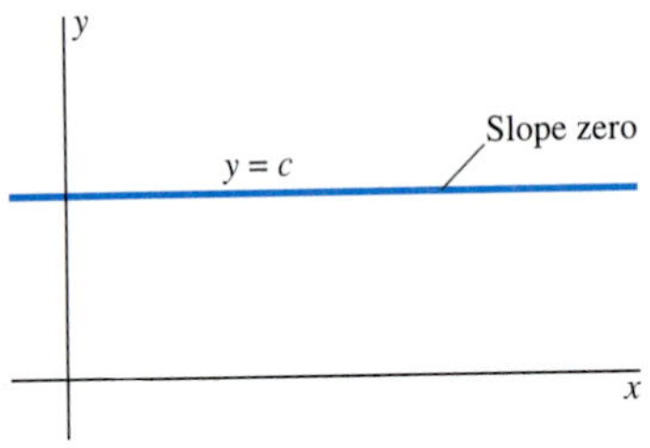

Fig. 3.2.2 The derivative of a constant-valued function is zero (Theorem 1).

> **Theorem 1** ***Derivative of a Constant***
>
> If $f(x) = c$ (a constant) for all x, then $f'(x) = 0$ for all x. That is,
>
> $$\frac{dc}{dx} = D_x c = 0. \tag{4}$$

PROOF Because $f(x + h) = f(x) = c$, we see that

$$f'(x) = \lim_{h \to 0} \frac{f(x + h) - f(x)}{h} = \lim_{h \to 0} \frac{c - c}{h} = \lim_{h \to 0} \frac{0}{h} = 0. \qquad \blacksquare$$

The Power Rule

As motivation for the next rule, consider the following list of derivatives, all of which have already appeared in the text (or as problems). The first two are special cases of the formula $D_x(ax^2 + bx + c) = 2ax + b$.

$$D_x x = 1$$

$$D_x x^2 = 2x = 2 \cdot x^1$$

$$D_x x^3 = 3 \cdot x^2 \qquad \text{(Problem 37, Section 2.2)}$$

$$D_x \frac{1}{x} = D_x x^{-1} = -\frac{1}{x^2} = - = -1 \cdot x^{-2} \qquad \text{(Problem 38, Section 2.2)}$$

$$D_x \frac{1}{x^2} = D_x x^{-2} = -\frac{2}{x^3} = -2 \cdot x^{-3} \qquad \text{(Problem 39, Section 2.2)}$$

$$D_x \sqrt{x} = D_x x^{1/2} = \frac{1}{2\sqrt{x}} = \frac{1}{2} \cdot x^{-1/2} \qquad \text{(Example 13, Section 2.2)}$$

Each of these formulas fits the simple pattern

$$D_x x^n = nx^{n-1}. \qquad \textbf{(5)}$$

Equation (5), inferred from the preceding list of derivatives, as yet is only a conjecture. But many discoveries in mathematics are made by detecting such patterns and then proving that they hold universally.

Eventually, we shall see that the formula in Eq. (5), called the **power rule,** is valid for all real numbers n. At this time we give a proof only for the case in which the exponent n is a *positive integer.*

Theorem 2 ***Power Rule for a Positive Integer n***

If n is a positive integer and $f(x) = x^n$, then

$$f'(x) = nx^{n-1}. \qquad \textbf{(6)}$$

PROOF For a positive integer n, the identity

$$b^n - a^n = (b - a)(b^{n-1} + b^{n-2}a + b^{n-3}a^2 + \cdots + ba^{n-2} + a^{n-1})$$

is easy to verify by multiplication. Thus, if $b \neq a$, then

$$\frac{b^n - a^n}{b - a} = b^{n-1} + b^{n-2}a + b^{n-3}a^2 + \cdots + ba^{n-2} + a^{n-1}.$$

Because there are n terms on the right-hand side, this tells us that

$$\lim_{b \to a} \frac{b^n - a^n}{b - a} = na^{n-1}$$

by various limit laws. Now let $b = x + h$ and $a = x$, so that $h = b - a$. Then $h \to 0$ as $b \to a$, and hence

$$f'(x) = \lim_{h \to 0} \frac{(x + h)^n - x^n}{h} = nx^{n-1}. \tag{7}$$

This establishes Theorem 2. ■

We need not always use the same symbols x and n for the independent variable and the constant exponent in the power rule. For instance,

$$D_t t^m = mt^{m-1} \quad \text{and} \quad D_z z^k = kz^{k-1}.$$

If it is perfectly clear what the independent variable is, the subscript may be dropped from D_x (or D_t, or D_z), as in Example 1.

EXAMPLE 1 $Dx^7 = 7x^6, \qquad Dt^{17} = 17t^{16}, \qquad Dz^{100} = 100z^{99}.$ ■

The Derivative of a Linear Combination

To use the power rule to differentiate polynomials, we need to know how to differentiate *linear combinations.* A **linear combination** of the functions f and g is a function of the form $af + bg$, where a and b are constants. It follows from the sum and product laws for limits that

$$\lim_{x \to c} [af(x) + bg(x)] = a\left(\lim_{x \to c} f(x)\right) + b\left(\lim_{x \to c} g(x)\right) \tag{8}$$

provided that both limits on the right in Eq. (8) exist. The formula in Eq. (8) is called the **linearity property** of the limit operation. It implies an analogous linearity property of differentiation.

Theorem 3 ***Derivative of a Linear Combination***

If f and g are differentiable functions and a and b are fixed real numbers, then

$$D_x[af(x) + bg(x)] = a[D_x f(x)] + b[D_x g(x)]. \tag{9}$$

With $u = f(x)$ and $v = g(x)$, this takes the form

$$\frac{d(au + bv)}{dx} = a\frac{du}{dx} + b\frac{dv}{dx}. \tag{9'}$$

PROOF The linearity property of limits immediately gives

$$\begin{aligned} D_x[af(x) + bg(x)] &= \lim_{h \to 0} \frac{[af(x + h) + bg(x + h)] - [af(x) + bg(x)]}{h} \\ &= a\left(\lim_{h \to 0} \frac{f(x + h) - f(x)}{h}\right) + b\left(\lim_{h \to 0} \frac{g(x + h) - g(x)}{h}\right) \\ &= a[D_x f(x)] + b[D_x g(x)], \end{aligned}$$

as desired. ■

Now take $a = c$ and $b = 0$ in Eq. (9). The result is

$$D_x[cf(x)] = cD_x f(x); \tag{10}$$

alternatively,

$$\frac{d(cu)}{dx} = c\frac{du}{dx}, \tag{10'}$$

Thus *the derivative of a constant multiple of a function is the same constant multiple of its derivative.*

EXAMPLE 2

(a) $D_x(16x^6) = 16 \cdot 6x^5 = 96x^5$.

(b) If $f(z) = 7z^3$, then $f'(z) = 21z^2$.

(c) $\dfrac{d}{du}(99u^{100}) = 9900u^{99}$. ■

Next, take $a = b = 1$ in Eq. (9). We find that

$$D_x[f(x) + g(x)] = [D_x f(x)] + [D_x g(x)]. \tag{11}$$

In differential notation,

$$\frac{d(u+v)}{dx} = \frac{du}{dx} + \frac{dv}{dx}. \tag{11'}$$

Thus *the derivative of the sum of two functions is the sum of their derivatives.* Similarly, for differences we have

$$\frac{d(u-v)}{dx} = \frac{du}{dx} - \frac{dv}{dx}. \tag{12}$$

It's easy to see that these rules generalize to sums and differences of more than two functions. For example, repeated application of Eq. (11) to the sum of a finite number of differentiable functions gives

$$\frac{d(u_1 + u_2 + \cdots + u_n)}{dx} = \frac{du_1}{dx} + \frac{du_2}{dx} + \cdots + \frac{du_n}{dx}. \tag{13}$$

EXAMPLE 3

$$\begin{aligned} D_x(36 + 26x + 7x^5 - 5x^9) &= 0 + 26 \cdot 1 + 7 \cdot 5x^4 - 5 \cdot 9x^8 \\ &= 26 + 35x^4 - 45x^8. \end{aligned}$$ ■

The Derivative of a Polynomial

When we apply Eqs. (10) and (13) and the power rule to the polynomial

$$p(x) = a_n x^n + a_{n-1}x^{n-1} + \cdots + a_2 x^2 + a_1 x + a_0,$$

we find the derivative *as fast as we can write it:*

$$p'(x) = na_n x^{n-1} + (n-1)a_{n-1}x^{n-2} + \cdots + 3a_3x^2 + 2a_2x + a_1. \tag{14}$$

With this result, it becomes a routine matter to write an equation for a line tangent to the graph of a polynomial.

EXAMPLE 4 Write an equation for the straight line that is tangent to the graph of $y = 2x^3 - 7x^2 + 3x + 4$ at the point $(1, 2)$.

Solution We compute the derivative as in Eq. (14):

$$\frac{dy}{dx} = 2 \cdot 3x^2 - 7 \cdot 2x + 3 + 0 = 6x^2 - 14x + 3.$$

We substitute $x = 1$ in dy/dx and find that the slope of the tangent line at $(1, 2)$ is $m = -5$. So the point-slope equation of the tangent line is

$$y - 2 = -5(x - 1); \quad \text{that is,}$$

$$y = -5x + 7.$$

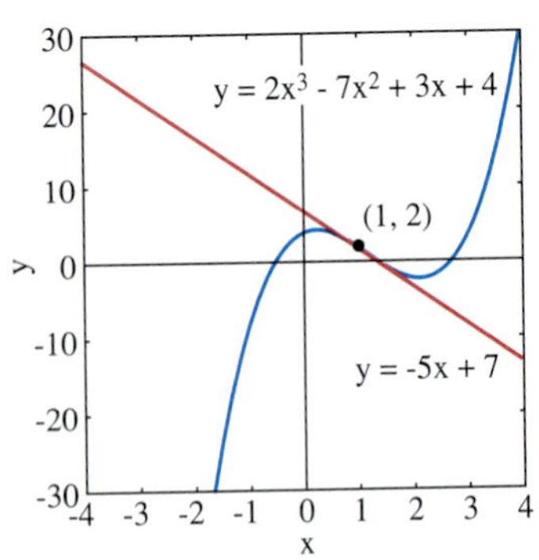

Fig. 3.2.3 The graph $y = 2x^3 - 7x^2 + 3x + 4$ and its tangent line $y = -5x + 7$

A calculator- or computer-generated picture like Fig. 3.2.3 provides suggestive visual evidence of the correctness of this tangent line computation. ■

EXAMPLE 5 The volume V (in cubic centimeters) of a given sample of water varies with changing temperature T. For T between 0°C and 30°C, the relation is given almost exactly by the formula

$$V = V_0[1 - (6.427 \times 10^{-5})T + (8.505 \times 10^{-6})T^2 - (6.790 \times 10^{-8})T^3],$$

where V_0 is the volume of the water (*not* ice) sample at 0°C. Suppose that $V_0 = 10^5$ cm³. Find both the volume and the rate of change of volume with respect to temperature when $T = 20$°C.

Solution Substitution of $T = 20$ in the given volume formula yields $V \approx$ 100,157 cm³, so the sample would expand by 157 cm³ if heated from 0°C to 20°C. The rate of change of volume with respect to temperature is given by

$$\frac{dV}{dT} = 10^5 \cdot [-6.427 \times 10^{-5} + (17.010 \times 10^{-6})T - (20.370 \times 10^{-8})T^2],$$

and substitution of $T = 20$ yields

$$\left.\frac{dV}{dT}\right|_{T=20} \approx 19.45 \quad (\text{cm}^3/^\circ\text{C}).$$

Thus we should expect the volume of the water sample to increase by slightly more than 19 cm³ if it is heated by 1°C from 20°C to 21°C. In fact, direct substitution into the original volume formula gives

$$V(21) - V(20) \approx 19.88.$$

Finally, we note that the average rate of change of V with respect to T on the interval $19.5 \leqq T \leqq 20.5$ centered at $T = 20$ is

$$\frac{\Delta V}{\Delta T} = \frac{V(20.5) - V(19.5)}{20.5 - 19.5} \approx 19.44 \quad (\text{cm}^3/^\circ\text{C}),$$

which is very close to the derivative dV/dT at $T = 20$. ■

The Product Rule and the Quotient Rule

It might be natural to conjecture that the derivative of a product $f(x)g(x)$ is the product of the derivatives. This is *false*! For example, if $f(x) = g(x) = x$, then

$$D_x[_f(x)g(x)] = D_x\, x^2 = 2x.$$

But

$$[D_x\, f(x)]\cdot[D_x\, g(x)] = (D_x\, x)\cdot(D_x\, x) = 1\cdot 1 = 1.$$

In general, the derivative of a product is *not* merely the product of the derivatives. Theorem 4 tells us what it *is*.

Theorem 4 *The Product Rule*

If f and g are differentiable at x, then fg is differentiable at x, and

$$D_x[f(x)g(x)] = f'(x)g(x) + f(x)g'(x). \tag{15}$$

With $u = f(x)$ and $v = g(x)$, this **product rule** takes the form

$$\frac{d(uv)}{dx} = u\frac{dv}{dx} + v\frac{du}{dx}. \tag{15$'$}$$

When it is clear what the independent variable is, we can write the product rule even more briefly:

$$(uv)' = u'v + uv'. \tag{15$''$}$$

PROOF We use an "add and subtract" device.

$$\begin{aligned}
D_x[f(x)g(x)] &= \lim_{h\to 0}\frac{f(x+h)g(x+h) - f(x)g(x)}{h}\\
&= \lim_{h\to 0}\frac{f(x+h)g(x+h) - f(x)g(x+h) + f(x)g(x+h) - f(x)g(x)}{h}\\
&= \lim_{h\to 0}\frac{f(x+h)g(x+h) - f(x)g(x+h)}{h} + \lim_{h\to 0}\frac{f(x)g(x+h) - f(x)g(x)}{h}\\
&= \left(\lim_{h\to 0}\frac{f(x+h) - f(x)}{h}\right)\left(\lim_{h\to 0} g(x+h)\right) + f(x)\left(\lim_{h\to 0}\frac{g(x+h) - g(x)}{h}\right)\\
&= f'(x)g(x) + f(x)g'(x).
\end{aligned}$$

In this proof we used the sum law and product law for limits, the definitions of $f'(x)$ and $g'(x)$, and the fact that

$$\lim_{h\to 0} g(x+h) = g(x).$$

This last equation holds because g is differentiable and therefore continuous at x (as we will see in Theorem 2 in Section 3.4). ■

In words, the product rule says that *the derivative of the product of two functions is formed by multiplying the derivative of each by the other and then adding the results.*

EXAMPLE 6 Find the derivative of

$$f(x) = (1 - 4x^3)(3x^2 - 5x + 2)$$

without first multiplying out the two factors.

Solution $D_x[(1 - 4x^3)(3x^2 - 5x + 2)]$

$$= [D_x(1 - 4x^3)](3x^2 - 5x + 2) + (1 - 4x^3)[D_x(3x^2 - 5x + 2)]$$
$$= (-12x^2)(3x^2 - 5x + 2) + (1 - 4x^3)(6x - 5)$$
$$= -60x^4 + 80x^3 - 24x^2 + 6x - 5.$$

We can apply the product rule repeatedly to find the derivative of a product of three or more differentiable functions $u_1, u_2, \ldots, u_n$ of x. For example,

$$\begin{aligned} D[u_1u_2u_3] &= (u_1u_2)'u_3 + (u_1u_2)u_3' \\ &= (u_1'u_2 + u_1u_2')u_3 + u_1u_2u_3' \\ &= u_1'u_2u_3 + u_1u_2'u_3 + u_1u_2u_3'. \end{aligned}$$

Note that the derivative of each factor in the original product is multiplied by the other two factors and then the three resulting products are added. This is, indeed, the general result:

$$\begin{aligned} D(u_1u_2\cdots u_n) = {} & u_1'u_2u_3\cdots u_{n-1}u_n + u_1u_2'u_3\cdots u_{n-1}u_n \\ & +\cdots+\cdots+ u_1u_2u_3\cdots u_{n-1}u_n', \end{aligned} \tag{16}$$

where the sum in Eq. (16) has one term corresponding to each of the n factors in the product $u_1u_2\cdots u_n$. It is easy to establish this **extended product rule** (see Problem 62) one step at a time—next with $n = 4$, then with $n = 5$, and so forth.

Our next result tells us how to find the derivative of the reciprocal of a function if we know the derivative of the function itself.

The Reciprocal Rule

If f is differentiable at x and $f(x) \neq 0$, then

$$D_x\left[\frac{1}{f(x)}\right] = -\frac{f'(x)}{[f(x)]^2}. \tag{17}$$

With $u = f(x)$, the reciprocal rule takes the form

$$\frac{d}{dx}\left(\frac{1}{u}\right) = -\frac{1}{u^2}\cdot\frac{du}{dx}. \tag{17$'$}$$

If there can be no doubt what the independent variable is, we can write

$$\left(\frac{1}{u}\right)' = -\frac{u'}{u^2}. \tag{17$''$}$$

PROOF As in the proof of Theorem 4, we use the limit laws, the definition of the derivative, and the fact that a function is continuous wherever it is differentiable (by Theorem 2 of Section 3.4). Moreover, note that $f(x + h) \neq 0$ for h near zero because $f(x) \neq 0$ and f is continuous at x (see Problem 16 in Appendix D). Therefore

$$D_x\left[\frac{1}{f(x)}\right] = \lim_{h\to 0}\frac{1}{h}\left(\frac{1}{f(x+h)} - \frac{1}{f(x)}\right) = \lim_{h\to 0}\frac{f(x) - f(x+h)}{hf(x+h)f(x)}$$

$$= -\left(\lim_{h\to 0}\frac{1}{f(x+h)f(x)}\right)\left(\lim_{h\to 0}\frac{f(x+h) - f(x)}{h}\right) = -\frac{f'(x)}{[f(x)]^2}.$$

■

EXAMPLE 7 With $f(x) = x^2 + 1$ in Eq. (17), we get

$$D_x\left(\frac{1}{x^2+1}\right) = -\frac{D_x(x^2+1)}{(x^2+1)^2} = -\frac{2x}{(x^2+1)^2}.$$

■

We now combine the reciprocal rule with the power rule for positive integral exponents to establish the power rule for negative integral exponents.

Theorem 5 ***Power Rule for a Negative Integer n***
If n is a negative integer, then $D_x\, x^n = nx^{n-1}$.

Proof Let $m = -n$, so that m is a positive integer. Then

$$D_x x^n = D_x\left(\frac{1}{x^m}\right) = -\frac{D_x(x^m)}{(x^m)^2} = -\frac{mx^{m-1}}{x^{2m}} = (-m)x^{-m-1} = nx^{n-1}.$$

■

This proof also shows that the rule of Theorem 5 holds exactly when the function being differentiated is defined: when $x \neq 0$.

EXAMPLE 8

$$D_x\left(\frac{5x^4 - 6x + 7}{2x^2}\right) = D_x\left(\frac{5}{2}x^2 - 3x^{-1} + \frac{7}{2}x^{-2}\right)$$

$$= \frac{5}{2}(2x) - 3(-x^{-2}) + \frac{7}{2}(-2x^{-3}) = 5x + \frac{3}{x^2} - \frac{7}{x^3}.$$

The key here was to “divide out” before differentiating. ■

Now we apply the product rule and reciprocal rule to get a rule for differentiation of the quotient of two functions.

Theorem 6 ***The Quotient Rule***
If f and g are differentiable at x and $g(x) \neq 0$, then f/g is differentiable at x and

$$D_x\left[\frac{f(x)}{g(x)}\right] = \frac{f'(x)g(x) - f(x)g'(x)}{[g(x)]^2}. \quad \textbf{(18)}$$

With $u = f(x)$ and $v = g(x)$, this rule takes the form

$$\frac{d}{dx}\left(\frac{u}{v}\right) = \frac{v\dfrac{du}{dx} - u\dfrac{dv}{dx}}{v^2}. \quad \textbf{(18')}$$

If it is clear what the independent variable is, we can write the quotient rule in the form

$$\left(\frac{u}{v}\right)' = \frac{u'v - uv'}{v^2}. \tag{18''}$$

PROOF We apply the product rule to the factorization

$$\frac{f(x)}{g(x)} = f(x) \cdot \frac{1}{g(x)}.$$

This gives

$$\begin{aligned} D_x\left[\frac{f(x)}{g(x)}\right] &= [D_x f(x)] \cdot \frac{1}{g(x)} + f(x) \cdot D_x\left[\frac{1}{g(x)}\right] \\ &= \frac{f'(x)}{g(x)} + f(x) \cdot \left(-\frac{g'(x)}{[g(x)]^2}\right) = \frac{f'(x)g(x) - f(x)g'(x)}{[g(x)]^2}. \end{aligned}$$ ■

Note that the numerator in Eq. (18) is *not* the derivative of the product of f and g. And the minus sign means that the *order* of terms in the numerator is important.

EXAMPLE 9 Find $z'(t) = dz/dt$ if z is given by

$$z = \frac{1 - t^3}{1 + t^4}.$$

Solution Here, primes denote derivatives with respect to t. With t (rather than x) as the independent variable, the quotient rule gives

$$\begin{aligned} \frac{dz}{dt} &= \frac{(1 - t^3)'(1 + t^4) - (1 - t^3)(1 + t^4)'}{(1 + t^4)^2} \\ &= \frac{(-3t^2)(1 + t^4) - (1 - t^3)(4t^3)}{(1 + t^4)^2} = \frac{t^6 - 4t^3 - 3t^2}{(1 + t^4)^2}. \end{aligned}$$

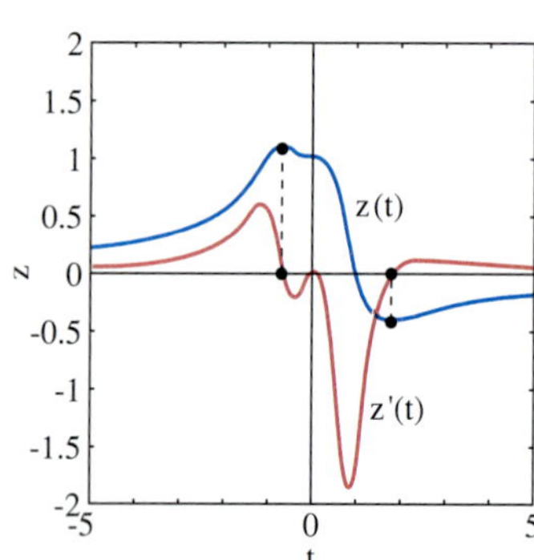

Fig. 3.2.4 Graphs of the function $z(t)$ of Example 9 and its derivative $z'(t)$

Figure 3.2.4 shows computer-generated graphs of the function $z(t)$ and its derivative $z'(t)$. Observe that $z(t)$ is increasing on intervals where $z'(t)$ is positive and is decreasing on intervals where $z'(t)$ is negative (thus corroborating our computation of the derivative). A quick computer or calculator graph of a function and its alleged derivative will often reveal an error if one has been made. ■

3.2 PROBLEMS

Apply the differentiation rules of this section to find the derivatives of the functions in Problems 1 through 40.

1. $f(x) = 3x^2 - x + 5$

2. $g(t) = 1 - 3t^2 - 2t^4$

3. $f(x) = (2x + 3)(3x - 2)$

4. $g(x) = (2x^2 - 1)(x^3 + 2)$

5. $h(x) = (x + 1)^3$

6. $g(t) = (4t - 7)^2$

7. $f(y) = y(2y - 1)(2y + 1)$

8. $f(x) = 4x^4 - \dfrac{1}{x^2}$

9. $g(x) = \dfrac{1}{x + 1} - \dfrac{1}{x - 1}$

10. $f(t) = \dfrac{1}{4 - t^2}$

11. $h(x) = \dfrac{3}{x^2 + x + 1}$

12. $f(x) = \dfrac{1}{1 - \dfrac{2}{x}}$

13. $g(t) = (t^2 + 1)(t^3 + t^2 + 1)$

14. $f(x) = (2x^3 - 3)(17x^4 - 6x + 2)$

15. $g(z) = \dfrac{1}{2z} - \dfrac{1}{3z^2}$

16. $f(x) = \dfrac{2x^3 - 3x^2 + 4x - 5}{x^2}$

17. $g(y) = 2y(3y^2 - 1)(y^2 + 2y + 3)$

18. $f(x) = \dfrac{x^2 - 4}{x^2 + 4}$

19. $g(t) = \dfrac{t - 1}{t^2 + 2t + 1}$

20. $u(x) = \dfrac{1}{(x + 2)^2}$

21. $v(t) = \dfrac{1}{(t - 1)^3}$

22. $h(x) = \dfrac{2x^3 + x^2 - 3x + 17}{2x - 5}$

23. $g(x) = \dfrac{3x}{x^3 + 7x - 5}$

24. $f(t) = \dfrac{1}{\left(t + \dfrac{1}{t}\right)^2}$

25. $g(x) = \dfrac{\dfrac{1}{x} - \dfrac{2}{x^2}}{\dfrac{2}{x^3} - \dfrac{3}{x^4}}$

26. $f(x) = \dfrac{x^3 - \dfrac{1}{x^2 + 1}}{x^4 + \dfrac{1}{x^2 + 1}}$

27. $y(x) = x^3 - 6x^5 + \frac{3}{2}x^{-4} + 12$

28. $x(t) = \dfrac{3}{t} - \dfrac{4}{t^2} - 5$

29. $y(x) = \dfrac{5 - 4x^2 + x^5}{x^3}$

30. $u(x) = \dfrac{2x - 3x^2 + 2x^4}{5x^2}$

31. $y(x) = 3x - \dfrac{1}{4x^2}$

32. $f(z) = \dfrac{1}{z(z^2 + 2z + 2)}$

33. $y(x) = \dfrac{x}{x - 1} + \dfrac{x + 1}{3x}$

34. $u(t) = \dfrac{1}{1 - 4t^{-2}}$

35. $y(x) = \dfrac{x^3 - 4x + 5}{x^2 + 9}$

36. $w(z) = z^2\left(2z^3 - \dfrac{3}{4z^4}\right)$

37. $y(x) = \dfrac{2x^2}{3x - \dfrac{4}{5x^4}}$

38. $z(t) = \dfrac{4}{(t^2 - 3)^2}$

39. $y(x) = \dfrac{x^2}{x + 1}$

40. $h(w) = \dfrac{w + 10}{w^2}$

In Problems 41 through 50, write an equation of the line tangent to the curve $y = f(x)$ at the given point P on the curve. Express the answer in the form $ax + by = c$.

41. $y = x^3$; $P(2, 8)$

42. $y = 3x^2 - 4$; $P(1, -1)$

43. $y = \dfrac{1}{x - 1}$; $P(2, 1)$

44. $y = 2x - \dfrac{1}{x}$; $P(0.5, -1)$

45. $y = x^3 + 3x^2 - 4x - 5$; $P(1, -5)$

46. $y = \left(\dfrac{1}{x} - \dfrac{1}{x^2}\right)^{-1}$; $P(2, 4)$

47. $y = \dfrac{3}{x^2} - \dfrac{4}{x^3}$; $P(-1, 7)$

48. $y = \dfrac{3x - 2}{3x + 2}$; $P(2, 0.5)$

49. $y = \dfrac{3x^2}{x^2 + x + 1}$; $P(-1, 3)$

50. $y = \dfrac{6}{1 - x^2}$; $P(2, -2)$

51. Apply the formula in Example 5 to answer the following two questions. (a) If 1000 cm^3 of water at 0°C is heated, does it initially expand or contract? (b) What is the rate (in cm^3/°C) at which it initially contracts or expands?

52. Susan's weight in pounds is given by the formula $W = (2 \times 10^9)/R^2$, where R is her distance in miles from the center of the earth. What is the rate of change of W with respect to R when $R = 3960$ mi? If Susan climbs a mountain, beginning at sea level, at what rate in ounces per (vertical) mile does her weight initially decrease?

53. The conical tank shown in Fig. 3.2.5 has radius 160 cm and height 800 cm. Water is running out of a small hole in the bottom of the tank. When the height h of water in the tank is 600 cm, what is the rate of change of its volume V with respect to h?

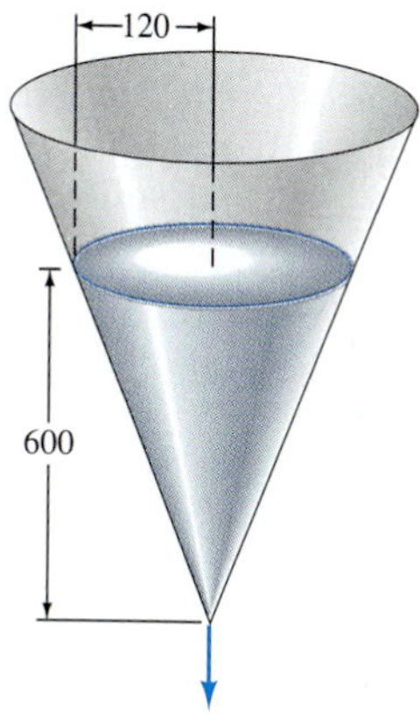

Fig. 3.2.5 The leaky tank of Problem 53

54. Find the x- and y-intercepts of the straight line that is tangent to the curve $y = x^3 + x^2 + x$ at the point $(1, 3)$ (Fig. 3.2.6).

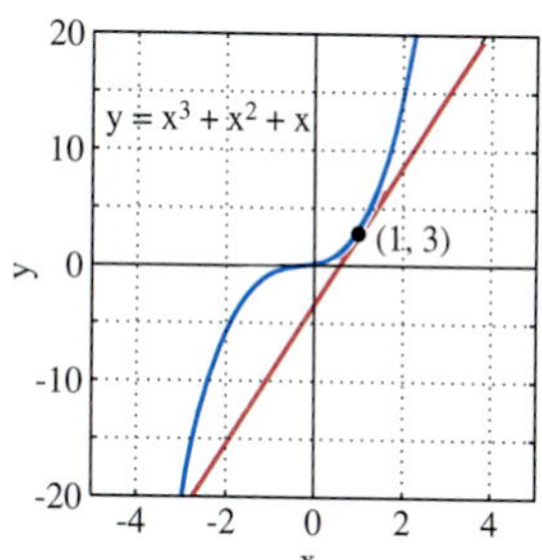

Fig. 3.2.6 The tangent line of Problem 54

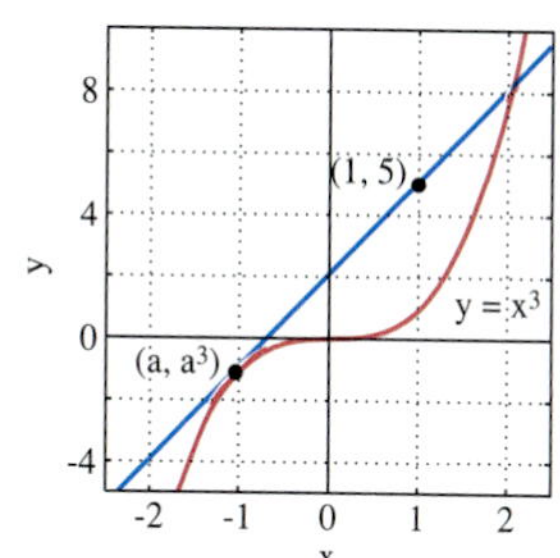

Fig. 3.2.7 The tangent line of Problem 55

55. Find an equation for the straight line that passes through the point $(1, 5)$ and is tangent to the curve $y = x^3$. [*Suggestion:* Denote by (a, a^3) the point of tangency, as indicated in Fig 3.2.7. Find by inspection small integral solutions of the resulting cubic equation in a.]

56. Find *two* lines through the point $(2, 8)$ that are tangent to the curve $y = x^3$. (See the suggestion for Problem 55.)

57. Prove that no straight line can be tangent to the curve $y = x^2$ at two different points.

58. Find the two straight lines of slope -2 that are tangent to the curve $y = 1/x$.

59. Let $n \geqq 2$ be a fixed but unspecified integer. Find the x-intercept of the line that is tangent to the curve $y = x^n$ at the point $P(x_0, y_0)$.

60. Prove that the curve $y = x^5 + 2x$ has no horizontal tangents. What is the smallest slope that a line tangent to this curve can have?

61. Apply Eq. (16) with $n = 3$ and $u_1 = u_2 = u_3 = f(x)$ to show that

$$D_x([f(x)]^3) = 3[f(x)]^2 \cdot f'(x).$$

62. (a) First write $u_1u_2u_3u_4 = (u_1u_2u_3)u_4$ to verify Eq. (16) for $n = 4$. (b) Then write $u_1u_2u_3u_4u_5 = (u_1u_2u_3u_4)u_5$ and apply the result in part (a) to verify Eq. (16) for $n = 5$.

63. Apply Eq. (16) to show that

$$D_x([f(x)]^n) = n[f(x)]^{n-1} \cdot f'(x)$$

if n is a positive integer and $f'(x)$ exists.

64. Use the result of Problem 63 to compute $D_x[(x^2 + x + 1)^{100}]$.

65. Find $g'(x)$ given $g(x) = (x^3 - 17x + 35)^{17}$.

66. Find constants $a, b, c,$ and d such that the graph of

$$f(x) = ax^3 + bx^2 + cx + d$$

has horizontal tangent lines at the points $(0, 1)$ and $(1, 0)$.

In connection with Problems 67 through 71, Figs. 3.2.8 through 3.2.11 show the curves

$$y = \frac{x^n}{1 + x^2}$$

for $n = 0, 1, 2,$ and 3.

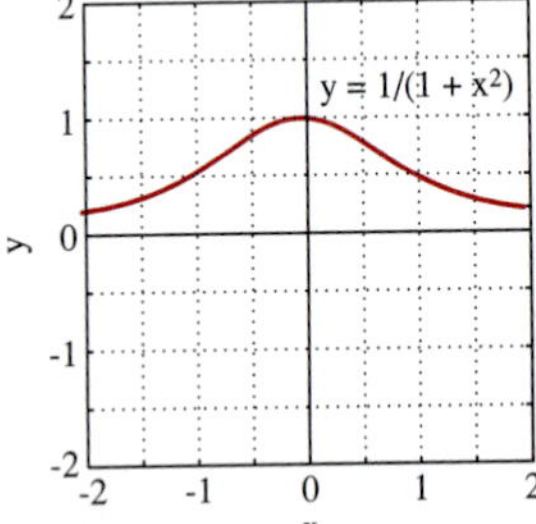

Fig. 3.2.8 The graph of $y = \dfrac{1}{1 + x^2}$

Fig. 3.2.9 The graph of 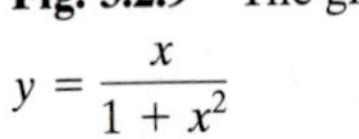 $y = \dfrac{x}{1 + x^2}$

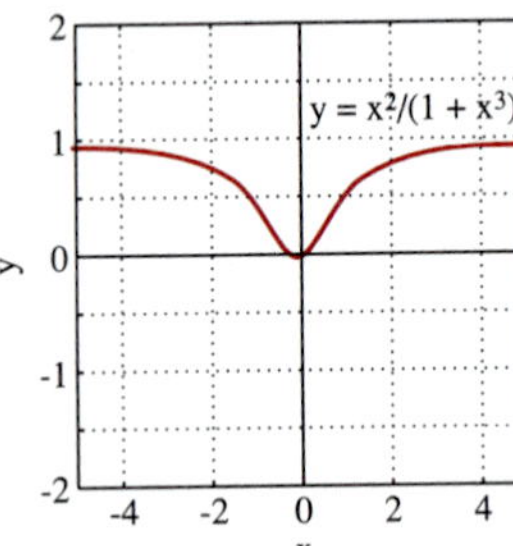

Fig. 3.2.10 The graph of $y = \dfrac{x^2}{1 + x^2}$

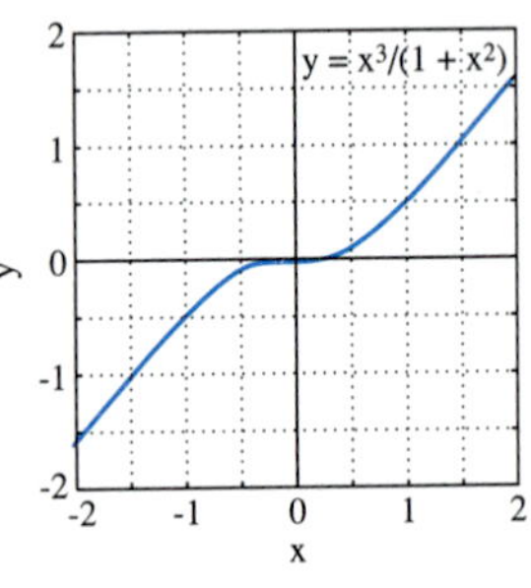

Fig. 3.2.11 The graph of $y = \dfrac{x^3}{1 + x^2}$

67. Show that for $n = 0$ and $n = 2$, the curve has only a single point where the tangent line is horizontal (Figs. 3.2.8 and 3.2.10).

68. When $n = 1$, there are two points on the curve where the tangent line is horizontal (Fig. 3.2.9). Find them.

69. Show that for $n \geqq 3$, $(0, 0)$ is the only point on the graph of

$$y = \frac{x^n}{1 + x^2}$$

at which the tangent line is horizontal (Fig. 3.2.11).

70. Figure 3.2.12 shows the graph of the derivative $f'(x)$ of the function

$$f(x) = \frac{x^3}{1 + x^2}.$$

There appear to be two points on the graph of $y = f(x)$ at which the tangent line has slope 1. Find them.

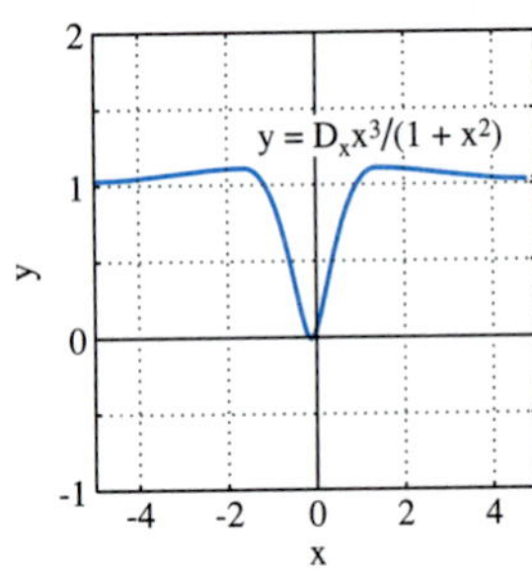

Fig. 3.2.12 The graph of $y = D_x\left(\dfrac{x^3}{1 + x^2}\right)$ of Problems 70 and 71

71. It appears in Fig. 3.2.12 that there are three points on the curve $y = f'(x)$ at which the tangent line is horizontal. Find them.

3.2 PROJECT: A LITER OF COLD WATER

Much of life on earth (as we know it) depends critically on the thermal expansion of water as described in Example 5 of this section. If a given sample of water—we'll suppose that it's a 1-kg sample, although this isn't quite its precise mass—has a volume of exactly 1000 cm^3 when measured at 0°C, then its volume when measured at Celsius temperature T (between 0°C and 30°C) is almost exactly

$$V(T) = 1000 - (0.06247)T + (0.0085053)T^2 - (0.0000679)T^3 \quad (\text{cm}^3). \tag{1}$$

1. Investigate graphically the minimum value V_m of $V(T)$ for $0 \leqq T \leqq 30$, and determine (accurate to two decimal places) the temperature T_m at which this minimal volume occurs. You should find that $T_m \approx 4$°C. Thus water slightly above the freezing point of 0°C is denser than water at the freezing point and therefore sinks to the bottom as a cooling lake freezes. Ice at 0°C (or colder) is less dense than water at 0°C, so such ice floats on the surface. The resulting fact that ice at the surface traps "warm" water at the bottom of a lake—which otherwise might freeze solid—permits the survival of life forms that can withstand cold water but not freezing.
2. Let p denote the last digit of your student I.D. number. Find the temperature between 0°C and 30°C at which the volume of our 1-kg sample of water is $1000 + \frac{1}{3}p$ cm^3.
3. Consider a sample of a hypothetical liquid (not real water) whose volume is given by Eq. (1) on the entire interval $0 \leqq T \leqq 100$. Then investigate graphically the maximum volume of this sample for temperatures between 0°C and 100°C.

3.3 THE CHAIN RULE

We saw in Section 3.2 how to differentiate polynomials and rational functions. But we often need to differentiate *powers* of such functions. For instance, if

$$y = [g(x)]^3, \tag{1}$$

then the extended product rule [Eq. (16) in Section 3.2] yields

$$\begin{aligned}\frac{dy}{dx} &= D_x[g(x) \cdot g(x) \cdot g(x)] \\ &= g'(x) \cdot g(x) \cdot g(x) + g(x) \cdot g'(x) \cdot g(x) + g(x) \cdot g(x) \cdot g'(x),\end{aligned}$$

so collecting terms gives

$$\frac{dy}{dx} = 3[g(x)]^2 \cdot g'(x). \tag{2}$$

Is it a surprise that the derivative of $[g(x)]^3$ is not simply $3[g(x)]^2$, which you might expect in analogy with the (correct) formula $D_x x^3 = 3x^2$? There is an additional factor $g'(x)$ whose origin can be explained by writing $y = [g(x)]^3$ in the form

$$y = u^3 \quad \text{with} \quad u = g(x).$$

Then

$$\frac{dy}{dx} = D_x[g(x)]^3,$$

$$\frac{dy}{du} = 3u^2 = 3[g(x)]^2, \text{ and} \tag{3}$$

$$\frac{du}{dx} = g'(x),$$

so the derivative formula in Eq. (2) takes the form

$$\frac{dy}{dx} = \frac{dy}{du} \cdot \frac{du}{dx}. \tag{4}$$

Equation (4), the **chain rule,** holds for *any* two differentiable functions $y = f(u)$ and $u = g(x)$. The formula in Eq. (2) is simply the special case of Eq. (4) with $f(u) = u^3$.

EXAMPLE 1 If

$$y = (3x + 5)^{17},$$

it would be impractical to write the binomial expansion of the seventeenth power of $3x + 5$ before differentiating. The computer algebra commands in Fig. 3.3.1 yield a polynomial in x having 18 terms, some of which have 15-digit coefficients:

$$(3x + 5)^{17} = 762939453125 + 7781982421875x + \cdots$$
$$+ 186911613281250x^9 + \cdots + 129140163x^{17}.$$

Mathematica	`Expand[(3x+5)^17]`
Maple	`expand((3*x+5)^17)`

Fig. 3.3.1 CAS commands to expand $(3x + 5)^{17}$

(Each ellipsis replaces seven omitted terms.) But if we simply write

$$y = u^{17} \quad \text{with} \quad u = 3x + 5,$$

then

$$\frac{dy}{du} = 17u^{16} \quad \text{and} \quad \frac{du}{dx} = 3.$$

Hence the chain rule yields

$$D_x[(3x + 5)^{17}] = \frac{dy}{dx} = \frac{dy}{du} \cdot \frac{du}{dx} = (17u^{16}) \cdot (3)$$
$$= 17(3x + 5)^{16} \cdot 3 = 51(3x + 5)^{16}.$$

The formula in Eq. (4) is one that, once learned, is unlikely to be forgotten. Although dy/du and du/dx are *not* fractions—they are merely symbols representing the derivatives $f'(u)$ and $g'(x)$—it is much as though they were fractions, with the du in the first factor "canceling" the du in the second. Of course, such "cancellation" no more proves the chain rule than canceling d's proves that

$$\frac{dy}{dx} = \frac{y}{x} \quad \text{(an absurdity).}$$

It is nevertheless an excellent way to remember the chain rule. Such manipulations with differentials are so suggestive (even when invalid) that they played a substantial role in the early development of calculus in the seventeenth and eighteenth centuries. Many formulas were thereby produced that were later proved valid (as were some formulas that were incorrect).

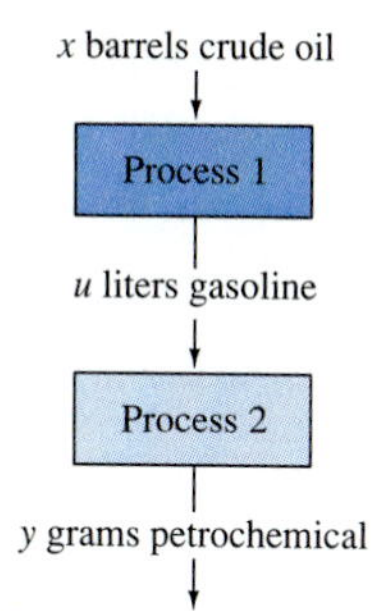

Fig. 3.3.2 The two-process oil refinery (Example 2)

EXAMPLE 2 For a physical interpretation of the chain rule, imagine an oil refinery that first makes u liters of gasoline from x barrels of crude oil. Then, in a second process, the refinery makes y grams of a marketable petrochemical from the u liters of gasoline. (The two processes are illustrated in Fig. 3.3.2.) Then y is a function of u and u is a function of x, so the final output y is a function also of the input x. Consider the *units* in which the *derivatives* of these functions are measured:

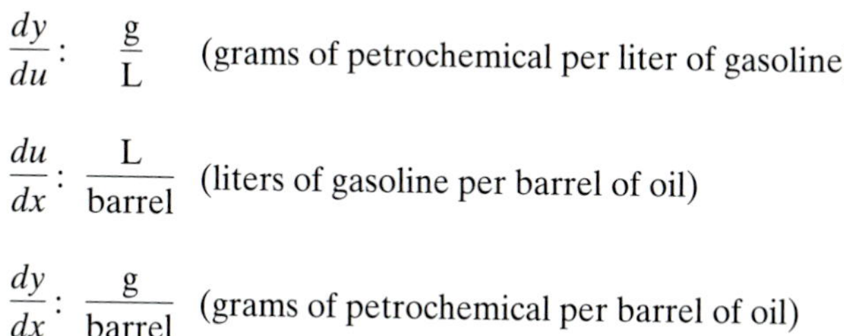

$$\frac{dy}{du}: \quad \frac{\text{g}}{\text{L}} \quad \text{(grams of petrochemical per liter of gasoline)}$$

$$\frac{du}{dx}: \quad \frac{\text{L}}{\text{barrel}} \quad \text{(liters of gasoline per barrel of oil)}$$

$$\frac{dy}{dx}: \quad \frac{\text{g}}{\text{barrel}} \quad \text{(grams of petrochemical per barrel of oil)}$$

When we include the units in the chain rule equation

$$\frac{dy}{dx} = \frac{dy}{du}\frac{du}{dx},$$

we get

$$\frac{dy}{dx}\frac{\text{g}}{\text{barrel}} = \left(\frac{dy}{du}\frac{\text{g}}{\not{\text{L}}}\right)\cdot\left(\frac{du}{dx}\frac{\not{\text{L}}}{\text{barrel}}\right) = \left(\frac{dy}{du}\cdot\frac{du}{dx}\right)\frac{\text{g}}{\text{barrel}}.$$

The handy cancellation of units seems to confirm the validity of the chain rule (at least in this application). For example, if we get 3 g of petrochemical per liter of gasoline and 75 L of gasoline per barrel of oil, how could we fail to get $225 = 3 \cdot 75$ g of petrochemical per barrel of oil? ■

The Chain Rule in Function Notation

Although Eq. (4) is a memorable statement of the chain rule in differential notation, it has the disadvantage of not specifying the values of the variables at which the derivatives are evaluated. This problem can be solved by the use of function notation for the derivatives. Let us write

$$y = f(u), \quad u = g(x), \quad y = h(x) = f(g(x)). \tag{5}$$

Then

$$\frac{du}{dx} = g'(x), \quad \frac{dy}{dx} = h'(x),$$

and

$$\frac{dy}{du} = f'(u) = f'(g(x)). \tag{6}$$

Substitution of these derivatives into Eq. (4) recasts the chain rule in the form

$$h'(x) = f'(g(x)) \cdot g'(x). \tag{7}$$

This version of the chain rule gives the derivative of the *composition* $h = f \circ g$ of two functions f and g in terms of *their* derivatives.

Theorem 1 ***The Chain Rule***

Suppose that g is differentiable at x and that f is differentiable at $g(x)$. Then the composition $h = f \circ g$ defined by $h(x) = f(g(x))$ is differentiable at x, and its derivative is

$$h'(x) = D_x[f(g(x))] = f'(g(x)) \cdot g'(x). \tag{8}$$

Note: Although the derivative of $h = f \circ g$ is a product of the derivatives of f and g, these two derivatives are evaluated at *different* points. The derivative f' is evaluated at $g(x)$, whereas g' is evaluated at x. For a particular number $x = a$, Eq. (7) tells us that

$$h'(a) = f'(b) \cdot g'(a) \qquad \text{where} \quad b = g(a). \tag{9}$$

That is, if $y = f(u)$ and $u = g(x)$, then

$$\left.\frac{dy}{dx}\right|_{x=a} = \left(\left.\frac{dy}{du}\right|_{u=b}\right) \cdot \left(\left.\frac{du}{dx}\right|_{x=a}\right)$$

in differential notation.

EXAMPLE 3 If $h(x) = f(g(x))$ where

$$f(u) = u^5 + u^3 \quad \text{and} \quad g(x) = x^4 + x^2,$$

then

$$f'(u) = 5u^4 + 3u^2 \quad \text{and} \quad g'(x) = 4x^3 + 2x.$$

Therefore $g(1) = 2$, $g'(1) = 6$, and $f'(2) = 5 \cdot 2^4 + 3 \cdot 2^2 = 92$, so

$$h'(1) = f'(g(1)) \cdot g'(1) = f'(2) \cdot g'(1) = 92 \cdot 6 = 552.$$

■

Outline of the Proof of the Chain Rule

To outline a proof of the chain rule, suppose that we are given differentiable functions $y = f(u)$ and $u = g(x)$ and want to compute the derivative

$$\frac{dy}{dx} = \lim_{\Delta x \to 0} \frac{\Delta y}{\Delta x} = \lim_{\Delta x \to 0} \frac{f(g(x + \Delta x)) - f(g(x))}{\Delta x}. \tag{10}$$

The differential form of the chain rule suggests the factorization

$$\frac{\Delta y}{\Delta x} = \frac{\Delta y}{\Delta u} \frac{\Delta u}{\Delta x}. \tag{11}$$

The product law of limits then gives

$$\frac{dy}{dx} = \lim_{\Delta x \to 0} \frac{\Delta y}{\Delta u}\frac{\Delta u}{\Delta x} = \left(\lim_{\Delta u \to 0} \frac{\Delta y}{\Delta u}\right)\left(\lim_{\Delta x \to 0} \frac{\Delta u}{\Delta x}\right) = \frac{dy}{du}\frac{du}{dx}. \tag{12}$$

This will suffice to prove the chain rule *provided that*

$$\Delta u = g(x + \Delta x) - g(x) \tag{13}$$

is a *nonzero* quantity that approaches zero as $\Delta x \to 0$. Certainly $\Delta u \to 0$ as $\Delta x \to 0$, because g is differentiable and therefore continuous. But it is still quite possible that Δu is zero for some—even all—nonzero values of Δx. In such a case, the factorization in Eq. (11) would include the *invalid* step of division by zero. Thus our proof is incomplete. A complete proof of the chain rule is given in Appendix E.

The Generalized Power Rule

If we substitute $g(x) = u$ and $g'(x) = du/dx$ into Eq. (8), we get the hybrid form

$$D_x[f(u)] = f'(u)\frac{du}{dx} \tag{14}$$

of the chain rule that typically is the most useful for purely computational purposes. Recall that the subscript x in D_x specifies that $f(u)$ is being differentiated with respect to x, not with respect to u.

Let us set $f(u) = u^n$ in Eq. (14), where n is an integer. Because $f'(u) = nu^{n-1}$, we thereby obtain

$$D_x u^n = nu^{n-1}\frac{du}{dx}, \tag{15}$$

the *chain rule version* of the power rule. If $u = g(x)$ is a differentiable function, then Eq. (15) implies that

$$D_x[g(x)]^n = n[g(x)]^{n-1} \cdot g'(x). \tag{16}$$

[If $n - 1 < 0$, we must add the proviso that $g(x) \neq 0$ in order for the right-hand side in Eq. (16) to be meaningful.] We refer to this chain rule version of the power rule as the **generalized power rule.**

EXAMPLE 4 To differentiate

$$y = \frac{1}{(2x^3 - x + 7)^2},$$

we first write

$$y = (2x^3 - x + 7)^{-2}$$

in order to apply the generalized power rule, Eq. (16), with $n = -2$. This gives

$$\begin{aligned}\frac{dy}{dx} &= (-2)(2x^3 - x + 7)^{-3}D_x(2x^3 - x + 7)\\ &= (-2)(2x^3 - x + 7)^{-3}(6x^2 - 1) = \frac{2(1 - 6x^2)}{(2x^3 - x + 7)^3}.\end{aligned}$$

EXAMPLE 5 Find the derivative of the function

$$h(z) = \left(\frac{z-1}{z+1}\right)^5.$$

Solution The key to applying the generalized power rule is observing *what* the given function is a power *of.* Here,

$$h(z) = u^5, \quad \text{where} \quad u = \frac{z-1}{z+1},$$

and z, not x, is the independent variable. Hence we apply first Eq. (15) and then the quotient rule to get

$$\begin{aligned} h'(z) = 5u^4\frac{du}{dz} &= 5\left(\frac{z-1}{z+1}\right)^4 D_z\left(\frac{z-1}{z+1}\right) \\ &= 5\left(\frac{z-1}{z+1}\right)^4 \cdot \frac{(1)(z+1)-(z-1)(1)}{(z+1)^2} \\ &= 5\left(\frac{z-1}{z+1}\right)^4 \cdot \frac{2}{(z+1)^2} = \frac{10(z-1)^4}{(z+1)^6}. \end{aligned}$$

■

The importance of the chain rule goes far beyond the power function differentiations illustrated in Examples 1, 4, and 5. We shall learn in later sections how to differentiate exponential, logarithmic, and trigonometric functions. Each time we learn a new differentiation formula—for the derivative $f'(x)$ of a new function $f(x)$—the formula in Eq. (14) immediately provides us with the chain rule version of that formula,

$$D_x f(u) = f'(u) D_x u.$$

The step from the power rule $D_x x^n = nx^{n-1}$ to the generalized power rule $D_x u^n = nu^{n-1} D_x u$ is our first instance of this general phenomenon.

Rate-of-Change Applications

Suppose that the physical or geometric quantity p depends on the quantity q, which in turn depends on time t. Then the *dependent variable* p is a function both of the *intermediate* variable q and of the *independent* variable t. Hence the derivatives that appear in the chain rule formula

$$\frac{dp}{dt} = \frac{dp}{dq}\frac{dq}{dt}$$

are rates of change (as in Section 3.1) of these variables with respect to one another. For instance, suppose that a spherical balloon is being inflated or deflated. Then its volume V and its radius r are changing with time t, and

$$\frac{dV}{dt} = \frac{dV}{dr}\frac{dr}{dt}.$$

Remember that a positive derivative signals an increasing quantity and that a negative derivative signals a decreasing quantity.

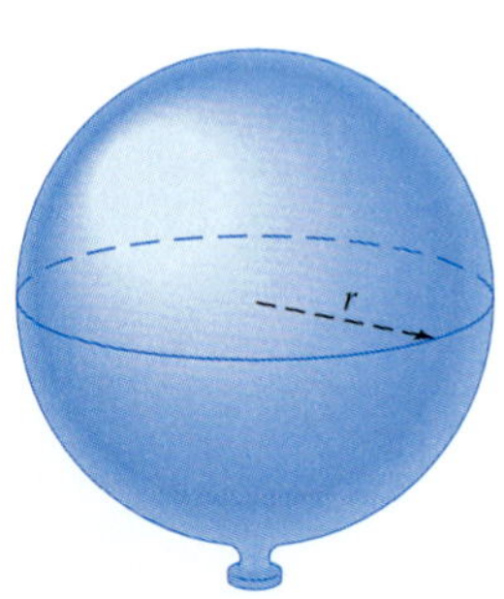

Fig. 3.3.3 The spherical balloon with volume $V = \frac{4}{3}\pi r^3$

EXAMPLE 6 A spherical balloon is being inflated (Fig. 3.3.3). The radius r of the balloon is increasing at the rate of 0.2 cm/s when $r = 5$ cm. At what rate is the volume V of the balloon increasing at that instant?

Solution Given $dr/dt = 0.2$ cm/s when $r = 5$ cm, we want to find dV/dt at that instant. Because the volume of the balloon is

$$V = \frac{4}{3}\pi r^3,$$

we see that $dV/dr = 4\pi r^2$. So the chain rule gives

$$\frac{dV}{dt} = \frac{dV}{dr}\cdot\frac{dr}{dt} = 4\pi r^2\frac{dr}{dt} = 4\pi(5)^2(0.2) \approx 62.83 \qquad (\text{cm}^3/\text{s})$$

at the instant when $r = 5$ cm. ■

In Example 6 we didn't need to know r explicitly as a function of t. But suppose we are told that after t seconds the radius (in centimeters) of an inflating balloon is $r = 3 + (0.2)t$ (until the balloon bursts). Then the volume of this balloon is

$$V = \frac{4}{3}\pi r^3 = \frac{4}{3}\pi\left(3 + \frac{t}{5}\right)^3,$$

so dV/dt is given explicitly as a function of t by

$$\frac{dV}{dt} = \frac{4}{3}\pi(3)\left(3 + \frac{t}{5}\right)^2\left(\frac{1}{5}\right) = \frac{4}{5}\pi\left(3 + \frac{t}{5}\right)^2.$$

EXAMPLE 7 Imagine a spherical raindrop that is falling through water vapor in the air. Suppose that the vapor adheres to the surface of the raindrop in such a way that the time rate of increase of the mass M of the droplet is proportional to the surface area S of the droplet. If the initial radius of the droplet is, in effect, zero and the radius is $r = 1$ mm after 20 s, when is the radius 3 mm?

Solution We are given

$$\frac{dM}{dt} = kS,$$

where k is some constant that depends upon atmospheric conditions. Now

$$M = \frac{4}{3}\pi\rho r^3 \quad \text{and} \quad S = 4\pi r^2,$$

where ρ is the density of water. Hence the chain rule gives

$$4\pi k r^2 = kS = \frac{dM}{dt} = \frac{dM}{dr}\cdot\frac{dr}{dt};$$

that is,

$$4\pi k r^2 = 4\pi\rho r^2\frac{dr}{dt}.$$

This implies that

$$\frac{dr}{dt} = \frac{k}{\rho},$$

a constant. So the radius of the droplet grows at a *constant* rate. Thus if it takes 20 s for r to grow to 1 mm, it will take 1 min for r to grow to 3 mm. ■

3.3 PROBLEMS

Find dy/dx in Problems 1 through 12.

1. $y = (3x + 4)^5$

2. $y = (2 - 5x)^3$

3. $y = \dfrac{1}{3x - 2}$

4. $y = \dfrac{1}{(2x + 1)^3}$

5. $y = (x^2 + 3x + 4)^3$

6. $y = (7 - 2x^3)^{-4}$

7. $y = (2 - x)^4(3 + x)^7$

8. $y = (x + x^2)^5(1 + x^3)^2$

9. $y = \dfrac{x + 2}{(3x - 4)^3}$

10. $y = \dfrac{(1 - x^2)^3}{(4 + 5x + 6x^2)^2}$

11. $y = [1 + (1 + x)^3]^4$

12. $y = [x + (x + x^2)^{-3}]^{-5}$

In Problems 13 through 20, express the derivative dy/dx in terms of x.

13. $y = (u + 1)^3$ and $u = \dfrac{1}{x^2}$

14. $y = \dfrac{1}{2u} - \dfrac{1}{3u^2}$ and $u = 2x + 1$

15. $y = (1 + u^2)^3$ and $u = (4x - 1)^2$

16. $y = u^5$ and $u = \dfrac{1}{3x - 2}$

17. $y = u(1 - u)^3$ and $u = \dfrac{1}{x^4}$

18. $y = \dfrac{u}{u + 1}$ and $u = \dfrac{x}{x + 1}$

19. $y = u^2(u - u^4)^3$ and $u = \dfrac{1}{x^2}$

20. $y = \dfrac{u}{(2u + 1)^4}$ and $u = x - \dfrac{2}{x}$

In Problems 21 through 26, identify a function u of x and an integer $n \neq 1$ such that $f(x) = u^n$. Then compute $f'(x)$.

21. $f(x) = (2x - x^2)^3$

22. $f(x) = \dfrac{1}{2 + 5x^3}$

23. $f(x) = \dfrac{1}{(1 - x^2)^4}$

24. $f(x) = (x^2 - 4x + 1)^3$

25. $f(x) = \left(\dfrac{x + 1}{x - 1}\right)^7$

26. $f(x) = \dfrac{(x^2 + x + 1)^4}{(x + 1)^4}$

Differentiate the functions given in Problems 27 through 36.

27. $g(y) = y + (2y - 3)^5$

28. $h(z) = z^2(z^2 + 4)^3$

29. $F(s) = \left(s - \dfrac{1}{s^2}\right)^3$

30. $G(t) = \left(t^2 + 1 + \dfrac{1}{t}\right)^2$

31. $f(u) = (1 + u)^3(1 + u^2)^4$

32. $g(w) = (w^2 - 3w + 4)(w + 4)^5$

33. $h(v) = \left[v - \left(1 - \dfrac{1}{v}\right)^{-1}\right]^{-2}$

34. $p(t) = \left(\dfrac{1}{t} + \dfrac{1}{t^2} + \dfrac{1}{t^3}\right)^{-4}$

35. $F(z) = \dfrac{1}{(3 - 4z + 5z^5)^{10}}$

36. $G(x) = \{1 + [x + (x^2 + x^3)^4]^5\}^6$

In Problems 37 through 44, dy/dx can be found in two ways—either using the chain rule or not using it. Use both techniques to find dy/dx, then compare the answers. (They should agree!)

37. $y = (x^3)^4 = x^{12}$

38. $y = x = \left(\dfrac{1}{x}\right)^{-1}$

39. $y = (x^2 - 1)^2 = x^4 - 2x^2 + 1$

40. $y = (1 - x)^3 = 1 - 3x + 3x^2 - x^3$

41. $y = (x + 1)^4 = x^4 + 4x^3 + 6x^2 + 4x + 1$

42. $y = (x + 1)^{-2} = \dfrac{1}{x^2 + 2x + 1}$

43. $y = (x^2 + 1)^{-1} = \dfrac{1}{x^2 + 1}$

44. $y = (x^2 + 1)^2 = (x^2 + 1)(x^2 + 1)$

We shall see in Section 3.7 that $D_x[\sin x] = \cos x$ (provided that x is in radian measure). Use this fact and the chain rule to find the derivatives of the functions in Problems 45 through 48.

45. $f(x) = \sin(x^3)$

46. $g(t) = (\sin t)^3$

47. $g(z) = (\sin 2z)^3$

48. $k(u) = \sin(1 + \sin u)$

49. A pebble dropped into a lake creates an expanding circular ripple (Fig. 3.3.4). Suppose that the radius of the circle is increasing at the rate of 2 in./s. At what rate is its area increasing when its radius is 10 in.?

50. The area of a circle is decreasing at the rate of 2π cm^2/s. At what rate is the radius of the circle decreasing when its area is 75π cm^2?

51. Each edge x of a square is increasing at the rate of 2 in./s. At what rate is the area A of the square increasing when each edge is 10 in.?

52. Each edge of an equilateral triangle is increasing at 2 cm/s (Fig. 3.3.5). At what rate is the area of the triangle increasing when each edge is 10 cm?

53. A cubical block of ice is melting in such a way that each edge decreases steadily by 2 in. every hour. At what rate is its volume decreasing when each edge is 10 in. long?

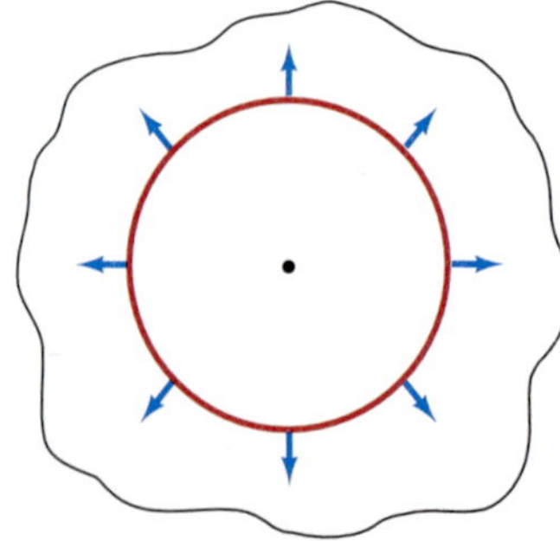

Fig. 3.3.4 Expanding circular ripple in a lake (Problem 49)

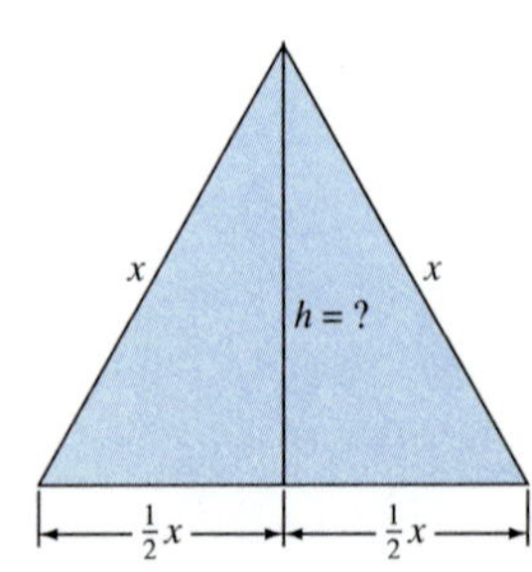

Fig. 3.3.5 The triangle of Problem 52 with area $A = \frac{1}{2}xh$

54. Find $f'(-1)$, given $f(y) = h(g(y))$, $h(2) = 55$, $g(-1) = 2$, $h'(2) = -1$, and $g'(-1) = 7$.

55. Given: $G(t) = f(h(t))$, $h(1) = 4$, $f'(4) = 3$, and $h'(1) = -6$. Find $G'(1)$.

56. Suppose that $f(0) = 0$ and that $f'(0) = 1$. Calculate the derivative of $f(f(f(x)))$ at $x = 0$.

57. Air is being pumped into a spherical balloon in such a way that its radius r is increasing at the rate of $dr/dt = 1$ cm/s. What is the time rate of increase, in cubic centimeters per second, of the balloon's volume when $r = 10$ cm?

58. Suppose that the air is being pumped into the balloon of Problem 57 at the constant rate of 200π cm^3/s. What is the time rate of increase of the radius r when $r = 5$ cm?

59. Air is escaping from a spherical balloon at the constant rate of 300π cm^3/s. What is the radius of the balloon when its radius is decreasing at the rate of 3 cm/s?

60. A spherical hailstone is losing mass by melting uniformly over its surface as it falls. At a certain time, its radius is 2 cm and its volume is decreasing at the rate of 0.1 cm^3/s. How fast is its radius decreasing at that time?

61. A spherical snowball is melting in such a way that the rate of decrease of its volume is proportional to its surface area. At 10 A.M. its volume is 500 in.^{3} and at 11 A.M. its volume is 250 in.^{3}. When does the snowball finish melting? (See Example 7.)

62. A cubical block of ice with edges 20 in. long begins to melt at 8 A.M. Each edge decreases at a constant rate thereafter and each is 8 in. long at 4 P.M. What was the rate of change of the block's volume at noon?

63. Suppose that u is a function of v, that v is a function of w, that w is a function of x, and that all these functions are differentiable. Explain why it follows from the chain rule that

$$\frac{du}{dx} = \frac{du}{dv} \cdot \frac{dv}{dw} \cdot \frac{dw}{dx}.$$

64. Let f be a differentiable function such that $f(1) = 1$. If $F(x) = f(x^n)$ and $G(x) = [f(x)]^n$ (where n is a fixed integer), show that $F(1) = G(1)$ and that $F'(1) = G'(1)$.

3.4 DERIVATIVES OF ALGEBRAIC FUNCTIONS

We saw in Section 3.3 that the chain rule yields the differentiation formula

$$D_x\, u^n = nu^{n-1}\frac{du}{dx} \tag{1}$$

if $u = f(x)$ is a differentiable function and the exponent n is an integer. We shall see in Theorem 1 of this section that this **generalized power rule** holds not only when the exponent is an integer, but also when it is a rational number $r = p/q$ (where p and q are integers and $q \neq 0$). Recall that rational powers are defined in terms of integral roots and powers as follows:

$$u^{p/q} = \sqrt[q]{u^p} = \left(\sqrt[q]{u}\right)^p.$$

We first consider the case of a rational power of the independent variable x:

$$y = x^{p/q}, \tag{2}$$

where p and q are integers with q positive. We show independently in Section 7.4 that $g(x) = x^{p/q}$ is differentiable wherever its derivative does not involve division by zero or an even root of a negative number. Assuming this fact, let us take the qth power of each side in Eq. (2) to obtain

$$y^q = x^p \tag{3}$$

[because $(x^{p/q})^q = x^p$]. Note that Eq. (3) is an identity—the functions y^q and x^p of x are identical where defined. Therefore their derivatives with respect to x must also be identical. That is,

$$D_x(y^q) = D_x(x^p);$$

$$qy^{q-1}\frac{dy}{dx} = px^{p-1}.$$

To differentiate the left-hand side, we used Eq. (1) with $u = y$ and $n = q$. Finally, we solve for the derivative

$$\frac{dy}{dx} = \frac{px^{p-1}}{qy^{q-1}} = \frac{p}{q}x^{p-1}y^{1-q} = \frac{p}{q}x^{p-1}(x^{p/q})^{1-q}$$
$$= \frac{p}{q}x^{p-1}x^{p/q}x^{-p} = \frac{p}{q}x^{(p/q)-1}.$$

Thus we have shown that

$$D_x\left[x^{p/q}\right] = \frac{p}{q}x^{(p/q)-1};$$

that is, that the power rule

$$D_x\, x^r = rx^{r-1} \tag{4}$$

holds if the exponent $r = p/q$ is a rational number (subject to the provisos previously mentioned).

Using Eq. (4), we can differentiate a simple "radical" (or "root") function by first rewriting it as a power with a fractional exponent.

EXAMPLE 1

(a) $D_x[\sqrt{x}] = D_x\left[x^{1/2}\right] = \dfrac{1}{2}x^{-1/2} = \dfrac{1}{2\sqrt{x}}.$

(b) If $y = \sqrt{x^3} = x^{3/2}$, then $\dfrac{dy}{dx} = \dfrac{3}{2}x^{1/2} = \dfrac{3}{2}\sqrt{x}.$

(c) If $g(t) = \dfrac{1}{\sqrt[3]{t^2}} = t^{-2/3}$, then $g'(t) = -\dfrac{2}{3}t^{-5/3} = -\dfrac{2}{3\sqrt[3]{t^5}}.$ ■

REMARK In parts (a) and (b) of Example 1 it is necessary that $x \geqq 0$ in order that $\sqrt{x}$ be defined. In part (a) it is moreover necessary that $x \neq 0$; if $x = 0$ then the formula

$$D_x[\sqrt{x}] = \frac{1}{2\sqrt{x}}$$

would involve division by zero. Figure 3.4.1 shows the graphs of the function $f(x) = \sqrt{x}$ and its derivative $f'(x) = 1/(2\sqrt{x})$ for $x > 0$. Note that $f'(x) \to \infty$ as $x \to 0^+$, further emphasizing the fact that $f(x) = \sqrt{x}$ is not differentiable at $x = 0$.

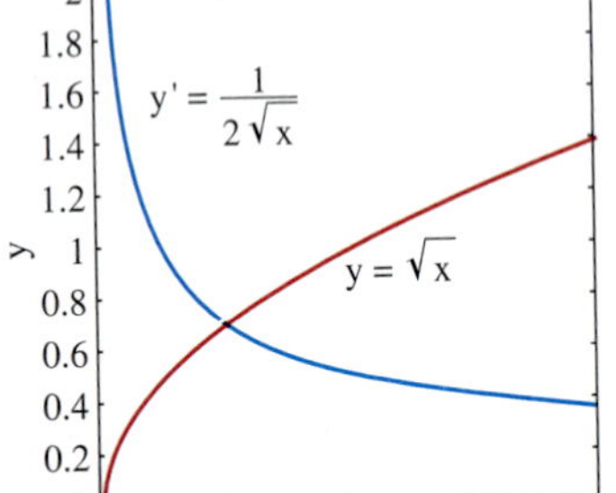

Fig. 3.4.1 The graphs of $f(x) = \sqrt{x}$ and $f'(x) = \dfrac{1}{2\sqrt{x}}$

The Generalized Power Rule

For the more general form of the power rule, let

$$y = u^r$$

where u is a differentiable function of x and $r = p/q$ is rational. Then

$$\frac{dy}{du} = ru^{r-1}$$

by Eq. (4), so the chain rule gives

$$\frac{dy}{dx} = \frac{dy}{du}\frac{du}{dx} = ru^{r-1}\frac{du}{dx}.$$

Thus

$$D_x\, u^r = ru^{r-1}\frac{du}{dx}, \tag{5}$$

which is the generalized power rule for rational exponents.

Theorem 1 ***Generalized Power Rule***

If r is a rational number, then

$$D_x[f(x)]^r = r[f(x)]^{r-1}\cdot f'(x) \tag{6}$$

wherever the function f is differentiable and the right-hand side is defined.

For the right-hand side in Eq. (6) to be "defined" means that $f'(x)$ exists, there is no division by zero, and no even root of a negative number appears.

EXAMPLE 2

$$D_x\left[\sqrt{4 - x^2}\right] = D_x\left[(4 - x^2)^{1/2}\right] = \tfrac{1}{2}(4 - x^2)^{-1/2}\cdot D_x(4 - x^2)$$
$$= \tfrac{1}{2}(4 - x^2)^{-1/2}\cdot(-2x);$$

$$D_x\left[\sqrt{4 - x^2}\right] = -\frac{x}{\sqrt{4 - x^2}} \tag{7}$$

except where $x = \pm 2$ (division by zero) or where $|x| > 2$ (square root of a negative number). Thus Eq. (7) holds if $-2 < x < 2$. In writing derivatives of algebraic functions, we ordinarily omit such disclaimers unless they are pertinent to some specific purpose at hand. But note in Fig. 3.4.2 that if $f(x) = \sqrt{4 - x^2}$, then $f'(x) \to +\infty$ as $x \to -2^+$ and $f'(x) \to -\infty$ as $x \to +2^-$. ■

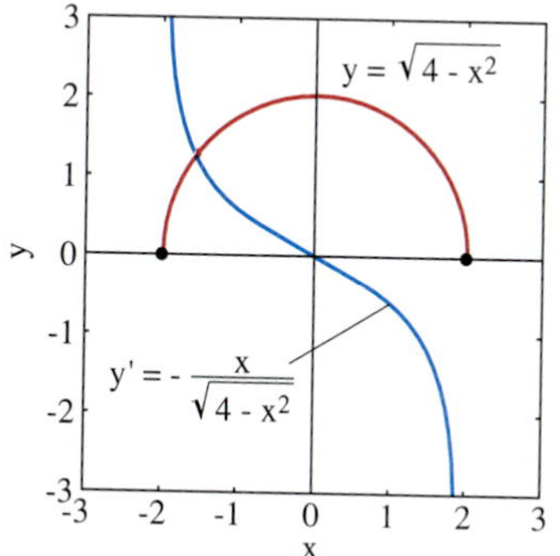

Fig. 3.4.2 The graphs of $f(x) = \sqrt{4 - x^2}$ and $f'(x) = \dfrac{-x}{\sqrt{4 - x^2}}$

A template for the application of the generalized power rule is

$$D_x([***]^n) = n[***]^{n-1}D_x[***],$$

where $***$ represents a function of x and (as we now know) n can be either an integer or a fraction (a quotient of integers).

But to differentiate a *power of a function,* we must first recognize *what function* it is a power *of.* So to differentiate a function involving roots (or radicals), we first "prepare" it for an application of the generalized power rule by rewriting it as a power function with fractional exponent. Examples 3 through 6 illustrate this technique.

EXAMPLE 3 If $y = 5\sqrt{x^3} - \dfrac{2}{\sqrt[3]{x}}$, then

$$y = 5x^{3/2} - 2x^{-1/3}, \quad \text{so}$$

$$\frac{dy}{dx} = 5\cdot\left(\frac{3}{2}x^{1/2}\right) - 2\cdot\left(-\frac{1}{3}x^{-4/3}\right)$$

$$= \frac{15}{2}x^{1/2} + \frac{2}{3}x^{-4/3} = \frac{15}{2}\sqrt{x} + \frac{2}{3\sqrt[3]{x^4}}.$$ ■

EXAMPLE 4 With $f(x) = 3 - 5x$ and $r = 7$, the generalized power rule yields

$$\begin{aligned} D_x[(3-5x)^7] &= 7(3-5x)^6 D_x(3-5x) \\ &= 7(3-5x)^6(-5) = -35(3-5x)^6. \end{aligned}$$ ■

EXAMPLE 5 With $f(x) = 2x^2 - 3x + 5$ and $r = \frac{1}{2}$, the generalized power rule yields

$$\begin{aligned} D_x\sqrt{2x^2-3x+5} &= D_x(2x^2-3x+5)^{1/2} \\ &= \frac{1}{2}(2x^2-3x+5)^{-1/2}D_x(2x^2-3x+5) \\ &= \frac{4x-3}{2\sqrt{2x^2-3x+5}}. \end{aligned}$$ ■

EXAMPLE 6 If

$$x = \left[5t + \sqrt[3]{(3t-1)^4}\right]^{10},$$

then Eq. (5) with $u = 5t + (3t-1)^{4/3}$ and with independent variable t gives

$$\begin{aligned} \frac{dx}{dt} &= 10u^9\frac{du}{dt} \\ &= 10\left[5t + (3t-1)^{4/3}\right]^9 D_t\left[5t + (3t-1)^{4/3}\right] \\ &= 10\left[5t + (3t-1)^{4/3}\right]^9 \left[D_t(5t) + D_t(3t-1)^{4/3}\right] \\ &= 10\left[5t + (3t-1)^{4/3}\right]^9 \left[5 + \frac{4}{3}(3t-1)^{1/3}\cdot 3\right]; \\ \frac{dx}{dt} &= 10\left[5t + (3t-1)^{4/3}\right]^9 \left[5 + 4(3t-1)^{1/3}\right]. \end{aligned}$$ ■

Example 6 illustrates the fact that we apply the chain rule (or generalized power rule) by working from the *outside* to the *inside*. At each step the derivative of the outside function is multiplied by the derivative of the inside function. We continue until no "inside function" remains undifferentiated. Does the process remind you of peeling an onion, one layer at a time, until its core is reached?

Differentiability and Vertical Tangent Lines

Whereas polynomials and rational functions are both continuous and differentiable wherever they are defined, simple algebraic functions can be continuous at points where their derivatives do not exist.

EXAMPLE 7 If

$$f(x) = |x| = \sqrt{x^2}$$

denotes the absolute value function, then for $x \neq 0$ we find that

$$f'(x) = D_x[(x^2)^{1/2}] = \frac{1}{2}(x^2)^{-1/2}(2x) = \frac{x}{\sqrt{x^2}} = \frac{x}{|x|} = \begin{cases} -1 & \text{if } x < 0, \\ +1 & \text{if } x > 0. \end{cases}$$

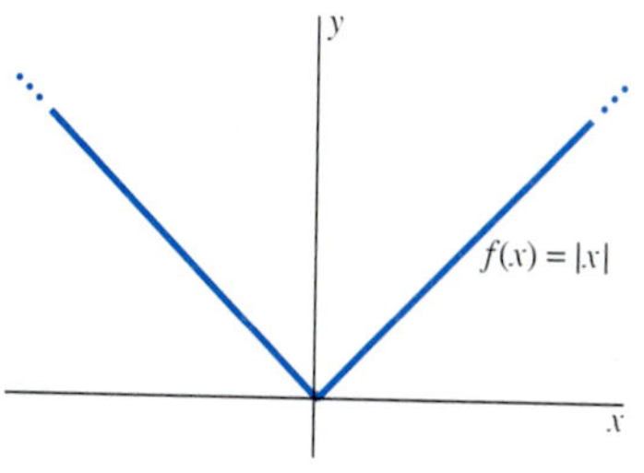

Fig. 3.4.3 The graph of $f(x) = |x|$

Thus f is differentiable at every point except possibly for the origin $x = 0$. In fact, the graph of $f(x) = |x|$ in Fig. 3.4.3 makes it clear that the difference quotient

$$\frac{f(x) - f(0)}{x - 0} = \frac{|x|}{x}$$

has left-hand limit -1 and right-hand limit $+1$ at $x = 0$. Thus the absolute value function is not differentiable at the isolated point $x = 0$, where the graph $y = |x|$ has a "corner point" rather than a tangent line. (Can you think of a continuous function whose graph has *infinitely many* such corner points?) ■

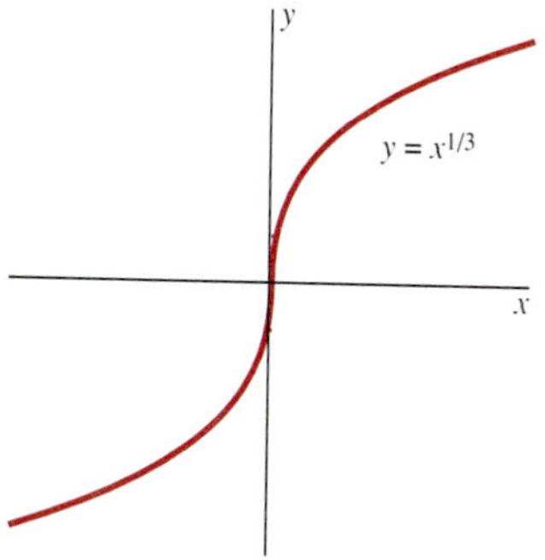

Fig. 3.4.4 The graph of the cube root function

EXAMPLE 8 Figure 3.4.4 shows the graph of the cube-root function

$$y = \sqrt[3]{x} = x^{1/3}$$

and illustrates another way in which a function can fail to be differentiable at an isolated point. Its derivative,

$$\frac{dy}{dx} = \frac{1}{3}x^{-2/3} = \frac{1}{3\sqrt[3]{x^2}},$$

increases without bound as $x \to 0$ but does not exist at $x = 0$. Therefore, the definition of tangent line does not apply to this graph at $(0, 0)$. Nevertheless, from the figure it seems appropriate to regard the vertical line $x = 0$ as the line tangent to the curve $y = x^{1/3}$ at the point $(0, 0)$. ■

Definition ***Vertical Tangent Line***

The curve $y = f(x)$ has a **vertical tangent line** at the point $(a, f(a))$ provided that f is continuous at a and

$$|f'(x)| \to +\infty \quad \text{as} \quad x \to a. \tag{8}$$

The requirement that f be continuous at $x = a$ implies that $f(a)$ must be defined. Thus it would be pointless to ask about a line (vertical or not) tangent to the curve $y = 1/x$ where $x = 0$.

If f is defined (and differentiable) on only one side of $x = a$, we mean in Eq. (8) that $|f'(x)| \to +\infty$ as x approaches a from that side.

EXAMPLE 9 Find the points on the curve

$$y = f(x) = x\sqrt{1 - x^2}, \quad -1 \leqq x \leqq 1,$$

at which the tangent line is either horizontal or vertical.

Solution We differentiate using first the product rule and then the chain rule:

$$\begin{aligned} f'(x) &= (1 - x^2)^{1/2} + \frac{x}{2}(1 - x^2)^{-1/2}(-2x) \\ &= (1 - x^2)^{-1/2}[(1 - x^2) - x^2] = \frac{1 - 2x^2}{\sqrt{1 - x^2}}. \end{aligned}$$

Now $f'(x) = 0$ only when the numerator $1 - 2x^2$ is zero—that is, when $x = \pm 1/\sqrt{2}$. Because $f(\pm 1/\sqrt{2}) = \pm 1/2$, the curve has a horizontal tangent line at each of the two points $(1/\sqrt{2}, 1/2)$ and $(-1/\sqrt{2}, -1/2)$.

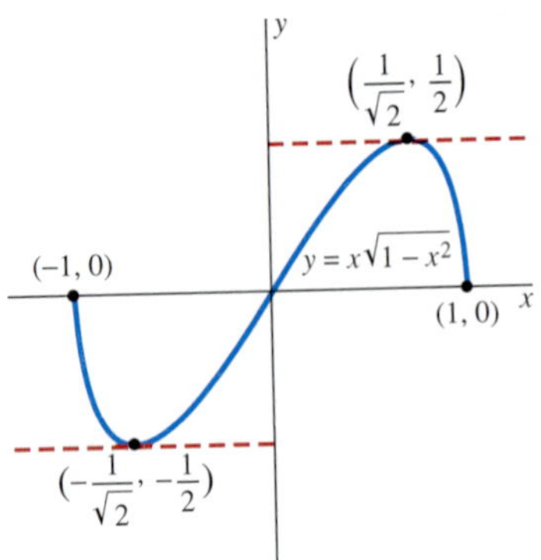

Fig. 3.4.5 The graph of $f(x) = x\sqrt{1-x^2}$, $-1 \leqq x \leqq 1$ (Example 9)

We also observe that the denominator $\sqrt{1-x^2}$ approaches zero as $x \to -1^+$ and as $x \to +1^-$. Because $f(\pm 1) = 0$, we see that the curve has a vertical tangent line at each of the two points $(1, 0)$ and $(-1, 0)$. The graph of f is shown in Fig. 3.4.5. ■

EXAMPLE 10 Figure 3.4.6 shows the graph of the function $f(x) = 1 - \sqrt[5]{x^2}$, which appears to have a sharp "cusp" (rather than a corner) at the point $(0, 1)$. Because the absolute value of the derivative $f'(x) = -\frac{2}{5}x^{-3/5}$ approaches $+\infty$ as $x \to 0$, the curve $y = f(x)$ has a vertical tangent at that point. ■

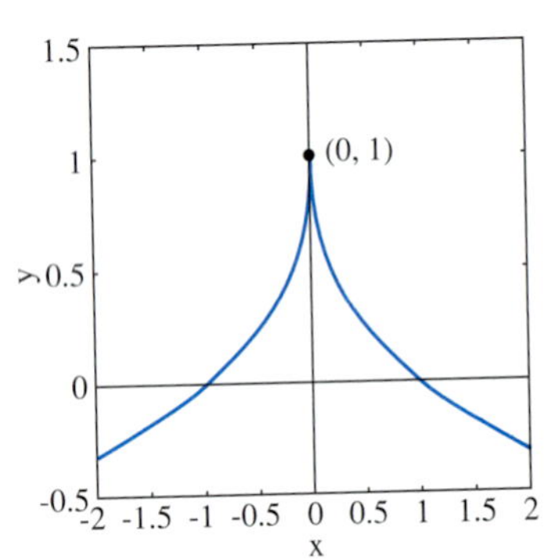

Fig. 3.4.6 The graph of $y = 1 - \sqrt[5]{x^2}$ with a cusp at $(0, 1)$

Whereas the preceding examples show that a function can be continuous without being differentiable, the following theorem says that a function *is* continuous wherever it is differentiable. Thus differentiability of a function is a stronger condition than continuity alone.

Theorem 2 ***Differentiability Implies Continuity***

Suppose that the function f is defined in a neighborhood of a. If f is differentiable at a, then f is continuous at a.

PROOF Because $f'(a)$ exists, the product law for limits yields

$$\begin{aligned}\lim_{x\to a}[f(x)-f(a)] &= \lim_{x\to a}(x-a)\cdot\frac{f(x)-f(a)}{x-a}\\ &= \left(\lim_{x\to a}(x-a)\right)\left(\lim_{x\to a}\frac{f(x)-f(a)}{x-a}\right)\\ &= 0\cdot f'(a) = 0.\end{aligned}$$

Thus $\lim_{x\to a} f(x) = f(a)$, so f is continuous at a. ■

3.4 PROBLEMS

Differentiate the functions given in Problems 1 through 44.

1. $f(x) = 4\sqrt{x^5} + \dfrac{2}{\sqrt{x}}$

2. $g(t) = 9\sqrt[3]{t^4} - \dfrac{3}{\sqrt[3]{t}}$

3. $f(x) = \sqrt{2x+1}$

4. $h(z) = \dfrac{1}{\sqrt[3]{7-6z}}$

5. $f(x) = \dfrac{6-x^2}{\sqrt{x}}$

6. $\phi(u) = \dfrac{7+2u-3u^4}{\sqrt[3]{u^2}}$

7. $f(x) = (2x+3)^{3/2}$

8. $g(x) = (3x+4)^{4/3}$

9. $f(x) = (3-2x^2)^{-3/2}$

10. $f(y) = (4-3y^3)^{-2/3}$

11. $f(x) = \sqrt{x^3+1}$

12. $g(z) = \dfrac{1}{(z^4+3)^2}$

13. $f(x) = \sqrt{2x^2+1}$

14. $f(t) = \dfrac{t}{\sqrt{1+t^4}}$

15. $f(t) = \sqrt{2t^3}$

16. $g(t) = \sqrt{\dfrac{1}{3t^5}}$

17. $f(x) = (2x^2 - x + 7)^{3/2}$

18. $g(z) = (3z^2 - 4)^{97}$

19. $g(x) = \dfrac{1}{(x-2x^3)^{4/3}}$

20. $f(t) = [t^2 + (1+t)^4]^5$

21. $f(x) = x\sqrt{1-x^2}$

22. $g(x) = \sqrt{\dfrac{2x+1}{x-1}}$

23. $f(t) = \sqrt{\dfrac{t^2+1}{t^2-1}}$

24. $h(y) = \left(\dfrac{y+1}{y-1}\right)^{17}$

25. $f(x) = \left(x - \dfrac{1}{x}\right)^3$

26. $g(z) = \dfrac{z^2}{\sqrt{1+z^2}}$

27. $f(v) = \dfrac{\sqrt{v+1}}{v}$

28. $h(x) = \left(\dfrac{x}{1+x^2}\right)^{5/3}$

29. $f(x) = \sqrt[3]{1-x^2}$

30. $g(x) = \sqrt{x+\sqrt{x}}$

31. $f(x) = x(3-4x)^{1/2}$

32. $g(t) = \dfrac{t-(1+t^2)^{1/2}}{t^2}$

33. $f(x) = (1-x^2)(2x+4)^{4/3}$

34. $f(x) = (1-x)^{1/2}(2-x)^{1/3}$

35. $g(t) = \left(1 + \dfrac{1}{t}\right)^2 (3t^2+1)^{1/2}$

36. $f(x) = x(1+2x+3x^2)^{10}$

37. $f(x) = \dfrac{2x-1}{(3x+4)^5}$

38. $h(z) = (z - 1)^4(z + 1)^6$

39. $f(x) = \dfrac{(2x + 1)^{1/2}}{(3x + 4)^{1/3}}$

40. $f(x) = (1 - 3x^4)^5(4 - x)^{1/3}$

41. $h(y) = \dfrac{\sqrt{1 + y} + \sqrt{1 - y}}{\sqrt[3]{y^5}}$

42. $f(x) = \sqrt{1 - \sqrt[3]{x}}$

43. $g(t) = \sqrt{t + \sqrt{t + \sqrt{t}}}$

44. $f(x) = x^3\sqrt{1 - \dfrac{1}{x^2 + 1}}$

For each curve given in Problems 45 through 50, find all points on the graph where the tangent line is either horizontal or vertical.

45. $y = x^{2/3}$

46. $y = x\sqrt{4 - x^2}$

47. $y = x^{1/2} - x^{3/2}$

48. $y = \dfrac{1}{\sqrt{9 - x^2}}$

49. $y = \dfrac{x}{\sqrt{1 - x^2}}$

50. $y = \sqrt{(1 - x^2)(4 - x^2)}$

In Problems 51 through 56, first write an equation of the line tangent to the given curve $y = f(x)$ at the indicated point P. Then illustrate your result with a graphing calculator or computer by graphing both the curve and the tangent line on the same screen.

51. $y = 2\sqrt{x}$, at the point P where $x = 4$

52. $y = 3\sqrt[3]{x}$, at the point P where $x = 8$

53. $y = 3\sqrt[3]{x^2}$, at the point P where $x = -1$

54. $y = 2\sqrt{1 - x}$, at the point P where $x = \frac{3}{4}$

55. $y = x\sqrt{4 - x}$, at the point P where $x = 0$

56. $y = (1 - x)\sqrt{x}$, at the point P where $x = 4$

In Problems 57 through 62 (Figs. 3.4.7 through 3.4.12), match the graph $y = f(x)$ of a function with the graph $y = f'(x)$ of its derivative among those shown in Figs. 3.4.13(a) through 3.4.13(f).

57. Figure 3.4.7

58. Figure 3.4.8

59. Figure 3.4.9

60. Figure 3.4.10

61. Figure 3.4.11

62. Figure 3.4.12

63. The period of oscillation P (in seconds) of a simple pendulum of length L (in feet) is given by $P = 2\pi\sqrt{L/g}$, where $g = 32$ ft/s^2. Find the rate of change of P with respect to L when $P = 2$.

64. Find the rate of change of the volume $V = \frac{4}{3}\pi r^3$ of a sphere of radius r with respect to its surface area $A = 4\pi r^2$ when $r = 10$.

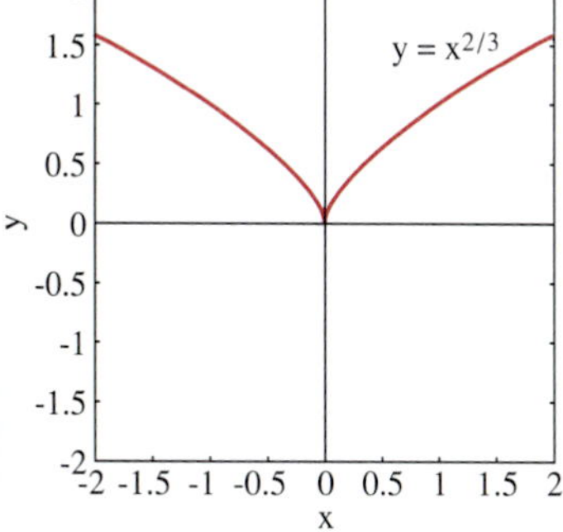

Fig. 3.4.7 $y = x^{2/3}$ (Problem 57)

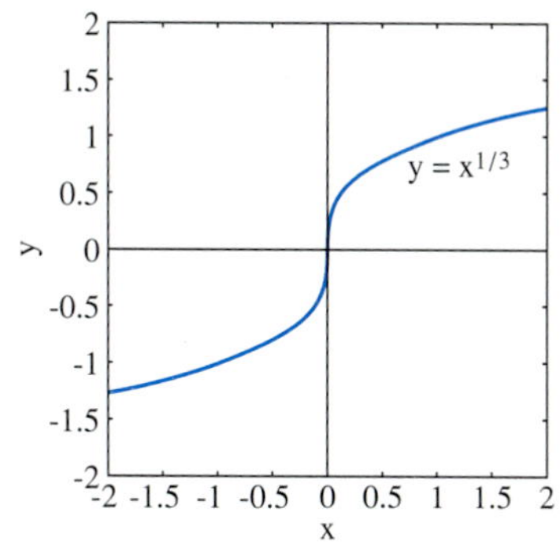

Fig. 3.4.8 $y = x^{1/3}$ (Problem 58)

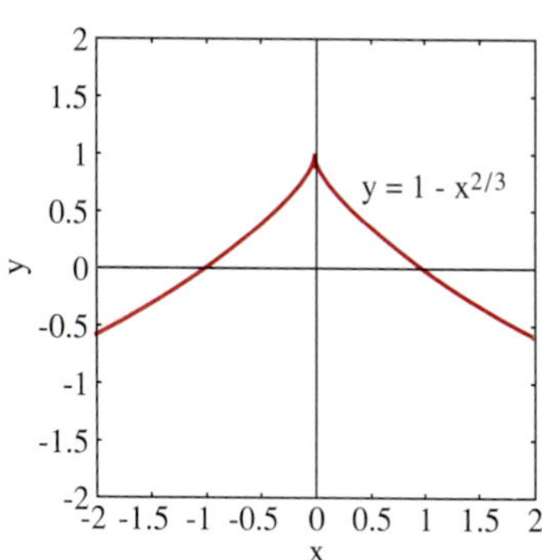

Fig. 3.4.9 $y = 1 - x^{2/3}$ (Problem 59)

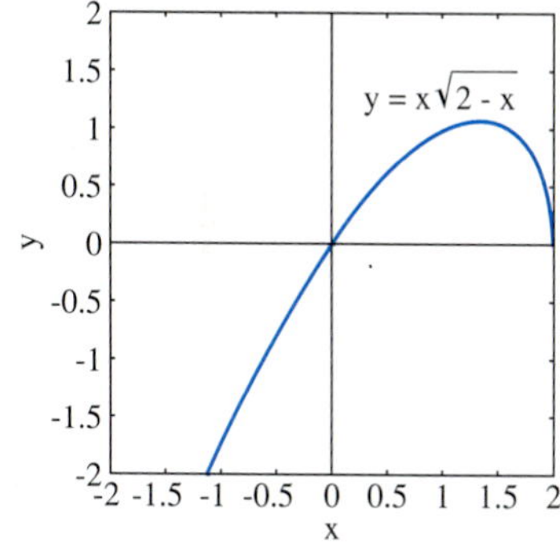

Fig. 3.4.10 $y = x\sqrt{2 - x}$ (Problem 60)

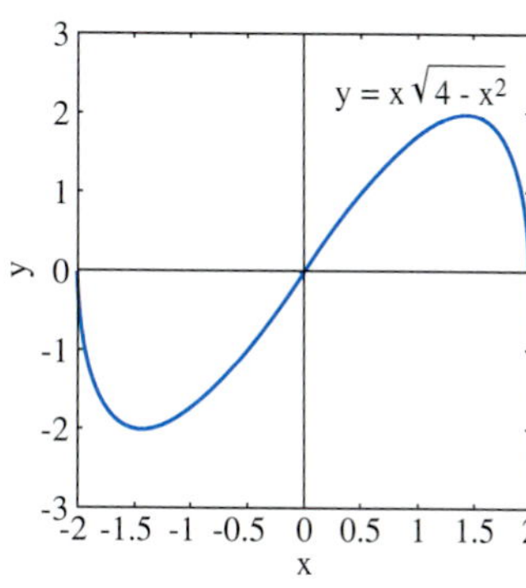

Fig. 3.4.11 $y = x\sqrt{4 - x^2}$ (Problem 61)

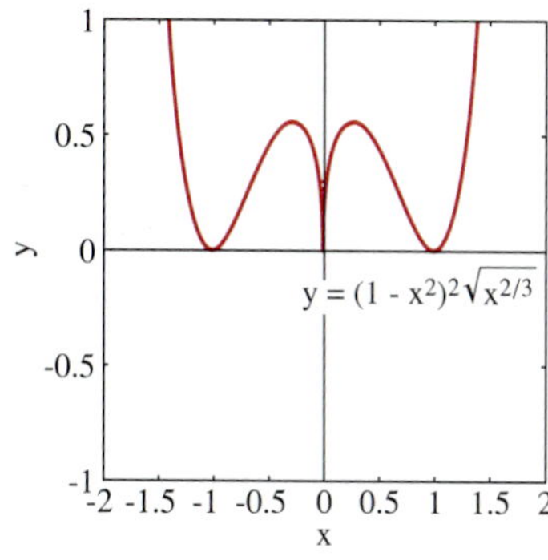

Fig. 3.4.12 $y = (1 - x^2)^2\sqrt{x^{2/3}}$ (Problem 62)

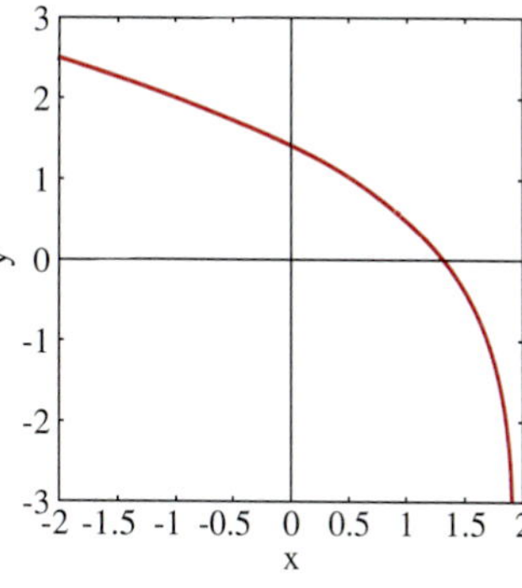

Fig. 3.4.13 (a)

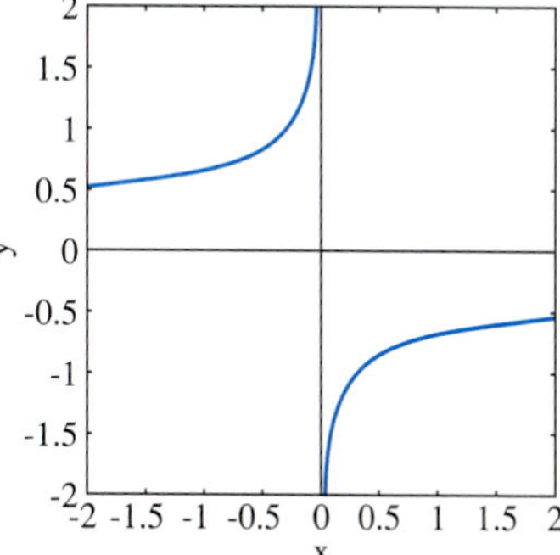

Fig. 3.4.13 (b)

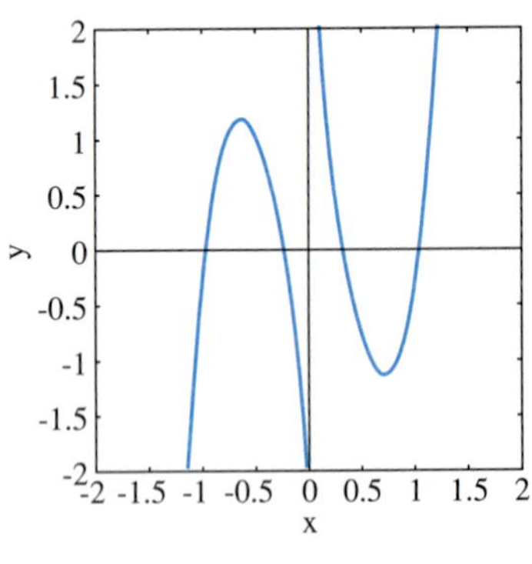

Fig. 3.4.13 (c)

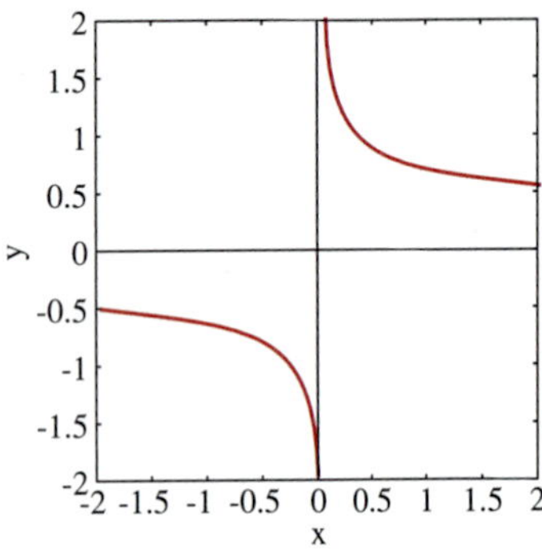

Fig. 3.4.13 (d)

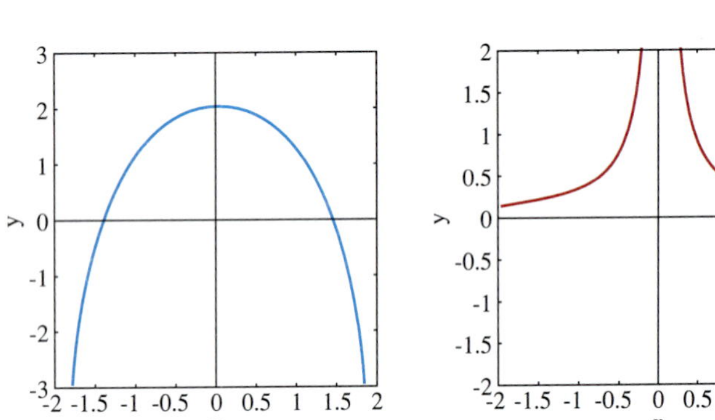

Fig. 3.4.13 (e)

Fig. 3.4.13 (f)

65. Find the two points on the circle $x^2 + y^2 = 1$ at which the slope of the tangent line is -2 (Fig. 3.4.14).

66. Find the two points on the circle $x^2 + y^2 = 1$ at which the slope of the tangent line is 3.

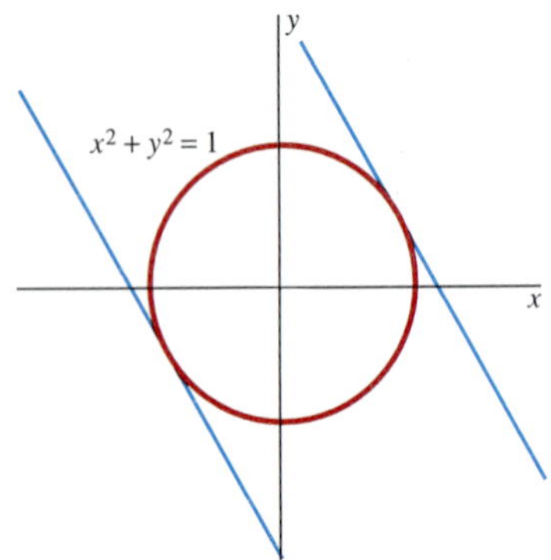

Fig. 3.4.14 The two tangent lines of Problem 65

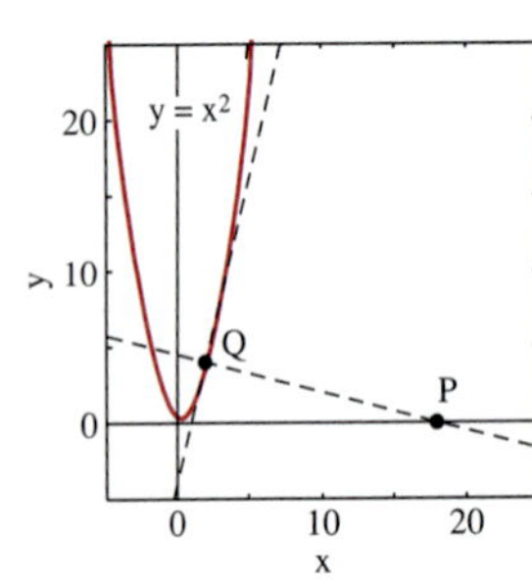

Fig. 3.4.15 The tangent and normal of Problem 67

67. Find a line through the point $P(18, 0)$ that is normal to the tangent line to the parabola $y = x^2$ at some point $Q(a, a^2)$ (see Fig. 3.4.15). (*Suggestion:* You will obtain a cubic equation in the unknown a. Find by inspection a small integral root r. The cubic polynomial is then the product of $a - r$ and a quadratic polynomial; you can find the latter by division of $a - r$ into the cubic.)

68. Find three distinct lines through the point $P(3, 10)$ that are normal to the parabola $y = x^2$ (Fig. 3.4.16). (See the suggestion for Problem 67. This problem will require a certain amount of calculator-aided computation.)

69. Find two distinct lines through the point $P(0, \frac{5}{2})$ that are normal to the curve $y = x^{2/3}$ (Fig. 3.4.17).

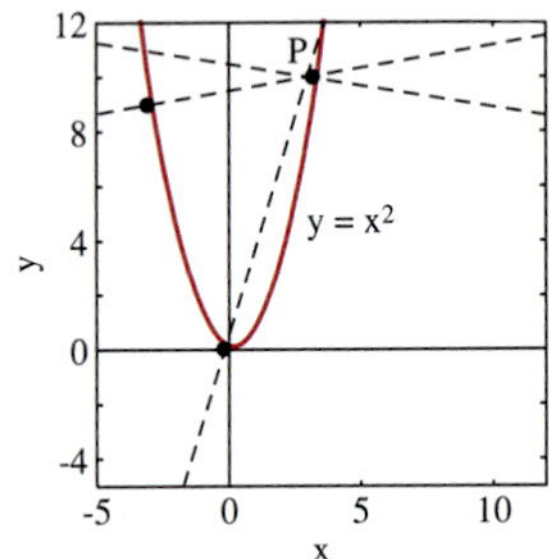

Fig. 3.4.16 The three normal lines of Problem 68

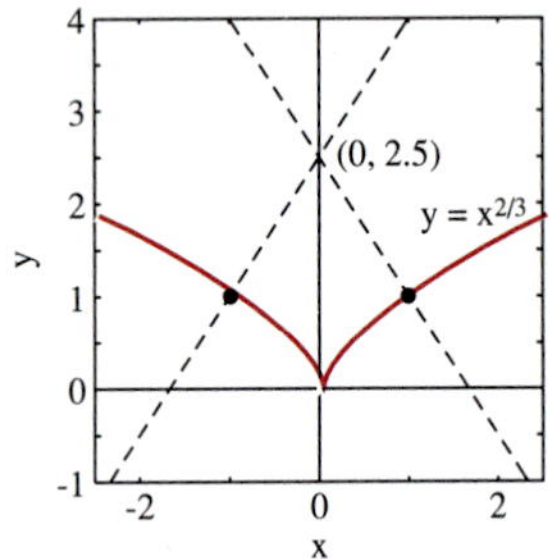

Fig. 3.4.17 The two normal lines of Problem 69

70. Verify that the line tangent to the circle $x^2 + y^2 = a^2$ at the point P is perpendicular to the radius OP (Fig. 3.4.18).

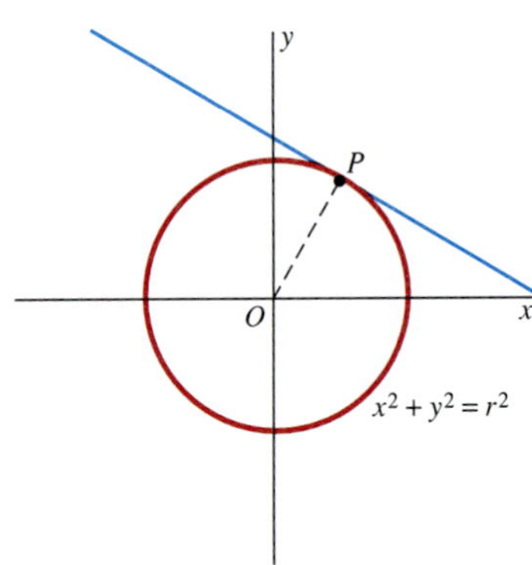

Fig. 3.4.18 The circle, radius, and tangent line of Problem 70

71. Consider the cubic equation $x^3 = 3x + 8$. If we differentiate each side with respect to x, we obtain $3x^2 = 3$, which has the two solutions $x = 1$ and $x = -1$. But neither of these is a solution of the original cubic equation. What went wrong? After all, in several examples and theorems of this section, we *appeared* to differentiate both sides of an equation. Explain carefully why differentiation of both sides in Eq. (3) is valid and why the differentiation in this problem is not.

3.5 MAXIMA AND MINIMA OF FUNCTIONS ON CLOSED INTERVALS

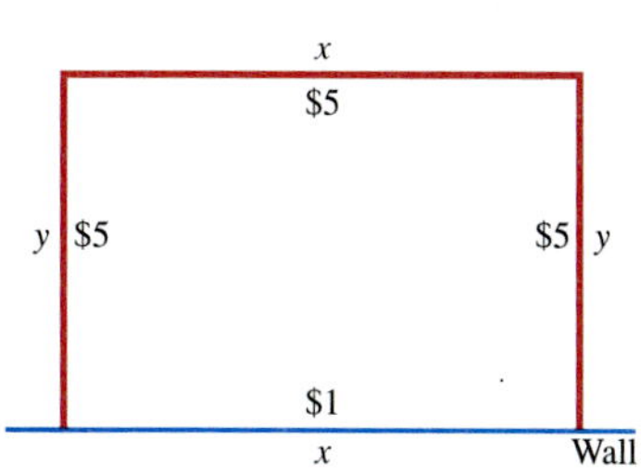

Fig. 3.5.1 The animal pen

In applications we often need to find the maximum (largest) or minimum (smallest) value that a specified quantity can attain. The animal pen problem posed in Section 1.1 is a simple yet typical example of an applied maximum-minimum problem. There we investigated the animal pen shown in Fig. 3.5.1, with the indicated dollar-per-foot cost figures for its four sides. We showed that if $180 is allocated for material to construct this pen, then its area $A = f(x)$ is given as a function of its base length x by

$$f(x) = \tfrac{3}{5}x(30 - x), \qquad 0 \leqq x \leqq 30. \tag{1}$$

Hence the question of the largest possible area of the animal pen is equivalent to the purely mathematical problem of finding the maximum value attained by the function $f(x) = \frac{3}{5}x(30 - x)$ on the closed interval $[0, 30]$.

Definition *Maximum and Minimum Values*

If c is in the closed interval $[a, b]$, then $f(c)$ is called the **minimum value** of $f(x)$ on $[a, b]$ if $f(c) \leqq f(x)$ for all x in $[a, b]$. Similarly, if d is in $[a, b]$, then $f(d)$ is called the **maximum value** of $f(x)$ on $[a, b]$ if $f(d) \geqq f(x)$ for all x in $[a, b]$.

Thus if $f(c)$ is the minimum value and $f(d)$ the maximum value of $f(x)$ on $[a, b]$, then

$$f(c) \leqq f(x) \leqq f(d) \tag{2}$$

for all x in $[a, b]$, and hence $f(x)$ attains no value smaller than $f(c)$ or larger than $f(d)$. In geometric terms, $(c, f(c))$ is a *low point* and $(d, f(d))$ is a *high point* on the curve $y = f(x)$, $a \leqq x \leqq b$, as illustrated in Figs. 3.5.2 and 3.5.3.

Theorem 1 (proved in Appendix F) says that a continuous function f on a closed interval $[a, b]$ attains a minimum value $f(c)$ and a maximum value $f(d)$, so the inequalities in (2) hold: The curve $y = f(x)$ over $[a, b]$ has both a lowest point and a highest point.

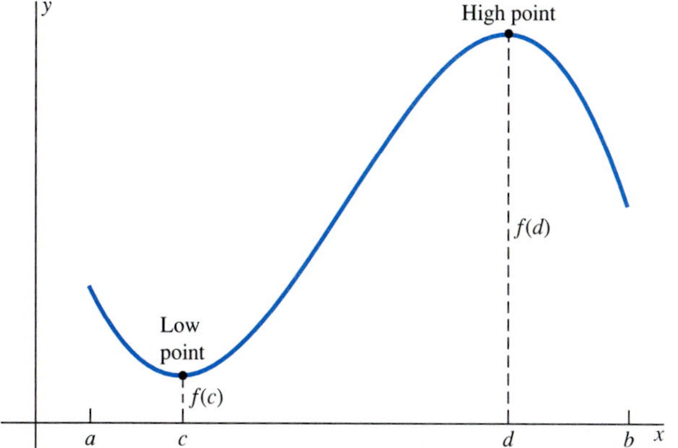

Fig. 3.5.2 $f(c)$ is the minimum value and $f(d)$ is the maximum value of $f(x)$ on $[a, b]$.

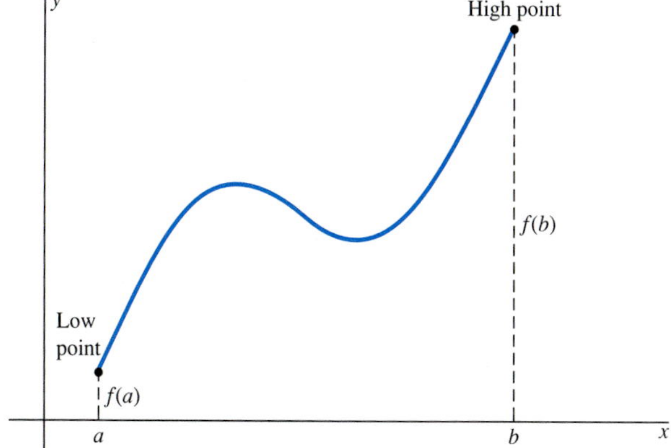

Fig. 3.5.3 Maximum and minimum values can occur at the endpoints of an interval. Here $f(a)$ is the minimum value and $f(b)$ is the maximum value of $f(x)$ on $[a, b]$.

Theorem 1 ***Maximum and Minimum Value Property***

If the function f is continuous on the closed interval $[a, b]$, then there exist numbers c and d in $[a, b]$ such that $f(c)$ is the minimum value, and $f(d)$ the maximum value, of f on $[a, b]$.

In short, a continuous function defined on a closed and bounded interval attains both a minimum value and a maximum value at points of the interval. Hence we see it is the *continuity* of the function

$$f(x) = \tfrac{3}{5}x(30 - x)$$

on the *closed* interval $[0, 30]$ that guarantees that the maximum value of f exists and is attained at some point of the interval $[0, 30]$.

Suppose that the function f is defined on the interval I. Examples 1 and 2 show that if *either* f is not continuous *or* I is not closed, then f may fail to attain maximum and minimum values at points of I. Thus both hypotheses in Theorem 1 are necessary.

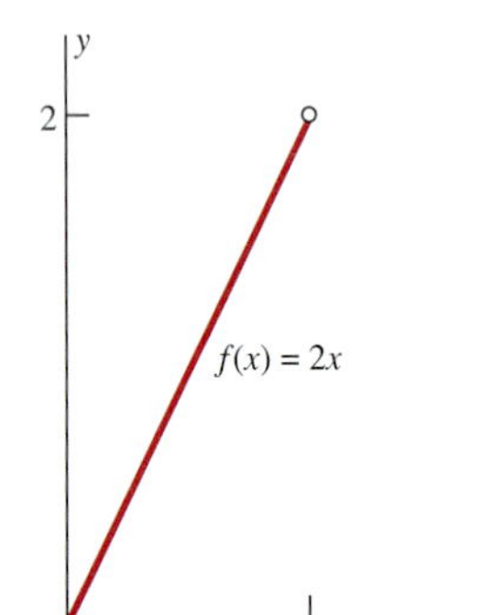

Fig. 3.5.4 The graph of the function of Example 1

EXAMPLE 1 Let the continuous function $f(x) = 2x$ be defined only for $0 \leqq x < 1$, so that its domain of definition is not a closed interval. From the graph shown in Fig. 3.5.4, it is clear that f attains its minimum value 0 at $x = 0$. But $f(x) = 2x$ attains *no* maximum value at any point of $[0, 1)$. The only possible candidate for a maximum value would be the value 2 at $x = 1$, but $f(1)$ is not defined. ■

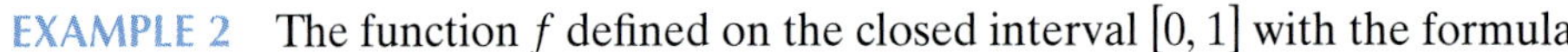

EXAMPLE 2 The function f defined on the closed interval $[0, 1]$ with the formula

$$f(x) = \begin{cases} \dfrac{1}{x} & \text{if } 0 < x \leqq 1, \\ 1 & \text{if } x = 0 \end{cases}$$

is not continuous on $[0, 1]$ because $\lim_{x \to 0^+} (1/x)$ does not exist (Fig. 3.5.5). This function does attain its minimum value of 1 at $x = 0$ and also at $x = 1$. But it attains no maximum value on $[0, 1]$, because $1/x$ can be made arbitrarily large by choosing x positive and very close to zero. ■

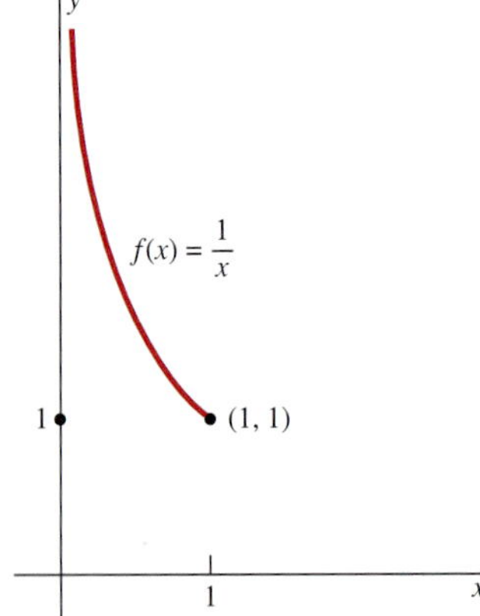

Fig. 3.5.5 The graph of the function of Example 2

For a variation on Example 2, the function $g(x) = 1/x$ with domain the *open* interval $(0, 1)$ attains neither a maximum nor a minimum there.

Local Maxima and Minima

Once we know that the continuous function f *does* attain minimum and maximum values on the closed interval $[a, b]$, the remaining question is this: Exactly *where* are these values located? We solved the animal pen problem in Section 2.1 on the basis of the following assumption, motivated by geometry: The function $f(x) = \frac{3}{5}x(30 - x)$ attains its maximum value on $[0, 30]$ at an interior point of that interval, a point at which the tangent line is horizontal. Theorems 2 and 3 of this section provide a rigorous basis for the method we used there.

We say that the value $f(c)$ is a **local maximum value** of the function f if $f(x) \leqq f(c)$ for all x sufficiently near c. More precisely, if this inequality holds for all x that are simultaneously in the domain of f and in some open interval containing c, then $f(c)$ is a local maximum of f. Similarly, we say that the value $f(c)$ is a **local minimum value** of f if $f(x) \geqq f(c)$ for all x sufficiently near c.

Local maximum

Local minimum

Fig. 3.5.6 Local extrema

As Fig. 3.5.6 shows, a local maximum is a point such that no nearby points on the graph are higher, and a local minimum is one such that no nearby points on the graph are lower. A **local extremum** of f is a value of f that is either a local maximum or a local minimum.

Theorem 2 ***Local Maxima and Minima***

If f is differentiable at c and is defined on an open interval containing c and if $f(c)$ is either a local maximum value or a local minimum value of f, then $f'(c) = 0$.

Thus a local extremum of a *differentiable* function on an *open* interval can occur only at a point where the derivative is zero and therefore where the line tangent to the graph is horizontal.

Proof of Theorem 2 Suppose, for instance, that $f(c)$ is a local maximum value of f. The assumption that $f'(c)$ exists means that the right-hand and left-hand limits

$$\lim_{h\to 0^+} \frac{f(c+h)-f(c)}{h} \quad \text{and} \quad \lim_{h\to 0^-} \frac{f(c+h)-f(c)}{h}$$

both exist and are equal to $f'(c)$.

If $h > 0$, then

$$\frac{f(c+h)-f(c)}{h} \leqq 0,$$

because $f(c) \geqq f(c+h)$ for all small positive values of h. Hence, by a one-sided version of the squeeze law for limits (in Section 2.2), this inequality is preserved when we take the limit as $h \to 0$. We thus find that

$$f'(c) = \lim_{h\to 0^+} \frac{f(c+h)-f(c)}{h} \leqq \lim_{h\to 0^+} 0 = 0.$$

Similarly, in the case $h < 0$, we find that

$$\frac{f(c+h)-f(c)}{h} \geqq 0.$$

Therefore,

$$f'(c) = \lim_{h\to 0^-} \frac{f(c+h)-f(c)}{h} \geqq \lim_{h\to 0^-} 0 = 0.$$

Because both $f'(c) \leqq 0$ and $f'(c) \geqq 0$, we conclude that $f'(c) = 0$. This establishes Theorem 2. ■

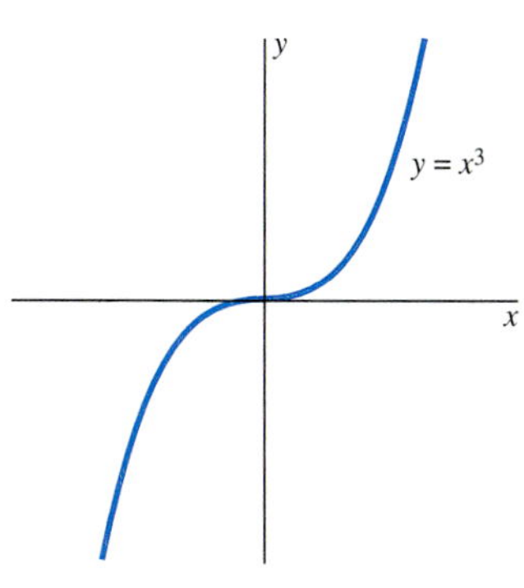

Fig. 3.5.7 There is no extremum at $x = 0$ even though the derivative is zero there.

Beware The converse of Theorem 2 is false. That is, the fact that $f'(c) = 0$ is *not enough* to imply that $f(c)$ is a local extremum. For example, consider the function $f(x) = x^3$. Its derivative $f'(x) = 3x^2$ is zero at $x = 0$. But a glance at its graph (Fig. 3.5.7) shows us that $f(0)$ is *not* a local extremum of f.

Thus the equation $f'(c) = 0$ is a *necessary* condition for $f(c)$ to be a local maximum or minimum value for a function f that is differentiable on an open interval. It is not a *sufficient* condition. The reason: $f'(x)$ may well be zero at points other than local maxima and minima. We give sufficient conditions for local maxima and minima in Chapter 4.

Absolute Maxima and Minima

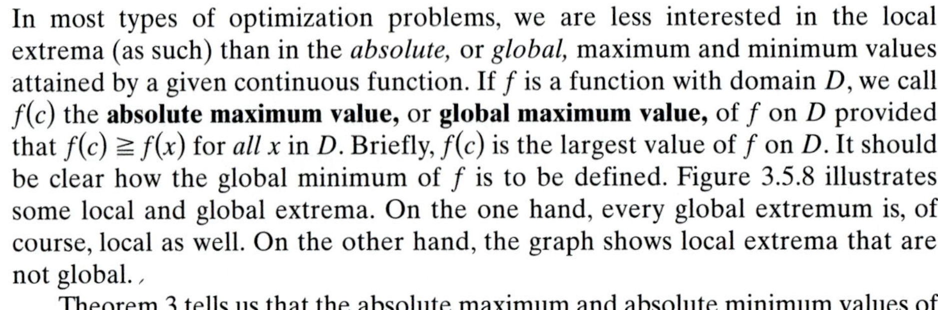

In most types of optimization problems, we are less interested in the local extrema (as such) than in the *absolute,* or *global,* maximum and minimum values attained by a given continuous function. If f is a function with domain D, we call $f(c)$ the **absolute maximum value,** or **global maximum value,** of f on D provided that $f(c) \geqq f(x)$ for *all* x in D. Briefly, $f(c)$ is the largest value of f on D. It should be clear how the global minimum of f is to be defined. Figure 3.5.8 illustrates some local and global extrema. On the one hand, every global extremum is, of course, local as well. On the other hand, the graph shows local extrema that are not global.

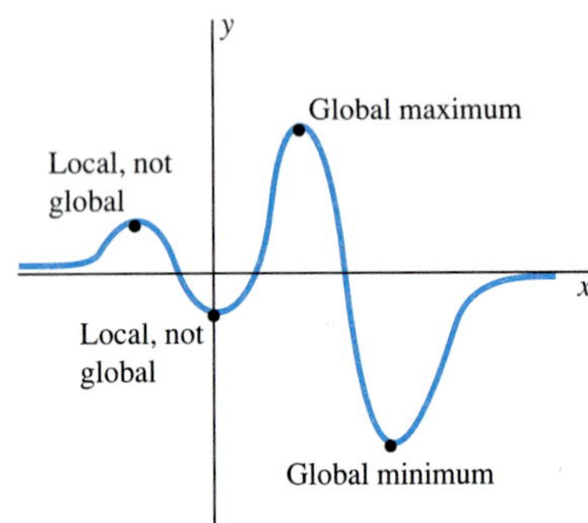

Fig. 3.5.8 Some extrema are global; others are merely local.

Theorem 3 tells us that the absolute maximum and absolute minimum values of the continuous function f on the closed interval $[a, b]$ occur either at one of the endpoints a or b or at a *critical point* of f. The number c in the domain of f is called a **critical point** of f if either

- ▼ $f'(c) = 0$, or
- ▼ $f'(c)$ does not exist.

> **Theorem 3** ***Absolute Maxima and Minima***
>
> Suppose that $f(c)$ is the absolute maximum (or absolute minimum) value of the continuous function f on the closed interval $[a, b]$. Then c is either a critical point of f or one of the endpoints a and b.

Proof This result follows almost immediately from Theorem 2. If c is not an endpoint of $[a, b]$, then $f(c)$ is a local extremum of f on the open interval (a, b). In this case Theorem 2 implies that $f'(c) = 0$, provided that f is differentiable at c. ■

As a consequence of Theorem 3, we can find the (absolute) maximum and minimum values of the function f on the closed interval $[a, b]$ as follows:

1. *Locate* the critical points of f: those points where $f'(x) = 0$ and those points where $f'(x)$ does not exist.
2. *List* the values of x that yield *possible* extrema of f: the two endpoints a and b and those critical points that lie in $[a, b]$.
3. *Evaluate* $f(x)$ at each point in this list of possible extrema.
4. *Inspect* these values of $f(x)$ to see which is the smallest and which is the largest.

The largest of the values in step 4 is the absolute maximum value of f; the smallest, the absolute minimum. We call this procedure the **closed-interval maximum-minimum method.**

EXAMPLE 3 For our final discussion of the animal pen problem, let us apply the closed-interval maximum-minimum method to find the maximum and minimum values of the differentiable function

$$f(x) = \tfrac{3}{5}x(30 - x) = \tfrac{3}{5}(30x - x^2)$$

on the closed interval $[0, 30]$.

Solution The derivative of f is

$$f'(x) = \tfrac{3}{5}(30 - 2x),$$

which is zero only at the point $x = 15$ in $[0, 30]$. Including the two endpoints, our list of the only values of x that can yield extrema of f consists of 0, 15, and 30. We evaluate f at each:

$$f(0) = 0, \quad \longleftarrow \text{ absolute minimum}$$

$$f(15) = 135, \quad \longleftarrow \text{ absolute maximum}$$

$$f(30) = 0. \quad \longleftarrow \text{ absolute minimum}$$

Thus the maximum value of $f(x)$ on $[0, 30]$ is 135 (attained at $x = 15$), and the minimum value is 0 (attained both at $x = 0$ and at $x = 30$). ■

EXAMPLE 4 Find the maximum and minimum values of

$$f(x) = 2x^3 - 3x^2 - 12x + 15$$

on the closed interval $[0, 3]$.

Solution The derivative of f is

$$f'(x) = 6x^2 - 6x - 12 = 6(x - 2)(x + 1).$$

So the critical points of f are the solutions of the equation

$$6(x - 2)(x + 1) = 0$$

and the numbers c for which $f'(c)$ does not exist. There are none of the latter, so the critical points of f occur at $x = -1$ and $x = 2$. The first of these is not in the domain of f; we discard it, and thus the only critical point of f in $[0, 3]$ is $x = 2$. Including the two endpoints, our list of all values of x that yield a possible maximum or minimum value of f consists of 0, 2, and 3. We evaluate the function f at each:

$$f(0) = 15, \quad \longleftarrow \text{ absolute maximum}$$

$$f(2) = -5, \quad \longleftarrow \text{ absolute minimum}$$

$$f(3) = 6.$$

Therefore the maximum value of f on $[0, 3]$ is $f(0) = 15$, and its minimum value is $f(2) = -5$. ■

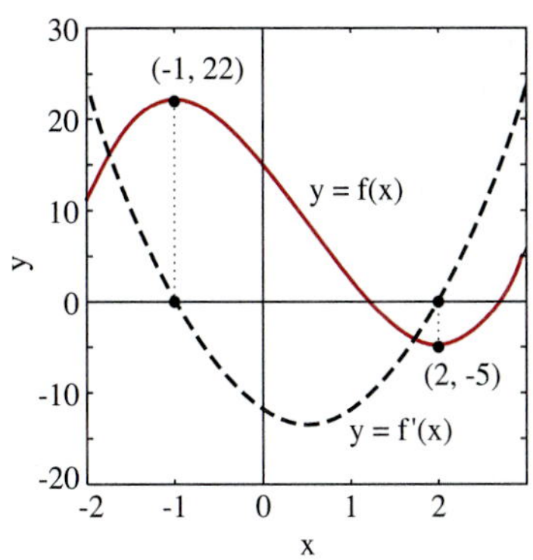

Fig. 3.5.9 The critical points of the differentiable function $f(x)$ are the zeros of $f'(x)$.

If in Example 4 we had asked for the maximum and minimum values of $f(x)$ on the interval $[-2, 3]$ (instead of the interval $[0, 3]$), then we would have included *both* critical points $x = -1$ and $x = 2$ in our list of possibilities. The resulting values of f would have been

$$f(-2) = 11,$$

$$f(-1) = 22, \quad \longleftarrow \text{ absolute maximum}$$

$$f(2) = -5, \quad \longleftarrow \text{ absolute minimum}$$

$$f(3) = 6.$$

Figure 3.5.9 shows both the curve $y = f(x)$ and the graph of its derivative. Note the vertical line segments joining high and low points on $y = f(x)$ with x-intercepts of $dy/dx = f'(x)$. Thus the figure illustrates the following fact:

The critical points of a differentiable function $f(x)$ are the zeros of its derivative $f'(x)$.

On the basis of this principle, we can approximate a critical point of f graphically by "zooming in" on a zero of f'.

In Example 4 the function f was differentiable everywhere. Examples 5 and 6 illustrate the case of an extremum at a critical point where the function is not differentiable.

EXAMPLE 5 Find the maximum and minimum values of $f(x) = 3 - |x - 2|$ on the interval $[1, 4]$.

Solution If $x \leqq 2$, then $x - 2 \leqq 0$, so

$$f(x) = 3 - (2 - x) = x + 1.$$

If $x \geqq 2$, then $x - 2 \geqq 0$, so

$$f(x) = 3 - (x - 2) = 5 - x.$$

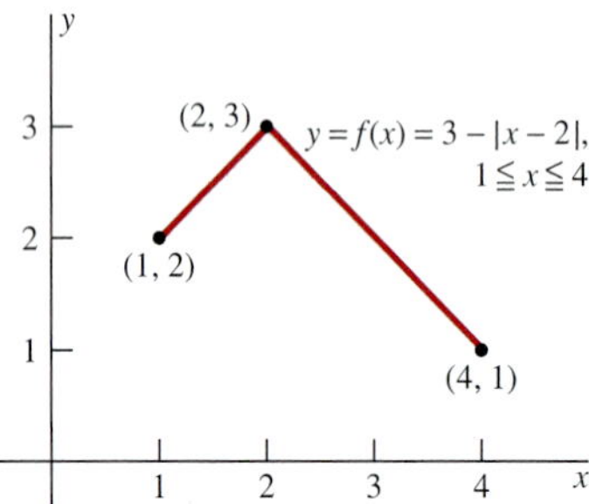

Fig. 3.5.10 Graph of the function of Example 5

Consequently, the graph of f looks like the one shown in Fig. 3.5.10. The only critical point of f in $[1, 4]$ is the point $x = 2$, because $f'(x)$ takes on only the two values $+1$ and -1 (and so is never zero), and $f'(2)$ does not exist. (Why not?) Evaluation of f at this critical point and at the two endpoints yields

$$f(1) = 2,$$

$$f(2) = 3, \quad \longleftarrow \quad \text{absolute maximum}$$

$$f(4) = 1. \quad \longleftarrow \quad \text{absolute minimum}$$

■

EXAMPLE 6 Find the maximum and minimum values of

$$f(x) = 5x^{2/3} - x^{5/3}$$

on the closed interval $[-1, 4]$.

Solution Differentiation of f yields

$$f'(x) = \frac{10}{3}x^{-1/3} - \frac{5}{3}x^{2/3} = \frac{5}{3}x^{-1/3}(2 - x) = \frac{5(2 - x)}{3x^{1/3}}.$$

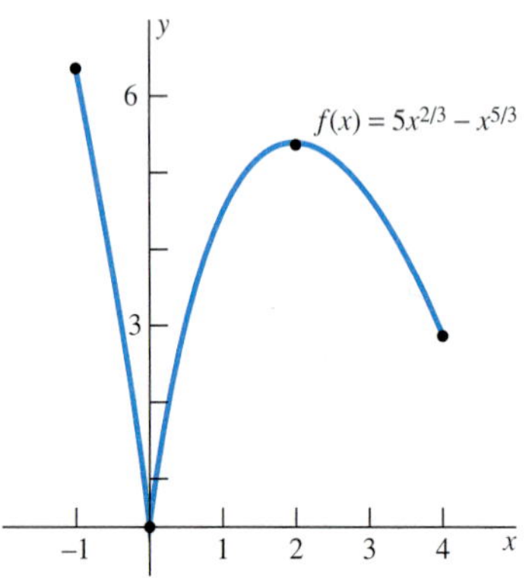

Fig. 3.5.11 Graph of the function of Example 6

Hence f has two critical points in the interval: $x = 2$, where $f'(x) = 0$, and $x = 0$, where $f'(x)$ does not exist (the graph of f has a vertical tangent at $(0, 0)$). When we evaluate f at these two critical points and at the two endpoints, we get

$$f(-1) = 6, \quad \leftarrow \text{absolute maximum}$$

$$f(0) = 0 \quad \leftarrow \text{absolute minimum}$$

$$f(2) = 5 \cdot 2^{2/3} - 2^{5/3} \approx 4.76,$$

$$f(4) = 5 \cdot 4^{2/3} - 4^{5/3} \approx 2.52.$$

Thus the maximum value $f(-1) = 6$ occurs at an endpoint. The minimum value $f(0) = 0$ occurs at a point where f is not differentiable. ■

By using a graphics calculator or computer with graphics capabilities, you can verify that the graph of the function f of Example 6 is that shown in Fig. 3.5.11. But in the usual case of a continuous function that has only finitely many critical points in a given closed interval, the closed-interval maximum-minimum method suffices to determine its maximum and minimum values without requiring any detailed knowledge of the graph of the function.

3.5 PROBLEMS

In Problems 1 through 10, state whether the given function attains a maximum value or a minimum value (or both) on the given interval. (*Suggestion:* Begin by sketching a graph of the function.)

1. $f(x) = 1 - x;\quad [-1, 1)$

2. $f(x) = 2x + 1;\quad [-1, 1)$

3. $f(x) = |x|;\quad (-1, 1)$

4. $f(x) = \dfrac{1}{\sqrt{x}};\quad (0, 1]$

5. $f(x) = |x - 2|;\quad (1, 4]$

6. $f(x) = 5 - x^2;\quad [-1, 2)$

7. $f(x) = x^3 + 1;\quad [-1, 1]$

8. $f(x) = \dfrac{1}{x^2 + 1};\quad (-\infty, \infty)$

9. $f(x) = \dfrac{1}{x(1 - x)};\quad [2, 3]$

10. $f(x) = \dfrac{1}{x(1 - x)};\quad (0, 1)$

In Problems 11 through 40, find the maximum and minimum values attained by the given function on the indicated closed interval.

11. $f(x) = 3x - 2;\quad [-2, 3]$

12. $f(x) = 4 - 3x;\quad [-1, 5]$

13. $h(x) = 4 - x^2;\quad [1, 3]$

14. $f(x) = x^2 + 3;\quad [0, 5]$

15. $g(x) = (x - 1)^2;\quad [-1, 4]$

16. $h(x) = x^2 + 4x + 7;\quad [-3, 0]$

17. $f(x) = x^3 - 3x;\quad [-2, 4]$

18. $g(x) = 2x^3 - 9x^2 + 12x;\quad [0, 4]$

19. $h(x) = x + \dfrac{4}{x};\quad [1, 4]$

20. $f(x) = x^2 + \dfrac{16}{x};\quad [1, 3]$

21. $f(x) = 3 - 2x;\quad [-1, 1]$

22. $f(x) = x^2 - 4x + 3;\quad [0, 2]$

23. $f(x) = 5 - 12x - 9x^2;\quad [-1, 1]$

24. $f(x) = 2x^2 - 4x + 7;\quad [0, 2]$

25. $f(x) = x^3 - 3x^2 - 9x + 5;\quad [-2, 4]$

26. $f(x) = x^3 + x;\quad [-1, 2]$

27. $f(x) = 3x^5 - 5x^3;\quad [-2, 2]$

28. $f(x) = |2x - 3|;\quad [1, 2]$

29. $f(x) = 5 + |7 - 3x|;\quad [1, 5]$

30. $f(x) = |x + 1| + |x - 1|;\quad [-2, 2]$

31. $f(x) = 50x^3 - 105x^2 + 72x;\quad [0, 1]$

32. $f(x) = 2x + \dfrac{1}{2x};\quad [1, 4]$

33. $f(x) = \dfrac{x}{x + 1};\quad [0, 3]$

34. $f(x) = \dfrac{x}{x^2 + 1};\quad [0, 3]$

35. $f(x) = \dfrac{1 - x}{x^2 + 3};\quad [-2, 5]$

36. $f(x) = 2 - \sqrt[3]{x};\quad [-1, 8]$

37. $f(x) = x\sqrt{1 - x^2};\quad [-1, 1]$

38. $f(x) = x\sqrt{4 - x^2};\quad [0, 2]$

39. $f(x) = x(2 - x)^{1/3};\quad [1, 3]$

40. $f(x) = x^{1/2} - x^{3/2};\quad [0, 4]$

41. Suppose that $f(x) = Ax + B$ is a linear function and that $A \neq 0$. Explain why the maximum and minimum values of f on a closed interval $[a, b]$ must occur at the endpoints of the interval.

42. Suppose that f is continuous on $[a, b]$ and differentiable on (a, b) and that $f'(x)$ is never zero at any point of (a, b). Explain why the maximum and minimum values of f must occur at the endpoints of the interval $[a, b]$.

43. Explain why every real number is a critical point of the greatest integer function $f(x) = \lfloor x \rfloor$.

44. Prove that every quadratic function

$$f(x) = ax^2 + bx + c \qquad (a \neq 0)$$

has exactly one critical point on the real line.

45. Explain why the cubic polynomial function

$$f(x) = ax^3 + bx^2 + cx + d \qquad (a \neq 0)$$

can have either two, one, or no critical points on the real line. Produce examples that illustrate each of the three cases.

46. Define $f(x)$ to be the distance from x to the nearest integer. What are the critical points of f?

In Problems 47 through 52, match the given graph of the function with the graph of its derivative f' from those in Fig. 3.5.12, parts (a) through (f).

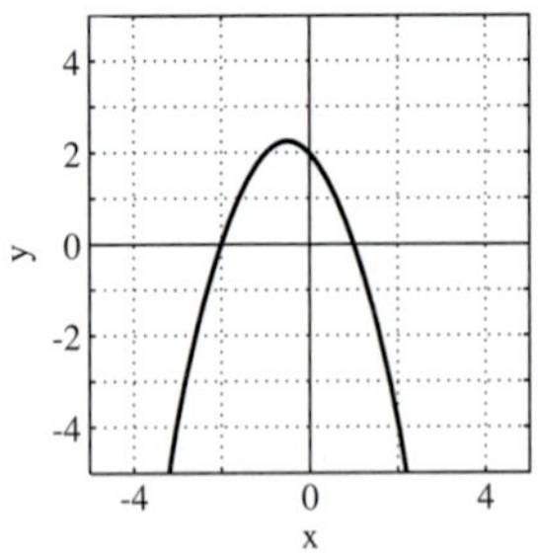

Fig. 3.5.12 (a)

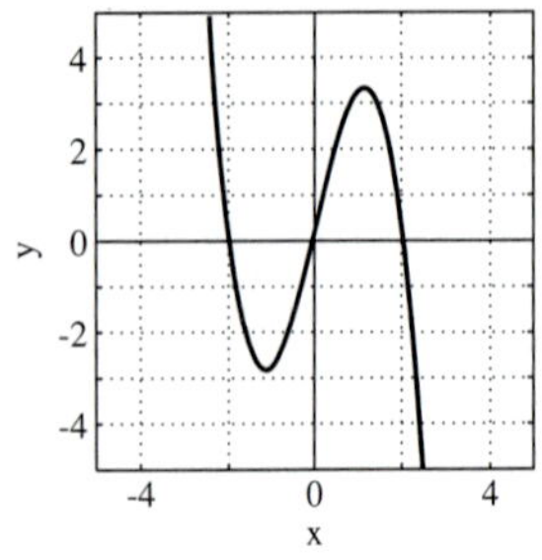

Fig. 3.5.12 (b)

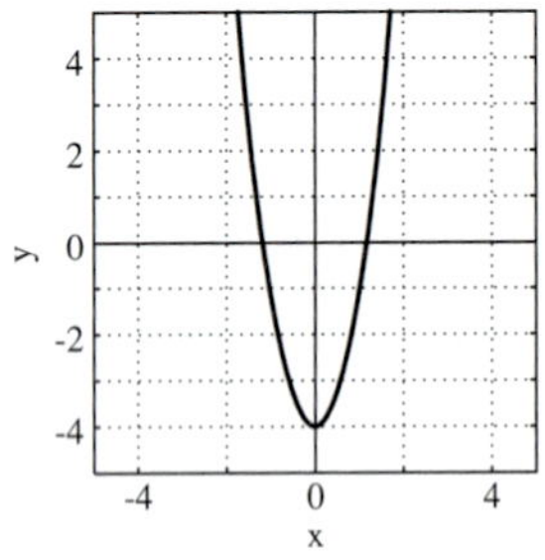

Fig. 3.5.12 (c)

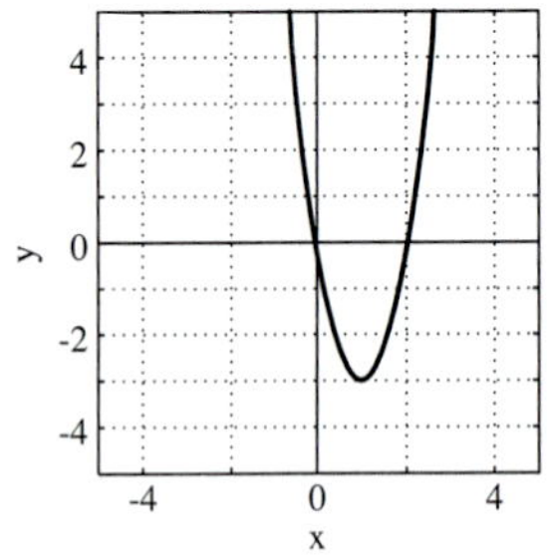

Fig. 3.5.12 (d)

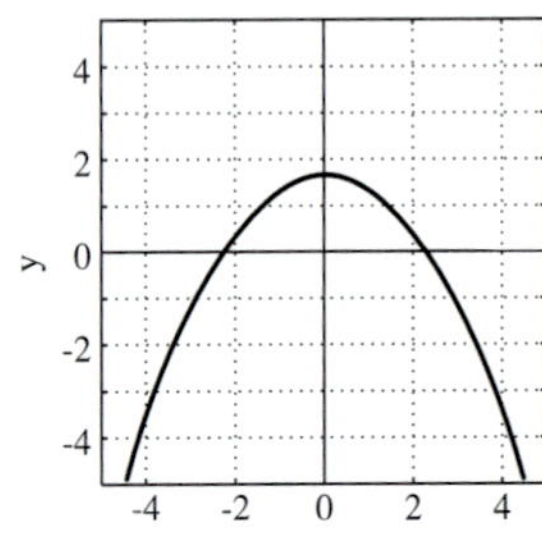

Fig. 3.5.12 (e)

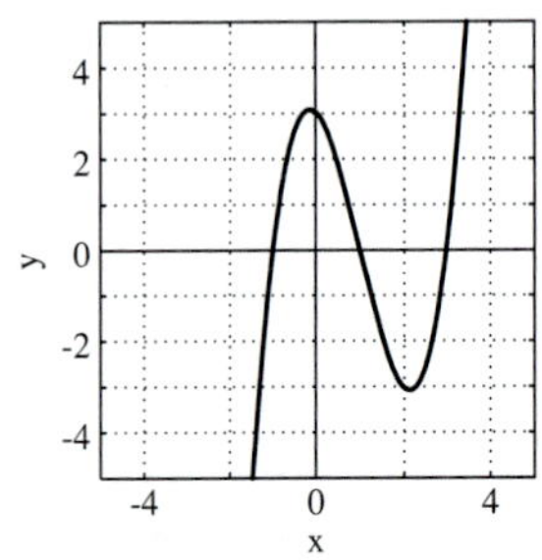

Fig. 3.5.12 (f)

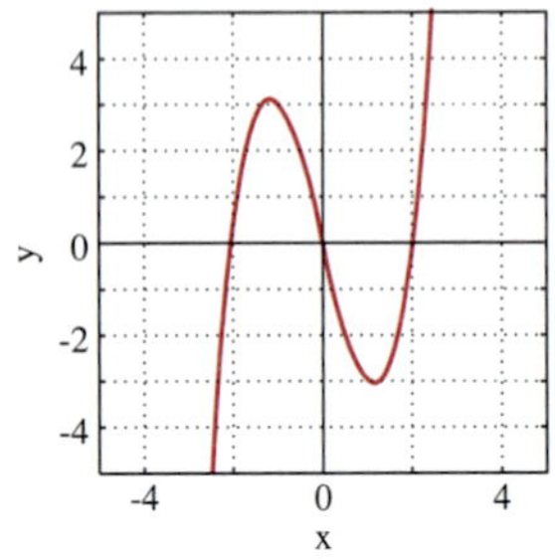

47. Figure 3.5.13 **48.** Figure 3.5.14

49. Figure 3.5.15 **50.** Figure 3.5.16

51. Figure 3.5.17 **52.** Figure 3.5.18

Fig. 3.5.13

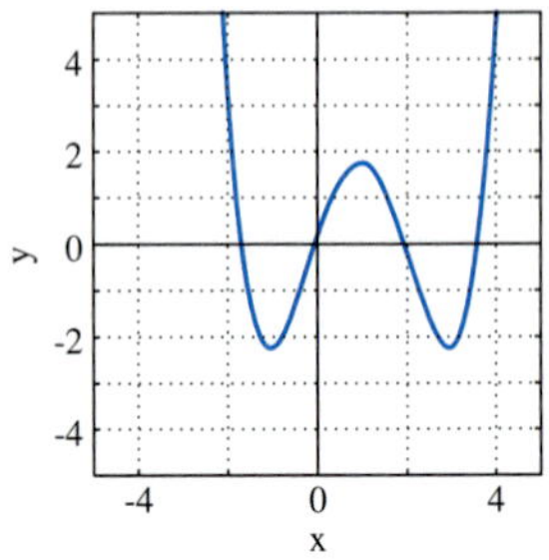

Fig. 3.5.14

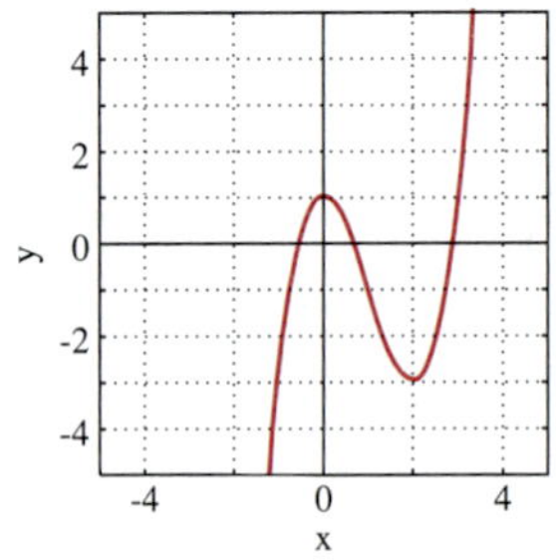

Fig. 3.5.15

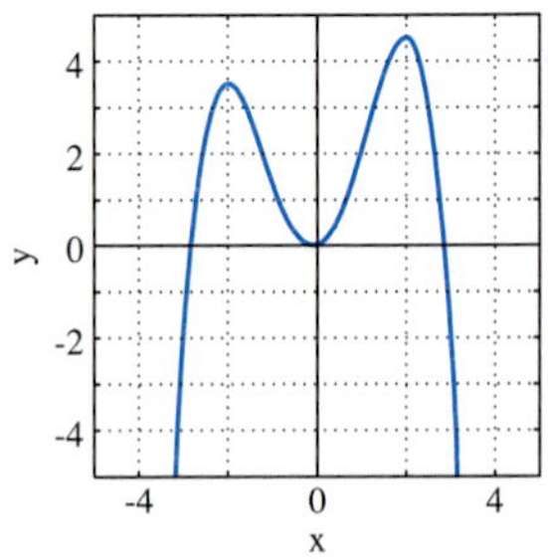

Fig. 3.5.16

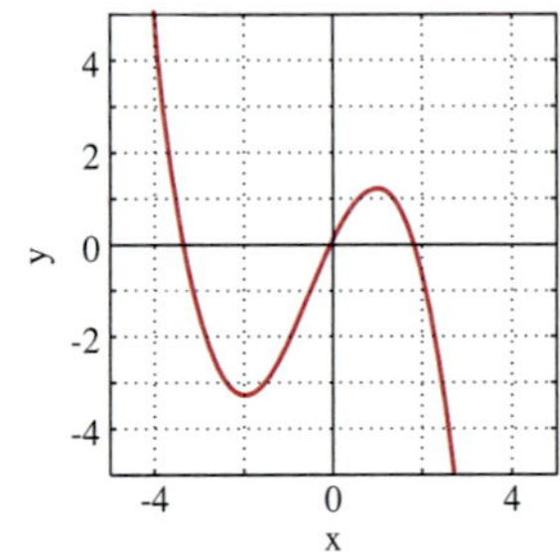

Fig. 3.5.17

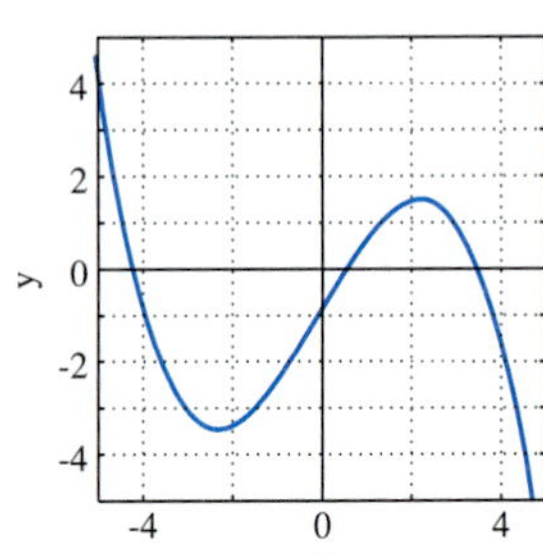

Fig. 3.5.18

3.5 PROJECT: ZOOMING IN ON ZEROS OF THE DERIVATIVE

Figure 3.5.19 shows the graph of the function

$$f(x) = 4x^4 - 11x^2 - 5x - 3$$

and its derivative

$$f'(x) = 16x^3 - 22x - 5$$

on the interval $[-3, 3]$. The maximum value of $f(x)$ on $[-2, 2]$ is $f(-2) = 27$ at the left endpoint. The lowest point on the curve $y = f(x)$ and the corresponding zero of the derivative $dy/dx = f'(x)$ lie within the small boxes in the figure.

If we attempt to zoom in on the lowest point—without changing the "range factors" or "aspect ratio" of the viewing windows—we get the picture shown in

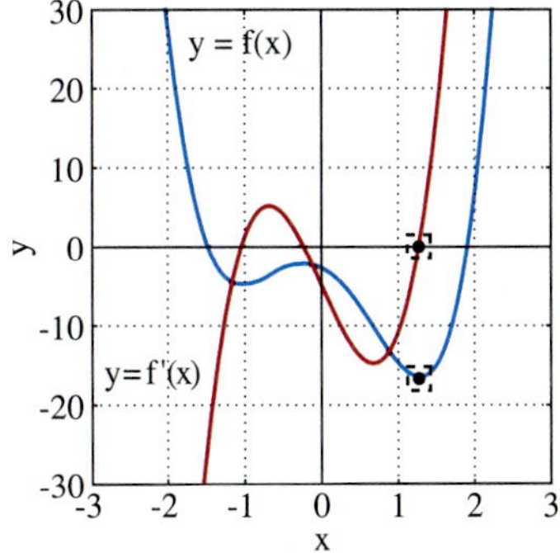

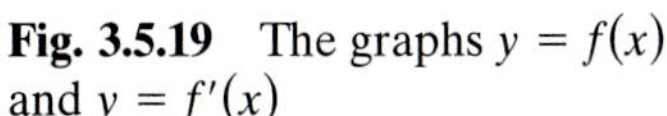

Fig. 3.5.19 The graphs $y = f(x)$ and $y = f'(x)$

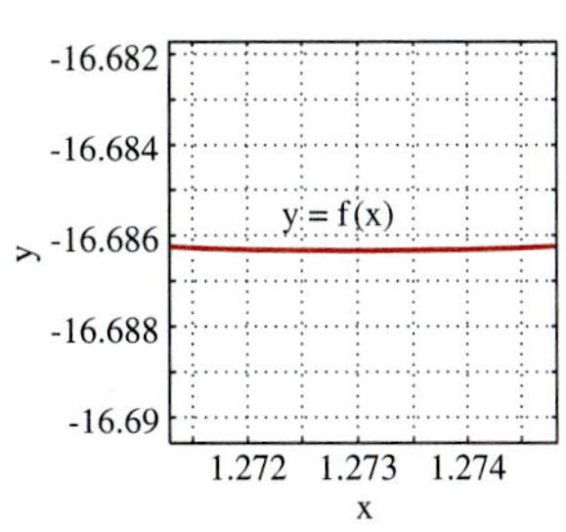

Fig. 3.5.20 Zooming in on the minimum shown in Fig. 3.5.19

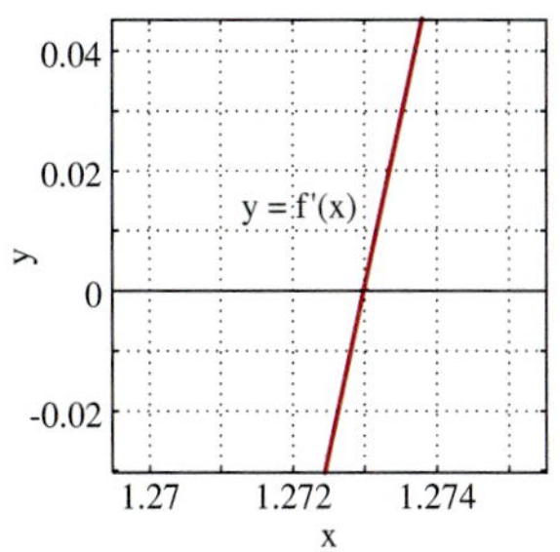

Fig. 3.5.21 Zooming in instead on the zero of $f'(x)$ shown in Fig. 3.5.19.

Fig. 3.5.20, where it is difficult to locate the lowest point precisely. The reason is this: After sufficient magnification, the graph is indistinguishable from its tangent line, which is horizontal at a local maximum or minimum point.

Consequently, it is much better to zoom in on the corresponding zero of the derivative $f'(x)$. We can then locate the indicated critical point with much greater precision (Fig. 3.5.21). Here it is clear that the minimum value attained by $f(x)$ on $[-2, 2]$ is approximately $f(1.273) \approx -16.686$.

In Problems 1 through 8, find the maximum and minimum values of the given function on the indicated closed interval by zooming in on the zeros of the derivative.

1. $f(x) = x^3 + 3x^2 - 7x + 10$; $[-2, 2]$

2. $f(x) = x^3 + 3x^2 - 7x + 10$; $[-4, 2]$

3. $f(x) = x^4 - 3x^3 + 7x - 5$; $[-3, 3]$

4. $f(x) = x^4 - 5x^3 + 17x - 5$; $[-3, 3]$

5. $f(x) = x^4 - 5x^3 + 17x - 5$; $[0, 2]$

6. $f(x) = x^5 - 5x^4 - 15x^3 + 17x^2 + 23x$; $[-1, 1]$

7. $f(x) = x^5 - 5x^4 - 15x^3 + 17x^2 + 23x$; $[-3, 3]$

8. $f(x) = x^5 - 5x^4 - 15x^3 + 17x^2 + 23x$; $[0, 10]$

3.6 APPLIED MAXIMUM-MINIMUM PROBLEMS

This section is devoted to applied maximum-minimum problems (like the animal pen problem of Section 1.1) for which the closed-interval maximum-minimum method of Section 3.5 can be used. When confronting such a problem, we face an important first step: We must determine the quantity to be maximized or minimized. This quantity will be the dependent variable in our analysis of the problem.

This dependent variable must then be expressed as a function of an independent variable, one that "controls" the values of the dependent variable. If the domain of values of the independent variable—those that are pertinent to the applied problem—is a closed interval, then we may proceed with the closed-interval maximum-minimum method. This plan of attack can be summarized in the following steps.

1. *Find the quantity to be maximized or minimized.* This quantity, which you should describe with a word or short phrase and label with a descriptive letter, will be the dependent variable. Because it is a *dependent* variable, it depends on something else; that quantity will be the independent variable. Here we call the independent variable x.

2. *Express the dependent variable as a function of the independent variable.* Use the information in the problem to write the dependent variable as a function of x. Always draw a figure and *label the variables;* this is generally the best way to find the relationship between the dependent and independent variables. Use auxiliary variables if they help, but not too many, for you must eventually eliminate them. You *must* express the dependent variable as a function of the *single* independent variable x and various constants before you can compute any derivatives. Find the domain of this function as well as its formula. Force the domain to be a closed and bounded interval if possible—if the natural domain is an open interval, adjoin the endpoints if you can.
3. *Apply calculus to find the critical points.* Compute the derivative f' of the function f that you found in step 2. Use the derivative to find the critical points—where $f'(x) = 0$ and where $f'(x)$ does not exist. If f is differentiable everywhere, then its only critical points occur where $f'(x) = 0$.
4. *Identify the extrema.* Evaluate f at each critical point in its domain *and* at the two endpoints. The values you obtain will tell you which is the absolute maximum and which is the absolute minimum. Of course, either or both extreme values may occur at more than one point.
5. *Answer the question posed in the original problem.* In other words, interpret your results. The answer to the original problem may be something other than merely the largest (or smallest) value of f. Give a precise answer to the specific question originally asked.

Observe how we follow this five-step process in Example 1.

EXAMPLE 1 A farmer has 200 yd of fence with which to construct three sides of a rectangular pen; an existing long, straight wall will form the fourth side. What dimensions will maximize the area of the pen?

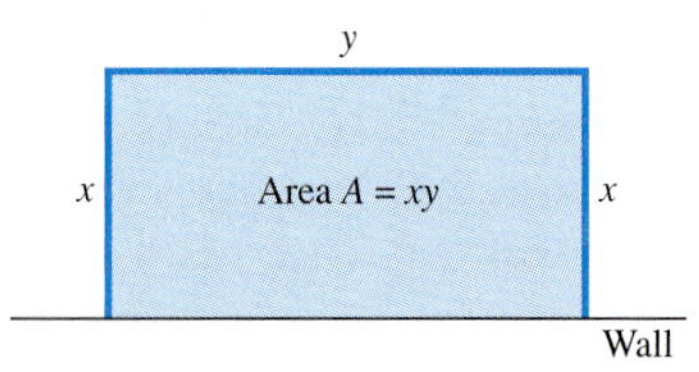

Fig. 3.6.1 The rectangular pen of Example 1

Solution We want to maximize the area A of the pen shown in Fig. 3.6.1. To get a formula for the *dependent* variable A, we observe that the area of a rectangle is the product of its base and its height. So we let x denote the length of each of the two sides of the pen perpendicular to the wall. We also let y denote the length of the side parallel to the wall. Then the area of the rectangle is given by the *formula*

$$A = xy.$$

Now we need to write A as a *function* of either x or y. Because all 200 yd of fence are to be used,

$$2x + y = 200, \quad \text{so} \quad y = 200 - 2x. \tag{1}$$

(We chose to express y in terms of x merely because the algebra is slightly simpler.) Next, we substitute this value of y into the formula $A = xy$ to obtain

$$A(x) = x(200 - 2x) = 200x - 2x^2. \tag{2}$$

This equation expresses the dependent variable A as a function of the independent variable x.

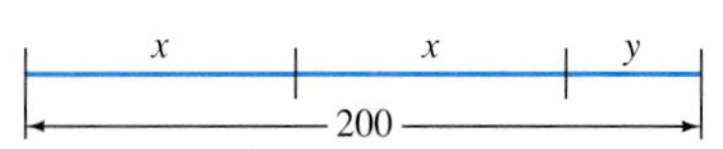

Fig. 3.6.2 The relation in Eq. (1) between x and y (Example 1)

Before proceeding, we must find the domain of the function A. It is clear from Fig. 3.6.2 that $0 < x < 100$. But to apply the closed-interval maximum-minimum method, we need a closed interval. In this example, we may adjoin the endpoints to $(0, 100)$ to get the *closed* interval $[0, 100]$. The values $x = 0$ and $x = 100$ correspond to "degenerate" pens of area zero. Because zero cannot be the maximum value of A, there is no harm in thus enlarging the domain of the function A.

Now we compute the derivative of the function A in Eq. (2):

$$\frac{dA}{dx} = 200 - 4x.$$

Because A is differentiable, its only critical points occur when

$$\frac{dA}{dx} = 0;$$

that is, when

$$200 - 4x = 0.$$

So $x = 50$ is the only interior critical point. Including the endpoints, the extrema of A can occur only at $x = 0, 50$, and 100. We evaluate A at each:

$$A(0) = 0,$$

$$A(50) = 5000, \quad \longleftarrow \quad \text{absolute maximum}$$

$$A(100) = 0.$$

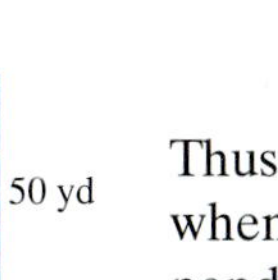

Fig. 3.6.3 The pen with maximal area of Example 1

Thus the maximal area is $A(50) = 5000$ (yd^2). From Eq. (1) we find that $y = 100$ when $x = 50$. Therefore, for the pen to have maximal area, each of the two sides perpendicular to the wall should be 50 yd long and the side parallel to the wall should be 100 yd long (Fig. 3.6.3). ■

EXAMPLE 2 A piece of sheet metal is rectangular, 5 ft wide and 8 ft long. Congruent squares are to be cut from its four corners. The resulting piece of metal is to be folded and welded to form an open-topped box (Fig. 3.6.4). How should this be done to get a box of largest possible volume?

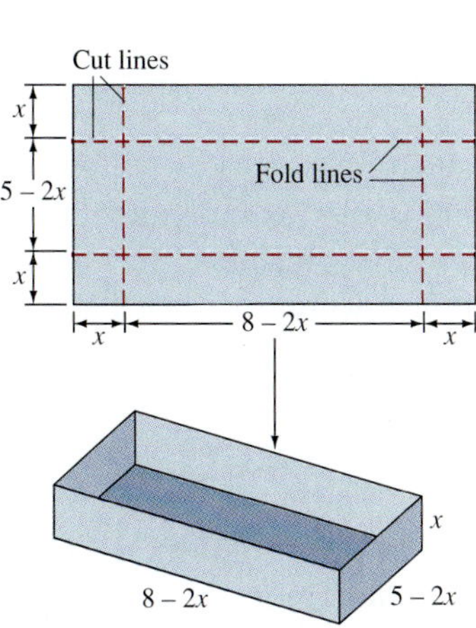

Fig. 3.6.4 Making the box of Example 2

Solution The quantity to be maximized—the dependent variable—is the volume V of the box to be constructed. The shape and thus the volume of the box are determined by the length x of the edge of each corner square removed. Hence x is a natural choice for the independent variable.

To write the volume V as a function of x, note that the finished box will have height x and its base will measure $8 - 2x$ ft by $5 - 2x$ ft. Hence its volume is given by

$$V(x) = x(5 - 2x)(8 - 2x) = 4x^3 - 26x^2 + 40x.$$

The procedure described in this example will produce an actual box only if $0 < x < 2.5$ (Fig. 3.6.5). But we make the domain the *closed* interval $[0, 2.5]$ to ensure that a maximum of $V(x)$ exists and to use the closed-interval maximum-minimum method. The values $x = 0$ and $x = 2.5$ correspond to "degenerate" boxes of zero volume, so adjoining these points to $(0, 2.5)$ will affect neither the location of the absolute maximum nor its value.

Now we compute the derivative of V:

$$V'(x) = 12x^2 - 52x + 40 = 4(3x - 10)(x - 1).$$

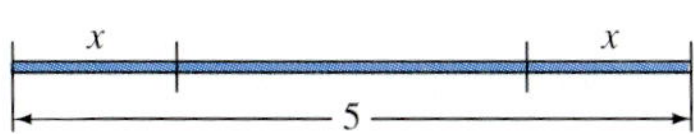

Fig. 3.6.5 The 5-ft width of the metal sheet (Example 2)

The only critical points of the differentiable function V occur where

$$V'(x) = 0;$$

that is, where

$$4(3x - 10)(x - 1) = 0.$$

The solutions of this equation are $x = 1$ and $x = \frac{10}{3}$. We discard the latter because it does not lie in the domain $[0, 2.5]$ of V. So we examine these values of V:

$$V(0) = 0,$$

$$V(1) = 18, \qquad \leftarrow \text{absolute maximum}$$

$$V(2.5) = 0.$$

Thus the maximum value of $V(x)$ on $[0, 2.5]$ is $V(1) = 18$. The answer to the question posed is this: The squares cut from the corners should be of edge length 1 ft each. The resulting box will measure 6 ft by 3 ft by 1 ft, and its volume will be 18 ft^3 (Fig. 3.6.6). ■

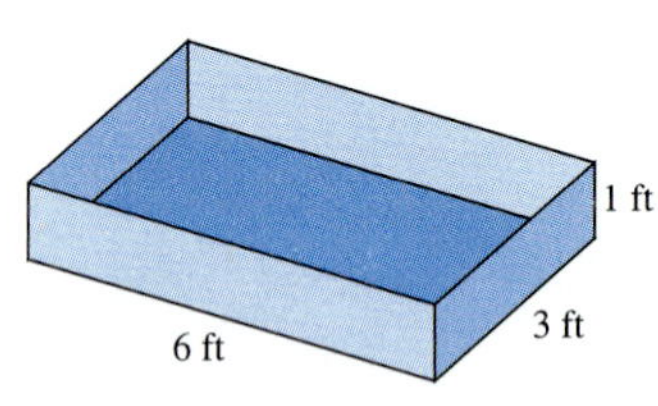

Fig. 3.6.6 The box with maximal volume of Example 2

For our next application of the closed-interval maximum-minimum method, let us consider a typical problem in business management. Suppose that x units of computer diskettes are to be manufactured at a *total cost* of $C(x)$ dollars. We make the simple (but not always valid) assumption that the cost function $C(x)$ is the sum of two terms:

1. A constant term a representing the *fixed cost* of acquiring and maintaining production facilities (overhead), and
2. A variable term representing the *additional cost* of making x units at, for example, b dollars each.

Then

the **total cost** is the **sum** of the **fixed cost** and the **additional cost,**

so the *cost function* $C(x)$ is given by

$$C(x) = a + bx. \tag{3}$$

We also assume that the number of units that can be sold is a linear function of the selling price p, so that $x = m - np$, where m and n are positive constants. The minus sign indicates that an increase in selling price will result in a decrease in sales. If we solve this last equation for p, we get the *price function*

$$p(x) = A - Bx \tag{4}$$

(A and B are also constants).

The quantity to be maximized is profit, given here by the *profit function* $P(x)$, which is equal to the sales revenue minus the production costs. Thus

$$P(x) = xp(x) - C(x). \tag{5}$$

EXAMPLE 3 Suppose that the cost of publishing a small book is \$10000 to set up the (annual) press run plus \$8 for each book printed. The publisher sold 8000 copies last year at \$13 each, but sales dropped to 5000 copies this year when the price was raised to \$15 per copy. Assume that up to 10000 copies can be printed in a single press run. How many copies should be printed, and what should be the selling price of each copy, to maximize the year's profit on this book?

Solution The dependent variable to be maximized is the profit P. As independent variable we choose the number x of copies to be printed; also, $0 \leqq x \leqq 10000$. The given cost information then implies that

$$C(x) = 10000 + 8x.$$

Now we substitute into Eq. (4) the data $x = 8000$ when $p = 13$ as well as the data $x = 5000$ when $p = 15$. We obtain the equations

$$A - 7000B = 13, \qquad A - 5000B = 15.$$

When we solve these equations simultaneously, we find that $A = 20$ and $B = 0.001$. Hence the price function is

$$p(x) = 20 - \frac{x}{1000},$$

and thus the profit function is

$$P(x) = x\left(20 - \frac{x}{1000}\right) - (10000 + 8x).$$

We expand and collect terms to obtain

$$P(x) = 12x - \frac{x^2}{1000} - 10000, \qquad 0 \leqq x \leqq 10000.$$

Now

$$\frac{dP}{dx} = 12 - \frac{x}{500},$$

and the only critical points of the differentiable function P occur when

$$\frac{dP}{dx} = 0;$$

that is, when

$$12 - \frac{x}{500} = 0; \qquad x = 12 \cdot 500 = 6000.$$

We check P at this value of x as well as the values of $P(x)$ at the endpoints to find the maximum profit:

$$P(0) = -10000,$$

$$P(6000) = 26000, \leftarrow \text{ absolute maximum}$$

$$P(10000) = 10000.$$

Therefore, the maximum possible annual profit of \$26000 results from printing 6000 copies of the book. Each copy should be sold for \$14, because

$$p(6000) = 20 - \frac{6000}{1000} = 14.$$

■

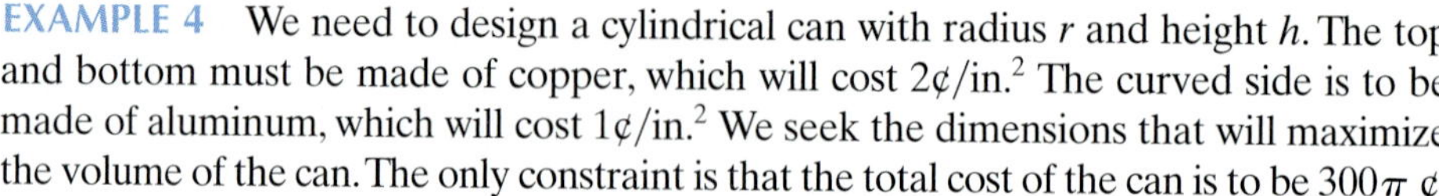

EXAMPLE 4 We need to design a cylindrical can with radius r and height h. The top and bottom must be made of copper, which will cost $2¢/\text{in.}^2$ The curved side is to be made of aluminum, which will cost $1¢/\text{in.}^2$ We seek the dimensions that will maximize the volume of the can. The only constraint is that the total cost of the can is to be 300π ¢.

Solution We need to maximize the volume V of the can, which we can compute if we know its radius r and its height h (Fig. 3.6.7). With these dimensions, we find that

$$V = \pi r^2 h, \tag{6}$$

but we need to express V as a function of r alone (or as a function of h alone).

Both the circular top and bottom of the can have area πr^2 in.^{2}, so the area of copper to be used is $2\pi r^2$ and its cost is $4\pi r^2$ cents. The area of the curved side of the can is $2\pi rh$ in.^{2}, so the area of aluminum used is the same, and the aluminum costs $2\pi rh$ cents.

We obtain the total cost of the can by adding the cost of the copper to the cost of the aluminum. This sum must be 300π ¢, and therefore

$$4\pi r^2 + 2\pi rh = 300\pi. \tag{7}$$

Fig. 3.6.7 The cylindrical can of Example 4

We eliminate h in Eq. (6) by solving Eq. (7) for h:

$$h = \frac{300\pi - 4\pi r^2}{2\pi r} = \frac{1}{r}(150 - 2r^2). \tag{8}$$

Hence

$$V = V(r) = (\pi r^2)\frac{1}{r}(150 - 2r^2) = 2\pi(75r - r^3). \tag{9}$$

To determine the domain of definition of V, we note from Eq. (7) that $4\pi r^2 < 300\pi$, so $r < \sqrt{75}$ for the desired can; with $r = \sqrt{75} = 5\sqrt{3}$, we get a degenerate can with height $h = 0$. With $r = 0$, we obtain *no* value of h in Eq. (8) and therefore no can, but $V(r)$ is nevertheless continuous at $r = 0$. Consequently, we can take the closed interval $[0, 5\sqrt{3}]$ to be the domain of V.

Calculating the derivative yields

$$V'(r) = 2\pi(75 - 3r^2) = 6\pi(25 - r^2).$$

Because $V(r)$ is a polynomial, $V'(r)$ exists for all values of r, so we obtain all critical points by solving the equation

$$V'(r) = 0;$$

that is,

$$6\pi(25 - r^2) = 0.$$

We discard the solution -5, as it does not lie in the domain of V. Thus we obtain only the single critical point $r = 5$ in $[0, 5\sqrt{3}]$. Now

$$V(0) = 0,$$

$$V(5) = 500\pi, \quad \leftarrow \text{absolute maximum}$$

$$V(5\sqrt{3}) = 0.$$

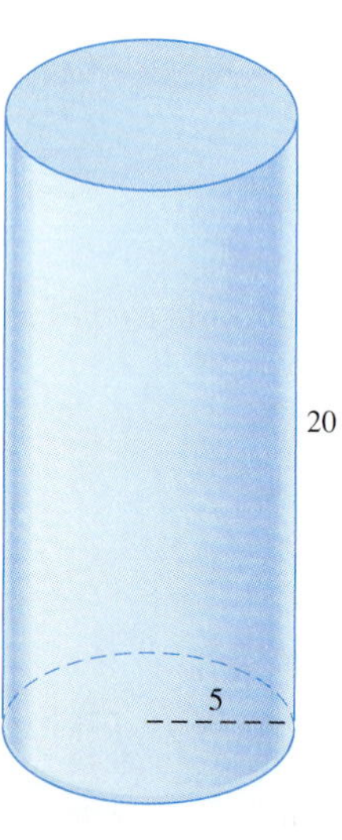

Fig. 3.6.8 The can of maximal volume in Example 4

Thus the can of maximum volume has radius $r = 5$ in., and Eq. (8) yields its height to be $h = 20$ in. Figure 3.6.8 shows such a can. ∎

EXAMPLE 5 (A SAWMILL PROBLEM) Suppose that you need to cut a beam with maximal rectangular cross section from a circular log of radius 1 ft. (This is the geometric problem of finding the rectangle of greatest area that can be inscribed in a circle of radius 1.) What are the shape and cross-sectional area of such a beam?

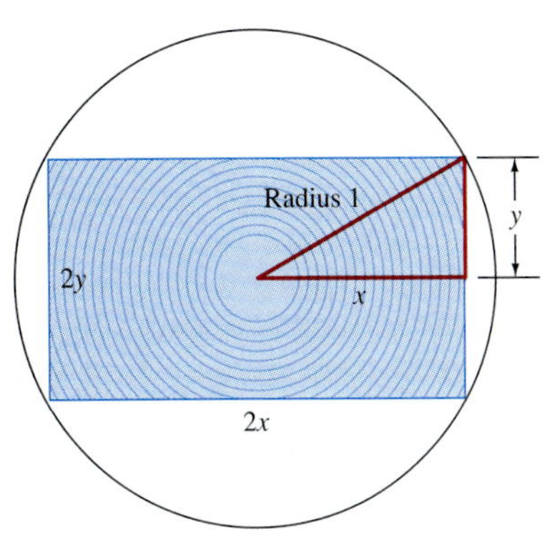

Fig. 3.6.9 A sawmill problem—Example 5

Solution Let x and y denote half the base and half the height, respectively, of the inscribed rectangle (Fig. 3.6.9). Apply the Pythagorean theorem to the small right triangle in the figure. This yields the equation

$$x^2 + y^2 = 1, \quad \text{so} \quad y = \sqrt{1 - x^2}.$$

The area of the inscribed rectangle is $A = (2x)(2y) = 4xy$. You may now express A as a function of x alone:

$$A(x) = 4x\sqrt{1 - x^2}.$$

The practical domain of definition of A is $(0, 1)$, and there is no harm (and much advantage) in adjoining the endpoints, so you take $[0, 1]$ to be the domain. Next,

$$\frac{dA}{dx} = 4 \cdot (1 - x^2)^{1/2} + 2x(1 - x^2)^{-1/2}(-2x) = \frac{4 - 8x^2}{(1 - x^2)^{1/2}}.$$

You observe that $A'(1)$ does not exist, but this causes no trouble, because differentiability at the endpoints is not assumed in Theorem 3 of Section 3.5. Hence you need only solve the equation

$$A'(x) = 0;$$

that is,

$$\frac{4 - 8x^2}{\sqrt{1 - x^2}} = 0.$$

A fraction can be zero only when its numerator is zero and its denominator is *not,* so $A'(x) = 0$ when $4 - 8x^2 = 0$. Thus you find the only critical point of A in the open interval $(0, 1)$ to be $x = \frac{1}{2}\sqrt{2}$ (and $2x = 2y = \sqrt{2}$). You evaluate A here and at the two endpoints to find that

$$A(0) = 0,$$

$$A\left(\frac{1}{2}\sqrt{2}\right) = 2, \leftarrow \text{absolute maximum}$$

$$A(1) = 0.$$

Therefore the beam with rectangular cross section of maximal area is square, with edges $\sqrt{2}$ ft long and with cross-sectional area 2 ft^2. ■

In Problem 43 we ask you to maximize the total cross-sectional area of the four planks that can be cut from the four pieces of log that remain after cutting the square beam (Fig. 3.6.10).

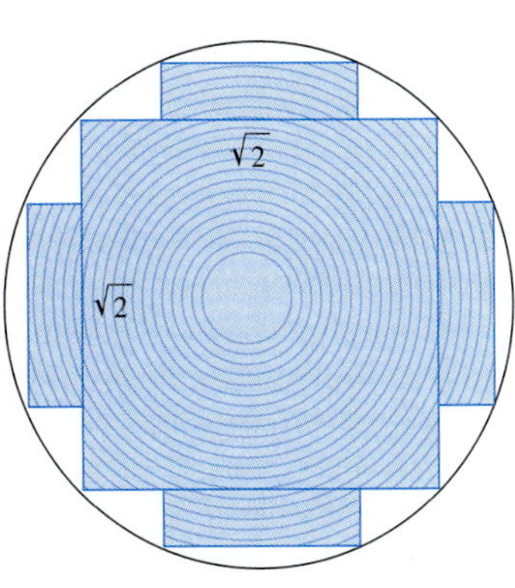

Fig. 3.6.10 Cut four more beams after cutting one large beam.

Plausibility You should always check your answers for *plausibility*. In Example 5, the cross-sectional area of the log from which the beam is to be cut is $\pi \approx 3.14$ ft^2. The beam of maximal cross-sectional area 2 ft^2 thus uses a little less than 64% of the log. This *is* plausible. Had the fraction been an extremely inefficient 3% or a wildly optimistic 98%, you should have searched for an error in arithmetic, algebra, calculus, or logic (as you would had the fraction been −14% or 150%). Check the results of Examples 1 through 4 for plausibility.

Dimensions Another way to check answers is to use *dimensional analysis.* Work the problem with unspecified constants in place of the actual numbers. In Example 5, it would be good practice to find the beam of maximal rectangular cross section that can be cut from a circular log of radius R rather than radius 1 ft. You can always substitute the given value $R = 1$ at the conclusion of the solution. A brief solution to this problem might go as follows:

Dimensions of beam: base $2x$, height $2y$.

Area of beam: $A = 4xy$.

Draw a radius of the log from its center to one corner of the rectangular beam, as in Fig. 3.6.11. This radius has length R, so the Pythagorean theorem gives

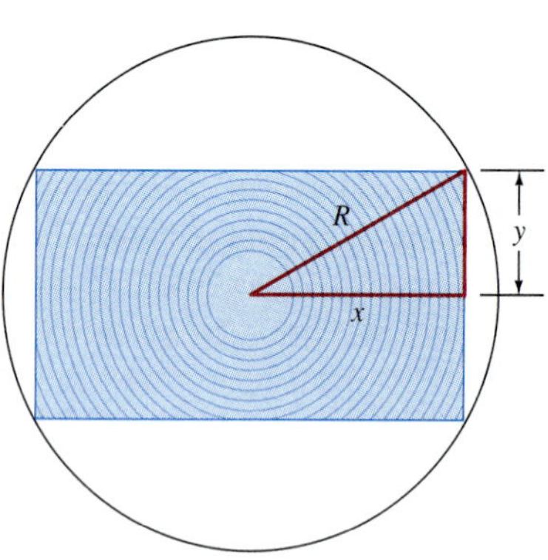

Fig. 3.6.11 The log with radius R

$$x^2 + y^2 = R^2; \quad y = \sqrt{R^2 - x^2}.$$

Area of beam:

$$A(x) = 4x\sqrt{R^2 - x^2}, \quad 0 \leqq x \leqq R.$$

$$A'(x) = 4(R^2 - x^2)^{1/2} + 2x(R^2 - x^2)^{-1/2}(-2x) = \frac{4R^2 - 8x^2}{\sqrt{R^2 - x^2}}.$$

$A'(x)$ does not exist when $x = R$, but that's an endpoint; we'll check it separately.

$A'(x) = 0$ when $x = \frac{1}{2}R\sqrt{2}$ (ignore the negative root; it's not in the domain of A).

$$A(0) = 0,$$

$$A\left(\frac{1}{2}R\sqrt{2}\right) = 2R^2, \quad \longleftarrow \text{ absolute maximum}$$

$$A(R) = 0.$$

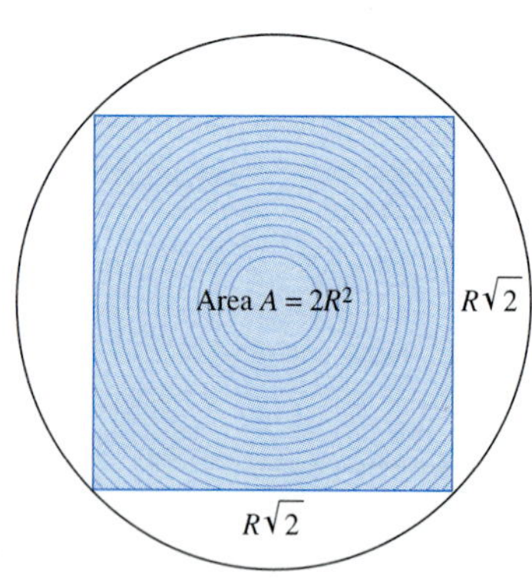

Fig. 3.6.12 The inscribed square beam with maximal cross-sectional area

Figure 3.6.12 shows the dimensions of the inscribed rectangle of maximal area.

Now you can check the results for dimensional accuracy. The value of x that maximizes A is a length (R) multiplied by a pure (dimensionless) numerical constant $(\frac{1}{2}\sqrt{2})$, so x has the dimensions of length—that's correct; had it been anything else, you would need to search for the error. Moreover, the maximum cross-sectional area of the beam is $2R^2$, the product of a pure number and the square of a length, thus having the dimensions of area. This, too, is correct.

EXAMPLE 6 We consider the reflection of a ray of light by a mirror M as in Fig. 3.6.13, which shows a ray traveling from point A to point B via reflection off M at the point P. We assume that the location of the point of reflection is such that the total distance $d_1 + d_2$ traveled by the light ray will be minimized. This is an application of *Fermat's principle of least time* for the propagation of light. The problem is to find P.

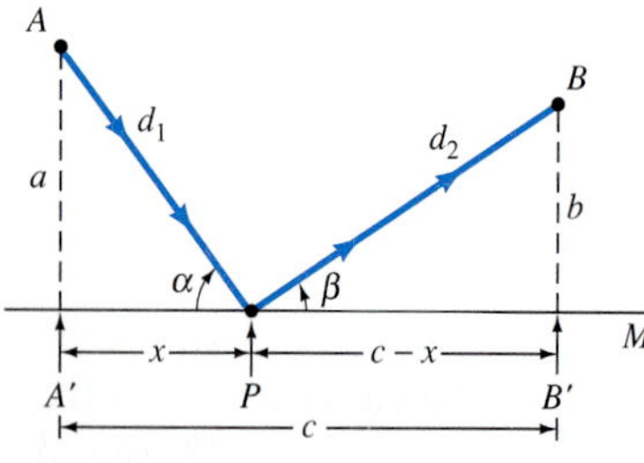

Fig. 3.6.13 Reflection at P of a light ray by a mirror M (Example 6)

Solution Drop perpendiculars from A and B to the plane of the mirror M. Denote the feet of these perpendiculars by A' and B' (Fig. 3.6.13). Let a, b, c, and x denote the lengths of the segments AA', BB', $A'B'$, and $A'P$, respectively. Then $c - x$ is the length of the segment PB'. By the Pythagorean theorem, the distance to be minimized is then

$$d_1 + d_2 = f(x) = \sqrt{a^2 + x^2} + \sqrt{b^2 + (c - x)^2}. \tag{10}$$

We may choose as the domain of f the interval $[0, c]$, because the minimum of f must occur somewhere within that interval. (To see why, examine the picture you get if x is *not* in that interval.)

Then

$$f'(x) = \frac{x}{\sqrt{a^2 + x^2}} + \frac{(c - x)(-1)}{\sqrt{b^2 + (c - x)^2}}. \tag{11}$$

Because

$$f'(x) = \frac{x}{d_1} - \frac{c - x}{d_2}, \tag{12}$$

we find that any horizontal tangent to the graph of f must occur over the point x determined by the equation

$$\frac{x}{d_1} = \frac{c - x}{d_2}. \tag{13}$$

At such a point, $\cos\alpha = \cos\beta$, where α is the angle of the incident light ray and β is the angle of the reflected ray (Fig. 3.6.13). Both α and β lie between 0 and $\pi/2$, and thus we find that $\alpha = \beta$. In other words, the angle of incidence is equal to the angle of reflection, a familiar principle from physics. ■

The computation in Example 6 has an alternative interpretation that is interesting, if somewhat whimsical. Figure 3.6.14 shows a feedlot 200 ft long with a water trough along one edge and a feed bin located on an adjacent edge. A cow enters the gate at the point A, 90 ft from the water trough. She walks straight to point P, gets a drink from the trough, and then walks straight to the feed bin at point B, 60 ft from the trough. If the cow knew calculus, what point P along the water trough would she select to minimize the total distance she walks?

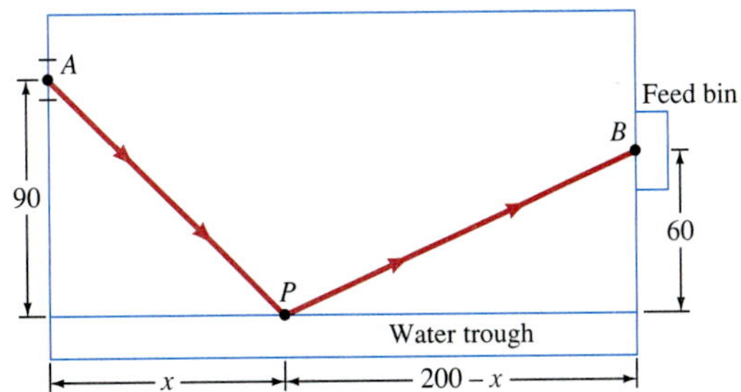

Fig. 3.6.14 The feedlot

In comparing Figs. 3.6.13 and 3.6.14, we see that the cow's problem is to minimize the distance function f in Eq. (10) with the numerical values $a = 90$, $b = 60$, and $c = 200$. When we substitute these values and

$$d_1 = \sqrt{a^2 + x^2} \quad \text{and} \quad d_2 = \sqrt{b^2 + (c - x)^2}$$

in Eq. (13), we get

$$\frac{x}{\sqrt{8100 + x^2}} = \frac{200 - x}{\sqrt{3600 + (200 - x)^2}}.$$

We square both sides, clear the equation of fractions, and simplify. The result is

$$\begin{aligned} x^2[3600 + (200 - x)^2] &= (200 - x)^2(8100 + x^2); \\ 3600x^2 &= 8100(200 - x)^2; \quad \text{(Why?)} \\ 60x &= 90(200 - x); \\ 150x &= 18000; \\ x &= 120. \end{aligned}$$

Thus the cow should proceed directly to the point P located 120 ft along the water trough. ■

These examples indicate that the closed-interval maximum-minimum method is applicable to a wide range of problems. Indeed, applied optimization problems that seem as different as light rays and cows may have essentially identical mathematical models. This is only one illustration of the power of generality that calculus exploits so effectively.

3.6 PROBLEMS

1. Find two positive real numbers x and y such that their sum is 50 and their product is as large as possible.

2. Find the maximum possible area of a rectangle of perimeter 200 m.

3. A rectangle with sides parallel to the coordinate axes has one vertex at the origin, one on the positive x-axis, one on the positive y-axis, and its fourth vertex in the first quadrant on the line with equation $2x + y = 100$ (Fig. 3.6.15). What is the maximum possible area of such a rectangle?

4. A farmer has 600 m of fencing with which to enclose a rectangular pen adjacent to a long existing wall. He will use the wall for one side of the pen and the available fencing for the remaining three sides. What is the maximum area that can be enclosed in this way?

5. A rectangular box has a square base with edges at least 1 in. long. It has no top, and the total area of its five sides

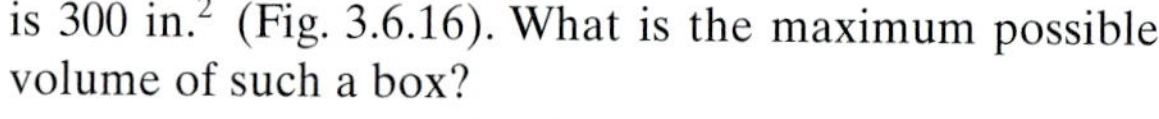

is 300 in.^{2} (Fig. 3.6.16). What is the maximum possible volume of such a box?

6. If x is in the interval $[0, 1]$, then $x - x^2$ is not negative. What is the maximum value that $x - x^2$ can have on that interval? In other words, what is the greatest amount by which a real number can exceed its square?

7. The sum of two positive numbers is 48. What is the smallest possible value of the sum of their squares?

8. A rectangle of fixed perimeter 36 is rotated around one of its sides, thus sweeping out a figure in the shape of a right circular cylinder (Fig. 3.6.17). What is the maximum possible volume of that cylinder?

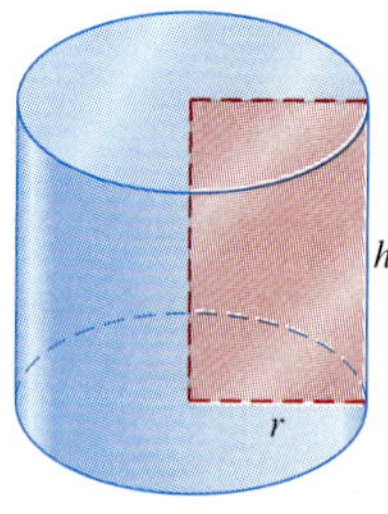

Fig. 3.6.17 The rectangle and cylinder of Problem 8

(x, y)

2x + y = 100

x

Fig. 3.6.15 The rectangle of Problem 3

y

x

x

Fig. 3.6.16 A box with square base and volume $V = x^2y$ (Problems 5, 17, and 20)

9. The sum of two nonnegative real numbers is 10. Find the minimum possible value of the sum of their cubes.

10. Suppose that the strength of a rectangular beam is proportional to the product of the width and the square of the height of its cross section. What shape beam should be cut from a cylindrical log of radius r to achieve the greatest possible strength?

11. A farmer has 600 yd of fencing with which to build a rectangular corral. Some of the fencing will be used to construct two internal divider fences, both parallel to the same two sides of the corral (Fig. 3.6.18). What is the maximum possible total area of such a corral?

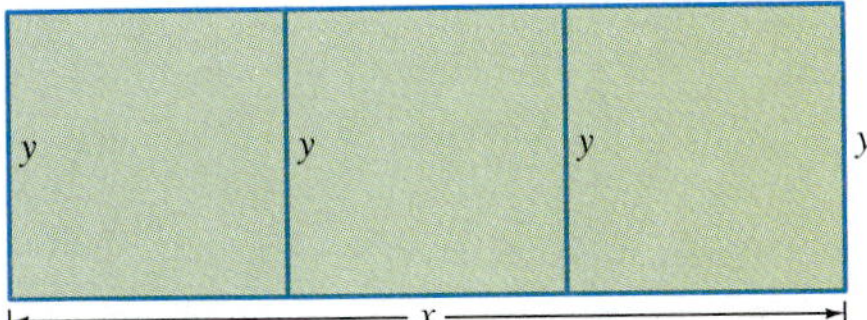

Fig. 3.6.18 The divided corral of Problem 11

12. Find the maximum possible volume of a right circular cylinder if its total surface area—including both circular ends—is 150π.
13. Find the maximum possible area of a rectangle with diagonals of length 16.
14. A rectangle has a line of fixed length L reaching from one vertex to the midpoint of one of the far sides (Fig. 3.6.19). What is the maximum possible area of such a rectangle?

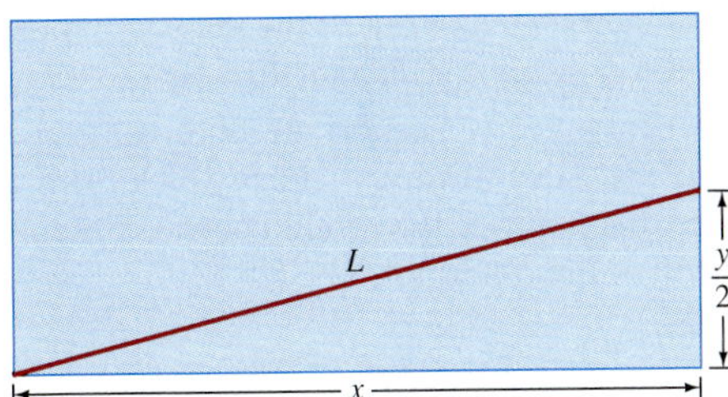

Fig. 3.6.19 The rectangle of Problem 14

15. The volume V (in cubic centimeters) of 1 kg of water at temperature T between 0°C and 30°C is very closely approximated by

$$V = 999.87 - (0.06426)T + (0.0085043)T^2 - (0.0000679)T^3.$$

At what temperature does water have its maximum density?

16. What is the maximum possible area of a rectangle with a base that lies on the x-axis and with two upper vertices that lie on the graph of the equation $y = 4 - x^2$ (Fig. 3.6.20)?
17. A rectangular box has a square base with edges at least 1 cm long. Its total surface area is 600 cm^2. What is the largest possible volume that such a box can have?
18. You must make a cylindrical can with a bottom but no top from 300π in.^{2} of sheet metal. No sheet metal will be wasted; you are allowed to order a circular piece of any size for its base and any appropriate rectangular piece to make into its curved side as long as the given conditions are met. What is the greatest possible volume of such a can?
19. Three large squares of tin, each with edges 1 m long, have four small, equal squares cut from their corners. All twelve resulting small squares are to be the same size (Fig. 3.6.21). The three large cross-shaped pieces are then folded and welded to make boxes with no tops, and the twelve small squares are used to make two small cubes. How should this be done to maximize the total volume of all five boxes?

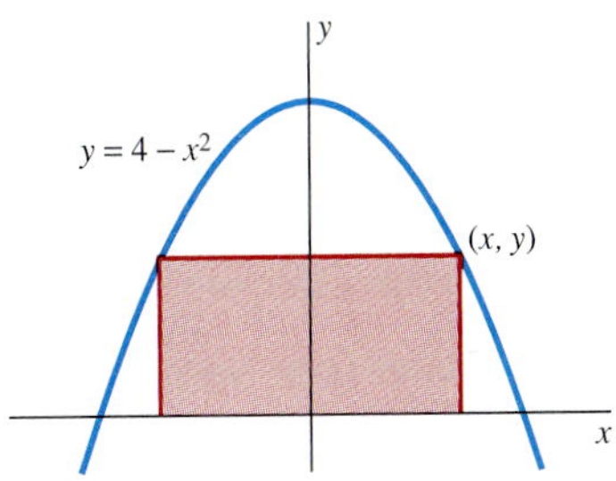

Fig. 3.6.20 The rectangle of Problem 16

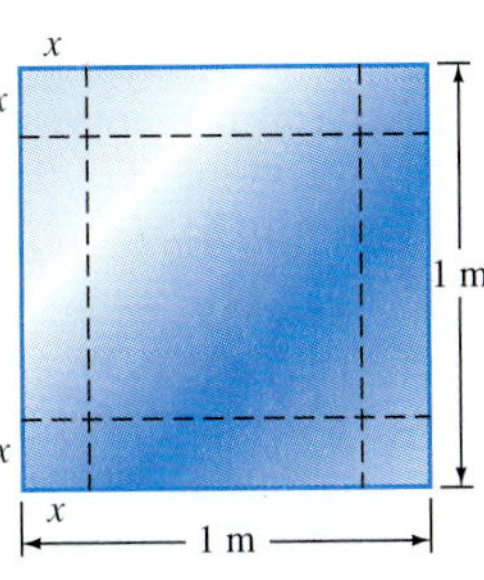

Fig. 3.6.21 One of the three 1-m squares of Problem 19

20. Suppose that you are to make a rectangular box with a square base from two different materials. The material for the top and four sides of the box costs $\$1/\text{ft}^2$; the material for the base costs $\$2/\text{ft}^2$. Find the dimensions of the box of greatest possible volume if you are allowed to spend \$144 for the material to make it.
21. A piece of wire 80 in. long is cut into at most two pieces. Each piece is bent into the shape of a square. How should this be done to minimize the sum of the area(s) of the square(s)? to maximize it?
22. A wire of length 100 cm is cut into two pieces. One piece is bent into a circle, the other into a square. Where should the cut be made to maximize the sum of the areas of the square and the circle? to minimize that sum?
23. A farmer has 600 m of fencing with which she plans to enclose a rectangular pasture adjacent to a long existing wall. She plans to build one fence parallel to the wall, two to form the ends of the enclosure, and a fourth (parallel to the ends of the enclosure) to divide it equally. What is the maximum area that can be enclosed?
24. A zookeeper needs to add a rectangular outdoor pen to an animal house with a corner notch, as shown in Fig. 3.6.22. If

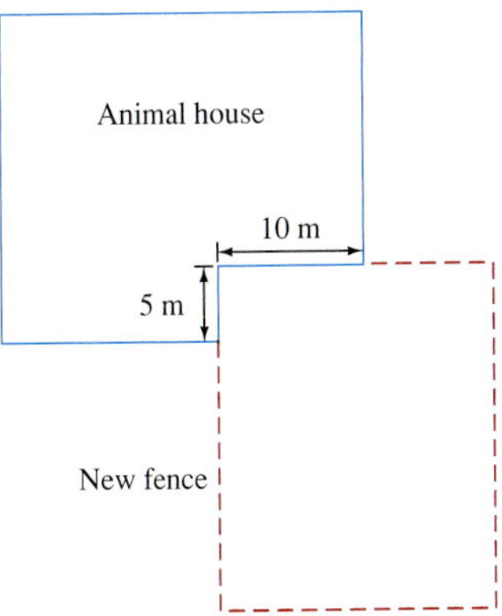

Fig. 3.6.22 The rectangular pen of Problem 24

85 m of new fence is available, what dimensions of the pen will maximize its area? No fence will be used along the walls of the animal house.

25. Suppose that a post office can accept a package for mailing only if the sum of its length and its girth (the circumference of its cross section) is at most 100 in. What is the maximum volume of a rectangular box with square cross section that can be mailed?

26. Repeat Problem 25, but use a cylindrical package; its cross section is circular.

27. A printing company has eight presses, each of which can print 3600 copies per hour. It costs \$5.00 to set up each press for a run and $10 + 6n$ dollars to run n presses for 1 h. How many presses should be used to print 50000 copies of a poster most profitably?

28. A farmer wants to hire workers to pick 900 bushels of beans. Each worker can pick 5 bushels per hour and is paid \$1.00 per bushel. The farmer must also pay a supervisor \$10 per hour while the picking is in progress, and he has additional miscellaneous expenses of \$8 per worker. How many workers should he hire to minimize the total cost? What will then be the cost per bushel picked?

29. The heating and cooling costs for a certain uninsulated house are \$500/yr, but with $x \leqq 10$ in. of insulation, the costs are $1000/(2 + x)$ dollars/yr. It costs \$150 for each inch (thickness) of insulation installed. How many inches of insulation should be installed to minimize the *total* (initial plus annual) costs over a 10-yr period? What will then be the annual savings resulting from this optimal insulation?

30. A concessionaire had been selling 5000 burritos each game night at 50¢ each. When she raised the price to 70¢ each, sales dropped to 4000 per night. Assume a linear relationship between price and sales. If she has fixed costs of \$1000 per night and each burrito costs her 25¢, what price will maximize her nightly profit?

31. A commuter train carries 600 passengers each day from a suburb to a city. The cost to ride the train is \$1.50 per person. Market research reveals that 40 fewer people would ride the train for each 5¢ increase in the fare, 40 more for each 5¢ decrease. What fare should be charged to get the largest possible revenue?

32. Find the shape of the cylinder of maximal volume that can be inscribed in a sphere of radius R (Fig. 3.6.23). Show that the ratio of the height of the cylinder to its radius is $\sqrt{2}$ and that the ratio of the volume of the sphere to that of the maximal cylinder is $\sqrt{3}$.

33. Find the dimensions of the right circular cylinder of greatest volume that can be inscribed in a right circular cone of radius R and height H (Fig. 3.6.24).

34. Figure 3.6.25 shows a circle of radius 1 in which a trapezoid is inscribed. The longer of the two parallel sides of the trapezoid coincides with a diameter of the circle. What is the maximum possible area of such a trapezoid? (*Suggestion:* A positive quantity is maximized when its square is maximized.)

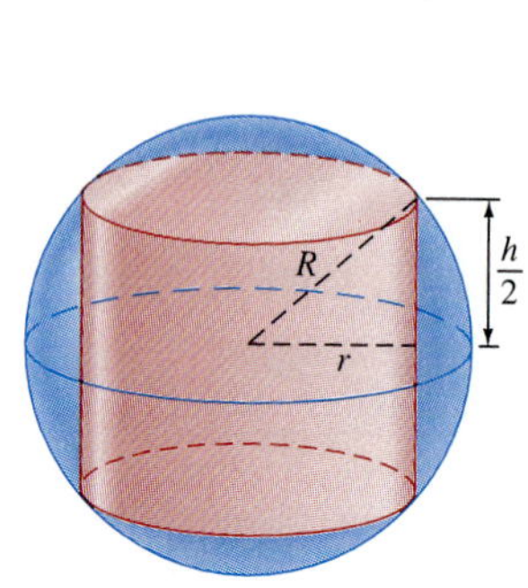

Fig. 3.6.23 The sphere and cylinder of Problem 32

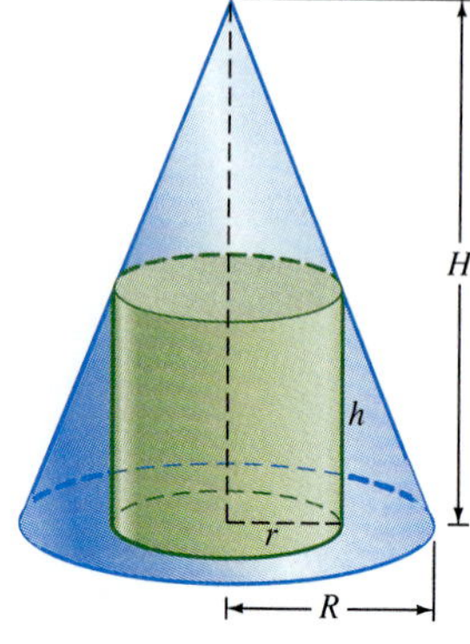

Fig. 3.6.24 The cone and cylinder of Problem 33

35. Show that the rectangle of maximal perimeter that can be inscribed in a circle is a square.

36. Find the dimensions of the rectangle (with sides parallel to the coordinate axes) of maximal area that can be inscribed in the ellipse shown in Fig. 3.6.26.

37. A right circular cone of radius r and height h has slant height $L = \sqrt{r^2 + h^2}$. What is the maximum possible volume of a cone with slant height 10?

38. Two vertical poles 10 ft apart are both 10 ft tall. Find the length of the shortest rope that can reach from the top of one pole to a point on the ground between them and then to the top of the other pole.

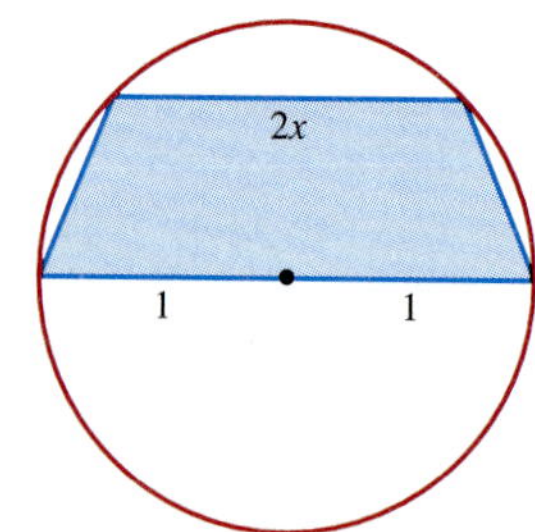

Fig. 3.6.25 The circle and trapezoid of Problem 34

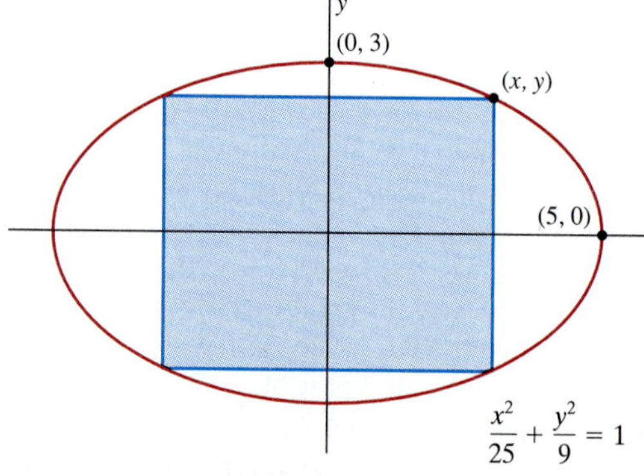

Fig. 3.6.26 The ellipse and rectangle of Problem 36

39. The sum of two nonnegative real numbers is 16. Find the maximum possible value and the minimum possible value of the sum of their cube roots.

40. A straight wire 60 cm long is bent into an L shape. What is the shortest possible distance between the two ends of the bent wire?

41. What is the shortest possible distance from a point on the parabola $y = x^2$ to the point $(0, 1)$?

42. Given: There is exactly one point on the graph of $y = \sqrt[3]{3x - 4}$ that is closest to the origin. Find it. (*Suggestion:* See Fig. 3.6.27, and solve the equation you obtain by inspection.)

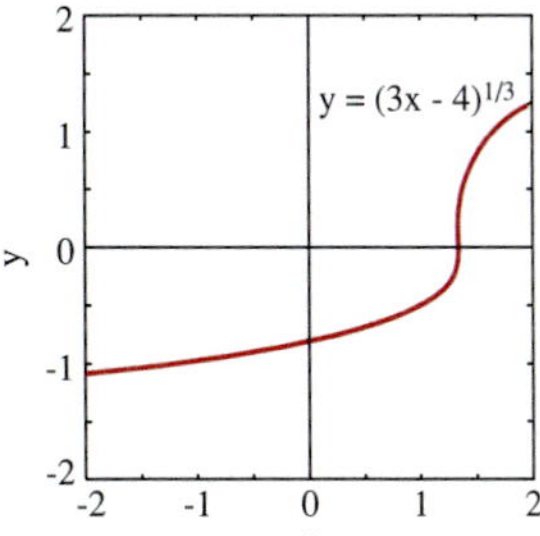

Fig. 3.6.27 The curve of Problem 42

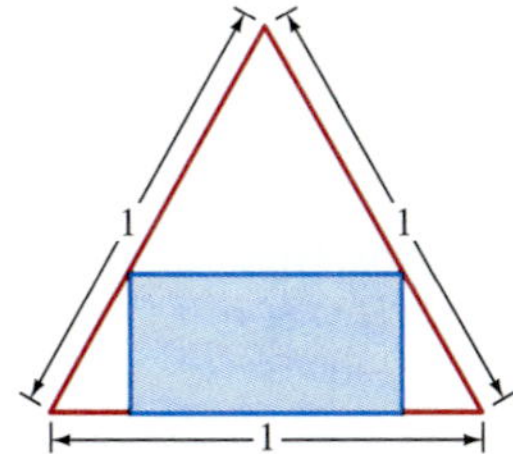

Fig. 3.6.28 The rectangle and equilateral triangle of Problem 44

43. Find the dimensions that maximize the cross-sectional area of the four planks that can be cut from the four pieces of the circular log of Example 5—the pieces that remain after a square beam has been cut from the log (see Fig. 3.6.10).

44. Find the maximal area of a rectangle inscribed in an equilateral triangle with edges of length 1, as in Fig. 3.6.28.

45. A small island is 2 km off shore in a large lake. A woman on the island can row her boat 10 km/h and can run at a speed of 20 km/h. If she rows to the closest point of the straight shore, she will land 6 km from a village on the shore. Where should she land to reach the village most quickly by a combination of rowing and running?

46. A factory is located on one bank of a straight river that is 2000 m wide. On the opposite bank but 4500 m downstream is a power station from which the factory draws its electricity. Assume that it costs three times as much per meter to lay an underwater cable as to lay an aboveground cable. What path should a cable connecting the power station to the factory take to minimize the cost of laying the cable?

47. A company has plants that are located (in an appropriate coordinate system) at the points $A(0, 1)$, $B(0, -1)$, and $C(3, 0)$ (Fig. 3.6.29). The company plans to construct a distribution center at the point $P(x, 0)$. What value of x would minimize the sum of the distances from P to A, B, and C?

48. Light travels at speed c in air and at a slower speed v in water. (The constant c is approximately 3×10^{10} cm/s; the ratio $n = c/v$, known as the **index of refraction,** depends on the color of the light but is approximately 1.33 for water.)

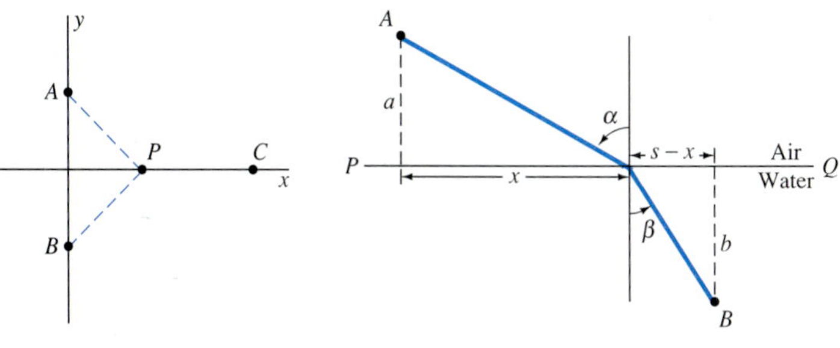

Fig. 3.6.29 The locations in Problem 47

Fig. 3.6.30 Snell's law gives the path of refracted light (Problem 48)

Figure 3.6.30 shows the path of a light ray traveling from point A in air to point B in water, with what appears to be a sudden change in direction as the ray moves through the air-water interface. (a) Write the time T required for the ray to travel from A to B in terms of the variable x and the constants a, b, c, s, and v, all of which have been defined or are shown in the figure. (b) Show that the equation $T'(x) = 0$ for minimizing T is equivalent to the condition

$$\frac{\sin \alpha}{\sin \beta} = \frac{c}{v} = n.$$

This is **Snell's law:** The ratio of the sines of the angles of incidence and refraction is equal to the index of refraction.

Refraction of light at an air-water interface

49. The mathematics of Snell's law (Problem 48) is applicable to situations other than the refraction of light. Figure 3.6.31 shows an east-west geologic fault that separates two towns at points A and B. Assume that A is a miles north of the fault, that B is b miles south of the fault, and that B is L miles east of A. We want to build a road from A to B. Because of differences in terrain, the cost of construction is C_1 (in millions of dollars per mile) north of the fault and C_2 south

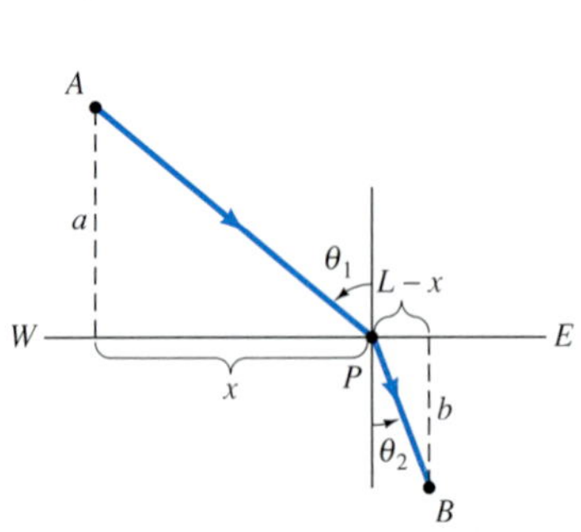

Fig. 3.6.31 Building a road from A to B (Problem 49)

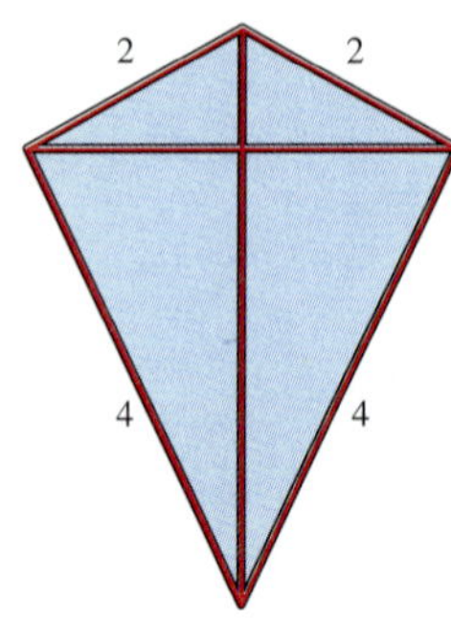

Fig. 3.6.32 The kite frame (Problem 52)

of it. Where should the point P be placed to minimize the total cost of road construction? (a) Using the notation in the figure, show that the cost is minimized when $C_1 \sin\theta_1 = C_2 \sin\theta_2$. (b) Take $a = b = C_1 = 1, C_2 = 2$, and $L = 4$. Show that the equation in part (a) is equivalent to

$$f(x) = 3x^4 - 24x^3 + 51x^2 - 32x + 64 = 0.$$

To approximate the desired solution of this equation, calculate $f(0), f(1), f(2), f(3)$, and $f(4)$. You should find that $f(3) > 0 > f(4)$. Interpolate between $f(3)$ and $f(4)$ to approximate the desired root of this equation.

50. The sum of the volumes of two cubes is 2000 in.^{3} What should their edges x and y be to maximize the sum of their surface areas? to minimize it?

51. The sum of the surface areas of a cube and a sphere is 1000 in.^{2} What should their dimensions be to minimize the sum of their volumes? to maximize it?

52. Your brother has six pieces of wood with which to make the kite frame shown in Fig. 3.6.32. The four outer pieces with the indicated lengths have already been cut. How long should the lengths of the inner struts be to maximize the area of the kite?

The following three problems deal with alternative methods of constructing a tent.

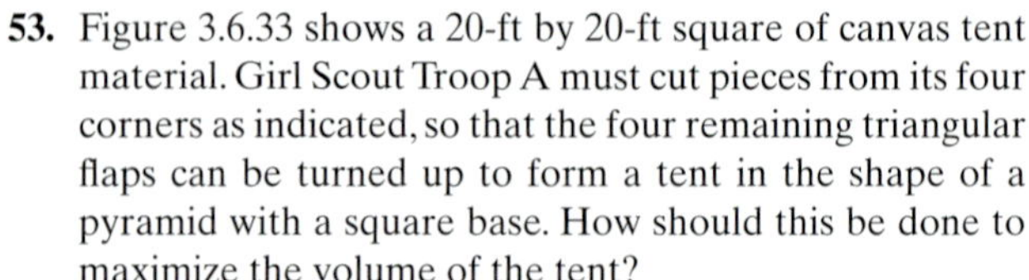

53. Figure 3.6.33 shows a 20-ft by 20-ft square of canvas tent material. Girl Scout Troop A must cut pieces from its four corners as indicated, so that the four remaining triangular flaps can be turned up to form a tent in the shape of a pyramid with a square base. How should this be done to maximize the volume of the tent?

Let A denote the area of the base of the tent and h its height. With x as indicated in the figure, show that the volume $V = \frac{1}{3}Ah$ of the tent is given by

$$V(x) = \tfrac{4}{3}x^2\sqrt{100 - 20x}, \quad 0 \leqq x \leqq 5.$$

Maximize V by graphing $V(x)$ and $V'(x)$ and zooming in on the zero of $V'(x)$.

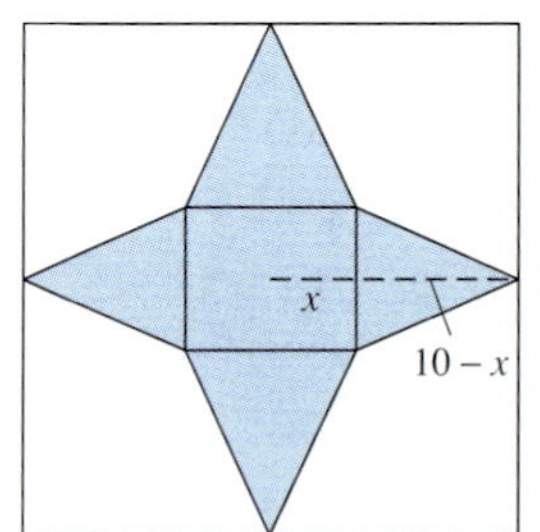

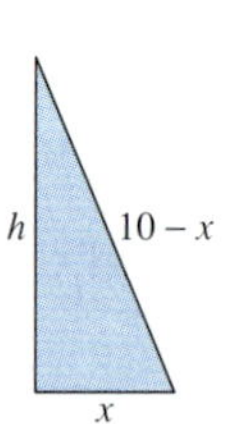

Fig. 3.6.33 The canvas square—first attempt

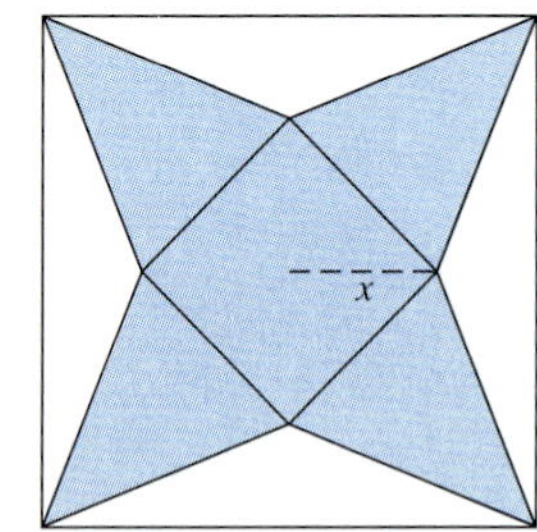

Fig. 3.6.34 The canvas square—second attempt

54. Girl Scout Troop B must make a tent in the shape of a pyramid with a square base from a similar 20-ft by 20-ft square of canvas but in the manner indicated in Fig. 3.6.34. With x as indicated in the figure, show that the volume of the tent is given by

$$V(x) = \tfrac{2}{3}x^2\sqrt{200 - 20x}, \quad 0 \leqq x \leqq 10.$$

Maximize V graphically as in Problem 53.

55. Solve Problems 53 and 54 analytically to verify that the maximal volume in Problem 54 is exactly $2\sqrt{2}$ times the maximal volume in Problem 53. It pays to think before making a tent!

3.6 PROJECT A: MAKING A CANDY BOX WITH LID

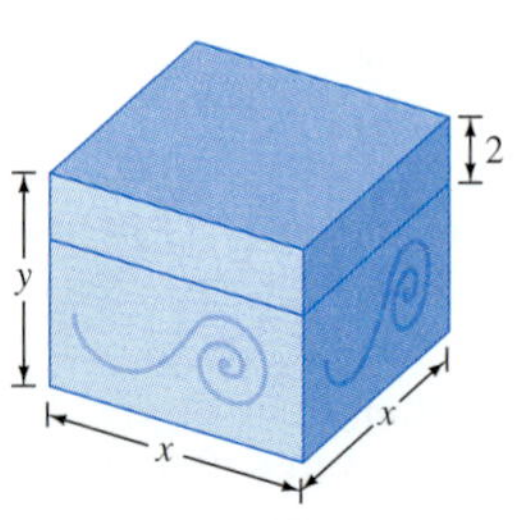

Fig. 3.6.35 The square-based candy box with lid

A candy maker wants to package jelly beans in boxes each having a volume of 1000 in.^{3} Each of these jelly bean boxes is to be an open-topped rectangular box with a square base having edge length x between 6 and 12 in. (Fig. 3.6.35). In addition, the box is also to have a square lid with a 2-in. rim. Thus the box-with-lid actually consists of two open-topped boxes—the x by x by y candy box itself with height $y \geqq 2$ in., and the x by x by 2 lid with height 2 in. (we assume that the lid fits *very* snugly). The candy maker's problem is to determine the dimensions x and y that will minimize the total area A (and hence the cost) of the two open-topped boxes that comprise a single candy box with its lid.

Solve the candy maker's problem. Begin by expressing the total area of the two open-topped boxes as a function A of the base edge length x. Show that the equation $A'(x) = 0$ simplifies to the cubic equation

$$x^3 + 2x^2 - 1000 = 0.$$

But instead of attempting to solve this equation directly, graph $A(x)$ and $A'(x)$ on the same set of coordinate axes and zoom in on the zero of $A'(x)$.

For your own personal candy box problem—or is it to be a home for your pet ferret?—suppose that the box itself is to have volume $V = 400 + 50n$ in.^{3}, where n is the last nonzero digit in your student I.D. number. Its lid still has a 2-in. rim. If you make your box-with-lid out of nice foil-covered cardboard costing \$1/ft^2, what dimensions will minimize the total cost of the material needed? What is this minimal cost?

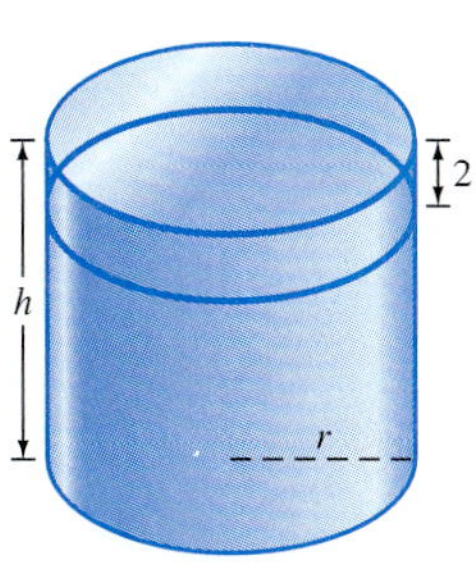

Fig. 3.6.36 The cylindrical candy box with lid

Suppose instead that you want to make a cylindrical box as indicated in Fig. 3.6.36. Now the base box and its lid are both open-topped circular cylinders. Find the dimensions r and h that minimize the cost of this cylindrical box-with-lid (with the same volume $V = 400 + 50n$ in.^{3} as before, and with its lid having a 2-in. rim as before). Which costs less to make—the rectangular box or the cylindrical box?

3.6 PROJECT B: POWER LINE DESIGN

You are the city engineer for the town of Villabuena, which encircles a beautiful park. This park is a square one mile on each side. The local power company needs to run a power line from the northwest corner A of the park to the southeast corner B (see Fig. 3.6.37). Local ordinances prohibit overhead lines within the park, but it is permitted to run underground cable through the park. It is your job to determine the optimal design of the power line, the design that will minimize the cost of the power line.

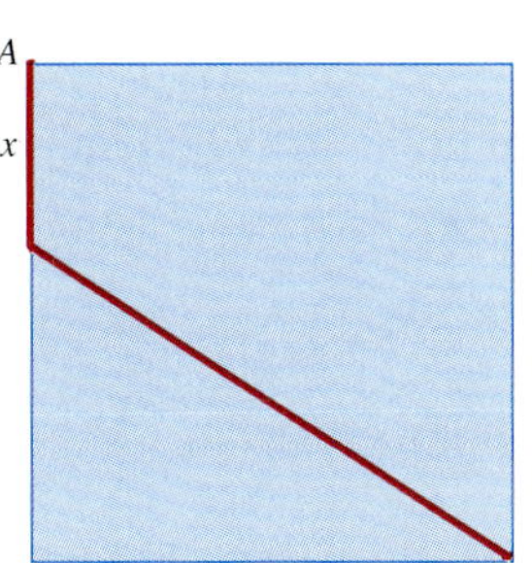

Fig. 3.6.37 The one-mile-square park in central Villabuena

You can see that one reasonable design for the power line is to stretch overhead lines along one edge of the park for a distance x, then lay underground cable straight to point B. Suppose that you already have a bid on the underground cable; it will cost \$100,000/mi. You await the final bid on the overhead cable, but the city council wants you to report at the next meeting, before the bid comes in, exactly how to design the least expensive cable. The overhead cable will cost k hundred thousand dollars per mile, where $0 \leqq k \leqq 1$. Your design will, of course, depend on the bid that you await, but you must present a clear report tonight to the city council that explains how the path of the power line depends on the bid (i.e., upon the value of k) and how you made your calculations. Fortunately, all of the city council members have studied calculus. Experiment with various values of k to determine the best design for your power line, depending on the value of k, to minimize the cost of this project, and write your report to the city council.

3.7 DERIVATIVES OF TRIGONOMETRIC FUNCTIONS

In this section we begin our study of the calculus of trigonometric functions, focusing first on the sine and cosine functions. The definitions and the elementary properties of trigonometric functions are reviewed in Appendix C.

When we write $\sin\theta$ (or $\cos\theta$), we mean the sine (or cosine) of an angle of θ radians (rad). Recall the fundamental relation between radian measure and degree measure of angles:

$$\pi \text{ radians} = 180 \text{ degrees}. \tag{1}$$

Upon division of both sides of this equation by π and 180, respectively, and abbreviating the units, we get the conversion relations

$$1 \text{ rad} = \frac{180}{\pi} \text{ deg} \quad \text{and} \quad 1 \text{ deg} = \frac{\pi}{180} \text{ rad}.$$

Radians	Degrees
0	0
$\pi/6$	30
$\pi/4$	45
$\pi/3$	60
$\pi/2$	90
$2\pi/3$	120
$3\pi/4$	135
$5\pi/6$	150
π	180
$3\pi/2$	270
2π	360
4π	720

Fig. 3.7.1 Some radian-degree conversions

Figure 3.7.1 shows radian-degree conversions for some frequently occurring angles.

The derivatives of the sine and cosine functions depend on the limits

$$\lim_{\theta \to 0} \frac{\sin \theta}{\theta} = 1, \qquad \lim_{\theta \to 0} \frac{1 - \cos \theta}{\theta} = 0 \tag{2}$$

that we established in Section 2.3. The addition formulas

$$\begin{aligned} \cos(x + y) &= \cos x \cos y - \sin x \sin y, \\ \sin(x + y) &= \sin x \cos y + \cos x \sin y \end{aligned} \tag{3}$$

are needed as well.

Theorem 1 ***Derivatives of Sines and Cosines***

The functions $f(x) = \sin x$ and $g(x) = \cos x$ are differentiable for all x, and

$$D_x \sin x = \cos x, \tag{4}$$

$$D_x \cos x = -\sin x. \tag{5}$$

PROOF To differentiate $f(x) = \sin x$, we begin with the definition of the derivative,

$$f'(x) = \lim_{h \to 0} \frac{f(x + h) - f(x)}{h} = \lim_{h \to 0} \frac{\sin(x + h) - \sin x}{h}.$$

Next we apply the addition formula for the sine and the limit laws to get

$$\begin{aligned} f'(x) &= \lim_{h \to 0} \frac{(\sin x \cos h + \sin h \cos x) - \sin x}{h} \\ &= \lim_{h \to 0} \left[(\cos x) \frac{\sin h}{h} - (\sin x) \frac{1 - \cos h}{h} \right] \\ &= (\cos x) \left(\lim_{h \to 0} \frac{\sin h}{h} \right) - (\sin x) \left(\lim_{h \to 0} \frac{1 - \cos h}{h} \right). \end{aligned}$$

The limits in Eq. (2) now yield

$$f'(x) = (\cos x)(1) - (\sin x)(0) = \cos x,$$

which proves Eq. (4). The proof of Eq. (5) is quite similar (see Problem 72). ■

Examples 1 through 4 illustrate the applications of Eqs. (4) and (5) in conjunction with the general differentiation formulas of Sections 3.2, 3.3, and 3.4 to differentiate various combinations of trigonometric and other functions.

EXAMPLE 1 The product rule yields

$$\begin{aligned} D_x(x^2 \sin x) &= (D_x x^2)(\sin x) + (x^2)(D_x \sin x) \\ &= 2x \sin x + x^2 \cos x. \end{aligned}$$

■

EXAMPLE 2 If $y = \dfrac{\cos x}{1 - \sin x}$, then the quotient rule yields

$$\begin{aligned}\frac{dy}{dx} &= \frac{(D_x \cos x)(1 - \sin x) - (\cos x)[D_x(1 - \sin x)]}{(1 - \sin x)^2}\\ &= \frac{(-\sin x)(1 - \sin x) - (\cos x)(-\cos x)}{(1 - \sin x)^2}\\ &= \frac{-\sin x + \sin^2 x + \cos^2 x}{(1 - \sin x)^2} = \frac{-\sin x + 1}{(1 - \sin x)^2};\end{aligned}$$

$$\frac{dy}{dx} = \frac{1}{1 - \sin x}.$$

EXAMPLE 3 If $x = \cos^3 t$ and $u = \cos t$—so that $x = u^3$—then the chain rule yields

$$\frac{dx}{dt} = \frac{dx}{du}\frac{du}{dt} = (3u^2)(-\sin t) = (3\cos^2 t)(-\sin t) = -3\cos^2 t \sin t.$$

EXAMPLE 4 If $g(t) = (2 - 3\cos t)^{3/2}$, then the chain rule yields

$$\begin{aligned}g'(t) &= \frac{3}{2}(2 - 3\cos t)^{1/2} D_t(2 - 3\cos t)\\ &= \frac{3}{2}(2 - 3\cos t)^{1/2}(3\sin t) = \frac{9}{2}(2 - 3\cos t)^{1/2}\sin t.\end{aligned}$$

EXAMPLE 5 Write an equation of the line tangent to the curve $y = \cos^2 x$ at the point P on the graph where $x = 0.5$. Approximations are allowed.

Solution The y-coordinate of P is $y(0.5) = (\cos 0.5)^2 \approx (0.8776)^2 \approx 0.7702$. Because

$$\frac{dy}{dx} = -2\cos x \sin x,$$

the slope of the tangent line at P is

$$m = \left.\frac{dy}{dx}\right|_{x=0.5} = -2(\cos 0.5)(\sin 0.5) \approx -0.8415.$$

Then the point-slope formula gives the (approximate) equation

$$y - 0.7702 = -(0.8415)(x - 0.5);$$

that is, $y = -(0.8415)x + 1.1909$, as the desired equation of the tangent line at P. Figure 3.7.2 shows the result of checking this computation by graphing both the curve $y = \cos^2 x$ and the line with this equation.

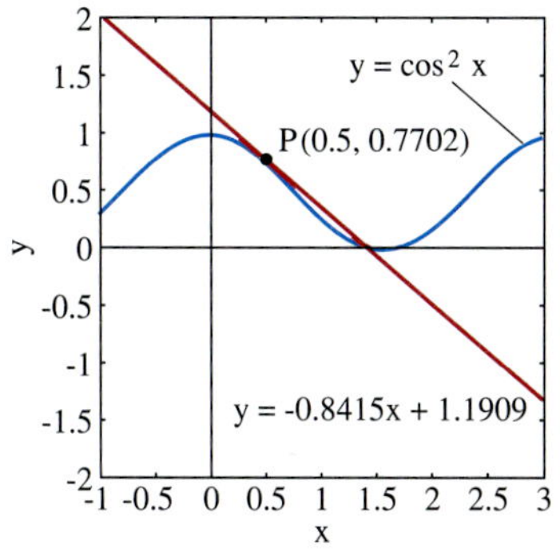

Fig. 3.7.2 The curve $y = \cos^2 x$ and its tangent line at the point P where $x = 0.5$

The Remaining Trigonometric Functions

It is easy to differentiate the other four trigonometric functions, because they can be expressed in terms of the sine and cosine functions:

$$\tan x = \frac{\sin x}{\cos x}, \quad \cot x = \frac{\cos x}{\sin x},$$
$$\sec x = \frac{1}{\cos x}, \quad \csc x = \frac{1}{\sin x}. \tag{6}$$

Each of these formulas is valid except where a zero denominator is encountered. Thus $\tan x$ and $\sec x$ are undefined when x is an odd integral multiple of $\pi/2$, and $\cot x$ and $\csc x$ are undefined when x is an integral multiple of π. The graphs of the six trigonometric functions appear in Fig. 3.7.3. There we show the sine and its reciprocal, the cosecant, in the same coordinate plane; we also pair the cosine with the secant but show the tangent and cotangent functions separately.

The functions in Eq. (6) can be differentiated by using the quotient rule and the derivatives of the sine and cosine functions. For example,

$$\tan x = \frac{\sin x}{\cos x},$$

so

$$D_x \tan x = \frac{(D_x \sin x)(\cos x) - (\sin x)(D_x \cos x)}{(\cos x)^2}$$
$$= \frac{(\cos x)(\cos x) - (\sin x)(-\sin x)}{\cos^2 x} = \frac{\cos^2 x + \sin^2 x}{\cos^2 x} = \frac{1}{\cos^2 x};$$
$$D_x \tan x = \sec^2 x.$$

As an exercise, you should derive in similar fashion the differentiation formulas in Eqs. (8) through (10) of Theorem 2.

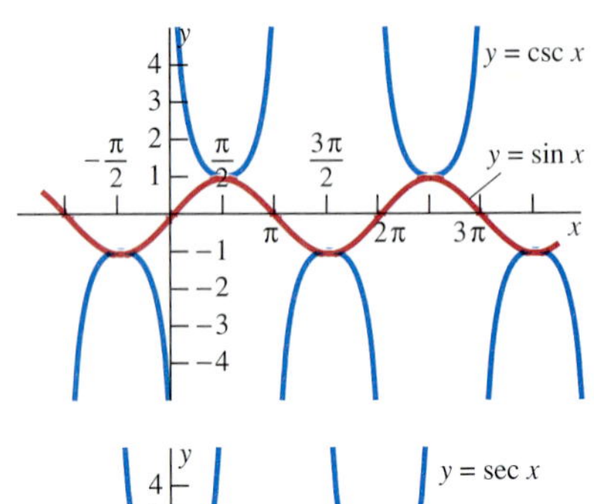

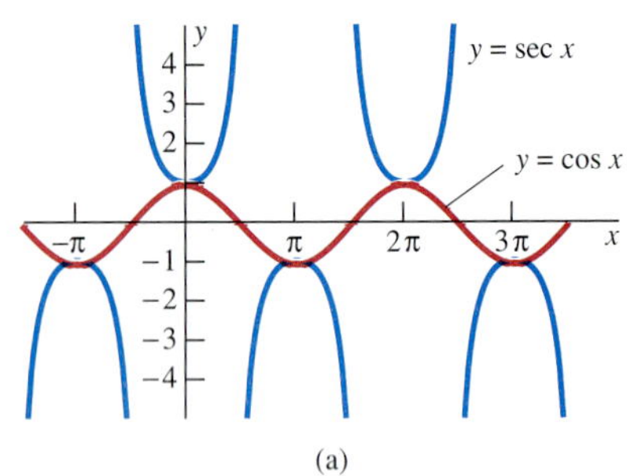

(a)

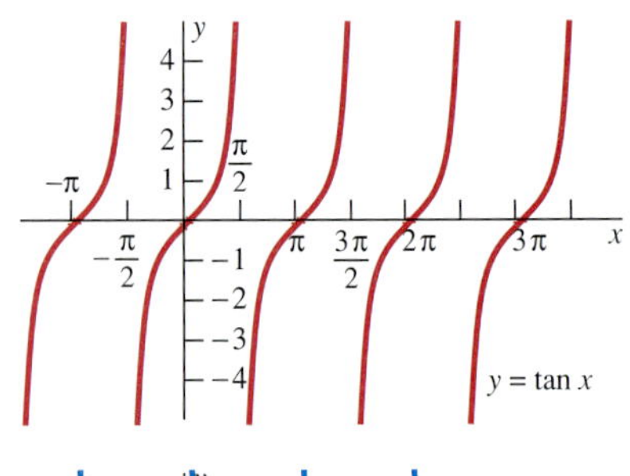

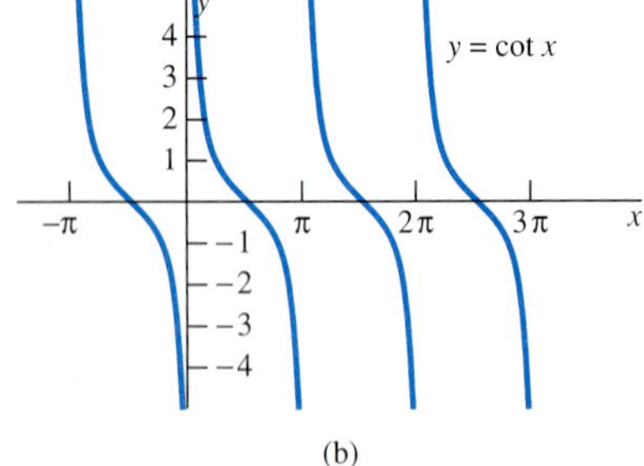

(b)

Fig. 3.7.3 Graphs of the six trigonometric functions

Theorem 2 ***Derivatives of Trigonometric Functions***

The functions $f(x) = \tan x$, $g(x) = \cot x$, $p(x) = \sec x$, and $q(x) = \csc x$ are differentiable wherever they are defined, and

$$D_x \tan x = \sec^2 x, \tag{7}$$

$$D_x \cot x = -\csc^2 x, \tag{8}$$

$$D_x \sec x = \sec x \tan x, \tag{9}$$

$$D_x \csc x = -\csc x \cot x. \tag{10}$$

The patterns in the formulas of Theorem 2 and in Eqs. (4) and (5) make them easy to remember. The formulas in Eqs. (5), (8), and (10) are the "cofunction analogues" of those in Eqs. (4), (7), and (9), respectively. Note that the derivative formulas for the three cofunctions are those involving minus signs.

EXAMPLE 6

$$D_x(x \tan x) = (D_x x)(\tan x) + (x)(D_x \tan x)$$
$$= (1)(\tan x) + (x)(\sec^2 x) = \tan x + x \sec^2 x.$$

$$D_t(\cot^3 t) = D_t(\cot t)^3 = 3(\cot t)^2 D_t \cot t$$
$$= 3(\cot t)^2(-\csc^2 t) = -3 \csc^2 t \cot^2 t.$$

$$D_z\left(\frac{\sec z}{\sqrt{z}}\right) = \frac{(D_z \sec z)(\sqrt{z}) - (\sec z)(D_z \sqrt{z})}{(\sqrt{z})^2}$$

$$= \frac{(\sec z \tan z)(\sqrt{z}) - (\sec z)\left(\frac{1}{2}z^{-1/2}\right)}{z}$$

$$= \frac{1}{2}z^{-3/2}(2z \tan z - 1)\sec z.$$

Chain Rule Formulas

Recall from Eq. (14) in Section 3.3 that the chain rule gives

$$D_x[g(u)] = g'(u)\frac{du}{dx} \tag{11}$$

for the derivative of the composition $g(u(x))$ of two differentiable functions g and u. This formula yields a *chain rule version* of each new differentiation formula that we learn.

If we apply Eq. (11) first with $g(u) = \sin u$, then with $g(u) = \cos u$, and so on, we get the chain rule versions of the trigonometric differentiation formulas:

$$D_x \sin u = (\cos u)\frac{du}{dx}, \tag{12}$$

$$D_x \cos u = (-\sin u)\frac{du}{dx}, \tag{13}$$

$$D_x \tan u = (\sec^2 u)\frac{du}{dx}, \tag{14}$$

$$D_x \cot u = (-\csc^2 u)\frac{du}{dx}, \tag{15}$$

$$D_x \sec u = (\sec u \tan u)\frac{du}{dx}, \tag{16}$$

$$D_x \csc u = (-\csc u \cot u)\frac{du}{dx}. \tag{17}$$

The cases in which $u = kx$ (where k is a constant) are worth mentioning. For example,

$$D_x \sin kx = k \cos kx \quad \text{and} \quad D_x \cos kx = -k \sin kx. \tag{18}$$

The formulas in (18) provide an explanation of why radian measure is more appropriate than degree measure. Because it follows from Eq. (1) that an angle of degree measure x has radian measure $\pi x/180$, the "sine of an angle of x degrees" is a *new* and *different* function with the formula

$$\sin x^\circ = \sin \frac{\pi x}{180},$$

expressed on the right-hand side in terms of the standard (radian-measure) sine function. Hence the first formula in (18) yields

$$D_x \sin x^\circ = \frac{\pi}{180}\cos\frac{\pi x}{180},$$

so

$$D_x \sin x^\circ \approx (0.01745) \cos x^\circ.$$

The necessity of using the approximate value 0.01745 here—and indeed its very presence—is one reason why radians instead of degrees are used in the calculus of trigonometric functions: When we work with radians, we don't need such approximations.

EXAMPLE 7 If $y = 2 \sin 10t + 3 \cos \pi t$, then

$$\frac{dy}{dt} = 20 \cos 10t - 3\pi \sin \pi t.$$

EXAMPLE 8 $D_x(\sin^2 3x \cos^4 5x)$

$$= \left[D_x(\sin 3x)^2\right](\cos^4 5x) + (\sin^2 3x)\left[D_x(\cos 5x)^4\right]$$

$$= 2(\sin 3x)(D_x \sin 3x)\cdot(\cos^4 5x) + (\sin^2 3x)\cdot 4(\cos 5x)^3\cdot(D_x \cos 5x)$$

$$= 2(\sin 3x)(3 \cos 3x)(\cos^4 5x) + (\sin^2 3x)(4 \cos^3 5x)(-5 \sin 5x)$$

$$= 6 \sin 3x \cos 3x \cos^4 5x - 20 \sin^2 3x \sin 5x \cos^3 5x.$$

EXAMPLE 9 Differentiate $f(x) = \cos\sqrt{x}$.

Solution If $u = \sqrt{x}$, then $du/dx = 1/(2\sqrt{x})$, so Eq. (13) yields

$$D_x \cos\sqrt{x} = D_x \cos u = (-\sin u)\frac{du}{dx}$$

$$= -(\sin\sqrt{x})\frac{1}{2\sqrt{x}} = -\frac{\sin\sqrt{x}}{2\sqrt{x}}.$$

Alternatively, we can carry out this computation without introducing the auxiliary variable u:

$$D_x \cos\sqrt{x} = (-\sin\sqrt{x})\cdot D_x(\sqrt{x}) = -\frac{\sin\sqrt{x}}{2\sqrt{x}}.$$

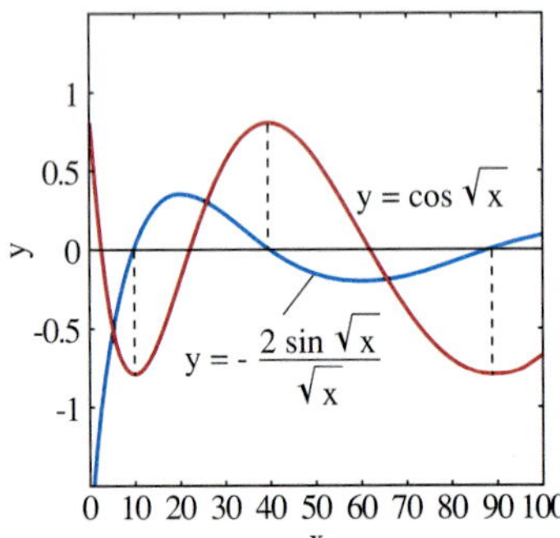

Fig. 3.7.4 The curve $y = \cos\sqrt{x}$ and the constant multiple $y = -(2\sin\sqrt{x})/\sqrt{x}$ of its derivative

In Fig. 3.7.4 we have plotted both the curve $y = y(x) = \cos\sqrt{x}$ and (to show the vertical scale more clearly) the constant multiple

$$y = 4y'(x) = -\frac{2 \sin\sqrt{x}}{\sqrt{x}}$$

of its derivative. Note the correspondence in this figure between the local maxima and minima of the function $y(x) = \cos\sqrt{x}$ and the zeros of its derivative $y'(x)$ (which are the same as the zeros of $4y'(x)$).

EXAMPLE 10 Differentiate

$$y = \sin^2(2x - 1)^{3/2} = \left[\sin(2x - 1)^{3/2}\right]^2.$$

Solution Here, $y = u^2$, where $u = \sin(2x - 1)^{3/2}$, so

$$\frac{dy}{dx} = 2u\frac{du}{dx} = 2\left[\sin(2x - 1)^{3/2}\right]\cdot D_x\left[\sin(2x - 1)^{3/2}\right]$$

$$= 2\left[\sin(2x-1)^{3/2}\right]\left[\cos(2x-1)^{3/2}\right]\cdot D_x(2x-1)^{3/2}$$
$$= 2\left[\sin(2x-1)^{3/2}\right]\left[\cos(2x-1)^{3/2}\right]\tfrac{3}{2}(2x-1)^{1/2}\cdot 2$$
$$= 6(2x-1)^{1/2}\left[\sin(2x-1)^{3/2}\right]\left[\cos(2x-1)^{3/2}\right].$$

EXAMPLE 11

$$D_x \tan 2x^3 = (\sec^2 2x^3)\cdot D_x(2x^3) = 6x^2 \sec^2 2x^3.$$

$$\begin{aligned} D_t \cot^3 2t &= D_t(\cot 2t)^3 = 3(\cot 2t)^2\cdot D_t(\cot 2t) \\ &= (3\cot^2 2t)(-\csc^2 2t)\cdot D_t(2t) \\ &= -6\csc^2 2t \cot^2 2t. \end{aligned}$$

$$D_y \sec\sqrt{y} = (\sec\sqrt{y}\tan\sqrt{y})\cdot D_y\sqrt{y} = \frac{\sec\sqrt{y}\tan\sqrt{y}}{2\sqrt{y}}.$$

$$\begin{aligned} D_z\sqrt{\csc x} &= D_z(\csc z)^{1/2} = \frac{1}{2}(\csc z)^{-1/2}\cdot D_z(\csc z) \\ &= \frac{1}{2}(\csc z)^{-1/2}(-\csc z\cot z) = -\frac{1}{2}(\cot z)\sqrt{\csc z}. \end{aligned}$$

Examples 12 and 13 illustrate the applications of trigonometric functions to rate-of-change and maximum-minimum problems.

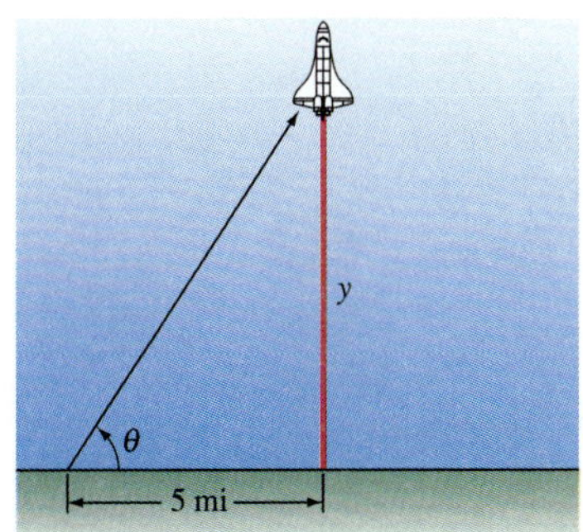

Fig. 3.7.5 Tracking an ascending rocket (Example 12)

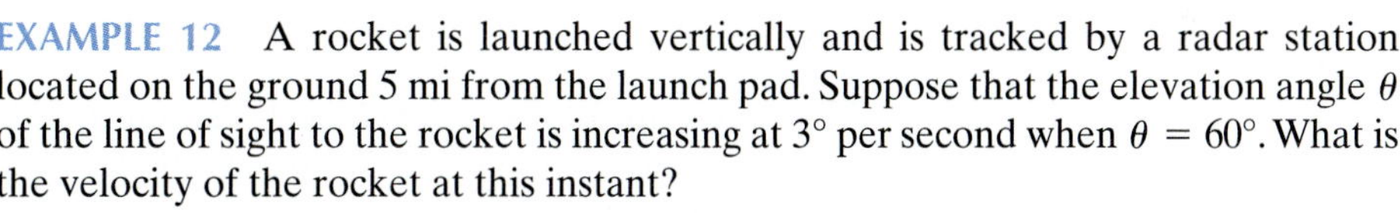

EXAMPLE 12 A rocket is launched vertically and is tracked by a radar station located on the ground 5 mi from the launch pad. Suppose that the elevation angle θ of the line of sight to the rocket is increasing at $3°$ per second when $\theta = 60°$. What is the velocity of the rocket at this instant?

Solution First we convert the given data from degrees into radians. Because there are $\pi/180$ rad in $1°$, the rate of increase of θ becomes

$$\frac{3\pi}{180} = \frac{\pi}{60} \quad (\text{rad/s})$$

at the instant when

$$\theta = \frac{60\pi}{180} = \frac{\pi}{3} \quad (\text{rad}).$$

From Fig. 3.7.5 we see that the height y (in miles) of the rocket is

$$y = 5\tan\theta.$$

Hence its velocity is

$$\frac{dy}{dt} = \frac{dy}{d\theta}\cdot\frac{d\theta}{dt} = 5(\sec^2\theta)\frac{d\theta}{dt}.$$

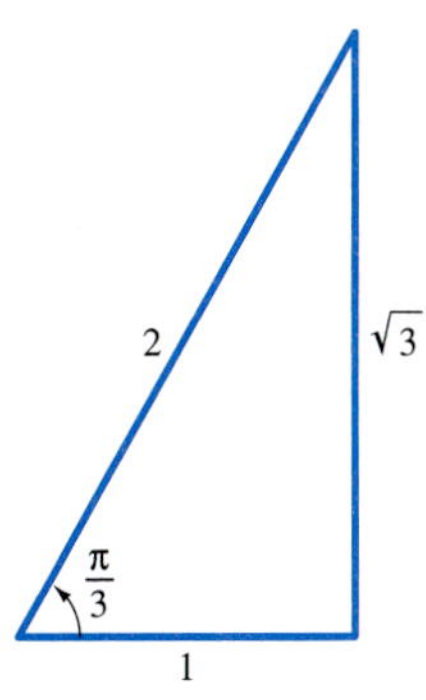

Fig. 3.7.6 $\sec\frac{\pi}{3} = 2$ (Example 12)

Because $\sec(\pi/3) = 2$ (Fig. 3.7.6), the velocity of the rocket is

$$\frac{dy}{dt} = 5\cdot 2^2\cdot\frac{\pi}{60} = \frac{\pi}{3} \quad (\text{mi/s}),$$

about 3770 mi/h, at the instant when $\theta = 60°$.

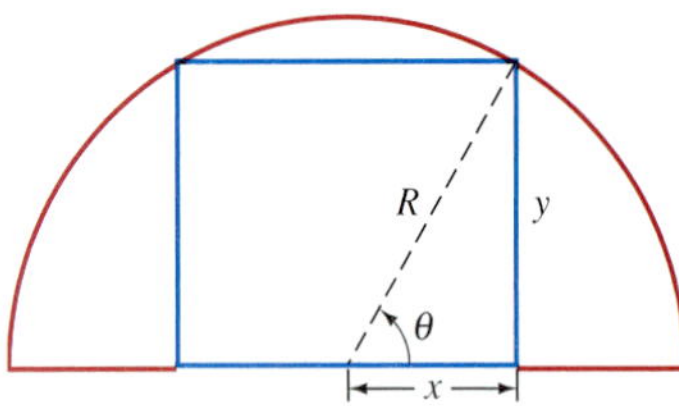

Fig. 3.7.7 The rectangle of Example 13

EXAMPLE 13 A rectangle is inscribed in a semicircle of radius R (Fig. 3.7.7). What is the maximum possible area of such a rectangle?

Solution If we denote the length of *half* the base of the rectangle by x and its height by y, then its area is $A = 2xy$. We see in Fig. 3.7.7 that the right triangle has hypotenuse R, the radius of the circle. So

$$x = R\cos\theta \quad \text{and} \quad y = R\sin\theta. \tag{19}$$

Each value of θ between 0 and $\pi/2$ corresponds to a possible inscribed rectangle. The values $\theta = 0$ and $\theta = \pi/2$ will yield degenerate rectangles.

We substitute the data in Eq. (19) into the formula $A = 2xy$ to obtain the area,

$$A = A(\theta) = 2(R\cos\theta)(R\sin\theta) = 2R^2\cos\theta\sin\theta, \tag{20}$$

as a function of θ on the closed interval $[0, \pi/2]$. To find the critical points, we differentiate:

$$\frac{dA}{d\theta} = 2R^2(-\sin\theta\sin\theta + \cos\theta\cos\theta) = 2R^2(\cos^2\theta - \sin^2\theta).$$

Because $dA/d\theta$ always exists, we have critical points only if

$$\begin{aligned}\cos^2\theta - \sin^2\theta &= 0;\\ \sin^2\theta &= \cos^2\theta;\\ \tan^2\theta &= 1;\\ \tan\theta &= \pm 1.\end{aligned}$$

The only value of θ in $[0, \pi/2]$ such that $\tan\theta = \pm 1$ is $\theta = \pi/4$.

Upon evaluation of $A(\theta)$ at each of the possible values $\theta = 0$, $\theta = \pi/4$, and $\theta = \pi/2$ (the endpoints and the critical point), we find that

$$\begin{aligned}A(0) &= 0,\\ A\left(\frac{\pi}{4}\right) &= 2R^2\left(\frac{1}{\sqrt{2}}\right)\left(\frac{1}{\sqrt{2}}\right) = R^2, \leftarrow \text{absolute maximum}\\ A\left(\frac{\pi}{2}\right) &= 0.\end{aligned}$$

Thus the largest inscribed rectangle has area R^2, and its dimensions are $2x = R\sqrt{2}$ and $y = R/\sqrt{2}$. ■

3.7 PROBLEMS

Differentiate the functions given in Problems 1 through 20.

1. $f(x) = 3\sin^2 x$

2. $f(x) = 2\cos^4 x$

3. $f(x) = x\cos x$

4. $f(x) = \sqrt{x}\sin x$

5. $f(x) = \dfrac{\sin x}{x}$

6. $f(x) = \dfrac{\cos x}{\sqrt{x}}$

7. $f(x) = \sin x\cos^2 x$

8. $f(x) = \cos^3 x\sin^2 x$

9. $g(t) = (1 + \sin t)^4$

10. $g(t) = (2 - \cos^2 t)^3$

11. $g(t) = \dfrac{1}{\sin t + \cos t}$

12. $g(t) = \dfrac{\sin t}{1 + \cos t}$

13. $f(x) = 2x\sin x - 3x^2\cos x$

14. $f(x) = x^{1/2}\cos x - x^{-1/2}\sin x$

15. $f(x) = \cos 2x\sin 3x$

16. $f(x) = \cos 5x\sin 7x$

17. $g(t) = t^3\sin^2 2t$

18. $g(t) = \sqrt{t}\cos^3 3t$

19. $g(t) = (\cos 3t + \cos 5t)^{5/2}$

20. $g(t) = \dfrac{1}{\sqrt{\sin^2 t + \sin^2 3t}}$

Find dy/dx in Problems 21 through 40.

21. $y = \sin^2\sqrt{x}$

22. $y = \dfrac{\cos 2x}{x}$

23. $y = x^2\cos(3x^2 - 1)$

24. $y = \sin^3 x^4$

25. $y = \sin 2x\cos 3x$

26. $y = \dfrac{x}{\sin 3x}$

27. $y = \dfrac{\cos 3x}{\sin 5x}$ **28.** $y = \sqrt{\cos\sqrt{x}}$

29. $y = \sin^2 x^2$ **30.** $y = \cos^3 x^3$

31. $y = \sin 2\sqrt{x}$ **32.** $y = \cos 3\sqrt[3]{x}$

33. $y = x \sin x^2$ **34.** $y = x^2 \cos\left(\dfrac{1}{x}\right)$

35. $y = \sqrt{x}\sin\sqrt{x}$ **36.** $y = (\sin x - \cos x)^2$

37. $y = \sqrt{x}\,(x - \cos x)^3$ **38.** $y = \sqrt{x}\sin\sqrt{x + \sqrt{x}}$

39. $y = \cos(\sin x^2)$ **40.** $y = \sin(1 + \sqrt{\sin x})$

Find dx/dt in Problems 41 through 60.

41. $x = \tan t^7$ **42.** $x = \sec t^7$

43. $x = (\tan t)^7$ **44.** $x = (\sec 2t)^7$

45. $x = t^7 \tan 5t$ **46.** $x = \dfrac{\sec t^5}{t}$

47. $x = \sqrt{t}\sec\sqrt{t}$ **48.** $x = \sec\sqrt{t}\tan\sqrt{t}$

49. $x = \csc\left(\dfrac{1}{t^2}\right)$ **50.** $x = \cot\left(\dfrac{1}{\sqrt{t}}\right)$

51. $x = \dfrac{\sec 5t}{\tan 3t}$ **52.** $x = \sec^2 t - \tan^2 t$

53. $x = t \sec t \csc t$ **54.** $x = t^3 \tan^3 t^3$

55. $x = \sec(\sin t)$ **56.** $x = \cot(\sec 7t)$

57. $x = \dfrac{\sin t}{\sec t}$ **58.** $x = \dfrac{\sec t}{1 + \tan t}$

59. $x = \sqrt{1 + \cot 5t}$ **60.** $x = \sqrt{\csc\sqrt{t}}$

In Problems 61 through 64, write an equation of the line that is tangent to the given curve $y = f(x)$ at the point P with the given x-coordinate. Then check the plausibility of your result by plotting both the curve and the line you found on the same screen.

61. $y = x \cos x;\quad x = \pi$ **62.** $y = \cos^2 x;\quad x = \pi/4$

63. $y = \dfrac{4}{\pi}\tan\dfrac{\pi x}{4};\quad x = 1$ **64.** $y = \dfrac{3}{\pi}\sin^2\dfrac{\pi x}{3};\quad x = 5$

In Problems 65 through 68, find all points on the given curve $y = f(x)$ where the tangent line is horizontal.

65. $y = \cos 2x$ **66.** $y = x - 2\sin x$

67. $y = \sin x \cos x$ **68.** $y = \dfrac{1}{3\sin^2 x + 2\cos^2 x}$

69. Figure 3.7.8 shows the graph $y = x - 2\cos x$ and two lines of slope 1 both tangent to this graph. Write equations of these two lines.

70. Figure 3.7.9 shows the graph

$$y = \frac{16 + \sin x}{3 + \sin x}$$

and its two horizontal tangent lines. Write equations of these two lines.

71. Derive the differentiation formulas in Eqs. (8) through (10).

72. Use the definition of the derivative to show directly that $g'(x) = -\sin x$ if $g(x) = \cos x$.

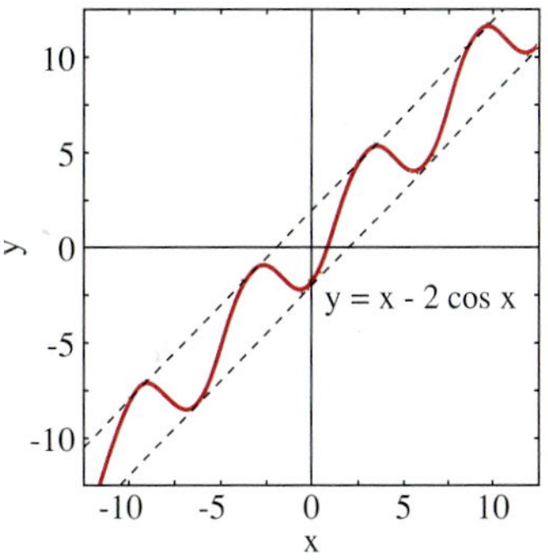

Fig. 3.7.8 The curve $y = x - 2\cos x$ and two tangent lines each having slope 1

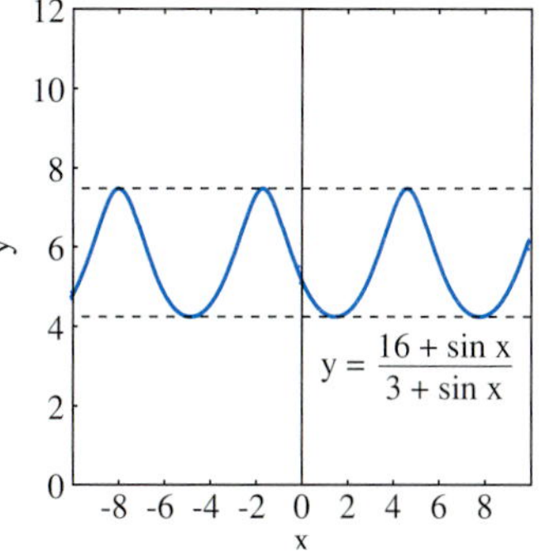

Fig. 3.7.9 The curve $y = \dfrac{16 + \sin x}{3 + \sin x}$ and its two horizontal tangent lines

73. If a projectile is fired from ground level with initial velocity v_0 and inclination angle α and if air resistance can be ignored, then its range—the horizontal distance it travels—is

$$R = \frac{1}{16} v_0^2 \sin\alpha \cos\alpha$$

(Fig. 3.7.10). What value of α maximizes R?

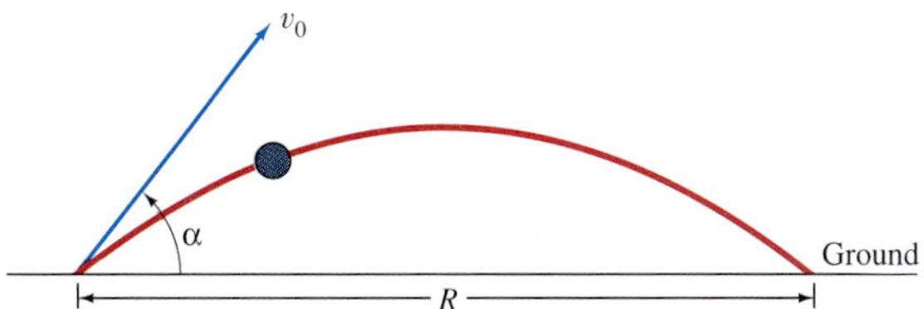

Fig. 3.7.10 The projectile of Problem 73

74. A weather balloon that is rising vertically is observed from a point on the ground 300 ft from the spot directly beneath the balloon (Fig. 3.7.11). At what rate is the balloon rising when the angle between the ground and the observer's line of sight is 45° and is increasing at 1° per second?

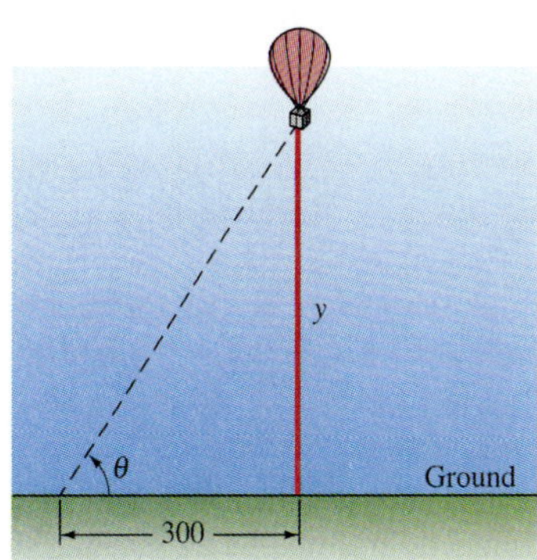

Fig. 3.7.11 The weather balloon of Problem 74

75. A rocket is launched vertically upward from a point 2 mi west of an observer on the ground. What is the speed of the rocket when the angle of elevation (from the horizontal) of the observer's line of sight to the rocket is 50° and is increasing at 5° per second?

76. A plane flying at an altitude of 25000 ft has a defective airspeed indicator. To determine her speed, the pilot sights a fixed point on the ground. At the moment the angle of depression (from the horizontal) of her line of sight is 65°, she notes that this angle is increasing at 1.5° per second (Fig. 3.7.12). What is the speed of the plane?

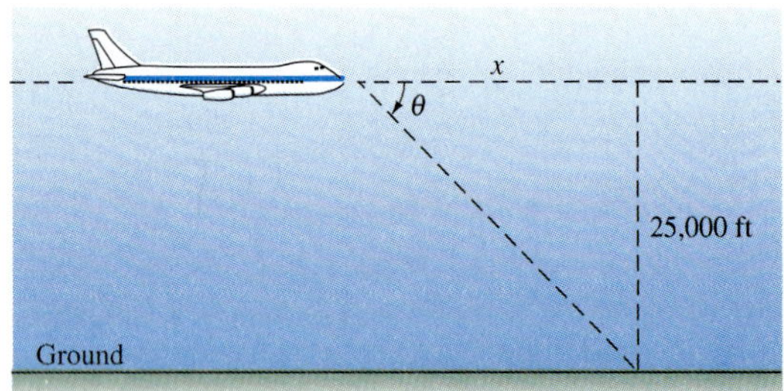

Fig. 3.7.12 The airplane of Problem 76

77. An observer on the ground sights an approaching plane flying at constant speed and at an altitude of 20000 ft. From his point of view, the plane's angle of elevation is increasing at 0.5° per second when the angle is 60°. What is the speed of the plane?

78. Find the largest possible area A of a rectangle inscribed in the unit circle $x^2 + y^2 = 1$ by maximizing A as a function of the angle θ indicated in Fig. 3.7.13.

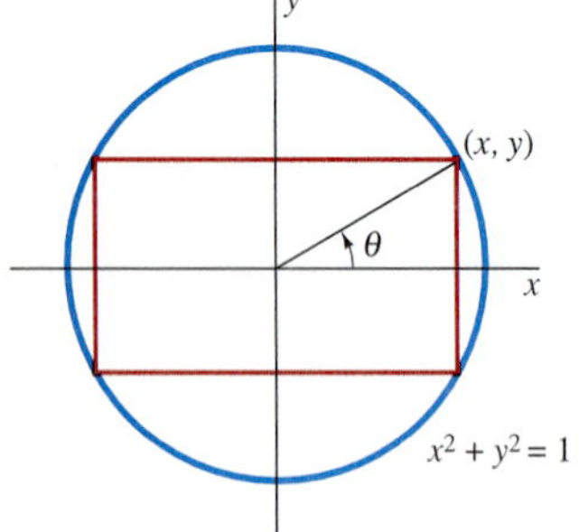

Fig. 3.7.13 A rectangle inscribed in the unit circle (Problem 78)

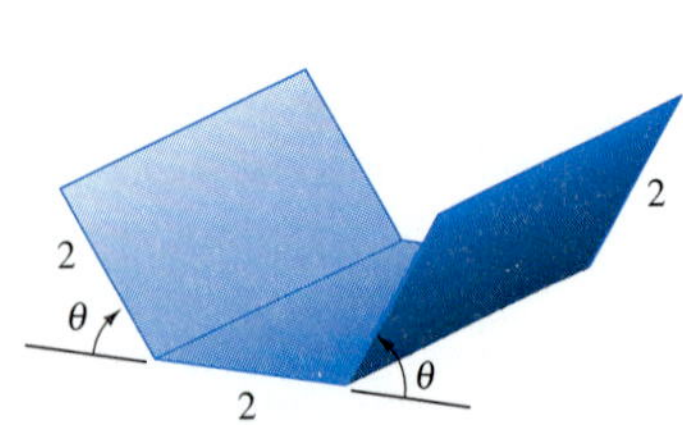

Fig. 3.7.14 The water trough of Problem 79

79. A water trough is to be made from a long strip of tin 6 ft wide by bending up at an angle θ a 2-ft strip on each side (Fig. 3.7.14). What angle θ would maximize the cross-sectional area, and thus the volume, of the trough?

80. A circular patch of grass of radius 20 m is surrounded by a walkway, and a light is placed atop a lamppost at the circle's center. At what height should the light be placed to illuminate the walkway most strongly? The intensity of illumination I of a surface is given by $I = (k \sin \theta)/D^2$, where D is the distance from the light source to the surface, θ is the angle at which light strikes the surface, and k is a positive constant.

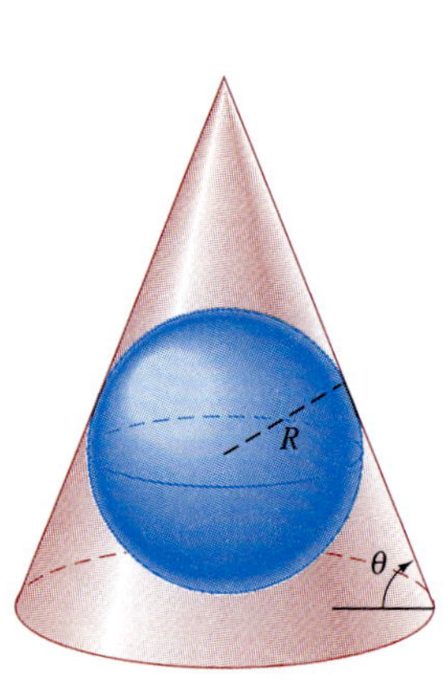

Fig. 3.7.15 Finding the smallest cone containing a fixed sphere (Problem 81)

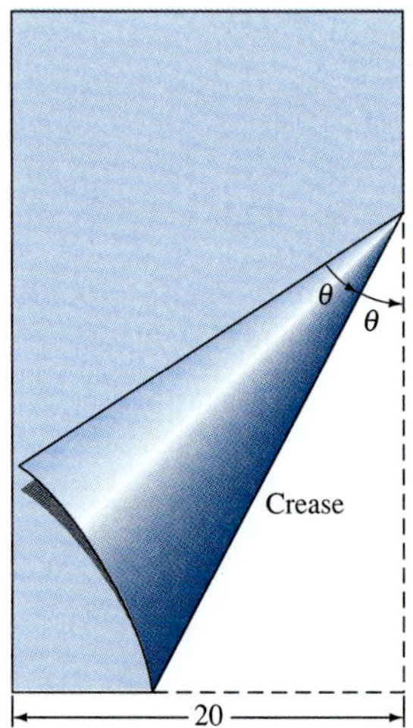

Fig. 3.7.16 Fold a piece of paper; make the crease of minimal length (Problem 82).

81. Find the minimum possible volume V of a cone in which a sphere of given radius R is inscribed. Minimize V as a function of the angle θ indicated in Fig. 3.7.15.

82. A very long rectangular piece of paper is 20 cm wide. The bottom right-hand corner is folded along the crease shown in Fig. 3.7.16, so that the corner just touches the left-hand side of the page. How should this be done so that the crease is as short as possible?

83. Find the maximum possible area A of a trapezoid inscribed in a semicircle of radius 1, as shown in Fig. 3.7.17. Begin by expressing A as a function of the angle θ shown there.

84. A logger must cut a six-sided beam from a circular log of diameter 30 cm so that its cross section is as shown in Fig. 3.7.18. The beam is symmetrical, with only two different internal angles α and β. Show that the cross section is maximal when it is in the shape of a regular hexagon, with equal sides and angles (corresponding to $\alpha = \beta = 2\pi/3$). Note that $\alpha + 2\beta = 2\pi$. (Why?)

85. Consider a circular arc of length s with its endpoints on the x-axis (Fig. 3.7.19). Show that the area A bounded by this arc and the x-axis is maximal when the circular arc is in the shape of a semicircle. (Suggestion: Express A in terms of the angle θ subtended by the arc at the center of the circle, as shown in Fig. 3.7.19. Show that A is maximal when $\theta = \pi$.)

86. A hiker starting at a point P on a straight road wants to reach a forest cabin that is 2 km from a point Q 3 km

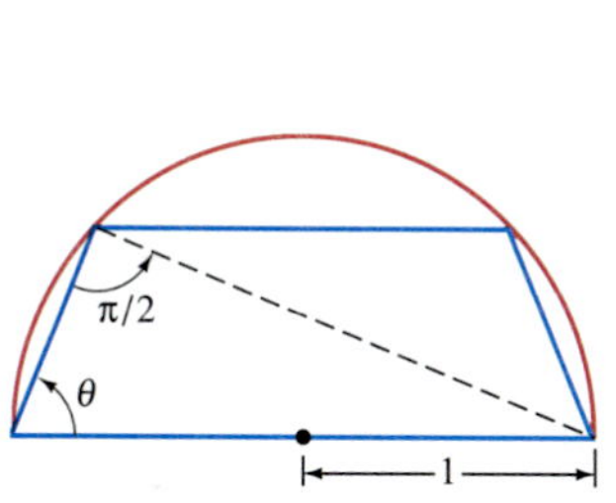

Fig. 3.7.17 A trapezoid inscribed in a circle (Problem 83)

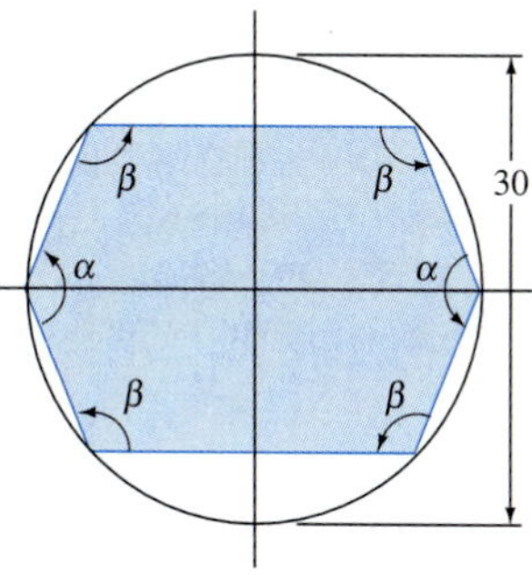

Fig. 3.7.18 A hexagonal beam cut from a circular log (Problem 84)

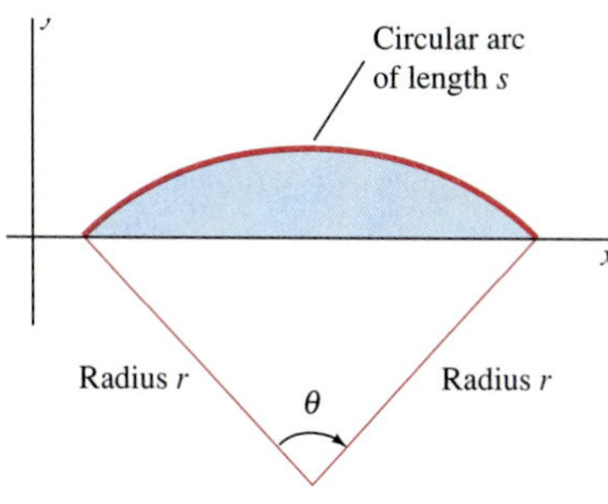

Fig. 3.7.19 Finding the maximum area bounded by a circular arc and its chord (Problem 85)

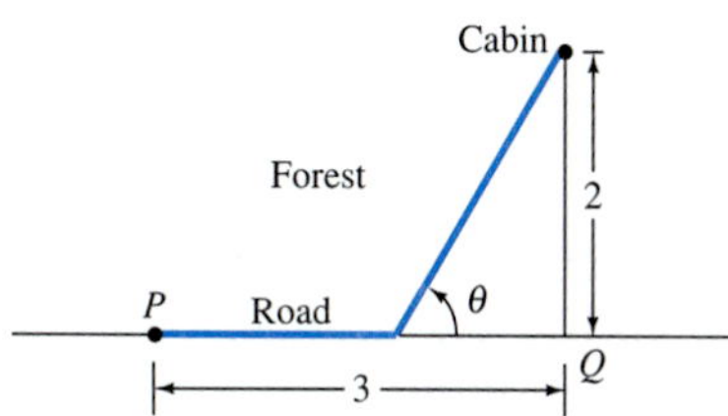

Fig. 3.7.20 Finding the quickest path to the cabin in the forest (Problem 86).

down the road from P (Fig. 3.7.20). She can walk 8 km/h along the road but only 3 km/h through the forest. She wants to minimize the time required to reach the cabin. How far down the road should she walk before setting off through the forest straight for the cabin? (*Suggestion:* Use the angle θ between the road and the path she takes through the forest as the independent variable.)

87. Show that the function (graphed in Fig. 3.7.21)

$$f(x) = \begin{cases} x\sin\dfrac{1}{x} & \text{if } x \neq 0, \\ 0 & \text{if } x = 0 \end{cases}$$

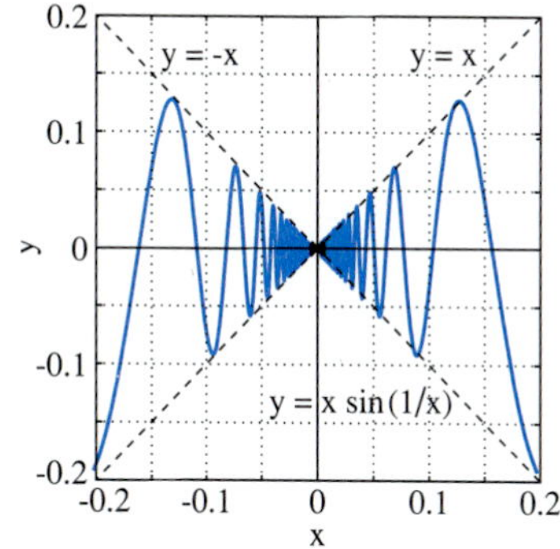

Fig. 3.7.21 The graph of $y = x\sin\dfrac{1}{x}$ near $x = 0$

(see Example 4 in Section 2.3) is *not* differentiable at $x = 0$. (*Suggestion:* Show that whether $z = 1$ or $z = -1$, there are arbitrarily small values of h such that $[f(h) - f(0)]/h = z$. Then use the *definition* of the derivative.)

88. Let

$$f(x) = \begin{cases} x^2\sin\dfrac{1}{x} & \text{if } x \neq 0, \\ 0 & \text{if } x = 0 \end{cases}$$

(the graph of f appears in Figs. 3.7.22 and 3.7.23). Apply the definition of the derivative to show that f is differentiable at $x = 0$ and that $f'(0) = 0$.

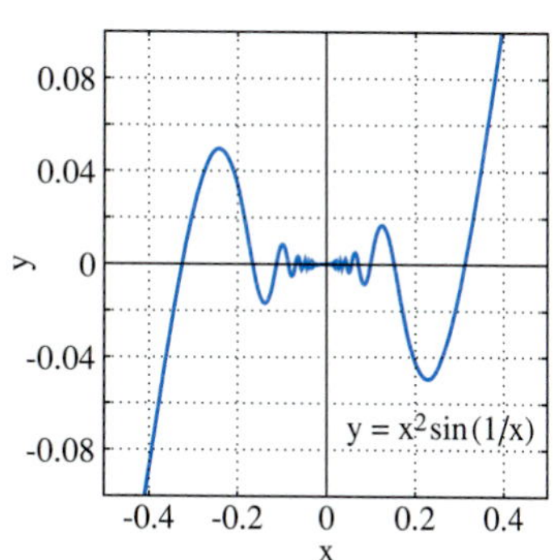

Fig. 3.7.22 The graph of $y = x^2\sin\dfrac{1}{x}$ (Problem 88)

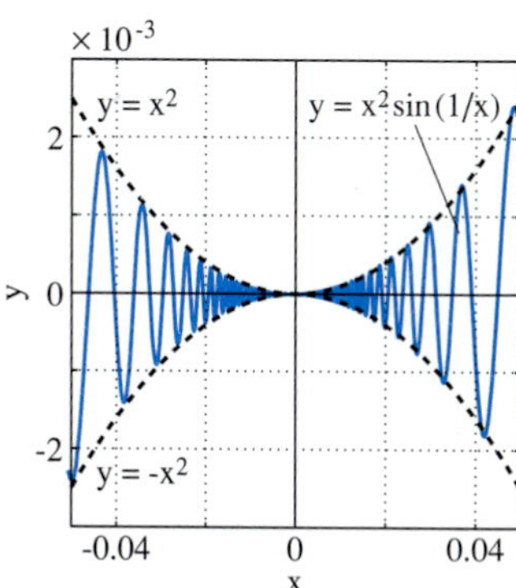

Fig. 3.7.23 The graph in Fig. 3.7.22 magnified (Problem 88)

3.8 IMPLICIT DIFFERENTIATION AND RELATED RATES

An equation in two variables x and y may have one or more solutions for y in terms of x or for x in terms of y. These solutions are functions that are said to be **implicitly defined** by the equation. Here we discuss the differentiation of such functions and the use of their derivatives in solving rate-of-change problems.

For example, the equation

$$y^2 - x = 0$$

implicitly defines two continuous functions of x:

$$y = \sqrt{x} \quad \text{and} \quad y = -\sqrt{x}.$$

Each has domain the half-line $x \geqq 0$. The graphs of these two functions are the upper and lower branches of the parabola shown in Fig. 3.8.1. The whole parabola is not the graph of any single function of x because no vertical line can meet the graph of such a function at more than one point.

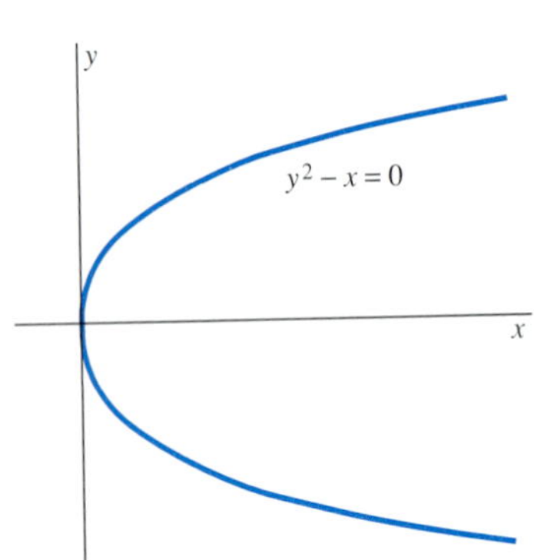

Fig. 3.8.1 The parabola $y^2 - x = 0$

The equation of the unit circle, $x^2 + y^2 = 1$, implicitly defines four functions (among others):

$$\begin{aligned}
y &= +\sqrt{1 - x^2} && \text{for } x \text{ in } [-1, 1],\\
y &= -\sqrt{1 - x^2} && \text{for } x \text{ in } [-1, 1],\\
x &= +\sqrt{1 - y^2} && \text{for } y \text{ in } [-1, 1], \text{ and}\\
x &= -\sqrt{1 - y^2} && \text{for } y \text{ in } [-1, 1].
\end{aligned}$$

The graphs of these four functions are highlighted against four copies of the unit circle shown in Fig. 3.8.2.

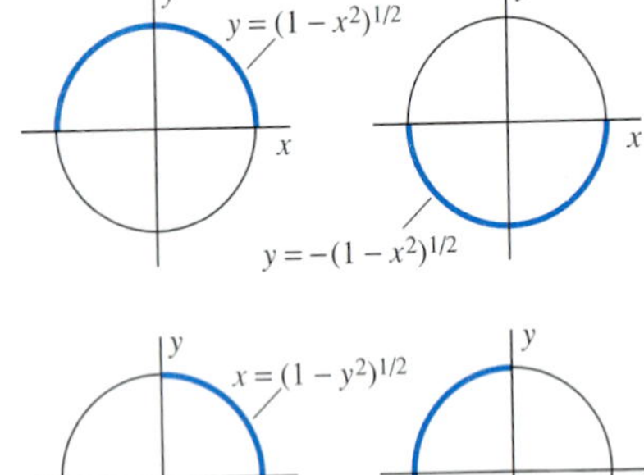

Fig. 3.8.2 Continuous functions defined implicitly by $x^2 + y^2 = 1$

Whereas the equations $y^2 = x$ and $x^2 + y^2 = 1$ are readily solved for y in terms of x, an equation such as $x^3 + y^3 = 3xy$ or $\sin(x + 2y) = 2x \cos y$ may be difficult or impossible to solve for an implicitly defined function $y(x)$. Nevertheless, the derivative dy/dx can be calculated without first finding y in terms of x. Here's how: We can use the chain rule to differentiate both sides of the given equation, thinking of x as the independent variable. We then solve the resulting equation for the derivative $y'(x) = dy/dx$ of the implicitly defined function $y(x)$. This process is called **implicit differentiation.** In the examples and problems of this section, we proceed on the assumption that our implicitly defined functions are differentiable at almost all points in their domains. (The functions with the graphs shown in Fig. 3.8.2 are *not* differentiable at the endpoints of their domains.)

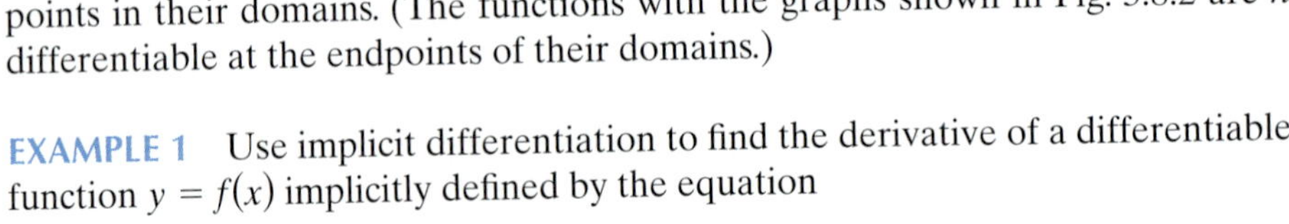

EXAMPLE 1 Use implicit differentiation to find the derivative of a differentiable function $y = f(x)$ implicitly defined by the equation

$$x^2 + y^2 = 1.$$

Solution The equation $x^2 + y^2 = 1$ is to be regarded as an *identity* that implicitly defines $y = y(x)$ as a function of x. Because $x^2 + [y(x)]^2$ is then a function of x, it has the same derivative as the constant function 1 on the right-hand side of the identity. Thus we may differentiate both sides of the *identity* $x^2 + y^2 = 1$ with respect to x and equate the results. We obtain

$$2x + 2y\frac{dy}{dx} = 0.$$

In this step, it is essential to remember that y is a function of x, so the chain rule yields $D_x(y^2) = 2y\,D_x y$.

Then we solve for

$$\frac{dy}{dx} = -\frac{x}{y}. \tag{1}$$

It may be surprising to see a formula for dy/dx containing both x and y, but such a formula can be just as useful as one containing only x. For example, the formula in Eq. (1) tells us that the slope of the line tangent to the circle $x^2 + y^2 = 1$ at the point $\left(\frac{3}{5}, \frac{4}{5}\right)$ is

$$\left.\frac{dy}{dx}\right|_{\left(\frac{3}{5},\frac{4}{5}\right)} = -\frac{\frac{3}{5}}{\frac{4}{5}} = -\frac{3}{4}.$$

The circle and this line are shown in Fig. 3.8.3. ■

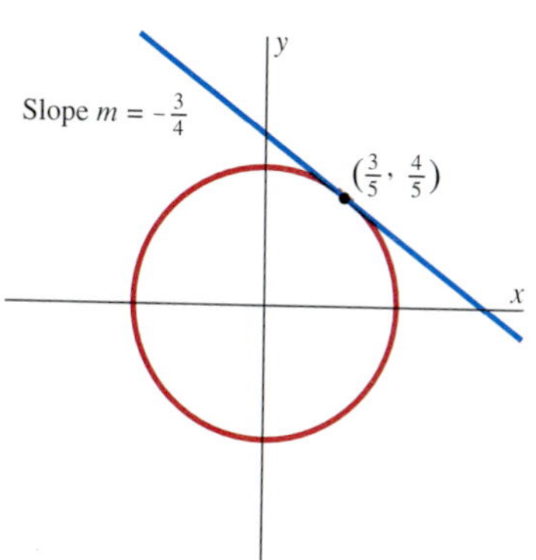

Fig. 3.8.3 The tangent line of Example 1

Note If we solve for $y = \pm\sqrt{1 - x^2}$ in Example 1, then

$$\frac{dy}{dx} = \frac{-x}{\pm\sqrt{1 - x^2}} = -\frac{x}{y},$$

in agreement with Eq. (1). Thus Eq. (1) simultaneously gives us the derivatives of both the functions $y = +\sqrt{1 - x^2}$ and $y = -\sqrt{1 - x^2}$ implicitly defined by the equation $x^2 + y^2 = 1$.

EXAMPLE 2 The *folium of Descartes* is the graph of the equation

$$x^3 + y^3 = 3xy. \tag{2}$$

This curve was first proposed by René Descartes as a challenge to Pierre de Fermat (1601–1665) to find its tangent line at an arbitrary point. The graph of the folium appears in Fig. 3.8.4. The graph very nearly coincides with the straight line $x + y = 1$ when $|x|$ and $|y|$ are large, and it produces a loop shaped like a laurel leaf (thus the name *folium*) in the first quadrant. We indicate in Problem 30 of Section 10.4 how this graph can be constructed. Here we want to find the slope of its tangent line.

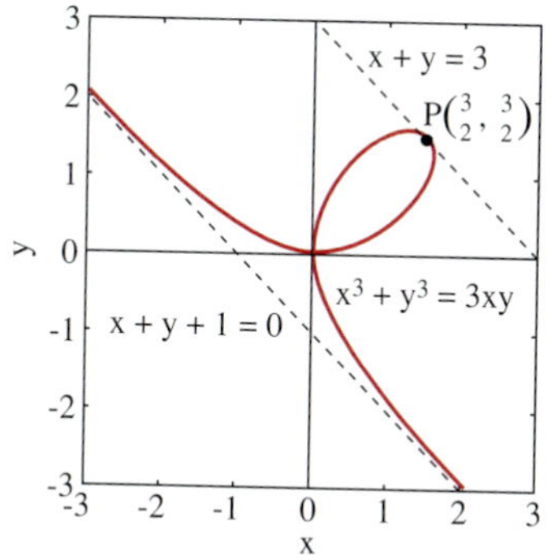

Fig. 3.8.4 The curve $x^3 + y^3 = 3xy$, its asymptote, and its tangent line at the point $P\left(\frac{3}{2}, \frac{3}{2}\right)$

Solution Equation (2) is a cubic equation in x, so it is inconvenient to solve for y in terms of x. Indeed,

$$y = \frac{3x\sqrt[3]{2}}{\sqrt[3]{-27x^3 + \sqrt{729x^6 - 2916x^3}}} + \frac{\sqrt[3]{-27x^3 + \sqrt{729x^6 - 2916x^3}}}{3\sqrt[3]{2}}$$

is one of *three* solutions for y in terms of x. Fermat invented an approach specific to this problem, but we can readily differentiate each side of Eq. (2) with respect to x, remembering that y is a function of x, and thus using the chain rule to differentiate y^3 and the product rule to differentiate $3xy$. This yields

$$3x^2 + 3y^2\frac{dy}{dx} = 3y + 3x\frac{dy}{dx}.$$

We can now collect coefficients (those involving dy/dx and those not) and solve for the derivative:

$$(3y^2 - 3x)\frac{dy}{dx} = 3y - 3x^2;$$

$$\frac{dy}{dx} = \frac{y - x^2}{y^2 - x}. \tag{3}$$

For instance, at the point $P(\frac{3}{2}, \frac{3}{2})$ of the folium, the slope of the tangent line is

$$\left.\frac{dy}{dx}\right|_{\left(\frac{3}{2}, \frac{3}{2}\right)} = \frac{\frac{3}{2} - (\frac{3}{2})^2}{(\frac{3}{2})^2 - \frac{3}{2}} = -1.$$

This result agrees with our intuition about the figure, because the evident symmetry of the folium around the line $y = x$ suggests that the tangent line at P should, indeed, have slope -1. The equation of this tangent line is

$$y - \tfrac{3}{2} = -(x - \tfrac{3}{2});$$

that is, $x + y = 3$. ■

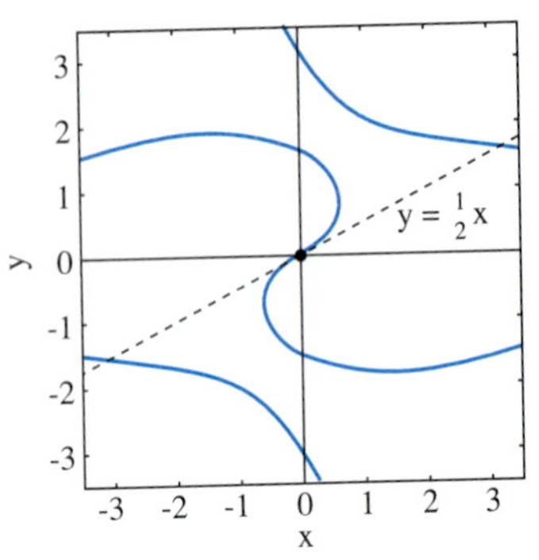

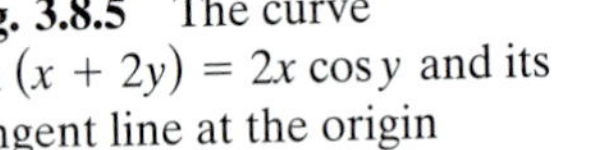

Fig. 3.8.5 The curve $\sin(x + 2y) = 2x \cos y$ and its tangent line at the origin

EXAMPLE 3 Figure 3.8.5 shows a computer plot of the graph of the equation

$$\sin(x + 2y) = 2x \cos y. \tag{4}$$

Write the equation of the line tangent to this curve at the origin $(0, 0)$.

Solution When we differentiate each side in (4) with respect to the independent variable x, regarding y as a function of x, we get

$$[\cos(x + 2y)] \cdot \left(1 + \frac{dy}{dx}\right) = 2 \cos y - (2x \sin y)\frac{dy}{dx}. \tag{5}$$

We could collect coefficients and solve for the derivative dy/dx. But because we need only the slope $y'(0)$ at the origin, let us instead substitute $x = y = 0$ in Eq. (5). Noting that $\cos(0) = 1$ and $\sin(0) = 0$, we get the equation

$$1 + 2y'(0) = 2,$$

from which we see that $y'(0) = \frac{1}{2}$. The resulting tangent line $y = \frac{1}{2}x$ plotted in Fig. 3.8.5 "looks right" and thus corroborates the results of our calculations. ■

Related Rates

A **related-rates** problem involves two or more quantities that vary with time and an equation that expresses some relationship between these quantities. Typically, the values of these quantities at some instant are given, together with all their time rates of change but one. The problem is usually to find the time rate of change that is *not* given, at some instant specified in the problem. Implicit differentiation, with respect to time t, of the equation that relates the given quantities will produce an equation that relates the *rates of change* of the given quantities. This is the key to solving a related-rates problem.

EXAMPLE 4 Suppose that $x(t)$ and $y(t)$ are the x- and y-coordinates at time t of a point moving around the circle with equation

$$x^2 + y^2 = 25. \tag{6}$$

Let us use the chain rule to differentiate both sides of this equation *with respect to time t*. This produces the equation

$$2x\frac{dx}{dt} + 2y\frac{dy}{dt} = 0. \tag{7}$$

If the values of x, y, and dx/dt are known at a certain instant t, then Eq. (7) can be solved for the value of dy/dt. It is not necessary to know x and y as functions of t. Indeed, it is common for a related-rates problem to contain insufficient information to express x and y as functions of t.

For instance, suppose that we are given $x = 3$, $y = 4$, and $dx/dt = 12$ at a certain instant. Substitution of these values into Eq. (7) yields

$$2\cdot 3\cdot 12 + 2\cdot 4\cdot\frac{dy}{dt} = 0,$$

so we find that $dy/dx = -9$ at the same instant. ■

EXAMPLE 5 A rocket that is launched vertically is tracked by a radar station located on the ground 3 mi from the launch site. What is the vertical speed of the rocket at the instant that its distance from the radar station is 5 mi and this distance is increasing at the rate of 5000 mi/h?

Solution Figure 3.8.6 illustrates this situation. We denote the altitude of the rocket (in miles) by y and its distance from the radar station by z. We are given

$$\frac{dz}{dt} = 5000 \quad \text{when} \quad z = 5.$$

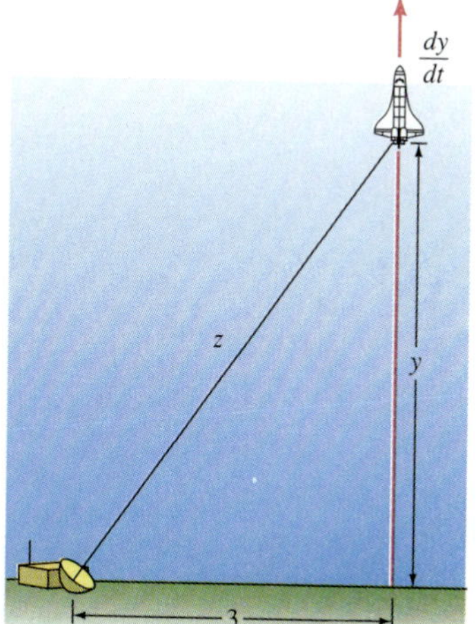

Fig. 3.8.6 The rocket of Example 5

We want to find dy/dt (in miles per hour) at this instant.

We apply the Pythagorean theorem to the right triangle in the figure and obtain

$$y^2 + 9 = z^2$$

as a relation between y and z. From this we see that $y = 4$ when $z = 5$. Implicit differentiation then gives

$$2y\frac{dy}{dt} = 2z\frac{dz}{dt}.$$

We substitute the data $y = 4$, $z = 5$, and $dz/dt = 5000$. Thus we find that

$$\frac{dy}{dt} = 6250 \quad \text{(mi/h)}$$

at the instant in question. ■

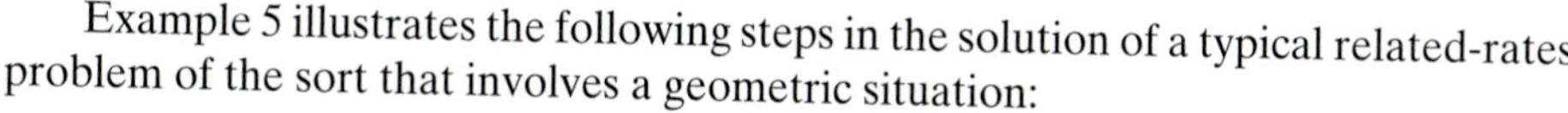

Example 5 illustrates the following steps in the solution of a typical related-rates problem of the sort that involves a geometric situation:

1. Draw a diagram and label as variables the various changing quantities involved in the problem.
2. Record the values of the variables and their rates of change, as given in the problem.
3. Use the diagram to determine an equation that relates the important variables in the problem.
4. Differentiate this equation implicitly with respect to time t.
5. Substitute the given numerical data in the resulting equation, and then solve for the unknown.

Warning The most common error to be avoided is the premature substitution of the given data before differentiating implicitly. If we had substituted $z = 5$ to begin with in Example 5, our equation would have been $y^2 + 9 = 25$, and implicit differentiation would have given the absurd result $dy/dt = 0$.

We use similar triangles (rather than the Pythagorean theorem) in Example 6 to discover the needed relation between the variables.

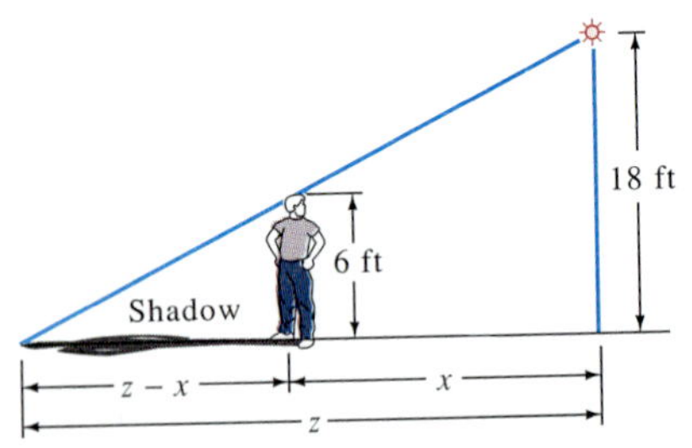

Fig. 3.8.7 The moving shadow of Example 6

EXAMPLE 6 A man 6 ft tall walks with a speed of 8 ft/s away from a street light that is atop an 18-ft pole. How fast is the tip of his shadow moving along the ground when he is 100 ft from the light pole?

Solution Let x be the man's distance from the pole and z the distance from the tip of his shadow to the base of the pole (Fig. 3.8.7). Although x and z are functions of time t, we do not attempt to find explicit formulas for either.

We are given $dx/dt = 8$ (in feet per second); we want to find dz/dt when $x = 100$ (ft). We equate ratios of corresponding sides of the two similar triangles of Fig. 3.8.7 and find that

$$\frac{z}{18} = \frac{z - x}{6}.$$

It follows that

$$2z = 3x,$$

and implicit differentiation gives

$$2\frac{dz}{dt} = 3\frac{dx}{dt}.$$

We substitute $dx/dt = 8$ and find that

$$\frac{dz}{dt} = \frac{3}{2}\cdot\frac{dx}{dt} = \frac{3}{2}\cdot 8 = 12.$$

So the tip of the man's shadow is moving at 12 ft/s. ■

Example 6 is somewhat unusual in that the answer is independent of the man's distance from the light pole—the given value $x = 100$ is superfluous because the tip of the man's shadow is moving at constant speed. Example 7 is a related-rates problem with two relationships between the variables, which is not quite so unusual.

EXAMPLE 7 Two radar stations at A and B, with B 6 km east of A, are tracking a ship. At a certain instant, the ship is 5 km from A, and this distance is increasing at the rate of 28 km/h. At the same instant, the ship is also 5 km from B, but this distance is increasing at only 4 km/h. Where is the ship, how fast is it moving, and in what direction is it moving?

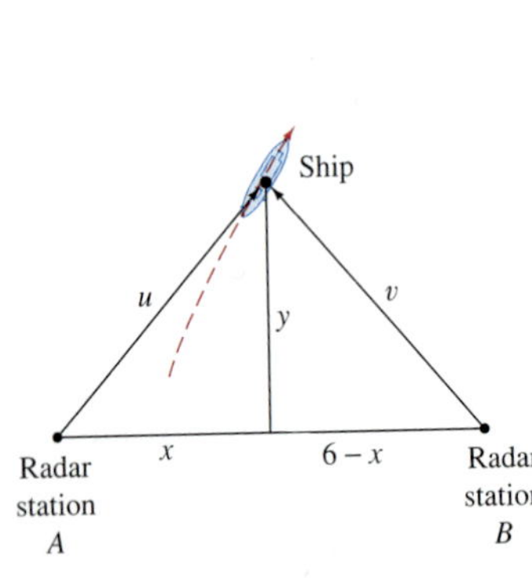

Fig. 3.8.8 Radar stations tracking a ship (Example 7)

Solution With the distances indicated in Fig. 3.8.8, we find—again with the aid of the Pythagorean theorem—that

$$x^2 + y^2 = u^2 \quad \text{and} \quad (6 - x)^2 + y^2 = v^2. \tag{8}$$

We are given these data: $u = v = 5$, $du/dt = 28$, and $dv/dt = 4$ at the instant in question. Because the ship is equally distant from A and B, it is clear that $x = 3$. Thus $y = 4$. Hence the ship is 3 km east and 4 km north of A.

We differentiate implicitly the two equations in (8), and we obtain

$$2x\frac{dx}{dt} + 2y\frac{dy}{dt} = 2u\frac{du}{dt} \quad \text{and} \quad -2(6-x)\frac{dx}{dt} + 2y\frac{dy}{dt} = 2v\frac{dv}{dt}.$$

When we substitute the numerical data given and data deduced, we find that

$$3\frac{dx}{dt} + 4\frac{dy}{dt} = 140 \quad \text{and} \quad -3\frac{dx}{dt} + 4\frac{dy}{dt} = 20.$$

These equations are easy to solve: $dx/dt = dy/dt = 20$. Therefore, the ship is sailing northeast at a speed of

$$\sqrt{20^2 + 20^2} = 20\sqrt{2} \quad \text{(km/h)}$$

—*if* the figure is correct! A mirror along the line AB will reflect *another* ship, 3 km east and 4 km *south* of A, sailing *southeast* at a speed of $20\sqrt{2}$ km/h.

The lesson? Figures are important, helpful, often essential—but potentially misleading. Avoid taking anything for granted when you draw a figure. In this example there would be no real problem, for each radar station could determine whether the ship was generally to the north or to the south. ■

3.8 PROBLEMS

In Problems 1 through 4, first find the derivative dy/dx by implicit differentiation. Then solve the original equation for y explicitly in terms of x and differentiate to find dy/dx. Finally verify that your two results are the same by substitution of the explicit expression for $y(x)$ in the implicit form of the derivative.

1. $x^2 - y^2 = 1$ **2.** $xy = 1$

3. $16x^2 + 25y^2 = 400$ **4.** $x^3 + y^3 = 1$

In Problems 5 through 14, find dy/dx by implicit differentiation.

5. $\sqrt{x} + \sqrt{y} = 1$ **6.** $x^4 + x^2y^2 + y^4 = 48$

7. $x^{2/3} + y^{2/3} = 1$ **8.** $(x-1)y^2 = x + 1$

9. $x^2(x - y) = y^2(x + y)$ **10.** $x^5 + y^5 = 5x^2y^2$

11. $x \sin y + y \sin x = 1$ **12.** $\cos(x + y) = \sin x \sin y$

13. $\cos^3 x + \cos^3 y = \sin(x + y)$

14. $xy = \tan xy$

In Problems 15 through 28, use implicit differentiation to find an equation of the line tangent to the given curve at the given point.

15. $x^2 + y^2 = 25$; $(3, -4)$

16. $xy = -8$; $(4, -2)$

17. $x^2y = x + 2$; $(2, 1)$

18. $x^{1/4} + y^{1/4} = 4$; $(16, 16)$

19. $xy^2 + x^2y = 2$; $(1, -2)$

20. $\dfrac{1}{x+1} + \dfrac{1}{y+1} = 1$; $(1, 1)$

21. $12(x^2 + y^2) = 25xy$; $(3, 4)$

22. $x^2 + xy + y^2 = 7$; $(3, -2)$

23. $\dfrac{1}{x^3} + \dfrac{1}{y^3} = 2$; $(1, 1)$

24. $(x^2 + y^2)^3 = 8x^2y^2$; $(1, -1)$

25. $x^{2/3} + y^{2/3} = 5$; $(8, 1)$ (Fig. 3.8.9)

26. $x^2 - xy + y^2 = 19$; $(3, -2)$ (Fig. 3.8.10)

27. $(x^2 + y^2)^2 = 50xy$; $(2, 4)$ (Fig. 3.8.11)

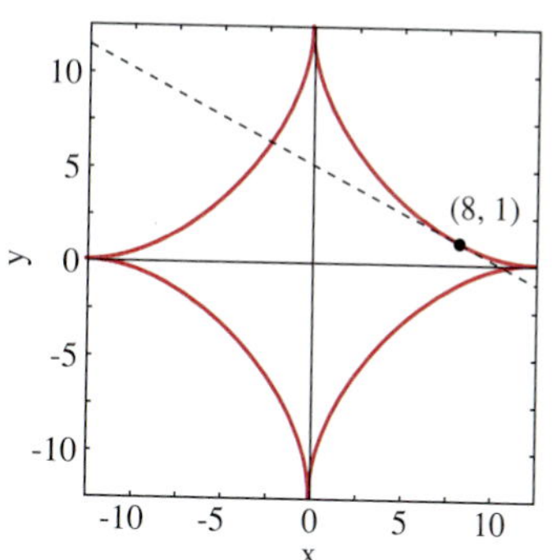

Fig. 3.8.9 Problem 25

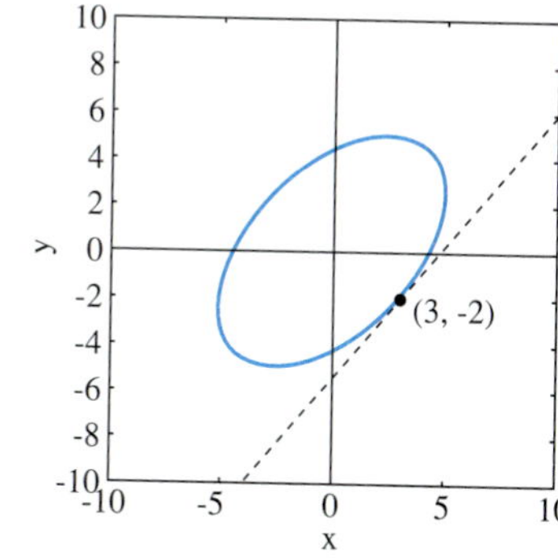

Fig. 3.8.10 Problem 26

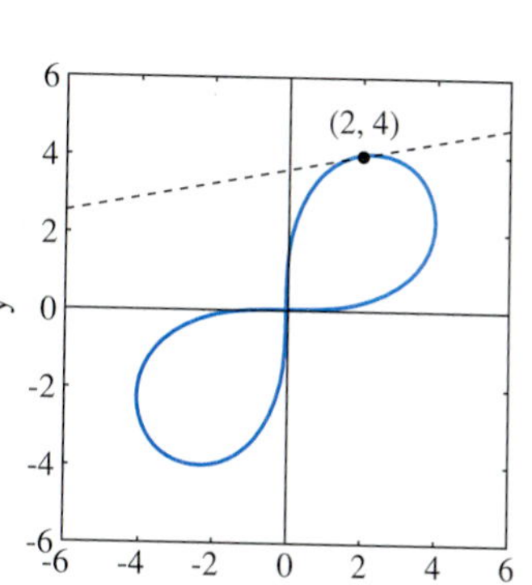

Fig. 3.8.11 Problem 27

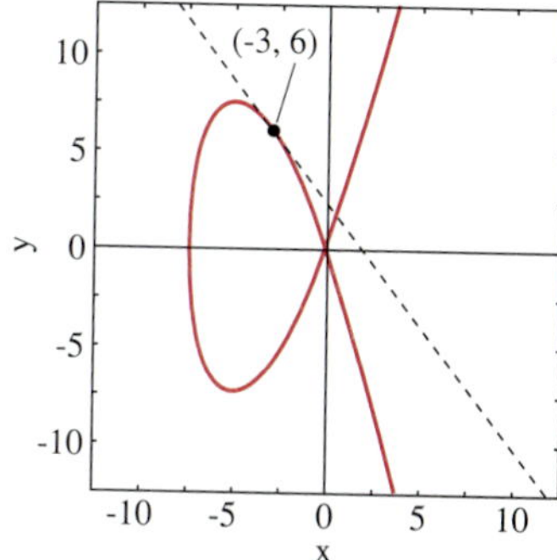

Fig. 3.8.12 Problem 28

28. $y^2 = x^2(x + 7)$; $(-3, 6)$ (Fig. 3.8.12)

29. The curve $x^3 + y^3 = 9xy$ is similar in shape and appearance to the folium of Descartes in Fig. 3.8.4. Find (a) the equation of its tangent line at the point $(2, 4)$ and (b) the equation of its tangent line with slope -1.

30. (a) Factor the left-hand side of the equation

$$2x^2 - 5xy + 2y^2 = 0$$

to show that its graph consists of two straight lines through the origin. Hence the derivative $y'(x)$ has only two possible numerical values (the slopes of these two lines). (b) Calculate dy/dx by implicit differentiation of the equation in part (a). Verify that the expression you obtain yields the proper slope for each of the straight lines of part (a).

31. Find all points on the graph of $x^2 + y^2 = 4x + 4y$ at which the tangent line is horizontal.

32. Find the first-quadrant points of the folium of Example 2 at which the tangent line is either horizontal $(dy/dx = 0)$ or vertical [where $dx/dy = 1/(dy/dx) = 0$].

33. The graph of the equation $x^2 - xy + y^2 = 9$ is the rotated ellipse shown in Fig. 3.8.13. Find the lines tangent to this curve at the two points where it intersects the x-axis, and show that these lines are parallel.

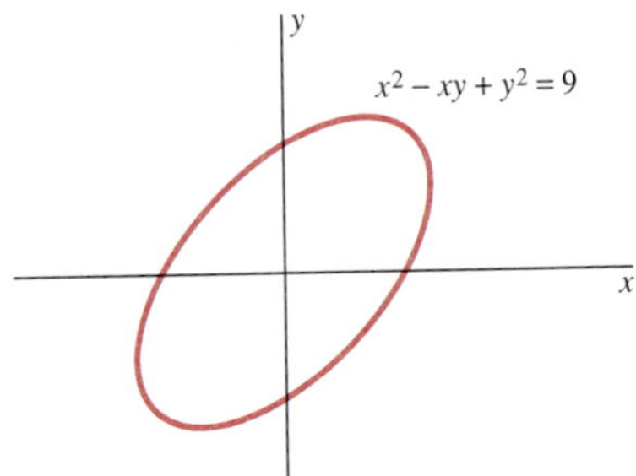

Fig. 3.8.13 The rotated ellipse of Problem 33

34. Find the points on the curve of Problem 33 where the tangent line is horizontal $(dy/dx = 0)$ and those where it is vertical $(dx/dy = 0)$.

35. The graph in Fig. 3.8.14 is a *lemniscate* with equation $(x^2 + y^2)^2 = x^2 - y^2$. Find by implicit differentiation the four points on the lemniscate where the tangent line is horizontal. Then find the two points where the tangent line is vertical—that is, where $dx/dy = 1/(dy/dx) = 0$.

36. Water is being collected from a block of ice with a square base (Fig. 3.8.15). The water is produced because the ice is melting in such a way that each edge of the base of the block is decreasing at 2 in./h while the height of the block is decreasing at 3 in./h. What is the rate of flow of water into the collecting pan when the base has edge length 20 in. and the height of the block is 15 in.? Make the simplifying assumption that water and ice have the same density.

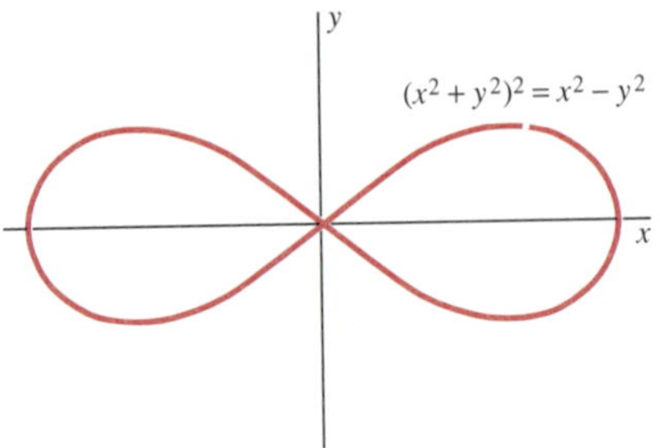

Fig. 3.8.14 The lemniscate of Problem 35

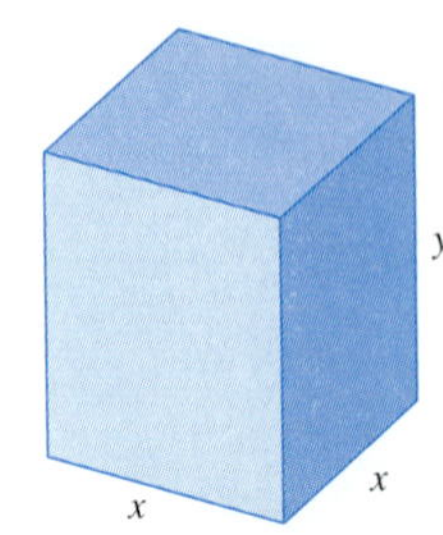

Fig. 3.8.15 The ice block of Problem 36

37. Sand is being emptied from a hopper at the rate of 10 ft^3/s. The sand forms a conical pile whose height is always twice its radius (Fig. 3.8.16). At what rate is the radius of the pile increasing when its height is 5 ft?

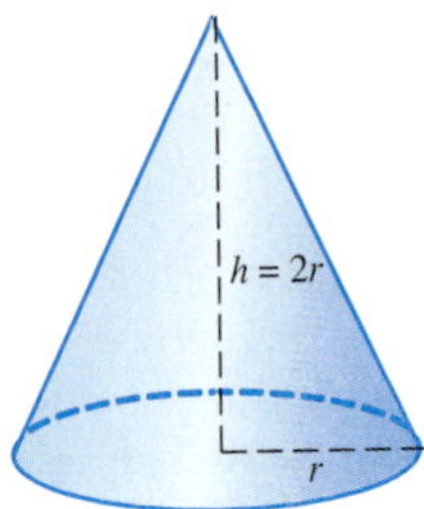

Fig. 3.8.16 The conical sand pile of Problem 37 with volume $V = \frac{1}{3}\pi r^2 h$

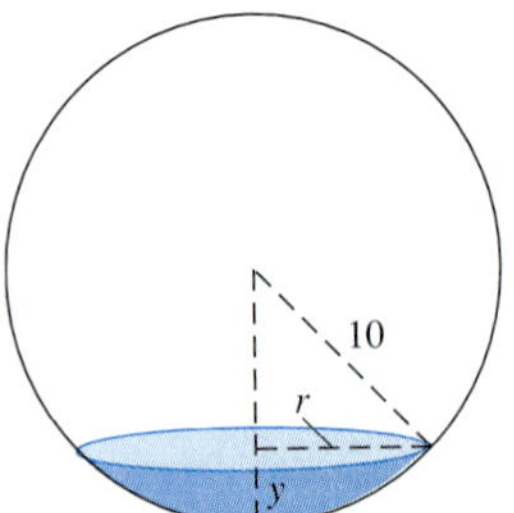

Fig. 3.8.17 The spherical tank of Problem 38

38. Suppose that water is being emptied from a spherical tank of radius 10 ft (Fig. 3.8.17). If the depth of the water in the tank is 5 ft and is decreasing at the rate of 3 ft/s, at what rate is the radius r of the top surface of the water decreasing?

39. A circular oil slick of uniform thickness is caused by a spill of 1 m^3 of oil. The thickness of the oil slick is decreasing at the rate of 0.1 cm/h. At what rate is the radius of the slick increasing when the radius is 8 m?

40. Suppose that an ostrich 5 ft tall is walking at a speed of 4 ft/s directly toward a street light 10 ft high. How fast is the tip of the ostrich's shadow moving along the ground? At what rate is the ostrich's shadow decreasing in length?

41. The width of a rectangle is half its length. At what rate is its area increasing if its width is 10 cm and is increasing at 0.5 cm/s?

42. At what rate is the area of an equilateral triangle increasing if its base is 10 cm long and is increasing at 0.5 cm/s?

43. A gas balloon is being filled at the rate of 100π cm^3 of gas per second. At what rate is the radius of the balloon increasing when its radius is 10 cm?

44. The volume V (in cubic inches) and pressure p (in pounds per square inch) of a certain gas satisfy the equation $pV = 1000$. At what rate is the volume of the sample changing if the pressure is 100 lb/in.^{2} and is increasing at the rate of 2 lb/in.^{2} per second?

45. Figure 3.8.18 shows a kite in the air at an altitude of 400 ft. The kite is being blown horizontally at the rate of 10 ft/s away from the person holding the kite string at ground level. At what rate is the string being payed out when 500 ft of string is already out? (Assume that the string forms a straight line.)

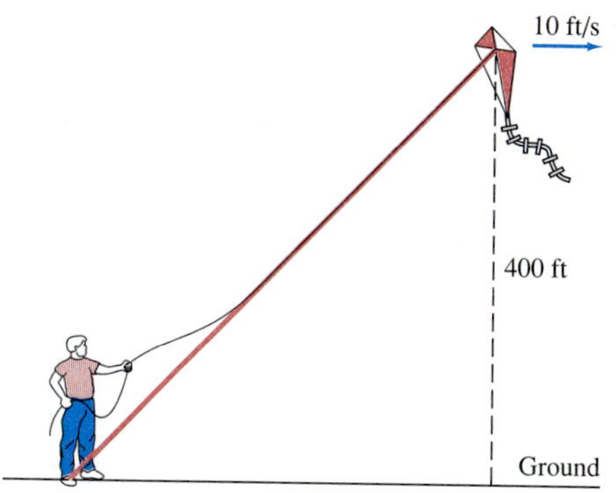

Fig. 3.8.18 The kite of Problem 45

46. A weather balloon that is rising vertically is being observed from a point on the ground 300 ft from the spot directly beneath the balloon. At what rate is the balloon rising when the angle between the ground and the observer's line of sight is 45° and is increasing at 1° per second?

47. An airplane flying horizontally at an altitude of 3 mi and at a speed of 480 mi/h passes directly above an observer on the ground. How fast is the distance from the observer to the airplane increasing 30 s later?

48. Figure 3.8.19 shows a spherical tank of radius a partly filled with water. The maximum depth of water in the tank is y. A formula for the volume V of water in the tank—a formula you can derive after you study Chapter 5—is $V = \frac{1}{3}\pi y^2(3a - y)$. Suppose that water is being drained from a spherical tank of radius 5 ft at the rate of 100 gal/min. Find the rate at which the depth y of water is decreasing when (a) $y = 7$ (ft); (b) $y = 3$ (ft). (*Note:* One gallon of water occupies a volume of approximately 0.1337 ft^3.)

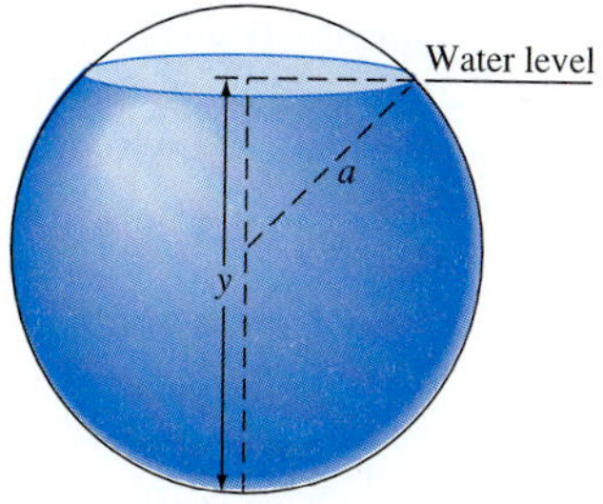

Fig. 3.8.19 The spherical water tank of Problem 48

49. Repeat Problem 48, this time for a tank that is hemispherical, flat side on top, with radius 10 ft.

50. A swimming pool is 50 ft long and 20 ft wide. Its depth varies uniformly from 2 ft at the shallow end to 12 ft at the deep end (Fig. 3.8.20). Suppose that the pool is being filled at the rate of 1000 gal/min. At what rate is the depth of water at the deep end increasing when the depth there is 6 ft? (*Note:* One gallon of water occupies a volume of approximately 0.1337 ft^3.)

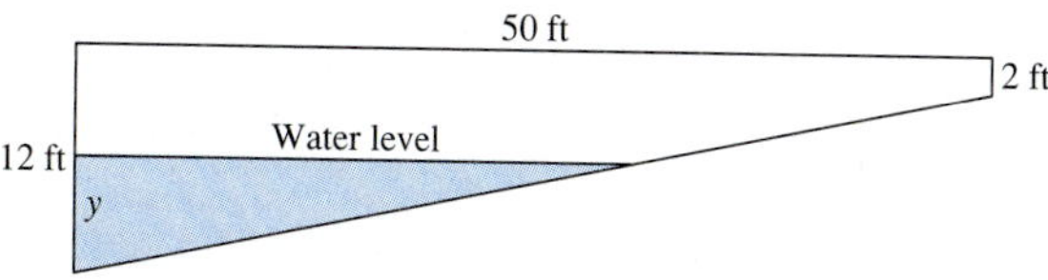

Fig. 3.8.20 Cross section of the swimming pool of Problem 50

51. A ladder 41 ft long that was leaning against a vertical wall begins to slip. Its top slides down the wall while its bottom moves along the level ground at a constant speed of 4 ft/s. How fast is the top of the ladder moving when it is 9 ft above the ground?

52. The base of a rectangle is increasing at 4 cm/s while its height is decreasing at 3 cm/s. At what rate is its area changing when its base is 20 cm and its height is 12 cm?

53. The height of a cone is decreasing at 3 cm/s while its radius is increasing at 2 cm/s. When the radius is 4 cm and the height is 6 cm, is the volume of the cone increasing or decreasing? At what rate is the volume changing then?

54. A square is expanding. When each edge is 10 in., its area is increasing at 120 in.^{2}/s. At what rate is the length of each edge changing then?

55. A rocket that is launched vertically is tracked by a radar station located on the ground 4 mi from the launch site. What is the vertical speed of the rocket at the instant its distance from the radar station is 5 mi and this distance is increasing at the rate of 3600 mi/h?

56. Two straight roads intersect at right angles. At 10 A.M. a car passes through the intersection headed due east at 30 mi/h. At 11 A.M. a truck heading due north at 40 mi/h passes through the intersection. Assume that the two vehicles maintain the given speeds and directions. At what rate are they separating at 1 P.M.?

57. A 10-ft ladder is leaning against a wall. The bottom of the ladder begins to slide away from the wall at a speed of 1 mi/h. (a) Find the rate at which the top of the ladder is moving when it is 4 ft from the ground. If the top of the ladder maintained contact with the wall, find the speed with which it would be moving when it is (b) 1 in. above the ground; (c) 1 mm above the ground. Do you believe your answers? The key to the apparent paradox is that when the top of the ladder is about 1.65 ft high, it disengages altogether from the wall and thereafter moves away from it.

58. Two ships are sailing toward a very small island. One ship, the *Pinta,* is east of the island and is sailing due west at 15 mi/h. The other ship, the *Niña,* is north of the island and is sailing due south at 20 mi/h. At a certain time the *Pinta* is 30 mi from the island and the *Niña* is 40 mi from the island. At what rate are the two ships drawing closer together at that time?

59. At time $t = 0$, a single-engine military jet is flying due east at 12 mi/min. At the same altitude and 208 mi directly ahead of the military jet, still at time $t = 0$, a commercial jet is flying due north at 8 mi/min. When are the two planes closest to each other? What is the minimum distance between them?

60. A ship with a long anchor chain is anchored in 11 fathoms of water. The anchor chain is being wound in at the rate of 10 fathoms/min, causing the ship to move toward the spot directly above the anchor resting on the seabed. The hawsehole—the point of contact between ship and chain—is located 1 fathom above the water line. At what speed is the ship moving when there are exactly 13 fathoms of chain still out?

61. A water tank is in the shape of a cone with vertical axis and vertex downward. The tank has radius 3 ft and is 5 ft high. At first the tank is full of water, but at time $t = 0$ (in seconds), a small hole at the vertex is opened and the water begins to drain. When the height of the water in the tank has dropped to 3 ft, the water is flowing out at 2 ft^3/s. At what rate, in feet per second, is the water level dropping then?

62. A spherical tank of radius 10 ft is being filled with water at the rate of 200 gal/min. How fast is the water level rising when the maximum depth of water in the tank is 5 ft? See Problem 48 for a useful formula and a helpful note.

63. A water bucket is shaped like the frustum of a cone with height 2 ft, base radius 6 in., and top radius 12 in. Water is leaking from the bucket at 10 in.^{3}/min. At what rate is the water level falling when the depth of water in the bucket is 1 ft? (*Note:* The volume V of a conical frustum with height h and base radii a and b is

$$V = \frac{\pi h}{3}(a^2 + ab + b^2).$$

Such a frustum is shown in Fig. 3.8.21.)

64. Suppose that the radar stations A and B of Example 7 are now 12.6 km apart. At a certain instant, a ship is 10.4 km from A and its distance from A is increasing at 19.2 km/h. At the same instant, its distance from B is 5 km and is decreasing at 0.6 km/h. Find the location, speed, and direction of motion of the ship.

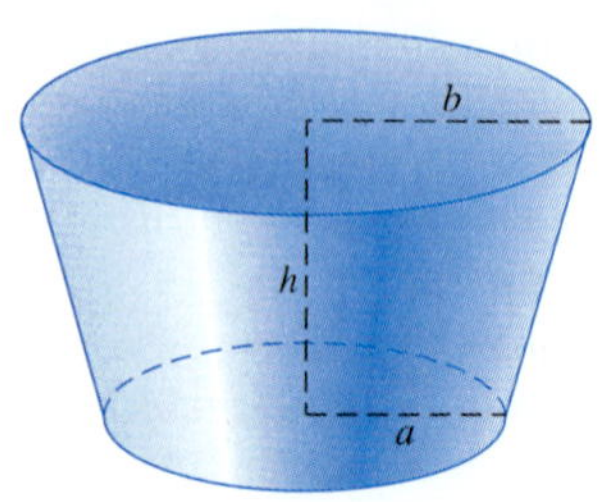

Fig. 3.8.21 The volume of this conical frustum is given in Problem 63.

65. An airplane climbing at an angle of 45° passes directly over a ground radar station at an altitude of 1 mi. A later reading shows that the distance from the radar station to the plane is 5 mi and is increasing at 7 mi/min. What is the speed of the plane then (in miles per hour)? (*Suggestion:* You may find the law of cosines useful—see Appendix C.)

66. The water tank of Problem 62 is completely full when a plug at its bottom is removed. According to *Torricelli's law,* the water drains in such a way that $dV/dt = -k\sqrt{y}$, where V is the volume of water in the tank and k is a positive empirical constant. (a) Find dy/dt as a function of the depth y. (b) Find the depth of water when the water level is falling the *least* rapidly. (You will need to compute the derivative of dy/dt with respect to y.)

67. A person 6 ft tall walks at 5 ft/s along one edge of a road 30 ft wide. On the other edge of the road is a light atop a pole 18 ft high. How fast is the length of the person's shadow (on the horizontal ground) increasing when the person is 40 ft from the point directly across the road from the pole?

68. A highway patrol officer's radar unit is parked behind a billboard 200 ft from a long straight stretch of U.S. 17. Down the highway, 200 ft from the point on the highway closest to the officer, is an emergency call box. The officer points the radar gun at the call box. A minivan passes the call box and, at that moment, the radar unit indicates that the *distance between the officer and the minivan* is increasing at 45 mi/h—that is, 66 ft/s. The posted speed limit is 55 mi/h. Does the officer have any reason to apprehend the driver of the minivan?

3.9 SUCCESSIVE APPROXIMATIONS AND NEWTON'S METHOD

The solution of equations has always been a central task of mathematics. More than two millennia ago, mathematicians of ancient Babylon discovered the method of "completing the square," which leads to the *quadratic formula* for an exact solution of any second-degree equation $ax^2 + bx + c = 0$. Early in the sixteenth century, several Italian mathematicians (Cardan, del Ferro, Ferrari, Tartaglia) discovered formulas for the exact solutions of third- and fourth-degree equations. (Because they are quite complicated, these formulas are seldom used today except in computer algebra systems.) And in 1824 a brilliant young Norwegian mathematician, Niels Henrik Abel* (1802–1829), published a proof that there is *no* general formula giving the solution of an arbitrary polynomial equation of degree 5 (or higher) in terms of algebraic combinations of its coefficients. Thus the exact solution (for all its roots) of an equation such as

$$f(x) = x^5 - 3x^3 + x^2 - 23x + 19 = 0 \tag{1}$$

may be quite difficult or even—as a practical matter—impossible to find. In such a case it may be necessary to resort to *approximate methods.*

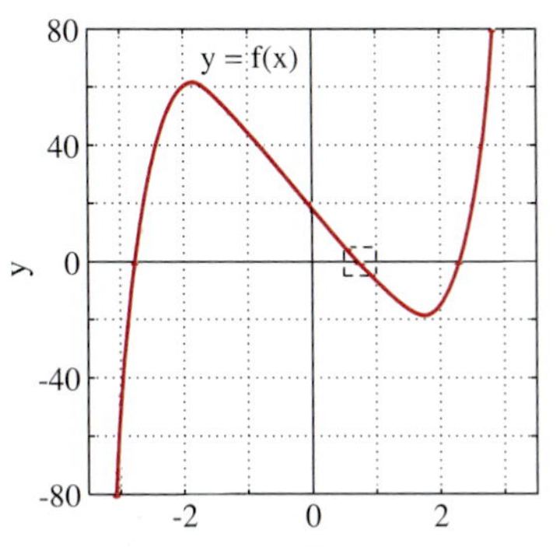

Fig. 3.9.1 The graph $y = f(x)$ in Eq. (1)

For example, the graph of $y = f(x)$ in Fig. 3.9.1 indicates that Eq. (1) has three real solutions (and hence two complex ones as well). The indicated small rectangle $0.5 \leqq x \leqq 1$, $-5 \leqq y \leqq 5$ encloses one of these solutions. If we use this small rectangle as a new "viewing window" with a computer or graphics calculator, then we see that this solution is near 0.8 (Fig. 3.9.2). A few additional magnifications might yield greater accuracy, showing that the solution is approximately 0.801.

Graphical methods are good for three- or four-place approximations. Here we shall discuss an analytical method developed by Isaac Newton that can rapidly provide much more accurate approximations.

Iteration and the Babylonian Square Root Method

What it means to solve even so simple an equation as

$$x^2 - 2 = 0 \tag{2}$$

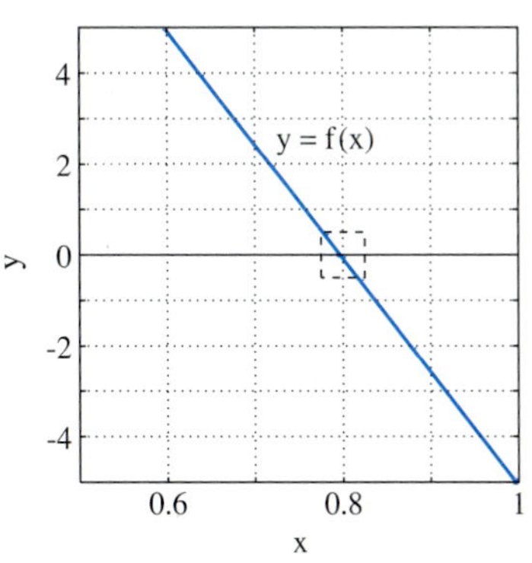

Fig. 3.9.2 Magnification of Fig. 3.9.1 near a solution

is open to question. The positive exact solution is $x = \sqrt{2}$. But the number $\sqrt{2}$ is irrational and hence cannot be expressed as a terminating or repeating decimal. Thus if we mean by a *solution* an exact decimal value for x, then even Eq. (2) can be solved only approximately.

The ancient Babylonians devised an effective way to generate a sequence of better and better approximations to $\sqrt{A}$, the square root of a given positive number A. Here is the *Babylonian square root method:* We begin with a first guess x_0 for the value of $\sqrt{A}$. For $\sqrt{2}$, we might guess $x_0 = 1.5$. If x_0 is too large—that is, if $x_0 > \sqrt{A}$—then

$$\frac{A}{x_0} < \frac{A}{\sqrt{A}} = \sqrt{A},$$

so A/x_0 is too small an estimate of $\sqrt{A}$. Similarly, if x_0 is too small (if $x_0 < \sqrt{A}$), then A/x_0 is too large an estimate of $\sqrt{A}$; that is, $A/x_0 > \sqrt{A}$.

Thus in each case one of the two numbers x_0 and A/x_0 is an underestimate of $\sqrt{A}$ and the other is an overestimate. The Babylonian idea was that we should get a better estimate of $\sqrt{A}$ by *averaging* x_0 and A/x_0. This yields a better approximation,

$$x_1 = \frac{1}{2}\left(x_0 + \frac{A}{x_0}\right), \tag{3}$$

* For the complete story of Abel's remarkable achievements in his brief lifetime, see Oystein Ore's very readable biography *Niels Henrik Abel* (The University of Minnesota and Chelsea Publishing Company, 1974).

to $\sqrt{A}$. But why not repeat this process? We can average x_1 and A/x_1 to get a second approximation x_2, average x_2 and A/x_2 to get x_3, and so on. By repeating this process, we generate a sequence of numbers

$$x_1, \quad x_2, \quad x_3, \quad x_4, \ \ldots$$

that we have every right to expect will consist of better and better approximations to $\sqrt{A}$.

Specifically, having calculated the nth approximation x_n, we calculate the next one by means of the *iterative formula*

$$x_{n+1} = \frac{1}{2}\left(x_n + \frac{A}{x_n}\right). \tag{4}$$

In other words, we plow each approximation to $\sqrt{A}$ back into the right-hand side in Eq. (4) to calculate the next approximation. This is an *iterative* process—the words *iteration* and *iterative* are derived from the Latin *iterare,* "to plow again."

Suppose we find that after sufficiently many steps in this iteration, $x_{n+1} \approx x_n$ accurate to the number of decimal places we are retaining in our computations. Then Eq. (4) yields

$$x_n \approx x_{n+1} = \frac{1}{2}\left(x_n + \frac{A}{x_n}\right) = \frac{1}{2x_n}(x_n^2 + A),$$

so $2x_n^2 \approx x_n^2 + A$, and hence $x_n^2 \approx A$ to some degree of accuracy.

EXAMPLE 1 With $A = 2$ we begin with the crude first guess $x_0 = 1$ at the value of $\sqrt{2}$. Then successive applications of the formula in Eq. (4) yield

$$x_1 = \frac{1}{2}\left(1 + \frac{2}{1}\right) = \frac{3}{2} = 1.5,$$

$$x_2 = \frac{1}{2}\left(\frac{3}{2} + \frac{2}{3/2}\right) = \frac{17}{12} \approx 1.416666667,$$

$$x_3 = \frac{1}{2}\left(\frac{17}{12} + \frac{2}{17/12}\right) = \frac{577}{408} \approx 1.414215686,$$

$$x_4 = \frac{1}{2}\left(\frac{577}{408} + \frac{2}{577/408}\right) = \frac{665857}{470832} \approx 1.414213562,$$

rounding results to nine decimal places. It happens that x_4 gives $\sqrt{2}$ accurate to all nine places! ■

The Babylonian iteration defined in Eq. (4) is a method for generating a sequence of approximations to the positive root $r = \sqrt{A}$ of the particular equation $x^2 - A = 0$. We turn next to a method that gives such a sequence of approximations for more general equations.

Newton's Method

Newton's method is an iterative method for generating a sequence $x_1, x_2, x_3, \ldots$ of approximations to a solution r of a given equation written in the general form

$$f(x) = 0. \tag{5}$$

We hope that this sequence of approximations will "converge" to the root r in the sense of the following definition.

Definition *Convergence of Approximations*

We say that the sequence of approximations $x_1, x_2, x_3, \ldots$ **converges** to the number r provided that we can make x_n as close to r as we please merely by choosing n sufficiently large. More precisely, for any given $\epsilon > 0$, there exists a positive integer N such that $|x_n - r| < \epsilon$ for all $n \geqq N$.

As a practical matter such convergence means, as illustrated in Example 1, that for any positive integer k, x_n and r will agree to k or more decimal places once n becomes sufficiently large.

The idea is that we begin with an *initial guess* x_0 that roughly approximates a solution r of the equation $f(x) = 0$. This initial guess may, for example, be obtained by inspection of the graph of $y = f(x)$, perhaps obtained from a computer or graphics calculator. We use x_0 to calculate an approximation x_1, use x_1 to calculate a better approximation x_2, use x_2 to calculate a still better approximation x_3, and so on.

Here is the general step in the process. Having reached the nth approximation x_n, we use the tangent line at $(x_n, f(x_n))$ to construct the next approximation x_{n+1} to the solution r as follows: Begin at the point x_n on the x-axis. Go vertically up (or down) to the point $(x_n, f(x_n))$ on the curve $y = f(x)$. Then follow the tangent line L there to the point where L meets the x-axis (Fig. 3.9.3). That point will be x_{n+1}.

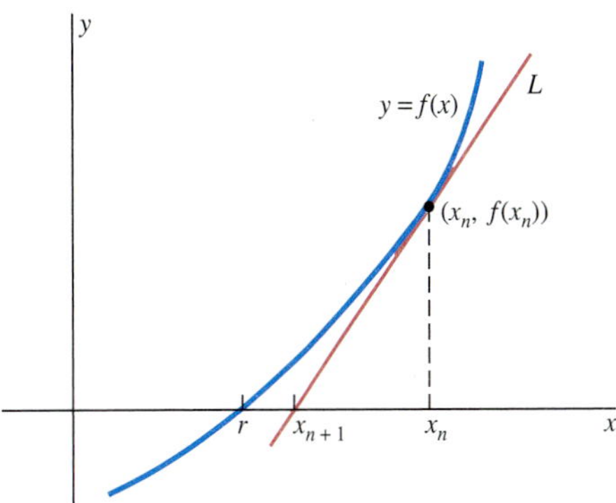

Fig. 3.9.3 Geometry of the formula of Newton's method

Here is a formula for x_{n+1}. We obtain it by computing the slope of the line L in two ways: from the derivative and from the two-point definition of slope. Thus

$$f'(x_n) = \frac{f(x_n) - 0}{x_n - x_{n+1}},$$

and we easily solve for

$$x_{n+1} = x_n - \frac{f(x_n)}{f'(x_n)}. \qquad \textbf{(6)}$$

This equation is the **iterative formula** of Newton's method, so called because in about 1669, Newton introduced an algebraic procedure (rather than the geometric construction illustrated in Fig. 3.9.3) that is equivalent to the iterative use of Eq. (6). Newton's first example was the cubic equation $x^3 - 2x - 5 = 0$, for which he found the root $r \approx 2.0946$ (as we ask you to do in Problem 18).

Suppose now that we want to apply Newton's method to solve the equation

$$f(x) = 0 \tag{7}$$

to an accuracy of k decimal places (k digits to the right of the decimal correct or correctly rounded). Remember that an equation must be written precisely in the form of Eq. (7) in order to use the formula in Eq. (6). If we reach the point in our iteration at which x_n and x_{n+1} agree to k decimal places, it then follows that

$$x_n \approx x_{n+1} = x_n - \frac{f(x_n)}{f'(x_n)}; \qquad 0 \approx -\frac{f(x_n)}{f'(x_n)}; \qquad f(x_n) \approx 0.$$

Thus we have found an approximate root $x_n \approx x_{n+1}$ of Eq. (7). In practice, then, we retain k decimal places in our computations and persist until $x_n \approx x_{n+1}$ to this degree of accuracy. (We do not consider here the possibility of round-off error, an important topic in numerical analysis.)

EXAMPLE 2 Use Newton's method to find $\sqrt{2}$ accurate to nine decimal places.

Solution More generally, consider the square root of the positive number A as the positive root of the equation

$$f(x) = x^2 - A = 0.$$

Because $f'(x) = 2x$, Eq. (6) gives the iterative formula

$$x_{n+1} = x_n - \frac{x_n^2 - A}{2x_n} = \frac{1}{2}\left(x_n + \frac{A}{x_n}\right). \tag{8}$$

Thus we have derived the Babylonian iterative formula as a special case of Newton's method. The use of Eq. (8) with $A = 2$ therefore yields exactly the values of x_1, x_2, x_3, and x_4 that we computed in Example 1, and after performing another iteration we find that

$$x_5 = \frac{1}{2}\left(x_4 + \frac{2}{x_4}\right) \approx 1.414213562,$$

which agrees with x_4 to nine decimal places. The very rapid convergence here is an important characteristic of Newton's method. As a general rule (with some exceptions), each iteration doubles the number of decimal places of accuracy. ■

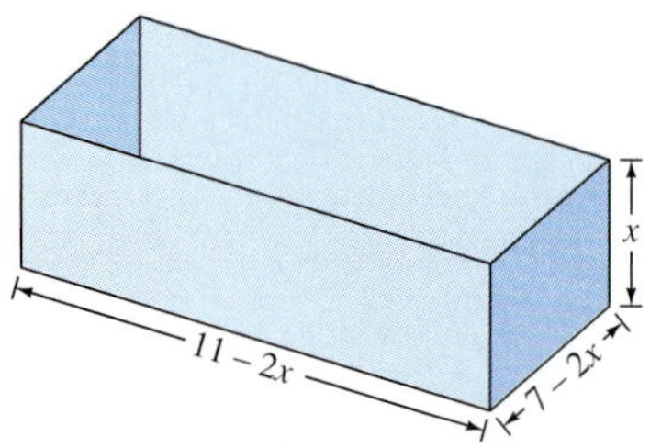

Fig. 3.9.4 The tray of Example 3

EXAMPLE 3 Figure 3.9.4 shows an open-topped tray constructed by the method of Example 2 in Section 3.6. We begin with a 7-in. by 11-in. rectangle of sheet metal. We cut a square with edge length x from each of its four corners and then fold up the resulting flaps to obtain a rectangular tray with volume

$$V(x) = x(7 - 2x)(11 - 2x) = 4x^3 - 36x^2 + 77x, \qquad 0 \leqq x \leqq 3.5. \tag{9}$$

In Section 3.6 we inquired about the maximum possible volume of such a tray. Here we want to find instead the value(s) of x that will yield a tray with volume 40 in.^{3}; we will find x by solving the equation

$$V(x) = 4x^3 - 36x^2 + 77x = 40.$$

To solve this equation for x, first we write an equation of the form in Eq. (7):

$$f(x) = 4x^3 - 36x^2 + 77x - 40 = 0. \tag{10}$$

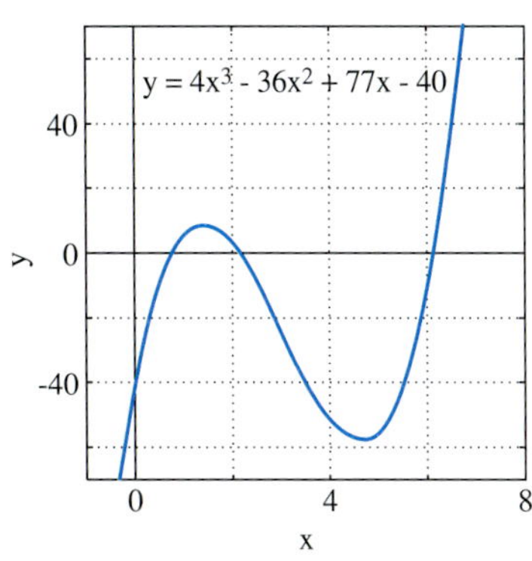

Fig. 3.9.5 The graph of $f(x)$ in Eq. (10) of Example 3

Figure 3.9.5 shows the graph of f. We see three solutions: a root r_1 between 0 and 1, a root r_2 slightly greater than 2, and a root r_3 slightly larger than 6. Because

$$f'(x) = 12x^2 - 72x + 77,$$

Newton's iterative formula in Eq. (6) takes the form

$$x_{n+1} = x_n - \frac{f(x_n)}{f'(x_n)} = x_n - \frac{4x_n^3 - 36x_n^2 + 77x_n - 40}{12x_n^2 - 72x_n + 77}. \tag{11}$$

Beginning with the initial guess $x_0 = 1$ (because it's reasonably close to r_1), Eq. (11) gives

$$x_1 = 1 - \frac{4\cdot 1^3 - 36\cdot 1^2 + 77\cdot 1 - 40}{12\cdot 1^2 - 72\cdot 1 + 77} \approx 0.7059,$$

$$x_2 \approx 0.7736,$$

$$x_3 \approx 0.7780,$$

$$x_4 \approx 0.7780.$$

Thus we obtain the root $r_1 \approx 0.7780$, retaining only four decimal places.

If we had begun with a different initial guess, the sequence of Newton iterates might well have converged to a different root of the equation $f(x) = 0$. The approximate solution obtained therefore depends on the initial guess. For example, with $x_0 = 2$ and, later, with $x_0 = 6$, the iteration in Eq. (11) produces the two sequences

$x_0 = 2$	$x_0 = 6$
$x_1 \approx 2.1053$	$x_1 \approx 6.1299$
$x_2 \approx 2.0993$	$x_2 \approx 6.1228$
$x_3 \approx 2.0992$	$x_3 \approx 6.1227$
$x_4 \approx 2.0992$	$x_4 \approx 6.1227.$

Thus the other two roots of Eq. (10) are $r_2 \approx 2.0992$ and $r_3 \approx 6.1277$.

With $x = r_1 \approx 0.7780$, the tray in Fig. 3.9.4 has the approximate dimensions 9.4440 in. by 5.4440 in. by 0.7780 in. With $x = r_2 \approx 2.0992$, its approximate dimensions are 6.8015 in. by 2.8015 in. by 2.0992 in. But the third root $r_3 \approx 6.1227$ would *not* lead to a tray that is physically possible. (Why not?) Thus the *two* values of x that yield trays with volume 40 in. are $x \approx 0.7780$ and $x \approx 2.0992$. ■

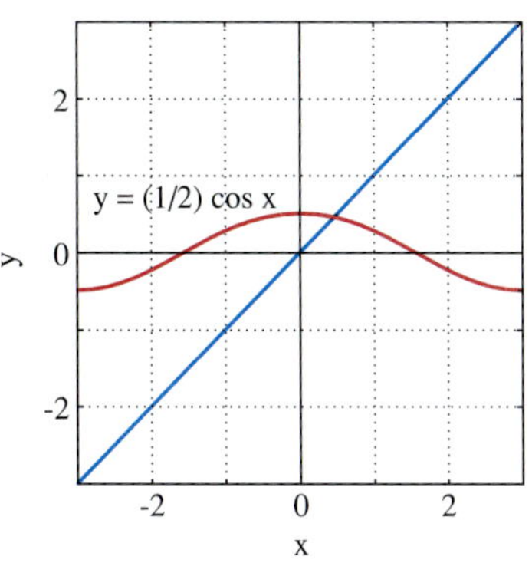

Fig. 3.9.6 Solving the equation $x = \frac{1}{2}\cos x$ (Example 4)

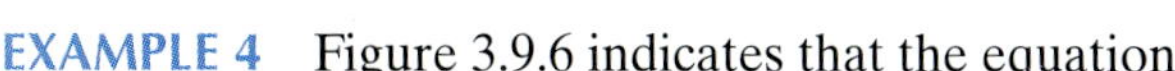

EXAMPLE 4 Figure 3.9.6 indicates that the equation

$$x = \frac{1}{2}\cos x \tag{12}$$

has a solution r near 0.5. To apply Newton's method to approximate r, we rewrite Eq. (12) in the form

$$f(x) = 2x - \cos x = 0.$$

Because $f'(x) = 2 + \sin x$, the iterative formula of Newton's method is

$$x_{n+1} = x_n - \frac{2x_n - \cos x_n}{2 + \sin x_n}.$$

Beginning with $x_0 = 0.5$ and retaining five decimal places, this formula yields

$$x_1 \approx 0.45063, \qquad x_2 \approx 0.45018, \qquad x_3 \approx 0.45018.$$

Thus the root is 0.45018 to five decimal places. ■

Newton's method is one for which "the proof is in the pudding." If it works, it's obvious that it does, and everything's fine. When Newton's method fails, it may do so spectacularly. For example, suppose that we want to solve the equation

$$x^{1/3} = 0.$$

Here $r = 0$ is the only solution. The iterative formula in Eq. (6) becomes

$$x_{n+1} = x_n - \frac{(x_n)^{1/3}}{\frac{1}{3}(x_n)^{-2/3}} = x_n - 3x_n = -2x_n.$$

If we begin with $x_0 = 1$, Newton's method yields $x_1 = -2, x_2 = +4, x_3 = -8$, and so on. Figure 3.9.7 indicates why our "approximations" are not converging.

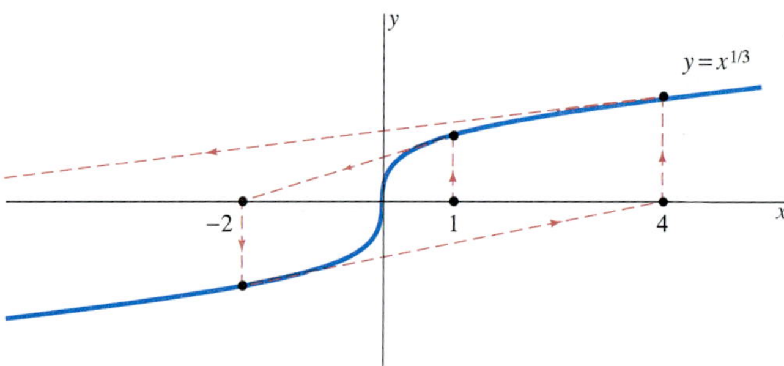

Fig. 3.9.7 A failure of Newton's method

When Newton's method fails, a graph will typically indicate the reason why. Then the use of an alternative method such as repeated tabulation or successive magnification is appropriate.

Newton's Method with Calculators and Computers

With calculators and computers that permit user-defined functions, Newton's method is very easy to set up and apply repeatedly. It is helpful to interpret Newton's iteration

$$x_{n+1} = x_n - \frac{f(x_n)}{f'(x_n)}$$

as follows. Having first defined the functions f and f', we then define the "iteration function"

$$g(x) = x - \frac{f(x)}{f'(x)}.$$

Newton's method is then equivalent to the following procedure. Begin with an initial estimate x_0 of the solution of the equation

$$f(x) = 0.$$

Calculate successive approximations $x_1, x_2, x_3, \ldots$ to the exact solution by means of the iteration

$$x_{n+1} = g(x_n).$$

That is, apply the function g to each approximation to get the next.

Figure 3.9.8 shows a TI graphics calculator prepared to solve the equation

$$f(x) = x^3 - 3x^2 + 1 = 0.$$

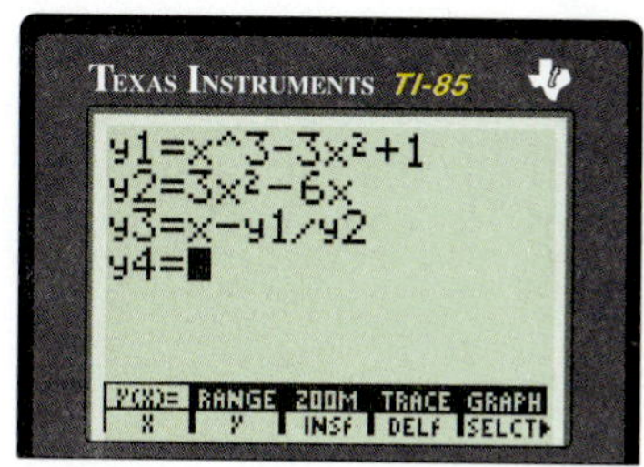

Fig. 3.9.8 Preparing to solve the equation $x^3 - 3x^2 + 1 = 0$

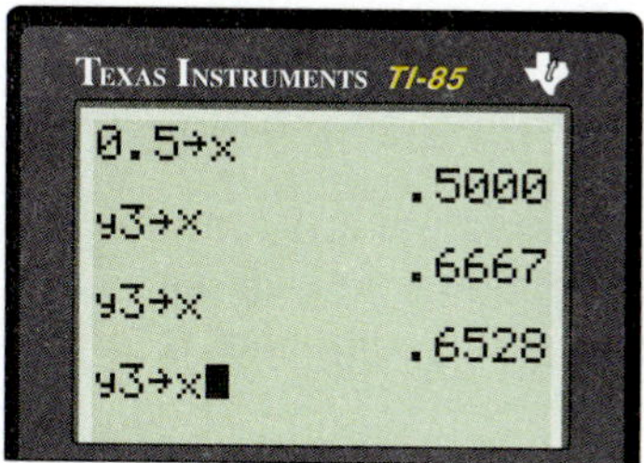

Fig. 3.9.9 Solving the equation $x^3 - 3x^2 + 1 = 0$

Then we need only store the initial guess, **0.5→X**, and next enter repeatedly the command **Y3→X**, as indicated in Fig. 3.9.9.

Figure 3.9.10 shows an HP calculator prepared to carry out the same iteration. The functions **F(X)**, **D(X)** (for $f'(x)$), and **G(X)** are each defined by pressing the **DEFINE** key. Then it is necessary only to **ENTER** the initial guess x_0 and press the **G** key repeatedly to generate the desired successive appropriations.

With *Maple* or *Mathematica*, you can define the functions f and g and then repeatedly enter the command $x = g(x)$, as shown in Fig. 3.9.11.

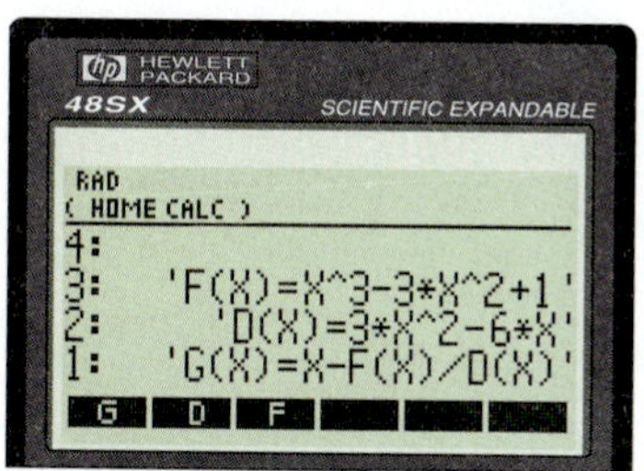

Fig. 3.9.10 Preparing to solve the equation $x^3 - 3x^2 + 1 = 0$

Mathematica Command	*Maple* Command	Result
f [x_] := x^3 − 3x^2 + 1	f := x −> x^3 − 3 * x^2 + 1;	
g [x_] := x − f [x] /f′ [x]	g := x −> x − f(x)/D(f)(x);	
x = 0.5	x := 0.5;	0.500000
x = g [x]	x := g(x);	0.666667
x = g [x]	x := g(x);	0.652778
x = g [x]	x := g(x);	0.652704
x = g [x]	x := g(x);	0.652704

Fig. 3.9.11 *Mathematica* and *Maple* implementations of Newton's method

Newton's Method and Computer Graphics

Newton's method and similar iterative techniques are often used to generate vividly colored "fractal patterns," in which the same or similar structures are replicated on smaller and smaller scales at successively higher levels of magnification. To describe one method of such replication, we replace the real numbers in our Newton's method computations with *complex* numbers. We illustrate this idea with the cubic equation

$$f(x) = x^3 - 3x^2 + 1 = 0. \tag{13}$$

In the project at the end of this section, we ask you to approximate the three solutions

$$r_1 \approx -0.53, \qquad r_2 \approx 0.65, \qquad r_3 \approx 2.88$$

of this equation.

Fig. 3.9.12 $-2 \leqq x \leqq 4$, $-2.25 \leqq y \leqq 2.25$

Fig. 3.9.13 $1.6 \leqq x \leqq 2.4$, $-0.3 \leqq y \leqq 0.3$

Fig. 3.9.14 $1.64 \leqq x \leqq 1.68$, $-0.015 \leqq y \leqq 0.015$

Fig. 3.9.15 $1.648 \leqq x \leqq 1.650$, $-0.00075 \leqq y \leqq 0.00075$

First, recall that a *complex number* is a number of the form $a + bi$, where $i = \sqrt{-1}$, so $i^2 = -1$. The real numbers a and b are called the *real part* and the *imaginary part*, respectively, of $a + bi$. You add, multiply, and divide complex numbers as if they were binomials, with real and imaginary parts "collected" as in the computations

$$(3 + 4i) + (5 - 7i) = (3 + 5) + (4 - 7)i = 8 - 3i,$$

$$\begin{aligned}(2 + 5i)(3 - 4i) &= 2(3 - 4i) + 5i(3 - 4i)\\ &= 6 - 8i + 15i - 20i^2 = 26 + 7i,\end{aligned}$$

and

$$\frac{2 + 5i}{3 + 4i} = \frac{2 + 5i}{3 + 4i} \cdot \frac{3 - 4i}{3 - 4i} = \frac{26 + 7i}{9 - 16i^2} = \frac{26 + 7i}{25} = 1.04 + (0.28)i.$$

The use of the *conjugate* $3 - 4i$ of the denominator $3 + 4i$ in the last computation is a very common technique for writing a complex fraction in the standard form $a + bi$. (The **conjugate** of $x + yi$ is $x - yi$; it follows that the conjugate of $x - yi$ is $x + yi$.)

Now let us substitute the complex number $z = x + iy$ into the cubic polynomial

$$f(z) = z^3 - 3z^2 + 1$$

of Eq. (13) and into its derivative $f'(z) = 3z^2 - 6z$. We find that

$$\begin{aligned} f(z) &= (x + iy)^3 - 3(x + iy)^2 + 1\\ &= (x^3 - 3xy^2 - 3x^2 + 3y^2 + 1) + (3x^2y - y^3 - 6xy)i \end{aligned} \tag{14}$$

and

$$f'(z) = 3(x + iy)^2 - 6(x + iy) = (3x^2 - 3y^2 - 6x) + (6xy - 6y)i. \tag{15}$$

Consequently, there is nothing to prevent us from applying Newton's method to Eq. (13) with complex numbers. Beginning with a *complex* initial guess $z_0 = x_0 + iy_0$, we can substitute Eqs. (14) and (15) into Newton's iterative formula

$$z_{n+1} = z_n - \frac{f(z_n)}{f'(z_n)} \tag{16}$$

to generate the complex sequence $\{z_n\}$, which may yet converge to a (real) solution of Eq. (13).

With this preparation, we can now explain how Fig. 3.9.12 was generated. A computer was programmed to carry out Newton's iteration repeatedly, beginning with many thousands of initial guesses $z_0 = x_0 + iy_0$ that "fill" the rectangle $-2 \leqq x \leqq 4$, $-2.25 \leqq y \leqq 2.25$ in the complex plane. The initial point $z_0 = x_0 + iy_0$ was then color-coded according to the root (if any) to which the corresponding sequence $\{z_n\}$ converged:

Color z_0 green if $\{z_n\}$ converges to the root $r_1 \approx -0.53$;

Color z_0 red if $\{z_n\}$ converges to the root $r_2 \approx 0.65$;

Color z_0 yellow if $\{z_n\}$ converges to the root $r_3 \approx 2.88$.

Thus we use different colors to distinguish different "*Newton basins* of attraction" for the equation we are investigating. It is not surprising that a red region containing the root r_2 appears in the middle of Fig. 3.9.12, separating a green region to the left that contains r_1 and a yellow region to the right that contains r_3. But why would yellow lobes protrude from the green region into the red region and green lobes protrude from the yellow region into the red one? To see what's happening near these lobes, we generated some blowups.

Figure 3.9.13 shows a blowup of the rectangle $1.6 \leqq x \leqq 2.4$, $-0.3 \leqq y \leqq 0.3$ containing the green lobe indicated in Fig. 3.9.12. Figure 3.9.14 ($1.64 \leqq x \leqq 1.68$, $-0.015 \leqq y \leqq 0.015$) and Fig. 3.9.15 ($1.648 \leqq x \leqq 1.650$, $-0.00075 \leqq y \leqq 0.00075$) are further magnifications. The rectangle shown in Fig. 3.9.15 corresponds to less than one millionth of a square inch of Fig. 3.9.12.

Fig. 3.9.16 Newton basis for the twelfth-degree polynomial

At every level of magnification, each green lobe has smaller yellow lobes protruding into the surrounding red region, and each of these yellow lobes has still smaller green lobes protruding from it, and so on ad infinitum (just like the proverbial little fleas that are bitten by still smaller fleas, and so on ad infinitum).

Figure 3.9.16 shows the Newton basins picture for the twelfth-degree polynomial equation

$$\begin{aligned} f(x) = x^{12} - 14x^{10} + 183x^8 - 612x^6 \\ - 2209x^4 - 35374x^2 + 38025 = 0, \end{aligned} \tag{17}$$

Fig. 3.9.17 The flower at the center of Fig. 3.9.16

which has as its solution the twelve complex numbers

$$1, \quad 1 \pm 2i, \quad -1, \quad -1 \pm 2i,$$

$$3, \quad 3 \pm 2i, \quad -3, \quad -3 \pm 2i.$$

Twelve different colors are used to distinguish the Newton basins of these twelve solutions of Eq. (17).

Where the fractal common boundary appears to separate basins of different colors, it is studded with "flowers" like the one at the center of Fig. 3.9.16, which is magnified in Fig. 3.9.17. Each of these flowers has ten "leaves" (in the remaining ten colors). Each of these leaves has "buds" like the one shown in Fig. 3.9.18. Each of these buds is encircled with flowers that have leaves that have buds that are encircled with flowers—and so on ad infinitum.

Fig. 3.9.18 A bud on a petal of the flower in Fig. 3.9.17

3.9 PROBLEMS

In Problems 1 through 20, use Newton's method to find the solution of the given equation $f(x) = 0$ in the indicated interval $[a, b]$ accurate to four decimal places. You may choose the initial guess either on the basis of a calculator graph or by interpolation between the values $f(a)$ and $f(b)$.

1. $x^2 - 5 = 0$; $[2, 3]$ (to find the positive square root of 5)

2. $x^3 - 2 = 0$; $[1, 2]$ (to find the cube root of 2)

3. $x^5 - 100 = 0$; $[2, 3]$ (to find the fifth root of 100)

4. $x^{3/2} - 10 = 0$; $[4, 5]$ (to find $10^{2/3}$)

5. $x^2 + 3x - 1 = 0$; $[0, 1]$

6. $x^3 + 4x - 1 = 0$; $[0, 1]$

7. $x^6 + 7x^2 - 4 = 0$; $[-1, 0]$

8. $x^3 + 3x^2 + 2x = 10$; $[1, 2]$

9. $x - \cos x = 0$; $[0, 2]$

10. $x^2 - \sin x = 0$; $[0.5, 1.0]$

11. $4x - \sin x = 4$; $[1, 2]$

12. $5x + \cos x = 5$; $[0, 1]$

13. $x^5 + x^4 = 100$; $[2, 3]$

14. $x^5 + 2x^4 + 4x = 5$; $[0, 1]$

15. $x + \tan x = 0$; $[2, 3]$

16. $x + \tan x = 0$; $[11, 12]$

17. $x^3 - 10 = 0$; $[2, 3]$

18. $x^3 - 2x - 5 = 0$; $[2, 3]$ (Newton's own example)

19. $x^5 - 5x - 10 = 0$; $[1, 2]$

20. $x^5 = 32$; $[0, 5]$

21. (a) Show that Newton's method applied to the equation $x^3 - a = 0$ yields the formula

$$x_{n+1} = \frac{1}{3}\left(2x_n + \frac{a}{x_n^2}\right)$$

for approximating the cube root of a. (b) Use this formula to find $\sqrt[3]{2}$ accurate to five decimal places.

22. (a) Show that Newton's method yields the formula

$$x_{n+1} = \frac{1}{k}\left[(k-1)x_n + \frac{a}{(x_n)^{k-1}}\right]$$

for approximating the kth root of the positive number a. (b) Use this formula to find $\sqrt[10]{100}$ accurate to five decimal places.

23. Equation (12) has the special form $x = G(x)$, where $G(x) = \frac{1}{2}\cos x$. For an equation of this form, the iterative formula $x_{n+1} = G(x_n)$ produces a sequence of approximations that *sometimes* converges to a root. In the case of Eq. (12), this *repeated substitution* formula is simply $x_{n+1} = \frac{1}{2}\cos x_n$. Begin with $x_0 = 0.5$ as in Example 4 and retain five decimal places in your computation of the solution of Eq. (12). (*Check:* You should find that $x_8 \approx 0.45018$.)

24. The equation $x^4 = x + 1$ has a solution between $x = 1$ and $x = 2$. Use the initial guess $x_0 = 1.5$ and the method of repeated substitution (see Problem 23) to discover that one of the solutions of this equation is approximately 1.220744. Iterate using the formula

$$x_{n+1} = (x_n + 1)^{1/4}.$$

Then compare the result with what happens when you iterate using the formula

$$x_{n+1} = (x_n)^4 - 1.$$

25. The equation $x^3 - 3x^2 + 1 = 0$ has a solution between $x = 0$ and $x = 1$. To apply the method of repeated substitution (see Problem 23) to this equation, you may write it either in the form

$$x = 3 - \frac{1}{x^2}$$

or in the form

$$x = (3x^2 - 1)^{1/3}.$$

If you begin with $x_0 = 0.5$ in the hope of finding the nearby solution (approximately 0.6527) of the original equation by using each of the preceding iterative formulas, you will observe some of the drawbacks of the method. Describe what goes wrong.

26. Show that Newton's method applied to the equation

$$\frac{1}{x} - a = 0$$

yields the iterative formula

$$x_{n+1} = 2x_n - a(x_n)^2$$

and thus provides a method for approximating the reciprocal $1/a$ without performing any divisions. Such a method is useful because, in most high-speed computers, the operation of division is more time-consuming than even several additions and multiplications.

27. Prove that the equation $x^5 + x = 1$ has exactly one real solution. Then use Newton's method to find it with four places correct to the right of the decimal.

In Problems 28 through 30, use Newton's method to find all real roots of the given equation with four digits correct to the right of the decimal. (*Suggestion:* To determine the number of roots and their approximate locations, graph the left- and right-hand sides of each equation and observe where the graphs cross.)

28. $x^2 = \cos x$

29. $x = 2\sin x$

30. $\cos x = -\frac{1}{5}x$ (There are exactly three solutions, as indicated in Fig. 3.9.19.)

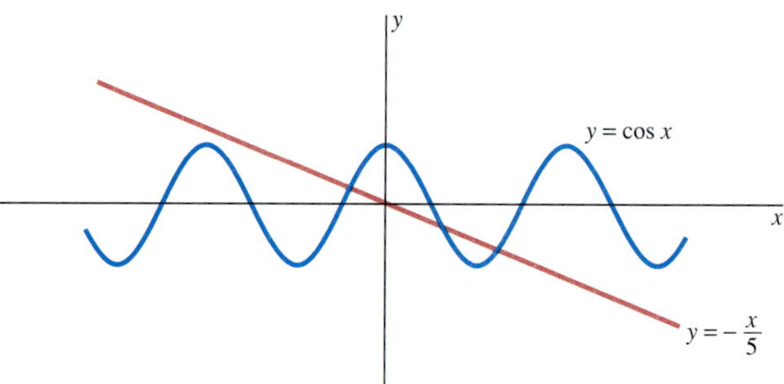

Fig. 3.9.19 Solving the equation in Problem 30

31. Prove that the equation $x^7 - 3x^3 + 1 = 0$ has at least one solution. Then use Newton's method to find one solution to three-place accuracy.

32. Use Newton's method to approximate $\sqrt[3]{5}$ to four-place accuracy.

33. Use Newton's method to find the value of x for which $x^3 = \cos x$.

34. Use Newton's method to find the smallest positive value of x for which $x = \tan x$.

35. In Problem 49 of Section 3.6, we dealt with the problem of minimizing the cost of building a road to two points on opposite sides of a geologic fault. This problem led to the equation

$$f(x) = 3x^4 - 24x^3 + 51x^2 - 32x + 64 = 0.$$

Use Newton's method to find, to four-place accuracy, the root of this equation that lies in the interval [3, 4].

36. The moon of Planet Gzyx has an elliptical orbit with eccentricity 0.5, and its period of revolution around the planet is 100 days. If the moon is at the position $(a, 0)$ when $t = 0$, then (Fig. 3.9.20) the central angle after t days is given by *Kepler's equation*

$$\frac{2\pi t}{100} = \theta - \frac{1}{2}\sin\theta.$$

Use Newton's method to solve for θ when $t = 17$ (days). Take $\theta_0 = 1.5$ (rad) and calculate the first two approximations θ_1 and θ_2. Express θ_2 in degrees as well.

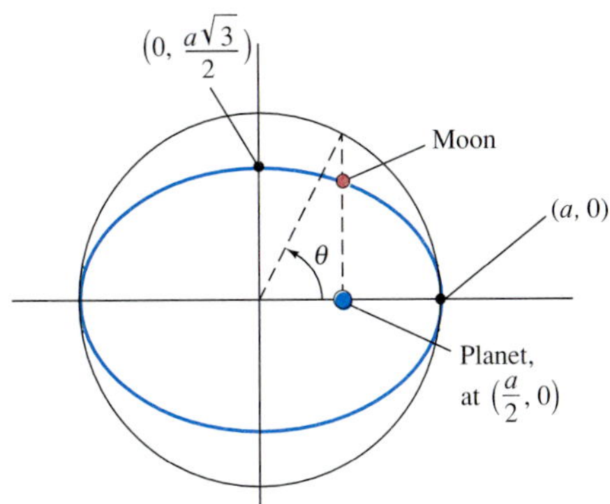

Fig. 3.9.20 The elliptical orbit of Problem 36

37. A great problem of Archimedes was that of using a plane to cut a sphere into two segments with volumes in a given (preassigned) ratio. Archimedes showed that the volume of a segment of height h of a sphere of radius a is $V = \frac{1}{3}\pi h^2(3a - h)$. If a plane at distance x from the center of a sphere of radius 1 cuts the sphere into two segments, one with twice the volume of the other, show that $3x^3 - 9x + 2 = 0$. Then use Newton's method to find x accurate to four decimal places.

38. The equation $f(x) = x^3 - 4x + 1 = 0$ has three distinct real roots. Approximate their locations by evaluating f at $x = -3, -2, -1, 0, 1, 2,$ and 3. Then use Newton's method to approximate each of the three roots to four-place accuracy.

39. The equation $x + \tan x = 0$ is important in a variety of applications—for example, in the study of the diffusion of heat. It has a sequence $\alpha_1, \alpha_2, \alpha_3, \ldots$ of positive roots, with the nth root slightly larger than $(n - 0.5)\pi$. Use Newton's method to compute α_1 and α_2 to three-place accuracy.

40. Investigate the cubic equation

$$4x^3 - 42x^2 - 19x - 28 = 0.$$

Perhaps you can see graphically that it has only a single real solution. Find it (accurate to four decimal places). First try the initial guess $x_0 = 0$; be prepared for at least 25 iterations. Then try initial guesses $x_0 = 10$ and $x_0 = 100$.

41. A 15-ft ladder and a 20-ft ladder lean in opposite directions against the vertical walls of a hall (Fig. 3.9.21). The ladders cross at a height of 5 ft. You must find the width w of the hall. First, let x and y denote the heights of the tops of the ladders on the walls and u and v the lengths shown in the figure, so that $w = u + v$. Use similar triangles to show that

$$x = 5\left(1 + \frac{u}{v}\right), \qquad y = 5\left(1 + \frac{v}{u}\right).$$

Then apply the Pythagorean theorem to show that $t = u/v$ satisfies the equation

$$t^4 + 2t^3 - 7t^2 - 2t - 1 = 0.$$

Finally, use Newton's method to find first the possible values of t, and then those of w, accurate to four decimal places.

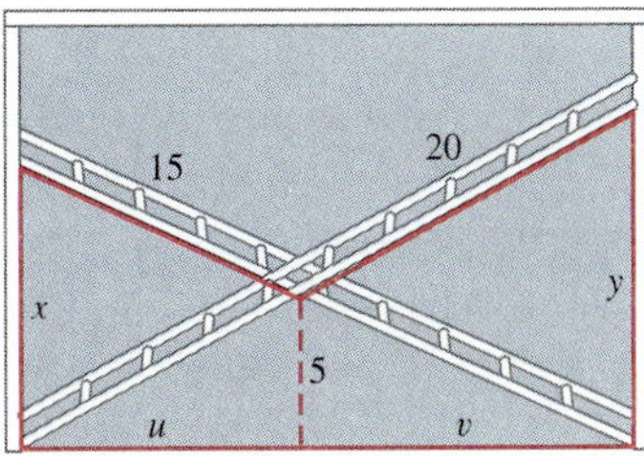

Fig. 3.9.21 The crossing ladders of Problem 41

3.9 PROJECT: HOW DEEP DOES A FLOATING BALL SINK?

Figure 3.9.22 shows a large cork ball of radius $a = 1$ floating in water of density 1. If the ball's density ρ is one-fourth that of water, $\rho = \frac{1}{4}$, then Archimedes' law of buoyancy implies that the ball floats in such a way that one-fourth of its total volume is submerged. Because the volume of the ball is $4\pi/3$, it follows that the volume of the part of the ball beneath the waterline is given by

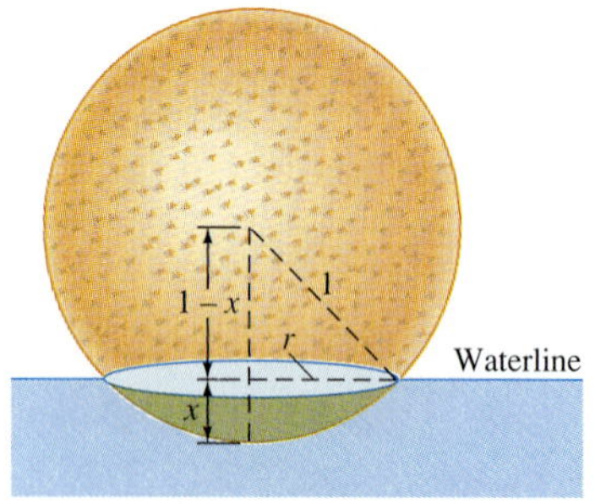

Fig. 3.9.22 The floating cork ball

$$V = \rho \cdot \frac{4\pi}{3} = \frac{1}{4} \cdot \frac{4\pi}{3} = \frac{\pi}{3}. \tag{1}$$

The shape of the submerged part of the ball is that of a **spherical segment** with a circular flat top. The volume of a spherical segment of *top radius* r and *depth* $h = x$ (as in Fig. 3.9.22) is given by the formula

$$V = \frac{\pi x}{6}(3r^2 + x^2). \tag{2}$$

This formula is also due to Archimedes and holds for any depth x, whether the spherical segment is smaller or larger than a hemisphere. For instance, note that with $r = 0$ and $x = 2a$ it gives $V = \frac{4}{3}\pi a^3$, the volume of an entire sphere of radius a.

For a preliminary investigation, proceed as follows to find the depth x to which the ball sinks in the water. Equate the two expressions for V in Eqs. (1) and (2), then use the right triangle in Fig. 3.9.22 to eliminate r. You should find that x must be a solution of the cubic equation

$$f(x) = x^3 - 3x^2 + 1 = 0. \tag{3}$$

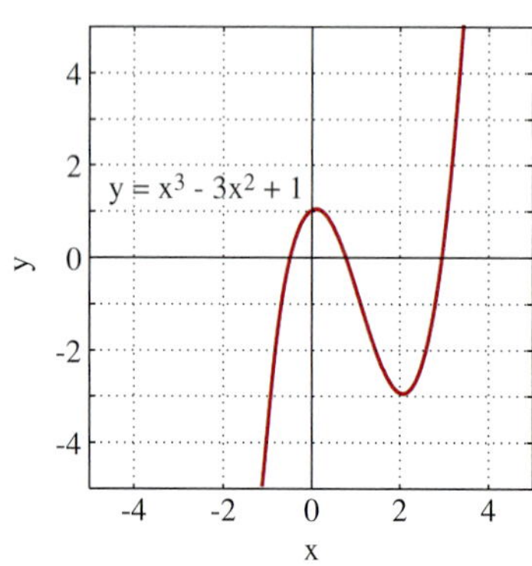

Fig. 3.9.23 Graph for the cork-ball equation

As the graph $y = f(x)$ in Fig. 3.9.23 indicates, this equation has three real solutions—one in $(-1, 0)$, one in $(0, 1)$, and one in $(2, 3)$. The solution between 0 and 1 gives the actual depth x to which the ball sinks (why?). You can find x using Newton's method.

YOUR INVESTIGATION For your very own floating ball to investigate, let its density ρ in Eq. (1) be given by

$$\rho = \frac{10 + k}{20}$$

where k denotes the last nonzero digit in the sum of the final four digits of your student I.D. number. Your objective is to find the depth to which this ball sinks in the water. Begin by deriving the cubic equation that you need to solve, explaining each step carefully. Then find all of its solutions accurate to at least four decimal places. Include in your report a sketch of a spherical ball with the waterline located accurately (to scale) in the position corresponding to your result for the desired depth.

CHAPTER 3 REVIEW: *Formulas, Concepts, Definitions*

Differentiation Formulas

$$D_x(cu) = c\frac{du}{dx}$$

$$D_x(u + v) = \frac{du}{dx} + \frac{dv}{dx}$$

$$D_x(uv) = u\frac{dv}{dx} + v\frac{du}{dx}$$

$$D_x\frac{u}{v} = \frac{v\dfrac{du}{dx} - u\dfrac{dv}{dx}}{v^2}$$

$$D_x\, g(u) = g'(u)\frac{du}{dx}$$

$$D_x(u^r) = ru^{r-1}\frac{du}{dx}$$

$$D_x \sin u = (\cos u)\frac{du}{dx}$$

$$D_x \cos u = (-\sin u)\frac{du}{dx}$$

$$D_x \tan u = (\sec^2 u)\frac{du}{dx}$$

$$D_x \cot u = (-\csc^2 u)\frac{du}{dx}$$

$$D_x \sec u = (\sec u \tan u)\frac{du}{dx}$$

$$D_x \csc u = (-\csc u \cot u)\frac{du}{dx}$$

Use the following list as a guide to concepts that you may need to review.

1. Definition of the derivative
2. Average rate of change of a function
3. Instantaneous rate of change of a function
4. Position function; velocity and acceleration
5. Differential, function, and operator notation for derivatives
6. The power rule
7. The binomial formula
8. Linearity of differentiation
9. The product rule
10. The reciprocal rule
11. The quotient rule
12. The chain rule
13. The generalized power rule
14. Vertical tangent lines
15. Local maxima and minima
16. $f'(c) = 0$ as a necessary condition for local extrema
17. Absolute (or global) extrema
18. Critical points
19. The closed-interval maximum-minimum method
20. Steps in the solution of applied maximum-minimum problems
21. Derivatives of the sine and cosine functions
22. Derivatives of the other four trigonometric functions
23. Implicitly defined functions
24. Implicit differentiation
25. Solving related-rates problems
26. Newton's method

CHAPTER 3 *Miscellaneous Problems*

Find dy/dx in Problems 1 through 35.

1. $y = x^2 + \dfrac{3}{x^2}$

2. $y^2 = x^2$

3. $y = \sqrt{x} + \dfrac{1}{\sqrt[3]{x}}$

4. $y = (x^2 + 4x)^{5/2}$

5. $y = (x - 1)^7(3x + 2)^9$

6. $y = \dfrac{x^4 + x^2}{x^2 + x + 1}$

7. $y = \left(3x - \dfrac{1}{2x^2}\right)^4$

8. $y = x^{10}\sin 10x$

9. $xy = 9$

10. $y = \sqrt{\dfrac{1}{5x^6}}$

11. $y = \dfrac{1}{\sqrt{(x^3 - x)^3}}$

12. $y = \sqrt[3]{2x + 1}\sqrt[5]{3x - 2}$

13. $y = \dfrac{1}{1 + u^2}$ where $u = \dfrac{1}{1 + x^2}$

14. $x^3 = \sin^2 y$

15. $y = (\sqrt{x} + \sqrt[3]{2x})^{7/3}$

16. $y = \sqrt{3x^5 - 4x^2}$

17. $y = \dfrac{u + 1}{u - 1}$, where $u = \sqrt{x + 1}$

18. $y = \sin(2\cos 3x)$

19. $x^2y^2 = x + y$

20. $y = \sqrt{1 + \sin\sqrt{x}}$

21. $y = \sqrt{x + \sqrt{2x + \sqrt{3x}}}$

22. $y = \dfrac{x + \sin x}{x^2 + \cos x}$

23. $\sqrt[3]{x} + \sqrt[3]{y} = 4$

24. $x^3 + y^3 = xy$

25. $y = (1 + 2u)^3$, where $u = \dfrac{1}{(1 + x)^3}$

26. $y = \cos^2(\sin^2 x)$

27. $y = \sqrt{\dfrac{\sin^2 x}{1 + \cos x}}$

28. $y = (1 + \sqrt{x})^3(1 - 2\sqrt[3]{x})^4$

29. $y = \dfrac{\cos 2x}{\sqrt{\sin 3x}}$

30. $x^3 - x^2y + xy^2 - y^3 = 4$

31. $y = \sin^3 2x \cos^2 3x$

32. $y = [1 + (2 + 3x)^{-3/2}]^{2/3}$

33. $y = \sin^5\left(x + \dfrac{1}{x}\right)$

34. $\sqrt{x + y} = \sqrt[3]{x - y}$

35. $y = \cos^3(\sqrt[3]{x^4 + 1})$

In Problems 36 through 39, write an equation of the line tangent to the given curve at the indicated point.

36. $y = \dfrac{x + 1}{x - 1}$; $(0, -1)$

37. $x = \sin 2y$; $(1, \pi/4)$

38. $x^2 - 3xy + 2y^2 = 0$; $(2, 1)$

39. $y^3 = x^2 + x$; $(0, 0)$

40. If a hemispherical bowl with radius 1 ft is filled with water to a depth of x in., then the volume of water in the bowl is

$$V = \frac{\pi}{3}(36x^2 - x^3) \quad (\text{in.}^3).$$

If the water flows out a hole at the bottom of the bowl at the rate of 36π in.^{3}/s, how fast is x decreasing when $x = 6$ in.?

41. Falling sand forms a conical sandpile. Its height h always remains twice its radius r while both are increasing. If sand is falling onto the pile at the rate of 25π ft^3/min, how fast is r increasing when $r = 5$ ft?

Find the limits in Problems 42 through 47.

42. $\lim_{x \to 0} \frac{x - \tan x}{\sin x}$

43. $\lim_{x \to 0} x \cot 3x$

44. $\lim_{x \to 0} \frac{\sin 2x}{\sin 5x}$

45. $\lim_{x \to 0} x^2 \csc 2x \cot 2x$

46. $\lim_{x \to 0} x^2 \sin \frac{1}{x^2}$

47. $\lim_{x \to 0^+} \sqrt{x} \sin \frac{1}{x}$

In Problems 48 through 53, identify two functions f and g such that $h(x) = f(g(x))$. Then apply the chain rule to find $h'(x)$.

48. $h(x) = \sqrt[3]{x + x^4}$

49. $h(x) = \frac{1}{\sqrt{x^2 + 25}}$

50. $h(x) = \sqrt{\frac{x}{x^2 + 1}}$

51. $h(x) = \sqrt[3]{(x - 1)^5}$

52. $h(x) = \frac{(x + 1)^{10}}{(x - 1)^{10}}$

53. $h(x) = \cos(x^2 + 1)$

54. The period T of oscillation (in seconds) of a simple pendulum of length L (in feet) is given by $T = 2\pi\sqrt{L/32}$. What is the rate of change of T with respect to L when $L = 4$ ft?

55. What is the rate of change of the volume $V = \frac{4}{3}\pi r^3$ of a sphere with respect to its surface area $A = 4\pi r^2$?

56. What is an equation for the straight line through $(1, 0)$ that is tangent to the graph of

$$h(x) = x + \frac{1}{x}$$

at a point in the first quadrant?

57. A rocket is launched vertically upward from a point 3 mi west of an observer on the ground. What is the speed of the rocket when the angle of elevation (from the horizontal) of the observer's line of sight to the rocket is 50° and is increasing at 5° per second?

58. An oil field containing 20 wells has been producing 4000 barrels of oil daily. For each new well drilled, the daily production of each well decreases by 5 barrels. How many new wells should be drilled to maximize the total daily production of the oil field?

59. A triangle is inscribed in a circle of radius R. One side of the triangle coincides with a diameter of the circle. In terms of R, what is the maximum possible area of such a triangle?

60. Five rectangular pieces of sheet metal measure 210 cm by 336 cm each. Equal squares are to be cut from all their corners, and the resulting five cross-shaped pieces of metal are to be folded and welded to form five boxes without tops. The 20 little squares that remain are to be assembled in groups of four into five larger squares, and these five larger squares are to be assembled into a cubical box with no top. What is the maximum possible total volume of the six boxes that are constructed in this way?

61. A mass of clay of volume V is formed into two spheres. For what distribution of clay is the total surface area of the two spheres a maximum? a minimum?

62. A right triangle has legs of lengths 3 m and 4 m. What is the maximum possible area of a rectangle inscribed in the triangle in the "obvious" way—with one corner at the triangle's right angle, two adjacent sides of the rectangle lying on the triangle's legs, and the opposite corner on the hypotenuse?

63. What is the maximum possible volume of a right circular cone inscribed in a sphere of radius R?

64. A farmer has 400 ft of fencing with which to build a rectangular corral. He will use some or even all of an existing straight wall 100 ft long as part of the perimeter of the corral. What is the maximum area that can be enclosed?

65. In one simple model of the spread of a contagious disease among members of a population of M people, the incidence of the disease, measured as the number of new cases per day, is given in terms of the number x of individuals already infected by

$$R(x) = kx(M - x) = kMx - kx^2,$$

where k is a positive constant. How many individuals in the population are infected when the incidence R is the greatest?

66. Three sides of a trapezoid have length L, a constant. What should be the length of the fourth side if the trapezoid is to have maximal area?

67. A box with no top must have a base twice as long as it is wide, and the total surface area of the box is to be 54 ft^2. What is the maximum possible volume of such a box?

68. A small right circular cone is inscribed in a larger one (Fig. 3.MP.1). The larger cone has fixed radius R and fixed altitude H. What is the largest fraction of the volume of the larger cone that the smaller one can occupy?

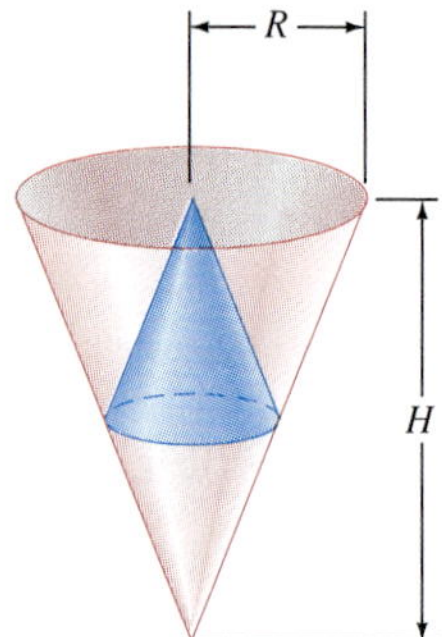

Fig. 3.MP.1 A small cone inscribed in a larger one (Problem 68)

69. Two vertices of a trapezoid are at $(-2, 0)$ and $(2, 0)$, and the other two lie on the semicircle $x^2 + y^2 = 4$, $y \geqq 0$. What is the maximum possible area of the trapezoid? [*Note:* The area of a trapezoid with bases b_1 and b_2 and height h is $A = h(b_1 + b_2)/2$.]

70. Suppose that f is a differentiable function defined on the whole real number line $\boldsymbol{R}$ and that the graph of f contains a point $Q(x, y)$ closest to the point $P(x_0, y_0)$ not on the graph. Show that

$$f'(x) = -\frac{x - x_0}{y - y_0}$$

at Q. Conclude that the segment PQ is perpendicular to the line tangent to the curve at Q. (*Suggestion:* Minimize the square of the distance PQ.)

71. Use the result of Problem 70 to show that the minimum distance from the point (x_0, y_0) to a point of the straight line $Ax + By + C = 0$ is

$$\frac{|Ax_0 + By_0 + C|}{\sqrt{A^2 + B^2}}.$$

72. A race track is to be built in the shape of two parallel and equal straightaways connected by semicircles on each end (Fig. 3.MP.2). The length of the track, one lap, is to be exactly 4 km. What should its design be to maximize the rectangular area within it?

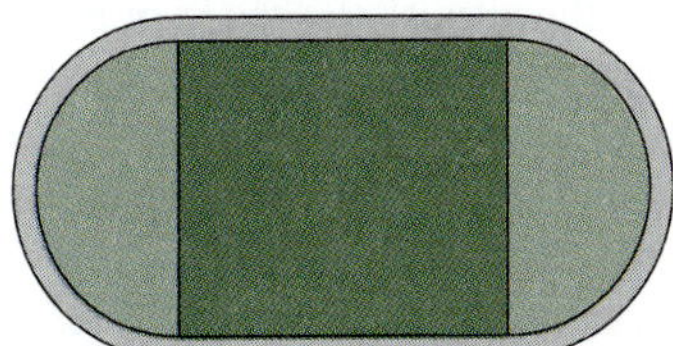

Fig. 3.MP.2 Design the race track to maximize the rectangular area (Problem 72).

73. Two towns are located near the straight shore of a lake. Their nearest distances to points on the shore are 1 mi and 2 mi, respectively, and these points on the shore are 6 mi apart. Where should a fishing pier be located to minimize the total amount of paving necessary to build a straight road from each town to the pier?

74. A hiker finds herself in a forest 2 km from a long straight road. She wants to walk to her cabin, which is 10 km away in the forest and also 2 km from the road (Fig. 3.MP.3). She can walk at a rate of 8 km/h along the road but only 3 km/h through the forest. So she decides to walk first to the road, then along the road, and finally through the forest to the cabin. What angle θ (shown in the figure) would minimize the total time required for the hiker to reach her cabin? How much time is saved in comparison with the straight route through the forest?

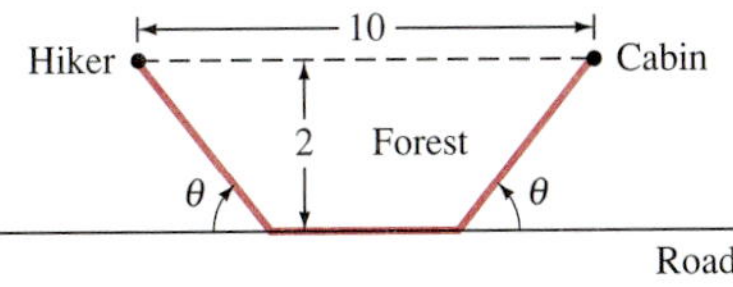

Fig. 3.MP.3 The hiker's quickest path to the cabin (Problem 74)

75. When an arrow is shot from the origin with initial velocity v and initial angle of inclination α (from the horizontal x-axis, which represents the ground), then its trajectory is the curve

$$y = mx - \frac{16}{v^2}(1 + m^2)x^2,$$

where $m = \tan\alpha$. (a) Find the maximum height reached by the arrow in terms of m and v. (b) For what value of m (and hence, for what α) does the arrow travel the greatest horizontal distance?

76. A projectile is fired with initial velocity v and angle of elevation θ from the base of a plane inclined at 45° from the horizontal (Fig. 3.MP.4). The range of the projectile, as measured up this slope, is given by

$$R = \frac{v^2\sqrt{2}}{16}(\cos\theta\sin\theta - \cos^2\theta).$$

What value of θ maximizes R?

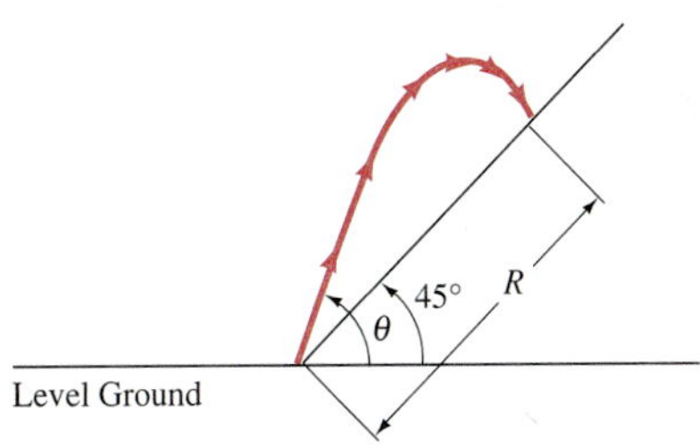

Fig. 3.MP.4 A projectile fired uphill (Problem 76)

In Problems 77 through 88, use Newton's method to find the solution of the given equation $f(x) = 0$ in the indicated interval $[a, b]$ accurate to four decimal places.

77. $x^2 - 7 = 0$; $[2, 3]$ (to find the positive square root of 7)

78. $x^3 - 3 = 0$; $[1, 2]$ (to find the cube root of 3)

79. $x^5 - 75 = 0$; $[2, 3]$ (to find the fifth root of 75)

80. $x^{4/3} - 10 = 0$; $[5, 6]$ (to approximate $10^{3/4}$)

81. $x^3 - 3x - 1 = 0$; $[-1, 0]$

82. $x^3 - 4x - 1 = 0$; $[-1, 0]$

83. $x^6 + 7x^2 - 4 = 0$; $[0, 1]$

84. $x^3 - 3x^2 + 2x + 10 = 0$; $[-2, -1]$

85. $x + \cos x = 0$; $[-2, 0]$

86. $x^2 + \sin x = 0$; $[-1.0, -0.5]$

87. $4x - \sin x + 4 = 0$; $[-2, -1]$

88. $5x - \cos x + 5 = 0$; $[-1, 0]$

89. Find the depth to which a wooden ball with radius 2 ft sinks in water if its density is one-third that of water. A useful formula appears in Problem 37 of Section 3.9.

90. The equation $x^2 + 1 = 0$ has no real solutions. Try finding a solution by using Newton's method and report what happens. Use the initial estimate $x_0 = 2$.

91. At the beginning of Section 3.9 we mentioned the fifth-degree equation

$$x^5 - 3x^3 + x^2 - 23x + 19 = 0;$$

its graph appears in Fig. 3.9.1. The graph makes it clear that this equation has exactly three real solutions. Find all of them, to four-place accuracy, using Newton's method.

92. The equation

$$\tan x = \frac{1}{x}$$

has a sequence $\alpha_1, \alpha_2, \alpha_3, \ldots$ of positive roots, with α_n slightly larger than $(n-1)\pi$. Use Newton's method to approximate α_1 and α_2 to three-place accuracy.

93. Criticize the following "proof" that $3 = 2$. Begin by writing

$$x^3 = x \cdot x^2 = x^2 + x^2 + \cdots + x^2 \quad (x \text{ summands}).$$

Differentiate to obtain

$$3x^2 = 2x + 2x + \cdots + 2x \quad (\text{still } x \text{ summands}).$$

Thus $3x^2 = 2x^2$, and "therefore" $3 = 2$.

If we substitute $z = x + h$ into the definition of the derivative, the result is

$$f'(x) = \lim_{z \to x} \frac{f(z) - f(x)}{z - x}.$$

Use this formula in Problems 94 and 95, together with the formula

$$a^3 - b^3 = (a - b)(a^2 + ab + b^2)$$

for factoring the difference of two cubes.

94. Show that

$$D_x x^{3/2} = \lim_{z \to x} \frac{z^{3/2} - x^{3/2}}{z - x} = \frac{3}{2} x^{1/2}.$$

(*Suggestion:* Factor the numerator as a difference of cubes and the denominator as a difference of squares.)

95. Prove that

$$D_x x^{2/3} = \lim_{z \to x} \frac{z^{2/3} - x^{2/3}}{z - x} = \frac{2}{3} x^{-1/3}.$$

(*Suggestion:* Factor the numerator as a difference of squares and the denominator as a difference of cubes.)

96. A rectangular block with square base is being squeezed in such a way that its height y is decreasing at the rate of 2 cm/min while its volume remains constant. At what rate is the edge x of its base increasing when $x = 30$ cm and $y = 20$ cm?

97. Air is being pumped into a spherical balloon at the constant rate of 10 in.^{3}/s. At what rate is the surface area of the balloon increasing when its radius is 5 in.?

98. A ladder 10 ft long is leaning against a wall. If the bottom of the ladder slides away from the wall at the constant rate of 1 mi/h, how fast (in miles per hour) is the top of the ladder moving when it is 0.01 ft above the ground? (Assume that the top of the ladder maintains contact with the wall.)

99. A water tank in the shape of an inverted cone, axis vertical and vertex downward, has a top radius of 5 ft and height 10 ft. Water is flowing out of the tank through a hole at the vertex at the rate of 50 ft^3/min. What is the time rate of change of the water depth at the instant when the water is 6 ft deep?

100. Plane A is flying west toward an airport at an altitude of 2 mi. Plane B is flying south toward the same airport at an altitude of 3 mi. When both planes are 2 mi (ground distance) from the airport, the speed of plane A is 500 mi/h and the distance between the two planes is decreasing at 600 mi/h. What is the speed of plane B then?

101. A water tank is shaped in such a way that the volume of water in the tank is $V = 2y^{3/2}$ in.^{3} when its depth is y inches. If water flows out through a hole at the bottom of the tank at the rate of $3\sqrt{y}$ in.^{3}/min, at what rate does the water level in the tank fall? What is a practical application for such a water tank?

102. Water is being poured into the conical tank of Problem 99 at the rate of 50 ft^3/min and is draining through the hole at the bottom at the rate of $10\sqrt{y}$ ft^3/min, where y is the depth of water in the tank. (a) At what rate is the water level rising when the water is 5 ft deep? (b) Suppose that the tank is initially empty, water is poured in at 25 ft^3/min, and water continues to drain at $10\sqrt{y}$ ft^3/min. What is the maximum depth attained by the water?

103. Let L be a straight line passing through the fixed point $P(x_0, y_0)$ and tangent to the parabola $y = x^2$ at the point $Q(a, a^2)$. (a) Show that $a^2 - 2ax_0 + y_0 = 0$. (b) Apply the quadratic formula to show that if $y_0 < (x_0)^2$ (that is, if P lies below the parabola), then there are two possible values for a and thus two lines through P that are tangent to the parabola. (c) Similarly, show that if $y_0 > (x_0)^2$ (P lies above the parabola), then no line through P can be tangent to the parabola.

CHAPTER 4

ADDITIONAL APPLICATIONS OF THE DERIVATIVE

G. W. Leibniz (1646–1716)

Gottfried Wilhelm Leibniz entered the University of Leipzig at age 15, studied philosophy and law, graduated at 17, and received his doctorate in philosophy at 21. Upon completion of his academic work, Leibniz entered the political and governmental service of the Elector of Mainz (Germany). His serious study of mathematics did not begin until 1672 (at age 26) when he was sent to Paris on a diplomatic mission. During the next four years there he conceived the principal features of calculus. For this work he is remembered (with Newton) as a co-discoverer of the subject. Newton's discoveries had come slightly earlier (in the late 1660s), but Leibniz's were the first to be published, beginning in 1684. Despite an unfortunate priority dispute between supporters of Newton and supporters of Leibniz that raged for more than a century, it is clear now that the discoveries were made independently.

Throughout his life, Leibniz sought a universal language incorporating notation and terminology that would provide *all* educated people with the powers of clear and correct reasoning in all subjects. But only in mathematics did he largely accomplish this goal. His differential notation for calculus is arguably the best example of a system of notation chosen so as to mirror perfectly the basic operations and processes of the subject. Indeed, it can be said that Leibniz's notation for calculus brings within the range of ordinary students problems that once required the ingenuity of an Archimedes or a Newton. For this reason, Leibniz's approach to calculus was dominant during the eighteenth century, even though Newton's somewhat different approach may have been closer to our modern understanding of the subject.

The origin of differential notation was an infinitesimal right triangle with legs dx and dy and with hypotenuse a tiny segment of the curve $y = f(x)$. Leibniz later described the moment he first visualized this "characteristic triangle" as a burst of light that was the inception of his calculus. Indeed, he sometimes referred to his calculus as "my method of the Characteristic Triangle."

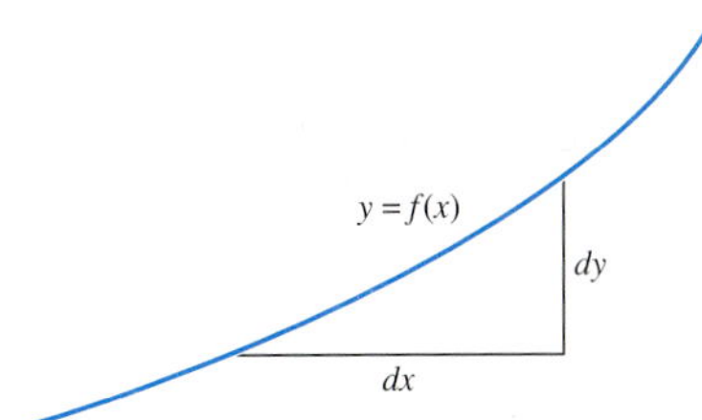

Leibniz's characteristic triangle

The following excerpt shows the opening paragraphs of Leibniz's first published article (in the 1684 *Acta Eruditorum*) in which the differential notation initially appeared. In the fifth line of the second paragraph, the product rule for differentiation is expressed as

$$d(xv) = x\,dv + v\,dx.$$

MENSIS OCTOBRIS A. M DC LXXXIV. 467

NOVA METHODUS PRO MAXIMIS ET MInimis, itemque tangentibus, quæ nec fractas, nec irrationales quantitates moratur, & singulare pro illis calculi genus, per G.G.L.

Sit axis AX, & curvæ plures, ut VV, WW, YY, ZZ, quarum ordinatæ, ad axem normales, VX, WX, YX, ZX, quæ vocentur respective, v, vv, y, z, & ipsa AX abscissa ab axe, vocetur x. Tangentes sint VB, WC, YD, ZE axi occurrentes respective in punctis B, C, D, E. Jam recta aliqua pro arbitrio assumta vocetur dx, & recta quæ sit ad dx, ut v (vel vv, vel y, vel z) est ad VB (vel WC, vel YD, vel ZE) vocetur dv (vel dvv, vel dy vel dz) sive differentia ipsarum v (vel ipsarum vv, aut y, aut z) His positis calculi regulæ erunt tales:

Sit a quantitas data constans, erit da æqualis 0, & d ax erit æqu. a dx: si sit y æqu. v (seu ordinata quævis curvæ YY, æqualis cuivis ordinatæ respondenti curvæ VV) erit dy æqu. dv. Jam *Additio & Subtractio*: si sit z -y + vv + x æqu. v, erit dz-y+vv+x seu dv, æqu. dz -dy + dvv + dx. *Multiplicatio*, dxv æqu. xdv + vdx, seu posito y æqu. xv, fiet dy æqu. xdv + vdx. In arbitrio enim est vel formulam, ut xv, vel compendio pro ea literam, ut y, adhibere. Notandum & x & dx eodem modo in hoc calculo tractari, ut y & dy, vel aliam literam indeterminatam cum sua differentiali. Notandum etiam non dari semper regressum a differentiali Æquatione, nisi cum quadam cautio-

4.1 INTRODUCTION

We learned in Chapter 3 how to differentiate a wide variety of algebraic and trigonometric functions. We saw that derivatives have such diverse applications as maximum-minimum problems, related-rates problems, and the solution of equations by Newton's method. The further applications of differentiation that we discuss in this chapter all depend ultimately on a single fundamental question. Suppose that $y = f(x)$ is a differentiable function defined on the closed interval $[a, b]$ of length $\Delta x = b - a$. Then the *increment* Δy in the value of $f(x)$ as x changes from $x = a$ to $x = b = a + \Delta x$ is

$$\Delta y = f(b) - f(a). \tag{1}$$

The question is this: How is the increment Δy related to the derivative—the rate of change—of the function f at the points of the interval $[a, b]$?

An *approximate* answer is given in Section 4.2. If the function continued throughout the interval with the same rate of change $f'(a)$ as it had at $x = a$, then the change in its value would be $f'(a)(b - a) = f'(a)\,\Delta x$. This observation motivates the tentative approximation

$$\Delta y \approx f'(a)\,\Delta x. \tag{2}$$

A precise answer to the preceding question is provided by the mean value theorem of Section 4.3. This theorem implies that the exact increment is given by

$$\Delta y = f'(c)\,\Delta x \tag{3}$$

for some number c in (a, b). The mean value theorem is the central theoretical result of differential calculus and is the key to many of the more advanced applications of derivatives.

4.2 INCREMENTS, DIFFERENTIALS, AND LINEAR APPROXIMATION

Sometimes we need a quick and simple estimate of the change in $f(x)$ that results from a given change in x. We write y for $f(x)$ and suppose first that the change in the independent variable is the *increment* Δx, so that x changes from its original value to the new value $x + \Delta x$. The change in the value of y is the **increment** Δy, computed by subtracting the old value of y from its new value:

$$\Delta y = f(x + \Delta x) - f(x). \tag{1}$$

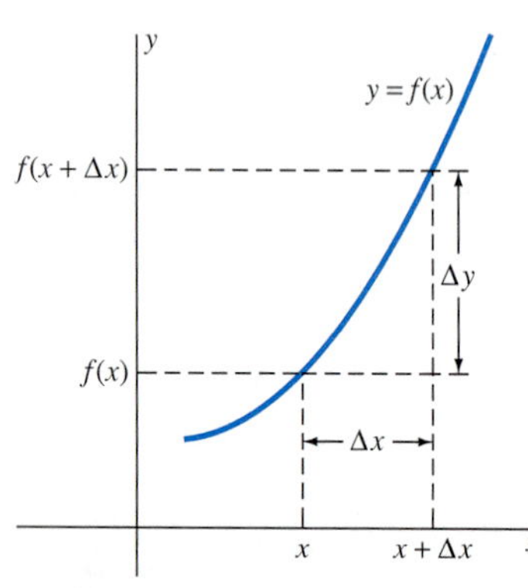

Fig. 4.2.1 The increments Δx and Δy

The increments Δx and Δy are represented geometrically in Fig. 4.2.1.

Now we compare the actual increment Δy with the change that *would* occur in the value of y *if* it continued to change at the *fixed* rate $f'(x)$ while the value of the independent variable changes from x to $x + \Delta x$. This hypothetical change in y is the **differential**

$$dy = f'(x)\,\Delta x. \tag{2}$$

As Fig. 4.2.2 shows, dy is the change in height of a point that moves along the tangent line at the point $(x, f(x))$ rather than along the curve $y = f(x)$.

If we think of x as fixed, then Eq. (2) shows that the differential dy is a *linear* function of the increment Δx. For this reason, dy is called the **linear approximation** to the increment Δy. We can approximate $f(x + \Delta x)$ by substituting dy for Δy:

$$f(x + \Delta x) = y + \Delta y \approx y + dy.$$

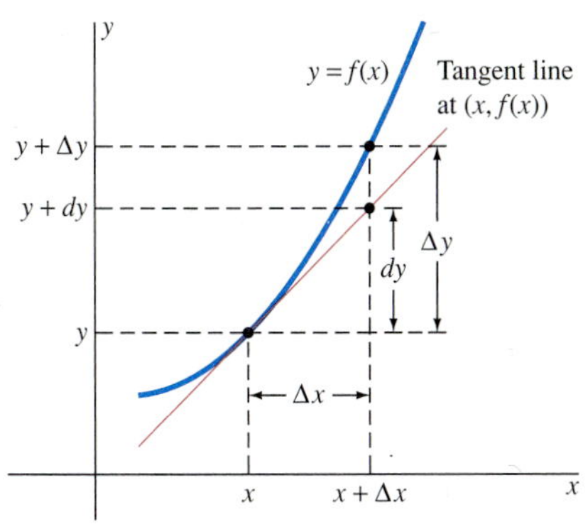

Fig. 4.2.2 The estimate dy of the actual increment Δy

Because $y = f(x)$ and $dy = f'(x)\,\Delta x$, this gives the **linear approximation formula**

$$f(x + \Delta x) \approx f(x) + f'(x)\,\Delta x. \tag{3}$$

The point is that this approximation is a "good" one, at least when Δx is relatively small. If we combine Eqs. (1), (2), and (3), we see that

$$\Delta y \approx f'(x)\,\Delta x = dy. \tag{4}$$

Thus the differential $dy = f'(x)\,\Delta x$ is a good approximation to the increment $\Delta y = f(x + \Delta x) - f(x)$.

If we replace x with a in Eq. (3), we get the approximation

$$f(a + \Delta x) \approx f(a) + f'(a)\,\Delta x. \tag{5}$$

If we now write $\Delta x = x - a$, so that $x = a + \Delta x$, the result is

$$f(x) \approx f(a) + f'(a)\cdot(x - a). \tag{6}$$

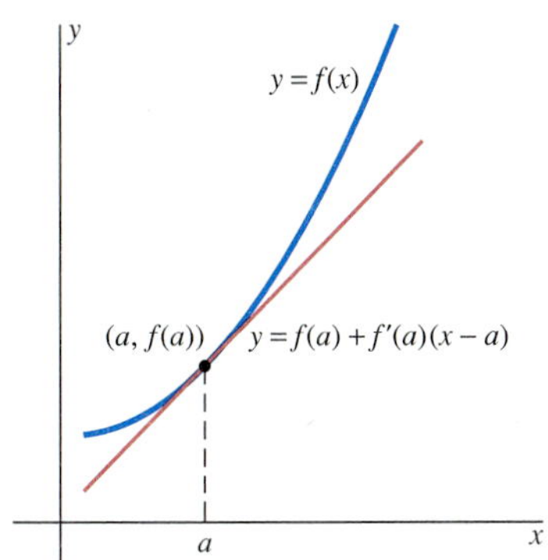

Fig. 4.2.3 The graph of the linear approximation $L(x) = f(a) + f'(a)\cdot(x - a)$ is the line tangent to $y = f(x)$ at the point $(a, f(a))$.

Because the right-hand side

$$L(x) = f(a) + f'(a)\cdot(x - a) \tag{7}$$

in Eq. (6) is a linear function of x, we call it the **linear approximation** $L(x)$ **to the function** $f(x)$ **near the point** $x = a$. As illustrated in Fig. 4.2.3, the graph $y = L(x)$ is the straight line tangent to the graph $y = f(x)$ at the point $(a, f(a))$.

EXAMPLE 1 Find the linear approximation to the function $f(x) = \sqrt{1 + x}$ near the point $a = 0$.

Solution Note that $f(0) = 1$ and that

$$f'(x) = \frac{1}{2}(1 + x)^{-1/2} = \frac{1}{2\sqrt{1 + x}},$$

so $f'(0) = \frac{1}{2}$. Hence Eq. (6) with $a = 0$ yields

$$f(x) \approx f(0) + f'(0)\cdot(x - 0) = 1 + \tfrac{1}{2}x = L(x).$$

Thus the desired linear approximation is

$$\sqrt{1 + x} \approx 1 + \tfrac{1}{2}x. \tag{8}$$

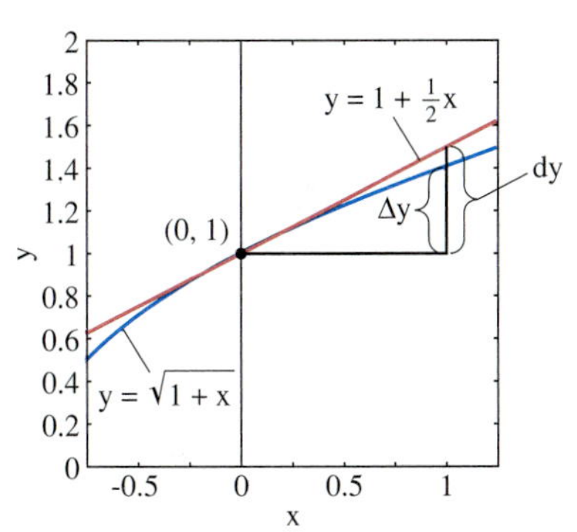

Fig. 4.2.4 The function $f(x) = \sqrt{1 + x}$ and its linear approximation $L(x) = 1 + \frac{1}{2}x$ near $a = 0$

Figure 4.2.4 illustrates the close approximation near $x = 0$ of the nonlinear function $f(x) = \sqrt{1 + x}$ by its linear approximation $L(x) = 1 + \frac{1}{2}x$. ■

IMPORTANT It is evident in Fig. 4.2.4 that the value of the linear approximation $L(x) = 1 + \frac{1}{2}x$ is closer to the actual value of the function $f(x) = \sqrt{1 + x}$ when x is closer to $a = 0$. For instance, the approximate values

$$\sqrt{1.1} \approx 1 + \tfrac{1}{2}(0.1) = 1.05 \quad (\text{using } x = 0.1 \text{ in (8)})$$

and

$$\sqrt{1.03} \approx 1 + \tfrac{1}{2}(0.03) = 1.015 \quad (\text{using } x = 0.03 \text{ in (8)})$$

are accurate to two and three decimal places (rounded), respectively. But

$$\sqrt{3} \approx 1 + \tfrac{1}{2} \cdot 2 = 2,$$

using $x = 2$, is a very poor approximation to $\sqrt{3} \approx 1.732$.

The approximation $\sqrt{1 + x} \approx 1 + \frac{1}{2}x$ is a special case of the approximation

$$(1 + x)^k \approx 1 + kx \tag{9}$$

(k is a constant, x is near zero), an approximation with numerous applications. The derivation of (9) is similar to that in Example 1 (see Problem 39).

EXAMPLE 2 Use the linear approximation formula to approximate $(122)^{2/3}$. Note that

$$(125)^{2/3} = \left[(125)^{1/3}\right]^2 = 5^2 = 25.$$

Solution We need to approximate a particular value of $x^{2/3}$, so our strategy is to apply Eq. (6) with $f(x) = x^{2/3}$. We first note that $f'(x) = \frac{2}{3}x^{-1/3}$. We choose $a = 125$, because we know the *exact* values

$$f(125) = (125)^{2/3} = 25 \quad \text{and} \quad f'(125) = \tfrac{2}{3}(125)^{-1/3} = \tfrac{2}{15}$$

and because 125 is relatively close to 122. Then the linear approximation in (6) to $f(x) = x^{2/3}$ near $a = 125$ takes the form

$$f(x) \approx f(125) + f'(125) \cdot (x - 125);$$

that is,

$$x^{2/3} \approx 25 + \tfrac{2}{15}(x - 125).$$

With $x = 122$ we get

$$(122)^{2/3} \approx 25 + \tfrac{2}{15}(-3) = 24.6.$$

Thus $(122)^{2/3}$ is approximately 24.6. The actual value of $(122)^{2/3}$ is about 24.5984, so the formula in (6) gives a relatively good approximation in this case. ■

EXAMPLE 3 A hemispherical bowl of radius 10 in. is filled with water to a depth of x inches. The volume V of water in the bowl (in cubic inches) is given by the formula

$$V = \frac{\pi}{3}(30x^2 - x^3) \tag{10}$$

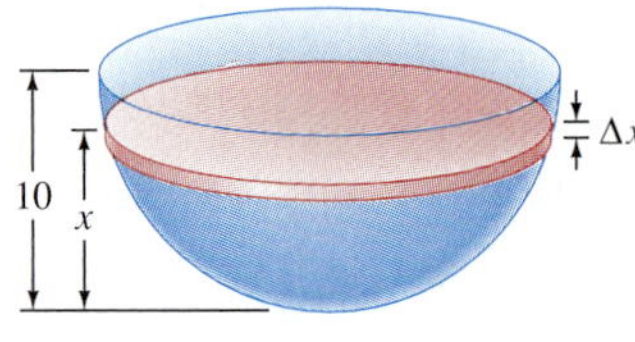

Fig. 4.2.5 The bowl of Example 3

(Fig. 4.2.5). (You will be able to derive this formula after you study Chapter 6.) Suppose that you measure the depth of water in the bowl to be 5 in. with a maximum possible measured error of $\frac{1}{16}$ in. Estimate the maximum error in the calculated volume of water in the bowl.

Solution The error in the calculated volume $V(5)$ is the difference

$$\Delta V = V(x) - V(5)$$

between the actual volume $V(x)$ and the calculated volume. We do not know the depth x of water in the bowl. We are given only that the difference

$$\Delta x = x - 5$$

between the actual and the measured depths is numerically at most $\frac{1}{16}$ in.: $|\Delta x| \leqq \frac{1}{16}$. Because Eq. (10) yields

$$V'(x) = \frac{\pi}{3}(60x - 3x^2) = \pi(20x - x^2),$$

the linear approximation

$$\Delta V \approx dV = V'(5)\,\Delta x$$

at $x = 5$ gives

$$\Delta V \approx \pi(20 \cdot 5 - 5^2)\,\Delta x = 75\pi\,\Delta x.$$

With the common practice in science of writing $\Delta x = \pm\frac{1}{16}$ to signify that $-\frac{1}{16} \leqq \Delta x \leqq \frac{1}{16}$, this gives

$$\Delta V \approx (75\pi)(\pm\tfrac{1}{16}) \approx \pm 14.73 \quad (\text{in.}^3).$$

The formula in Eq. (10) gives the calculated volume $V(5) \approx 654.50$ in.^{3}, but we now see that this may be in error by almost 15 in.^{3} in either direction. ■

The **absolute error** in a measured or approximated value is defined to be the remainder when the approximate value is subtracted from the true value. The **relative error** is the ratio of the absolute error to the true value. Thus in Example 3 a relative error in the measured depth x of

$$\frac{\Delta x}{x} = \frac{\frac{1}{16}}{5} = 0.0125 = 1.25\%$$

leads to a relative error in the estimated volume of

$$\frac{dV}{V} \approx \frac{14.73}{654.50} \approx 0.0225 = 2.25\%.$$

The relationship between these two relative errors is of some interest. The formulas for dV and V in Example 3 give

$$\frac{dV}{V} = \frac{\pi(20x - x^2)\,\Delta x}{\frac{1}{3}\pi(30x^2 - x^3)} = \frac{3(20 - x)}{30 - x} \cdot \frac{\Delta x}{x}.$$

When $x = 5$, this gives

$$\frac{dV}{V} = (1.80)\frac{\Delta x}{x}.$$

Hence, to approximate the volume of water in the bowl with a relative error of at most 0.5%, for instance, we would need to measure the depth with a relative error of at most $(0.5\%)/1.8$, thus with a relative error of less than 0.3%.

The Error in Linear Approximation

Now we consider briefly the question of the difference between the values of a function $f(x)$ and its linear approximation $L(x)$ near the point $x = a$. If we let $\Delta x = x - a$ and write

$$y = f(x), \qquad f(a + \Delta x) = f(a) + \Delta y,$$

and

$$L(x) = f(x) + f'(a)\cdot\Delta x = f(a) + dy,$$

it then follows that the error in the linear approximation is given by

$$f(x) - L(x) = \Delta y - dy, \tag{11}$$

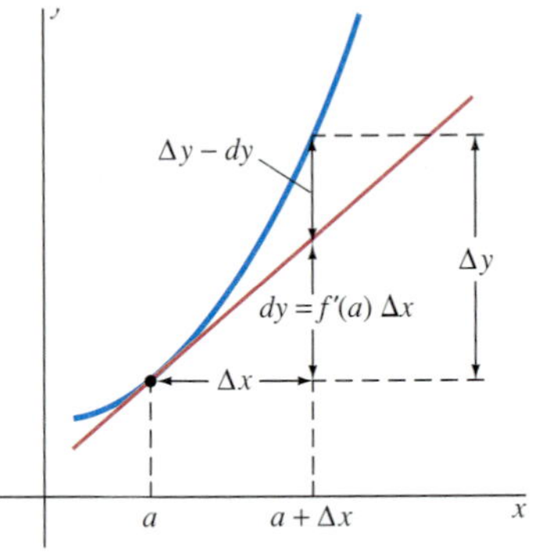

Fig. 4.2.6 The error $\Delta y - dy$ in the linear approximation $\Delta y \approx f'(a)\,\Delta x = dy$

as illustrated in Fig. 4.2.6. It appears in the figure that, the smaller Δx is, the closer are the corresponding points on the curve $y = f(x)$ and its tangent line $y = L(x)$. Because Eq. (11) implies that the difference in the heights of two such points is equal to $\Delta y - dy$, the figure suggests that $\Delta y - dy$ approaches zero as $\Delta x \to 0$.

But even more is true: As $\Delta x \to 0$, the difference $\Delta y - dy$ is small *even in comparison* with Δx. To see why, note that

$$\begin{aligned}\frac{\Delta y - dy}{\Delta x} &= \frac{f(a + \Delta x) - f(a) - f'(a)\,\Delta x}{\Delta x}\\ &= \frac{f(a + \Delta x) - f(a)}{\Delta x} - f'(a).\end{aligned}$$

Therefore the definition of the derivative $f'(a)$—as the limit of the last quotient here—implies that

$$\lim_{\Delta x \to 0} \frac{\Delta y - dy}{\Delta x} = 0.$$

Consequently, the **error**

$$\Delta y - dy = \Delta x \cdot \frac{\Delta y - dy}{\Delta x}$$

in the linear approximation $\Delta y \approx dy = f'(a)\,\Delta x$ is the product of two quantities that both approach zero as $\Delta x \to 0$. If Δx is "very small," so that the quotient $(\Delta y - dy)/\Delta x$ also is "very small," then we might well describe their product $\Delta y - dy$ as "*very* very small."

EXAMPLE 4 If $y = f(x) = x^3$, then simple computations (with $\Delta x = x - a$) give

$$\begin{aligned}\Delta y &= f(a + \Delta x) - f(a)\\ &= (a + \Delta x)^3 - a^3 = 3a^2\Delta x + 3a(\Delta x)^2 + (\Delta x)^3\end{aligned}$$

and

$$dy = f'(a)\,\Delta x = 3a^2\,\Delta x.$$

Hence

$$\Delta y - dy = 3a(\Delta x)^2 + (\Delta x)^3.$$

If $a = 1$ and $\Delta x = 0.1$, for instance, then these formulas yield

$$\Delta y = 0.331, \qquad dy = 0.3, \qquad \text{and} \qquad \Delta y - dy = 0.031,$$

thereby illustrating the smallness in the error $\Delta y - dy$ in the linear approximation in comparison with the values of Δy and dy. ■

Example 5 indicates how we sometimes can use a graphing calculator or computer to specify *how accurate* a linear approximation is—in terms of its accuracy throughout an entire interval containing the point $x = a$. In concrete situations we often want to determine an interval throughout which the linear approximation provides a specified accuracy.

EXAMPLE 5 Find an interval on which the approximation

$$\sqrt{1+x} \approx 1 + \tfrac{1}{2}x \tag{12}$$

of Example 1 is accurate to within 0.1.

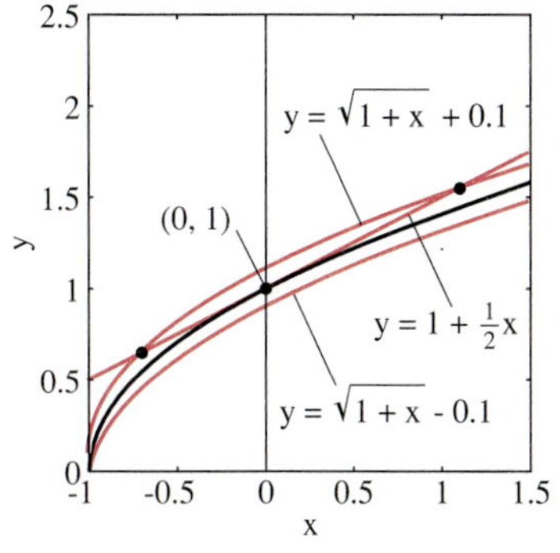

Fig. 4.2.7 The function $f(x) = \sqrt{1+x}$ on the interval $-1 < x < 1.5$

Solution Accuracy to within 0.1 means that the two functions in (12) differ by less than 0.1:

$$\left|\sqrt{1+x} - (1 + \tfrac{1}{2}x)\right| < 0.1,$$

which is equivalent to

$$\sqrt{1+x} - 0.1 < 1 + \tfrac{1}{2}x < \sqrt{1+x} + 0.1.$$

Thus we want the graph of the linear approximation $y = 1 + \frac{1}{2}x$ to lie between the two curves obtained by shifting the graph $y = \sqrt{1+x}$ vertically up and down by the amount 0.1. Figure 4.2.7 shows the graphs of all these curves on the interval $-1 < x < 1.5$. The points at which the linear approximation $y = 1 + \frac{1}{2}x$ emerges from the band of width 0.2 around the graph $y = \sqrt{1+x}$ are marked, and we see that a smaller interval around $x = 0$ is needed to confine the linear approximation within the desired range. Indeed, the zoom shown in Fig. 4.2.8 indicates that the approximation in (12) is accurate to within 0.1 for every x in the interval $-0.6 < x < 0.9$. ■

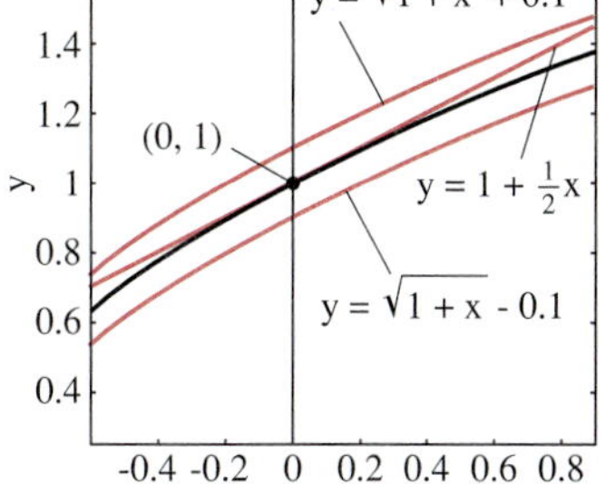

Fig. 4.2.8 The function $f(x) = \sqrt{1+x}$ on the smaller interval $-0.6 < x < 0.9$

Differentials

The linear approximation formula in (3) is often written with dx in place of Δx:

$$f(x + dx) \approx f(x) + f'(x)\,dx. \tag{13}$$

In this case dx is an independent variable, called the **differential** of x, and x is fixed. Thus the differentials of x and y are defined to be

$$dx = \Delta x \quad \text{and} \quad dy = f'(x)\,\Delta x = f'(x)\,dx. \tag{14}$$

From this definition it follows immediately that

$$\frac{dy}{dx} = \frac{f'(x)\,dx}{dx} = f'(x),$$

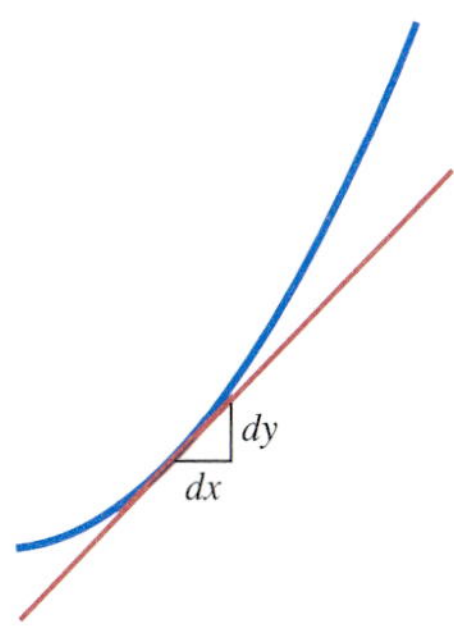

Fig. 4.2.9 The slope of the tangent line as the ratio of the infinitesimals dy and dx

in perfect accord with the notation we have been using. Indeed, Leibniz originated differential notation by visualizing "infinitesimal" increments dx and dy (Fig. 4.2.9), with their ratio dy/dx being the slope of the tangent line. The key to Leibniz's independent discovery of differential calculus in the 1670s was his insight that if dx and dy are sufficiently small, then the segment of the curve $y = f(x)$ and the straight line segment joining (x, y) and $(x + dx, y + dy)$ are virtually indistinguishable. This insight is illustrated by the successive magnifications in Figs. 4.2.10 through 4.2.12 of the curve $y = x^2$ near the point $(1, 1)$.

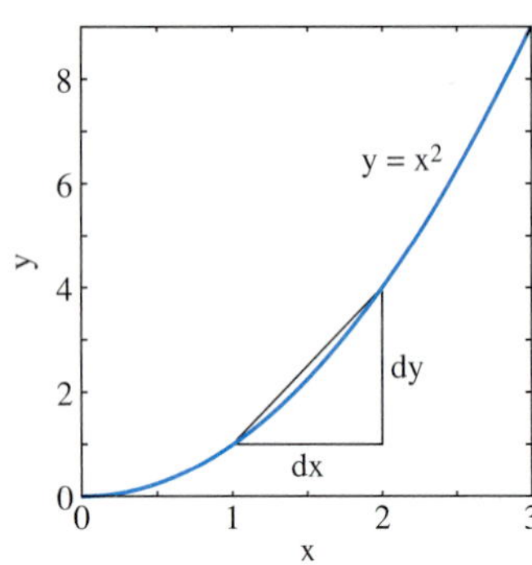

Fig. 4.2.10 $dx = 1$

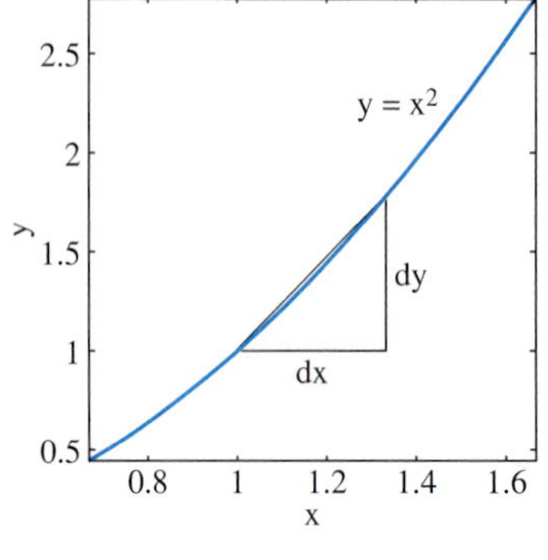

Fig. 4.2.11 $dx = \frac{1}{3}$

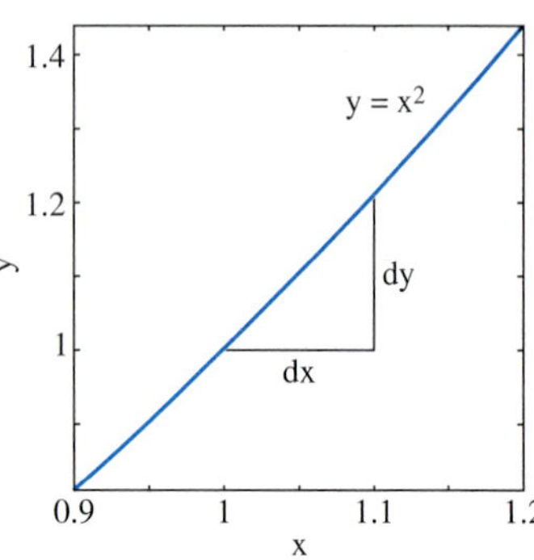

Fig. 4.2.12 $dx = \frac{1}{10}$

Differential notation provides us with a convenient way to write derivative formulas. Suppose that $z = f(u)$, so that $dz = f'(u)\,du$. For particular choices of the function f, we get the formulas

$$d(u^n) = nu^{n-1}\,du, \tag{15}$$

$$d(\sin u) = (\cos u)\,du, \tag{16}$$

and so on. Thus we can write differentiation rules in differential form without having to identify the independent variable. The sum, product, and quotient rules take the respective forms

$$d(u + v) = du + dv, \tag{17}$$

$$d(uv) = u\,dv + v\,du, \quad \text{and} \tag{18}$$

$$d\left(\frac{u}{v}\right) = \frac{v\,du - u\,dv}{v^2}. \tag{19}$$

If $z = f(u)$ and $u = g(x)$, we may substitute $du = g'(x)\,dx$ into the formula $dz = f'(u)\,du$. This gives

$$dz = f'\big(g(x)\big)\cdot g'(x)\,dx. \tag{20}$$

This is the differential form of the chain rule

$$D_x f\big(g(x)\big) = f'\big(g(x)\big)\cdot g'(x).$$

Thus the chain rule appears here as though it were the result of mechanical manipulations of the differential notation. This compatibility with the chain rule is one reason for the extraordinary usefulness of differential notation in calculus.

EXAMPLE 6

(a) If $y = 3x^2 - 2x^{3/2}$, then $dy = (6x - 3\sqrt{x})\,dx$.

(b) If $u = \sin^2 t - \cos 2t$, then

$$du = (2\sin t\cos t + 2\sin 2t)\,dt = 3\sin 2t\,dt$$

(using the trigonometric identity $\sin 2t = 2\sin t\cos t$). ■

Example 5 indicates how we sometimes can use a graphing calculator or computer to specify *how accurate* a linear approximation is—in terms of its accuracy throughout an entire interval containing the point $x = a$. In concrete situations we often want to determine an interval throughout which the linear approximation provides a specified accuracy.

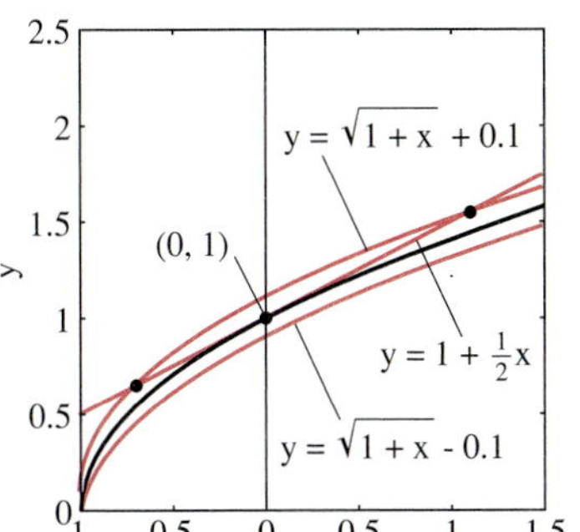

Fig. 4.2.7 The function $f(x) = \sqrt{1+x}$ on the interval $-1 < x < 1.5$

EXAMPLE 5 Find an interval on which the approximation

$$\sqrt{1+x} \approx 1 + \tfrac{1}{2}x \qquad \textbf{(12)}$$

of Example 1 is accurate to within 0.1.

Solution Accuracy to within 0.1 means that the two functions in (12) differ by less than 0.1:

$$\left|\sqrt{1+x} - (1 + \tfrac{1}{2}x)\right| < 0.1,$$

which is equivalent to

$$\sqrt{1+x} - 0.1 < 1 + \tfrac{1}{2}x < \sqrt{1+x} + 0.1.$$

Thus we want the graph of the linear approximation $y = 1 + \frac{1}{2}x$ to lie between the two curves obtained by shifting the graph $y = \sqrt{1+x}$ vertically up and down by the amount 0.1. Figure 4.2.7 shows the graphs of all these curves on the interval $-1 < x < 1.5$. The points at which the linear approximation $y = 1 + \frac{1}{2}x$ emerges from the band of width 0.2 around the graph $y = \sqrt{1+x}$ are marked, and we see that a smaller interval around $x = 0$ is needed to confine the linear approximation within the desired range. Indeed, the zoom shown in Fig. 4.2.8 indicates that the approximation in (12) is accurate to within 0.1 for every x in the interval $-0.6 < x < 0.9$. ■

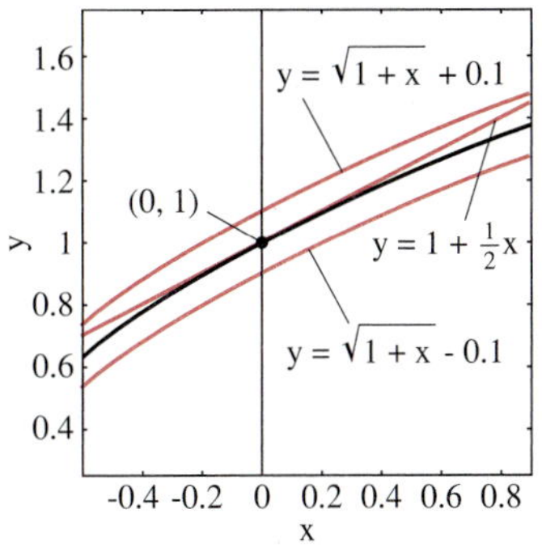

Fig. 4.2.8 The function $f(x) = \sqrt{1+x}$ on the smaller interval $-0.6 < x < 0.9$

Differentials

The linear approximation formula in (3) is often written with dx in place of Δx:

$$f(x + dx) \approx f(x) + f'(x)\,dx. \qquad \textbf{(13)}$$

In this case dx is an independent variable, called the **differential** of x, and x is fixed. Thus the differentials of x and y are defined to be

$$dx = \Delta x \quad \text{and} \quad dy = f'(x)\,\Delta x = f'(x)\,dx. \qquad \textbf{(14)}$$

From this definition it follows immediately that

$$\frac{dy}{dx} = \frac{f'(x)\,dx}{dx} = f'(x),$$

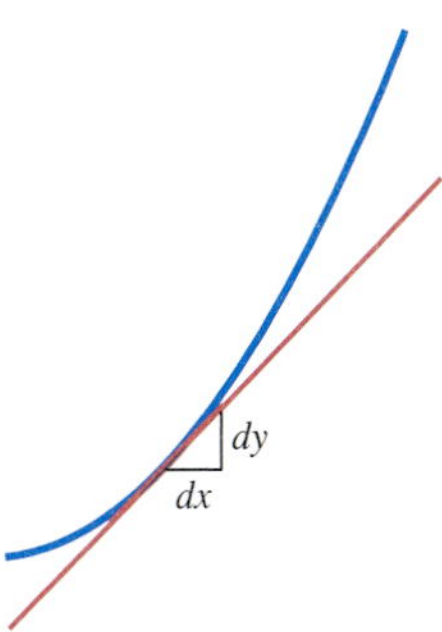

Fig. 4.2.9 The slope of the tangent line as the ratio of the infinitesimals dy and dx

in perfect accord with the notation we have been using. Indeed, Leibniz originated differential notation by visualizing "infinitesimal" increments dx and dy (Fig. 4.2.9), with their ratio dy/dx being the slope of the tangent line. The key to Leibniz's independent discovery of differential calculus in the 1670s was his insight that if dx and dy are sufficiently small, then the segment of the curve $y = f(x)$ and the straight line segment joining (x, y) and $(x + dx, y + dy)$ are virtually indistinguishable. This insight is illustrated by the successive magnifications in Figs. 4.2.10 through 4.2.12 of the curve $y = x^2$ near the point $(1, 1)$.

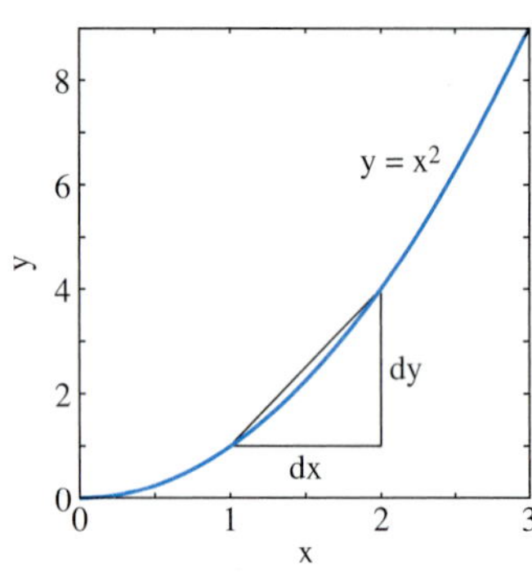

Fig. 4.2.10 $dx = 1$

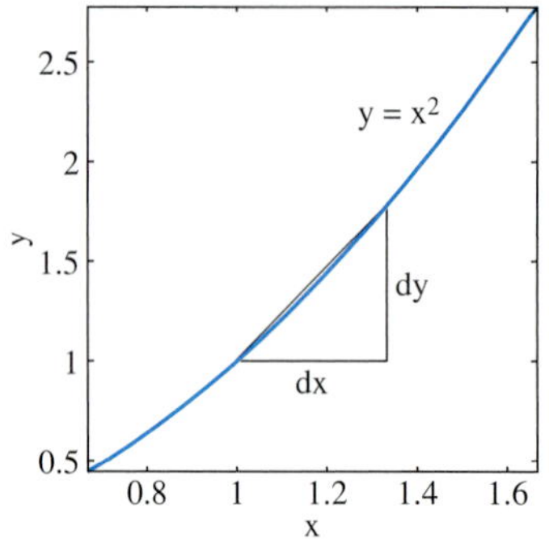

Fig. 4.2.11 $dx = \frac{1}{3}$

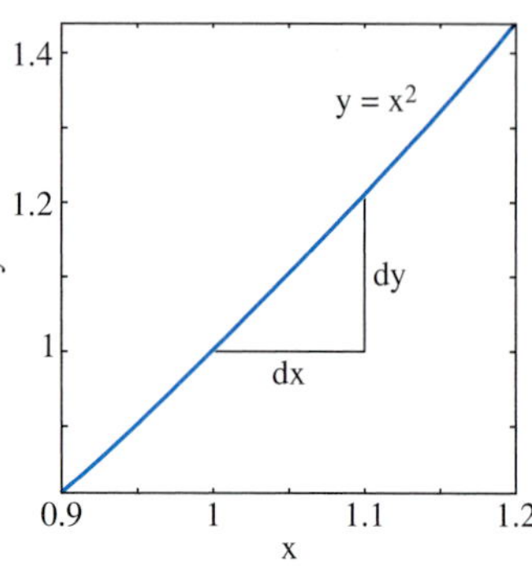

Fig. 4.2.12 $dx = \frac{1}{10}$

Differential notation provides us with a convenient way to write derivative formulas. Suppose that $z = f(u)$, so that $dz = f'(u)\,du$. For particular choices of the function f, we get the formulas

$$d(u^n) = nu^{n-1}\,du, \tag{15}$$

$$d(\sin u) = (\cos u)\,du, \tag{16}$$

and so on. Thus we can write differentiation rules in differential form without having to identify the independent variable. The sum, product, and quotient rules take the respective forms

$$d(u + v) = du + dv, \tag{17}$$

$$d(uv) = u\,dv + v\,du, \quad \text{and} \tag{18}$$

$$d\left(\frac{u}{v}\right) = \frac{v\,du - u\,dv}{v^2}. \tag{19}$$

If $z = f(u)$ and $u = g(x)$, we may substitute $du = g'(x)\,dx$ into the formula $dz = f'(u)\,du$. This gives

$$dz = f'\big(g(x)\big)\cdot g'(x)\,dx. \tag{20}$$

This is the differential form of the chain rule

$$D_x f\big(g(x)\big) = f'\big(g(x)\big)\cdot g'(x).$$

Thus the chain rule appears here as though it were the result of mechanical manipulations of the differential notation. This compatibility with the chain rule is one reason for the extraordinary usefulness of differential notation in calculus.

EXAMPLE 6

(a) If $y = 3x^2 - 2x^{3/2}$, then $dy = (6x - 3\sqrt{x})\,dx$.

(b) If $u = \sin^2 t - \cos 2t$, then

$$du = (2\sin t\cos t + 2\sin 2t)\,dt = 3\sin 2t\,dt$$

(using the trigonometric identity $\sin 2t = 2\sin t\cos t$). ■

4.2 PROBLEMS

In Problems 1 through 16, write dy in terms of x and dx.

1. $y = 3x^2 - \dfrac{4}{x^2}$

2. $y = 2\sqrt{x} - \dfrac{3}{\sqrt[3]{x}}$

3. $y = x - \sqrt{4 - x^3}$

4. $y = \dfrac{1}{x - \sqrt{x}}$

5. $y = 3x^2(x - 3)^{3/2}$

6. $y = \dfrac{x}{x^2 - 4}$

7. $y = x(x^2 + 25)^{1/4}$

8. $y = \dfrac{1}{(x^2 - 1)^{4/3}}$

9. $y = \cos\sqrt{x}$

10. $y = x^2 \sin x$

11. $y = \sin 2x \cos 2x$

12. $y = \cos^3 3x$

13. $y = \dfrac{\sin 2x}{3x}$

14. $y = \dfrac{\cos x}{\sqrt{x}}$

15. $y = \dfrac{1}{1 - x \sin x}$

16. $y = (1 + \cos 2x)^{3/2}$

In Problems 17 through 24, find—as in Example 1—the linear approximation $L(x)$ to the given function $f(x)$ near the point $a = 0$.

17. $f(x) = \dfrac{1}{1 - x}$

18. $f(x) = \dfrac{1}{\sqrt{1 + x}}$

19. $f(x) = (1 + x)^2$

20. $f(x) = (1 - x)^3$

21. $f(x) = (1 - 2x)^{3/2}$

22. $f(x) = \dfrac{1}{(1 + 3x)^{2/3}}$

23. $f(x) = \sin x$

24. $f(x) = \cos x$

In Problems 25 through 34, use—as in Example 2—a linear approximation $L(x)$ to an appropriate function $f(x)$, with an appropriate value of a, to estimate the given number.

25. $\sqrt[3]{25}$

26. $\sqrt{102}$

27. $\sqrt[4]{15}$

28. $\sqrt{80}$

29. $65^{-2/3}$

30. $80^{3/4}$

31. $\cos 43°$

32. $\sin 32°$

33. $\sin 88°$

34. $\cos 62°$

In Problems 35 through 38, compute the differential of each side of the given equation, regarding x and y as dependent variables (as though both were functions of some third, unspecified, variable). Then solve for dy/dx.

35. $x^2 + y^2 = 1$

36. $x^{2/3} + y^{2/3} = 4$

37. $x^3 + y^3 = 3xy$

38. $x \sin y = 1$

39. Assuming that $D_x x^k = kx^{k-1}$ for any real constant k (which we shall establish in Chapter 7), derive the linear approximation formula $(1 + x)^k \approx 1 + kx$ for x near zero.

In Problems 40 through 47, use linear approximations to estimate the change in the given quantity.

40. The circumference of a circle, if its radius is increased from 10 in. to 10.5 in.

41. The area of a square, if its edge length is decreased from 10 in. to 9.8 in.

42. The surface area of a sphere, if its radius is increased from 5 in. to 5.2 in. (Fig. 4.2.13)

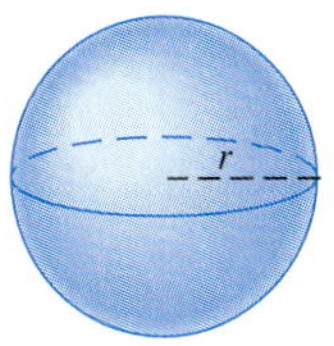

Fig. 4.2.13 The sphere of Problem 42: area $A = 4\pi r^2$, volume $V = \frac{4}{3}\pi r^3$

Fig. 4.2.14 The cylinder of Problem 43: volume $V = \pi r^2 h$

43. The volume of a cylinder, if both its height and its radius are decreased from 15 cm to 14.7 cm (Fig. 4.2.14)

44. The volume of the conical sandpile of Fig. 4.2.15, if its radius is 14 in. and its height is increased from 7 in. to 7.1 in.

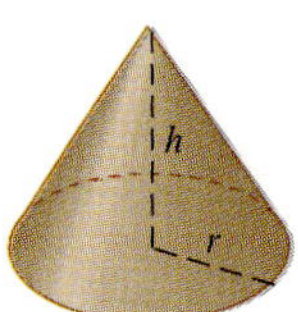

Fig. 4.2.15 The conical sandpile of Problem 44: volume $V = \frac{1}{3}\pi r^2 h$

45. The range $R = \frac{1}{16}v^2 \sin 2\theta$ of a shell fired at inclination angle $\theta = 45°$, if its initial velocity v is increased from 80 ft/s to 81 ft/s

46. The range $R = \frac{1}{16}v^2 \sin 2\theta$ of a projectile fired with initial velocity $v = 80$ ft/s, if its initial inclination angle θ is increased from 45° to 46°

47. The wattage $W = RI^2$ of a floodlight with resistance $R = 10$ ohms, if the current I is increased from 3 amperes to 3.1 amperes

48. The equatorial radius of the earth is approximately 3960 mi. Suppose that a wire is wrapped tightly around the earth at the equator. Approximately how much must this wire be lengthened if it is to be strung all the way around the earth on poles 10 ft above the ground? Use the linear approximation formula!

49. The radius of a spherical ball is measured as 10 in., with a maximum error of $\frac{1}{16}$ in. What is the maximum resulting error in its calculated volume?

50. With what accuracy must the radius of the ball of Problem 49 be measured to ensure an error of at most 1 in.^{3} in its calculated volume?

51. The radius of a hemispherical dome is measured as 100 m with a maximum error of 1 cm (Fig. 4.2.16). What is the maximum resulting error in its calculated surface area?

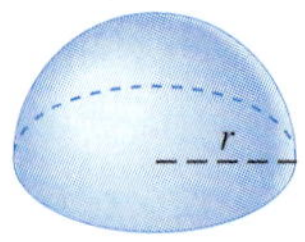

Fig. 4.2.16 The hemisphere of Problem 51: curved surface area $A = 2\pi r^2$

52. With what accuracy must the radius of a hemispherical dome be measured to ensure an error of at most 0.01% in its calculated surface area?

In Problems 53 through 60, a function $f(x)$ and a point $x = a$ are given. Determine graphically an open interval I centered at a so that the function $f(x)$ and its linear approximation $L(x)$ differ by less than the given value ϵ at each point of I.

53. $f(x) = x^2, \quad a = 1, \quad \epsilon = 0.2$

54. $f(x) = \sqrt{x}, \quad a = 1, \quad \epsilon = 0.1$

55. $f(x) = \dfrac{1}{x}, \quad a = 2, \quad \epsilon = 0.01$

56. $f(x) = \sqrt[3]{x}, \quad a = 8, \quad \epsilon = 0.01$

57. $f(x) = \sin x, \quad a = 0, \quad \epsilon = 0.05$

58. $f(x) = \cos x, \quad a = \pi/2, \quad \epsilon = 0.05$

59. $f(x) = \sin x, \quad a = \pi/4, \quad \epsilon = 0.02$

60. $f(x) = \tan x, \quad a = \pi/4, \quad \epsilon = 0.02$

4.3 INCREASING AND DECREASING FUNCTIONS AND THE MEAN VALUE THEOREM

The significance of the *sign* of the first derivative of a function is simple but crucial:

> f is increasing on an interval where $f'(x) > 0$;
> f is decreasing on an interval where $f'(x) < 0$.

Geometrically, this means that where $f'(x) > 0$, the graph of $y = f(x)$ is rising as you scan it from left to right. Where $f'(x) < 0$, the graph is falling. We can clarify the terms *increasing* and *decreasing* as follows.

Definition *Increasing and Decreasing Functions*

The function f is **increasing** on the interval $I = (a, b)$ provided that

$$f(x_1) < f(x_2)$$

for all pairs of numbers x_1 and x_2 in I for which $x_1 < x_2$. The function f is **decreasing** on I provided that

$$f(x_1) > f(x_2)$$

for all pairs of numbers x_1 and x_2 for which $x_1 < x_2$.

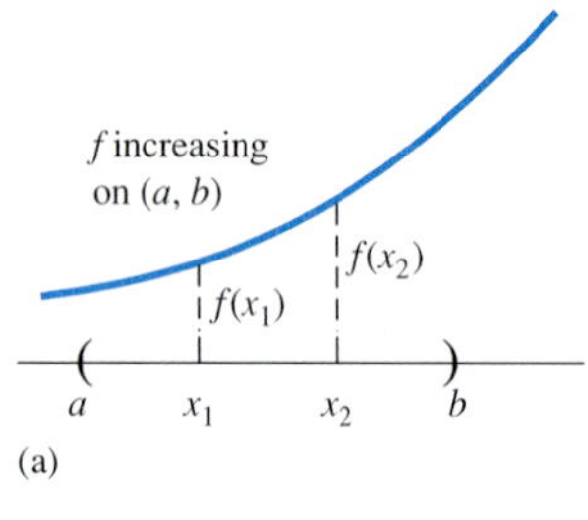

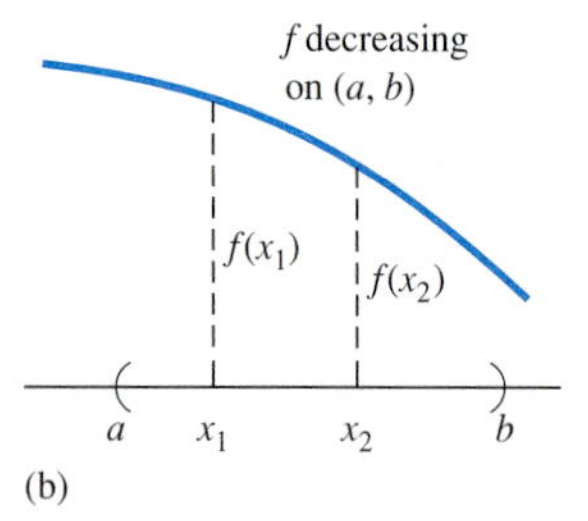

Fig. 4.3.1 (a) An increasing function and (b) a decreasing function

Figure 4.3.1 illustrates this definition. In short, the function f is increasing on $I = (a, b)$ if the values of $f(x)$ increase as x increases (Fig. (4.3.1a)); f is decreasing on I if the values of $f(x)$ decrease as x increases (Fig. (4.3.1b)).

EXAMPLE 1 As illustrated in Fig. 4.3.2, the simple function $f(x) = x^2$ is decreasing on the interval $(-\infty, 0)$ and increasing on the interval $(0, +\infty)$. This follows immediately from the elementary fact that $u^2 < v^2$ if $0 < u < v$. Because $f'(x) = 2x$, we also see immediately that $f'(x) < 0$ on the interval $(-\infty, 0)$ and that $f'(x) > 0$ on the interval $(0, +\infty)$. But for more general functions, the mean value theorem of this section is needed to establish the precise relationship between the sign of the derivative of a function and its increasing-decreasing behavior. ■

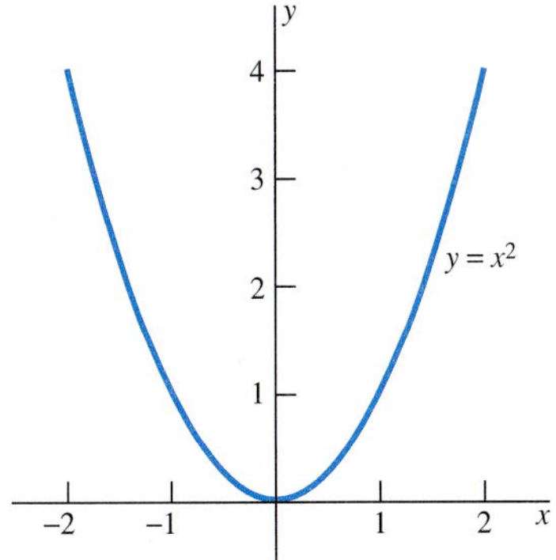

Fig. 4.3.2 $f(x) = x^2$ is decreasing for $x < 0$, increasing for $x > 0$.

REMARK We speak of a function as increasing or decreasing *on an interval,* not at a single point. Nevertheless, if we consider the sign of f', the derivative of f, at a single point, we get a useful intuitive picture of the significance of the sign of the derivative. This is because the derivative $f'(x)$ is the slope of the tangent line at the point $(x, f(x))$ on the graph of f. If $f'(x) > 0$, then the tangent line has positive slope. Therefore, it rises as you scan from left to right. Intuitively, a rising tangent would seem to correspond to a rising graph and thus to an increasing function. Similarly, we expect to see a falling graph where $f'(x)$ is negative (Fig. 4.3.3). One caution: To determine whether a function f is increasing or decreasing, we must examine the sign of f' on a whole interval, not merely at a single point (see Problem 59).

The Mean Value Theorem

Although pictures of rising and falling graphs are suggestive, they provide no actual *proof* of the significance of the sign of the derivative. To establish rigorously the connection between a graph's rising and falling and the sign of the derivative of the graphed function, we need the *mean value theorem,* stated later in this section. This theorem is the principal theoretical tool of differential calculus, and we shall see that it has many important applications.

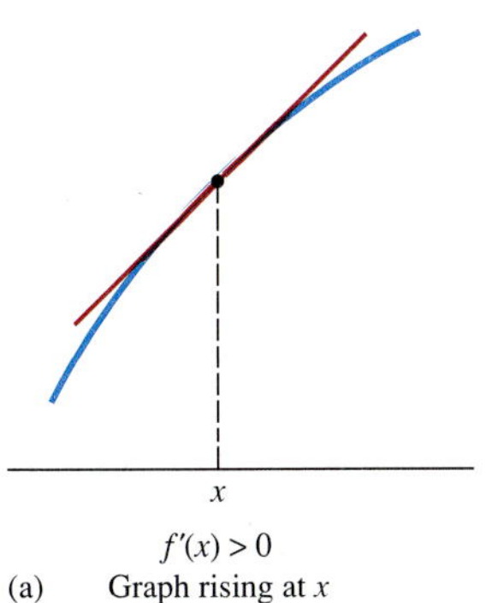

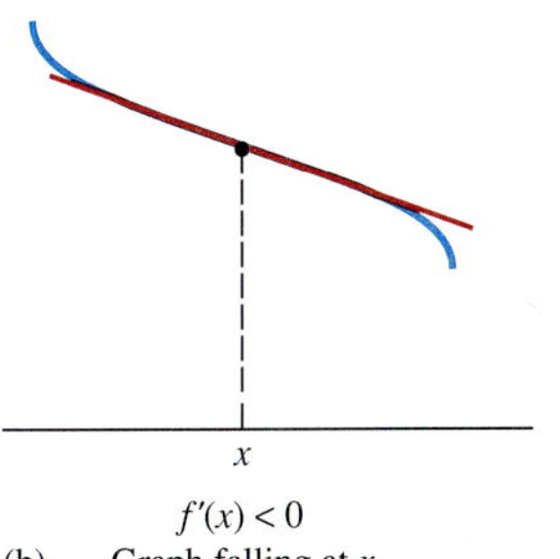

Fig. 4.3.3 (a) A graph rising at x and (b) a graph falling at x

A QUESTION As an introduction to the mean value theorem, we pose the following question. Suppose that P and Q are two points on the surface of the sea, with Q lying generally to the east of P (Fig. 4.3.4). Is it possible to sail a boat from P to Q, always sailing roughly east, without *ever* (even for an instant) sailing in the exact direction from P to Q? That is, can we sail from P to Q without our instantaneous line of motion ever being parallel to the line PQ?

The mean value theorem answers this question: No. There will always be at least one instant when we are sailing parallel to the line PQ, no matter which path we choose.

To paraphrase: Let the path of the sailboat be the graph of a differentiable function $y = f(x)$ with endpoints $P(a, f(a))$ and $Q(b, f(b))$. Then we say that there must be some point on this graph where the tangent line (corresponding to the instantaneous line of motion of the boat) to the curve is parallel to the line PQ that joins the curve's endpoints. This is a *geometric interpretation* of the mean value theorem.

THE GEOMETRIC FORMULATION The slope of the line tangent at the point $(c, f(c))$ (Fig. 4.3.5) is $f'(c)$, whereas the slope of the line PQ is

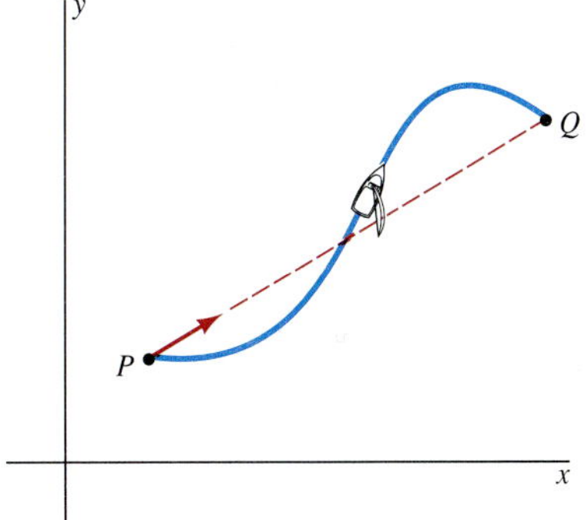

Fig. 4.3.4 Can you sail from P to Q without ever sailing—even for an instant—in the direction PQ (the direction of the arrow)?

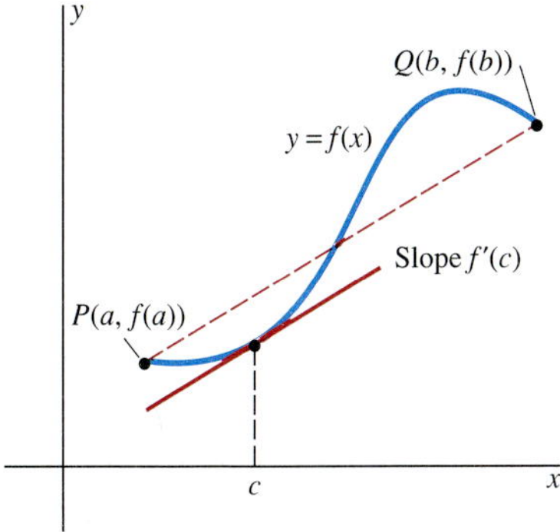

Fig. 4.3.5 The sailboat problem in mathematical terminology

$$\frac{f(b) - f(a)}{b - a}.$$

We may think of this last quotient as the average (or *mean*) value of the slope of the curve $y = f(x)$ over the interval $[a, b]$. The mean value theorem guarantees that there is a point c in (a, b) for which the line tangent to $y = f(x)$ at $(c, f(c))$ is indeed parallel to the line PQ. In the language of algebra, there's a number c in (a, b) such that

$$f'(c) = \frac{f(b) - f(a)}{b - a}. \qquad \textbf{(1)}$$

A Preliminary Result We first state a "lemma" to expedite the proof of the mean value theorem. This lemma is called *Rolle's theorem,* after Michel Rolle (1652–1719), who discovered it in 1690. In his youth Rolle studied the emerging subject of calculus but later renounced it. He argued that the subject was based on logical fallacies, and he is remembered today only for the single theorem that bears his name. It is ironic that his theorem plays an important role in the rigorous proofs of several calculus theorems.

Rolle's Theorem

Suppose that the function f is continuous on the closed interval $[a, b]$ and is differentiable in its interior (a, b). If $f(a) = 0 = f(b)$, then there exists some number c in (a, b) such that $f'(c) = 0$.

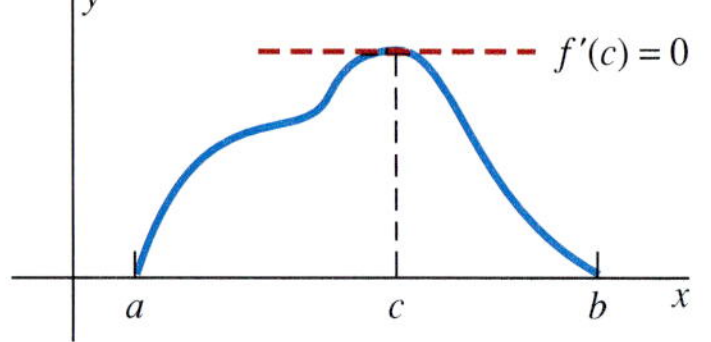

Fig. 4.3.6 The idea of the proof of Rolle's theorem

Figure 4.3.6 illustrates the first case in the following proof of Rolle's theorem. The idea of the proof is this: Suppose that the smooth graph $y = f(x)$ starts $(x = a)$ at height zero and ends $(x = b)$ at height zero. Then if it goes up, it must come back down. But where it stops going up and starts coming back down, its tangent line must be horizontal. Therefore the derivative is zero at that point.

Proof of Rolle's Theorem Because f is continuous on $[a, b]$, it must attain both a maximum and a minimum value on $[a, b]$ (by the maximum value property of Section 3.5). If f has any positive values, consider its maximum value $f(c)$. Now c is not an endpoint of $[a, b]$, because $f(a) = 0$ and $f(b) = 0$. Therefore c is a point of (a, b). But we know that f is differentiable at c. So it follows from Theorem 2 of Section 3.5 that $f'(c) = 0$.

Similarly, if f has any negative values, we can consider its minimum value $f(c)$ and conclude much as before that $f'(c) = 0$.

If f has neither positive nor negative values, then f is identically zero on $[a, b]$, and it follows that $f'(c) = 0$ for *every* c in (a, b).

Thus we see that the conclusion of Rolle's theorem is justified in every case. ■

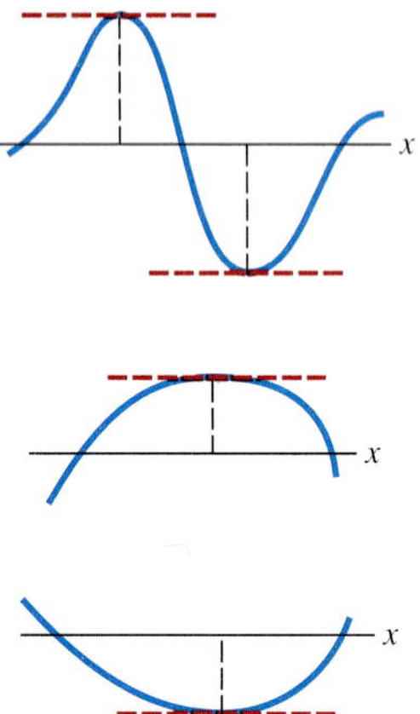

Fig. 4.3.7 The existence of the horizontal tangent is a consequence of Rolle's theorem.

An important consequence of Rolle's theorem is that between each pair of zeros of a differentiable function, there is *at least one* point at which the tangent line is horizontal. Some possible pictures of the situation are indicated in Fig. 4.3.7.

EXAMPLE 2 Suppose that $f(x) = x^{1/2} - x^{3/2}$ on $[0, 1]$. Find a number c that satisfies the conclusion of Rolle's theorem.

Solution Note that f is continuous on $[0, 1]$ and differentiable on $(0, 1)$. Because the term $x^{1/2}$ is present, f is *not* differentiable at $x = 0$, but this is irrelevant. Also, $f(0) = 0 = f(1)$, so all of the hypotheses of Rolle's theorem are satisfied. Finally,

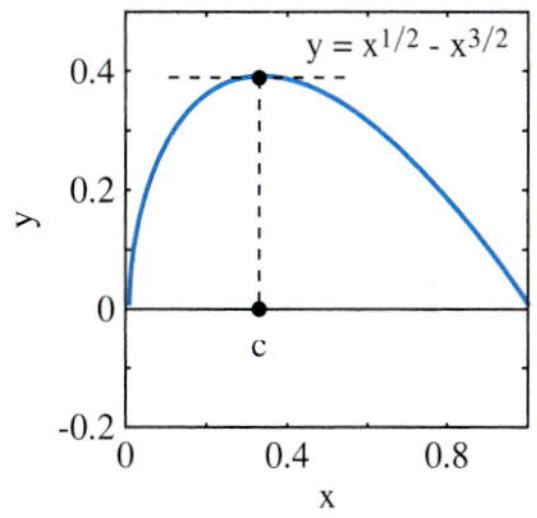

Fig. 4.3.8 The number c of Example 2

$$f'(x) = \frac{1}{2}x^{-1/2} - \frac{3}{2}x^{1/2} = \frac{1}{2}x^{-1/2}(1 - 3x),$$

so we see that $f'(c) = 0$ for $c = \frac{1}{3}$. An accurate graph of f on $[0, 1]$, including c and the horizontal tangent line, is shown in Fig. 4.3.8. ■

EXAMPLE 3 Suppose that $f(x) = 1 - x^{2/3}$ on $[-1, 1]$. Then f satisfies the hypotheses of Rolle's theorem *except* for the fact that $f'(0)$ does not exist. It is clear from the graph of f that there is *no* point at which the tangent line is horizontal (Fig. 4.3.9). Indeed,

$$f'(x) = -\frac{2}{3}x^{-1/3} = -\frac{2}{3\sqrt[3]{x}},$$

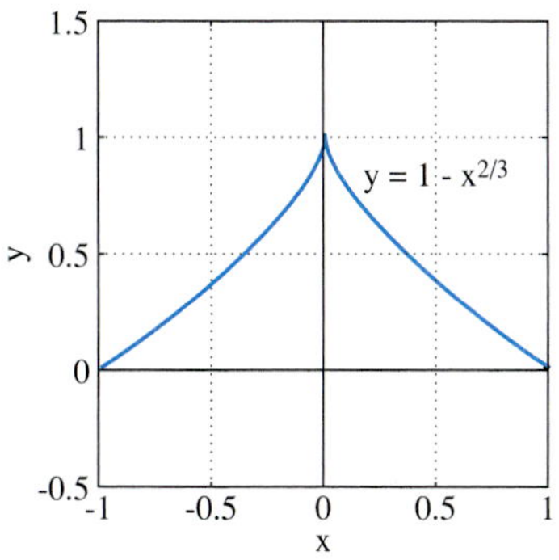

Fig. 4.3.9 The function $f(x) = 1 - x^{2/3}$ of Example 3

so $f'(x) \neq 0$ for $x \neq 0$, and we see that $|f'(x)| \to \infty$ as $x \to 0$. Hence the graph of f has a vertical tangent line—rather than a horizontal one—at the point $(0, 1)$. Thus the conclusion of Rolle's theorem—like that of any theorem—may fail to hold if any of its hypotheses are not satisfied. ■

Now we are ready to state formally and prove the mean value theorem.

The Mean Value Theorem

Suppose that the function f is continuous on the closed interval $[a, b]$ and differentiable on the open interval (a, b). Then

$$f(b) - f(a) = f'(c)\cdot(b - a) \qquad \textbf{(2)}$$

for some number c in (a, b).

COMMENT Because Eq. (2) is equivalent to Eq. (1), the conclusion of the mean value theorem is that there must be at least one point on the curve $y = f(x)$ at which the tangent line is parallel to the line joining its endpoints $P(a, f(a))$ and $Q(b, f(b))$.

MOTIVATION FOR THE PROOF OF THE MEAN VALUE THEOREM We consider the auxiliary function ϕ suggested by Fig. 4.3.10. By definition, the value of $\phi(x)$ is the vertical height difference over x of the point $(x, f(x))$ on the curve and the corresponding point on the line PQ. It appears that a point on the curve $y = f(x)$ at which the

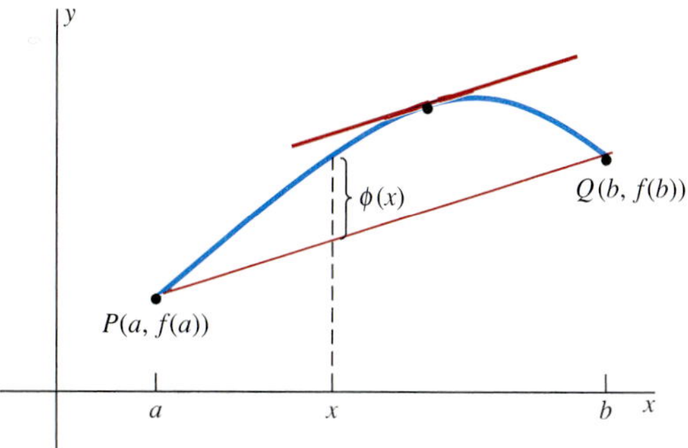

Fig. 4.3.10 The construction of the auxiliary function ϕ

tangent line is parallel to PQ corresponds to a maximum or minimum of ϕ. It's also clear that $\phi(a) = 0 = \phi(b)$, so Rolle's theorem can be applied to the function ϕ on $[a, b]$. So our plan for proving the mean value theorem is this: First, we obtain a formula for the function ϕ. Second, we locate the point c such that $\phi'(c) = 0$. Finally, we show that this number c is exactly the number needed to satisfy the conclusion of the mean value theorem in Eq. (2).

PROOF OF THE MEAN VALUE THEOREM Because the line PQ passes through $P(a, f(a))$ and has slope

$$m = \frac{f(b) - f(a)}{b - a},$$

the point-slope formula for the equation of a straight line gives us the following equation for PQ:

$$y = y_{\text{line}} = f(a) + m(x - a).$$

Thus

$$\phi(x) = y_{\text{curve}} - y_{\text{line}} = f(x) - f(a) - m(x - a).$$

You may verify by direct substitution that $\phi(a) = 0 = \phi(b)$. And, because ϕ is continuous on $[a, b]$ and differentiable on (a, b), we may apply Rolle's theorem to it. Thus there is a point c somewhere in the open interval (a, b) at which $\phi'(c) = 0$. But

$$\phi'(x) = f'(x) - m = f'(x) - \frac{f(b) - f(a)}{b - a}.$$

Because $\phi'(c) = 0$, we conclude that

$$0 = f'(c) - \frac{f(b) - f(a)}{b - a}.$$

That is,

$$f(b) - f(a) = f'(c) \cdot (b - a).$$ ■

The proof of the mean value theorem is an application of Rolle's theorem, whereas Rolle's theorem is the special case of the mean value theorem in which $f(a) = 0 = f(b)$.

EXAMPLE 4 Suppose that we drive from Kristiansand, Norway, to Oslo—a road distance of almost exactly 350 km—in exactly 4 h, from time $t = 0$ to time $t = 4$. Let $f(t)$ denote the distance we have traveled at time t and assume that f is a differentiable function. Then the mean value theorem implies that

$$350 = f(4) - f(0) = f'(c) \cdot (4 - 0) = 4f'(c)$$

and thus that

$$f'(c) = \tfrac{350}{4} = 87.5$$

at some instant c in $(0, 4)$. But $f'(c)$ is our *instantaneous* velocity at time $t = c$, and 87.5 km/h is our *average* velocity for the trip. Thus the mean value theorem implies that we must have an instantaneous velocity of exactly 87.5 km/h at least once during the trip. ■

The argument in Example 4 is quite general—during any trip, the instantaneous velocity must at *some* instant equal the average velocity for the whole trip. For instance, it follows that if two toll stations are 70 mi apart and you drive between the two in exactly 1 h, then at some instant you must have been speeding in excess of the posted limit of 65 mi/h. Speeding tickets have been issued by the Pennsylvania State Police to speeders on the Pennsylvania Turnpike on exactly such evidence!

Consequences of the Mean Value Theorem

The first of three important consequences of the mean value theorem is the *non*trivial converse of the trivial fact that the derivative of a constant function is identically zero. That is, we prove that there can be *no* exotic function that is nonconstant but has a derivative that is identically zero. In Corollaries 1 through 3 we assume, as in Rolle's theorem and the mean value theorem, that f and g are continuous on the closed interval $[a, b]$ and differentiable on (a, b).

Corollary 1 ***Functions with Zero Derivative***

If $f'(x) \equiv 0$ on (a, b) (that is, $f'(x) = 0$ for all x in (a, b)), then f is a constant function on $[a, b]$. In other words, there exists a constant C such that $f(x) \equiv C$.

Proof Apply the mean value theorem to the function f on the interval $[a, x]$, where x is a fixed but arbitrary point of the interval $(a, b]$. We find that

$$f(x) - f(a) = f'(c) \cdot (x - a)$$

for some number c between a and x. But $f'(x)$ is always zero on the interval (a, b), so $f'(c) = 0$. Thus $f(x) - f(a) = 0$, and therefore $f(x) = f(a)$.

But this last equation holds for *all* x in $(a, b]$. Therefore, $f(x) = f(a)$ for all x in $(a, b,]$ and, indeed, for all x in $[a, b]$. That is, $f(x)$ has the constant value $C = f(a)$. This establishes Corollary 1. ■

Corollary 1 is usually applied in a different but equivalent form, which we state and prove next.

Corollary 2 ***Functions with Equal Derivatives***

Suppose that $f'(x) = g'(x)$ for all x in the open interval (a, b). Then f and g differ by a constant on $[a, b]$. That is, there exists a constant K such that

$$f(x) = g(x) + K$$

for all x in $[a, b]$.

Proof Given the hypotheses, let $h(x) = f(x) - g(x)$. Then

$$h'(x) = f'(x) - g'(x) = 0$$

for all x in (a, b). So, by Corollary 1, $h(x)$ is a constant K on $[a, b]$. That is, $f(x) - g(x) = K$ for all x in $[a, b]$; therefore,

$$f(x) = g(x) + K$$

for all x in $[a, b]$. This establishes Corollary 2. ■

EXAMPLE 5 If $f'(x) = 2\cos x$ and $f(0) = 5$, what is the function $f(x)$?

Solution From our knowledge of the derivatives of trigonometric functions, we know that one function with derivative $g'(x) = 2\cos x$ is

$$g(x) = 2\sin x.$$

Hence Corollary 2 implies that there exists a constant K such that

$$f(x) = g(x) + K = 2\sin x + K$$

on any given interval $[a, b]$ containing zero. But we can find the value of K by substituting $x = 0$:

$$f(0) = 2\sin 0 + K;$$
$$5 = 2 \cdot 0 + K;$$

so $K = 5$. Thus the function f is

$$f(x) = 2\sin x + 5.$$

The following consequence of the mean value theorem verifies the remarks about increasing and decreasing functions with which we opened this section.

Corollary 3 ***Increasing and Decreasing Functions***
If $f'(x) > 0$ for all x in (a, b), then f is an increasing function on $[a, b]$. If $f'(x) < 0$ for all x in (a, b), then f is a decreasing function on $[a, b]$.

PROOF Suppose, for example, that $f'(x) > 0$ for all x in (a, b). We need to show the following: If u and v are points of $[a, b]$ with $u < v$, then $f(u) < f(v)$. We apply the mean value theorem to f, but on the closed interval $[u, v]$. This is legitimate because $[u, v]$ is contained in $[a, b]$, so f satisfies the hypotheses of the mean value theorem on $[u, v]$ as well as on $[a, b]$. The result is that

$$f(v) - f(u) = f'(c) \cdot (v - u)$$

for some number c in (u, v). Because $v > u$ and because, by hypothesis, $f'(c) > 0$, it follows that

$$f(v) - f(u) > 0; \qquad \text{that is,} \quad f(u) < f(v),$$

as we wanted to show. The proof is similar when $f'(x)$ is negative on (a, b).

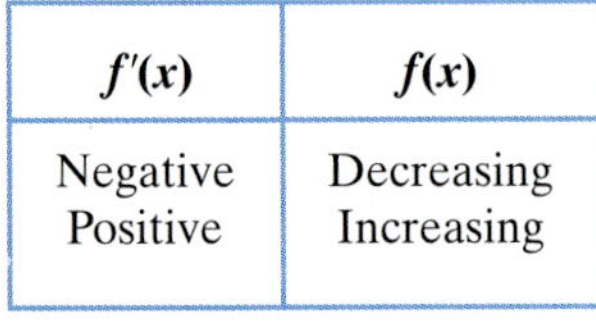

$f'(x)$	$f(x)$
Negative Positive	Decreasing Increasing

Fig. 4.3.11 Corollary 3

The meaning of Corollary 3 is summarized in Fig. 4.3.11. Figure 4.3.12 shows a graph $y = f(x)$ labeled in accord with this correspondence between the sign of the derivative $f'(x)$ and the increasing or decreasing behavior of the function $f(x)$.

EXAMPLE 6 Where is the function $f(x) = x^2 - 4x + 5$ increasing, and where is it decreasing?

Solution The derivative of f is $f'(x) = 2x - 4$. Clearly $f'(x) > 0$ if $x > 2$, whereas $f'(x) < 0$ if $x < 2$. Hence f is decreasing on $(-\infty, 2)$ and increasing on $(2, +\infty)$, as we see in Fig. 4.3.13.

EXAMPLE 7 Show that the equation $x^3 + x - 1 = 0$ has exactly one [real] solution.

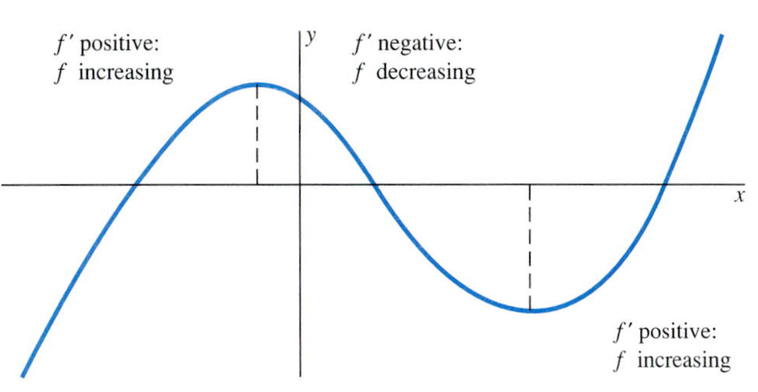

Fig. 4.3.12 The significance of the sign of $f'(x)$

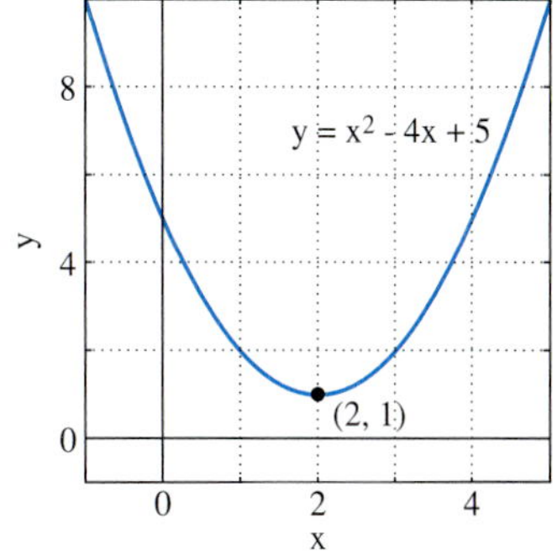

Fig. 4.3.13 The parabola of Example 6

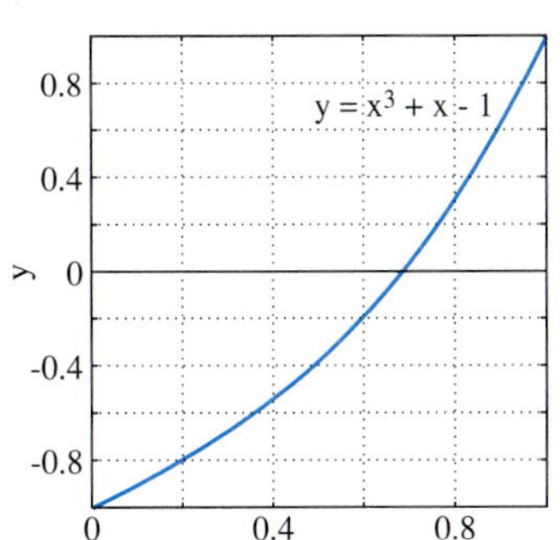

Fig. 4.3.14 The increasing function of Example 7

Solution We need a *function* in order to apply the tools of this section, so we let $f(x) = x^3 + x - 1$. Because $f(0) = -1 < 0$, $f(1) = +1 > 0$, and f is continuous (everywhere), the intermediate value property guarantees that $f(x)$ has at *least* one real zero in $[0, 1]$ (Fig. 4.3.14). But it might have more; because we cannot plot $y = f(x)$ on its entire domain $(-\infty, +\infty)$, we do not *know* that the equation $f(x) = 0$ has no additional real solutions. We need conclusive evidence that this is true.

But

$$f'(x) = 3x^2 + 1,$$

and it is evident that $f'(x) > 0$ for all x. So Corollary 3 implies that the graph of $y = f(x)$ is increasing on the whole real number line. Therefore the equation $f(x) = 0$ can have at most one solution. (Why?)

In summary, the equation $f(x) = 0$ has both at *least* one solution and at *most* one solution. Therefore, it has exactly one [real] solution. ■

EXAMPLE 8 Determine the open intervals on the x-axis on which the function

$$f(x) = 3x^4 - 4x^3 - 12x^2 + 5$$

is increasing and those on which it is decreasing.

Solution The derivative of f is

$$\begin{aligned} f'(x) &= 12x^3 - 12x^2 - 24x \\ &= 12x(x^2 - x - 2) = 12x(x + 1)(x - 2). \end{aligned} \qquad (3)$$

The critical points $x = -1, 0$, and 2 separate the x-axis into the four open intervals $(-\infty, -1)$, $(-1, 0)$, $(0, 2)$, and $(2, +\infty)$ (Fig. 4.3.14). The derivative $f'(x)$ does not change sign within any of these intervals, because

- The factor $x + 1$ in Eq. (3) changes sign only at $x = -1$,
- The factor $12x$ changes sign only at $x = 0$, and
- The factor $x - 2$ changes sign only at $x = 2$.

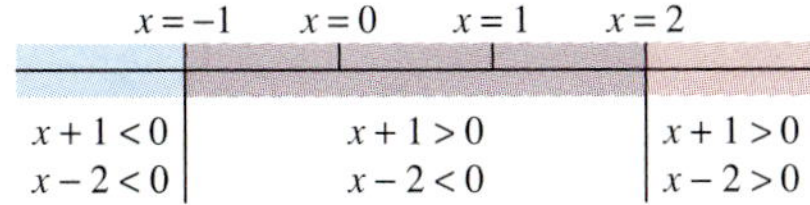

Fig. 4.3.15 The signs of $x + 1$ and $x - 2$ (Example 8)

Let's illustrate two methods of determining the sign of $f'(x)$ on each of the four intervals (Fig. 4.3.15).

METHOD 1 The second, third, and fourth columns of the next table record the signs of the factors in Eq. (3) on each of the four intervals listed in the first column. The signs of $f'(x)$ shown in the fifth column are then obtained by multiplication. The sixth column lists the resulting increasing or decreasing behavior of f on the four intervals.

Interval	$x+1$	$12x$	$x-2$	$f'(x)$	f
$(-\infty,-1)$	Neg.	Neg.	Neg.	Neg.	Decreasing
$(-1,0)$	Pos.	Neg.	Neg.	Pos.	Increasing
$(0,2)$	Pos.	Pos.	Neg.	Neg.	Decreasing
$(2,+\infty)$	Pos.	Pos.	Pos.	Pos.	Increasing

Method 2 Because the derivative $f'(x)$ does not change sign within any of the four intervals, we need only calculate its value at a single point in each interval. Whatever the sign at that point may be, it is the sign of $f'(x)$ throughout that interval.

In $(-\infty,-1)$: $f'(-2) = -96 < 0$; f is decreasing.

In $(-1,0)$: $f'(-0.5) = 7.5 > 0$; f is increasing.

In $(0,2)$: $f'(1) = -24 < 0$; f is decreasing.

In $(2,+\infty)$: $f'(3) = 144 > 0$; f is increasing.

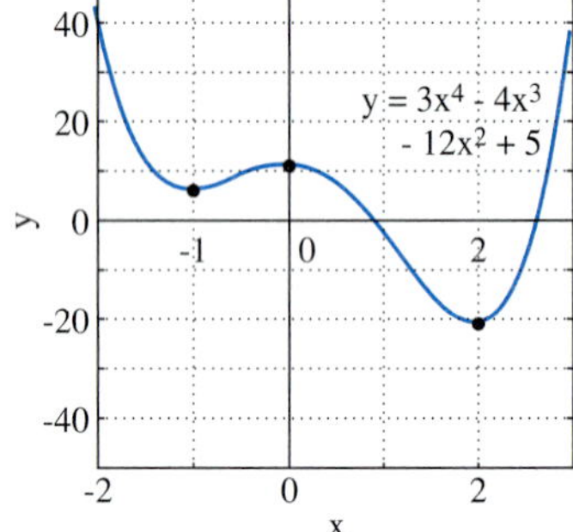

Fig. 4.3.16 The critical points of the polynomial of Example 8

The second method is especially convenient if the derivative is complicated, but an appropriate calculator for computation of its values is available.

Finally, note that the results we have obtained in each method are consistent with the graph of $y = f(x)$ shown in Fig. 4.3.16. ■

EXAMPLE 9 The graph in Fig. 4.3.17 suggests that $\sin x < x$ for $x > 0$. To show that this is so, we consider the function

$$f(x) = x - \sin x.$$

Because $|\sin x| \leqq 1$ for *all* x, $\sin x < x$ if $x > 1$, so we need to show only that $\sin x < x$ for x in the interval $(0, 1]$. But

$$f'(x) = 1 - \cos x > 0$$

for such x, because $\cos x < 1$ if $0 < x < 2\pi$, and so $\cos x < 1$ for $0 < x \leqq 1$. Hence $f(x)$ is an increasing function on $(0, 1]$. But $f(0) = 0$, so it follows that

$$f(x) = x - \sin x > 0$$

for all x in $(0, 1]$. Consequently $\sin x < x$ for x in $(0, 1]$. By our earlier remark, it now follows that $\sin x < x$ for all $x > 0$. ■

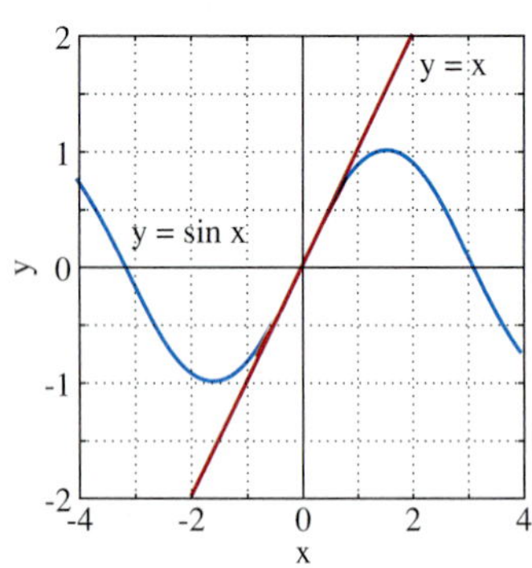

Fig. 4.3.17 x and $\sin x$ (Example 9)

4.3 PROBLEMS

For the functions in Problems 1 through 6, first determine (as in Example 8) the open intervals on the x-axis on which each function is increasing and those on which it is decreasing. Then use this information to match the function to its graph, one of the six shown in Fig. 4.3.18.

1. $f(x) = 4 - x^2$

2. $f(x) = x^2 - 2x - 1$

3. $f(x) = x^2 + 4x + 1$

4. $f(x) = \frac{1}{4}x^3 - 3x$

5. $f(x) = \frac{1}{3}x^3 - \frac{1}{2}x^2 - 2x + 1$

6. $f(x) = 2x - \frac{1}{6}x^2 - \frac{1}{9}x^3$

In Problems 7 through 10, the derivative $f'(x)$ and the value $f(0)$ are given. Use the method of Example 5 to find the function $f(x)$.

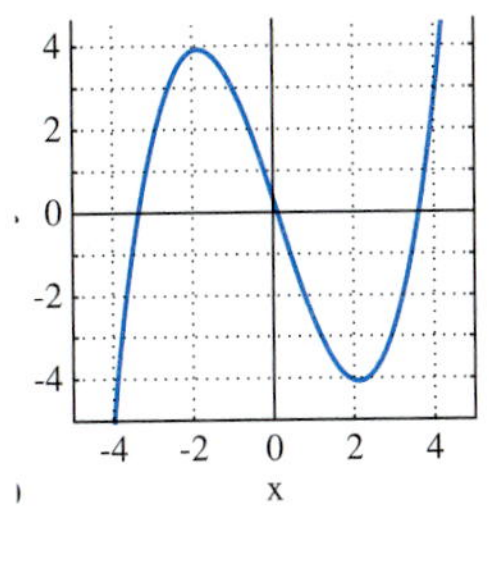
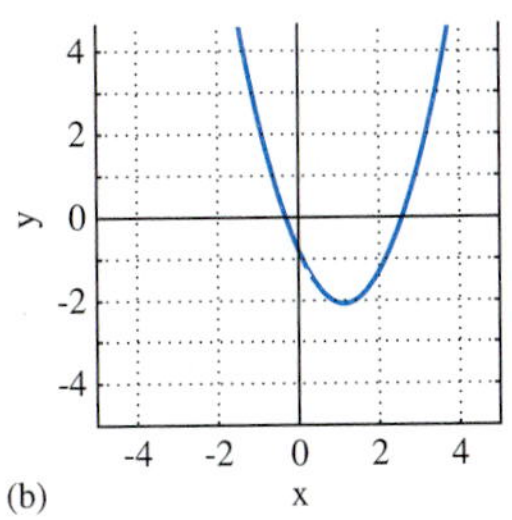
(b)
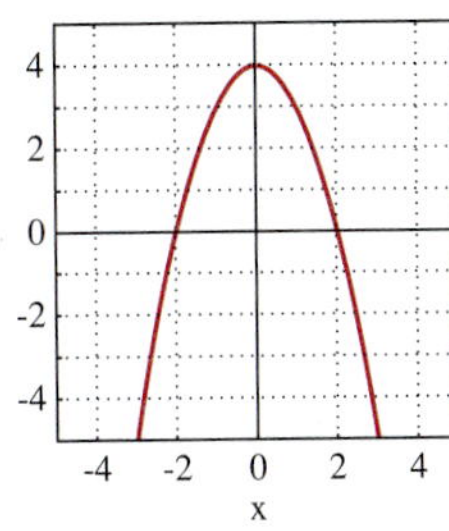
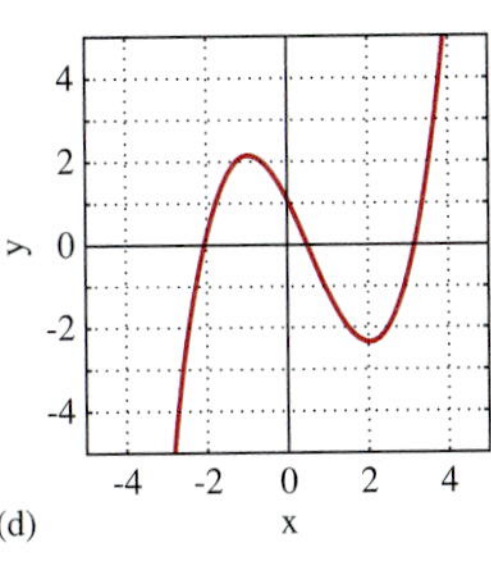
(d)
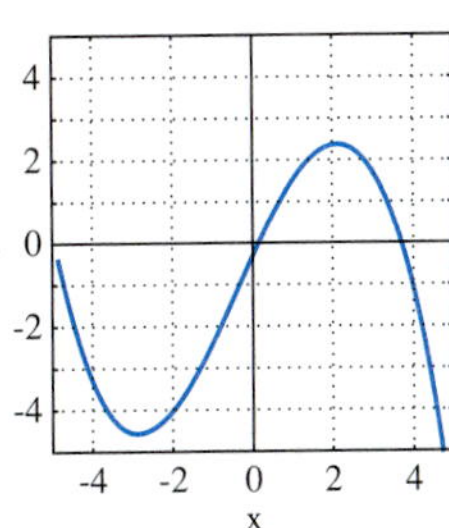
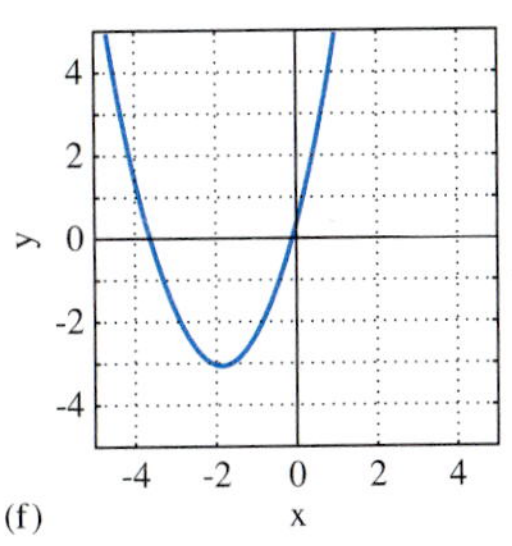
(f)

Fig. 4.3.18 Problems 1 through 6

7. $f'(x) = 4x;\ f(0) = 5$ **8.** $f'(x) = 3\sqrt{x};\ f(0) = 4$

9. $f'(x) = \dfrac{1}{x^2};\ f(1) = 1$ **10.** $f'(x) = \dfrac{2}{\sqrt{x}};\ f(0) = 3$

In Problems 11 through 24, determine (as in Example 8) the open intervals on the x-axis on which the function is increasing as well as those on which it is decreasing. If you have a graphics calculator or computer, plot the graph $y = f(x)$ to see whether it agrees with your results.

11. $f(x) = 3x + 2$ **12.** $f(x) = 4 - 5x$

13. $f(x) = 8 - 2x^2$ **14.** $f(x) = 4x^2 + 8x + 13$

15. $f(x) = 6x - 2x^2$ **16.** $f(x) = x^3 - 12x + 17$

17. $f(x) = x^4 - 2x^2 + 1$

18. $f(x) = \dfrac{x}{x+1}$ (*Note:* $f'(x)$ can't change sign at $x = -1$. Why?)

19. $f(x) = 3x^4 + 4x^3 - 12x^2$

20. $f(x) = x\sqrt{x^2 + 1}$

21. $f(x) = 8x^{1/3} - x^{4/3}$

22. $f(x) = 2x^3 + 3x^2 - 12x + 5$

23. $f(x) = \dfrac{(x-1)^2}{x^2 - 3}$ (See the note for Problem 18.)

24. $f(x) = x^2 + \dfrac{16}{x^2}$ (See the note for Problem 18.)

In Problems 25 through 28, show that the given function satisfies the hypotheses of Rolle's theorem on the indicated interval $[a, b]$, and find all numbers x in (a, b) that satisfy the conclusion of that theorem.

25. $f(x) = x^2 - 2x;\ [0, 2]$ **26.** $f(x) = 9x^2 - x^4;\ [-3, 3]$

27. $f(x) = \dfrac{1 - x^2}{1 + x^2};\ [-1, 1]$ **28.** $f(x) = 5x^{2/3} - x^{5/3};\ [0, 5]$

In Problems 29 through 31, show that the given function f does not satisfy the conclusion of Rolle's theorem on the indicated interval. Which of the hypotheses does it fail to satisfy?

29. $f(x) = 1 - |x|;\ [-1, 1]$

30. $f(x) = 1 - (2 - x)^{2/3};\ [1, 3]$

31. $f(x) = x^4 + x^2;\ [0, 1]$

In Problems 32 through 36, show that the given function f satisfies the hypotheses of the mean value theorem on the indicated interval, and find all numbers c in that interval that satisfy the conclusion of that theorem.

32. $f(x) = x^3;\ [-1, 1]$

33. $f(x) = 3x^2 + 6x - 5;\ [-2, 1]$

34. $f(x) = \sqrt{x - 1};\ [2, 5]$

35. $f(x) = (x - 1)^{2/3};\ [1, 2]$

36. $f(x) = x + \dfrac{1}{x};\ [1, 2]$

In Problems 37 through 40, show that the given function f satisfies neither the hypotheses nor the conclusion of the mean value theorem on the indicated interval.

37. $f(x) = |x - 2|;\ [1, 4]$

38. $f(x) = 1 + |x - 1|;\ [0, 3]$

39. $f(x) = [\![x]\!]$ (the greatest integer function); $[-1, 1]$

40. $f(x) = 3x^{2/3};\ [-1, 1]$

In Problems 41 through 44, show that the given equation has exactly one solution in the indicated interval.

41. $x^5 + 2x - 3 = 0;\ [0, 1]$

42. $x^{10} = 1000;\ [1, 2]$

43. $x^4 - 3x = 20;\ [2, 3]$ **44.** $\sin x = 3x - 1;\ [-1, 1]$

45. A car is driving along a rural road where the speed limit is 70 mi/h. At 3:00 P.M. its odometer (measuring distance traveled) reads 8075 mi. At 3:18 P.M. it reads 8100 mi. Prove that the driver violated the speed limit at some instant between 3:00 and 3:18 P.M.

46. Suppose that a car's speedometer reads 50 mi/h at 3:25 P.M. and 65 mi/h at 3:35 P.M. Prove that at some instant in this 10-minute time interval the car's acceleration was exactly 90 mi/h².

47. Points A and B along Interstate Highway 80 in Nebraska are 60 miles apart. Two cars pass point A at 9:00 A.M., then pass point B at 10:00 A.M. Show that at some instant between 9:00 and 10:00 A.M. the two cars have the same velocity. (*Suggestion:* Consider the difference $h(t) = f(t) - g(t)$ between the position functions of the two cars.)

48. Show that the function $f(x) = x^{2/3}$ does not satisfy the hypotheses of the mean value theorem on $[-1, 27]$ but that nevertheless there is a number c in $(-1, 27)$ such that

$$f'(c) = \frac{f(27) - f(-1)}{27 - (-1)}.$$

49. Prove that the function

$$f(x) = (1 + x)^{3/2} - \tfrac{3}{2}x - 1$$

is increasing on $(0, +\infty)$. Explain carefully how you could conclude that

$$(1 + x)^{3/2} > 1 + \tfrac{3}{2}x$$

for all $x > 0$.

50. Suppose that f' is a constant function on the interval $[a, b]$. Prove that f must be a linear function (a function whose graph is a straight line).

51. Suppose that $f'(x)$ is a polynomial of degree $n - 1$ on the interval $[a, b]$. Prove that $f(x)$ must be a polynomial of degree n on $[a, b]$.

52. Suppose that there are k different points of $[a, b]$ at which the differentiable function f vanishes (is zero). Prove that f' must vanish on at least $k - 1$ points of $[a, b]$.

53. (a) Apply the mean value theorem to $f(x) = \sqrt{x}$ on $[100, 101]$ to show that

$$\sqrt{101} = 10 + \frac{1}{2\sqrt{c}}$$

for some number c in $(100, 101)$. (b) Show that if $100 < c < 101$, then $10 < \sqrt{c} < 10.5$, and use this fact to conclude from part (a) that $10.0475 < \sqrt{101} < 10.0500$.

54. Prove that the equation $x^7 + x^5 + x^3 + 1 = 0$ has exactly one real solution.

55. (a) Show that $D_x \tan^2 x = D_x \sec^2 x$ on the open interval $(-\pi/2, \pi/2)$. (b) Conclude that there exists a constant C such that $\tan^2 x = \sec^2 x + C$ for all x in $(-\pi/2, \pi/2)$. Then evaluate C.

56. Explain why the mean value theorem does not apply to the function $f(x) = |x|$ on the interval $[-1, 2]$.

57. Suppose that the function f is differentiable on the interval $[-1, 2]$ and that $f(-1) = -1$ and $f(2) = 5$. Prove that there is a point on the graph of f at which the tangent line is parallel to the line with the equation $y = 2x$.

58. Let $f(x) = x^4 - x^3 + 7x^2 + 3x - 11$. Prove that the graph of f has at least one horizontal tangent line.

59. Let the function g be defined as follows:

$$g(x) = \begin{cases} \dfrac{x}{2} + x^2 \sin \dfrac{1}{x} & \text{if } x \neq 0, \\ 0 & \text{if } x = 0. \end{cases}$$

(a) Show that $g'(0) = \frac{1}{2} > 0$. (b) Sketch the graph of g near $x = 0$. Is g increasing on any open interval containing $x = 0$? (*Answer:* No.)

60. Suppose that f is increasing on every closed interval $[a, b]$ provided that $2 \leqq a < b$. Prove that f is increasing on the unbounded open interval $(2, +\infty)$. Note that the principle you discover was used implicitly in Example 6 of this section.

APPROXIMATIONS

Problems 61 through 63 illustrate the use of the mean value theorem to approximate numerical values of functions.

61. Use the method of Example 9 with $f(x) = \cos x$ and $g(x) = 1 - \frac{1}{2}x^2$ to show that

$$\cos x > 1 - \tfrac{1}{2}x^2$$

for all $x > 0$ (Fig. 4.3.19).

62. (a) Use the method of Example 9 and the result of Problem 61 to show that

$$\sin x > x - \tfrac{1}{6}x^3$$

for all $x > 0$ (Fig. 4.3.20). (b) Use the results of Example 9 and part (a) to calculate the sine of a 5° angle accurate to three decimal places.

63. (a) Use the result of Problem 62(a) to show that

$$\cos x < 1 - \tfrac{1}{2}x^2 + \tfrac{1}{24}x^4$$

for all $x > 0$. (b) Use the result of Problem 61 and part (a) to calculate the cosine of a 10° angle accurate to three decimal places.

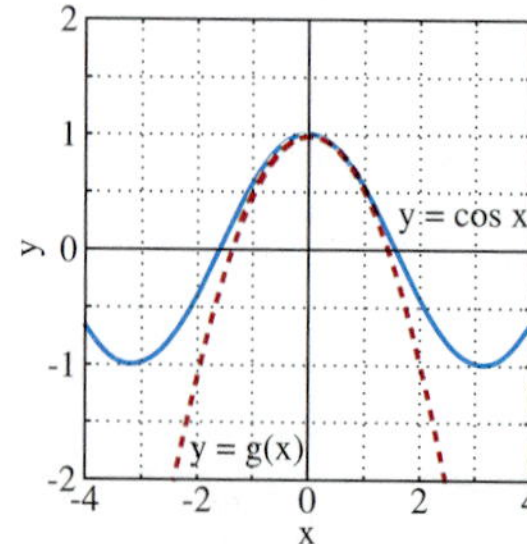

Fig. 4.3.19 $\cos x$ and $g(x) = 1 - \frac{1}{2}x^2$ (Problem 61)

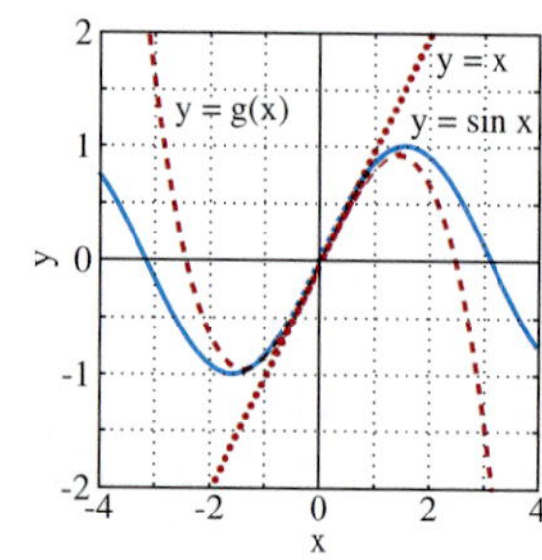

Fig. 4.3.20 x, $\sin x$, and $g(x) = x - \frac{1}{6}x^3$ (Problem 62)

4.4 THE FIRST DERIVATIVE TEST AND APPLICATIONS

In Section 3.5 we discussed maximum and minimum values of a function defined on a closed and bounded interval $[a, b]$. Now we consider extreme values of functions defined on more general domains, including open or unbounded intervals as well as closed and bounded intervals.

The distinction between *absolute* and *local* extrema is important here. Let c be a point of the domain D of the function f. Then recall from Section 3.5 that $f(c)$ is the **(absolute) maximum value** of $f(x)$ on D provided that $f(c) \geqq f(x)$ for all x in D, whereas the value $f(c)$ is a **local maximum value** of $f(x)$ if it is the maximum value of $f(x)$ on some open interval containing c. Similarly, $f(c)$ is the **(absolute) minimum value** of $f(x)$ on D provided that $f(c) \leqq f(x)$ for all x in D; $f(c)$ is a **local minimum value** of $f(x)$ if it is the minimum value of $f(x)$ on some open interval containing c. Thus a local maximum value is one that is as large as or greater than any nearby value of $f(x)$, and a local minimum value is one that is as small as or less than any nearby value. Figure 4.4.1 shows a typical example of a function that has neither an absolute maximum nor an absolute minimum value. But each of the two local extrema pictured there is an (absolute) extreme value on a sufficiently small open interval.

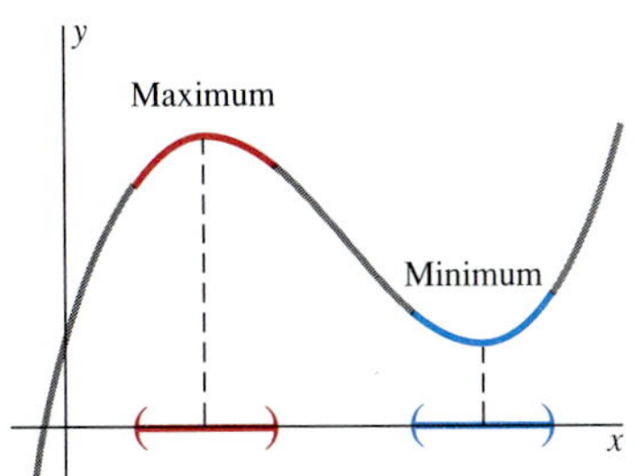

Fig. 4.4.1 Local extrema are absolute extrema on sufficiently small intervals.

REMARK Absolute extreme values are sometimes called **global** extreme values, and local extreme values are sometimes called **relative** extreme values.

Theorem 2 of Section 3.5 tells us that any extremum of the differentiable function f on an open interval I must occur at a *critical point* where the derivative vanishes:

$$f'(x) = 0.$$

But the mere fact that $f'(c) = 0$ does *not*, by itself, imply that the critical value $f(c)$ is an extreme value of f. Figures 4.4.2 through 4.4.5 illustrate different possibilities for the nature of $f(c)$: as a local or global maximum or minimum value, or neither.

A Test for Local Extrema

What we need is a way to test whether, at the critical point $x = c$, the value $f(c)$ is actually a maximum or a minimum value of $f(x)$, either local or global. Figure 4.4.6 shows how such a test might be developed. Suppose that the function f is continuous

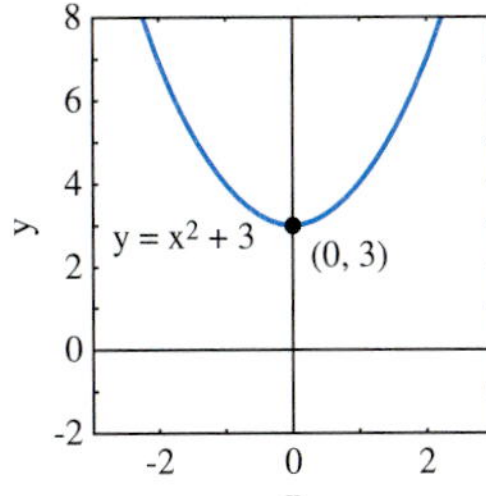

Fig. 4.4.2 The graph of $f(x) = x^2 + 3$. The local minimum value $f(0) = 3$ is also the global minimum value of $f(x)$.

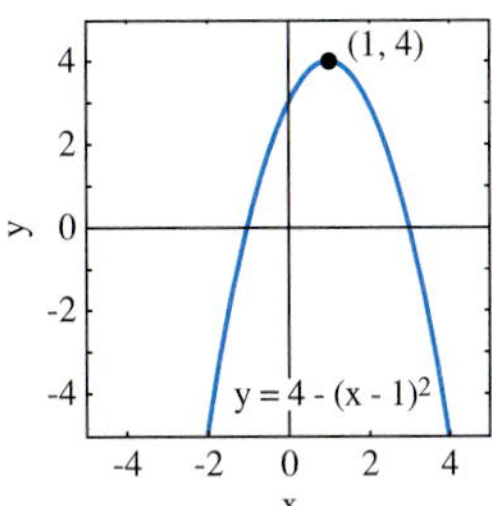

Fig. 4.4.3 The graph of $f(x) = 4 - (x - 1)^2$. The local maximum value $f(1) = 4$ is also the global maximum value of $f(x)$.

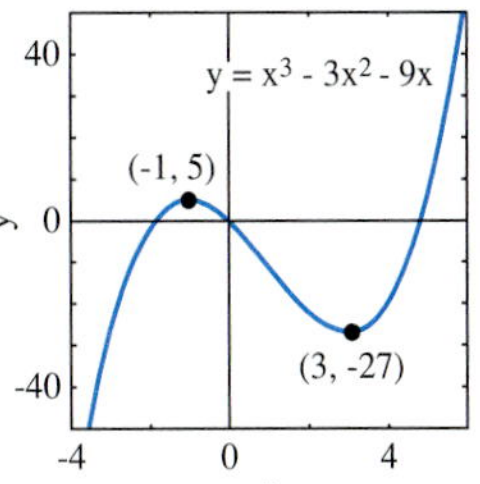

Fig. 4.4.4 The graph of $f(x) = x^3 - 3x^2 - 9x$. The local minimum value $f(3) = -27$ clearly is not the global minimum value. Similarly, the local maximum value $f(-1) = 5$ is not the global maximum value.

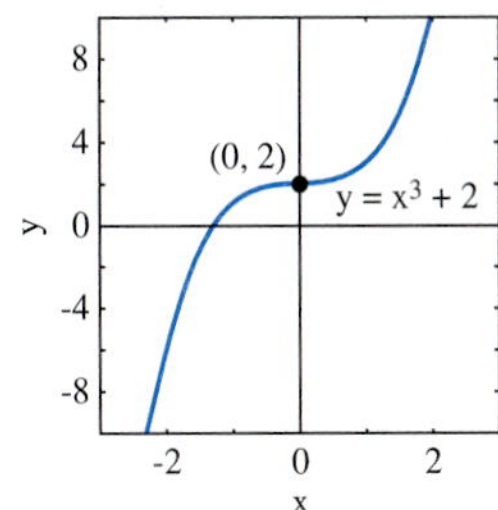

Fig. 4.4.5 The graph of $f(x) = x^3 + 2$. The critical value $f(0) = 2$ is neither a global nor a local extreme value of $f(x)$.

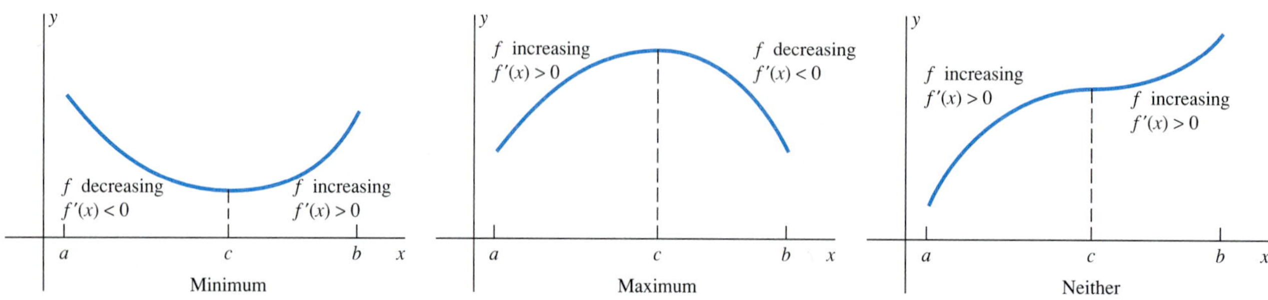

Fig. 4.4.6 The first derivative test

at c and that c is an **interior point** of the domain of f—that is, f is defined on some open interval that contains c. If f is decreasing immediately to the left of c and increasing immediately to the right, then $f(c)$ should be a local minimum value of $f(x)$. But if f is increasing immediately to the left of c and decreasing immediately to its right, then $f(c)$ should be a local maximum. If f is increasing on both sides or decreasing on both sides, then $f(c)$ should be neither a maximum value nor a minimum value of $f(x)$.

Moreover, we know from Corollary 3 in Section 4.3 that the *sign* of the derivative $f'(x)$ determines where $f(x)$ is decreasing and where it is increasing:

- f is decreasing where $f'(x) < 0$;
- f is increasing where $f'(x) > 0$.

In the following test for local extrema, we say that

- $f'(x) < 0$ *to the left of* c if $f'(x) < 0$ on some interval (a, c) of numbers immediately to the left of c, and that
- $f'(x) > 0$ *to the right of* c if $f'(x) > 0$ on some interval (c, b) of numbers immediately to the right of c,

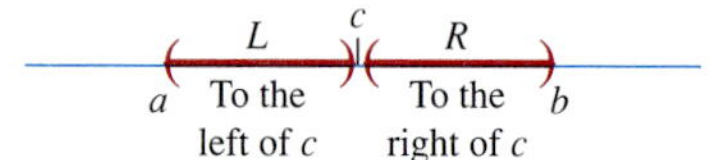

Fig. 4.4.7 Open intervals to the left and right of the point c

and so forth (see Fig. 4.4.7). Theorem 1 tells us how to use the *signs* of $f'(x)$ to the left and right of the point c to determine whether $f(x)$ has a local maximum or a local minimum value at $x = c$.

Theorem 1 *The First Derivative Test for Local Extrema*

Suppose that the function f is continuous on the interval I and also is differentiable there except possibly at the interior point c of I.

1. If $f'(x) < 0$ to the left of c and $f'(x) > 0$ to the right of c, then $f(c)$ is a *local minimum value* of $f(x)$ on I.
2. If $f'(x) > 0$ to the left of c and $f'(x) < 0$ to the right of c, then $f(c)$ is a *local maximum value* of $f(x)$ on I.
3. If $f'(x) > 0$ both to the left of c and to the right of c, *or* if $f'(x) < 0$ both to the left of c and to the right of c, then $f(c)$ is *neither* a maximum nor a minimum value of $f(x)$.

Comment Thus $f(c)$ is a local extremum if the first derivative $f'(x)$ *changes sign* as x increases through c, and the direction of this sign change determines whether $f(c)$ is a local maximum or a local minimum. A good way to remember the first derivative test for local extrema is simply to visualize Fig. 4.4.6.

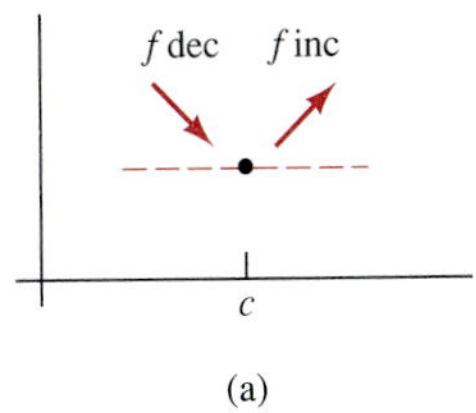

(a)

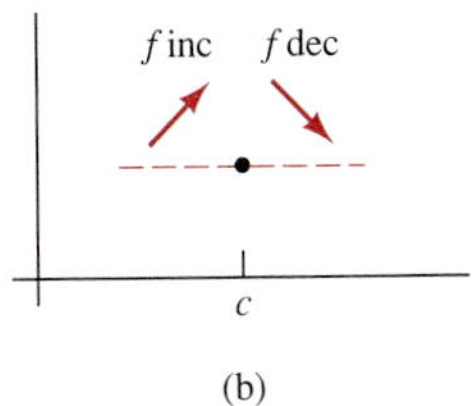

(b)

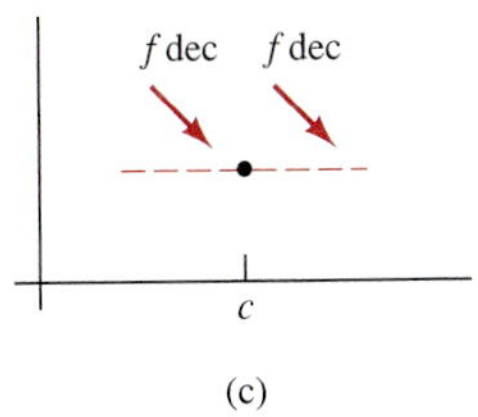

(c)

Fig. 4.4.8 The three cases in the first derivative test

Proof of Theorem 1 We will prove only part 1; the other two parts have similar proofs. Suppose that the hypotheses of Theorem 1 hold: that f is continuous on the interval I, that c is an interior point of I, and that f is differentiable on I except possibly at $x = c$. Then there exist two intervals (a, c) and (c, b), each wholly contained in I, such that $f'(x) < 0$ on (a, c) and $f'(x) > 0$ on (c, b).

Suppose that x is in (a, b). Then there are three cases to consider. First, if $x < c$, then x is in (a, c) and f is decreasing on $(a, c]$, so $f(x) > f(c)$. Second, if $x > c$, then x is in (c, b) and f is increasing on $[c, b)$, so again $f(x) > f(c)$. Finally, if $x = c$, then $f(x) = f(c)$. Thus, for each x in (a, b), $f(x) \geqq f(c)$. Therefore, by definition, $f(c)$ is a local minimum value of $f(x)$. ■

The idea of this proof is illustrated in Fig. 4.4.8. Part (a) shows f decreasing to the left of c and increasing to the right, so there must be a local minimum at $x = c$. Part (b) shows f increasing to the left of c and decreasing to the right, so $f(c)$ is a local maximum value of $f(x)$. In part (c), the derivative has the same sign on each side of c, and so there can be no extremum of any sort at $x = c$.

Remark Figures 4.4.9 through 4.4.13 illustrate various applications of Theorem 1, in which the interval I is the entire real number line $\boldsymbol{R}$. In Fig. 4.4.9 through 4.4.11, the origin $c = 0$ is a critical point because $f'(0) = 0$. In Figs. 4.4.12 and 4.4.13, $c = 0$ is a critical point because $f'(0)$ does not exist.

Classification of Critical Points

Suppose that we have found the critical points of a function. Then we can attempt to classify them—as local maxima, local minima, or neither—by applying the first derivative test at each point in turn. Example 1 illustrates a procedure for this purpose.

EXAMPLE 1 Find and classify the critical points of the function

$$f(x) = 2x^3 - 3x^2 - 36x + 7.$$

Solution The derivative is

$$f'(x) = 6x^2 - 6x - 36 = 6(x + 2)(x - 3), \tag{1}$$

so the critical points [where $f'(x) = 0$] are $x = -2$ and $x = 3$. These two points separate the x-axis into the three open intervals $(-\infty, -2)$, $(-2, 3)$, and $(3, +\infty)$. The derivative $f'(x)$ cannot change sign within any of these intervals. One reason is that the

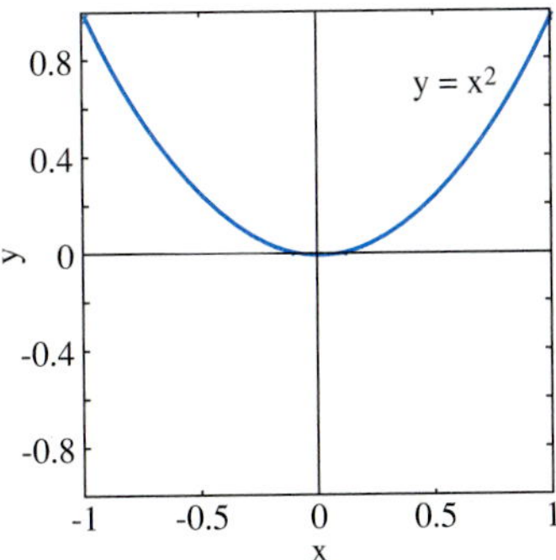

Fig. 4.4.9 $f(x) = x^2$, $f'(x) = 2x$: a local minimum at $x = 0$

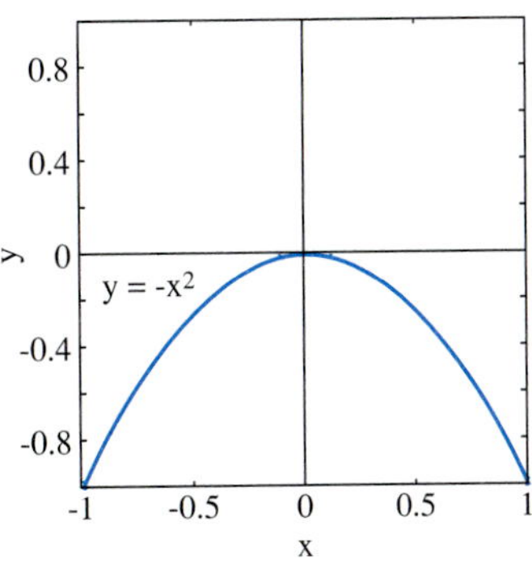

Fig. 4.4.10 $f(x) = -x^2$, $f'(x) = -2x$: a local maximum at $x = 0$

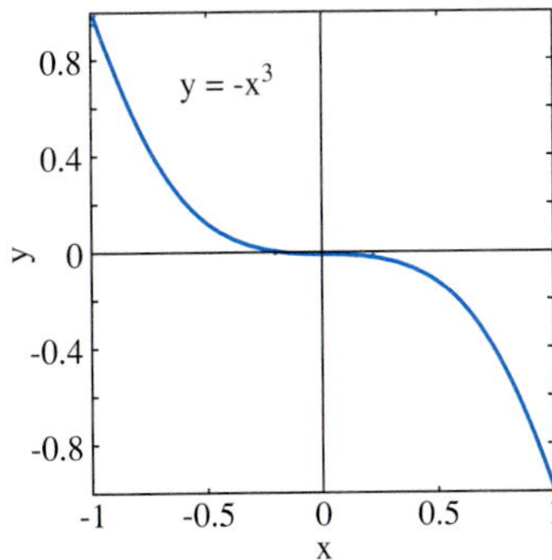

Fig. 4.4.11 $f(x) = -x^3$, $f'(x) = -3x^2$: no extremum at $x = 0$

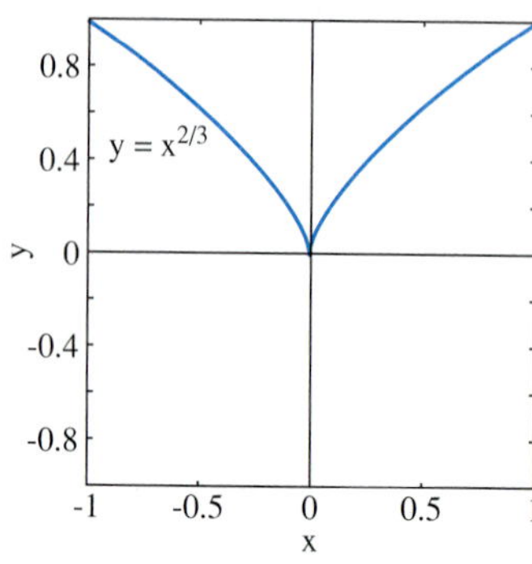

Fig. 4.4.12 $f(x) = x^{2/3}$, $f'(x) = \frac{2}{3}x^{-1/3}$: a local minimum at $x = 0$

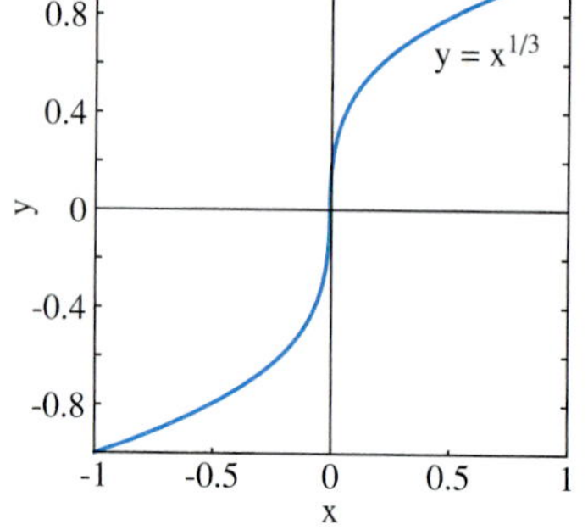

Fig. 4.4.13 $f(x) = x^{1/3}$, $f'(x) = \frac{1}{3}x^{-2/3}$: no extremum at $x = 0$

factor $x + 2$ in Eq. (1) changes sign only at -2, whereas the factor $x - 3$ changes sign only at 3 (Fig. 4.4.14). As in Example 7 of Section 4.3, we illustrate here two methods of determining the signs of $f'(x)$ on the intervals $(-\infty, -2)$, $(-2, 3)$, and $(3, +\infty)$.

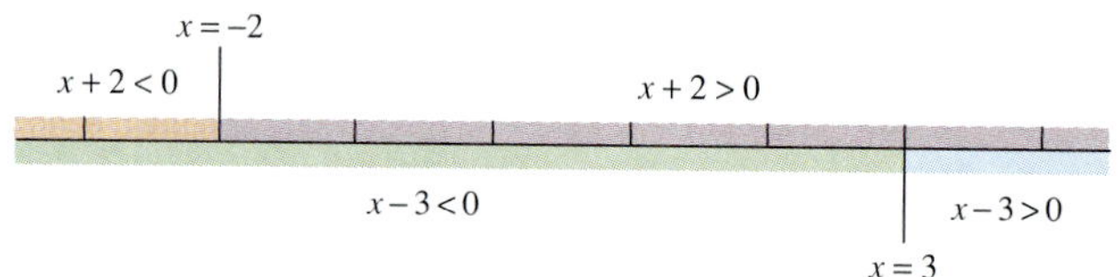

Fig. 4.4.14 The signs of $x + 2$ and $x - 3$ (Example 1)

Method 1 The second and third columns of the following table record (from Fig. 4.4.14) the signs of the factors $x + 2$ and $x - 3$ in Eq. (1) on the three intervals listed in the first column. The signs of $f'(x)$ in the fourth column are then obtained by multiplication.

Interval	$x + 2$	$x - 3$	$f'(x)$
$(-\infty, -2)$	Neg.	Neg.	Pos.
$(-2, 3)$	Pos.	Neg.	Neg.
$(3, +\infty)$	Pos.	Pos.	Pos.

Method 2 Because the derivative $f'(x)$ does not change sign within any of the three intervals, we need to calculate its value only at a single point in each interval:

$$\text{In } (-\infty, -2): \quad f'(-3) = 36 > 0; \quad f' \text{ is positive;}$$

$$\text{In } (-2, 3): \quad f'(0) = -36 < 0; \quad f' \text{ is negative;}$$

$$\text{In } (3, +\infty): \quad f'(4) = 36 > 0; \quad f' \text{ is positive.}$$

Figure 4.4.15 summarizes our information about the signs of $f'(x)$. Because $f'(x)$ is positive to the left and negative to the right of the critical point $x = -2$, the first derivative test implies that $f(-2) = 51$ is a local maximum value. Because $f'(x)$ is negative to the left and positive to the right of $x = 3$, it follows that $f(3) = -74$ is a local minimum value. The graph of $y = f(x)$ in Fig. 4.4.16 confirms this classification of the critical points $x = -2$ and $x = 3$. ■

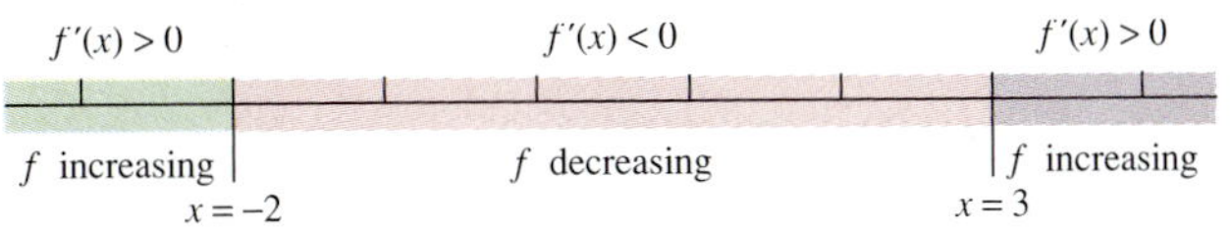

Fig. 4.4.15 The three intervals of Example 1

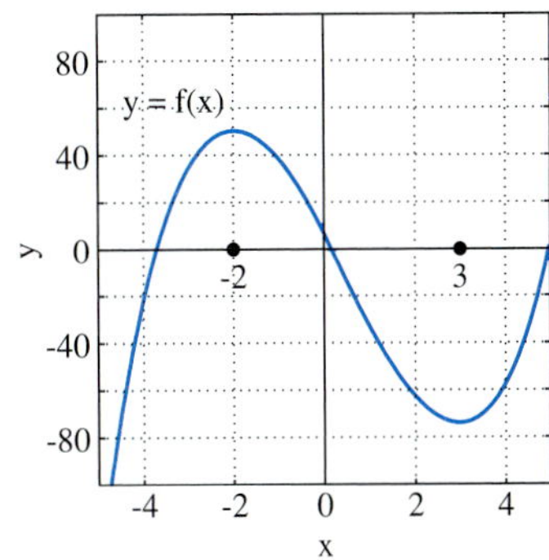

Fig. 4.4.16 $y = f(x)$ (Example 1)

Open-Interval Maximum-Minimum Problems

In Section 3.6 we discussed applied maximum-minimum problems in which the values of the dependent variable are given by a function defined on a closed and bounded interval. Sometimes, though, the function f describing the variable to be maximized (or minimized) is defined on an *open* interval (a, b), possibly an *unbounded* open interval such as $(1,+\infty)$ or $(-\infty,+\infty)$, and we cannot "close" the interval by adjoining endpoints. Typically, the reason is that $|f(x)| \to +\infty$ as x approaches a or b. But if f has only a single critical point in (a, b), then the first derivative test can tell us that $f(c)$ is the desired extreme value and can even determine whether it is a maximum or a minimum value of $f(x)$.

EXAMPLE 2 Find the (absolute) minimum value of

$$f(x) = x + \frac{4}{x} \quad \text{for} \quad 0 < x < +\infty.$$

Solution The derivative is

$$f'(x) = 1 - \frac{4}{x^2} = \frac{x^2 - 4}{x^2}. \tag{2}$$

The roots of the equation

$$f'(x) = \frac{x^2 - 4}{x^2} = 0$$

are $x = -2$ and $x = 2$. But $x = -2$ is not in the open interval $(0,+\infty)$, so we have only the critical point $x = 2$ to consider.

We see immediately from Eq. (2) that

- $f'(x) < 0$ to the left of $x = 2$ (because $x^2 < 4$ there), and
- $f'(x) > 0$ to the right of $x = 2$ (because $x^2 > 4$ there).

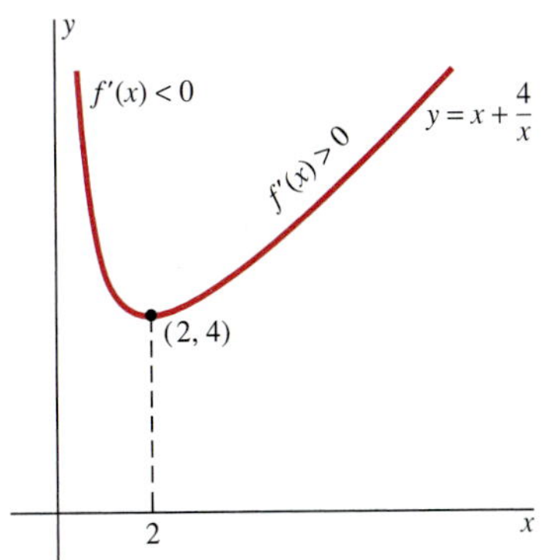

Fig. 4.4.17 The graph of the function of Example 2

Therefore, the first derivative test implies that $f(2) = 4$ is a local minimum value. We note also that $f(x) \to +\infty$ as either $x \to 0^+$ or as $x \to +\infty$. Hence the graph of f must resemble Fig. 4.4.17, and we see that $f(2) = 4$ is in fact the absolute minimum value of $f(x)$ on the entire interval $(0,+\infty)$. ■

EXAMPLE 3 We must make a cylindrical can with volume 125 in.^{3} (about 2 L) by cutting its top and bottom from squares of metal and forming its curved side by

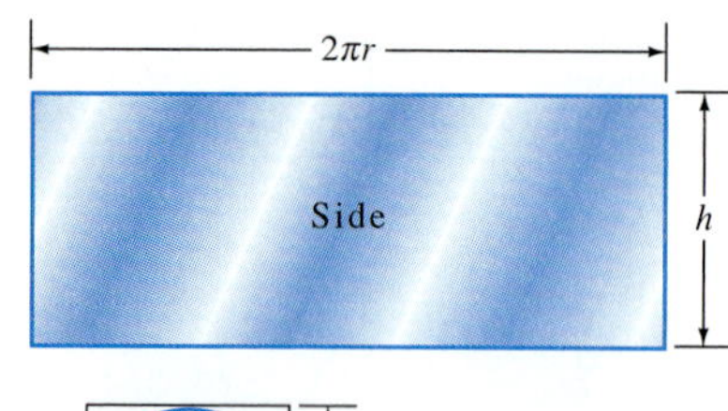

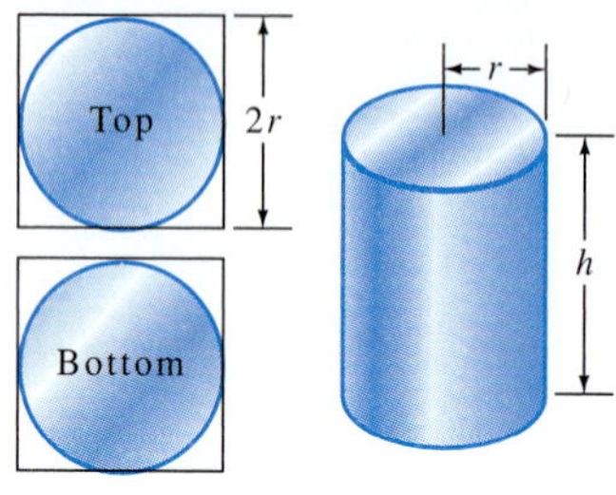

Fig. 4.4.18 The parts to make the cylindrical can of Example 3

bending a rectangular sheet of metal to match its ends. What radius r and height h of the can will minimize the total amount of material required for the rectangle and the two squares?

Solution We assume that the corners cut from the two squares, shown in Fig. 4.4.18, are wasted but that there is no other waste. As the figure shows, the area of the total amount of sheet metal required is

$$A = 8r^2 + 2\pi rh.$$

The volume of the resulting can is then

$$V = \pi r^2 h = 125,$$

so $h = 125/(\pi r^2)$. Hence A is given as a function of r by

$$A(r) = 8r^2 + 2\pi r \cdot \frac{125}{\pi r^2} = 8r^2 + \frac{250}{r}, \qquad 0 < r < +\infty.$$

The domain of A is the unbounded open interval $(0, +\infty)$ because r can have any positive value, so $A(r)$ is defined for every number r in $(0, +\infty)$. But $A(r) \to +\infty$ as $r \to 0^+$ and as $r \to +\infty$. So we cannot use the closed-interval maximum-minimum method. But we *can* use the first derivative test.

The derivative of $A(r)$ is

$$\frac{dA}{dr} = 16r - \frac{250}{r^2} = \frac{16}{r^2}\left(r^3 - \frac{125}{8}\right). \tag{3}$$

Thus the only critical point in $(0, +\infty)$ is where $r^3 = \frac{125}{8}$; that is,

$$r = \sqrt[3]{\tfrac{125}{8}} = \tfrac{5}{2} = 2.5.$$

We see immediately from Eq. (3) that

- $dA/dr < 0$ to the left of $r = \frac{5}{2}$ because $r^3 < \frac{125}{8}$ there, and
- $dA/dr > 0$ to the right, where $r^3 > \frac{125}{8}$.

Therefore, the first derivative test implies that a local minimum value of $A(r)$ on $(0, +\infty)$ is

$$A(\tfrac{5}{2}) = 8 \cdot (\tfrac{5}{2})^2 + \frac{250}{\frac{5}{2}} = 150.$$

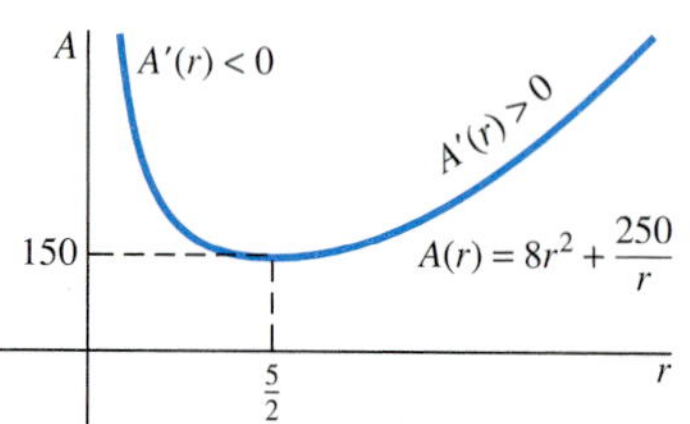

Fig. 4.4.19 Graph of the function of Example 3

Considering that $A(r) \to +\infty$ as $x \to 0^+$ and as $x \to +\infty$, we see that the graph of $A(r)$ on $(0, +\infty)$ looks like Fig. 4.4.19. This clinches the fact that $A(\frac{5}{2}) = 150$ is the *absolute* minimum value of $A(r)$. Therefore we minimize the amount of material required by making a can with radius $r = 2.5$ in. and height

$$h = \frac{125}{\pi(2.5)^2} = \frac{20}{\pi} \approx 6.37 \quad \text{(in.)}.$$

The total amount of material used is 150 in.^{2} ■

EXAMPLE 4 Find the length of the longest rod that can be carried horizontally around the corner from a hall 2 m wide into one that is 4 m wide.

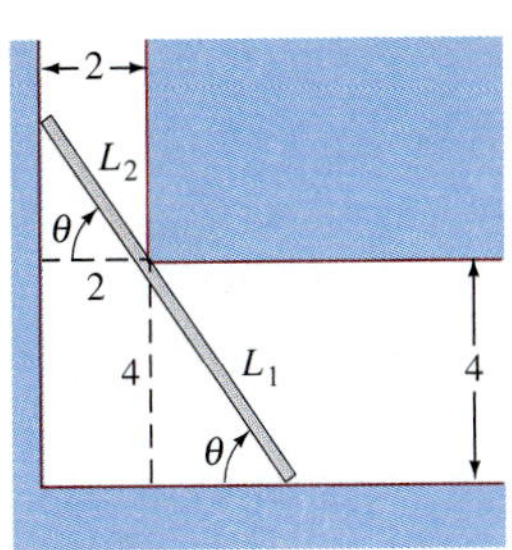

Fig. 4.4.20 Carrying a rod around a corner (Example 4)

Solution The desired length is the *minimum* length $L = L_1 + L_2$ of the rod being carried around the corner in Fig. 4.4.20. We see from the two similar triangles in the figure that

$$\frac{4}{L_1} = \sin\theta \quad \text{and} \quad \frac{2}{L_2} = \cos\theta,$$

so

$$L_1 = 4\csc\theta \quad \text{and} \quad L_2 = 2\sec\theta.$$

Therefore, the length $L = L_1 + L_2$ of the rod is given as a function of θ by

$$L(\theta) = 4\csc\theta + 2\sec\theta$$

on the open interval $(0, \pi/2)$. Note that $L(\theta) \to +\infty$ as either $\theta \to 0^+$ or as $\theta \to (\pi/2)^-$. (Why?)

The derivative of $L(\theta)$ is

$$\begin{aligned}\frac{dL}{d\theta} &= -4\csc\theta\cot\theta + 2\sec\theta\tan\theta \\ &= -\frac{4\cos\theta}{\sin^2\theta} + \frac{2\sin\theta}{\cos^2\theta} = \frac{2\sin^3\theta - 4\cos^3\theta}{\sin^2\theta\cos^2\theta} \\ &= \frac{(2\cos\theta)(\tan^3\theta - 2)}{\sin^2\theta}.\end{aligned} \tag{4}$$

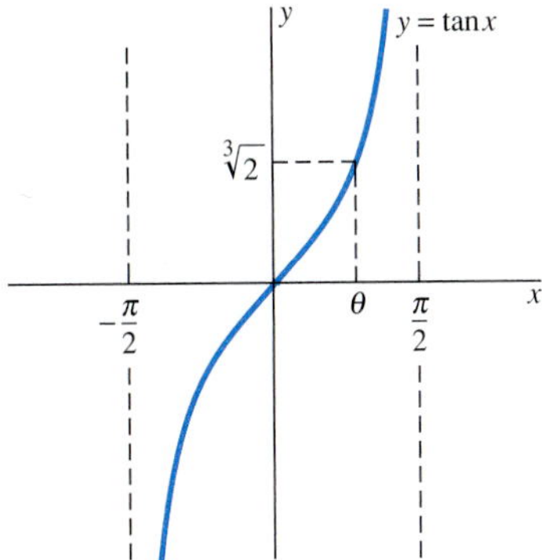

Fig. 4.4.21 $y = \tan x$ (Example 4)

Hence $dL/d\theta = 0$ exactly when

$$\tan\theta = \sqrt[3]{2}, \quad \text{so} \quad \theta \approx 0.90 \quad \text{(rad)}.$$

We now see from Eq. (4) and from the graph of the tangent function (Fig. 4.4.21) that

- $dL/d\theta < 0$ to the left of $\theta \approx 0.90$, where $\tan\theta < \sqrt[3]{2}$, so $\tan^3\theta < 2$, and
- $dL/d\theta > 0$ to the right, where $\tan^3\theta > 2$.

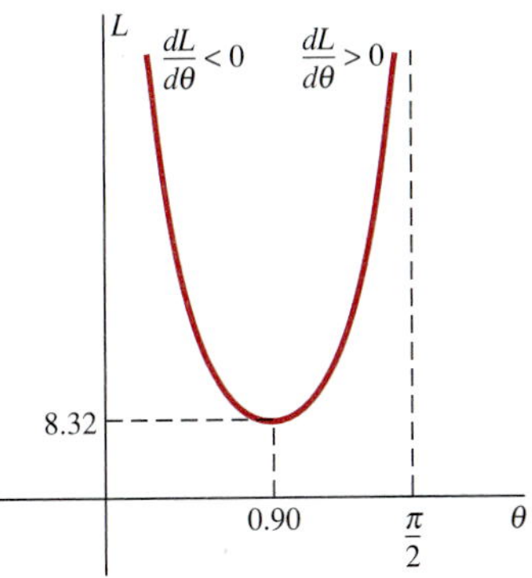

Fig. 4.4.22 The graph of $L(\theta)$ (Example 4)

Hence the graph of L resembles Fig. 4.4.22. This means that the absolute minimum value of L—and therefore the maximum length of the rod in question—is about

$$L(0.90) = 4\csc(0.90) + 2\sec(0.90),$$

approximately 8.32 m. ■

The method we used in Examples 2 and 4 to establish absolute extrema illustrates the following global version of the first derivative test.

Theorem 2 ***The First Derivative Test for Global Extrema***

Suppose that f is defined on an open interval I, either bounded or unbounded, and that f is differentiable at each point of I except possibly at the critical point c where f is continuous.

1. If $f'(x) < 0$ for all x in I with $x < c$ and $f'(x) > 0$ for all x in I with $x > c$, then $f(c)$ is the absolute minimum value of $f(x)$ on I.

2. If $f'(x) > 0$ for all x in I with $x < c$ and $f'(x) < 0$ for all x in I with $x > c$, then $f(c)$ is the absolute maximum value of $f(x)$ on I.

The proof of this theorem is essentially the same as that of Theorem 1.

REMARK When the function $f(x)$ has only one critical point c in an open interval I, Theorem 2 may apply to tell us either that $f(c)$ is the absolute minimum or that it is the absolute maximum of $f(x)$ on I. But it is good practice to verify your conclusion by sketching the graph as we did in Examples 2 through 4.

4.4 PROBLEMS

Apply the first derivative test to classify each of the critical points of the functions in Problems 1 through 16 (local or global, maximum or minimum, or not an extremum). If you have a graphics calculator or computer, plot $y = f(x)$ to see whether the appearance of the graph corresponds to your classification of the critical points.

1. $f(x) = x^2 - 4x + 5$ **2.** $f(x) = 6x - x^2$

3. $f(x) = x^3 - 3x^2 + 5$ **4.** $f(x) = x^3 - 3x + 5$

5. $f(x) = x^3 - 3x^2 + 3x + 5$

6. $f(x) = 2x^3 + 3x^2 - 36x + 17$

7. $f(x) = 10 + 60x + 9x^2 - 2x^3$

8. $f(x) = 27 - x^3$ **9.** $f(x) = x^4 - 2x^2$

10. $f(x) = 3x^5 - 5x^3$ **11.** $f(x) = x + \dfrac{1}{x}$

12. $f(x) = x + \dfrac{9}{x}$ **13.** $f(x) = x^2 + \dfrac{2}{x}$

14. $f(x) = x^2 + \dfrac{8}{x}$ **15.** $f(x) = 3 - x^{2/3}$

16. $f(x) = 4 + x^{1/3}$

In Problems 17 through 26, find and classify the critical points of the given function in the indicated open interval. You may find it useful to construct a table of signs as in Example 1.

17. $f(x) = \sin^2 x$; $(0, 3)$ **18.** $f(x) = \cos^2 x$; $(-1, 3)$

19. $f(x) = \sin^3 x$; $(-3, 3)$ **20.** $f(x) = \cos^4 x$; $(0, 4)$

21. $f(x) = \tan^2 x$; $(-1, 1)$ **22.** $f(x) = \tan^3 x$; $(-1, 1)$

23. $f(x) = \sin x - x \cos x$; $(-5, 5)$

24. $f(x) = \cos x + x \sin x$; $(-3, 3)$

25. $f(x) = 2 \tan x - \tan^2 x$; $(0, 1)$

26. $f(x) = (1 - 2 \sin x)^2$; $(0, 2)$

Problems 27 through 50 are applied maximum-minimum problems. Use the first derivative test to verify your answers.

27. Determine two real numbers with difference 20 and minimum possible product.

28. A long rectangular sheet of metal is to be made into a rain gutter by turning up two sides at right angles to the remaining center strip (Fig. 4.4.23). The rectangular cross section of the gutter is to have area 18 in.^{2} Find the minimum possible width of the sheet.

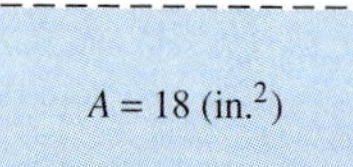

Fig. 4.4.23 The rectangular cross section of the gutter of Problem 28

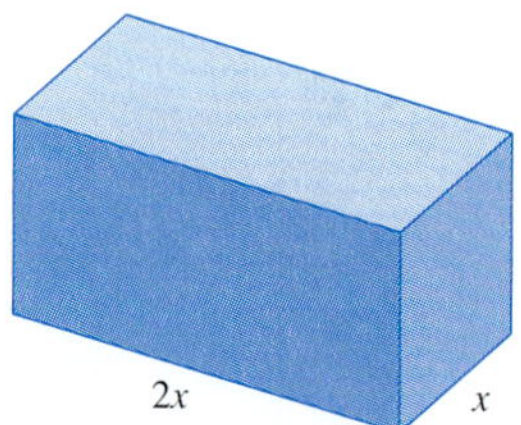

Fig. 4.4.24 The box of Problem 30

29. Find the point (x, y) on the line $2x + y = 3$ that is closest to the point $(3, 2)$.

30. You must construct a closed rectangular box with volume 576 in.^{3} and with its bottom twice as long as it is wide (Fig. 4.4.24). Find the dimensions of the box that will minimize its total surface area.

31. Repeat Problem 30, but use an open-topped rectangular box with volume 972 in.^{3}

32. An open-topped cylindrical pot is to have volume 125 in.^{3} What dimensions will minimize the total amount of material used in making this pot (Fig. 4.4.25)? Neglect the thickness of the material and possible wastage.

33. An open-topped cylindrical pot is to have volume 250 cm^3 (Fig. 4.4.25). The material for the bottom of the pot costs 4¢/cm^2; that for its curved side costs 2¢/cm^2. What dimensions will minimize the total cost of this pot?

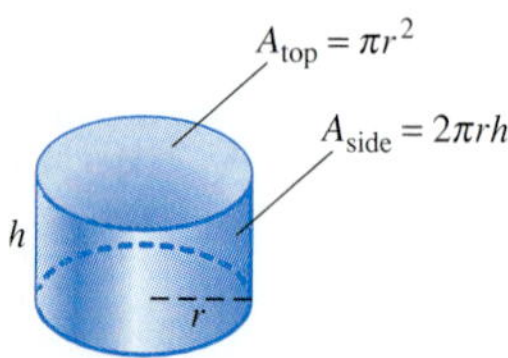

Fig. 4.4.25 The cylinder of Problems 32, 33, 38, and 39

34. Find the point (x, y) on the parabola $y = 4 - x^2$ that is closest to the point $(3, 4)$. (*Suggestions:* The cubic equation that you should obtain has a small integer as one of its roots. Minimize the *square* of the distance.)

35. Show that the rectangle with area 100 and minimum perimeter is a square.

36. Show that the rectangular solid with a square base, volume 1000, and minimum total surface area is a cube.

37. A box with a square base and an open top is to have volume 62.5 in.^{3} Neglect the thickness of the material used to make the box, and find the dimensions that will minimize the amount of material used.

38. You need a tin can in the shape of a right circular cylinder of volume 16π cm^3 (Fig. 4.4.25). What radius r and height h would minimize its total surface area (including top and bottom)?

39. The metal used to make the top and bottom of a cylindrical can (Fig. 4.4.25) costs 4¢/in.^{2}; the metal used for the sides costs 2¢/in.^{2} The volume of the can must be exactly 100 in.^{3} What dimensions of the can would minimize its total cost?

40. Each page of a book will contain 30 in.^{2} of print, and each page must have 2-in. margins at top and bottom and 1-in. margins at each side. What is the minimum possible area of such a page?

41. What point or points on the curve $y = x^2$ are nearest the point $(0, 2)$? (*Suggestion:* The square of a distance is minimized exactly when the distance itself is minimized.)

42. What is the length of the shortest line segment lying wholly in the first quadrant tangent to the graph of $y = 1/x$ and with its endpoints on the coordinate axes?

43. A rectangle has area 64 cm^2. A straight line is to be drawn from one corner of the rectangle to the midpoint of one of the two more distant sides. What is the minimum possible length of such a line?

44. An oil can is to have volume 1000 in.^{3} and is to be shaped like a cylinder with a flat bottom but capped by a hemisphere (Fig. 4.4.26). Neglect the thickness of the material of the can, and find the dimensions that will minimize the total amount of material needed to construct it.

45. Find the exact length L of the longest rod that can be carried horizontally around a corner from a corridor 2 m wide into one 4 m wide. Do this by *minimizing* the length of the rod in Fig. 4.4.27 by minimizing the square of that length as a function of x.

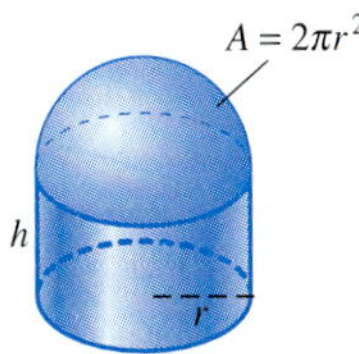

Fig. 4.4.26 The oil can of Problem 44

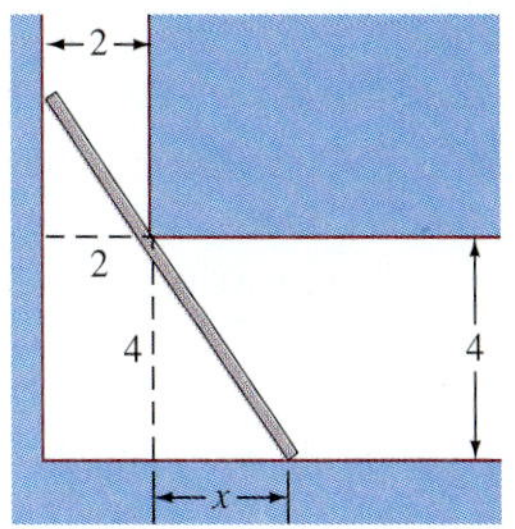

Fig. 4.4.27 Carrying a rod around a corner (Problem 45)

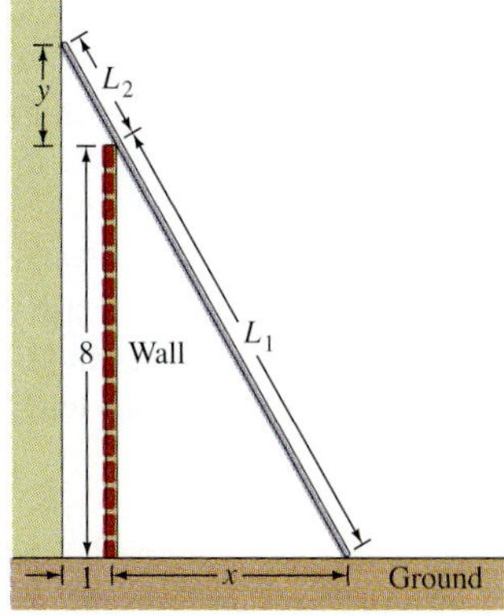

Fig. 4.4.28 The ladder of Problem 46

46. Find the length of the shortest ladder that will reach from the ground, over a wall 8 ft high, to the side of a building 1 ft behind the wall. That is, minimize the length $L = L_1 + L_2$ shown in Fig. 4.4.28.

47. A sphere with fixed radius a is inscribed in a pyramid with a square base so that the sphere touches the base of the pyramid and also each of its four sides. Show that the minimum possible volume of the pyramid is $8/\pi$ times the volume of the sphere. (*Suggestion:* Use the two right triangles in Fig. 4.4.29 to show that the volume of the pyramid is

$$V = V(y) = \frac{4a^2y^2}{3(y - 2a)}.$$

This can be done easily with the aid of the angle θ and *without* the formula for $\tan(\theta/2)$. Don't forget the domain of $V(y)$.)

48. Two noisy discothèques, one four times as noisy as the other, are located on opposite ends of a block 1000 ft long. What is the quietest point on the block between the two discos? The intensity of noise at a point away from its

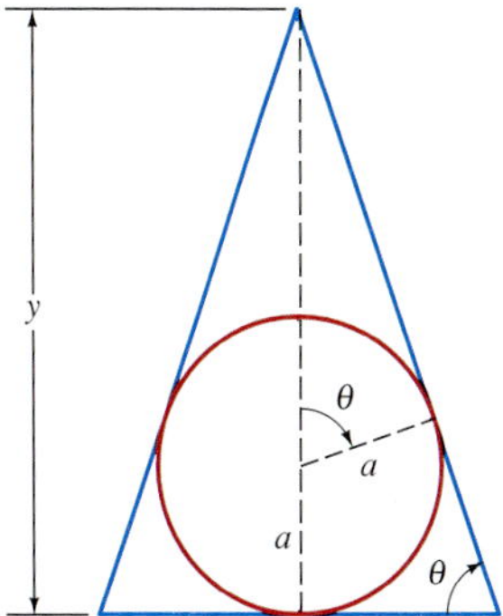

Fig. 4.4.29 Cross section through the centers of the sphere and pyramid of Problem 47

source is proportional to the noisiness and inversely proportional to the square of the distance from the source.

49. A floored tent with fixed volume V is to be shaped like a pyramid with a square base and congruent sides (Fig. 4.4.30). What height y and base edge $2x$ would minimize its total surface area (including its floor)?

50. Suppose that the distance from the building to the wall in Problem 46 is a and that the height of the wall is b. Show that the minimal length of the ladder is

$$L_{\min} = (a^{2/3} + b^{2/3})^{3/2}.$$

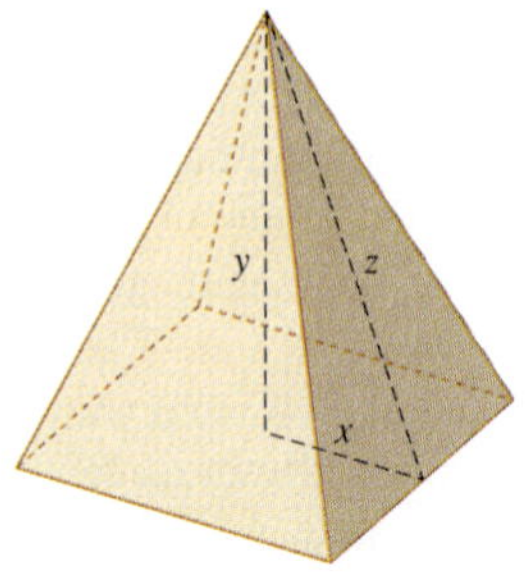

Fig. 4.4.30 The tent of Problem 49

4.4 PROJECT: CONSTRUCTING A BOX AT MINIMAL COST

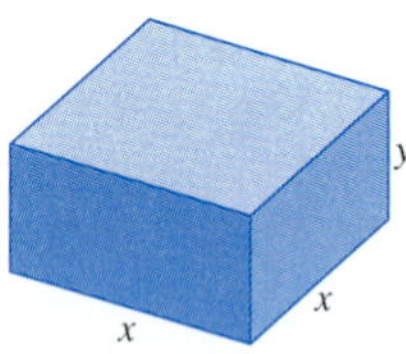

Fig. 4.4.31 A box with square base

Figure 4.4.31 shows a rectangular box with square base. Suppose that its volume $V = x^2y$ is to be 1000 in.[3] We want to make this box at minimal cost. Each of the six faces costs a cents per square inch, and gluing each of the 12 edges costs b cents per inch of edge length.

1. Show first that the total cost C is given as a function of x by

$$C(x) = 2ax^2 + 8bx + \frac{4000a}{x} + \frac{4000b}{x^2}. \quad \textbf{(1)}$$

2. Suppose that $a = b = 1$. Find the dimensions of the box of minimal cost by zooming in on the appropriate solution of $C'(x) = 0$.

3. Repeat Problem 2 with a and b being the last two distinct nonzero digits of your student I.D. number.

4. After doing Problems 2 and 3, you should smell a rat. Is it possible that the shape of the optimal box is independent of the values of a and b? Show that the equation $C'(x) = 0$ simplifies to the fourth-degree equation

$$ax^4 + 2bx^3 - 1000ax - 2000b = 0. \quad \textbf{(2)}$$

Then solve this equation (for x) by using a symbolic algebra system such as *Derive, Maple,* or *Mathematica,* if one is available. If not, you may still be able to solve Eq. (2) by hand—begin by factoring x^3 from the first two terms and 1000 from the last two.

5. Suppose that the top and bottom of the box in Fig. 4.4.31 costs p cents per square inch and that the four sides cost q cents per square inch [the twelve edges still cost b cents per (linear) inch]. For instance, let p, q, and b be the last three nonzero digits of your student I.D. number. Then determine graphically the dimensions of the box with volume 1000 in.[3] and minimal cost.

4.5 SIMPLE CURVE SKETCHING

We can construct a reasonably accurate graph of the polynomial function

$$f(x) = a_nx^n + a_{n-1}x^{n-1} + \cdots + a_2x^2 + a_1x + a_0 \quad \textbf{(1)}$$

by assembling the following information.

1. **The critical points of** f—that is, the points on the graph at which the tangent line is horizontal, so that $f'(x) = 0$.
2. **The increasing/decreasing behavior of** f—that is, the intervals on which f is increasing and those on which it is decreasing.
3. **The behavior of** f **"at" infinity**—that is, the behavior of $f(x)$ as $x \to +\infty$ and as $x \to -\infty$.

The same information is often the key to understanding the structure of a graph that has been plotted with a calculator or computer.

Behavior at Infinity

To carry out the task in item 3, we write $f(x)$ in the form

$$f(x) = x^n\left(a_n + \frac{a_{n-1}}{x} + \cdots + \frac{a_1}{x^{n-1}} + \frac{a_0}{x^n}\right).$$

Thus we conclude that the behavior of $f(x)$ as $x \to +\infty$ is much the same as that of its *leading term* $a_n x^n$, because all the terms that have powers of x in the denominator approach zero as $x \to +\infty$. In particular, if $a_n > 0$, then

$$\lim_{x\to\infty} f(x) = +\infty, \tag{2}$$

meaning that $f(x)$ increases without bound as $x \to +\infty$. Also

$$\lim_{x\to-\infty} f(x) = \begin{cases} +\infty & \text{if } n \text{ is even,} \\ -\infty & \text{if } n \text{ is odd.} \end{cases} \tag{3}$$

If $a_n < 0$, simply reverse the signs on the right-hand sides in Eqs. (2) and (3). It follows that the graph of any (nonconstant) polynomial function exhibits one of the four "behaviors as $x \to \pm\infty$" that are illustrated in Fig. 4.5.1.

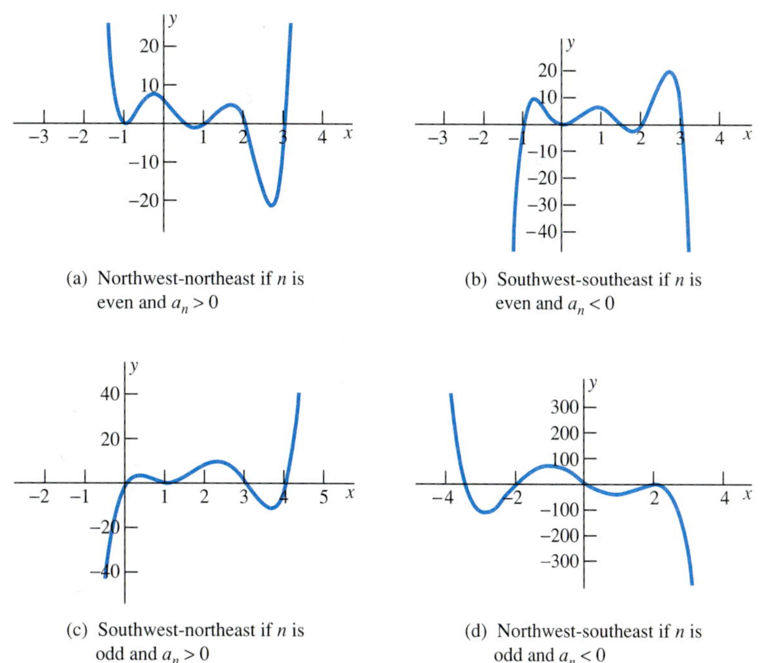

Fig. 4.5.1 The behavior of polynomial graphs as $x \to \pm\infty$

Critical Points

Every polynomial, such as $f(x)$ in Eq. (1), is differentiable everywhere. So the critical points of $f(x)$ are the roots of the polynomial equation $f'(x) = 0$—that is, solutions of

$$na_n x^{n-1} + (n-1)a_{n-1}x^{n-2} + \cdots + 2a_2 x + a_1 = 0. \tag{4}$$

Sometimes we can find all (real) solutions of such an equation by factoring, but most often in practice we must resort to numerical methods aided by calculator or computer.

Increasing and Decreasing Functions

Suppose that we have somehow found all the (real) solutions $c_1, c_2, \ldots, c_k$ of Eq. (4). Then these solutions are the critical points of f. If they are arranged in increasing order, as in Fig. 4.5.2, then they separate the x-axis into the finite number of open intervals

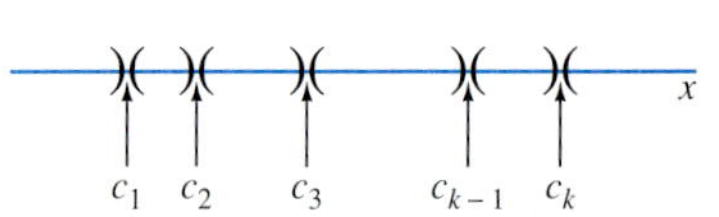

Fig. 4.5.2 The zeros of $f'(x)$ divide the x-axis into intervals on which $f'(x)$ does not change sign.

$$(-\infty, c_1), \quad (c_1, c_2), \quad (c_2, c_3), \quad \ldots, \quad (c_{k-1}, c_k), \quad (c_k, +\infty)$$

that also appear in the figure. The intermediate value property applied to $f'(x)$ tells us that $f'(x)$ can change sign only at the critical points of f, so $f'(x)$ has only one sign on each of these open intervals. It is typical for $f'(x)$ to be negative on some intervals and positive on others. Moreover, it's easy to find the sign of $f'(x)$ on any one such interval I: We simply substitute *any* convenient number in I into $f'(x)$.

Once we know the sign of $f'(x)$ on each of these intervals, we know where f is increasing and where it is decreasing. We then apply the first derivative test to find which of the critical values are local maxima, which are local minima, and which are neither—merely places where the tangent line is horizontal. With this information, the knowledge of the behavior of f as $x \to \pm\infty$, and the fact that f is continuous, we can sketch its graph. We plot the critical points $(c_i, f(c_i))$ and connect them with a smooth curve that is consistent with our other data.

It may also be helpful to plot the y-intercept $(0, f(0))$ and also any x-intercepts that are easy to find. But we recommend (until inflection points are introduced in Section 4.6) that you plot *only* these points—critical points and intercepts—and rely otherwise on the increasing and decreasing behavior of f.

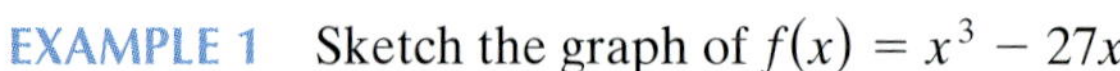

EXAMPLE 1 Sketch the graph of $f(x) = x^3 - 27x$.

Solution Because the leading term is x^3, we see that

$$\lim_{x\to+\infty} f(x) = +\infty \quad \text{and} \quad \lim_{x\to-\infty} f(x) = -\infty.$$

Moreover, because

$$f'(x) = 3x^2 - 27 = 3(x+3)(x-3), \tag{5}$$

we see that the critical points at which $f'(x) = 0$ are $x = -3$ and $x = 3$. The corresponding points on the graph of f are $(-3, 54)$ and $(3, -54)$. The critical points separate the x-axis into the three open intervals $(-\infty, -3)$, $(-3, 3)$, and $(3, +\infty)$ (Fig. 4.5.3). To determine the increasing or decreasing behavior of f on these intervals, let's substitute a number in each interval into the derivative in Eq. (5):

On $(-\infty, -3)$: $\quad f'(-4) = (3)(-1)(-7) = 21 > 0$; f is increasing;

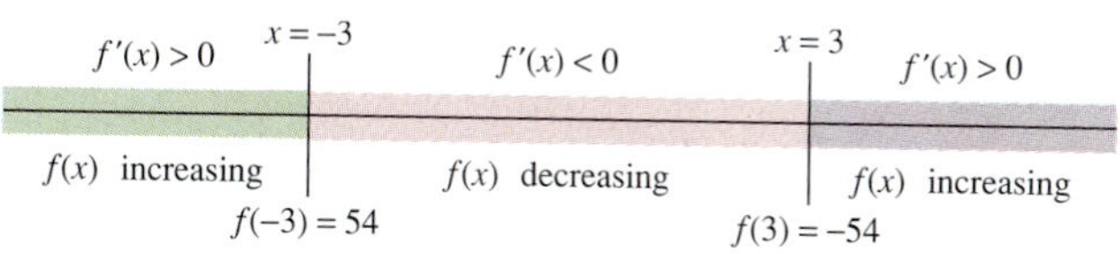

Fig. 4.5.3 The three open intervals of Example 1

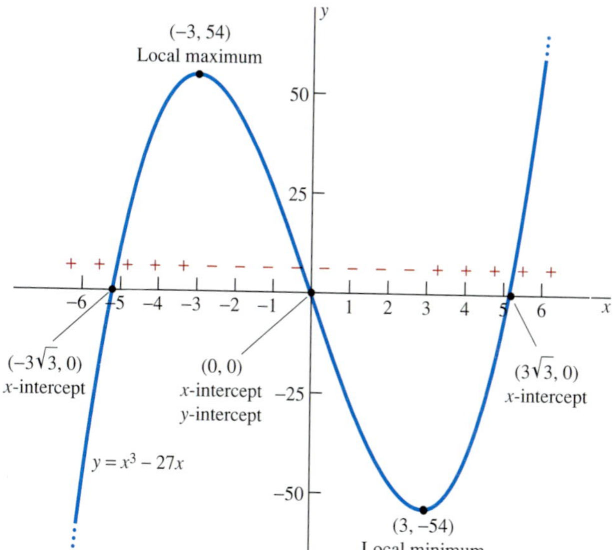

Fig. 4.5.4 Graph of the function of Example 1

On $(-3, 3)$: $\quad f'(0) = (3)(3)(-3) = -27 < 0;\ f$ is decreasing;

On $(3, +\infty)$: $\quad f'(4) = (3)(7)(1) = 21 > 0;\quad f$ is increasing.

We plot the critical points and the intercepts $(0, 0)$, $(3\sqrt{3}, 0)$, and $(-3\sqrt{3}, 0)$. Then we use the information about where f is increasing or decreasing to connect these points with a smooth curve. Remembering that there are horizontal tangents at the two critical points, we obtain the graph shown in Fig. 4.5.4.

In the figure we use plus and minus signs to mark the sign of $f'(x)$ in each interval. This makes it clear that $(-3, 54)$ is a local maximum and that $(3, -54)$ is a local minimum. The limits we found at the outset show that neither is global. ■

EXAMPLE 2 Sketch the graph of $f(x) = 8x^5 - 5x^4 - 20x^3$.

Solution Because

$$f'(x) = 40x^4 - 20x^3 - 60x^2 = 20x^2(x + 1)(2x - 3), \tag{6}$$

the critical points at which $f'(x) = 0$ are $x = -1$, $x = 0$, and $x = \frac{3}{2}$. These three critical points separate the x-axis into the four open intervals shown in Fig. 4.5.5. This time, let's determine the increasing or decreasing behavior of f by recording the

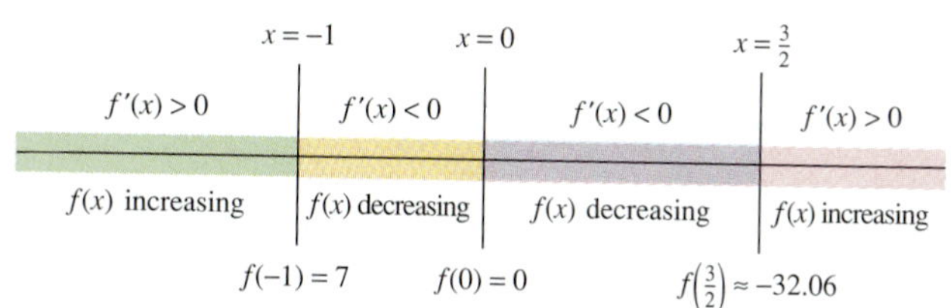

Fig. 4.5.5 The four open intervals of Example 2

signs of the factors in Eq. (6) on each of the subintervals shown in Fig. 4.5.5. In this way we get the following table:

Interval	$x + 1$	$20x^2$	$2x - 3$	$f'(x)$	f
$(-\infty, -1)$	Neg.	Pos.	Neg.	Pos.	Increasing
$(-1, 0)$	Pos.	Pos.	Neg.	Neg.	Decreasing
$\left(0, \frac{3}{2}\right)$	Pos.	Pos.	Neg.	Neg.	Decreasing
$\left(\frac{3}{2}, +\infty\right)$	Pos.	Pos.	Pos.	Pos.	Increasing

The points on the graph that correspond to the critical points are $(-1, 7)$, $(0, 0)$, and $(1.5, -32.0625)$.

We write $f(x)$ in the form

$$f(x) = x^3(8x^2 - 5x - 20)$$

in order to use the quadratic formula to find the intercepts. They turn out to be $(-1.30, 0)$, $(1.92, 0)$ (the abscissas are given only approximately), and the origin $(0, 0)$. The latter is also the y-intercept. We apply the first derivative test with the increasing or decreasing behavior shown in the table. It follows that $(-1, 7)$ is a local maximum, $(1.5, -32.0625)$ is a local minimum, and $(0, 0)$ is neither. The graph resembles the one shown in Fig. 4.5.6. ■

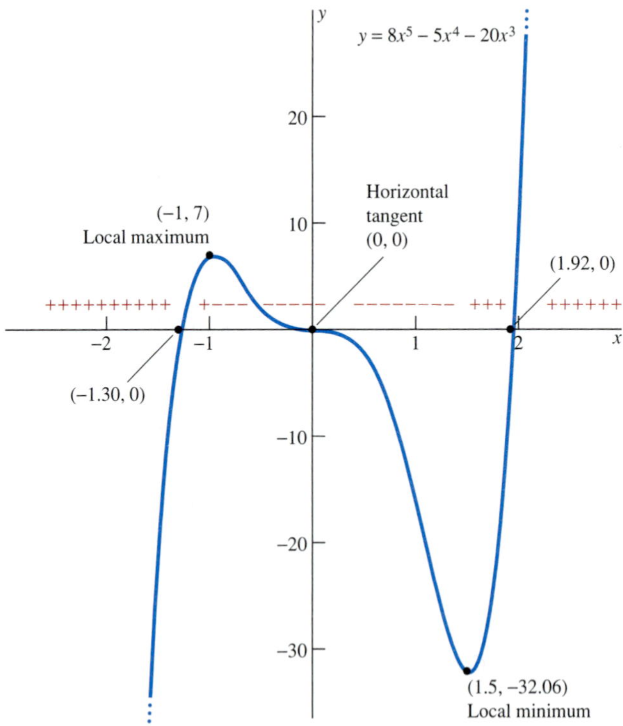

Fig. 4.5.6 Graph of the function of Example 2

In Example 3, the function is not a polynomial. Nevertheless, the methods of this section suffice for sketching its graph.

EXAMPLE 3 Sketch the graph of

$$f(x) = x^{2/3}(x^2 - 2x - 6) = x^{8/3} - 2x^{5/3} - 6x^{2/3}.$$

Solution The derivative of f is

$$f'(x) = \frac{8}{3}x^{5/3} - \frac{10}{3}x^{2/3} - \frac{12}{3}x^{-1/3}$$
$$= \frac{2}{3}x^{-1/3}(4x^2 - 5x - 6) = \frac{2(4x+3)(x-2)}{3x^{1/3}}. \qquad \textbf{(7)}$$

The tangent line is horizontal at the two critical points $x = -\frac{3}{4}$ and $x = 2$, where the numerator in the last fraction of Eq. (7) is zero (and the denominator is not). Moreover, because of the presence of the factor $x^{1/3}$ in the denominator, $|f'(x)| \to +\infty$ as $x \to 0$. Thus $x = 0$ (a critical point because f is not differentiable there) is a point at which the tangent line is vertical. These three critical points separate the x-axis into the four open intervals shown in Fig. 4.5.7. We determine the increasing or decreasing behavior of f by substituting a number from each interval in $f'(x)$ (Eq. (7)).

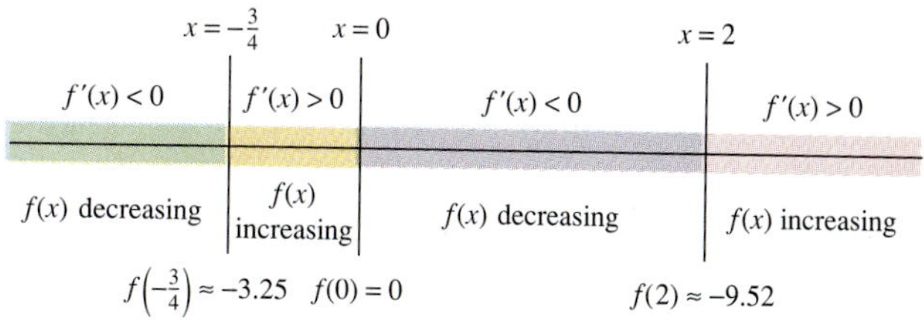

Fig. 4.5.7 The four open intervals of Example 3

On $(-\infty, -\frac{3}{4})$: $f'(-1) = \dfrac{2\cdot(-1)(-3)}{3\cdot(-1)} < 0$; f is decreasing;

On $(-\frac{3}{4}, 0)$: $f'(-\frac{1}{2}) = \dfrac{2\cdot(+1)(-\frac{5}{2})}{3\cdot(-\frac{1}{2})^{1/3}} > 0$; f is increasing;

On $(0, 2)$: $f'(1) = \dfrac{2\cdot(+7)(-1)}{3\cdot(+1)} < 0$; f is decreasing;

On $(2, +\infty)$: $f'(3) = \dfrac{2\cdot(+15)(+1)}{3\cdot(+3)^{1/3}} > 0$; f is increasing.

The three critical points $x = -\frac{3}{4}$, $x = 0$, and $x = 2$ give the points $(-0.75, -3.25)$, $(0, 0)$, and $(2, -9.52)$ on the graph (using approximations where appropriate).

The first derivative test now shows local minima at $(-0.75, -3.25)$ and at $(2, -9.52)$; there is a local maximum at $(0, 0)$. Although $f'(0)$ does not exist, f is continuous at $x = 0$, so it is continuous everywhere.

We use the quadratic formula to find the x-intercepts. In addition to the origin, they occur where $x^2 - 2x - 6 = 0$, and thus they are located at $(1 - \sqrt{7}, 0)$ and at $(1 + \sqrt{7}, 0)$. We then plot the approximations $(-1.65, 0)$ and $(3.65, 0)$. Finally, we note that $f(x) \to +\infty$ as $x \to \pm\infty$. So the graph has the shape shown in Fig. 4.5.8. ■

Curve Sketching and Solution of Equations

An important application of curve-sketching techniques is the solution of an equation of the form

$$f(x) = 0. \qquad \textbf{(8)}$$

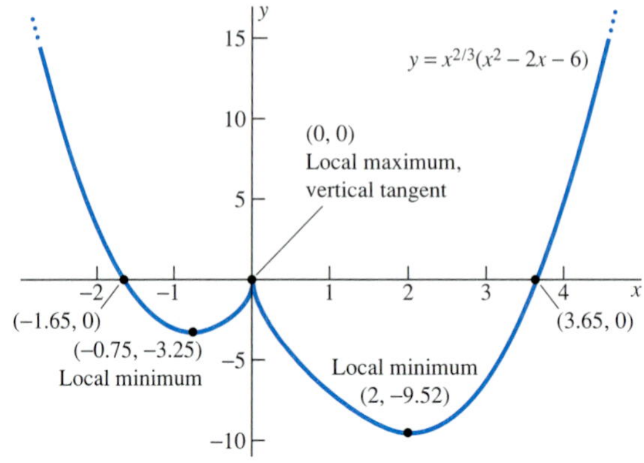

Fig. 4.5.8 The technique is effective for nonpolynomial functions, as in Example 3

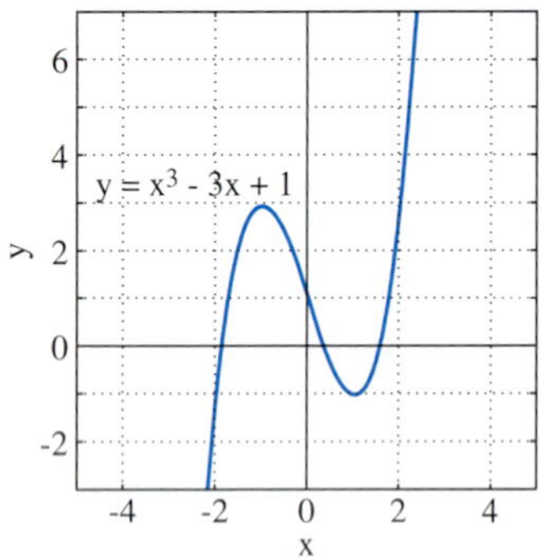

Fig. 4.5.9 $y = x^3 - 3x + 1$

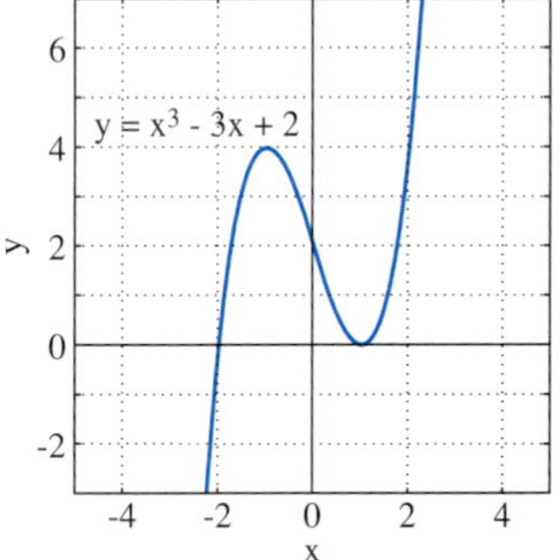

Fig. 4.5.10 $y = x^3 - 3x + 2$

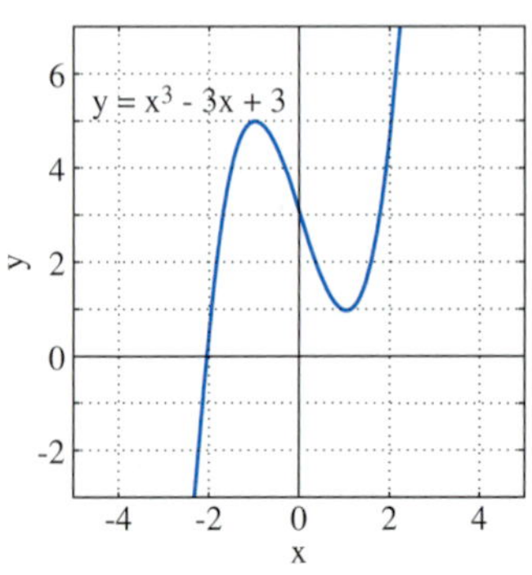

Fig. 4.5.11 $y = x^3 - 3x + 3$

The real (as opposed to complex) solutions of this equation are simply the x-intercepts of the graph of $y = f(x)$. Hence by sketching this graph with reasonable accuracy—either "by hand" or with a calculator or computer—we can glean information about the number of real solutions of Eq. (8) as well as their approximate locations.

For example, Figs. 4.5.9 through 4.5.11 show the graphs of the cubic polynomials on the left-hand sides of the equations

$$x^3 - 3x + 1 = 0, \tag{9}$$

$$x^3 - 3x + 2 = 0, \tag{10}$$

$$x^3 - 3x + 3 = 0. \tag{11}$$

Note that the polynomials differ only in their constant terms.

It is clear from Fig. 4.5.9 that Eq. (9) has three real solutions, one in each of the intervals $[-2, -1]$, $[0, 1]$, and $[1, 2]$. These solutions could be approximated graphically by successive magnification or analytically by Newton's method. [As we have previously mentioned, there are even formulas—*Cardan's formulas*—for the exact solution of an arbitrary cubic equation, but they are unwieldy and are seldom used except in computer algebra programs. For example, they yield (via *Mathematica*)

$$x_1 = \left(\frac{-1 + i\sqrt{3}}{2}\right)^{-1/3} + \left(\frac{-1 + i\sqrt{3}}{2}\right)^{1/3},$$

$$x_2 = \left(\frac{-1 + i\sqrt{3}}{2}\right)^{4/3} + \left(\frac{-1 + i\sqrt{3}}{2}\right)^{5/3}, \quad \text{and}$$

$$x_3 = \left(\frac{-1 + i\sqrt{3}}{2}\right)^{2/3} + \left(\frac{-1 + i\sqrt{3}}{2}\right)^{7/3}$$

for the three solutions of Eq. (9). Despite the appearance of the imaginary number $i = \sqrt{-1}$ in these three expressions, Fig. 4.5.9—with its three x-intercepts—indicates that all three solutions simplify to ordinary real numbers.]

It appears in Fig. 4.5.10 that Eq. (10) has the two real solutions $x = 1$ and $x = -2$. Once we verify that $x = 1$ is a solution, then it follows from the *factor theorem* of algebra that $x - 1$ is a factor of $x^3 - 3x + 2$. The other factor can be found by division (long or synthetic) of $x - 1$ into $x^3 - 3x + 2$; the quotient is $x^2 + x - 2$. Thus we see that

$$x^3 - 3x + 2 = (x - 1)(x^2 + x - 2) = (x - 1)^2(x + 2).$$

Hence $x = 1$ is a "double root" and $x = -2$ is a "single root" of Eq. (10), thereby accounting for the three solutions that a cubic equation "ought to have."

We see in Fig. 4.5.11 that Eq. (11) has only one real solution. It is given approximately by $x \approx -2.1038$. Problem 55 asks you to divide $x + 2.0138$ into $x^3 - 3x + 3$ to obtain a factorization of the form

$$x^3 - 3x + 3 \approx (x + 2.1038)(x^2 + bx + c). \quad \textbf{(12)}$$

The quadratic equation $x^2 + bx + c = 0$ has two complex conjugate solutions, which are the other two solutions of Eq. (11).

Calculator and Computer Graphing

With a graphing calculator or computer, we may construct the graph of a given function with a few keystrokes. Nevertheless, the viewpoint of this section may be useful in analyzing and understanding what we see on the screen.

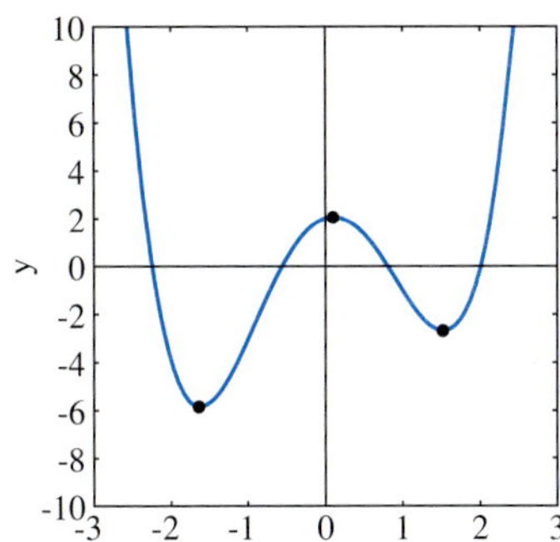

Fig. 4.5.12 $y = x^4 - 5x^2 + x + 2$

EXAMPLE 4 Figure 4.5.12 shows a computer-generated graph of the function

$$f(x) = x^4 - 5x^2 + x + 2. \quad \textbf{(13)}$$

Three critical points are visible, separating the x-axis into two intervals on which the function f increases and two on which it decreases. To find these critical points, we need to solve the cubic equation

$$f'(x) = 4x^3 - 10x + 1 = 0. \quad \textbf{(14)}$$

For this purpose we could graph the derivative $f'(x)$ and zoom in on its solutions, or we could use Newton's method to approximate these solutions accurately, or simply use the "solve" command on our calculator or computer. The approximate solutions of Eq. (14) thus found are -1.6289, 0.1004, and 1.5285. The corresponding numerical values of y obtained by substitution in Eq. (13) are -5.8554, 2.0501, and -2.6947. Thus the three critical points that we see on the graph in Fig. 4.5.12 are $(-1.6289, -5.8554)$, $(0.1004, 2.0501)$, and $(1.5285, -2.6947)$. The function f is decreasing on the intervals $-\infty < x < -1.6289$ and $0.1004 < x < 1.5285$ and increasing on the intervals $-1.6289 < x < 0.1004$ and $1.5285 < x < \infty$. ■

4.5 PROBLEMS

In Problems 1 through 4, use behavior "at infinity" to match the given function with its graph in Fig. 4.5.13.

1. $f(x) = x^3 - 5x + 2$
2. $f(x) = x^4 - 3x^2 + x - 2$
3. $f(x) = -\frac{1}{3}x^5 - 3x^2 + 3x + 2$
4. $f(x) = -\frac{1}{3}x^6 + 2x^5 - 3x^4 + \frac{1}{2}x + 5$

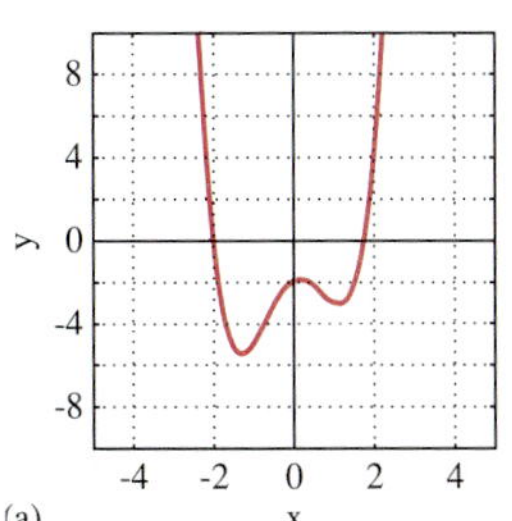

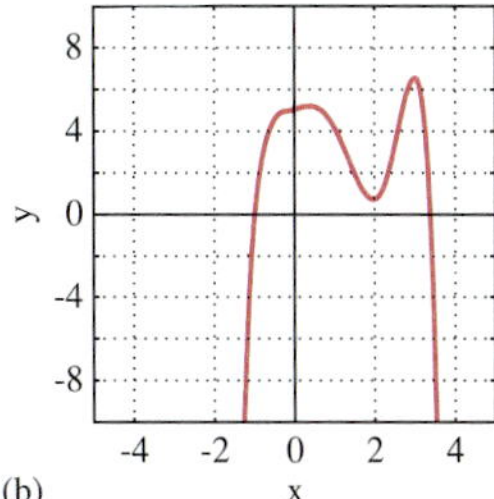

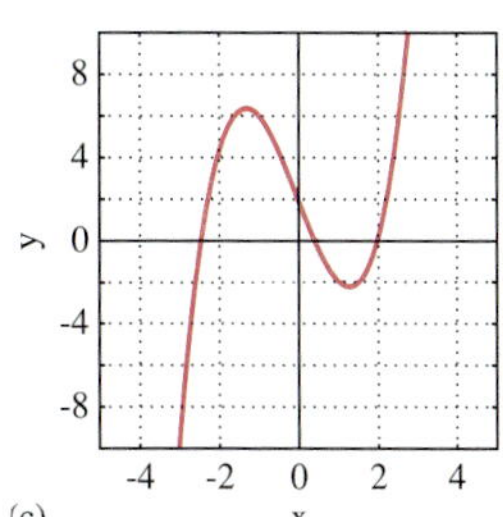

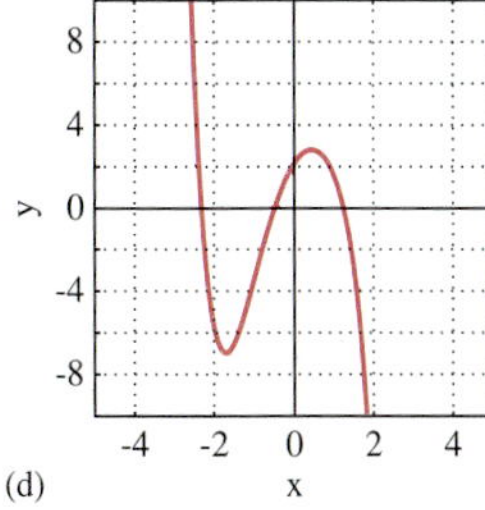

Fig. 4.5.13 Problems 1 through 4

In Problems 5 through 14 a function $y = f(x)$ and its computer-generated graph are given. Find both the critical points and the increasing/decreasing intervals for $f(x)$.

5. $y = 2x^2 - 10x - 7$ (Fig. 4.5.14)

6. $y = 27 + 12x - 4x^2$ (Fig. 4.5.15)

7. $y = 4x^3 - 3x^2 - 90x + 23$ (Fig. 4.5.16)

8. $y = 85 + 70x - 11x^2 - 4x^3$ (Fig. 4.5.17)

9. $y = 3x^4 + 4x^3 - 36x^2 + 40$ (Fig. 4.5.18)

10. $y = 125 + 120x^2 - 2x^3 - 9x^4$ (Fig. 4.5.19)

11. $y = 3x^5 - 100x^3 + 960x$ (Fig. 4.5.20)

12. $y = 2x^6 - 87x^4 + 600x^2 + 3000$ (Fig. 4.5.21)

13. $y = 3x^7 - 84x^5 + 448x^3$ (Fig. 4.5.22)

14. $y = 3x^8 - 52x^6 + 216x^4 - 500$ (Fig. 4.5.23)

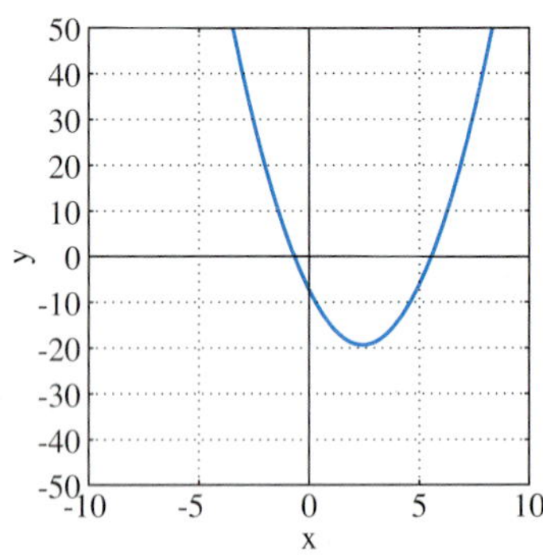

Fig. 4.5.14 Problem 5

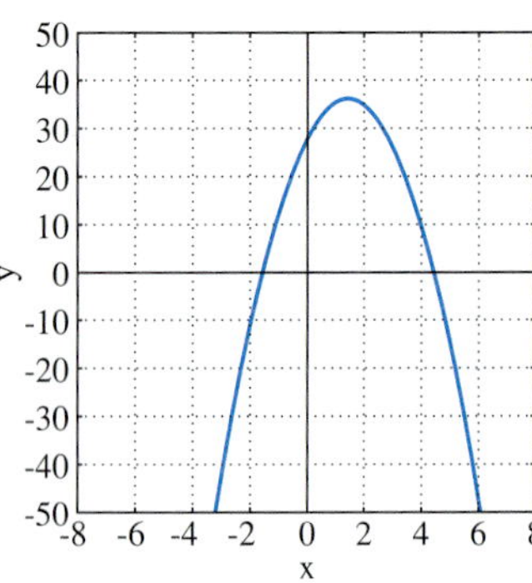

Fig. 4.5.15 Problem 6

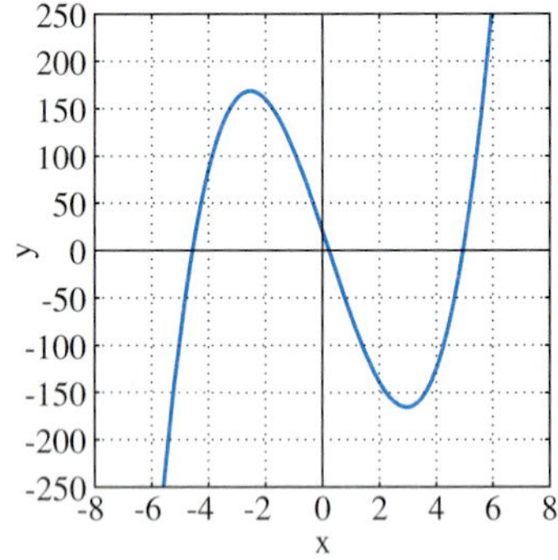

Fig. 4.5.16 Problem 7

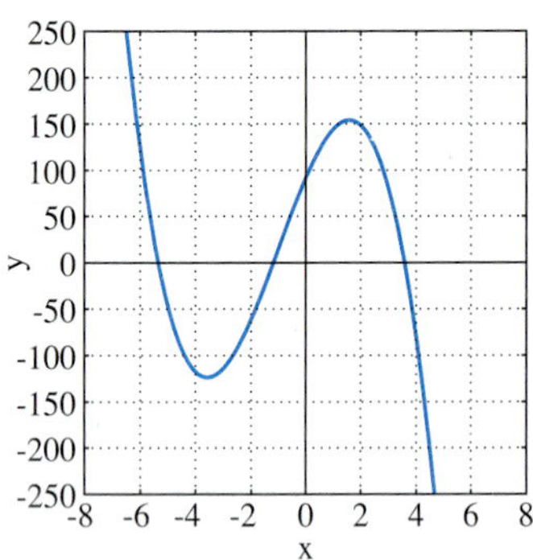

Fig. 4.5.17 Problem 8

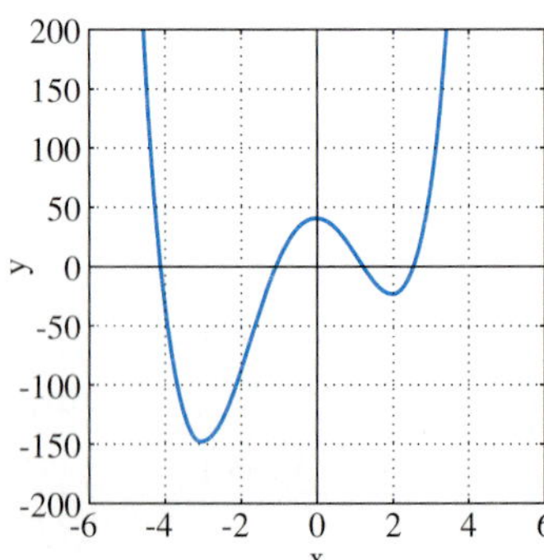

Fig. 4.5.18 Problem 9

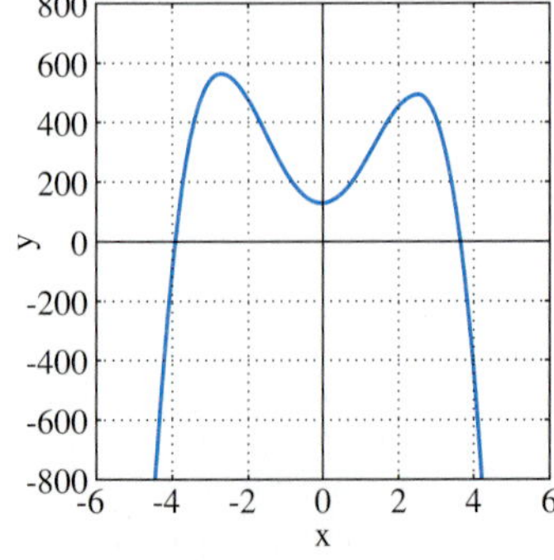

Fig. 4.5.19 Problem 10

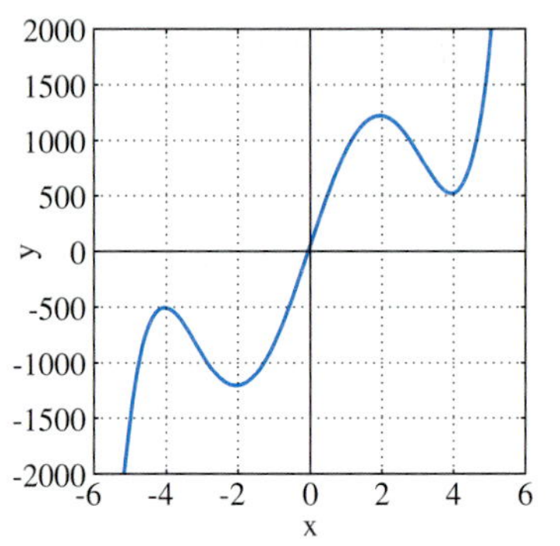

Fig. 4.5.20 Problem 11

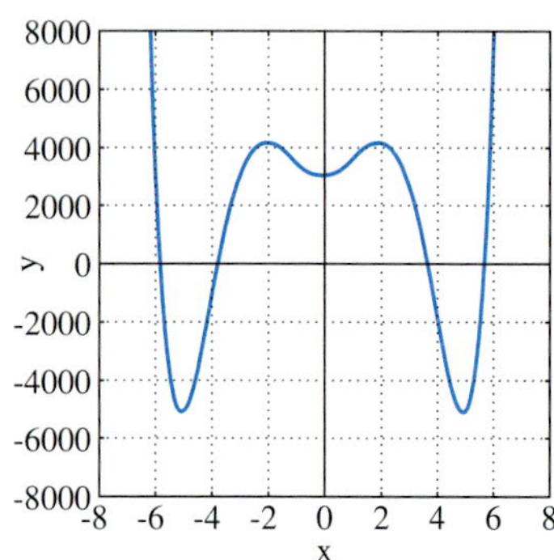

Fig. 4.5.21 Problem 12

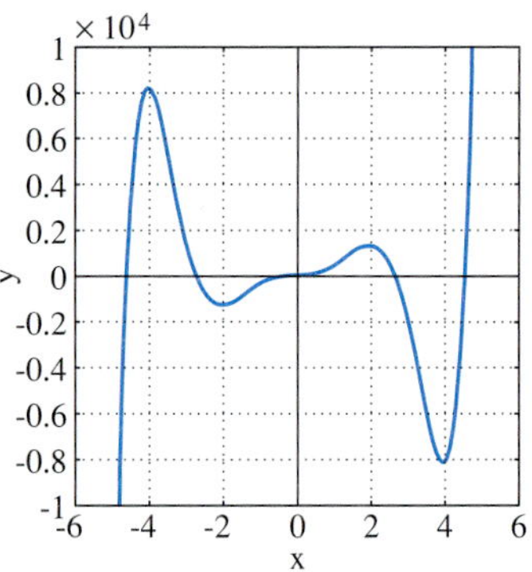

Fig. 4.5.22 Problem 13

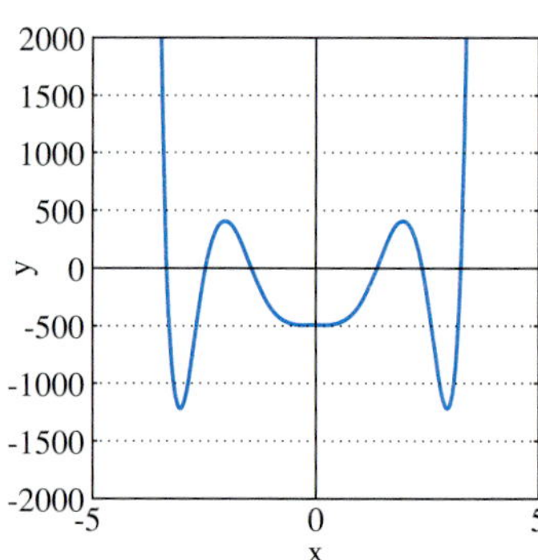

Fig. 4.5.23 Problem 14

In Problems 15 through 48, find the intervals on which the function f is increasing and those on which it is decreasing. Sketch the graph of $y = f(x)$ and label the local maxima and minima. Global extrema should be so identified.

15. $f(x) = 3x^2 - 6x + 5$

16. $f(x) = 5 - 8x - 2x^2$

17. $f(x) = x^3 - 12x$

18. $f(x) = x^3 + 3x$

19. $f(x) = x^3 - 6x^2 + 9x$

20. $f(x) = x^3 + 6x^2 + 9x$

21. $f(x) = x^3 + 3x^2 + 9x$

22. $f(x) = x^3 - 27x$

23. $f(x) = (x - 1)^2(x + 2)^2$

24. $f(x) = (x - 2)^2(2x + 3)^2$

25. $f(x) = 3\sqrt{x} - x\sqrt{x}$

26. $f(x) = x^{2/3}(5 - x)$

27. $f(x) = 3x^5 - 5x^3$

28. $f(x) = x^4 + 4x^3$

29. $f(x) = x^4 - 8x^2 + 7$

30. $f(x) = \dfrac{1}{x}$

31. $f(x) = 2x^2 - 3x - 9$

32. $f(x) = 6 - 5x - 6x^2$

33. $f(x) = 2x^3 + 3x^2 - 12x$

34. $f(x) = x^3 + 4x$

35. $f(x) = 50x^3 - 105x^2 + 72x$

36. $f(x) = x^3 - 3x^2 + 3x - 1$

37. $f(x) = 3x^4 - 4x^3 - 12x^2 + 8$

38. $f(x) = x^4 - 2x^2 + 1$ **39.** $f(x) = 3x^5 - 20x^3$

40. $f(x) = 3x^5 - 25x^3 + 60x$

41. $f(x) = 2x^3 + 3x^2 + 6x$ **42.** $f(x) = x^4 - 4x^3$

43. $f(x) = 8x^4 - x^8$ **44.** $f(x) = 1 - x^{1/3}$

45. $f(x) = x^{1/3}(4 - x)$ **46.** $f(x) = x^{2/3}(x^2 - 16)$

47. $f(x) = x(x - 1)^{2/3}$ **48.** $f(x) = x^{1/3}(2 - x)^{2/3}$

Problems 49 through 54 give the values of the function $f(x)$ at its critical points, together with the graph $y = f'(x)$ of its derivative. Use this information to construct a sketch of the graph $y = f(x)$ of the function.

49. $f(-3) = 78,\quad f(2) = -47;$ Fig. 4.5.24

50. $f(-2) = 106,\quad f(4) = -110;$ Fig. 4.5.25

51. $f(-3) = -66,\quad f(2) = 59;$ Fig. 4.5.26

52. $f(-3) = -130,\quad f(0) = 5,\quad f(1) = -2;$ Fig. 4.5.27

53. $f(-2) = -107,\quad f(1) = 82,\quad f(3) = 18;$ Fig. 4.5.28

54. $f(-3) = 5336,\quad f(0) = 17,\quad f(2) = 961,\quad f(4) = -495;$ Fig. 4.5.29

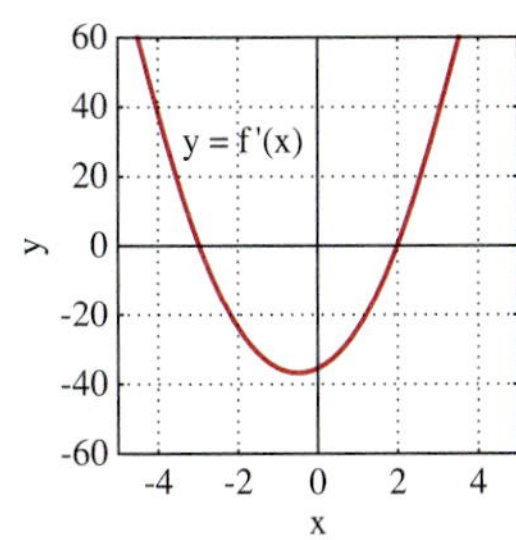

Fig. 4.5.24 $y = f'(x)$ of Problem 49

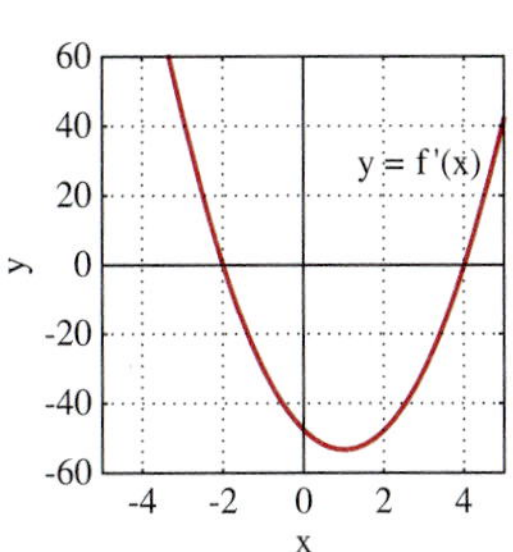

Fig. 4.5.25 $y = f'(x)$ of Problem 50

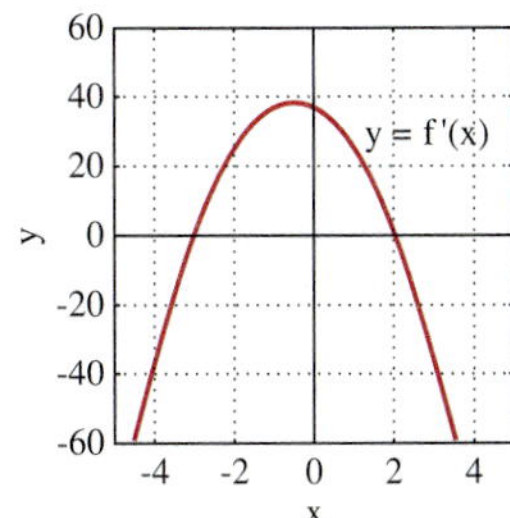

Fig. 4.5.26 $y = f'(x)$ of Problem 51

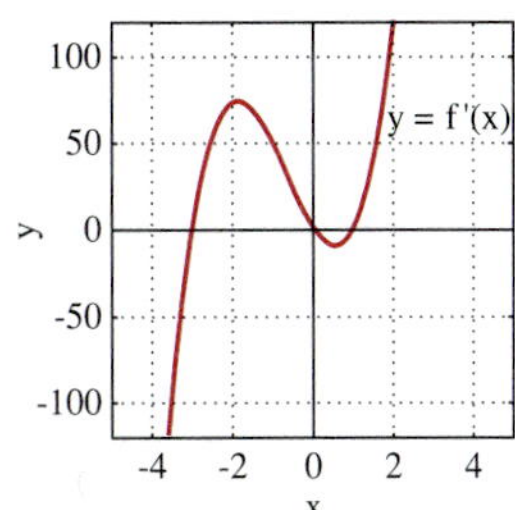

Fig. 4.5.27 $y = f'(x)$ of Problem 52

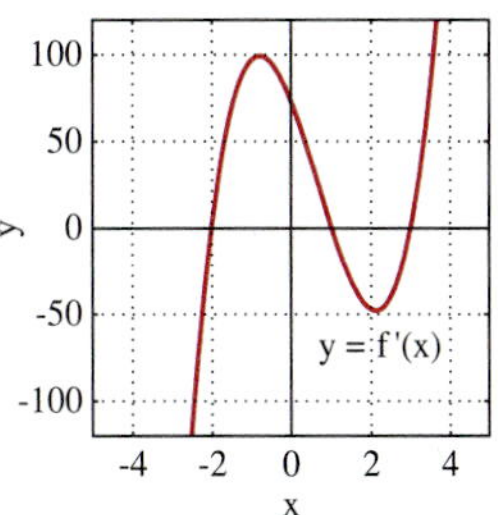

Fig. 4.5.28 $y = f'(x)$ of Problem 53

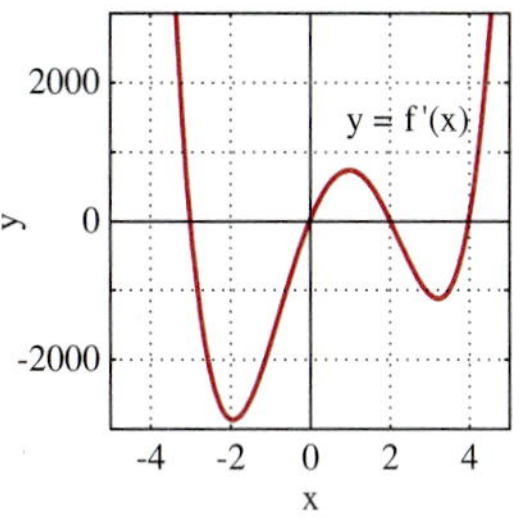

Fig. 4.5.29 $y = f'(x)$ of Problem 54

55. (a)Verify the approximate solution $x \approx -2.0138$ of Eq. (11). (b) Divide $x^3 - 3x + 3$ by $x + 2.1038$ to obtain the factorization in Eq. (12). (c) Use the quotient in part (d) to find (approximately) the complex conjugate pair of solutions of Eq. (11).

56. Explain why Figs. 4.5.9 and 4.5.10 imply that the cubic equation $x^3 - 3x + q = 0$ has exactly one real solution if $|q| > 2$ but has three distinct real solutions if $|q| < 2$. What is the situation if $q = -2$?

57. The computer-generated graph in Fig. 4.5.30 shows how the curve

$$y = [x(x - 1)(2x - 1)]^2$$

looks on any "reasonable" scale with integral units of measurement on the y-axis. Use the methods of this section to show that the graph really has the appearance shown in Fig. 4.5.31 (the values on the y-axis are in thousandths), with critical points at $0, \frac{1}{2}, \frac{1}{6}(3 \pm \sqrt{3})$, and 1.

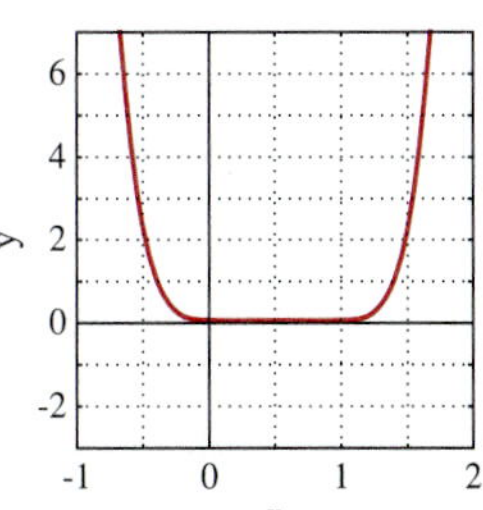

Fig. 4.5.30 The graph $y = [x(x - 1)(2x - 1)]^2$ on a "reasonable scale" (Problem 57)

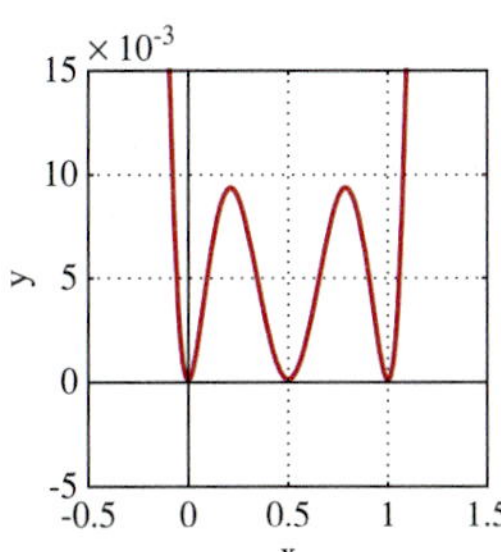

Fig. 4.5.31 The graph $y = [x(x - 1)(2x - 1)]^2$ on a finer scale: $-0.005 \leqq y \leqq 0.005$ (Problem 57)

PROJECT 4.5: SOME EXOTIC GRAPHS

These projects require the use of a graphing calculator or a computer. Feel free to solve any necessary equations either graphically or using a "solve" key or command.

Problem 1 First show that, on a "reasonable" scale with integral units of measurement on the y-axis, the graph of the polynomial

$$f(x) = [\tfrac{1}{6}\, x(9x - 5)(x - 1)]^4$$

looks much like Fig. 4.5.30, with a seemingly flat section. Then produce a plot that reveals the true structure of the graph, as in Fig. 4.5.31. Finally, find the approximate coordinates of the local maximum and minimum points on the graph $y = f(x)$.

Problem 2 For an investigation of your own personal polynomial graph with a seeming "flat spot," repeat Problem 1 with the function

$$f(x) = [\tfrac{1}{50}\, x(ax - b)(x - 1)]^2$$

where a is the largest and b is the smallest nonzero digit in your student I.D. number.

Problem 3 First show graphically that the quartic (fourth-degree) equation

$$f(x) = x^4 - 55x^3 + 505x^2 + 11000x - 110000 = 0$$

has four distinct solutions. Then let p be the smallest nonzero digit in your student I.D. number and show that the equation

$$g(x) = x^4 - 55x^3 + (505 + p)x^2 + 11000x - 110000 = 0,$$

which differs only (slightly) in the coefficient of x^2, has only two distinct real solutions. Then find the local maximum and minimum points on each graph.

4.6 HIGHER DERIVATIVES AND CONCAVITY

We saw in Section 4.3 that the sign of the first derivative f' of a differentiable function f indicates whether the graph of f is rising or is falling. Here we shall see that the sign of the *second* derivative of f, the derivative of f', indicates which way the curve $y = f(x)$ is *bending,* upward or downward.

Higher Derivatives

The **second derivative** of f is the derivative of f', is denoted by f'', and its value at x is

$$f''(x) = D_x\big(f'(x)\big) = D_x\big(D_x f(x)\big) = D_x^2 f(x).$$

(The superscript 2 is not an exponent but only an indication that the operator D_x is to be applied twice.) The derivative of f'' is the **third derivative** f''' of f, and

$$f'''(x) = D_x\big(f''(x)\big) = D_x\big(D_x^2 f(x)\big) = D_x^3 f(x).$$

The third derivative is also denoted by $f^{(3)}$. More generally, the result of beginning with the function f and differentiating n times in succession is the n**th derivative** $f^{(n)}$ of f, with $f^{(n)}(x) = D_x^n f(x)$.

If $y = f(x)$, then the first n derivatives are written in operator notation as

$$D_x y, \quad D_x^2 y, \quad D_x^3 y, \quad \dots, \quad D_x^n y,$$

in dependent-independent variable notation as

$$y'(x), \quad y''(x), \quad y'''(x), \quad \dots, \quad y^{(n)}(x),$$

and in differential notation as

$$\frac{dy}{dx}, \quad \frac{d^2y}{dx^2}, \quad \frac{d^3y}{dx^3}, \quad \ldots, \quad \frac{d^ny}{dx^n}.$$

The history of the curious use of superscripts in differential notation for higher derivatives involves the metamorphosis

$$\frac{d}{dx}\left(\frac{dy}{dx}\right) \to \frac{d}{dx}\frac{dy}{dx} \to \frac{(d)^2y}{(dx)^2} \to \frac{d^2y}{dx^2}.$$

EXAMPLE 1 Find the first four derivatives of

$$f(x) = 2x^3 + \frac{1}{x^2} + 16x^{7/2}.$$

Solution Write

$$f(x) = 2x^3 + x^{-2} + 16x^{7/2}. \quad \text{Then}$$

$$f'(x) = 6x^2 - 2x^{-3} + 56x^{5/2} = 6x^2 - \frac{2}{x^3} + 56x^{5/2},$$

$$f''(x) = 12x + 6x^{-4} + 140x^{3/2} = 12x + \frac{6}{x^4} + 140x^{3/2},$$

$$f'''(x) = 12 - 24x^{-5} + 210x^{1/2} = 12 - \frac{24}{x^5} + 210\sqrt{x}, \quad \text{and}$$

$$f^{(4)}(x) = 120x^{-6} + 105x^{-1/2} = \frac{120}{x^6} + \frac{105}{\sqrt{x}}.$$

■

Example 2 shows how to find higher derivatives of implicitly defined functions.

EXAMPLE 2 Find the second derivative $y''(x)$ of the function $y = y(x)$ that is defined implicitly by the equation

$$x^2 - xy + y^2 = 9.$$

Solution A first implicit differentiation of the given equation *with respect to x* gives

$$2x - y - x\frac{dy}{dx} + 2y\frac{dy}{dx} = 0,$$

so

$$\frac{dy}{dx} = \frac{y - 2x}{2y - x}.$$

We obtain d^2y/dx^2 by differentiating implicitly, again with respect to x, using the quotient rule. After that, we substitute the expression we just found for dy/dx:

$$\frac{d^2y}{dx^2} = D_x\left(\frac{y - 2x}{2y - x}\right) = \frac{\left(\frac{dy}{dx} - 2\right)(2y - x) - (y - 2x)\left(2\frac{dy}{dx} - 1\right)}{(2y - x)^2}$$

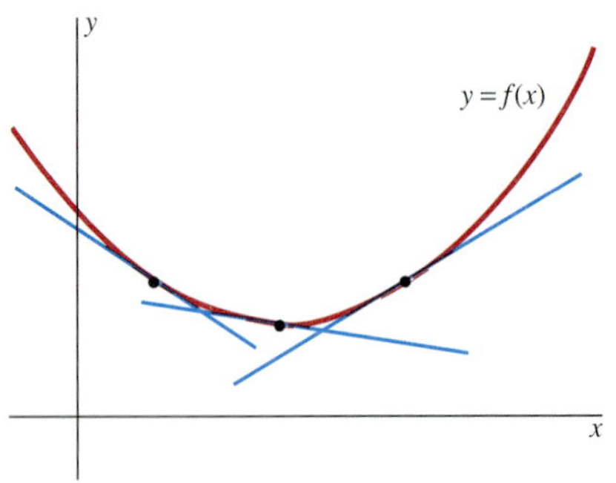

Fig. 4.6.1 The graph is bending upward.

$$= \frac{3x\dfrac{dy}{dx} - 3y}{(2y - x)^2} = \frac{3x\dfrac{y - 2x}{2y - x} - 3y}{(2y - x)^2}.$$

Thus

$$\frac{d^2y}{dx^2} = -\frac{6(x^2 - xy + y^2)}{(2y - x)^3}.$$

We now substitute the original equation, $x^2 - xy + y^2 = 9$, for one final simplification:

$$\frac{d^2y}{dx^2} = -\frac{54}{(2y - x)^3}.$$

The somewhat unexpected final simplification is always available when the original equation is symmetric in x and y.

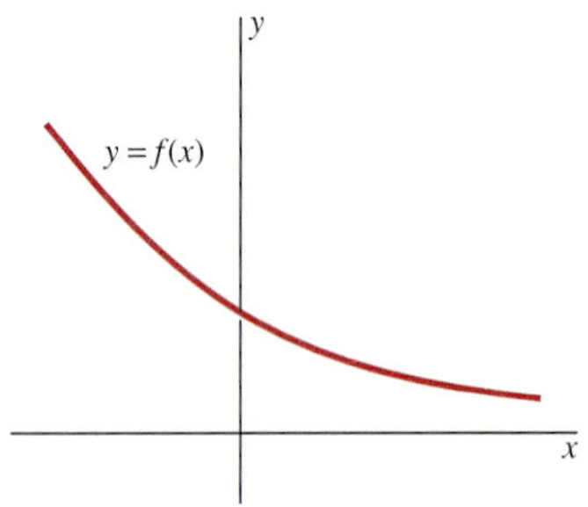

Fig. 4.6.2 Another graph bending upward

The Sign of the Second Derivative

Now we shall investigate the significance of the *sign* of the second derivative. If $f''(x) > 0$ on the interval I, then the first derivative f' is an increasing function on I, because *its* derivative $f''(x)$ is positive. Thus as we scan the graph $y = f(x)$ from left to right, we see the tangent line turning counterclockwise (Fig. 4.6.1). We describe this situation by saying that the curve $y = f(x)$ is **bending upward.** Note that a curve can bend upward without rising, as in Fig. 4.6.2.

If $f''(x) < 0$ on the interval I, then the first derivative f' is decreasing on I, so the tangent line turns clockwise as x increases. We say in this case that the curve $y = f(x)$ is **bending downward.** Figures 4.6.3 and 4.6.4 show two ways this can happen.

The two cases are summarized in the brief table in Fig. 4.6.5.

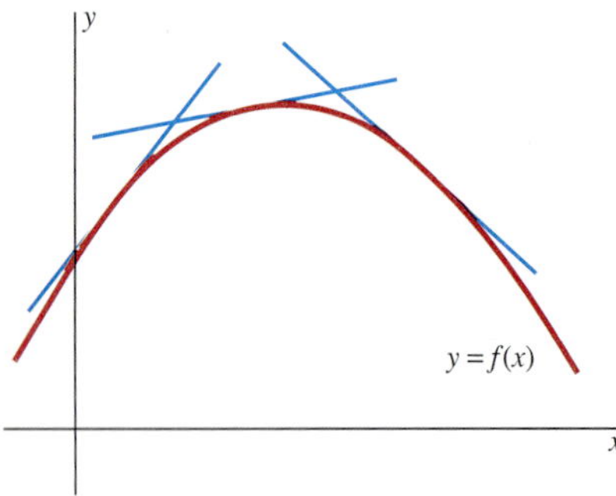

Fig. 4.6.3 A graph bending downward

EXAMPLE 3 Figure 4.6.6 shows the graph of the function

$$f(x) = x^3 - 3x^2 + 3.$$

Because

$$f'(x) = 3x^2 - 6x \quad \text{and} \quad f''(x) = 6x - 6 = 6(x - 1),$$

we see that

$$f''(x) < 0 \quad \text{for} \quad x < 1,$$
$$f''(x) > 0 \quad \text{for} \quad x > 1.$$

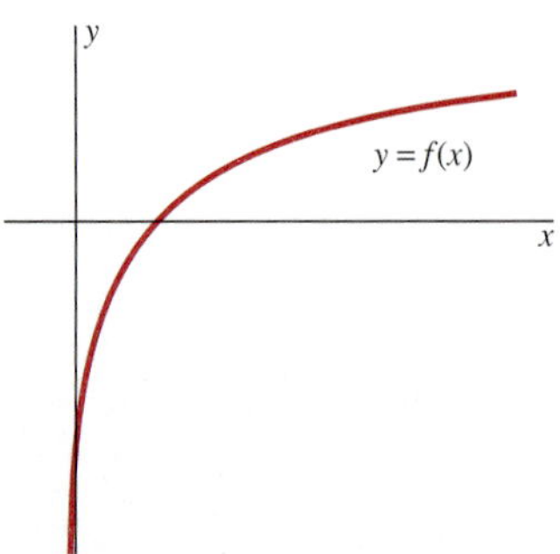

Fig. 4.6.4 Another graph bending downward

$f''(x)$	$y = f(x)$
Negative Positive	Bending downward Bending upward

Fig. 4.6.5 Significance of the sign of $f''(x)$ on an interval

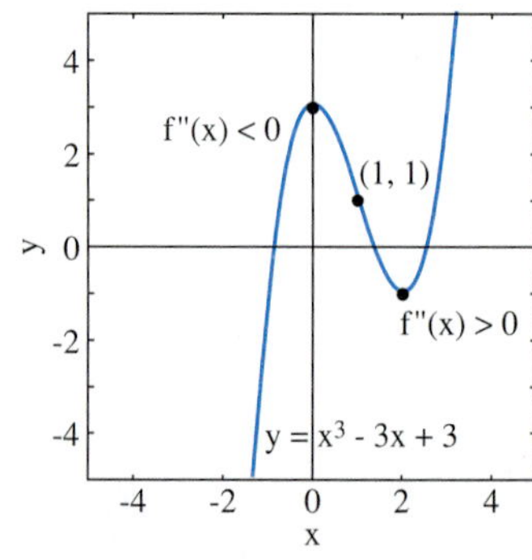

Fig. 4.6.6 The graph of $y = x^3 - 3x^2 + 3$ (Example 3)

Observe in the figure that the curve bends downward on $(-\infty, 1)$ but bends upward on $(1, +\infty)$, consistent with the correspondences in Fig. 4.6.5. ■

The Second Derivative Test

We know from Section 3.5 that a local extremum of a differentiable function f can occur only at a critical point where $f'(c) = 0$, so the tangent line at the point $(c, f(c))$ on the curve $y = f(x)$ is horizontal. But the example $f(x) = x^3$, for which $x = 0$ is a critical point but not an extremum (Fig. 4.6.7), shows that the *necessary condition* $f'(c) = 0$ is *not* a sufficient condition from which to conclude that $f(c)$ is an extreme value of the function f.

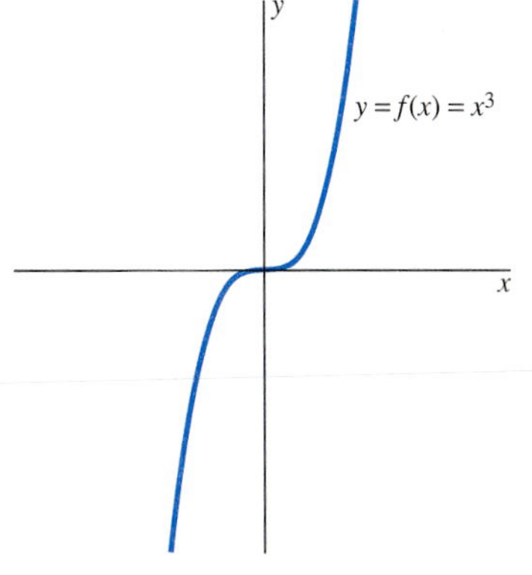

Fig. 4.6.7 Although $f'(0) = 0$, $f(0)$ is not an extremum

Now suppose not only that $f'(c) = 0$, but also that the curve $y = f(x)$ is bending upward on some open interval that contains the critical point $x = c$. It is apparent from Fig. 4.6.8(a) that $f(c)$ is a local minimum value. Similarly, $f(c)$ is a local maximum value if $f'(c) = 0$ while $y = f(x)$ is bending downward on some open interval containing c (Fig. 4.6.8b). But the *sign* of the second derivative $f''(x)$ tells us whether $y = f(x)$ is bending upward or downward and therefore provides us with a *sufficient* condition for a local extremum.

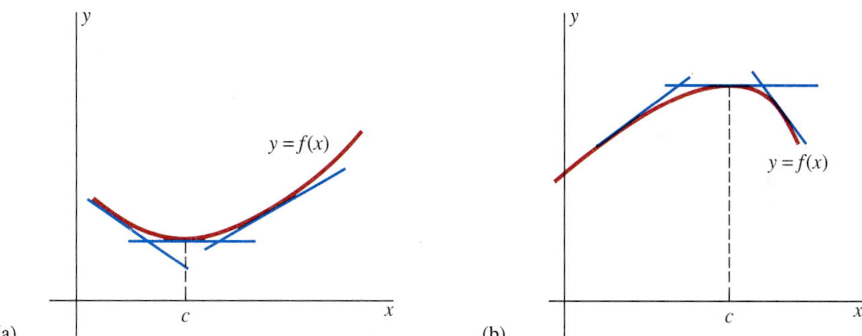

Fig. 4.6.8 The second derivative test (Theorem 1). (a) $f''(x) > 0$; tangent turning counterclockwise; graph bending upward; local minimum at $x = c$. (b) $f''(x) < 0$: tangent turning clockwise; graph bending downward; local maximum at $x = c$.

Theorem 1 *Second Derivative Test*

Suppose that the function f is twice differentiable on the open interval I containing the critical point c at which $f'(c) = 0$. Then

1. If $f''(x) > 0$ on I, then $f(c)$ is the minimum value of $f(x)$ on I.
2. If $f''(x) < 0$ on I, then $f(c)$ is the maximum value of $f(x)$ on I.

PROOF We will prove only part 1. If $f''(x) > 0$ on I, then it follows that the first derivative f' is an increasing function on I. Because $f'(c) = 0$, we may conclude that $f'(x) < 0$ for $x < c$ in I and that $f'(x) > 0$ for $x > c$ in I. Consequently, the first derivative test of Section 4.4 implies that $f(c)$ is the minimum value of $f(x)$ on I. ■

$f''(x)$	$f(c)$
Positive Negative	Minimum Maximum

Fig. 4.6.9 Significance of the sign of $f''(x)$ on an interval containing the critical point c

REMARK 1 Rather than memorizing verbatim the conditions in parts 1 and 2 of Theorem 1 (summarized in Fig. 4.6.9), it is easier and more reliable to remember the second derivative test by visualizing continuously turning tangent lines (Fig. 4.6.8).

REMARK 2 Theorem 1 implies that the function f has a local minimum at the critical point c if $f''(x) > 0$ on some open interval about c but a local maximum if $f''(x) < 0$ near c. But the hypothesis on $f''(x)$ in Theorem 1 is *global* in that $f''(x)$ is assumed to have the same sign at *every* point of the open interval I that contains the critical point c. There is a strictly *local* version of the second derivative test that involves only the sign of $f''(c)$ at the critical point c (rather than on a whole open interval). According to Problem 90, if $f'(c) = 0$, then $f(c)$ is a local minimum value of f if $f''(c) > 0$ but a local maximum if $f''(c) < 0$.

Remark 3 The second derivative test says nothing about what happens if $f''(c) = 0$ at the critical point c. Consider the three functions $f(x) = x^4$, $f(x) = -x^4$, and $f(x) = x^3$. For each, $f'(0) = 0$ and $f''(0) = 0$. But their graphs, shown in Fig. 4.6.10, demonstrate that anything can happen at such a point—maximum, minimum, or neither.

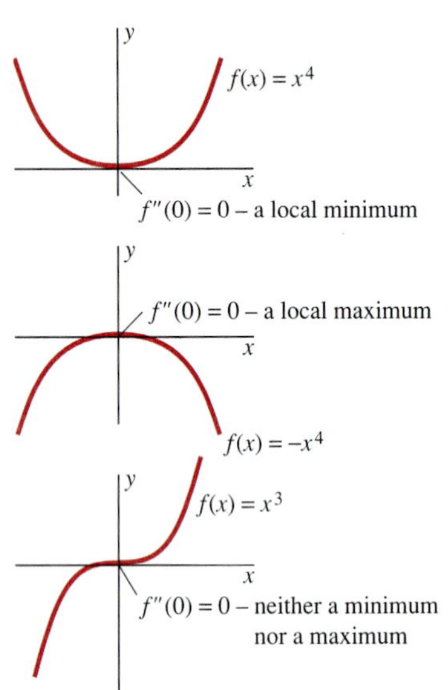

Fig. 4.6.10 No conclusion is possible if $f'(c) = 0 = f''(c)$.

Remark 4 Suppose that we want to maximize or minimize the function f on the open interval I, and we find that f has only one critical point in I, a number c at which $f'(c) = 0$. If $f''(x)$ has the same sign at all points of I, then Theorem 1 implies that $f(c)$ is an absolute extremum of f on I—a minimum if $f''(x) > 0$ and a maximum if $f''(x) < 0$. This absolute interpretation of the second derivative test can be useful in applied open-interval maximum-minimum problems.

EXAMPLE 3 (continued) Consider again the function $f(x) = x^3 - 3x^2 + 3$, for which

$$f'(x) = 3x(x-2) \quad \text{and} \quad f''(x) = 6(x-1).$$

Then f has the two critical points $x = 0$ and $x = 2$, as marked in Fig. 4.6.6. Because $f''(x) < 0$ for x near zero, the second derivative test implies that $f(0) = 3$ is a local maximum value of f. And because $f''(x) > 0$ for x near 2, it follows that $f(2) = -1$ is a local minimum value. ■

EXAMPLE 4 An open-topped rectangular box with square base has volume 500 cm^3. Find the dimensions that minimize the total area A of its base and four sides.

Solution We denote by x the edge length of the square base and by y the height of the box (Fig. 4.6.11). The volume of the box is

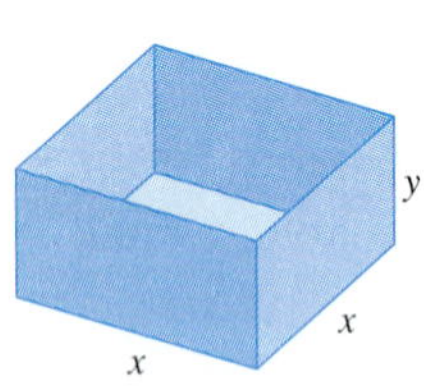

Fig. 4.6.11 The open-topped box of Example 4

$$V = x^2 y = 500, \tag{1}$$

and the total area of its base and four sides is

$$A = x^2 + 4xy. \tag{2}$$

When we solve Eq. (1) for $y = 500/x^2$ and substitute this into Eq. (2), we get the area function

$$A(x) = x^2 + \frac{2000}{x}, \qquad 0 < x < +\infty.$$

The domain of A is the open and unbounded interval $(0, +\infty)$ because x can take on any positive value; to make the box volume 500, simply choose $y = 500/x^2$. But x cannot be zero or negative.

The first derivative of $A(x)$ is

$$A'(x) = 2x - \frac{2000}{x^2} = \frac{2(x^3 - 1000)}{x^2}. \tag{3}$$

The equation $A'(x) = 0$ yields $x^3 = 1000$, so the only critical point of A in $(0, +\infty)$ is $x = 10$. To investigate this critical point, we calculate the second derivative,

$$A''(x) = 2 + \frac{4000}{x^3}. \tag{4}$$

Because it is clear that $A''(x) > 0$ on $(0, +\infty)$, it follows from the second derivative test and Remark 4 that $A(10) = 300$ is the absolute minimum value of $A(x)$ on $(0, +\infty)$. Finally, because $y = 500/x^2$, $y = 5$ when $x = 10$. Therefore, this absolute minimum corresponds to a box with base 10 cm by 10 cm and height 5 cm. ■

Concavity and Inflection Points

A comparison of Fig. 4.6.1 with Fig. 4.6.3 suggests that the question of whether the curve $y = f(x)$ is bending upward or downward is closely related to the question of whether it lies above or below its tangent lines. The latter question refers to the important property of *concavity.*

Definition *Concavity*

Suppose that the function f is differentiable at the point a and that L is the line tangent to the graph $y = f(x)$ at the point $(a, f(a))$. Then the function f (or its graph) is said to be

1. **Concave upward** at a if, on some open interval containing a, the graph of f lies *above* L.
2. **Concave downward** at a if, on some open interval containing a, the graph of f lies *below* L.

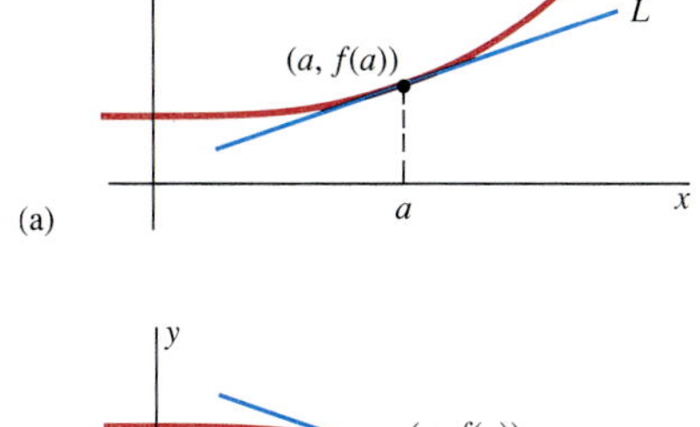

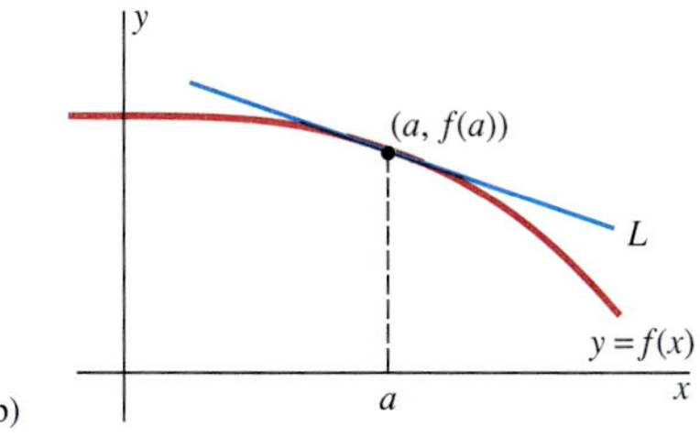

Fig. 4.6.12 (a) At $x = a$, f is a concave upward. (b) At $x = a$, f is concave downward.

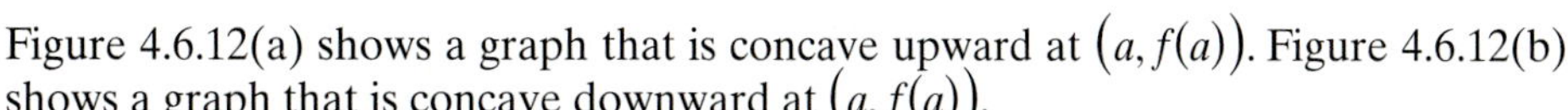

Figure 4.6.12(a) shows a graph that is concave upward at $(a, f(a))$. Figure 4.6.12(b) shows a graph that is concave downward at $(a, f(a))$.

Theorem 2 establishes the connection between concavity and the sign of the second derivative. That connection is the one suggested by our discussion of bending.

Theorem 2 *Test for Concavity*

Suppose that the function f is twice differentiable on the open interval I.

1. If $f''(x) > 0$ on I, then f is concave upward at each point of I.
2. If $f''(x) < 0$ on I, then f is concave downward at each point of I.

A proof of Theorem 2 based on the second derivative test is given at the end of this section.

Note The significance of the sign of the *first* derivative must not be confused with the significance of the sign of the *second* derivative. The possibilities illustrated in Figs. 4.6.13 through 4.6.16 show that the signs of f' and f'' are independent of each other.

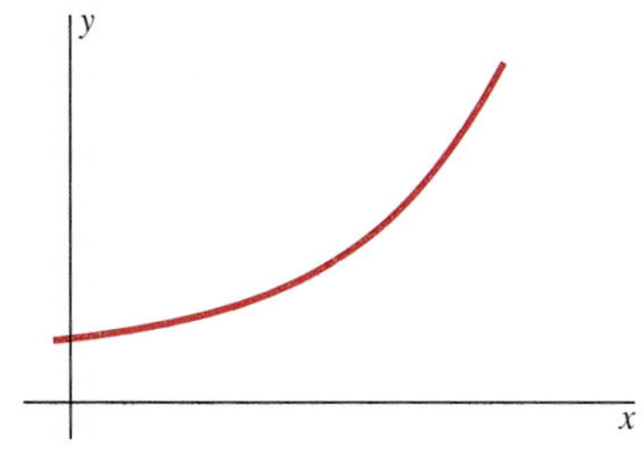

Fig. 4.6.13 $f'(x) > 0$, f increasing; $f''(x) > 0$, f concave upward

EXAMPLE 3 (continued again) For the function $f(x) = x^3 - 3x + 3$, the second derivative changes sign from positive to negative at the point $x = 1$. Observe in Fig. 4.6.6 that the corresponding point $(1, 1)$ on the graph of f is where the curve changes from bending downward to bending upward. ■

The test for concavity in Theorem 2 says nothing about the case in which $f''(x) = 0$. A point where the second derivative is zero *may* or *may not* be a point

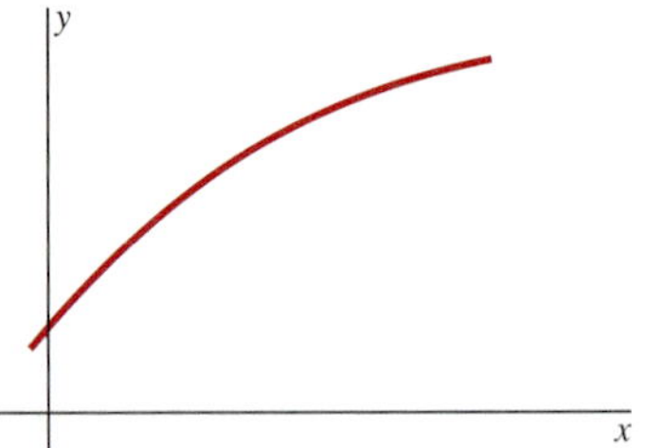

Fig. 4.6.14 $f'(x) > 0$, f increasing; $f''(x) < 0$; f concave downward

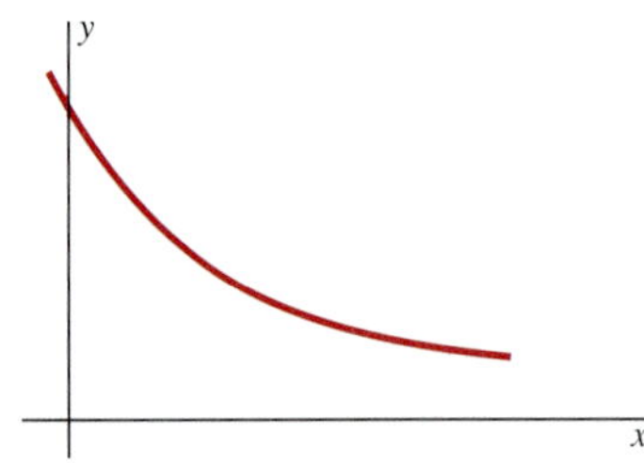

Fig. 4.6.15 $f'(x) < 0$, f decreasing; $f''(x) > 0$, f concave upward

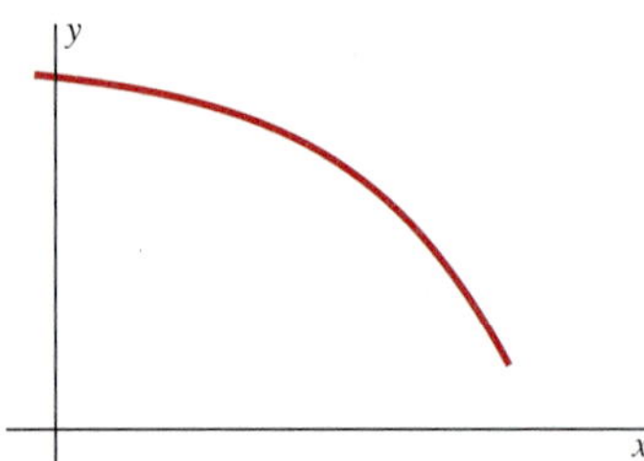

Fig. 4.6.16 $f'(x) < 0$, f decreasing; $f''(x) < 0$, f concave downward

where the function changes from concave upward on one side to concave downward on the other. But a point like $(1, 1)$ in Fig. 4.6.6, where the concavity *does* change in this manner, is called an *inflection point* of the graph of f. More precisely, the point $x = a$ where f is continuous is an **inflection point** of the function f provided that f is concave upward on one side of $x = a$ and concave downward on the other side. We also refer to $(a, f(a))$ as an inflection point on the graph of f.

Theorem 3 *Inflection Point Test*

Suppose that the function f is continuous and f'' exists on an open interval containing the point a. Then a is an inflection point of f provided that $f''(x) < 0$ on one side of a and $f''(x) > 0$ on the other side.

The fact that a point where the second derivative changes sign is an inflection point follows from Theorem 2 and the definition of an inflection point.

Remark At the inflection point itself, either

- $f''(a) = 0$, or
- $f''(a)$ does not exist.

Thus we find *inflection points* of f by examining the *critical points* of f'. Some of the possibilities are indicated in Fig. 4.6.17. We mark the intervals of upward concavity and downward concavity by small cups opening upward and downward, respectively.

EXAMPLE 5 Figure 4.6.18 shows the graph of $f(x) = \sin^2 x$ on $[0, \pi]$. Two evident inflection points are marked. Find their coordinates.

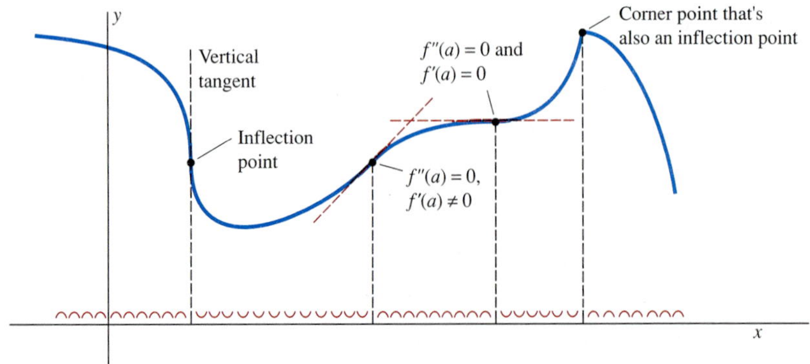

Fig. 4.6.17 Some inflection points

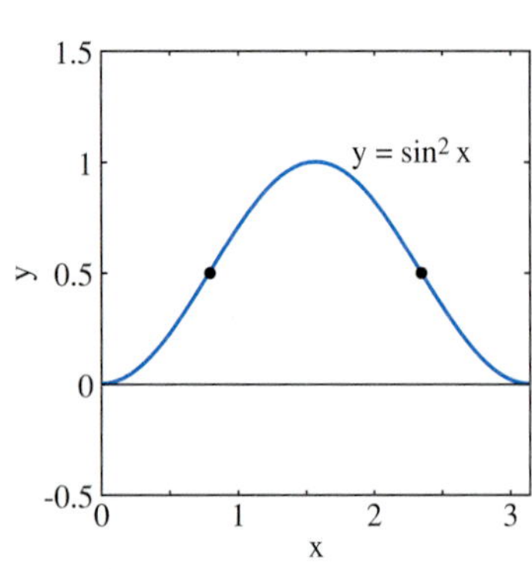

Fig. 4.6.18 $y = \sin^2 x$ (Example 5)

Solution We calculate

$$f'(x) = 2\sin x\cos x$$

and

$$f''(x) = 2\cos^2 x - 2\sin^2 x = 2\cos 2x.$$

Then

$$f''(x) = 2\cos 2x = 0$$

at $x = \pi/4$ and at $x = 3\pi/4$ in the interval $[0, \pi]$. Now

$$f''(x) = 2\cos 2x > 0 \quad \text{if} \quad 0 < x < \frac{\pi}{4}, \qquad \text{so} \qquad 0 < 2x < \frac{\pi}{2};$$

$$f''(x) = 2\cos 2x < 0 \quad \text{if} \quad \frac{\pi}{4} < x < \frac{3\pi}{4}, \qquad \text{so} \qquad \frac{\pi}{2} < 2x < \frac{3\pi}{2};$$

$$f''(x) = 2\cos 2x > 0 \quad \text{if} \quad \frac{3\pi}{4} < x < \pi, \qquad \text{so} \qquad \frac{3\pi}{2} < 2x < 2\pi.$$

It follows that $f(x) = \sin^2 x$ has inflection points at $x = \pi/4$ and $x = 3\pi/4$. The corresponding points marked on the graph in Fig. 4.6.18 are $(\pi/4, 1/2)$ and $(3\pi/4, 1/2)$. ■

Inflection Points and Curve Sketching

Let the function f be twice differentiable for all x. Just as the critical points where $f'(x) = 0$ separate the x-axis into open intervals on which $f'(x)$ does not change sign, the *possible* inflection points where $f''(x) = 0$ separate the x-axis into open intervals on which $f''(x)$ does not change sign. On each of these intervals, the curve $y = f(x)$ either is bending downward $[f''(x) < 0]$ or is bending upward $[f''(x) > 0]$. We can determine the sign of $f''(x)$ in each of these intervals in either of two ways:

1. Evaluation of $f''(x)$ at a typical point of each interval. The sign of $f''(x)$ at that particular point is the sign of $f''(x)$ throughout the interval.
2. Construction of a table of signs of the factors of $f''(x)$. Then the sign of $f''(x)$ on each interval can be deduced from the table.

These are the same two methods we used in Sections 4.4 and 4.5 to determine the sign of $f'(x)$. We use the first method in Example 6 and the second in Example 7.

EXAMPLE 6 Sketch the graph of $f(x) = 8x^5 - 5x^4 - 20x^3$, indicating local extrema, inflection points, and concave structure.

Solution We sketched this curve in Example 2 of Section 4.5; see Fig. 4.5.5 for the graph. In that example we found the first derivative to be

$$f'(x) = 40x^4 - 20x^3 - 60x^2 = 20x^2(x + 1)(2x - 3),$$

so the critical points are $x = -1$, $x = 0$, and $x = \frac{3}{2}$. The second derivative is

$$f''(x) = 160x^3 - 60x^2 - 120x = 20x(8x^2 - 3x - 6).$$

When we compute $f''(x)$ at each critical point, we find that

$$f''(-1) = -100 < 0, \qquad f''(0) = 0, \qquad \text{and} \qquad f''(\tfrac{3}{2}) = 225 > 0.$$

Continuity of f'' ensures that $f''(x) < 0$ near the critical point $x = -1$ and that $f''(x) > 0$ near the critical point $x = \frac{3}{2}$. The second derivative test therefore tells us that f has a local maximum at $x = -1$ and a local minimum at $x = \frac{3}{2}$. We cannot determine from the second derivative test the behavior of f at $x = 0$.

Because $f''(x)$ exists everywhere, the possible inflection points are the solutions of the equation

$$f''(x) = 0; \quad \text{that is,} \quad 20x(8x^2 - 3x - 6) = 0.$$

Clearly, one solution is $x = 0$. To find the other two, we use the quadratic formula to solve the equation

$$8x^2 - 3x - 6 = 0.$$

This gives

$$x = \tfrac{1}{16}\left(3 \pm \sqrt{201}\right),$$

so $x \approx 1.07$ and $x \approx -0.70$ are possible inflection points along with $x = 0$.

These three possible inflection points separate the x-axis into the intervals indicated in Fig. 4.6.19. We check the sign of $f''(x)$ on each.

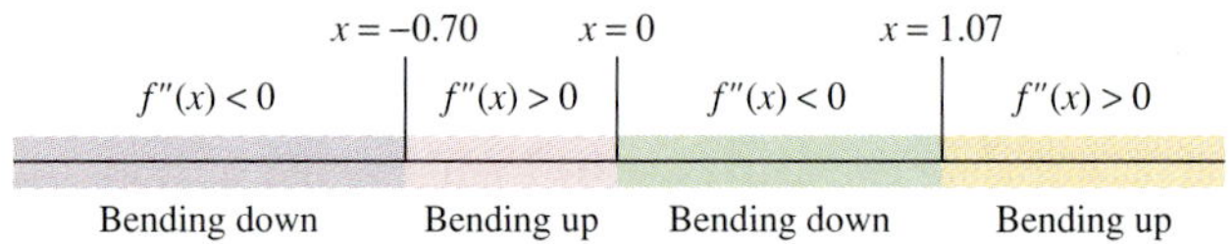

Fig. 4.6.19 Intervals of concavity of Example 6

On $(-\infty, -0.70)$: $f''(-1) = -100 < 0$; f is concave downward;

On $(-0.70, 0)$: $f''(-\frac{1}{2}) = 25 > 0$; f is concave upward;

On $(0, 1.07)$: $f''(1) = -20 < 0$; f is concave downward;

On $(1.07, +\infty)$: $f''(2) = 800 > 0$; f is concave upward.

Thus we see that the direction of concavity of f changes at each of the three points $x \approx -0.70$, $x = 0$, and $x \approx 1.07$. These three points are indeed inflection points. This information is shown in the graph of f sketched in Fig 4.6.20. ■

EXAMPLE 7 Sketch the graph of $f(x) = 4x^{1/3} + x^{4/3}$. Indicate local extrema, inflection points, and concave structure.

Solution First,

$$f'(x) = \frac{4}{3}x^{-2/3} + \frac{4}{3}x^{1/3} = \frac{4(x + 1)}{3x^{2/3}},$$

so the critical points are $x = -1$ (where the tangent line is horizontal) and $x = 0$ (where it is vertical). Next,

$$f''(x) = -\frac{8}{9}x^{-5/3} + \frac{4}{9}x^{-2/3} = \frac{4(x - 2)}{9x^{5/3}},$$

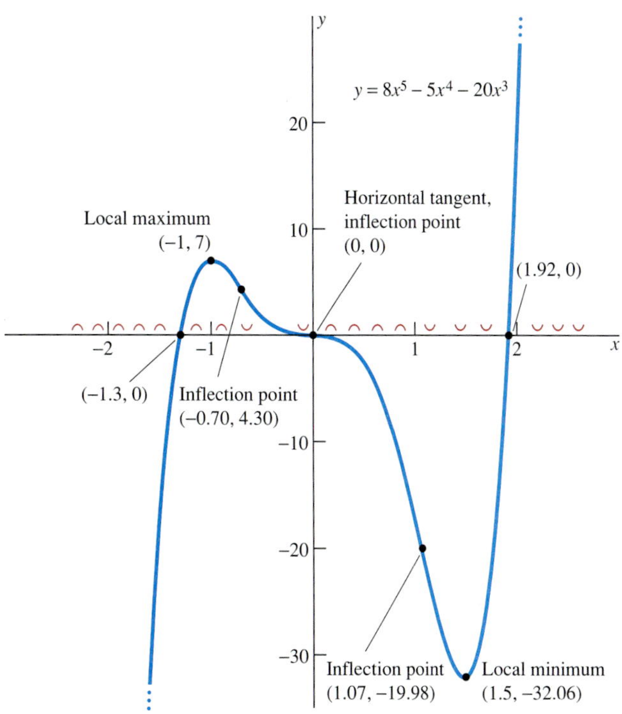

Fig. 4.6.20 The graph of the function of Example 6

so the possible inflection points are $x = 2$ (where $f''(x) = 0$) and $x = 0$ (where $f''(x)$ does not exist).

To determine where f is increasing and where it is decreasing, we construct the following table.

Interval	$x + 1$	$x^{2/3}$	$f'(x)$	f
$(-\infty, -1)$	Neg.	Pos.	Neg.	Decreasing
$(-1, 0)$	Pos.	Pos.	Pos.	Increasing
$(0, +\infty)$	Pos.	Pos.	Pos.	Increasing

Thus f is decreasing when $x < -1$ and increasing when $x > -1$ (Fig. 4.6.21).

To determine the concavity of f, we construct a table to find the sign of $f''(x)$ on each of the intervals separated by its zeros.

Interval	$x^{5/3}$	$x - 2$	$f''(x)$	f
$(-\infty, 0)$	Neg.	Neg.	Pos.	Concave upward
$(0, 2)$	Pos.	Neg.	Neg.	Concave downward
$(2, +\infty)$	Pos.	Pos.	Pos.	Concave upward

The table shows that f is concave downward on $(0, 2)$ and concave upward for $x < 0$ and for $x > 2$ (Fig. 4.6.22).

We note that $f(x) \to +\infty$ as $x \to \pm\infty$, and we mark with plus signs the intervals on the x-axis where f is increasing, minus signs where it is decreasing, cups opening upward

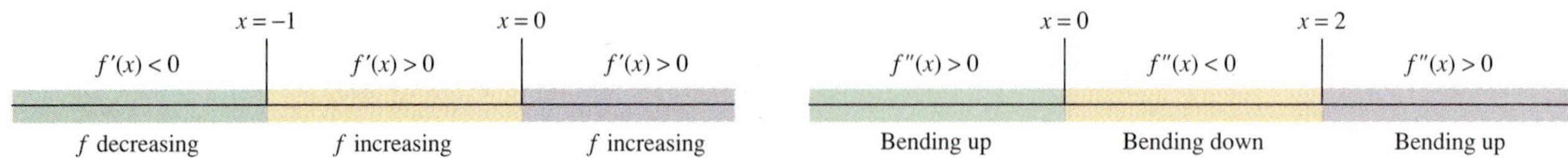

Fig. 4.6.21 Increasing and decreasing intervals of Example 7

Fig. 4.6.22 Intervals of concavity of Example 7

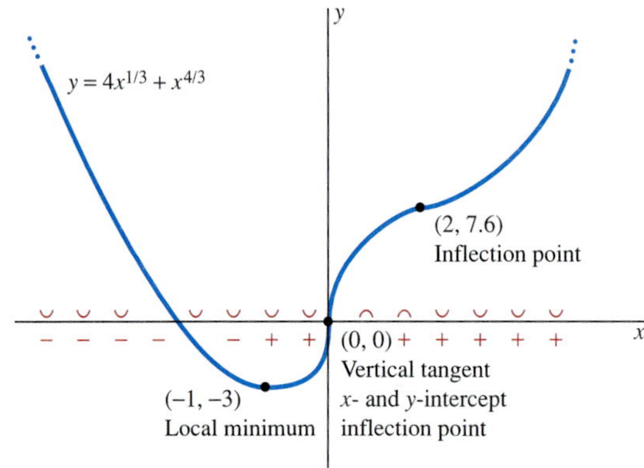

Fig. 4.6.23 The graph of the function of Example 7

where f is concave upward, and cups opening downward where f is concave downward. We plot (at least approximately) the points on the graph of f that correspond to the zeros and discontinuities of f' and f''; these are $(-1,-3)$, $(0,0)$, and $(2, 6\sqrt[3]{2})$. Finally, we use all this informtion to draw the smooth curve shown in Fig. 4.6.23. ■

Proof of Theorem 2 We will prove only part 1—the proof of part 2 is similar. Given a fixed point a of the open interval I where $f''(x) > 0$, we want to show that the graph $y = f(x)$ lies above the tangent line at $(a, f(a))$. The tangent line in question has the equation

$$y = T(x) = f(a) + f'(a)\cdot(x - a). \tag{5}$$

Consider the auxiliary function

$$g(x) = f(x) - T(x) \tag{6}$$

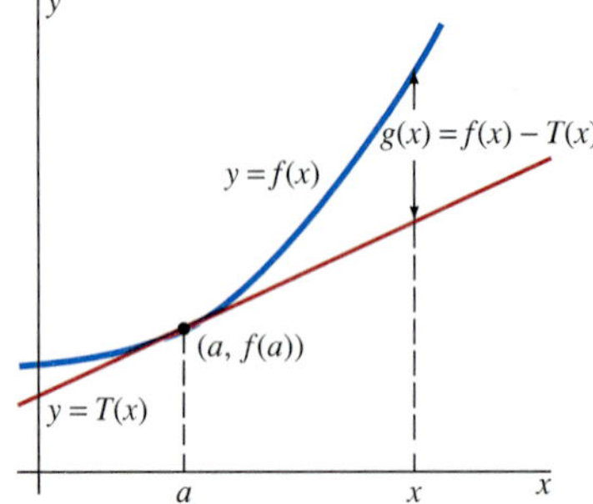

Fig. 4.6.24 Illustrating the proof of Theorem 2

illustrated in Fig. 4.6.24. Note first that $g(a) = g'(a) = 0$, so $x = a$ is a critical point of g. Moreover, Eq. (5) implies that $T'(x) \equiv f'(a)$ and that $T''(x) \equiv 0$, so

$$g''(x) = f''(x) - T''(x) = f''(x) > 0$$

at each point of I. Therefore, the second derivative test implies that $g(a) = 0$ is the minimum value of $g(x) = f(x) - T(x)$ on I. It follows that the curve $y = f(x)$ lies above the tangent line $y = T(x)$. ■

4.6 PROBLEMS

Calculate the first three derivatives of the functions given in Problems 1 through 15.

1. $f(x) = 2x^4 - 3x^3 + 6x - 17$

2. $f(x) = 2x^5 + x^{3/2} - \dfrac{1}{2x}$

3. $f(x) = \dfrac{2}{(2x - 1)^2}$

4. $g(t) = t^2 + \sqrt{t + 1}$

5. $g(t) = (3t - 2)^{4/3}$

6. $f(x) = x\sqrt{x + 1}$

7. $h(y) = \dfrac{y}{y + 1}$

8. $f(x) = (1 + \sqrt{x})^3$

9. $g(t) = \dfrac{1}{2\sqrt{t}} - \dfrac{3}{\sqrt[3]{1-t}}$ **10.** $h(z) = \dfrac{z^2}{z^2+4}$

11. $f(x) = \sin 3x$ **12.** $f(x) = \cos^2 2x$

13. $f(x) = \sin x \cos x$ **14.** $f(x) = x^2 \cos x$

15. $f(x) = \dfrac{\sin x}{x}$

In Problems 16 through 22, calculate dy/dx and d^2y/dx^2, assuming that y is defined implicitly as a function of x by the given equation.

16. $x^2 + y^2 = 4$ **17.** $x^2 + xy + y^2 = 3$

18. $x^{1/3} + y^{1/3} = 1$ **19.** $y^3 + x^2 + x = 5$

20. $\dfrac{1}{x} + \dfrac{1}{y} = 1$ **21.** $\sin y = xy$

22. $\sin^2 x + \cos^2 y = 1$

In Problems 23 through 30, find the exact coordinates of the inflection points and critical points marked on the given graph.

23. The graph of $f(x) = x^3 - 3x^2 - 45x$ (Fig. 4.6.25)

24. The graph of $f(x) = 2x^3 - 9x^2 - 108x + 200$ (Fig. 4.6.26)

25. The graph of $f(x) = 4x^3 - 6x^2 - 189x + 137$ (Fig. 4.6.27)

26. The graph of $f(x) = -40x^3 - 171x^2 + 2550x + 4150$ (Fig. 4.6.28)

27. The graph of $f(x) = x^4 - 54x^2 + 237$ (Fig. 4.6.29)

28. The graph of $f(x) = x^4 - 10x^3 - 250$ (Fig. 4.6.30)

29. The graph of $f(x) = 3x^5 - 20x^4 + 1000$ (Fig. 4.6.31)

30. The graph of $f(x) = 3x^5 - 160x^3$ (Fig. 4.6.32)

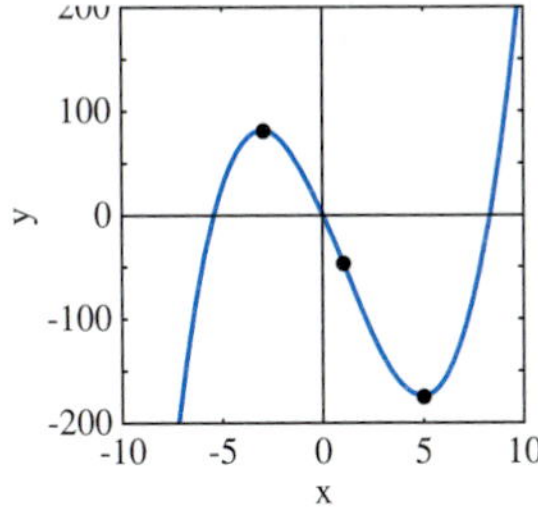

Fig. 4.6.25 The graph of $f(x) = x^3 - 3x^2 - 45x$ (Problem 23)

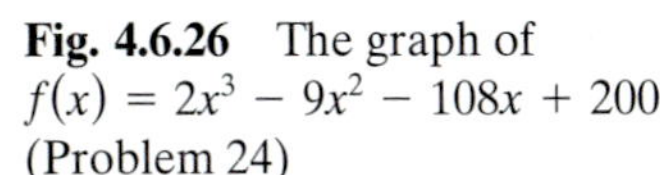

Fig. 4.6.26 The graph of $f(x) = 2x^3 - 9x^2 - 108x + 200$ (Problem 24)

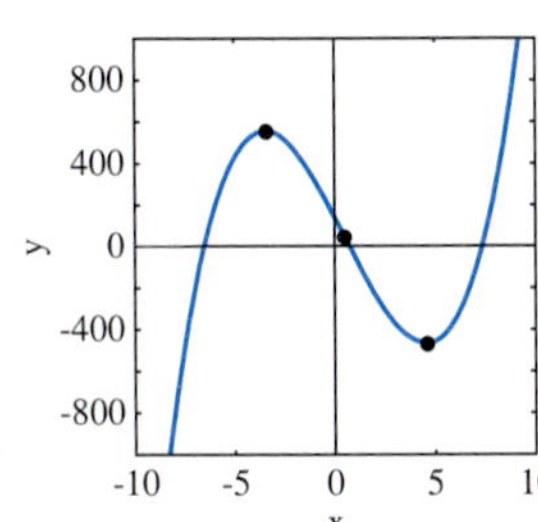

Fig. 4.6.27 The graph of $f(x) = 4x^3 - 6x^2 - 189x + 137$ (Problem 25)

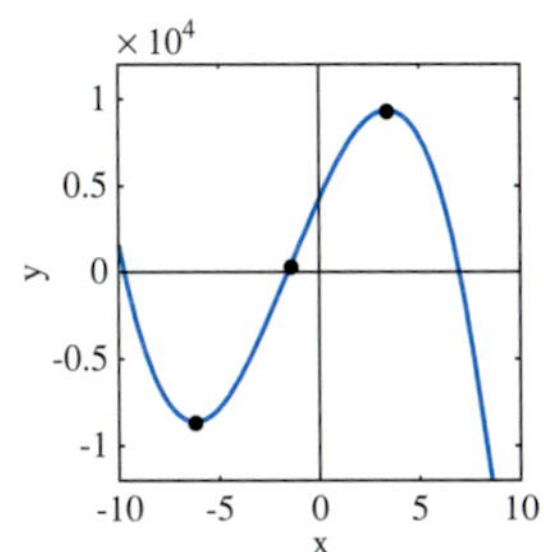

Fig. 4.6.28 The graph of $f(x) = -40x^3 - 171x^2 + 2550x + 4150$ (Problem 26)

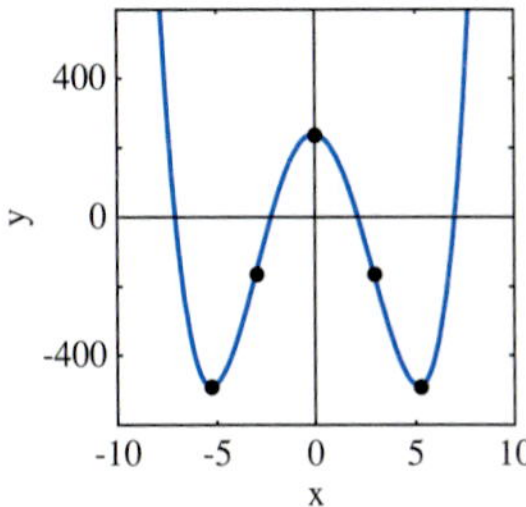

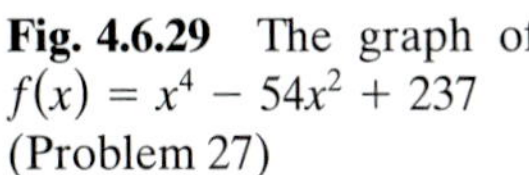

Fig. 4.6.29 The graph of $f(x) = x^4 - 54x^2 + 237$ (Problem 27)

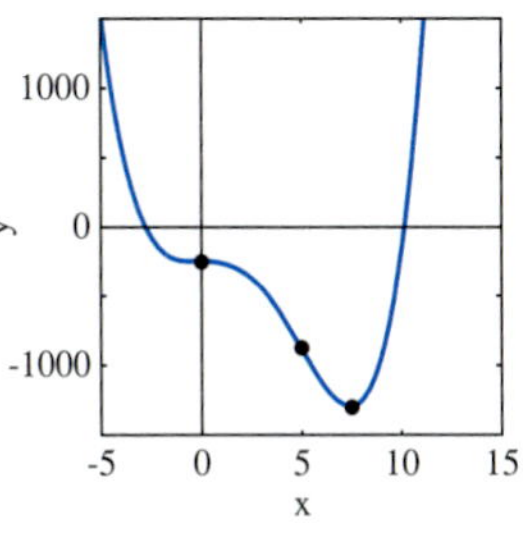

Fig. 4.6.30 The graph of $f(x) = x^4 - 10x^3 - 250$ (Problem 28)

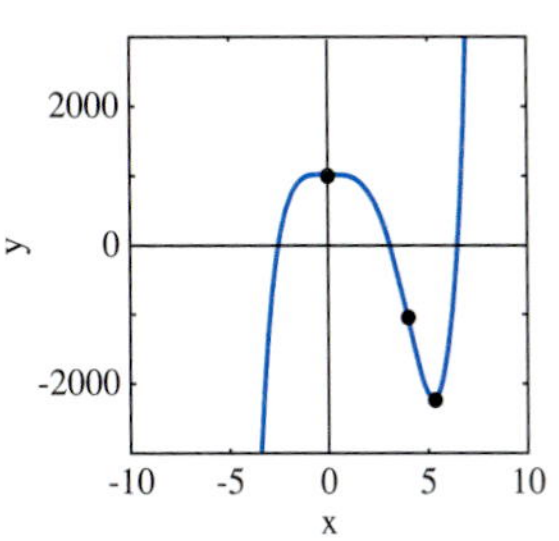

Fig. 4.6.31 The graph of $f(x) = 3x^5 - 20x^4 + 1000$ (Problem 29)

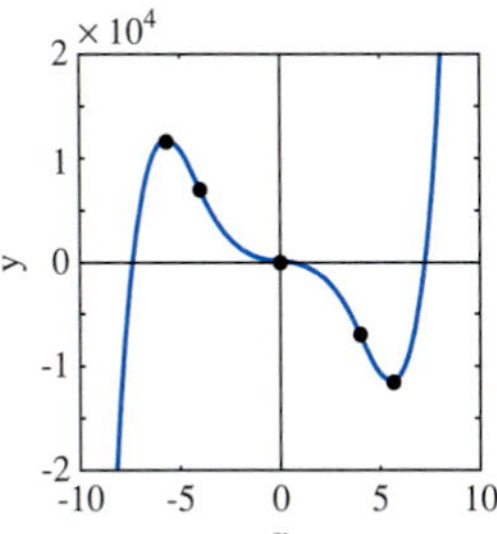

Fig. 4.6.32 The graph of $f(x) = 3x^5 - 160x^3$ (Problem 30)

Apply the second derivative test to find the local maxima and local minima of the functions given in Problems 31 through 50, and apply the inflection point test to find all inflection points.

31. $f(x) = x^2 - 4x + 3$ **32.** $f(x) = 5 - 6x - x^2$

33. $f(x) = x^3 - 3x + 1$ **34.** $f(x) = x^3 - 3x^2$

35. $f(x) = x^3$ **36.** $f(x) = x^4$

37. $f(x) = x^5 + 2x$ **38.** $f(x) = x^4 - 8x^2$

39. $f(x) = x^2(x-1)^2$ **40.** $f(x) = x^3(x+2)^2$

41. $f(x) = \sin x$ on $(0, 2\pi)$

42. $f(x) = \cos x$ on $(-\pi/2, \pi/2)$

43. $f(x) = \tan x$ on $(-\pi/2, \pi/2)$

44. $f(x) = \sec x$ on $(-\pi/2, \pi/2)$

45. $f(x) = \cos^2 x$ on $(-\pi/2, 3\pi/2)$

46. $f(x) = \sin^3 x$ on $(-\pi, \pi)$

47. $f(x) = \sin x + \cos x$ on $(0, 2\pi)$

48. $f(x) = \cos x - \sin x$ on $(0, 2\pi)$

49. $f(x) = \sin x + 2\cos x$ on $(0, 2\pi)$

50. $f(x) = 3\sin x - 4\cos x$ on $(0, 2\pi)$

In Problems 51 through 62, rework the indicated problem from Section 4.4, this time using the second derivative test to verify that you have found the desired absolute maximum or minimum value.

51. Problem 27 **52.** Problem 28

53. Problem 29 **54.** Problem 30

55. Problem 31 **56.** Problem 32

57. Problem 33 **58.** Problem 36

59. Problem 37 **60.** Problem 38

61. Problem 39 **62.** Problem 40

Sketch the graphs of the functions in Problems 63 through 76, indicating all critical points and inflection points. Apply the second derivative test at each critical point. Show the correct concave structure and indicate the behavior of $f(x)$ as $x \to \pm\infty$.

63. $f(x) = 2x^3 - 3x^2 - 12x + 3$

64. $f(x) = 3x^4 - 4x^3 - 5$

65. $f(x) = 6 + 8x^2 - x^4$ **66.** $f(x) = 3x^5 - 5x^3$

67. $f(x) = 3x^4 - 4x^3 - 12x^2 - 1$

68. $f(x) = 3x^5 - 25x^3 + 60x$

69. $f(x) = x^3(x - 1)^4$ **70.** $f(x) = (x - 1)^2(x + 2)^3$

71. $f(x) = 1 + x^{1/3}$ **72.** $f(x) = 2 - (x - 3)^{1/3}$

73. $f(x) = (x + 3)\sqrt{x}$ **74.** $f(x) = x^{2/3}(5 - 2x)$

75. $f(x) = (4 - x)\sqrt[3]{x}$ **76.** $f(x) = x^{1/3}(6 - x)^{2/3}$

Problems 77 through 82 show the graph of a function $f(x)$. Match each with the graph of its second derivative $f''(x)$ in Fig. 4.6.33.

77. See Fig. 4.6.34. **78.** See Fig. 4.6.35.

79. See Fig. 4.6.36. **80.** See Fig. 4.6.37.

81. See Fig. 4.6.38. **82.** See Fig. 4.6.39.

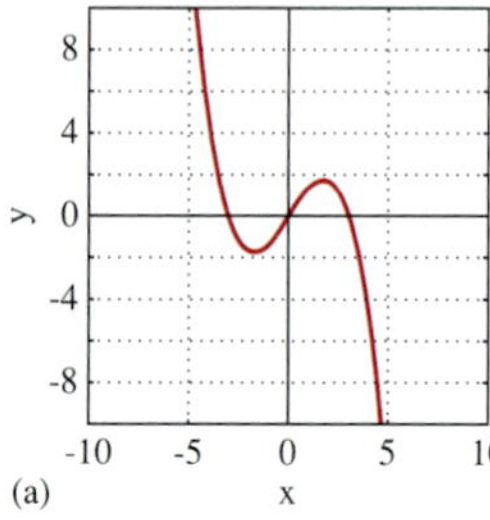

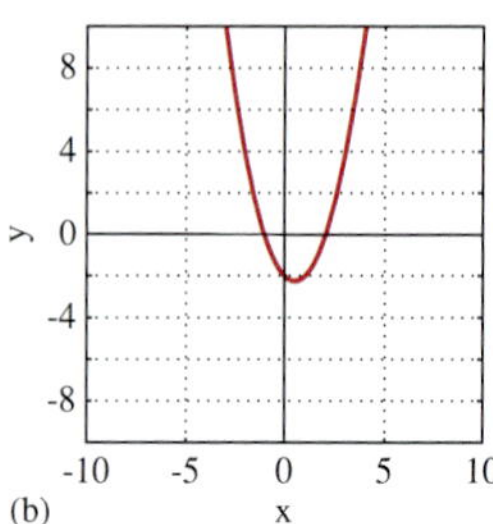

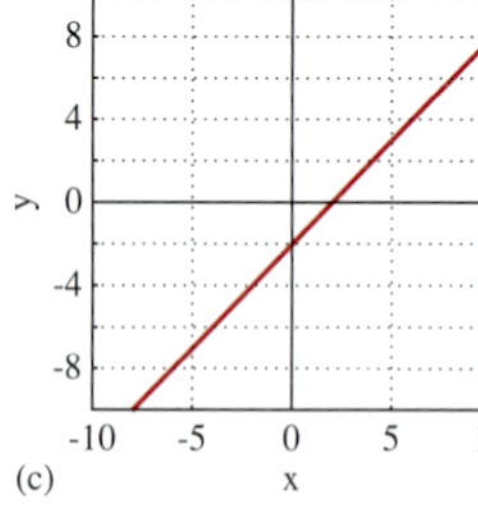

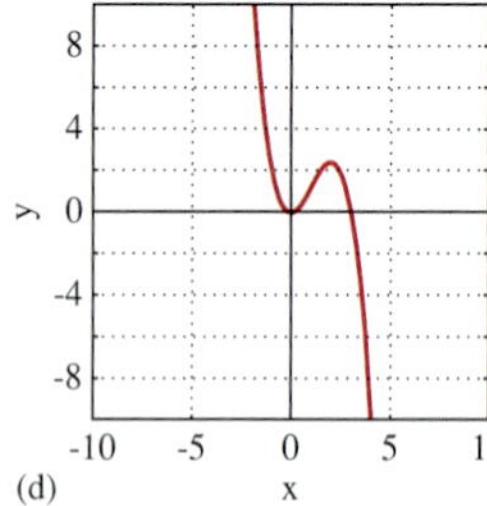

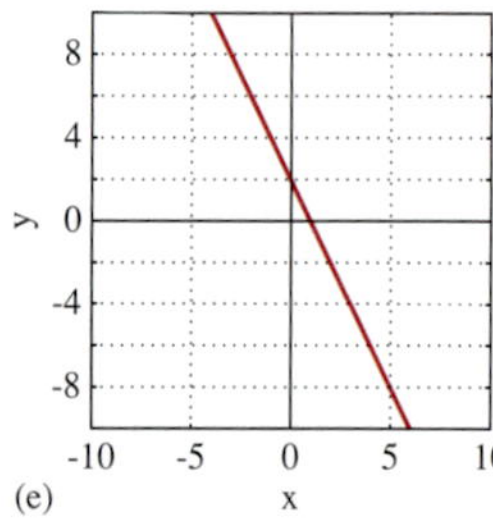

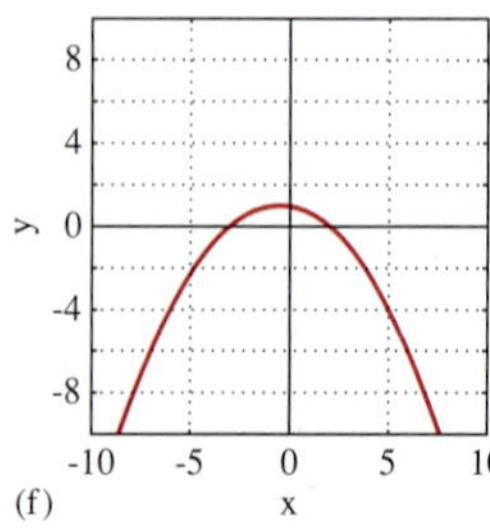

Fig. 4.6.33

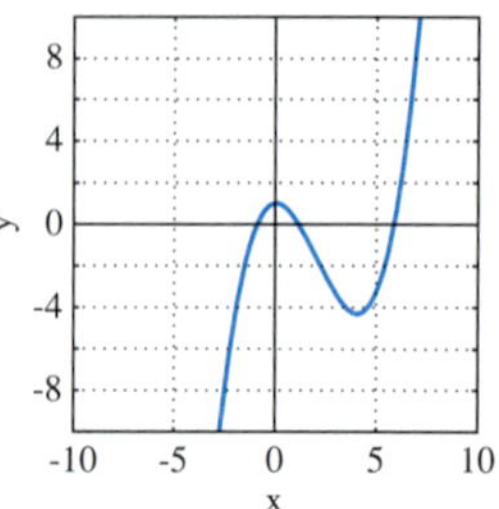

Fig. 4.6.34

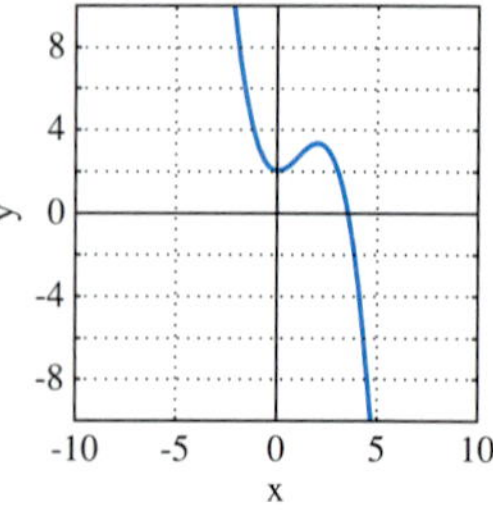

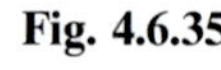

Fig. 4.6.35

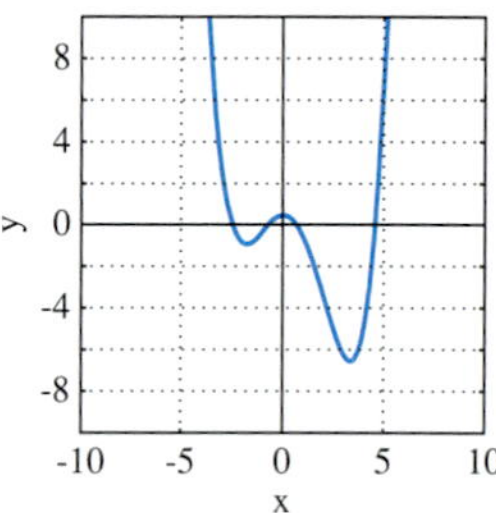

Fig. 4.6.36

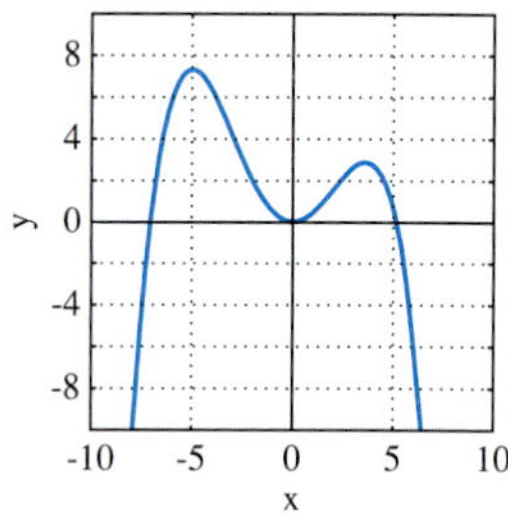

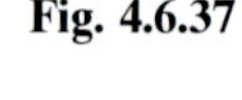

Fig. 4.6.37

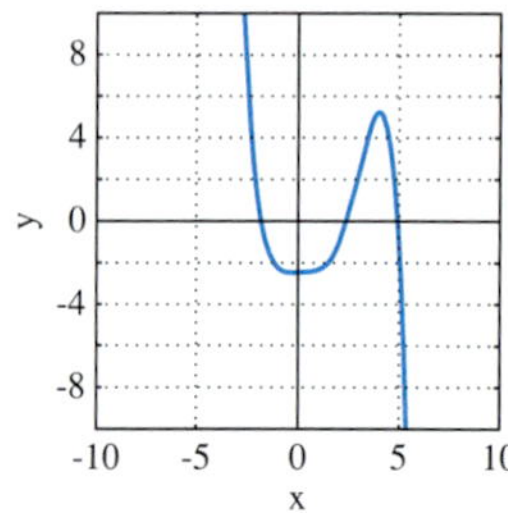

Fig. 4.6.38

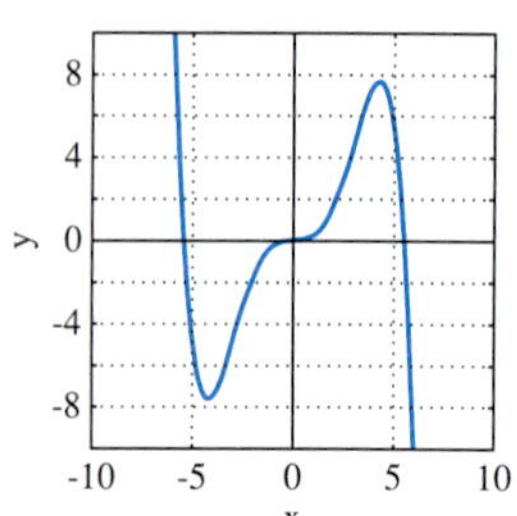

Fig. 4.6.39

83. (a) Show first that the nth derivative of $f(x) = x^n$ is

$$f^{(n)}(x) \equiv n! = n\cdot(n-1)\cdot(n-2)\cdots 3\cdot 2\cdot 1.$$

(b) Conclude that if $f(x)$ is a polynomial of degree n, then $f^{(k)}(x) \equiv 0$ if $k > n$.

84. (a) Calculate the first four derivatives of $f(x) = \sin x$. (b) Explain why it follows that $D_x^{n+4} \sin x = D_x^n \sin x$ if n is a positive integer.

85. Suppose that $z = g(y)$ and that $y = f(x)$. Show that

$$\frac{d^2z}{dx^2} = \frac{d^2z}{dy^2}\left(\frac{dy}{dx}\right)^2 + \frac{dz}{dy}\cdot\frac{d^2y}{dx^2}.$$

86. Prove that the graph of a quadratic polynomial has no inflection points.

87. Prove that the graph of a cubic polynomial has exactly one inflection point.

88. Prove that the graph of a polynomial function of degree 4 has either no inflection point or exactly two inflection points.

89. Suppose that the pressure p (in atmospheres), volume V (in cubic centimeters), and temperature T (in kelvins) of

n moles of carbon dioxide (CO_2) satisfies van der Waals' equation

$$\left(p + \frac{n^2 a}{V^2}\right)(V - nb) = nRT,$$

where a, b, and R are empirical constants. The following experiment was carried out to find the values of these constants.

One mole of CO_2 was compressed at the constant temperature $T = 304$ K. The measured pressure-volume (pV) data were then plotted as in Fig. 4.6.40, with the pV curve showing an inflection point coinciding with a horizontal tangent at $V = 128.1$, $p = 72.8$. Use this information to calculate a, b, and R. (*Suggestion:* Solve van der Waals' equation for p and then calculate dp/dV and d^2p/dV^2.)

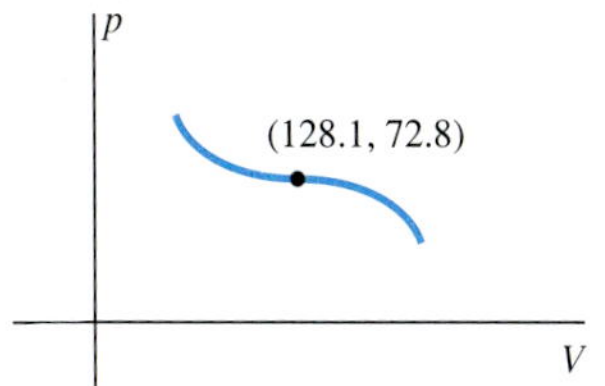

Fig. 4.6.40 A problem involving van der Waals' equation

90. Suppose that the function f is differentiable on an open interval containing the point c at which $f'(c) = 0$ and that the second derivative

$$f''(c) = \lim_{h \to 0} \frac{f'(c+h) - f'(c)}{h} = \lim_{h \to 0} \frac{f'(c+h)}{h}$$

exists. (a) First assume that $f''(c) > 0$. Reason that $f'(c + h)$ and h have the same sign if $h \neq 0$ is sufficiently small. Hence apply the first derivative test to show in this case that $f(c)$ is a local minimum value of f. (b) Show similarly that $f(c)$ is a local maximum value of f if $f''(c) < 0$.

4.6 PROJECT: INVISIBLE CRITICAL POINTS AND INFLECTION POINTS

The investigations described next deal with curves having critical and inflection points that are not clearly visible on their initially plotted graphs, because the "action" takes place on different scales on the x- and y-axes.

Investigation A Let the coefficients a, b, and c be defined as

$$a = 3011 + 20n,$$

$$b = 3022 + 40n, \quad \text{and}$$

$$c = 1010 + 20n,$$

where n denotes the last nonzero digit of your student I.D. number, and consider the cubic curve

$$y = 1000x^3 - ax^2 + bx + c. \tag{1}$$

The sizes of the coefficients suggest that y be measured in thousands while x is measured in units. The graph of this cubic curve is plotted on these scales in Fig. 4.6.41, and it appears to exhibit a single combination critical/inflection point somewhere near the point $(1, 2000)$. Nevertheless, the curve in (1) has "two real wiggles like a good cubic should." Find them!

In particular, find the local maximum and minimum points and the inflection point (or points!) on this curve, giving their coordinates accurate to four decimal places (to the right of the decimal point). Then produce a graph that plainly exhibits all these points—and mark and label each. As you zoom in with a calculator or computer, you'll need to control carefully the successive viewing windows.

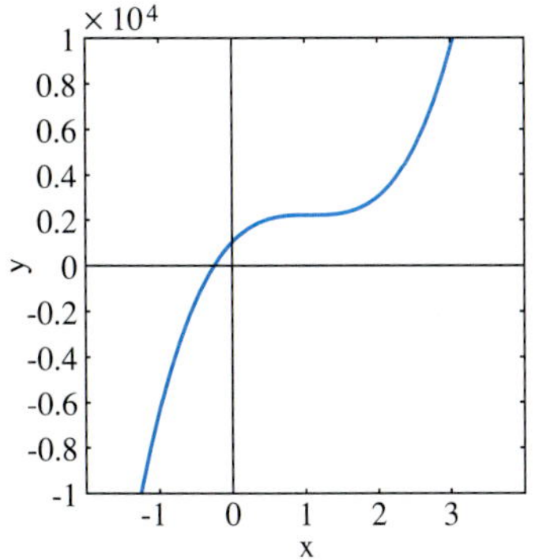

Fig. 4.6.41 The cubic graph of Investigation A

Investigation B Your task now is to investigate the curve

$$f(x) = [x(1 - x)(ax - b)(cx - d)]^4, \tag{2}$$

where a, b, c, and d are positive integers with $a > b > c > d$ and $a/b \neq c/d$. For instance, you might pick the four largest digits of your student I.D. number.

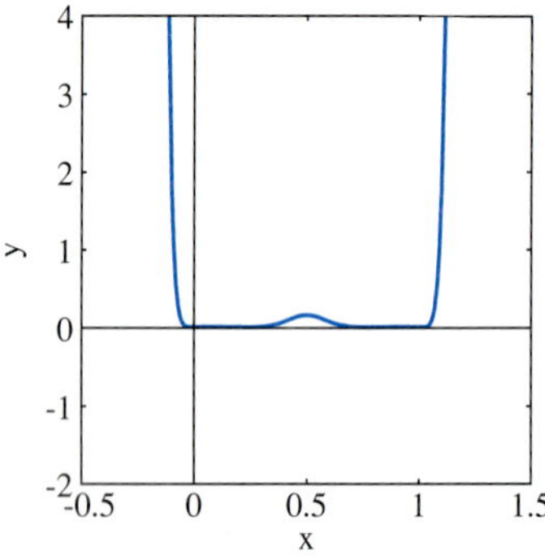

Fig. 4.6.42 The graph of Investigation B

Figure 4.6.42 shows the graph of $f(x)$ with $a = 9, b = 7, c = 4$, and $d = 1$. At first glance it appears that there is a local maximum near $x = \frac{1}{2}$, with "flat spots" along the x-axis to the left and to the right. But no nonconstant polynomial graph can actually have a "flat spot" where $y = 0$ on an open interval on the x-axis. (Why not?) Indeed—subject to the conditions we have imposed on a, b, c, and d—our graph really has four local minima, three local maxima, and six inflection points in the interval $0 \leqq x \leqq 1$. Find the approximate coordinates of all thirteen of these points, and show the graph of Eq. (2) on a scale that makes all these points evident. You can either zoom in graphically on the solutions of the equations $f'(x) = 0$ and $f''(x) = 0$, or use the **SOLVE** command on your calculator or computer to find these solutions.

4.7 CURVE SKETCHING AND ASYMPTOTES

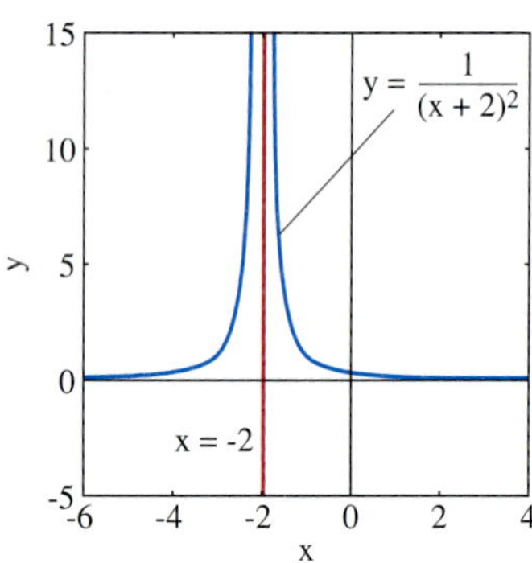

Fig. 4.7.1 $\dfrac{1}{(x+2)^2} \to \infty$ as $x \to -2$, and the line $x = -2$ is a vertical asymptote.

We now extend the limit concept to include infinite limits and limits at infinity. This extension will add a powerful weapon to our arsenal of curve-sketching techniques, the notion of an *asymptote* to a curve—a straight line that the curve approaches arbitrarily closely in a sense we soon make precise.

Recall from Section 2.3 that $f(x)$ is said to **increase without bound,** or **become infinite,** as x approaches a, and we write

$$\lim_{x \to a} f(x) = +\infty, \tag{1}$$

provided that $f(x)$ can be made arbitrarily large by choosing x sufficiently close (but not equal) to a. The statement that $f(x)$ **decreases without bound,** or **becomes negatively infinite,** as $x \to a$, written

$$\lim_{x \to a} f(x) = -\infty, \tag{2}$$

has an analogous definition.

EXAMPLE 1 It is apparent that

$$\lim_{x \to -2} \frac{1}{(x+2)^2} = +\infty$$

because, as $x \to -2$, $(x+2)^2$ is positive and approaches zero (see Fig. 4.7.1). By contrast,

$$\lim_{x \to -2} \frac{x}{(x+2)^2} = -\infty$$

because, as $x \to -2$, the denominator $(x+2)^2$ is still positive and approaches zero, but the numerator x is negative (see Fig. 4.7.2). Thus when x is very close to -2, we have in $x/(x+2)^2$ a negative number close to -2 divided by a very small positive number. Hence the quotient becomes a negative number of large magnitude ■

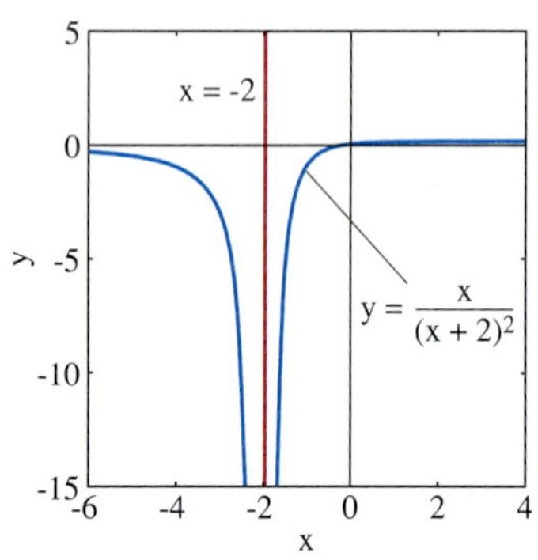

Fig. 4.7.2 $\dfrac{x}{(x+2)^2} \to -\infty$ as $x \to -2$, and the line $x = -2$ is a vertical asymptote

One-sided versions of Eqs. (1) and (2) are valid also. For instance, if n is an *odd* positive integer, then it is apparent that

$$\lim_{x \to 2^-} \frac{1}{(x-2)^n} = -\infty \quad \text{and} \quad \lim_{x \to 2^+} \frac{1}{(x-2)^n} = +\infty,$$

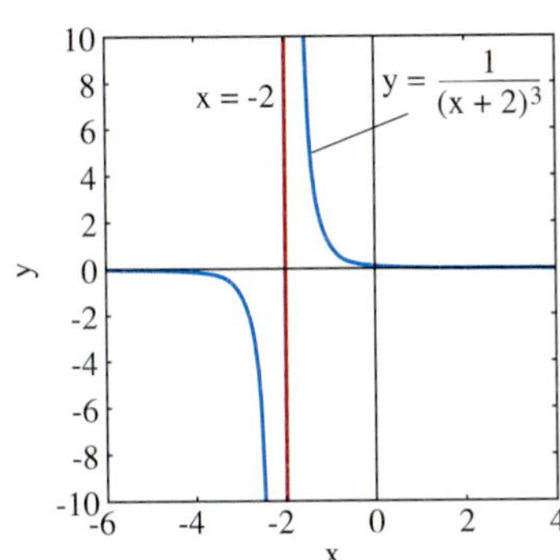

Fig. 4.7.3 $\dfrac{1}{(x+2)^3}$ has infinite one-sided limits as $x \to -2$, and the line $x = -2$ is a vertical asymptote

because $(x-2)^n$ is negative when x is to the left of 2 and positive when x is to the right of 2. The case $n = 3$ is illustrated in Fig. 4.7.3.

Vertical Asymptotes

The vertical lines at $x = -2$ in Figs. 4.7.1 through 4.7.3 are examples of *vertical asymptotes* associated with infinite limits. The line $x = a$ is a **vertical asymptote** of the curve $y = f(x)$ provided that either

$$\lim_{x \to a^-} f(x) = \pm\infty \tag{3a}$$

or

$$\lim_{x \to a^+} f(x) = \pm\infty \tag{3b}$$

or both. It is usually the case that both one-sided limits, rather than only one, are infinite. If so, we write

$$\lim_{x \to a} f(x) = \pm\infty. \tag{3c}$$

The geometry of a vertical asymptote is illustrated by the graphs in Figs. 4.7.1 through 4.7.3. In each case, as $x \to -2$ and $f(x) \to \pm\infty$, the point $(x, f(x))$ on the curve approaches the vertical asymptote $x = -2$ and the shape and direction of the curve are better and better approximated by the asymptote.

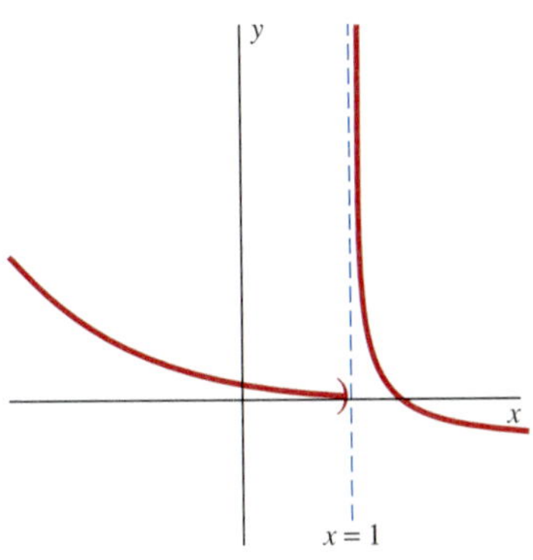

Fig. 4.7.4 A "right-hand only" vertical asymptote

Figure 4.7.4 shows the graph of a function whose left-hand limit is zero at $x = 1$. But the right-hand limit there is $+\infty$, which explains why the line $x = 1$ is a vertical asymptote for this graph. The right-hand limit in Fig. 4.7.5 doesn't even exist, but because the left-hand limit at $x = 1$ is $-\infty$, the vertical line at $x = 1$ is again a vertical asymptote.

A vertical asymptote typically appears in the case of a rational function $f(x) = p(x)/q(x)$ at a point $x = a$ where $q(a) = 0$ but $p(a) \neq 0$. (See Examples 4 through 8 later in this section.)

Limits at Infinity

In Section 4.5 we mentioned infinite limits at infinity in connection with the behavior of a polynomial as $x \to \pm\infty$. There is also such a thing as a *finite* limit at infinity. We say that $f(x)$ **approaches the number** L **as** x **increases without bound** and write

$$\lim_{x \to +\infty} f(x) = L \tag{4}$$

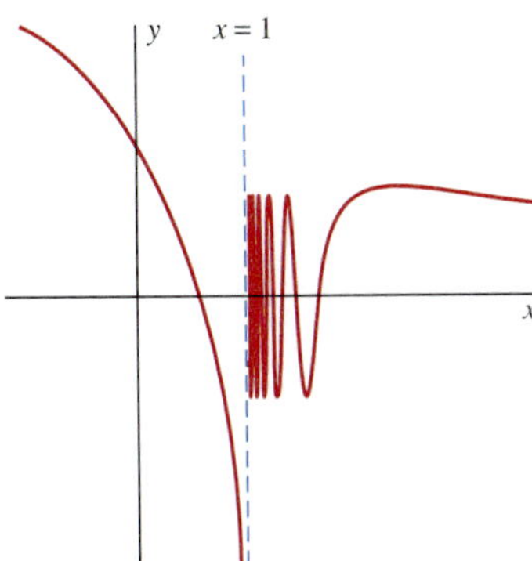

Fig. 4.7.5 The behavior of the graph to its *left* produces the vertical asymptote

provided that $|f(x) - L|$ can be made arbitrarily small (close to zero) merely by choosing x sufficiently large. That is, given $\epsilon > 0$, there exists $M > 0$ such that

$$x > M \quad \text{implies} \quad |f(x) - L| < \epsilon. \tag{5}$$

The statement that

$$\lim_{x \to -\infty} f(x) = L$$

has a definition of similar form—merely replace the condition $x > M$ with the condition $x < -M$.

The analogues for limits at infinity of the limit laws of Section 2.2 all hold, including, in particular, the sum, product, and quotient laws. In addition, it is not difficult to show that if

$$\lim_{x \to +\infty} f(x) = L \quad \text{and} \quad \lim_{x \to +\infty} g(x) = \pm\infty,$$

then

$$\lim_{x\to+\infty}\frac{f(x)}{g(x)}=0.$$

It follows from this result that

$$\lim_{x\to+\infty}\frac{1}{x^k}=0 \tag{6}$$

for any choice of the positive rational number k.

Using Eq. (6) and the limit laws, we can easily evaluate limits at infinity of rational functions. The general method is this: First divide each term in both the numerator and the denominator by the highest power of x that appears in any of the terms. Then apply the limit laws.

EXAMPLE 2 Find

$$\lim_{x\to+\infty} f(x) \quad \text{if} \quad f(x)=\frac{3x^3-x}{2x^3+7x^2-4}.$$

Solution We begin by dividing each term in the numerator and denominator by x^3:

$$\lim_{x\to+\infty}\frac{3x^3-x}{2x^3+7x^2-4}=\lim_{x\to+\infty}\frac{3-\dfrac{1}{x^2}}{2+\dfrac{7}{x}-\dfrac{4}{x^3}}$$

$$=\frac{\displaystyle\lim_{x\to+\infty}\left(3-\frac{1}{x^2}\right)}{\displaystyle\lim_{x\to+\infty}\left(2+\frac{7}{x}-\frac{4}{x^3}\right)}=\frac{3-0}{2+0-0}=\frac{3}{2}.$$

The same computation, but with $x\to-\infty$, also gives the result

$$\lim_{x\to-\infty} f(x)=\frac{3}{2}.$$ ■

EXAMPLE 3 Find $\lim_{x\to+\infty}(\sqrt{x+a}-\sqrt{x})$.

Solution We use the familiar "divide and multiply" technique with the conjugate of $\sqrt{x+a}-\sqrt{x}$:

$$\lim_{x\to+\infty}(\sqrt{x+a}-\sqrt{x})=\lim_{x\to+\infty}(\sqrt{x+a}-\sqrt{x})\cdot\frac{\sqrt{x+a}+\sqrt{x}}{\sqrt{x+a}+\sqrt{x}}$$

$$=\lim_{x\to+\infty}\frac{a}{\sqrt{x+a}+\sqrt{x}}=0.$$ ■

Horizontal Asymptotes

In geometric terms, the statement

$$\lim_{x\to+\infty} f(x)=L$$

means that the point $(x, f(x))$ on the curve $y=f(x)$ approaches the horizontal line $y=L$ as $x\to+\infty$. In particular, with the numbers M and ϵ of the condition in Eq. (5), the part of the curve for which $x>M$ lies between the horizontal lines $y=L-\epsilon$

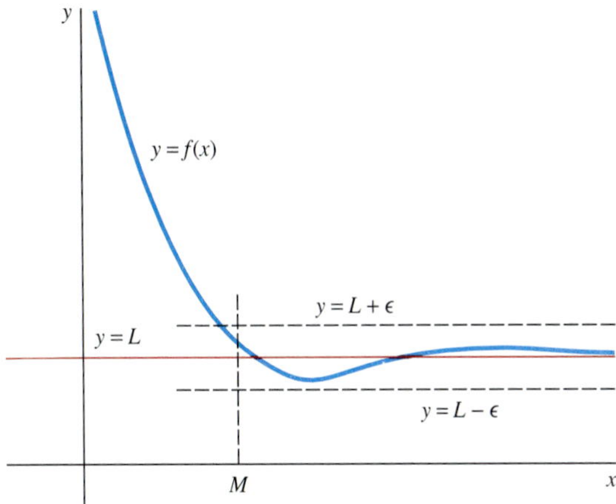

Fig. 4.7.6 Geometry of the definition of horizontal asymptote

and $y = L + \epsilon$ (Fig. 4.7.6). Therefore we say that the line $y = L$ is a **horizontal asymptote** of the curve $y = f(x)$ if either

$$\lim_{x\to+\infty} f(x) = L \quad \text{or} \quad \lim_{x\to-\infty} f(x) = L.$$

EXAMPLE 4 Sketch the graph of $f(x) = x/(x - 2)$. Indicate any horizontal or vertical asymptotes.

Solution First we note that $x = 2$ is a vertical asymptote because $|f(x)| \to +\infty$ as $x \to 2$. Also,

$$\lim_{x\to\pm\infty} \frac{x}{x-2} = \lim_{x\to\pm\infty} \frac{1}{1 - \dfrac{2}{x}} = \frac{1}{1-0} = 1.$$

So the line $y = 1$ is a horizontal asymptote. The first two derivatives of f are

$$f'(x) = -\frac{2}{(x-2)^2} \quad \text{and} \quad f''(x) = \frac{4}{(x-2)^3}.$$

Neither $f'(x)$ nor $f''(x)$ is zero anywhere, so the function f has no critical points and no inflection points. Because $f'(x) < 0$ for $x \neq 2$, we see that $f(x)$ is decreasing on the open intervals $(-\infty, 2)$ and $(2, +\infty)$. And because $f''(x) < 0$ for $x < 2$ and $f''(x) > 0$ for $x > 2$, the graph of f is concave downward on $(-\infty, 2)$ and concave upward on $(2, +\infty)$. The graph of f appears in Fig. 4.7.7. ■

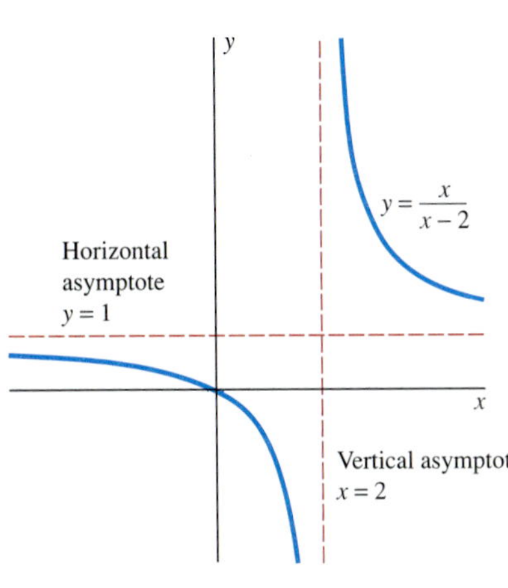

Fig. 4.7.7 The graph for Example 4

EXAMPLE 5 Let's reexamine the function

$$f(x) = \frac{x}{(x+2)^2}$$

whose graph was shown in Fig. 4.7.2. We note that

$$\lim_{x\to\infty} \frac{x}{(x+2)^2} = \lim_{x\to\infty} \frac{\dfrac{1}{x}}{\left(1 + \dfrac{2}{x}\right)^2} = 0,$$

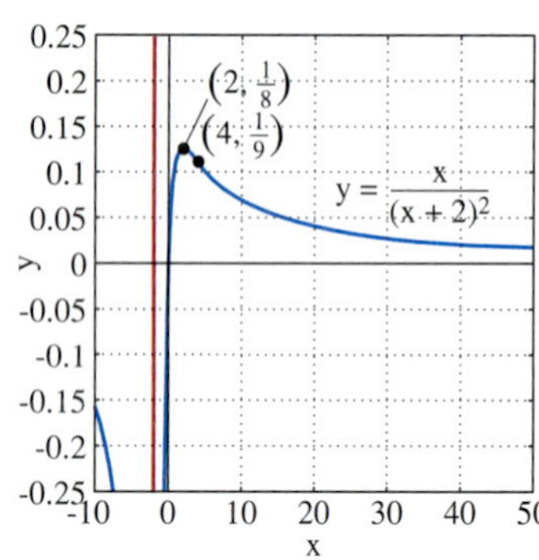

Fig. 4.7.8 $\frac{x}{(x+2)^2} \to 0$ as $x \to \infty$, so the x-axis $y = 0$ is a horizontal asymptote

so the x-axis $y = 0$ is a horizontal asymptote of the graph $y = f(x)$. We must change the viewing window to see clearly the behavior of this curve for $x > 0$. With the window $-10 < x < 50$, $-0.25 < y < 0.25$ of Fig. 4.7.8 we see that $f(x)$ appears to attain a local maximum value near the point where $x = 2$ before approaching zero as $x \to \infty$. Indeed, upon differentiating f and simplifying the result, we see that

$$f'(x) = \frac{2 - x}{(x + 2)^3},$$

so the indicated maximum point on the curve is $(2, \frac{1}{8})$. The second derivative of f is

$$f''(x) = \frac{2(x - 4)}{(x + 2)^4},$$

and it follows that the inflection point apparent in Fig. 4.7.8 is at $(4, \frac{1}{9})$. ■

Curve-Sketching Strategy

The curve-sketching techniques of Sections 4.5 and 4.6, together with those of this section, can be summarized as a list of steps. If you follow these steps, loosely rather than rigidly, you will obtain a qualitatively accurate sketch of the graph of a given function f.

1. Solve the equation $f'(x) = 0$ and also find where $f'(x)$ does not exist. This gives the critical points of f. Note whether the tangent line is horizontal, vertical, or nonexistent at each critical point.
2. Determine the intervals on which f is increasing and those on which it is decreasing.
3. Solve the equation $f''(x) = 0$ and also find where $f''(x)$ does not exist. These points are the *possible* inflection points of the graph.
4. Determine the intervals on which the graph of f is concave upward and those on which it is concave downward.
5. Find the y-intercepts and the x-intercepts (if any) of the graph.
6. Plot and label the critical points, possible inflection points, and intercepts.
7. Determine the asymptotes (if any), discontinuities (if any), and *especially* the behavior of $f(x)$ and $f'(x)$ near discontinuities of f. Also determine the behavior of $f(x)$ as $x \to +\infty$ and as $x \to -\infty$.
8. Finally, join the plotted points with a curve that is consistent with the information you have gathered. Remember that corner points are rare and that straight sections of graph are even rarer.

You may follow these steps in any convenient order and omit any that present major computational difficulties. Many problems require fewer than all eight steps; see Example 4. But Example 6 requires all these steps.

EXAMPLE 6 Sketch the graph of

$$f(x) = \frac{2 + x - x^2}{(x - 1)^2}.$$

Solution We notice immediately that

$$\lim_{x \to 1} f(x) = +\infty,$$

because the numerator approaches 2 as $x \to 1$ and the denominator approaches zero through *positive* values. So the line $x = 1$ is a vertical asymptote. Also,

$$\lim_{x\to\pm\infty}\frac{2+x-x^2}{(x-1)^2}=\lim_{x\to\pm\infty}\frac{\dfrac{2}{x^2}+\dfrac{1}{x}-1}{\left(1-\dfrac{1}{x}\right)^2}=-1,$$

so the line $y=-1$ is a horizontal asymptote (in both the positive and the negative directions).

Next we apply the quotient rule and simplify to find that

$$f'(x)=\frac{x-5}{(x-1)^3}.$$

Thus the only critical point in the domain of f is $x=5$, and we plot the point $(5, f(5))=(5, -\frac{9}{8})$ on a convenient coordinate plane and mark the horizontal tangent there. To determine the increasing or decreasing behavior of f, we use both the critical point $x=5$ and the point $x=1$ (where f' is not defined) to separate the x-axis into open intervals. Here are the results.

Interval	$(x-1)^3$	$x-5$	$f'(x)$	f
$(-\infty, 1)$	Neg.	Neg.	Pos.	Increasing
$(1, 5)$	Pos.	Neg.	Neg.	Decreasing
$(5, +\infty)$	Pos.	Pos.	Pos.	Increasing

After some simplifications, we find the second derivative to be

$$f''(x)=\frac{2(7-x)}{(x-1)^4}.$$

The only possible inflection point is at $x=7$, corresponding to the point $(7, -\frac{10}{9})$ on the graph. We use both $x=7$ and $x=1$ (where f'' is undefined) to separate the x-axis into open intervals. The concave structure of the graph can be deduced with the aid of the next table.

Interval	$(x-1)^4$	$7-x$	$f''(x)$	f
$(-\infty, 1)$	Pos.	Pos.	Pos.	Concave upward
$(1, 7)$	Pos.	Pos.	Pos.	Concave upward
$(7, +\infty)$	Pos.	Neg.	Neg.	Concave downward

The y-intercept of f is $(0, 2)$, and the equation $2+x-x^2=0$ readily yields the x-intercepts $(-1, 0)$ and $(2, 0)$. We plot these intercepts, sketch the asymptotes, and finally sketch the graph with the aid of the two tables; their information now is symbolized along the x-axis in Fig. 4.7.9. ■

Slant Asymptotes

Not all asymptotes are horizontal or vertical—some are inclined. The nonvertical line $y=mx+b$ is an **asymptote** for the curve $y=f(x)$ provided that either

$$\lim_{x\to+\infty}[f(x)-(mx+b)]=0 \tag{7a}$$

or

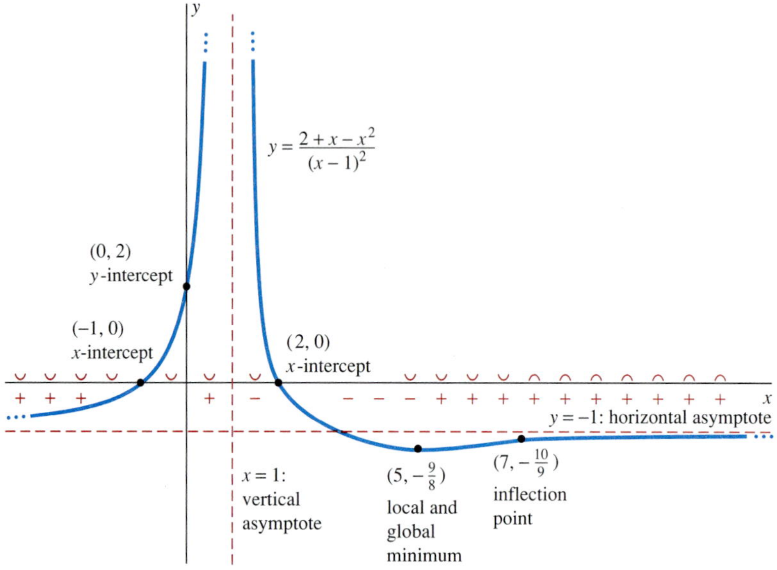

Fig. 4.7.9 Graphing the function of Example 6

$$\lim_{x \to -\infty} [f(x) - (mx + b)] = 0 \tag{7b}$$

(or both). These conditions mean that as $x \to +\infty$ or as $x \to -\infty$ (or both), the vertical distance between the point $\big(x, f(x)\big)$ on the curve and the point $(x, mx + b)$ on the line approaches zero.

Suppose that $f(x) = p(x)/q(x)$ is a rational function for which the degree of $p(x)$ is greater by 1 than the degree of $q(x)$. Then, by long division of $q(x)$ into $p(x)$, we find that $f(x)$ has the form

$$f(x) = mx + b + g(x)$$

where $m \neq 0$ and

$$\lim_{x \to \pm\infty} g(x) = 0.$$

Thus the nonvertical line $y = mx + b$ is an asymptote of the graph of $y = f(x)$. Such an asymptote is called a **slant** asymptote.

EXAMPLE 7 Sketch the graph of

$$f(x) = \frac{x^2 + x - 1}{x - 1}.$$

Solution The long division suggested previously takes the form

$$\begin{array}{r} x + 2 \\ x - 1 \overline{)x^2 + x - 1} \\ \underline{x^2 - x} \\ 2x - 1 \\ \underline{2x - 2} \\ 1 \end{array}$$

Thus

$$f(x) = x + 2 + \frac{1}{x - 1}.$$

So $y = x + 2$ is a slant asymptote of the curve. Also,

$$\lim_{x \to 1} |f(x)| = +\infty,$$

so $x = 1$ is a vertical asymptote. The first two derivatives of f are

$$f'(x) = 1 - \frac{1}{(x-1)^2} = \frac{x(x-2)}{(x-1)^2}$$

and

$$f''(x) = \frac{2}{(x-1)^3}.$$

It follows that f has critical points at $x = 0$ and at $x = 2$ but no inflection points. The sign of f' tells us that f is increasing on $(-\infty, 0)$ and on $(2, +\infty)$, decreasing on $(0, 1)$ and on $(1, 2)$. Examination of $f''(x)$ reveals that f is concave downward on $(-\infty, 1)$ and concave upward on $(1, +\infty)$. In particular, $f(0) = 1$ is a local maximum value and $f(2) = 5$ is a local minimum value. The graph of f looks much like the one in Fig. 4.7.10. ■

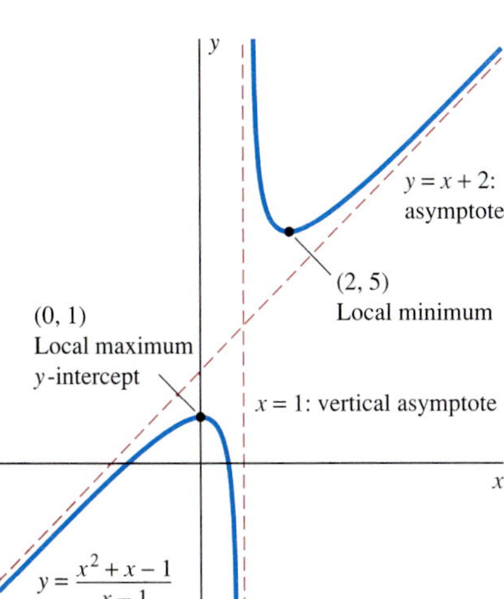

Fig. 4.7.10 A function with slant asymptote $y = x + 2$ (Example 7)

Calculator/Computer Graphing

Instead of using concepts of calculus to construct a graph from scratch, we can go the other way. That is, we can *begin* with a graph plotted by a calculator or computer, then use calculus to analyze the graph and refine our understanding of it. In Sections 1.3 and 1.4 we discussed the fact that a calculator or computer graph can sometimes be misleading or incomplete. But now we can use calculus—and in particular the computation of critical points and inflection points—to make sure that the machine-generated graph exhibits all of its important features. Moreover, with graphing and automatic solution techniques, we can investigate graphs of functions that would be too complicated to study without a calculator or computer.

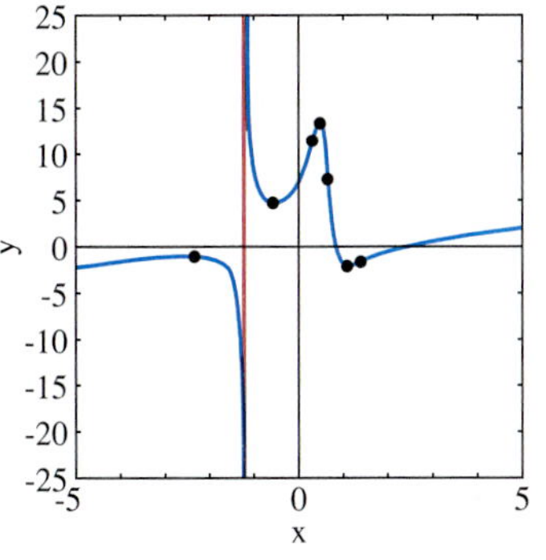

Fig. 4.7.11 $y = \dfrac{x^4 - 5x^2 - 5x + 7}{2x^3 - 2x + 1}$

EXAMPLE 8 Figure 4.7.11 shows a computer-generated graph of the function

$$f(x) = \frac{x^4 - 5x^2 - 5x + 7}{2x^3 - 2x + 1}. \tag{8}$$

It appears to have a vertical asymptote somewhere near $x = -1$. To test this hypothesis, we need to know where the denominator in Eq. (8) is zero. The graph of this denominator, shown in Fig. 4.7.12, indicates that the equation $2x^3 - 2x + 1 = 0$ has a single real solution near $x = -1.2$. We could zoom in graphically to show that the corresponding vertical asymptote is still closer to $x = -1.19$, and a calculator or computer **SOLVE** command would yield the solution $x \approx -1.1915$ accurate to four decimal places.

Noting that the degree of the numerator in Eq. (8) exceeds that of the denominator, we find by long division that

$$f(x) = \frac{1}{2}x + \frac{-4x^2 - \frac{11}{2}x + 7}{2x^3 - 2x + 1}.$$

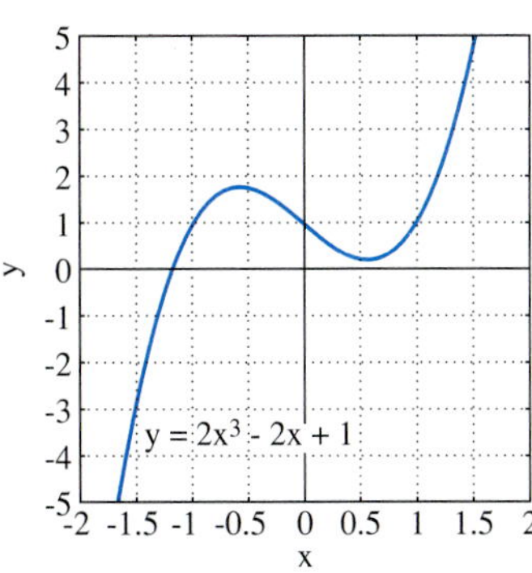

Fig. 4.7.12 Graph of the denominator in (8)

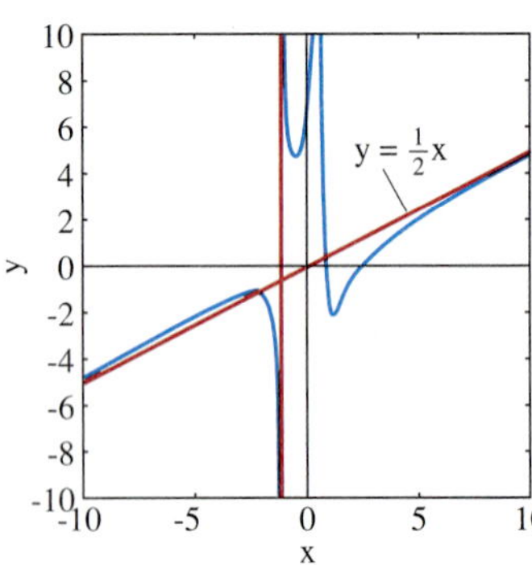

Fig. 4.7.13 Now we see both the vertical asymptote and the slant asymptote $y = \frac{1}{2}x$.

Thus the graph $y = f(x)$ has the slant asymptote $y = \frac{1}{2}x$ (Fig. 4.7.13).

To investigate the critical points of $f(x)$, we calculate the derivative

$$f'(x) = \frac{2x^6 + 4x^4 + 24x^3 - 32x^2 - 10x + 9}{(2x^3 - 2x + 1)^2}. \tag{9}$$

The critical points of $f(x)$ are the zeros of the numerator of $f'(x)$, together with the zero of the denominator that yields the vertical asymptote. The graph of the numerator, shown in Fig. 4.7.14, indicates that the equation

$$2x^6 + 4x^4 + 24x^3 - 32x^2 - 10x + 9 = 0$$

has four real solutions, near the points $x = -2.3, -0.6, 0.5$, and 1.1. We could zoom in closer to each of these solutions, or we could use a calculator or computer **SOLVE** command to get the approximations $x \approx -2.3440, -0.5775, 0.4673$, and 1.0864 that agree with the overall structure of the graph shown in Fig. 4.7.11, where four critical points with horizontal tangent lines are apparent.

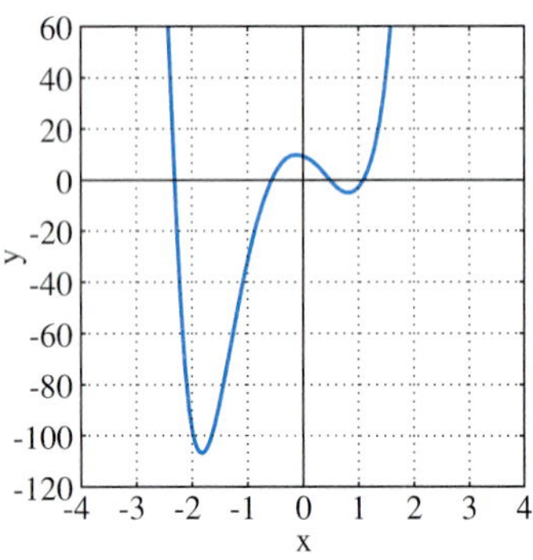

Fig. 4.7.14 Graph of the numerator in (9)

The leftmost critical point $x \approx -2.3440$ deserves closer examination. In Fig. 4.7.11 it appears to be just to the left of the point where the left branch of the graph $y = f(x)$ crosses the slant asymptote $y = \frac{1}{2}x$. The zoom shown in Fig. 4.7.15 bears out this observation.

Finally, an examination of the original graph $y = f(x)$ in Fig. 4.7.11 suggests the approximate locations of three inflection points in the first quadrant. But if the graph is to approach the slant asymptote as $x \to -\infty$, then Fig. 4.7.15 suggests the presence of a fourth inflection point somewhere to the left of the leftmost critical point. (Why?) To investigate this possibility, we calculate the second derivative

$$f''(x) = \frac{2(-16x^6 - 66x^5 + 120x^4 + 34x^3 - 18x^2 - 42x + 13)}{(2x^3 - 2x + 1)^3}. \tag{10}$$

The inflection points of $y = f(x)$ have x-coordinates given by the zeros of the numerator in Eq. (10). The graph of this numerator, shown in Fig. 4.7.16, indicates that the equation

$$2(-16x^6 - 66x^5 + 120x^4 + 34x^3 - 18x^2 - 42x + 13) = 0$$

has four real solutions—a negative one near -5.5 as well as three positive solutions between 0 and 2 that correspond to the visually apparent first-quadrant inflection

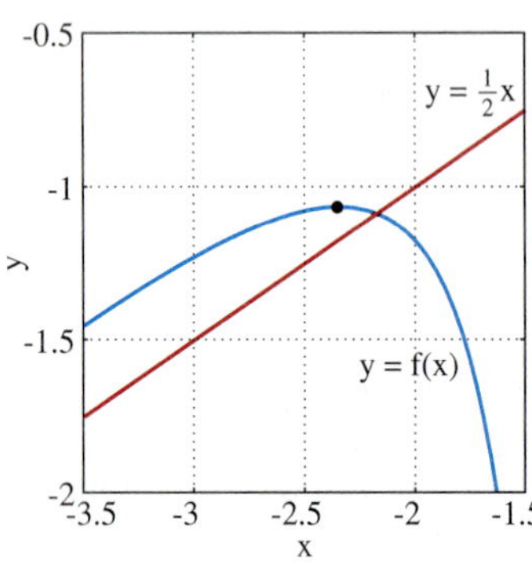

Fig. 4.7.15 Near the leftmost critical point

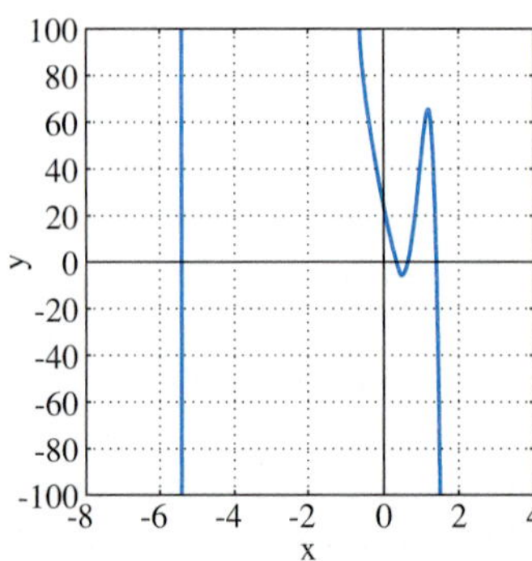

Fig. 4.7.16 Graph of the numerator in (10)

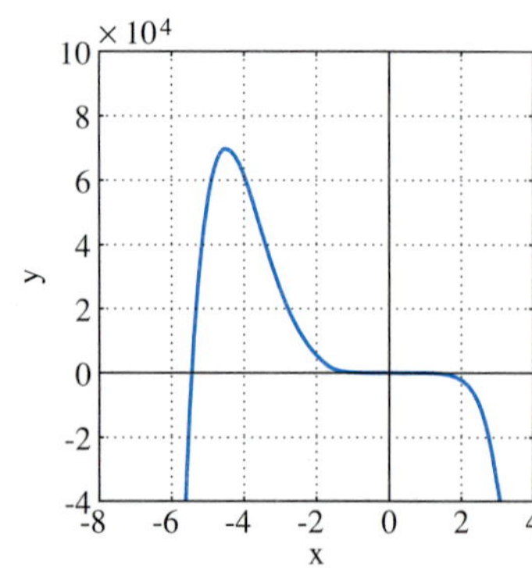

Fig. 4.7.17 Larger view of the graph of the numerator in (10)

points in Fig. 4.7.11. We could zoom in closer to each of these solutions, or we could use a calculator or computer **SOLVE** command to get the approximations $-5.4303, 0.3152, 0.6503$, and 1.3937. The larger view shown in Fig. 4.7.17 convinces us that we've found *all* the inflection points of $y = f(x)$. In particular, we see that $y = f(x)$ is concave upward to the right of the inflection point $x \approx -5.4303$, where the denominator in Eq. (10) is negative (why?), and is concave downward just to its left (consistent with what we see in Fig. 4.7.15).

This thorough analysis of the graph of the function f of Eq. (8) involves a certain amount of manual labor—just to calculate and simplify the derivatives in Eqs. (9) and (10) unless we use a computer algebra system for this task—but would be very challenging without the use of a graphing calculator or computer. ■

4.7 PROBLEMS

Investigate the limits in Problems 1 through 16.

1. $\lim_{x\to+\infty} \dfrac{x}{x+1}$

2. $\lim_{x\to-\infty} \dfrac{x^2+1}{x^2-1}$

3. $\lim_{x\to1} \dfrac{x^2+x-2}{x-1}$

4. $\lim_{x\to1} \dfrac{x^2-x-2}{x-1}$

5. $\lim_{x\to+\infty} \dfrac{2x^2-1}{x^2-3x}$

6. $\lim_{x\to-\infty} \dfrac{x^2+3x}{x^3-5}$

7. $\lim_{x\to-1} \dfrac{x^2+2x+1}{(x+1)^2}$

8. $\lim_{x\to+\infty} \dfrac{5x^3-2x+1}{7x^3+4x^2-2}$

9. $\lim_{x\to4} \dfrac{x-4}{\sqrt{x}-2}$

10. $\lim_{x\to+\infty} \dfrac{2x+1}{x-x\sqrt{x}}$

11. $\lim_{x\to-\infty} \dfrac{8-\sqrt[3]{x}}{2+x}$

12. $\lim_{x\to+\infty} \dfrac{2x^2-17}{x^3-2x+27}$

13. $\lim_{x\to+\infty} \sqrt{\dfrac{4x^2-x}{x^2+9}}$

14. $\lim_{x\to-\infty} \dfrac{\sqrt[3]{x^3-8x+1}}{3x-4}$

15. $\lim_{x\to-\infty} \left(\sqrt{x^2+2x}-x\right)$

16. $\lim_{x\to-\infty} \left(2x-\sqrt{4x^2-5x}\right)$

Apply your knowledge of limits and asymptotes to match each function in Problems 17 through 28 with its graph-with-asymptotes in one of the twelve parts of Fig. 4.7.18.

17. $f(x) = \dfrac{1}{x-1}$

18. $f(x) = \dfrac{1}{1-x}$

19. $f(x) = \dfrac{1}{(x-1)^2}$

20. $f(x) = -\dfrac{1}{(1-x)^2}$

21. $f(x) = \dfrac{1}{x^2-1}$

22. $f(x) = \dfrac{1}{1-x^2}$

23. $f(x) = \dfrac{x}{x^2-1}$

24. $f(x) = \dfrac{x}{1-x^2}$

25. $f(x) = \dfrac{x}{x-1}$

26. $f(x) = \dfrac{x^2}{x^2-1}$

27. $f(x) = \dfrac{x^2}{x-1}$

28. $f(x) = \dfrac{x^3}{x^2-1}$

Sketch by hand the graph of each function in Problems 29 through 54. Identify and label all extrema, inflection points, intercepts, and asymptotes. Show the concave structure clearly as well as the behavior of the graph for $|x|$ large and for x near any discontinuities of the function.

29. $f(x) = \dfrac{2}{x-3}$

30. $f(x) = \dfrac{4}{5-x}$

31. $f(x) = \dfrac{3}{(x+2)^2}$

32. $f(x) = -\dfrac{4}{(3-x)^2}$

33. $f(x) = \dfrac{1}{(2x-3)^3}$

34. $f(x) = \dfrac{x+1}{x-1}$

35. $f(x) = \dfrac{x^2}{x^2+1}$

36. $f(x) = \dfrac{2x}{x^2+1}$

37. $f(x) = \dfrac{1}{x^2-9}$

38. $f(x) = \dfrac{x}{4-x^2}$

39. $f(x) = \dfrac{1}{x^2+x-6}$

40. $f(x) = \dfrac{2x^2+1}{x^2-2x}$

41. $f(x) = x + \dfrac{1}{x}$

42. $f(x) = 2x + \dfrac{1}{x^2}$

43. $f(x) = \dfrac{x^2}{x-1}$

44. $f(x) = \dfrac{2x^3-5x^2+4x}{x^2-2x+1}$

45. $f(x) = \dfrac{1}{(x-1)^2}$

46. $f(x) = \dfrac{1}{x^2-4}$

47. $f(x) = \dfrac{x}{x+1}$

48. $f(x) = \dfrac{1}{(x+1)^3}$

49. $f(x) = \dfrac{1}{x^2-x-2}$

50. $f(x) = \dfrac{1}{(x-1)(x+1)^2}$

51. $f(x) = \dfrac{x^2-4}{x}$

52. $f(x) = \dfrac{x}{x^2-1}$

53. $f(x) = \dfrac{x^3-4}{x^2}$

54. $f(x) = \dfrac{x^2+1}{x-2}$

In Problems 55 through 60, you can determine by inspection the x-intercepts as well as the vertical and horizontal asymptotes of the curve $y = f(x)$. First sketch the graph by hand, using this information, and without calculating any derivatives. Then use a calculator or computer to locate accurately the critical and inflection points of $f(x)$. Finally, use a calculator or computer to produce graphs that display the major features of the curve.

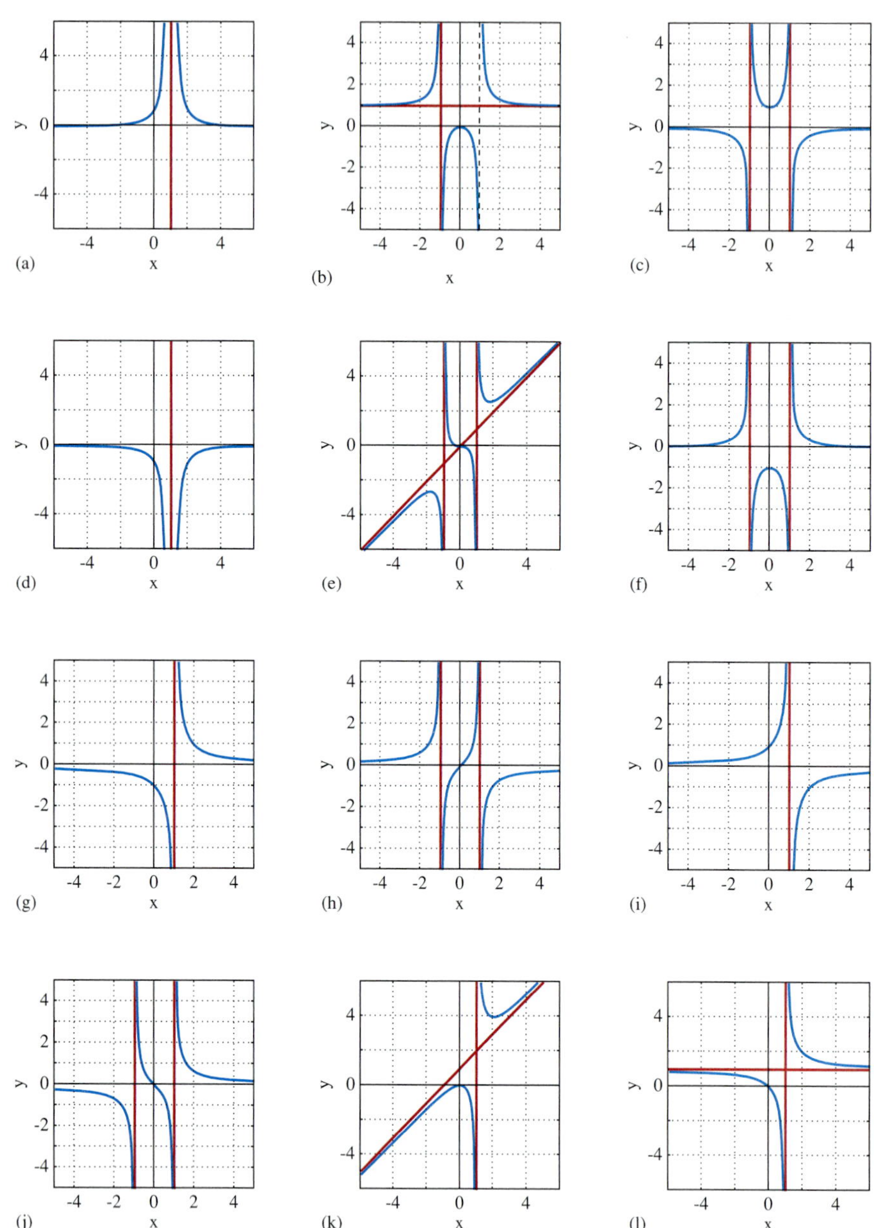

Fig. 4.7.18 Problems 17 through 28

55. $f(x) = \dfrac{(x+1)(x-3)}{x^2(x-2)}$ **56.** $f(x) = \dfrac{(x+1)^2(x-3)}{x^2(x-4)}$

57. $f(x) = \dfrac{(x+1)^2(x-3)}{x^3(x-2)}$ **58.** $f(x) = \dfrac{(x+1)^2(x-3)^2}{x^3(x-2)}$

59. $f(x) = \dfrac{(x+1)^2(x-3)^2}{x^3(x-2)^2}$ **60.** $f(x) = \dfrac{(x+1)(x-3)^4}{x^3(x-2)^3}$

In Problems 61 through 68, begin with a calculator- or computer-generated graph of the curve $y = f(x)$. Then use a calculator or computer to locate accurately the vertical asymptotes and the critical and inflection points of $f(x)$. Finally, use a calculator or computer to produce graphs that display the major features of the curve, including any vertical, horizontal, and slant asymptotes.

61. $f(x) = \dfrac{x^2}{x^3 - 3x^2 + 1}$ **62.** $f(x) = \dfrac{x^2}{x^3 - 3x^2 + 5}$

63. $f(x) = \dfrac{x^4 - 4x + 5}{x^3 - 3x^2 + 5}$ **64.** $f(x) = \dfrac{x^4 - 4x + 1}{2x^3 - 3x + 2}$

65. $f(x) = \dfrac{x^5 - 4x^2 + 1}{2x^4 - 3x + 2}$ **66.** $f(x) = \dfrac{x^5 - 4x^3 + 2}{2x^4 - 5x + 5}$

67. $f(x) = \dfrac{x^6 - 4x^3 + 5x}{2x^5 - 5x^3 + 5}$ **68.** $f(x) = \dfrac{2x^6 - 5x^4 + 6}{3x^5 - 5x^4 + 4}$

69. Suppose that

$$f(x) = x^2 + \frac{2}{x}.$$

Note that

$$\lim_{x \to \pm\infty} [f(x) - x^2] = 0,$$

so the curve $y = f(x)$ approaches the parabola $y = x^2$ as $x \to \pm\infty$. Use this observation to make an accurate sketch of the graph of f.

70. Use the method of Problem 69 to make an accurate sketch of the graph of

$$f(x) = x^3 - \frac{12}{x - 1}.$$

4.7 PROJECT: LOCATING SPECIAL POINTS ON EXOTIC GRAPHS

The investigations described here deal with fairly exotic curves having critical and inflection points that are not clearly visible on their graphs if plotted on a "natural" scale. The reason is that different scales on the x- and y-axes are required to see the unusual behavior in question. In both projects you are to begin with a graph that you generate with calculator or computer, then analyze the curve—locating accurately all critical and inflection points—in order to plot additional graphs that demonstrate clearly all of the major features of the curve.

INVESTIGATION A Choose in advance a single-digit integer n (perhaps the final nonzero digit of your student I.D. number). Then your task is to analyze the structure of the curve

$$y = x^7 + 5x^6 - 11x^5 - 21x^4 + 31x^3 - 57x^2 - (101 + 2n)x + (89 - 3n).$$

Find the local maximum and minimum points and the inflection point (or points) on this curve, giving their coordinates accurate to four decimal places. To display all these points, you probably will need to produce separate plots with different scales, showing different parts of this curve. In the end, use all the information accumulated to produce a careful hand sketch (*not* to scale) displaying all the maxima, minima, and inflection points with their (approximate) coordinates labeled.

INVESTIGATION B Explore in the detail structure of the graph of the function

$$f(x) = -1{,}234{,}567{,}890 + 2{,}695{,}140{,}459x^2 + 605{,}435{,}400x^3$$
$$- 411{,}401{,}250x^4 - 60{,}600{,}000x^5 + 25{,}000{,}000x^6.$$

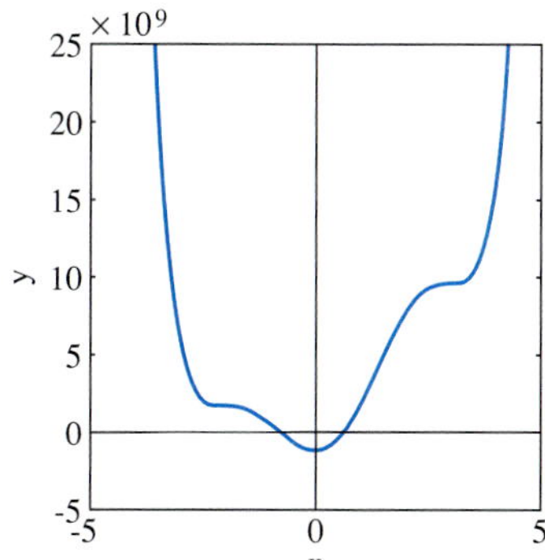

Fig. 4.7.19 The "big picture" in Investigation B

The graph $y = f(x)$ is shown in Fig. 4.7.19. At a glance, it might appear that we have only three critical points—a local minimum near the origin and two critical points that are also inflection points, as well as two more inflection points that are not critical points. Settle the matter. How many of each, in fact, are there? Find and exhibit all of them in a graph; your graph may be a neat hand sketch and need not be to scale.

CHAPTER 4 REVIEW: *Definitions, Concepts, Results*

Use the following list as a guide to concepts that you may need to review.

1. Increment Δy
2. Differential dy
3. Linear approximation formula
4. Differentiation rules in differential form
5. Increasing functions and decreasing functions
6. Significance of the sign of the first derivative
7. Rolle's theorem

8. The mean value theorem
9. Consequences of the mean value theorem
10. First derivative test
11. Open-interval maximum-minimum problems
12. Graphs of polynomials
13. Calculation of higher derivatives
14. Concave-upward functions and concave-downward functions
15. Test for concavity
16. Second derivative test
17. Inflection points
18. Inflection point test
19. Infinite limits
20. Vertical asymptotes
21. Limits as $x \to \pm\infty$
22. Horizontal asymptotes
23. Slant asymptotes
24. Curve-sketching techniques

CHAPTER 4 *Miscellaneous Problems*

In Problems 1 through 6, write dy in terms of x and dx.

1. $y = (4x - x^2)^{3/2}$ **2.** $y = 8x^3\sqrt{x^2 + 9}$

3. $y = \dfrac{x+1}{x-1}$ **4.** $y = \sin x^2$

5. $y = x^2 \cos\sqrt{x}$ **6.** $y = \dfrac{x}{\sin 2x}$

In Problems 7 through 16, estimate the indicated number by linear approximation.

7. $\sqrt{6401}$ (Note that $80^2 = 6400$.)

8. $\dfrac{1}{1.000007}$

9. $(2.0003)^{10}$ (Note that $2^{10} = 1024$.)

10. $\sqrt[3]{999}$ (Note that $10^3 = 1000$.)

11. $\sqrt[3]{1005}$ **12.** $\sqrt[3]{62}$

13. $26^{3/2}$ **14.** $\sqrt[5]{30}$

15. $\sqrt[4]{17}$ **16.** $\sqrt[10]{1000}$

In Problems 17 through 22, estimate by linear approximation the change in the indicated quantity.

17. The volume $V = s^3$ of a cube, if its side length s is increased from 5 in. to 5.1 in.

18. The area $A = \pi r^2$ of a circle, if its radius r is decreased from 10 cm to 9.8 cm

19. The volume $V = \frac{4}{3}\pi r^3$ of a sphere, if its radius r is increased from 5 cm to 5.1 cm

20. The volume $V = 1000/p$ in.^{3} of a gas, if the pressure p is decreased from 100 lb/in.^{2} to 99 lb/in.^{2}

21. The period of oscillation $T = 2\pi\sqrt{L/32}$ of a pendulum, if its length L is increased from 2 ft to 25 in. (Time T is in seconds, L is in feet.)

22. The lifetime $L = 10^{30}/E^{13}$ of a light bulb with applied voltage E volts (V), if the voltage is increased from 110 V to 111 V. Compare your result with the exact change in the function L.

If the mean value theorem applies to the function f on the interval $[a, b]$, it ensures the existence of a solution c in the interval (a, b) of the equation

$$f'(c) = \frac{f(b) - f(a)}{b - a}.$$

Problems 23 through 28 give a function f and an interval $[a, b]$. Verify that the hypotheses of the mean value theorem are satisfied for f on $[a, b]$. Then use the given equation to find the value of the number c.

23. $f(x) = x - \dfrac{1}{x}$; $[1, 3]$

24. $f(x) = x^3 + x - 4$; $[-2, 3]$

25. $f(x) = x^3$; $[-1, 2]$ **26.** $f(x) = x^3$; $[-2, 1]$

27. $f(x) = \frac{11}{5}x^5$; $[-1, 2]$ **28.** $f(x) = \sqrt{x}$; $[0, 4]$

Sketch the graphs of the functions in Problems 29 through 33. Indicate the local maxima and minima of each function and the intervals on which the function is increasing or decreasing. Show the concave structure of the graph and identify all inflection points.

29. $f(x) = x^2 - 6x + 4$ **30.** $f(x) = 2x^3 - 3x^2 - 36x$

31. $f(x) = 3x^5 - 5x^3 + 60x$

32. $f(x) = (3 - x)\sqrt{x}$ **33.** $f(x) = (1 - x)\sqrt[3]{x}$

34. Show that the equation $x^5 + x = 5$ has exactly one real solution.

Calculate the first three derivatives of the functions in Problems 35 through 44.

35. $f(x) = x^3 - 2x$ **36.** $f(x) = (x + 1)^{100}$

37. $g(t) = \dfrac{1}{t} - \dfrac{1}{2t + 1}$ **38.** $h(y) = \sqrt{3y - 1}$

39. $f(t) = 2t^{3/2} - 3t^{4/3}$ **40.** $g(x) = \dfrac{1}{x^2 + 9}$

41. $h(t) = \dfrac{t + 2}{t - 2}$ **42.** $f(z) = \sqrt[3]{z} + \dfrac{3}{\sqrt[5]{z}}$

43. $g(x) = \sqrt[3]{5 - 4x}$ **44.** $g(t) = \dfrac{8}{(3 - t)^{3/2}}$

In Problems 45 through 52, calculate dy/dx and d^2y/dx^2 under the assumption that y is defined implicitly as a function of x by the given equation.

45. $x^{1/3} + y^{1/3} = 1$

46. $2x^2 - 3xy + 5y^2 = 25$

47. $y^5 - 4y + 1 = \sqrt{x}$

48. $\sin xy = xy$

49. $x^2 + y^2 = 5xy + 5$

50. $x^5 + xy^4 = 1$

51. $y^3 - y = x^2 y$

52. $(x^2 - y^2)^2 = 4xy$

Sketch the graphs of the functions in Problems 53 through 72, indicating all critical points, inflection points, and asymptotes. Show the concave structure clearly.

53. $f(x) = x^4 - 32x$

54. $f(x) = 18x^2 - x^4$

55. $f(x) = x^6 - 2x^4$

56. $f(x) = x\sqrt{x - 3}$

57. $f(x) = x\sqrt[3]{4 - x}$

58. $f(x) = \dfrac{x - 1}{x + 2}$

59. $f(x) = \dfrac{x^2 + 1}{x^2 - 4}$

60. $f(x) = \dfrac{x}{x^2 - x - 2}$

61. $f(x) = \dfrac{2x^2}{x^2 - x - 2}$

62. $f(x) = \dfrac{x^3}{x^2 - 1}$

63. $f(x) = 3x^4 - 4x^3$

64. $f(x) = x^4 - 2x^2$

65. $f(x) = \dfrac{x^2}{x^2 - 1}$

66. $f(x) = x^3 - 12x$

67. $f(x) = -10 + 6x^2 - x^3$

68. $f(x) = \dfrac{x}{1 + x^2}$; note that

$$f'(x) = -\frac{(x - 1)(x + 1)}{(x^2 + 1)^2}$$

and that

$$f''(x) = \frac{2x(x^2 - 3)}{(x^2 + 1)^3}.$$

69. $f(x) = x^3 - 3x$

70. $f(x) = x^4 - 12x^2$

71. $f(x) = x^3 + x^2 - 5x + 3$

72. $f(x) = \dfrac{1}{x} + \dfrac{1}{x^2}$

73. The function

$$f(x) = \frac{1}{x^2 + 2x + 2}$$

has a maximum value, and only one. Find it.

74. You need to manufacture a cylindrical pot, without a top, with a volume of 1 ft^3. The cylindrical part of the pot is to be made of aluminum, the bottom of copper. Copper is five times as expensive as aluminum. What dimensions would minimize the total cost of the pot?

75. An open-topped rectangular box is to have a volume of 4500 cm^3. If its bottom is a rectangle whose length is twice its width, what dimensions would minimize the total area of the bottom and four sides of the box?

76. A small rectangular box must be made with a volume of 324 in.^3 Its bottom is square and costs twice as much (per square inch) as its top and four sides. What dimensions would minimize the total cost of the material needed to make this box?

77. You must make a small rectangular box with a volume of 400 in.^3 Its bottom is a rectangle whose length is twice its width. The bottom costs 7¢/in.^2; the top and four sides of the box cost 5¢/in.^2 What dimensions would minimize the cost of the box?

78. Suppose that $f(x)$ is a cubic polynomial with exactly three distinct real zeros. Prove that the two zeros of $f'(x)$ are real and distinct.

79. Suppose that it costs $1 + (0.0003)v^{3/2}$ dollars per mile to operate a truck at v miles per hour. If there are additional costs (such as the driver's pay) of \$10/hr, what speed would minimize the total cost of a 1000-mi trip?

80. The numbers $a_1, a_2, \dots, a_n$ are fixed. Find a simple formula for the number x such that the sum of the squares of the distances of x from the n fixed numbers is as small as possible.

81. Sketch the curve $y^2 = x(x - 1)(x - 2)$, showing that it consists of two pieces—one bounded and the other unbounded—and has two horizontal tangent lines, three vertical tangent lines, and two inflection points. (*Suggestion:* Note that the curve is symmetric around the x-axis. Begin by determining the intervals on which the product $x(x - 1)(x - 2)$ is positive. Compute dy/dx and d^2y/dx^2 by implicit differentiation.)

82. Farmer Rogers wants to fence in a rectangular plot of area 2400 ft^2. She also wants to use additional fencing to build an internal divider fence parallel to two of the boundary sections (Fig. 4.MP.1). What is the minimum total length of fencing that this project will require? Verify that your answer yields the global minimum.

Fig. 4.MP.1 The fencing of Problem 82

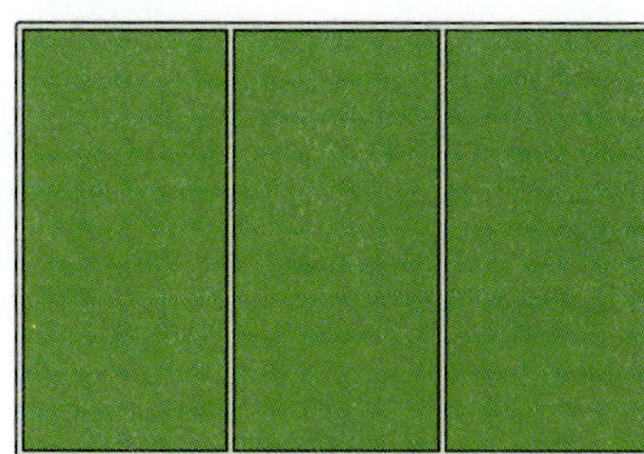

Fig. 4.MP.2 The fencing of Problem 83

83. Farmer Simmons wants to fence in a rectangular plot of area 1800 ft^2. He also wants to use additional fencing to build two internal divider fences, both parallel to the same two outer boundary sections (Fig. 4.MP.2). What is the minimum total length of fencing that this project will require? Verify that your answer yields the global minimum.

84. Farmer Taylor wants to fence in a rectangular plot of area 2250 m^2. She also wants to use additional fencing to build three internal divider fences, all parallel to the same two outer boundary sections. What is the minimum total length of fencing that this project will require? Verify that your answer yields the global minimum.

85. Farmer Upshaw wants to fence in a rectangular plot of area A ft^2. He also wants to use additional fencing to build n (a fixed but unspecified positive integer) internal divider fences, all parallel to the same two outer boundary sections. What is the minimum total length of fencing that this project will require? Verify that your answer yields the global minimum.

86. What is the length of the shortest line segment that lies in the first quadrant with its endpoints on the coordinate axes and is also tangent to the graph of $y = 1/x^2$? Verify that your answer yields the global minimum.

87. A right triangle is formed in the first quadrant by a line segment that is tangent to the graph of $y = 1/x^2$ and whose endpoints lie on the coordinate axes. Is there a maximum possible area of such a triangle? Is there a minimum? Justify your answers.

88. A right triangle is formed in the first quadrant by a line segment that is tangent to the graph of $y = 1/x$ and whose endpoints lie on the coordinate axes. Is there a maximum possible area of such a triangle? Is there a minimum? Justify your answers.

89. A rectangular box (with a top) is to have volume 288 in.^{3}, and its base is to be exactly three times as long as it is wide. What is the minimum possible surface area of such a box? Verify that your answer yields the global minimum.

90. A rectangular box (with a top) is to have volume 800 in.^{3}, and its base is to be exactly four times as long as it is wide. What is the minimum possible surface area of such a box? Verify that your answer yields the global minimum.

91. A rectangular box (with a top) is to have volume 225 cm^3, and its base is to be exactly five times as long as it is wide. What is the minimum possible surface area of such a box? Verify that your answer yields the global minimum.

92. A rectangular box (with a top) is to have volume V and its base is to be exactly n times as long as it is wide (n is a fixed but unspecified positive integer). What is the minimum possible surface area of such a box? Verify that your answer yields the global minimum.

93. The graph of $f(x) = x^{1/3}(1 - x)^{2/3}$ is shown in Fig. 4.MP.3. Recall from Section 4.7 that this graph has a slant asymptote with equation $y = mx + b$ provided that

$$\lim_{x \to +\infty} \left[f(x) - (mx + b)\right] = 0$$

or that

$$\lim_{x \to -\infty} \left[f(x) - (mx + b)\right] = 0.$$

(The values of m and b may be different in the two cases $x \to +\infty$ and $x \to -\infty$.) The graph here appears to have such an asymptote as $x \to +\infty$. Find m by evaluating

$$\lim_{x \to +\infty} \frac{f(x)}{x}.$$

Then find b by evaluating

$$\lim_{x \to +\infty} [f(x) - mx].$$

Finally, find m and b for the case $x \to -\infty$.

94. You are at the southernmost point of a circular lake of radius 1 mi. Your plan is to swim a straight course to another point on the shore of the lake, then jog to the northernmost point. You can jog twice as fast as you can swim. What route gives the minimum time required for your journey?

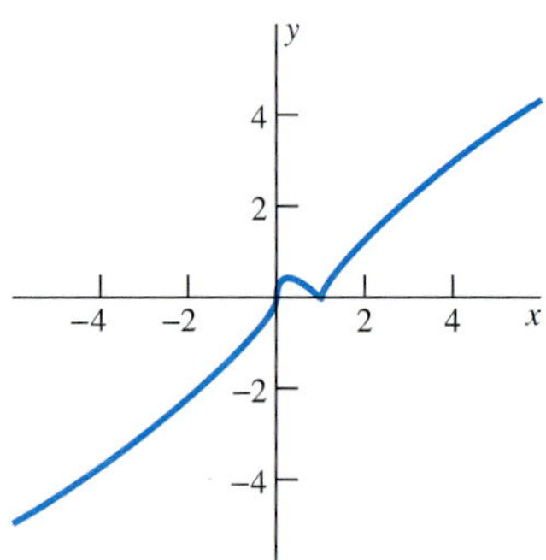

Fig. 4.MP.3 The graph of $y = f(x)$ of Problem 93

CHAPTER 5

THE INTEGRAL

Archimedes (287–212 B.C.)

Archimedes of Syracuse was the greatest mathematician of the ancient era from the fifth century B.C. to the second century A.D., when the seeds of modern mathematics sprouted in Greek communities located mainly on the shores of the Mediterranean Sea. He was famous in his own time for mechanical inventions—the so-called Archimedean screw for pumping water, lever-and-pulley devices ("give me a place to stand and I can move the earth"), a planetarium that duplicated the motions of heavenly bodies so accurately as to show eclipses of the sun and moon, and machines of war that terrified Roman soldiers in the siege of Syracuse, during which Archimedes was killed. But it is said that for Archimedes himself these inventions were merely the "diversions of geometry at play," and his writings are devoted to mathematical investigations.

Archimedes carried out many area and volume computations that now use integral calculus—ranging from areas of circles, spheres, and segments of conic sections to volumes of cones, spheres, ellipsoids, and paraboloids. It had been proved earlier in Euclid's *Elements* that the area A of a circle is proportional to the square of its radius r, so $A = \pi r^2$ for some proportionality constant π. But it was Archimedes who accurately approximated the numerical value of π, showing that it lies between the value $3\frac{1}{7}$ memorized by elementary school children and the lower bound $3\frac{10}{71}$. Euclid had also proved that the volume V of a sphere of radius r is given by $V = \mu r^3$ (μ constant), but it was Archimedes who discovered (and proved) that $\mu = 4\pi/3$. He also discovered the now-familiar volume formulas $V = \pi r^2 h$ and $V = \frac{1}{3}\pi r^2 h$ for the cylinder and the cone, respectively, of base radius r and height h.

It was long suspected that Archimedes had not originally discovered his area and volume formulas by means of the limit-based arguments that he used to establish them rigorously. In 1906 an Archimedean treatise entitled *The Method* was rediscovered virtually by accident after having been lost since ancient times. In it he described a "method of discovery" based on using infinitesimals much as they were employed during the invention and exploration of calculus in the seventeenth and eighteenth centuries.

To commemorate his sphere and cylinder formulas, Archimedes requested that on his tombstone be carved a sphere inscribed in a circular cylinder. If the height of the cylinder is $h = 2r$, can you verify that the total surface areas A_C and A_S of the cylinder and sphere, and their volumes V_C and V_S, are related by Archimedes' formulas

$$A_S = \frac{2}{3}A_C \quad \text{and} \quad V_S = \frac{2}{3}V_C?$$

Thus the volumes and surface areas of the sphere and cylinder have the same 2:3 ratio.

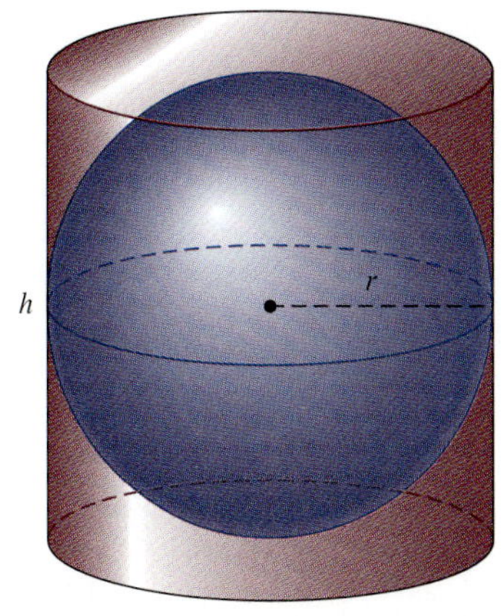

5.1 INTRODUCTION

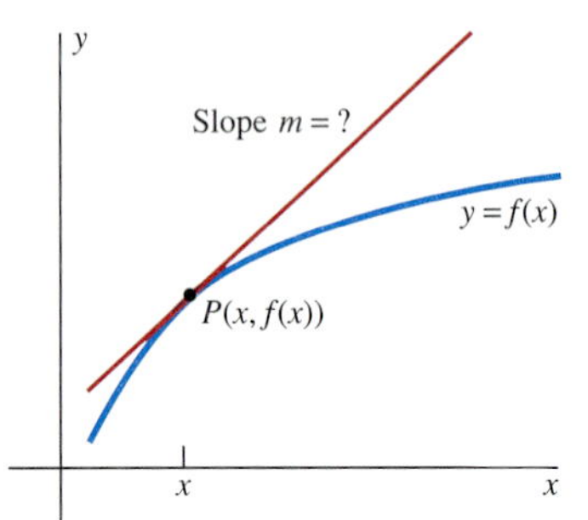

Fig. 5.1.1 The tangent-line problem motivates differential calculus.

Chapters 1 through 4 dealt with **differential calculus,** which is one of two closely related parts of *the* calculus. Differential calculus is centered on the concept of the *derivative.* Recall that the original motivation for the derivative was the problem of defining what it means for a straight line to be tangent to the graph of a function and calculating the slopes of such lines (Fig. 5.1.1). By contrast, the importance of the derivative stems from its applications to diverse problems that may, upon initial inspection, seem to have little connection with tangent lines.

Integral calculus is based on the concept of the *integral.* The definition of the integral is motivated by the problem of defining and calculating the area of the region that lies between the graph of a positive-valued function f and the x-axis over a given closed interval $[a, b]$. The area of the region R of Fig. 5.1.2 is given by the *integral* of f from a to b, denoted by the symbol

$$\int_a^b f(x)\,dx. \tag{1}$$

But the integral, like the derivative, is important due to its applications in many problems that may appear unrelated to its original motivation—problems involving motion and velocity, population growth, volume, arc length, surface area, and center of gravity, among others.

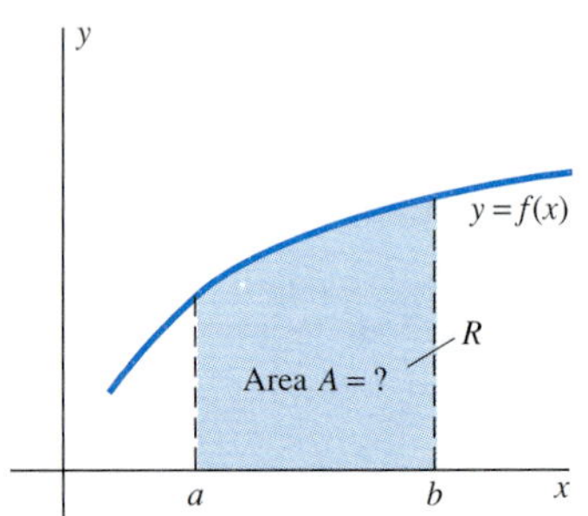

Fig. 5.1.2 The area problem motivates integral calculus.

The principal theorem of this chapter is the *fundamental theorem of calculus,* in Section 5.6. It provides a vital connection between the operations of differentiation and integration, one that provides an effective method for computing values of integrals. It turns out that instead of finding the derivative of the function $f(x)$ in Eq. (1), we need instead to find a new function $F(x)$ whose derivative is $f(x)$:

$$F'(x) = f(x). \tag{2}$$

Thus we need to do "differentiation in reverse." We therefore begin in Section 5.2 with an investigation of *antidifferentiation.*

5.2 ANTIDERIVATIVES AND INITIAL VALUE PROBLEMS

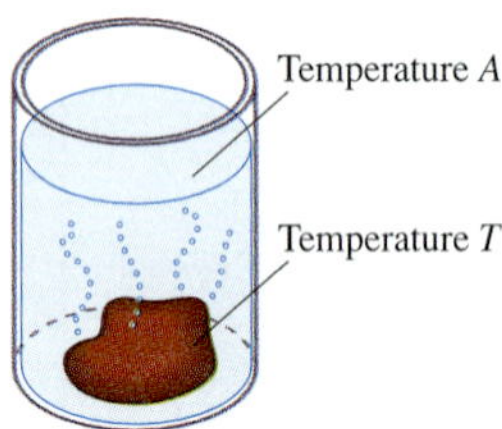

Fig. 5.2.1 Newton's law of cooling (Eq. (1)) describes the cooling of a hot rock in cold water.

The language of change is the natural language for the statement of most scientific laws and principles. For example, Newton's law of cooling says that the *rate of change* of the temperature T of a body is proportional to the difference between T and the temperature of the surrounding medium (Fig. 5.2.1). That is,

$$\frac{dT}{dt} = -k(T - A), \tag{1}$$

where k is a positive constant and A, normally assumed to be constant, is the surrounding temperature. Similarly, the *rate of change* of a population P with constant birth and death rates is proportional to the size of the population:

$$\frac{dP}{dt} = kP \qquad (k \text{ constant}). \tag{2}$$

Torricelli's law of draining (Fig. 5.2.2) implies that the *rate of change* of the depth y of water in a draining tank is proportional to the *square root* of y; that is,

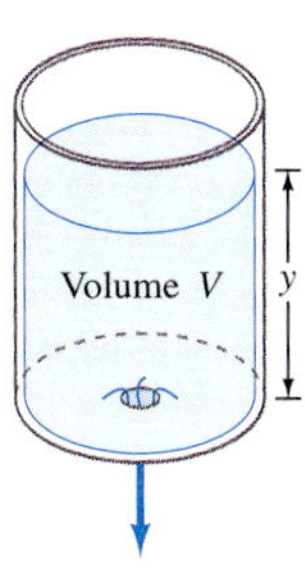

Fig. 5.2.2 Torricelli's law of draining (Eq. (3)) describes the draining of a cylindrical water tank.

$$\frac{dy}{dt} = -k\sqrt{y} \qquad (k \text{ constant}). \tag{3}$$

Mathematical models of real-world situations frequently involve equations that contain *derivatives* of unknown functions. Such equations, including (1) through (3), are called **differential equations.**

Antiderivatives

The simplest kind of differential equation has the form

$$\frac{dy}{dx} = f(x), \tag{4}$$

where f is a given (known) function and the function $y(x)$ is unknown. The process of finding a function from its derivative is the opposite of differentiation and is therefore called **antidifferentiation.** If we can find a function $y(x)$ whose derivative is $f(x)$,

$$y'(x) = f(x),$$

then we call $y(x)$ an *antiderivative* of $f(x)$.

Definition *Antiderivative*

An **antiderivative** of the function f is a function F such that

$$F'(x) = f(x)$$

wherever $f(x)$ is defined.

The table in Fig. 5.2.3 shows some examples of functions, each paired with one of its antiderivatives. Figure 5.2.4 illustrates the operations of differentiation and antidifferentiation, beginning with the same function f and going in opposite directions. Figure 5.2.5 illustrates differentiation "undoing" the result of antidifferentiation—the derivative of an antiderivative of $f(x)$ is the original function $f(x)$.

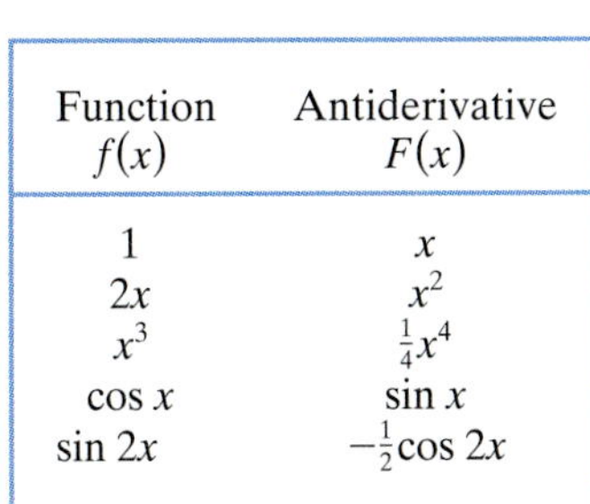

Function $f(x)$	Antiderivative $F(x)$
1	x
$2x$	x^2
x^3	$\frac{1}{4}x^4$
$\cos x$	$\sin x$
$\sin 2x$	$-\frac{1}{2}\cos 2x$

Fig. 5.2.3 Some antiderivatives

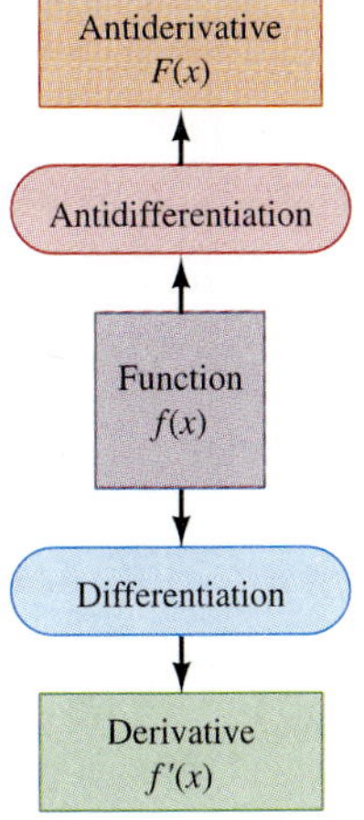

Fig. 5.2.4 Differentiation and antidifferentiation are opposites.

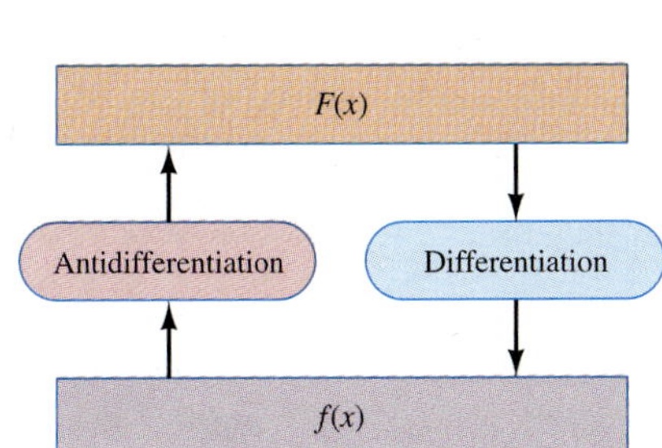

Fig. 5.2.5 Differentiation undoes the result of antidifferentiation.

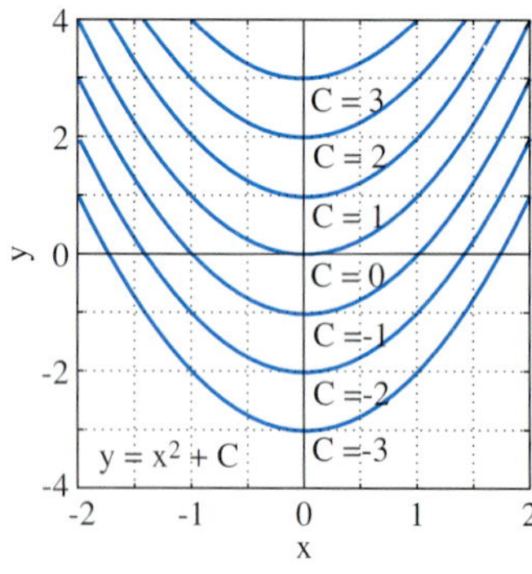

Fig. 5.2.6 Graph of $y = x^2 + C$ for various values of C

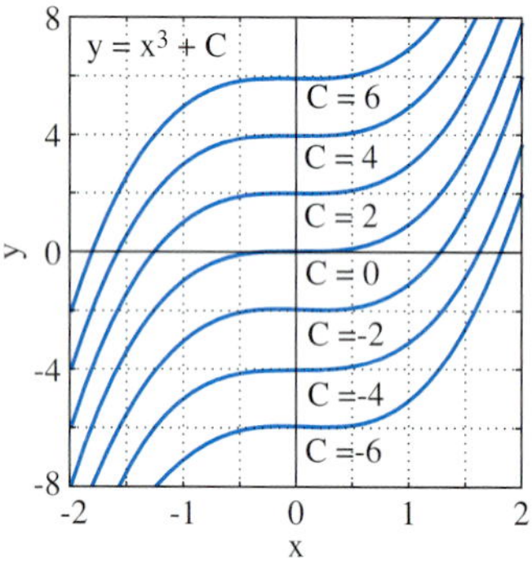

Fig. 5.2.7 Graph of $y = x^3 + C$ for various values of C

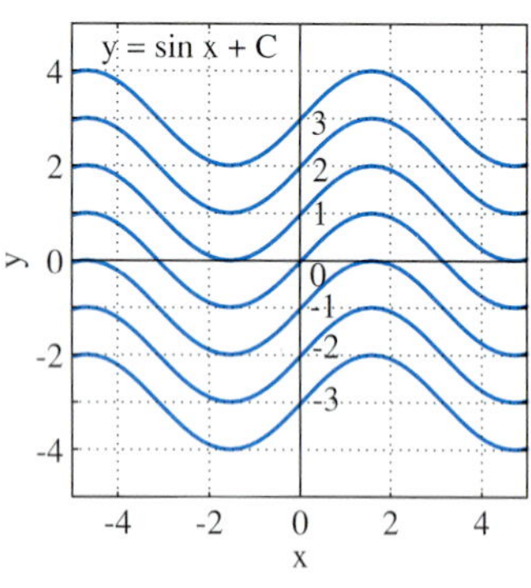

Fig. 5.2.8 Graph of $y = \sin x + C$ for various values of C

EXAMPLE 1 Given: the function $f(x) = 3x^2$. Then $F(x) = x^3$ is an antiderivative of $f(x) = 3x^2$, as are the functions

$$G(x) = x^3 + 17, \quad H(x) = x^3 + \pi, \quad \text{and} \quad K(x) = x^3 - \sqrt{2}.$$

Indeed, $J(x) = x^3 + C$ is an antiderivative of $f(x) = 3x^2$ for *any* choice of the *constant C*. ■

Thus a single function has *many* antiderivatives, whereas a function can have *only one* derivative. If $F(x)$ is an antiderivative of $f(x)$, then so is $F(x) + C$ for any choice of the constant C. The converse of this statement is more subtle: If $F(x)$ is one antiderivative of $f(x)$ *on the interval I*, then *every* antiderivative of $f(x)$ on I is of the form $F(x) + C$. This follows directly from Corollary 2 of the mean value theorem in Section 4.3, according to which two functions with the same derivative on an interval differ only by a constant on that interval.

Thus the graphs of any two antiderivatives $F(x) + C_1$ and $F(x) + C_2$ of the same function $f(x)$ on the same interval I are "parallel" in the sense illustrated in Figs. 5.2.6 through 5.2.8. There we see that the constant C is the vertical distance between the curves $y = F(x)$ and $y = F(x) + C$ for each x in I. This is the geometric interpretation of Theorem 1.

Theorem 1 ***The Most General Antiderivative***

If $F'(x) = f(x)$ at each point of the open interval I, then every antiderivative G of f on I has the form

$$G(x) = F(x) + C, \tag{5}$$

where C is a constant.

Thus if F is any single antiderivative of f on the interval I, then the *most general* antiderivative of f on I has the form $F(x) + C$, as given in Eq. (5). The collection of *all* antiderivatives of the function $f(x)$ is called the **indefinite integral** of f with respect to x and is denoted by

$$\int f(x)\,dx.$$

On the basis of Theorem 1, we write

$$\int f(x)\,dx = F(x) + C, \tag{6}$$

where $F(x)$ is any particular antiderivative of $f(x)$. Therefore

$$\int f(x)\,dx = F(x) + C \quad \text{if and only if} \quad F'(x) = f(x).$$

The integral symbol $\int$ is made like an elongated capital S. It is, in fact, a medieval S, used by Leibniz as an abbreviation for the Latin word *summa* ("sum"). We think of the combination $\int \ldots dx$ as a single symbol; we fill in the "blank" with the formula of the function whose antiderivative we seek. We may regard the differential dx as specifying the independent variable x both in the function $f(x)$ and in its antiderivatives.

EXAMPLE 2 The entries in Fig. 5.2.3 yield the indefinite integrals

$$\int 1\,dx = x + C,$$

$$\int 2x\,dx = x^2 + C,$$

$$\int x^3\,dx = \tfrac{1}{4}x^4 + C,$$

$$\int \cos x\,dx = \sin x + C, \quad \text{and}$$

$$\int \sin 2x\,dx = -\tfrac{1}{2}\cos 2x + C.$$

You can verify each such formula by differentiating the right-hand side. Indeed, this is the *sure-fire* way to check any antidifferentiation: To verify that $F(x)$ is an antiderivative of $f(x)$, compute $F'(x)$ to see whether you obtain $f(x)$. For instance, the differentiation

$$D_x(-\tfrac{1}{2}\cos 2x + C) = -\tfrac{1}{2}(-2\sin 2x) + 0 = \sin 2x$$

is sufficient to verify the fifth formula of this example. ■

The differential dx in Eq. (6) specifies that the independent variable is x. But we can describe a specific antidifferentiation in terms of *any* independent variable that is convenient. For example, the indefinite integrals

$$\int 3t^2\,dt = t^3 + C, \qquad \int 3y^2\,dy = y^3 + C, \qquad \text{and} \qquad \int 3u^2\,du = u^3 + C$$

mean exactly the same thing as

$$\int 3x^2\,dx = x^3 + C.$$

Using Integral Formulas

Every differentiation formula yields immediately—by "reversal" of the differentiation—a corresponding indefinite integral formula. The now-familiar derivatives of power functions and trigonometric functions yield the integral formulas stated in Theorem 2.

Theorem 2 ***Some Integral Formulas***

$$\int x^k\,dx = \frac{x^{k+1}}{k+1} + C \quad (\text{if } k \neq -1), \tag{7}$$

$$\int \cos kx\,dx = \frac{1}{k}\sin kx + C, \tag{8}$$

$$\int \sin kx\,dx = -\frac{1}{k}\cos kx + C, \tag{9}$$

$$\int \sec^2 kx\,dx = \frac{1}{k}\tan kx + C, \tag{10}$$

$$\int \csc^2 kx\,dx = -\frac{1}{k}\cot kx + C, \tag{11}$$

$$\int \sec kx \tan kx\,dx = \frac{1}{k}\sec kx + C, \quad \text{and} \tag{12}$$

$$\int \csc kx \cot kx\,dx = -\frac{1}{k}\csc kx + C. \tag{13}$$

Remark Be sure you see why there is a minus sign in Eq. (9) but none in Eq. (8)!

Recall that the operation of differentiation is *linear,* meaning that

$$D_x[cF(x)] = cF'(x) \qquad \text{(where } c \text{ is a constant)}$$

and

$$D_x[F(x) \pm G(x)] = F'(x) \pm G'(x).$$

It follows in the notation of antidifferentiation that

$$\int cf(x)\,dx = c\int f(x)\,dx \qquad (c \text{ is a constant}) \tag{14}$$

and

$$\int [f(x) \pm g(x)]\,dx = \int f(x)\,dx \pm \int g(x)\,dx. \tag{15}$$

We can summarize these two equations by saying that antidifferentiation is **linear.** In essence, then, we antidifferentiate a sum of functions by antidifferentiating each function individually. This is *termwise* (or *term-by-term*) antidifferentiation. Moreover, a constant coefficient in any such term is merely "carried through" the antidifferentiation.

EXAMPLE 3 Find

$$\int \left(x^3 + 3\sqrt{x} - \frac{4}{x^2}\right)dx.$$

Solution Just as in differentiation, we prepare for antidifferentiation by writing roots and reciprocals as powers with fractional or negative exponents. Thus

$$\int \left(x^3 + 3\sqrt{x} - \frac{4}{x^2}\right)dx = \int (x^3 + 3x^{1/2} - 4x^{-2})\,dx$$

$$= \int x^3\,dx + 3\int x^{1/2}\,dx - 4\int x^{-2}\,dx \qquad \text{[using Eqs. (14) and (15)]}$$

$$= \frac{x^4}{4} + 3\cdot\frac{x^{3/2}}{\frac{3}{2}} - 4\cdot\frac{x^{-1}}{-1} + C \qquad \text{[using Eq. (7)]}$$

$$= \frac{1}{4}x^4 + 2x\sqrt{x} + \frac{4}{x} + C.$$

There's only one "$+C$" because the sure-fire check verifies that $\frac{1}{4}x^4 + 2x^{3/2} + 4x^{-1}$ is a particular antiderivative. Hence any other antiderivative differs from this one by only a (single) constant C. ■

EXAMPLE 4

$$\int (2\cos 3t + 5\sin 4t)\,dt$$

$$= 2\int \cos 3t\,dt + 5\int \sin 4t\,dt \qquad \text{[using Eqs. (14) and (15)]}$$

$$= 2\left(\tfrac{1}{3}\sin 3t\right) + 5\left(-\tfrac{1}{4}\cos 4t\right) + C \qquad \text{[using Eqs. (8) and (9)]}$$

$$= \tfrac{2}{3}\sin 3t - \tfrac{5}{4}\cos 4t + C.$$

■

Equation (7) is the power rule "in reverse." The generalized power rule in reverse is

$$\int u^k\,du = \frac{u^{k+1}}{k+1} + C \qquad (\text{if } k \neq -1), \tag{16}$$

where

$$u = g(x) \quad \text{and} \quad du = g'(x)\,dx.$$

EXAMPLE 5 With $u = x + 5$ (so that $du = dx$), Eq. (16) yields

$$\int (x+5)^{10}\,dx = \int u^{10}\,du = \tfrac{1}{11}u^{11} + C = \tfrac{1}{11}(x+5)^{11} + C.$$

Note that, after making the substitution $u = x + 5$ and integrating with respect to u, our final step is to express the resulting antiderivative in terms of the original variable x. ■

EXAMPLE 6 We want to find

$$\int \frac{20}{(4-5x)^3}\,dx.$$

We plan to use Eq. (16) with $u = 4 - 5x$. But we must get the differential $du = -5\,dx$ into the act. The "constant multiplier rule" of Eq. (14) permits us to do this:

$$\int \frac{20}{(4-5x)^3}\,dx = 20\int (4-5x)^{-3}\,dx$$

$$= \frac{20}{-5}\int (4-5x)^{-3}(-5\,dx) \tag{17}$$

$$= -4\int u^{-3}\,du \qquad (u = 4 - 5x,\ du = -5\,dx)$$

$$= -4\cdot\frac{u^{-2}}{-2} + C \quad [\text{Eq. (7) with } k = -3].$$

Thus

$$\int \frac{20}{(4-5x)^3}\,dx = \frac{2}{(4-5x)^2} + C.$$

The key step occurs in Eq. (17). There we, in effect, multiplied by the *constant* -5 inside the integral and compensated for that by dividing by -5 outside the integral. At the end it was necessary to replace u with $4 - 5x$ to express the antiderivative in terms of the original independent variable x. ■

Simple Differential Equations

The technique of antidifferentiation can often be used to solve a differential equation of the special form

$$\frac{dy}{dx} = f(x) \tag{18}$$

in which the dependent variable y does not appear on the right-hand side. To *solve* the differential equation in (18) is simply to find a function $y(x)$ that satisfies Eq. (18)—a function whose derivative is the given function $f(x)$. Hence the **general solution** of Eq. (18) is the indefinite integral

$$y(x) = \int f(x)\,dx + C \tag{19}$$

of the function $f(x)$.

EXAMPLE 7 The general solution of the differential equation

$$\frac{dy}{dx} = 3x^2$$

is given by

$$y(x) = \int 3x^2\,dx = x^3 + C.$$ ■

A differential equation of the form in Eq. (18) may appear in conjunction with an **initial condition,** a condition of the form

$$y(x_0) = y_0. \tag{20}$$

This condition specifies the value $y = y_0$ that the solution function $y(x)$ must have at $x = x_0$. Once we have found the general solution in Eq. (19), we can determine the value of the constant C by substituting the information that $y = y_0$ when $x = x_0$. With this specific value of C, Eq. (19) then gives the **particular solution** of the differential equation in Eq. (18) that satisfies the initial condition in Eq. (20). The combination

$$\frac{dy}{dx} = f(x), \quad y(x_0) = y_0 \tag{21}$$

of a differential equation with an initial condition is called an **initial value problem.**

EXAMPLE 8 Solve the initial value problem

$$\frac{dy}{dx} = 2x + 3, \quad y(1) = 2. \tag{22}$$

Solution By Eq. (19) the *general solution* of the differential equation $dy/dx = 2x + 3$ is given by

$$y(x) = \int (2x + 3)\,dx = x^2 + 3x + C.$$

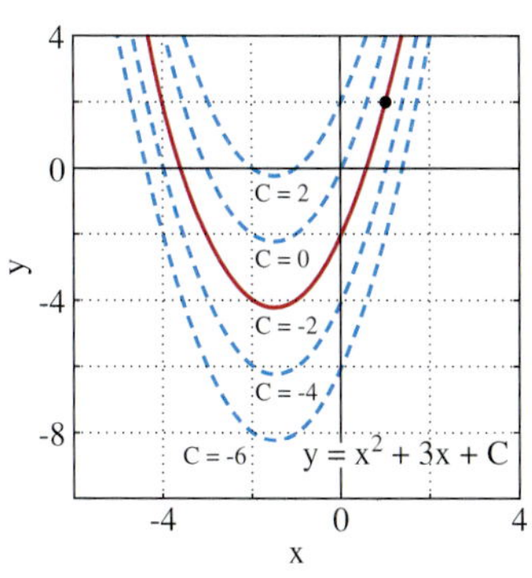

Fig. 5.2.9 General solutions $y = x^2 + 3x + C$ of the differential equation in (22) (Example 8)

Figure 5.2.9 shows the graph $y = x^2 + 3x + C$ for various values of C. The particular solution we seek corresponds to the curve in Fig. 5.2.9 that passes through the point $(1, 2)$, thereby satisfying the initial condition

$$y(1) = (1)^2 + 3 \cdot (1) + C = 2.$$

It follows that $4 + C = 2$, and hence that $C = -2$. So the desired *particular solution* is given by

$$y(x) = x^2 + 3x - 2. \quad \blacksquare$$

REMARK The method used in Example 8 may be described as "integrating both sides of a differential equation" with respect to x:

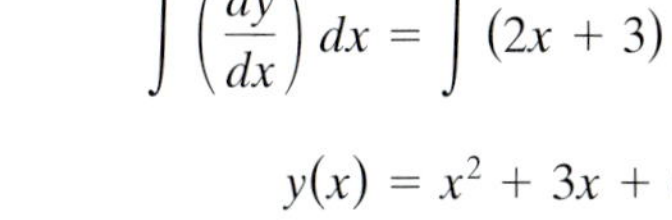

$$\int \left(\frac{dy}{dx}\right) dx = \int (2x + 3)\,dx;$$

$$y(x) = x^2 + 3x + C.$$

Rectilinear Motion

Antidifferentiation enables us, in many important cases, to analyze the motion of a particle (or "mass point") in terms of the forces acting on it. If the particle moves in *rectilinear motion* along a straight line—the x-axis, for instance—under the influence of a given (possibly variable) force, then (as in Section 3.1) the motion of the particle is described by its **position function**

$$x = x(t), \tag{23}$$

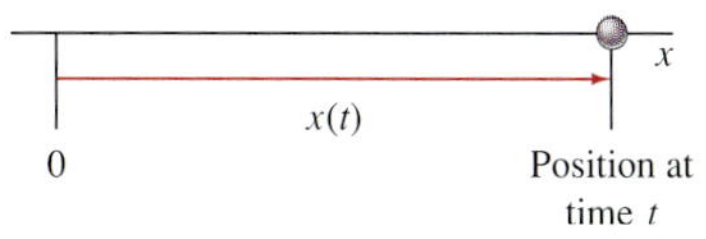

Fig. 5.2.10 The position function $x(t)$ of a particle moving along the x-axis

which gives its x-coordinate at time t (Fig. 5.2.10). The particle's **velocity** $v(t)$ is the time derivative of its position function,

$$v(t) = \frac{dx}{dt}, \tag{24}$$

and its **acceleration** $a(t)$ is the time derivative of its velocity:

$$a(t) = \frac{dv}{dt} = \frac{d^2x}{dt^2}. \tag{25}$$

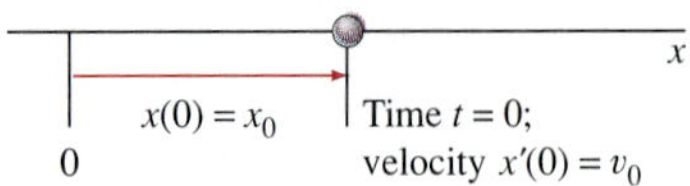

Fig. 5.2.11 Initial data for linear motion

In a typical situation, the following information is given (Fig. 5.2.11):

$$\begin{array}{ll} a(t) & \text{the particle's acceleration;} \\ x(0) = x_0 & \text{its } \textit{initial position}; \\ v(0) = v_0 & \text{its } \textit{initial velocity}. \end{array} \tag{26}$$

In principle, we can then proceed as follows to find the particle's position function $x(t)$. First we solve the initial value problem

$$\frac{dv}{dt} = a(t), \quad v(0) = v_0 \tag{27}$$

for the velocity function $v(t)$. Knowing $v(t)$, we then solve the initial value problem

$$\frac{dx}{dt} = v(t), \quad x(0) = x_0 \tag{28}$$

for the particle's position function $x(t)$. Thus we determine $x(t)$ from the acceleration and initial data given in Eq. (26) by solving two successive initial value problems. For this purpose we can use the integral versions

$$v(t) = \int a(t)\,dt \tag{29}$$

and

$$x(t) = \int v(t)\,dt \tag{30}$$

of the derivative formulas in (24) and (25).

EXAMPLE 9 A particle starts from rest (that is, with initial velocity zero) at the point $x = 10$ and moves along the x-axis with acceleration function $a(t) = 12t$. Find its resulting position function $x(t)$.

Solution First we must solve the initial value problem

$$\frac{dv}{dt} = a(t) = 12t, \quad v(0) = 0$$

to find the velocity function $v(t)$. Using Eq. (29), we get

$$v(t) = \int a(t)\,dt = \int 12t\,dt = 6t^2 + C_1.$$

[We write C_1 because we anticipate the appearance of a second constant when we integrate again to find $x(t)$.] Then substitution of the initial data $t = 0$, $v = 0$ yields

$$0 = 6 \cdot 0^2 + C_1 = C_1,$$

so it follows that $v(t) = 6t^2$. Next we must solve the initial value problem

$$\frac{dx}{dt} = v(t) = 6t^2, \quad x(0) = 10$$

for $x(t)$. Using (30), we get

$$x(t) = \int v(t)\,dt = \int 6t^2\,dt = 2t^3 + C_2.$$

Then substitution of the initial data $t = 0, x = 10$ yields

$$10 = 2 \cdot 0^3 + C_2 = C_2,$$

so it follows finally that the particle's position function is

$$x(t) = 2t^3 + 10.$$ ■

Constant Acceleration

The solution of the initial value problems in Eqs. (27) and (28) is simplest when the given acceleration a is *constant*. We begin with

$$\frac{dv}{dt} = a \qquad (a \text{ is a constant})$$

and antidifferentiate:

$$v(t) = \int a\,dt.$$

So

$$v(t) = at + C_1. \tag{31}$$

To evaluate the constant C_1, we substitute the initial datum $v(0) = v_0$; this gives

$$v_0 = a \cdot 0 + C_0 = C_1.$$

Therefore Eq. (31) becomes

$$v(t) = at + v_0. \tag{32}$$

Because $x'(t) = v(t)$, a second antidifferentiation yields

$$\begin{aligned} x(t) &= \int v(t)\,dt \\ &= \int (at + v_0)\,dt; \\ x(t) &= \tfrac{1}{2}at^2 + v_0 t + C_2. \end{aligned} \tag{33}$$

Now substitution of the initial datum $x(0) = x_0$ gives

$$x_0 = \tfrac{1}{2}a \cdot (0)^2 + v_0 \cdot (0) + C_2 = C_2$$

in Eq. (33). Thus the position function of the particle is

$$x(t) = \tfrac{1}{2}at^2 + v_0 t + x_0. \tag{34}$$

WARNING Equations (32) and (34) are valid only in the case of *constant* acceleration a. They do not apply when acceleration varies.

EXAMPLE 10 The skid marks made by an automobile indicate that its brakes were fully applied for a distance of 160 ft before it came to a stop. Suppose that the car in question has a constant deceleration of 20 ft/s^2 under the conditions of the skid. How fast was the car traveling when its brakes were first applied?

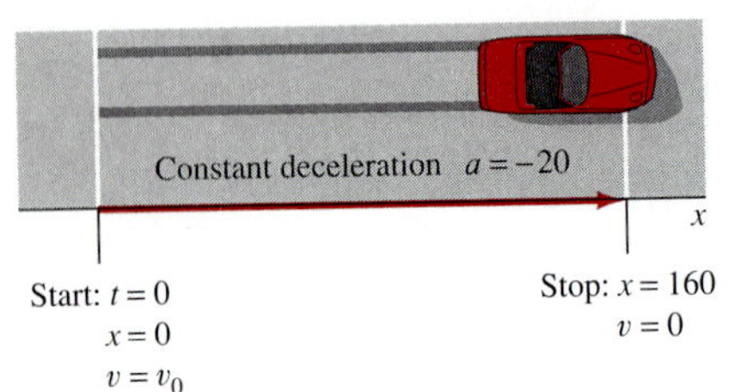

Fig. 5.2.12 Skid marks 160 ft long (Example 10)

Solution The introduction of a convenient coordinate system is often crucial to the successful solution of a physical problem. Here we take the x-axis to be positively oriented in the direction of motion of the car. We choose the origin so that $x_0 = 0$ when $t = 0$, the time when the brakes were first applied (Fig. 5.2.12). In this coordinate system, the car's velocity $v(t)$ is a decreasing function of time t (in seconds), so its acceleration is $a = -20$ (ft/s^2) rather than $a = +20$. Hence we begin with the constant acceleration equation

$$\frac{dv}{dt} = -20.$$

Antidifferentiation as in Eq. (29) gives

$$v(t) = \int (-20)\,dt = -20t + C_1.$$

Even though the initial velocity is unknown and not given, the initial data $t = 0$, $v = v_0$ still yield $C_1 = v_0$. So the car's velocity function is

$$v(t) = -20t + v_0. \tag{35}$$

A second antidifferentiation, as in Eq. (30), gives

$$x(t) = \int (-20t + v_0)\,dt = -10t^2 + v_0 t + C_2.$$

Substitution of the initial data $t = 0, x_0 = 0$ yields $C_2 = 0$, so the position function of the car is

$$x(t) = -10t^2 + v_0 t. \tag{36}$$

The fact that the skid marks are 160 ft long tells us that $x = 160$ when the car comes to a stop; that is,

$$x = 160 \quad \text{when} \quad v = 0.$$

Substitution of these values into the velocity and position equations [Eqs. (35) and (36)] then yields the two simultaneous equations

$$-20t + v_0 = 0, \qquad -10t^2 + v_0 t = 160.$$

We now solve these for v_0 and t to find the initial velocity v_0 *and* the duration t of the car's skid. If we multiply the first equation by $-t$ and add the result to the second equation, we find that $10t^2 = 160$, so $t = 4$ when the car first comes to a stop. It follows that the velocity of the car was

$$v_0 = 20 \cdot 4 = 80 \quad \text{(ft/s)},$$

or about 55 mi/h, when the brakes were first applied. ■

Vertical Motion with Constant Gravitational Acceleration

Stroboscopic photograph of a ball falling with constant acceleration due to gravity.

One common application of Eqs. (32) and (34) involves vertical motion near the surface of the earth. A particle in such motion is subject to a *downward* acceleration a, which is almost exactly constant if only small vertical distances are involved. The magnitude of this constant is denoted by g, approximately 32 ft/s^2 or 9.8 m/s^2. (If you need more accurate values for g, use 32.174 ft/s^2 in the fps system or 9.80665 m/s^2 in the mks system.)

If we neglect air resistance, we may assume that this acceleration due to gravity is the only outside influence on the moving particle. Because we deal with vertical motion here, it is natural to choose the y-axis as the coordinate system for the position of the particle and to place "ground level" where $y = 0$ (Fig. 5.2.13). If we choose the upward direction to be the positive direction, then the effect of gravity on the particle is to *decrease* its height and also to *decrease* its velocity $v = dy/dt$. Then the acceleration of the particle is

$$a = \frac{dv}{dt} = -g = -32 \quad (\text{ft/s}^2).$$

Equations (32) and (34) then become

$$v(t) = -32t + v_0 \tag{37}$$

and

$$y(t) = -16t^2 + v_0 t + y_0. \tag{38}$$

Here y_0 is the initial height of the particle in feet, v_0 is its initial velocity in feet per second, and time t is measured in seconds.

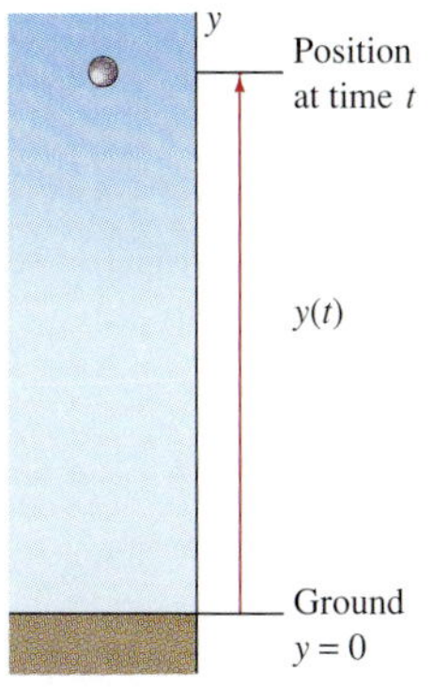

Fig. 5.2.13 The position function $y(t)$ of a particle moving vertically

EXAMPLE 11 Suppose that a bolt was fired vertically upward from a crossbow at ground level and that it struck the ground 20 s later. If air resistance may be neglected, find the initial velocity of the bolt and the maximum altitude that it reached.

Solution We set up the coordinate system illustrated in Fig. 5.2.14, with ground level corresponding to $y = 0$, with the bolt fired at time $t = 0$ (in seconds), and with the positive direction being the upward direction. Units on the y-axis are in feet.

We are given that $y = 0$ when $t = 20$. We lack any information about the initial velocity v_0. But we may use Eqs. (37) and (38) because we have set up a coordinate system in which the acceleration due to gravity acts in the negative direction. Thus

$$y(t) = -16t^2 + v_0 t + y_0 = -16t^2 + v_0 t$$

and

$$v(t) = -32t + v_0.$$

We use the information that $y = 0$ when $t = 20$ in the first equation:

$$0 = -16 \cdot 20^2 + 20v_0, \quad \text{and thus} \quad v_0 = 16 \cdot 20 = 320 \quad (\text{ft/s}).$$

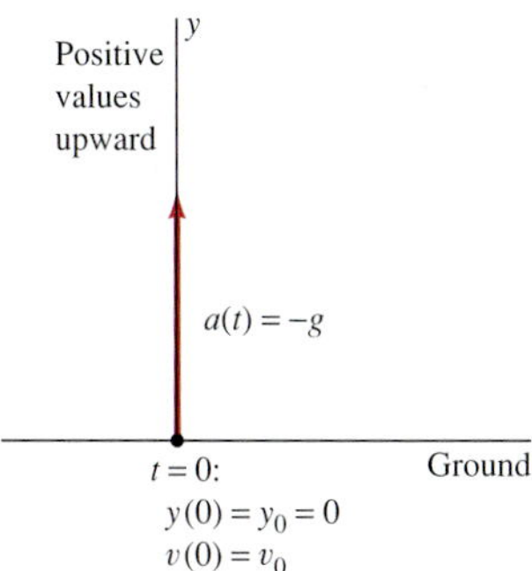

Fig. 5.2.14 A bolt fired straight upward from a crossbow (Example 11)

To find the maximum altitude of the bolt, we maximize $y(t)$ by finding the value of t for which its derivative is zero. In other words, the bolt reaches its maximum altitude when its velocity is zero,

$$-32t + v_0 = 0,$$

so at maximum altitude, $t = v_0/32 = 10$. At that time, the bolt has reached its maximum altitude of

$$y_{\text{max}} = y(10) = -16 \cdot 10^2 + 320 \cdot 10 = 1600 \quad \text{(ft)}.$$

The result seems contrary to experience. We must conclude that air resistance cannot always be neglected, particularly not in problems involving long journeys at high velocity. ■

5.2 PROBLEMS

Evaluate the indefinite integrals in Problems 1 through 30.

1. $\displaystyle\int (3x^2 + 2x + 1)\,dx$ **2.** $\displaystyle\int (3t^4 + 5t - 6)\,dt$

3. $\displaystyle\int (1 - 2x^2 + 3x^3)\,dx$ **4.** $\displaystyle\int \left(-\frac{1}{t^2}\right) dt$

5. $\displaystyle\int \left(\frac{3}{x^3} + 2x^{3/2} - 1\right) dx$ **6.** $\displaystyle\int \left(x^{5/2} - \frac{5}{x^4} - \sqrt{x}\right) dx$

7. $\displaystyle\int \left(\tfrac{3}{2}t^{1/2} + 7\right) dt$ **8.** $\displaystyle\int \left(\frac{2}{x^{3/4}} - \frac{3}{x^{2/3}}\right) dx$

9. $\displaystyle\int \left(\sqrt[3]{x^2} + \frac{4}{\sqrt[4]{x^5}}\right) dx$ **10.** $\displaystyle\int \left(2x\sqrt{x} - \frac{1}{\sqrt{x}}\right) dx$

11. $\displaystyle\int (4x^3 - 4x + 6)\,dx$ **12.** $\displaystyle\int \left(\frac{1}{4}t^5 - \frac{5}{t^2}\right) dt$

13. $\displaystyle\int 7\,dx$ **14.** $\displaystyle\int \left(4\sqrt[3]{x^2} - \frac{6}{\sqrt[3]{x}}\right) dx$

15. $\displaystyle\int (x + 1)^4\,dx$ **16.** $\displaystyle\int (t + 1)^{10}\,dt$

17. $\displaystyle\int \frac{1}{(x - 10)^7}\,dx$ **18.** $\displaystyle\int \sqrt{z + 1}\,dz$

19. $\displaystyle\int \sqrt{x}(1 - x)^2\,dx$ **20.** $\displaystyle\int \sqrt[3]{x}\,(x + 1)^3\,dx$

21. $\displaystyle\int \frac{2x^4 - 3x^3 + 5}{7x^2}\,dx$ **22.** $\displaystyle\int \frac{(3x + 4)^2}{\sqrt{x}}\,dx$

23. $\displaystyle\int (9t + 11)^5\,dt$ **24.** $\displaystyle\int \frac{1}{(3z + 10)^7}\,dz$

25. $\displaystyle\int \frac{7}{(x + 77)^2}\,dx$ **26.** $\displaystyle\int \frac{3}{\sqrt{(x - 1)^3}}\,dx$

27. $\displaystyle\int (5\cos 10x - 10\sin 5x)\,dx$

28. $\displaystyle\int (2\cos \pi x + 3\sin \pi x)\,dx$

29. $\displaystyle\int (3\cos \pi t + \cos 3\pi t)\,dt$ **30.** $\displaystyle\int (4\sin 2\pi t - 2\sin 4\pi t)\,dt$

31. Verify by differentiation that the integral formulas

$$\int \sin x \cos x\,dx = \frac{1}{2}\sin^2 x + C_1 \quad \text{and}$$

$$\int \sin x \cos x\,dx = -\frac{1}{2}\cos^2 x + C_2$$

are both valid. Reconcile these seemingly different results. What is the relation between the constants C_1 and C_2?

32. Show that the obviously different functions

$$F_1(x) = \frac{1}{1 - x} \quad \text{and} \quad F_2(x) = \frac{x}{1 - x}$$

are both antiderivatives of $f(x) = 1/(1 - x)^2$. What is the relation between $F_1(x)$ and $F_2(x)$?

33. Use the identities

$$\sin^2 x = \frac{1 - \cos 2x}{2} \quad \text{and} \quad \cos^2 x = \frac{1 + \cos 2x}{2}$$

to find the antiderivatives

$$\int \sin^2 x\,dx \quad \text{and} \quad \int \cos^2 x\,dx.$$

34. (a) First explain why $\int \sec^2 x\,dx = \tan x + C$. (b) Then use the identity $1 + \tan^2 x = \sec^2 x$ to find the antiderivative

$$\int \tan^2 x\,dx.$$

Solve the initial value problems in 35 through 46.

35. $\dfrac{dy}{dx} = 2x + 1;\ \ y(0) = 3$ **36.** $\dfrac{dy}{dx} = (x - 2)^3;\ \ y(2) = 1$

37. $\dfrac{dy}{dx} = \sqrt{x};\ \ y(4) = 0$ **38.** $\dfrac{dy}{dx} = \dfrac{1}{x^2};\ \ y(1) = 5$

39. $\dfrac{dy}{dx} = \dfrac{1}{\sqrt{x + 2}};\ \ y(2) = -1$

40. $\dfrac{dy}{dx} = \sqrt{x + 9};\ \ y(-4) = 0$

41. $\dfrac{dy}{dx} = 3x^3 + \dfrac{2}{x^2};\ \ y(1) = 1$

42. $\dfrac{dy}{dx} = x^4 - 3x + \dfrac{3}{x^3};\ \ y(1) = -1$

43. $\dfrac{dy}{dx} = (x - 1)^3;\ \ y(0) = 2$

44. $\dfrac{dy}{dx} = \sqrt{x + 5};\ \ y(4) = -3$

45. $\dfrac{dy}{dx} = \dfrac{1}{\sqrt{x - 13}};\ \ y(17) = 2$

46. $\dfrac{dy}{dx} = (2x + 3)^{3/2};\ \ y(3) = 100$

Stroboscopic photograph of a ball falling with constant acceleration due to gravity.

Vertical Motion with Constant Gravitational Acceleration

One common application of Eqs. (32) and (34) involves vertical motion near the surface of the earth. A particle in such motion is subject to a *downward* acceleration a, which is almost exactly constant if only small vertical distances are involved. The magnitude of this constant is denoted by g, approximately 32 ft/s^2 or 9.8 m/s^2. (If you need more accurate values for g, use 32.174 ft/s^2 in the fps system or 9.80665 m/s^2 in the mks system.)

If we neglect air resistance, we may assume that this acceleration due to gravity is the only outside influence on the moving particle. Because we deal with vertical motion here, it is natural to choose the y-axis as the coordinate system for the position of the particle and to place "ground level" where $y = 0$ (Fig. 5.2.13). If we choose the upward direction to be the positive direction, then the effect of gravity on the particle is to *decrease* its height and also to *decrease* its velocity $v = dy/dt$. Then the acceleration of the particle is

$$a = \frac{dv}{dt} = -g = -32 \quad (\text{ft/s}^2).$$

Equations (32) and (34) then become

$$v(t) = -32t + v_0 \tag{37}$$

and

$$y(t) = -16t^2 + v_0 t + y_0. \tag{38}$$

Here y_0 is the initial height of the particle in feet, v_0 is its initial velocity in feet per second, and time t is measured in seconds.

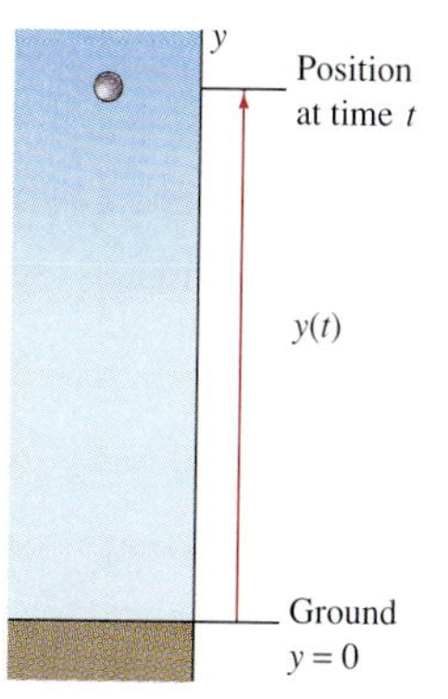

Fig. 5.2.13 The position function $y(t)$ of a particle moving vertically

EXAMPLE 11 Suppose that a bolt was fired vertically upward from a crossbow at ground level and that it struck the ground 20 s later. If air resistance may be neglected, find the initial velocity of the bolt and the maximum altitude that it reached.

Solution We set up the coordinate system illustrated in Fig. 5.2.14, with ground level corresponding to $y = 0$, with the bolt fired at time $t = 0$ (in seconds), and with the positive direction being the upward direction. Units on the y-axis are in feet.

We are given that $y = 0$ when $t = 20$. We lack any information about the initial velocity v_0. But we may use Eqs. (37) and (38) because we have set up a coordinate system in which the acceleration due to gravity acts in the negative direction. Thus

$$y(t) = -16t^2 + v_0 t + y_0 = -16t^2 + v_0 t$$

and

$$v(t) = -32t + v_0.$$

We use the information that $y = 0$ when $t = 20$ in the first equation:

$$0 = -16 \cdot 20^2 + 20v_0, \quad \text{and thus} \quad v_0 = 16 \cdot 20 = 320 \quad (\text{ft/s}).$$

To find the maximum altitude of the bolt, we maximize $y(t)$ by finding the value of t for which its derivative is zero. In other words, the bolt reaches its maximum altitude when its velocity is zero,

$$-32t + v_0 = 0,$$

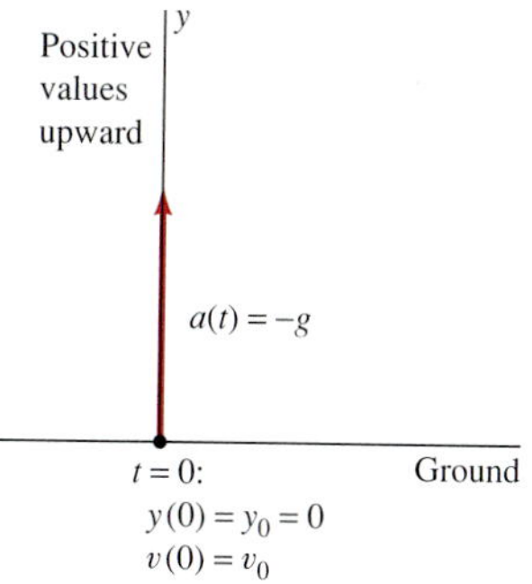

Fig. 5.2.14 A bolt fired straight upward from a crossbow (Example 11)

so at maximum altitude, $t = v_0/32 = 10$. At that time, the bolt has reached its maximum altitude of

$$y_{\max} = y(10) = -16 \cdot 10^2 + 320 \cdot 10 = 1600 \quad \text{(ft)}.$$

The result seems contrary to experience. We must conclude that air resistance cannot always be neglected, particularly not in problems involving long journeys at high velocity. ■

5.2 PROBLEMS

Evaluate the indefinite integrals in Problems 1 through 30.

1. $\int (3x^2 + 2x + 1)\,dx$ **2.** $\int (3t^4 + 5t - 6)\,dt$

3. $\int (1 - 2x^2 + 3x^3)\,dx$ **4.** $\int \left(-\frac{1}{t^2}\right) dt$

5. $\int \left(\frac{3}{x^3} + 2x^{3/2} - 1\right) dx$ **6.** $\int \left(x^{5/2} - \frac{5}{x^4} - \sqrt{x}\right) dx$

7. $\int \left(\frac{3}{2}t^{1/2} + 7\right) dt$ **8.** $\int \left(\frac{2}{x^{3/4}} - \frac{3}{x^{2/3}}\right) dx$

9. $\int \left(\sqrt[3]{x^2} + \frac{4}{\sqrt[4]{x^5}}\right) dx$ **10.** $\int \left(2x\sqrt{x} - \frac{1}{\sqrt{x}}\right) dx$

11. $\int (4x^3 - 4x + 6)\,dx$ **12.** $\int \left(\frac{1}{4}t^5 - \frac{5}{t^2}\right) dt$

13. $\int 7\,dx$ **14.** $\int \left(4\sqrt[3]{x^2} - \frac{6}{\sqrt[3]{x}}\right) dx$

15. $\int (x + 1)^4\,dx$ **16.** $\int (t + 1)^{10}\,dt$

17. $\int \frac{1}{(x - 10)^7}\,dx$ **18.** $\int \sqrt{z + 1}\,dz$

19. $\int \sqrt{x}(1 - x)^2\,dx$ **20.** $\int \sqrt[3]{x}\,(x + 1)^3\,dx$

21. $\int \frac{2x^4 - 3x^3 + 5}{7x^2}\,dx$ **22.** $\int \frac{(3x + 4)^2}{\sqrt{x}}\,dx$

23. $\int (9t + 11)^5\,dt$ **24.** $\int \frac{1}{(3z + 10)^7}\,dz$

25. $\int \frac{7}{(x + 77)^2}\,dx$ **26.** $\int \frac{3}{\sqrt{(x - 1)^3}}\,dx$

27. $\int (5\cos 10x - 10\sin 5x)\,dx$

28. $\int (2\cos \pi x + 3\sin \pi x)\,dx$

29. $\int (3\cos \pi t + \cos 3\pi t)\,dt$ **30.** $\int (4\sin 2\pi t - 2\sin 4\pi t)\,dt$

31. Verify by differentiation that the integral formulas

$$\int \sin x \cos x\,dx = \frac{1}{2}\sin^2 x + C_1 \quad \text{and}$$

$$\int \sin x \cos x\,dx = -\frac{1}{2}\cos^2 x + C_2$$

are both valid. Reconcile these seemingly different results. What is the relation between the constants C_1 and C_2?

32. Show that the obviously different functions

$$F_1(x) = \frac{1}{1 - x} \quad \text{and} \quad F_2(x) = \frac{x}{1 - x}$$

are both antiderivatives of $f(x) = 1/(1 - x)^2$. What is the relation between $F_1(x)$ and $F_2(x)$?

33. Use the identities

$$\sin^2 x = \frac{1 - \cos 2x}{2} \quad \text{and} \quad \cos^2 x = \frac{1 + \cos 2x}{2}$$

to find the antiderivatives

$$\int \sin^2 x\,dx \quad \text{and} \quad \int \cos^2 x\,dx.$$

34. (a) First explain why $\int \sec^2 x\,dx = \tan x + C$. (b) Then use the identity $1 + \tan^2 x = \sec^2 x$ to find the antiderivative

$$\int \tan^2 x\,dx.$$

Solve the initial value problems in 35 through 46.

35. $\frac{dy}{dx} = 2x + 1;\ y(0) = 3$ **36.** $\frac{dy}{dx} = (x - 2)^3;\ y(2) = 1$

37. $\frac{dy}{dx} = \sqrt{x};\ y(4) = 0$ **38.** $\frac{dy}{dx} = \frac{1}{x^2};\ y(1) = 5$

39. $\frac{dy}{dx} = \frac{1}{\sqrt{x + 2}};\ y(2) = -1$

40. $\frac{dy}{dx} = \sqrt{x + 9};\ y(-4) = 0$

41. $\frac{dy}{dx} = 3x^3 + \frac{2}{x^2};\ y(1) = 1$

42. $\frac{dy}{dx} = x^4 - 3x + \frac{3}{x^3};\ y(1) = -1$

43. $\frac{dy}{dx} = (x - 1)^3;\ y(0) = 2$

44. $\frac{dy}{dx} = \sqrt{x + 5};\ y(4) = -3$

45. $\frac{dy}{dx} = \frac{1}{\sqrt{x - 13}};\ y(17) = 2$

46. $\frac{dy}{dx} = (2x + 3)^{3/2};\ y(3) = 100$

In Problems 47 through 52, a particle moves along the x-axis with the given acceleration function $a(t)$, initial position $x(0)$, and initial velocity $v(0)$. Find the particle's position function $x(t)$.

47. $a(t) = 12t - 4;\quad x(0) = 0, v(0) = -10$

48. $a(t) = 10 - 30t;\quad x(0) = 5, v(0) = -5$

49. $a(t) = 2t^2;\quad x(0) = -7, v(0) = 3$

50. $a(t) = 15\sqrt{t};\quad x(0) = 5, v(0) = 7$

51. $a(t) = \sin t;\quad x(0) = 0, v(0) = 0$

52. $a(t) = 8\cos 2t;\quad x(0) = -2, v(0) = 4$

In Problems 53 through 56, a particle starts at the origin and travels along the x-axis with the velocity function $v(t)$ whose graph is shown in Figs. 5.2.15 through 5.2.18. Sketch the graph of the resulting position function $x(t)$ for $0 \leqq t \leqq 10$.

53. Figure 5.2.15

54. Figure 5.2.16

55. Figure 5.2.17

56. Figure 5.2.18

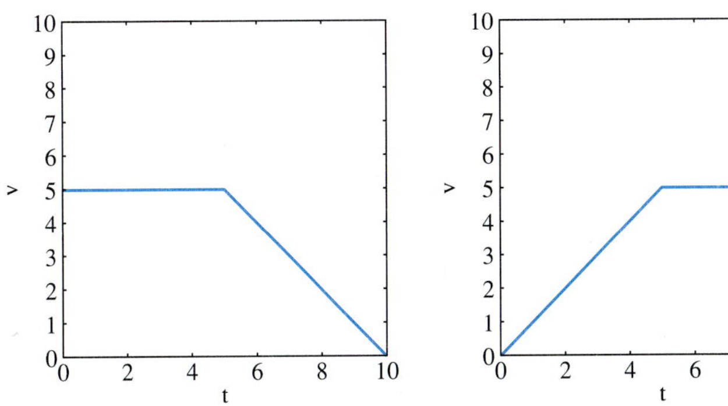

Fig. 5.2.15 Graph of the velocity function $v(t)$ of Problem 53

Fig. 5.2.16 Graph of the velocity function $v(t)$ of Problem 54

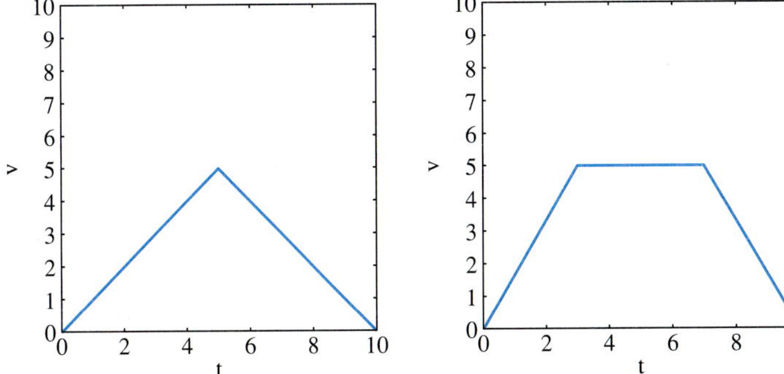

Fig. 5.2.17 Graph of the velocity function $v(t)$ of Problem 55

Fig. 5.2.18 Graph of the velocity function $v(t)$ of Problem 56

Problems 57 through 73 deal with vertical motion near the surface of the earth (with air resistance considered negligible). Use $g = 32$ ft/s^2 for the magnitude of the gravitational acceleration.

57. You throw a ball straight upward from the ground with initial velocity 96 ft/s. How high does the ball rise, and how long does it remain aloft?

58. When Alex shot a marble straight upward from ground level with his slingshot, it reached a maximum height of 400 ft. What was the marble's initial velocity?

59. Laura drops a stone into a well; it hits bottom 3 s later. How deep is the well?

60. Fran throws a rock straight upward alongside a tree (Fig. 5.2.19). The rock rises until it is even with the top of the tree and then falls back to the ground; it remains aloft for 4 s. How tall is the tree?

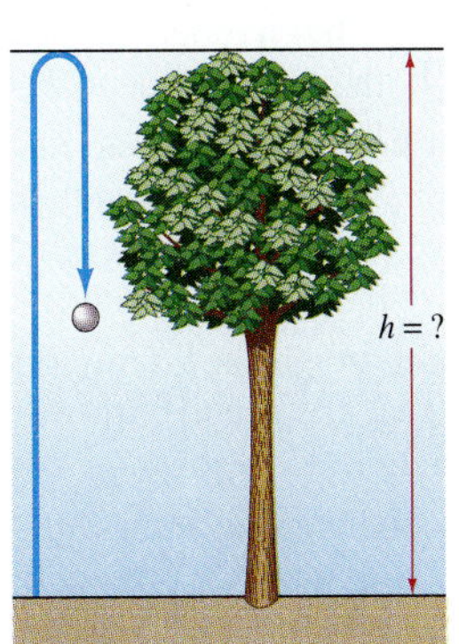

Fig. 5.2.19 The tree of Problem 60

Fig. 5.2.20 The building of Problem 61

61. Mickey throws a ball upward with an initial velocity of 48 ft/s from the top of a building 160 ft high. The ball soon falls to the ground at the base of the building (Fig. 5.2.20). How long does the ball remain aloft, and with what speed does it strike the ground?

62. A ball is dropped from the top of a building 576 ft high. With what velocity should a second ball be thrown straight downward 3 s later so that the two balls hit the ground simultaneously?

63. A ball is dropped from the top of the Empire State Building, 960 ft above 34th Street. How long does it take for the ball to reach the street, and with what velocity does it strike the street?

64. Lynda shoots an arrow straight upward from the ground with initial velocity 320 ft/s. (a) How high is the arrow after exactly 3 s have elapsed? (b) At what time is the arrow exactly 1200 ft above the ground? (c) How many seconds after its release does the arrow strike the ground?

65. Bill throws a stone upward from the ground. The stone reaches a maximum height of 225 ft. What was its initial velocity?

66. Sydney drops a rock into a well in which the water surface is 98 m below the ground. How long does it take the rock to reach the water surface? How fast is the rock moving as it penetrates the water surface?

67. Gloria drops a tennis ball from the top of a building 400 ft high. How long does it take the ball to reach the ground? With what velocity does it strike the ground?

68. Kosmo throws a baseball straight downward from the top of a tall building. The initial speed of the ball is 25 ft/s. It hits the ground with a speed of 153 ft/s. How tall is the building?

69. A ball is thrown straight upward from ground level with an initial speed of 160 ft/s. What is the maximum height that the ball attains?

70. Carolyn drops a sandbag from the top of a tall building h feet high. At the same time Jon throws a ball upward from ground level from a point directly below the sandbag. What velocity should be given the ball so that it meets the sandbag at the halfway point, where both have altitude $h/2$?

71. Kelly throws a baseball straight downward with an initial speed of 40 ft/s from the top of the Washington Monument, 555 ft high. How long does it take the baseball to reach the ground, and with what speed does it strike the ground?

72. A rock is dropped from an initial height of h feet above the surface of the earth. Show that the speed with which the rock strikes the surface is $\sqrt{2gh}$.

73. A bomb is dropped from a balloon hovering at an altitude of 800 ft. From directly below the balloon, a projectile is fired straight upward toward the bomb exactly 2 s after the bomb is released. With what initial speed should the projectile be fired in order to hit the bomb at an altitude of exactly 400 ft?

74. A car's brakes are applied when the car is moving at 60 mi/h (exactly 88 ft/s). The brakes provide a constant deceleration of 40 ft/s^2. How far does the car travel before coming to a stop?

75. A car traveling at 60 mi/h (exactly 88 ft/s) skids 176 ft after its brakes are applied. The deceleration provided by the braking system is constant. What is its value?

76. A spacecraft is in free fall toward the surface of the moon at a speed of 1000 mi/h. Its retrorockets, when fired, provide a deceleration of 20000 mi/h^2. At what height above the surface should the astronauts activate the retrorockets to ensure a "soft touchdown" ($v = 0$ at impact)? (See Fig. 5.2.21.) Ignore the effect of the moon's gravitational field.

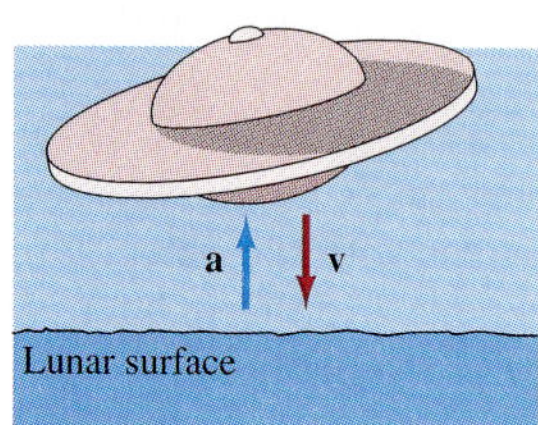

Fig. 5.2.21 The spacecraft of Problem 76

77. (a) What initial velocity v_0 must you use to throw a ball to a maximum height of 144 ft? (b) Now suppose that you throw a ball upward with the same initial velocity v_0 on the moon, where the surface gravitational acceleration is only 5.2 ft/s^2. How high will it go, and how long will it remain aloft?

78. Arthur C. Clarke's *The Wind from the Sun* (1963) describes *Diana,* a spacecraft propelled by the solar wind. Its 2-mi^2 aluminized sail provides it with an acceleration of $(0.001)g = 0.032$ ft/s^2. If the *Diana* starts from rest and travels in a straight line, calculate its distance x traveled (in miles) and its velocity v (in mi/h) after 1 min, 1 h, and 1 day.

79. A driver involved in an accident claims he was going only 25 mi/h. When police tested his car, they found that when the brakes were applied at 25 mi/h, the car skidded only 45 ft before coming to a stop. The driver's skid marks at the accident scene measured 210 ft. Assuming the same (constant) deceleration, calculate the speed he was traveling prior to the accident.

5.3 ELEMENTARY AREA COMPUTATIONS

The indefinite integrals of Section 5.2 stem from the concept of antidifferentiation. The most fundamental type of integral is the one mentioned in Section 5.1, associated with the concept of area. It is called the *definite integral,* or simply *the* integral. Surprisingly, the quite different concepts of area and antidifferentiation have a close and deep relationship. This fact, discovered and exploited by Newton and Leibniz late in the seventeenth century, is the reason the same word, *integral,* is used in both contexts.

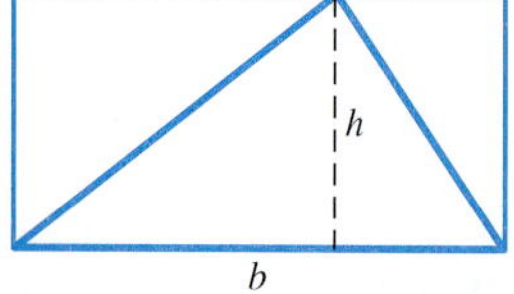

Fig. 5.3.1 The formula for the area of a triangle, $A = \frac{1}{2}bh$, follows with the aid of this figure.

The Concept of Area

Perhaps everyone's first contact with the concept of area is the formula $A = bh$, which gives the area A of a rectangle as the product of its base length b and its height h. We next learn that the area of a triangle is half the product of its base and height. This follows because any triangle can be split into two right triangles, and every right triangle is exactly half a rectangle (Fig. 5.3.1).

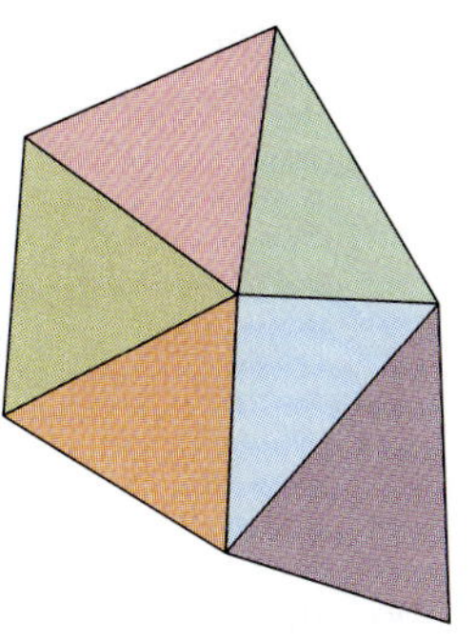

Fig. 5.3.2 Every polygon can be represented as the union of nonoverlapping triangles.

Given the formula $A = \frac{1}{2}bh$ for the area of a triangle, we can—in principle—find the area of any polygonal figure (a bounded plane set bounded by a closed "curve" consisting of a finite number of straight line segments). The reason is that any polygonal figure can be divided into nonoverlapping triangles (Fig. 5.3.2), and the area of the polygonal figure is then the sum of the areas of these triangles. This approach to area dates back several thousand years to the ancient civilizations of Egypt and Babylonia.

The ancient Greeks began the investigation of areas of *curvilinear* figures in the fourth and fifth centuries B.C. Given a plane region R whose area they sought, they worked both with a polygon P *inscribed* in R (Fig. 5.3.3) and with a polygon Q *circumscribed* about R (Fig. 5.3.4). If the polygons P and Q have sufficiently many sides, all short, then it would appear that their areas $a(P)$ and $a(Q)$ closely approximate the area of the region R. Moreover, error control is possible: We see that

$$a(P) < a(R) < a(Q) \tag{1}$$

because R contains the polygon P but is contained in the polygon Q.

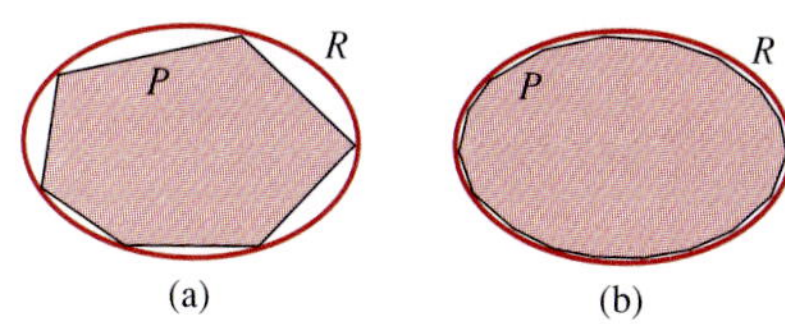

Fig. 5.3.3 (a) A six-sided polygon P inscribed in R; (b) a many-sided inscribed polygon P more closely approximating the area of R

The inequalities in (1) bracket the desired area $a(R)$. Suppose, for instance, that calculations based on triangular dissections (as in Fig. 5.3.2) yield $a(P) = 7.341$ and $a(Q) = 7.343$. Then the resulting inequality

$$7.341 < a(R) < 7.343$$

implies that $a(R) \approx 7.34$, accurate to two decimal places.

Our primary objective here is to describe a systematic technique by which to approximate the area of an appropriate curvilinear region using easily calculated polygonal areas.

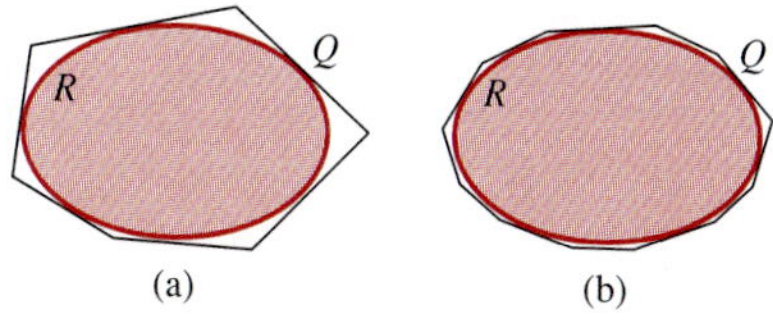

Fig. 5.3.4 (a) A six-sided polygon Q circumscribed around R; (b) a many-sided circumscribed polygon Q more closely approximating the area of R

Areas Under Graphs

We consider the type of region that is determined by a continuous positive-valued function f defined on a closed interval $[a, b]$. Suppose that we want to calculate the area A of the region R that lies *below* the curve $y = f(x)$ and *above* the interval $[a, b]$ on the x-axis (Fig. 5.3.5). The region R is bounded on the left by the vertical line $x = a$ and on the right by the vertical line $x = b$.

We divide the base interval $[a, b]$ into subintervals, all with the same length. Above each subinterval lies a vertical strip (Fig. 5.3.6), and the area of A is the sum of the areas of these strips.

On each of these base subintervals, we erect a rectangle that approximates the corresponding vertical strip. We may choose either an "inscribed" or a "circumscribed" rectangle (both possibilities are illustrated in Fig. 5.3.6), or even a rectangle that is intermediate between the two. These rectangles then make up a polygon that approximates the region R, and therefore the sum of the areas of these rectangles *approximates* the desired area A.

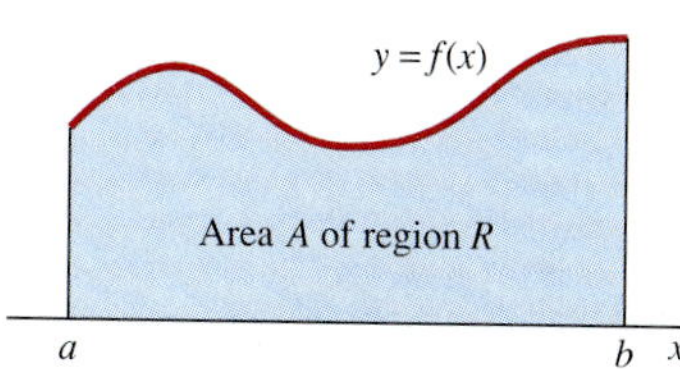

Fig. 5.3.5 The area under the graph of $y = f(x)$ from $x = a$ to $x = b$

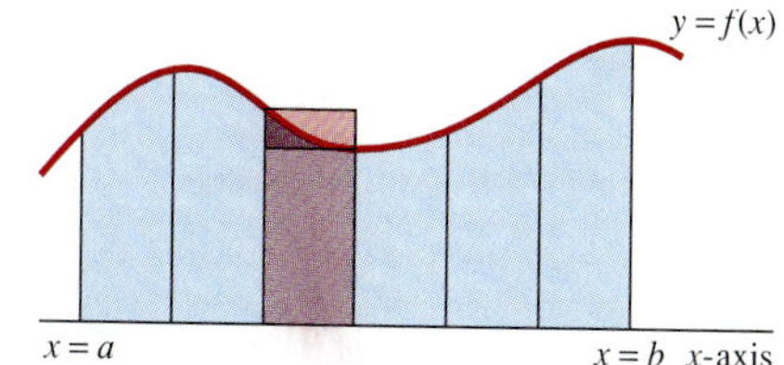

Fig. 5.3.6 Vertical strips determined by a division of $[a, b]$ into equal-length subintervals

For example, suppose that we want to approximate the area A of the region R that lies below the parabola $y = x^2$ above the interval $[0, 3]$. The computer plots in Fig. 5.3.7 show successively

- 5 inscribed and 5 circumscribed rectangles;
- 10 inscribed and 10 circumscribed rectangles;
- 20 inscribed and 20 circumscribed rectangles;
- 40 inscribed and 40 circumscribed rectangles.

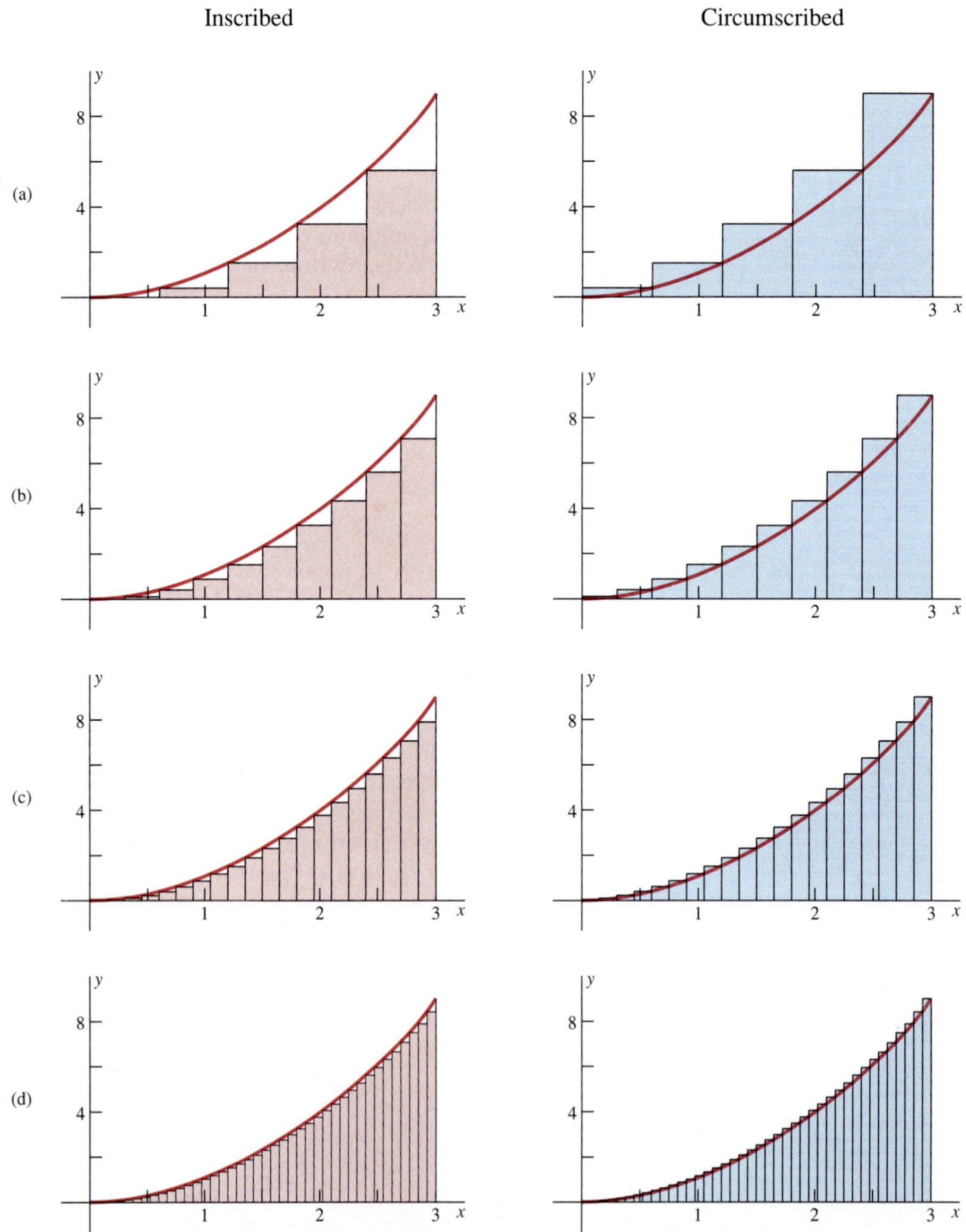

Fig. 5.3.7 (a) Five inscribed and circumscribed polygons; (b) ten inscribed and circumscribed polygons; (c) twenty inscribed and circumscribed polygons; (d) forty inscribed and circumscribed polygons

Each collection of inscribed rectangles gives an *underestimate* of A, whereas each collection of circumscribed rectangles gives an *overestimate* of A. The "curvilinear triangles" (by which the rectangular polygons in Fig. 5.3.7 undershoot the region R) constitute the *errors* in these estimates. The more rectangles we use, the more accurate the approximation. Thus, to approximate accurately the area of such a region R, we need an effective way to calculate and sum the areas of collections of rectangles like those in Fig. 5.3.7.

EXAMPLE 1 As in Fig. 5.3.7, let R denote the region that lies below the graph of $f(x) = x^2$ and above the interval $[0, 3]$. Calculate the underestimate and the overestimate of the area A of R by using 5 rectangles each of width $\frac{3}{5}$. Then repeat the computations using 10 rectangles each of width $\frac{3}{10}$.

Fig. 5.3.8 Five subintervals, each of length $\frac{3}{5}$ (Example 1)

Solution First suppose that $n = 5$ rectangles are used. Let $\underline{A}_5$ denote the underestimate and $\overline{A}_5$ the overestimate obtained by using 5 rectangles based on the 5 subintervals of length $\frac{3}{5}$ (Fig. 5.3.8). From Fig. 5.3.7(a) we see that the heights of the 5 inscribed rectangles (the first of which is degenerate—its height is zero) are the values of the function $f(x) = x^2$ at the 5 *left-hand* endpoints $0, \frac{3}{5}, \frac{6}{5}, \frac{9}{5}$, and $\frac{12}{5}$. Because the base of each rectangle has length $\frac{3}{5}$, we see also that

$$\begin{aligned}\underline{A}_5 &= \tfrac{3}{5}\cdot\left[(0)^2 + \left(\tfrac{3}{5}\right)^2 + \left(\tfrac{6}{5}\right)^2 + \left(\tfrac{9}{5}\right)^2 + \left(\tfrac{12}{5}\right)^2\right]\\ &= \tfrac{3}{5}\cdot\left(0 + \tfrac{9}{25} + \tfrac{36}{25} + \tfrac{81}{25} + \tfrac{144}{25}\right) = 6.48.\end{aligned}$$

The heights of the 5 circumscribed rectangles are the values of $f(x) = x^2$ at the 5 right-hand endpoints $\frac{3}{5}, \frac{6}{5}, \frac{9}{5}, \frac{12}{5}$, and 3, so the corresponding overestimate is

$$\begin{aligned}\overline{A}_5 &= \tfrac{3}{5}\cdot\left[\left(\tfrac{3}{5}\right)^2 + \left(\tfrac{6}{5}\right)^2 + \left(\tfrac{9}{5}\right)^2 + \left(\tfrac{12}{5}\right)^2 + \left(\tfrac{15}{5}\right)^2\right]\\ &= \tfrac{3}{5}\cdot\left(\tfrac{9}{25} + \tfrac{36}{25} + \tfrac{81}{25} + \tfrac{144}{25} + \tfrac{225}{25}\right) = 11.88.\end{aligned}$$

These are crude approximations to the actual area A. On the basis of this information alone, our best estimate of A might well be the average of the under- and overestimates:

$$\frac{\underline{A}_5 + \overline{A}_5}{2} = \frac{6.48 + 11.88}{2} = 9.18.$$

Let us see if doubling the number of subintervals to $n = 10$ increases the accuracy significantly. Looking at Fig. 5.3.7(b), we see that the heights of the 10 inscribed rectangles are the values of $f(x) = x^2$ at the 10 *left-hand* endpoints $0, \frac{3}{10}, \frac{6}{10}, \frac{9}{10}, \frac{12}{10}, \frac{15}{10}, \frac{18}{10}, \frac{21}{10}, \frac{24}{10}$, and $\frac{27}{10}$ of the subintervals in Fig. 5.3.9. The base of each rectangle has length $\frac{3}{10}$, so the resulting underestimate is

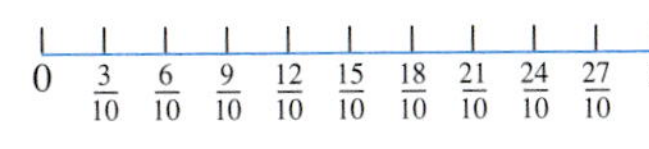

Fig. 5.3.9 Ten subintervals, each of length $\frac{3}{10}$ (Example 1)

$$\begin{aligned}\underline{A}_{10} &= \tfrac{3}{10}\cdot\left[(0)^2 + \left(\tfrac{3}{10}\right)^2 + \left(\tfrac{6}{10}\right)^2 + \left(\tfrac{9}{10}\right)^2 + \left(\tfrac{12}{10}\right)^2 + \left(\tfrac{15}{10}\right)^2 + \left(\tfrac{18}{10}\right)^2 + \left(\tfrac{21}{10}\right)^2 + \left(\tfrac{24}{10}\right)^2 + \left(\tfrac{27}{10}\right)^2\right]\\ &= \tfrac{3}{10}\cdot\left(0 + \tfrac{9}{100} + \tfrac{36}{100} + \tfrac{81}{100} + \tfrac{144}{100} + \tfrac{225}{100} + \tfrac{324}{100} + \tfrac{441}{100} + \tfrac{576}{100} + \tfrac{729}{100}\right)\\ &= \tfrac{7695}{1000} = 7.695.\end{aligned}$$

Similarly, the sum of the areas of the 10 circumscribed rectangles in Fig. 5.3.7(b) is the overestimate

$$\begin{aligned}\overline{A}_{10} &= \tfrac{3}{10}\cdot\left[\left(\tfrac{3}{10}\right)^2 + \left(\tfrac{6}{10}\right)^2 + \left(\tfrac{9}{10}\right)^2 + \left(\tfrac{12}{10}\right)^2 + \left(\tfrac{15}{10}\right)^2 + \left(\tfrac{18}{10}\right)^2 + \left(\tfrac{21}{10}\right)^2 + \left(\tfrac{24}{10}\right)^2 + \left(\tfrac{27}{10}\right)^2 + \left(\tfrac{30}{10}\right)^2\right].\\ &= \tfrac{10395}{1000} = 10.395.\end{aligned}$$

At this point, our best estimate of the actual area A might be the average

$$\frac{\underline{A}_{10} + \overline{A}_{10}}{2} = \frac{7.695 + 10.395}{2} = 9.045.$$ ■

We used a computer to calculate more refined underestimates and overestimates of the area A under the graph $y = x^2$ over $[0, 3]$, with 20, 40, 80, 160, and, finally, 320 rectangles. The results (rounded to four decimal places) are shown in Fig. 5.3.10. The average values in the final column of the table suggest that $A \approx 9$.

Number of rectangles	Underestimate	Overestimate	Average
5	6.4800	11.8800	9.1800
10	7.6950	10.3950	9.0450
20	8.3363	9.6863	9.0113
40	8.6653	9.3403	9.0028
80	8.8320	9.1695	9.0007
160	8.9158	9.0846	9.0002
320	8.9579	9.0422	9.0000

Fig. 5.3.10 Estimate of the area under $y = x^2$ over $[0,3]$

Summation Notation

For more convenient computation of area estimates, as in Example 1, we need a concise notation for sums of many numbers. The symbol $\sum_{i=1}^{n} a_i$ is used to abbreviate the sum of the n numbers $a_1, a_2, a_3, \dots, a_n$:

$$\sum_{i=1}^{n} a_i = a_1 + a_2 + a_3 + \cdots + a_n. \tag{2}$$

The symbol Σ (capital Greek sigma) indicates the sum of the **terms** a_i as the **summation index** i takes successive *integer* values from 1 to n. For instance, the sum of the squares of the first 10 positive integers is

$$\begin{aligned}\sum_{i=1}^{10} i^2 &= 1^2 + 2^2 + 3^2 + 4^2 + 5^2 + 6^2 + 7^2 + 8^2 + 9^2 + 10^2 \\ &= 1 + 4 + 9 + 16 + 25 + 36 + 49 + 64 + 81 + 100 = 385.\end{aligned}$$

The particular symbol used for the summation index is immaterial:

$$\sum_{i=1}^{10} i^2 = \sum_{k=1}^{10} k^2 = \sum_{r=1}^{10} r^2 = 385.$$

EXAMPLE 2

$$\sum_{k=1}^{7} (k + 1) = 2 + 3 + 4 + 5 + 6 + 7 + 8 = 35,$$

$$\sum_{n=1}^{6} 2^n = 2 + 4 + 8 + 16 + 32 + 64 = 126, \quad \text{and}$$

$$\sum_{j=1}^{5} \frac{(-1)^{j+1}}{j^2} = 1 - \tfrac{1}{4} + \tfrac{1}{9} - \tfrac{1}{16} + \tfrac{1}{25} = \tfrac{3019}{3600} \approx 0.8386.$$

■

The simple rules of summation

$$\sum_{i=1}^{n} ca_i = c\sum_{i=1}^{n} a_i \tag{3}$$

and

$$\sum_{i=1}^{n} (a_i + b_i) = \left(\sum_{i=1}^{n} a_i\right) + \left(\sum_{i=1}^{n} b_i\right) \tag{4}$$

are easy to verify by writing out each sum in full.

Note that if $a_i = a$ (a constant) for $i = 1, 2, \dots, n$, then Eq. (4) yields

$$\sum_{i=1}^{n} (a + b_i) = \sum_{i=1}^{n} a + \sum_{i=1}^{n} b_i = \underbrace{(a + a + \cdots + a)}_{n \text{ terms}} + \sum_{i=1}^{n} b_i,$$

and hence

$$\sum_{i=1}^{n} (a + b_i) = na + \sum_{i=1}^{n} b_i. \tag{5}$$

In particular,

$$\sum_{i=1}^{n} 1 = n. \tag{6}$$

The sum of the kth powers of the first n positive integers,

$$\sum_{i=1}^{n} i^k = 1^k + 2^k + 3^k + \cdots + n^k,$$

commonly occurs in area computations. The values of this sum for $k = 1, 2,$ and 3 are given by the following formulas (see Problems 43 and 44):

$$\sum_{i=1}^{n} i = \frac{n(n+1)}{2} = \tfrac{1}{2}n^2 + \tfrac{1}{2}n, \tag{7}$$

$$\sum_{i=1}^{n} i^2 = \frac{n(n+1)(2n+1)}{6} = \tfrac{1}{3}n^3 + \tfrac{1}{2}n^2 + \tfrac{1}{6}n, \tag{8}$$

$$\sum_{i=1}^{n} i^3 = \frac{n^2(n+1)^2}{4} = \tfrac{1}{4}n^4 + \tfrac{1}{2}n^3 + \tfrac{1}{4}n^2. \tag{9}$$

EXAMPLE 3 The sum of the first 10 positive integers is given by Eq. (7) with $n = 10$:

$$1 + 2 + 3 + \cdots + 10 = \sum_{i=1}^{10} i = \frac{10 \cdot 11}{2} = 55.$$

The sum of their squares and cubes are given by Eqs. (8) and (9):

$$1^2 + 2^2 + 3^2 + \cdots + 10^2 = \sum_{i=1}^{10} i^2 = \frac{10 \cdot 11 \cdot 21}{6} = 385$$

and

$$1^3 + 2^3 + 3^3 + \cdots + 10^3 = \sum_{i=1}^{10} i^3 = \frac{10^2 \cdot 11^2}{4} = 3025.$$

■

EXAMPLE 4 Consider the sum

$$\sum_{i=1}^{10}(7i^2 - 5i) = 2 + 18 + 48 + \cdots + 522 + 650.$$

Using the rules in Eqs. (3) and (4) as well as Eqs. (7) and (8), we find that

$$\sum_{i=1}^{10} = 7\sum_{i=1}^{10} i^2 - 5\sum_{i=1}^{10} i = 7\cdot\frac{10\cdot 11\cdot 21}{6} - 5\cdot\frac{10\cdot 11}{2} = 2420.$$

EXAMPLE 5 We can use Eq. (8) to simplify the evaluation of the sum for $\underline{A}_{10}$ in Example 1, as follows:

$$\begin{aligned}\underline{A}_{10} &= \tfrac{3}{10}\cdot\left[\left(\tfrac{0}{10}\right)^2 + \left(\tfrac{3}{10}\right)^2 + \left(\tfrac{6}{10}\right)^2 + \cdots + \left(\tfrac{27}{10}\right)^2\right] = \tfrac{3}{10}\sum_{i=0}^{9}\left(\tfrac{3}{10}\right)^2 i^2 \\ &= \tfrac{3}{10}\cdot\left(\tfrac{3}{10}\right)^2\cdot\left[1^2 + 2^2 + 3^2 + \cdots + 9^2\right] = \left(\tfrac{3}{10}\right)^3\sum_{i=1}^{9} i^2 \\ &= \frac{27}{1000}\cdot\frac{9\cdot 10\cdot 19}{6} = \frac{7695}{1000} = 7.695.\end{aligned}$$

EXAMPLE 6 Evaluate the limit

$$\lim_{n\to+\infty}\frac{1 + 2 + 3 + \cdots + n}{n^2}.$$

Solution Using Eq. (7), we obtain

$$\begin{aligned}\lim_{n\to+\infty}\frac{1 + 2 + 3 + \cdots + n}{n^2} &= \lim_{n\to+\infty}\frac{\frac{1}{2}n(n+1)}{n^2} \\ &= \lim_{n\to+\infty}\frac{n+1}{2n} = \lim_{n\to+\infty}\left(\frac{1}{2} + \frac{1}{2n}\right) = \frac{1}{2},\end{aligned}$$

because the term $1/(2n)$ has limit zero as $n \to +\infty$.

Area Sums

Figure 5.3.11 shows the region R that lies below the graph of the positive-valued *increasing* function f and above the interval $[a, b]$. To approximate the area A of R, we have chosen a fixed integer n and divided the interval $[a, b]$ into n subintervals

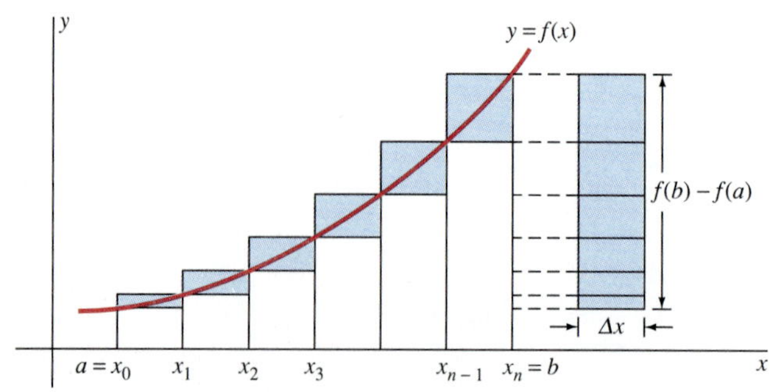

Fig. 5.3.11 The area under $y = f(x)$ over the interval $[a, b]$

$$[x_0, x_1], \quad [x_1, x_2], \quad [x_2, x_3], \quad \dots, \quad [x_{n-1}, x_n],$$

all with the same length

$$\Delta x = \frac{b - a}{n}. \tag{10}$$

On each of the subintervals we have erected one inscribed rectangle and one circumscribed rectangle.

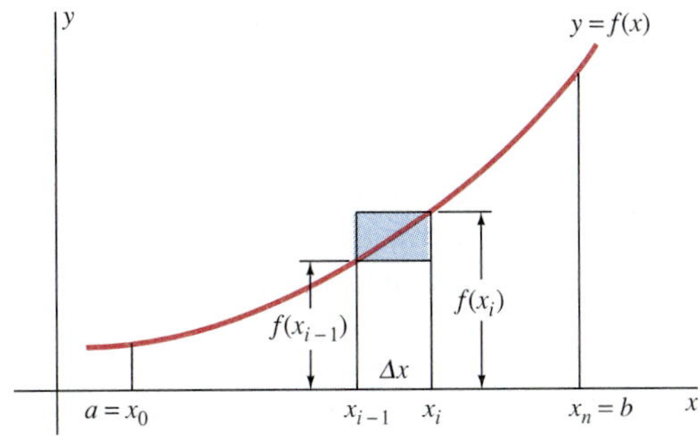

Fig. 5.3.12 Inscribed and circumscribed rectangles on the ith subinterval $[x_{i-1}, x_i]$

As indicated in Fig. 5.3.12, the inscribed rectangle over the ith subinterval $[x_{i-1}, x_i]$ has height $f(x_{i-1})$, whereas the ith circumscribed rectangle has height $f(x_i)$. Because the base of each rectangle has length Δx, the areas of the rectangles are

$$f(x_{i-1})\,\Delta x \quad \text{and} \quad f(x_i)\,\Delta x, \tag{11}$$

respectively. Adding the areas of the inscribed rectangles for $i = 1, 2, 3, \dots, n$, we get the underestimate

$$\underline{A}_n = \sum_{i=1}^{n} f(x_{i-1})\,\Delta x \tag{12}$$

of the actual area A. Similarly, the sum of the areas of the circumscribed rectangles is the overestimate

$$\overline{A}_n = \sum_{i=1}^{n} f(x_i)\,\Delta x. \tag{13}$$

The inequality $\underline{A}_n \leqq A \leqq \overline{A}_n$ then yields

$$\sum_{i=1}^{n} f(x_{i-1})\,\Delta x \leqq A \leqq \sum_{i=1}^{n} f(x_i)\,\Delta x. \tag{14}$$

The inequalities in (14) would be reversed if $f(x)$ were decreasing (rather than increasing) on $[a, b]$. (Why?)

Areas as Limits

An illustration such as Fig. 5.3.7 suggests that if the number n of subintervals is very large, so that Δx is small, then the areas $\underline{A}_n$ and $\overline{A}_n$ of the inscribed and circumscribed polygons will differ by very little. Hence both will be very close to the actual area A of the region R. We can also see this because if f either is increasing or is decreasing on the whole interval $[a, b]$, then the small rectangles in Fig. 5.3.11 (representing the difference between $\overline{A}_n$ and $\underline{A}_n$) can be reassembled in a "stack," as indicated on the right in the figure. It follows that

$$|\overline{A}_n - \underline{A}_n| = |f(b) - f(a)|\,\Delta x. \tag{15}$$

But $\Delta x = (b - a)/n \to 0$ as $n \to \infty$. Thus the difference between the left-hand and right-hand sums in (14) is approaching zero as $n \to \infty$, whereas A does not change as $n \to \infty$. It follows that the area of the region R is given by

$$A = \lim_{n\to\infty} \sum_{i=1}^{n} f(x_{i-1})\,\Delta x = \lim_{n\to\infty} \sum_{i=1}^{n} f(x_i)\,\Delta x. \tag{16}$$

The meaning of these limits is simply that A can be found with any desired accuracy by calculating either sum in Eq. (16) with a sufficiently large number n of subintervals. In applying Eq. (16), recall that

$$\Delta x = \frac{b - a}{n}. \tag{17}$$

Also note that

$$x_i = a + i\,\Delta x \tag{18}$$

for $i = 0, 1, 2, \ldots, n$, because x_i is i "steps" of length Δx to the right of $x_0 = a$.

EXAMPLE 7 We can now compute exactly the area we approximated in Example 1—the area of the region under the graph of $f(x) = x^2$ over the interval $[0, 3]$. If we divide $[0, 3]$ into n subintervals all of the same length, then Eqs. (17) and (18) give

$$\Delta x = \frac{3}{n} \quad \text{and} \quad x_i = 0 + i \cdot \frac{3}{n} = \frac{3i}{n}$$

for $i = 0, 1, 2, \ldots, n$. Therefore

$$\sum_{i=1}^{n} f(x_i)\,\Delta x = \sum_{i=1}^{n} (x_i)^2\,\Delta x = \sum_{i=1}^{n} \left(\frac{3i}{n}\right)^2 \left(\frac{3}{n}\right) = \frac{27}{n^3} \sum_{i=1}^{n} i^2.$$

Then Eq. (8) for $\sum i^2$ yields

$$\sum_{i=1}^{n} f(x_i)\,\Delta x = \frac{27}{n^3}\left(\frac{1}{3}n^3 + \frac{1}{2}n^2 + \frac{1}{6}n\right) = 27\left(\frac{1}{3} + \frac{1}{2n} + \frac{1}{6n^2}\right).$$

When we take the limit as $n \to \infty$, Eq. (16) gives

$$A = \lim_{n\to\infty} 27\left(\frac{1}{3} + \frac{1}{2n} + \frac{1}{6n^2}\right) = 9,$$

because the terms $1/(2n)$ and $1/(6n^2)$ approach zero as $n \to \infty$. Thus our earlier inference from the data in Fig. 5.3.10 was correct: $A = 9$ *exactly.* ■

EXAMPLE 8 Find the area under the graph of $f(x) = 100 - 3x^2$ from $x = 1$ to $x = 5$.

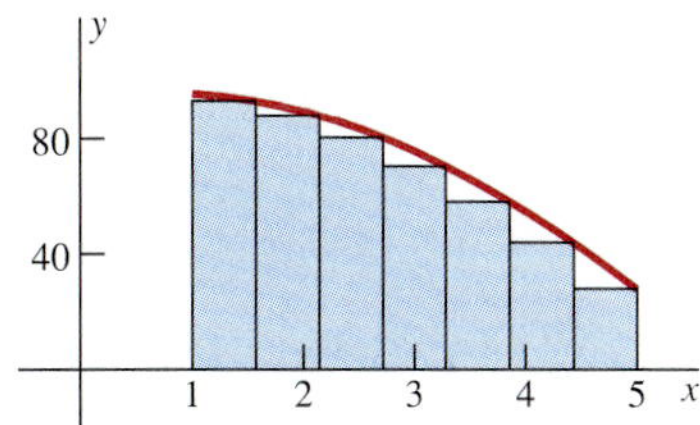

Fig. 5.3.13 The region of Example 8

Solution As shown in Fig. 5.3.13, the sum $\sum f(x_i)\,\Delta x$ gives the area of the inscribed rectangular polygon. With $a = 1$ and $b = 5$, Eqs. (17) and (18) give

$$\Delta x = \frac{4}{n} \quad \text{and} \quad x_i = 1 + i \cdot \frac{4}{n} = 1 + \frac{4i}{n}.$$

Therefore

$$\begin{aligned}
\sum_{i=1}^{n} f(x_i)\,\Delta x &= \sum_{i=1}^{n} \left[100 - 3 \cdot \left(1 + \frac{4i}{n}\right)^2 \right] \left(\frac{4}{n}\right) \\
&= \sum_{i=1}^{n} \left[97 - \frac{24i}{n} - \frac{48i^2}{n^2} \right] \left(\frac{4}{n}\right) \\
&= \frac{388}{n} \sum_{i=1}^{n} 1 - \frac{96}{n^2} \sum_{i=1}^{n} i - \frac{192}{n^3} \sum_{i=1}^{n} i^2 \\
&= \frac{388}{n} \cdot n - \frac{96}{n^2}\left(\frac{1}{2}n^2 + \frac{1}{2}n\right) - \frac{192}{n^3}\left(\frac{1}{3}n^3 + \frac{1}{2}n^2 + \frac{1}{6}n\right) \\
&= 276 - \frac{144}{n} - \frac{32}{n^2}.
\end{aligned}$$

[We have applied Eqs. (6) through (8).] Consequently, the second limit in Eq. (16) yields

$$A = \lim_{n \to \infty} \left(276 - \frac{144}{n} - \frac{32}{n} \right) = 276$$

for the desired area. ■

Historical Note—The Number π

Mathematicians of ancient times tended to employ inscribed and circumscribed triangles rather than rectangles for area approximations. In the third century B.C., Archimedes, the greatest mathematician of antiquity, used such an approach to derive the famous estimate

$$\frac{223}{71} = 3\frac{10}{71} < \pi < 3\frac{1}{7} = \frac{22}{7}.$$

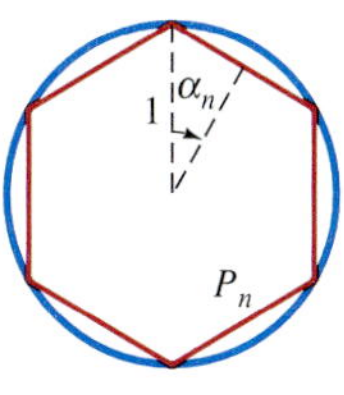

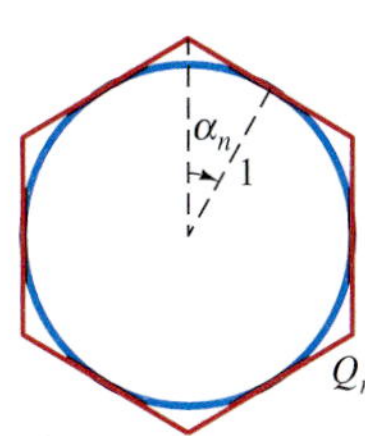

Fig. 5.3.14 Estimating π by using inscribed and circumscribed regular polygons and the unit circle

Because the area of a circle of radius r is πr^2, the number π may be *defined* to be the area of the unit circle of radius $r = 1$. We will approximate π, then, by approximating the area of the unit circle.

Let P_n and Q_n be n-sided regular polygons, with P_n inscribed in the unit circle and Q_n circumscribed around it (Fig. 5.3.14). Because both polygons are regular, all their sides and angles are equal, so we need to find the area of only *one* of the triangles that we've shown making up P_n and *one* of those making up Q_n.

Let α_n be the central angle subtended by *half* of one of the sides. The angle α_n is the same whether we work with P_n or with Q_n. In degrees,

$$\alpha_n = \frac{360°}{2n} = \frac{180°}{n}.$$

We can read various dimensions and proportions from Fig. 5.3.14. For example, we see that the area $a(P_n) = A_n$ of P_n is given by

n	$a(P_n)$	$a(Q_n)$
6	2.598076	3.464102
12	3.000000	3.215390
24	3.105829	3.159660
48	3.132629	3.146086
96	3.139350	3.142715
180	3.140955	3.141912
360	3.141433	3.141672
720	3.141553	3.141613
1440	3.141583	3.141598
2880	3.141590	3.141594
5760	3.141592	3.141593

Fig. 5.3.15 Data for estimating π (rounded to six-place accuracy)

$$\underline{A}_n = a(P_n) = n \cdot 2 \cdot \frac{1}{2} \sin \alpha_n \cos \alpha_n = \frac{n}{2} \sin 2\alpha_n = \frac{n}{2} \sin\left(\frac{360^\circ}{n}\right) \tag{19}$$

and that the area of Q_n is

$$\overline{A}_n = a(Q_n) = n \cdot 2 \cdot \frac{1}{2} \tan \alpha_n = n \tan\left(\frac{180^\circ}{n}\right). \tag{20}$$

We substituted selected values of n into Eqs. (19) and (20) to obtain the entries of the table in Fig. 5.3.15. Because $\underline{A}_n \leqq \pi \leqq \overline{A}_n$ for all n, we see that $\pi \approx 3.14159$ to five decimal places. Archimedes' reasoning was *not* circular—he used a direct method for computing the sines and cosines in Eqs. (19) and (20) that does not depend upon a priori knowledge of the value of π. *

5.3 PROBLEMS

Write each of the sums in Problems 1 through 8 in expanded notation.

1. $\sum_{i=1}^{5} 3^i$

2. $\sum_{i=1}^{6} \sqrt{2i}$

3. $\sum_{j=1}^{5} \frac{1}{j+1}$

4. $\sum_{j=1}^{6} (2j-1)$

5. $\sum_{k=1}^{6} \frac{1}{k^2}$

6. $\sum_{k=1}^{6} \frac{(-1)^{k+1}}{k^2}$

7. $\sum_{n=1}^{5} x^n$

8. $\sum_{n=1}^{5} (-1)^{n+1} x^{2n-1}$

Write the sums in Problems 9 through 18 in summation notation.

9. $1 + 4 + 9 + 16 + 25$

10. $1 - 2 + 3 - 4 + 5 - 6$

11. $1 + \frac{1}{2} + \frac{1}{3} + \frac{1}{4} + \frac{1}{5}$

12. $1 + \frac{1}{4} + \frac{1}{9} + \frac{1}{16} + \frac{1}{25}$

13. $\frac{1}{2} + \frac{1}{4} + \frac{1}{8} + \frac{1}{16} + \frac{1}{32} + \frac{1}{64}$

14. $\frac{1}{3} - \frac{1}{9} + \frac{1}{27} - \frac{1}{81} + \frac{1}{243}$

15. $\frac{2}{3} + \frac{4}{9} + \frac{8}{27} + \frac{16}{81} + \frac{32}{243}$

16. $1 + \sqrt{2} + \sqrt{3} + 2 + \sqrt{5} + \sqrt{6} + \sqrt{7} + 2\sqrt{2} + 3$

17. $x + \frac{x^2}{2} + \frac{x^3}{3} + \cdots + \frac{x^{10}}{10}$

18. $x - \frac{x^3}{3} + \frac{x^5}{5} - \frac{x^7}{7} + \cdots - \frac{x^{19}}{19}$

Use Eqs. (6) through (9) to find the sums in Problems 19 through 28.

19. $\sum_{i=1}^{10} (4i - 3)$

20. $\sum_{j=1}^{8} (5 - 2j)$

21. $\sum_{i=1}^{10} (3i^2 + 1)$

22. $\sum_{k=1}^{6} (2k - 3k^2)$

23. $\sum_{r=1}^{8} (r-1)(r+2)$

24. $\sum_{i=1}^{5} (i^3 - 3i + 2)$

25. $\sum_{i=1}^{6} (i^3 - i^2)$

26. $\sum_{k=1}^{10} (2k-1)^2$

27. $\sum_{i=1}^{100} i^2$

28. $\sum_{i=1}^{100} i^3$

Use the method of Example 6 to evaluate the limits in Problems 29 and 30.

29. $\lim_{n \to \infty} \frac{1^2 + 2^2 + 3^2 + \cdots + n^2}{n^3}$

30. $\lim_{n \to \infty} \frac{1^3 + 2^3 + 3^3 + \cdots + n^3}{n^4}$

Use Eqs. (6) through (9) to derive concise formulas in terms of n for the sums in Problems 31 and 32.

31. $\sum_{i=1}^{n} (2i - 1)$

32. $\sum_{i=1}^{n} (2i-1)^2$

In Problems 33 through 42, let R denote the region that lies below the graph of $y = f(x)$ over the interval $[a, b]$ on the x-axis. Use the method of Example 1 to calculate both an underestimate $\underline{A}_n$ and an overestimate $\overline{A}_n$ for the area A of R, based on a division of $[a, b]$ into n subintervals all with the same length $\Delta x = (b - a)/n$.

33. $f(x) = x$ on $[0, 1]$; $n = 5$

34. $f(x) = x$ on $[1, 3]$; $n = 5$

35. $f(x) = 2x + 3$ on $[0, 3]$; $n = 6$

36. $f(x) = 13 - 3x$ on $[0, 3]$; $n = 6$ (Fig. 5.3.16)

37. $f(x) = x^2$ on $[0, 1]$; $n = 5$

*See Chapter 2 of C. H. Edwards, Jr., *The Historical Development of the Calculus* (New York: Springer-Verlag, 1979).

Fig. 5.3.16 Problem 36

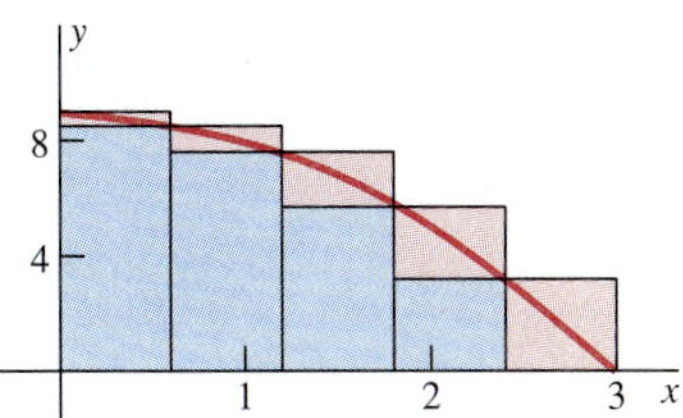

Fig. 5.3.17 Problem 39

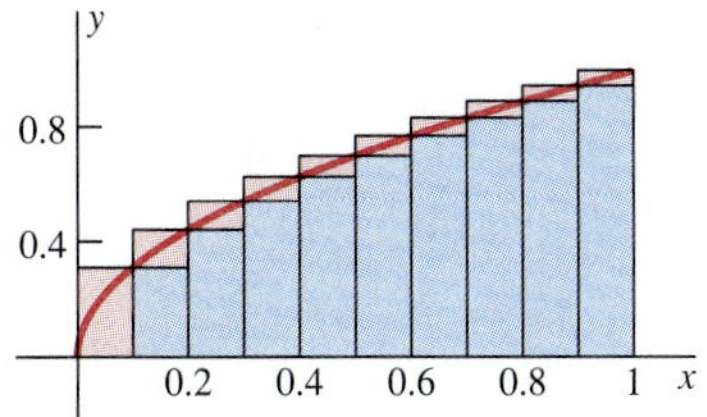

Fig. 5.3.18 Problem 42

38. $f(x) = x^2$ on $[1, 3]$; $n = 5$

39. $f(x) = 9 - x^2$ on $[0, 3]$; $n = 5$ (Fig. 5.3.17)

40. $f(x) = 9 - x^2$ on $[1, 3]$; $n = 8$

41. $f(x) = x^3$ on $[0, 1]$; $n = 10$

42. $f(x) = \sqrt{x}$ on $[0, 1]$; $n = 10$ (Fig. 5.3.18)

43. Derive Eq. (7) by adding the equations

$$\sum_{i=1}^{n} i = 1 + 2 + 3 + \cdots + n$$

and

$$\sum_{i=1}^{n} i = n + (n - 1) + (n - 2) + \cdots + 2 + 1.$$

44. Write the n equations obtained by substituting the values $k = 1, 2, 3, \dots, n$ into the identity

$$(k + 1)^3 - k^3 = 3k^2 + 3k + 1.$$

Add these n equations and use their sum to deduce Eq. (8) from Eq. (7).

In Problems 45 through 50, first calculate (in terms of n) the sum

$$\sum_{i=1}^{n} f(x_i)\,\Delta x$$

to approximate the area A of the region under $y = f(x)$ above the interval $[a, b]$. Then find A exactly (as in Examples 7 and 8) by taking the limit as $n \to \infty$.

45. $f(x) = x$ on $[0, 1]$

46. $f(x) = x^2$ on $[0, 2]$

47. $f(x) = x^3$ on $[0, 3]$

48. $f(x) = x + 2$ on $[0, 2]$

49. $f(x) = 5 - 3x$ on $[0, 1]$

50. $f(x) = 9 - x^2$ on $[0, 3]$

51. As in Fig. 5.3.19, the region under the graph of $f(x) = hx/b$ is a triangle with base b and height h. Use Eq. (7) to verify—with the notation of Eq. (16)—that

$$\lim_{n\to\infty} \sum_{i=1}^{n} f(x_i)\,\Delta x = \frac{1}{2}bh,$$

in agreement with the familiar formula for the area of a triangle.

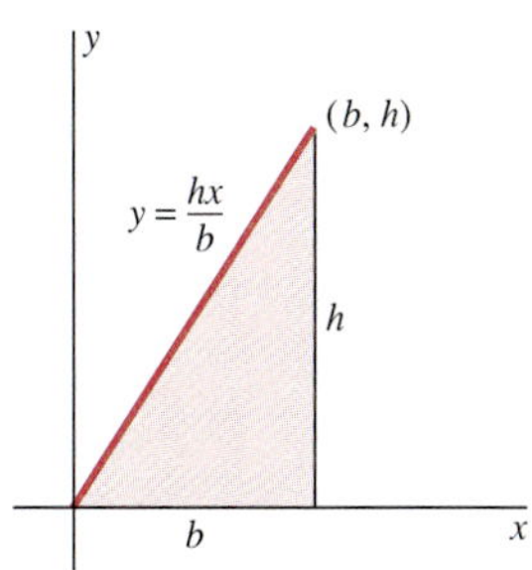

Fig. 5.3.19 Problem 51

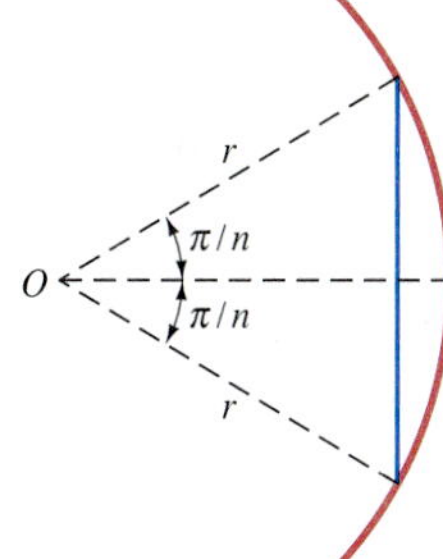

Fig. 5.3.20 Problem 52

In Problems 52 and 53, let A denote the area and C the circumference of a circle of radius r and let A_n and C_n denote the area and perimeter, respectively, of a regular n-sided polygon inscribed in this circle.

52. Figure 5.3.20 shows one side of the n-sided polygon subtending an angle $2\pi/n$ at the center O of the circle. Show that

$$A_n = nr^2 \sin\left(\frac{\pi}{n}\right)\cos\left(\frac{\pi}{n}\right) \quad \text{and that} \quad C_n = 2nr\sin\left(\frac{\pi}{n}\right).$$

53. Deduce that $A = \frac{1}{2}rC$ by taking the limit of A_n/C_n as $n \to \infty$. Then, under the assumption that $A = \pi r^2$, deduce that $C = 2\pi r$. Thus the familiar circumference formula for a circle follows from the familiar area formula for a circle.

5.4 RIEMANN SUMS AND THE INTEGRAL

Suppose that f is a positive-valued and increasing function defined on a set of real numbers that includes the interval $[a, b]$. In Section 5.3 we used inscribed and circumscribed rectangles to set up the sums

$$\sum_{i=1}^{n} f(x_{i-1})\,\Delta x \quad \text{and} \quad \sum_{i=1}^{n} f(x_i)\,\Delta x \tag{1}$$

that approximate the area A under the graph of $y = f(x)$ from $x = a$ to $x = b$. Recall that the notation in Eq. (1) is based on a division of the interval $[a, b]$ into n subintervals, all with the same length $\Delta x = (b - a)/n$, and that $[x_{i-1}, x_i]$ denotes the ith subinterval.

The approximating sums in Eq. (1) are both of the form

$$\sum_{i=1}^{n} f(x_i^{\star})\,\Delta x, \tag{2}$$

where $x_i^{\star}$ denotes a selected point of the ith subinterval $[x_{i-1}, x_i]$ (Fig. 5.4.1). Sums of the form in (2) appear as approximations in a wide range of applications and form the basis for the definition of the integral. Motivated by our discussion of area in Section 5.3, we want to define the integral of f from a to b as some sort of limit, as $\Delta x \to 0$, of sums such as the one in (2). Our goal is to begin with a fairly general function f and define a computable real number I (the *integral* of f) that—in the special case when f is continuous and positive-valued on $[a, b]$—will equal the area under the graph of $y = f(x)$.

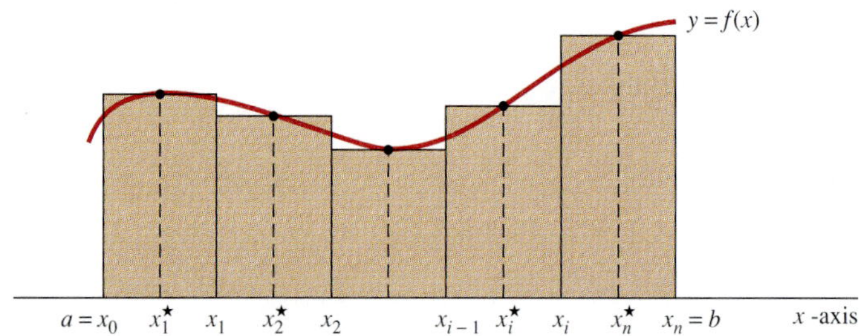

Fig. 5.4.1 The Riemann sum in Eq. (2) as a sum of areas of rectangles

Riemann Sums

We begin with a function f defined on $[a, b]$ that is *not* necessarily either continuous or positive-valued. A **partition** P of $[a, b]$ is a collection of subintervals

$$[x_0, x_1], \quad [x_1, x_2], \quad [x_2, x_3], \quad \dots, \quad [x_{n-1}, x_n]$$

of $[a, b]$ such that

$$a = x_0 < x_1 < x_2 < x_3 < \cdots < x_{n-1} < x_n = b,$$

as in Fig. 5.4.1. To get a sum such as that in (2), we need a point $x_i^{\star}$ in the ith subinterval for each i, $1 \leqq i \leqq n$. A collection of points

$$S = \{x_1^{\star}, x_2^{\star}, x_3^{\star}, \dots, x_n^{\star}\}$$

with $x_i^{\star}$ in $[x_{i-1}, x_i]$ for each i (Fig. 5.4.2) is called a **selection** for the partition P.

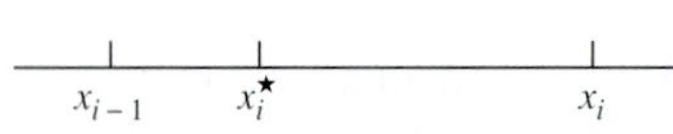

Fig. 5.4.2 The selected point $x_i^{\star}$ in the ith subinterval $[x_{i-1}, x_i]$

Definition *Riemann Sum*

Let f be a function defined on the interval $[a, b]$. If P is a partition of $[a, b]$ and S is a selection for P, then the **Riemann sum** for f determined by P and S is

$$R = \sum_{i=1}^{n} f(x_i^\star)\,\Delta x_i. \tag{3}$$

We also say that this Riemann sum is **associated with** the partition P.

The German mathematician G.F.B. Riemann (1826–1866) provided a rigorous definition of the integral. Various special types of "Riemann sums" had appeared in area and volume computations since the time of Archimedes, but it was Riemann who framed the preceding definition in its full generality.

The point $x_i^\star$ in Eq. (3) is simply a selected point of the ith subinterval $[x_{i-1}, x_i]$. That is, it can be *any* point of this subinterval. But when computing Riemann sums, we usually choose the points of the selection S in some systematic manner, as illustrated in Fig. 5.4.3. There we show different Riemann sums for the function $f(x) = 2x^3 - 6x^2 + 5$ on the interval $[0, 3]$. Figure 5.4.3(a) shows rectangles associated with the *left-endpoint sum*

$$R_{\text{left}} = \sum_{i=1}^{n} f(x_{i-1})\,\Delta x, \tag{4}$$

in which each $x_i^\star$ is selected to be x_{i-1}, the *left endpoint* of the ith subinterval $[x_{i-1}, x_i]$ of length $\Delta x = (b - a)/n$. Figure 5.4.3(b) shows rectangles associated with the *right-endpoint sum*

$$R_{\text{right}} = \sum_{i=1}^{n} f(x_i)\,\Delta x, \tag{5}$$

in which each $x_i^\star$ is selected to be x_i, the *right endpoint* of $[x_{i-1}, x_i]$. In each figure, some of the rectangles are inscribed and other are circumscribed.

Figure 5.4.3(c) shows rectangles associated with the *midpoint sum*

$$R_{\text{mid}} = \sum_{i=1}^{n} f(m_i)\,\Delta x, \tag{6}$$

in which

$$x_i^\star = m_i = \frac{x_{i-1} + x_i}{2},$$

the *midpoint* of the ith subinterval $[x_{i-1}, x_i]$. The dashed lines in Fig. 5.4.3(c) represent the ordinates of f at these midpoints.

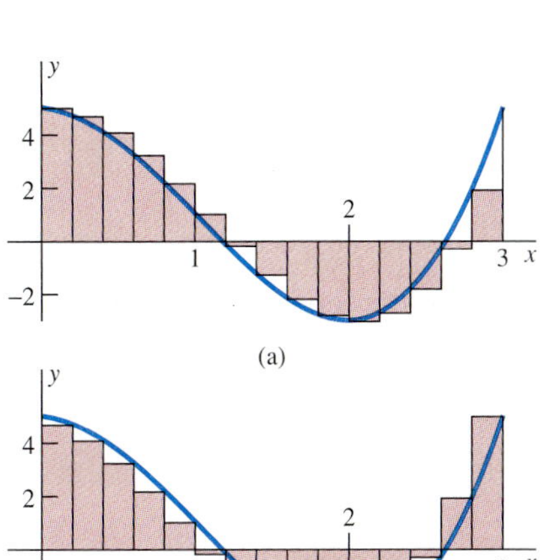

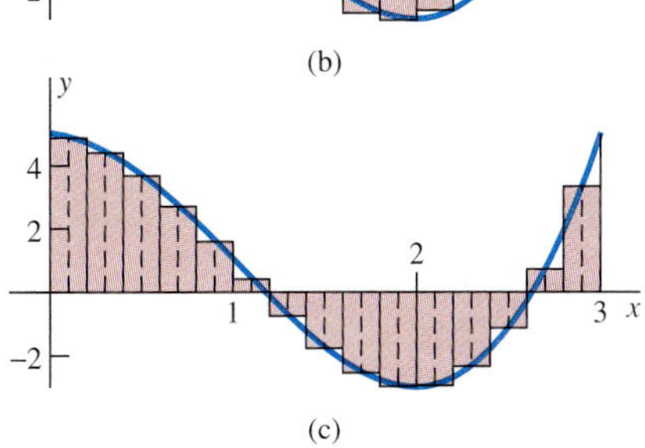

Fig. 5.4.3 Riemann sums for $f(x) = 2x^3 - 6x^2 + 5$ on $[0, 3]$: (a) Left-endpoint sum; (b) Right-endpoint sum; (c) Midpoint sum

EXAMPLE 1 In Example 1 of Section 5.3 we calculated left- and right-endpoint sums for $f(x) = x^2$ on $[0, 3]$ with $n = 10$ subintervals. We now do this more concisely by using summation notation, and we also calculate the analogous midpoint sum. Figure 5.4.4 shows a typical approximating rectangle for each of these sums. With $a = 0$, $b = 3$, and $\Delta x = (b - a)/n = \frac{3}{10}$, we see that the ith subdivision point is

$$x_i = a + i \cdot \Delta x = \frac{3}{10}\,i.$$

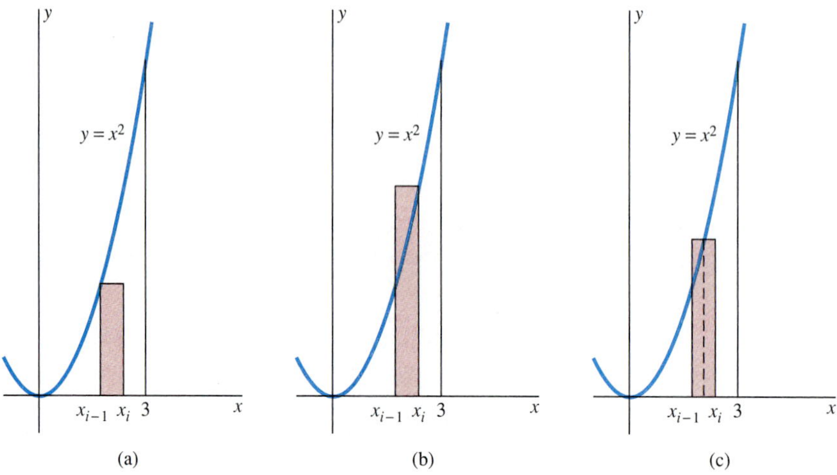

Fig. 5.4.4 Example 1: (a) The case $x_i^\star = x_{i-1}$; (b) The case $x_i^\star = x_i$; (c) The case $x_i^\star = m_i$

The ith subinterval as well as its midpoint

$$m_i = \frac{x_{i-1} + x_i}{2} = \frac{1}{2}\left(\frac{3i-3}{10} + \frac{3i}{10}\right) = \frac{3}{20}(2i-1)$$

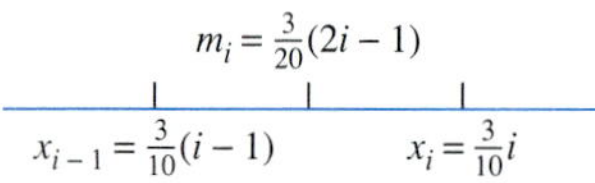

Fig. 5.4.5 The ith subinterval of Example 1

are shown in Fig. 5.4.5. With $x_i^\star = x_{i-1} = \frac{3}{10}(i-1)$, we obtain the left-endpoint sum in Eq. (4),

$$\begin{aligned} R_{\text{left}} &= \sum_{i=1}^{n} f(x_{i-1})\,\Delta x = \sum_{i=1}^{10} \left[\tfrac{3}{10}(i-1)\right]^2\left(\tfrac{3}{10}\right) \\ &= \tfrac{27}{1000}\cdot(0^2 + 1^2 + 2^2 + \cdots + 9^2) \\ &= \tfrac{7695}{1000} = 7.695 \qquad \text{[using Eq. (8) of Section 5.3].} \end{aligned}$$

With $x_i^\star = x_i = \frac{3}{10}i$, we get the right-endpoint sum in Eq. (5),

$$\begin{aligned} R_{\text{right}} &= \sum_{i=1}^{n} f(x_i)\,\Delta x = \sum_{i=1}^{10} \left[\tfrac{3}{10}i\right]^2\left(\tfrac{3}{10}\right) \\ &= \tfrac{27}{1000}\cdot(1^2 + 2^2 + 3^2 + \cdots + 10^2) \\ &= \tfrac{10395}{1000} = 10.395 \quad \text{[using Eq. (8) of Section 5.3].} \end{aligned}$$

Finally, with $x_i^\star = m_i = \frac{3}{20}(2i-1)$, we get the midpoint sum in Eq. (6),

$$\begin{aligned} R_{\text{mid}} &= \sum_{i-1}^{n} f(m_i)\,\Delta x = \sum_{i=1}^{10} \left[\tfrac{3}{20}(2i-1)\right]^2\left(\tfrac{3}{10}\right) \\ &= \tfrac{27}{4000}\cdot(1^2 + 3^2 + 5^2 + \cdots + 17^2 + 19^2) = \tfrac{35910}{4000} = 8.9775. \end{aligned}$$

The midpoint sum is much closer than either endpoint sum to the actual value 9 (of the area under the graph of $y = x^2$ over $[0, 3]$) that we found in Example 7 of Section 5.3. ■

EXAMPLE 2 Figure 5.4.6 illustrates Riemann sums for $f(x) = \sin x$ on $[0, \pi]$ based on $n = 3$ subintervals: $[0, \pi/3]$, $[\pi/3, 2\pi/3]$, and $[2\pi/3, \pi]$, of length $\Delta x = \pi/3$, and with midpoints $\pi/6$, $\pi/2$, and $5\pi/6$. The left-endpoint sum is

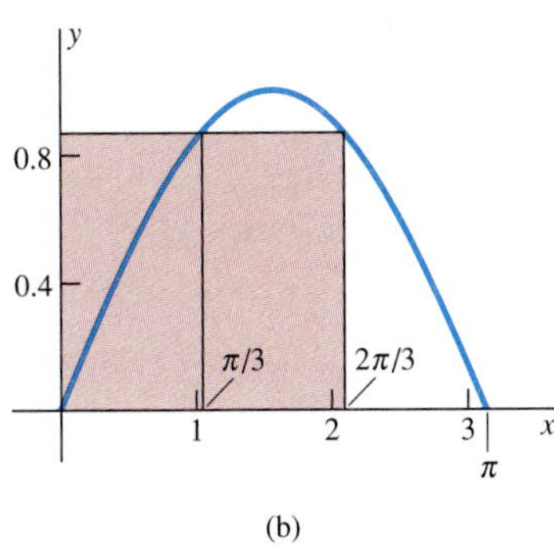

$$R_{\text{left}} = (\Delta x)\cdot\left(\sum_{i=1}^{n} f(x_{i-1})\right) = \frac{\pi}{3}\cdot\left(\sin 0 + \sin\frac{\pi}{3} + \sin\frac{2\pi}{3}\right)$$
$$= \frac{\pi}{3}\cdot\left(0 + \frac{\sqrt{3}}{2} + \frac{\sqrt{3}}{2}\right) = \frac{\pi\sqrt{3}}{3} \approx 1.81.$$

It is clear from the figure that the right-endpoint sum has the same value. The corresponding midpoint sum is

$$R_{\text{mid}} = \frac{\pi}{3}\cdot\left(\sin\frac{\pi}{6} + \sin\frac{\pi}{2} + \sin\frac{5\pi}{6}\right) = \frac{\pi}{3}\cdot\left(\frac{1}{2} + 1 + \frac{1}{2}\right) = \frac{2\pi}{3} \approx 2.09.$$

(We will soon be able to show that the area under one arch of the sine curve is exactly 2.) ■

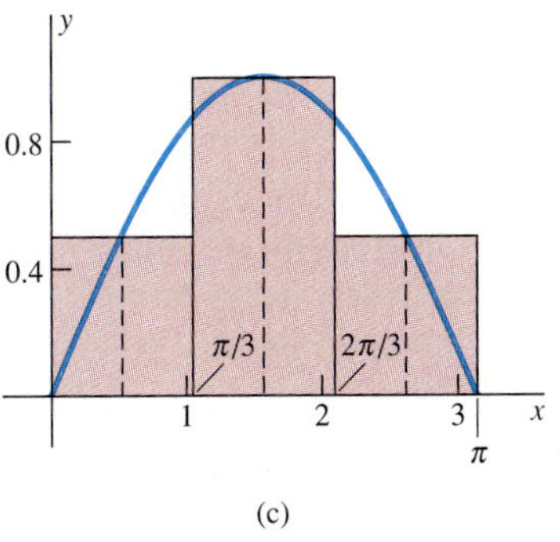

The Integral as a Limit

In the case of a function f that has both positive and negative values on $[a, b]$, it is necessary to consider the *signs* indicated in Fig. 5.4.7 when we interpret geometrically the Riemann sum in Eq. (3). On each subinterval $[x_{i-1}, x_i]$, we have a rectangle with width Δx and "height" $f(x_i^\star)$. If $f(x_i^\star) > 0$, then this rectangle stands *above* the x-axis; if $f(x_i^\star) < 0$, it lies *below* the x-axis. The Riemann sum R is then the sum of the **signed** areas of these rectangles—that is, the sum of the areas of those rectangles that lie above the x-axis *minus* the sum of the areas of those that lie below the x-axis.

(c)

Fig. 5.4.6 Approximating the area under $y = \sin x$ on $[0, \pi]$ (Example 2): (a) Left-endpoint sum; (b) Right-endpoint sum; (c) Midpoint sum

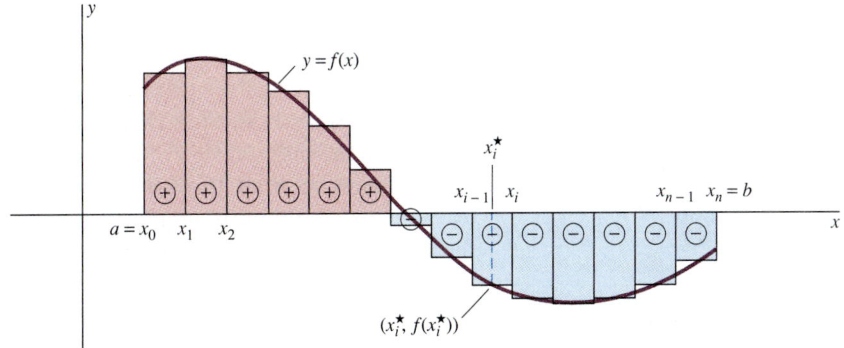

Fig. 5.4.7 A geometric representation of the Riemann sum in Eq. (3)

If the widths Δx_i of these rectangles are all very small, then it appears that the corresponding Riemann sum R will closely approximate the area from $x = a$ to $x = b$ under $y = f(x)$ and above the x-axis, *minus* the area that lies above the graph and below the x-axis. This suggests that the integral of f from a to b should be defined by taking the limit of the Riemann sums as the widths Δx_i all approach zero:

$$I = \lim_{\Delta x_i \to 0} \sum_{i=1}^{n} f(x_i^\star)\,\Delta x_i. \tag{7}$$

The formal definition of the integral is obtained by saying precisely what it means for this limit to exist. The **norm** of the partition P is the largest of the lengths $\Delta x_i = x_i - x_{i-1}$ of the subintervals in P and is denoted by $|P|$. Briefly, Eq. (7) means that if $|P|$ is sufficiently small, then *all* Riemann sums associated with the partition P are close to the number I.

Definition *The Definite Integral*

The **definite integral of the function f from a to b** is the number

$$I = \lim_{|P| \to 0} \sum_{i=1}^{n} f(x_i^{\star})\,\Delta x_i, \tag{8}$$

provided that this limit exists, in which case we say that f is **integrable** on $[a, b]$. Equation (8) means that, for each number $\epsilon > 0$, there exists a number $\delta > 0$ such that

$$\left| I - \sum_{i=1}^{n} f(x_i^{\star})\,\Delta x_i \right| < \epsilon$$

for every Riemann sum associated with any partition P of $[a, b]$ for which $|P| < \delta$.

The customary notation for the integral of f from a to b, due to the German mathematician and philosopher G. W. Leibniz, is

$$I = \int_a^b f(x)\,dx = \lim_{|P| \to 0} \sum_{i=1}^{n} f(x_i^{\star})\,\Delta x_i. \tag{9}$$

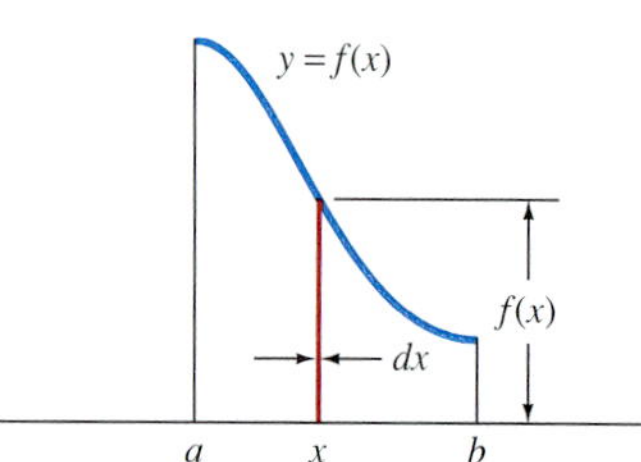

Fig. 5.4.8 Origin of Leibniz's notation for the integral

Considering I to be the area under $y = f(x)$ from a to b, Leibniz first thought of a narrow strip with height $f(x)$ and (infinitesimally small) width dx (as in Fig. 5.4.8), so that its area would be the product $f(x)\,dx$. He regarded the integral as a sum of areas of such strips and denoted this sum by the elongated capital S (for *summa*) that appears as the integral sign in Eq. (9).

We shall see that this integral notation is not only highly suggestive but also exceedingly useful in manipulations with integrals. The numbers a and b are called the **lower limit** and **upper limit,** respectively, of the integral; they are the endpoints of the interval of integration. The function $f(x)$ that appears between the integral sign and dx is called the **integrand.** The symbol dx that follows the integrand in Eq. (9) should, for the time being, be thought of as simply an indication of what the independent variable is. Like the index of summation, the independent variable x is a "dummy variable"—it may be replaced with any other variable without affecting the meaning of Eq. (9). Thus if f is integrable on $[a, b]$, we can write

$$\int_a^b f(x)\,dx = \int_a^b f(t)\,dt = \int_a^b f(u)\,du.$$

The definition given of the definite integral applies only if $a < b$, but it is convenient to include the cases $a = b$ and $a > b$ as well. The integral is defined in these cases as follows:

$$\int_a^a f(x)\,dx = 0 \tag{10}$$

and

$$\int_a^b f(x)\,dx = -\int_b^a f(x)\,dx, \tag{11}$$

provided that the right-hand integral exists. Thus *interchanging the limits of integration reverses the sign of the integral.*

Just as not all functions are differentiable, not every function is integrable. Suppose that c is a point of $[a, b]$ such that $f(x) \to +\infty$ as $x \to c$. If $[x_{k-1}, x_k]$ is the subinterval of the partition P that contains c, then the Riemann sum in Eq. (3) can be made arbitrarily large by choosing $x_k^\star$ to be sufficiently close to c. For our purposes, however, we need to know only that every continuous function is integrable. The following theorem is proved in Appendix G.

Theorem 1 ***Existence of the Integral***
If the function f is continuous on $[a, b]$, then f is integrable on $[a, b]$.

Although we omit the details, it is not difficult to show that the definition of the integral can be reformulated in terms of sequences of Riemann sums, as follows.

Theorem 2 ***The Integral as a Limit of a Sequence***
The function f is integrable on $[a, b]$ with integral I if and only if

$$\lim_{n\to\infty} R_n = I \tag{12}$$

for every sequence $\{R_n\}_1^\infty$ of Riemann sums associated with a sequence of partitions $\{P_n\}_1^\infty$ of $[a, b]$ such that $|P_n| \to 0$ as $n \to +\infty$.

Riemann Sum Computations

The reformulation in Theorem 2 of the definition of the integral is helpful because it is easier to visualize a specific sequence of Riemann sums than to visualize the vast totality of all possible Riemann sums. In the case of a continuous function f (known to be integrable by Theorem 1), the situation can be simplified even more by using only Riemann sums associated with partitions consisting of subintervals all with the same length

$$\Delta x_1 = \Delta x_2 = \cdots = \Delta x_n = \frac{b-a}{n} = \Delta x.$$

Such a partition of $[a, b]$ into equal-length subintervals is called a **regular partition** of $[a, b]$.

Any Riemann sum associated with a regular partition can be written in the form

$$\sum_{i=1}^{n} f(x_i^\star)\,\Delta x \tag{13}$$

where the absence of a subscript in Δx signifies that the sum is associated with a regular partition. In such a case the conditions $|P| \to 0$, $\Delta x \to 0$, and $n \to +\infty$ are equivalent, so the integral of a *continuous* function can be defined quite simply:

$$\int_a^b f(x)\,dx = \lim_{n\to\infty} \sum_{i=1}^{n} f(x_i^\star)\,\Delta x = \lim_{\Delta x\to 0} \sum_{i=1}^{n} f(x_i^\star)\,\Delta x. \tag{14}$$

Therefore, because we are concerned for the most part only with integrals of *continuous* functions, in our subsequent discussions we will employ only regular partitions.

EXAMPLE 3 Use Riemann sums to compute

$$\int_a^b x\,dx,$$

where $a < b$.

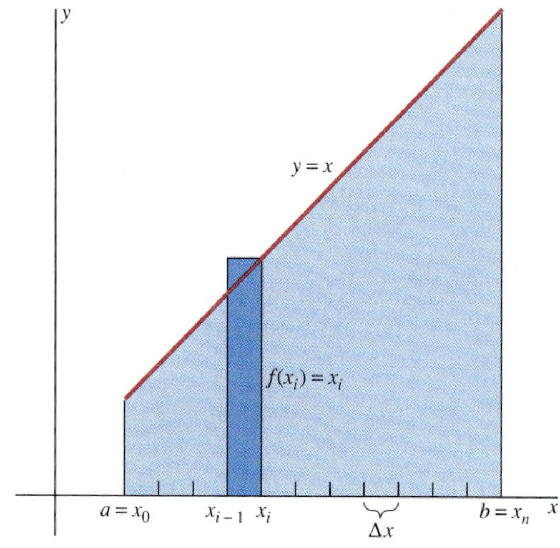

Fig. 5.4.9 Calculating the area under $y = x$ from $x = a$ to $x = b$

Solution We take $f(x) = x$ and $x_i^{\star} = x_i$ (see Fig. 5.4.9), where

$$\Delta x = \frac{b-a}{n} \quad \text{and} \quad x_i = a + i \cdot \Delta x.$$

The Riemann sum in (13) is then

$$\begin{aligned}\sum_{i=1}^{n} f(x_i)\,\Delta x &= \sum_{i=1}^{n} (a + i\cdot\Delta x)\,\Delta x = (a\,\Delta x)\sum_{i=1}^{n} 1 + (\Delta x)^2 \sum_{i=1}^{n} i \\ &= a\cdot\frac{b-a}{n}\cdot n + \left(\frac{b-a}{n}\right)^2 \cdot \frac{n}{2}(n+1) \\ &= a\cdot(b-a) + (b-a)^2\cdot\left(\frac{1}{2} + \frac{1}{2n}\right).\end{aligned}$$

[The second line follows from Eqs. (6) and (7) of Section 5.3.] Because $1/(2n) \to 0$ as $n \to +\infty$, it follows that

$$\begin{aligned}\int_a^b x\,dx &= \lim_{n\to\infty} \sum_{i=1}^{n} f(x_i)\,\Delta x \\ &= \lim_{n\to\infty} \left[a(b-a) + (b-a)^2\left(\frac{1}{2} + \frac{1}{2n}\right)\right] \\ &= a(b-a) + \frac{1}{2}(b-a)^2 = (b-a)\left(a + \frac{1}{2}b - \frac{1}{2}a\right),\end{aligned}$$

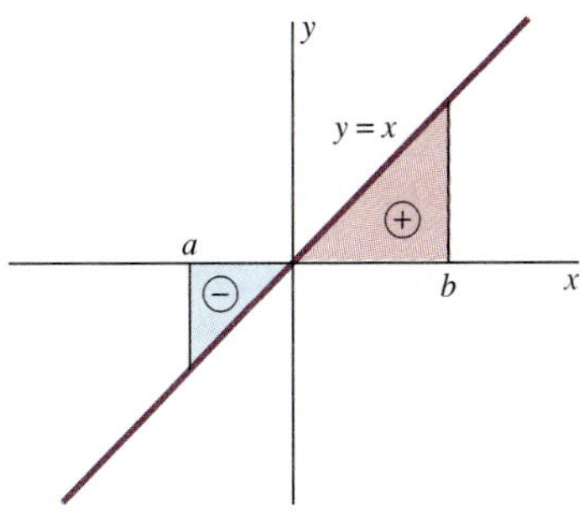

Fig. 5.4.10 Example 3 with $a < 0 < b$

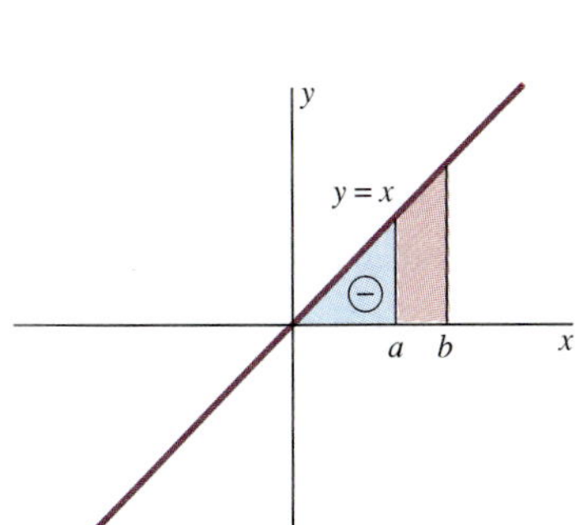

Fig. 5.4.11 Example 3 with $0 < a < b$

and therefore

$$\int_a^b x\,dx = \frac{1}{2}b^2 - \frac{1}{2}a^2. \tag{15}$$

■

Remark 1 If $0 < a < b$, then $A = \int_a^b x\,dx$ is the area of the trapezoid shown in Fig. 5.4.9. Then Eq. (15) says that

$$A = (b - a)\cdot\tfrac{1}{2}(a + b) = w\cdot\overline{h},$$

where $w = b - a$ is the width and $\overline{h} = \frac{1}{2}(a + b)$ is the average height of the trapezoid.

Remark 2 Figures 5.4.10 and 5.4.11 illustrate two different cases in Example 3. In each case Eq. (15) agrees with the sum of the indicated *signed* areas. The minus sign in Fig. 5.4.10 represents the fact that area *beneath* the x-axis is measured with a *negative* number.

5.4 PROBLEMS

In Problems 1 through 10, express the given limit as a definite integral over the indicated interval $[a, b]$. Assume that $[x_{i-1}, x_i]$ denotes the ith subinterval of a subdivision of $[a, b]$ into n subintervals, all with the same length $\Delta x = (b - a)/n$, and that $m_i = \frac{1}{2}(x_{i-1} + x_i)$ is the midpoint of the ith subinterval.

1. $\displaystyle\lim_{n\to\infty}\sum_{i=1}^{n}(2x_i - 1)\,\Delta x$ over $[1, 3]$

2. $\displaystyle\lim_{n\to\infty}\sum_{i=1}^{n}(2 - 3x_{i-1})\,\Delta x$ over $[-3, 2]$

3. $\displaystyle\lim_{n\to\infty}\sum_{i=1}^{n}(x_i^2 + 4)\,\Delta x$ over $[0, 10]$

4. $\displaystyle\lim_{n\to\infty}\sum_{i=1}^{n}(x_i^3 - 3x_i^2 + 1)\,\Delta x$ over $[0, 3]$

5. $\displaystyle\lim_{n\to\infty}\sum_{i=1}^{n}\sqrt{m_i}\,\Delta x$ over $[4, 9]$

6. $\displaystyle\lim_{n\to\infty}\sum_{i=1}^{n}\sqrt{25 - x_i^2}\,\Delta x$ over $[0, 5]$

7. $\displaystyle\lim_{n\to\infty}\sum_{i=1}^{n}\frac{1}{\sqrt{1 + m_i}}\,\Delta x$ over $[3, 8]$

8. $\displaystyle\lim_{n\to\infty}\sum_{i=1}^{n}(\cos 2x_{i-1})\,\Delta x$ over $[0, \pi/2]$

9. $\displaystyle\lim_{n\to\infty}\sum_{i=1}^{n}(\sin 2\pi m_i)\,\Delta x$ over $[0, 1/2]$

10. $\displaystyle\lim_{n\to\infty}\sum_{i=1}^{n}(\tan x_i)\,\Delta x$ over $[0, \pi/4]$

In Problems 11 through 20, compute the Riemann sum

$$\sum_{i=1}^{n} f(x_i^\star)\,\Delta x$$

for the indicated function and a regular partition of the given interval into n subintervals. Use $x_i^\star = x_i$, the right-hand endpoint of the ith subinterval $[x_{i-1}, x_i]$.

11. $f(x) = x^2$ on $[0, 1]$; $n = 5$

12. $f(x) = x^3$ on $[0, 1]$; $n = 5$

13. $f(x) = \dfrac{1}{x}$ on $[1, 6]$; $n = 5$

14. $f(x) = \sqrt{x}$ on $[0, 5]$; $n = 5$

15. $f(x) = 2x + 1$ on $[1, 4]$; $n = 6$

16. $f(x) = x^2 + 2x$ on $[1, 4]$; $n = 6$

17. $f(x) = x^3 - 3x$ on $[1, 4]$; $n = 5$

18. $f(x) = 1 + 2\sqrt{x}$ on $[2, 3]$; $n = 5$

19. $f(x) = \cos x$ on $[0, \pi]$; $n = 6$

20. $f(x) = \sin \pi x$ on $[0, 1]$; $n = 6$

21. through **30**. Repeat Problems 11 through 20, except with $x_i^\star = x_{i-1}$, the left-hand endpoint.

31. through **40.** Repeat Problems 11 through 20, except with $x_i^\star = (x_{i-1} + x_i)/2$, the midpoint of the ith subinterval.

41. Work Problem 13 with $x_i^\star = (3x_{i-1} + 2x_i)/5$.

42. Work Problem 14 with $x_i^\star = (x_{i-1} + 2x_i)/3$.

In Problems 43 through 48, evaluate the given integral by computing

$$\lim_{n\to\infty} \sum_{i=1}^{n} f(x_i)\,\Delta x$$

for a regular partition of the given interval of integration.

43. $\displaystyle\int_0^2 x^2\,dx$ **44.** $\displaystyle\int_0^4 x^3\,dx$

45. $\displaystyle\int_0^3 (2x + 1)\,dx$ **46.** $\displaystyle\int_1^5 (4 - 3x)\,dx$

47. $\displaystyle\int_0^3 (3x^2 + 1)\,dx$ **48.** $\displaystyle\int_0^4 (x^3 - x)\,dx$

49. Show by the method of Example 3 that

$$\int_0^b x\,dx = \frac{1}{2}b^2$$

if $b > 0$.

50. Show by the method of Example 3 that

$$\int_0^b x^3\,dx = \frac{1}{4}b^4$$

if $b > 0$.

51. Let $f(x) = x$, and let $\{x_0, x_1, x_2, \dots, x_n\}$ be an arbitrary partition of the closed interval $[a, b]$. For each i $(1 \leqq i \leqq n)$, let $x_i^\star = (x_{i-1} + x_i)/2$. Then show that

$$\sum_{i=1}^{n} x_i^\star\,\Delta x_i = \frac{1}{2}b^2 - \frac{1}{2}a^2.$$

Explain why this computation proves that

$$\int_a^b x\,dx = \frac{b^2 - a^2}{2}.$$

52. Suppose that f is a function continuous on $[a, b]$ and that k is a constant. Use Riemann sums to prove that

$$\int_a^b kf(x)\,dx = k\int_a^b f(x)\,dx.$$

53. Suppose that $f(x) \equiv c$, a constant. Use Riemann sums to prove that

$$\int_a^b c\,dx = c(b - a).$$

(*Suggestion:* First consider the case $a < b$.)

54. Suppose that the function f is defined on the interval $[0, 1]$ as follows:

$$f(x) = \begin{cases} \dfrac{1}{x} & \text{if } 0 < x \leqq 1, \\ 0 & \text{if } x = 0. \end{cases}$$

Show that the integral $\int_0^1 f(x)\,dx$ does not exist. [*Suggestion:* Show that, whatever n may be, the first term in the Riemann sum $\Sigma_{i=1}^n f(x_i^\star)\,\Delta x$ can be made arbitrarily large by the choice of the first selected point $x_1^\star$.] Why does this not contradict Theorem 1?

55. Suppose that the function f is defined as follows:

$$f(x) = \begin{cases} 0 & \text{if } x \text{ is rational,} \\ 1 & \text{if } x \text{ is irrational.} \end{cases}$$

Show that the integral $\int_0^1 f(x)\,dx$ does not exist. [*Suggestion:* Show that, whatever n may be, the Riemann sum $\Sigma_{i=1}^n f(x_i^\star)\,\Delta x$ has the value 0 for one possible selection of points $\{x_i^\star\}$, but the value 1 for another possible selection.] Why does this not contradict Theorem 1?

5.4 PROJECT: CALCULATOR/COMPUTER RIEMANN SUMS

Suppose that we want to use Riemann sums to approximate numerically the integral $\int_0^2 \sqrt{x}\,dx$. Its exact value is known (see Section 5.5) to be given by

$$\int_0^2 \sqrt{x}\,dx = \tfrac{2}{3}\sqrt{8} \approx 1.8856. \tag{1}$$

If we use a midpoint sum with $n = 10$ subintervals, then $\Delta x = 0.2$ and $f(0.1)$ is the first function value used. (Why?) Using a TI graphing calculator, we first define the function $f(x) = \sqrt{x}$ by entering

```
Y1=√X
```

under the `Y=` menu. Then we return to the home screen and enter successively the commands

`0→A:2→B`	The two endpoints
`0.2→H`	The subinterval length with $n = 10$
`A+H/2→X`	The initial x-value selected
`X-H→X`	Back up one step before starting
`0→S`	Initialize the accumulating sum
`X+H→X:S+H*Y1→S`	Add on the successive rectangular areas

When we now press the **ENTER** key $n = 10$ times, we get the final value $1.8905 \approx 1.89$ for the midpoint sum, correct when rounded to two decimal places.

Only one line—the third command—needs to be changed to calculate either a left-endpoint sum or a right-endpoint sum. For a left-endpoint sum use

`A→X`	The initial x-value selected

and for a right-endpoint sum use

`A+H→X`	The initial x-value selected

Of course, `H` $= \Delta x = (b - a)/n$ must be entered (in the second command) consistent with the desired number n of subintervals. Many calculators have "repeat entry" features that enable you to re-enter (without retyping) the commands that are to be used again without change.

Programmable calculators have "if-then" and "for-next" loop-programming features for writing self-contained programs for calculating Riemann sums—see the calculus projects WWW site that supports this text. Most computer algebra systems include a `SUM` function that can be used to calculate Riemann sums directly. Suppose that the integrand function $f(x)$ has been defined appropriately. Then all the single-line commands listed in Fig. 5.4.12 calculate (in various systems) the right-endpoint sum with n subintervals. The alterations that produce left-endpoint and midpoint sums should be apparent.

Computer	Command
Derive	`SUM(f(a+(b-a)*i/n),i,1,n)*(b-a)/n`
Maple	`sum(f(a+(b-a)*i/n),i=1..n)*(b-a)/n`
Mathematica	`Sum(f[x],{x,a+(b-a)/n,b,(b-a)/n}]*(b-a)/n`

Fig. 5.4.12 Computer algebra system commands for approximating the integral of f from $x = a$ to $x = b$ with n subintervals

INVESTIGATION 1 Approximate the integral in (1) using Riemann sums with $n = 20$, $n = 50$, and $n = 100$ subintervals. Can you calculate in this way the value of the integral accurate to three decimal places (when rounded)?

INVESTIGATION 2 Verify the approximations to the integral

$$\int_0^3 x^2\,dx$$

that are listed in Fig. 5.3.10. Include a column of midpoint-sum approximations. Which appears to give a more accurate approximation to this integral for a given number of subintervals: the average of the left- and right-endpoint sums or the midpoint sum?

INVESTIGATION 3 Use midpoint sums and left- and right-endpoint sum averages to corroborate the claim that

$$\int_0^{\pi} \sin x \, dx = 2.$$

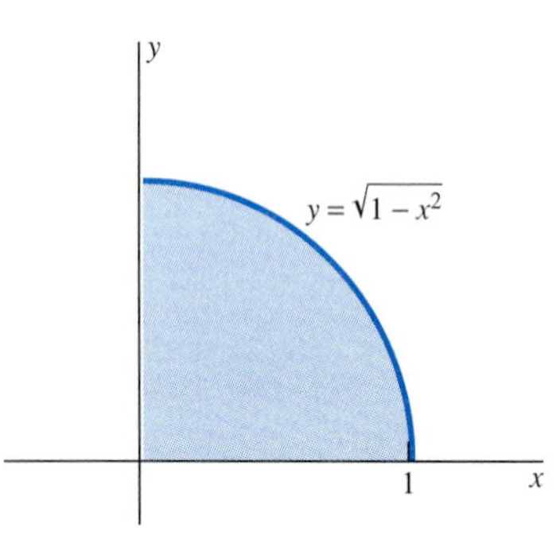

Fig. 5.4.13 Investigation 4

INVESTIGATION 4 First explain why Fig. 5.4.13 and the circle area formula $A = \pi r^2$ imply that

$$\int_0^1 4\sqrt{1 - x^2} \, dx = \pi.$$

Then use midpoint-sum approximations to this integral to approximate the number π. Begin with $n = 25$ subintervals, then successively double n. How large must n be for you to obtain the familiar approximation $\pi \approx 3.1416$?

5.5 EVALUATION OF INTEGRALS

The evaluation of integrals by using Riemann sums, as in Section 5.4, is tedious and time consuming. Fortunately, we will seldom find it necessary to evaluate an integral in this way. In 1666 Isaac Newton, while still a student at Cambridge University, discovered a much more efficient way to evaluate an integral. A few years later, Gottfried Wilhelm Leibniz, working with a different approach, discovered this method independently.

Newton's key idea was that to evaluate the *number*

$$\int_a^b f(x) \, dx,$$

we should first introduce the *function* $A(x)$ defined as follows:

$$A(x) = \int_a^x f(t) \, dt. \tag{1}$$

The independent variable x appears as the *upper limit* of the integral in Eq. (1); the dummy variable t is used in the integrand merely to avoid confusion. If f is positive-valued and continuous and if $x > a$, then $A(x)$ is the area below the curve $y = f(x)$ above the interval $[a, x]$ (Fig. 5.5.1).

It is apparent from Fig. 5.5.1 that $A(x)$ increases as x increases. When x increases by Δx, A increases by the area ΔA of the narrow strip in Fig. 5.5.1 with base $[x, x + \Delta x]$. If Δx is very small, then the area of this strip is very close to the area $f(x)\,\Delta x$ of the rectangle with base $[x, x + \Delta x]$ and height $f(x)$. Thus

$$\Delta A \approx f(x)\,\Delta x; \qquad \frac{\Delta A}{\Delta x} \approx f(x). \tag{2}$$

Moreover, the figure makes it plausible that we get equality in the limit as $\Delta x \to 0$:

$$\frac{dA}{dx} = \lim_{\Delta x \to 0} \frac{\Delta A}{\Delta x} = f(x).$$

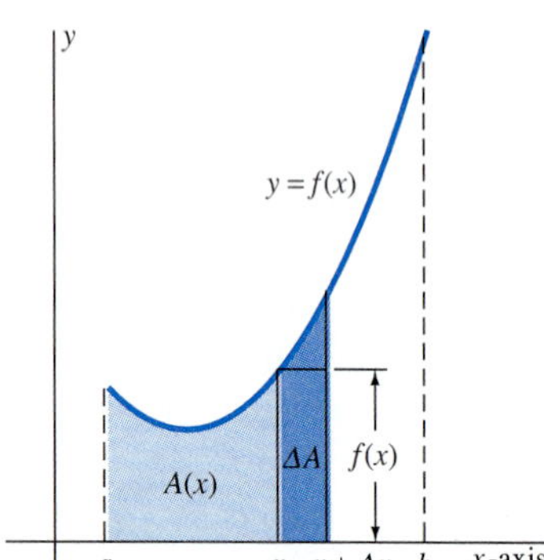

Fig. 5.5.1 The area function $A(x)$

That is,

$$A'(x) = f(x), \tag{3}$$

so *the derivative of the area function $A(x)$ is the curve's height function $f(x)$*. In other words, Eq. (3) implies that $A(x)$ is an *antiderivative* of $f(x)$.

Figure 5.5.2 shows a physical interpretation of Eq. (3). A paint roller is laying down a 1-mm-thick coat of paint to cover the region under the curve $y = f(t)$. The paint roller is of adjustable length—as it rolls with a speed of 1 mm/s from left to right, one end traces the x-axis and the other end traces the curve $y = f(t)$. At any time t, the volume V of paint the roller has laid down equals the area of the region already painted:

$$V = A(t) \quad (\text{mm}^3).$$

Then Eq. (3) yields

$$\frac{dV}{dt} = A'(t) = f(t).$$

Thus the instantaneous rate at which the roller is depositing paint is equal to the current length of the roller.

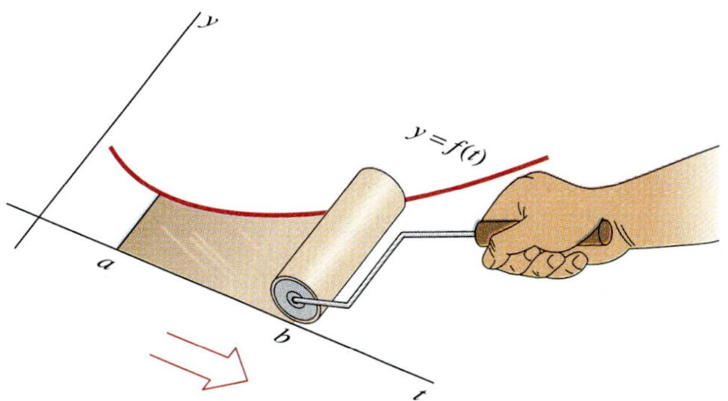

Fig. 5.5.2 The adjustable-length paint roller

The Evaluation Theorem

Equation (3) says that the area function $A(x)$ illustrated in Fig. 5.5.1 is *one* antiderivative of the given function $f(x)$. Now suppose that $G(x)$ is any other antiderivative of $f(x)$—perhaps one found by the methods of Section 5.2. Then

$$A(x) = G(x) + C, \tag{4}$$

because (by the second corollary to the mean value theorem) two antiderivatives of the same function (on an interval) can differ only by a constant. Also,

$$A(a) = \int_a^a f(t)\,dt = 0 \tag{5}$$

and

$$A(b) = \int_a^b f(t)\,dt = \int_a^b f(x)\,dx \tag{6}$$

by Eq. (1). So it follows that

$$\int_a^b f(x)\,dx = A(b) - A(a) = [G(b) + C] - [G(a) + C],$$

and thus

$$\int_a^b f(x)\,dx = G(b) - G(a). \tag{7}$$

Our intuitive discussion has led us to the statement of Theorem 1.

Theorem 1 ***Evaluation of Integrals***

If G is an antiderivative of the continuous function f on the interval $[a, b]$, then

$$\int_a^b f(x)\,dx = G(b) - G(a). \tag{7}$$

In Section 5.6 we will fill in the details of the preceding discussion, thus giving a rigorous proof of Theorem 1 (which is part of the fundamental theorem of calculus). Here we concentrate on the computational applications of this theorem. The difference $G(b) - G(a)$ is customarily abbreviated as $[G(x)]_a^b$, so Theorem 1 implies that

$$\int_a^b f(x)\,dx = \Big[G(x)\Big]_a^b = G(b) - G(a) \tag{8}$$

if G is any antiderivative of the continuous function f on the interval $[a, b]$. Thus if we can find an antiderivative G of f, we can quickly evaluate the integral without having to resort to the paraphernalia of limits of Riemann sums.

If $G'(x) = f(x)$, then (as in Section 5.2) we write

$$\int f(x)\,dx = G(x) + C \tag{9}$$

for the indefinite integral of f. With the indefinite integral $\int f(x)\,dx$ in place of the antiderivative $G(x)$, Eq. (8) takes the form

$$\int_a^b f(x)\,dx = \left[\int f(x)\,dx\right]_a^b. \tag{10}$$

This is the connection between the indefinite integral and the definite integral to which we have alluded in the earlier sections of Chapter 5.

EXAMPLE 1 Because

$$\int x^n\,dx = \frac{x^{n+1}}{n+1} + C \qquad (\text{if } n \neq -1),$$

it follows that

$$\int_a^b x^n\,dx = \left[\frac{x^{n+1}}{n+1}\right]_a^b = \frac{b^{n+1} - a^{n+1}}{n+1}$$

if $n \neq -1$. For instance,

$$\int_0^3 x^2\,dx = \left[\tfrac{1}{3}x^3\right]_0^3 = \tfrac{1}{3}\cdot 3^3 - \tfrac{1}{3}\cdot 0^3 = 9.$$

Contrast the immediacy of this result with the complexity of the computations of Example 6 in Section 5.3.

EXAMPLE 2 Because

$$\int \cos x\,dx = \sin x + C,$$

it follows that

$$\int_a^b \cos x\,dx = \Big[\sin x\Big]_a^b = \sin b - \sin a.$$

Similarly,

$$\int_a^b \sin x\,dx = \Big[-\cos x\Big]_a^b = \cos a - \cos b.$$

In particular, as we mentioned in Example 2 of Section 5.4,

$$\int_0^\pi \sin x\,dx = \Big[-\cos x\Big]_0^\pi = (-\cos\pi) - (-\cos 0) = (+1) - (-1) = 2.$$

EXAMPLE 3

$$\int_0^2 x^5\,dx = \left[\tfrac{1}{6}x^6\right]_0^2 = \tfrac{64}{6} - 0 = \tfrac{32}{3}.$$

$$\int_1^9 (2x - x^{-1/2} - 3)\,dx = \left[x^2 - 2x^{1/2} - 3x\right]_1^9 = 52.$$

$$\int_0^1 (2x+1)^3\,dx = \left[\tfrac{1}{8}(2x+1)^4\right]_0^1 = \tfrac{1}{8}\cdot(81 - 1) = 10.$$

$$\int_0^{\pi/2} \sin 2x\,dx = \left[-\tfrac{1}{2}\cos 2x\right]_0^{\pi/2} = -\tfrac{1}{2}(\cos\pi - \cos 0) = 1.$$

We have not shown the details of finding the antiderivatives, but you can (and should) check each of these results by showing that the derivative of the function within the evaluation brackets on the right is equal to the integrand on the left. In Example 4 we show the details.

EXAMPLE 4 Evaluate $\displaystyle\int_1^5 \sqrt{3x+1}\,dx$.

Solution We apply the antiderivative form of the generalized power rule,

$$\int u^k\,du = \frac{u^{k+1}}{k+1} + C \qquad (k \neq -1),$$

with

$$u = 3x + 1, \qquad du = 3\,dx.$$

This gives

$$\int (3x+1)^{1/2}\,dx = \tfrac{1}{3}\int (3x+1)^{1/2}(3\,dx) = \tfrac{1}{3}\int u^{1/2}\,du$$

$$= \tfrac{1}{3}\cdot\frac{u^{3/2}}{\frac{3}{2}} + C = \tfrac{2}{9}(3x+1)^{3/2} + C$$

for the indefinite integral, so it follows from Eq. (10) that

$$\int_1^5 \sqrt{3x+1}\,dx = \left[\tfrac{2}{9}(3x+1)^{3/2}\right]_1^5$$

$$= \tfrac{2}{9}(16^{3/2} - 4^{3/2}) = \tfrac{2}{9}(4^3 - 2^3) = \tfrac{112}{9}.$$

If the derivative $F'(x)$ of the function $F(x)$ is continuous, then the evaluation theorem, with $F'(x)$ in place of $f(x)$ and $F(x)$ in place of $G(x)$, yields

$$\int_a^b F'(x)\,dx = \Big[F(x)\Big]_a^b = F(b) - F(a). \tag{11}$$

Here is an immediate application.

EXAMPLE 5 Suppose that an animal population $P(t)$ initially numbers $P(0) = 100$ and that its rate of growth after t months is given by

$$P'(t) = 10 + t + (0.06)t^2.$$

What is the population after 10 months?

Solution By Eq. (11), we know that

$$P(10) - P(0) = \int_0^{10} P'(t)\,dt = \int_0^{10} [10 + t + (0.06)t^2]\,dt$$

$$= \left[10t + \tfrac{1}{2}t^2 + (0.02)t^3\right]_0^{10} = 170.$$

Thus $P(10) = 100 + 170 = 270$ individuals.

EXAMPLE 6 Evaluate

$$\lim_{n\to\infty} \sum_{i=1}^{n} \frac{2i}{n^2}$$

by recognizing this limit as the value of an integral.

Solution If we write

$$\sum_{i=1}^{n} \frac{2i}{n^2} = \sum_{i=1}^{n} \left(\frac{2i}{n}\right)\left(\frac{1}{n}\right),$$

we recognize that we have a Riemann sum for the function $f(x) = 2x$ associated with a partition of the interval $[0, 1]$ into n equal-length subintervals. The ith point of subdivision is $x_i = i/n$, and $\Delta x = 1/n$. Hence it follows from the definition of the integral and from the evaluation theorem that

$$\lim_{n\to\infty} \sum_{i=1}^{n} \frac{2i}{n^2} = \lim_{n\to\infty} \sum_{i=1}^{n} 2x_i\,\Delta x = \lim_{n\to\infty} \sum_{i=1}^{n} f(x_i)\,\Delta x$$

$$= \int_0^1 f(x)\,dx = \int_0^1 2x\,dx.$$

Therefore

$$\lim_{n\to\infty} \sum_{i=1}^{n} \frac{2i}{n^2} = \Big[x^2\Big]_0^1 = 1.$$ ■

Basic Properties of Integrals

Problems 59 through 62 outline elementary proofs of the integral properties that are stated next. We assume throughout that each function mentioned is integrable on $[a, b]$.

Integral of a Constant

$$\int_a^b c\,dx = c(b - a).$$

This property is intuitively obvious because the area represented by the integral is simply a rectangle with base $b - a$ and height c (Fig. 5.5.3).

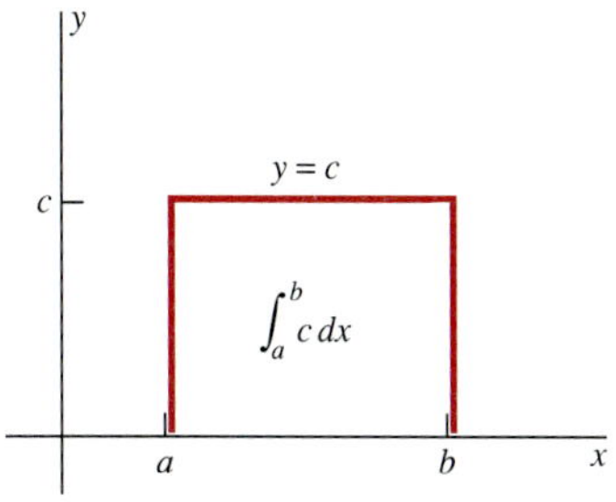

y
y = c
c
$\int_a^b c\,dx$
a
b
x

Fig. 5.5.3 The integral of a constant is the area of a rectangle.

Constant Multiple Property

$$\int_a^b cf(x)\,dx = c\int_a^b f(x)\,dx.$$

Thus a constant can be "moved across" the integral sign. For example,

$$\int_0^{\pi/2} 2\sin x\,dx = 2\int_0^{\pi/2} \sin x\,dx = 2\Big[-\cos x\Big]_0^{\pi/2} = 2.$$

Sum Property

$$\int_a^b [f(x) + g(x)]\,dx = \int_a^b f(x)\,dx + \int_a^b g(x)\,dx.$$

Thus if both of the functions f and g are integrable on $[a, b]$, then *the integral of their sum is equal to the sum of their integrals.* This fact sometimes permits a "divide and conquer" strategy for the calculation of integrals:

$$\int_0^{\pi}\left(3\sqrt{x} + \cos\frac{x}{2}\right)dx = \int_0^{\pi} 3\sqrt{x}\,dx + \int_0^{\pi} \cos\frac{x}{2}\,dx$$

$$= \left[2x^{3/2}\right]_0^{\pi} + \left[2\sin\frac{x}{2}\right]_0^{\pi} = 2\pi^{3/2} + 2.$$

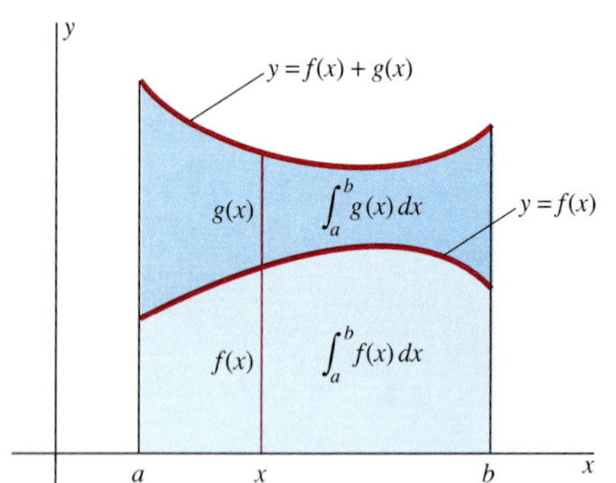

Fig. 5.5.4 The integral of the sum of two positive-valued functions

Figure 5.5.4 illustrates geometrically the sum property of integrals. The proof of the sum property illustrates a Riemann sums approach that can be adapted to all of the properties under discussion here. Let us think of partitions of the interval $[a, b]$ into subintervals all having the same length Δx. If the functions f, g, and $f + g$ are all integrable, then Theorem 2 in Section 5.4 gives

$$\int_a^b [f(x) + g(x)]\,dx = \lim_{\Delta x \to 0} \sum_{i=1}^{n} [f(x_i^{\star}) + g(x_i^{\star})]\,\Delta x$$

$$= \lim_{\Delta x \to 0} \left[\sum_{i=1}^{n} f(x_i^{\star})\,\Delta x + \sum_{i=1}^{n} g(x_i^{\star})\,\Delta x\right]$$

$$= \left[\lim_{\Delta x \to 0} \sum_{i=1}^{n} f(x_i^{\star})\,\Delta x\right] + \left[\lim_{\Delta x \to 0} \sum_{i=1}^{n} g(x_i^{\star})\,\Delta x\right]$$

$$= \int_a^b f(x)\,dx + \int_a^b g(x)\,dx.$$

Interval Union Property

If $a < c < b$, then

$$\int_a^b f(x)\,dx = \int_a^c f(x)\,dx + \int_c^b f(x)\,dx.$$

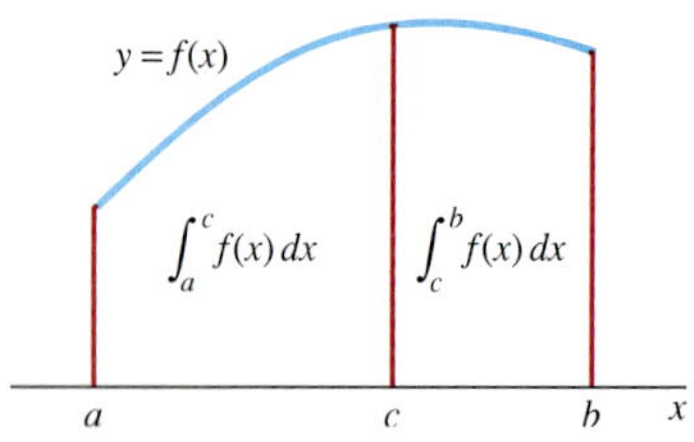

Fig. 5.5.5 The way the interval union property works

Figure 5.5.5 indicates the plausibility of the interval union property.

EXAMPLE 7 If $f(x) = 2\,|x|$, then

$$f(x) = \begin{cases} -2x & \text{if } x \leqq 0, \\ 2x & \text{if } x \geqq 0. \end{cases}$$

The graph of f is shown in Fig. 5.5.5. An antiderivative of $f(x)$ is not evident, but the interval union property allows us to split the integral of f on $[-1, 3]$ into two easily calculated integrals:

$$\int_{-1}^{3} 2\,|x|\,dx = \int_{-1}^{0} (-2x)\,dx + \int_0^3 (2x)\,dx$$

$$= \left[-x^2\right]_{-1}^{0} + \left[x^2\right]_0^3 = [0 - (-1)] + [9 - 0] = 10.$$

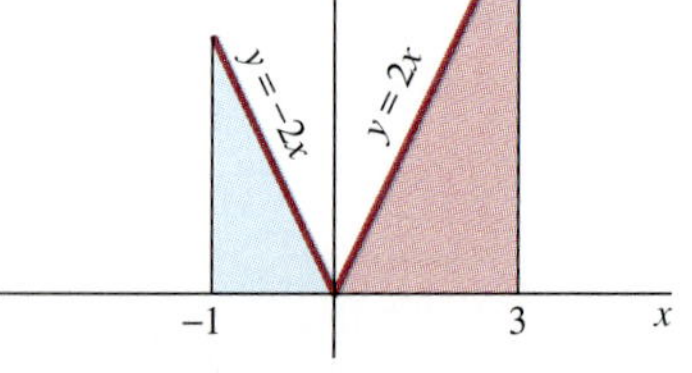

Fig. 5.5.6 The area under the graph of $y = 2|x|$

Does the result agree with Fig. 5.5.6? ■

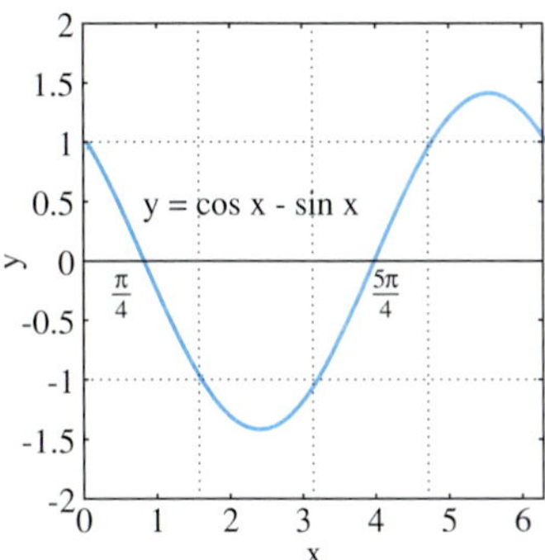

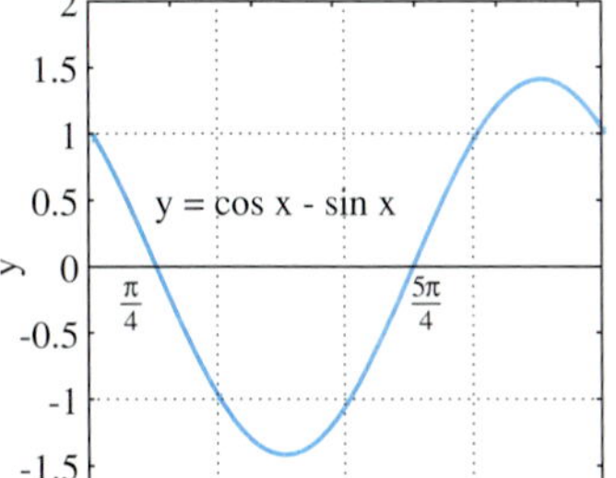

Fig. 5.5.7 $y = \cos x - \sin x$

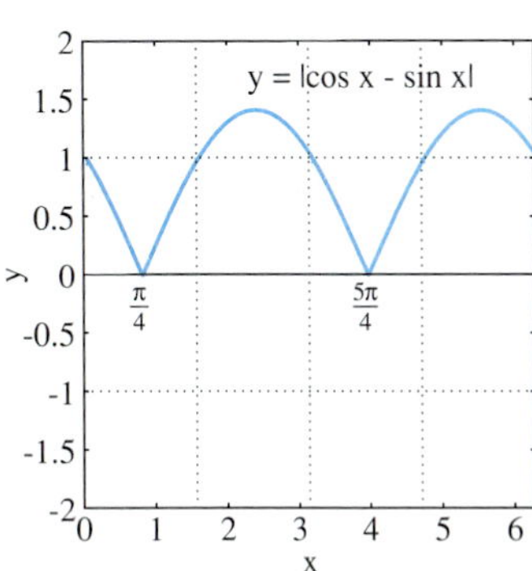

Fig. 5.5.8 $y = |\cos x - \sin x|$

EXAMPLE 8 Evaluate the integral $\displaystyle\int_0^{2\pi} |\cos x - \sin x|\,dx$.

Solution Figure 5.5.7 shows the graph of the function $f(x) = \cos x - \sin x$ and Fig. 5.5.8 shows the graph of its absolute value $|f(x)| = |\cos x - \sin x|$ that we want to integrate. We readily see that $f(x) = 0$ at $x = \pi/4$ and $x = 5\pi/4$, so

$$|f(x)| = \begin{cases} \cos x - \sin x & \text{if } 0 \leqq x < \pi/4, \\ \sin x - \cos x & \text{if } \pi/4 \leqq x < 5\pi/4, \\ \cos x - \sin x & \text{if } 5\pi/4 \leqq x \leqq 2\pi. \end{cases}$$

The interval union property therefore gives

$$\begin{aligned} &\int_0^{2\pi} |\cos x - \sin x|\,dx \\ &\quad = \int_0^{\pi/4} (\cos x - \sin x)\,dx + \int_{\pi/4}^{5\pi/4} (\sin x - \cos x)\,dx + \int_{5\pi/4}^{2\pi} (\cos x - \sin x)\,dx \\ &\quad = \Big[\sin x + \cos x\Big]_0^{\pi/4} + \Big[-\cos x - \sin x\Big]_{\pi/4}^{5\pi/4} + \Big[\sin x + \cos x\Big]_{5\pi/4}^{2\pi} \\ &\quad = \left(\tfrac{1}{2}\sqrt{2} + \tfrac{1}{2}\sqrt{2}\right) - (0 + 1) + \left(\tfrac{1}{2}\sqrt{2} + \tfrac{1}{2}\sqrt{2}\right) \\ &\qquad\qquad - \left(-\tfrac{1}{2}\sqrt{2} - \tfrac{1}{2}\sqrt{2}\right) + (0 + 1) - \left(-\tfrac{1}{2}\sqrt{2} - \tfrac{1}{2}\sqrt{2}\right) \\ &\quad = 4\sqrt{2}. \end{aligned}$$

Comparison Properties

(1) If $f(x) \leqq g(x)$ for all x in $[a, b]$, then

$$\int_a^b f(x)\,dx \leqq \int_a^b g(x)\,dx.$$

(2) If $m \leqq f(x) \leqq M$ for all x in $[a, b]$, then

$$m(b - a) \leqq \int_a^b f(x)\,dx \leqq M(b - a).$$

Fig. 5.5.9 Plausibility of the comparison property

The first comparison property says that the larger function has the larger integral. The plausibility of the second comparison property is indicated in Fig. 5.5.9. Note that m and M need not necessarily be the minimum and maximum values of $f(x)$ on $[a, b]$.

EXAMPLE 9 Figure 5.5.10 shows the graphs

$$y = \sqrt{1 + x}, \qquad y = \sqrt{1 + \sqrt{x}}, \qquad \text{and} \qquad y = 1.2 + (0.3)x,$$

and we see that

$$\sqrt{1 + x} \leqq \sqrt{1 + \sqrt{x}} \leqq 1.2 + (0.3)x \tag{12}$$

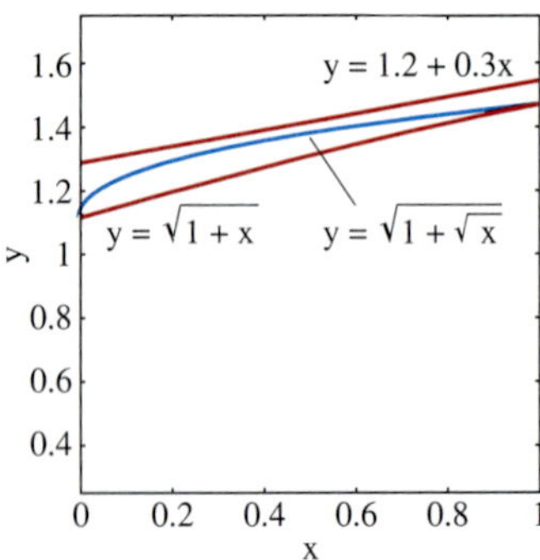

Fig. 5.5.10 Bounding the graph of $f(x) = \sqrt{1 + \sqrt{x}}$

for x in $[0, 1]$. Indeed, the fact that $x \leqq \sqrt{x}$ for x in $[0, 1]$ implies that $\sqrt{1 + x} \leqq \sqrt{1 + \sqrt{x}}$ there. The graph $y = 1.2 + (0.3)x$ that lies above $y = \sqrt{1 + \sqrt{x}}$ was discovered empirically using a graphing calculator. At any rate, the inequalities in (12) and the first comparison property of integrals imply that

$$\int_0^1 \sqrt{1 + x}\, dx \leqq \int_0^1 \sqrt{1 + \sqrt{x}}\, dx \leqq \int_0^1 [1.2 + (0.3)x]\, dx; \quad \text{thus}$$

$$\left[\tfrac{2}{3}(1 + x)^{3/2}\right]_0^1 \leqq \int_0^1 \sqrt{1 + \sqrt{x}}\, dx \leqq \left[(1.2)x + (0.15)x^2\right]_0^1 = 1.35.$$

Now $\frac{2}{3}(2^{3/2} - 1) \approx 1.2190$, so we see finally that

$$1.21 \leqq \int_0^1 \sqrt{1 + \sqrt{x}}\, dx \leqq 1.35. \tag{13}$$

It turns out (using the methods of Section 5.9) that the actual value of $\int_0^1 \sqrt{1 + \sqrt{x}}\, dx$ is 1.29 rounded to two decimal places—quite close to the average 1.28 of the upper and lower bounds in (13). ■

The properties of integrals stated here are frequently used in computing and will be applied in the proof of the fundamental theorem of calculus in Section 5.6.

5.5 PROBLEMS

Apply the evaluation theorem to evaluate the integrals in Problems 1 through 36.

1. $\displaystyle\int_0^1 (3x^2 + 2\sqrt{x} + 3\sqrt[3]{x})\, dx$

2. $\displaystyle\int_1^3 \frac{6}{x^2}\, dx$

3. $\displaystyle\int_0^1 x^3(1 + x)^2\, dx$

4. $\displaystyle\int_{-2}^{-1} \frac{1}{x^4}\, dx$

5. $\displaystyle\int_0^1 (x^4 - x^3)\, dx$

6. $\displaystyle\int_1^2 (x^4 - x^3)\, dx$

7. $\displaystyle\int_{-1}^0 (x + 1)^3\, dx$

8. $\displaystyle\int_1^3 \frac{x^4 + 1}{x^2}\, dx$

9. $\displaystyle\int_0^4 \sqrt{x}\, dx$

10. $\displaystyle\int_1^4 \frac{1}{\sqrt{x}}\, dx$

11. $\displaystyle\int_{-1}^2 (3x^2 + 2x + 4)\, dx$

12. $\displaystyle\int_0^1 x^{99}\, dx$

13. $\displaystyle\int_{-1}^1 x^{99}\, dx$

14. $\displaystyle\int_0^4 (7x^{5/2} - 5x^{3/2})\, dx$

15. $\displaystyle\int_1^3 (x - 1)^5\, dx$

16. $\displaystyle\int_1^2 (x^2 + 1)^3\, dx$

17. $\displaystyle\int_{-1}^0 (2x + 1)^3\, dx$

18. $\displaystyle\int_1^3 \frac{10}{(2x + 3)^2}\, dx$

19. $\displaystyle\int_1^8 x^{2/3}\, dx$

20. $\displaystyle\int_1^9 (1 + \sqrt{x})^2\, dx$

21. $\displaystyle\int_0^1 (x^2 - 3x + 4)\, dx$

22. $\displaystyle\int_0^4 \sqrt{3t}\, dt$

23. $\displaystyle\int_1^9 \left(\sqrt{x} - \frac{2}{\sqrt{x}}\right) dx$

24. $\displaystyle\int_2^3 \frac{du}{u^2}$ $\left(\text{Note the abbreviation for } \dfrac{1}{u^2}\, du.\right)$

25. $\displaystyle\int_1^4 \frac{x^2 - 1}{\sqrt{x}}\, dx$

26. $\displaystyle\int_1^4 (t^2 - 2)\sqrt{t}\, dt$

27. $\displaystyle\int_4^7 \sqrt{3x + 4}\, dx$

28. $\displaystyle\int_0^{\pi/2} \cos 2x\, dx$

29. $\displaystyle\int_0^{\pi/4} \sin x \cos x\, dx$

30. $\displaystyle\int_0^{\pi} \sin^2 x \cos x\, dx$

31. $\displaystyle\int_0^{\pi} \sin 5x\, dx$

32. $\displaystyle\int_0^2 \cos \pi t\, dt$

33. $\displaystyle\int_0^{\pi/2} \cos 3x\, dx$

34. $\displaystyle\int_0^5 \sin \frac{\pi x}{10}\, dx$

35. $\displaystyle\int_0^2 \cos \frac{\pi x}{4}\, dx$

36. $\displaystyle\int_0^{\pi/8} \sec^2 2t\, dt$

In Problems 37 through 42, evaluate the given limit by first recognizing the indicated sum as a Riemann sum associated with a regular partition of $[0, 1]$ and then evaluating the corresponding integral.

37. $\displaystyle\lim_{n\to\infty} \sum_{i=1}^{n} \left(\frac{2i}{n} - 1\right)\frac{1}{n}$

38. $\displaystyle\lim_{n\to\infty} \sum_{i=1}^{n} \frac{i^2}{n^3}$

39. $\displaystyle\lim_{n\to\infty} \frac{1 + 2 + 3 + \cdots + n}{n^2}$

40. $\lim_{n\to\infty} \dfrac{1^3 + 2^3 + 3^3 + \cdots + n^3}{n^4}$

41. $\lim_{n\to\infty} \dfrac{\sqrt{1} + \sqrt{2} + \sqrt{3} + \cdots + \sqrt{n}}{n\sqrt{n}}$

42. $\lim_{n\to\infty} \sum_{i=1}^{n} \dfrac{1}{n} \sin \dfrac{\pi i}{n}$

Problems 43 through 48 give an integral $\int_a^b f(x)\,dx$. First sketch the graph $y = f(x)$ on the interval $[a, b]$. Then, interpreting the integral as the area of a region, evaluate it using known area formulas for rectangles, triangles, and circles.

43. $\int_{-2}^{2} |1 - x|\,dx$

44. $\int_{-3}^{3} |3x - 2|\,dx$

45. $\int_{0}^{5} (2 - |x|)\,dx$

46. $\int_{0}^{6} |5 - |2x||\,dx$

47. $\int_{0}^{5} \sqrt{25 - x^2}\,dx$

48. $\int_{0}^{6} \sqrt{6x - x^2}\,dx$ (*Suggestion:* Complete the square.)

In Problems 49 through 54, use properties of integrals to establish each inequality without evaluating the integrals involved.

49. $1 \leqq \int_0^1 \sqrt{1 + x^2}\,dx \leqq \int_0^1 \sqrt{1 + x}\,dx$

50. $\int_1^2 \sqrt{1 + x}\,dx \leqq \int_1^2 \sqrt{1 + x^3}\,dx \leqq 3$

51. $\int_0^1 \dfrac{1}{1 + \sqrt{x}}\,dx \leqq \int_0^1 \dfrac{1}{1 + x^2}\,dx$

52. $\int_2^5 \dfrac{1}{1 + x^5}\,dx \leqq \int_2^5 \dfrac{1}{1 + x^2}\,dx$

53. $\int_0^2 \sin\sqrt{x}\,dx \leqq 2$

54. $\dfrac{\pi}{8} \leqq \int_0^{\pi/4} \dfrac{1}{1 + \cos^2 x}\,dx \leqq \dfrac{\pi}{6}$

In Problems 55 through 58, use the second comparison property of integrals to estimate—giving both a lower bound and an upper bound as in Problem 54—the value of the given integral.

55. $\int_0^1 \dfrac{1}{1 + x}\,dx$

56. $\int_4^9 \dfrac{1}{1 + \sqrt{x}}\,dx$

57. $\int_0^{\pi/6} \cos^2 x\,dx$

58. $\int_0^{\pi/4} \sqrt{16 + 2\sin^2 x}\,dx$

59. Use Riemann sums—as in the proof of the sum property of integrals—to establish the constant multiple property.

60. Use Riemann sums to establish the first comparison property of integrals.

61. Deduce the second comparison property of integrals from the first comparison property.

62. Use sequences of Riemann sums to establish the interval union property of the integral. Note that if R_n' and R_n'' are Riemann sums for f on the intervals $[a, c]$ and $[c, b]$, respectively, then $R_n = R_n' + R_n''$ is a Riemann sum for f on $[a, b]$.

63. Suppose that a tank initially contains 1000 gal of water and that the rate of change of its volume after the tank drains for t min is $V'(t) = (0.8)t - 40$ (in gallons per minute). How much water does the tank contain after it has been draining for a half-hour?

64. Suppose that the population of Juneau in 1970 was 125 (in thousands) and that its rate of growth t years later was $P'(t) = 8 + (0.5)t + (0.03)t^2$ (in thousands per year). What was its population in 1990?

65. Figure 5.5.11 shows the graph of $f(x) = 1/x$ on the interval $[1, 2]$, the line joining its endpoints $(1, 1)$ and $(2, \frac{1}{2})$, and its tangent line at the point $(\frac{3}{2}, \frac{2}{3})$. Use this construction to estimate the value of the integral

$$\int_1^2 \frac{1}{x}\,dx$$

(whose exact value is known to be $\ln 2 \approx 0.693$).

66. Figure 5.5.12 shows the graph of $f(x) = 1/(1 + x^2)$ on the interval $[0, 1]$, the line $y = L(x)$ joining its endpoints $(0, 1)$ and $(1, \frac{1}{2})$, and the line $y = L(x) + 0.07$. First graph $f(x) - L(x)$ to verify that the latter line lies above $y = f(x)$ on the interval $[0, 1]$. Then use this construction to estimate the value of the integral

$$\int_0^1 \frac{1}{1 + x^2}\,dx$$

(whose exact value is known to be $\frac{1}{4}\pi \approx 0.785$).

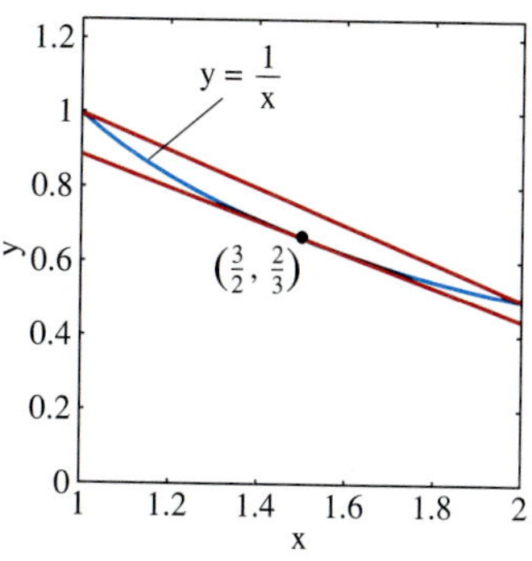

Fig. 5.5.11 Bounding the graph of $f(x) = \dfrac{1}{x}$

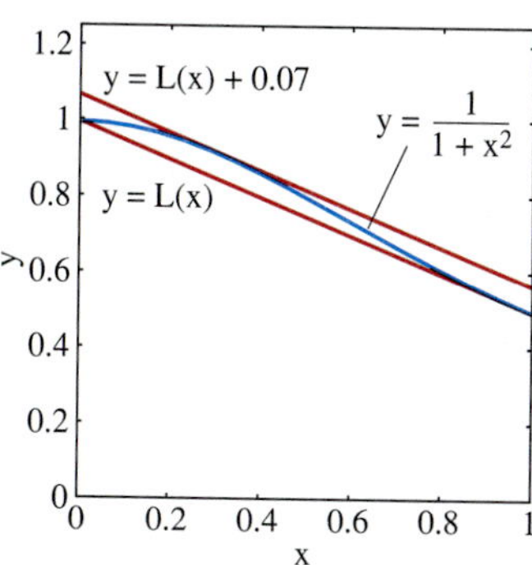

Fig. 5.5.12 Bounding the graph of $f(x) = \dfrac{1}{1 + x^2}$

5.6 AVERAGE VALUES AND THE FUNDAMENTAL THEOREM OF CALCULUS

Newton and Leibniz are generally credited with the invention of calculus in the latter part of the seventeenth century. Actually, others had earlier calculated areas essentially equivalent to integrals and tangent line slopes essentially equivalent to derivatives. The great accomplishments of Newton and Leibniz were the discovery and computational exploitation of the inverse relationship between differentiation and integration. This relationship is embodied in the *fundamental theorem of calculus.* One part of this theorem is the evaluation theorem of Section 5.5: To evaluate

$$\int_a^b f(x)\,dx,$$

it suffices to find an antiderivative of f on $[a, b]$. The other part of the fundamental theorem tells us that doing so is usually possible, at least in theory: Every continuous function has an antiderivative.

The Average Value of a Function

The concept of the *average value* of a function is useful for the proof of the fundamental theorem and has numerous important applications in its own right. The ordinary (arithmetic) **average** of n given numbers $a_1, a_2, \ldots, a_n$ is defined to be

$$\overline{a} = \frac{a_1 + a_2 + \cdots + a_n}{n} = \frac{1}{n}\sum_{i=1}^{n} a_i. \tag{1}$$

But a function f defined on an interval generally has infinitely many values $f(x)$, so we cannot simply divide the sum of all these values by their number to find the average value of $f(x)$. We introduce the proper notion with a discussion of average temperature.

EXAMPLE 1 Let the measured temperature T during a particular 24-h day at a certain location by given by the function

$$T = f(t), \qquad 0 \leqq t \leqq 24$$

(with the 24-h clock running from $t = 0$ at one midnight to $t = 24$ at the following midnight). Thus, for example, the temperatures $f(1), f(2), \ldots, f(24)$ are recorded at 1-h intervals during the day. We might define the average temperature $\overline{T}$ for the day as the (ordinary arithmetic) average of the hourly temperatures:

$$\overline{T} = \frac{1}{24}\sum_{i=1}^{24} f(t_i),$$

where $t_i = i$. If we divided the day into n equal subintervals rather than into 24 1-h intervals, we would obtain the more general average

$$\overline{T} = \frac{1}{n}\sum_{i=1}^{n} f(t_i).$$

The larger n is, the closer we would expect $\overline{T}$ to be to the "true" average temperature for the entire day. It is therefore plausible to define the true average temperature by letting n increase without bound. This gives

$$\overline{T} = \lim_{n\to\infty}\frac{1}{n}\sum_{i=1}^{n} f(t_i).$$

The right-hand side resembles a Riemann sum, and we can make it into a Riemann sum by introducing the factor

$$\Delta t = \frac{b-a}{n},$$

where $a = 0$ and $b = 24$. Then

$$\overline{T} = \lim_{n\to\infty} \frac{1}{n} \cdot \frac{n}{b-a} \sum_{i=1}^{n} f(t_i) \cdot \frac{b-a}{n} = \lim_{n\to\infty} \frac{1}{b-a} \sum_{i=1}^{n} f(t_i) \cdot \frac{b-a}{n}$$

$$= \frac{1}{b-a} \lim_{n\to\infty} \sum_{i=1}^{n} f(t_i)\,\Delta t = \frac{1}{b-a} \int_a^b f(t)\,dt.$$

Thus

$$\overline{T} = \frac{1}{24} \int_0^{24} f(t)\,dt \tag{2}$$

under the assumption that f is continuous, so the Riemann sums converge to the integral as $n \to \infty$. ■

The final result in Eq. (2) is the *integral of the function divided by the length of the interval.* Example 1 motivates the following definition.

Definition *Average Value of a Function*

Suppose that the function f is integrable on $[a, b]$. Then the **average value** $\overline{y}$ of $y = f(x)$ for x in the interval $[a, b]$ is

$$\overline{y} = \frac{1}{b-a} \int_a^b f(x)\,dx. \tag{3}$$

We can rewrite Eq. (3) in the form

$$\int_a^b f(x)\,dx = \overline{y} \cdot (b-a). \tag{4}$$

If f is positive-valued on $[a, b]$, then Eq. (4) implies that the area under $y = f(x)$ over $[a, b]$ is equal to the area of a rectangle with base length $b - a$ and height $\overline{y}$ (Fig. 5.6.1).

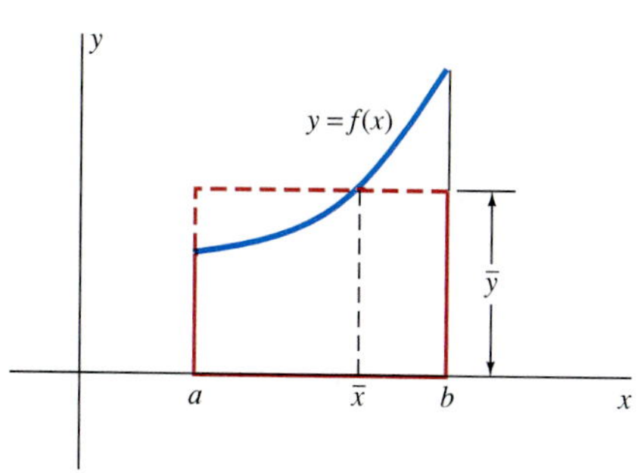

Fig. 5.6.1 A rectangle illustrating the average value of a function

EXAMPLE 2 The average value of $f(x) = x^2$ for x in $[0, 2]$ is

$$\overline{y} = \frac{1}{2} \int_0^2 x^2\,dx = \frac{1}{2} \left[\frac{1}{3} x^3 \right]_0^2 = \frac{4}{3}.$$ ■

EXAMPLE 3 The mean daily temperature in degrees Fahrenheit in Athens, Georgia, t months after July 15 is closely approximated by

$$T = 61 + 18 \cos \frac{\pi t}{6} = f(t). \tag{5}$$

Find the average temperature between September 15 ($t = 2$) and December 15 ($t = 5$).

Solution Equation (3) gives

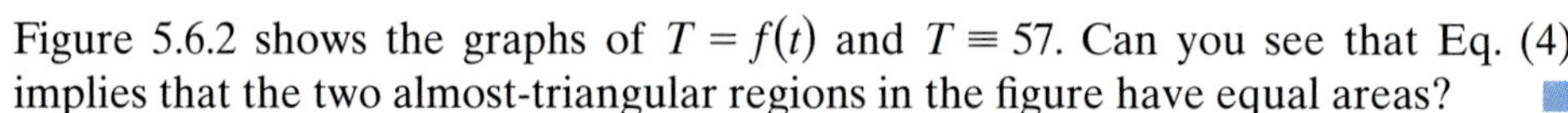

$$\overline{T} = \frac{1}{5-2}\int_2^5 \left(61 + 18\cos\frac{\pi t}{6}\right)dt = \frac{1}{3}\left[61t + \frac{6\cdot 18}{\pi}\sin\frac{\pi t}{6}\right]_2^5 \approx 57°\text{F}.$$

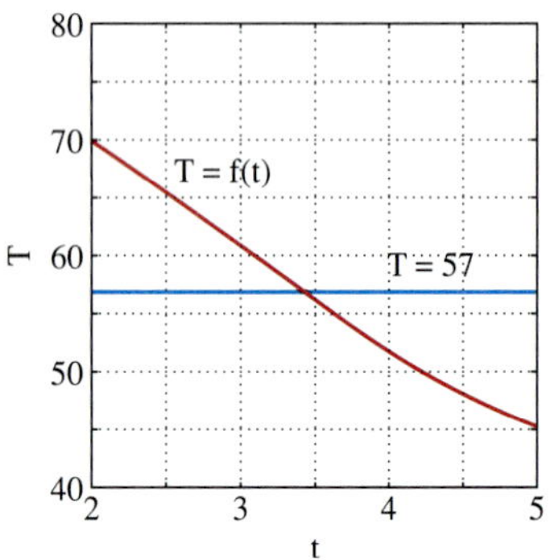

Fig. 5.6.2 The temperature function $T = f(t)$ of Example 3

Figure 5.6.2 shows the graphs of $T = f(t)$ and $T \equiv 57$. Can you see that Eq. (4) implies that the two almost-triangular regions in the figure have equal areas? ■

Theorem 1 tells us that every continuous function on a closed interval *attains* its average value at some point of the interval.

Theorem 1 ***Average Value Theorem***
If f is continuous on $[a, b]$, then

$$f(\overline{x}) = \frac{1}{b-a}\int_a^b f(x)\,dx \qquad \textbf{(6)}$$

for some number $\overline{x}$ in $[a, b]$.

Proof Let $m = f(c)$ be the minimum value of $f(x)$ on $[a, b]$ and let $M = f(d)$ be its maximum value there. Then, by the comparison property of Section 5.5,

$$m = f(c) \leqq \overline{y} = \frac{1}{b-a}\int_a^b f(x)\,dx \leqq f(d) = M.$$

Because f is continuous, we can now apply the intermediate value property. The number $\overline{y}$ is between the two values m and M of f, and consequently, $\overline{y}$ itself must be a value of f. Specifically, $\overline{y} = f(\overline{x})$ for some number $\overline{x}$ between a and b. This yields Eq. (6). ■

EXAMPLE 4 If $v(t)$ denotes the velocity function of a sports car accelerating during the time interval $a \leqq t \leqq b$, then the car's average velocity is given by

$$\overline{v} = \frac{1}{b-a}\int_a^b v(t)\,dt.$$

The average value theorem implies that $\overline{v} = v(\overline{t})$ for some number $\overline{t}$ in $[a, b]$. Thus $\overline{t}$ is an instant at which the car's instantaneous velocity is equal to its average velocity over the entire time interval. ■

The Fundamental Theorem

We state the fundamental theorem of calculus in two parts. The first part is the fact that every function f that is continuous on an interval I has an antiderivative on I. In particular, an antiderivative of f can be obtained by integrating f in a certain way. Intuitively, when $f(x) > 0$, we let $F(x)$ denote the area under the graph of f from a fixed point a of I to x, a point of I with $x > a$. We shall prove that $F'(x) = f(x)$. We show the construction of the function F in Fig. 5.6.3. More precisely, we define the function F as follows:

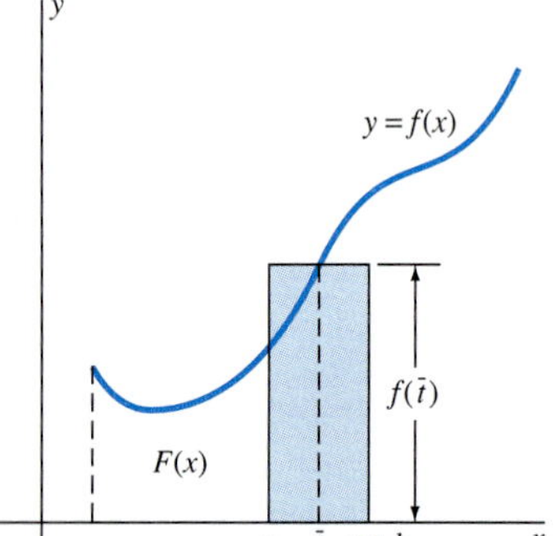

Fig. 5.6.3 The area function F is an antiderivative of f.

$$F(x) = \int_a^x f(t)\,dt,$$

where we use the dummy variable t in the integrand to avoid confusion with the upper limit x. The proof that $F'(x) = f(x)$ will be independent of the supposition that $x > a$.

The Fundamental Theorem of Calculus

Suppose that f is continuous on the closed interval $[a, b]$.

PART 1: If the function F is defined on $[a, b]$ by

$$F(x) = \int_a^x f(t)\,dt, \tag{7}$$

then F is an antiderivative of f. That is, $F'(x) = f(x)$ for x in $[a, b]$.

PART 2: If G is any antiderivative of f on $[a, b]$, then

$$\int_a^b f(x)\,dx = \Big[G(x)\Big]_a^b = G(b) - G(a). \tag{8}$$

PROOF OF PART 1 By the definition of the derivative,

$$F'(x) = \lim_{h\to 0} \frac{F(x+h) - F(x)}{h} = \lim_{h\to 0} \frac{1}{h}\left(\int_a^{x+h} f(t)\,dt - \int_a^x f(t)\,dt\right).$$

But

$$\int_a^{x+h} f(t)\,dt = \int_a^x f(t)\,dt + \int_x^{x+h} f(t)\,dt$$

by the interval union property of Section 5.5. Thus

$$F'(x) = \lim_{h\to 0} \frac{1}{h}\int_x^{x+h} f(t)\,dt.$$

The average value theorem tells us that

$$\frac{1}{h}\int_x^{x+h} f(t)\,dt = f(\bar{t})$$

for some number $\bar{t}$ in $[x, x+h]$. Finally, we note that $\bar{t} \to x$ as $h \to 0$. Thus, because f is continuous, we see that

$$F'(x) = \lim_{h\to 0} \frac{1}{h}\int_x^{x+h} f(t)\,dt = \lim_{h\to 0} f(\bar{t}) = \lim_{\bar{t}\to x} f(\bar{t}) = f(x).$$

Hence the function F in Eq. (7) is indeed an antiderivative of f. ■

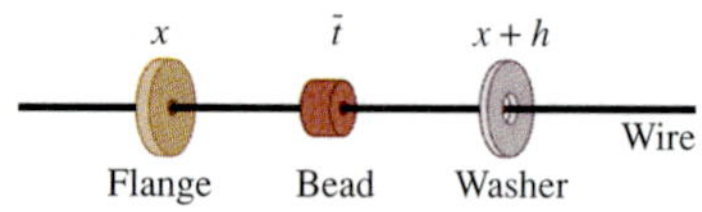

Fig. 5.6.4 The bead at $\bar{t}$ trapped between the washer at $x + h$ and the flange at x

Remark Figure 5.6.4 indicates why $\bar{t}$ must approach x as $h \to 0$. As the moving washer at $x + h$ approaches the fixed flange at x, the bead $\bar{t}$ between them has nowhere else to go.

Proof of Part 2 Here we apply part 1 to give a proof of the evaluation theorem in Section 5.5. If G is *any* antiderivative of f, then—because it and the function F of part 1 are both antiderivatives of f on the interval $[a, b]$—we know that

$$G(x) = F(x) + C$$

on $[a, b]$ for some constant C. To evaluate C, we substitute $x = a$ and obtain

$$C = G(a) - F(a) = G(a),$$

because

$$F(a) = \int_a^a f(t)\,dt = 0.$$

Hence $G(x) = F(x) + G(a)$. In other words,

$$F(x) = G(x) - G(a)$$

for all x in $[a, b]$. With $x = b$ this gives

$$G(b) - G(a) = F(b) = \int_a^b f(x)\,dx,$$

which establishes Eq. (8) ■

Sometimes the fundamental theorem of calculus is interpreted to mean that differentiation and integration are *inverse processes.* Part 1 can be written in the form

$$\frac{d}{dx}\left(\int_a^x f(t)\,dt\right) = f(x) \tag{9}$$

if f is continuous on an open interval containing a and x. That is, if we first integrate the function f (with *variable* upper limit of integration x) and then differentiate with respect to x, the result is the function f again. So differentiation "cancels" the effect of integration of continuous functions.

Moreover, part 2 of the fundamental theorem can be written in the form

$$\int_a^x G'(t)\,dt = G(x) - G(a) \tag{10}$$

if we assume that G' is continuous. If so, this equation means that if we first differentiate the function G and then integrate the result from a to x, the result can differ from the original function G by, at worst, the *constant* $G(a)$. If a is chosen so that $G(a) = 0$, this means that integration "cancels" the effect of differentiation.

Computational Applications

Examples 1 through 4 of Section 5.5 illustrate the use of part 2 of the fundamental theorem in the evaluation of integrals. Additional examples appear in this section, in the problems following this section, and in Section 5.7. Example 5 illustrates the necessity of splitting an integral into a sum of integrals when its integrand has different antiderivative formulas on different intervals.

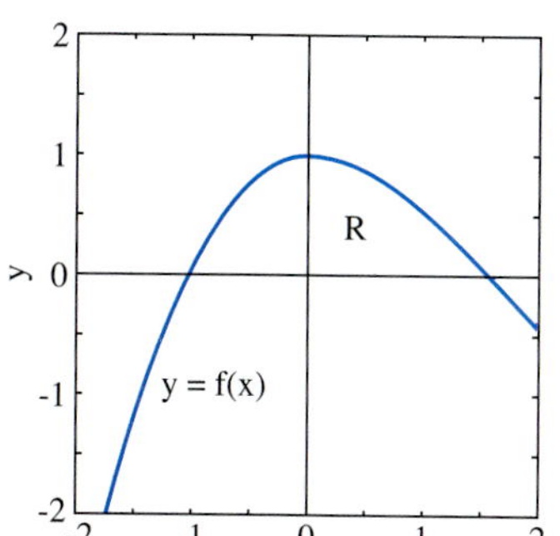

Fig. 5.6.5 The region of Example 5

EXAMPLE 5 Figure 5.6.5 shows the graph of the function f defined by

$$f(x) = \begin{cases} \cos x & \text{if } x \geqq 0, \\ 1 - x^2 & \text{if } x \leqq 0. \end{cases}$$

Find the area A of the region R bounded above by the graph of $y = f(x)$ and below by the x-axis.

Solution The x-intercepts shown in the figure are $x = -1$ (where $1 - x^2 = 0$ and $x < 0$) and $x = \pi/2$ (where $\cos x = 0$ and $x > 0$). Hence

$$A = \int_{-1}^{\pi/2} f(x)\,dx = \int_{-1}^{0} (1 - x^2)\,dx + \int_{0}^{\pi/2} \cos x\,dx$$

$$= \left[x - \frac{1}{3}x^3 \right]_{-1}^{0} + \left[\sin x \right]_{0}^{\pi/2} = \frac{2}{3} + 1 = \frac{5}{3}.$$

EXAMPLE 6 Figure 5.6.6 shows the graph of

$$f(x) = x^3 - x^2 - 6x.$$

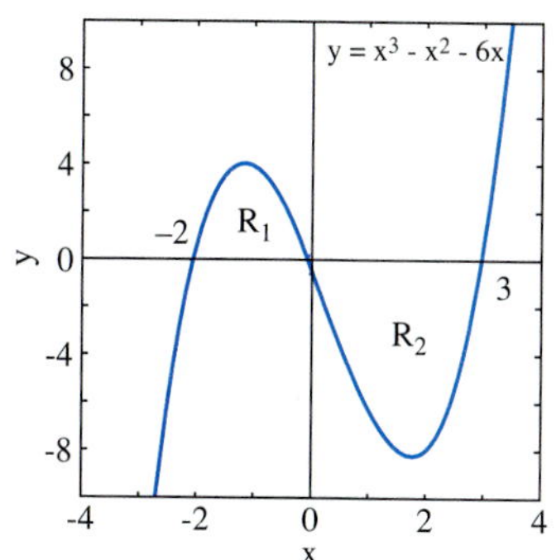

Fig. 5.6.6 The graph $y = x^3 - x^2 - 6x$ of Example 6

Find the area A of the entire region R bounded by the graph of f and the x-axis.

Solution The region R consists of the two regions R_1 and R_2 and extends from $x = -2$ to $x = 3$. The area of R_1 is

$$A_1 = \int_{-2}^{0} (x^3 - x^2 - 6x)\,dx = \left[\tfrac{1}{4}x^4 - \tfrac{1}{3}x^3 - 3x^2 \right]_{-2}^{0} = \tfrac{16}{3}.$$

But on the interval $(0, 3)$, the function $f(x)$ is negative-valued, so to get the (positive) area A_2 of R_2, we must integrate the *negative* of f:

$$A_2 = \int_{0}^{3} (-x^3 + x^2 + 6x)\,dx = \left[-\tfrac{1}{4}x^4 + \tfrac{1}{3}x^3 + 3x^2 \right]_{0}^{3} = \tfrac{63}{4}.$$

Consequently the area of the entire region R is

$$A = A_1 + A_2 = \tfrac{16}{3} + \tfrac{63}{4} = \tfrac{253}{12} \approx 21.08.$$

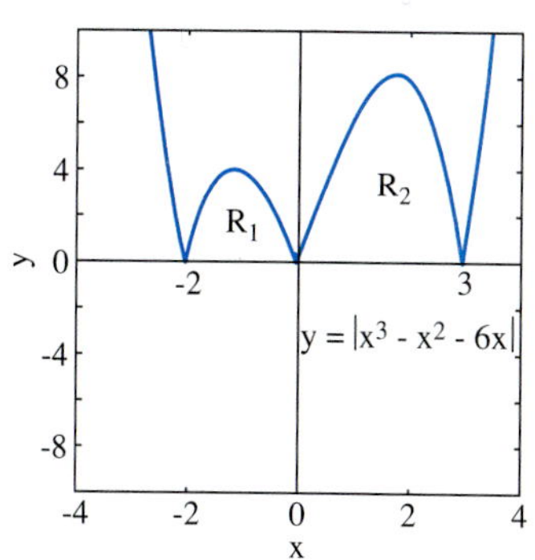

Fig. 5.6.7 The graph $y = |x^3 - x^2 - 6x|$ of Example 6

In effect, we have integrated the *absolute value* of $f(x)$:

$$A = \int_{-2}^{3} |f(x)|\,dx$$

$$= \int_{-2}^{0} (x^3 - x^2 - 6x)\,dx + \int_{0}^{3} (-x^3 + x^2 + 6x)\,dx = \tfrac{253}{12}.$$

Compare the graph of $y = |f(x)|$ in Fig. 5.6.7 with that of $y = f(x)$ in Fig. 5.6.6.

EXAMPLE 7 Evaluate

$$\int_{-1}^{2} |x^3 - x|\,dx.$$

Solution We note that $x^3 - x \geqq 0$ on $[-1, 0]$, that $x^3 - x \leqq 0$ on $[0, 1]$, and that $x^3 - x \geqq 0$ on $[1, 2]$. So we write

$$\int_{-1}^{2} |x^3 - x|\,dx = \int_{-1}^{0} (x^3 - x)\,dx + \int_{0}^{1} (x - x^3)\,dx + \int_{1}^{2} (x^3 - x)\,dx$$

$$= \left[\tfrac{1}{4}x^4 - \tfrac{1}{2}x^2\right]_{-1}^{0} + \left[\tfrac{1}{2}x^2 - \tfrac{1}{4}x^4\right]_{0}^{1} + \left[\tfrac{1}{4}x^4 - \tfrac{1}{2}x^2\right]_{1}^{2}$$

$$= \tfrac{1}{4} + \tfrac{1}{4} + [2 - (-\tfrac{1}{4})] = \tfrac{11}{4} = 2.75.$$

Part 1 of the fundamental theorem of calculus says that the derivative of an integral with respect to its upper limit is equal to the value of the integrand at the upper limit. For example, if

$$y(x) = \int_0^x t^3 \sin t\,dt,$$

then

$$\frac{dy}{dx} = x^3 \sin x.$$

Example 8 is a bit more complicated in that the upper limit of the integral is a nontrivial function of the independent variable.

EXAMPLE 8 Find $h'(x)$ given

$$h(x) = \int_0^{x^2} t^3 \sin t\,dt.$$

Solution Let $y = h(x)$ and $u = x^2$. Then

$$y = \int_0^u t^3 \sin t\,dt,$$

so

$$\frac{dy}{du} = u^3 \sin u$$

by the fundamental theorem of calculus. Then the chain rule yields

$$h'(x) = \frac{dy}{dx} = \frac{dy}{du} \cdot \frac{du}{dx} = (u^3 \sin u)(2x) = 2x^7 \sin x^2.$$

Initial Value Problems

Note that if

$$y(x) = \int_a^x f(t)\,dt, \tag{11}$$

then $y(a) = 0$. Hence $y(x)$ is a solution of the initial value problem

$$\frac{dy}{dx} = f(x), \qquad y(a) = 0. \tag{12}$$

To get a solution of the initial value problem

$$\frac{dy}{dx} = f(x), \qquad y(a) = b, \tag{13}$$

we need only add the desired initial value:

$$y(x) = b + \int_a^x f(t)\,dt. \tag{14}$$

EXAMPLE 9 Express as an integral the solution of the initial value problem

$$\frac{dy}{dx} = \sec x, \qquad y(2) = 3. \tag{15}$$

Solution With $a = 2$ and $b = 3$, Eq. (14) gives

$$y(x) = 3 + \int_2^x \sec t\,dt. \tag{16}$$

With our present knowledge, we cannot antidifferentiate $\sec t$, but for a particular value of x the integral in Eq. (16) can be approximated using Riemann sums. For instance, with $x = 4$ a calculator with an **INTEGRATE** key gives

$$\int_2^4 \sec t\,dt \approx -2.5121.$$

Hence the value of the solution in Eq. (16) at $x = 4$ is

$$y(4) \approx 3 - 2.5121 = 0.4879.$$

■

5.6 PROBLEMS

In Problems 1 through 12, find the average value of the given function on the specified interval.

1. $f(x) = x^4$; $[0, 2]$ **2.** $g(x) = \sqrt{x}$; $[1, 4]$

3. $h(x) = 3x^2\sqrt{x^3 + 1}$; $[0, 2]$

4. $f(x) = 8x$; $[0, 4]$ **5.** $g(x) = 8x$; $[-4, 4]$

6. $h(x) = x^2$; $[-4, 4]$ **7.** $f(x) = x^3$; $[0, 5]$

8. $g(x) = x^{-1/2}$; $[1, 4]$ **9.** $f(x) = \sqrt{x + 1}$; $[0, 3]$

10. $g(x) = \sin 2x$; $[0, \pi/2]$ **11.** $f(x) = \sin 2x$; $[0, \pi]$

12. $g(t) = \cos 2\pi t$; $[-\frac{1}{2}, \frac{1}{2}]$

Evaluate the integrals in Problems 13 through 28.

13. $\displaystyle\int_{-1}^{3} dx$ (Here dx stands for $1\,dx$.)

14. $\displaystyle\int_1^2 (y^5 - 1)\,dy$ **15.** $\displaystyle\int_1^4 \frac{dx}{\sqrt{9x^3}}$

16. $\displaystyle\int_{-1}^{1} (x^3 + 2)^2\,dx$ **17.** $\displaystyle\int_1^3 \frac{3t - 5}{t^4}\,dt$

18. $\displaystyle\int_{-2}^{-1} \frac{x^2 - x + 3}{\sqrt[3]{x}}\,dx$ **19.** $\displaystyle\int_0^{\pi} \sin x \cos x\,dx$

20. $\displaystyle\int_{-1}^{2} |x|\,dx$ **21.** $\displaystyle\int_1^2 \left(t - \frac{1}{2t}\right)^2 dt$

22. $\displaystyle\int_{-1}^{1} \frac{x^2 - 4}{x + 2}\,dx$ **23.** $\displaystyle\int_0^{\sqrt{\pi}} x \cos x^2\,dx$

24. $\displaystyle\int_0^2 |x - \sqrt{x}|\,dx$ **25.** $\displaystyle\int_{-2}^{2} |x^2 - 1|\,dx$

26. $\displaystyle\int_0^{\pi/3} \sin 3x\,dx$ **27.** $\displaystyle\int_2^7 \sqrt{x + 2}\,dx$

28. $\displaystyle\int_5^{10} \frac{dx}{\sqrt{x - 1}}$

In Problems 29 through 32, the graph of f and the x-axis divide the xy-plane into several regions, some of which are bounded. Find the total area of the bounded regions in each problem.

29. $f(x) = 1 - x^4$ if $x \leqq 0$; $f(x) = 1 - x^3$ if $x \geqq 0$ (Fig. 5.6.8)

30. $f(x) = (\pi/2)^2 \sin x$ on $[0, \pi/2]$; $f(x) = x(\pi - x)$ on $[\pi/2, \pi]$ (Fig. 5.6.9)

31. $f(x) = x^3 - 9x$ (Fig. 5.6.10)

32. $f(x) = x^3 - 2x^2 - 15x$ (Fig. 5.6.11)

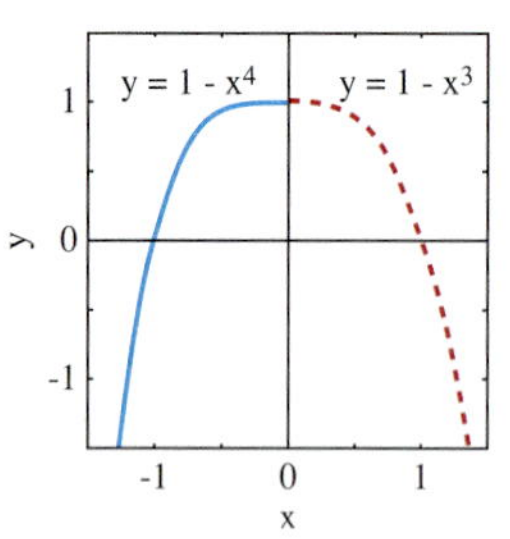

Fig. 5.6.8 Problem 29

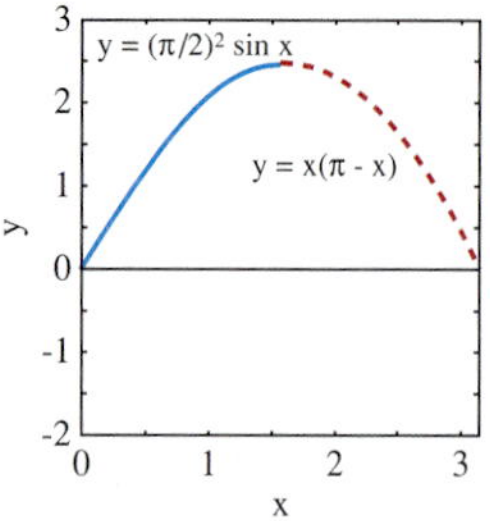

Fig. 5.6.9 Problem 30

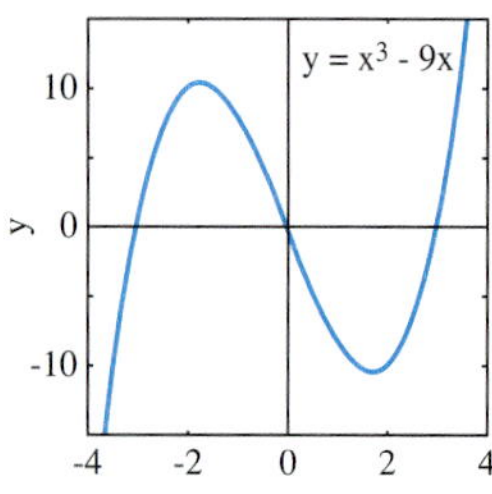

Fig. 5.6.10 Problem 31

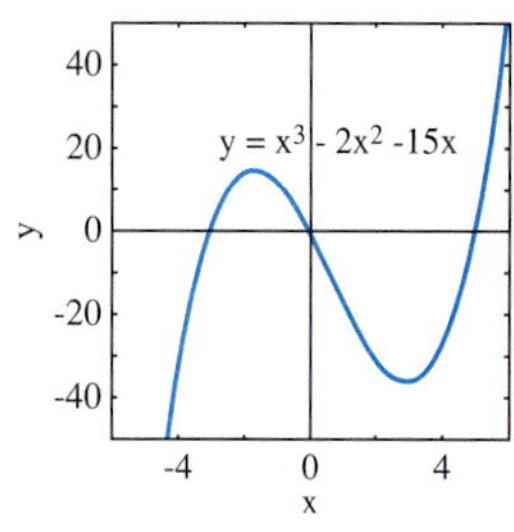

Fig. 5.6.11 Problem 32

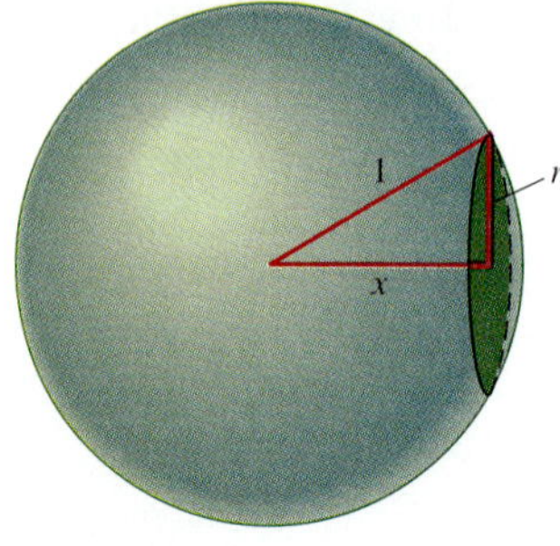
Fig. 5.6.12 The sphere of Problem 38

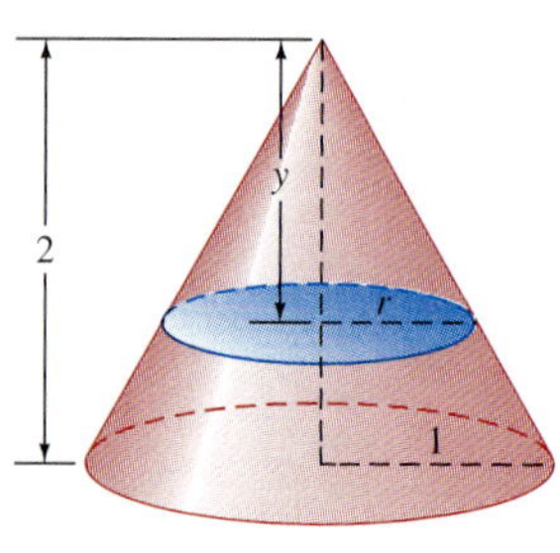
Fig. 5.6.13 The cone of Problem 39

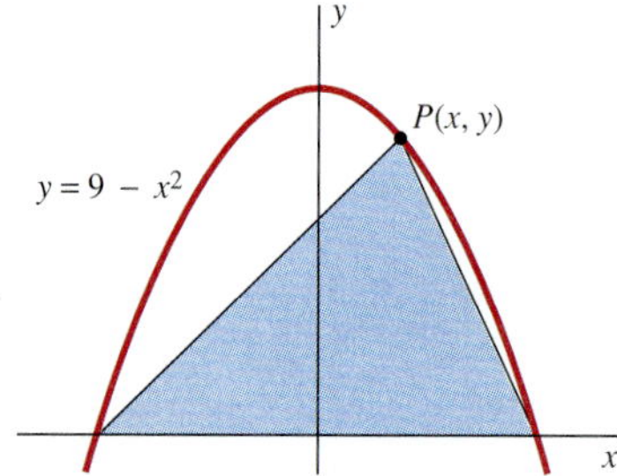

Fig. 5.6.14 The typical triangle of Problem 41

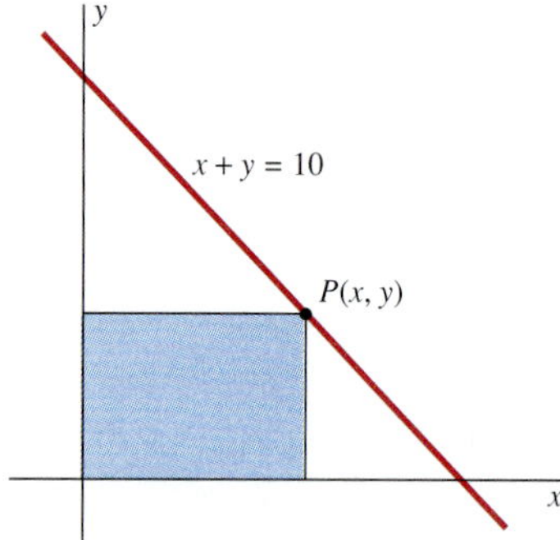

Fig. 5.6.15 The typical rectangle of Problem 42

33. Rosanne drops a ball from a height of 400 ft. Find the ball's average height and its average velocity between the time it is dropped and the time it strikes the ground.

34. Find the average value of the animal population $P(t) = 100 + 10t + (0.02)t^2$ over the time interval $[0, 10]$.

35. Suppose that a 5000-L water tank takes 10 min to drain and that after t minutes, the amount of water remaining in the tank is $V(t) = 50(10 - t)^2$ liters. What is the average amount of water in the tank during the time it drains?

36. On a certain day the temperature t hours past midnight was

$$T(t) = 80 + 10 \sin\left(\frac{\pi}{12}(t - 10)\right).$$

What was the average temperature between noon and 6 P.M.?

37. Suppose that a heated rod lies along the interval $0 \leqq x \leqq 10$. If the temperature at points of the rod is given by $T(x) = 4x(10 - x)$, what is the rod's average temperature?

38. Figure 5.6.12 shows a cross section at distance x from the center of a sphere of radius 1. Find the average area of the cross section for $0 \leqq x \leqq 1$.

39. Figure 5.6.13 shows a cross section at distance y from the vertex of a cone with base radius 1 and height 2. Find the average area of this cross section for $0 \leqq y \leqq 2$.

40. A sports car starts from rest ($x = 0$, $t = 0$) and experiences constant acceleration $x''(t) = a$ for T seconds. Find, in terms of a and T, (a) its final and average velocities and (b) its final and average positions.

41. (a) Figure 5.6.14 shows a triangle inscribed in the region that lies between the x-axis and the curve $y = 9 - x^2$. Express the area of this triangle as a function $A(x)$ of the x-coordinate of its upper vertex P. (b) Find the average area $\overline{A}$ of $A(x)$ for x in the interval $[-3, 3]$. (c) Sketch a triangle as in Fig. 5.6.14 that has the area $\overline{A}$ found in part (b). How many different such triangles are there?

42. (a) Figure 5.6.15 shows a rectangle inscribed in the first-quadrant region that lies between the x-axis and the line $y = 10 - x$. Express the area of this rectangle as a function $A(x)$ of the x-coordinate of its vertex P on the line. (b) Find the average area $\overline{A}$ of $A(x)$ for x in the interval $[0, 10]$. (c) Sketch a rectangle as in Fig. 5.6.15 that has the area $\overline{A}$ found in part (b). How many different such rectangles are there?

43. Figure 5.6.16 shows a rectangle inscribed in the semicircular region that lies between the x-axis and the graph

$y = \sqrt{16 - x^2}$. Express the area of the rectangle as a function $A(x)$ of the x-coordinate of its vertex P on the line. (b) Find the average area $\overline{A}$ of $A(x)$ for x in the interval $[0, 4]$. (c) Sketch a rectangle as in Fig. 5.6.16 that has the area $\overline{A}$ found in part (b). How many different such rectangles are there?

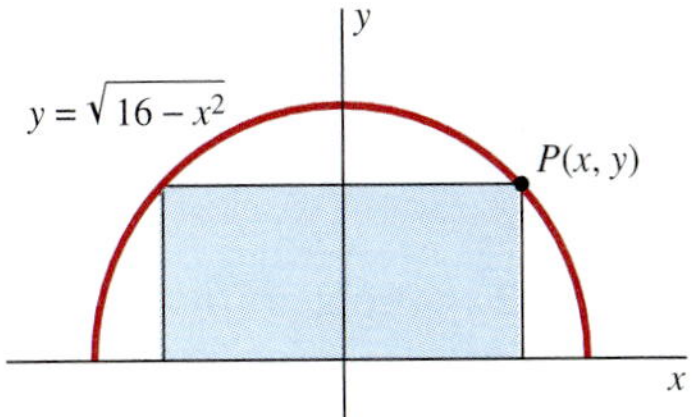

Fig. 5.6.16 The typical rectangle of Problem 43

44. Repeat Problem 43 in the case that the rectangle has two vertices on the x-axis and two on the parabola $y = 16 - x^2$ (rather than on the semicircle $y = \sqrt{16 - x^2}$). You may need to use a calculator or computer to find the base of a rectangle whose area is the average $\overline{A}$ of $A(x)$ for x in $[0, 4]$.

In Problems 45 through 49, apply the fundamental theorem of calculus to find the derivative of the given function.

45. $f(x) = \displaystyle\int_{-1}^{x} (t^2 + 1)^{17}\,dt$ **46.** $g(t) = \displaystyle\int_{0}^{t} \sqrt{x^2 + 25}\,dx$

47. $h(z) = \displaystyle\int_{2}^{z} \sqrt[3]{u - 1}\,du$ **48.** $A(x) = \displaystyle\int_{1}^{x} \frac{1}{t}\,dt$

49. $f(x) = \displaystyle\int_{x}^{10} \left(t + \frac{1}{t}\right) dt$

In Problems 50 through 53, $G(x)$ is the integral of the given function $f(t)$ over the specified interval of the form $[a, x]$, $x > a$. Apply part 1 of the fundamental theorem of calculus to find $G'(x)$.

50. $f(t) = \dfrac{t}{t^2 + 1}$; $[2, x]$ **51.** $f(t) = \sqrt{t + 4}$; $[0, x]$

52. $f(t) = \sin^3 t$; $[0, x]$ **53.** $f(t) = \sqrt{t^3 + 1}$; $[1, x]$

In Problems 54 through 60, differentiate the function by first writing $f(x)$ in the form $g(u)$, where u denotes the upper limit of integration.

54. $f(x) = \displaystyle\int_{0}^{x^2} \sqrt{1 + t^3}\,dt$ **55.** $f(x) = \displaystyle\int_{2}^{3x} \sin t^2\,dt$

56. $f(x) = \displaystyle\int_{0}^{\sin x} \sqrt{1 - t^2}\,dt$ **57.** $f(x) = \displaystyle\int_{0}^{x^2} \sin t\,dt$

58. $f(x) = \displaystyle\int_{1}^{\sin x} (t^2 + 1)^3\,dt$ **59.** $f(x) = \displaystyle\int_{1}^{x^2+1} \frac{dt}{t}$

60. $f(x) = \displaystyle\int_{1}^{x^5} \sqrt{1 + t^2}\,dt$

Use integrals (as in Example 9) to solve the initial value problems in 61 through 64.

61. $\dfrac{dy}{dx} = \dfrac{1}{x}$, $y(1) = 0$ **62.** $\dfrac{dy}{dx} = \dfrac{1}{1 + x^2}$, $y(1) = \dfrac{\pi}{4}$

63. $\dfrac{dy}{dx} = \sqrt{1 + x^2}$, $y(5) = 10$

64. $\dfrac{dy}{dx} = \tan x$, $y(1) = 2$

65. The fundamental theorem of calculus *seems* to say that

$$\int_{-1}^{1} \frac{dx}{x^2} = \left[-\frac{1}{x}\right]_{-1}^{1} = -2,$$

in apparent contradiction to the fact that $1/x^2$ is always positive. What's wrong here?

66. Prove that the average rate of change

$$\frac{f(b) - f(a)}{b - a}$$

of the differentiable function f on $[a, b]$ is equal to the average value of its derivative on $[a, b]$.

67. The graph $y = f(x), 0 \leqq x \leqq 10$ is shown in Fig. 5.6.17. Let

$$g(x) = \int_{0}^{x} f(t)\,dt.$$

(a) Find the values $g(0)$, $g(2)$, $g(4)$, $g(6)$, $g(8)$, and $g(10)$. (b) Find the intervals on which $g(x)$ is increasing and those on which it is decreasing. (c) Find the global maximum and minimum values of $g(x)$ for $0 \leqq x \leqq 10$. (d) Sketch a rough graph of $y = g(x)$.

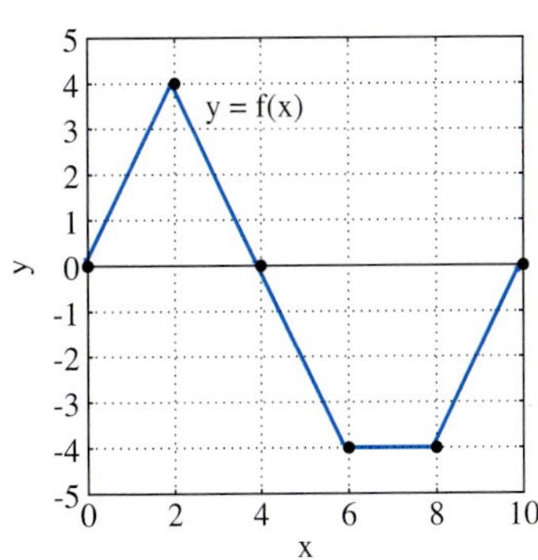

Fig. 5.6.17 Problem 67

68. Repeat Problem 67, except use the graph of the function f shown in Fig. 5.6.18.

69. Figure 5.6.19 shows the graph of the function $f(x) = x \sin x$ on the interval $[0, 4\pi]$. Let

$$g(x) = \int_{0}^{x} f(t)\,dt.$$

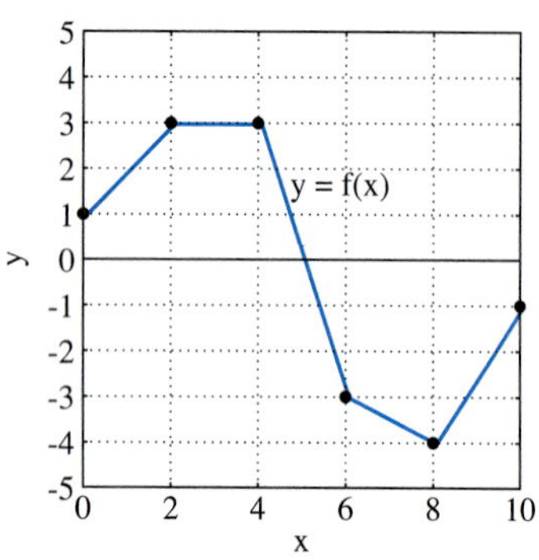

Fig. 5.6.18 Problem 68

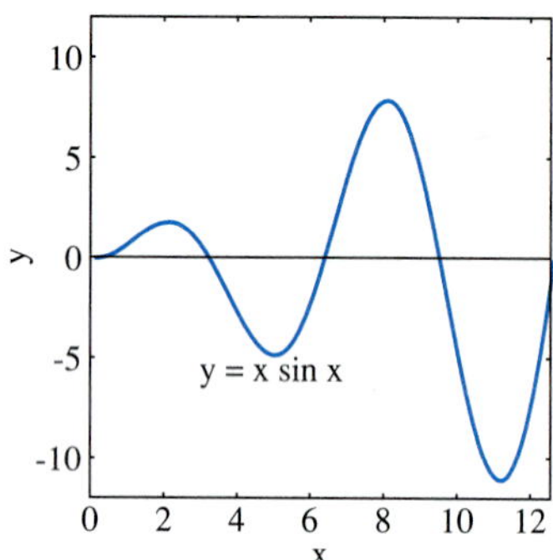

Fig. 5.6.19 Problem 69

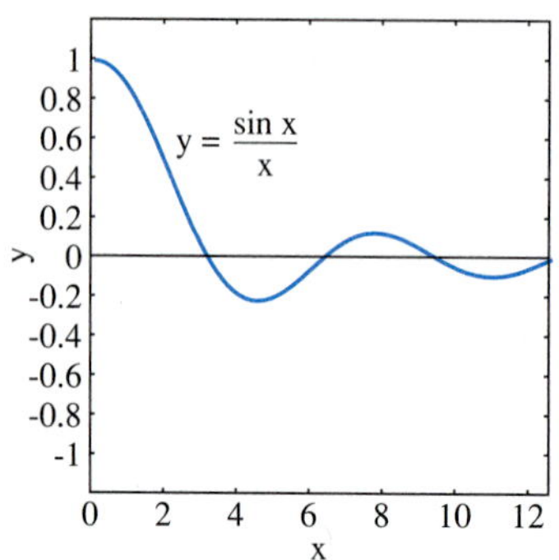

Fig. 5.6.20 Problem 70

(a) Find the values of x at which $g(x)$ has local maximum and minimum values on the interval $[0, 4\pi]$. (b) Where does $g(x)$ attain its global maximum and minimum values on $[0, 4\pi]$? (c) Which points on the graph $y = f(x)$ correspond to inflection points on the graph $y = g(x)$? (d) Sketch a rough graph of $y = g(x)$.

70. Repeat Problem 69, except use the function

$$f(x) = \frac{\sin x}{x}$$

on the interval $[0, 4\pi]$ (as shown in Fig. 5.6.20). Take $f(0) = 1$ because $(\sin x)/x \to 1$ as $x \to 0$.

5.7 INTEGRATION BY SUBSTITUTION

The fundamental theorem of calculus in the form

$$\int_a^b f(x)\,dx = \left[\int f(x)\,dx\right]_a^b \tag{1}$$

implies that we can readily evaluate the definite integral on the left if we can find the indefinite integral (that is, antiderivative) on the right. We now discuss a powerful method of antidifferentiation that amounts to "the chain rule in reverse." This method is a generalization of the "generalized power rule in reverse,"

$$\int u^n\,du = \frac{u^{n+1}}{n+1} + C \quad (n \neq -1), \tag{2}$$

that we introduced in Section 5.2.

Equation (2) is an abbreviation for the formula

$$\int [g(x)]^n g'(x)\,dx = \frac{[g(x)]^{n+1}}{n+1} + C \quad (n \neq -1) \tag{3}$$

that results when we write

$$u = g(x), \quad du = g'(x)\,dx.$$

So to apply Eq. (2) to a given integral, we must be able to visualize the integrand as a *product* of a power of a differentiable function $g(x)$ and its derivative $g'(x)$.

EXAMPLE 1 With

$$u = 2x + 1, \qquad du = 2\,dx,$$

we see that

$$\int (2x+1)^5 \cdot 2\,dx = \int u^5\,du = \frac{u^6}{6} + C = \frac{1}{6}(2x+1)^6 + C. \quad \blacksquare$$

EXAMPLE 2

$$\begin{aligned} \int 2x\sqrt{1+x^2}\,dx &= \int (1+x^2)^{1/2} \cdot 2x\,dx \\ &= \int u^{1/2}\,du \qquad [u = 1 + x^2,\ du = 2x\,dx] \\ &= \frac{u^{3/2}}{\frac{3}{2}} + C = \tfrac{2}{3}(1+x^2)^{3/2} + C. \quad \blacksquare \end{aligned}$$

Equation (3) is the special case $f(u) = u^n$ of the general integral formula

$$\int f(g(x)) \cdot g'(x)\,dx = \int f(u)\,du. \tag{4}$$

The right-hand side of Eq. (4) results when we make the formal substitution

$$u = g(x), \qquad du = g'(x)\,dx$$

into the left-hand side.

One of the beauties of differential notation is that Eq. (4) not only is plausible but is, in fact, true—with the understanding that u is to be replaced with $g(x)$ after the indefinite integration on the right-hand side of Eq. (4) has been carried out. Indeed, Eq. (4) is merely an indefinite integral version of the chain rule. For if $F'(x) = f(x)$, then

$$D_x F(g(x)) = F'(g(x)) \cdot g'(x) = f(g(x)) \cdot g'(x)$$

by the chain rule, so

$$\begin{aligned} \int f(g(x)) \cdot g'(x)\,dx &= \int F'(g(x)) \cdot g'(x)\,dx = F(g(x)) + C \\ &= F(u) + C \qquad [u = g(x)] \\ &= \int f(u)\,du. \end{aligned}$$

Equation (4) is the basis for the powerful technique of indefinite **integration by substitution.** It may be used whenever the integrand function is recognizable in the form $f'(g(x)) \cdot g'(x)$.

EXAMPLE 3 Find

$$\int x^2\sqrt{x^3 + 9}\,dx.$$

Solution Note that x^2 is, to within a *constant* factor, the derivative of $x^3 + 9$. We can therefore substitute

$$u = x^3 + 9, \qquad du = 3x^2\,dx. \tag{5}$$

The constant factor 3 can be supplied if we compensate by multiplying the integral by $\frac{1}{3}$. This gives

$$\begin{aligned}\int x^2\sqrt{x^3 + 9}\,dx &= \frac{1}{3}\int (x^2 + 9)^{1/2}\cdot 3x^2\,dx = \frac{1}{3}\int u^{1/2}\,du \\ &= \frac{1}{3}\cdot\frac{u^{3/2}}{\frac{3}{2}} + C = \frac{2}{9}u^{3/2} + C = \frac{2}{9}(x^3 + 9)^{3/2} + C.\end{aligned}$$

An alternative way to carry out the substitution in (5) is to solve

$$du = 3x^2\,dx \quad \text{for} \quad x^2\,dx = \tfrac{1}{3}\,du$$

and then write

$$\int (x^3 + 9)^{1/2}\,x^2\,dx = \int u^{1/2}\cdot\tfrac{1}{3}\,du = \tfrac{1}{3}\int u^{1/2}\,du,$$

concluding the computation as before. ■

The following three steps in the solution of Example 3 are worth special mention:

1. The differential dx along with the rest of the integrand is "transformed," or replaced, in terms of u and du.
2. Once the integration has been performed, the constant C of integration is added.
3. A final resubstitution is necessary to write the answer in terms of the original variable x.

Substitution in Trigonometric Integrals

By now we know that every differentiation formula yields—upon "reversal"—a corresponding antidifferentiation formula. The familiar formulas for the derivatives of the six trigonometric functions thereby yield the following indefinite-integral formulas:

$$\int \cos u\,du = \sin u + C \tag{6}$$

$$\int \sin u\,du = -\cos u + C \tag{7}$$

$$\int \sec^2 u\,du = \tan u + C \tag{8}$$

$$\int \csc^2 u\,du = -\cot u + C \tag{9}$$

$$\int \sec u\tan u\,du = \sec u + C \tag{10}$$

$$\int \csc u\cot u\,du = -\csc u + C \tag{11}$$

Any of these integrals can appear as the integral $\int f(u)\,du$ that results from an appropriate *u-substitution* in a given integral.

EXAMPLE 4

$$\int \sin(3x+4)\,dx = \int (\sin u)\cdot\tfrac{1}{3}\,du \qquad (u = 3x+4,\ du = 3\,dx)$$
$$= \tfrac{1}{3}\int \sin u\,du = -\tfrac{1}{3}\cos u + C$$
$$= -\tfrac{1}{3}\cos(3x+4) + C.$$

EXAMPLE 5

$$\int 3x\cos(x^2)\,dx = 3\int (\cos x^2)\cdot x\,dx$$
$$= 3\int (\cos u)\cdot\tfrac{1}{2}\,du \qquad (u = x^2,\ du = 2x\,dx)$$
$$= \tfrac{3}{2}\int \cos u\,du = \tfrac{3}{2}\sin u + C = \tfrac{3}{2}\sin(x^2) + C.$$

EXAMPLE 6

$$\int \sec^2 3x\,dx = \int (\sec^2 u)\cdot\tfrac{1}{3}\,du \qquad (u = 3x,\ du = 3\,dx)$$
$$= \tfrac{1}{3}\tan u + C = \tfrac{1}{3}\tan 3x + C.$$

EXAMPLE 7 Evaluate

$$\int 2\sin^3 x\cos x\,dx.$$

Solution None of the integrals in Eqs. (6) through (11) appears to "fit," but the substitution

$$u = \sin x, \qquad du = \cos x\,dx$$

yields

$$\int 2\sin^3 x\cos x\,dx = 2\int u^3\,du = 2\cdot\frac{u^4}{4} + C = \frac{1}{2}\sin^4 x + C.$$

Substitution in Definite Integrals

The method of integration by substitution can be used with definite integrals as well as with indefinite integrals. Only one additional step is required—evaluation of the final antiderivative at the original limits of integration.

EXAMPLE 8 The substitution used in Example 3 gives

$$\int_0^3 x^2\sqrt{x^3+9}\,dx = \int_{-}^{-} u^{1/2}\cdot\tfrac{1}{3}\,du \qquad (u = x^3+9,\ du = 3x^2\,dx)$$

$$= \frac{1}{3}\left[\frac{2}{3}u^{3/2}\right]_{-}^{-} = \frac{2}{9}\left[(x^3+9)^{3/2}\right]_0^3 \quad \text{(resubstitute)}$$

$$= \frac{2}{9}(216 - 27) = 42.$$

The limits on u were left "blank" because they weren't calculated—there was no need to know them, because we planned to resubstitute for u in terms of the original variable x before using the original limits of integration.

But sometimes it is more convenient to determine the limits of integration with respect to the new variable u. With the substitution $u = x^4 + 9$, $du = 3x^2\,dx$, we see that

- $u = 9$ when $x = 0$ (lower limit);
- $u = 36$ when $x = 3$ (upper limit).

Use of these limits on u (rather than resubstitution in terms of x) gives

$$\int_0^3 x^2\sqrt{x^3+9}\,dx = \frac{1}{3}\int_9^{36} u^{1/2}\,du = \frac{1}{3}\left[\frac{2}{3}u^{3/2}\right]_9^{36} = 42.$$

■

Theorem 1 says that the "natural" way of transforming an integral's limits under a u-substitution, like the work just done, is in fact correct.

Theorem 1 ***Definite Integration by Substitution***

Suppose that the function g has a continuous derivative on $[a, b]$ and that f is continuous on the set $g([a, b])$. Let $u = g(x)$. Then

$$\int_a^b f(g(x)) \cdot g'(x)\,dx = \int_{g(a)}^{g(b)} f(u)\,du. \tag{12}$$

Remark Thus we get the new limits on u by applying the substitution function $u = g(x)$ to the old limits on x. Then:

- The new lower limit is $g(a)$, and
- The new upper limit is $g(b)$,

whether or not $g(b)$ is greater than $g(a)$.

Choose an antiderivative F of f, so $F' = f$. Then, by the chain rule,

$$D_x[F(g(x))] = F'(g(x)) \cdot g'(x) = f(g(x)) \cdot g'(x).$$

Therefore

$$\int_a^b f(g(x)) \cdot g'(x)\,dx = \Big[F(g(x))\Big]_a^b = F(g(b)) - F(g(a))$$

$$= \Big[F(u)\Big]_{u=g(a)}^{g(b)} = \int_{g(a)}^{g(b)} f(u)\,du.$$

Note how we used the fundamental theorem to obtain the first and last equalities in this argument. ■

Whether it is simpler to apply Theorem 1 and transform to new u-limits or to resubstitute $u = g(x)$ and use the old x-limits depends on the specific problem at hand. Examples 9 and 10 illustrate the technique of transforming to new limits.

EXAMPLE 9 Evaluate

$$\int_3^5 \frac{x\,dx}{(30 - x^2)^2}.$$

Solution Note that $30 - x^2$ is nonzero on $[3, 5]$, so the integrand is continuous there. We substitute

$$u = 30 - x^2, \quad du = -2x\,dx,$$

and observe that:

$$\text{If } x = 3, \quad \text{then } u = 21 \quad \text{(lower limit);}$$

$$\text{If } x = 5, \quad \text{then } u = 5 \quad \text{(upper limit).}$$

Hence our substitution gives

$$\int_3^5 \frac{x\,dx}{(30 - x^2)^2} = \int_{21}^5 \frac{-\frac{1}{2}\,du}{u^2} = -\frac{1}{2}\left[-\frac{1}{u}\right]_{21}^5 = -\frac{1}{2}\left(-\frac{1}{5} + \frac{1}{21}\right) = \frac{8}{105}.$$ ■

EXAMPLE 10 Evaluate

$$\int_0^{\pi/4} \sin^3 2t \cos 2t\,dt.$$

Solution We substitute

$$u = \sin 2t, \quad \text{so} \quad du = 2\cos 2t\,dt.$$

Then $u = 0$ when $t = 0$; $u = 1$ when $t = \pi/4$. Hence

$$\int_0^{\pi/4} \sin^3 2t \cos 2t\,dt = \tfrac{1}{2}\int_0^1 u^3\,du = \tfrac{1}{2}\left[\tfrac{1}{4}u^4\right]_0^1 = \tfrac{1}{8}.$$ ■

5.7 PROBLEMS

In Problems 1 through 10, use the indicated substitution to evaluate the given integral.

1. $\int (3x - 5)^{17}\,dx;\quad u = 3x - 5$
2. $\int \frac{1}{(4x + 7)^6}\,dx;\quad u = 4x + 7$
3. $\int x\sqrt{x^2 + 9}\,dx;\quad u = x^2 + 9$
4. $\int \frac{x^2}{\sqrt[3]{2x^3 - 1}}\,dx;\quad u = 2x^3 - 1$
5. $\int \sin 5x\,dx;\quad u = 5x$
6. $\int \cos kx\,dx;\quad u = kx$
7. $\int x\sin(2x^2)\,dx;\quad u = 2x^2$
8. $\int \sqrt{x}\cos(x^{3/2})\,dx;\quad u = x^{3/2}$
9. $\int (1 - \cos x)^5 \sin x\,dx;\quad u = 1 - \cos x$
10. $\int \frac{\cos 3x}{\sqrt{5 + 2\sin 3x}}\,dx;\quad u = 5 + 2\sin 3x$

Evaluate the indefinite integrals in Problems 11 through 40.

11. $\int (x + 1)^6\,dx$
12. $\int (2 - x)^5\,dx$

13. $\int (4 - 3x)^7\,dx$

14. $\int \sqrt{2x + 1}\,dx$

15. $\int \frac{dx}{\sqrt{7x + 5}}$

16. $\int \frac{dx}{(3 - 5x)^2}$

17. $\int \sin(\pi x + 1)\,dx$

18. $\int \cos\frac{\pi t}{3}\,dt$

19. $\int \sec 2\theta \tan 2\theta\,d\theta$

20. $\int \csc^2 5x\,dx$

21. $\int x\sqrt{x^2 - 1}\,dx$

22. $\int 3t(1 - 2t^2)^{10}\,dt$

23. $\int x\sqrt{2 - 3x^2}\,dx$

24. $\int \frac{t\,dt}{\sqrt{2t^2 + 1}}$

25. $\int x^3\sqrt{x^4 + 1}\,dx$

26. $\int \frac{x^2\,dx}{\sqrt[3]{x^3 + 1}}$

27. $\int x^2 \cos(2x^3)\,dx$

28. $\int t\sec^2(t^2)\,dt$

29. $\int \frac{x^2\,dx}{(x^3 + 5)^4}$

30. $\int y^2\sqrt[3]{2 - 4y^3}\,dy$

31. $\int \cos^3 x \sin x\,dx$

32. $\int \sin^5 3z \cos 3z\,dz$

33. $\int \tan^3\theta \sec^2\theta\,d\theta$

34. $\int \sec^3\theta \tan\theta\,d\theta$

35. $\int \frac{\cos\sqrt{x}}{\sqrt{x}}\,dx$ (*Suggestion:* Try $u = \sqrt{x}$.)

36. $\int \frac{dx}{\sqrt{x}(1 + \sqrt{x})^2}$

37. $\int (x^2 + 2x + 1)^4(x + 1)\,dx$

38. $\int \frac{(x + 2)\,dx}{(x^2 + 4x + 3)^3}$

39. $\int (2 + t^2)\sqrt[3]{6t + t^3}\,dt$

40. $\int \frac{2 - x^2}{(x^3 - 6x + 1)^5}\,dx$

Evaluate the definite integrals in Problems 41 through 54.

41. $\int_1^2 \frac{dt}{(t + 1)^3}$

42. $\int_0^4 \frac{dx}{\sqrt{2x + 1}}$

43. $\int_0^4 x\sqrt{x^2 + 9}\,dx$

44. $\int_1^4 \frac{(1 + \sqrt{x})^4}{\sqrt{x}}\,dx$ (*Suggestion:* Try $u = 1 + \sqrt{x}$.)

45. $\int_0^8 t\sqrt{t + 1}\,dt$ (*Suggestion:* Try $u = t + 1$.)

46. $\int_0^{\pi/2} \sin x \cos x\,dx$

47. $\int_0^{\pi/6} \sin 2x \cos^3 2x\,dx$

48. $\int_0^{\sqrt{\pi}} t\sin\frac{t^2}{2}\,dt$

49. $\int_0^{\pi/2} (1 + 3\sin\theta)^{3/2}\cos\theta\,d\theta$
(*Suggestion:* Try $u = 1 + 3\sin\theta$.)

50. $\int_0^{\pi/2} \sec^2\frac{x}{2}\,dx$

51. $\int_0^4 x\sqrt{4 - x}\,dx$ (*Suggestion:* Try $u = 4 - x$.)

52. $\int_0^{\pi/2} (\cos x)\sqrt{\sin x}\,dx$ (*Suggestion:* Try $u = \sin x$.)

53. $\int_0^1 t^3 \sin \pi t^4\,dt$

54. $\int_{\pi^2/4}^{\pi^2} \frac{\sin\sqrt{x}\cos\sqrt{x}}{\sqrt{x}}\,dx$

Use the half-angle identities

$$\cos^2\theta = \frac{1 + \cos 2\theta}{2} \quad\text{and}\quad \sin^2\theta = \frac{1 - \cos 2\theta}{2}$$

to evaluate the integrals in Problems 55 through 58.

55. $\int \sin^2 x\,dx$

56. $\int \cos^2 x\,dx$

57. $\int_0^{\pi} \sin^2 3t\,dt$

58. $\int_0^1 \cos^2 \pi t\,dt$

Use the identity $1 + \tan^2\theta = \sec^2\theta$ to evaluate the integrals in Problems 59 and 60.

59. $\int \tan^2 x\,dx$

60. $\int_0^{\pi/12} \tan^2 3t\,dt$

61. Substitute $\sin^3 x = (\sin x)(1 - \cos^2 x)$ to show that

$$\int \sin^3 x\,dx = \tfrac{1}{3}\cos^3 x - \cos x + C.$$

62. Evaluate

$$\int_0^{\pi/2} \cos^3 x\,dx$$

by the method of Problem 61.

63. Substitute first $u = \sin\theta$ and then $u = \cos\theta$ to obtain

$$\int \sin\theta\cos\theta\,d\theta = \tfrac{1}{2}\sin^2\theta + C_1 = -\tfrac{1}{2}\cos^2\theta + C_2.$$

Reconcile these results. What is the relation between the constants C_1 and C_2? *Suggestion:* Compare the graphs (on the same screen) of

$$f(x) = \tfrac{1}{2}\sin^2\theta \quad\text{and}\quad g(x) = -\tfrac{1}{2}\cos^2\theta.$$

64. Substitute first $u = \tan\theta$ and then $u = \sec\theta$ to obtain

$$\int \sec^2\theta\tan\theta\,d\theta = \tfrac{1}{2}\tan^2\theta + C_1 = \tfrac{1}{2}\sec^2\theta + C_2.$$

Reconcile these results. What is the relation between the constants C_1 and C_2? *Suggestion:* Compare the graphs (on the same screen) of

$$f(x) = \tfrac{1}{2}\tan^2\theta \quad \text{and} \quad g(x) = \tfrac{1}{2}\sec^2\theta.$$

65. (a) Verify by differentiation that

$$\int \frac{dx}{(1-x)^2} = \frac{x}{1-x} + C_1.$$

(b) Substitute $u = 1 - x$ to show that

$$\int \frac{dx}{(1-x)^2} = \frac{1}{1-x} + C_2.$$

(c) Reconcile the results of parts (a) and (b). *Suggestion:* Compare the graphs (on the same screen) of

$$f(x) = \frac{x}{1-x} \quad \text{and} \quad g(x) = \frac{1}{1-x}.$$

66. (a) Substitute $u = x^2$ and apply part (a) of Problem 65 to show that

$$\int \frac{x\,dx}{(1-x^2)^2} = \frac{x^2}{2(1-x^2)} + C_1.$$

(b) Substitute $u = 1 - x^2$ to show that

$$\int \frac{x\,dx}{(1-x^2)^2} = \frac{1}{2(1-x^2)} + C_2.$$

(c) Reconcile the results of parts (a) and (b). *Suggestion:* Compare the graphs (on the same screen) of

$$f(x) = \frac{x^2}{2(1-x^2)} \quad \text{and} \quad g(x) = \frac{1}{2(1-x^2)}.$$

Problems 67 through 70 deal with even and odd functions. An **even** function f is a function such that

$$f(-x) = f(x)$$

for all x. This means that the graph of $y = f(x)$ is symmetric under reflection across the y-axis (Fig. 5.7.1). Examples of even functions include $f(x) = \cos x$, 1, x^2, x^4, and x^6. An **odd** function f is a function such that

$$f(-x) = -f(x)$$

for all x. This means that the graph of $y = f(x)$ is symmetric under reflections first across the y-axis, then across the x-axis (Fig. 5.7.2). Examples of odd functions are $f(x) = \sin x$, x, x^3, and x^5. Think about the indicated reflections with the (even) cosine function (in Fig. 5.7.3) and the (odd) sine function (in Fig. 5.7.4).

67. See Fig. 5.7.5. If the continuous function f is odd, substitute $u = -x$ into the integral

$$\int_{-a}^{0} f(x)\,dx \quad \text{to show that} \quad \int_{-a}^{a} f(x)\,dx = 0.$$

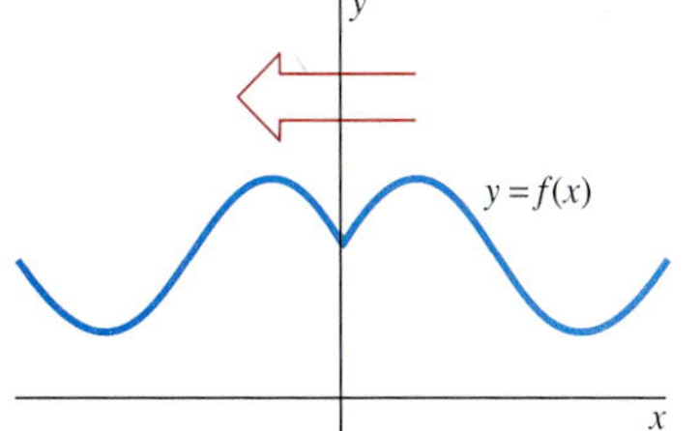

Fig. 5.7.1 The graph of the even function $y = f(x)$ is invariant under reflection through the y-axis.

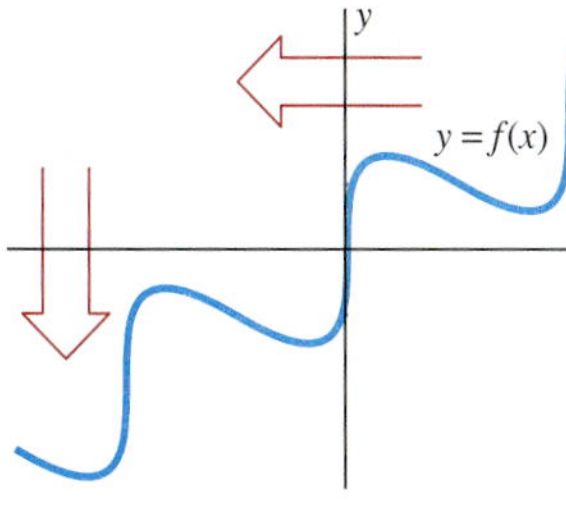

Fig. 5.7.2 The graph of the odd function $y = f(x)$ is invariant under successive reflections through both axes.

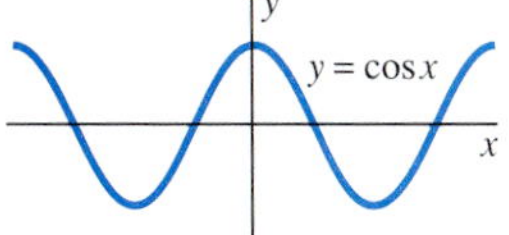

Fig. 5.7.3 The cosine function is even.

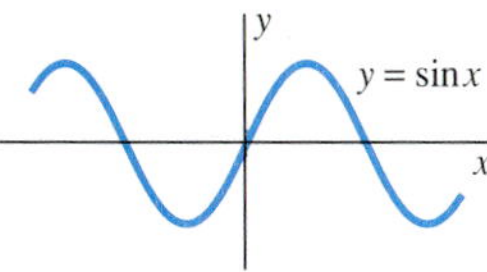

Fig. 5.7.4 The sine function is odd.

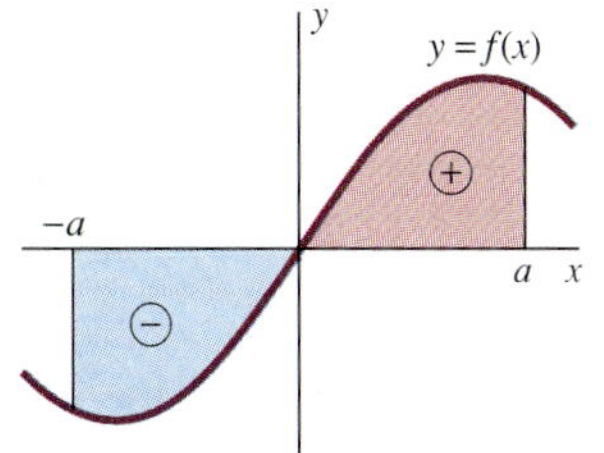

Fig. 5.7.5 Areas cancel when f is odd (Problem 67).

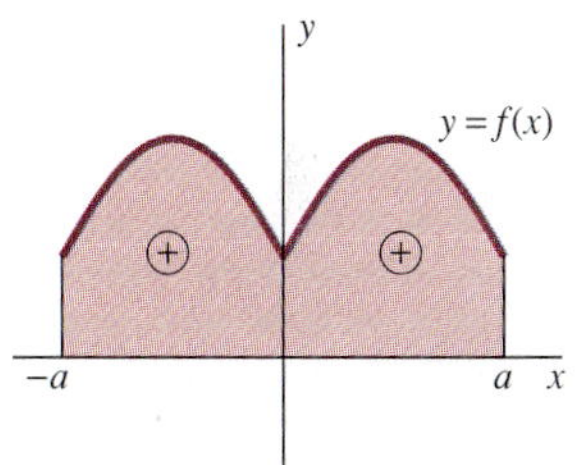

Fig. 5.7.6 Areas add when f is even (Problem 68).

68. See Fig. 5.7.6. If the continuous function f is even, use the method of Problem 67 to show that

$$\int_{-a}^{a} f(x)\,dx = 2\int_{0}^{a} f(x)\,dx.$$

69. Explain without extensive computation why it is evident that

$$\int_{-1}^{1}\left[\tan x + \frac{\sqrt[3]{x}}{(1+x^2)^7} - x^{17}\cos x\right]dx = 0.$$

70. Explain without extensive computation why it is evident that

$$\int_{-5}^{5}\left(3x^2 - x^{10}\sin x + x^5\sqrt{1+x^4}\right)dx = 2\Big[x^3\Big]_0^5 = 250.$$

5.8 AREAS OF PLANE REGIONS

In Section 5.3 we discussed the area A under the graph of a positive-valued continuous function f on the interval $[a, b]$. This discussion motivated our definition in Section 5.4 of the integral of f from a to b as the limit of Riemann sums. An important result was that

$$A = \int_a^b f(x)\,dx, \tag{1}$$

by definition.

Here we consider the problem of finding the areas of more general regions in the coordinate plane. Regions such as those illustrated in Fig. 5.8.1 may be bounded by the graphs of *two* (or more) different functions.

Let f and g be continuous functions such that $f(x) \geqq g(x)$ for all x in the interval $[a, b]$. We are interested in the area A of the region R in Fig. 5.8.2, which lies *between* the graphs of $y = f(x)$ and $y = g(x)$ for x in $[a, b]$. Thus R is bounded by

- The curve $y = f(x)$, the upper boundary of R, by
- The curve $y = g(x)$, the lower boundary of R, *and* by
- The vertical lines $x = a$ and $x = b$ (if needed).

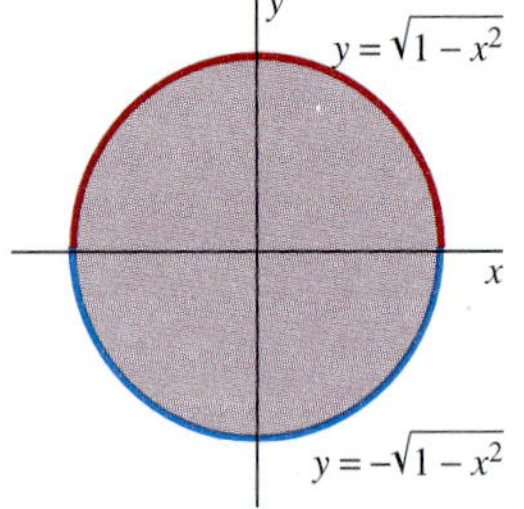

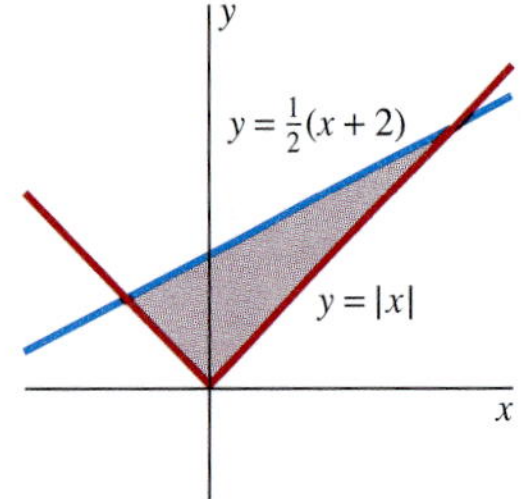

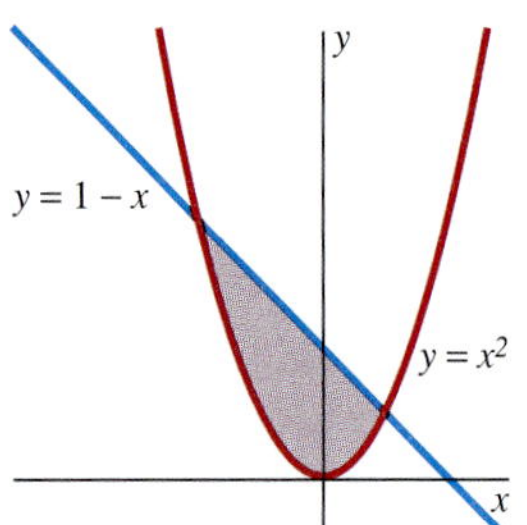

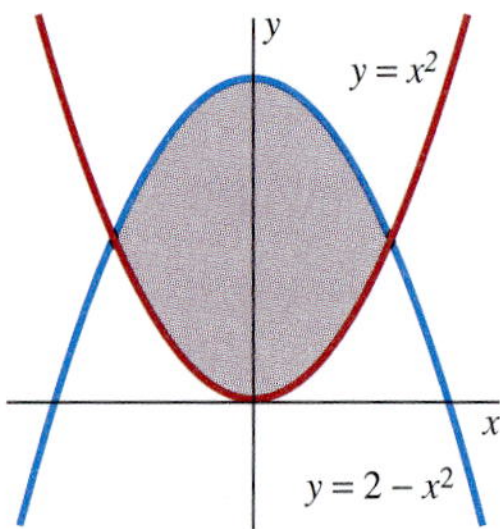

Fig. 5.8.1 Plane regions bounded by pairs of curves.

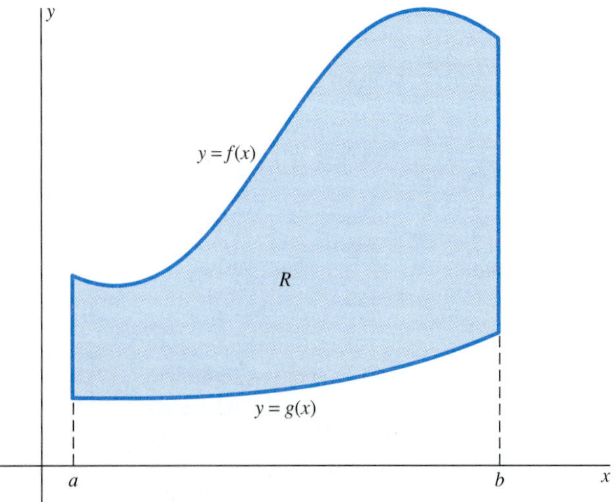

Fig. 5.8.2 A region between two graphs

To approximate A, we consider a partition of $[a, b]$ into n subintervals, all with the same length $\Delta x = (b - a)/n$. If ΔA_i denotes the area of the region between the graphs of f and g over the ith subinterval $[x_{i-1}, x_i]$, and if $x_i^\star$ is a selected number chosen in that subinterval (all this for $i = 1, 2, 3, \dots, n$), then ΔA_i is approximately equal to the area of a rectangle with height $f(x_i^\star) - g(x_i^\star)$ and width Δx (Fig. 5.8.3). Hence

$$\Delta A_i \approx [f(x_i^\star) - g(x_i^\star)]\,\Delta x;$$

so

$$A = \sum_{i=1}^{n} \Delta A_i \approx \sum_{i=1}^{n} [f(x_i^\star) - g(x_i^\star)]\,\Delta x.$$

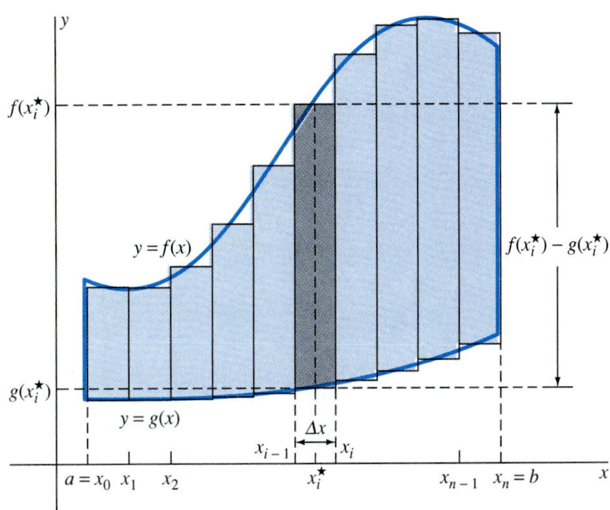

Fig. 5.8.3 A partition of $[a, b]$ divides R into vertical strips that we approximate with rectangular strips.

We introduce the *height* function $h(x) = f(x) - g(x)$ and observe that A is approximated by a Riemann sum for $h(x)$ associated with our partition of $[a, b]$:

$$A \approx \sum_{i=1}^{n} h(x_i^\star)\,\Delta x.$$

Both intuition and reason suggest that this approximation can be made arbitrarily accurate by choosing n to be sufficiently large [and hence $\Delta x = (b - a)/n$ to be sufficiently small]. We therefore conclude that

$$A = \lim_{\Delta x \to 0} \sum_{i=1}^{n} h(x_i^\star)\,\Delta x = \int_a^b h(x)\,dx = \int_a^b [f(x) - g(x)]\,dx.$$

Because our discussion is based on an intuitive concept rather than on a precise logical definition of area, it does not constitute a proof of this area formula. It does, however, provide justification for the following *definition* of the area in question.

Definition ***The Area Between Two Curves***

Let f and g be continuous with $f(x) \geqq g(x)$ for x in $[a, b]$. Then the **area** A of the region bounded by the curves $y = f(x)$ and $y = g(x)$ and by the vertical lines $x = a$ and $x = b$ is

$$A = \int_a^b [f(x) - g(x)]\,dx \qquad \textbf{(2)}$$

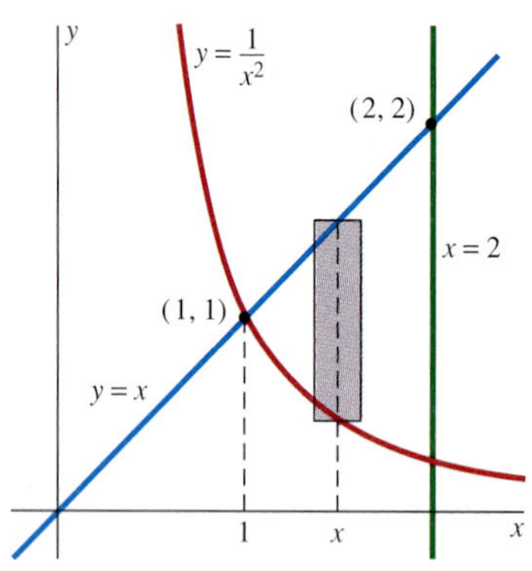

Fig. 5.8.4 The region of Example 1

EXAMPLE 1 Find the area of the region bounded by the lines $y = x$ and $x = 2$ and by the curve $y = 1/x^2$ (Fig. 5.8.4).

Solution Here the top curve is $y = f(x) = x$, the bottom curve is $y = g(x) = 1/x^2$, $a = 1$, and $b = 2$. The vertical line $x = 2$ is "needed" (to form the right-hand boundary of the region), whereas $x = 1$ is not. Equation (2) yields

$$A = \int_1^2 \left(x - \frac{1}{x^2}\right) dx = \left[\frac{1}{2}x^2 + \frac{1}{x}\right]_1^2 = \left(2 + \frac{1}{2}\right) - \left(\frac{1}{2} + 1\right) = 1.$$

Equation (1) is the special case of Eq. (2) in which $g(x)$ is identically zero on $[a, b]$. But if $f(x) \equiv 0$ and $g(x) \leqq 0$ on $[a, b]$, then Eq. (2) reduces to

$$A = -\int_a^b g(x)\,dx; \quad \text{that is,} \quad \int_a^b g(x)\,dx = -A.$$

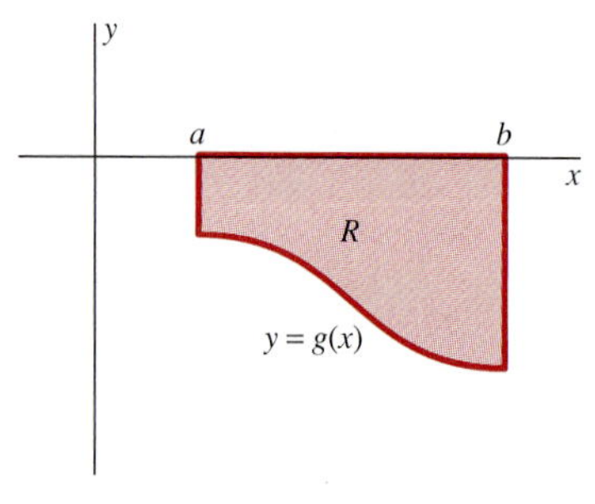

Fig. 5.8.5 The integral gives the negative of the geometric area for a region that lies below the x-axis.

In this case the region R lies beneath the x-axis (Fig. 5.8.5). Thus the integral from a to b of a negative-valued function is the *negative* of the area of the region bounded by its graph, the x-axis, and the vertical lines $x = a$ and $x = b$.

More generally, consider a continuous function f with a graph that crosses the x-axis at finitely many points $c_1, c_2, c_3, \ldots, c_k$ between a and b (Fig. 5.8.6). We write

$$\int_a^b f(x)\,dx = \int_a^{c_1} f(x)\,dx + \int_{c_1}^{c_2} f(x)\,dx + \cdots + \int_{c_k}^b f(x)\,dx.$$

Thus we see that

$$\int_a^b f(x)\,dx$$

is equal to the area below $y = f(x)$ and *above* the x-axis *minus* the area above $y = f(x)$ and *below* the x-axis.

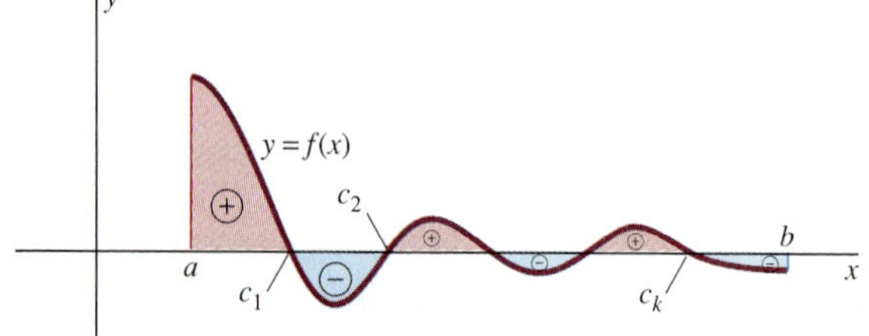

Fig. 5.8.6 The integral computes the area above the x-axis *minus* the area below the x-axis.

The following *heuristic* (suggestive, though nonrigorous) way of setting up integral formulas such as Eq. (2) can be useful. Consider the vertical strip of area that lies above the interval $[x, x + dx]$, shown shaded in Fig. 5.8.7, where we have written

$$y_{\text{top}} = f(x) \quad \text{and} \quad y_{\text{bot}} = g(x)$$

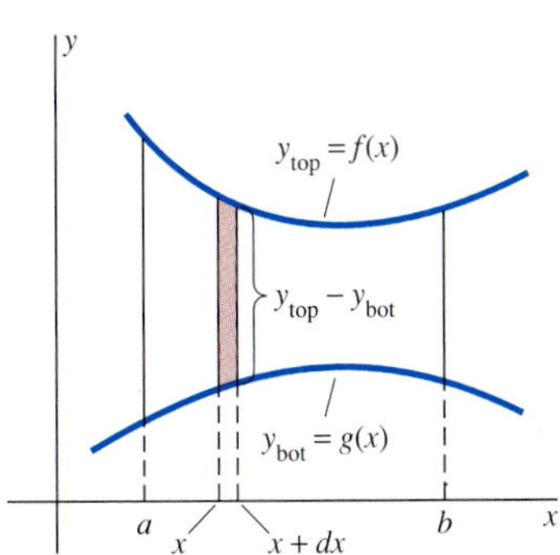

Fig. 5.8.7 Heuristic (suggestive but nonrigorous) approach to setting up area integrals

for the top and bottom boundary curves. We think of the length dx of the interval $[x, x + dx]$ as being so small that we can regard this strip as a rectangle with width dx and height $y_{\text{top}} - y_{\text{bot}}$. Its area is then

$$dA = (y_{\text{top}} - y_{\text{bot}})\,dx.$$

Think now of the region over $[a, b]$ that lies between $y_{\text{top}} = f(x)$ and $y_{\text{bot}} = g(x)$ as being made up of many such vertical strips. We can regard its area as a sum of areas of such rectangular strips. If we write $\int$ for *sum,* we get the formula

$$A = \int dA = \int_a^b (y_{\text{top}} - y_{\text{bot}})\,dx.$$

This heuristic approach bypasses the subscript notation associated with Riemann sums. Nevertheless, it *is not and should not be regarded as a complete* derivation of the last formula. It is best used only as a convenient memory device. For instance, in the figures accompanying examples, we shall often show a strip of width dx as a visual aid in properly setting up the correct integral.

EXAMPLE 2 Find the area A of the region R bounded by the line $y = x$ and the parabola $y = 6 - x^2$.

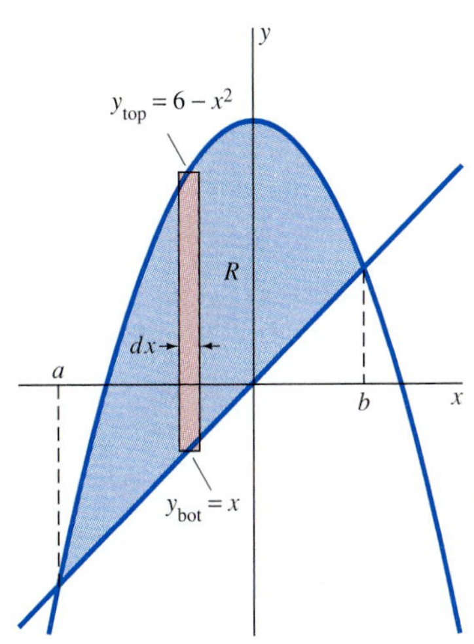

Fig. 5.8.8 The region R of Example 2

Solution The region R is shown in Fig. 5.8.8. We can use Eq. (2) and take $f(x) = 6 - x^2$ and $g(x) = x$. The limits a and b will be the x-coordinates of the two points of intersection of the line and the parabola; our first order of business is to find a and b. To do so, we equate $f(x)$ and $g(x)$ and solve the resulting equation for x:

$$x = 6 - x^2; \qquad x^2 + x - 6 = 0; \qquad (x - 2)(x + 3) = 0; \qquad x = -3, 2.$$

Thus $a = -3$ and $b = 2$, so Eq. (2) gives

$$A = \int_{-3}^{2} (6 - x^2 - x)\,dx = \left[6x - \tfrac{1}{3}x^3 - \tfrac{1}{2}x^2\right]_{-3}^{2}$$
$$= \left[6\cdot 2 - \tfrac{1}{3}\cdot 2^3 - \tfrac{1}{2}\cdot 2^2\right] - \left[6\cdot(-3) - \tfrac{1}{3}\cdot(-3)^3 - \tfrac{1}{2}\cdot(-3)^2\right] = \tfrac{125}{6}.$$ ■

Subdividing Regions Before Integrating

Example 3 shows that it is sometimes necessary to subdivide a region before applying Eq. (2), typically because the formula for either the top or the bottom boundary curve (or both) changes somewhere between $x = a$ and $x = b$.

EXAMPLE 3 Find the area of the region R bounded by the line $y = \frac{1}{2}x$ and the parabola $y^2 = 8 - x$.

Solution The region R is shown in Fig. 5.8.9. The points of intersection $(-8, -4)$ and $(4, 2)$ are found by equating $y = \frac{1}{2}x$ and $y = \pm\sqrt{8 - x}$ and then solving for x. The lower boundary of R is given by $y_{\text{bot}} = -\sqrt{8 - x}$ on $[-8, 8]$. But the upper boundary of R is given by

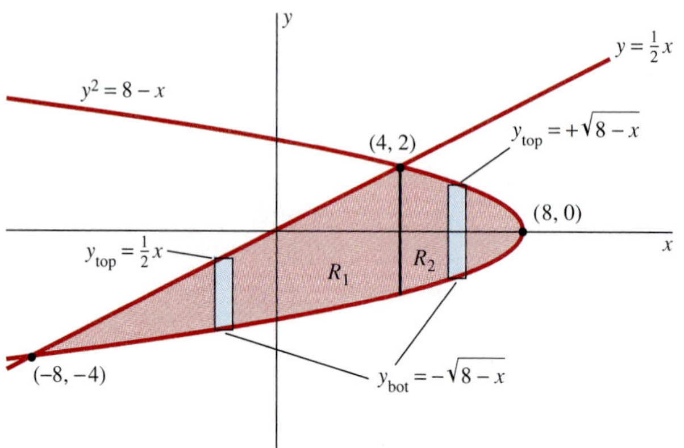

Fig. 5.8.9 In Example 3, we split region R into two regions R_1 and R_2.

$$y_{top} = \tfrac{1}{2}x \quad \text{on} \quad [-8, 4], \qquad y_{top} = +\sqrt{8-x} \quad \text{on} \quad [4, 8].$$

We must therefore divide R into the two regions R_1 and R_2, as indicated in Fig. 5.8.9. Then Eq. (2) gives

$$\begin{aligned} A &= \int_{-8}^{4} \left(\tfrac{1}{2}x + \sqrt{8-x}\right) dx + \int_{4}^{8} 2\sqrt{8-x}\, dx \\ &= \left[\tfrac{1}{4}x^2 - \tfrac{2}{3}(8-x)^{3/2}\right]_{-8}^{4} + \left[-\tfrac{4}{3}(8-x)^{3/2}\right]_{4}^{8} \\ &= \left[\left(\tfrac{16}{4} - \tfrac{16}{3}\right) - \left(\tfrac{64}{4} - \tfrac{128}{3}\right)\right] + \left[0 + \tfrac{32}{3}\right] = 36. \end{aligned}$$

■

Determining Area by Integrating with Respect to y

The region in Example 3 appears to be simpler if we consider it to be bounded by graphs of functions of y rather than by graphs of functions of x. Figure 5.8.10 shows a region R bounded by the curves $x = f(y)$ and $x = g(y)$, with $f(y) \geqq g(y)$ for y in $[c, d]$, and by the horizontal lines $y = c$ and $y = d$. To approximate the area A of R, we begin with a partition of $[c, d]$ into n subintervals, all with the same length $\Delta y = (d - c)/n$. We choose a point $y_i^\star$ in the ith subinterval $[y_{i-1}, y_i]$ for each i $(1 \leqq i \leqq n)$. The horizontal strip of R lying opposite $[y_{i-1}, y_i]$ is approximated by a rectangle with width Δy (measured vertically) and height $f(y_i^\star) - g(y_i^\star)$ (measured horizontally). Hence

$$A \approx \sum_{i=1}^{n} [f(y_i^\star) - g(y_i^\star)]\,\Delta y.$$

Recognition of the sum as a Riemann sum for the integral

$$\int_{c}^{d} [f(y) - g(y)]\, dy$$

motivates the following definition.

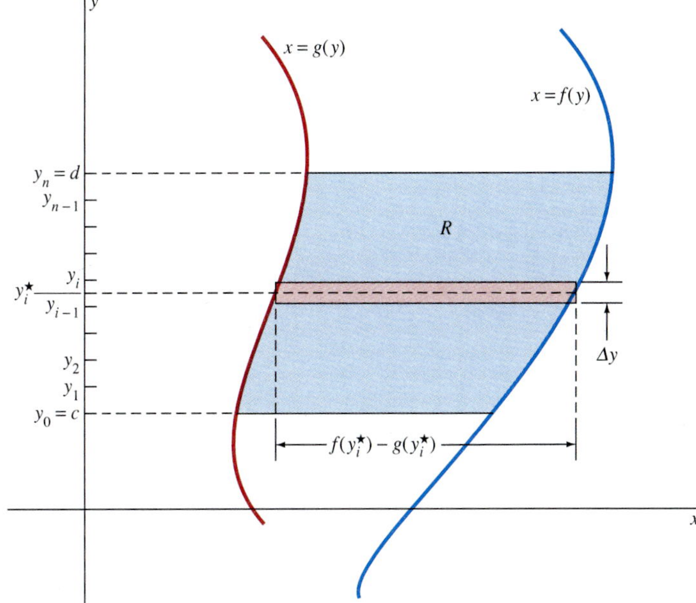

Fig. 5.8.10 Find area by using an integral with respect to y

Definition *The Area Between Two Curves*

Let f and g be continuous functions of y with $f(y) \geqq g(y)$ for y in $[c, d]$. Then the **area** A of the region bounded by the curves $x = f(y)$ and $x = g(y)$ and by the horizontal lines $y = c$ and $y = d$ is

$$A = \int_c^d [f(y) - g(y)]\,dy \tag{3}$$

In a more advanced course, we would now prove that Eqs. (2) and (3) yield the same area A for a region that can be described both in the manner shown in Fig. 5.8.2 and in the manner shown in Fig. 5.8.10.

Let us write

$$x_{\text{right}} = f(y) \quad \text{and} \quad x_{\text{left}} = g(y)$$

for the right and left boundary curves, respectively, of the region in Fig. 5.8.10. Then Eq. (3) takes the form

$$A = \int_c^d [x_{\text{right}} - x_{\text{left}}]\,dy.$$

Comparison of Example 3 with Example 4 illustrates the advantage of choosing the "right" variable of integration—the one that makes the resulting computations simpler.

EXAMPLE 4 Integrate with respect to y to find the area of the region R of Example 3.

Solution We see from Fig. 5.8.11 that Eq. (3) applies with $x_{\text{right}} = f(y) = 8 - y^2$ and $x_{\text{left}} = g(y) = 2y$ for y in $[-4, 2]$. This gives

$$A = \int_{-4}^{2} [(8 - y^2) - 2y]\,dy = \left[8y - \tfrac{1}{3}y^3 - y^2\right]_{-4}^{2} = 36.$$

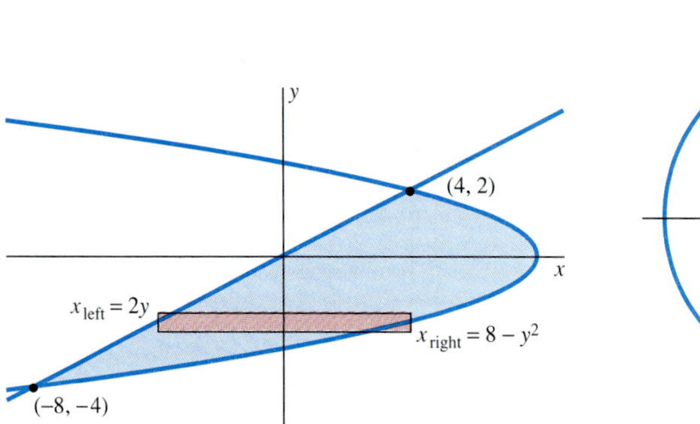

Fig. 5.8.11 Recomputation of the area of Example 3 (Example 4)

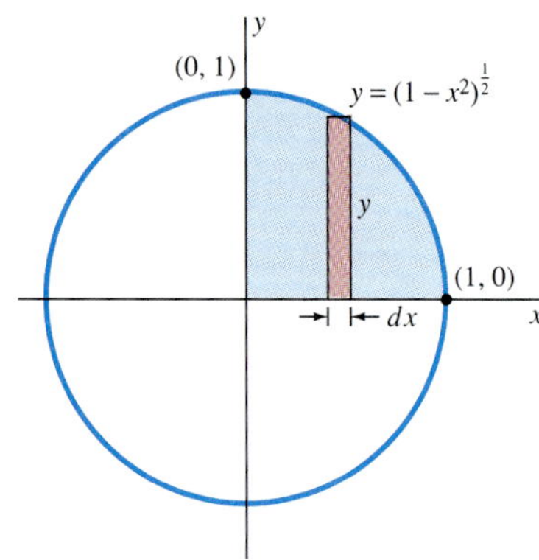

Fig. 5.8.12 The number π is four times the shaded area.

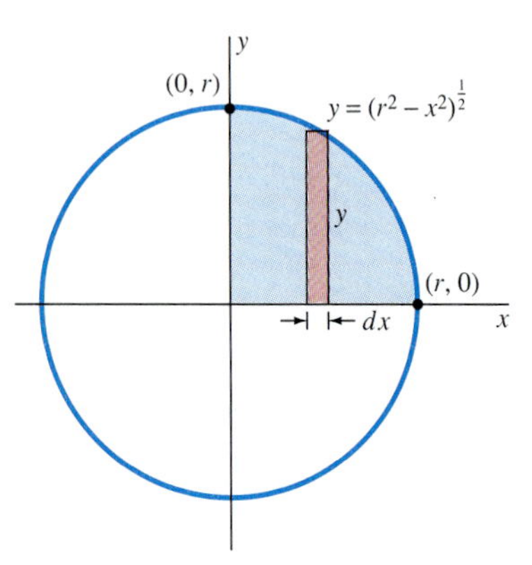

Fig. 5.8.13 The shaded area can be expressed as an integral.

EXAMPLE 5 Use calculus to derive the formula $A = \pi r^2$ for the area of a circle of radius r.

Solution We begin with the *definition* (in Section 5.3) of the number π as the area of the unit circle $x^2 + y^2 = 1$. Then, with the aid of Fig. 5.8.12, we write

$$\pi = 4\int_0^1 \sqrt{1 - x^2}\,dx, \tag{4}$$

because the integral in Eq. (4) is, by Eq. (1), the area of the first quadrant of the circle. We apply Eq. (1) to the first quadrant of the circle of radius r in Fig. 5.8.13, and we find the total area A of that circle to be

$$\begin{aligned} A &= 4\int_0^r \sqrt{r^2 - x^2}\,dx = 4r\int_0^r \sqrt{1 - (x/r)^2}\,dx \\ &= 4r\int_0^1 \sqrt{1 - u^2}\,r\,du \qquad \left(\textit{Substitution: } u = \frac{x}{r},\ dx = r\,du.\right) \\ &= 4r^2\int_0^1 \sqrt{1 - u^2}\,du. \end{aligned}$$

Therefore, by Eq. (4) (with u in place of x), $A = \pi r^2$.

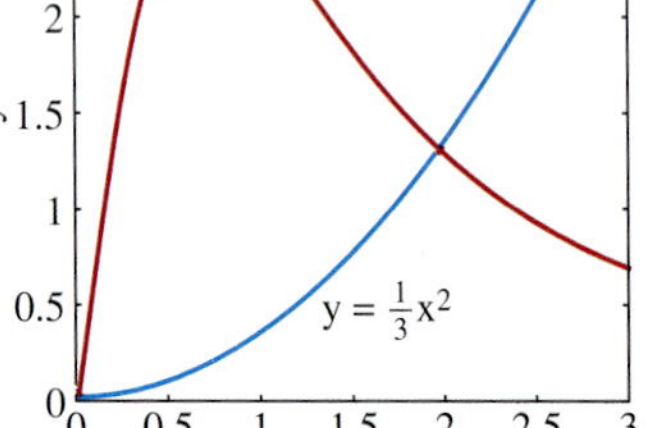

Fig. 5.8.14 The region of Example 6

EXAMPLE 6 Approximate the area A of the first-quadrant region shown in Fig. 5.8.14. This is the region bounded by the curves

$$y = \frac{7x}{(x^2 + 1)^{3/2}} \quad \text{and} \quad y = \frac{x^2}{3}.$$

Solution To find the exact coordinates of the first-quadrant intersection point in Fig. 5.8.14, we would need to solve the equation

$$\frac{7x}{(x^2+1)^{3/2}} = \frac{x^2}{3}.$$

We might begin by first canceling x, then cross-multiplying and squaring both sides. The result, as you can verify, simplifies to the equation

$$x^8 + 3x^6 + 3x^4 + x^2 - 441 = 0.$$

Now you see why we asked for the *approximate* area. Although it is impractical to solve this eighth-degree equation exactly, we can use a graphing calculator or computer to zoom in on the desired intersection point shown in Fig. 5.8.14. In this way we find that $x \approx 1.963$ is its approximate x-coordinate. (We could also use Newton's method or a calculator root-finder.) We can now proceed to approximate the desired area.

$$\begin{aligned} A &\approx \int_0^{1.963} \left(\frac{7x}{(x^2+1)^{3/2}} - \frac{x^2}{3} \right) dx \\ &= \int_0^{1.963} \frac{7x}{(x^2+1)^{3/2}}\,dx - \int_0^{1.963} \frac{x^2}{3}\,dx. \end{aligned}$$

We substitute $u = x^2 + 1$, $du = 2x\,dx$ to evaluate the first integral and get

$$A \approx \left[\frac{-7}{\sqrt{x^2+1}} - \frac{x^3}{9} \right]_0^{1.963} \approx -4.018 - (-7) = 2.982.$$

5.8 PROBLEMS

Find the areas of the regions shown in Problems 1 through 10.

1. (See Fig. 5.8.15.)
2. (See Fig. 5.8.16.)
3. (See Fig. 5.8.17.)
4. (See Fig. 5.8.18.)
5. (See Fig. 5.8.19.)
6. (See Fig. 5.8.20.)
7. (See Fig. 5.8.21.)
8. (See Fig. 5.8.22.)
9. (See Fig. 5.8.23.)
10. (See Fig. 5.8.24.)

Find the areas of the regions described in Problems 11 through 20.

11. The region R bounded below by the graph of $y = x^3$ and above by the graph of $y = x$ over the interval $[0, 1]$

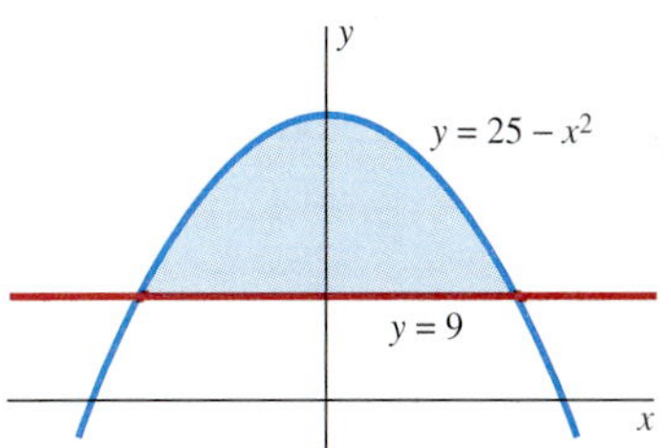

Fig. 5.8.15 Problem 1

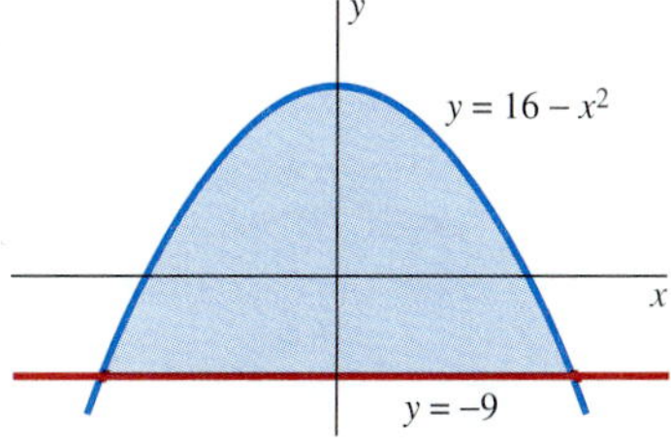

Fig. 5.8.16 Problem 2

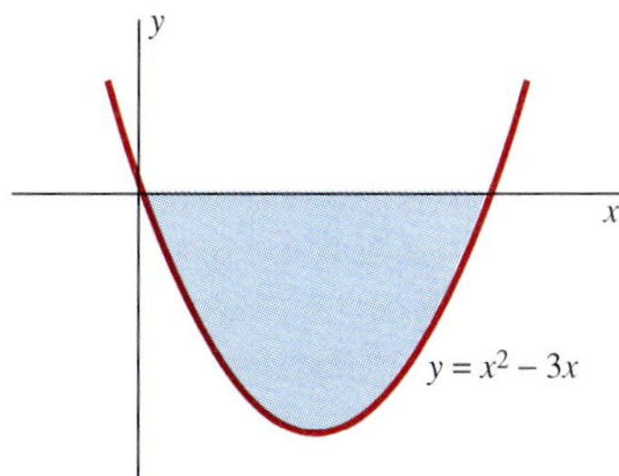

Fig. 5.8.17 Problem 3

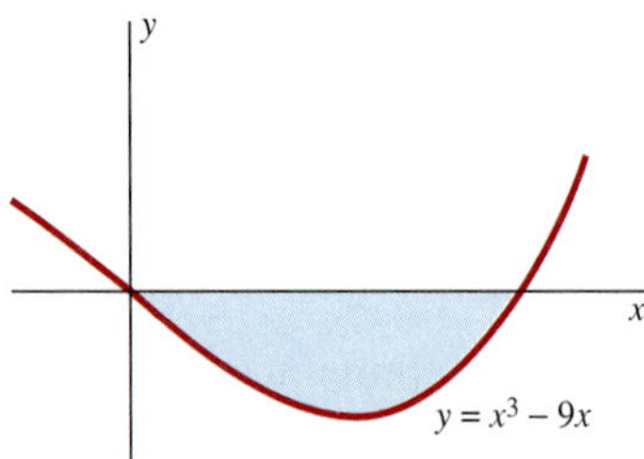

Fig. 5.8.18 Problem 4

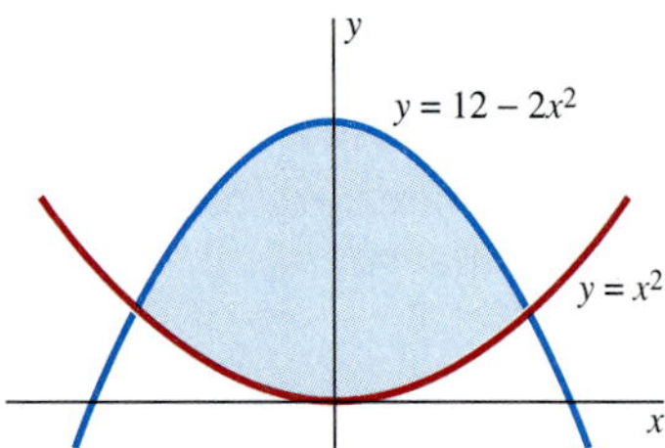

Fig. 5.8.19 Problem 5

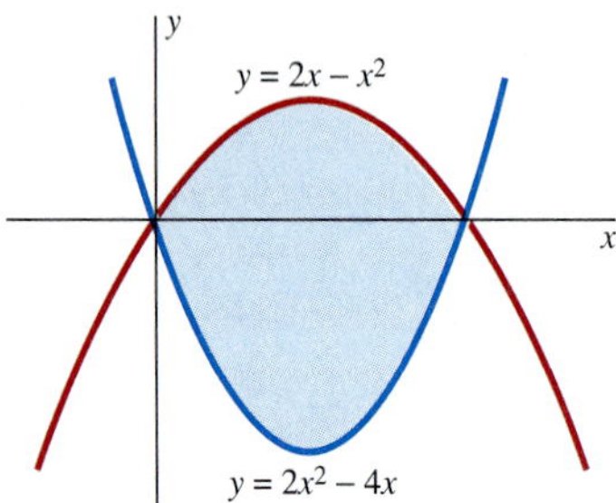

Fig. 5.8.20 Problem 6

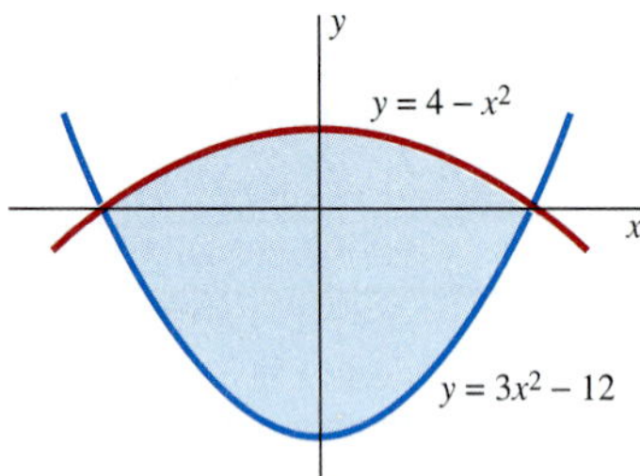

Fig. 5.8.21 Problem 7

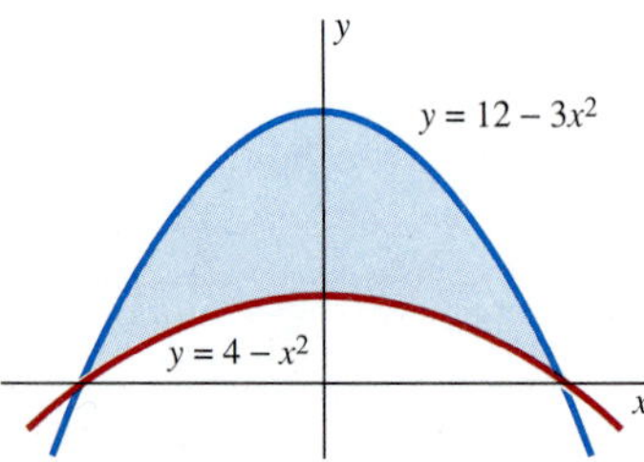

Fig. 5.8.22 Problem 8

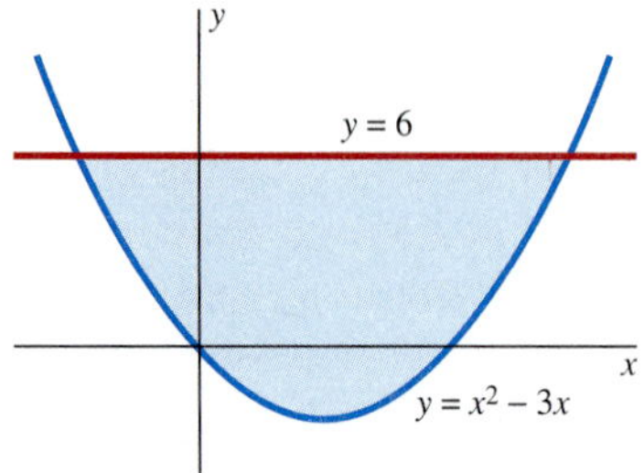

Fig. 5.8.23 Problem 9

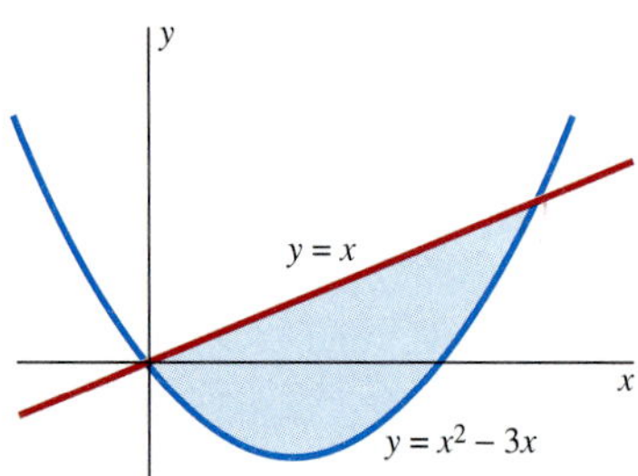

Fig. 5.8.24 Problem 10

12. The region R between the graph of $y = 1/(x+1)^2$ and the x-axis over the interval $[1, 3]$

13. The region R bounded above by the graph of $y = x^3$ and below by the graph of $y = x^4$ over the interval $[0, 1]$

14. The region R bounded above by the graph of $y = x^2$ and below by the horizontal line $y = -1$ over the interval $[-1, 2]$

15. The region R bounded above by the graph of $y = 1/(x+1)^3$ and below by the x-axis over the interval $[0, 2]$

16. The region R bounded above by the graph of $y = 4x - x^2$ and below by the x-axis

17. The region R bounded on the left by the graph of $x = y^2$ and on the right by the vertical line $x = 4$

18. The region R between the graphs of $y = x^4 - 4$ and $y = 3x^2$

19. The region R between the graphs of $x = 8 - y^2$ and $x = y^2 - 8$

20. The region R between the graphs of $y = x^{1/3}$ and $y = x^3$

In Problems 21 through 40, sketch the region bounded by the given curves, then find its area.

21. $y = x^2,\quad y = 2x$

22. $y = x^2,\quad y = 8 - x^2$

23. $x = y^2,\quad x = 25$

24. $x = y^2,\quad x = 32 - y^2$

25. $y = x^2,\quad y = 2x + 3$

26. $y = x^2,\quad y = 2x + 8$

27. $x = y^2,\quad x = y + 6$

28. $x = y^2,\quad x = 8 - 2y$

29. $y = x^2,\quad x = y^2$

30. $y = x^2,\quad y = x^3$

31. $x = 4y^2,\quad x + 12y + 5 = 0$

32. $y = x^2,\quad y = 3 + 5x - x^2$

33. $x = 3y^2,\quad x = 12y - y^2 - 5$

34. $y = x^2,\quad y = 4(x - 1)^2$

35. $x = y^2 - 2y - 2, \quad x = 4 + y - 2y^2$

36. $y = x^4, \quad y = 32 - x^4$ **37.** $y = x^3, \quad y = 32\sqrt{x}$

38. $y = x^3, \quad y = 2x - x^2$ **39.** $y = x^2, \quad y = \sqrt[3]{x^2}$

40. $y^2 = x, \quad y^2 = 2(x - 3)$

In Problems 41 and 42, first use a calculator or computer to graph the given curves $y = f(x)$ and $y = g(x)$. You should then be able to find the coordinates of the points of intersection that will be evident in your figure. Finally, find the area of the region bounded by the two curves. Problems 43 and 44 are similar, except that the two curves bound *two* regions; find the sum of the areas of these two regions.

41. $y = x^2 - x, \quad y = 1 - x^3$ **42.** $y = x^3 - x, \quad y = 1 - x^4$

43. $y = x^2, \quad y = x^3 - 2x$

44. $y = x^3, \quad y = 2x^3 + x^2 - 2x$

45. Evaluate

$$\int_{-3}^{3} (4x + 5)\sqrt{9 - x^2}\, dx$$

by writing this integral as a sum of two integrals and interpreting one of them in terms of a known (circular) area.

46. Evaluate

$$\int_{0}^{3} x\sqrt{81 - x^4}\, dx$$

by making a substitution of the form $u = x^p$ (you choose p) and then interpreting the result in terms of a known area.

47. The *ellipse* $x^2/a^2 + y^2/b^2 = 1$ is shown in Fig. 5.8.25. Use the method of Example 5 to show that the area of the region it bounded is $A = \pi ab$, a pleasing generalization of the area formula for the circle.

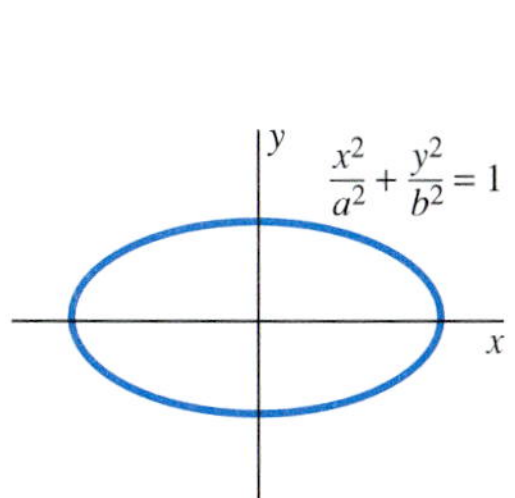

Fig. 5.8.25 The ellipse of Problem 47

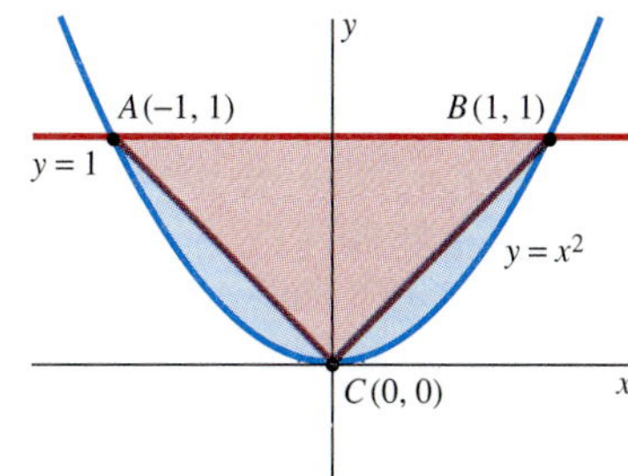

Fig. 5.8.26 The parabolic segment of Problem 48

48. Figure 5.8.26 shows a *parabolic segment* bounded by the parabola $y = x^2$ and the line $y = 1$. In the third century B.C., Archimedes showed that the area of a parabolic segment is four-thirds the area of the triangle ABC where AB is the "base" of the parabolic segment and C is its vertex (as in Fig. 5.8.26). Verify this for the indicated parabolic segment.

49. Let A and B be the points of intersection of the parabola $y = x^2$ and the line $y = x + 2$, and let C be the point on the parabola where the tangent line is parallel to the graph of $y = x + 2$. Show that the area of the parabolic segment cut from the parabola by the line (Fig. 5.8.27) is four-thirds the area of the triangle ABC.

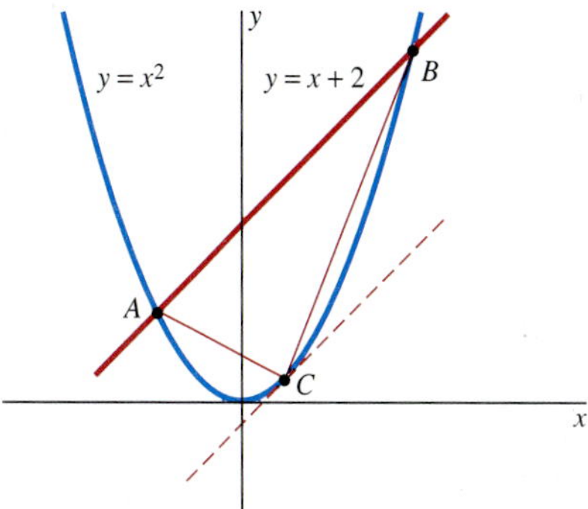

Fig. 5.8.27 The parabolic segment of Problem 49

50. Find the area of the unbounded region R shaded in Fig. 5.8.28—regard it as the limit as $b \to \infty$ of the region bounded by $y = 1/x^2$, $y = 0$, $x = 1$, and $x = b > 1$.

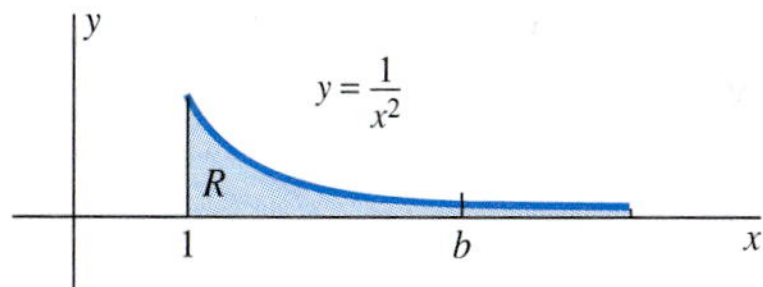

Fig. 5.8.28 The unbounded region of Problem 50

51. Find the total area of the bounded regions that are bounded by the x-axis and the curve $y = 2x^3 - 2x^2 - 12x$.

52. Suppose that the quadratic function

$$f(x) = px^2 + qx + r$$

is never negative on $[a, b]$. Show that the area under the graph of f from a to b is $A = \frac{1}{3}h[f(a) + 4f(m) + f(b)]$, where $h = (b - a)/2$ and $m = (a + b)/2$. (*Suggestion:* By a horizontal translation of this region, you may assume that $a = -h$, $m = 0$, and $b = h$.)

The curves defined in Problems 53 and 54 include the loops shown in Figs. 5.8.29 and 5.8.30. Find the area of the region bounded by each loop.

53. $y^2 = x(5 - x)^2$ **54.** $y^2 = x^2(x + 3)$

In Problems 55 through 58, use a calculator to approximate (graphically or otherwise) the points of intersection of the two given curves. Then integrate to find (approximately) the area of the region bounded by these curves.

55. $y = x^2, \quad y = \cos x$ **56.** $y = x^2 - 2x, \quad y = \sin x$

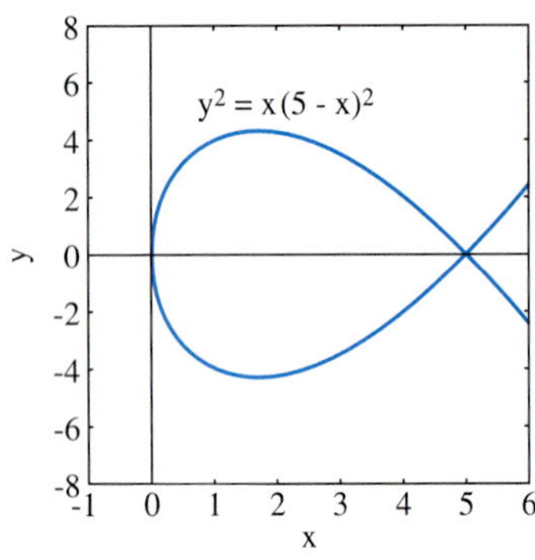

Fig. 5.8.29 The region of Problem 53

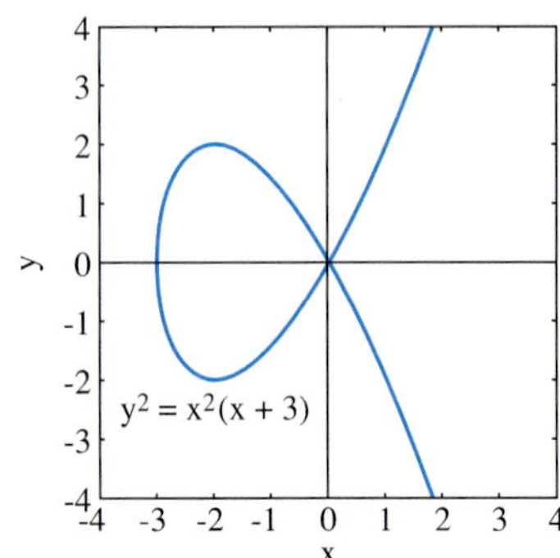

Fig. 5.8.30 The region of Problem 54

57. $y = x^2 - 1, \quad y = \dfrac{1}{1 + x^2}$

58. $y = x^4 - 16, \quad y = 2x - x^2$

59. Find a number $k > 0$ such that the area bounded by the curves $y = x^2$ and $y = k - x^2$ is 72.

60. Find a number $k > 0$ such that the line $y = k$ divides the region between the parabola $y = 100 - x^2$ and the x-axis into two regions having equal areas.

5.8 PROJECT: APPROXIMATE AREA CALCULATIONS

These problems require the use of a calculator or computer that provides a facility for the numerical approximation of an integral

$$\int_a^b f(x)\,dx,$$

in which the integrand function $f(x)$ and the limits a and b are given. The table in Fig. 5.8.31 lists the appropriate commands for common calculator and computer systems.

Calculator/Computer	Numerical Integration Command
Derive	**Author f(x), Calculus, Integrate, Enter Limits, Simplify**
HP-48	**'∫(a, b, f(x), x)'**
Mathematica	**NIntegrate[f[x], {x, a, b}]**
Maple	**evalf(Int(f(x), x = a..b))**
TI-85	**fnInt(a, b, f(x), x)**

Fig. 5.8.31 Calculator and computer commands for numerical integration

For your own versions of Problems 1, 2, and 3, choose a fixed integer n between 1 and 9 and use it throughout. For instance, n could be the last nonzero digit of your student I.D. number.

Fig. 5.8.32 The regions of Problem 1

Problem 1: Let the two functions f and g be defined as follows:

$$f(x) = x, \tag{1}$$

$$g(x) = x(x - n - 1)^2. \tag{2}$$

As illustrated in Fig. 5.8.32 (for the case $n = 4$), the graphs $y = f(x)$ and $y = g(x)$ bound two regions, R_1 and R_2. You are to find the sum A of the areas A_1 and A_2 of these two regions.

First solve manually (that is, by ordinary pencil-and-paper computations) for the x-coordinates of the three points of intersection of $y = f(x)$ and $y = g(x)$. Then use your calculator or computer to evaluate the two integrals required to find A numerically. Also, calculate A_1 and A_2 exactly (by using the fundamental theorem of calculus) and see whether your results agree.

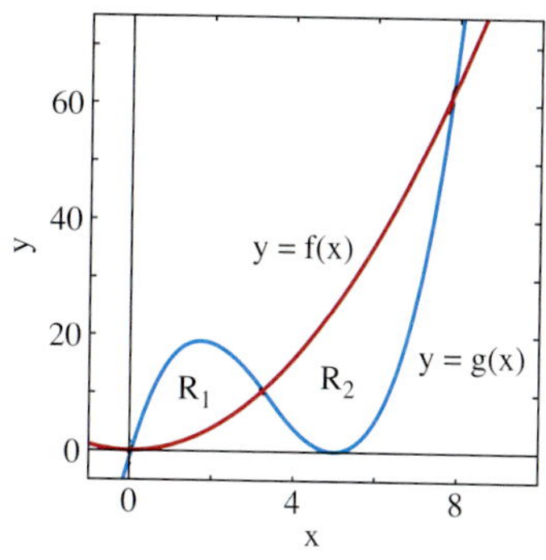

Fig. 5.8.33 The regions of Problem 2

PROBLEM 2 Repeat Problem 1, but replace the function $f(x)$ in Eq. (1) with

$$f(x) = x^2. \tag{3}$$

As before, you can solve manually for the three points of intersection of $y = f(x)$ and $y = g(x)$ (shown in Fig. 5.8.33 for the case $n = 4$), but you surely would not want to calculate the areas A_1 and A_2 manually. Instead, use your calculator or computer to evaluate the necessary integrals numerically.

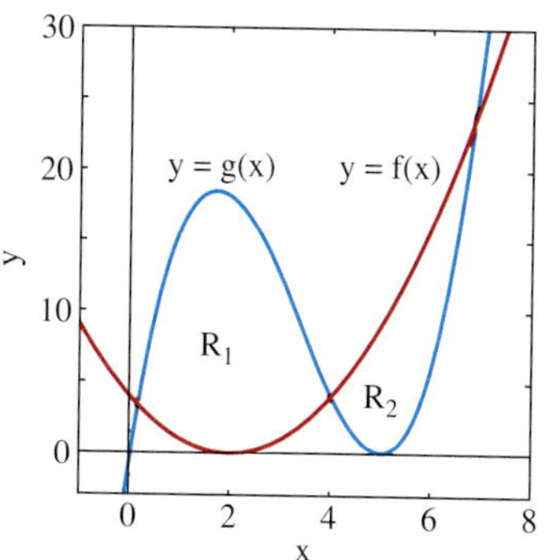

Fig. 5.8.34 The regions of Problem 3

PROBLEM 3 Repeat Problem 1, but replace the function $f(x)$ in Eq. (1) with

$$f(x) = \left(x - \frac{n}{2}\right)^2. \tag{4}$$

Once again a plot (Fig. 5.8.34 illustrates the case $n = 4$) shows two regions R_1 and R_2 bounded by $y = f(x)$ and $y = g(x)$. But now you would *not* be able to solve manually for the three points of intersection that provide the needed limits for the area integrals. Instead, use graphical solution, Newton's method, or the **SOLVE** command on your calculator or computer. Finally, find the numerical values of the areas A_1 and A_2 and their sum A.

5.9 NUMERICAL INTEGRATION

The fundamental theorem of calculus,

$$\int_a^b f(x)\,dx = \Big[\,G(x)\,\Big]_a^b,$$

can be used to evaluate an integral only if a convenient formula for the antiderivative G of f can be found. But there are simple functions with antiderivatives that are not elementary functions. An **elementary function** is one that can be expressed in terms of polynomial, trigonometric, exponential, or logarithmic functions by means of finite combinations of sums, differences, products, quotients, roots, and function composition.

The problem is that elementary functions can have nonelementary antiderivatives. For example, it is known that the elementary function $f(x) = (1 + x^2)^{1/3}$ has no elementary antiderivative. Consequently, we cannot use the fundamental theorem of calculus to evaluate an integral such as

$$\int_0^1 (1 + x^2)^{1/3}\,dx.$$

Here we discuss the use of Riemann sums to approximate numerically integrals that cannot conveniently be evaluated exactly, whether or not nonelementary functions are involved. Given a continuous function f on $[a, b]$ with an integral to be approximated, consider a partition of $[a, b]$ into n subintervals, each with the same length $\Delta x = (b - a)/n$. Then the value of any Riemann sum of the form

$$S = \sum_{i=1}^{n} f(x_i^\star)\,\Delta x \tag{1}$$

may be taken to be an approximation to the value of the integral $\int_a^b f(x)\,dx$.

With $x_i^\star = x_{i-1}$ and with $x_i^\star = x_i$ in Eq. (1), we get the *left-endpoint approximation* L_n and the *right-endpoint approximation* R_n to the definite integral $\int_a^b f(x)\,dx$ associated with the partition of $[a, b]$ into n equal-length subintervals. Thus

$$L_n = \sum_{i=1}^{n} f(x_{i-1})\,\Delta x \tag{2}$$

and

$$R_n = \sum_{i=1}^{n} f(x_i)\,\Delta x. \tag{3}$$

We can simplify the notation for L_n and R_n by writing y_i for $f(x_i)$ (Fig. 5.9.1).

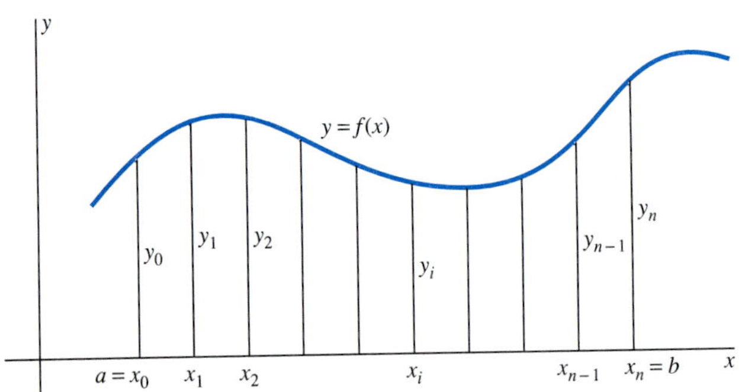

Fig. 5.9.1 $y_i = f(x_i)$

Definition ***Endpoint Approximations***

The **left-endpoint approximation** L_n and the **right-endpoint approximation** R_n to $\int_a^b f(x)\,dx$ with $\Delta x = (b - a)/n$ are

$$L_n = (\Delta x)(y_0 + y_1 + y_2 + \cdots + y_{n-1}) \tag{2$'$}$$

and

$$R_n = (\Delta x)(y_1 + y_2 + y_3 + \cdots + y_n). \tag{3$'$}$$

In Example 1 of Section 5.3 we calculated the left- and right-endpoint approximations to the integral

$$\int_0^3 x^2\,dx = 9 \tag{4}$$

with $n = 5$ and $n = 10$. The table in Fig. 5.9.2 shows values of L_n and R_n with larger values of n.

n	L_n	R_n	$\frac{1}{2}(L_n + R_n)$
5	6.4800	11.8800	9.1800
10	7.6950	10.3950	9.0450
20	8.3363	9.6863	9.0113
40	8.6653	9.3403	9.0028
80	8.8320	9.1695	9.0007
160	8.9258	9.0846	9.0002
320	9.9579	9.0422	9.0000

Fig. 5.9.2 Left- and right-endpoint approximations to the integral in Eq. (4)

The final column of this table gives the *average* of the endpoint sums L_n and R_n. It is apparent that (for a given value of n) this average is a considerably more accurate approximation to the integral than is either one-sided approximation by itself.

The Trapezoidal and Midpoint Approximations

The average $T_n = (L_n + R_n)/2$ of the left-endpoint and right-endpoint approximations is called the *trapezoidal approximation* to $\int_a^b f(x)\,dx$ associated with the partition of $[a, b]$ into n equal-length subintervals. Written in full,

$$\begin{aligned}
T_n &= \frac{1}{2}(L_n + R_n) \\
&= \frac{\Delta x}{2}\sum_{i=1}^{n}[f(x_{i-1}) + f(x_i)] \\
&= \frac{\Delta x}{2}\{[f(x_0) + f(x_1)] + [f(x_1) + f(x_2)] + [f(x_2) + f(x_3)] + \cdots \\
&\qquad + [f(x_{n-2}) + f(x_{n-1})] + [f(x_{n-1}) + f(x_n)]\};
\end{aligned}$$

that is,

$$T_n = \frac{\Delta x}{2}[f(x_0) + 2f(x_1) + 2f(x_2) + \cdots + 2f(x_{n-2}) + 2f(x_{n-1}) + f(x_n)]. \tag{5}$$

Note the 1-2-2- $\cdots$ -2-2-1 pattern of the coefficients.

Definition *The Trapezoidal Approximation*

The **trapezoidal approximation** to

$$\int_a^b f(x)\,dx \quad \text{with} \quad \Delta x = \frac{b-a}{n}$$

is

$$T_n = \frac{\Delta x}{2}(y_0 + 2y_1 + 2y_2 + \cdots + 2y_{n-2} + 2y_{n-1} + y_n). \tag{6}$$

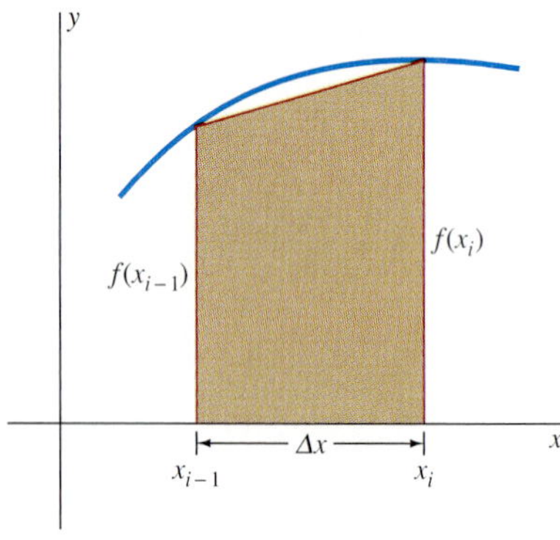

Fig. 5.9.3 The area of the trapezoid is $\frac{1}{2}[f(x_{i-1})+f(x_i)]\,\Delta x$.

Figure 5.9.3 shows where the trapezoidal approximation gets its name. The partition points $x_0, x_1, x_2, \ldots, x_n$ are used to build trapezoids from the x-axis to the graph of the function f. The trapezoid over the ith subinterval $[x_{i-1}, x_i]$ has height Δx, and its parallel bases have widths $f(x_{i-1})$ and $f(x_i)$. So its area is

$$\frac{\Delta x}{2}[f(x_{i-1}) + f(x_i)] = \frac{\Delta x}{2}(y_{i-1} + y_i).$$

Comparison of this with Eq. (6) shows that T_n is merely the sum of the areas of the n trapezoids shown in Fig. 5.9.4.

EXAMPLE 1 Calculate the trapezoidal approximation to the integral in Eq. (4) with $n = 6$ and $\Delta x = 0.5$.

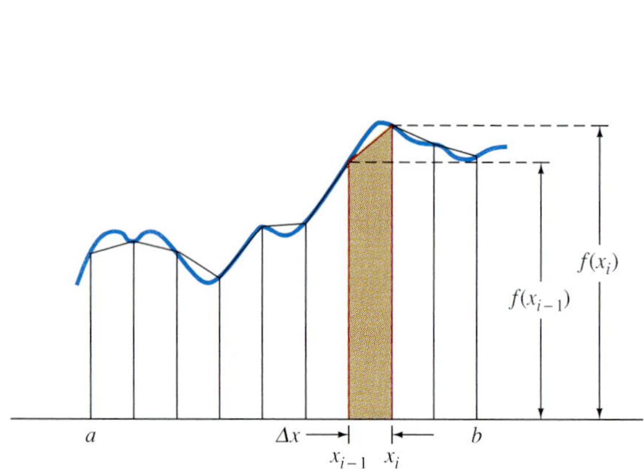

Fig. 5.9.4 Geometry of the trapezoidal approximation

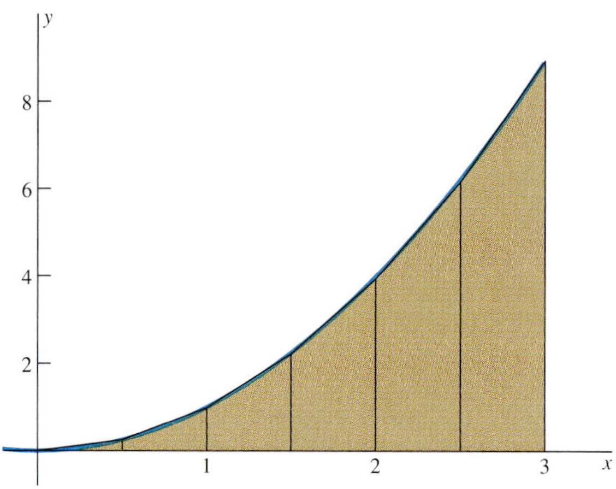

Fig. 5.9.5 The area under $y = x^2$ (Example 1)

Solution The trapezoids in Fig. 5.9.5 indicate why T_6 should be a much better approximation than either of the endpoint approximations L_6 or R_6. The table in Fig. 5.9.6 shows the values of $f(x) = x^2$ that are needed to compute L_6 and R_6. The 1-2-2-$\cdots$-2-2-1 coefficients appear in the final column. Using Eq. (6), we get

$$\begin{aligned} T_6 &= \frac{0.5}{2}[1\cdot(0) + 2\cdot(0.25) + 2\cdot(1) \\ &\qquad + 2\cdot(2.25) + 2\cdot(4) + 2\cdot(6.25) + 1\cdot(9)] \\ &= 9.125 \end{aligned}$$

(as compared with the actual value 9). ■

n	x_n	$f(x_n) = x_n^2$	Coefficients
0	0	0	1
1	0.5	0.25	2
2	1	1	2
3	1.5	2.25	2
4	2	4	2
5	2.5	6.25	2
6	3	9	1

Fig. 5.9.6 Data for Example 1

Another useful approximation to $\int_a^b f(x)\,dx$ is the *midpoint approximation* M_n. It is the Riemann sum obtained by choosing the point $x_i^\star$ in $[x_{i-1}, x_i]$ to be its midpoint $m_i = (x_{i-1} + x_i)/2$. Thus

$$M_n = \sum_{i=1}^{n} f(m_i)\,\Delta x = (\Delta x)[f(m_1) + f(m_2) + \cdots + f(m_n)]. \tag{7}$$

Because m_1 is the midpoint of $[x_0, x_1]$, it is sometimes convenient to write $y_{1/2}$ for $f(m_1)$, $y_{3/2}$ for $f(m_2)$, and so on (Fig. 5.9.7).

Definition *The Midpoint Approximation*

The **midpoint approximation** to

$$\int_a^b f(x)\,dx \quad \text{with} \quad \Delta x = \frac{b-a}{n}$$

is

$$M_n = (\Delta x)(y_{1/2} + y_{3/2} + y_{5/2} + \cdots + y_{n-(1/2)}). \tag{7$'$}$$

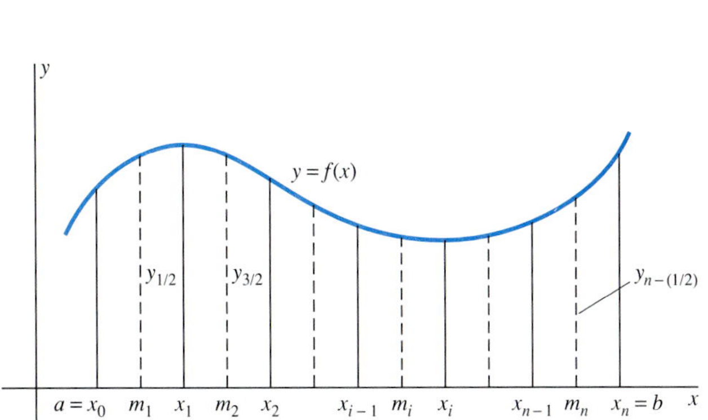

Fig. 5.9.7 The ordinates used in the midpoint approximation

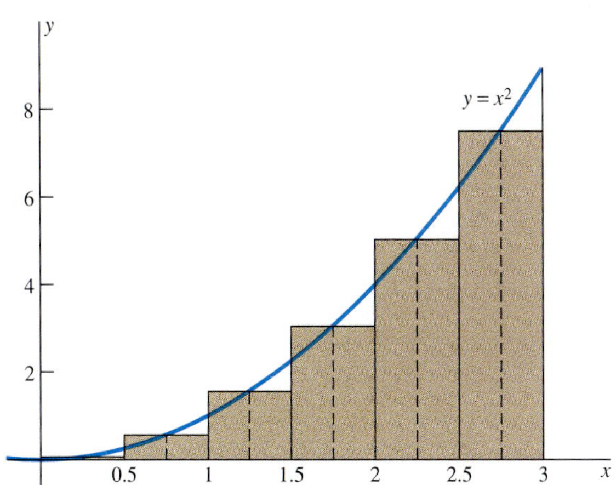

Fig. 5.9.8 Midpoint rectangles approximating the area under $y = x^2$ (Example 2)

EXAMPLE 2 Figure 5.9.8 illustrates the midpoint approximation to the integral

$$\int_0^3 x^2\,dx = 9$$

of Example 1, with $n = 6$ and $\Delta x = 0.5$, and the table in Fig. 5.9.9 shows the values of $f(x) = x^2$ needed to compute M_6. Using Eq. (7), we obtain

$$\begin{aligned} M_6 &= (0.5)\,[1\cdot(0.0625) + 1\cdot(0.5625) + 1\cdot(1.5625). \\ &\qquad + 1\cdot(3.0625) + 1\cdot(5.0625) + 1\cdot(7.5625)] \\ &= 8.9375. \end{aligned}$$

■

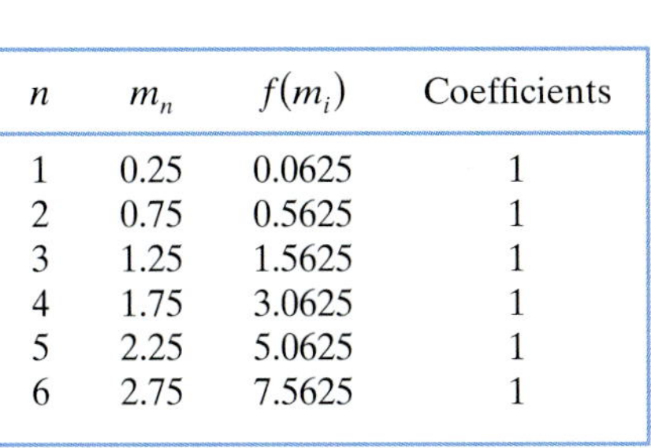

n	m_n	$f(m_i)$	Coefficients
1	0.25	0.0625	1
2	0.75	0.5625	1
3	1.25	1.5625	1
4	1.75	3.0625	1
5	2.25	5.0625	1
6	2.75	7.5625	1

Fig. 5.9.9 Data for Example 2

EXAMPLE 3 Figure 5.9.10 shows the graph of the function

$$f(x) = \frac{\sin x}{x}. \tag{8}$$

Recall from Section 2.5 that

$$\lim_{x\to 0} \frac{\sin x}{x} = 1.$$

Hence we *define* $f(0) = 1$; then there will be no difficulty at $x = 0$, even though both the numerator and denominator in Eq. (8) are zero. It happens that the function $f(x)$

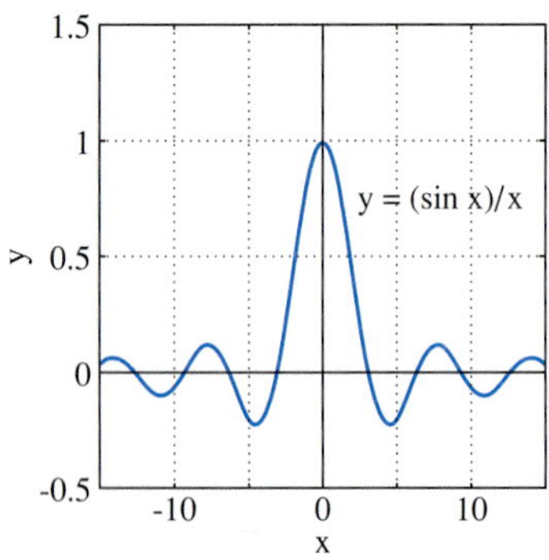
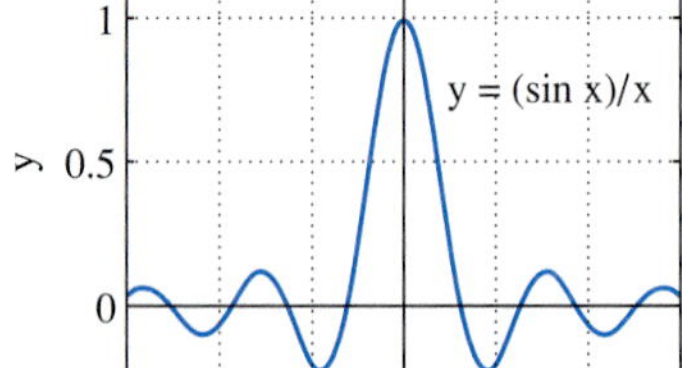

Fig. 5.9.10 The graph of $f(x) = \dfrac{\sin x}{x}$ (Example 3)

has *no* elementary antiderivative, so the fundamental theorem of calculus cannot be used to calculate the value of the integral

$$I = \int_0^1 \frac{\sin x}{x}\,dx. \tag{9}$$

But such integrals are important in the design of precise photographic lenses (among other applications), so we are well motivated to resort to approximating its value numerically.

With $n = 10$ and $\Delta x = 0.1$, the trapezoidal approximation is

$$T_{10} = \frac{0.1}{2}\left[1 \cdot 1 + 2 \cdot \frac{\sin 0.1}{0.1} + 2 \cdot \frac{\sin 0.2}{0.2} + 2 \cdot \frac{\sin 0.3}{0.3} + \cdots + 2 \cdot \frac{\sin 0.9}{0.9} + 1 \cdot \frac{\sin 1.0}{1.0}\right];$$

$$T_{10} \approx 0.94583.$$

The corresponding midpoint approximation is

$$M_{10} = (0.1)\left[1 \cdot \frac{\sin 0.05}{0.05} + 1 \cdot \frac{\sin 0.15}{0.15} + 1 \cdot \frac{\sin 0.25}{0.25} + \cdots + 1 \cdot \frac{\sin 0.85}{0.85} + 1 \cdot \frac{\sin 0.95}{0.95}\right];$$

$$M_{10} \approx 0.94621.$$

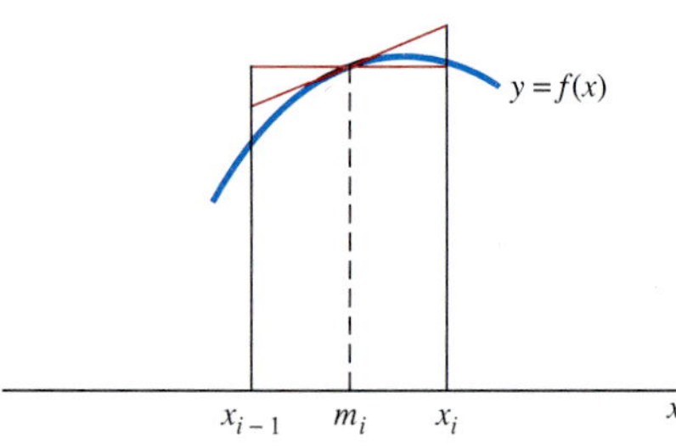

Fig. 5.9.11 The midpoint (or tangent) approximation

The actual value of the integral in Eq. (9) is $I \approx 0.94608$ (accurate to five decimal places). Thus both T_{10} and M_{10} give the correct value 0.946 when rounded to three places. But

- T_{10} underestimates I by about 0.00025, whereas
- M_{10} overestimates I by about 0.00013.

Thus in this example the midpoint approximation is somewhat more accurate than the trapezoidal approximation. ■

Simpson's Approximation

The midpoint approximation in Eq. (7) is sometimes called the **tangent-line approximation,** because the area of the rectangle with base $[x_{i-1}, x_i]$ and height $f(m_i)$ is also the area of another approximating figure. As shown in Fig. 5.9.11, we draw a segment tangent to the graph of f, tangent at the point $(m_i, f(m_i))$ of its graph, and use that segment for one side of a trapezoid (somewhat like the method of the trapezoidal approximation). The trapezoid and the rectangle mentioned earlier have the same area, and so the value of M_n is the sum of the areas of trapezoids like the one in Fig. 5.9.11.

The area of the trapezoid associated with the midpoint approximation is generally closer to the true value of

$$\int_{x_{i-1}}^{x_i} f(x)\,dx$$

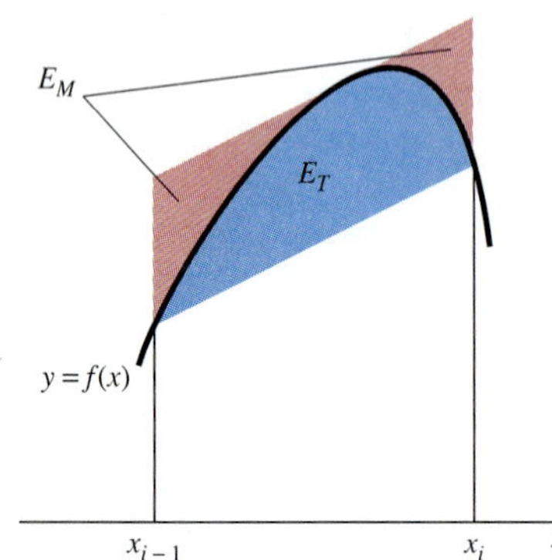

Fig. 5.9.12 Comparison of the midpoint approximation error E_M with the trapezoidal approximation error E_T

than is the area of the trapezoid associated with the trapezoidal approximation, as was the case in Example 3. Figure 5.9.12 shows this too, in that the midpoint error E_M (above the curve in this figure) is generally smaller than the trapezoidal error E_T (below the curve in this figure). Figure 5.9.12 also indicates that if $y = f(x)$ is concave downward, then M_n will be an overestimate and T_n will be an underestimate of $\int_a^b f(x)\,dx$. If the graph is concave upward, then the situation is reversed.

Such observations motivate the consideration of a *weighted average* of M_n and T_n, with M_n weighted more heavily than T_n, to improve our numerical estimates of the definite integral. The particular weighted average

$$S_{2n} = \tfrac{1}{3}(2M_n + T_n) = \tfrac{2}{3}M_n + \tfrac{1}{3}T_n \qquad \textbf{(10)}$$

is called *Simpson's approximation* to $\int_a^b f(x)\,dx$. The reason for the subscript $2n$ is that we associate S_{2n} with a partition of $[a, b]$ into an *even* number, $2n$, of equal-length subintervals with the endpoints

$$a = x_0 < x_1 < x_2 < \cdots < x_{2n-2} < x_{2n-1} < x_{2n} = b.$$

The midpoint and trapezoidal approximations associated with the n subintervals

$$[x_0, x_2], \quad [x_2, x_4], \quad [x_4, x_6], \quad \ldots, \quad [x_{2n-4}, x_{2n-2}], \quad [x_{2n-2}, x_{2n}],$$

all with the same length $2\,\Delta x$, can then be written in the respective forms

$$M_n = (2\,\Delta x)(y_1 + y_3 + y_5 + \cdots + y_{2n-1})$$

and

$$T_n = \frac{2\,\Delta x}{2}(y_0 + 2y_2 + 2y_4 + \cdots + 2y_{2n-2} + y_{2n}).$$

We substitute these formulas for M_n and T_n into Eq. (10) and find—after a bit of algebra—that

$$S_{2n} = \frac{\Delta x}{3}(y_0 + 4y_1 + 2y_2 + 4y_3 + 2y_4 + \cdots + 2y_{2n-2} + 4y_{2n-1} + y_{2n}). \qquad \textbf{(11)}$$

To be consistent with our other approximation formulas, we next rewrite Eq. (11) with n (rather than $2n$) to denote the total number of subintervals used.

Definition *Simpson's Approximation*

Simpson's approximation to $\int_a^b f(x)\,dx$ with $\Delta x = (b - a)/n$, associated with a partition of $[a, b]$ into an *even* number n of equal-length subintervals, is the sum S_n defined as

$$S_n = \frac{\Delta x}{3}(y_0 + 4y_1 + 2y_2 + 4y_3 + 2y_4 + \cdots + 2y_{n-2} + 4y_{n-1} + y_n). \qquad \textbf{(12)}$$

Remark Note the 1-4-2-4-2- · · · -2-4-2-4-1 pattern of coefficients in Simpson's approximation. This pattern is symmetric (ending in -2-4-1), as shown, if and only if n is *even.*

EXAMPLE 4 Simpson's approximation (with $n = 6$ and $\Delta x = 0.5$) to the integral

$$\int_0^3 x^2\,dx = 9$$

of Examples 1 and 2 is

$$S_6 = \frac{0.5}{3}[1\cdot(0)^2 + 4\cdot(0.5)^2 + 2\cdot(1)^2 + 4\cdot(1.5)^2 + 2\cdot(2)^2 + 4\cdot(2.5)^2 + 1\cdot(3)^2];$$

$$S_6 = 9 \qquad \text{(exactly)}.$$

Problem 29 explains why Simpson's approximation to this particular integral is *exact* rather than merely a good approximation.

EXAMPLE 5 Simpson's approximation (with $n = 10$ and $\Delta x = 0.1$) to the integral

$$\int_0^1 \frac{\sin x}{x}\,dx$$

of Example 3 is

$$S_{10} = \frac{0.1}{3}\left[1\cdot 1 + 4\cdot\frac{\sin 0.1}{0.1} + 2\cdot\frac{\sin 0.2}{0.2} + 4\cdot\frac{\sin 0.3}{0.3} + \cdots + 2\cdot\frac{\sin 0.8}{0.8} + 4\cdot\frac{\sin 0.9}{0.9} + 1\cdot\frac{\sin 1.0}{1.0}\right];$$

$$S_{10} \approx 0.94608,$$

which is accurate to *all five* decimal places shown.

Examples 4 and 5 illustrate the greater accuracy of Simpson's approximation in comparison with the midpoint and trapezoidal approximations.

The numerical methods of this section are especially useful for approximating integrals of functions that are available only in graphical or in tabular form. This is often the case with functions derived from empirical data or from experimental measurements.

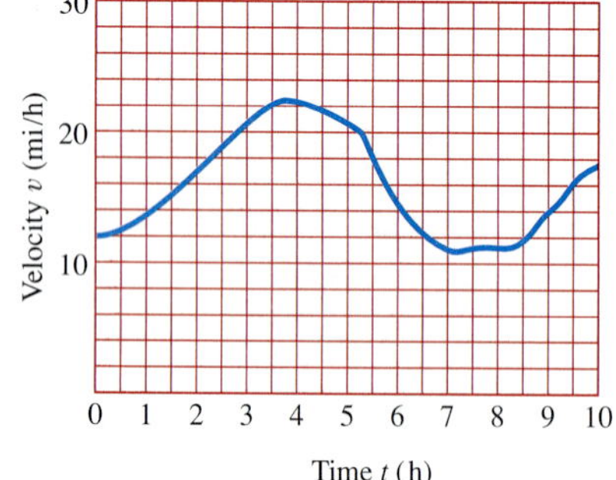

Fig. 5.9.13 Velocity graph for the submarine of Example 6

EXAMPLE 6 Suppose that the graph in Fig. 5.9.13 shows the velocity $v(t)$ recorded by instruments on board a submarine traveling under the polar ice cap directly toward the North Pole. Use the trapezoidal approximation and Simpson's approximation to estimate the distance $s = \int_a^b v(t)\,dt$ traveled by the submarine during the 10-h period from $t = 0$ to $t = 10$.

Solution We read the following data from the graph.

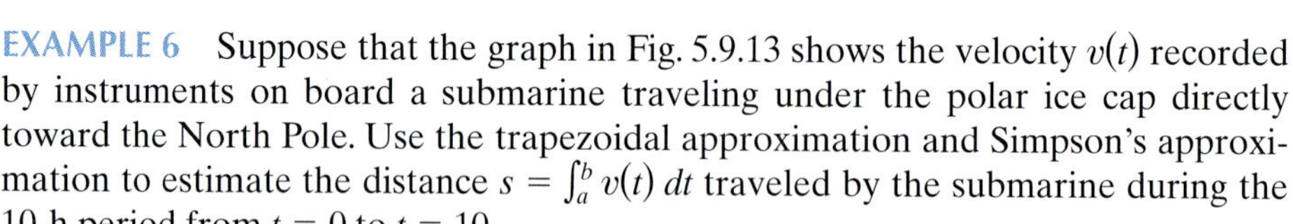

t	0	1	2	3	4	5	6	7	8	9	10	h
v	12	14	17	21	22	21	15	11	11	14	17	mi/h

Using the trapezoidal approximation with $n = 10$ and $\Delta x = 1$, we obtain

$$s = \int_0^{10} v(t)\,dt$$

$$\approx \tfrac{1}{2}[12 + 2(14 + 17 + 21 + 22 + 21 + 15 + 11 + 11 + 14) + 17]$$

$$= 160.5 \quad \text{(mi)}.$$

Using Simpson's approximation with $2n = 10$ and $\Delta x = 1$, we obtain

$$\begin{aligned} s &= \int_0^{10} v(t)\,dt \\ &\approx \tfrac{1}{3}[12 + 4\cdot 14 + 2\cdot 17 + 4\cdot 21 + 2\cdot 22 + 4\cdot 21 \\ &\qquad + 2\cdot 15 + 4\cdot 11 + 2\cdot 11 + 4\cdot 14 + 17] \\ &= 161 \quad \text{(mi)} \end{aligned}$$

as an estimate of the distance traveled by the submarine during this 10-h period. ■

Parabolic Approximations

Although we have defined Simpson's approximation S_{2n} as a weighted average of the midpoint and trapezoidal approximations, Simpson's approximation has an important interpretation in terms of **parabolic approximations** to the curve $y = f(x)$. Beginning with the partition of $[a, b]$ into $2n$ equal-length subintervals, we define the parabolic function

$$p_i(x) = A_i + B_i x + C_i x^2$$

on $[x_{2i-2}, x_{2i}]$ as follows: We choose the coefficients A_i, B_i, and C_i so that $p_i(x)$ agrees with $f(x)$ at the three points x_{2i-2}, x_{2i-1}, and x_{2i} (Fig. 5.9.14). This can be done by solving the three equations

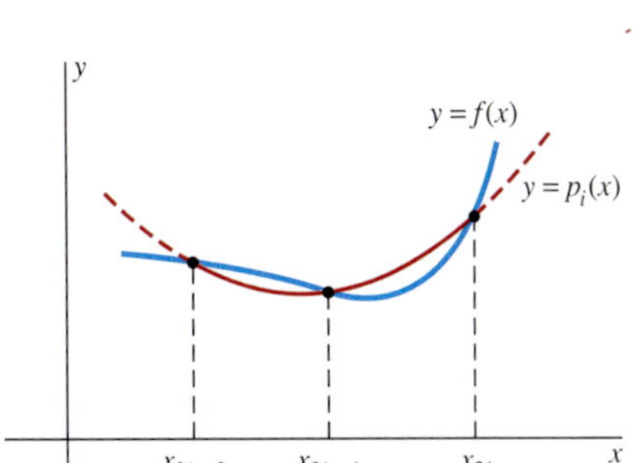

Fig. 5.9.14 The parabolic approximation $y = p_i(x)$ to $y = f(x)$ on $[x_{2i-2}, x_{2i}]$

$$\begin{aligned} A_i + B_i x_{2i-2} + C_i(x_{2i-2})^2 &= f(x_{2i-2}), \\ A_i + B_i x_{2i-1} + C_i(x_{2i-1})^2 &= f(x_{2i-1}), \\ A_i + B_i x_{2i} \quad + C_i(x_{2i})^2 &= f(x_{2i}) \end{aligned}$$

for the three unknowns A_i, B_i, and C_i. A routine (but tedious) algebraic computation—see Problem 52 of Section 5.8—shows that

$$\int_{x_{2i-2}}^{x_{2i}} p_i(x)\,dx = \frac{\Delta x}{3}\left(y_{2i-2} + 4y_{2i-1} + y_{2i}\right).$$

We now approximate $\int_a^b f(x)\,dx$ by replacing $f(x)$ with $p_i(x)$ on the interval $[x_{2i-2}, x_{2i}]$ for $i = 1, 2, 3, \dots, n$. This gives

$$\begin{aligned} \int_a^b f(x)\,dx &= \sum_{i=1}^{n} \int_{x_{2i-2}}^{x_{2i}} f(x)\,dx \approx \sum_{i=1}^{n} \int_{x_{2i-2}}^{x_{2i}} p_i(x)\,dx \\ &= \sum_{i=1}^{n} \frac{\Delta x}{3}\left(y_{2i-2} + 4y_{2i-1} + y_{2i}\right) \\ &= \frac{\Delta x}{3}\left(y_0 + 4y_1 + 2y_2 + 4y_3 + \cdots \right. \\ &\qquad \left. + 4y_{2n-3} + 2y_{2n-2} + 4y_{2n-1} + y_{2n}\right). \end{aligned}$$

Thus the parabolic approximation described here results in Simpson's approximation S_{2n} to $\int_a^b f(x)\,dx$.

Error Estimates

The trapezoidal approximation, the midpoint approximation, and Simpson's approximation are widely used for numerical integration, and there are *error estimates* that can be used to predict the maximum possible error in a particular approximation. The trapezoidal error ET_n, the midpoint error EM_n, and Simpson's error ES_n are defined by the equations

$$\int_a^b f(x)\,dx = T_n + ET_n, \tag{13}$$

$$\int_a^b f(x)\,dx = M_n + EM_n, \tag{14}$$

and

$$\int_a^b f(x)\,dx = S_n + ES_n \qquad (n \text{ even}) \tag{15}$$

Note that each of these formulas is of the form

$$\int_a^b f(x)\,dx = \{\text{approximation}\} + \{\text{error}\}.$$

The absolute value $|ET_n|$ is the difference between the value of the integral and the trapezoidal approximation with n subintervals (and similarly for $|EM_n|$ and $|ES_n|$). Theorems 1 and 2 are proved in numerical analysis textbooks.

Theorem 1 ***Trapezoidal and Midpoint Error Estimates***
Suppose that the second derivative f'' is continuous on $[a, b]$ and that $|f''(x)| \leqq K_2$ for $a \leqq x \leqq b$. Then

$$|ET_n| \leqq \frac{K_2(b-a)^3}{12n^2} \tag{16}$$

and

$$|EM_n| \leqq \frac{K_2(b-a)^3}{24n^2}. \tag{17}$$

Remark Comparing (16) and (17), we see that the maximal predicted midpoint error is *half* the predicted trapezoidal error. This is the reason for weighting M_n *twice* as heavily as T_n when we calculate Simpson's approximation using the formula $S_{2n} = \frac{2}{3}M_n + \frac{1}{3}T_n$ in (10).

Theorem 2 ***Simpson's Error Estimate***
Suppose that the fourth derivative $f^{(4)}$ is continuous on $[a, b]$ and that $|f^{(4)}(x)| \leqq K_4$ for $a \leqq x \leqq b$. If n is even, then

$$|ES_n| \leqq \frac{K_4(b-a)^5}{180n^4}. \tag{18}$$

Remark The factor n^4 in (18)—compared with the n^2 in (16) and (17)—explains the greater accuracy typical of Simpson's approximation. For instance, if $n = 10$, then

$n^2 = 100$ but $n^4 = 10000$, so the denominator in the error formula for Simpson's approximation is much larger.

EXAMPLE 7 In Chapter 7 we will see that the natural logarithm of the number 2 (given approximately by the LN key on your calculator) is the value of the integral

$$\ln 2 = \int_1^2 \frac{1}{x}\,dx,$$

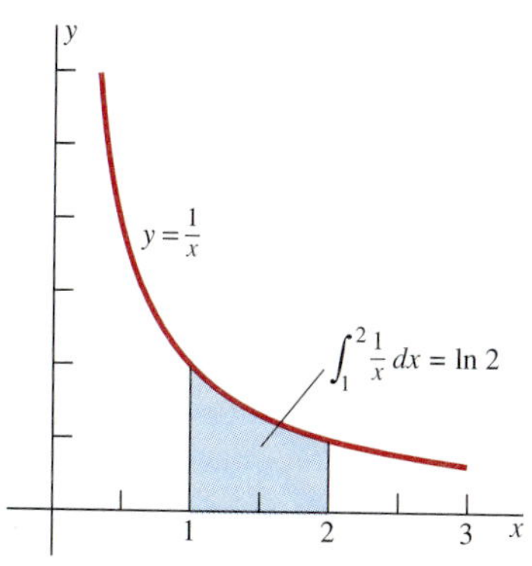

Fig. 5.9.15 The number ln 2 as an area

illustrated in Fig. 5.9.15. Estimate the errors in the trapezoidal, midpoint, and Simpson's approximations to this integral by using $n = 10$ subintervals. (The actual value of ln 2 is approximately 0.693147.)

Solution With $f(x) = 1/x$, we calculate

$$f'(x) = -\frac{1}{x^2}, \qquad f''(x) = \frac{2}{x^3},$$

$$f'''(x) = -\frac{6}{x^4}, \qquad f^{(4)}(x) = \frac{24}{x^5}.$$

The maximum values of all these derivatives for $1 \leqq x \leqq 2$ occur at $x = 1$, so we may take $K_2 = 2$ and $K_4 = 24$ in Eqs. (16), (17), and (18). From (16) and (17), we see that

$$|ET_{10}| \leqq \frac{2 \cdot 1^3}{12 \cdot 10^2} \approx 0.0016667 \quad \text{and} \quad |EM_{10}| \leqq \frac{2 \cdot 1^3}{24 \cdot 10^2} \approx 0.000833. \tag{19}$$

Hence we would expect both the trapezoidal approximation T_{10} and the midpoint approximation M_{10} to give ln 2 accurate to at least two decimal places. From (18) we see that

$$|ES_{10}| \leqq \frac{24 \cdot 1^5}{180 \cdot 10^4} \approx 0.000013, \tag{20}$$

so we would expect Simpson's approximation S_{10} to be accurate to at least four decimal places. When we calculate these approximations, we find that

$$T_{10} = (0.1)\left(\frac{1}{1} + \frac{2}{1.1} + \frac{2}{1.2} + \frac{2}{1.3} + \cdots + \frac{2}{1.9} + \frac{1}{2}\right) \approx 0.693771,$$

$$M_{10} = (0.1)\left(\frac{1}{1.05} + \frac{1}{1.15} + \frac{1}{1.25} + \cdots + \frac{1}{1.85} + \frac{1}{1.95}\right) \approx 0.692835,$$

and

$$S_{10} = (0.1)\left(\frac{1}{1} + \frac{4}{1.1} + \frac{2}{1.2} + \frac{4}{1.3} + \cdots + \frac{4}{1.9} + \frac{1}{2}\right) \approx 0.693150.$$

It follows that the values of the errors in these approximations (in comparison with the actual value $\ln 2 \approx 0.693147$) are

$$ET_{10} \approx -0.000624, \qquad EM_{10} \approx 0.000312, \qquad \text{and} \qquad ES_{10} \approx -0.000003.$$

Comparing these actual errors with the maximal predicted errors in (19) and (20), we see that our approximations are somewhat more accurate than predicted—M_{10}

actually is accurate to three decimal places and S_{10} is accurate to five. It is fairly typical of numerical integration that the trapezoidal, midpoint, and Simpson approximations are somewhat more accurate than the (worst-case) estimates provided by Theorems 1 and 2. ■

5.9 PROBLEMS

In Problems 1 through 6, calculate the trapezoidal approximation T_n to the given integral, and compare T_n with the exact value of the integral. Use the indicated number n of subintervals, and round answers to two decimal places.

1. $\int_0^4 x\,dx, \quad n = 4$

2. $\int_1^2 x^2\,dx, \quad n = 5$

3. $\int_0^1 \sqrt{x}\,dx, \quad n = 5$

4. $\int_1^3 \frac{1}{x^2}\,dx, \quad n = 4$

5. $\int_0^{\pi/2} \cos x\,dx, \quad n = 3$

6. $\int_0^{\pi} \sin x\,dx, \quad n = 4$

In Problems **7** through **12.** Calculate the midpoint approximations to the integrals in Problems 1 through 6, using the indicated number of subintervals. In each case compare M_n with the exact value of the integral.

In Problems 13 through 20, calculate both the trapezoidal approximation T_n and Simpson's approximation S_n to the given integral. Use the indicated number of subintervals and round answer to four decimal places. In Problems 13 through 16, also compare these approximations with the exact value of the integral.

13. $\int_1^3 x^2\,dx, \quad n = 4$

14. $\int_1^4 x^3\,dx, \quad n = 4$

15. $\int_2^4 \frac{1}{x^3}\,dx, \quad n = 4$

16. $\int_0^1 \sqrt{1 + x}\,dx, \quad n = 4$

17. $\int_0^2 \sqrt{1 + x^3}\,dx, \quad n = 6$

18. $\int_0^3 \frac{1}{1 + x^4}\,dx, \quad n = 6$

19. $\int_1^5 \sqrt[3]{1 + x^2}\,dx, \quad n = 8$

20. $\int_0^1 \frac{\tan x}{x}\,dx, \quad n = 10$

(*Note:* Make the integrand in Problem 20 continuous by assuming that its value at $x = 0$ is its limit there,

$$\lim_{x \to 0} \frac{\tan x}{x} = 1.)$$

In Problems 21 and 22, calculate (a) the trapezoidal approximation and (b) Simpson's approximation to

$$\int_a^b f(x)\,dx,$$

where f is the given tabulated function.

21.

x	$a = 1.00$	1.25	1.50	1.75	2.00	2.25	$2.50 = b$
$f(x)$	3.43	2.17	0.38	1.87	2.65	2.31	1.97

22.

x	$a = 0$	1	2	3	4	5	6	7	8	9	$10 = b$
$f(x)$	23	8	−4	12	35	47	53	50	39	29	5

23. Figure 5.9.16 shows the measured rate of water flow (in liters per minute) into a tank during a 10-min period. Using 10 subintervals in each case, estimate the total amount of water that flows into the tank during this period by using (a) the trapezoidal approximation and (b) Simpson's approximation.

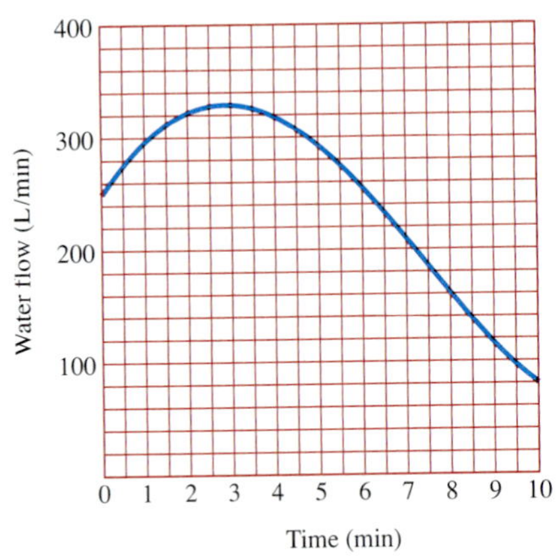

Fig. 5.9.16 Water-flow graph for Problem 23

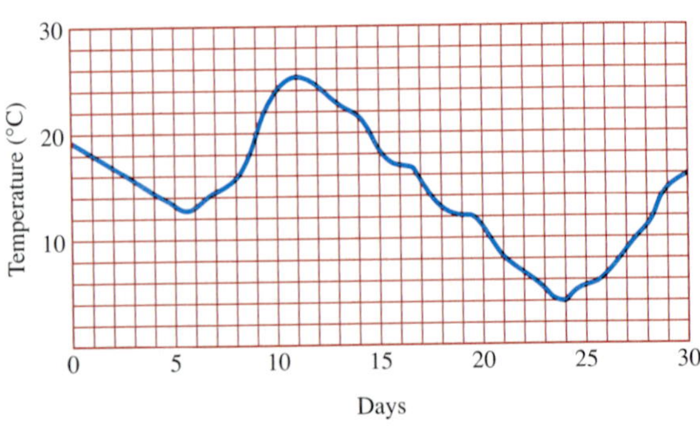

Fig. 5.9.17 Temperature graph for Problem 24

24. Figure 5.9.17 shows the daily mean temperature recorded during December at Big Frog, California. Using 10 subintervals in each case, estimate the average temperature during that month by using (a) the trapezoidal approximation and (b) Simpson's approximation.

25. Figure 5.9.18 shows a tract of land with measurements in feet. A surveyor has measured its width w at 50-ft intervals (the values of x shown in the figure), with the following results:

x	0	50	100	150	200	250	300
w	0	165	192	146	63	42	84

x	350	400	450	500	550	600
w	155	224	270	267	215	0

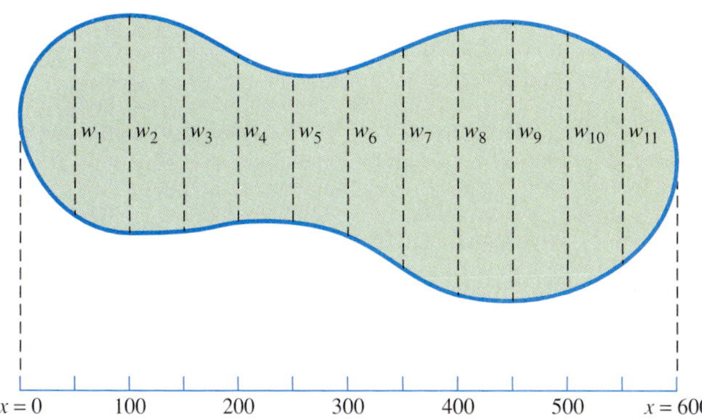

Fig. 5.9.18 The tract of land of Problem 25

Estimate the acreage of this tract using (a) the trapezoidal approximation and (b) Simpson's approximation. (*Note:* An acre is 4840 yd^2.)

26. The base for natural logarithms is the number e. In Chapter 7 we will see that

$$\int_1^e \frac{1}{x}\,dx = 1.$$

Approximate the integrals

$$\int_1^{2.7} \frac{1}{x}\,dx \quad \text{and} \quad \int_1^{2.8} \frac{1}{x}\,dx$$

with sufficient accuracy to show that $2.7 < e < 2.8$.

Problems 27 and 28 deal with the integral

$$\ln 2 = \int_1^2 \frac{1}{x}\,dx$$

of Example 7.

27. Use the trapezoidal error estimate to determine how large n must be to guarantee that T_n differs from $\ln 2$ by at most 0.0005.

28. Use the Simpson's error estimate to determine how large n must be to guarantee that S_n differs from $\ln 2$ by at most 0.000005.

29. Deduce the following from the error estimate for Simpson's approximation: If $p(x)$ is a polynomial of degree at most 3, then Simpson's approximation with $n = 2$ subintervals gives the exact value of the integral

$$\int_a^b p(x)\,dx.$$

30. Use the result of Problem 29 to calculate (without explicit integration) the area of the region shown in Fig. 5.9.19. (*Answer:* 1331/216.)

31. Whereas the carefully weighted average in Eq. (10) of the midpoint and trapezoidal approximations M_n and T_n gives Simpson's approximation S_{2n}, show that their equally weighted average gives the trapezoidal approximation with twice as many intervals; that is, $\frac{1}{2}(M_n + T_n) = T_{2n}$.

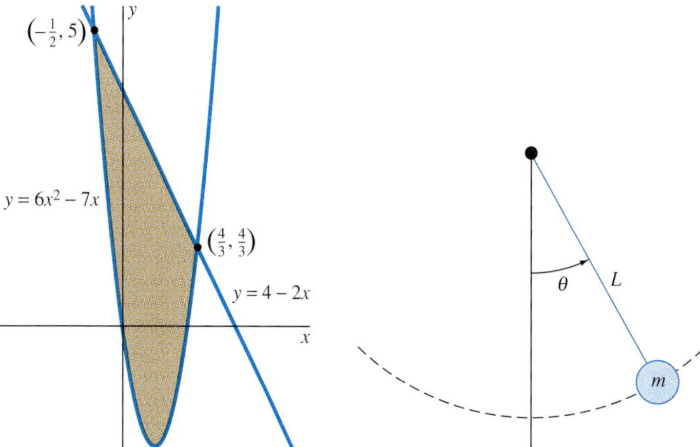

Fig. 5.9.19 The region of Problem 30

Fig. 5.9.20 The pendulum of Problem 32

32. Figure 5.9.20 shows a pendulum of length L. If this pendulum is released from rest at an angle α from the vertical, then it swings back and forth with its period T (for one complete oscillation) given by

$$T = 4\sqrt{\frac{L}{g}} \int_0^{\pi/2} \frac{1}{\sqrt{1 - k^2 \sin^2 x}}\,dx,$$

where $k = \sin(\alpha/2)$. Taking $L = 1$ m and $g = 9.8$ m/s^2, use Simpson's approximation with $n = 10$ subintervals to calculate the pendulum's period of oscillation if its initial angle α is (a) 10°; (b) 50°.

33. Note in Example 7 that the midpoint and trapezoidal approximations gave under- and overestimates, respectively, of the value $\int_1^2 (1/x)\,dx = \ln 2$. Show that this is a general phenomenon. That is, if both $f(x)$ and $f''(x)$ are positive for $a \leqq x \leqq b$, then

$$M_n < \int_a^b f(x)\,dx < T_n,$$

whereas if $f(x)$ is positive but $f''(x)$ is negative for $a \leqq x \leqq b$, then the inequalities are reversed.

5.9 PROJECT: APPROXIMATING ln 2 AND π BY NUMERICAL INTEGRATION

In the Section 5.4 project we discussed programs and computer algebra system commands that can be used to calculate the Riemann sums

L_n, the left-endpoint approximation,

R_n, the right-endpoint approximation, and

M_n, the midpoint approximation,

based on a partition of $[a, b]$ into n equal-length subintervals, to approximate the integral

$$\int_a^b f(x)\,dx. \tag{1}$$

The Riemann sums L_n, R_n, and M_n suffice, in turn, to calculate the trapezoidal and Simpson sums of this section. In particular, the trapezoidal approximation is given—using Eq. (5) of this section—by

$$T_n = \tfrac{1}{2}(L_n + R_n). \tag{2}$$

Once these sums are known, Simpson's approximation based on a subdivision of $[a, b]$ into $2n$ equal-length subintervals is given—using Eq. (10) of this section—by

$$S_{2n} = \tfrac{1}{3}(2M_n + T_n) = \tfrac{1}{6}(L_n + 4M_n + R_n). \tag{3}$$

Here, then, is a practical scheme for approximating accurately the integral in (1). Begin with a selected value of n, such as $n = 5$, and calculate the Riemann sums L_5, R_5, and M_5. Then use Eq. (3) to calculate the Simpson approximation S_{10}. Next, double the value of n and calculate similarly L_{10}, R_{10}, M_{10}, and finally S_{20}. A typical strategy is to continue in this manner, always doubling the current value of n for the next cycle of computations, until two successively calculated Simpson approximations agree to the desired number of decimal places of accuracy.

INVESTIGATION A: According to Example 7, the natural logarithm (corresponding to the **LN** or, in some cases, the **LOG** key on your calculator) of the number 2 is the value of the integral

$$\ln 2 = \int_1^2 \frac{1}{x}\,dx.$$

The value of ln 2 correct to 15 decimal places is

$$\ln 2 \approx 0.693147180559945.$$

See how many correct decimal places you can obtain in a reasonable period of time by using a Simpson's approximation procedure.

INVESTIGATION B: In Chapter 8 we will study the inverse tangent function $y = \arctan x$ (y is the angle between $-\pi/2$ and $\pi/2$ such that $\tan y = x$). There we will discover that the derivative of $y = \arctan x$ is

$$\frac{dy}{dx} = \frac{1}{1 + x^2}.$$

This implies that

$$\int_0^1 \frac{1}{1 + x^2}\,dx = \Big[\arctan x\Big]_0^1 = \arctan 1 - \arctan 0 = \frac{\pi}{4}.$$

It follows that the number π is the value of the integral

$$\pi = \int_0^1 \frac{4}{1 + x^2}\,dx.$$

The value of π to 15 decimal places is

$$\pi \approx 3.14159\;26535\;89793.$$

See how many correct decimal places you can obtain in a reasonable period of time by using a Simpson's approximation procedure.

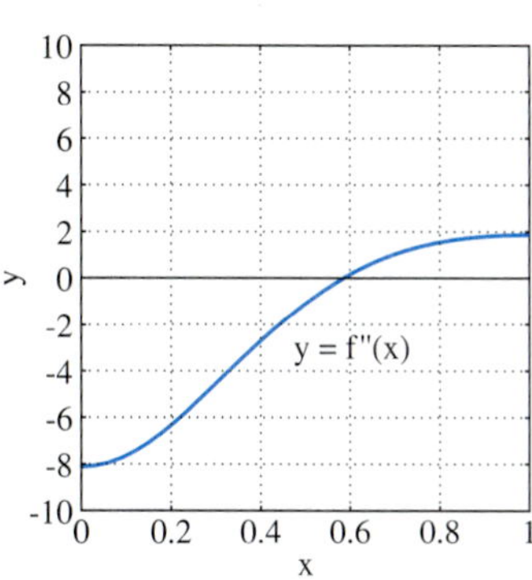

Fig. 5.9.21 Graph of the second derivative of $f(x) = \dfrac{4}{1 + x^2}$

INVESTIGATION C: Taking $f(x) = 4/(1 + x^2)$ as the integrand of the integral

$$\int_0^1 \frac{4}{1 + x^2}\,dx = \pi$$

of Investigation B, it would be somewhat lengthy to first calculate and then maximize by hand the derivatives $f''(x)$ and $f^{(4)}(x)$ as needed to apply the error estimates in Theorems 1 and 2 of this section. Instead, a computer algebra system readily yields

$$f''(x) = \frac{8(3x^2 - 1)}{(1 + x^2)^3} \quad \text{and} \quad f^{(4)}(x) = \frac{96(5x^4 - 10x^2 + 1)}{(1 + x^2)^5}.$$

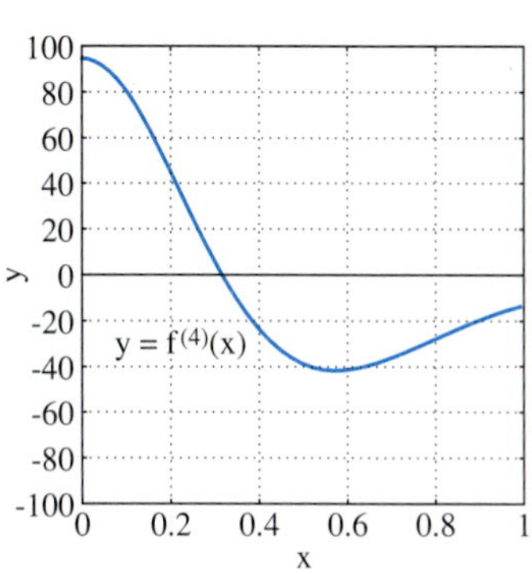

Fig. 5.9.22 Graph of the fourth derivative of $f(x) = \dfrac{4}{1 + x^2}$

Figures 5.9.21 and 5.9.22 show the graphs of these two derivatives on the interval $[0, 1]$. From these graphs it is clear that each of these derivatives attains its maximum absolute value at the left endpoint $x = 0$. Thus you can take

$$K_2 = 8 \quad \text{and} \quad K_4 = 96$$

in Theorems 1 and 2. Use this information to determine how large the integer n must be so that:

1. $|EM_n| < 5 \times 10^{-6}$, so the midpoint approximation M_n will give the number π accurate to five decimal places;
2. $|ES_n| < 5 \times 10^{-11}$, so the Simpson approximation S_n will give the number π accurate to ten decimal places;
3. The Simpson approximation S_n will give the number π accurate to 15 decimal places.

CHAPTER 5 REVIEW: *Definitions, Concepts, Results*

Use the following list as a guide to concepts that you may need to review.

1. Antidifferentiation and antiderivatives
2. The most general antiderivative of a function
3. Integral formulas
4. Initial value problems
5. Velocity and acceleration
6. Solution of problems involving constant acceleration
7. Properties of area
8. Summation notation
9. The area under the graph of f from a to b
10. Inscribed and circumscribed rectangular polygons
11. A partition of $[a, b]$
12. The norm of a partition
13. A Riemann sum associated with a partition
14. The definite integral of f from a to b
15. Existence of the integral of a continuous function
16. The integral as the limit of a sequence of Riemann sums
17. Regular partitions and

$$\lim_{\Delta x \to 0} \sum_{i=1}^{n} f(x_i^\star)\,\Delta x$$

18. Riemann sums
19. The constant multiple, sum, interval union, and comparison properties of integrals
20. Evaluation of definite integrals by using antiderivatives
21. The average value of $f(x)$ on the interval $[a, b]$
22. The average value theorem
23. The fundamental theorem of calculus
24. Indefinite integrals
25. The method of integration by substitution
26. Transforming the limits in integration by substitution
27. The area between $y = f(x)$ and $y = g(x)$ by integration with respect to x
28. The area between $x = f(y)$ and $x = g(y)$ by integration with respect to y
29. The right-endpoint and left-endpoint approximations
30. The trapezoidal approximation
31. The midpoint approximation
32. Simpson's approximation
33. Error estimates for the trapezoidal approximation and Simpson's approximation

CHAPTER 5 *Miscellaneous Problems*

Find the indefinite integrals in Problems 1 through 24. In Problems 13 through 24, use the indicated substitution.

1. $\displaystyle\int \frac{x^5 - 2x + 5}{x^3}\,dx$

2. $\displaystyle\int \sqrt{x}(1 + \sqrt{x})^3\,dx$

3. $\displaystyle\int (1 - 3x)^9\,dx$

4. $\displaystyle\int \frac{7}{(2x + 3)^3}\,dx$

5. $\displaystyle\int \sqrt[3]{9 + 4x}\,dx$

6. $\displaystyle\int \frac{24}{\sqrt{6x + 7}}\,dx$

7. $\displaystyle\int x^3(1 + x^4)^5\,dx$

8. $\displaystyle\int 3x^2\sqrt{4 + x^3}\,dx$

9. $\displaystyle\int x\sqrt[3]{1 - x^2}\,dx$

10. $\displaystyle\int \frac{3x}{\sqrt{1 + 3x^2}}\,dx$

11. $\displaystyle\int (7\cos 5x - 5\sin 7x)\,dx$

12. $\displaystyle\int 5\sin^3 4x\cos 4x\,dx$

13. $\displaystyle\int x^3\sqrt{1 + x^4}\,dx;\quad u = x^4$

14. $\displaystyle\int \sin^2 x\cos x\,dx;\quad u = \sin x$

15. $\displaystyle\int \frac{1}{\sqrt{x}(1 + \sqrt{x})^2}\,dx;\quad u = 1 + \sqrt{x}$

16. $\displaystyle\int \frac{1}{\sqrt{x}(1 + \sqrt{x})^2}\,dx;\quad u = \sqrt{x}$

17. $\displaystyle\int x^2\cos 4x^3\,dx;\quad u = 4x^3$

18. $\displaystyle\int x(x + 1)^{14}\,dx;\quad u = x + 1$

19. $\displaystyle\int x(x^2 + 1)^{14}\,dx;\quad u = x^2 + 1$

20. $\displaystyle\int x^3\cos x^4\,dx;\quad u = x^4$

21. $\displaystyle\int x\sqrt{4 - x}\,dx;\quad u = 4 - x$

22. $\displaystyle\int \frac{x + 2x^3}{(x^4 + x^2)^3}\,dx;\quad u = x^4 + x^2$

23. $\displaystyle\int \frac{2x^3}{\sqrt{1 + x^4}}\,dx;\quad u = x^4$

24. $\displaystyle\int \frac{2x + 1}{\sqrt{x^2 + x}}\,dx;\quad u = x^2 + x$

Solve the initial value problems in 25 through 30.

25. $\dfrac{dy}{dx} = 3x^2 + 2x;\quad y(0) = 5$

26. $\dfrac{dy}{dx} = 3\sqrt{x};\quad y(4) = 20$

27. $\dfrac{dy}{dx} = (2x+1)^5;\quad y(0) = 2$

28. $\dfrac{dy}{dx} = \dfrac{2}{\sqrt{x+5}};\quad y(4) = 3$

29. $\dfrac{dy}{dx} = \dfrac{1}{\sqrt[3]{x}};\quad y(1) = 1$

30. $\dfrac{dy}{dx} = 1 - \cos x;\quad y(0) = 0$

31. When its brakes are fully applied, a certain automobile has a constant deceleration of 22 ft/s^2. If its initial velocity is 90 mi/h, how long will it take to come to a stop? How many feet will it travel during that time?

32. In Hal Clement's novel *Mission of Gravity*, much of the action takes place in the polar regions of the planet Mesklin, where the acceleration of gravity is 22500 ft/s^2. A stone is dropped near the north pole of Mesklin from a height of 450 ft. How long does it remain aloft? With what speed does it strike the ground?

33. An automobile is traveling along the x-axis in the positive direction. At time $t = 0$ its brakes are fully applied, and the car experiences a constant deceleration of 40 ft/s^2 while skidding. The car skids 180 ft before coming to a stop. What was its initial velocity?

34. If a car starts from rest with an acceleration of 8 ft/s^2, how far has it traveled by the time it reaches a speed of 60 mi/h?

35. On the planet Zorg, a ball dropped from a height of 20 ft hits the ground in 2 s. If the ball is dropped from the top of a 200-ft building on Zorg, how long will it take to reach the ground? With what speed will it hit?

36. Suppose that you can throw a ball from the earth's surface straight upward to a maximum height of 144 ft. (a) How high could you throw it on the planet Zorg (Problem 35)? (b) How high could you throw it in the polar regions of Mesklin (Problem 32)?

37. Suppose that a car skids 44 ft if its velocity is 30 mi/h when the brakes are fully applied. Assuming the same constant deceleration, how far will it skid if its velocity is 60 mi/h when the brakes are fully applied?

38. The graph of the velocity of a model rocket fired at time $t = 0$ is shown in Fig. 5.MP.1. (a) At what time was the fuel exhausted? (b) At what time did the parachute open? (c) At what time did the rocket reach its maximum altitude? (d) At what time did the rocket land? (e) How high did the rocket go? (f) How high was the pole on which the rocket landed?

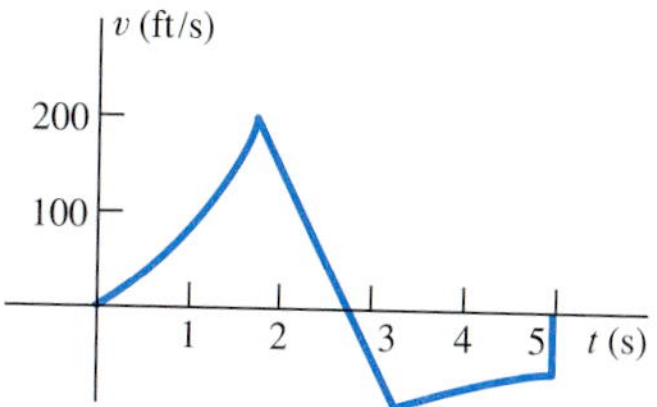

Fig. 5.MP.1 Rocket velocity graph for Problem 38

Find the sums in Problems 39 through 42.

39. $\displaystyle\sum_{i=1}^{100} 17$

40. $\displaystyle\sum_{k=1}^{100}\left(\frac{1}{k} - \frac{1}{k+1}\right)$

41. $\displaystyle\sum_{n=1}^{10} (3n-2)^2$

42. 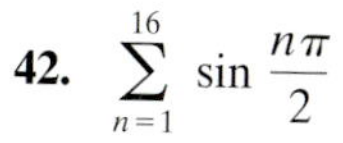 $\displaystyle\sum_{n=1}^{16} \sin\frac{n\pi}{2}$

In Problems 43 through 45, find the limit of the given Riemann sum associated with a regular partition of the indicated interval $[a, b]$. First express it as in integral from a to b; then evaluate that integral.

43. $\displaystyle\lim_{n\to\infty}\sum_{i=1}^{n}\frac{\Delta x}{\sqrt{x_i^\star}};\quad [1, 2]$

44. $\displaystyle\lim_{n\to\infty}\sum_{i=1}^{n}\left[(x_i^\star)^2 - 3x_i^\star\right]\Delta x;\quad [0, 3]$

45. $\displaystyle\lim_{n\to\infty}\sum_{i=1}^{n} 2\pi x_i^\star\sqrt{1 + (x_i^\star)^2}\,\Delta x;\quad [0, 1]$

46. Evaluate

$$\lim_{n\to\infty}\frac{1^{10} + 2^{10} + 3^{10} + \cdots + n^{10}}{n^{11}}$$

by expressing this limit as an integral over $[0, 1]$.

47. Use Riemann sums to prove that if $f(x) \equiv c$ (a constant), then

$$\int_a^b f(x)\,dx = c(b-a).$$

48. Use Riemann sums to prove that if f is continuous on $[a, b]$ and $f(x) \geqq 0$ for all x in $[a, b]$, then

$$\int_a^b f(x)\,dx \geqq 0.$$

49. Use the comparison property of integrals (Section 5.5) to prove that

$$\int_a^b f(x)\,dx > 0$$

if f is a continuous function with $f(x) > 0$ on $[a, b]$.

Evaluate the integrals in Problems 50 through 63.

50. $\displaystyle\int_0^1 (1 - x^2)^3\,dx$

51. $\displaystyle\int\left(\sqrt{2x} - \frac{1}{\sqrt{3x^3}}\right)dx$

52. $\displaystyle\int\frac{(1 + \sqrt[3]{x})^2}{\sqrt{x}}\,dx$

53. $\displaystyle\int\frac{4 - x^3}{2x^2}\,dx$

54. $\displaystyle\int_0^1 \frac{dt}{(3-2t)^2}$

55. $\displaystyle\int \sqrt{x}\cos x\sqrt{x}\,dx$

56. $\displaystyle\int_0^2 x^2\sqrt{9-x^3}\,dx$

57. $\displaystyle\int \frac{1}{t^2}\sin\frac{1}{t}\,dt$

58. $\displaystyle\int_1^2 \frac{2t+1}{\sqrt{t^2+t}}\,dt$

59. $\displaystyle\int \frac{\sqrt[3]{u}}{(1+u^{4/3})^3}\,du$

60. $\displaystyle\int_0^{\pi/4} \frac{\sin t}{\sqrt{\cos t}}\,dt$

61. $\displaystyle\int_1^4 \frac{(1+\sqrt{t})^2}{\sqrt{t}}\,dt$

62. $\displaystyle\int \frac{1}{u^2}\sqrt[3]{1-\frac{1}{u}}\,du$

63. $\displaystyle\int \frac{\sqrt{4x^2-1}}{x^4}\,dx$

Find the areas of the plane regions bounded by the curves given in Problems 64 through 70.

64. $y = x^3, \quad x = -1, \quad y = 1$

65. $y = x^4, \quad y = x^5$

66. $y^2 = x, \quad 3y^2 = x + 6$

67. $y = x^4, \quad y = 2 - x^2$

68. $y = x^4, \quad y = 2x^2 - 1$

69. $y = (x-2)^2, \quad y = 10 - 5x$

70. $y = x^{2/3}, \quad y = 2 - x^2$

71. Evaluate the integral

$$\int_0^2 \sqrt{2x - x^2}\,dx$$

by interpreting it as the area of a region.

72. Evaluate the integral

$$\int_1^5 \sqrt{6x - 5 - x^2}\,dx$$

by interpreting it as the area of a region.

73. Find a function f such that

$$x^2 = 1 + \int_1^x \sqrt{1 + [f(t)]^2}\,dt$$

for all $x > 1$. (*Suggestion:* Differentiate both sides of the equation with the aid of the fundamental theorem of calculus.)

74. Show that $G'(x) = \phi(h(x)) \cdot h'(x)$ if

$$G(x) = \int_a^{h(x)} \phi(t)\,dt.$$

75. Use right-endpoint and left-endpoint approximations to estimate

$$\int_0^1 \sqrt{1 + x^2}\,dx$$

with error not exceeding 0.05.

76. Calculate the trapezoidal approximation and Simpson's approximation to

$$\int_0^\pi \sqrt{1 - \cos x}\,dx$$

with six subintervals. For comparison, use an appropriate half-angle identity to calculate the exact value of this integral.

77. Calculate the midpoint approximation and trapezoidal approximation to

$$\int_1^2 \frac{1}{x + x^2}\,dx$$

with $n = 5$ subintervals. Then explain why the exact value of the integral lies between these two approximations.

In Problems 78 through 80, let $\{x_0, x_1, x_2, \dots, x_n\}$ be a partition of $[a, b]$, where $a < b$.

78. For $i = 1, 2, 3, \dots, n$, let $x_i^\star$ be given by

$$(x_i^\star)^2 = \tfrac{1}{3}\left[(x_{i-1})^2 + x_{i-1}x_i + (x_i)^2\right].$$

Show first that $x_{i-1} < x_i^\star < x_i$. Then use the algebraic identity

$$(c - d)(c^2 + cd + d^2) = c^3 - d^3$$

to show that

$$\sum_{i=1}^n (x_i^\star)^2\,\Delta x_i = \tfrac{1}{3}(b^3 - a^3).$$

Explain why this computation proves that

$$\int_a^b x^2\,dx = \tfrac{1}{3}(b^3 - a^3).$$

79. Let $x_i^\star = \sqrt{x_{i-1}x_i}$ for $i = 1, 2, 3, \dots, n$, and assume that $0 < a < b$. Show that

$$\sum_{i=1}^n \frac{\Delta x_i}{(x_i^\star)^2} = \frac{1}{a} - \frac{1}{b}.$$

Then explain why this computation proves that

$$\int_a^b \frac{dx}{x^2} = \frac{1}{a} - \frac{1}{b}.$$

80. Assume that $0 < a < b$. Define

$$\sqrt{x_i^\star} = \frac{\tfrac{2}{3}\left[(x_i)^{3/2} - (x_{i-1})^{3/2}\right]}{x_i - x_{i-1}}.$$

First show that $x_{i-1} < x_i^\star < x_i$. Then use this selection for the given partition to prove that

$$\int_a^b \sqrt{x}\,dx = \frac{2}{3}(b^{3/2} - a^{3/2}).$$

CHAPTER 6

APPLICATIONS OF THE INTEGRAL

G.F.B. Riemann (1826–1866)

The general concept of integration traces back to the area and volume computations of ancient times, but the integrals used by Newton and Leibniz were not defined with sufficient precision for full understanding. We owe to the German mathematician G. F. Bernhard Riemann the modern definition that uses "Riemann sums." The son of a Protestant minister, Riemann studied theology and philology at Göttingen University until finally gaining his father's permission to concentrate on mathematics. He transferred to Berlin University, where he received his Ph.D. in 1851. His work during the next decade justifies his place on everyone's short list of the most profound and creative mathematicians of all time. But in 1862 he was stricken ill. He never fully recovered and in 1866 died prematurely at the age of 39.

Riemann's mathematical investigations were as varied as they were deep, ranging from the basic concepts of functions and integrals to such areas as non-Euclidean (differential) geometry and the distribution of prime numbers. Recall that the positive integer p is *prime* if it cannot be factored into smaller integers. In a famous paper of 1859, Riemann analyzed the approximation

$$\pi(x) \approx \int_2^x \frac{dt}{\ln t} = \text{li}(x)$$

to the number $\pi(x)$ of those primes $p \leqq x$ (with $\ln x$ denoting the natural logarithm of x). There is a remarkable correspondence between the values of $\pi(x)$ and the "logarithmic integral" approximation $\text{li}(x)$:

x	1,000,000	10,000,000
$\text{li}(x)$	78,628	664,918
$\pi(x)$	78,498	664,579
error	0.165%	0.051%
x	100,000,000	1,000,000,000
$\text{li}(x)$	5,762,209	50,849,235
$\pi(x)$	5,761,455	50,847,534
error	0.013%	0.003%

Thirty years after Riemann's death, his ideas led ultimately to a proof that the percentage error in the approximation $\text{li}(x)$ to $\pi(x)$ approaches 0 as $x \to \infty$.

In his 1851 thesis, Riemann introduced a geometric way of visualizing "multivalued" functions such as the square root function with two values $\pm\sqrt{x}$. The following graph illustrates the *cube root* function. For each complex number $z = x + iy$ in the unit disk $x^2 + y^2 \leqq 1$, the three (complex) cube roots of z are plotted directly above z. Each root is plotted at a height equal to its real part, with color determined by its imaginary part. The result is the "Riemann surface" of the cube root function.

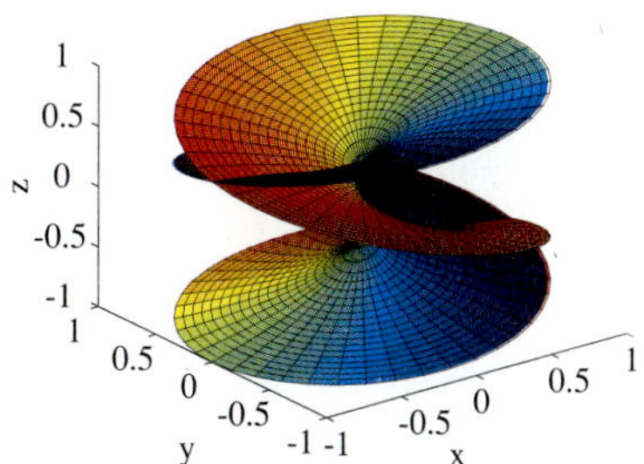

6.1 SETTING UP INTEGRAL FORMULAS

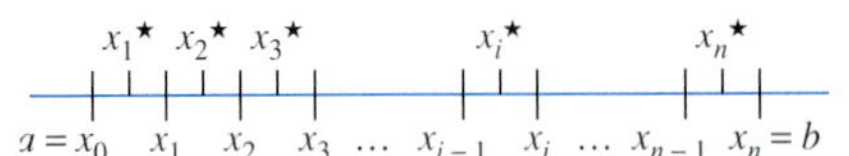

Fig. 6.1.1 A division (or partition) of $[a, b]$ into n equal-length subintervals

In Section 5.4 we defined the integral of the function f on the interval $[a, b]$ as a limit of Riemann sums. Specifically, let the interval $[a, b]$ be divided into n subintervals, all with the same length $\Delta x = (b - a)/n$ (Fig. 6.1.1). Then a selection of numbers $x_1^\star, x_2^\star, \dots, x_n^\star$ in these subintervals ($x_i^\star$ being a point of the ith subinterval $[x_{i-1}, x_i]$) produces a Riemann sum

$$\sum_{i=1}^{n} f(x_i^\star)\,\Delta x \tag{1}$$

whose value approximates the integral of f on $[a, b]$. The value of the integral is the limiting value (if any) of such sums as the subinterval length Δx approaches zero. That is,

$$\int_a^b f(x)\,dx = \lim_{\Delta x \to 0} \sum_{i=1}^{n} f(x_i^\star)\,\Delta x. \tag{2}$$

The wide applicability of the definite integral arises from the fact that many geometric and physical quantities can be approximated arbitrarily closely by Riemann sums. Such approximations lead to integral formulas for the computation of such quantities.

For example, suppose that $f(x)$ is positive-valued on $[a, b]$ and that our goal—as in Section 5.3—is to calculate the area A of the region that lies below the graph of $y = f(x)$ over the interval $[a, b]$. Beginning with the subdivision (or partition) of $[a, b]$ indicated in Fig. 6.1.1, let ΔA_i denote the area of the vertical "strip" that lies under $y = f(x)$ over the ith subinterval $[x_{i-1}, x_i]$. Then, as illustrated in Fig. 6.1.2, the "strip areas"

$$\Delta A_1, \Delta A_2, \dots, \Delta A_n$$

add up to the total area A:

$$A = \sum_{i=1}^{n} \Delta A_i. \tag{3}$$

But the ith strip is approximated by a rectangle with base $[x_{i-1}, x_i]$ and height $f(x_i^\star)$, so its area is given approximately by

$$\Delta A_i \approx f(x_i^\star)\,\Delta x. \tag{4}$$

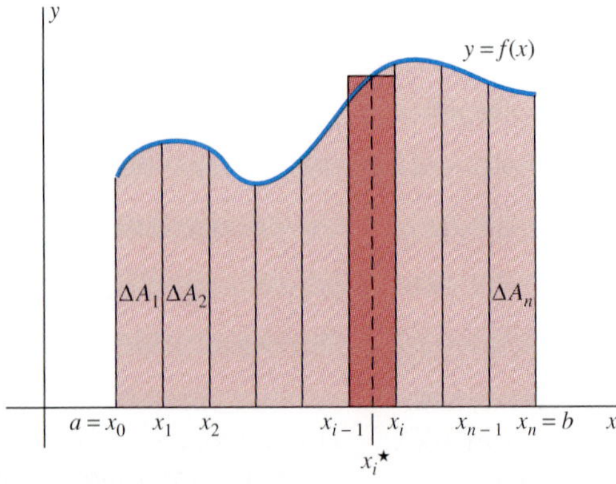

Fig. 6.1.2 Approximating an area by means of a Riemann sum

After we substitute Eq. (4) into Eq. (3), it becomes apparent that the total area A under the graph of f is given approximately by

$$A \approx \sum_{i=1}^{n} f(x_i^\star)\,\Delta x. \tag{5}$$

Note that the approximating sum on the right is a Riemann sum for f on $[a, b]$. Moreover,

1. It is intuitively evident that the Riemann sum in (5) approaches the actual area A as $n \to +\infty$ (which forces $\Delta x \to 0$);
2. By the definition of the integral, this Riemann sum approaches $\int_a^b f(x)\,dx$ as $n \to +\infty$.

These observations justify the *definition* of the area A by the formula

$$A = \int_a^b f(x)\,dx. \tag{6}$$

Other Quantities as Integrals

Our justification of the area formula in Eq. (6) illustrates an important general method of setting up integral formulas. Suppose that we want to calculate a certain quantity Q that is associated with an interval $[a, b]$ in such a way that subintervals of $[a, b]$ correspond to specific portions of Q (such as the portion of area lying above a particular subinterval). Then a subdivision of $[a, b]$ into n subintervals produces portions

$$\Delta Q_1, \Delta Q_2, \ldots, \Delta Q_n,$$

which add up to the quantity

$$Q = \sum_{i=1}^{n} \Delta Q_i. \tag{7}$$

Now suppose that we can find a function f such that the ith portion ΔQ_i is given approximately by

$$\Delta Q_i \approx f(x_i^\star)\,\Delta x \tag{8}$$

(for each i, $1 \leqq i \leqq n$) for a selected point $x_i^\star$ of the ith subinterval $[x_{i-1}, x_i]$ of $[a, b]$. Then substitution of Eq. (8) into Eq. (7) yields the Riemann sum approximation

$$Q \approx \sum_{i=1}^{n} f(x_i^\star)\,\Delta x \tag{9}$$

analogous to the approximation in Eq. (5). The right-hand sum in Eq. (9) is a Riemann sum that approaches the integral

$$\int_a^b f(x)\,dx \quad \text{as} \quad n \to +\infty.$$

If it is also evident—for geometric or physical reasons, for example—that this Riemann sum must approach the quantity Q as $n \to +\infty$, then Eq. (9) justifies our setting up the integral formula

$$Q = \int_a^b f(x)\,dx. \tag{10}$$

Because the right-hand side in Eq. (10) is frequently easy to calculate (by the fundamental theorem of calculus), this gives us a practical way of finding the exact numerical value of the quantity Q.

In addition to area, the following are some of the quantities that can be calculated by using integral formulas such as Eq. (10). (The variable x is replaced by t where appropriate.)

- The *mass* of a thin rod of variable density lying along the interval $a \leqq x \leqq b$
- The *profit* earned by a company between time $t = a$ and time $t = b$
- The *number* of people in a city who contract a certain disease between time $t = a$ and time $t = b$
- The *distance* traveled by a moving particle during the time interval $a \leqq t \leqq b$
- The *volume* of water flowing into a tank during the time interval $a \leqq t \leqq b$
- The *work* done by a variable force in moving a particle from the point $x = a$ to the point $x = b$

In each case it is evident that a *subinterval* of $[a, b]$ determines a specific portion ΔQ of the whole quantity Q that corresponds to the whole interval $[a, b]$. The question is this: What function f should be integrated from a to b? Examples 1 through 3 illustrate the process of finding the needed function f by approximating the portion ΔQ_i of the quantity Q that corresponds to the subinterval $[x_{i-1}, x_i]$. An approximation of the form

$$\Delta Q_i \approx f(x_i^\star)\,\Delta x \tag{8}$$

leads to the desired integral formula

$$Q = \int_a^b f(x)\,dx. \tag{10}$$

The integral in Eq. (10) results from the summation in Eq. (9) when we make the following replacements:

$$\sum_{i=1}^{n} \quad \text{becomes} \quad \int_a^b,$$
$$x_i^\star \quad \text{becomes} \quad x, \quad \text{and}$$
$$\Delta x \quad \text{becomes} \quad dx.$$

Fig. 6.1.3 The tank of Example 1

EXAMPLE 1 Suppose that water is pumped into the initially empty tank of Fig. 6.1.3. The rate of water flow into the tank at time t (in seconds) is $50 - t$ liters (L) per second. How much water flows into the tank during the first 30 s?

Solution We want to compute the amount Q of water that flows into the tank during the time interval $[0, 30]$. Think of a subdivision of $[0, 30]$ into n subintervals, all with the same length $\Delta t = 30/n$.

Next choose a point $t_i^\star$ in the ith subinterval $[t_{i-1}, t_i]$. If this subinterval is very short, then the rate of water flow between time t_{i-1} and time t_i remains approximately $50 - t_i^\star$ liters per second. So the amount ΔQ_i of water *in liters* that flows into the tank during this subinterval of time is obtained approximately by multiplying the *flow rate in liters per second* by the *duration of flow in seconds:*

$$\left[(50 - t_i^\star)\,\frac{\text{liters}}{\text{second}}\right]\cdot[\Delta t\ \text{second}],$$

and hence

$$\Delta Q_i \approx (50 - t_i^\star)\,\Delta t \quad \text{(liters)}.$$

Therefore, the total amount Q that we seek is given approximately by

$$Q = \sum_{i=1}^{n} \Delta Q_i \approx \sum_{i=1}^{n} (50 - t_i^\star)\,\Delta t \quad \text{(liters)}.$$

We recognize that the sum on the right is a Riemann sum, and—most important—we see that it is a Riemann sum for the function $f(t) = 50 - t$. Hence we may conclude that

$$Q = \lim_{n\to\infty} \sum_{i=1}^{n} (50 - t_i^\star)\Delta t = \int_0^{30} (50 - t)\,dt$$

$$= \left[50t - \frac{1}{2}t^2\right]_0^{30} = 1050 \quad \text{(liters)}.$$

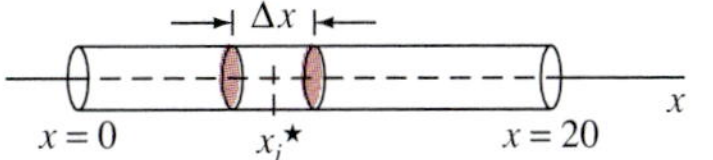

Fig. 6.1.4 The 20-cm rod of Example 2

EXAMPLE 2 Figure 6.1.4 shows a thin rod 20 cm long. Its (linear) *density* at the point x is $15 + 2x$ grams of mass per centimeter of the rod's length (g/cm). The rod's density thus varies from 15 g/cm at the left end $x = 0$ to 55 g/cm at the right end $x = 20$. Find the total mass M of this rod.

Solution Think of a subdivision of $[0, 20]$ into n subintervals of length $\Delta x = 20/n$ each. Figure 6.1.4 shows the short piece of the rod that corresponds to the typical ith subinterval $[x_{i-1}, x_i]$. If $x_i^\star$ is, say, the midpoint of $[x_{i-1}, x_i]$, then the density of this short piece is obtained approximately by multiplying its *density in grams per centimeter* by its *length in centimeters:*

$$\left[(15 + 2x_i^\star)\frac{\text{grams}}{\text{centimeter}}\right]\cdot[\Delta x\ \text{centimeters}].$$

That is,

$$\Delta M_i \approx (15 + 2x_i^\star)\,\Delta x \quad \text{(grams)}.$$

Therefore the total mass M of the entire rod is given approximately by

$$M = \sum_{i=1}^{n} \Delta M_i \approx \sum_{i=1}^{n} (15 + 2x_i^\star)\,\Delta x.$$

We recognize a Riemann sum on the right, as in Example 1, although this time for the function $f(x) = 15 + 2x$ on the interval $[0, 20]$. Hence we may conclude that

$$M = \lim_{n\to\infty} \sum_{i=1}^{n} (15 + 2x_i^\star)\,\Delta x = \int_0^{20} (15 + 2x)\,dx$$

$$= \left[15x + x^2\right]_0^{20} = 700 \quad \text{(g)}.$$

EXAMPLE 3 We deduced in Example 5 of Section 5.8 the area formula for a circle, $A = \pi r^2$, from the definition of the number π as the area of a circle of radius 1. Here we take a different approach and begin instead with the formula

$$C = 2\pi r \tag{11}$$

for the *circumference* of a circle of radius r.

Figure 6.1.5 shows a circular disk of radius r divided into annular rings (washers) by concentric circles whose radii are the points $x_1, x_2, \ldots, x_n = r$ of a subdivision of the x-interval $[0, r]$ into n equal-length subintervals. The ith annular ring corresponding to the ith subinterval $[x_{i-1}, x_i]$ is bounded by the circles of radii x_{i-1} and x_i. To approximate its area ΔA_i, we think of cutting this annular ring of the circular disk and straightening it into a long strip of width Δx (Fig. 6.1.6). The lengths of the top and bottom edges of this strip are simply the circumferences $2\pi x_{i-1}$ and $2\pi x_i$ of the two circles that once bounded the strip. It follows that

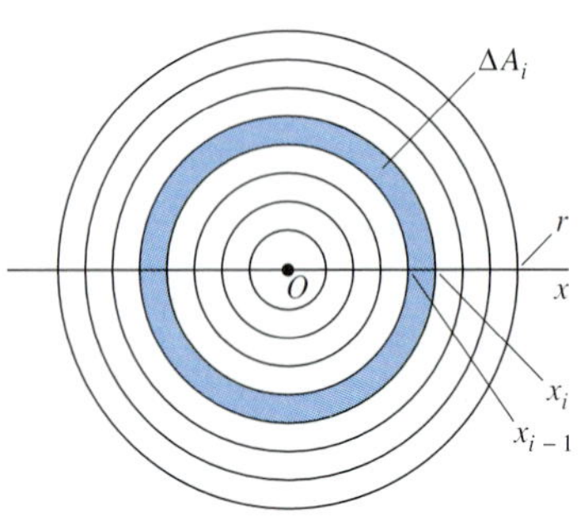

Fig. 6.1.5 The circle divided into annular rings (Example 3)

$$2\pi x_{i-1}\,\Delta x \leqq \Delta A_i \leqq 2\pi x_i\,\Delta x,$$

so

$$\Delta A_i \approx 2\pi x_i^\star\,\Delta x$$

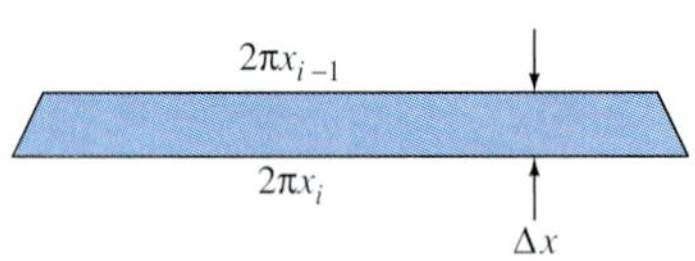

Fig. 6.1.6 An annular ring "straightened out" (Example 3)

for any selected $x_i^\star$ in $[x_{i-1}, x_i]$. Summing the areas of all these annular rings that make up the entire circular disk, we get the approximation

$$A = \sum_{i=1}^{n} \Delta A_i \approx \sum_{i=1}^{n} 2\pi x_i^\star\,\Delta x.$$

Recognizing on the right a Riemann sum for $f(x) = 2\pi x$, we conclude that the area of a circle of radius r is given by

$$A = \lim_{n\to\infty} \sum_{i=1}^{n} 2\pi x_i^\star\,\Delta x = \int_0^r 2\pi x\,dx = \Big[\pi x^2\Big]_0^r = \pi r^2. \quad ■$$

In setting up integral formulas as in Examples 1 through 3, the key is to recognize a Riemann sum when one is present.

EXAMPLE 4 Calculate Q if

$$Q = \lim_{n\to\infty} \sum_{i=1}^{n} m_i\sqrt{25 - m_i^2}\,\Delta x,$$

where (for each $i = 1, 2, \ldots, n$) m_i denotes the midpoint of the ith subinterval $[x_{i-1}, x_i]$ of a partition of the interval $[3, 4]$ into n subintervals, all with the same length $\Delta x = 1/n$.

Solution We recognize the given sum as a Riemann sum (with $x_i^\star = m_i$) for the function $f(x) = x\sqrt{25 - x^2}$. Hence

$$Q = \int_3^4 x\sqrt{25 - x^2}\,dx = \Big[-\tfrac{1}{3}(25 - x^2)^{3/2}\Big]_3^4 = \tfrac{1}{3}(16^{3/2} - 9^{3/2}) = \tfrac{37}{3}. \quad ■$$

Distance and Velocity

Consider a particle that travels with velocity $v = f(t)$ at time t along a (directed) straight line. We want to compute the **net distance** s it travels between time $t = a$ and time $t = b$—that is, the distance between its initial position at time $t = a$ and its final position at time $t = b$.

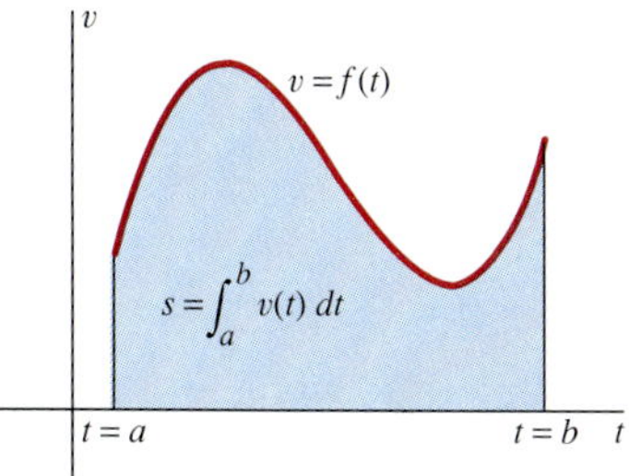

Fig. 6.1.7 Equation (13) means that the (net) distance traveled is equal to the (signed) area under the velocity curve.

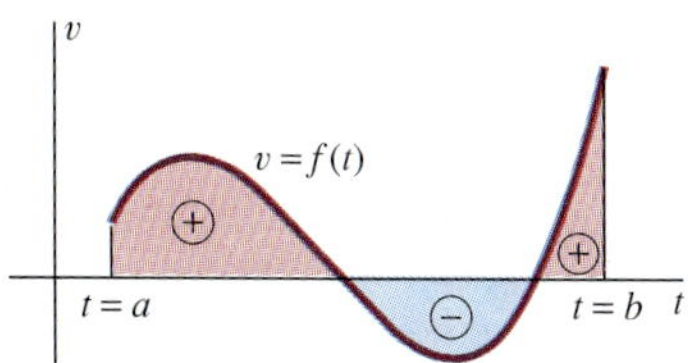

Fig. 6.1.8 Velocity curve of a particle that first travels forward, then backward, then forward again.

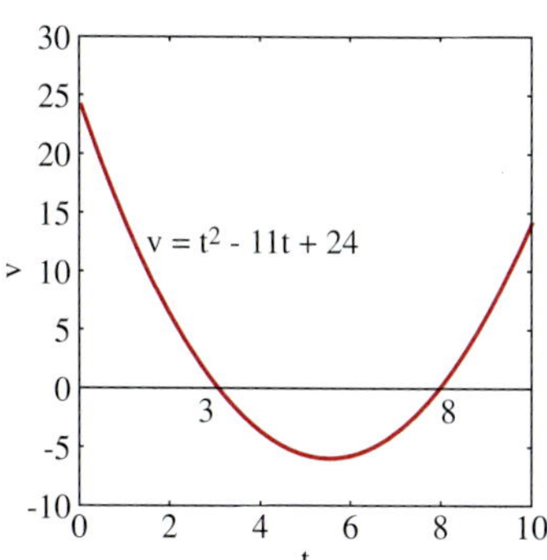

Fig. 6.1.9 Graph of the velocity function of Example 5

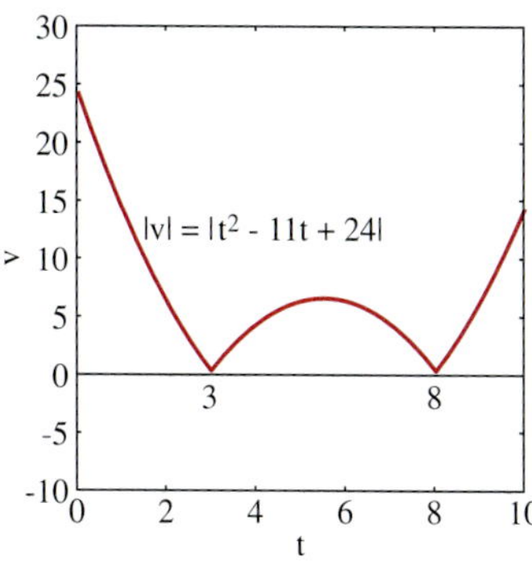

Fig. 6.1.10 Graph of the absolute value of the velocity function of Example 5

We begin with a partition of $[a, b]$ into n subintervals, all with the same length $\Delta t = (b - a)/n$. If $t_i^\star$ is an arbitrary point of the ith subinterval $[t_{i-1}, t_i]$, then during this subinterval the velocity of the particle is approximately $f(t_i^\star)$. Hence the distance Δx_i traveled by the particle between time $t = t_{i-1}$ and time $t = t_i$ is given approximately by

$$\Delta s_i \approx f(t_i^\star)\,\Delta t.$$

Consequently, the net distance s is given approximately by

$$s = \sum_{i=1}^{n} \Delta s_i \approx \sum_{i=1}^{n} f(t_i^\star)\,\Delta t. \tag{12}$$

We recognize the right-hand approximation as a Riemann sum for the function $f(t)$ on the interval $[a, b]$, and we reason that this approximation should approach the actual net distance s as $n \to +\infty$. Thus we conclude that

$$s = \int_a^b f(t)\,dt = \int_a^b v\,dt. \tag{13}$$

Note that the *net distance is the definite integral of the velocity* (Fig. 6.1.7).

If the velocity function $v = f(t)$ attains both positive and negative values between $t = a$ and $t = b$—as illustrated in Fig. 6.1.8—then forward and backward distances cancel when we compute the net distance s given by the integral formula in Eq. (13). The **total distance** traveled, irrespective of direction, may be found by evaluating the integral of the *absolute value* of the velocity. Thus the total distance is given by

$$\int_a^b |f(t)|\,dt = \int_a^b |v|\,dt. \tag{14}$$

This integral may be computed by integrating separately over the subintervals where v is positive and those where v is negative and then adding the absolute values of the results. This is exactly the same procedure we use to find the area between the graph of a function and the x-axis when the function takes on both positive and negative values.

EXAMPLE 5 Suppose that the velocity of a moving particle is $v(t) = t^2 - 11t + 24$ (ft/s). Find both the net distance and the total distance it travels between the times $t = 0$ and $t = 10$ (s).

Solution For net distance, we use Eq. (10) and find that

$$v = \int_0^{10} (t^2 - 11t + 24)\,dt = \left[\tfrac{1}{3}t^3 - \tfrac{11}{2}t^2 + 24t\right]_0^{10}$$
$$= \tfrac{1000}{3} - \tfrac{1100}{2} + 240 = \tfrac{70}{3} \quad \text{(ft)}.$$

To find the total distance traveled, we note from the graph of f in Fig. 6.1.9—or from the factorization $v(t) = (t - 3)(t - 8)$—that $v(t) > 0$ if $0 \leqq t < 3$, $v(t) < 0$ if $3 < t < 8$, and $v(t) > 0$ if $8 < t \leqq 10$. We need to integrate the absolute value of $v(t)$ shown in Fig. 6.1.10. Now

$$\int_0^3 (t^2 - 11t + 24)\,dt = \left[\tfrac{1}{3}t^3 - \tfrac{11}{2}t^2 + 24t\right]_0^3 = \tfrac{63}{2},$$

$$\int_3^8 (-t^2 + 11t - 24)\, dt = \left[-\tfrac{1}{3}t^3 + \tfrac{11}{2}t^2 - 24t \right]_3^8 = \tfrac{125}{6}, \quad \text{and}$$

$$\int_8^{10} (t^2 - 11t + 24)\, dt = \left[\tfrac{1}{3}t^3 - \tfrac{11}{2}t^2 + 24t \right]_8^{10} = \tfrac{38}{3}.$$

Thus the particle first travels 63/2 ft forward, then 125/6 ft backward, and finally 38/3 ft forward, for a total distance traveled of $\frac{63}{2} + \frac{125}{6} + \frac{38}{3} = 65$ ft. ■

6.1 PROBLEMS

In Problems 1 through 10, $x_i^\star$ denotes a selected point, and m_i the midpoint, of the ith subinterval $[x_{i-1}, x_i]$ of a partition of the indicated interval $[a, b]$ into n subintervals each of length Δx. Evaluate the given limit by computing the value of the appropriate related integral.

1. $\displaystyle\lim_{n\to\infty} \sum_{i=1}^{n} 2x_i^\star \,\Delta x;\quad a = 0, b = 1$

2. $\displaystyle\lim_{n\to\infty} \sum_{i=1}^{n} \frac{\Delta x}{(x_i^\star)^2};\quad a = 1, b = 2$

3. $\displaystyle\lim_{n\to\infty} \sum_{i=1}^{n} (\sin \pi x_i^\star)\,\Delta x;\quad a = 0, b = 1$

4. $\displaystyle\lim_{n\to\infty} \sum_{i=1}^{n} [3(x_i^\star)^2 - 1]\,\Delta x;\quad a = -1, b = 3$

5. $\displaystyle\lim_{n\to\infty} \sum_{i=1}^{n} x_i^\star\sqrt{(x_i^\star)^2 + 9}\,\Delta x;\quad a = 0, b = 4$

6. $\displaystyle\lim_{n\to\infty} \sum_{i=1}^{n} (x_i)^2\,\Delta x;\quad a = 2, b = 4$

7. $\displaystyle\lim_{n\to\infty} \sum_{i=1}^{n} (2m_i - 1)\,\Delta x;\quad a = -1, b = 3$

8. $\displaystyle\lim_{n\to\infty} \sum_{i=1}^{n} \sqrt{2m_i + 1}\,\Delta x;\quad a = 0, b = 4$

9. $\displaystyle\lim_{n\to\infty} \sum_{i=1}^{n} \frac{m_i}{\sqrt{(m_i)^2 + 16}}\,\Delta x;\quad a = -3, b = 0$

10. $\displaystyle\lim_{n\to\infty} \sum_{i=1}^{n} m_i \cos(m_i^2)\,\Delta x;\quad a = 0, b = \sqrt{\pi}$

The notation in Problems 11 through 14 is the same as in Problems 1 through 10. Express the given limit as an integral involving the function f.

11. $\displaystyle\lim_{n\to\infty} \sum_{i=1}^{n} 2\pi x_i^\star f(x_i^\star)\,\Delta x;\quad a = 1, b = 4$

12. $\displaystyle\lim_{n\to\infty} \sum_{i=1}^{n} [f(x_i^\star)]^2\,\Delta x;\quad a = -1, b = 1$

13. $\displaystyle\lim_{n\to\infty} \sum_{i=1}^{n} \sqrt{1 + [f(x_i^\star)]^2}\,\Delta x;\quad a = 0, b = 10$

14. $\displaystyle\lim_{n\to\infty} \sum_{i=1}^{n} 2\pi m_i\sqrt{1 + [f(m_i)]^2}\,\Delta x;\quad a = -2, b = 3$

In Problems 15 through 18, a rod coinciding with the interval $[a, b]$ on the x-axis (units in centimeters) has the specified density function $\rho(x)$ that gives its density (in grams per centimeter) at the point x. Find the mass M of the rod.

15. $a = 0, b = 100;\quad \rho(x) = \frac{1}{5}x$

16. $a = 0, b = 25;\quad \rho(x) = 60 - 2x$

17. $a = 0, b = 10;\quad \rho(x) = x(10 - x)$

18. $a = 0, b = 10;\quad \rho(x) = 10 \sin \dfrac{\pi x}{10}$

In Problems 19 through 30, compute both the net distance and the total distance traveled between time $t = a$ and time $t = b$ by a particle moving with the given velocity function $v = f(t)$ along a line.

19. $v = -32;\quad a = 0, b = 10$

20. $v = 2t + 10;\quad a = 1, b = 5$

21. $v = 4t - 25;\quad a = 0, b = 10$

22. $v = |2t - 5|;\quad a = 0, b = 5$

23. $v = 4t^3;\quad a = -2, b = 3$

24. $v = t - \dfrac{1}{t^2};\quad a = 0.1, b = 1$

25. $v = \sin 2t;\quad a = 0, b = \dfrac{\pi}{2}$

26. $v = \cos 2t;\quad a = 0, b = \dfrac{\pi}{2}$

27. $v = \cos \pi t;\quad a = -1, b = 1$

28. $v = \sin t + \cos t;\quad a = 0, b = \pi$

29. $v = t^2 - 9t + 14;\quad a = 0, b = 10$

30. $v = t^3 - 8t^2 + 15t;\quad a = 0, b = 6$

In Problems 31 through 34, use a calculator or computer to approximate both the net distance and the total distance traveled by a particle with the given velocity function $v(t)$ during the indicated time interval $[a, b]$. Begin by graphing $v = v(t)$ to estimate the intervals where $v(t) > 0$ and where $v(t) < 0$. You may then integrate numerically if your calculator or computer has this facility.

31. $v(t) = t^3 - 7t + 4;\quad a = 0, b = 3$

32. $v(t) = t^3 - 5t^2 + 10;\quad a = 0, b = 5$

33. $v(t) = t \sin t - \cos t;\quad a = 0, b = \pi$

34. $v(t) = \sin t + \sqrt{t} \cos t;\quad a = 0, b = 2\pi$

35. Suppose that the circular disk of Example 3 has mass density $\rho(x)$ (in grams per square centimeter) at distance x from the origin. Then the annular ring of Figs. 6.1.5 and 6.1.6 has density approximately $\rho(x_i^\star)$ at each point. Conclude that the mass M of this disk of radius r is given by

$$M = \int_0^r 2\pi x \rho(x)\,dx.$$

In Problems 36 and 37, use the result of Problem 35 to find the mass of a circular disk with the given radius r and density function ρ.

36. $r = 10,\quad \rho(x) = x$ **37.** $r = 5,\quad \rho(x) = 25 - x^2$

38. If a particle is thrown straight upward from the ground with an initial velocity of 160 ft/s, then its velocity after t seconds is $v = -32t + 160$ feet per second, and it attains its maximum height when $t = 5$ s (and $v = 0$). Use Eq. (13) to compute this maximum height. Check your answer by the methods of Section 5.2.

39. Suppose that the rate of water flow into an initially empty tank is $100 - 3t$ gallons per minute at time t (in minutes). How much water flows into the tank during the interval from $t = 10$ to $t = 20$ min?

40. Suppose that the birth rate in Calgary t years after 1970 was $16 + t$ thousands of births per year. Set up and evaluate an appropriate integral to compute the total number of births between 1970 and 1990.

41. Assume that Calgary (see Problem 40) had a death rate of $5 + \frac{1}{2}t$ thousands per year t years after 1970. If the population of the city was 375,000 in 1970, what was its population in 1990? Consider both births and deaths.

42. The average daily rainfall in Charleston is $r(t)$ inches per day at time t (in days), $0 \leqq t \leqq 365$. Begin with a partition of the interval $[0, 365]$ and derive the formula

$$R = \int_0^{365} r(t)\,dt$$

for the average total annual rainfall R.

43. Take the average daily rainfall of Problem 42 to be

$$r(t) = a - b\cos\frac{2\pi t}{365},$$

where a and b are constants to be determined. If the value of $r(t)$ on January 1 ($t = 0$) is 0.1 in. and the value of $r(t)$ on July 1 ($t = 182.5$) is 0.3 in., what is the average total annual rainfall in this locale?

44. Suppose that the rate of water flow into a tank is $r(t)$ liters per minute at time t (in minutes). Use the method of Example 1 to derive the formula

$$Q = \int_a^b r(t)\,dt$$

for the amount of water that flows into the tank between times $t = a$ and $t = b$.

45. Evaluate

$$\lim_{n\to\infty} \frac{\sqrt[3]{1} + \sqrt[3]{2} + \sqrt[3]{3} + \cdots + \sqrt[3]{n}}{n^{4/3}}$$

by first finding a function f such that the limit is equal to

$$\int_0^1 f(x)\,dx.$$

46. In this problem you are to derive the volume formula $V = \frac{4}{3}\pi r^3$ for a spherical ball of radius r, assuming as *known* the formula $S = 4\pi r^2$ for the surface area of a sphere of radius r. Assume it follows that the volume of a thin spherical shell of radius r and thickness t (Fig. 6.1.11) is given approximately by $\Delta V \approx S \cdot t = 4\pi r^2 t$. Then divide the spherical ball into concentric spherical shells, analogous to the concentric annular rings of Example 3. Finally, interpret the sum of the volumes of these spherical shells as a Riemann sum.

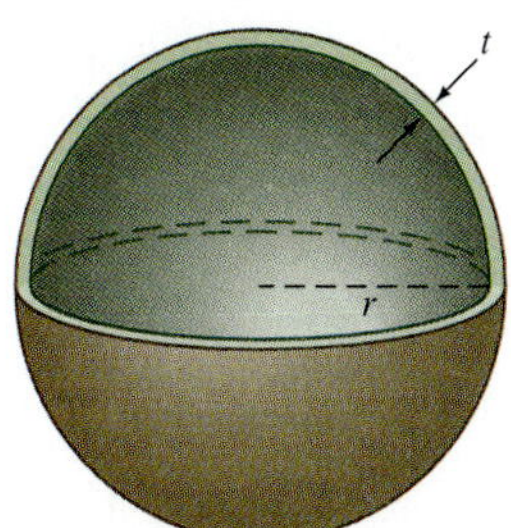

Fig. 6.1.11 A thin spherical shell of thickness t and inner radius r (Problem 46)

47. A spherical ball has radius 1 ft and, at distance x from its center, its density is $100(1 + x)$ lb/ft^3. Use Riemann sums to find a function $f(x)$ such that the weight of the ball is

$$W = \int_0^1 f(x)\,dx$$

(in pounds). Then compute W by evaluating this integral. (*Suggestion:* Given a partition $0 = x_0 < x_1 < x_2 < \cdots < x_n = 1$ of $[0, 1]$, estimate the weight ΔW_i of the spherical shell $x_{i-1} \leqq x \leqq x_i$ of the ball.)

6.2 VOLUMES BY THE METHOD OF CROSS SECTIONS

Here we use integrals to calculate the volumes of certain solids or regions in space. We begin with an intuitive idea of volume as a measure of solids, analogous to area as a measure of plane regions. In particular, we assume that every simply expressible bounded solid region R has a volume measured by a nonnegative number $v(R)$ such that

- If R consists of two nonoverlapping pieces, then $v(R)$ is the sum of *their* volumes;
- Two different solids have the same volume if they have the same size and shape.

The **method of cross sections** is a way of computing the volume of a solid that is described in terms of its cross sections (or "slices") in planes perpendicular to a fixed *reference line* (such as the x-axis or y-axis). For instance, Fig. 6.2.1 shows a solid R with volume $V = v(R)$ lying alongside the interval $[a, b]$ on the x-axis. That is, a plane perpendicular to the x-axis intersects the solid if and only if this plane meets the x-axis in a point of $[a, b]$. Let R_x denote the intersection of R with the perpendicular plane that meets the x-axis at the point x of $[a, b]$. We call R_x the (plane) **cross section** of the solid at x.

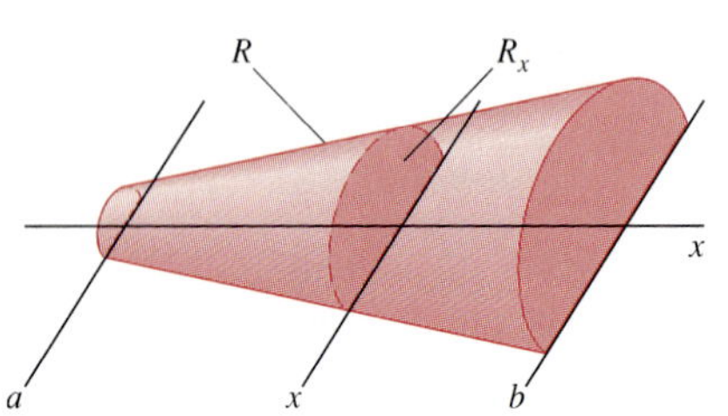

Fig. 6.2.1 R_x is the cross section of R in the plane perpendicular to the x-axis at x.

Volumes of Cylinders

This situation is especially simple if all the cross sections of R are congruent to one another and are parallel translations of one another. In this case the solid R is called a **cylinder** with **bases** R_a and R_b and **height** $h = b - a$. If R_a and R_b are circular disks, then R is the familiar **circular cylinder**. Recall that the volume formula for a circular cylinder of height h and circular base of radius r and area $A = \pi r^2$ is

$$V = \pi r^2 h = Ah.$$

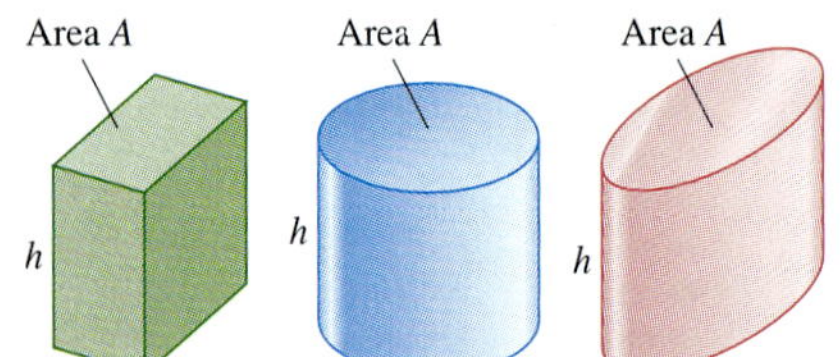

Fig. 6.2.2 Every cylinder of height h and base area A has volume $V = Ah$.

Figure 6.2.2 shows several (general) cylinders with bases of various shapes. The method of cross sections is based on the fact that the volume V of any cylinder—circular or not—is equal to the product of the cylinder's height h and the area A of its base:

$$V = Ah \qquad \text{(volume of a cylinder).} \tag{1}$$

More General Volumes

The volume of a more general solid, as in Fig. 6.2.1, can be approximated by using cylinders. For each x in $[a, b]$, let $A(x)$ denote the area of the cross section R_x of the solid R:

$$A(x) = \text{area}(R_x). \tag{2}$$

We shall assume that the shape of R is sufficiently simple that this **cross-sectional area function** A is continuous (and therefore integrable).

To set up an integral formula for $V = v(R)$, we begin with a partition of $[a, b]$ into n subintervals, all with the same length $\Delta x = (b - a)/n$. Let R_i denote the slab or slice of the solid R positioned alongside the ith subinterval $[x_{i-1}, x_i]$ (Fig. 6.2.3). We denote the volume of this ith slice of R by $\Delta V_i = v(R_i)$, so

$$V = \sum_{i=1}^{n} \Delta V_i.$$

To approximate ΔV_i, we select a typical point $x_i^\star$ in $[x_{i-1}, x_i]$ and consider the *cylinder* C_i whose height is Δx and whose base is the cross section $R_{x_i^\star}$ of R at $x_i^\star$. Figure 6.2.4 suggests that if Δx is small, then $v(C_i)$ is a good approximation to $\Delta V_i = v(R_i)$:

$$\Delta V_i \approx v(C_i) = \text{area}(R_{x_i^\star}) \cdot \Delta x = A(x_i^\star)\,\Delta x,$$

a consequence of Eq. (1) with $A = A(x_i^\star)$ and $h = \Delta x$.

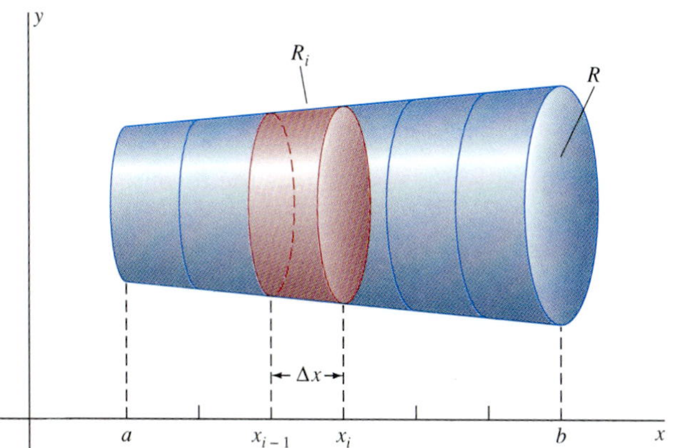

Fig. 6.2.3 Planes through the partition points $x_0, x_1, x_2, \ldots, x_n$ partition the solid R into slabs $R_1, R_2, \ldots, R_n$.

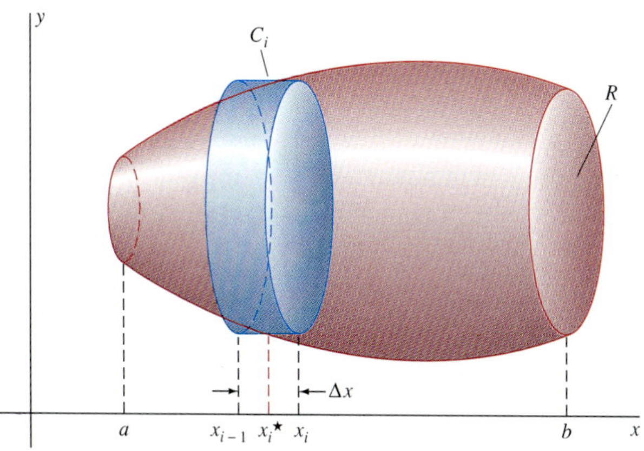

Fig. 6.2.4 The slab R_i is approximated by the cylinder C_i of volume $A(x_i^\star)\,\Delta x$.

Then we add the volumes of these approximating cylinders for $i = 1, 2, 3, \ldots, n$. We find that

$$V = \sum_{i=1}^{n} \Delta V_i \approx \sum_{i=1}^{n} A(x_i^\star)\,\Delta x.$$

We recognize the approximating sum on the right to be a Riemann sum that approaches $\int_a^b A(x)\,dx$ as $n \to +\infty$. This justifies the following *definition* of the volume of a solid R in terms of its cross-sectional area function $A(x)$.

Definition ***Volume by Cross Sections***

If the solid R lies alongside the interval $[a, b]$ on the x-axis and has continuous cross-sectional area function $A(x)$, then its volume $V = v(R)$ is

$$V = \int_a^b A(x)\,dx. \tag{3}$$

Equation (3) is known as **Cavalieri's principle,** after the Italian mathematician Bonaventura Cavalieri (1598–1647), who systematically exploited the fact that the volume of a solid is determined by the areas of its cross sections perpendicular to a given reference line.

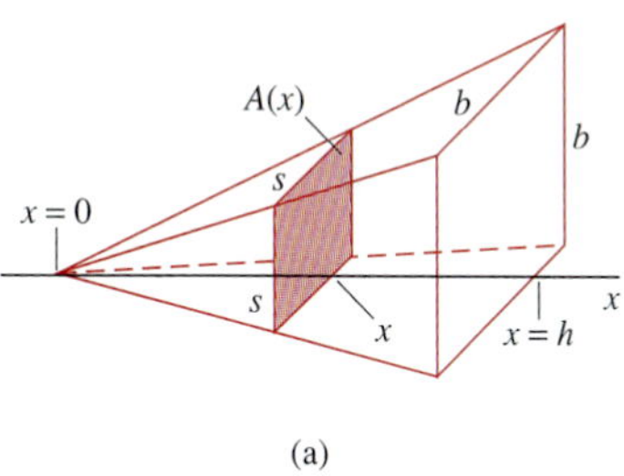

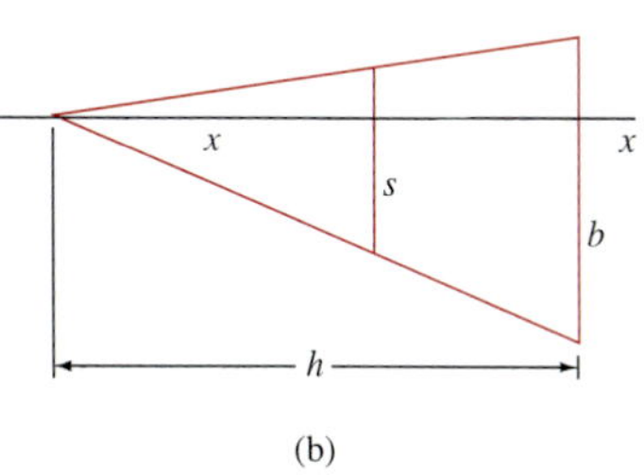

Fig. 6.2.5 The squared-base pyramid of Example 1

EXAMPLE 1 Figure 6.2.5(a) shows a square-based pyramid oriented so that its height h corresponds to the interval $[0, h]$ on the x-axis. Its *base* is a b-by-b square, and each cross section perpendicular to the x-axis is also a square. To find the area $A(x)$ of the s-by-s cross section at x, we equate height-to-length ratios in the similar triangles of Fig. 6.2.5(b):

$$\frac{s}{x} = \frac{b}{h}, \quad \text{so} \quad s = \frac{b}{h}x.$$

Therefore

$$A(x) = s^2 = \frac{b^2}{h^2}x^2,$$

and Eq. (3)—with $[0, h]$ as the interval of integration—gives

$$V = \int_0^h A(x)\,dx = \int_0^h \frac{b^2}{h^2}x^2\,dx = \left[\frac{b^2}{h^2}\cdot\frac{x^3}{3}\right]_{x=0}^{x=h} = \frac{1}{3}b^2h.$$

With $A = b^2$ denoting the area of the base, our result takes the form

$$V = \tfrac{1}{3}Ah$$

for the volume of a pyramid. ■

Cross Sections Perpendicular to the y-Axis

In the case of a solid R lying alongside the interval $[c, d]$ on the y-axis, we denote by $A(y)$ the area of the solid's cross section R_y in the plane perpendicular to the y-axis at the point y of $[c, d]$ (Fig. 6.2.6). A similar discussion, beginning with a partition of $[c, d]$, leads to the volume formula

$$V = \int_c^d A(y)\,dy. \tag{4}$$

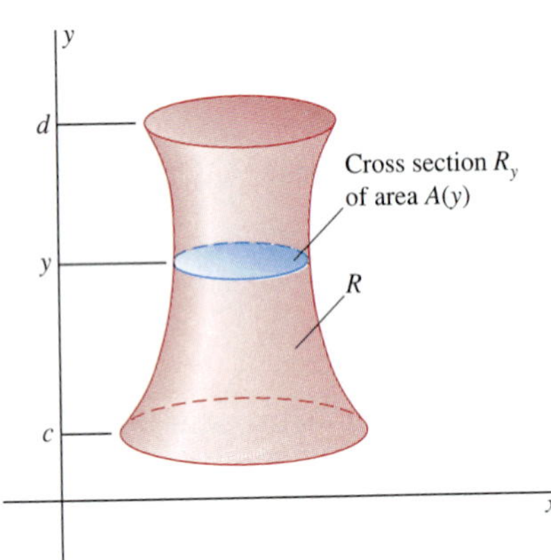

Fig. 6.2.6 $A(y)$ is the area of the cross section R_y in the plane perpendicular to the y-axis at the point y.

Solids of Revolution—Disks and Washers

An important special case of Eq. (3) gives the volume of a **solid of revolution.** For example, consider the solid R obtained by revolving around the x-axis the region under the graph of $y = f(x)$ over the interval $[a, b]$, where $f(x) \geqq 0$. Such a region and the resulting solid of revolution are shown in Fig. 6.2.7.

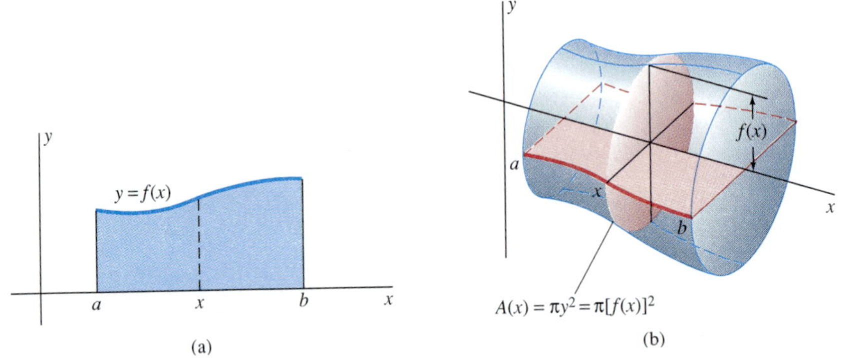

Fig. 6.2.7 (a) A region from which we can determine the volume of a (b) solid of revolution around the x-axis.

Because the solid R is obtained by revolution, each cross section of R at x is a circular *disk* of radius $y = f(x)$. The cross-sectional area function is then $A(x) = \pi y^2 = \pi[f(x)]^2$, so Eq. (3) yields

$$V = \int_a^b \pi y^2\,dx = \int_a^b \pi[f(x)]^2\,dx \tag{5}$$

for the **volume of a solid of revolution around the x-axis.**

Note In the expression $\pi y^2\,dx$, the differential dx tells us that the independent variable is x. We *must* express the dependent variable y in terms of x in order to perform the indicated integration.

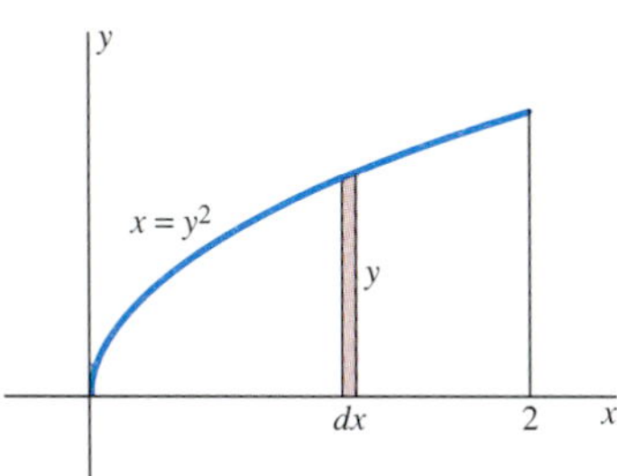

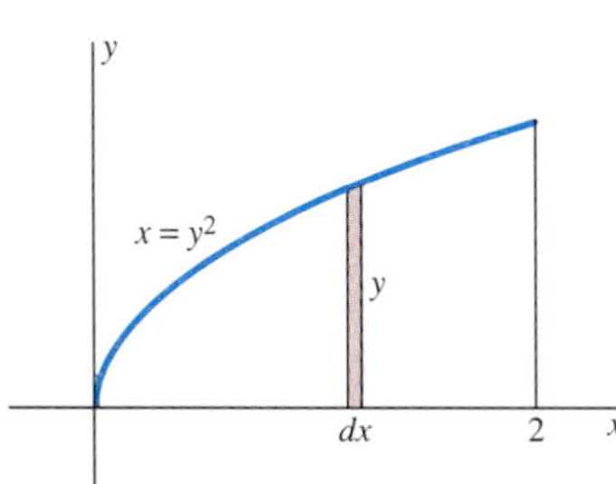

Fig. 6.2.8 The parabolic region of Example 2

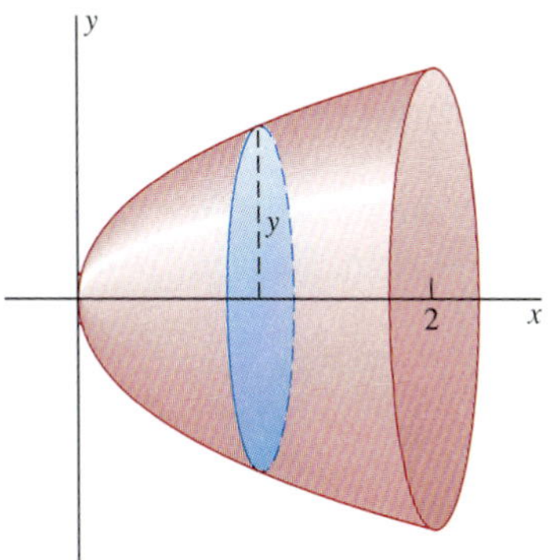

Fig. 6.2.9 The solid paraboloid of Example 2

EXAMPLE 2 Figure 6.2.8 shows the region that lies below the parabola $y^2 = x$ and above the x-axis over the interval $[0, 2]$. Find the volume V of the solid paraboloid (Fig. 6.2.9) obtained by revolving this region around the x-axis.

Solution Because $y^2 = x$ on the parabola, the cross-sectional area function in Eq. (5) is given in terms of x by

$$A(x) = \pi y^2 = \pi x.$$

Hence integration immediately gives

$$V = \int_0^2 \pi x\,dx = \left[\tfrac{1}{2}\pi x^2\right]_0^2 = 2\pi. \qquad \blacksquare$$

EXAMPLE 3 Use the method of cross sections to verify the familiar formula $V = \frac{4}{3}\pi R^3$ for the volume of a sphere of radius R.

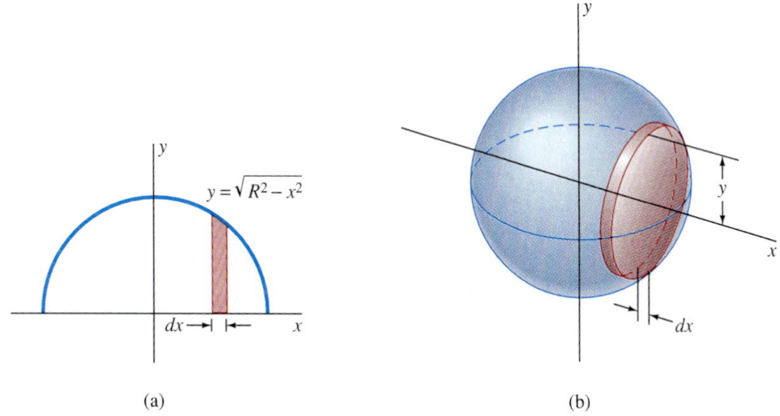

Fig. 6.2.10 (a) A semicircular region that we rotate (b) to generate a sphere (Example 3)

Solution We think of the sphere as the solid of revolution obtained by revolving the semicircular plane region in Fig. 6.2.10 around the x-axis. This is the region bounded above by the semicircle

$$y = \sqrt{R^2 - x^2}, \qquad -R \leqq x \leqq R$$

and below by the interval $[-R, R]$ on the x-axis. To use Eq. (5), we take

$$f(x) = \sqrt{R^2 - x^2}, \qquad a = -R, \quad \text{and} \quad b = R.$$

This gives

$$V = \int_{-R}^{R} \pi\left(\sqrt{R^2 - x^2}\right)^2 dx = \pi\int_{-R}^{R} (R^2 - x^2)\,dx$$

$$= \pi\left[R^2x - \tfrac{1}{3}x^3\right]_{-R}^{R} = \tfrac{4}{3}\pi R^3. \qquad \blacksquare$$

Fig. 6.2.11 A region lying between the y-axis and the curve $x = g(y)$, $c \leqq y \leqq d$, is rotated around the y-axis.

Revolution Around the y-Axis

Figure 6.2.11 shows a solid of revolution around the y-axis. The region being revolved is bounded by the y-axis and the curve $x = g(y)$, $c \leqq y \leqq d$ (as well as the lines $y = c$ and $y = d$). In this case the *horizontal* circular cross section has radius x, and thus the cross-sectional area at y is πx^2, where $x = g(y)$. Hence the cross-sectional area function is $A(y) = \pi x^2 = \pi[g(y)]^2$. We therefore obtain the formula

$$V = \int_c^d \pi x^2\,dy = \int_c^d \pi[g(y)]^2\,dy \tag{6}$$

(contrast it with Eq. (5)) for the **volume of a solid of revolution around the y-axis.**

Note In the expression $\pi x^2\,dy$, the differential dy tells us that the independent variable is y. So here we must express the dependent variable x in terms of y before integrating.

EXAMPLE 4 Use the method of cross sections to verify the familiar formula $V = \frac{1}{3}\pi r^2 h$ for the volume of a right circular cone with base radius r and height h.

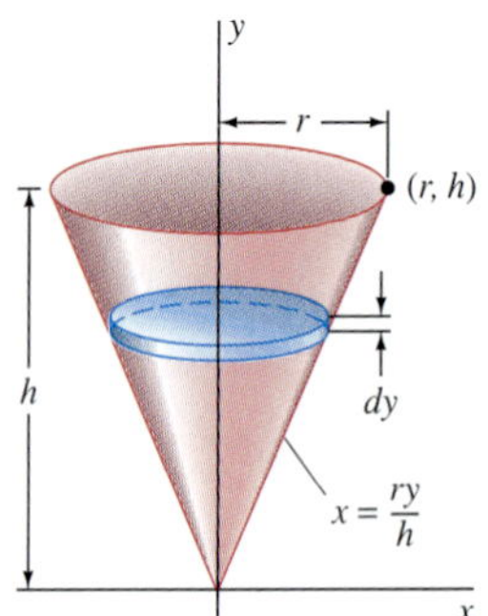

Fig. 6.2.12 Generating a cone by rotation (Example 4)

Solution Figure 6.2.12 depicts the cone as the solid of revolution obtained by revolving around the y-axis the triangle with vertices $(0, 0)$, $(0, h)$, and (r, h). The similar triangles in Fig. 6.2.13 yield the equation $x/y = r/h$, so the radius of the circular cross section perpendicular to the y-axis at the point y is $x = ry/h$. Then Eq. (6), with $g(y) = ry/h$, gives

$$\begin{aligned} V &= \int_a^b A(y)\,dy = \int_a^b \pi x^2\,dy = \int_0^h \pi\left(\frac{ry}{h}\right)^2 dy \\ &= \frac{\pi r^2}{h^2}\int_0^h y^2\,dy = \frac{1}{3}\pi r^2 h = \frac{1}{3}Ah, \end{aligned}$$

where $A = \pi r^2$ is the area of the base of the cone. ■

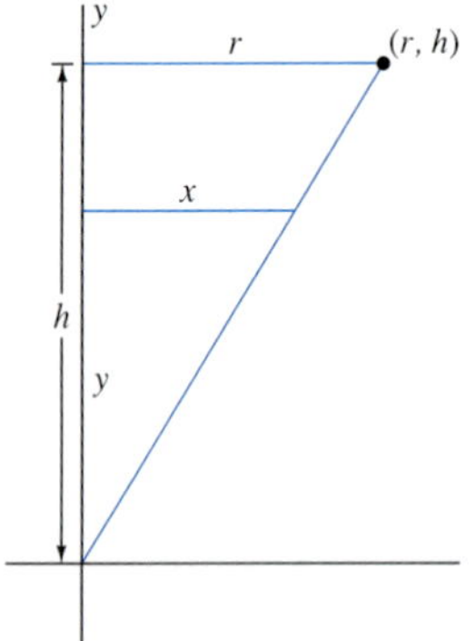

Fig. 6.2.13 Finding the radius x of the circular cross section (Example 4)

Revolving the Region Between Two Curves

Sometimes we need to calculate the volume of a solid generated by revolution of a plane region that lies between two given curves. Suppose that $f(x) \geqq g(x) \geqq 0$ for x in the interval $[a, b]$ and that the solid R is generated by revolving around the x-axis the region between $y = f(x)$ and $y = g(x)$. Then the cross section at x is an **annular ring** (or **washer**) bounded by two circles (Fig. 6.2.14). The ring has inner radius $r_{\text{in}} = g(x)$ and outer radius $r_{\text{out}} = f(x)$, so the formula for the cross-sectional area at x is

$$A(x) = \pi(r_{\text{out}})^2 - \pi(r_{\text{in}})^2 = \pi[(y_{\text{top}})^2 - (y_{\text{bot}})^2] = \pi\{[f(x)]^2 - [g(x)]^2\},$$

where we write $y_{\text{top}} = f(x)$ and $y_{\text{bot}} = g(x)$ for the top and bottom curves of the plane region. Therefore, Eq. (3) yields

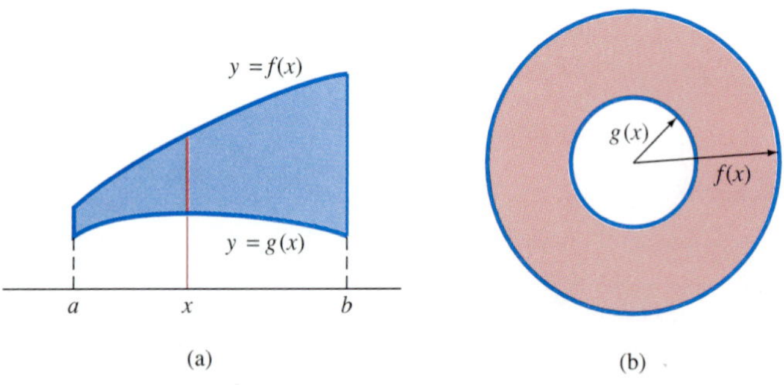

Fig. 6.2.14 (a) The region between two positive graphs (b) is rotated around the x-axis. Cross sections are annular rings.

$$V = \int_a^b \pi[(y_{\text{top}})^2 - (y_{\text{bot}})^2]\,dx = \int_a^b \pi\{[f(x)]^2 - [g(x)]^2\}\,dx \tag{7}$$

for the volume V of the solid.

Similarly, if $f(y) \geqq g(y) \geqq 0$ for $c \leqq y \leqq d$, then the volume of the solid obtained by revolving around the y-axis the region between $x_{\text{right}} = f(y)$ and $x_{\text{left}} = g(y)$ is

$$V = \int_c^d \pi[(x_{\text{right}})^2 - (x_{\text{left}})^2]\,dy = \int_c^d \pi\{[f(y)]^2 - [g(y)]^2\}\,dy. \tag{8}$$

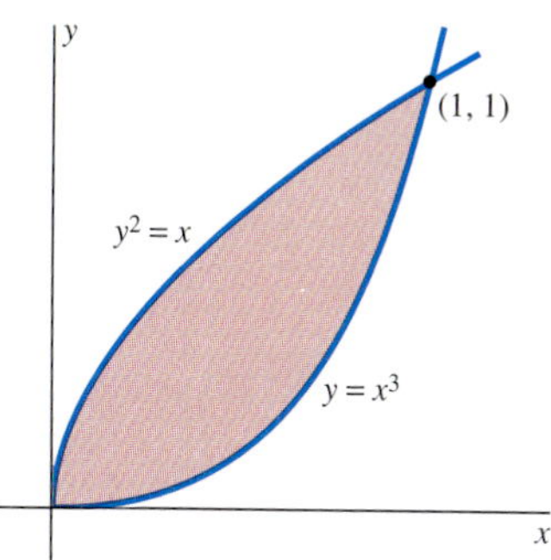

Fig. 6.2.15 The plane region of Example 5

EXAMPLE 5 Consider the plane region shown in Fig. 6.2.15, bounded by the curves $y^2 = x$ and $y = x^3$, which intersect at the points $(0, 0)$ and $(1, 1)$. If this region is revolved around the x-axis (Fig. 6.2.16), then Eq. (7) with

$$y_{\text{top}} = \sqrt{x}, \qquad y_{\text{bot}} = x^3$$

gives

$$V = \int_0^1 \pi[(\sqrt{x})^2 - (x^3)^2]\,dx = \int_0^1 \pi(x - x^6)\,dx$$

$$= \pi\left[\tfrac{1}{2}x^2 - \tfrac{1}{7}x^7\right]_0^1 = \tfrac{5}{14}\pi$$

for the volume of revolution.

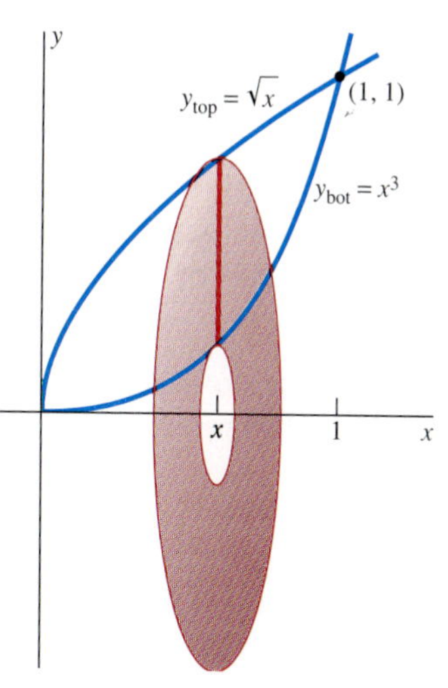

Fig. 6.2.16 Revolution around the x-axis (Example 5)

If the same region is revolved around the y-axis (Fig. 6.2.17), then each cross section perpendicular to the y-axis is an annular ring with outer radius $x_{\text{right}} = y^{1/3}$ and inner radius $x_{\text{left}} = y^2$. Hence Eq. (8) gives the volume of revolution generated by this region as

$$V = \int_0^1 \pi[(y^{1/3})^2 - (y^2)^2]\,dy = \int_0^1 \pi(y^{2/3} - y^4)\,dy$$

$$= \pi\left[\tfrac{3}{5}y^{5/3} - \tfrac{1}{5}y^5\right]_0^1 = \tfrac{2}{5}\pi.$$

■

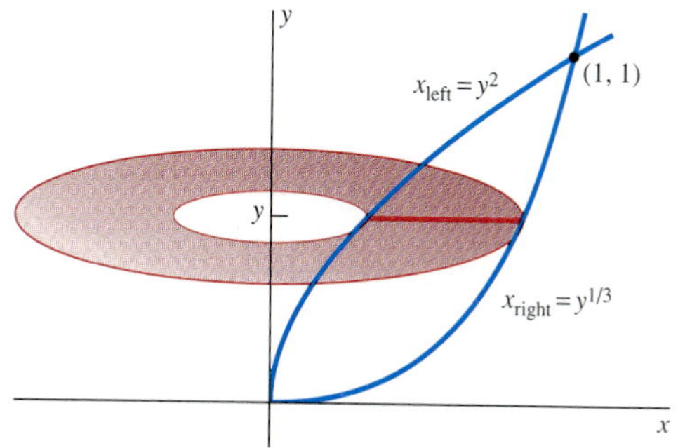

Fig. 6.2.17 Revolution around the y-axis (Example 5)

EXAMPLE 6 Suppose that the plane region of Example 5 (Fig. 6.2.15) is revolved around the vertical line $x = -1$ (Fig. 6.2.18). Then each cross section of the resulting solid is an annular ring with outer radius

$$r_{\text{out}} = 1 + x_{\text{right}} = 1 + y^{1/3}$$

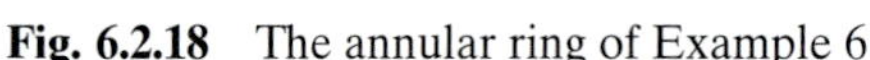

Fig. 6.2.18 The annular ring of Example 6

and inner radius

$$r_{\text{in}} = 1 + x_{\text{left}} = 1 + y^2.$$

The area of such a cross section is

$$A(y) = \pi(1 + y^{1/3})^2 - \pi(1 + y^2)^2 = \pi(2y^{1/3} + y^{2/3} - 2y^2 - y^4),$$

so the volume of the resulting solid of revolution is

$$V = \int_0^1 \pi(2y^{1/3} + y^{2/3} - 2y^2 - y^4)\,dy$$

$$= \pi\left[\tfrac{3}{2}\,y^{4/3} + \tfrac{3}{5}\,y^{5/3} - \tfrac{2}{3}\,y^3 - \tfrac{1}{5}\,y^5\right]_0^1 = \tfrac{37}{30}\,\pi.$$

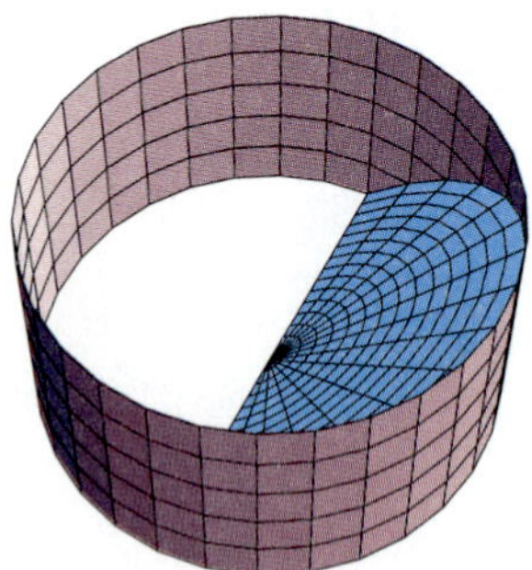

Fig. 6.2.19 The wedge and cylinder of Example 7

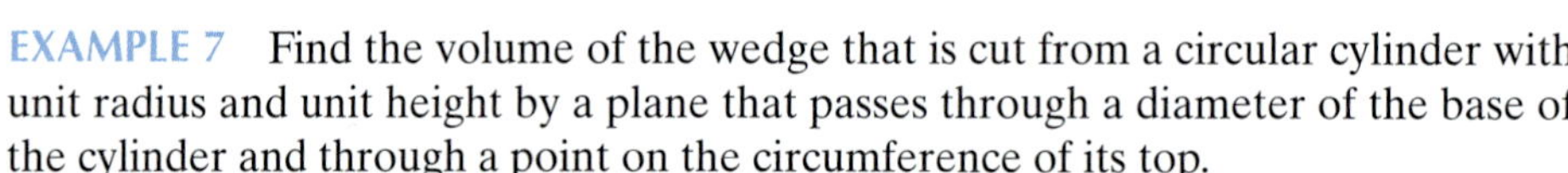

EXAMPLE 7 Find the volume of the wedge that is cut from a circular cylinder with unit radius and unit height by a plane that passes through a diameter of the base of the cylinder and through a point on the circumference of its top.

Solution The cylinder and wedge are shown in Fig. 6.2.19. To form such a wedge, fill a cylindrical glass with cider and then drink slowly until half the bottom of the glass is exposed; the remaining cider forms the wedge.

We choose as reference line and x-axis the line through the "edge of the wedge"—the original diameter of the base of the cylinder. We can verify with similar triangles that each cross section of the wedge perpendicular to the diameter is an isosceles right triangle. One of these triangles is shown in Fig. 6.2.20. We denote by y the equal base and height of this triangle.

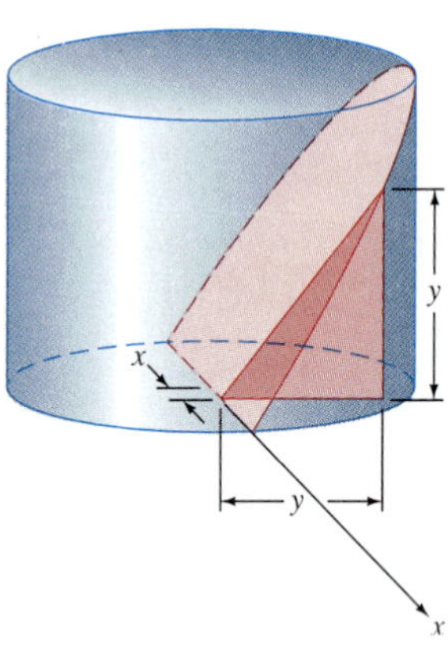

Fig. 6.2.20 A cross section of the wedge—an isosceles triangle (Example 7)

To determine the cross-sectional area function $A(x)$, we must express y in terms of x. Figure 6.2.21 shows the unit circular base of the original cylinder. We apply the Pythagorean theorem to the right triangle in this figure and find that $y = \sqrt{1 - x^2}$. Hence

$$A(x) = \tfrac{1}{2}\,y^2 = \tfrac{1}{2}\,(1 - x^2),$$

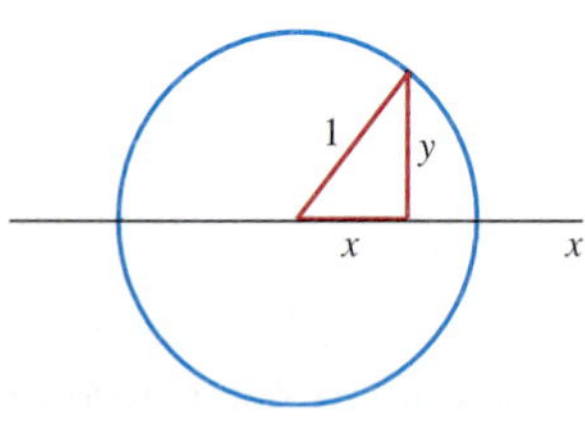

Fig. 6.2.21 The base of the cylinder of Example 7

so Eq. (3) gives

$$V = \int_{-1}^{1} A(x)\,dx = 2\int_0^1 A(x)\,dx \qquad \text{(by symmetry)}$$

$$= 2\int_0^1 \tfrac{1}{2}(1 - x^2)\,dx = \left[x - \tfrac{1}{3}x^3\right]_0^1 = \tfrac{2}{3}$$

for the volume of the wedge. ■

COMMENT It is a useful habit to check answers for plausibility whenever convenient. For example, we may compare a given solid with one whose volume is known. Because the volume of the original cylinder in Example 7 is π, we have found that the wedge occupies the fraction

$$\frac{V_{\text{wedge}}}{V_{\text{cyl}}} = \frac{\frac{2}{3}}{\pi} \approx 21\%$$

of the volume of the cylinder. A glance at Fig. 6.2.19 indicates that this is plausible. An error in our computations could well have given an unbelievable answer.

HISTORICAL NOTE The wedge of Example 7 has an ancient history. Its volume was first calculated in the third century B.C. by Archimedes, who also derived the formula $V = \frac{4}{3}\pi r^3$ for the volume of a sphere of radius r. His work on the wedge is found in a manuscript that was discovered in 1906 after having been lost for centuries. Archimedes used a method of exhaustion for volume similar to that discussed for areas in Section 5.3. For more information, see pp. 73–74 of C. H. Edwards, *The Historical Development of the Calculus* (New York: Springer-Verlag, 1979).

6.2 PROBLEMS

In Problems 1 through 24, find the volume of the solid that is generated by rotating around the indicated axis the plane region bounded by the given curves.

1. $y = x^2, y = 0, x = 1$; the x-axis

2. $y = \sqrt{x}, y = 0, x = 4$; the x-axis

3. $y = x^2$, $y = 4$, $x = 0$ (first quadrant only); the y-axis (Fig. 6.2.22)

4. $y = 1/x, y = 0, x = 0.1, x = 1$; the x-axis (Fig. 6.2.23)

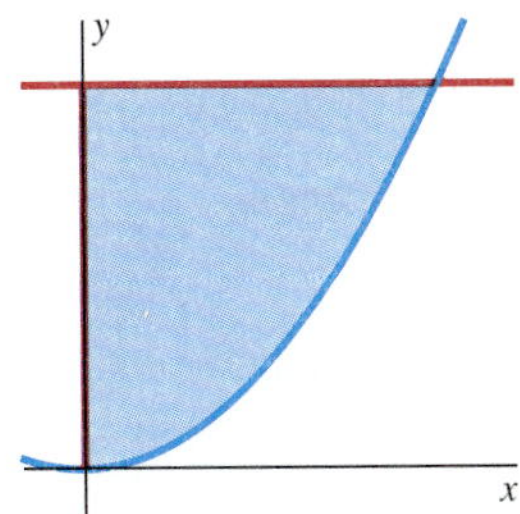

Fig. 6.2.22 Problem 3

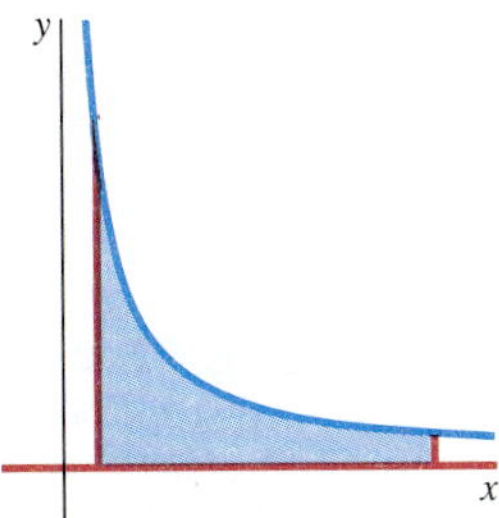

Fig. 6.2.23 Problem 4

5. $y = \sin x$ on $[0, \pi], y = 0$; the x-axis

6. $y = 9 - x^2, y = 0$; the x-axis

7. $y = x^2, x = y^2$; the x-axis (Fig. 6.2.24)

8. $y = x^2, y = 4x$; the line $x = 5$ (Fig. 6.2.25)

9. $y = x^2, y = 8 - x^2$; the x-axis

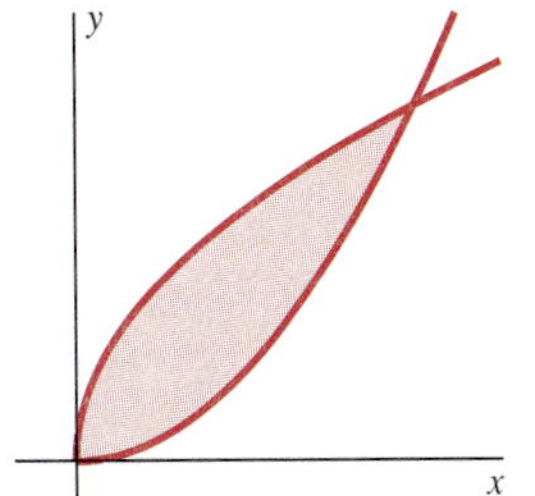

Fig. 6.2.24 Problem 7

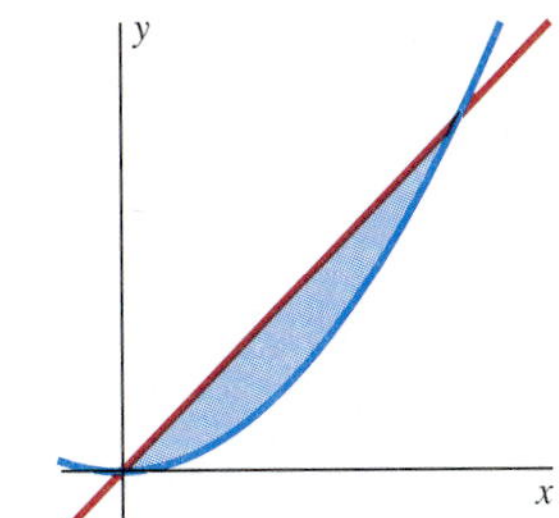

Fig. 6.2.25 Problem 8

10. $x = y^2, x = y + 6$; the y-axis

11. $y = 1 - x^2, y = 0$; the x-axis (Fig. 6.2.26)

12. $y = x - x^3, y = 0\ (0 \leqq x \leqq 1)$; the x-axis (Fig. 6.2.27)

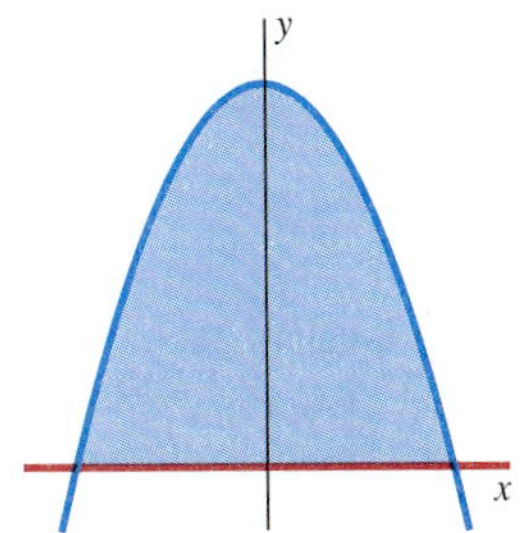

Fig. 6.2.26 Problem 11

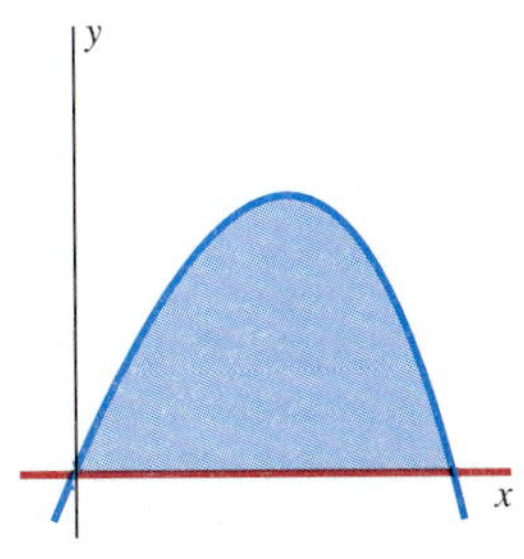

Fig. 6.2.27 Problem 12

13. $y = 1 - x^2, y = 0$; the y-axis

14. $y = 6 - x^2, y = 2$; the x-axis

15. $y = 6 - x^2, y = 2$; the y-axis (Fig. 6.2.28)

16. $y = 1 - x^2, y = 0$; the vertical line $x = 2$

17. $y = x - x^3$, $y = 0$ $(0 \leqq x \leqq 1)$; the horizontal line $y = -1$

18. $y = 4, x = 0, y = x^2$; the x-axis

19. $y = 4, x = 0, y = x^2$; the y-axis

20. $x = 16 - y^2, x = 0, y = 0$ (first quadrant only); the x-axis (Fig. 6.2.29)

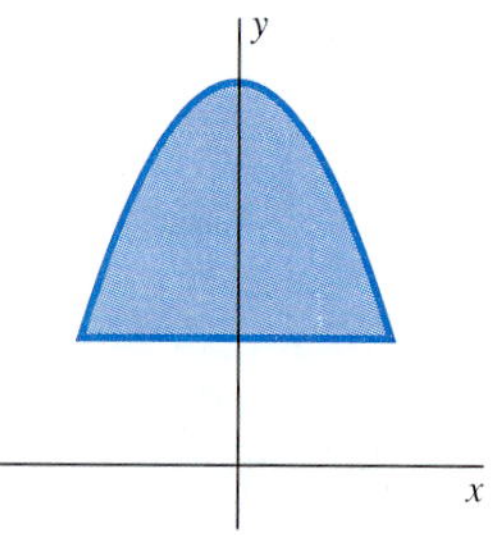

Fig. 6.2.28 Problem 15

Fig. 6.2.29 Problem 20

21. $y = x^2, x = y^2$; the line $y = -2$

22. $y = x^2, y = 8 - x^2$; the line $y = -1$

23. $y = x^2, x = y^2$; the line $x = 3$

24. $y = x^2, y = 8 - x^2$; the line $x = 4$

In Problems 25 through 30, find the volume obtained by revolving the region R around the x-axis. You may use the trigonometric identities

$$\cos^2 x = \frac{1 + \cos 2x}{2} \quad \text{and} \quad \sin^2 x = \frac{1 - \cos 2x}{2}$$

to help you evaluate some of the integrals.

25. R is the region between the graph $y = \sin x$ and the x-axis for $0 \leqq x \leqq \pi$.

26. R is the region between the graph $y = \cos\left(\frac{1}{2}\pi x\right)$ and the x-axis for $-1 \leqq x \leqq 1$.

27. R is the region between the curves $y = \sin x$ and $y = \cos x$ for $0 \leqq x \leqq \pi/4$.

28. R is the region between $x = -\pi/3$ and $x = \pi/3$ that is bounded by the curves $y = \cos x$ and $y = 1/2$.

29. R is bounded by the curve $y = \tan x$ and the lines $y = 0$ and $x = \pi/4$.

30. R is bounded by the curve $y = \tan x$ and the lines $x = 0$ and $y = 1$.

In Problems 31 through 34, first use a calculator or computer to approximate (graphically or otherwise) the points of intersection of the two given curves. Let R be the region bounded by these curves. Integrate to approximate the volume of the solid obtained by revolving the region R around the x-axis.

31. $y = x^3 + 1$, $y = 3x^2$

32. $y = x^4$, $y = x + 4$

33. $y = x^2$, $y = \cos x$

34. $y = \sin x$, $y = (x - 1)^2$

35. The region R shown in Fig. 6.2.30 is bounded by the parabolas $y^2 = 2(x - 3)$ and $y^2 = x$. Find the volume of the solid generated by rotating R around the x-axis.

36. Find the volume of the ellipsoid generated by rotating around the x-axis the region bounded by the ellipse with equation

$$\left(\frac{x}{a}\right)^2 + \left(\frac{y}{b}\right)^2 = 1$$

(Fig. 6.2.31).

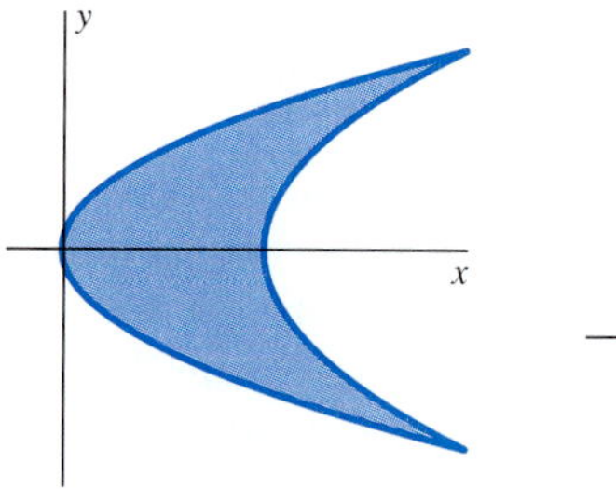

Fig. 6.2.30 The region of Problem 35

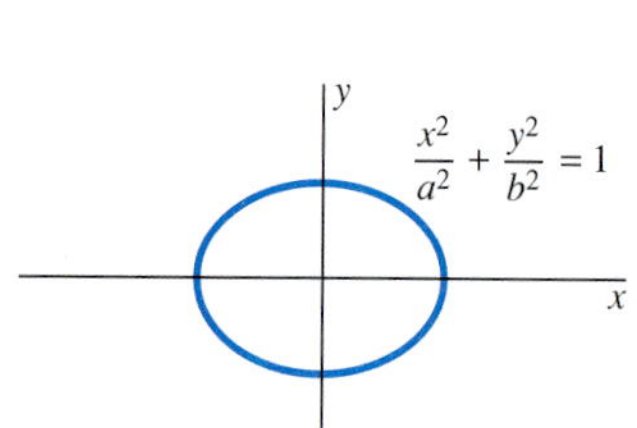

Fig. 6.2.31 The ellipse of Problems 36 and 37

37. Repeat Problem 36, except rotate the elliptical region around the y-axis.

38. Find the volume of the unbounded solid generated by rotating the unbounded region of Fig. 6.2.32 around the x-axis. This is the region between the graph of $y = 1/x^2$ and the x-axis for $x \geqq 1$. (*Method:* Compute the volume from $x = 1$ to $x = b$, where $b > 1$. Then find the limit of this volume as $b \to +\infty$.)

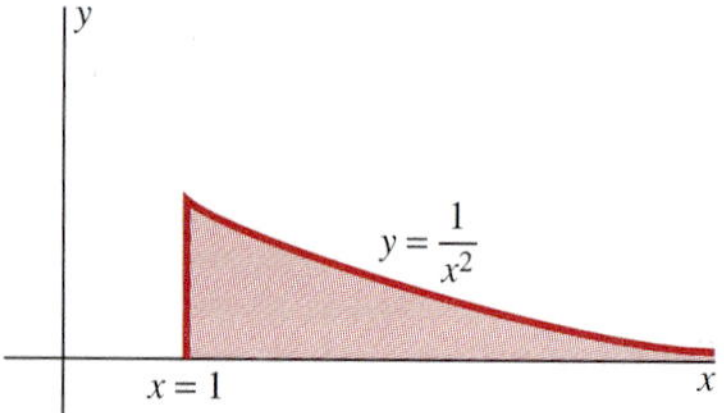

Fig. 6.2.32 The unbounded plane region of Problem 38

39. An observatory (Fig. 6.2.33) is shaped like a solid whose base is a circular disk with diameter AB of length $2a$ (Fig. 6.2.34). Find the volume of this solid if each cross section perpendicular to AB is a square.

40. The base of a certain solid is a circular disk with diameter AB of length $2a$. Find the volume of the solid if each cross section perpendicular to AB is a semicircle.

41. The base of a certain solid is a circular disk with diameter AB of length $2a$. Find the volume of the solid if each cross section perpendicular to AB is an equilateral triangle.

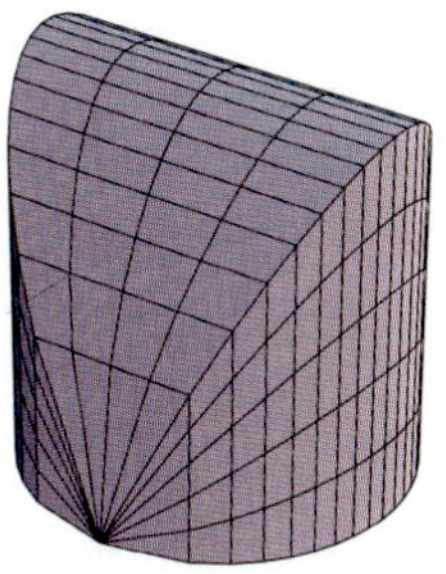

Fig. 6.2.33 The observatory of Problem 39

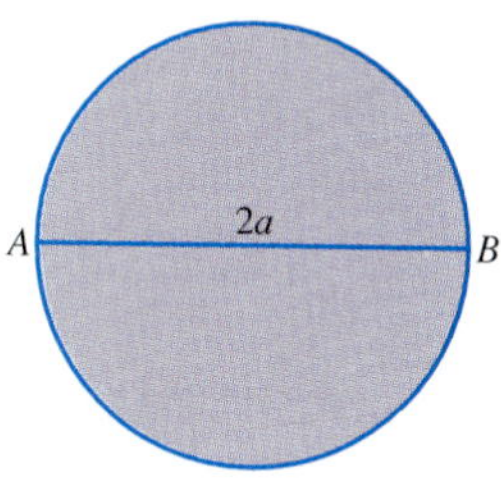

Fig. 6.2.34 The circular base of the observatory (Problem 39)

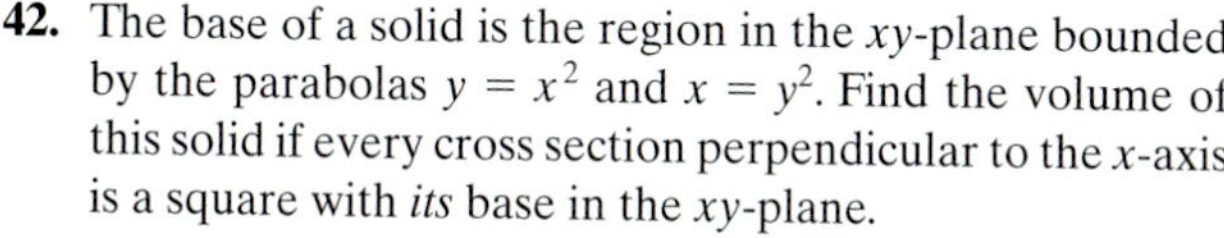

42. The base of a solid is the region in the xy-plane bounded by the parabolas $y = x^2$ and $x = y^2$. Find the volume of this solid if every cross section perpendicular to the x-axis is a square with *its* base in the xy-plane.

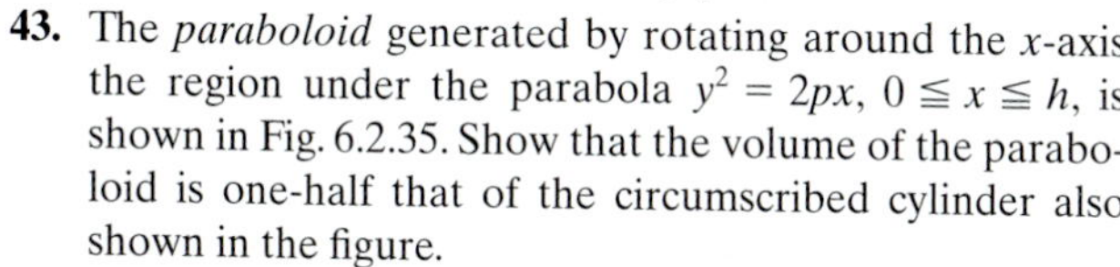

43. The *paraboloid* generated by rotating around the x-axis the region under the parabola $y^2 = 2px$, $0 \leqq x \leqq h$, is shown in Fig. 6.2.35. Show that the volume of the paraboloid is one-half that of the circumscribed cylinder also shown in the figure.

44. A pyramid has height h and rectangular base with area A. Show that its volume is $V = \frac{1}{3}Ah$. (*Suggestion:* Note that each cross section parallel to the base is a rectangle.)

45. Repeat Problem 44, except make the base a triangle with area A.

46. Find the volume that remains after a hole of radius 3 is bored through the center of a solid sphere of radius 5 (Fig. 6.2.36).

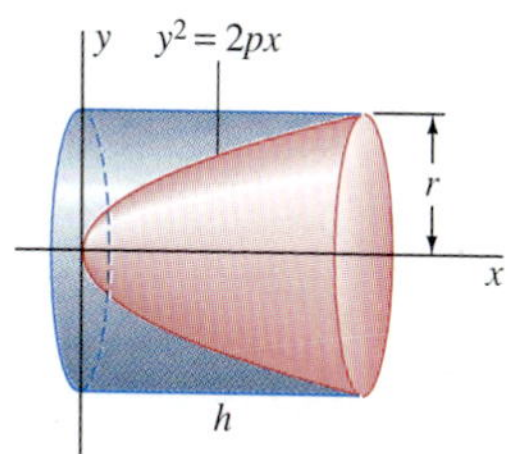

Fig. 6.2.35 The paraboloid and cylinder of Problem 43

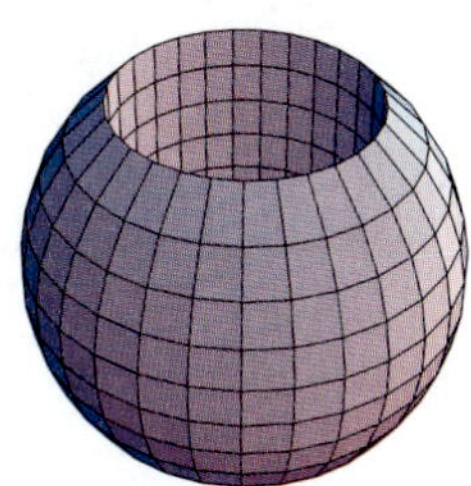

Fig. 6.2.36 The sphere-with-hole of Problem 46

47. Two horizontal circular cylinders both have radius a, and their axes intersect at right angles. Find the volume of their solid of intersection (Figs. 6.2.37 and 6.2.38, where $a = 1$). Is it clear to you that each horizontal cross section of the solid is a square?

48. Figure 6.2.39 shows a "spherical segment" of height h that is cut off from a sphere of radius r by a horizontal plane. Show that its volume is

$$V = \tfrac{1}{3}\pi h^2(3r - h).$$

49. A doughnut-shaped solid, called a *torus* (Fig. 6.2.40), is generated by revolving around the y-axis the circular disk

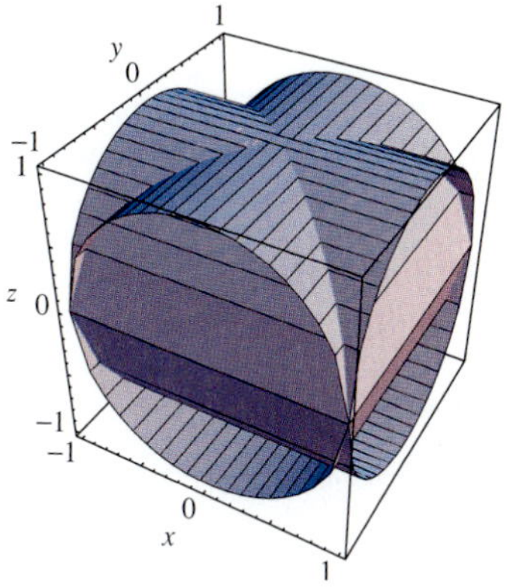

Fig. 6.2.37 The intersecting cylinders of Problem 47

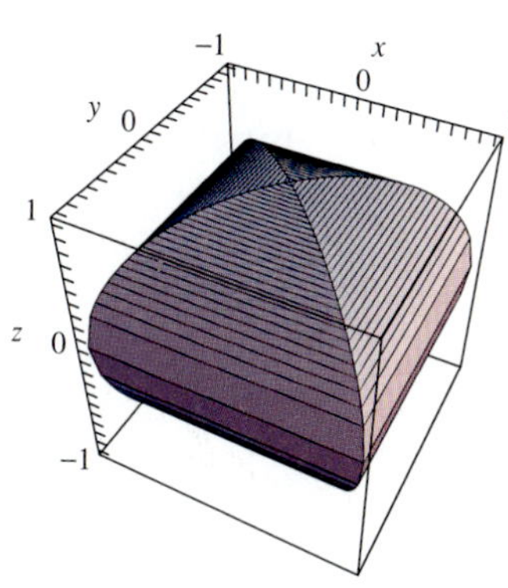

Fig. 6.2.38 The solid of intersection (Problem 47)

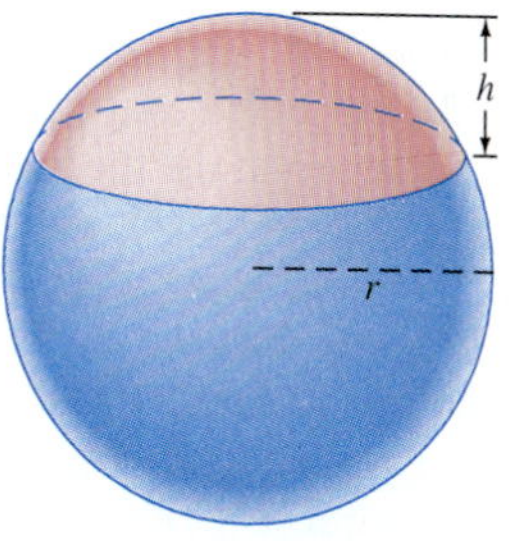

Fig. 6.2.39 A spherical segment (Problem 48)

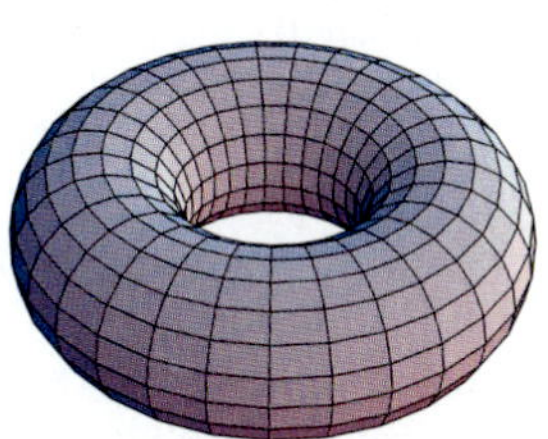

Fig. 6.2.40 The torus of Problem 49

$(x - b)^2 + y^2 \leqq a^2$ centered at the point $(b, 0)$, where $0 < a < b$. Show that the volume of this torus is $V = 2\pi^2a^2b$. (*Suggestion:* Note that each cross section perpendicular to the y-axis is an annular ring, and recall that

$$\int_0^a \sqrt{a^2 - y^2}\,dy = \tfrac{1}{4}\pi a^2$$

because the integral represents the area of a quarter-circle of radius a.)

50. The summit of a hill is 100 ft higher than the surrounding level terrain, and each horizontal cross section of the hill is circular. The following table gives the radius r (in feet) for selected values of the height h (in feet) above the surrounding terrain. Use Simpson's approximation to estimate the volume of the hill.

h	0	25	50	75	100
r	60	55	50	35	0

51. *Newton's Wine Barrel* Consider a barrel with the shape of the solid generated by revolving around the x-axis the region under the parabola

$$y = R - kx^2, \ -\tfrac{1}{2}h \leqq x \leqq \tfrac{1}{2}h$$

(Fig. 6.2.41). (a) Show that the radius of each end of the barrel is $r = R - \delta$, where $4\delta = kh^2$. (b) Then show that the volume of the barrel is

$$V = \tfrac{1}{3}\pi h(2R^2 + r^2 - \tfrac{2}{5}\delta^2).$$

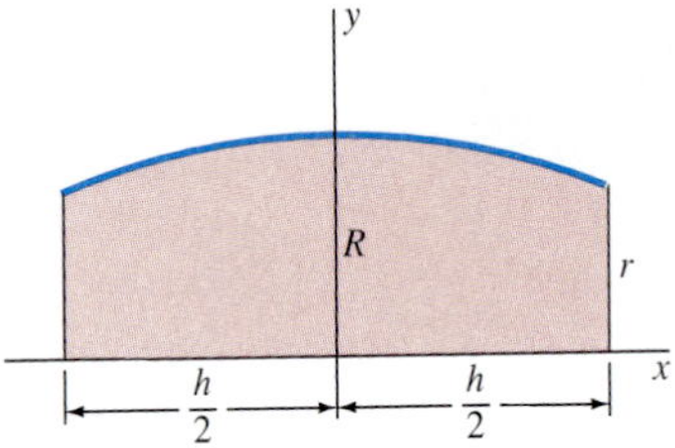

Fig. 6.2.41 The region of Problem 51

52. *The Clepsydra, or Water Clock* Consider a water tank whose side surface is generated by rotating the curve $y = kx^4$ around the y-axis (k is a positive constant). (a) Compute $V(y)$, the volume of water in the tank as a function of the depth y. (b) Suppose that water drains from the tank through a small hole at its bottom. Use the chain rule and Torricelli's law of draining (in the form $dV/dt = -c\sqrt{y}$) to show that the water level in this tank falls at a *constant* rate. How could such a tank be used as a clock?

53. A contractor wants to bid on the job of leveling a 60-ft hill. It will cost $\$3.30/\text{yd}^3$ of material in the hill to be removed. The following table, based on surveying data, shows areas of horizontal cross sections of the hill at 10-ft height intervals. Use (a) the trapezoidal approximation and (b) Simpson's approximation to estimate how much this job should cost. Round each answer to the nearest hundred dollars.

Height x (ft)	0	10	20	30	40	50	60
Area (ft^2)	1513	882	381	265	151	50	0

54. Water evaporates from an open bowl at a rate proportional to the area of the surface of the water. Show that whatever the shape of the bowl, the water level will drop at a constant rate.

55. A frustum of a right circular cone has height h and volume V. Its base is a circular disk with radius R and its top is a circular disk with radius r (Fig. 6.2.42). Apply the method of cross sections to show that

$$V = \tfrac{1}{3}\pi h(r^2 + rR + R^2).$$

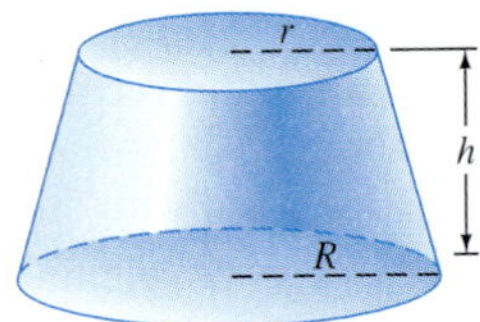

Fig. 6.2.42 A frustum of a cone (Problem 55)

56. Find the volume of the solid of intersection of two spheres of radius a, if the center of each lies on the surface of the other.

6.2 PROJECT: APPROXIMATING VOLUMES OF SOLIDS OF REVOLUTION

We can derive the volume formula

$$V = \int_a^b \pi[f(x)]^2\,dx \tag{1}$$

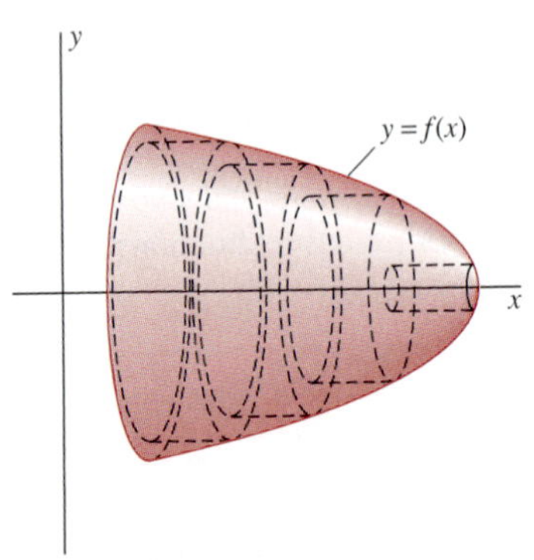

Fig. 6.2.43 Using cylinders to approximate a solid of revolution

of this section by using circular cylinders to approximate the volume of the solid obtained by revolving around the x-axis the region that lies under the curve $y = f(x)$ over the interval $[a, b]$. For instance, in the right-endpoint approximation

$$R_n = \sum_{i=1}^{n} \pi[f(x_i)]^2\,\Delta x,$$

where $\Delta x = (b - a)/n$, the ith term $\pi[f(x_i)]^2\,\Delta x$ is the volume of the cylinder with radius $r = f(x_i)$ and height $h = \Delta x$ that approximates the slice of the solid corresponding to the subinterval $[x_{i-1}, x_i]$ (Fig. 6.2.43).

In the third century B.C. Archimedes regarded the sphere of radius r as a solid of revolution in deriving his famous volume formula $V = \frac{4}{3}\pi r^3$. A major difference between his method and ours is that his derivation used conical frusta rather than

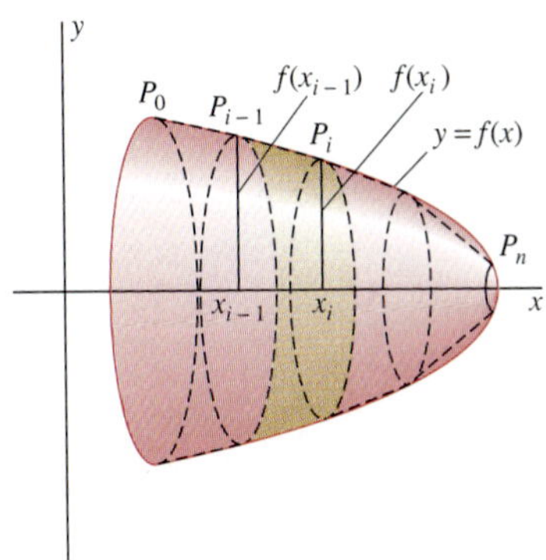

Fig. 6.2.44 Using conical frusta to approximate a solid of revolution

cylinders. Figure 6.2.44 shows the approximating solid obtained by revolving around the x-axis the polygonal arc $P_0P_1P_2 \cdots P_n$, where P_i denotes the point $(x_i, f(x_i))$ on the curve $y = f(x)$. The approximating slice that corresponds to the ith subinterval $[x_{i-1}, x_i]$ is the *conical frustum* highlighted in Fig. 6.2.44. By using the volume formula of Problem 55, we therefore get the "frustum approximation"

$$F_n = \sum_{i=1}^{n} \frac{\pi}{3}\left\{[f(x_{i-1})]^2 + f(x_{i-1})f(x_i) + [f(x_i)]^2\right\}\Delta x$$

to the volume integral of Eq. (1).

In Problems 1 through 4, use a programmable calculator, computer, or computer algebra system (as in the projects for Sections 5.4 and 5.9) to calculate the right- and left-endpoint approximations R_n and L_n, the trapezoidal approximation $T_n = \frac{1}{2}(R_n + L_n)$, and the frustum approximation F_n to the volume integral of Eq. (5). In each case compare the accuracy of these approximations for $n = 10, 20, \ldots,$ subintervals.

1. Let $f(x) = x$ on $[0, 1]$, so that $V = \pi/3$. (Why?)

2. Let $f(x) = \sqrt{1 - x^2}$ on $[0, 1]$. Explain why V is then the volume $2\pi/3$ of a hemisphere of radius 1.

3. Let $f(x) = \sin x$ on $[0, \pi]$. Use the identity $\sin^2 x = \frac{1}{2}(1 - \cos 2x)$ to show that $V = \frac{1}{2}\pi^2$.

4. Let $f(x) = \sec x$ on $[0, \pi/4]$, so that $V = \pi$. (Why?)

6.3 VOLUMES BY THE METHOD OF CYLINDRICAL SHELLS

The method of cross sections of Section 6.2 is a technique of approximating a solid by a stack of thin slabs or slices. In the case of a solid of revolution, these slices are circular disks or annular rings. The **method of cylindrical shells** is a second way of computing volumes of solids of revolution. It is a technique of approximating a solid of revolution by a collection of thin cylindrical shells, and it frequently leads to simpler computations than does the method of cross sections.

Volume of a Cylindrical Shell

A **cylindrical shell** is a region bounded by two concentric circular cylinders of the same height h. If, as in Fig. 6.3.1, the inner cylinder has radius r_1 and the outer one has radius r_2, then $\bar{r} = (r_1 + r_2)/2$ is the **average radius** of the cylindrical shell and $t = r_2 - r_1$ is its **thickness.** We then get the volume of the cylindrical shell by subtracting the volume of the inner cylinder from that of the outer one:

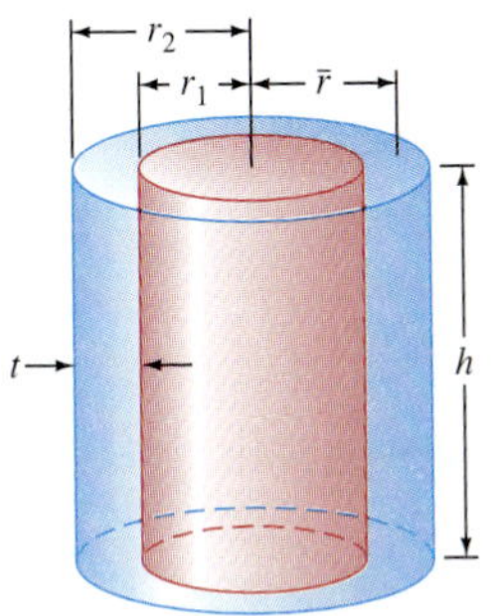

Fig. 6.3.1 A cylindrical shell

$$V = \pi r_2^2 h - \pi r_1^2 h = 2\pi \frac{r_1 + r_2}{2}(r_2 - r_1)h = 2\pi \bar{r} t h. \qquad \textbf{(1)}$$

In words, the volume of the shell is the product of 2π, its average radius, its thickness, and its height. Thus the volume of the shell is closely approximated by multiplying its surface area by its thickness.

More General Volumes

Now suppose that we want to find the volume V of revolution generated by revolving around the y-axis the region under $y = f(x)$ from $x = a$ to $x = b$. We assume, as indicated in Fig. 6.3.2(a), that $0 \leqq a < b$ and that $f(x)$ is continuous and nonnegative on $[a, b]$. The solid will then resemble the one shown in Fig. 6.3.2(b).

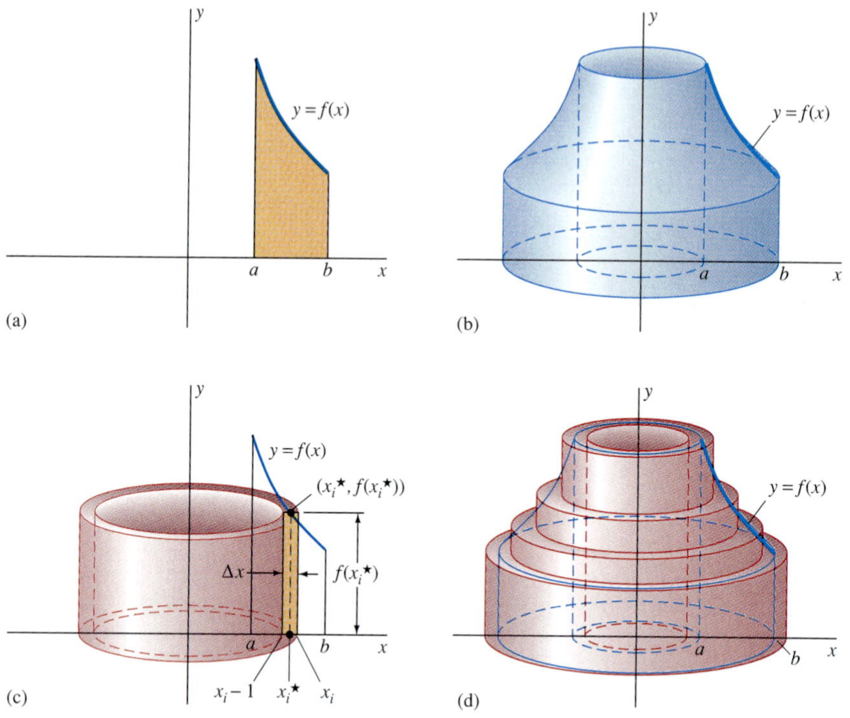

Fig. 6.3.2 A solid of revolution—note the hole through its center—and a way to approximate it with nested cylindrical shells

To find V, we begin with a partition of $[a, b]$ into n subintervals, all with the same length $\Delta x = (b - a)/n$. Let $x_i^\star$ denote the midpoint of the ith subinterval $[x_{i-1}, x_i]$. Consider the rectangle in the xy-plane with base $[x_{i-1}, x_i]$ and height $f(x_i^\star)$. This cylindrical shell approximates the solid with volume ΔV_i that is obtained by revolving the region under $y = f(x)$ and over $[x_{i-1}, x_i]$, and thus Eq. (1) gives

$$\Delta V_i \approx 2\pi x_i^\star f(x_i^\star)\,\Delta x.$$

We add the volumes of the n cylindrical shells determined by the partition of $[a, b]$. This sum should approximate V because—as Fig. 6.3.2(d) suggests—the union of these shells physically approximates the solid of revolution. Thus we obtain the approximation

$$V = \sum_{i=1}^{n} \Delta V_i \approx \sum_{i=1}^{n} 2\pi x_i^\star f(x_i^\star)\,\Delta x.$$

This approximation to the volume V is a Riemann sum that approaches the integral

$$\int_a^b 2\pi x f(x)\,dx \quad \text{as} \quad \Delta x \to 0,$$

so it appears that the volume of the solid of revolution is given by

$$V = \int_a^b 2\pi x f(x)\,dx. \tag{2}$$

A complete discussion would require a proof that this formula gives the same volume as that *defined* by the method of cross sections in Section 6.2 (see Appendix H).

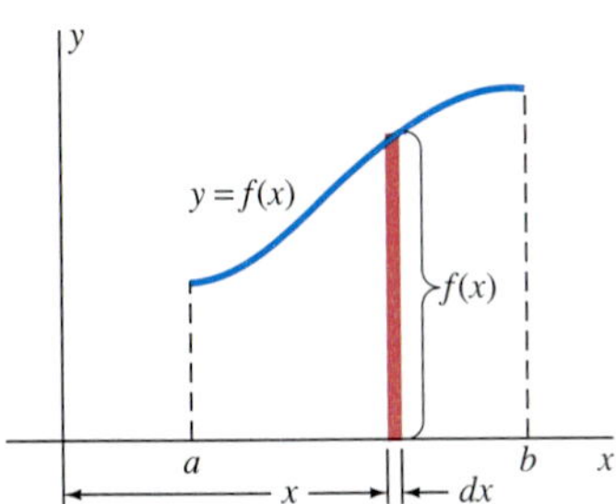

Fig. 6.3.3 Heuristic device for setting up Eq. (2)

It is more reliable to learn how to set up integral formulas than merely to memorize such formulas. A useful heuristic (suggestive but nonrigorous) device for setting up Eq. (2) is to picture the very narrow rectangular strip of area shown in Fig. 6.3.3. When this strip is revolved around the y-axis, it produces a thin cylindrical shell of radius x, height $y = f(x)$, and thickness dx (Fig. 6.3.4). So, if its volume is denoted by dV, we may write

$$dV = 2\pi x \cdot f(x) \cdot dx = 2\pi x f(x)\,dx.$$

This is easy to remember if you visualize Fig. 6.3.5.

We think of V as a sum of very many such volumes, nested concentrically around the axis of revolution and forming the solid itself. We can then write

$$V = \int_{\star}^{\star\star} dV = \int_a^b 2\pi xy\,dx = \int_a^b 2\pi x f(x)\,dx.$$

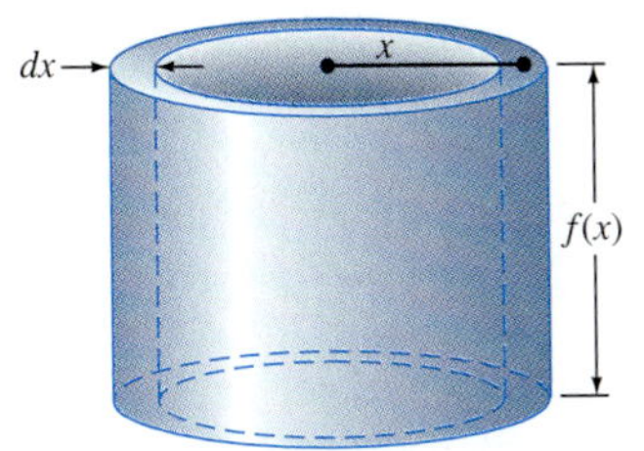

Fig. 6.3.4 Cylindrical shell of infinitesimal thickness

Do not forget to express y (and any other dependent variable) in terms of the independent variable x (identified here by the differential dx) before you integrate.

EXAMPLE 1 Find the volume V of the solid generated by revolving around the y-axis the region under $y = 3x^2 - x^3$ from $x = 0$ to $x = 3$ (Fig. 6.3.6).

Solution Here it would be impractical to use the method of cross sections, because a cross section perpendicular to the y-axis is an annular ring, and finding its inner and outer radii would require us to solve the equation $y = 3x^2 - x^3$ for x in terms of y. We prefer to avoid this troublesome task, and Eq. (2) provides us with an alternative: We take $f(x) = 3x^2 - x^3$, $a = 0$, and $b = 3$. It immediately follows that

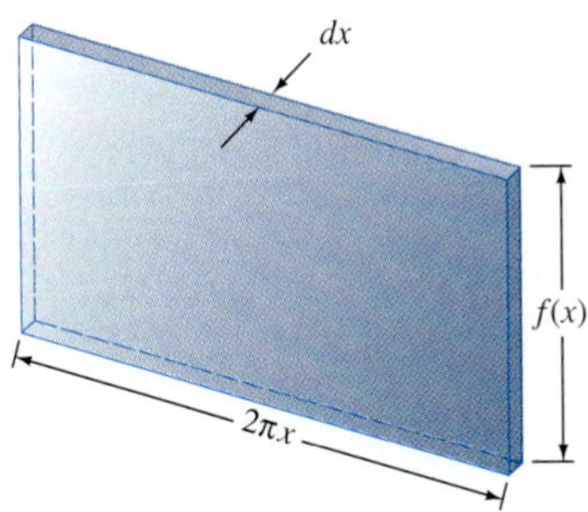

Fig. 6.3.5 Infinitesimal cylindrical shell, flattened out

$$V = \int_0^3 2\pi x(3x^2 - x^3)\,dx = 2\pi\int_0^3 (3x^3 - x^4)\,dx$$

$$= 2\pi\left[\tfrac{3}{4}x^4 - \tfrac{1}{5}x^5\right]_0^3 = \tfrac{243}{10}\pi.$$

EXAMPLE 2 Find the volume V of the solid that remains after you bore a circular hole of radius a through the center of a solid sphere of radius $b > a$ (Fig. 6.3.7).

Solution We think of the sphere of radius b as generated by revolving the right half of the circular disk $x^2 + y^2 = b^2$ around the y-axis, and we think of the hole as vertical and with its centerline lying on the y-axis. Then the upper *half* of the solid in question is generated by revolving around the y-axis the region shaded in Fig. 6.3.8. This is the region below the graph of $y = \sqrt{b^2 - x^2}$ (and above the x-axis) from $x = a$ to $x = b$. The volume of the entire sphere-with-hole is then double that of the upper half, and Eq. (2) gives

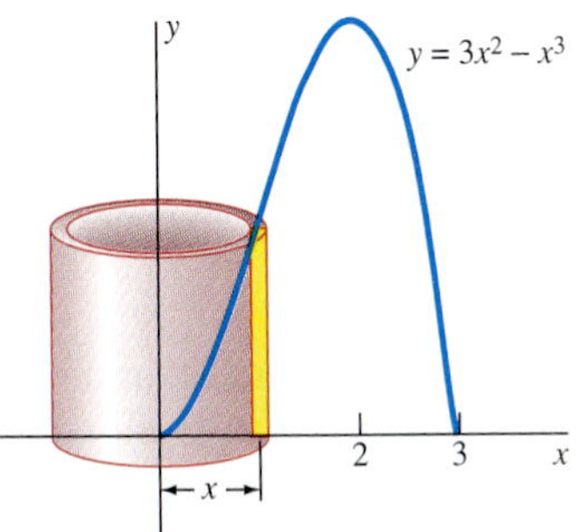

Fig. 6.3.6 The region of Example 1: Rotate it around the y-axis.

$$V = 2\int_a^b 2\pi x(b^2 - x^2)^{1/2}\,dx = 4\pi\left[-\tfrac{1}{3}(b^2 - x^2)^{3/2}\right]_a^b,$$

so

$$V = \tfrac{4}{3}\pi(b^2 - a^2)^{3/2}.$$

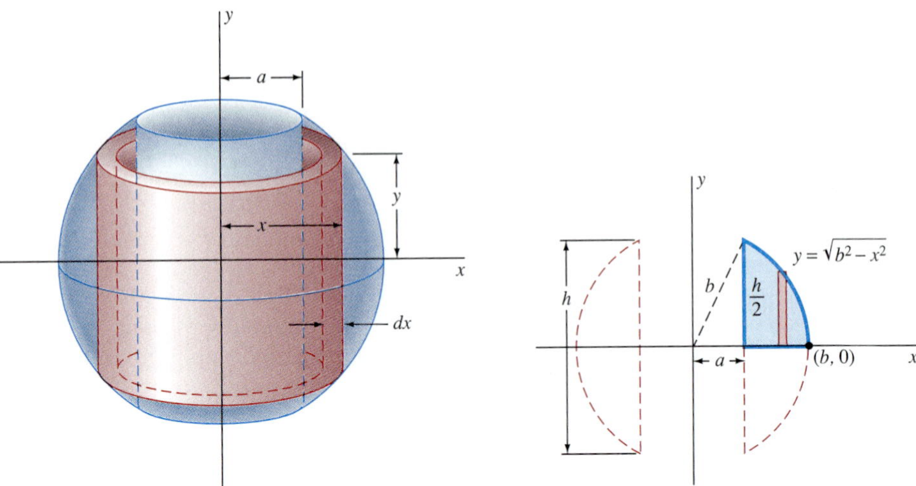

Fig. 6.3.7 The sphere-with-hole of Example 2

Fig. 6.3.8 Middle cross section of the sphere-with-hole (Example 2)

A way to check an answer such as this is to test it in some extreme cases. If $a = 0$ and $b = r$, which corresponds to drilling no hole at all through a sphere of radius r, then our result reduces to the volume $V = \frac{4}{3}\pi r^3$ of the entire sphere. If $a = b$, which corresponds to using a drill bit as large as the sphere, then $V = 0$; this, too, is correct.

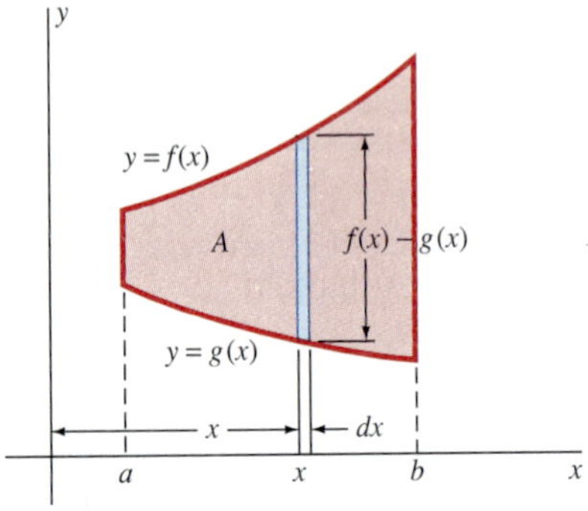

Fig. 6.3.9 The region A between the graphs of f and g over $[a, b]$ is to be rotated around the y-axis.

Revolving the Region Between Two Curves

Now let A denote the region between the curves $y = f(x)$ and $y = g(x)$ over the interval $[a, b]$, where $0 \leqq a < b$ and $g(x) \leqq f(x)$ for x in $[a, b]$. Such a region is shown in Fig. 6.3.9. When A is rotated around the y-axis, it generates a solid of revolution. Suppose that we want to find the volume V of this solid. A development similar to that of Eq. (2) leads to the approximation

$$V \approx \sum_{i=1}^{n} 2\pi x_i^{\star}[f(x_i^{\star}) - g(x_i^{\star})]\,\Delta x,$$

from which we may conclude that

$$V = \int_a^b 2\pi x[f(x) - g(x)]\,dx. \tag{3}$$

Thus

$$V = \int_a^b 2\pi x\,[y_{\text{top}} - y_{\text{bot}}]\,dx, \tag{3'}$$

where $y_{\text{top}} = f(x)$ and $y_{\text{bot}} = g(x)$.

The method of cylindrical shells is also an effective way to compute volumes of solids of revolution around the x-axis. Figure 6.3.10 shows the region A bounded by the curves $x = f(y)$ and $x = g(y)$ for $c \leqq y \leqq d$ and by the horizontal lines $y = c$ and $y = d$. Let V be the volume obtained by revolving the region A around the x-axis. To compute V, we begin with a partition of $[c, d]$ into n subintervals, all of the same length $\Delta y = (d - c)/n$. Let $y_i^{\star}$ denote the midpoint of the ith subinterval $[y_{i-1}, y_i]$ of the partition. Then the volume of the cylindrical shell with average radius $y_i^{\star}$, height $f(y_i^{\star}) - g(y_i^{\star})$, and thickness Δy is

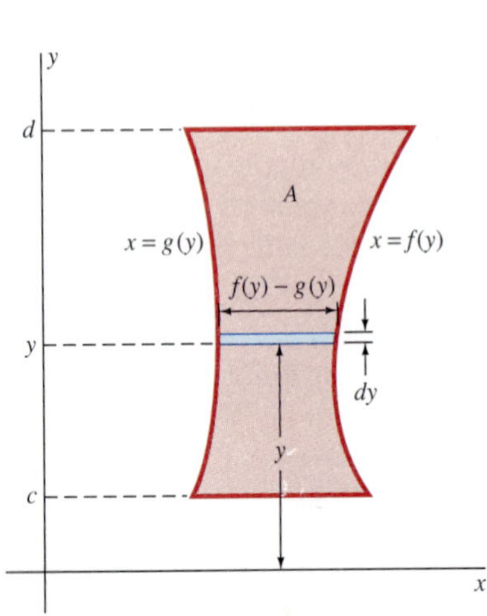

Fig. 6.3.10 The region A is to be rotated around the x-axis.

$$\Delta V_i = 2\pi y_i^\star[f(y_i^\star) - g(y_i^\star)]\,\Delta y.$$

We add the volumes of these cylindrical shells and thus obtain the approximation

$$V \approx \sum_{i=1}^{n} 2\pi y_i^\star[f(y_i^\star) - g(y_i^\star)]\,\Delta y.$$

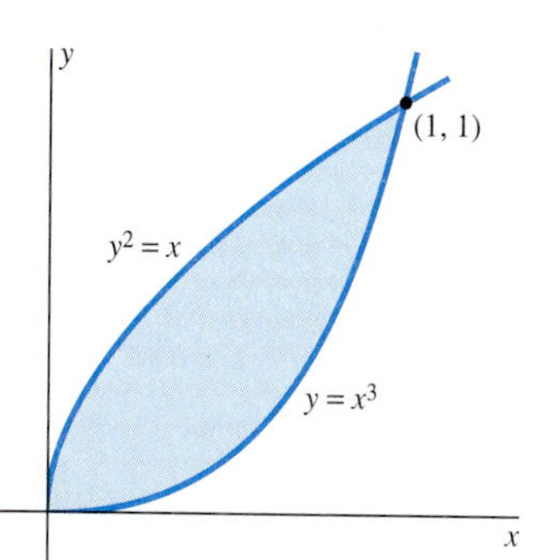

Fig. 6.3.11 The region of Example 3

We recognize the right-hand side to be a Riemann sum for an integral with respect to y from c to d and so conclude that the volume of the solid of revolution is given by

$$V = \int_c^d 2\pi y[f(y) - g(y)]\,dy. \tag{4}$$

Thus

$$V = \int_c^d 2\pi y\,[x_{\text{right}} - x_{\text{left}}]\,dy, \tag{4'}$$

where $x_{\text{right}} = f(y)$ and $x_{\text{left}} = g(y)$.

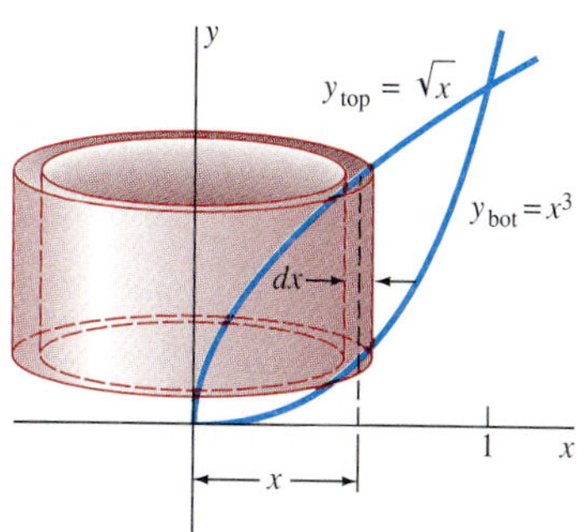

Fig. 6.3.12 Revolution around the y-axis (Example 3)

Note To use Eqs. (3′) and (4′), the integrand must be expressed in terms of the variable of integration specified by the differential.

EXAMPLE 3 Consider the region in the first quadrant bounded by the curves $y^2 = x$ and $y = x^3$ (Fig. 6.3.11). Use the method of cylindrical shells to compute the volume of the solids obtained by revolving this region first around the y-axis and then around the x-axis.

Solution It is best to use cylindrical shells, as in Figs. 6.3.12 and 6.3.13, rather than memorized formulas, to set up the appropriate integrals. Thus the volume of revolution around the y-axis (Fig. 6.3.12) is given by

$$V = \int_0^1 2\pi x(y_{\text{top}} - y_{\text{bot}})\,dx = \int_0^1 2\pi x(\sqrt{x} - x^3)\,dx$$

$$= \int_0^1 2\pi(x^{3/2} - x^4)\,dx = 2\pi\left[\tfrac{2}{5}x^{5/2} - \tfrac{1}{5}x^5\right]_0^1 = \tfrac{2}{5}\pi.$$

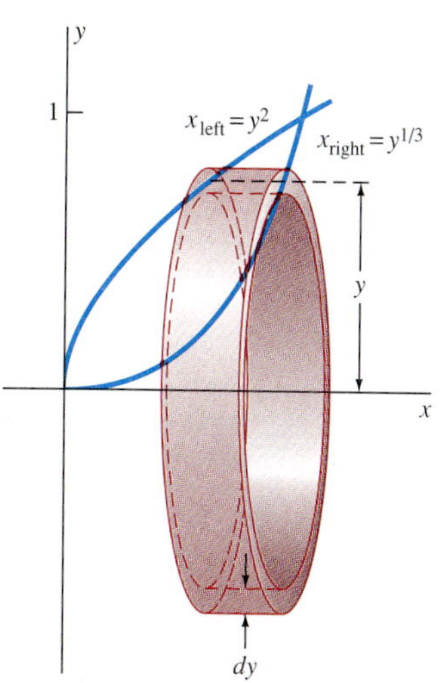

Fig. 6.3.13 Revolution around the x-axis (Example 3)

The volume of revolution around the x-axis (Fig. 6.3.13) is given by

$$V = \int_0^1 2\pi y(x_{\text{right}} - x_{\text{left}})\,dy = \int_0^1 2\pi y(y^{1/3} - y^2)\,dy$$

$$= \int_0^1 2\pi(y^{4/3} - y^3)\,dy = 2\pi\left[\tfrac{3}{7}y^{7/3} - \tfrac{1}{4}y^4\right]_0^1 = \tfrac{5}{14}\pi.$$

The answers are the same, of course, as those we obtained by using the method of cross sections in Example 5 of Section 6.2. ■

EXAMPLE 4 Suppose that the region of Example 3 is rotated around the vertical line $x = -1$ (Fig. 6.3.14). Then the area element

$$dA = (y_{\text{top}} - y_{\text{bot}})\,dx = (\sqrt{x} - x^3)\,dx$$

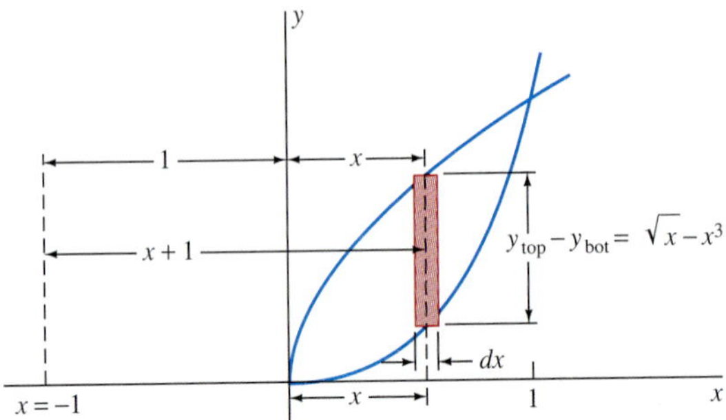

Fig. 6.3.14 Revolution around the vertical line $x = -1$ (Example 4)

is revolved through a circle of radius $r = 1 + x$. Hence the volume of the resulting cylindrical shell is

$$\begin{aligned} dV &= 2\pi r\, dA = 2\pi(1 + x)(x^{1/2} - x^3)\, dx \\ &= 2\pi(x^{1/2} + x^{3/2} - x^3 - x^4)\, dx. \end{aligned}$$

The volume of the resulting solid of revolution is then

$$\begin{aligned} V &= \int_0^1 2\pi(x^{1/2} + x^{3/2} - x^3 - x^4)\, dx \\ &= 2\pi\left[\tfrac{2}{3}x^{3/2} + \tfrac{2}{5}x^{5/2} - \tfrac{1}{4}x^4 - \tfrac{1}{5}x^5\right]_0^1 = \tfrac{37}{30}\pi, \end{aligned}$$

as we found by using the method of cross sections in Example 6 of Section 6.2. ■

We may observe finally that the method of cylindrical shells is summarized by the heuristic formula

$$V = \int_{\star}^{\star\star} 2\pi r\, dA,$$

where dA denotes the area of an infinitesimal strip that is revolved through a circle of radius r to generate a thin cylindrical shell. The stars indicate limits of integration that you need to find.

6.3 PROBLEMS

In Problems 1 through 28, use the method of cylindrical shells to find the volume of the solid generated by rotating around the indicated axis the region bounded by the given curves.

1. $y = x^2$, $y = 0$, $x = 2$; the y-axis
2. $x = y^2$, $x = 4$; the y-axis
3. $y = 25 - x^2$, $y = 0$; the y-axis (Fig. 6.3.15)
4. $y = 2x^2$, $y = 8$; the y-axis (Fig. 6.3.16)
5. $y = x^2$, $y = 8 - x^2$; the y-axis
6. $x = 9 - y^2$, $x = 0$; the x-axis
7. $x = y$, $x + 2y = 3$, $y = 0$; the x-axis (Fig. 6.3.17)
8. $y = x^2$, $y = 2x$; the line $y = 5$

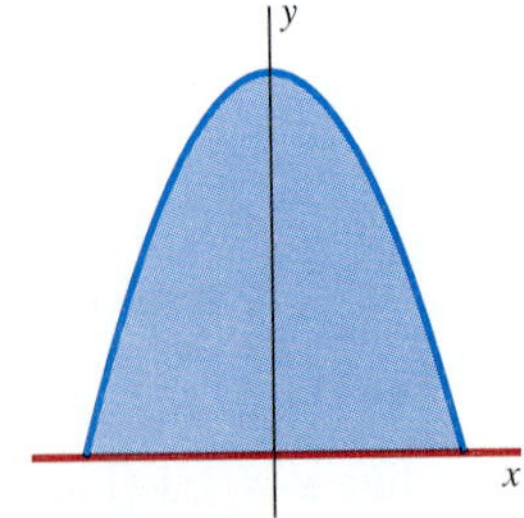

Fig. 6.3.15 Problem 3

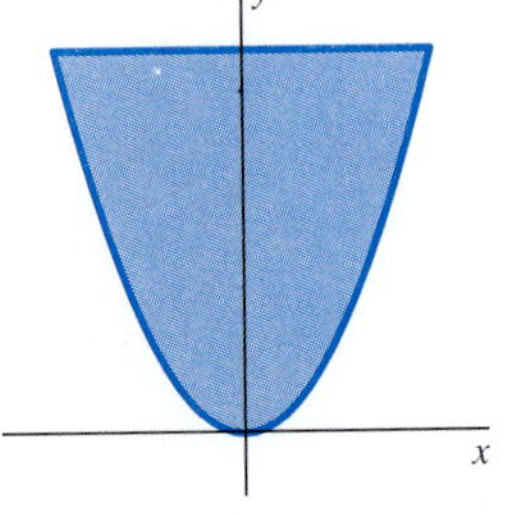

Fig. 6.3.16 Problem 4

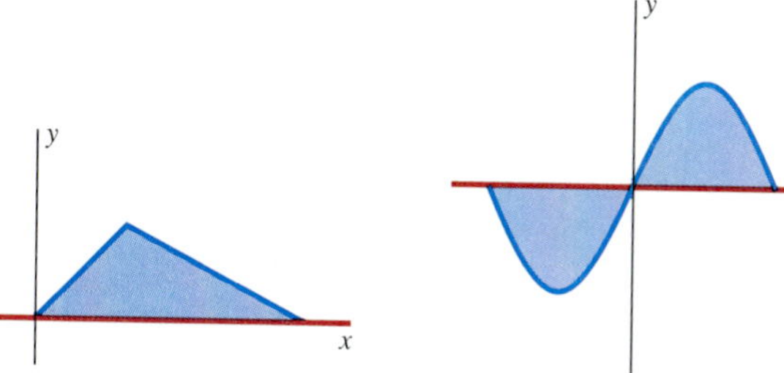

Fig. 6.3.17 Problem 7

Fig. 6.3.18 Problem 11

9. $y = 2x^2,\ y^2 = 4x$; the x-axis

10. $y = 3x - x^2,\ y = 0$; the y-axis

11. $y = 4x - x^3,\ y = 0$; the y-axis (Fig. 6.3.18)

12. $x = y^3 - y^4,\ x = 0$; the line $y = -2$ (Fig. 6.3.19)

13. $y = x - x^3,\ y = 0\ (0 \leqq x \leqq 1)$; the y-axis

14. $x = 16 - y^2,\ x = 0,\ y = 0\ (0 \leqq y \leqq 4)$; the x-axis

15. $y = x - x^3,\ y = 0\ (0 \leqq x \leqq 1)$; the line $x = 2$ (Fig. 6.3.20)

16. $y = x^3,\ y = 0,\ x = 2$; the y-axis (Fig. 6.3.21)

17. $y = x^3,\ y = 0,\ x = 2$; the line $x = 3$

18. $y = x^3,\ y = 0,\ x = 2$; the x-axis

19. $y = x^2,\ y = 0,\ x = -1,\ x = 1$; the line $x = 2$

20. $y = x^2,\ y = x\ (0 \leqq x \leqq 1)$; the y-axis

21. $y = x^2,\ y = x\ (0 \leqq x \leqq 1)$; the x-axis

22. $y = x^2,\ y = x\ (0 \leqq x \leqq 1)$; the line $y = 2$

23. $y = x^2,\ y = x\ (0 \leqq x \leqq 1)$; the line $x = -1$

24. $x = y^2,\ x = 2 - y^2$; the x-axis (Fig. 6.3.22)

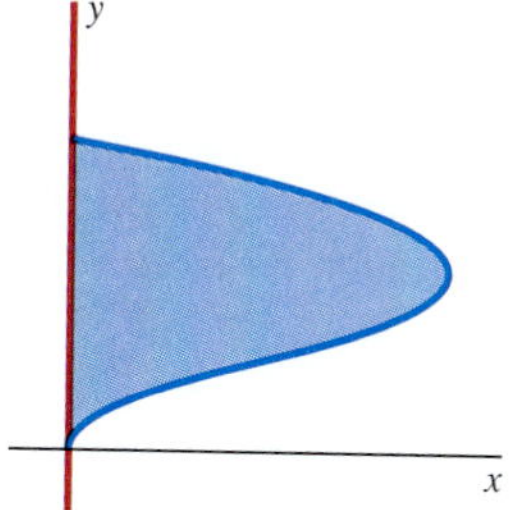

Fig. 6.3.19 Problem 12

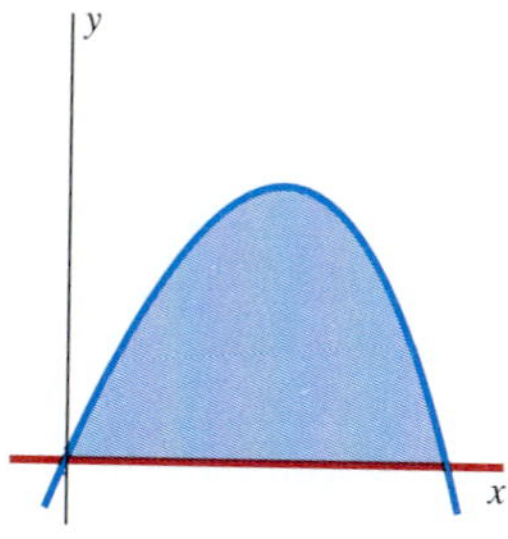

Fig. 6.3.20 Problem 15

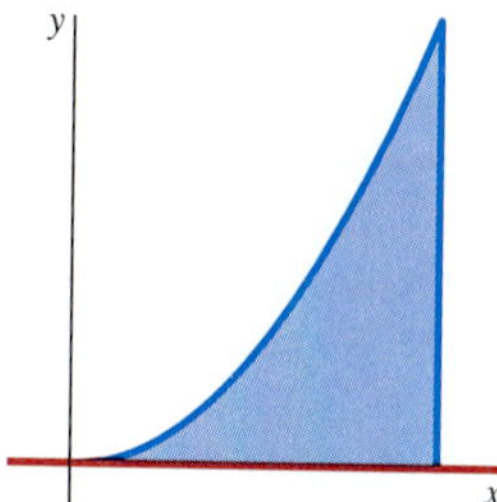

Fig. 6.3.21 Problem 16

Fig. 6.3.22 Problem 24

25. $x = y^2,\ x = 2 - y^2$; the line $y = 1$

26. $y = 4x - x^2,\ y = 0$; the y-axis

27. $y = 4x - x^2,\ y = 0$; the line $x = -1$

28. $y = x^2,\ x = y^2$; the line $y = -1$

In Problems 29 through 34, first use a calculator or computer to approximate (graphically or otherwise) the points of intersection of the two given curves. Let R be the region bounded by these curves. Integrate to approximate the volume of the solid obtained by revolving the region R around the y-axis. In Problems 31 through 34 you will find helpful the integral formula

$$\int u \cos u \, du = \cos u + u \sin u + C,$$

which you can verify by differentiation of the right-hand side.

29. $y = x^3 + 1,\quad y = 6x - x^2$ (R lies to the right of the y-axis)

30. $y = x^4,\quad y = 10x - 5$

31. $y = \cos x,\quad y = x^2$

32. $y = \cos x,\quad y = (x - 1)^2$

33. $y = \cos x,\quad y = 3x^2 - 6x + 2$

34. $y = 3 \cos x,\quad y = -\cos 4x$ (R lies between $x = -2$ and $x = 2$)

35. Verify the formula for the volume of a right circular cone by using the method of cylindrical shells. Apply the method to the figure generated by rotating the triangular region with vertices $(0, 0)$, $(r, 0)$, and $(0, h)$ around the y-axis.

36. Use the method of cylindrical shells to compute the volume of the paraboloid of Problem 43 in Section 6.2.

37. Use the method of cylindrical shells to find the volume of the ellipsoid obtained by revolving the elliptical region bounded by the graph of the equation

$$\left(\frac{x}{a}\right)^2 + \left(\frac{y}{b}\right)^2 = 1$$

around the y-axis.

38. Use the method of cylindrical shells to derive the formula given in Problem 48 of Section 6.2 for the volume of a spherical segment.

39. Use the method of cylindrical shells to compute the volume of the torus in Problem 49 in Section 6.2. [*Suggestion:* Substitute u for $x - b$ in the integral given by the formula in Eq. (2).]

40. (a) Find the volume of the solid generated by revolving the region bounded by the curves $y = x^2$ and $y = x + 2$ around the line $x = -2$. (b) Repeat part (a), but revolve the region around the line $x = 3$.

41. Find the volume of the solid generated by revolving the circular disk $x^2 + y^2 \leqq a^2$ around the vertical line $x = a$.

42. (a) Verify by differentiation that

$$\int x \sin x \, dx = \sin x - x \cos x + C.$$

(b) Find the volume of the solid obtained by rotating around the y-axis the area under $y = \sin x$ from $x = 0$ to $x = \pi$.

43. We found in Example 2 that the volume remaining after a hole of radius a is bored through the center of a sphere of radius $b > a$ is

$$V = \frac{4}{3}\pi(b^2 - a^2)^{3/2}.$$

(a) Express the volume V in this formula *without* use of the hole radius a; use instead the hole height h. (*Suggestion:* Use the right triangle in Fig. 6.3.8.) (b) What is remarkable about the answer to part (a)?

44. The plane region R is bounded above and on the right by the graph of $y = 25 - x^2$, on the left by the y-axis, and below by the x-axis. A paraboloid is generated by revolving R around the y-axis. Then a vertical hole of radius 3 and centered along the y-axis is bored through the paraboloid. Find the volume of the solid that remains by using (a) the method of cross sections and (b) the method of cylindrical shells.

45. The loop of the curve $y^2 = x(5 - x)^2$ bounds the region shown in Fig. 6.3.23. Find the volume of the solid obtained when this region is revolved around (a) the x-axis; (b) the y-axis; (c) the line $x = 5$.

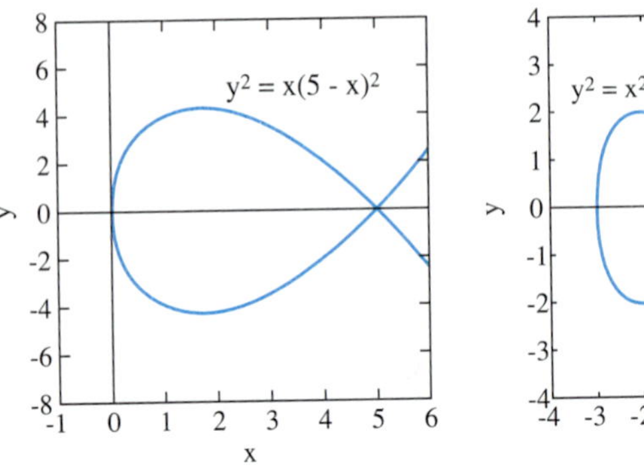

Fig. 6.3.23 The region of Problem 45

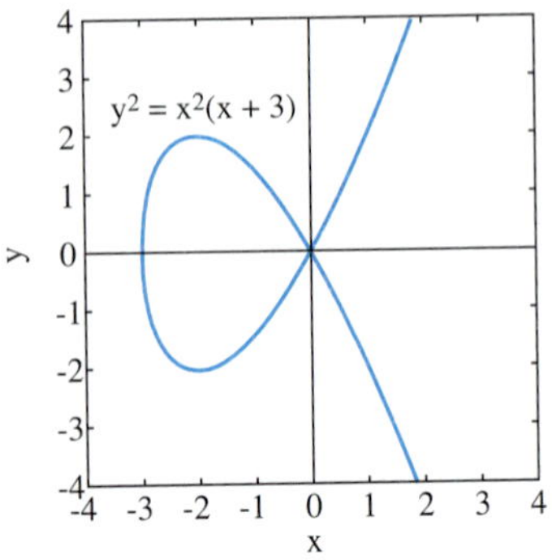

Fig. 6.3.24 The region of Problem 46

46. The loop of the curve $y^2 = x^2(x + 3)$ bounds the region shown in Fig. 6.3.24. Find the volume of the solid obtained when this region is revolved around (a) the x-axis; (b) the y-axis; (c) the line $x = -3$. (*Suggestion:* If useful, substitute $u = x + 3$ before integrating.)

6.3 PROJECT: DESIGN YOUR OWN RING!

This project deals with the custom-made gold wedding band pictured in Fig. 6.3.25. Its shape is obtained by revolving the region A shown in Fig. 6.3.26 around the vertical axis shown there. The resulting wedding band has

- Inner radius R,
- Minimum thickness T, and
- Width W.

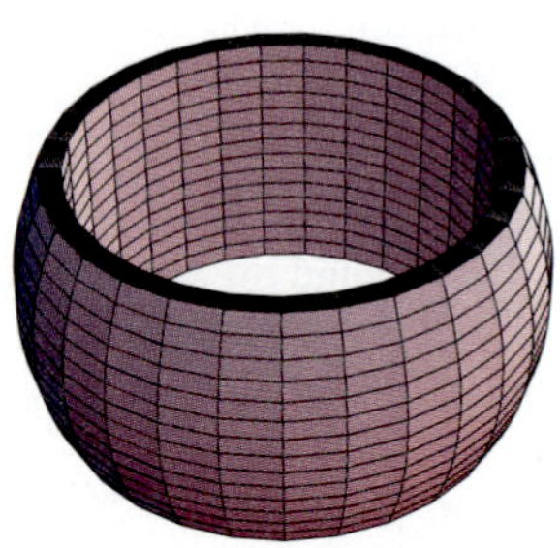

Fig. 6.3.25 Wedding band

The curved boundary of the region A is an arc of a circle whose center lies on the axis of revolution. For a typical wedding band, R might be anywhere from 6 to 12 mm, T might be 0.5 to 1.5 mm, and W might be 4 to 10 mm.

If a customer asks the price of a wedding band with given dimensions R, T, and W, the jeweler must first calculate the volume of the desired band to determine how much gold will be required to make it. Use the methods of this section to show that the volume V is given by the formula

$$V = \frac{\pi W}{6}(W^2 + 12RT + 6T^2). \tag{1}$$

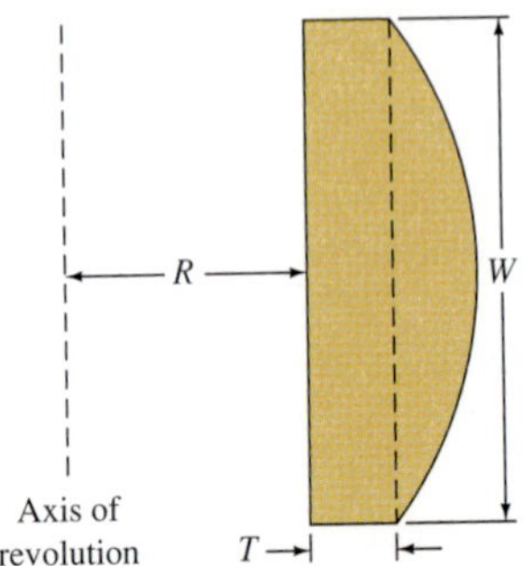

Fig. 6.3.26 Cross section of the wedding band

If these dimensions are measured in millimeters, then V is given in cubic millimeters. (There are 1000 mm³ in 1 cm³.)

Suppose that the jeweler plans to charge the customer \$1000 per *troy ounce* of alloy (90% gold, 10% silver) used to make the ring. (The profit on the sale, covering the jeweler's time and overhead in making the ring, is fairly substantial because the price of gold is generally under \$400/oz and that of silver, under \$6/oz.) The inner radius R of the wedding band is determined by the measurement of the customer's finger (in millimeters; there are exactly 25.4 mm per inch). Suppose that the jeweler makes all wedding bands with $T = 1$ (mm). Then, for a given acceptable cost C (in dollars), the customer wants to know the maximum width W of the wedding band he or she can afford.

PROJECT: Measure your own ring finger to determine R (you can measure its circumference C with a piece of string and then divide by 2π). Then choose a cost figure C in the \$100 to \$500 price range. Use Eq. (1) with $T = 1$ to find the width C of a band that costs C dollars (at \$1000/oz). You will need to know that the density of the gold-silver alloy is 18.4 gm/cm^3 and that 1 lb contains 12 troy ounces and 453.59 gm. Use a graphics calculator or a calculator with a **SOLVE** key to solve the resulting cubic equation in W.

6.4 ARC LENGTH AND SURFACE AREA OF REVOLUTION

If you plan to hike the Continental Divide Trail, you will need to know the length of this curved path so you'll know how much equipment to take. Here we investigate how to find the length of a curved path and the closely related idea of finding the surface area of a curved surface.

A **smooth arc** is the graph of a smooth function defined on a closed interval; a **smooth function** f on $[a, b]$ is a function whose derivative f' is continuous on $[a, b]$. The continuity of f' rules out the possibility of corner points on the graph of f, points where the direction of the tangent line changes abruptly. The graphs of $f(x) = |x|$ and $g(x) = x^{2/3}$ are shown in Fig. 6.4.1; neither is smooth because each has a corner point at the origin.

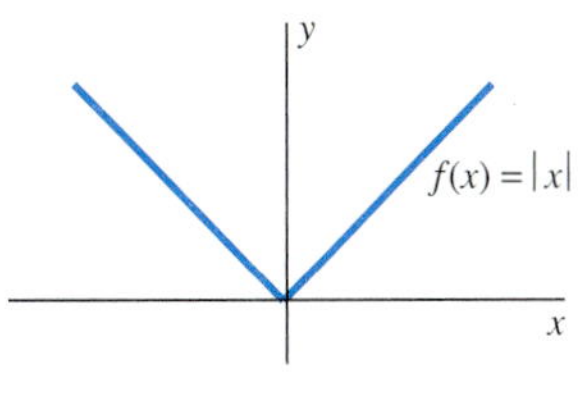

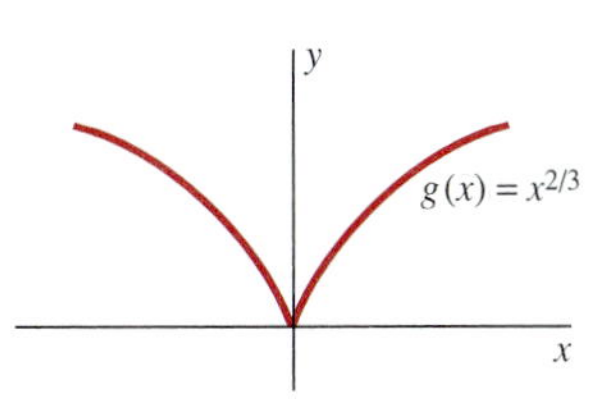

Fig. 6.4.1 Graphs that have corner points

The Length of a Curve

To investigate the length of a smooth arc, we begin with the length of a straight line segment, which is simply the distance between its endpoints. Then, given a smooth arc C, we pose the following question: If C were a thin wire and we straightened it without stretching it, how long would the resulting straight wire be? The answer is what we call the *length* of C.

To approximate the length s of the smooth arc C, we can inscribe in C a polygonal arc—one made up of straight line segments—and then calculate the length of this polygonal arc. We proceed in the following way, under the assumption that C is the graph of a smooth function f defined on the closed interval $[a, b]$. Consider a partition of $[a, b]$ into n subintervals, all with the same length Δx. Let P_i denote the point $(x_i, f(x_i))$ on the arc C corresponding to the ith subdivision point x_i. Our polygonal arc "inscribed in C" is then the union of the line segments $P_0P_1, P_1P_2, P_2P_3, \dots, P_{n-1}P_n$. So an approximation to the length x of C is

$$s \approx \sum_{i=1}^{n} |P_{i-1}P_i|, \tag{1}$$

the sum of the lengths of these line segments (Fig. 6.4.2). Our plan is to take the limit of this sum as $n \to \infty$: We want to evaluate

$$s = \lim_{n\to\infty} \sum_{i=1}^{n} |P_{i-1}P_i|.$$

The length of the typical line segment $P_{i-1}P_i$ is

$$|P_{i-1}P_i| = \left[(x_i - x_{i-1})^2 + (f(x_i) - f(x_{i-1}))^2 \right]^{1/2}.$$

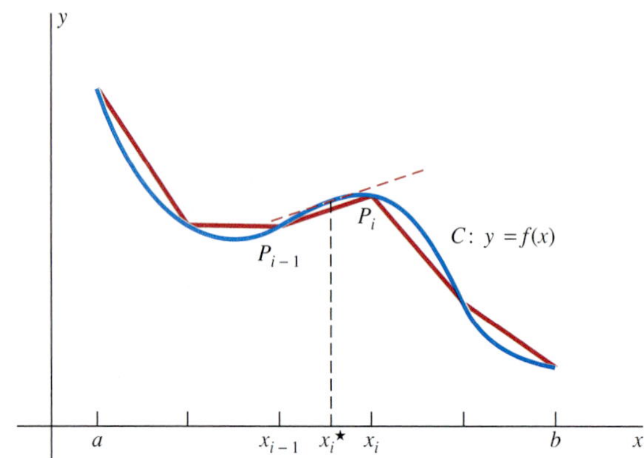

Fig. 6.4.2 A polygonal arc inscribed in the smooth curve C

We apply the mean value theorem to the function f on the interval $[x_{i-1}, x_i]$ and thereby conclude the existence of a point $x_i^\star$ in this interval such that

$$f(x_i) - f(x_{i-1}) = f'(x_i^\star) \cdot (x_i - x_{i-1}).$$

Hence

$$\begin{aligned} |P_{i-1}P_i| &= \left[1 + \left(\frac{f(x_i) - f(x_{i-1})}{x_i - x_{i-1}}\right)^2\right]^{1/2} \cdot (x_i - x_{i-1}) \\ &= \sqrt{1 + [f'(x_i^\star)]^2}\,\Delta x, \end{aligned}$$

where $\Delta x = x_i - x_{i-1}$.

We next substitute this expression for $|P_{i-1}P_i|$ into Eq. (1) and get the approximation

$$s \approx \sum_{i=1}^{n} \sqrt{1 + [f'(x_i^\star)]^2}\,\Delta x.$$

This sum is a Riemann sum for the function $\sqrt{1 + [f'(x)]^2}$ on $[a, b]$, and therefore—because f' is continuous—such sums approach the integral

$$\int_a^b \sqrt{1 + [f'(x)]^2}\,dx$$

as $\Delta x \to 0$. But our approximation ought to approach, as well, the actual length s as $\Delta x \to 0$. On this basis we *define* the **length** s of the smooth arc C to be

$$s = \int_a^b \sqrt{1 + [f'(x)]^2}\,dx = \int_a^b \sqrt{1 + \left(\frac{dy}{dx}\right)^2}\,dx. \tag{2}$$

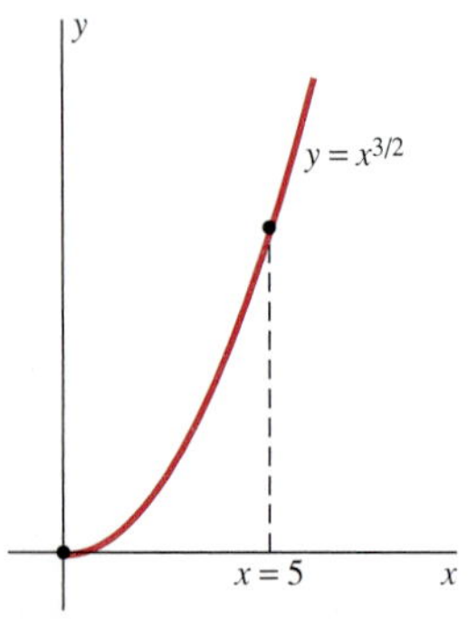

Fig. 6.4.3 The semicubical parabola of Example 1

EXAMPLE 1 Find the length of the so-called semicubical parabola (it's not really a parabola) $y = x^{3/2}$ on $[0, 5]$ (Fig. 6.4.3).

Solution We first compute the integrand in Eq. (2):

$$\sqrt{1 + \left(\frac{dy}{dx}\right)^2} = \sqrt{1 + (\tfrac{3}{2}x^{1/2})^2} = \sqrt{1 + \tfrac{9}{4}x} = \tfrac{1}{2}(4 + 9x)^{1/2}.$$

Hence the length of the arc $y = x^{3/2}$ over the interval $[0, 5]$ is

$$s = \int_0^5 \tfrac{1}{2}(4 + 9x)^{1/2}\,dx = \left[\tfrac{1}{27}(4 + 9x)^{3/2}\right]_0^5 = \tfrac{335}{27} \approx 12.41.$$

As a plausibility check, the endpoints of the arc are $(0, 0)$ and $(5, 5\sqrt{5})$, so the straight line segment connecting these points has length $5\sqrt{6} \approx 12.25$. This is, as it should be, somewhat less than the calculated length of the arc. ■

EXAMPLE 2 A manufacturer needs to make corrugated metal sheets 36 in. wide with cross sections in the shape of the curve

$$y = \tfrac{1}{2}\sin \pi x, \qquad 0 \leqq x \leqq 36$$

(Fig. 6.4.4). How wide must the original flat sheets be for the manufacturer to produce these corrugated sheets?

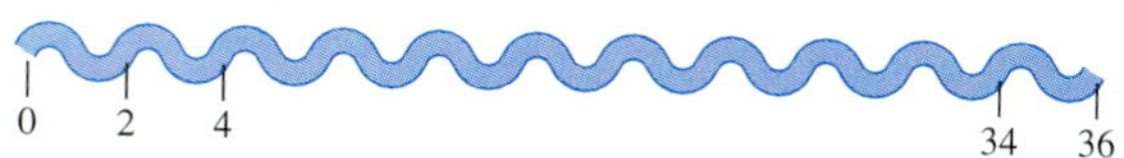

Fig. 6.4.4 The corrugated metal sheet in the shape of $y = \frac{1}{2}\sin \pi x$ (Example 2)

Solution If

$$f(x) = \tfrac{1}{2}\sin \pi x, \quad \text{then} \quad f'(x) = \tfrac{1}{2}\pi \cos \pi x.$$

Hence Eq. (2) yields the arc length of the graph of f over $[0, 36]$:

$$s = \int_0^{36} \sqrt{1 + (\tfrac{1}{2}\pi)^2 \cos^2 \pi x}\;dx = 36\int_0^1 \sqrt{1 + (\tfrac{1}{2}\pi)^2 \cos^2 \pi x}\;dx.$$

These integrals cannot be evaluated in terms of elementary functions. For this reason, we cannot apply the fundamental theorem of calculus. So we estimate their values with the aid of Simpson's approximation (Section 5.9). Both with $n = 6$ and with $n = 12$ subintervals we find that

$$\int_0^1 \sqrt{1 + (\tfrac{1}{2}\pi)^2 \cos^2 \pi x}\;dx \approx 1.46$$

inches. Therefore the manufacturer should use flat sheets of approximate width $36 \cdot 1.46 \approx 52.6$ in. ■

Arc Length by Integration with Respect to y

In the case of a smooth arc given as a graph $x = g(y)$ for y in $[c, d]$, a similar discussion beginning with a subdivision of $[c, d]$ leads to the formula

$$s = \int_c^d \sqrt{1 + [g'(y)]^2}\;dy = \int_c^d \sqrt{1 + \left(\frac{dx}{dy}\right)^2}\;dy \tag{3}$$

for its length. We can compute the length of a more general curve, such as a circle, by dividing it into a finite number of smooth arcs and then applying to each of these arcs whichever of Eqs. (2) and (3) is required.

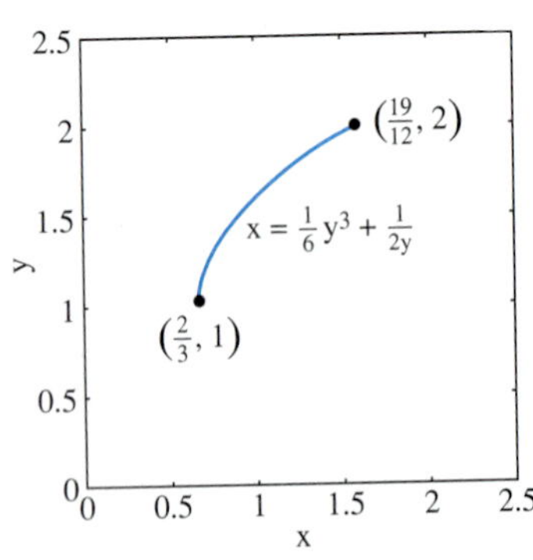

Fig. 6.4.5 The curve of Example 3 (although we can calculate its arc length without even visualizing it)

EXAMPLE 3 Find the length s of the curve (Fig. 6.4.5)

$$x = \frac{1}{6}y^3 + \frac{1}{2y}, \qquad 1 \leqq y \leqq 2.$$

Solution Here y is the natural independent variable, so we use the arc-length formula in Eq. (3). First we calculate

$$1 + \left(\frac{dx}{dy}\right)^2 = 1 + \left(\frac{1}{2}y^2 - \frac{1}{2y^2}\right)^2 = 1 + \frac{1}{4}y^4 - \frac{1}{2} + \frac{1}{4y^4}$$
$$= \frac{1}{4}y^4 + \frac{1}{2} + \frac{1}{4y^4} = \left(\frac{1}{2}y^2 + \frac{1}{2y^2}\right)^2.$$

Thus we can "get out from under the radical" in Eq. (3):

$$s = \int_c^d \sqrt{1 + \left(\frac{dx}{dy}\right)^2}\, dy = \int_1^2 \left(\frac{1}{2}y^2 + \frac{1}{2y^2}\right) dy$$
$$= \left[\frac{1}{6}y^3 - \frac{1}{2y}\right]_1^2 = \frac{17}{12}.$$

A Symbolic Device

We can employ a convenient symbolic device to remember both Eqs. (2) and (3) simultaneously. We think of two nearby points $P(x, y)$ and $Q(x + dx, y + dy)$ on the smooth arc C and denote by ds the length of the arc that joins P and Q. Imagine that P and Q are so close together that ds is, for all practical purposes, equal to the length of the straight line segment PQ. Then the Pythagorean theorem applied to the small right triangle in Fig. 6.4.6 gives

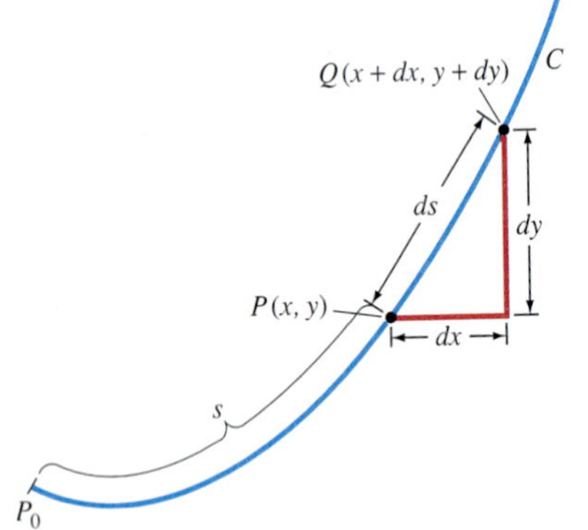

Fig. 6.4.6 Heuristic development of the arc-length formula

$$ds = \sqrt{(dx)^2 + (dy)^2} \tag{4}$$
$$= \sqrt{1 + \left(\frac{dy}{dx}\right)^2}\, dx \tag{4$'$}$$
$$= \sqrt{1 + \left(\frac{dx}{dy}\right)^2}\, dy. \tag{4$''$}$$

Thinking of the entire length s of C as the sum of small pieces such as ds, we write

$$s = \int_{\star}^{\star\star} ds. \tag{5}$$

Then formal (symbolic) substitution of the expressions in Eqs. (4′) and (4″) for ds in Eq. (5) yields Eqs. (2) and (3); only the limits of integration remain to be determined.

Cones and Conical Frusta

A **surface of revolution** is a surface obtained by revolving an arc or curve around an axis that lies in the same plane as the arc. The surface of a cylinder or of a sphere and the curved surface of a cone are important as examples of surfaces of revolution.

Our basic approach to finding the area of such a surface is this: First we inscribe a polygonal arc in the curve to be revolved. We then regard the area of the surface generated by revolving the polygonal arc to be an approximation to the surface generated by revolving the original curve. Because a surface generated by revolving a polygonal arc around an axis consists of frusta (sections) of cones, we can calculate its area in a reasonably simple way.

This approach to surface area originated with Archimedes. For example, he used this method to establish the formula $A = 4\pi r^2$ for the surface area of a sphere of radius r.

We will need the formula

$$A = 2\pi \bar{r} L \tag{6}$$

for the curved surface area of a cone with average radius $\bar{r} = \frac{1}{2}(r_1 + r_2)$ and *slant height* L (Fig. 6.4.7). Equation (6) follows from the formula

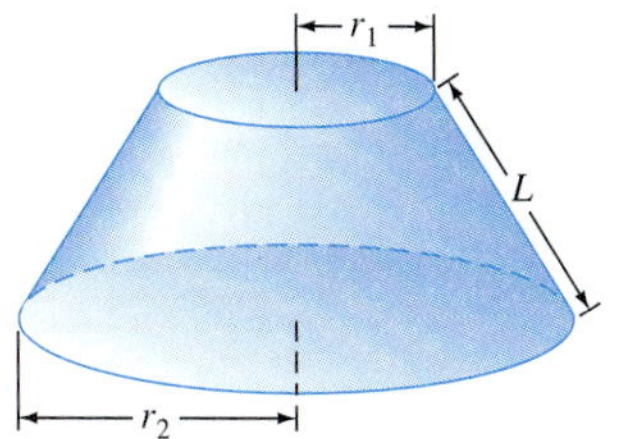

Fig. 6.4.7 A frustum of a cone; the slant height is L.

$$A = \pi r L \tag{7}$$

for the area of a conical surface with base radius r and slant height L (Fig. 6.4.8). It is easy to derive Eq. (7) by "unrolling" the conical surface onto a sector of a circle of radius L, because the area of this sector is

$$A = \frac{2\pi r}{2\pi L} \cdot \pi L^2 = \pi r L.$$

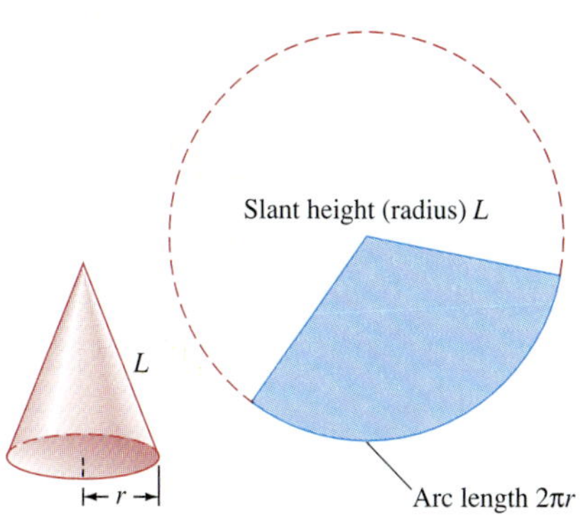

Fig. 6.4.8 Surface area of a cone: Cut along L, then unroll the cone onto the circular sector.

To derive Eq. (6) from Eq. (7), we think of the frustum as the lower section of a cone with slant height $L_2 = L + L_1$ (Fig. 6.4.9). Then subtraction of the area of the upper conical section from that of the entire cone gives

$$A = \pi r_2 L_2 - \pi r_1 L_1 = \pi r_2 (L + L_1) - \pi r_1 L_1 = \pi (r_2 - r_1) L_1 + \pi r_2 L$$

for the area of the frustum. But the similar right triangles in Fig. 6.4.9 yield the proportion

$$\frac{r_1}{L_1} = \frac{r_2}{L_2} = \frac{r_2}{L + L_1},$$

from which we find that $(r_2 - r_1)L_1 = r_1 L$. Hence the area of the frustum is

$$A = \pi r_1 L + \pi r_2 L = 2\pi \bar{r} L,$$

where $\bar{r} = \frac{1}{2}(r_1 + r_2)$. Thus we have verified Eq. (6).

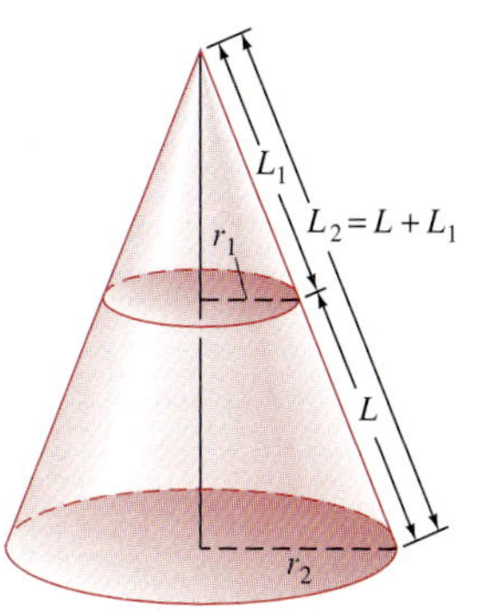

Fig. 6.4.9 Derivation of Eq. (6)

Surface Area of Revolution

Suppose that the surface S has area A and is generated by revolving around the x-axis the smooth arc $y = f(x)$, $a \leqq x \leqq b$; suppose also that $f(x)$ is never negative on $[a, b]$. To approximate A, we begin with a partition of $[a, b]$ into n subintervals, each of length Δx. As in our discussion of arc length leading to Eq. (2), let P_i denote the point $(x_i, f(x_i))$ on the arc. Then, as before, the line segment $P_{i-1}P_i$ has length

$$L_i = |P_{i-1}P_i| = \sqrt{1 + [f'(x_i^\star)]^2}\,\Delta x$$

for some point $x_i^\star$ in the ith subinterval $[x_{i-1}, x_i]$.

The conical frustum obtained by revolving the segment $P_{i-1}P_i$ around the x-axis has slant height L_i and, as shown in Fig. 6.4.10, average radius

$$\bar{r}_i = \tfrac{1}{2}\,[f(x_{i-1}) + f(x_i)].$$

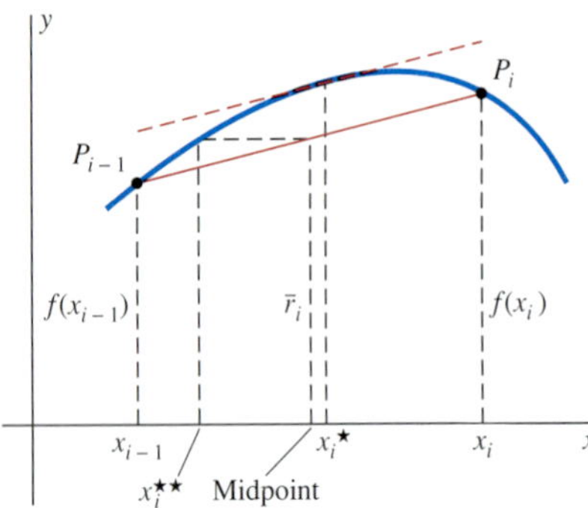

Fig. 6.4.10 Approximation of a surface area of revolution by the surface of a frustum of a cone

Because $\bar{r}_i$ lies between the values $f(x_{i-1})$ and $f(x_i)$, the intermediate value property of continuous functions (Section 2.4) yields a point $x_i^{\star\star}$ in $[x_{i-1}, x_i]$ such that $\bar{r}_i = f(x_i^{\star\star})$. By Eq. (6), the area of this conical frustum is therefore

$$2\pi\bar{r}_i L_i = 2\pi f(x_i^{\star\star})\sqrt{1 + [f'(x_i^{\star})]^2}\ \Delta x.$$

We add the areas of these conical frusta for $i = 1, 2, 3, \dots, n$. This gives the approximation

$$A \approx \sum_{i=1}^{n} 2\pi f(x_i^{\star\star})\sqrt{1 + [f'(x_i^{\star})]^2}\ \Delta x.$$

If $x_i^{\star}$ and $x_i^{\star\star}$ were the *same* point of the ith subinterval $[x_{i-1}, x_i]$, then this approximation would be a Riemann sum for the integral

$$\int_a^b 2\pi f(x)\sqrt{1 + [f'(x)]^2}\ dx.$$

Even though the numbers $x_i^{\star}$ and $x_i^{\star\star}$ are generally not equal, it still follows (from a result in Appendix H) that our approximation approaches the integral above as $\Delta x \to 0$. Intuitively, this is easy to believe; after all, as $\Delta x \to 0$, the difference between $x_i^{\star}$ and $x_i^{\star\star}$ also approaches zero.

We therefore *define* the **area** A of the surface generated by revolving around the x-axis the smooth arc $y = f(x)$, $a \leqq x \leqq b$, by the formula

$$A = \int_a^b 2\pi f(x)\sqrt{1 + [f'(x)]^2}\ dx. \tag{8}$$

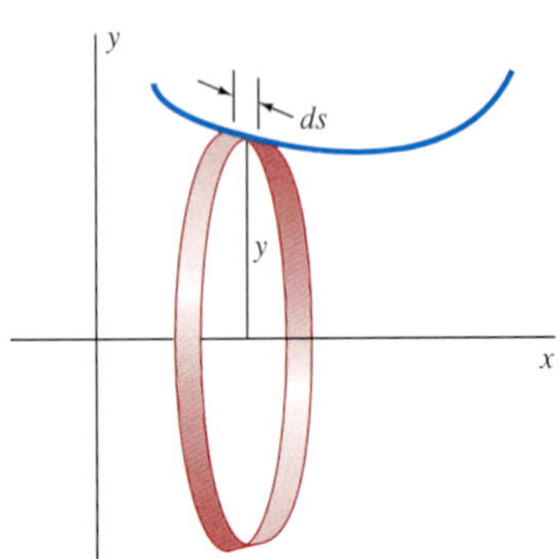

Fig. 6.4.11 The tiny arc ds generates a ribbon with circumference $2\pi y$ when it is revolved around the x-axis.

If we write y for $f(x)$ and ds for $\sqrt{1 + (dy/dx)^2}\ dx$, as in Eq. (4′), then we can abbreviate Eq. (8) as

$$A = \int_a^b 2\pi y\ ds \qquad (x\text{-axis}). \tag{9}$$

This abbreviated formula is conveniently remembered by thinking of $dA = 2\pi y\,ds$ as the area of the narrow frustum obtained by revolving the tiny arc ds around the x-axis in a circle of radius y (Fig. 6.4.11).

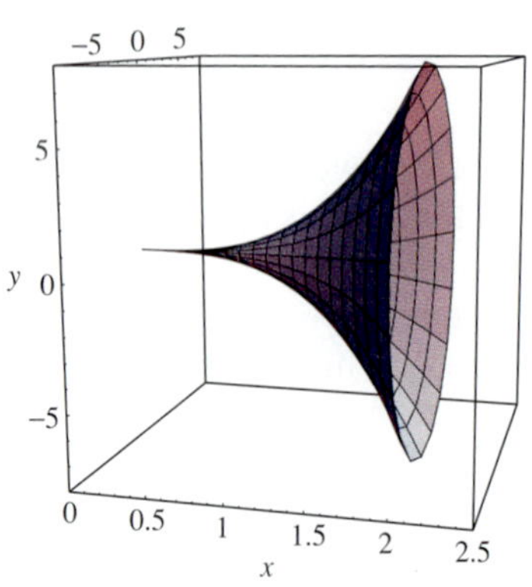

Fig. 6.4.12 The "horn" generated by revolving the curve $y = x^3$, $0 \leqq x \leqq 2$, around the x-axis

EXAMPLE 4 Figure 6.4.12 shows the horn-shaped surface generated by revolving the curve $y = x^3$, $0 \leqq x \leqq 2$, around the x-axis. Find its surface area of revolution.

Solution Substituting $y = x^3$ and

$$ds = \sqrt{1 + [y'(x)]^2}\ dx = \sqrt{1 + 9x^4}\ dx$$

in Eq. (9), we get

$$A = \int_0^2 2\pi x^3(1 + 9x^4)^{1/2}\ dx \qquad (\text{let } u = 1 + 9x^4)$$

$$= \left[\frac{\pi}{27}(1 + 9x^4)^{3/2}\right]_0^2 = \frac{\pi}{27}(145^{3/2} - 1) \approx 203.04. \qquad ■$$

If the smooth arc being revolved around the x-axis is given instead by $x = g(y)$, $c \leqq y \leqq d$, then an approximation based on a subdivision of $[c, d]$ leads to the area formula

$$A = \int_c^d 2\pi y\sqrt{1 + [g'(y)]^2}\, dy. \tag{10}$$

We can obtain Eq. (10) by making the formal substitution $ds = \sqrt{1 + (dx/dy)^2}\, dy$ of Eq. (4″) into the abbreviated formula in Eq. (9) for surface area of revolution and then replacing a and b with the correct limits of integration.

Revolution Around the y-Axis

Now let us consider the surface generated by revolving a smooth arc around the y-axis rather than around the x-axis. In Fig. 6.4.13 we see that the average radius of the narrow frustum obtained by revolving the tiny arc ds is now x instead of y. This suggests the abbreviated formula

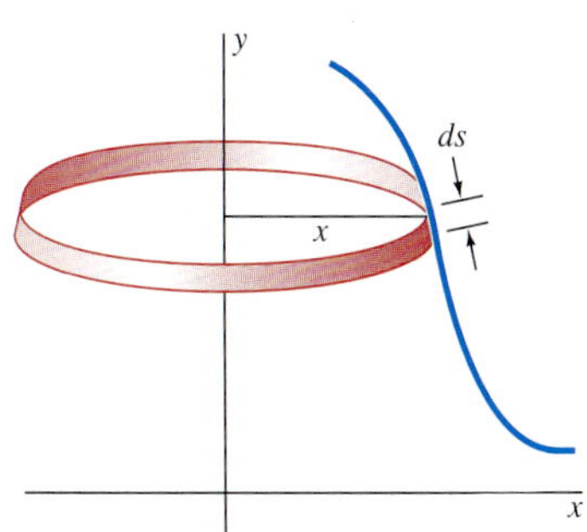

Fig. 6.4.13 The tiny arc ds generates a ribbon with circumference $2\pi x$ when it is revolved around the y-axis.

$$A = \int_a^b 2\pi x\, ds \qquad (y\text{-axis}) \tag{11}$$

for a surface area of revolution around the y-axis. If the smooth arc is given by $y = f(x)$, $a \leqq x \leqq b$, then the symbolic substitution $ds = \sqrt{1 + (dy/dx)^2}\, dx$ gives

$$A = \int_a^b 2\pi x\sqrt{1 + [f'(x)]^2}\, dx. \tag{12}$$

But if the smooth arc is presented in the form $x = g(y)$, $c \leqq y \leqq d$, then the symbolic substitution of $ds = \sqrt{1 + (dx/dy)^2}\, dy$ into Eq. (11) gives

$$A = \int_c^d 2\pi g(y)\sqrt{1 + [g'(y)]^2}\, dy. \tag{13}$$

Equations (12) and (13) may be verified by using approximations similar to the one leading to Eq. (8).

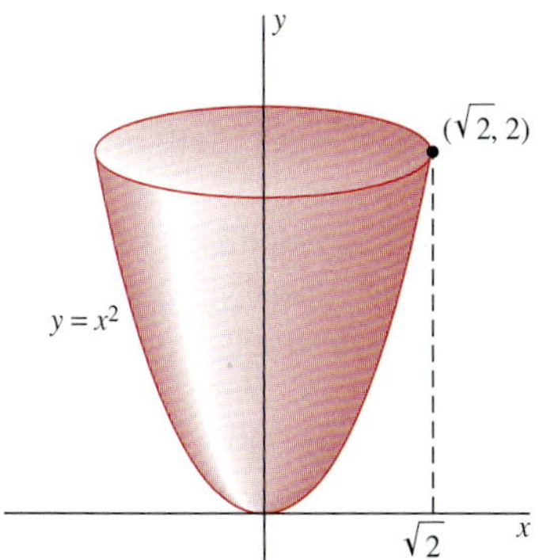

Fig. 6.4.14 The paraboloid of Example 5

EXAMPLE 5 Find the area of the paraboloid shown in Fig. 6.4.14, which is obtained by revolving the parabolic arc $y = x^2$, $0 \leqq x \leqq \sqrt{2}$, around the y-axis.

Solution Following the suggestion that precedes the example, we get

$$A = \int_{\star}^{\star\star} 2\pi x\, ds = \int_a^b 2\pi x\sqrt{1 + \left(\frac{dy}{dx}\right)^2}\, dx$$

$$= \int_0^{\sqrt{2}} 2\pi x\sqrt{1 + (2x)^2}\, dx$$

$$= \int_0^{\sqrt{2}} \frac{\pi}{4}(1 + 4x^2)^{1/2} \cdot 8x\, dx = \left[\frac{\pi}{6}(1 + 4x^2)^{3/2}\right]_0^{\sqrt{2}} = \frac{13}{3}\pi.$$

Surface Area Summary

In conclusion, we have four formulas for areas of surfaces of revolution, summarized in the table in Fig. 6.4.15. Which of these formulas is appropriate for computing the area of a given surface depends on two factors:

1. Whether the smooth arc that generates the surface is presented in the form $y = f(x)$ or in the form $x = g(y)$, and
2. Whether this arc is to be revolved around the x-axis or around the y-axis.

Description of curve C		Axis of revolution: x-axis		Axis of revolution: y-axis	
	$y = f(x)$, $a \leqq x \leqq b$	$\int_a^b 2\pi f(x)\sqrt{1 + [f'(x)]^2}\,dx$	**(8)**	$\int_a^b 2\pi x\sqrt{1 + [f'(x)]^2}\,dx$	**(12)**
	$x = g(y)$, $c \leqq y \leqq d$	$\int_c^d 2\pi y\sqrt{1 + [g'(y)]^2}\,dy$	**(10)**	$\int_c^d 2\pi g(y)\sqrt{1 + [g'(y)]^2}\,dy$	**(13)**

Fig. 6.4.15 Area formulas for surfaces of revolution

Memorizing the four formulas in the table in Fig. 6.4.15 is unnecessary. We suggest that you instead remember the abbreviated formulas in Eqs. (9) and (11) in conjunction with Figs. 6.4.11 and 6.4.13 and make either the substitution

$$y = f(x), \qquad ds = \sqrt{1 + \left(\frac{dy}{dx}\right)^2}\,dx$$

or the substitution

$$x = g(y), \qquad ds = \sqrt{1 + \left(\frac{dx}{dy}\right)^2}\,dy,$$

depending on whether the smooth arc is presented as a function of x or as a function of y. It may also be helpful to note that all four of these surface-area formulas have the form

$$A = \int_{\star}^{\star\star} 2\pi r\,ds, \tag{14}$$

where r denotes the radius of the circle around which the arc length element ds is revolved.

As in earlier sections, we again caution you to identify the independent variable by examining the differential and to express every dependent variable in terms of the independent variable before you antidifferentiate. That is, express everything, including ds, either in terms of x (and dx) or in terms of y (and dy).

The decision of which abbreviated formula—Eq. (9) or Eq. (11)—to use is determined by the axis of revolution. In contrast, the decision of whether the variable of integration should be x or y depends on how the smooth arc is given: as a function of x or as a function of y. In some problems, either x or y may be used as the variable of integration, but the integral is usually much simpler to evaluate if you make the correct choice. Experience is very helpful here. Right now, try Example 5 with independent variable y.

6.4 PROBLEMS

In Problems 1 through 10, set up and simplify the integral that gives the length of the given smooth arc. Do not evaluate the integral.

1. $y = x^2$, $0 \leqq x \leqq 1$

2. $y = x^{5/2}$, $1 \leqq x \leqq 3$

3. $y = 2x^3 - 3x^2$, $0 \leqq x \leqq 2$

4. $y = x^{4/3}, \quad -1 \leqq x \leqq 1$

5. $y = 1 - x^2, \quad 0 \leqq x \leqq 100$

6. $x = 4y - y^2, \quad 0 \leqq y \leqq 1$

7. $x = y^4, \quad -1 \leqq y \leqq 2$

8. $x^2 = y, \quad 1 \leqq y \leqq 4$

9. $xy = 1, \quad 1 \leqq x \leqq 2$

10. $x^2 + y^2 = 4, \quad 0 \leqq x \leqq 2$

In Problems 11 through 20, set up and simplify the integral that gives the surface area of revolution generated by rotation of the given smooth arc around the given axis. Do not evaluate the integral.

11. $y = x^2, \quad 0 \leqq x \leqq 4;$ the x-axis

12. $y = x^2, \quad 0 \leqq x \leqq 4;$ the y-axis

13. $y = x - x^2, \quad 0 \leqq x \leqq 1;$ the x-axis

14. $y = x^2, \quad 0 \leqq x \leqq 1;$ the line $y = 4$

15. $y = x^2, \quad 0 \leqq x \leqq 1;$ the line $x = 2$

16. $y = x - x^3, \quad 0 \leqq x \leqq 1;$ the x-axis

17. $y = \sqrt{x}, \quad 1 \leqq x \leqq 4;$ the x-axis

18. $y = \sqrt{x}, \quad 1 \leqq x \leqq 4;$ the y-axis

19. $y = x^{3/2}, \quad 1 \leqq x \leqq 4;$ the line $x = -1$

20. $y = x^{5/2}, \quad 1 \leqq x \leqq 4;$ the line $y = -2$

Find the lengths of the smooth arcs in Problems 21 through 28.

21. $y = \frac{2}{3}(x^2 + 1)^{3/2}$ from $x = 0$ to $x = 2$

22. $x = \frac{2}{3}(y - 1)^{3/2}$ from $y = 1$ to $y = 5$

23. $y = \dfrac{1}{6}x^3 + \dfrac{1}{2x}$ from $x = 1$ to $x = 3$

24. $x = \dfrac{1}{8}y^4 + \dfrac{1}{4y^2}$ from $y = 1$ to $y = 2$

25. $8x^2y - 2x^6 = 1$ from $\left(1, \frac{3}{8}\right)$ to $\left(2, \frac{129}{32}\right)$

26. $12xy - 4y^4 = 3$ from $\left(\frac{7}{12}, 1\right)$ to $\left(\frac{67}{24}, 2\right)$

27. $y^3 = 8x^2$ from $(1, 2)$ to $(8, 8)$

28. $(y - 3)^2 = 4(x + 2)^3$ from $(-1, 5)$ to $(2, 19)$

In Problems 29 through 35, find the area of the surface of revolution generated by revolving the given curve around the indicated axis.

29. $y = \sqrt{x}, 0 \leqq x \leqq 1;$ the x-axis

30. $y = x^3, 1 \leqq x \leqq 2;$ the x-axis

31. $y = \dfrac{1}{5}x^5 + \dfrac{1}{12x^3}, 1 \leqq x \leqq 2;$ the y-axis

32. $x = \dfrac{1}{8}y^4 + \dfrac{1}{4y^2}, 1 \leqq y \leqq 2;$ the x-axis

33. $y^3 = 3x, 0 \leqq x \leqq 9;$ the y-axis

34. $y = \frac{2}{3}x^{3/2}, 1 \leqq x \leqq 2;$ the y-axis (*Suggestion:* Make the substitution $u = 1 + x$.)

35. $y = (2x - x^2)^{1/2}, 0 \leqq x \leqq 2;$ the x-axis

36. Prove that the length of one arch of the sine curve $y = \sin x$ is equal to half the circumference of the ellipse $2x^2 + y^2 = 2$. (*Suggestion:* Substitute $x = \cos\theta$ into the arc length integral for the ellipse.)

37. Use Simpson's approximation with $n = 6$ subintervals to estimate the length of the sine arch of Problem 36.

38. Use Simpson's approximation with $n = 10$ subintervals to estimate the length of the parabola $y = x^2$ from $x = 0$ to $x = 1$.

39. Verify Eq. (6) for the area of a conical frustum. Think of the frustum as being generated by revolving around the y-axis the line segment from $(r_1, 0)$ to (r_2, h).

40. By considering a sphere of radius r to be a surface of revolution, derive the formula $A = 4\pi r^2$ for its surface area.

41. Find the total length of the *astroid* shown in Fig. 6.4.16. The equation of its graph is $x^{2/3} + y^{2/3} = 1$.

42. Find the area of the surface generated by revolving the astroid of Problem 41 around the y-axis (Fig. 6.4.17).

43. Figure 6.4.18 shows a *spherical zone* of height h—it is cut out of the sphere by two parallel planes that intersect the sphere. Show that the surface area of this zone is $A = 2\pi rh$, where r is the radius of the sphere and h (the *height* of the zone) is the distance between the two planes. Note that A depends only on the height of the zone, and not (otherwise) on the specific location of the two planes relative to the sphere.

44. Figure 6.4.19 shows a loop of the curve $32y^2 = x^2(4 - x^2)$. Find the surface area generated by revolving this loop around the x-axis.

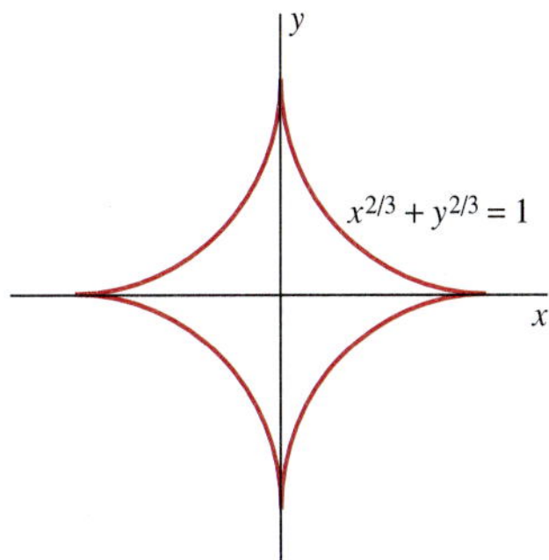

Fig. 6.4.16 The astroid of Problem 41

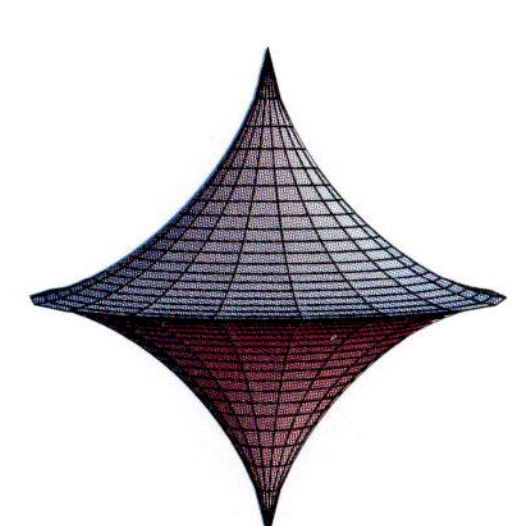

Fig. 6.4.17 The surface of Problem 42

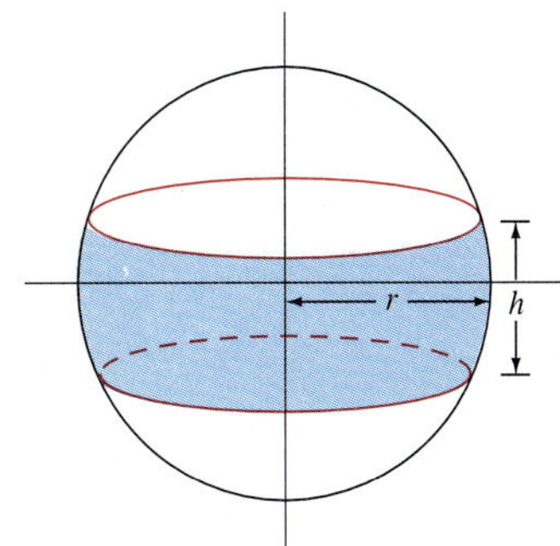

Fig. 6.4.18 The spherical zone of Problem 43

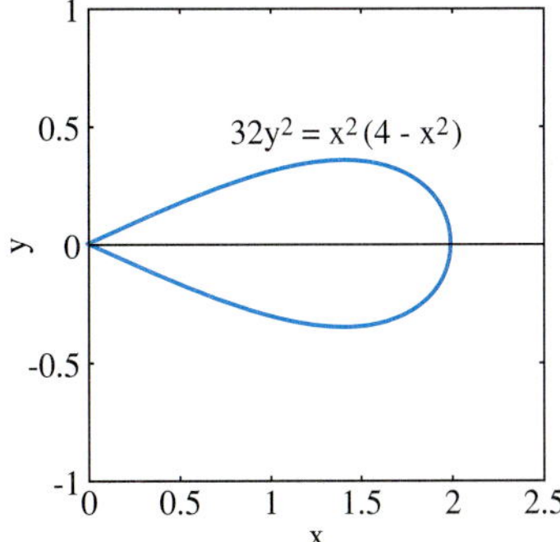

Fig. 6.4.19 The loop of Problem 44

6.4 PROJECT: APPROXIMATING ARC LENGTH AND SURFACE AREA

Arc-length integrals that appear in applications are seldom easy to evaluate exactly. This project uses numerical integration techniques (as described in the projects for Sections 5.4 and 5.9) to approximate such integrals.

In Problems 1 through 10, calculate the right- and left-endpoint approximations R_n and L_n, the trapezoidal approximation $T_n = (R_n + L_n)/2$, and/or Simpson's approximation S_n. In each case, use $n = 10, 20, 40, \ldots,$ subintervals, doubling the number of subintervals until satisfied with the accuracy you obtain.

1 through **5.** Find the arc lengths of the graphs described in Problems 1 through 5 of this section.

6 through **10.** Find the areas of the surfaces described in Problems 11 through 15 of this section.

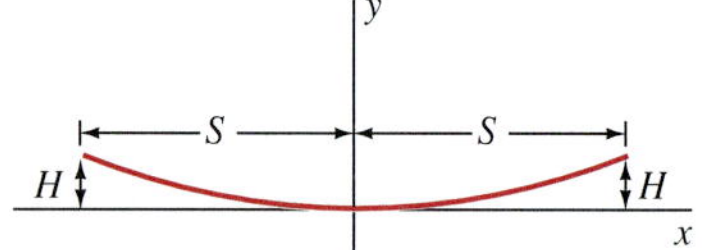

Fig. 6.4.20 The parabolic supporting cable of a suspension bridge

11. Figure 6.4.20 shows a cable for a suspension bridge, which has the shape of a parabola with equation $y = kx^2$. The suspension bridge has total span $2S$, and the height of the cable (relative to its lowest point) is H at each end. Show that the total length of this cable is

$$L = 2\int_0^S \left(1 + \frac{4H^2}{S^4}x^2\right)^{1/2} dx.$$

12. Italian engineers have proposed a single-span suspension bridge across the Strait of Messina (8 km wide) between Italy and Sicily. The plans include suspension towers 380 m high at each end. Use the integral in Problem 11 to approximate the length L of the parabolic suspension cable for this proposed bridge. Assuming that the given dimensions are exact, approximate the integral with sufficient accuracy to determine L to the nearest meter.

6.5 SEPARABLE DIFFERENTIAL EQUATIONS

We remarked in Section 5.2 that mathematical models of changing real-world phenomena frequently involve differential equations—that is, equations containing *derivatives* of unknown functions. A *first-order* **differential equation** is a differential equation that can be written in the form

$$\frac{dy}{dx} = F(x, y), \tag{1}$$

where x denotes the independent variable and $y = y(x)$ is the unknown function. A **solution** of Eq. (1) is a function $y = y(x)$ such that $y'(x) = F(x, y(x))$ for all x in some appropriate interval I.

EXAMPLE 1 If $y = x^{3/2}$, then

$$\frac{dy}{dx} = \frac{3}{2}x^{1/2} = \frac{3}{2}\cdot\frac{x^{3/2}}{x} = \frac{3y}{2x}$$

if $x > 0$. Hence the function $y(x) = x^{3/2}$ is a solution (on the interval $x > 0$) of the differential equation

$$\frac{dy}{dx} = \frac{3y}{2x}.$$

Thus to verify that $y = y(x)$ is a solution of Eq. (1), we first compute dy/dx and then show that it is equal to $F(x, y)$. ■

The differential equation in (1) is said to be **separable** provided that its right-hand side is the product of a function of x and a function of y. If so, the equation takes the special form

$$\frac{dy}{dx} = g(x)\phi(y). \tag{2}$$

In this case the variables x and y can be "separated"—isolated on opposite sides of the equation—by writing informally the differential equation

$$f(y)\,dy = g(x)\,dx. \tag{3}$$

We understand Eq. (3) to be concise notation for the equation

$$f(y)\frac{dy}{dx} = g(x), \tag{4}$$

which we get upon multiplying each side of Eq. (2) by $f(y) = 1/\phi(y)$.

Method of Solution

One of the most important applications of integration is the solution of differential equations. We now show that the solution of the differential equation in (4) reduces to the evaluation of two integrals. When we integrate both sides of Eq. (4) with respect to x, we get

$$\int f(y(x))\frac{dy}{dx}\,dx = \int g(x)\,dx + C, \tag{5}$$

because two antiderivatives of the same function can differ only by a constant. If we make the substitution

$$y = y(x), \qquad dy = \frac{dy}{dx}\,dx$$

in the left-hand side of Eq. (5), we get

$$\int f(y)\,dy = \int g(x)\,dx + C, \tag{6}$$

just as though we integrated each side of $f(y)\,dy = g(x)\,dx$ with respect to its own variable. If the indefinite integrals

$$F(y) = \int f(y)\,dy \quad \text{and} \quad G(x) = \int g(x)\,dx \tag{7}$$

can be evaluated, then Eq. (6) takes the form

$$F(y) = G(x) + C. \tag{8}$$

Finally, we can hope to solve Eq. (8) algebraically for an *explicit* solution $y = y(x)$ of the original differential equation. If not, it may be equally satisfactory to solve Eq. (8) for $x = x(y)$. If this, too, is impossible, we will generally be satisfied with the solution in the form in Eq. (8). We call that form an *implicit solution* of the original differential equation because Eq. (8) presents the solution in implicitly defined form.

Observe that $y(x)$ will involve the arbitrary constant C. When a solution of a differential equation contains an arbitrary constant, we call it a **general solution** of a differential equation. A general solution actually describes an infinite collection of different solutions of a given differential equation, because different values for the constant C yield different *particular* solutions (solutions containing no arbitrary constants).

EXAMPLE 2 Find a general solution of the differential equation

$$\frac{dy}{dx} = \sqrt{xy} \qquad (x, y > 0). \tag{9}$$

Solution When we separate the variables and integrate as in Eq. (6), we get

$$\int y^{-1/2}\,dy = \int x^{1/2}\,dx + C; \qquad 2y^{1/2} = \tfrac{2}{3}x^{3/2} + C. \tag{10}$$

We can now solve for y to obtain the general solution

$$y(x) = \left(\tfrac{1}{3}x^{3/2} + \tfrac{1}{2}C\right)^2 \tag{11}$$

in explicit form. ◼

Slope Fields and Solution Curves

To investigate the possible behavior of solutions of a differential equation of the form $dy/dx = f(x, y)$, we may think of the differential equation in a very geometric way: At various points (x, y) of the xy-plane, the value of $f(x, y)$ determines a slope dy/dx. A *solution* of this differential equation is a differentiable function whose graph has slope $dy/dx = f(x, y)$ at each point (x, y) on its graph. The *graph* of a solution of a differential equation is sometimes called a **solution curve** of the equation. In geometric terms, a solution curve of the differential equation $dy/dx = f(x, y)$ is a curve in the plane whose tangent line at each point (x, y) has slope $m = f(x, y)$.

Suppose that we draw a short line segment with slope $m = f(x, y)$ through each of a representative collection of points (x, y) in the xy-plane. (Computer programs that make this task almost effortless are available.) The set of all these line segments is called a **slope field,** or a **direction field,** for the equation $dy/dx = f(x, y)$. We can attempt to sketch (or visualize) a solution curve that threads its way through the slope field in such a way that the curve is tangent to each of the short line segments that it intersects.

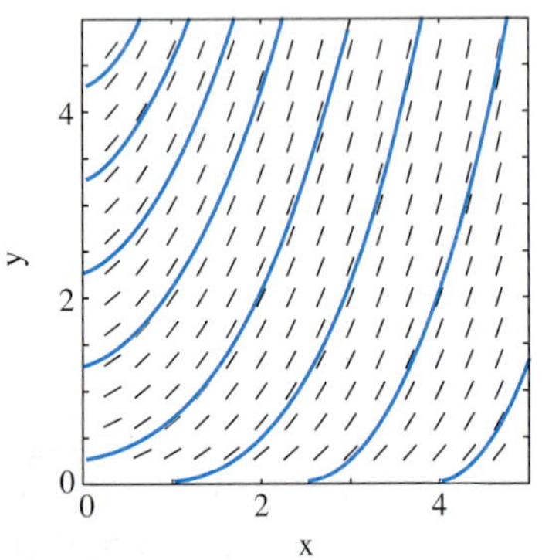

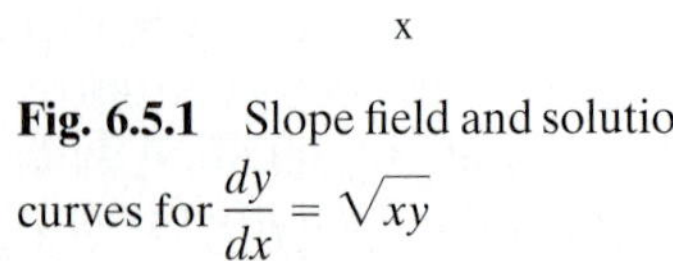
Fig. 6.5.1 Slope field and solution curves for $\frac{dy}{dx} = \sqrt{xy}$

Figure 6.5.1 shows a slope field and typical solution curves for the differential equation $dy/dx = \sqrt{xy}$ of Example 2. As the figure suggests, these solution curves fill the first quadrant of the xy-plane. Through any point (x_0, y_0) of the first quadrant there passes exactly one solution curve—the one corresponding to the value of the constant

$$C = 2(y_0)^{1/2} - \tfrac{2}{3}(x_0)^{3/2}$$

obtained by substituting $x = x_0$ and $y = y_0$ into the second equation in (10).

A specific solution of a differential equation—such as one obtained by specifying the value of the constant C in a general solution—is called a **particular solution** of the differential equation. Often we seek a particular solution that satisfies an **initial condition** of the form

$$y(x_0) = y_0. \tag{12}$$

In this case we want to solve the **initial value problem**

$$\frac{dy}{dx} = F(x, y), \quad y(x_0) = y_0, \tag{13}$$

consisting of a differential equation *and* an initial condition.

It is proved in differential equations texts that if $F(x, y)$ is a sufficiently well-behaved function of x and y, then the initial value problem in (13) has exactly one solution for x near x_0. If this differential equation is also separable, then we can attempt to find this solution as follows:

- First separate the variables and integrate to find a general solution.
- Then impose the initial condition $y(x_0) = y_0$ to evaluate the constant C and thereby determine the desired particular solution.

EXAMPLE 3 Solve the initial value problems

(a) $\dfrac{dy}{dx} = \sqrt{xy}, \quad y(0) = 1;$

(b) $\dfrac{dy}{dx} = \sqrt{xy}, \quad y(2) = 2.$

Solution In Example 2 we found the implicit solution

$$2y^{1/2} = \tfrac{2}{3}x^{3/2} + C \tag{10}$$

that yields the general solution

$$y(x) = \left(\tfrac{1}{3}x^{3/2} + \tfrac{1}{2}C\right)^2$$

of the differential equation $dy/dx = \sqrt{xy}$.

To solve the initial value problems in part (a), we substitute $x = 0$ and $y = 1$ into Eq. (10). This yields $C = 2$. Hence the desired particular solution is given by

$$y(x) = \left(\tfrac{1}{3}x^{3/2} + 1\right)^2 = \tfrac{1}{9}(x^{3/2} + 3)^2.$$

To solve the initial value problem in part (b), we substitute $x = 2$ and $y = 2$ into Eq. (10):

$$2 \cdot 2^{1/2} = \tfrac{2}{3} \cdot 2^{3/2} + C = \tfrac{4}{3} \cdot 2^{1/2} + C.$$

It follows that $C = \frac{2}{3}\sqrt{2}$, so the desired particular solution is

$$y(x) = \left(\tfrac{1}{3}x^{3/2} + \tfrac{1}{3}\sqrt{2}\right)^2 = \tfrac{1}{9}\left(x^{3/2} + \sqrt{2}\right)^2.$$

The solution curves of parts (a) and (b) are shown in Fig. 6.5.2. ∎

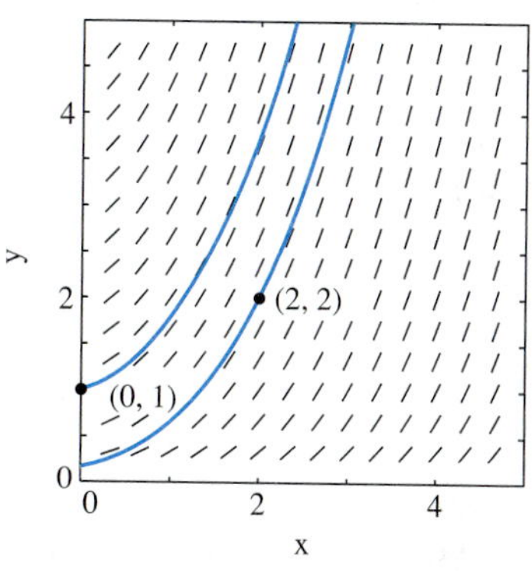

Fig. 6.5.2 The solution curves of Example 3

Population Growth

A differential equation of the form

$$\frac{dP}{dt} = kP^{\alpha} \tag{14}$$

is frequently used to model the growth of a population of $P(t)$ individuals at time t—if we can regard the continuous function $P(t)$ as a sufficiently accurate approximation to the actual discrete population. The proportionality constant k ordinarily must be determined by experiment, whereas the value of the exponent α depends on the assumptions made about the way in which the population changes. For example, the case of *constant* birth and death rates corresponds to the value $\alpha = 1$ and yields the *natural* population growth equation $dP/dt = kP$. In Example 4 we take $\alpha = 0.5$ (mainly to make the calculations simpler).

EXAMPLE 4 Suppose that a lake is stocked initially with $P(0) = 100$ fish and that the fish population $P(t)$ thereafter satisfies the differential equation

$$\frac{dP}{dt} = k\sqrt{P} \qquad (k \text{ constant}). \tag{15}$$

If after $t = 6$ mo there are 169 fish in the lake, how many will there be after 1 yr (that is, when $t = 12$)?

Solution We will use the datum $P(6) = 169$ to find the value of k. But first we must solve the differential equation. We separate the variables in Eq. (15) and integrate:

$$\int P^{-1/2}\,dP = \int k\,dt + C;$$
$$2\sqrt{P} = kt + C. \tag{16}$$

Now $P = 100$ when $t = 0$. Simultaneous substitution of these values in Eq. (16) gives $C = 2\sqrt{100} = 20$, so

$$2\sqrt{P} = kt + 20.$$

Next, substitution of $t = 6$ and $P = 169$ yields the value $k = 1$, so $2\sqrt{P} = t + 20$. Hence the fish population after t months is given by

$$P(t) = \tfrac{1}{4}(t + 20)^2.$$

Finally, the number of fish in the lake after 1 yr is

$$P(12) = \tfrac{1}{4} \cdot 32^2 = 256.$$

Torricelli's Law

Suppose that a water tank has a hole with area a at its bottom and that water is draining from the hole. Denote by $y(t)$ the depth (in feet) of water in the tank at time t (in seconds) and by $V(t)$ the volume of water (in cubic feet) in the tank then. It is plausible—and true under ideal conditions—that the velocity of the stream of water exiting through the hole is

$$v = \sqrt{2gy} \qquad (g \approx 32 \text{ ft/s}^2), \tag{17}$$

which is the velocity that a drop of water would acquire in falling freely from the water surface to the hole. This is *Torricelli's law of draining*.

As indicated in Fig. 6.5.3, the amount of water that leaves through the bottom hole during a short time interval dt amounts to a cylinder with base area a and height $v\,dt$. Hence the resulting change dV in the volume of water in the tank is given by

$$dV = -av\,dt = -a\sqrt{2gy}\,dt. \tag{18}$$

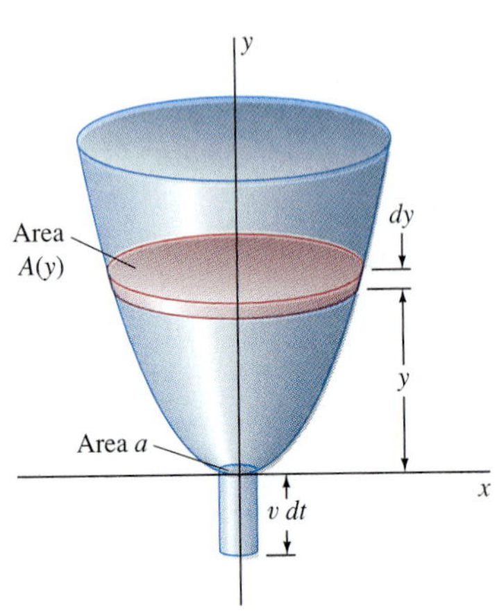

Fig. 6.5.3 Derivation of Torricelli's law

But if $A(y)$ denotes the horizontal cross-sectional area of the tank at height y above the hole, then (with dy negative)

$$dV = A(y)\,dy, \tag{19}$$

as usual. Comparing Eqs. (18) and (19), we see that $y(t)$ satisfies the differential equation

$$A(y)\frac{dy}{dt} = -a\sqrt{2gy}. \tag{20}$$

In some applications this is a very convenient form of Torricelli's law (see Section 5.2). In other situations you may prefer to work with the differential equation in (18) in the form

$$\frac{dV}{dt} = -a\sqrt{2gy} \tag{21}$$

or, if the area of the bottom hole is unknown, in the form

$$\frac{dV}{dt} = -c\sqrt{y}, \tag{22}$$

where $c = a\sqrt{2g}$ is a positive constant.

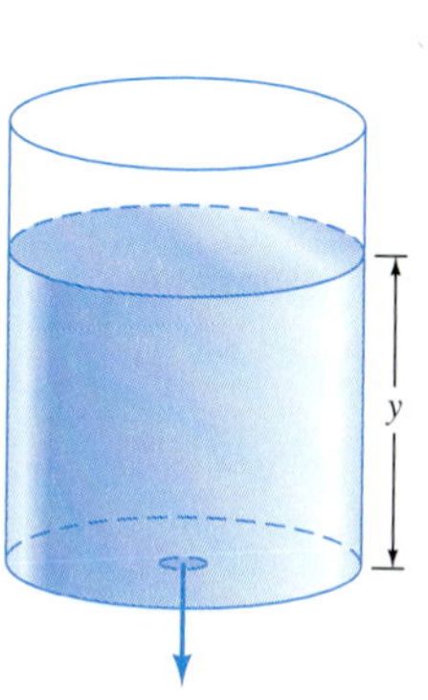

Fig. 6.5.4 A cylindrical water tank has constant cross-sectional area.

Cylindrical water tanks are common (Fig. 6.5.4). In this case the cross-sectional area function in Eq. (20) is constant: $A(y) \equiv A$. Consequently, Eq. (20) reduces to the simple differential equation

$$\frac{dy}{dt} = -k\sqrt{y}, \tag{23}$$

where $k = (a/A)\sqrt{2g}$ is a positive constant that frequently is determined from given tank-draining data (rather than from a knowledge of the areas a and A).

EXAMPLE 5 The water in a draining cylindrical tank is 10 ft deep at noon. At 1:00 P.M. it is 5 ft deep. When will the tank be empty?

Solution Separation of the variables in Eq. (23) gives

$$\frac{dy}{\sqrt{y}} = -k\,dt,$$

and then integration yields

$$2\sqrt{y} = -kt + C.$$

Substitution of the initial data $y = 10$ when $t = 0$ (noon) gives $C = 2\sqrt{10}$, so

$$2\sqrt{y} = -kt + 2\sqrt{10}. \tag{24}$$

Then substitution of the additional data $y = 5$ when $t = 1$ (1:00 P.M.) gives $k = 2\sqrt{10} - 2\sqrt{5}$. Substitution of this value in Eq. (24) and division by 2 yields

$$\sqrt{y} = \left(\sqrt{5} - \sqrt{10}\right)t + \sqrt{10}. \tag{25}$$

Finally, the tank is empty when $y = 0$ in Eq. (25) and thus when

$$t = \frac{\sqrt{10}}{\sqrt{10} - \sqrt{5}} \approx 3.414,$$

about 3 h 25 min. So we see that—whereas a natural (but naive and ***wrong***) guess might have been 2:00 P.M. (one more hour for the remaining 5 ft of water to drain)—the tank actually is not empty until about 3:25 P.M. You should use Eq. (25) to show that the actual water depth in the tank at 2:00 P.M. is about 1.72 ft, and the depth at 3:00 P.M. is about 2 in., so it takes about 25 min for the last 2 in. of water in the tank to drain! ■

EXAMPLE 6 A hemispherical tank has top radius 4 ft and, at time $t = 0$, is full of water. At that moment a circular hole of diameter 1 in. is opened in the bottom of the tank. How long will it take for all the water to drain from the tank?

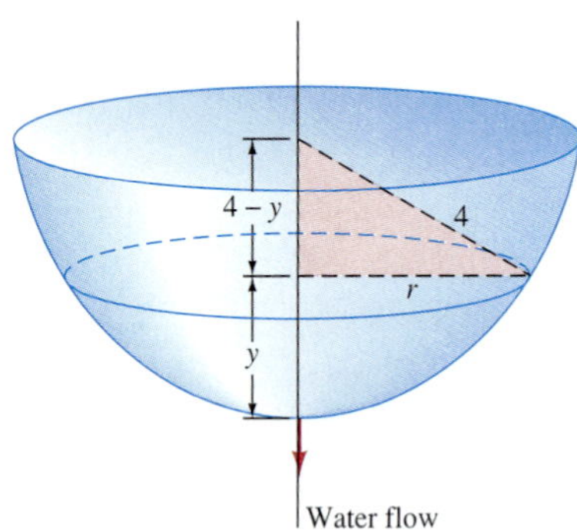

Fig. 6.5.5 Draining a hemispherical tank

Solution From the right triangle in Fig. 6.5.5, we see that

$$A(y) = \pi r^2 = \pi[16 - (4 - y)^2] = \pi(8y - y^2).$$

With $g = 32$ ft/s^2, Eq. (20) takes the form

$$\pi(8y - y^2)\frac{dy}{dt} = -\pi\left(\frac{1}{24}\right)^2\sqrt{64y};$$

$$\int(8y^{1/2} - y^{3/2})\,dy = -\int\frac{1}{72}\,dt + C;$$

$$\frac{16}{3}y^{3/2} - \frac{2}{5}y^{5/2} = -\frac{1}{72}t + C.$$

Now $y(0) = 4$, so

$$C = \tfrac{16}{3}\cdot 4^{3/2} - \tfrac{2}{5}\cdot 4^{5/2} = \tfrac{448}{15}.$$

The tank is empty when $y = 0$—that is, when

$$t = 72\cdot\tfrac{448}{15} \approx 2150\ \ (\mathrm{s}),$$

about 35 min 50 s. So it takes slightly less than 36 min for the tank to drain. ■

6.5 PROBLEMS

Find general solutions (implicit if necessary, explicit if possible) of the differential equations in Problems 1 through 10.

1. $\dfrac{dy}{dx} = 2x\sqrt{y}$

2. $\dfrac{dy}{dx} = 2xy^2$

3. $\dfrac{dy}{dx} = x^2y^2$

4. $\dfrac{dy}{dx} = (xy)^{3/2}$

5. $\dfrac{dy}{dx} = 2x\sqrt{y - 1}$

6. $\dfrac{dy}{dx} = 4x^3(y - 4)^2$

7. $\dfrac{dy}{dx} = \dfrac{1 + \sqrt{x}}{1 + \sqrt{y}}$

8. $\dfrac{dy}{dx} = \dfrac{x + x^3}{y + y^3}$

9. $\dfrac{dy}{dx} = \dfrac{x^2 + 1}{x^2(3y^2 + 1)}$

10. $\dfrac{dy}{dx} = \dfrac{(x^3 - 1)y^3}{x^2(2y^3 - 3)}$

Solve the initial value problems in Problems 11 through 20.

11. $\dfrac{dy}{dx} = y^2,\ \ y(0) = 1$

12. $\dfrac{dy}{dx} = \sqrt{y},\ \ y(0) = 4$

13. $\dfrac{dy}{dx} = \dfrac{1}{4y^3},\ \ y(0) = 1$

14. $\dfrac{dy}{dx} = \dfrac{1}{x^2y},\ \ y(1) = 2$

15. $\dfrac{dy}{dx} = \sqrt{xy^3},\ \ y(0) = 4$

16. $\dfrac{dy}{dx} = \dfrac{x}{y},\ \ y(3) = 5$

17. $\dfrac{dy}{dx} = -\dfrac{x}{y},\ \ y(12) = -5$

18. $y^2\dfrac{dy}{dx} = x^2 + 2x + 1,\ \ y(1) = 2$

19. $\dfrac{dy}{dx} = 3x^2y^2 - y^2,\ \ y(0) = 1$

20. $\dfrac{dy}{dx} = 2xy^3(2x^2 + 1),\ \ y(1) = 1$

21. Suppose that the fish population $P(t)$ in a lake is attacked by disease at time $t = 0$, with the result that

$$\frac{dP}{dt} = -k\sqrt{P} \qquad (k > 0)$$

thereafter. If there were initially 900 fish in the lake and 441 were left after 6 weeks, how long would it take all the fish in the lake to die?

22. Prove that the solution of the initial value problem

$$\frac{dP}{dt} = k\sqrt{P}, \qquad P(0) = P_0 \qquad (P_0 > 0)$$

is given by

$$P(t) = \left(\tfrac{1}{2}kt + \sqrt{P_0}\right)^2.$$

23. Suppose that the population of Fremont satisfies the differential equation of Problem 22. (a) If $P = 100{,}000$ in 1970 and $P = 121{,}000$ in 1980, what will the population be in 2000? (b) When will the population reach 200,000?

24. Consider a breed of rabbits whose population $P(t)$ satisfies the initial value problem

$$\frac{dP}{dt} = kP^2, \qquad P(0) = P_0,$$

where k is a positive constant. Derive the solution

$$P(t) = \frac{P_0}{1 - kP_0 t}.$$

25. In Problem 24, suppose that $P_0 = 2$ and that there are 4 rabbits after 3 months. What happens in the next 3 months?

26. Suppose that a motorboat is traveling at $v = 40$ ft/s when its motor is cut off at time $t = 0$. Thereafter its deceleration due to water resistance is given by $dv/dt = -kv^2$, where k is a positive constant. (a) Solve this differential equation to show that the speed of the boat at time $t > 0$ is $v = 40/(1 + 40kt)$ feet per second. (b) If the boat's speed after 10 s is 20 ft/s, how long does it take to slow to 5 ft/s?

27. A tank shaped like a vertical cylinder initially contains water to a depth of 9 ft (Fig. 6.5.6). A bottom plug is pulled at time $t = 0$ (t in hours). After 1 h the depth has dropped to 4 ft. How long will it take all the water to drain from this tank?

28. Suppose that the tank of Problem 27 has a radius of 3 ft and that its bottom hole is circular with radius 1 in. How long will it take for the water, initially 9 ft deep, to drain completely?

29. A water tank is in the shape of a right circular cone with its axis vertical and its vertex at the bottom. The tank is 16 ft high and the radius of its top is 5 ft. At time $t = 0$, a plug at its vertex is removed and the tank, initially full of water, begins to drain. After 1 h the water in the tank is 9 ft deep. When will the tank be empty (Fig. 6.5.7)?

30. Suppose that a cylindrical tank (axis vertical) initially containing V_0 liters of water drains through a bottom hole in T minutes. Use Torricelli's law to show that the volume of water in the tank after $t \leqq T$ minutes is $V(t) = V_0[1 - (t/T)]^2$.

31. The shape of a water tank is obtained by revolving the curve $y = x^{4/3}$ around the y-axis (units on the coordinate axes are in feet). A plug at the bottom is removed at 12 noon, when the water depth in the tank is 12 ft. At 1 P.M. the water depth is 6 ft. When will the tank be empty?

32. The shape of a water tank is obtained by revolving the parabola $y = x^2$ around the y-axis (units on the coordinate axes are in feet; see Fig. 6.5.8). The water depth is 4 ft at 12 noon; at that time, the plug in the circular hole at the bottom of the tank is removed. At 1 P.M. the water depth is 1 ft. (a) Find the water depth $y(t)$ after t hours. (b) When will the tank be empty? (c) What is the radius of the circular hole at the bottom?

33. A cylindrical tank of length 5 ft and radius 3 ft is situated with its axis horizontal. If a circular bottom hole of radius 1 in. is opened and the tank is initially half full of xylene, how long will it take the liquid to drain completely?

34. A spherical tank of radius 4 ft is full of mercury when a circular bottom hole of radius 1 in. is opened. How long will it be before all of the mercury drains from the tank?

35. *The Clepsydra, or Water Clock* A 12-h water clock is to be designed with the dimensions shown in Fig. 6.5.9, shaped like the surface obtained by revolving the curve $y = f(x)$ around the y-axis. What equation should this curve have, *and* what radius should the bottom hole have, so that the water level will fall at the *constant* rate of 4 in./h?

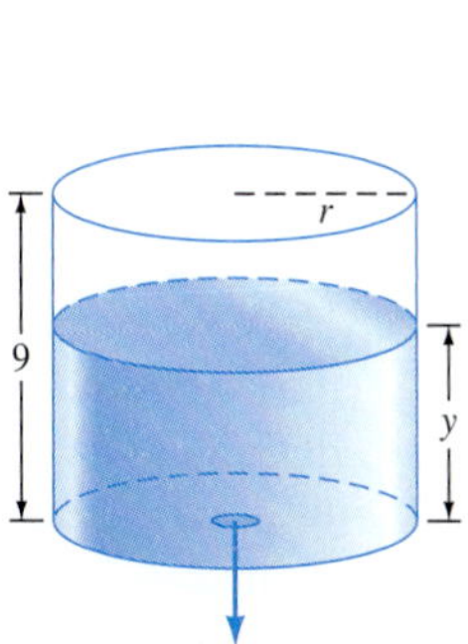

Fig. 6.5.6 The cylindrical tank of Problem 27

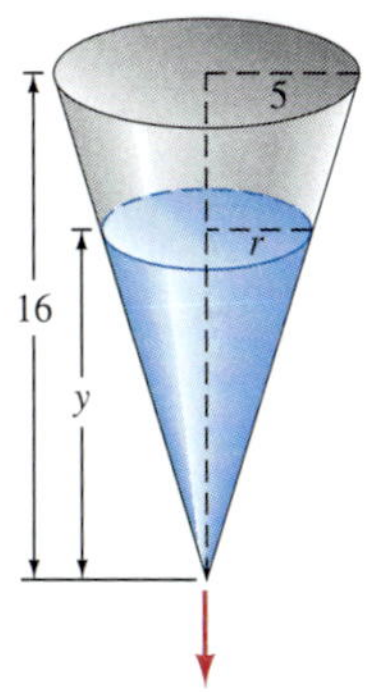

Fig. 6.5.7 The conical tank of Problem 29

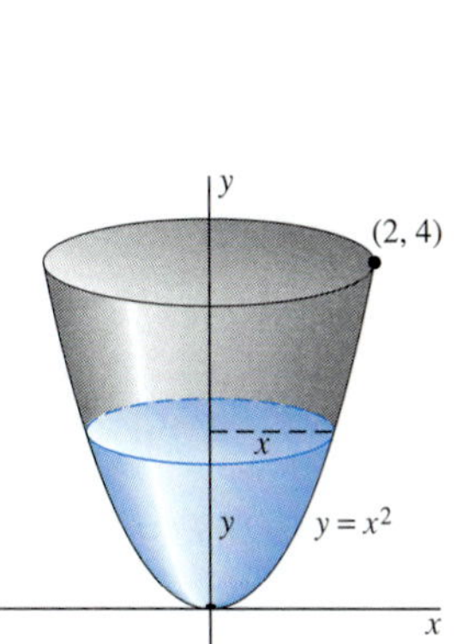

Fig. 6.5.8 The tank of Problem 32

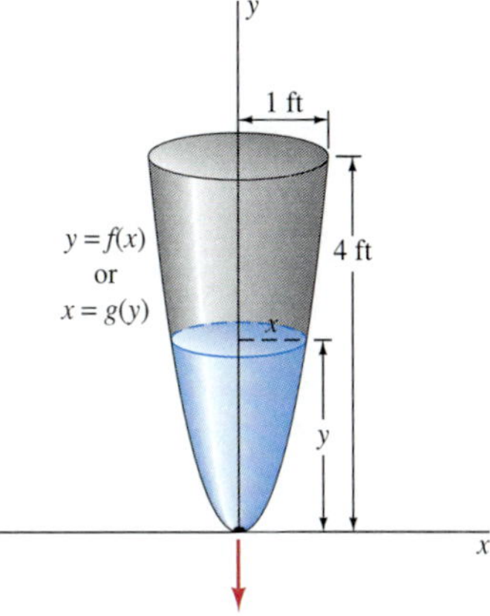

Fig. 6.5.9 The clepsydra of Problem 35

6.6 FORCE AND WORK

The concept of *work* is introduced to measure the cumulative effect of a force in moving a body from one position to another. In the simplest case, a particle is moved along a straight line by the action of a *constant* force. The work done by such a force is defined to be the product of the force and the distance through which it acts. Thus if the constant force has magnitude F and the particle is moved through the distance d, then the work done by the force is given by

$$W = F \cdot d. \tag{1}$$

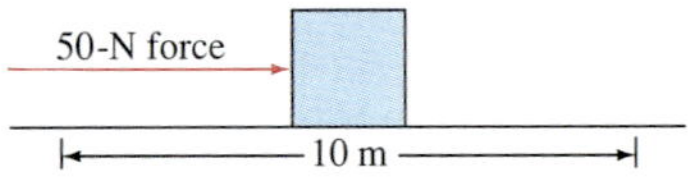

50-N force

10 m

Fig. 6.6.1 A 50-N force does 500 N·m of work in pushing a box 10 m.

EXAMPLE 1 If a constant horizontal force of 50 newtons (N) is applied to a heavy box to push it a distance of 10 m along a rough floor (Fig. 6.6.1), then the work done by the force is

$$W = 50 \cdot 10 = 500$$

newton-meters (N·m). Note the units; because of the definition of work, units of work are always products of force units and distance units. For another example, to lift a weight of 75 lb a vertical distance of 5 ft, a constant force of 75 lb must be applied. The work done by this force is

$$W = 75 \cdot 5 = 375$$

foot-pounds (ft·lb). ■

Work Done by a Variable Force

Here we use the integral to generalize the definition of work to the case in which a particle is moved along a straight line by a *variable* force. Given a **force function** $F(x)$ defined at each point x of the straight line segment $[a, b]$, we want to define the work W done by this variable force in pushing the particle from the point $x = a$ to the point $x = b$ (Fig. 6.6.2).

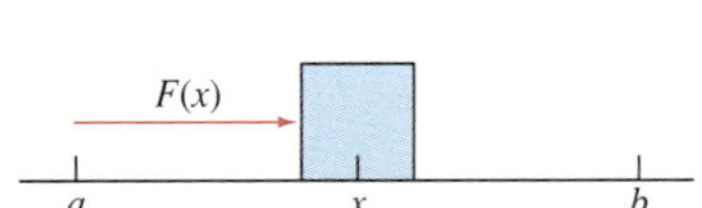

Fig. 6.6.2 A variable force pushing a particle from a to b

We begin with the usual partition of the interval $[a, b]$ into n subintervals, all with the same length $\Delta x = (b - a)/n$. For each i ($1 \leqq i \leqq n$), let $x_i^\star$ be an arbitrary point of the ith subinterval $[x_{i-1}, x_i]$. The key idea is to approximate the actual work ΔW_i done by the *variable force* $F(x)$ in moving the particle from x_{i-1} to x_i by the work $F(x_i^\star)\,\Delta x$ (force × distance) done in moving a particle the distance Δx from x_{i-1} to x_i (Fig. 6.6.3). Thus

$$\Delta W_i \approx F(x_i^\star)\,\Delta x. \tag{2}$$

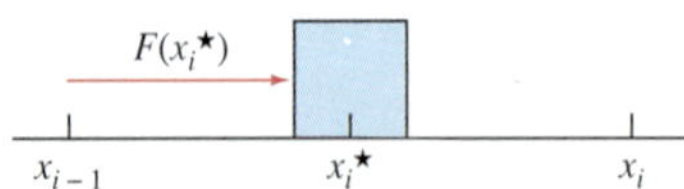

Fig. 6.6.3 The *constant* force $F(x_i^\star)$ acting through the ith subinterval

We approximate the total work W by summing from $i = 1$ to $i = n$:

$$W = \sum_{i=1}^{n} \Delta W_i \approx \sum_{i=1}^{n} F(x_i^\star)\,\Delta x. \tag{3}$$

But the final sum in Eq. (3) is a Riemann sum for $F(x)$ on the interval $[a, b]$, and as $n \to +\infty$ (and $\Delta x \to 0$), such sums approach the *integral* of $F(x)$ from $x = a$ to $x = b$. We therefore are motivated to *define* the **work** W done by the force $F(x)$ in moving the particle from $x = a$ to $x = b$ to be

$$W = \int_a^b F(x)\,dx. \tag{4}$$

The following heuristic way of setting up Eq. (4) is useful in obtaining integrals for work problems. Imagine that dx is so small a number that the value of $F(x)$ does not change appreciably on the tiny interval from x to $x + dx$. Then the work done by the force F in moving a particle from x to $x + dx$ should be very close to

$$dW = F(x)\,dx.$$

The natural additive property of work then implies that we could obtain the total work W by adding these tiny elements of work:

$$W = \int_{\star}^{\star\star} dW = \int_a^b F(x)\,dx.$$

Elastic Springs

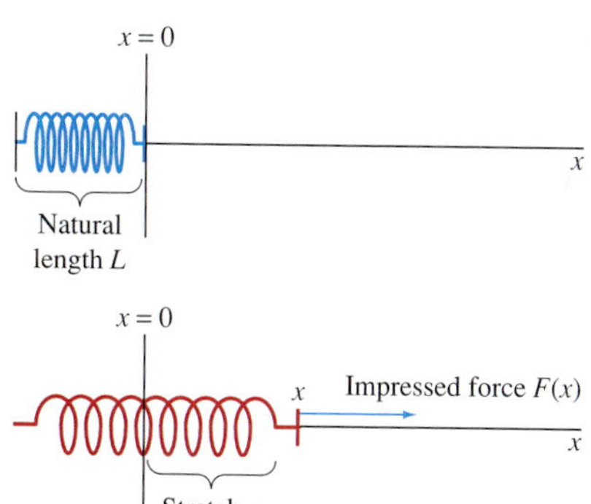

Fig. 6.6.4 The stretch x is proportional to the impressed force F.

Consider a spring whose left end is held fixed and whose right end is free to move along the x-axis. We assume that the right end is at the origin $x = 0$ when the spring has its **natural length**—that is, when the spring is in its rest position, neither compressed nor stretched by outside forces.

According to **Hooke's law** for elastic springs, the force $F(x)$ that must be exerted on the spring to hold its right end at the point x is proportional to the displacement x of the right end from its rest position. That is,

$$F(x) = kx, \tag{5}$$

where k is a positive constant. The constant k, called the **spring constant,** is a characteristic of the particular spring under study.

Figure 6.6.4 shows the arrangement of such a spring along the x-axis. The right end of the spring is held at position x on the x-axis by a force $F(x)$. The figure shows the situation for $x > 0$, so the spring is stretched. The force that the spring exerts on its right-hand end is directed to the left, so—as the figure indicates—the external force $F(x)$ must act to the right. The right is the positive direction here, so $F(x)$ must be a positive number. Because x and $F(x)$ have the same sign, k must also be positive. You can check that k is positive as well for $x < 0$.

EXAMPLE 2 Suppose that a spring has a natural length of 1 ft and that a force of 10 lb is required to hold it compressed to a length of 6 in. How much work is done in stretching the spring from its natural length to a total length of 2 ft?

Solution To move the free end from $x = 0$ (the natural-length position) to $x = 1$ (stretched by 1 ft), we must exert a variable force $F(x)$ determined by Hooke's law. We are given that $F = -10$ (lb) when $x = -0.5$ (ft), so Eq. (5), $F = kx$, implies that the spring constant for this spring is $k = 20$ (lb/ft). Thus $F(x) = 20x$, and so—using Eq. (4)—we find that the work done in stretching this spring in the manner given is

$$W = \int_0^1 20x\,dx = \Big[10x^2\Big]_0^1 = 10 \quad (\text{ft} \cdot \text{lb}).$$ ■

Work in the sense of physics is different than work in the sense of physiology. At this moment the weightlifter is doing no work in the physics sense because he is holding the weight still.

*Work Done Against Gravity

According to Newton's law of gravitation, the force that must be exerted on a body to hold it at a distance r from the center of the earth is inversely proportional to r^2 (if $r \geqq R$, the radius of the earth). In other words, if $F(r)$ denotes the holding force, then

$$F(r) = \frac{k}{r^2} \tag{6}$$

for some positive constant k. The value of this force at the surface of the earth, where $r = R \approx 4000$ mi (about 6370 km), is called the **weight** of the body.

Given the weight $F(R)$ of a particular body, we can find the corresponding value of k by using Eq. (6):

$$k = R^2 \cdot F(R).$$

The work that must be done to lift the body vertically from the surface to a distance $R_1 > R$ from the center of the earth is then

$$W = \int_R^{R_1} \frac{k}{r^2}\, dr. \tag{7}$$

If distance is measured in miles and force in pounds, then this integral gives the work in mile-pounds. This is a very unconventional unit of work. We shall multiply by 5280 (ft/mi) to convert any such result into foot-pounds.

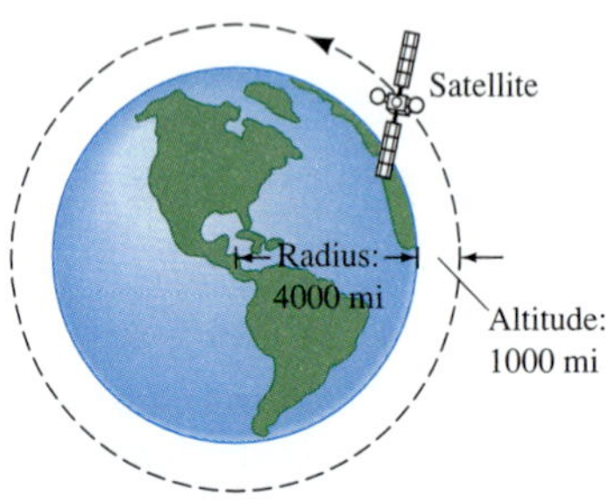

Fig. 6.6.5 A satellite in orbit 1000 mi above the surface of the earth (Example 3)

EXAMPLE 3 (*SATELLITE LAUNCH*) How much work must be done to lift a 1000-lb satellite vertically from the earth's surface to an orbit 1000 mi above the surface? See Fig. 6.6.5, and take $R = 4000$ (mi) to be the radius of the earth.

Solution Because $F = 1000$ (lb) when $r = R = 4000$ (mi), we find from Eq. (6) that

$$k = 4000^2 \cdot 1000 = 16 \times 10^9 \quad (\text{mi}^2 \cdot \text{lb}).$$

Then by Eq. (7), the work done is

$$W = \int_{4000}^{5000} \frac{k}{r^2}\, dr = \left[-\frac{k}{r} \right]_{4000}^{5000}$$

$$= (16 \times 10^9) \cdot \left(\frac{1}{4000} - \frac{1}{5000} \right) = 8 \times 10^5 \quad (\text{mi} \cdot \text{lb}).$$

We multiply by 5280 (ft/mi) and write the answer as

$$4.224 \times 10^9 = 4{,}224{,}000{,}000 \quad (\text{ft} \cdot \text{lb}).$$

We can instead express the answer to Example 3 in terms of the power that the launch rocket must provide. **Power** is the rate at which work is done. For instance, 1 **horsepower** (hp) is defined to be 33000 ft·lb/min. If the ascent to orbit takes 15 min and if only 2% of the power generated by the rocket is effective in lifting the satellite (the rest is used to lift the rocket and its fuel), we can convert the answer in Example 3 to horsepower. The *average* power that the rocket engine must produce during the 15-min ascent is

$$P = \frac{50 \cdot (4.224 \times 10^9)}{15 \cdot 33000} \approx 426{,}667 \quad (\text{hp}).$$

The factor of 50 in the numerator comes from the 2% "efficiency" of the rocket: The total power must be multiplied by $1/(0.02) = 50$.

Work Done in Filling a Tank

Examples 2 and 3 are applications of Eq. (4) for calculating the work done by a variable force in moving a particle a certain distance. Another common type of force-work problem involves the summation of work done by constant forces that act through different distances. For example, consider the problem of pumping a fluid from ground level—where we take $y = 0$—up into an above-ground tank (Fig. 6.6.6).

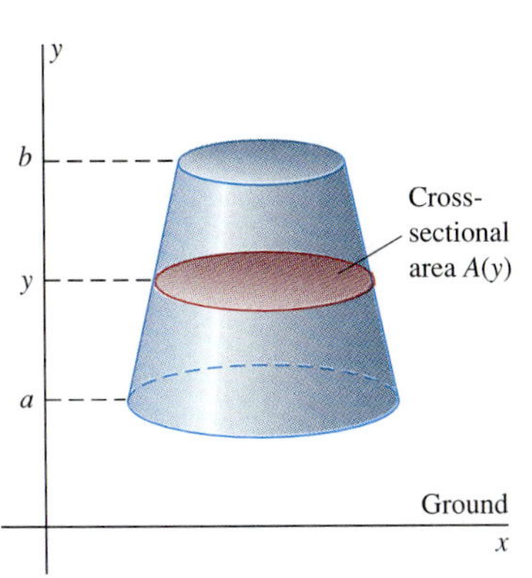

Fig. 6.6.6 An above-ground tank

It is convenient to think of the tank as being filled in thin, horizontal layers of fluid, each lifted from the ground to its final position in the tank. No matter how the fluid actually behaves as the tank is filled, this simple way of thinking about the process gives us a way to compute the work done in the filling process. But when we think of filling the tank in this way, we must allow for the fact that different layers of fluid are lifted different distances to reach their final positions in the tank.

Suppose that the bottom of the tank is at height $y = a$ and that its top is at height $y = b > a$. Let $A(y)$ be the cross-sectional area of the tank at height y. Consider a partition of $[a, b]$ into n subintervals, all with the same length Δy. Then the volume of the horizontal slice (Fig. 6.6.7) of the tank that corresponds to the ith subinterval $[y_{i-1}, y_i]$ is

$$\Delta V_i = \int_{y_{i-1}}^{y_i} A(y)\,dy = A(y_i^\star)\,\Delta y$$

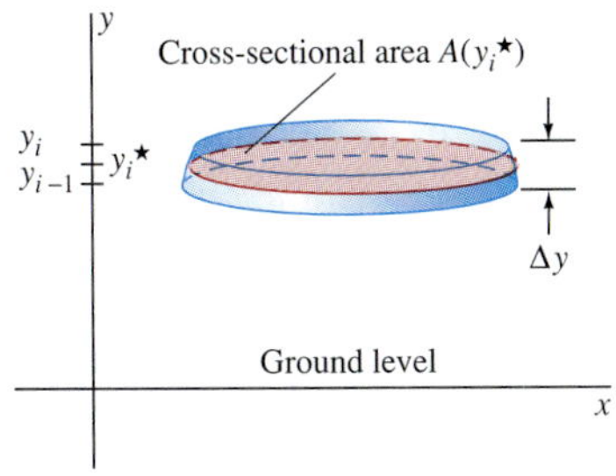

Fig. 6.6.7 A thin horizontal slice of fluid with volume $\Delta V = A(y_i^\star)\,\Delta y$. Each particle of this slice must be lifted (from the ground at $y = 0$) a distance between y_{i-1} and y_i.

for some number $y_i^\star$ in $[y_{i-1}, y_i]$; this is a consequence of the average value theorem for integrals (Section 5.6). If ρ is the density of the fluid (in pounds per cubic foot, for example), then the force required to lift this slice from the ground to its final position in the tank is simply the (constant) weight of the slice:

$$F_i = \rho\Delta V_i = \rho A(y_i^\star)\,\Delta y.$$

But what about the distance through which this force must act? The fluid in question is lifted from ground level to the level of the subinterval $[y_{i-1}, y_i]$, so every particle of the fluid is lifted at least the distance y_{i-1} and at most the distance y_i (remember, the fluid begins its journey at ground level, where $y = 0$). Hence the work ΔW_i needed to lift this ith slice of fluid satisfies the inequalities

$$F_i y_{i-1} \leqq \Delta W_i \leqq F_i y_i;$$

that is,

$$\rho y_{i-1} A(y_i^\star)\,\Delta y \leqq \Delta W_i \leqq \rho y_i A(y_i^\star)\,\Delta y.$$

Now we add these inequalities for $i = 1, 2, 3, \dots, n$ and find thereby that the total work $W = \sum W_i$ satisfies the inequalities

$$\sum_{i=1}^{n} \rho y_{i-1} A(y_i^\star)\,\Delta y \leqq W \leqq \sum_{i-1}^{n} \rho y_i A(y_i^\star)\,\Delta y.$$

If the three points y_{i-1}, y_i, and $y_i^\star$ of $[y_{i-1}, y_i]$ were the same, then both the last two sums would be Riemann sums for the function $f(y) = \rho y A(y)$ on $[a, b]$. Although the three points are not the same, it still follows—from a result stated in Appendix H—that both sums approach

$$\int_a^b \rho y A(y)\,dy \quad \text{as} \quad \Delta y \to 0.$$

The squeeze law of limits therefore gives the formula

$$W = \int_a^b \rho y A(y)\, dy. \tag{8}$$

This is the work W done in pumping fluid of density ρ from the ground into a tank that has horizontal cross-sectional area $A(y)$ and is located between heights $y = a$ and $y = b$ above the ground.

A quick heuristic way to set up Eq. (8), and many variants of it, is to think of a thin, horizontal slice of fluid with volume $dV = A(y)\, dy$ and weight $\rho\, dV = \rho A(y)\, dy$. The work required to lift this slice a distance y is

$$dW = y \cdot \rho\, dV = \rho y A(y)\, dy,$$

so the total work required to fill the tank is

$$W = \int_{\star}^{\star\star} dW = \int_a^b \rho y A(y)\, dy,$$

because the horizontal slices lie between $y = a$ and $y = b$.

The great pyramid of Khufu

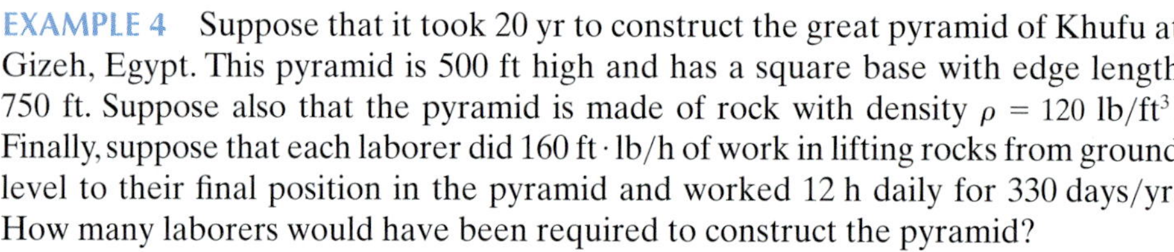

EXAMPLE 4 Suppose that it took 20 yr to construct the great pyramid of Khufu at Gizeh, Egypt. This pyramid is 500 ft high and has a square base with edge length 750 ft. Suppose also that the pyramid is made of rock with density $\rho = 120\ \text{lb/ft}^3$. Finally, suppose that each laborer did 160 ft·lb/h of work in lifting rocks from ground level to their final position in the pyramid and worked 12 h daily for 330 days/yr. How many laborers would have been required to construct the pyramid?

Solution We assume a constant labor force throughout the 20-yr construction period. We think of the pyramid as being made up of thin, horizontal slabs of rock, each slab lifted (just like a slice of liquid) from ground level to its ultimate height. Hence we can use Eq. (8) to compute the work W required.

Figure 6.6.8 shows a vertical cross section of the pyramid. The horizontal cross section at height y is a square with edge length s. We see from the similar triangles in Fig. 6.6.8 that

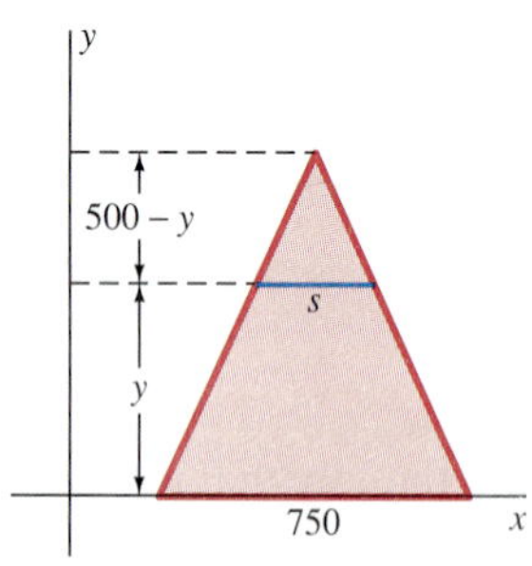

Fig. 6.6.8 Vertical cross section of Khufu's pyramid

$$\frac{s}{750} = \frac{500 - y}{500}, \quad \text{so} \quad s = \frac{3}{2}(500 - y).$$

Hence the cross-sectional area at height y is

$$A(y) = \tfrac{9}{4}(500 - y)^2.$$

Equation (8) therefore gives

$$\begin{aligned} W &= \int_0^{500} 120 \cdot y \cdot \tfrac{9}{4}\,(500 - y)^2\, dy \\ &= 270 \int_0^{500} (250{,}000y - 1000y^2 + y^3)\, dy \\ &= 270 \left[125{,}000y^2 - \tfrac{1000}{3} y^3 + \tfrac{1}{4} y^4 \right]_0^{500}, \end{aligned}$$

so $W \approx 1.406 \times 10^{12}$ ft·lb.

Because each laborer does

$$160 \cdot 12 \cdot 330 \cdot 20 \approx 1.267 \times 10^7 \quad \text{ft} \cdot \text{lb}$$

of work, the construction of the pyramid would—under our assumptions—have required

$$\frac{1.406 \times 10^{12}}{1.267 \times 10^7} \approx 111{,}000$$

laborers. ■

Emptying a Tank

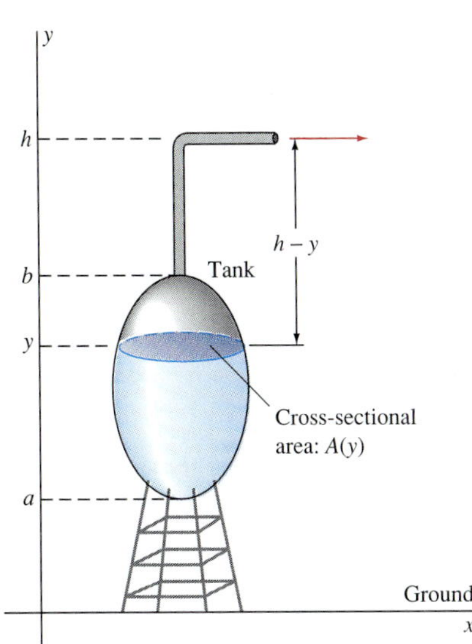

Fig. 6.6.9 Pumping liquid from a tank to a higher level

Suppose now that the tank shown in Fig. 6.6.9 is already filled with a liquid of density ρ lb/ft^3 and that we want to pump all this liquid from the tank up to the level $y = h$ above the top of the tank. We imagine a thin, horizontal slice of liquid at height y. If its thickness is dy, then its volume is $dV = A(y)\,dy$, so its weight is $\rho\,dV = \rho A(y)\,dy$. This slice must be lifted the distance $h - y$, so the work done to lift the slice is

$$dW = (h - y)\rho\,dV = \rho(h - y)A(y)\,dy.$$

Hence the total amount of work done on all the liquid originally in the tank is

$$W = \int_a^b \rho(h - y)A(y)\,dy. \tag{9}$$

Problem 14 asks you to use Riemann sums to set up this integral.

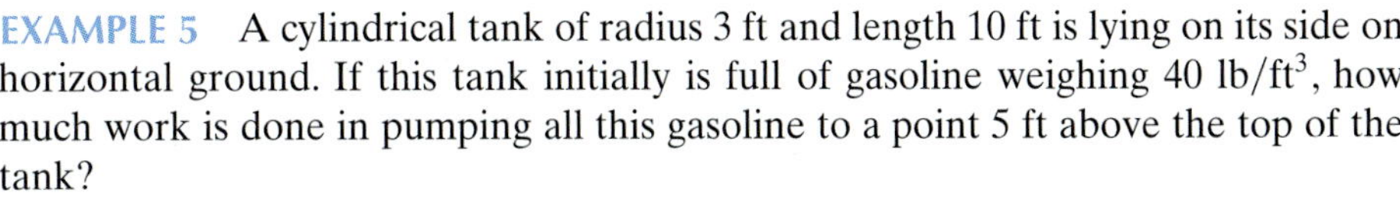

EXAMPLE 5 A cylindrical tank of radius 3 ft and length 10 ft is lying on its side on horizontal ground. If this tank initially is full of gasoline weighing 40 lb/ft^3, how much work is done in pumping all this gasoline to a point 5 ft above the top of the tank?

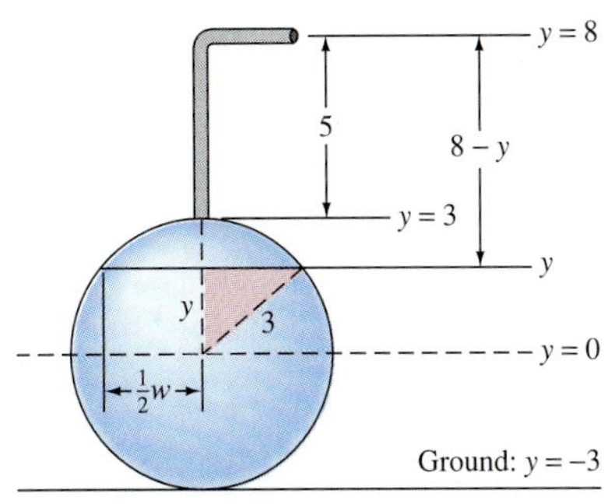

Fig. 6.6.10 End view of the cylindrical tank of Example 5

Solution Figure 6.6.10 shows an end view of the tank. To exploit its circular symmetry, we choose $y = 0$ at the *center* of the circular vertical section, so the tank lies between $y = -3$ and $y = 3$. A horizontal cross section of the tank that meets the y-axis is a rectangle of length 10 ft and width w. From the right triangle in Fig. 6.6.10, we see that

$$\tfrac{1}{2}w = \sqrt{9 - y^2},$$

so the area of this cross section is

$$A(y) = 10w = 20\sqrt{9 - y^2}.$$

This cross section must be lifted from its initial position y to the final position $5 + 3 = 8$, so it is to be lifted the distance $8 - y$. Thus Eq. (9) with $\rho = 40$, $a = -3$, and $b = 3$ yields

$$\begin{aligned} W &= \int_{-3}^{3} 40 \cdot (8 - y) \cdot 20\sqrt{9 - y^2}\,dy \\ &= 6400\int_{-3}^{3} \sqrt{9 - y^2}\,dy - 800\int_{-3}^{3} y\sqrt{9 - y^2}\,dy. \end{aligned}$$

We attack the two integrals separately. First,

$$\int_{-3}^{3} y\sqrt{9 - y^2}\, dy = \left[-\tfrac{1}{3}(9 - y^2)^{3/2} \right]_{-3}^{3} = 0.$$

Second,

$$\int_{-3}^{3} \sqrt{9 - y^2}\, dy = \tfrac{1}{2}\pi \cdot 3^2 = \tfrac{9}{2}\pi,$$

because the integral is simply the area of a semicircle of radius 3. Hence

$$W = 6400 \cdot \tfrac{9}{2}\pi = 28800\pi,$$

approximately 90478 ft·lb. ■

Remark As in Example 5, you may use as needed in the problems the integral

$$\int_{0}^{a} \sqrt{a^2 - x^2}\, dx = \tfrac{1}{4}\pi a^2, \tag{10}$$

which corresponds to the area of a quarter-circle of radius a.

Force Exerted by a Liquid

The **pressure** p at depth h in a liquid is the force per unit area exerted by the liquid at that depth. Pressure is given by

$$p = \rho h, \tag{11}$$

where ρ is the (weight) density of the liquid. For example, at a depth of 10 ft in water, for which $\rho = 62.4\ \text{lb/ft}^3$, the pressure is $62.4 \cdot 10 = 624\ \text{lb/ft}^2$. Hence if a thin, flat plate of area 5 ft^2 is suspended in a horizontal position at a depth of 10 ft in water, then the water exerts a downward force of $624 \cdot 5 = 3120$ lb on the top face of the plate and an equal upward force on its bottom face.

It is an important fact that at a given depth in a liquid, the pressure is the same in all directions. But if a flat plate is submerged in a vertical position in the liquid, then the pressure on the face of the plate is *not* constant, because by Eq. (11) the pressure increases with increasing depth. Consequently, the total force exerted on a vertical plate must be computed by integration.

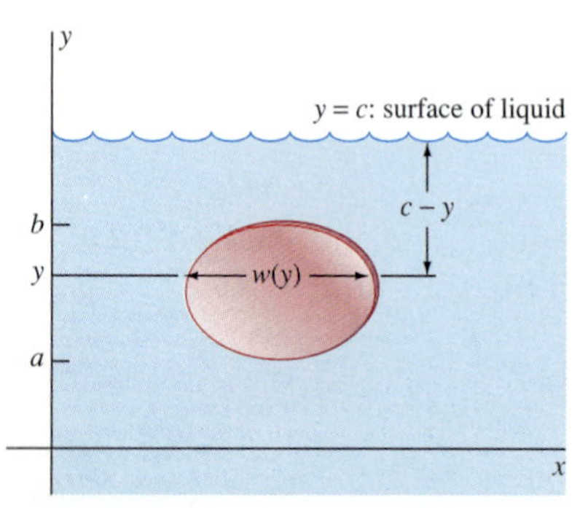

Fig. 6.6.11 A thin plate suspended verticaly in a liquid

Consider a thin, vertical, flat plate submerged in a liquid of density ρ (Fig. 6.6.11). The surface of the liquid is at the line $y = c$, and the plate lies alongside the interval $a \leqq y \leqq b$. The width of the plate at depth $c - y$ is some function of y, which we denote by $w(y)$.

To compute the total force F exerted by the liquid on either face of this plate, we begin with a partition of $[a, b]$ into n subintervals, all with the same length Δy, and denote by $y_i^\star$ the midpoint of the subinterval $[y_{i-1}, y_i]$. The horizontal strip of the plate opposite this ith subinterval is approximated by a rectangle of width $w(y_i^\star)$and height Δy, and its average depth in the liquid is $c - y_i^\star$. Hence the force ΔF_i exerted by the liquid on this horizontal strip is given approximately by

$$\Delta F_i \approx \rho(c - y_i^\star)w(y_i^\star)\,\Delta y. \tag{12}$$

The total force on the entire plate is given approximately by

$$F = \sum_{i=1}^{n} \Delta F_i \approx \sum_{i=1}^{n} \rho(c - y_i^{\star})w(y_i^{\star})\,\Delta y.$$

We obtain the exact value of F by taking the limit of such Riemann sums as $\Delta y \to 0$:

$$F = \int_a^b \rho(c - y)w(y)\,dy. \tag{13}$$

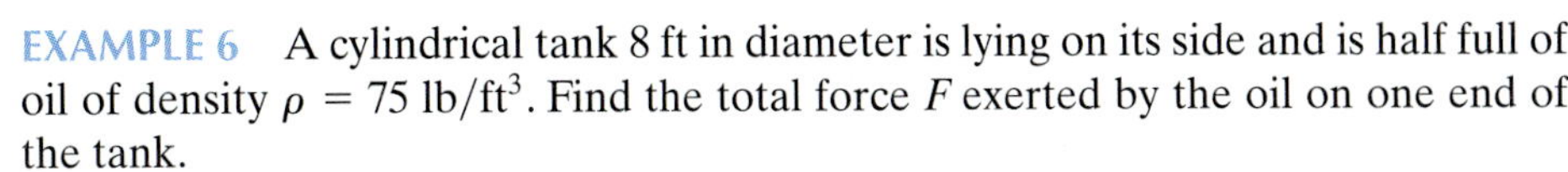

EXAMPLE 6 A cylindrical tank 8 ft in diameter is lying on its side and is half full of oil of density $\rho = 75\ \text{lb/ft}^3$. Find the total force F exerted by the oil on one end of the tank.

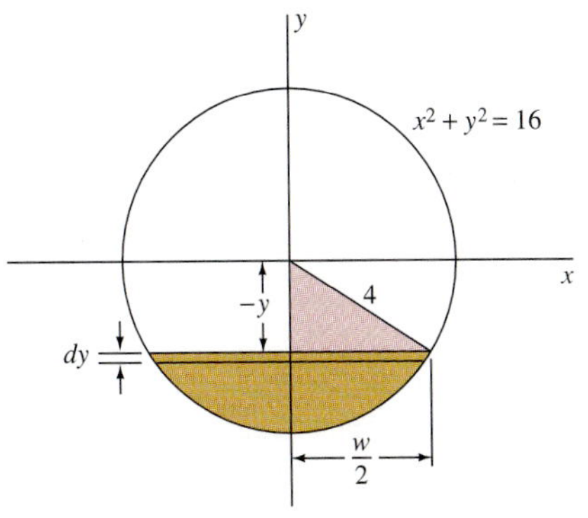

Fig. 6.6.12 View of one end of the tank of Example 6

Solution We locate the y-axis as indicated in Fig. 6.6.12, so that the surface of the oil is at the level $y = 0$. The oil lies alongside the interval $-4 \leqq y \leqq 0$. We see from the right triangle in the figure that the width of the oil at depth $-y$ (and thus at location y) is

$$w(y) = 2\sqrt{16 - y^2}.$$

Hence Eq. (13) gives

$$F = \int_{-4}^{0} 75(-y)(2\sqrt{16 - y^2})\,dy = 75\left[\tfrac{2}{3}(16 - y^2)^{3/2}\right]_{-4}^{0} = 3200 \quad \text{(lb)}.$$ ■

6.6 PROBLEMS

In Problems 1 through 5, find the work done by the given force $F(x)$ in moving a particle along the x-axis from $x = a$ to $x = b$.

1. $F(x) = 10;\quad a = -2, b = 1$

2. $F(x) = 3x - 1;\quad a = 1, b = 5$

3. $F(x) = \dfrac{10}{x^2};\quad a = 1, b = 10$

4. $F(x) = -3\sqrt{x};\quad a = 0, b = 4$

5. $F(x) = \sin \pi x;\quad a = -1, b = 1$

6. A spring has a natural length of 1 m, and a force of 10 N is required to hold it stretched to a total length of 2 m. How much work is done in compressing this spring from its natural length to a length of 60 cm?

7. A spring has a natural length of 2 ft, and a force of 15 lb is required to hold it compressed at a length of 18 in. How much work is done in stretching this spring from its natural length to a length of 3 ft?

8. Apply Eq. (4) to compute the amount of work done in lifting a 100-lb weight a height of 10 ft, assuming that this work is done against the constant force of gravity.

9. Compute the amount of work (in foot-pounds) done in lifting a 1000-lb weight from an orbit 1000 mi above the earth's surface to one 2000 mi above the earth's surface. Use the value of k given in Example 3.

10. A cylindrical tank of radius 5 ft and height 10 ft is resting on the ground with its axis vertical. Use Eq. (8) to compute the amount of work done in filling this tank with water pumped in from ground level. (Use $\rho = 62.4\ \text{lb/ft}^3$ for the weight density of water.)

11. A conical tank is resting on its base, which is at ground level, and its axis is vertical. The tank has radius 5 ft and height 10 ft (Fig. 6.6.13). Compute the work done in filling this tank with water ($\rho = 62.4\ \text{lb/ft}^3$) pumped in from ground level.

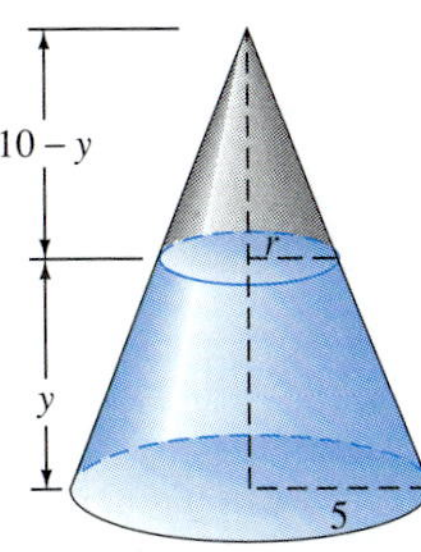

Fig. 6.6.13 The conical tank of Problem 11

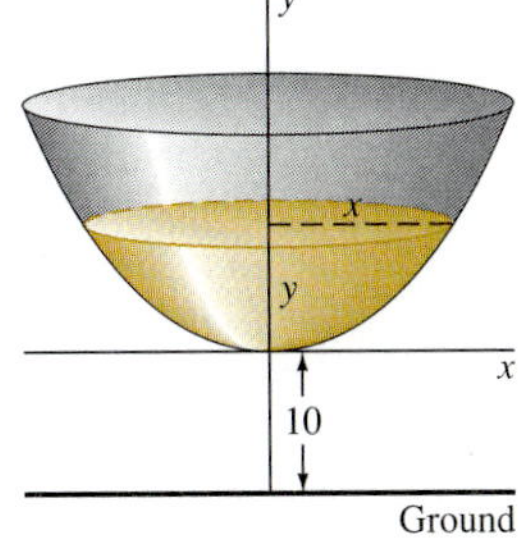

Fig. 6.6.14 The cup-shaped tank of Problem 13

12. Repeat Problem 11, except that now the tank is upended: Its vertex is at ground level and its base is 10 ft above the ground.

13. A tank whose lowest point is 10 ft above the ground has the shape of a cup obtained by rotating the parabola $x^2 = 5y$, $0 \leqq y \leqq 5$, around the y-axis (Fig. 6.6.14). The units on the coordinate axes are in feet. How much work is done in filling this tank with oil of density 50 lb/ft^3 if the oil is pumped in from ground level?

14. Suppose that the tank of Fig. 6.6.9 is filled with fluid of density ρ and that all this fluid must be pumped from the tank to the level $y = h$ above the top of the tank. Use Riemann sums, as in the derivation of Eq. (8), to obtain the formula

$$W = \int_a^b \rho(h - y)A(y)\,dy$$

for the work required to do so.

15. Use the formula in Problem 14 to find the amount of work done in pumping the water in the tank of Problem 10 to a height of 5 ft above the top of the tank.

16. Gasoline at a service station is stored in a cylindrical tank buried on its side, with the highest part of the tank 5 ft below the surface. The tank is 6 ft in diameter and 10 ft long. The density of gasoline is 45 lb/ft^3. Assume that the filler cap of each automobile gas tank is 2 ft above the ground (Fig. 6.6.15). (a) How much work is done in emptying all the gasoline from this tank, initially full, into automobiles? (b) Recall that 1 hp is equivalent to 33000 ft · lb/min. For electrical conversions, 1 kW (1000 W) is the same as 1.341 hp. The charge for use of electricity generated by a power company is typically about 7.2¢/kWh. Assume that the electrical motor in the gas pump at this station is 30% efficient. How much does it cost to pump all the gasoline from this tank into automobiles?

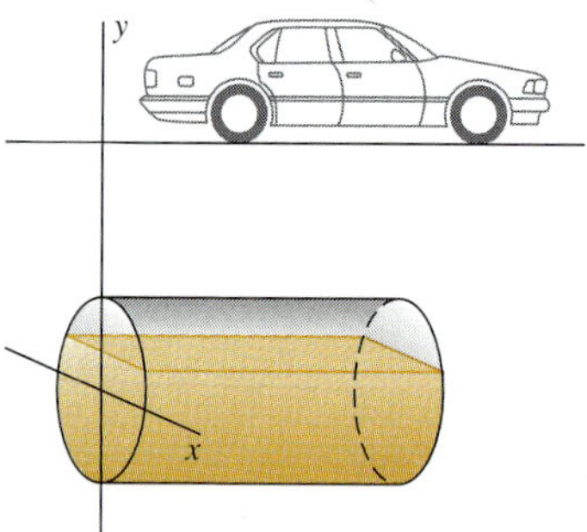

Fig. 6.6.15 The gasoline tank of Problem 16

17. Consider a spherical water tank whose radius is 10 ft and whose center is 50 ft above the ground. How much work is required to fill this tank by pumping water up from ground level? (*Suggestion:* It may simplify your computations to take $y = 0$ at the center of the tank and to think of the distance each horizontal slice of water must be lifted.)

18. A hemispherical tank of radius 10 ft is located with its flat side down atop a tower 60 ft high (Fig. 6.6.16). How much work is required to fill this tank with oil of density 50 lb/ft^3 if the oil is to be pumped into the tank from ground level?

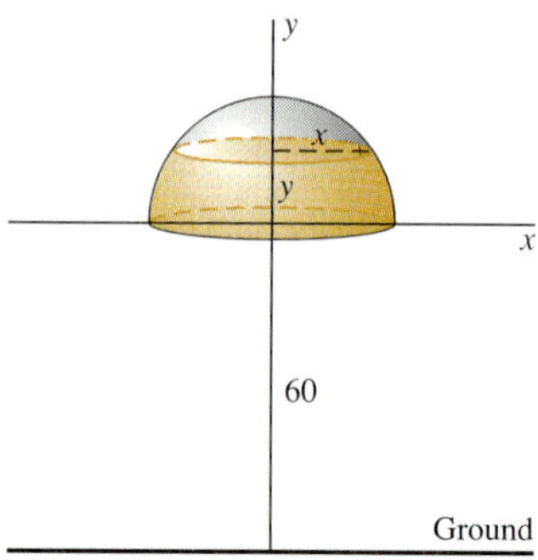

Fig. 6.6.16 The hemispherical tank of Problem 18

19. Water is being drawn from a well 100 ft deep, using a bucket that scoops up 100 lb of water. The bucket is pulled up at the rate of 2 ft/s, but it has a hole in the bottom through which water leaks out at the rate of 0.5 lb/s. How much work is done in pulling the bucket to the top of the well? Neglect the weight of the bucket, the weight of the rope, and the work done in overcoming friction. (*Suggestion:* Take $y = 0$ at the level of the water surface in the well, so that $y = 100$ at ground level. Let $\{y_0, y_1, y_2, \dots, y_n\}$ be a partition of $[0, 100]$ into n equal-length subintervals. Estimate the amount of work ΔW_i required to raise the bucket from y_{i-1} to y_i. Then set up the sum $W = \sum \Delta W_i$ and proceed to the appropriate integral by letting $n \to +\infty$.)

20. A rope that is 100 ft long and weighs 0.25 lb per linear foot hangs from the edge of a very tall building. How much work is required to pull this rope to the top of the building?

21. Suppose that we plug the hole in the leaky bucket of Problem 19. How much work do we do in lifting the mended bucket, full of water, to the surface, using the rope of Problem 20? Ignore friction and the weight of the bucket, but allow for the weight of the rope.

22. Consider a volume V of gas in a cylinder fitted with a piston at one end, where the pressure p of the gas is a function $p(V)$ of its volume (Fig. 6.6.17). Let A be the area of the face of the piston. Then the force exerted on the piston by gas in the cylinder is $F = pA$. Assume that the gas expands from volume V_1 to volume V_2. Show that the work done by the force F is then given by

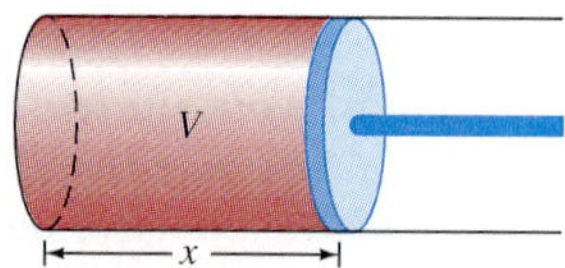

Fig. 6.6.17 A cylinder fitted with a piston (Problem 22)

$$W = \int_{V_1}^{V_2} p(V)\, dV.$$

(*Suggestion:* If x is the length of the cylinder (from its fixed end to the face of the piston), then $F = A \cdot p(Ax)$. Apply Eq. (4) and substitute $V = Ax$ into the resulting integral.)

23. The pressure p and volume V of the steam in a small steam engine satisfy the condition $pV^{1.4} = c$ (where c is a constant). In one cycle, the steam expands from a volume $V_1 = 50$ in.^{3} to $V_2 = 500$ in.^{3} with an initial pressure of 200 lb/in.^{2} Use the formula in Problem 22 to compute the work, in foot-pounds, done by this engine in each such cycle.

24. A tank in the shape of a hemisphere of radius 60 is resting on its flat base with the curved surface on top. It is filled with alcohol of density 40 lb/ft^3. How much work is done in pumping all the alcohol to the level of the top of the tank?

25. A tank has the shape of the surface generated by rotating around the y-axis the graph of $y = x^4$, $0 \leqq x \leqq 1$. The tank is initially full of oil of density 60 lb/ft^3. The units on the coordinate axes are in feet. How much work is done in pumping all the oil to the level of the top of the tank?

26. A cylindrical tank of radius 3 ft and length 20 ft is lying on its side on horizontal ground. Gasoline weighing 40 lb/ft^3 is at ground level and is to be pumped into the tank. Find the work required to fill the tank.

27. The base of a spherical storage tank of radius 12 ft is at ground level. Find the amount of work done in filling the tank with oil of density 50 lb/ft^3 if all the oil is initially at ground level.

28. A 20-lb monkey is attached to a 50-ft chain that weighs 0.5 lb per (linear) foot. The other end of the chain is attached to the 40-ft-high ceiling of the monkey's cage (Fig. 6.6.18). Find the amount of work the monkey does in climbing up her chain to the ceiling.

29. Tom is flying his kite at a height of 500 ft above the ground. Suppose that the kite string weighs $\frac{1}{16}$ oz per (linear) foot and that its string is stretched in a straight line at a 45° angle to the ground. How much work was done by the wind in lifting the string from ground level up to its flying position?

30. The center of a spherical tank of radius R is at a distance $H > R$ above the ground. A liquid of weight density ρ is at ground level. Show that the work required to pump the initially empty tank full of this liquid is the same as that to lift the full tank the distance H.

31. A water trough 10 ft long has a square cross section that is 2 ft wide. If the trough is full of water (density $\rho =$ 62.4 lb/ft^3), find the force exerted by the water on one end of the trough.

32. Repeat Problem 31 for a trough whose cross section is an equilateral triangle with edges 2 ft long.

33. Repeat Problem 31 for a trough whose cross section is a trapezoid 3 ft high, 2 ft wide at the bottom, and 4 ft wide at the top.

34. Find the force on one end of the cylindrical tank of Example 6 if the tank is filled with oil of density 50 lb/ft^3. Remember that

$$\int_0^a \sqrt{a^2 - y^2}\, dy = \tfrac{1}{4}\pi a^2,$$

because the integral represents the area of a quarter-circle of radius a.

Problems 35 through 38 describe a gate in the vertical face of a dam. Find the total force of water on this gate if its top is 10 ft beneath the surface of the water.

35. A square of edge length 5 ft whose top is parallel to the water surface

36. A circle of radius 3 ft

37. An isosceles triangle 5 ft high and 8 ft wide at the top

38. A semicircle of radius 4 ft whose top edge is its diameter (also parallel to the water surface)

39. Suppose that the dam of Fig. 6.6.19 is $L = 200$ ft long and $T = 30$ ft thick at its base. Find the force of water on the dam if the water is 100 ft deep and the *slanted* end of the dam faces the water.

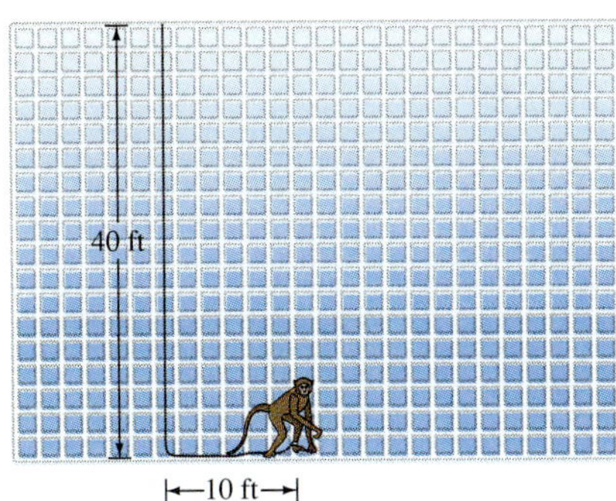

Fig. 6.6.18 The monkey of Problem 28

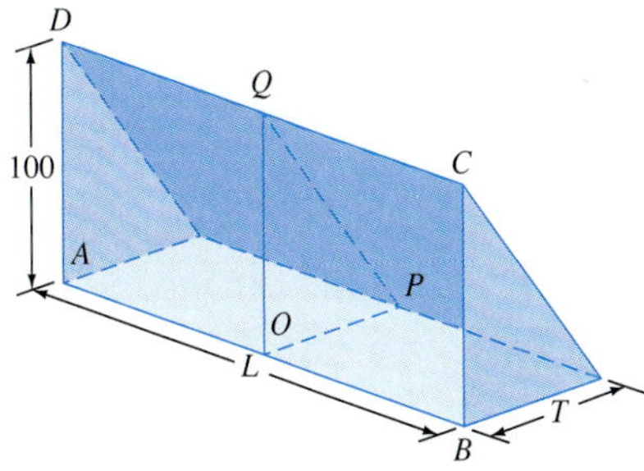

Fig. 6.6.19 View of a model of a dam (Problem 39)

CHAPTER 6 REVIEW: *Definitions, Concepts, Results*

Use the following list as a guide to concepts that you may need to review.

1. The general method of setting up an integral formula for a quantity by approximating it and then recognizing the approximation to be a Riemann sum that corresponds to the desired integral: If the interval $[a, b]$ is partitioned into n subintervals of equal length $\Delta x = (b - a)/n$ and if $x_i^\star$ denotes a point of the ith subinterval, then

$$\lim_{n\to\infty} \sum_{i=1}^{n} f(x_i^\star)\,\Delta x = \int_a^b f(x)\,dx.$$

2. Net distance traveled as the integral of velocity:

$$s = \int_a^b v(t)\,dt$$

3. The method of cross sections for computing volumes:

$$V = \int_a^b A(x)\,dx,$$

where $A(x)$ denotes the area of a slice with infinitesimal thickness dx

4. Determining the volume of a solid of revolution by the method of cross sections, with the cross sections being either disks or annular rings
5. Determining the volume of a solid of revolution by the method of cylindrical shells:

$$V = \int_\star^{\star\star} 2\pi r\,dA,$$

where r denotes the radius of the circle through which the area element dA is revolved

6. The arc length of a smooth arc described either in the form $y = f(x)$, $a \leqq x \leqq b$ or in the form $x = g(y)$, $c \leqq y \leqq d$:

$$s = \int_\star^{\star\star} ds \qquad \left(ds = \sqrt{(dx)^2 + (dy)^2}\right),$$

where

$$ds = \sqrt{1 + [f'(x)]^2}\,dx \quad \text{for} \quad y = f(x),$$

$$ds = \sqrt{1 + [g'(y)]^2}\,dy \quad \text{for} \quad x = g(y)$$

7. Determining the area of the surface of revolution generated by revolving a smooth arc, given in the form $y = f(x)$ or in the form $x = g(y)$, around either the x-axis or the y-axis:

$$A = \int_\star^{\star\star} 2\pi r\,ds,$$

where r denotes the radius of the circle through which the arc length element ds is revolved

8. The work done by a force function in moving a particle along a straight line segment:

$$W = \int_a^b F(x)\,dx$$

if the force $F(x)$ acts from $x = a$ to $x = b$

9. Hooke's law and the work done in stretching or compressing an elastic spring
10. Work done against the varying force of gravity
11. Work done in filling a tank or in pumping the liquid in a tank to another level:

$$W = \int_a^b \rho h(y) A(y)\,dy,$$

where $h(y)$ denotes the vertical distance that the horizontal fluid slice of volume $dV = A(y)\,dy$, at the height y, must be lifted

12. The force exerted by a liquid on the face of a submerged vertical plate:

$$F = \int_\star^{\star\star} \rho h\,dA,$$

where h denotes the depth of the horizontal area element dA beneath the surface of the fluid of weight density ρ

13. Solution of the separable differential equation

$$\frac{dy}{dx} = g(x)\phi(y)$$

by integration:

$$\int f(y)\,dy = \int g(x)\,dx + C,$$

where $f(y) = 1/\phi(y)$

CHAPTER 6 *Miscellaneous Problems*

In Problems 1 through 3, find both the net distance and the total distance traveled between times $t = a$ and $t = b$ by a particle moving along a line with the given velocity function $v = f(t)$.

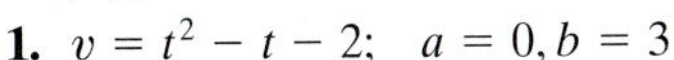

1. $v = t^2 - t - 2;\quad a = 0, b = 3$

2. $v = |t^2 - 4|;\quad a = 1, b = 4$

3. $v = \pi \sin \frac{1}{2}\pi(2t - 1);\quad a = 0, b = \frac{3}{2}$

In Problems 4 through 8, a solid extends along the x-axis from $x = a$ to $x = b$, and its cross-sectional area at x is $A(x)$. Find its volume.

4. $A(x) = x^3;\quad a = 0, b = 1$

5. $A(x) = \sqrt{x};\quad a = 1, b = 4$

6. $A(x) = x^3;\quad a = 1, b = 2$

7. $A(x) = \pi(x^2 - x^4);\quad a = 0, b = 1$

8. $A(x) = x^{100};\quad a = -1, b = 1$

9. Suppose that rainfall begins at time $t = 0$ and that the rate after t hours is $(t + 6)/12$ inches per hour. How many inches of rain falls during the first 12 h?

10. The base of a certain solid is the region in the first quadrant bounded by the curves $y = x^3$ and $y = 2x - x^2$. Find the solid's volume if each cross section perpendicular to the x-axis is a square with one edge in the base of the solid.

11. Find the volume of the solid generated by revolving around the x-axis the first-quadrant region of Problem 10.

12. Find the volume of the solid generated by revolving the region bounded by $y = 2x^4$ and $y = x^2 + 1$ around (a) the x-axis; (b) the y-axis.

13. A wire made of copper (density 8.5 g/cm^3) is shaped like a helix that spirals around the x-axis from $x = 0$ to $x = 20$. Each cross section of this wire perpendicular to the x-axis is a circular disk of radius 0.25 cm. What is the total mass of the wire?

14. Derive the formula $V = \frac{1}{3}\pi h(r_1^2 + r_1 r_2 + r_2^2)$ for the volume of a frustum of a cone with height h and base radii r_1 and r_2.

15. Suppose that the point P lies on a line perpendicular to the xy-plane at the origin O, with $|OP| = h$. Consider the "elliptical cone" that consists of all points on line segments from P to points on and within the ellipse with equation

$$\left(\frac{x}{a}\right)^2 + \left(\frac{y}{b}\right)^2 = 1.$$

Show that the volume of this elliptical cone is $V = \frac{1}{3}\pi abh$.

16. Figure 6.MP.1 shows the region R bounded by the ellipse $(x/a)^2 + (y/b)^2 = 1$ and by the line $x = a - h$, where $0 < h < a$. Revolution of R around the x-axis generates a "segment of an ellipsoid" of radius r, height h, and volume V. Show that

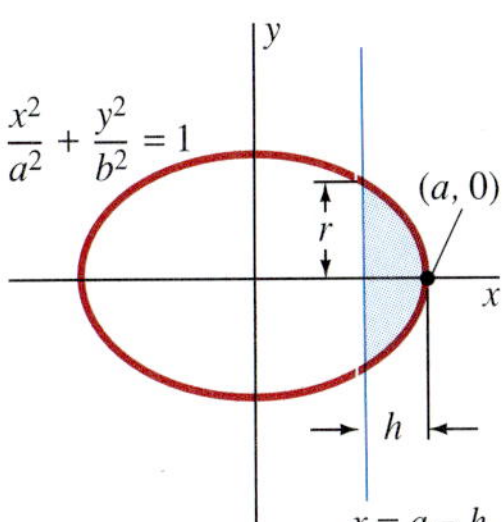

Fig. 6.MP.1 A segment of an ellipse (Problem 16)

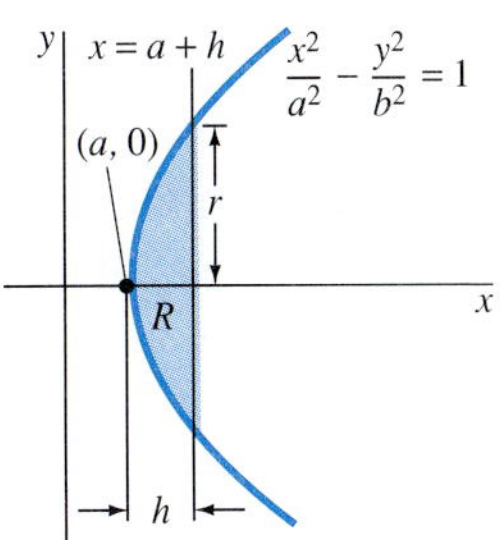

Fig. 6.MP.2 The region R of Problem 17

$$r^2 = \frac{b^2(2ah - h^2)}{a^2} \quad \text{and that} \quad V = \frac{1}{3}\pi r^2 h \frac{3a - h}{2a - h}.$$

17. Figure 6.MP.2 shows the region R bounded by the hyperbola $(x/a)^2 - (y/b)^2 = 1$ and the line $x = a + h$, where $h > 0$. Revolution of R around the x-axis generates a "segment of a hyperboloid" of radius r, height h, and volume V. Show that

$$r^2 = \frac{b^2(2ah + h^2)}{a^2} \quad \text{and that} \quad V = \frac{1}{3}\pi r^2 h \frac{3a + h}{2a + h}.$$

In Problems 18 through 20, the function $f(x)$ is nonnegative and continuous for $x \geqq 1$. When the region lying under $y = f(x)$ from $x = 1$ to $x = t$ is revolved around the indicated axis, the volume of the resulting solid is $V(t)$. Find the function $f(x)$.

18. $V(t) = \pi\left(1 - \dfrac{1}{t}\right);\quad$ the x-axis

19. $V(t) = \frac{1}{6}\pi[(1 + 3t)^2 - 16];\quad$ the x-axis

20. $V(t) = \frac{2}{9}\pi[(1 + 3t^2)^{3/2} - 8];\quad$ the y-axis

21. Use the integral formula in Problem 42 of Section 6.3 to find the volume of the solid generated by revolving around the y-axis the first-quadrant region bounded by $y = x$ and $y = \sin\left(\frac{1}{2}\pi x\right)$.

22. Use the method of cylindrical shells to find the volume of the solid generated by revolving around the line $x = -2$ the region bounded by $y = x^2$ and $y = x + 2$.

23. Find the length of the curve $y = \frac{1}{3}x^{3/2} - x^{1/2}$ from $x = 1$ to $x = 4$.

24. Find the area of the surface generated by revolving the curve of Problem 23 around (a) the x-axis; (b) the y-axis.

25. Find the length of the curve $x = \frac{3}{8}(y^{4/3} - 2y^{2/3})$ from $y = 1$ to $y = 8$.

26. Find the area of the surface generated by revolving the curve of Problem 25 around (a) the x-axis; (b) the y-axis.

27. Find the area of the surface generated by revolving the curve of Problem 23 around the line $x = 1$.

28. If $-r \leqq a < b \leqq r$, then a "spherical zone" of "height" $h = b - a$ is generated by revolving around the x-axis the circular arc $y = \sqrt{r^2 - x^2}$, $a \leqq x \leqq b$. Show that the area of this spherical zone is $A = 2\pi rh$, the same as that of a cylinder of radius r and height h.

29. Apply the result of Problem 28 to show that the surface area of a sphere of radius r is $A = 4\pi r^2$.

30. Let R denote the region bounded by the curves $y = 2x^3$ and $y^2 = 4x$. Find the volumes of the solids obtained by revolving the region R around (a) the x-axis; (b) the y-axis; (c) the line $y = -1$; (d) the line $x = 2$. In each case use both the method of cross sections and the method of cylindrical shells.

In Problems 31 through 42, find the general solution of the given differential equation. If an initial condition is given, find the corresponding particular solution.

31. $\dfrac{dy}{dx} = 2x + \cos x;\quad y(0) = 0$

32. $\dfrac{dy}{dx} = 3\sqrt{x} + \dfrac{1}{\sqrt{x}};\quad y(1) = 10$

33. $\dfrac{dy}{dx} = (y + 1)^2$

34. $\dfrac{dy}{dx} = \sqrt{y + 1}$

35. $\dfrac{dy}{dx} = 3x^2y^2;\quad y(0) = 1$

36. $\dfrac{dy}{dx} = \sqrt[3]{xy};\quad y(1) = 1$

37. $x^2y^2\dfrac{dy}{dx} = 1$

38. $\sqrt{xy}\,\dfrac{dy}{dx} = 1$

39. $\dfrac{dy}{dx} = y^2 \cos x;\quad y(0) = 1$

40. $\dfrac{dy}{dx} = \sqrt{y} \sin x;\quad y(0) = 4$

41. $\dfrac{dy}{dx} = \dfrac{y^2(1 - \sqrt{x})}{x^2(1 - \sqrt{y})}$

42. $\dfrac{dy}{dx} = \dfrac{\sqrt{y}(x + 1)^3}{\sqrt{x}(y + 1)^3}$

43. Find the natural length L of a spring if five times as much work is required to stretch it from a length of 2 ft to a length of 5 ft as is required to stretch it from a length of 2 ft to a length of 3 ft.

44. A steel beam weighing 1000 lb hangs from a 50-ft cable that weighs 5 lb per linear foot. How much work is done in winding in 25 ft of the cable with a windlass?

45. A spherical tank of radius R (in feet) is initially full of oil of density ρ lb/ft^3. Find the total work done in pumping all the oil from the sphere to a height of $2R$ above the top of the tank.

46. How much work is done by a colony of ants in building a conical anthill of height and diameter 1 ft, using sand initially at ground level and with a density of 150 lb/ft^3?

47. The gravitational attraction below the earth's surface is directly proportional to the distance from the center of the earth. Suppose that a straight cylindrical hole of radius 1 ft is dug from the earth's surface to its center. Assume that the earth has radius 3960 mi and uniform density 350 lb/ft^3. How much work, in foot-pounds, is done in lifting a 1-lb weight from the bottom of this hole to its top?

48. How much work is done in digging the hole of Problem 47—that is, in lifting all the material it initially contained to the earth's surface?

49. Suppose that a dam is shaped like a trapezoid of height 100 ft, 300 ft wide at the top and 200 ft wide at the bottom. When the water level behind the dam is even with its top, what is the total force that the water exerts on the dam?

50. Suppose that a dam has the same top and bottom widths as the dam of Problem 49 and the same vertical height of 100 ft, but that its face toward the water is slanted at an angle of 30° from the vertical. What is the total force of water pressure on this dam?

51. For $c > 0$, the graphs of $y = c^2x^2$ and $y = c$ bound a plane region. Revolve this region around the horizontal line $y = -1/c$ to form a solid. For what value of c is the volume of this solid maximal? minimal?

CHAPTER 7

EXPONENTIAL AND LOGARITHMIC FUNCTIONS

John Napier (1550–1617)

John Napier was the eighth baron of Merchiston (near Edinburgh), Scotland. As a landowner, he participated vigorously in local affairs and actively managed crops and cattle on his lands around Merchiston Castle. He is said to have regarded as his most important and lasting contribution his book *A Plaine Discovery of the Whole Revelation of Saint John* (1593), in which (among other things) he predicted that the world would come to an end in the year 1786. But the world—still here some two centuries later—remembers Napier as the inventor of logarithms. The late sixteenth century was an age of numerical computation, as developments in astronomy and navigation called for increasingly accurate and lengthy trigonometric computations. With his theological work behind him, in 1594 Napier began the 20-year labor that was to revolutionize the practical art of numerical computation. The result of this monumental project was his 1614 book *Mirifici Logarithmorum Canonis Descriptio* (Description of the Wonderful Canon of Logarithms), which contained his logarithmic tables along with a brief introduction and guide to their use.

Students today first learn of the "common" base-10 logarithm $\log_{10} x$ as the power to which 10 must be raised to get the number x. Because

$$\log_{10} xy = \log_{10} x + \log_{10} y,$$

multiplying the two numbers x and y effectively reduces to the simpler operation of adding their logarithms. The "logarithm" $\text{Nlog}(x)$ that Napier used was somewhat different. Whereas $\log_{10} 1 = 0$, Napier's definition implies that $\text{Nlog}(10^7) = 0$ with the previous equation replaced by a rule of the form $\text{Nlog}(xy) = \text{Nlog}(x) + \text{Nlog}(y) - C$. Hence the use of Napier's original table of logarithms would involve continual addition or subtraction of the particular constant $C = 10^7 \times \ln 10^7$ (where ln denotes the modern "natural" logarithm).

In 1615 the English mathematics professor Henry Briggs visited Napier in Scotland. Their discussions led to Briggs's construction of a table of "improved" logarithms, ones for which the logarithm of 1 is 0 and the logarithm of 10 is 1. Within a decade after Napier's death, 10-place tables of these common logarithms had been completed. These tables played a central role in scientific computation during the next three centuries.

On a slide rule of the type used universally by calculus students prior to the advent of electronic pocket calculators in the 1970s, numbers are inscribed on the scales at distances proportional to their logarithms. Consequently, the logarithms of x on a fixed scale and of y on a sliding scale are readily added to determine the product xy.

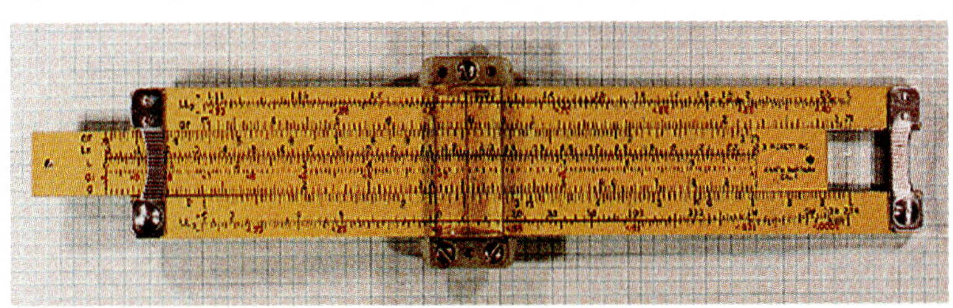

7.1 EXPONENTIALS, LOGARITHMS, AND INVERSE FUNCTIONS

Until now, our study of calculus has concentrated largely on trigonometric and power functions and on the algebraic functions obtained from power functions by forming finite combinations and compositions. This chapter is devoted to the calculus of exponential and logarithmic functions. This introductory section gives a brief and somewhat informal overview of these nonalgebraic functions, partly from the perspective of precalculus mathematics. In addition to reviewing the laws of exponents and the laws of logarithms, our purpose here is to provide some intuitive background for the systematic treatment of exponential and logarithmic functions that begins in Section 7.2.

Exponential Functions

An *exponential function* is a function of the form

$$f(x) = a^x \tag{1}$$

where $a > 0$. Note that the exponent x is the variable here and that the number a, called the *base,* is a constant. Thus

- An exponential function $f(x) = a^x$ is a constant raised to a variable power, whereas
- The power function $p(x) = x^k$ is a variable raised to a constant power.

In elementary algebra a *rational* power of the positive real number a is defined in terms of integral roots and powers. If n is a positive integer, then

$$a^n = a \cdot a \cdot a \cdots a \quad (n \text{ factors})$$

and

$$a^{-n} = \frac{1}{a^n}.$$

Next we learn that if $r = p/q$ where p and q are integers (with q positive), then the rational power a^r is defined by

$$a^{p/q} = \sqrt[q]{a^p} = \left(\sqrt[q]{a}\right)^p.$$

The following **laws of exponents** are then established for all *rational* exponents r and s:

$$\begin{aligned} a^{r+s} &= a^r \cdot a^s, \qquad & (a^r)^s &= a^{r \cdot s}, \\ a^{-r} &= \frac{1}{a^r}, \qquad & (ab)^r &= a^r \cdot b^r. \end{aligned} \tag{2}$$

Moreover, recall that

$$a^0 = 1 \tag{3}$$

for every positive real number a.

EXAMPLE 1 A typical calculation using the laws of exponents is

$$\frac{(2^2)^3(3)^{-4}}{(2)^{-2}} = \frac{2^6 \cdot 2^2}{3^4} = \frac{2^8}{3^4} = \frac{256}{81}.$$

Some fractional powers of 2 are

$$2^{3/2} = \sqrt{2^3} = \sqrt{8}, \quad 2^{5/3} = \sqrt[3]{2^5} = \sqrt[3]{32},$$

and

$$2^{0.7} = 2^{7/10} = \sqrt[10]{2^7} = \sqrt[10]{128} \approx 1.6245.$$

The following example illustrates the fact that applications often call for irrational exponents as well as rational exponents.

EXAMPLE 2 Consider a bacteria population $P(t)$ that begins (at time $t = 0$) with initial population $P(0) = 1$ (million) and doubles every hour thereafter. The growing population is given at 1-hour intervals as in the following table.

t	1	2	3	4	5	(hours)
P	2	4	8	16	32	(millions)

It is evident that $P(n) = 2^n$ if n is an integer. Now let's make the plausible assumption that the population increases by the same factor in any two time intervals of the same length—for example, if it grows by 10% in any one eight-minute interval, then it grows by 10% in *any* eight-minute interval. If q is a positive integer and k denotes the factor by which the population increases during a time interval of length $\Delta t = 1/q$, then the population is given at successive time intervals of length $1/q$ as in the next table.

t	$\frac{1}{q}$	$\frac{2}{q}$	$\frac{3}{q}$	$\cdots$	$\frac{q}{q} = 1$	
P	k	k^2	k^3	$\cdots$	$k^q = 2$	(Why?)

We therefore see that $k = 2^{1/q}$. If p is another positive integer, then during p/q hours the population P will increase p times by the factor $k = 2^{1/q}$, so it follows that

$$P\left(\frac{p}{q}\right) = k^p = (2^{1/q})^p = 2^{p/q}.$$

Thus the bacteria population after t hours is given (in millions) by

$$P(t) = 2^t$$

if the exponent t is a rational number. But because time is not restricted to rational values alone, we surely ought to conclude that $P(t) = 2^t$ for *all* $t \geqq 0$.

INVESTIGATION But what do we mean by an expression involving an irrational exponent, such as $2^{\sqrt{2}}$ or 2^{π}? To find the value of 2^{π}, we might work with (rational) finite decimal approximations to the irrational number $\pi = 3.1415926\ldots$. For example, a calculator gives

$$2^{3.1} = 2^{31/10} = \left(\sqrt[10]{2}\right)^{31} \approx 8.5742.$$

t	2^t
3.1	8.5742
3.14	8.8152
3.141	8.8214
3.1415	8.8244
3.14159	8.8250
3.141592	8.8250
3.1415926	8.8250
↓	↓
π	2^π

Fig. 7.1.1 Investigating 2^π

The approximate values shown in the table in Fig. 7.1.1 indicate that the bacteria population in Example 2 after π hours is

$$P(\pi) = 2^\pi \approx 8.8250 \quad \text{(million).}$$

Because any irrational number can be approximated arbitrarily closely by rational numbers, the preceding investigation suggests that the value of a^x with irrational exponent x can be regarded as a limit of the form

$$a^x = \lim_{r \to x} a^r \qquad (r \text{ rational}). \tag{4}$$

Indeed, we will see in this chapter that, for any fixed base $a > 0$, the exponential function $f(x) = a^x$ can be defined for *all* x in such a way that:

- The laws of exponents in Eq. (2) hold for irrational exponents as well as for rational exponents;
- The function $f(x) = a^x$ is continuous, indeed differentiable.

The limit in Eq. (4) will then follow from the continuity of f.

On a calculator, the ∧ key (sometimes the y^x key) is ordinarily used to calculate values of exponential functions. For instance, Fig. 7.1.2 shows the result of graphing the function defined by **y = 2∧x**. We see an increasing function that appears to be positive-valued for all x. Indeed, these observations are predictable. If $a > 1$ and r is a positive rational number, note first that $a^r > 1$. (Why?) If s is a second positive rational number with $r < s$, then the laws of exponents give

$$a^r < a^r \cdot a^{s-r} = a^s.$$

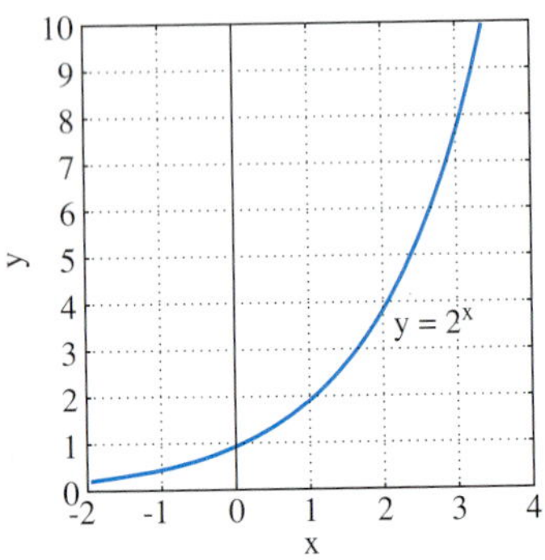

Fig. 7.1.2 The graph $y = 2^x$

Thus $a^r < a^s$ whenever $0 < r < s$, so the exponential function $f(x) = a^x$ is certainly an *increasing* function if only positive rational values of the exponent x are considered. A graphing calculator or computer actually plots only finitely many points (x, a^x), but the curve plotted in Fig. 7.1.2 looks "continuous" because these points are plotted too closely together for the eye to distinguish them. By contrast, the graph in Fig. 7.1.3 is shown with a dotted curve to suggest that it is densely filled with "holes" corresponding to the missing points (x, a^x) with x irrational. We will show in Section 7.4 that the holes in this graph can be filled to obtain the graph of an increasing and continuous function f that is defined for all x and for which $f(r) = a^r$ for every rational number r. We therefore write $f(x) = a^x$ for all x and call f the **exponential function with (positive) base** a.

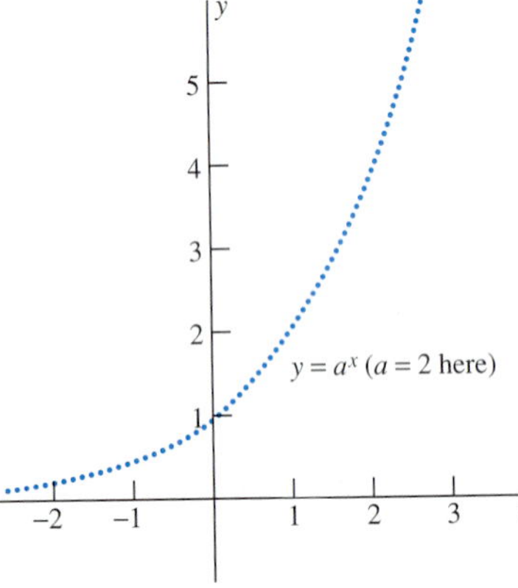

Fig. 7.1.3 The graph of $y = a^x$ has "holes" if only rational values of x are used.

As illustrated in Fig. 7.1.4, the function $f(x) = a^x$ with $a > 1$ increases rapidly as $x > 0$ increases, and the graphs of $y = a^x$ look qualitatively similar for different values of $a > 1$ as the base. The steep rate of increase of $f(x) = a^x$ is a characteristic feature of exponential functions. Figures 7.1.5 and 7.1.6 compare the graphs $y = 2^x$ and $y = x^2$.

Derivatives of Exponential Functions

To compute the derivative of the exponential function $f(x) = a^x$, we begin with the definition of the derivative and then use the first law of exponents in Eq. (2) to simplify. This gives

$$f'(x) = D_x a^x = \lim_{h \to 0} \frac{f(x+h) - f(x)}{h} = \lim_{h \to 0} \frac{a^{x+h} - a^x}{h}$$

$$= \lim_{h \to 0} \frac{a^x a^h - a^x}{h} \qquad \text{(by the laws of exponents)}$$

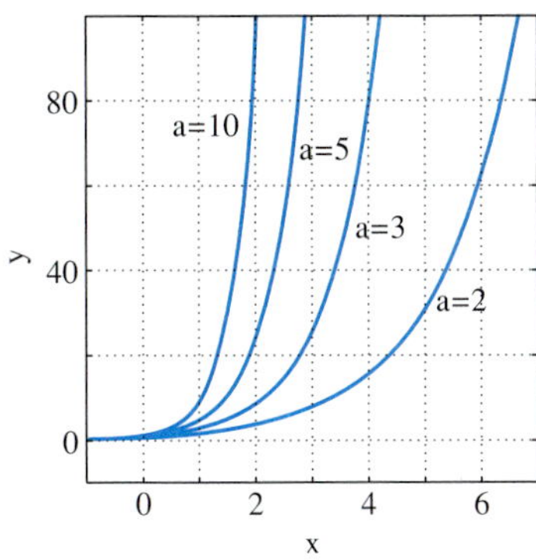

Fig. 7.1.4 $y = a^x$ for $a = 2, 3, 5, 10$

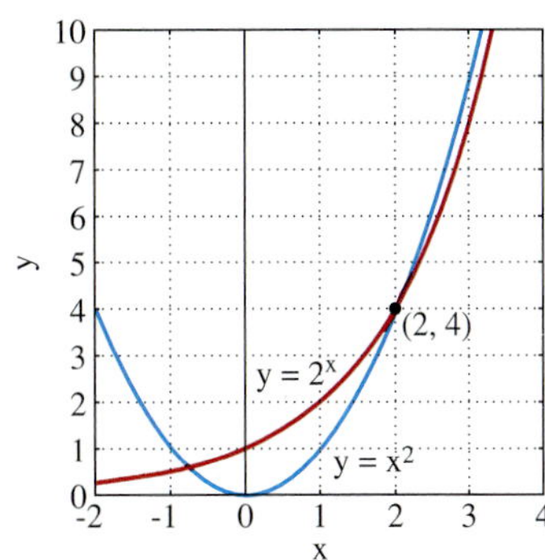

Fig. 7.1.5 Here the graphs $y = 2^x$ and $y = x^2$ look similar for x near 2.

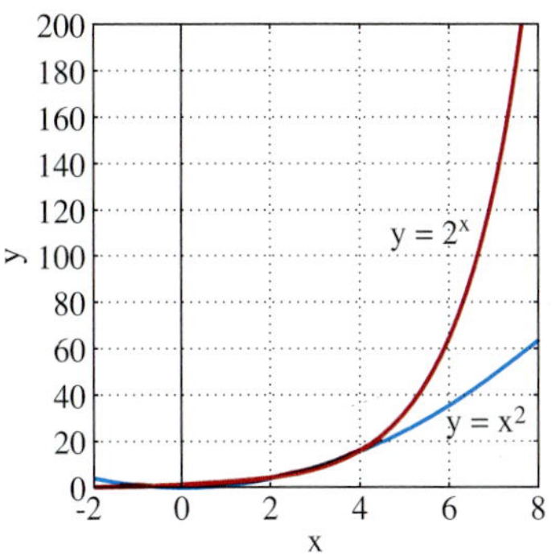

Fig. 7.1.6 But here we see 2^x increasing much more rapidly than x^2

$$= a^x\left(\lim_{h\to 0}\frac{a^h - 1}{h}\right) \qquad \text{(because } a^x \text{ is "constant" with respect to } h\text{).}$$

Assuming that $f(x) = a^x$ is differentiable, it follows that the limit

$$m(a) = \lim_{h\to 0}\frac{a^h - 1}{h} \tag{5}$$

exists. Although its value $m(a)$ depends on a, the limit is a constant with regard to x. Thus we find that the derivative of a^x is a *constant multiple* of a^x itself:

$$D_x\, a^x = m(a)\cdot a^x. \tag{6}$$

Because $a^0 = 1$, we see from Eq. (6) that the constant $m(a)$ is the slope of the line tangent to the curve $y = a^x$ at the point $(0, 1)$, where $x = 0$.

The numerical data shown in Fig. 7.1.7 suggest that $m(2) \approx 0.693$ and that $m(3) \approx 1.099$. The tangent lines with these slopes are shown in Fig. 7.1.8. Thus it appears that

$$D_x\, 2^x \approx (0.693)\cdot 2^x \quad \text{and} \quad D_x\, 3^x \approx (1.099)\cdot 3^x. \tag{7}$$

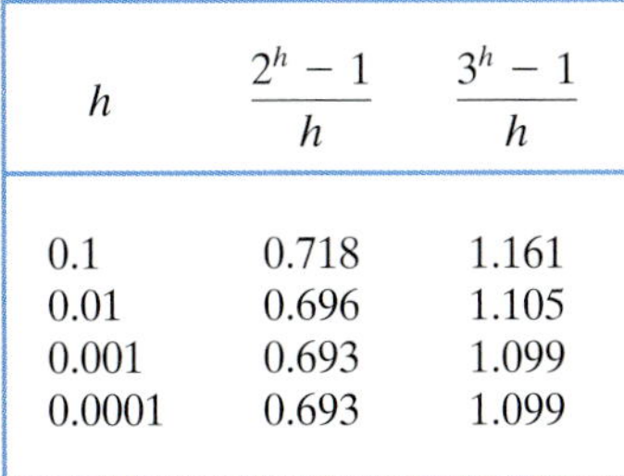

h	$\dfrac{2^h - 1}{h}$	$\dfrac{3^h - 1}{h}$
0.1	0.718	1.161
0.01	0.696	1.105
0.001	0.693	1.099
0.0001	0.693	1.099

Fig. 7.1.7 Investigating the values of $m(2)$ and $m(3)$

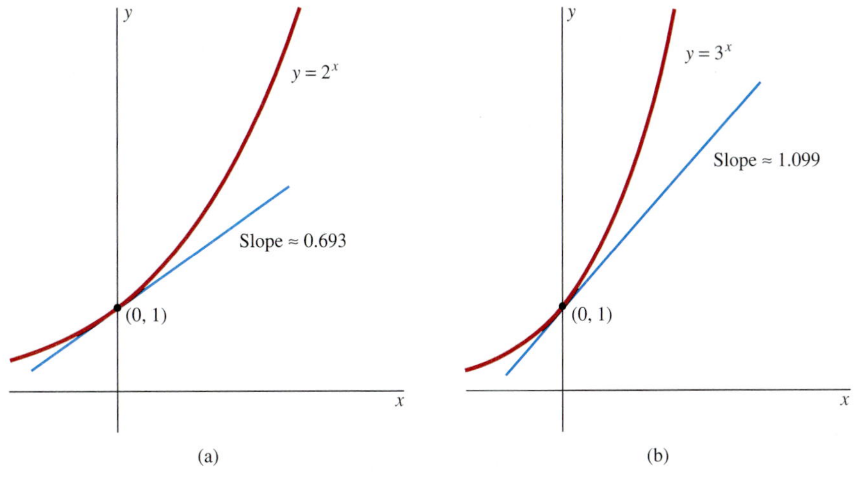

Fig. 7.1.8 The graphs (a) $y = 2^x$ and (b) $y = 3^x$

We would like somehow to avoid awkward numerical factors like those in Eq. (7). It seems plausible that the value $m(a)$ defined in Eq. (5) is a continuous function of a. If so, then because $m(2) < 1$ and $m(3) > 1$, the intermediate value property implies that $m(e) = 1$ (exactly) for some number e between 2 and 3. If we use this particular number e as the base, then it follows from Eq. (6) that the derivative of the resulting exponential function $f(x) = e^x$ is

$$D_x\, e^x = e^x. \tag{8}$$

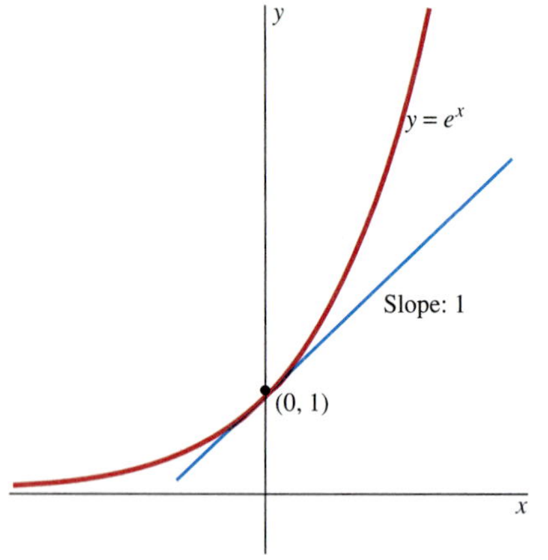

Fig. 7.1.9 The graph $y = e^x$

Thus the function e^x is its own derivative. We call $f(x) = e^x$ the **natural exponential function.** Its graph is shown in Fig. 7.1.9.

We will see in Section 7.3 that the number e is given by the limit

$$e = \lim_{n\to\infty}\left(1 + \frac{1}{n}\right)^n.$$

Let us investigate this limit numerically. With a calculator we obtain the values in the table of Fig. 7.1.10. The evidence suggests (but does not prove) that $e \approx 2.718$ to three places. This number e is one of the most important special numbers in mathematics. It is known to be irrational; its value accurate to 15 places is

$$e \approx 2.71828\,1828\,459045.$$

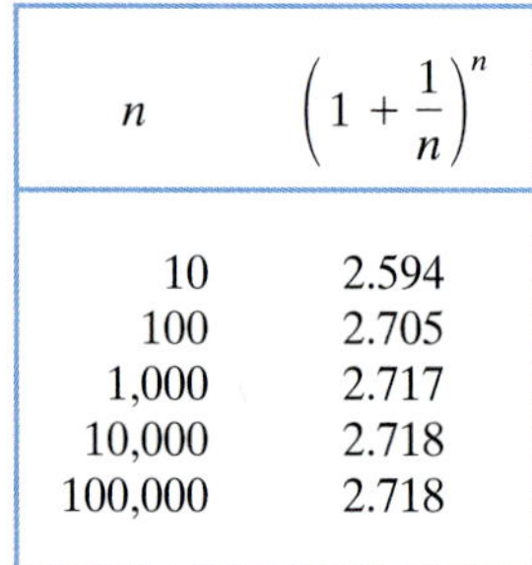

n	$\left(1 + \frac{1}{n}\right)^n$
10	2.594
100	2.705
1,000	2.717
10,000	2.718
100,000	2.718

Fig. 7.1.10 Numerical estimate of the number e

The chain rule version of Eq. (8) is the differentiation formula

$$D_x\, e^u = e^u \frac{du}{dx}, \tag{9}$$

where u denotes a differentiable function of x. In particular,

$$D_x\, e^{kx} = ke^{kx}$$

if k is a constant. For instance, $D_x\, e^{-x} = -e^{-x}$ and $D_x e^{2x} = 2e^{2x}$.

EXAMPLE 3

(a) If $f(x) = x^2e^{-x}$, then the product rule gives

$$\begin{aligned} f'(x) &= (D_x\, x^2)e^{-x} + x^2(D_x e^{-x}) \\ &= 2xe^{-x} + x^2(-e^{-x}) = (2x - x^2)e^{-x}. \end{aligned}$$

(b) If $y = \dfrac{e^{2x}}{2x+1}$, then the quotient rule gives

$$\begin{aligned} \frac{dy}{dx} &= \frac{(D_x\, e^{2x})(2x+1) - (e^{2x})D_x(2x+1)}{(2x+1)^2} \\ &= \frac{(2e^{2x})(2x+1) - (e^{2x})(2)}{(2x+1)^2} = \frac{4xe^{2x}}{(2x+1)^2}. \end{aligned}$$

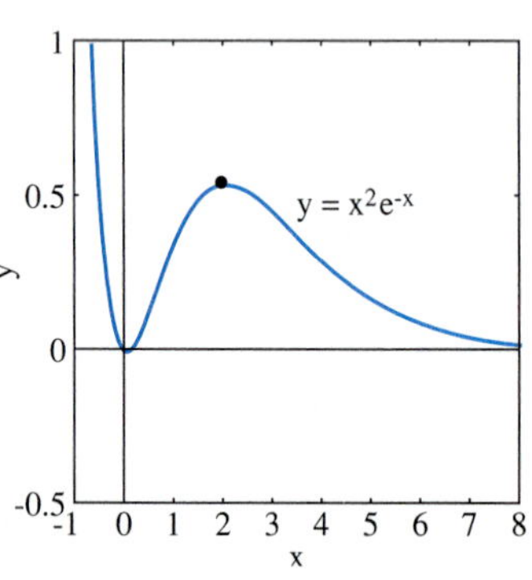

Fig. 7.1.11 The graph of Example 4

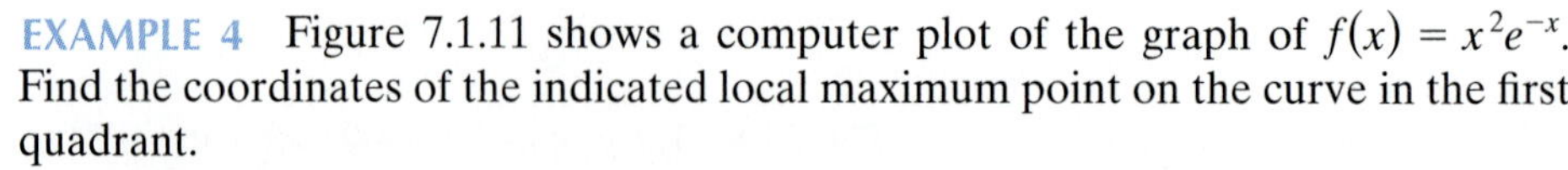

EXAMPLE 4 Figure 7.1.11 shows a computer plot of the graph of $f(x) = x^2e^{-x}$. Find the coordinates of the indicated local maximum point on the curve in the first quadrant.

Solution The calculation in part (a) of Example 3 yields $f'(x) = 0$ when

$$(2x - x^2)e^{-x} = \frac{x(2-x)}{e^x} = 0,$$

so the only critical points of f are at $x = 0$ and $x = 2$. Thus the indicated first-quadrant critical point on the curve is $(2, f(2)) = (2, 4e^{-2}) \approx (2, 0.5413)$. ■

Logarithms and Inverse Functions

In precalculus courses, the **base *a* logarithm function** $\log_a x$ is introduced as the "opposite" of the exponential function $f(x) = a^x$ with base $a > 1$. That is, $\log_a x$ is the power to which a must be raised to get x. Thus

$$y = \log_a x \quad \text{if and only if} \quad a^y = x. \tag{10}$$

With $a = 10$, this is the base 10 *common logarithm* $\log_{10} x$.

EXAMPLE 5

$$\begin{aligned}
\log_{10} 1000 &= 3 &&\text{because} & 1000 &= 10^3;\\
\log_{10}(0.1) &= -1 &&\text{because} & 0.1 &= 10^{-1};\\
\log_2 16 &= 4 &&\text{because} & 16 &= 2^4;\\
\log_3 9 &= 2 &&\text{because} & 9 &= 3^2.
\end{aligned}$$ ■

If $y = \log_a x$, then $a^y = x > 0$. Hence it follows that

$$a^{\log_a x} = x \tag{11a}$$

and

$$\log_a(a^y) = y. \tag{11b}$$

Thus the base a exponential and logarithmic functions are natural opposites, in the sense that each undoes the result of applying the other. Apply both in succession—in either order—and you're back where you started (Fig. 7.1.12). Example 6 gives other familiar pairs of functions that are *inverses* of each other.

EXAMPLE 6 The following are pairs of *inverse functions:*

$$f(x) = x + 1 \quad \text{and} \quad g(x) = x - 1.$$

Adding 1 and subtracting 1 are inverse operations; doing either undoes the other. Next, doubling and halving are inverse operations:

$$f(x) = 2x \quad \text{and} \quad g(x) = \frac{x}{2}.$$

A function can be its own inverse:

$$f(x) = \frac{1}{x} \quad \text{and} \quad g(x) = \frac{1}{x}.$$

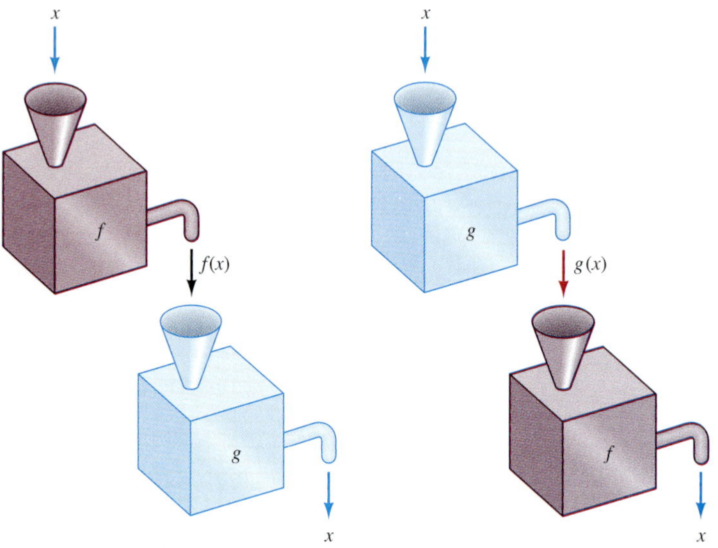

Fig. 7.1.12 Inverse functions f and g. Each undoes the effect of the other.

Squaring and taking the square root are inverse operations when only nonnegative numbers are involved:

$$\begin{cases} f\colon [0,\infty) \to [0,\infty) \\ \text{by } f(x) = x^2 \end{cases} \quad \text{and} \quad \begin{cases} g\colon [0,\infty) \to [0,\infty) \\ \text{by } g(x) = \sqrt{x}. \end{cases}$$ ■

Like $f(x) = a^x$ and $g(x) = \log_a x$, each pair f and g of functions given in Example 5 has the property that

$$f(g(x)) = x \quad \text{and} \quad g(f(x)) = x \tag{12}$$

for all values of x in the domains of g and f, respectively.

Definition *Inverse Functions*

The two functions f and g are **inverse functions,** or are **inverses** of each other, provided that

- The range of values of each function is the domain of definition of the other, and
- The relations in (12) hold for all x in the domains of g and f, respectively.

Care is required when we specify the domains of f and g to ensure that the condition in (12) is satisfied. For example, the rule $f(x) = x^2$ makes sense for all real x, but it is not the same as the function

$$f(x) = x^2, \qquad x \geqq 0,$$

which is the inverse of $g(x) = \sqrt{x}$, $x \geqq 0$. But the functions

$$f(x) = x^3 \quad \text{and} \quad g(x) = \sqrt[3]{x}$$

are inverses of each other with no restriction on their natural domains.

Because $a^x > 0$ for all x (as illustrated in Fig. 7.1.3), it follows that $\log_a x$ is defined *only* for $x > 0$. Because interchanging x and y in $a^y = x$ yields $y = a^x$, it follows from Eq. (10) that the graph of $y = \log_a x$ is the reflection in the line $y = x$ of the graph of

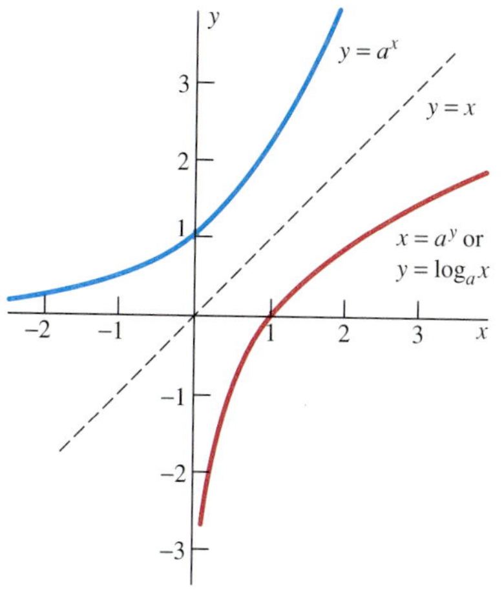

Fig. 7.1.13 The graph of $x = a^y$ is the graph of the inverse function $\log_a x$ of the exponential function a^x. The case $a > 1$ is shown here.

$y = a^x$ and therefore has the shape shown in Fig. 7.1.13. Because $a^0 = 1$, it also follows that

$$\log_a 1 = 0,$$

so the intercepts in the figure are independent of the choice of a.

We can use the inverse-function relationship between $\log_a x$ and a^x to deduce, from the laws of exponents in Eq. (2), the following **laws of logarithms:**

$$\log_a xy = \log_a x + \log_a y, \qquad \log_a \frac{1}{x} = -\log_a x,$$
$$\log_a \frac{x}{y} = \log_a x - \log_a y, \qquad \log_a x^y = y \log_a x. \tag{13}$$

We will verify these laws of logarithms in Section 7.2.

EXAMPLE 7 A typical calculation using the laws of logarithms is

$$\log_{10}\left(\tfrac{3000}{17}\right) = \log_{10}3000 - \log_{10}17 = \log_{10}(3\cdot 10^3) - \log_{10}17$$
$$= \log_{10}3 + 3\log_{10}10 - \log_{10}17 \approx 0.47712 + 3\cdot 1 - 1.23045;$$
$$\log_{10}\tfrac{3000}{17} \approx 2.24667.$$

We used a calculator to find the approximations $\log_{10}3 \approx 0.47712$ and $\log_{10}17 \approx 1.23045$. ■

Derivatives of Inverse Functions

Our interest in inverse-function pairs at this point stems from the following general principle: When we know the derivative of either of two inverse functions, then we can use the inverse-function relationship between them to *discover* the derivative of the other of the two functions. Theorem 1 is usually proved in an advanced calculus course.

Theorem 1 ***Differentiation of an Inverse Function***

Suppose that the function f is defined on the open interval I and that $f'(x) > 0$ for all x in I. Then f has an inverse function g, the function g is differentiable, and

$$g'(x) = \frac{1}{f'(g(x))} \tag{14}$$

for all x in the domain of g.

COMMENT 1 This theorem is also true when the condition $f'(x) > 0$ is replaced with the condition $f'(x) < 0$. Assuming that g is differentiable, the derivative formula in Eq. (14) can be derived by differentiating each side in the inverse-function relation

$$f(g(x)) = x.$$

When we differentiate each side, using the fact that this equation is actually an identity on some interval and using the chain rule on the left-hand side, the result is

$$f'(g(x)) \cdot g'(x) = 1.$$

When we solve this equation for $g'(x)$, the result is Eq. (14).

Comment 2 Equation (14) is easy to remember in differential notation. Let us write $x = f(y)$ and $y = g(x)$. Then $dy/dx = g'(x)$ and $dx/dy = f'(y)$. So Eq. (14) becomes the seemingly inevitable formula

$$\frac{dy}{dx} = \frac{1}{\dfrac{dx}{dy}}. \tag{15}$$

In using Eq. (15), it is important to remember that dy/dx is to be evaluated at x but that dx/dy is to be evaluated at the corresponding value of y; namely, $y = g(x)$.

The Natural Logarithm

The logarithm function $g(x) = \log_e x$ with base e is called the **natural logarithm function.** It is commonly denoted (on many calculator keys, for instance) by the special symbol ln:

$$\ln x = \log_e x. \tag{16}$$

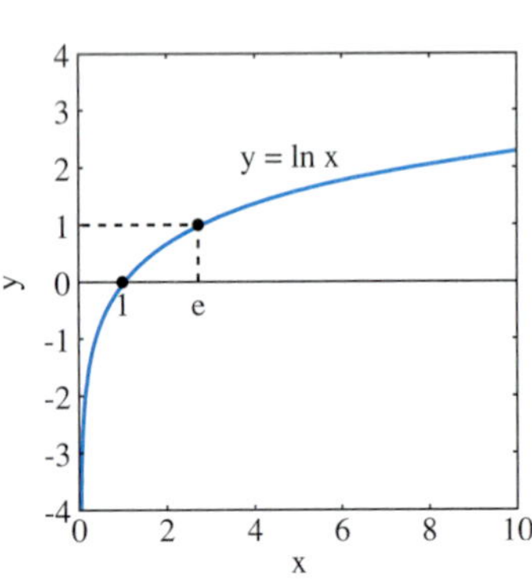

Fig. 7.1.14 The graph of the natural logarithm function

Because $e^x > 0$ for all x, it follows that $\ln x$ is defined only for $x > 0$. The graph $y = \ln x$ is shown in Fig. 7.1.14. We note that $\ln 1 = 0$ (so the graph has x-intercept $x = 1$), that $\ln e = 1$ (because $\ln e = \log_e e$), and that $\ln x$ appears to increase quite slowly when x is large.

The natural logarithm function is the inverse of the natural exponential function e^x, so

$$e^{\ln x} = x \qquad \text{for all} \quad x > 0 \tag{17a}$$

and

$$\ln(e^x) = x \qquad \text{for all } x. \tag{17b}$$

Because $D_x e^x = e^x > 0$ for all x, it therefore follows that the function $\ln x$ is differentiable. To find the derivative of the function

$$u = \ln x,$$

let us follow the prescription in Comment 1 (following Theorem 1) and begin with Eq. (17a) in the form

$$e^u = x, \quad \text{where} \quad u = \ln x.$$

Because this last equation is actually an identity (for $x > 0$), the derivative of the left-hand side with respect to x is also identically equal (for $x > 0$) to the derivative of the right-hand side with respect to x:

$$D_x(e^u) = D_x x.$$

With the aid of the chain rule in Eq. (9), we differentiate and find that

$$e^u \frac{du}{dx} = 1,$$

so it follows that

$$\frac{du}{dx} = \frac{1}{e^u} = \frac{1}{e^{\ln x}} = \frac{1}{x}.$$

Hence the derivative du/dx of the natural logarithm function $u = \ln x$ is given by

$$D_x \ln x = \frac{1}{x}. \tag{18}$$

Thus $\ln x$ is the hitherto missing function whose derivative is $1/x$.

Just as with exponentials, the derivative of a logarithm function with a base other than e involves an inconvenient numerical factor. For instance, in Section 7.4 we will see that

$$D_x \log_{10} x \approx \frac{0.4343}{x}. \tag{19}$$

The contrast between Eqs. (18) and (19) illustrates one way in which base e logarithms are "natural."

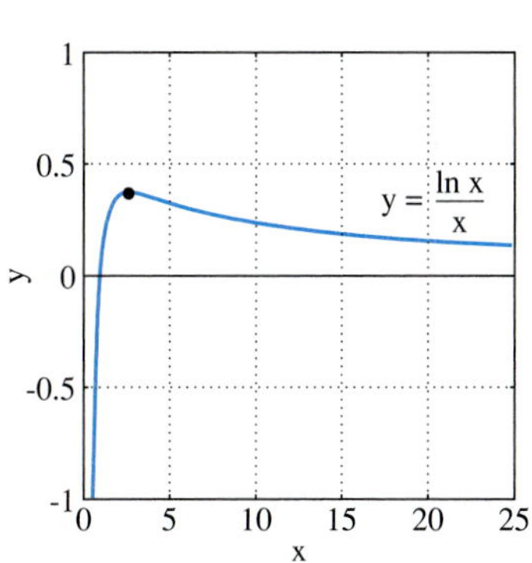

Fig. 7.1.15 The graph of Example 8

EXAMPLE 8 Figure 7.1.15 shows the graph of the function

$$f(x) = \frac{\ln x}{x}.$$

Find the coordinates of the indicated first-quadrant critical point on this curve.

Solution Equation (18) and the quotient rule yield

$$f'(x) = \frac{(D_x \ln x)(x) - (\ln x)(D_x x)}{x^2} = \frac{\frac{1}{x}\cdot x - (\ln x)\cdot 1}{x^2} = \frac{1 - \ln x}{x^2}.$$

Hence the only critical point of f occurs when $\ln x = 1$; that is, when $x = e$. Thus the critical point indicated in Fig. 7.1.15 is $(e, 1/e) \approx (2.718, 0.368)$. ■

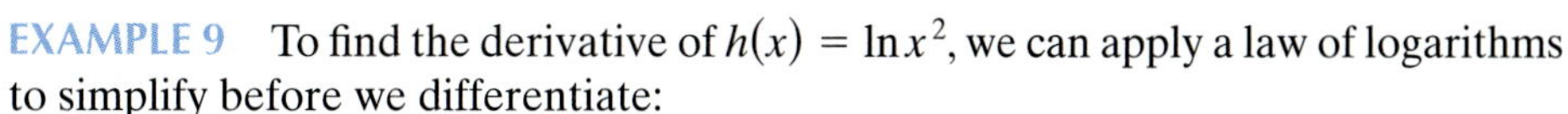

EXAMPLE 9 To find the derivative of $h(x) = \ln x^2$, we can apply a law of logarithms to simplify before we differentiate:

$$h(x) = \ln x^2 = 2 \ln x,$$

and hence

$$h'(x) = 2 \cdot D_x(\ln x) = \frac{2}{x}.$$ ■

EXAMPLE 10 Recall the bacteria population in Example 2, given by $P(t) = 2^t$ (millions) after t months. Find the rate of growth of this population when three months have elapsed, thus when $P(3) = 8$ million bacteria are present.

Solution We need to calculate the derivative $P'(t)$. For this purpose, let's write $2 = e^{\ln 2}$ (by the definition of the natural logarithm), so

$$P(t) = 2^t = (e^{\ln 2})^t = e^{t \ln 2}.$$

Then the differentiation formula $D_t e^{kt} = ke^{kt}$ with $k = \ln 2$ yields

$$P'(t) = (\ln 2)e^{t \ln 2} = 2^t \ln 2.$$

Therefore the rate of growth of the bacteria population at time $t = 3$ is

$$P'(3) = 2^3 \ln 2 = 8 \ln 2 \approx 8 \cdot (0.69315) \approx 5.54 \quad \text{(millions per month).}$$

Summary

In the somewhat informal discussion in this section, we have introduced

- The number $e \approx 2.71828$,
- The natural exponential function e^x, and
- The natural logarithm function $\ln x$.

Our investigation of the derivatives of e^x and $\ln x$—given in Eqs. (8) and (18)—should be regarded as provisional, pending a more complete discussion of these new functions in Sections 7.2 and 7.3. In any case, this brief preview of the connections between exponential functions and logarithmic functions should help you better understand the systematic development in those two sections.

We shall see in subsequent sections of this chapter that the functions e^x and $\ln x$ play a vital role in the quantitative analysis of a wide range of natural phenomena—including population growth, radioactive decay, the spread of epidemics, growth of investments, diffusion of pollutants, and motion with the effect of resistance taken into account.

7.1 PROBLEMS

Use the laws of exponents to simplify the expressions in Problems 1 and 2. Write each answer as an integer.

1. (a) $2^3 \cdot 2^4$; (b) $3^2 \cdot 3^3$; (c) $(2^2)^3$; (d) $2^{(2^3)}$; (e) $3^5 \cdot 3^{-5}$
2. (a) $10^{10} \cdot 10^{-10}$; (b) $(2^{12})^{1/3}$; (c) $(3^6)^{1/2}$; (d) $4^5 \cdot 2^{-6}$; (e) $6^5 \cdot 3^{-5}$

Note that $\log_{10} 100 = \log_{10} 10^2 = 2 \log_{10} 10 = 2$. Use a similar technique to simplify and evaluate (without the use of a calculator) the expressions in Problems 3 and 4.

3. (a) $\log_2 16$; (b) $\log_3 27$; (c) $\log_5 125$
4. (a) $\log_7 49$; (b) $\log_{10} 1000$; (c) $\log_{12} 144$

Use the laws of logarithms to express the natural logarithms in Problems 5 through 8 in terms of the three numbers $\ln 2$, $\ln 3$, and $\ln 5$.

5. (a) $\ln 8$; (b) $\ln 9$; (c) $\ln 6$
6. (a) $\ln 15$; (b) $\ln 72$; (c) $\ln 200$
7. (a) $\ln \frac{8}{27}$; (b) $\ln \frac{12}{25}$
8. (a) $\ln \frac{27}{40}$; (b) $\ln \frac{1}{90}$
9. Which is larger, $2^{(3^4)}$ or $(2^3)^4$?
10. Evaluate $\log_{0.5} 16$.
11. Find by inspection two values of x such that $x^2 = 2^x$. (Examine Fig. 7.1.5 if you wish.)
12. Suppose that the graphs of $f(x) = x^2$ and $g(x) = 2^x$ in Fig. 7.1.6 are drawn on a scale such that the unit of measurement (on both coordinate axes) is 1 cm. Then show that at a distance of 1/2 meter to the right of the origin, the height of the graph $y = f(x)$ is 25 meters above the x-axis but the height of $y = g(x)$ is over 10 billion kilometers.

In Problems 13 through 16, graph the given functions on the same calculator or computer screen. Then describe any relationships you notice among their graphs.

13. $y = 2^x$, $y = e^x$, $y = 3^x$
14. $y = 2^{-x}$, $y = e^{-x}$, $y = 3^{-x}$
15. $y = 4^x$, $y = (\frac{1}{4})^x$
16. $y = (0.2)^x$, $y = (0.4)^x$, $y = (0.6)^x$

In Problems 17 through 26, solve for x without using a calculator.

17. $2^x = 64$
18. $10^{-x} = 0.001$
19. $10^{-x} = 100$
20. $(3^x)^2 = 81$
21. $x^x = x^2$ (Find *all* solutions.)
22. $\log_x 16 = 2$
23. $\log_3 x = 4$
24. $e^{5x} = 7$
25. $3e^x = 3$
26. $2e^{-7x} = 5$

In Problems 27 through 44, find dy/dx.

27. $y = xe^x$

28. $y = x^3 e^x$

29. $y = \sqrt{x}\, e^x$

30. $y = \dfrac{e^x}{x}$

31. $y = \dfrac{e^x}{x^2}$

32. $y = \dfrac{e^x}{\sqrt{x}}$

33. $y = x \ln x$

34. $y = x^2 \ln x$

35. $y = \sqrt{x} \ln x$

36. $y = \dfrac{\ln x}{\sqrt{x}}$

37. $y = \dfrac{x}{e^x}$

38. $y = e^x \ln x$

39. $y = \ln x^3$ [This means $\ln(x^3)$.]

40. $y = \ln \sqrt[3]{x}$

41. $y = \sqrt{e^x - e^{-x}}$

42. $y = e^{2x}\sqrt{x+1}$

43. $y = e^{3x} \cos 4x$

44. $y = e^{3x} \ln 4x$

In Problems 45 through 48, write an equation of line tangent to the given curve at the indicated point.

45. $y = xe^{2x}$ at the point $(1, e^2)$

46. $y = e^{2x} \cos x$ at the point $(0, 1)$

47. $y = x^3 \ln x$ at the point $(1, 0)$

48. $y = \dfrac{\ln x}{x^2}$ at the point (e, e^{-2})

In Problems 49 and 50, calculate the first five or six derivatives of the given function $f(x)$. Then deduce a likely formula for the nth derivative $f^{(n)}(x)$.

49. $f(x) = e^{2x}$

50. $f(x) = xe^x$

Like the functions $f(x) = x^2 e^{-x}$ and $f(x) = (\ln x)/x$ graphed in Figs. 7.1.11 and 7.1.15, respectively, each of the functions in Problems 51 and 52 has an absolute maximum for $x > 0$. Find it.

51. $f(x) = (x - 1)e^{-x}$

52. $f(x) = (x - 3)e^{-x/2}$

Like the function graphed in Fig. 7.1.16, each of the functions in Problems 53 and 54 has both an absolute maximum value and an absolute minimum value for $x > 0$. Find them.

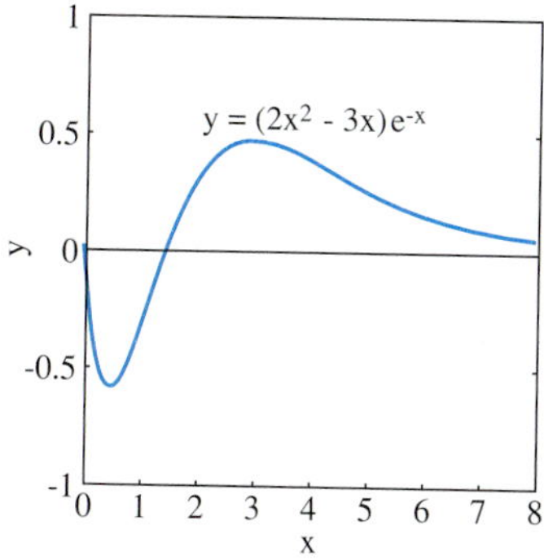

Fig. 7.1.16 The graph for Problem 53

53. $f(x) = (2x^2 - 3x)e^{-x}$

54. $f(x) = (6x^2 - 5x)e^{-x}$

55. Figure 7.1.17 shows the graph of the function $f(x) = e^{-x/6} \sin x$, together with the graphs of its "envelope curves" $y = e^{-x/6}$ and $y = -e^{-x/6}$. Find the first local maximum point and the first local minimum point on the graph of f for $x > 0$.

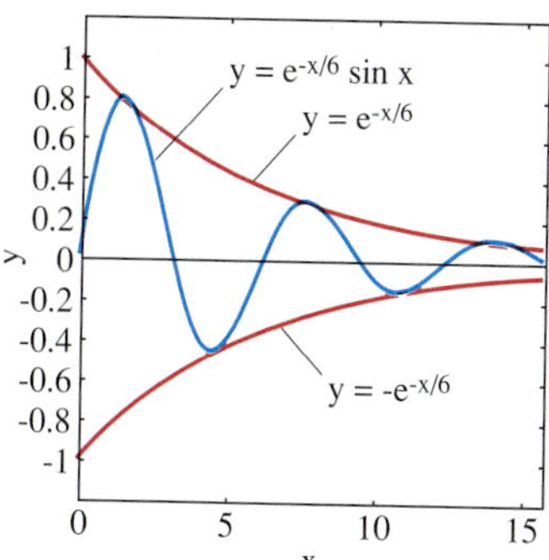

Fig. 7.1.17 The graph for Problems 55 and 56

56. Find the first two points of tangency of the curve $y = e^{-x/6} \sin x$ with the two envelope curves shown in Fig. 7.1.17. Are these the same as the two local extreme points found in Problem 55?

57. Suppose that a population of bacteria originally (at time $t = 0$) numbers 2 million and triples every hour, so the population is $P(t) = 2 \cdot 3^t$ (million) bacteria after t hours. (a) Find how long it takes this bacteria population to quadruple (to 8 million). (b) What is the rate of growth of this population after 4 hours?

58. Suppose that a rabbit population starts (at time $t = 0$) with 4 rabbits and doubles every three months, so the population is $P(t) = 4 \cdot 2^{t/3}$ rabbits after t months. (a) Find how long it takes this rabbit population to triple. (b) What is the rate of growth of this population after 1 year?

59. If you deposit \$1000 in a savings account that earns interest compounded continuously with an annual yield of 8%, then the amount in the account after t years is $A(t) = 1000 \cdot (1.08)^t$. (a) How long will it take your account to double its value? (b) At what rate is the amount in this account growing after five years?

60. Suppose that a nuclear accident has left the radiation level on campus at three times the safe level for human occupation. If thereafter it is observed that the radiation level declines by 10% each month, then the radiation level after t months will be given by $R(t) = 3 \cdot (0.9)^t$ (in "safe units"). How long will it take for the radiation level to decline to 1 safe unit, so students and faculty can return to campus?

61. The **LOG** key on the typical calculator denotes the base 10 logarithm $f(x) = \log_{10} x$. (a) Use the definition of the derivative to show that

$$f'(1) = \lim_{h \to 0} \log_{10}(1 + h)^{1/h}.$$

(b) Investigate the limit in (a) numerically to show that $f'(1) \approx 0.4343$.

62. The object of this problem is to differentiate the base 10 logarithm function of Problem 61. (a) First use the known formula for $D_x e^u$ to show that $D_x 10^x = 10^x \ln 10$. (b) Conclude from the chain rule that

$$D_x 10^u = 10^u(\ln 10)\frac{du}{dx}.$$

(c) Substitute $u = \log_{10} x$ in the inverse function identity $10^{\log_{10} x} = x$ and then differentiate using the result of part (b) to conclude that

$$D_x \log_{10} x = \frac{1}{x \ln 10} \approx \frac{0.4343}{x},$$

consistent with the result (for $x = 1$) of Problem 61.

63. Show that the number $\log_2 3$ is irrational. (*Suggestion:* Assume to the contrary that $\log_2 3 = p/q$, where p and q are positive integers, and then express the consequence of this assumption in exponential form. Under what circumstances can an integral power of 2 equal an integral power of 3?)

7.1 PROJECT: DISCOVERING THE NUMBER *e* FOR YOURSELF

You can investigate numerically the limit

$$m(a) = \lim_{h \to 0} \frac{a^h - 1}{h}$$

in Eq. (5) by using a calculator or computer with which the function

$$\phi(h) = \frac{a^h - 1}{h}$$

(with a fixed) can be defined and rapidly calculated for successively smaller values of h.

For instance, calculate $\phi(h)$ with $a = 2$ and with $a = 3$ for $h = 0.1, 0.01, 0.001, \ldots$, and conclude that

$$m(2) \approx 0.6931 < 1,$$

whereas

$$m(3) \approx 1.0986 > 1.$$

It follows that the mysterious number e for which

$$m(e) = 1$$

lies somewhere between 2 and 3.

Indeed, interpolation between the values $m(2) \approx 0.6931$ and $m(3) \approx 1.0986$ suggests that $e \approx 2.7$ or $e \approx 2.8$. Investigate the values $m(2.7)$ and $m(2.8)$ to verify the entries shown in Fig. 7.1.18.

Continue in this way to close in on the number e. Don't quit until you're convinced that $e \approx 2.718$ accurate to three decimal places.

a	$m(a)$
2	0.6931
2.7	0.9933
↓	↓
e	1.0000
↑	↑
2.8	1.0296
3	1.0986

Fig. 7.1.18 Closing in on the number e

7.2 THE NATURAL LOGARITHM

Section 7.1 was an informal overview of exponential and logarithmic functions. We now present a more systematic development of the properties of these functions.

It is simplest to make the definition of the natural logarithm our starting point. Guided by the results in Section 7.1, we want to define $\ln x$ for $x > 0$ in such a way that

$$\ln 1 = 0 \quad \text{and} \quad D_x \ln x = \frac{1}{x}. \tag{1}$$

To do so, we recall part 1 of the fundamental theorem of calculus (Section 5.6), according to which

$$D_x \int_a^x f(t)\,dt = f(x)$$

if f is continuous on an interval that contains a and x. In order that $\ln x$ satisfy the equations in (1), we take $a = 1$ and $f(t) = 1/t$.

Definition ***The Natural Logarithm***

The **natural logarithm** $\ln x$ of the positive number x is defined to be

$$\ln x = \int_1^x \frac{1}{t}\,dt. \tag{2}$$

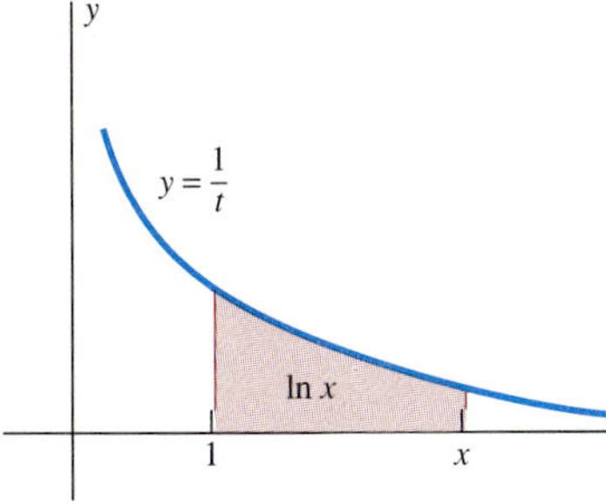

Fig. 7.2.1 The natural logarithm function defined by means of an integral

Note that $\ln x$ is *not* defined for $x \leqq 0$. Geometrically, $\ln x$ is the area under the graph of $y = 1/t$ from $t = 1$ to $t = x$ if $x > 1$ (Fig. 7.2.1), the negative of this area if $0 < x < 1$, and 0 if $x = 1$. That $D_x \ln x = 1/x$ follows immediately from the fundamental theorem of calculus. And by Theorem 2 in Section 3.4, that the function $\ln x$ is differentiable for $x > 0$ implies that it is continuous for $x > 0$.

The Graph $y = \ln x$

Because $D_x \ln x = 1/x$ is positive for $x > 0$, we see that $\ln x$ must be an increasing function. Also, because its second derivative

$$D_x^2 \ln x = D_x\left(\frac{1}{x}\right) = -\frac{1}{x^2}$$

is negative for $x > 0$, it follows from Theorem 2 in Section 4.6 that the graph of $y = \ln x$ is everywhere concave downward. Later in this section we shall use the laws of logarithms to show that

$$\lim_{x \to 0^+} \ln x = -\infty \tag{3}$$

and that

$$\lim_{x \to +\infty} \ln x = +\infty. \tag{4}$$

When we assemble all these facts, we see that the graph of the equation $y = \ln x$ has the shape shown in Fig. 7.2.2.

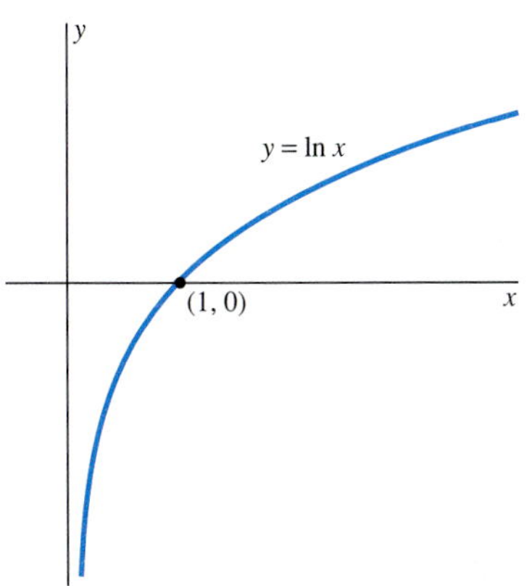

Fig. 7.2.2 The graph of the natural logarithm function

Derivatives and Integrals Involving Logarithms

When we combine the formula $D_x \ln x = 1/x$ with the chain rule, we obtain the differentiation formula

$$D_x \ln u = \frac{D_x u}{u} = \frac{1}{u} \cdot \frac{du}{dx}. \tag{5}$$

Here, $u = u(x)$ denotes a positive-valued differentiable function of x.

EXAMPLE 1 If $y = \ln(x^2 + 1)$, then Eq. (5) with $u = x^2 + 1$ gives

$$\frac{dy}{dx} = \frac{D_x(x^2 + 1)}{x^2 + 1} = \frac{2x}{x^2 + 1}.$$

If $y = \ln\sqrt{x^2 + 1}$, then with $u = \sqrt{x^2 + 1}$ we obtain

$$\frac{dy}{dx} = \frac{D_x\sqrt{x^2 + 1}}{\sqrt{x^2 + 1}} = \frac{1}{\sqrt{x^2 + 1}} \cdot \frac{x}{\sqrt{x^2 + 1}} = \frac{x}{x^2 + 1}.$$

These two derivatives are consistent with the fact that

$$\ln\sqrt{x^2 + 1} = \tfrac{1}{2}\ln(x^2 + 1).$$

EXAMPLE 2 To differentiate $f(x) = (\ln x)^2$, we can use the chain rule with $u = \ln x$. This gives

$$f'(x) = D_x(\ln x)^2 = D_x(u^2) = 2u\frac{du}{dx} = 2(\ln x)D_x \ln x = \frac{2}{x}\ln x.$$

The function $f(x) = \ln|x|$ is defined for all $x \neq 0$. If $x > 0$, then $|x| = x$ and, in this case,

$$f'(x) = D_x \ln x = \frac{1}{x}.$$

But if $x < 0$, then $|x| = -x$, so

$$f'(x) = D_x \ln(-x) = \frac{-1}{-x} = \frac{1}{x}.$$

[To compute the derivative of $\ln(-x)$, we used Eq. (5) with $u = -x$.] Thus we have shown that

$$D_x \ln|x| = \frac{1}{x} \qquad (x \neq 0) \tag{6}$$

whether x is positive or is negative.

When we combine Eq. (6) with the chain rule, we obtain the formula

$$D_x \ln|u| = \frac{D_x u}{u} = \frac{1}{u} \cdot \frac{du}{dx}, \tag{7}$$

which is valid wherever the function $u = u(x)$ is both differentiable and nonzero.

Equation (7) is equivalent to the integral formula

$$\int \frac{u'(x)}{u(x)}\,dx = \ln|u(x)| + C,$$

or simply

$$\int \frac{du}{u} = \ln|u| + C. \tag{8}$$

EXAMPLE 3 If $x > 0$, then Eq. (8) with $u = 4x + 3$, $du = 4\,dx$ yields

$$\int \frac{dx}{4x+3} = \frac{1}{4}\int \frac{4\,dx}{4x+3} = \frac{1}{4}\int \frac{du}{u}$$

$$= \frac{1}{4}\ln|u| + C = \frac{1}{4}\ln|4x+3| + C = \frac{1}{4}\ln(4x+3) + C.$$

■

EXAMPLE 4 With $u = \cos x$, Eq. (7) yields

$$D_x \ln|\cos x| = \frac{D_x \cos x}{\cos x} = \frac{-\sin x}{\cos x} = -\tan x.$$

It follows that

$$\int \tan x\,dx = -\ln|\cos x| + C = \ln|\sec x| + C.$$

■

EXAMPLE 5 With $u = x^2 - 1$, Eq. (8) gives

$$\int \frac{2x}{x^2+1}\,dx = \ln|x^2 - 1| + C = \begin{cases} \ln(x^2-1) + C & \text{if } |x| > 1, \\ \ln(1 - x^2) + C & \text{if } |x| < 1. \end{cases}$$

Note the dependence on whether $|x| > 1$ or $|x| < 1$. For example,

$$\int_2^4 \frac{2x}{x^2-1}\,dx = \Big[\ln(x^2-1)\Big]_2^4 = \ln 15 - \ln 3 = \ln 5,$$

but

$$\int_{1/3}^{3/4} \frac{2x}{x^2-1}\,dx = \Big[\ln(1-x^2)\Big]_{1/3}^{3/4} = \ln\frac{7}{16} - \ln\frac{8}{9} = \ln\frac{63}{128}.$$

■

The Laws of Logarithms

We now use our ability to differentiate logarithms to establish rigorously the laws of logarithms.

Theorem 1 *Laws of Logarithms*

If x and y are positive numbers and r is a rational number, then

$$\ln xy = \ln x + \ln y; \qquad \textbf{(9)}$$

$$\ln\left(\frac{1}{x}\right) = -\ln x; \qquad \textbf{(10)}$$

$$\ln\left(\frac{x}{y}\right) = \ln x - \ln y; \qquad \textbf{(11)}$$

$$\ln(x^r) = r\ln x. \qquad \textbf{(12)}$$

The restriction that r is rational is removed in Section 7.4.

PROOF OF EQUATION (9) We temporarily fix y, so that we may regard x as the independent variable and y as a constant in what follows. Then

$$D_x \ln xy = \frac{D_x(xy)}{xy} = \frac{y}{xy} = \frac{1}{x} = D_x \ln x.$$

Thus $\ln xy$ and $\ln x$ have the same derivative with respect to x. We antidifferentiate both and conclude that

$$\ln xy = \ln x + C$$

for some constant C. To evaluate C, we substitute $x = 1$ into both sides of the last equation. The fact that $\ln 1 = 0$ then implies that $C = \ln y$, and this is enough to establish Eq. (9). ■

PROOF OF EQUATION (10) We differentiate $\ln(1/x)$:

$$D_x\left(\ln \frac{1}{x}\right) = \frac{-\dfrac{1}{x^2}}{\dfrac{1}{x}} = -\frac{1}{x} = D_x(-\ln x).$$

Thus $\ln(1/x)$ and $-\ln x$ have the same derivative. Hence antidifferentiation gives

$$\ln\left(\frac{1}{x}\right) = -\ln x + C,$$

where C is a constant. We substitute $x = 1$ into this last equation. Because $\ln 1 = 0$, it follows that $C = 0$, and this proves Eq. (10). ■

PROOF OF EQUATION (11) Because $x/y = x \cdot (1/y)$, Eq. (11) follows immediately from Eqs. (9) and (10). ■

PROOF OF EQUATION (12) We know that $D_x x^r = rx^{r-1}$ if r is rational. So

$$D_x(\ln x^r) = \frac{rx^{r-1}}{x^r} = \frac{r}{x} = D_x(r \ln x).$$

Antidifferentiation then gives

$$\ln(x^r) = r \ln x + C$$

for some constant C. As before, substitution of $x = 1$ then gives $C = 0$, which proves Eq. (12). We show in Section 7.4 that Eq. (12) holds whether or not r is rational. ■

The proofs of Eqs. (9), (10), and (12) are all quite similar—we differentiate the left-hand side, apply the fact that two functions with the same derivative (on an interval) differ by a constant C (on that interval), and evaluate C by using the fact that $\ln 1 = 0$.

The laws of logarithms can often be used to simplify an expression prior to differentiating it, as in Examples 6 and 7.

EXAMPLE 6 If $f(x) = \ln(3x + 1)^{17}$, then

$$f'(x) = D_x \ln(3x + 1)^{17} = D_x[17 \ln(3x + 1)] = 17 \cdot \frac{3}{3x + 1} = \frac{51}{3x + 1}.$$ ■

EXAMPLE 7 Find dy/dx given

$$y = \ln \frac{\sqrt{x^2+1}}{\sqrt[3]{x^3+1}}.$$

Solution Immediate differentiation would require us to use the quotient rule and the chain rule (*several* times), all the while working with a complicated fraction. Our work is greatly eased if we first use the laws of logarithms to simplify the formula for y:

$$y = \ln[(x^2+1)^{1/2}] - \ln[(x^3+1)^{1/3}] = \tfrac{1}{2}\ln(x^2+1) - \tfrac{1}{3}\ln(x^3+1).$$

Finding dy/dx is now easy:

$$\frac{dy}{dx} = \frac{1}{2}\cdot\frac{2x}{x^2+1} - \frac{1}{3}\cdot\frac{3x^2}{x^3+1} = \frac{x}{x^2+1} - \frac{x^2}{x^3+1}.$$

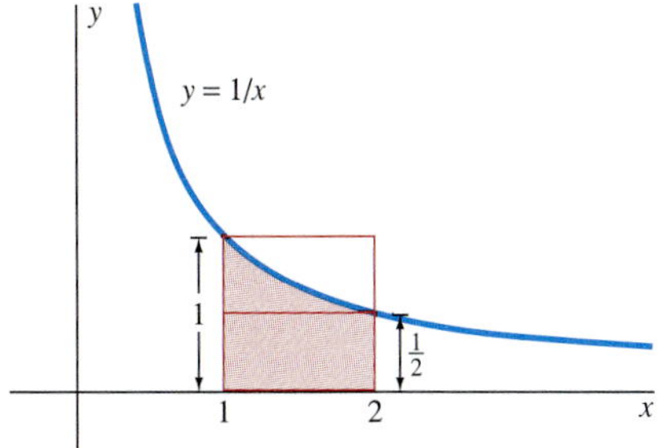

Fig. 7.2.3 Using rectangles to estimate ln 2

Order of Magnitude and Limits Involving $\ln x$

Now we establish the limits of $\ln x$ as $x \to 0^+$ and as $x \to +\infty$, as stated in Eqs. (3) and (4). Because $\ln 2$ is the area under $y = 1/x$ from $x = 1$ to $x = 2$, we can inscribe and circumscribe a pair of rectangles (Fig. 7.2.3) and conclude that

$$\tfrac{1}{2} < \ln 2 < 1.$$

Given a large positive number x, let n be the largest integer such that $x > 2^n$. Then, because $\ln x$ is an increasing function, the law of logarithms in Eq. (12) gives

$$\ln x > \ln(2^n) = n\ln 2 > \frac{n}{2}.$$

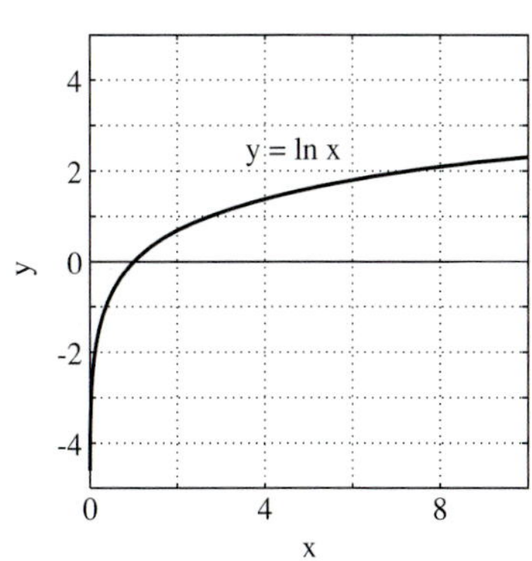

Fig. 7.2.4 $y = \ln x$ for $x \leqq 10$

Because $n \to \infty$ as $x \to \infty$, it follows that we can make $\ln x$ as large as we please simply by choosing x to be sufficiently large. This proves Eq. (4):

$$\lim_{x\to+\infty} \ln x = +\infty.$$

To prove Eq. (3), we use the law of logarithms in Eq. (10) to write

$$\lim_{x\to 0^+} \ln x = -\lim_{x\to 0^+} \ln\left(\frac{1}{x}\right) = -\lim_{y\to+\infty} \ln y = -\infty$$

by taking $y = 1/x$ and applying the result in Eq. (4).

The graphs of the natural logarithm function shown in Figs. 7.2.4 through 7.2.6 indicate that, although $\ln x \to +\infty$ as $x \to +\infty$, the logarithm of x increases *very* slowly. For instance,

$$\ln(1000) \approx 6.908, \quad \ln(1{,}000{,}000) \approx 13.812, \quad \text{and} \quad \ln(1{,}000{,}000{,}000) \approx 20.723.$$

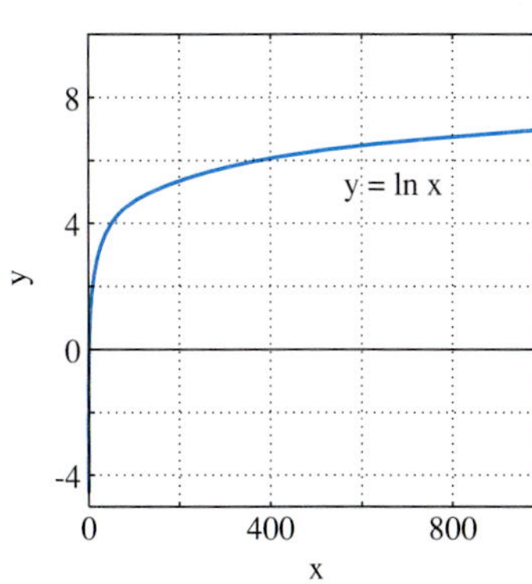

Fig. 7.2.5 $y = \ln x$ for $x \leqq 1000$

Indeed, the function $\ln x$ increases more slowly than does any positive integral power of x. By this we mean that

$$\lim_{x\to\infty} \frac{\ln x}{x^n} = 0 \qquad \textbf{(13)}$$

if n is a fixed positive integer. To prove this, note first that

$$\ln x = \int_1^x \frac{dt}{t} \leqq \int_1^x \frac{dt}{\sqrt{t}} = 2\left(\sqrt{x} - 1\right) < 2\sqrt{x}$$

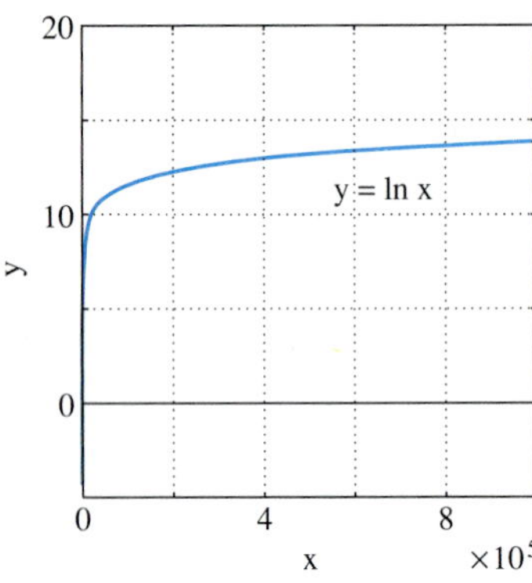

Fig. 7.2.6 $y = \ln x$ for $x \leqq 1{,}000{,}000$

because $1/t \leqq 1/\sqrt{t}$ if $t \geqq 1$. Hence if $x > 1$, then

$$0 < \frac{\ln x}{x^n} \leqq \frac{\ln x}{x} < \frac{2}{\sqrt{x}}.$$

The last expression on the right approaches zero as $x \to +\infty$. Equation (13) follows by the squeeze law for limits. Moreover, this argument shows that the positive integer n in Eq. (13) may be replaced with any rational number $k \geqq 1$. Indeed, in Problem 71 we ask you to deduce that

$$\lim_{x\to\infty} \frac{\ln x}{x^k} = 0 \tag{13$'$}$$

for *any* positive rational number k (even if $k < 1$). This limit is a precise way of saying that, *as* $x \to \infty$, $\ln x$ *approaches* $+\infty$ *more slowly than does any (fixed) positive power of* x.

Another interpretation of Eq. (13′) is that, when x is large, the *order of magnitude* of $\ln x$ is less than that of x^k. This fact may at first seem surprising when the value of k is small. For instance, Fig. 7.2.7 might suggest to the unwary that $\ln x$ is greater (rather than less) than $x^{1/10}$ when x is large. But Eq. (13) implies that the graph of $y = x^{1/10}$ must eventually overtake and recross the graph of $y = \ln x$.

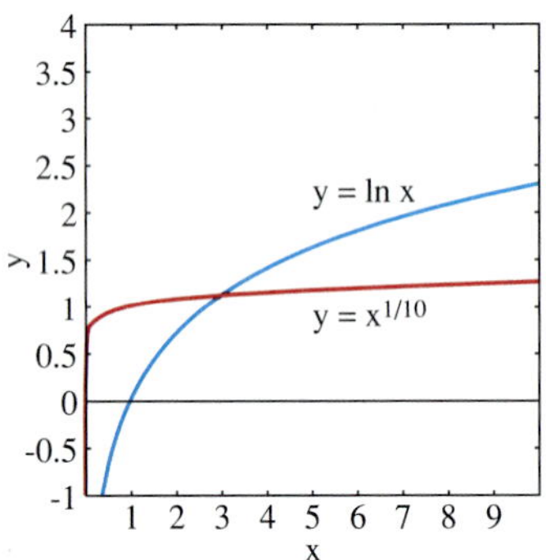

Fig. 7.2.7 Comparing $y = \ln x$ with $y = x^{1/10}$

The Number e

Because $\ln x$ is an increasing function, the intermediate value property implies that the curve $y = \ln x$ crosses the horizontal line $y = 1$ precisely once. The abscissa of the point of intersection is the important number $e \approx 2.71828$ mentioned in Section 7.1 (see Fig. 7.2.8).

Definition of e

The number e is the unique real number such that

$$\ln e = 1. \tag{14}$$

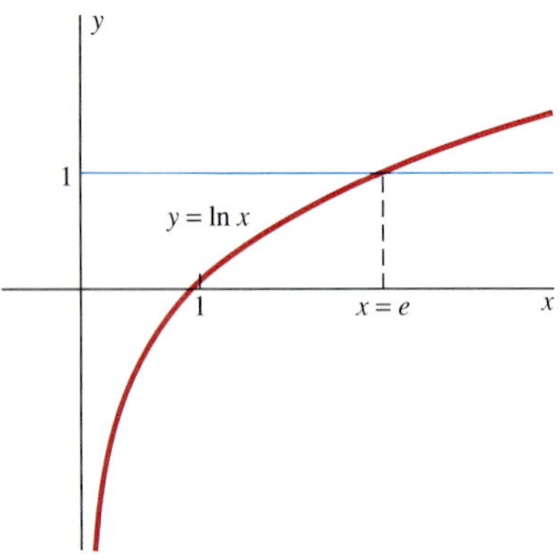

Fig. 7.2.8 The fact that $\ln e = 1$ is expressed graphically here.

The number e has been used to denote the number whose natural logarithm is 1 ever since this number was introduced by the Swiss mathematician Leonhard Euler (1707–1783), who used e for "exponential."

With a graphing calculator or computer you can zoom in on the intersection of the graphs $y = \ln x$ and $y = 1$ to verify the first few decimal places of e. For instance, the viewing window $2.71 \leqq x \leqq 2.72$, $0.99 \leqq y \leqq 1.01$ of Fig. 7.2.9 suffices to verify that $e = 2.718$ rounded to three decimal places.

EXAMPLE 8 Sketch the graph of

$$f(x) = \frac{\ln x}{x}, \qquad x > 0.$$

Solution First we compute the derivative

$$f'(x) = \frac{\frac{1}{x}\cdot x - (\ln x)\cdot 1}{x^2} = \frac{1 - \ln x}{x^2}.$$

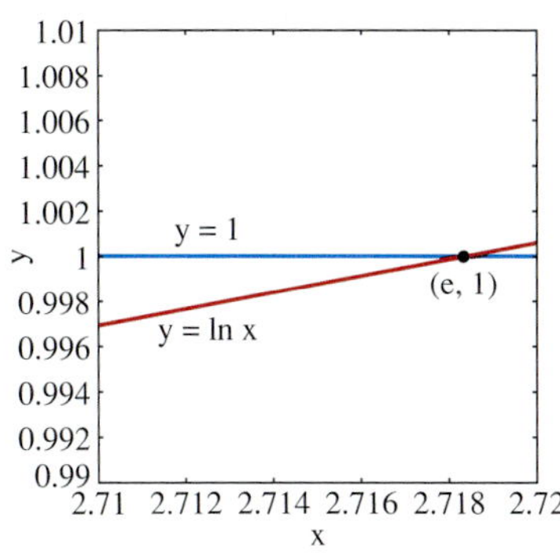

Fig. 7.2.9 Zooming in on the number e

Thus the only critical point occurs where $\ln x = 1$—that is, where $x = e$. Because $f'(x) > 0$ if $x < e$ (the same as $\ln x < 1$) and $f'(x) < 0$ if $x > e$ (the same as $\ln x > 1$), we see that f is increasing if $x < e$ and decreasing if $x > e$. Hence f has a global maximum at $x = e$.

The second derivative of f is

$$f''(x) = \frac{-\dfrac{1}{x}\cdot x^2 - (1 - \ln x)\cdot 2x}{x^4} = \frac{2\ln x - 3}{x^3},$$

so the only inflection point of the graph is where $\ln x = \frac{3}{2}$—that is, at $x = e^{3/2} \approx 4.48$.

Because

$$\lim_{x\to 0^+} \frac{\ln x}{x} = -\infty \quad \text{and} \quad \lim_{x\to\infty} \frac{\ln x}{x} = 0,$$

consequences of Eqs. (3) and (13), respectively, we conclude that the graph of f looks like Fig. 7.2.10. ■

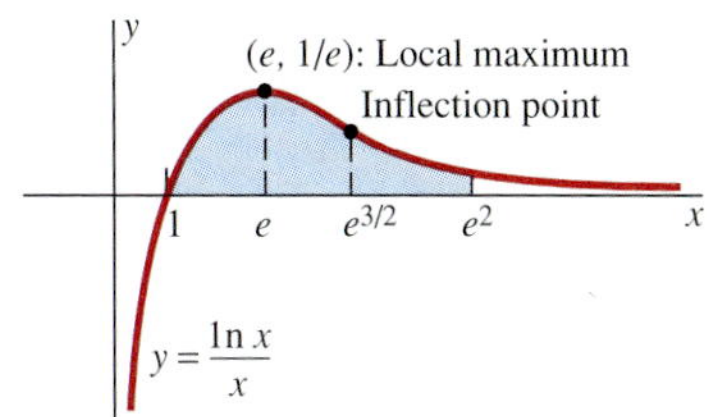

Fig. 7.2.10 The graph of Example 8

EXAMPLE 8 (continued) Now that we know what the graph $y = (\ln x)/x$ looks like, let's find the area A of the region that's shaded in Fig. 7.2.10, lying between the graph and the x-axis over the interval $1 \leqq x \leqq e^2$. With the aid of the substitution

$$u = \ln x, \qquad du = \frac{1}{x}\,dx$$

for which $u = \ln 1 = 0$ when $x = 1$ and $u = \ln e^2 = 2$ when $x = e^2$, we find that

$$A = \int_1^{e^2} \frac{\ln x}{x}\,dx = \int_0^2 u\,du = \left[\tfrac{1}{2}u^2\right]_0^2 = 2. \qquad ■$$

Logarithms and Experimental Data

Certain empirical data can be explained by assuming that the observed dependent variable is a **power function** of the independent variable x. In other words, y is described by a mathematical model of the form

$$y = kx^m,$$

where k and m are constants. If so, the laws of logarithms imply that

$$\ln y = \ln k + m\ln x.$$

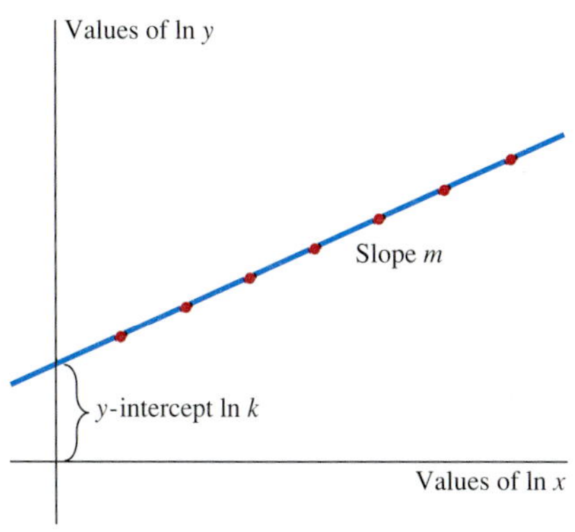

Fig. 7.2.11 Plotting the logarithms of data may reveal a hidden relationship.

An experimenter can then plot values of $\ln y$ against values of $\ln x$. If the power-function model is valid, the resulting data points will lie on a straight line of slope m and with y-intercept $\ln k$ (Fig. 7.2.11). A simple graph will exhibit plainly whether the data lie on a straight line. If they do, then we can measure the slope and y-intercept of the line and thereby determine the values of k and m.

EXAMPLE 9 *(PLANETARY MOTION)* The table in Fig. 7.2.12 gives the period of revolution T and the major semiaxis a of the elliptical orbit of each of the first six planets around the sun, together with the logarithms of these numbers. If we plot

Planet	T(in days)	a (in 10^6 km)	ln T	ln a
Mercury	87.97	58	4.48	4.06
Venus	224.70	108	5.41	4.68
Earth	365.26	149	5.90	5.00
Mars	686.98	228	6.53	5.43
Jupiter	4332.59	778	8.37	6.66
Saturn	10,759.20	1426	9.28	7.26

Fig. 7.2.12 Data for Example 9

$\ln T$ against $\ln a$, it is immediately apparent that the resulting points lie on a straight line of slope $m = \frac{3}{2}$. Hence T and a satisfy an equation of the form $T = ka^{3/2}$, so

$$T^2 = Ca^3.$$

This means that the square of the period T is proportional to the cube of the major semiaxis a. This is Kepler's third law of planetary motion, which Johannes Kepler (1571–1630) discovered empirically in 1619. ■

7.2 PROBLEMS

Differentiate the functions given in Problems 1 through 18.

1. $f(x) = \ln(3x - 1)$

2. $f(x) = \ln(4 - x^2)$

3. $f(x) = \ln\sqrt{1 + 2x}$

4. $f(x) = \ln[(1 + x)^2]$

5. $f(x) = \ln\sqrt[3]{x^3 - x}$

6. $f(x) = \ln(\sin^2 x)$

7. $f(x) = \cos(\ln x)$

8. $f(x) = (\ln x)^3$

9. $f(x) = \dfrac{1}{\ln x}$

10. $f(x) = \ln(\ln x)$

11. $f(x) = \ln(x\sqrt{x^2 + 1})$

12. $g(t) = t^{3/2}\ln(t + 1)$

13. $f(x) = \ln(\cos x)$

14. $f(x) = \ln(2\sin x)$

15. $f(t) = t^2\ln(\cos t)$

16. $f(x) = \sin(\ln 2x)$

17. $g(t) = t(\ln t)^2$

18. $g(t) = \sqrt{t}\,[\cos(\ln t)]^2$

In Problems 19 through 28, apply laws of logarithms to simplify the given function; then write its derivative.

19. $f(x) = \ln[(2x + 1)^3(x^2 - 4)^4]$

20. $f(x) = \ln\sqrt{\dfrac{1 - x}{1 + x}}$

21. $f(x) = \ln\sqrt{\dfrac{4 - x^2}{9 + x^2}}$

22. $f(x) = \ln\dfrac{\sqrt{4x - 7}}{(3x - 2)^3}$

23. $f(x) = \ln\dfrac{x + 1}{x - 1}$

24. $f(x) = x^2\ln\dfrac{1}{2x + 1}$

25. $g(t) = \ln\dfrac{t^2}{t^2 + 1}$

26. $f(x) = \ln\dfrac{\sqrt{x + 1}}{(x - 1)^3}$

27. $f(x) = \ln\dfrac{\sin x}{x}$

28. $f(x) = \ln\dfrac{\sin x}{\cos x}$

In Problems 29 through 32, find dy/dx by implicit differentiation.

29. $y = x\ln y$

30. $y = (\ln x)(\ln y)$

31. $xy = \ln(\sin y)$

32. $xy + x^2(\ln y)^2 = 4$

Evaluate the indefinite integrals in Problems 33 through 50.

33. $\displaystyle\int \frac{dx}{2x - 1}$

34. $\displaystyle\int \frac{dx}{3x + 5}$

35. $\displaystyle\int \frac{x}{1 + 3x^2}\,dx$

36. $\displaystyle\int \frac{x^2}{4 - x^3}\,dx$

37. $\displaystyle\int \frac{x + 1}{2x^2 + 4x + 1}\,dx$

38. $\displaystyle\int \frac{\cos x}{1 + \sin x}\,dx$

39. $\displaystyle\int \frac{1}{x}(\ln x)^2\,dx$

40. $\displaystyle\int \frac{1}{x\ln x}\,dx$

41. $\displaystyle\int \frac{1}{x + 1}\,dx$

42. $\displaystyle\int \frac{x}{1 - x^2}\,dx$

43. $\displaystyle\int \frac{2x + 1}{x^2 + x + 1}\,dx$

44. $\displaystyle\int \frac{x + 1}{x^2 + 2x + 3}\,dx$

45. $\displaystyle\int \frac{\ln x}{x}\,dx$

46. $\displaystyle\int \frac{\ln(x^3)}{x}\,dx$

47. $\displaystyle\int \frac{\sin 2x}{1 - \cos 2x}\,dx$

48. $\displaystyle\int \frac{dx}{x(\ln x)^2}$

49. $\displaystyle\int \frac{x^2 - 2x}{x^3 - 3x^2 + 1}\,dx$

50. $\displaystyle\int \frac{dx}{\sqrt{x}(1 + \sqrt{x})}$ (*Suggestion:* Let $y = 1 + \sqrt{x}$.)

Apply Eq. (13) to evaluate the limits in Problems 51 through 56.

51. $\displaystyle\lim_{x\to\infty} \frac{\ln\sqrt{x}}{x}$

52. $\displaystyle\lim_{x\to\infty} \frac{\ln(x^3)}{x^2}$

53. $\displaystyle\lim_{x\to\infty} \frac{\ln x}{\sqrt{x}}$ (*Suggestion:* Substitute $x = u^2$.)

54. $\lim_{x\to 0^+} x \ln x$ (*Suggestion:* Substitute $x = 1/u$.)

55. $\lim_{x\to 0^+} \sqrt{x}\ln x$ **56.** $\lim_{x\to\infty} \dfrac{(\ln x)^2}{x}$

57. Use the method of Example 4 to deduce that

$$\int \cot x\, dx = \ln|\sin x| + C.$$

58. Find a formula for $f^{(n)}(x)$, given $f(x) = \ln x$ (n denotes a positive integer).

59. The heart rate R (in beats per minute) and weight W (in pounds) of various mammals were measured, with the results shown in Fig. 7.2.13. Use the method of Example 9 to find a relation between the two of the form $R = kW^m$.

W	25	67	127	175	240	975
R	131	103	88	81	75	53

Fig. 7.2.13 Data for Problem 59

60. During the adiabatic expansion of a certain diatomic gas, its volume V (in liters) and pressure p (in atmospheres) were measured, with the results shown in Fig. 7.2.14. Use the method of Example 9 to find a relation between V and p of the form $p = kV^m$.

V	1.46	2.50	3.51	5.73	7.26
p	28.3	13.3	8.3	4.2	3.0

Fig. 7.2.14 Data for Problem 60

In Problems 61 and 62, graph (on a single calculator or computer screen) the functions on either side of the given equation to locate its single positive solution. Then determine numerically the value of this solution accurate to three decimal places.

61. $\ln x = 5 - x$ **62.** $\ln x = \dfrac{1}{x}$

In Problems 63 through 66, graph (on a single calculator or computer screen) the functions on both sides of the given equation to determine how many positive solutions the equation has. Then determine numerically the value of each solution accurate to three decimal places.

63. $2\ln x = x - 2$ **64.** $4\ln x = (x-3)^2$

65. $2\ln x = 3\sin 2x$ **66.** $2\ln x = 5\cos x$

67. Find graphically the coordinates (accurate to three decimal places) of the intersection point of the graphs $y = \ln x$ and $y = x^{1/10}$ shown in Fig. 7.2.7.

68. Determine a viewing rectangle that reveals the second intersection point (with $x > 10$) of the graphs $y = \ln x$ and $y = x^{1/10}$. Then determine graphically the first three digits of the larger solution x of the equation $\ln x = x^{1/10}$ (thus writing this solution in the form $p.qr \times 10^k$).

69. Approximate numerically the area of the region that lies beneath the curve $y = 5\ln x - 2x + 3$ and above the x-axis. You will first need to estimate graphically the x-intercepts of this curve, then integrate numerically (perhaps using the numerical integration facility of your calculator or computer).

70. Approximate numerically the area of the region bounded by the curves $y = 10\ln x$ and $y = (x-5)^2$. You will first need to estimate graphically the x-coordinates of the intersection points of the two curves, then integrate numerically (perhaps using the numerical integration facility of your calculator or computer).

71. Substitute $y = x^p$ and then apply Eq. (13) to show that

$$\lim_{x\to\infty} \frac{\ln x}{x^p} = 0 \quad \text{if } 0 < p < 1.$$

72. Deduce from the result of Problem 71 that

$$\lim_{x\to\infty} \frac{(\ln x)^k}{x} = 0 \quad \text{if } k > 0.$$

73. Substitute $y = 1/x$ and then apply Eq. (13) to show that

$$\lim_{x\to 0^+} x^k \ln x = 0 \quad \text{if } k > 0.$$

Use the limits in Problems 71 through 73 to help you sketch the graphs, for $x > 0$, of the functions given in Problems 74 through 77.

74. $y = x\ln x$ **75.** $y = x^2 \ln x$

76. $y = \sqrt{x}\ln x$ **77.** $y = \dfrac{\ln x}{\sqrt{x}}$

78. Problem 26 of Section 5.9 asks you to show by numerical integration that

$$\int_1^{2.7} \frac{dx}{x} < 1 < \int_1^{2.8} \frac{dx}{x}.$$

Explain carefully why this result proves that $2.7 < e < 2.7$.

79. If n moles of an ideal gas expand at *constant* temperature T, then the pressure and volume satisfy the ideal-gas equation $pV = nRT$ (n and R are constants). With the aid of Problem 22 in Section 6.6, show that the work W done by the gas in expanding from volume V_1 to volume V_2 is

$$W = nRT \ln \frac{V_2}{V_1}.$$

80. "Gabriel's horn" is obtained by revolving around the x-axis the curve $y = 1/x$, $x \geqq 1$ (Fig. 7.2.15). Let A_b denote its surface area from $x = 1$ to $x = b$. Show that $A_b \geqq 2\pi \ln b$, so—as a consequence—$A_b \to +\infty$ as

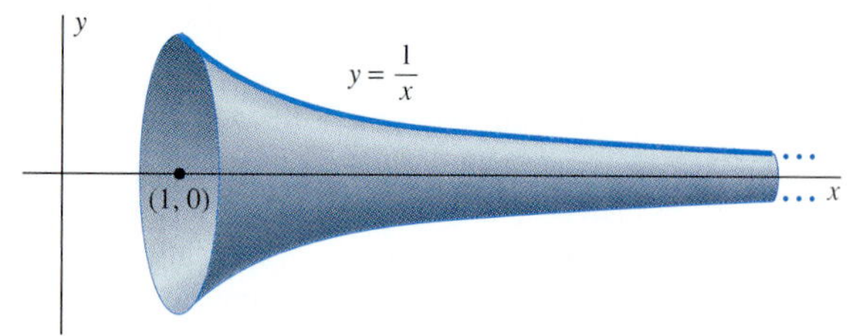

Fig. 7.2.15 Gabriel's horn (Problem 80)

$b \to +\infty$. Thus the surface area of Gabriel's horn is infinite. Is its volume finite or infinite?

81. According to the prime number theorem, which was conjectured by the great German mathematician Carl Friedrich Gauss in 1792 (when he was 15 years old) but not proved until 1896 (independently, by Jacques Hadamard and C. J. de la Vallée Poussin), the number of prime numbers between the large positive numbers a and b ($a < b$) is given to a close approximation by the integral

$$\int_a^b \frac{1}{\ln x}\,dx.$$

The midpoint and trapezoidal approximations with $n = 1$ subinterval provide an underestimate and an overestimate of the value of this integral. (Why?) Calculate these estimates with $a = 90000$ and $b = 100{,}000$. The actual number of prime numbers in this range is 879.

7.2 PROJECT: DISCOVERING THE NUMBER e BY NUMERICAL INTEGRATION

The fact that the number e satisfies (by definition) the equation

$$\int_1^e \frac{dx}{x} = 1 \tag{1}$$

suggests the possibility of using numerical integration to investigate the value of e. You can thereby verify that

$$e \approx 2.71828.$$

Your objective in this project is to bracket the number e between closer and closer numerical approximations—attempting to determine the upper limit b so that the value of the integral

$$\int_1^b \frac{dx}{x} \tag{2}$$

is as close as possible to the target value 1.

You may use whatever computational technology is available. For example, with $b = 2$ and $n = 50$, a Simpson's approximation will yield the sum 0.6931; with $b = 3$ and $n = 50$, you'll obtain the sum 1.0986. These results show that $b = 2$ is too small, and that $b = 3$ is too large, to yield 1 for the value of the integral in (2). Thus $2 < e < 3$.

Interpolation between 0.6931 and 1.0986 suggests that you try the upper limits $b = 2.6$, $b = 2.7$, and perhaps $b = 2.8$. In fact, you should obtain the sums 0.9933 and 1.0296, respectively, when you use $b = 2.7$ and $b = 2.8$ (and $n = 100$). Thus it follows that $2.7 < e < 2.8$.

As you bracket e more closely, you will need to increase the number of decimal places displayed and the number of subintervals used in Simpson's approximation to keep pace with your increasing accuracy. Continue until you have bracketed e between two seven-place approximations that both round to 2.71828.

7.3 THE EXPONENTIAL FUNCTION

We saw in Section 7.2 that the natural logarithm function $\ln x$ is continuous and increasing for $x > 0$ and that it attains arbitrarily large positive and negative values (because of the limits in Eqs. (3) and (4) of Section 7.2). It follows that $\ln x$ has an

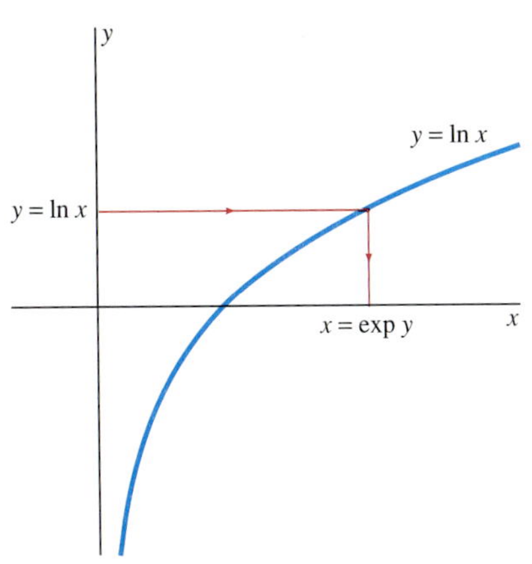

Fig. 7.3.1 To get $x = \exp y$, move straight over from y to the graph $y = \ln x$, then move straight down (or up) to x.

inverse function that is defined for all x. To see this, let y be any (fixed) real number. If a and b are positive numbers such that $\ln a < y < \ln b$, then the intermediate value property gives a number $x > 0$, with x between a and b, such that $\ln x = y$. Because $\ln x$ is an increasing function, there is only *one* such number x such that $\ln x = y$ (Fig. 7.3.1). Because y determines precisely one such value x, we see that x is a *function* of y.

This function x of y is the inverse function of the natural logarithm function, and it is called the *natural exponential function*. It is commonly denoted by exp (for *exp*onential), so

$$x = \exp y \quad \text{provided that} \quad y = \ln x.$$

Interchanging x and y yields the following definition.

Definition ***The Natural Exponential Function***

The **natural exponential function** exp is defined for all x as follows:

$$\exp x = y \quad \text{if and only if} \quad \ln y = x. \tag{1}$$

Thus $\exp x$ is simply that (positive) number whose natural logarithm is x. It is an immediate consequence of Eq. (1) that

$$\ln(\exp x) = x \quad \text{for all } x \tag{2}$$

and that

$$\exp(\ln y) = y \quad \text{for all } y > 0. \tag{3}$$

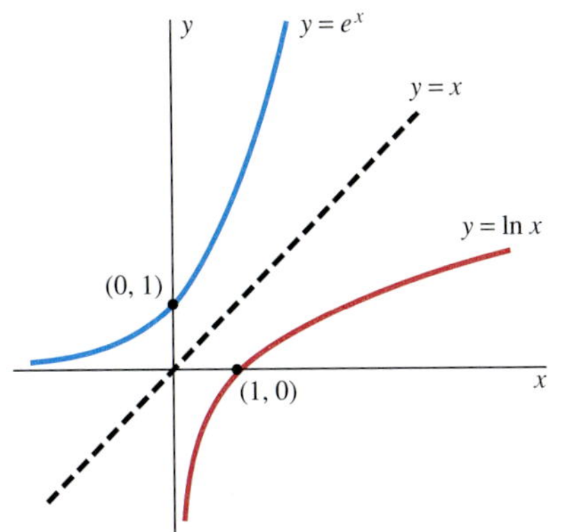

Fig. 7.3.2 The graphs $y = e^x$ and $y = \ln x$ are reflections of each other across the 45° line $y = x$.

As in the case of the graphs of $y = a^x$ and $y = \log_a x$ discussed informally in Section 7.1, the fact that $\exp x$ and $\ln x$ are inverse functions implies that the graphs of $y = \exp x$ and $y = \ln x$ are reflections of each other across the line $y = x$ (Fig. 7.3.2). Therefore the graph of the exponential function looks like the one shown in Fig. 7.3.3. In particular, $\exp x$ is positive-valued for all x, and

$$\exp 0 = 1, \tag{4}$$

$$\lim_{x\to\infty} \exp x = +\infty, \quad \text{and} \tag{5}$$

$$\lim_{x\to-\infty} \exp x = 0. \tag{6}$$

These facts follow from the equation $\ln 1 = 0$ and from the limits in Eqs. (3) and (4) of Section 7.2.

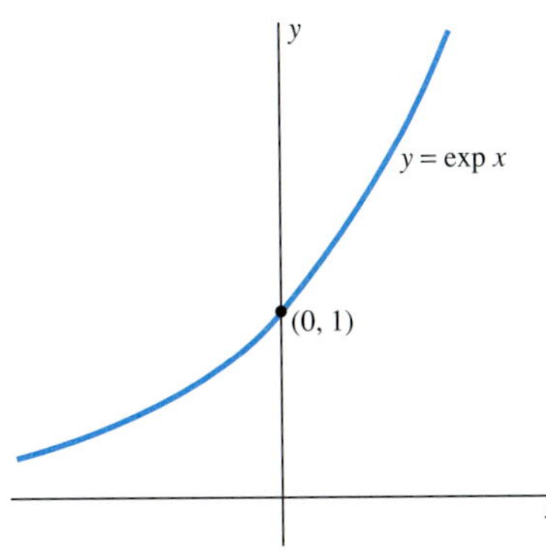

Fig. 7.3.3 The graph of the natural exponential function, exp.

Exponentials and Powers of e

Recall from Section 7.2 that we defined the number $e \approx 2.71828$ as the number whose natural logarithm is 1. If r is any rational number, it follows that

$$\ln(e^r) = r \ln e = r.$$

But Eq. (1) implies that $\ln(e^r) = r$ if and only if

$$\exp r = e^r.$$

Thus $\exp x$ is equal to e^x (e raised to the power x) if x is a rational number. We therefore *define* e^x for irrational as well as rational values of x by

$$e^x = \exp x. \tag{7}$$

This is our first instance of powers with irrational exponents.

Equation (7) is the reason for calling exp the natural exponential function. With this notation, Eqs. (1) through (3) become

$$e^x = y \quad \text{if and only if} \quad \ln y = x, \tag{8}$$

$$\ln(e^x) = x \quad \text{for all } x, \text{ and} \tag{9}$$

$$e^{\ln x} = x \quad \text{for all } x > 0. \tag{10}$$

To justify Eq. (7), we should show rigorously that powers of e satisfy the laws of exponents. We can do this immediately.

Theorem 1 ***Laws of Exponents***

If x and y are real numbers and r is rational, then

$$e^x e^y = e^{x+y}, \tag{11}$$

$$e^{-x} = \frac{1}{e^x}, \text{ and} \tag{12}$$

$$(e^x)^r = e^{rx}. \tag{13}$$

PROOF The laws of logarithms and Eq. (9) give

$$\ln(e^x e^y) = \ln(e^x) + \ln(e^y) = x + y = \ln(e^{x+y}).$$

Then Eq. (11) follows from the fact that ln is an increasing function and therefore is one-to-one—if $x_1 \neq x_2$, then $\ln x_1 \neq \ln x_2$. Similarly,

$$\ln([e^x]^r) = r\ln(e^x) = rx = \ln(e^{rx}).$$

So Eq. (13) follows in the same way. The proof of Eq. (12) is almost identical. We will see in Section 7.4 that the restriction that r is rational in Eq. (13) is unnecessary; that is,

$$(e^x)^y = e^{xy}$$

for *all* real numbers x and y. ■

Derivatives and Integrals of Exponentials

Because e^x is the inverse of the differentiable and increasing function $\ln x$, it follows from Theorem 1 of Section 7.1 that e^x is differentiable and therefore also continuous. We may thus differentiate both sides of the equation (actually, the *identity*)

$$\ln(e^x) = x$$

with respect to x. Let $u = e^x$. Then the preceding equation becomes

$$\ln u = x,$$

and the derivatives also must be equal:

$$\frac{1}{u} \cdot \frac{du}{dx} = 1 \qquad (\text{because } u > 0).$$

So

$$\frac{du}{dx} = u = e^x;$$

that is,

$$D_x\, e^x = e^x, \tag{14}$$

as we indicated in Section 7.1.

If u denotes a differentiable function of x, then Eq. (14) in combination with the chain rule gives

$$D_x\, e^u = e^u \frac{du}{dx}. \tag{15}$$

The corresponding integration formula is

$$\int e^u\, du = e^u + C. \tag{16}$$

The special case of Eq. (15) with $u = kx$ (k constant) is worth noting:

$$D_x\, e^{kx} = ke^{kx}.$$

For instance, $D_x\, e^{5x} = 5e^{5x}$.

EXAMPLE 1 Find dy/dx, given $y = e^{\sqrt{x}}$.

Solution With $u = \sqrt{x}$, Eq. (15) gives

$$\frac{dy}{dx} = e^{\sqrt{x}} D_x(\sqrt{x}) = e^{\sqrt{x}}\left(\frac{1}{2}x^{-1/2}\right) = \frac{e^{\sqrt{x}}}{2\sqrt{x}}.$$

■

EXAMPLE 2 If $y = x^2 e^{-2x^3}$, then Eq. (15) and the product rule yield

$$\begin{aligned}\frac{dy}{dx} &= (D_x\, x^2) \cdot e^{-2x^3} + x^2 \cdot D_x\left(e^{-2x^3}\right) = 2xe^{-2x^3} + x^2e^{-2x^3} D_x(-2x^3)\\ &= 2xe^{-2x^3} + x^2e^{-2x^3}(-6x^2).\end{aligned}$$

Therefore

$$\frac{dy}{dx} = (2x - 6x^4)e^{-2x^3}.$$

■

EXAMPLE 3 Find $\displaystyle\int xe^{-3x^2}\, dx$.

Solution We substitute $u = -3x^2$, so $du = -6x\,dx$. Then we have $x\,dx = -\frac{1}{6}\,du$, and hence we obtain

$$\int xe^{-3x^2}\,dx = -\frac{1}{6}\int e^u\,du = -\frac{1}{6}e^u + C = -\frac{1}{6}e^{-3x^2} + C.$$ ■

Order of Magnitude

The exponential function is remarkable for its high rate of increase with increasing x. In fact, e^x increases more rapidly as $x \to +\infty$ than *any* fixed power of x. In the language of limits,

$$\lim_{x\to\infty} \frac{x^k}{e^x} = 0 \qquad \text{for any fixed } k > 0. \tag{17}$$

Alternatively,

$$\lim_{x\to\infty} \frac{e^x}{x^k} = +\infty \qquad \text{for any fixed } k > 0. \tag{17'}$$

Because we have not yet defined x^k for irrational values of k, we prove Eq. (17) for the case in which k is rational. Once we know that (for $x > 1$) the power function x^k is an increasing function of k for all k, then the general case will follow.

We begin by taking logarithms. We find that

$$\ln\left(\frac{e^x}{x^k}\right) = x - k\ln x = \left(\frac{x}{\ln x} - k\right)\ln x.$$

Because we deduce (from Eq. (13) of Section 7.2) that $x/(\ln x) \to +\infty$ as $x \to +\infty$, this makes it clear that

$$\lim_{x\to\infty} \ln\left(\frac{e^x}{x^k}\right) = +\infty.$$

Hence $e^x/x^k \to +\infty$ as $x \to +\infty$, so we have proved both Eqs. (17) and (17′). The table in Fig. 7.3.4 illustrates the limit in Eq. (17) for the case $k = 5$. Although both $x^5 \to +\infty$ and $e^x \to +\infty$ as $x \to +\infty$, we see that e^x increases so much more rapidly than x^5 that $x^5/e^x \to 0$.

x	x^5	e^x	x^5/e^x
10	1.00×10^5	2.20×10^4	4.54×10^0
20	3.20×10^6	4.85×10^8	6.60×10^{-3}
30	2.43×10^7	1.07×10^{13}	2.27×10^{-6}
40	1.02×10^8	2.35×10^{17}	4.35×10^{-10}
50	3.13×10^8	5.18×10^{21}	6.03×10^{-14}
	$\downarrow$	$\downarrow$	$\downarrow$
	∞	∞	0

Fig. 7.3.4 Orders of magnitude of x^5 and e^x

If we write Eq. (17) in the form

$$\lim_{x\to\infty} \frac{e^{-x}}{1/x^k} = 0,$$

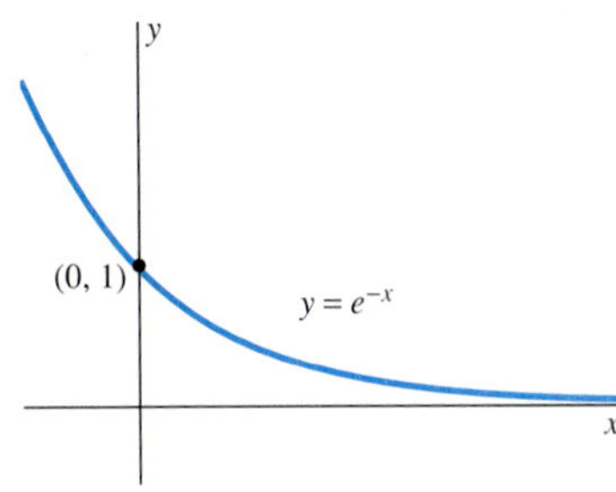

Fig. 7.3.5 The graph of $y = e^{-x}$ has the positive x-axis as an aysmptote.

we see that not only does e^{-x} approach zero as $x \to +\infty$ (Fig. 7.3.5), it does so more rapidly than any power reciprocal $1/x^k$.

We can use Eqs. (17) and (17′) to evaluate certain limits with the limit laws. For example,

$$\lim_{x\to\infty} \frac{2e^x - 3x}{3e^x + x^3} = \lim_{x\to\infty} \frac{2 - 3xe^{-x}}{3 + x^3e^{-x}} = \frac{2 - 0}{3 + 0} = \frac{2}{3}.$$

EXAMPLE 4 Sketch the graph of $f(x) = xe^{-x}$.

Solution We see from Eq. (17) that $f(x) \to 0$ as $x \to +\infty$, whereas $f(x) \to -\infty$ as $x \to -\infty$. Because

$$f'(x) = e^{-x} - xe^{-x} = e^{-x}(1 - x),$$

the only critical point of f is $x = 1$, where $y = e^{-1} \approx 0.37$. Moreover,

$$f''(x) = -e^{-x}(1 - x) + e^{-x}(-1) = e^{-x}(x - 2),$$

so the only inflection point occurs at $x = 2$, where $y = 2e^{-2} \approx 0.27$. Hence the graph of f resembles Fig. 7.3.6. ■

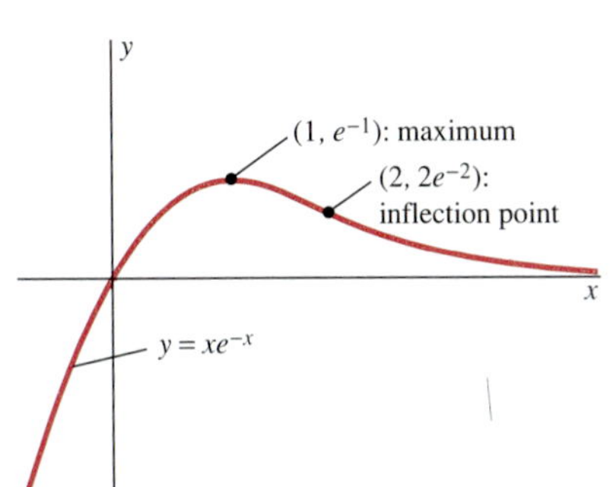

Fig. 7.3.6 Graph of the function of Example 4

The Number e as a Limit

We now establish the following limit expression for the exponential function:

$$e^x = \lim_{n\to\infty}\left(1 + \frac{x}{n}\right)^n. \tag{18}$$

We begin by differentiating $\ln t$, using the definition of the derivative in combination with the fact that we already know that the derivative is $1/t$. Thus

$$\frac{1}{t} = D_t \ln t = \lim_{h\to 0} \frac{\ln(t + h) - \ln t}{h} = \lim_{h\to 0} \frac{1}{h}\ln\left(\frac{t + h}{t}\right)$$

$$= \lim_{h\to 0} \ln\left[\left(1 + \frac{h}{t}\right)^{1/h}\right] \quad \text{(by laws of logarithms)}$$

$$= \ln\left[\lim_{h\to 0}\left(1 + \frac{h}{t}\right)^{1/h}\right] \quad \text{(by continuity of the logarithm function).}$$

The substitution $n = 1/h$ allows us to write

$$\frac{1}{t} = \ln\left[\lim_{n\to\infty}\left(1 + \frac{1}{nt}\right)^n\right].$$

Then the substitution $x = 1/t$ gives

$$x = \ln\left[\lim_{n\to\infty}\left(1 + \frac{x}{n}\right)^n\right].$$

Now Eq. (18) follows, because $x = \ln y$ implies that $e^x = y$. With $x = 1$, we obtain also the following important expression of e as a limit:

$$e = \lim_{n\to\infty}\left(1 + \frac{1}{n}\right)^n. \tag{19}$$

7.3 PROBLEMS

Differentiate the functions in Problems 1 through 30.

1. $f(x) = e^{2x}$
2. $f(x) = e^{3x-1}$
3. $f(x) = e^{x^2} = \exp(x^2)$
4. $f(x) = e^{4-x^3}$
5. $f(x) = e^{1/x^2}$
6. $f(x) = x^2 e^{x^3}$
7. $g(t) = te^{\sqrt{t}}$
8. $g(t) = (e^{2t} + e^{3t})^7$
9. $g(t) = (t^2 - 1)e^{-t}$
10. $g(t) = \sqrt{e^t - e^{-t}}$
11. $g(t) = e^{\cos t}$
12. $f(x) = xe^{\sin x}$
13. $f(x) = \cos(1 - e^{-x})$
14. $f(x) = \sin^2(e^{-x})$
15. $f(x) = \ln(x + e^{-x})$
16. $f(x) = e^x \cos 2x$
17. $f(x) = e^{-2x} \sin 3x$
18. $g(t) = \ln(te^{t^2})$
19. $g(t) = 3(e^t - \ln t)^5$
20. $g(t) = \sin(e^t)\cos(e^{-t})$
21. $f(x) = \dfrac{2 + 3x}{e^{4x}}$
22. $g(t) = \dfrac{1 + e^t}{1 - e^t}$
23. $g(t) = \dfrac{1 - e^{-t}}{t}$
24. $f(x) = e^{-1/x}$
25. $f(x) = \dfrac{1 - x}{e^x}$
26. $f(x) = e^{\sqrt{x}} + e^{-\sqrt{x}}$
27. $f(x) = e^{e^x}$
28. $f(x) = \sqrt{e^{2x} + e^{-2x}}$
29. $f(x) = \sin(2e^x)$
30. $f(x) = \cos(e^x + e^{-x})$

In Problems 31 through 35, find dy/dx by implicit differentiation.

31. $xe^y = y$
32. $\sin(e^{xy}) = x$
33. $e^x + e^y = e^{xy}$
34. $x = ye^y$
35. $e^{x-y} = xy$

Find the antiderivatives indicated in Problems 36 through 53.

36. $\displaystyle\int e^{3x}\,dx$
37. $\displaystyle\int e^{1-2x}\,dx$
38. $\displaystyle\int xe^{x^2}\,dx$
39. $\displaystyle\int x^2 e^{3x^3-1}\,dx$
40. $\displaystyle\int \sqrt{x}\,e^{2x\sqrt{x}}\,dx$
41. $\displaystyle\int \frac{e^{2x}}{1 + e^{2x}}\,dx$
42. $\displaystyle\int (\cos x)e^{\sin x}\,dx$
43. $\displaystyle\int (\sin 2x)e^{1-\cos 2x}\,dx$
44. $\displaystyle\int (e^x + e^{-x})^2\,dx$
45. $\displaystyle\int \frac{x + e^{2x}}{x^2 + e^{2x}}\,dx$
46. $\displaystyle\int e^{2x+3}\,dx$
47. $\displaystyle\int te^{-t^2/2}\,dt$
48. $\displaystyle\int x^2 e^{1-x^3}\,dx$
49. $\displaystyle\int \frac{e^{\sqrt{x}}}{\sqrt{x}}\,dx$
50. $\displaystyle\int \frac{e^{1/t}}{t^2}\,dt$
51. $\displaystyle\int \frac{e^x}{1 + e^x}\,dx$
52. $\displaystyle\int \exp(x + e^x)\,dx$
53. $\displaystyle\int \sqrt{x}\exp\left(-\sqrt{x^3}\right)dx$

Apply Eq. (18) to evaluate (in terms of the exponential function) the limits in Problems 54 through 58.

54. $\displaystyle\lim_{n\to\infty}\left(1 - \frac{1}{n}\right)^n$
55. $\displaystyle\lim_{n\to\infty}\left(1 + \frac{2}{n}\right)^n$
56. $\displaystyle\lim_{n\to\infty}\left(1 + \frac{2}{3n}\right)^n$
57. $\displaystyle\lim_{h\to 0}(1 + h)^{1/h}$
58. $\displaystyle\lim_{h\to 0}(1 + 2h)^{1/h}$ (*Suggestion*: Substitute $k = 2h$.)

Evaluate the limits in Problems 59 through 62 by applying the fact that

$$\lim_{x\to\infty} x^k e^{-x} = 0.$$

59. $\displaystyle\lim_{x\to\infty}\frac{e^x}{x}$
60. $\displaystyle\lim_{x\to\infty}\frac{e^x}{\sqrt{x}}$
61. $\displaystyle\lim_{x\to\infty}\frac{e^{\sqrt{x}}}{x}$
62. $\displaystyle\lim_{x\to\infty} x^2 e^{-x}$

In Problems 63 through 65, sketch the graph of the given equation. Show and label all extrema, inflection points, and asymptotes; show the concave structure clearly.

63. $y = x^2 e^{-x}$
64. $y = x^3 e^{-x}$
65. $y = \exp(-x^2)$
66. Find the area under the graph of $y = e^x$ from $x = 0$ to $x = 1$.
67. Find the volume generated by revolving the region of Problem 66 around the x-axis.
68. Let R be the plane figure bounded below by the x-axis, above by the graph of $y = \exp(-x^2)$, and on the sides by the vertical lines at $x = 0$ and $x = 1$ (Fig. 7.3.7). Find the volume generated by rotating R around the y-axis.

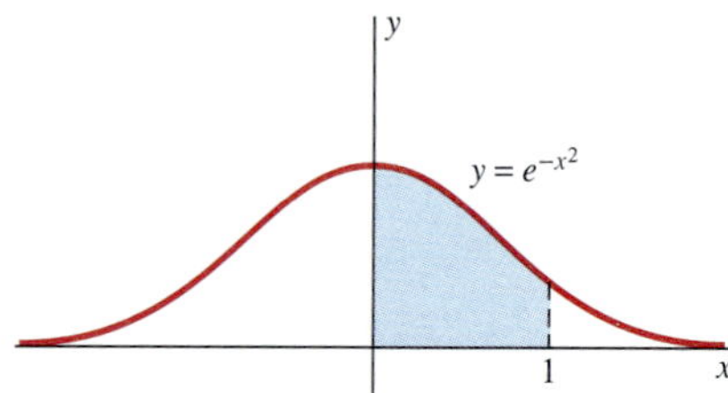

Fig. 7.3.7 The region of Problem 68

69. Find the length of the curve $y = \frac{1}{2}(e^x + e^{-x})$ from $x = 0$ to $x = 1$.
70. Find the area of the surface generated by revolving around the x-axis the curve of Problem 69 (Fig. 7.3.8).

In Problems 71 and 72, graph (on a single calculator or computer screen) the functions on both sides of the given equation to locate its single positive solution. Then determine numerically the value of this solution accurate to three decimal places.

71. $e^{-x} = x - 1$
72. $e^{-x} = \ln x$

In Problems 73 through 76, graph (on a single calculator or computer screen) the functions on both sides of the given equation to determine how many positive solutions it has. Then determine numerically the value of each solution accurate to three decimal places.

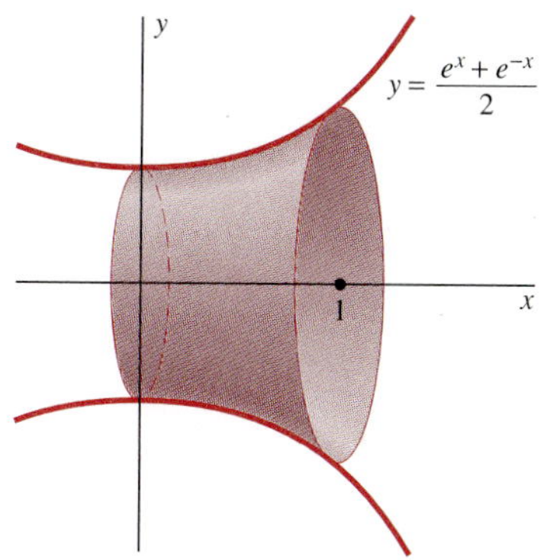

Fig. 7.3.8 The surface of Problem 70

73. $e^x = 5x - 2$

74. $e^x = 7(x-1)^2$

75. $e^x = 3\sin 2x$

76. $4e^x = 47\cos 5x$

77. Find graphically the coordinates (accurate to three decimal places) of the intersection point of the graphs $y = e^x$ and $y = x^{10}$ indicated in Fig. 7.3.9.

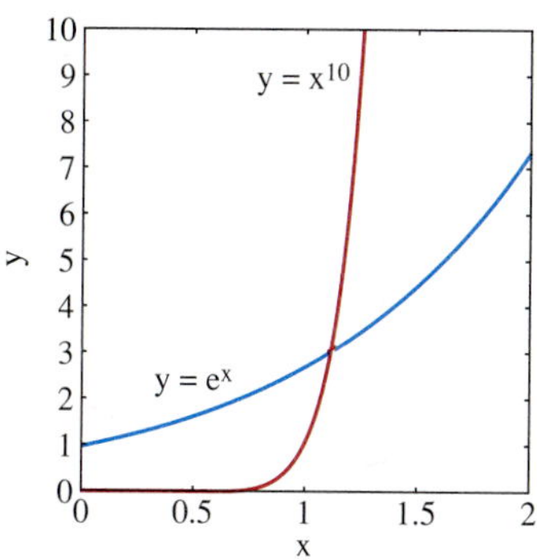

Fig. 7.3.9 Comparing $y = e^x$ and $y = x^{10}$

78. Determine a viewing rectangle that reveals a second intersection point (with $x > 10$) of the graphs $y = e^x$ and $y = x^{10}$. Then determine graphically the first three digits of the larger solution x of the equation $e^x = x^{10}$ (thus writing this solution in the form $p.qr \times 10^k$).

79. Approximate numerically the area of the region bounded by the curves $y = e^x$ and $y = 10(1 + 5x - x^2)$. You will first need to estimate graphically the x-coordinates of the intersection points of the two curves, then integrate numerically (perhaps using the numerical integration facility of your calculator or computer).

80. If a chemical plant releases an amount A of pollutant into a canal at time $t = 0$, then the resulting concentration of pollutant at time t in the water at a town on the canal a distance x_0 downstream is

$$C(t) = \frac{A}{\sqrt{k\pi t}} \exp\left(-\frac{x_0^2}{4kt}\right),$$

where k is a constant. Show that the maximum concentration at the town is

$$C_{\max} = \frac{A}{x_0}\sqrt{\frac{2}{\pi e}}.$$

81. Sketch the graph of $f(x) = x^n e^{-x}$ for $x \geqq 0$ (n is a fixed but arbitrary positive integer). In particular, show that the maximum value of $f(x)$ is $f(n) = n^n e^{-n}$.

82. Approximate the number e as follows. First apply Simpson's approximation with $n = 2$ subintervals to the integral

$$\int_0^1 e^x\,dx = e - 1$$

to obtain the approximation $5e - 4\sqrt{e} - 7 \approx 0$. Then solve for e.

83. Suppose that $f(x) = x^n e^{-x}$, where n is a fixed but arbitrary positive integer. Conclude from Problem 81 that the numbers $f(n-1)$ and $f(n+1)$ are each less than $f(n) = n^n e^{-n}$. Deduce from this that

$$\left(1 + \frac{1}{n}\right)^n < e < \left(1 - \frac{1}{n}\right)^{-n}.$$

Substitute $n = 1024$ to show that $2.716 < e < 2.720$. Note that $1024 = 2^{10}$, so a^{1024} can be computed easily with almost any calculator by entering a and then squaring ten times in succession.

84. Suppose that the quadratic equation $am^2 + bm + c = 0$ has the two real roots m_1 and m_2, and suppose that C_1 and C_2 are arbitrary constants. Show that the function

$$y = y(x) = C_1 e^{m_1 x} + C_2 e^{m_2 x}$$

satisfies the differential equation $ay'' + by' + cy = 0$.

85. Use the result of Problem 84 to find a solution $y = y(x)$ of the differential equation $y'' + y' - 2y = 0$ such that $y(0) = 5$ and $y'(0) = 2$.

7.3 PROJECT: DISCOVERING THE NUMBER e AS A LIMIT

In this brief calculator project you are to use the limit

$$e = \lim_{n\to\infty}\left(1 + \frac{1}{n}\right)^n \tag{1}$$

to investigate numerically the number e. Assuming that the limit in Eq. (1) exists, you can "accelerate" the convergence to the limit by calculating the quantity $(1 + 1/n)^n$

INTRODUCTIO
IN ANALYSIN
INFINITORUM.
AUCTORE
LEONHARDO EULERO,
Professore Regio BEROLINENSI, & Academiæ Imperialis Scientiarum PETROPOLITANÆ Socio.
TOMUS PRIMUS.
LAUSANNÆ,
Apud MARCUM-MICHAELEM BOUSQUET & Socios.
MDCCXLVIII.

Title page of the book in which Euler introduced the number e

only for each power $n = 2^k$ instead of for every positive integer n. That is, consider the sequence $1, 2, 4, 8, \ldots, 2^k, \ldots$ instead of the sequence $1, 2, 3, 4, \ldots$ of all positive integers:

$$e = \lim_{k\to\infty}\left(1 + \frac{1}{2^k}\right)^{2^k}. \tag{2}$$

This method has the advantage that the terms

$$s_k = \left(1 + \frac{1}{2^k}\right)^{2^k} \tag{3}$$

approaching e can be calculated by "successive squaring," because

$$(x^2)^2 = x^4, \quad (x^4)^2 = x^8, \quad (x^8)^2 = x^{16},$$

and so forth. Thus to calculate the (2^k)th power $x^{(2^k)}$ of the number x, you need only enter x and then press the **x^2** key k times in succession. (On some calculators, such as the TI-85, you must first press **x^2** and then **ENTER** each time.)

Consequently, you can calculate the number

$$s_k = \left(1 + \frac{1}{2^k}\right)^{2^k} \tag{3}$$

by the following steps:

1. Calculate 2^k (no squaring here).
2. Calculate the reciprocal $1/2^k$.
3. Add 1 to get the sum $1 + (1/2^k)$.
4. Square the result k times in succession.

Use your calculator to do this with $k = 2, 4, 6, \ldots, 18, 20$. Construct a table showing each result s_k accurate to five decimal places. When you finish, you will have verified (once again?) that $e \approx 2.71828$.

7.4 GENERAL EXPONENTIAL AND LOGARITHMIC FUNCTIONS

The natural exponential function e^x and the natural logarithm function $\ln x$ are often called the exponential and logarithm with *base e*. We now define general exponential and logarithm functions, with the forms a^x and $\log_a x$, whose base is a positive number $a \neq 1$. But it is now convenient to reverse the order of treatment from that of Sections 7.2 and 7.3, so we first consider the general exponential function.

If r is a rational number, then one of the laws of exponents (Eq. (13) of Section 7.3) gives

$$a^r = (e^{\ln a})^r = e^{r \ln a}.$$

We therefore *define* arbitrary powers (rational *and* irrational) of the positive number a in this way:

$$a^x = e^{x \ln a} \tag{1}$$

for all x. Thus

$$3^{\sqrt{2}} = e^{\sqrt{2}\ln 3} \approx e^{1.5537} \approx 4.7289$$

and

$$(0.5)^{-\pi} = e^{-\pi \ln(0.5)} \approx e^{2.1776} \approx 8.8251.$$

Then $f(x) = a^x$ is called the **exponential function with base** a. Note that $a^x > 0$ for all x and that $a^0 = e^0 = 1$ for all $a > 0$.

The *laws of exponents* for general exponentials follow almost immediately from the definition in Eq. (1) and from the laws of exponents for the natural exponential function:

$$a^x a^y = a^{x+y}, \tag{2}$$

$$a^{-x} = \frac{1}{a^x}, \quad \text{and} \tag{3}$$

$$(a^x)^y = a^{xy} \tag{4}$$

for all x and y. To prove Eq. (2), we write

$$a^x a^y = e^{x \ln a} e^{y \ln a} = e^{(x \ln a)+(y \ln a)} = e^{(x+y)\ln a} = a^{x+y}.$$

To derive Eq. (4), note first from Eq. (1) that $\ln a^x = x \ln a$. Then

$$(a^x)^y = e^{y \ln(a^x)} = e^{xy \ln a} = a^{xy}.$$

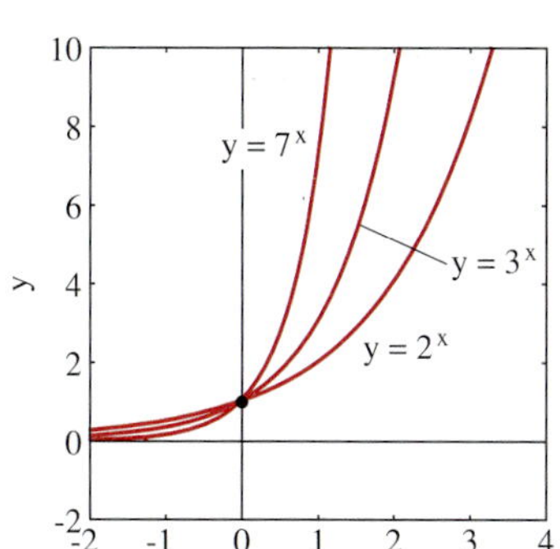

Fig. 7.4.1 If $a > 1$ then $\lim_{x\to-\infty} a^x = 0$, $\lim_{x\to+\infty} a^x = +\infty$

This follows for all real numbers x and y, so the restriction that r is rational in the formula $(e^x)^r = e^{rx}$ (see Eq. (13) of Section 7.3) has now been removed.

If $a > 1$, so that $\ln a > 0$, then Eqs. (5) and (6) in Section 7.3 immediately give us the results

$$\lim_{x\to\infty} a^x = +\infty \quad \text{and} \quad \lim_{x\to-\infty} a^x = 0. \tag{5}$$

Because

$$D_x a^x = D_x(e^{x \ln a}) = (\ln a)e^{x \ln a} = a^x \ln a \tag{6}$$

is positive for all x if $a > 1$, we see that—in this case—$f(x) = a^x$ is an *increasing* function of x. The graph $y = a^x$ then resembles that of the natural exponential function $y = e^x$, but its relative steepness depends on the magnitude of a (Fig. 7.4.1).

If $0 < a < 1$, then $\ln a < 0$. In this case it therefore follows from Eq. (6) that $f(x) = a^x$ is a *decreasing* function, and the values of the two limits in (5) are interchanged (Fig. 7.4.2). Whether $a > 1$ or $0 < a < 1$, it follows from Eq. (6) that $f''(x) = a^x(\ln a)^2 > 0$ for all x, so the graphs in both Figs. 7.4.1 and 7.4.2 are concave upward for all x.

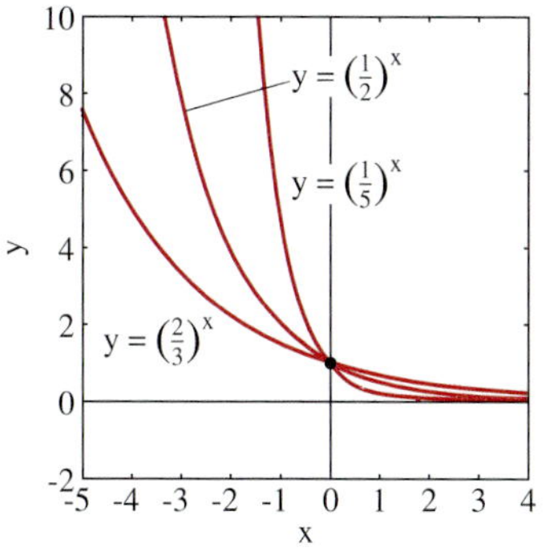

Fig. 7.4.2 If $0 < a < 1$ then $\lim_{x\to-\infty} a^x = +\infty$, $\lim_{x\to+\infty} a^x = 0$

Derivatives and Integrals

If $u = u(x)$ is a differentiable function of x, then Eq. (6) combined with the chain rule gives

$$D_x a^u = (a^u \ln a)\frac{du}{dx}. \tag{7}$$

The corresponding integral formula is

$$\int a^u\,du = \frac{a^u}{\ln a} + C. \tag{8}$$

But rather than using these general formulas, it usually is simpler to rely solely on the definition in Eq. (1), as in Examples 1, 2, and 3.

EXAMPLE 1 To differentiate $f(x) = 3^{x^2}$, we may first write

$$3^{x^2} = (e^{\ln 3})^{x^2} = e^{x^2 \ln 3}.$$

Then

$$D_x\, 3^{x^2} = D_x\, e^{x^2 \ln 3} = e^{x^2 \ln 3} D_x(x^2 \ln 3) = 3^{x^2}(\ln 3)(2x).$$

EXAMPLE 2 Find $\displaystyle\int \frac{10^{\sqrt{x}}}{\sqrt{x}}\,dx.$

Solution We first write $10^{\sqrt{x}} = (e^{\ln 10})^{\sqrt{x}} = e^{\sqrt{x}\ln 10}$. Then

$$\begin{aligned}
\int \frac{10^{\sqrt{x}}}{\sqrt{x}}\,dx &= \int \frac{e^{\sqrt{x}\ln 10}}{\sqrt{x}}\,dx \\
&= \int \frac{2e^u}{\ln 10}\,du \qquad \left(u = \sqrt{x}\ln 10,\ du = \frac{\ln 10}{2\sqrt{x}}\,dx\right) \\
&= \frac{2e^u}{\ln 10} + C = \frac{2}{\ln 10}10^{\sqrt{x}} + C.
\end{aligned}$$

EXAMPLE 3 The function

$$P(t) = 3\cdot(1.07)^t$$

describes a population that starts with $P(0) = 3$ (million) bacteria at time $t = 0$ and increases in number by 7% every hour. After 10 hours the population is

$$P(10) = 3\cdot(1.07)^{10} \approx 5.90 \quad \text{(millions)},$$

so the population has almost doubled. The derivative of $P(t)$ is

$$\begin{aligned}
P'(t) &= D_t[3\cdot(1.07)^t] = 3\cdot D_t(e^{t\ln(1.07)}) \\
&= 3\cdot[\ln(1.07)]e^{t\ln(1.07)} = 3(\ln 1.07)(1.07)^t,
\end{aligned}$$

so at time $t = 10$ the rate of growth of this bacteria population is

$$P'(10) = 3(\ln 1.07)(1.07)^{10} \approx 17.70 \quad \text{(millions per hour)}.$$

Whether or not the exponent r is rational, the **general power function** $f(x) = x^r$ is now defined for $x > 0$ by

$$x^r = e^{r\ln x}.$$

We may now prove the power rule of differentiation for an *arbitrary* (constant) exponent as follows:

$$D_x x^r = D_x(e^{r \ln x}) = e^{r \ln x} D_x(r \ln x) = x^r \cdot \frac{r}{x} = r x^{r-1}.$$

For example, we now know that

$$D_x x^{\pi} = \pi x^{\pi - 1} \approx (3.14159) x^{2.14159}.$$

General Logarithm Functions

If $a > 1$, then the general exponential function a^x is continuous and increasing for all x and attains all positive values. (This follows by an argument similar to that in the first paragraph of Section 7.3, using Eq. (6) and the limits in Eq. (5).) It therefore has an inverse function that is defined for all $x > 0$. This inverse function of a^x is called the **logarithm function with base** a and is denoted by $\log_a x$. Thus

$$y = \log_a x \quad \text{if and only if} \quad x = a^y. \tag{9}$$

The logarithm function with base e is the natural logarithm function: $\log_e x = \ln x$.

The following *laws of logarithms* are easy to derive from the laws of exponents in Eqs. (2) through (4):

$$\log_a xy = \log_a x + \log_a y, \tag{10}$$

$$\log_a \left(\frac{1}{x}\right) = -\log_a x, \tag{11}$$

$$\log_a x^y = y \log_a x. \tag{12}$$

These formulas hold for any positive base $a \neq 1$ and for all positive values of x and y; in Eq. (12), y may be negative or zero as well.

Logarithms with one base are related to logarithms with another base, and the relationship is most easily expressed by the formula

$$(\log_a b)(\log_b c) = \log_a c. \tag{13}$$

This formula holds for all values of a, b, and c for which it makes sense—the bases a and b are positive numbers other than 1 and c is positive. The proof of this formula is outlined in Problem 57. Equation (13) should be easy to remember—it is as though some arcane cancellation law applies.

If we take $c = a$ in Eq. (13), this gives

$$(\log_a b)(\log_b a) = 1, \tag{14}$$

which in turn, with $b = e$, gives

$$\ln a = \frac{1}{\log_a e}. \tag{15}$$

If we replace a with e, b with a, and c with x in Eq. (13), we obtain

$$(\log_e a)(\log_a x) = \log_e x,$$

so

$$\log_a x = \frac{\log_e x}{\log_e a} = \frac{\ln x}{\ln a}. \tag{16}$$

On most calculators, the **LOG** key denotes common (base 10) logarithms: $\log x = \log_{10} x$. In contrast, in many programming languages, such as BASIC, and some symbolic algebra programs, such as *Mathematica,* only the natural logarithm appears explicitly—as `LOG(X)` (in BASIC) and as `Log[x]` (in *Mathematica*). To get $\log_{10} x$, we write `LOG(X)/LOG(10)` and `Log[10,x]`, respectively.

Differentiation of both sides of Eq. (16) yields

$$D_x \log_a x = \frac{1}{x \ln a} = \frac{\log_a e}{x}. \tag{17}$$

For example,

$$D_x \log_{10} x = \frac{\log_{10} e}{x} \approx \frac{0.4343}{x}.$$

If we now reason as we did to obtain Eq. (6) of Section 7.2, the chain rule yields the general formula

$$D_x \log_a |u| = \frac{1}{u \ln a} \cdot \frac{du}{dx} = \frac{\log_a e}{u} \cdot \frac{du}{dx} \qquad (u \neq 0) \tag{18}$$

if u is a differentiable function of x. For example,

$$D_x \log_2 \sqrt{x^2+1} = \frac{1}{2} D_x \log_2 (x^2+1) = \frac{1}{2} \cdot \frac{\log_2 e}{x^2+1} \cdot 2x \approx \frac{(1.4427)x}{x^2+1}.$$

Here we used the fact that $\log_2 e = 1/(\ln 2)$ by Eq. (15).

Logarithmic Differentiation

The derivatives of certain functions are most conveniently found by first differentiating their logarithms. This process—called **logarithmic differentiation**—involves the following steps for finding $f'(x)$:

1. Given: $\quad y = f(x).$
2. Take *natural* logarithms; then simplify, using laws of logarithms: $\quad \ln y = \ln f(x).$
3. Differentiate with respect to x: $\quad \dfrac{1}{y} \cdot \dfrac{dy}{dx} = D_x[\ln f(x)].$
4. Multiply both sides by $y = f(x)$. $\quad \dfrac{dy}{dx} = f(x) D_x[\ln f(x)].$

Remark If $f(x)$ is not positive-valued everywhere, then steps 1 and 2 should be replaced with $y = |f(x)|$ and $\ln y = \ln |f(x)|$, respectively. The differentiation in step 3 then leads to the result $dy/dx = f(x) D_x[\ln |f(x)|]$ in step 4. In practice, we need not be overly concerned in advance with the sign of $f(x)$, because the appearance of what seems to be the logarithm of a negative quantity will signal the fact that absolute values should be used.

EXAMPLE 4 Find dy/dx, given

$$y = \frac{\sqrt{(x^2+1)^3}}{\sqrt[3]{(x^3+1)^4}}.$$

Solution The laws of logarithms give

$$\ln y = \ln \frac{(x^2+1)^{3/2}}{(x^3+1)^{4/3}} = \frac{3}{2}\ln(x^2+1) - \frac{4}{3}\ln(x^3+1).$$

Then differentiation with respect to x gives

$$\frac{1}{y}\cdot\frac{dy}{dx} = \frac{3}{2}\cdot\frac{2}{x^2+1} - \frac{4}{3}\cdot\frac{3x^2}{x^3+1} = \frac{3x}{x^2+1} - \frac{4x^2}{x^3+1}.$$

Finally, to solve for dy/dx, we multiply both sides by

$$y = \frac{(x^2+1)^{3/2}}{(x^3+1)^{4/3}},$$

and we obtain

$$\frac{dy}{dx} = \left(\frac{3x}{x^2+1} - \frac{4x^2}{x^3+1}\right)\cdot\frac{(x^2+1)^{3/2}}{(x^3+1)^{4/3}}.$$

EXAMPLE 5 Find dy/dx, given $y = x^{x+1}$ for $x > 0$.

Solution If $y = x^{x+1}$, then

$$\ln y = \ln(x^{x+1}) = (x+1)\ln x;$$

$$\frac{1}{y}\cdot\frac{dy}{dx} = (1)(\ln x) + (x+1)\left(\frac{1}{x}\right) = 1 + \frac{1}{x} + \ln x.$$

Multiplication by $y = x^{x+1}$ gives

$$\frac{dy}{dx} = \left(1 + \frac{1}{x} + \ln x\right)x^{x+1}.$$

7.4 PROBLEMS

In Problems 1 through 24, find the derivative of the given function $f(x)$.

1. $f(x) = 10^x$
2. $f(x) = 2^{1/x^2}$
3. $f(x) = \dfrac{3^x}{4^x}$
4. $f(x) = \log_{10}\cos x$
5. $f(x) = 7^{\cos x}$
6. $f(x) = 2^x 3^{x^2}$
7. $f(x) = 2^{x\sqrt{x}}$
8. $f(x) = \log_{100}10^x$
9. $f(x) = 2^{\ln x}$
10. $f(x) = 7^{8^x}$
11. $f(x) = 17^x$
12. $f(x) = 2^{\sqrt{x}}$
13. $f(x) = 10^{1/x}$
14. $f(x) = 3^{\sqrt{1-x^2}}$
15. $f(x) = 2^{2^x}$
16. $f(x) = \log_2 x$
17. $f(x) = \log_3\sqrt{x^2+4}$
18. $f(x) = \log_{10}(e^x)$
19. $f(x) = \log_3(2^x)$
20. $f(x) = \log_{10}(\log_{10}x)$
21. $f(x) = \log_2(\log_3 x)$
22. $f(x) = \pi^x + x^\pi + \pi^\pi$
23. $f(x) = \exp(\log_{10}x)$
24. $f(x) = \pi^{x^3}$

Evaluate the integrals given in Problems 25 through 32.

25. $\displaystyle\int 3^{2x}\,dx$
26. $\displaystyle\int x\cdot 10^{-x^2}\,dx$
27. $\displaystyle\int \frac{2^{\sqrt{x}}}{\sqrt{x}}\,dx$
28. $\displaystyle\int \frac{10^{1/x}}{x^2}\,dx$
29. $\displaystyle\int x^2 7^{x^3+1}\,dx$
30. $\displaystyle\int \frac{1}{x\log_{10}x}\,dx$
31. $\displaystyle\int \frac{\log_2 x}{x}\,dx$
32. $\displaystyle\int (2^x)3^{(2^x)}\,dx$

In Problems 33 through 50, find dy/dx by logarithmic differentiation.

33. $y = \sqrt{(x^2 - 4)\sqrt{2x + 1}}$

34. $y = \dfrac{(3 - x^2)^{1/2}}{(x^4 + 1)^{1/4}}$ **35.** $y = 2^x$

36. $y = x^x$ **37.** $y = x^{\ln x}$

38. $y = (1 + x)^{1/x}$

39. $y = \left[\dfrac{(x + 1)(x + 2)}{(x^2 + 1)(x^2 + 2)}\right]^{1/3}$

40. $y = \sqrt{x + 1}\sqrt[3]{x + 2}\sqrt[4]{x + 3}$

41. $y = (\ln x)^{\sqrt{x}}$ **42.** $y = (3 + 2^x)^x$

43. $y = \dfrac{(1 + x^2)^{3/2}}{(1 + x^3)^{4/3}}$ **44.** $y = (x + 1)^x$

45. $y = (x^2 + 1)^{x^2}$ **46.** $y = \left(1 + \dfrac{1}{x}\right)^x$

47. $y = (\sqrt{x})^{\sqrt{x}}$ **48.** $y = x^{\sin x}$

49. $y = x^{(e^x)}$ **50.** $y = (\ln x)^{\ln x}, \quad x > 1$

51. Find the highest point on the curve $f(x) = x \cdot 2^{-x}$ for $x > 0$.

52. Approximate the area of the first-quadrant region bounded by the curves $y = 2^{-x}$ and $y = (x - 1)^2$. One of the points of intersection of these two curves should be evident, but you will need to approximate the other point.

53. Approximate the volume of the solid generated by rotation of the region of Problem 52 around the x-axis.

54. Approximate the area of the first-quadrant region bounded by the curves $y = 3^{2-x}$ and $y = (3x - 4)^2$. You will need to approximate the two points of intersection of these two curves.

55. Find $f'(x)$ given $f(x) = x^{\cos x}$ for $x \geqq 0$. Then plot the curves $y = f(x)$ and $y = f'(x)$ for $0 \leqq x \leqq 5\pi$ and verify that the local extrema of $f(x)$ correspond to the zeros of $f'(x)$.

56. Figure 7.4.3 shows the graph of the function $f(x) = (\ln x)^{\ln x}$ for $x > 1$. Approximate graphically the coordinates of the apparent local minimum point on the curve. Are your results consistent with the derivative $f'(x)$ asked for in Problem 50?

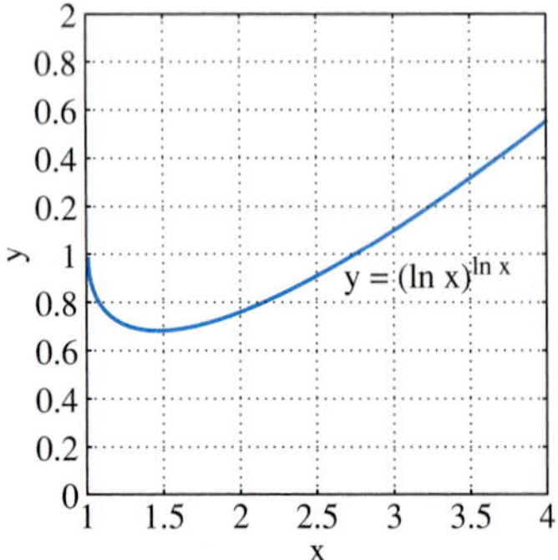

Fig. 7.4.3 The graph in Problem 56

57. Prove Eq. (13). (*Suggestion:* Let $x = \log_a b$, $y = \log_b c$, and $z = \log_a c$. Then show that $a^z = a^{xy}$, and conclude that $z = xy$.)

58. Suppose that u and v are differentiable functions of x. Show by logarithmic differentiation that

$$D_x(u^v) = v(u^{v-1})\frac{du}{dx} + (u^v \ln u)\frac{dv}{dx}.$$

Interpret the two terms on the right in relation to the special cases in which (a) u is a constant; (b) v is a constant.

59. Suppose that $a > 0$. By examining $\ln(a^{1/x})$, show that

$$\lim_{x \to \infty} a^{1/x} = 1.$$

It follows that

$$\lim_{n \to \infty} a^{1/n} = 1.$$

(The inference is that n is a positive integer.) Test this conclusion by entering some positive numbers in your calculator and then pressing the square root key repeatedly. Make a table of the results of two such experiments.

60. Show that $\lim_{n \to \infty} n^{1/n} = 1$ by showing that $\lim_{x \to \infty} x^{1/x} = 1$. (The inference is that x is an arbitrary positive real number, whereas n is a positive integer.) Use the method of Problem 59.

61. By examining $\ln(x^x/e^x)$, show that

$$\lim_{x \to \infty} \frac{x^x}{e^x} = +\infty.$$

Thus x^x increases even faster than does the exponential function e^x as $x \to +\infty$.

62. Consider the function

$$f(x) = \frac{1}{1 + 2^{1/x}} \qquad \text{for} \quad x \neq 0.$$

Show that both the left-hand and right-hand limits of $f(x)$ at $x = 0$ exist but are unequal.

63. Find dy/dx if $y = \log_x 2$.

64. Suppose that $y = uvw/pqr$, where u, v, w, p, q, and r are nonzero differentiable functions of x. Show by logarithmic differentiation that

$$\frac{dy}{dx} = y \cdot \left(\frac{1}{u} \cdot \frac{du}{dx} + \frac{1}{v} \cdot \frac{dv}{dx} + \frac{1}{w} \cdot \frac{dw}{dx} - \frac{1}{p} \cdot \frac{dp}{dx} - \frac{1}{q} \cdot \frac{dq}{dx} - \frac{1}{r} \cdot \frac{dr}{dx}\right).$$

Is the generalization—for an arbitrary finite number of factors in numerator and denominator—obvious?

7.4 PROJECT: GOING WHERE NO ONE HAS GONE BEFORE

This project investigates the equation

$$2^x = x^{10}. \tag{1}$$

1. The graphs of $y = 2^x$ and $y = x^{10}$ (Fig. 7.4.4) *suggest* that Eq. (1) has two solutions—one positive and the other negative. If a graphics calculator or computer is available, find these two solutions (accurate to three or four decimal places) by successive magnification (the method of "zooming").
2. Figure 7.4.5 seems to indicate that x^{10} leaves 2^x forever behind as $x \to +\infty$. Show, however, that Eq. (1) has the same positive solutions as the equation
$$\frac{\ln x}{x} = \frac{\ln 2}{10}.$$
Hence conclude from the graph of $y = (\ln x)/x$ (Example 8 of Section 7.2) that Eq. (1) has precisely *two* positive solutions.
3. Tabulate values of 2^x and x^{10} for $x = 10, 20, 30, 40, 50$, and 60, and thereby verify that the missing positive solution is somewhere between $x = 50$ and $x = 60$ (Fig. 7.4.6). If you attempt to locate this solution by successive magnifications, you may well have the feeling of going boldly where no one has gone before!
4. Use Newton's method to approximate (with four-place accuracy) all three solutions of Eq. (1).

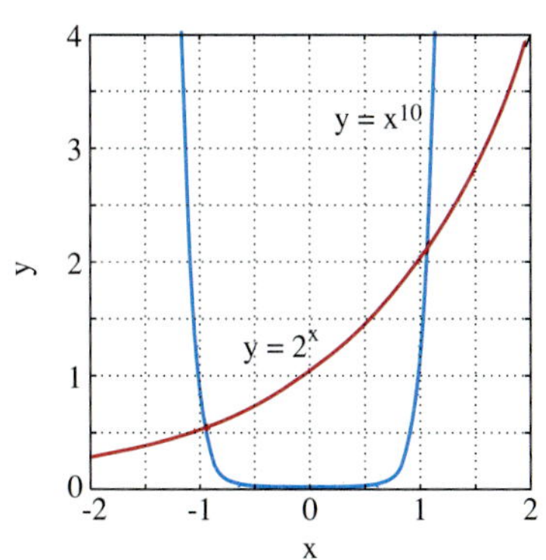

Fig. 7.4.4 $y = 2^x$ and $y = x^{10}$ for $-2 \leqq x \leqq 2$

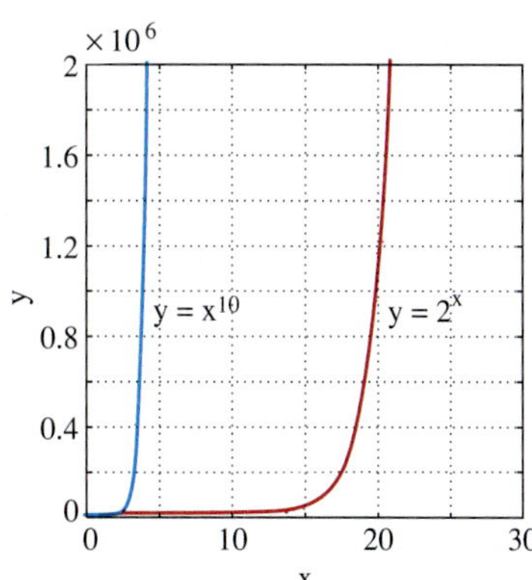

Fig. 7.4.5 $y = 2^x$ and $y = x^{10}$ for $0 \leqq x \leqq 30$

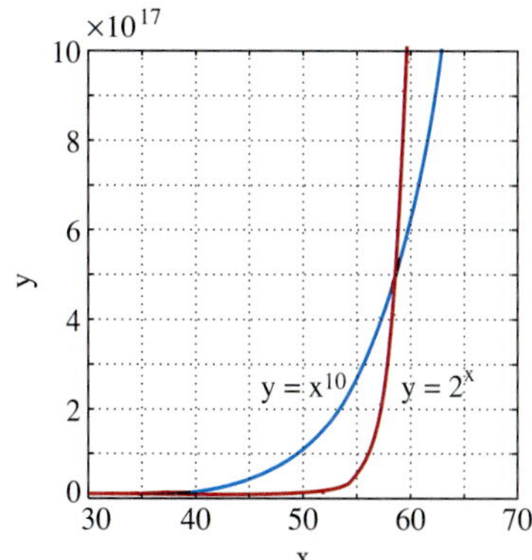

Fig. 7.4.6 $y = 2^x$ and $y = x^{10}$ for $30 \leqq x \leqq 70$

7.5 NATURAL GROWTH AND DECAY

In this section we use the natural exponential function to model the "natural growth" of populations. Consider a population that numbers $P(t)$ persons or other animals, bacteria, dollars, molecules—any sort of entity—at time t. We assume that this population has a constant *birth rate* β and a constant *death rate* δ. Roughly speaking, this means that during any 1-yr period, βP births and δP deaths occur.

Because P changes during the course of a year, some allowance must be made for changes in the number of births and the number of deaths. We think of a very brief time interval from t to $t + \Delta t$. For very small values of Δt, the value of $P = P(t)$ will change by such a small amount during the time interval $[t, t + \Delta t]$ that we can regard $P(t)$ as almost constant. We require that the number of births and deaths during this time interval be given with sufficient accuracy by the approximations

$$\begin{aligned} &\text{Number of births:} \quad \text{approximately } \beta P(t)\,\Delta t; \\ &\text{Number of deaths:} \quad \text{approximately } \delta P(t)\,\Delta t. \end{aligned} \tag{1}$$

More precisely, we assume that the ratio to Δt of the error in each of these approximations approaches zero as $\Delta t \to 0$.

On this basis we wish to deduce the form of the function $P(t)$ that describes the population in question. Our strategy begins with finding the **time rate of change** of $P(t)$. Hence we consider the increment

$$\Delta P = P(t + \Delta t) - P(t)$$

of P during the time interval $[t, t + \Delta t]$. Because ΔP is simply the number of births minus the number of deaths, we find from Eq. (1) that

$$\begin{aligned} \Delta P &= P(t + \Delta t) - P(t) \\ &\approx \beta P(t)\,\Delta t - \delta P(t)\,\Delta t \qquad \text{(the number of births minus the number of deaths).} \end{aligned}$$

Therefore

$$\frac{\Delta P}{\Delta t} = \frac{P(t + \Delta t) - P(t)}{\Delta t} \approx (\beta - \delta)P(t).$$

The quotient on the left-hand side approaches the derivative $P'(t)$ as $\Delta t \to 0$, and, by the assumption following (1), it also approaches the right-hand side, $(\beta - \delta)P(t)$. Hence, when we take the limit as $\Delta t \to 0$, we get the differential equation

$$P'(t) = (\beta - \delta)P(t);$$

that is,

$$\frac{dP}{dt} = kP, \qquad \text{where} \quad k = \beta - \delta. \tag{2}$$

This differential equation may be regarded as a *mathematical model* of the changing population.

The Natural Growth Equation

With $x(t)$ in place of $P(t)$ in Eq. (2), we have the differential equation

$$\frac{dx}{dt} = kx, \tag{3}$$

which serves as the mathematical model for an extraordinarily wide range of natural phenomena. It is easily solved if we first "separate the variables" and then integrate:

$$\begin{aligned} \frac{dx}{x} &= k\,dt; \\ \int \frac{dx}{x} &= \int k\,dt; \\ \ln x &= kt + C. \end{aligned}$$

Next, we apply the exponential function to both sides of the last equation to solve for x:

$$x = e^{\ln x} = e^{kt+C} = e^{kt}e^{C} = Ae^{kt}.$$

Here, $A = e^{C}$ is a constant that remains to be determined. But we see that A is simply the value $x_0 = x(0)$ of $x(t)$ when $t = 0$, and thus $A = x_0$.

Theorem 1 ***The Natural Growth Equation***

The solution of the initial value problem

$$\frac{dx}{dt} = kx, \qquad x(0) = x_0 \tag{4}$$

is

$$x(t) = x_0 e^{kt}. \tag{5}$$

As a consequence, Eq. (3) is often called the **exponential growth equation,** or the **natural growth equation.** We see from Eq. (5) that, with $x_0 > 0$, the solution $x(t)$ is an increasing function if $k > 0$ and a decreasing function if $k < 0$. (The situation $k < 0$ is sometimes called *exponential decay.*) These two cases are illustrated in Figs. 7.5.1 and 7.5.2, respectively. The remainder of this section concerns examples of natural phenomena for which this differential equation serves as a mathematical model.

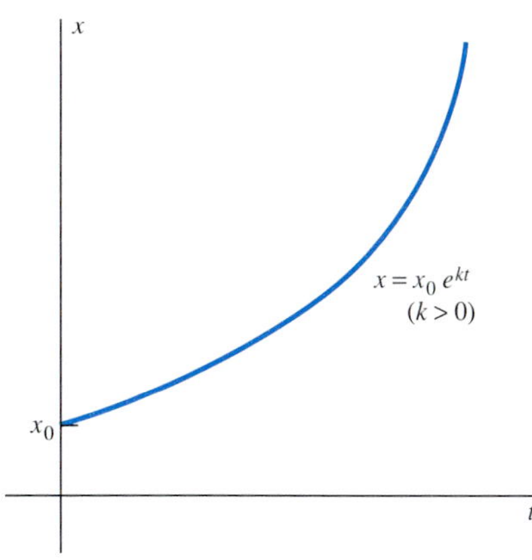

Fig. 7.5.1 Solution of the exponential growth equation for $k > 0$

Population Growth

By comparing Eqs. (2), (3), and (5), we see that a population $P(t)$ with constant birth rate β and constant death rate δ is given by

$$P(t) = P_0 e^{kt}, \tag{6}$$

where $P_0 = P(0)$ and $k = \beta - \delta$. If t is measured in years, then k is called the **annual growth rate,** which can be positive, negative, or zero. Its value is often given as a percentage (its decimal value multiplied by 100). If k is close to zero, this value is fairly close to the annual percentage increase (or decrease) of the population each year.

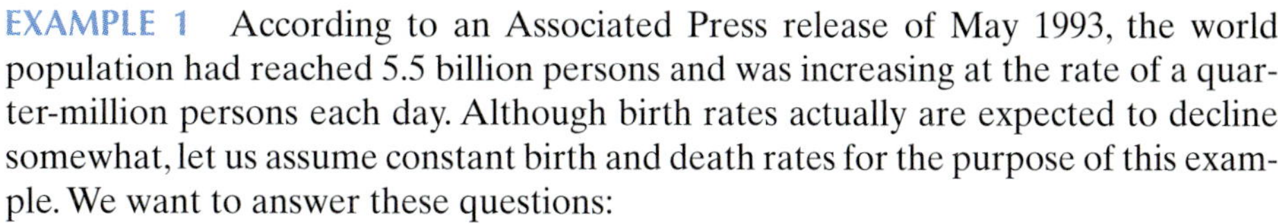

EXAMPLE 1 According to an Associated Press release of May 1993, the world population had reached 5.5 billion persons and was increasing at the rate of a quarter-million persons each day. Although birth rates actually are expected to decline somewhat, let us assume constant birth and death rates for the purpose of this example. We want to answer these questions:

1. What is the annual growth rate k?
2. What will be the world population 10 years thereafter, in the year 2003?
3. How long will the world population take to double?
4. How long will it take for the world population to increase tenfold (thereby exceeding the 55 billion that some demographers believe to be the maximum for which the planet can provide food)?

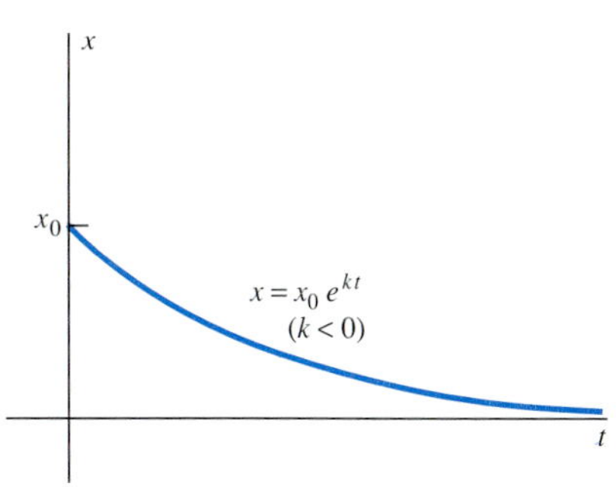

Fig. 7.5.2 Solution of the exponential growth equation—now actually a *decay* equation—for the case $k < 0$

Solution We measure the world population $P(t)$ in billions and measure time t in years. We take $t = 0$ to correspond to 1993, so $P_0 = 5.5$. The fact that P is increasing by 250,000, or 0.00025 billion, persons per day at time $t = 0$ means that

The population of the world is growing rapidly, but how rapidly?

$$P'(0) = (0.00025)(365.25) \approx 0.0913$$

billion per year. From Eq. (2) we now obtain

$$k = \left[\frac{1}{P}\cdot\frac{dP}{dt}\right]_{t=0} = \frac{P'(0)}{P(0)} = \frac{0.0913}{5.5} \approx 0.0166.$$

Thus the world population was growing at the rate of about 1.66% per year in 1993.

We use this value of k to conclude that the world population at time t should be

$$P(t) = (5.5)e^{(0.0166)t}.$$

For example, $t = 10$ yields

$$P(10) = (5.5)e^{(0.0166)(10)} \approx 6.49 \text{ (billion)}$$

for the population in the year 2003.

To find when the population will double to 11 billion, we solve the equation

$$11 = (5.5)e^{(0.0166)t}$$

for t. We first take the natural logarithm of each side, then solve for

$$t = \frac{\ln 2}{0.0166} \approx 42 \text{ (yr)},$$

which corresponds to the year 2035. Finally, under the assumptions made here, the world population will reach 55 billion when

$$55 = (5.5)e^{(0.0166)t}; \qquad t = \frac{\ln 10}{0.0166} \approx 139;$$

that is, in the year 2132. ■

Radioactive Decay

Consider a sample of material that contains $N(t)$ atoms of a certain radioactive isotope at time t. Many experiments have confirmed that a constant fraction of these radioactive atoms will spontaneously decay (into atoms of another element or another isotope of the same element) during each given unit of time. Consequently, the sample behaves exactly like a population with a constant death rate but with no births occurring. To write a model for $N(t)$, we use Eq. (2) with N in place of P, $k > 0$ in place of δ, and $\beta = 0$. We thus obtain the differential equation

$$\frac{dN}{dt} = -kN. \tag{7}$$

From the solution in Eq. (5) of Eq. (3), with k replaced with $-k$, we conclude that

$$N(t) = N_0 e^{-kt}, \tag{8}$$

where $N_0 = N(0)$, the number of radioactive atoms of the original isotope present in the sample at time $t = 0$.

The value of the *decay constant* k depends on the particular isotope. If k is large, then the isotope decays rapidly. If k is near zero, the isotope decays quite slowly and thus may be a relatively persistent factor in its environment. The decay constant k is often specified in terms of another empirical parameter that is more convenient, the *half-life* of the isotope. The **half-life** τ of a sample of a radioactive isotope is the time required for *half* of that sample to decay. To find the relationship between k and τ, we set

$$t = \tau \quad \text{and} \quad N = \tfrac{1}{2}N_0$$

in Eq. (8), so that

$$\tfrac{1}{2}N_0 = N_0 e^{-kt}. \tag{9}$$

Solving for τ, we find that

$$\tau = \frac{\ln 2}{k}. \tag{10}$$

Note that the concept of half-life is meaningful—the value of τ depends *only* on k and thus depends only on the particular isotope involved. It does *not* depend on the amount of that isotope present.

Radiocarbon Dating

The method of *radiocarbon dating* is based on the fact that the radioactive carbon isotope ^{14}C has a known half-life of about 5700 yr. Living organic matter maintains a constant level of ^{14}C by "breathing" air (or by consuming organic matter that does so). But air contains ^{14}C along with the much more common, stable isotope ^{12}C of carbon, mostly in the gas CO_2. Thus all living organisms maintain the same percentage of ^{14}C as in air, because organic processes seem to make no distinction between the two isotopes. But when an organism dies, it ceases to metabolize carbon, and the process of radioactive decay begins to deplete its ^{14}C content. The fraction of ^{14}C in the air remains roughly constant because new ^{14}C is continuously generated by the bombardment of nitrogen atoms in the upper atmosphere by cosmic rays, and this generation has long been in steady-state equilibrium with the loss of ^{14}C through radioactive decay.

EXAMPLE 2 A specimen of charcoal found at Stonehenge contains 63% as much ^{14}C as does a sample of present-day charcoal. What is the age of the sample?

Solution We take $t = 0$ (in years) as the time of death of the tree from which the charcoal was made. From Eq. (9), we know that

$$\tfrac{1}{2}N_0 = N_0 e^{-5700k},$$

so

$$k = \frac{\ln 2}{\tau} = \frac{\ln 2}{5700} \approx 0.0001216.$$

We are given that $N = (0.63)N_0$ at present, so we solve the equation

$$(0.63)N_0 = N_0 e^{-kt}$$

with this value of k. We thus find that

$$t = -\frac{\ln(0.63)}{0.0001216} \approx 3800 \quad \text{(yr)}.$$

Therefore, the sample is about 3800 yr old. If it is connected in any way with the builders of Stonehenge, our computations suggest that this observatory, monument, or temple—whichever it may be—dates from almost 1800 B.C. ■

EXAMPLE 3 According to one cosmological theory, there were equal amounts of the uranium isotopes ^{235}U and ^{238}U at the creation of the universe in the "big bang." At present there are 137.7 ^{238}U atoms for each ^{235}U atom. Using the known half-lives

$$4.51 \text{ billion yr for } {}^{238}\text{U},$$

$$0.71 \text{ billion yr for } {}^{235}\text{U},$$

calculate the age of the universe.

Solution Let $N_8(t)$ and $N_5(t)$ be the numbers of ^{238}U and ^{235}U atoms, respectively, at time t, in billions of years after the creation of the universe. Then

$$N_8(t) = N_0 e^{-kt} \quad \text{and} \quad N_5(t) = N_0 e^{-ct},$$

where N_0 is the initial number of atoms of each isotope. Also,

$$k = \frac{\ln 2}{4.51} \quad \text{and} \quad c = \frac{\ln 2}{0.71},$$

a consequence of Eq. (10). We divide the equation for N_8 by the equation for N_5 and find that when t has the value corresponding to "now,"

$$137.7 = \frac{N_8}{N_5} = e^{(c-k)t}.$$

Finally, we solve this equation for t:

$$t = \frac{\ln(137.7)}{\left(\dfrac{1}{0.71} - \dfrac{1}{4.51}\right)\ln 2} \approx 5.99.$$

Thus we estimate the age of the universe to be about 6 billion years, which is roughly on the same order of magnitude as recent estimates of 10 to 15 billion years (based on astronomical observations of the rate of expansion of the universe). ■

Continuously Compounded Interest

Consider a savings account that is opened with an initial deposit of A_0 dollars and earns interest at the annual rate r. If there are $A(t)$ dollars in the account at time t and the interest is compounded at time $t + \Delta t$, this means that $rA(t)\,\Delta t$ dollars in interest are added to the account then. So

$$A(t + \Delta t) = A(t) + rA(t)\,\Delta t,$$

and thus

$$\frac{\Delta A}{\Delta t} = \frac{A(t + \Delta t) - A(t)}{\Delta t} = rA(t).$$

Continuous compounding of interest results from taking the limit as $\Delta t \to 0$, so

$$\frac{dA}{dt} = rA. \tag{11}$$

This is an exponential growth equation with solution

$$A(t) = A_0 e^{rt}. \tag{12}$$

EXAMPLE 4 If $A_0 = \$1000$ is invested at an annual interest rate of 6% compounded continuously, then $r = 0.06$, and Eq. (12) gives

$$A(1) = 1000e^{(0.06)(1)} = \$1061.84$$

for the value of the investment after one year. Thus the *effective annual interest rate* is 6.184%. The more often interest is compounded, the more rapidly savings grow, but bank advertisements sometimes overemphasize this advantage. For instance, 6% compounded *monthly* multiplies your investment by

$$1 + \frac{0.06}{12} = 1.005$$

at the end of each month, so an initial investment of \$1000 would grow in one year to

$$(1000)(1.005)^{12} = \$1061.68,$$

only 16¢ less than would be yielded by continuous compounding. ■

*Drug Elimination

The amount $A(t)$ of a certain drug in the human bloodstream, as measured by the excess above the natural level of the drug in the bloodstream, typically declines at a rate proportional to that excess amount. That is,

$$\frac{dA}{dt} = -\lambda A, \quad \text{so} \quad A(t) = A_0 e^{-\lambda t}. \tag{13}$$

The parameter λ is called the *elimination constant* of the drug, and $T = 1/\lambda$ is called the *elimination time.*

EXAMPLE 5 The elimination time for alcohol varies from one person to another. If a person's "sobering time" $T = 1/\lambda$ is 2.5 h, how long will the excess concentration of alcohol in the bloodstream require to fall from 0.10% to 0.02%?

Solution We assume that the normal concentration of alcohol in the blood is zero, so any amount is an excess amount. In this problem, we have $\lambda = 1/2.5 = 0.4$, so Eq. (13) yields

$$0.02 = (0.10)e^{-(0.4)t}.$$

Thus

$$t = -\frac{\ln(0.2)}{0.4} \approx 4.02 \quad \text{(h)}.$$

■

*Sales Decline

According to marketing studies, if advertising for a particular product is halted and other market conditions—such things as number and promotion of competing products, their prices, and so on—remain unchanged, then the sales of the unadvertised product will decline at a rate that is proportional at any time t to the current sales S. That is,

$$\frac{dS}{dt} = -\lambda S, \quad \text{so} \quad S(t) = S_0 e^{-\lambda t}. \tag{14}$$

Here S_0 denotes the initial value of the sales, which we take to be sales in the last month of advertising. If we take months as the units for time t, then $S(t)$ gives the number of sales t months after advertising is halted, and λ might be called the *sales decay constant.*

*Linguistics

Consider a basic list of N_0 words in use in a given language at time $t = 0$. Let $N(t)$ denote the number of these words that are still in use at time t—those that have neither disappeared from the language nor been replaced. According to one theory in linguistics, the rate of decrease of N is proportional to N. That is,

$$\frac{dN}{dt} = -\lambda N, \quad \text{so} \quad N(t) = N_0 e^{-\lambda t}. \tag{15}$$

If t is measured in millennia (as is standard in linguistics), then $k = e^{-\lambda}$ is the fraction of the words in the original list that survive for 1000 yr.

7.5 PROBLEMS

1. *Continuously Compounded Interest* Suppose that \$1000 is deposited in a savings account that pays 8% annual interest compounded continuously. At what rate (in dollars per year) is it earning interest after 5 yr? after 20 yr?
2. *Population Growth* Coopersville had a population of 25000 in 1970 and a population of 30000 in 1980. Assume that its population will continue to grow exponentially at a constant rate. What population can the Coopersville city planners expect in the year 2010?
3. *Population Growth* In a certain culture of bacteria, the number of bacteria increased sixfold in 10 h. Assuming natural growth, how long did it take for their number to double?
4. *Radiocarbon Dating* Carbon extracted from an ancient skull recently unearthed contained only one-sixth as much radioactive ^{14}C as carbon extracted from present-day bone. How old is the skull?
5. *Radiocarbon Dating* Carbon taken from a relic purported to date from A.D. 30 contained 4.6×10^{10} atoms of ^{14}C per gram. Carbon extracted from a present-day specimen of the same substance contained 5.0×10^{10} atoms of ^{14}C per gram. Compute the approximate age of the relic. What is your opinion regarding its authenticity?
6. *Continuously Compounded Interest* An amount A of money is invested for t years at an annual interest rate r compounded n times over these years at equal intervals. (a) Explain why the amount accrued after t years is
$$A_{r,n} = A \cdot \left(1 + \frac{rt}{n}\right)^n.$$
(b) Conclude from the limit in Eq. (18) of Section 7.3 that
$$\lim_{n \to \infty} A_{r,n} = Ae^{rt},$$
in agreement with Eq. (12) of this section.
7. *Continuously Compounded Interest* If an investment of A_0 dollars returns A_1 dollars after 1 yr, the **effective annual interest rate** r is defined by the equation
$$A_1 = (1 + r)A_0.$$
Banks sometimes advertise that they increase the effective interest rates on their customers' savings accounts by increasing the frequency of compounding. Calculate the effective annual interest rate if a 9% interest rate is compounded (a) quarterly; (b) monthly; (c) weekly; (d) daily; (e) continuously.

8. *Continuously Compounded Interest* Upon the birth of their first child, a couple deposited \$5000 in a savings account that pays 6% annual interest compounded continuously. The interest payments are allowed to accumulate. How much will the account contain when the child is ready to go to college at age 18?

9. *Continuously Compounded Interest* You discover in your attic an overdue library book on which your great-great-great-grandfather owed a fine of 30¢ exactly 100 years ago. If an overdue fine grows exponentially at a 5% annual interest rate compounded continuously, how much would you have to pay if you returned the book today?

10. *Drug Elimination* Suppose that sodium pentobarbitol will anesthetize a dog when its bloodstream contains at least 45 mg of this drug per kilogram of the dog's body weight. Suppose also that sodium pentobarbitol is eliminated exponentially from a dog's bloodstream, with a half-life of 5 h. What single dose should be administered to anesthetize a 50-kg dog for 1 h?

11. *Sales Decline* Moonbeam Motors has discontinued advertising of its sports-utility vehicle. The company plans to resume advertising when sales have declined to 75% of their initial rate. If after 1 week without advertising, sales have declined to 95% of their initial rate, when should the company expect to resume advertising?

12. *Linguistics* The English language evolves in such a way that 77% of all words disappear (or are replaced) every 1000 yr. Of a basic list of words used by Chaucer in A.D. 1400, what percentage should we expect to find still in use today?

13. *Radioactive Decay* The half-life of radioactive cobalt is 5.27 yr. Suppose that a nuclear accident has left the level of cobalt radiation in a certain region at 100 times the level acceptable for human habitation. How long will it be before the region is again habitable? (Ignore the likely presence of other radioactive substances.)

14. *Radioactive Decay* Suppose that a rare mineral deposit formed in an ancient cataclysm—such as the collision of a meteorite with the earth—originally contained the uranium isotope ^{238}U (which has a half-life of 4.51×10^9 yr) but none of the lead isotope ^{207}Pb, the end product of the radioactive decay of ^{238}U. If the ratio of ^{238}U atoms to ^{207}Pb atoms in the mineral deposit today is 0.9, when did the cataclysm occur?

15. *Radioactive Decay* A certain moon rock contains equal numbers of potassium atoms and argon atoms. Assume that all the argon is present because of radioactive decay of potassium (its half-life is about 1.28×10^9 yr) and that 1 out of every 9 potassium atom disintegrations yields an argon atom. What is the age of the rock, measured from the time it contained only potassium?

16. *Newton's Law of Cooling* If a body is cooling in a medium with constant temperature A, then—according to Newton's law of cooling (Section 7.6)—the rate of change of the body's temperature T is proportional to $T - A$. We want to cool a pitcher of buttermilk initially at 25°C by setting it out on the front porch, where the temperature is 0°C. If the temperature of the buttermilk drops to 15°C after 20 min, when will it be at 5°C?

17. When sugar is dissolved in water, the amount A of sugar that remains undissolved after t minutes satisfies the differential equation $dA/dt = -kA$ $(k > 0)$. If 25% of the sugar dissolves in 1 min, how long does it take for half the sugar to dissolve?

18. The intensity I of light at a depth x meters below the surface of a lake satisfies the differential equation $dI/dx = -(1.4)I$. (a) At what depth is the intensity half the intensity I_0 at the surface (where $x = 0$)? (b) What is the intensity at a depth of 10 m (as a fraction of I_0)? (c) At what depth will the intensity be 1% of its value at the surface?

19. The barometric pressure p (in inches of mercury) at an altitude x miles above sea level satisfies the differential equation $dp/dx = -(0.2)p$; $p(0) = 29.92$. (a) Calculate the barometric pressure at 10000 ft and again at 30000 ft. (b) Without prior conditioning, few people can survive when the pressure drops to less than 15 in. of mercury. How high is that?

20. An accident at a nuclear power plant has left the surrounding area polluted with a radioactive element that decays at a rate proportional to its current amount $A(t)$. The initial radiation level is 10 times the maximum amount S that is safe, and 100 days later it is still 7 times that amount. (a) Set up and solve a differential equation to find $A(t)$. (b) How long (to the nearest day after the original accident) will it be before it is safe for people to return to the area?

21. Suppose that a nuclear accident was contained to a single room of a nuclear research laboratory but has left that room contaminated with polonium-210, which has a half-life of 140 days. If the initial contamination of the room is five times the amount safe for human exposure, how long should laboratory workers wait before entering the room to clean it up?

22. Suppose that the national government's annual budget is currently \$2 trillion, but only \$1.85 trillion in taxes is being collected annually (so the current deficit is \$150 billion per year). Suppose that both the annual budget and the annual tax revenues increase exponentially. If revenues increase at 3% annually, what annual percentage increase in the national budget will yield a balanced budget seven years in the future? You can take either a symbolic approach or a graphical approach (in which case you want to determine the budget's rate of increase so that the budget and revenue graphs intersect seven years from now).

23. Dinosaurs became extinct late in the Cretaceous Era, about 70 million years ago. Suppose that you find a dinosaur bone containing exactly one atom of ^{14}C. Obtain an *underestimate* of the weight of the dinosaur. You will need to look up Avogadro's number and perhaps read about what it tells you.

7.5 PROJECT: THE RULE OF 72—TRUE OR FALSE?

Suppose that you inherit a lump sum A_0 and invest the entire amount in a guaranteed savings account that pays $r\%$ interest once annually. Then the amount A_n in the account after n years is given by

$$A_n = A_0\left(1 + \frac{r}{100}\right)^n.$$

According to a country banker's "rule of thumb," the number of years you must wait for your money to double is given approximately by $n \approx 72/r$.

Your first task is to investigate the accuracy of this "rule of 72." For this purpose, construct a table of values of the "doubling time" n as a function of r (for $r = 1, 2, 3, \ldots, 20$, perhaps). For what range of values of the simple interest rate r is the value $n \approx 72/r$ correct to the nearest year?

Next, answer these two questions:

1. For what value of r (if any) is the formula $n = 72/r$ precisely correct? You may estimate r graphically or numerically.
2. If your interest rate r is compounded continuously rather than once annually, with what constant should the number 72 be replaced in the banker's rule of thumb? And is there a value k such that $n = k/r$ for all (reasonable) values of r?

Finally, a rule of thumb for you to discover: Determine how long $1000 placed in your child's personal account at birth would take to grow to a million dollars. If the account draws $r\%$ annually, you would divide r into *what* constant to determine the approximate age at which your child will be a millionaire?

*7.6 LINEAR FIRST-ORDER EQUATIONS AND APPLICATIONS

A **first-order differential equation** is one in which only the first derivative (not higher derivatives) of the dependent variable appears. It is a **linear** first-order differential equation if it can be written in the form

$$\frac{dx}{dt} = ax + b, \tag{1}$$

where a and b denote functions of the independent variable t. Here we discuss applications in the special case in which the coefficients a and b are *constants.*

Equation (1) is separable, so we can separate the variables as in Section 6.5 and immediately integrate. Assuming that $ax + b > 0$, we get

$$\int \frac{a\,dx}{ax + b} = \int a\,dt; \qquad \ln(ax + b) = at + C.$$

Then application of the natural exponential function to both sides gives

$$ax + b = e^{at+C} = Ke^{at},$$

where $K = e^C$. When we substitute $t = 0$ and denote the resulting value of x by x_0, we find that $K = ax_0 + b$. So

$$ax + b = (ax_0 + b)e^{at}.$$

Finally, we solve this equation for the solution $x = x(t)$ of Eq. (1):

$$x(t) = \left(x_0 + \frac{b}{a}\right)e^{at} - \frac{b}{a}. \tag{2}$$

Here we have assumed that $ax + b > 0$, but Eq. (2) also gives the correct solution in the case $ax + b < 0$ (see Problem 17). In Problem 21 we outline a method by which Eq. (1) may be solved when the coefficients a and b are functions of t rather than constants. The solution in Eq. (2) for the constant-coefficient case will, however, be sufficient for the following applications.

Population Growth with Immigration

Consider a population $P(t)$ with constant birth and death rates (β and δ, respectively), as in Section 7.5, but also with a constant immigration rate of I persons per year entering the country. To account for the immigration, our derivation of Eq. (2) in Section 7.5 must be amended as follows:

$$\begin{aligned} P(t + \Delta t) - P(t) &= \{\text{births}\} - \{\text{deaths}\} + \{\text{immigrants}\} \\ &\approx \beta P(t)\,\Delta t - \delta P(t)\,\Delta t + I\,\Delta t, \end{aligned}$$

so

$$\frac{P(t + \Delta t) - P(t)}{\Delta t} \approx (\beta - \delta)P(t) + I.$$

We take limits as $\Delta t \to 0$ and thus obtain the linear first-order differential equation

$$\frac{dP}{dt} = kP + I \tag{3}$$

with constant coefficients $k = \beta - \delta$ and I. According to Eq. (2), the solution of Eq. (3) is

$$P(t) = P_0 e^{kt} + \frac{I}{k}(e^{kt} - 1). \tag{4}$$

The first term on the right-hand side is the effect of natural population growth, and the second term is the effect of immigration.

EXAMPLE 1 In 1993 the U.S. population was approximately $P_0 = 260$ million and was increasing at the rate of 0.8% annually. In addition, a net immigration into the country at the rate of about 900,000 people per year was occurring. Let's examine the effects of these growth and immigration rates, assuming that they hold constant for the next 20 years. With $k = 0.008$ and $I = 0.9$ (counting people by the million), the differential equation in (3) is

$$\frac{dP}{dt} = (0.008)P + 0.9,$$

and its solution with $P(0) = 260$ is given by

$$P(t) = 260e^{(0.008)t} + (112.5)(e^{(0.008)t} - 1).$$

The two summands here correspond to the two terms on the right-hand side in Eq. (4). The predicted population for the year 2013 is $P(20) \approx 324.6$ million. Of the

20-year predicted U.S. population increase of $324.6 - 260 = 64.6$ million, the amount due to the 0.8% growth rate alone (as if there were no immigration) is

$$260e^{(0.008)(20)} - 260 \approx 45.1 \quad \text{(million)},$$

and the remaining amount due ultimately to immigration is

$$(112.5)\left(e^{(0.008)(20)} - 1\right) \approx 19.5 \quad \text{(million)}.$$

Savings Account with Continuous Deposits

Consider the savings account of Section 7.5 that contains A_0 dollars initially and earns interest at the annual rate r compounded continuously. Now suppose that deposits are added to this account at the rate of Q dollars per year. To simplify the mathematical model, we assume that these deposits are made continuously rather than (for instance) monthly. We may then regard the amount $A(t)$ in the account at time t as a "population" of dollars, with a natural (annual) growth rate r and "immigration" (deposits) at the rate of Q dollars annually. Then by merely changing the notation in Eqs. (3) and (4), we get the differential equation

$$\frac{dA}{dt} = rA + Q, \tag{5}$$

which has the solution

$$A(t) = A_0 e^{rt} + \frac{Q}{r}(e^{rt} - 1). \tag{6}$$

EXAMPLE 2 Suppose that you wish to arrange for your daughter, at the time of her birth, to have \$80000 available for her college expenses when she is 18. You plan to do so by making frequent, small—essentially continuous—deposits in a mutual fund at the rate of Q dollars per year. This fund will accumulate 9% annual interest compounded continuously. What should Q be so that you achieve your goal?

Solution With $A_0 = 0$ and $r = 0.09$, we want the value of Q so that Eq. (6) yields the result

$$A(18) = 80000.$$

That is, we must find Q so that

$$80000 = \frac{Q}{0.09}\left(e^{(0.09)(18)} - 1\right).$$

Solving this equation, we find that $Q \approx 1776.42$. Thus you should deposit \$1776.42 per year, or about \$148.04 per month, to have \$80000 in the fund after 18 years. You may wish to verify that your total deposits will be \$31975.60 and that the total interest accumulated will be \$48024.40.

Cooling and Heating

According to **Newton's law of cooling** (or heating), the time rate of change of the temperature T of a body is proportional to the difference between T and the temperature A of its surroundings. We will assume that A is constant. We may translate this law into the language of differential equations by writing

$$\frac{dT}{dt} = -k(T - A). \tag{7}$$

Here k is a positive constant; the minus sign is required to make $T'(t)$ negative when T exceeds A, and it correctly makes $T'(t)$ positive when A exceeds T.

From Eq. (2) we obtain the solution of Eq. (7):

$$T(t) = A + (T_0 - A)e^{-kt}, \tag{8}$$

where $T_0 = T(0)$.

EXAMPLE 3 A 5-lb roast initially at 50°F is put into a 375°F oven when $t = 0$. The temperature $T(t)$ of the roast is 125°F when $t = 75$ (min). When will the roast be medium-rare, a temperature of 150°F?

Solution Although we could simply substitute $A = 375$ and $T_0 = 50$ into Eq. (8), let us instead solve explicitly the differential equation

$$\frac{dT}{dt} = -k(T - 375) = k(375 - T)$$

that we get from Eq. (7). We separate the variables and integrate to obtain

$$\int \frac{dT}{375 - T} = \int k\,dt; \qquad -\ln(375 - T) = kt + C.$$

When $t = 0$, $T = T_0 = 50$. Substitution of this datum now yields the value $C = -\ln 325$, so

$$-\ln(375 - T) = kt - \ln 325; \qquad 375 - T = 325e^{-kt};$$

and hence

$$T = T(t) = 375 - 325e^{-kt}.$$

We also know that $T = 125$ when $t = 75$. It follows that

$$k = \tfrac{1}{75} \ln \tfrac{325}{250} \approx 0.0035.$$

So all we need to do is solve the equation

$$150 = 375 - 325e^{-kt}.$$

We find that t is approximately 105, so the roast should remain in the oven for about another 30 min. ■

Diffusion of Information and Spread of Disease

Let $N(t)$ denote the number of people (in a fixed population P) who by time t have heard a certain news item spread by the mass media. Under certain common conditions, the time rate of increase of N will be proportional to the number of people who have not yet heard the news. Thus

$$\frac{dN}{dt} = k(P - N). \tag{9}$$

If $N(0) = 0$, the solution of Eq. (9) is

$$N(t) = P \cdot (1 - e^{-kt}). \tag{10}$$

If P and some later value $N(t_1)$ are known, we can then solve for k and thereby determine $N(t)$ for all t. Problem 15 illustrates this situation.

Different infectious diseases spread in different ways. A simple model may be built on the assumption that some infectious diseases spread like information—in a fixed population P, the rate of increase of the number $N(t)$ of people infected with the disease is proportional to the number $P - N$ who are not yet infected. Then N satisfies the differential equation in (9). See Problems 24 and 25 for applications.

Elimination of Pollutants

In Example 4, we envision a lake that has been polluted, perhaps by factories operating on its shores. Suppose that the pollution is halted, perhaps by a legal order or by improved technology. We ask how long it will take for natural processes to reduce the pollutant concentration in the lake to an acceptable level.

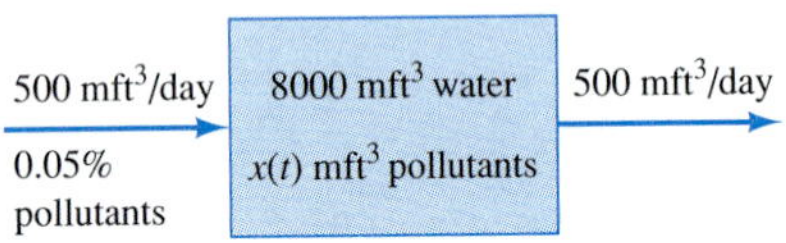

Fig. 7.6.1 The polluted lake of Example 4; volumes are in millions of cubic feet (mft^3)

EXAMPLE 4 Consider a lake with a volume of 8 billion ft^3 and an initial pollutant concentration of 0.25%. An inflowing river brings in 500 million ft^3 (mft^3) of water daily with a (low) pollutant concentration of 0.05%, and an outflowing river removes 500 mft^3 of the lake water daily (Fig. 7.6.1). We make the simplifying assumption that the water in the lake, including that removed by the second river, is perfectly mixed at all times. If so, how long will it take to reduce the pollutant concentration in the lake to 0.10%?

Solution Let $x(t)$ denote the amount of pollutants in the lake after t days, measured in millions of cubic feet. The volume of the lake is 8000 mft^3, and the initial amount $x(0)$ of pollutants is

$$x_0 = (0.25\%)(8000) = (0.0025)(8000) = 20 \quad (\text{mft}^3).$$

We want to know when

$$x(t) = (0.10\%)(8000) - (0.0010)(8000) = 8 \quad (\text{mft}^3).$$

We construct a mathematical model of this situation by estimating the increment Δx in x during a short time interval of duration Δt days. In that time 500 Δt mft^3 of water flows out of the lake. Thus

$$\begin{aligned}\Delta x &= \{\text{pollutant in}\} - \{\text{pollutant out}\}\\ &\approx (0.0005)(500)\,\Delta t - \frac{x(t)}{8000}\cdot 500\,\Delta t = \frac{1}{4}\,\Delta t - \frac{x}{16}\,\Delta t.\end{aligned}$$

So

$$\frac{\Delta x}{\Delta t} \approx \frac{1}{4} - \frac{x}{16}.$$

It follows that

$$\frac{dx}{dt} = \frac{1}{4} - \frac{x}{16}.$$

With $x_0 = 20$, Eq. (2) gives the solution

$$x(t) = 4 + 16e^{-t/16}.$$

We can find the value of t at which $x(t) = 8$ by solving the equation $8 = 4 + 16e^{-t/16}$. This gives

$$t = 16 \ln 4 \approx 22.2 \quad \text{(days)}.$$

7.6 PROBLEMS

In Problems 1 through 10, use the method of derivation of Eq. (2), rather than the equation itself, to find the solution of the given initial value problem.

1. $\dfrac{dy}{dx} = y + 1;\ \ y(0) = 1$ **2.** $\dfrac{dy}{dx} = 2 - y;\ \ y(0) = 3$

3. $\dfrac{dy}{dx} = 2y - 3;\ \ y(0) = 2$

4. $\dfrac{dy}{dx} = \dfrac{1}{4} - \dfrac{y}{16};\ \ y(0) = 20$

5. $\dfrac{dx}{dt} = 2(x - 1);\ \ x(0) = 0$

6. $\dfrac{dx}{dt} = 2 - 3x;\ \ x(0) = 4$

7. $\dfrac{dx}{dt} = 5(x + 2);\ \ x(0) = 25$

8. $\dfrac{dx}{dt} = -3 - 4x;\ \ x(0) = -5$

9. $\dfrac{dv}{dt} = 10(10 - v);\ \ v(0) = 0$

10. $\dfrac{dv}{dt} = -5(10 - v);\ \ v(0) = -10$

11. Zembla had a population of 1.5 million in 1990. Assume that this country's population is growing continuously at a 4% annual rate and that Zembla absorbs 50000 newcomers per year. What will its population be in the year 2010?

12. When a cake is removed from an oven, the temperature of the cake is 210°F. The cake is left to cool at room temperature, which is 70°F. After 30 min the temperature of the cake is 140°F. When will it be 100°F?

13. Payments are made continuously on a mortgage of original amount P_0 dollars at the constant rate of c dollars per month. Let $P(t)$ denote the balance (amount still owed) after t months, and let r denote the monthly interest rate paid by the mortgage holder. (For example, $r = 0.06/12 = 0.005$ if the annual interest rate is 6%.) Derive the differential equation

$$\frac{dP}{dt} = rP - c, \qquad P(0) = P_0.$$

14. Your cousin must pay off an auto loan of \$3600 continuously over a period of 36 months. Apply the result of Problem 13 to determine the monthly payment required if the annual interest rate is (a) 12%; (b) 18%.

15. A rumor about phenylethylamine in the drinking water began to spread one day in a city with a population of 100,000. Within a week, 10000 people had heard this rumor. Assuming that the rate of increase of the number of people who have heard the rumor is proportional to the number who have not yet heard it, how long will it be until half the population of the city has heard the rumor?

16. A tank contains 1000 L of a solution consisting of 50 kg of salt dissolved in water. Pure water is pumped into the tank at the rate of r L/s, and the mixture—kept uniform by stirring—is pumped out at the same rate. After how many seconds will only 10 kg of salt remain in the tank?

17. Derive the solution in Eq. (2) of Eq. (1) under the assumption that $ax + b < 0$.

18. Suppose that a body moves through a resisting medium with resistance proportional to its velocity v, so $dv/dt = -kv$. (a) Show that its velocity $v(t)$ and position $x(t)$ at time t are given by

$$v(t) = v_0 e^{-kt} \quad \text{and} \quad x(t) = x_0 + \frac{v_0}{k}(1 - e^{-kt}).$$

(b) Conclude that the body travels only a *finite* distance v_0/k.

19. A motorboat is moving at 40 ft/s when its motor suddenly quits; 10 s later the boat has slowed to 20 ft/s. Assume, as in Problem 18, that the resistance it encounters while it coasts is proportional to its velocity. How far will the motorboat coast in all?

20. The acceleration of a Lamborghini is proportional to the difference between 250 km/h and the velocity of this sports car. If this car can accelerate from rest to 100 km/h in 10 s, how long will it take the car to accelerate from rest to 200 km/h?

21. Consider the linear first-order differential equation

$$\frac{dx}{dt} + p(t)x(t) = q(t)$$

with variable coefficients. Let $P(t)$ be an antiderivative of $p(t)$. Multiply both sides of the given equation by $e^{P(t)}$, and note that the left-hand side of the resulting equation is $D_t[e^{P(t)}x(t)]$. Conclude by antidifferentiation that

$$x(t) = e^{-P(t)}\left[\int e^{P(t)}q(t)\,dt + C\right].$$

22. Use the method of Problem 21 to derive the solution

$$x(t) = x_0e^{-at} + b\frac{e^{ct} - e^{-at}}{a + c}$$

of the differential equation $dx/dt + ax = be^{ct}$ (under the assumption that $a + c \neq 0$).

23. A 30-yr-old engineer accepts a position with a starting salary of \$30000/yr. Her salary S increases exponentially, with

$$S(t) = 30e^{(0.05)t}$$

thousand dollars after t years. Meanwhile, 12% of her salary is deposited continuously in a retirement account, which accumulates interest at an annual rate of 6% compounded continuously. (a) Estimate ΔA in terms of Δt to derive this equation for the amount $A(t)$ in her retirement account at time t:

$$\frac{dA}{dt} - (0.06)A = (3.6)e^{(0.05)t}.$$

(b) Apply the result of Problem 22 to compute $A(40)$, the amount available for her retirement at age 70.

24. Pottstown has a fixed population of 10000 people. On January 1, 1000 people have the flu; on April 1, 2000 people have it. Assume that the rate of increase of the number $N(t)$ who have the flu is proportional to the number who don't have it. How many will have the disease on October 1?

25. Let $x(t)$ denote the number of people in Athens, Georgia, of population 100,000, who have the Tokyo flu. The rate of change of $x(t)$ is proportional to the number of those in Athens who do not yet have the disease. Suppose that 20000 have the flu on March 1 and that 60000 have it on March 16. (a) Set up and solve a differential equation to find $x(t)$. (b) On what date will the number of people infected with the disease reach 80000? (c) What happens in the long run?

CHAPTER 7 REVIEW: *Definitions, Concepts, Results*

Use this list as a guide to concepts that you may need to review.

1. The laws of exponents
2. The laws of logarithms
3. The definition of the natural logarithm function
4. The graph of $y = \ln x$
5. The definition of the number e
6. The definition of the natural exponential function
7. The inverse function relationship between $\ln x$ and e^x
8. The graphs of $y = e^x$ and $y = e^{-x}$
9. Differentiation of $\ln u$ and e^u, where u is a differentiable function of x
10. The order of magnitude of $(\ln x)/x^k$ and x^k/e^x as $x \to +\infty$
11. The number e as a limit
12. The definition of general exponential and logarithm functions
13. Differentiation of a^u and $\log_a u$
14. Logarithmic differentiation
15. Solution of the differential equation $dx/dt = kx$
16. The natural growth equation
17. Radioactive decay and radiocarbon dating
18. Solution of a linear first-order differential equation with constant coefficients
19. Solution of separable first-order differential equations
20. Evaluating the constant of integration in an initial value problem

CHAPTER 7 *Miscellaneous Problems*

Differentiate the functions given in Problems 1 through 24.

1. $f(x) = \ln 2\sqrt{x}$

2. $f(x) = e^{-2\sqrt{x}}$

3. $f(x) = \ln(x - e^x)$

4. $f(x) = 10^{\sqrt{x}}$

5. $f(x) = \ln(2^x)$

6. $f(x) = \log_{10}(\sin x)$

7. $f(x) = x^3e^{-1/x^2}$

8. $f(x) = x(\ln x)^2$

9. $f(x) = (\ln x)[\ln(\ln x)]$

10. $f(x) = \exp(10^x)$

11. $f(x) = 2^{\ln x}$

12. $f(x) = \ln\left(\dfrac{e^x + e^{-x}}{e^x - e^{-x}}\right)$

13. $f(x) = e^{(x+1)/(x-1)}$

14. $f(x) = \ln\left(\sqrt{1 + x}\,\sqrt[3]{2 + x^2}\right)$

15. $f(x) = \ln\left(\dfrac{x - 1}{3 - 4x^2}\right)^{3/2}$

16. $f(x) = \sin(\ln x)$

17. $f(x) = \exp\left(\sqrt{1 + \sin^2 x}\right)$

18. $f(x) = \dfrac{x}{(\ln x)^2}$

19. $f(x) = \ln(3^x \sin x)$

20. $f(x) = (\ln x)^x$

21. $f(x) = x^{1/x}$

22. $f(x) = x^{\sin x}$ **23.** $f(x) = (\ln x)^{\ln x}$

24. $f(x) = (\sin x)^{\cos x}$

Evaluate the indefinite integrals in Problems 25 through 36.

25. $\displaystyle\int \frac{dx}{1-2x}$ **26.** $\displaystyle\int \frac{\sqrt{x}}{1+x^{3/2}}\,dx$

27. $\displaystyle\int \frac{3-x}{1+6x-x^2}\,dx$ **28.** $\displaystyle\int \frac{e^x - e^{-x}}{e^x + e^{-x}}\,dx$

29. $\displaystyle\int \frac{\sin x}{2+\cos x}\,dx$ **30.** $\displaystyle\int \frac{e^{-1/x^2}}{x^3}\,dx$

31. $\displaystyle\int \frac{10^{\sqrt{x}}}{\sqrt{x}}\,dx$ **32.** $\displaystyle\int \frac{1}{x(\ln x)^2}\,dx$

33. $\displaystyle\int e^x\sqrt{1+e^x}\,dx$ **34.** $\displaystyle\int \frac{1}{x}\sqrt{1+\ln x}\,dx$

35. $\displaystyle\int 2^x 3^x\,dx$ **36.** $\displaystyle\int \frac{dx}{x^{1/3}(1+x^{2/3})}$

Solve the initial value problems in Problems 37 through 44.

37. $\dfrac{dx}{dt} = 2t;\quad x(0) = 17$ **38.** $\dfrac{dx}{dt} = 2x;\quad x(0) = 17$

39. $\dfrac{dx}{dt} = e^t;\quad x(0) = 2$ **40.** $\dfrac{dx}{dt} = e^x;\quad x(0) = 2$

41. $\dfrac{dx}{dt} = 3x - 2;\quad x(0) = 3$

42. $\dfrac{dx}{dt} = x^2t^2;\quad x(0) = -1$

43. $\dfrac{dx}{dt} = x\cos t;\quad x(0) = \sqrt{2}$

44. $\dfrac{dx}{dt} = \sqrt{x};\quad x(1) = 0$

Sketch the graphs of the equations given in Problems 45 through 49.

45. $y = e^{-x}\sqrt{x}$ **46.** $y = x - \ln x$

47. $y = \sqrt{x} - \ln x$ **48.** $y = x(\ln x)^2$

49. $y = e^{-1/x}$

50. Find the length of the curve $y = \frac{1}{2}x^2 - \frac{1}{4}\ln x$ from $x = 1$ to $x = e$.

51. A grain warehouse holds B bushels of grain, which is deteriorating such that only $B \cdot 2^{-t/12}$ bushels will be salable after t months. Meanwhile, the grain's market price is increasing linearly: After t months it will be $2 + \frac{1}{12}t$ dollars per bushel. After how many months should the grain be sold to maximize the revenue obtained?

52. You have borrowed \$1000 at 10% annual interest, compounded continuously, to plant timber on a tract of land. Your agreement is to repay the loan, plus interest, when the timber is cut and sold. If the cut timber can be sold after t years for $800\exp\left(\frac{1}{2}\sqrt{t}\right)$ dollars, when should you cut and sell to maximize the profit?

53. Blood samples from 1000 students are to be tested for a certain disease known to occur in 1% of the population. Each test costs \$5, so it would cost \$5000 to test the samples individually. Suppose, however, that "lots" made up of x samples each are formed by pooling halves of individual samples and that these lots are tested first (for \$5 each). Only when a lot tests positive—the probability of this is $1 - (0.99)^x$—will the x samples used to make up this lot be tested individually. (a) Show that the total expected number of tests is

$$f(x) = \frac{1000}{x}\left[(1)(0.99)^x + (x+1)(1-(0.99)^x)\right]$$

$$= 1000 + \frac{1000}{x} - 1000\cdot(0.99)^x$$

if $x \geqq 2$. (b) Show that the value of x that minimizes $f(x)$ is a root of the equation

$$x = \frac{(0.99)^{-x/2}}{\sqrt{\ln(100/99)}}.$$

Because the denominator is approximately 0.1, it may be convenient to solve instead the simpler equation $x = 10\cdot(0.99)^{-x/2}$. (c) From the results in parts (a) and (b), compute the minimum (expected) cost of using this batch method to test the original 1000 samples.

54. Deduce from Problem 73 in Section 7.2 that

$$\lim_{x\to 0^+} x^x = 1.$$

55. Show that

$$\lim_{x\to 0}\frac{\ln(1+x)}{x} = 1$$

by considering the value of $D_x \ln x$ for $x = 1$. Thus show that $\ln(1+x) \approx x$ if x is very close to zero.

56. (a) Prove that

$$\lim_{h\to 0}\frac{a^h - 1}{h} = \ln a$$

by considering the definition of the derivative of a^x at $x = 0$. (b) Substitute $h = 1/n$ to obtain

$$\ln a = \lim_{n\to\infty} n(a^{1/n} - 1).$$

(c) Approximate $\ln 2$ by taking $n = 1024 = 2^{10}$ and using only the square root key (10 times) on a pocket calculator.

57. Suppose that the fish population $P(t)$ in a lake is attacked by disease at time $t = 0$, with the result that

$$\frac{dP}{dt} = -3\sqrt{P}$$

thereafter. Time t is measured in weeks. Initially there are $P_0 = 900$ fish in the lake. How long will it take for all the fish to die?

58. A race car sliding along a level surface is decelerated by frictional forces proportional to its speed. Suppose that it decelerates initially at 2 m/s^2 and travels a total distance of 1800 m. What was its initial velocity? (See Problem 18 of Section 7.6.)

59. A home mortgage of $120,000 is to be paid off continuously over a period of 25 yr. Apply the result of Problem 13 in Section 7.6 to determine the monthly payment if the annual interest rate, compounded continuously, is (a) 8%; (b) 12%.

60. A powerboat weighs 32000 lb, and its motor provides a thrust of 5000 lb. Assume that the water resistance is 100 lb for each foot per second of the boat's speed. Then the velocity $v(t)$ (in ft/s) of the boat at time t (in seconds) satisfies the differential equation

$$1000\frac{dv}{dt} = 5000 - 100v.$$

Find the maximum velocity that the boat can attain if it starts from rest.

61. The temperature inside my freezer is $-16°$C and the room temperature is a constant 20°C. At 11 P.M. one evening the power goes off due to an ice storm. At 6 A.M. the next morning I see that the temperature in the freezer has risen to $-10°$C. At what time will the temperature in the freezer reach the critical value of 0°C if the power remains off?

62. Suppose that the action of fluorocarbons depletes the ozone in the upper atmosphere by 0.25% annually, so that the amount A of ozone in the upper atmosphere satisfies the differential equation

$$\frac{dA}{dt} = -\frac{1}{400}A \quad (t \text{ in years}).$$

(a) What percentage of the original amount A_0 of upper-atmospheric ozone will remain 25 yr from now? (b) How long will it take for the amount of upper-atmospheric ozone to be reduced to half its initial amount?

63. A car starts from rest and travels along a straight road. Its engine provides a constant acceleration of a feet per second per second. Air resistance and road friction cause a deceleration of ρ feet per second per second for every foot per second of the car's velocity. (a) Show that the car's velocity after t seconds is

$$v(t) = \frac{a}{\rho}(1 - e^{-\rho t}).$$

(b) If $a = 17.6$ ft/s^2 and $\rho = 0.1$, find v when $t = 10$ s, and find the limiting velocity as $t \to +\infty$. Give each answer in *miles per hour* as well as feet per second.

64. Immediately after an accident in a nuclear power plant, the level of radiation there was 10 times the safe limit. After 6 mo it dropped to 9 times the safe limit. Assuming exponential decay, how long (in years) after the accident will the radiation level drop to the safe limit?

65. Figure 7.MP.1 shows the graphs of $f(x) = x^{1/2}$, $g(x) = \ln x$, and $h(x) = x^{1/3}$ plotted on the interval $[0.2, 10]$. You can see that the graph of f remains above the graph of $\ln x$, whereas the graph of h dips below the graph of $\ln x$. But because $\ln x$ increases *less* rapidly than any positive power of x, the graph of h must eventually cross the graph of $\ln x$ and rise above it. Finally, it is easy to believe that, for a suitable choice of p between 2 and 3, the graph of $j(x) = x^{1/p}$ never dips below the graph of $\ln x$ but does drop down just far enough to be tangent to the graph of $\ln x$ at a certain point. (a) Show that $f(x) > \ln x$ for all $x > 0$ by finding the global minimum value of $f(x) - \ln x$ on the interval $(0, +\infty)$. (b) Use Newton's method to find the value at which $h(x)$ crosses the graph of $\ln x$ and rises above it—the value of x *not* shown in Fig. 7.MP.1. (c) Find the value of p for which the graph of $j(x)$ is tangent to the graph of $\ln x$ at the point $(q, \ln q)$.

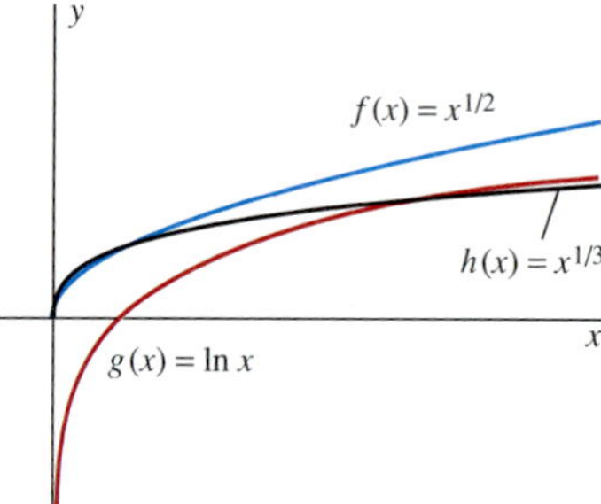

Fig. 7.MP.1 The three functions of Problem 65

CHAPTER 8

FURTHER CALCULUS OF TRANSCENDENTAL FUNCTIONS

John Bernoulli (1667–1748)

In the eighteenth century the remarkable Swiss family Bernoulli was to mathematics what the Bach family was to music. Eight different Bernoullis were sufficiently prominent that more than two centuries later, they rate entries in the *Dictionary of Scientific Biography*. The brothers James (1654–1705) and John Bernoulli played crucial roles in the early development of Leibniz's version of the calculus based on infinitely small differentials, which in continental European science predominated over Newton's version based more explicitly on limits of ratios. It was James Bernoulli who introduced the word *integral* in suggesting the name *calculus integralis* (instead of Leibniz's original *calculus summatorius*) for the subject inverse to the *calculus differentialis*.

John Bernoulli first studied mathematics under his older brother James at the university in Basel, Switzerland, but soon they were on an equal footing in mathematical understanding. In 1691 John Bernoulli visited Paris and there met the young Marquis de l'Hôpital (1661–1704), who was anxious to learn the secrets of the new infinitesimal calculus. In return for a generous monthly stipend, Bernoulli agreed to tutor the wealthy Marquis and continued the lessons (and the financial arrangement) by mail after his return to Basel. The result of this correspondence was the first differential calculus textbook, published by l'Hôpital in 1696. This text is remembered mainly for its inclusion of a result of Bernoulli that is known as *l'Hôpital's rule*. This result concerns the limit as $x \to a$ of a quotient $f(x)/g(x)$ whose numerator and denominator both approach zero as $x \to a$, so mere substitution of the value $x = a$ would give the "indeterminate form" $0/0$. The definition of the derivative involves such a limit, so indeterminate forms pervade the subject of calculus.

Both James and John Bernoulli worked on (and solved) the *catenary problem,* which asks for the shape of a hanging cable suspended between two fixed points, assuming that it is inelastic (unstretchable) but perfectly flexible. The Bernoullis showed that, in terms of hyperbolic functions, such a hanging cable takes the shape of a curve of the form

$$y = a\cosh\left(\frac{x}{a}\right).$$

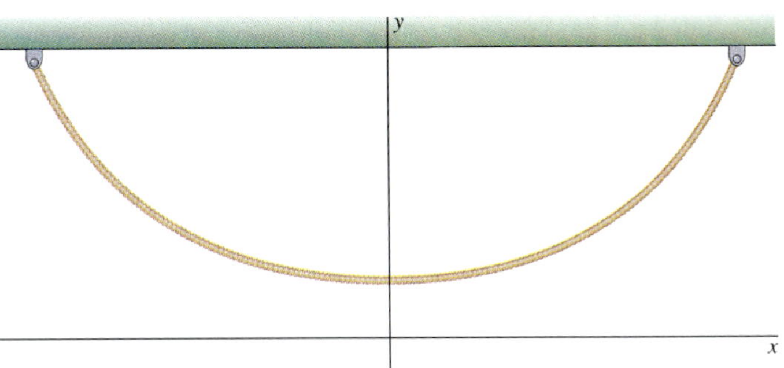

What is the shape of a hanging cable?

8.1 INTRODUCTION

The function f is called an **algebraic function** provided that $y = f(x)$ satisfies an equation of the form

$$a_n(x)y^n + a_{n-1}(x)y^{n-1} + \cdots + a_1(x)y + a_0(x) = 0, \qquad \textbf{(1)}$$

where the coefficients $a_0(x), a_1(x), \ldots, a_n(x)$ are polynomials in x. For example, because the equation $y^2 - p(x) = 0$ has the form in (1), the square root of the polynomial $p(x)$—that is, $f(x) = \sqrt{p(x)}$—is an algebraic function. The equation $q(x)y - p(x) = 0$ also has the necessary form, so a rational function [a quotient of polynomials; here, $y(x) = p(x)/q(x)$] is also an algebraic function.

A function that is *not* algebraic is said to be **transcendental.** The natural logarithm function $\ln x$ and the natural exponential function e^x are transcendental functions, as are the six familiar trigonometric functions. In this chapter we shall study the remaining transcendental functions of elementary character—the inverse trigonometric functions and the hyperbolic functions. These functions have extensive scientific applications and provide the basis for certain important methods of integration (discussed in Chapter 9). In Sections 8.3 and 8.4 we also study certain limit expressions ("indeterminate forms") that typically involve transcendental functions.

8.2 INVERSE TRIGONOMETRIC FUNCTIONS

Recall that the function f is said to be **one-to-one** on its domain of definition D if, given x_1 and x_2 in D, $x_1 \neq x_2$ implies that $f(x_1) \neq f(x_2)$: "Different inputs give different outputs." (To prove that f is one-to-one, it is usually easier to prove the contrapositive, that if $f(x_1) = f(x_2)$, then $x_1 = x_2$.) What is important here is that if f is one-to-one on its domain of definition, then it has an inverse function f^{-1}. This inverse function is *defined* by

$$f^{-1}(x) = y \quad \text{if and only if} \quad f(y) = x. \qquad \textbf{(1)}$$

For example, from Chapter 7 we are familiar with the pair of inverse functions

$$f(x) = e^x \quad \text{and} \quad f^{-1}(x) = \ln x.$$

From a geometric viewpoint, Eq. (1) implies that the graphs $y = f(x)$ and $y = f^{-1}(x)$ are reflections across the 45° line $y = x$, like the familiar graphs $y = e^x$ and $y = \ln x$ in Fig. 8.2.1.

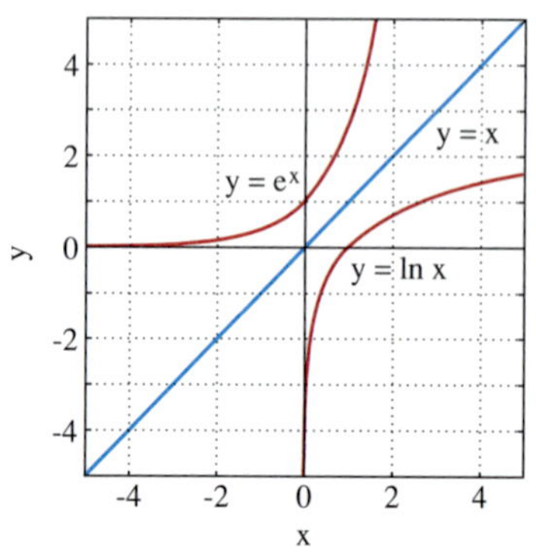

Fig. 8.2.1 The graphs $y = e^x$ and $y = \ln x$ are reflections across the line $y = x$.

The Inverse Tangent Function

Here we want to define the inverses of the trigonometric functions, beginning with the inverse tangent function. We must, however, confront the fact that the trigonometric functions fail to be one-to-one because the period of each of the six is π or 2π. For example, $\tan x = 1$ if x is $\pi/4$ *or* $\pi/4$ plus any integral multiple of π. These many values of x, all with tangent equal to 1, correspond to the multiple points of intersection of the graph $y = \tan x$ and the horizontal line $y = 1$ in Fig. 8.2.2.

Figure 8.2.3 is the reflection of Fig. 8.2.2 across the 45° line $y = x$. The multiple intersections of $x = \tan y$ and the vertical line $x = 1$ indicate that we must make a choice in order to define $\tan^{-1} 1$. That is, we *cannot* define $y = \tan^{-1} x$, the inverse of the tangent function, by saying simply that y is the number such that $\tan y = x$. There are *many* such values of y, and we must specify which particular one is to be used.

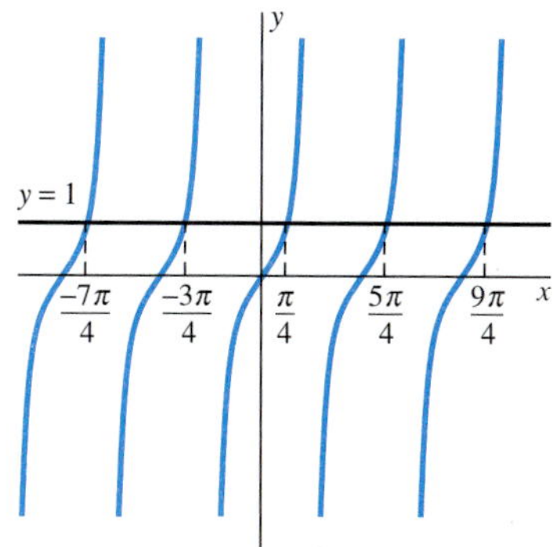

Fig. 8.2.2 The tangent function takes on every real number value infinitely often.

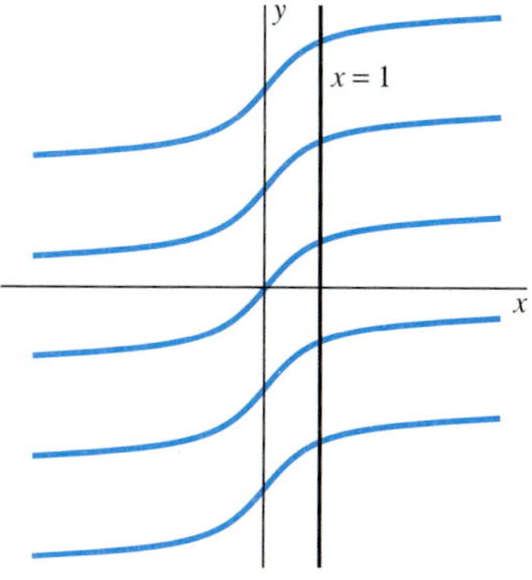

Fig. 8.2.3 Simply reflecting the graph of $y = \tan x$ across the 45°-line $y = x$ does not produce the graph of a function.

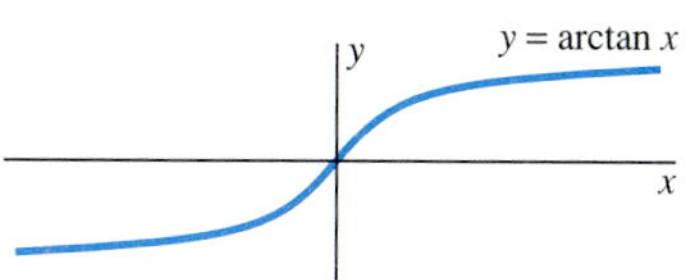

Fig. 8.2.4 The inverse tangent function has domain the set of all real numbers.

[Note that the symbol -1 in the notation $\tan^{-1}x$ is not an exponent—it does *not* mean $(\tan x)^{-1}$.]

We do this by suitably restricting the domain of the tangent function. Because the function $\tan x$ is increasing on $(-\pi/2, \pi/2)$ and its range of values is $(-\infty, +\infty)$, for each x in $(-\infty, +\infty)$ there is *one* number y in $(-\pi/2, \pi/2)$ such that $\tan y = x$. This observation leads to the following definition of the *inverse tangent* (or *arctangent*) function, denoted by $\tan^{-1}x$ or $\arctan x$.

Definition ***The Inverse Tangent Function***

The **inverse tangent** (or **arctangent**) **function** is defined as follows:

$$y = \tan^{-1}x \quad \text{if and only if} \quad \tan y = x \quad \text{and} \quad -\pi/2 < y < \pi/2, \tag{2}$$

where x is an arbitrary real number.

Because the tangent function attains all real values, $\tan^{-1}x$ is defined for all real numbers x; $\tan^{-1}x$ is that number y in the interval $(-\pi/2, \pi/2)$ whose tangent is x. The graph of $y = \tan^{-1}x$ is the reflection of the graph of $y = \tan x$, $-\pi/2 < x < \pi/2$, across the line $y = x$ (Fig. 8.2.4).

It follows from Eq. (2) that

$$\tan(\tan^{-1}x) = x \quad \text{for all } x \quad \text{and} \tag{3a}$$

$$\tan^{-1}(\tan x) = x \quad \text{if } -\pi/2 < x < \pi/2. \tag{3b}$$

Because the derivative of $\tan x$ is positive for all x in the interval $(-\pi/2, \pi/2)$, it follows from Theorem 1 in Section 7.1 that $\tan^{-1}x$ is differentiable for all x. We can therefore differentiate both sides of the identity in Eq. (3a). First we write that identity in the form

$$\tan y = x,$$

where $y = \tan^{-1}x$. Then the chain rule yields

$$(\sec^2 y)\frac{dy}{dx} = 1;$$

$$\frac{dy}{dx} = \frac{1}{\sec^2 y} = \frac{1}{1 + \tan^2 y} = \frac{1}{1 + x^2}.$$

Thus

$$D_x \tan^{-1}x = \frac{1}{1 + x^2}, \tag{4}$$

and if u is any differentiable function of x, then the chain rule yields

$$D_x \tan^{-1}u = \frac{1}{1 + u^2} \cdot \frac{du}{dx}. \tag{5}$$

The definition of the inverse cotangent function is similar to that of the inverse tangent function, except that we begin by restricting the cotangent function to the interval $(0, \pi)$, where it is a decreasing function attaining all real values. Thus the **inverse cotangent** (or **arccotangent**) **function** is defined as

$$y = \cot^{-1} x \quad \text{if and only if} \quad \cot y = x \quad \text{and} \quad 0 < y < \pi, \tag{6}$$

where x is any real number. Then differentiation of both sides of the identity $\cot(\cot^{-1} x) = x$ leads, as in the derivation of Eq. (4), to

$$D_x \cot^{-1} x = -\frac{1}{1 + x^2}.$$

If u is a differentiable function of x, then the chain rule gives

$$D_x \cot^{-1} u = -\frac{1}{1 + u^2} \cdot \frac{du}{dx}. \tag{7}$$

The inverse cotangent function is of little use in antidifferentiation problems because, if

$$\int f(x)\,dx = \cot^{-1} x + C, \quad \text{then} \quad \int f(x)\,dx = -\tan^{-1} x + C.$$

EXAMPLE 1 A mountain climber on one edge of a deep canyon 800 ft wide sees a large rock fall from the opposite edge at time $t = 0$. As he watches the rock plummet downward, his eyes first move slowly, then faster, then more slowly again. Let α be the angle of depression of his line of sight below the horizontal. At what angle α would the rock *seem* to be moving the most rapidly? That is, when would $d\alpha/dt$ be maximal?

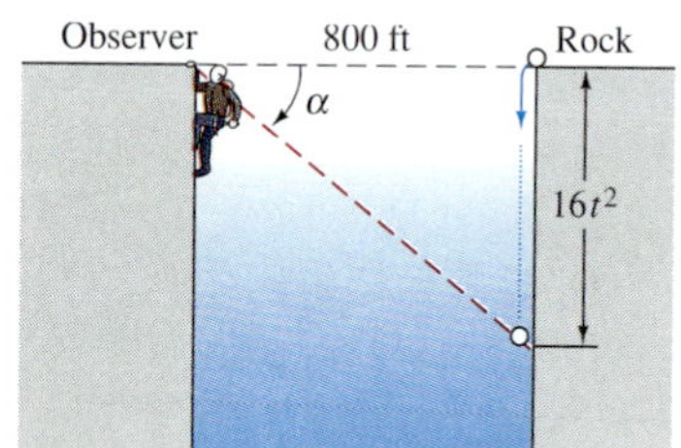

Fig. 8.2.5 The falling rock of Example 1.

Solution From our study of constant acceleration in Section 5.2, we know that the rock will fall $16t^2$ feet in the first t seconds. Refering to Fig. 8.2.5, we see that the value of α at time t will be

$$\alpha = \alpha(t) = \tan^{-1}\left(\frac{16t^2}{800}\right) = \tan^{-1}\left(\frac{t^2}{50}\right).$$

Hence

$$\frac{d\alpha}{dt} = \frac{1}{1 + \left(\dfrac{t^2}{50}\right)^2} \cdot \frac{2t}{50} = \frac{100t}{t^4 + 2500}.$$

To find when $d\alpha/dt$ is maximal, we find when *its* derivative is zero:

$$\frac{d}{dt}\left(\frac{d\alpha}{dt}\right) = \frac{100(t^4 + 2500) - (100t)(4t^3)}{(t^4 + 2500)^2} = \frac{100(2500 - 3t^4)}{(t^4 + 2500)^2}.$$

So $d^2\alpha/dt^2$ is zero when $3t^4 = 2500$—that is, when

$$t = \sqrt[4]{\frac{2500}{3}} \approx 5.37 \quad \text{(s)}.$$

This is the value of t when $d\alpha/dt$ is maximal, and at this time we have $t^2 = 50/\sqrt{3}$. So the angle at this time is

$$\alpha = \arctan\left(\frac{1}{50}\cdot\frac{50}{\sqrt{3}}\right) = \arctan\left(\frac{1}{\sqrt{3}}\right) = \frac{\pi}{6}.$$

The *apparent* speed of the falling rock is greatest when the climber's line of sight is 30° below the horizontal. The actual speed of the rock is then $32t$ with $t \approx 5.37$ and thus is about 172 ft/s. ■

The Inverse Sine Function

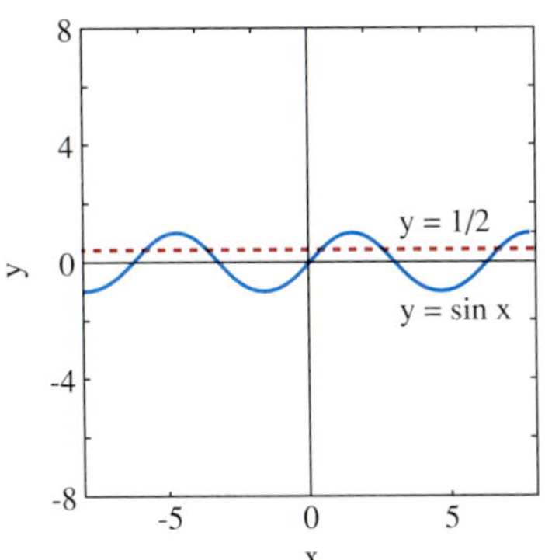

Fig. 8.2.6 Multiple values of x such that $\sin x = \frac{1}{2}$

Figure 8.2.6 shows the graph of the sine function and the horizontal line $y = \frac{1}{2}$. Because this line meets the graph of the sine function in more than one point (indeed, at infinitely many points), the sine function takes on the value $\frac{1}{2}$ for many different values of x. For example, $\sin x = \frac{1}{2}$ if x is *either* $\pi/6$ plus any integral multiple of 2π *or* $5\pi/6$ plus any integral multiple of 2π.

Figure 8.2.7 is the reflection of Fig. 8.2.6 across the 45° line $y = x$. The multiple intersections of $x = \sin y$ and the vertical line $x = \frac{1}{2}$ indicate that we must make a choice in order to define $\sin^{-1}\left(\frac{1}{2}\right)$. That is, we *cannot* define $y = \sin^{-1}x$, the inverse of the sine function, by saying merely that y is the number such that $\sin y = x$. There are *many* such values of y, and we must specify which particular one is to be used.

We do this by suitably restricting the domain of the sine function. Because the function $\sin x$ is increasing on $[-\pi/2, \pi/2]$ and its range of values is $[-1, 1]$, for each x in $[-1, 1]$ there is *one* number y in $[-\pi/2, \pi/2]$ such that $\sin y = x$. This observation leads to the following definition of the *inverse sine* (or *arcsine*) function, denoted by $\sin^{-1}x$ or by $\arcsin x$.

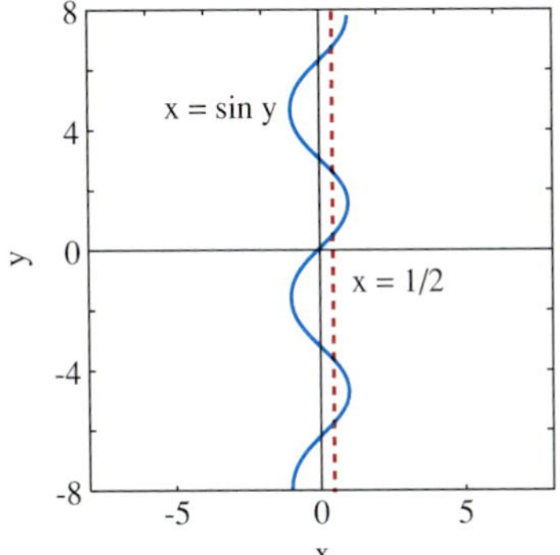

Fig. 8.2.7 There are many possible choices for $y = \sin^{-1}(\frac{1}{2})$.

Definition *The Inverse Sine Function*

The **inverse sine** (or **arcsine**) **function** is defined as follows:

$$y = \sin^{-1}x \quad \text{if and only if} \quad \sin y = x \quad \text{and} \quad -\pi/2 \leqq y \leqq \pi/2, \tag{8}$$

where $-1 \leqq x \leqq 1$.

Thus if x is between -1 and $+1$ (inclusive), then $\sin^{-1}x$ is that number y between $-\pi/2$ and $\pi/2$ such that $\sin y = x$. Even more briefly, $\arcsin x$ is the angle (in radians) nearest zero whose sine is x. For instance,

$$\sin^{-1}1 = \frac{\pi}{2}, \quad \sin^{-1}0 = 0, \quad \sin^{-1}(-1) = -\frac{\pi}{2},$$

and $\sin^{-1}2$ does not exist.

Because interchanging x and y in the equation $\sin y = x$ yields $y = \sin x$, it follows from Eq. (8) that the graph of $y = \sin^{-1}x$ is the reflection of the graph of $y = \sin x$, $-\pi/2 \leqq x \leqq \pi/2$, across the line $y = x$ (Fig. 8.2.8).

It also follows from Eq. (8) that

$$\sin(\sin^{-1}x) = x \quad \text{if} \quad -1 \leqq x \leqq 1 \quad \text{and} \tag{9a}$$

$$\sin^{-1}(\sin x) = x \quad \text{if} \quad -\pi/2 \leqq x \leqq \pi/2. \tag{9b}$$

Because the derivative of $\sin x$ is positive for $-\pi/2 < x < \pi/2$, it follows from Theorem 1 of Section 7.1 that $\sin^{-1}x$ is differentiable on $(-1, 1)$. We can therefore differentiate both sides of the identity in (9a), but we begin by writing it in the form

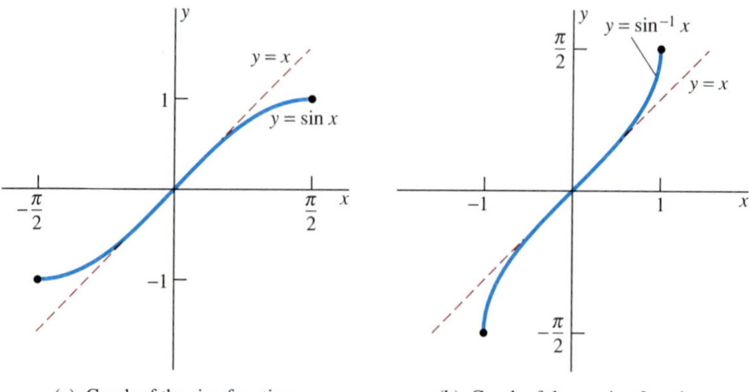

(a) Graph of the sine function
(b) Graph of the arcsine function

Fig. 8.2.8 The graphs $y = \sin x$ and $y = \sin^{-1} x$ are reflections of each other across the line $y = x$.

$$\sin y = x,$$

where $y = \sin^{-1} x$. Then differentiation with respect to x gives

$$(\cos y)\,\frac{dy}{dx} = 1,$$

so

$$\frac{dy}{dx} = \frac{1}{\cos y} = \frac{1}{\sqrt{1 - \sin^2 y}} = \frac{1}{\sqrt{1 - x^2}}.$$

We are correct in taking the positive square root in this computation because $\cos y > 0$ for $-\pi/2 < y < \pi/2$. Thus

$$D_x \sin^{-1} x = \frac{1}{\sqrt{1 - x^2}} \tag{10}$$

provided that $-1 < x < 1$. Combining this result with the chain rule, we get

$$D_x \sin^{-1} u = \frac{1}{\sqrt{1 - u^2}} \cdot \frac{du}{dx} \tag{11}$$

if u is a differentiable function with values in the interval $(-1, 1)$.

EXAMPLE 2 If $y = \sin^{-1} x^2$, then Eq. (11) with $u = x^2$ yields

$$\frac{dy}{dx} = \frac{1}{\sqrt{1 - (x^2)^2}} \cdot 2x = \frac{2x}{\sqrt{1 - x^4}}.$$

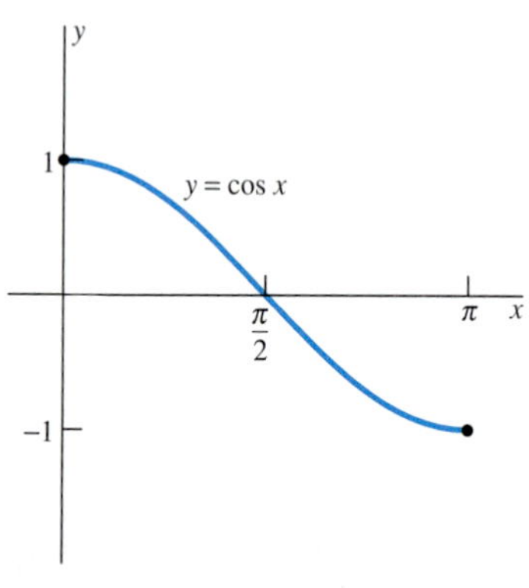

Fig. 8.2.9 The cosine function is decreasing on the interval $0 \leqq x \leqq \pi$

The definition of the inverse cosine function is similar to that of the inverse sine function, except that we begin by restricting the cosine function to the interval $[0, \pi]$, where it is a decreasing function (Fig. 8.2.9). Thus the **inverse cosine** (or **arccosine**) **function** is defined by means of the rule

$$y = \cos^{-1} x \quad \text{if and only if} \quad \cos y = x \quad \text{and} \quad 0 \leqq y \leqq \pi, \tag{12}$$

where $-1 \leqq x \leqq 1$. Thus $\cos^{-1} x$ is the angle in $[0, \pi]$ whose cosine is x. For instance,

$$\cos^{-1} 1 = 0, \qquad \cos^{-1} 0 = \frac{\pi}{2}, \qquad \cos^{-1}(-1) = \pi.$$

We may compute the derivative of $\cos^{-1} x$, also written $\arccos x$, by differentiation of both sides of the identity

$$\cos(\cos^{-1} x) = x \qquad (-1 < x < 1).$$

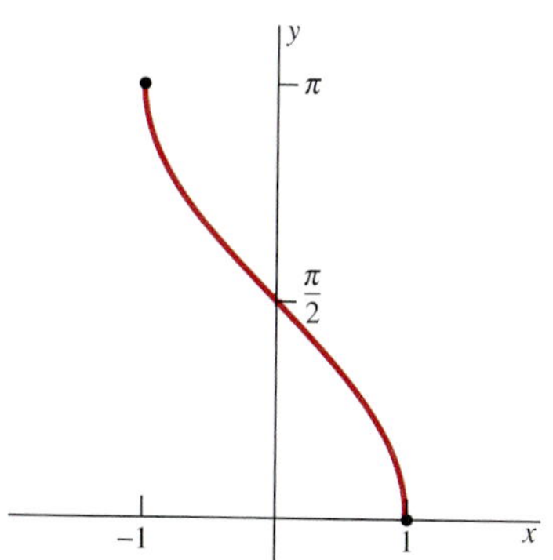

Fig. 8.2.10 The graph $y = \cos^{-1} x$ of the arccosine function

The computations are similar to those for $D_x \sin^{-1} x$ and lead to the result

$$D_x \cos^{-1} x = -\frac{1}{\sqrt{1 - x^2}}.$$

And if u denotes a differentiable function of x, the chain rule then gives

$$D_x \cos^{-1} u = -\frac{1}{\sqrt{1 - u^2}} \cdot \frac{du}{dx}. \qquad \textbf{(13)}$$

Figure 8.2.10 shows the graph of $y = \cos^{-1} x$ as the reflection of the graph of $y = \cos x$, $0 \leqq x \leqq \pi$, across the line $y = x$.

The Inverse Secant Function

Figure 8.2.11 shows that the secant function is increasing on each of the intervals $[0, \pi/2)$ and $(\pi/2, \pi]$. On the union of these two intervals, the secant function attains all real values y such that $|y| \geqq 1$. We can therefore define the inverse secant function, denoted by $\sec^{-1} x$ or by $\operatorname{arcsec} x$, by restricting the secant function to the union of the two intervals $[0, \pi/2)$ and $(\pi/2, \pi]$.

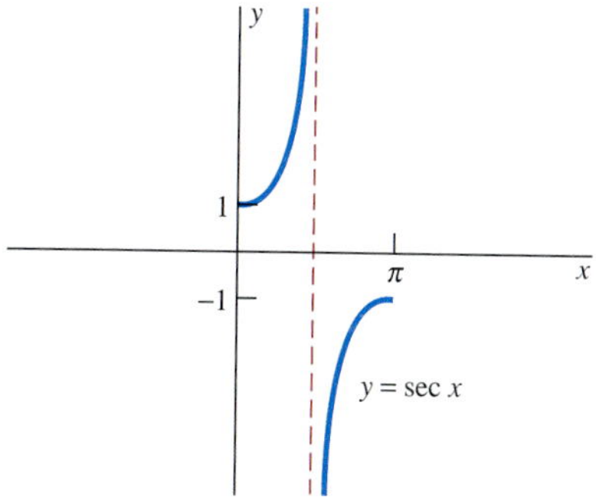

Fig. 8.2.11 Restriction of the secant function to the union of the two intervals $[0, \pi/2)$ and $(\pi/2, \pi]$

Definition *The Inverse Secant Function*

The **inverse secant** (or **arcsecant**) **function** is defined as follows:

$$y = \sec^{-1} x \quad \text{if and only if} \quad \sec y = x \quad \text{and} \quad 0 \leqq y \leqq \pi, \qquad \textbf{(14)}$$

where $|x| \geqq 1$.

Remark Some textbooks offer alternative definitions of the inverse secant based on different intervals of definition of $\sec x$ (see Problem 63). The definition given here, however, satisfies the condition that

$$\sec^{-1} x = \cos^{-1} \frac{1}{x} \qquad (\text{if } |x| > 1),$$

which is convenient for calculator-computer calculations (see Problem 62). Moreover, our definition of $\sec^{-1} x$ is the same as that used in computer algebra systems such as *Maple* and *Mathematica* (see Problem 61).

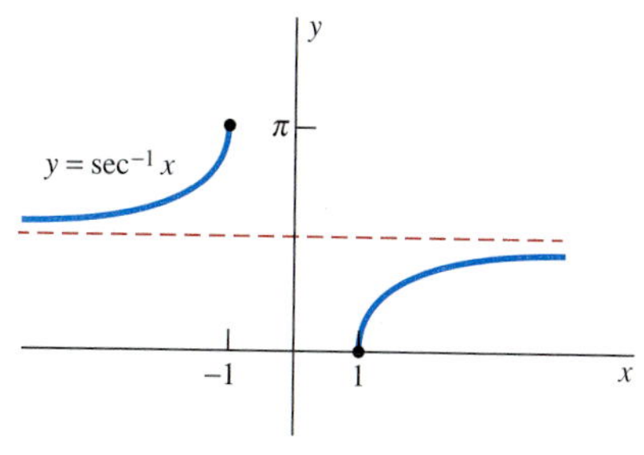

Fig. 8.2.12 The graph of $y = \operatorname{arcsec} x = \sec^{-1} x$

The graph of $y = \sec^{-1} x$ is the reflection of the graph of $y = \sec x$, suitably restricted to the intervals $0 \leqq x < \pi/2$ and $\pi/2 < x \leqq \pi$, across the line $y = x$ (Fig. 8.2.12). It follows from the definition of the inverse secant that

$$\sec(\sec^{-1} x) = x \qquad \text{if } |x| \geqq 1, \tag{15a}$$

$$\sec^{-1}(\sec x) = x \qquad \text{for } x \text{ in } [0, \pi/2) \cup (\pi/2, \pi]. \tag{15b}$$

Following the now-familiar pattern, we find $D_x \sec^{-1} x$ by differentiating both sides of Eq. (15a) in the form

$$\sec y = x,$$

where $y = \sec^{-1} x$. This yields

$$(\sec y \tan y)\frac{dy}{dx} = 1,$$

so

$$\frac{dy}{dx} = \frac{1}{\sec y \tan y} = \frac{1}{\pm x\sqrt{x^2 - 1}},$$

because $\tan y = \pm\sqrt{\sec^2 y - 1} = \pm\sqrt{x^2 - 1}$.

To obtain the correct choice of sign here, note what happens in the two cases $x > 1$ and $x < -1$. In the first case, $0 < y < \pi/2$ and $\tan y > 0$, so we choose the plus sign. If $x < -1$, then $\pi/2 < y < \pi$ and $\tan y < 0$, so we take the minus sign. Thus

$$\frac{dy}{dx} = \frac{1}{|x|\sqrt{x^2 - 1}} \qquad (|x| > 1). \tag{16}$$

If u is a differentiable function of x with values that exceed 1 in magnitude, then by the chain rule we have

$$D_x \sec^{-1} u = \frac{1}{|u|\sqrt{u^2 - 1}} \cdot \frac{du}{dx}. \tag{17}$$

EXAMPLE 3 The function $f(x) = \sec^{-1} e^x$ is defined if $x > 0$, because then $e^x > 1$. Then by Eq. (17),

$$D_x \sec^{-1} e^x = \frac{e^x}{|e^x|\sqrt{e^{2x} - 1}} = \frac{1}{\sqrt{e^{2x} - 1}}$$

because $|e^x| = e^x$ for all x. ■

The **inverse cosecant** (or **arccosecant**) **function** is the inverse of the function $y = \csc x$, where x is restricted to the union of the intervals $[-\pi/2, 0)$ and $(0, \pi/2]$. Thus

$$y = \csc^{-1} x \quad \text{if and only if} \quad \csc y = x \quad \text{and} \quad -\pi/2 \leqq y \leqq \pi/2, \tag{18}$$

where $|x| \geqq 1$. Its derivative formula, which has a derivation similar to that of the inverse secant function, is

$$D_x \csc^{-1} u = -\frac{1}{|u|\sqrt{u^2 - 1}} \cdot \frac{du}{dx}. \tag{19}$$

Summary

The following table summarizes the domains, ranges, and derivatives of the six inverse trigonometric functions.

Function	Domain of Definition	Range of Values	Derivative
$\sin^{-1}x$	$-1 \leqq x \leqq 1$	$-\pi/2 \leqq y \leqq \pi/2$	$\dfrac{1}{\sqrt{1-x^2}}$
$\cos^{-1}x$	$-1 \leqq x \leqq 1$	$0 \leqq y \leqq \pi$	$-\dfrac{1}{\sqrt{1-x^2}}$
$\tan^{-1}x$	$-\infty < x < +\infty$	$-\pi/2 < y < \pi/2$	$\dfrac{1}{1+x^2}$
$\cot^{-1}x$	$-\infty < x < +\infty$	$0 < y < \pi$	$-\dfrac{1}{1+x^2}$
$\sec^{-1}x$	$\lvert x\rvert \geqq 1$	$0 \leqq y < \pi/2, \pi/2 < y \leqq \pi$	$\dfrac{1}{\lvert x\rvert\sqrt{x^2-1}}$
$\csc^{-1}x$	$\lvert x\rvert \geqq 1$	$-\pi/2 \leqq y < 0, 0 < y \leqq \pi/2$	$-\dfrac{1}{\lvert x\rvert\sqrt{x^2-1}}$

It is worth noting that

- $\sin^{-1}x$ has the range $[-\pi/2, \pi/2]$ and $\tan^{-1}x$ has the range $(-\pi/2, \pi/2)$, whereas
- $\cos^{-1}x$ has the range $[0, \pi]$ and $\sec^{-1}x$ has the range $[0, \frac{1}{2}\pi) \cup (\frac{1}{2}\pi, \pi]$.

Observe also the "difference only in sign" of the derivatives of function/cofunction pairs of inverse functions.

Integrals Involving Inverse Trigonometric Functions

The derivatives of the six inverse trigonometric functions are all simple *algebraic* functions. As a consequence, inverse trigonometric functions typically occur when we integrate algebraic functions. Moreover, as mentioned earlier, the derivatives of $\cos^{-1}x$, $\cot^{-1}x$, and $\csc^{-1}x$ differ only in sign from the derivatives of their respective cofunctions. For this reason, only the arctangent, arcsine, and arcsecant functions are necessary for integration, and only these three are in common use. That is, you need commit to memory the integral formulas only for the latter three functions. They follow immediately from Eqs. (5), (11), and (17) and may be written in the forms shown next:

$$\int \frac{du}{1+u^2} = \tan^{-1}u + C, \tag{20}$$

$$\int \frac{du}{\sqrt{1-u^2}} = \sin^{-1}u + C, \tag{21}$$

$$\int \frac{du}{u\sqrt{u^2-1}} = \sec^{-1}\lvert u\rvert + C. \tag{22}$$

It is easy to verify that the absolute value on the right-hand side in Eq. (22) follows from the one in Eq. (17) (see Problem 57). And remember that because $\sec^{-1}\lvert u\rvert$ is undefined unless $\lvert u\rvert \geqq 1$, the definite integral

$$\int_a^b \frac{du}{u\sqrt{u^2-1}}$$

is meaningful only when both of the limits a and b are at least 1 or both are at most -1.

EXAMPLE 4 It follows immediately from Eq. (20) that

$$\int_0^1 \frac{dx}{1+x^2} = \Big[\tan^{-1}x\Big]_0^1 = \tan^{-1}1 - \tan^{-1}0 = \frac{\pi}{4}.$$ ■

EXAMPLE 5 The substitution $u = 3x$, $du = 3\,dx$ gives

$$\int \frac{1}{1+9x^2}\,dx = \frac{1}{3}\int \frac{3}{1+(3x)^2}\,dx$$
$$= \frac{1}{3}\int \frac{du}{1+u^2} = \frac{1}{3}\tan^{-1}u + C = \frac{1}{3}\tan^{-1}3x + C.$$ ■

EXAMPLE 6 The substitution $u = \frac{1}{2}x$, $du = \frac{1}{2}\,dx$ gives

$$\int \frac{1}{\sqrt{4-x^2}}\,dx = \int \frac{1}{2\sqrt{1-(x/2)^2}}\,dx$$
$$= \int \frac{1}{\sqrt{1-u^2}}\,du = \arcsin u + C = \arcsin\left(\frac{x}{2}\right) + C.$$ ■

EXAMPLE 7 The substitution $u = x\sqrt{2}$, $du = \sqrt{2}\,dx$ gives

$$\int_1^{\sqrt{2}} \frac{1}{x\sqrt{2x^2-1}}\,dx = \int_{\sqrt{2}}^2 \frac{1}{u\sqrt{u^2-1}}\,du$$
$$= \Big[\sec^{-1}|u|\Big]_{\sqrt{2}}^2 = \sec^{-1}2 - \sec^{-1}\sqrt{2}$$
$$= \frac{\pi}{3} - \frac{\pi}{4} = \frac{\pi}{12}.$$ ■

8.2 PROBLEMS

Find the values indicated in Problems 1 through 4.

1. (a) $\sin^{-1}\left(\frac{1}{2}\right)$; (b) $\sin^{-1}\left(-\frac{1}{2}\right)$; (c) $\sin^{-1}\left(\frac{1}{2}\sqrt{2}\right)$; (d) $\sin^{-1}\left(-\frac{1}{2}\sqrt{3}\right)$

2. (a) $\cos^{-1}\left(\frac{1}{2}\right)$; (b) $\cos^{-1}\left(-\frac{1}{2}\right)$; (c) $\cos^{-1}\left(\frac{1}{2}\sqrt{2}\right)$; (d) $\cos^{-1}\left(-\frac{1}{2}\sqrt{3}\right)$

3. (a) $\tan^{-1}0$; (b) $\tan^{-1}1$; (c) $\tan^{-1}(-1)$; (d) $\tan^{-1}\sqrt{3}$

4. (a) $\sec^{-1}1$; (b) $\sec^{-1}(-1)$; (c) $\sec^{-1}2$; (d) $\sec^{-1}\left(-\sqrt{2}\right)$

Differentiate the functions in Problems 5 through 26.

5. $f(x) = \sin^{-1}(x^{100})$
6. $f(x) = \arctan(e^x)$
7. $f(x) = \sec^{-1}(\ln x)$
8. $f(x) = \ln(\tan^{-1}x)$
9. $f(x) = \arcsin(\tan x)$
10. $f(x) = x\arctan x$
11. $f(x) = \sin^{-1}e^x$
12. $f(x) = \arctan\sqrt{x}$
13. $f(x) = \cos^{-1}x + \sec^{-1}\left(\dfrac{1}{x}\right)$
14. $f(x) = \cot^{-1}\left(\dfrac{1}{x^2}\right)$
15. $f(x) = \csc^{-1}x^2$
16. $f(x) = \arccos\left(\dfrac{1}{\sqrt{x}}\right)$
17. $f(x) = \dfrac{1}{\arctan x}$
18. $f(x) = (\arcsin x)^2$
19. $f(x) = \tan^{-1}(\ln x)$
20. $f(x) = \operatorname{arcsec}\sqrt{x^2+1}$
21. $f(x) = \tan^{-1}e^x + \cot^{-1}e^{-x}$
22. $f(x) = \exp(\arcsin x)$
23. $f(x) = \sin(\arctan x)$
24. $f(x) = \sec(\sec^{-1}e^x)$
25. $f(x) = \dfrac{\arctan x}{(1+x^2)^2}$
26. $f(x) = (\sin^{-1}2x^2)^{-2}$

In Problems 27 through 30, find dy/dx by implicit differentiation. Then find the line tangent to the graph of the equation at the indicated point P.

27. $\tan^{-1}x + \tan^{-1}y = \dfrac{\pi}{2}$; $P(1, 1)$

28. $\sin^{-1}x + \sin^{-1}y = \dfrac{\pi}{2}$; $P\left(\frac{1}{2}, \frac{1}{2}\sqrt{3}\right)$

29. $(\sin^{-1}x)(\sin^{-1}y) = \dfrac{\pi^2}{16}$; $P\left(\frac{1}{2}\sqrt{2}, \frac{1}{2}\sqrt{2}\right)$

30. $(\sin^{-1}x)^2 + (\sin^{-1}y)^2 = \dfrac{5\pi^2}{36}$; $P\left(\frac{1}{2}, \frac{1}{2}\sqrt{3}\right)$

Evaluate or antidifferentiate, as appropriate, in Problems 31 through 55.

31. $\displaystyle\int_0^1 \frac{dx}{1 + x^2}$

32. $\displaystyle\int_0^{1/2} \frac{dx}{\sqrt{1 - x^2}}$

33. $\displaystyle\int_{\sqrt{2}}^2 \frac{dx}{x\sqrt{x^2 - 1}}$

34. $\displaystyle\int_{-2}^{-2/\sqrt{3}} \frac{dx}{x\sqrt{x^2 - 1}}$

35. $\displaystyle\int_0^3 \frac{dx}{9 + x^2}$

36. $\displaystyle\int_0^{\sqrt{12}} \frac{dx}{\sqrt{16 - x^2}}$

37. $\displaystyle\int \frac{dx}{\sqrt{1 - 4x^2}}$

38. $\displaystyle\int \frac{dx}{9x^2 + 4}$

39. $\displaystyle\int \frac{dx}{x\sqrt{x^2 - 25}}$

40. $\displaystyle\int \frac{dx}{x\sqrt{4x^2 - 9}}$

41. $\displaystyle\int \frac{e^x}{1 + e^{2x}}\,dx$

42. $\displaystyle\int \frac{x^2}{x^6 + 25}\,dx$

43. $\displaystyle\int \frac{dx}{x\sqrt{x^6 - 25}}$

44. $\displaystyle\int \frac{\sqrt{x}}{1 + x^3}\,dx$

45. $\displaystyle\int \frac{dx}{\sqrt{x(1 - x)}}$

46. $\displaystyle\int \frac{\sec x \tan x}{1 + \sec^2 x}\,dx$

47. $\displaystyle\int \frac{x^{49}}{1 + x^{100}}\,dx$

48. $\displaystyle\int \frac{x^4}{\sqrt{1 - x^{10}}}\,dx$

49. $\displaystyle\int \frac{1}{x[1 + (\ln x)^2]}\,dx$

50. $\displaystyle\int \frac{\arctan x}{1 + x^2}\,dx$

51. $\displaystyle\int_0^1 \frac{1}{1 + (2x - 1)^2}\,dx$

52. $\displaystyle\int_0^1 \frac{x^3}{1 + x^4}\,dx$

53. $\displaystyle\int_1^e \frac{dx}{x\sqrt{1 - (\ln x)^2}}$

54. $\displaystyle\int_1^2 \frac{dx}{x\sqrt{x^2 - 1}}$

55. $\displaystyle\int_1^3 \frac{dx}{2\sqrt{x}(1 + x)}$ (*Suggestion:* Let $u = x^{1/2}$.)

56. Conclude from the formula $D_x \cos^{-1}x = -D_x \sin^{-1}x$ that $\sin^{-1}x + \cos^{-1}x = \pi/2$ if $0 \leqq x \leqq 1$.

57. The integral formula in Eq. (22) is equivalent to

$$D_u \sec^{-1}|u| = \frac{1}{u\sqrt{u^2 - 1}} \quad \text{if } |u| > 1. \tag{22$'$}$$

This is the same as Eq. (17) if $u > 1$. If $u < -1$, substitute $x = -u$ in $y = \sec^{-1}|u|$ and use the chain rule $dy/du = (dy/dx)(dx/du)$ to verify Eq. (22′).

In Problems 58 through 60, substitute $u = ax$ (assuming that $a > 0$) to derive the given integral formula.

58. $\displaystyle\int \frac{1}{\sqrt{a^2 - u^2}}\,du = \sin^{-1}\left(\frac{u}{a}\right) + C \quad (u < a).$

59. $\displaystyle\int \frac{1}{a^2 + u^2}\,du = \frac{1}{a}\tan^{-1}\left(\frac{u}{a}\right) + C.$

60. $\displaystyle\int \frac{1}{u\sqrt{u^2 - a^2}}\,du = \frac{1}{a}\sec^{-1}\left|\frac{u}{a}\right| + C \quad (u > a).$

61. If $f(x) = \sec^{-1}x$, then both of the computer algebra systems *Mathematica* and *Maple* give

$$f'(x) = \frac{1}{x^2\sqrt{1 - \dfrac{1}{x^2}}}$$

for the derivative of f. Verify carefully that (if either $x < -1$ or $x > 1$) this result is equivalent to the derivative formula for $\sec^{-1}x$ given in this section.

62. Show that

$$D_x \sec^{-1}x = D_x \cos^{-1}\frac{1}{x} \quad \text{if } |x| > 1,$$

and conclude that

$$\sec^{-1}x = \cos^{-1}\frac{1}{x} \quad \text{if } |x| > 1.$$

This fact can be used to find arcsecant values on a calculator that has a key for the arccosine function, usually written **INV** **COS** or **COS⁻¹**, but no arcsecant key.

63. Some calculus textbooks define the inverse secant function as that function g such that $y = g(x)$ if and only if $\sec y = x$ with y in either $[0, \pi/2)$ or $[\pi, 3\pi/2)$ (the latter instead of the interval $(\pi/2, \pi]$ used in this text). In contrast with Fig. 8.2.12, show that the graph of this "alternative arcsecant function" is as shown in Fig. 8.2.13. Then show that its derivative is given by

$$g'(x) = \frac{1}{x\sqrt{x^2 - 1}}$$

(with no absolute value on the right).

64. (a) Deduce from the addition formula for tangents (Problem 28 in Appendix C) that

$$\arctan x + \arctan y = \arctan\frac{x + y}{1 - xy}$$

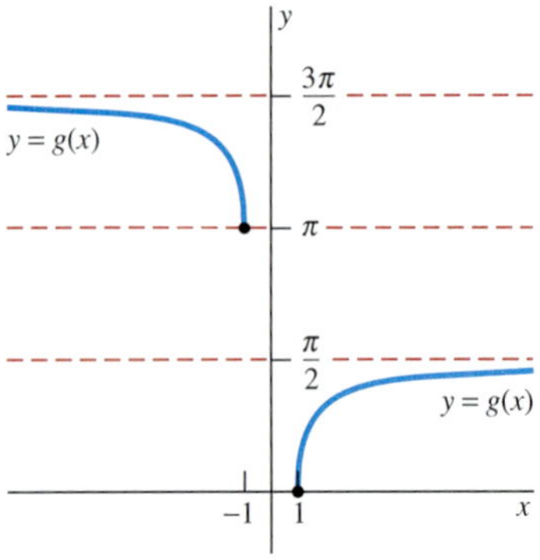

Fig. 8.2.13 Graph of the alternative inverse secant function of Problem 63

provided that $xy < 1$. (b) Apply part (a) to show that each of the following numbers is equal to $\pi/4$: (i) $\arctan\left(\frac{1}{2}\right) + \arctan\left(\frac{1}{3}\right)$; (ii) $2\arctan\left(\frac{1}{3}\right) + \arctan\left(\frac{1}{7}\right)$; (iii) $\arctan\left(\frac{120}{119}\right) - \arctan\left(\frac{1}{239}\right)$; (iv) $4\arctan\left(\frac{1}{5}\right) - \arctan\left(\frac{1}{239}\right)$.

65. A billboard to be built *parallel* to a highway will be 12 m high with its bottom 4 m above the eye level of the average passing motorist. How far from the highway should the billboard be placed to maximize the vertical angle it subtends at the motorist's eyes?

66. Use inverse trigonometric functions to prove that the vertical angle subtended by a rectangular painting on a wall is greatest when the painting is hung with its center at the level of the observer's eyes.

67. Show that the circumference of a circle of radius a is $2\pi a$ by finding the length of the circular arc

$$y = \sqrt{a^2 - x^2}$$

from $x = 0$ to $x = a/\sqrt{2}$ and then multiplying by 8.

68. Find the volume generated by revolving around the y-axis the area under $y = 1/(1 + x^4)$ from $x = 0$ to $x = 1$.

69. The unbounded region R is bounded on the left by the y-axis, below by the x-axis, and above by the graph of $y = 1/(1 + x^2)$. Show that the area of R is finite by evaluating

$$\lim_{a\to\infty} \int_0^a \frac{dx}{1 + x^2}.$$

70. A building 250 ft high is equipped with an external elevator. The elevator starts at the top at time $t = 0$ and descends at the constant rate of 25 ft/s. You are watching the elevator from a window that is 100 ft above the ground and in a building 50 ft from the elevator. At what height does the elevator appear to you to be moving the fastest?

71. Suppose that the function f is defined for all x such that $|x| > 1$ and has the property that

$$f'(x) = \frac{1}{x\sqrt{x^2 - 1}}$$

for all such x. (a) Explain why there exist two constants A and B such that

$$f(x) = \operatorname{arcsec} x + A \qquad \text{if } x > 1;$$

$$f(x) = -\operatorname{arcsec} x + B \qquad \text{if } x < -1.$$

(b) Determine the values of A and B so that $f(2) = 1 = f(-2)$. Then sketch the graph of $y = f(x)$.

In some computing languages the arctangent is the only inverse trigonometric function that is programmed directly, so it is necessary to express $\sin^{-1} x$ and $\sec^{-1} x$ in terms of $\tan^{-1} x$. In Problems 72 and 73 verify each given identity by differentiating both sides. What else must be done?

72. $\sin^{-1} x = \tan^{-1}\left(\dfrac{x}{\sqrt{1 - x^2}}\right)$.

73. (a) $\sec^{-1} x = \tan^{-1}\sqrt{x^2 - 1}$ if $x > 1$;
(b) $\sec^{-1} x = \pi - \tan^{-1}\sqrt{x^2 - 1}$ if $x < -1$.

In Problems 74 through 76, estimate the absolute maximum value of $f(x)$ for $x > 0$. You may want to begin by locating the pertinent critical point graphically.

74. $f(x) = x^{-1/2}\tan^{-1} x$

75. $f(x) = e^{-x/10}\tan^{-1} x$

76. $f(x) = e^{-x/100}\sec^{-1} x$

8.3 INDETERMINATE FORMS AND L'HÔPITAL'S RULE

An *indeterminate form* is a certain type of expression with a limit that is not evident by inspection. There are several types of indeterminate forms. If

$$\lim_{x\to a} f(x) = 0 = \lim_{x\to a} g(x),$$

then we say that the quotient $f(x)/g(x)$ has the **indeterminate form** $0/0$ at $x = a$. For example, to differentiate the trigonometric functions (Section 3.7), we needed to know that

$$\lim_{x\to 0} \frac{\sin x}{x} = 1. \quad \textbf{(1)}$$

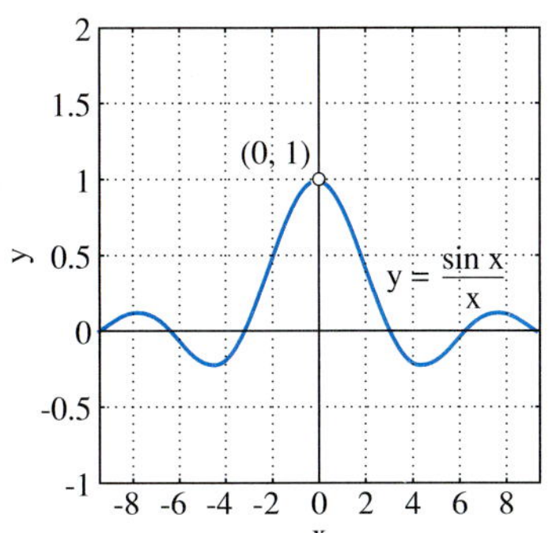

Fig. 8.3.1 Visual evidence that the quotient $(\sin x)/x$ is near 1 when x is near zero

Figure 8.3.1 corroborates the fact that $(\sin x)/x$ is close to 1 when x is close to zero.

The quotient $(\sin x)/x$ in Eq. (1) has the indeterminate form $0/0$ at $x = 0$ because the functions $f(x) = \sin x$ and $g(x) = x$ both approach zero as $x \to 0$. Hence the quotient law of limits cannot be used to evaluate this limit. We therefore needed a special geometric argument (see Section 2.3) to find the limit in Eq. (1). Something similar happens whenever we compute a derivative, because the quotient

$$\frac{f(x) - f(a)}{x - a},$$

whose limit as $x \to a$ is the derivative $f'(a)$, has the indeterminate form $0/0$ at $x = a$.

We can sometimes find the limit of an indeterminate form by performing a special algebraic manipulation or construction, as in our earlier computation of derivatives. Often, however, it is more convenient to apply a rule that appeared in the first calculus textbook ever published, by the Marquis de l'Hôpital, in 1696. L'Hôpital was a French nobleman who had hired the Swiss mathematician John Bernoulli as his calculus tutor, and "l'Hôpital's rule" is actually the work of Bernoulli.

Theorem 1 ***L'Hôpital's Rule***

Suppose that the functions f and g are differentiable for $x \neq a$ in some open interval containing the point a and that $g'(x)$ is nonzero there. Suppose also that

$$\lim_{x\to a} f(x) = 0 = \lim_{x\to a} g(x).$$

Then

$$\lim_{x\to a} \frac{f(x)}{g(x)} = \lim_{x\to a} \frac{f'(x)}{g'(x)}, \quad \textbf{(2)}$$

provided that the limit on the right either exists (as a finite real number) or is $+\infty$ or $-\infty$.

In essence, l'Hôpital's rule says that if $f(x)/g(x)$ has the indeterminate form $0/0$ at $x = a$, then—subject to a few mild restrictions—this quotient has the same limit at $x = a$ as does the quotient $f'(x)/g'(x)$ of *derivatives*. The proof of l'Hôpital's rule is discussed at the end of this section.

EXAMPLE 1 Find $\displaystyle\lim_{x\to 0} \frac{e^x - 1}{\sin 2x}$.

Solution The fraction whose limit we seek has the indeterminate form $0/0$ at $x = 0$. The numerator and denominator are clearly differentiable in some open interval containing the point $x = 0$, and the derivative of the denominator is certainly nonzero for $x \neq 0$ if the interval is small enough (specifically, if $0 < |x| < \pi/4$). So l'Hôpital's rule applies, and

$$\lim_{x\to 0} \frac{e^x - 1}{\sin 2x} = \lim_{x\to 0} \frac{e^x}{2\cos 2x} = \frac{e^0}{2\cos 0} = \frac{1}{2}$$

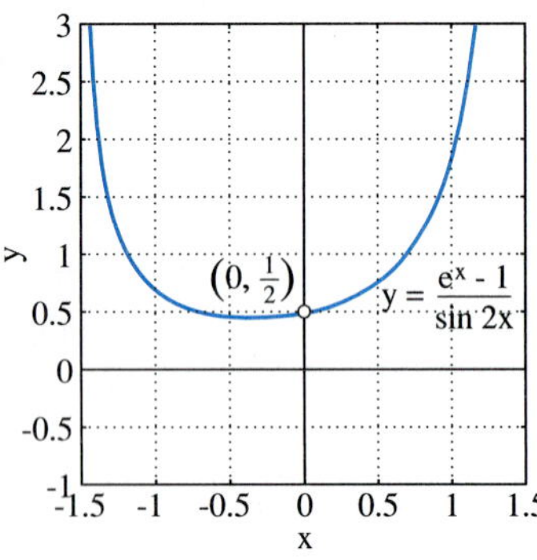

Fig. 8.3.2 Visual evidence that the quotient $\dfrac{e^x - 1}{\sin 2x}$ is near $\frac{1}{2}$ when x is near 0

because (by continuity) both e^x and $\cos 2x$ approach 1 as $x \to 0$. Figure 8.3.2 corroborates this limit. ■

If the quotient $f'(x)/g'(x)$ is itself indeterminate, then l'Hôpital's rule may be applied a second (or third, ...) time, as in Example 2. When the rule is applied repeatedly, however, the conditions for its applicability must be checked at each stage.

EXAMPLE 2 Find $\displaystyle\lim_{x\to 1} \frac{1 - x + \ln x}{1 + \cos \pi x}$.

Solution

$$\begin{aligned}
\lim_{x\to 1} \frac{1 - x + \ln x}{1 + \cos \pi x} &= \lim_{x\to 1} \frac{-1 + \dfrac{1}{x}}{-\pi \sin \pi x} && \text{(still of the form 0/0)} \\
&= \lim_{x\to 1} \frac{x - 1}{\pi x \sin \pi x} && \text{(algebraic simplification)} \\
&= \lim_{x\to 1} \frac{1}{\pi \sin \pi x + \pi^2 x \cos \pi x} && \text{(l'Hôpital's rule again)} \\
&= -\frac{1}{\pi^2} && \text{(by inspection).}
\end{aligned}$$

Because the final limit exists, so do the previous ones; the existence of the final limit in Eq. (2) implies the existence of the first. ■

When you need to apply l'Hôpital's rule repeatedly in this way, you need only keep differentiating the numerator and denominator separately until at least one of them has a nonzero finite limit. At that point you can recognize the limit of the quotient by inspection, as in the final step in Example 2.

EXAMPLE 3 Find $\displaystyle\lim_{x\to 0} \frac{\sin x}{x + x^2}$.

Solution If we simply apply l'Hôpital's rule twice in succession, the result is the *incorrect* computation

$$\begin{aligned}
\lim_{x\to 0} \frac{\sin x}{x + x^2} &= \lim_{x\to 0} \frac{\cos x}{1 + 2x} \\
&= \lim_{x\to 0} \frac{-\sin x}{2} = 0. \qquad \text{(Wrong!)}
\end{aligned}$$

The answer is wrong because $(\cos x)/(1 + 2x)$ is *not* an indeterminate form. Thus l'Hôpital's rule cannot be applied to it. The *correct* computation is

$$\lim_{x\to 0} \frac{\sin x}{x + x^2} = \lim_{x\to 0} \frac{\cos x}{1 + 2x} = \frac{\displaystyle\lim_{x\to 0} \cos x}{\displaystyle\lim_{x\to 0} (1 + 2x)} = \frac{1}{1} = 1.$$ ■

The point of Example 3 is to issue a warning: Verify the hypotheses of l'Hôpital's rule *before* you apply it. It is an oversimplification to say that l'Hôpital's rule works when you need it and doesn't work when you don't, but there is still much truth in this statement.

Indeterminate Forms Involving ∞

L'Hôpital's rule has several variations. In addition to the fact that the limit in Eq. (2) is allowed to be infinite, the real number a in l'Hôpital's rule may be replaced with either $+\infty$ or $-\infty$. For example,

$$\lim_{x\to+\infty} \frac{f(x)}{g(x)} = \lim_{x\to+\infty} \frac{f'(x)}{g'(x)} \tag{3}$$

provided that the other hypotheses are satisfied for all x in some open interval of the form $(c, +\infty)$. In particular, to use Eq. (3), we must first verify that

$$\lim_{x\to\infty} f(x) = 0 = \lim_{x\to\infty} g(x)$$

and that the right-hand limit in Eq. (3) exists. The proof of this version of l'Hôpital's rule is outlined in Problem 70.

L'Hôpital's rule may also be used when $f(x)/g(x)$ has the **indeterminate form** ∞/∞. This means that

$$\lim_{x\to a} f(x) \quad \text{is either} \quad +\infty \text{ or } -\infty$$

and

$$\lim_{x\to a} g(x) \quad \text{is either} \quad +\infty \text{ or } -\infty.$$

The proof of this extension of the rule is difficult and is omitted here. [For a proof, see, for example, A. E. Taylor and W. R. Mann, *Advanced Calculus,* 3rd ed. (New York: John Wiley, 1983), p. 107.]

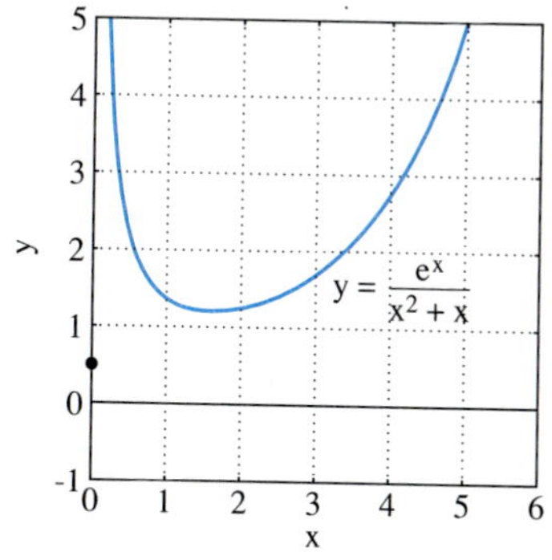

Fig. 8.3.3 Graph of the function $f(x) = \dfrac{e^x}{x^2 + x}$ of Example 4. The fact that $f(x) \to \infty$ as $x \to 0$ is evident by inspection. The l'Hôpital's computation of the example verifies that $f(x) \to \infty$ as $x \to \infty$ as well (consistent with the fact that exponentials have larger magnitude than polynomials as $x \to \infty$).

EXAMPLE 4 Find $\displaystyle\lim_{x\to+\infty} \frac{e^x}{x^2 + x}$.

Solution Both quotients $e^x/(x^2 + x)$ and $e^x/(2x + 1)$ have the indeterminate form ∞/∞, so two applications of l'Hôpital's rule yield

$$\lim_{x\to+\infty} \frac{e^x}{x^2 + x} = \lim_{x\to+\infty} \frac{e^x}{2x + 1} = \lim_{x\to+\infty} \frac{e^x}{2} = +\infty.$$

Remember that l'Hôpital's rule also "allows" the final result to be an infinite limit. ■

Remark If x is large, then finding the behavior of the graph of $y = f(x)/g(x)$ may require the computation of

$$\lim_{x\to+\infty} \frac{f(x)}{g(x)}.$$

See Figs. 8.3.3 and 8.3.4 for illustrations of the limits in Examples 4 and 5.

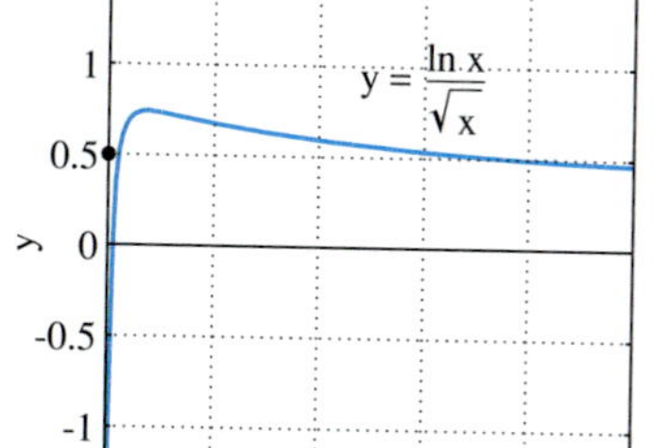

Fig. 8.3.4 Does the curve $y = \dfrac{\ln x}{\sqrt{x}}$ actually have the x-axis as an asymptote? The limit of Example 5 implies that it does (consistent with the fact that logarithms have smaller magnitude than powers as $x \to \infty$).

EXAMPLE 5

$$\lim_{x\to+\infty} \frac{\ln x}{\sqrt{x}} = \lim_{x\to+\infty} \frac{\frac{1}{x}}{\frac{1}{2}x^{-1/2}} = \lim_{x\to+\infty} \frac{2}{\sqrt{x}} = 0.$$ ■

Proof of L'Hôpital's Rule

Suppose that the functions f and g of Theorem 1 are not merely differentiable but have continuous derivatives near $x = a$ and that $g'(a) \neq 0$. Then

$$\lim_{x \to a} \frac{f'(x)}{g'(x)} = \frac{\lim_{x \to a} f'(x)}{\lim_{x \to a} g'(x)} = \frac{f'(a)}{g'(a)} \tag{4}$$

by the quotient law for limits. In this case l'Hôpital's rule in Eq. (2) reduces to the limit

$$\lim_{x \to a} \frac{f(x)}{g(x)} = \frac{f'(a)}{g'(a)}, \tag{5}$$

which is a weak form of the rule. It actually is this weak form that is typically applied in single-step applications of l'Hôpital's rule.

EXAMPLE 6 In Example 1 we had

$$f(x) = e^x - 1, \qquad g(x) = \sin 2x$$

so

$$f'(x) = e^x, \qquad g'(x) = 2\cos 2x,$$

and $g'(0) = 2 \neq 0$. With $a = 0$, Eq. (5) therefore gives

$$\lim_{x \to 0} \frac{e^x - 1}{\sin 2x} = \lim_{x \to 0} \frac{f(x)}{g(x)} = \frac{f'(0)}{g'(0)} = \frac{1}{2}.$$

■

Theorem 2 ***L'Hôpital's Rule (weak form)***

Suppose that the functions f and g are differentiable at $x = a$, that

$$f(a) = 0 = g(a),$$

and that $g'(a) \neq 0$. Then

$$\lim_{x \to a} \frac{f(x)}{g(x)} = \frac{f'(a)}{g'(a)}. \tag{4}$$

PROOF We begin with the right-hand side of Eq. (4) and work toward the left-hand side.

$$\begin{aligned}
\frac{f'(a)}{g'(a)} &= \frac{\lim_{x \to a} \dfrac{f(x) - f(a)}{x - a}}{\lim_{x \to a} \dfrac{g(x) - g(a)}{x - a}} && \text{(the definition of the derivative)} \\
&= \lim_{x \to a} \frac{\dfrac{f(x) - f(a)}{x - a}}{\dfrac{g(x) - g(a)}{x - a}} && \text{(the quotient law of limits)} \\
&= \lim_{x \to a} \frac{f(x) - f(a)}{g(x) - g(a)} && \text{(algebraic simplification)} \\
&= \lim_{x \to a} \frac{f(x)}{g(x)}
\end{aligned}$$

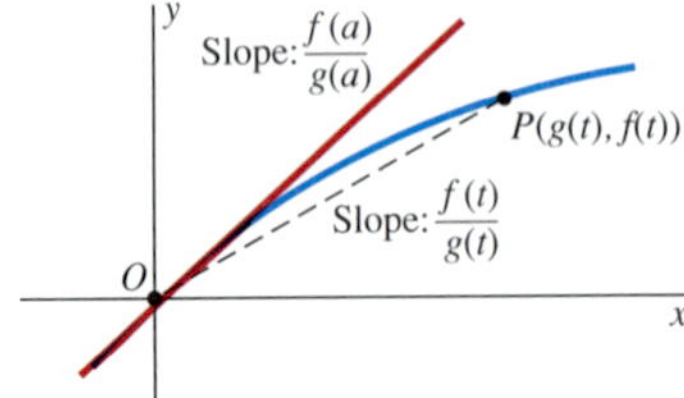

Fig. 8.3.5 Suppose that the point $P(g(t), f(t))$ traces a continuous curve that passes through the origin O when $t = a$. Then the secant line OP approaches the tangent line at O as $t \to a$, so its slope $f(t)/g(t)$ approaches the slope $f(a)/g(a)$ of the tangent line at O.

[because $f(a) = 0 = g(a)$]. ■

Figure 8.3.5 illustrates the meaning and proof of Theorem 2. Appendix I includes a proof of the strong form of l'Hôpital's rule, the form stated in Theorem 1.

8.3 PROBLEMS

Find the limits in Problems 1 through 48.

1. $\lim_{x\to 1} \frac{x-1}{x^2-1}$

2. $\lim_{x\to\infty} \frac{3x-4}{2x-5}$

3. $\lim_{x\to\infty} \frac{2x^2-1}{5x^2+3x}$

4. $\lim_{x\to 0} \frac{e^{3x}-1}{x}$

5. $\lim_{x\to 0} \frac{\sin x^2}{x}$

6. $\lim_{x\to 0^+} \frac{1-\cos\sqrt{x}}{x}$

7. $\lim_{x\to 1} \frac{x-1}{\sin x}$

8. $\lim_{x\to 0} \frac{1-\cos x}{x^3}$

9. $\lim_{x\to 0} \frac{e^x-x-1}{x^2}$

10. $\lim_{z\to\pi/2} \frac{1+\cos 2z}{1-\sin 2z}$

11. $\lim_{u\to 0} \frac{u\arctan u}{1-\cos u}$

12. $\lim_{x\to 0} \frac{x-\arctan x}{x^3}$

13. $\lim_{x\to\infty} \frac{\ln x}{\sqrt[10]{x}}$

14. $\lim_{r\to\infty} \frac{e^r}{(r+1)^4}$

15. $\lim_{x\to 10} \frac{\ln(x-9)}{x-10}$

16. $\lim_{t\to\infty} \frac{t^2+1}{t\ln t}$

17. $\lim_{x\to 0} \frac{e^x+e^{-x}-2}{x\sin x}$

18. $\lim_{x\to(\pi/2)^-} \frac{\tan x}{\ln(\cos x)}$

19. $\lim_{x\to 0} \frac{\sin 3x}{\tan 5x}$

20. $\lim_{x\to 0} \frac{e^x-e^{-x}}{x}$

21. $\lim_{x\to 1} \frac{x^3-1}{x^2-1}$

22. $\lim_{x\to 2} \frac{x^3-8}{x^4-16}$

23. $\lim_{x\to\infty} \frac{x+\sin x}{3x+\cos x}$

24. $\lim_{x\to\infty} \frac{\sqrt{x^2+4}}{x}$

25. $\lim_{x\to 0} \frac{2^x-1}{3^x-1}$

26. $\lim_{x\to\infty} \frac{2^x}{3^x}$

27. $\lim_{x\to\infty} \frac{\sqrt{x^2-1}}{\sqrt{4x^2-x}}$

28. $\lim_{x\to\infty} \frac{\sqrt{x^3+x}}{\sqrt{2x^3-4}}$

29. $\lim_{x\to 0} \frac{\ln(1+x)}{x}$

30. $\lim_{x\to\infty} \frac{\ln(\ln x)}{x\ln x}$

31. $\lim_{x\to 0} \frac{2e^x-x^2-2x-2}{x^3}$

32. $\lim_{x\to 0} \frac{\sin x-\tan x}{x^3}$

33. $\lim_{x\to 0} \frac{2-e^x-e^{-x}}{2x^2}$

34. $\lim_{x\to 0} \frac{e^{3x}-e^{-3x}}{2x}$

35. $\lim_{x\to\pi/2} \frac{2x-\pi}{\tan 2x}$

36. $\lim_{x\to\pi/2} \frac{\sec x}{\tan x}$

37. $\lim_{x\to 2} \frac{x-2\cos\pi x}{x^2-4}$

38. $\lim_{x\to 1/2} \frac{2x-\sin\pi x}{4x^2-1}$

39. $\lim_{x\to 0} \frac{\arctan 2x}{\arctan 3x}$

40. $\lim_{x\to\infty} \frac{\arctan 2x}{\arctan 3x}$

41. $\lim_{x\to 0} \frac{\exp(x^3)-1}{x-\sin x}$

42. $\lim_{x\to 0} \frac{\sqrt{1+3x}-1}{x}$

43. $\lim_{x\to 0} \frac{\sqrt[3]{1+4x}-1}{x}$

44. $\lim_{x\to 0} \frac{\sqrt{3+2x}-\sqrt{3+x}}{x}$

45. $\lim_{x\to 0} \frac{\sqrt[3]{1+x}-\sqrt[3]{1-x}}{x}$

46. $\lim_{x\to\pi/4} \frac{1-\tan x}{4x-\pi}$

47. $\lim_{x\to 0} \frac{\ln(1+x^2)}{e^x-\cos x}$

48. $\lim_{x\to 2} \frac{x^5-5x^2-12}{x^{10}-500x-24}$

Sketch the graphs of the curves in Problems 49 through 60. Even if you use a graphing calculator or computer, apply l'Hôpital's rule as necessary to verify the apparent behavior of the curve as x approaches a point where the function has an indeterminate form.

49. $y = \frac{\sin^2 x}{x}$

50. $y = \frac{\sin^2 x}{x^2}$

51. $y = \frac{\sin x}{x-\pi}$

52. $y = \frac{\cos x}{2x-\pi}$

53. $y = \frac{1-\cos x}{x^2}$

54. $y = \frac{x-\sin x}{x^3}$

55. $y = xe^{-x}$

56. $y = e^{-x}\sqrt{x}$

57. $y = xe^{-\sqrt{x}}$

58. $y = x^2e^{-2x}$

59. $y = \frac{\ln x}{x}$

60. $y = \frac{\ln x}{\sqrt{x}+\sqrt[3]{x}}$

61. Apply l'Hôpital's rule to prove that

$$\lim_{x\to\infty} \frac{x^n}{e^x} = 0$$

if n is an arbitrary positive integer, thereby verifying that exponentials have larger magnitude than powers as $x \to +\infty$.

62. Apply l'Hôpital's rule to prove that

$$\lim_{x\to\infty} \frac{\ln x}{x^k} = 0$$

if k is an arbitrary positive real number, thereby verifying that logarithms have smaller magnitude than powers as $x \to +\infty$.

63. Suppose that n is a fixed positive integer larger than 1. Show that the curve $y = x^n e^{-x}$ has a single local maximum and two inflection points for $x > 0$ and has the x-axis as an asymptote.

64. Suppose that k is an arbitrary positive real number. Show that the curve $y = x^{-k}\ln x$ has a single local maximum and a single inflection point for $x > 0$ and has the x-axis as an asymptote.

Apply l'Hôpital's rule and the fundamental theorem of calculus to evaluate the limits in Problems 65 and 66.

65. $\lim_{x\to 0} \frac{E(x)}{x}$, where $E(x) = \int_0^x e^{-t^2}\,dt.$

66. $\lim_{x\to 0} \frac{S(x)}{x^3}$, where $S(x) = \int_0^x \sin^2 t\,dt.$

67. Suppose that $f'(x)$ is continuous. Show that

$$\lim_{h\to 0} \frac{f(x+h) - f(x-h)}{2h} = f'(x).$$

The *symmetric difference quotient* on the left can be used (with h very small) to approximate the derivative numerically, and it turns out to be a better approximation than the one-sided difference quotient $[f(x+h) - f(x)]/h$.

68. Suppose that $f''(x)$ is continuous. Show that

$$\lim_{h\to 0} \frac{f(x+h) - 2f(x) + f(x-h)}{h^2} = f''(x).$$

The *second difference quotient* on the left can be used (with h very small) to approximate the second derivative numerically.

69. In his calculus textbook of 1696, l'Hôpital used a limit similar to

$$\lim_{x\to 1} \frac{\sqrt{2x - x^4} - \sqrt[3]{x}}{1 - x^{4/3}}$$

to illustrate his rule. Evaluate this limit.

70. Establish the 0/0 version of l'Hôpital's rule for the case $a = \infty$. [*Suggestion:* Let $F(t) = f(1/t)$ and $G(t) = g(1/t)$. Then show that

$$\lim_{x\to\infty} \frac{f(x)}{g(x)} = \lim_{t\to 0^+} \frac{F(t)}{G(t)} = \lim_{t\to 0^+} \frac{F'(t)}{G'(t)} = \lim_{x\to\infty} \frac{f'(x)}{g'(x)},$$

using l'Hôpital's rule for the case $a = 0$.]

8.3 PROJECT: GRAPHICAL INVESTIGATION OF INDETERMINATE FORMS

If one accepts at face value the graph

$$y = \frac{\sin x}{x}$$

shown in Fig. 8.3.1, it seems clear that the point (x, y) on the curve is near the point $(0, 1)$ when the number x is close to zero. As noted previously, this observation corroborates the fundamental trigonometric limit

$$\lim_{x\to 0} \frac{\sin x}{x} = 1. \tag{1}$$

Your objective in this project is to explore graphically the meaning of one number or point being "close" to another number or point. Recall from Section 2.2 the formal definition of the limit: We say that

$$\lim_{x\to a} = L \tag{2}$$

provided that, given any $\epsilon > 0$, there exists a number $\delta > 0$ such that

$$0 < |x - a| < \delta \quad \text{implies that} \quad |f(x) - L| < \epsilon. \tag{3}$$

The geometric meaning of the implication in Eq. (3) is this:

> Suppose that two horizontal lines $y = L + \epsilon$ and $y = L - \epsilon$ above and below the point (a, L) are drawn. Then it is possible to draw two vertical lines $x = a - \delta$ and $x = a + \delta$ on either side of (a, L) such that the portion of the graph $y = f(x)$ (with $x \neq a$) that lies between the two vertical lines lies also between the two horizontal lines.

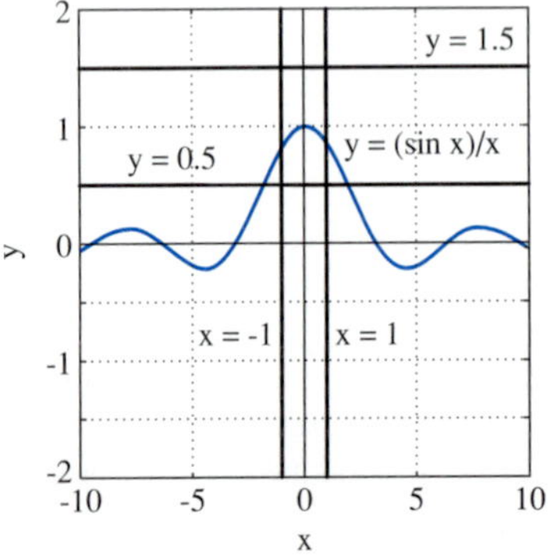

Fig. 8.3.6 The value $\delta = 1$ "works" if $\epsilon = \frac{1}{2}$.

The question is this: Given a specific value of $\epsilon > 0$, what value of δ makes this so? With a graphics calculator or computer graphing program, we can first plot the horizontal lines $y = L + \epsilon$ and $y = L - \epsilon$ and then experiment with different locations of the vertical lines $x = a - \delta$ and $x = a + \delta$ to find a value of δ that "works" with the given value of ϵ.

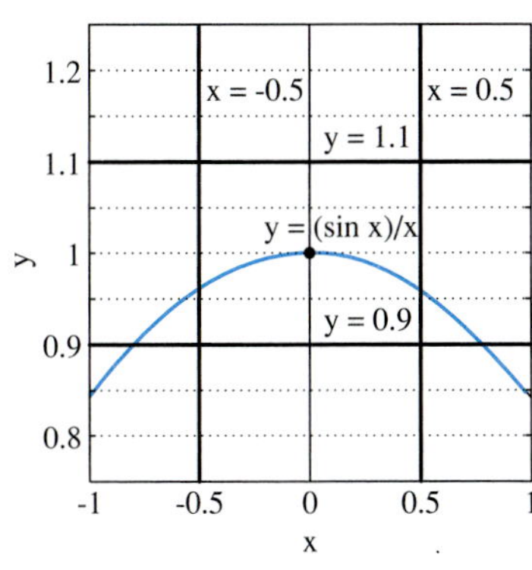

Fig. 8.3.7 The value $\delta = \frac{1}{2}$ "works" if $\epsilon = \frac{1}{10}$.

Thus we see in Fig. 8.3.6 that every point on $y = (\sin x)/x$ between the vertical lines $x = -1$ and $x = 1$ also lies between the two horizontal lines $y = 0.5$ and $y = 1.5$. So for the limit in Eq. (1) (for which $a = 0$ and $L = 1$ in Eq. (3)), if $\epsilon = 1/2$, then we can choose $\delta = 1$.

The smaller the value of $\epsilon > 0$, the smaller the value of $\delta > 0$ we should expect to need. We see in Fig. 8.3.7 that if $\epsilon = 1/10$, then we can choose $\delta = 1/2$.

Problems 1 through 6 specify a limit $\lim_{x \to a} f(x)$. First inspect the graph $y = f(x)$ near $x = a$ to determine the apparent value L of the limit. Then experiment with horizontal and vertical lines to find values of $\delta > 0$ that satisfy the definition of the limit for the indicated values of $\epsilon > 0$. Finally, apply l'Hôpital's rule to verify that $\lim_{x \to a} f(x) = L$.

1. $\displaystyle\lim_{x \to 0} \frac{e^x - 1}{x}$; $\epsilon = 1, 0.5, 0.1$

2. $\displaystyle\lim_{x \to 0} \frac{1 - \cos x}{x^2}$; $\epsilon = 0.25, 0.1, 0.05$

3. $\displaystyle\lim_{x \to 0} \frac{\sqrt{1 + x} - 1}{x}$; $\epsilon = 0.25, 0.1, 0.05$

4. $\displaystyle\lim_{x \to 4} \frac{\sqrt{x} - 2}{x - 4}$; $\epsilon = 0.1, 0.05, 0.01$

5. $\displaystyle\lim_{x \to 0} \frac{\tan x - \sin x}{x^3}$; $\epsilon = 0.25, 0.1, 0.05$

6. $\displaystyle\lim_{x \to 0} \frac{\sin^{-1} x - \tan^{-1} x}{x^3}$; $\epsilon = 0.25, 0.1, 0.05$

8.4 ADDITIONAL INDETERMINATE FORMS

We saw in Section 8.3 that l'Hôpital's rule can be applied to the indeterminate forms $0/0$ and ∞/∞. There are other indeterminate forms; although l'Hôpital's rule cannot applied directly to these other forms, it may be possible to convert them into the form $0/0$ or into the form ∞/∞. If so, it may be possible to apply l'Hôpital's rule.

Suppose that

$$\lim_{x \to a} f(x) = 0 \quad \text{and} \quad \lim_{x \to a} g(x) = \infty.$$

Then we say that the product $f(x) \cdot g(x)$ has the **indeterminate form** $0 \cdot \infty$ at $x = a$. To find the limit of $f(x) \cdot g(x)$ at $x = a$, we can change the problem to one of the forms $0/0$ or ∞/∞ in this way:

$$f(x) \cdot g(x) = \frac{f(x)}{1/g(x)} = \frac{g(x)}{1/f(x)}.$$

Now l'Hôpital's rule may be applied if its other hypotheses are satisfied, as illustrated in Example 1.

EXAMPLE 1 Find $\displaystyle\lim_{x \to \infty} x \ln\left(\frac{x - 1}{x + 1}\right)$.

Solution We are dealing with the indeterminate form $0 \cdot \infty$, so we write

$$\lim_{x \to \infty} x \ln\left(\frac{x - 1}{x + 1}\right) = \lim_{x \to \infty} \frac{\ln\left(\dfrac{x - 1}{x + 1}\right)}{\dfrac{1}{x}}.$$

The right-hand limit has the form 0/0, so we can apply l'Hôpital's rule. First we note that

$$D_x \ln\left(\frac{x-1}{x+1}\right) = \frac{2}{x^2-1}.$$

Thus

$$\lim_{x\to\infty} x \ln\left(\frac{x-1}{x+1}\right) = \lim_{x\to\infty} \frac{\dfrac{2}{x^2-1}}{-\dfrac{1}{x^2}}$$

$$= \lim_{x\to\infty} \frac{-2x^2}{x^2-1} = \lim_{x\to\infty} \frac{-2}{1-\dfrac{1}{x^2}} = -2.$$

Hence the curve

$$y = x \ln \frac{x-1}{x+1}, \qquad x > 1,$$

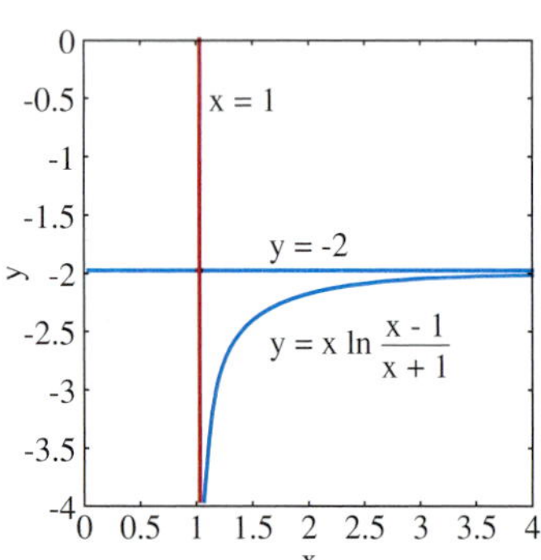

Fig. 8.4.1 Visual corroboration of the limit in Example 1

has the line $y = -2$ as a horizontal asymptote as $x \to +\infty$ (see Fig. 8.4.1). It also has the line $x = 1$ as a vertical asymptote as $x \to 1^+$. (Why?) ■

If

$$\lim_{x\to a} f(x) = +\infty = \lim_{x\to a} g(x),$$

then we say that $f(x) - g(x)$ has the **indeterminate form** $\infty - \infty$. To then evaluate

$$\lim_{x\to a} [f(x) - g(x)],$$

we try by algebraic manipulation to convert $f(x) - g(x)$ into a form of type 0/0 or ∞/∞ so that it may be possible to apply l'Hôpital's rule. If $f(x)$ or $g(x)$ is expressed as a fraction, we can sometimes do this by finding a common denominator. In most cases, however, subtler methods are required. Example 2 illustrates the technique of finding a common denominator. Example 3 demonstrates a factoring technique that can be effective.

EXAMPLE 2

$$\lim_{x\to 0}\left(\frac{1}{x} - \frac{1}{\sin x}\right) = \lim_{x\to 0} \frac{(\sin x) - x}{x \sin x} \qquad \text{(form } 0/0\text{)}$$

$$= \lim_{x\to 0} \frac{(\cos x) - 1}{\sin x + x\cos x} \qquad \text{(still } 0/0\text{)}$$

$$= \lim_{x\to 0} \frac{-\sin x}{2\cos x - x \sin x} = 0.$$ ■

EXAMPLE 3

$$\lim_{x\to+\infty} \left(\sqrt{x^2+3x} - x\right) = \lim_{x\to+\infty} x\left(\sqrt{1+\frac{3}{x}} - 1\right) \qquad \text{(form } \infty \cdot 0\text{)}$$

$$= \lim_{x\to+\infty} \frac{\sqrt{1+\dfrac{3}{x}}-1}{\dfrac{1}{x}} \qquad \text{(form 0/0 now)}$$

$$= \lim_{x\to+\infty} \frac{\dfrac{1}{2}\left(1+\dfrac{3}{x}\right)^{-1/2}\left(-\dfrac{3}{x^2}\right)}{-\dfrac{1}{x^2}}$$

$$= \lim_{x\to+\infty} \frac{\dfrac{3}{2}}{\sqrt{1+\dfrac{3}{x}}} = \frac{3}{2}.$$

Thus the curve $y = \sqrt{x^2+3x} - x$, $x > 0$, has the line $y = \frac{3}{2}$ as a horizontal asymptote as $x \to +\infty$ (see Fig. 8.4.2). ■

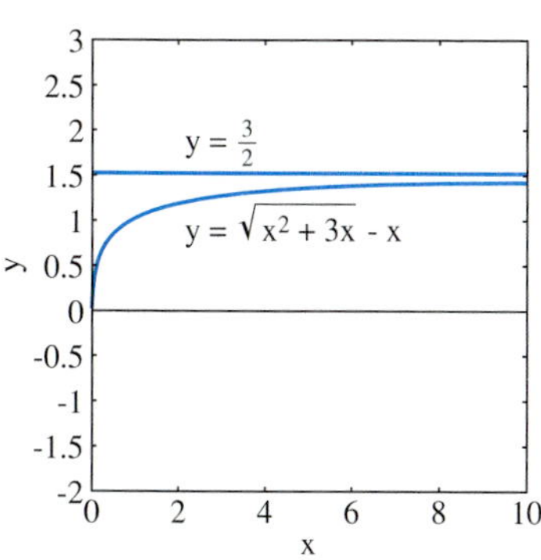

Fig. 8.4.2 Visual corroboration of the limit in Example 3

The Indeterminate Forms 0^0, ∞^0, and 1^∞

Suppose that we need to find the limit of a quantity

$$y = [f(x)]^{g(x)},$$

where the limits of f and g as $x \to a$ are such that one of the **indeterminate forms** 0^0, ∞^0, or 1^∞ is produced. We first compute the natural logarithm

$$\ln y = \ln\left([f(x)]^{g(x)}\right) = g(x)\ln f(x).$$

For each of the three indeterminate forms mentioned here, $g(x)\ln f(x)$ has the form $0\cdot\infty$, so we can use our earlier methods to find $L = \lim_{x\to a} \ln y$ (assuming that $f(x) > 0$ near $x = a$, so that $y > 0$). Then

$$\lim_{x\to a} [f(x)]^{g(x)} = \lim_{x\to a} y = \lim_{x\to a} \exp(\ln y) = \exp(\lim_{x\to a} \ln y) = e^L,$$

because the exponential function is continuous. Thus we have the following four steps for finding the limit of $[f(x)]^{g(x)}$ as $x \to a$:

1. Let $y = [f(x)]^{g(x)}$.
2. Simplify $\ln y = g(x)\ln f(x)$.
3. Evaluate $L = \lim_{x\to a} \ln y$.
4. Conclude that $\lim_{x\to a} [f(x)]^{g(x)} = e^L$.

EXAMPLE 4 Find $\lim_{x\to 0} (\cos x)^{1/x^2}$.

Solution Here we have the indeterminate form 1^∞. If we let $y = (\cos x)^{1/x^2}$, then

$$\ln y = \ln[(\cos x)^{1/x^2}] = \frac{\ln \cos x}{x^2}.$$

As $x \to 0$, $\cos x \to 1$, and so $\ln \cos x \to 0$; we are now dealing with the indeterminate form $0/0$. Hence two applications of l'Hôpital's rule yield

$$\lim_{x\to 0} \ln y = \lim_{x\to 0} \frac{\ln \cos x}{x^2} = \lim_{x\to 0} \frac{-\tan x}{2x} \qquad (0/0 \text{ form})$$

$$= \lim_{x\to 0} \frac{-\sec^2 x}{2} = -\frac{1}{2}.$$

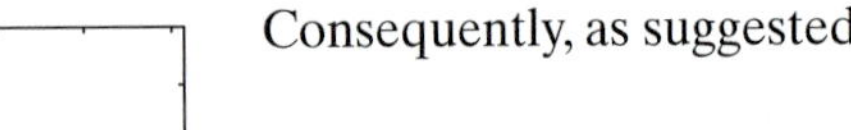

Consequently, as suggested by Fig. 8.4.3,

$$\lim_{x\to 0} (\cos x)^{1/x^2} = e^{-1/2} = \frac{1}{\sqrt{e}}.$$

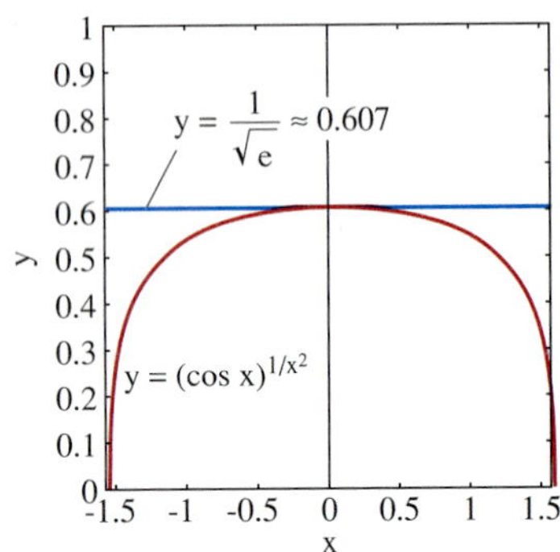

Fig. 8.4.3 Visual corroboration of the limit in Example 4

EXAMPLE 5 Find $\lim_{x\to 0^+} x^{\tan x}$.

Solution This has the indeterminate form 0^0. If $y = x^{\tan x}$, then

$$\ln y = (\tan x)(\ln x) = \frac{\ln x}{\cot x}.$$

Now we have the indeterminate form ∞/∞, and l'Hôpital's rule yields

$$\lim_{x\to 0^+} \ln y = \lim_{x\to 0^+} \frac{\ln x}{\cot x} = \lim_{x\to 0^+} \frac{\frac{1}{x}}{-\csc^2 x} = -\lim_{x\to 0^+} \frac{\sin^2 x}{x}$$

$$= -\lim_{x\to 0^+} \left(\frac{\sin x}{x}\right)(\sin x) = (-1)\cdot 0 = 0.$$

Therefore, $\lim_{x\to 0^+} x^{\tan x} = e^0 = 1$. The graph of the curve $y = x^{\tan x}, 0 < x < \pi/2$ in Fig. 8.4.4 provides corroboration of this limit. We note also a local minimum on the curve near $x = 0.4$ (see Problem 45).

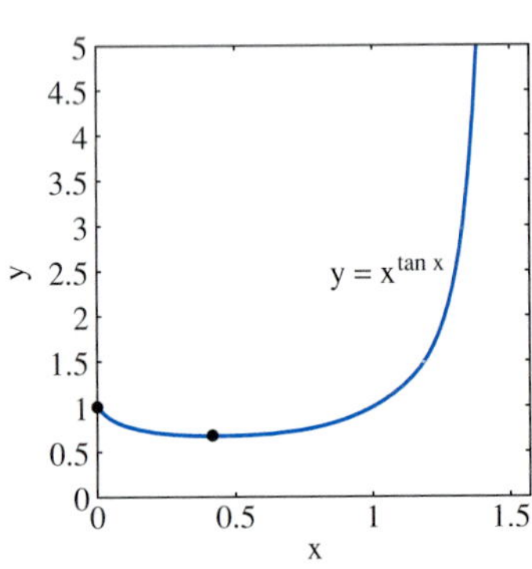

Fig. 8.4.4 Visual corroboration of the limit in Example 5

Although $a^0 = 1$ for any *nonzero* constant a, the form 0^0 is indeterminate—the limit is not necessarily 1 (see Problem 52). But the form 0^∞ is not indeterminate; its limit is zero. For example,

$$\lim_{x\to 0^+} x^{1/x} = 0.$$

8.4 PROBLEMS

Find the limits in Problems 1 through 34.

1. $\lim_{x\to 0} x \cot x$

2. $\lim_{x\to 0} \left(\frac{1}{x} - \cot x\right)$

3. $\lim_{x\to 0} \frac{1}{x} \ln\left(\frac{7x+8}{4x+8}\right)$

4. $\lim_{x\to 0^+} (\sin x)(\ln \sin x)$

5. $\lim_{x\to 0} x^2 \csc^2 x$

6. $\lim_{x\to\infty} e^{-x} \ln x$

7. $\lim_{x\to\infty} x(e^{1/x} - 1)$

8. $\lim_{x\to 2} \left(\frac{1}{x-2} - \frac{1}{\ln(x-1)}\right)$

9. $\lim_{x\to 0^+} x \ln x$

10. $\lim_{x\to\pi/2} (\tan x)(\cos 3x)$

11. $\lim_{x\to\pi} (x - \pi) \csc x$

12. $\lim_{x\to\infty} (x - \sin x) \exp(-x^2)$

13. $\lim_{x\to 0^+} \left(\frac{1}{\sqrt{x}} - \frac{1}{\sin x}\right)$

14. $\lim_{x\to 0} \left(\frac{1}{x} - \frac{1}{e^x - 1}\right)$

15. $\lim_{x\to 1^+} \left(\frac{x}{x^2 + x - 2} - \frac{1}{x-1}\right)$

16. $\lim_{x\to\infty} (\sqrt{x+1} - \sqrt{x})$

17. $\lim_{x\to 0} \left(\frac{1}{x} - \frac{1}{\ln(1+x)}\right)$

18. $\lim_{x\to\infty} (\sqrt{x^2 + x} - \sqrt{x^2 - x})$

19. $\lim_{x\to\infty} (\sqrt[3]{x^3 + 2x + 5} - x)$

20. $\lim_{x\to 0^+} x^x$

21. $\lim_{x\to 0^+} x^{\sin x}$

22. $\lim_{x\to\infty} \left(1 + \frac{1}{x}\right)^x$

23. $\lim_{x\to\infty} (\ln x)^{1/x}$

24. $\lim_{x\to\infty} \left(1 - \frac{1}{x^2}\right)^x$

25. $\lim_{x\to 0} \left(\frac{\sin x}{x}\right)^{1/x^2}$

26. $\lim_{x\to 0^+} (1 + 2x)^{1/(3x)}$

27. $\lim_{x\to\infty} \left(\cos \frac{1}{x^2}\right)^{x^4}$

28. $\lim_{x\to 0^+} (\sin x)^{\sec x}$ **29.** $\lim_{x\to 0+} (x + \sin x)^x$

30. $\lim_{x\to \pi/2} (\tan x - \sec x)$ **31.** $\lim_{x\to 1} x^{1/(1-x)}$

32. $\lim_{x\to 1+} (x - 1)^{\ln x}$ **33.** $\lim_{x\to 2+} \left(\dfrac{1}{\sqrt{x^2 - 4}} - \dfrac{1}{x - 2}\right)$

34. $\lim_{x\to\infty} (\sqrt[5]{x^5 - 3x^4 + 17} - x)$

Figures 8.4.5 through 8.4.7 illustrate the graphs of some of the functions defined for $x > 0$ in Problems 35 through 42. In each of these problems,

(a) First use your own calculator or computer to graph the given function $f(x)$ with an x-range sufficient to suggest its behavior both as $x \to 0^+$ and as $x \to +\infty$.

(b) Then apply l'Hôpital's rule as necessary to verify this suspected behavior near zero and $+\infty$.

(c) Finally, estimate graphically and/or numerically the maximum value attained by $f(x)$ for $x \geqq 0$. If possible, find this maximum value exactly.

35. $f(x) = x^{1/x}$ **36.** $f(x) = x^{(1/x^2)}$

37. $f(x) = (x^2)^{1/x}$ **38.** $f(x) = x^{-x}$

39. $f(x) = (1 + x^2)^{1/x}$ **40.** $f(x) = \left(1 + \dfrac{1}{x^2}\right)^x$

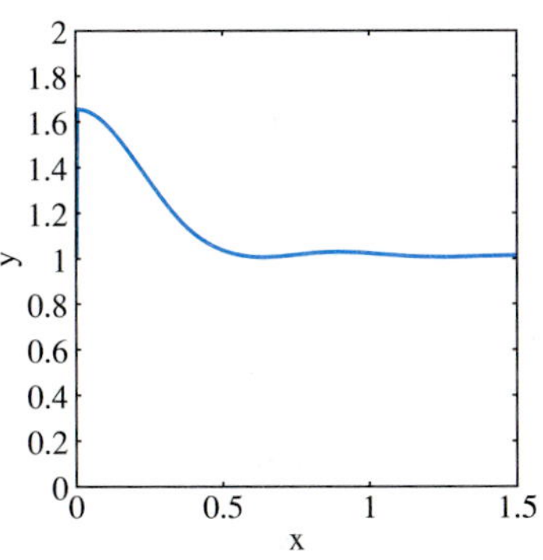

Fig. 8.4.5

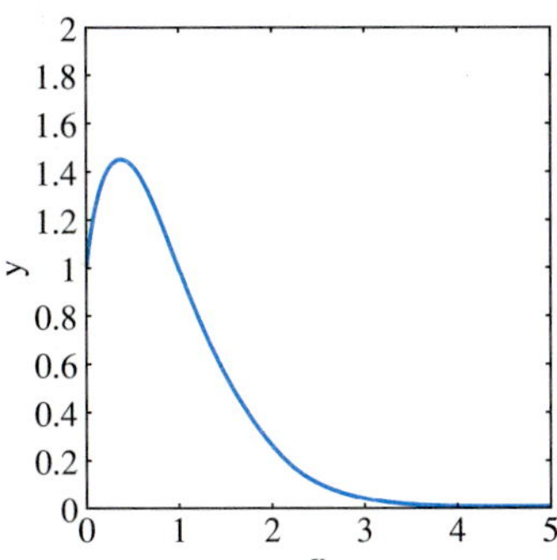

Fig. 8.4.6

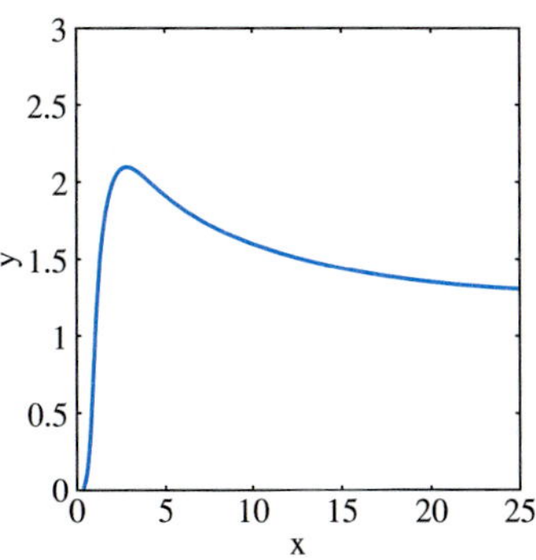

Fig. 8.4.7

41. $f(x) = (x + \sin x)^{1/x}$ **42.** $f(x) = (e^{1/x^2})^{(\cos x - 1)}$

Use l'Hôpital's rule to establish the limits in Problems 43 and 44.

43. $\lim_{h\to 0} (1 + hx)^{1/h} = e^x$ **44.** $\lim_{n\to\infty} \left(1 + \dfrac{x}{n}\right)^n = e^x$

45. Estimate graphically or numerically the location of the local minimum point on the graph $y = x^{\tan x}$ shown in Fig. 8.4.4.

46. Let n be a fixed positive integer, and let $p(x)$ be the polynomial

$$p(x) = x^n + a_1x^{n-1} + a_2x^{n-2} + \cdots + a_{n-1}x + a_n;$$

the numbers $a_1, a_2, \ldots, a_n$ are fixed real numbers. Prove that

$$\lim_{x\to\infty} ([p(x)]^{1/n} - x) = \frac{a_1}{n}.$$

47. As we shall see in Problem 52 of Section 9.6, the surface area of the ellipsoid obtained by revolving the ellipse

$$\frac{x^2}{a^2} + \frac{y^2}{b^2} = 1 \quad (a > b > 0)$$

around the x-axis is

$$A = 2\pi ab\left[\frac{b}{a} + \frac{a}{c}\sin^{-1}\left(\frac{c}{a}\right)\right],$$

where $c = \sqrt{a^2 - b^2}$. Use l'Hôpital's rule to show that

$$\lim_{b\to a} A = 4\pi a^2,$$

the surface area of a sphere of radius a.

48. Consider a long, thin rod that has heat diffusivity k and coincides with the x-axis. Suppose that at time $t = 0$ the temperature at x is $A/(2\epsilon)$ if $-\epsilon \leqq x \leqq \epsilon$ and is zero if $|x| > \epsilon$. Then it turns out that the temperature $T(x, t)$ of the rod at the point x at time $t > 0$ is given by

$$T(x, t) = \frac{A}{\epsilon\sqrt{4\pi kt}}\int_0^{\epsilon} \exp\left(-\frac{(x - u)^2}{4kt}\right) du.$$

Use l'Hôpital's rule to show that

$$\lim_{\epsilon\to 0} T(x,t) = \frac{A}{\sqrt{4\pi kt}}\exp\left(-\frac{x^2}{4kt}\right).$$

This is the temperature resulting from an initial "hot spot" at the origin.

49. Graph the function $f(x) = |\ln x|^{1/x}$ for $x > 0$, and determine its behavior as $x \to 0^+$ and as $x \to +\infty$. Estimate graphically and/or numerically the locations of any critical points or inflection points on the graph of f.

50. Graph the function $f(x) = |\ln x|^{1/|\ln x|}$ for $x > 0$, and determine its behavior as $x \to 0^+$, as $x \to +\infty$, and as x approaches 1 from either side.

51. Graph the function $f(x) = |\ln x|^{|\ln x|}$ for $x > 0$ and determine its behavior as $x \to 0^+$, as $x \to +\infty$, and as x approaches 1 from either side. Explore both graphically (by zooming) and symbolically (by differentiating) the question of whether f is differentiable at $x = 1$.

52. Let α be a fixed real number. (a) Evaluate (in terms of α) the 0^0 indeterminate form

$$\lim_{x\to 0}\left[\exp\left(-\frac{1}{x^2}\right)\right]^{\alpha x^2}.$$

(Note that l'Hôpital's rule is not needed.) Thus the indeterminate form 0^0 may have as its limit any positive real number. Explain why. (b) Can the limit of a 0^0 indeterminate form be zero, negative, or infinite? Explain.

8.5 HYPERBOLIC FUNCTIONS AND INVERSE HYPERBOLIC FUNCTIONS

The **hyperbolic cosine** and the **hyperbolic sine** of the real number x are denoted by $\cosh x$ and $\sinh x$ and are defined to be

$$\cosh x = \frac{e^x + e^{-x}}{2} \quad\text{and}\quad \sinh x = \frac{e^x - e^{-x}}{2}. \tag{1}$$

These particular combinations of familiar exponentials are useful in certain applications of calculus and are also helpful in evaluating certain integrals. The other four hyperbolic functions—the hyperbolic tangent, cotangent, secant, and cosecant—are defined in terms of $\cosh x$ and $\sinh x$ by analogy with trigonometry:

$$\begin{aligned} \tanh x &= \frac{\sinh x}{\cosh x} = \frac{e^x - e^{-x}}{e^x + e^{-x}}, \\ \coth x &= \frac{\cosh x}{\sinh x} = \frac{e^x + e^{-x}}{e^x - e^{-x}} \qquad (x \neq 0); \end{aligned} \tag{2}$$

$$\begin{aligned} \operatorname{sech} x &= \frac{1}{\cosh x} = \frac{2}{e^x + e^{-x}}, \\ \operatorname{csch} x &= \frac{1}{\sinh x} = \frac{2}{e^x - e^{-x}} \qquad (x \neq 0). \end{aligned} \tag{3}$$

The trigonometric terminology and notation for these hyperbolic functions stems from the fact that these functions satisfy a list of identities that, apart from an occasional difference of sign, much resemble familiar trigonometric identities:

$$\cosh^2 x - \sinh^2 x = 1; \tag{4}$$

$$1 - \tanh^2 x = \operatorname{sech}^2 x; \tag{5}$$

$$\coth^2 x - 1 = \operatorname{csch}^2 x; \tag{6}$$

$$\sinh(x + y) = \sinh x \cosh y + \cosh x \sinh y; \tag{7}$$

$$\cosh(x + y) = \cosh x \cosh y + \sinh x \sinh y; \tag{8}$$

$$\sinh 2x = 2 \sinh x \cosh x; \tag{9}$$

$$\cosh 2x = \cosh^2 x + \sinh^2 x; \tag{10}$$

$$\cosh^2 x = \tfrac{1}{2}(\cosh 2x + 1); \tag{11}$$

$$\sinh^2 x = \tfrac{1}{2}(\cosh 2x - 1). \tag{12}$$

The identities in Eqs. (4), (7), and (8) follow directly from the definitions of $\cosh x$ and $\sinh x$, as in Example 1.

EXAMPLE 1 To establish the "fundamental identity" in Eq. (4), we simply substitute the definitions of $\cosh x$ and $\sinh x$ on the left-hand side and write

$$\begin{aligned}\cosh^2 x - \sinh^2 x &= \tfrac{1}{4}(e^x + e^{-x})^2 - \tfrac{1}{4}(e^x - e^{-x})^2 \\ &= \tfrac{1}{4}(e^{2x} + 2 + e^{-2x}) - \tfrac{1}{4}(e^{2x} - 2 + e^{-2x}) = 1.\end{aligned}$$

The other identities listed previously may be derived from Eqs. (4), (7), and (8) in ways that parallel the derivations of the corresponding trigonometric identities. ■

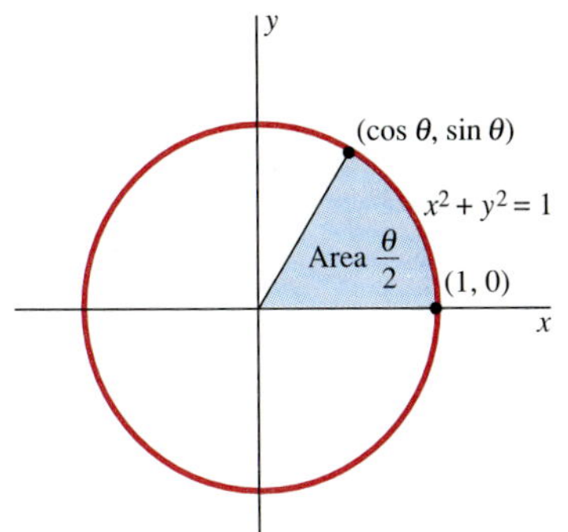

Fig. 8.5.1 Relation of the ordinary cosine and sine functions to the circle $x^2 + y^2 = 1$

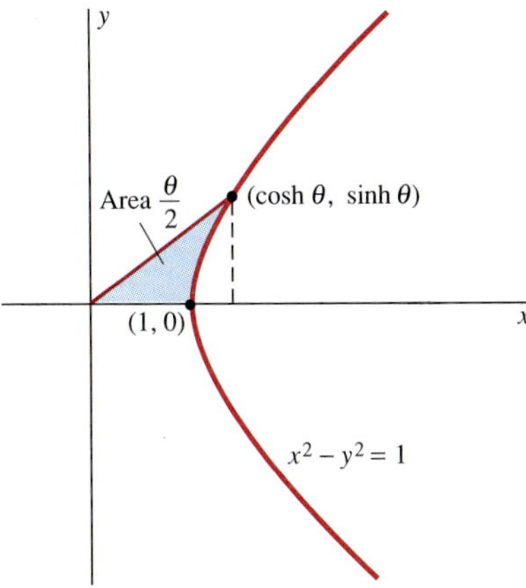

Fig. 8.5.2 Relation of the hyperbolic cosine and hyperbolic sine functions to the hyperbola $x^2 - y^2 = 1$

The trigonometric functions are sometimes called the *circular functions* because the point $(\cos\theta, \sin\theta)$ lies on the circle $x^2 + y^2 = 1$ for all θ (Fig. 8.5.1). Similarly, the identity in Eq. (4) tells us that the point $(\cosh\theta, \sinh\theta)$ lies on the hyperbola $x^2 - y^2 = 1$, and this is how the name *hyperbolic function* originated (Fig. 8.5.2).

The graphs of $y = \cosh x$ and $y = \sinh x$ are easy to construct. Add (for cosh) or subtract (for sinh) the ordinates of the graphs of $y = \frac{1}{2}e^x$ and $y = \frac{1}{2}e^{-x}$. The graphs of the other four hyperbolic functions can then be constructed by dividing ordinates. The graphs of all six are shown in Fig. 8.5.3.

These graphs show a striking difference between the hyperbolic functions and the ordinary trigonometric functions: None of the hyperbolic functions is periodic. They do, however, have even-odd properties, as the circular functions do. Like cosine and secant, the two functions cosh and sech are even, because

$$\cosh(-x) = \cosh x \quad \text{and} \quad \operatorname{sech}(-x) = \operatorname{sech} x$$

for all x. The other four hyperbolic functions, like the sine and tangent functions, are odd:

$$\sinh(-x) = -\sinh x, \qquad \tanh(-x) = -\tanh x,$$

and so on.

Derivatives and Integrals of Hyperbolic Functions

The formulas for the derivatives of the hyperbolic functions parallel those for the trigonometric functions, with occasional sign differences. For example,

$$D_x \cosh x = D_x\left(\tfrac{1}{2}e^x + \tfrac{1}{2}e^{-x}\right) = \tfrac{1}{2}e^x - \tfrac{1}{2}e^{-x} = \sinh x.$$

The chain rule then gives

$$D_x \cosh u = (\sinh u)\frac{du}{dx} \tag{13}$$

if u is a differentiable function of x. The other five differentiation formulas are

$$D_x \sinh u = (\cosh u)\frac{du}{dx}, \tag{14}$$

$$D_x \tanh u = (\operatorname{sech}^2 u)\frac{du}{dx}, \tag{15}$$

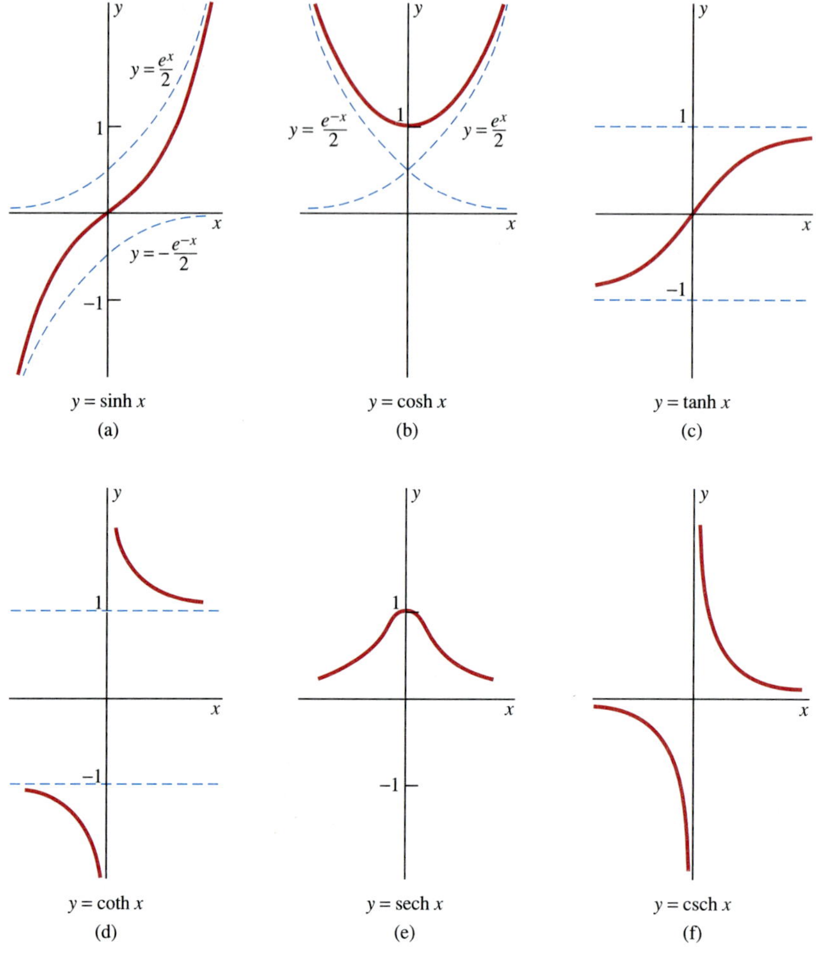

Fig. 8.5.3 Graphs of the six hyperbolic functions

$$D_x \coth u = (-\operatorname{csch}^2 u)\frac{du}{dx}, \tag{16}$$

$$D_x \operatorname{sech} u = (-\operatorname{sech} u \tanh u)\frac{du}{dx}, \tag{17}$$

$$D_x \operatorname{csch} u = (-\operatorname{csch} u \coth u)\frac{du}{dx}. \tag{18}$$

Equation (14) is derived exactly as Eq. (13) is. Then Eqs. (15) through (18) follow from Eqs. (13) and (14) with the aid of the quotient rule and the identities in Eqs. (5) and (6). As indicated in Example 2, the differentiation of hyperbolic functions using Eqs. (13) through (18) is very similar to the differentiation of trigonometric functions.

EXAMPLE 2

(a) $D_x \cosh 2x = 2 \sinh 2x$.

(b) $D_x \sinh^2 x = 2 \sinh x \cosh x$.

(c) $D_x (x \tanh x) = \tanh x + x \operatorname{sech}^2 x$.

(d) $D_x \operatorname{sech}(x^2) = -2x \operatorname{sech}(x^2) \tanh(x^2)$. ■

The antiderivative versions of the differentiation formulas in Eqs. (13) through (18) are the following integral formulas:

$$\int \sinh u\, du = \cosh u + C, \tag{19}$$

$$\int \cosh u\, du = \sinh u + C, \tag{20}$$

$$\int \operatorname{sech}^2 u\, du = \tanh u + C, \tag{21}$$

$$\int \operatorname{csch}^2 u\, du = -\coth u + C, \tag{22}$$

$$\int \operatorname{sech} u \tanh u\, du = -\operatorname{sech} u + C, \tag{23}$$

$$\int \operatorname{csch} u \coth u\, du = -\operatorname{csch} u + C. \tag{24}$$

The integrals in Example 3 illustrate the fact that simple hyperbolic integrals may be treated in much the same way as simple trigonometric integrals.

EXAMPLE 3

(a) With $u = 3x$, we have

$$\int \cosh 3x\, dx = \int (\cosh u)\left(\tfrac{1}{3}\, du\right) = \tfrac{1}{3} \sinh u + C = \tfrac{1}{3} \sinh 3x + C.$$

(b) With $u = \sinh x$, we have

$$\int \sinh x \cosh x\, dx = \int u\, du = \tfrac{1}{2} u^2 + C = \tfrac{1}{2} \sinh^2 x + C.$$

(c) Using Eq. (12), we find that

$$\int \sinh^2 x\, dx = \int \tfrac{1}{2} (\cosh 2x - 1)\, dx = \tfrac{1}{4} \sinh 2x - \tfrac{1}{2} x + C.$$

(d) Finally, using Eq. (5), we see that

$$\int_0^1 \tanh^2 x\, dx = \int_0^1 (1 - \operatorname{sech}^2 x)\, dx = \Big[x - \tanh x \Big]_0^1$$

$$= 1 - \tanh 1 = 1 - \frac{e - e^{-1}}{e + e^{-1}} = \frac{2}{e^2 + 1} \approx 0.238406.$$

■

Inverse Hyperbolic Functions

Figure 8.5.3 shows that

- The functions $\sinh x$ and $\tanh x$ are increasing for all x;
- The functions $\coth x$ and $\operatorname{csch} x$ are decreasing and defined for all $x \neq 0$;

- The function $\cosh x$ is increasing on the half-line $x \geqq 0$; and
- The function $\operatorname{sech} x$ is decreasing on the half-line $x \geqq 0$.

It follows that each of the six hyperbolic functions can be "inverted" on the indicated domain where it is either increasing or decreasing. The resulting inverse hyperbolic functions and their domains of definition are listed in the following table:

Inverse Hyperbolic Function	Defined For
$\sinh^{-1} x$	All x
$\cosh^{-1} x$	$x \geqq 1$
$\tanh^{-1} x$	$\lvert x \rvert < 1$
$\coth^{-1} x$	$\lvert x \rvert > 1$
$\operatorname{sech}^{-1} x$	$0 < x \leqq 1$
$\operatorname{csch}^{-1} x$	$x \neq 0$

EXAMPLE 4 Find the numerical value of $\tanh^{-1}\left(\frac{1}{2}\right)$.

Solution If $y = \tanh^{-1} x$, then

$$\begin{aligned}
\tanh y &= x; \\
\frac{e^y - e^{-y}}{e^y + e^{-y}} &= x \quad \text{[by Eq. (2)]}; \\
e^y - e^{-y} &= xe^y + xe^{-y}; \\
(1 - x)e^y &= (1 + x)e^{-y}; \\
e^{2y} &= \frac{1 + x}{1 - x}; \\
y &= \frac{1}{2} \ln \frac{1 + x}{1 - x}.
\end{aligned}$$

Hence, with $x = \frac{1}{2}$, we find that $\tanh^{-1}\left(\frac{1}{2}\right) = \frac{1}{2} \ln 3 \approx 0.549306$. ■

Scientific calculators ordinarily are used to find values of hyperbolic and inverse hyperbolic functions. Many calculators give values only of $\sinh^{-1}$, $\cosh^{-1}$, and $\tanh^{-1}$. Values of the other three inverse hyperbolic functions can then be found by using the identities

$$\operatorname{sech}^{-1} x = \cosh^{-1}\left(\frac{1}{x}\right), \tag{25}$$

$$\operatorname{csch}^{-1} x = \sinh^{-1}\left(\frac{1}{x}\right), \quad \text{and} \tag{26}$$

$$\coth^{-1} x = \tanh^{-1}\left(\frac{1}{x}\right). \tag{27}$$

For example, $\coth^{-1} 2 = \tanh^{-1}\left(\frac{1}{2}\right) \approx 0.549306$.

Derivatives of Inverse Hyperbolic Functions

Here are the derivatives of the six inverse hyperbolic functions:

$$D_x \sinh^{-1} x = \frac{1}{\sqrt{1 + x^2}}, \tag{28}$$

$$D_x \cosh^{-1} x = \frac{1}{\sqrt{x^2 - 1}}, \tag{29}$$

$$D_x \tanh^{-1} x = \frac{1}{1 - x^2}, \tag{30}$$

$$D_x \coth^{-1} x = \frac{1}{1 - x^2}, \tag{31}$$

$$D_x \operatorname{sech}^{-1} x = -\frac{1}{x\sqrt{1 - x^2}}, \tag{32}$$

$$D_x \operatorname{csch}^{-1} x = -\frac{1}{|x|\sqrt{1 + x^2}}. \tag{33}$$

We can derive these formulas by the standard method of finding the derivative of the inverse of a function when the derivative of the function itself is known. The only requirement is that the inverse function is known in advance to be differentiable.

EXAMPLE 5 To differentiate $\tanh^{-1} x$, we begin with the inverse function relation

$$\tanh(\tanh^{-1} x) = x$$

and substitute $u = \tanh^{-1} x$. Then, because this equation is actually an identity,

$$D_x \tanh u = D_x x = 1,$$

so

$$(\operatorname{sech}^2 u)\frac{du}{dx} = 1.$$

Thus, because $u = \tanh^{-1} x$,

$$D_x \tanh^{-1} x = \frac{du}{dx} = \frac{1}{\operatorname{sech}^2 u} = \frac{1}{1 - \tanh^2 u}$$

$$= \frac{1}{1 - \tanh^2(\tanh^{-1} x)} = \frac{1}{1 - x^2}.$$

This establishes Eq. (30). We can use similar methods to verify the formulas for the derivatives of the other five hyperbolic functions. ■

The hyperbolic functions are defined in terms of the natural exponential e^x, so it's no surprise that their inverses may be expressed in terms of $\ln x$ (see Example 3). In fact,

$$\sinh^{-1} x = \ln\left(x + \sqrt{x^2 + 1}\right) \quad \text{for all } x; \tag{34}$$

$$\cosh^{-1} x = \ln\left(x + \sqrt{x^2 - 1}\right) \quad \text{for all } x \geqq 1; \tag{35}$$

$$\tanh^{-1} x = \frac{1}{2}\ln\left(\frac{1 + x}{1 - x}\right) \quad \text{for } |x| < 1; \tag{36}$$

$$\coth^{-1}x = \frac{1}{2}\ln\left(\frac{x+1}{x-1}\right) \qquad \text{for } |x| > 1; \tag{37}$$

$$\operatorname{sech}^{-1}x = \ln\left(\frac{1+\sqrt{1-x^2}}{x}\right) \qquad \text{if } 0 < x \leqq 1; \tag{38}$$

$$\operatorname{csch}^{-1}x = \ln\left(\frac{1}{x} + \frac{\sqrt{1+x^2}}{|x|}\right) \qquad \text{if } x \neq 0. \tag{39}$$

Each of these identities may be established by showing that each side has the same derivative and also that the two sides agree for at least one value of x in every interval of their respective domains.

EXAMPLE 6 To establish the identity in Eq. (34), we begin by differentiating each side:

$$D_x \ln\left(x + \sqrt{x^2+1}\right) = \frac{1 + \dfrac{x}{\sqrt{x^2+1}}}{x + \sqrt{x^2+1}} = \frac{1}{\sqrt{x^2+1}} = D_x \sinh^{-1} x.$$

Thus

$$\sinh^{-1}x = \ln\left(x + \sqrt{x^2+1}\right) + C.$$

But $\sinh^{-1}(0) = 0 = \ln(0 + \sqrt{0+1})$. This implies that $C = 0$ and thus establishes Eq. (34). It is not quite as easy to show that $C = 0$ in the proofs of Eqs. (37) and (39); see Problems 64 and 65. ■

Equations (34) through (39) may be used to calculate the values of inverse hyperbolic functions. This is convenient if you own a calculator whose repertoire does not include the inverse hyperbolic functions or if you are programming in a language such as BASIC, most forms of which do not include these functions.

Integrals Involving Inverse Hyperbolic Functions

The principal applications of inverse hyperbolic functions are to the evaluation of algebraic integrals. The differentiation formulas in Eqs. (28) through (33) may, in the usual way, be written as the following integral formulas:

$$\int \frac{du}{\sqrt{u^2+1}} = \sinh^{-1}u + C, \tag{40}$$

$$\int \frac{du}{\sqrt{u^2-1}} = \cosh^{-1}u + C, \tag{41}$$

$$\int \frac{du}{1-u^2} = \tanh^{-1}u + C \qquad \text{if } |u| < 1, \tag{42a}$$

$$\int \frac{du}{1-u^2} = \coth^{-1}u + C \qquad \text{if } |u| > 1, \tag{42b}$$

$$\int \frac{du}{1-u^2} = \tfrac{1}{2}\ln\left|\frac{1+u}{1-u}\right| + C, \tag{42c}$$

$$\int \frac{du}{u\sqrt{1-u^2}} = -\operatorname{sech}^{-1}|u| + C, \tag{43}$$

$$\int \frac{du}{u\sqrt{1+u^2}} = -\operatorname{csch}^{-1}|u| + C. \tag{44}$$

The distinction between the two cases $|u| < 1$ and $|u| > 1$ in Eqs. (42a) and (42b) results from the fact that the inverse hyperbolic tangent is defined for $|x| < 1$, whereas the inverse hyperbolic cotangent is defined for $|x| > 1$.

EXAMPLE 7 The substitution $u = 2x$, $dx = \frac{1}{2}\,du$ yields

$$\int \frac{dx}{\sqrt{4x^2+1}} = \frac{1}{2}\int \frac{du}{\sqrt{u^2+1}} = \frac{1}{2}\sinh^{-1}2x + C.$$

EXAMPLE 8

$$\begin{aligned}\int_0^{1/2} \frac{dx}{1-x^2} &= \Big[\tanh^{-1}x\Big]_0^{1/2}\\ &= \frac{1}{2}\left[\ln\left|\frac{1+x}{1-x}\right|\right]_0^{1/2} = \frac{1}{2}\ln 3 \approx 0.549306.\end{aligned}$$

EXAMPLE 9

$$\begin{aligned}\int_2^{5} \frac{dx}{1-x^2} &= \Big[\coth^{-1}x\Big]_2^{5} = \frac{1}{2}\left[\ln\left|\frac{1+x}{1-x}\right|\right]_2^{5}\\ &= \frac{1}{2}\left[\ln\left(\frac{6}{4}\right) - \ln 3\right] = -\frac{1}{2}\ln 2 \approx -0.346574.\end{aligned}$$

8.5 PROBLEMS

Find the derivatives of the functions in Problems 1 through 14.

1. $f(x) = \cosh(3x-2)$

2. $f(x) = \sinh\sqrt{x}$

3. $f(x) = x^2\tanh\left(\frac{1}{x}\right)$

4. $f(x) = \operatorname{sech} e^{2x}$

5. $f(x) = \coth^3 4x$

6. $f(x) = \ln\sinh 3x$

7. $f(x) = e^{\operatorname{csch} x}$

8. $f(x) = \cosh\ln x$

9. $f(x) = \sin(\sinh x)$

10. $f(x) = \tan^{-1}(\tanh x)$

11. $f(x) = \sinh x^4$

12. $f(x) = \sinh^4 x$

13. $f(x) = \dfrac{1}{x + \tanh x}$

14. $f(x) = \cosh^2 x - \sinh^2 x$

Evaluate the integrals in Problems 15 through 28.

15. $\displaystyle\int x\sinh x^2\,dx$

16. $\displaystyle\int \cosh^2 3u\,du$

17. $\displaystyle\int \tanh^2 3x\,dx$

18. $\displaystyle\int \frac{\operatorname{sech}\sqrt{x}\tanh\sqrt{x}}{\sqrt{x}}\,dx$

19. $\displaystyle\int \sinh^2 2x\cosh 2x\,dx$

20. $\displaystyle\int \tanh 3x\,dx$

21. $\displaystyle\int \frac{\sinh x}{\cosh^3 x}\,dx$

22. $\displaystyle\int \sinh^4 x\,dx$

23. $\displaystyle\int \coth x\operatorname{csch}^2 x\,dx$

24. $\displaystyle\int \operatorname{sech} x\,dx$

25. $\displaystyle\int \frac{\sinh x}{1+\cosh x}\,dx$

26. $\displaystyle\int \frac{\sinh\ln x}{x}\,dx$

27. $\displaystyle\int \frac{1}{(e^x+e^{-x})^2}\,dx$

28. $\displaystyle\int \frac{e^x+e^{-x}}{e^x-e^{-x}}\,dx$

Find the derivatives of the functions in Problems 29 through 38.

29. $f(x) = \sinh^{-1}2x$

30. $f(x) = \cosh^{-1}(x^2+1)$

31. $f(x) = \tanh^{-1}\sqrt{x}$

32. $f(x) = \coth^{-1}\sqrt{x^2+1}$

33. $f(x) = \operatorname{sech}^{-1}\left(\frac{1}{x}\right)$

34. $f(x) = \operatorname{csch}^{-1}e^x$

35. $f(x) = (\sinh^{-1}x)^{3/2}$

36. $f(x) = \sinh^{-1}(\ln x)$

37. $f(x) = \ln(\tanh^{-1}x)$

38. $f(x) = \dfrac{1}{\tanh^{-1}3x}$

Use inverse hyperbolic functions to evaluate the integrals in Problems 39 through 48.

39. $\displaystyle\int \frac{dx}{\sqrt{x^2+9}}$ **40.** $\displaystyle\int \frac{dy}{\sqrt{4y^2-9}}$

41. $\displaystyle\int_{1/2}^{1} \frac{dx}{4-x^2}$ **42.** $\displaystyle\int_{5}^{10} \frac{dx}{4-x^2}$

43. $\displaystyle\int \frac{dx}{x\sqrt{4-9x^2}}$ **44.** $\displaystyle\int \frac{dx}{x\sqrt{x^2+25}}$

45. $\displaystyle\int \frac{e^x}{\sqrt{e^{2x}+1}}\,dx$ **46.** $\displaystyle\int \frac{x}{\sqrt{x^4-1}}\,dx$

47. $\displaystyle\int \frac{1}{\sqrt{1-e^{2x}}}\,dx$ **48.** $\displaystyle\int \frac{\cos x}{\sqrt{1+\sin^2 x}}\,dx$

49. Apply the definitions in Eq. (1) to prove the identity in Eq. (7).

50. Derive the identities in Eqs. (5) and (6) from the identity in Eq. (4).

51. Deduce the identities in Eqs. (10) and (11) from the identity in Eq. (8).

52. Suppose that A and B are constants. Show that the function $x(t) = A\cosh kt + B\sinh kt$ is a solution of the differential equation

$$\frac{d^2x}{dt^2} = k^2x(t).$$

53. Find the length of the curve $y = \cosh x$ over the interval $[0, a]$.

54. Find the volume of the solid obtained by revolving around the x-axis the area under $y = \sinh x$ from $x = 0$ to $x = \pi$.

55. Show that the area $A(\theta)$ of the shaded sector in Fig. 8.5.1 is $\theta/2$. This corresponds to the fact that the area of the sector of the unit circle between the positive x-axis and the radius to the point $(\cos\theta, \sin\theta)$ is $\theta/2$. [*Suggestion:* Note first that

$$A(\theta) = \tfrac{1}{2}\cosh\theta\sinh\theta - \int_1^{\cosh\theta} \sqrt{x^2-1}\,dx.$$

Then use the fundamental theorem of calculus to show that $A'(\theta) = \frac{1}{2}$ for all θ.]

56. Evaluate the following limits: (a) $\displaystyle\lim_{x\to 0}\frac{\sinh x}{x}$; (b) $\displaystyle\lim_{x\to\infty}\tanh x$; (c) $\displaystyle\lim_{x\to\infty}\frac{\cosh x}{e^x}$.

57. Use the method of Example 4 to find the numerical value of $\sinh^{-1}1$.

58. Apply Eqs. (34) and (39) to verify the identity

$$\operatorname{csch}^{-1}x = \sinh^{-1}\left(\frac{1}{x}\right) \quad \text{if } x \neq 0.$$

59. Establish the formula for $D_x \sinh^{-1}x$ in Eq. (28).

60. Establish the formula for $D_x \operatorname{sech}^{-1}x$ in Eq. (32).

61. Prove Eq. (36) by differentiating both sides.

62. Establish Eq. (34) by solving the equation

$$x = \sinh y = \frac{e^y - e^{-y}}{2}$$

for y in terms of x.

63. Establish Eq. (37) by solving the equation

$$x = \coth y = \frac{e^y + e^{-y}}{e^y - e^{-y}}$$

for y in terms of x.

64. (a) Differentiate both sides of Eq. (37) to show that they differ by a constant C. (b) Then prove that $C = 0$ by using the definition of $\coth x$ to show that $\coth^{-1}2 = \frac{1}{2}\ln 3$.

65. (a) Differentiate both sides of Eq. (39) to show that they differ by a constant C. (b) Then prove that $C = 0$ by using the definition of $\operatorname{csch} x$ to show that $\operatorname{csch}^{-1}1 = \ln\left(1+\sqrt{2}\right)$.

66. Estimate (graphically or numerically) the points of intersection of the curves $y = x + 2$ and $y = \cosh x$. Then approximate the area of the region bounded by these two curves.

In Problems 67 and 68, show first that $f(x) \to 0$ as $x \to +\infty$. Then estimate (graphically or numerically) the absolute maximum value of $f(x)$ for $x > 0$. Differentiate $f(x)$ to verify that you have an approximate critical point.

67. $f(x) = e^{-2x}\tanh x$ **68.** $f(x) = e^{-x}\sinh^{-1}x$

Problems 69 and 70 deal with the hanging cable illustrated in Fig. 8.5.4. If the cable is flexible and has uniform density, then elementary principles of physics can be used to show that its shape function $y = y(x)$ satisfies the differential equation

$$\frac{d^2y}{dx^2} = k\sqrt{1+\left(\frac{dy}{dx}\right)^2}, \tag{45}$$

where k is a constant determined by the density and tension of the cable.

69. Verify that the function

$$y(x) = y_0 + \frac{1}{k}(-1 + \cosh kx) \tag{46}$$

satisfies the differential equation in (45) and also satisfies the initial conditions $y(0) = y_0$, $y'(0) = 0$. A curve

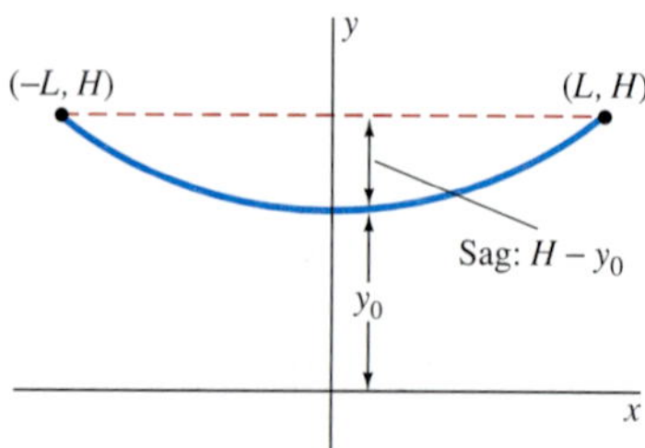

Fig. 8.5.4 A flexible uniform cable suspended between two points at equal heights

with this shape is called a *catenary,* from the Latin word *catena* (chain).

70. A high-voltage line is to be strung between two 50-ft towers 200 ft apart. (a) If the line sags 20 ft at its middle (where $x = 0$)—so $y_0 = 30$ (ft) in Eq. (46)—estimate graphically or numerically the value of the parameter k. (b) Then approximate the total length of the high-voltage line.

8.5 PROJECT: YOUR FIRST SKY-DIVE

You have long wanted to try sky-diving, and in this project you will plan—very carefully—your first sky-dive.

Suppose that while falling through the air you are affected by two forces: a downward force equal to your weight W (in pounds) and an upward force of air resistance that is proportional to the *square* of your velocity v (in feet per second). Then Newton's law $F = ma$ provides the differential equation

$$m\frac{dv}{dt} = W - kv^2 \tag{1}$$

if the distance $x(t)$ you have fallen after t seconds is measured downward (so $v = dx/dt > 0$ as you are falling). Your mass is given in fps units by $m = W/g$, where $g = 32\ \text{ft/s}^2$ denotes the acceleration of gravity. If we divide by m in Eq. (1), the result is

$$\frac{dv}{dt} = g - \rho v^2 \tag{2}$$

where $\rho = k/m$. Ultimately you want to know how long it will take to reach the ground if you jump from a given height and what your velocity will be t seconds after your jump begins—especially when it ends. The first two of the following steps will provide the basis for your calculations.

1. Assume that you step out of a hovering hot-air balloon with initial velocity $v(0) = 0$. Separate the variables in Eq. (2) and thus write

$$\int \frac{1}{g - \rho v^2}\, dv = \int 1\, dt = t + C.$$

Then substitute $v = u\sqrt{g/\rho}$ and apply Eq. (42a) in this section to deduce that your velocity after t seconds is given by

$$v(t) = \frac{dx}{dt} = \sqrt{\frac{g}{\rho}}\tanh\left(t\sqrt{\rho g}\right). \tag{3}$$

2. Take $x = 0$ at the altitude of the hovering balloon (so that $x(0) = 0$) and take the downward direction as the positive direction. Integrate both sides in Eq. (3) to deduce that the distance you fall in t seconds is given by

$$x(t) = \frac{1}{\rho}\ln\cosh\left(t\sqrt{\rho g}\right). \tag{4}$$

In the remaining computations, suppose that you step out of the hovering balloon at an altitude of $h = 1000n$ feet, where n is the largest digit in your student I.D. number.

3. Suppose that $k = 0.005$ in Eq. (1) and that your parachute fails to open, so you free-fall all the way to the ground. Determine the time required to reach the ground and the velocity (both in ft/s and in mi/h) with which you hit. Do you have a chance of survival? *Suggestion:* Instead of setting $x = h$ in Eq. (4) and trying to solve for t, note first that Eq. (3) implies that you reach a limiting *terminal velocity* of $v_\tau = \sqrt{g/\rho}$ as t increases indefinitely. (Why?) Then use Eqs. (3) and (4) to construct a table of values of v and x at successive one-second intervals. How many seconds does it take you to reach terminal velocity (accurate to two decimal places)? Assume then that you fall the remaining distance to the ground at *constant* velocity v_τ, so how long it takes is easy to calculate.

4. Now suppose that you open your parachute immediately upon exit from the balloon and thence fall all the way with $k = 0.4$ in Eq. (1). Answer the same questions as in step 3. You should land safely (if on open ground rather than in a thornbush).

5. Finally, suppose that your parachute opens, but in a defective manner, so you fall all the way to the ground with $k = 0.05$ in Eq. (1). Might you survive the experience, given a haystack on the ground fortunately located at your point of impact?

Challenge: Suppose that you free-fall halfway to the ground as in Problem 3, then descend the remaining distance using the parachute of Problem 4. Answer the usual questions. Note that Eqs. (3) and (4) depend on the initial conditions $x(0) = v(0) = 0$ and therefore do not apply during the parachute portion of your descent. Thus you will need to begin again from scratch.

CHAPTER 8 REVIEW: *Definitions and Formulas*

Use the following list as a guide to concepts that you may need to review.

1. The definitions of the six inverse trigonometric functions
2. The derivatives of the inverse trigonometric functions
3. The integral formulas corresponding to the derivatives of the inverse sine, inverse tangent, and inverse secant functions
4. L'Hôpital's rule and the indeterminate forms $0/0$, ∞/∞, $0 \cdot \infty$, $\infty - \infty$, 0^0, ∞^0, and 1^∞
5. The definitions and derivatives of the hyperbolic functions
6. The definition and derivatives of the inverse hyperbolic functions
7. The use of inverse hyperbolic functions to evaluate certain algebraic integrals

CHAPTER 8 *Miscellaneous Problems*

Differentiate the functions in Problems 1 through 20.

1. $f(x) = \sin^{-1} 3x$

2. $f(x) = \tan^{-1} 7x$

3. $g(t) = \sec^{-1} t^2$

4. $g(t) = \tan^{-1} e^t$

5. $f(x) = \sin^{-1}(\cos x)$

6. $f(x) = \sinh^{-1} 2x$

7. $g(t) = \cosh^{-1} 10t$

8. $h(u) = \tanh^{-1}\left(\dfrac{1}{u}\right)$

9. $f(x) = \sin^{-1}\left(\dfrac{1}{x^2}\right)$

10. $f(x) = \tan^{-1}\left(\dfrac{1}{x}\right)$

11. $f(x) = \arcsin \sqrt{x}$

12. $f(x) = x \sec^{-1} x^2$

13. $f(x) = \tan^{-1}(1 + x^2)$

14. $f(x) = \sin^{-1}\sqrt{1 - x^2}$

15. $f(x) = e^x \sinh e^x$

16. $f(x) = \ln \cosh x$

17. $f(x) = \tanh^2 3x + \operatorname{sech}^2 3x$

18. $f(x) = \sinh^{-1}\sqrt{x^2 - 1}$

19. $f(x) = \cosh^{-1}\sqrt{x^2 + 1}$

20. $f(x) = \tanh^{-1}(1 - x^2)$

Evaluate the integrals in Problems 21 through 40.

21. $\displaystyle\int \frac{dx}{\sqrt{1 - 4x^2}}$

22. $\displaystyle\int \frac{dx}{1 + 4x^2}$

23. $\displaystyle\int \frac{dx}{\sqrt{4 - x^2}}$

24. $\displaystyle\int \frac{dx}{4 + x^2}$

25. $\displaystyle\int \frac{e^x}{\sqrt{1 - e^{2x}}}\,dx$

26. $\displaystyle\int \frac{x}{1 + x^4}\,dx$

27. $\displaystyle\int \frac{1}{\sqrt{9 - 4x^2}}\,dx$

28. $\displaystyle\int \frac{1}{9 + 4x^2}\,dx$

29. $\displaystyle\int \frac{x^2}{1+x^6}\,dx$

30. $\displaystyle\int \frac{\cos x}{1+\sin^2 x}\,dx$

31. $\displaystyle\int \frac{1}{x\sqrt{4x^2-1}}\,dx$

32. $\displaystyle\int \frac{1}{x\sqrt{x^4-1}}\,dx$

33. $\displaystyle\int \frac{1}{\sqrt{e^{2x}-1}}\,dx$

34. $\displaystyle\int x^2\cosh x^3\,dx$

35. $\displaystyle\int \frac{\sinh\sqrt{x}}{\sqrt{x}}\,dx$

36. $\displaystyle\int \operatorname{sech}^2(3x-2)\,dx$

37. $\displaystyle\int \frac{\arctan x}{1+x^2}\,dx$

38. $\displaystyle\int \frac{1}{\sqrt{4x^2-1}}\,dx$

39. $\displaystyle\int \frac{1}{\sqrt{4x^2+9}}\,dx$

40. $\displaystyle\int \frac{x}{\sqrt{x^4+1}}\,dx$

Find the limits in Problems 41 through 55.

41. $\displaystyle\lim_{x\to 2}\frac{x-2}{x^2-4}$

42. $\displaystyle\lim_{x\to 0}\frac{\sin 2x}{x}$

43. $\displaystyle\lim_{x\to \pi}\frac{1+\cos x}{(x-\pi)^2}$

44. $\displaystyle\lim_{x\to 0}\frac{x-\sin x}{x^3}$

45. $\displaystyle\lim_{t\to 0}\frac{\arctan t-\sin t}{t^3}$

46. $\displaystyle\lim_{x\to \infty}\frac{\ln(\ln x)}{\ln x}$

47. $\displaystyle\lim_{x\to 0}(\cot x)\ln(1+x)$

48. $\displaystyle\lim_{x\to 0^+}(e^{1/x}-1)\tan x$

49. $\displaystyle\lim_{x\to 0}\left(\frac{1}{x^2}-\frac{1}{1-\cos x}\right)$

50. $\displaystyle\lim_{x\to \infty}\left(\frac{x^2}{x+2}-\frac{x^3}{x^2+3}\right)$

51. $\displaystyle\lim_{x\to \infty}\left(\sqrt{x^2-x-1}-\sqrt{x}\right)$

52. $\displaystyle\lim_{x\to \infty}x^{1/x}$

53. $\displaystyle\lim_{x\to \infty}(e^{2x}-2x)^{1/x}$

54. $\displaystyle\lim_{x\to \infty}[1-\exp(-x^2)]^{1/x^2}$

55. $\displaystyle\lim_{x\to \infty}x\left[\left(1+\frac{1}{x}\right)^x-e\right]$ (*Suggestion:* Let $u=1/x$, and take the limit as $u\to 0^+$.)

56. According to Problem 53 of Section 9.6, the surface area of the ellipsoid obtained by revolving around the x-axis the ellipse with equation

$$\left(\frac{x}{a}\right)^2+\left(\frac{y}{b}\right)^2=1 \quad (0<a<b)$$

is

$$A=2\pi ab\left[\frac{b}{a}+\frac{a}{c}\ln\left(\frac{b+c}{a}\right)\right],$$

where $c=\sqrt{b^2-a^2}$. Use l'Hôpital's rule to show that

$$\lim_{b\to a}A=4\pi a^2,$$

the surface area of a sphere of radius a.

57. Find the volume generated by revolving around the y-axis the region under $y=1/\sqrt{1-x^4}$ from $x=0$ to $x=1/\sqrt{2}$.

58. Find the volume generated by revolving around the y-axis the region under $y=1/\sqrt{x^4+1}$ from $x=0$ to $x=1$.

59. Use Eqs. (35) through (38) of Section 8.5 to show that (a) $\coth^{-1}x=\tanh^{-1}\left(\dfrac{1}{x}\right)$; (b) $\operatorname{sech}^{-1}x=\cosh^{-1}\left(\dfrac{1}{x}\right)$.

60. Show that $x''(t)=k^2x(t)$ if

$$x(t)=A\cosh kt+B\sinh kt,$$

where A and B are constants. Determine A and B if (a) $x(0)=1, x'(0)=0$; (b) $x(0)=0, x'(0)=1$.

61. Use Newton's method to find the least positive solution of the equation $\cos x\cosh x=1$. Begin by sketching the graphs of $y=\cos x$ and $y=\operatorname{sech} x$.

62. (a) Verify by differentiation that

$$\int \sec x\,dx=\sinh^{-1}(\tan x)+C.$$

(b) Show similarly that

$$\int \operatorname{sech} x\,dx=\tan^{-1}(\sinh x)+C.$$

CHAPTER 9

TECHNIQUES OF INTEGRATION

Leonhard Euler (1707–1783)

The most prolific mathematician in all history was Leonhard Euler, who was born in 1707 in Basel, Switzerland, the home of the Bernoulli family of mathematicians. His father preferred a theological career for his son, but young Euler learned mathematics from John Bernoulli and thereby found his true vocation. During his lifetime Euler published more than 500 books and papers. His work continued unabated even after he lost his sight in 1766. Upon his death in 1783, he left behind more than 300 additional manuscripts whose publication continued in a steady flow for another half century. His collected works fill approximately 75 substantial volumes.

No other mathematician of the past more directly affects the modern student of mathematics, because it was largely Euler who shaped the notation and terminology still used today in teaching high school algebra and trigonometry as well as calculus. His *Introductio in Analysin Infinitorium* (Introduction to Infinitesimal Analysis) is the earliest mathematics textbook whose exposition would (in translation from the original Latin) be accessible to a modern student. Here are some now-familiar notations whose use was popularized and standardized by Euler:

e	for the base of natural logarithms;
i	for the square root of -1;
Σ	for the summation symbol;
$f(x)$	for function notation;
π	for the area of the unit circle;

and the trigonometric abbreviations sin, cos, tang, cot, sec, and cosec, which are close to their current forms. It was Euler's *Introductio* that once and for all based calculus squarely on the function concept. His 1755 and 1768 calculus treatises provide the original source for much of the content and methods of modern calculus courses and texts.

Euler originally discovered so many of the standard formulas and identities of mathematics that it is customary to attribute a formula to the first mathematician *after* Euler to *re*discover it. But the identity $e^{ix} = \cos x + i \sin x$ relating the exponential and trigonometric functions is still known as Euler's formula. Substitution of $x = \pi$ yields the relation $e^{i\pi} + 1 = 0$, which links five of the most important constants in mathematics.

The photograph—part of a page from Chapter VII of the *Introductio*—shows the first appearance in public print of the number $e \approx 2.71828$. Immediately following its definition as the sum of the infinite series $\Sigma\,(1/n!)$, Euler gives the numerical value of e accurate to 23 decimal places.

90 *DE QUANTITATUM EXPONENTIALIUM*

LIB. I. (116) inventam, $a = 1 + \frac{1}{1} + \frac{1}{1.2} + \frac{1}{1.2.3} + \frac{1}{1.2.3.4} + \&c.$, qui termini, ſi in fractiones decimales convertantur atque actu addantur, præbebunt hunc valorem pro $a =$ 2,71828182845904523536028, cujus ultima adhuc nota veritati eſt conſentanea. Quod ſi jam ex hac baſi Logarithmi conſtruantur, ii vocari ſolent Logarithmi *naturales* ſeu *hyperbolici*, quoniam quadratura hyperbolæ per iſtiuſmodi Logarithmos exprimi poteſt. Ponamus autem brevitatis gratia pro numero hoc 2,718281828459 &c. conſtanter litteram e, quæ ergo denotabit baſin Logarithmorum naturalium ſeu hyperbolicorum, cui reſpondet valor litteræ $k = 1$; ſive hæc littera e quoque exprimet ſummam hujus Seriei $1 + \frac{1}{1} + \frac{1}{1.2} + \frac{1}{1.2.3} + \frac{1}{1.2.3.4} + \&c.$ in infinitum.

9.1 INTRODUCTION

We have seen in the preceding three chapters that many geometric and physical quantities can be expressed as definite integrals. The fundamental theorem of calculus reduces the problem of calculating the definite integral

$$\int_a^b f(x)\,dx$$

to that of finding an antiderivative $G(x)$ of $f(x)$. Once this is accomplished, then

$$\int_a^b f(x)\,dx = \Big[\, G(x) \,\Big]_a^b = G(b) - G(a).$$

But as yet we have relied largely on trial-and-error methods for finding the required antiderivative $G(x)$. In some cases a knowledge of elementary derivative formulas, perhaps in combination with a simple substitution, allows us to integrate a given function. This approach can, however, be inefficient and time-consuming, especially in view of the following surprising fact: Some simple-looking integrals, such as

$$\int e^{-x^2}\,dx, \qquad \int \frac{\sin x}{x}\,dx, \qquad \text{and} \qquad \int \sqrt{1+x^4}\,dx,$$

cannot be evaluated in terms of finite combinations of the familiar algebraic and elementary transcendental functions. For example, the antiderivative

$$H(x) = \int_0^x e^{-t^2}\,dt$$

of $\exp(-x^2)$ has no finite expression in terms of elementary functions. Any attempt to find such an expression therefore will inevitably be unsuccessful.

The presence of such integrals indicates that we cannot hope to reduce integration to a routine process such as differentiation. In fact, finding antiderivatives is an art, the mastery of which depends on experience and practice. Nevertheless, there are a number of techniques whose systematic use can substantially reduce our dependence on chance and intuition alone. This chapter deals with some of these systematic techniques of integration.

9.2 INTEGRAL TABLES AND SIMPLE SUBSTITUTIONS

Integration would be a simple matter if we had a list of integral formulas (an *integral table*) in which we could locate any integral that we needed to evaluate. But the diversity of integrals that we encounter is too great for such an all-inclusive integral table to be practical. It is more sensible to print or memorize a short table of integrals of the sort seen frequently and to learn techniques by which the range of applicability of this short table can be extended. We begin with the list of integrals in Fig. 9.2.1, which are familiar from earlier chapters. Each formula is equivalent to one of the basic derivative formulas.

A table of 113 integral formulas appears on the inside back cover of this book. Even more extensive integral tables are readily available. For example, the volume

$$\int u^n\,du = \frac{u^{n+1}}{n+1} + C \qquad [n \neq -1] \quad (1)$$

$$\int \frac{du}{u} = \ln|u| + C \quad (2)$$

$$\int e^u\,du = e^u + C \quad (3)$$

$$\int \cos u\,du = \sin u + C \quad (4)$$

$$\int \sin u\,du = -\cos u + C \quad (5)$$

$$\int \sec^2 u\,du = \tan u + C \quad (6)$$

$$\int \csc^2 u\,du = -\cot u + C \quad (7)$$

$$\int \sec u \tan u\,du = \sec u + C \quad (8)$$

$$\int \csc u \cot u\,du = -\csc u + C \quad (9)$$

$$\int \frac{du}{\sqrt{1-u^2}} = \sin^{-1} u + C \quad (10)$$

$$\int \frac{du}{1+u^2} = \tan^{-1} u + C \quad (11)$$

$$\int \frac{du}{u\sqrt{u^2-1}} = \sec^{-1}|u| + C \quad (12)$$

Fig. 9.2.1 A short table of integrals

of *Standard Mathematical Tables and Formulas,* edited by William H. Beyer and published by the CRC Press, Inc. (Boca Raton, Florida), contains over 700 integral formulas. But even such a lengthy table can be expected to include only a small fraction of the integrals we may need to evaluate. Thus it is necessary to learn techniques for deriving new formulas and for transforming a given integral either into one that's already familiar or into one that appears in an accessible table.

The principal such technique is the *method of substitution,* which we first considered in Section 5.7. Recall that if

$$\int f(u)\,du = F(u) + C,$$

then

$$\int f(g(x)) \cdot g'(x)\,dx = F(g(x)) + C.$$

Thus the substitution

$$u = g(x), \qquad du = g'(x)\,dx$$

transforms the integral

$$\int f(g(x)) \cdot g'(x)\,dx \quad \text{into the simpler integral} \quad \int f(u)\,du.$$

The key to making this simplification lies in spotting the composition $f(g(x))$ in the given integrand. For this integrand to be converted into a function of u alone, the remaining factor must be a constant multiple of the derivative $g'(x)$ of the "inside function" $g(x)$. In this case we replace $f(g(x))$ with the simpler $f(u)$ and $g'(x)\,dx$ with the simpler du. Chapters 6 through 8 contain numerous illustrations of this method of substitution, and the problems at the end of this section provide an opportunity to review it.

EXAMPLE 1 Find $\displaystyle\int \frac{1}{x}(1+\ln x)^5\,dx$.

Solution We need to spot *both* the inner function $g(x)$ and its derivative $g'(x)$. If we choose $g(x) = 1 + \ln x$, then $g'(x) = 1/x$. Hence the given integral is of the form discussed above with $f(u) = u^5$, $u = 1 + \ln x$, and $du = dx/x$. Therefore

$$\int \frac{1}{x}(1+\ln x)^5\,dx = \int u^5\,du = \frac{1}{6}u^6 + C = \frac{1}{6}(1+\ln x)^6 + C.$$

EXAMPLE 2 Find $\displaystyle\int \frac{x}{1+x^4}\,dx$.

Solution Here it is not so clear what the inside function is. But, looking at the integral formula in Eq. (11) (Fig. 9.2.1), we try the substitution $u = x^2$, $du = 2x\,dx$. We take advantage of the factor $x\,dx = \frac{1}{2}\,du$ that is available in the integrand and compute as follows:

$$\int \frac{x}{1+x^4}\,dx = \frac{1}{2}\int \frac{du}{1+u^2} = \frac{1}{2}\tan^{-1} u + C = \frac{1}{2}\tan^{-1} x^2 + C.$$

Note that the substitution $u = x^2$ would have been of no use had the integrand been either $1/(1 + x^4)$ or $x^2/(1 + x^4)$. ■

Example 2 illustrates how to make a substitution that converts a given integral into a familiar one. This is a kind of *pattern matching.* Often an integral that does not appear in any integral table can be transformed into one that does by using the techniques of this chapter. In Example 3 we employ an appropriate substitution to match the given integral with the standard integral formula

$$\int \frac{u^2}{\sqrt{a^2 - u^2}}\,du = \frac{a^2}{2}\sin^{-1}\left(\frac{u}{a}\right) - \frac{u}{2}\sqrt{a^2 - u^2} + C, \qquad \textbf{(13)}$$

which is formula (56) (inside the back cover).

EXAMPLE 3 Find $\displaystyle\int \frac{x^2}{\sqrt{25 - 16x^2}}\,dx.$

Solution So that $25 - 16x^2$ will be equal to $a^2 - u^2$ in Eq. (13), we take $a = 5$ and $u = 4x$. Then $du = 4\,dx$, and so $dx = \frac{1}{4}\,du$. This gives

$$\begin{aligned}\int \frac{x^2}{\sqrt{25 - 16x^2}}\,dx &= \int \frac{(\frac{1}{4}u)^2}{\sqrt{25 - u^2}}\cdot\frac{1}{4}\,du = \frac{1}{64}\int \frac{u^2}{\sqrt{25 - u^2}}\,du \\ &= \frac{1}{64}\left[\frac{25}{2}\sin^{-1}\left(\frac{u}{5}\right) - \frac{u}{2}\sqrt{25 - u^2}\right] + C \\ &= \frac{25}{128}\sin^{-1}\left(\frac{4x}{5}\right) - \frac{x}{32}\sqrt{25 - 16x^2} + C.\end{aligned}$$ ■

In Section 9.6 we will see how to derive integral formulas such as that in Eq. (13).

Computer Algebra Systems

Computer systems such as *Derive, Maple,* and *Mathematica* have integral formulas stored internally and can perform pattern matching substitutions like the one used in Example 3. For instance, the *Mathematica* command

```
Integrate[ x∧2 / Sqrt[25 - 16 x∧2], x]
```

and the *Maple* command

```
int( x∧2 / sqrt (25 - 16*x∧2), x);
```

as well as the corresponding *Derive* command all produce precisely the same result as that found in Example 3 (except without adding the arbitrary constant of integration, which computer algebra systems generally omit).

Sometimes different methods (whether manual, table, or computer methods) produce integrals that appear to differ. For instance, *Derive* and *Maple* yield the same antiderivative $\frac{1}{2}\tan^{-1}x^2$ of $x/(1 + x^4)$ found in Example 2, whereas *Mathematica* returns the function $-\frac{1}{2}\tan^{-1}x^{-2}$ as the result. Naturally we wonder whether $\frac{1}{2}\tan^{-1}x^2 \equiv -\frac{1}{2}\tan^{-1}x^{-2}$. Figure 9.2.2 shows that the answer is No! In Problem 55 we ask you to reconcile these apparently different antiderivatives of the same function.

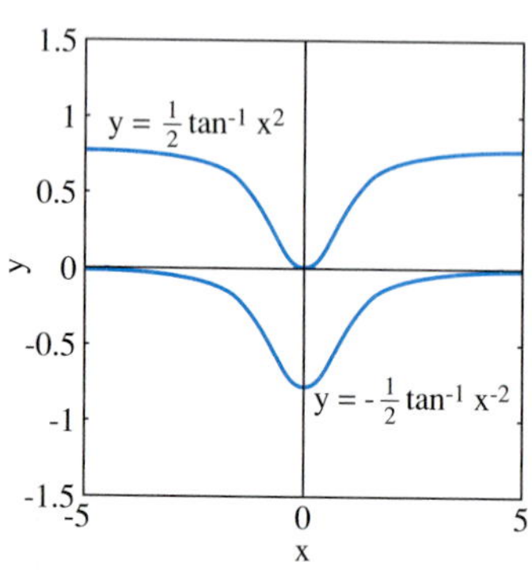

Fig. 9.2.2 The graphs $y = \frac{1}{2}\tan^{-1}x^2$ and $y = -\frac{1}{2}\tan^{-1}x^{-2}$ do not agree. What *is* the relationship between them?

If computer algebra systems can match patterns with tables of integrals stored in computer memory, you may wonder why manual integration techniques should still be learned. One answer is that a hand computation may yield an integral in a simpler or more convenient form than would a computer result. For instance, *Derive, Maple,* and *Mathematica* all yield a result of the form

$$\int \frac{1}{x}(1+\ln x)^5\,dx = \ln x + \frac{5}{2}(\ln x)^2 + \frac{10}{3}(\ln x)^3 + \frac{5}{2}(\ln x)^4 + (\ln x)^5 + \frac{1}{6}(\ln x)^6$$

that looks considerably less appealing than the hand result $\frac{1}{6}(1+\ln x)^6$ of Example 1. Is the relationship between the two obvious? See Problem 54.

9.2 PROBLEMS

Evaluate the integrals in Problems 1 through 30.

1. $\int (2-3x)^4\,dx$

2. $\int \frac{1}{(1+2x)^2}\,dx$

3. $\int x^2\sqrt{2x^3-4}\,dx$

4. $\int \frac{5t}{5+2t^2}\,dt$

5. $\int \frac{2x}{\sqrt[3]{2x^2+3}}\,dx$

6. $\int x\sec^2 x^2\,dx$

7. $\int \frac{\cot\sqrt{y}\,\csc\sqrt{y}}{\sqrt{y}}\,dy$

8. $\int \sin\pi(2x+1)\,dx$

9. $\int (1+\sin\theta)^5\cos\theta\,d\theta$

10. $\int \frac{\sin 2x}{4+\cos 2x}\,dx$

11. $\int e^{-\cot x}\csc^2 x\,dx$

12. $\int \frac{e^{\sqrt{x+4}}}{\sqrt{x+4}}\,dx$

13. $\int \frac{(\ln t)^{10}}{t}\,dt$

14. $\int \frac{t}{\sqrt{1-9t^2}}\,dt$

15. $\int \frac{1}{\sqrt{1-9t^2}}\,dt$

16. $\int \frac{e^{2x}}{1+e^{2x}}\,dx$

17. $\int \frac{e^{2x}}{1+e^{4x}}\,dx$

18. $\int \frac{e^{\arctan x}}{1+x^2}\,dx$

19. $\int \frac{3x}{\sqrt{1-x^4}}\,dx$

20. $\int \sin^3 2x\cos 2x\,dx$

21. $\int \tan^4 3x\sec^2 3x\,dx$

22. $\int \frac{1}{1+4t^2}\,dt$

23. $\int \frac{\cos\theta}{1+\sin^2\theta}\,d\theta$

24. $\int \frac{\sec^2\theta}{1+\tan\theta}\,d\theta$

25. $\int \frac{(1+\sqrt{x})^4}{\sqrt{x}}\,dx$

26. $\int t^{-1/3}\sqrt{t^{2/3}-1}\,dt$

27. $\int \frac{1}{(1+t^2)\arctan t}\,dt$

28. $\int \frac{\sec 2x\tan 2x}{(1+\sec 2x)^{3/2}}\,dx$

29. $\int \frac{1}{\sqrt{e^{2x}-1}}\,dx$

30. $\int \frac{x}{\sqrt{\exp(2x^2)-1}}\,dx$

In Problems 31 through 35, evaluate the given integral by making the indicated substitution.

31. $\int x^2\sqrt{x-2}\,dx;\quad u=x-2$

32. $\int \frac{x^2}{\sqrt{x+3}}\,dx;\quad u=x+3$

33. $\int \frac{x}{\sqrt{2x+3}}\,dx;\quad u=2x+3$

34. $\int x\sqrt[3]{x-1}\,dx;\quad u=x-1$

35. $\int \frac{x}{\sqrt[3]{x+1}}\,dx;\quad u=x+1$

In Problems 36 through 50, evaluate the given integral. First make a substitution that transforms it into a standard form. The standard forms with the given formula numbers are inside the back cover of this book. If a computer algebra system is available, compare and reconcile (if necessary) the result found using the integral table formula with a machine result.

36. $\int \frac{1}{100+9x^2}\,dx;$ formula (17)

37. $\int \frac{1}{100-9x^2}\,dx;$ formula (18)

38. $\int \sqrt{9-4x^2}\,dx;$ formula (54)

39. $\int \sqrt{4+9x^2}\,dx;$ formula (44)

40. $\int \frac{1}{\sqrt{16x^2+9}}\,dx;$ formula (45)

41. $\int \frac{x^2}{\sqrt{16x^2+9}}\,dx;$ formula (49)

42. $\int \frac{x^2}{\sqrt{25+16x^2}}\,dx;$ formula (49)

43. $\int x^2\sqrt{25-16x^2}\,dx;$ formula (57)

44. $\int x\sqrt{4-x^4}\,dx;$ formula (54)

45. $\int e^x\sqrt{9+e^{2x}}\,dx;$ formula (44)

46. $\int \frac{\cos x}{(\sin^2 x)\sqrt{1+\sin^2 x}}\,dx;$ formula (50)

47. $\int \frac{\sqrt{x^4-1}}{x}\,dx;$ formula (47)

48. $\int \frac{e^{3x}}{\sqrt{25+16e^{2x}}}\,dx;$ formula (49)

49. $\int \frac{(\ln x)^2}{x}\sqrt{1+(\ln x)^2}\,dx;$ formula (48)

50. $\int x^8\sqrt{4x^6-1}\,dx;$ formula (48)

51. The substitution $u = x^2, x = \sqrt{u},\ dx = du/(2\sqrt{u})$ appears to lead to this result:

$$\int_{-1}^{1} x^2\,dx = \frac{1}{2}\int_{1}^{1} \sqrt{u}\,du = 0.$$

Do you believe this result? If not, why not?

52. Use the fact that $x^2 + 4x + 5 = (x+2)^2 + 1$ to evaluate

$$\int \frac{1}{x^2 + 4x + 5}\,dx.$$

53. Use the fact that $1 - (x-1)^2 = 2x - x^2$ to evaluate

$$\int \frac{1}{\sqrt{2x - x^2}}\,dx.$$

54. Use the binomial expansion

$$(1+t)^6 = 1 + 6t + 15t^2 + 20t^3 + 15t^4 + 6t^5 + t^6$$

to reconcile the result of Example 1 with the machine result listed at the end of this section. Are the two results precisely equal?

55. Establish the precise relationship between the two functions

$$\frac{1}{2}\tan^{-1}x^2 \quad \text{and} \quad -\frac{1}{2}\tan^{-1}x^{-2}$$

graphed in Fig. 9.2.2. Are both actually antiderivatives of $x/(1+x^4)$?

56. With $u = x$ and $a = 1$, formula (44) inside the back cover yields

$$\int \sqrt{x^2+1}\,dx = \tfrac{1}{2}x\sqrt{x^2+1} + \tfrac{1}{2}\ln\left|x + \sqrt{x^2+1}\right|,$$

whereas both *Maple* and *Mathematica* give

$$\int \sqrt{x^2+1}\,dx = \tfrac{1}{2}x\sqrt{x^2+1} + \tfrac{1}{2}\sinh^{-1}x.$$

Consult Section 8.5 to reconcile these two results.

9.2 PROJECT: COMPARING DIFFERENT ANSWERS

According to formula (44) on the inside back cover of this book,

$$\int \sqrt{x^2+1}\,dx = G(x) + C, \tag{1}$$

where

$$G(x) = \frac{x}{2}\sqrt{x^2+1} + \frac{1}{2}\ln\left(x + \sqrt{x^2+1}\right). \tag{2}$$

According to Serge Lang's *A First Course in Calculus* (5th ed., New York: Springer-Verlag, 1991, p. 376), this same indefinite integral is given by

$$\int \sqrt{x^2+1}\,dx = H(x) + C, \tag{3}$$

where

$$H(x) = \tfrac{1}{8}\left[\left(x + \sqrt{x^2+1}\right)^2 + 4\ln\left(x + \sqrt{x^2+1}\right) - \left(x + \sqrt{x^2+1}\right)^{-2}\right]. \tag{4}$$

Your mission in this project is to determine whether the functions $G(x)$ and $H(x)$ in Eqs. (2) and (4) are, in fact, both antiderivatives of $f(x) = \sqrt{x^2+1}$. We list possible ways to investigate the relationships among the functions f, G, and H. You should explore several different approaches.

1. If you have a graphics calculator or a computer with a graphing utility, plot both $y = G(x)$ and $y = H(x)$. If the functions G and H are both antiderivatives of f, how should their graphs be related?
2. If your calculator or computer can graph the *derivative* of a user-defined function, plot the graphs of f and the derivatives G' and H'. Does the visual evidence convince you that $G' = H' = f$?

3. Even if your calculator or computer cannot plot derivatives directly, you can still plot $f(x)$ and the quotients

$$\frac{G(x+h)-G(x-h)}{2h} \quad \text{and} \quad \frac{H(x+h)-H(x-h)}{2h},$$

which—with $h = 0.001$—should closely approximate the derivatives $G'(x)$ and $H'(x)$, respectively. (Why?)

4. With a calculator or computer that can approximate integrals numerically—either with an appropriate function key or by using programs like those in the Section 5.9 projects—you can determine whether

$$\int_a^b f(x)\,dx = G(b) - G(a) = H(b) - H(a), \tag{5}$$

as the fundamental theorem implies if $G' = H' = f$. It should be fairly convincing if you can verify Eq. (5) numerically with several different pairs of limits, such as $a = 1, b = 5$ and $a = 7, b = 11$.

5. Even without a graphics calculator, you can compute numerical values of $G(x)$, $H(x)$, $G'(x)$, $H'(x)$, and $f(x)$ for several selected values of x. Do the numerical results imply that $G(x) = H(x)$ or that $G'(x) = H'(x) = f(x)$?

6. With a computer algebra system, such as *Derive, Maple,* or *Mathematica,* you can calculate the derivatives $G'(x)$ and $H'(x)$ symbolically to determine whether both are equal to $f(x)$. Perhaps you can even do this the old-fashioned way, by using nothing but paper and pencil. And you might investigate in the same way whether $G(x) = H(x)$.

9.3 INTEGRATION BY PARTS

One reason for transforming a given integral into another is to make it easier to evaluate. There are two general ways to accomplish this. We have seen the first, integration by substitution. The second is *integration by parts.*

The formula for integration by parts is a simple consequence of the product rule for derivatives,

$$D_x(uv) = v\frac{du}{dx} + u\frac{dv}{dx}.$$

If we write this formula in the form

$$u(x)v'(x) = D_x[u(x)v(x)] - v(x)u'(x), \tag{1}$$

then antidifferentiation gives

$$\int u(x)v'(x)\,dx = u(x)v(x) - \int v(x)u'(x)\,dx. \tag{2}$$

This is the formula for **integration by parts**. With $du = u'(x)\,dx$ and $dv = v'(x)\,dx$, Eq. (2) becomes

$$\int u\,dv = uv - \int v\,du. \tag{3}$$

To apply the integration by parts formula to a given integral, we must first factor its integrand into two "parts," u and dv, the latter including the differential dx. We try to choose the parts in accordance with two principles:

1. The antiderivative $v = \int dv$ is easy to find.
2. The new integral $\int v\,du$ is easier to compute than the original integral $\int u\,dv$.

An effective strategy is to choose for dv the most complicated factor that can readily be integrated. Then we differentiate the other part, u, to find du.

We begin with two examples in which we have little flexibility in choosing the parts u and dv.

EXAMPLE 1 Find $\int \ln x\,dx$.

Solution Here there is little alternative to the natural choice $u = \ln x$ and $dv = dx$. It is helpful to systematize the procedure of integration by parts by writing u, dv, du, and v in a rectangular array like this:

$$\begin{aligned}&\text{Let} \quad u = \ln x \quad \text{and} \quad dv = dx.\\ &\text{Then} \quad du = \frac{1}{x}\,dx \quad \text{and} \quad v = x.\end{aligned}$$

The first line in the array specifies the choice of u and dv; the second line is computed from the first. Then Eq. (3) gives

$$\int \ln x\,dx = x\ln x - \int dx = x\ln x - x + C.$$

COMMENT 1 The constant of integration appears only at the last step. We know that once we have found one antiderivative, any other may be obtained by adding a constant C to the one we have found.

COMMENT 2 In computing $v = \int dv$, we ordinarily take the constant of integration to be zero. Had we written $v = x + C_1$ in Example 1, the answer would have been

$$\begin{aligned}\int \ln x\,dx &= (x + C_1)\ln x - \int \left(1 + \frac{C_1}{x}\right)dx\\ &= x\ln x + C_1\ln x - (x + C_1\ln x) + C = x\ln x - x + C\end{aligned}$$

as before, so introducing the extra constant C_1 has no effect. See Fig. 9.3.1.

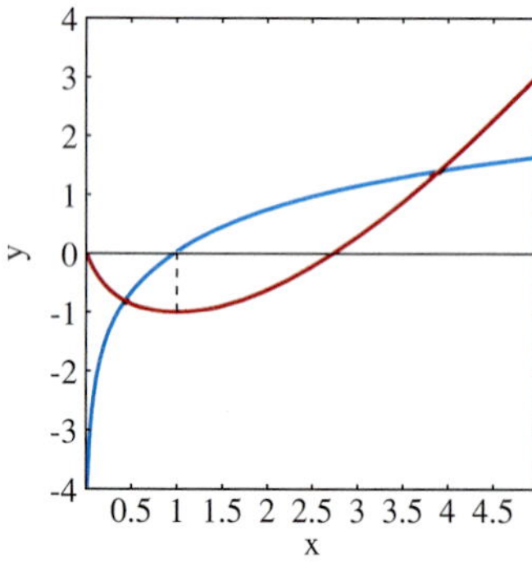

Fig. 9.3.1 Graphs of the functions $\ln x$ and $x\ln x - x$ of Example 1. If you did not recognize either graph, but noted that the zero of one corresponds to a critical point of the other, which would you conclude is the antiderivative?

EXAMPLE 2 Find $\int \arcsin x\,dx$.

Solution Again, there is only one plausible choice for u and dv:

$$\begin{aligned}&\text{Let} \quad u = \arcsin x \quad \text{and} \quad dv = dx.\\ &\text{Then} \quad du = \frac{dx}{\sqrt{1 - x^2}} \quad \text{and} \quad v = x.\end{aligned}$$

Then Eq. (3) gives

$$\begin{aligned}\int \arcsin x\,dx &= x\arcsin x - \int \frac{x}{\sqrt{1 - x^2}}\,dx\\ &= x\arcsin x + \sqrt{1 - x^2} + C.\end{aligned}$$

EXAMPLE 3 Find $\int xe^{-x}\,dx$.

Solution Here we appear to have some flexibility. Suppose that we try

$$u = e^{-x}, \qquad dv = x\,dx$$

so that

$$du = -e^{-x}\,dx, \qquad v = \tfrac{1}{2}x^2.$$

Then integration by parts gives

$$\int xe^{-x}\,dx = \tfrac{1}{2}x^2e^{-x} + \tfrac{1}{2}\int x^2e^{-x}\,dx.$$

The new integral on the right looks more troublesome than the one we started with! Let us begin anew.

$$\begin{array}{llll} \text{Let} & u = x & \text{and} & dv = e^{-x}\,dx. \\ \text{Then} & du = dx & \text{and} & v = -e^{-x}. \end{array}$$

Now integration by parts gives

$$\int xe^{-x}\,dx = -xe^{-x} + \int e^{-x}\,dx = -xe^{-x} - e^{-x} + C.$$

■

Integration by parts can be applied to definite integrals as well as to indefinite integrals. We integrate Eq. (1) from $x = a$ to $x = b$ and apply the fundamental theorem of calculus. This gives

$$\begin{aligned} \int_a^b u(x)\,v'(x)\,dx &= \int_a^b D_x\,[u(x)v(x)]\,dx - \int_a^b v(x)u'(x)\,dx \\ &= \Big[u(x)v(x)\Big]_a^b - \int_a^b v(x)u'(x)\,dx. \end{aligned}$$

In the notation of Eq. (3), this equation would be written

$$\int_{x=a}^{x=b} u\,dv = \Big[uv\Big]_a^b - \int_{x=a}^{x=b} v\,du, \tag{4}$$

although we must not forget that u and v are functions of x. For example, with $u = x$ and $dv = e^{-x}\,dx$, as in Example 3, we obtain

$$\int_0^1 xe^{-x}\,dx = \Big[-xe^{-x}\Big]_0^1 + \int_0^1 e^{-x}\,dx = -e^{-1} + \Big[-e^{-x}\Big]_0^1 = 1 - \frac{2}{e}.$$

EXAMPLE 4 Find $\int x^2e^{-x}\,dx$.

Solution If we choose $u = x^2$, then $du = 2x\,dx$, so we will reduce the exponent of x by this choice.

$$\begin{array}{llll} \text{Let} & u = x^2 & \text{and} & dv = e^{-x}\,dx. \\ \text{Then} & du = 2x\,dx & \text{and} & v = -e^{-x}. \end{array}$$

Then integration by parts gives

$$\int x^2 e^{-x}\,dx = -x^2 e^{-x} + 2\int x e^{-x}\,dx.$$

We apply integration by parts a second time to the right-hand integral and obtain the result

$$\int x e^{-x}\,dx = -x e^{-x} - e^{-x}$$

of Example 3. Substitution then yields

$$\int x^2 e^{-x}\,dx = -x^2 e^{-x} - 2x e^{-x} - 2e^{-x} + C = -(x^2 + 2x + 2)e^{-x} + C.$$

In effect, we have annihilated the original factor x^2 by integrating by parts twice in succession. See Fig. 9.3.2. ■

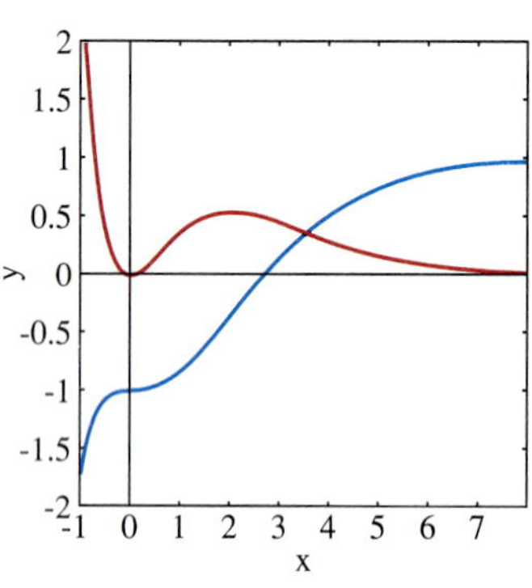

Fig. 9.3.2 Graphs of the functions $x^2 e^{-x}$ and $-(x^2 + 2x + 2)e^{-x} + 1$. Noting only that the zero of one corresponds to a critical point of the other, which do you conclude is the antiderivative?

EXAMPLE 5 Find $\int e^{2x} \sin 3x\,dx$.

Solution This is another example in which repeated integration by parts succeeds, but with a twist:

Let $\quad u = \sin 3x, \qquad dv = e^{2x}\,dx.$

Then $\quad du = 3\cos 3x\,dx, \qquad v = \frac{1}{2}e^{2x}.$

Therefore

$$\int e^{2x}\sin 3x\,dx = \tfrac{1}{2}e^{2x}\sin 3x - \tfrac{3}{2}\int e^{2x}\cos 3x\,dx.$$

At first it might appear that little progress has been made, for the integral on the right is as difficult to evaluate as the one on the left. We ignore this objection and try again, applying integration by parts to the new integral:

Let $\quad u = \cos 3x, \qquad dv = e^{2x}\,dx.$

Then $\quad du = -3\sin 3x\,dx, \qquad v = \frac{1}{2}e^{2x}.$

Now we find that

$$\int e^{2x}\cos 3x\,dx = \tfrac{1}{2}e^{2x}\cos 3x + \tfrac{3}{2}\int e^{2x}\sin 3x\,dx.$$

By substituting this result into the previous equation, we discover that

$$\int e^{2x}\sin 3x\,dx = \tfrac{1}{2}e^{2x}\sin 3x - \tfrac{3}{4}e^{2x}\cos 3x - \tfrac{9}{4}\int e^{2x}\sin 3x\,dx.$$

So we are back where we started. Or *are* we? In fact we are *not,* because we can *solve* this last equation for the desired integral. We add the right-hand integral here to both sides of the last equation. This gives

$$\tfrac{13}{4}\int e^{2x}\sin 3x\,dx = \tfrac{1}{4}e^{2x}(2\sin 3x - 3\cos 3x) + C_1,$$

so

$$\int e^{2x}\sin 3x\,dx = \tfrac{1}{13}e^{2x}(2\sin 3x - 3\cos 3x) + C.$$

See Fig. 9.3.3. ■

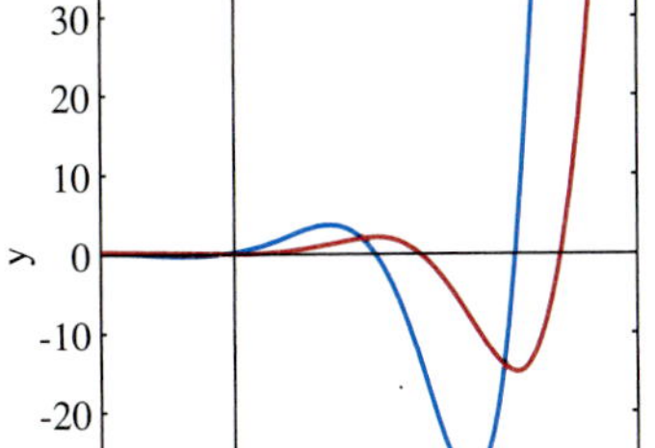

Fig. 9.3.3 Graphs of the functions $e^{2x}\sin 3x$ and $\frac{1}{13}e^{2x}(2\sin 3x - 3\cos 3x)$. Noting only that the zeros of one correspond to the critical points of the other, which do you conclude is the antiderivative?

EXAMPLE 6 Find a reduction formula for $\int \sec^n x\,dx$.

Solution The idea is that n is a (large) positive integer and that we want to express the given integral in terms of the integral of a lower power of $\sec x$. The easiest power of $\sec x$ to integrate is $\sec^2 x$, so we proceed as follows:

$$\text{Let}\quad u = \sec^{n-2} x, \qquad dv = \sec^2 x\,dx.$$
$$\text{Then}\quad du = (n-2)\sec^{n-2} x\tan x\,dx, \quad v = \tan x.$$

This gives

$$\begin{aligned}\int \sec^n x\,dx &= \sec^{n-2} x\tan x - (n-2)\int \sec^{n-2} x\tan^2 x\,dx\\ &= \sec^{n-2} x\tan x - (n-2)\int (\sec^{n-2} x)(\sec^2 x - 1)\,dx.\end{aligned}$$

Hence

$$\int \sec^n x\,dx = \sec^{n-2} x\tan x - (n-2)\int \sec^n x\,dx + (n-2)\int \sec^{n-2} x\,dx.$$

We solve this equation for the original integral and find that

$$\int \sec^n x\,dx = \frac{\sec^{n-2} x\tan x}{n-1} + \frac{n-2}{n-1}\int \sec^{n-2} x\,dx. \tag{5}$$

This is the desired reduction formula. For example, if we take $n = 3$ in this formula, we find that

$$\begin{aligned}\int \sec^3 x\,dx &= \tfrac{1}{2}\sec x\tan x + \tfrac{1}{2}\int \sec x\,dx\\ &= \tfrac{1}{2}\sec x\tan x + \tfrac{1}{2}\ln|\sec x + \tan x| + C.\end{aligned} \tag{6}$$

In the last step we used the integral formula

$$\int \sec x\,dx = \ln|\sec x + \tan x| + C, \tag{7}$$

which is tricky to derive systematically (see Section 9.4) but is easy to verify by differentiation:

$$\begin{aligned}D_x(\ln|\sec x + \tan x|) &= \frac{D_x(\sec x + \tan x)}{\sec x + \tan x}\\ &= \frac{\sec x\tan x + \sec^2 x}{\sec x + \tan x} = \frac{(\sec x)(\tan x + \sec x)}{\sec x + \tan x} = \sec x.\end{aligned}$$

The reason for using the reduction formula in Eq. (5) is that—if n is a positive integer—repeated application must yield either Eq. (7) or the elementary integral

$$\int \sec^2 x\,dx = \tan x + C.$$ ■

EXAMPLE 7 With $n = 4$ in Eq. (5) we get

$$\int \sec^4 x\,dx = \tfrac{1}{3}\sec^2 x\tan x + \tfrac{2}{3}\int \sec^2 x\,dx$$
$$= \tfrac{1}{3}\sec^2 x\tan x + \tfrac{2}{3}\tan x + C, \qquad \textbf{(8)}$$

and with $n = 5$ we get

$$\int \sec^5 x\,dx = \tfrac{1}{4}\sec^3 x\tan x + \tfrac{3}{4}\int \sec^3 x\,dx$$
$$= \tfrac{1}{4}\sec^3 x\tan x + \tfrac{3}{8}\sec x\tan x + \tfrac{3}{8}\ln|\sec x + \tan x| + C, \qquad \textbf{(9)}$$

using Eq. (6) in the last step. ■

9.3 PROBLEMS

Use integration by parts to compute the integrals in Problems 1 through 34.

1. $\int xe^{2x}\,dx$

2. $\int x^2e^{2x}\,dx$

3. $\int t\sin t\,dt$

4. $\int t^2\sin t\,dt$

5. $\int x\cos 3x\,dx$

6. $\int x\ln x\,dx$

7. $\int x^3\ln x\,dx$

8. $\int e^{3z}\cos 3z\,dz$

9. $\int \arctan x\,dx$

10. $\int \frac{\ln x}{x^2}\,dx$

11. $\int \sqrt{y}\ln y\,dy$

12. $\int x\sec^2 x\,dx$

13. $\int (\ln t)^2\,dt$

14. $\int t(\ln t)^2\,dt$

15. $\int x\sqrt{x+3}\,dx$

16. $\int x^3\sqrt{1-x^2}\,dx$

17. $\int x^5\sqrt{x^3+1}\,dx$

18. $\int \sin^2\theta\,d\theta$

19. $\int \csc^3\theta\,d\theta$

20. $\int \sin(\ln t)\,dt$

21. $\int x^2\arctan x\,dx$

22. $\int \ln(1+x^2)\,dx$

23. $\int \sec^{-1}\sqrt{x}\,dx$

24. $\int x\tan^{-1}\sqrt{x}\,dx$

25. $\int \tan^{-1}\sqrt{x}\,dx$

26. $\int x^2\cos 4x\,dx$

27. $\int x\csc^2 x\,dx$

28. $\int x\arctan x\,dx$

29. $\int x^3\cos x^2\,dx$

30. $\int e^{-3x}\sin 4x\,dx$

31. $\int \frac{\ln x}{x\sqrt{x}}\,dx$

32. $\int \frac{x^7}{(1+x^4)^{3/2}}\,dx$

33. $\int x\cosh x\,dx$

34. $\int e^x\cosh x\,dx$

In Problems 35 through 38, first make a substitution of the form $t = x^k$ and then integrate by parts.

35. $\int x^3\sin x^2\,dx$

36. $\int x^7\cos x^4\,dx$

37. $\int \exp(-\sqrt{x})\,dx$

38. $\int x^2\sin x^{3/2}\,dx$

In Problems 39 through 42, use the method of cylindrical shells to calculate the volume of the solid obtained by revolving the region R around the y-axis.

39. R is bounded below by the x-axis and above by the curve $y = \cos x$, $-\pi/2 \leqq x \leqq \pi/2$.

40. R is bounded below by the x-axis and above by the curve $y = \sin x$, $0 \leqq x \leqq \pi$.

41. R is bounded below by the x-axis, on the right by the line $x = e$, and above by the curve $y = \ln x$.

42. R is bounded below by the x-axis, on the left by the y-axis, on the right by the line $x = 1$, and above by the curve $y = e^{-x}$.

In Problems 43 through 45, first estimate graphically or numerically the points of intersection of the two given curves, then approximate the volume of the solid that is generated when the region bounded by these two curves is revolved around the y-axis.

43. $y = x^2$ and $y = \cos x$

44. $y = 10x - x^2$ and $y = e^x - 1$

45. $y = x^2 - 2x$ and $y = \ln(x+1)$

46. Use integration by parts to evaluate

$$\int 2x\arctan x\,dx,$$

with $dv = 2x\,dx$, but let $v = x^2 + 1$ rather than $v = x^2$. Is there a reason why v should not be chosen in this way?

47. Use integration by parts to evaluate $\int xe^x \cos x\,dx$.

48. Use integration by parts to evaluate $\int \sin 3x \cos x\,dx$.

Derive the reduction formulas given in Problems 49 through 54. Throughout, n denotes a positive integer with an appropriate side condition (such as $n \geqq 1$ or $n \geqq 2$).

49. $\int x^n e^x\,dx = x^n e^x - n\int x^{n-1}e^x\,dx$

50. $\int x^n e^{-x^2}\,dx = -\frac{1}{2}x^{n-1}e^{-x^2} + \frac{n-1}{2}\int x^{n-2}e^{-x^2}\,dx$

51. $\int (\ln x)^n\,dx = x(\ln x)^n - n\int (\ln x)^{n-1}\,dx$

52. $\int x^n \cos x\,dx = x^n \sin x - n\int x^{n-1}\sin x\,dx$

53. $\int \sin^n x\,dx = -\frac{\sin^{n-1}x\cos x}{n} + \frac{n-1}{n}\int \sin^{n-2}x\,dx$

54. $\int \cos^n x\,dx = \frac{\cos^{n-1}x\sin x}{n} + \frac{n-1}{n}\int \cos^{n-2}x\,dx$

Use appropriate reduction formulas from the preceding list to evaluate the integrals in Problems 55 through 57.

55. $\int_0^1 x^3 e^x\,dx$ **56.** $\int_0^1 x^5 e^{-x^2}\,dx$

57. $\int_1^e (\ln x)^3\,dx$

58. Apply the reduction formula in Problem 53 to show that for each positive integer n,

$$\int_0^{\pi/2} \sin^{2n}x\,dx = \frac{\pi}{2}\cdot\frac{1}{2}\cdot\frac{3}{4}\cdot\frac{5}{6}\cdots\frac{2n-1}{2n}$$

and

$$\int_0^{\pi/2} \sin^{2n+1}x\,dx = \frac{2}{3}\cdot\frac{4}{5}\cdot\frac{6}{7}\cdot\frac{8}{9}\cdots\frac{2n}{2n+1}.$$

59. Derive the formula

$$\int \ln(x+10)\,dx = (x+10)\ln(x+10) - x + C$$

in three different ways: (a) by substituting $u = x + 10$ and applying the result of Example 1; (b) by integrating by parts with $u = \ln(x+10)$ and $dv = dx$, noting that

$$\frac{x}{x+10} = 1 - \frac{10}{x+10};$$

and (c) by integrating by parts with $u = \ln(x+10)$ and $dv = dx$, but with $v = x + 10$.

60. Derive the formula

$$\int x^3 \tan^{-1}x\,dx = \tfrac{1}{4}(x^4-1)\tan^{-1}x - \tfrac{1}{12}x^3 + \tfrac{1}{4}x + C$$

by integrating by parts with $u = \tan^{-1}x$ and $v = \frac{1}{4}(x^4-1)$.

61. Let $J_n = \int_0^1 x^n e^{-x}\,dx$ for each integer $n \geqq 0$. (a) Show that

$$J_0 = 1 - \frac{1}{e} \quad\text{and that}\quad J_n = nJ_{n-1} - \frac{1}{e}$$

for $n \geqq 1$. (b) Deduce by mathematical induction that

$$J_n = n! - \frac{n!}{e}\sum_{k=0}^{n}\frac{1}{k!}$$

for each integer $n \geqq 0$. (c) Explain why $J_n \to 0$ as $n \to +\infty$. (d) Conclude that

$$e = \lim_{n\to\infty}\sum_{k=0}^{n}\frac{1}{k!}.$$

62. Let m and n be positive integers. Derive the reduction formula

$$\int x^m(\ln x)^n\,dx = \frac{x^{m+1}}{m+1}(\ln x)^n - \frac{n}{m+1}\int x^m(\ln x)^{n-1}\,dx.$$

63. An advertisement for a symbolic algebra program claims that an engineer worked for three weeks on the integral

$$\int (k\ln x - 2x^3 + 3x^2 + b)^4\,dx,$$

which deals with turbulence in an aerospace application. The advertisement said that the engineer never got the same answer twice during the three weeks. Explain how you could use the reduction formula of Problem 62 to find the engineer's integral (but don't actually do it). Can you see any reason why it should have taken three weeks?

64. Figure 9.3.4 shows the region bounded by the x-axis and the graph of $y = \frac{1}{2}x^2\sin x$, $0 \leqq x \leqq \pi$. Use Formulas (42) and (43) (inside the back cover)—which are derived by integration by parts—to find (a) the area of this region; (b) the volume obtained by revolving this region around the y-axis.

65. The top shown in Fig. 9.3.5 has the shape of the solid obtained by revolving the region of Problem 64 around the x-axis. Find the volume of this top.

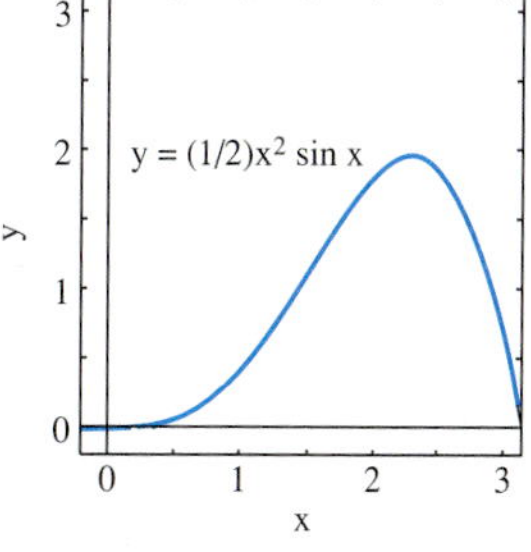

Fig. 9.3.4 The region of Problem 64

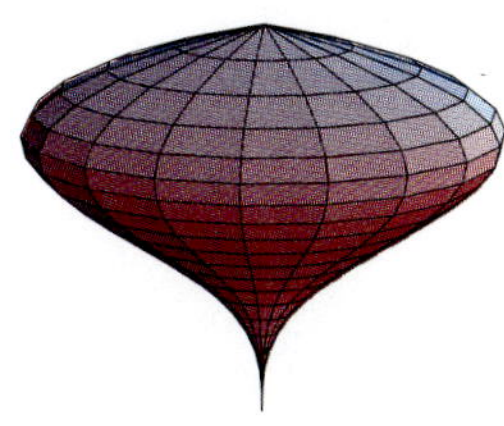

Fig. 9.3.5 The top of Problem 65

9.4 TRIGONOMETRIC INTEGRALS

Here we discuss the evaluation of certain integrals in which the integrand is either a power of a trigonometric function or the product of two such powers. Such integrals are among the most common trigonometric integrals in applications of calculus.

To evaluate the integrals

$$\int \sin^2 u\,du \quad \text{and} \quad \int \cos^2 u\,du$$

that appear in numerous applications, we use the **half-angle identities**

$$\sin^2\theta = \tfrac{1}{2}(1 - \cos 2\theta), \tag{1}$$

$$\cos^2\theta = \tfrac{1}{2}(1 + \cos 2\theta) \tag{2}$$

of Eqs. (11) and (10) in Appendix C.

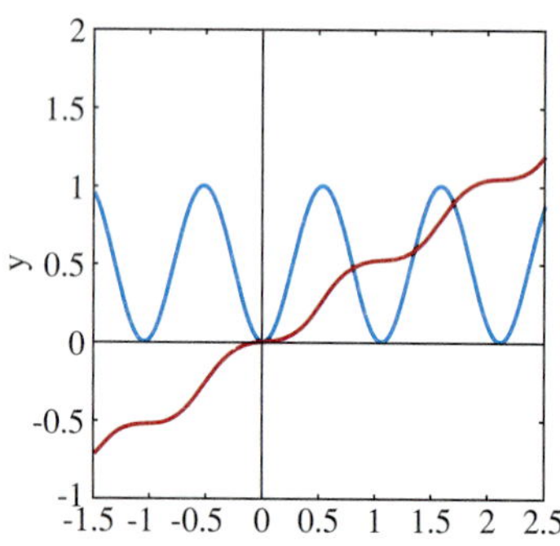

Fig. 9.4.1 Graphs of the functions $\sin^2 3x$ and $\frac{1}{12}(6x - \sin 6x)$ of Example 1. The zeros of which correspond to the critical points of the other? So which is the antiderivative?

EXAMPLE 1 Find $\int \sin^2 3x\,dx$ (see Fig. 9.4.1).

Solution The identity in Eq. (1)—with $3x$ in place of x—yields

$$\int \sin^2 3x\,dx = \int \tfrac{1}{2}(1 - \cos 6x)\,dx$$
$$= \tfrac{1}{2}\left(x - \tfrac{1}{6}\sin 6x\right) + C = \tfrac{1}{12}(6x - \sin 6x) + C.$$ ■

To integrate $\tan^2 x$ and $\cot^2 x$, we use the identities

$$1 + \tan^2 x = \sec^2 x \quad \text{and} \quad 1 + \cot^2 x = \csc^2 x. \tag{3}$$

The first of these follows from the fundamental identity $\sin^2 x + \cos^2 x = 1$ upon division of both sides by $\cos^2 x$. To obtain the second formula in (3), we divide both sides of the fundamental identity by $\sin^2 x$.

EXAMPLE 2 Compute the antiderivative $\int \cot^2 3x\,dx$.

Solution By using the second identity in (3), we obtain

$$\int \cot^2 3x\,dx = \int (\csc^2 3x - 1)\,dx = \int (\csc^2 u - 1)\left(\tfrac{1}{3}\,du\right) \qquad (u = 3x)$$
$$= \tfrac{1}{3}(-\cot u - u) + C = -\tfrac{1}{3}\cot 3x - x + C.$$ ■

Integrals of Products of Sines and Cosines

The substitution $u = \sin x$, $du = \cos x\,dx$ gives

$$\int \sin^3 x \cos x\,dx = \int u^3\,du = \tfrac{1}{4}u^4 + C = \tfrac{1}{4}\sin^4 x + C.$$

This substitution, or the similar substitution $u = \cos x$, $du = -\sin x\,dx$, can be used to evaluate an integral of the form

$$\int \sin^m x \cos^n x\,dx \tag{4}$$

in the first of the following two cases:

Case 1: At least one of the two numbers m and n is an *odd positive integer.* If so, then the other may be any real number.

Case 2: Both m and n are *nonnegative even integers*.

Suppose, for example, that $m = 2k + 1$ is an odd positive integer. Then we isolate one $\sin x$ factor and use the identity $\sin^2 x = 1 - \cos^2 x$ to express the remaining $\sin^{m-1} x$ factor in terms of $\cos x$, as follows:

$$\begin{aligned}\int \sin^m x \cos^n x\, dx &= \int \sin^{m-1} x \cos^n x \sin x\, dx = \int (\sin^2 x)^k \cos^n x \sin x\, dx \\ &= \int (1 - \cos^2 x)^k \cos^n x \sin x\, dx.\end{aligned}$$

Now the substitution $u = \cos x$, $du = -\sin x\, dx$ yields

$$\int \sin^m x \cos^n x\, dx = -\int (1 - u^2)^k u^n\, du.$$

The exponent $k = (m - 1)/2$ is a nonnegative integer because m is an odd positive integer. Thus the factor $(1 - u^2)^k$ of the integrand is a polynomial in the variable u, and so its product with u^n is easy to integrate.

In essence, this method consists of peeling off one copy of $\sin x$ (if m is odd) and then converting the remaining sines into cosines. If n is odd, then we can split off one copy of $\cos x$ and convert the remaining cosines into sines.

EXAMPLE 3

(a)
$$\begin{aligned}\int \sin^3 x \cos^2 x\, dx &= \int (1 - \cos^2 x) \cos^2 x \sin x\, dx \\ &= \int (u^4 - u^2)\, du \qquad (u = \cos x) \\ &= \tfrac{1}{5} u^5 - \tfrac{1}{3} u^3 + C = \tfrac{1}{5}\cos^5 x - \tfrac{1}{3}\cos^3 x + C.\end{aligned}$$

(b)
$$\begin{aligned}\int \cos^5 x\, dx &= \int (1 - \sin^2 x)^2 \cos x\, dx \\ &= \int (1 - u^2)^2\, du \qquad (u = \sin x) \\ &= \int (1 - 2u^2 + u^4)\, du = u - \tfrac{2}{3}u^3 + \tfrac{1}{5}u^5 + C \\ &= \sin x - \tfrac{2}{3}\sin^3 x + \tfrac{1}{5}\sin^5 x + C.\end{aligned}$$

■

In case 2 of the sine-cosine integral in Eq. (4), with both m and n nonnegative even integers, we use the half-angle formulas in Eqs. (1) and (2) to halve the even powers of $\sin x$ and $\cos x$. If we repeat this process with the resulting powers of $\cos 2x$ (if necessary), we get integrals involving odd powers, and we have seen how to handle these in case 1.

EXAMPLE 4 Use of Eqs. (1) and (2) gives

$$\int \sin^2 x \cos^2 x\, dx = \int \tfrac{1}{2}(1 - \cos 2x)\, \tfrac{1}{2}(1 + \cos 2x)\, dx$$

$$= \frac{1}{4}\int (1 - \cos^2 2x)\, dx = \frac{1}{4}\int \left[1 - \frac{1}{2}(1 + \cos 4x)\right] dx$$

$$= \frac{1}{8}\int (1 - \cos 4x)\, dx = \frac{1}{8}x - \frac{1}{32}\sin 4x + C.$$

In the third step we used Eq. (2) with $\theta = 2x$. ■

EXAMPLE 5 Here we apply Eq. (2), first with $\theta = 3x$ and then with $\theta = 6x$.

$$\int \cos^4 3x\, dx = \int \frac{1}{4}(1 + \cos 6x)^2\, dx$$

$$= \frac{1}{4}\int (1 + 2\cos 6x + \cos^2 6x)\, dx$$

$$= \frac{1}{4}\int \left(\frac{3}{2} + 2\cos 6x + \frac{1}{2}\cos 12x\right) dx$$

$$= \frac{3}{8}x + \frac{1}{12}\sin 6x + \frac{1}{96}\sin 12x + C.$$ ■

Integrals of Products of Secants and Tangents

To integrate $\tan x$, the substitution

$$u = \cos x, \qquad du = -\sin x\, dx$$

gives

$$\int \tan x\, dx = \int \frac{\sin x}{\cos x}\, dx = -\int \frac{1}{u}\, du = -\ln|u| + C,$$

and thus

$$\int \tan x\, dx = -\ln|\cos x| + C = \ln|\sec x| + C. \tag{5}$$

In Eq. (5) we used the fact that $|\sec x| = 1/|\cos x|$. Similarly,

$$\int \cot x\, dx = \ln|\sin x| + C = -\ln|\csc x| + C. \tag{6}$$

The first person to integrate $\sec x$ may well have spent much time doing so. Here is one of several methods. First we "prepare" the function for integration:

$$\sec x = \frac{1}{\cos x} = \frac{\cos x}{\cos^2 x} = \frac{\cos x}{1 - \sin^2 x}.$$

Now

$$\frac{1}{1+z} + \frac{1}{1-z} = \frac{2}{1-z^2}.$$

Similarly, working backward, we have

$$\frac{2\cos x}{1 - \sin^2 x} = \frac{\cos x}{1 + \sin x} + \frac{\cos x}{1 - \sin x}.$$

Therefore

$$\int \sec x\,dx = \frac{1}{2}\int\left(\frac{\cos x}{1+\sin x}+\frac{\cos x}{1-\sin x}\right)dx$$
$$= \frac{1}{2}\left(\ln|1+\sin x| - \ln|1-\sin x|\right) + C.$$

It's customary to simplify this result:

$$\int \sec x\,dx = \frac{1}{2}\ln\left|\frac{1+\sin x}{1-\sin x}\right| + C = \frac{1}{2}\ln\left|\frac{(1+\sin x)^2}{1-\sin^2 x}\right| + C$$
$$= \ln\left|\frac{(1+\sin x)^2}{\cos^2 x}\right|^{1/2} + C = \ln\left|\frac{1+\sin x}{\cos x}\right| + C$$
$$= \ln|\sec x + \tan x| + C.$$

Once we know that

$$\int \sec x\,dx = \ln|\sec x + \tan x| + C, \tag{7}$$

we can always "derive" this result quickly by using an unmotivated trick:

$$\int \sec x\,dx = \int(\sec x)\frac{\tan x + \sec x}{\sec x + \tan x}\,dx$$
$$= \int \frac{\sec x\tan x + \sec^2 x}{\sec x + \tan x}\,dx = \ln|\sec x + \tan x| + C.$$

A similar technique yields

$$\int \csc x\,dx = -\ln|\csc x + \cot x| + C. \tag{8}$$

EXAMPLE 6 The substitution $u = \frac{1}{2}x$, $du = \frac{1}{2}dx$ gives

$$\int_0^{\pi/2} \sec\frac{x}{2}\,dx = 2\int_0^{\pi/4}\sec u\,du$$
$$= 2\Big[\ln|\sec u + \tan u|\Big]_0^{\pi/4} = 2\ln(1+\sqrt{2}) \approx 1.76275.$$ ■

An integral of the form

$$\int \tan^m x \sec^n x\,dx \tag{9}$$

can be routinely evaluated in either of the following two cases:

- *Case 1*: m is an *odd positive integer.*
- *Case 2*: n is an *even positive integer*.

In case 1, we split off the factor $\sec x\tan x$ to form, along with dx, the differential $\sec x \tan x\,dx$ of $\sec x$. We then use the identity $\tan^2 x = \sec^2 x - 1$ to convert the remaining even power of $\tan x$ into powers of $\sec x$. This prepares the integrand for the substitution $u = \sec x$.

EXAMPLE 7

$$\begin{aligned}\int \tan^3 x \sec^3 x\,dx &= \int (\sec^2 x - 1)\sec^2 x \sec x \tan x\,dx \\ &= \int (u^4 - u^2)\,du \qquad (u = \sec x) \\ &= \tfrac{1}{5}u^5 - \tfrac{1}{3}u^3 + C = \tfrac{1}{5}\sec^5 x - \tfrac{1}{3}\sec^3 x + C.\end{aligned}$$

■

To evaluate the integral in Eq. (9) in case 2, we split off $\sec^2 x$ to form, along with dx, the differential of $\tan x$. We then use the identity $\sec^2 x = 1 + \tan^2 x$ to convert the remaining even power of $\sec x$ into powers of $\tan x$. This prepares the integrand for the substitution $u = \tan x$.

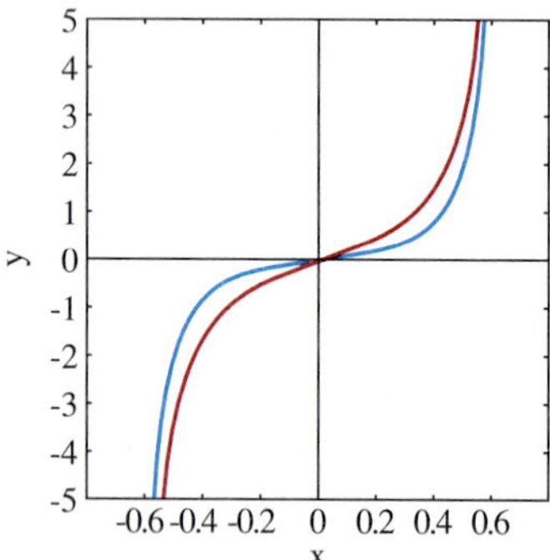

Fig. 9.4.2 After an initial attempt to evaluate $\int \sec^6 2x\,dx$ using the two methods of Example 8, a computer was used to plot the two alleged antiderivatives. Why does this figure indicate the presence of an error in the calculations? What is the relationship between any two antiderivatives of a given function?

EXAMPLE 8 *Method 1.* Use of the secant-tangent form of the fundamental identity of trigonometry gives

$$\begin{aligned}\int \sec^6 2x\,dx &= \int (1 + \tan^2 2x)^2 \sec^2 2x\,dx \\ &= \tfrac{1}{2}\int (1 + \tan^2 2x)^2 (2\sec^2 2x)\,dx \\ &= \tfrac{1}{2}\int (1 + u^2)^2\,du \qquad (u = \tan 2x) \\ &= \tfrac{1}{2}\int (1 + 2u^2 + u^4)\,du = \tfrac{1}{2}u + \tfrac{1}{3}u^3 + \tfrac{1}{10}u^5 + C \\ &= \tfrac{1}{2}\tan 2x + \tfrac{1}{3}\tan^3 2x + \tfrac{1}{10}\tan^5 2x + C.\end{aligned} \tag{10}$$

Method 2. Alternatively, we could apply the reduction formula

$$\int \sec^n x\,dx = \frac{\sec^{n-2} x \tan x}{n-1} + \frac{n-2}{n-1}\int \sec^{n-2} x\,dx$$

of Section 9.3, first with $n = 6$ and then with $n = 4$. This gives

$$\begin{aligned}\int \sec^6 2x\,dx &= \tfrac{1}{2}\int \sec^6 u\,du \qquad (u = 2x) \\ &= \tfrac{1}{2}\left(\tfrac{1}{5}\sec^4 u \tan u + \tfrac{4}{5}\int \sec^4 u\,du\right) \\ &= \tfrac{1}{10}\sec^4 u \tan u + \tfrac{2}{5}\left(\tfrac{1}{3}\sec^2 u \tan u + \tfrac{2}{3}\int \sec^2 u\,du\right) \\ &= \tfrac{1}{10}\sec^4 2x \tan 2x + \tfrac{2}{15}\sec^2 2x \tan 2x + \tfrac{4}{15}\tan 2x + C.\end{aligned} \tag{11}$$

See Figs. 9.4.2 and 9.4.3. ■

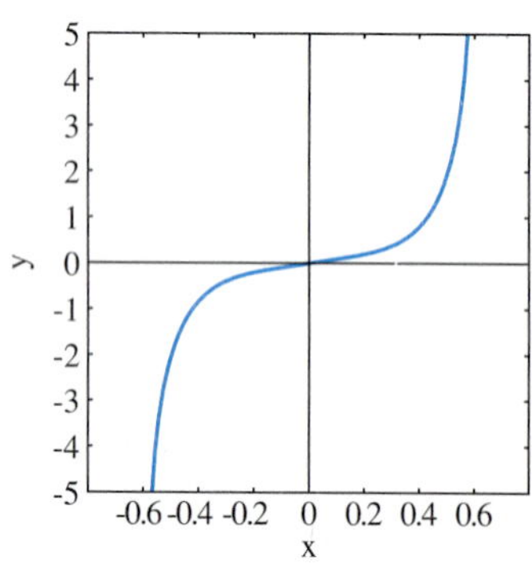

Fig. 9.4.3 After the error made initially was found and corrected, this figure resulted when the two antiderivatives in (10) and (11) were plotted simultaneously (each with $C = 0$). It indicates that the two antiderivative formulas are, in fact, equivalent.

Similar methods are effective with integrals of the form

$$\int \csc^m x \cot^n x\,dx,$$

because the cotangent and cosecant functions satisfy analogous differentiation formulas and trigonometric identities:

$$1 + \tan^2 x = \sec^2 x, \qquad D_x \tan x = \sec^2 x, \qquad D_x \sec x = \sec x \tan x$$

and

$$1 + \cot^2 x = \csc^2 x, \quad D_x \cot x = -\csc^2 x, \quad D_x \csc x = -\csc x \cot x.$$

The method of case 1 succeeds with the integral

$$\int \tan^n x \, dx$$

only when n is an odd positive integer, but there is another approach that works equally well whether n is even or odd. We split off the factor $\tan^2 x$ and replace it with $\sec^2 x - 1$:

$$\begin{aligned}\int \tan^n x \, dx &= \int (\tan^{n-2} x)(\sec^2 x - 1) \, dx \\ &= \int \tan^{n-2} x \sec^2 x \, dx - \int \tan^{n-2} x \, dx.\end{aligned}$$

We integrate what we can and find that

$$\int \tan^n x \, dx = \frac{\tan^{n-1} x}{n-1} - \int \tan^{n-2} x \, dx. \tag{12}$$

Equation (12) is another example of a reduction formula. Its use effectively reduces the original exponent from n to $n - 2$. If we apply Eq. (12) repeatedly, we eventually reduce the integral to either

$$\int \tan^2 x \, dx = \int (\sec^2 x - 1) \, dx = \tan x - x + C$$

or

$$\int \tan x \, dx = \ln |\sec x| + C.$$

EXAMPLE 9 Two applications of Eq. (12) give

$$\begin{aligned}\int \tan^6 x \, dx &= \tfrac{1}{5} \tan^5 x - \int \tan^4 x \, dx \\ &= \tfrac{1}{5} \tan^5 x - \left(\tfrac{1}{3} \tan^3 x - \int \tan^2 x \, dx \right) \\ &= \tfrac{1}{5} \tan^5 x - \tfrac{1}{3} \tan^3 x + \tan x - x + C.\end{aligned}$$ ■

Finally, in the case of an integral involving an unusual mixture of trigonometric functions—tangents and cosecants, for example—expressing the integrand entirely in terms of sines and cosines may yield an expression that's easy to integrate.

9.4 PROBLEMS

Evaluate the integrals in Problems 1 through 44.

1. $\int \sin^2 2x \, dx$

2. $\int \cos^2 5x \, dx$

3. $\int \sec^2 \frac{x}{2} \, dx$

4. $\int \tan^2 \frac{x}{2} \, dx$

5. $\int \tan 3x \, dx$

6. $\int \cot 4x \, dx$

7. $\int \sec 3x \, dx$

8. $\int \csc 2x \, dx$

9. $\displaystyle\int \frac{1}{\csc^2 x}\,dx$ **10.** $\displaystyle\int \sin^2 x \cot^2 x\,dx$

11. $\displaystyle\int \sin^3 x\,dx$ **12.** $\displaystyle\int \sin^4 x\,dx$

13. $\displaystyle\int \sin^2\theta \cos^3\theta\,d\theta$ **14.** $\displaystyle\int \sin^3 t \cos^3 t\,dt$

15. $\displaystyle\int \cos^5 x\,dx$ **16.** $\displaystyle\int \frac{\sin t}{\cos^3 t}\,dt$

17. $\displaystyle\int \frac{\sin^3 x}{\sqrt{\cos x}}\,dx$ **18.** $\displaystyle\int \sin^3\phi \cos^4\phi\,d\phi$

19. $\displaystyle\int \sin^5 2z \cos^2 2z\,dz$ **20.** $\displaystyle\int \sin^{3/2} x \cos^3 x\,dx$

21. $\displaystyle\int \frac{\sin^3 4x}{\cos^2 4x}\,dx$ **22.** $\displaystyle\int \cos^6 4\theta\,d\theta$

23. $\displaystyle\int \sec^4 t\,dt$ **24.** $\displaystyle\int \tan^3 x\,dx$

25. $\displaystyle\int \cot^3 2x\,dx$ **26.** $\displaystyle\int \tan\theta \sec^4\theta\,d\theta$

27. $\displaystyle\int \tan^5 2x \sec^2 2x\,dx$ **28.** $\displaystyle\int \cot^3 x \csc^2 x\,dx$

29. $\displaystyle\int \csc^6 2t\,dt$ **30.** $\displaystyle\int \frac{\sec^4 t}{\tan^2 t}\,dt$

31. $\displaystyle\int \frac{\tan^3\theta}{\sec^4\theta}\,d\theta$ **32.** $\displaystyle\int \frac{\cot^3 x}{\csc^2 x}\,dx$

33. $\displaystyle\int \frac{\tan^3 t}{\sqrt{\sec t}}\,dt$ **34.** $\displaystyle\int \frac{1}{\cos^4 2x}\,dx$

35. $\displaystyle\int \frac{\cot\theta}{\csc^3\theta}\,d\theta$ **36.** $\displaystyle\int \sin^2 3\alpha \cos^2 3\alpha\,d\alpha$

37. $\displaystyle\int \cos^3 5t\,dt$ **38.** $\displaystyle\int \tan^4 x\,dx$

39. $\displaystyle\int \cot^4 3t\,dt$ **40.** $\displaystyle\int \tan^2 2t \sec^4 2t\,dt$

41. $\displaystyle\int \sin^5 2t \cos^{3/2} 2t\,dt$ **42.** $\displaystyle\int \cot^3\xi \csc^{3/2}\xi\,d\xi$

43. $\displaystyle\int \frac{\tan x + \sin x}{\sec x}\,dx$ **44.** $\displaystyle\int \frac{\cot x + \csc x}{\sin x}\,dx$

In Problems 45 through 48, find the area of the region bounded by the two given curves.

45. The x-axis and the curve $y = \sin^3 x$, from $x = 0$ to $x = \pi$

46. $y = \cos^2 x$ and $y = \sin^2 x$, from $x = -\pi/4$ to $x = \pi/4$

47. $y = \sin x \cos x$ and $y = \sin^2 x$, from $x = \pi/4$ to $x = \pi$

48. $y = \cos^3 x$ and $y = \sin^3 x$, from $x = \pi/4$ to $x = 5\pi/4$

In Problems 49 and 50, first graph the integrand function and guess the value of the integral. Then verify your guess by actually evaluating the integral.

49. $\displaystyle\int_0^{2\pi} \sin^3 x \cos^2 x\,dx$ **50.** $\displaystyle\int_0^{\pi} \sin^5 2x\,dx$

In Problems 51 through 54, find the volume of the solid generated by revolving the given region R around the x-axis.

51. R is bounded by the x-axis and the curve $y = \sin^2 x$, $0 \leqq x \leqq \pi$.

52. R is the region of Problem 46.

53. R is bounded by $y = 2$ and $y = \sec x$ for $-\pi/3 \leqq x \leqq \pi/3$.

54. R is bounded by $y = 4\cos x$ and $y = \sec x$ for $-\pi/3 \leqq x \leqq \pi/3$.

55. Let R denote the region that lies between the curves $y = \tan^2 x$ and $y = \sec^2 x$ for $0 \leqq x \leqq \pi/4$. Find: (a) the area of R; (b) the volume of the solid obtained by revolving R around the x-axis.

56. Find the length of the graph of $y = \ln(\cos x)$ from $x = 0$ to $x = \pi/4$.

57. Find

$$\int \tan x \sec^4 x\,dx$$

in two different ways. Then show that your two results are equivalent.

58. Find

$$\int \cot^3 x\,dx$$

in two different ways. Then show that your two results are equivalent.

Problems 59 through 62 are applications of the trigonometric identities

$$\sin A \sin B = \tfrac{1}{2}\left[\cos(A - B) - \cos(A + B)\right],$$

$$\sin A \cos B = \tfrac{1}{2}\left[\sin(A - B) + \sin(A + B)\right],$$

$$\cos A \cos B = \tfrac{1}{2}\left[\cos(A - B) + \cos(A + B)\right].$$

59. Find $\displaystyle\int \sin 3x \cos 5x\,dx$.

60. Find $\displaystyle\int \sin 2x \sin 4x\,dx$.

61. Find $\displaystyle\int \cos x \cos 4x\,dx$.

62. Suppose that m and n are positive integers with $m \neq n$. Show that (a) $\int_0^{2\pi} \sin mx \sin nx\,dx = 0$; (b) $\int_0^{2\pi} \cos mx \sin nx\,dx = 0$; (c) $\int_0^{2\pi} \cos mx \cos nx\,dx = 0$.

63. Substitute $\sec x \csc x = (\sec^2 x)/(\tan x)$ to derive the formula

$$\int \sec x \csc x\,dx = \ln|\tan x| + C.$$

64. Show that

$$\csc x = \frac{1}{2\sin(\frac{1}{2}x)\cos(\frac{1}{2}x)},$$

then apply the result of Problem 63 to derive the formula

$$\int \csc x \, dx = \ln\left|\tan \frac{x}{2}\right| + C.$$

65. Substitute $x = \frac{1}{2}\pi - u$ into the integral formula of Problem 64 to show that

$$\int \sec x \, dx = \ln\left|\cot\left(\frac{\pi}{4} - \frac{x}{2}\right)\right| + C.$$

66. Use appropriate trigonometric identities to deduce from the result of Problem 65 that

$$\int \sec x \, dx = \ln|\sec x + \tan x| + C.$$

67. Show first that the reduction formula in Eq. (12) gives

$$\int \tan^4 x \, dx = \tfrac{1}{3}\tan^3 x - \tan x + x.$$

Then compare this result with the alleged antiderivative

$$\int \tan^4 x \, dx = \tfrac{1}{12}(\sec^3 x)(9x \cos x + 3x \cos 3x - 4 \sin 3x)$$

given by some versions of *Mathematica.*

68. Compare the result given in Example 9 with the integral

$$\int \tan^6 x \, dx$$

as given by your favorite computer algebra system.

9.5 RATIONAL FUNCTIONS AND PARTIAL FRACTIONS

We now discuss methods with which every rational function can be integrated in terms of elementary functions. Recall that a rational function $R(x)$ is a function that can be expressed as the quotient of two polynomials. That is,

$$R(x) = \frac{P(x)}{Q(x)}, \tag{1}$$

where $P(x)$ and $Q(x)$ are polynomials. The **method of partial fractions** is an *algebraic* technique that decomposes $R(x)$ into a sum of terms:

$$R(t) = \frac{P(x)}{Q(x)} = p(x) + F_1(x) + F_2(x) + \cdots + F_k(x), \tag{2}$$

where $p(x)$ is a polynomial and each expression $F_i(x)$ is a fraction that can be integrated with little difficulty.

EXAMPLE 1 We can verify (by finding a common denominator on the right) that

$$\frac{x^3 - 1}{x^3 + x} = 1 - \frac{1}{x} + \frac{x - 1}{x^2 + 1}. \tag{3}$$

It follows that

$$\int \frac{x^3 - 1}{x^3 + x} \, dx = \int \left(1 - \frac{1}{x} + \frac{x}{x^2 + 1} - \frac{1}{x^2 + 1}\right) dx$$
$$= x - \ln|x| + \tfrac{1}{2}\ln(x^2 + 1) - \tan^{-1} x + C.$$

The key to this simple integration lies in finding the decomposition given in Eq. (3). The existence of such a decomposition and the technique of finding it are what the method of partial fractions is about. See Fig. 9.5.1. ■

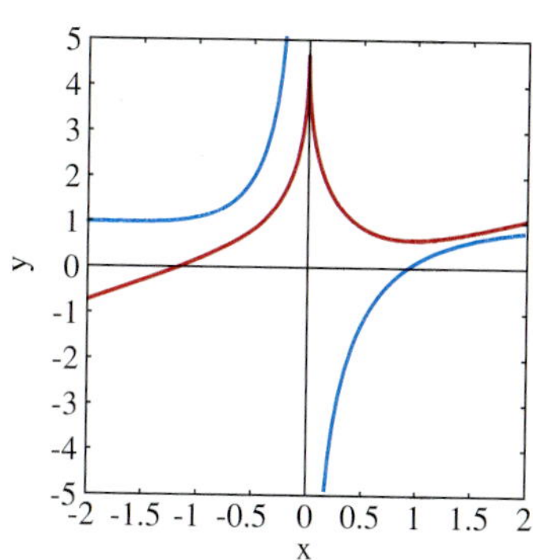

Fig. 9.5.1 Graphs of the function $f(x) = (x^3 - 1)/(x^3 + x)$ of Example 1 and its indefinite integral with $C = 0$. Which is which?

According to a theorem proved in advanced algebra, every rational function can be written in the form in Eq. (2) with each $F_i(x)$ being a fraction either of the form

$$\frac{A}{(ax + b)^n} \tag{4}$$

or of the form

$$\frac{Bx + C}{(ax^2 + bx + c)^n} \tag{5}$$

(where A, B, C, a, b, and c are constants). Here the quadratic polynomial $ax^2 + bx + c$ is **irreducible**: It is not a product of linear factors with real coefficients. This is the same as saying that the equation $ax^2 + bx + c = 0$ has no real roots, and the quadratic formula tells us that this is the case exactly when its discriminant is negative: $b^2 - 4ac < 0$.

Fractions of the forms in Eqs. (4) and (5) are called **partial fractions**, and the sum in Eq. (2) is called the **partial fraction decomposition** of $R(x)$. Thus Eq. (3) gives the partial fraction decomposition of $(x^3 - 1)/(x^3 + x)$. A partial fraction of the form in Eq. (4) may be integrated immediately, and we will see in Section 9.7 how to integrate one of the form in Eq. (5).

The first step in finding the partial fraction decomposition of $R(x)$ is to find the polynomial $p(x)$ in Eq. (2). It turns out that $p(x) \equiv 0$ provided that the degree of the numerator $P(x)$ is *less than* that of the denominator $Q(x)$; such a rational function $R(x) = P(x)/Q(x)$ is said to be **proper**. If $R(x)$ is not proper, then $p(x)$ may be found by dividing $Q(x)$ into $P(x)$, as in Example 2.

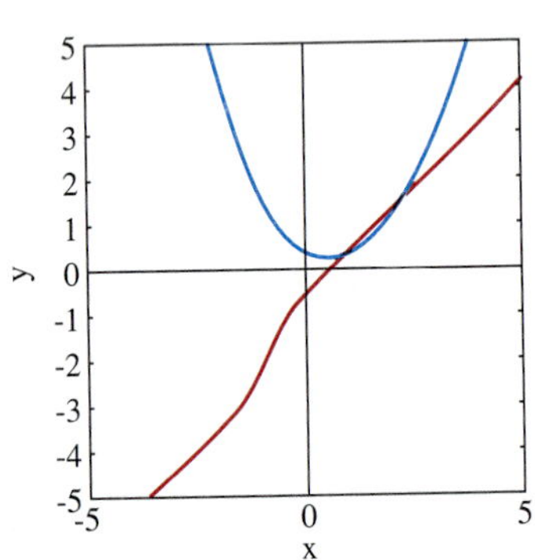

Fig. 9.5.2 Graphs of the function $f(x) = (x^3 + x^2 + x - 1) / (x^2 + 2x + 2)$ of Example 2 and its indefinite integral with $C = 0$. Which is which?

EXAMPLE 2 Find $\displaystyle\int \frac{x^3 + x^2 + x - 1}{x^2 + 2x + 2}\,dx$. See Fig. 9.5.2.

Solution Long division of denominator into numerator may be carried out as follows:

$$\begin{array}{rll}
 & x - 1 \longleftarrow & p(x) \quad \text{(quotient)} \\
x^2 + 2x + 2 \,\big) & x^3 + x^2 + x - 1 & \\
 & \underline{x^3 + 2x^2 + 2x} & \\
 & -x^2 - x - 1 & \\
 & \underline{-x^2 - 2x - 2} & \\
 & x + 1 \longleftarrow & r(x) \quad \text{(remainder)}
\end{array}$$

As in simple arithmetic,

$$\text{“fraction} = \text{quotient} + \frac{\text{remainder}}{\text{divisor}}.\text{”}$$

Thus

$$\frac{x^3 + x^2 + x - 1}{x^2 + 2x + 2} = (x - 1) + \frac{x + 1}{x^2 + 2x + 2},$$

and hence

$$\begin{aligned}
\int \frac{x^3 + x^2 + x - 1}{x^2 + 2x + 2}\,dx &= \int \left(x - 1 + \frac{x + 1}{x^2 + 2x + 2} \right) dx \\
&= \tfrac{1}{2}x^2 - x + \tfrac{1}{2}\ln(x^2 + 2x + 2) + C.
\end{aligned}$$

■

By using long division as in Example 2, any rational function $R(x)$ can be written as a sum of a polynomial $p(x)$ and a *proper* rational fraction,

$$R(x) = p(x) + \frac{r(x)}{Q(x)}.$$

To see how to integrate an arbitrary rational function, we therefore need only see how to find the partial fraction decomposition of a proper rational fraction.

To obtain such a decomposition, the first step is to factor the denominator $Q(x)$ into a product of linear factors (those of the form $ax + b$) and irreducible quadratic factors (those of the form $ax^2 + bx + c$ with $b^2 - 4ac < 0$). This is always possible in principle but may be difficult in practice. But once we have found the factorization of $Q(x)$, we can obtain the partial fraction decomposition by routine algebraic methods (described next). Each linear or irreducible quadratic factor of $Q(x)$ leads to one or more partial fractions of the forms in Eqs. (4) and (5).

Linear Factors

Let $R(x) = P(x)/Q(x)$ be a *proper* rational fraction, and suppose that the linear factor $ax + b$ occurs n times in the complete factorization of $Q(x)$. That is, $(ax + b)^n$ is the highest power of $ax + b$ that divides "evenly" into $Q(x)$. In this case we call n the **multiplicity** of the factor $ax + b$.

Rule 1 *Linear Factor Partial Fractions*

The part of the partial fraction decomposition of $R(x)$ that corresponds to the linear factor $ax + b$ of multiplicity n is a sum of n partial fractions, specifically

$$\frac{A_1}{ax+b} + \frac{A_2}{(ax+b)^2} + \cdots + \frac{A_n}{(ax+b)^n}, \tag{6}$$

where $A_1, A_2, \ldots, A_n$ are constants.

If *all* the factors of $Q(x)$ are linear, then the partial fraction decomposition of $R(x)$ is a sum of expressions like the one in (6). The situation is especially simple if each of these linear factors is *nonrepeated*—that is, if each has multiplicity $n = 1$. In this case, the expression in (6) reduces to its first term, and the partial fraction decomposition of $R(x)$ is a sum of such terms. The solutions in Examples 3 and 4 illustrate how the constant numerators can be determined.

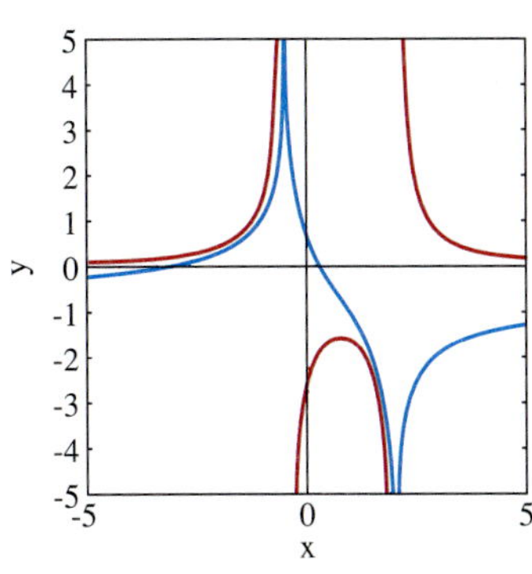

Fig. 9.5.3 Graphs of the function $f(x) = 5/[(2x + 1)(x - 2)]$ of Example 3 and its indefinite integral with $C = 0$. Which is which?

EXAMPLE 3 Find $\displaystyle\int \frac{5}{(2x+1)(x-2)}\,dx$. See Fig. 9.5.3.

Solution The linear factors in the denominator are distinct, so we seek a partial fraction decomposition of the form

$$\frac{5}{(2x+1)(x-2)} = \frac{A}{2x+1} + \frac{B}{x-2}.$$

To find the constants A and B, we multiply both sides of this *identity* by the left-hand (common) denominator $(2x + 1)(x - 2)$. The result is

$$5 = A(x-2) + B(2x+1) = (A+2B)x + (-2A+B).$$

Next we equate coefficients of x and coefficients of 1 on the left-hand and right-hand sides of this equation. This yields the simultaneous equations

$$A + 2B = 0,$$

$$-2A + B = 5,$$

which we readily solve for $A = -2$, $B = 1$. Hence

$$\frac{5}{(2x+1)(x-2)} = \frac{-2}{2x+1} + \frac{1}{x-2},$$

and therefore

$$\int \frac{5}{(2x+1)(x-2)}\,dx = -\ln|2x+1| + \ln|x-2| + C$$
$$= \ln\left|\frac{x-2}{2x+1}\right| + C.$$

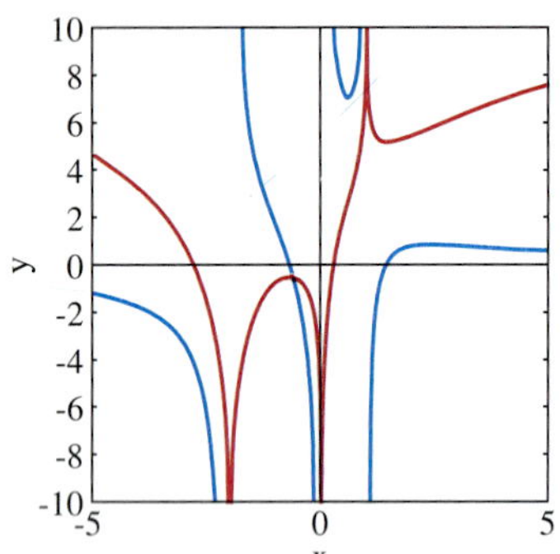

Fig. 9.5.4 Graphs of the function $f(x) = (4x^2 - 3x - 4)/(x^3 + x^2 - 2x)$ of Example 4 and its indefinite integral with $C = 0$. Which is which?

EXAMPLE 4 Find $\displaystyle\int \frac{4x^2 - 3x - 4}{x^3 + x^2 - 2x}\,dx$. See Fig. 9.5.4.

Solution The rational function to be integrated is proper, so we immediately factor its denominator:

$$x^3 + x^2 - 2x = x(x^2 + x - 2) = x(x-1)(x+2).$$

We are dealing with three nonrepeated linear factors, so the partial fraction decomposition has the form

$$\frac{4x^2 - 3x - 4}{x^3 + x^2 - 2x} = \frac{A}{x} + \frac{B}{x-1} + \frac{C}{x+2}.$$

To find the constants A, B, and C, we multiply both sides of this equation by the common denominator $x(x-1)(x+2)$ and find thereby that

$$4x^2 - 3x - 4 = A(x-1)(x+2) + Bx(x+2) + Cx(x-1). \tag{7}$$

Then we collect coefficients of like powers of x on the right:

$$4x^2 - 3x - 4 = (A + B + C)x^2 + (A + 2B - C)x + (-2A).$$

Because two polynomials are (identically) equal only if the coefficients of corresponding powers of x are the same, we conclude that

$$\begin{aligned} A + B + C &= 4, \\ A + 2B - C &= -3, \\ -2A \qquad\qquad &= -4. \end{aligned}$$

We solve these simultaneous equations and thus find that $A = 2$, $B = -1$, and $C = 3$.

There is an alternative way to find A, B, and C that is especially effective in the case of nonrepeated linear factors. Substitute the values of $x = 0$, $x = 1$, and $x = -2$ (the zeros of the linear factors of the denominator) into Eq. (7). Substitution of $x = 0$ into Eq. (7) immediately gives $-4 = -2A$, so $A = 2$. Substitution of $x = 1$ into Eq. (7) gives $-3 = 3B$, so $B = -1$. Substitution of $x = -2$ gives $18 = 6C$, so $C = 3$.

With these values of $A = 2$, $B = -1$, and $C = 3$, however obtained, we find that

$$\int \frac{4x^2 - 3x - 4}{x^3 + x^2 - 2x}\,dx = \int \left(\frac{2}{x} - \frac{1}{x-1} + \frac{3}{x+2}\right) dx$$
$$= 2\ln|x| - \ln|x-1| + 3\ln|x+2| + C.$$

Laws of logarithms allow us to write this antiderivative in the more compact form

$$\int \frac{4x^2 - 3x - 4}{x^3 + x^2 - 2x}\,dx = \ln\left|\frac{x^2(x+2)^3}{x-1}\right| + C.$$

EXAMPLE 5 Find $\displaystyle\int \frac{x^3 - 4x - 1}{x(x-1)^3}\,dx.$

Solution Here we have the linear factor x of multiplicity 1 but also the linear factor $x - 1$ of multiplicity 3. According to rule 1, the partial fraction decomposition of the integrand has the form

$$\frac{x^3 - 4x - 1}{x(x-1)^3} = \frac{A}{x} + \frac{B}{x-1} + \frac{C}{(x-1)^2} + \frac{D}{(x-1)^3}.$$

To find the constants A, B, C, and D, we multiply both sides of this equation by the least common denominator $x(x-1)^3$. We find that

$$x^3 - 4x - 1 = A(x-1)^3 + Bx(x-1)^2 + Cx(x-1) + Dx.$$

We expand and then collect coefficients of like powers of x on the right-hand side. This yields

$$\begin{aligned} x^3 - 4x + 1 = (A + B)x^3 &+ (-3A - 2B + C)x^2 \\ &+ (3A + B - C + D)x - A. \end{aligned}$$

Then we equate coefficients of like powers of x on each side of this equation. We get the four simultaneous equations

$$\begin{aligned} A + B \quad &= 1, \\ -3A - 2B + C \quad &= 0, \\ 3A + B - C + D &= -4, \\ -A \quad &= -1. \end{aligned}$$

The last equation gives $A = 1$, and then the first equation gives $B = 0$. Next, the second equation gives $C = 3$. When we substitute these values into the third equation, we finally get $D = -4$. Hence

$$\begin{aligned} \int \frac{x^3 - 4x - 1}{x(x-1)^3}\,dx &= \int \left(\frac{1}{x} + \frac{3}{(x-1)^2} - \frac{4}{(x-1)^3} \right) dx \\ &= \ln|x| - \frac{3}{x-1} + \frac{2}{(x-1)^2} + C. \end{aligned}$$

Quadratic Factors

Suppose that $R(x) = P(x)/Q(x)$ is a proper rational fraction and that the irreducible quadratic factor $ax^2 + bx + c$ occurs n times in the factorization of $Q(x)$. That is, $(ax^2 + bx + c)^n$ is the highest power of $ax^2 + bx + c$ that divides evenly into $Q(x)$. As before, we call n the **multiplicity** of the quadratic factor $ax^2 + bx + c$.

Rule 2 *Quadratic Factor Partial Fractions*

The part of the partial fraction decomposition of $R(x)$ that corresponds to the irreducible quadratic factor $ax^2 + bx + c$ of multiplicity n is a sum of n partial fractions. It has the form

$$\frac{B_1 x + C_1}{ax^2 + bx + c} + \frac{B_2 x + C_2}{(ax^2 + bx + c)^2} + \cdots + \frac{B_n x + C_n}{(ax^2 + bx + c)^n}, \qquad \textbf{(8)}$$

where $B_1, B_2, \ldots, B_n, C_1, C_2, \ldots,$ and C_n are constants.

If $Q(x)$ has both linear and irreducible quadratic factors, then the partial fraction decomposition of $R(x)$ is simply the sum of the expressions of the form in (6) that correspond to the linear factors plus the sum of the expressions of the form in (8) that correspond to the quadratic factors. In the case of an irreducible quadratic factor of multiplicity $n = 1$, the expression in (8) reduces to its first term alone.

The most important case is that of a nonrepeated quadratic factor of the sum of squares form $x^2 + k^2$ (where k is a positive constant). The corresponding partial fraction $(Bx + C)/(x^2 + k^2)$ is readily integrated by using the familiar integrals

$$\int \frac{x}{x^2 + k^2}\,dx = \frac{1}{2}\ln(x^2 + k^2) + C,$$

$$\int \frac{1}{x^2 + k^2}\,dx = \frac{1}{k}\arctan\frac{x}{k} + C.$$

We will discuss in Section 9.7 the integration of more general partial fractions involving irreducible quadratic factors.

EXAMPLE 6 Find $\displaystyle\int \frac{5x^3 - 3x^2 + 2x - 1}{x^4 + x^2}\,dx.$

Solution The denominator $x^4 + x^2 = x^2(x^2 + 1)$ is the product of an irreducible quadratic factor and a repeated linear factor. The partial fraction decomposition of the integrand takes the form

$$\frac{5x^3 - 3x^2 + 2x - 1}{x^4 + x^2} = \frac{A}{x} + \frac{B}{x^2} + \frac{Cx + D}{x^2 + 1}.$$

We multiply both sides by $x^4 + x^2$ and obtain

$$\begin{aligned} 5x^3 - 3x^2 + 2x - 1 &= Ax(x^2 + 1) + B(x^2 + 1) + (Cx + D)x^2 \\ &= (A + C)x^3 + (B + D)x^2 + Ax + B. \end{aligned}$$

As before, we equate coefficients of like powers of x. This yields the four simultaneous equations

$$\begin{aligned} A \quad + C \quad\quad &= 5, \\ B \quad\quad + D &= -3, \\ A \quad\quad\quad\quad &= 2, \\ B \quad\quad\quad &= -1. \end{aligned}$$

These equations are easily solved for $A = 2$, $B = -1$, $C = 3$, and $D = -2$. Thus

$$\begin{aligned} \int \frac{5x^3 - 3x^2 + 2x - 1}{x^4 + x^2}\,dx &= \int\left(\frac{2}{x} - \frac{1}{x^2} + \frac{3x - 2}{x^2 + 1}\right)dx \\ &= 2\ln|x| + \frac{1}{x} + \frac{3}{2}\int\frac{2x\,dx}{x^2 + 1} - 2\int\frac{dx}{x^2 + 1} \\ &= 2\ln|x| + \frac{1}{x} + \frac{3}{2}\ln(x^2 + 1) - 2\tan^{-1}x + C. \quad ■ \end{aligned}$$

REMARK Numerical solution for the coefficients in a partial fraction decomposition frequently is more tedious than in Examples 3 through 6. But computer algebra systems can do this work automatically. For instance, if we write

```
f := (5*x^3 - 3*x^2 + 2*x -1)/(x^4 + x^2)
```

then either the *Mathematica* command `Apart[ f ]` or the *Maple* command

```
convert( f, parfrac, x )
```

quickly produces the partial fractions decomposition found in Example 6.

*Applications to Differential Equations

Example 7 illustrates the use of partial fractions to solve certain types of separable differential equations.

EXAMPLE 7 Suppose that at time $t = 0$, half of a population of 100,000 people have heard a certain rumor and that the number $P(t)$ of those who have heard it is then increasing at the rate of 1000 people per day. If $P(t)$ satisfies the differential equation

$$\frac{dP}{dt} = kP(100 - P) \tag{9}$$

(with P measured in thousands of people and t in days), determine how many people will have heard the rumor after $t = 30$ days.

Solution The differential equation in (9) is the model for a simple but widely used assumption: The rate at which the rumor spreads is proportional to the number of contacts between those who already know the rumor and those who have not yet heard it. That is, dP/dt is proportional both to P and to $100 - P$; thus dP/dt is proportional to the product of P and $100 - P$.

To find the constant k of proportionality, we substitute the given values $P(0) = 50$ and $P'(0) = 1$ in Eq. (9). This yields the equation

$$1 = k \cdot 50 \cdot (100 - 50).$$

It follows that $k = 0.0004$, so the differential equation in (9) takes the form

$$\frac{dP}{dt} = (0.0004) \cdot P \cdot (100 - P). \tag{10}$$

Separation of the variables leads to

$$\int \frac{dP}{P(100 - P)} = \int 0.0004\, dt.$$

Then the partial fraction decomposition

$$\frac{100}{P(100 - P)} = \frac{1}{P} + \frac{1}{100 - P}$$

yields

$$\begin{aligned} \int \left(\frac{1}{P} + \frac{1}{100 - P} \right) dP &= \int 0.04\, dt; \\ \ln P - \ln(100 - P) &= (0.04)t + C. \end{aligned} \tag{11}$$

Substitution of the initial data $P = 50$ when $t = 0$ now gives $C = 0$, so Eq. (11) takes the form

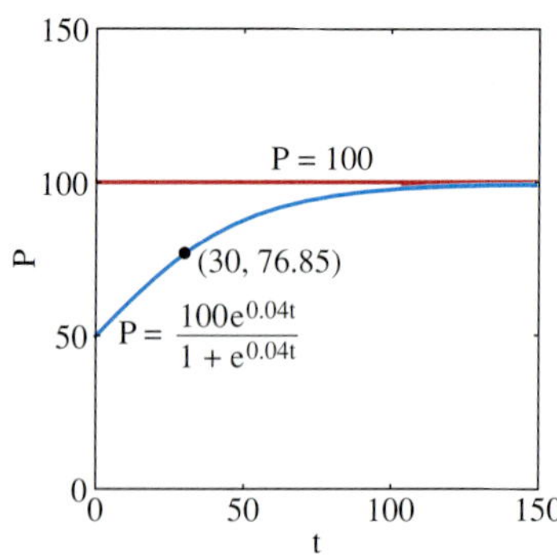

Fig. 9.5.5 Graph of the population function in Eq. (12). It is apparent that $P(t) \to 100$ as $t \to +\infty$. Thus everyone eventually hears the rumor. Can you verify this directly by taking the limit in Eq. (12) as $t \to +\infty$?

$$\ln \frac{P}{100 - P} = (0.04)t, \quad \text{so} \quad \frac{P}{100 - P} = e^{(0.04)t}.$$

We readily solve this last equation for the solution

$$P(t) = \frac{100e^{(0.04)t}}{1 + e^{(0.04)t}}. \tag{12}$$

Hence the number of people who have heard the rumor after 30 days is $P(30) \approx 76.85$ thousand people. See Fig. 9.5.5.

The method of Example 7 can be used to solve any differential equation of the form

$$\frac{dx}{dt} = k(x - a)(x - b), \tag{13}$$

where a, b, and k are constants. As Problems 63 through 68 indicate, this differential equation serves as a mathematical model for a wide variety of natural phenomena.

9.5 PROBLEMS

Find the integrals in Problems 1 through 36.

1. $\displaystyle\int \frac{x^2}{x + 1}\,dx$

2. $\displaystyle\int \frac{x^3}{2x - 1}\,dx$

3. $\displaystyle\int \frac{1}{x^2 - 3x}\,dx$

4. $\displaystyle\int \frac{x}{x^2 + 4x}\,dx$

5. $\displaystyle\int \frac{1}{x^2 + x - 6}\,dx$

6. $\displaystyle\int \frac{x^3}{x^2 + x - 6}\,dx$

7. $\displaystyle\int \frac{1}{x^3 + 4x}\,dx$

8. $\displaystyle\int \frac{1}{(x + 1)(x^2 + 1)}\,dx$

9. $\displaystyle\int \frac{x^4}{x^2 + 4}\,dx$

10. $\displaystyle\int \frac{1}{(x^2 + 1)(x^2 + 4)}\,dx$

11. $\displaystyle\int \frac{x - 1}{x + 1}\,dx$

12. $\displaystyle\int \frac{2x^3 - 1}{x^2 + 1}\,dx$

13. $\displaystyle\int \frac{x^2 + 2x}{(x + 1)^2}\,dx$

14. $\displaystyle\int \frac{2x - 4}{x^2 - x}\,dx$

15. $\displaystyle\int \frac{1}{x^2 - 4}\,dx$

16. $\displaystyle\int \frac{x^4}{x^2 + 4x + 4}\,dx$

17. $\displaystyle\int \frac{x + 10}{2x^2 + 5x - 3}\,dx$

18. $\displaystyle\int \frac{x + 1}{x^3 - x^2}\,dx$

19. $\displaystyle\int \frac{x^2 + 1}{x^3 + 2x^2 + x}\,dx$

20. $\displaystyle\int \frac{x^2 + x}{x^3 - x^2 - 2x}\,dx$

21. $\displaystyle\int \frac{4x^3 - 7x}{x^4 - 5x^2 + 4}\,dx$

22. $\displaystyle\int \frac{2x^2 + 3}{x^4 - 2x^2 + 1}\,dx$

23. $\displaystyle\int \frac{x^2}{(x + 2)^3}\,dx$

24. $\displaystyle\int \frac{x^2 + x}{(x^2 - 4)(x + 4)}\,dx$

25. $\displaystyle\int \frac{1}{x^3 + x}\,dx$

26. $\displaystyle\int \frac{6x^3 - 18x}{(x^2 - 1)(x^2 - 4)}\,dx$

27. $\displaystyle\int \frac{x + 4}{x^3 + 4x}\,dx$

28. $\displaystyle\int \frac{4x^4 + x + 1}{x^5 + x^4}\,dx$

29. $\displaystyle\int \frac{x}{(x + 1)(x^2 + 1)}\,dx$

30. $\displaystyle\int \frac{x^2 + 2}{(x^2 + 1)^2}\,dx$

31. $\displaystyle\int \frac{x^2 - 10}{2x^4 + 9x^2 + 4}\,dx$

32. $\displaystyle\int \frac{x^2}{x^4 - 1}\,dx$

33. $\displaystyle\int \frac{x^3 + x^2 + 2x + 3}{x^4 + 5x^2 + 6}\,dx$

34. $\displaystyle\int \frac{x^2 + 4}{(x^2 + 1)^2(x^2 + 2)}\,dx$

35. $\displaystyle\int \frac{x^4 + 3x^2 - 4x + 5}{(x - 1)^2(x^2 + 1)}\,dx$

36. $\displaystyle\int \frac{2x^3 + 5x^2 - x + 3}{(x^2 + x - 2)^2}\,dx$

In Problems 37 through 40, make a preliminary substitution before using the method of partial fractions.

37. $\displaystyle\int \frac{e^{4t}}{(e^{2t} - 1)^3}\,dt$

38. $\displaystyle\int \frac{\cos\theta}{\sin^2\theta - \sin\theta - 6}\,d\theta$

39. $\displaystyle\int \frac{1 + \ln t}{t(3 + 2\ln t)^2}\,dt$

40. $\displaystyle\int \frac{\sec^2 t}{\tan^3 t + \tan^2 t}\,dt$

In Problems 41 through 44, find the area of the region R between the curve and the x-axis over the given interval.

41. $y = \dfrac{x - 9}{x^2 - 3x}$, $\quad 1 \leqq x \leqq 2$

42. $y = \dfrac{x + 5}{3 + 2x - x^2}$, $\quad 0 \leqq x \leqq 2$

43. $y = \dfrac{3x - 15 - 2x^2}{x^3 - 9x}$, $1 \leqq x \leqq 2$

44. $y = \dfrac{x^2 + 10x + 16}{x^3 + 8x^2 + 16x}$, $2 \leqq x \leqq 5$

In Problems 45 through 48, find the volume of the solid obtained by revolving the region R around the y-axis.

45. The region R of Problem 41

46. The region R of Problem 42

47. The region R of Problem 43

48. The region R of Problem 44

In Problems 49 and 50, find the volume of the solid obtained by revolving the region R around the x-axis.

49. The region R of Problem 41

50. The region R of Problem 42

51. The plane region R shown in Fig. 9.5.6 is bounded by the curve

$$y^2 = \frac{1 - x}{1 + x}x^2, \qquad 0 \leqq x \leqq 1.$$

Find the volume generated by revolving R around the x-axis.

52. Figure 9.5.7 shows the region bounded by the curve

$$y^2 = \frac{(1 - x)^2}{(1 + x)^2}x^4, \qquad 0 \leqq x \leqq 1.$$

Find the volume generated by revolving this region around: (a) the x-axis; (b) the y-axis.

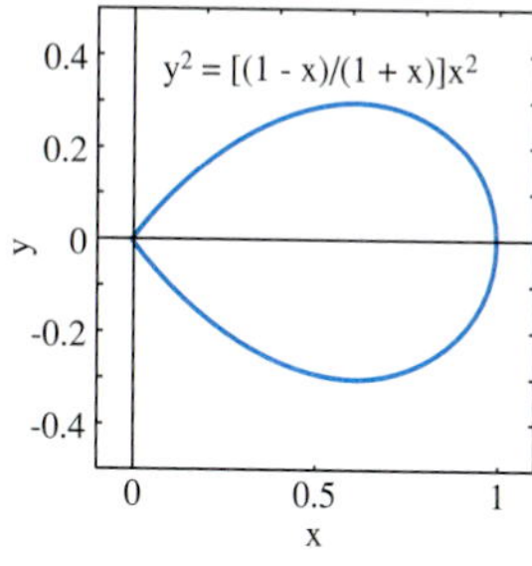

Fig. 9.5.6 The region of Problem 51

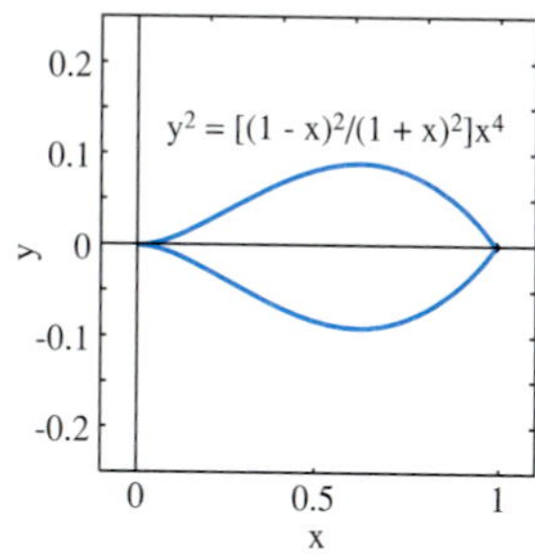

Fig. 9.5.7 The region of Problem 52

In Problems 53 through 56, write the general form of a partial fraction decomposition of the given rational function $f(x)$ (with coefficients A, B, C, . . . , remaining to be determined). Then use a computer algebra system (as in the remark following Example 6) to find the numerical values of the coefficients in the decomposition. Finally, find the indefinite integral $\int f(x)\,dx$ both by hand and using the computer algebra system, and resolve any apparent discrepancy between the two results.

53. $f(x) = \dfrac{98(x^3 - 50x + 100)}{x^2(x^2 - 12x + 35)}$

54. $f(x) = \dfrac{16(2x^3 + 77x - 99)}{(x^2 + 10x + 21)^2}$

55. $f(x) = \dfrac{324(x^3 + 8)}{(x^2 - x - 6)(x^2 + x - 20)^2}$

56. $f(x) = \dfrac{500(4x^4 - 23x^2 + 16)}{(x^2 - 4)^2(x - 3)^2}$

Solve the initial value problems in 57 through 62.

57. $\dfrac{dx}{dt} = x - x^2$, $x(0) = 2$

58. $\dfrac{dx}{dt} = 10x - x^2$, $x(0) = 1$

59. $\dfrac{dx}{dt} = 1 - x^2$, $x(0) = 3$

60. $\dfrac{dx}{dt} = 9 - 4x^2$, $x(0) = 0$

61. $\dfrac{dx}{dt} = x^2 + 5x + 6$, $x(0) = 5$

62. $\dfrac{dx}{dt} = 2x^2 + x - 15$, $x(0) = 10$

63. Suppose that the population $P(t)$ (in millions) of Ruritania satisfies the differential equation

$$\frac{dP}{dt} = k \cdot P \cdot (200 - P) \qquad (k \text{ constant}).$$

Its population in 1940 was 100 million and was then growing at the rate of 1 million per year. Predict this country's population for the year 2000.

64. Suppose that a community contains 15000 people who are susceptible to Michaud's syndrome, a contagious disease. At time $t = 0$ the number $N(t)$ of people who have caught Michaud's syndrome is 5000 and is increasing at the rate of 500 per day. Assume that $N'(t)$ is proportional to the product of the numbers of those who have caught the disease and those who have not. How long will it take for another 5000 people to contract Michaud's syndrome?

65. As the salt KNO_3 dissolves in methanol, the number $x(t)$ of grams of the salt in solution after t seconds satisfies the differential equation

$$\frac{dx}{dt} = (0.8)x - (0.004)x^2.$$

(a) If $x = 50$ when $t = 0$, how long will it take an additional 50 g of the salt to dissolve? (b) What is the maximum amount of the salt that will ever dissolve in the methanol?

66. A population $P(t)$ (t in months) of squirrels satisfies the differential equation

$$\frac{dP}{dt} = (0.001)P^2 - kP \qquad (k \text{ constant}).$$

If $P(0) = 100$ and $P'(0) = 8$, how long will it take for this population to double to 200 squirrels?

67. Consider an animal population $P(t)$ (t in years) that satisfies the differential equation

$$\frac{dP}{dt} = kP^2 - (0.01)P \qquad (k \text{ constant}).$$

Suppose also that $P(0) = 200$ and that $P'(0) = 2$. (a) When will $P = 1000$? (b) When will doomsday occur for this population?

68. Suppose that the number $x(t)$ (t in months) of alligators in a swamp satisfies the differential equation

$$\frac{dx}{dt} = (0.0001)x^2 - (0.01)x.$$

(a) If initially there are 25 alligators, solve this equation to determine what happens to this alligator population in the long run. (b) Repeat part (a), but use 150 alligators initially.

9.5 PROJECT: BOUNDED POPULATION GROWTH

The differential equation in (9) of Example 7 is a **logistic equation,** an equation of the form

$$\frac{dP}{dt} = kP(M - P) \qquad (k, M \text{ constants}). \tag{1}$$

The logistic equation models many animal (including human) populations more accurately than does the natural growth equation $dP/dt = kP$ that we studied in Section 7.5. For instance, think of an environment that can support a population of at most M individuals. We might then think of $M - P$ as the potential for further expansion when the population is P. The hypothesis that the rate of change dP/dt is therefore proportional to $M - P$ as well as to P itself then yields Eq. (1) with the proportionality constant k. The classic example of such a limited-environment situation is a fruit fly population in a closed container.

The object of this project is to investigate the behavior of populations that can be modeled by the logistic equation.

1. First separate the variables in Eq. (1) and then use partial fractions as in Example 7 to derive the solution

$$P(t) = \frac{MP_0}{P_0 + (M - P_0)e^{-kMt}} \tag{2}$$

that satisfies the initial condition $P(0) = P_0$. If k and M are positive constants, then

$$\lim_{t\to\infty} P(t) = M. \tag{3}$$

Hence the parameter M in (1) is the *limiting population.*

2. During the period from 1790 to 1930, the U.S. population $P(t)$ (t in years) grew from 3.9 million to 123.2 million. Throughout this period, $P(t)$ remained close to the solution of the initial value problem

$$\frac{dP}{dt} = (0.03135)P - (0.0001589)P^2, \qquad P(0) = 3.9.$$

Has this differential equation continued since 1930 to predict accurately the U.S. population? If so, what is the limiting population of the United States? (This problem is based on a computation by the Belgian demographer Verhulst, who in 1845 used the 1790–1840 U.S. population data to predict accurately the U.S. population through the year 1930.)

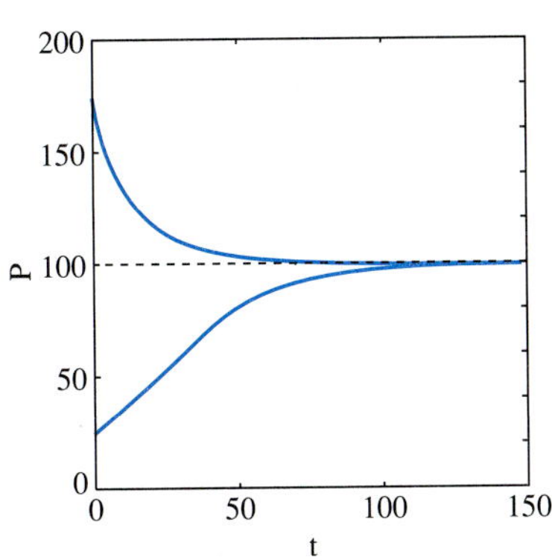

Fig. 9.5.8 Typical solutions of the logistic equation

3. For your very own logistic equation, choose nonzero integers r and s (for example, the last two nonzero digits of your student I.D. number), and then take

$$M = 10r, \qquad k = \frac{s}{100M}$$

in Eq. (1). Then plot the corresponding solution in Eq. (2) with several different values of the initial population P_0. What determines whether the graph $P = P(t)$ looks like the upper curve or the lower curve in Fig. 9.5.8?

4. Now, with the same fixed value of the limiting population M as in step 3, plot solution curves with both larger and smaller values of k. What appears to be the relationship between the size of k and the rate at which the solution curve approaches its horizontal asymptote $P = M$?

9.6 TRIGONOMETRIC SUBSTITUTION

The method of *trigonometric substitution* can be very effective in dealing with integrals when the integrands contain algebraic expressions such as $(a^2 - u^2)^{1/2}$, $(u^2 - a^2)^{3/2}$, and $1/(a^2 + u^2)^2$. There are three basic trigonometric substitutions:

If the Integral Involves	Then Substitute	And Use the Identity
$a^2 - u^2$	$u = a \sin\theta$	$1 - \sin^2\theta = \cos^2\theta$
$a^2 + u^2$	$u = a \tan\theta$	$1 + \tan^2\theta = \sec^2\theta$
$u^2 - a^2$	$u = a \sec\theta$	$\sec^2\theta - 1 = \tan^2\theta$

What we mean by the substitution $u = a\sin\theta$ is, more precisely, the *inverse* trigonometric substitution

$$\theta = \sin^{-1}\frac{u}{a}, \qquad -\frac{\pi}{2} \leqq \theta \leqq \frac{\pi}{2},$$

where $|u| \leqq a$. Suppose, for example, that an integral contains the expression $(a^2 - u^2)^{1/2}$. Then this substitution yields

$$(a^2 - u^2)^{1/2} = (a^2 - a^2\sin^2\theta)^{1/2} = (a^2\cos^2\theta)^{1/2} = a\cos\theta.$$

We chose the nonnegative square root in the last step because $\cos\theta \geqq 0$ for $\pi/2 \leqq \theta \leqq \pi/2$. Thus the troublesome factor $(a^2 - u^2)^{1/2}$ becomes $a\cos\theta$ and, meanwhile, $du = a\cos\theta\, d\theta$. If the trigonometric integral that results from the substitution can be evaluated by earlier methods of this chapter, the result will ordinarily involve $\theta = \sin^{-1}(u/a)$ and trigonometric functions of θ. The final step will be to express the answer in terms of the original variable. For this purpose the values of the various trigonometric functions can be read from the right triangle in Fig. 9.6.1, which contains an angle θ such that $\sin\theta = u/a$ (if u is negative, then θ is negative).

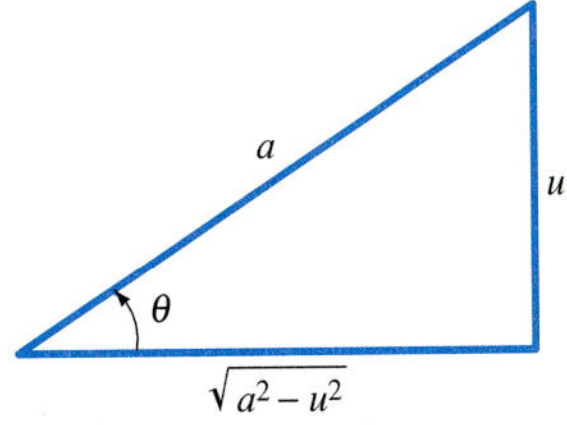

Fig. 9.6.1 The reference triangle for the substitution $u = a\sin\theta$

EXAMPLE 1 Evaluate $\displaystyle\int \frac{x^3}{\sqrt{1 - x^2}}\, dx$, where $|x| < 1$.

Solution Here $a = 1$ and $u = x$, so we substitute

$$x = \sin\theta, \qquad dx = \cos\theta\, d\theta.$$

This gives

$$\int \frac{x^3}{\sqrt{1-x^2}}\,dx = \int \frac{\sin^3\theta\cos\theta}{\sqrt{1-\sin^2\theta}}\,d\theta$$
$$= \int \sin^3\theta\,d\theta = \int (\sin\theta)(1-\cos^2\theta)\,d\theta$$
$$= \tfrac{1}{3}\cos^3\theta - \cos\theta + C.$$

Figure 9.6.2, in which $\sin\theta = x$, reminds us that

$$\cos\theta = (1-\sin^2\theta)^{1/2} = \sqrt{1-x^2},$$

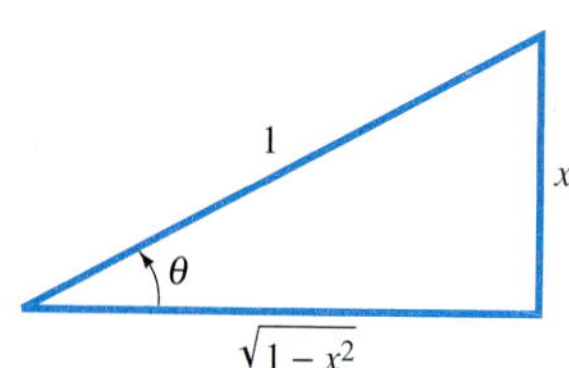

Fig. 9.6.2 Reference triangle for the substitution $x = \sin\theta$

so our final result in terms of x is

$$\int \frac{x^3}{\sqrt{1-x^2}}\,dx = \tfrac{1}{3}(1-x^2)^{3/2} - \sqrt{1-x^2} + C.$$ ■

Example 2 illustrates the use of trigonometric substitution to find integrals like those in formulas (44) through (62) on the inside back cover.

EXAMPLE 2 Find $\int \sqrt{a^2-u^2}\,du$, where $|u| \leqq a$.

Solution The substitution $u = a\sin\theta$, $du = a\cos\theta\,d\theta$ gives

$$\int \sqrt{a^2-u^2}\,du = \int \sqrt{a^2-a^2\sin^2\theta}\,(a\cos\theta)\,d\theta$$
$$= \int a^2\cos^2\theta\,d\theta = \tfrac{1}{2}a^2\int (1+\cos 2\theta)\,d\theta$$
$$= \tfrac{1}{2}a^2(\theta + \tfrac{1}{2}\sin 2\theta) + C = \tfrac{1}{2}a^2(\theta + \sin\theta\cos\theta) + C.$$

(We used the identity $\sin 2\theta = 2\sin\theta\cos\theta$ in the last step.) Now from Fig. 9.6.1 we see that

$$\sin\theta = \frac{u}{a} \quad \text{and} \quad \cos\theta = \frac{\sqrt{a^2-u^2}}{a}.$$

Hence

$$\int \sqrt{a^2-u^2}\,du = \frac{1}{2}a^2\left(\sin^{-1}\frac{u}{a} + \frac{u}{a}\cdot\frac{\sqrt{a^2-u^2}}{a}\right) + C$$
$$= \frac{u}{2}\sqrt{a^2-u^2} + \frac{a^2}{2}\sin^{-1}\frac{u}{a} + C.$$

Thus we have obtained formula (54) (from the inside back cover). ■

What we mean by the substitution $u = a\tan\theta$ in an integral that contains $a^2 + u^2$ is the substitution

$$\theta = \tan^{-1}\frac{u}{a}, \qquad -\frac{\pi}{2} < \theta < \frac{\pi}{2}.$$

In this case

$$\sqrt{a^2 + u^2} = \sqrt{a^2 + a^2 \tan^2 \theta} = \sqrt{a^2 \sec^2 \theta} = a \sec \theta,$$

under the assumption that $a > 0$. We take the positive square root in the last step here because $\sec \theta > 0$ for $-\pi/2 < \theta < \pi/2$. The values of the various trigonometric functions of θ under this substitution can be read from the right triangle of Fig. 9.6.3, which shows a (positive or negative) acute angle θ such that $\tan \theta = u/a$.

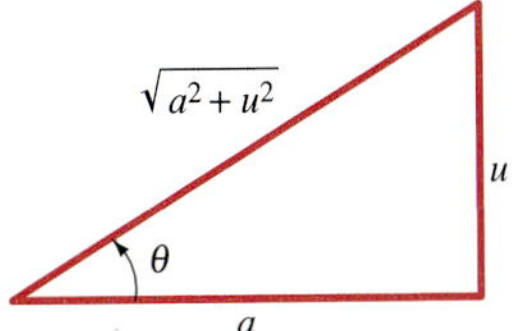

Fig. 9.6.3 The reference triangle for the substitution $u = a \tan \theta$

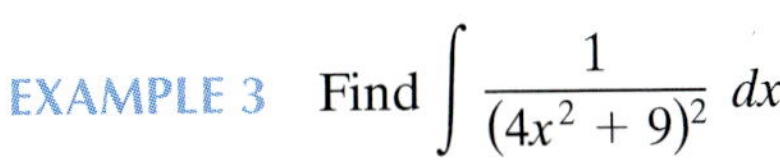

EXAMPLE 3 Find $\displaystyle\int \frac{1}{(4x^2 + 9)^2}\,dx$.

Solution The factor $4x^2 + 9$ corresponds to $u^2 + a^2$ with $u = 2x$ and $a = 3$. Hence the substitution $u = a \tan \theta$ amounts to

$$2x = 3 \tan \theta, \qquad x = \tfrac{3}{2} \tan \theta, \qquad dx = \tfrac{3}{2} \sec^2 \theta \, d\theta.$$

This gives

$$\begin{aligned}
\int \frac{1}{(4x^2 + 9)^2}\,dx &= \int \frac{\frac{3}{2}\sec^2 \theta}{(9 \tan^2 \theta + 9)^2}\,d\theta \\
&= \frac{3}{2} \int \frac{\sec^2 \theta}{(9 \sec^2 \theta)^2}\,d\theta = \frac{1}{54} \int \frac{1}{\sec^2 \theta}\,d\theta \\
&= \frac{1}{54} \int \cos^2 \theta \, d\theta = \frac{1}{108}(\theta + \sin \theta \cos \theta) + C.
\end{aligned}$$

(The integration in the last step is the same as in Example 2.) Now $\theta = \tan^{-1}(2x/3)$, and the triangle of Fig. 9.6.4 gives

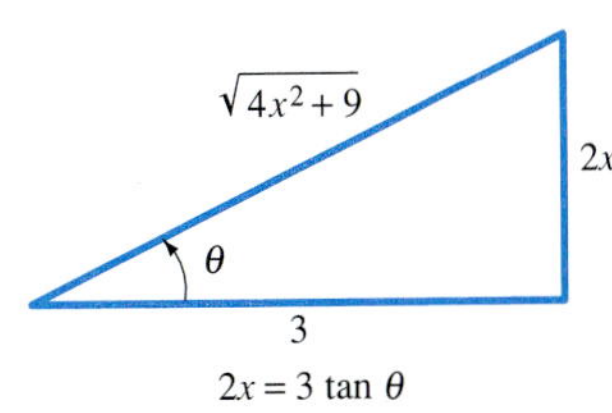

Fig. 9.6.4 The reference triangle for Example 3

$$\sin \theta = \frac{2x}{\sqrt{4x^2 + 9}}, \qquad \cos \theta = \frac{3}{\sqrt{4x^2 + 9}}.$$

Hence

$$\begin{aligned}
\int \frac{1}{(4x^2 + 9)^2}\,dx &= \frac{1}{108}\left[\tan^{-1}\left(\frac{2x}{3}\right) + \frac{2x}{\sqrt{4x^2 + 9}} \cdot \frac{3}{\sqrt{4x^2 + 9}}\right] + C \\
&= \frac{1}{108} \tan^{-1}\left(\frac{2x}{3}\right) + \frac{x}{18(4x^2 + 9)} + C.
\end{aligned}$$

■

What we mean by the substitution $u = a \sec \theta$ in an integral that contains $u^2 - a^2$ is the substitution

$$\theta = \sec^{-1} \frac{u}{a}, \qquad 0 \leqq \theta \leqq \pi,$$

where $|u| \geqq a > 0$ (because of the domain and range of the inverse secant function). Then

$$\sqrt{u^2 - a^2} = \sqrt{a^2 \sec^2 \theta - a^2} = \sqrt{a^2 \tan^2 \theta} = \pm\, a \tan \theta.$$

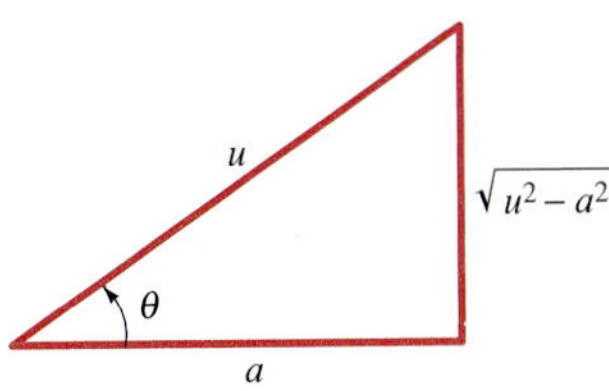

Fig. 9.6.5 The reference triangle for the substitution $u = a \sec \theta$

Here we must take the plus sign if $u > a$, so that $0 < \theta < \pi/2$ and $\tan \theta > 0$. If $u < -a$, so that $\pi/2 < \theta < \pi$ and $\tan \theta < 0$, we take the minus sign. In either case the values of the various trigonometric functions of θ can be read from the right triangle in Fig. 9.6.5.

EXAMPLE 4 Find $\displaystyle\int \frac{\sqrt{x^2 - 25}}{x}\,dx$, where $x > 5$.

Solution We substitute $x = 5\sec\theta$, $dx = 5\sec\theta\tan\theta\,d\theta$. Then

$$\sqrt{x^2 - 25} = \sqrt{25(\sec^2\theta - 1)} = 5\tan\theta,$$

because $x > 5$ implies that $0 < \theta < \pi/2$, so $\tan\theta > 0$. Hence this substitution gives

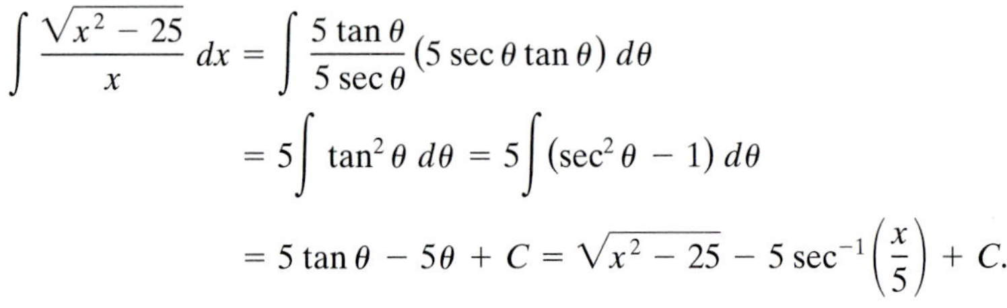

$$\begin{aligned}\int \frac{\sqrt{x^2 - 25}}{x}\,dx &= \int \frac{5\tan\theta}{5\sec\theta}(5\sec\theta\tan\theta)\,d\theta\\ &= 5\int \tan^2\theta\,d\theta = 5\int(\sec^2\theta - 1)\,d\theta\\ &= 5\tan\theta - 5\theta + C = \sqrt{x^2 - 25} - 5\sec^{-1}\left(\frac{x}{5}\right) + C.\end{aligned}$$

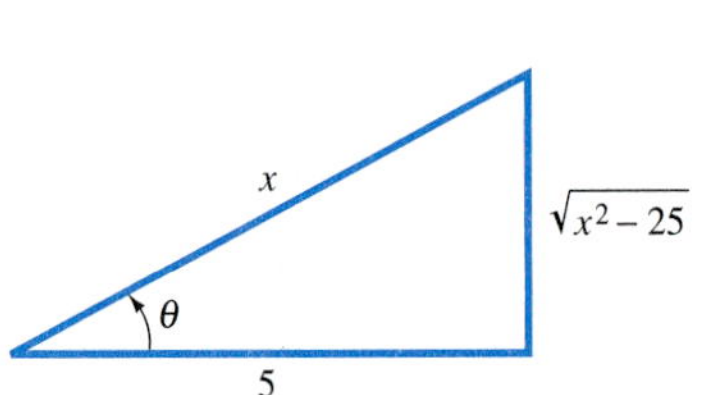

Fig. 9.6.6 The reference triangle for the substitution $x = 5\sec\theta$

The substitutions in the last step may be read from the reference triangle in Fig. 9.6.6. ■

Hyperbolic substitutions may be used in a similar way—and with the same effect—as trigonometric substitutions. The three basic hyperbolic substitutions, which are not ordinarily memorized, are listed here for reference.

If the Integral Involves	Then Substitute	And Use the Identity
$a^2 - u^2$	$u = a\tanh\theta$	$1 - \tanh^2\theta = \operatorname{sech}^2\theta$
$a^2 + u^2$	$u = a\sinh\theta$	$1 + \sinh^2\theta = \cosh^2\theta$
$u^2 - a^2$	$u = a\cosh\theta$	$\cosh^2\theta - 1 = \sinh^2\theta$

EXAMPLE 5 Find $\displaystyle\int \frac{1}{\sqrt{x^2 - 1}}\,dx$, where $x > 1$.

Solution For purposes of comparison, we evaluate this integral both by trigonometric substitution and by hyperbolic substitution. The trigonometric substitution

$$x = \sec\theta,\qquad dx = \sec\theta\tan\theta\,d\theta,\qquad \tan\theta = \sqrt{x^2 - 1}$$

gives

$$\begin{aligned}\int \frac{1}{\sqrt{x^2 - 1}}\,dx &= \int \frac{\sec\theta\tan\theta}{\tan\theta}\,d\theta = \int \sec\theta\,d\theta\\ &= \ln|\sec\theta + \tan\theta| + C \qquad \text{[Eq. (7), Section 9.4]}\\ &= \ln\left|x + \sqrt{x^2 - 1}\right| + C.\end{aligned}$$

Using instead the hyperbolic substitution $x = \cosh\theta$, $dx = \sinh\theta\,d\theta$, we have

$$\sqrt{x^2 - 1} = \sqrt{\cosh^2\theta - 1} = \sinh\theta.$$

We take the positive square root here, because $x > 1$ implies that $\theta = \cosh^{-1}x > 0$ and thus that $\sinh\theta > 0$. Hence

$$\int \frac{1}{\sqrt{x^2 - 1}}\, dx = \int \frac{\sinh\theta}{\sinh\theta}\, d\theta = \int 1\, d\theta = \theta + C = \cosh^{-1}x + C.$$

The two results appear to differ, but Eq. (35) in Section 8.5 shows that they are equivalent.

9.6 PROBLEMS

Use trigonometric substitutions to evaluate the integrals in Problems 1 through 36.

1. $\int \frac{1}{\sqrt{16 - x^2}}\, dx$

2. $\int \frac{1}{\sqrt{4 - 9x^2}}\, dx$

3. $\int \frac{1}{x^2\sqrt{4 - x^2}}\, dx$

4. $\int \frac{1}{x^2\sqrt{x^2 - 25}}\, dx$

5. $\int \frac{x^2}{\sqrt{16 - x^2}}\, dx$

6. $\int \frac{x^2}{\sqrt{9 - 4x^2}}\, dx$

7. $\int \frac{1}{(9 - 16x^2)^{3/2}}\, dx$

8. $\int \frac{1}{(25 + 16x^2)^{3/2}}\, dx$

9. $\int \frac{\sqrt{x^2 - 1}}{x^2}\, dx$

10. $\int x^3\sqrt{4 - x^2}\, dx$

11. $\int x^3\sqrt{9 + 4x^2}\, dx$

12. $\int \frac{x^3}{\sqrt{x^2 + 25}}\, dx$

13. $\int \frac{\sqrt{1 - 4x^2}}{x}\, dx$

14. $\int \frac{1}{\sqrt{1 + x^2}}\, dx$

15. $\int \frac{1}{\sqrt{9 + 4x^2}}\, dx$

16. $\int \sqrt{1 + 4x^2}\, dx$

17. $\int \frac{x^2}{\sqrt{25 - x^2}}\, dx$

18. $\int \frac{x^3}{\sqrt{25 - x^2}}\, dx$

19. $\int \frac{x^2}{\sqrt{1 + x^2}}\, dx$

20. $\int \frac{x^3}{\sqrt{1 + x^2}}\, dx$

21. $\int \frac{x^2}{\sqrt{4 + 9x^2}}\, dx$

22. $\int (1 - x^2)^{3/2}\, dx$

23. $\int \frac{1}{(1 + x^2)^{3/2}}\, dx$

24. $\int \frac{1}{(4 - x^2)^2}\, dx$

25. $\int \frac{1}{(4 - x^2)^3}\, dx$

26. $\int \frac{1}{(4x^2 + 9)^3}\, dx$

27. $\int \sqrt{9 + 16x^2}\, dx$

28. $\int (9 + 16x^2)^{3/2}\, dx$

29. $\int \frac{\sqrt{x^2 - 25}}{x}\, dx$

30. $\int \frac{\sqrt{9x^2 - 16}}{x}\, dx$

31. $\int x^2\sqrt{x^2 - 1}\, dx$

32. $\int \frac{x^2}{\sqrt{4x^2 - 9}}\, dx$

33. $\int \frac{1}{(4x^2 - 1)^{3/2}}\, dx$

34. $\int \frac{1}{x^2\sqrt{4x^2 - 9}}\, dx$

35. $\int \frac{\sqrt{x^2 - 5}}{x^2}\, dx$

36. $\int (4x^2 - 5)^{3/2}\, dx$

Use hyperbolic substitutions to evaluate the integrals in Problems 37 through 41.

37. $\int \frac{1}{\sqrt{25 + x^2}}\, dx$

38. $\int \sqrt{1 + x^2}\, dx$

39. $\int \frac{\sqrt{x^2 - 4}}{x^2}\, dx$

40. $\int \frac{1}{\sqrt{1 + 9x^2}}\, dx$

41. $\int x^2\sqrt{1 + x^2}\, dx$

42. Use the result of Example 2 to show that the area bounded by the ellipse

$$\frac{x^2}{a^2} + \frac{y^2}{b^2} = 1$$

of Fig. 9.6.7 is given by $A = \pi ab$. (The special case $b = a$ is the familiar circular area formula $A = \pi a^2$.)

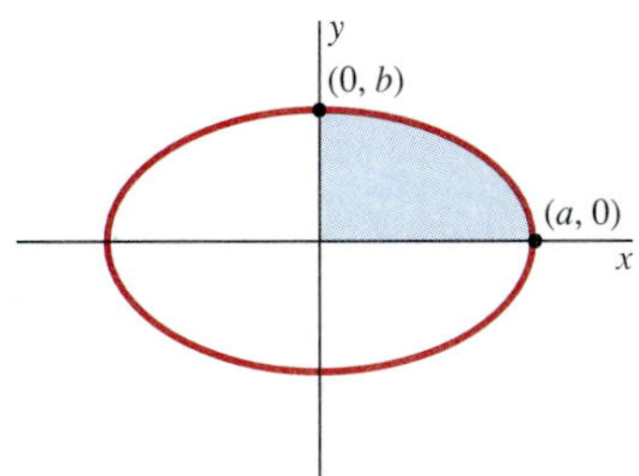

Fig. 9.6.7 The ellipse of Problem 42

43. Derive the formula $A = \frac{1}{2}a^2\theta$ for the area of a circular sector with radius a and central angle θ by calculating and adding the areas of the right triangle OAC and the region ABC of Fig. 9.6.8.

44. Compute the arc length of the parabola $y = x^2$ over the interval $[0, 1]$.

45. Compute the area of the surface obtained by revolving around the x-axis the parabolic arc of Problem 44.

46. Show that the length of one arch of the sine curve $y = \sin x$ is equal to half the circumference of the ellipse $x^2 + \frac{1}{2}y^2 = 1$. (*Suggestion:* Substitute $x = \cos\theta$ into the arc-length integral for the ellipse.) See Fig. 9.6.9.

47. Compute the arc length of the curve $y = \ln x$ over the interval $[1, 2]$.

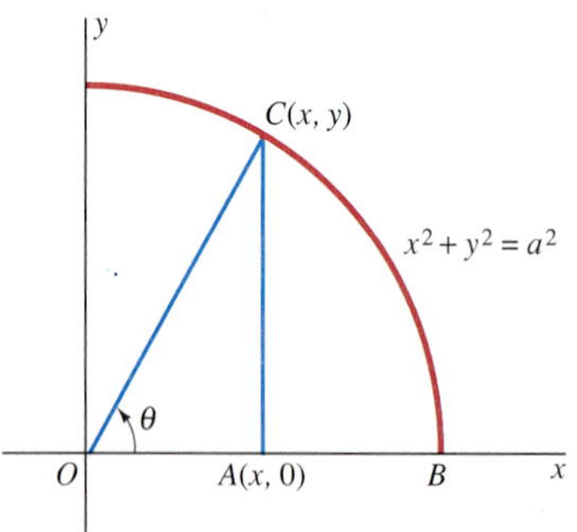

Fig. 9.6.8 The circular sector of Problem 43

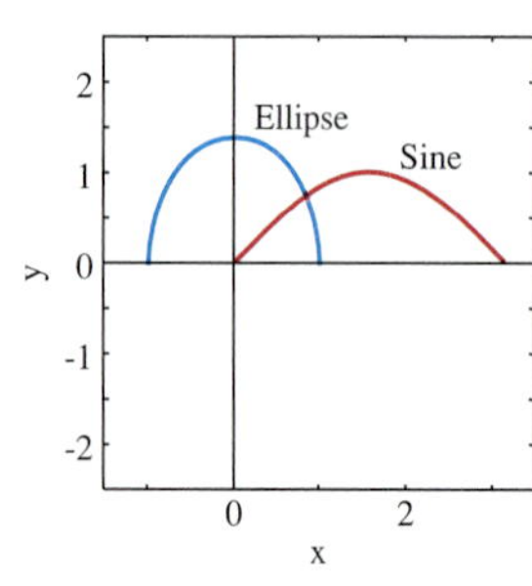

Fig. 9.6.9 Two arcs with the same length (Problem 46)

48. Compute the area of the surface obtained by revolving around the y-axis the curve of Problem 47.

49. A torus (see Fig. 9.6.10) is obtained by revolving around the y-axis the circle

$$(x - b)^2 + y^2 = a^2 \qquad (0 < a \leqq b).$$

Show that the surface area of the torus is $4\pi^2 ab$.

50. Find the area under the curve $y = \sqrt{9 + x^2}$ over the interval $[0, 4]$.

51. Find the area of the surface obtained by revolving around the x-axis the curve $y = \sin x, 0 \leqq x \leqq \pi$ (see Fig. 9.6.11).

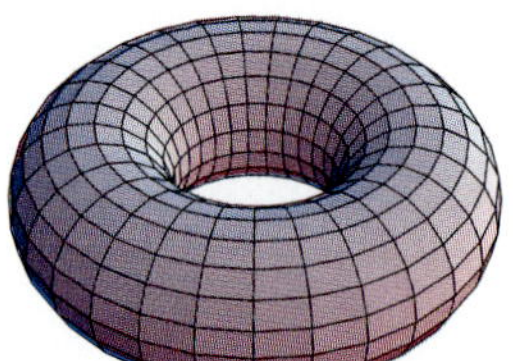

Fig. 9.6.10 The torus of Problem 49

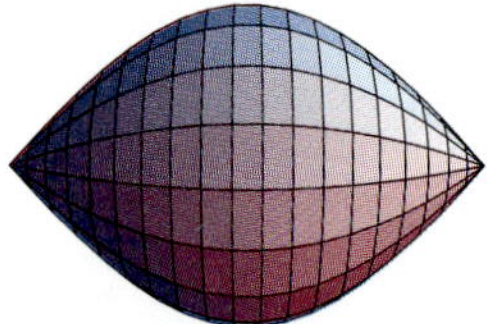

Fig. 9.6.11 The pointed football of Problem 51

52. An ellipsoid of revolution is obtained by revolving the ellipse $x^2/a^2 + y^2/b^2 = 1$ around the x-axis. Suppose that $a > b$. Show that the ellipsoid has surface area

$$A = 2\pi ab\left[\frac{b}{a} + \frac{a}{c}\sin^{-1}\left(\frac{c}{a}\right)\right],$$

where $c = \sqrt{a^2 - b^2}$. Assume that $a \approx b$, so that $c \approx 0$ and $\sin^{-1}(c/a) \approx c/a$. Conclude that $A \approx 4\pi a^2$.

53. Suppose that $b > a$ for the ellipsoid of revolution of Problem 52. Show that its surface area is then

$$A = 2\pi ab\left[\frac{b}{a} + \frac{a}{c}\ln\left(\frac{b + c}{a}\right)\right],$$

where $c = \sqrt{b^2 - a^2}$. Use the fact that $\ln(1 + x) \approx x$ if $x \approx 0$, and thereby conclude that $A \approx 4\pi a^2$ if $a \approx b$.

54. A road is to be built from the point $(2, 1)$ to the point $(5, 3)$, following the path of the parabola

$$y = -1 + 2\sqrt{x - 1}.$$

Calculate the length of this road (the units on the coordinate axes are in miles). (*Suggestion:* Substitute $x = \sec^2\theta$ into the arc-length integral.)

55. Suppose that the cost of the road in Problem 54 is $\sqrt{x}$ million dollars per mile. Calculate the total cost of the road.

56. A kite is flying at a height of 500 ft and at a horizontal distance of 100 ft from the string-holder on the ground. The kite string weighs 1/16 oz/ft and is hanging in the shape of the parabola $y = x^2/20$ that joins the string-holder at $(0, 0)$ to the kite at $(100, 500)$ (Fig. 9.6.12). Calculate the work (in foot-pounds) done in lifting the kite string from the ground to its present position.

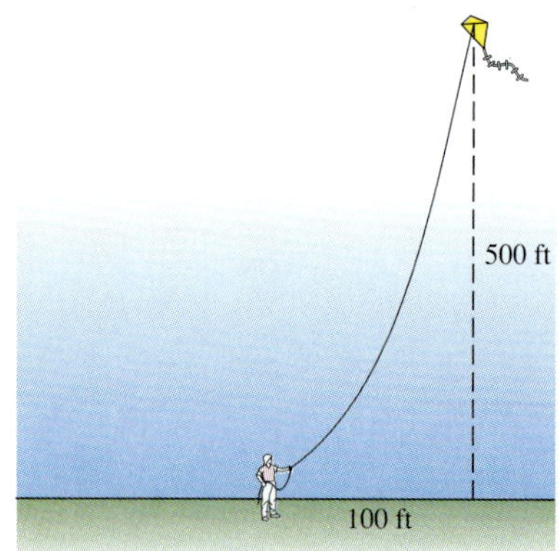

Fig. 9.6.12 The kite string of Problem 56

9.7 INTEGRALS CONTAINING QUADRATIC POLYNOMIALS

Many integrals containing a square root or negative power of a quadratic polynomial $ax^2 + bx + c$ can be simplified by the process of *completing the square.* For example,

$$x^2 + 2x + 2 = (x + 1)^2 + 1,$$

and hence the substitution $u = x + 1, du = dx$ yields

$$\int \frac{1}{x^2 + 2x + 2}\,dx = \int \frac{1}{u^2 + 1}\,du = \tan^{-1}u + C = \tan^{-1}(x + 1) + C.$$

In general, the objective is to convert $ax^2 + bx + c$ into either a sum or difference of two squares—either $u^2 \pm a^2$ or $a^2 - u^2$—so that the method of trigonometric substitution can then be used. To see how this works in practice, suppose first that $a = 1$, so that the quadratic in question has the form $x^2 + bx + c$. The sum $x^2 + bx$ of the first two terms can be completed to a perfect square by adding $b^2/4$, the square of half the coefficient of x, and in turn subtracting $b^2/4$ from the constant term c. This gives

$$\begin{aligned} x^2 + bx + c &= \left(x^2 + bx + \frac{b^2}{4}\right) + \left(c - \frac{b^2}{4}\right) \\ &= \left(x + \frac{b}{2}\right)^2 + \left(c - \frac{b^2}{4}\right). \end{aligned}$$

With $u = x + \frac{1}{2}b$, this result is of the form $u^2 + A^2$ or $u^2 - A^2$, depending on the sign of $c - \frac{1}{4}b^2$. If the coefficient a of x^2 is not 1, we first factor it out and proceed as before:

$$ax^2 + bx + c = a\left(x^2 + \frac{b}{a}x + \frac{c}{a}\right).$$

EXAMPLE 1 Find $\displaystyle\int \frac{1}{9x^2 + 6x + 5}\,dx.$

Solution The first step is to complete the square:

$$\begin{aligned} 9x^2 + 6x + 5 &= 9(x^2 + \tfrac{2}{3}x) + 5 = 9(x^2 + \tfrac{2}{3}x + \tfrac{1}{9}) - 1 + 5 \\ &= 9(x + \tfrac{1}{3})^2 + 4 = (3x + 1)^2 + 2^2. \end{aligned}$$

Hence

$$\begin{aligned} \int \frac{1}{9x^2 + 6x + 5}\,dx &= \int \frac{1}{(3x + 1)^2 + 4}\,dx \\ &= \frac{1}{3}\int \frac{1}{u^2 + 4}\,du \qquad (u = 3x + 1) \\ &= \frac{1}{6}\int \frac{\frac{1}{2}}{(\frac{1}{2}u)^2 + 1}\,dx \\ &= \frac{1}{6}\int \frac{1}{v^2 + 1}\,dv \qquad (v = \tfrac{1}{2}u) \\ &= \frac{1}{6}\tan^{-1}v + C = \frac{1}{6}\tan^{-1}\left(\frac{u}{2}\right) + C \\ &= \frac{1}{6}\tan^{-1}\left(\frac{3x + 1}{2}\right) + C. \end{aligned}$$

■

EXAMPLE 2 Find $\displaystyle\int \frac{1}{\sqrt{9 + 16x - 4x^2}}\,dx.$

Solution First we complete the square:

$$\begin{aligned} 9 + 16x - 4x^2 &= 9 - 4(x^2 - 4x) \\ &= 9 - 4(x^2 - 4x + 4) + 16 = 25 - 4(x - 2)^2. \end{aligned}$$

Hence

$$\begin{aligned}\int \frac{1}{\sqrt{9 + 16x - 4x^2}}\,dx &= \int \frac{1}{\sqrt{25 - 4(x-2)^2}}\,dx\\ &= \frac{1}{5}\int \frac{1}{\sqrt{1 - \frac{4}{25}(x-2)^2}}\,dx\\ &= \frac{1}{2}\int \frac{1}{\sqrt{1 - u^2}}\,du \qquad \left[u = \tfrac{2}{5}(x-2)\right]\\ &= \frac{1}{2}\sin^{-1} u + C = \frac{1}{2}\sin^{-1}\left[\tfrac{2}{5}(x-2)\right] + C.\end{aligned}$$

An alternative approach is to make the trigonometric substitution

$$2(x-2) = 5\sin\theta, \qquad 2\,dx = 5\cos\theta\,d\theta$$

immediately after completing the square. This yields

$$\begin{aligned}\int \frac{1}{\sqrt{9 + 16x - 4x^2}}\,dx &= \int \frac{1}{\sqrt{25 - 4(x-2)^2}}\,dx\\ &= \int \frac{\frac{5}{2}\cos\theta}{\sqrt{25 - 25\sin^2\theta}}\,d\theta\\ &= \frac{1}{2}\int 1\,d\theta = \frac{1}{2}\theta + C\\ &= \frac{1}{2}\arcsin\frac{2(x-2)}{5} + C.\end{aligned}$$

Some integrals that contain a quadratic expression can be split into two simpler integrals. Examples 3 and 4 illustrate this technique.

EXAMPLE 3 Find $\displaystyle\int \frac{2x + 3}{9x^2 + 6x + 5}\,dx.$

Solution Because $D_x(9x^2 + 6x + 5) = 18x + 6$, this would be a simpler integral if the numerator $2x + 3$ were a constant multiple of $18x + 6$. Our strategy is to write

$$2x + 3 = A\cdot(18x + 6) + B$$

so that we can split the given integral into a sum of two integrals, one of which has numerator $18x + 6$ in its integrand. By matching coefficients in

$$2x + 3 = 18Ax + (6A + B),$$

we find that $A = \frac{1}{9}$ and $B = \frac{7}{3}$. Hence

$$\int \frac{2x + 3}{9x^2 + 6x + 5}\,dx = \frac{1}{9}\int \frac{18x + 6}{9x^2 + 6x + 5}\,dx + \frac{7}{3}\int \frac{1}{9x^2 + 6x + 5}\,dx.$$

The first integral on the right is a logarithm, and the second is given in Example 1. Thus

$$\int \frac{2x + 3}{9x^2 + 6x + 5}\,dx = \frac{1}{9}\ln(9x^2 + 6x + 5) + \frac{7}{18}\tan^{-1}\left(\frac{3x + 1}{2}\right) + C.$$

Alternatively, we could first complete the square in the denominator. The substitution $u = 3x + 1, x = \frac{1}{3}(u - 1), dx = \frac{1}{3}\,du$ then gives

$$\begin{aligned}\int \frac{2x+3}{(3x+1)^2+4}\,dx &= \int \frac{\frac{2}{3}(u-1)+3}{u^2+4}\cdot\frac{1}{3}\,du\\ &= \frac{1}{9}\int \frac{2u}{u^2+4}\,du + \frac{7}{9}\int\frac{1}{u^2+4}\,du\\ &= \frac{1}{9}\ln(u^2+4) + \frac{7}{18}\tan^{-1}\left(\frac{u}{2}\right) + C\\ &= \frac{1}{9}\ln(9x^2+6x+5) + \frac{7}{18}\tan^{-1}\left(\frac{3x+1}{2}\right) + C.\end{aligned}$$

■

EXAMPLE 4 Find $\displaystyle\int \frac{2+6x}{(3+2x-x^2)^2}\,dx$ given $|x-1| < 2$.

Solution Because $D_x(3 + 2x - x^2) = 2 - 2x$, we first write

$$\begin{aligned}&\int \frac{2+6x}{(3+2x-x^2)^2}\,dx\\ &= -3\int\frac{2-2x}{(3+2x-x^2)^2}\,dx + 8\int\frac{1}{(3+2x-x^2)^2}\,dx.\end{aligned}$$

Then let $u = 3 + 2x - x^2$, $du = (2 - 2x)\,dx$ in the first integral to obtain

$$-3\int\frac{2-2x}{(3+2x-x^2)^2}\,dx = -3\int\frac{du}{u^2} = \frac{3}{u} + C_1 = \frac{3}{3+2x-x^2} + C_1.$$

Therefore

$$\int\frac{2+6x}{(3+2x-x^2)^2}\,dx = \frac{3}{3+2x-x^2} + 8\int\frac{1}{(3+2x-x^2)^2}\,dx. \tag{1}$$

(We can drop the constant C_1 because it can be absorbed by the constant C we will obtain when we evaluate the remaining integral.) To evaluate the remaining integral, we complete the square:

$$3 + 2x - x^2 = 4 - (x^2 - 2x + 1) = 4 - (x-1)^2.$$

Because $|x - 1| < 2$, this suggests the substitution

$$x - 1 = 2\sin\theta, \qquad dx = 2\cos\theta\,d\theta,$$

with which

$$3 + 2x - x^2 = 4 - 4\sin^2\theta = 4\cos^2\theta.$$

This substitution yields

$$\begin{aligned}8\int\frac{1}{(3+2x-x^2)^2}\,dx = 8\int\frac{2\cos\theta}{(4\cos^2\theta)^2}\,d\theta &= \int\sec^3\theta\,d\theta\\ &= \tfrac{1}{2}\sec\theta\tan\theta + \tfrac{1}{2}\int\sec\theta\,d\theta \qquad \text{[Section 9.3, Eq. (6)]}\\ &= \tfrac{1}{2}\sec\theta\tan\theta + \tfrac{1}{2}\ln|\sec\theta + \tan\theta| + C\end{aligned}$$

$$= \frac{x-1}{3+2x-x^2} + \tfrac{1}{2}\ln\left|\frac{x+1}{\sqrt{3+2x-x^2}}\right| + C. \qquad \textbf{(2)}$$

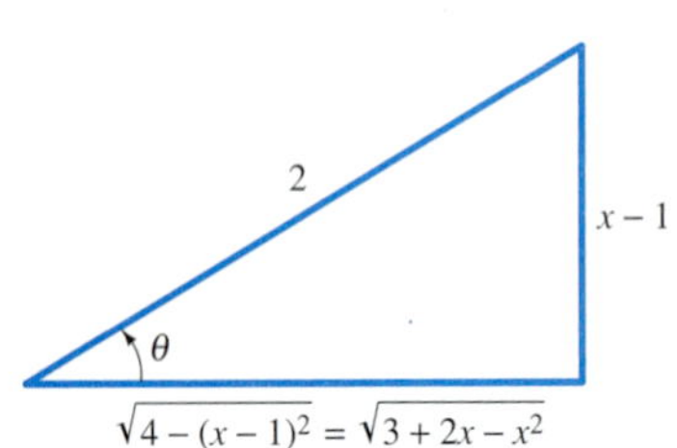

Fig. 9.7.1 The reference triangle for Example 4.

In the last step we read the values of $\sec\theta$ and $\tan\theta$ from the right triangle in Fig. 9.7.1. When we substitute Eq. (2) into Eq. (1), we finally obtain the result

$$\int \frac{2+6x}{(3+2x-x^2)^2}\,dx = \frac{x+2}{3+2x-x^2} + \frac{1}{2}\ln\left|\frac{x+1}{\sqrt{3+2x-x^2}}\right| + C.$$

The method of Example 4 can be used to evaluate a general integral of the form

$$\int \frac{Ax+B}{(ax^2+bx+c)^n}\,dx, \qquad \textbf{(3)}$$

where n is a positive integer. By splitting such an integral into two simpler ones and then completing the square in the quadratic expression in the denominator, the problem of evaluating the integral in Eq. (3) can be reduced to that of computing

$$\int \frac{1}{(a^2 \pm u^2)^n}\,du. \qquad \textbf{(4)}$$

If the sign in the denominator of Eq. (4) is the plus sign, then the substitution $u = a\tan\theta$ transforms the integral into the form

$$\int \cos^m\theta\,d\theta$$

(see Problem 35). This integral can be handled by the methods of Section 9.4 or by using the reduction formula

$$\int \cos^k\theta\,d\theta = \frac{1}{k}\cos^{k-1}\theta\sin\theta + \frac{k-1}{k}\int \cos^{k-2}\theta\,d\theta$$

of Problem 54 in Section 9.3.

If the sign of the denominator in Eq. (4) is the minus sign, then the substitution $u = a\sin\theta$ transforms the integral into the form

$$\int \sec^m\theta\,d\theta$$

(see Problem 36). This integral may be evaluated with the aid of the reduction formula

$$\int \sec^k\theta\,d\theta = \frac{1}{k-1}\sec^{k-2}\theta\tan\theta + \frac{k-2}{k-1}\int \sec^{k-2}\theta\,d\theta$$

[Eq. (5) of Section 9.3].

9.7 PROBLEMS

Evaluate the antiderivatives in Problems 1 through 34.

1. $\displaystyle\int \frac{1}{x^2+4x+5}\,dx$

2. $\displaystyle\int \frac{2x+5}{x^2+4x+5}\,dx$

3. $\displaystyle\int \frac{5-3x}{x^2+4x+5}\,dx$

4. $\displaystyle\int \frac{x+1}{(x^2+4x+5)^2}\,dx$

5. $\displaystyle\int \frac{1}{\sqrt{3-2x-x^2}}\,dx$

6. $\displaystyle\int \frac{x+3}{\sqrt{3-2x-x^2}}\,dx$

7. $\displaystyle\int x\sqrt{3-2x-x^2}\,dx$

8. $\displaystyle\int \frac{1}{4x^2+4x-3}\,dx$

9. $\displaystyle\int \frac{3x+2}{4x^2+4x-3}\,dx$

10. $\displaystyle\int \sqrt{4x^2+4x-3}\,dx$

11. $\displaystyle\int \frac{1}{x^2+4x+13}\,dx$

12. $\displaystyle\int \frac{1}{\sqrt{2x-x^2}}\,dx$

13. $\displaystyle\int \frac{1}{3+2x-x^2}\,dx$

14. $\displaystyle\int x\sqrt{8+2x-x^2}\,dx$

15. $\displaystyle\int \frac{2x-5}{x^2+2x+2}\,dx$

16. $\displaystyle\int \frac{2x-1}{4x^2+4x-15}\,dx$

17. $\displaystyle\int \frac{x}{\sqrt{5+12x-9x^2}}\,dx$

18. $\displaystyle\int (3x-2)\sqrt{9x^2+12x+8}\,dx$

19. $\displaystyle\int (7-2x)\sqrt{9+16x-4x^2}\,dx$

20. $\displaystyle\int \frac{2x+3}{\sqrt{x^2+2x+5}}\,dx$

21. $\displaystyle\int \frac{x+4}{(6x-x^2)^{3/2}}\,dx$

22. $\displaystyle\int \frac{x-1}{(x^2+1)^2}\,dx$

23. $\displaystyle\int \frac{2x+3}{(4x^2+12x+13)^2}\,dx$

24. $\displaystyle\int \frac{x^3}{(1-x^2)^4}\,dx$

25. $\displaystyle\int \frac{3x-1}{x^2+x+1}\,dx$

26. $\displaystyle\int \frac{3x-1}{(x^2+x+1)^2}\,dx$

27. $\displaystyle\int \frac{1}{(x^2-4)^2}\,dx$

28. $\displaystyle\int (x-x^2)^{3/2}\,dx$

29. $\displaystyle\int \frac{x^2+1}{x^3+x^2+x}\,dx$

30. $\displaystyle\int \frac{x^2+2}{(x^2+1)^2}\,dx$

31. $\displaystyle\int \frac{2x^2+3}{x^4-2x^2+1}\,dx$

32. $\displaystyle\int \frac{x^2+4}{(x^2+1)^2(x^2+2)}\,dx$

33. $\displaystyle\int \frac{3x+1}{(x^2+2x+5)^2}\,dx$

34. $\displaystyle\int \frac{x^3-2x}{x^2+2x+2}\,dx$

35. Show that the substitution $u = a\tan\theta$ gives

$$\int \frac{1}{(a^2+u^2)^n}\,du = \frac{1}{a^{2n-1}}\int \cos^{2n-2}\theta\,d\theta.$$

36. Show that the substitution $u = a\sin\theta$ gives

$$\int \frac{1}{(a^2-u^2)^n}\,du = \frac{1}{a^{2n-1}}\int \sec^{2n-1}\theta\,d\theta.$$

In Problems 37 through 39, the region R lies between the curve $y = 1/(x^2-2x+5)$ and the x-axis from $x=0$ to $x=5$.

37. Find the area of the region R.

38. Find the volume of the solid generated by revolving R around the y-axis.

39. Find the volume of the solid generated by revolving R around the x-axis.

In Problems 40 through 42 the region R lies between the curve $y = 1/(4x^2-20x+29)$ and the x-axis from $x=1$ to $x=4$.

40. Find the area of the region R.

41. Find the volume of the solid generated by revolving R around the y-axis.

42. Find the volume of the solid generated by revolving R around the x-axis.

43. Your task is to build a road that joins the points $(0,0)$ and $(3,2)$ and follows the path of the circle with equation $(4x+4)^2+(4y-19)^2 = 377$. Find the length of this road. (Units on the coordinate axes are measured in miles.)

44. Suppose that the road of Problem 43 costs $10/(1+x)$ million dollars per mile. (a) Calculate its total cost. (b) With the same cost per mile, calculate the total cost of a straight line road from $(0,0)$ to $(3,2)$. You should find that it is *more* expensive than the *longer* circular road!

In Problems 45 through 47, factor the denominator by first noting by inspection a root r of the denominator and then employing long division by $x-r$. Finally, use the method of partial fractions to aid in finding the indicated antiderivative.

45. $\displaystyle\int \frac{3x+2}{x^3+x^2-2}\,dx$

46. $\displaystyle\int \frac{1}{x^3+8}\,dx$

47. $\displaystyle\int \frac{x^4+2x^2}{x^3-1}\,dx$

48. (a) Find constants a and b such that

$$x^4+1 = (x^2+ax+1)(x^2+bx+1).$$

(b) Prove that

$$\int_0^1 \frac{x^2+1}{x^4+1}\,dx = \frac{\pi}{2\sqrt{2}}.$$

(*Suggestion:* If u and v are positive numbers and $uv=1$, then $\arctan u + \arctan v = \frac{1}{2}\pi$.)

49. Factor x^4+x^2+1 with the aid of ideas suggested in Problem 48. Then evaluate

$$\int \frac{2x^3+3x}{x^4+x^2+1}\,dx.$$

50. Evaluate the integral to show that

$$\int_0^1 \frac{16(x-1)}{x^4-2x^3+4x-4}\,dx = \pi.$$

This integral was (in effect) used by D. Bailey, P. Borwein, and S. Plouffe as a starting point in their recent determination of the 5 billionth hexagesimal digit of the number π (it's a 9). (*Suggestion:* Long divide to verify that x^2-2 is a factor of the denominator and to find the other factor.)

In Problems 51 through 54, write the general form of a partial fraction decomposition of the given rational function $f(x)$ (with coefficients $A, B, C, \ldots$ remaining to be determined). Then use a computer algebra system (as in the remark following Example 6 in Section 9.5) to find the numerical values of the coefficients in the decomposition. Finally, find the indefinite integral $\int f(x)\,dx$ both by hand and by using the computer algebra system, and resolve any apparent discrepancy between the two results.

51. $\displaystyle\int \frac{7x^4 + 28x^3 + 50x^2 + 67x + 23}{(x-1)(x^2+2x+2)^2}\,dx$

52. $\displaystyle\int \frac{35 + 84x + 55x^2 - x^3 + 5x^4 - 4x^5}{(x^2+1)^2(x^2+6x+10)}\,dx$

53. $\displaystyle\int \frac{32x^5 + 16x^4 + 19x^3 - 98x^2 - 107x - 15}{(x^2-2x-15)(4x^2+4x+5)^2}\,dx$

54. $\displaystyle\int \frac{63x^5 + 302x^4 + 480x^3 + 376x^2 - 240x - 300}{(x^2+6x+10)^2(4x^2+4x+5)^2}\,dx$

9.8 IMPROPER INTEGRALS

To show the existence of the definite integral, until now we have relied on the existence theorem stated in Section 5.4. This theorem guarantees the existence of the definite integral

$$\int_a^b f(x)\,dx$$

provided that the function f is *continuous* on the closed and bounded interval $[a, b]$. Certain applications of calculus, however, lead naturally to the formulation of integrals in which either

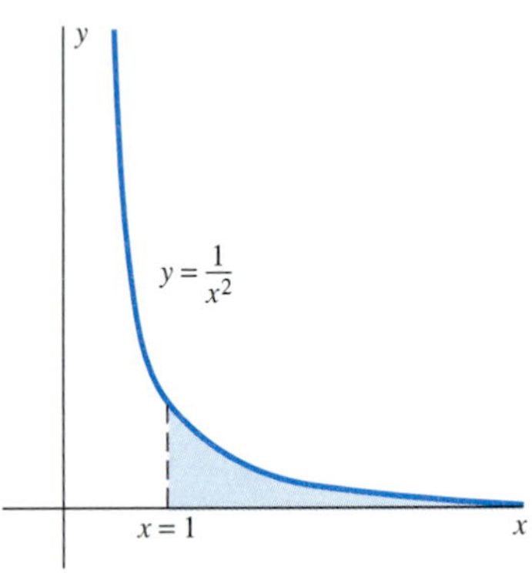

Fig. 9.8.1 The shaded area cannot be measured by using our earlier techniques.

1. The interval of integration is not bounded; it has one of the forms

$$[a, +\infty), \qquad (-\infty, a], \qquad \text{or} \qquad (-\infty, +\infty); \qquad \text{or}$$

2. The integrand has an infinite discontinuity at some point c in the interval $[a, b]$:

$$\lim_{x\to c} f(x) = \pm\infty.$$

An example of case 1 is the integral

$$\int_1^\infty \frac{1}{x^2}\,dx.$$

A geometric interpretation of this integral is the area of the unbounded region (shaded in Fig. 9.8.1) that lies between the curve $y = 1/x^2$ and the x-axis and to the right of the vertical line $x = 1$. An example of case 2 is the integral

$$\int_0^1 \frac{1}{\sqrt{x}}\,dx.$$

This integral may be interpreted as the area of the unbounded region (shaded in Fig. 9.8.2) that lies under the curve $y = 1/\sqrt{x}$ from $x = 0$ to $x = 1$.

Such integrals are called **improper integrals**. The natural interpretation of an improper integral is the area of an unbounded region. It is perhaps surprising that such an area can nevertheless be finite, and here we shall show how to find such areas—that is, how to evaluate improper integrals.

To see why improper integrals require special care, let us consider the integral

$$\int_{-1}^1 \frac{1}{x^2}\,dx.$$

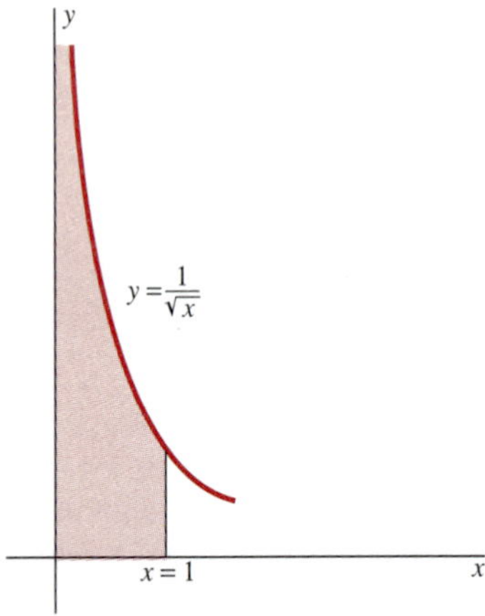

Fig. 9.8.2 Another area that must be measured with an improper integral.

This integral is improper because its integrand $f(x) = 1/x^2$ is unbounded as $x \to 0$, and thus f is not continuous at $x = 0$. If we thoughtlessly applied the fundamental theorem of calculus, we would obtain

$$\int_{-1}^{1} \frac{1}{x^2}\,dx = \left[-\frac{1}{x}\right]_{-1}^{1} = (-1) - (+1) = -2. \qquad \text{(Wrong!)}$$

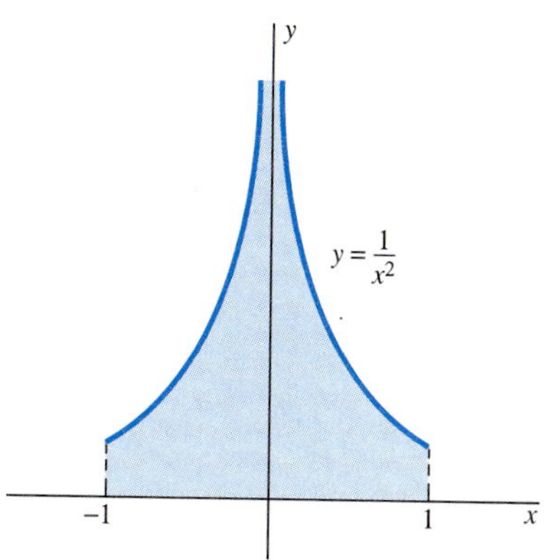

Fig. 9.8.3 The area under $y = 1/x^2$, $-1 \leqq x \leqq 1$

The negative answer is obviously incorrect, because the area shown in Fig. 9.8.3 lies above the x-axis and hence the integral cannot be negative. This simple example emphasizes that we cannot ignore the hypotheses—*continuous* function and *bounded closed* interval—of the fundamental theorem of calculus.

Infinite Limits of Integration

Suppose that the function f is continuous and nonnegative on the unbounded interval $[a, +\infty)$. Then, for any fixed $t > a$, the area $A(t)$ of the region under $y = f(x)$ from $x = a$ to $x = t$ (shaded in Fig. 9.8.4) is given by the (ordinary) definite integral

$$A(t) = \int_a^t f(x)\,dx.$$

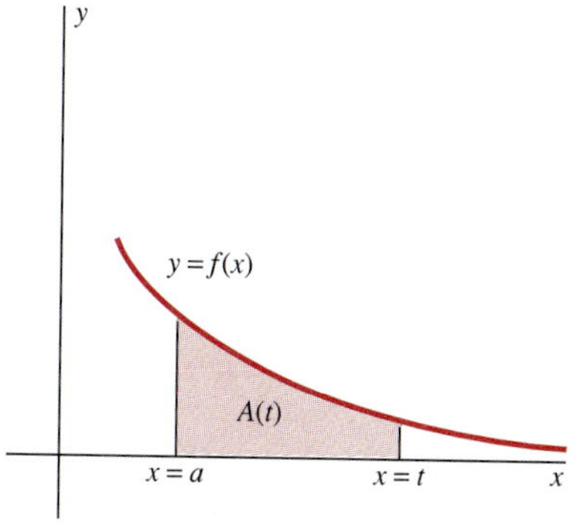

Fig. 9.8.4 The shaded area $A(t)$ exists provided that f is continuous.

Suppose now that we let $t \to +\infty$ and find that the limit of $A(t)$ exists. Then we may regard this limit as the area of the unbounded region that lies under $y = f(x)$ and over $[a, +\infty)$. For f continuous on $[a, +\infty)$, we therefore *define*

$$\int_a^{\infty} f(x)\,dx = \lim_{t\to\infty} \int_a^t f(x)\,dx \tag{1}$$

provided that this limit exists (as a finite real number). If this limit does exist, we say that the improper integral on the left **converges**; if the limit does not exist, we say that the improper integral **diverges**. If $f(x)$ is nonnegative on $[a, +\infty)$, then the limit in Eq. (1) either exists or is infinite, and in the latter case we write

$$\int_a^{\infty} f(x)\,dx = +\infty$$

and say that the improper integral **diverges to infinity.**

If the function f has both positive and negative values on $[a, +\infty)$, then the improper integral can diverge *by oscillation*—that is, without diverging to infinity. This occurs with $\int_0^{\infty} \sin x\,dx$, because it is easy to verify that $\int_0^t \sin x\,dx$ is zero if t is an even integral multiple of π but is 2 if t is an odd integral multiple of π. Thus $\int_0^t \sin x\,dx$ oscillates between 0 and 2 as $t \to +\infty$, and so the limit in Eq. (1) does not exist.

We handle an infinite lower limit of integration similarly: We define

$$\int_{-\infty}^{b} f(x)\,dx = \lim_{t\to-\infty} \int_t^b f(x)\,dx \tag{2}$$

provided that the limit exists. If the function f is continuous on the whole real line, we define

$$\int_{-\infty}^{\infty} f(x)\,dx = \int_{-\infty}^{c} f(x)\,dx + \int_c^{\infty} f(x)\,dx \tag{3}$$

for any convenient choice of c, provided that both improper integrals on the right-hand side converge. Note that the integral in Eq. (3) is *not* necessarily equal to

$$\lim_{t\to\infty}\int_{-t}^{t} f(x)\,dx$$

(see Problem 52).

It makes no difference what value of c is used in Eq. (3), because if $c < d$, then

$$\begin{aligned}\int_{-\infty}^{c} f(x)\,dx + \int_{c}^{\infty} f(x)\,dx &= \int_{-\infty}^{c} f(x)\,dx + \int_{c}^{d} f(x)\,dx + \int_{d}^{\infty} f(x)\,dx \\ &= \int_{-\infty}^{d} f(x)\,dx + \int_{d}^{\infty} f(x)\,dx,\end{aligned}$$

under the assumption that the limits involved all exist.

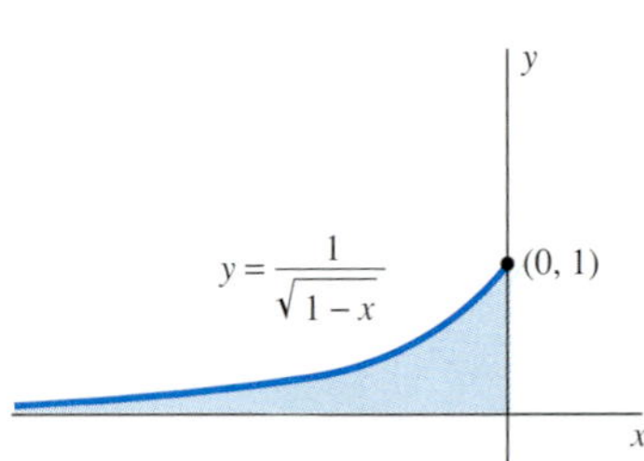

Fig. 9.8.5 The unbounded region represented by the improper integral in Example 1(b)

EXAMPLE 1 Investigate the improper integrals

(a) $\displaystyle\int_{1}^{\infty} \frac{1}{x^2}\,dx$ and **(b)** $\displaystyle\int_{-\infty}^{0} \frac{1}{\sqrt{1-x}}\,dx.$

Solution

(a) $$\int_{1}^{\infty} \frac{1}{x^2}\,dx = \lim_{t\to\infty}\int_{1}^{t} \frac{1}{x^2}\,dx = \lim_{t\to\infty}\left[-\frac{1}{x}\right]_1^t = \lim_{t\to\infty}\left(-\frac{1}{t}+1\right) = 1.$$

Thus this improper integral converges, and this is the area of the region shaded in Fig. 9.8.1.

(b) $$\begin{aligned}\int_{-\infty}^{0} \frac{1}{\sqrt{1-x}}\,dx &= \lim_{t\to-\infty}\int_{t}^{0} \frac{1}{\sqrt{1-x}}\,dx \\ &= \lim_{t\to-\infty}\Big[-2\sqrt{1-x}\Big]_t^0 = \lim_{t\to-\infty}\left(2\sqrt{1-t}-2\right) = +\infty.\end{aligned}$$

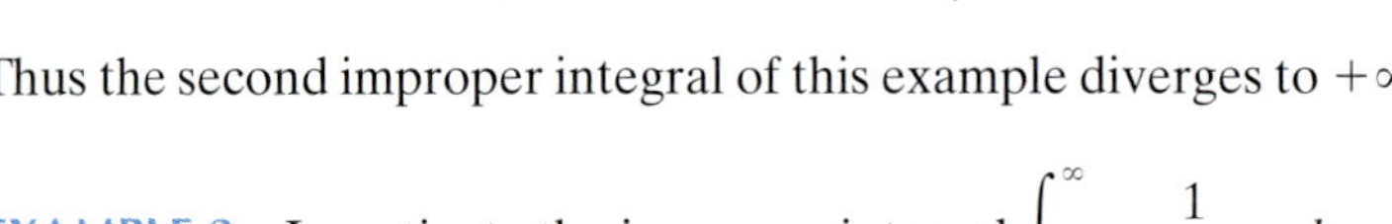

Thus the second improper integral of this example diverges to $+\infty$ (Fig. 9.8.5). ■

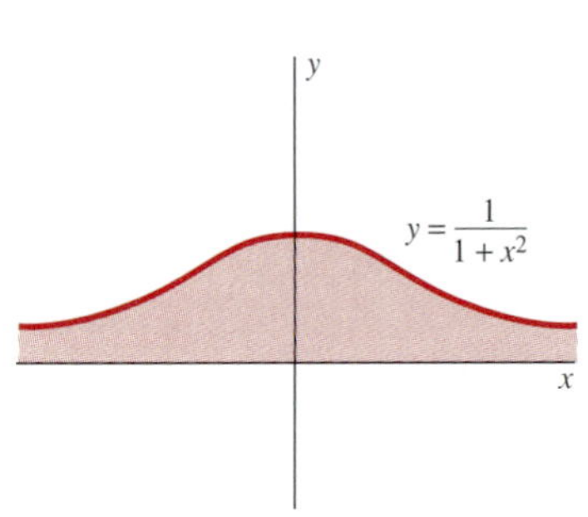

Fig. 9.8.6 The area measured by the integral in Example 2

EXAMPLE 2 Investigate the improper integral $\displaystyle\int_{-\infty}^{\infty} \frac{1}{1+x^2}\,dx.$

Solution The choice $c = 0$ in Eq. (3) gives

$$\begin{aligned}\int_{-\infty}^{\infty} \frac{1}{1+x^2}\,dx &= \int_{-\infty}^{0} \frac{1}{1+x^2}\,dx + \int_{0}^{\infty} \frac{1}{1+x^2}\,dx \\ &= \lim_{s\to-\infty}\int_{s}^{0} \frac{1}{1+x^2}\,dx + \lim_{t\to\infty}\int_{0}^{t} \frac{1}{1+x^2}\,dx \\ &= \lim_{s\to-\infty}\Big[\tan^{-1}x\Big]_s^0 + \lim_{t\to\infty}\Big[\tan^{-1}x\Big]_0^t \\ &= \lim_{s\to-\infty}\left(-\tan^{-1}s\right) + \lim_{t\to\infty}\left(\tan^{-1}t\right) = \frac{\pi}{2}+\frac{\pi}{2} = \pi.\end{aligned}$$

The shaded region in Fig. 9.8.6 is a geometric interpretation of the integral of Example 2. ■

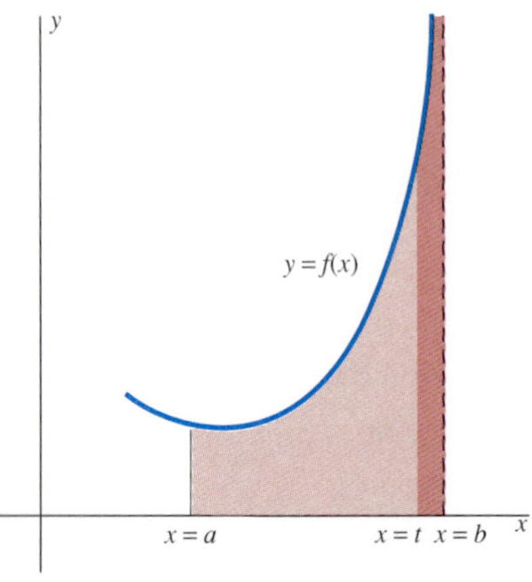

Fig. 9.8.7 An improper integral of the second type: $f(x) \to \infty$ as $x \to b^-$

Infinite Integrands

Suppose that the function f is continuous and nonnegative on $[a, b)$ but that $f(x) \to +\infty$ as $x \to b^-$. The graph of such a function appears in Fig. 9.8.7. The area

$A(t)$ of the region lying under $y = f(x)$ from $x = a$ to $x = t < b$ is the value of the (ordinary) definite integral

$$A(t) = \int_a^t f(x)\,dx.$$

If the limit of $A(t)$ exists as $t \to b^-$, then this limit may be regarded as the area of the (unbounded) region under $y = f(x)$ from $x = a$ to $x = b$. For f continuous on $[a, b)$, we therefore *define*

$$\int_a^b f(x)\,dx = \lim_{t\to b^-} \int_a^t f(x)\,dx, \tag{4}$$

provided that this limit exists (as a finite number), in which case we say that the improper integral on the left **converges**; if the limit does not exist, we say that the integral **diverges**. If

$$\int_a^b f(x)\,dx = \lim_{t\to b^-} \int_a^t f(x)\,dx = \infty,$$

then we say that the improper integral **diverges to infinity**.

If f is continuous on $(a, b]$ but the limit of $f(x)$ as $x \to a^+$ is infinite, then we *define*

$$\int_a^b f(x)\,dx = \lim_{t\to a^+} \int_t^b f(x)\,dx, \tag{5}$$

provided that the limit exists. If f is continuous at every point of $[a, b]$ except for the point c in (a, b) and one or more both one-sided limits of f at c are infinite, then we *define*

$$\int_a^b f(x)\,dx = \int_a^c f(x)\,dx + \int_c^b f(x)\,dx \tag{6}$$

provided that both improper integrals on the right converge.

EXAMPLE 3 Investigate the improper integrals

(a) $\displaystyle\int_0^1 \frac{1}{\sqrt{x}}\,dx$ and **(b)** $\displaystyle\int_1^2 \frac{1}{(x-2)^2}\,dx.$

Solution

(a) The integrand $1/\sqrt{x}$ becomes infinite as $x \to 0^+$, so

$$\begin{aligned}\int_0^1 \frac{1}{\sqrt{x}}\,dx &= \lim_{t\to 0^+} \int_t^1 \frac{1}{\sqrt{x}}\,dx \\ &= \lim_{t\to 0^+} \Big[2\sqrt{x}\Big]_t^1 = \lim_{t\to 0^+} 2(1 - \sqrt{t}) = 2.\end{aligned}$$

Thus the area of the unbounded region shown in Fig. 9.8.2 is 2.

(b) Here the integrand becomes infinite as x approaches the right-hand endpoint, so

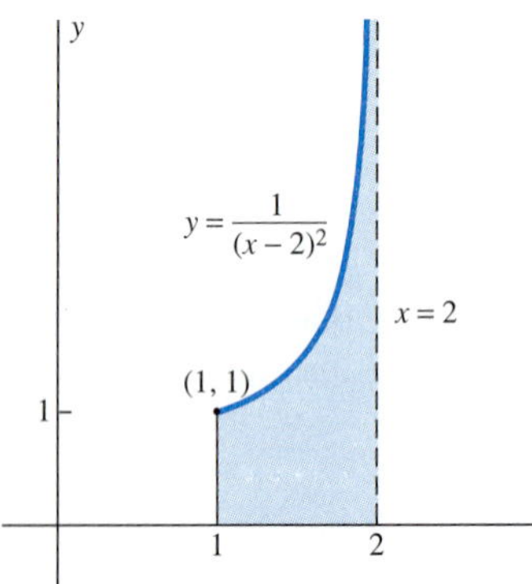

Fig. 9.8.8 The unbounded region represented by the improper integral in Example 3(b)

$$\int_1^2 \frac{1}{(x-2)^2}\,dx = \lim_{t\to 2^-} \int_1^t \frac{1}{(x-2)^2}\,dx$$
$$= \lim_{t\to 2^-} \left[-\frac{1}{x-2} \right]_1^t = \lim_{t\to 2^-} \left(-1 - \frac{1}{t-2} \right) = +\infty.$$

Hence this improper integral diverges to infinity (Fig. 9.8.8). It follows that the improper integral

$$\int_1^3 \frac{1}{(x-2)^2}\,dx = \int_1^2 \frac{1}{(x-2)^2}\,dx + \int_2^3 \frac{1}{(x-2)^2}\,dx$$

also diverges, because not both of the right-hand improper integrals converge. (You can verify that the second one also diverges to $+\infty$.) ■

EXAMPLE 4 Investigate the improper integral

$$\int_0^2 \frac{1}{(2x-1)^{2/3}}\,dx.$$

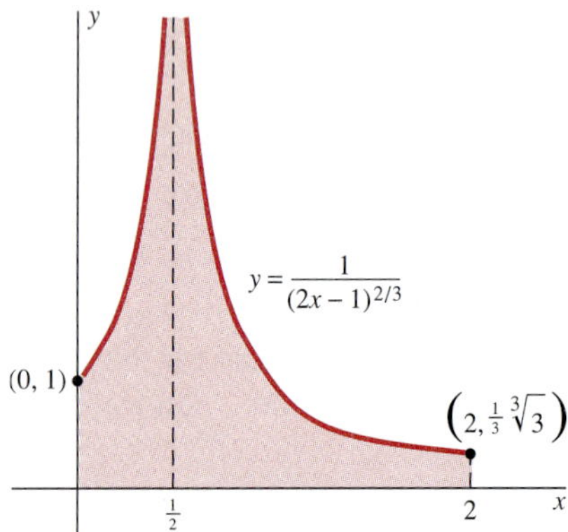

Fig. 9.8.9 The region of Example 4

Solution This improper integral corresponds to the region shaded in Fig. 9.8.9. The integrand has an infinite discontinuity at the point $c = \frac{1}{2}$ within the interval of integration, so we write

$$\int_0^2 \frac{1}{(2x-1)^{2/3}}\,dx = \int_0^{1/2} \frac{1}{(2x-1)^{2/3}}\,dx + \int_{1/2}^2 \frac{1}{(2x-1)^{2/3}}\,dx$$

and investigate separately the two improper integrals on the right. We find that

$$\int_0^{1/2} \frac{1}{(2x-1)^{2/3}}\,dx = \lim_{t\to(1/2)^-} \int_0^t \frac{1}{(2x-1)^{2/3}}\,dx$$
$$= \lim_{t\to(1/2)^-} \left[\tfrac{3}{2}(2x-1)^{1/3} \right]_0^t$$
$$= \lim_{t\to(1/2)^-} \tfrac{3}{2}\left[(2t-1)^{1/3} - (-1)^{1/3}\right] = \tfrac{3}{2},$$

and

$$\int_{1/2}^2 \frac{1}{(2x-1)^{2/3}}\,dx = \lim_{t\to(1/2)^+} \int_t^2 \frac{1}{(2x-1)^{2/3}}\,dx$$
$$= \lim_{t\to(1/2)^+} \left[\tfrac{3}{2}(2x-1)^{1/3} \right]_t^2$$
$$= \lim_{t\to(1/2)^+} \tfrac{3}{2}\left[3^{1/3} - (2t-1)^{1/3}\right] = \tfrac{3}{2}\sqrt[3]{3}.$$

Therefore,

$$\int_0^2 \frac{1}{(2x-1)^{2/3}}\,dx = \tfrac{3}{2}\left(1 + \sqrt[3]{3}\right).$$ ■

*Special Functions and Improper Integrals

Special functions in advanced mathematics frequently are defined by means of improper integrals. An important example is the **gamma function** $\Gamma(t)$ that the prolific Swiss mathematician Leonhard Euler (1707–1783) introduced to "interpolate" values of the factorial function

$$n! = 1 \cdot 2 \cdot 3 \cdots (n-1) \cdot n.$$

The gamma function is defined for all real numbers $t > 0$ by the improper integral

$$\Gamma(t) = \int_0^\infty x^{t-1} e^{-x}\,dx. \tag{7}$$

Thus for a fixed positive number t, the value $\Gamma(t)$ is the area under the curve $y = x^{t-1}e^{-x}$ from $x = 0$ to ∞ (Fig. 9.8.10). It turns out that the improper integral in Eq. (7) converges for all $t > 0$. The following example treats the illustrative cases $t = 1$ and $t = 2$.

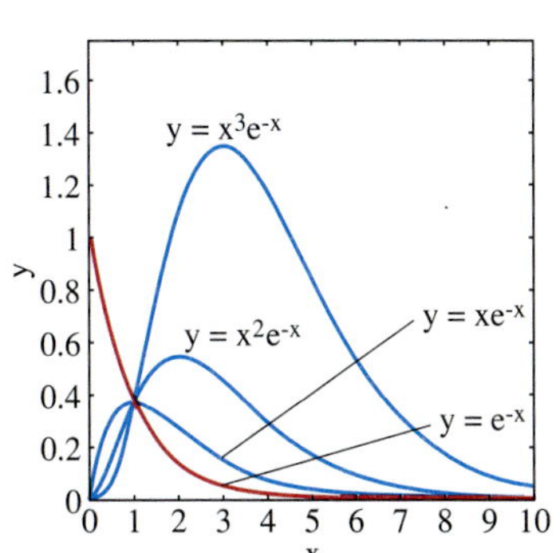

Fig. 9.8.10 The graphs $y = x^{t-1}e^{-x}$ for $t = 1, 2, 3, 4$

EXAMPLE 5 If $t = 1$, then

$$\Gamma(1) = \int_0^\infty e^{-x}\,dx = \lim_{b\to\infty}\int_0^b e^{-x}\,dx$$

$$= \lim_{b\to\infty}\Big[-e^{-x}\Big]_0^b = \lim_{b\to\infty}(1 - e^{-b}) = 1.$$

If $t = 2$, then integration by parts with $u = x$, $dv = e^{-x}\,dx$ yields

$$\Gamma(2) = \int_0^\infty xe^{-x}\,dx = \lim_{b\to\infty}\int_0^b xe^{-x}\,dx$$

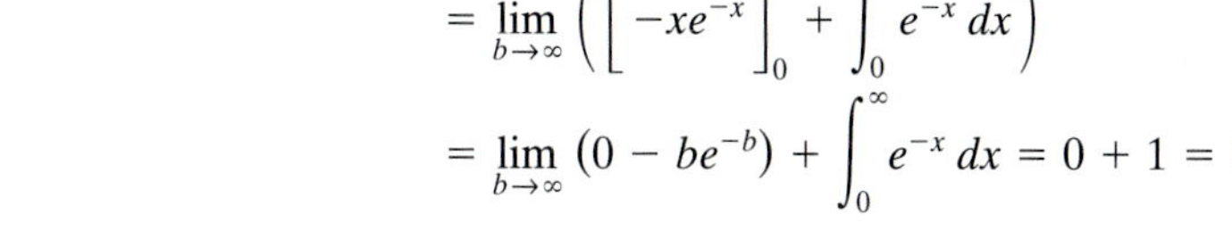

$$= \lim_{b\to\infty}\left(\Big[-xe^{-x}\Big]_0^b + \int_0^b e^{-x}\,dx\right)$$

$$= \lim_{b\to\infty}(0 - be^{-b}) + \int_0^\infty e^{-x}\,dx = 0 + 1 = 1$$

as well. ■

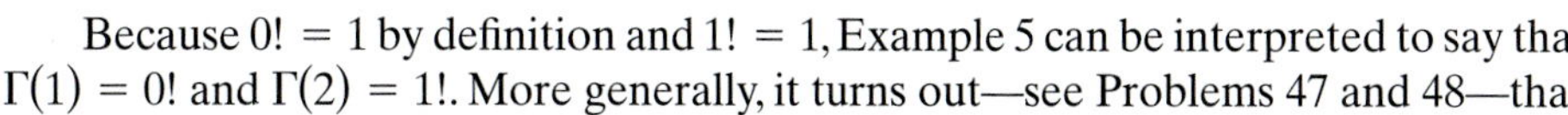

Because $0! = 1$ by definition and $1! = 1$, Example 5 can be interpreted to say that $\Gamma(1) = 0!$ and $\Gamma(2) = 1!$. More generally, it turns out—see Problems 47 and 48—that

$$\Gamma(n+1) = n! \tag{8}$$

for every nonnegative integer n. But remember, the gamma function is defined for all positive *real numbers*. In Problem 57 we ask you to show that

$$\Gamma(\tfrac{1}{2}) = 2\int_0^\infty e^{-x^2}\,dx. \tag{9}$$

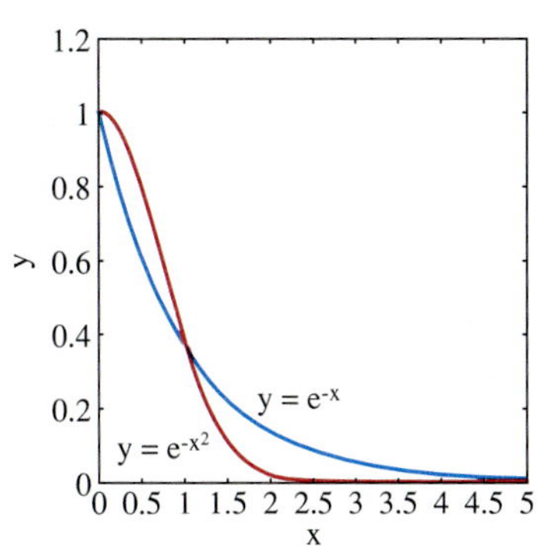

Fig. 9.8.11 The graphs $y = e^{-x}$ and $y = e^{-x^2}$

Figure 9.8.11 shows the graph $y = e^{-x^2}$ in comparison with $y = e^{-x}$ for $x \geqq 0$. Because $e^{-x^2} < e^{-x}$ for $x > 1$ and the improper integral

$$\int_0^\infty e^{-x}\,dx$$

converges by Example 5, it is plausible that the area under $y = e^{-x^2}$ for $x \geqq 0$ is finite, and hence that the improper integral in Eq. (9) converges as well. Indeed, we will see in Section 14.4 that $\Gamma(\frac{1}{2}) = \sqrt{\pi}$.

*Escape Velocity

We saw in Section 6.6 how to compute the work W_r required to lift a body of mass m from the surface of a planet of mass M and radius R to a distance $r > R$ from the center of the planet. According to Eq. (7) there, the answer is

$$W_r = \int_R^r \frac{GMm}{x^2}\,dx.$$

So the work required to move the mass m "infinitely far" from the planet is

$$W = \lim_{r\to\infty} W_r = \int_R^\infty \frac{GMm}{x^2}\,dx = \lim_{r\to\infty}\left[-\frac{GMm}{x}\right]_R^r = \frac{GMm}{R}.$$

Suppose that the mass is projected with initial velocity v straight upward from the planet's surface, as in Jules Verne's novel *From the Earth to the Moon* (1865), in which a spacecraft was fired from an immense cannon. Then the initial kinetic energy $\frac{1}{2}mv^2$ is available to supply this work—by conversion into potential energy. From the equation

$$\frac{1}{2}mv^2 = \frac{GMm}{R},$$

we find that

$$v = \sqrt{\frac{2GM}{R}}.$$

Substitution of appropriate numerical values for the constants G, M, and R yields the value $v \approx 11175$ mi/s (about 25000 mi/h) for the *escape velocity* from the earth.

*Present Value of a Perpetuity

Consider a perpetual annuity, under which you and your heirs (and theirs, ad infinitum) will be paid A dollars annually. The question we pose is this: What is the fair market value of such an annuity? What should you pay to purchase it?

If the annual interest rate r is continuously compounded, then a dollar deposited in a savings account would grow to e^{rt} dollars in t years. Hence e^{-rt} dollars deposited now would yield \$1 after t years. Consequently, the **present value** of the amount you (and your heirs) will receive between time $t = 0$ (the present) and time $t = T > 0$ is defined to be

$$P_T = \int_0^T Ae^{-rt}\,dt.$$

Hence the total present value of the perpetual annuity is

$$P = \lim_{T\to\infty} P_T = \int_0^\infty Ae^{-rt}\,dt = \lim_{T\to\infty}\left[-\frac{A}{r}e^{-rt}\right]_0^T = \frac{A}{r}.$$

Thus $A = rP$. For instance, at an annual interest rate of 8% ($r = 0.08$), you should be able to purchase for $P = (\$50000)/(0.08) = \$625{,}000$ a perpetuity that pays you (and your heirs) an annual sum of \$50000 forever.

9.8 PROBLEMS

Determine whether the improper integrals in Problems 1 through 38 converge. Evaluate those that do converge.

1. $\displaystyle\int_2^\infty \frac{1}{x\sqrt{x}}\,dx$

2. $\displaystyle\int_1^\infty \frac{1}{x^{2/3}}\,dx$

3. $\displaystyle\int_0^4 \frac{1}{x\sqrt{x}}\,dx$

4. $\displaystyle\int_0^8 \frac{1}{x^{2/3}}\,dx$

5. $\displaystyle\int_1^\infty \frac{1}{x+1}\,dx$

6. $\displaystyle\int_3^\infty \frac{1}{\sqrt{x+1}}\,dx$

7. $\displaystyle\int_5^\infty \frac{1}{(x-1)^{3/2}}\,dx$

8. $\displaystyle\int_0^4 \frac{1}{\sqrt{4-x}}\,dx$

9. $\displaystyle\int_0^9 \frac{1}{(9-x)^{3/2}}\,dx$

10. $\displaystyle\int_0^3 \frac{1}{(x-3)^2}\,dx$

11. $\displaystyle\int_{-\infty}^{-2} \frac{1}{(x+1)^3}\,dx$

12. $\displaystyle\int_{-\infty}^{0} \frac{1}{\sqrt{4-x}}\,dx$

13. $\displaystyle\int_{-1}^{8} \frac{1}{\sqrt[3]{x}}\,dx$

14. $\displaystyle\int_{-4}^{4} \frac{1}{(x+4)^{2/3}}\,dx$

15. $\displaystyle\int_{2}^{\infty} \frac{1}{\sqrt[3]{x-1}}\,dx$

16. $\displaystyle\int_{-\infty}^{\infty} \frac{x}{(x^2+4)^{3/2}}\,dx$

17. $\displaystyle\int_{-\infty}^{\infty} \frac{x}{x^2+4}\,dx$

18. $\displaystyle\int_{0}^{\infty} e^{-(x+1)}\,dx$

19. $\displaystyle\int_{0}^{1} \frac{e^{\sqrt{x}}}{\sqrt{x}}\,dx$

20. $\displaystyle\int_{0}^{2} \frac{x}{x^2-1}\,dx$

21. $\displaystyle\int_{0}^{\infty} xe^{-3x}\,dx$

22. $\displaystyle\int_{-\infty}^{2} e^{2x}\,dx$

23. $\displaystyle\int_{0}^{\infty} xe^{-x^2}\,dx$

24. $\displaystyle\int_{-\infty}^{\infty} |x|e^{-x^2}\,dx$

25. $\displaystyle\int_{0}^{\infty} \frac{1}{1+x^2}\,dx$

26. $\displaystyle\int_{0}^{\infty} \frac{x}{1+x^2}\,dx$

27. $\displaystyle\int_{0}^{\infty} \cos x\,dx$

28. $\displaystyle\int_{0}^{\infty} \sin^2 x\,dx$

29. $\displaystyle\int_{1}^{\infty} \frac{\ln x}{x}\,dx$

30. $\displaystyle\int_{2}^{\infty} \frac{1}{x\ln x}\,dx$

31. $\displaystyle\int_{2}^{\infty} \frac{1}{x(\ln x)^2}\,dx$

32. $\displaystyle\int_{1}^{\infty} \frac{\ln x}{x^2}\,dx$

33. $\displaystyle\int_{0}^{\pi/2} \frac{\cos x}{\sqrt{\sin x}}\,dx$

34. $\displaystyle\int_{0}^{\pi/2} \frac{\sin x}{(\cos x)^{4/3}}\,dx$

35. $\displaystyle\int_{0}^{1} \ln x\,dx$

36. $\displaystyle\int_{0}^{1} \frac{\ln x}{x}\,dx$

37. $\displaystyle\int_{0}^{1} \frac{\ln x}{x^2}\,dx$

38. $\displaystyle\int_{0}^{\infty} e^{-x}\cos x\,dx$

In Problems 39 through 42, the given integral is improper both because the interval of integration is unbounded and because the integrand is unbounded near zero. Investigate its convergence by expressing it as a sum of two integrals—one from 0 to 1, the other from 1 to ∞. Evaluate those integrals that converge.

39. $\displaystyle\int_{0}^{\infty} \frac{1}{x+x^2}\,dx$

40. $\displaystyle\int_{0}^{\infty} \frac{1}{x^2+x^4}\,dx$

41. $\displaystyle\int_{0}^{\infty} \frac{1}{x^{1/2}+x^{3/2}}\,dx$

42. $\displaystyle\int_{0}^{\infty} \frac{1}{x^{2/3}+x^{4/3}}\,dx$

In Problems 43 through 46, find all real number values of k for which the given improper integral converges. Evaluate the integral for those values of k.

43. $\displaystyle\int_{0}^{1} \frac{1}{x^k}\,dx$

44. $\displaystyle\int_{1}^{\infty} \frac{1}{x^k}\,dx$

45. $\displaystyle\int_{0}^{1} x^k\ln x\,dx$

46. $\displaystyle\int_{1}^{\infty} \frac{1}{x(\ln x)^k}\,dx$

47. Beginning with the definition of the gamma function in Eq. (7), integrate by parts to show that

$$\Gamma(x+1) = x\Gamma(x)$$

for every positive real number x.

48. Explain how to apply the result of Problem 47 n times in succession to show that if n is a positive integer, then $\Gamma(n+1) = n!\Gamma(1) = n!$.

Problems 49 through 51 deal with Gabriel's horn, the surface obtained by revolving the curve $y = 1/x$, $x \geqq 1$, around the x-axis (Fig. 9.8.12).

49. Show that the area under the curve $y = 1/x$, $x \geqq 1$, is infinite.

50. Show that the volume of revolution enclosed by Gabriel's horn is finite, and compute it.

51. Show that the surface area of Gabriel's horn is infinite. (*Suggestion:* Let A_t denote the surface area from $x = 1$ to $x = t > 1$. Prove that $A_t > 2\pi \ln t$.) In any case, the implication is that we could fill Gabriel's horn with a finite amount of paint (Problem 50), but no finite amount suffices to paint its surface.

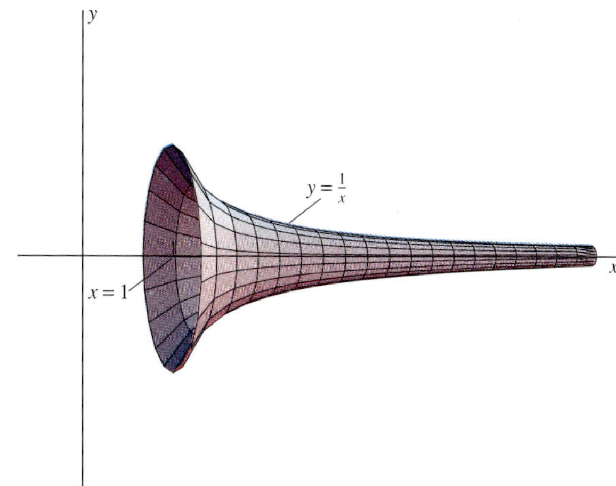

Fig. 9.8.12 Gabriel's horn (Problems 49 through 51)

52. Show that

$$\int_{-\infty}^{\infty} \frac{1+x}{1+x^2}\,dx$$

diverges, but that

$$\lim_{t\to\infty} \int_{-t}^{t} \frac{1+x}{1+x^2}\,dx = \pi.$$

53. Use the substitution $x = e^{-u}$ and the fact that $\Gamma(n+1) = n!$ (Problem 48) to prove that if m and n are fixed but arbitrary positive integers, then

$$\int_0^1 x^m (\ln x)^n \, dx = \frac{n!(-1)^n}{(m+1)^{n+1}}.$$

54. Consider a perpetual annuity under which you and your heirs will be paid at the rate of $10 + t$ thousand dollars per year t years hence. Thus you will receive \$20000 ten years from now, your heirs will receive \$110,000 one hundred years from now, and so on. Assuming a constant annual interest rate of 10%, show that the present value of this perpetuity is

$$P = \int_0^\infty (10 + t)e^{-t/10} \, dt,$$

and then evaluate this improper integral.

55. A "semi-infinite" uniform rod occupies the nonnegative x-axis ($x \geqq 0$) and has linear density δ; that is, a segment of length dx has mass $\delta \, dx$. Show that the force of gravitational attraction that the rod exerts on a point mass m at $(-a, 0)$ is

$$F = \int_0^\infty \frac{Gm\delta}{(a+x)^2} \, dx = \frac{Gm\delta}{a}.$$

56. A rod of linear density δ occupies the entire y-axis. A point mass m is located at $(a, 0)$ on the x-axis, as indicated in Fig. 9.8.13. Show that the total (horizontal) gravitational attraction that the rod exerts on m is

$$F = \int_{-\infty}^\infty \frac{Gm\delta \cos\theta}{r^2} \, dy = \frac{2Gm\delta}{a},$$

where $r^2 = a^2 + y^2$ and $\cos\theta = a/r$.

57. Verify Eq. (9) by substituting $x = u^2$ in the integral whose value is $\Gamma(\frac{1}{2})$.

58. Given the fact that

$$\int_0^\infty e^{-x^2} \, dx = \frac{1}{2}\sqrt{\pi},$$

find the volume of the unbounded solid that is obtained by revolving around the x-axis the unbounded region R between the x-axis and the curve $y = e^{-x^2}$ for $x \geqq 0$.

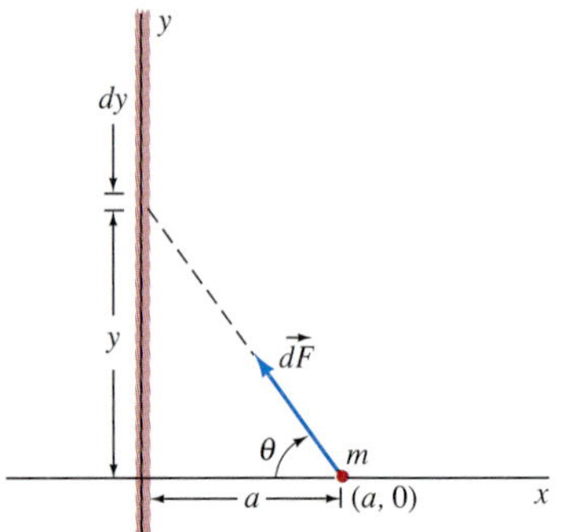

Fig. 9.8.13 Gravitational attraction exerted on a point mass by an infinite rod (Problem 56)

59. Find the volume of the unbounded solid that is obtained by revolving around the y-axis the region R of Problem 58.

60. Recall from Problem 47 that $\Gamma(x+1) = x\Gamma(x)$ if $x > 0$. Suppose that n is a positive integer. Use Eq. (9) to establish that

$$\Gamma\left(n + \frac{1}{2}\right) = \frac{1 \cdot 3 \cdot 5 \cdots (2n-1)}{2^n}\sqrt{\pi}.$$

61. (a) Suppose that $k > 1$. Use integration by parts to show that

$$\int_0^\infty x^k \exp(-x^2) \, dx = \frac{k-1}{2}\int_0^\infty x^{k-2} \exp(-x^2) \, dx.$$

(b) Suppose that n is a positive integer. Prove that

$$\int_0^\infty x^{n-1} \exp(-x^2) \, dx = \frac{1}{2}\Gamma\left(\frac{n}{2}\right).$$

62. Suppose that you win the Florida lottery and decide to use part of your winnings to purchase a perpetual annuity that will pay you and your heirs \$10000 per year (forever). Assuming an annual interest rate of 6%, what is a fair price for an insurance company to charge you for such an annuity?

9.8 PROJECT: NUMERICAL APPROXIMATION OF IMPROPER INTEGRALS

The improper integral

$$\int_0^\infty \exp(-x^2) \, dx$$

is important in applications ranging from probability and statistics (political polling, for example) to traffic flow and the theory of heat. Because the function $f(x) = \exp(-x^2)$ has no elementary antiderivative as a finite combination of familiar functions, a simple and direct evaluation of the limit

$$\int_0^\infty \exp(-x^2)\,dx = \lim_{b\to\infty} \int_0^b \exp(-x^2)\,dx \tag{1}$$

by using only the fundamental theorem of calculus is not feasible. But the fact that $\exp(-x^2) \leqq e^{-x}$ for $x \geqq 1$ implies that the improper integral in Eq. (1) converges rather than diverges to infinity. (Can you explain why?)

We will see in Section 14.4 that the exact value of the integral in Eq. (1) is given by

$$\int_0^\infty \exp(-x^2)\,dx = \frac{\sqrt{\pi}}{2} \approx 0.88622693. \tag{2}$$

To verify this value numerically, we used the **INTEGRATE** key on an HP-48SX calculator to compute values of the integral

$$\int_0^b \exp(-x^2)\,dx$$

with successively larger values of the upper limit b. The results appear in Fig. 9.8.14.

b	$\int_0^b e^{-x^2}dx$
1	0.74682 413
2	0.88208 139
3	0.88620 735
4	0.88622 691
5	0.88622 693
10	0.88622 693
100	0.88622 693

Fig. 9.8.14 Verifying the numerical value in Eq. (2)

As an alternative to the **INTEGRATE** command on a scientific calculator or computer algebra system, you could use the numerical integration methods described in the project of Section 5.4. If you use a method such as Simpson's approximation, it is a good idea to increase the accuracy by doubling, say, the number of subintervals with each increase in the value b of the upper limit.

Use one of the techniques indicated here to verify numerically the values specified in Problems 1 through 6.

1. $\displaystyle\int_0^\infty x^5 e^{-x}\,dx = 120$

2. $\displaystyle\int_0^\infty \frac{\sin x}{x}\,dx = \frac{\pi}{2}$

3. $\displaystyle\int_0^\infty \frac{1}{x^2+2}\,dx = \frac{\pi}{2\sqrt{2}}$

4. $\displaystyle\int_0^\infty \exp(-x^2)\cos 2x\,dx = \frac{\pi}{2e}$

5. $\displaystyle\int_0^\infty \frac{1-e^{-3x}}{x}\,dx = \frac{\ln 10}{2}$

6. $\displaystyle\int_0^\infty \sin(x^2)\,dx = \frac{\sqrt{\pi}}{2\sqrt{2}}$

SUMMARY *Strategy and Tactics for Integration*

When you confront the problem of evaluating a particular integral, you must first decide which of the several methods of this chapter to try. There are only two *general* methods of integration:

1. Integration by substitution (Section 9.2), and
2. Integration by parts (Section 9.3).

These are the analogues for integration of the chain rule and product rule, respectively, for differentiation.

Look first at the given integral to see whether you can spot a substitution that would transform it into an elementary or familiar integral or one likely to be found in an integral table. In the case of an integral

$$\int f(x)q(x)\,dx,$$

whose integral is an unfamiliar product of two functions, one of which is easily differentiated and the other easily integrated, then an attempt to integrate by parts is indicated.

Beyond these two general methods, the chapter deals with a number of *special* methods. In the case of an integral that is obviously trigonometric, $\int \text{trig}(x)\,dx$, the simple "spin-off" methods of Section 9.4 may succeed. Remember that reduction formulas [such as Eq. (5) and Problems 53 and 54 of Section 9.3] are available for integrating an integral power of a single trigonometric function.

Any integral of a rational function—that is, an integral of the form

$$\int \frac{p(x)}{q(x)}\,dx,$$

where the integrand is a quotient of polynomials—can be evaluated by the method of partial fractions (Section 9.5). If the degree of the numerator is not less than that of the denominator—that is, if the rational function is not proper—first use long division to express it as the sum of a polynomial (easily integrated) and a proper rational fraction. Then decompose the latter into partial fractions. Partial fractions corresponding to linear factors are easily integrated, and those corresponding to irreducible quadratic factors can be integrated by completing the square and making (if necessary) a trigonometric substitution. As we explained in Section 9.7, the trigonometric integrals that result can always be evaluated.

In the case of an integral involving $\sqrt{ax^2 + bx + c}$, first complete the square (Section 9.7) and then rationalize the integral by making an appropriate trigonometric substitution (Section 9.6). This will leave you with a trigonometric integral.

Some additional special substitutions are introduced in the Miscellaneous Problems that follow. Notable among these is the substitution

$$u = \tan\frac{\theta}{2},$$

which transforms any integral $\int R(\sin\theta, \cos\theta)\,d\theta$ of a rational function of $\sin\theta$ and $\cos\theta$ into an integral of a rational function of u. The latter integral can then be evaluated by the method of partial fractions.

A final comment: Computer algebra systems are increasingly used for the evaluation of integrals such as those studied in this chapter. Nevertheless, the availability of these systems is no panacea. For instance, such computer systems are likely to be stumped by the integral

$$\int (1 + \ln x)\sqrt{1 + (x \ln x)^2}\,dx.$$

But you probably notice that the substitution

$$u = x \ln x, \quad du = (1 + \ln x)\,dx$$

transforms this integral into the integral

$$\int \sqrt{1 + u^2}\,du,$$

which is amenable to trigonometric substitution (and can be found in almost any integral table). Thus the human factor remains—thankfully—essential.

CHAPTER 9 *Miscellaneous Problems*

Evaluate the integrals in Problems 1 through 100.

1. $\displaystyle\int \frac{1}{(1+x)\sqrt{x}}\,dx$ (*Suggestion:* Let $x = u^2$.)

2. $\displaystyle\int \frac{\sec^2 t}{1 + \tan t}\,dt$

3. $\displaystyle\int \sin x \sec x\,dx$

4. $\displaystyle\int \frac{\csc x \cot x}{1 + \csc^2 x}\,dx$

5. $\displaystyle\int \frac{\tan\theta}{\cos^2\theta}\,d\theta$

6. $\displaystyle\int \csc^4 x\,dx$

7. $\displaystyle\int x \tan^2 x\,dx$

8. $\displaystyle\int x^2 \cos^2 x\,dx$

9. $\displaystyle\int x^5\sqrt{2 - x^3}\,dx$

10. $\displaystyle\int \frac{1}{\sqrt{x^2+4}}\,dx$

11. $\displaystyle\int \frac{x^2}{\sqrt{25+x^2}}\,dx$

12. $\displaystyle\int (\cos x)\sqrt{4 - \sin^2 x}\,dx$

13. $\displaystyle\int \frac{1}{x^2 - x + 1}\,dx$

14. $\displaystyle\int \sqrt{x^2 + x + 1}\,dx$

15. $\displaystyle\int \frac{5x + 31}{3x^2 - 4x + 11}\,dx$

16. $\displaystyle\int \frac{x^4+1}{x^2+1}\,dx$

17. $\displaystyle\int \sqrt{x^4 + x^7}\,dx$

18. $\displaystyle\int \frac{\sqrt{x}}{1+x}\,dx$ (*Suggestion:* Let $x = u^2$.)

19. $\displaystyle\int \frac{\cos x}{\sqrt{4 - \sin^2 x}}\,dx$

20. $\displaystyle\int \frac{\cos 2x}{\cos x}\,dx$

21. $\displaystyle\int \frac{\tan x}{\ln(\cos x)}\,dx$

22. $\displaystyle\int \frac{x^7}{\sqrt{1 - x^4}}\,dx$

23. $\displaystyle\int \ln(1 + x)\,dx$

24. $\displaystyle\int x \sec^{-1} x\,dx$

25. $\displaystyle\int \sqrt{x^2+9}\,dx$

26. $\displaystyle\int \frac{x^2}{\sqrt{4 - x^2}}\,dx$

27. $\displaystyle\int \sqrt{2x - x^2}\,dx$

28. $\displaystyle\int \frac{4x - 2}{x^3 - x}\,dx$

29. $\displaystyle\int \frac{x^4}{x^2 - 2}\,dx$

30. $\displaystyle\int \frac{\sec x \tan x}{\sec x + \sec^2 x}\,dx$

31. $\displaystyle\int \frac{x}{(x^2 + 2x + 2)^2}\,dx$

32. $\displaystyle\int \frac{x^{1/3}}{x^{1/2} + x^{1/4}}\,dx$ (*Suggestion:* Let $x = u^{12}$.)

33. $\displaystyle\int \frac{1}{1 + \cos 2\theta}\,d\theta$

34. $\displaystyle\int \frac{\sec x}{\tan x}\,dx$

35. $\displaystyle\int \sec^3 x \tan^3 x\,dx$

36. $\displaystyle\int x^2 \tan^{-1} x\,dx$

37. $\displaystyle\int x(\ln x)^3\,dx$

38. $\displaystyle\int \frac{1}{x\sqrt{1+x^2}}\,dx$

39. $\displaystyle\int e^x\sqrt{1 + e^{2x}}\,dx$

40. $\displaystyle\int \frac{x}{\sqrt{4x - x^2}}\,dx$

41. $\displaystyle\int \frac{1}{x^3\sqrt{x^2 - 9}}\,dx$

42. $\displaystyle\int \frac{x}{(7x+1)^{17}}\,dx$

43. $\displaystyle\int \frac{4x^2 + x + 1}{4x^3 + x}\,dx$

44. $\displaystyle\int \frac{4x^3 - x + 1}{x^3 + 1}\,dx$

45. $\displaystyle\int \tan^2 x \sec x\,dx$

46. $\displaystyle\int \frac{x^2 + 2x + 2}{(x+1)^3}\,dx$

47. $\displaystyle\int \frac{x^4 + 2x + 2}{x^5 + x^4}\,dx$

48. $\displaystyle\int \frac{8x^2 - 4x + 7}{(x^2+1)(4x+1)}\,dx$

49. $\displaystyle\int \frac{3x^5 - x^4 + 2x^3 - 12x^2 - 2x + 1}{(x^3 - 1)^2}\,dx$

50. $\displaystyle\int \frac{x}{x^4 + 4x^2 + 8}\,dx$

51. $\displaystyle\int (\ln x)^6\,dx$

52. $\displaystyle\int \frac{(1 + x^{2/3})^{3/2}}{x^{1/3}}\,dx$ (*Suggestion:* Let $x = u^3$.)

53. $\displaystyle\int \frac{(\arcsin x)^2}{\sqrt{1 - x^2}}\,dx$

54. $\displaystyle\int \frac{1}{x^{3/2}(1 + x^{1/3})}\,dx$ (*Suggestion:* Let $x = u^6$.)

55. $\displaystyle\int \tan^3 z\,dz$

56. $\displaystyle\int \sin^2\omega \cos^4\omega\,d\omega$

57. $\displaystyle\int \frac{xe^{x^2}}{1 + e^{2x^2}}\,dx$

58. $\displaystyle\int \frac{\cos^3 x}{\sqrt{\sin x}}\,dx$

59. $\displaystyle\int x^3 e^{-x^2}\,dx$

60. $\displaystyle\int \sin\sqrt{x}\,dx$

61. $\int \frac{\arcsin x}{x^2}\,dx$

62. $\int \sqrt{x^2 - 9}\,dx$

63. $\int x^2\sqrt{1 - x^2}\,dx$

64. $\int x\sqrt{2x - x^2}\,dx$

65. $\int \frac{x - 2}{4x^2 + 4x + 1}\,dx$

66. $\int \frac{2x^2 - 5x - 1}{x^3 - 2x^2 - x + 2}\,dx$

67. $\int \frac{e^{2x}}{e^{2x} - 1}\,dx$

68. $\int \frac{\cos x}{\sin^2 x - 3\sin x + 2}\,dx$

69. $\int \frac{2x^3 + 3x^2 + 4}{(x + 1)^4}\,dx$

70. $\int \frac{\sec^2 x}{\tan^2 x + 2\tan x + 2}\,dx$

71. $\int \frac{x^3 + x^2 + 2x + 1}{x^4 + 2x^2 + 1}\,dx$

72. $\int \sin x \cos 3x\,dx$

73. $\int x^5\sqrt{x^3 - 1}\,dx$

74. $\int \ln(x^2 + 2x)\,dx$

75. $\int \frac{\sqrt{1 + \sin x}}{\sec x}\,dx$

76. $\int \frac{1}{x^{2/3}(1 + x^{2/3})}\,dx$

77. $\int \frac{\sin x}{\sin 2x}\,dx$

78. $\int \sqrt{1 + \cos t}\,dt$

79. $\int \sqrt{1 + \sin t}\,dt$

80. $\int \frac{\sec^2 t}{1 - \tan^2 t}\,dt$

81. $\int \ln(x^2 + x + 1)\,dx$

82. $\int e^x \sin^{-1}(e^x)\,dx$

83. $\int \frac{\arctan x}{x^2}\,dx$

84. $\int \frac{x^2}{\sqrt{x^2 - 25}}\,dx$

85. $\int \frac{x^3}{(x^2 + 1)^2}\,dx$

86. $\int \frac{1}{x\sqrt{6x - x^2}}\,dx$

87. $\int \frac{3x + 2}{(x^2 + 4)^{3/2}}\,dx$

88. $\int x^{3/2}\ln x\,dx$

89. $\int \frac{\sqrt{1 + \sin^2 x}}{\sec x \csc x}\,dx$

90. $\int \frac{\exp(\sqrt{\sin x})}{(\sec x)\sqrt{\sin x}}\,dx$

91. $\int xe^x \sin x\,dx$

92. $\int x^2 e^{x^{3/2}}\,dx$

93. $\int \frac{\arctan x}{(x - 1)^3}\,dx$

94. $\int \ln(1 + \sqrt{x})\,dx$

95. $\int \frac{2x + 3}{\sqrt{3 + 6x - 9x^2}}\,dx$

96. $\int \frac{1}{\sqrt{e^{2x} - 1}}\,dx$

97. $\int \frac{x^4}{(x - 1)^2}\,dx$ (*Suggestion:* Let $u = x - 1$.)

98. $\int x^{3/2}\tan^{-1}(\sqrt{x})\,dx$

99. $\int \operatorname{arcsec}(\sqrt{x})\,dx$

100. $\int x\sqrt{\frac{1 - x^2}{1 + x^2}}\,dx$

101. Find the area of the surface generated by revolving the curve $y = \cosh x, 0 \leqq x \leqq 1$, around the x-axis.

102. Find the length of the curve $y = e^{-x}, 0 \leqq x \leqq 1$.

103. (a) Find the area A_t of the surface generated by revolving the curve $y = e^{-x}$, $0 \leqq x \leqq t$, around the x-axis. (b) Find $\lim_{t\to\infty} A_t$.

104. (a) Find the area A_t of the surface generated by revolving the curve $y = 1/x$, $1 \leqq x \leqq t$, around the x-axis. (b) Find $\lim_{t\to\infty} A_t$.

105. Find the area of the surface generated by revolving the curve $y = \sqrt{x^2 - 1}, 1 \leqq x \leqq 2$, around the x-axis.

106. (a) Derive the reduction formula

$$\int x^m(\ln x)^n\,dx = \frac{1}{m + 1}x^{m+1}(\ln x)^n - \frac{n}{m + 1}\int x^m(\ln x)^{n-1}\,dx.$$

(b) Evaluate $\int_1^e x^3(\ln x)^3\,dx$.

107. Derive the reduction formula

$$\int \sin^m x \cos^n x\,dx =$$

$$-\frac{1}{m + n}\sin^{m-1}x\cos^{n+1}x + \frac{m - 1}{m + n}\int \sin^{m-2}x\cos^n x\,dx.$$

108. Use the reduction formulas of Problem 107 here and Problem 54 of Section 9.3 to evaluate

$$\int_0^{\pi/2} \sin^6 x \cos^5 x\,dx.$$

109. Find the area bounded by the curve $y^2 = x^5(2 - x)$, $0 \leqq x \leqq 2$. (*Suggestion:* Substitute $x = 2\sin^2\theta$; then use the result of Problem 58 of Section 9.3.)

110. Show that

$$0 < \int_0^1 \frac{t^4(1 - t)^4}{1 + t^2}\,dt$$

and that

$$\int_0^1 \frac{t^4(1 - t)^4}{1 + t^2}\,dt = \frac{22}{7} - \pi.$$

111. Evaluate

$$\int_0^1 t^4(1 - t)^4\,dt;$$

then apply the results of Problem 110 to conclude that

$$\frac{22}{7} - \frac{1}{630} < \pi < \frac{22}{7} - \frac{1}{1260}.$$

Thus $3.1412 < \pi < 3.1421$.

112. Find the length of the curve $y = \frac{4}{5}x^{5/4}, 0 \leqq x \leqq 1$.

113. Find the length of the curve $y = \frac{4}{3}x^{3/4}, 1 \leqq x \leqq 4$.

114. An initially empty water tank is shaped like a cone whose axis is vertical. Its vertex is at the bottom; the cone is 9 ft deep and has a top radius of 4.5 ft. Beginning at time $t = 0$, water is poured into this tank at 50 ft^3/min. Meanwhile, water leaks from a hole at the tank's bottom at the rate of $10\sqrt{y}$ cubic feet per minute, where y is the depth of water in the tank. (This is consistent with Torricelli's law of draining.) How long does it take to fill the tank?

115. (a) Evaluate

$$\int \frac{1}{1 + e^x + e^{-x}}\,dx.$$

(b) Explain why your substitution in part (a) suffices to integrate any rational function of e^x.

116. (a) The equation $x^3 + x + 1 = 0$ has at least one real root r. Use Newton's method to find it, accurate to at least two places. (b) Use long division to find (approximately) the irreducible quadratic factor of $x^3 + x + 1$. (c) Use the factorization obtained in part (b) to evaluate (approximately)

$$\int_0^1 \frac{1}{x^3 + x + 1}\,dx.$$

117. Evaluate $\displaystyle\int \frac{1}{1 + e^x}\,dx$.

118. The integral

$$\int \frac{1 + 2x^2}{x^5(1 + x^2)^3}\,dx = \int \frac{x + 2x^3}{(x^4 + x^2)^3}\,dx$$

would require that you solve 11 equations in 11 unknowns if you were to use the method of partial fractions to evaluate it. Use the substitution $u = x^4 + x^2$ to evaluate it much more simply.

119. Evaluate

$$\int \sqrt{\tan\theta}\,d\theta.$$

(*Suggestion:* First substitute $u = \tan\theta$. Then substitute $u = x^2$. Finally, use the method of partial fractions; see Problem 48 of Section 9.7.)

120. Prove that if $p(x)$ is a polynomial, then the substitution $u^n = (ax + b)/(cx + d)$ transforms the integral

$$\int p(x)\left(\frac{ax + b}{cx + d}\right)^{1/n} dx$$

into the integral of a rational function of u. (The substitution indicated here is called a *rationalizing substitution;* its name comes from the fact that it converts the integrand into a *rational* function of u.)

In Problems 121 through 129, use a rationalizing substitution as indicated in Problem 120.

121. $\displaystyle\int x^3\sqrt{3x - 2}\,dx$

122. $\displaystyle\int x^3\sqrt[3]{x^2 + 1}\,dx$

123. $\displaystyle\int \frac{x^3}{(x^2 - 1)^{4/3}}\,dx$

124. $\displaystyle\int x^2(x - 1)^{3/2}\,dx$

125. $\displaystyle\int \frac{x^5}{\sqrt{x^3 + 1}}\,dx$

126. $\displaystyle\int x^7\sqrt[3]{x^4 + 1}\,dx$

127. $\displaystyle\int \sqrt{\frac{1 + x}{1 - x}}\,dx$

128. $\displaystyle\int \frac{x}{\sqrt{x + 1}}\,dx$

129. $\displaystyle\int \frac{\sqrt[3]{x + 1}}{x}\,dx$

130. Substitute $x = u^2$ to find

$$\int \sqrt{1 + \sqrt{x}}\,dx.$$

131. Substitute $u^2 = 1 + e^{2x}$ to find

$$\int \sqrt{1 + e^{2x}}\,dx.$$

132. Find the area A of the surface obtained by revolving the curve $y = \frac{2}{3}x^{3/2}$, $3 \leqq x \leqq 8$, around the x-axis. (*Suggestion:* Substitute $x = u^2$ into the surface area integral. *Note:* $A \approx 732.39$.)

133. Find the area bounded by one loop of the curve

$$y^2 = x^2(1 - x), \qquad 0 \leqq x \leqq 1.$$

134. Find the area bounded by the loop of the curve

$$y^2 = x^2\left(\frac{1 - x}{1 + x}\right), \qquad 0 \leqq x \leqq 1.$$

More General Trigonometric Integrals

As a last resort, any trigonometric integral can be transformed into an integral

$$\int R(\sin\theta, \cos\theta)\,d\theta \tag{1}$$

of sines and cosines. If the integrand in Eq. (1) is a quotient of polynomials in the variables $\sin\theta$ and $\cos\theta$, then the special substitution

$$u = \tan\frac{\theta}{2} \tag{2}$$

suffices for its evaluation.

To carry out the substitution indicated in Eq. (2), we must express $\sin\theta$, $\cos\theta$, and $d\theta$ in terms of u and du. Note first that

$$\theta = 2\tan^{-1}u, \quad \text{so} \quad d\theta = \frac{2\,du}{1 + u^2}. \tag{3}$$

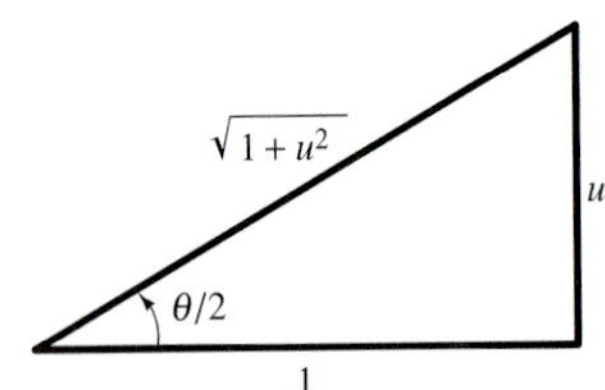

Fig. 9.MP.1 The special rationalizing substitution $u = \tan\dfrac{\theta}{2}$

We see from the triangle in Fig. 9.MP.1 that

$$\sin\frac{\theta}{2} = \frac{u}{\sqrt{1+u^2}}, \qquad \cos\frac{\theta}{2} = \frac{1}{\sqrt{1+u^2}}.$$

Hence

$$\sin\theta = 2\sin\frac{\theta}{2}\cos\frac{\theta}{2} = \frac{2u}{1+u^2}, \tag{4}$$

$$\cos\theta = \cos^2\frac{\theta}{2} - \sin^2\frac{\theta}{2} = \frac{1-u^2}{1+u^2}. \tag{5}$$

These substitutions will convert the integral in Eq. (1) into an integral of a rational function of u. The latter can then be evaluated by the methods of Section 9.5.

EXAMPLE

$$\int \frac{1}{5+3\cos\theta}\,d\theta = \int \frac{1}{5+3\cdot\dfrac{1-u^2}{1+u^2}}\cdot\frac{2}{1+u^2}\,du$$

$$= \int \frac{2}{8+2u^2}\,du = \int \frac{1}{4+u^2}\,du$$

$$= \frac{1}{2}\tan^{-1}\frac{u}{2} + C = \frac{1}{2}\tan^{-1}\left(\frac{1}{2}\tan\frac{\theta}{2}\right) + C.$$

■

In Problems 135 through 142, use the rationalizing substitution given in Eqs. (2) through (5).

135. $\displaystyle\int \frac{1}{1+\cos\theta}\,d\theta$

136. $\displaystyle\int \frac{1}{5+4\cos\theta}\,d\theta$

137. $\displaystyle\int \frac{1}{1+\sin\theta}\,d\theta$

138. $\displaystyle\int \frac{1}{(1-\cos\theta)^2}\,d\theta$

139. $\displaystyle\int \frac{1}{\sin\theta+\cos\theta}\,d\theta$

140. $\displaystyle\int \frac{1}{2+\sin\phi+\cos\phi}\,d\phi$

141. $\displaystyle\int \frac{\sin\theta}{2+\cos\theta}\,d\theta$

142. $\displaystyle\int \frac{\sin\theta-\cos\theta}{\sin\theta+\cos\theta}\,d\theta$

143. (a) Substitute $u = \tan(\theta/2)$ to show that

$$\int \sec\theta\,d\theta = \ln\left|\frac{1+\tan\dfrac{\theta}{2}}{1-\tan\dfrac{\theta}{2}}\right| + C.$$

(b) Use the trigonometric identity

$$\tan\frac{\theta}{2} = \sqrt{\frac{1-\cos\theta}{1+\cos\theta}}$$

to derive our earlier formula

$$\int \sec\theta\,d\theta = \ln|\sec\theta + \tan\theta| + C$$

from the solution in part (a).

144. (a) Use the method of Problem 143 to show that

$$\int \csc\theta\,d\theta = \ln\left|\tan\frac{\theta}{2}\right| + C.$$

(b) Use trigonometric identities to derive the formula

$$\int \csc\theta\,d\theta = \ln|\csc\theta - \cot\theta| + C$$

from part (a).

CHAPTER 10

POLAR COORDINATES AND PLANE CURVES

Pierre de Fermat (1601–1665)

Pierre de Fermat exemplifies the distinguished tradition of great amateurs in mathematics. Like his contemporary René Descartes, he was educated as a lawyer. But unlike Descartes, Fermat actually practiced law as his profession and served in the regional parliament, devoting only his leisure time to the study of mathematics and ancient manuscripts.

In a margin of one such manuscript (by the Greek mathematician Diophantus) was found a handwritten note that has remained an enigma ever since. Fermat asserts that for *no* integer $n > 2$ do positive integers x, y, and z exist such that $x^n + y^n = z^n$. For instance, although $15^2 + 8^2 = 17^2$, the sum of two (positive integer) cubes cannot be a cube. "I have found an admirable proof of this," Fermat wrote, "but this margin is too narrow to contain it." Despite the publication of many incorrect proofs, "Fermat's last theorem" remained unproved for three and one-half centuries. But in a June 1993 lecture, the British mathematician Andrew Wiles of Princeton University announced a long and complex proof of Fermat's last theorem. With some gaps in this proof now repaired, experts in the field agree that Fermat's last *conjecture* is, finally, a *theorem*.

Descartes and Fermat shared in the discovery of analytic geometry. But whereas Descartes typically used geometrical methods to solve algebraic equations (see the Chapter 1 opening), Fermat concentrated on the investigation of geometric curves defined by algebraic equations. For instance, he showed that the graph of an equation of the form $Ax^2 + Bxy + Cy^2 + Dx + Ey + F = 0$ is generally one of the "conic sections" described in this chapter.

The brilliantly colored left-hand photograph is a twentieth-century example of a geometric object defined by means of algebraic operations. Starting with the point $P(a, b)$ in the xy-plane, we interpret P as the complex number $c = a + bi$ and define the sequence $\{z_n\}$ of points of the complex plane iteratively (as in Section 3.9) by the equations

$$z_0 = c, \qquad z_{n+1} = z_n^2 + c.$$

If this sequence of points remains inside the circle $x^2 + y^2 = 4$ for all n, then the original point $P(a, b)$ is colored black. Otherwise, the color assigned to P is determined by the speed with which this sequence "escapes" that circular disk. The set of all black points is the famous *Mandelbrot set,* discovered in 1980 by the French mathematician Benoit Mandelbrot.

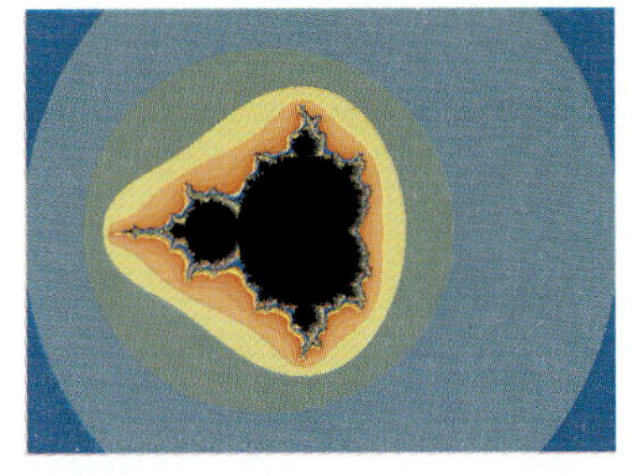

The object in the right-hand figure is a subset of that in the left-hand figure.

10.1 ANALYTIC GEOMETRY AND THE CONIC SECTIONS

Plane analytic geometry, a central topic of this chapter, is the use of algebra and calculus to study the properties of curves in the xy-plane. The ancient Greeks used deductive reasoning and the methods of axiomatic Euclidean geometry to study lines, circles, and the **conic sections** (parabolas, ellipses, and hyperbolas). The properties of conic sections have played an important role in diverse scientific applications since the seventeenth century, when Kepler discovered—and Newton explained—the fact that the orbits of planets and other bodies in the solar system are conic sections.

The French mathematicians Descartes and Fermat, working almost independently of each other, initiated analytic geometry in 1637. The central idea of analytic geometry is the correspondence between an equation $F(x, y) = 0$ and its **locus** (typically, a curve), the set of all those points (x, y) in the plane with coordinates that satisfy this equation.

A central idea of analytic geometry is this: Given a geometric locus or curve, its properties can be derived algebraically or analytically from its defining equation $F(x, y) = 0$. For example, suppose that the equation of a given curve turns out to be the linear equation

$$Ax + By = C, \tag{1}$$

where A, B, and C are constants with $B \neq 0$. This equation may be written in the form

$$y = mx + b, \tag{2}$$

where $m = -A/B$ and $b = C/B$. But Eq. (2) is the slope-intercept equation of the straight line with slope m and y-intercept b. Hence the given curve is this straight line. We use this approach in Example 1 to show that a specific geometrically described locus is a particular straight line.

EXAMPLE 1 Prove that the set of all points equidistant from the points $(1, 1)$ and $(5, 3)$ is the perpendicular bisector of the line segment that joins these two points.

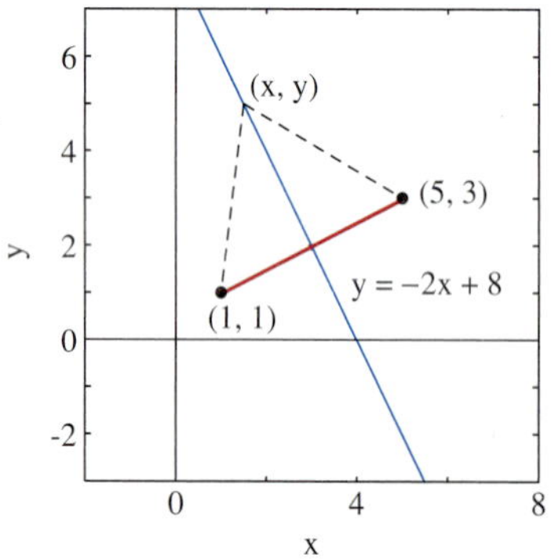

Fig. 10.1.1 The perpendicular bisector of Example 1

Solution The typical point $P(x, y)$ in Fig. 10.1.1 is equally distant from $(1, 1)$ and $(5, 3)$ if and only if

$$\begin{aligned}(x-1)^2 + (y-1)^2 &= (x-5)^2 + (y-3)^2;\\ x^2 - 2x + 1 + y^2 - 2y + 1 &= x^2 - 10x + 25 + y^2 - 6y + 9;\\ 2x + y &= 8;\\ y &= -2x + 8. \end{aligned} \tag{3}$$

Thus the given locus is the straight line in Eq. (3) whose slope is -2. The straight line through $(1, 1)$ and $(5, 3)$ has equation

$$y - 1 = \frac{1}{2}(x - 1) \tag{4}$$

and thus has slope $\frac{1}{2}$. Because the product of the slopes of these two lines is -1, it follows (from Theorem 2 in Section 1.2) that these lines are perpendicular. If we solve Eqs. (3) and (4) simultaneously, we find that the intersection of these lines is, indeed, the midpoint $(3, 2)$ of the given line segment. Thus the locus described is the perpendicular bisector of this line segment. ■

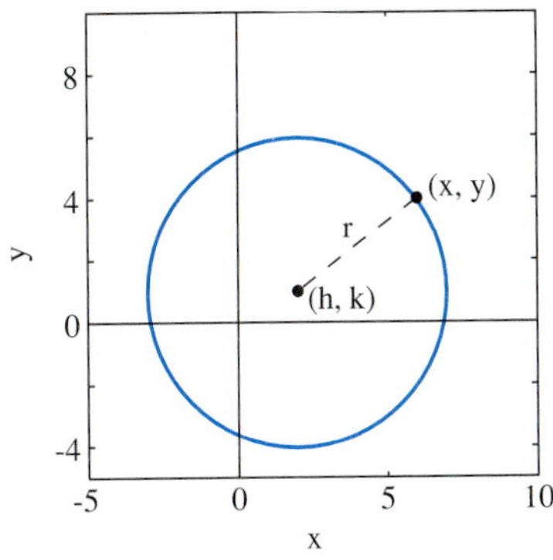

Fig. 10.1.2 The circle with center (h, k) and radius r

The circle shown in Fig. 10.1.2 has center (h, k) and radius r. It may be described geometrically as the set or locus of all points $P(x, y)$ whose distance from (h, k) is r. The distance formula then gives

$$(x - h)^2 + (y - k)^2 = r^2 \tag{5}$$

as the equation of this circle. In particular, if $h = k = 0$, then Eq. (5) takes the simple form

$$x^2 + y^2 = r^2. \tag{6}$$

We can see directly from this equation, without further reference to the definition of *circle*, that a circle centered at the origin has the following symmetry properties:

- *Symmetry around the x-axis*: The equation of the curve is unchanged when y is replaced with $-y$.
- *Symmetry around the y-axis*: The equation of the curve is unchanged when x is replaced with $-x$.
- *Symmetry with respect to the origin*: The equation of the curve is unchanged when x is replaced with $-x$ and y is replaced with $-y$.
- *Symmetry around the* 45° *line* $y = x$: The equation is unchanged when x and y are interchanged.

The relationship between Eqs. (5) and (6) is an illustration of the *translation principle* stated informally in Section 1.3. Imagine a translation (or "slide") of the plane that moves the point (x, y) to the new position $(x + h, y + k)$. Under such a translation, a curve C is moved to a new curve. The equation of the new curve is easy to obtain from the old equation—we simply replace x with $x - h$ and y with $y - k$. Conversely, we can recognize a translated circle from its equation: Any equation of the form

$$x^2 + y^2 + Ax + By + C = 0 \tag{7}$$

can be rewritten in the form

$$(x - h)^2 + (y - k)^2 = p$$

by completing squares, as in Example 2 of Section 1.3. Thus the graph of Eq. (7) is either a circle (if $p > 0$), a single point (if $p = 0$), or no points at all (if $p < 0$). We use this approach in Example 2 to discover that the locus described is a particular circle.

EXAMPLE 2 Determine the locus of a point $P(x, y)$ if its distance $|AP|$ from $A(7, 1)$ is twice its distance $|BP|$ from $B(1, 4)$.

Solution The points A, B, and P appear in Fig. 10.1.3, along with a curve through P that represents the given locus. From

$$|AP|^2 = 4|BP|^2 \quad (\text{because } |AP| = 2\,|BP|),$$

we get the equation

$$(x - 7)^2 + (y - 1)^2 = 4[(x - 1)^2 + (y - 4)^2].$$

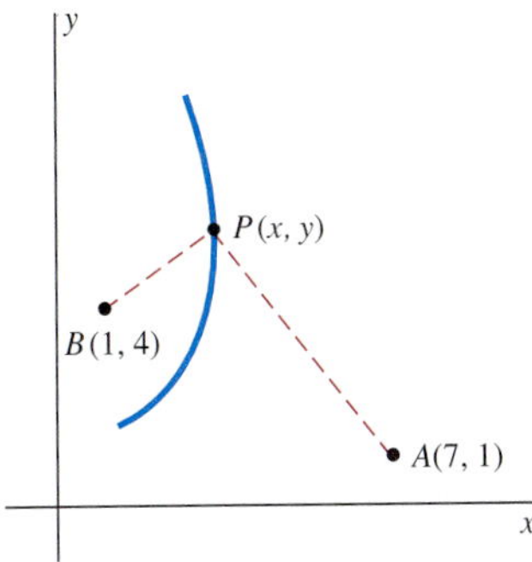

Fig. 10.1.3 The locus of Example 2

Hence

$$3x^2 + 3y^2 + 6x - 30y + 18 = 0;$$

$$x^2 + y^2 + 2x - 10y = -6;$$
$$(x + 1)^2 + (y - 5)^2 = 20.$$

Thus the locus is a circle with center $(-1, 5)$ and radius $r = \sqrt{20} = 2\sqrt{5}$. ■

Conic Sections

Conic sections are so named because they are the curves formed by a plane intersecting a cone. The cone used is a right circular cone with two *nappes* extending infinitely far in both directions (Fig. 10.1.4). There are three types of conic sections, as illustrated in Fig. 10.1.5. If the cutting plane is parallel to some generator of the cone (a line that, when revolving around an axis, forms the cone), then the curve of intersection is a *parabola.* If the plane is not parallel to a generator, then the curve of intersection is either a single closed curve—an *ellipse*—or a *hyperbola* with two *branches.*

Fig. 10.1.4 A cone with two nappes

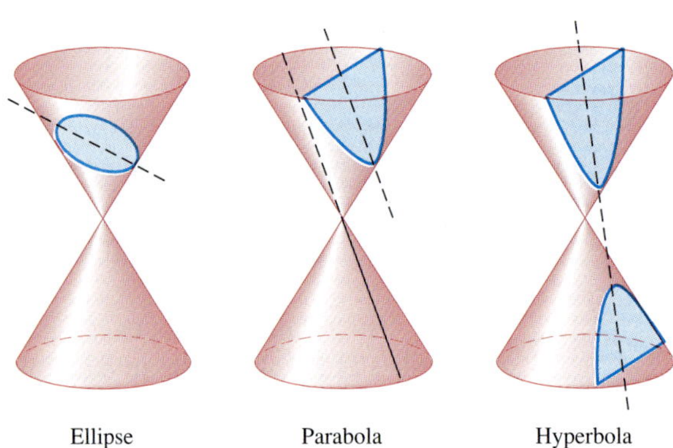
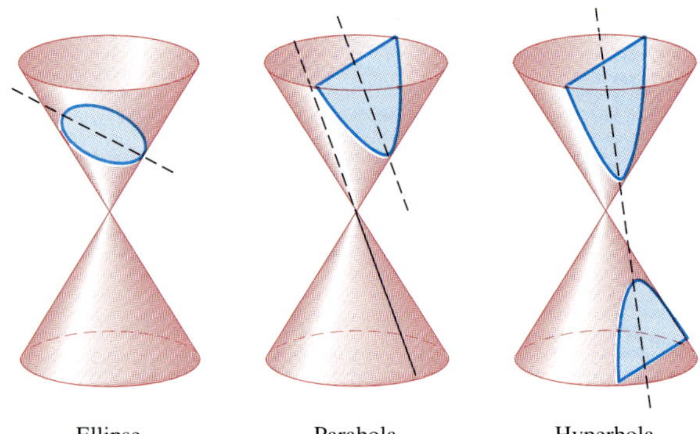

Fig. 10.1.5 The conic sections

In Appendix K we use methods of three-dimensional geometry to show that if an appropriate xy-coordinate system is set up in the intersecting plane, then the equations of the three conic sections take the following forms:

$$\text{Parabola:} \qquad y^2 = kx; \tag{8}$$

$$\text{Ellipse:} \qquad \frac{x^2}{a^2} + \frac{y^2}{b^2} = 1; \tag{9}$$

$$\text{Hyperbola:} \qquad \frac{x^2}{a^2} - \frac{y^2}{b^2} = 1. \tag{10}$$

In Sections 10.6 through 10.8 we discuss these conic sections on the basis of definitions that are two-dimensional—they do not require the three-dimensional setting of a cone and an intersecting plane. Example 3 illustrates one such approach to the conic sections.

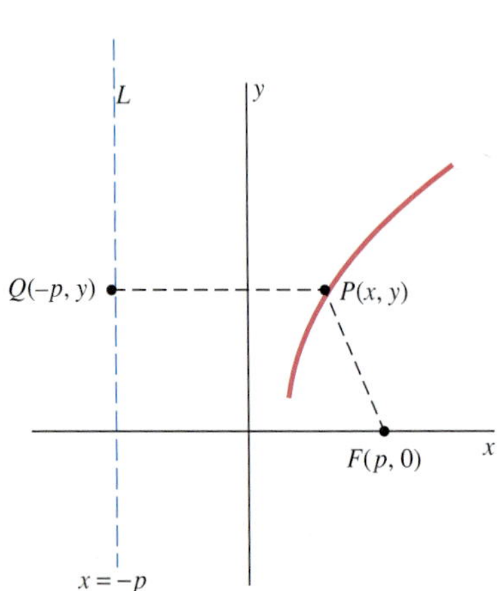

Fig. 10.1.6 The locus of Example 3

EXAMPLE 3 Let e be a given positive number (*not* to be confused with the natural logarithm base; in the context of conic sections, e stands for *eccentricity*). Determine the locus of a point $P(x, y)$ if its distance from the fixed point $F(p, 0)$ is e times its distance from the vertical line L whose equation is $x = -p$ (Fig. 10.1.6).

Solution Let PQ be the perpendicular from P to the line L. Then the condition

$$|PF| = e|PQ|$$

takes the analytic form

$$\sqrt{(x-p)^2+y^2} = e|x-(-p)|.$$

That is,

$$(x^2-2px+p^2)+y^2 = e^2(x^2+2px+p^2),$$

so

$$x^2(1-e^2)-2p(1+e^2)x+y^2 = -p^2(1-e^2). \tag{11}$$

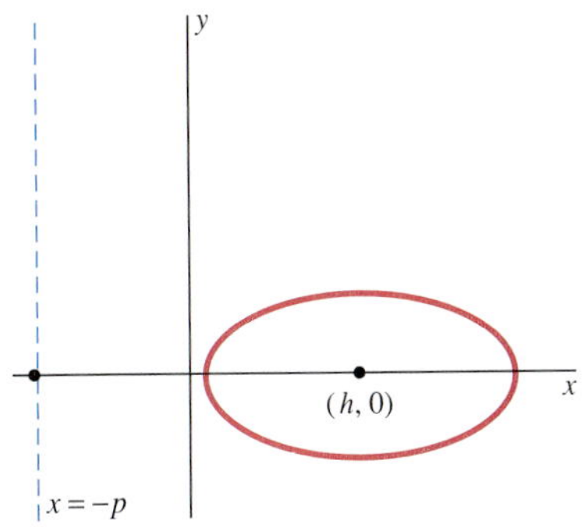

Fig. 10.1.7 An ellipse: $e < 1$ (Example 3)

▼ *Case* 1: $e = 1$. Then Eq. (11) reduces to

$$y^2 = 4px. \tag{12}$$

We see upon comparison with Eq. (8) that the locus of P is a *parabola* if $e = 1$.

▼ *Case* 2: $e < 1$. Dividing both sides of Eq. (11) by $1-e^2$, we get

$$x^2 - 2p\cdot\frac{1+e^2}{1-e^2}x + \frac{y^2}{1-e^2} = -p^2.$$

We now complete the square in x. The result is

$$\left(x - p\cdot\frac{1+e^2}{1-e^2}\right)^2 + \frac{y^2}{1-e^2} = p^2\left[\left(\frac{1+e^2}{1-e^2}\right)^2 - 1\right] = a^2.$$

This equation has the form

$$\frac{(x-h)^2}{a^2} + \frac{y^2}{b^2} = 1, \tag{13}$$

where

$$h = +p\cdot\frac{1+e^2}{1-e^2} \quad\text{and}\quad b^2 = a^2(1-e^2). \tag{14}$$

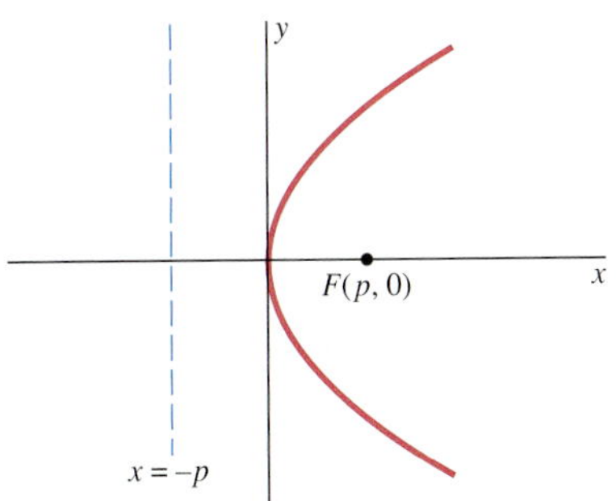

Fig. 10.1.8 A parabola: $e = 1$ (Example 3)

When we compare Eqs. (9) and (13), we see that if $e < 1$, then the locus of P is an *ellipse* with $(0, 0)$ translated to $(h, 0)$, as illustrated in Fig. 10.1.7.

▼ *Case* 3: $e > 1$. In this case, Eq. (11) reduces to a translated version of Eq. (10), so the locus of P is a *hyperbola*. The details, which are similar to those in case 2, are left for Problem 35.

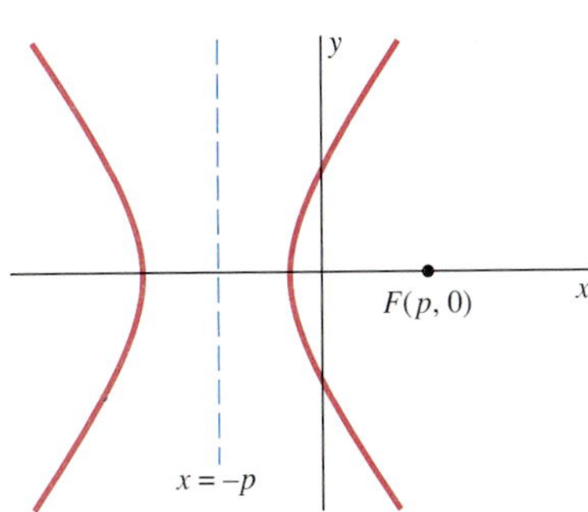

Fig. 10.1.9 A hyperbola: $e > 1$ (Example 3)

Thus the locus in Example 3 is a *parabola* if $e = 1$, an *ellipse* if $e < 1$, and a *hyperbola* if $e > 1$. The number e is called the **eccentricity** of the conic section. The point $F(p, 0)$ is commonly called its **focus** in the parabolic case. Figure 10.1.8 shows the parabola of case 1, and Fig. 10.1.9 illustrates the hyperbola of case 3. ■

If we begin with Eqs. (8) through (10), we can derive the general characteristics of the three conic sections shown in Figs. 10.1.7 through 10.1.9. For example, in the case of the parabola of Eq. (8) with $k > 0$, the curve passes through the origin, $x \geqq 0$ at each of the curve's points, $y \to \pm\infty$ as $x \to \infty$, and the graph is symmetric around the x-axis (because the curve is unchanged when y is replaced with $-y$).

In the case of the ellipse of Eq. (9), the graph must be symmetric around both coordinate axes. At each point (x, y) of the graph, we must have $|x| \leqq a$ and $|y| \leqq b$. The graph intersects the axes at the four points $(\pm a, 0)$ and $(0, \pm b)$.

Finally, the hyperbola of Eq. (10), or its alternative form

$$y = \pm \frac{b}{a}\sqrt{x^2 - a^2},$$

is symmetric around both coordinate axes. It meets the x-axis at the two points $(\pm a, 0)$ and has one branch consisting of points with $x \geqq a$ and has another branch where $x \leqq -a$. Also, $|y| \to \infty$ as $|x| \to \infty$.

10.1 PROBLEMS

In Problems 1 through 6, write an equation of the specified straight line.

1. The line through the point $(1,-2)$ that is parallel to the line with equation $x + 2y = 5$
2. The line through the point $(-3, 2)$ that is perpendicular to the line with equation $3x - 4y = 7$
3. The line that is tangent to the circle $x^2 + y^2 = 25$ at the point $(3,-4)$
4. The line that is tangent to the curve $y^2 = x + 3$ at the point $(6,-3)$
5. The line that is perpendicular to the curve $x^2 + 2y^2 = 6$ at the point $(2,-1)$
6. The perpendicular bisector of the line segment with endpoints $(-3, 2)$ and $(5,-4)$

In Problems 7 through 16, find the center and radius of the circle described in the given equation.

7. $x^2 + 2x + y^2 = 4$
8. $x^2 + y^2 - 4y = 5$
9. $x^2 + y^2 - 4x + 6y = 3$
10. $x^2 + y^2 + 8x - 6y = 0$
11. $4x^2 + 4y^2 - 4x = 3$
12. $4x^2 + 4y^2 + 12y = 7$
13. $2x^2 + 2y^2 - 2x + 6y = 13$
14. $9x^2 + 9y^2 - 12x = 5$
15. $9x^2 + 9y^2 + 6x - 24y = 19$
16. $36x^2 + 36y^2 - 48x - 108y = 47$

In Problems 17 through 20, show that the graph of the given equation consists either of a single point or of no points.

17. $x^2 + y^2 - 6x - 4y + 13 = 0$
18. $2x^2 + 2y^2 + 6x + 2y + 5 = 0$
19. $x^2 + y^2 - 6x - 10y + 84 = 0$
20. $9x^2 + 9y^2 - 6x - 6y + 11 = 0$

In Problems 21 through 24, write the equation of the specified circle.

21. The circle with center $(-1,-2)$ that passes through the point $(2, 3)$
22. The circle with center $(2,-2)$ that is tangent to the line $y = x + 4$
23. The circle with center $(6, 6)$ that is tangent to the line $y = 2x - 4$
24. The circle that passes through the points $(4, 6)$, $(-2,-2)$, and $(5,-1)$

In Problems 25 through 30, derive the equation of the set of all points $P(x, y)$ that satisfy the given condition. Then sketch the graph of the equation.

25. The point $P(x, y)$ is equally distant from the two points $(3, 2)$ and $(7, 4)$.
26. The distance from P to the point $(-2, 1)$ is half the distance from P to the point $(4,-2)$.
27. The point P is three times as far from the point $(-3, 2)$ as it is from the point $(5, 10)$.
28. The distance from P to the line $x = -3$ is equal to its distance from the point $(3, 0)$.
29. The sum of the distances from P to the points $(4, 0)$ and $(-4, 0)$ is 10.
30. The sum of the distances from P to the points $(0, 3)$ and $(0,-3)$ is 10.
31. Find all the lines through the point $(2, 1)$ that are tangent to the parabola $y = x^2$.
32. Find all lines through the point $(-1, 2)$ that are normal to the parabola $y = x^2$.
33. Find all lines that are normal to the curve $xy = 4$ and simultaneously are parallel to the line $y = 4x$.
34. Find all lines that are tangent to the curve $y = x^3$ and are also parallel to the line $3x - y = 5$.
35. Suppose that $e > 1$. Show that Eq. (11) of this section can be written in the form

$$\frac{(x-h)^2}{a^2} - \frac{y^2}{b^2} = 1,$$

thus showing that its graph is a hyperbola. Find a, b, and h in terms of p and e.

10.2 POLAR COORDINATES

A familiar way to locate a point in the coordinate plane is by specifying its rectangular coordinates (x, y)—that is, by giving its abscissa x and ordinate y relative to given perpendicular axes. In some problems it is more convenient to locate a point by means of its *polar coordinates*. The polar coordinates give its position relative to a fixed reference point O (the **pole**) and to a given ray (the **polar axis**) beginning at O.

For convenience, we begin with a given xy-coordinate system and then take the origin as the pole and the nonnegative x-axis as the polar axis. Given the pole O and the polar axis, the point P with **polar coordinates** r and θ, written as the ordered pair (r, θ), is located as follows. First find the terminal side of the angle θ, given in radians, where θ is measured counterclockwise (if $\theta > 0$) from the x-axis (the polar axis) as its initial side. If $r \geqq 0$, then P is on the terminal side of this angle at the distance r from the origin. If $r < 0$, then P lies on the ray opposite the terminal side at the distance $|r| = -r > 0$ from the pole (Fig. 10.2.1). The **radial coordinate** r can be described as the *directed* distance of P from the pole along the terminal side of the angle θ. Thus, if r is positive, the point P lies in the same quadrant as θ, whereas if r is negative, then P lies in the opposite quadrant. If $r = 0$, the angle θ does not matter; the polar coordinates $(0, \theta)$ represent the origin whatever the **angular coordinate** θ might be. The origin, or pole, is the only point for which $r = 0$.

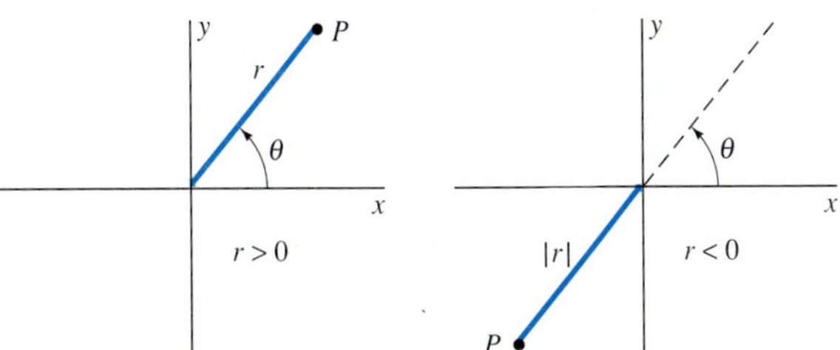

Fig. 10.2.1 The difference between the two cases $r > 0$ and $r < 0$

EXAMPLE 1 Polar coordinates differ from rectangular coordinates in that any point has more than one representation in polar coordinates. For example, the polar coordinates (r, θ) and $(-r, \theta + \pi)$ represent the same point P, as shown in Fig. 10.2.2. More generally, this point P has the polar coordinates $(r, \theta + n\pi)$ for any even integer n *and* the coordinates $(-r, \theta + n\pi)$ for any odd integer n. Thus the polar coordinate pairs

$$\left(2, \frac{\pi}{3}\right), \quad \left(-2, \frac{4\pi}{3}\right), \quad \left(2, \frac{7\pi}{3}\right), \quad \text{and} \quad \left(-2, -\frac{2\pi}{3}\right)$$

all represent the same point P in Fig. 10.2.3. [The rectangular coordinates of P are $(1, \sqrt{3})$.] ■

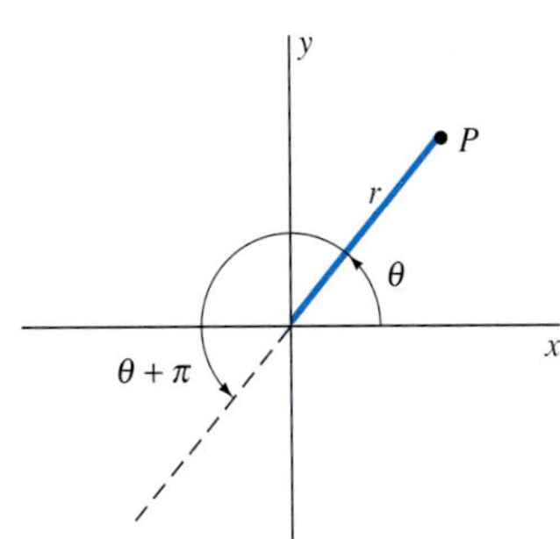

Fig. 10.2.2 The polar coordinates (r, θ) and $(-r, \theta + \pi)$ represent the same point P (Example 1).

To convert polar coordinates into rectangular coordinates, we use the basic relations

$$x = r\cos\theta, \quad y = r\sin\theta \tag{1}$$

that we read from the right triangle in Fig. 10.2.4. Converting in the opposite direction, we have

$$r^2 = x^2 + y^2, \qquad \tan\theta = \frac{y}{x} \quad \text{if } x \neq 0. \tag{2}$$

Some care is required in making the correct choice of θ in the formula $\tan\theta = y/x$. If $x > 0$, then (x, y) lies in either the first or the fourth quadrant, so $-\pi/2 < \theta < \pi/2$, which is the range of the inverse tangent function. Hence if $x > 0$, then $\theta = \arctan(y/x)$. But if $x < 0$, then (x, y) lies in the second or the third quadrant. In this case a proper choice for the angle is $\theta = \pi + \arctan(y/x)$. In any event, the signs of x and y in Eq. (1) with $r > 0$ indicate the quadrant in which θ lies.

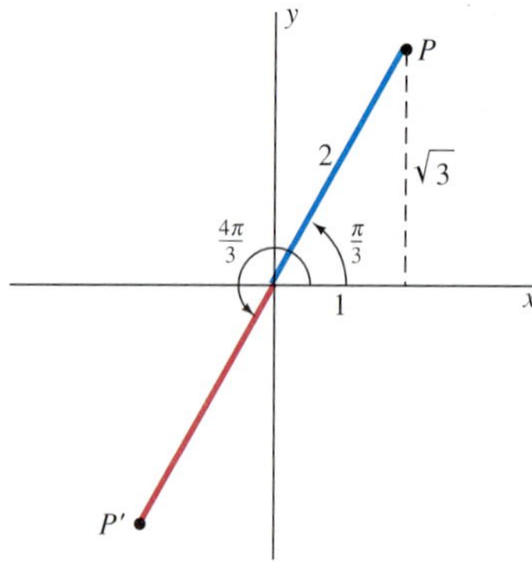

Fig. 10.2.3 The point P of Example 1 can be described in many different ways using polar coordinates.

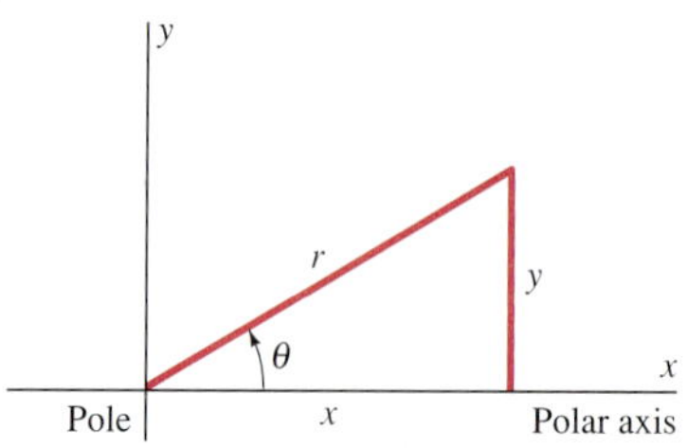

Fig. 10.2.4 Read Eqs. (1) and (2)—conversions between polar and rectangular coordinates—from this figure.

Polar Coordinate Equations

Some curves have equations in polar coordinates that are simpler than their equations in rectangular coordinates, an important reason for the usefulness of polar coordinates. The **graph** of an equation in the polar coordinates variables r and θ is the set of all those points P such that P has some pair of polar coordinates (r, θ) that satisfy the given equation. The graph of a polar equation $r = f(\theta)$ can be constructed by computing a table of values of r against θ and then plotting the corresponding points (r, θ) on polar coordinates graph paper.

EXAMPLE 2 One reason for the importance of polar coordinates is that many real-world problems involve circles, and the polar coordinates equation (or *polar equation*) of the circle with center $(0, 0)$ and radius $a > 0$ (Fig. 10.2.5) is very simple:

$$r = a. \tag{3}$$

Note that if we begin with the rectangular coordinates equation $x^2 + y^2 = a^2$ of this circle and transform it using the first relation in (2), we get the polar coordinates equation $r^2 = a^2$. Then Eq. (3) results upon taking positive square roots. ■

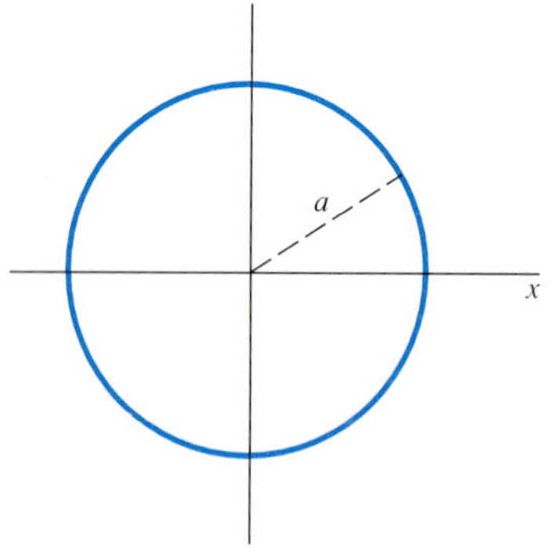

Fig. 10.2.5 The circle $r = a$ centered at the origin (Example 2).

EXAMPLE 3 Construct the polar coordinates graph of the equation $r = 2\sin\theta$.

Solution Figure 10.2.6 shows a table of values of r as a function of θ. The corresponding points (r, θ) are plotted in Fig. 10.2.7, using the rays at multiples of $\pi/6$ and the circles (centered at the pole) of radii 1 and 2 to locate these points. A visual inspection of the smooth curve connecting these points suggests that it is a circle of radius 1. Let us assume for the moment that this is so. Note then that the point $P(r, \theta)$ moves *once around this circle counterclockwise* as θ increases from 0 to π and then moves around this circle a *second time* as θ increases from π to 2π. The reason is the negative values of r for θ between π and 2π give—in this example—the same geometric points as do the positive values of r for θ between 0 and π. (Why?) ■

The verification that the graph of $r = 2\sin\theta$ is the indicated circle illustrates the general procedure for transferring back and forth between polar and rectangular coordinates, using the relations in (1) and (2).

EXAMPLE 4 To transform the equation $r = 2\sin\theta$ of Example 3 into rectangular coordinates, we first multiply both sides by r to get

$$r^2 = 2r\sin\theta.$$

θ	r
0	0.00
$\pi/6$	1.00
$\pi/3$	1.73
$\pi/2$	2.00
$2\pi/3$	1.73
$5\pi/6$	1.00
π	0.00
$7\pi/6$	−1.00
$4\pi/3$	−1.73
$3\pi/2$	−2.00
$5\pi/3$	−1.73
$11\pi/6$	−1.00
2π	0.00
	(data rounded)

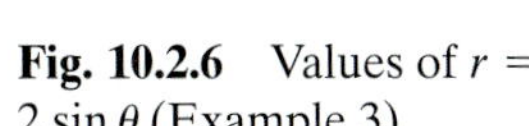

Fig. 10.2.6 Values of $r = 2 \sin \theta$ (Example 3)

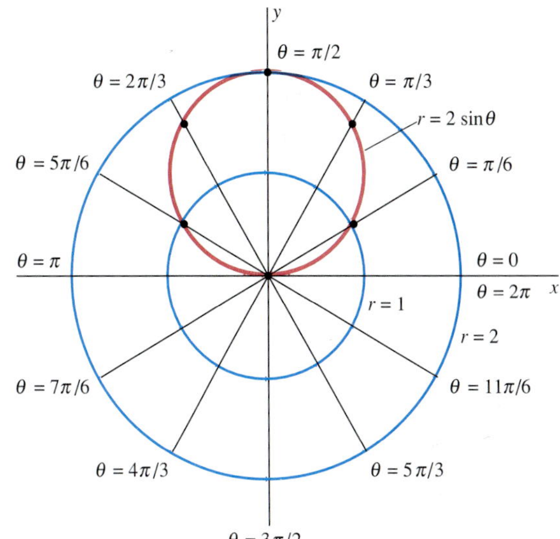

Fig. 10.2.7 The graph of the polar equation $r = 2 \sin \theta$ (Example 3)

Equations (1) and (2) now give

$$x^2 + y^2 = 2y.$$

Finally, after we complete the square in y, we have

$$x^2 + (y - 1)^2 = 1,$$

the rectangular coordinates equation (or *rectangular equation*) of a circle whose center is $(0, 1)$ and whose radius is 1. ■

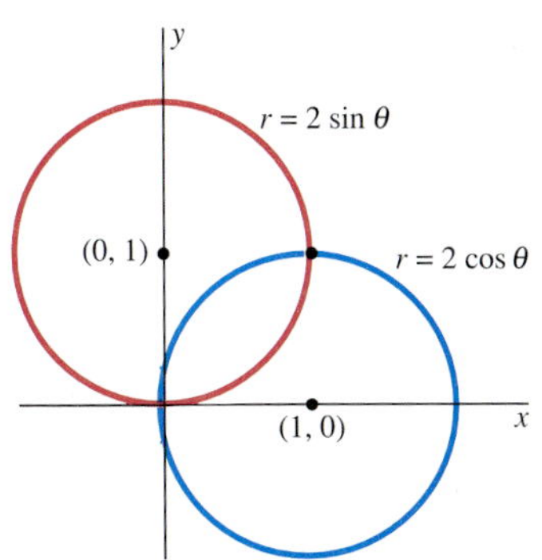

Fig. 10.2.8 The graphs of the circles whose equations appear in Eq. (4) with $a = 1$

More generally, the graphs of the equations

$$r = 2a \sin \theta \quad \text{and} \quad r = 2a \cos \theta \tag{4}$$

are circles of radius a centered, respectively, at the points $(0, a)$ and $(a, 0)$. This is illustrated (with $a = 1$) in Fig. 10.2.8.

By substituting the equations in (1), we can transform the rectangular equation $ax + by = c$ of a straight line into

$$ar \cos \theta + br \sin \theta = c.$$

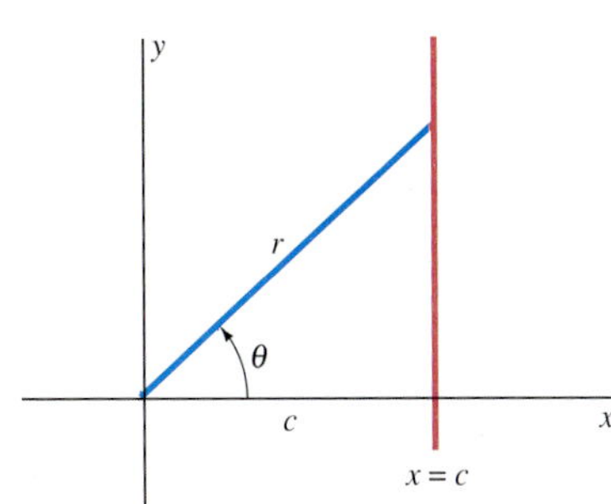

Fig. 10.2.9 Finding the polar equation of the vertical line $x = c$

Let us take $a = 1$ and $b = 0$. Then we see that the polar equation of the vertical line $x = c$ is $r = c \sec \theta$, as we can deduce directly from Fig. 10.2.9.

EXAMPLE 5 Sketch the graph of the polar equation $r = 2 + 2 \sin \theta$.

Solution If we scan the second column of the table in Fig. 10.2.6, mentally adding 2 to each entry for r, we see that

- r increases from 2 to 4 as θ increases from 0 to $\pi/2$;
- r decreases from 4 to 2 as θ increases from $\pi/2$ to π;
- r decreases from 2 to 0 as θ increases from π to $3\pi/2$;
- r increases from 0 to 2 as θ increases from $3\pi/2$ to 2π.

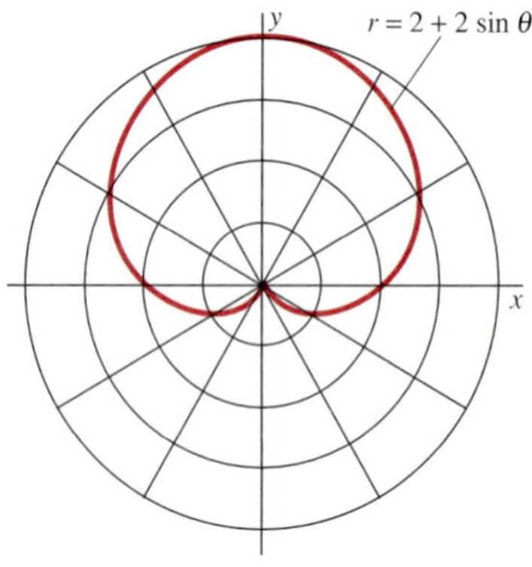

Fig. 10.2.10 A cardioid (Example 5)

This information tells us that the graph resembles the curve shown in Fig. 10.2.10. This heart-shaped graph is called a **cardioid**. The graphs of the equations

$$r = a(1 \pm \sin\theta) \quad \text{and} \quad r = a(1 \pm \cos\theta)$$

are all cardioids, differing only in size (determined by a), axis of symmetry (horizontal or vertical), and the direction in which the cusp at the pole points. ■

EXAMPLE 6 Sketch the graph of the equation $r = 2\cos 2\theta$.

Solution Rather than constructing a table of values of r as a function of θ and then plotting individual points, let us begin with a *rectangular coordinates graph* of r as a function of θ. In Fig. 10.2.11, we see that $r = 0$ if θ is an odd integral multiple of $\pi/4$ and that r is alternately positive and negative on successive intervals of length $\pi/2$ from one odd integral multiple of $\pi/4$ to the next.

Now let's think about how r changes as θ increases, beginning at $\theta = 0$. As θ increases from 0 to $\pi/4$, r decreases in value from 2 to 0, and so we draw the first portion (labeled "1") of the polar curve in Fig. 10.2.12. As θ increases from $\pi/4$ to $3\pi/4$, r first decreases from 0 to -2 and then increases from -2 to 0. Because r is now negative, we draw the second and third portions (labeled "2" and "3") of the polar curve in the third and fourth quadrants (rather than in the first and second quadrants) in Fig. 10.2.12. Continuing in this fashion, we draw the fourth through eighth portions of the polar curve, with those portions where r is negative in the quadrants opposite those in which θ lies. The arrows on the resulting polar curve in Fig. 10.2.12 indicate the direction of motion of the point $P(r, \theta)$ along the curve as θ increases. The whole graph consists of four loops, each of which begins and ends at the pole. ■

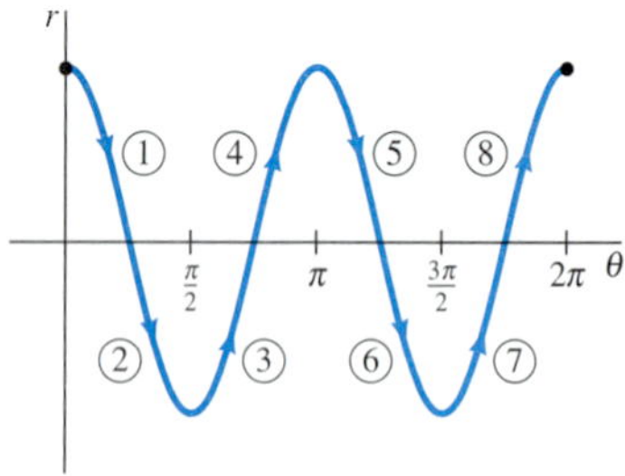

Fig. 10.2.11 The rectangular coordinates graph of $r = 2\cos 2\theta$ as a function of θ. Numbered portions of the graph correspond to numbered portions of the polar-coordinates graph in Fig. 10.2.12.

The curve in Example 6 is called a *four-leaved rose*. The equations $r = a\cos n\theta$ and $r = a\sin n\theta$ represent "roses" with $2n$ "leaves," or loops, if n is even and $n \geqq 2$ but with n loops if n is odd and $n \geqq 3$.

The four-leaved rose exhibits several types of symmetry. The following are some *sufficient* conditions for symmetry in polar coordinates:

- *For symmetry around the x-axis:* The equation is unchanged when θ is replaced with $-\theta$.
- *For symmetry around the y-axis*: The equation is unchanged when θ is replaced with $\pi - \theta$.
- *For symmetry with respect to the origin*: The equation is unchanged when r is replaced with $-r$.

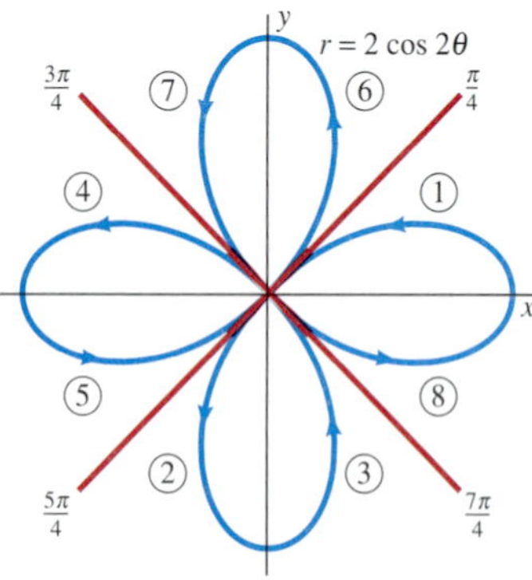

Fig. 10.2.12 A four-leaved rose (Example 6)

Because $\cos 2\theta = \cos(-2\theta) = \cos 2(\pi - \theta)$, the equation $r = 2\cos 2\theta$ of the four-leaved rose satisfies the first two symmetry conditions, and therefore its graph is symmetric around both the x-axis and the y-axis. Thus it is also symmetric around the origin. Nevertheless, this equation does *not* satisfy the third condition, the one for symmetry around the origin. This illustrates that although the symmetry conditions given are *sufficient* for the symmetries described, they are not *necessary* conditions.

EXAMPLE 7 Figure 10.2.13 shows the lemniscate with equation

$$r^2 = -4\sin 2\theta.$$

To see why it has loops only in the second and fourth quadrants, we examine a table of values of $-4\sin 2\theta$.

θ	2θ	$-4\sin 2\theta$
$0 < \theta < \frac{1}{2}\pi$	$0 < 2\theta < \pi$	Negative
$\frac{1}{2}\pi < \theta < \pi$	$\pi < 2\theta < 2\pi$	Positive
$\pi < \theta < \frac{3}{2}\pi$	$2\pi < 2\theta < 3\pi$	Negative
$\frac{3}{2}\pi < \theta < 2\pi$	$3\pi < 2\theta < 4\pi$	Positive

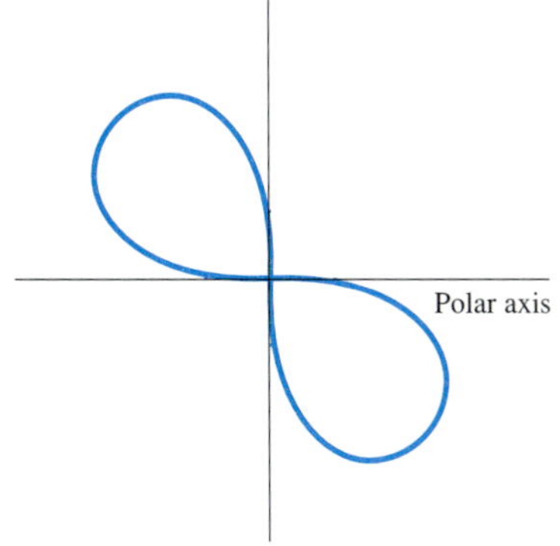

Fig. 10.2.13 The lemniscate $r^2 = -4\sin 2\theta$ (Example 7)

When θ lies in the first or the third quadrant, the quantity $-4\sin 2\theta$ is negative, so the equation $r^2 = -4\sin 2\theta$ cannot be satisfied for any real values of r. ■

Example 6 illustrates a peculiarity of graphs of polar equations, caused by the fact that a single point has multiple representations in polar coordinates. The point with polar coordinates $(2, \pi/2)$ clearly lies on the four-leaved rose, but these coordinates do *not* satisfy the equation $r = 2\cos 2\theta$. This means that a point may have one pair of polar coordinates that satisfy a given equation and others that do not. Hence we must be careful to understand this: The graph of a polar equation consists of all those points with *at least one* polar coordinates representation that satisfies the given equation.

Another result of the multiplicity of polar coordinates is that the simultaneous solution of two polar equations does not always give all the points of intersection of their graphs. For instance, consider the circles $r = 2\sin\theta$ and $r = 2\cos\theta$ shown in Fig. 10.2.8. The origin is clearly a point of intersection of these two circles. Its polar representation $(0, \pi)$ satisfies the equation $r = 2\sin\theta$, and its representation $(0, \pi/2)$ satisfies the other equation, $r = 2\cos\theta$. But the origin has no *single* polar representation that satisfies both equations simultaneously! If we think of θ as increasing uniformly with time, then the corresponding moving points on the two circles pass through the origin at different times. Hence the origin cannot be discovered as a point of intersection of the two circles by solving their equations simultaneously—try it.

As a consequence of the phenomenon illustrated by this example, the only way we can be certain of finding *all* points of intersection of two curves in polar coordinates is to graph both curves.

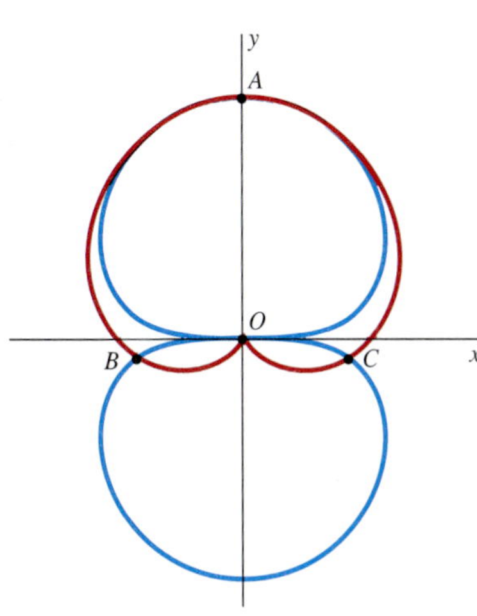

Fig. 10.2.14 The cardioid $r = 1 + \sin\theta$ and the figure eight $r^2 = 4\sin\theta$ meet in four points (Example 8).

EXAMPLE 8 Find all points of intersection of the graphs of the equations $r = 1 + \sin\theta$ and $r^2 = 4\sin\theta$.

Solution The graph of $r = 1 + \sin\theta$ is a scaled-down version of the cardioid of Example 5. In Problem 52 we ask you to show that the graph of $r^2 = 4\sin\theta$ is the figure eight curve shown with the cardioid in Fig. 10.2.14. The figure shows four points of intersection: A, B, C, and O. Can we find all four with algebra?

Given the two equations, we begin by eliminating r. Because

$$(1 + \sin\theta)^2 = r^2 = 4\sin\theta,$$

it follows that

$$\begin{aligned} \sin^2\theta - 2\sin\theta + 1 &= 0; \\ (\sin\theta - 1)^2 &= 0; \end{aligned}$$

and thus that $\sin\theta = 1$. So θ must be an angle of the form $\frac{1}{2}\pi + 2n\pi$, where n is an integer. All points on the cardioid and all points on the figure eight curve are produced by letting θ range from 0 to 2π, so $\theta = \pi/2$ will produce all the solutions that we can possibly obtain by algebraic methods. The only such point is $A(2, \pi/2)$, and the other three points of intersection are detected only when the two equations are graphed. ■

10.2 PROBLEMS

1. Plot the points with the given polar coordinates, then find the rectangular coordinates of each.

(a) $(1, \pi/4)$ (b) $(-2, 2\pi/3)$
(c) $(1, -\pi/3)$ (d) $(3, 3\pi/2)$
(e) $(2, -\pi/4)$ (f) $(-2, -7\pi/6)$
(g) $(2, 5\pi/6)$

2. Find two polar coordinates representations, one with $r > 0$ and the other with $r < 0$, for the points with the given rectangular coordinates.

(a) $(-1, -1)$ (b) $(\sqrt{3}, -1)$
(c) $(2, 2)$ (d) $(-1, \sqrt{3})$
(e) $(\sqrt{2}, -\sqrt{2})$ (f) $(-3, \sqrt{3})$

In Problems 3 through 10, express the given rectangular equations in polar form.

3. $x = 4$ **4.** $y = 6$
5. $x = 3y$ **6.** $x^2 + y^2 = 25$
7. $xy = 1$ **8.** $x^2 - y^2 = 1$
9. $y = x^2$ **10.** $x + y = 4$

In Problems 11 through 18, express the given polar equation in rectangular coordinates.

11. $r = 3$ **12.** $\theta = 3\pi/4$
13. $r = -5\cos\theta$ **14.** $r = \sin 2\theta$
15. $r = 1 - \cos 2\theta$ **16.** $r = 2 + \sin\theta$
17. $r = 3\sec\theta$ **18.** $r^2 = \cos 2\theta$

For the curves described in Problems 19 through 28, write equations in both rectangular and polar coordinates.

19. The vertical line through $(2, 0)$
20. The horizontal line through $(1, 3)$
21. The line with slope -1 through $(2, -1)$
22. The line with slope 1 through $(4, 2)$
23. The line through the points $(1, 3)$ and $(3, 5)$
24. The circle with center $(3, 0)$ that passes through the origin
25. The circle with center $(0, -4)$ that passes through the origin
26. The circle with center $(3, 4)$ and radius 5
27. The circle with center $(1, 1)$ that passes through the origin
28. The circle with center $(5, -2)$ that passes through the point $(1, 1)$

In Problems 29 through 32, transform the given polar coordinates equation into a rectangular coordinates equation, then match the equation with its graph among those in Figs. 10.2.15 through 10.2.18.

29. $r = -4\cos\theta$ **30.** $r = 5\cos\theta + 5\sin\theta$
31. $r = -4\cos\theta + 3\sin\theta$ **32.** $r = 8\cos\theta - 15\sin\theta$

The graph of a polar equation of the form $r = a + b\cos\theta$ (or $r = a + b\sin\theta$) is called a *limaçon* (from the French word for *snail*). In Problems 33 through 36, match the given polar coordinates equation with its graph among the limaçons in Figs. 10.2.19 through 10.2.22.

33. $r = 8 + 6\cos\theta$ **34.** $r = 7 + 7\cos\theta$
35. $r = 5 + 9\cos\theta$ **36.** $r = 3 + 11\cos\theta$

37. Show that the graph of the polar equation $r = a\cos\theta + b\sin\theta$ is a circle if $ab \neq 0$. Express the center (h, k) and radius r of this circle in terms of a and b.

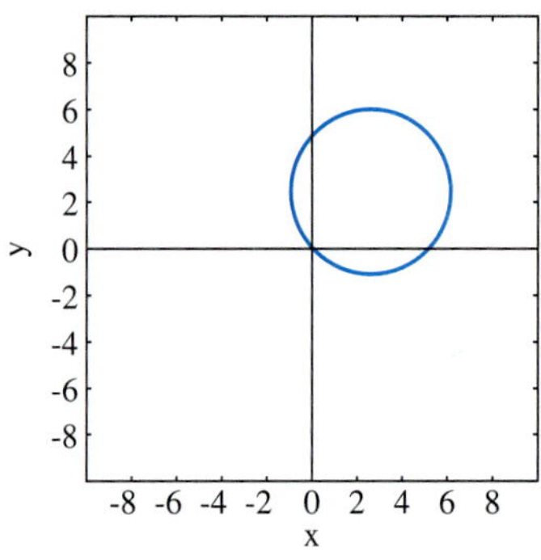

Fig. 10.2.15

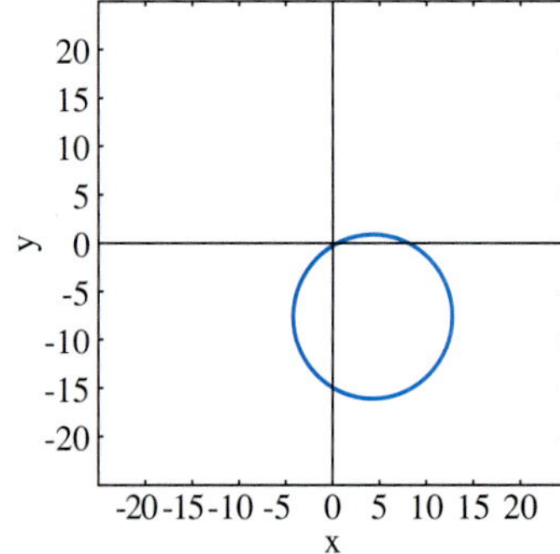

Fig. 10.2.16

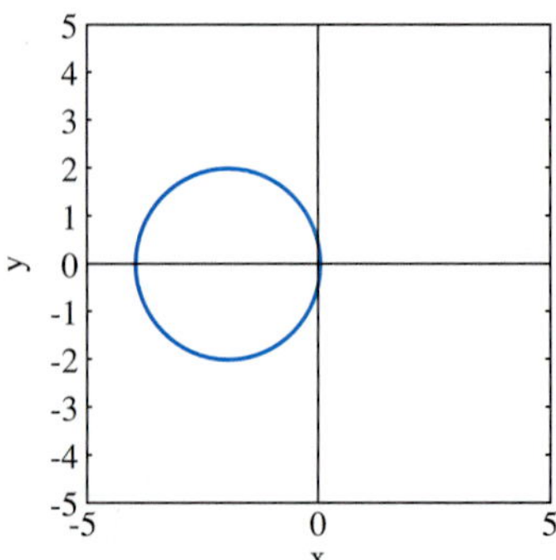

Fig. 10.2.17

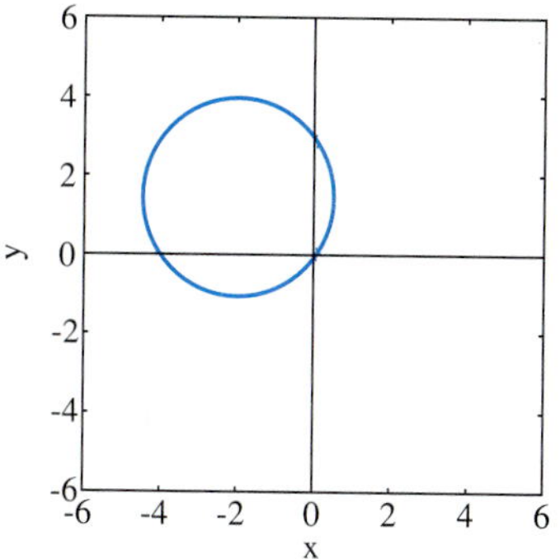

Fig. 10.2.18

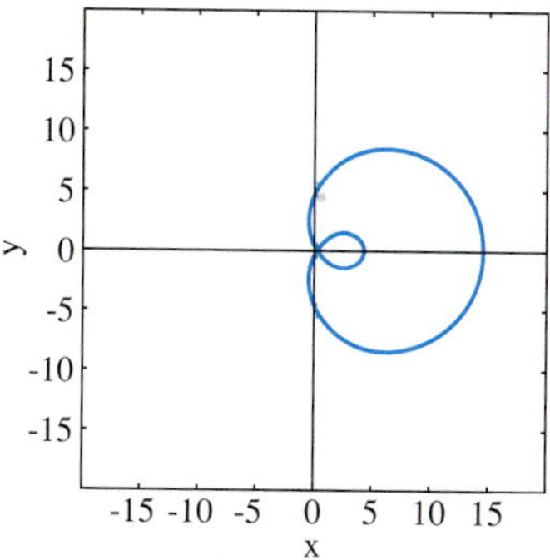

Fig. 10.2.19

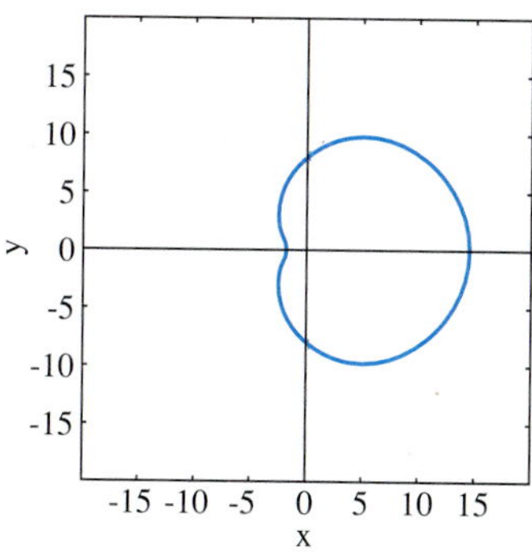

Fig. 10.2.20

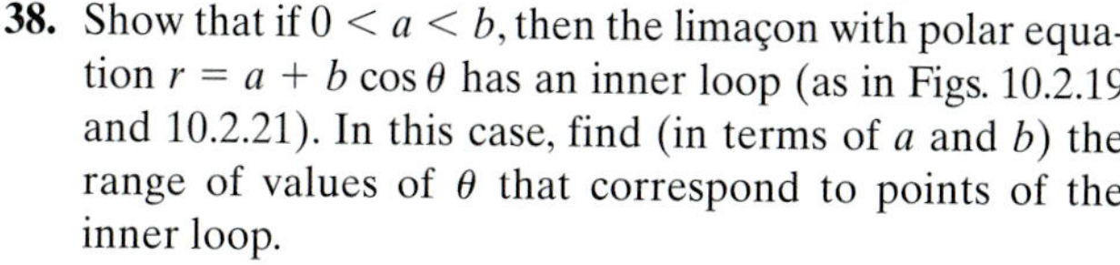

38. Show that if $0 < a < b$, then the limaçon with polar equation $r = a + b\cos\theta$ has an inner loop (as in Figs. 10.2.19 and 10.2.21). In this case, find (in terms of a and b) the range of values of θ that correspond to points of the inner loop.

Sketch the graphs of the polar equations in Problems 39 through 52. Indicate any symmetries around either coordinate axis or the origin.

39. $r = 2\cos\theta$ (circle)

40. $r = 2\sin\theta + 2\cos\theta$ (circle)

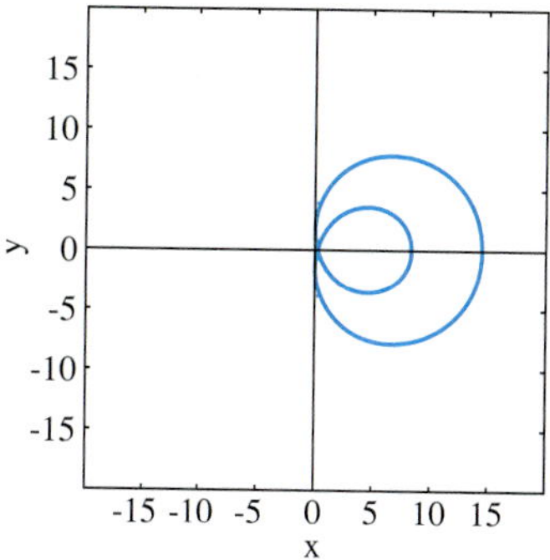

Fig. 10.2.21

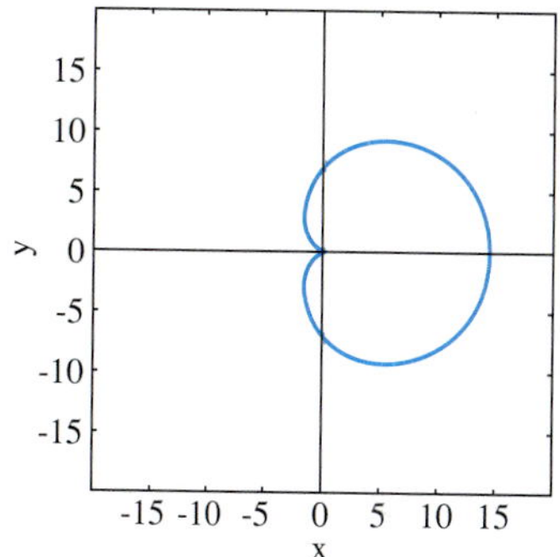

Fig. 10.2.22

41. $r = 1 + \cos\theta$ (cardioid)

42. $r = 1 - \sin\theta$ (cardioid)

43. $r = 2 + 4\sin\theta$ (limaçon)

44. $r = 4 + 2\cos\theta$ (limaçon)

45. $r^2 = 4\sin 2\theta$ (lemniscate)

46. $r^2 = 4\cos 2\theta$ (lemniscate)

47. $r = 2\sin 2\theta$ (four-leaved rose)

48. $r = 3\sin 3\theta$ (three-leaved rose)

49. $r = 3\cos 3\theta$ (three-leaved rose)

50. $r = 3\theta$ (spiral of Archimedes)

51. $r = 2\sin 5\theta$ (five-leaved rose)

52. $r^2 = 4\sin\theta$ (figure eight)

In Problems 53 through 58, find all points of intersection of the curves with the given polar equations.

53. $r = 1, \quad r = \cos\theta$

54. $r = \sin\theta, \quad r^2 = 3\cos^2\theta$

55. $r = \sin\theta, \quad r = \cos 2\theta$

56. $r = 1 + \cos\theta, \quad r = 1 - \sin\theta$

57. $r = 1 - \cos\theta, \quad r^2 = 4\cos\theta$

58. $r^2 = 4\sin\theta, \quad r^2 = 4\cos\theta$

59. (a) The straight line L passes through the point with polar coordinates (p, α) and is perpendicular to the line segment joining the pole and the point (p, α). Write the polar coordinates equation of L. (b) Show that the rectangular coordinates equation of L is

$$x \cos \alpha + y \sin \alpha = p.$$

60. Find a rectangular coordinates equation of the cardioid with polar equation $r = 1 - \cos \theta$.

61. Use polar coordinates to identify the graph of the rectangular coordinates equation $a^2(x^2 + y^2) = (x^2 + y^2 - by)^2$.

62. Plot the polar equations

$$r = 1 + \cos \theta \quad \text{and} \quad r = -1 + \cos \theta$$

on the same coordinate plane. Comment on the results.

10.2 PROJECT: CALCULATOR/COMPUTER-GENERATED POLAR COORDINATES GRAPHS

Even if your graphing calculator or computer has no polar graphing facility, the graph of the polar coordinates equation $r = f(\theta)$ can be plotted in rectangular coordinates by using the equations

$$x = r \cos \theta = f(\theta) \cos \theta, \tag{1a}$$

$$y = r \sin \theta = f(\theta) \sin \theta. \tag{1b}$$

Then, as θ ranges from 0 to 2π (or, in some cases, through a larger domain), the point (x, y) traces the polar graph $r = f(\theta)$.

For instance, with

$$r = \cos\left(\frac{7\theta}{4}\right), \tag{2}$$

we get the curve in Fig. 10.2.23 having 14 overlapping loops. By contrast, with $r = \cos\left(\frac{7}{3}\theta\right)$, we get the curve in Fig. 10.2.24, having only seven loops.

If (as required by many graphing devices) we use t instead of θ and substitute Eq. (2) in (1), we get the equations

$$x = \cos\left(\frac{7t}{4}\right) \cos t, \qquad y = \cos\left(\frac{7t}{4}\right) \sin t. \tag{3}$$

With a TI graphing calculator set in **Par** mode, we then graph the equations

```
X1T=cos(7T/4)*cos(T),    Y1T=cos(7T/4)*sin(T).
```

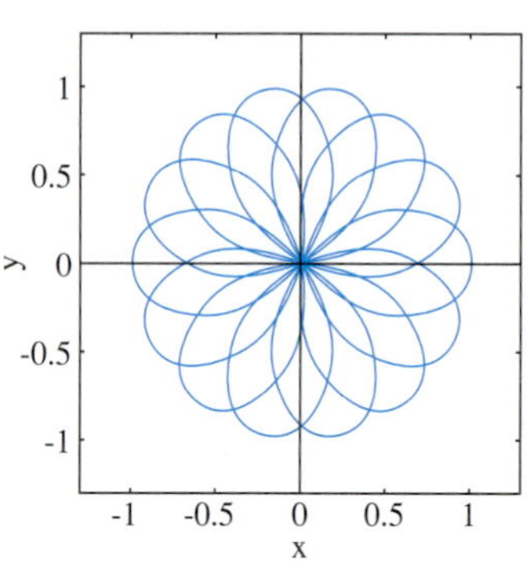

Fig. 10.2.23 $r = \cos\left(\frac{7\theta}{4}\right)$

The corresponding *Mathematica* and *Maple* commands are (respectively)

```
ParametricPlot[ {Cos[7t/4] Cos[t], Cos[7t/4] Sin[t]},
                {t, 0, 8 Pi} ];
```

and

```
plot([cos(7*t/4)*cos(t), cos(7*t/4)*sin(t), t = 0..8*Pi]);
```

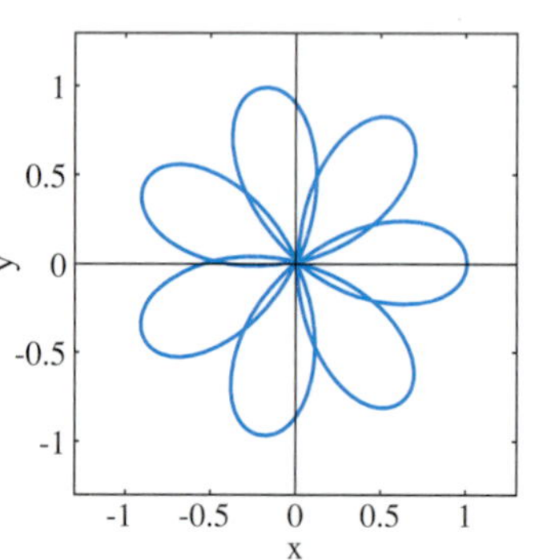

Fig. 10.2.24 $r = \cos\left(\frac{7\theta}{3}\right)$

The t-range $[0, 8\pi]$ is selected because 8π is the smallest complete revolution $2n\pi$ such that $\cos\left(\frac{7}{4}[t + 2n\pi]\right) = \cos\left(\frac{7}{4}t\right)$, so that the curve repeats itself for larger values of t.

INVESTIGATION 1 Plot the polar coordinates curves $r = \cos(p\theta/q)$ and $r = \sin(p\theta/q)$ with various nonzero integers p and q. Select the t-interval so that the entire graph is plotted. What determines the number of loops in the resulting graph?

INVESTIGATION 2 Plot the polar coordinates curve $r = (a + b \cos m\theta)(c + d \sin n\theta)$ with various values of the coefficients a, b, c, and d and the positive integers m and n. You might begin with the special case $a = 1, b = 0$, or the special case $c = 1, d = 0$,

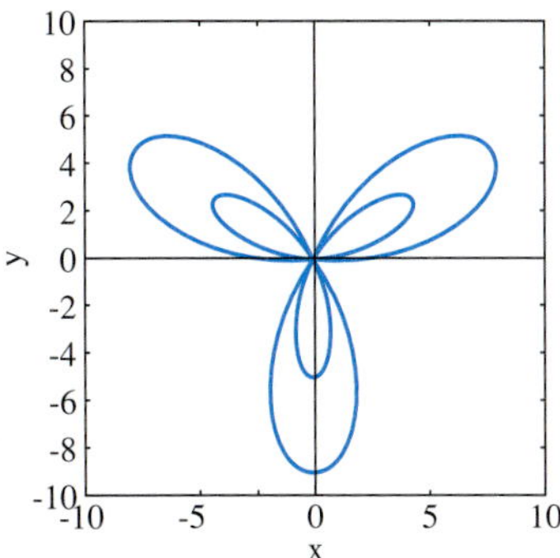

Fig. 10.2.25 $r = 2 + 7\sin 3\theta$

or the special case $a = c = 0$. Figures 10.2.25 and 10.2.26 illustrate just two of the possibilities.

INVESTIGATION 3 The simple "butterfly" shown in Fig. 10.2.27 is the graph of the polar coordinates equation

$$r = e^{\cos\theta} - 2\cos 4\theta.$$

Now plot the polar coordinates equation

$$r = e^{\cos\theta} - 2\cos 4\theta + \sin^5\left(\tfrac{1}{12}\theta\right)$$

for $0 \leqq \theta \leqq 24\pi$. The incredibly beautiful curve that results was discovered by Temple H. Fay. His article "The Butterfly Curve" (*American Mathematical Monthly*, May 1989, p. 442) is well worth a trip to the library.

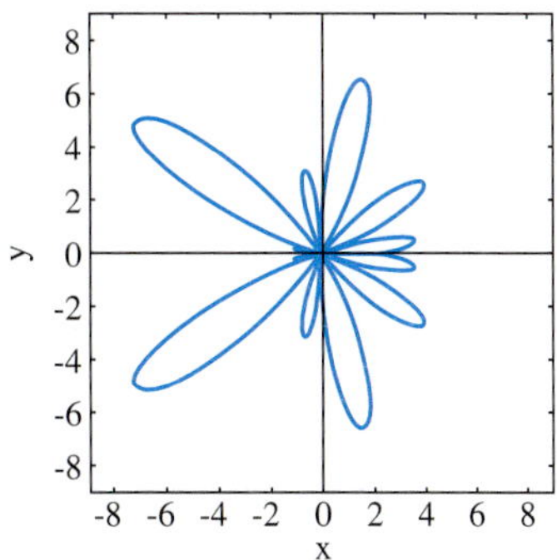

Fig. 10.2.26
$r = (2 + 7\sin 3\theta)\cos 5\theta$

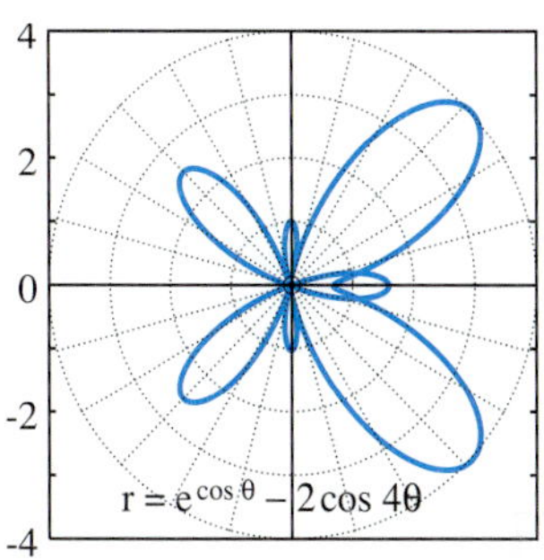

Fig. 10.2.27 The butterfly curve

10.3 AREA COMPUTATIONS IN POLAR COORDINATES

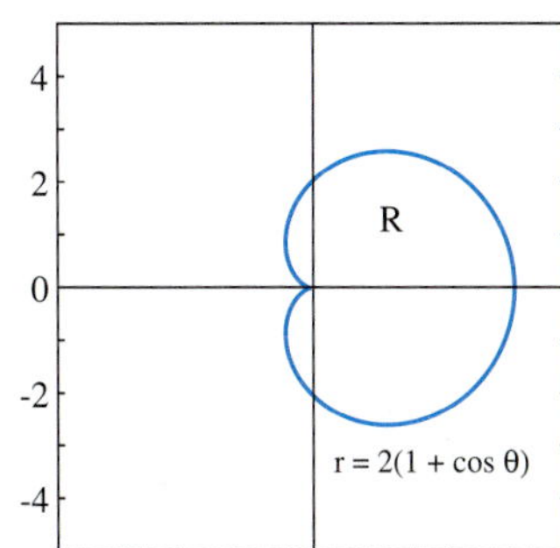

Fig. 10.3.1 What is the area of the region R bounded by the cardioid $r = 2(1 + \cos\theta)$?

The graph of the polar coordinates equation $r = f(\theta)$ may bound an area, as does the cardioid $r = 2(1 + \cos\theta)$—see Fig. 10.3.1. To calculate the area of this region, we may find it convenient to work directly in polar coordinates rather than to change to rectangular coordinates.

To see how to set up an area integral in polar coordinates, we consider the region R of Fig. 10.3.2. This region is bounded by the two radial lines $\theta = \alpha$ and $\theta = \beta$ and by the curve $r = f(\theta)$, $\alpha \leqq \theta \leqq \beta$. To approximate the area A of R, we begin with a partition

$$\alpha = \theta_0 < \theta_1 < \theta_2 < \cdots < \theta_n = \beta$$

of the interval $[\alpha, \beta]$ into n subintervals, all with the same length $\Delta\theta = (\beta - \alpha)/n$. We select a point $\theta_i^\star$ in the ith subinterval $[\theta_{i-1}, \theta_i]$ for $i = 1, 2, \dots, n$.

Let ΔA_i denote the area of the sector bounded by the lines $\theta = \theta_{i-1}$ and $\theta = \theta_i$ and by the curve $r = f(\theta)$. We see from Fig. 10.3.2 that for small values of $\Delta\theta$, ΔA_i is approximately equal to the area of the *circular* sector that has radius $r_i^\star = f(\theta_i^\star)$ and is bounded by the same lines. That is,

$$\Delta A_i \approx \tfrac{1}{2}(r_i^\star)^2\,\Delta\theta = \tfrac{1}{2}[f(\theta_i^\star)]^2\,\Delta\theta.$$

We add the areas of these sectors for $i = 1, 2, \dots, n$ and thereby find that

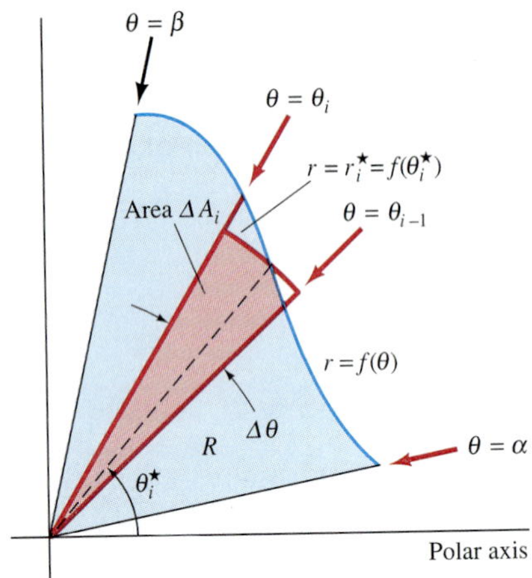

Fig. 10.3.2 We obtain the area formula from Riemann sums.

$$A = \sum_{i=1}^{n} \Delta A_i \approx \sum_{i=1}^{n} \tfrac{1}{2}[f(\theta_i^\star)]^2\,\Delta\theta.$$

The right-hand sum is a Riemann sum for the integral

$$\int_\alpha^\beta \tfrac{1}{2}[f(\theta)]^2\,d\theta.$$

Hence, if f is continuous, the value of this integral is the limit, as $\Delta\theta \to 0$, of the preceding sum. We therefore conclude that the *area A of the region R bounded by the lines $\theta = \alpha$ and $\theta = \beta$ and the curve $r = f(\theta)$* is

$$A = \int_\alpha^\beta \tfrac{1}{2}[f(\theta)]^2\,d\theta. \qquad \textbf{(1)}$$

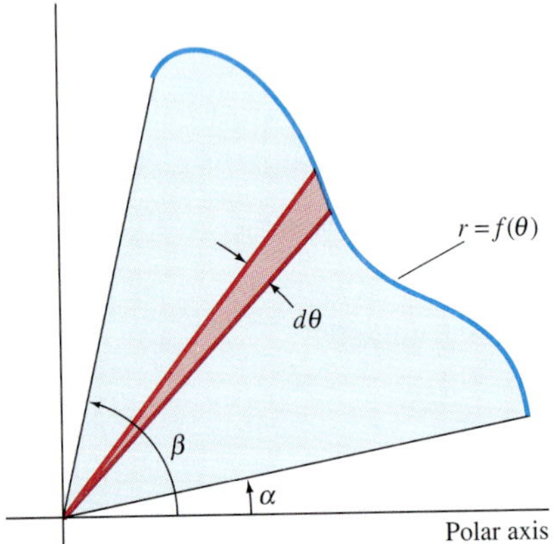

Fig. 10.3.3 Nonrigorous derivation of the area formula in polar coordinates

The infinitesimal sector shown in Fig. 10.3.3, with radius r, central angle $d\theta$, and area $dA = \frac{1}{2}r^2\,d\theta$, serves as a useful device for remembering Eq. (1) in the abbreviated form

$$A = \int_\alpha^\beta \tfrac{1}{2}r^2\,d\theta. \qquad \textbf{(2)}$$

EXAMPLE 1 Find the area of the region bounded by the limaçon with equation $r = 3 + 2\cos\theta, 0 \leqq \theta \leqq 2\pi$ (Fig. 10.3.4).

Solution We could apply Eq. (2) with $\alpha = 0$ and $\beta = 2\pi$. Here, instead, we will make use of symmetry. We will calculate the area of the upper half of the region and then double the result. Note that the infinitesimal sector shown in Fig. 10.3.4 sweeps out the upper half of the limaçon as θ increases from 0 to π (Fig. 10.3.5). Hence

$$A = 2\int_\alpha^\beta \tfrac{1}{2}r^2\,d\theta = \int_0^\pi (3 + 2\cos\theta)^2\,d\theta$$

$$= \int_0^\pi (9 + 12\cos\theta + 4\cos^2\theta)\,d\theta.$$

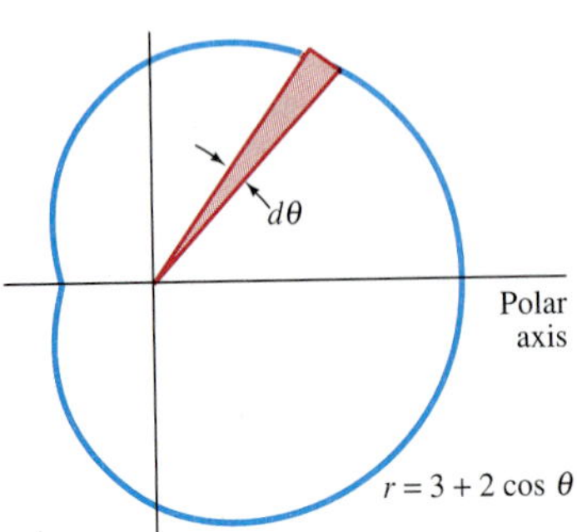

Fig. 10.3.4 The limaçon of Example 1

Because

$$4\cos^2\theta = 4\cdot\frac{1 + \cos 2\theta}{2} = 2 + 2\cos 2\theta,$$

we now get

$$A = \int_0^\pi (11 + 12\cos\theta + 2\cos 2\theta)\,d\theta$$

$$= \Big[11\theta + 12\sin\theta + \sin 2\theta\Big]_0^\pi = 11\pi.$$

■

EXAMPLE 2 Find the area bounded by each loop of the limaçon with equation $r = 1 + 2\cos\theta$ (Fig. 10.3.6).

Solution The equation $1 + 2\cos\theta = 0$ has two solutions for θ in the interval $[0, 2\pi]$: $\theta = 2\pi/3$ and $\theta = 4\pi/3$. The upper half of the outer loop of the limaçon

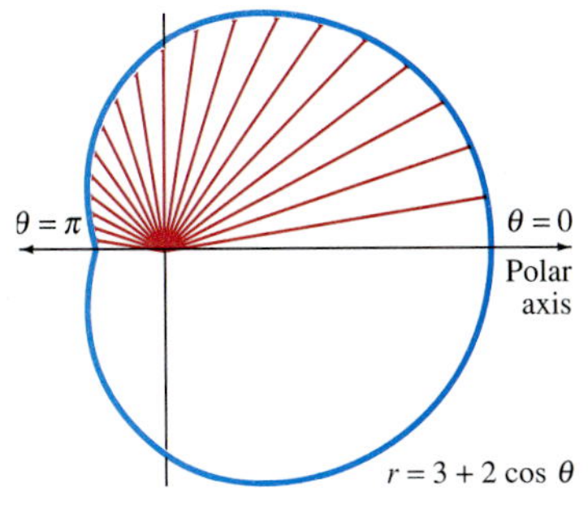

Fig. 10.3.5 Infinitesimal sectors from $\theta = 0$ to $\theta = \pi$ (Example 1)

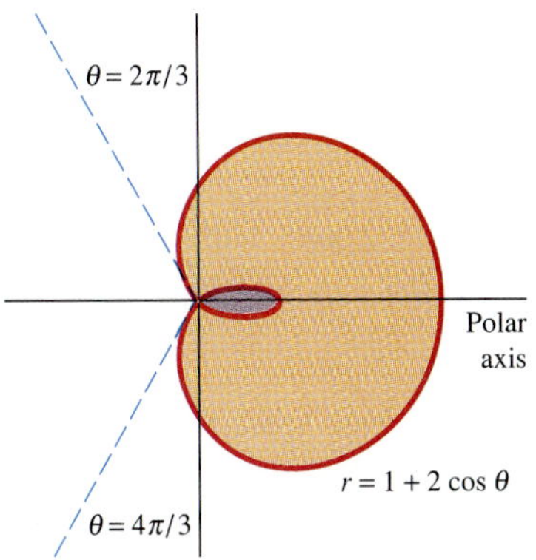

Fig. 10.3.6 The limaçon of Example 2

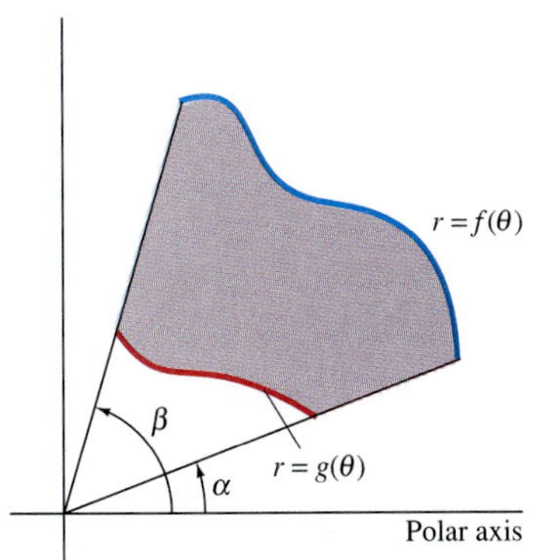

Fig. 10.3.7 The area between the graphs of f and g

corresponds to values of θ between 0 and $2\pi/3$, where r is positive. Because the curve is symmetric around the x-axis, we can find the total area A_1 bounded by the outer loop by integrating from 0 to $2\pi/3$ and then doubling. Thus

$$\begin{aligned} A_1 &= 2\int_0^{2\pi/3} \tfrac{1}{2}(1 + 2\cos\theta)^2\, d\theta = \int_0^{2\pi/3} (1 + 4\cos\theta + 4\cos^2\theta)\, d\theta \\ &= \int_0^{2\pi/3} (3 + 4\cos\theta + 2\cos 2\theta)\, d\theta \\ &= \Big[3\theta + 4\sin\theta + \sin 2\theta \Big]_0^{2\pi/3} = 2\pi + \tfrac{3}{2}\sqrt{3}. \end{aligned}$$

The inner loop of the limaçon corresponds to values of θ between $2\pi/3$ and $4\pi/3$, where r is negative. Hence the area bounded by the inner loop is

$$\begin{aligned} A_2 &= \int_{2\pi/3}^{4\pi/3} \tfrac{1}{2}(1 + 2\cos\theta)^2\, d\theta \\ &= \tfrac{1}{2}\Big[3\theta + 4\sin\theta + \sin 2\theta \Big]_{2\pi/3}^{4\pi/3} = \pi - \tfrac{3}{2}\sqrt{3}. \end{aligned}$$

The area of the region lying *between* the two loops of the limaçon is then

$$A = A_1 - A_2 = 2\pi + \tfrac{3}{2}\sqrt{3} - (\pi - \tfrac{3}{2}\sqrt{3}) = \pi + 3\sqrt{3}.$$

The Area Between Two Polar Curves

Now consider two curves $r = f(\theta)$ and $r = g(\theta)$, with $f(\theta) \geqq g(\theta) \geqq 0$ for $\alpha \leqq \theta \leqq \beta$. Then we can find the area of the region bounded by these curves and the rays (radial lines) $\theta = \alpha$ and $\theta = \beta$ (Fig. 10.3.7) by subtracting the area bounded by the inner curve from that bounded by the outer curve. That is, the area A between the two curves is given by

$$\begin{aligned} A &= \int_\alpha^\beta \tfrac{1}{2}[f(\theta)]^2\, d\theta - \int_\alpha^\beta \tfrac{1}{2}[g(\theta)]^2\, d\theta \\ &= \tfrac{1}{2}\int_\alpha^\beta \{[f(\theta)]^2 - [g(\theta)]^2\}\, d\theta. \end{aligned} \qquad \textbf{(3)}$$

With r_{outer} for the outer curve and r_{inner} for the inner curve, we get the abbreviated formula

$$A = \tfrac{1}{2}\int_a^\beta \Big[(r_{\text{outer}})^2 - (r_{\text{inner}})^2 \Big]\, d\theta \qquad \textbf{(4)}$$

for the area of the region shown in Fig. 10.3.8.

EXAMPLE 3 Find the area A of the region that lies within the limaçon $r = 1 + 2\cos\theta$ and outside the circle $r = 2$.

Solution The circle and limaçon are shown in Fig. 10.3.9, with the area A between them shaded. The points of intersection of the circle and limaçon are given by

$$1 + 2\cos\theta = 2, \quad \text{so} \quad \cos\theta = \tfrac{1}{2},$$

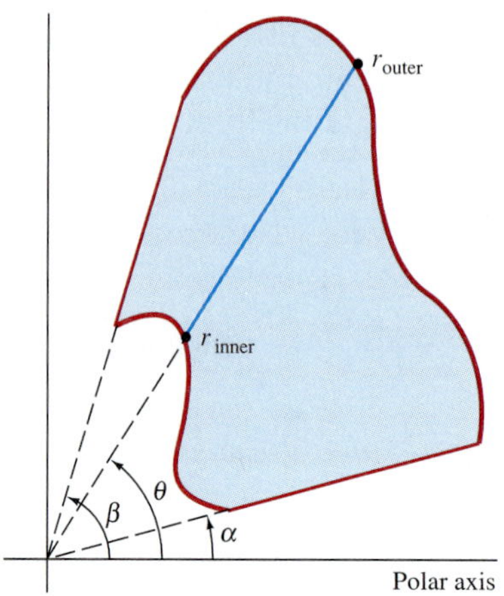

Fig. 10.3.8 The radial line segment illustrates the radii r_{inner} and r_{outer} of Eq. (4).

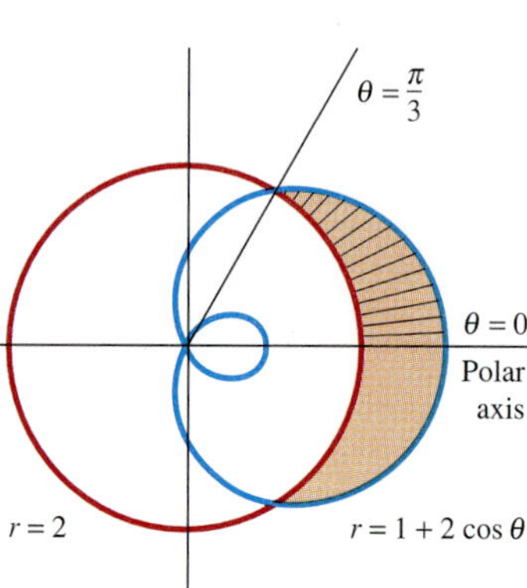

Fig. 10.3.9 The region of Example 3

and the figure shows that we should choose the solutions $\theta = \pm\pi/3$. These two values of θ are the needed limits of integration. When we use Eq. (3), we find that

$$\begin{aligned} A &= \tfrac{1}{2}\int_{-\pi/3}^{\pi/3} [(1 + 2\cos\theta)^2 - 2^2]\,d\theta \\ &= \int_0^{\pi/3} (4\cos\theta + 4\cos^2\theta - 3)\,d\theta \qquad \text{(by symmetry)} \\ &= \int_0^{\pi/3} (4\cos\theta + 2\cos 2\theta - 1)\,d\theta \\ &= \Big[4\sin\theta + \sin 2\theta - \theta\Big]_0^{\pi/3} = \frac{15\sqrt{3} - 2\pi}{6}. \end{aligned}$$

10.3 PROBLEMS

In Problems 1 through 6, sketch the plane region bounded by the given polar curve $r = f(\theta)$, $\alpha \leqq \theta \leqq \beta$, and the rays $\theta = \alpha$, $\theta = \beta$.

1. $r = \theta$, $0 \leqq \theta \leqq \pi$ **2.** $r = \theta$, $0 \leqq \theta \leqq 2\pi$

3. $r = 1/\theta$, $\pi \leqq \theta \leqq 3\pi$ **4.** $r = 1/\theta$, $3\pi \leqq \theta \leqq 5\pi$

5. $r = e^{-\theta}$, $0 \leqq \theta \leqq \pi$

6. $r = e^{-\theta}$, $\pi/2 \leqq \theta \leqq 3\pi/2$

In Problems 7 through 16, find the area bounded by the given curve.

7. $r = 2\cos\theta$

8. $r = 4\sin\theta$

9. $r = 1 + \cos\theta$

10. $r = 2 - 2\sin\theta$ (Fig. 10.3.10)

11. $r = 2 - \cos\theta$

12. $r = 3 + 2\sin\theta$ (Fig. 10.3.11)

13. $r = -4\cos\theta$ **14.** $r = 5(1 + \sin\theta)$

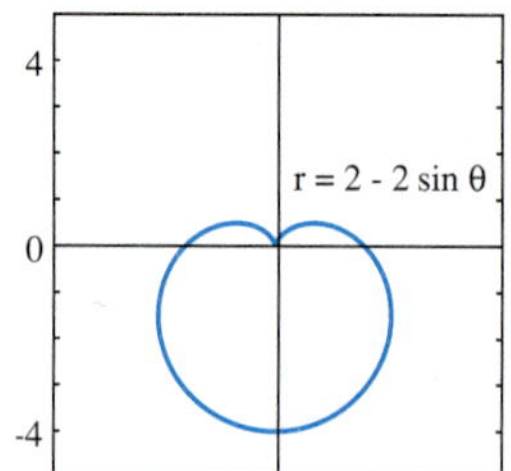

Fig. 10.3.10 The cardioid of Problem 10

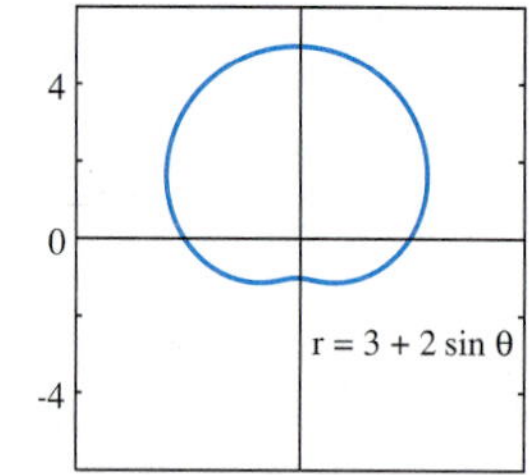

Fig. 10.3.11 The limaçon of Problem 12

15. $r = 3 - \cos\theta$ **16.** $r = 2 + \sin\theta + \cos\theta$

In Problems 17 through 24, find the area bounded by one loop of the given curve.

17. $r = 2\cos 2\theta$

18. $r = 3\sin 3\theta$ (Fig. 10.3.12)

19. $r = 2\cos 4\theta$ (Fig. 10.3.13)

20. $r = \sin 5\theta$ (Fig. 10.3.14)

21. $r^2 = 4\sin 2\theta$

22. $r^2 = 4\cos 2\theta$ (Fig. 10.3.15)

23. $r^2 = 4\sin\theta$

24. $r = 6\cos 6\theta$

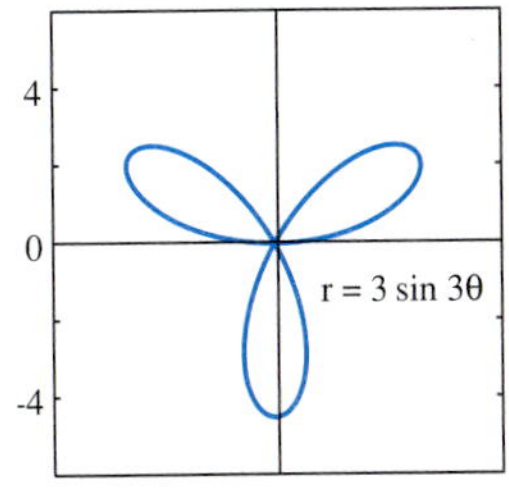

Fig. 10.3.12 The three-leaved rose of Problem 18

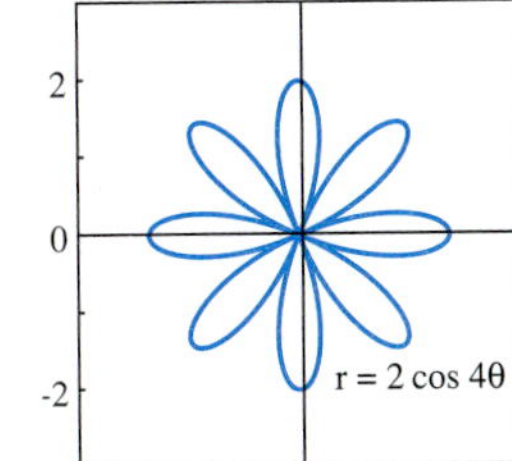

Fig. 10.3.13 The eight-leaved rose of Problem 19

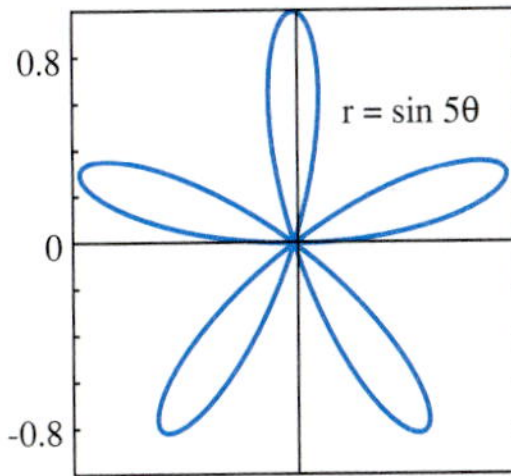

Fig. 10.3.14 The five-leaved rose of Problem 20

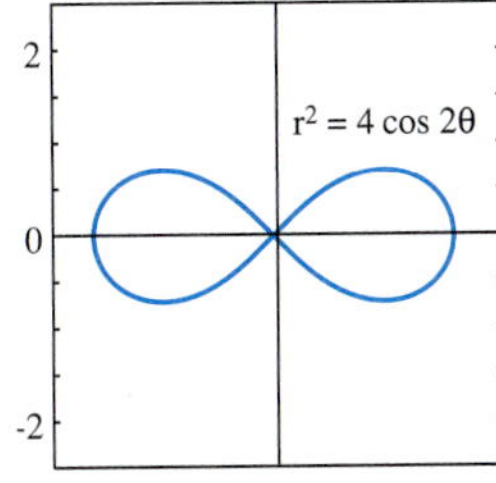

Fig. 10.3.15 The lemniscate of Problem 22

In Problems 25 through 36, find the area of the region described.

25. Inside $r = 2\sin\theta$ and outside $r = 1$

26. Inside both $r = 4\cos\theta$ and $r = 2$

27. Inside both $r = \cos\theta$ and $r = \sqrt{3}\sin\theta$

28. Inside $r = 2 + \cos\theta$ and outside $r = 2$

29. Inside $r = 3 + 2\cos\theta$ and outside $r = 4$

30. Inside $r^2 = 2\cos 2\theta$ and outside $r = 1$

31. Inside $r^2 = \cos 2\theta$ and $r^2 = \sin 2\theta$ (Fig. 10.3.16)

32. Inside the large loop and outside the small loop of $r = 1 - 2\sin\theta$ (Fig. 10.3.17)

33. Inside $r = 2(1 + \cos\theta)$ and outside $r = 1$

34. Inside the figure eight curve $r^2 = 4\cos\theta$ and outside $r = 1 - \cos\theta$

35. Inside both $r = 2\cos\theta$ and $r = 2\sin\theta$

36. Inside $r = 2 + 2\sin\theta$ and outside $r = 2$

37. Find the area of the circle $r = \sin\theta + \cos\theta$ by integration in polar coordinates (Fig. 10.3.18). Check your answer by writing the equation of the circle in rectangular coordinates, finding its radius, and then using the familiar formula for the area of a circle.

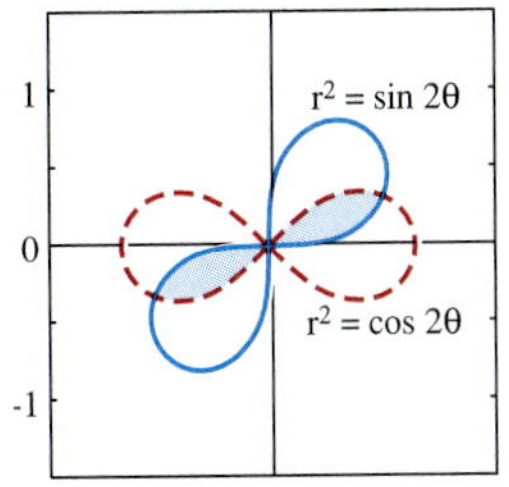

Fig. 10.3.16 Problem 31

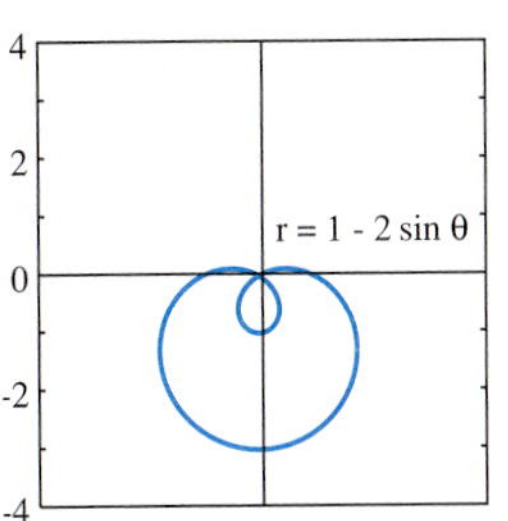

Fig. 10.3.17 Problem 32

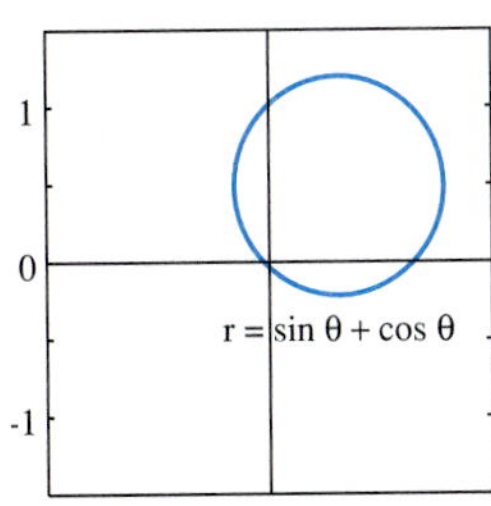

Fig. 10.3.18 The circle $r = \sin\theta + \cos\theta$ (Problem 37)

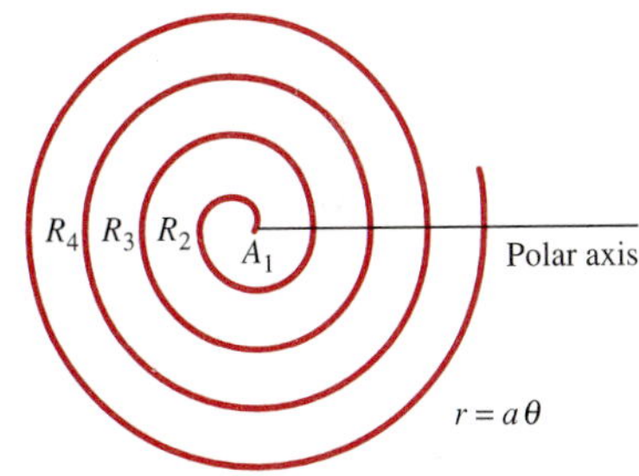

Fig. 10.3.19 The spiral of Archimedes (Problem 39)

38. Find the area of the region that lies interior to all three circles $r = 1$, $r = 2\cos\theta$, and $r = 2\sin\theta$.

39. The *spiral of Archimedes,* shown in Fig. 10.3.19, has the simple equation $r = a\theta$ (a is a constant). Let A_n denote the area bounded by the nth turn of the spiral, where $2(n-1)\pi \leqq \theta \leqq 2n\pi$, and by the portion of the polar axis joining its endpoints. For each $n \geqq 2$, let $R_n = A_n - A_{n-1}$ denote the area between the $(n-1)$th and the nth turns. Then derive the following results of Archimedes:

(a) $A_1 = \frac{1}{3}\pi(2\pi a)^2$; (b) $A_2 = \frac{7}{12}\pi(4\pi a)^2$;

(c) $R_2 = 6A_1$; (d) $R_{n+1} = nR_2$ for $n \geqq 2$.

40. Two circles each have radius a and each circle passes through the center of the other. Find the area of the region that lies within both circles.

41. A polar curve of the form $r = ae^{-k\theta}$ is called a *logarithmic spiral,* and the portion given by $2(n-1)\pi \leqq \theta \leqq 2n\pi$ is called the nth *turn* of this spiral. Figure 10.3.20 shows the first five turns of the logarithmic spiral $r = e^{-\theta/10}$, and the area of the region lying between the second and third turns is shaded. Find:

(a) The area of the region that lies between the first and second turns.

(b) The area of the region between the $(n-1)$th and nth turns for $n > 1$.

42. Figure 10.3.21 shows the first turn of the logarithmic spiral $r = 2e^{-\theta/10}$ together with the two circles, both cen-

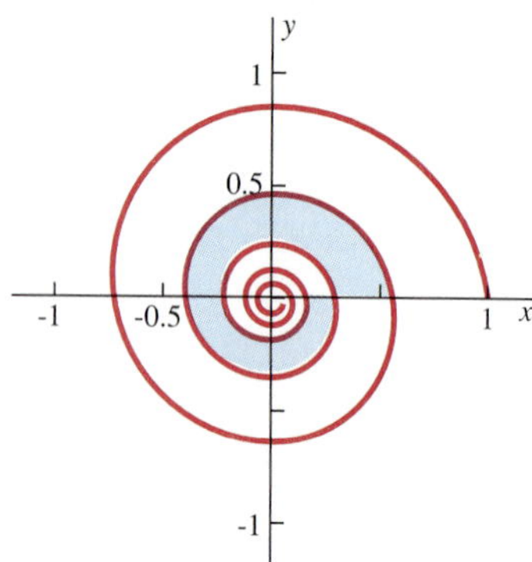

Fig. 10.3.20 The logarithmic spiral of Problem 41

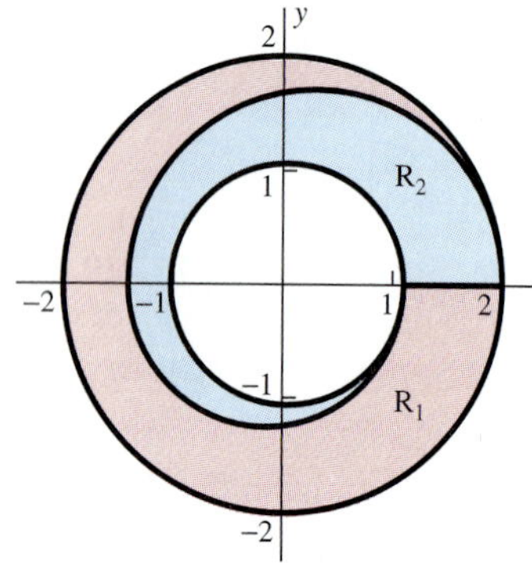

Fig. 10.3.21 The two regions of Problem 42

tered at $(0, 0)$, through the endpoints of the spiral. Find the areas of the two shaded regions and verify that their sum is the area of the annular region between the two circles.

43. The shaded region R in Fig. 10.3.22 is bounded by the cardioid $r = 1 + \cos\theta$, the spiral $r = e^{-\theta/5}$, $0 \leqq \theta \leqq \pi$, and the spiral $r = e^{\theta/5}$, $-\pi \leqq \theta \leqq 0$. Graphically estimate the points of intersection of the cardioid and the spirals, and then approximate the area of the region R.

44. The shaded region R in Fig. 10.3.23 lies inside both the cardioid $r = 3 + 3\sin\theta$ and the polar curve $r = 3 + \cos 4\theta$. Graphically estimate the points of intersection of the two curves; then approximate the area of the region R.

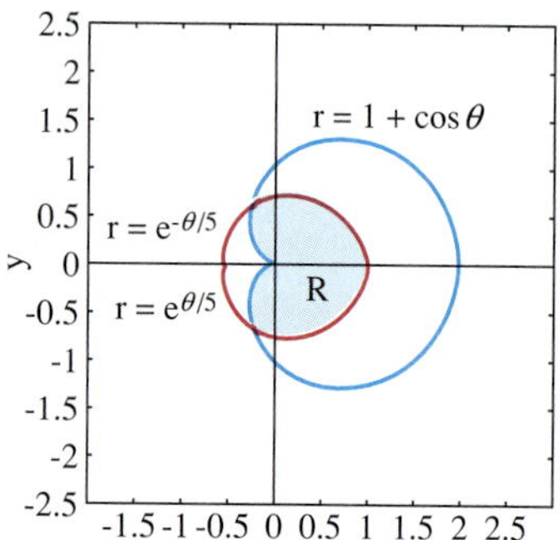

Fig. 10.3.22 The region of Problem 43

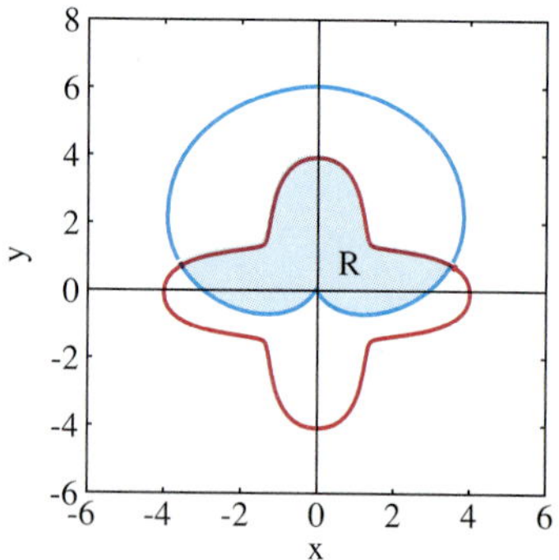

Fig. 10.3.23 The region of Problem 44

10.4 PARAMETRIC CURVES

Until now we have encountered *curves* mainly as graphs of equations. An equation of the form $y = f(x)$ or of the form $x = g(y)$ determines a curve by giving one of the coordinate variables explicitly as a function of the other. An equation of the form $F(x, y) = 0$ may also determine a curve, but then each variable is given implicitly as a function of the other.

Another important type of curve is the trajectory of a point moving in the coordinate plane. The motion of the point can be described by giving its position $(x(t), y(t))$ at time t. Such a description involves expressing both the rectangular coordinates variables x and y as functions of a third variable, or *parameter*, t rather than as functions of each other. In this context, a **parameter** is an independent variable (not a constant, as is sometimes meant in popular usage). This approach motivates the following definition.

Definition ***Parametric Curve***

A **parametric curve** C in the plane is a pair of functions

$$x = f(t), \quad y = g(t), \tag{1}$$

that give x and y as continuous functions of the real number t (the parameter) in some interval I.

Each value of the parameter t determines a point $(f(t), g(t))$, and the set of all such points is the **graph** of the curve C. Often the distinction between the curve—the pair of **coordinate functions** f and g—and the graph is not made. Therefore we may refer interchangeably to the curve and to its graph. The two equations in (1) are called the **parametric equations** of the curve.

The graph of a parametric curve may be sketched by plotting enough points to indicate its likely shape. In some cases we can eliminate the parameter t and thus obtain an equation in x and y. This equation may give us more information about the shape of the curve.

EXAMPLE 1 Determine the graph of the curve

$$x = \cos t, \quad y = \sin t, \quad 0 \leqq t \leqq 2\pi. \tag{2}$$

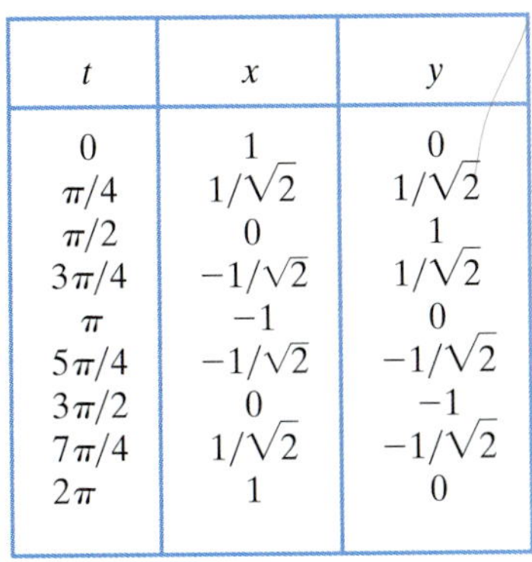

t	x	y
0	1	0
$\pi/4$	$1/\sqrt{2}$	$1/\sqrt{2}$
$\pi/2$	0	1
$3\pi/4$	$-1/\sqrt{2}$	$1/\sqrt{2}$
π	-1	0
$5\pi/4$	$-1/\sqrt{2}$	$-1/\sqrt{2}$
$3\pi/2$	0	-1
$7\pi/4$	$1/\sqrt{2}$	$-1/\sqrt{2}$
2π	1	0

Fig. 10.4.1 A table of values for Example 1

Solution Figure 10.4.1 shows a table of values of x and y that correspond to multiples of $\pi/4$ for the parameter t. These values give the eight points highlighted in Fig. 10.4.2, all of which lie on the unit circle. This suggests that the graph is, in fact, the unit circle. To verify this, we note that the fundamental identity of trigonometry gives

$$x^2 + y^2 = \cos^2 t + \sin^2 t = 1,$$

so every point of the graph lies on the circle with equation $x^2 + y^2 = 1$. Conversely, the point of the circle with angular (polar) coordinate t is the point $(\cos t, \sin t)$ of the graph. Thus the graph is precisely the unit circle. ■

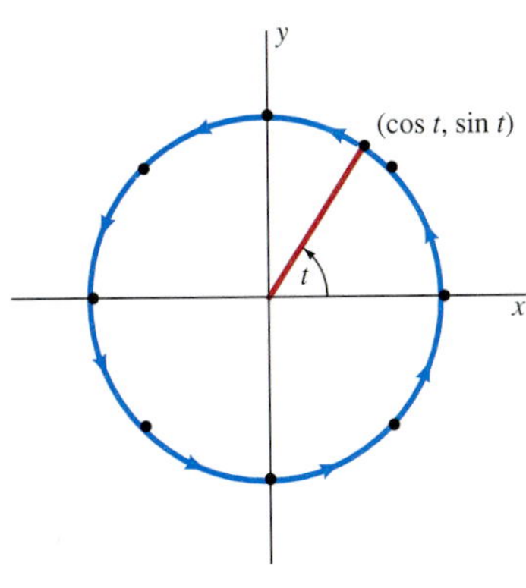

Fig. 10.4.2 The graph of the parametric functions of Example 1

What is lost in the process in Example 1 is the information about how the graph is produced as t goes from 0 to 2π. But this is easy to determine by inspection. As t travels from 0 to 2π, the point $(\cos t, \sin t)$ begins at $(1, 0)$ and travels counterclockwise around the circle, ending at $(1, 0)$ when $t = 2\pi$.

A given figure in the plane may be the graph of different curves. To speak more loosely, a given curve may have different **parametrizations**.

EXAMPLE 2 The graph of the parametric curve

$$x = \frac{1 - t^2}{1 + t^2}, \quad y = \frac{2t}{1 + t^2}, \quad -\infty < t < +\infty$$

also lies on the unit circle, because we find that $x^2 + y^2 = 1$ here as well. If t begins at 0 and increases, then the point $P(x(t), y(t))$ begins at $(1, 0)$ and travels along the upper half of the circle. If t begins at 0 and decreases, then the point $P(x(t), y(t))$ travels along the lower half of the circle. As t approaches either $+\infty$ or $-\infty$, the point P approaches the point $(-1, 0)$. Thus the graph consists of the unit circle with the single point $(-1, 0)$ deleted. A slight modification of the curve of Example 1,

$$x = \cos t, \quad y = \sin t, \quad -\pi < t < \pi,$$

is a different parametrization of this circle. ■

EXAMPLE 3 Eliminate the parameter to determine the graph of the parametric curve

$$x = t - 1, \quad y = 2t^2 - 4t + 1, \quad 0 \leqq t \leqq 2.$$

Solution We substitute $t = x + 1$ (from the equation for x) into the equation for y. This yields

$$y = 2(x + 1)^2 - 4(x + 1) + 1 = 2x^2 - 1$$

for $-1 \leqq x \leqq 1$. Thus the graph of the given curve is a portion of the parabola $y = 2x^2 - 1$ (Fig. 10.4.3). As t increases from 0 to 2, the point $(t - 1, 2t^2 - 4t + 1)$ travels along the parabola from $(-1, 1)$ to $(1, 1)$. ■

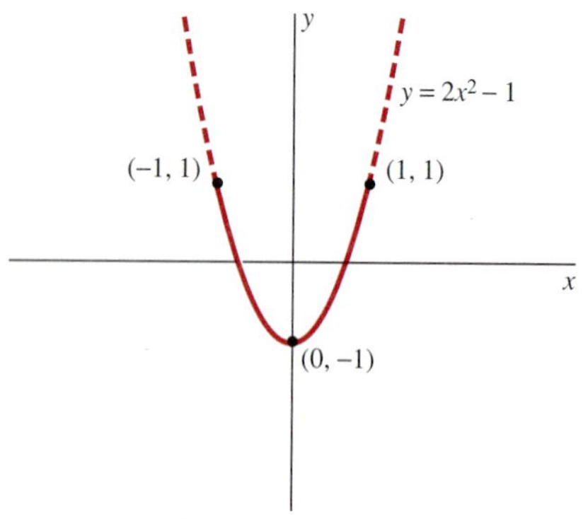

Fig. 10.4.3 The curve of Example 3 is part of a parabola.

REMARK The parabolic arc of Example 3 can be reparametrized with

$$x = \sin t, \quad y = 2\sin^2 t - 1.$$

Now, as t increases, the point $(\sin t, 2\sin^2 t - 1)$ travels back and forth along the parabola between the two points $(-1, 1)$ and $(1, 1)$, rather like the bob of a pendulum.

The parametric curve of Example 3 is one in which we can eliminate the parameter and thus obtain an explicit equation $y = f(x)$. Moreover, any explicitly presented curve $y = f(x)$ can be viewed as a parametric curve by writing

$$x = t, \quad y = f(t),$$

with the parameter t taking on values in the original domain of f. By contrast, the circle of Example 1 illustrates a parametric curve whose graph is not the graph of any single function. (Why not?) Example 4 exhibits another way in which parametric curves can differ from graphs of functions—they can have self-intersections.

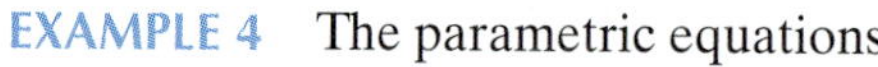
EXAMPLE 4 The parametric equations

$$x = \cos at, \quad y = \sin bt$$

(with a and b constant) define the *Lissajous curves* that typically appear on oscilloscopes in physics and electronics laboratories. The Lissajous curve with $a = 3$ and $b = 5$ is shown in Fig. 10.4.4. You probably would not want to calculate and plot by hand enough points to produce a Lissajous curve. Figure 10.4.4 was plotted with a computer program that generated it almost immediately. But it is perhaps more instructive to watch a slower graphing calculator plot a parametric curve like this, because the curve is traced by a point that moves on the screen as the parameter t increases (from 0 to 2π in this case). ■

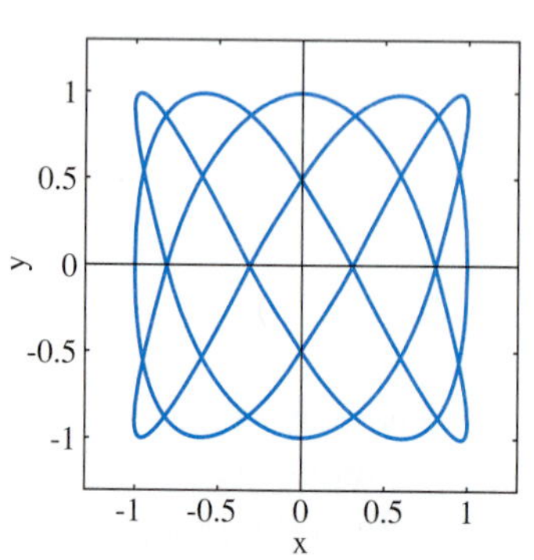

Fig. 10.4.4 The Lissajous curve with $a = 3, b = 5$

The use of parametric equations $x = x(t)$, $y = y(t)$ is most advantageous when elimination of the parameter is either impossible or would lead to an equation $y = f(x)$ that is considerably more complicated than the original parametric equations. This often happens when the curve is a geometric locus or the path of a point moving under specified conditions.

EXAMPLE 5 The curve traced by a point P on the edge of a rolling circle is called a **cycloid**. The circle rolls along a straight line without slipping or stopping. (You will see a cycloid if you watch a patch of bright paint on the tire of a bicycle that crosses your path.) Find parametric equations for the cycloid if the line along which the circle rolls is the x-axis, the circle is above the x-axis but always tangent to it, and the point P begins at the origin.

Solution Evidently the cycloid consists of a series of arches. We take as parameter t the angle (in radians) through which the circle has turned since it began with P at the origin O. This is the angle TCP in Fig. 10.4.5.

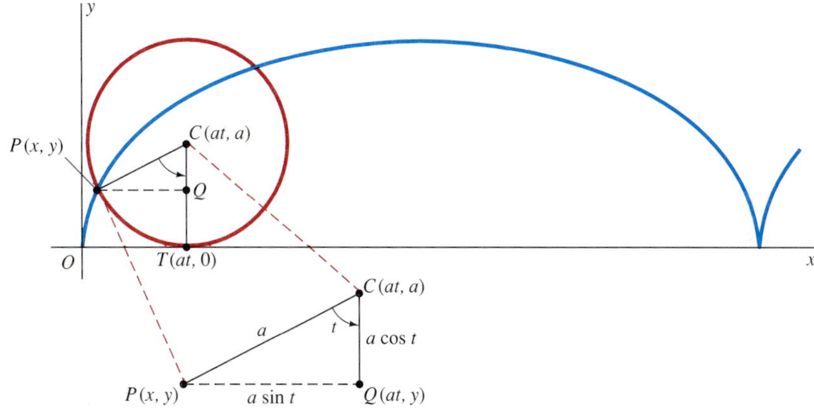

Fig. 10.4.5 The cycloid and the right triangle CPQ (Example 5)

The distance the circle has rolled is $|OT|$, so this is also the length of the circumference subtended by the angle TCP. Thus $|OT| = at$ if a is the radius of the circle, so the center C of the rolling circle has coordinates (at, a) when the angle TCP is t. The right triangle CPQ in Fig. 10.4.5 provides us with the relations

$$at - x = a\sin t \quad \text{and} \quad a - y = a\cos t.$$

Therefore the cycloid—the path of the moving point P—has parametric equations

$$x = a(t - \sin t), \qquad y = a(1 - \cos t). \tag{3}$$

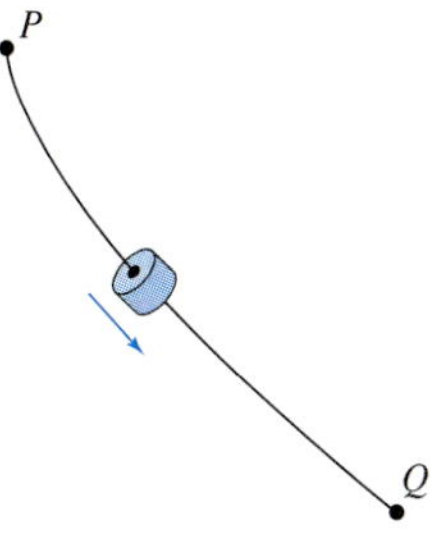

Fig. 10.4.6 A bead sliding down a wire—the brachistochrone problem

Historical Note Figure 10.4.6 shows a bead sliding down a frictionless wire from point P to point Q. The *brachistochrone problem* asks what shape the wire should be to minimize the bead's time of descent from P to Q. In June 1696, John Bernoulli proposed the brachistochrone problem as a public challenge, with a six-month deadline (later extended to Easter 1697, at Leibniz's request). Isaac Newton, then retired from academic life and serving as Warden of the Mint in London, received Bernoulli's challenge on January 29, 1697. The very next day he communicated his own solution—the curve of minimal descent time is an arc of an inverted cycloid—to the Royal Society of London.

Lines Tangent to Parametric Curves

The parametric curve $x = f(t)$, $y = g(t)$ is called **smooth** if the derivatives $f'(t)$ and $g'(t)$ are continuous and never simultaneously zero. In some neighborhood of each point of its graph, a smooth parametric curve can be described in one or possibly both of the forms $y = F(x)$ and $x = G(y)$. To see why this is so, suppose (for example) that $f'(t) > 0$ on the interval I. Then $f(t)$ is an increasing function on I and therefore has an inverse function $t = \phi(x)$ there. If we substitute $t = \phi(x)$ into the equation $y = g(t)$, then we get

$$y = g(\phi(x)) = F(x).$$

We can use the chain rule to compute the slope dy/dx of the line tangent to a smooth parametric curve at a given point. Differentiation of $y = F(x)$ with respect to t yields

$$\frac{dy}{dt} = \frac{dy}{dx} \cdot \frac{dx}{dt},$$

so

$$\frac{dy}{dx} = \frac{dy/dt}{dx/dt} = \frac{g'(t)}{f'(t)} \tag{4}$$

at any point where $f'(t) \neq 0$. The tangent line is vertical at any point where $f'(t) = 0$ but $g'(t) \neq 0$.

Equation (4) gives $y' = dy/dx$ as a function of t. Another differentiation with respect to t, again with the aid of the chain rule, results in the formula

$$\frac{dy'}{dt} = \frac{dy'}{dx} \cdot \frac{dx}{dt},$$

so

$$\frac{d^2y}{dx^2} = \frac{dy'}{dx} = \frac{dy'/dt}{dx/dt}. \tag{5}$$

EXAMPLE 6 Calculate dy/dx and d^2y/dx^2 for the cycloid with the parametric equations in (3).

Solution We begin with

$$x = a(t - \sin t), \qquad y = a(1 - \cos t). \tag{3}$$

Then Eq. (4) gives

$$\frac{dy}{dx} = \frac{dy/dt}{dx/dt} = \frac{a \sin t}{a(1 - \cos t)} = \frac{\sin t}{1 - \cos t}. \tag{6}$$

This derivative is zero when t is an odd integral multiple of π, so the tangent line is horizontal at the midpoint of each arch of the cycloid. The endpoints of the arches correspond to even integral multiples of π, where both the numerator and the denominator in Eq. (6) are zero. These are isolated points (called *cusps*) at which the cycloid fails to be a smooth curve. See Fig. 10.4.7.

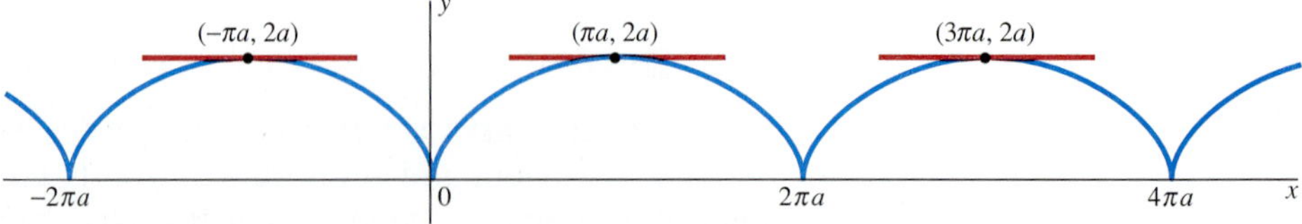

Fig. 10.4.7 Horizontal tangents and cusps of the cycloid

Next, Eq. (5) yields

$$\frac{d^2y}{dx^2} = \frac{(\cos t)(1 - \cos t) - (\sin t)(\sin t)}{(1 - \cos t)^2 \cdot a(1 - \cos t)} = -\frac{1}{a(1 - \cos t)^2}.$$

Because $d^2y/dx^2 < 0$ for all t (except for the isolated even integral multiples of π), this shows that each arch of the cycloid is concave downward (Fig. 10.4.7). ■

Polar Curves as Parametric Curves

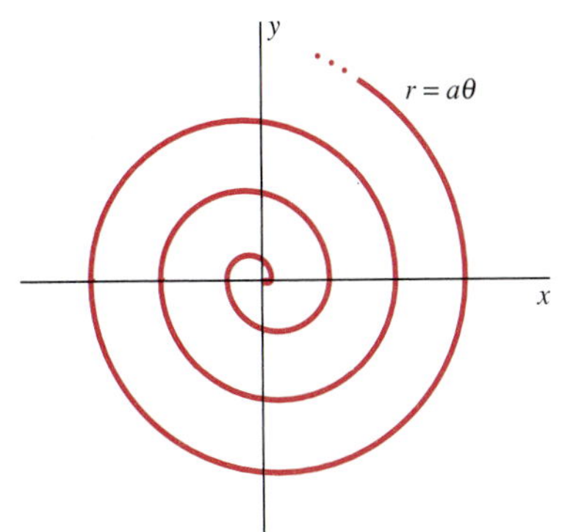

Fig. 10.4.8 The spiral of Archimedes (Example 7)

A curve given in polar coordinates by the equation $r = f(\theta)$ can be regarded as a parametric curve with parameter θ. To see this, we recall that the equations $x = r\cos\theta$ and $y = r\sin\theta$ allow us to change from polar to rectangular coordinates. We replace r with $f(\theta)$, and this gives the parametric equations

$$x = f(\theta)\cos\theta, \quad y = f(\theta)\sin\theta, \tag{7}$$

which express x and y in terms of the parameter θ.

EXAMPLE 7 The *spiral of Archimedes* has the polar coordinates equation $r = a\theta$ (Fig. 10.4.8). The equations in (7) give the spiral the parametrization

$$x = a\theta\cos\theta, \quad y = a\theta\sin\theta.$$ ■

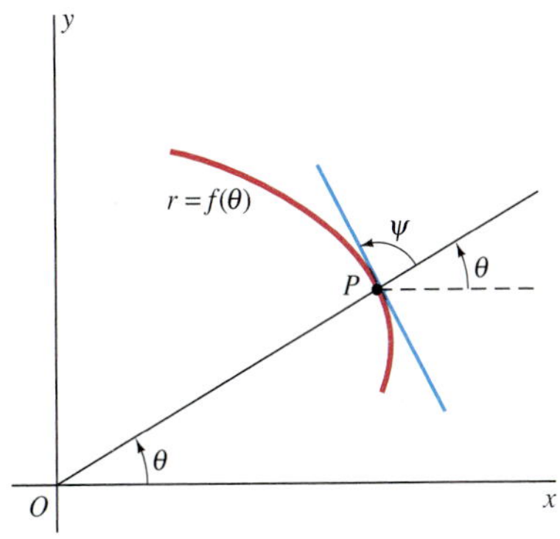

Fig. 10.4.9 The interpretation of the angle ψ [see Eq. (10)]

The slope dy/dx can be computed in terms of polar coordinates as well as rectangular coordinates. Given a polar coordinates curve $r = f(\theta)$, we use the parametrization

$$x = f(\theta)\cos\theta, \quad y = f(\theta)\sin\theta$$

shown in (7). Then Eq. (4), with θ in place of t, gives

$$\frac{dy}{dx} = \frac{dy/d\theta}{dx/d\theta} = \frac{f'(\theta)\sin\theta + f(\theta)\cos\theta}{f'(\theta)\cos\theta - f(\theta)\sin\theta}, \tag{8}$$

or, alternatively, denoting $f'(\theta)$ by r',

$$\frac{dy}{dx} = \frac{r'\sin\theta + r\cos\theta}{r'\cos\theta - r\sin\theta}. \tag{9}$$

Equation (9) has the following useful consequence. Let ψ denote the angle between the tangent line at P and the radius OP (extended) from the origin (Fig. 10.4.9). Then

$$\cot\psi = \frac{1}{r}\cdot\frac{dr}{d\theta} \quad (0 \leqq \psi \leqq \pi). \tag{10}$$

In Problem 32 we indicate how Eq. (10) can be derived from Eq. (9).

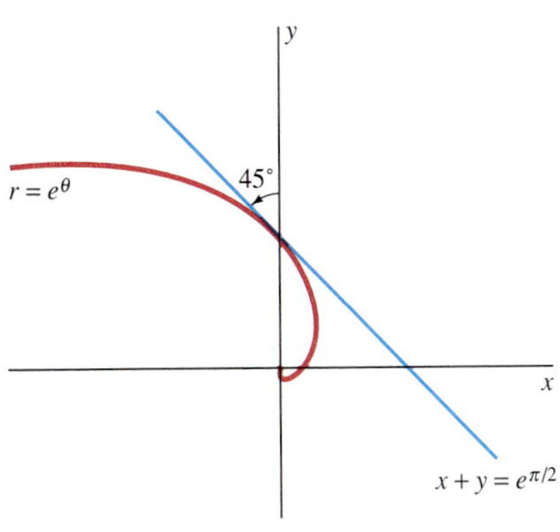

Fig. 10.4.10 The angle ψ is always 45° for the logarithmic spiral (Example 8)

EXAMPLE 8 Consider the *logarithmic spiral* with polar equation $r = e^{\theta}$. Show that $\psi = \pi/4$ at every point of the spiral, and write an equation of its tangent line at the point $(e^{\pi/2}, \pi/2)$.

Solution Because $dr/d\theta = e^{\theta}$, Eq. (10) tells us that $\cot\psi = e^{\theta}/e^{\theta} = 1$. Thus $\psi = \pi/4$. When $\theta = \pi/2$, Eq. (9) gives

$$\frac{dy}{dx} = \frac{e^{\pi/2}\sin(\pi/2) + e^{\pi/2}\cos(\pi/2)}{e^{\pi/2}\cos(\pi/2) - e^{\pi/2}\sin(\pi/2)} = -1.$$

But when $\theta = \pi/2$, we have $x = 0$ and $y = e^{\pi/2}$. It follows that an equation of the desired tangent line is

$$y - e^{\pi/2} = -x; \quad \text{that is,} \quad x + y = e^{\pi/2}.$$

The line and the spiral appear in Fig. 10.4.10. ■

10.4 PROBLEMS

In Problems 1 through 12, eliminate the parameter and then sketch the curve.

1. $x = t + 1, \quad y = 2t - 1$
2. $x = t^2 + 1, \quad y = 2t^2 - 1$
3. $x = t^2, \quad y = t^3$
4. $x = \sqrt{t}, \quad y = 3t - 2$
5. $x = t + 1, \quad y = 2t^2 - t - 1$
6. $x = t^2 + 3t, \quad y = t - 2$
7. $x = e^t, \quad y = 4e^{2t}$
8. $x = 2e^t, \quad y = 2e^{-t}$
9. $x = 5\cos t, \quad y = 3\sin t$
10. $x = \sinh t, \quad y = \cosh t$
11. $x = 2\cosh t, \quad y = 3\sinh t$
12. $x = \sec t, \quad y = \tan t$

In Problems 13 through 16, first eliminate the parameter and sketch the curve. Then describe the motion of the point $(x(t), y(t))$ as t varies in the given interval.

13. $x = \sin 2\pi t, \quad y = \cos 2\pi t; \quad 0 \leqq t \leqq 1$
14. $x = 3 + 2\cos t, \quad y = 5 - 2\sin t; \quad 0 \leqq t \leqq 2\pi$
15. $x = \sin^2 \pi t, \quad y = \cos^2 \pi t; \quad 0 \leqq t \leqq 2$
16. $x = \cos t, \quad y = \sin^2 t; \quad -\pi \leqq t \leqq \pi$

In Problems 17 through 20, (a) first write the equation of the line tangent to the given parametric curve at the point that corresponds to the given value of t, and (b) then calculate d^2y/dx^2 to determine whether the curve is concave upward or concave downward at this point.

17. $x = 2t^2 + 1, \ y = 3t^3 + 2; \ t = 1$
18. $x = \cos^3 t, \ y = \sin^3 t; \ t = \pi/4$
19. $x = t\sin t, \ y = t\cos t; \ t = \pi/2$
20. $x = e^t, \ y = e^{-t}; \ t = 0$

In Problems 21 through 24, find the angle ψ between the radius OP and the tangent line at the point P that corresponds to the given value of θ.

21. $r = \exp(\theta\sqrt{3}), \quad \theta = \pi/2$
22. $r = 1/\theta, \quad \theta = 1$
23. $r = \sin 3\theta, \quad \theta = \pi/6$
24. $r = 1 - \cos\theta, \quad \theta = \pi/3$

In Problems 25 through 28, find:

(a) The points on the curve where the tangent line is horizontal.

(b) The slope of each tangent line at any point where the curve intersects the x-axis.

25. $x = t^2, y = t^3 - 3t$ (Fig. 10.4.11)
26. $x = \sin t, y = \sin 2t$ (Fig. 10.4.12)

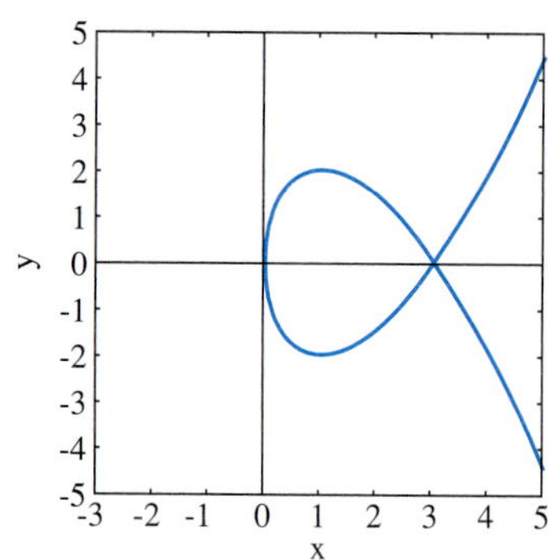

Fig. 10.4.11 The curve of Problem 25

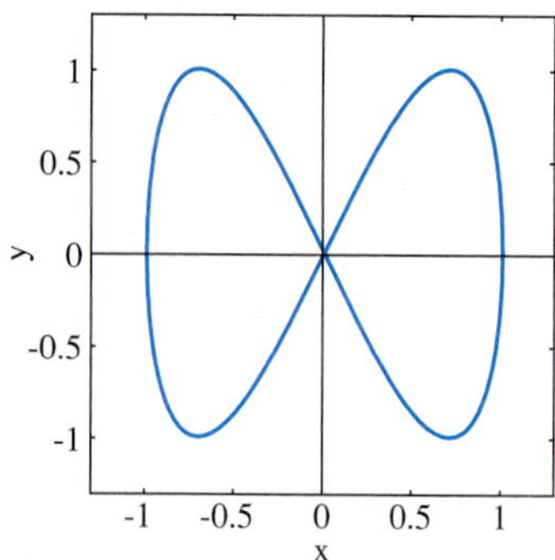

Fig. 10.4.12 The curve of Problem 26

27. $r = 1 + \cos\theta$
28. $r^2 = 4\cos 2\theta$ (see Fig. 10.3.15)
29. The curve C is determined by the parametric equations $x = e^{-t}$, $y = e^{2t}$. Calculate dy/dx and d^2y/dx^2 directly

from these parametric equations. Conclude that C is concave upward at every point. Then sketch C.

30. The graph of the folium of Descartes with rectangular equation $x^3 + y^3 = 3xy$ appears in Fig. 10.4.13. Parametrize its loop as follows: Let P be the point of intersection of the line $y = tx$ with the loop; then solve for the coordinates x and y of P in terms of t.

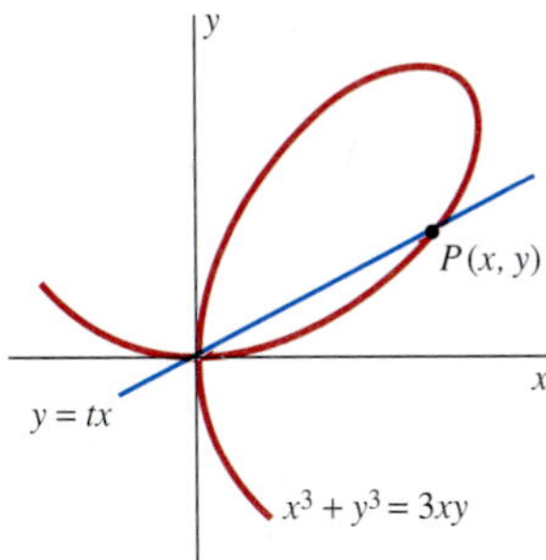

Fig. 10.4.13 The loop of the folium of Descartes (Problem 30)

31. Parametrize the parabola $y^2 = 4px$ by expressing x and y as functions of the slope m of the tangent line at the point $P(x, y)$ of the parabola.

32. Let P be a point of the curve with polar equation $r = f(\theta)$, and let ψ be the angle between the extended radius OP and the tangent line at P. Let α be the angle of inclination of this tangent line, measured counterclockwise from the horizontal. Then $\psi = \alpha - \theta$. Verify Eq. (10) by substituting $\tan\alpha = dy/dx$ from Eq. (9) and $\tan\theta = y/x = (\sin\theta)/(\cos\theta)$ into the identity

$$\cot\psi = \frac{1}{\tan(\alpha - \theta)} = \frac{1 + \tan\alpha\tan\theta}{\tan\alpha - \tan\theta}.$$

33. Let P_0 be the highest point of the circle of Fig. 10.4.5—the circle that generates the cycloid of Example 5. Show that the line through P_0 and the point P of the cycloid (the point P is shown in Fig. 10.4.5) is tangent to the cycloid at P. This fact gives a geometric construction of the line tangent to the cycloid.

34. A circle of radius b rolls without slipping inside a circle of radius $a > b$. The path of a point fixed on the circumference of the rolling circle is called a *hypocycloid* (Fig. 10.4.14). Let P begin its journey at $A(a, 0)$ and let t be the angle AOC, where O is the center of the large circle and C is the center of the rolling circle. Show that the coordinates of P are given by the parametric equations

$$x = (a - b)\cos t + b\cos\left(\frac{a - b}{b}t\right),$$

$$y = (a - b)\sin t - b\sin\left(\frac{a - b}{b}t\right).$$

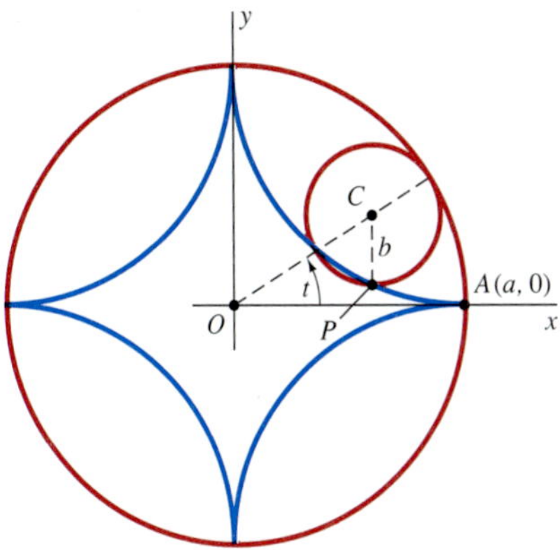

Fig. 10.4.14 The hypocycloid of Problem 34

35. If $b = a/4$ in Problem 34, show that the parametric equations of the hypocycloid reduce to

$$x = a\cos^3 t, \qquad y = a\sin^3 t.$$

36. (a) Prove that the hypocycloid of Problem 35 is the graph of the equation

$$x^{2/3} + y^{2/3} = a^{2/3}.$$

(b) Find all points of this hypocycloid where its tangent line is either horizontal or vertical, and find the intervals on which it is concave upward and those on which it is concave downward. (c) Sketch this hypocycloid.

37. Consider a point P on the spiral of Archimedes, the curve shown in Fig. 10.4.15 with polar equation $r = a\theta$. Archimedes viewed the path of P as compounded of two motions, one with speed a directly away from the origin O and another a circular motion with unit angular speed around O. This suggests Archimedes' result that the line PQ in the figure is tangent to the spiral at P. Prove that this is indeed true.

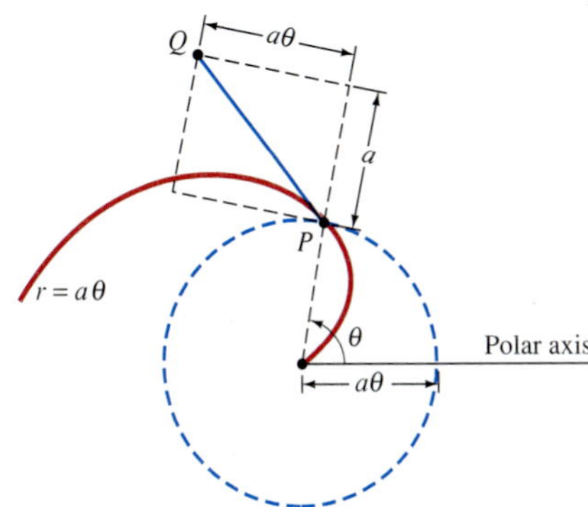

Fig. 10.4.15 The segment PQ is tangent to the spiral (a result of Archimedes; see Problem 37).

38. (a) Deduce from Eq. (7) that if t is not an integral multiple of 2π, then the slope of the tangent line at the corresponding point of the cycloid is $\cot(t/2)$. (b) Conclude that at the cusp of the cycloid where t is an integral multiple of 2π, the cycloid has a vertical tangent line.

39. A *loxodrome* is a curve $r = f(\theta)$ such that the tangent line at P and the radius OP in Fig. 10.4.9 make a constant angle. Use Eq. (10) to prove that every loxodrome is of the form $r = Ae^{k\theta}$, where A and k are constants. Thus every loxodrome is a logarithmic spiral similar to the one considered in Example 8.

40. Let a curve be described in polar coordinates by $r = f(\theta)$, where f is continuous. If $f(\alpha) = 0$, then the origin is the point of the curve corresponding to $\theta = \alpha$. Deduce from the parametrization $x = f(\theta)\cos\theta,\ y = f(\theta)\sin\theta$ that the line tangent to the curve at this point makes the angle α with the positive x-axis. For example, the cardioid $r = f(\theta) = 1 - \sin\theta$ shown in Fig. 10.4.16 is tangent to the y-axis at the origin. And, indeed, $f(\pi/2) = 0$: The y-axis is the line $\theta = \alpha = \pi/2$.

41. Use the technique of Problem 30 to parametrize the first-quadrant loop of the folium-like curve $x^5 + y^5 = 5x^2y^2$.

42. A line segment of length $2a$ has one endpoint constrained to lie on the x-axis and the other endpoint constrained to lie on the y-axis, but its endpoints are free to move along those axes. As they do so, its midpoint sweeps out a locus in the xy-plane. Obtain a rectangular coordinates equation of this locus and thereby identify this curve.

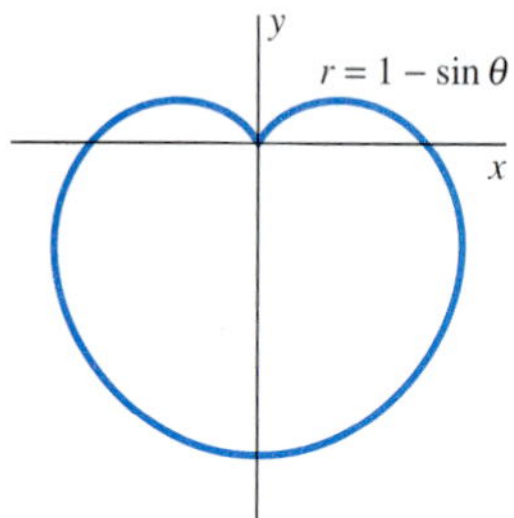

Fig. 10.4.16 The cardioid of Problem 40

10.4 PROJECT: CALCULATOR/COMPUTER GRAPHING OF PARAMETRIC CURVES

The most fun in plotting parametric curves comes from trying your own hand—especially if you have a graphics calculator or computer with graphing utility to do the real work. Try various values of the constants $a, b, p, \ldots$ in the investigations. When sines and cosines are involved, the interval $0 \leqq t \leqq 2\pi$ is a reasonable one to try first. You will need to experiment with different viewing windows to find one that shows the whole curve (or the most interesting part of it).

Investigation A Given: $x = at - b\sin t$, $y = a - b\cos t$. This *trochoid* is traced by a point P on a solid wheel of radius a as it rolls along the x-axis; the distance of P from the center of the wheel is $b > 0$. (The graph is a cycloid if $a = b$.) Try both cases $a > b$ and $a < b$.

Investigation B Given:

$$x = (a - b)\cos t + b\cos\left(\frac{a - b}{b}t\right),$$

$$y = (a - b)\sin t - b\sin\left(\frac{a - b}{b}t\right).$$

This is a *hypocycloid*—the path of a point P on a circle of radius b that rolls along the inside of a circle of radius $a > b$. Figure 10.4.14 shows the case $b = a/4$.

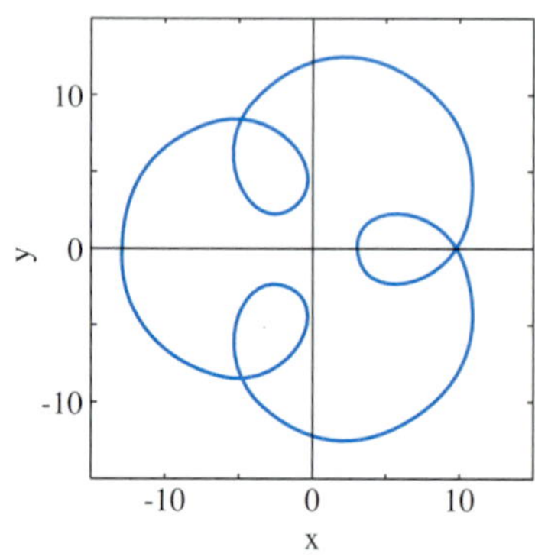

Fig. 10.4.17 The epitrochoid with $a = 8$, $b = 5$ (Investigation D)

Investigation C Given:

$$x = (a + b)\cos t - b\cos\left(\frac{a + b}{b}t\right),$$

$$y = (a + b)\sin t - b\sin\left(\frac{a + b}{b}t\right).$$

This is an *epicycloid* traced by a point P on a circle of radius b that rolls along the outside of a circle of radius $a > b$.

INVESTIGATION D Given:

$$x = a\cos t - b\cos\frac{at}{2}, \qquad y = a\sin t - b\sin\frac{at}{2}.$$

This is an *epitrochoid*—it is to the epicycloid what a trochoid is to a cycloid. With $a = 8$ and $b = 5$, you should get the curve shown in Fig. 10.4.17.

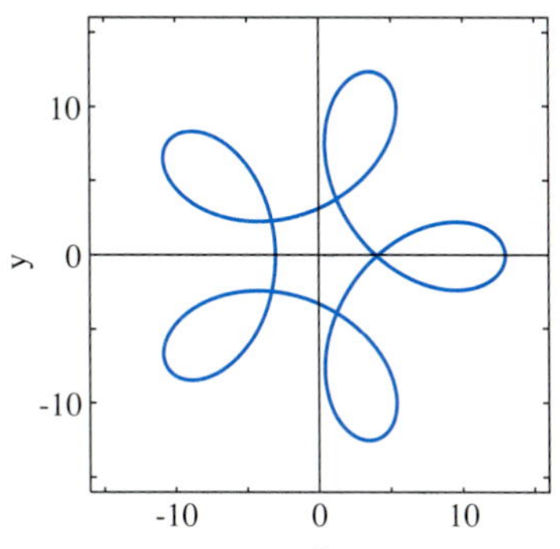

Fig. 10.4.18 The hypotrochoid with $a = 8$, $b = 5$ (Investigation E)

INVESTIGATION E Given:

$$x = a\cos t + b\cos\frac{at}{2}, \qquad y = a\sin t - b\sin\frac{at}{2}.$$

With $a = 8$ and $b = 5$, this *hypotrochoid* looks like the curve shown in Fig. 10.4.18.

INVESTIGATION F Given: $x = \cos at$, $y = \sin bt$. These are Lissajous curves (Example 4). The Lissajous curve with $a = 3$ and $b = 5$ is shown in Fig. 10.4.4.

INVESTIGATION G Consider the parametric equations

$$x = a\cos t - b\cos pt, \qquad y = c\sin t - d\sin qt.$$

The values $a = 16$, $b = 5$, $c = 12$, $d = 3$, $p = \frac{47}{3}$, and $q = \frac{44}{3}$ yield the "slinky curve" shown in Fig. 10.4.19. Experiment with various combinations of constants to see whether you can produce a prettier picture.

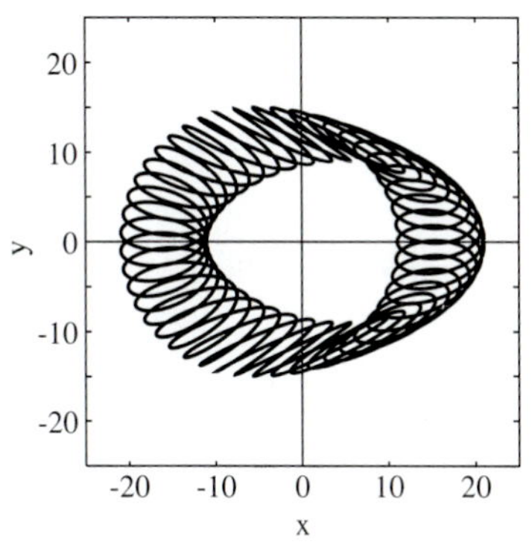

Fig. 10.4.19 The slinky curve of Investigation G

INVESTIGATION H Given a polar coordinates equation $r = f(\theta)$, it's frequently easiest to plot by converting to parametric form:

$$x = f(\theta)\cos\theta, \qquad y = f(\theta)\sin\theta,$$

so that the parameter is θ. Given

$$r(\theta) = 2\cos 3\theta - \sin 11\theta + \sin 2\theta,$$

plot this polar equation for $0 \leqq \theta \leqq 2\pi$. Then vary some of the coefficients or otherwise modify this function to see whether you can produce a more attractive curve.

10.5 INTEGRAL COMPUTATIONS WITH PARAMETRIC CURVES

In Chapter 6 we discussed the computation of a variety of geometric quantities associated with the graph $y = f(x)$ of a nonnegative function on the interval $[a, b]$. They included the following.

- The area under the curve:

$$A = \int_a^b y\,dx. \tag{1}$$

- The volume of revolution around the x-axis:

$$V_x = \int_a^b \pi y^2\,dx. \tag{2a}$$

- The volume of revolution around the y-axis:

$$V_y = \int_a^b 2\pi xy\,dx. \tag{2b}$$

▼ The arc length of the curve:

$$s = \int_0^s ds = \int_a^b \sqrt{1 + (dy/dx)^2}\,dx. \tag{3}$$

▼ The area of the surface of revolution around the x-axis:

$$S_x = \int_{x=a}^{b} 2\pi y\,ds. \tag{4a}$$

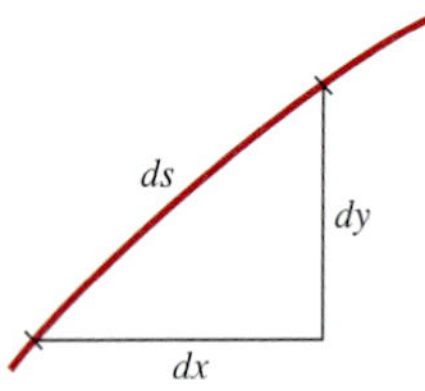

Fig. 10.5.1 Nearly a right triangle for dx and dy close to zero

▼ The area of the surface of revolution around the y-axis:

$$S_y = \int_{x=a}^{b} 2\pi x\,ds. \tag{4b}$$

We substitute $y = f(x)$ into each of these integrals before we integrate from $x = a$ to $x = b$.

We now want to compute these same quantities for a smooth parametric curve

$$x = f(t), \qquad y = g(t), \qquad \alpha \leqq t \leqq \beta. \tag{5}$$

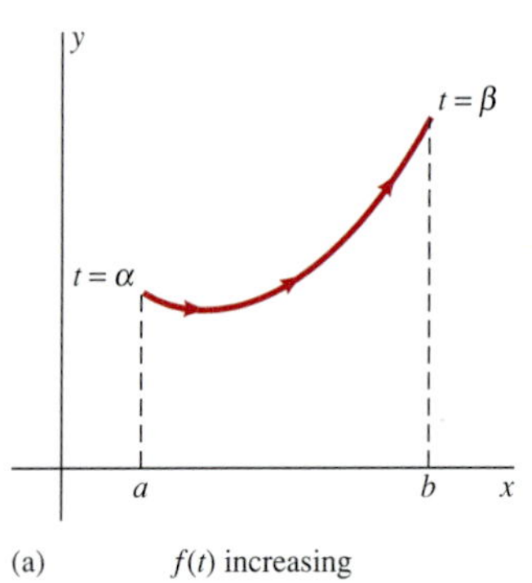

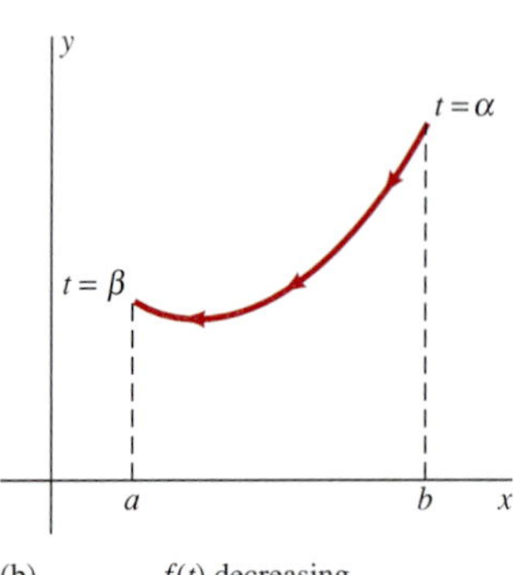

Fig. 10.5.2 Tracing a parametrized curve: (a) $f(t)$ increasing; (b) $f(t)$ decreasing

The area, volume, arc length, and surface integrals in Eqs. (1) through (4) can then be evaluated by making the formal substitutions

$$\begin{aligned} x &= f(t), & y &= g(t), \\ dx &= f'(t)\,dt, & dy &= g'(t)\,dt, \quad \text{and} \\ ds &= \sqrt{[f'(t)]^2 + [g'(t)]^2}\,dt. \end{aligned} \tag{6}$$

The infinitesimal "right triangle" in Fig. 10.5.1 serves as a convenient device for remembering the latter substitution for ds. The Pythagorean theorem then leads to the symbolic manipulation

$$ds = \sqrt{dx^2 + dy^2} = \sqrt{\left(\frac{dx}{dt}\right)^2 + \left(\frac{dy}{dt}\right)^2}\,dt = \sqrt{[f'(t)]^2 + [g'(t)]^2}\,dt. \tag{7}$$

It simplifies the discussion to assume that the graph of the parametric curve in (5) resembles Fig. 10.5.2, in which $y = g(t) \geqq 0$ and $x = f(t)$ is either increasing on the entire interval $\alpha \leqq t \leqq \beta$ or is decreasing there. The two parts of Fig. 10.5.2 illustrate the two possibilities—whether as t increases the curve is traced in the positive x-direction from left to right, or in the negative x-direction from right to left. How and whether to take this direction of motion into account depends on which integral we are computing.

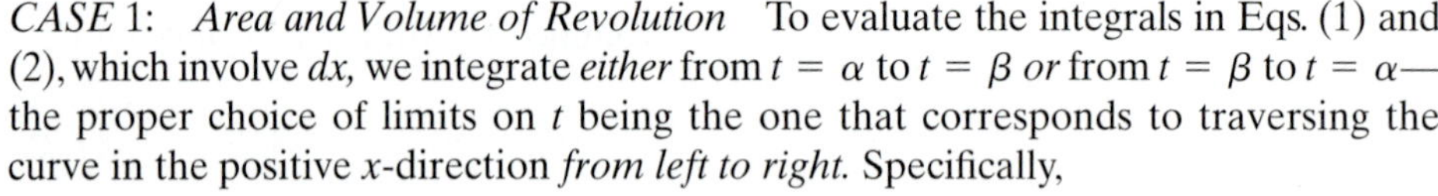

CASE 1: *Area and Volume of Revolution* To evaluate the integrals in Eqs. (1) and (2), which involve dx, we integrate *either* from $t = \alpha$ to $t = \beta$ *or* from $t = \beta$ to $t = \alpha$—the proper choice of limits on t being the one that corresponds to traversing the curve in the positive x-direction *from left to right.* Specifically,

$$A = \int_{\alpha}^{\beta} g(t)f'(t)\,dt \quad \text{if } f(\alpha) < f(\beta),$$

whereas

$$A = \int_{\beta}^{\alpha} g(t)f'(t)\,dt \quad \text{if } f(\beta) < f(\alpha).$$

The validity of this method of evaluating the integrals in Eqs. (1) and (2) follows from Theorem 1 of Section 5.7, on integration by substitution.

CASE 2: Arc Length and Surface Area To evaluate the integrals in Eqs. (3) and (4), which involve ds rather than dx, we integrate from $t = \alpha$ to $t = \beta$ irrespective of the direction of motion along the curve. To see why this is so, recall from Eq. (4) of Section 10.4 that $dy/dx = g'(t)/f'(t)$ if $f'(t) \neq 0$ on $[\alpha, \beta]$. Hence

$$s = \int_{a}^{b} \sqrt{1 + \left(\frac{dy}{dx}\right)^2}\,dx = \int_{f^{-1}(a)}^{f^{-1}(b)} \sqrt{1 + \left[\frac{g'(t)}{f'(t)}\right]^2}\,f'(t)\,dt.$$

Assuming that $f'(t) > 0$ if $f(\alpha) = a$ and $f(\beta) = b$, whereas $f'(t) < 0$ if $f(\alpha) = b$ and $f(\beta) = a$, it follows in either event that

$$s = \int_{\alpha}^{\beta} \sqrt{1 + \left[\frac{g'(t)}{f'(t)}\right]^2}\,|f'(t)|\,dt,$$

and so

$$s = \int_{\alpha}^{\beta} \sqrt{[f'(t)]^2 + [g'(t)]^2}\,dt = \int_{\alpha}^{\beta} \sqrt{\left(\frac{dx}{dt}\right)^2 + \left(\frac{dy}{dt}\right)^2}\,dt. \qquad \textbf{(8)}$$

This formula, derived under the assumption that $f'(t) \neq 0$ on $[\alpha, \beta]$, may be taken to be the *definition* of arc length for an arbitrary smooth parametric curve. Similarly, the area of a surface of revolution is defined for smooth parametric curves as the result of first making the substitutions of Eq. (6) into Eq. (4a) or (4b) and then integrating from $t = \alpha$ to $t = \beta$.

EXAMPLE 1 Use the parametrization $x = a\cos t$, $y = a\sin t$ of the circle with center $(0, 0)$ and radius a to find the volume V and surface area S of the sphere obtained by revolving this circle around the x-axis.

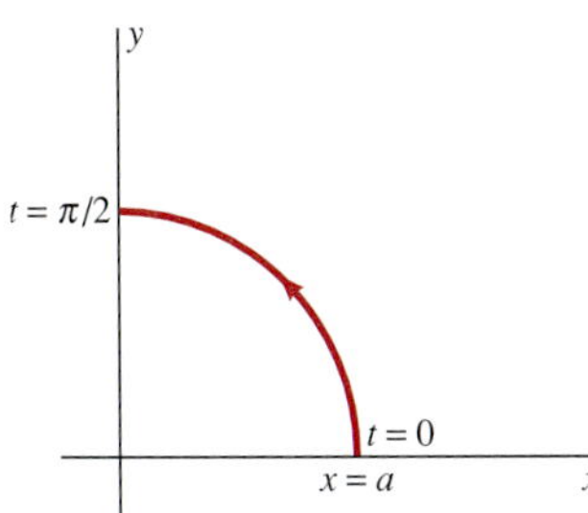

Fig. 10.5.3 The quarter-circle of Example 1

Solution Half the sphere is obtained by revolving the first quadrant of the circle (Fig. 10.5.3). The left-to-right direction along the curve is from $t = \pi/2$ to $t = 0$, and $dx = -a\sin t\,dt$, so Eq. (2a) gives

$$\begin{aligned} V &= 2\int_{t=\pi/2}^{0} \pi y^2\,dx \\ &= 2\int_{\pi/2}^{0} \pi(a\sin t)^2(-a\sin t\,dt) = 2\pi a^3\int_{0}^{\pi/2} (1 - \cos^2 t)\sin t\,dt \\ &= 2\pi a^3\left[-\cos t + \frac{1}{3}\cos^3 t\right]_0^{\pi/2} = \frac{4}{3}\pi a^3. \end{aligned}$$

The arc-length differential for the parametrized curve is

$$ds = \sqrt{(-a\sin t)^2 + (a\cos t)^2} = a\,dt.$$

Hence Eq. (4a) gives

$$\begin{aligned} S &= 2\int_{t=0}^{\pi/2} 2\pi y\,ds = 2\int_0^{\pi/2} 2\pi(a\sin t)(a\,dt) \\ &= 4\pi a^2\int_0^{\pi/2} \sin t\,dt = 4\pi a^2\Big[-\cos t\Big]_0^{\pi/2} = 4\pi a^2. \end{aligned}$$

Of course, the results of Example 1 are familiar. In contrast, Example 2 requires the methods of this section.

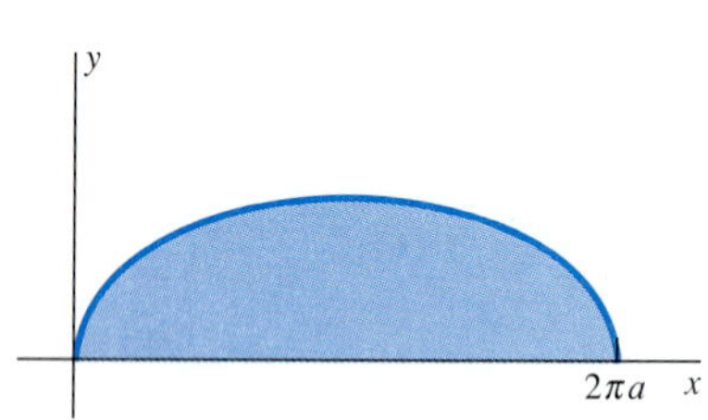

Fig. 10.5.4 The cycloidal arch of Example 2

EXAMPLE 2 Find the area under, and the arc length of, the cycloidal arch of Fig. 10.5.4. Its parametric equations are

$$x = a(t - \sin t), \qquad y = a(1 - \cos t), \qquad 0 \leqq t \leqq 2\pi.$$

Solution Because $dx = a(1 - \cos t)\,dt$ and the left-to-right direction along the curve is from $t = 0$ to $t = 2\pi$, Eq. (1) gives

$$\begin{aligned} A &= \int_{t=0}^{2\pi} y\,dx \\ &= \int_0^{2\pi} a(1-\cos t)\cdot a(1-\cos t)\,dt = a^2\int_0^{2\pi}(1-\cos t)^2\,dt \end{aligned}$$

for the area. Now we use the half-angle identity

$$1 - \cos t = 2\sin^2\left(\frac{t}{2}\right)$$

and a consequence of Problem 58 in Section 9.3:

$$\int_0^{\pi} \sin^{2n} u\,du = \pi\cdot\frac{1}{2}\cdot\frac{3}{4}\cdot\frac{5}{6}\cdots\frac{2n-1}{2n}.$$

We thereby get

$$\begin{aligned} A &= 4a^2\int_0^{2\pi}\sin^4\left(\frac{t}{2}\right)dt = 8a^2\int_0^{\pi}\sin^4 u\,du \qquad \left(u = \frac{t}{2}\right) \\ &= 8a^2\cdot\pi\cdot\frac{1}{2}\cdot\frac{3}{4} = 3\pi a^2 \end{aligned}$$

for the area under one arch of the cycloid. The arc-length differential is

$$\begin{aligned} ds &= \sqrt{a^2(1-\cos t)^2 + (a\sin t)^2}\,dt \\ &= a\sqrt{2(1-\cos t)}\,dt = 2a\sin\left(\frac{t}{2}\right)dt, \end{aligned}$$

so Eq. (3) gives

$$s = \int_0^{2\pi} 2a\sin\frac{t}{2}\,dt = \left[-4a\cos\frac{t}{2}\right]_0^{2\pi} = 8a$$

for the length of one arch of the cycloid. ■

Parametric Polar Coordinates

Suppose that a parametric curve is determined by giving its polar coordinates

$$r = r(t), \qquad \theta = \theta(t), \qquad \alpha \leqq t \leqq \beta$$

as functions of the parameter t. Then this curve is described in rectangular coordinates by the parametric equations

$$x(t) = r(t)\cos\theta(t), \qquad y(t) = r(t)\sin\theta(t), \qquad \alpha \leqq t \leqq \beta,$$

giving x and y as functions of t. The latter parametric equations may then be used in the integral formulas in Eqs. (1) through (4).

To compute ds, we first calculate the derivatives

$$\frac{dx}{dt} = (\cos\theta)\frac{dr}{dt} - (r\sin\theta)\frac{d\theta}{dt}, \qquad \frac{dy}{dt} = (\sin\theta)\frac{dr}{dt} + (r\cos\theta)\frac{d\theta}{dt}.$$

Upon substituting these expressions for dx/dt and dy/dt into Eq. (8) and making algebraic simplifications, we find that the arc-length differential in parametric polar coordinates is

$$ds = \sqrt{\left(\frac{dr}{dt}\right)^2 + \left(r\frac{d\theta}{dt}\right)^2}\,dt. \tag{9}$$

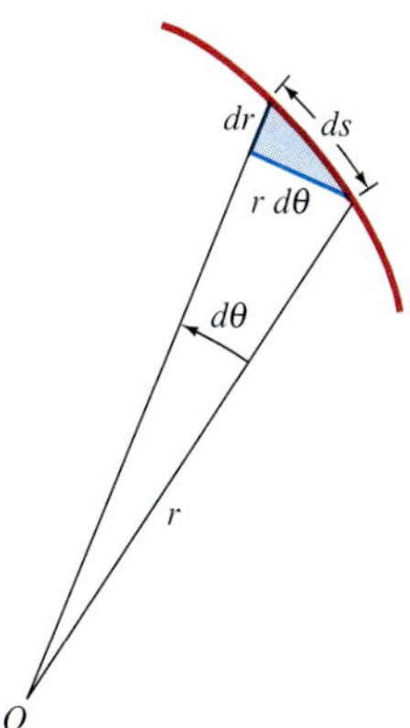

Fig. 10.5.5 The differential triangle in polar coordinates

In the case of a curve with the explicit polar coordinates equation $r = f(\theta)$, we may use θ itself as the parameter. Then Eq. (9) takes the simpler form

$$ds = \sqrt{\left(\frac{dr}{d\theta}\right)^2 + r^2}\,d\theta. \tag{10}$$

This formula is easy to remember with the aid of the tiny "almost-triangle" shown in Fig. 10.5.5.

EXAMPLE 3 Find the perimeter (arc length) s of the cardioid with polar equation $r = 1 + \cos\theta$ (Fig. 10.5.6). Find also the surface area S generated by revolving the cardioid around the x-axis.

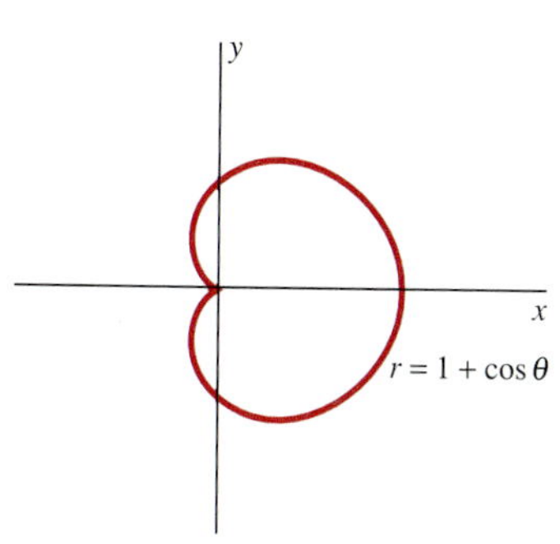

Fig. 10.5.6 The cardioid of Example 3

Solution Because $dr/d\theta = -\sin\theta$, Eq. (10) and the identity

$$1 + \cos\theta = 2\cos^2\left(\frac{\theta}{2}\right) \tag{11}$$

give

$$\begin{aligned} ds &= \sqrt{(-\sin\theta)^2 + (1+\cos\theta)^2}\,d\theta = \sqrt{2(1+\cos\theta)}\,d\theta \\ &= \sqrt{4\cos^2\left(\frac{\theta}{2}\right)}\,d\theta = \left|2\cos\left(\frac{\theta}{2}\right)\right|\,d\theta. \end{aligned}$$

Hence $ds = 2\cos(\theta/2)\,d\theta$ on the upper half of the cardioid, where $0 \leqq \theta \leqq \pi$, and thus $\cos(\theta/2) \geqq 0$. Therefore

$$s = 2\int_0^{\pi} 2\cos\frac{\theta}{2}\,d\theta = 8\left[\sin\frac{\theta}{2}\right]_0^{\pi} = 8.$$

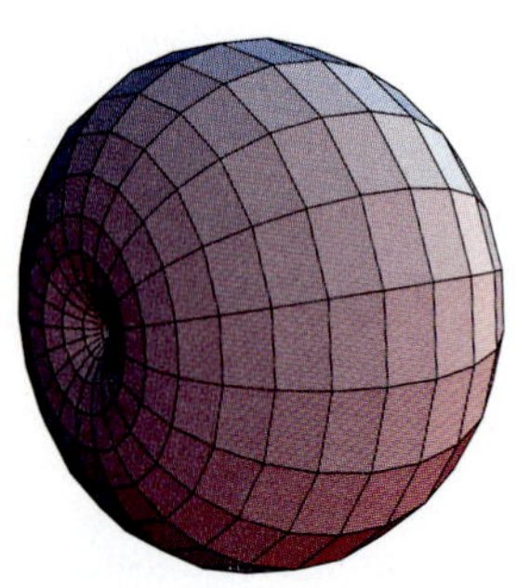

Fig. 10.5.7 The surface generated by rotating the cardioid around the x-axis

The surface area of revolution around the x-axis (Fig. 10.5.7) is given by

$$\begin{aligned} S &= \int_{\theta=0}^{\pi} 2\pi y\,ds \\ &= \int_{\theta=0}^{\pi} 2\pi(r\sin\theta)\,ds = \int_0^{\pi} 2\pi(1+\cos\theta)(\sin\theta)\cdot 2\cos\left(\frac{\theta}{2}\right)d\theta \\ &= 16\pi\int_0^{\pi}\cos^4\frac{\theta}{2}\sin\frac{\theta}{2}\,d\theta = 16\pi\left[-\frac{2}{5}\cos^5\frac{\theta}{2}\right]_0^{\pi} = \frac{32\pi}{5}, \end{aligned}$$

using the identity

$$\sin\theta = 2\sin\left(\frac{\theta}{2}\right)\cos\left(\frac{\theta}{2}\right)$$

as well as the identity in Eq. (11).

10.5 PROBLEMS

In Problems 1 through 6, find the area of the region that lies between the given parametric curve and the x-axis.

1. $x = t^3, \quad y = 2t^2 + 1; \quad -1 \leqq t \leqq 1$

2. $x = e^{3t}, \quad y = e^{-t}; \quad 0 \leqq t \leqq \ln 2$

3. $x = \cos t, \quad y = \sin^2 t; \quad 0 \leqq t \leqq \pi$

4. $x = 2 - 3t, \quad y = e^{2t}; \quad 0 \leqq t \leqq 1$

5. $x = \cos t, \quad y = e^t; \quad 0 \leqq t \leqq \pi$

6. $x = 1 - e^t, \quad y = 2t + 1; \quad 0 \leqq t \leqq 1$

In Problems 7 through 10, find the volume obtained by revolving around the x-axis the region described in the given problem.

7. Problem 1 **8.** Problem 2

9. Problem 3 **10.** Problem 5

In Problems 11 through 16, find the arc length of the given curve.

11. $x = 2t, \quad y = \frac{2}{3}t^{3/2}; \quad 5 \leqq t \leqq 12$

12. $x = \frac{1}{2}t^2, \quad y = \frac{1}{3}t^3; \quad 0 \leqq t \leqq 1$

13. $x = \sin t - \cos t, \quad y = \sin t + \cos t; \quad \frac{1}{4}\pi \leqq t \leqq \frac{1}{2}\pi$

14. $x = e^t \sin t, \quad y = e^t \cos t; \quad 0 \leqq t \leqq \pi$

15. $r = e^{\theta/2}; \quad 0 \leqq \theta \leqq 4\pi$ **16.** $r = \theta; \quad 2\pi \leqq \theta \leqq 4\pi$

In Problems 17 through 22, find the area of the surface of revolution generated by revolving the given curve around the indicated axis.

17. $x = 1 - t, \quad y = 2\sqrt{t}, \quad 1 \leqq t \leqq 4;$ the x-axis

18. $x = 2t^2 + t^{-1}, \quad y = 8\sqrt{t}, \quad 1 \leqq t \leqq 2;$ the x-axis

19. $x = t^3, \quad y = 2t + 3, \quad -1 \leqq t \leqq 1;$ the y-axis

20. $x = 2t + 1, \quad y = t^2 + t, \quad 0 \leqq t \leqq 3;$ the y-axis

21. $r = 4\sin\theta, \quad 0 \leqq \theta \leqq \pi;$ the x-axis

22. $r = e^{\theta}, \quad 0 \leqq \theta \leqq \frac{1}{2}\pi;$ the y-axis

23. Find the volume generated by revolving around the x-axis the region under the cycloidal arch of Example 2.

24. Find the area of the surface generated by revolving around the x-axis the cycloidal arch of Example 2.

25. Use the parametrization $x = a\cos t$, $y = b\sin t$ to find (a) the area bounded by the ellipse $x^2/a^2 + y^2/b^2 = 1$; (b) the volume of the ellipsoid generated by revolving this ellipse around the x-axis.

26. Find the area bounded by the loop of the parametric curve $x = t^2$, $y = t^3 - 3t$ of Problem 25 in Section 10.4.

27. Use the parametrization $x = t\cos t$, $y = t\sin t$ of the Archimedean spiral to find the arc length of the first full turn of this spiral (corresponding to $0 \leqq t \leqq 2\pi$).

28. The circle $(x - b)^2 + y^2 = a^2$ with radius $a < b$ and center $(b, 0)$ can be parametrized by

$$x = b + a\cos t, \quad y = a\sin t, \quad 0 \leqq t \leqq 2\pi.$$

Find the surface area of the torus obtained by revolving this circle around the y-axis (Fig. 10.5.8).

29. The *astroid* (four-cusped hypocycloid) has equation $x^{2/3} + y^{2/3} = a^{2/3}$ (Fig. 10.4.14) and the parametrization

$$x = a\cos^3 t, \quad y = a\sin^3 t, \quad 0 \leqq t \leqq 2\pi.$$

Find the area of the region bounded by the astroid.

30. Find the total length of the astroid of Problem 29.

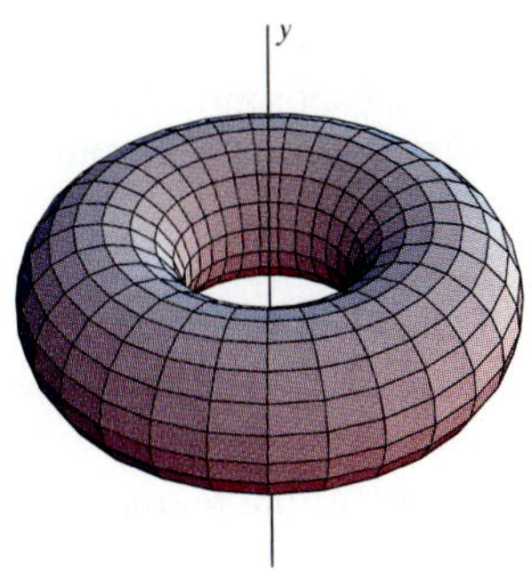

Fig. 10.5.8 The torus of Problem 28

31. Find the area of the surface obtained by revolving the astroid of Problem 29 around the x-axis.

32. Find the area of the surface generated by revolving the lemniscate $r^2 = 2a^2 \cos 2\theta$ around the y-axis (Fig. 10.5.9). (*Suggestion:* Use Eq. (9); note that $r\,dr = -2a^2 \sin 2\theta\, d\theta$.)

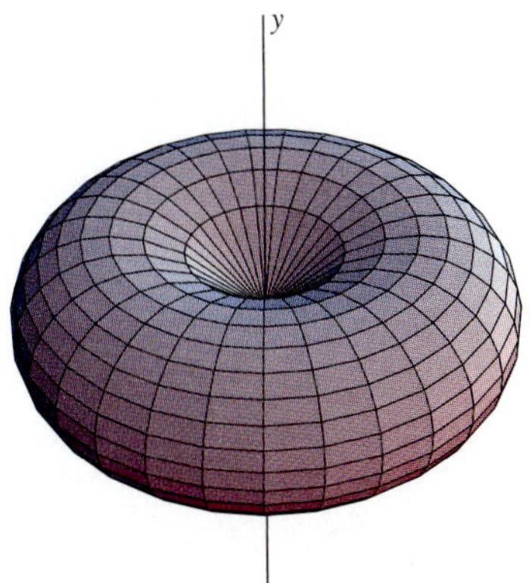

Fig. 10.5.9 The surface generated by rotating the lemniscate of Problem 32 around the y-axis

33. Figure 10.5.10 shows the graph of the parametric curve

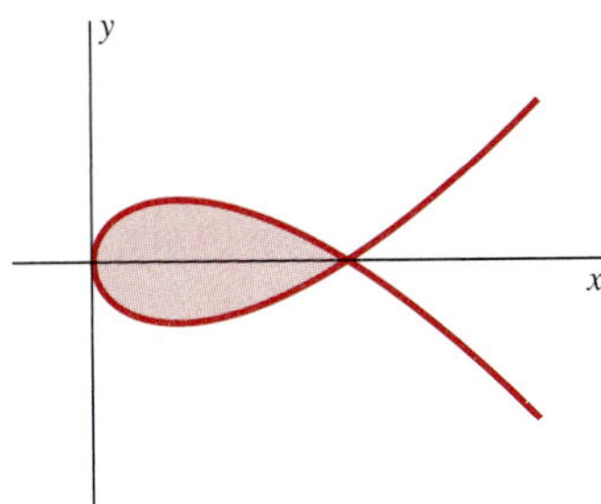

Fig. 10.5.10 The parametric curve of Problems 33 through 36

$$x = t^2\sqrt{3}, \quad y = 3t - \tfrac{1}{3}t^3.$$

The shaded region is bounded by the part of the curve for which $-3 \leqq t \leqq 3$. Find its area.

34. Find the arc length of the loop of the curve of Problem 33.

35. Find the volume of the solid obtained by revolving around the x-axis the shaded region in Fig. 10.5.10.

36. Find the surface area of revolution generated by revolving around the x-axis the loop of Fig. 10.5.10.

37. (a) With reference to Problem 30 and Fig. 10.4.13 in Section 10.4, show that the arc length of the first-quadrant loop of the folium of Descartes is

$$s = 6\int_0^1 \frac{\sqrt{1 + 4t^2 - 4t^3 - 4t^5 + 4t^6 + t^8}}{(1 + t^3)^2}\, dt.$$

(b) Use a programmable calculator or a computer to approximate this length.

38. Find the surface area generated by rotating around the y-axis the cycloidal arch of Example 2. (*Suggestion:* $\sqrt{x^2} = x$ only if $x \geqq 0$.)

39. Find the volume generated by rotating around the y-axis the region under the cycloidal arch of Example 2.

40. Suppose that after a string is wound clockwise around a circle of radius a, its free end is at the point $A(a, 0)$ (see Fig. 10.5.11). Now the string is unwound, always stretched tight so the unwound portion TP is tangent to the circle at T. The locus of the string's free endpoint P is called the **involute** of the circle.

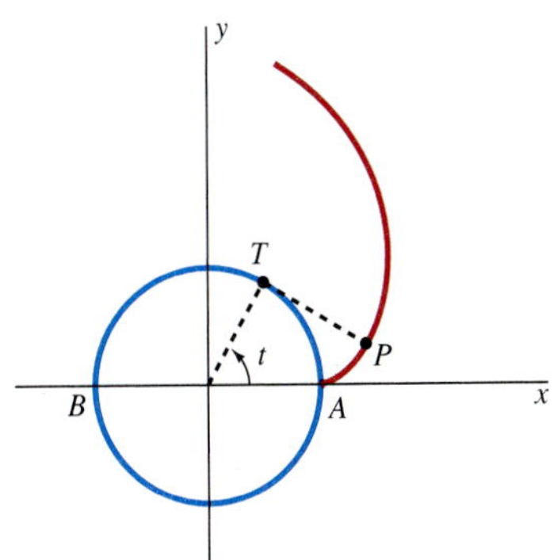

Fig. 10.5.11 The involute of a circle

(a) Show that the parametric equations of the involute (in terms of the angle t of Fig. 10.5.11) are

$$x = a(\cos t + t \sin t), \quad y = a(\sin t - t \cos t).$$

(b) Find the length of the involute from $t = 0$ to $t = \pi$.

41. Suppose that the circle of Problem 40 is a water tank and the "string" is a rope of length πa. It is anchored at the point B opposite A. Figure 10.5.12 depicts the total area that can be grazed by a cow tied to the free end of the rope. Find this total area. (The three labeled arcs of the curve in the figure represent, respectively, an involute APQ generated as the cow unwinds the rope in the counterclockwise direction, a semicircle QR of radius πa centered at B, and an involute RSA generated as the cow winds the rope around the tank proceeding in the counterclockwise direction from B to A. These three arcs form a closed curve that resembles a cardioid, and the cow can reach every point that lies inside this curve and outside the original circle.)

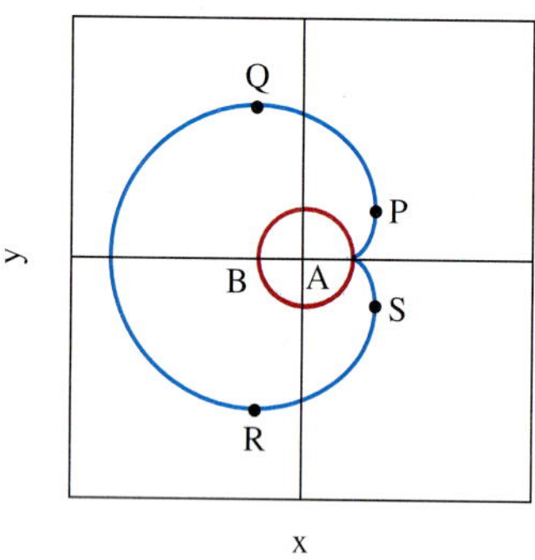

Fig. 10.5.12 The area that the cow of Problem 41 can graze

42. Now suppose that the rope of the previous problem has length $2\pi a$ and is anchored at the point A before being wound completely around the tank. Now find the total area that the cow can graze. Figure 10.5.13 shows an involute APQ, a semicircle QR of radius $2\pi a$ centered at A, and an involute RSA. The cow can reach every point that lies inside the outer curve and outside the original circle.

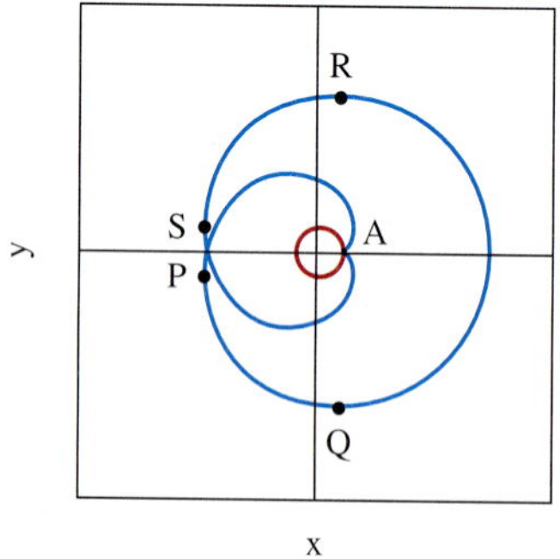

Fig. 10.5.13 The area that the cow of Problem 42 can graze

In Problems 43 through 54, use a graphing calculator or computer algebra system as appropriate. Approximate (by integrating numerically) the desired quantity if it cannot be calculated exactly.

43. Find the total arc length of the three-leaved rose $r = 3\sin 3\theta$ of Fig. 10.3.12.

44. Find the total surface area generated by rotating around the y-axis the three-leaved rose of Problem 43.

45. Find the total length of the four-leaved rose $r = 2\cos 2\theta$ of Fig. 10.2.12.

46. Find the total surface area generated by revolving around the x-axis the four-leaved rose of Problem 45.

47. Find the total arc length of the limaçon (both loops) $r = 5 + 9\cos\theta$ of Fig. 10.2.19.

48. Find the total surface area generated by revolving around the x-axis the limaçon of Problem 47.

49. Find the total arc length (all seven loops) of the polar curve $r = \cos\left(\frac{7}{3}\theta\right)$ of Fig. 10.2.24.

50. Find the total arc length of the figure eight curve $x = \sin t$, $y = \sin 2t$ of Fig. 10.4.12.

51. Find the total surface area and volume generated by revolving around the x-axis the figure eight curve of Problem 50.

52. Find the total surface area and volume generated by revolving around the y-axis the figure eight curve of Problem 50.

53. Find the total arc length of the Lissajous curve $x = \cos 3t$, $y = \sin 5t$ of Fig. 10.4.4.

54. Find the total arc length of the epitrochoid $x = 8\cos t - 5\cos 4t$, $y = 8\sin t - 5\sin 4t$ of Fig. 10.4.17.

55. Frank A. Farris of Santa Clara University, while designing a computer laboratory exercise for his calculus students, discovered an extremely lovely curve with the parametrization

$$x(t) = \cos t + \tfrac{1}{2}\cos 7t + \tfrac{1}{3}\sin 17t,$$

$$y(t) = \sin t + \tfrac{1}{2}\sin 7t + \tfrac{1}{3}\cos 17t.$$

For information on what these equations represent, see his article "Wheels on Wheels on Wheels—Surprising Symmetry" in the June 1996 issue of *Mathematics Magazine*. Plot these equations so you can enjoy this extraordinary figure, then numerically integrate to approximate the length of its graph. What kind of symmetry does the graph have? Is this predictable from the coefficients of t in the parametric equations?

10.5 PROJECT: MOON ORBITS AND RACE TRACKS

This project calls for the use of numerical integration techniques (using a calculator or computer) to approximate the parametric arc-length integral

$$s = \int_a^b \sqrt{[x'(t)]^2 + [y'(t)]^2}\, dt. \tag{1}$$

Consider the ellipse with equation

$$\frac{x^2}{a^2} + \frac{y^2}{b^2} = 1 \quad (a > b) \tag{2}$$

and *eccentricity* $\epsilon = \sqrt{1 - (a/b)^2}$ (see Section 10.6). Substitute the parametrization

$$x = a \cos t, \quad y = b \sin t \tag{3}$$

into Eq. (1) to show that the perimeter of the ellipse is given by the *elliptic integral*

$$p = 4a \int_0^{\pi/2} \sqrt{1 - \epsilon^2 \cos^2 t}\, dt. \tag{4}$$

This integral is known to be nonelementary if $0 < \epsilon < 1$. A common simple approximation to it is

$$p \approx \pi(A + R), \tag{5}$$

where

$$A = \frac{1}{2}(a + b) \quad \text{and} \quad R = \sqrt{\frac{a^2 + b^2}{2}}$$

denote the arithmetic mean and root-square mean, respectively, of the semiaxes a and b of the ellipse.

INVESTIGATION A As a warm-up, consider the ellipse whose major and minor semiaxes a and b are, respectively, the largest and smallest nonzero digits of your student I.D. number. For this ellipse, compare the arc-length estimates given by (5) and by numerical evaluation of the integral in Eq. (4).

INVESTIGATION B If we ignore the perturbing effects of the sun and the planets other than the earth, the orbit of the moon is an almost perfect ellipse with the earth at one focus. Assume that this ellipse has major semiaxis $a = 384{,}403$ km (exactly) and eccentricity $\epsilon = 0.0549$ (exactly). Approximate the perimeter p of this ellipse [using Eq. (4)] to the nearest meter.

INVESTIGATION C Suppose that you are designing an elliptical auto racetrack. Choose semiaxes for *your* racetrack so that its perimeter will be somewhere between a half mile and two miles. Your task is to construct a table with *time* and *speed* columns that an observer can use to determine the average speed of a particular car as it circles the track. The times listed in the first column should correspond to speeds up to perhaps 150 mi/h. The observer clocks a car's circuit of the track and locates its time for the lap in the first column of the table. The corresponding figure in the second column then gives the car's average speed (in miles per hour) for that circuit of the track. Your report should include a convenient table to use in this way—so you can successfully sell it to racetrack patrons attending the auto races.

10.6 THE PARABOLA

The case $e = 1$ of Example 3 in Section 10.1 is motivation for this formal definition.

> **Definition** ***The Parabola***
>
> A **parabola** is the set of all points P in the plane that are equidistant from a fixed point F (called the **focus** of the parabola) and a fixed line L (called the parabola's **directrix**) not containing F.

If the focus of the parabola is $F(p, 0)$ and its directrix is the vertical line $x = -p$, $p > 0$, then it follows from Eq. (12) of Section 10.1 that the equation of this parabola is

$$y^2 = 4px. \qquad \textbf{(1)}$$

When we replace x with $-x$ in the equation and in the discussion that precedes it, we get the equation of the parabola whose focus is $(-p, 0)$ and whose directrix is the vertical line $x = p$. The new parabola has equation

$$y^2 = -4px. \qquad \textbf{(2)}$$

The old and new parabolas appear in Fig. 10.6.1.

We could also interchange x and y in Eq. (1). This would give the equation of a parabola whose focus is $(0, p)$ and whose directrix is the horizontal line $y = -p$. This parabola opens upward, as in Fig. 10.6.2(a); its equation is

$$x^2 = 4py. \qquad \textbf{(3)}$$

Finally, we replace y with $-y$ in Eq. (3). This gives the equation

$$x^2 = -4py \qquad \textbf{(4)}$$

of a parabola opening downward, with focus $(0, -p)$ and with directrix $y = p$, as in Fig. 10.6.2(b).

Each of the parabolas discussed so far is symmetric around one of the coordinate axes. The line around which a parabola is symmetric is called the **axis** of the parabola. The point of a parabola midway between its focus and its directrix is called the **vertex** of the parabola. The vertex of each parabola that we discussed in connection with Eqs. (1) through (4) is the origin $(0, 0)$.

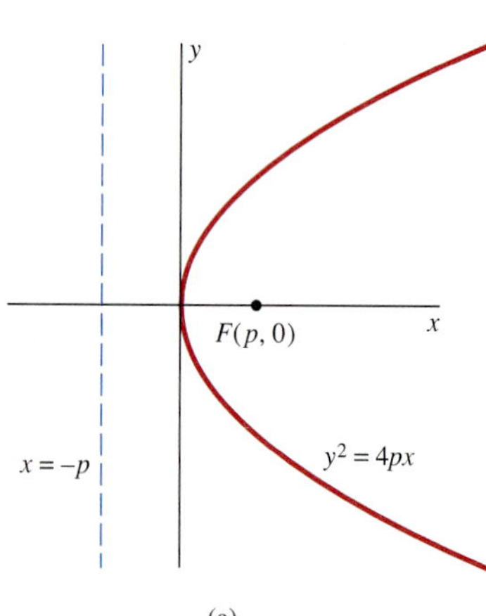

(a)

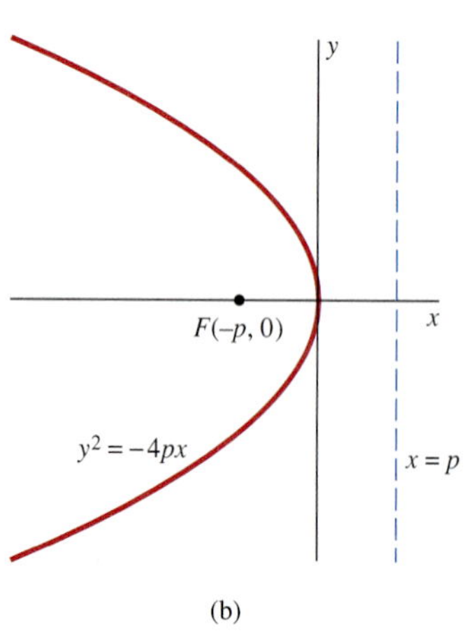

(b)

Fig. 10.6.1 Two parabolas with vertical directrices

EXAMPLE 1 Determine the focus, directrix, axis, and vertex of the parabola $x^2 = 12y$.

Solution We write the given equation as $x^2 = 4 \cdot (3y)$. In this form it matches Eq. (3) with $p = 3$. Hence the focus of the given parabola is $(0, 3)$ and its directrix is the horizontal line $y = -3$. The y-axis is its axis of symmetry, and the parabola opens upward from its vertex at the origin. ■

Suppose that we begin with the parabola of Eq. (1) and translate it in such a way that its vertex moves to the point (h, k). Then the translated parabola has equation

$$(y - k)^2 = 4p(x - h). \qquad \textbf{(1a)}$$

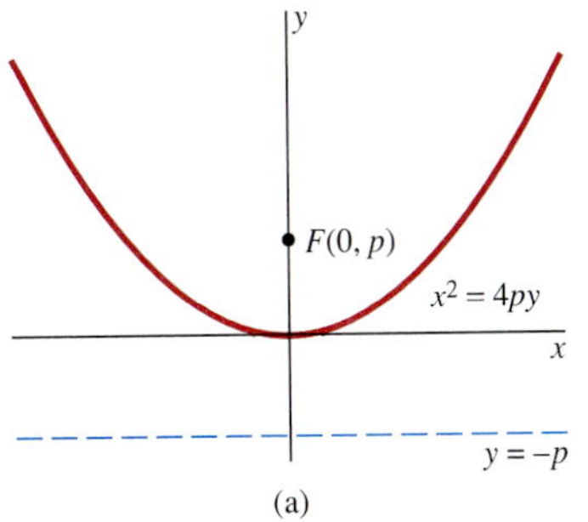

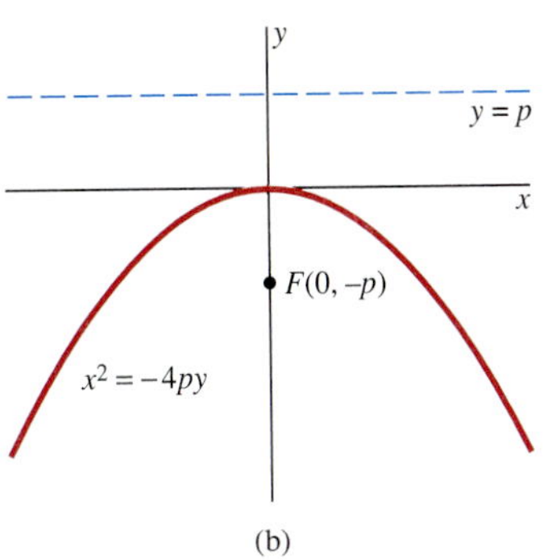

Fig. 10.6.2 Two parabolas with horizontal directrices: (a) opening upward; (b) opening downward

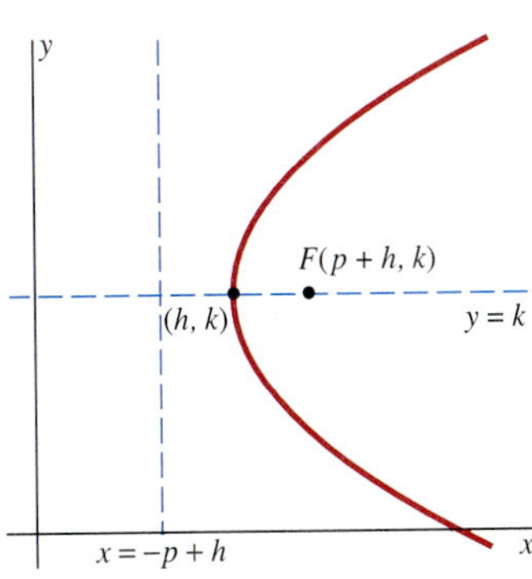

Fig. 10.6.3 A translation of the parabola $y^2 = 4px$

The new parabola has focus $F(p + h, k)$, and its directrix is the vertical line $x = -p + h$ (Fig. 10.6.3). Its axis is the horizontal line $y = k$.

We can obtain the translates of the other three parabolas in Eqs. (2) through (4) in the same way. If the vertex is moved from the origin to the point (h, k), then the three equations take these forms:

$$(y - k)^2 = -4p(x - h), \tag{2a}$$

$$(x - h)^2 = 4p(y - k), \quad \text{and} \tag{3a}$$

$$(x - h)^2 = -4p(y - k). \tag{4a}$$

Equations (1a) and (2a) both take the general form

$$y^2 + Ax + By + C = 0 \quad (A \neq 0), \tag{5}$$

whereas Eqs. (3a) and (4a) both take the general form

$$x^2 + Ax + By + C = 0 \quad (B \neq 0). \tag{6}$$

What is significant about Eqs. (5) and (6) is what they have in common: Both are linear in one of the coordinate variables and quadratic in the other. In fact, we can reduce *any* such equation to one of the standard forms in Eqs. (1a) through (4a) by completing the square in the coordinate variable that appears quadratically. This means that the graph of any equation of the form of either Eqs. (5) or (6) is a parabola. The features of the parabola can be read from the standard form of its equation, as in Example 2.

EXAMPLE 2 Determine the graph of the equation

$$4y^2 - 8x - 12y + 1 = 0.$$

Solution This equation is linear in x and quadratic in y. We divide through by the coefficient of y^2 and then collect on one side of the equation all terms that include y:

$$y^2 - 3y = 2x - \tfrac{1}{4}.$$

Then we complete the square in the variable y and thus find that

$$y^2 - 3y + \tfrac{9}{4} = 2x - \tfrac{1}{4} + \tfrac{9}{4} = 2x + 2 = 2(x + 1).$$

The final step is to write in the form $4p(x - h)$ the terms on the right-hand side that include x:

$$\left(y - \tfrac{3}{2}\right)^2 = 4 \cdot \tfrac{1}{2} \cdot (x + 1).$$

This equation has the form of Eq. (1a) with $p = \frac{1}{2}$, $h = -1$, and $k = \frac{3}{2}$. Thus the graph is a parabola that opens to the right from its vertex at $\left(-1, \frac{3}{2}\right)$. Its focus is at $\left(-\frac{1}{2}, \frac{3}{2}\right)$, its directrix is the vertical line $x = -\frac{3}{2}$, and its axis is the horizontal line $y = \frac{3}{2}$. It appears in Fig. 10.6.4. ■

Applications of Parabolas

The parabola $y^2 = 4px$, $(p > 0)$ is shown in Fig. 10.6.5 along with an incoming ray of light traveling to the left and parallel to the x-axis. This light ray strikes the parabola at the point $Q(a, b)$ and is reflected toward the x-axis, which it meets at the point

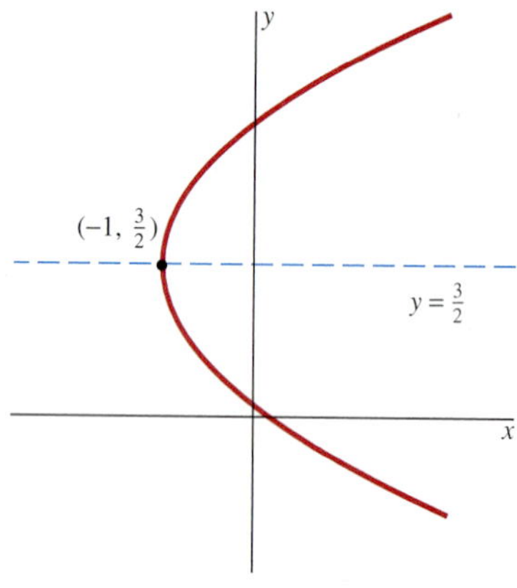

Fig. 10.6.4 The parabola of Example 2

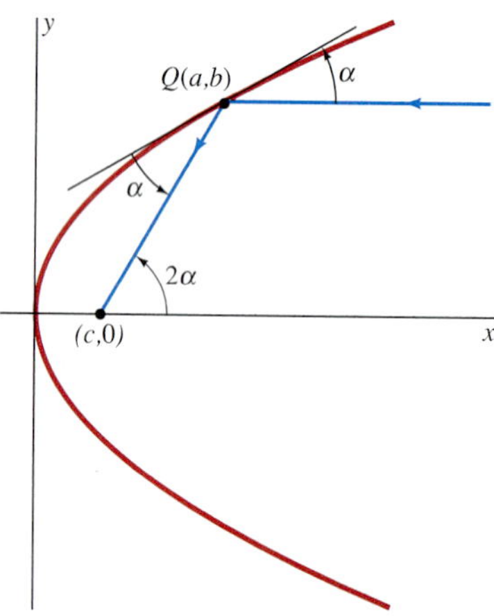

Fig. 10.6.5 The reflection property of the parabola: $\alpha = \beta$

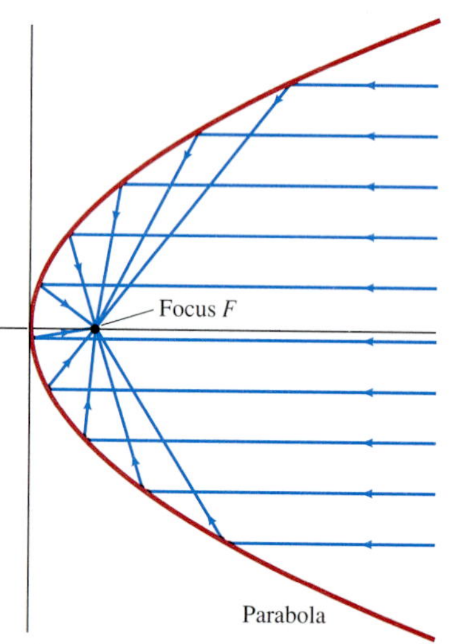

Fig. 10.6.6 Incident rays parallel to the axis reflect through the focus.

$(c, 0)$. The light ray's angle of reflection must equal its angle of incidence, which is why both of these angles—measured with respect to the tangent line L at Q—are labeled α in the figure. The angle vertical to the angle of incidence is also equal to α. Hence, because the incoming ray is parallel to the x-axis, the angle the reflected ray makes with the x-axis at $(c, 0)$ is 2α.

Using the points Q and $(c, 0)$ to compute the slope of the reflected light ray, we find that

$$\frac{b}{a - c} = \tan 2\alpha = \frac{2\tan\alpha}{1 - \tan^2\alpha}.$$

(The second equality follows from a trigonometric identity in Problem 64 of Section 8.2.) But the angle α is related to the slope of the tangent line L at Q. To find that slope, we begin with

$$y = 2\sqrt{px} = 2(px)^{1/2}$$

and compute

$$\frac{dy}{dx} = \left(\frac{p}{x}\right)^{1/2}.$$

Hence the slope of L is both $\tan\alpha$ and dy/dx evaluated at (a, b); that is,

$$\tan\alpha = \left(\frac{p}{a}\right)^{1/2}.$$

Therefore

$$\frac{b}{a - c} = \frac{2\tan\alpha}{1 - \tan^2\alpha} = \frac{2\sqrt{\dfrac{p}{a}}}{1 - \dfrac{p}{a}} = \frac{2\sqrt{pa}}{a - p} = \frac{b}{a - p},$$

because $b = 2\sqrt{pa}$. Hence $c = p$. The surprise is that c is independent of a and b and depends only on the equation $y^2 = 4px$ of the parabola. Therefore *all* incoming light rays parallel to the x-axis will be reflected to the single point $F(p, 0)$. This is why F is called the *focus* of the parabola.

This **reflection property** of the parabola is exploited in the design of parabolic mirrors. Such a mirror has the shape of the surface obtained by revolving a parabola around its axis of symmetry. Then a beam of incoming light rays parallel to the axis will be focused at F, as shown in Fig. 10.6.6. The reflection property can also be used in reverse—rays emitted at the focus are reflected in a beam parallel to the axis, thus keeping the light beam intense. Moreover, applications are not limited to light rays alone; parabolic mirrors are used in visual and radio telescopes, radar antennas, searchlights, automobile headlights, microphone systems, satellite ground stations, and solar heating devices.

Galileo discovered early in the seventeenth century that the trajectory of a projectile fired from a gun is a parabola (under the assumptions that air resistance can be ignored and that the gravitational acceleration remains constant). Suppose that a projectile is fired with initial velocity v_0 at time $t = 0$ from the origin and at an angle α of inclination from the horizontal x-axis. Then the initial velocity of the projectile splits into the components

$$v_{0x} = v_0\cos\alpha \quad \text{and} \quad v_{0y} = v_0\sin\alpha,$$

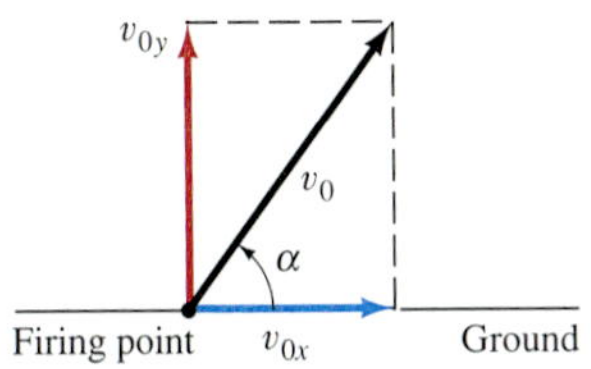

Fig. 10.6.7 Resolution of the initial velocity v_0 into its horizontal and vertical components

as indicated in Fig. 10.6.7. The fact that the projectile continues to move horizontally with *constant* speed v_{0x}, together with Eq. (34) of Section 5.2, implies that its x- and y-coordinates after t seconds are

$$x = (v_0 \cos\alpha)t, \tag{7}$$

$$y = -\tfrac{1}{2}gt^2 + (v_0 \sin\alpha)t. \tag{8}$$

By substituting $t = x/(v_0 \cos\alpha)$ from Eq. (7) into Eq. (8) and then completing the square, we can derive (as in Problem 24) an equation of the form

$$y - M = -4p(x - \tfrac{1}{2}R)^2. \tag{9}$$

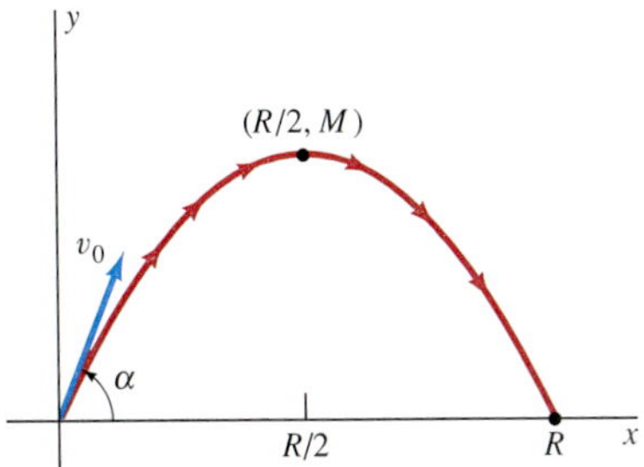

Fig. 10.6.8 The trajectory of the projectile, showing its maximum altitude M and its range R

Here,

$$M = \frac{v_0^2 \sin^2\alpha}{2g} \tag{10}$$

is the maximum height attained by the projectile, and

$$R = \frac{v_0^2 \sin 2\alpha}{g} \tag{11}$$

is its **range**, the horizontal distance the projectile will travel before it returns to the ground. Thus its trajectory is the parabola shown in Fig. 10.6.8.

10.6 PROBLEMS

In Problems 1 through 5, find the equation and sketch the graph of the parabola with vertex V and focus F.

1. $V(0, 0)$, $F(3, 0)$

2. $V(0, 0)$, $F(0, -2)$

3. $V(2, 3)$, $F(2, 1)$

4. $V(-1, -1)$, $F(-3, -1)$

5. $V(2, 3)$, $F(0, 3)$

In Problems 6 through 10, find the equation and sketch the graph of the parabola with the given focus and directrix.

6. $F(1, 2)$, $x = -1$

7. $F(0, -3)$, $y = 0$

8. $F(1, -1)$, $x = 3$

9. $F(0, 0)$, $y = -2$

10. $F(-2, 1)$, $x = -4$

In Problems 11 through 18, sketch the parabola with the given equation. Show and label its vertex, focus, axis, and directrix.

11. $y^2 = 12x$

12. $x^2 = -8y$

13. $y^2 = -6x$

14. $x^2 = 7y$

15. $x^2 - 4x - 4y = 0$

16. $y^2 - 2x + 6y + 15 = 0$

17. $4x^2 + 4x + 4y + 13 = 0$

18. $4y^2 - 12y + 9x = 0$

19. Prove that the point of the parabola $y^2 = 4px$ closest to its focus is its vertex.

20. Find an equation of the parabola that has a vertical axis and passes through the points $(2, 3)$, $(4, 3)$, and $(6, -5)$.

21. Show that an equation of the line tangent to the parabola $y^2 = 4px$ at the point (x_0, y_0) is $2px - y_0 y + 2px_0 = 0$. Conclude that the tangent line intersects the x-axis at the point $(-x_0, 0)$. This fact provides a quick method for constructing a line tangent to a parabola at a given point.

22. A comet's orbit is a parabola with the sun at its focus. When the comet is $100\sqrt{2}$ million miles from the sun, the line from the sun to the comet makes an angle of 45° with the axis of the parabola (Fig. 10.6.9). What will be the minimum distance between the comet and the sun? (*Suggestion:* Write the equation of the parabola with the origin at the focus, then use the result of Problem 19.)

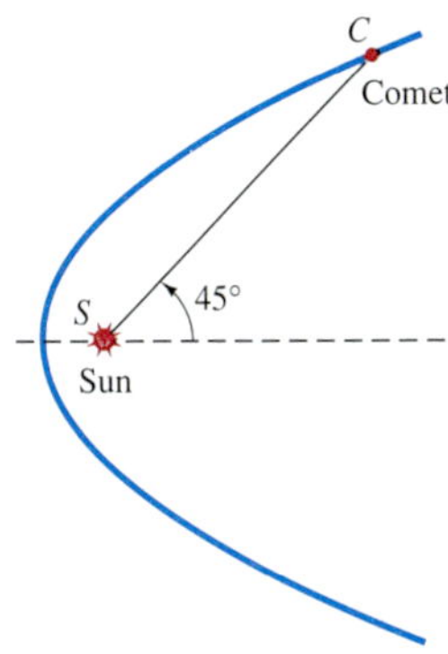

Fig. 10.6.9 The comet of Problem 22 in parabolic orbit around the sun

23. Suppose that the angle of Problem 22 increases from 45° to 90° in 3 days. How much longer will the comet take to reach its point of closest approach to the sun? Assume that the line segment from the sun to the comet sweeps out area at a constant rate (Kepler's second law).

24. Use Eqs. (7) and (8) to derive Eq. (9) with the values of M and R given in Eqs. (10) and (11).

25. Deduce from Eq. (11) that, given a fixed initial velocity v_0, the maximum range of the projectile is $R_{\max} = v_0^2/g$ and is attained when $\alpha = 45°$.

In Problems 26 through 28, assume that a projectile is fired with initial velocity $v_0 = 50$ m/s from the origin and at an angle of inclination α. Use $g = 9.8$ m/s^2.

26. If $\alpha = 45°$, find the range of the projectile and the maximum height it attains.

27. For what value or values of α is the range $R = 125$ m?

28. Find the range of the projectile and the length of time it remains above the ground if (a) $\alpha = 30°$; (b) $\alpha = 60°$.

29. The book *Elements of Differential and Integral Calculus* by William Granville, Percey Smith, and William Longley (Ginn and Company: Boston, 1929) lists a number of "curves for reference"; the curve with equation $\sqrt{x} + \sqrt{y} = \sqrt{a}$ is called a parabola. Verify that the curve in question actually is a parabola, or show that it is not.

30. The 1992 edition of the study guide for the national actuarial examinations has a problem similar to this one: Every point on the plane curve K is equally distant from the point $(-1,-1)$ and the line $x + y = 1$, and K has equation

$$x^2 + Bxy + Cy^2 + Dx + Ey + F = 0.$$

Which is the value of D: -2, 2, 4, 6, or 8?

10.7 THE ELLIPSE

An ellipse is a conic section with eccentricity e less than 1, as in Example 3 of Section 10.1.

Definition ***The Ellipse***

Suppose that $e < 1$, and let F be a fixed point and L a fixed line not containing F. The **ellipse** with **eccentricity** e, **focus** F, and **directrix** L is the set of all points P such that the distance $|PF|$ is e times the (perpendicular) distance from P to the line L.

The equation of the ellipse is especially simple if F is the point $(c, 0)$ on the x-axis and L is the vertical line $x = c/e^2$. The case $c > 0$ is shown in Fig. 10.7.1. If Q is the point $(c/e^2, y)$, then PQ is the perpendicular from $P(x, y)$ to L. The condition $|PF| = e|PQ|$ then gives

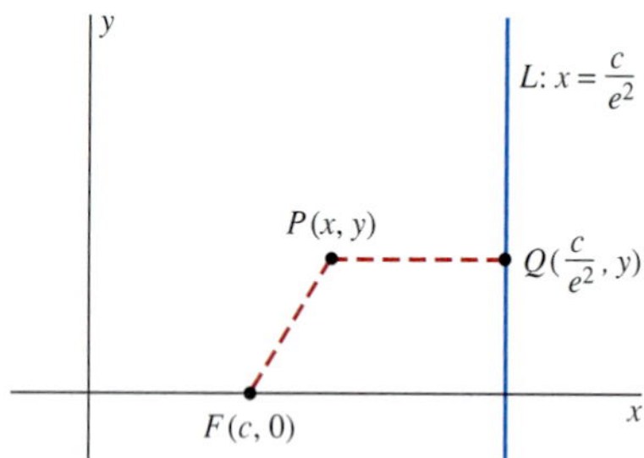

Fig. 10.7.1 Ellipse: focus F, directrix L, eccentricity e

$$(x - c)^2 + y^2 = e^2\left(x - \frac{c}{e^2}\right)^2;$$

$$x^2 - 2cx + c^2 + y^2 = e^2x^2 - 2cx + \frac{c^2}{e^2};$$

$$x^2(1 - e^2) + y^2 = c^2\left(\frac{1}{e^2} - 1\right) = \frac{c^2}{e^2}(1 - e^2).$$

Thus

$$x^2(1 - e^2) + y^2 = a^2(1 - e^2),$$

where

$$a = \frac{c}{e}. \tag{1}$$

We divide both sides of the next-to-last equation by $a^2(1 - e^2)$ and get

$$\frac{x^2}{a^2} + \frac{y^2}{a^2(1 - e^2)} = 1.$$

Finally, with the aid of the fact that $e < 1$, we may let

$$b^2 = a^2(1 - e^2) = a^2 - c^2. \tag{2}$$

Then the equation of the ellipse with focus $(c, 0)$ and directrix $x = c/e^2 = a/e$ takes the simple form

$$\frac{x^2}{a^2} + \frac{y^2}{b^2} = 1. \tag{3}$$

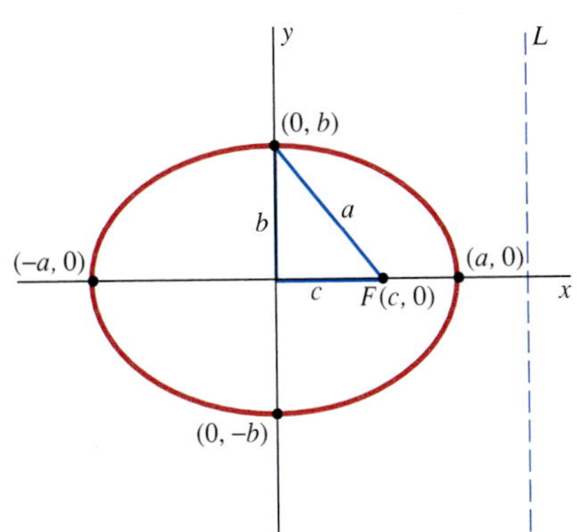

Fig. 10.7.2 The parts of an ellipse

We see from Eq. (3) that this ellipse is symmetric around both coordinate axes. Its x-intercepts are $(\pm a, 0)$ and its y-intercepts are $(0, \pm b)$. The points $(\pm a, 0)$ are called the **vertices** of the ellipse, and the line segment joining them is called its **major axis**. The line segment joining $(0, b)$ and $(0, -b)$ is called the **minor axis** [note from Eq. (2) that $b < a$]. The alternative form

$$a^2 = b^2 + c^2 \tag{4}$$

of Eq. (2) is the Pythagorean relation for the right triangle of Fig. 10.7.2. Indeed, visualization of this triangle is an excellent way to remember Eq. (4). The numbers a and b are the lengths of the major and minor **semiaxes**, respectively.

Because $a = c/e$, the directrix of the ellipse in Eq. (3) is $x = a/e$. If we had begun instead with the focus $(-c, 0)$ and directrix $x = -a/e$, we would still have obtained Eq. (3), because only the squares of a and c are involved in its derivation. Thus the ellipse in Eq. (3) has *two* foci, $(c, 0)$ and $(-c, 0)$, and *two* directrices, $x = a/e$ and $x = -a/e$ (Fig. 10.7.3).

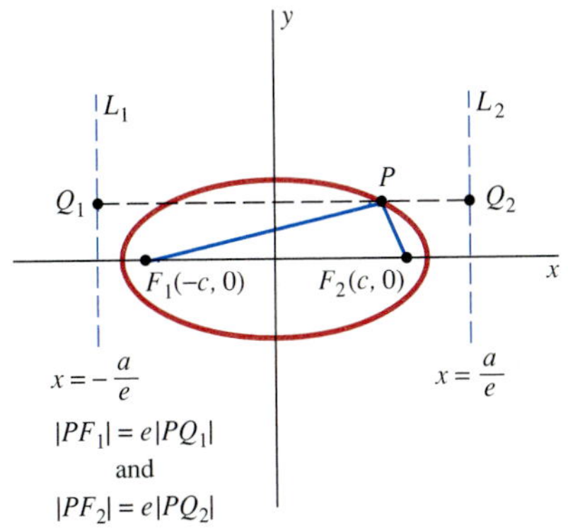

Fig. 10.7.3 The ellipse as a conic section: two foci, two directrices

The larger the eccentricity $e < 1$, the more elongated the ellipse. (Remember that $e = 1$ is the eccentricity of every parabola.) But if $e = 0$, then Eq. (2) gives $b = a$, so Eq. (3) reduces to the equation of a circle of radius a. Thus a circle is an ellipse of eccentricity zero. Compare the three cases shown in Fig. 10.7.4.

EXAMPLE 1 Find an equation of the ellipse with foci $(\pm 3, 0)$ and vertices $(\pm 5, 0)$.

Solution We are given $c = 3$ and $a = 5$, so Eq. (2) gives $b = 4$. Thus Eq. (3) gives

$$\frac{x^2}{25} + \frac{y^2}{16} = 1$$

for the desired equation. This ellipse is shown in Fig. 10.7.5. ■

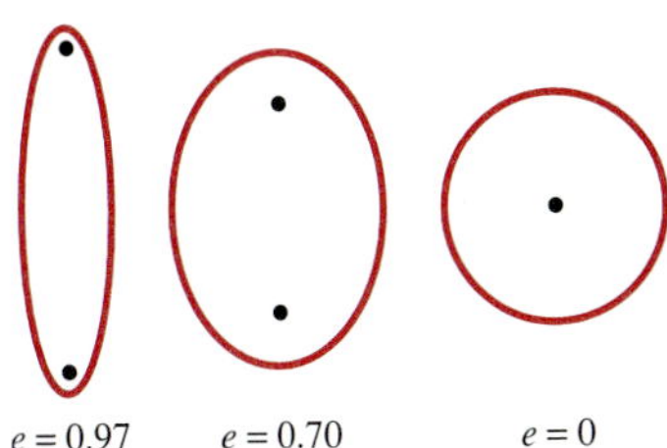

Fig. 10.7.4 The relation between the eccentricity of an ellipse and its shape

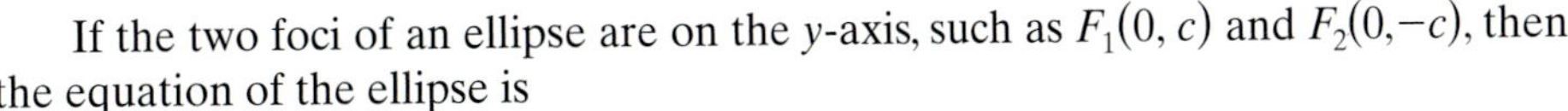

If the two foci of an ellipse are on the y-axis, such as $F_1(0, c)$ and $F_2(0, -c)$, then the equation of the ellipse is

$$\frac{x^2}{b^2} + \frac{y^2}{a^2} = 1, \tag{5}$$

and it is still true that $a^2 = b^2 + c^2$, as in Eq. (4). But now the major axis of length $2a$ is vertical and the minor axis of length $2b$ is horizontal. The derivation of Eq. (5) is similar to that of Eq. (3); see Problem 23. Figure 10.7.6 shows the case of an ellipse whose major axis is vertical. The vertices of such an ellipse are at $(0, \pm a)$; they are always the endpoints of the major axis.

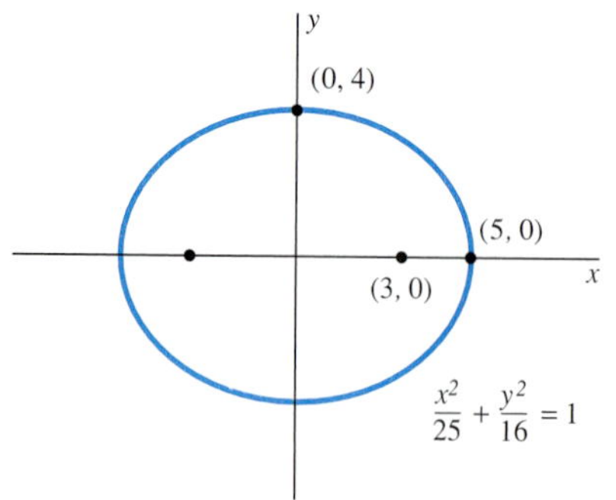

Fig. 10.7.5 The ellipse of Example 1

In practice there is little chance of confusing Eqs. (3) and (5). The equation or the given data will make clear whether the major axis of the ellipse is horizontal or vertical. Just use the equation to read the ellipse's intercepts. The two intercepts that are farthest from the origin are the endpoints of the major axis; the other two are the endpoints of the minor axis. The two foci lie on the major axis, each at distance c from the center of the ellipse—which will be the origin if the equation of the ellipse has the form of either Eq. (3) or Eq. (5).

EXAMPLE 2 Sketch the graph of the equation

$$\frac{x^2}{16} + \frac{y^2}{25} = 1.$$

Solution The x-intercepts are $(\pm 4, 0)$; the y-intercepts are $(0, \pm 5)$. So the major axis is vertical. We take $a = 5$ and $b = 4$ in Eq. (4) and find that $c = 3$. The foci are thus at $(0, \pm 3)$. Hence this ellipse has the appearance of the one shown in Fig. 10.7.7. ◼

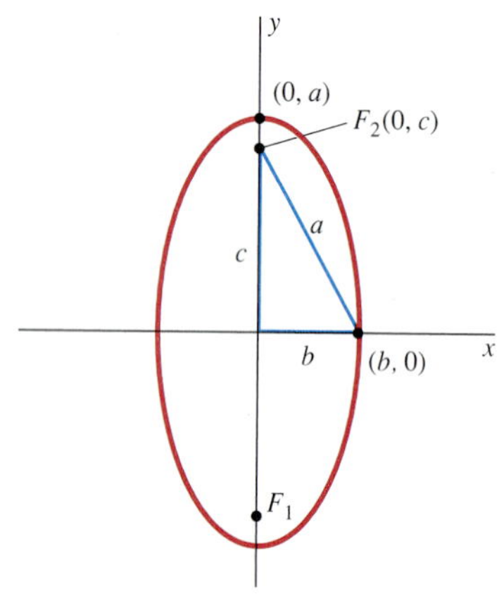

Fig. 10.7.6 An ellipse with vertical major axis

Any equation of the form

$$Ax^2 + Cy^2 + Dx + Ey + F = 0, \tag{6}$$

in which the coefficients A and C of the squared variables are *both nonzero* and *have the same sign,* may be reduced to the form

$$A(x - h)^2 + C(y - k)^2 = G$$

by completing the square in x and y. We may assume that A and C are both positive. Then if $G < 0$, there are no points that satisfy Eq. (6), and the graph is the empty set. If $G = 0$, then there is exactly one point on the locus—the single point (h, k). And if $G > 0$, we can divide both sides of the last equation by G and get an equation that resembles one of these two:

$$\frac{(x - h)^2}{a^2} + \frac{(y - k)^2}{b^2} = 1, \tag{7a}$$

$$\frac{(x - h)^2}{b^2} + \frac{(y - k)^2}{a^2} = 1. \tag{7b}$$

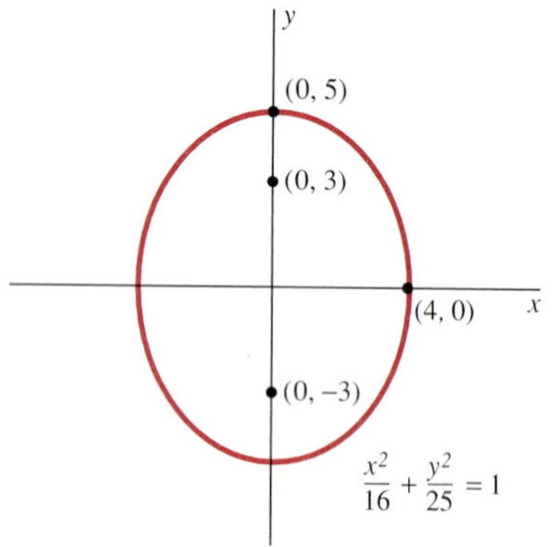

Fig. 10.7.7 The ellipse of Example 2

Which equation should you choose? Select the one that is consistent with the condition $a \geqq b > 0$. Finally, note that either of the equations in (7) is the equation of a translated ellipse. Thus, apart from the exceptional cases already noted, the graph of Eq. (6) is an ellipse if $AC > 0$.

EXAMPLE 3 Determine the graph of the equation

$$3x^2 + 5y^2 - 12x + 30y + 42 = 0.$$

Solution We collect terms containing x and terms containing y and complete the square in each variable. This gives

$$3(x^2 - 4x) + 5(y^2 + 6y) = -42;$$

$$3(x^2 - 4x + 4) + 5(y^2 + 6y + 9) = 15;$$

$$\frac{(x - 2)^2}{5} + \frac{(y + 3)^2}{3} = 1.$$

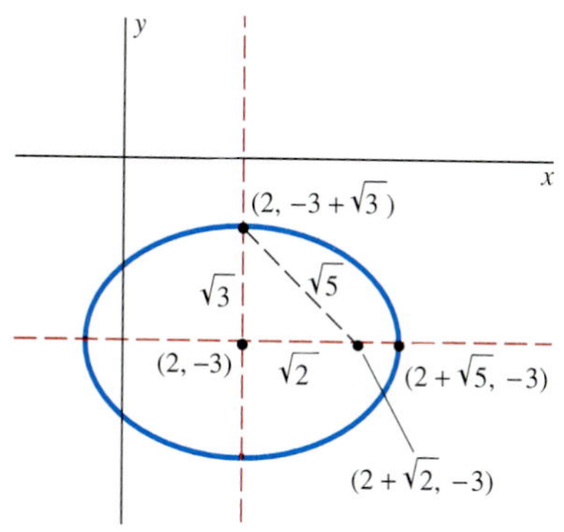

Fig. 10.7.8 The ellipse of Example 3

Thus the given equation is that of a translated ellipse with center at $(2, -3)$. Its horizontal major semiaxis has length $a = \sqrt{5}$ and its minor semiaxis has length $b = \sqrt{3}$ (Fig. 10.7.8). The distance from the center to each focus is $c = \sqrt{2}$ and the eccentricity is $e = c/a = \sqrt{2/5}$. ■

Applications of Ellipses

EXAMPLE 4 The orbit of the earth is an ellipse with the sun at one focus. The planet's maximum distance from the center of the sun is 94.56 million miles and its minimum distance is 91.44 million miles. What are the major and minor semiaxes of the earth's orbit, and what is its eccentricity?

Solution As Fig. 10.7.9 shows, we have

$$a + c = 94.56 \quad \text{and} \quad a - c = 91.44,$$

with units in millions of miles. We conclude from these equations that $a = 93.00$, that $c = 1.56$, and then that

$$b = \sqrt{(93.00)^2 - (1.56)^2} \approx 92.99$$

million miles. Finally,

$$e = \frac{1.56}{93.00} \approx 0.017,$$

a number relatively close to zero. This means that the earth's orbit is nearly circular. Indeed, the major and minor semiaxes are so nearly equal that, on any usual scale, the earth's orbit would appear to be a perfect circle. But the difference between uniform circular motion and the earth's actual motion has some important aspects, including the facts that the sun is 1.56 million miles off center and that the orbital speed of the earth is not constant. ■

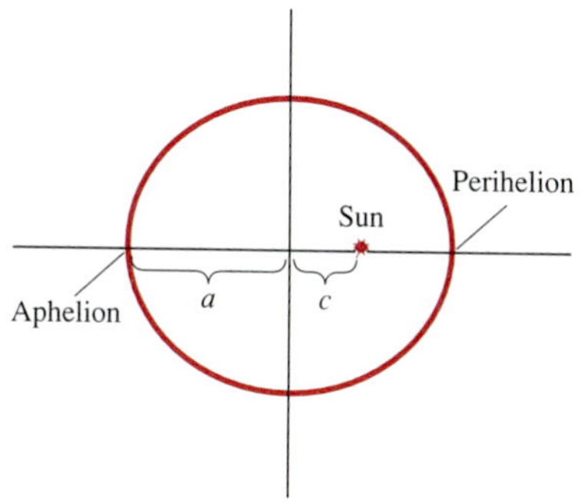

Fig. 10.7.9 The orbit of the earth with its eccentricity exaggerated (Example 4)

EXAMPLE 5 One of the most famous comets is Halley's comet, named for Edmund Halley (1656–1742), a disciple of Newton. By studying the records of the paths of earlier comets, Halley deduced that the comet of 1682 was the same one that had been sighted in 1607, in 1531, in 1456, and in 1066 (an omen at the Battle of Hastings). In 1682 Halley predicted that this comet would return in 1759, in 1835, and in 1910; he was correct each time. The period of Halley's comet is about 76 years—it can vary one or two years in either direction because of perturbations of its orbit by the planet Jupiter. The orbit of Halley's comet is an ellipse with the sun at one focus. In terms of astronomical units (1 AU is the mean distance from the earth to the sun), the major and minor semiaxes of this elliptical orbit are 18.09 AU and 4.56 AU, respectively. What are the maximum and minimum distances from the sun of Halley's comet?

Solution We are given that $a = 18.09$ (all distance measurements are in astronomical units) and that $b = 4.56$, so

$$c = \sqrt{(18.09)^2 - (4.56)^2} \approx 17.51.$$

Hence its maximum distance from the sun is $a + c \approx 35.60$ AU and its minimum distance is $a - c \approx 0.58$ AU. The eccentricity of its orbit is

$$e = \frac{17.51}{18.09} \approx 0.97,$$

a very eccentric orbit (but see Problem 21). ■

The *reflection property* of the ellipse states that the tangent line at a point P of an ellipse makes equal angles with the two lines PF_1 and PF_2 from P to the two foci of the ellipse (Fig. 10.7.10). This property is the basis of the "whispering gallery" phenomenon, which has been observed in the so-called whispering gallery of the U.S. Senate. Suppose that the ceiling of a large room is shaped like half an ellipsoid obtained by revolving an ellipse around its major axis. Sound waves, like light waves, are reflected with equal angles of incidence and reflection. Thus if two diplomats are holding a quiet conversation near one focus of the ellipsoidal surface, a reporter standing near the other focus—perhaps 50 feet away—would be able to eavesdrop on their conversation even if the conversation were inaudible to others in the same room.

Fig. 10.7.10 The reflection property: $\alpha = \beta$

Some billiard tables are manufactured in the shape of an ellipse. The foci of such tables are plainly marked for the convenience of enthusiasts of this unusual game.

A more serious application of the reflection property of ellipses is the nonsurgical kidney stone treatment called *shockwave lithotripsy.* An ellipsoidal reflector with a transducer (an energy transmitter) at one focus is positioned outside the patient's body so that the offending kidney stone is located at the other focus. The stone then is pulverized by reflected shockwaves emanating from the transducer. (For further details, see the COMAP *Newsletter* 20, November 1986.)

An alternative definition of the ellipse with foci F_1 and F_2 and major axis of length $2a$ is this: It is the locus of a point P such that the sum of the distances $|PF_1|$ and $|PF_2|$ is the constant $2a$ (see Problem 26). This fact gives us a convenient way to draw the ellipse by using two tacks placed at F_1 and F_2, a string of length $2a$, and a pencil (Fig. 10.7.11).

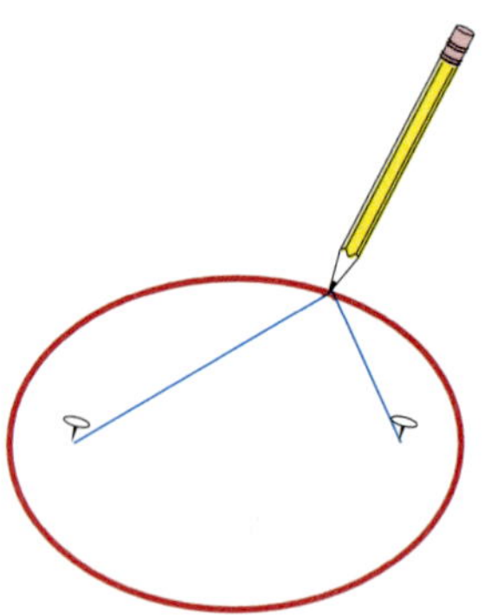

Fig. 10.7.11 One way to draw an ellipse

10.7 PROBLEMS

In Problems 1 through 15, find an equation of the ellipse specified.

1. Vertices $(\pm 4, 0)$ and $(0, \pm 5)$

2. Foci $(\pm 5, 0)$, major semiaxis 13

3. Foci $(0, \pm 8)$, major semiaxis 17

4. Center $(0, 0)$, vertical major axis 12, minor axis 8

5. Foci $(\pm 3, 0)$, eccentricity $\frac{3}{4}$

6. Foci $(0, \pm 4)$, eccentricity $\frac{2}{3}$

7. Center $(0, 0)$, horizontal major axis 20, eccentricity $\frac{1}{2}$

8. Center $(0, 0)$, horizontal minor axis 10, eccentricity $\frac{1}{2}$

9. Foci $(\pm 2, 0)$, directrices $x = \pm 8$

10. Foci $(0, \pm 4)$, directrices $y = \pm 9$

11. Center $(2, 3)$, horizontal axis 8, vertical axis 4

12. Center $(1, -2)$, horizontal major axis 8, eccentricity $\frac{3}{4}$

13. Foci $(-2, 1)$ and $(4, 1)$, major axis 10

14. Foci $(-3, 0)$ and $(-3, 4)$, minor axis 6

15. Foci $(-2, 2)$ and $(4, 2)$, eccentricity $\frac{1}{3}$

Sketch the graphs of the equations in Problems 16 through 20. Indicate centers, foci, and lengths of axes.

16. $4x^2 + y^2 = 16$ **17.** $4x^2 + 9y^2 = 144$

18. $4x^2 + 9x^2 = 24x$

19. $9x^2 + 4y^2 - 32y + 28 = 0$

20. $2x^2 + 3y^2 + 12x - 24y + 60 = 0$

21. (a) The orbit of the comet Kahoutek is an ellipse of extreme eccentricity $e = 0.999925$; the sun is at one focus of this ellipse. The minimum distance between the sun and Kahoutek is 0.13 AU. What is the maximum distance between Kahoutek and the sun? (b) The orbit of the comet Hyakutake is an ellipse of extreme eccentricity $e = 0.999643856$; the sun is at one focus of this ellipse. The minimum distance between the sun and Hyakutake is 0.2300232 AU. What is the maximum distance between Hyakutake and the sun?

22. The orbit of the planet Mercury is an ellipse of eccentricity $e = 0.206$. Its maximum and minimum distances from the sun are 0.467 and 0.307 AU, respectively. What are the major and minor semiaxes of the orbit of Mercury? Does "nearly circular" accurately describe the orbit of Mercury?

23. Derive Eq. (5) for an ellipse whose foci lie on the y-axis.

24. Show that the line tangent to the ellipse

$$\frac{x^2}{a^2} + \frac{y^2}{b^2} = 1$$

at the point $P(x_0, y_0)$ of that ellipse has equation

$$\frac{x_0 x}{a^2} + \frac{y_0 y}{b^2} = 1.$$

25. Use the result of Problem 24 to establish the reflection property of the ellipse. [*Suggestion:* Let m be the slope of the line normal to the ellipse at $P(x_0, y_0)$ and let m_1 and m_2 be the slopes of the lines PF_1 and PF_2, respectively, from P to the two foci F_1 and F_2 of the ellipse. Show that

$$\frac{m - m_1}{1 + m_1 m} = \frac{m_2 - m}{1 + m_2 m};$$

then use the identity for $\tan(A - B)$.]

26. Given $F_1(-c, 0)$ and $F_2(c, 0)$ with $a > c > 0$, prove that the ellipse

$$\frac{x^2}{a^2} + \frac{y^2}{b^2} = 1$$

(with $b^2 = a^2 - c^2$) is the locus of those points P such that $|PF_1| + |PF_2| = 2a$.

27. Find an equation of the ellipse with horizontal and vertical axes that passes through the points $(-1, 0)$, $(3, 0)$, $(0, 2)$, and $(0, -2)$.

28. Derive an equation for the ellipse with foci $(3, -3)$ and $(-3, 3)$ and major axis of length 10. Note that the foci of this ellipse lie on neither a vertical line nor a horizontal line.

10.8 THE HYPERBOLA

A hyperbola is a conic section defined in the same way as is an ellipse, except that the eccentricity e of a hyperbola is greater than 1.

Definition ***The Hyperbola***

Suppose that $e > 1$, and let F be a fixed point and L a fixed line not containing F. Then the **hyperbola** with **eccentricity** e, **focus** F, and **directrix** L is the set of all points P such that the distance $|PF|$ is e times the (perpendicular) distance from P to the line L.

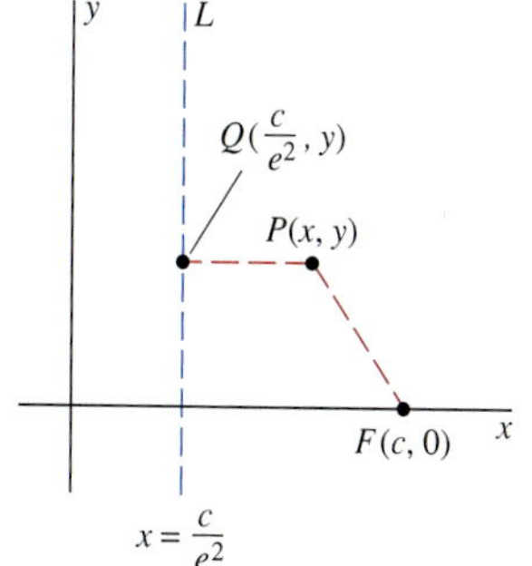

Fig. 10.8.1 The definition of the hyperbola

As with the ellipse, the equation of a hyperbola is simplest if F is the point $(c, 0)$ on the x-axis and L is the vertical line $x = c/e^2$. The case $c > 0$ is shown in Fig. 10.8.1. If Q is the point $(c/e^2, y)$, then PQ is the perpendicular from $P(x, y)$ to L. The condition $|PF| = e|PQ|$ gives

$$(x - c)^2 + y^2 = e^2\left(x - \frac{c}{e^2}\right)^2;$$

$$x^2 - 2cx + c^2 + y^2 = e^2x^2 - 2cx + \frac{c^2}{e^2};$$

$$(e^2 - 1)x^2 - y^2 = c^2\left(1 - \frac{1}{e^2}\right) = \frac{c^2}{e^2}(e^2 - 1).$$

Thus

$$(e^2 - 1)x^2 - y^2 = a^2(e^2 - 1),$$

where

$$a = \frac{c}{e}. \tag{1}$$

If we divide both sides of the next-to-last equation by $a^2(e^2 - 1)$, we get

$$\frac{x^2}{a^2} - \frac{y^2}{a^2(e^2 - 1)} = 1.$$

To simplify this equation, we let

$$b^2 = a^2(e^2 - 1) = c^2 - a^2. \tag{2}$$

This is permissible because $e > 1$. So the equation of the hyperbola with focus $(c, 0)$ and directrix $x = c/e^2 = a/e$ takes the form

$$\frac{x^2}{a^2} - \frac{y^2}{b^2} = 1. \tag{3}$$

The minus sign on the left-hand side is the only difference between the equation of a hyperbola and that of an ellipse. Of course, Eq. (2) also differs from the relation

$$b^2 = a^2(1 - e^2) = a^2 - c^2$$

for the case of the ellipse.

The hyperbola of Eq. (3) is clearly symmetric around both coordinate axes and has x-intercepts $(\pm a, 0)$. But it has no y-intercept. If we rewrite Eq. (3) in the form

$$y = \pm\frac{b}{a}\sqrt{x^2 - a^2}, \tag{4}$$

then we see that there are points on the graph only if $|x| \geqq a$. Hence the hyperbola has two **branches**, as shown in Fig. 10.8.2. We also see from Eq. (4) that $|y| \to \infty$ as $|x| \to \infty$.

The x-intercepts $V_1(-a, 0)$ and $V_2(a, 0)$ are the **vertices** of the hyperbola, and the line segment joining them is its **transverse axis** (Fig. 10.8.3). The line segment joining $W_1(0, -b)$ and $W_2(0, b)$ is its **conjugate axis**. The alternative form

$$c^2 = a^2 + b^2 \tag{5}$$

of Eq. (2) is the Pythagorean relation for the right triangle shown in Fig. 10.8.2.

The lines $y = \pm bx/a$ that pass through the **center** $(0, 0)$ and the opposite vertices of the rectangle in Fig. 10.8.3 are **asymptotes** of the two branches of the hyperbola in both directions. That is, if

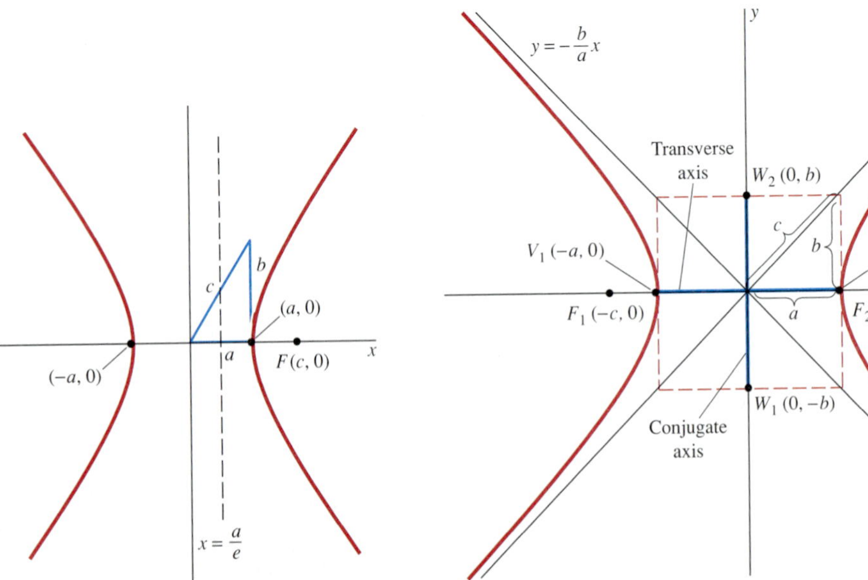

Fig. 10.8.2 A hyperbola has two *branches.*

Fig. 10.8.3 The parts of a hyperbola

$$y_1 = \frac{bx}{a} \quad \text{and} \quad y_2 = \frac{b}{a}\sqrt{x^2 - a^2},$$

then

$$\lim_{x\to\infty} (y_1 - y_2) = 0 = \lim_{x\to-\infty} (y_1 - (-y_2)). \tag{6}$$

To verify the first limit, note that

$$\begin{aligned}\lim_{x\to\infty} \frac{b}{a}\left(x - \sqrt{x^2 - a^2}\right) &= \lim_{x\to\infty} \frac{b}{a} \cdot \frac{(x - \sqrt{x^2 - a^2})(x + \sqrt{x^2 - a^2})}{x + \sqrt{x^2 - a^2}} \\ &= \lim_{x\to\infty} \frac{b}{a} \cdot \frac{a^2}{x + \sqrt{x^2 - a^2}} = 0.\end{aligned}$$

Just as in the case of the ellipse, the hyperbola with focus $(c, 0)$ and directrix $x = a/e$ also has focus $(-c, 0)$ and directrix $x = -a/e$ (Fig. 10.8.3). Because $c = ae$ by Eq. (1), the foci $(\pm ae, 0)$ and the directrices $x = \pm a/e$ take the same forms in terms of a and e for both the hyperbola ($e > 1$) and the ellipse ($e < 1$).

If we interchange x and y in Eq. (3), we obtain

$$\frac{y^2}{a^2} - \frac{x^2}{b^2} = 1. \tag{7}$$

This hyperbola has foci at $(0, \pm c)$. The foci as well as this hyperbola's transverse axis lie on the y-axis. Its asymptotes are $y = \pm ax/b$, and its graph generally resembles the one in Fig. 10.8.4.

When we studied the ellipse, we saw that its orientation—whether the major axis is horizontal or vertical—is determined by the relative sizes of a and b. In the case of the hyperbola, the situation is quite different because the relative sizes of a and b make no such difference: They affect only the slopes of the asymptotes. The direction in which the hyperbola opens—horizontal as in Fig. 10.8.3 or vertical as in Fig. 10.8.4—is determined by the signs of the terms that contain x^2 and y^2.

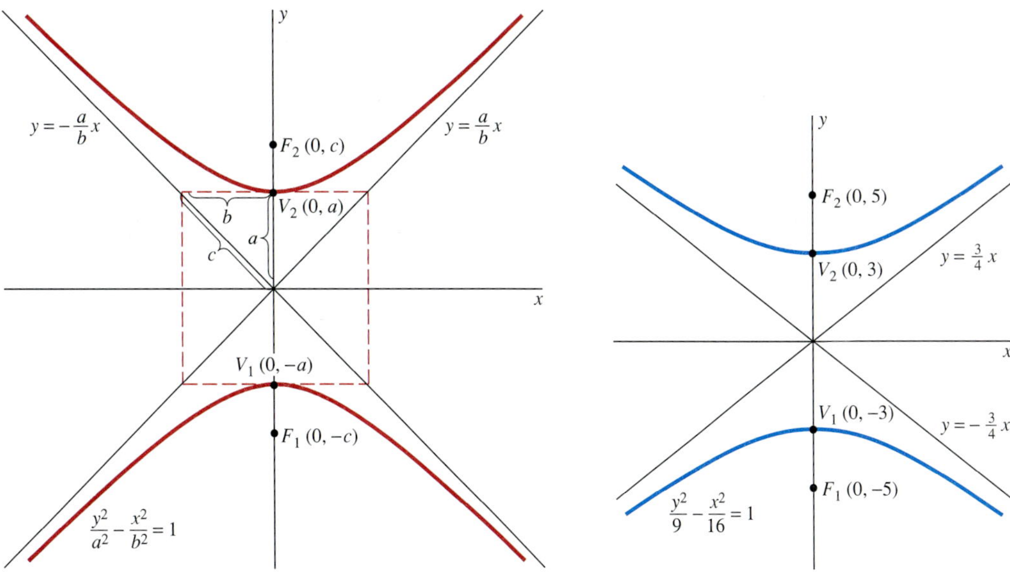

Fig. 10.8.4 The hyperbola of Eq. (7) has horizontal directrices.

Fig. 10.8.5 The hyperbola of Example 1

EXAMPLE 1 Sketch the graph of the hyperbola with equation

$$\frac{y^2}{9} - \frac{x^2}{16} = 1.$$

Solution This is an equation of the form in Eq. (7), so the hyperbola opens vertically. Because $a = 3$ and $b = 4$, we find that $c = 5$ by using Eq. (5): $c^2 = a^2 + b^2$. Thus the vertices are $(0, \pm 3)$, the foci are the two points $(0, \pm 5)$, and the asymptotes are the two lines $y = \pm 3x/4$. This hyperbola appears in Fig. 10.8.5. ■

EXAMPLE 2 Find an equation of the hyperbola with foci $(\pm 10, 0)$ and asymptotes $y = \pm 4x/3$.

Solution Because $c = 10$, we have

$$a^2 + b^2 = 100 \quad \text{and} \quad \frac{b}{a} = \frac{4}{3}.$$

Thus $b = 8$ and $a = 6$, and the standard equation of the hyperbola is

$$\frac{x^2}{36} - \frac{y^2}{64} = 1.$$ ■

As we noted in Section 10.7, any equation of the form

$$Ax^2 + Cy^2 + Dx + Ey + F = 0 \tag{8}$$

with both A and C nonzero can be reduced to the form

$$A(x - h)^2 + B(y - k)^2 = G$$

by completing the square in x and y. Now suppose that the coefficients A and C of the quadratic terms have *opposite signs*. For example, suppose that $A = p^2$ and $B = -q^2$. The last equation then becomes

$$p^2(x - h)^2 - q^2(y - k)^2 = G. \tag{9}$$

If $G = 0$, then factorization of the difference of squares on the left-hand side yields the equations

$$p(x - h) + q(y - k) = 0 \quad \text{and} \quad p(x - h) - q(y - k) = 0$$

of two straight lines through (h, k) with slopes $m = \pm p/q$. If $G \neq 0$, then division of Eq. (9) by G gives an equation that looks either like

$$\frac{(x - h)^2}{a^2} - \frac{(y - k)^2}{b^2} = 1 \qquad (\text{if } G > 0)$$

or like

$$\frac{(y - k)^2}{a^2} - \frac{(x - h)^2}{b^2} = 1 \qquad (\text{if } G < 0).$$

Thus if $AC < 0$ in Eq. (8), the graph is either a pair of intersecting straight lines or a hyperbola.

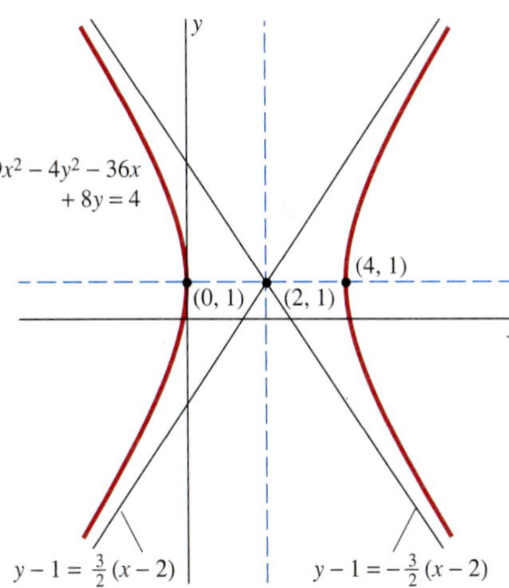

Fig. 10.8.6 The hyperbola of Example 3, a translate of the hyperbola $x^2/4 - y^2/9 = 1$

EXAMPLE 3 Determine the graph of the equation

$$9x^2 - 4y^2 - 36x + 8y = 4.$$

Solution We collect the terms that contain x and those that contain y, and we then complete the square in each variable. We find that

$$9(x - 2)^2 - 4(y - 1)^2 = 36,$$

so

$$\frac{(x-2)^2}{4} - \frac{(y-1)^2}{9} = 1.$$

Hence the graph is a hyperbola with a horizontal transverse axis and center $(2, 1)$. Because $a = 2$ and $b = 3$, we find that $c = \sqrt{13}$. The vertices of the hyperbola are $(0, 1)$ and $(4, 1)$, and its foci are the two points $(2 \pm \sqrt{13}, 1)$. Its asymptotes are the two lines

$$y - 1 = \pm\tfrac{3}{2}(x - 2),$$

translates of the asymptotes $y = \pm 3x/2$ of the hyperbola $\frac{1}{4}x^2 - \frac{1}{9}y^2 = 1$. Figure 10.8.6 shows the graph of the translated hyperbola. ■

Applications of Hyperbolas

The *reflection property* of the hyperbola takes the same form as that for the ellipse. If P is a point on a hyperbola, then the two lines PF_1 and PF_2 from P to the two foci make equal angles with the tangent line at P. In Fig. 10.8.7 this means that $\alpha = \beta$.

For an important application of this reflection property, consider a mirror that is shaped like one branch of a hyperbola and is reflective on its outer (convex) surface. An incoming light ray aimed toward one focus will be reflected toward the other focus (Fig. 10.8.8). Figure 10.8.9 indicates the design of a reflecting telescope that

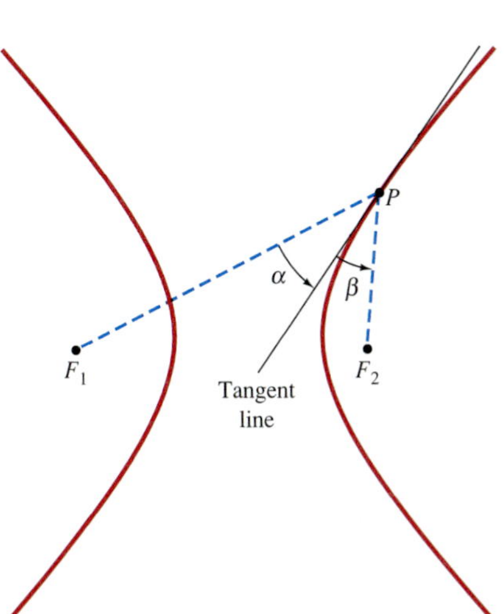

Fig. 10.8.7 The reflection property of the hyperbola

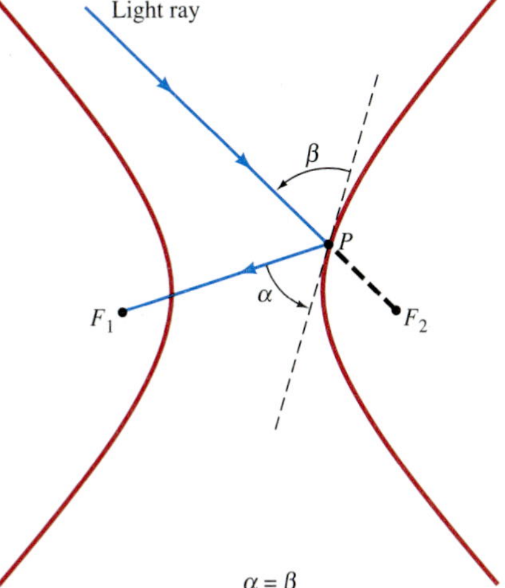

Fig. 10.8.8 How a hyperbolic mirror reflects a ray aimed at one focus: $\alpha = \beta$ again

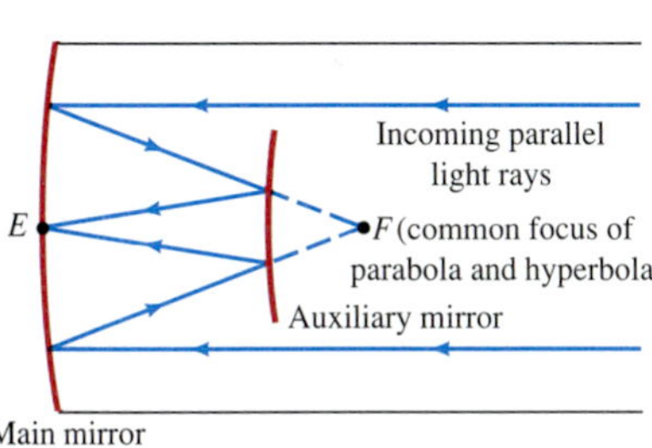

Fig. 10.8.9 One type of reflecting telescope: main mirror parabolic, auxiliary mirror hyperbolic

makes use of the reflection properties of the parabola and the hyperbola. The parallel incoming light rays first are reflected by the parabola toward its focus at F. Then they are intercepted by an auxiliary hyperbolic mirror with foci at E and F and reflected into the eyepiece located at E.

Example 4 illustrates how hyperbolas are used to determine the position of ships at sea.

EXAMPLE 4 A ship lies in the Labrador Sea due east of Wesleyville, point A, on the long north-south coastline of Newfoundland. Simultaneous radio signals are transmitted by radio stations at A and at St. John's, point B, which is on the coast 200 km due south of A. The ship receives the signal from A 500 microseconds (μs) before it receives the signal from B. Assume that the speed of radio signals is 300 m/μs. How far out at sea is the ship?

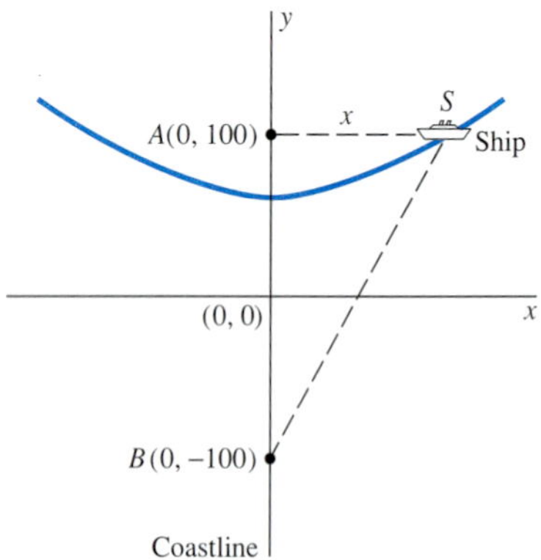

Fig. 10.8.10 A navigation problem (Example 4)

Solution The situation is diagrammed in Fig. 10.8.10. The difference between the distances of the ship at S from A and B is

$$|SB| - |SA| = 500 \cdot 300 = 150{,}000$$

meters; that is, 150 km. Thus (by Problem 24) the ship lies on a hyperbola with foci A and B. From Fig. 10.8.10 we see that $c = 100$, so $a = \frac{1}{2} \cdot 150 = 75$, and thus

$$b = \sqrt{c^2 - a^2} = \sqrt{100^2 - 75^2} = 25\sqrt{7}.$$

In the coordinate system of Fig. 10.8.10, the hyperbola has equation

$$\frac{y^2}{75^2} - \frac{x^2}{7 \cdot 25^2} = 1.$$

We substitute $y = 100$ because the ship is due east of A. Thus we find that the ship's distance from the coastline is $x = \frac{175}{3} \approx 58.3$ km. ■

10.8 PROBLEMS

In Problems 1 through 14, find an equation of the hyperbola described there.

1. Foci $(\pm 4, 0)$, vertices $(\pm 1, 0)$
2. Foci $(0, \pm 3)$, vertices $(0, \pm 2)$
3. Foci $(\pm 5, 0)$, asymptotes $y = \pm 3x/4$
4. Vertices $(\pm 3, 0)$, asymptotes $y = \pm 3x/4$
5. Vertices $(0, \pm 5)$, asymptotes $y = \pm x$
6. Vertices $(\pm 3, 0)$, eccentricity $e = \frac{5}{3}$
7. Foci $(0, \pm 6)$, eccentricity $e = 2$
8. Vertices $(\pm 4, 0)$ and passing through $(8, 3)$
9. Foci $(\pm 4, 0)$, directrices $x = \pm 1$
10. Foci $(0, \pm 9)$, directrices $y = \pm 4$
11. Center $(2, 2)$, horizontal transverse axis of length 6, eccentricity $e = 2$
12. Center $(-1, 3)$, vertices $(-4, 3)$ and $(2, 3)$, foci $(-6, 3)$ and $(4, 3)$
13. Center $(1, -2)$, vertices $(1, 1)$ and $(1, -5)$, asymptotes $3x - 2y = 7$ and $3x + 2y = -1$
14. Focus $(8, -1)$, asymptotes $3x - 4y = 13$ and $3x + 4y = 5$

Sketch the graphs of the equations given in Problems 15 through 20; indicate centers, foci, and asymptotes.

15. $x^2 - y^2 - 2x + 4y = 4$
16. $x^2 - 2y^2 + 4x = 0$
17. $y^2 - 3x^2 - 6y = 0$
18. $x^2 - y^2 - 2x + 6y = 9$
19. $9x^2 - 4y^2 + 18x + 8y = 31$
20. $4y^2 - 9x^2 - 18x - 8y = 41$
21. Show that the graph of the equation

$$\frac{x^2}{15 - c} - \frac{y^2}{c - 6} = 1$$

is (a) a hyperbola with foci $(\pm 3, 0)$ if $6 < c < 15$ and (b) an ellipse if $c < 6$. (c) Identify the graph in the case $c > 15$.

22. Establish that the line tangent to the hyperbola

$$\frac{x^2}{a^2} - \frac{y^2}{b^2} = 1$$

at the point $P(x_0, y_0)$ has equation

$$\frac{x_0 x}{a^2} - \frac{y_0 y}{b^2} = 1.$$

23. Use the result of Problem 22 to establish the reflection property of the hyperbola. (See the suggestion for Problem 25 of Section 10.7.)

24. Suppose that $0 < a < c$, and let $b = \sqrt{c^2 - a^2}$. Show that the hyperbola $x^2/a^2 - y^2/b^2 = 1$ is the locus of a point P such that the *difference* between the distances $|PF_1|$ and $|PF_2|$ is equal to $2a$ (F_1 and F_2 are the foci of the hyperbola).

25. Derive an equation for the hyperbola with vertices $(\pm 3/\sqrt{2}, \pm 3/\sqrt{2})$ and foci $(\pm 5, \pm 5)$. Use the difference definition of a hyperbola implied by Problem 24.

26. Two radio signaling stations at A and B lie on an east-west line, with A 100 mi west of B. A plane is flying west on a line 50 mi north of the line AB. Radio signals are sent (traveling at 980 ft/μs) simultaneously from A and B, and the one sent from B arrives at the plane 400 μs before the one sent from A. Where is the plane?

27. Two radio signaling stations are located as in Problem 26 and transmit radio signals that travel at the same speed. But now we know only that the plane is generally somewhere north of the line AB, that the signal from B arrives 400 μs before the one sent from A, and that the signal sent from A and reflected by the plane takes a total of 600 μs to reach B. Where is the plane?

CHAPTER 10 REVIEW: *Concepts and Definitions*

Use the following list as a guide to additional concepts that you may need to review.

1. Conic sections
2. The relationship between rectangular and polar coordinates
3. The graph of an equation in polar coordinates
4. The area formula in polar coordinates
5. Definition of a parametric curve and a smooth parametric curve
6. The slope of the line tangent to a smooth parametric curve (both in rectangular and in polar coordinates)
7. Integral computations with parametric curves [Eqs. (1) through (4) of Section 10.5]
8. Arc length of a parametric curve

Review of Conic Sections

The parabola with focus $(p, 0)$ and directrix $x = -p$ has eccentricity $e = 1$ and equation $y^2 = 4px$. The accompanying table compares the properties of an ellipse and a hyperbola, each with foci $(\pm c, 0)$ and major axis of length $2a$.

	Ellipse	Hyperbola
Eccentricity	$e = \frac{c}{a} < 1$	$e = \frac{c}{a} > 1$
a, b, c relation	$a^2 = b^2 + c^2$	$c^2 = a^2 + b^2$
Equation	$\frac{x^2}{a^2} + \frac{y^2}{b^2} = 1$	$\frac{x^2}{a^2} - \frac{y^2}{b^2} = 1$
Vertices	$(\pm a, 0)$	$(\pm a, 0)$
y-intercepts	$(0, \pm b)$	None
Directrices	$x = \pm\frac{a}{e}$	$x = \pm\frac{a}{e}$
Asymptotes	None	$y = \pm\frac{bx}{a}$

CHAPTER 10 *Miscellaneous Problems*

Sketch the graphs of the equations in Problems 1 through 30. In Problems 1 through 18, if the graph is a conic section, label its center, foci, and vertices.

1. $x^2 + y^2 - 2x - 2y = 2$

2. $x^2 + y^2 = x + y$

3. $x^2 + y^2 - 6x + 2y + 9 = 0$

4. $y^2 = 4(x + y)$

5. $x^2 = 8x - 2y - 20$

6. $x^2 + 2y^2 - 2x + 8y + 8 = 0$

7. $9x^2 + 4y^2 = 36x$

8. $x^2 - y^2 = 2x - 2y - 1$

9. $y^2 - 2x^2 = 4x + 2y + 3$

10. $9y^2 - 4x^2 = 8x + 18y + 31$

11. $x^2 + 2y^2 = 4x + 4y - 12$

12. $y^2 - 6y + 4x + 5 = 0$

13. $9(x^2 - 2x + 1) = 4(y^2 + 9)$

14. $(x^2 - 4)(y^2 - 1) = 0$

15. $x^2 - 8x + y^2 - 2y + 16 = 0$

16. $(x - 1)^2 + 4(y - 2)^2 = 1$

17. $(x^2 - 4x + y^2 - 4y + 8)(x + y)^2 = 0$

18. $x = y^2 + 4y + 5$

19. $r = -2\cos\theta$

20. $\cos\theta + \sin\theta = 0$

21. $r = \dfrac{1}{\sin\theta - \cos\theta}$

22. $r\sin^2\theta = \cos\theta$

23. $r = 3\csc\theta$

24. $r = 2(\cos\theta - 1)$

25. $r^2 = 4\cos\theta$

26. $r\theta = 1$

27. $r = 3 - 2\sin\theta$

28. $r = \dfrac{1}{1 + \cos\theta}$

29. $r = \dfrac{4}{2 + \cos\theta}$

30. $r = \dfrac{4}{1 - 2\cos\theta}$

In Problems 31 through 38, find the area of the region described.

31. Inside both $r = 2\sin\theta$ and $r = 2\cos\theta$

32. Inside $r^2 = 4\cos\theta$

33. Inside $r = 3 - 2\sin\theta$ and outside $r = 4$

34. Inside $r^2 = 2\sin 2\theta$ and outside $r = 2\sin\theta$

35. Inside $r = 2\sin 2\theta$ and outside $r = \sqrt{2}$

36. Inside $r = 3\cos\theta$ and outside $r = 1 + \cos\theta$

37. Inside $r = 1 + \cos\theta$ and outside $r = \cos\theta$

38. Between the loops of $r = 1 - 2\sin\theta$

In Problems 39 through 43, eliminate the parameter and sketch the curve.

39. $x = 2t^3 - 1, \quad y = 2t^3 + 1$

40. $x = \cosh t, \quad y = \sinh t$

41. $x = 2 + \cos t, \quad y = 1 - \sin t$

42. $x = \cos^4 t, \quad y = \sin^4 t$

43. $x = 1 + t^2, \quad y = t^3$

In Problems 44 through 48, write an equation of the line tangent to the given curve at the indicated point.

44. $x = t^2, \quad y = t^3; \quad t = 1$

45. $x = 3\sin t, \quad y = 4\cos t; \quad t = \pi/4$

46. $x = e^t, \quad y = e^{-t}; \quad t = 0$

47. $r = \theta; \quad \theta = \pi/2$

48. $r = 1 + \sin\theta; \quad \theta = \pi/3$

In Problems 49 through 52, find the area of the region between the given curve and the x-axis.

49. $x = 2t + 1, \quad y = t^2 + 3; \quad -1 \leqq t \leqq 2$

50. $x = e^t, \quad y = e^{-t}; \quad 0 \leqq t \leqq 10$

51. $x = 3\sin t, \quad y = 4\cos t; \quad 0 \leqq t \leqq \pi/2$

52. $x = \cosh t, \quad y = \sinh t; \quad 0 \leqq t \leqq 1$

In Problems 53 through 57, find the arc length of the given curve.

53. $x = t^2, \quad y = t^3; \quad 0 \leqq t \leqq 1$

54. $x = \ln(\cos t), \quad y = t; \quad 0 \leqq t \leqq \pi/4$

55. $x = 2t, \quad y = t^3 + \dfrac{1}{3t}; \quad 1 \leqq t \leqq 2$

56. $r = \sin\theta; \quad 0 \leqq \theta \leqq \pi$

57. $r = \sin^2(\theta/3); \quad 0 \leqq \theta \leqq \pi$

In Problems 58 through 62, find the area of the surface generated by revolving the given curve around the x-axis.

58. $x = t^2 + 1, \quad y = 3t; \quad 0 \leqq t \leqq 2$

59. $x = 4\sqrt{t}, \quad y = \dfrac{t^3}{3} + \dfrac{1}{2t^2}; \quad 1 \leqq t \leqq 4$

60. $r = \cos\theta$

61. $r = e^{\theta/2}; \quad 0 \leqq \theta \leqq \pi$

62. $x = e^t\cos t, \quad y = e^t\sin t; \quad 0 \leqq t \leqq \pi/2$

63. Consider the rolling circle of radius a that was used to generate the cycloid in Example 5 of Section 10.4. Suppose that this circle is the rim of a disk, and let Q be a point of this disk at distance $b < a$ from its center. Find parametric equations for the curve traced by Q as the circle rolls along the x-axis. Assume that Q begins at the point $(0, a - b)$. Sketch this curve, which is called a **trochoid.**

64. If the smaller circle of Problem 34 in Section 10.4 rolls around the *outside* of the larger circle, the path of the point P is called an **epicycloid.** Show that it has parametric equations

$$x = (a + b)\cos t - b\cos\left(\frac{a + b}{b}t\right),$$

$$y = (a + b)\sin t - b\sin\left(\frac{a + b}{b}t\right).$$

65. Suppose that $b = a$ in Problem 64. Show that the epicycloid is then the cardioid $r = 2a(1 - \cos\theta)$ translated a units to the right.

66. Find the area of the surface generated by revolving the lemniscate $r^2 = 2a^2\cos 2\theta$ around the x-axis.

67. Find the volume generated by revolving around the y-axis the area under the cycloid

$$x = a(t - \sin t), \qquad y = a(1 - \cos t), \qquad 0 \leqq t \leqq 2\pi.$$

68. Show that the length of one arch of the hypocycloid of Problem 34 in Section 10.4 is $s = 8b(a - b)/a$.

69. Find a polar coordinates equation of the circle that passes through the origin and is centered at the point with polar coordinates (p, α).

70. Find a simple equation of the parabola whose focus is the origin and whose directrix is the line $y = x + 4$. Recall from Miscellaneous Problem 71 of Chapter 3 that the distance from the point (x_0, y_0) to the line with equation $Ax + By + C = 0$ is

$$\frac{|Ax_0 + By_0 + C|}{\sqrt{A^2 + B^2}}.$$

71. A **diameter** of an ellipse is a chord through its center. Find the maximum and minimum lengths of diameters of the ellipse with equation

$$\frac{x^2}{a^2} + \frac{y^2}{b^2} = 1.$$

72. Use calculus to prove that the ellipse of Problem 71 is normal to the coordinate axes at each of its four vertices.

73. The parabolic arch of a bridge has base width b and height h at its center. Write its equation, choosing the origin on the ground at the left end of the arch.

74. Use methods of calculus to find the points of the ellipse

$$\frac{x^2}{a^2} + \frac{y^2}{b^2} = 1$$

that are nearest to and farthest from (a) the center $(0, 0)$; (b) the focus $(c, 0)$.

75. Consider a line segment QR that contains a point P such that $|QP| = a$ and $|PR| = b$. Suppose that Q is constrained to move on the y-axis, whereas R must remain on the x-axis. Prove that the locus of P is an ellipse.

76. Suppose that $a > 0$ and that F_1 and F_2 are two fixed points in the plane with $|F_1F_2| > 2a$. Imagine a point P that moves in such a way that $|PF_2| = 2a + |PF_1|$. Prove that the locus of P is one branch of a hyperbola with foci F_1 and F_2. Then—as a consequence—explain how to construct points on a hyperbola by drawing appropriate circles centered at its foci.

77. Let Q_1 and Q_2 be two points on the parabola $y^2 = 4px$. Let P be the point of the parabola at which the tangent line is parallel to Q_1Q_2. Prove that the horizontal line through P bisects the segment Q_1Q_2.

78. Determine the locus of a point P such that the product of its distances from the two fixed points $F_1(-a, 0)$ and $F_1(a, 0)$ is a^2.

79. Find the eccentricity of the conic section with equation $3x^2 - y^2 + 12x + 9 = 0$.

80. Find the area bounded by the loop of the *strophoid*

$$r = \sec\theta - 2\cos\theta$$

shown in Fig. 10.MP.1.

81. Find the area bounded by the loop of the *folium of Descartes* with equation $x^3 + y^3 = 3xy$ shown in Fig. 10.MP.2. (*Suggestion:* Change to polar coordinates and then substitute $u = \tan\theta$ to evaluate the area integral.)

82. Use the method of Problem 81 to find the area bounded by the first-quadrant loop (similar to the folium of Problem 51) of the curve $x^5 + y^5 = 5x^2y^2$.

83. The graph of a conic section in the xy-plane has intercepts at $(5, 0)$, $(-5, 0)$, $(0, 4)$, and $(0, -4)$. Deduce all the information you can about this conic. Can you determine whether it is a parabola, a hyperbola, or an ellipse? What if you also know that the graph of this conic is normal to the y-axis at $(0, 4)$?

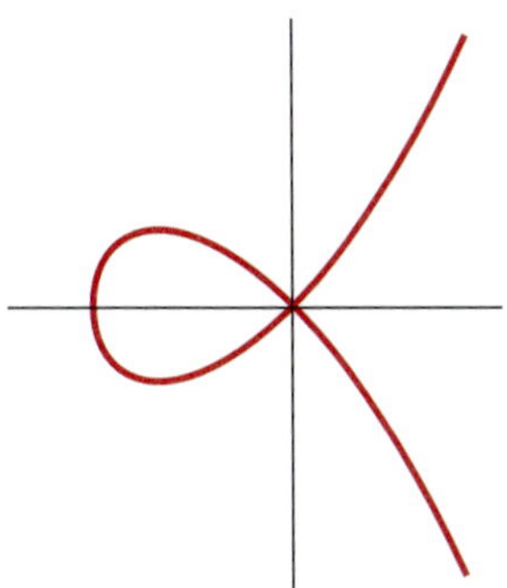

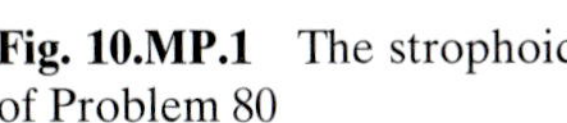

Fig. 10.MP.1 The strophoid of Problem 80

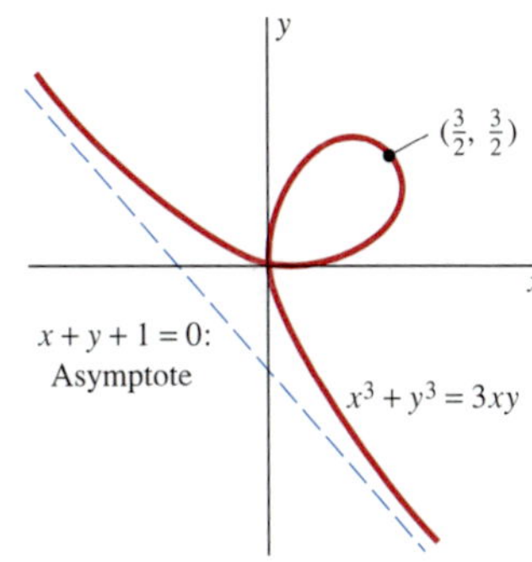

Fig. 10.MP.2 The folium of Descartes $x^3 + y^3 = 3xy$ (Problem 81)

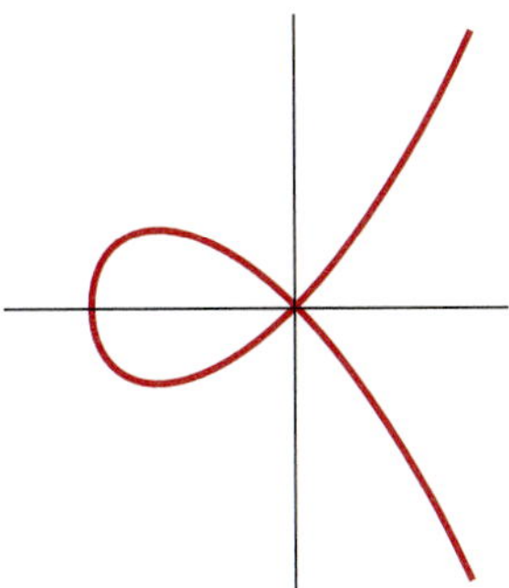

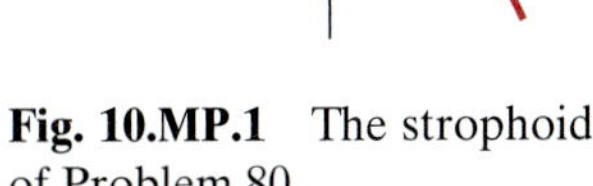

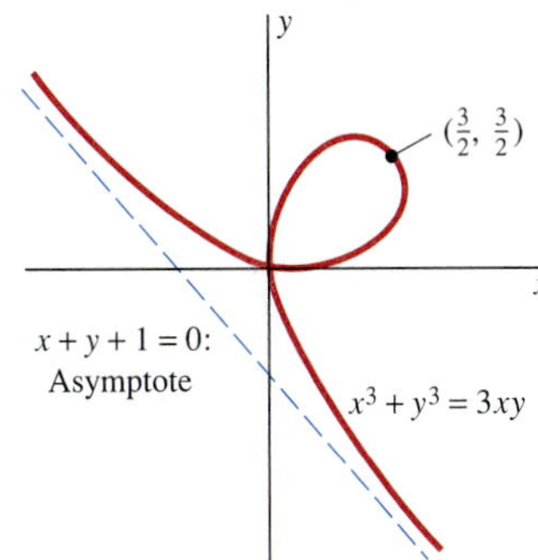

CHAPTER 11

INFINITE SERIES

Srinivasa Ramanujan (1887–1920)

On a cold January day in 1913, the eminent Cambridge mathematics professor G. H. Hardy received a letter from an unknown 25-year-old clerk in the accounting department of a government office in Madras, India. Its author, Srinivasa Ramanujan, had no university education, he admitted—he had flunked out—but "after leaving school I have employed the spare time at my disposal to work at Mathematics...I have not trodden through the conventional regular course...but am striking out a new path for myself." The ten pages that followed listed in neat handwritten script approximately fifty formulas, most dealing with integrals and infinite series that Ramanujan had discovered, and asked Hardy's advice whether they contained anything of value. The formulas were of such exotic and unlikely appearance that Hardy at first suspected a hoax, but he and his colleague J. E. Littlewood soon realized that they were looking at the work of an extraordinary mathematical genius.

Thus began one of the most romantic episodes in the history of mathematics. In April 1914 Ramanujan arrived in England a poor, self-taught Indian mathematical amateur called to collaborate as an equal with the most sophisticated professional mathematicians of the day. For the next three years a steady stream of remarkable discoveries poured forth from his pen. But in 1917 he fell seriously ill, apparently with tuberculosis. The following year he returned to India in an attempt to regain his health but never recovered, and he died in 1920 at the age of 32. Up to the very end he worked feverishly to record his final discoveries. He left behind notebooks outlining work whose completion has occupied prominent mathematicians throughout the twentieth century.

With the possible exception of Euler, no one before or since has exhibited Ramanujan's virtuosity with infinite series. An example of his discoveries is the infinite series

$$\frac{1}{\pi} = \frac{\sqrt{8}}{9801}\sum_{n=0}^{\infty}\frac{(4n)!}{(n!)^4}\cdot\frac{(1103+26390n)}{396^{4n}},$$

whose first term yields the familiar approximation $\pi \approx 3.14159$, and with each additional term giving π to roughly eight more decimal places of accuracy. For instance, just four terms of Ramanujan's series are needed to calculate the 30-place approximation

$$\pi \approx 3.14159\ 26535\ 89793\ 23846\ 26433\ 83279$$

that suffices for virtually any imaginable "practical" application.

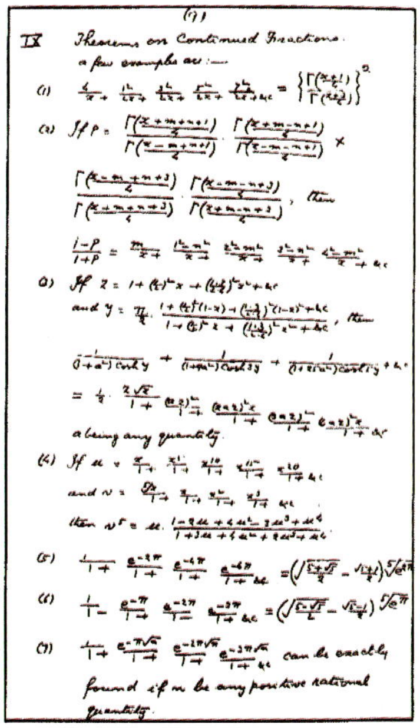

A typical page of Ramanujan's letter to Hardy, listing formulas Ramanujan had discovered, but with no hint of proof or derivation.

11.1 INTRODUCTION

In the fifth century B.C., the Greek philosopher Zeno proposed the following paradox: For a runner to travel a given distance, the runner must first travel halfway, then half the remaining distance, then half the distance that yet remains, and so on ad infinitum. But, Zeno argued, it is clearly impossible for a runner to accomplish infinitely many such tasks in a finite period of time, so motion from one point to another is impossible.

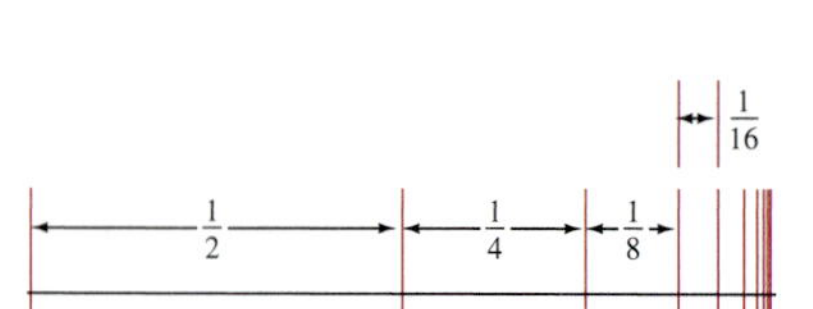

Fig. 11.1.1 Subdivision of an interval to illustrate Zeno's paradox

Zeno's paradox suggests the infinite subdivision of $[0, 1]$ indicated in Fig. 11.1.1. There is one subinterval of length $1/2^n$ for each integer $n = 1, 2, 3, \ldots$. If the length of the interval is the sum of the lengths of the subintervals into which it is divided, then it would appear that

$$1 = \frac{1}{2} + \frac{1}{4} + \frac{1}{8} + \frac{1}{16} + \cdots + \frac{1}{2^n} + \cdots,$$

with infinitely many terms somehow adding up to 1. But the formal infinite sum

$$1 + 2 + 3 + \cdots + n + \cdots$$

of all the positive integers seems meaningless—it does not appear to add up to *any* (finite) value.

The question is this: What, if anything, do we mean by the sum of an *infinite* collection of numbers? This chapter explores conditions under which an *infinite* sum

$$a_1 + a_2 + a_3 + \cdots + a_n + \cdots,$$

known as an *infinite series,* is meaningful. We discuss methods for computing the sum of an infinite series and applications of the algebra and calculus of infinite series. Infinite series are important in science and mathematics because many functions either arise most naturally in the form of infinite series or have infinite series representations (such as the Taylor series of Section 11.4) that are useful for numerical computations.

11.2 INFINITE SEQUENCES

An **infinite sequence** of real numbers is an ordered, unending list

$$a_1, a_2, a_3, a_4, \ldots, a_n, a_{n+1}, \ldots \qquad \textbf{(1)}$$

of numbers. That this list is *ordered* implies that it has a first term a_1, a second term a_2, a third term a_3, and so forth. That the sequence is unending, or *infinite,* implies that (for every n) the ***n*th term** a_n has a successor a_{n+1}. Thus, as indicated by the final ellipsis in Eq. (1), an infinite sequence never ends and—despite the fact that we write explicitly only a finite number of terms—it actually has an infinite number of terms. Concise notation for the infinite sequence in (1) is

$$\{a_n\}_{n=1}^{\infty}, \qquad \{a_n\}_1^{\infty}, \qquad \text{or simply} \qquad \{a_n\}. \qquad \textbf{(2)}$$

Frequently an infinite sequence $\{a_n\}$ of numbers can be described "all at once" by a single function f that gives the successive terms of the sequence as successive values of the function:

$$a_n = f(n) \qquad \text{for } n = 1, 2, 3, \ldots. \qquad \textbf{(3)}$$

Here $a_n = f(n)$ is simply a *formula for the nth term* of the sequence. Conversely, if the sequence $\{a_n\}$ is given in advance, we can regard Eq. (3) as the definition of the function f having the set of positive integers as its domain of definition. Ordinarily we will use the subscript notation a_n in preference to the function notation $f(n)$.

EXAMPLE 1 The following table exhibits several particular infinite sequences. Each is described in three ways: in the concise sequential notation $\{a_n\}$ of (2), by writing the formula as in Eq. (3) for its nth term, and in extended list notation as in (1). Note that n need not begin with initial value 1.

$\left\{\frac{1}{n}\right\}_1^\infty$	$a_n = \frac{1}{n}$	$1, \frac{1}{2}, \frac{1}{3}, \frac{1}{4}, \dots, \frac{1}{n}, \dots$
$\left\{\frac{1}{10^n}\right\}_0^\infty$	$a_n = \frac{1}{10^n}$	$1, \frac{1}{10}, \frac{1}{100}, \frac{1}{1000}, \dots, \frac{1}{10^n}, \dots$
$\left\{\sqrt{3n-7}\right\}_3^\infty$	$a_n = \sqrt{3n-7}$	$\sqrt{2}, \sqrt{5}, \sqrt{8}, \sqrt{11}, \dots, \sqrt{3n-7}, \dots$
$\left\{\sin\frac{n\pi}{2}\right\}_1^\infty$	$a_n = \sin\frac{n\pi}{2}$	$1, 0, -1, 0, \dots, \sin\frac{n\pi}{2}, \dots$
$\{3+(-1)^n\}_1^\infty$	$a_n = 3+(-1)^n$	$2, 4, 2, 4, \dots, 3+(-1)^n, \dots$

■

Sometimes it is inconvenient or impossible to give an explicit formula for the nth term of a particular sequence. The following example illustrates how sequences can be defined in other ways.

EXAMPLE 2 Here we give the first ten terms of each sequence.

(a) The sequence of prime integers (those positive integers n having precisely two divisors, 1 and n with $n > 1$):

$$2, 3, 5, 7, 11, 13, 17, 19, 23, 29, \dots$$

(b) The sequence whose nth term is the nth decimal digit of the number

$$\pi = 3.14159265358979323846\dots:$$

$$1, 4, 1, 5, 9, 2, 6, 5, 3, 5, \dots$$

(c) The **Fibonacci sequence** $\{F_n\}$, which may be defined by

$$F_1 = 1, \quad F_2 = 1, \quad \text{and} \quad F_{n+1} = F_n + F_{n-1} \quad \text{for } n \geqq 2.$$

Thus each term after the second is the sum of the preceding two terms:

$$1, 1, 2, 3, 5, 8, 13, 21, 34, 55, \dots$$

This is an example of a *recursively defined sequence* in which each term (after the first few) is given by a formula involving its predecessors. The thirteenth-century Italian mathematician Fibonacci asked the following question: If we start with a single pair of rabbits that gives birth to a new pair after two months, and each such new pair does the same, how many pairs of rabbits will we have after n months? See Problems 55 and 56.

(d) If the amount $A_0 = 100$ dollars is invested in a savings account that draws 10% interest compounded annually, then the amount A_n in the account at the end of n years is defined (for $n \geqq 1$) by the *iterative formula* $A_n = (1.10)A_{n-1}$ (rounded to the nearest number of cents) in terms of the preceding amount:

$$110.00, 121.00, 133.10, 146.41, 161.05, 177.16, 194.87, 214.36, 235.79, 259.37, \ldots$$ ■

Limits of Sequences

The limit of a sequence is defined in much the same way as the limit of an ordinary function (Section 2.2).

Definition *Limit of a Sequence*

We say that the sequence $\{a_n\}$ **converges** to the real number L, or has the **limit** L, and we write

$$\lim_{n\to\infty} a_n = L, \tag{4}$$

provided that a_n can be made as close to L as we please merely by choosing n to be sufficiently large. That is, given any number $\epsilon > 0$, there exists an integer N such that

$$|a_n - L| < \epsilon \qquad \text{for all } n \geqq N. \tag{5}$$

If the sequence $\{a_n\}$ does *not* converge, then we say that $\{a_n\}$ **diverges.**

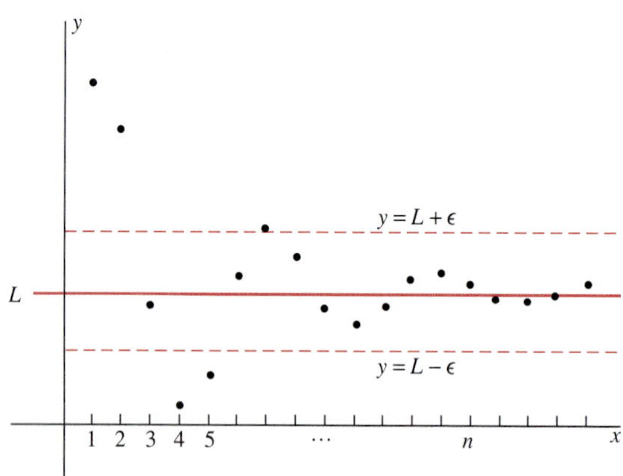

Fig. 11.2.1 The point (n, a_n) approaches the line $y = L$ as $n \to +\infty$.

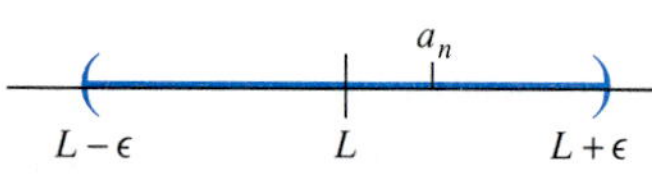

Fig. 11.2.2 The inequality $|a_n - L| < \epsilon$ means that a_n lies somewhere between $L - \epsilon$ and $L + \epsilon$.

Figure 11.2.1 illustrates geometrically the definition of the limit of a sequence. Because

$$|a_n - L| < \epsilon \quad \text{means that} \quad L - \epsilon < a_n < L + \epsilon,$$

the condition in (5) means that if $n \geqq N$, then the point (n, a_n) lies between the horizontal lines $y = L - \epsilon$ and $y = L + \epsilon$. Alternatively, if $n \geqq N$, then the number a_n lies between the points $L - \epsilon$ and $L + \epsilon$ on the real line (Fig. 11.2.2).

EXAMPLE 3 Suppose that we want to establish rigorously the intuitively evident fact that the sequence $\{1/n\}_1^\infty$ converges to zero,

$$\lim_{n\to\infty} \frac{1}{n} = 0. \tag{6}$$

Because $L = 0$ here, we need only convince ourselves that to each positive number ϵ there corresponds an integer N such that

$$\left|\frac{1}{n}\right| = \frac{1}{n} < \epsilon \qquad \text{if } n \geqq N.$$

But evidently it suffices to choose any fixed integer $N > 1/\epsilon$. Then $n \geqq N$ implies immediately that

$$\frac{1}{n} \leqq \frac{1}{N} < \epsilon,$$

as desired (Fig. 11.2.3). ■

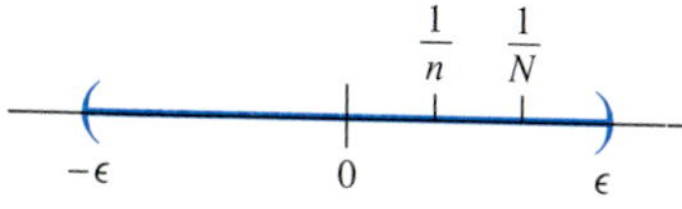

Fig. 11.2.3 If $N > \dfrac{1}{\epsilon}$ and $n \geqq N$ then $0 < \dfrac{1}{n} \leqq \dfrac{1}{N} < \epsilon$.

EXAMPLE 4 (a) The sequence $\{(-1)^n\}$ diverges because its successive terms "oscillate" between the two values $+1$ and -1. Hence $(-1)^n$ cannot approach any single value as $n \to \infty$. (b) The terms of the sequence $\{n^2\}$ increase without bound as $n \to \infty$. Thus the sequence $\{n^2\}$ diverges. In this case, we might also say that $\{n^2\}$ diverges *to infinity*. ■

Using Limit Laws

The limit laws in Section 2.2 for limits of functions have natural analogues for limits of sequences. Their proofs are based on techniques similar to those used in Appendix D.

Theorem 1 ***Limit Laws for Sequences***

If the limits

$$\lim_{n\to\infty} a_n = A \quad \text{and} \quad \lim_{n\to\infty} b_n = B$$

exist (so A and B are real numbers), then

1. $\displaystyle\lim_{n\to\infty} ca_n = A$ (c any real number);
2. $\displaystyle\lim_{n\to\infty} (a_n + b_n) = A + B$;
3. $\displaystyle\lim_{n\to\infty} a_n b_n = AB$;
4. $\displaystyle\lim_{n\to\infty} \frac{a_n}{b_n} = \frac{A}{B}$.

In part (4) we must assume that $B \neq 0$ and that $b_n \neq 0$ for all sufficiently large values of n.

Theorem 2 ***Substitution Law for Sequences***

If $\lim_{n\to\infty} a_n = A$ and the function f is continuous at $x = A$, then

$$\lim_{n\to\infty} f(a_n) = f(A).$$

Theorem 3 ***Squeeze Law for Sequences***

If $a_n \leqq b_n \leqq c_n$ for all n and

$$\lim_{n\to\infty} a_n = L = \lim_{n\to\infty} c_n,$$

then $\lim_{n\to\infty} b_n = L$ as well.

These theorems can be used to compute limits of many sequences formally, without recourse to the definition. For example, Eq. (6) and the product law of limits yield

$$\lim_{n\to\infty} \frac{1}{n^k} = 0 \tag{7}$$

for every positive integer k.

EXAMPLE 5 Equation (7) and the limit laws give (after dividing numerator and denominator by the highest power of n that is present)

$$\lim_{n\to\infty} \frac{7n^2}{5n^2 - 3} = \lim_{n\to\infty} \frac{7}{5 - \dfrac{3}{n^2}}$$

$$= \frac{\lim_{n\to\infty} 7}{\left(\lim_{n\to\infty} 5\right) - 3 \cdot \left(\lim_{n\to\infty} \dfrac{1}{n^2}\right)} = \frac{7}{5 - 3 \cdot 0} = \frac{7}{5}.$$ ■

EXAMPLE 6 Show that $\lim_{n\to\infty} \dfrac{\cos n}{n} = 0.$

Solution This follows from the squeeze law and the fact that $1/n \to 0$ as $n \to \infty$, because

$$-\frac{1}{n} \leqq \frac{\cos n}{n} \leqq \frac{1}{n}$$

for every positive integer n. ■

REMARK With a typical graphing calculator (in "dot plot" mode) or computer algebra system (using its "list plot" facility), one can plot the points (n, a_n) in the xy-plane corresponding to a given sequence $\{a_n\}$. Figure 11.2.4 shows such a plot for the sequence of Example 6 and provides visual evidence of its convergence to zero.

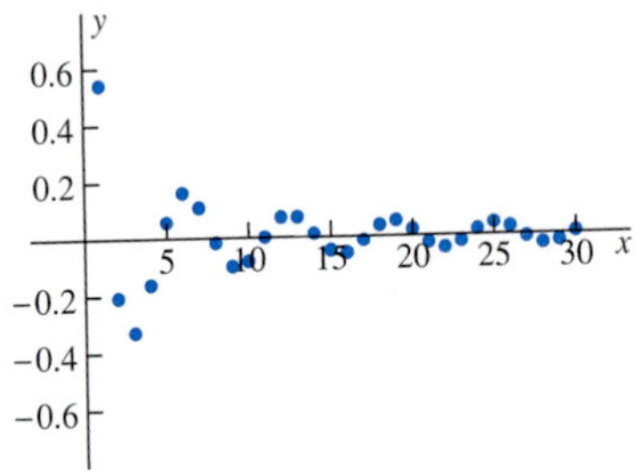

Fig. 11.2.4 The points $(n, (\cos n)/n)$ for $n = 1, 2, \ldots, 30$

EXAMPLE 7 Show that if $a > 0$, then $\lim_{n\to\infty} \sqrt[n]{a} = 1.$

Solution We apply the substitution law with $f(x) = a^x$ and $A = 0$. Because $1/n \to 0$ as $n \to \infty$ and f is continuous at $x = 0$, this gives

$$\lim_{n\to\infty} a^{1/n} = a^0 = 1.$$ ■

EXAMPLE 8 The limit laws and the continuity of $f(x) = \sqrt{x}$ at $x = 4$ yield

$$\lim_{n\to\infty}\sqrt{\frac{4n-1}{n+1}} = \left(\lim_{n\to\infty}\frac{4-\dfrac{1}{n}}{1+\dfrac{1}{n}}\right)^{1/2} = \sqrt{4} = 2.$$

EXAMPLE 9 Show that if $|r| < 1$, then $\lim_{n\to\infty} r^n = 0$.

Solution Because $|r^n| = |(-r)^n|$, we may assume that $0 < r < 1$. Then $1/r = 1 + a$ for some number $a > 0$, so the binomial formula yields

$$\frac{1}{r^n} = (1+a)^n = 1 + na + \{\text{positive terms}\} > 1 + na;$$

$$0 < r^n < \frac{1}{1+na}.$$

Now $1/(1+na) \to 0$ as $n \to \infty$. Therefore the squeeze law implies that $r^n \to 0$ as $n \to \infty$.

Figure 11.2.5 shows the graph of a function f such that $\lim_{x\to\infty} f(x) = L$. If the sequence $\{a_n\}$ is defined by the formula $a_n = f(n)$ for each positive integer n, then all the points $(n, f(n))$ lie on the graph of $y = f(x)$. It therefore follows from the definition of the limit of a function that $\lim_{n\to\infty} a_n = L$ as well.

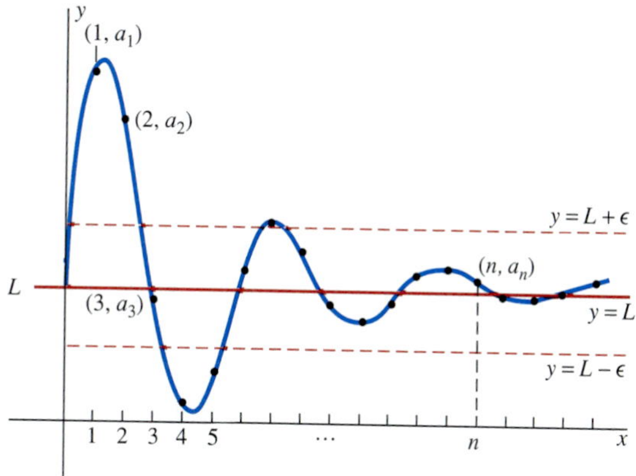

Fig. 11.2.5 If $\lim_{x\to\infty} f(x) = L$ and $a_n = f(n)$, then $\lim_{n\to\infty} a_n = L$.

Theorem 4 ***Limits of Functions and Sequences***

If $a_n = f(n)$ for each positive integer n, then

$$\lim_{x\to\infty} f(x) = L \quad \text{implies that} \quad \lim_{n\to\infty} a_n = L. \tag{8}$$

The converse of the statement in (8) is generally false. For example, take $f(x) = \sin \pi x$ and, for each positive integer n, let $a_n = f(n) = \sin n\pi$. Then

$$\lim_{n\to\infty} a_n = \lim_{n\to\infty} \sin n\pi = 0, \quad \text{but}$$

$$\lim_{x\to\infty} f(x) = \lim_{x\to\infty} \sin \pi x \quad \text{does not exist.}$$

Because of (8) we can use **l'Hôpital's rule for sequences:** If $a_n = f(n)$, $b_n = g(n)$, and $f(x)/g(x)$ has the indeterminate form ∞/∞ as $x \to \infty$, then

$$\lim_{n\to\infty} \frac{a_n}{b_n} = \lim_{x\to\infty} \frac{f(x)}{g(x)} = \lim_{x\to\infty} \frac{f'(x)}{g'(x)}, \tag{9}$$

provided that f and g satisfy the other hypotheses of l'Hôpital's rule, including the important assumption that the right-hand limit exists.

EXAMPLE 10 Show that $\displaystyle\lim_{n\to\infty} \frac{\ln n}{n} = 0.$

Solution The function $(\ln x)/x$ is defined for all $x \geqq 1$ and agrees with the given sequence $\{(\ln n)/n\}$ when $x = n$, a positive integer. Because $(\ln x)/x$ has the indeterminate form ∞/∞ as $x \to \infty$, l'Hôpital's rule gives

$$\lim_{n\to\infty} \frac{\ln n}{n} = \lim_{x\to\infty} \frac{\ln x}{x} = \lim_{x\to\infty} \frac{\frac{1}{x}}{1} = 0. \qquad \blacksquare$$

EXAMPLE 11 Show that $\displaystyle\lim_{n\to\infty} \sqrt[n]{n} = 1.$

Solution First we note that

$$\ln \sqrt[n]{n} = \ln n^{1/n} = \frac{\ln n}{n} \to 0 \qquad \text{as } n \to \infty,$$

by Example 10. By the substitution law with $f(x) = e^x$, this gives

$$\lim_{n\to\infty} n^{1/n} = \lim_{n\to\infty} \exp(\ln n^{1/n}) = e^0 = 1. \qquad \blacksquare$$

EXAMPLE 12 Find $\displaystyle\lim_{n\to\infty} \frac{3n^3}{e^{2n}}.$

Solution We apply l'Hôpital's rule repeatedly, although we must be careful at each intermediate step to verify that we still have an indeterminate form. Thus we find that

$$\lim_{n\to\infty} \frac{3n^3}{e^{2n}} = \lim_{x\to\infty} \frac{3x^3}{e^{2x}} = \lim_{x\to\infty} \frac{9x^2}{2e^{2x}} = \lim_{x\to\infty} \frac{18x}{4e^{2x}} = \lim_{x\to\infty} \frac{18}{8e^{2x}} = 0. \qquad \blacksquare$$

Bounded Monotonic Sequences

The set of all *rational* numbers has by itself all of the most familiar elementary algebraic properties of the entire real number system. To guarantee the existence of irrational numbers, we must assume in addition a "completeness property" of the real numbers. Otherwise, the real line might have "holes" where the irrational numbers ought to be. One way of stating this completeness property is in terms of the convergence of an important type of sequence, a bounded monotonic sequence.

The sequence $\{a_n\}_1^\infty$ is said to be **increasing** if

$$a_1 \leqq a_2 \leqq a_3 \leqq \cdots \leqq a_n \leqq \cdots$$

and **decreasing** if

$$a_1 \geqq a_2 \geqq a_3 \geqq \cdots \geqq a_n \geqq \cdots.$$

The sequence $\{a_n\}$ is **monotonic** if it is either increasing or decreasing. The sequence $\{a_n\}$ is **bounded** if there is a number M such that $|a_n| \leqq M$ for all n. The following assertion may be taken to be an axiom for the real number system.

Bounded Monotonic Sequence Property

Every bounded monotonic infinite sequence converges—that is, has a finite limit.

Suppose, for example, that the increasing sequence $\{a_n\}_1^\infty$ is bounded above by a number M, meaning that $a_n \leqq M$ for all $n \geqq 1$. Because the sequence is also bounded below (by a_1, for instance), the bounded monotonic sequence property implies that

$$\lim_{n\to\infty} a_n = A \quad \text{for some real number } A \leqq M,$$

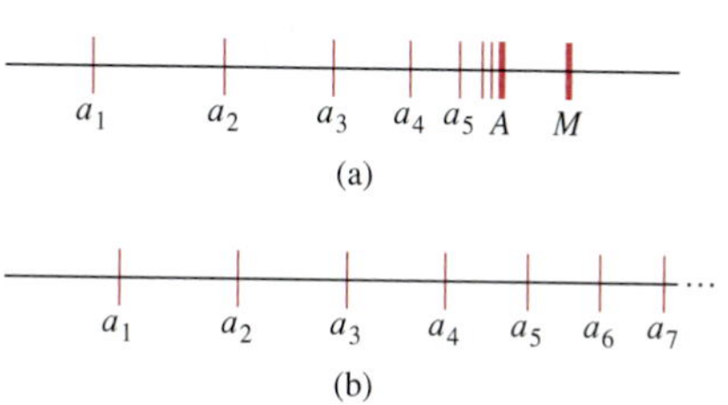

Fig. 11.2.6 (a) If the increasing sequence $\{a_n\}$ is bounded above by M, then its terms "pile up" at some point $A \leqq M$. (b) If the sequence is unbounded, then its terms "keep going" and diverge to infinity.

as in Fig. 11.2.6(a). If the increasing sequence $\{a_n\}$ is not bounded above, then it follows that

$$\lim_{n\to\infty} a_n = +\infty$$

as in Fig. 11.2.6(b) (see Problem 52). Figure 11.2.7 illustrates the graph of a typical bounded increasing sequence, with the heights of the points (n, a_n) steadily rising toward A.

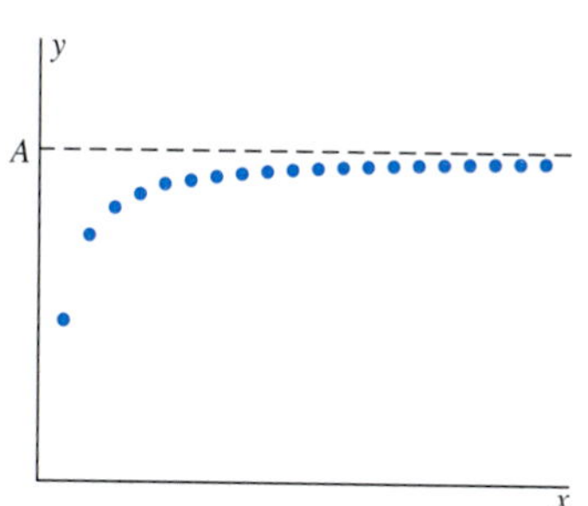

Fig. 11.2.7 Graph of a bounded increasing sequence with limit A

EXAMPLE 13 Investigate the sequence $\{a_n\}$ that is defined recursively by

$$a_1 = \sqrt{6}, \qquad a_{n+1} = \sqrt{6 + a_n} \qquad \text{for } n \geqq 1. \tag{10}$$

Solution The first four terms of this sequence are

$$\sqrt{6}, \quad \sqrt{6+\sqrt{6}}, \quad \sqrt{6+\sqrt{6+\sqrt{6}}}, \quad \sqrt{6+\sqrt{6+\sqrt{6+\sqrt{6}}}}. \tag{11}$$

If the sequence $\{a_n\}$ converges, then its limit A would seem to be the natural interpretation of the infinite expression

$$\sqrt{6+\sqrt{6+\sqrt{6+\sqrt{6+\cdots}}}}.$$

A calculator gives 2.449, 2.907, 2.984, and 2.997 for the approximate values of the terms in (11). This suggests that the sequence may be bounded above by $M = 3$. Indeed, if we assume that a particular term a_n satisfies the inequality $a_n < 3$, then it follows that

$$a_{n+1} = \sqrt{6 + a_n} < \sqrt{6 + 3} = 3;$$

that is, $a_{n+1} < 3$ as well. Can you see that this implies that *all* terms of the sequence are less than 3? (If there were a first term not less than 3, then its predecessor would be less than 3, and we would have a contradiction. This is a "proof by mathematical induction.")

To apply the bounded monotonic sequence property in order to conclude that the sequence $\{a_n\}$ converges, it remains to show that it is an increasing sequence. But

$$(a_{n+1})^2 - (a_n)^2 = (6 + a_n) - (a_n)^2 = (2 + a_n)(3 - a_n) > 0$$

because $a_n < 3$. Because all terms of the sequence are positive (why?), it therefore follows that $a_{n+1} > a_n$ for all $n \geqq 1$, as desired.

Now that we know that the limit A of the sequence $\{a_n\}$ exists, we can write

$$A = \lim_{n\to\infty} a_{n+1} = \lim_{n\to\infty} \sqrt{6 + a_n} = \sqrt{6 + A},$$

and thus $A^2 = 6 + A$. The roots of this quadratic equation are -2 and 3. Because $A > 0$ (why?), we conclude that $A = \lim_{n\to\infty} a_n = 3$, and so

$$\sqrt{6 + \sqrt{6 + \sqrt{6 + \sqrt{6 + \cdots}}}} = 3. \tag{12}$$

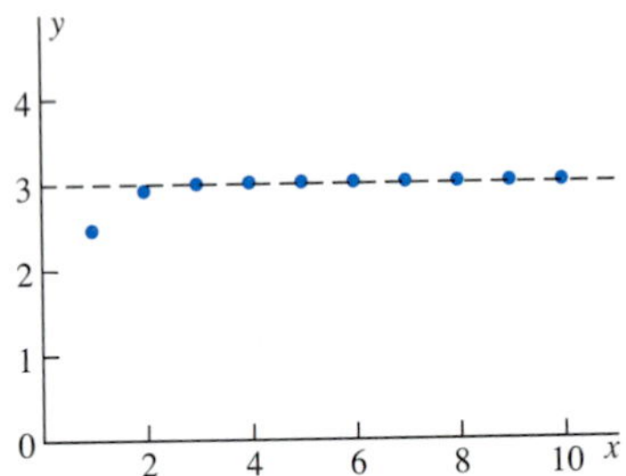

Fig. 11.2.8 Graph of the sequence of Example 13

The graph in Fig. 11.2.8 of the first ten terms of the sequence $\{a_n\}$ shows that the convergence to its limit 3 is quite rapid. ■

To indicate what the bounded monotonic sequence property has to do with the "completeness property" of the real numbers, in Problem 63 we outline a proof, using this property, of the existence of the number $\sqrt{2}$. In Problems 61 and 62, we outline a proof of the equivalence of the bounded monotonic sequence property and another common statement of the completeness of the real number system—the *least upper bound property.*

11.2 PROBLEMS

In Problems 1 through 8, find a pattern in the sequence with given terms a_1, a_2, a_3, a_4, and (assuming that it continues as indicated) write a formula for the general term a_n of the sequence.

1. $1, 4, 9, 16, \dots$

2. $2, 7, 12, 17, \dots$

3. $\frac{1}{3}, \frac{1}{9}, \frac{1}{27}, \frac{1}{81}, \dots$

4. $1, -\frac{1}{2}, \frac{1}{4}, -\frac{1}{8}, \dots$

5. $\frac{1}{2}, \frac{1}{5}, \frac{1}{8}, \frac{1}{11}, \dots$

6. $\frac{1}{2}, \frac{1}{5}, \frac{1}{10}, \frac{1}{17}, \dots$

7. $0, 2, 0, 2, \dots$

8. $10, 5, 10, 5, \dots$

In Problems 9 through 42, determine whether the sequence $\{a_n\}$ converges, and find its limit if it does converge.

9. $a_n = \dfrac{2n}{5n - 3}$

10. $a_n = \dfrac{1 - n^2}{2 + 3n^2}$

11. $a_n = \dfrac{n^2 - n + 7}{2n^3 + n^2}$

12. $a_n = \dfrac{n^3}{10n^2 + 1}$

13. $a_n = 1 + \left(\frac{9}{10}\right)^n$

14. $a_n = 2 - \left(-\frac{1}{2}\right)^n$

15. $a_n = 1 + (-1)^n$

16. $a_n = \dfrac{1 + (-1)^n}{\sqrt{n}}$

17. $a_n = \dfrac{1 + (-1)^n\sqrt{n}}{\left(\frac{3}{2}\right)^n}$

18. $a_n = \dfrac{\sin n}{3^n}$

19. $a_n = \dfrac{\sin^2 n}{\sqrt{n}}$

20. $a_n = \sqrt{\dfrac{2 + \cos n}{n}}$

21. $a_n = n \sin \pi n$

22. $a_n = n \cos \pi n$

23. $a_n = \pi^{-(\sin n)/n}$

24. $a_n = 2^{\cos \pi n}$

25. $a_n = \dfrac{\ln n}{\sqrt{n}}$

26. $a_n = \dfrac{\ln 2n}{\ln 3n}$

27. $a_n = \dfrac{(\ln n)^2}{n}$

28. $a_n = n \sin\left(\dfrac{1}{n}\right)$

29. $a_n = \dfrac{\tan^{-1} n}{n}$

30. $a_n = \dfrac{n^3}{e^{n/10}}$

31. $a_n = \dfrac{2^n + 1}{e^n}$

32. $a_n = \dfrac{\sinh n}{\cosh n}$

33. $a_n = \left(1 + \dfrac{1}{n}\right)^n$

34. $a_n = (2n + 5)^{1/n}$

35. $a_n = \left(\dfrac{n - 1}{n + 1}\right)^n$

36. $a_n = (0.001)^{-1/n}$

37. $a_n = \sqrt[n]{2^{n+1}}$

38. $a_n = \left(1 - \dfrac{2}{n^2}\right)^n$

39. $a_n = \left(\dfrac{2}{n}\right)^{3/n}$

40. $a_n = (-1)^n(n^2 + 1)^{1/n}$

41. $a_n = \left(\dfrac{2 - n^2}{3 + n^2}\right)^n$

42. $a_n = \dfrac{\left(\frac{2}{3}\right)^n}{1 - \sqrt[n]{n}}$

In Problems 43 through 50, investigate the given sequence $\{a_n\}$ numerically or graphically. Formulate a reasonable guess for the value of its limit. Then apply limit laws to verify that your guess is correct.

43. $a_n = \dfrac{n - 2}{n + 13}$

44. $a_n = \dfrac{2n + 3}{5n - 17}$

45. $a_n = \sqrt{\dfrac{4n^2 + 7}{n^2 + 3n}}$

46. $a_n = \left(\dfrac{n^3 - 5}{8n^3 + 7n}\right)^{1/3}$

47. $a_n = e^{-1/\sqrt{n}}$

48. $a_n = n \sin \dfrac{2}{n}$

49. $a_n = 4 \tan^{-1} \dfrac{n - 1}{n + 1}$

50. $a_n = 3 \sin^{-1} \sqrt{\dfrac{3n - 1}{4n + 1}}$

51. Prove that if $\lim_{n\to\infty} a_n = A \neq 0$, then the sequence $\{(-1)^n a_n\}$ diverges.

52. Prove that if the increasing sequence $\{a_n\}$ is not bounded, then $\lim_{n\to\infty} a_n = +\infty$. (It's largely a matter of saying precisely what this means.)

53. Suppose that $A > 0$. Given $x_1 \neq 0$ but otherwise arbitrary, define the sequence $\{x_n\}$ recursively by

$$x_{n+1} = \frac{1}{2}\cdot\left(x_n + \frac{A}{x_n}\right) \qquad \text{if } n \geqq 1.$$

Prove that if $L = \lim_{n\to\infty} x_n$ exists, then $L = \pm\sqrt{A}$.

54. Suppose that A is a fixed real number. Given $x_1 \neq 0$ but otherwise arbitrary, define the sequence $\{x_n\}$ recursively by

$$x_{n+1} = \frac{1}{3}\cdot\left(2x_n + \frac{A}{(x_n)^2}\right) \qquad \text{if } n \geqq 1.$$

Prove that if $L = \lim_{n\to\infty} x_n$ exists, then $L = \sqrt[3]{A}$.

55. (a) Suppose that every newborn pair of rabbits becomes productive after two months and thereafter gives birth to a new pair of rabbits every month. If we begin with a single newborn pair of rabbits, denote by F_n the total number of pairs of rabbits we have after n months. Explain carefully why $\{F_n\}$ is the Fibonacci sequence of Example 2. (b) If, instead, every newborn pair of rabbits becomes productive after three months, denote by $\{G_n\}$ the number of pairs of rabbits we have after n months. Give a recursive definition of the sequence $\{G_n\}$ and calculate its first ten terms.

56. Let $\{F_n\}$ be the Fibonacci sequence of Example 2, and assume that

$$\tau = \lim_{n\to\infty} \frac{F_{n+1}}{F_n}$$

exists. (It does.) Show that $\tau = \frac{1}{2}(1 + \sqrt{5})$. (*Suggestion:* Write $a_n = F_n/F_{n-1}$ and show that $a_{n+1} = 1 + (1/a_n)$.)

57. Let the sequence $\{a_n\}$ be defined recursively as follows:

$$a_1 = 2; \qquad a_{n+1} = \tfrac{1}{2}(a_n + 4) \qquad \text{for } n \geqq 1.$$

(a) Prove by induction on n that $a_n < 4$ for each n and that $\{a_n\}$ is an increasing sequence. (b) Find the limit of this sequence.

58. Investigate as in Example 13 the sequence $\{a_n\}$ that is defined recursively by

$$a_1 = \sqrt{2}, \qquad a_{n+1} = \sqrt{2 + a_n} \qquad \text{for } n \geqq 1.$$

In particular, show that

$$\sqrt{2 + \sqrt{2 + \sqrt{2 + \sqrt{2 + \cdots}}}} = 2.$$

Verify the results stated in Problems 59 and 60.

59. $\sqrt{20 + \sqrt{20 + \sqrt{20 + \sqrt{20 + \cdots}}}} = 5.$

60. $\sqrt{90 + \sqrt{90 + \sqrt{90 + \sqrt{90 + \cdots}}}} = 10.$

Problems 61 and 62 deal with the *least upper bound* property of the real numbers: If the nonempty set S of real numbers has an upper bound, then S has a least upper bound. The number M is an **upper bound** for the set S if $x \leqq M$ for all x in S. The upper bound L of S is a **least upper bound** for S if no number smaller than L is an upper bound for S. You can easily show that if the set S has least upper bounds L_1 and L_2, then $L_1 = L_2$; in other words, if a least upper bound for a set exists, then it is unique.

61. Prove that the least upper bound property implies the bounded monotonic sequence property. (*Suggestion:* If $\{a_n\}$ is a bounded increasing sequence and A is the least upper bound of the set $\{a_n : n \geqq 1\}$ of terms of the sequence, you can prove that $A = \lim_{n\to\infty} a_n$.)

62. Prove that the bounded monotonic sequence property implies the least upper bound property. (*Suggestion:* For each positive integer n, let a_n be the least integral multiple of $1/10^n$ that is an upper bound of the set S. Prove that $\{a_n\}$ is a bounded decreasing sequence and then that $A = \lim_{n\to\infty} a_n$ is a least upper bound for S.)

63. For each positive integer n, let a_n be the largest integral multiple of $1/10^n$ such that $a_n^2 \leqq 2$. (a) Prove that $\{a_n\}$ is a bounded increasing sequence, so $A = \lim_{n\to\infty} a_n$ exists. (b) Prove that if $A^2 > 2$, then $a_n^2 > 2$ for n sufficiently large. (c) Prove that if $A^2 < 2$, then $a_n^2 < B$ for some number $B < 2$ and all sufficiently large n. (d) Conclude that $A^2 = 2$.

64. Investigate the sequence $\{a_n\}$, where

$$a_n = [\![n + \tfrac{1}{2} + \sqrt{n}]\!].$$

You may need a computer or programmable calculator to discover what is remarkable about this sequence.

11.2 PROJECT: NESTED RADICALS AND CONTINUED FRACTIONS

This project is an investigation of the relation

$$\sqrt{q + p\sqrt{q + p\sqrt{q + p\sqrt{q + \cdots}}}} = p + \cfrac{q}{p + \cfrac{q}{p + \cfrac{q}{p + \cdots}}}, \tag{1}$$

where p and q are the last two nonzero digits in your student I.D. number. We ask not only whether Eq. (1) could possibly be true, but also what it means.

1. Define the sequence $\{a_n\}$ recursively by

$$a_1 = \sqrt{q} \quad \text{and} \quad a_{n+1} = \sqrt{q + pa_n} \qquad \text{for } n \geqq 1.$$

To investigate the convergence of this sequence, approximate successive terms numerically by entering the TI graphing calculator commands

```
√q→A
√(q + p*A) → A        (re-enter repeatedly)
```

or equivalent computer algebra system commands. Does the sequence appear to converge? Assuming that it does, write the first several terms of the sequence symbolically, and conclude that $A = \lim_{n\to\infty} a_n$ is a natural intepretation of the *nested radical* on the left-hand side in Eq. (1). Finally, show that A is the positive solution of the quadratic equation $x^2 - px - q = 0$. Does the quadratic formula then yield a result consistent with your numerical evidence?

2. Define the sequence $\{b_n\}$ recursively by

$$b_1 = p \quad \text{and} \quad b_{n+1} = p + \frac{q}{b_n} \qquad \text{for } n \geqq 1.$$

To investigate its convergence, approximate successive terms numerically by entering the TI graphing calculator commands

```
p →B
p + q/B → B        (re-enter repeatedly)
```

or equivalent computer algebra system commands. Does the sequence appear to converge? Assuming that it does, write the first several terms of the sequence symbolically, and conclude that $B = \lim_{n\to\infty} b_n$ is a natural interpretation of the *continued fraction* on the right-hand side in Eq. (1). Finally, show that B is also the positive solution of the quadratic equation $x^2 - px - q = 0$, and thereby conclude that Eq. (1) is indeed true.

11.3 INFINITE SERIES AND CONVERGENCE

An **infinite series** is an expression of the form

$$\sum_{n=1}^{\infty} a_n = a_1 + a_2 + a_3 + \cdots + a_n + \cdots, \tag{1}$$

where $\{a_n\}$ is an infinite sequence of real numbers. The number a_n is called the ***n*th term** of the series. The symbol $\sum_{n=1}^{\infty} a_n$ is simply an abbreviation for the right-hand side of Eq. (1). In this section we discover what is meant by the **sum** of an infinite series.

EXAMPLE 1 Consider the infinite series

$$\sum_{n=1}^{\infty} \frac{1}{2^n} = \frac{1}{2} + \frac{1}{4} + \frac{1}{8} + \frac{1}{16} + \cdots + \frac{1}{2^n} + \cdots, \tag{2}$$

which was mentioned in Section 11.1; its nth term is $a_n = 1/2^n$. Although we cannot literally add an infinite number of terms, we can add any finite number of the terms in Eq. (2). For instance, the sum of the first five terms is

$$\frac{1}{2} + \frac{1}{4} + \frac{1}{8} + \frac{1}{16} + \frac{1}{32} = \frac{31}{32} = 0.96875.$$

n	Sum of first n terms
5	0.96875000
10	0.99902344
15	0.99996948
20	0.99999905
25	0.99999997

Fig. 11.3.1 Sums of terms in the infinite series of Example 1

We could add five more terms, then five more, and so forth. The table in Fig. 11.3.1 shows what happens. It appears that the sums get closer and closer to 1 as we add more and more terms. If indeed this is so, then it is natural to say that the sum of the (whole) infinite series in Eq. (2) is 1 and hence to write

$$\sum_{n=1}^{\infty} \frac{1}{2^n} = \frac{1}{2} + \frac{1}{4} + \frac{1}{8} + \frac{1}{16} + \cdots + \frac{1}{2^n} + \cdots = 1.$$ ■

Motivated by Example 1, we introduce the *partial sums* of the general infinite series in Eq. (1). The **nth partial sum** S_n of the series is the sum of its first n terms:

$$S_n = a_1 + a_2 + a_3 + \cdots + a_n = \sum_{k=1}^{n} a_k. \tag{3}$$

Thus each infinite series has not only an infinite sequence of terms, but also an **infinite sequence of partial sums** $S_1, S_2, S_3, \ldots, S_n, \ldots,$ where

$$S_1 = a_1,$$
$$S_2 = a_1 + a_2,$$
$$S_3 = a_1 + a_2 + a_3,$$
$$\vdots$$
$$S_{10} = a_1 + a_2 + a_3 + a_4 + a_5 + a_6 + a_7 + a_8 + a_9 + a_{10},$$

and so forth. We define the sum of the infinite series to be the limit of its sequence of partial sums, provided that this limit exists.

Definition *The Sum of an Infinite Series*

We say that the infinite series

$$\sum_{n=1}^{\infty} a_n \qquad \textbf{converges} \text{ (or is } \textbf{convergent}\text{)}$$

with **sum** S provided that the limit of its sequence of partial sums,

$$S = \lim_{n\to\infty} S_n, \tag{4}$$

exists (and is finite). Otherwise we say that the series **diverges** (or is **divergent**). If a series diverges, then it has no sum.

Thus an infinite series is a limit of finite sums,

$$S = \sum_{n=1}^{\infty} a_n = \lim_{N\to\infty} \sum_{n=1}^{N} a_n,$$

provided that this limit exists.

EXAMPLE 1 (continued) Show that the series

$$\sum_{n=1}^{\infty} \left(\frac{1}{2}\right)^n = \frac{1}{2} + \frac{1}{4} + \frac{1}{8} + \frac{1}{16} + \cdots$$

converges, and find its sum.

Solution The first four partial sums are

$$S_1 = \frac{1}{2}, \quad S_2 = \frac{3}{4}, \quad S_3 = \frac{7}{8}, \quad \text{and} \quad S_4 = \frac{15}{16}.$$

It seems likely that $S_n = (2^n - 1)/2^n$, and indeed this follows easily by induction because

$$S_{n+1} = S_n + \frac{1}{2^{n+1}} = \frac{2^n - 1}{2^n} + \frac{1}{2^{n+1}} = \frac{2^{n+1} - 2 + 1}{2^{n+1}} = \frac{2^{n+1} - 1}{2^{n+1}}.$$

Hence the sum of the given series is

$$S = \lim_{n\to\infty} S_n = \lim_{n\to\infty} \frac{2^n - 1}{2^n} = \lim_{n\to\infty} \left(1 - \frac{1}{2^n}\right) = 1.$$

The graph in Fig. 11.3.2 illustrates the convergence of the partial sums to the number 1. ■

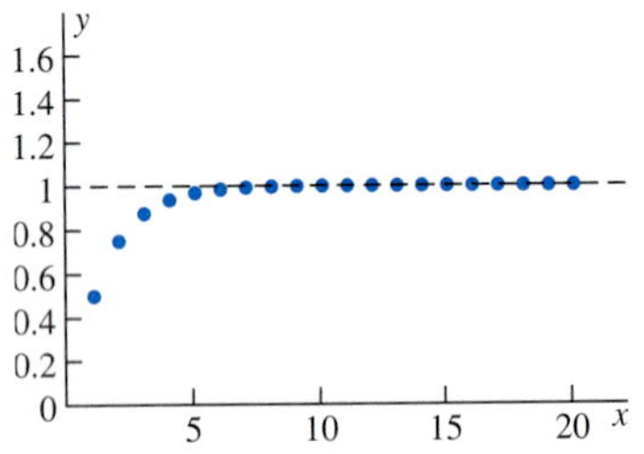

Fig. 11.3.2 Graph of the first 20 partial sums of the infinite series in Example 1

EXAMPLE 2 Show that the series

$$\sum_{n=1}^{\infty} (-1)^{n+1} = 1 - 1 + 1 - 1 + \cdots$$

diverges.

Solution The sequence of partial sums of this series is

$$1, 0, 1, 0, 1, \ldots,$$

which has no limit. Therefore the series diverges. ■

EXAMPLE 3 Show that the infinite series

$$\sum_{n=1}^{\infty} \frac{1}{n(n+1)}$$

converges, and find its sum.

Solution We need a formula for the nth partial sum S_n so that we can evaluate its limit as $n \to \infty$. To find such a formula, we begin with the observation that the nth term of the series is

$$a_n = \frac{1}{n(n+1)} = \frac{1}{n} - \frac{1}{n+1}.$$

(In more complicated cases, such as those in Problems 50 through 55, such a decomposition can be obtained by the method of partial fractions.) It follows that the sum of the first n terms of the given series is

$$\begin{aligned} S_n &= \left(1 - \frac{1}{2}\right) + \left(\frac{1}{2} - \frac{1}{3}\right) + \left(\frac{1}{3} - \frac{1}{4}\right) + \left(\frac{1}{4} - \frac{1}{5}\right) + \cdots + \left(\frac{1}{n} - \frac{1}{n+1}\right) \\ &= 1 - \frac{1}{n+1} = \frac{n}{n+1}. \end{aligned}$$

Hence

$$\sum_{n=1}^{\infty} \frac{1}{n(n+1)} = \lim_{n\to\infty} \frac{n}{n+1} = 1.$$ ■

The sum for S_n in Example 3, called a *telescoping sum*, provides a way to find the sums of certain series. The series in Examples 1 and 2 are examples of a more common and more important type of series, the *geometric series.*

Definition *Geometric Series*

The series $\sum_{n=0}^{\infty} a_n$ is said to be a **geometric series** if each term after the first is a fixed multiple of the term immediately before it. That is, there is a number r, called the **ratio** of the series, such that

$$a_{n+1} = ra_n \qquad \text{for all } n \geqq 0.$$

If we write $a = a_0$ for the initial constant term, then $a_1 = ar$, $a_2 = ar^2$, $a_3 = ar^3$, and so forth. Thus every geometric series takes the form

$$a + ar + ar^2 + ar^3 + \cdots = \sum_{n=0}^{\infty} ar^n. \tag{5}$$

Note that the summation begins at $n = 0$ (rather than at $n = 1$). It is therefore convenient to regard the sum

$$S_n = a(1 + r + r^2 + r^3 + \cdots + r^n)$$

of the first $n + 1$ terms as the nth partial sum of the series.

EXAMPLE 4 The infinite series

$$\sum_{n=0}^{\infty} \frac{2}{3^n} = 2 + \frac{2}{3} + \frac{2}{9} + \cdots + \frac{2}{3^n} + \cdots$$

is a geometric series whose first term is $a = 2$ and whose ratio is $r = \frac{1}{3}$. ■

Theorem 1 *The Sum of a Geometric Series*

If $|r| < 1$, then the geometric series in Eq. (5) converges, and its sum is

$$S = \sum_{n=0}^{\infty} ar^n = \frac{a}{1-r}. \tag{6}$$

If $|r| \geqq 1$ and $a \neq 0$, then the geometric series diverges.

Proof If $r = 1$, then $S_n = (n+1)a$, so the series certainly diverges if $a \neq 0$. If $r = -1$ and $a \neq 0$, then the series diverges by an argument like that in Example 2. So we may suppose that $|r| \neq 1$. Then the elementary identity

$$1 + r + r^2 + r^3 + \cdots + r^n = \frac{1 - r^{n+1}}{1 - r}$$

follows if we multiply each side by $1 - r$. Hence the nth partial sum of the geometric series is

$$S_n = a(1 + r + r^2 + r^3 + \cdots + r^n) = a\left(\frac{1}{1-r} - \frac{r^{n+1}}{1-r}\right).$$

If $|r| < 1$, then $r^{n+1} \to 0$ as $n \to \infty$, by Example 9 in Section 11.2. So in this case the geometric series converges to

$$S = \lim_{n\to\infty} a \cdot \left(\frac{1}{1-r} - \frac{r^{n+1}}{1-r}\right) = \frac{a}{1-r}.$$

But if $|r| > 1$, then $\lim_{n\to\infty} r^{n+1}$ does not exist, so $\lim_{n\to\infty} S_n$ does not exist. This establishes the theorem. ■

EXAMPLE 5 With $a = 1$ and $r = -\frac{2}{3}$, we find that

$$1 - \frac{2}{3} + \frac{4}{9} - \frac{8}{27} + \cdots = \sum_{n=0}^{\infty}\left(-\frac{2}{3}\right)^n = \frac{1}{1-(-\frac{2}{3})} = \frac{3}{5}.$$

The graph in Fig. 11.3.3 shows the partial sums of this series approaching its sum $\frac{3}{5}$ alternately from above and below. ■

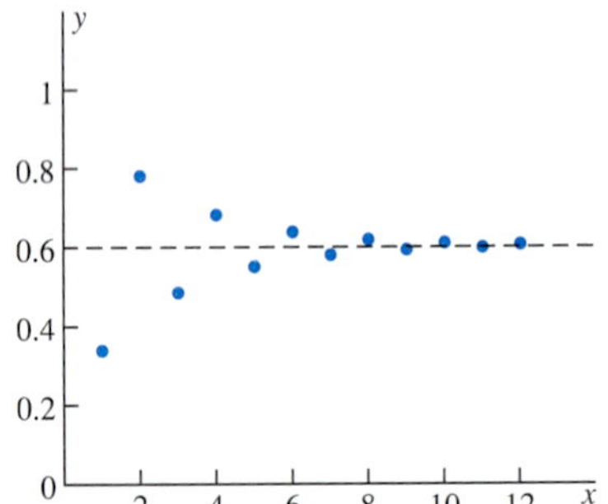

Fig. 11.3.3 Graph of the first dozen partial sums of the infinite series in Example 5

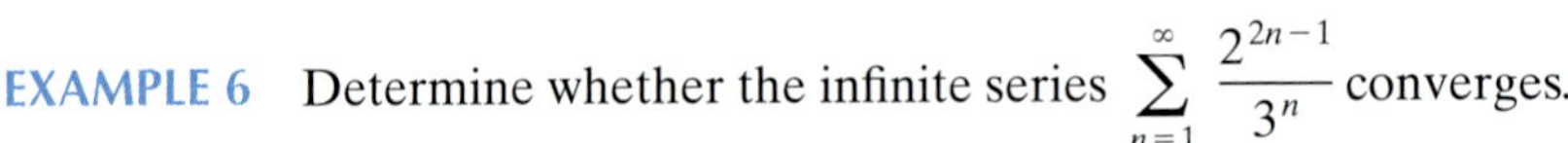

EXAMPLE 6 Determine whether the infinite series $\sum_{n=1}^{\infty} \frac{2^{2n-1}}{3^n}$ converges.

Solution If we write this series in the form

$$\sum_{n=1}^{\infty} \frac{2^{2n-1}}{3^n} = \frac{2}{3} + \frac{8}{9} + \frac{32}{27} + \frac{128}{81} + \cdots = \frac{2}{3}\left(1 + \frac{4}{3} + \frac{16}{9} + \frac{64}{27} + \cdots\right),$$

then we recognize it as a geometric series with $a = \frac{2}{3}$ and $r = \frac{4}{3}$. Because $r > 1$, the second part of Theorem 1 implies that this series diverges. ■

Theorem 2 implies that the operations of addition and of multiplication by a constant can be carried out term by term in the case of *convergent series.* Because the sum of an infinite series is the limit of its sequence of partial sums, this theorem follows immediately from the limit laws for sequences (Theorem 1 of Section 11.2).

Theorem 2 ***Termwise Addition and Multiplication***

If the series $A = \sum a_n$ and $B = \sum b_n$ converge to the indicated sums and c is a constant, then the series $\sum(a_n + b_n)$ and $\sum ca_n$ also converge, with sums

1. $\sum(a_n + b_n) = A + B$;
2. $\sum ca_n = cA$.

The geometric series in Eq. (6) may be used to find the rational number represented by a given infinite repeating decimal.

EXAMPLE 7

$$0.55555\ldots = \frac{5}{10} + \frac{5}{100} + \frac{5}{1000} + \cdots = \frac{5}{10}\left(1 + \frac{1}{10} + \frac{1}{100} + \cdots\right)$$

$$= \sum_{n=0}^{\infty} \frac{5}{10}\left(\frac{1}{10}\right)^n = \frac{\frac{5}{10}}{1-\frac{1}{10}} = \frac{5}{10}\cdot\frac{10}{9} = \frac{5}{9}.$$

In a more complicated situation, we may need to use the termwise algebra of Theorem 2:

$$0.7282828\ldots = \frac{7}{10} + \frac{28}{10^3} + \frac{28}{10^5} + \frac{28}{10^7} + \cdots$$

$$= \frac{7}{10} + \frac{28}{10^3}\left(1 + \frac{1}{10^2} + \frac{1}{10^4} + \cdots\right)$$

$$= \frac{7}{10} + \frac{28}{1000}\sum_{n=0}^{\infty}\left(\frac{1}{100}\right)^n = \frac{7}{10} + \frac{28}{1000}\left(\frac{1}{1-\frac{1}{100}}\right)$$

$$= \frac{7}{10} + \frac{28}{1000}\cdot\frac{100}{99} = \frac{7}{10} + \frac{28}{990} = \frac{721}{990}.$$

This technique can be used to show that every repeated infinite decimal represents a rational number. Consequently, the decimal expansions of irrational numbers such as π, e, and $\sqrt{2}$ must be nonrepeating as well as infinite. Conversely, if p and q are integers with $q \neq 0$, then long division of q into p yields a repeating decimal expansion for the rational number p/q because such a division can yield at each stage only q possible different remainders. ■

EXAMPLE 8 Suppose that Paul and Mary toss a fair six-sided die in turn until one of them wins by getting the first six. If Paul tosses first, calculate the probability that he will win the game.

Solution Because the die is fair, the probability that Paul gets a six on the first round is $\frac{1}{6}$. The probability that he gets the game's first six on the second round is $(\frac{5}{6})^2(\frac{1}{6})$—the product of the probability $(\frac{5}{6})^2$ that neither Paul nor Mary rolls a six in the first round and the probability $\frac{1}{6}$ that Paul rolls a six in the second round. Paul's probability p of getting the first six in the game is the *sum* of his probabilities of getting it in the first round, in the second round, in the third round, and so on. Hence

$$p = \frac{1}{6} + \left(\frac{5}{6}\right)^2\left(\frac{1}{6}\right) + \left(\frac{5}{6}\right)^2\left(\frac{5}{6}\right)^2\left(\frac{1}{6}\right) + \cdots$$

$$= \frac{1}{6}\left[1 + \left(\frac{5}{6}\right)^2 + \left(\frac{5}{6}\right)^4 + \cdots\right]$$

$$= \frac{1}{6}\cdot\frac{1}{1-(\frac{5}{6})^2} = \frac{1}{6}\cdot\frac{36}{11} = \frac{6}{11}.$$

Because he has the advantage of tossing first, Paul has more than the fair probability $\frac{1}{2}$ of getting the first six and thus winning the game. ■

Theorem 3 is often useful in showing that a given series does *not* converge.

Theorem 3 ***The nth-Term Test for Divergence***

If either

$$\lim_{n\to\infty} a_n \neq 0$$

or this limit does not exist, then the infinite series $\sum a_n$ diverges.

Proof We want to show under the stated hypothesis that the series $\sum a_n$ diverges. It suffices to show that *if* the series $\sum a_n$ does converge, then $\lim_{n\to\infty} a_n = 0$. So suppose that $\sum A_n$ converges with sum $S = \lim_{n\to\infty} S_n$, where

$$S_n = a_1 + a_2 + a_3 + \cdots + a_n$$

is the nth partial sum of the series. Because $a_n = S_n - S_{n-1}$,

$$\lim_{n\to\infty} a_n = \lim_{n\to\infty} (S_n - S_{n-1}) = (\lim_{n\to\infty} S_n) - (\lim_{n\to\infty} S_{n-1}) = S - S = 0.$$

Consequently, if $\lim_{n\to\infty} a_n \neq 0$, then the series $\sum a_n$ diverges. ■

Remark It is important to remember also the *contrapositive* of the nth-term divergence test: *If the infinite series* $\sum a_n$ *converges with sum S, then its sequence* $\{a_n\}$ *of terms converges to* 0. Thus we have *two* sequences associated with the single infinite series $\sum a_n$: its sequence $\{a_n\}$ of *terms* and its sequence $\{S_n\}$ of *partial sums.* And (assuming that the series converges to S) these two sequences have generally different limits:

$$\lim_{n\to\infty} a_n = 0 \quad \text{and} \quad \lim_{n\to\infty} S_n = S.$$

EXAMPLE 9 The series

$$\sum_{n=1}^{\infty} (-1)^{n-1} n^2 = 1 - 4 + 9 - 16 + 25 - \cdots$$

diverges because $\lim_{n\to\infty} a_n$ does not exist, whereas the series

$$\sum_{n=1}^{\infty} \frac{n}{3n+1} = \frac{1}{4} + \frac{2}{7} + \frac{3}{10} + \frac{4}{13} + \cdots$$

diverges because

$$\lim_{n\to\infty} \frac{n}{3n+1} = \frac{1}{3} \neq 0.$$ ■

Warning The converse of Theorem 3 is *false*! The condition

$$\lim_{n\to\infty} a_n = 0$$

is necessary *but not sufficient* to guarantee convergence of the series

$$\sum_{n=1}^{\infty} a_n.$$

That is, a series may satisfy the condition $a_n \to 0$ as $n \to \infty$ and yet diverge. An important example of a divergent series with terms that approach zero is the **harmonic series**

$$\sum_{n=1}^{\infty} \frac{1}{n} = 1 + \frac{1}{2} + \frac{1}{3} + \frac{1}{4} + \frac{1}{5} + \cdots. \tag{7}$$

Theorem 4

The harmonic series diverges.

PROOF The nth term of the harmonic series in Eq. (7) is $a_n = 1/n$, and Fig. 11.3.4 shows the graph of the related function $f(x) = 1/x$ on the interval $1 \leqq x \leqq n + 1$. For each integer k, $1 \leqq k \leqq n$, we have erected on the subinterval $[k, k + 1]$ a rectangle with height $f(k) = 1/k$. All of these n rectangles have base length 1, and their respective heights are the successive terms $1, 1/2, 1/3, \dots, 1/n$ of the harmonic series. Hence the sum of their areas is the nth partial sum

$$S_n = 1 + \frac{1}{2} + \frac{1}{3} + \frac{1}{4} + \cdots + \frac{1}{n}$$

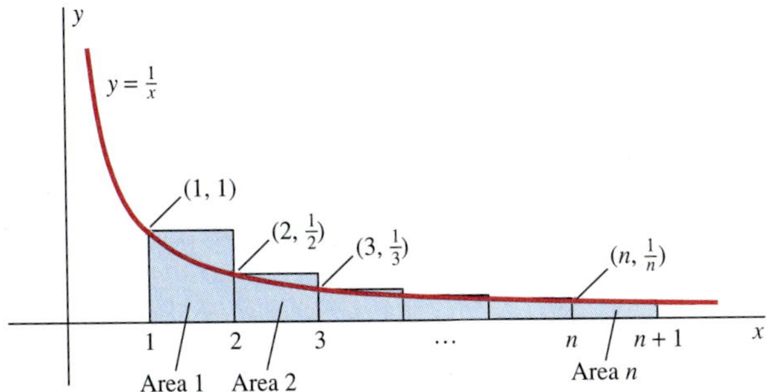

Fig. 11.3.4 Idea of the proof of Theorem 4

of the series. Because these rectangles circumscribe the area under the curve $y = 1/x$ from $x = 1$ to $x = n + 1$, we therefore see that S_n must exceed this area. That is,

$$S_n > \int_1^{n+1} \frac{1}{x}\,dx = \Big[\ln x\Big]_1^{n+1} = \ln(n + 1).$$

But $\ln(n + 1)$ takes on arbitrarily large positive values with increasing n. Because $S_n > \ln(n + 1)$, it follows that the partial sums of the harmonic series also take on arbitrarily large positive values. Now the terms of the harmonic series are positive, so its sequence of partial sums is increasing. We may therefore conclude that $S_n \to +\infty$ as $n \to +\infty$, and hence that the harmonic series diverges. ■

If the sequence of partial sums of the series $\sum a_n$ diverges to infinity, then we say that the series **diverges to infinity,** and we write

$$\sum_{n=1}^{\infty} a_n = \infty.$$

The series $\sum (-1)^{n+1}$ of Example 2 is a series that diverges but does not diverge to infinity. In the nineteenth century it was common to say that such a series was divergent by *oscillation*; today we say merely that it diverges.

Our proof of Theorem 4 shows that

$$\sum_{n=1}^{\infty} \frac{1}{n} = \infty.$$

But the partial sums of the harmonic series diverge to infinity very slowly. If N_A denotes the smallest integer such that

$$\sum_{n=1}^{N_A} \frac{1}{n} \geqq A,$$

then with the aid of a programmable calculator you can verify that $N_5 = 83$. With the aid of a computer and refinements of estimates like those in the proof of Theorem 4, one can show that

$$N_{10} = 12367,$$

$$N_{20} = 272{,}400{,}600,$$

$$N_{100} \approx 1.5 \times 10^{43}, \quad \text{and}$$

$$N_{1000} \approx 1.1 \times 10^{434}.$$

Thus you would need to add more than a quarter of a billion terms of the harmonic series to get a partial sum that exceeds 20. At this point each of the next few terms would be approximately $0.000000004 = 4 \times 10^{-9}$. The number of terms you'd have to add to reach 1000 is far greater than the estimated number of elementary particles in the entire universe (10^{80}). If you enjoy such large numbers, see the article "Partial sums of infinite series, and how they grow," by R. P. Boas, Jr., in *American Mathematical Monthly* **84** (1977): 237–248.

Theorem 5 says that if two infinite series have the same terms from some point on, then either both series converge or both series diverge. The proof is left for Problem 63.

Theorem 5 ***Series That Are Eventually the Same***

If there exists a positive integer k such that $a_n = b_n$ for all $n > k$, then the series $\sum a_n$ and $\sum b_n$ either both converge or both diverge.

It follows that a *finite* number of terms can be changed, deleted from, or adjoined to an infinite series without altering its convergence or divergence (although the *sum* of a convergent series will generally be changed by such alternations). In particular, taking $b_n = 0$ for $n \leqq k$ and $b_n = a_n$ for $n > k$, we see that the series

$$\sum_{n=1}^{\infty} a_n = a_1 + a_2 + a_3 + \cdots + a_k + a_{k+1} + \cdots$$

and the series

$$\sum_{n=k+1}^{\infty} a_n = a_{k+1} + a_{k+2} + a_{k+3} + a_{k+4} + \cdots$$

that is obtained by deleting its first k terms either both converge or both diverge.

11.3 PROBLEMS

In Problems 1 through 37, determine whether the given infinite series converges or diverges. If it converges, find its sum.

1. $1 + \frac{1}{3} + \frac{1}{9} + \cdots + \frac{1}{3^n} + \cdots$

2. $1 + e^{-1} + e^{-2} + e^{-3} + \cdots + e^{-n} + \cdots$

3. $1 + 3 + 5 + 7 + \cdots + (2n - 1) + \cdots$

4. $\frac{1}{2} + \frac{1}{\sqrt{2}} + \frac{1}{\sqrt[3]{2}} + \cdots + \frac{1}{\sqrt[n]{2}} + \cdots$

5. $1 - 2 + 4 - 8 + 16 - \cdots + (-2)^n + \cdots$

6. $1 - \frac{1}{4} + \frac{1}{16} - \cdots + \left(-\frac{1}{4}\right)^n + \cdots$

7. $4 + \frac{4}{3} + \frac{4}{9} + \frac{4}{27} + \cdots + \frac{4}{3^n} + \cdots$

8. $\frac{1}{3} + \frac{2}{9} + \frac{4}{27} + \frac{8}{81} + \cdots + \frac{2^{n-1}}{3^n} + \cdots$

9. $1 + (1.01) + (1.01)^2 + (1.01)^3 + \cdots + (1.01)^n + \cdots$

10. $1 + \dfrac{1}{\sqrt{2}} + \dfrac{1}{\sqrt[3]{3}} + \cdots + \dfrac{1}{\sqrt[n]{n}} + \cdots$

11. $\displaystyle\sum_{n=0}^{\infty} \frac{(-1)^n n}{n+1}$

12. $\displaystyle\sum_{n=1}^{\infty} \left(\frac{e}{10}\right)^n$

13. $\displaystyle\sum_{n=0}^{\infty} (-1)^n \left(\frac{3}{e}\right)^n$

14. $\displaystyle\sum_{n=0}^{\infty} \frac{3^n - 2^n}{4^n}$

15. $\displaystyle\sum_{n=1}^{\infty} (\sqrt{2})^{1-n}$

16. $\displaystyle\sum_{n=1}^{\infty} \left(\frac{2}{n} - \frac{1}{2^n}\right)$

17. $\displaystyle\sum_{n=1}^{\infty} \frac{n}{10n+17}$

18. $\displaystyle\sum_{n=1}^{\infty} \frac{\sqrt{n}}{\ln(n+1)}$

19. $\displaystyle\sum_{n=1}^{\infty} (5^{-n} - 7^{-n})$

20. $\displaystyle\sum_{n=0}^{\infty} \frac{1}{1 + \left(\frac{9}{10}\right)^n}$

21. $\displaystyle\sum_{n=1}^{\infty} \left(\frac{e}{\pi}\right)^n$

22. $\displaystyle\sum_{n=1}^{\infty} \left(\frac{\pi}{e}\right)^n$

23. $\displaystyle\sum_{n=0}^{\infty} \left(\tfrac{100}{99}\right)^n$

24. $\displaystyle\sum_{n=0}^{\infty} \left(\tfrac{99}{100}\right)^n$

25. $\displaystyle\sum_{n=0}^{\infty} \frac{1 + 2^n + 3^n}{5^n}$

26. $\displaystyle\sum_{n=0}^{\infty} \frac{1 + 2^n + 5^n}{3^n}$

27. $\displaystyle\sum_{n=0}^{\infty} \frac{7 \cdot 5^n + 3 \cdot 11^n}{13^n}$

28. $\displaystyle\sum_{n=1}^{\infty} \sqrt[n]{2}$

29. $\displaystyle\sum_{n=1}^{\infty} \left[\left(\tfrac{7}{11}\right)^n - \left(\tfrac{3}{5}\right)^n\right]$

30. $\displaystyle\sum_{n=1}^{\infty} \frac{2n}{\sqrt{4n^2+3}}$

31. $\displaystyle\sum_{n=1}^{\infty} \frac{n^2 - 1}{3n^2 + 1}$

32. $\displaystyle\sum_{n=1}^{\infty} \sin^n 1$

33. $\displaystyle\sum_{n=1}^{\infty} \tan^n 1$

34. $\displaystyle\sum_{n=1}^{\infty} (\arcsin 1)^n$

35. $\displaystyle\sum_{n=1}^{\infty} (\arctan 1)^n$

36. $\displaystyle\sum_{n=1}^{\infty} \arctan n$

37. $\displaystyle\sum_{n=2}^{\infty} \frac{1}{n \ln n}$ (*Suggestion:* Mimic the proof of Theorem 4 to show divergence.)

38. Use the method of Example 6 to verify that
(a) $0.666\,666\,666\ldots = \frac{2}{3}$; (b) $0.111\,111\,111\ldots = \frac{1}{9}$;
(c) $0.249\,999\,999\ldots = \frac{1}{4}$; (d) $0.999\,999\,999\ldots = 1$.

In Problems 39 through 43, find the rational number represented by the given repeating decimal.

39. $0.4747\,4747\ldots$

40. $0.2525\,2525\ldots$

41. $0.123\,123\,123\ldots$

42. $0.3377\,3377\,3377\ldots$

43. $3.14159\,14159\,14159\ldots$

In Problems 44 through 49, find the set of all those values of x for which the given series is a convergent geometric series, then express the sum of the series as a function of x.

44. $\displaystyle\sum_{n=1}^{\infty} (2x)^n$

45. $\displaystyle\sum_{n=1}^{\infty} \left(\frac{x}{3}\right)^n$

46. $\displaystyle\sum_{n=1}^{\infty} (x-1)^n$

47. $\displaystyle\sum_{n=1}^{\infty} \left(\frac{x-2}{3}\right)^n$

48. $\displaystyle\sum_{n=1}^{\infty} \left(\frac{x^2}{x^2+1}\right)^n$

49. $\displaystyle\sum_{n=1}^{\infty} \left(\frac{5x^2}{x^2+16}\right)^n$

In Problems 50 through 55, express the nth partial sum of the infinite series as a telescoping sum (as in Example 3) and thereby find the sum of the series if it converges.

50. $\displaystyle\sum_{n=1}^{\infty} \frac{1}{4n^2 - 1}$

51. $\displaystyle\sum_{n=1}^{\infty} \frac{1}{9n^2 + 3n - 2}$

52. $\displaystyle\sum_{n=1}^{\infty} \ln \frac{n+1}{n}$

53. $\displaystyle\sum_{n=1}^{\infty} \frac{1}{16n^2 - 8n - 3}$

54. $\displaystyle\sum_{n=1}^{\infty} \frac{1}{n(n+2)}$

55. $\displaystyle\sum_{n=2}^{\infty} \frac{1}{n^2 - 1}$

In Problems 56 through 60, use a computer algebra system to find the partial fraction decomposition of the general term, then apply the methods of Problems 50 through 55 to sum the series.

56. $\displaystyle\sum_{n=1}^{\infty} \frac{2n+1}{n^2(n+1)^2}$

57. $\displaystyle\sum_{n=1}^{\infty} \frac{6n^2 + 2n - 1}{n(n+1)(4n^2 - 1)}$

58. $\displaystyle\sum_{n=1}^{\infty} \frac{2}{n(n+1)(n+2)}$

59. $\displaystyle\sum_{n=1}^{\infty} \frac{6}{n(n+1)(n+2)(n+3)}$

60. $\displaystyle\sum_{n=3}^{\infty} \frac{6n}{n^4 - 5n^2 + 4}$

61. Prove: If $\sum a_n$ diverges and c is a nonzero constant, then $\sum ca_n$ diverges.

62. Suppose that $\sum a_n$ converges and that $\sum b_n$ diverges. Prove that $\sum(a_n + b_n)$ diverges.

63. Let S_n and T_n denote the nth partial sums of $\sum a_n$ and $\sum b_n$, respectively. Suppose that k is a fixed positive integer and that $a_n = b_n$ for all $n \geqq k$. Show that $S_n - T_n = S_k - T_k$ for all $n > k$. Hence prove Theorem 5.

64. A ball has *bounce coefficient* $r < 1$ if, when it is dropped from a height h, it bounces back to a height of rh (Fig. 11.3.5). Suppose that such a ball is dropped from the initial height a and subsequently bounces infinitely many times. Use a geometric series to show that the total up-and-down distance it travels in all its bouncing is

$$D = a \cdot \frac{1+r}{1-r}.$$

Note that D is *finite*.

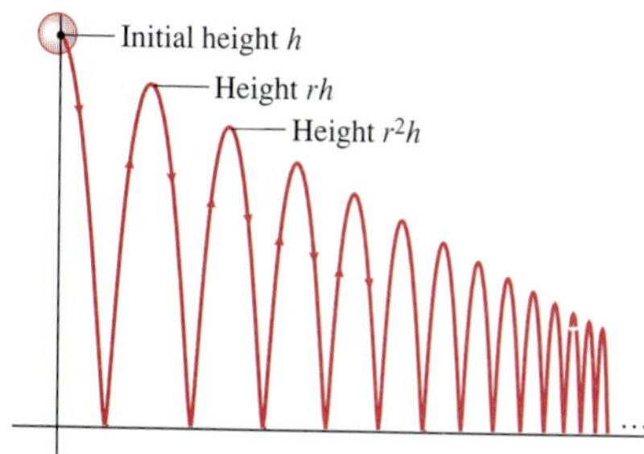

Fig. 11.3.5 Successive bounces of the ball of Problems 64 and 65

65. A ball with bounce coefficient $r = 0.64$ (see Problem 64) is dropped from an initial height of $a = 4$ ft. Use a geometric series to compute the total time required for it to complete its infinitely many bounces. The time required for a ball to drop h feet (from rest) is $\sqrt{2h/g}$ seconds, where $g = 32$ ft/s^2.

66. Suppose that the government spends \$1 billion and that each recipient of a fraction of this wealth spends 90% of the dollars that he or she receives. In turn, the secondary recipients spend 90% of the dollars they receive, and so on. How much total spending thereby results from the original injection of \$1 billion into the economy?

67. A tank initially contains a mass M_0 of air. Each stroke of a vacuum pump removes 5% of the air in the container. Compute (a) the mass M_n of air remaining in the tank after n strokes of the pump; (b) $\lim_{n\to\infty} M_n$.

68. Paul and Mary toss a fair coin in turn until one of them wins the game by getting the first head. Calculate for each the probability that he or she wins the game.

69. Peter, Paul, and Mary toss a fair coin in turn until one of them wins by getting the first head. Calculate for each the probability that he or she wins the game. Check your answer by verifying that the sum of the three probabilities is 1.

70. Peter, Paul, and Mary roll a fair die in turn until one of them wins by getting the first six. Calculate for each the probability that he or she wins the game. Check your answer by verifying that the sum of the three probabilities is 1.

71. A pane of a certain type of glass reflects half the incident light, absorbs one-fourth, and transmits one-fourth. A window is made of two panes of this glass separated by a small space (Fig. 11.3.6). What fraction of the incident light I is transmitted by the double window?

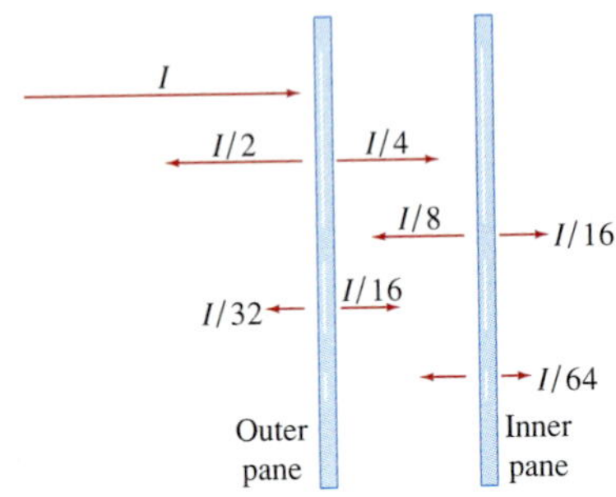

Fig. 11.3.6 The double-pane window of Problem 71

72. Criticize the following evaluation of the sum of an infinite series:

Let $x = 1 - 2 + 4 - 8 + 16 - 32 + 64 - \cdots$.

Then $2x = 2 - 4 + 8 - 16 + 32 - 64 + \cdots$.

Add the equations to obtain $3x = 1$. Thus $x = \frac{1}{3}$, and "therefore"

$$1 - 2 + 4 - 8 + 16 - 32 + 64 - \cdots = \tfrac{1}{3}.$$

11.3 PROJECT: NUMERICAL SUMMATION AND GEOMETRIC SERIES

With a modern calculator, the calculation of a new term of an infinite series and its addition to the preceding partial sum to get the new partial sum is a one-liner. Suppose, for instance, that we want to check numerically the fact that

$$\sum_{n=0}^{\infty} \left(\frac{1}{5}\right)^n = \frac{5}{4}.$$

First we enter the ratio $R = \frac{1}{5}$, the initial index $N = 0$, and the initial partial sum $S = 0$ of this geometric series:

```
1/5→R : 0→N : 1→S
```

If we next type

```
N+1→N : R∧N→T : S+T→S
```

then each press of the **ENTER** key executes these instructions in order, and thus calculates the next index N, the next term T, and the next partial sum S. Thus just six key presses yield the successive partial sums 1.2000, 1.2400, 1.2480, 1.2496, 1.2499, and 1.2500.

Any computer algebra system includes a sum function that can be used to calculate partial sums directly. If $a(n)$ denotes the nth term of Σa_n, then the typical commands

```
sum( a(k), k=1..n )            Maple
Sum[ a[k], {k,1,n} ]           Mathematica
sum (seq( a(k), n, 1, k ))     TI-92
```

calculate the nth partial sum of the infinite series.

INVESTIGATION A Calculate partial sums of the geometric series

$$\sum_{n=0}^{\infty} r^n$$

with $r = 0.2, 0.5, 0.75, 0.9$, and 0.99. For each value of r, calculate the partial sums S_n with $n = 10, 20, 30, \dots$, continuing until two successive results agree to four or five decimal places. (For $r = 0.9$ and 0.99, you may decide to use $n = 100, 200, 300, \dots$.) How does the apparent rate of convergence—as measured by the number of terms required for the desired accuracy—depend on the value of r?

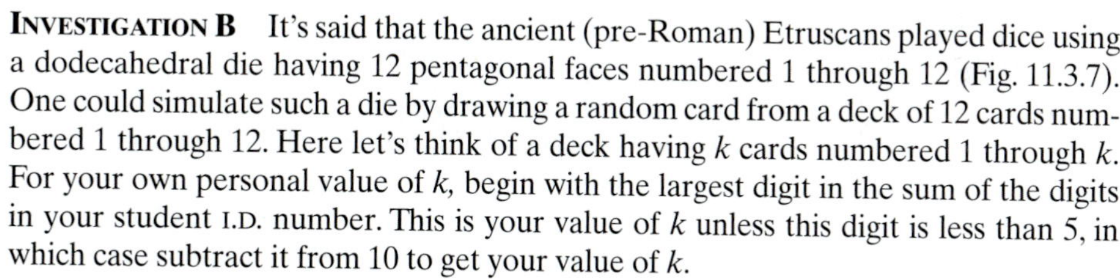

INVESTIGATION B It's said that the ancient (pre-Roman) Etruscans played dice using a dodecahedral die having 12 pentagonal faces numbered 1 through 12 (Fig. 11.3.7). One could simulate such a die by drawing a random card from a deck of 12 cards numbered 1 through 12. Here let's think of a deck having k cards numbered 1 through k. For your own personal value of k, begin with the largest digit in the sum of the digits in your student I.D. number. This is your value of k unless this digit is less than 5, in which case subtract it from 10 to get your value of k.

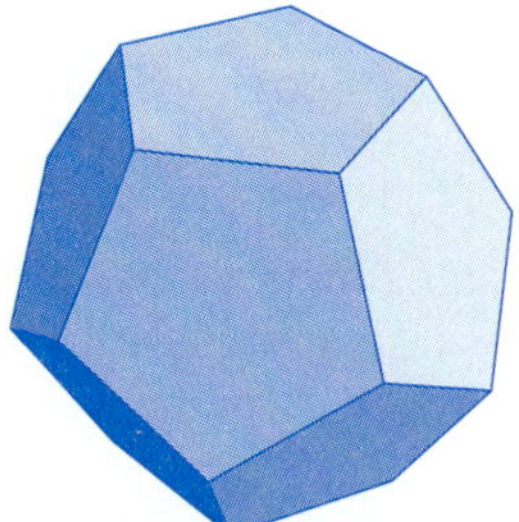

Fig. 11.3.7 The 12-sided dodecahedron

(a) John and Mary draw alternately from a shuffled deck of k cards. The first one to draw an ace—the card numbered 1—wins. Assume that John draws first. Use the geometric series sum formula to calculate (both as a rational number and as a four-place decimal) the probability J that John wins, and similarly the probability M that Mary wins. Check that $J + M = 1$.

(b) Now John, Mary, and Paul draw alternately from the deck of k cards. Calculate separately their respective probabilities of winning, given that John draws first and Mary draws second. Check that $J + M + P = 1$.

11.4 TAYLOR SERIES AND TAYLOR POLYNOMIALS

The infinite series we studied in Section 11.3 have *constant* terms, and the sum of such a series (assuming it converges) is a *number.* In contrast, much of the practical importance of infinite series derives from the fact that many functions have useful representations as infinite series with *variable* terms.

EXAMPLE 1 If we write $r = x$ for the ratio in a geometric series, then Theorem 1 in Section 11.3 gives the infinite series representation

$$\frac{1}{1-x} = \sum_{n=0}^{\infty} x^n = 1 + x + x^2 + x^3 + \cdots \tag{1}$$

of the function $f(x) = 1/(1 - x)$. That is, for each fixed number x with $|x| < 1$, the infinite series in Eq. (1) converges to the number $1/(1 - x)$. The nth partial sum

$$S_n(x) = 1 + x + x^2 + x^3 + \cdots + x^n \tag{2}$$

of the geometric series in Eq. (1) is now an nth-degree *polynomial* that approximates the function $f(x) = 1/(1 - x)$. The convergence of the infinite series for $|x| < 1$ suggests that the approximation

$$\frac{1}{1-x} \approx 1 + x + x^2 + x^3 + \cdots + x^n \tag{3}$$

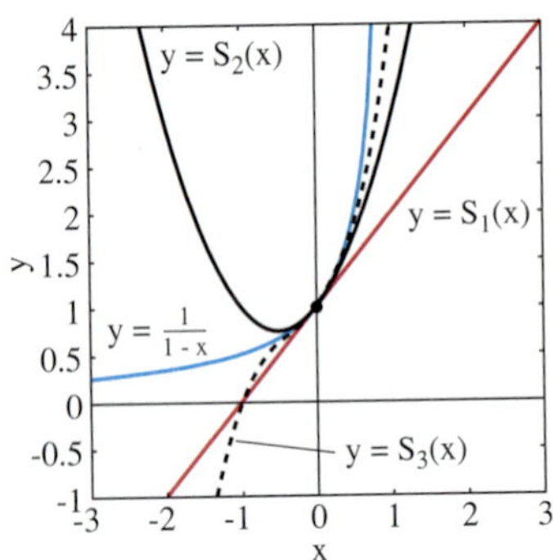

Fig. 11.4.1 The first three polynomials approximating $f(x) = 1/(1 - x)$ near $x = 0$.

should then be accurate if n is sufficiently large. Figure 11.4.1 shows the graphs of $1/(1 - x)$ and the three approximations $S_1(x)$, $S_2(x)$, and $S_3(x)$. It appears that the approximations are more accurate when n is larger and when x is closer to zero. ■

REMARK The approximation in (3) could be used to calculate numerical quotients with a calculator that has only $+$, $-$, $\times$ keys (but no $\div$ key). For instance,

$$\begin{aligned}\frac{329}{73} &= \frac{3.29}{0.73} = 3.29 \times \frac{1}{1 - 0.27}\\ &\approx (3.29)[1 + (0.27) + (0.27)^2 + \cdots + (0.27)^{10}]\\ &\approx (3.29)(1.36986); \quad \text{thus}\\ \frac{329}{73} &\approx 4.5068,\end{aligned}$$

accurate to four decimal places. This is a simple illustration of the use of polynomial approximation for numerical computation.

The definitions of the various elementary transcendental functions leave it unclear how to compute their values precisely, except at a few isolated points. For example,

$$\ln x = \int_1^x \frac{1}{t}\,dt \quad (x > 0)$$

by definition, so obviously $\ln 1 = 0$, but no other value of $\ln x$ is obvious. The natural exponential function is the inverse of $\ln x$, so it is clear that $e^0 = 1$, but it is not at all clear how to compute e^x for $x \neq 0$. Indeed, even such an innocent-looking expression as $\sqrt{x}$ is not computable (precisely and in a finite number of steps) unless x happens to be the square of a rational number.

But *any* value of a polynomial

$$P(x) = c_0 + c_1x + c_2x^2 + \cdots + c_nx^n$$

with known coefficients $c_0, c_1, c_2, \ldots, c_n$ is easy to calculate—as in the preceding remark, only addition and multiplication are required. One goal of this section is to use the fact that polynomial values are so readily computable to help us calculate approximate values of functions such as $\ln x$ and e^x.

Polynomial Approximations

Suppose that we want to calculate (or, at least, closely approximate) a specific value $f(x_0)$ of a given function f. It would suffice to find a polynomial $P(x)$ with a graph that is very close to that of f on some interval containing x_0. For then we could use the value $P(x_0)$ as an approximation to the actual value of $f(x_0)$. Once we know how to find such an approximating polynomial $P(x)$, our next question would be how accurately $P(x_0)$ approximates the desired value $f(x_0)$.

The simplest example of polynomial approximation is the linear approximation

$$f(x) \approx f(a) + f'(a)(x - a)$$

obtained by writing $\Delta x = x - a$ in the linear approximation formula, Eq. (3) of Section 4.2. The graph of the first-degree polynomial

$$P_1(x) = f(a) + f'(a)(x - a) \tag{4}$$

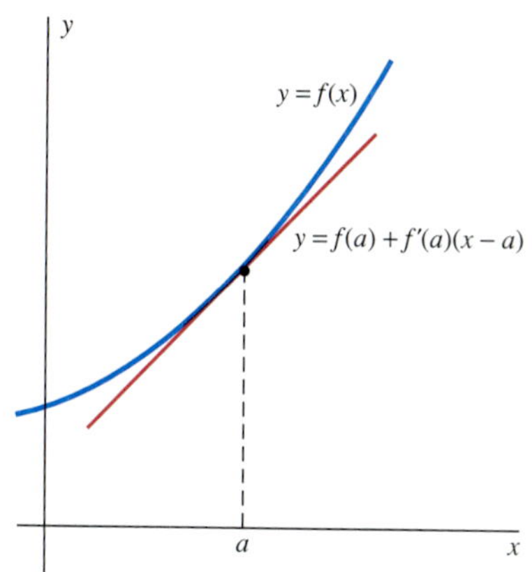

Fig. 11.4.2 The tangent line at $(a, f(a))$ is the best linear approximation to $y = f(x)$ near a.

is the line tangent to the curve $y = f(x)$ at the point $(a, f(a))$; see Fig. 11.4.2. This first-degree polynomial agrees with f and with its first derivative at $x = a$. That is,

$$P_1(a) = f(a) \quad \text{and} \quad P_1'(a) = f'(a).$$

EXAMPLE 2 Suppose that $f(x) = \ln x$ and that $a = 1$. Then $f(1) = 0$ and $f'(1) = 1$, so $P_1(x) = x - 1$. Hence we expect that $\ln x \approx x - 1$ for x near 1. With $x = 1.1$, we find that

$$P_1(1.1) = 0.1000, \quad \text{whereas} \quad \ln(1.1) \approx 0.0953.$$

The error in this approximation is about 5%.

To better approximate $\ln x$ near $x = 1$, let us look for a second-degree polynomial $P_2(x) = c_0 + c_1x + c_2x^2$ that has not only the same value and the same first derivative as does f at $x = 1$, but also has the same second derivative there: $P_2''(1) = f''(1) = -1$. To satisfy these conditions, we must have

$$P_2(1) = c_2 + c_1 + c_0 = 0,$$

$$P_2{}'(1) = 2c_2 + c_1 = 1, \quad \text{and}$$

$$P_2{}''(1) = 2c_2 = -1.$$

When we solve these equations, we find that $c_0 = -\frac{3}{2}$, $c_1 = 2$, and $c_2 = -\frac{1}{2}$, so

$$P_2(x) = -\tfrac{1}{2}x^2 + 2x - \tfrac{3}{2}.$$

With $x = 1.1$, we find that $P_2(1.1) = 0.0950$, which is accurate to three decimal places because $\ln(1.1) \approx 0.0953$. The graph of $y = P_2(x) = -\frac{1}{2}x^2 + 2x - \frac{3}{2}$ is a parabola through $(1, 0)$ with the same value, slope, *and curvature* there as $y = \ln x$ (Fig. 11.4.3).

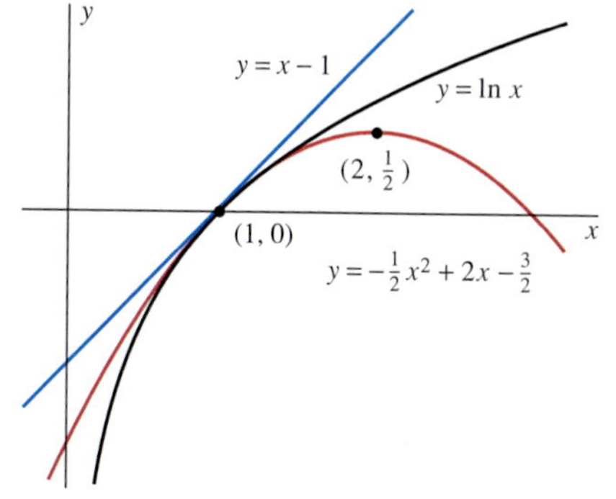

Fig. 11.4.3 The linear and parabolic approximations to $y = \ln x$ near the point $(1, 0)$ (Example 2)

The tangent line and the parabola used in the computations of Example 2 illustrate one general approach to polynomial approximation. To approximate the function $f(x)$ near $x = a$, we look for an nth-degree polynomial

$$P_n(x) = c_0 + c_1x + c_2x^2 + \cdots + c_nx^n$$

such that its value at a and the value of its first n derivatives at a agree with the corresponding values of f. That is, we require that

$$\begin{aligned} P_n(a) &= f(a), \\ P_n'(a) &= f'(a), \\ P_n''(a) &= f''(a), \\ &\vdots \\ P_n^{(n)}(a) &= f^{(n)}(a). \end{aligned} \tag{5}$$

We can use these $n + 1$ conditions to evaluate the values of the $n + 1$ coefficients $c_0, c_1, c_2, \ldots, c_n$.

The algebra involved is much simpler, however, if we begin with $P_n(x)$ expressed as an nth-degree polynomial in powers of $x - a$ rather than in powers of x:

$$P_n(x) = c_0 + c_1(x - a) + c_2(x - a)^2 + \cdots + c_n(x - a)^n. \tag{6}$$

Then substitution of $x = a$ into Eq. (3) yields

$$c_0 = P_n(a) = f(a)$$

by the first condition in Eq. (5). Substitution of $x = a$ into

$$P_n'(x) = c_1 + 2c_2(x - a) + 3c_3(x - a)^2 + \cdots + nc_n(x - a)^{n-1}$$

yields

$$c_1 = P_n'(a) = f'(a)$$

by the second condition in Eq. (5). Next, substitution of $x = a$ into

$$P_n''(x) = 2c_2 + 3 \cdot 2c_3(x - a) + \cdots + n(n - 1)c_n(x - a)^{n-2}$$

yields $2c_2 = P_n''(a) = f''(a)$, so

$$c_2 = \tfrac{1}{2} f''(a).$$

We continue this process to find $c_3, c_4, \ldots, c_n$. In general, the constant term in the kth derivative $P_n^{(k)}(x)$ is $k!c_k$, because it is the kth derivative of the kth-degree term $c_k(x - a)^k$ in $P_n(x)$:

$$P_n^{(k)}(x) = k!\,c_k + \{\text{powers of } x - a\}.$$

[Recall that $k! = 1 \cdot 2 \cdot 3 \cdots (k - 1) \cdot k$ denotes the *factorial* of the positive integer k, read "k factorial."] So when we substitute $x = a$ into $P_n^{(k)}(x)$, we find that

$$k!\,c_k = P_n^{(k)}(a) = f^{(k)}(a)$$

and thus that

$$c_k = \frac{f^{(k)}(a)}{k!} \tag{7}$$

for $k = 1, 2, 3, \ldots, n$.

Indeed, Eq. (7) holds also for $k = 0$ if we use the universal convention that $0! = 1$ and agree that the zeroth derivative $g^{(0)}$ of the function g is just g itself. With such conventions, our computations establish the following theorem.

Theorem 1 ***The nth-Degree Taylor Polynomial***

Suppose that the first n derivatives of the function $f(x)$ exist at $x = a$. Let $P_n(x)$ be the nth-degree polynomial

$$\begin{aligned} P_n(x) &= \sum_{k=0}^{n} \frac{f^{(k)}(a)}{k!}(x - a)^k \\ &= f(a) + f'(a)(x - a) + \frac{f''(a)}{2!}(x - a)^2 + \cdots + \frac{f^{(n)}(a)}{n!}(x - a)^n. \end{aligned} \tag{8}$$

Then the values of $P_n(x)$ and its first n derivatives agree, at $x = a$, with the values of f and its first n derivatives there. That is, the equations in (5) all hold.

The polynomial in Eq. (8) is called the **nth-degree Taylor polynomial of the function f at the point $x = a$.** Note that $P_n(x)$ is a polynomial in powers of $x - a$ rather than in powers of x. To use $P_n(x)$ effectively for the approximation of $f(x)$ near a, we must be able to compute the value $f(a)$ and the values of its derivatives $f'(a), f''(a)$, and so on, all the way to $f^{(n)}(a)$.

The line $y = P_1(x)$ is simply the line tangent to the curve $y = f(x)$ at the point $(a, f(a))$. Thus $y = f(x)$ and $y = P_1(x)$ have the same slope at this point. Now recall from Section 4.6 that the second derivative measures the way the curve $y = f(x)$ is bending as it passes through $(a, f(a))$. Therefore let us call $f''(a)$ the "concavity" of $y = f(x)$ at $(a, f(a))$. Then, because $P_2''(a) = f''(a)$, it follows that $y = P_2(x)$ has the same value, the same slope, *and* the same concavity at $(a, f(a))$ as does $y = f(x)$. Moreover, $P_3(x)$ and $f(x)$ will also have the same rate of change of concavity at $(a, f(a))$. Such observations suggest that the larger n is, the more closely the nth-degree Taylor polynomial will approximate $f(x)$ for x near a.

EXAMPLE 3 Find the nth-degree Taylor polynomial of $f(x) = \ln x$ at $a = 1$.

Solution The first few derivatives of $f(x) = \ln x$ are

$$f'(x) = \frac{1}{x}, \quad f''(x) = -\frac{1}{x^2}, \quad f^{(3)}(x) = \frac{2}{x^3}, \quad f^{(4)}(x) = -\frac{3!}{x^4}, \quad f^{(5)}(x) = \frac{4!}{x^5}.$$

The pattern is clear:

$$f^{(k)}(x) = (-1)^{k-1}\frac{(k-1)!}{x^k} \qquad \text{for } k \geqq 1.$$

Hence $f^{(k)}(1) = (-1)^{k-1}(k-1)!$, so Eq. (8) gives

$$P_n(x) = (x-1) - \frac{1}{2}(x-1)^2 + \frac{1}{3}(x-1)^3 - \frac{1}{4}(x-1)^4 + \cdots + \frac{(-1)^{n-1}}{n}(x-1)^n.$$

With $n = 2$, we obtain the quadratic polynomial

$$P_2(x) = (x-1) - \tfrac{1}{2}(x-1)^2 = -\tfrac{1}{2}x^2 + 2x - \tfrac{3}{2},$$

the same as in Example 2. With the third-degree Taylor polynomial

$$P_3(x) = (x-1) - \tfrac{1}{2}(x-1)^2 + \tfrac{1}{3}(x-1)^3,$$

we can go one step further in approximating $\ln(1.1) = 0.095310\ldots \approx 0.0953$. The value

$$P_3(1.1) = (0.1) - \tfrac{1}{2}(0.1)^2 + \tfrac{1}{3}(0.1)^3 \approx 0.095333 \approx 0.0953$$

is accurate to four decimal places (rounded). In Fig. 11.4.4 we see that, the higher the degree and the closer x is to 1, the more accurate the approximation $\ln x \approx P_n(x)$ appears to be. ■

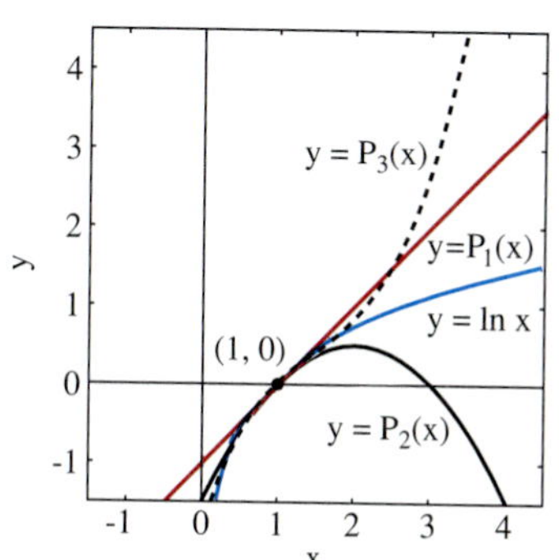

Fig. 11.4.4 The first three Taylor polynomials approximating $f(x) = \ln x$ near $x = 1$

In the common case $a = 0$, the nth-degree Taylor polynomial in Eq. (8) reduces to

$$P_n(x) = f(0) + f'(0) \cdot x + \frac{f''(0)}{2!}x^2 + \cdots + \frac{f^{(n)}(0)}{n!}x^n. \tag{9}$$

EXAMPLE 4 Find the nth-degree Taylor polynomial for $f(x) = e^x$ at $a = 0$.

Solution This is the easiest of all Taylor polynomials to compute, because $f^{(k)}(x) = e^x$ for all $k \geqq 0$. Hence $f^{(k)}(0) = 1$ for all $k \geqq 0$, so Eq. (9) yields

$$P_n(x) = 1 + x + \frac{x^2}{2!} + \frac{x^3}{3!} + \cdots + \frac{x^n}{n!}.$$ ■

The first few Taylor polynomials of the natural exponential function at $a = 0$ are therefore

$$P_0(x) = 1,$$

$$P_1(x) = 1 + x,$$

$$P_2(x) = 1 + x + \tfrac{1}{2}x^2,$$

$$P_3(x) = 1 + x + \tfrac{1}{2}x^2 + \tfrac{1}{6}x^3,$$

$$P_4(x) = 1 + x + \tfrac{1}{2}x^2 + \tfrac{1}{6}x^3 + \tfrac{1}{24}x^4,$$

$$P_5(x) = 1 + x + \tfrac{1}{2}x^2 + \tfrac{1}{6}x^3 + \tfrac{1}{24}x^4 + \tfrac{1}{120}x^5.$$

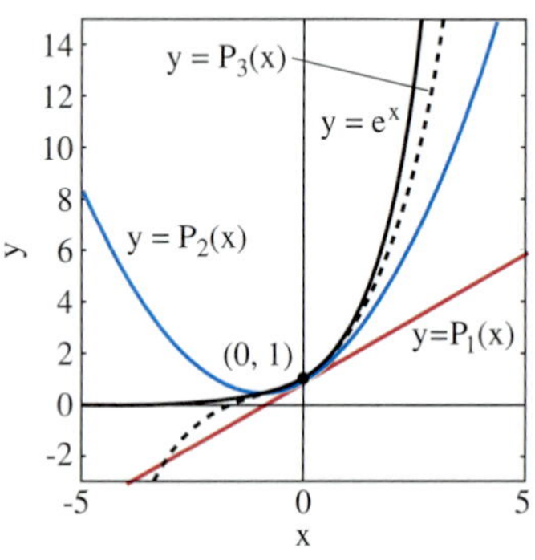

Fig. 11.4.5 The first three Taylor polynomials approximating $f(x) = e^x$ near $x = 0$

Figure 11.4.5 shows the graphs of $P_1(x)$, $P_2(x)$, and $P_3(x)$. The table in Fig. 11.4.6 shows how these polynomials approximate $f(x) = e^x$ for $x = 0.1$ and for $x = 0.5$. At least for these two values of x, the closer x is to $a = 0$, the more rapidly $P_n(x)$ appears to approach $f(x)$ as n increases.

$x = 0.1$

n	$P_n(x)$	e^x	$e^x - P_n(x)$
0	1.00000	1.10517	0.10517
1	1.10000	1.10517	0.00517
2	1.10500	1.10517	0.00017
3	1.10517	1.10517	0.00000
4	1.10517	1.10517	0.00000

$x = 0.5$

n	$P_n(x)$	e^x	$e^x - P_n(x)$
0	1.00000	1.64872	0.64872
1	1.50000	1.64872	0.14872
2	1.62500	1.64872	0.02372
3	1.64583	1.64872	0.00289
4	1.64844	1.64872	0.00028
5	1.64879	1.64872	0.00002

Fig. 11.4.6 Approximating $y = e^x$ with Taylor polynomials at $a = 0$

Taylor's Formula

The closeness with which the polynomial $P_n(x)$ approximates the function $f(x)$ is measured by the difference

$$R_n(x) = f(x) - P_n(x),$$

for which

$$f(x) = P_n(x) + R_n(x). \tag{10}$$

This difference $R_n(x)$ is called the ***n*th-degree remainder for $f(x)$ at $x = a$**. It is the *error* made if the value $f(x)$ is replaced with the approximation $P_n(x)$.

The theorem that lets us estimate the error, or remainder, $R_n(x)$ is called **Taylor's formula,** after Brook Taylor (1685–1731), a follower of Newton who introduced Taylor polynomials in an article published in 1715. The particular expression for $R_n(x)$ that we give next is called the *Lagrange form* for the remainder because it first appeared in 1797 in a book written by the French mathematician Joseph Louis Lagrange (1736–1813).

Theorem 2 ***Taylor's Formula***

Suppose that the $(n + 1)$th derivative of the function f exists on an interval containing the points a and b. Then

$$\begin{aligned} f(b) = f(a) + f'(a)(b - a) + \frac{f''(a)}{2!}(b - a)^2 \\ + \frac{f^{(3)}(a)}{3!}(b - a)^3 + \cdots + \frac{f^{(n)}(a)}{n!}(b - a)^n \\ + \frac{f^{(n+1)}(z)}{(n + 1)!}(b - a)^{n+1} \end{aligned} \tag{11}$$

for some number z between a and b.

Remark With $n = 0$, Eq. (11) reduces to the equation

$$f(b) = f(a) + f'(z)(b - a),$$

the conclusion of the mean value theorem (Section 4.3). Thus Taylor's formula is a far-reaching generalization of the mean value theorem of differential calculus.

A proof of Taylor's formula is given in Appendix J. If we replace b with x in Eq. (11), we get the ***n*th-degree Taylor formula with remainder at $x = a$**,

$$\begin{aligned} f(x) = f(a) + f'(a)(x - a) + \frac{f''(a)}{2!}(x - a)^2 + \frac{f^{(3)}(a)}{3!}(x - a)^3 \\ + \cdots + \frac{f^{(n)}(a)}{n!}(x - a)^n + \frac{f^{(n+1)}(z)}{(n + 1)!}(x - a)^{n+1}, \end{aligned} \tag{12}$$

where z is some number between a and x. Thus the nth-degree remainder term is

$$R_n(x) = \frac{f^{(n+1)}(z)}{(n + 1)!}(x - a)^{n+1}, \tag{13}$$

which is easy to remember—it's the same as the *last* term of $P_{n+1}(x)$, except that $f^{(n+1)}(a)$ is replaced with $f^{(n+1)}(z)$.

EXAMPLE 3 (continued) To estimate the accuracy of the approximation

$$\ln 1.1 \approx 0.095333,$$

we substitute $x = 1$ into the formula

$$f^{(k)}(x) = (-1)^{k-1}\frac{(k-1)!}{x^k}$$

for the kth derivative of $f(x) = \ln x$ and get

$$f^{(k)}(1) = (-1)^{k-1}(k-1)!.$$

Hence the third-degree Taylor formula *with remainder* at $x = 1$ is

$$\ln x = (x-1) - \frac{1}{2}(x-1)^2 + \frac{1}{3}(x-1)^3 - \frac{3!}{4!\,z^4}(x-1)^4$$

with z between $a = 1$ and x. With $x = 1.1$, this gives

$$\ln(1.1) \approx 0.095333 - \frac{(0.1)^4}{4z^4},$$

where $1 < z < 1.1$. The value $z = 1$ gives the largest possible magnitude $(0.1)^4/4 = 0.000025$ of the remainder term. It follows that

$$0.0953083 < \ln(1.1) < 0.0953334,$$

so we can conclude that $\ln(1.1) = 0.0953$ to four-place accuracy. ■

Taylor Series

If the function f has derivatives of all orders, then we can write Taylor's formula [Eq. (11)] with any degree n that we please. Ordinarily, the exact value of z in the Taylor remainder term in Eq. (13) is unknown. Nevertheless, we can sometimes use Eq. (13) to show that the remainder approaches zero as $n \to \infty$:

$$\lim_{n\to\infty} R_n(x) = 0 \tag{14}$$

for some particular *fixed* value of x. Then Eq. (10) gives

$$f(x) = \lim_{n\to\infty}[P_n(x) + R_n(x)] = \lim_{n\to\infty} P_n(x) = \lim_{n\to\infty}\sum_{k=0}^{n}\frac{f^{(k)}(a)}{k!}(x-a)^k;$$

that is,

$$f(x) = \sum_{k=0}^{\infty}\frac{f^{(k)}(x)}{k!}(x-a)^k. \tag{15}$$

The infinite series

$$\sum_{n=0}^{\infty}\frac{f^{(n)}(a)}{n!}(x-a)^n = f(a) + f'(a)(x-a) + \frac{f''(a)}{2!}(x-a)^2 + \cdots + \frac{f^{(n)}(a)}{n!}(x-a)^n + \cdots \tag{16}$$

is called the **Taylor series** of the function f at $x = a$. Its partial sums are the successive Taylor polynomials of f at $x = a$.

We can write the Taylor series of a function f without knowing that it converges. But if the limit in Eq. (14) can be established, then it follows as in Eq. (15) that the Taylor series in Eq. (16) actually converges to $f(x)$. If so, then we can approximate the value of $f(x)$ accurately by calculating the value of a Taylor polynomial of f of sufficiently high degree.

EXAMPLE 5 In Example 4 we noted that if $f(x) = e^x$, then $f^{(k)}(x) = e^x$ for all integers $k \geqq 0$. Hence the Taylor formula

$$f(x) = f(0) + f'(0)\cdot x + \frac{f''(0)}{2!}x^2 + \cdots + \frac{f^{(n)}(0)}{n!}x^n + \frac{f^{(n+1)}(z)}{(n+1)!}x^{n+1}$$

at $a = 0$ gives

$$e^x = 1 + x + \frac{x^2}{2!} + \frac{x^3}{3!} + \cdots + \frac{x^n}{n!} + \frac{e^z x^{n+1}}{(n+1)!} \tag{17}$$

for some z between 0 and x. Thus the remainder term $R_n(x)$ satisfies the inequalities

$$0 < |R_n(x)| < \frac{|x|^{n+1}}{(n+1)!} \quad \text{if } x < 0,$$

$$0 < |R_n(x)| < \frac{e^x x^{n+1}}{(n+1)!} \quad \text{if } x > 0.$$

Therefore, the fact that

$$\lim_{n\to\infty} \frac{x^n}{n!} = 0 \tag{18}$$

for all x (see Problem 55) implies that $\lim_{n\to\infty} R_n(x) = 0$ for all x. This means that the Taylor series for e^x converges to e^x for all x, and we may write

$$e^x = \sum_{n=0}^{\infty} \frac{x^n}{n!} = 1 + x + \frac{x^2}{2!} + \frac{x^3}{3!} + \frac{x^4}{4!} + \cdots. \tag{19}$$

The series in Eq. (19) is the most famous and most important of all Taylor series.

With $x = 1$, Eq. (19) yields a numerical series

$$e = \sum_{n=0}^{\infty} \frac{1}{n!} = 1 + \frac{1}{1!} + \frac{1}{2!} + \frac{1}{3!} + \frac{1}{4!} + \cdots \tag{20}$$

for the number e itself. The 10th and 20th partial sums of this series give the approximations

$$e \approx 1 + \frac{1}{1!} + \frac{1}{2!} + \cdots + \frac{1}{10!} \approx 2.7182818$$

and

$$e \approx 1 + \frac{1}{1!} + \frac{1}{2!} + \cdots + \frac{1}{20!} \approx 2.71828182845904523 5,$$

both of which are accurate to the number of decimal places shown. ■

EXAMPLE 6 To find the Taylor series at $a = 0$ for $f(x) = \cos x$, we first calculate the derivatives

$$\begin{aligned} f(x) &= \cos x, & f'(x) &= -\sin x, \\ f''(x) &= -\cos x, & f^{(3)}(x) &= \sin x, \\ f^{(4)}(x) &= \cos x, & f^{(5)}(x) &= -\sin x, \\ &\vdots & &\vdots \\ f^{(2n)}(x) &= (-1)^n \cos x, & f^{(2n+1)}(x) &= (-1)^{n+1} \sin x. \end{aligned}$$

It follows that

$$f^{(2n)}(0) = (-1)^n \quad \text{but} \quad f^{(2n+1)}(0) = 0,$$

so the Taylor polynomials and Taylor series of $f(x) = \cos x$ include only terms of *even* degree. The Taylor formula of degree $2n$ for $\cos x$ at $a = 0$ is

$$\cos x = 1 - \frac{x^2}{2!} + \frac{x^4}{4!} - \cdots + (-1)^n \frac{x^{2n}}{(2n)!} + (-1)^{n+1} \frac{\cos z}{(2n+2)!} x^{2n+2},$$

where z is between 0 and x. Because $|\cos z| \leqq 1$ for all z, it follows from Eq. (18) that the remainder term approaches zero as $n \to \infty$ *for all* x. Hence the desired Taylor series of $f(x) = \cos x$ at $a = 0$ converges to $\cos x$ for all x, so we may write

$$\cos x = \sum_{n=0}^{\infty} \frac{(-1)^n x^{2n}}{(2n)!} = 1 - \frac{x^2}{2!} + \frac{x^4}{4!} - \frac{x^6}{6!} + \cdots. \tag{21}$$

■

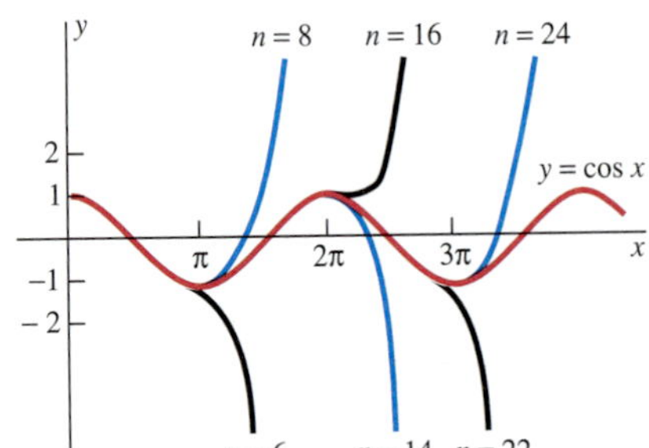

Fig. 11.4.7 Approximating cos x with nth-degree Taylor polynomials

In Problem 41 we ask you to show similarly that the Taylor series at $a = 0$ of $f(x) = \sin x$ is

$$\sin x = \sum_{n=0}^{\infty} \frac{(-1)^n x^{2n+1}}{(2n+1)!} = x - \frac{x^3}{3!} + \frac{x^5}{5!} - \frac{x^7}{7!} + \cdots. \tag{22}$$

Figures 11.4.7 and 11.4.8 illustrate the increasingly better approximations to $\cos x$ and $\sin x$ that we get by using more and more terms of the series in Eqs. (21) and (22).

The case $a = 0$ of Taylor's series is called the **Maclaurin series** of the function $f(x)$,

$$\sum_{n=0}^{\infty} \frac{f^{(n)}(0)}{n!} x^n = f(0) + f'(0) \cdot x + \frac{f''(0)}{2!} x^2 + \frac{f^{(3)}(0)}{3!} x^3 + \cdots. \tag{23}$$

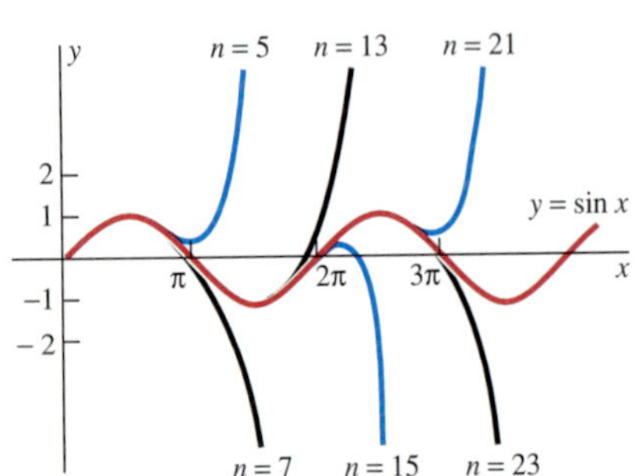

Fig. 11.4.8 Approximating sin x with nth-degree Taylor polynomials

Colin Maclaurin (1698–1746) was a Scottish mathematician who used this series as a basic tool in a calculus book he published in 1742. The three Maclaurin series

$$e^x = \sum_{n=0}^{\infty} \frac{x^n}{n!} = 1 + x + \frac{x^2}{2!} + \frac{x^3}{3!} + \frac{x^4}{4!} + \cdots. \tag{19}$$

$$\cos x = \sum_{n=0}^{\infty} \frac{(-1)^n x^{2n}}{(2n)!} = 1 - \frac{x^2}{2!} + \frac{x^4}{4!} - \frac{x^6}{6!} + \cdots, \quad \text{and} \tag{21}$$

$$\sin x = \sum_{n=0}^{\infty} \frac{(-1)^n x^{2n+1}}{(2n+1)!} = x - \frac{x^3}{3!} + \frac{x^5}{5!} - \frac{x^7}{7!} + \cdots \tag{22}$$

(which actually were discovered by Newton) bear careful examination and comparison. Observe that

- The terms in the *even* cosine series are the *even*-degree terms in the exponential series but with alternating signs.
- The terms in the *odd* sine series are the *odd*-degree terms in the exponential series but with alternating signs.

These series are *identities* that hold for all values of x. Consequently, new series can be derived by substitution, as in Examples 7 and 8.

EXAMPLE 7 The substitution of $x = -t^2$ into Eq. (19) yields

$$e^{-t^2} = 1 - t^2 + \frac{t^4}{2!} - \frac{t^6}{3!} + \cdots + (-1)^n \frac{t^{2n}}{n!} + \cdots.$$

EXAMPLE 8 The substitution of $x = 2t$ into Eq. (22) gives

$$\sin 2t = 2t - \frac{4}{3}t^3 + \frac{4}{15}t^5 - \frac{8}{315}t^7 + \cdots.$$

*The Number π

In Section 5.3 we described how Archimedes used polygons inscribed in and circumscribed about the unit circle to show that $3\frac{10}{71} < \pi < 3\frac{1}{7}$. With the aid of electronic computers, π has been calculated to well over a *billion* decimal places. We describe now some of the methods that have been used for such computations. [For a chronicle of humanity's perennial fascination with the number π, see Peter Beckmann, *A History of π* (New York: St. Martin's Press), 1971.]

We begin with the elementary algebraic identity

$$\frac{1}{1+x} = 1 - x + x^2 - x^3 + \cdots + (-1)^{k-1}x^{k-1} + \frac{(-1)^k x^k}{1+x}, \tag{24}$$

which can be verified by multiplying both sides by $1 + x$. We substitute t^2 for x and $n + 1$ for k and thus find that

$$\frac{1}{1+t^2} = 1 - t^2 + t^4 - t^6 + \cdots + (-1)^n t^{2n} + \frac{(-1)^{n+1}t^{2n+2}}{1+t^2}.$$

Because $D_t \tan^{-1} t = 1/(1 + t^2)$, integration of both sides of this last equation from $t = 0$ to $t = x$ gives

$$\tan^{-1} x = x - \frac{x^3}{3} + \frac{x^5}{5} - \frac{x^7}{7} + \cdots + (-1)^n \frac{x^{2n+1}}{2n+1} + R_{2n+1}, \tag{25}$$

where

$$|R_{2n+1}| = \left| \int_0^x \frac{t^{2n+2}}{1+t^2}\,dx \right| \leqq \left| \int_0^x t^{2n+2}\,dx \right| = \frac{|x|^{2n+3}}{2n+3}. \tag{26}$$

This estimate of the error makes it clear that

$$\lim_{n\to\infty} R_n = 0$$

if $|x| \leqq 1$. Hence we obtain the Taylor series for the inverse tangent function:

$$\tan^{-1} x = \sum_{n=0}^{\infty} (-1)^n \frac{x^{2n+1}}{2n+1} = x - \frac{x^3}{3} + \frac{x^5}{5} - \frac{x^7}{7} + \cdots, \tag{27}$$

valid for $-1 \leqq x \leqq 1$.

If we substitute $x = 1$ into Eq. (27), we obtain *Leibniz's series*

$$\frac{\pi}{4} = 1 - \frac{1}{3} + \frac{1}{5} - \frac{1}{7} + \cdots.$$

Although this is a beautiful series, it is not an effective way to compute π. But the error estimate in Eq. (26) shows that we can use Eq. (25) to calculate $\tan^{-1} x$ if $|x|$ is small. For example, if $x = \frac{1}{5}$, then the fact that

$$\frac{1}{9 \cdot 5^9} \approx 0.0000000057 < 0.0000001$$

implies that the approximation

$$\tan^{-1}\left(\tfrac{1}{5}\right) \approx \tfrac{1}{5} - \tfrac{1}{3}\left(\tfrac{1}{5}\right)^3 + \tfrac{1}{5}\left(\tfrac{1}{5}\right)^5 - \tfrac{1}{7}\left(\tfrac{1}{5}\right)^7 \approx 0.197396$$

is accurate to six decimal places.

Accurate inverse tangent calculations lead to accurate computations of the number π. For example, we can use the addition formula for the tangent function to show (Problem 52) that

$$\frac{\pi}{4} = 4 \tan^{-1}\left(\frac{1}{5}\right) - \tan^{-1}\left(\frac{1}{239}\right). \tag{28}$$

Historical Note In 1706, John Machin (?–1751) used Eq. (28) to calculate the first 100 decimal places of π. (In Problem 54 we ask you to use it to show that $\pi = 3.14159$ to five decimal places.) In 1844 the lightning-fast mental calculator Zacharias Dase (1824–1861) of Germany computed the first 200 decimal places of π, using the related formula

$$\frac{\pi}{4} = \tan^{-1}\left(\frac{1}{2}\right) + \tan^{-1}\left(\frac{1}{5}\right) + \tan^{-1}\left(\frac{1}{8}\right). \tag{29}$$

You might enjoy verifying this formula (see Problem 53). A recent computation of 1 million decimal places of π used the formula

$$\frac{\pi}{4} = 12 \tan^{-1}\left(\frac{1}{18}\right) + 8 \tan^{-1}\left(\frac{1}{57}\right) - 5 \tan^{-1}\left(\frac{1}{239}\right).$$

For derivations of this formula and others like it, with further discussion of the computations of the number π, see the article "An algorithm for the calculation of π" by George Miel in the *American Mathematical Monthly* **86** (1979), pp. 694–697. Although no practical application is ever likely to require more than ten or twelve decimal places of π, these computations provide dramatic evidence of the power of Taylor's formula. Moreover, the number π continues to serve as a challenge both to human ingenuity and to the accuracy and efficiency of modern electronic computers. For an account of how investigations of the Indian mathematical genius Srinivasa Ramanujan (1887–1920) have led recently to the computation of over a billion decimal places of π, see the article "Ramanujan and pi," Jonathan M. Borwein and Peter B. Borwein, *Scientific American* (Feb. 1988), pp. 112–117.

11.4 PROBLEMS

In Problems 1 through 10, find Taylor's formula for the given function f at $a = 0$. Find both the Taylor polynomial $P_n(x)$ of the indicated degree n and the remainder term $R_n(x)$.

1. $f(x) = e^{-x}, \quad n = 5$

2. $f(x) = \sin x, \quad n = 4$

3. $f(x) = \cos x, \quad n = 4$

4. $f(x) = \dfrac{1}{1-x}, \quad n = 4$

5. $f(x) = \sqrt{1+x}, \quad n = 3$

6. $f(x) = \ln(1+x), \quad n = 4$

7. $f(x) = \tan x, \quad n = 3$

8. $f(x) = \arctan x, \quad n = 2$

9. $f(x) = \sin^{-1} x, \quad n = 2$

10. $f(x) = x^3 - 3x^2 + 5x - 7, \quad n = 4$

In Problems 11 through 20, find the Taylor polynomial with remainder by using the given values of a and n.

11. $f(x) = e^x; \ a = 1, \quad n = 4$

12. $f(x) = \cos x; \ a = \pi/4, \quad n = 3$

13. $f(x) = \sin x; \ a = \pi/6, \quad n = 3$

14. $f(x) = \sqrt{x}; \ a = 100, \quad n = 3$

15. $f(x) = \dfrac{1}{(x-4)^2}; \ a = 5, \quad n = 5$

16. $f(x) = \tan x; \ a = \pi, \quad n = 4$

17. $f(x) = \cos x; \ a = \pi, \quad n = 4$

18. $f(x) = \sin x; \ a = \pi/2, \quad n = 4$

19. $f(x) = x^{3/2}; \ a = 1, \quad n = 4$

20. $f(x) = \dfrac{1}{\sqrt{1-x}}; \ a = 0, \quad n = 4$

In Problems 21 through 28, find the Maclaurin series of the given function f by substitution in one of the known series in Eqs. (19), (21), and (22).

21. $f(x) = e^{-x}$

22. $f(x) = e^{2x}$

23. $f(x) = e^{-3x}$

24. $f(x) = \exp(x^3)$

25. $f(x) = \sin 2x$

26. $f(x) = \sin \dfrac{x}{2}$

27. $f(x) = \sin x^2$

28. $f(x) = \sin^2 x = \frac{1}{2}(1 - \cos 2x)$

In Problems 29 through 40, find the Taylor series [Eq. (16)] of the given function at the indicated point a.

29. $f(x) = \ln(1+x), \quad a = 0$

30. $f(x) = \dfrac{1}{1-x}, \quad a = 0$

31. $f(x) = e^{-x}, \quad a = 0$

32. $f(x) = \sin x, \quad a = \pi/2$

33. $f(x) = \ln x, \quad a = 1$

34. $f(x) = e^{2x}, \quad a = 0$

35. $f(x) = \cos x, \quad a = \pi/4$

36. $f(x) = \dfrac{1}{(1-x)^2}, \quad a = 0$

37. $f(x) = \dfrac{1}{x}, \quad a = 1$

38. $f(x) = \cos x, \quad a = \pi/2$

39. $f(x) = \sin x, \quad a = \pi/4$

40. $f(x) = \sqrt{1+x}, \quad a = 0$

41. Derive, as in Example 5, the Taylor series in Eq. (22) of $f(x) = \sin x$ at $a = 0$.

42. Granted that it is valid to differentiate the sine and cosine Taylor series in a term-by-term manner, use these series to verify that $D_x \cos x = -\sin x$ and $D_x \sin x = \cos x$.

43. Use the differentiation formulas $D_x \sinh x = \cosh x$ and $D_x \cosh x = \sinh x$ to derive the Maclaurin series

$$\cosh x = \sum_{n=0}^{\infty} \frac{x^{2n}}{(2n)!} \quad \text{and} \quad \sinh x = \sum_{n=0}^{\infty} \frac{x^{2n+1}}{(2n+1)!}$$

for the hyperbolic sine and cosine functions. What is their relationship to the Maclaurin series of the ordinary sine and cosine functions?

44. Derive the Maclaurin series stated in Problem 43 by substituting the known Maclaurin series for the exponential function in the definitions

$$\cosh x = \frac{e^x + e^{-x}}{2} \quad \text{and} \quad \sinh x = \frac{e^x - e^{-x}}{2}$$

of the hyperbolic functions.

The sum commands listed for several computer algebra systems in the Section 11.3 Project can be used to calculate Taylor polynomials efficiently. For instance, when the TI graphing calculator definitions

```
Y1 = sin(X)
Y2 = sum(seq((-1)∧(N-1)*X∧(2N-1)/(2N-
     1)!,N,1,7))
```

are graphed, the result is Fig. 11.4.9, showing that the 13th-degree Taylor polynomial $P_{13}(x)$ approximates $\sin x$ rather closely if $-3\pi/2 < x < 3\pi/2$ but not outside this range. By plotting several successive Taylor polynomials of a function $f(x)$ simultaneously, we can get a visual sense of the way in

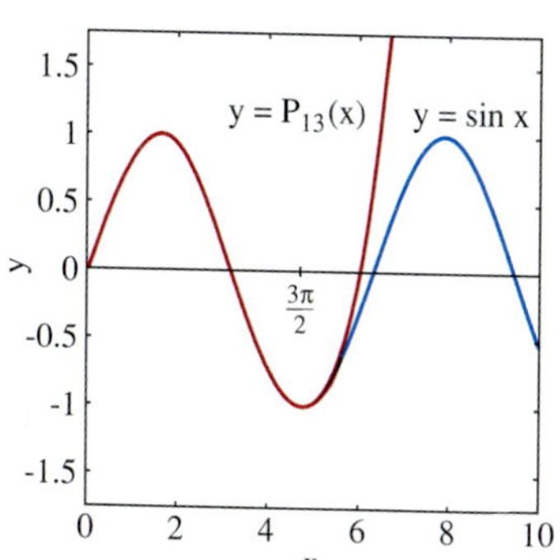

Fig. 11.4.9 Graphs of $\sin x$ and its 13th-degree Taylor polynomial $P_{13}(x)$

which they approximate the function. Do this for each function given in Problems 45 through 50.

45. $f(x) = e^{-x}$

46. $f(x) = \sin x$

47. $f(x) = \cos x$

48. $f(x) = \ln(1 + x)$

49. $f(x) = \dfrac{1}{1 + x}$

50. $f(x) = \dfrac{1}{1 - x^2}$

51. Let the function

$$f(x) = \sum_{n=0}^{\infty} \frac{(-1)^n x^n}{(2n)!} = 1 - \frac{x}{2!} + \frac{x^2}{4!} - \frac{x^3}{6!} + \cdots$$

be defined by replacing x with $\sqrt{x}$ in the Maclaurin series for $\cos x$. Plot partial sums of this series to verify graphically that $f(x)$ agrees with the function $g(x)$ defined by

$$g(x) = \begin{cases} \cos\sqrt{x} & \text{if } x \geqq 0, \\ \cosh\sqrt{|x|} & \text{if } x < 0. \end{cases}$$

52. Beginning with $\alpha = \tan^{-1}\left(\frac{1}{5}\right)$, use the addition formula

$$\tan(A + B) = \frac{\tan A + \tan B}{1 - \tan A \tan B}$$

to show in turn that (a) $\tan 2\alpha = \frac{5}{12}$; (b) $\tan 4\alpha = \frac{120}{119}$; (c) $\tan(\pi/4 - 4\alpha) = -\frac{1}{239}$. Finally, show that part (c) implies Eq. (28).

53. Apply the addition formula for the tangent function to verify Eq. (29).

54. Every young person deserves the thrill, just once, of calculating personally the first several decimal places of the number π. The seemingly random nature of this decimal expansion demands an explanation; how, indeed, are the digits 3.14159 26535 89793 ... determined? For a partial answer, set your calculator to display nine decimal places. Then add enough terms of the arctangent series in Eq. (27) with $x = \frac{1}{5}$ to calculate $\arctan\left(\frac{1}{5}\right)$ accurate to nine places. Next, calculate the value of $\arctan\left(\frac{1}{239}\right)$ similarly. Finally, substitute these numerical results in Eq. (28) and solve for π. How many accurate decimal places do you get?

55. Prove that

$$\lim_{n\to\infty} \frac{x^n}{n!} = 0$$

if x is a real number. (*Suggestion:* Choose an integer k such that $k > |2x|$, and let $L = |x|^k/k!$. Then show that

$$\frac{|x|^n}{n!} < \frac{L}{2^{n-k}}$$

if $n > k$.)

56. Suppose that $0 < x \leqq 1$. Integrate both sides of the identity

$$\frac{1}{1+t} = 1 - t + t^2 - t^3 + \cdots + (-1)^n t^n + \frac{(-1)^{n+1} t^{n+1}}{1+t}$$

from $t = 0$ to $t = x$ to show that

$$\ln(1 + x) = x - \frac{x^2}{2} + \frac{x^3}{3} - \cdots + (-1)^n \frac{x^{n+1}}{n+1} + R_n,$$

where $\lim_{n\to\infty} R_n = 0$. Hence conclude that

$$\ln(1 + x) = \sum_{n=1}^{\infty} (-1)^{n+1} \frac{x^n}{n}$$

if $0 \leqq x \leqq 1$.

57. Criticize the following "proof" that $2 = 1$. Substitution of $x = 1$ into the result in Problem 56 yields the fact that

$$\ln 2 = 1 - \tfrac{1}{2} + \tfrac{1}{3} - \tfrac{1}{4} + \cdots.$$

If

$$S = 1 + \tfrac{1}{2} + \tfrac{1}{3} + \tfrac{1}{4} + \cdots,$$

then

$$\ln 2 = S - 2\cdot\left(\tfrac{1}{2} + \tfrac{1}{4} + \tfrac{1}{6} + \tfrac{1}{8} + \cdots\right) = S - S = 0.$$

Hence $2 = e^{\ln 2} = e^0 = 1$.

58. Deduce from the result of Problem 56 first that

$$\ln(1 - x) = -\sum_{n=1}^{\infty} \frac{x^n}{n} = -x - \frac{x^2}{3} - \frac{x^3}{3} - \cdots$$

and then that

$$\ln\frac{1+x}{1-x} = \sum_{n \text{ odd}} \frac{2x^n}{n} = 2\left(x + \frac{x^3}{3} + \frac{x^5}{5} + \cdots\right)$$

if $0 \leqq x \leqq 1$.

59. Approximate the number $\ln 2 \approx 0.69315$ first by substituting $x = 1$ in the Maclaurin series of Problem 56, and then by substituting $x = \frac{1}{3}$ (Why?) in the second series of Problem 58. Which approach appears to require the fewest terms to yield the value of $\ln 2$ accurate to a given number of decimal places?

11.4 PROJECT: CALCULATING LOGARITHMS ON A DESERTED ISLAND

You are stranded for life on a desert island with only a very basic calculator that does not calculate natural logarithms. So to get modern science going on this miserable island, you need to use the Problem 58 infinite series for $\ln[(1 + x)/(1 - x)]$ to produce a simple table of logarithms (with five-place accuracy, say), giving $\ln x$ at least for the integers $x = 1, 2, 3, \dots, 9$, and 10.

The most direct way might be to use the series for $\ln[(1 + x)/(1 - x)]$ to calculate first $\ln 2$, $\ln 3$, $\ln 5$, and $\ln 7$. Then use the law of logarithms $\ln xy = \ln x + \ln y$ to fill in the other entries in the table by simple addition of logarithms already computed. Unfortunately, larger values of x result in series that are more slowly convergent. So you could save yourself time and work by exercising some ingenuity: Calculate from scratch some four *other* logarithms from which you can build up the rest. For example, if you know $\ln 2$ and $\ln 1.25$, then $\ln 10 = \ln 1.25 + 3\ln 2$. (Why?) Be as ingenious as you wish. Can you complete your table of ten logarithms by initially calculating directly (using the series) *fewer* than four logarithms?

For a finale, calculate somehow (from scratch, and accurate to five rounded decimal places) the natural logarithm $\ln(pq.rs)$, where p, q, r, and s denote the last four *nonzero* digits in your student I.D. number.

11.5 THE INTEGRAL TEST

A Taylor series (as in Section 11.4) is a special type of infinite series with *variable* terms. We saw that Taylor's formula can sometimes be used—as in the case of the exponential, sine, and cosine series—to establish the convergence of such a series.

But given an infinite series $\sum a_n$ with *constant* terms, it is the exception rather than the rule when a simple formula for the nth partial sum of that series can be found and used directly to determine whether the series converges or diverges. There are, however, several *convergence tests* that use the *terms* of an infinite series rather than its partial sums. Such a test, when successful, will tell us whether the series converges. Once we know that the series $\sum a_n$ does converge, it is then a separate matter to find its sum S. It may be necessary to approximate S by adding sufficiently many terms; in this case we shall need to know how many terms are required for the desired accuracy.

Here and in Section 11.6, we concentrate our attention on **positive-term series**—that is, series with terms that are all positive. If $a_n > 0$ for all n, then

$$S_1 < S_2 < S_3 < \cdots < S_n < \cdots,$$

so the sequence $\{S_n\}$ of partial sums of the series is increasing. Hence there are just two possibilities. If the sequence $\{S_n\}$ is *bounded*—there exists a number M such that $S_n < M$ for all n—then the bounded monotonic sequence property (Section 11.2) implies that $S = \lim_{n\to\infty} S_n$ exists, so the series $\sum a_n$ *converges*. Otherwise, it diverges to infinity (by Problem 52 in Section 11.2).

A similar alternative holds for improper integrals. Suppose that the function f is continuous and positive-valued for $x \geqq 1$. Then it follows (from Problem 51) that the improper integral

$$\int_1^\infty f(x)\,dx = \lim_{b\to\infty}\int_1^b f(x)\,dx \tag{1}$$

either converges (the limit is a real number) or diverges to infinity (the limit is $+\infty$). This analogy between positive-term series and improper integrals of positive functions is the key to the **integral test.** We compare the behavior of the series $\sum a_n$ with that of the improper integral in Eq. (1), where f is an appropriately chosen function. [Among other things, we require that $f(n) = a_n$ for all n.]

Theorem 1 *The Integral Test*

Suppose that $\sum a_n$ is a positive-term series and that f is a positive-valued, decreasing, continuous function for $x \geqq 1$. If $f(n) = a_n$ for all integers $n \geqq 1$, then the series and the improper integral

$$\sum_{n=1}^{\infty} a_n \quad \text{and} \quad \int_1^{\infty} f(x)\,dx$$

either both converge or both diverge.

PROOF Because f is a decreasing function, the rectangular polygon with area

$$S_n = a_1 + a_2 + a_3 + \cdots + a_n$$

shown in Fig. 11.5.1 contains the region under $y = f(x)$ from $x = 1$ to $x = n + 1$. Hence

$$\int_1^{n+1} f(x)\,dx \leqq S_n. \tag{2}$$

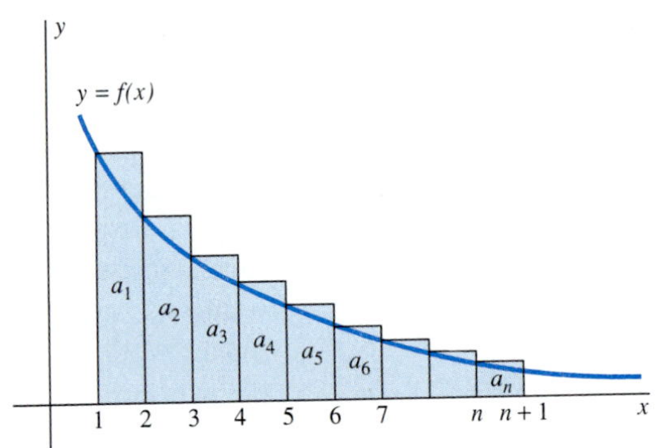

Fig. 11.5.1 Underestimating the partial sums with an integral

Similarly, the rectangular polygon with area

$$S_n - a_1 = a_2 + a_3 + a_4 + \cdots + a_n$$

shown in Fig. 11.5.2 is contained in the region under $y = f(x)$ from $x = 1$ to $x = n$. Hence

$$S_n - a_1 \leqq \int_1^{n} f(x)\,dx. \tag{3}$$

Suppose first that the improper integral $\int_1^{\infty} f(x)\,dx$ diverges (necessarily to $+\infty$). Then

$$\lim_{n\to\infty} \int_1^{n+1} f(x)\,dx = +\infty,$$

so it follows from (2) that $\lim_{n\to\infty} S_n = +\infty$ as well, and hence the infinite series $\sum a_n$ likewise diverges.

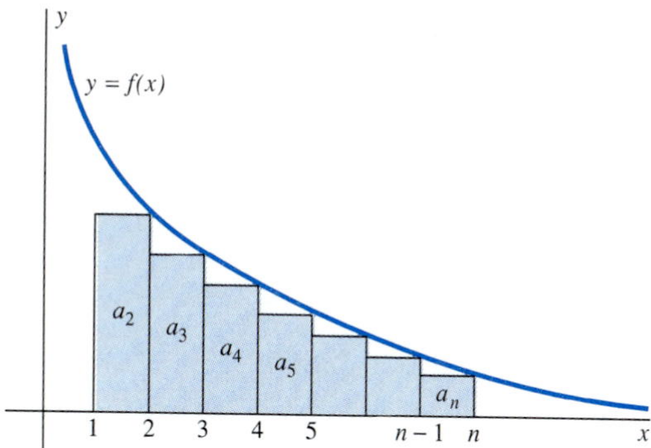

Fig. 11.5.2 Overestimating the partial sums with an integral

Now suppose instead that the improper integral $\int_1^\infty f(x)\,dx$ converges and has the (finite) value I. Then (3) implies that

$$S_n \leqq a_1 + \int_1^n f(x)\,dx \leqq a_1 + I,$$

so the increasing sequence $\{S_n\}$ is bounded. Thus the infinite series

$$\sum_{n=1}^{\infty} a_n = \lim_{n\to\infty} S_n$$

converges as well. Hence we have shown that the infinite series and the improper integral either both converge or both diverge. ■

EXAMPLE 1 We used a version of the integral test to prove in Section 11.3 that the harmonic series

$$\sum_{n=1}^{\infty} \frac{1}{n} = 1 + \frac{1}{2} + \frac{1}{3} + \frac{1}{4} + \cdots$$

diverges. Using the test as stated in Theorem 1 is a little simpler: We note that $f(x) = 1/x$ is positive, continuous, and decreasing for $x \geqq 1$ and that $f(n) = 1/n$ for each positive integer n. Now

$$\int_1^\infty \frac{1}{x}\,dx = \lim_{b\to\infty} \int_1^b \frac{1}{x}\,dx = \lim_{b\to\infty} \Big[\ln x\Big]_1^b = \lim_{b\to\infty} (\ln b - \ln 1) = +\infty.$$

Thus the improper integral diverges and therefore so does the harmonic series. ■

The harmonic series is the case $p = 1$ of the ***p*-series**

$$\sum_{n=1}^{\infty} \frac{1}{n^p} = 1 + \frac{1}{2^p} + \frac{1}{3^p} + \cdots + \frac{1}{n^p} + \cdots. \tag{4}$$

Whether the p-series converges or diverges depends on the value of p.

EXAMPLE 2 Show that the p-series converges if $p > 1$ but diverges if $0 < p \leqq 1$.

Solution The case $p = 1$ has already been settled in Example 1. If $p > 0$ but $p \neq 1$, then the function $f(x) = 1/x^p$ satisfies the conditions of the integral test, and

$$\int_1^\infty \frac{1}{x^p}\,dx = \lim_{b\to\infty}\int_1^b \frac{1}{x^p}\,dx = \lim_{b\to\infty}\left[-\frac{1}{(p-1)x^{p-1}}\right]_1^b$$
$$= \lim_{b\to\infty}\frac{1}{p-1}\left(1-\frac{1}{b^{p-1}}\right).$$

If $p > 1$, then

$$\int_1^\infty \frac{1}{x^p}\,dx = \frac{1}{p-1} < \infty,$$

so the integral and the series both converge. But if $0 < p < 1$, then

$$\int_1^\infty \frac{1}{x^p}\,dx = \lim_{b\to\infty}\frac{1}{1-p}\left(b^{1-p}-1\right) = \infty,$$

and in this case the integral and the series both diverge. ■

As specific examples, the series

$$\sum_{n=1}^\infty \frac{1}{n^2} = 1 + \frac{1}{2^2} + \frac{1}{3^2} + \cdots + \frac{1}{n^2} + \cdots$$

converges ($p = 2 > 1$), whereas the series

$$\sum_{n=1}^\infty \frac{1}{\sqrt{n}} = 1 + \frac{1}{\sqrt{2}} + \frac{1}{\sqrt{3}} + \cdots + \frac{1}{\sqrt{n}} + \cdots$$

diverges ($p = \frac{1}{2} \leqq 1$).

Now suppose that the positive-term series $\sum a_n$ converges by the integral test and that we wish to approximate its sum by adding sufficiently many of its initial terms. The difference between the sum S of the series and its nth partial sum S_n is the **remainder**

$$R_n = S - S_n = a_{n+1} + a_{n+2} + a_{n+3} + \cdots. \qquad \textbf{(5)}$$

This remainder is the error made when the sum is estimated by using in its place the partial sum S_n.

Theorem 2 ***The Integral Test Remainder Estimate***

Suppose that the infinite series and improper integral

$$\sum_{n=1}^\infty a_n \quad \text{and} \quad \int_1^\infty f(x)\,dx$$

satisfy the hypotheses of the integral test, and suppose in addition that both converge. Then

$$\int_{n+1}^\infty f(x)\,dx \leqq R_n \leqq \int_n^\infty f(x)\,dx, \qquad \textbf{(6)}$$

where R_n is the remainder given in Eq. (5).

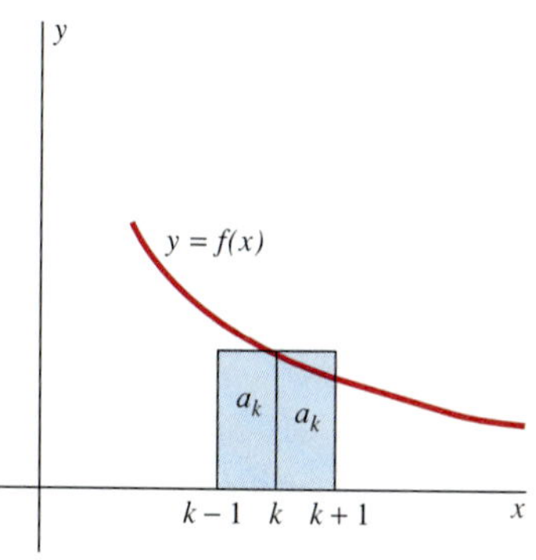

Fig. 11.5.3 Establishing the integral test remainder estimate

Proof We see from Fig. 11.5.3 that

$$\int_k^{k+1} f(x)\,dx \leqq a_k \leqq \int_{k-1}^{k} f(x)\,dx$$

for $k = n+1, n+2, \ldots$. We add these inequalities for all such values of k, and the result is the inequality in (6) because

$$\sum_{k=n+1}^{\infty} \int_k^{k+1} f(x)\,dx = \int_{n+1}^{\infty} f(x)\,dx$$

and

$$\sum_{k=n+1}^{\infty} \int_{k-1}^{k} f(x)\,dx = \int_{n}^{\infty} f(x)\,dx. \quad \blacksquare$$

If we substitute $R_n = S - S_n$, then it follows from (6) that the sum S of the series satisfies the inequality

$$S_n + \int_{n+1}^{\infty} f(x)\,dx \leqq S \leqq S_n + \int_{n}^{\infty} f(x)\,dx. \tag{7}$$

If the nth partial sum S_n is known and the difference

$$\int_n^{n+1} f(x)\,dx$$

between the two integrals is small, then (7) provides an accurate estimate of the sum S of the infinite series.

EXAMPLE 3 We will see in Section 11.8 that the exact sum of the p-series with $p = 2$ is $\pi^2/6$, thus giving the beautiful formula

$$\frac{\pi^2}{6} = 1 + \frac{1}{2^2} + \frac{1}{3^2} + \frac{1}{4^2} + \cdots. \tag{8}$$

Use this series to approximate the number π by applying the integral test remainder estimate, first with $n = 50$, then with $n = 200$.

Solution Obviously we take $f(x) = 1/x^2$ in the remainder estimate. Because

$$\int_n^{\infty} \frac{1}{x^2}\,dx = \lim_{b\to\infty} \left[-\frac{1}{x}\right]_n^b = \lim_{b\to\infty} \left(\frac{1}{n} - \frac{1}{b}\right) = \frac{1}{n},$$

Eq. (7) gives

$$S_n + \frac{1}{n+1} \leqq \frac{\pi^2}{6} \leqq S_n + \frac{1}{n}, \tag{9}$$

where

$$S_n = 1 + \frac{1}{2^2} + \frac{1}{3^2} + \cdots + \frac{1}{n^2}$$

is the nth partial sum of the series in Eq. (8). Upon multiplying by 6 and taking square roots, Eq. (9) gives the inequality

$$\sqrt{6\left(S_n + \frac{1}{n+1}\right)} \leqq \pi \leqq \sqrt{6\left(S_n + \frac{1}{n}\right)}. \qquad \textbf{(10)}$$

You could add the first 50 terms in Eq. (8) one by one in a few minutes using a simple four-function calculator, but this kind of arithmetic is precisely the task for which a modern calculator or computer algebra system is designed. A one-line instruction like the calculator command `sum(series(1/n^2,n,1,50))` yields

$$S_{50} = \sum_{n=1}^{50} \frac{1}{n^2} \approx 1.62513273.$$

Then, using (9) for illustration rather than (10), we calculate

$$1.62513273 + \frac{1}{51} \leqq \frac{\pi^2}{6} \leqq 1.62513273 + \frac{1}{50};$$

$$1.64474057 \leqq \frac{\pi^2}{6} \leqq 1.64513273;$$

$$3.14140788 \leqq \pi \leqq 3.14178236.$$

Finally, rounding down on the left and up on the right (why?), we conclude that $3.1414 < \pi < 3.1418$. The average of these two bounds is the traditional four-place approximation $\pi \approx 3.1416$.

The 200th partial sum of the series in Eq. (8) is

$$S_{200} = \sum_{n=1}^{200} \frac{1}{n^2} \approx 1.63994655.$$

Substituting this sum and $n = 200$ in (10), we get

$$3.14158082 \leqq \pi \leqq 3.14160457.$$

This proves that $\pi \approx 3.1416$ rounded accurate to four decimal places. ■

EXAMPLE 4 Show that the series

$$\sum_{n=2}^{\infty} \frac{1}{n(\ln n)^2} \qquad \textbf{(11)}$$

converges, and determine how many terms you would need to add to find its sum accurate to within 0.01. That is, how large must n be for the remainder to satisfy the inequality $R_n < 0.01$?

Solution We begin the sum at $n = 2$ because $\ln 1 = 0$. Let $f(x) = 1/[x(\ln x)^2]$. Then

$$\int_n^{\infty} \frac{1}{x(\ln x)^2}\,dx = \lim_{b\to\infty}\left[-\frac{1}{\ln x}\right]_n^b = \lim_{b\to\infty}\left(\frac{1}{\ln n} - \frac{1}{\ln b}\right) = \frac{1}{\ln n}.$$

Substitution of $n = 2$ shows that the series in (11) converges (by the integral test). Our calculations and the right-hand inequality in (6) now give $R_n < 1/(\ln n)$, so we need

$$\frac{1}{\ln n} \leqq 0.01, \qquad \ln n \geqq 100, \qquad n \geqq e^{100} \approx 2.7 \times 10^{43}.$$

A computer that could calculate a billion (10^9) terms per second would require about 8.5×10^{26} years—far longer than the expected lifetime of the universe—to sum this many terms. But you can check that accuracy to only one decimal place—that is, $R_n < 0.05$—would require only about $n = 4.85 \times 10^8$ (fewer than a half billion) terms, well within the range of a powerful desktop computer. ■

11.5 PROBLEMS

In Problems 1 through 30, use the integral test to test the given series for convergence.

1. $\sum_{n=1}^{\infty} \frac{n}{n^2+1}$ **2.** $\sum_{n=1}^{\infty} \frac{n}{e^{n^2}}$

3. $\sum_{n=1}^{\infty} \frac{1}{\sqrt{n+1}}$ **4.** $\sum_{n=1}^{\infty} \frac{1}{(n+1)^{4/3}}$

5. $\sum_{n=1}^{\infty} \frac{1}{n^2+1}$ **6.** $\sum_{n=1}^{\infty} \frac{1}{n(n+1)}$

7. $\sum_{n=2}^{\infty} \frac{1}{n \ln n}$ **8.** $\sum_{n=1}^{\infty} \frac{\ln n}{n}$

9. $\sum_{n=1}^{\infty} \frac{1}{2^n}$ **10.** $\sum_{n=1}^{\infty} \frac{n}{e^n}$

11. $\sum_{n=1}^{\infty} \frac{n^2}{e^n}$ **12.** $\sum_{n=1}^{\infty} \frac{1}{17n-13}$

13. $\sum_{n=1}^{\infty} \frac{\ln n}{n^2}$ **14.** $\sum_{n=1}^{\infty} \frac{n+1}{n^2}$

15. $\sum_{n=1}^{\infty} \frac{n}{n^4+1}$ **16.** $\sum_{n=1}^{\infty} \frac{1}{n^3+n}$

17. $\sum_{n=1}^{\infty} \frac{2n+5}{n^2+5n+17}$ **18.** $\sum_{n=1}^{\infty} \ln\left(\frac{n+1}{n}\right)$

19. $\sum_{n=1}^{\infty} \ln\left(1+\frac{1}{n^2}\right)$ **20.** $\sum_{n=1}^{\infty} \frac{2^{1/n}}{n^2}$

21. $\sum_{n=1}^{\infty} \frac{n}{4n^2+5}$ **22.** $\sum_{n=1}^{\infty} \frac{n}{(4n^2+5)^{3/2}}$

23. $\sum_{n=2}^{\infty} \frac{1}{n\sqrt{\ln n}}$ **24.** $\sum_{n=2}^{\infty} \frac{1}{n(\ln n)^3}$

25. $\sum_{n=1}^{\infty} \frac{1}{4n^2+9}$ **26.** $\sum_{n=1}^{\infty} \frac{n+1}{n+100}$

27. $\sum_{n=1}^{\infty} \frac{n}{n^4+2n^2+1}$ **28.** $\sum_{n=1}^{\infty} \frac{1}{(n+1)^3}$

29. $\sum_{n=1}^{\infty} \frac{\arctan n}{n^2+1}$ **30.** $\sum_{n=3}^{\infty} \frac{1}{n(\ln n)[\ln(\ln n)]}$

In Problems 31 through 34, tell why the integral test does *not* apply to the given series.

31. $\sum_{n=1}^{\infty} \frac{(-1)^n}{n}$ **32.** $\sum_{n=1}^{\infty} e^{-n} \sin n$

33. $\sum_{n=1}^{\infty} \frac{2+\sin n}{n^2}$ **34.** $\sum_{n=1}^{\infty} \left(\frac{\sin n}{n}\right)^4$

In Problems 35 through 38, determine the values of p for which the given series converges.

35. $\sum_{n=1}^{\infty} \frac{1}{p^n}$ **36.** $\sum_{n=1}^{\infty} \frac{n}{(n^2+1)^p}$

37. $\sum_{n=2}^{\infty} \frac{1}{n(\ln n)^p}$ **38.** $\sum_{n=3}^{\infty} \frac{1}{n(\ln n)[\ln(\ln n)]^p}$

In Problems 39 through 42, find the least positive integer n such that the remainder R_n in Theorem 2 is less than E.

39. $\sum_{n=1}^{\infty} \frac{1}{n^2}$; $E = 0.0001$ **40.** $\sum_{n=1}^{\infty} \frac{1}{n^2}$; $E = 0.00005$

41. $\sum_{n=1}^{\infty} \frac{1}{n^3}$; $E = 0.00005$ **42.** $\sum_{n=1}^{\infty} \frac{1}{n^6}$; $E = 2 \times 10^{-11}$

In Problems 43 through 46, find the sum of the given series accurate to the indicated number k of decimal places. Begin by finding the smallest value of n such that the remainder satisfies the inequality $R_n < 5 \times 10^{-(k+1)}$. Then use a calculator to compute the partial sum S_n and round off appropriately.

43. $\sum_{n=1}^{\infty} \frac{1}{n^{3/2}}$; $k = 2$ **44.** $\sum_{n=1}^{\infty} \frac{1}{n^3}$; $k = 3$

45. $\sum_{n=1}^{\infty} \frac{1}{n^5}$; $k = 5$ **46.** $\sum_{n=1}^{\infty} \frac{1}{n^7}$; $k = 7$

In Problems 47 and 48, use a computer algebra system (if necessary) to determine the values of p for which the given infinite series converges.

47. $\sum_{n=1}^{\infty} \frac{\ln n}{n^p}$ **48.** $\sum_{n=1}^{\infty} \frac{1}{p^{\ln n}}$

49. Deduce from the inequalities in (2) and (3) with the function $f(x) = 1/x$ that

$$\ln n \leqq 1 + \frac{1}{2} + \frac{1}{3} + \cdots + \frac{1}{n} \leqq 1 + \ln n$$

for $n = 1, 2, 3, \ldots$. If a computer adds 1 million terms of the harmonic series per second, how long will it take for the partial sum to reach 50?

50. (a) Let

$$c_n = 1 + \frac{1}{2} + \frac{1}{3} + \cdots + \frac{1}{n} - \ln n$$

for $n = 1, 2, 3, \ldots$. Deduce from Problem 49 that $0 \leqq c_n \leqq 1$ for all n. (b) Note that

$$\int_n^{n+1} \frac{1}{x}\,dx \geqq \frac{1}{n+1}.$$

Conclude that the sequence $\{c_n\}$ is decreasing. Therefore, the sequence $\{c_n\}$ converges. The number

$$\gamma = \lim_{n\to\infty} c_n = \lim_{n\to\infty}\left(1 + \frac{1}{2} + \frac{1}{3} + \cdots + \frac{1}{n} - \ln n\right) \approx 0.57722$$

is known as **Euler's constant.**

51. Suppose that the function f is continuous and positive-valued for $x \geqq 1$. Let

$$b_n = \int_1^n f(x)\,dx$$

for $n = 1, 2, 3, \ldots$. (a) Suppose that the increasing sequence $\{b_n\}$ is bounded, so that $B = \lim_{n\to\infty} b_n$ exists. Prove that

$$\int_1^\infty f(x)\,dx = B.$$

(b) Prove that if the sequence $\{b_n\}$ is not bounded, then

$$\int_1^\infty f(x)\,dx = +\infty.$$

11.5 PROJECT: THE NUMBER π, ONCE AND FOR ALL

When we replace the parameter p in the p-series $\sum 1/n^p$ with the variable x, we get one of the most important transcendental functions in higher mathematics, the **Riemann zeta function**

$$\zeta(x) = \sum_{n=1}^{\infty} \frac{1}{n^x} = 1 + \frac{1}{2^x} + \frac{1}{3^x} + \frac{1}{4^x} + \cdots.$$

Remark One can substitute a complex number $x = a + bi$ in the zeta function. Now that Fermat's last theorem has been proved, the most famous unsolved conjecture in mathematics is the **Riemann hypothesis**—that $\zeta(a + bi) = 0$ implies that $a = \frac{1}{2}$; that is, that the only complex zeros of the Riemann zeta function have real part $\frac{1}{2}$. (The smallest such example is approximately $\frac{1}{2} + 14.13475i$.) The truth of the Riemann hypothesis would have profound implications in number theory, including information about the distribution of the prime numbers.

It has been known since the time of Euler that if k is an *even* positive integer, then the value of $\zeta(k)$ is a rational multiple of π^k. But despite great efforts by many of the best mathematicians of the past two centuries, little is known about the value of $\zeta(k)$ if k is an *odd* positive integer. Indeed, it was only in the last decade that $\zeta(3)$ was shown to be irrational.

In Problems 1 through 4, use the given value of the zeta function and the integral test remainder estimate (as in Example 3 of this section) with the given value of n to determine how accurately the value of the number π is thereby determined. Knowing that

$$\pi \approx 3.14159\,26535\,89793\,23846,$$

write each final answer in the form $\pi \approx 3.abcde\ldots$, giving precisely those digits that are correct or correctly rounded.

1. $\zeta(2) = \dfrac{\pi^2}{6}$ with $n = 25$.

2. $\zeta(4) = \dfrac{\pi^4}{90}$ with $n = 20$.

3. $\zeta(6) = \dfrac{\pi^6}{945}$ with $n = 15$.

4. $\zeta(8) = \dfrac{\pi^8}{9450}$ with $n = 10$.

5. Finally, use one of the preceding four problems and your own careful choice of n to show that $\pi \approx 3.14159\,2654$ with all digits correct or correctly rounded.

11.6 COMPARISON TESTS FOR POSITIVE-TERM SERIES

With the integral test we attempt to determine whether an infinite series converges by comparing it with an improper integral. The methods of this section involve comparing the terms of the *positive-term* series $\sum a_n$ with those of another positive-term series $\sum b_n$ whose convergence or divergence is known. We have already developed two families of *reference series* for the role of the known series $\sum b_n$; they are the geometric series of Section 11.3 and the p-series of Section 11.5. They are well adapted for our new purposes because their convergence or divergence is quite easy to determine. Recall that the geometric series $\sum r^n$ converges if $|r| < 1$ and diverges if $|r| \geqq 1$, and that the p-series $\sum 1/n^p$ converges if $p > 1$ and diverges if $0 < p \leqq 1$.

Let $\sum a_n$ and $\sum b_n$ be positive-term series. Then we say that the series $\sum b_n$ **dominates** the series $\sum a_n$ provided that $a_n \leqq b_n$ for all n. Theorem 1 says that the positive-term series $\sum a_n$ converges if it is dominated by a convergent series and diverges if it dominates a positive-term divergent series.

Theorem 1 ***Comparison Test***

Suppose that $\sum a_n$ and $\sum b_n$ are positive-term series. Then

1. $\sum a_n$ converges if $\sum b_n$ converges and $a_n \leqq b_n$ for all n;

2. $\sum a_n$ diverges if $\sum b_n$ diverges and $a_n \geqq b_n$ for all n.

PROOF Denote the nth partial sums of the series $\sum a_n$ and $\sum b_n$ by S_n and T_n, respectively. Then $\{S_n\}$ and $\{T_n\}$ are increasing sequences. To prove part (1), suppose that $\sum b_n$ converges, so $T = \lim_{n\to\infty} T_n$ exists (so that T is a real number). Then the fact that $a_n \leqq b_n$ for all n implies that $S_n \leqq T_n \leqq T$ for all n. Thus the sequence $\{S_n\}$ of partial sums of $\sum a_n$ is bounded and increasing and therefore converges. Thus $\sum a_n$ converges.

Part (2) is merely a restatement of part (1). If the series $\sum a_n$ converged, then the fact that $\sum a_n$ dominates $\sum b_n$ would imply—by part (1), with a_n and b_n interchanged—that $\sum b_n$ converged. But $\sum b_n$ diverges, so it follows that $\sum a_n$ must also diverge. ■

We know by Theorem 5 of Section 11.3 that the convergence or divergence of an infinite series is not affected by the addition or deletion of a finite number of terms. Consequently, the conditions $a_n \leqq b_n$ and $a_n \geqq b_n$ in the two parts of the comparison test really need to hold only for all $n \geqq k$, where k is some fixed positive integer. Thus we can say that the positive-term series $\sum a_n$ converges if it is "eventually dominated" by the convergent positive-term series $\sum b_n$.

EXAMPLE 1 Because

$$\frac{1}{n(n+1)(n+2)} < \frac{1}{n^3}$$

for all $n \geqq 1$, the series

$$\sum_{n=1}^{\infty} \frac{1}{n(n+1)(n+2)} = \frac{1}{1\cdot 2\cdot 3} + \frac{1}{2\cdot 3\cdot 4} + \frac{1}{3\cdot 4\cdot 5} + \cdots$$

is dominated by the series $\sum 1/n^3$, which is a convergent p-series with $p = 3$. Both are positive-term series, and hence the series $\sum 1/[n(n+1)(n+2)]$ converges by part (1) of the comparison test. ■

EXAMPLE 2 Because

$$\frac{1}{\sqrt{2n-1}} > \frac{1}{\sqrt{2n}}$$

for all $n \geqq 1$, the positive-term series

$$\sum_{n=1}^{\infty} \frac{1}{\sqrt{2n-1}} = 1 + \frac{1}{\sqrt{3}} + \frac{1}{\sqrt{5}} + \frac{1}{\sqrt{7}} + \cdots$$

dominates the series

$$\sum_{n=1}^{\infty} \frac{1}{\sqrt{2n}} = \frac{1}{\sqrt{2}} \sum_{n=1}^{\infty} \frac{1}{n^{1/2}}.$$

But $\sum 1/n^{1/2}$ is a divergent p-series with $p = \frac{1}{2}$, and a constant nonzero multiple of a divergent series diverges. So part (2) of the comparison test implies that the series $\sum 1/\sqrt{2n-1}$ also diverges. ■

EXAMPLE 3 Test the series

$$\sum_{n=0}^{\infty} \frac{1}{n!} = 1 + \frac{1}{1!} + \frac{1}{2!} + \frac{1}{3!} + \cdots$$

for convergence.

Solution We note first that if $n \geqq 1$, then

$$\begin{aligned} n! &= n(n-1)(n-2)\cdots 3\cdot 2\cdot 1 \\ &\geqq 2\cdot 2\cdot 2 \cdots 2\cdot 2\cdot 1 \quad \text{(the same number of factors);} \end{aligned}$$

that is, $n! \geqq 2^{n-1}$ for $n \geqq 1$. Thus

$$\frac{1}{n!} \leqq \frac{1}{2^{n-1}} \qquad \text{for } n \geqq 1,$$

so the series

$$\sum_{n=0}^{\infty} \frac{1}{n!} \quad \text{is dominated by the series} \quad 1 + \sum_{n=1}^{\infty} \frac{1}{2^{n-1}} = 1 + \sum_{n=0}^{\infty} \frac{1}{2^n},$$

which is a convergent geometric series (after the first term). Both are positive-term series, so by the comparison test the given series converges. We saw in Section 11.4 that the sum of the series is the number e, so

$$e = 1 + \frac{1}{1!} + \frac{1}{2!} + \frac{1}{3!} + \cdots + \frac{1}{n!} + \cdots.$$

Indeed, this series provides perhaps the simplest way of showing that

$$e \approx 2.71828\ 18284\ 59045\ 23536.$$

■

Limit Comparison of Terms

Suppose that $\sum a_n$ is a positive-term series such that $a_n \to 0$ as $n \to \infty$. Then, in connection with the nth-term divergence test of Section 11.3, the series $\sum a_n$ has at least a *chance* of converging. How do we choose an appropriate positive-term series $\sum b_n$

with which to compare it? A good idea is to express b_n as a *simple* function of n, simpler than a_n but such that a_n and b_n approach zero at the same rate as $n \to \infty$. If the formula for a_n is a fraction, we can try discarding all but the terms of largest magnitude in its numerator and denominator to form b_n. For example, if

$$a_n = \frac{3n^2 + n}{n^4 + \sqrt{n}},$$

then we reason that n is small in comparison with $3n^2$, and that $\sqrt{n}$ is small in comparison with n^4, when n is quite large. This suggests that we choose $b_n = 3n^2/n^4 = 3/n^2$. The series $\sum 3/n^2$ converges ($p = 2$), but when we attempt to compare $\sum a_n$ and $\sum b_n$, we find that $a_n \geqq b_n$ (rather than $a_n \leqq b_n$). Consequently, the comparison test does not apply immediately—the fact that $\sum a_n$ dominates a convergent series does *not* imply that $\sum a_n$ itself converges. Theorem 2 provides a convenient way of handling such a situation.

Theorem 2 ***Limit Comparison Test***

Suppose that $\sum a_n$ and $\sum b_n$ are positive-term series. If the limit

$$L = \lim_{n\to\infty} \frac{a_n}{b_n}$$

exists and $0 < L < +\infty$, then either both series converge or both series diverge.

Proof Choose two fixed positive numbers P and Q such that $P < L < Q$. Then $P < a_n/b_n < Q$ for n sufficiently large, and so

$$Pb_n < a_n < Qb_n$$

for all sufficiently large values of n. If $\sum b_n$ converges, then $\sum a_n$ is eventually dominated by the convergent series $\sum Qb_n = Q\sum b_n$, so part (1) of the comparison test implies that $\sum a_n$ also converges. If $\sum b_n$ diverges, then $\sum a_n$ eventually dominates the divergent series $\sum Pb_n = P\sum b_n$, so part (2) of the comparison test implies that $\sum a_n$ also diverges. Thus the convergence of either series implies the convergence of the other. ■

EXAMPLE 4 With

$$a_n = \frac{3n^2 + n}{n^4 + \sqrt{n}} \quad \text{and} \quad b_n = \frac{1}{n^2}$$

(motivated by the discussion preceding Theorem 2), we find that

$$\lim_{n\to\infty} \frac{a_n}{b_n} = \lim_{n\to\infty} \frac{3n^4 + n^3}{n^4 + \sqrt{n}} = \lim_{n\to\infty} \frac{3 + \dfrac{1}{n}}{1 + \dfrac{1}{n^{7/2}}} = 3.$$

Because $\sum 1/n^2$ is a convergent p-series ($p = 2$), the limit comparison test tells us that the series

$$\sum_{n=1}^{\infty} \frac{3n^2 + n}{n^4 + \sqrt{n}}$$

also converges. ■

EXAMPLE 5 Test for convergence: $\sum_{n=1}^{\infty} \frac{1}{2n + \ln n}$.

Solution Because $\lim_{n\to\infty} (\ln n)/n = 0$ (by l'Hôpital's rule), $\ln n$ is very small in comparison with $2n$ when n is large. We therefore take $a_n = 1/(2n + \ln n)$ and, ignoring the constant coefficient 2, we take $b_n = 1/n$. Then we find that

$$\lim_{n\to\infty} \frac{a_n}{b_n} = \lim_{n\to\infty} \frac{n}{2n + \ln n} = \lim_{n\to\infty} \frac{1}{2 + \dfrac{\ln n}{n}} = \frac{1}{2}.$$

Because the harmonic series $\sum 1/n = \sum b_n$ diverges, it follows that the given series $\sum a_n$ also diverges. ■

It is important to remember that if $L = \lim_{n\to\infty} (a_n/b_n)$ is either zero or infinite, then the limit comparison test does not apply. (See Problem 52 for a discussion of what conclusions may sometimes be drawn in these cases.) Note, for example, that if $a_n = 1/n^2$ and $b_n = 1/n$, then $\lim_{n\to\infty} (a_n/b_n) = 0$. But in this case $\sum a_n$ converges, whereas $\sum b_n$ diverges.

Estimating Remainders

Suppose that $0 \leqq a_n \leqq b_n$ for all n and we know that $\sum b_n$ converges, so the comparison test implies that $\sum a_n$ converges as well. Let us write $s = \sum a_n$ and $S = \sum b_n$. If a numerical estimate is available for the remainder

$$R_n = S - S_n = b_{n+1} + b_{n+2} + \cdots$$

in the dominating series $\sum b_n$, then we can use it to estimate the remainder

$$r_n = s - s_n = a_{n+1} + a_{n+2} + \cdots$$

in the series $\sum a_n$. The reason is that $0 \leqq a_n \leqq b_n$ (for all n) implies that $0 \leqq r_n \leqq R_n$. We can apply this fact if, for instance, we have used the integral test remainder estimate to calculate an upper bound for R_n—which is, then, an upper bound for r_n as well.

EXAMPLE 6 The series

$$\sum_{n=1}^{\infty} a_n = \sum_{n=1}^{\infty} \frac{1}{n^3 + \sqrt{n}}$$

converges because it is dominated by the convergent p-series

$$\sum_{n=1}^{\infty} b_n = \sum_{n=1}^{\infty} \frac{1}{n^3}.$$

It therefore follows by the integral test remainder estimate (Section 11.5) that

$$0 < r_n \leqq R_n \leqq \int_n^{\infty} \frac{1}{x^3}\,dx = \lim_{b\to\infty} \left[-\frac{1}{2x^2} \right]_n^b = \frac{1}{2n^2}.$$

Now a calculator gives

$$s_{100} = \sum_{n=1}^{100} \frac{1}{n^3 + \sqrt{n}} \approx 0.680284 \quad \text{and} \quad R_{100} \leqq \frac{1}{2 \cdot 100^2} = 0.00005.$$

It follows that $0.680284 \leqq s \leqq 0.680334$. In particular,

$$\sum_{n=1}^{\infty} \frac{1}{n^3 + \sqrt{n}} \approx 0.6803$$

rounded accurate to four decimal places. ■

Rearrangement and Grouping

We close our discussion of positive-term series with the observation that the sum of a convergent *positive*-term series is not altered by grouping or rearranging its terms. For example, let $\sum a_n$ be a convergent positive-term series and consider

$$\sum_{n=1}^{\infty} b_n = (a_1 + a_2 + a_3) + a_4 + (a_5 + a_6) + \cdots.$$

That is, the new series has terms

$$b_1 = a_1 + a_2 + a_3,$$

$$b_2 = a_4,$$

$$b_3 = a_5 + a_6,$$

and so on. Then every partial sum T_n of $\sum b_n$ is equal to some partial sum $S_{n'}$ of $\sum a_n$. Because $\{S_n\}$ is an increasing sequence with limit $S = \sum a_n$, it follows easily that $\{T_n\}$ is an increasing sequence with the same limit. Thus $\sum b_n = S$ as well. The argument is more subtle if terms of $\sum a_n$ are moved "out of place," as in

$$\sum_{n=1}^{\infty} b_n = a_1 + a_2 + a_4 + a_3 + a_6 + a_8 + a_5 + a_{10} + a_{12} + \cdots,$$

but the same conclusion holds: Any rearrangement of a convergent *positive*-term series also converges, and it converges to the same sum.

Similarly, it is easy to prove that any grouping or rearrangement of a divergent positive-term series also diverges. But these observations all fail in the case of an infinite series with both positive and negative terms. For example, the series $\sum(-1)^n$ diverges, but it has the convergent grouping

$$(-1 + 1) + (-1 + 1) + (-1 + 1) + \cdots = 0 + 0 + 0 + \cdots = 0.$$

It follows from Problem 56 of Section 11.4 that

$$\ln 2 = 1 - \frac{1}{2} + \frac{1}{3} - \frac{1}{4} + \frac{1}{5} - \cdots,$$

but the rearrangement

$$1 + \frac{1}{3} - \frac{1}{2} + \frac{1}{5} + \frac{1}{7} - \frac{1}{4} + \frac{1}{9} + \frac{1}{11} - \frac{1}{6} + \cdots$$

converges instead to $\frac{3}{2}\ln 2$. This series for $\ln 2$ even has rearrangements that converge to zero and others that diverge to $+\infty$ (see Problem 64 of Section 11.7).

11.6 PROBLEMS

Use comparison tests to determine whether the infinite series in Problems 1 through 36 converge or diverge.

1. $\sum_{n=1}^{\infty} \frac{1}{n^2+n+1}$

2. $\sum_{n=1}^{\infty} \frac{n^3+1}{n^4+2}$

3. $\sum_{n=1}^{\infty} \frac{1}{n+\sqrt{n}}$

4. $\sum_{n=1}^{\infty} \frac{1}{n+n^{3/2}}$

5. $\sum_{n=1}^{\infty} \frac{1}{1+3^n}$

6. $\sum_{n=1}^{\infty} \frac{10n^2}{n^4+1}$

7. $\sum_{n=2}^{\infty} \frac{10n^2}{n^3-1}$

8. $\sum_{n=1}^{\infty} \frac{n^2-n}{n^4+2}$

9. $\sum_{n=1}^{\infty} \frac{1}{\sqrt{37n^3+3}}$

10. $\sum_{n=1}^{\infty} \frac{1}{\sqrt{n^2+1}}$

11. $\sum_{n=1}^{\infty} \frac{\sqrt{n}}{n^2+n}$

12. $\sum_{n=1}^{\infty} \frac{1}{3+5^n}$

13. $\sum_{n=2}^{\infty} \frac{1}{\ln n}$

14. $\sum_{n=1}^{\infty} \frac{1}{n-\ln n}$

15. $\sum_{n=1}^{\infty} \frac{\sin^2 n}{n^2+1}$

16. $\sum_{n=1}^{\infty} \frac{\cos^2 n}{3^n}$

17. $\sum_{n=1}^{\infty} \frac{n+2^n}{n+3^n}$

18. $\sum_{n=1}^{\infty} \frac{1}{2^n+3^n}$

19. $\sum_{n=2}^{\infty} \frac{1}{n^2 \ln n}$

20. $\sum_{n=1}^{\infty} \frac{1}{n^{1+\sqrt{n}}}$

21. $\sum_{n=1}^{\infty} \frac{\ln n}{n^2}$

22. $\sum_{n=1}^{\infty} \frac{\arctan n}{n}$

23. $\sum_{n=1}^{\infty} \frac{\sin^2(1/n)}{n^2}$

24. $\sum_{n=1}^{\infty} \frac{e^{1/n}}{n}$

25. $\sum_{n=1}^{\infty} \frac{\ln n}{e^n}$

26. $\sum_{n=1}^{\infty} \frac{n^2+2}{n^3+3n}$

27. $\sum_{n=1}^{\infty} \frac{n^{3/2}}{n^2+4}$

28. $\sum_{n=1}^{\infty} \frac{1}{n \cdot 2^n}$

29. $\sum_{n=1}^{\infty} \frac{3}{4+\sqrt{n}}$

30. $\sum_{n=1}^{\infty} \frac{n^2+1}{e^n(n+1)^2}$

31. $\sum_{n=1}^{\infty} \frac{2n^2-1}{n^2 \cdot 3^n}$

32. $\sum_{n=1}^{\infty} \frac{1}{\sqrt[3]{2n^4+1}}$

33. $\sum_{n=1}^{\infty} \frac{2+\sin n}{n^2}$

34. $\sum_{n=1}^{\infty} \frac{\ln n}{n^3}$

35. $\sum_{n=1}^{\infty} \frac{(n+1)^n}{n^{n+1}}$ $\left(\textit{Suggestion: } \lim_{n\to\infty}\left(1+\frac{1}{n}\right)^n = e\right).$

36. $\sum_{n=1}^{\infty} \left(\frac{\sin n}{n}\right)^4$

In Problems 37 through 40, calculate the sum of the first ten terms of the series, then estimate the error made in using this partial sum to approximate the sum of the series.

37. $\sum_{n=1}^{\infty} \frac{1}{n^2+1}$

38. $\sum_{n=1}^{\infty} \frac{1}{3^n+1}$

39. $\sum_{n=1}^{\infty} \frac{\cos^2 n}{n^2}$

40. $\sum_{n=2}^{\infty} \frac{1}{(n+1)(\ln n)^2}$

In Problems 41 through 44, first determine the smallest positive integer n such that the remainder satisfies the inequality $R_n < 0.005$. Then use a calculator or computer to approximate the sum of the series accurate to two decimal places.

41. $\sum_{n=1}^{\infty} \frac{1}{n^3+1}$

42. $\sum_{n=1}^{\infty} \frac{n}{(n+1)2^n}$

43. $\sum_{n=1}^{\infty} \frac{\cos^4 n}{n^4}$

44. $\sum_{n=1}^{\infty} \frac{1}{n^{2+(1/n)}}$

45. Show that if $\sum a_n$ is a convergent positive-term series, then the series $\sum \sin(a_n)$ also converges.

46. (a) Prove that $\ln n < n^{1/8}$ for all sufficiently large values of n. (b) Explain why part (a) shows that the series $\sum 1/(\ln n)^8$ diverges.

47. Prove that if $\sum a_n$ is a convergent positive-term series, then $\sum (a_n/n)$ converges.

48. Suppose that $\sum a_n$ is a convergent positive-term series and that $\{c_n\}$ is a sequence of positive numbers with limit zero. Prove that $\sum a_n c_n$ converges.

49. Use the result of Problem 48 to prove that if $\sum a_n$ and $\sum b_n$ are convergent positive-term series, then $\sum a_n b_n$ converges.

50. Prove that the series

$$\sum_{n=1}^{\infty} \frac{1}{1+2+3+\cdots+n}$$

converges.

51. Use the result of Problem 50 in Section 11.5 to prove that the series

$$\sum_{n=1}^{\infty} \frac{1}{1+\frac{1}{2}+\frac{1}{3}+\cdots+\frac{1}{n}}$$

diverges.

52. Adapt the proof of the limit comparison test to prove the following two results. (a) Suppose that $\sum a_n$ and $\sum b_n$ are positive-term series and that $\sum b_n$ converges. If

$$L = \lim_{n\to\infty} \frac{a_n}{b_n} = 0,$$

then $\sum a_n$ converges. (b) Suppose that $\sum a_n$ and $\sum b_n$ are positive-term series and that $\sum b_n$ diverges. If

$$L = \lim_{n\to\infty} \frac{a_n}{b_n} = +\infty,$$

then $\sum a_n$ diverges.

11.7 ALTERNATING SERIES AND ABSOLUTE CONVERGENCE

In Sections 11.5 and 11.6 we considered only positive-term series. Now we discuss infinite series that have both positive terms and negative terms. An important example is a series with terms that are alternately positive and negative. An **alternating series** is an infinite series of the form

$$\sum_{n=1}^{\infty} (-1)^{n+1} a_n = a_1 - a_2 + a_3 - a_4 + a_5 - \cdots \tag{1}$$

or of the form $\sum_{n=1}^{\infty} (-1)^n a_n$, where $a_n > 0$ for all n. For example, both the *alternating harmonic series*

$$\sum_{n=1}^{\infty} \frac{(-1)^{n+1}}{n} = 1 - \frac{1}{2} + \frac{1}{3} - \frac{1}{4} + \frac{1}{5} - \cdots$$

and the geometric series

$$\sum_{n=0}^{\infty} \left(-\frac{1}{2}\right)^n = 1 - \frac{1}{2} + \frac{1}{4} - \frac{1}{8} + \frac{1}{16} - \cdots$$

are alternating series. Theorem 1 shows that both of these series converge because the sequence of absolute values of their terms is decreasing and has limit zero.

Theorem 1 ***Alternating Series Test***

If the alternating series in Eq. (1) satisfies the two conditions

1. $a_n \geqq a_{n+1} > 0$ for all n and
2. $\lim_{n\to\infty} a_n = 0$,

then the infinite series converges.

PROOF We first consider the even-numbered partial sums $S_2, S_4, S_6, \ldots, S_{2n}, \ldots$. We may write

$$S_{2n} = (a_1 - a_2) + (a_3 - a_4) + \cdots + (a_{2n-1} - a_{2n}).$$

Because $a_k - a_{k+1} \geqq 0$ for all k, the sequence $\{S_{2n}\}$ is increasing. Also, because

$$S_{2n} = a_1 - (a_2 - a_3) - \cdots - (a_{2n-2} - a_{2n-1}) - a_{2n},$$

$S_{2n} \leqq a_1$ for all n. So the increasing sequence $\{S_{2n}\}$ is bounded above. Hence the limit

$$S = \lim_{n\to\infty} S_{2n}$$

exists by the bounded monotonic sequence property of Section 11.2. It remains only for us to verify that the odd-numbered partial sums $S_1, S_3, S_5, \ldots$ also converge to S. But $S_{2n+1} = S_{2n} + a_{2n+1}$ and $\lim_{n\to\infty} a_{2n+1} = 0$, so

$$\lim_{n\to\infty} S_{2n+1} = \left(\lim_{n\to\infty} S_{2n}\right) + \left(\lim_{n\to\infty} a_{2n+1}\right) = S.$$

Thus $\lim_{n\to\infty} S_n = S$, and therefore the series in Eq. (1) converges. ■

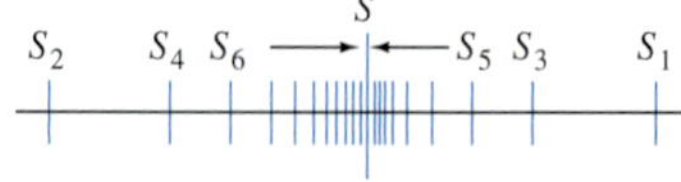

Fig. 11.7.1 The even partial sums $\{S_{2n}\}$ increase and the odd partial sums $\{S_{2n+1}\}$ decrease.

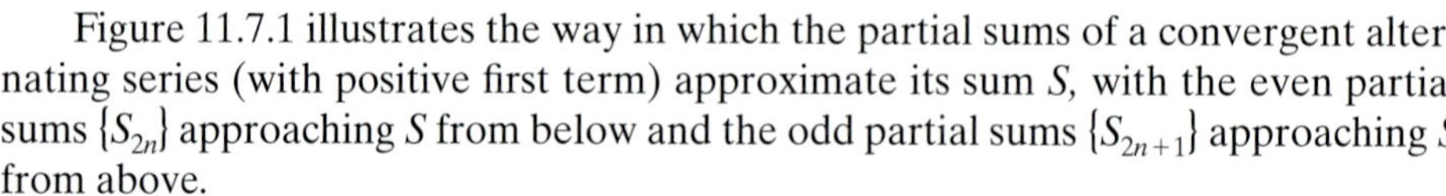

Figure 11.7.1 illustrates the way in which the partial sums of a convergent alternating series (with positive first term) approximate its sum S, with the even partial sums $\{S_{2n}\}$ approaching S from below and the odd partial sums $\{S_{2n+1}\}$ approaching S from above.

EXAMPLE 1 The series

$$\sum_{n=1}^{\infty} \frac{(-1)^{n+1}}{2n-1} = 1 - \frac{1}{3} + \frac{1}{5} - \frac{1}{7} + \frac{1}{9} - \cdots$$

satisfies the conditions of Theorem 1 and therefore converges. The alternating series test does not tell us the sum of this series, but we saw in Section 11.4 that its sum is $\pi/4$. The graph in Fig. 11.7.2 of the partial sums of this series illustrates the typical convergence of an alternating series, with its partial sums approaching its sum alternately from above and below. ■

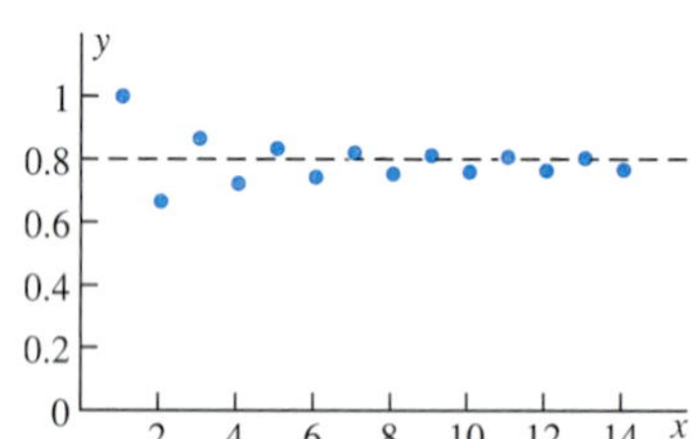

Fig. 11.7.2 Graph of the first 14 partial sums of the alternating series in Example 1

EXAMPLE 2 The series

$$\sum_{n=1}^{\infty} \frac{(-1)^{n+1}n}{2n-1} = 1 - \frac{2}{3} + \frac{3}{5} - \frac{4}{7} + \frac{5}{9} - \cdots$$

is an alternating series, and it is easy to verify that

$$a_n = \frac{n}{2n-1} > \frac{n+1}{2n+1} = a_{n+1}$$

for all $n \geqq 1$. But

$$\lim_{n\to\infty} a_n = \frac{1}{2} \neq 0,$$

so the alternating series test *does not apply.* (This fact alone does not imply that the series in question diverges—many series in Section 11.5 and 11.6 converge even though the alternating series test does not apply. But the series of this example diverges by the nth-term divergence test.) ■

If a series converges by the alternating series test, then Theorem 2 shows how to approximate its sum with any desired degree of accuracy—*if* you have a computer fast enough to add a large number of its terms.

Theorem 2 ***Alternating Series Remainder Estimate***

Suppose that the series $\sum(-1)^{n+1}a_n$ satisfies the conditions of the alternating series test and therefore converges. Let S denote the sum of the series. Denote by $R_n = S - S_n$ the error made in replacing S with the nth partial sum S_n of the series. Then this **remainder** R_n has the same sign as the next term $(-1)^{n+2}a_{n+1}$ of the series, and

$$0 \leqq |R_n| < a_{n+1}. \tag{2}$$

In particular, the *sum S of a convergent alternating series lies between any two consecutive partial sums.* This follows from the proof of Theorem 1, where we saw that $\{S_{2n}\}$ is an increasing sequence and that $\{S_{2n+1}\}$ is a decreasing sequence, both converging to S. The resulting inequalities

$$S_{2n-1} > S > S_{2n} = S_{2n-1} - a_{2n}$$

and

$$S_{2n} < S < S_{2n+1} = S_{2n} + a_{2n+1}$$

(see Fig. 11.7.3) imply the inequality in (2).

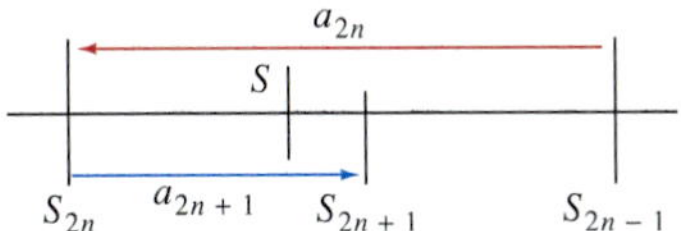

Fig. 11.7.3 Illustrating the proof of the alternating series remainder estimate

REMARK The inequality in (2) means the following. Suppose that you are given an *alternating series* that satisfies the conditions of Theorem 2 and has sum S. Then, if S is replaced with a partial sum S_n, the error made is numerically less than the first term a_{n+1} not retained and has the same sign as this first neglected term. **Important:** This error estimate *does not* apply to other types of series.

EXAMPLE 3 We saw in Section 11.4 that

$$e^x = \sum_{n=0}^{\infty} \frac{x^n}{n!}$$

for all x and thus that

$$\frac{1}{e} = e^{-1} = 1 - 1 + \frac{1}{2!} - \frac{1}{3!} + \frac{1}{4!} - \cdots.$$

Use this alternating series to compute e^{-1} accurate to four decimal places.

Solution To attain four-place accuracy, we want the error to be less than a half unit in the fourth place. Thus we want

$$|R_n| < \frac{1}{(n+1)!} \leqq 0.00005.$$

If we use a calculator to compute the reciprocals of the factorials of the first several integers, we find that the least value of n for which this inequality holds is $n = 7$. Then

$$e^{-1} = 1 - \frac{1}{1!} + \frac{1}{2!} - \frac{1}{3!} + \frac{1}{4!} - \frac{1}{5!} + \frac{1}{6!} - \frac{1}{7!} + R_7 \approx 0.367857 + R_7.$$

(Relying on a common "+2 rule of thumb," we are carrying six decimal places because we want four-place accuracy in the final answer.) Now the first neglected term $1/8!$ is positive, so the inequality in (2) gives

$$0 < R_7 < \frac{1}{8!} < 0.000025.$$

Therefore

$$S_7 \approx 0.367857 < e^{-1} < S_7 + 0.000025 \approx 0.367882.$$

The two bounds here both round to $e^{-1} \approx 0.3679$. Although this approximation is accurate to four decimal places, its reciprocal

$$e = 1/e^{-1} \approx 1/(0.3679) \approx 2.7181 \approx 2.718$$

gives the number e accurate to only three decimal places. ■

Absolute Convergence

The series

$$\sum_{n=1}^{\infty} \frac{(-1)^{n+1}}{n} = 1 - \frac{1}{2} + \frac{1}{3} - \frac{1}{4} + \frac{1}{5} - \cdots$$

converges, but if we simply replace each term with its absolute value, we get the *divergent* series

$$1 + \frac{1}{2} + \frac{1}{3} + \frac{1}{4} + \frac{1}{5} + \cdots.$$

In contrast, the *convergent* series

$$\sum_{n=0}^{\infty} \frac{(-1)^n}{2^n} = 1 - \frac{1}{2} + \frac{1}{4} - \frac{1}{8} + \cdots = \frac{2}{3}$$

has the property that the associated positive-term series

$$1 + \frac{1}{2} + \frac{1}{4} + \frac{1}{8} + \cdots = 2$$

also converges. Theorem 3 tells us that if a series of *positive* terms converges, then we may insert minus signs in front of any of the terms—every other one, for instance—and the resulting series will also converge.

Theorem 3 ***Absolute Convergence Implies Convergence***

If the series $\sum |a_n|$ converges, then so does the series $\sum a_n$.

Proof Suppose that the series $\sum |a_n|$ converges. Note that

$$0 \leqq a_n + |a_n| \leqq 2|a_n|$$

for all n. Let $b_n = a_n + |a_n|$. It then follows from the comparison test that the positive-term series $\sum b_n$ converges, because it is dominated by the convergent series $\sum 2|a_n|$. It is easy to verify, too, that the termwise difference of two convergent series also converges. Hence we now see that the series

$$\sum a_n = \sum (b_n - |a_n|) = \sum b_n - \sum |a_n|$$

converges. ■

Thus we have another convergence test, one not limited to positive-term series or to alternating series: Given the series $\sum a_n$, test the series $\sum |a_n|$ for convergence. If the latter converges, then so does the former. (But the converse is *not* true!) This phenomenon motivates us to make the following definition.

Definition *Absolute Convergence*

The series $\sum a_n$ is said to **converge absolutely** (and is called **absolutely convergent**) provided that the series

$$\sum |a_n| = |a_1| + |a_2| + |a_3| + \cdots + |a_n| + \cdots$$

converges.

Thus we have explained the title of Theorem 3, and we can rephrase the theorem as follows: *If a series converges absolutely, then it converges.* The two examples preceding Theorem 3 show that a convergent series may either converge absolutely or fail to do so:

$$1 - \frac{1}{2} + \frac{1}{4} - \frac{1}{8} + \frac{1}{16} - \cdots$$

is an absolutely convergent series because

$$1 + \frac{1}{2} + \frac{1}{4} + \frac{1}{8} + \frac{1}{16} + \cdots$$

converges, whereas

$$1 - \frac{1}{2} + \frac{1}{3} - \frac{1}{4} + \frac{1}{5} - \cdots$$

is a series that, though convergent, is *not* absolutely convergent. A series that converges but does not converge absolutely is said to be **conditionally convergent.** Consequently, the terms *absolutely convergent, conditionally convergent,* and *divergent* are simultaneously all-inclusive and mutually exclusive: Any given numerical series belongs to exactly one of those three classes.

There is some advantage in the application of Theorem 3, because to apply it we test the *positive*-term series $\sum |a_n|$ for convergence—and we have a variety of tests, such as comparison tests or the integral test, designed for use on positive-term series.

Note also that absolute convergence of the series $\sum a_n$ means that a *different* series $\sum |a_n|$ converges, and the two sums will generally differ. For example, with $a_n = (-\frac{1}{3})^n$, the formula for the sum of a geometric series gives

$$\sum_{n=0}^{\infty} a_n = \sum_{n=0}^{\infty} \left(-\frac{1}{3}\right)^n = \frac{1}{1 - (-\frac{1}{3})} = \frac{3}{4},$$

whereas

$$\sum_{n=0}^{\infty} |a_n| = \sum_{n=0}^{\infty} \left(\frac{1}{3}\right)^n = \frac{1}{1 - \frac{1}{3}} = \frac{3}{2}.$$

EXAMPLE 4 Discuss the convergence of the series

$$\sum_{n=1}^{\infty} \frac{\cos n}{n^2} = \cos 1 + \frac{\cos 2}{4} + \frac{\cos 3}{9} + \cdots.$$

Solution Let $a_n = (\cos n)/n^2$. Then

$$|a_n| = \frac{|\cos n|}{n^2} \leqq \frac{1}{n^2}$$

for all $n \geqq 1$. Hence the positive-term series $\sum |a_n|$ converges by the comparison test, because it is dominated by the convergent p-series $\sum(1/n^2)$. Thus the given series is absolutely convergent, and it therefore converges by Theorem 3. ■

One reason for the importance of absolute convergence is the fact (proved in advanced calculus) that the terms of an absolutely convergent series may be regrouped or rearranged without changing the sum of the series. As we suggested at the end of Section 11.6, this is *not* true of conditionally convergent series.

The Ratio Test and the Root Test

Our next two convergence tests involve a way of measuring the rate of growth or decrease of the sequence $\{a_n\}$ of terms of a series to determine whether $\sum a_n$ converges absolutely or diverges.

Theorem 4 ***The Ratio Test***

Suppose that the limit

$$\rho = \lim_{n\to\infty} \left|\frac{a_{n+1}}{a_n}\right| \tag{3}$$

either exists or is infinite. Then the infinite series $\sum a_n$ of nonzero terms

1. Converges absolutely if $\rho < 1$;
2. Diverges if $\rho > 1$.

If $\rho = 1$, the ratio test is inconclusive.

Proof If $\rho < 1$, choose a (fixed) number r with $\rho < r < 1$. Then Eq. (3) implies that there exists an integer N such that $|a_{n+1}| \leqq r|a_n|$ for all $n \geqq N$. It follows that

$$|a_{N+1}| \leqq r|a_N|,$$

$$|a_{N+2}| \leqq r|a_{N+1}| \leqq r^2|a_N|,$$

$$|a_{N+3}| \leqq r|a_{N+2}| \leqq r^3|a_N|,$$

and in general that

$$|a_{N+k}| \leqq r^k|a_N| \qquad \text{for } k \geqq 0.$$

Hence the series

$$|a_N| + |a_{N+1}| + |a_{N+2}| + \cdots$$

is dominated by the geometric series

$$|a_N|(1 + r + r^2 + r^3 + \cdots),$$

and the latter converges because $|r| < 1$. Thus the series $\sum |a_n|$ converges, so the series $\sum a_n$ converges absolutely.

If $\rho > 1$, then Eq. (3) implies that there exists a positive integer N such that $|a_{n+1}| > |a_n|$ for all $n \geqq N$. It follows that $|a_n| > |a_N| > 0$ for all $n > N$. Thus the sequence $\{a_n\}$ cannot approach zero as $n \to \infty$, and consequently, by the nth-term divergence test, the series $\sum a_n$ diverges. ■

To see that $\sum a_n$ may either converge or diverge if $\rho = 1$, consider the divergent series $\sum(1/n)$ and the convergent series $\sum(1/n^2)$. You should verify that, for both series, the value of the ratio ρ is 1.

EXAMPLE 5 Consider the series

$$\sum_{n=1}^{\infty} \frac{(-1)^n 2^n}{n!} = -2 + \frac{4}{2!} - \frac{8}{3!} + \frac{16}{4!} - \cdots.$$

Then

$$\rho = \lim_{n\to\infty} \left|\frac{a_{n+1}}{a_n}\right| = \lim_{n\to\infty} \left|\frac{\dfrac{(-1)^{n+1}2^{n+1}}{(n+1)!}}{\dfrac{(-1)^n 2^n}{n!}}\right| = \lim_{n\to\infty} \frac{2}{n+1} = 0.$$

Because $\rho < 1$, the series converges absolutely. ■

EXAMPLE 6 Test for convergence: $\displaystyle\sum_{n=1}^{\infty} \frac{n}{2^n}$.

Solution We have

$$\rho = \lim_{n\to\infty} \left|\frac{a_{n+1}}{a_n}\right| = \lim_{n\to\infty} \frac{\dfrac{n+1}{2^{n+1}}}{\dfrac{n}{2^n}} = \lim_{n\to\infty} \frac{n+1}{2n} = \frac{1}{2}.$$

Because $\rho < 1$, this series converges (absolutely). ■

EXAMPLE 7 Test for convergence: $\displaystyle\sum_{n=1}^{\infty} \frac{3^n}{n^2}$.

Solution Here we have

$$\rho = \lim_{n\to\infty} \left|\frac{a_{n+1}}{a_n}\right| = \lim_{n\to\infty} \frac{\dfrac{3^{n+1}}{(n+1)^2}}{\dfrac{3^n}{n^2}} = \lim_{n\to\infty} \frac{3n^2}{(n+1)^2} = 3.$$

In this case $\rho > 1$, so the given series diverges. ■

Theorem 5 ***The Root Test***

Suppose that the limit

$$\rho = \lim_{n\to\infty} \sqrt[n]{|a_n|} \tag{4}$$

exists or is infinite. Then the infinite series $\sum a_n$

1. Converges absolutely if $\rho < 1$;
2. Diverges if $\rho > 1$.

If $\rho = 1$, the root test is inconclusive.

PROOF If $\rho < 1$, choose a (fixed) number r such that $\rho < r < 1$. Then $|a_n|^{1/n} < r$, and hence $|a_n| < r^n$, for n sufficiently large. Thus the series $\sum |a_n|$ is eventually dominated by the convergent geometric series $\sum r^n$. Therefore $\sum |a_n|$ converges, and so the series $\sum a_n$ converges absolutely.

If $\rho > 1$, then $|a_n|^{1/n} > 1$, and hence $|a_n| > 1$, for n sufficiently large. Therefore the nth-term test for divergence implies that the series $\sum a_n$ diverges. ■

The ratio test is generally simpler to apply than the root test, and therefore it is ordinarily the one to try first. But there are certain series for which the root test succeeds and the ratio test fails, as in Example 8.

EXAMPLE 8 Consider the series

$$\sum_{n=0}^{\infty} \frac{1}{2^{n+(-1)^n}} = \frac{1}{2} + \frac{1}{1} + \frac{1}{8} + \frac{1}{4} + \frac{1}{32} + \frac{1}{16} + \cdots.$$

Then $a_{n+1}/a_n = 2$ if n is even, whereas $a_{n+1}/a_n = \frac{1}{8}$ if n is odd. So the limit required for the ratio test does not exist. But

$$\lim_{n\to\infty} |a_n|^{1/n} = \lim_{n\to\infty} \left|\frac{1}{2^{n+(-1)^n}}\right|^{1/n} = \lim_{n\to\infty} \frac{1}{2}\left|\frac{1}{2^{(-1)^n/n}}\right| = \frac{1}{2},$$

so the given series converges by the root test. (Its convergence also follows from the fact that it is a rearrangement of the positive-term convergent geometric series $\sum 1/2^n$.) ■

11.7 PROBLEMS

Determine whether the alternating series in Problems 1 through 20 converge or diverge.

1. $\sum_{n=1}^{\infty} \frac{(-1)^{n+1}}{n^2}$

2. $\sum_{n=1}^{\infty} \frac{(-1)^{n+1}}{\sqrt{n^2+1}}$

3. $\sum_{n=1}^{\infty} \frac{(-1)^n n}{3n+2}$

4. $\sum_{n=1}^{\infty} \frac{(-1)^n n}{3n^2+2}$

5. $\sum_{n=1}^{\infty} \frac{(-1)^{n+1} n}{\sqrt{n^2+2}}$

6. $\sum_{n=1}^{\infty} \frac{(-1)^{n+1} n^2}{\sqrt{n^5+5}}$

7. $\sum_{n=2}^{\infty} \frac{(-1)^{n+1} n}{\ln n}$

8. $\sum_{n=1}^{\infty} \frac{(-1)^n \ln n}{\sqrt{n}}$

9. $\sum_{n=1}^{\infty} \frac{(-1)^n n}{2^n}$

10. $\sum_{n=1}^{\infty} n \cdot \left(-\frac{2}{3}\right)^{n+1}$

11. $\sum_{n=1}^{\infty} \frac{(-1)^n n}{\sqrt{2^n+1}}$

12. $\sum_{n=1}^{\infty} \left(-\frac{n\pi}{10}\right)^{n+1}$

13. $\sum_{n=1}^{\infty} \frac{1}{n^{2/3}} \sin\left(\frac{n\pi}{2}\right)$

14. $\sum_{n=1}^{\infty} \frac{\cos n\pi}{n^{3/2}}$

15. $\sum_{n=1}^{\infty} (-1)^n \sin\left(\frac{1}{n}\right)$

16. $\sum_{n=1}^{\infty} (-1)^n n \sin\left(\frac{\pi}{n}\right)$

17. $\sum_{n=1}^{\infty} \frac{(-1)^{n+1}}{\sqrt[n]{2}}$

18. $\sum_{n=1}^{\infty} \frac{(-1.01)^{n+1}}{n^4}$

19. $\sum_{n=1}^{\infty} \frac{(-1)^{n+1}}{\sqrt[n]{n}}$

20. $\sum_{n=1}^{\infty} \frac{(-1)^{n+1} n!}{(2n)!}$

Determine whether the series in Problems 21 through 42 converge absolutely, converge conditionally, or diverge.

21. $\sum_{n=1}^{\infty} \frac{(-1)^{n+1}}{2^n}$

22. $\sum_{n=1}^{\infty} \frac{1}{n^2+1}$

23. $\sum_{n=1}^{\infty} \frac{(-1)^n \ln n}{n}$

24. $\sum_{n=1}^{\infty} \frac{1}{n^n}$

25. $\sum_{n=1}^{\infty} \left(\frac{10}{n}\right)^n$

26. $\sum_{n=1}^{\infty} \frac{3^n}{n! n}$

27. $\sum_{n=0}^{\infty} \frac{(-10)^n}{n!}$

28. $\sum_{n=1}^{\infty} \frac{(-1)^{n+1} n!}{n^n}$

29. $\sum_{n=1}^{\infty} (-1)^{n+1}\left(\frac{n}{n+1}\right)^n$ **30.** $\sum_{n=1}^{\infty} \frac{n!n^2}{(2n)!}$

31. $\sum_{n=1}^{\infty} \left(\frac{\ln n}{n}\right)^n$ **32.** $\sum_{n=0}^{\infty} \frac{(-1)^n 2^{3n}}{7^n}$

33. $\sum_{n=0}^{\infty} (-1)^n(\sqrt{n+1} - \sqrt{n})$

34. $\sum_{n=1}^{\infty} n \cdot \left(\frac{3}{4}\right)^n$ **35.** $\sum_{n=1}^{\infty} \left[\ln\left(\frac{1}{n}\right)\right]^n$

36. $\sum_{n=0}^{\infty} \frac{(n!)^2}{(2n)!}$ **37.** $\sum_{n=1}^{\infty} \frac{(-1)^{n+1}3^n}{n(2^n+1)}$

38. $\sum_{n=1}^{\infty} \frac{(-1)^{n+1}\arctan n}{n}$ **39.** $\sum_{n=1}^{\infty} \frac{(-1)^{n+1}n!}{1\cdot 3\cdot 5\cdots(2n-1)}$

40. $\sum_{n=1}^{\infty} (-1)^{n+1}\frac{1\cdot 3\cdot 5\cdots(2n-1)}{1\cdot 4\cdot 7\cdots(3n-2)}$

41. $\sum_{n=1}^{\infty} \frac{(n+2)!}{3^n(n!)^2}$ **42.** $\sum_{n=1}^{\infty} \frac{(-1)^{n+1}n^n}{3^{n^2}}$

In Problems 43 through 48, sum the indicated number of terms of the given alternating series. Then apply the alternating series remainder estimate to estimate the error in approximating the sum of the series with this partial sum. Finally, approximate the sum of the series, writing precisely the number of decimal places that thereby are guaranteed to be correct (after rounding).

43. $\sum_{n=1}^{\infty} \frac{(-1)^{n+1}}{n^3}$, 5 terms **44.** $\sum_{n=1}^{\infty} \frac{(-1)^{n+1}}{3^n}$, 8 terms

45. $\sum_{n=1}^{\infty} \frac{(-1)^{n+1}}{n!}$, 6 terms **46.** $\sum_{n=1}^{\infty} \frac{(-1)^{n+1}}{n^n}$, 7 terms

47. $\sum_{n=1}^{\infty} \frac{(-1)^{n+1}}{n}$, 12 terms

48. $\sum_{n=1}^{\infty} \frac{(-1)^{n+1}}{n^2}$, 15 terms

In Problems 49 through 54, sum enough terms (tell how many) to approximate the sum of series, writing the sum rounded to the indicated number of correct decimal places.

49. $\sum_{n=1}^{\infty} \frac{(-1)^{n+1}}{n^4}$, 3 decimal places

50. $\sum_{n=1}^{\infty} \frac{(-1)^{n+1}}{n^5}$, 4 decimal places

51. $\frac{1}{\sqrt{e}} = \sum_{n=0}^{\infty} \frac{(-1)^n}{n!2^n}$, 4 decimal places

52. $\cos 1 = \sum_{n=0}^{\infty} \frac{(-1)^n}{(2n)!}$, 5 decimal places

53. $\sin 60° = \sum_{n=0}^{\infty} \frac{(-1)^n}{(2n+1)!}\left(\frac{\pi}{3}\right)^{2n+1}$, 5 decimal places

54. $\ln(1.1) = \sum_{n=1}^{\infty} \frac{(-1)^{n+1}}{n\cdot 10^n}$, 7 decimal places

In Problems 55 and 56, show that the indicated alternating series $\sum(-1)^{n+1}a_n$ satisfies the condition that $a_n \to 0$ as $n \to \infty$, but nevertheless diverges. Tell why the alternating series test does not apply. It may be informative to graph the first 10 or 20 partial sums.

55. $a_n = \begin{cases} \dfrac{1}{n} & \text{if } n \text{ is odd} \\ \dfrac{1}{n^2} & \text{if } n \text{ is even.} \end{cases}$

56. $a_n = \begin{cases} \dfrac{1}{\sqrt{n}} & \text{if } n \text{ is odd,} \\ \dfrac{1}{n^3} & \text{if } n \text{ is even.} \end{cases}$

57. Give an example of a pair of convergent series $\sum a_n$ and $\sum b_n$ such that $\sum a_n b_n$ diverges.

58. Prove that $\sum |a_n|$ diverges if the series $\sum a_n$ diverges.

59. Prove that

$$\lim_{n\to\infty} \frac{a^n}{n!} = 0$$

(for any real number a) by applying the ratio test to show that the infinite series $\sum a^n/n!$ converges.

60. (a) Suppose that r is a (fixed) number such that $|r| < 1$. Use the ratio test to prove that the series $\sum_{n=0}^{\infty} nr^n$ converges. Let S denote its sum. (b) Show that

$$(1 - r)S = \sum_{n=1}^{\infty} r^n.$$

Show how to conclude that

$$\sum_{n=0}^{\infty} nr^n = \frac{r}{(1-r)^2}.$$

61. Let

$$H_n = \sum_{k=1}^{n} \frac{1}{k} \quad\text{and}\quad S_n = \sum_{k=1}^{n} \frac{(-1)^{k+1}}{k}$$

denote the nth partial sums of the harmonic and alternating harmonic series, respectively. (a) Show that $S_{2n} = H_{2n} - H_n$ for all $n \geqq 1$. (b) Problem 50 in Section 11.5 says that

$$\lim_{n\to\infty} (H_n - \ln n) = \gamma$$

(where $\gamma \approx 0.57722$ denotes Euler's constant). Explain why it follows that

$$\lim_{n\to\infty} (H_{2n} - \ln 2n) = \gamma.$$

(c) Conclude from parts (a) and (b) that $\lim_{n\to\infty} S_{2n} = \ln 2$. Thus

$$\ln 2 = 1 - \frac{1}{2} + \frac{1}{3} - \frac{1}{4} + \frac{1}{5} - \frac{1}{6} + \cdots.$$

62. Suppose that $\sum a_n$ is a conditionally convergent infinite series. For each n, let

$$a_n^+ = \frac{a_n + |a_n|}{2} \quad \text{and} \quad a_n^- = \frac{a_n - |a_n|}{2}.$$

(a) Explain why $\sum a_n^+$ consists of the positive terms of $\sum a_n$ and why $\sum a_n^-$ consists of the negative terms of $\sum a_n$. (b) Given a real number r, show that some rearrangement of the conditionally convergent series $\sum a_n$ converges to r. (*Suggestion*: If r is positive, for instance, begin with the first partial sum of the positive series $\sum a_n^+$ that exceeds r. Then add just enough terms of the negative series $\sum a_n^-$ so that the cumulative sum is less than r. Next add just enough terms of the positive series that the cumulative sum is greater than r, and continue in this way to define the desired rearrangement.) Why does it follow that this rearranged infinite series converges to r?

63. Use the method of Problem 62 to write the first dozen terms of a rearrangement of the alternating harmonic series (Problem 61) that converges to 1 rather than to ln 2.

64. Describe a way to rearrange the terms of the alternating harmonic series to obtain (a) a rearranged series that converges to -2; (b) a rearranged series that diverges to $+\infty$.

65. Here is another rearrangement of the alternating harmonic series of Problem 61:

$$\begin{aligned} 1 &- \frac{1}{2} - \frac{1}{4} - \frac{1}{6} - \frac{1}{8} \\ &+ \frac{1}{3} - \frac{1}{10} - \frac{1}{12} - \frac{1}{14} - \frac{1}{16} \\ &+ \frac{1}{5} - \frac{1}{18} - \frac{1}{20} - \frac{1}{22} - \frac{1}{24} \\ &+ \frac{1}{7} - \frac{1}{26} - \frac{1}{28} - \frac{1}{30} - \frac{1}{32} + \cdots. \end{aligned}$$

Use a computer to collect evidence about the value of its sum.

11.8 POWER SERIES

The most important infinite series representations of functions are those whose terms are constant multiples of (successive) integral powers of the independent variable x—that is, series that resemble "infinite polynomials." For example, we discussed in Section 11.4 the geometric series

$$\frac{1}{1-x} = 1 + x + x^2 + x^3 + \cdots \quad (|x| < 1) \tag{1}$$

and the Taylor series

$$e^x = \sum_{n=0}^{\infty} \frac{x^n}{n!} = 1 + x + \frac{x^2}{2!} + \frac{x^3}{3!} + \frac{x^4}{4!} + \cdots, \tag{2}$$

$$\cos x = \sum_{n=0}^{\infty} \frac{(-1)^n x^{2n}}{(2n)!} = 1 - \frac{x^2}{2!} + \frac{x^4}{4!} - \frac{x^6}{6!} + \cdots, \text{ and} \tag{3}$$

$$\sin x = \sum_{n=0}^{\infty} \frac{(-1)^n x^{2n+1}}{(2n+1)!} = x - \frac{x^3}{3!} + \frac{x^5}{5!} - \frac{x^7}{7!} + \cdots. \tag{4}$$

There we used Taylor's formula to show that the series in Eqs. (2) through (4) converge, for all x, to the functions e^x, $\cos x$, and $\sin x$, respectively. Here we investigate the convergence of a "power series" without knowing in advance the function (if any) to which it converges.

All the infinite series in Eqs. (1) through (4) have the form

$$\sum_{n=0}^{\infty} a_n x^n = a_0 + a_1 x + a_2 x^2 + \cdots + a_n x^n + \cdots \tag{5}$$

with the constant *coefficients* $a_0, a_1, a_2, \ldots$. An infinite series of this form is called a **power series** in (powers of) x. To have the initial terms of the two sides of Eq. (5) agree, we adopt here the convention that $x^0 = 1$ even if $x = 0$.

Convergence of Power Series

The partial sums of the power series in Eq. (5) are the *polynomials*

$$s_1(x) = a_0 + a_1x, \quad s_2(x) = a_0 + a_1x + a_2x^2, \quad s_3(x) = a_0 + a_1x + a_2x^2 + a_3x^3,$$

and so forth. The nth partial sum is an nth-degree polynomial. When we ask *where* the power series converges, we seek those values of x for which the limit

$$s(x) = \lim_{n\to\infty} s_n(x)$$

exists. The sum $s(x)$ of a power series is then a function of x that is defined wherever the series converges.

The power series in Eq. (5) obviously converges when $x = 0$. In general, it will converge for some nonzero values of x and diverge for others. Because of the way in which powers of x are involved, the ratio test of Section 11.7 is particularly effective in determining the values of x for which a given power series converges.

Assume that the limit

$$\rho = \lim_{n\to\infty} \left| \frac{a_{n+1}}{a_n} \right| \tag{6}$$

exists. This is the limit that we need if we want to apply the ratio test to the series $\sum a_n$ of constants. To apply the ratio test to the power series in Eq. (5), we write $u_n = a_nx^n$ and compute the limit

$$\lim_{n\to\infty} \left| \frac{u_{n+1}}{u_n} \right| = \lim_{n\to\infty} \left| \frac{a_{n+1}x^{n+1}}{a_nx^n} \right| = \rho\,|x|. \tag{7}$$

If $\rho = 0$, then $\sum a_nx^n$ converges absolutely for all x. If $\rho = +\infty$, then $\sum a_nx^n$ diverges for all $x \neq 0$. If ρ is a positive real number, we see from Eq. (7) that $\sum a_nx^n$ converges absolutely for all x such that $\rho \cdot |x| < 1$—that is, when

$$|x| < R = \frac{1}{\rho} = \lim_{n\to\infty} \left| \frac{a_n}{a_{n+1}} \right|. \tag{8}$$

In this case the ratio test also implies that $\sum a_nx^n$ diverges if $|x| > R$ but is inconclusive when $x = \pm R$. We have therefore proved Theorem 1, under the additional hypothesis that the limit in Eq. (6) exists. In Problems 69 and 70 we outline a proof that does not require this additional hypothesis.

Theorem 1 ***Convergence of Power Series***

If $\sum a_nx^n$ is a power series, then either

1. The series converges absolutely for all x, or
2. The series converges only when $x = 0$, or
3. There exists a number $R > 0$ such that $\sum a_nx^n$ converges absolutely if $|x| < R$ and diverges if $|x| > R$.

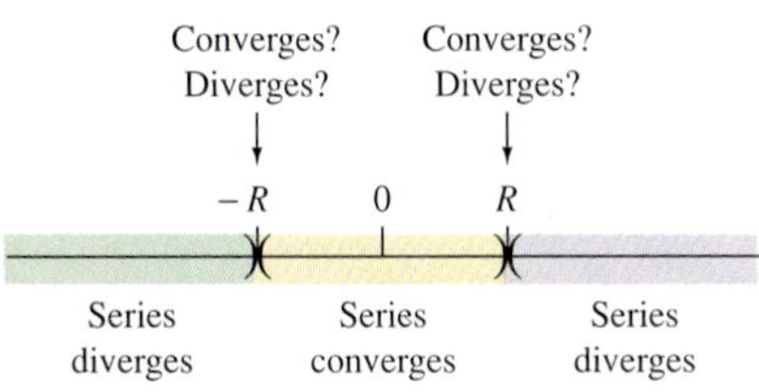

Fig. 11.8.1 The interval of convergence if $0 < R = \lim_{n\to\infty} \left| \dfrac{a_n}{a_{n+1}} \right| < \infty$

The number R of case 3 is called the **radius of convergence** of the power series $\sum a_nx^n$. We write $R = \infty$ in case 1 and $R = 0$ in case 2. The set of all real numbers x for which the series converges is called its **interval of convergence** (Fig. 11.8.1); note that this set *is* an interval. If $0 < R < \infty$, then the interval of convergence is one of the intervals

$$(-R, R), \quad (-R, R], \quad [-R, R), \quad \text{or} \quad [-R, R].$$

When we substitute either of the endpoints $x = \pm R$ into the series $\sum a_n x^n$, we obtain an infinite series with constant terms whose convergence must be determined separately. Because these will be numerical series, the earlier tests of this chapter are appropriate.

EXAMPLE 1 Find the interval of convergence of the series

$$\sum_{n=1}^{\infty} \frac{x^n}{n \cdot 3^n}.$$

Solution With $u_n = x^n/(n \cdot 3^n)$, we find that

$$\lim_{n\to\infty} \left|\frac{u_{n+1}}{u_n}\right| = \lim_{n\to\infty} \left| \frac{\dfrac{x^{n+1}}{(n+1)\cdot 3^{n+1}}}{\dfrac{x^n}{n \cdot 3^n}} \right| = \lim_{n\to\infty} \frac{n|x|}{3(n+1)} = \frac{|x|}{3}.$$

Now $|x|/3 < 1$ provided that $|x| < 3$, so the ratio test implies that the given series converges absolutely if $|x| < 3$ and diverges if $|x| > 3$. When $x = 3$, we have the divergent harmonic series $\Sigma(1/n)$, and when $x = -3$ we have the convergent alternating series $\Sigma(-1)^n/n$. Thus the interval of convergence of the given power series is $[-3, 3)$. We see in Fig. 11.8.2 the dramatic difference between convergence at $x = -3$ and divergence at $x = +3$. ■

Fig. 11.8.2 Graphs of the partial sums $S_2(x)$ and $S_6(x)$ of the power series in Example 1

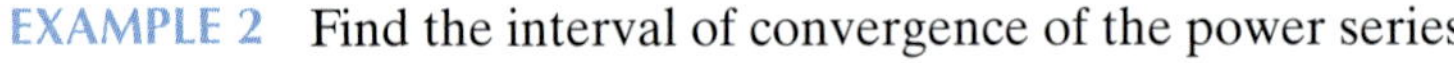

EXAMPLE 2 Find the interval of convergence of the power series

$$\sum_{n=0}^{\infty} \frac{(-2)^n x^n}{(2n)!} = 1 - \frac{2x}{2!} + \frac{4x^2}{4!} - \frac{8x^3}{6!} + \frac{16x^4}{8!} - \cdots.$$

Solution With $u_n = (-2)^n x^n/(2n)!$ we find that

$$\lim_{n\to\infty} \left|\frac{u_{n+1}}{u_n}\right| = \lim_{n\to\infty} \left| \frac{\dfrac{(-2)^{n+1}x^{n+1}}{(2n+1)!}}{\dfrac{(-2)^n x^n}{(2n)!}} \right| = \lim_{n\to\infty} \frac{2|x|}{(2n+1)(2n+2)} = 0$$

for all x [using the fact that $(2n+2)! = (2n)!(2n+1)(2n+2)$]. Hence the ratio test implies that the given power series converges for all x, and its interval of convergence is therefore $(-\infty, +\infty)$, the entire real line. ■

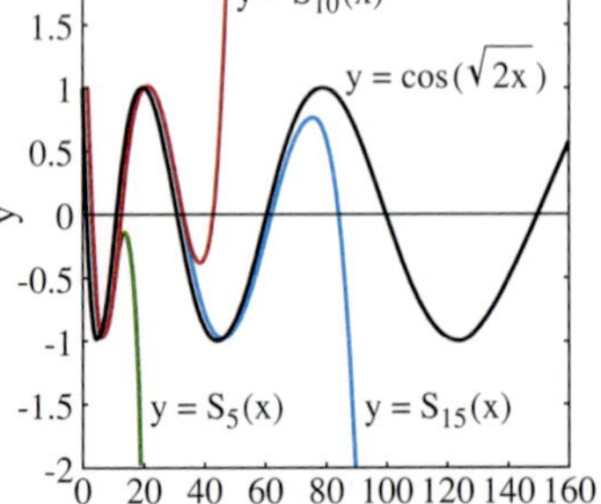

Fig. 11.8.3 Graphs of the partial sums $S_5(x)$, $S_{10}(x)$, and $S_{15}(x)$ and the sum $s(x) = \cos\sqrt{2x}$ (for $x > 0$) of the power series in Example 2

REMARK The power series of Example 2 results upon substitution of $\sqrt{2x}$ for x in the Taylor series for $\cos x$ [Eq. (3)]. But only for $x > 0$ does the sum $s(x)$ of the series exhibit the oscillatory character of the function $\cos\sqrt{2x}$ (Fig. 11.8.3). For $x < 0$, the power series converges to the quite different (and nonoscillatory) function $\cosh\sqrt{|2x|}$.

EXAMPLE 3 Find the interval of convergence of the series $\sum_{n=1}^{\infty} n^n x^n$.

Solution With $u_n = n^n x^n$, we find that

$$\lim_{n\to\infty}\left|\frac{u_{n+1}}{u_n}\right| = \lim_{n\to\infty}\left|\frac{(n+1)^{n+1}x^{n+1}}{n^n x^n}\right| = \lim_{n\to\infty}(n+1)\left(1+\frac{1}{n}\right)^n |x| = +\infty$$

for all $x \neq 0$, because

$$\lim_{n\to\infty}\left(1+\frac{1}{n}\right)^n = e.$$

Thus the given series diverges for all $x \neq 0$, and its interval of convergence consists of the single point $x = 0$. ■

EXAMPLE 4 Use the ratio test to verify that the Taylor series for $\cos x$ in Eq. (3) converges for all x.

Solution With $u_n = (-1)^n x^{2n}/(2n)!$ we find that

$$\lim_{n\to\infty}\left|\frac{u_{n+1}}{u_n}\right| = \lim_{n\to\infty}\left|\frac{\dfrac{(-1)^{n+1}x^{2n+2}}{(2n+2)!}}{\dfrac{(-1)^n x^{2n}}{(2n)!}}\right| = \lim_{n\to\infty}\frac{x^2}{(2n+1)(2n+2)} = 0$$

for all x, so the series converges for all x. ■

IMPORTANT In Example 4, the ratio test tells us only that the series for $\cos x$ converges to *some* number, *not* necessarily the particular number $\cos x$. The argument of Section 11.4, using Taylor's formula with remainder, is required to establish that the sum of the series is actually $\cos x$.

Power Series in Powers of $x - c$

An infinite series of the form

$$\sum_{n=0}^{\infty} a_n(x-c)^n = a_0 + a_1(x-c) + a_2(x-c)^2 + \cdots, \qquad \textbf{(9)}$$

where c is a constant, is called a **power series in** (powers of) $x - c$. By the same reasoning that led us to Theorem 1, with x^n replaced with $(x-c)^n$ throughout, we conclude that either

1. The series in Eq. (9) converges absolutely for all x, or
2. The series converges only when $x - c = 0$—that is, when $x = c$—or
3. There exists a number $R > 0$ such that the series in Eq. (9) converges absolutely if $|x - c| < R$ and diverges if $|x - c| > R$.

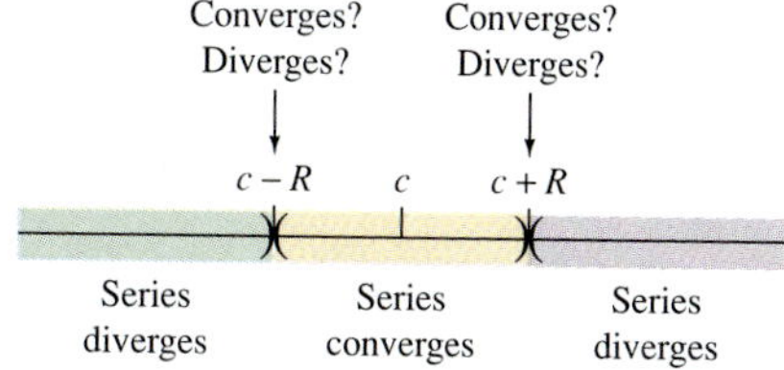

Fig. 11.8.4 The interval of convergence of $\sum_{n=0}^{\infty} a_n(x-c)^n$

As in the case of a power series with $c = 0$, the number R is called the **radius of convergence** of the series, and the **interval of convergence** of the series $\sum a_n(x-c)^n$ is the set of all numbers x for which it converges (Fig. 11.8.4). As before, when $0 < R < \infty$, the convergence of the series at the endpoints $x = c - R$ and $x = c + R$ of its interval of convergence must be checked separately.

EXAMPLE 5 Determine the interval of convergence of the series

$$\sum_{n=1}^{\infty}\frac{(-1)^n(x-2)^n}{n\cdot 4^n}.$$

Solution We let $u_n = (-1)^n(x-2)^n/(n \cdot 4^n)$. Then

$$\lim_{n\to\infty}\left|\frac{u_{n+1}}{u_n}\right| = \lim_{n\to\infty}\left|\frac{\dfrac{(-1)^{n+1}(x-2)^{n+1}}{(n+1)\cdot 4^{n+1}}}{\dfrac{(-1)^n(x-2)^n}{n\cdot 4^n}}\right|$$

$$= \lim_{n\to\infty}\frac{|x-4|}{4}\cdot\frac{n}{n+1} = \frac{|x-2|}{4}.$$

Hence the given series converges when $|x-2| < 4$, so the radius of convergence is $R = 4$. Because $c = 2$, the series converges when $-2 < x < 6$ and diverges if either $x < -2$ or $x > 6$. When $x = -2$, the series reduces to the divergent harmonic series, and when $x = 6$ it reduces to the convergent alternating series $\Sigma(-1)^n/n$. Thus the interval of convergence of the given power series is $(-2, 6]$. ■

Power Series Representations of Functions

Power series are important tools for computing (or approximating) values of functions. Suppose that the series $\sum a_n x^n$ converges to the value $f(x)$; that is,

$$f(x) = a_0 + a_1x + a_2x^2 + \cdots + a_nx^n + \cdots$$

for each x in the interval of convergence of the power series. Then we call $\sum a_n x^n$ a **power series representation** of $f(x)$. For example, the geometric series $\sum x^n$ in Eq. (1) is a power series representation of the function $f(x) = 1/(1-x)$ on the interval $(-1, 1)$.

We saw in Section 11.4 how Taylor's formula with remainder can often be used to find a power series representation of a given function. Recall that the nth-degree Taylor's formula for $f(x)$ at $x = a$ is

$$f(x) = f(a) + f'(a)(x-a) + \frac{f''(a)}{2!}(x-a)^2 + \frac{f^{(3)}(a)}{3!}(x-a)^3 + \cdots + \frac{f^{(n)}(a)}{n!}(x-a)^n + R_n(x). \tag{10}$$

The remainder $R_n(x)$ is given by

$$R_n(x) = \frac{f^{(n+1)}(z)}{(n+1)!}(x-a)^{n+1},$$

where z is some number between a and x. If we let $n \to \infty$ in Eq. (10), we obtain Theorem 2.

Theorem 2 ***Taylor Series Representations***

Suppose that the function f has derivatives of all orders on some interval containing a and also that

$$\lim_{n\to\infty} R_n(x) = 0 \tag{11}$$

for each x in that interval. Then

$$f(x) = \sum_{n=0}^{\infty}\frac{f^{(n)}(a)}{n!}(x-a)^n \tag{12}$$

for each x in the interval.

The power series in Eq. (12) is the **Taylor series** of the function f **at** $x = a$ (or *in powers of* $x - a$, or *with center* a). If $a = 0$, we obtain the power series

$$f(x) = \sum_{n=0}^{\infty} \frac{f^{(n)}(0)}{n!} x^n = f(0) + f'(0)x + \frac{f''(0)}{2!} x^2 + \cdots, \tag{13}$$

commonly called the **Maclaurin series** of f. Thus the power series in Eqs. (2) through (4) are the Maclaurin series of the functions e^x, $\cos x$, and $\sin x$, respectively.

EXAMPLE 6 New power series can be constructed from old ones. For instance, upon replacing x with $-x$ in the Maclaurin series for e^x, we obtain

$$e^{-x} = 1 - x + \frac{x^2}{2!} - \frac{x^3}{3!} + \cdots + (-1)^n \frac{x^n}{n!} + \cdots.$$

Let us now add the series for e^x and e^{-x} and divide by 2. This gives

$$\cosh x = \frac{e^x + e^{-x}}{2} = \frac{1}{2}\left(1 + x + \frac{x^2}{2!} + \frac{x^3}{3!} + \frac{x^4}{4!} + \cdots\right)$$
$$+ \frac{1}{2}\left(1 - x + \frac{x^2}{2!} - \frac{x^3}{3!} + \frac{x^4}{4!} - \cdots\right),$$

so

$$\cosh x = 1 + \frac{x^2}{2!} + \frac{x^4}{4!} + \frac{x^6}{6!} + \cdots.$$

Similarly,

$$\sinh x = x + \frac{x^3}{3!} + \frac{x^5}{5!} + \frac{x^7}{7!} + \cdots.$$

Note the strong resemblance to Eqs. (3) and (4), the series for $\cos x$ and $\sin x$, respectively.

Upon replacing x with $-x^2$ in the series for e^x, we obtain

$$e^{-x^2} = \sum_{n=0}^{\infty} (-1)^n \frac{x^{2n}}{n!} = 1 - x^2 + \frac{x^4}{2!} - \frac{x^6}{3!} + \cdots.$$

Because this power series converges to $\exp(-x^2)$ for all x, it must be the Maclaurin series for $\exp(-x^2)$ (see Problem 66). Think how tedious it would be to compute the derivatives of $\exp(-x^2)$ needed to write its Maclaurin series directly from Eq. (13). ■

EXAMPLE 7 Sometimes a function is originally defined by means of a power series. One of the most important "higher transcendental functions" of applied mathematics is the Bessel function $J_0(x)$ of order zero defined by

$$J_0(x) = \sum_{n=0}^{\infty} \frac{(-1)^n x^{2n}}{2^{2n}(n!)^2} = 1 - \frac{x^2}{4} + \frac{x^4}{64} - \frac{x^6}{2304} + \cdots.$$

Only terms of even degree appear, so let us write $u_n = (-1)^n x^{2n}/[2^{2n}(n!)^2]$. Then

$$\lim_{n\to\infty} \left|\frac{u_{n+1}}{u_n}\right| = \lim_{n\to\infty} \left| \frac{\dfrac{(-1)^{n+1} x^{2n+2}}{2^{2n+2}[(n+1)!]^2}}{\dfrac{(-1)^n x^{2n}}{2^{2n}(n!)^2}} \right| = \lim_{n\to\infty} \frac{x^2}{4(n+1)^2} = 0$$

for all x, so the ratio test implies that $J_0(x)$ is defined on the whole real line. The series for $J_0(x)$ resembles somewhat the cosine series, but the graph of $J_0(x)$ exhibits *damped* oscillations (Fig. 11.8.5). Bessel functions are important in such applications as the distribution of temperature in a cylindrical steam pipe and distribution of thermal neutrons in a cylindrical reactor. ■

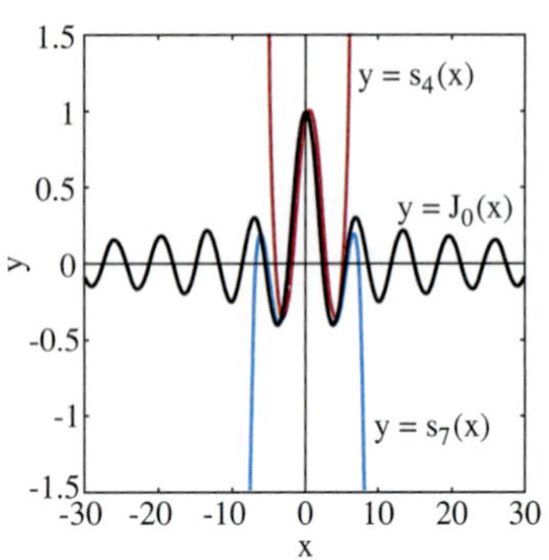

Fig. 11.8.5 Graphs of the Bessel function $J_0(x)$ and its partial sums $s_4(x)$ and $s_7(x)$ (polynomials of degrees 8 and 14)

The Binomial Series

One of the most famous and useful of all series, the *binomial series,* was discovered by Newton in the 1660s and is given in Example 8. It is the infinite series generalization of the (finite) binomial formula of elementary algebra.

EXAMPLE 8 Suppose that α is a nonzero real number. Show that the Maclaurin series of $f(x) = (1 + x)^\alpha$ is

$$\begin{aligned}(1+x)^\alpha &= 1 + \sum_{n=1}^{\infty} \frac{\alpha(\alpha-1)(\alpha-2)\cdots(\alpha-n+1)}{n!}x^n \\ &= 1 + \alpha x + \frac{\alpha(\alpha-1)}{2!}x^2 + \frac{\alpha(\alpha-1)(\alpha-2)}{3!}x^3 + \cdots. \end{aligned} \quad \textbf{(14)}$$

Also determine the interval of convergence of this **binomial series.**

Solution To derive the series itself, we simply list all the derivatives of $f(x) = (1 + x)^\alpha$, including its "zeroth" derivative:

$$\begin{aligned} f(x) &= (1+x)^\alpha \\ f'(x) &= \alpha(1+x)^{\alpha-1} \\ f''(x) &= \alpha(\alpha-1)(1+x)^{\alpha-2} \\ f^{(3)}(x) &= \alpha(\alpha-1)(\alpha-2)(1+x)^{\alpha-3}, \\ &\vdots \\ f^{(n)}(x) &= \alpha(\alpha-1)(\alpha-2)\cdots(\alpha-n+1)(1+x)^{\alpha-n}. \end{aligned}$$

Thus

$$f^{(n)}(0) = \alpha(\alpha-1)(\alpha-2)\cdots(\alpha-n+1).$$

If we substitute this value of $f^{(n)}(0)$ into the Maclaurin series formula in Eq. (13), we get the binomial series in Eq. (14).

To determine the interval of convergence of the binomial series, we let

$$u_n = \frac{\alpha(\alpha-2)(\alpha-2)\cdots(\alpha-n+1)}{n!}x^n.$$

We find that

$$\lim_{n\to\infty}\left|\frac{u_{n+1}}{u_n}\right| = \lim_{n\to\infty}\left|\frac{\dfrac{\alpha(\alpha-1)(\alpha-2)\cdots(\alpha-n)x^{n+1}}{(n+1)!}}{\dfrac{\alpha(\alpha-1)(\alpha-2)\cdots(\alpha-n+1)x^n}{n!}}\right|$$

$$= \lim_{n\to\infty} \left| \frac{(\alpha - n)x}{n + 1} \right| = |x|.$$

Hence the ratio test shows that the binomial series converges absolutely if $|x| < 1$ and diverges if $|x| > 1$. Its convergence at the endpoints $x = \pm 1$ depends on the value of α; we shall not pursue this problem. Problem 67 outlines a proof that the sum of the binomial series actually is $(1 + x)^\alpha$ if $|x| < 1$.

If $\alpha = k$, a positive integer, then the coefficient of x^n is zero for $n > k$, and the binomial series reduces to the binomial formula

$$(1 + x)^k = \sum_{n=0}^{k} \frac{k!}{n!(k - n)!} x^n.$$

Otherwise Eq. (14) is an infinite series. For example, with $\alpha = \frac{1}{2}$, we obtain

$$\begin{aligned} \sqrt{1 + x} &= 1 + \frac{\frac{1}{2}}{1!}x + \frac{(\frac{1}{2})(-\frac{1}{2})}{2!}x^2 + \frac{(\frac{1}{2})(-\frac{1}{2})(-\frac{3}{2})}{3!}x^3 \\ &\quad + \frac{(\frac{1}{2})(-\frac{1}{2})(-\frac{3}{2})(-\frac{5}{2})}{4!}x^4 + \cdots \\ &= 1 + \tfrac{1}{2}x - \tfrac{1}{8}x^2 + \tfrac{1}{16}x^3 - \tfrac{5}{128}x^4 + \cdots. \end{aligned} \tag{15}$$

If we replace x with $-x$ and take $\alpha = -\frac{1}{2}$, we get the series

$$\begin{aligned} \frac{1}{\sqrt{1 - x}} &= 1 + \frac{-\frac{1}{2}}{1!}(-x) + \frac{(-\frac{1}{2})(-\frac{3}{2})}{2!}(-x)^2 \\ &\quad + \cdots + \frac{1\cdot 3\cdot 5 \cdots (2n - 1)}{n!\cdot 2^n} x^n + \cdots, \end{aligned}$$

which in summation notation takes the form

$$\frac{1}{\sqrt{1 - x}} = 1 + \sum_{n=1}^{\infty} \frac{1\cdot 3\cdot 5 \cdots (2n - 1)}{2\cdot 4\cdot 6 \cdots (2n)} x^n. \tag{16}$$

We will find this series quite useful in Example 12 and in Problem 68.

Differentiation and Integration of Power Series

Sometimes it is inconvenient to compute the repeated derivatives of a function in order to find its Taylor series. An alternative method of finding new power series is by the differentiation and integration of known power series.

Suppose that a power series representation of the function $f(x)$ is known. Then Theorem 3 (we leave its proof to advanced calculus) implies that the function $f(x)$ may be differentiated by separately differentiating the individual terms in its power series. That is, the power series obtained by termwise differentiation converges to the derivative $f'(x)$. Similarly, a function can be integrated by termwise integration of its power series.

Theorem 3 ***Termwise Differentiation and Integration***

Suppose that the function f has a power series representation

$$f(x) = \sum_{n=0}^{\infty} a_n x^n = a_0 + a_1 x + a_2 x^2 + a_3 x^3 + \cdots$$

with nonzero radius of convergence R. Then f is differentiable on $(-R, R)$ and

$$f'(x) = \sum_{n=1}^{\infty} na_n x^{n-1} = a_1 + 2a_2 x + 3a_3 x^2 + 4a_4 x^3 + \cdots. \tag{17}$$

Also,

$$\int_0^x f(t)\,dt = \sum_{n=0}^{\infty} \frac{a_n x^{n+1}}{n+1} = a_0 x + \frac{1}{2}a_1 x^2 + \frac{1}{3}a_2 x^3 + \cdots \tag{18}$$

for each x in $(-R, R)$. Moreover, the power series in Eqs. (17) and (18) have the same radius of convergence R.

Remark 1 Although we omit the proof of Theorem 3, we observe that the radius of convergence of the series in Eq. (17) is

$$R = \lim_{n\to\infty} \left| \frac{na_n}{(n+1)a_{n+1}} \right| = \left(\lim_{n\to\infty} \frac{n}{n+1} \right) \cdot \left(\lim_{n\to\infty} \left| \frac{a_n}{a_{n+1}} \right| \right) = \lim_{n\to\infty} \left| \frac{a_n}{a_{n+1}} \right|.$$

Thus, by Eq. (8), the power series for $f(x)$ and the power series for $f'(x)$ have the same radius of convergence (under the assumption that the preceding limit exists).

Remark 2 Theorem 3 has this important consequence: If both the power series $\sum a_n x^n$ and $\sum b_n x^n$ converge and, for all x with $|x| < R$ $(R > 0)$, $\sum a_n x^n = \sum b_n x^n$, then $a_n = b_n$ for all n. In particular, the Taylor series of a function is its unique power series representation (if any). See Problem 66.

EXAMPLE 9 Termwise differentiation of the geometric series for

$$f(x) = \frac{1}{1-x}$$

yields

$$\frac{1}{(1-x)^2} = D_x\left(\frac{1}{1-x}\right) = D_x(1 + x + x^2 + x^3 + \cdots)$$
$$= 1 + 2x + 3x^2 + 4x^3 + \cdots.$$

Thus

$$\frac{1}{(1-x)^2} = \sum_{n=1}^{\infty} nx^{n-1} = \sum_{n=0}^{\infty} (n+1)x^n.$$

The series converges to $1/(1-x)^2$ if $-1 < x < 1$. ■

EXAMPLE 10 Replacement of x with $-t$ in the geometric series of Example 9 gives

$$\frac{1}{1+t} = 1 - t + t^2 - t^3 + \cdots + (-1)^n t^n + \cdots.$$

Because $D_t \ln(1+t) = 1/(1+t)$, termwise integration from $t = 0$ to $t = x$ now gives

$$\ln(1+x) = \int_0^x \frac{1}{1+t}\,dt$$

$$= \int_0^x (1 - t + t^2 - \cdots + (-1)^n t^n + \cdots)\, dt;$$

$$\ln(1 + x) = x - \frac{1}{2}x^2 + \frac{1}{3}x^3 - \frac{1}{4}x^4 + \cdots + \frac{(-1)^{n-1}}{n}x^n + \cdots \tag{19}$$

if $|x| < 1$. ■

EXAMPLE 11 Find a power series representation for the arctangent function.

Solution Because $D_t \tan^{-1} t = 1/(1 + t^2)$, termwise integration of the series

$$\frac{1}{1 + t^2} = 1 - t^2 + t^4 - t^6 + t^8 - \cdots$$

gives

$$\tan^{-1} x = \int_0^x \frac{1}{1 + t^2}\, dt = \int_0^x (1 - t^2 + t^4 - t^6 + t^8 - \cdots)\, dt.$$

Therefore

$$\tan^{-1} x = x - \tfrac{1}{3}x^3 + \tfrac{1}{5}x^5 - \tfrac{1}{7}x^7 + \tfrac{1}{9}x^9 - \cdots \tag{20}$$

if $-1 \leqq x \leqq 1$. ■

EXAMPLE 12 Find a power series representation for the arcsine function.

Solution First we substitute t^2 for x in Eq. (16). This yields

$$\frac{1}{\sqrt{1 - t^2}} = 1 + \sum_{n=1}^{\infty} \frac{1 \cdot 3 \cdot 5 \cdots (2n - 1)}{2 \cdot 4 \cdot 6 \cdots (2n)} t^{2n}$$

if $|t| < 1$. Because $D_t \sin^{-1} t = 1/\sqrt{1 - t^2}$, termwise integration of this series from $t = 0$ to $t = x$ gives

$$\sin^{-1} x = \int_0^x \frac{1}{\sqrt{1 - t^2}}\, dt = x + \sum_{n=1}^{\infty} \frac{1 \cdot 3 \cdot 5 \cdots (2n - 1)}{2 \cdot 4 \cdot 6 \cdots (2n)} \cdot \frac{x^{2n+1}}{2n + 1} \tag{21}$$

if $|x| < 1$. Problem 68 shows how to use this series to derive the series

$$\frac{\pi^2}{6} = 1 + \frac{1}{2^2} + \frac{1}{3^2} + \frac{1}{4^2} + \cdots + \frac{1}{n^2} + \cdots,$$

which we used in Example 3 of Section 11.5 to approximate the number π. ■

11.8 PROBLEMS

Find the interval of convergence of each power series in Problems 1 through 30.

1. $\sum_{n=1}^{\infty} nx^n$

2. $\sum_{n=1}^{\infty} \frac{x^n}{\sqrt{n}}$

3. $\sum_{n=1}^{\infty} \frac{nx^n}{2^n}$

4. $\sum_{n=1}^{\infty} \frac{(-1)^n x^n}{n^{1/2} 5^n}$

5. $\sum_{n=1}^{\infty} n! x^n$

6. $\sum_{n=1}^{\infty} \frac{(-1)^n x^n}{n^n}$

7. $\sum_{n=1}^{\infty} \frac{3^n x^n}{n^3}$

8. $\sum_{n=1}^{\infty} \frac{(-4)^n x^n}{\sqrt{2n + 1}}$

9. $\sum_{n=1}^{\infty} (-1)^n n^{1/2} (2x)^n$

10. $\sum_{n=1}^{\infty} \frac{n^2 x^n}{3n - 1}$

11. $\sum_{n=1}^{\infty} \frac{(-1)^n n x^n}{2^n (n + 1)^3}$

12. $\sum_{n=1}^{\infty} \frac{n^{10} x^n}{10^n}$

13. $\sum_{n=1}^{\infty} \frac{(\ln n)x^n}{3^n}$ **14.** $\sum_{n=2}^{\infty} \frac{(-1)^n 4^n x^n}{n \ln n}$

15. $\sum_{n=0}^{\infty} (5x-3)^n$ **16.** $\sum_{n=1}^{\infty} \frac{(2x-1)^n}{n^4+16}$

17. $\sum_{n=1}^{\infty} \frac{2^n(x-3)^n}{n^2}$

18. $\sum_{n=1}^{\infty} \frac{n!}{n^n} x^n$ (Do not test the endpoints; the series diverges at each.)

19. $\sum_{n=1}^{\infty} \frac{(2n)!}{n!} x^n$

20. $\sum_{n=1}^{\infty} \frac{1 \cdot 3 \cdot 5 \cdots (2n+1)}{n!} x^n$ (Do not test the endpoints; the series diverges at each.)

21. $\sum_{n=1}^{\infty} \frac{n^3(x+1)^n}{3^n}$ **22.** $\sum_{n=1}^{\infty} \frac{(-1)^{n+1}(x-2)^n}{n^2}$

23. $\sum_{n=1}^{\infty} \frac{(3-x)^n}{n^3}$ **24.** $\sum_{n=1}^{\infty} \frac{(-1)^{n+1}10^n}{n!}(x-10)^n$

25. $\sum_{n=1}^{\infty} \frac{n!}{2^n}(x-5)^n$ **26.** $\sum_{n=1}^{\infty} \frac{(-1)^{n+1}}{n \cdot 10^n}(x-2)^n$

27. $\sum_{n=0}^{\infty} x^{(2^n)}$ **28.** $\sum_{n=0}^{\infty} \left(\frac{x^2+1}{5}\right)^n$

29. $\sum_{n=1}^{\infty} \frac{(-1)^n x^n}{1 \cdot 3 \cdot 5 \cdots (2n-1)}$ **30.** $\sum_{n=1}^{\infty} \frac{1 \cdot 3 \cdot 5 \cdots (2n-1)}{2 \cdot 5 \cdot 8 \cdots (3n-1)} x^n$

In Problems 31 through 42, use power series established in this section to find a power series representation of the given function. Then determine the radius of convergence of the resulting series.

31. $f(x) = \frac{x}{1-x}$ **32.** $f(x) = \frac{1}{10+x}$

33. $f(x) = x^2 e^{-3x}$ **34.** $f(x) = \frac{x}{9-x^2}$

35. $f(x) = \sin(x^2)$

36. $f(x) = \cos^2 2x = \frac{1}{2}(1 + \cos 4x)$

37. $f(x) = \sqrt[3]{1-x}$ **38.** $f(x) = (1+x^2)^{3/2}$

39. $f(x) = (1+x)^{-3}$ **40.** $f(x) = \frac{1}{\sqrt{9+x^3}}$

41. $f(x) = \frac{\ln(1+x)}{x}$ **42.** $f(x) = \frac{x - \arctan x}{x^3}$

In Problems 43 through 48, find a power series representation for the given function $f(x)$ by using termwise integration.

43. $f(x) = \int_0^x \sin t^3 \, dt$ **44.** $f(x) = \int_0^x \frac{\sin t}{t} \, dt$

45. $f(x) = \int_0^x \exp(-t^3) \, dt$ **46.** $f(x) = \int_0^x \frac{\arctan t}{t} \, dt$

47. $f(x) = \int_0^x \frac{1 - \exp(-t^2)}{t^2} \, dt$

48. $\tanh^{-1} x = \int_0^x \frac{1}{1-t^2} \, dt$

Beginning with the geometric series $\sum_{n=0}^{\infty} x^n$ as in Example 9, differentiate termwise to find the sums (for $|x| < 1$) of the power series in Problems 49 through 51.

49. $\sum_{n=1}^{\infty} nx^n$ **50.** $\sum_{n=1}^{\infty} n(n-1)x^n$ **51.** $\sum_{n=1}^{\infty} n^2 x^n$

52. Use the power series of the preceding problems to sum the numerical series

$$\sum_{n=1}^{\infty} \frac{n}{2^n} \quad \text{and} \quad \sum_{n=1}^{\infty} \frac{n^2}{3^n}.$$

53. Verify by termwise differentiation of its Maclaurin series that the exponential function $y = e^x$ satisfies the differential equation $dy/dx = y$. (Thus the exponential series arises naturally as a power series that is its own termwise derivative.)

54. Verify by termwise differentiation of their Maclaurin series that the sine function $y = \sin x$ and the cosine function $y = \cos x$ both satisfy the differential equation

$$\frac{d^2y}{dx^2} + y = 0.$$

55. Introduce the usual notation $i = \sqrt{-1}$ for the complex square root of -1 that lies on the positive y-axis in the complex plane. You can verify that $i^2 = -1$, $i^3 = -i$, $i^4 = 1$, and so forth, because the arithmetic of complex numbers obeys the same axioms of arithmetic as that of the real numbers. Assuming that the substitution $x = i\theta$ in the exponential series makes sense (it does), collect real and imaginary terms in the resulting series to derive formally Euler's formula

$$e^{i\theta} = \cos\theta + i \sin\theta.$$

56. In elementary mathematics one sees various definitions (some circular!) of the trigonometric functions. But the now-familiar Maclaurin series for the sine and cosine functions arise naturally and independently (for example, as in the computation in Problem 55). So an alternative approach is to begin by defining $\cos x$ and $\sin x$ by means of their Maclaurin series. For instance, never having heard of sine, cosine, or the number π, we might define the function

$$S(x) = \sum_{n=1}^{\infty} \frac{(-1)^{n-1} x^{2n-1}}{(2n-1)!}$$

and verify using the ratio test that this series converges for all x. Use a computer algebra system to plot graphs of high-degree partial sums $s_n(x)$ of this series. Does it appear that the function $S(x)$ appears to have a zero somewhere near the number 3? Solve the equation $s_n(x) = 0$ numerically (for some large values of n) to verify that this least positive zero of the sine function is approximately 3.14159 (and thus the famous number π makes a fresh new appearance).

57. The Bessel function of order 1 is defined by

$$J_1(x) = \sum_{n=0}^{\infty} \frac{(-1)^n x^{2n+1}}{2^{2n+1} n!(n+1)!} = \frac{x}{2} - \frac{x^3}{16} + \frac{x^5}{384} - \cdots.$$

Verify that this series converges for all x and that the derivative of the Bessel function of order zero is given by $J_0'(x) = -J_1(x)$. Are the graphs in Fig. 11.8.6 consistent with this latter fact?

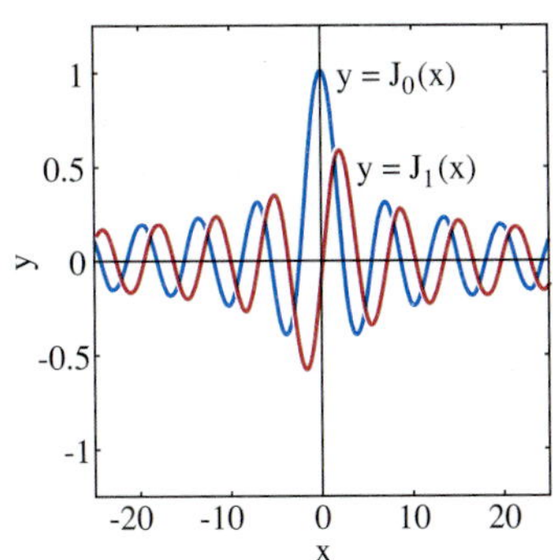

Fig. 11.8.6 Graphs of the Bessel functions $J_0(x)$ and $J_1(x)$. Note that their zeros are interlaced, like the zeros of the cosine and sine functions.

58. Verify by termwise integration that

$$\int x J_0(x)\,dx = x J_1(x) + C.$$

59. Bessel's equation of order n is the second-order differential equation

$$x^2 y'' + x y' + (x^2 - n^2) y = 0.$$

Verify by termwise differentiation that $y = J_0(x)$ satisfies Bessel's equation of order zero.

60. Verify that $y = J_1(x)$ satisfies Bessel's equation of order 1.

61. First use the sine series to find the Taylor series of $f(x) = (\sin x)/x$. Then use a graphing calculator or computer to illustrate the approximation of $f(x)$ by its Taylor polynomials with center $a = 0$.

62. First find the Taylor series of the function

$$g(x) = \int_0^x \frac{\sin t}{t}\,dt.$$

Then determine where this power series converges. Finally, use a graphing calculator or computer to illustrate the approximation of $g(x)$ by its Taylor polynomials with center $a = 0$.

63. Deduce from the arctangent series (Example 11) that

$$\pi = \frac{6}{\sqrt{3}} \sum_{n=0}^{\infty} \frac{(-1)^n}{2n+1} \left(\frac{1}{3}\right)^n.$$

Then use this alternating series to show that $\pi = 3.14$ accurate to two decimal places.

64. Substitute the Maclaurin series for $\sin x$, then assume the validity of termwise integration of the resulting series, to derive the formula

$$\int_0^{\infty} e^{-t} \sin xt\,dt = \frac{x}{1+x^2} \qquad (|x| < 1).$$

Use the fact from Section 9.8 that

$$\int_0^{\infty} t^n e^{-t}\,dt = \Gamma(n+1) = n!.$$

65. (a) Deduce from the Maclaurin series for e^t that

$$\frac{1}{x^x} = \sum_{n=0}^{\infty} \frac{(-1)^n}{n!} (x \ln x)^n.$$

(b) Assuming the validity of termwise integration of the series in part (a), use the integral formula of Problem 53 in Section 9.8 to conclude that

$$\int_0^1 \frac{1}{x^x}\,dx = \sum_{n=1}^{\infty} \frac{1}{n^n}.$$

66. Suppose that $f(x)$ is represented by the power series

$$\sum_{n=0}^{\infty} a_n x^n$$

for all x in some open interval centered at $x = 0$. Show by repeated differentiation of the series, substituting $x = 0$ after each differentiation, that $a_n = f^{(n)}(0)/n!$ for all $n \geqq 0$. Thus the only power series in x that represents a function at and near $x = 0$ is its Maclaurin series.

67. (a) Consider the binomial series

$$f(x) = \sum_{n=0}^{\infty} \frac{\alpha(\alpha-1)(\alpha-2)\cdots(\alpha-n+1)}{n!} x^n,$$

which converges (to *something*) if $|x| < 1$. Compute the derivative $f'(x)$ by termwise differentiation, and show that it satisfies the differential equation $(1+x)f'(x) = \alpha f(x)$.
(b) Solve the differential equation in part (a) to obtain $f(x) = C(1+x)^\alpha$ for some constant C. Finally, show that $C = 1$. Thus the binomial series converges to $(1+x)^\alpha$ if $|x| < 1$.

68. (a) Show by direct integration that

$$\int_0^1 \frac{\arcsin x}{\sqrt{1-x^2}}\,dx = \frac{\pi^2}{8}.$$

(b) Use the result of Problem 58 in Section 9.3 to show that

$$\int_0^1 \frac{x^{2n+1}}{\sqrt{1-x^2}}\,dx = \frac{2\cdot 4\cdot 6\cdots(2n)}{1\cdot 3\cdot 5\cdots(2n+1)}.$$

(c) Substitute the series of Example 12 for $\arcsin x$ into the integral of part (a); then use the integral of part (b) to integrate termwise. Conclude that

$$\int_0^1 \frac{\arcsin x}{\sqrt{1-x^2}}\,dx = 1 + \frac{1}{3^2} + \frac{1}{5^2} + \frac{1}{7^2} + \cdots.$$

(d) Note that

$$\sum_{n=1}^{\infty} \frac{1}{n^2} = \sum_{n=1}^{\infty} \frac{1}{(2n-1)^2} + \sum_{n=1}^{\infty} \frac{1}{(2n)^2}.$$

Use this information and parts (a) and (c) to show that

$$\sum_{n=1}^{\infty} \frac{1}{n^2} = \frac{\pi^2}{6}.$$

69. Prove that if the power series $\sum a_n x^n$ converges for some $x = x_0 \neq 0$, then it converges absolutely for all x such that $|x| < |x_0|$. (*Suggestion:* Conclude from the fact that $\lim_{n\to\infty} a_n x_0^n = 0$ that $|a_n x^n| \leqq |x/x_0|^n$ for all n sufficiently large. Thus the series $\sum |a_n x^n|$ is eventually dominated by the geometric series $\sum |x/x_0|^n$, which converges if $|x| < |x_0|$.)

70. Suppose that the power series $\sum a_n x^n$ converges for some but not all nonzero values of x. Let S be the set of real numbers for which the series converges absolutely. (a) Conclude from Problem 69 that the set S is bounded above. (b) Let λ be the least upper bound of the set S (see Problem 61 of Section 11.2). Then show that $\sum a_n x^n$ converges absolutely if $|x| < \lambda$ and diverges if $|x| > \lambda$. Explain why this proves Theorem 1 without the additional hypothesis that $\lim_{n\to\infty} |a_{n+1}/a_n|$ exists.

11.9 POWER SERIES COMPUTATIONS

Power series often are used to approximate numerical values of functions and integrals. *Alternating* power series (such as the sine and cosine series) are especially common and useful. Recall the alternating series remainder (or "error") estimate of Theorem 2 in Section 11.7. It applies to a convergent alternating series $\sum(-1)^{n+1}a_n$ whose terms are decreasing (so $a_n > a_{n+1}$ for every n). If we write

$$\sum_{k=1}^{\infty} (-1)^{k+1} a_k = (a_1 - a_2 + a_3 - \cdots \pm a_n) + E, \tag{1}$$

then $E = \mp a_{n+1} \pm a_{n+2} \mp a_{n+3} \pm \cdots$ is the error made when the series is *truncated*—the terms following $(-1)^{n+1}a_n$ are simply chopped off and discarded, and the n-term partial sum is used in place of the actual sum of the whole series. The remainder estimate then says that the error E has the same sign as the first term not retained and is less in magnitude than this first neglected term; that is, $|E| < a_{n+1}$.

EXAMPLE 1 Use the first four terms of the binomial series

$$\sqrt{1+x} = 1 + \tfrac{1}{2}x - \tfrac{1}{8}x^2 + \tfrac{1}{16}x^3 - \tfrac{5}{128}x^4 + \cdots \tag{2}$$

to estimate the number $\sqrt{105}$ and to estimate the accuracy in the approximation.

Solution If $x > 0$, then the binomial series is, after the first term, an alternating series. To match the pattern on the left-hand side in Eq. (2), we first write

$$\sqrt{105} = \sqrt{100+5} = 10\sqrt{1+\tfrac{5}{100}} = 10\sqrt{1+0.05}.$$

Then with $x = 0.05$ the series in Eq. (2) gives

$$\begin{aligned}\sqrt{105} &= 10[1 + \tfrac{1}{2}(0.05) - \tfrac{1}{8}(0.05)^2 + \tfrac{1}{16}(0.05)^3 + E]\\ &= 10[1.02469531 + E] = 10.2469531 + 10E.\end{aligned}$$

Note that the error $10E$ in our approximation $\sqrt{105} \approx 10.2469531$ is 10 times the error E in the truncated series itself. It follows from the remainder estimate that E is negative and that

$$|10E| < 10 \cdot \tfrac{5}{128}\,(0.05)^4 \approx 0.0000024.$$

Therefore

$$10.2469531 - 0.0000024 = 10.2469507 < \sqrt{105} < 10.2469531,$$

so it follows that $\sqrt{105} \approx 10.24695$ rounded accurate to five decimal places. ■

REMARK Suppose that we had been asked in advance to approximate $\sqrt{105}$ accurate to five decimal places. A convenient way to do this is to continue writing terms of the series until it is clear that they have become too small in magnitude to affect the fifth decimal place. A good rule of thumb is to use two more decimal places in the computations than are required in the final answer. Thus we use seven decimal places in this case and get

$$\begin{aligned}\sqrt{105} &= 10\cdot(1 + 0.05)^{1/2} \\ &\approx 10\cdot(1 + 0.025 - 0.0003125 + 0.0000078 - 0.0000002 + \cdots) \\ &\approx 10.246951 \approx 10.24695.\end{aligned}$$

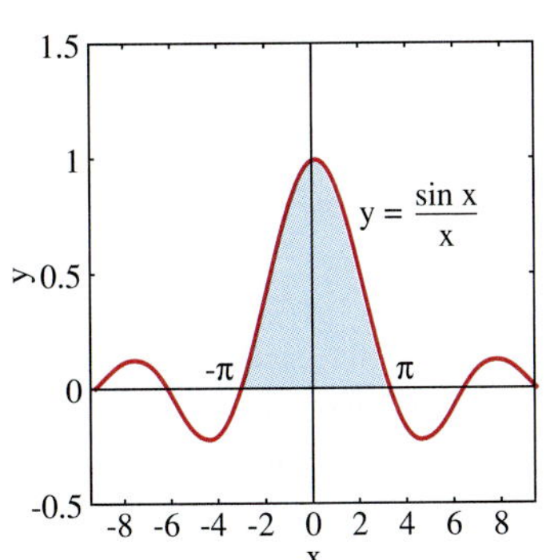

Fig. 11.9.1 The graph $y = \dfrac{\sin x}{x}$ of Example 2

EXAMPLE 2 Figure 11.9.1 shows the graph of the function $f(x) = (\sin x)/x$. Approximate (accurate to three decimal places) the area

$$A = \int_{-\pi}^{\pi} \frac{\sin x}{x}\,dx = 2\int_0^{\pi} \frac{\sin x}{x}\,dx \tag{3}$$

of the shaded region lying under the "principal arch" from $x = -\pi$ to π.

Solution When we substitute the Taylor series for $\sin x$ in Eq. (3) and integrate termwise, we get

$$\begin{aligned}A &= 2\int_0^{\pi} \frac{1}{x}\left(x - \frac{x^3}{3!} + \frac{x^5}{5!} - \frac{x^7}{7!} + \cdots\right)dx \\ &= 2\int_0^{\pi}\left(1 - \frac{x^2}{3!} + \frac{x^4}{5!} - \frac{x^6}{7!} + \cdots\right)dx \\ &= 2\left[x - \frac{x^3}{3!3} + \frac{x^5}{5!5} - \frac{x^7}{7!7} + \cdots\right]_0^{\pi},\end{aligned}$$

and thus

$$A = 2\pi - \frac{2\pi^3}{3!3} + \frac{2\pi^5}{5!5} - \frac{2\pi^7}{7!7} + \frac{2\pi^9}{9!9} - \frac{2\pi^{11}}{11!11} + \cdots.$$

Following the "+2 rule of thumb" and retaining five decimal places, we calculate

$$A = 6.28319 - 3.44514 + 1.02007 - 0.17122 + 0.01825 - 0.00134 + 0.00007 - \cdots.$$

The sum of the first six terms gives $A \approx 3.70381$. Because we are summing an alternating series, the error in this approximation is positive and less than the next term 0.00007. Neglecting possible roundoff in the last place, we would conclude that $3.70381 < A < 3.70388$. Thus $A \approx 3.704$ rounded accurate to three decimal places.

The Algebra of Power Series

Theorem 1, which we state without proof, implies that power series may be added and multiplied much like polynomials. The guiding principle is that of collecting coefficients of like powers of x.

Theorem 1 ***Adding and Multiplying Power Series***

Let $\sum a_n x^n$ and $\sum b_n x^n$ be power series with nonzero radii of convergence. Then

$$\sum_{n=0}^{\infty} a_n x^n + \sum_{n=0}^{\infty} b_n x^n = \sum_{n=0}^{\infty} (a_n + b_n) x^n \tag{4}$$

and

$$\left(\sum_{n=0}^{\infty} a_n x^n\right)\left(\sum_{n=0}^{\infty} b_n x^n\right) = \sum_{n=0}^{\infty} c_n x^n$$

$$= a_0 b_0 + (a_0 b_1 + a_1 b_0)x + (a_0 b_2 + a_1 b_1 + a_2 b_0)x^2 + \cdots, \tag{5}$$

where

$$c_n = a_0 b_n + a_1 b_{n-1} + a_2 b_{n-2} + \cdots + a_{n-1} b_1 + a_n b_0. \tag{6}$$

The series in Eqs. (4) and (5) converge for any x that lies interior to the intervals of convergence of both $\sum a_n x^n$ and $\sum b_n x^n$.

Thus if $\sum a_n x^n$ and $\sum b_n x^n$ are power series representations of the functions $f(x)$ and $g(x)$, respectively, then the product power series $\sum c_n x^n$ found by "ordinary multiplication" and collection of terms is a power series representation of the product function $f(x)g(x)$. This fact can also be used to divide one power series by another, *provided* that the quotient is known to have a power series representation.

EXAMPLE 3 Assume that the tangent function has a power series representation $\tan x = \sum a_n x^n$ (it does). Use the Maclaurin series for $\sin x$ and $\cos x$ to find a_0, a_1, a_2, and a_3.

Solution We multiply series to obtain

$$\sin x = \tan x \cos x$$

$$= (a_0 + a_1 x + a_2 x^2 + a_3 x^3 + \cdots)\left(1 - \frac{x^2}{2} + \frac{x^4}{24} - \cdots\right).$$

If we multiply each term in the first factor by each term in the second, then collect coefficients of like powers, the result is

$$\sin x = a_0 + a_1 x + \left(a_2 - \frac{1}{2}a_0\right)x^2 + \left(a_2 - \frac{1}{2}a_1\right)x^3 + \cdots.$$

But because

$$\sin x = x - \tfrac{1}{6}x^3 + \tfrac{1}{120}x^5 - \cdots,$$

comparison of coefficients gives the equations

$$\begin{aligned} a_0 \qquad\qquad\qquad &= 0, \\ a_1 \qquad\qquad &= 1, \\ -\tfrac{1}{2}a_0 \qquad + a_2 \qquad &= 0, \\ -\tfrac{1}{2}a_1 \qquad + a_3 &= -\tfrac{1}{6}. \end{aligned}$$

Thus we find that $a_0 = 0$, $a_1 = 1$, $a_2 = 0$, and $a_3 = \frac{1}{3}$. So

$$\tan x = x + \tfrac{1}{3}x^3 + \cdots.$$

Things are not always as they first appear. A computer algebra system gives the continuation

$$\tan x = x + \tfrac{1}{3}x^3 + \tfrac{2}{15}x^5 + \tfrac{17}{315}x^7 + \tfrac{62}{2835}x^9 + \tfrac{1382}{155{,}925}x^{11} + \cdots \tag{7}$$

of the tangent series. For the general form of the nth coefficient, see K. Knopp's *Theory and Application of Infinite Series* (New York: Hafner Press, 1971), p. 204. You may also check that the first few terms agree with the result of ordinary division of the Maclaurin series for $\cos x$ into the Maclaurin series for $\sin x$:

$$1 - \tfrac{1}{2}x^2 + \tfrac{1}{24}x^4 - \cdots \;\overline{)\,x - \tfrac{1}{3}x^3 + \tfrac{1}{120}x^5 - \cdots}\quad \text{quotient: } x + \tfrac{1}{3}x^3 + \tfrac{2}{15}x^5 + \cdots.$$

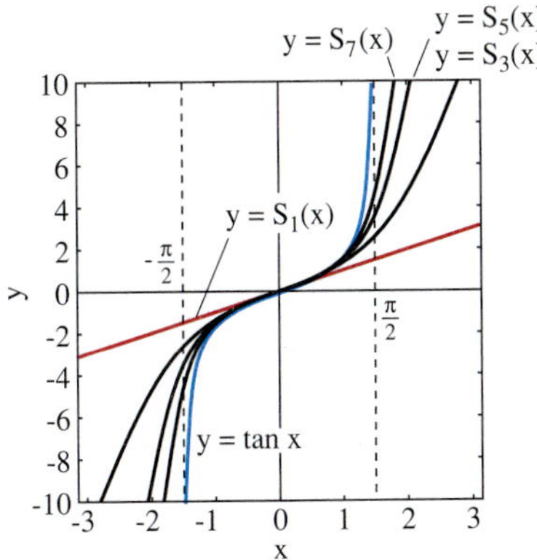

Fig. 11.9.2 The graphs of $y = \tan x$ and the first four partial sums of the power series in (7)

Figure 11.9.2 shows the approximation of the tangent function (on $-\pi/2 < x < \pi/2$) by the first four odd-degree polynomial partial sums corresponding to the terms exhibited in Eq. (7). Evidently these polynomial approximations have difficulty "keeping up" with $\tan x$ as it approaches $\pm\infty$ as $x \to \pm\pi/2$. ■

Power Series and Indeterminate Forms

According to Theorem 3 of Section 11.8, a power series is differentiable and therefore continuous within its interval of convergence. It follows that

$$\lim_{x \to c} \sum_{n=0}^{\infty} a_n(x - c)^n = a_0. \tag{8}$$

Examples 4 and 5 illustrate the use of this simple observation to find the limit of the indeterminate form $f(x)/g(x)$. The technique is first to substitute power series representations for $f(x)$ and $g(x)$.

EXAMPLE 4 Find $\displaystyle\lim_{x \to 0} \frac{\sin x - \arctan x}{x^2 \ln(1 + x)}$.

Solution The power series of Eqs. (4), (19), and (20) in Section 11.8 give

$$\begin{aligned} \sin x - \arctan x &= \left(x - \tfrac{1}{6}x^3 + \tfrac{1}{120}x^5 - \cdots\right) - \left(x - \tfrac{1}{3}x^3 + \tfrac{1}{5}x^5 - \cdots\right) \\ &= \tfrac{1}{6}x^3 - \tfrac{23}{120}x^5 + \cdots \end{aligned}$$

and

$$x^2 \ln(1 + x) = x^2 \cdot \left(x - \tfrac{1}{2}x^2 + \tfrac{1}{3}x^3 + \cdots\right) = x^3 - \tfrac{1}{2}x^4 + \tfrac{1}{3}x^5 - \cdots.$$

Hence

$$\lim_{x\to 0}\frac{\sin x - \arctan x}{x^2 \ln(1+x)} = \lim_{x\to 0}\frac{\frac{1}{6}x^3 - \frac{23}{120}x^5 + \cdots}{x^3 - \frac{1}{2}x^4 + \cdots}$$

$$= \lim_{x\to 0}\frac{\frac{1}{6} - \frac{23}{120}x^2 + \cdots}{1 - \frac{1}{2}x + \cdots} = \frac{1}{6}.$$

■

EXAMPLE 5 Find $\displaystyle\lim_{x\to 1}\frac{\ln x}{x-1}$.

Solution We first replace x with $x - 1$ in the power series for $\ln(1 + x)$ used in Example 4. [Equation (8) makes it clear that this method requires all series to have center c if we are taking limits as $x \to c$.] This gives us

$$\ln x = (x-1) - \tfrac{1}{2}(x-1)^2 + \tfrac{1}{3}(x-1)^3 - \cdots.$$

Hence

$$\lim_{x\to 1}\frac{\ln x}{x-1} = \lim_{x\to 1}\frac{(x-1) - \frac{1}{2}(x-1)^2 + \frac{1}{3}(x-1)^3 - \cdots}{x-1}$$

$$= \lim_{x\to 1}\left[1 - \frac{1}{2}(x-1) + \frac{1}{3}(x-1)^2 - \cdots\right] = 1.$$

■

The method of Examples 4 and 5 provides a useful alternative to l'Hôpital's rule, especially when repeated differentiation of numerator and denominator is inconvenient or too time-consuming.

Numerical and Graphical Error Estimation

The following examples show how to investigate the accuracy in a power series partial sum approximation for a specified interval of values of x. We will take the statement that a given approximation is "accurate to p decimal places" to mean that its error E is numerically less than half a unit in the pth decimal place; that is, that $|E| < 0.5 \times 10^{-p}$. For instance, four-place accuracy means that $|E| < 0.00005$. (Note that $p = 4$ is the number of zeros here.) Nevertheless, we should remember that in some cases a result accurate to within a half unit in the pth place may round "the wrong way," so that the result rounded to p places may still be in error by a unit in the pth decimal place (as in Problem 12).

EXAMPLE 6 Consider the polynomial approximation

$$\sin x \approx x - \frac{x^3}{3!} + \frac{x^5}{5!} - \cdots + (-1)^{n+1}\frac{x^{2n-1}}{(2n-1)!} \tag{9}$$

obtained by truncating the alternating Taylor series of the sine function. Questions:

(a) How accurate is the cubic approximation $P_3(x) \approx x - x^3/3!$ for angles from $0°$ to $10°$? Use this approximation to estimate $\sin 10°$.

(b) How many terms in (9) are needed to guarantee six-place accuracy in calculating $\sin x$ for angles from $0°$ to $45°$? Use the corresponding polynomial to approximate $\sin 30°$ and $\sin 40°$.

(c) For what values of x does the fifth-degree approximation yield five-place accuracy?

Solution (a) Of course, we must substitute x in radians in (9), so we deal here with values of x in the interval $0 \leqq x \leqq \pi/18$. For any such x, the error E is negative (why?) and is bounded by the magnitude of the next term:

$$|E| < \frac{x^5}{5!} \leqq \frac{(\pi/18)^5}{5!} \approx 0.00000135 < 0.000005.$$

We count five zeros on the right, and thus we have five-place accuracy. For instance, substitution of $x = \pi/18$ in the cubic polynomial $P_3(x)$ gives

$$\sin 10^\circ = \sin\left(\frac{\pi}{18}\right) \approx \frac{\pi}{18} - \frac{1}{3!}\cdot\left(\frac{\pi}{18}\right)^3$$

$$\approx 0.1736468 \approx 0.17365.$$

This five-place approximation $\sin 10^\circ \approx 0.17365$ is correct, because the actual seven-place value of $\sin 10^\circ$ is $0.1736482 \approx 0.17365$.

(b) For any x in the interval $0 \leqq x \leqq \pi/4$, the error E made if we use the polynomial value in (9) in place of the actual value $\sin x$ is bounded by the first neglected term,

$$|E| < \frac{x^{2n+1}}{(2n+1)!} \leqq \frac{(\pi/4)^{2n+1}}{(2n+1)!}.$$

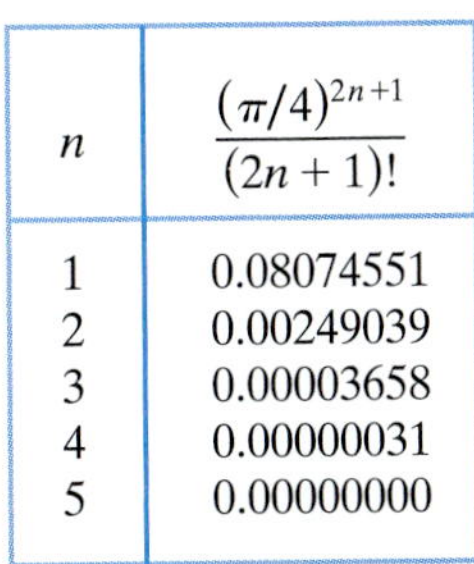

n	$\dfrac{(\pi/4)^{2n+1}}{(2n+1)!}$
1	0.08074551
2	0.00249039
3	0.00003658
4	0.00000031
5	0.00000000

Fig. 11.9.3 Estimating the error in Example 6(b)

The table in Fig. 11.9.3 shows calculator values for $n = 1, 2, 3, \ldots$ of this maximal error (rounded to eight decimal places). For six-place accuracy, we want $|E| < 0.0000005$, so we see that $n = 4$ will suffice. We therefore use the seventh-degree Taylor polynomial

$$P_7(x) = x - \frac{x^3}{3!} + \frac{x^5}{5!} - \frac{x^7}{7!} \tag{10}$$

to approximate $\sin x$ for $0 \leqq x \leqq \pi/4$. With $x = \pi/6$, we get

$$\sin 30^\circ \approx \frac{\pi}{6} - \frac{(\pi/6)^3}{3!} + \frac{(\pi/6)^5}{5!} - \frac{(\pi/6)^7}{7!} \approx 0.49999999 \approx \frac{1}{2},$$

as expected. Substitution of $x = 2\pi/9$ in Eq. (10) similarly gives $\sin 40^\circ \approx 0.64278750$, whereas the actual eight-place value of $\sin 40^\circ$ is $0.64278761 \approx 0.642788$.

(c) The fifth-degree approximation

$$\sin x \approx P_5(x) = x - \frac{x^3}{3!} + \frac{x^5}{5!} \tag{11}$$

gives five-place accuracy when x is such that the error E satisfies the inequality

$$|E| < \frac{|x|^7}{7!} = \frac{|x|^7}{5040} \leqq 0.000005;$$

that is, when $|x| \leqq [(5040)\cdot(0.000005)]^{1/7} \approx 0.5910$ (radians). In degrees, this corresponds to angles between -33.86° and $+33.86^\circ$. In Fig. 11.9.4 the graph of $y = x^7/7!$ in the viewing window $-1 \leqq x \leqq 1$, $-0.00001 \leqq y \leqq 0.00001$ provides visual corroboration of this analysis—we see clearly that $x^7/7!$ remains between -0.000005 and 0.000005 when x is between -0.59 and 0.59. ■

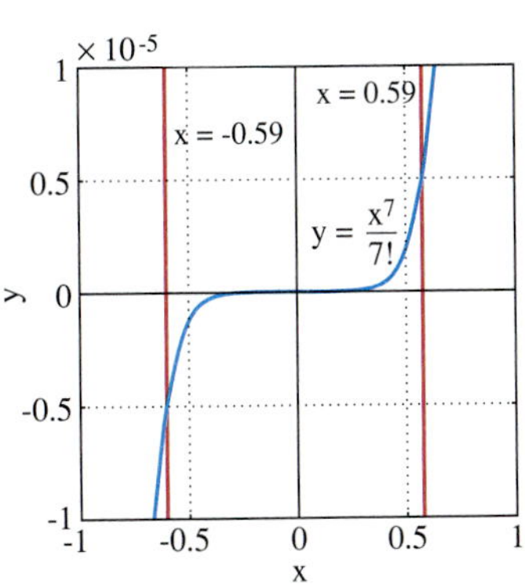

Fig. 11.9.4 The graph of the maximal error $y = \dfrac{x^7}{7!}$ in Example 6(c)

EXAMPLE 7 Suppose now that we want to approximate $f(x) = \sin x$ with three-place accuracy on the whole interval from 0° to 90°. Now it makes sense to begin with a Taylor series centered at the midpoint $x = \pi/4$ of the interval. Because the function $f(x)$ and its successive derivatives are $\sin x$, $\cos x$, $-\sin x$, $-\cos x$, and so forth, their values at $x = \pi/4$ are $\frac{1}{2}\sqrt{2}$, $\frac{1}{2}\sqrt{2}$, $-\frac{1}{2}\sqrt{2}$, $-\frac{1}{2}\sqrt{2}$, and so forth. Consequently Taylor's formula with remainder (Section 11.4) for $f(x) = \sin x$ centered at $x = \pi/4$ takes the form

$$\sin x = \frac{\sqrt{2}}{2}\cdot\left[1 + \left(x - \frac{\pi}{4}\right) - \frac{1}{2!}\left(x - \frac{\pi}{4}\right)^2 - \frac{1}{3!}\left(x - \frac{\pi}{4}\right)^3 + \cdots \pm \frac{1}{n!}\left(x - \frac{\pi}{4}\right)^n\right] + E(x), \tag{12}$$

where

$$|E(x)| = \left|\frac{f^{(n+1)}(z)}{(n+1)!}\left(x - \frac{\pi}{4}\right)^{n+1}\right| \leqq \frac{1}{(n+1)!}\left|x - \frac{\pi}{4}\right|^{n+1} \tag{13}$$

for some z in the interval $0 \leqq x \leqq \pi/2$. Observe that the corresponding Taylor series is not alternating—if $x > \pi/4$, it has instead a "+ + − − + + − −" pattern of signs—but we can still use the remainder estimate in Eq. (13). For three-place accuracy, we need to choose n so that $y = E(x)$ remains within the viewing window $-0.0005 \leqq y \leqq 0.0005$ on the whole interval $0 \leqq x \leqq \pi/2$. Looking at the graphs plotted in Fig. 11.9.5, we see that this is so if $n = 5$ but not if $n = 4$. The desired approximation is therefore

$$\sin x \approx \frac{\sqrt{2}}{2}\cdot\left[1 + \left(x - \frac{\pi}{4}\right) - \frac{1}{2!}\left(x - \frac{\pi}{4}\right)^2 - \frac{1}{3!}\left(x - \frac{\pi}{4}\right)^3 + \frac{1}{4!}\left(x - \frac{\pi}{4}\right)^4 + \frac{1}{5!}\left(x - \frac{\pi}{4}\right)^5\right].$$

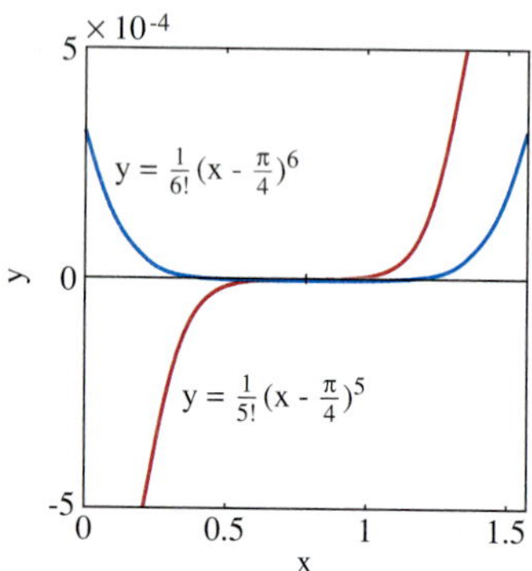

Fig. 11.9.5 Comparing errors in Example 7

For instance, substituting $x = 0$, we get $\sin 0° \approx 0.00020 \approx 0.000$ as desired, and $x = \pi/2$ gives $\sin 90° \approx 1.00025 \approx 1.000$. ■

11.9 PROBLEMS

In Problems 1 through 10, use an infinite series to approximate the indicated number accurate to three decimal places.

1. $\sqrt[3]{65}$ **2.** $\sqrt[4]{630}$

3. $\sin(0.5)$ **4.** $e^{-0.2}$

5. $\tan^{-1}(0.5)$ **6.** $\ln(1.1)$

7. $\sin\left(\dfrac{\pi}{10}\right)$ **8.** $\cos\left(\dfrac{\pi}{20}\right)$

9. $\sin 10°$ **10.** $\cos 5°$

In Problems 11 through 22, use power series to approximate the values of the given integrals accurate to four decimal places.

11. $\displaystyle\int_0^1 \frac{\sin x}{x}\,dx$ **12.** $\displaystyle\int_0^1 \frac{\sin x}{\sqrt{x}}\,dx$

13. $\displaystyle\int_0^{1/2} \frac{\arctan x}{x}\,dx$ **14.** $\displaystyle\int_0^1 \sin x^2\,dx$

15. $\displaystyle\int_0^{1/10} \frac{\ln(1+x)}{x}\,dx$ **16.** $\displaystyle\int_0^{1/2} \frac{1}{\sqrt{1+x^4}}\,dx$

17. $\displaystyle\int_0^{1/2} \frac{1-e^{-x}}{x}\,dx$ **18.** $\displaystyle\int_0^{1/2} \sqrt{1+x^3}\,dx$

19. $\displaystyle\int_0^1 e^{-x^2}\,dx$ **20.** $\displaystyle\int_0^1 \frac{1-\cos x}{x^2}\,dx$

21. $\displaystyle\int_0^{1/2} \sqrt[3]{1+x^2}\,dx$ **22.** $\displaystyle\int_0^{1/2} \frac{x}{\sqrt{1+x^3}}\,dx$

In Problems 23 through 28, use power series rather than l'Hôpital's rule to evaluate the given limit.

23. $\displaystyle\lim_{x\to 0} \frac{1+x-e^x}{x^2}$ **24.** $\displaystyle\lim_{x\to 0} \frac{x-\sin x}{x^3\cos x}$

25. $\displaystyle\lim_{x\to 0} \frac{1-\cos x}{x(e^x-1)}$ **26.** $\displaystyle\lim_{x\to 0} \frac{e^x-e^{-x}-2x}{x-\arctan x}$

27. $\displaystyle\lim_{x\to 0} \left(\frac{1}{x} - \frac{1}{\sin x}\right)$ **28.** $\displaystyle\lim_{x\to 1} \frac{\ln(x^2)}{x-1}$

In Problems 29 through 32, calculate the indicated number with the required accuracy using Taylor's formula for an appropriate function centered at the given point $x = a$.

29. $\sin 80°$; $a = \pi/4$, four decimal places

30. $\cos 35°$; $a = \pi/4$, four decimal places

31. $\cos 47°$; $a = \pi/4$, six decimal places

32. $\sin 58°$; $a = \pi/3$, six decimal places

In Problems 33 through 36, determine the number of decimal places of accuracy the given approximation formula yields for $|x| \leqq 0.1$.

33. $e^x \approx 1 + x + \frac{1}{2}x^2 + \frac{1}{6}x^3 + \frac{1}{24}x^4$

34. $\sin x \approx x - \frac{1}{6}x^3 + \frac{1}{120}x^5$

35. $\ln(1 + x) \approx x - \frac{1}{2}x^2 + \frac{1}{3}x^2 - \frac{1}{4}x^4$

36. $\sqrt{1 + x} \approx 1 + \frac{1}{2}x - \frac{1}{8}x^2$

37. Show that the approximation in Problem 33 gives the value of e^x accurate to within 0.001 if $|x| \leqq 0.5$. Then calculate $\sqrt[3]{e}$ accurate to two decimal places.

38. For what values of x is the approximation $\sin x \approx x - \frac{1}{6}x^3$ accurate to five decimal places?

39. (a) Show that the values of the cosine function for angles between 40° and 50° can be calculated with five-place accuracy using the approximation

$$\cos x \approx \frac{\sqrt{2}}{2}\left[1 - \left(x - \frac{\pi}{4}\right) - \frac{1}{2}\left(x - \frac{\pi}{4}\right)^2 + \frac{1}{6}\left(x - \frac{\pi}{4}\right)^3\right].$$

(b) Show that this approximation yields eight-place accuracy for angles between 44° and 46°.

40. Extend the approximation in Problem 39 to one that yields the values of $\cos x$ accurate to five decimal places for angles between 30° and 60°.

In Problems 41 through 44, use termwise integration of an appropriate power series to approximate the indicated area or volume accurate to two decimal places.

41. Figure 11.9.1 shows the region that lies between the graph of $y = (\sin x)/x$ and the x-axis from $x = -\pi$ to $x = \pi$. Substitute $\sin^2 x = \frac{1}{2}(1 - \cos 2x)$ to approximate the volume of the solid that is generated by revolving this region around the x-axis.

42. Approximate the area of the region that lies between the graph of $y = (1 - \cos x)/x^2$ and the x-axis from $x = -2\pi$ to $x = 2\pi$ (Fig. 11.9.6).

43. Approximate the volume of the solid generated by rotating the region of Problem 42 around the y-axis.

44. Approximate the volume of the solid generated by rotating the region of Problem 42 around the x-axis.

45. Derive the geometric series by long division of $1 - x$ into 1.

46. Derive the series for $\tan x$ listed in Example 3 by long division of the Maclaurin series of $\cos x$ into the Maclaurin series of $\sin x$.

47. Derive the geometric series representation of $1/(1 - x)$ by finding $a_0, a_1, a_2, \ldots$ such that

$$(1 - x)(a_0 + a_1x + a_2x^2 + a_3x^3 + \cdots) = 1.$$

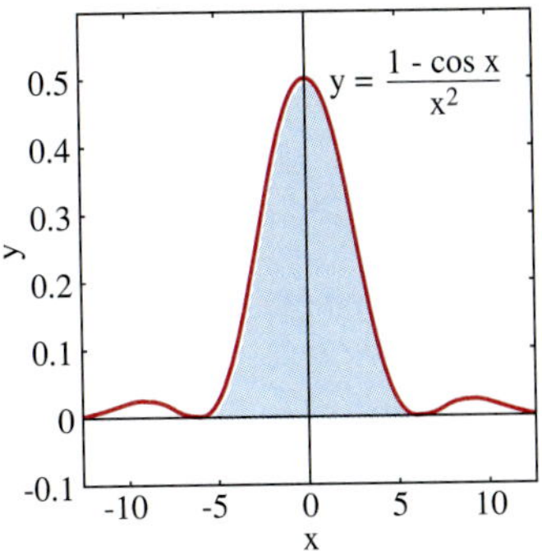

Fig. 11.9.6 The region of Problem 42

48. Derive the first five coefficients in the binomial series for $\sqrt{1 + x}$ by finding a_0, a_1, a_2, a_3, and a_4 such that

$$(a_0 + a_1x + a_2x^2 + a_3x^3 + a_4x^4 + \cdots)^2 = 1 + x.$$

49. Use the method of Example 4 to find the coefficients a_0, a_1, a_2, a_3, and a_4 in the series

$$\sec x = \frac{1}{\cos x} = \sum_{n=0}^{\infty} a_n x^n.$$

50. Multiply the geometric series for $1/(1 - x)$ and the series for $\ln(1 - x)$ to show that if $|x| < 1$, then

$$\frac{\ln(1 - x)}{1 - x} = -x - \left(1 + \tfrac{1}{2}\right)x^2 - \left(1 + \tfrac{1}{2} + \tfrac{1}{3}\right)x^3$$
$$-\left(1 + \tfrac{1}{2} + \tfrac{1}{3} + \tfrac{1}{4}\right)x^4 + \cdots.$$

51. Take as known the logarithmic series

$$\ln(1 + x) = x - \tfrac{1}{2}x^2 + \tfrac{1}{3}x^3 - \tfrac{1}{4}x^4 + \cdots.$$

Find the first four coefficients in the series for e^x by finding a_0, a_1, a_2, and a_3 such that

$$1 + x = e^{\ln(1+x)} = \sum_{n=0}^{\infty} a_n\left(x - \tfrac{1}{2}x^2 + \tfrac{1}{3}x^3 - \tfrac{1}{4}x^4 + \cdots\right)^n.$$

This is exactly how the power series for e^x was first discovered (by Newton)!

52. Use the method of Example 3 to show that

$$\frac{x}{\sin x} = 1 + \frac{1}{6}x^2 + \frac{7}{360}x^4 + \cdots.$$

53. Show that long division of power series gives

$$\frac{2 + x}{1 + x + x^2} = 2 - x - x^2 + 2x^3 - x^4 - x^5 + 2x^6$$
$$- x^7 - x^8 + 2x^9 - x^{10} - x^{11} + \cdots.$$

Show also that the radius of convergence of this series is $R = 1$.

54. Use the series in Problem 53 to approximate with two-place accuracy the value of the integral

$$\int_0^{1/2} \frac{x+2}{x^2+x+1}\,dx.$$

Compare your estimate with the exact result given by a computer algebra system.

Use the power series in Problem 53 to approximate with two-place accuracy the rather formidable integrals in Problems 55 and 56. Compare your estimates with the exact values given by a computer algebra system.

55. $\displaystyle\int_0^{1/2} \frac{1}{1+x^2+x^4}\,dx$ **56.** $\displaystyle\int_0^{1/2} \frac{1}{1+x^4+x^8}\,dx$

In Problems 57 and 58, graph the given function and several of its Taylor polynomials of the indicated degrees.

57. $f(x) = \dfrac{\sin x}{x}$; degrees $n = 2, 4, 6, \dots$.

58. $f(x) = \displaystyle\int_0^x \frac{\sin t}{t}\,dt$; degrees $n = 3, 5, 7, \dots$.

59. Use known power series to evaluate $\displaystyle\lim_{x\to 0} \frac{\sin x - \tan x}{\sin^{-1}x - \tan^{-1}x}$.

60. Substitute series such as

$$\sin(\tan x) = x + \frac{x^3}{6} - \frac{x^5}{40} - \frac{55x^7}{1008} + \cdots$$

provided by a computer algebra system to evaluate

$$\lim_{x\to 0} \frac{\sin(\tan x) - \tan(\sin x)}{\sin^{-1}(\tan^{-1}x) - \tan^{-1}(\sin^{-1}x)}.$$

61. (a) First use the parametrization $x(t) = a\cos t$, $y(t) = b\sin t$, $0 \leqq t \leqq 2\pi$ of the ellipse $(x/a)^2 + (y/b)^2 = 1$ to show that its perimeter (arc length) p is given by

$$p = 4a\int_0^{\pi/2} \sqrt{1-\epsilon^2\cos^2 t}\,dt$$

where $\epsilon = \sqrt{1-(b/a)^2}$ is the *eccentricity* of the ellipse. This so-called *elliptic integral* is nonelementary and so must be approximated numerically. (b) Use the binomial series to expand the integrand in the perimeter formula in part (a). Then integrate termwise—using formula 113 from the table of integrals on the inside back cover—to show that the perimeter of the ellipse is given in terms of its major semiaxis and eccentricity by the power series

$$p = 2\pi a\left(1 - \frac{1}{4}\epsilon^2 - \frac{3}{64}\epsilon^4 - \frac{5}{256}\epsilon^6 - \frac{175}{16384}\epsilon^8 - \cdots\right).$$

62. The *arithmetic mean* of the major and minor semiaxes of the ellipse of Problem 61 is $A = \frac{1}{2}(a+b)$; their *root-square mean* is $R = \sqrt{\frac{1}{2}(a^2+b^2)}$. Substitute $b = a\sqrt{1-\epsilon^2}$ and use the binomial series to derive the expansions

$$A = a\left(1 - \frac{1}{4}\epsilon^2 - \frac{1}{16}\epsilon^4 - \frac{1}{32}\epsilon^6 - \frac{5}{256}\epsilon^8 - \cdots\right)$$

and

$$R = a\left(1 - \frac{1}{4}\epsilon^2 - \frac{1}{32}\epsilon^4 - \frac{1}{128}\epsilon^6 - \frac{5}{2048}\epsilon^8 - \cdots\right).$$

Something wonderful happens when you average these two series; show that

$$\tfrac{1}{2}(A+R) = a\left(1 - \frac{1}{4}\epsilon^2 - \frac{3}{64}\epsilon^4 - \frac{5}{256}\epsilon^6 - \frac{180}{16384}\epsilon^8 - \cdots\right),$$

and then note that the first four terms of the series within the parentheses here are the same as in the ellipse perimeter series of Problem 61(b). Conclude that the perimeter p of the ellipse is given by

$$p = \pi(A+R) + \frac{5\pi a}{8192}\epsilon^8 + \cdots. \qquad \textbf{(14)}$$

If ϵ is quite small—as in a nearly circular ellipse—then the difference between the exact value of p and the simple approximation

$$p \approx \pi(A+R) = \pi\left(\tfrac{1}{2}(a+b) + \sqrt{\tfrac{1}{2}(a^2+b^2)}\right)$$

is extremely small. For instance, suppose that the orbit of the moon around the earth is an ellipse with major semi-axis a exactly 238,857 miles long and eccentricity ϵ exactly 0.0549. Then use Eq. (14) and a computer algebra system with extended-precision arithmetic to find the perimeter of the moon's orbit accurate *to the nearest inch;* give your answer in miles-feet-inches format.

11.9 PROJECT: CALCULATING TRIGONOMETRIC FUNCTIONS ON A DESERTED ISLAND

Again (as in the 11.4 project), you're stranded for life on a desert island with only a very basic calculator that doesn't know about transcendental functions. Now your task is to use the (alternating) sine and cosine series to construct a table presenting (with five-place accuracy) the sines, cosines, and tangents of angles from 0° to 90° in increments of 5°.

To begin with, you can find the sine, cosine, and tangent of an angle of 45° from the familiar 1–1–$\sqrt{2}$ right triangle. Then you can find the values of these functions at an angle of 60° from an equilateral triangle. Once you know all about 45° and 60° angles, you can use the sine and cosine addition formulas

$$\sin(\alpha \pm \beta) = \sin\alpha\cos\beta \pm \cos\alpha\sin\beta$$

and

$$\cos(\alpha \pm \beta) = \cos\alpha\cos\beta \mp \sin\alpha\sin\beta$$

and/or equivalent forms to find the sine, cosine, and tangent of such angles as 15°, 30°, 75°, and 90°.

But algebra and simple trigonometric identities will probably never give you the sine or cosine or an angle of 5°. For this you will need to use the power series for sine and cosine. Sum enough terms (and then some) so you know your result is accurate to nine decimal places. Then fill in all the entries in your table, rounding them to five places. Tell—honestly—whether your entries agree with those your *real* calculator gives.

Finally, explain what strategy you would use to complete a similar table of values of trigonometric functions with angles by increments of 1° rather than 5°.

CHAPTER 11 REVIEW: *Definitions, Concepts, Results*

Use the following list as a guide to concepts that you may need to review.

1. Definition of the limit of a sequence
2. The limit laws for sequences
3. The bounded monotonic sequence property
4. Definition of the sum of an infinite series
5. Formula for the sum of a geometric series
6. The nth-term test for divergence
7. Divergence of the harmonic series
8. The nth-degree Taylor polynomial of the function f at the point $x = a$
9. Taylor's formula with remainder
10. The Taylor series of the elementary transcendental functions
11. The integral test
12. Convergence of p-series
13. The comparison and limit comparison tests
14. The alternating series test
15. Absolute convergence: definition *and* the fact that it implies convergence
16. The ratio test
17. The root test
18. Power series; radius of convergence and interval of convergence
19. The binomial series
20. Termwise differentiation and integration of power series
21. The use of power series to approximate values of functions and integrals
22. The sum and product of two power series
23. The use of power series to evaluate indeterminate forms

CHAPTER 11 *Miscellaneous Problems*

In Problems 1 through 15, determine whether the sequence $\{a_n\}$ converges, and if so, find its limit.

1. $a_n = \dfrac{n^2+1}{n^2+4}$

2. $a_n = \dfrac{8n-7}{7n-8}$

3. $a_n = 10 - (0.99)^n$

4. $a_n = n\sin\pi n$

5. $a_n = \dfrac{1+(-1)^n\sqrt{n}}{n+1}$

6. $a_n = \sqrt{\dfrac{1+(-0.5)^n}{n+1}}$

7. $a_n = \dfrac{\sin 2n}{n}$

8. $a_n = 2^{-(\ln n)/n}$

9. $a_n = (-1)^{\sin(n\pi/2)}$

10. $a_n = \dfrac{(\ln n)^3}{n^2}$

11. $a_n = \dfrac{1}{n}\sin\dfrac{1}{n}$

12. $a_n = \dfrac{n - e^n}{n + e^n}$

13. $a_n = \dfrac{\sinh n}{n}$

14. $a_n = \left(1 + \dfrac{2}{n}\right)^{2n}$

15. $a_n = (2n^2 + 1)^{1/n}$

Determine whether each infinite series in Problems 16 through 30 converges or diverges.

16. $\displaystyle\sum_{n=1}^{\infty} \frac{(n^2)!}{n^n}$

17. $\displaystyle\sum_{n=1}^{\infty} \frac{(-1)^{n+1}\ln n}{n}$

18. $\displaystyle\sum_{n=0}^{\infty} \frac{3^n}{2^n + 4^n}$

19. $\displaystyle\sum_{n=0}^{\infty} \frac{n!}{e^{n^2}}$

20. $\displaystyle\sum_{n=1}^{\infty} \frac{1}{n^{3/2}}\sin\frac{1}{n}$

21. $\displaystyle\sum_{n=0}^{\infty} \frac{(-2)^n}{3^n + 1}$

22. $\displaystyle\sum_{n=1}^{\infty} 2^{-(2/n^2)}$

23. $\displaystyle\sum_{n=2}^{\infty} \frac{(-1)^n n}{(\ln n)^3}$

24. $\displaystyle\sum_{n=1}^{\infty} \frac{(-1)^n}{10^{1/n}}$

25. $\displaystyle\sum_{n=1}^{\infty} \frac{\sqrt{n} + \sqrt[3]{n}}{n^2 + n^3}$

26. $\displaystyle\sum_{n=1}^{\infty} \frac{(-1)^{n+1}}{n^{[1+(1/n)]}}$

27. $\displaystyle\sum_{n=1}^{\infty} \frac{(-1)^{n+1}\arctan n}{\sqrt{n}}$

28. $\displaystyle\sum_{n=1}^{\infty} n\sin\frac{1}{n}$

29. $\displaystyle\sum_{n=3}^{\infty} \frac{1}{n(\ln n)(\ln \ln n)}$

30. $\displaystyle\sum_{n=3}^{\infty} \frac{1}{n(\ln n)(\ln \ln n)^2}$

Find the interval of convergence of the power series in Problems 31 through 40.

31. $\displaystyle\sum_{n=0}^{\infty} \frac{2^n x^n}{n!}$

32. $\displaystyle\sum_{n=0}^{\infty} \frac{(3x)^n}{2^{n+1}}$

33. $\displaystyle\sum_{n=1}^{\infty} \frac{(x-1)^n}{n\cdot 3^n}$

34. $\displaystyle\sum_{n=0}^{\infty} \frac{(2x-3)^n}{4^n}$

35. $\displaystyle\sum_{n=1}^{\infty} \frac{(-1)^n x^n}{4n^2 - 1}$

36. $\displaystyle\sum_{n=0}^{\infty} \frac{(2x-1)^n}{n^2 + 1}$

37. $\displaystyle\sum_{n=0}^{\infty} \frac{n!x^{2n}}{10^n}$

38. $\displaystyle\sum_{n=2}^{\infty} \frac{x^n}{\ln n}$

39. $\displaystyle\sum_{n=0}^{\infty} \frac{1 + (-1)^n}{2(n!)} x^n$

40. $\displaystyle\sum_{n=1}^{\infty} \left(1 + \frac{1}{n}\right)^n (x-1)^n$

Find the set of all values of x for which the series in Problems 41 through 43 converge.

41. $\displaystyle\sum_{n=1}^{\infty} (x - n)^n$

42. $\displaystyle\sum_{n=1}^{\infty} (\ln x)^n$

43. $\displaystyle\sum_{n=0}^{\infty} \frac{e^{nx}}{n!}$

44. Find the rational number that has repeated decimal expansion 2.7 1828 1828 1828 . . . .

45. Give an example of two convergent numerical series $\sum a_n$ and $\sum b_n$ such that the series $\sum a_n b_n$ diverges.

46. Prove that if $\sum a_n$ is a convergent positive-term series, then $\sum a_n^2$ converges.

47. Let the sequence $\{a_n\}$ be defined recursively as follows:

$$a_1 = 1; \quad a_{n+1} = 1 + \frac{1}{1 + a_n} \quad \text{if } n \geqq 1.$$

The limit of the sequence $\{a_n\}$ is the value of the *continued fraction*

$$1 + \cfrac{1}{2 + \cfrac{1}{2 + \cfrac{1}{2 + \cfrac{1}{2 + \cdots}}}}.$$

Assuming that $A = \lim_{n\to\infty} a_n$ exists, prove that $A = \sqrt{2}$.

48. Let $\{F_n\}_1^{\infty}$ be the Fibonacci sequence of Example 2 in Section 11.2. (a) Prove that $0 < F_n \leqq 2^n$ for all $n \geqq 1$, and hence conclude that the power series

$$F(x) = \sum_{n=1}^{\infty} F_n x^n$$

converges if $|x| < \frac{1}{2}$. (b) Show that $(1 - x - x^2)F(x) = x$, so

$$F(x) = \frac{x}{1 - x - x^2}.$$

49. We say that the *infinite product* indicated by

$$\prod_{n=1}^{\infty} (1 + a_n) = (1 + a_1)(1 + a_2)(1 + a_3)\cdots$$

converges provided that the infinite series

$$S = \sum_{n=1}^{\infty} \ln(1 + a_n)$$

converges, in which case the value of the infinite product is, by definition, e^S. Use the integral test to prove that

$$\prod_{n=1}^{\infty}\left(1 + \frac{1}{n}\right)$$

diverges.

50. Prove that the infinite product (see Problem 49)

$$\prod_{n=1}^{\infty}\left(1 + \frac{1}{n^2}\right)$$

converges, and use the integral test remainder estimate to approximate its value. The actual value of this infinite product is known to be

$$\frac{\sinh \pi}{\pi} \approx 3.67607\,79103\,74977\,72069\,56975.$$

In Problems 51 through 55, use infinite series to approximate the indicated number accurate to three decimal places.

51. $\sqrt[5]{1.5}$

52. $\ln(1.2)$

53. $\displaystyle\int_0^{0.5} e^{-x^2}\,dx$

54. $\displaystyle\int_0^{0.5} \sqrt[3]{1+x^4}\,dx$

55. $\displaystyle\int_0^1 \frac{1-e^{-x}}{x}\,dx$

56. Substitute the Maclaurin series for $\sin x$ into that for e^x to obtain

$$e^{\sin x} = 1 + x + \tfrac{1}{2}x^2 - \tfrac{1}{8}x^4 + \cdots.$$

57. Substitute the Maclaurin series for the cosine and then integrate termwise to derive the formula

$$\int_0^\infty e^{-t^2}\cos 2xt\,dt = \frac{\sqrt{\pi}}{2}e^{-x^2}.$$

Use the reduction formula

$$\int_0^\infty t^{2n}e^{-t^2}\,dt = \frac{2n-1}{2}\int_0^\infty t^{2n-2}e^{-t^2}\,dt$$

derived in Problem 50 of Section 9.3. The validity of this improper termwise integration is subject to verification.

58. Prove that

$$\tanh^{-1}x = \int_0^x \frac{1}{1-t^2}\,dt = \sum_{n=0}^\infty \frac{x^{2n+1}}{2n+1}$$

if $|x| < 1$.

59. Prove that

$$\sinh^{-1}x = \int_0^x \frac{1}{\sqrt{1+t^2}}\,dt = \sum_{n=0}^\infty (-1)^n \frac{1\cdot 3\cdot 5\cdots(2n-1)}{2\cdot 4\cdot 6\cdots(2n)}\cdot\frac{x^{2n+1}}{2n+1}$$

if $|x| < 1$.

60. Suppose that $\tan y = \sum a_n y^n$. Determine a_0, a_1, a_2, and a_3 by substituting the inverse tangent series [Eq. (27) of Section 11.4] into the equation

$$x = \tan(\tan^{-1}x) = \sum_{n=0}^\infty a_n(\tan^{-1}x)^n.$$

61. According to *Stirling's series,* the value of $n!$ for large n is given to a close approximation by

$$n! \approx \sqrt{2\pi n}\left(\frac{n}{e}\right)^n e^{\mu(n)},$$

where

$$\mu(n) = \frac{1}{12n} - \frac{1}{360n^3} + \frac{1}{1260n^5}.$$

Substitute $\mu(n)$ into Maclaurin's series for e^x to show that

$$e^{\mu(n)} = 1 + \frac{1}{12n} + \frac{1}{288n^2} - \frac{139}{51840n^3} + \cdots.$$

Can you show that the next term in the last series is $-571/(2{,}488{,}320n^4)$?

62. Define

$$T(n) = \int_0^{\pi/4} \tan^n x\,dx$$

for $n \geqq 0$. (a) Show by "reduction" of the integral that

$$T(n+2) = \frac{1}{n+1} - T(n)$$

for $n \geqq 0$. (b) Conclude that $T(n) \to 0$ as $n \to \infty$. (c) Show that $T_0 = \pi/4$ and that $T_1 = \frac{1}{2}\ln 2$. (d) Prove by induction on n that

$$T(2n) = (-1)^{n+1}\left(1 - \frac{1}{3} + \frac{1}{5} - \cdots \pm \frac{1}{2n-1} - \frac{\pi}{4}\right).$$

(e) Conclude from parts (b) and (d) that

$$1 - \frac{1}{3} + \frac{1}{5} - \frac{1}{7} + \cdots = \frac{\pi}{4}.$$

(f) Prove by induction on n that

$$T(2n+1) = \frac{1}{2}(-1)^n\left(1 - \frac{1}{2} + \frac{1}{3} - \cdots \pm \frac{1}{n} - \ln 2\right).$$

(g) Conclude from parts (b) and (f) that

$$1 - \frac{1}{2} + \frac{1}{3} - \frac{1}{4} + \cdots = \ln 2.$$

63. Prove as follows that the number e is irrational. First suppose to the contrary that $e = p/q$, where p and q are positive integers. Note that $q > 1$. Write

$$\frac{p}{q} = e = \frac{1}{0!} + \frac{1}{1!} + \frac{1}{2!} + \frac{1}{3!} + \cdots + \frac{1}{q!} + R_q,$$

where $0 < R_q < 3/(q+1)!$. (Why?) Then show that multiplication of both sides of this equation by $q!$ would lead to the contradiction that one side of the result is an integer but the other side is not.

64. Evaluate the infinite product (see Problem 49)

$$\prod_{n=2}^\infty \frac{n^2}{n^2-1}$$

by finding an explicit formula for

$$\prod_{n=2}^{k} \frac{n^2}{n^2-1} \qquad (k \geqq 2)$$

and then taking the limit as $k \to \infty$.

65. Find a continued fraction representation (see Problem 47)

$$a_0 + \cfrac{1}{a_1 + \cfrac{1}{a_2 + \cfrac{1}{a_3 + \cfrac{1}{a_4 + \cdots}}}}.$$

of $\sqrt{5}$.

66. Evaluate

$$1 + \frac{1}{2} - \frac{2}{3} + \frac{1}{4} + \frac{1}{5} - \frac{2}{6} + \frac{1}{7} + \frac{1}{8} - \frac{2}{9} + \frac{1}{10} + \cdots.$$

CHAPTER 12

VECTORS, CURVES, AND SURFACES IN SPACE

J.W. Gibbs (1839–1903)

The study of vector quantities with both direction and magnitude (such as force and velocity) dates back at least to Newton's *Principia Mathematica.* But the vector notation and terminology used in science and mathematics today was largely "born in the USA." The modern system of vector analysis was created independently (and almost simultaneously) in the 1880s by the American mathematical physicist Josiah Willard Gibbs and the British electrical engineer Oliver Heaviside (1850–1925). Gibbs' first vector publication—*Elements of Vector Analysis* (1881)—appeared slightly earlier than Heaviside's (an 1885 paper) and was more systematic and complete in its exposition of the foundations of the subject.

Gibbs, the son of a Yale university professor, grew up in New Haven and attended Yale as an undergraduate. He studied both Latin and mathematics and remained at Yale for graduate work in engineering. In 1863 he received one of the first Ph.D. degrees awarded in the United States (and apparently the very first in engineering). After several years of postdoctoral study of mathematics and physics in France and Germany, he returned to Yale, where he served for more than three decades as professor of mathematical physics.

The educational careers of Gibbs and Heaviside were as different as their vectors were similar. At 16 Heaviside quit school, and at 18 he began work as a telegraph operator. Starting with this very practical introduction to electricity, he published in 1872 the first of a series of electrical papers leading to his famous three-volume treatise *Electromagnetic Theory.* The first volume of this treatise appeared in 1893 and included Heaviside's own systematic presentation of modern vector analysis.

Gibbs introduced the **i, j, k** notation now standard for three-dimensional vectors, adapting it from the algebra of "quaternions," in which the Irish mathematician William Rowan Hamilton (1805–1865) earlier had used *i, j,* and *k* to denote three distinct square roots of -1. Gibbs was the first to define clearly both the scalar (dot) product $\mathbf{a} \cdot \mathbf{b}$ and the vector (cross) product $\mathbf{a} \times \mathbf{b}$ of the vectors **a** and **b**. He observed that $\mathbf{b} \times \mathbf{a} = -\mathbf{a} \times \mathbf{b}$, in contrast with the commutativity of multiplication of ordinary numbers. In the Section 12.5 Project, you will see that the vector product is a key to the analysis of the "curve" of a baseball pitch.

12.1 VECTORS IN THE PLANE

A physical quantity such as length, temperature, or mass can be specified in terms of a single real number, its *magnitude*. Such a quantity is called a **scalar**. Other physical quantities, such as force and velocity, possess both magnitude and *direction*; these quantities are called **vector quantities,** or simply **vectors**.

For example, to specify the velocity of a moving point in the coordinate plane, we must give both the rate at which the point moves (its speed) and the direction of that motion. The *velocity vector* of the moving point incorporates both pieces of information—direction and speed. It is convenient to represent this velocity vector by an arrow, with its initial point located at the current position of the moving point on its trajectory (Fig. 12.1.1).

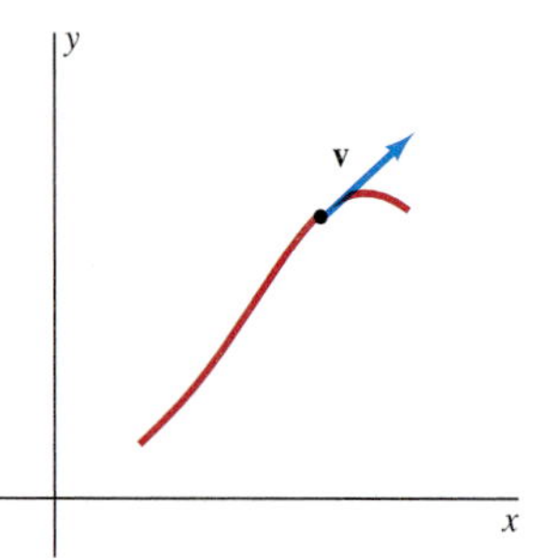

Fig. 12.1.1 A velocity vector may be represented by an arrow.

Although the arrow, a directed line segment, carries the desired information—both magnitude (the segment's length) and direction—it is a pictorial representation rather than a quantitative object. The following formal definition of a vector captures the essence of magnitude in combination with direction.

Definition *Vector*

A **vector v** in the Cartesian plane is an ordered pair of real numbers that has the form $\langle a, b\rangle$. We write $\mathbf{v} = \langle a, b\rangle$ and call a and b the **components** of the vector **v**.

The directed line segment $\overrightarrow{OP}$ from the origin O to the point $P(a, b)$ is one geometric representation of the vector **v** (see Fig. 12.1.2). For this reason, the vector $\mathbf{v} = \langle a, b\rangle$ is called the **position vector** of the point $P(a, b)$. In fact, the relationship between $\mathbf{v} = \langle a, b\rangle$ and $P(a, b)$ is so close that, in certain contexts, it is convenient to confuse the two deliberately—to regard **v** and P as the same mathematical object.

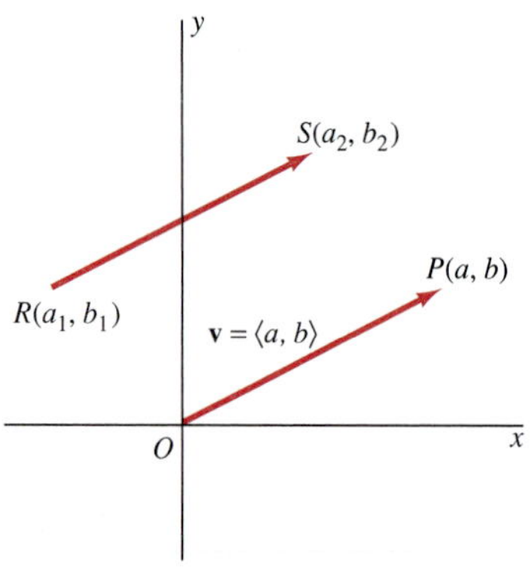

Fig. 12.1.2 The position vector **v** of the point P and another representation $\overrightarrow{RS}$ of **v**.

The directed line segment from the point $Q(a_1, b_1)$ to the point $R(a_2, b_2)$ has the same direction and magnitude as the directed line segment from the origin $O(0, 0)$ to the point $P(a_2 - a_1, b_2 - b_1)$, and consequently they represent the same vector $\mathbf{v} = \overrightarrow{OP} = \overrightarrow{QR}$. With this observation, it is easy to find the components of the vector with arbitrary initial point Q and arbitrary terminal point R.

REMARK When discussing vectors, we often use the term *scalar* to refer to an ordinary numerical quantity, one that is *not* a vector. In printed work we use **bold** type to distinguish the names of vectors from those of other mathematical objects, such as the scalars a and b that are the components of the vector $\mathbf{v} = \langle a, b\rangle$. In handwritten work a suitable alternative is to place an arrow—or just a bar—over every symbol that denotes a vector. Thus you may write $\vec{\text{v}} = \langle \text{a}, \text{b}\rangle$ or $\bar{\text{v}} = \langle \text{a}, \text{b}\rangle$. There is no need for an arrow or a bar over a vector $\langle a, b\rangle$ already identified by angle brackets, so none should be used there.

A directed line segment has both length and direction. The **length** of the vector $\mathbf{v} = \langle a, b\rangle$ is denoted by $v = |\mathbf{v}|$ and is defined as follows:

$$v = |\mathbf{v}| = |\langle a, b\rangle| = \sqrt{a^2 + b^2}. \tag{1}$$

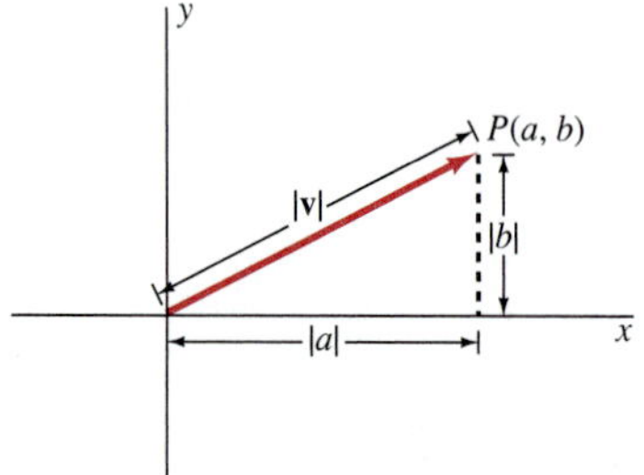

Fig. 12.1.3 The length $v = |\mathbf{v}|$ of the vector **v**.

The notation $v = |\mathbf{v}|$ is used because the length of a vector is in many ways analogous to the absolute value of a real number (Fig. 12.1.3).

EXAMPLE 1 The length of the vector $\mathbf{v} = \langle 1, -2\rangle$ is

$$v = |\langle 1, -2\rangle| = \sqrt{(1)^2 + (-2)^2} = \sqrt{5}.$$ ■

The only vector with length zero is the **zero vector** with both components zero, denoted by $\mathbf{0} = \langle 0, 0 \rangle$. The zero vector is unique in that it has no specific direction. Every nonzero vector has a specified direction; the vector represented by the arrow $\overrightarrow{OP}$ from the origin O to another point P in the plane has direction specified (for instance) by the counterclockwise angle from the positive x-axis to $\overrightarrow{OP}$.

What is important about the vector $\mathbf{v} = \langle a, b \rangle$ represented by $\overrightarrow{OP}$ often is not *where* it is, but how long it is and which way it points. If the directed line segment $\overrightarrow{RS}$ with endpoints $R(a_1, b_1)$ and $S(a_2, b_2)$ has the same length and direction as $\overrightarrow{OP}$, then we say that $\overrightarrow{RS}$ **represents** (or is a **representation** of) the vector $\mathbf{v}$. Thus a single vector has many representatives (Fig. 12.1.4).

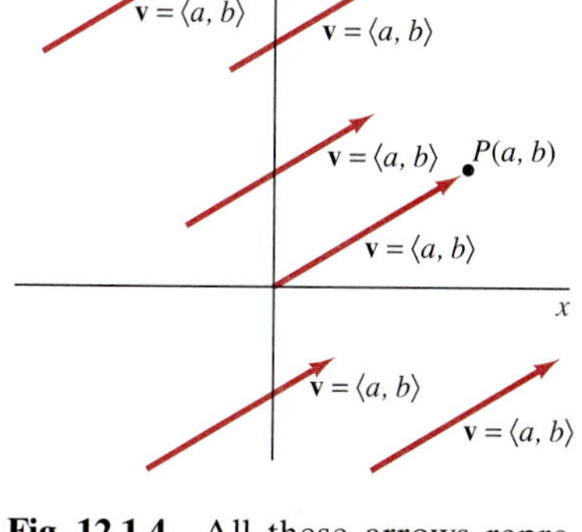

Fig. 12.1.4 All these arrows represent the same vector $\mathbf{v} = \langle a, b \rangle$.

Algebraic Operations with Vectors

The operations of addition and multiplication of real numbers have analogues for vectors. We shall define each of these operations of *vector algebra* in terms of components of vectors and then give a geometric interpretation in terms of arrows.

Definition ***Equality of Vectors***

The two vectors $\mathbf{u} = \langle u_1, u_2 \rangle$ and $\mathbf{v} = \langle v_1, v_2 \rangle$ are **equal** provided that $u_1 = v_1$ and $u_2 = v_2$.

In other words, two vectors are equal if and only if *corresponding components* are the same. Moreover, two directed line segments $\overrightarrow{PQ}$ and $\overrightarrow{RS}$ represent the same vector provided that they have the same length and direction. This will be the case provided that the segments $\overrightarrow{PQ}$ and $\overrightarrow{RS}$ are opposite sides of a parallelogram (Fig. 12.1.5).

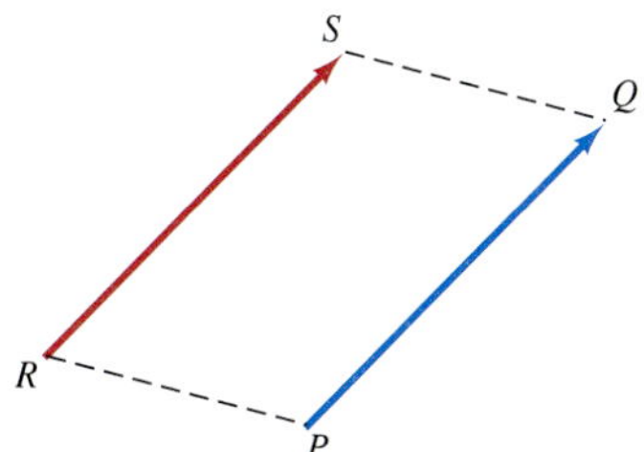

Fig. 12.1.5 Parallel directed segments representing equal vectors

Definition ***Addition of Vectors***

The **sum u + v** of the two vectors $\mathbf{u} = \langle u_1, u_2 \rangle$ and $\mathbf{v} = \langle v_1, v_2 \rangle$ is the vector

$$\mathbf{u} + \mathbf{v} = \langle u_1 + v_1, u_2 + v_2 \rangle. \tag{2}$$

Thus we add vectors by adding corresponding components—that is, by *componentwise addition*. The geometric interpretation of vector addition is the **triangle law of addition,** illustrated in Fig. 12.1.6, where the labeled lengths indicate why this interpretation is valid. An equivalent interpretation is the **parallelogram law of addition,** illustrated in Fig. 12.1.7.

EXAMPLE 2 The sum of the vectors $\mathbf{u} = \langle 4, 3 \rangle$ and $\mathbf{v} = \langle -5, 2 \rangle$ is the vector

$$\mathbf{u} + \mathbf{v} = \langle 4, 3 \rangle + \langle -5, 2 \rangle = \langle 4 + (-5), 3 + 2 \rangle = \langle -1, 5 \rangle.$$

It is natural to write $2\mathbf{u} = \mathbf{u} + \mathbf{u}$. But if $\mathbf{u} = \langle u_1, u_2 \rangle$, then

$$2\mathbf{u} = \mathbf{u} + \mathbf{u} = \langle u_1, u_2 \rangle + \langle u_1, u_2 \rangle = \langle 2u_1, 2u_2 \rangle.$$

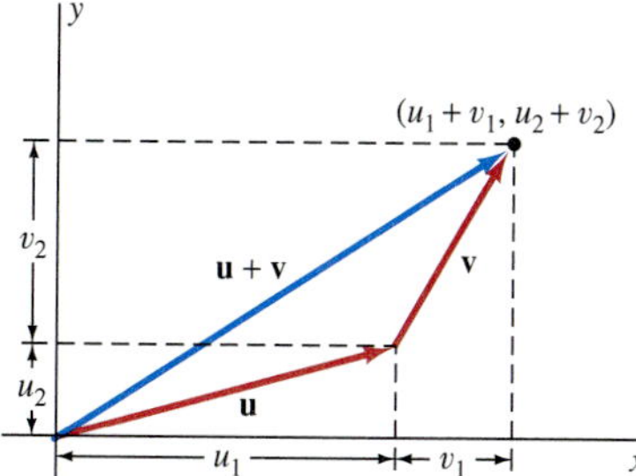

Fig. 12.1.6 The triangle law is a geometric interpretation of vector addition.

This suggests that multiplication of a vector by a scalar (real number) also is defined in a componentwise manner.

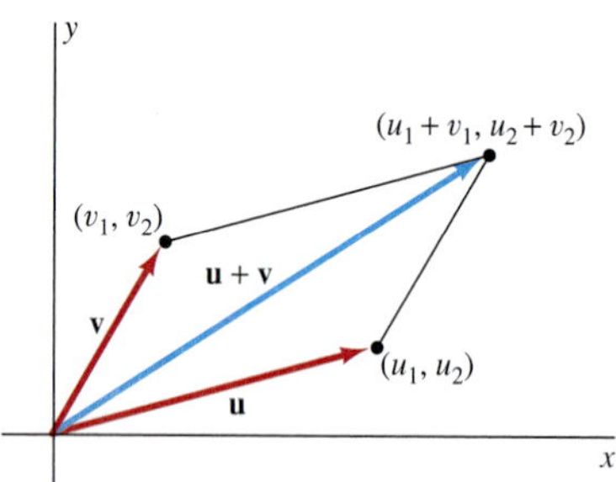

Fig. 12.1.7 The parallelogram law for vector addition

> **Definition** *Multiplication of a Vector by a Scalar*
>
> If $\mathbf{u} = \langle u_1, u_2 \rangle$ and c is a real number, then the **scalar multiple** $c\mathbf{u}$ is the vector
>
> $$c\mathbf{u} = \langle cu_1, cu_2 \rangle. \tag{3}$$

Note that

$$|c\mathbf{u}| = \sqrt{(cu_1)^2 + (cu_2)^2} = |c|\sqrt{(u_1)^2 + (u_2)^2} = |c| \cdot |\mathbf{u}|.$$

Thus the length of $|c\mathbf{u}|$ is $|c|$ times the length of $\mathbf{u}$. The **negative** of the vector $\mathbf{u}$ is the vector

$$-\mathbf{u} = (-1)\mathbf{u} = \langle -u_1, -u_2 \rangle,$$

with the same length as $\mathbf{u}$ but the opposite direction. We say that the two nonzero vectors $\mathbf{u}$ and $\mathbf{v}$ have

- The **same direction** if $\mathbf{u} = c\mathbf{v}$ for some $c > 0$;
- **Opposite directions** if $\mathbf{u} = c\mathbf{v}$ for some $c < 0$.

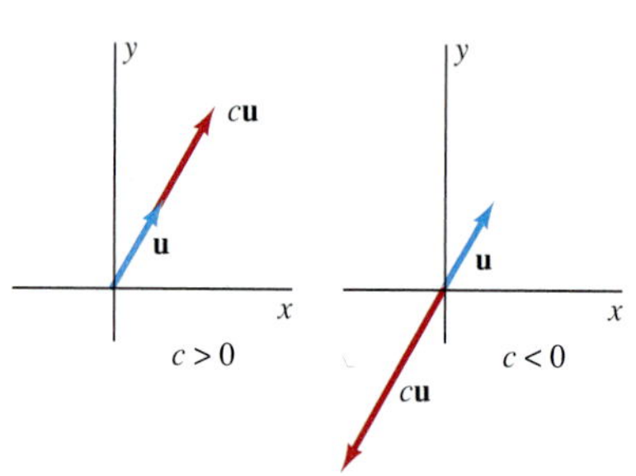

Fig. 12.1.8 The vector $c\mathbf{u}$ may have the same direction as $\mathbf{u}$ or the opposite direction, depending on the sign of c.

The geometric interpretation of scalar multiplication is that $c\mathbf{u}$ is the vector with length $|c| \cdot |\mathbf{u}|$, with the same direction as $\mathbf{u}$ if $c > 0$ but with the opposite direction if $c < 0$ (Fig. 12.1.8).

The **difference** $\mathbf{u} - \mathbf{v}$ of the vectors $\mathbf{u} = \langle u_1, u_2 \rangle$ and $\mathbf{v} = \langle v_1, v_2 \rangle$ is defined to be

$$\mathbf{u} - \mathbf{v} = \mathbf{u} + (-\mathbf{v}) = \langle u_1 - v_1, u_2 - v_2 \rangle. \tag{4}$$

If we think of $\langle u_1, u_2 \rangle$ and $\langle v_1, v_2 \rangle$ as position vectors of the points P and Q, respectively, then $\mathbf{u} - \mathbf{v}$ may be represented by the arrow $\overrightarrow{QP}$ from Q to P. We may therefore write

$$\mathbf{u} - \mathbf{v} = \overrightarrow{OP} - \overrightarrow{OQ} = \overrightarrow{QP},$$

as illustrated in Fig. 12.1.9.

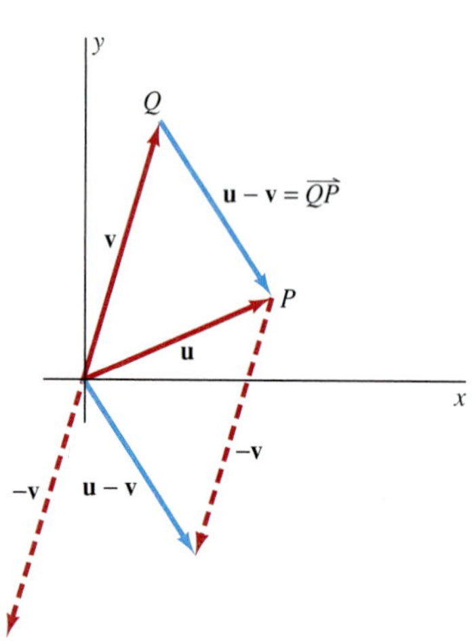

Fig. 12.1.9 Geometric interpretation of the difference $\mathbf{u} - \mathbf{v}$

EXAMPLE 3 Suppose that $\mathbf{u} = \langle 4, -3 \rangle$ and $\mathbf{v} = \langle -2, 3 \rangle$. Find $|\mathbf{u}|$ and the vectors $\mathbf{u} + \mathbf{v}$, $\mathbf{u} - \mathbf{v}$, $3\mathbf{u} - 2\mathbf{v}$, and $2\mathbf{u} + 4\mathbf{v}$.

Solution

$$|\mathbf{u}| = \sqrt{4^2 + (-3)^2} = \sqrt{25} = 5.$$

$$\mathbf{u} + \mathbf{v} = \langle 4 + (-2), -3 + 3 \rangle = \langle 2, 0 \rangle.$$

$$\mathbf{u} - \mathbf{v} = \langle 4 - (-2), -3 - 3 \rangle = \langle 6, -6 \rangle.$$

$$3\mathbf{u} = \langle 3 \cdot 4, 3 \cdot (-3) \rangle = \langle 12, -9 \rangle.$$

$$-2\mathbf{v} = \langle -2 \cdot (-2), -2 \cdot 3 \rangle = \langle 4, -6 \rangle.$$

$$2\mathbf{u} + 4\mathbf{v} = \langle 2 \cdot 4 + 4 \cdot (-2), 2 \cdot (-3) + 4 \cdot 3 \rangle = \langle 0, 6 \rangle.$$

■

The familiar algebraic properties of real numbers carry over to the following analogous properties of vector addition and scalar multiplication. Let **a**, **b**, and **c** be vectors and r and s real numbers. Then

1. $\mathbf{a} + \mathbf{b} = \mathbf{b} + \mathbf{a}$,
2. $\mathbf{a} + (\mathbf{b} + \mathbf{c}) = (\mathbf{a} + \mathbf{b}) + \mathbf{c}$,
3. $r(\mathbf{a} + \mathbf{b}) = r\mathbf{a} + r\mathbf{b}$, **(5)**
4. $(r + s)\mathbf{a} = r\mathbf{a} + s\mathbf{a}$,
5. $(rs)\mathbf{a} = r(s\mathbf{a}) = s(r\mathbf{a})$.

You can easily verify these identities by working with components. For example, if $\mathbf{a} = \langle a_1, a_2 \rangle$ and $\mathbf{b} = \langle b_1, b_2 \rangle$, then

$$\begin{aligned} r(\mathbf{a} + \mathbf{b}) &= r\langle a_1 + b_1, a_2 + b_2 \rangle = \langle r(a_1 + b_1), r(a_2 + b_2) \rangle \\ &= \langle ra_1 + rb_1, ra_2 + rb_2 \rangle = \langle ra_1, ra_2 \rangle + \langle rb_1, rb_2 \rangle = r\mathbf{a} + r\mathbf{b}. \end{aligned}$$

The proofs of the other four identities in Eq. (5) are left as exercises.

The Unit Vectors i and j

A **unit** vector is a vector of length 1. If $\mathbf{a} = \langle a_1, a_2 \rangle \neq \mathbf{0}$, then

$$\mathbf{u} = \frac{\mathbf{a}}{|\mathbf{a}|} \tag{6}$$

is the unit vector with the same direction as **a,** because

$$|\mathbf{u}| = \sqrt{\left(\frac{a_1}{|\mathbf{a}|}\right)^2 + \left(\frac{a_2}{|\mathbf{a}|}\right)^2} = \frac{1}{|\mathbf{a}|}\sqrt{a_1^2 + a_2^2} = 1.$$

For example, if $\mathbf{a} = \langle 3, -4 \rangle$, then $|\mathbf{a}| = 5$. Thus $\left\langle \frac{3}{5}, -\frac{4}{5} \right\rangle$ is a unit vector that has the same direction as **a**.

Two particular unit vectors play a special role, the vectors

$$\mathbf{i} = \langle 1, 0 \rangle \quad \text{and} \quad \mathbf{j} = \langle 0, 1 \rangle.$$

The first points in the positive x-direction; the second points in the positive y-direction (Fig. 12.1.10). Together they provide a useful alternative notation for vectors. If $\mathbf{a} = \langle a_1, a_2 \rangle$, then

$$\mathbf{a} = \langle a_1, 0 \rangle + \langle 0, a_2 \rangle = a_1\langle 1, 0 \rangle + a_2\langle 0, 1 \rangle = a_1\mathbf{i} + a_2\mathbf{j}. \tag{7}$$

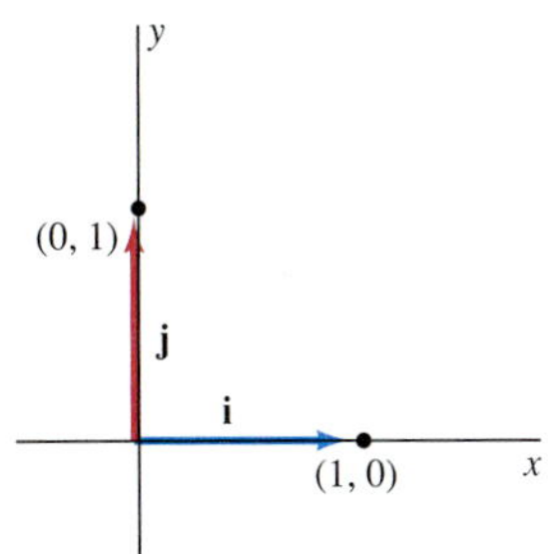

Fig. 12.1.10 The vectors **i** and **j**

Thus every vector in the plane is a **linear combination** of **i** and **j**. The usefulness of this notation is based on the fact that such linear combinations of **i** and **j** may be manipulated as if they were ordinary sums. For example, if

$$\mathbf{a} = a_1\mathbf{i} + a_2\mathbf{j} \quad \text{and} \quad \mathbf{b} = b_1\mathbf{i} + b_2\mathbf{j},$$

then

$$\mathbf{a} + \mathbf{b} = (a_1\mathbf{i} + a_2\mathbf{j}) + (b_1\mathbf{i} + b_2\mathbf{j}) = (a_1 + b_1)\mathbf{i} + (a_2 + b_2)\mathbf{j}.$$

Also,

$$c\mathbf{a} = c(a_1\mathbf{i} + a_2\mathbf{j}) = (ca_1)\mathbf{i} + (ca_2)\mathbf{j}.$$

EXAMPLE 4 Suppose that $\mathbf{a} = 2\mathbf{i} - 3\mathbf{j}$ and $\mathbf{b} = 3\mathbf{i} + 4\mathbf{j}$. Express $5\mathbf{a} - 3\mathbf{b}$ in terms of **i** and **j**.

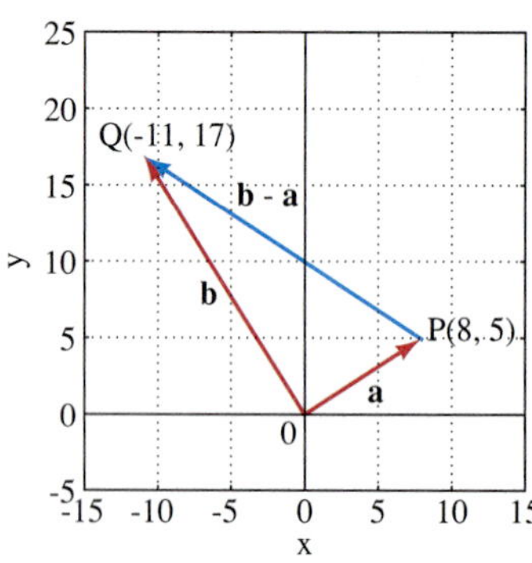

Fig. 12.1.11 The vectors **a, b,** and **b** − **a** of Example 5

Solution

$$\begin{aligned} 5\mathbf{a} - 3\mathbf{b} &= 5\cdot(2\mathbf{i} - 3\mathbf{j}) - 3\cdot(3\mathbf{i} + 4\mathbf{j}) \\ &= (10 - 9)\mathbf{i} + (-15 - 12)\mathbf{j} = \mathbf{i} - 27\mathbf{j}. \end{aligned}$$

EXAMPLE 5 When the vectors $\mathbf{a} = 8\mathbf{i} + 5\mathbf{j}$ and $\mathbf{b} = -11\mathbf{i} + 17\mathbf{j}$ are plotted carefully (Fig. 12.1.11), they look as though they might be perpendicular. Determine whether this is so.

Solution If the vectors **a** and **b** are regarded as position vectors of the points $P(8, 5)$ and $Q(-11, 17)$, then their difference $\mathbf{c} = \mathbf{b} - \mathbf{a} = -19\mathbf{i} + 12\mathbf{j}$ represents the third side $\overrightarrow{PQ}$ of the triangle OPQ (Fig. 12.1.11). According to the Pythagorean theorem, this triangle is a right triangle with hypotenuse PQ if and only if $|\mathbf{c}|^2 = |\mathbf{a}|^2 + |\mathbf{b}|^2$. But

$$|\mathbf{c}|^2 = (-19)^2 + 12^2 = 505 \quad \text{whereas} \quad |\mathbf{a}|^2 + |\mathbf{b}|^2 = [8^2 + 5^2] + [(-11)^2 + 17^2] = 499.$$

It follows that the vectors **a** and **b** are not perpendicular.

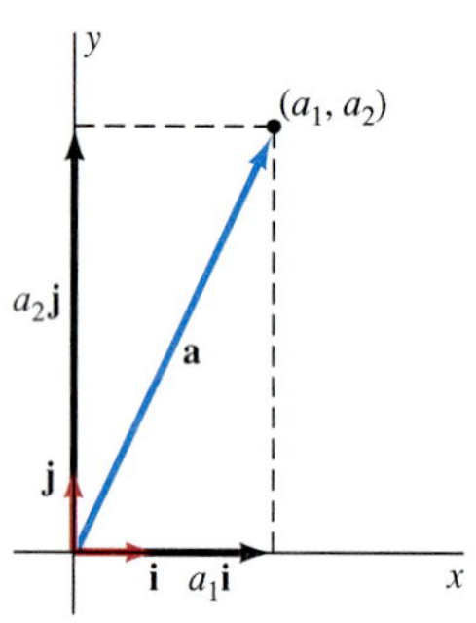

Fig. 12.1.12 Resolution of $\mathbf{a} = \langle a_1, a_2 \rangle$ into its horizontal and vertical components

Equation (7) expresses the vector $\mathbf{a} = \langle a_1, a_2 \rangle$ as the sum of a horizontal vector $a_1\mathbf{i}$ and a vertical vector $a_2\mathbf{j}$, as Fig. 12.1.12 shows. The decomposition or *resolution* of a vector into its horizontal and vertical components is an important technique in the study of vector quantities. For example, a force **F** may be decomposed into its horizontal and vertical components $F_1\mathbf{i}$ and $F_2\mathbf{j}$, respectively. The physical effect of the single force **F** is the same as the combined effect of the separate forces $F_1\mathbf{i}$ and $F_2\mathbf{j}$. (This is an instance of the empirically verifiable parallelogram law of addition of forces.) Because of this decomposition, many two-dimensional problems can be reduced to one-dimensional problems, the latter solved, and the two results combined (again by vector methods) to give the solution of the original problem.

EXAMPLE 6 A 100-lb weight is suspended from the ceiling by means of two perpendicular flexible cables of equal length (Fig. 12.1.13). Find the tension (in pounds) in each cable.

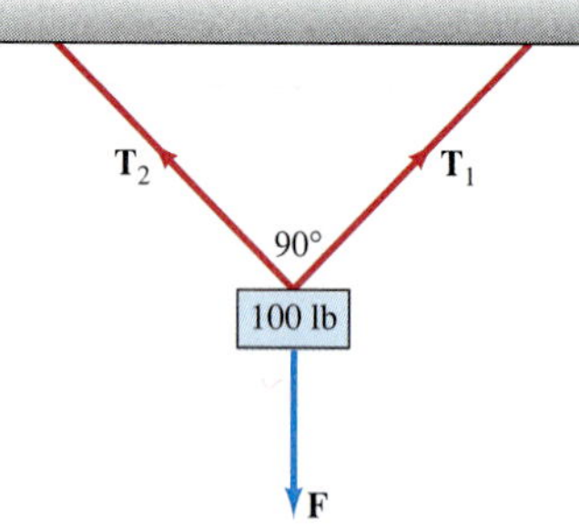

Fig. 12.1.13 The suspended weight of Example 6

Solution Each cable is inclined at an angle of 45° from the horizontal, so it follows readily upon calculating horizontal and vertical components that the indicated tension force vectors $\mathbf{T}_1$ and $\mathbf{T}_2$ are given by

$$\mathbf{T}_1 = (T_1 \cos 45°)\mathbf{i} + (T_1 \sin 45°)\mathbf{j} \quad \text{and} \quad \mathbf{T}_2 = (-T_2 \cos 45°)\mathbf{i} + (T_2 \sin 45°)\mathbf{j},$$

where $T_1 = |\mathbf{T}_1|$ and $T_2 = |\mathbf{T}_2|$ are the tension forces we seek. The downward force of gravity acting on the weight is given by $\mathbf{F} = -100\mathbf{j}$. So that the weight will hang motionless, the three forces must "balance," so that $\mathbf{T}_1 + \mathbf{T}_2 + \mathbf{F} = \mathbf{0}$; that is,

$$[(T_1 \cos 45°)\mathbf{i} + (T_1 \sin 45°)\mathbf{j}] + [(-T_2 \cos 45°)\mathbf{i} + (T_2 \sin 45°)\mathbf{j}] = 100\mathbf{j}.$$

When we equate the components of **i** in this equation and separately equate the components of **j**, we get the two scalar equations

$$T_1 \cos 45° - T_2 \cos 45° = 0 \quad \text{and} \quad T_1 \sin 45° + T_2 \sin 45° = 100.$$

The first of these scalar equations implies that $T_1 = T_2 = T$, and the second yields $T = 100/(2 \sin 45°) = 50\sqrt{2} \approx 70.71$ (pounds) for the tension in each cable.

12.1 PROBLEMS

In Problems 1 through 4, find a vector $\mathbf{v} = \langle a, b \rangle$ that is represented by the directed line segment $\overrightarrow{RS}$. Then sketch both $\overrightarrow{RS}$ and the position vector of the point $P(a, b)$.

1. $R(1, 2)$, $S(3, 5)$ **2.** $R(-2, -3)$, $S(1, 4)$

3. $R(5, 10)$, $S(-5, -10)$ **4.** $R(-10, 20)$, $S(15, -25)$

In Problems 5 through 8, find the sum $\mathbf{w} = \mathbf{u} + \mathbf{v}$ and illustrate it geometrically.

5. $\mathbf{u} = \langle 1, -2 \rangle$, $\mathbf{v} = \langle 3, 4 \rangle$ **6.** $\mathbf{u} = \langle 4, 2 \rangle$, $\mathbf{v} = \langle -2, 5 \rangle$

7. $\mathbf{u} = 3\mathbf{i} + 5\mathbf{j}$, $\mathbf{v} = 2\mathbf{i} - 7\mathbf{j}$

8. $\mathbf{u} = 7\mathbf{i} + 5\mathbf{j}$, $\mathbf{v} = -10\mathbf{i}$

In Problems 9 through 16, find $|\mathbf{a}|$, $|-2\mathbf{b}|$, $|\mathbf{a} - \mathbf{b}|$, $\mathbf{a} + \mathbf{b}$, and $3\mathbf{a} - 2\mathbf{b}$.

9. $\mathbf{a} = \langle 1, -2 \rangle$, $\mathbf{b} = \langle -3, 2 \rangle$

10. $\mathbf{a} = \langle 3, 4 \rangle$, $\mathbf{b} = \langle -4, 3 \rangle$

11. $\mathbf{a} = \langle -2, -2 \rangle$, $\mathbf{b} = \langle -3, -4 \rangle$

12. $\mathbf{a} = -2\langle 4, 7 \rangle$, $\mathbf{b} = -3\langle -4, -2 \rangle$

13. $\mathbf{a} = \mathbf{i} + 3\mathbf{j}$, $\mathbf{b} = 2\mathbf{i} - 5\mathbf{j}$

14. $\mathbf{a} = 2\mathbf{i} - 5\mathbf{j}$, $\mathbf{b} = \mathbf{i} - 6\mathbf{j}$

15. $\mathbf{a} = 4\mathbf{i}$, $\mathbf{b} = -7\mathbf{j}$

16. $\mathbf{a} = -\mathbf{i} - \mathbf{j}$, $\mathbf{b} = 2\mathbf{i} + 2\mathbf{j}$

In Problems 17 through 20, find a unit vector $\mathbf{u}$ with the same direction as the given vector $\mathbf{a}$. Express $\mathbf{u}$ in terms of $\mathbf{i}$ and $\mathbf{j}$. Also find a unit vector $\mathbf{v}$ with the direction opposite that of $\mathbf{a}$.

17. $\mathbf{a} = \langle -3, -4 \rangle$ **18.** $\mathbf{a} = \langle 5, -12 \rangle$

19. $\mathbf{a} = 8\mathbf{i} + 15\mathbf{j}$ **20.** $\mathbf{a} = 7\mathbf{i} - 24\mathbf{j}$

In Problems 21 through 24, find the vector $\mathbf{a}$, expressed in terms of $\mathbf{i}$ and $\mathbf{j}$, that is represented by the arrow $\overrightarrow{PQ}$ in the plane.

21. $P = (3, 2)$, $Q = (3, -2)$

22. $P = (-3, 5)$, $Q = (-3, 6)$

23. $P = (-4, 7)$, $Q = (4, -7)$

24. $P = (1, -1)$, $Q = (-4, -1)$

In Problems 25 through 28, determine whether the given vectors $\mathbf{a}$ and $\mathbf{b}$ are perpendicular.

25. $\mathbf{a} = \langle 6, 0 \rangle$, $\mathbf{b} = \langle 0, -7 \rangle$ **26.** $\mathbf{a} = 3\mathbf{j}$, $\mathbf{b} = 3\mathbf{i} - \mathbf{j}$

27. $\mathbf{a} = 2\mathbf{i} - \mathbf{j}$, $\mathbf{b} = 4\mathbf{j} + 8\mathbf{i}$

28. $\mathbf{a} = 8\mathbf{i} + 10\mathbf{j}$, $\mathbf{b} = 15\mathbf{i} - 12\mathbf{j}$

In Problems 29 and 30, express $\mathbf{i}$ and $\mathbf{j}$ in terms of $\mathbf{a}$ and $\mathbf{b}$.

29. $\mathbf{a} = 2\mathbf{i} + 3\mathbf{j}$, $\mathbf{b} = 3\mathbf{i} + 4\mathbf{j}$

30. $\mathbf{a} = 5\mathbf{i} - 9\mathbf{j}$, $\mathbf{b} = 4\mathbf{i} - 7\mathbf{j}$

In Problems 31 and 32, write $\mathbf{c}$ in the form $r\mathbf{a} + s\mathbf{b}$, where r and s are scalars.

31. $\mathbf{a} = \mathbf{i} + \mathbf{j}$, $\mathbf{b} = \mathbf{i} - \mathbf{j}$, $\mathbf{c} = 2\mathbf{i} - 3\mathbf{j}$

32. $\mathbf{a} = 3\mathbf{i} + 2\mathbf{j}$, $\mathbf{b} = 8\mathbf{i} + 5\mathbf{j}$, $\mathbf{c} = 7\mathbf{i} + 9\mathbf{j}$

33. Find a vector that has the same direction as $5\mathbf{i} - 7\mathbf{j}$ and is (a) three times its length; (b) one-third its length.

34. Find a vector that has the opposite direction from $-3\mathbf{i} + 5\mathbf{j}$ and is (a) four times its length; (b) one-fourth its length.

35. Find a vector of length 5 with (a) the same direction as $7\mathbf{i} - 3\mathbf{j}$; (b) the direction opposite that of $8\mathbf{i} + 5\mathbf{j}$.

36. For what numbers c are the vectors $\langle c, 2 \rangle$ and $\langle c, -8 \rangle$ perpendicular?

37. For what numbers c are the vectors $2c\mathbf{i} - 4\mathbf{j}$ and $3\mathbf{i} + c\mathbf{j}$ perpendicular?

38. Given the three points $A(2, 3)$, $B(-5, 7)$, and $C(1, -5)$, verify by direct computation of the vectors and their sum that $\overrightarrow{AB} + \overrightarrow{BC} + \overrightarrow{CA} = \mathbf{0}$.

In Problems 39 through 42, give a componentwise proof of the indicated property of vector algebra. Take $\mathbf{a} = \langle a_1, a_2 \rangle$, $\mathbf{b} = \langle b_1, b_2 \rangle$, and $\mathbf{c} = \langle c_1, c_2 \rangle$ throughout.

39. $\mathbf{a} + (\mathbf{b} + \mathbf{c}) = (\mathbf{a} + \mathbf{b}) + \mathbf{c}$

40. $(r + s)\mathbf{a} = r\mathbf{a} + s\mathbf{a}$

41. $(rs)\mathbf{a} = r(s\mathbf{a})$

42. If $\mathbf{a} + \mathbf{b} = \mathbf{a}$, then $\mathbf{b} = \mathbf{0}$.

43. Find the tension in each cable of Example 6 if the angle between them is 120°.

In Problems 44 through 46, a given weight (in pounds) is suspended by two cables as shown in the figure. Find the tension in each cable.

44. Figure 12.1.14

45. Figure 12.1.15

46. Figure 12.1.16

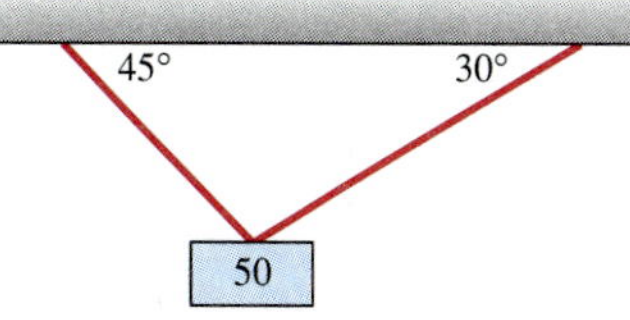

Fig. 12.1.14

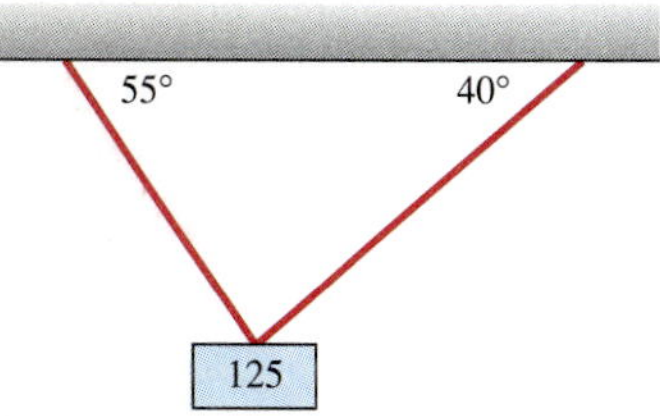

Fig. 12.1.15

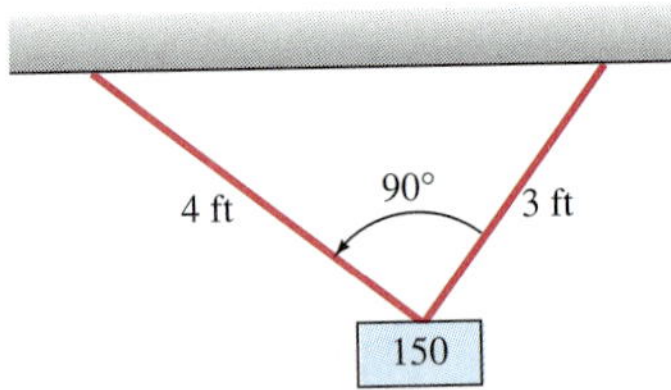

Fig. 12.1.16

In Problems 47 through 49, assume the following fact: If an airplane flies with velocity vector $\mathbf{v}_a$ relative to the air and the velocity of the wind is $\mathbf{w}$, then the velocity vector of the plane relative to the ground is $\mathbf{v}_g = \mathbf{v}_a + \mathbf{w}$ (Fig. 12.1.17). The velocity $\mathbf{v}_a$ is called the *apparent velocity vector,* and the vector $\mathbf{v}_g$ is called the *true velocity vector*.

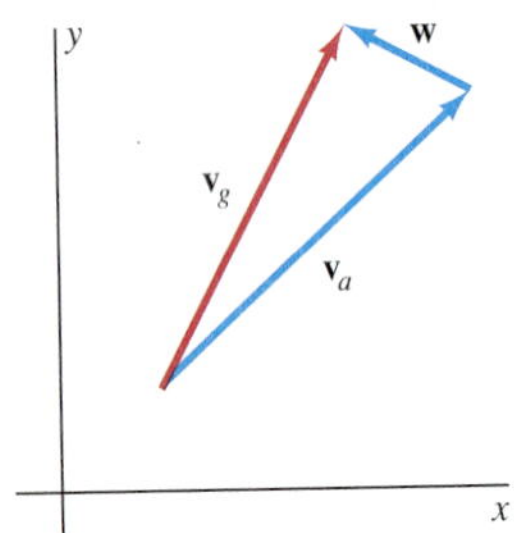

Fig. 12.1.17 The vectors of Problems 47 through 49

- Apparent velocity: $\mathbf{v}_a$
- Wind velocity: $\mathbf{w}$
- True velocity: $\mathbf{v}_g = \mathbf{v}_a + \mathbf{w}$

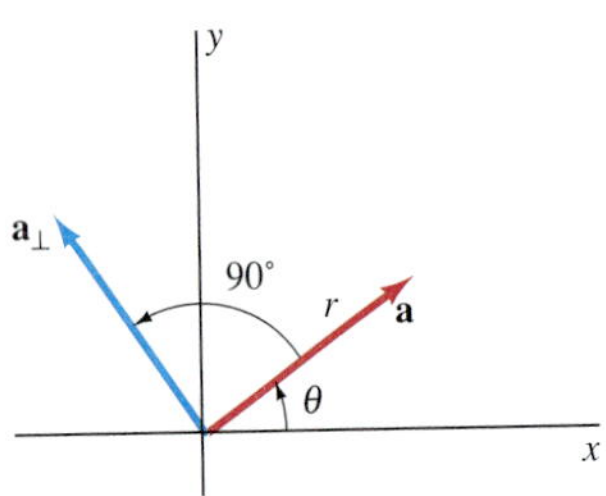

Fig. 12.1.18 Rotate **a** counterclockwise 90° to obtain $\mathbf{a}_\perp$ (Problem 55).

47. Suppose that the wind is blowing from the northeast at 50 mi/h and that the pilot wishes to fly due east at 500 mi/h. What should the plane's apparent velocity vector be?

48. Repeat Problem 47 with the phrase *due east* replaced with *due west*.

49. Repeat Problem 47 in the case that the pilot wishes to fly northwest at 500 mi/h.

50. Given any three points A, B, and C in the plane, show that $\overrightarrow{AB} + \overrightarrow{BC} + \overrightarrow{CA} = \mathbf{0}$. (*Suggestion:* Picture the triangle ABC.)

51. If **a** and **b** are the position vectors of the points P and Q in the plane and M is the point with position vector $\mathbf{v} = \frac{1}{2}(\mathbf{a} + \mathbf{b})$, show that M is the midpoint of the line segment PQ. Is it sufficient to show that the vectors $\overrightarrow{PM}$ and $\overrightarrow{QM}$ are equal and opposite?

52. In the triangle ABC, let M and N be the midpoints of AB and AC, respectively. Show that $\overrightarrow{MN} = \frac{1}{2}\overrightarrow{BC}$. Conclude that the line segment joining the midpoints of two sides of a triangle is parallel to the third side. How are their lengths related?

53. Prove that the diagonals of a parallelogram $ABCD$ bisect each other. (*Suggestion:* If M and N are the midpoints of the diagonals AC and BD, respectively, and O is the origin, show that $\overrightarrow{OM} = \overrightarrow{ON}$.)

54. Use vectors to prove that the midpoints of the four sides of an arbitrary quadrilateral are the vertices of a parallelogram.

55. Figure 12.1.18 shows the vector $\mathbf{a}_\perp$ obtained by rotating the vector $\mathbf{a} = a_1\mathbf{i} + a_2\mathbf{j}$ through a counterclockwise angle of 90°. Show that

$$\mathbf{a}_\perp = -a_2\mathbf{i} + a_1\mathbf{j}.$$

(*Suggestion:* Begin by writing $\mathbf{a} = (r\cos\theta)\mathbf{i} + (r\sin\theta)\mathbf{j}$.)

12.2 RECTANGULAR COORDINATES AND THREE-DIMENSIONAL VECTORS

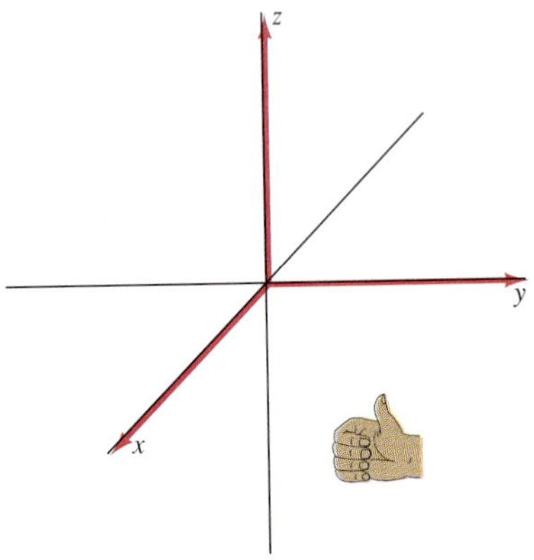

In the first eleven chapters we discussed many aspects of the calculus of functions of a *single* variable. The geometry of such functions is two-dimensional, because the graph of a function of a single variable is a curve in the coordinate plane. Most of the remaining chapters deal with the calculus of functions of *several* (two or more) independent variables. The geometry of functions of two variables is three-dimensional, because the graphs of such functions are generally surfaces in space.

Rectangular coordinates in the plane may be generalized to rectangular coordinates in space. A point in space is determined by giving its location relative to three mutually perpendicular **coordinate axes** that pass through the origin O. We shall usually draw the x-, y-, and z-axes as shown in Fig. 12.2.1, sometimes with arrows indicating the positive direction along each axis; the positive x-axis will always be labeled x, and similarly for the positive y- and z-axes. With this configuration of axes, our rectangular coordinate system is said to be **right-handed:** If you curl the fingers of your right hand in the direction of a 90° rotation from the positive x-axis to the positive y-axis, then your thumb will point in the direction of the positive z-axis. If the x- and y-axes were interchanged, then the coordinate system would be left-handed.

Fig. 12.2.1 The right-handed coordinate system

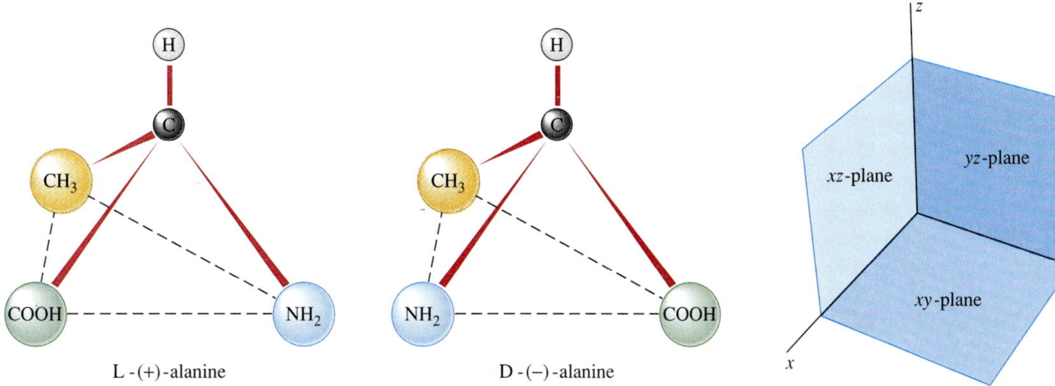

Fig. 12.2.2 The stereoisomers of the amino acid alanine are physically and biologically different even through they have the same molecular formula.

Fig. 12.2.3 The coordinate planes in space

These two coordinate systems are different in that it is impossible to bring one into coincidence with the other by means of rotations and translations. This is why the L- and D-alanine molecules shown in Fig. 12.2.2 are different; you can metabolize the left-handed ("levo") version but not the right-handed ("dextro") version. In this book we shall discuss right-handed coordinate systems exclusively and always draw the x-, y-, and z-axes with the right-handed orientation shown in Fig. 12.2.1.

The three coordinate axes taken in pairs determine the three **coordinate planes** (Fig. 12.2.3):

1. The (horizontal) xy-plane, where $z = 0$;
2. The (vertical) yz-plane, where $x = 0$; and
3. The (vertical) xz-plane, where $y = 0$.

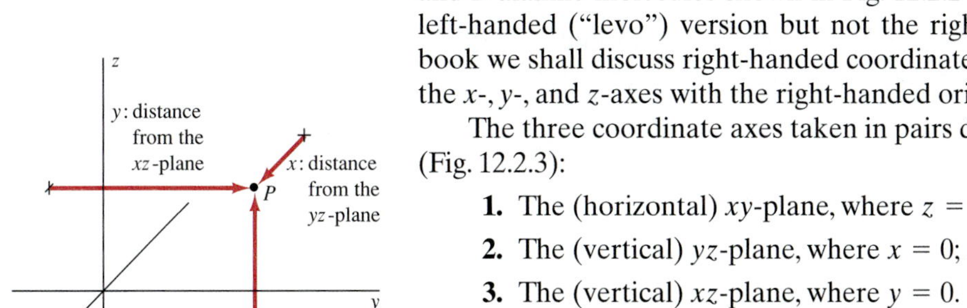
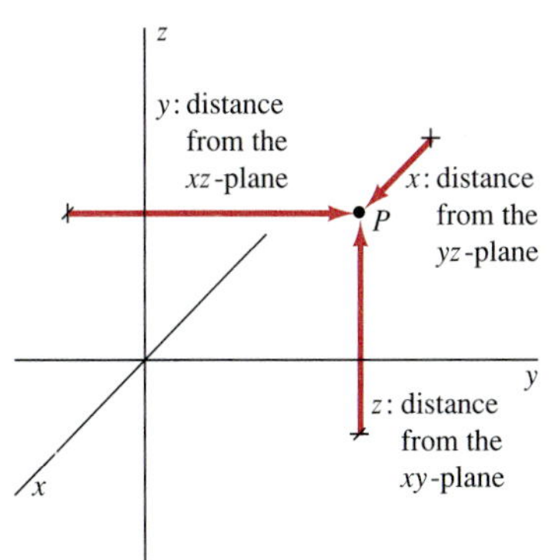

Fig. 12.2.4 Locating the point P in rectangular coordinates

The point P in space is said to have **rectangular coordinates** (x, y, z) if

1. x is its signed distance from the yz-plane,
2. y is its signed distance from the xz-plane, and
3. z is its signed distance from the xy-plane

(see Fig. 12.2.4). In this case we may describe the location of P simply by calling it "the point $P(x, y, z)$." There is a natural one-to-one correspondence between ordered triples (x, y, z) of real numbers and points P in space; this correspondence is called a **rectangular coordinate system** in space. In Fig. 12.2.5 the point P is located in the **first octant**—the eighth of space in which all three rectangular coordinates are positive.

Fig. 12.2.5 Completing the box to show P with the illusion of the third dimension

If we apply the Pythagorean theorem to the right triangles P_1QR and P_1RP_2 in Fig. 12.2.6, we get

$$\begin{aligned}|P_1P_2|^2 &= |RP_2|^2 + |P_1R|^2 = |RP_2|^2 + |QR|^2 + |P_1Q|^2 \\ &= (x_1 - x_2)^2 + (y_1 - y_2)^2 + (z_1 - z_2)^2.\end{aligned}$$

Thus the **distance formula** for the **distance** $|P_1P_2|$ between the points P_1 and P_2 is

$$|P_1P_2| = \sqrt{(x_1 - x_2)^2 + (y_1 - y_2)^2 + (z_1 - z_2)^2}. \tag{1}$$

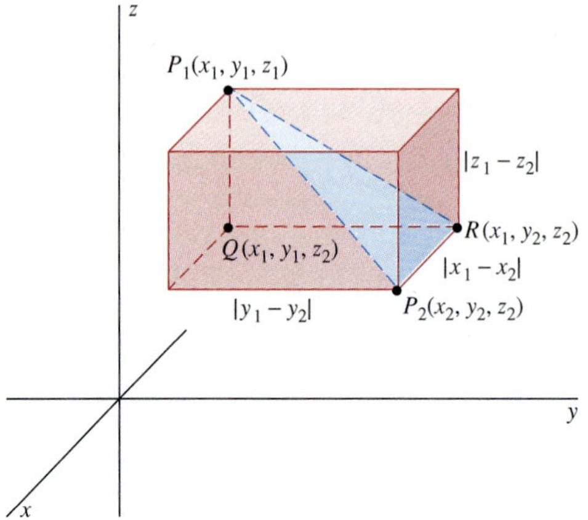

Fig. 12.2.6 The distance between P_1 and P_2 is the length of the long diagonal of the box.

EXAMPLE 1 The distance between the points $A(1, 3, -2)$ and $B(4, -3, 1)$ is

$$|AB| = \sqrt{(4-1)^2 + (-3-3)^2 + (1+2)^2} = \sqrt{54} \approx 7.348.$$

You can apply the distance formula in Eq. (1) to show that the **midpoint** M of the line segment joining $P_1(x_1, y_1, z_1)$ and $P_2(x_2, y_2, z_2)$ is

$$M\left(\frac{x_1 + x_2}{2}, \frac{y_1 + y_2}{2}, \frac{z_1 + z_2}{2}\right) \tag{2}$$

(see Problem 63).

The **graph** of an equation in three variables x, y, and z is the set of all points in space with rectangular coordinates that satisfy that equation. In general, the graph of an equation in three variables is a *two-dimensional surface* in $\mathbf{R}^3$ (three-dimensional space with rectangular coordinates).

EXAMPLE 2 Given a fixed point $C(h, k, l)$ and a number $r > 0$, find an equation of the sphere with radius r and center C.

Solution By definition, the sphere is the set of all points $P(x, y, z)$ such that the distance from P to C is r. That is, $|CP| = r$, and thus $|CP|^2 = r^2$. Therefore

$$(x-h)^2 + (y-k)^2 + (z-l)^2 = r^2. \tag{3}$$

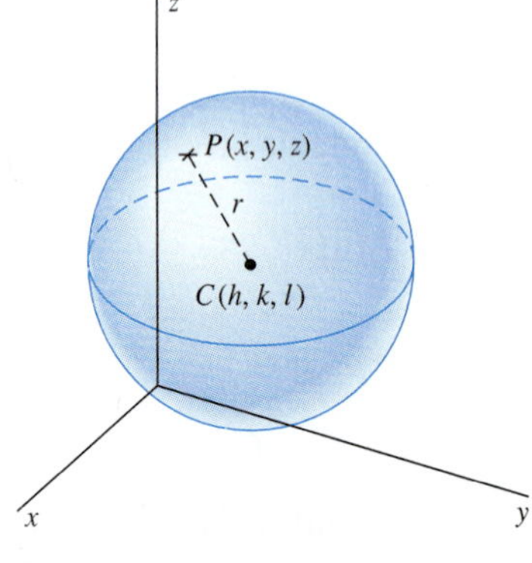

Fig. 12.2.7 The sphere with center (h, k, l) and radius r

Equation (3) is worth remembering as the equation of the **sphere with radius** r and **center** $C(h, k, l)$ shown in Fig. 12.2.7. Moreover, given an equation of the form

$$x^2 + y^2 + z^2 + Ax + By + Cz + D = 0,$$

we can attempt—by completing the square in each variable—to write it in the form of Eq. (3) and thereby show that its graph is a sphere.

EXAMPLE 3 Determine the graph of the equation

$$x^2 + y^2 + z^2 + 4x + 2y - 6z - 2 = 0.$$

Solution We complete the square in each variable. The equation then takes the form

$$(x^2 + 4x + 4) + (y^2 + 2y + 1) + (z^2 - 6z + 9) = 2 + (4 + 1 + 9) = 16;$$

that is,

$$(x + 2)^2 + (y + 1)^2 + (z - 3)^2 = 4^2.$$

Thus the graph of the given equation is the sphere with radius 4 and center $(-2, -1, 3)$.

Vectors in Space

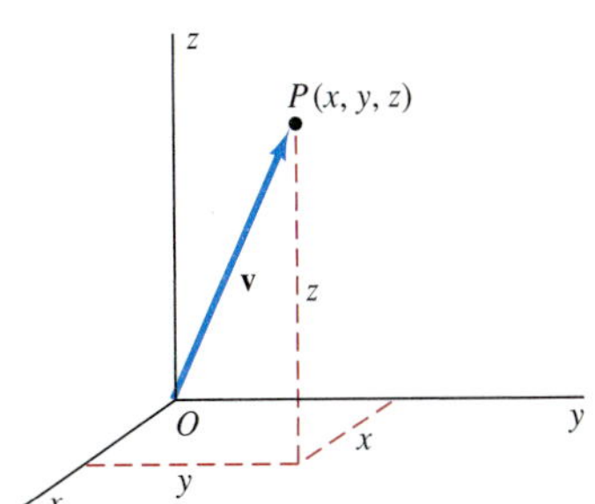

Fig. 12.2.8 The arrow $\overrightarrow{OP}$ represents the position vector $\mathbf{v} = \langle x, y, z \rangle$.

The discussion of vectors in space parallels the discussion in Section 12.1 of vectors in the plane. The difference is that a vector in space has three components rather than two. The point $P(x, y, z)$ has **position vector** $\mathbf{v} = \overrightarrow{OP} = \langle x, y, z \rangle$, which is represented by the directed line segment (or arrow) $\overrightarrow{OP}$ from the origin O to the point P (as well as by any parallel translate of this arrow—see Fig. 12.2.8). The distance formula in Eq. (1) gives

$$|\mathbf{v}| = \sqrt{x^2 + y^2 + z^2} \tag{4}$$

for the **length** (or **magnitude**) of the vector $\mathbf{v} = \langle x, y, z \rangle$.

Given two points $A(a_1, a_2, a_3)$ and $B(b_1, b_2, b_3)$ in space, the directed line segment $\overrightarrow{AB}$ in Fig. 12.2.9 represents the vector

$$\mathbf{v} = \langle b_1 - a_1, b_2 - a_2, b_3 - a_3 \rangle.$$

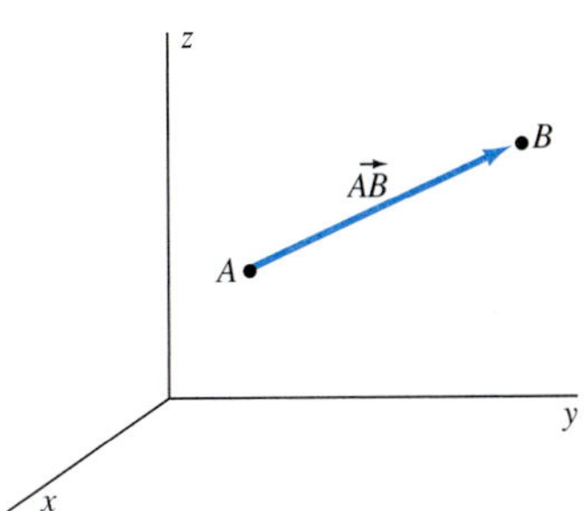

Fig. 12.2.9 The arrow $\overrightarrow{AB}$ represents the vector $\mathbf{v} = \langle b_1 - a_1, b_2 - a_2, b_3 - a_3 \rangle$.

Its length is the distance between the two points A and B:

$$|\mathbf{v}| = |\overrightarrow{AB}| = \sqrt{(b_1 - a_1)^2 + (b_2 - a_2)^2 + (b_3 - a_3)^2}.$$

What it means for two vectors in space to be equal is essentially the same as in the case of two-dimensional vectors: The vectors $\mathbf{a} = \langle a_1, a_2, a_3 \rangle$ and $\mathbf{b} = \langle b_1, b_2, b_3 \rangle$ are **equal** provided that $a_1 = b_1$, $a_2 = b_2$, and $a_3 = b_3$. That is, two vectors are equal exactly when corresponding components are equal.

We define addition and scalar multiplication of vectors exactly as we did in Section 12.1, taking into account that the vectors now have three components rather than two: The **sum** of the vectors $\mathbf{a} = \langle a_1, a_2, a_3 \rangle$ and $\mathbf{b} = \langle b_1, b_2, b_3 \rangle$ is the vector

$$\mathbf{a} + \mathbf{b} = \langle a_1 + b_1, a_2 + b_2, a_3 + b_3 \rangle. \tag{5}$$

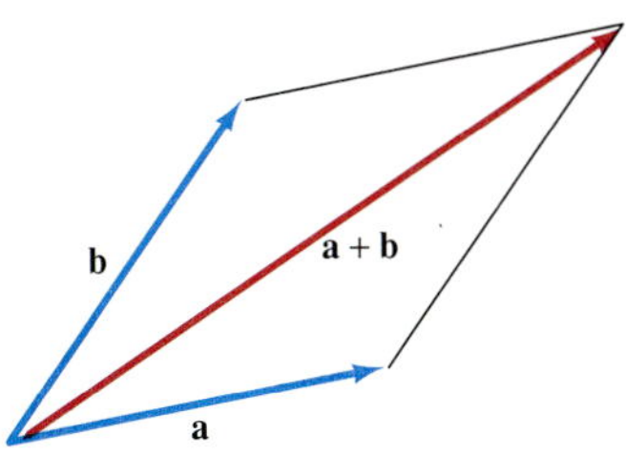

Fig. 12.2.10 The parallelogram law for addition of vectors

Because **a** and **b** lie in a plane (although not necessarily the xy-plane) if their initial points coincide, addition of vectors obeys the same **parallelogram law** as in the two-dimensional case (Fig. 12.2.10).

If c is a real number, then the **scalar multiple** $c\mathbf{a}$ is the vector

$$c\mathbf{a} = \langle ca_1, ca_2, ca_3 \rangle. \tag{6}$$

The length of $c\mathbf{a}$ is $|c|$ times the length of $\mathbf{a}$, and $c\mathbf{a}$ has the same direction as $\mathbf{a}$ if $c > 0$ but the opposite direction if $c < 0$. The following algebraic properties of vector addition and scalar multiplication are easy to establish; they follow from computations with components, exactly as in Section 12.1:

$$\begin{aligned} \mathbf{a} + \mathbf{b} &= \mathbf{b} + \mathbf{a}, \\ \mathbf{a} + (\mathbf{b} + \mathbf{c}) &= (\mathbf{a} + \mathbf{b}) + \mathbf{c}, \\ r(\mathbf{a} + \mathbf{b}) &= r\mathbf{a} + r\mathbf{b}, \\ (r + s)\mathbf{a} &= r\mathbf{a} + s\mathbf{a}, \\ (rs)\mathbf{a} &= r(s\mathbf{a}) = s(r\mathbf{a}). \end{aligned} \tag{7}$$

EXAMPLE 4 If $\mathbf{a} = \langle 3, 4, 12 \rangle$ and $\mathbf{b} = \langle -4, 3, 0 \rangle$, then

$$\begin{aligned} \mathbf{a} + \mathbf{b} &= \langle 3 - 4, 4 + 3, 12 + 0 \rangle = \langle -1, 7, 12 \rangle, \\ |\mathbf{a}| &= \sqrt{3^2 + 4^2 + 12^2} = \sqrt{169} = 13, \\ 2\mathbf{a} &= \langle 2 \cdot 3, 2 \cdot 4, 2 \cdot 12 \rangle = \langle 6, 8, 24 \rangle, \quad \text{and} \\ 2\mathbf{a} - 3\mathbf{b} &= \langle 6 + 12, 8 - 9, 24 - 0 \rangle = \langle 18, -1, 24 \rangle. \end{aligned}$$

A **unit vector** is a vector of length 1. We can express any vector in space (or *space vector*) in terms of the three **basic unit vectors**

$$\mathbf{i} = \langle 1, 0, 0 \rangle, \quad \mathbf{j} = \langle 0, 1, 0 \rangle, \quad \mathbf{k} = \langle 0, 0, 1 \rangle.$$

When located with their initial points at the origin, these basic unit vectors form a right-handed triple of vectors pointing in the positive directions along the three coordinate axes (Fig. 12.2.11).

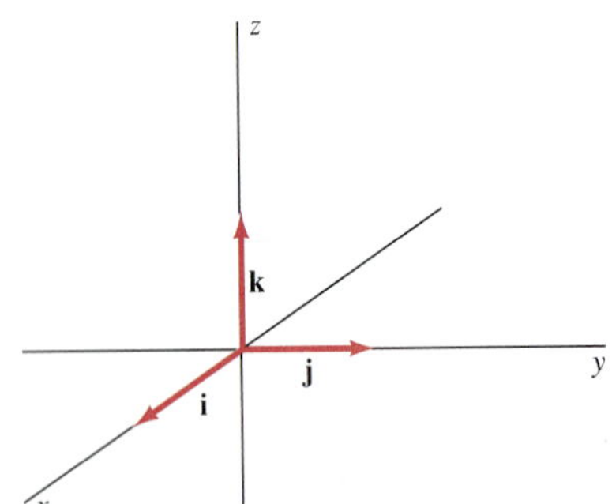

Fig. 12.2.11 The basic unit vectors **i**, **j**, and ***k***

The space vector $\mathbf{a} = \langle a_1, a_2, a_3 \rangle$ can be written as

$$\mathbf{a} = a_1\mathbf{i} + a_2\mathbf{j} + a_3\mathbf{k},$$

a linear combination of the basic unit vectors. As in the two-dimensional case, the usefulness of this representation is that algebraic operations involving vectors may be carried out simply by collecting coefficients of **i**, **j**, and **k**.

EXAMPLE 5 Given the vectors $\mathbf{a} = \langle 3, -4, 2 \rangle$ and $\mathbf{b} = \langle 5, 2, -5 \rangle$, we can write

$$\mathbf{a} = 3\mathbf{i} - 4\mathbf{j} + 2\mathbf{k} \quad \text{and} \quad \mathbf{b} = 5\mathbf{i} + 2\mathbf{j} - 7\mathbf{k}$$

in order to calculate

$$\begin{aligned} 7\mathbf{a} + 5\mathbf{b} &= 7 \cdot (3\mathbf{i} - 4\mathbf{j} + 2\mathbf{k}) + 5 \cdot (5\mathbf{i} + 2\mathbf{j} - 7\mathbf{k}) \\ &= (21 + 25)\mathbf{i} + (-28 + 10)\mathbf{j} + (14 - 35)\mathbf{k} \\ &= 46\mathbf{i} - 18\mathbf{j} - 25\mathbf{k} = \langle 46, -18, -25 \rangle. \end{aligned}$$

The Dot Product of Two Vectors

The **dot product** of the two vectors

$$\mathbf{a} = a_1\mathbf{i} + a_2\mathbf{j} + a_3\mathbf{k} \quad \text{and} \quad \mathbf{b} = b_1\mathbf{i} + b_2\mathbf{j} + b_3\mathbf{k}$$

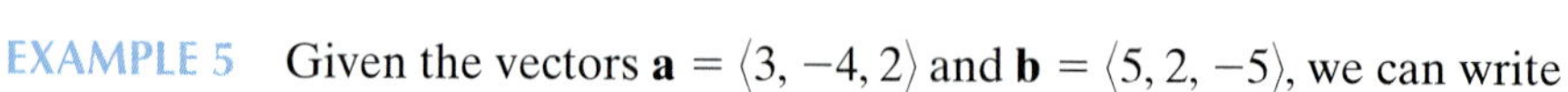

is the number obtained when we multiply corresponding components of **a** and **b** and add the results. That is,

$$\mathbf{a} \cdot \mathbf{b} = a_1 b_1 + a_2 b_2 + a_3 b_3. \tag{8}$$

Thus the dot product of two vectors is the *sum of the products of their corresponding components.* In the case of plane vectors $\mathbf{a} = \langle a_1, a_2 \rangle$ and $\mathbf{b} = \langle b_1, b_2 \rangle$, we simply dispense with third components and write $\mathbf{a} \cdot \mathbf{b} = a_1 b_1 + a_2 b_2$.

EXAMPLE 6 To apply the definition to calculate the dot product of the two vectors $\mathbf{a} = \langle 3, 4, 12 \rangle$ and $\mathbf{b} = \langle -4, 3, 0 \rangle$, we simply follow the pattern in Eq. (8):

$$\mathbf{a} \cdot \mathbf{b} = (3)(-4) + (4)(3) + (12)(0) = -12 + 12 + 0 = 0.$$

And if $\mathbf{c} = \langle 4, 5, -3 \rangle$, then

$$\mathbf{a} \cdot \mathbf{c} = (3)(4) + (4)(5) + (12)(-3) = 12 + 20 - 36 = -4.$$

IMPORTANT The dot product of two *vectors* is a *scalar*—that is, an ordinary real number. For this reason, the dot product is often called the **scalar product**. Example 6 illustrates the fact that the scalar product of two nonzero vectors (with positive lengths) may be zero, or even a negative number.

The following **properties of the dot product** show that dot products of vectors behave in many ways in analogy to the ordinary algebra of real numbers.

$$\begin{aligned} \mathbf{a} \cdot \mathbf{a} &= |\mathbf{a}|^2, \\ \mathbf{a} \cdot \mathbf{b} &= \mathbf{b} \cdot \mathbf{a}, \\ \mathbf{a} \cdot (\mathbf{b} + \mathbf{c}) &= \mathbf{a} \cdot \mathbf{b} + \mathbf{a} \cdot \mathbf{c}, \\ (r\mathbf{a}) \cdot \mathbf{b} &= r(\mathbf{a} \cdot \mathbf{b}) = \mathbf{a} \cdot (r\mathbf{b}). \end{aligned} \tag{9}$$

Each of the properties in (9) can be established by working with components of the vectors involved. For instance, to establish the second equation, suppose that $\mathbf{a} = \langle a_1, a_2, a_3 \rangle$ and $\mathbf{b} = \langle b_1, b_2, b_3 \rangle$. Then

$$\mathbf{a} \cdot \mathbf{b} = a_1 b_2 + a_2 b_2 + a_3 b_3 = b_1 a_1 + b_2 a_2 + b_3 a_3 = \mathbf{b} \cdot \mathbf{a}.$$

This derivation makes it clear that the commutative law for the dot product is a consequence of the commutative law $ab = ba$ for multiplication of ordinary real numbers.

Example 6 shows that the *algebraic definition* of the dot product is easy to apply in routine calculations. But what does it mean? The significance and **meaning of the dot product** lie in its *geometric interpretation.*

Let the vectors **a** and **b** be represented as position vectors by the directed segments $\overrightarrow{OP}$ and $\overrightarrow{OQ}$, respectively. Then the angle θ between **a** and **b** is the angle at O in triangle OPQ of Fig. 12.2.12. We say that **a** and **b** are **parallel** if $\theta = 0$ or if $\theta = \pi$ and that **a** and **b** are **perpendicular** if $\theta = \pi/2$. For convenience, we regard the zero vector $\mathbf{0} = \langle 0, 0, 0 \rangle$ as both parallel to *and* perpendicular to *every* vector.

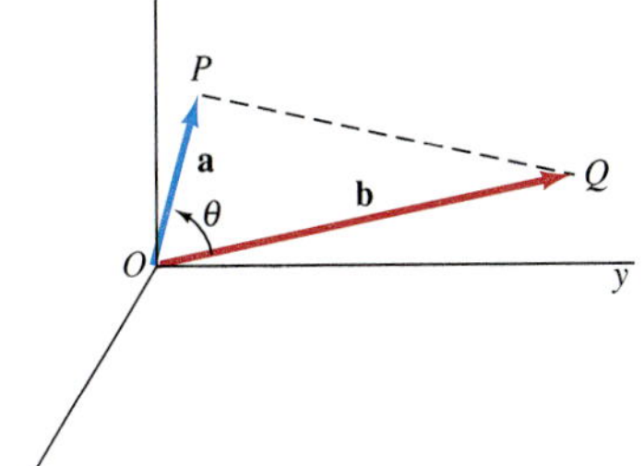

Fig. 12.2.12 The angle θ between the vectors **a** and **b**

Theorem 1 *Interpretation of the Dot Product*

If θ is the angle between the vectors **a** and **b,** then

$$\mathbf{a} \cdot \mathbf{b} = |\mathbf{a}|\,|\mathbf{b}| \cos\theta. \tag{10}$$

PROOF If either $\mathbf{a} = \mathbf{0}$ or $\mathbf{b} = \mathbf{0}$, then Eq. (10) follows immediately. If the vectors $\mathbf{a}$ and $\mathbf{b}$ are parallel, then $\mathbf{b} = t\mathbf{a}$ with either $t > 0$ and $\theta = 0$ or $t < 0$ and $\theta = \pi$. In either case, both sides in Eq. (10) reduce to $t|\mathbf{a}|^2$, so again the conclusion of Theorem 1 follows.

We turn to the general case in which the vector $\mathbf{a} = \overrightarrow{OP}$ and $\mathbf{b} = \overrightarrow{OQ}$ are nonzero and nonparallel. Then

$$\begin{aligned} |\overrightarrow{QP}| &= |\mathbf{a} - \mathbf{b}|^2 = (\mathbf{a} - \mathbf{b})\cdot(\mathbf{a} - \mathbf{b}) \\ &= \mathbf{a}\cdot\mathbf{a} - \mathbf{a}\cdot\mathbf{b} - \mathbf{b}\cdot\mathbf{a} + \mathbf{b}\cdot\mathbf{b} = |\mathbf{a}|^2 + |\mathbf{b}|^2 - 2\mathbf{a}\cdot\mathbf{b}. \end{aligned}$$

But $c = |\overrightarrow{QP}|$ is the side of triangle OPQ (Fig. 12.2.12) that is opposite the angle θ included between the sides $a = |\mathbf{a}|$ and $b = |\mathbf{b}|$. Hence the law of cosines (Appendix L) gives

$$\begin{aligned} |\overrightarrow{QP}|^2 &= c^2 = a^2 + b^2 - 2ab\cos\theta \\ &= |\mathbf{a}|^2 + |\mathbf{b}|^2 - 2|\mathbf{a}|\,|\mathbf{b}|\cos\theta. \end{aligned}$$

Finally, comparison of these two expressions for $|\overrightarrow{QP}|^2$ yields Eq. (10). ■

This theorem tells us that the angle θ between the nonzero vectors $\mathbf{a}$ and $\mathbf{b}$ can be found by using the equation

$$\cos\theta = \frac{\mathbf{a}\cdot\mathbf{b}}{|\mathbf{a}|\,|\mathbf{b}|}. \tag{11}$$

For instance, given the vectors $\mathbf{a} = \langle 8, 5\rangle$ and $\mathbf{b} = \langle -11, 17\rangle$ of Example 5 in Section 12.1, we calculate

$$\cos\theta = \frac{\langle 8,5\rangle\cdot\langle -11,17\rangle}{|\langle 8,5\rangle|\,|\langle -11,17\rangle|} = \frac{(8)(-11) + (5)(17)}{\sqrt{8^2+5^2}\sqrt{(-11)^2+17^2}} = \frac{-3}{\sqrt{89}\sqrt{410}}.$$

It follows that $\theta = \arccos\left(-3/\sqrt{89}\sqrt{410}\right) \approx 1.5865$ (*radians*) $\approx 90.90° \neq 90°$, so we see again that the vectors $\mathbf{a}$ and $\mathbf{b}$ are not perpendicular.

More generally, the two nonzero vectors $\mathbf{a}$ and $\mathbf{b}$ are perpendicular if and only if they make a right angle, so that $\theta = \pi/2$. By Eq. (11), this in turn is so if and only if $\mathbf{a}\cdot\mathbf{b} = 0$. Hence we have a quick computational check for perpendicularity of vectors.

Corollary ***Test for Perpendicular Vectors***

The two nonzero vectors $\mathbf{a}$ and $\mathbf{b}$ are perpendicular if and only if $\mathbf{a}\cdot\mathbf{b} = 0$.

EXAMPLE 7 (a) To show that the plane vectors $\mathbf{a} = \langle 8, 5\rangle$ and $\mathbf{b} = \langle -11, 17\rangle$ of Example 5 in Section 12.1 were not perpendicular, we need only have calculated their dot product $\mathbf{a}\cdot\mathbf{b} = -88 + 85 = -3$ and observed that its value is not zero. (b) Given the space vectors $\mathbf{a} = \langle 8, 5, -1\rangle$ and $\mathbf{b} = \langle -11, 17, -3\rangle$, we find that

$$\mathbf{a}\cdot\mathbf{b} = (8)(-11) + (5)(17) + (-1)(-3) = -88 + 85 + 3 = 0.$$

We may therefore conclude that $\mathbf{a}$ and $\mathbf{b}$ *are* perpendicular. ■

B(5, −4, 3)
∠A
∠B
A(2, −1, 0)
∠C
C(1, −3, 2)

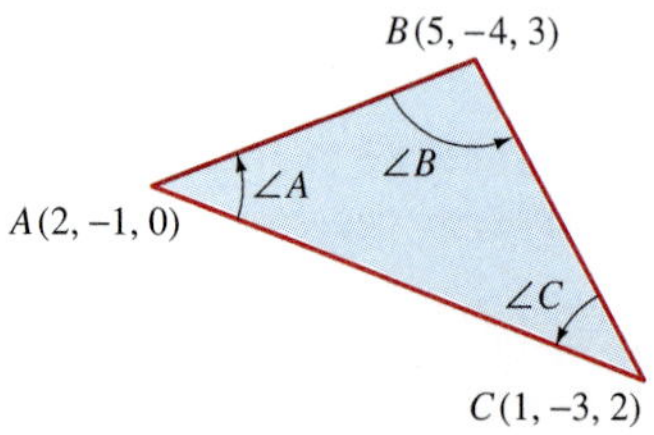

Fig. 12.2.13 The triangle of Example 8

EXAMPLE 8 Find the angles shown in the triangle of Fig. 12.2.13 with vertices at $A(2, -1, 0)$, $B(5, -4, 3)$, and $C(1, -3, 2)$.

Solution We apply Eq. (10) with $\theta = \angle A$, $\mathbf{a} = \overrightarrow{AB} = \langle 3, -3, 3 \rangle$, and $\mathbf{b} = \overrightarrow{AC} = \langle -1, -2, 2 \rangle$. This yields

$$\angle A = \cos^{-1}\left(\frac{\overrightarrow{AB} \cdot \overrightarrow{AC}}{|\overrightarrow{AB}||\overrightarrow{AC}|} \right) = \cos^{-1}\left(\frac{\langle 3, -3, 3 \rangle \cdot \langle -1, -2, 2 \rangle}{\sqrt{27}\sqrt{9}} \right)$$

$$= \cos^{-1}\left(\frac{9}{\sqrt{27}\sqrt{9}} \right) \approx 0.9553 \text{ (rad)} \approx 54.74^\circ.$$

Similarly,

$$\angle B = \cos^{-1}\left(\frac{\overrightarrow{BA} \cdot \overrightarrow{BC}}{|\overrightarrow{BA}||\overrightarrow{BC}|} \right) = \cos^{-1}\left(\frac{\langle -3, 3, -3 \rangle \cdot \langle -4, 1, -1 \rangle}{\sqrt{27}\sqrt{18}} \right)$$

$$= \cos^{-1}\left(\frac{18}{\sqrt{27}\sqrt{18}} \right) \approx 0.6155 \text{ (rad)} \approx 35.26^\circ.$$

Then $\angle C = 180^\circ - \angle A - \angle B = 90^\circ$. As a check, note that

$$\overrightarrow{CA} \cdot \overrightarrow{CB} = \langle 1, 2, -2 \rangle \cdot \langle 4, -1, 1 \rangle = 0.$$

So the angle at C is, indeed, a right angle. ■

Direction Angles and Projections

The **direction angles** of the nonzero vector $\mathbf{a} = \langle a_1, a_2, a_3 \rangle$ are the angles α, β, and γ that it makes with the vectors **i**, **j**, and **k**, respectively (Fig. 12.2.14). The cosines of these angles, $\cos\alpha$, $\cos\beta$, and $\cos\gamma$, are called the **direction cosines** of the vector **a**. When we replace **b** in Eq. (11) with **i, j,** and **k** in turn, we find that

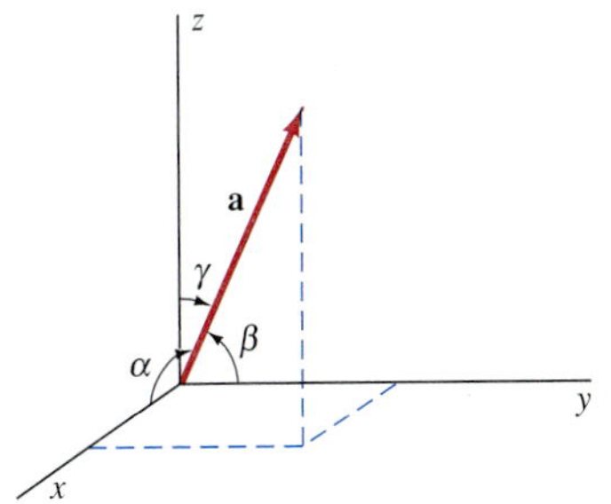

Fig. 12.2.14 The direction angles of the vector **a**

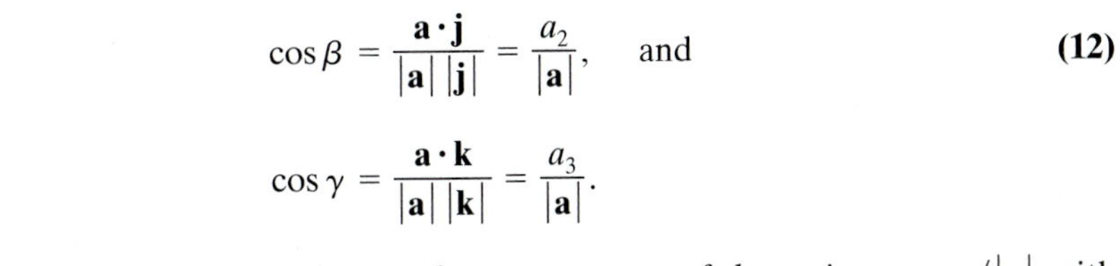

$$\cos\alpha = \frac{\mathbf{a} \cdot \mathbf{i}}{|\mathbf{a}|\,|\mathbf{i}|} = \frac{a_1}{|\mathbf{a}|},$$

$$\cos\beta = \frac{\mathbf{a} \cdot \mathbf{j}}{|\mathbf{a}|\,|\mathbf{j}|} = \frac{a_2}{|\mathbf{a}|}, \quad \text{and} \tag{12}$$

$$\cos\gamma = \frac{\mathbf{a} \cdot \mathbf{k}}{|\mathbf{a}|\,|\mathbf{k}|} = \frac{a_3}{|\mathbf{a}|}.$$

That is, the direction cosines of **a** are the components of the *unit vector* $\mathbf{a}/|\mathbf{a}|$ with the same direction as **a**. Consequently

$$\cos^2\alpha + \cos^2\beta + \cos^2\gamma = 1. \tag{13}$$

EXAMPLE 9 Find the direction angles of the vector $\mathbf{a} = 2\mathbf{i} + 3\mathbf{j} - \mathbf{k}$.

Solution Because $|\mathbf{a}| = \sqrt{14}$, the equations in (12) give

$$\alpha = \cos^{-1}\left(\frac{2}{\sqrt{14}} \right) \approx 57.69^\circ, \quad \beta = \cos^{-1}\left(\frac{3}{\sqrt{14}} \right) \approx 36.70^\circ,$$

$$\text{and} \quad \gamma = \cos^{-1}\left(\frac{-1}{\sqrt{14}} \right) \approx 105.50^\circ.$$ ■

Sometimes we need to find the component of one vector **a** in the direction of another *nonzero* vector **b**. Think of the two vectors located with the same initial point (Fig. 12.2.15). Then the (scalar) **component of a along b**, denoted by $\text{comp}_{\mathbf{b}}\mathbf{a}$, is numerically the length of the perpendicular projection of **a** onto the straight line determined by **b**. The number $\text{comp}_{\mathbf{b}}\mathbf{a}$ is positive if the angle θ between **a** and **b** is acute (so **a** and **b** point in the same general direction) and negative if $\theta > \pi/2$. Thus $\text{comp}_{\mathbf{b}}\mathbf{a} = |\mathbf{a}| \cos\theta$ in either case. Equation (10) then gives

$$\text{comp}_{\mathbf{b}}\mathbf{a} = \frac{|\mathbf{a}|\,|\mathbf{b}|\cos\theta}{|\mathbf{b}|} = \frac{\mathbf{a}\cdot\mathbf{b}}{|\mathbf{b}|}. \tag{14}$$

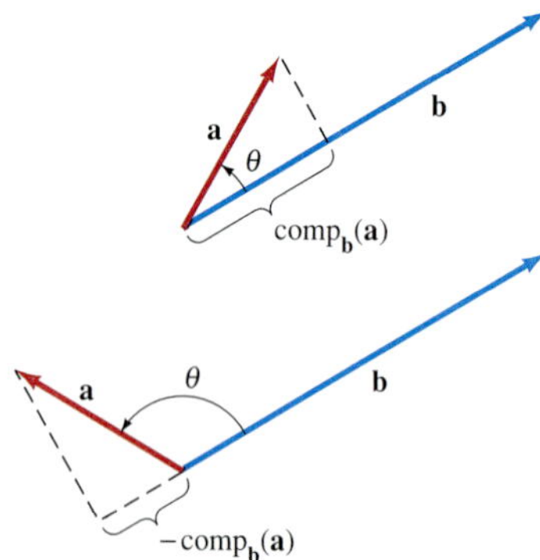

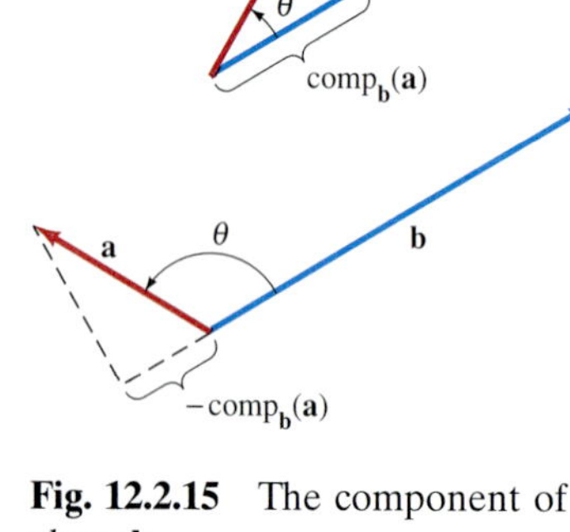

Fig. 12.2.15 The component of **a** along **b**

There is no need to memorize this formula, for—in practice—we can always read $\text{comp}_{\mathbf{b}}\mathbf{a} = |\mathbf{a}| \cos\theta$ from the figure and then apply Eq. (10) to eliminate $\cos\theta$. Note that $\text{comp}_{\mathbf{b}}\mathbf{a}$ is a scalar, not a vector.

EXAMPLE 10 Given $\mathbf{a} = \langle 4, -5, 3\rangle$ and $\mathbf{b} = \langle 2, 1, -2\rangle$, express **a** as the sum of a vector $\mathbf{a}_{\parallel}$ parallel to **b** and a vector $\mathbf{a}_{\perp}$ perpendicular to **b**.

Solution Our method of solution is motivated by the diagram in Fig. 12.2.16. We take

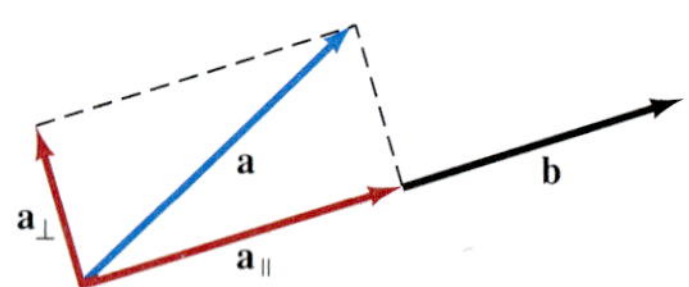

Fig. 12.2.16 Construction of $\mathbf{a}_{\parallel}$ and $\mathbf{a}_{\perp}$

$$\begin{aligned}\mathbf{a}_{\parallel} &= (\text{comp}_{\mathbf{b}}\mathbf{a})\frac{\mathbf{b}}{|\mathbf{b}|} = \frac{\mathbf{a}\cdot\mathbf{b}}{|\mathbf{b}|^2}\mathbf{b} = \frac{8-5-6}{9}\mathbf{b}\\ &= -\frac{1}{3}\langle 2, 1, -2\rangle = \left\langle -\frac{2}{3}, -\frac{1}{3}, \frac{2}{3}\right\rangle,\end{aligned}$$

and

$$\mathbf{a}_{\perp} = \mathbf{a} - \mathbf{a}_{\parallel} = \langle 4, -5, 3\rangle - \left\langle -\frac{2}{3}, -\frac{1}{3}, \frac{2}{3}\right\rangle = \left\langle \frac{14}{3}, -\frac{14}{3}, \frac{7}{3}\right\rangle.$$

The diagram makes our choice of $\mathbf{a}_{\parallel}$ plausible, and we have deliberately chosen $\mathbf{a}_{\perp}$ so that $\mathbf{a} = \mathbf{a}_{\parallel} + \mathbf{a}_{\perp}$. To verify that the vector $\mathbf{a}_{\parallel}$ is indeed parallel to **b,** we simply note that it is a scalar multiple of **b**. To verify that $\mathbf{a}_{\perp}$ is perpendicular to **b,** we compute the dot product

$$\mathbf{a}_{\perp}\cdot\mathbf{b} = \tfrac{28}{3} - \tfrac{14}{3} - \tfrac{14}{3} = 0.$$

Thus $\mathbf{a}_{\parallel}$ and $\mathbf{a}_{\perp}$ have the required properties. ■

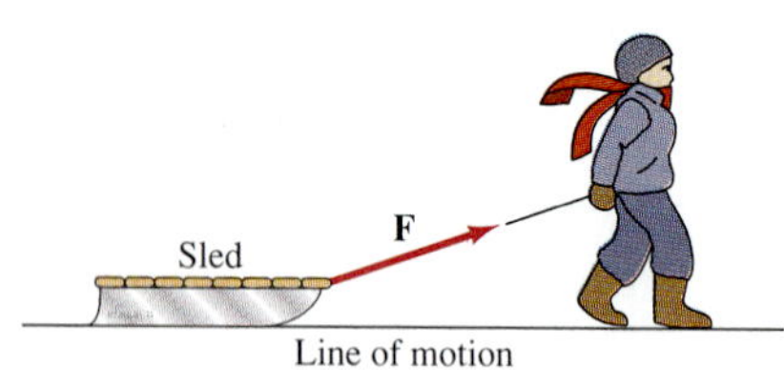

Fig. 12.2.17 The vector force **F** is constant but acts at an angle to the line of motion (Example 10).

One important application of vector components is to the definition and computation of *work*. Recall that the work W done by a constant force F exerted along the line of motion in moving a particle a distance d is given by $W = Fd$. But what if the force is a constant vector **F** pointing in some direction other than the line of motion, as when a child pulls a sled against the resistance of friction (Fig. 12.2.17)? Suppose that **F** moves a particle along the line segment from P to Q, and let $\mathbf{D} = \overrightarrow{PQ}$ be the resulting *displacement vector* of the object (Fig. 12.2.18). Then the **work** W done by the force **F** in moving the object along the line from P to Q is, by definition, the product of the component of **F** along **D** and the distance moved:

$$W = (\text{comp}_{\mathbf{D}}\mathbf{F})\,|\mathbf{D}|. \tag{15}$$

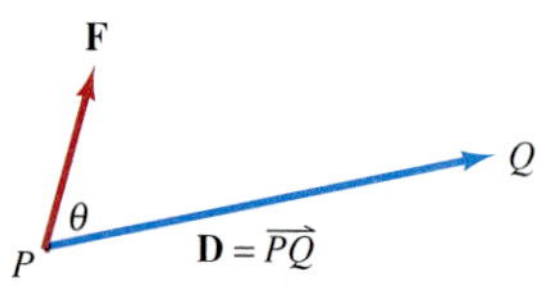

Fig. 12.2.18 The force vector **F** and displacement vector **D** in Eq. (16)

If we use Eq. (14) and substitute $\text{comp}_{\mathbf{D}}\mathbf{F} = (\mathbf{F}\cdot\mathbf{D})/|\mathbf{D}|$, we get

$$W = \mathbf{F}\cdot\mathbf{D} \tag{16}$$

for the work done by the constant force $\mathbf{F}$ in moving an object along the displacement vector $\mathbf{D} = \overrightarrow{PQ}$. This formula is the vector generalization of the scalar work formula $W = Fd$. Work is measured in foot-pounds (ft·lb) if distance is measured in feet and force in pounds. If metric units of meters (m) for distance and newtons (N) for force are used, then work is measured in joules (J).

EXAMPLE 11 Suppose that the force vector in Fig. 12.2.17 is inclined at an angle of 30° from the ground. If the child exerts a constant force of 20 lb, how much work is done in pulling the sled a distance of one mile?

Solution We are given that $|\mathbf{F}| = 20$ (lb) and $|\mathbf{D}| = 5280$ (ft). Because $\cos 30° = \frac{1}{2}\sqrt{3}$, Eq. (16) yields

$$W = \mathbf{F} \cdot \mathbf{D} = |\mathbf{F}|\,|\mathbf{D}|\cos 30° = (20)(5280)\left(\frac{1}{2}\sqrt{3}\right) \approx 91452 \quad (\text{ft}\cdot\text{lb}).$$

This may seem like a lot of work for a child to do. If the 1-mile trip takes an hour, then the child is generating *power* (work per unit time) at the rate of (91452 ft·lb)/(3600 s) $\approx$ 25.4 ft·lb/s. Because 1 horsepower (hp) is defined to be 550 ft·lb/s, the child's "power rating" is 25.4/550 $\approx \frac{1}{20}$ hp. By comparison, an adult in excellent physical condition can climb the 2570 steps of the staircase of the CN tower in Toronto in less than 40 minutes. On October 29, 1989, Brendon Keenory of Toronto set the world's record for the fastest stairclimb there with a time of 7 min 52 s. Assuming that he climbed 1672 ft and weighed 160 lb, he generated an average of more than 0.988 hp over this time interval. ■

12.2 PROBLEMS

In Problems 1 through 6, find (a) $2\mathbf{a} + \mathbf{b}$, (b) $3\mathbf{a} - 4\mathbf{b}$, (c) $\mathbf{a} \cdot \mathbf{b}$, (d) $|\mathbf{a} - \mathbf{b}|$, and (e) $\mathbf{a}/|\mathbf{a}|$.

1. $\mathbf{a} = \langle 2, 5, -4 \rangle$, $\mathbf{b} = \langle 1, -2, -3 \rangle$

2. $\mathbf{a} = \langle -1, 0, 2 \rangle$, $\mathbf{b} = \langle 3, 4, -5 \rangle$

3. $\mathbf{a} = \mathbf{i} + \mathbf{j} + \mathbf{k}$, $\mathbf{b} = \mathbf{j} - \mathbf{k}$

4. $\mathbf{a} = 2\mathbf{i} - 3\mathbf{j} + 5\mathbf{k}$, $\mathbf{b} = 5\mathbf{i} + 3\mathbf{j} - 7\mathbf{k}$

5. $\mathbf{a} = 2\mathbf{i} - \mathbf{j}$, $\mathbf{b} = \mathbf{j} - 3\mathbf{k}$

6. $\mathbf{a} = \mathbf{i} - 2\mathbf{j} + 3\mathbf{k}$, $\mathbf{b} = \mathbf{i} + 3\mathbf{j} - 2\mathbf{k}$

7 through **12.** Find, to the nearest degree, the angle between the vectors **a** and **b** in Problems 1 through 6.

13 through **18.** Find $\text{comp}_{\mathbf{a}}\mathbf{b}$ and $\text{comp}_{\mathbf{b}}\mathbf{a}$ for the vectors **a** and **b** given in Problems 1 through 6.

In Problems 19 through 24, write the equation of the indicated sphere.

19. Center $(3, 1, 2)$, radius 5

20. Center $(-2, 1, -5)$, radius $\sqrt{7}$

21. One diameter: the segment joining $(3, 5, -3)$ and $(7, 3, 1)$

22. Center $(4, 5, -2)$, passing through the point $(1, 0, 0)$

23. Center $(0, 0, 2)$, tangent to the xy-plane

24. Center $(3, -4, 3)$, tangent to the xz-plane

In Problems 25 through 28, find the center and radius of the sphere with the given equation.

25. $x^2 + y^2 + z^2 + 4x - 6y = 0$

26. $x^2 + y^2 + z^2 - 8x - 9y + 10z + 40 = 0$

27. $3x^2 + 3y^2 + 3z^2 - 18z - 48 = 0$

28. $2x^2 + 2y^2 + 2z^2 = 7x + 9y + 11z$

In Problems 29 through 38, describe the graph of the given equation in geometric terms, using plain, clear language.

29. $z = 0$

30. $x = 0$

31. $z = 10$

32. $xy = 0$

33. $xyz = 0$

34. $x^2 + y^2 + z^2 + 7 = 0$

35. $x^2 + y^2 + z^2 = 0$

36. $x^2 + y^2 + z^2 - 2x + 1 = 0$

37. $x^2 + y^2 + z^2 - 6x + 8y + 25 = 0$

38. $x^2 + y^2 = 0$

Two vectors are **parallel** provided that one is a scalar multiple of the other. Determine whether the vectors **a** and **b** in Problems 39 through 42 are parallel, perpendicular, or neither.

39. $\mathbf{a} = \langle 4, -2, 6 \rangle$ and $\mathbf{b} = \langle 6, -3, 9 \rangle$

40. $\mathbf{a} = \langle 4, -2, 6 \rangle$ and $\mathbf{b} = \langle 4, 2, 2 \rangle$

41. $\mathbf{a} = 12\mathbf{i} - 20\mathbf{j} + 16\mathbf{k}$ and $\mathbf{b} = -9\mathbf{i} + 15\mathbf{j} - 12\mathbf{k}$

42. $\mathbf{a} = 12\mathbf{i} - 20\mathbf{j} + 17\mathbf{k}$ and $\mathbf{b} = -9\mathbf{i} + 15\mathbf{j} + 24\mathbf{k}$

In Problems 43 and 44, determine whether the three given points lie on a single straight line.

43. $P(0, -2, 4)$, $Q(1, -3, 5)$, $R(4, -6, 8)$

44. $P(6, 7, 8)$, $Q(3, 3, 3)$, $R(12, 15, 18)$

In Problems 45 through 48, find (to the nearest degree) the three angles of the triangle with the given vertices.

45. $A(1,0,0)$, $B(0,1,0)$, $C(0,0,1)$

46. $A(1,0,0)$, $B(1,2,0)$, $C(1,2,3)$

47. $A(1,1,1)$, $B(3,-2,3)$, $C(3,4,6)$

48. $A(1,0,0)$, $B(0,1,0)$, $C(-1,-2,-2)$

In Problems 49 through 52, find the direction angles of the vector represented by $\overrightarrow{PQ}$.

49. $P(1,-1,0)$, $Q(3,4,5)$

50. $P(2,-3,5)$, $Q(1,0,-1)$

51. $P(-1,-2,-3)$, $Q(5,6,7)$

52. $P(0,0,0)$, $Q(5,12,13)$

In Problems 53 and 54, find the work W done by the force $\mathbf{F}$ in moving a particle in a straight line from P to Q.

53. $\mathbf{F} = \mathbf{i} - \mathbf{k}$; $P(0,0,0)$, $Q(3,1,0)$

54. $\mathbf{F} = 2\mathbf{i} - 3\mathbf{j} + 5\mathbf{k}$; $P(5,3,-4)$, $Q(-1,-2,5)$

55. Suppose that the force vector in Fig. 12.2.17 is inclined at an angle of 40° from the ground. If the child exerts a constant force of 40 N, how much heat energy (in calories) does the child expend in pulling the sled a distance of 1 km along the ground? (*Note:* 1 J of work requires an expenditure of 0.239 calories of energy.)

56. A 1000-lb dog sled has a coefficient of sliding friction of 0.2, so it requires a force with a horizontal component of 200 lb to keep it moving at a constant speed. Suppose that a dog-team harness is attached so that the team's force vector makes an angle of 5° with the horizontal. If the dog team pulls this sled at a speed of 10 mi/h, how much power (in horsepower) are the dogs generating? (*Note:* 1 hp is 550 ft·lb/s.)

57. Suppose that the horizontal and vertical components of the three vectors shown in Fig. 12.2.19 balance (the algebraic sum of the horizontal components is zero, as is the sum of the vertical components). How much work is done by the constant force $\mathbf{F}$ (parallel to the inclined plane) in pulling the weight mg up the inclined plane a vertical height h?

58. Prove the **Cauchy–Schwarz inequality:**

$$|\mathbf{a}\cdot\mathbf{b}| \leqq |\mathbf{a}|\,|\mathbf{b}|$$

for all pairs of vectors $\mathbf{a}$ and $\mathbf{b}$.

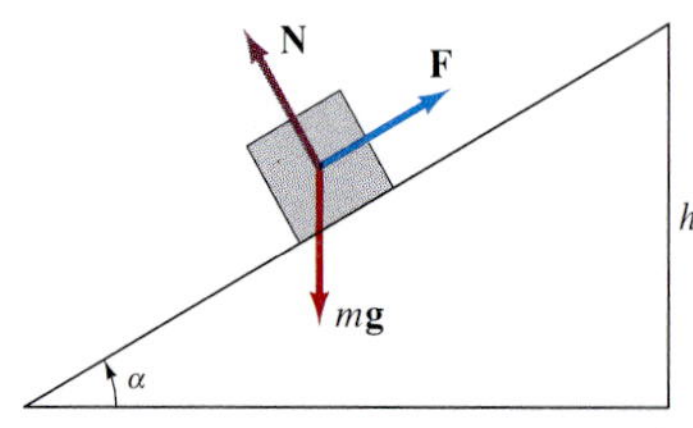

Fig. 12.2.19 The inclined plane of Problem 57

59. Given two arbitrary vectors $\mathbf{a}$ and $\mathbf{b}$, prove that they satisfy the **triangle inequality**

$$|\mathbf{a}+\mathbf{b}| \leqq |\mathbf{a}| + |\mathbf{b}|.$$

(*Suggestion:* Square both sides.)

60. Prove that if $\mathbf{a}$ and $\mathbf{b}$ are arbitrary vectors, then

$$|\mathbf{a}-\mathbf{b}| \geqq |\mathbf{a}| - |\mathbf{b}|.$$

(*Suggestion:* Write $\mathbf{a} = (\mathbf{a}-\mathbf{b}) + \mathbf{b}$; then apply the triangle inequality of Problem 59.)

61. Use the dot product to construct a nonzero vector $\mathbf{w} = \langle w_1, w_2, w_3\rangle$ perpendicular to both of the vectors $\mathbf{u} = \langle 1, 2, -3\rangle$ and $\mathbf{v} = \langle 2, 0, 1\rangle$.

62. The unit cube in the first octant in space has opposite vertices $O(0,0,0)$ and $P(1,1,1)$. Find the angle between the edge of the cube on the x-axis and the diagonal OP.

63. Prove that the point M given in Eq. (2) is indeed the midpoint of the segment P_1P_2. (*Note:* You must prove *both* that M is equally distant from P_1 and P_2 and that M lies on the segment P_1P_2.)

64. Given vectors $\mathbf{a}$ and $\mathbf{b}$, let $a = |\mathbf{a}|$ and $b = |\mathbf{b}|$. Prove that the vector

$$\mathbf{c} = \frac{b\mathbf{a} + a\mathbf{b}}{a+b}$$

bisects the angle between $\mathbf{a}$ and $\mathbf{b}$.

65. Let $\mathbf{a}$, $\mathbf{b}$, and $\mathbf{c}$ be three vectors in the xy-plane with $\mathbf{a}$ and $\mathbf{b}$ nonzero and nonparallel. Show that there exist scalars α and β such that $\mathbf{c} = \alpha\mathbf{a} + \beta\mathbf{b}$. (*Suggestion:* Begin by expressing $\mathbf{a}$, $\mathbf{b}$, and $\mathbf{c}$ in terms of $\mathbf{i}$, $\mathbf{j}$, and $\mathbf{k}$.)

66. Let $ax + by + c = 0$ be the equation of the line L in the xy-plane with normal vector $\mathbf{n}$. Let $P_0(x_0, y_0)$ be a point on this line and $P_1(x_1, y_1)$ be a point not on L. Prove that the perpendicular distance from P_1 to L is

$$d = \frac{|\mathbf{n}\cdot\overrightarrow{P_0P_1}|}{|\mathbf{n}|} = \frac{|ax_1 + by_1 + c|}{\sqrt{a^2+b^2}}.$$

67. Given the two points $A(3,-2,4)$ and $B(5,7,-1)$, write an equation in x, y, and z that says that the point $P(x,y,z)$ is equally distant from the points A and B. Then simplify this equation and give a geometric description of the set of all such points $P(x,y,z)$.

68. Given the fixed point $A(1,3,5)$, the point $P(x,y,z)$, and the vector $\mathbf{n} = \mathbf{i} - \mathbf{j} + 2\mathbf{k}$, use the dot product to help you write an equation in x, y, and z that says this: $\mathbf{n}$ and $\overrightarrow{AP}$ are perpendicular. Then simplify this equation and give a geometric description of all such points $P(x,y,z)$.

69. Prove that the points $(0,0,0)$, $(1,1,0)$, $(1,0,1)$, and $(0,1,1)$ are the vertices of a regular tetrahedron by showing that each of the six edges has length $\sqrt{2}$. Then use the dot product to find the angle between any two edges of the tetrahedron.

70. The methane molecule CH_4 is arranged with the four hydrogen atoms at the vertices of a regular tetrahedron and with the carbon atom at its center (Fig. 12.2.20). Suppose that the axes and scale are chosen so that the tetrahedron is that of Problem 69, with its center at $\left(\frac{1}{2}, \frac{1}{2}, \frac{1}{2}\right)$. Find the *bond angle* α between the lines from the carbon atom to two of the hydrogen atoms.

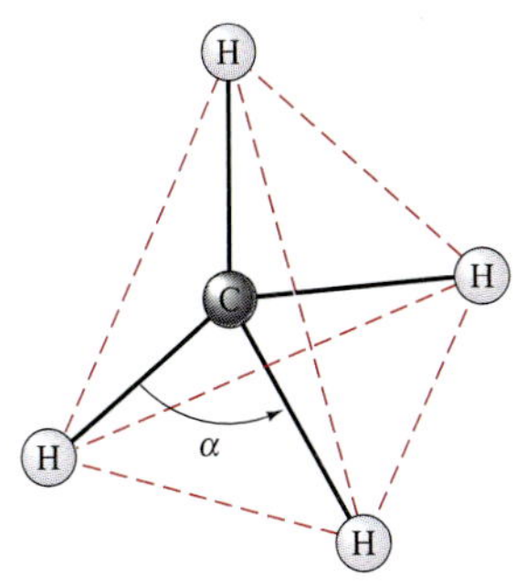

Fig. 12.2.20 The methane bond angle α of Problem 70

12.3 THE CROSS PRODUCT OF TWO VECTORS

We often need to find a vector that is perpendicular to each of two vectors **a** and **b** in space. A routine way of doing this is provided by the *cross product* $\mathbf{a} \times \mathbf{b}$ of the vectors **a** and **b**. This vector product is quite unlike the dot product $\mathbf{a} \cdot \mathbf{b}$ in that $\mathbf{a} \cdot \mathbf{b}$ is a *scalar,* whereas $\mathbf{a} \times \mathbf{b}$ is a *vector*. For this reason $\mathbf{a} \times \mathbf{b}$ is sometimes called the *vector product* of the two vectors **a** and **b**.

The **cross product** (or **vector product**) of the vectors $\mathbf{a} = \langle a_1, a_2, a_3 \rangle$ and $\mathbf{b} = \langle b_1, b_2, b_3 \rangle$ is defined algebraically by the formula

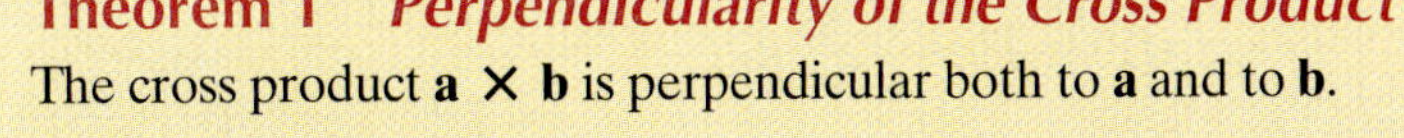

$$\mathbf{a} \times \mathbf{b} = \langle a_2b_3 - a_3b_2, a_3b_1 - a_1b_3, a_1b_2 - a_2b_1 \rangle. \quad (1)$$

Although this formula seems unmotivated, it has a redeeming feature: The product $\mathbf{a} \times \mathbf{b}$ is perpendicular both to **a** and to **b**, as suggested in Fig. 12.3.1.

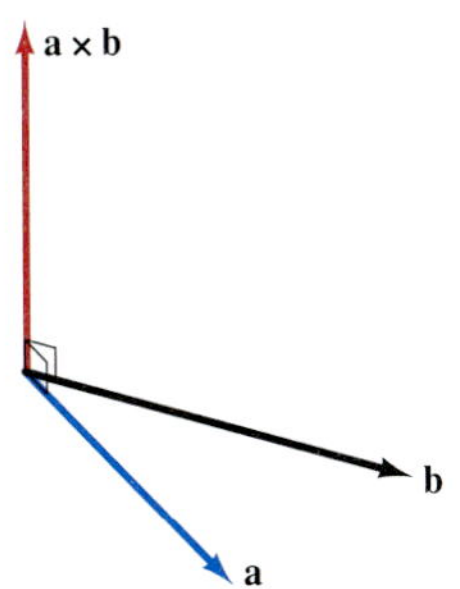

Fig. 12.3.1 The cross product $\mathbf{a} \times \mathbf{b}$ is perpendicular to both **a** and **b**.

Theorem 1 ***Perpendicularity of the Cross Product***
The cross product $\mathbf{a} \times \mathbf{b}$ is perpendicular both to **a** and to **b**.

PROOF We show that $\mathbf{a} \times \mathbf{b}$ is perpendicular to **a** by showing that the dot product of **a** and $\mathbf{a} \times \mathbf{b}$ is zero. With the components as in Eq. (1), we find that

$$\begin{aligned} \mathbf{a} \cdot (\mathbf{a} \times \mathbf{b}) &= a_1(a_2b_3 - a_3b_2) + a_2(a_3b_1 - a_1b_3) + a_3(a_1b_2 - a_2b_1) \\ &= a_1a_2b_3 - a_1a_3b_2 + a_2a_3b_1 - a_2a_1b_3 + a_3a_1b_2 - a_3a_2b_1 = 0. \end{aligned}$$

A similar computation shows that $\mathbf{b} \cdot (\mathbf{a} \times \mathbf{b}) = 0$ as well, so $\mathbf{a} \times \mathbf{b}$ is also perpendicular to the vector **b**. ■

You need not memorize Eq. (1), because there is an alternative version involving determinants that is easy both to remember and to use. Recall that a *determinant* of order 2 is defined as follows:

$$\begin{vmatrix} a_1 & a_2 \\ b_1 & b_2 \end{vmatrix} = a_1b_2 - a_2b_1. \quad (2)$$

EXAMPLE 1

$$\begin{vmatrix} 2 & -1 \\ 3 & 4 \end{vmatrix} = 2 \cdot 4 - (-1) \cdot 3 = 11.$$ ■

A determinant of order 3 can be defined in terms of determinants of order 2:

$$\begin{vmatrix} a_1 & a_2 & a_3 \\ b_1 & b_2 & b_3 \\ c_1 & c_2 & c_3 \end{vmatrix} = +a_1 \begin{vmatrix} b_2 & b_3 \\ c_2 & c_3 \end{vmatrix} - a_2 \begin{vmatrix} b_1 & b_3 \\ c_1 & c_3 \end{vmatrix} + a_3 \begin{vmatrix} b_1 & b_2 \\ c_1 & c_2 \end{vmatrix}. \tag{3}$$

Each element a_i of the first row is multiplied by the 2-by-2 "subdeterminant" obtained by deleting the row *and* column that contain a_i. Note in Eq. (3) that signs are attached to the a_i in accord with the checkerboard pattern

$$\begin{vmatrix} + & - & + \\ - & + & - \\ + & - & + \end{vmatrix}.$$

Equation (3) is an expansion of the 3-by-3 determinant along its first row. It can be expanded along any other row or column as well. For example, its expansion along its second column is

$$\begin{vmatrix} a_1 & a_2 & a_3 \\ b_1 & b_2 & b_3 \\ c_1 & c_2 & c_3 \end{vmatrix} = -a_2 \begin{vmatrix} b_1 & b_3 \\ c_1 & c_3 \end{vmatrix} + b_2 \begin{vmatrix} a_1 & a_3 \\ c_1 & c_3 \end{vmatrix} - c_2 \begin{vmatrix} a_1 & a_3 \\ b_1 & b_3 \end{vmatrix}.$$

In linear algebra it is shown that all such expansions yield the same value for the determinant.

Although we can expand a determinant of order 3 along any row or column, we shall use only expansions along the first row, as in Eq. (3) and in Example 2.

EXAMPLE 2

$$\begin{aligned} \begin{vmatrix} 1 & 3 & -2 \\ 2 & -1 & 4 \\ -3 & 7 & 5 \end{vmatrix} &= 1 \cdot \begin{vmatrix} -1 & 4 \\ 7 & 5 \end{vmatrix} - 3 \cdot \begin{vmatrix} 2 & 4 \\ -3 & 5 \end{vmatrix} + (-2) \cdot \begin{vmatrix} 2 & -1 \\ -3 & 7 \end{vmatrix} \\ &= 1 \cdot (-5 - 28) + (-3) \cdot (10 + 12) + (-2) \cdot (14 - 3) \\ &= -33 - 66 - 22 = -121. \end{aligned}$$

Equation (1) for the cross product of the vectors $\mathbf{a} = a_1\mathbf{i} + a_2\mathbf{j} + a_3\mathbf{k}$ and $\mathbf{b} = b_1\mathbf{i} + b_2\mathbf{j} + b_3\mathbf{k}$ is equivalent to

$$\mathbf{a} \times \mathbf{b} = \begin{vmatrix} a_2 & a_3 \\ b_2 & b_3 \end{vmatrix} \mathbf{i} - \begin{vmatrix} a_1 & a_3 \\ b_1 & b_3 \end{vmatrix} \mathbf{j} + \begin{vmatrix} a_1 & a_2 \\ b_1 & b_2 \end{vmatrix} \mathbf{k}. \tag{4}$$

This is easy to verify by expanding the 2-by-2 determinants on the right-hand side and noting that the three components of the right-hand side of Eq. (1) result. Motivated by Eq. (4), we write

$$\mathbf{a} \times \mathbf{b} = \begin{vmatrix} \mathbf{i} & \mathbf{j} & \mathbf{k} \\ a_1 & a_2 & a_3 \\ b_1 & b_2 & b_3 \end{vmatrix}. \tag{5}$$

The "symbolic determinant" in this equation is to be evaluated by expansion along its first row, just as in Eq. (3) and just as though it were an ordinary determinant with real number entries. The result of this expansion is the right-hand side of Eq. (4). The components of the *first* vector **a** in **a** × **b** form the *second* row of the 3-by-3 determinant, and the components of the *second* vector **b** form the *third* row. The order of the vectors **a** and **b** is important because, as we soon shall see, $\mathbf{a} \times \mathbf{b}$ is generally *not* equal to $\mathbf{b} \times \mathbf{a}$: The cross product is *not* commutative.

Equation (5) for the cross product is the form most convenient for computational purposes.

EXAMPLE 3 If $\mathbf{a} = 3\mathbf{i} - \mathbf{j} + 2\mathbf{k}$ and $\mathbf{b} = 2\mathbf{i} + 2\mathbf{j} - \mathbf{k}$, then

$$\mathbf{a} \times \mathbf{b} = \begin{vmatrix} \mathbf{i} & \mathbf{j} & \mathbf{k} \\ 3 & -1 & 2 \\ 2 & 2 & -1 \end{vmatrix} = \begin{vmatrix} -1 & 2 \\ 2 & -1 \end{vmatrix}\mathbf{i} - \begin{vmatrix} 3 & 2 \\ 2 & -1 \end{vmatrix}\mathbf{j} + \begin{vmatrix} 3 & -1 \\ 2 & 2 \end{vmatrix}\mathbf{k}$$
$$= (1 - 4)\mathbf{i} - (-3 - 4)\mathbf{j} + \big(6 - (-2)\big)\mathbf{k}.$$

Thus

$$\mathbf{a} \times \mathbf{b} = -3\mathbf{i} + 7\mathbf{j} + 8\mathbf{j}.$$

You might now pause to verify (by using the dot product) that the vector $-3\mathbf{i} + 7\mathbf{j} + 8\mathbf{k}$ is perpendicular both to **a** and to **b**. ■

If the vectors **a** and **b** share the same initial point, then Theorem 1 implies that $\mathbf{a} \times \mathbf{b}$ is perpendicular to the plane determined by **a** and **b** (Fig. 12.3.2). There are still two possible directions for $\mathbf{a} \times \mathbf{b}$, but if $\mathbf{a} \times \mathbf{b} \neq \mathbf{0}$, then the triple $\mathbf{a}, \mathbf{b}, \mathbf{a} \times \mathbf{b}$ is a *right-handed* triple in exactly the same sense as the triple **i, j, k**. Thus if the thumb of your right hand points in the direction of $\mathbf{a} \times \mathbf{b}$, then your fingers curl in the direction of rotation (less than 180°) from **a** to **b**.

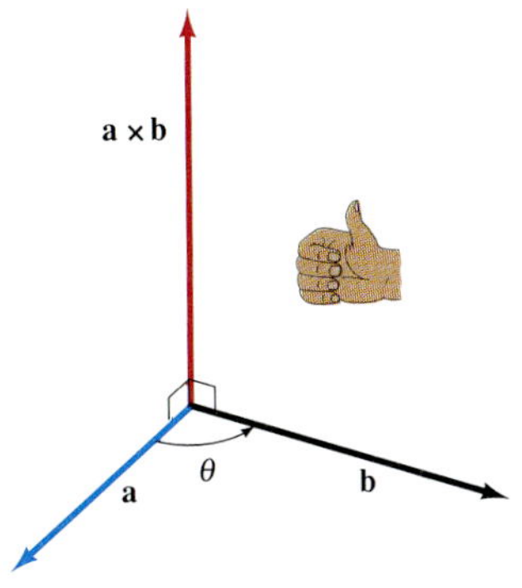

Fig. 12.3.2 The vectors **a**, **b**, and $\mathbf{a} \times \mathbf{b}$—in that order—form a right-handed triple.

Once we have established the direction of $\mathbf{a} \times \mathbf{b}$, we can describe the cross product in completely geometric terms by telling what the length $|\mathbf{a} \times \mathbf{b}|$ of the vector $\mathbf{a} \times \mathbf{b}$ is. This is given by the formula

$$|\mathbf{a} \times \mathbf{b}|^2 = |\mathbf{a}|^2|\mathbf{b}|^2 - (\mathbf{a} \cdot \mathbf{b})^2. \quad \textbf{(6)}$$

We can verify this vector identity routinely (though tediously) by writing $\mathbf{a} = \langle a_1, a_2, a_3 \rangle$ and $\mathbf{b} = \langle b_1, b_2, b_3 \rangle$, computing both sides of Eq. (6), and then noting that the results are equal (Problem 36).

Geometric Significance of the Cross Product

Equation (6) tells us what $|\mathbf{a} \times \mathbf{b}|$ is, but Theorem 2 reveals the geometric significance of the cross product.

Theorem 2 ***Length of the Cross Product***

Let θ be the angle between the nonzero vectors **a** and **b** (measured so that $0 \leqq \theta \leqq \pi$). Then

$$|\mathbf{a} \times \mathbf{b}| = |\mathbf{a}||\mathbf{b}| \sin \theta. \quad \textbf{(7)}$$

PROOF We begin with Eq. (6) and use the fact that $\mathbf{a} \cdot \mathbf{b} = |\mathbf{a}||\mathbf{b}| \cos \theta$. Thus

$$|\mathbf{a} \times \mathbf{b}|^2 = |\mathbf{a}|^2|\mathbf{b}|^2 - (\mathbf{a} \cdot \mathbf{b})^2 = |\mathbf{a}|^2|\mathbf{b}|^2 - (|\mathbf{a}||\mathbf{b}| \cos \theta)^2$$
$$= |\mathbf{a}|^2|\mathbf{b}|^2(1 - \cos^2 \theta) = |\mathbf{a}|^2|\mathbf{b}|^2 \sin^2 \theta.$$

Equation (7) now follows after we take the positive square root of both sides. (This is the correct root on the right-hand side because $\sin \theta \geqq 0$ for $0 \leqq \theta \leqq \pi$.) ■

Corollary ***Parallel Vectors***

Two nonzero vectors $\mathbf{a}$ and $\mathbf{b}$ are parallel ($\theta = 0$ or $\theta = \pi$) if and only if $\mathbf{a} \times \mathbf{b} = \mathbf{0}$.

In particular, the cross product of any vector with itself is the zero vector. Also, Eq. (1) shows immediately that the cross product of any vector with the zero vector is the zero vector itself. Thus

$$\mathbf{a} \times \mathbf{a} = \mathbf{a} \times \mathbf{0} = \mathbf{0} \times \mathbf{a} = \mathbf{0} \tag{8}$$

for every vector $\mathbf{a}$.

Equation (7) has an important geometric interpretation. Suppose that $\mathbf{a}$ and $\mathbf{b}$ are represented by adjacent sides of a parallelogram $PQRS$, with $\mathbf{a} = \overrightarrow{PQ}$ and $\mathbf{b} = \overrightarrow{PS}$ (Fig. 12.3.3). The parallelogram then has base of length $|\mathbf{a}|$ and height $|\mathbf{b}| \sin\theta$, so its area is

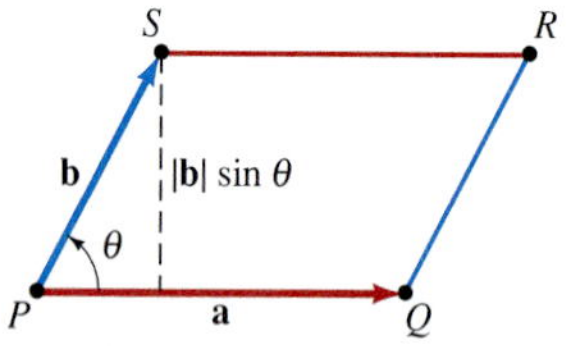

$$A = |\mathbf{a}||\mathbf{b}| \sin\theta = |\mathbf{a} \times \mathbf{b}|. \tag{9}$$

Fig. 12.3.3 The area of the parallelogram $PQRS$ is $|\mathbf{a} \times \mathbf{b}|$.

Thus *the length of the cross product* $\mathbf{a} \times \mathbf{b}$ *is numerically the same as the area of the parallelogram determined by* $\mathbf{a}$ *and* $\mathbf{b}$. It follows that the area of the triangle PQS in Fig. 12.3.4, whose area is half that of the parallelogram, is

$$\tfrac{1}{2}A = \tfrac{1}{2}|\mathbf{a} \times \mathbf{b}| = \tfrac{1}{2}\left|\overrightarrow{PQ} \times \overrightarrow{PS}\right|. \tag{10}$$

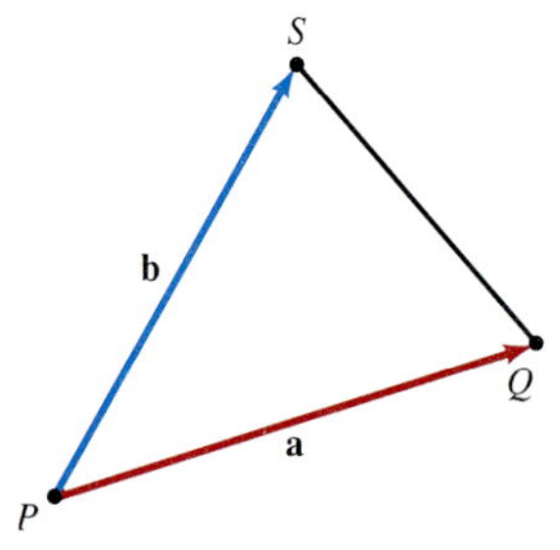

Fig. 12.3.4 The area of ΔPQS is $\frac{1}{2}|\mathbf{a} \times \mathbf{b}|$.

Equation (10) gives a quick way to compute the area of a triangle—even one in space—without the need of finding any of its angles.

EXAMPLE 4 Find the area of the triangle with vertices $A(3, 0, -10)$, $B(4, 2, 5)$, and $C(7, -2, 4)$.

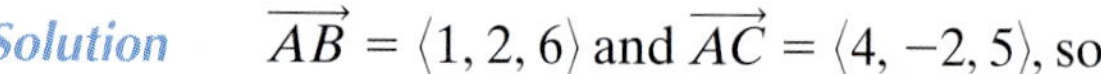

Solution $\overrightarrow{AB} = \langle 1, 2, 6\rangle$ and $\overrightarrow{AC} = \langle 4, -2, 5\rangle$, so

$$\overrightarrow{AB} \times \overrightarrow{AC} = \begin{vmatrix} \mathbf{i} & \mathbf{j} & \mathbf{k} \\ 1 & 2 & 6 \\ 4 & -2 & 5 \end{vmatrix} = 22\mathbf{i} + 19\mathbf{j} - 10\mathbf{k}.$$

Therefore, by Eq. (10), the area of triangle ABC is

$$\tfrac{1}{2}\sqrt{22^2 + 19^2 + (-10)^2} = \tfrac{1}{2}\sqrt{945} \approx 15.37.$$

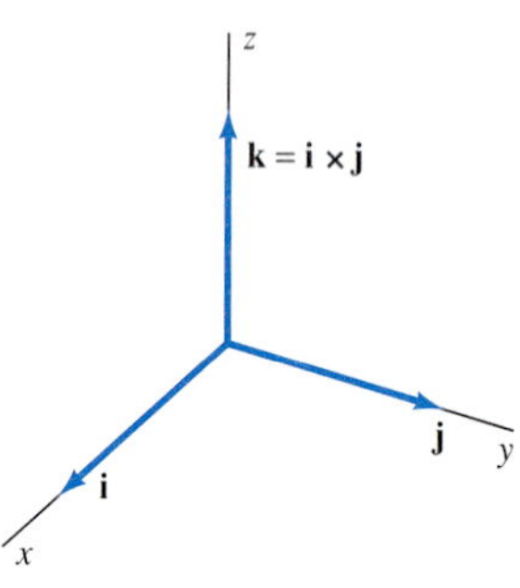

Fig. 12.3.5 The basic unit vectors in space

Now let $\mathbf{u}$, $\mathbf{v}$, $\mathbf{w}$ be a right-handed triple of mutually perpendicular *unit* vectors. The angle between any two of these is $\theta = \pi/2$, and $|\mathbf{u}| = |\mathbf{v}| = |\mathbf{w}| = 1$. Thus it follows from Eq. (7) that $\mathbf{u} \times \mathbf{v} = \mathbf{w}$. When we apply this observation to the basic unit vectors $\mathbf{i}$, $\mathbf{j}$, and $\mathbf{k}$ (Fig. 12.3.5), we see that

$$\mathbf{i} \times \mathbf{j} = \mathbf{k}, \quad \mathbf{j} \times \mathbf{k} = \mathbf{i}, \quad \text{and} \quad \mathbf{k} \times \mathbf{i} = \mathbf{j}. \tag{11a}$$

But

$$\mathbf{j} \times \mathbf{i} = -\mathbf{k}, \quad \mathbf{k} \times \mathbf{j} = -\mathbf{i}, \quad \text{and} \quad \mathbf{i} \times \mathbf{k} = -\mathbf{j}. \tag{11b}$$

These observations, together with the fact that

$$\mathbf{i} \times \mathbf{i} = \mathbf{j} \times \mathbf{j} = \mathbf{k} \times \mathbf{k} = \mathbf{0}, \tag{11c}$$

also follow directly from the original definition of the cross product [in the form in Eq. (5)]. The products in Eq. (11a) are easily remembered in terms of the sequence

$$\mathbf{i}, \quad \mathbf{j}, \quad \mathbf{k}, \quad \mathbf{i}, \quad \mathbf{j}, \quad \mathbf{k}, \quad \ldots.$$

The product of any two consecutive unit vectors, in the order in which they appear in this sequence, is the next in the sequence.

Note *The cross product is not commutative:* $\mathbf{i} \times \mathbf{j} \neq \mathbf{j} \times \mathbf{i}$. Instead, it is **anticommutative**: For any two vectors $\mathbf{a}$ and $\mathbf{b}$, $\mathbf{a} \times \mathbf{b} = -(\mathbf{b} \times \mathbf{a})$. This is the first part of Theorem 3.

Theorem 3 ***Algebraic Properties of the Cross Product***

If $\mathbf{a}$, $\mathbf{b}$, and $\mathbf{c}$ are vectors and k is a real number, then

1. $\mathbf{a} \times \mathbf{b} = -(\mathbf{b} \times \mathbf{a})$; **(12)**
2. $(k\mathbf{a}) \times \mathbf{b} = \mathbf{a} \times (k\mathbf{b}) = k(\mathbf{a} \times \mathbf{b})$; **(13)**
3. $\mathbf{a} \times (\mathbf{b} + \mathbf{c}) = (\mathbf{a} \times \mathbf{b}) + (\mathbf{a} \times \mathbf{c})$; **(14)**
4. $\mathbf{a} \cdot (\mathbf{b} \times \mathbf{c}) = (\mathbf{a} \times \mathbf{b}) \cdot \mathbf{c}$; **(15)**
5. $\mathbf{a} \times (\mathbf{b} \times \mathbf{c}) = (\mathbf{a} \cdot \mathbf{c})\mathbf{b} - (\mathbf{a} \cdot \mathbf{b})\mathbf{c}$. **(16)**

The proofs of Eqs. (12) through (15) are straightforward applications of the definition of the cross product in terms of components. See Problem 33 for an outline of the proof of Eq. (16).

We can find cross products of vectors expressed in terms of the basic unit vectors $\mathbf{i}$, $\mathbf{j}$, and $\mathbf{k}$ by means of computations that closely resemble those of ordinary algebra. We simply apply the algebraic properties summarized in Theorem 3 together with the relations in Eq. (11) giving the various products of the three unit vectors. We must be careful to preserve the order of factors, because vector multiplication is not commutative—although, of course, we should not hesitate to use Eq. (12).

EXAMPLE 5 $(\mathbf{i} - 2\mathbf{j} + 3\mathbf{k}) \times (3\mathbf{i} + 2\mathbf{j} - 4\mathbf{k})$

$$\begin{aligned}
&= 3(\mathbf{i} \times \mathbf{i}) + 2(\mathbf{i} \times \mathbf{j}) - 4(\mathbf{i} \times \mathbf{k}) - 6(\mathbf{j} \times \mathbf{i}) - 4(\mathbf{j} \times \mathbf{j}) \\
&\quad + 8(\mathbf{j} \times \mathbf{k}) + 9(\mathbf{k} \times \mathbf{i}) + 6(\mathbf{k} \times \mathbf{j}) - 12(\mathbf{k} \times \mathbf{k}) \\
&= 3 \cdot \mathbf{0} + 2\mathbf{k} - 4 \cdot (-\mathbf{j}) - 6 \cdot (-\mathbf{k}) - 4 \cdot \mathbf{0} \\
&\quad + 8\mathbf{i} + 9\mathbf{j} + 6 \cdot (-\mathbf{i}) - 12 \cdot \mathbf{0} \\
&= 2\mathbf{i} + 13\mathbf{j} + 8\mathbf{k}.
\end{aligned}$$

■

Scalar Triple Products

Let us examine the product $\mathbf{a} \cdot (\mathbf{b} \times \mathbf{c})$ that appears in Eq. (15). This expression would not make sense were the parentheses instead around $\mathbf{a} \cdot \mathbf{b}$, because $\mathbf{a} \cdot \mathbf{b}$ is a scalar, and thus we could not form the cross product of $\mathbf{a} \cdot \mathbf{b}$ with the vector $\mathbf{c}$. This means that we may omit the parentheses—the expression $\mathbf{a} \cdot \mathbf{b} \times \mathbf{c}$ is not ambiguous—but we keep them for simplicity. The dot product of the vectors $\mathbf{a}$ and $\mathbf{b} \times \mathbf{c}$ is a real number, called the **scalar triple product** of the vectors **a**, **b**, and **c**. Equation (15) implies the curious fact that we can interchange the operations · (dot) and × (cross) without affecting the value of the expression:

$$\mathbf{a} \cdot (\mathbf{b} \times \mathbf{c}) = (\mathbf{a} \times \mathbf{b}) \cdot \mathbf{c}$$

for all vectors **a**, **b**, and **c**.

To compute the scalar triple product in terms of components, write $\mathbf{a} = \langle a_1, a_2, a_3 \rangle$, $\mathbf{b} = \langle b_1, b_2, b_3 \rangle$, and $\mathbf{c} = \langle c_1, c_2, c_3 \rangle$. Then

$$\mathbf{b} \times \mathbf{c} = (b_2c_3 - b_3c_2)\mathbf{i} - (b_1c_3 - b_3c_1)\mathbf{j} + (b_1c_2 - b_2c_1)\mathbf{k},$$

so

$$\mathbf{a} \cdot (\mathbf{b} \times \mathbf{c}) = a_1(b_2c_3 - b_3c_2) - a_2(b_1c_3 - b_3c_1) + a_3(b_1c_2 - b_2c_1).$$

But the expression on the right is the value of the 3-by-3 determinant

$$\mathbf{a} \cdot (\mathbf{b} \times \mathbf{c}) = \begin{vmatrix} a_1 & a_2 & a_3 \\ b_1 & b_2 & b_3 \\ c_1 & c_2 & c_3 \end{vmatrix}. \tag{17}$$

This is the quickest way to compute the scalar triple product.

EXAMPLE 6 If $\mathbf{a} = 2\mathbf{i} - 3\mathbf{k}$, $\mathbf{b} = \mathbf{i} + \mathbf{j} + \mathbf{k}$, and $\mathbf{c} = 4\mathbf{j} - \mathbf{k}$, then

$$\begin{aligned} \mathbf{a} \cdot (\mathbf{b} \times \mathbf{c}) &= \begin{vmatrix} 2 & 0 & -3 \\ 1 & 1 & 1 \\ 0 & 4 & -1 \end{vmatrix} \\ &= +2 \cdot \begin{vmatrix} 1 & 1 \\ 4 & -1 \end{vmatrix} - 0 \cdot \begin{vmatrix} 1 & 1 \\ 0 & -1 \end{vmatrix} + (-3) \cdot \begin{vmatrix} 1 & 1 \\ 0 & 4 \end{vmatrix} \\ &= 2 \cdot (-5) + (-3) \cdot 4 = -22. \end{aligned}$$

■

The importance of the scalar triple product for applications depends on the following geometric interpretation. Let **a**, **b**, and **c** be three vectors with the same initial point. Figure 12.3.6 shows the parallelepiped determined by these vectors—that is, with arrows representing these vectors as adjacent edges. If the vectors **a**, **b**, and **c** are coplanar (lie in a single plane), then the parallelepiped is *degenerate* and its volume is zero. Theorem 4 holds whether or not the three vectors are coplanar, but it is most useful when they are not.

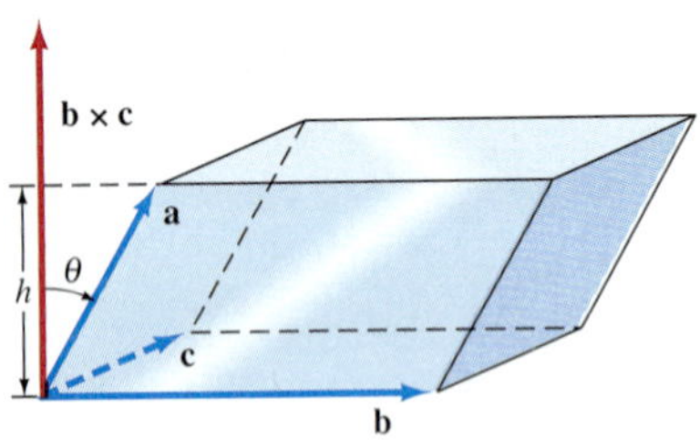

Fig. 12.3.6 The volume of the parallelepiped is $|\mathbf{a} \cdot (\mathbf{b} \times \mathbf{c})|$.

Theorem 4 *Scalar Triple Products and Volume*

The volume V of the parallelepiped determined by the vectors **a**, **b**, and **c** is the absolute value of the scalar triple product $\mathbf{a} \cdot (\mathbf{b} \times \mathbf{c})$; that is,

$$V = |\mathbf{a} \cdot (\mathbf{b} \times \mathbf{c})|. \tag{18}$$

PROOF If the three vectors are coplanar, then **a** and $\mathbf{b} \times \mathbf{c}$ are perpendicular, so $V = |\mathbf{a} \cdot (\mathbf{b} \times \mathbf{c})| = 0$. Assume, then, that they are not coplanar. By Eq. (9) the area of the base (determined by **b** and **c**) of the parallelepiped is $A = |\mathbf{b} \times \mathbf{c}|$.

Now let α be the *acute* angle between **a** and the line through $\mathbf{b} \times \mathbf{c}$ that is perpendicular to the base. Then the height of the parallelepiped is $h = |\mathbf{a}| \cos \alpha$. If θ is the angle between the vectors **a** and **b** × **c**, then either $\theta = \alpha$ or $\theta = \pi - \alpha$. Hence $\cos\alpha = |\cos\theta|$, so

$$V = Ah = |\mathbf{b} \times \mathbf{c}|\,|\mathbf{a}|\cos\alpha = |\mathbf{a}|\,|\mathbf{b} \times \mathbf{c}|\,|\cos\theta| = |\mathbf{a} \cdot (\mathbf{b} \times \mathbf{c})|.$$

Thus we have verified Eq. (18). ■

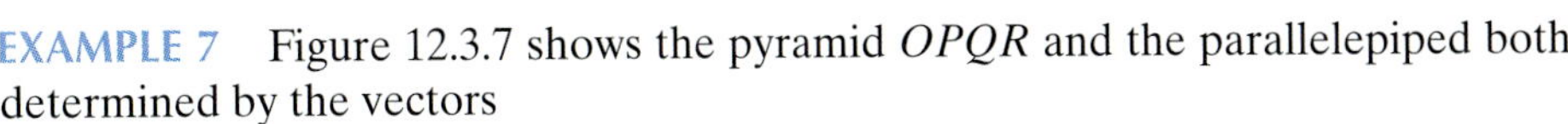

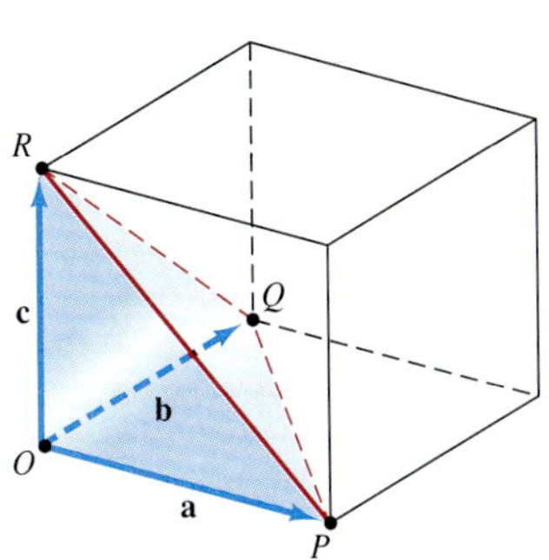

Fig. 12.3.7 The pyramid (and parallelepiped) of Example 7

EXAMPLE 7 Figure 12.3.7 shows the pyramid $OPQR$ and the parallelepiped both determined by the vectors

$$\mathbf{a} = \overrightarrow{OP} = \langle 3, 2, -1\rangle, \quad \mathbf{b} = \overrightarrow{OQ} = \langle -2, 5, 1\rangle, \quad \text{and} \quad \mathbf{c} = \overrightarrow{OR} = \langle 2, 1, 5\rangle.$$

The volume of the pyramid is $V = \frac{1}{3}Ah$, where h is its height and the area of its base OPQ is *half* the area of the corresponding base of the parallelepiped. It therefore follows from Eqs. (17) and (18) that V is one-sixth the volume of the parallelepiped:

$$V = \frac{1}{6}|\mathbf{a} \cdot (\mathbf{b} \times \mathbf{c})| = \frac{1}{6}\begin{vmatrix} 3 & 2 & -1 \\ -2 & 5 & 1 \\ 2 & 1 & 5 \end{vmatrix} = \frac{108}{6} = 18.$$ ■

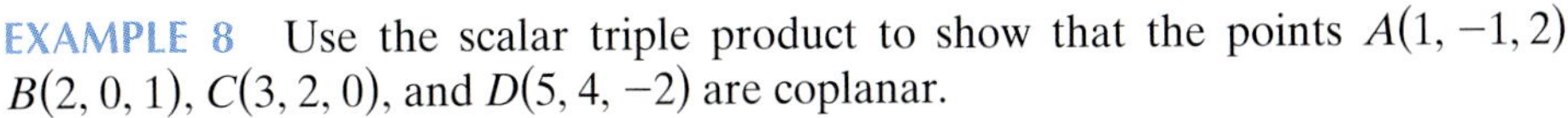

EXAMPLE 8 Use the scalar triple product to show that the points $A(1, -1, 2)$, $B(2, 0, 1)$, $C(3, 2, 0)$, and $D(5, 4, -2)$ are coplanar.

Solution It's enough to show that the vectors $\overrightarrow{AB} = \langle 1, 1, -1\rangle$, $\overrightarrow{AC} = \langle 2, 3, -2\rangle$, and $\overrightarrow{AD} = \langle 4, 5, -4\rangle$ are coplanar. But their scalar triple product is

$$\begin{vmatrix} 1 & 1 & -1 \\ 2 & 3 & -2 \\ 4 & 5 & -4 \end{vmatrix} = 1\cdot(-2) - 1\cdot 0 + (-1)\cdot(-2) = 0,$$

so Theorem 4 guarantees that the parallelepiped determined by these three vectors has volume zero. Hence the four given points are coplanar. ■

The cross product occurs quite often in scientific applications. For example, suppose that a body in space is free to rotate around the fixed point O. If a force **F** acts at a point P of the body, that force causes the body to rotate. This effect is measured by the **torque vector** $\boldsymbol{\tau}$ defined by the relation

$$\boldsymbol{\tau} = \mathbf{r} \times \mathbf{F},$$

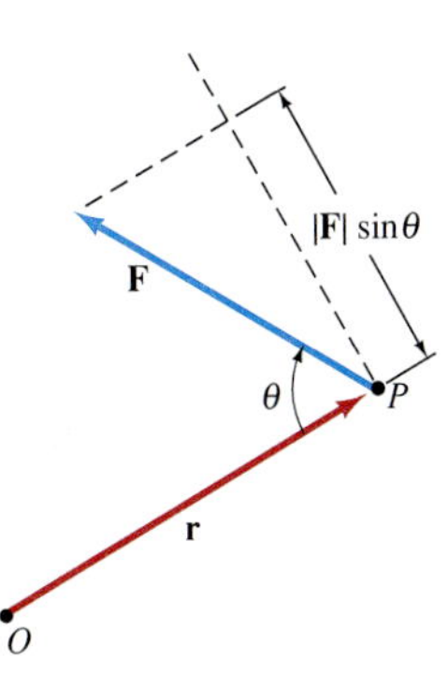

Fig. 12.3.8 The torque vector $\boldsymbol{\tau}$ is normal to both **r** and **F**.

where $\mathbf{r} = \overrightarrow{OP}$, the straight line through O determined by $\boldsymbol{\tau}$, is the axis of rotation, and the length

$$|\boldsymbol{\tau}| = |\mathbf{r}|\,|\mathbf{F}|\sin\theta$$

is the **moment** of the force **F** around this axis (Fig. 12.3.8).

Another application of the cross product involves the force exerted on a moving charged particle by a magnetic field. This force is important in particle accelerators, mass spectrometers, and television picture tubes; controlling the paths of the ions is accomplished through the interplay of electric and magnetic fields. In such circumstances, the force **F** on the particle due to a magnetic field depends on three things: the charge q of the particle, its velocity vector **v**, and the magnetic field vector **B** at the instantaneous location of the particle. And it turns out that

$$\mathbf{F} = (q\mathbf{v}) \times \mathbf{B}.$$

12.3 PROBLEMS

Find $\mathbf{a} \times \mathbf{b}$ in Problems 1 through 4.

1. $\mathbf{a} = \langle 5, -1, -2 \rangle$, $\mathbf{b} = \langle -3, 2, 4 \rangle$

2. $\mathbf{a} = \langle 3, -2, 0 \rangle$, $\mathbf{b} = \langle 0, 3, -2 \rangle$

3. $\mathbf{a} = \mathbf{i} - \mathbf{j} + 3\mathbf{k}$, $\mathbf{b} = -2\mathbf{i} + 3\mathbf{j} + \mathbf{k}$

4. $\mathbf{a} = 4\mathbf{i} + 2\mathbf{j} - 2\mathbf{k}$, $\mathbf{b} = 2\mathbf{i} - 5\mathbf{j} + 5\mathbf{k}$

In Problems 5 and 6, find the cross product of the given two-dimensional vectors $\mathbf{a} = \langle a_1, a_2 \rangle$ and $\mathbf{b} = \langle b_1, b_2 \rangle$ by first "extending" them to three-dimensional vectors $\mathbf{a} = \langle a_1, a_2, 0 \rangle$ and $\mathbf{b} = \langle b_1, b_2, 0 \rangle$.

5. $\mathbf{a} = \langle 2, -3 \rangle$ and $\mathbf{b} = \langle 4, 5 \rangle$

6. $\mathbf{a} = -5\mathbf{i} + 2\mathbf{j}$ and $\mathbf{b} = 7\mathbf{i} - 11\mathbf{j}$

In Problems 7 and 8, find two different unit vectors $\mathbf{u}$ and $\mathbf{v}$ both of which are perpendicular to both the given vectors $\mathbf{a}$ and $\mathbf{b}$.

7. $\mathbf{a} = \langle 3, 12, 0 \rangle$ and $\mathbf{b} = \langle 0, 4, 3 \rangle$

8. $\mathbf{a} = \mathbf{i} + 2\mathbf{j} + 3\mathbf{k}$ and $\mathbf{b} = 2\mathbf{i} + 3\mathbf{j} + 5\mathbf{k}$

9. Apply Eq. (5) to verify the equations in (11a).

10. Apply Eq. (5) to verify the equations in (11b).

11. Prove that the vector product is not associative by comparing $\mathbf{a} \times (\mathbf{b} \times \mathbf{c})$ with $(\mathbf{a} \times \mathbf{b}) \times \mathbf{c}$ in the case $\mathbf{a} = \mathbf{i}$, $\mathbf{b} = \mathbf{i} + \mathbf{j}$, and $\mathbf{c} = \mathbf{i} + \mathbf{j} + \mathbf{k}$.

12. Find nonzero vectors $\mathbf{a}$, $\mathbf{b}$, and $\mathbf{c}$ such that $\mathbf{a} \times \mathbf{b} = \mathbf{a} \times \mathbf{c}$ but $\mathbf{b} \neq \mathbf{c}$.

13. Suppose that the three vectors $\mathbf{a}$, $\mathbf{b}$, and $\mathbf{c}$ are mutually perpendicular. Prove that $\mathbf{a} \times (\mathbf{b} \times \mathbf{c}) = \mathbf{0}$.

14. Find the area of the triangle with vertices $P(1, 1, 0)$, $Q(1, 0, 1)$, and $R(0, 1, 1)$.

15. Find the area of the triangle with vertices $P(1, 3, -2)$, $Q(2, 4, 5)$, and $R(-3, -2, 2)$.

16. Find the volume of the parallelepiped with adjacent edges $\overrightarrow{OP}$, $\overrightarrow{OQ}$, and $\overrightarrow{OR}$, where P, Q, and R are the points given in Problem 14.

17. (a) Find the volume of the parallelepiped with adjacent edges $\overrightarrow{OP}$, $\overrightarrow{OQ}$, and $\overrightarrow{OR}$, where P, Q, and R are the points given in Problem 15. (b) Find the volume of the pyramid with vertices O, P, Q, and R.

18. Find a unit vector $\mathbf{n}$ perpendicular to the plane through the points P, Q, and R of Problem 15. Then find the distance from the origin to this plane by computing $\mathbf{n} \cdot \overrightarrow{OP}$.

In Problems 19 through 22, determine whether the four given points A, B, C, and D are coplanar. If not, find the volume of the pyramid with these four points as its vertices, given that its volume is one-sixth that of the parallelepiped spanned by $\overrightarrow{AB}$, $\overrightarrow{AC}$, and $\overrightarrow{AD}$.

19. $A(1, 3, -2)$, $B(3, 4, 1)$, $C(2, 0, -2)$, and $D(4, 8, 4)$

20. $A(13, -25, -37)$, $B(25, -14, -22)$, $C(24, -38, -25)$, and $D(26, 10, -19)$

21. $A(5, 2, -3)$, $B(6, 4, 0)$, $C(7, 5, 1)$, and $D(14, 14, 18)$

22. $A(25, 22, -33)$, $B(36, 34, -20)$, $C(27, 25, -29)$, and $D(34, 34, -12)$

23. Figure 12.3.9 shows a polygonal plot of land, with angles and lengths measured by a surveyor. First find the coordinates of each vertex. Then use the vector product [as in Eq. (10)] to calculate the area of the plot.

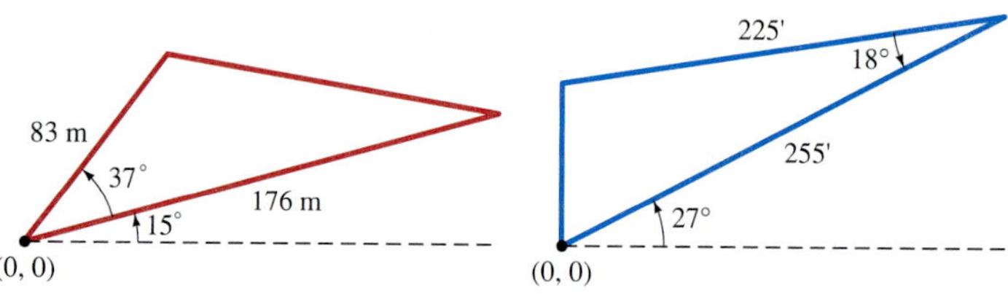

Fig. 12.3.9 Problem 23

Fig. 12.3.10 Problem 24

24. Repeat Problem 23 with the plot shown in Fig. 12.3.10.

25. Repeat Problem 23 with the plot shown in Fig. 12.3.11. (*Suggestion:* First divide the plot into two triangles.)

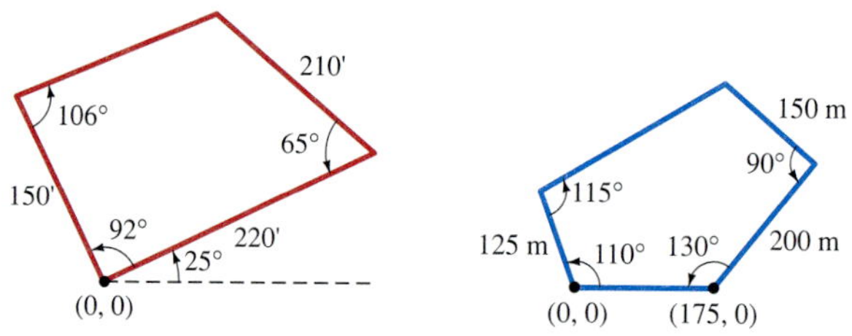

Fig. 12.3.11 Problem 25

Fig. 12.3.12 Problem 26

26. Repeat Problem 23 with the plot shown in Fig. 12.3.12.

27. Apply Eq. (5) to verify Eq. (12), the anticommutativity of the vector product.

28. Apply Eq. (17) to verify the identity for scalar triple products stated in Eq. (15).

29. Suppose that P and Q are points on a line L in space. Let A be a point not on L (Fig. 12.3.13). (a) Calculate in two ways the area of the triangle APQ to show that the perpendicular distance from A to the line L is

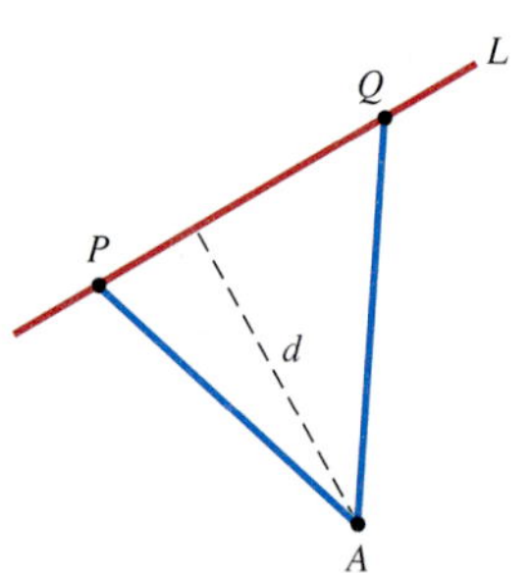

Fig. 12.3.13 Problem 29

$$d = \frac{|\overrightarrow{AP} \times \overrightarrow{AQ}|}{|\overrightarrow{PQ}|}.$$

(b) Use this formula to compute the distance from the point $(1, 0, 1)$ to the line through the two points $P(2, 3, 1)$ and $Q(-3, 1, 4)$.

30. Suppose that A is a point not on the plane determined by the three points P, Q, and R. Calculate in two ways the volume of the pyramid $APQR$ to show that the perpendicular distance from A to this plane is

$$d = \frac{|\overrightarrow{AP} \cdot (\overrightarrow{AQ} \times \overrightarrow{AR})|}{|\overrightarrow{PQ} \times \overrightarrow{PR}|}.$$

Use this formula to compute the distance from the point $(1, 0, 1)$ to the plane through the points $P(2, 3, 1)$, $Q(3, -1, 4)$, and $R(0, 0, 2)$.

31. Suppose that P_1 and Q_1 are two points on the line L_1 and that P_2 and Q_2 are two points on the line L_2. If the lines L_1 and L_2 are not parallel, then the perpendicular distance d between them is the projection of $\overrightarrow{P_1P_2}$ onto a vector $\mathbf{n}$ that is perpendicular to both $\overrightarrow{P_1Q_1}$ and $\overrightarrow{P_2Q_2}$. Prove that

$$d = \frac{|\overrightarrow{P_1P_2} \cdot (\overrightarrow{P_1Q_1} \times \overrightarrow{P_2Q_2})|}{|\overrightarrow{P_1Q_1} \times \overrightarrow{P_2Q_2}|}.$$

32. Use the following method to establish that the **vector triple product** $(\mathbf{a} \times \mathbf{b}) \times \mathbf{c}$ is equal to $(\mathbf{a} \cdot \mathbf{c})\mathbf{b} - (\mathbf{b} \cdot \mathbf{c})\mathbf{a}$.
(a) Let $\mathbf{I}$ be a unit vector in the direction of $\mathbf{a}$ and let $\mathbf{J}$ be a unit vector perpendicular to $\mathbf{I}$ and parallel to the plane of $\mathbf{a}$ and $\mathbf{b}$. Let $\mathbf{K} = \mathbf{I} \times \mathbf{J}$. Explain why there are scalars a_1, b_1, b_2, c_1, c_2, and c_3 such that

$$\mathbf{a} = a_1\mathbf{I}, \quad \mathbf{b} = b_1\mathbf{I} + b_2\mathbf{J}, \quad \text{and} \quad \mathbf{c} = c_1\mathbf{I} + c_2\mathbf{J} + c_2\mathbf{K}.$$

(b) Now show that

$$(\mathbf{a} \times \mathbf{b}) \times \mathbf{c} = -a_1b_2c_2\mathbf{I} + a_1b_2c_1\mathbf{J}.$$

(c) Finally, substitute for $\mathbf{I}$ and $\mathbf{J}$ in terms of $\mathbf{a}$ and $\mathbf{b}$.

33. By permutation of the vectors $\mathbf{a}$, $\mathbf{b}$, and $\mathbf{c}$, deduce from Problem 32 that

$$\mathbf{a} \times (\mathbf{b} \times \mathbf{c}) = (\mathbf{a} \cdot \mathbf{c})\mathbf{b} - (\mathbf{a} \cdot \mathbf{b})\mathbf{c}$$

[this is Eq. (16)].

34. Deduce from the orthogonality properties of the vector product that the vector $(\mathbf{a} \times \mathbf{b}) \times (\mathbf{c} \times \mathbf{d})$ can be written in the form $r_1\mathbf{a} + r_2\mathbf{b}$ and in the form $s_1\mathbf{c} + s_2\mathbf{d}$.

35. Consider the triangle in the xy-plane that has vertices $(x_1, y_1, 0)$, $(x_2, y_2, 0)$, and $(x_3, y_3, 0)$. Use the vector product to prove that the area of this triangle is *half* the *absolute value* of the determinant

$$\begin{vmatrix} 1 & x_1 & y_1 \\ 1 & x_2 & y_2 \\ 1 & x_3 & y_3 \end{vmatrix}.$$

36. Given the vectors $\mathbf{a} = \langle a_1, a_2, a_3 \rangle$ and $\mathbf{b} = \langle b_1, b_2, b_3 \rangle$, verify Eq. (6),

$$|\mathbf{a} \times \mathbf{b}|^2 = |\mathbf{a}|^2|\mathbf{b}|^2 - (\mathbf{a} \cdot \mathbf{b})^2,$$

by computing each side in terms of the components of $\mathbf{a}$ and $\mathbf{b}$.

12.4 LINES AND PLANES IN SPACE

Just as in the plane, a straight line in space is determined by any two points P_0 and P_1 that lie on it. We may write $\mathbf{v} = \overrightarrow{P_0P_1}$—meaning that the directed line segment $\overrightarrow{P_0P_1}$ represents the vector $\mathbf{v}$—to describe the "direction of the line." Thus, alternatively, a line in space can be specified by giving a point P_0 on it *and* a [nonzero] vector $\mathbf{v}$ that determines the direction of the line.

To investigate equations that describe lines in space, let us begin with a straight line L that passes through the point $P_0(x_0, y_0, z_0)$ and is parallel to the vector $\mathbf{v} = a\mathbf{i} + b\mathbf{j} + c\mathbf{k}$ (Fig. 12.4.1). Then another point $P(x, y, z)$ lies on the line L if and only if the vectors $\mathbf{v}$ and $\overrightarrow{P_0P}$ are parallel, in which case

$$\overrightarrow{P_0P} = t\mathbf{v} \tag{1}$$

for some real number t. If $\mathbf{r}_0 = \overrightarrow{OP_0}$ and $\mathbf{r} = \overrightarrow{OP}$ are the position vectors of the points P_0 and P, respectively, then $\overrightarrow{P_0P} = \mathbf{r} - \mathbf{r}_0$. Hence Eq. (1) gives the **vector equation**

$$\mathbf{r} = \mathbf{r}_0 + t\mathbf{v} \tag{2}$$

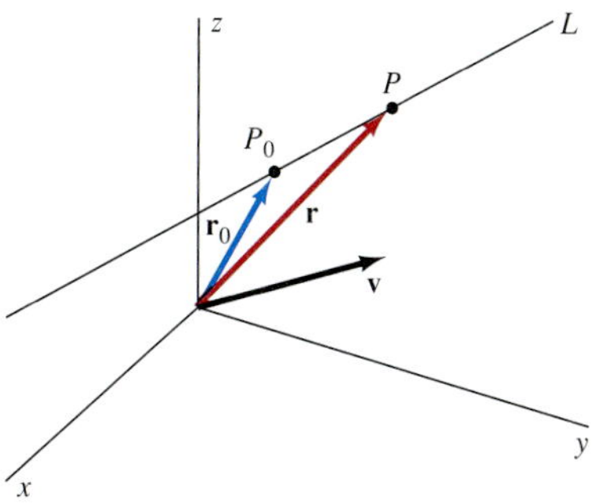

Fig. 12.4.1 Finding the equation of the line L that passes through the point P_0 and is parallel to the vector $\mathbf{v}$

describing the line L. As indicated in Fig. 12.4.1, $\mathbf{r}$ is the position vector of an *arbitrary* point P on the line L, and Eq. (2) gives $\mathbf{r}$ in terms of the parameter t, the position vector $\mathbf{r}_0$ of a *fixed* point P_0 on L, and the fixed vector $\mathbf{v}$ that determines the direction of L.

The left- and right-hand sides of Eq. (2) are equal, and each side is a vector. So corresponding components are also equal. When we write the resulting equations, we get a scalar description of the line L. Because $\mathbf{r}_0 = \langle x_0, y_0, z_0 \rangle$ and $\mathbf{r} = \langle x, y, z \rangle$, Eq. (2) thereby yields the three scalar equations

$$x = x_0 + at, \qquad y = y_0 + bt, \qquad z = z_0 + ct. \tag{3}$$

These are **parametric equations** of the line L that passes through the point (x_0, y_0, z_0) and is parallel to the vector $\mathbf{v} = \langle a, b, c \rangle$.

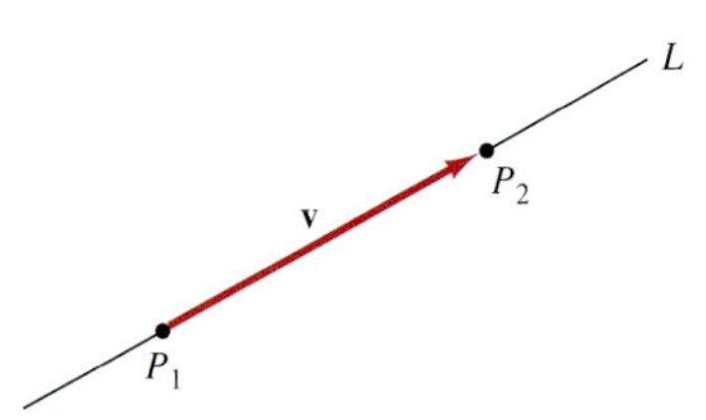

Fig. 12.4.2 The line L of Example 1

EXAMPLE 1 Write parametric equations of the line L that passes through the points $P_1(1, 2, 2)$ and $P_2(3, -1, 3)$ of Fig. 12.4.2.

Solution The line L is parallel to the vector

$$\mathbf{v} = \overrightarrow{P_1P_2} = (3\mathbf{i} - \mathbf{j} + 3\mathbf{k}) - (\mathbf{i} + 2\mathbf{j} + 2\mathbf{k}) = 2\mathbf{i} - 3\mathbf{j} + \mathbf{k},$$

so we take $a = 2$, $b = -3$, and $c = 1$. With P_1 as the fixed point, the equations in (3) give

$$x = 1 + 2t, \qquad y = 2 - 3t, \qquad z = 2 + t$$

as parametric equations of L. In contrast, with P_2 as the fixed point and with

$$-2\mathbf{v} = -4\mathbf{i} + 6\mathbf{j} - 2\mathbf{k}$$

as the direction vector, the equations in (3) yield the parametric equations

$$x = 3 - 4t, \qquad y = -1 + 6t, \qquad z = 3 - 2t.$$

Thus the parametric equations of a line are not unique. ■

Given two straight lines L_1 and L_2 with parametric equations

$$x = x_1 + a_1t, \qquad y = y_1 + b_1t, \qquad z = z_1 + c_1t \tag{4}$$

and

$$x = x_2 + a_2s, \qquad y = y_2 + b_2s, \qquad z = z_2 + c_2s, \tag{5}$$

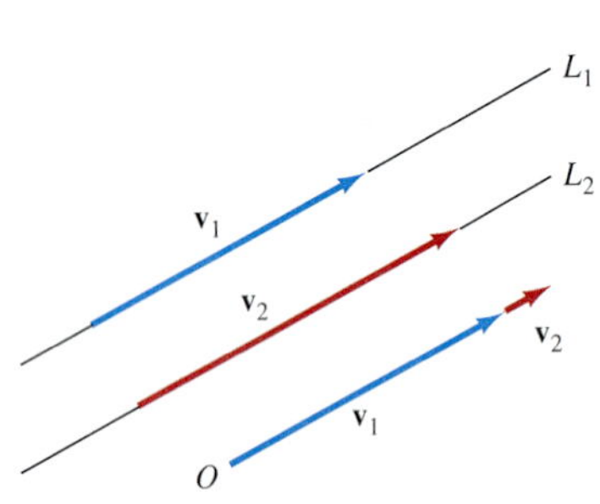

Fig. 12.4.3 Parallel lines

respectively, we can see at a glance whether L_1 and L_2 are parallel. Because L_1 is parallel to $\mathbf{v}_1 = \langle a_1, b_1, c_1 \rangle$ and L_2 is parallel to $\mathbf{v}_2 = \langle a_2, b_2, c_2 \rangle$, it follows that the lines L_1 and L_2 are parallel if and only if the vectors $\mathbf{v}_1$ and $\mathbf{v}_2$ are scalar multiples of each other (Fig. 12.4.3). If the two lines are not parallel, we can attempt to find a point of intersection by solving the equations

$$x_1 + a_1t = x_2 + a_2s \quad \text{and} \quad y_1 + b_1t = y_2 + b_2s$$

simultaneously for s and t. If these values of s and t also satisfy the equation $z_1 + c_1t = z_2 + c_2s$, then we have found a point of intersection. Its rectangular coordinates can be found by substituting the resulting value of t into Eq. (4) [or the resulting value of s into Eq. (5)]. Otherwise, the lines L_1 and L_2 do not intersect. Two nonparallel and nonintersecting lines in space are called **skew lines** (Fig. 12.4.4).

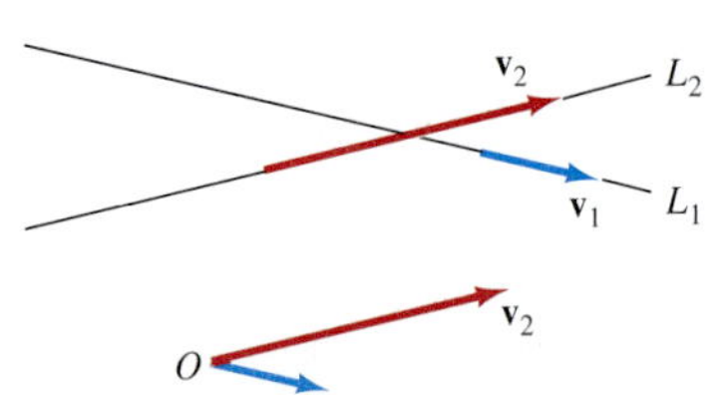

Fig. 12.4.4 Skew lines

EXAMPLE 2 The line L_1 with parametric equations

$$x = 1 + 2t, \qquad y = 2 - 3t, \qquad z = 2 + t$$

passes through the point $P_1(1, 2, 2)$ (discovered by substitution of $t = 0$) and is parallel to the vector $\mathbf{v}_1 = \langle 2, -3, 1 \rangle$. The line L_2 with parametric equations

$$x = 3 + 4t, \qquad y = 1 - 6t, \qquad z = 5 + 2t$$

passes through the point $P_2(3, 1, 5)$ and is parallel to the vector $\mathbf{v}_2 = \langle 4, -6, 2 \rangle$. Because $\mathbf{v}_2 = 2\mathbf{v}_1$, we see that L_1 and L_2 are parallel.

But are L_1 and L_2 actually different lines, or are we perhaps dealing with two different parametrizations of the same line? To answer this question, we note that $\overrightarrow{P_1P_2} = \langle 2, -1, 3 \rangle$ is not parallel to $\mathbf{v}_1 = \langle 2, -3, 1 \rangle$. Thus the point P_2 does not line on the line L_1, and hence the lines L_1 and L_2 are indeed distinct. ■

If all the coefficients a, b, and c in Eq. (3) are nonzero, then we can eliminate the parameter t. Simply solve each equation for t and then set the resulting expressions equal to each other. This gives

$$\frac{x - x_0}{a} = \frac{y - y_0}{b} = \frac{z - z_0}{c}. \tag{6}$$

These are called the **symmetric equations** of the line L. If one or more of a or b or c is zero, this means that L lies in a plane parallel to one of the coordinate planes, and in this case the line does not have symmetric equations. For example, if $c = 0$, then L lies in the horizontal plane $z = z_0$. Of course, it is still possible to write equations for L that don't include the parameter t; if $c = 0$, for instance, but a and b are nonzero, then we could describe the line L as the simultaneous solution of the equations

$$\frac{x - x_0}{a} = \frac{y - y_0}{b}, \qquad z = z_0.$$

EXAMPLE 3 Find both parametric and symmetric equations of the line L through the points $P_0(3, 1, -2)$ and $P_1(4, -1, 1)$. Find also the points at which L intersects the three coordinate planes.

Solution The line L is parallel to the vector $\mathbf{v} = \overrightarrow{P_0P_1} = \langle 1, -2, 3 \rangle$, so we take $a = 1$, $b = -2$, and $c = 3$. The equations in (3) then give the parametric equations

$$x = 3 + t, \qquad y = 1 - 2t, \qquad z = -2 + 3t$$

of L, and the equations in (6) give the symmetric equations

$$\frac{x - 3}{1} = \frac{y - 1}{-2} = \frac{z + 2}{3}.$$

To find the point at which L intersects the xy-plane, we set $z = 0$ in the symmetric equations. This gives

$$\frac{x - 3}{1} = \frac{y - 1}{-2} = \frac{2}{3},$$

and so $x = \frac{11}{3}$ and $y = -\frac{1}{3}$. Thus L meets the xy-plane at the point $\left(\frac{11}{3}, -\frac{1}{3}, 0\right)$. Similarly, $x = 0$ gives $(0, 7, -11)$ for the point where L meets the yz-plane, and $y = 0$ gives $\left(\frac{7}{2}, 0, -\frac{1}{2}\right)$ for its intersection with the xz-plane. ■

Planes in Space

A plane $\mathcal{P}$ in space is determined by a point $P_0(x_0, y_0, z_0)$ through which $\mathcal{P}$ passes and a line through P_0 that is normal to $\mathcal{P}$. Alternatively, we may be given P_0 on $\mathcal{P}$ and a vector $\mathbf{n} = \langle a, b, c \rangle$ normal to the plane $\mathcal{P}$. The point $P(x, y, z)$ lies on the plane $\mathcal{P}$ if and only if the vectors $\mathbf{n}$ and $\overrightarrow{P_0P}$ are perpendicular (Fig. 12.4.5), in which case $\mathbf{n} \cdot \overrightarrow{P_0P} = 0$. We write $\overrightarrow{P_0P} = \mathbf{r} - \mathbf{r}_0$, where $\mathbf{r}$ and $\mathbf{r}_0$ are the position vectors $\mathbf{r} = \overrightarrow{OP}$ and $\mathbf{r}_0 = \overrightarrow{OP_0}$ of the points P and P_0, respectively. Thus we obtain a **vector equation**

$$\mathbf{n} \cdot (\mathbf{r} - \mathbf{r}_0) = 0 \tag{7}$$

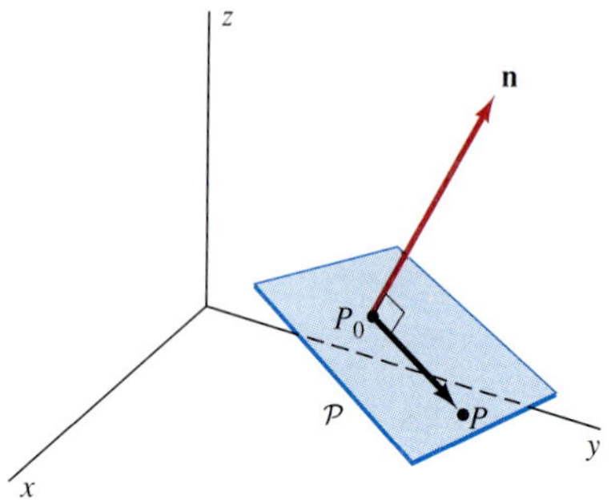

of the plane $\mathcal{P}$.

If we substitute $\mathbf{n} = \langle a, b, c \rangle$, $\mathbf{r} = \langle x, y, z \rangle$, and $\mathbf{r}_0 = \langle x_0, y_0, z_0 \rangle$ into Eq. (7), we thereby obtain a **scalar equation**

$$a(x - x_0) + b(y - y_0) + c(z - z_0) = 0 \tag{8}$$

of the plane through $P_0(x_0, y_0, z_0)$ with **normal vector** $\mathbf{n} = \langle a, b, c \rangle$.

Fig. 12.4.5 Because $\mathbf{n}$ is normal to $\mathcal{P}$, it follows that $\mathbf{n}$ is normal to $\overrightarrow{P_0P}$ for all points P in $\mathcal{P}$.

EXAMPLE 4 An equation of the plane through $P_0(-1, 5, 2)$ with normal vector $\mathbf{n} = \langle 1, -3, 2 \rangle$ is

$$1 \cdot (x + 1) + (-3) \cdot (y - 5) + 2 \cdot (z - 2) = 0;$$

that is, $x - 3y + 2z = -12$. ■

IMPORTANT The coefficients of x, y, and z in the last equation are the components of the normal vector. This is always the case, because we can write Eq. (8) in the form

$$ax + by + cz = d, \tag{9}$$

where $d = ax_0 + by_0 + cz_0$. Conversely, every *linear equation* in x, y, and z of the form in Eq. (9) represents a plane in space provided that the coefficients a, b, and c are not all zero. The reason is that if $c \neq 0$ (for instance), then we can choose x_0 and y_0 arbitrarily and solve the equation $ax_0 + by_0 + cz_0 = d$ for z_0. With these values, Eq. (9) takes the form

$$ax + by + cz = ax_0 + by_0 + cz_0;$$

that is,

$$a(x - x_0) + b(y - y_0) + c(z - z_0) = 0,$$

so this equation represents the plane through (x_0, y_0, z_0) with normal vector $\langle a, b, c \rangle$.

EXAMPLE 5 Find an equation for the plane through the three points $P(2, 4, -3)$, $Q(3, 7, -1)$, and $R(4, 3, 0)$.

Solution We want to use Eq. (8), so we first need a vector $\mathbf{n}$ that is normal to the plane in question. One easy way to obtain such a normal vector is by using the cross product. Let

$$\mathbf{n} = \overrightarrow{PQ} \times \overrightarrow{PR} = \begin{vmatrix} \mathbf{i} & \mathbf{j} & \mathbf{k} \\ 1 & 3 & 2 \\ 2 & -1 & 3 \end{vmatrix} = 11\mathbf{i} + \mathbf{j} - 7\mathbf{k}.$$

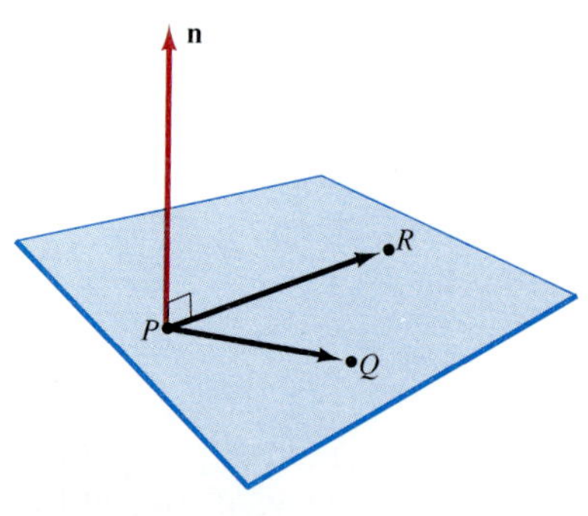

Fig. 12.4.6 The normal vector $\mathbf{n}$ as a cross product (Example 5)

Because $\overrightarrow{PQ}$ and $\overrightarrow{PR}$ are in the plane, their cross product $\mathbf{n}$ is normal to the plane (Fig. 12.4.6). Hence the plane has equation

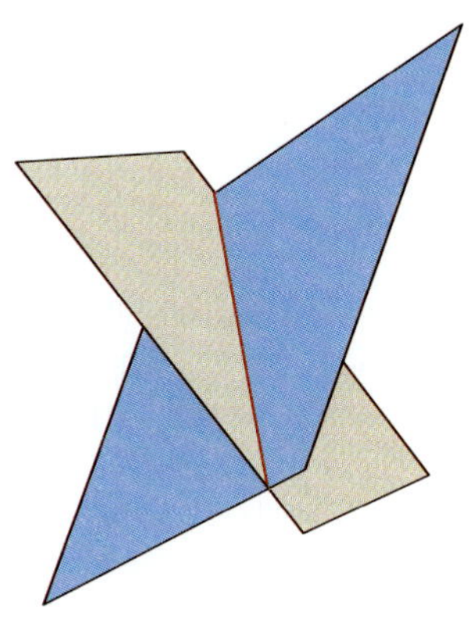

Fig. 12.4.7 The intersection of two nonparallel planes is a straight line.

$$11(x - 2) + (y - 4) - 7(z + 3) = 0.$$

After simplifying, we write the equation as

$$11x + y - 7z = 47.$$

Two planes with normal vectors **n** and **m** are said to be **parallel** provided that **n** and **m** are parallel. Otherwise, the two planes meet in a straight line (Fig. 12.4.7). We define the angle between the two planes to be the angle between their normal vectors **n** and **m**, as in Fig. 12.4.8.

EXAMPLE 6 Find the angle θ between the planes with equations

$$2x + 3y - z = 3 \quad \text{and} \quad 4x + 5y + z = 1.$$

Then write symmetric equations of their line of intersection L.

Solution The vectors $\mathbf{n} = \langle 2, 3, -1 \rangle$ and $\mathbf{m} = \langle 4, 5, 1 \rangle$ are normal to the two planes, so

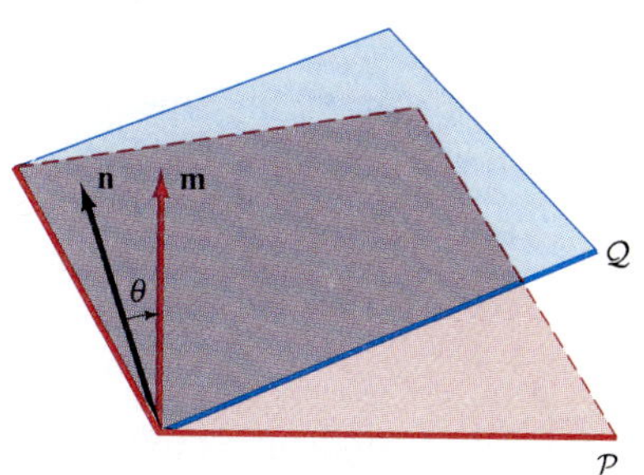

Fig. 12.4.8 Vectors **m** and **n** normal to the planes $\mathcal{P}$ and $\mathcal{Q}$, respectively

$$\cos\theta = \frac{\mathbf{n}\cdot\mathbf{m}}{|\mathbf{n}|\,|\mathbf{m}|} = \frac{22}{\sqrt{14}\sqrt{42}}.$$

Hence $\theta = \cos^{-1}\left(\frac{11}{21}\sqrt{3}\right) \approx 24.87°$.

To determine the line of intersection L of the two planes, we need first to find a point P_0 that lies on L. We can do this by substituting an arbitrarily chosen value of x into the equations of the given planes and then solving the resulting equations for y and z. With $x = 1$ we get the equations

$$2 + 3y - z = -3,$$

$$4 + 5y + z = 1.$$

The common solution is $y = -1$, $z = 2$. Thus the point $P_0(1, -1, 2)$ lies on the line L.

Next we need a vector **v** parallel to L. The vectors **n** and **m** normal to the two planes are both perpendicular to L, so their cross product is parallel to L. Alternatively, we can find a second point P_1 on L by substituting a second value of x into the equations of the given planes and solving for y and z, as before. With $x = 5$, we obtain the equations

$$10 + 3y - z = -3,$$

$$20 + 5y + z = 1,$$

with common solution $y = -4$, $z = 1$. Thus we obtain a second point $P_1(5, -4, 1)$ on L and thereby the vector

$$\mathbf{v} = \overrightarrow{P_0P_1} = \langle 4, -3, -1 \rangle$$

parallel to L. From Eq. (6) we now find symmetric equations

$$\frac{x - 1}{4} = \frac{y + 1}{-3} = \frac{z - 2}{-1}$$

of the line of intersection of the two given planes.

Finally, we may note that the symmetric equations of a line L present the line as an intersection of planes: We can rewrite the equations in (6) in the form

$$\begin{aligned} b(x - x_0) - a(y - y_0) &= 0, \\ c(x - x_0) - a(z - z_0) &= 0, \\ c(y - y_0) - b(z - z_0) &= 0. \end{aligned} \tag{10}$$

These are the equations of three planes that intersect in the line L. The first has normal vector $\langle b, -a, 0\rangle$, a vector parallel to the xy-plane. So the first plane is perpendicular to the xy-plane. Similarly, the second plane is perpendicular to the xz-plane and the third is perpendicular to the yz-plane.

The equations in (10) are symmetric equations of the line that passes through the point $P_0(x_0, y_0, z_0)$ and is parallel to $\mathbf{v} = \langle a, b, c\rangle$. Unlike the equations in (6), these equations are meaningful whether all the components a, b, and c are nonzero. They have a special form, though, if one of the three components is zero. If, say, $a = 0$, then the first two equations in (10) take the form $x = x_0$. The line is then the intersection of the two planes $x = x_0$ and $c(y - y_0) = b(z - z_0)$.

Lines, Planes, and Linear Mappings

Here we briefly discuss lines and planes from the viewpoint of elementary linear algebra. First, let L be a line through the origin O in space that is determined by the nonzero vector $\mathbf{v}$—that is, a point P lies on the line L provided that its position vector $\overrightarrow{OP}$ is a scalar multiple of $\mathbf{v}$. Then we may write

$$F(t) = t\mathbf{v} \tag{11}$$

to define a function or *mapping* that maps the real line $\boldsymbol{R}$ (considered as the t-axis) into 3-space $\boldsymbol{R}^3$. This mapping $F: \boldsymbol{R} \to \boldsymbol{R}^3$ is **linear,** meaning that

$$F(as + bt) = aF(s) + bF(t) \tag{12}$$

for all s and t in $\boldsymbol{R}$ and all scalars a and b. To see that this is so, we need only write

$$\begin{aligned} F(as + bt) &= (as + bt)\mathbf{v} \\ &= a(s\mathbf{v}) + b(t\mathbf{v}) = aF(s) + bF(t). \end{aligned}$$

Thus every line through the origin in space is the image of a linear mapping $F: \boldsymbol{R} \to \boldsymbol{R}^3$.

Similarly, every plane $\mathcal{P}$ through the origin in space is the image of a linear mapping $F: \boldsymbol{R}^2 \to \boldsymbol{R}^3$ of the plane $\boldsymbol{R}^2$ into space $\boldsymbol{R}^3$. To define the mapping F, we may write

$$F(\mathbf{t}) = t_1\mathbf{v}_1 + t_2\mathbf{v}_2 \tag{13}$$

where $\mathbf{t} = \langle t_1, t_2\rangle$ represents a typical point (t_1, t_2) of the plane and $\mathbf{v}_1 = \langle a_1, b_1, c_1\rangle$ and $\mathbf{v}_2 = \langle a_2, b_2, c_2\rangle$ are noncollinear (that is, nonparallel) vectors in the plane $\mathcal{P}$. In Problem 69 we ask you to verify that Eq. (13) defines a **linear mapping**; that is, that

$$F(a\mathbf{s} + b\mathbf{t}) = aF(\mathbf{s}) + bF(\mathbf{t}) \tag{14}$$

for any two scalars a and b and any two points $\mathbf{s} = \langle s_1, s_2\rangle$ and $\mathbf{t} = \langle t_1, t_2\rangle$ of $\boldsymbol{R}^2$.

Remark The mapping $F: \boldsymbol{R}^2 \to \boldsymbol{R}^3$ can still be defined by Eq. (13) if the nonzero vectors $\mathbf{v}_1$ and $\mathbf{v}_2$ are collinear; that is, if both are parallel to a single vector $\mathbf{v}$. But in this case the image of F is the *line* (rather than a plane) through the origin that consists of all scalar multiples of $\mathbf{v}$.

If we write $F(\mathbf{t}) = \langle x, y, z\rangle$ and $\langle s, t\rangle$ (rather than $\langle t_1, t_2\rangle$) for a typical point of $\boldsymbol{R}^2$, then the "scalar components" of the vector equation in (14) are the scalar equations

$$\begin{aligned} x &= a_1 s + b_1 t, \\ y &= a_2 s + b_2 t, \\ z &= a_3 s + b_3 t. \end{aligned} \tag{15}$$

These are **parametric equations** of the plane $\mathcal{P}$ through the origin that contains the points (a_1, b_1, c_1) and (a_2, b_2, c_2)—the coordinates (x, y, z) of a typical point of $\mathcal{P}$ are given in terms of the *parameters* s and t. Similarly, the equations

$$\begin{aligned} x &= a_1 s + b_1 t + c_1, \\ y &= a_2 s + b_2 t + c_2, \\ z &= a_3 s + b_3 t + c_3 \end{aligned} \tag{16}$$

describe parametrically a plane in space that passes through the point (c_1, c_2, c_3) (and may or may not also pass through the origin).

It is possible for two different triples of parametric equations—as in (15) or (16)—to describe the same plane in space. In a particular example, the methods of this section can be used to determine whether this is the case.

EXAMPLE 7 Determine whether the parametric equations

$$\begin{aligned} x &= 2s - t - 2 & \qquad x &= s - 2t + 1 \\ y &= s + 2t - 1 \quad \text{and} & y &= s - t \\ z &= s + t - 1 & z &= 2s + t - 3 \end{aligned} \tag{17}$$

represent the same plane in space.

Solution Substituting $s = t = 0$, we see that the first plane $\mathcal{P}$ in (17) passes through the point $A(-2, -1, -1)$. It is "spanned" by the vectors $\mathbf{v}_1 = \langle 2, 1, 1\rangle$ and $\mathbf{v}_2 = \langle -1, 2, 1\rangle$, and therefore has normal vector

$$\mathbf{n} = \mathbf{v}_1 \times \mathbf{v}_2 = \begin{vmatrix} \mathbf{i} & \mathbf{j} & \mathbf{k} \\ 2 & 1 & 1 \\ -1 & 2 & 1 \end{vmatrix} = -\mathbf{i} - 3\mathbf{j} + 5\mathbf{k}.$$

The second plane $\mathcal{Q}$ in (17) passes through the point $B(1, 0, -3)$. It is spanned by the vector $\mathbf{u}_1 = \langle 1, 1, 2\rangle$ and $\mathbf{u}_2 = \langle -2, -1, 1\rangle$ and therefore has normal vector

$$\mathbf{m} = \mathbf{u}_1 \times \mathbf{u}_2 = \begin{vmatrix} \mathbf{i} & \mathbf{j} & \mathbf{k} \\ 1 & 1 & 2 \\ -2 & -1 & 1 \end{vmatrix} = 3\mathbf{i} - 5\mathbf{j} + \mathbf{k}.$$

Because neither of the normal vectors $\mathbf{n}$ and $\mathbf{m}$ is a scalar multiple of the other, they are not parallel. It follows that the planes $\mathcal{P}$ and $\mathcal{Q}$ are not parallel, and hence cannot be identical. ■

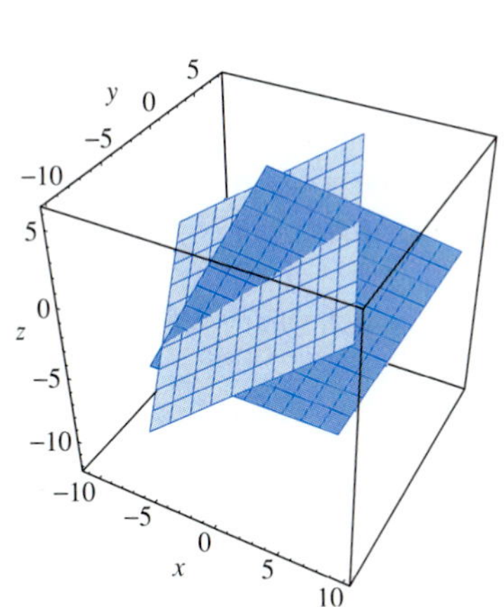

Fig. 12.4.9 Computer plot of the planes of Example 7

REMARK 1 Figure 12.4.9 corroborates the result of Example 7. It was plotted using typical computer algebra system syntax like the *Mathematica* command

```
ParametricPlot3D[{2s-t-2, s+2t-1, s+t-1}, {s,-3,3}, {t,-3,3}]
```

or the *Maple* command

```
plot3d([s-2*t+1, s-t, 2*s+t-3], s=-3..3, t=-3..3)
```

REMARK 2 If the two normal vectors **m** and **n** in Example 7 had turned out to be parallel, it would have remained for us to determine whether the two planes $\mathcal{P}$ and $\mathcal{Q}$ were identical or distinct (but parallel). For instance, we could settle this question by determining whether the point $A(-2, -1, -1)$ of $\mathcal{P}$ lies on $\mathcal{Q}$. If so, there must be values of s and t that satisfy the equations

$$s - 2t + 1 = -2, \qquad s - t = -1, \qquad 2s + t - 3 = -1.$$

(Why?) But if we solve the first two of these equations, we get the values $s = 1$ and $t = 2$, which do not satisfy the third equation. Therefore the point $A(-2, -1, -1)$ does *not* lie on $\mathcal{Q}$.

12.4 PROBLEMS

In Problems 1 through 4, write parametric equations of the straight line that passes through the point P and is parallel to the vector **v**.

1. $P(0, 0, 0)$, $\mathbf{v} = \mathbf{i} + 2\mathbf{j} + 3\mathbf{k}$

2. $P(3, -4, 5)$, $\mathbf{v} = -2\mathbf{i} + 7\mathbf{j} + 3\mathbf{k}$

3. $P(4, 13, -3)$, $\mathbf{v} = 2\mathbf{i} - 3\mathbf{k}$

4. $P(17, -13, -31)$, $\mathbf{v} = \langle -17, 13, 31 \rangle$

In Problems 5 through 8, write parametric equations of the straight line that passes through the points P_1 and P_2.

5. $P_1(0, 0, 0)$, $P_2(-6, 3, 5)$

6. $P_1(3, 5, 7)$, $P_2(6, -8, 10)$

7. $P_1(3, 5, 7)$, $P_2(6, 5, 4)$

8. $P_1(29, -47, 13)$, $P_2(73, 53, -67)$

In Problems 9 through 14, write both parametric and symmetric equations for the indicated straight line.

9. Through $P(2, 3, -4)$ and parallel to $\mathbf{v} = \langle 1, -1, -2 \rangle$

10. Through $P(2, 5, -7)$ and $Q(4, 3, 8)$

11. Through $P(1, 1, 1)$ and perpendicular to the xy-plane

12. Through the origin and perpendicular to the plane with equation $x + y + z = 1$

13. Through $P(2, -3, 4)$ and perpendicular to the plane with equation $2x - y + 3z = 4$

14. Through $P(2, -1, 5)$ and parallel to the line with parametric equations $x = 3t$, $y = 2 + t$, $z = 2 - t$

In Problems 15 through 20, determine whether the two lines L_1 and L_2 are parallel, skew, or intersecting. If they intersect, find the point of intersection.

15. L_1: $x - 2 = \frac{1}{2}(y + 1) = \frac{1}{3}(z - 3)$;
L_2: $\frac{1}{3}(x - 5) = \frac{1}{2}(y - 1) = z - 4$

16. L_1: $\frac{1}{4}(x - 11) = y - 6 = -\frac{1}{2}(z + 5)$;
L_2: $\frac{1}{6}(x - 13) = -\frac{1}{3}(y - 2) = \frac{1}{8}(z - 5)$

17. L_1: $x = 6 + 2t$, $y = 5 + 2t$, $z = 7 + 3t$;
L_2: $x = 7 + 3s$, $y = 5 + 3s$, $z = 10 + 5s$

18. L_1: $x = 14 + 3t$, $y = 7 + 2t$, $z = 21 + 5t$;
L_2: $x = 5 + 3s$, $y = 15 + 5s$, $z = 10 + 7s$

19. L_1: $\frac{1}{6}(x - 7) = \frac{1}{4}(y + 5) = -\frac{1}{8}(z - 9)$;
L_2: $-\frac{1}{9}(x - 11) = -\frac{1}{6}(y - 7) = \frac{1}{12}(z - 13)$

20. L_1: $x = 13 + 12t$, $y = -7 + 20t$, $z = 11 - 28t$;
L_2: $x = 22 + 9s$, $y = 8 + 15s$, $z = -10 - 21s$

In Problems 21 through 24, write an equation of the plane with normal vector **n** that passes through the point P.

21. $P(0, 0, 0)$, $\mathbf{n} = \langle 1, 2, 3 \rangle$

22. $P(3, -4, 5)$, $\mathbf{n} = \langle -2, 7, 3 \rangle$

23. $P(5, 12, 13)$, $\mathbf{n} = \mathbf{i} - \mathbf{k}$

24. $P(5, 12, 13)$, $\mathbf{n} = \mathbf{j}$

In Problems 25 through 32, write an equation of the indicated plane.

25. Through $P(5, 7, -6)$ and parallel to the xz-plane

26. Through $P(1, 0, -1)$ with normal vector $\mathbf{n} = \langle 2, 2, -1 \rangle$

27. Through $P(10, 4, -3)$ with normal vector $\mathbf{n} = \langle 7, 11, 0 \rangle$

28. Through $P(1, -3, 2)$ with normal vector $\mathbf{n} = \overrightarrow{OP}$

29. Through the origin and parallel to the plane with equation $3x + 4y = z + 10$

30. Through $P(5, 1, 4)$ and parallel to the plane with equation $x + y - 2z = 0$

31. Through the origin and the points $P(1, 1, 1)$ and $Q(1, -1, 3)$

32. Through the points $A(1, 0, -1)$, $B(3, 3, 2)$, and $C(4, 5, -1)$

In Problems 33 and 34, write an equation of the plane that contains both the point P and the line L.

33. $P(2, 4, 6)$; L: $x = 7 - 3t$, $y = 3 + 4t$, $z = 5 + 2t$

34. $P(13, -7, 29)$; L: $x = 17 - 9t$, $y = 23 + 14t$, $z = 35 - 41t$

In Problems 35 through 38, determine whether the line L and the plane $\mathcal{P}$ intersect or are parallel. If they intersect, find the point of intersection.

35. L: $x = 7 - 4t, y = 3 + 6t, z = 9 + 5t$;
$\mathcal{P}$: $4x + y + 2z = 17$

36. L: $x = 15 + 7t, y = 10 + 12t, z = 5 - 4t$;
$\mathcal{P}$: $12x - 5y + 6z = 50$

37. L: $x = 3 + 2t, y = 6 - 5t, z = 2 + 3t$;
$\mathcal{P}$: $3x + 2y - 4z = 1$

38. L: $x = 15 - 3t, y = 6 - 5t, z = 21 - 14t$;
$\mathcal{P}$: $23x + 29y - 31z = 99$

In Problems 39 through 42, find the angle between the planes with the given equations.

39. $x = 10$ and $x + y + z = 0$

40. $2x - y + z = 5$ and $x + y - z = 1$

41. $x - y - 2z = 1$ and $x - y - 2z = 5$

42. $2x + y + z = 4$ and $3x - y - z = 3$

In Problems 43 through 46, write both parametric and symmetric equations of the line of intersection of the indicated planes.

43. The planes of Problem 39

44. The planes of Problem 40

45. The planes of Problem 41

46. The planes of Problem 42

47. Write symmetric equations for the line through $P(3, 3, 1)$ that is parallel to the line of Problem 46.

48. Find an equation of the plane through $P(3, 3, 1)$ that is perpendicular to the planes $x + y = 2z$ and $2x + z = 10$.

49. Find an equation of the plane through $(1, 1, 1)$ that intersects the xy-plane in the same line as does the plane $3x + 2y - z = 6$.

50. Find an equation for the plane that passes through the point $P(1, 3, -2)$ and contains the line of intersection of the planes $x - y + z = 1$ and $x + y - z = 1$.

51. Find an equation of the plane that passes through the points $P(1, 0, -1)$ and $Q(2, 1, 0)$ and is parallel to the line of intersection of the planes $x + y + z = 5$ and $3x - y = 4$.

52. Prove that the lines $x - 1 = \frac{1}{2}(y + 1) = z - 2$ and $x - 2 = \frac{1}{3}(y - 2) = \frac{1}{2}(z - 4)$ intersect. Find an equation of the (only) plane that contains them both.

53. Prove that the line of intersection of the planes $x + 2y - z = 2$ and $3x + 2y + 2z = 7$ is parallel to the line $x = 1 + 6t, y = 3 - 5t, z = 2 - 4t$. Find an equation of the plane determined by these two lines.

54. Show that the perpendicular distance D from the point $P_0(x_0, y_0, z_0)$ to the plane $ax + by + cz = d$ is

$$D = \frac{|ax_0 + by_0 + cz_0 - d|}{\sqrt{a^2 + b^2 + c^2}}.$$

(*Suggestion:* The line that passes through P_0 and is perpendicular to the given plane has parametric equations $x = x_0 + at, y = y_0 + bt, z = z_0 + ct$. Let $P_1(x_1, y_1, z_1)$ be the point of this line, corresponding to $t = t_1$, at which it intersects the given plane. Solve for t_1, and then compute $D = |\overrightarrow{P_0P_1}|$.)

In Problems 55 and 56, use the formula of Problem 54 to find the distance between the given point and the given plane.

55. The origin and the plane $x + y + z = 10$

56. The point $P(5, 12, -13)$ and the plane with equation $3x + 4y + 5z = 12$

57. Prove that any two skew lines lie in parallel planes.

58. Use the formula of Problem 54 to show that the perpendicular distance D between the two parallel planes $ax + by + cz + d_1 = 0$ and $ax + by + cz + d_2 = 0$ is

$$D = \frac{|d_1 - d_2|}{\sqrt{a^2 + b^2 + c^2}}.$$

59. The line L_1 is described by the equations

$$x - 1 = 2y + 2, \quad z = 4.$$

The line L_2 passes through the points $P(2, 1, -3)$ and $Q(0, 8, 4)$. (a) Show that L_1 and L_2 are skew lines. (b) Use the results of Problems 57 and 58 to find the perpendicular distance between L_1 and L_2.

60. Find the shortest distance between points of the line L_1 with parametric equations

$$x = 7 + 2t, \quad y = 11 - 5t, \quad z = 13 + 6t$$

and the line L_2 of intersection of the planes $3x - 2y + 4z = 10$ and $5x + 3y - 2z = 15$.

In Problems 61 through 64, both the parametric equations of a line L and those of a plane $\mathcal{P}$ are given. Determine whether L and $\mathcal{P}$ are parallel, perpendicular, or neither. If the line and plane intersect, find both the point of intersection and the angle between the line L and the plane $\mathcal{P}$ (that is, the complement of the angle between L and a normal vector to $\mathcal{P}$).

61. L: $x = 11 + 3r, y = 17 - 9r, z = 5 + r$;
$\mathcal{P}$: $x = 5s + 4t, y = 7s - t, z = 3s + 2t$

62. L: $x = 16 + 13r, y = -18 - 23r, z = 23 + 4r$;
$\mathcal{P}$: $x = 9s + 3t, y = 7s + t, z = 11s - 4t$

63. L: $x = 12 + 5r, y = 11 + 10r, z = 21 + 20r$;
$\mathcal{P}$: $x = 2s + 5t, y = -s + 2t, z = 4s - 3t$

64. L: $x = 7 - 6r, y = 3 + 3r, z = 10 + 3r$;
$\mathcal{P}$: $x = 7s + 3t, y = 4s - 2t, z = -5s + 6t$

In Problems 65 through 68, the parametric equations of two planes $\mathcal{P}_1$ and $\mathcal{P}_2$ are given. Determine whether these two planes are equal, parallel, or neither. If $\mathcal{P}_1$ and $\mathcal{P}_2$ intersect, find the angle between them (that is, the angle between their normals) and write parametric equations of their line of intersection.

65. $\mathcal{P}_1$: $x = 5s + 7t, y = 3s + 8t, z = 2s + 5t$;
$\mathcal{P}_2$: $x = -1 + 12s + t, y = 7 + 11s + 12t, z = 4 + 7s + 7t$

66. $\mathcal{P}_1$: $x = 7 - s + t, y = 3 + 7s + 12t, z = 9 + 2s + 5t$;
$\mathcal{P}_2$: $x = 5 + 7t, y = 3 + 19s + 8t, z = 4 + 7s + 7t$

67. $\mathcal{P}_1$: $x = 7s + t, y = 8s + 12t, z = 7s + 5t$;
$\mathcal{P}_2$: $x = -3s + 4t, y = 7s + 4t, z = -2s + 11t$

68. $\mathcal{P}_1$: $x = 7 + 2s + 7t, y = 3 - s + 4t, z = 5 + 4s + 5t$;
$\mathcal{P}_2$: $x = 11 + 5s + 4t, y = 7 - s + 4t, z = 10 - 2s + 5t$

69. Show that the mapping in Eq. (13) defines a linear mapping; that is, show that Eq. (14) holds.

12.5 CURVES AND MOTION IN SPACE

In Section 10.4 we discussed parametric curves in the plane. Now think of a point that moves along a curve in three-dimensional space. We can describe this point's changing position by means of *parametric equations*

$$x = f(t), \quad y = g(t), \quad z = h(t) \tag{1}$$

that specify its coordinates as functions of time t. A **parametric curve** C in space is (by definition) simply a triple (f, g, h) of such *coordinate functions*. But often it is useful to refer informally to C as the trajectory in space that is traced out by a moving point with these coordinate functions. Space curves exhibit a number of interesting new phenomena that we did not see with plane curves.

Fig. 12.5.1 A tubular knot whose centerline is the parametric curve of Example 1

EXAMPLE 1 Figure 12.5.1 shows a common *trefoil knot* in space defined by the parametric equations

$$x(t) = (2 + \cos \tfrac{3}{2}t)\cos t, \quad y(t) = (2 + \cos \tfrac{3}{2}t)\sin t, \quad z(t) = \sin \tfrac{3}{2}t.$$

Actually, to enhance the three-dimensional appearance of this curve's shape, we have plotted in the figure a thin tubular surface whose centerline is the knot itself. The viewpoint for the computer plot is so chosen that we are looking down on the curve from a point on the positive z-axis. ■

EXAMPLE 2 Figure 12.5.2 shows simultaneously the circle

$$x(t) = 4\cos t, \quad y(t) = 4\sin t, \quad z(t) \equiv 0$$

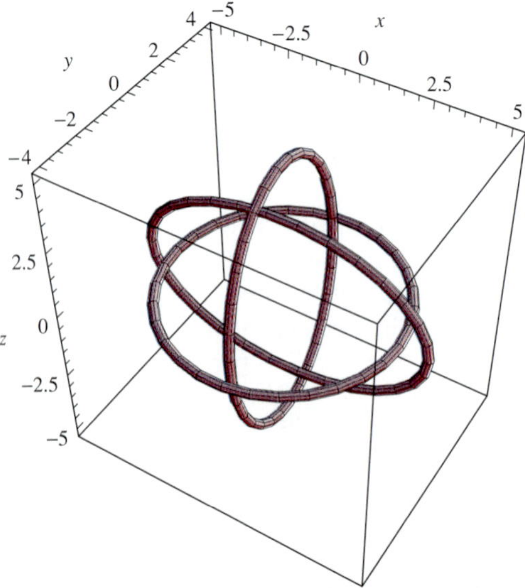

Fig. 12.5.2 The Borromean rings of Example 2

in the xy-plane, the ellipse

$$x(t) = 5\cos t, \quad y(t) \equiv 0, \quad z(t) = 3\sin t$$

in the xz-plane, and the ellipse

$$x(t) \equiv 0, \quad y(t) = 3\cos t, \quad z(t) = 5\sin t$$

in the yz-plane. Here, again, we actually have plotted thin tubular tori having these closed curves as centerlines. Can you see that any two of these curves are unlinked, but that the three together apparently cannot be "pulled apart"? ■

Vector-Valued Functions

The changing location of a point moving along the parametric curve in Eq. (1) can be described by giving its **position vector**

$$\mathbf{r}(t) = x(t)\mathbf{i} + y(t)\mathbf{j} + z(t)\mathbf{k} = \langle x(t), y(t), z(t)\rangle, \tag{2}$$

or simply

$$\mathbf{r} = x\mathbf{i} + y\mathbf{j} + z\mathbf{k} = \langle x, y, z\rangle,$$

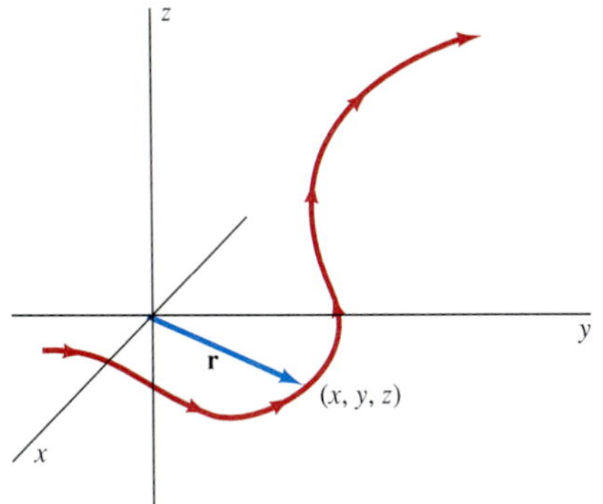

Fig. 12.5.3 The position vector $\mathbf{r} = \langle x, y, z\rangle$ of a moving particle in space

whose components are the coordinate functions of the moving point (Fig. 12.5.3). Equation (2) defines a **vector-valued function** that associates with the number t the vector $\mathbf{r}(t)$. In the case of a plane curve described by a two-dimensional position vector, we may suppress the third component in Eq. (2) and write $\mathbf{r}(t) = x(t)\mathbf{i} + y(t)\mathbf{j} = \langle x(t), y(t)\rangle$.

EXAMPLE 3 The position vector

$$\mathbf{r}(t) = \mathbf{i}\cos t + \mathbf{j}\sin t + t\mathbf{k} \tag{3}$$

describes the **helix** of Fig. 12.5.4. Because $x^2 + y^2 = \cos^2 t + \sin^2 t = 1$ for all t, the projection $\big(x(t), y(t)\big)$ into the xy-plane moves around and around the unit circle. Meanwhile, because $z = t$, the point $(\cos t, \sin t, t)$ steadily moves upward on the vertical cylinder in space that stands above and below the circle $x^2 + y^2 = 1$ in the xy-plane. The familiar corkscrew shape of the helix appears everywhere, from the coiled springs of an automobile to the *double helix* model of the DNA molecule that carries the genetic information of living cells (Fig. 12.5.5). ■

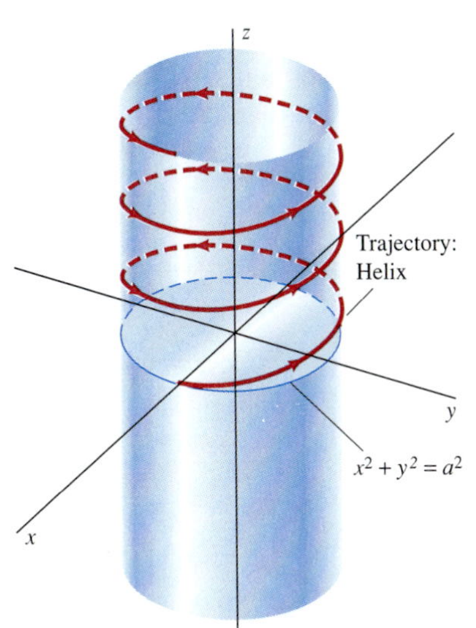

Fig. 12.5.4 The point of Example 3 moves in a helical path.

Much of the calculus of (ordinary) real-valued functions applies to vector-valued functions. To begin with, the **limit** of a vector-valued function $\mathbf{r} = \langle f, g, h\rangle$ is defined as follows:

$$\begin{aligned}\lim_{t\to a}\mathbf{r}(t) &= \langle \lim_{t\to a} f(t), \lim_{t\to a} g(t), \lim_{t\to a} h(t)\rangle \\ &= \mathbf{i}\big(\lim_{t\to a} f(t)\big) + \mathbf{j}\big(\lim_{t\to a} g(t)\big) + \mathbf{k}\big(\lim_{t\to a} h(t)\big),\end{aligned} \tag{4}$$

provided that the limits in the last three expressions exist. Thus we take limits of vector-valued functions by taking limits of their component functions.

We say that $\mathbf{r} = \mathbf{r}(t)$ is **continuous** at the number a provided that

$$\lim_{t\to a}\mathbf{r}(t) = \mathbf{r}(a).$$

This amounts to saying that $\mathbf{r}$ is continuous at a if and only if its component functions f, g, and h are continuous at a.

Fig. 12.5.5 The intertwined helices that model the DNA molecule served as a model for the DNA Tower in Kings Park, Perth, Australia. For a fascinating account of the discovery of the role of the helix as the genetic basis for life itself, see James D. Watson, *The Double Helix* (New York: Atheneum, 1968)

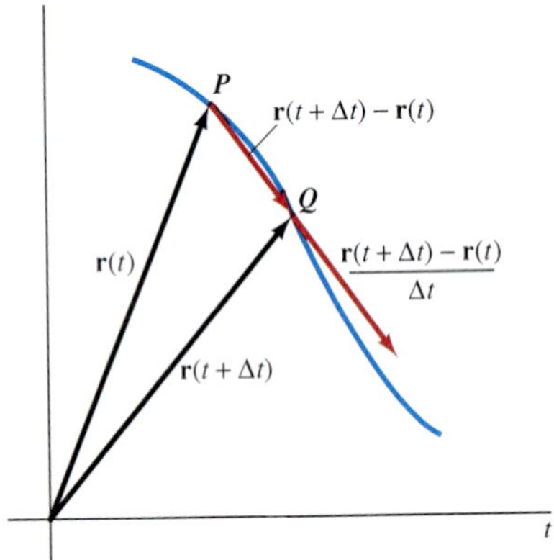

Fig. 12.5.6 Geometry of the derivative of a vector-valued function

The **derivative** $\mathbf{r}'(t)$ of the vector-valued function $\mathbf{r}(t)$ is defined in almost exactly the same way as the derivative of a real-valued function. Specifically,

$$\mathbf{r}'(t) = \lim_{\Delta t \to 0} \frac{\mathbf{r}(t + \Delta t) - \mathbf{r}(t)}{\Delta t}, \tag{5}$$

provided that this limit exists. Figures 12.5.6 and 12.5.7 correctly suggest that the **derivative vector**

$$\mathbf{r}'(t) = \frac{d\mathbf{r}}{dt} = D_t[\mathbf{r}(t)]$$

will be tangent to the curve C with position vector $\mathbf{r}(t)$. For this reason, we call $\mathbf{r}'(t)$ a **tangent vector** to the curve C at the point $\mathbf{r}(t)$ provided that $\mathbf{r}'(t)$ exists and is nonzero there. The **tangent line** to C at this point P with position vector $\mathbf{r}(t)$ is then the line through P determined by $\mathbf{r}'(t)$.

Our next result implies the simple *but important* fact that the derivative vector $\mathbf{r}'(t)$ can be calculated by **componentwise differentiation** of $\mathbf{r}(t)$—that is, by differentiating separately the component functions of $\mathbf{r}(t)$.

Theorem 1 *Componentwise Differentiation*

Suppose that

$$\mathbf{r}(t) = \langle f(t), g(t), h(t) \rangle = f(t)\mathbf{i} + g(t)\mathbf{j} + h(t)\mathbf{k},$$

where f, g, and h are differentiable functions. Then

$$\mathbf{r}'(t) = \langle f'(t), g'(t), h'(t) \rangle = f'(t)\mathbf{i} + g'(t)\mathbf{j} + h'(t)\mathbf{k}. \tag{6}$$

That is, if $\mathbf{r} = x\mathbf{i} + y\mathbf{j} + z\mathbf{k}$, then

$$\frac{d\mathbf{r}}{dt} = \frac{dx}{dt}\mathbf{i} + \frac{dy}{dt}\mathbf{j} + \frac{dz}{dt}\mathbf{k}.$$

Proof We take the limit in Eq. (5) simply by taking limits of components. We find that

$$\begin{aligned}
\mathbf{r}'(t) &= \lim_{\Delta t \to 0} \frac{\Delta \mathbf{r}}{\Delta t} = \lim_{\Delta t \to 0} \frac{\mathbf{r}(t + \Delta t) - \mathbf{r}(t)}{\Delta t} \\
&= \lim_{\Delta t \to 0} \frac{f(t + \Delta t)\mathbf{i} + g(t + \Delta t)\mathbf{j} + h(t + \Delta t)\mathbf{k} - f(t)\mathbf{i} - g(t)\mathbf{j} - h(t)\mathbf{k}}{\Delta t} \\
&= \left(\lim_{\Delta t \to 0} \frac{f(t + \Delta t) - f(t)}{\Delta t} \right)\mathbf{i} + \left(\lim_{\Delta t \to 0} \frac{g(t + \Delta t) - g(t)}{\Delta t} \right)\mathbf{j} \\
&\qquad + \left(\lim_{\Delta t \to 0} \frac{h(t + \Delta t) - h(t)}{\Delta t} \right)\mathbf{k} \\
&= f'(t)\mathbf{i} + g'(t)\mathbf{j} + h'(t)\mathbf{k}.
\end{aligned}$$

■

EXAMPLE 4 Find parametric equations of the line tangent to the helix C of Example 3 at the point $P(-1, 0, \pi)$ where $t = \pi$.

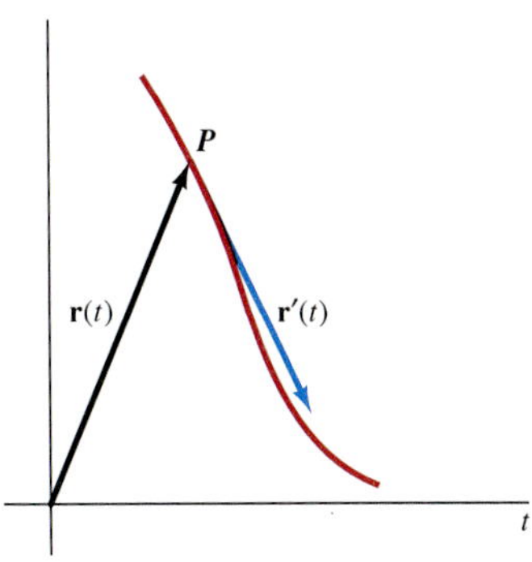

Fig. 12.5.7 The derivative vector is tangent to the curve at the point of evaluation.

Solution Componentwise differentiation of $\mathbf{r}(t) = \mathbf{i}\cos t + \mathbf{j}\sin t + t\mathbf{k}$ yields

$$\mathbf{r}'(t) = -\mathbf{i}\sin t + \mathbf{j}\cos t + \mathbf{k},$$

so the vector tangent to C at P is $\mathbf{v}(\pi) = \mathbf{r}'(\pi) = -\mathbf{j} + \mathbf{k} = \langle 0, -1, 1\rangle$. It follows that the parametric equations of the line tangent at P—with its own position vector $\mathbf{r}(\pi) + t\mathbf{r}'(\pi)$—are

$$x = -1, \quad y = -t, \quad z = \pi + t.$$

In particular, we see that this tangent line lies in the vertical plane $x = -1$. ■

Theorem 2 tells us that the formulas for computing derivatives of sums and products of vector-valued functions are formally similar to those for real-valued functions.

Theorem 2 ***Differentiation Formulas***

Let $\mathbf{u}(t)$ and $\mathbf{v}(t)$ be differentiable vector-valued functions. Let $h(t)$ be a differentiable real-valued function and let c be a (constant) scalar. Then

1. $D_t[\mathbf{u}(t) + \mathbf{v}(t)] = \mathbf{u}'(t) + \mathbf{v}'(t)$,
2. $D_t[c\mathbf{u}(t)] = c\mathbf{u}'(t)$,
3. $D_t[h(t)\mathbf{u}(t)] = h'(t)\mathbf{u}(t) + h(t)\mathbf{u}'(t)$,
4. $D_t[\mathbf{u}(t)\cdot\mathbf{v}(t)] = \mathbf{u}'(t)\cdot\mathbf{v}(t) + \mathbf{u}(t)\cdot\mathbf{v}'(t)$, and
5. $D_t[\mathbf{u}(t) \times \mathbf{v}(t)] = \mathbf{u}'(t) \times \mathbf{v}(t) + \mathbf{u}(t) \times \mathbf{v}'(t)$.

PROOF We'll prove part (4), working with two-dimensional vectors for simplicity, and leave the other parts as exercises. If

$$\mathbf{u}(t) = \langle f_1(t), f_2(t)\rangle \quad \text{and} \quad \mathbf{v}(t) = \langle g_1(t), g_2(t)\rangle,$$

then

$$\mathbf{u}(t)\cdot\mathbf{v}(t) = f_1(t)g_1(t) + f_2(t)g_2(t).$$

Hence the product rule for ordinary real-valued functions gives

$$\begin{aligned} D_t[\mathbf{u}(t)\cdot\mathbf{v}(t)] &= D_t[f_1(t)g_1(t) + f_2(t)g_2(t)] \\ &= [f_1'(t)g_1(t) + f_2'(t)g_2(t)] + [f_1(t)g_1'(t) + f_2(t)g_2'(t)] \\ &= \mathbf{u}'(t)\cdot\mathbf{v}(t) + \mathbf{u}(t)\cdot\mathbf{v}'(t). \end{aligned}$$ ■

REMARK The order of the factors in part (5) of Theorem 2 *must* be preserved because the cross product is not commutative.

EXAMPLE 5 The trajectory of the parametric curve $\mathbf{r}(t) = a\mathbf{i}\cos t + a\mathbf{j}\sin t$ is the circle of radius a centered at the origin in the xy-plane. Because $\mathbf{r}(t)\cdot\mathbf{r}(t) = a^2$, a constant, part 4 of Theorem 2 gives

$$0 \equiv \frac{d}{dt}(a^2) = \frac{d}{dt}[\mathbf{r}(t)\cdot\mathbf{r}(t)] = \mathbf{r}'(t)\cdot\mathbf{r}(t) + \mathbf{r}(t)\cdot\mathbf{r}'(t) = 2\mathbf{r}'(t)\cdot\mathbf{r}(t).$$

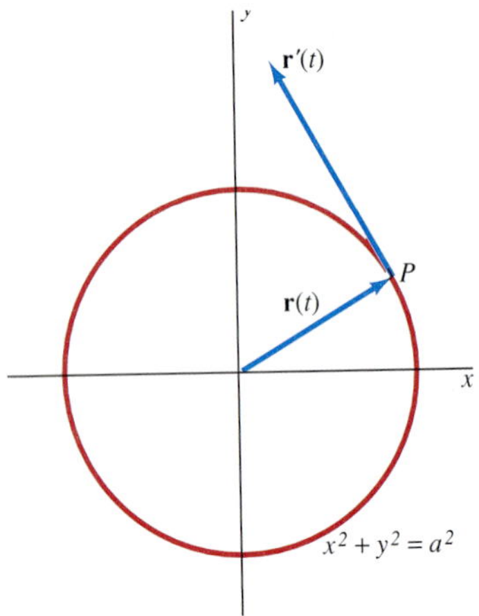

Fig. 12.5.8 The position and tangent vectors for the circle of Example 5

Because $\mathbf{r}'(t) \cdot \mathbf{r}(t) \equiv 0$, we see that (consistent with elementary geometry) the tangent vector $\mathbf{r}'(t)$ is perpendicular to the position vector $\mathbf{r}(t)$ at every point of the circle (Fig. 12.5.8). ■

Velocity and Acceleration Vectors

Looking at Fig. 12.5.6 and the definition of $\mathbf{r}'(t)$ in Eq. (5), we note that $|\mathbf{r}(t + \Delta t) - \mathbf{r}(t)|$ is the distance from the point with position vector $\mathbf{r}(t)$ to the point with position vector $\mathbf{r}(t + \Delta t)$. It follows that the quotient

$$\frac{|\mathbf{r}(t + \Delta t) - \mathbf{r}(t)|}{\Delta t}$$

is the average speed of a particle that travels from $\mathbf{r}(t)$ to $\mathbf{r}(t + \Delta t)$ in time Δt. Consequently, the limit in Eq. (5) yields both the direction of motion and the instantaneous speed of a particle moving along a curve with position vector $\mathbf{r}(t)$.

We therefore define the **velocity vector** $\mathbf{v}(t)$ at time t of a point moving along a curve with position vector $\mathbf{r}(t)$ as the derivative

$$\mathbf{v}(t) = \mathbf{r}'(t) = f'(t)\mathbf{i} + g'(t)\mathbf{j} + h'(t)\mathbf{k}; \tag{7a}$$

in differential notation,

$$\mathbf{v} = \frac{d\mathbf{r}}{dt} = \frac{dx}{dt}\mathbf{i} + \frac{dy}{dt}\mathbf{j} + \frac{dz}{dt}\mathbf{k}. \tag{7b}$$

Its **acceleration vector a** $= \mathbf{a}(t)$ is given by

$$\mathbf{a}(t) = \mathbf{v}'(t) = f''(t)\mathbf{i} + g''(t)\mathbf{j} + h''(t)\mathbf{k}; \tag{8a}$$

alternatively,

$$\mathbf{a} = \frac{d\mathbf{v}}{dt} = \frac{d^2x}{dt^2}\mathbf{i} + \frac{d^2y}{dt^2}\mathbf{j} + \frac{d^2z}{dt^2}\mathbf{k}. \tag{8b}$$

Thus, for motion in the plane or in space, just as for motion along a line,

> **velocity** is the **time derivative** of **position**;
> **acceleration** is the **time derivative** of **velocity**.

The **speed** $v(t)$ and **scalar acceleration** $a(t)$ of the moving point are the lengths of its velocity and acceleration vectors, respectively:

$$v(t) = |\mathbf{v}(t)| = \sqrt{\left(\frac{dx}{dt}\right)^2 + \left(\frac{dy}{dt}\right)^2 + \left(\frac{dz}{dt}\right)^2} \tag{9}$$

and

$$a(t) = |\mathbf{a}(t)| = \sqrt{\left(\frac{d^2x}{dt^2}\right)^2 + \left(\frac{d^2y}{dt^2}\right)^2 + \left(\frac{d^2z}{dt^2}\right)^2}. \tag{10}$$

Note The scalar acceleration $a = |d\mathbf{v}/dt|$ is generally *not* equal to the derivative dv/dt of the speed of a moving point. The difference between the two is discussed in Section 12.6.

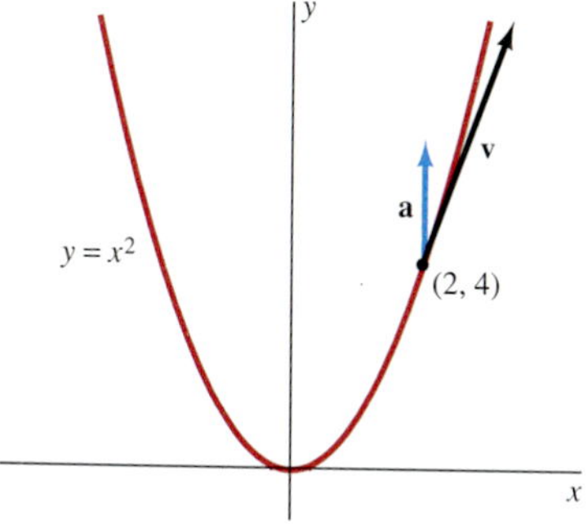

Fig. 12.5.9 The velocity and acceleration vectors at $t = 2$ (Example 6)

EXAMPLE 6 A particle moving along the parabola $y = x^2$ in the plane has position vector $\mathbf{r}(t) = t\mathbf{i} + t^2\mathbf{j}$. Find its velocity and acceleration vectors and its speed and scalar acceleration at the instant when $t = 2$.

Solution Because $\mathbf{r}(2) = 2\mathbf{i} + 4\mathbf{j}$, the location of the particle at time $t = 2$ is $(2, 4)$. Its velocity vector and speed are given by

$$\mathbf{v} = \mathbf{i} + 2t\mathbf{j} \quad \text{and} \quad v(t) = |\mathbf{v}(t)| = \sqrt{1 + 4t^2},$$

so $\mathbf{v}(2) = \mathbf{i} + 4\mathbf{j}$ (a vector) and $v(2) = \sqrt{17}$ (a scalar). Its acceleration is $\mathbf{a}(t) = \mathbf{v}'(t) = 2\mathbf{j}$ (a constant vector), so $\mathbf{a} = 2\mathbf{j}$ and $a = |\mathbf{a}| = 2$ (scalar acceleration) for all t, including the instant at which $t = 2$. Figure 12.5.9 shows the trajectory of the particle with its velocity and acceleration vectors $\mathbf{v}(2)$ and $\mathbf{a}(2)$ attached at its location $(2, 4)$ when $t = 2$. ■

EXAMPLE 7 Find the velocity, acceleration, speed, and scalar acceleration of a moving point P whose trajectory is the helix with position vector

$$\mathbf{r}(t) = (a\cos\omega t)\mathbf{i} + (a\sin\omega t)\mathbf{j} + bt\mathbf{k}. \tag{11}$$

Solution Equation (11) is a generalization of the position vector $\mathbf{r}(t) = \mathbf{i}\cos t + \mathbf{j}\sin t + t\mathbf{k}$ of the helix in Example 3. Here $x^2 + y^2 = a^2$, so the xy-projection $(a\cos\omega t, a\sin\omega t)$ of P lies on the circle of radius a centered at the origin. This projection moves around the circle with angular speed ω (radians per unit time). Meanwhile, the point P itself also is moving upward (if $b > 0$) on the vertical cylinder of radius a; the z-component of its velocity is $dz/dt = b$. Except for the radius of the cylinder, the picture looks the same as Fig. 12.5.4.

The derivative of the position vector in Eq. (11) is the velocity vector

$$\mathbf{v}(t) = (-a\omega\sin\omega t)\mathbf{i} + (a\omega\cos\omega t)\mathbf{j} + b\mathbf{k}. \tag{12}$$

Another differentiation gives its acceleration vector

$$\begin{aligned}\mathbf{a}(t) &= (-a\omega^2\cos\omega t)\mathbf{i} + (-a\omega^2\sin\omega t)\mathbf{j}\\ &= -a\omega^2(\mathbf{i}\cos\omega t + \mathbf{j}\sin\omega t).\end{aligned} \tag{13}$$

The speed of the moving point is a constant, because

$$v(t) = |\mathbf{v}(t)| = \sqrt{a^2\omega^2 + b^2}.$$

Note that the acceleration vector is a horizontal vector of length $a\omega^2$. Moreover, if we think of $\mathbf{a}(t)$ as attached to the moving point at the time t of evaluation—so that the initial point of $\mathbf{a}(t)$ is the terminal point of $\mathbf{r}(t)$—then $\mathbf{a}(t)$ points directly toward the point $(0, 0, bt)$ on the z-axis. ■

REMARK The helix of Example 7 is a typical trajectory of a charged particle in a constant magnetic field. Such a particle must satisfy both Newton's law $\mathbf{F} = m\mathbf{a}$ and the magnetic force law $\mathbf{F} = (q\mathbf{v}) \times \mathbf{B}$ mentioned in Section 12.3. Hence its velocity and acceleration vectors must satisfy the equation

$$(q\mathbf{v}) \times \mathbf{B} = m\mathbf{a}. \tag{14}$$

If the constant magnetic field is vertical, $\mathbf{B} = B\mathbf{k}$, then with the velocity vector of Eq. (12) we find that

$$q\mathbf{v} \times \mathbf{B} = q\begin{vmatrix} \mathbf{i} & \mathbf{j} & \mathbf{k} \\ -a\omega \sin \omega t & a\omega \cos \omega t & b \\ 0 & 0 & B \end{vmatrix} = qa\omega B\,(\mathbf{i}\cos \omega t + \mathbf{j}\sin \omega t).$$

The acceleration vector in Eq. (13) gives

$$m\mathbf{a} = -ma\omega^2(\mathbf{i}\cos\omega t + \mathbf{j}\sin\omega t).$$

When we compare the last two results, we see that the helix of Example 7 satisfies Eq. (14) provided that

$$qa\omega B = -ma\omega^2; \qquad \text{that is,} \quad \omega = -\frac{qB}{m}.$$

For example, this equation would determine the angular speed ω for the helical trajectory of electrons $(q < 0)$ in a cathode-ray tube placed in a constant magnetic field parallel to the axis of the tube (Fig. 12.5.10).

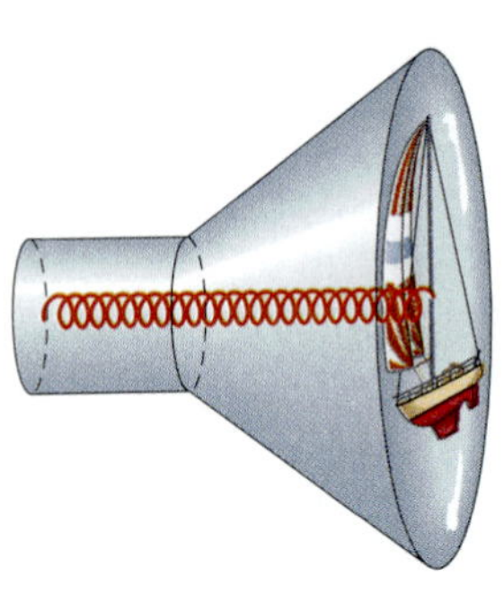

Fig. 12.5.10 A spiraling electron in a cathode-ray tube

Integration of Vector-Valued Functions

Integrals of vector-valued functions are defined by analogy with the definition of an integral of a real-valued function:

$$\int_a^b \mathbf{r}(t)\,dt = \lim_{\Delta t \to 0} \sum_{i=1}^{n} \mathbf{r}(t_i^\star)\,\Delta t,$$

where $t_i^\star$ is a point of the ith subinterval of a partition of $[a, b]$ into n subintervals, all with the same length $\Delta t = (b - a)/n$.

If $\mathbf{r}(t) = f(t)\mathbf{i} + g(t)\mathbf{j}$ is continuous on $[a, b]$, then—by taking limits componentwise—we get

$$\begin{aligned} \int_a^b \mathbf{r}(t)\,dt &= \lim_{\Delta t \to 0} \sum_{i=1}^{n} \mathbf{r}(t_i^\star)\,\Delta t \\ &= \mathbf{i}\left(\lim_{\Delta t \to 0} \sum_{i=1}^{n} f(t_i^\star)\,\Delta t\right) + \mathbf{j}\left(\lim_{\Delta t \to 0} \sum_{i=1}^{n} g(t_i^\star)\,\Delta t\right). \end{aligned}$$

This gives the result that

$$\int_a^b \mathbf{r}(t)\,dt = \mathbf{i}\left(\int_a^b f(t)\,dt\right) + \mathbf{j}\left(\int_a^b g(t)\,dt\right). \qquad \textbf{(15)}$$

Thus *a vector-valued function may be integrated componentwise.* The three-dimensional version of Eq. (15) is derived in the same way, merely including third components.

Now suppose that $\mathbf{R}(t)$ is an *antiderivative* of $\mathbf{r}(t)$, meaning that $\mathbf{R}'(t) = \mathbf{r}(t)$. That is, if $\mathbf{R}(t) = F(t)\mathbf{i} + G(t)\mathbf{j}$, then

$$\mathbf{R}'(t) = F'(t)\mathbf{i} + G'(t)\mathbf{j} = f(t)\mathbf{i} + g(t)\mathbf{j} = \mathbf{r}(t).$$

Then componentwise integration yields

$$\begin{aligned} \int_a^b \mathbf{r}(t)\,dt &= \mathbf{i}\left(\int_a^b f(t)\,dt\right) + \mathbf{j}\left(\int_a^b g(t)\,dt\right) = \mathbf{i}\Big[F(t)\Big]_a^b + \mathbf{j}\Big[G(t)\Big]_a^b \\ &= [F(b)\mathbf{i} + G(b)\mathbf{j}] - [F(a)\mathbf{i} + G(a)\mathbf{j}]. \end{aligned}$$

Thus the *fundamental theorem of calculus* for vector-valued functions takes the form

$$\int_a^b \mathbf{r}(t)\,dt = \Big[\mathbf{R}(t)\Big]_a^b = \mathbf{R}(b) - \mathbf{R}(a), \tag{16}$$

where $\mathbf{R}'(t) = \mathbf{r}(t)$.

Indefinite integrals of vector-valued functions may be computed as well. If $\mathbf{R}'(t) = \mathbf{r}(t)$, then every antiderivative of $\mathbf{r}(t)$ is of the form $\mathbf{R}(t) + \mathbf{C}$ for some constant vector $\mathbf{C}$. We therefore write

$$\int \mathbf{r}(t)\,dt = \mathbf{R}(t) + \mathbf{C} \quad \text{if} \quad \mathbf{R}'(t) = \mathbf{r}(t), \tag{17}$$

on the basis of a componentwise computation similar to the one leading to Eq. (16).

If $\mathbf{r}(t)$, $\mathbf{v}(t)$, and $\mathbf{a}(t)$ are the position, velocity, and acceleration vectors of a point moving in space, then the vector derivatives

$$\frac{d\mathbf{r}}{dt} = \mathbf{v} \quad \text{and} \quad \frac{d\mathbf{v}}{dt} = \mathbf{a}$$

imply the indefinite integrals

$$\mathbf{v}(t) = \int \mathbf{a}(t)\,dt \tag{18}$$

and

$$\mathbf{r}(t) = \int \mathbf{v}(t)\,dt. \tag{19}$$

Both of these integrals involve a *vector* constant of integration.

EXAMPLE 8 Suppose that a moving point has given initial position vector $\mathbf{r}(0) = 2\mathbf{i}$, initial velocity vector $\mathbf{v}(0) = \mathbf{i} - \mathbf{j}$, and acceleration vector $\mathbf{a}(t) = 2\mathbf{i} + 6t\mathbf{j}$. Find its position and velocity at time t.

Solution Equation (18) gives

$$\mathbf{v}(t) = \int \mathbf{a}(t)\,dt = \int (2\mathbf{i} + 6t\mathbf{j})\,dt = 2t\mathbf{i} + 3t^2\mathbf{j} + \mathbf{C}_1.$$

To evaluate the constant vector $\mathbf{C}_1$, we substitute $t = 0$ in this equation and find that $\mathbf{v}(0) = (0)\mathbf{i} + (0)\mathbf{j} + \mathbf{C}_1$, so $\mathbf{C}_1 = \mathbf{v}(0) = \mathbf{i} - \mathbf{j}$. Thus the velocity vector of the moving point at time t is

$$\mathbf{v}(t) = (2t\mathbf{i} + 3t^2\mathbf{j}) + (\mathbf{i} + \mathbf{j}) = (2t + 1)\mathbf{i} + (3t^2 - 1)\mathbf{j}.$$

A second integration, using Eq. (19), gives

$$\begin{aligned}\mathbf{r}(t) &= \int \mathbf{v}(t)\,dt \\ &= \int [(2t + 1)\mathbf{i} + (3t^2 - 1)\mathbf{j}]\,dt = (t^2 + t)\mathbf{i} + (t^3 - t)\mathbf{j} + \mathbf{C}_2.\end{aligned}$$

Again we substitute $t = 0$ and find that $\mathbf{C}_2 = \mathbf{r}(0) = 2\mathbf{i}$. Hence

$$\mathbf{r}(t) = (t^2 + t)\mathbf{i} + (t^3 - t)\mathbf{j} + 2\mathbf{i} = (t^2 + t + 2)\mathbf{i} + (t^3 - t)\mathbf{j}$$

is the position vector of the point at time t. ■

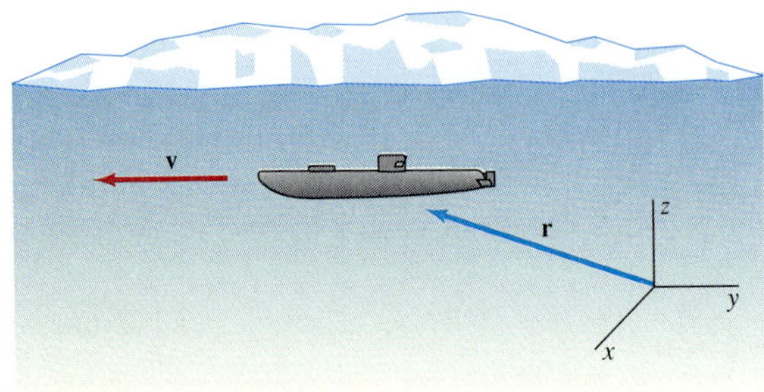

Fig. 12.5.11 A submarine beneath the polar icecap

Vector integration is the basis for at least one method of navigation. If a submarine is cruising beneath the icecap at the North Pole, as in Fig. 12.5.11, and thus can use neither visual nor radio methods to determine its position, there is an alternative. Build a sensitive gyroscope-accelerometer combination and install it in the submarine. The device continuously measures the sub's acceleration vector, beginning at the time $t = 0$ when its position $\mathbf{r}(0)$ and velocity $\mathbf{v}(0)$ are known. Because $\mathbf{v}'(t) = \mathbf{a}(t)$, Eq. (16) gives

$$\int_0^t \mathbf{a}(t)\,dt = \Big[\mathbf{v}(t)\Big]_0^t = \mathbf{v}(t) - \mathbf{v}(0),$$

so

$$\mathbf{v}(t) = \mathbf{v}(0) + \int_0^t \mathbf{a}(t)\,dt.$$

Thus the velocity at every time $t \geqq 0$ is known. Similarly, because $\mathbf{r}'(t) = \mathbf{v}(t)$, a second integration gives

$$\mathbf{r}(t) = \mathbf{r}(0) + \int_0^t \mathbf{v}(t)\,dt$$

for the position of the sub at every time $t \geqq 0$. On-board computers can be programmed to carry out these integrations (perhaps by using Simpson's approximation) and continuously provide captain and crew with the submarine's (almost) exact position and velocity.

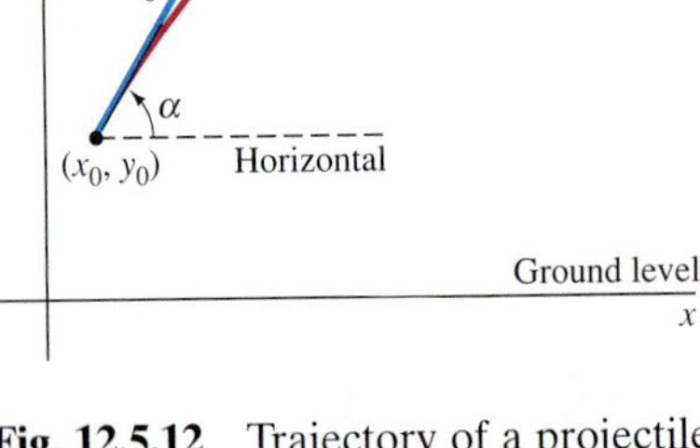

Fig. 12.5.12 Trajectory of a projectile launched at the angle α

Motion of Projectiles

Suppose that a projectile is launched from the point (x_0, y_0), with y_0 denoting its initial height above the surface of the earth. Let α be the angle of inclination from the horizontal of its initial velocity vector $\mathbf{v}_0$ (Fig. 12.5.12). Then its initial position vector is

$$\mathbf{r}_0 = x_0\mathbf{i} + y_0\mathbf{j}, \tag{20a}$$

and from Fig. 12.5.12 we see that

$$\mathbf{v}_0 = (v_0 \cos\alpha)\mathbf{i} + (v_0 \sin\alpha)\mathbf{j}, \tag{20b}$$

where $v_0 = |\mathbf{v}_0|$ is the initial speed of the projectile.

We suppose that the motion takes place sufficiently close to the surface that we may assume that the earth is flat and that gravity is perfectly uniform. Then, if we also ignore air resistance, the acceleration of the projectile is

$$\mathbf{a} = \frac{d\mathbf{v}}{dt} = -g\mathbf{j},$$

where $g = 32\ \text{ft/s}^2 \approx 9.8\ \text{m/s}^2$. Antidifferentiation gives

$$\mathbf{v}(t) = -gt\mathbf{j} + \mathbf{C}_1.$$

Put $t = 0$ in both sides of this last equation. This shows that $\mathbf{C}_0 = \mathbf{v}_0$ (as expected!) and thus that

$$\mathbf{v}(t) = \frac{d\mathbf{r}}{dt} = -gt\mathbf{j} + \mathbf{v}_0.$$

Another antidifferentiation gives

$$\mathbf{r}(t) = -\tfrac{1}{2}gt^2\mathbf{j} + \mathbf{v}_0 t + \mathbf{C}_2.$$

Now substitution of $t = 0$ yields $\mathbf{C}_2 = \mathbf{r}_0$, so the position vector of the projectile at time t is

$$\mathbf{r}(t) = -\tfrac{1}{2}gt^2\mathbf{j} + \mathbf{v}_0 t + \mathbf{r}_0. \tag{21}$$

Equations (20a) and (20b) now give

$$\mathbf{r}(t) = \left[(v_0 \cos\alpha)t + x_0\right]\mathbf{i} + \left[-\tfrac{1}{2}gt^2 + (v_0 \sin\alpha)t + y_0\right]\mathbf{j},$$

so parametric equations of the trajectory of the particle are

$$x(t) = (v_0 \cos\alpha)t + x_0, \tag{22}$$

$$y(t) = -\tfrac{1}{2}gt^2 + (v_0 \sin\alpha)t + y_0. \tag{23}$$

EXAMPLE 9 An airplane is flying horizontally at an altitude of 1600 ft to pass directly over snowbound cattle on the ground and release hay to land there. The plane's speed is a constant 150 mi/h (220 ft/s). At what angle of sight ϕ (between the horizontal and the direct line to the target) should a bale of hay be released in order to hit the target?

Solution See Fig. 12.5.13. We take $x_0 = 0$ where the bale of hay is released at time $t = 0$. Then $y_0 = 1600$ (ft), $v_0 = 220$ (ft/s), and $\alpha = 0$. Then Eqs. (22) and (23) take the forms

$$x(t) = 220t, \qquad y(t) = -16t^2 + 1600.$$

From the second of these equations, we find that $t = 10$ (s) when the bale of hay hits the ground ($y = 0$). It has then traveled a horizontal distance

$$x(10) = 220 \cdot 10 = 2200 \quad \text{(ft)}.$$

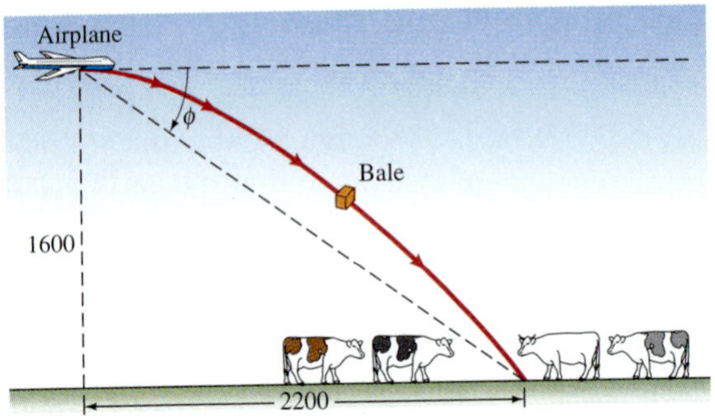

Fig. 12.5.13 Trajectory of the hay bale of Example 9

Hence the required angle of sight is

$$\phi = \tan^{-1}\left(\frac{1600}{2200}\right) \approx 36°.$$

EXAMPLE 10 A ball is thrown northward into the air from the origin in xyz-space (the xy-plane represents the ground and the positive y-axis points north). The initial velocity (vector) of the ball is

$$\mathbf{v}_0 = \mathbf{v}(0) = 80\mathbf{j} + 80\mathbf{k}.$$

The spin of the ball causes an eastward acceleration of 2 ft/s^2 in addition to gravitational acceleration. Thus the acceleration vector produced by the combination of gravity and spin is

$$\mathbf{a}(t) = 2\mathbf{i} - 32\mathbf{k}.$$

First find the velocity vector $\mathbf{v}(t)$ of the ball and its position vector $\mathbf{r}(t)$. Then determine where and with what speed the ball hits the ground (Fig. 12.5.14).

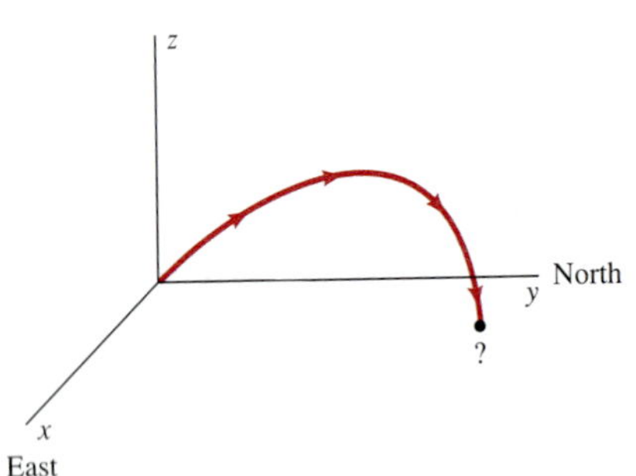

Fig. 12.5.14 The trajectory of the ball of Example 10

Solution When we antidifferentiate $\mathbf{a}(t)$, we get

$$\mathbf{v}(t) = \int \mathbf{a}(t)\,dt = \int (2\mathbf{i} - 32\mathbf{k})\,dt = 2t\mathbf{i} - 32t\mathbf{k} + \mathbf{c}_1.$$

We substitute $t = 0$ to find that $\mathbf{c}_1 = \mathbf{v}_0 = 80\mathbf{j} + 80\mathbf{k}$, so

$$\mathbf{v}(t) = 2t\mathbf{i} + 80\mathbf{j} + (80 - 32t)\mathbf{k}.$$

Another antidifferentiation yields

$$\mathbf{r}(t) = \int \mathbf{v}(t)\,dt = \int [2t\mathbf{i} + 80\mathbf{j} + (80 - 32t)\mathbf{k}]\,dt$$
$$= t^2\mathbf{i} + 80t\mathbf{j} + (80t - 16t^2)\mathbf{k} + \mathbf{c}_2,$$

and substitution of $t = 0$ gives $\mathbf{c}_2 = \mathbf{r}(0) = \mathbf{0}$. Hence the position vector of the ball is

$$\mathbf{r}(t) = t^2\mathbf{i} + 80t\mathbf{j} + (80t - 16t^2)\mathbf{k}.$$

The ball hits the ground when $z = 80t - 16t^2 = 0$; that is, when $t = 5$. Its position vector then is

$$\mathbf{r}(5) = 5^2\mathbf{i} + 80\cdot 5\mathbf{j} = 25\mathbf{i} + 400\mathbf{j},$$

so the ball has traveled 25 ft eastward and 400 ft northward. Its velocity vector at impact is

$$\mathbf{v}(5) = 2 \cdot 5\mathbf{i} + 80\mathbf{j} + (80 - 32 \cdot 5)\mathbf{k} = 10\mathbf{i} + 80\mathbf{j} - 80\mathbf{k},$$

so its speed when it hits the ground is

$$v(5) = |\mathbf{v}(5)| = \sqrt{10^2 + 80^2 + (-80)^2},$$

approximately 113.58 ft/s. Because the ball started with initial speed $v_0 = \sqrt{80^2 + 80^2} \approx 113.14$ ft/s, its eastward acceleration has slightly increased its terminal speed. ■

12.5 PROBLEMS

In Problems 1 through 4, also match the curves there defined with their three-dimensional plots in Figs. 12.5.15 through 12.5.18.

1. Show that the graph of the curve with parametric equations $x = t$, $y = \sin 5t$, $z = \cos 5t$ lies on the circular cylinder $y^2 + z^2 = 1$ centered along the x-axis.

2. Show that the graph of the curve with parametric equations $x = \sin t$, $y = \cos t$, $z = \cos 8t$ lies on the vertical circular cylinder $x^2 + y^2 = 1$.

3. Show that the graph of the curve with parametric equations $x = t \sin 6t$, $y = t \cos 6t$, $z = t$ lies on the cone $z = \sqrt{x^2 + y^2}$ with its vertex at the origin and opening upward.

4. Show that the graph of the curve with parametric equations $x = \cos t \sin 4t$, $y = \sin t \sin 4t$, $z = \cos 4t$ lies on the surface of the sphere $x^2 + y^2 + z^2 = 1$.

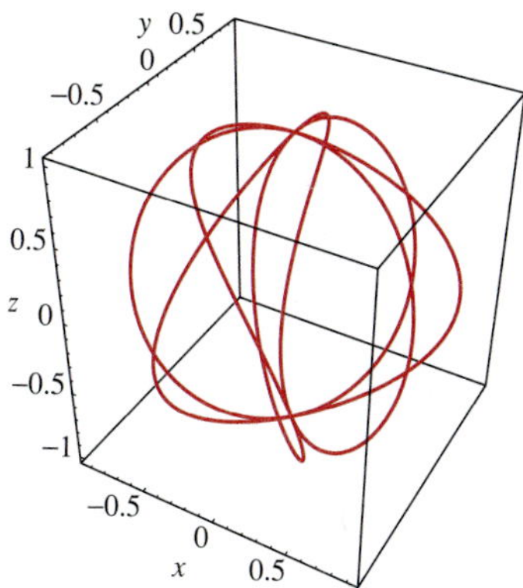

Fig. 12.5.15

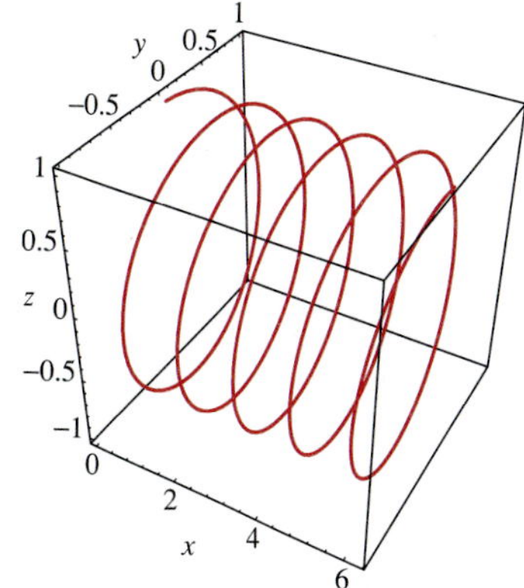

Fig. 12.5.17

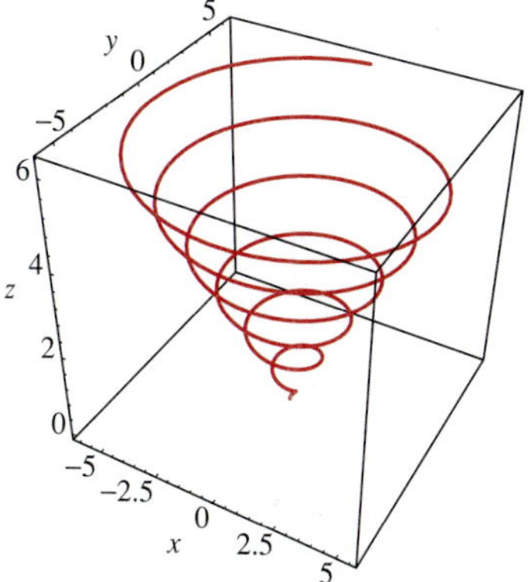

Fig. 12.5.16

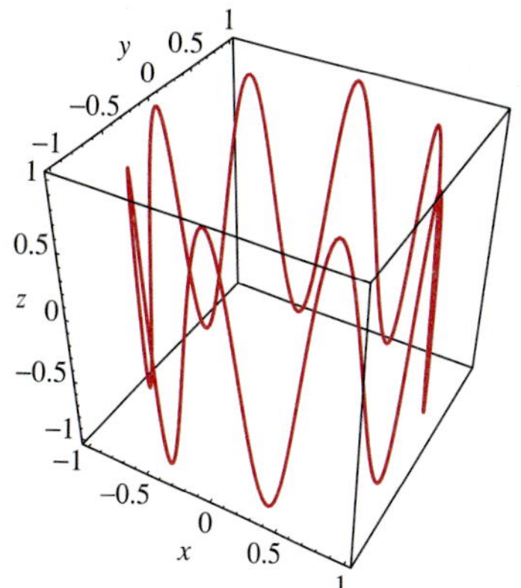

Fig. 12.5.18

In Problems 5 through 10, find the values of $\mathbf{r}'(t)$ and $\mathbf{r}''(t)$ for the given values of t.

5. $\mathbf{r}(t) = 3\mathbf{i} - 2\mathbf{j};\quad t = 1$ **6.** $\mathbf{r}(t) = t^2\mathbf{i} - t^3\mathbf{j};\quad t = 2$

7. $\mathbf{r}(t) = e^{2t}\mathbf{i} + e^{-t}\mathbf{j};\quad t = 0$

8. $\mathbf{r}(t) = \mathbf{i}\cos t + \mathbf{j}\sin t;\quad t = \pi/4$

9. $\mathbf{r}(t) = 3\mathbf{i}\cos 2\pi t + 3\mathbf{j}\sin 2\pi t;\quad t = 3/4$

10. $\mathbf{r}(t) = 5\mathbf{i}\cos t + 4\mathbf{j}\sin t;\quad t = \pi$

In Problems 11 through 16, the position vector $\mathbf{r}(t)$ of a particle moving in space is given. Find its velocity and acceleration vectors and its speed at time t.

11. $\mathbf{r}(t) = t\mathbf{i} + t^2\mathbf{j} + t^3\mathbf{k}$ **12.** $\mathbf{r}(t) = t^2(3\mathbf{i} + 4\mathbf{j} - 12\mathbf{k})$

13. $\mathbf{r}(t) = t\mathbf{i} + 3e^t\mathbf{j} + 4e^t\mathbf{k}$ **14.** $\mathbf{r}(t) = e^t\mathbf{i} + e^{2t}\mathbf{j} + e^{3t}\mathbf{k}$

15. $\mathbf{r}(t) = (3\cos t)\mathbf{i} + (3\sin t)\mathbf{j} - 4t\mathbf{k}$

16. $\mathbf{r}(t) = 12t\mathbf{i} + (5\sin 2t)\mathbf{j} - (5\cos 2t)\mathbf{k}$

Calculate the integrals in Problems 17 through 20.

17. $\displaystyle\int_0^{\pi/4} (\mathbf{i}\sin t + 2\mathbf{j}\cos t)\,dt$ **18.** $\displaystyle\int_1^e \left(\frac{1}{t}\mathbf{i} - \mathbf{j}\right)dt$

19. $\displaystyle\int_0^2 t^2(1 + t^3)^{3/2}\mathbf{i}\,dt$ **20.** $\displaystyle\int_0^1 (\mathbf{i}e^t - \mathbf{j}te^{-t^2})\,dt$

In Problems 21 through 24, apply Theorem 2 to compute the derivative $D_t[\mathbf{u}(t)\cdot\mathbf{v}(t)]$.

21. $\mathbf{u}(t) = 3t\mathbf{i} - \mathbf{j},\quad \mathbf{v}(t) = 2\mathbf{i} - 5t\mathbf{j}$

22. $\mathbf{u}(t) = t\mathbf{i} + t^2\mathbf{j},\quad \mathbf{v}(t) = t^2\mathbf{i} - t\mathbf{j}$

23. $\mathbf{u}(t) = \langle\cos t, \sin t\rangle,\quad \mathbf{v}(t) = \langle\sin t, -\cos t\rangle$

24. $\mathbf{u} = \langle t, t^2, t^3\rangle,\quad \mathbf{v} = \langle\cos 2t, \sin 2t, e^{-3t}\rangle$

In Problems 25 through 34, the acceleration vector $\mathbf{a}(t)$, the initial position $\mathbf{r}_0 = \mathbf{r}(0)$, and the initial velocity $\mathbf{v}_0 = \mathbf{v}(0)$ of a particle moving in xyz-space are given. Find its position vector $\mathbf{r}(t)$ at time t.

25. $\mathbf{a} = \mathbf{0};\quad \mathbf{r}_0 = \mathbf{i};\quad \mathbf{v}_0 = \mathbf{k}$

26. $\mathbf{a} = 2\mathbf{i};\quad \mathbf{r}_0 = 3\mathbf{j};\quad \mathbf{v}_0 = 4\mathbf{k}$

27. $\mathbf{a}(t) = 2\mathbf{i} - 4\mathbf{k};\quad \mathbf{r}_0 = \mathbf{0};\quad \mathbf{v}_0 = 10\mathbf{j}$

28. $\mathbf{a}(t) = \mathbf{i} - \mathbf{j} + 3\mathbf{k};\quad \mathbf{r}_0 = 5\mathbf{i};\quad \mathbf{v}_0 = 7\mathbf{j}$

29. $\mathbf{a}(t) = 2\mathbf{j} - 6t\mathbf{k};\quad \mathbf{r}_0 = 2\mathbf{i};\quad \mathbf{v}_0 = 5\mathbf{k}$

30. $\mathbf{a}(t) = 6t\mathbf{i} - 5\mathbf{j} + 12t^2\mathbf{k};\quad \mathbf{r}_0 = 3\mathbf{i} + 4\mathbf{j};\quad \mathbf{v}_0 = 4\mathbf{j} - 5\mathbf{k}$

31. $\mathbf{a}(t) = t\mathbf{i} + t^2\mathbf{j} + t^3\mathbf{k};\quad \mathbf{r}_0 = 10\mathbf{i};\quad \mathbf{v}_0 = 10\mathbf{j}$

32. $\mathbf{a}(t) = t\mathbf{i} + e^{-t}\mathbf{j};\quad \mathbf{r}_0 = 3\mathbf{i} + 4\mathbf{j};\quad \mathbf{v}_0 = 5\mathbf{k}$

33. $\mathbf{a}(t) = \mathbf{i}\cos t + \mathbf{j}\sin t;\quad \mathbf{r}_0 = \mathbf{j};\quad \mathbf{v}_0 = -\mathbf{i} + 5\mathbf{k}$

34. $\mathbf{a}(t) = 9(\mathbf{i}\sin 3t + \mathbf{j}\cos 3t) + 4\mathbf{k};\quad \mathbf{r}_0 = 3\mathbf{i} + 4\mathbf{j};$ $\mathbf{v}_0 = 2\mathbf{i} - 7\mathbf{k}$

35. The parametric equations of a moving point are

$$x(t) = 3\cos 2t,\quad y(t) = 3\sin 2t,\quad z(t) = 8t.$$

Find its velocity, speed, and acceleration at time $t = 7\pi/8$.

36. Use the equations in Theorem 2 to calculate

$$D_t[\mathbf{u}(t)\cdot\mathbf{v}(t)] \quad\text{and}\quad D_t[\mathbf{u}(t)\times\mathbf{v}(t)]$$

if $\mathbf{u}(t) = \langle t, t^2, t^3\rangle$ and $\mathbf{v}(t) = \langle e^t, \cos t, \sin t\rangle$.

37. Verify part 5 of Theorem 2 in the special case $\mathbf{u}(t) = \langle 0, 3, 4t\rangle$ and $\mathbf{v}(t) = \langle 5t, 0, -4\rangle$.

38. Prove part 5 of Theorem 2.

39. A point moves on a sphere centered at the origin. Show that its velocity vector is always tangent to the sphere.

40. A particle moves with constant speed along a curve in space. Show that its velocity and acceleration vectors are always perpendicular.

41. Find the maximum height reached by the ball in Example 10 and also its speed at that height.

42. The **angular momentum** $\mathbf{L}(t)$ and **torque** $\boldsymbol{\tau}(t)$ of a moving particle of mass m with position vector $\mathbf{r}(t)$ are defined to be

$$\mathbf{L}(t) = m\mathbf{r}(t)\times\mathbf{v}(t),\qquad \boldsymbol{\tau}(t) = m\mathbf{r}(t)\times\mathbf{a}(t).$$

Prove that $\mathbf{L}'(t) = \boldsymbol{\tau}(t)$. It follows that $\mathbf{L}(t)$ must be constant if $\boldsymbol{\tau} \equiv \mathbf{0}$; this is the law of conservation of angular momentum.

Problems 43 through 48 deal with a projectile fired from the origin (so $x_0 = y_0 = 0$) with initial speed v_0 and initial angle of inclination α. The **range** of the projectile is the horizontal distance it travels before it returns to the ground.

43. If $\alpha = 45°$, what value of v_0 gives a range of 1 mi?

44. If $\alpha = 60°$ and the range is $R = 1$ mi, what is the maximum height attained by the projectile?

45. Deduce from Eqs. (22) and (23) the fact that the range is

$$R = \tfrac{1}{16}v_0^2\sin\alpha\cos\alpha.$$

46. Given the initial speed v_0, find the angle α that maximizes the range. (*Suggestion:* Use the result of Problem 45.)

47. Suppose that $v_0 = 160$ (ft/s). Find the maximum height $y_{\max}$ and the range R of the projectile if (a) $\alpha = 30°$; (b) $\alpha = 45°$; (c) $\alpha = 60°$.

48. The projectile of Problem 47 is to be fired at a target 600 ft away, and there is a hill 300 ft high midway between the gun site and this target. At what initial angle of inclination should the projectile be fired?

49. A projectile is to be fired horizontally from the top of a 100-m cliff at a target 1 km from the base of the cliff. What should be the initial velocity of the projectile? (Use $g = 9.8\ \text{m/s}^2$.)

50. A bomb is dropped (initial speed zero) from a helicopter hovering at a height of 800 m. A projectile is fired from a gun located on the ground 800 m west of the point directly beneath the helicopter. The projectile is supposed to intercept the bomb at a height of exactly 400 m. If the projectile is fired at the same instant that the bomb is dropped, what should be its initial velocity and angle of inclination?

51. Suppose, more realistically, that the projectile of Problem 50 is fired 1 s after the bomb is dropped. What should be its initial velocity and angle of inclination?

52. An artillery gun with a muzzle velocity of 1000 ft/s is located atop a seaside cliff 500 ft high. At what initial inclination angle (or angles) should it fire a projectile in order to hit a ship at sea 20000 ft from the base of the cliff?

53. Suppose that the vector-valued functions $\mathbf{u}(t)$ and $\mathbf{v}(t)$ both have limits as $t \to a$. Prove

(a) $\lim_{t\to a} \big(\mathbf{u}(t) + \mathbf{v}(t)\big) = \lim_{t\to a} \mathbf{u}(t) + \lim_{t\to a} \mathbf{v}(t)$;

(b) $\lim_{t\to a} \big(\mathbf{u}(t)\cdot\mathbf{v}(t)\big) = \big(\lim_{t\to a} \mathbf{u}(t)\big)\cdot\big(\lim_{t\to a} \mathbf{v}(t)\big)$.

54. Suppose that both the vector-valued function $\mathbf{r}(t)$ and the real-valued function $h(t)$ are differentiable. Deduce the chain rule for vector-valued functions,

$$D_t[\mathbf{r}(h(t))] = h'(t)\mathbf{r}'(h(t)),$$

in componentwise fashion from the ordinary chain rule.

55. A point moves with constant speed, so its velocity vector $\mathbf{v}$ satisfies the condition

$$|\mathbf{v}|^2 = \mathbf{v}\cdot\mathbf{v} = C \quad \text{(a constant)}.$$

Prove that the velocity and acceleration vectors of the point are always perpendicular to each other.

56. A point moves on a circle whose center is at the origin. Use the dot product to show that the position and velocity vectors of the moving point are always perpendicular.

57. A point moves on the hyperbola $x^2 - y^2 = 1$ with position vector

$$\mathbf{r}(t) = \mathbf{i}\cosh\omega t + \mathbf{j}\sinh\omega t$$

(the number ω is a constant). Prove that the acceleration vector $\mathbf{a}(t)$ satisfies the equation $\mathbf{a}(t) = c\mathbf{r}(t)$, where c is a positive constant. What sort of external force would produce this kind of motion?

58. Suppose that a point moves on the ellipse

$$\frac{x^2}{a^2} + \frac{y^2}{b^2} = 1$$

with position vector $\mathbf{r}(t) = \mathbf{i}a\cos\omega t + \mathbf{j}b\sin\omega t$ (ω is a constant). Prove that the acceleration vector $\mathbf{a}$ satisfies the equation $\mathbf{a}(t) = c\mathbf{r}(t)$, where c is a negative constant. To what sort of external force $\mathbf{F}(t)$ does this motion correspond?

59. A point moves in the plane with constant acceleration vector $\mathbf{a} = a\mathbf{j}$. Prove that its path is a parabola or a straight line.

60. Suppose that a particle is subject to no force, so its acceleration vector $\mathbf{a}(t)$ is identically zero. Prove that the particle travels along a straight line at constant speed (Newton's first law of motion).

61. *Uniform Circular Motion* Consider a particle that moves counterclockwise around the circle with center $(0, 0)$ and radius r at a constant angular speed of ω radians per second (Fig. 12.5.19). If its initial position is $(r, 0)$, then its position vector is

$$\mathbf{r}(t) = \mathbf{i}r\cos\omega t + \mathbf{j}r\sin\omega t.$$

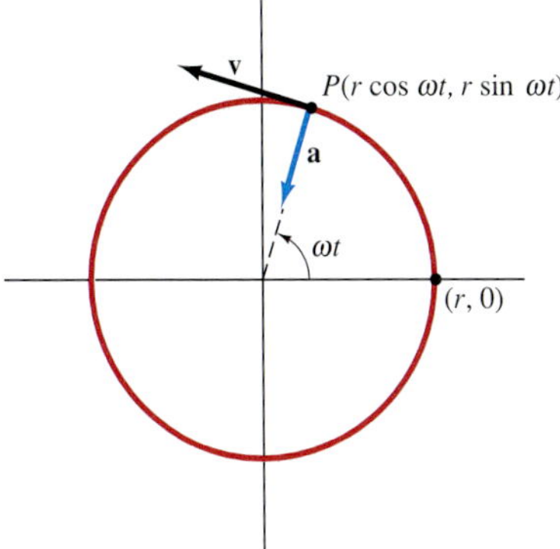

Fig. 12.5.19 Uniform circular motion (Problem 61)

(a) Show that the velocity vector of the particle is tangent to the circle and that the speed of the particle is

$$v(t) = |\mathbf{v}(t)| = r\omega.$$

(b) Show that the acceleration vector $\mathbf{a}$ of the particle is directed opposite to $\mathbf{r}$ and that

$$a(t) = |\mathbf{a}(t)| = r\omega^2.$$

62. Suppose that a particle is moving under the influence of a *central* force field $\mathbf{R} = k\mathbf{r}$, where k is a scalar function of x, y, and z. Conclude that the trajectory of the particle lies in a *fixed* plane through the origin.

63. A baseball is thrown with an initial velocity of 160 ft/s straight upward from the ground. It experiences a downward gravitational acceleration of 32 ft/s^2. Because of spin, it experiences also a (horizontal) northward acceleration of 0.1 ft/s^2; otherwise, the air has no effect on its motion. How far north of the throwing point will the ball land?

64. A baseball is hit with an initial velocity of 96 ft/s and an initial inclination angle of 15° from ground level straight down a foul line. Because of spin it experiences a horizontal acceleration of 2 ft/s^2 perpendicular to the foul line; otherwise, the air has no effect on its motion. When the ball hits the ground, how far is it from the foul line?

65. A projectile is fired northward (in the positive y-direction) out to sea from the top of a seaside cliff 384 ft high. The projectile's initial velocity vector is $\mathbf{v}_0 = 200\mathbf{j} + 160\mathbf{k}$. In addition to a downward (negative z-direction) gravitational acceleration of 32 ft/s^2, it experiences in flight an eastward (positive x-direction) acceleration of 2 ft/s^2 due to spin. (a) Find the projectile's velocity and position vectors t seconds after it is fired. (b) How long is the projectile in the air? (c) Where does the projectile hit the water ($z = 0$)? Give the answer by telling how far north out to sea and how far east along the coast its impact position is (d) What is the maximum height of the projectile above the water?

66. A gun fires a shell with a muzzle velocity of 150 m/s. While the shell is in the air, it experiences a downward (vertical) gravitational acceleration of 9.8 m/s^2 and an eastward (horizontal) Coriolis acceleration of 5 cm/s^2; air resistance may be ignored. The target is 1500 m due north of the gun, and both the gun and target are on level ground. Halfway between them is a hill 600 m high. Tell precisely how to aim the gun—both compass heading and inclination from the horizontal—so that the shell will clear the hill and hit the target.

12.5 PROJECT: DOES A PITCHED BASEBALL REALLY CURVE?

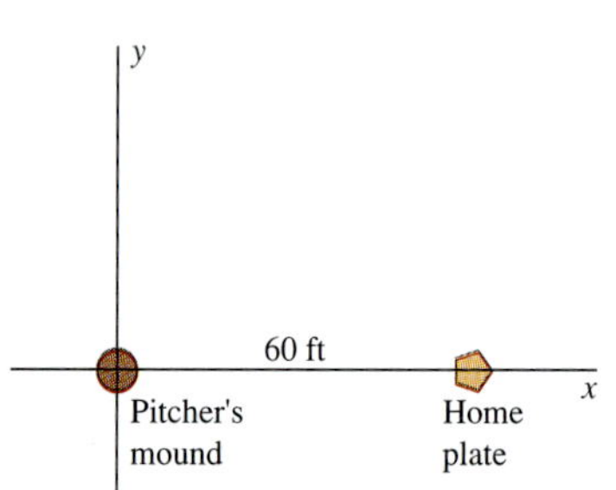

Fig. 12.5.20 The x-axis points toward home plate.

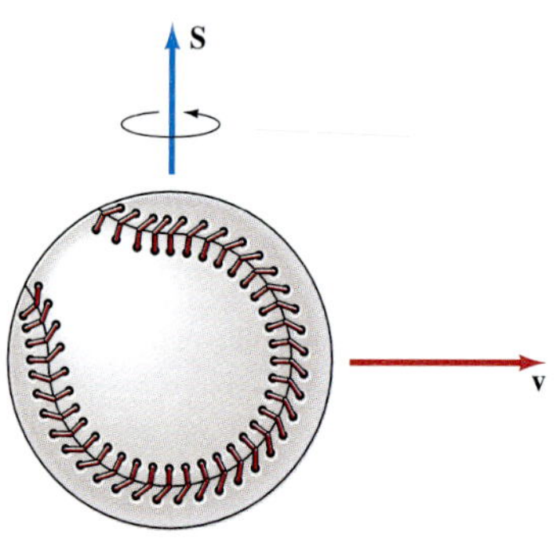

Fig. 12.5.21 The spin and velocity vectors

Have you ever wondered whether a baseball pitch really curves or whether it's some sort of optical illusion? In this project you'll use calculus to settle the matter.

Suppose that a pitcher throws a ball toward home plate (60 ft away, as in Fig. 12.5.20) and gives it a spin of S revolutions per second counterclockwise (as viewed from above) about a vertical axis through the center of the ball. This spin is described by the *spin vector* **S** that points along the axis of revolution in the right-handed direction and has length S (Fig. 12.5.21).

We know from studies of aerodynamics that this spin causes a difference in air pressure on the sides of the ball toward and away from this spin. Studies also show that this pressure difference results in a *spin acceleration*

$$\mathbf{a}_S = c\mathbf{S} \times \mathbf{v} \tag{1}$$

of the ball (where c is an empirical constant). The total acceleration of the ball is then

$$\mathbf{a} = (c\mathbf{S} \times \mathbf{v}) - g\mathbf{k}, \tag{2}$$

where $g \approx 32$ ft/s^2 is the gravitational acceleration. Here we will ignore any other effects of air resistance.

With the spin vector $\mathbf{S} = S\mathbf{k}$ pointing upward, as in Fig. 12.5.21, show first that

$$\mathbf{S} \times \mathbf{v} = -Sv_y\mathbf{i} + Sv_x\mathbf{j}, \tag{3}$$

where v_x is the component of **v** in the x-direction and v_y is the component of **v** in the y-direction.

For a ball pitched along the x-axis, v_x is much larger than v_y, and so the approximation $\mathbf{S} \times \mathbf{v} \approx Sv_x\mathbf{j}$ is sufficiently accurate for our purposes. We may then take the acceleration vector of the ball to be

$$\mathbf{a} = cSv_x\mathbf{j} - g\mathbf{k}. \tag{4}$$

Now suppose that the pitcher throws the ball from the initial position $x_0 = y_0 = 0$, $z_0 = 5$ (ft), with initial velocity vector

$$\mathbf{v}_0 = 120\mathbf{i} - 2\mathbf{j} + 4\mathbf{k} \tag{5}$$

(with components in feet per second, so $v_0 \approx 120$ ft/s, about 82 mi/h) and with a spin of $S = \frac{80}{3}$ rev/s. A reasonable value of c is

$$c = 0.005 \quad \text{ft/s}^2 \quad \text{per ft/s of velocity and rev/s of spin,}$$

although the precise value depends on whether the pitcher has (accidentally, of course) scuffed the ball or administered some foreign substance to it.

Show first that these values of the parameters yield

$$\mathbf{a} = 16\mathbf{j} - 32\mathbf{k}$$

for the ball's acceleration vector. Then integrate twice in succession to find the ball's position vector

$$\mathbf{r}(t) = x(t)\mathbf{i} + y(t)\mathbf{j} + z(t)\mathbf{k}.$$

Use your results to fill in the following table, giving the pitched ball's horizontal deflection y and height z (above the ground) at quarter-second intervals.

t (s)	x (ft)	y (ft)	z (ft)
0.0	0	0	5
0.25	30	?	?
0.50	60	?	?

Suppose that the batter gets a "fix" on the pitch by observing the ball during the first quarter-second and prepares to swing. After 0.25 s does the pitch still appear to be straight on target toward home plate at a height of 5 ft?

What happens to the ball during the final quarter-second of its approach to home plate—*after* the batter has begun to swing the bat? What were the ball's horizontal and vertical deflections during this brief period? What is your conclusion? Does the pitched ball really "curve"?

12.6 CURVATURE AND ACCELERATION

The speed of a moving point is closely related to the arc length of its trajectory. The arc-length formula for parametric curves in space (or *space curves*) is a natural generalization of the formula for parametric plane curves [Eq. (8) of Section 10.5]. The **arc length** s along the smooth curve with position vector

$$\mathbf{r}(t) = f(t)\mathbf{i} + g(t)\mathbf{j} + h(t)\mathbf{k} = x\mathbf{i} + y\mathbf{j} + z\mathbf{k} \tag{1}$$

from the point $\mathbf{r}(a)$ to the point $\mathbf{r}(b)$ is, by definition,

$$\begin{aligned} s &= \int_a^b \sqrt{[x'(t)]^2 + [y'(t)]^2 + [z'(t)]^2}\, dt \\ &= \int_a^b \sqrt{\left(\frac{dx}{dt}\right)^2 + \left(\frac{dy}{dt}\right)^2 + \left(\frac{dz}{dt}\right)^2}\, dt. \end{aligned} \tag{2}$$

We see from Eq. (9) in Section 12.5 that the integrand is the speed $v(t) = |\mathbf{r}'(t)|$ of the moving point with position vector $\mathbf{r}(t)$, so

$$s = \int_a^b v(t)\, dt. \tag{3}$$

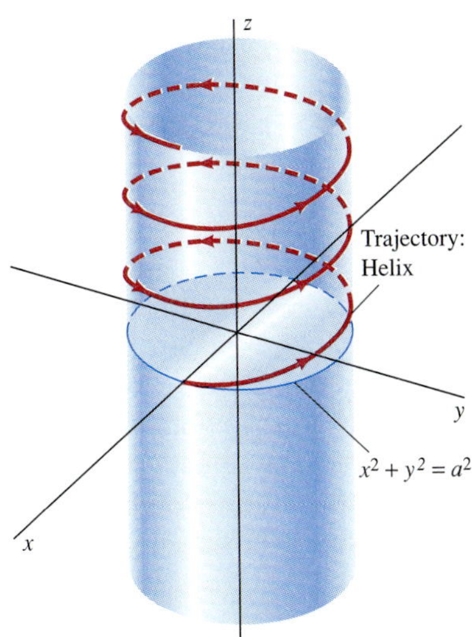

Fig. 12.6.1 The helix of Example 1

EXAMPLE 1 Find the arc length of one turn (from $t = 0$ to $t = 2\pi/\omega$) of the helix shown in Fig. 12.6.1. This helix has the parametric equations

$$x(t) = a\cos\omega t, \quad y(t) = a\sin\omega t, \quad z(t) = bt.$$

Solution We found in Example 7 of Section 12.5 that

$$v(t) = \sqrt{a^2\omega^2 + b^2}\,.$$

Hence Eq. (3) gives

$$s = \int_0^{2\pi/\omega} \sqrt{a^2\omega^2 + b^2}\, dt = \frac{2\pi}{\omega}\sqrt{a^2\omega^2 + b^2}\,.$$

For instance, if $a = b = \omega = 1$, then $s = 2\pi\sqrt{2}$, which is $\sqrt{2}$ times the circumference of the circle in the xy-plane over which the helix lies. ■

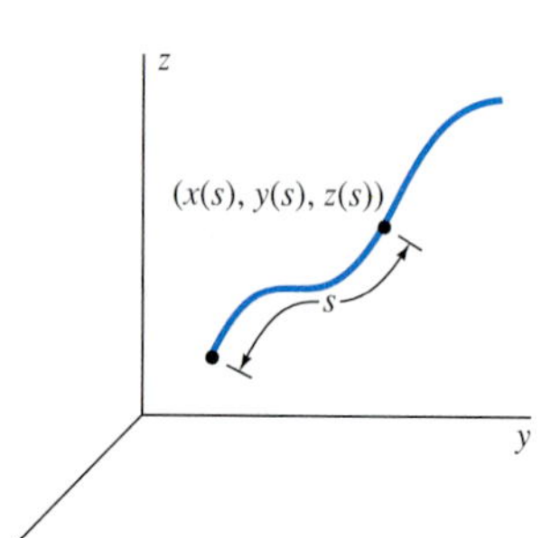

Fig. 12.6.2 A curve parametrized by arc length s

Let $s(t)$ denote the arc length along a smooth curve from its initial point $\mathbf{r}(a)$ to the variable point $\mathbf{r}(t)$ (Fig. 12.6.2). Then, from Eq. (3), we obtain the **arc-length function** $s(t)$ of the curve:

$$s(t) = \int_a^t v(\tau)\,d\tau. \tag{4}$$

The fundamental theorem of calculus then gives

$$\frac{ds}{dt} = v. \tag{5}$$

Thus *the speed of the moving point is the time rate of change of its arc-length function.* If $v(t) > 0$ for all t, then it follows that $s(t)$ is an increasing function of t and therefore has an inverse function $t(s)$. When we replace t with $t(s)$ in the curve's original parametric equations, we obtain the **arc-length parametrization**

$$x = x(s), \qquad y = y(s), \qquad z = z(s).$$

This gives the position of the moving point as a function of arc length measured along the curve from its initial point (see Fig. 12.6.2).

EXAMPLE 2 If we take $a = 5$, $b = 12$, and $\omega = 1$ for the helix of Example 1, then the velocity formula $v = (a^2\omega^2 + b^2)^{1/2}$ yields

$$v = \sqrt{5^2 \cdot 1^2 + 12^2} = \sqrt{169} = 13.$$

Hence Eq. (5) gives $ds/dt = 13$, so

$$s = 13t,$$

taking $s = 0$ when $t = 0$ and thereby measuring arc length from the natural starting point $(5, 0, 0)$. When we substitute $t = s/13$ and the numerical values of a, b, and ω into the original parametric equations of the helix, we get the arc-length parametrization

$$x(s) = 5\cos\frac{s}{13}, \qquad y(s) = 5\sin\frac{s}{13}, \qquad z(s) = \frac{12s}{13}$$

of the helix. ■

Curvature of Plane Curves

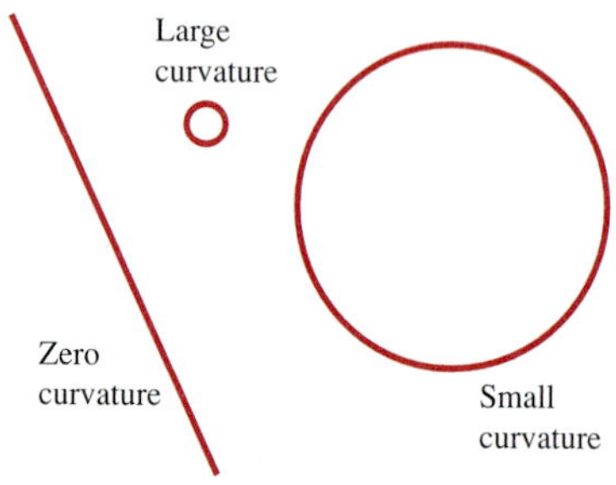

The word *curvature* has an intuitive meaning that we need to make precise. Most people would agree that a straight line does not curve at all, whereas a circle of small radius is more curved than a circle of large radius (Fig. 12.6.3). This judgment may be based on a feeling that curvature is "rate of change of direction." The direction of a curve is determined by its velocity vector, so you would expect the idea of curvature to have something to do with the rate at which the velocity vector is turning.

Large curvature

Zero curvature

Small curvature

Fig. 12.6.3 The intuitive idea of curvature

Let

$$\mathbf{r}(t) = x(t)\mathbf{i} + y(t)\mathbf{j}, \qquad a \leqq t \leqq b \tag{6}$$

be the position vector of a smooth *plane* curve with nonzero velocity vector $\mathbf{v}(t) = \mathbf{r}'(t)$. The curve's **unit tangent vector** at the point $\mathbf{r}(t)$ is the unit vector

$$\mathbf{T}(t) = \frac{\mathbf{v}(t)}{|\mathbf{v}(t)|} = \frac{\mathbf{v}(t)}{v(t)}, \tag{7}$$

where $v(t) = |\mathbf{v}(t)|$ is the speed. Now denote by ϕ the angle of inclination of $\mathbf{T}$, measured counterclockwise from the positive x-axis (Fig. 12.6.4). Then

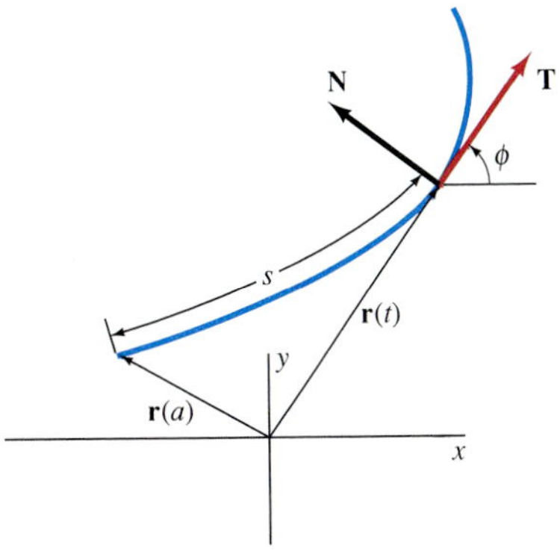

$$\mathbf{T} = \mathbf{i}\cos\phi + \mathbf{j}\sin\phi. \tag{8}$$

We can express the unit tangent vector $\mathbf{T}$ of Eq. (8) as a function of the arc-length parameter s indicated in Fig. 12.6.4. Then the rate at which $\mathbf{T}$ is turning is measured by the derivative

$$\frac{d\mathbf{T}}{ds} = \frac{d\mathbf{T}}{d\phi}\cdot\frac{d\phi}{ds} = (-\mathbf{i}\sin\phi + \mathbf{j}\cos\phi)\frac{d\phi}{ds}. \tag{9}$$

Note that

$$\left|\frac{d\mathbf{T}}{ds}\right| = \left|\frac{d\phi}{ds}\right| \tag{10}$$

Fig. 12.6.4 The unit tangent vector $\mathbf{T}$

because the vector on the right-hand side of Eq. (9) is a unit vector.

The **curvature** at a point of a plane curve, denoted by κ (lowercase Greek kappa), is therefore defined to be

$$\kappa = \left|\frac{d\phi}{ds}\right|, \tag{11}$$

the absolute value of the rate of change of the angle ϕ with respect to arc length s. We define the curvature κ in terms of $d\phi/ds$ rather than $d\phi/dt$ because the latter depends not only on the shape of the curve, but also on the speed of the moving point $\mathbf{r}(t)$. For a straight line the angle ϕ is a constant, so the curvature given by Eq. (11) is zero. If you imagine a point that is moving with constant speed along a curve, the curvature is greatest at points where ϕ changes the most rapidly, such as the points P and R on the curve of Fig. 12.6.5. The curvature is least at points such as Q and S, where ϕ is changing the least rapidly.

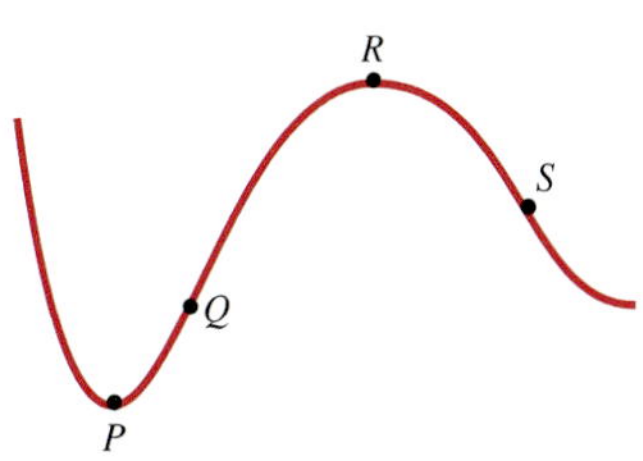

Fig. 12.6.5 The curvature is large at P and R, small at Q and S.

We need to derive a formula that is effective in computing the curvature of a smooth parametric plane curve $x = x(t)$, $y = y(t)$. First we note that

$$\phi = \tan^{-1}\left(\frac{dy}{dx}\right) = \tan^{-1}\left(\frac{y'(t)}{x'(t)}\right)$$

provided $x'(t) \neq 0$. Hence

$$\frac{d\phi}{dt} = \frac{y''x' - y'x''}{(x')^2} \div \left(1 + \left(\frac{y'}{x'}\right)^2\right) = \frac{x'y'' - x''y'}{(x')^2 + (y')^2},$$

where primes denote derivatives with respect to t. Because $v = ds/dt > 0$, Eq. (11) gives

$$\kappa = \left|\frac{d\phi}{ds}\right| = \left|\frac{d\phi}{dt}\cdot\frac{dt}{ds}\right| = \frac{1}{v}\left|\frac{d\phi}{dt}\right|;$$

thus

$$\kappa = \frac{|x'y'' - x''y'|}{[(x')^2 + (y')^2]^{3/2}} = \frac{|x'y'' - x''y'|}{v^3}. \tag{12}$$

At a point where $x'(t) = 0$, we know that $y'(t) \neq 0$, because the curve is smooth. Thus we will obtain the same result if we begin with the equation $\phi = \cot^{-1}(x'/y')$.

An explicitly described curve $y = f(x)$ may be regarded as a parametric curve $x = x$, $y = f(x)$. Then $x' = 1$ and $x'' = 0$, so Eq. (12)—with x in place of t as the parameter—becomes

$$\kappa = \frac{|y''|}{[1 + (y')^2]^{3/2}} = \frac{|d^2y/dx^2|}{[1 + (dy/dx)^2]^{3/2}}. \tag{13}$$

EXAMPLE 3 Show that the curvature at each point of a circle of radius a is $\kappa = 1/a$.

Solution With the familiar parametrization $x = a\cos t$, $y = a\sin t$ of such a circle centered at the origin, we let primes denote derivatives with respect to t and obtain

$$x' = -a\sin t, \qquad y' = a\cos t,$$

$$x'' = -a\cos t, \qquad y'' = -a\sin t.$$

Hence Eq. (12) gives

$$\kappa = \frac{|(-a\sin t)(-a\sin t) - (-a\cos t)(a\cos t)|}{[(-a\sin t)^2 + (a\cos t)^2]^{3/2}} = \frac{a^2}{a^3} = \frac{1}{a}.$$

Alternatively, we could have used Eq. (13). Our point of departure would then be the equation $x^2 + y^2 = a^2$ of the same circle, and we would compute y' and y'' by implicit differentiation (see Problem 27). ■

It follows immediately from Eqs. (8) and (9) that

$$\mathbf{T} \cdot \frac{d\mathbf{T}}{ds} = 0,$$

so the unit tangent vector $\mathbf{T}$ and its derivative vector $d\mathbf{T}/ds$ are perpendicular. The *unit* vector $\mathbf{N}$ that points in the direction of $d\mathbf{T}/ds$ is called the **principal unit normal vector** to the curve. Because $\kappa = |d\phi/ds| = |d\mathbf{T}/ds|$ by Eq. (10), it follows that

$$\frac{d\mathbf{T}}{ds} = \kappa\mathbf{N}. \tag{14}$$

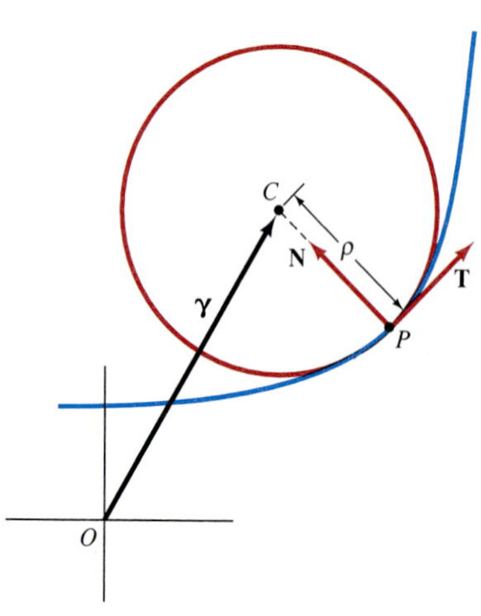

Fig. 12.6.6 Osculating circle, radius of curvature, and center of curvature

Intuitively, **N** *is the unit normal vector to the curve that points in the direction in which the curve is bending.*

Suppose that P is a point on a parametrized curve at which $\kappa \neq 0$. Consider the circle that is tangent to the curve at P and has the same curvature there. The center of the circle is to lie on the concave side of the curve—that is, on the side toward which the normal vector $\mathbf{N}$ points. This circle is called the **osculating circle** (or **circle of curvature**) of the curve at the given point because it touches the curve so closely there. (*Osculum* is the Latin word for *kiss.*) Let ρ be the radius of the osculating circle and let $\boldsymbol{\gamma}$ be the position vector of its center. Then $\boldsymbol{\gamma} = \overrightarrow{OC}$, where C is the center of the osculating circle (Fig. 12.6.6). Then ρ is called the **radius of curvature** of the curve at the point P and $\boldsymbol{\gamma}$ is called the (vector) **center of curvature** of the curve at P.

Example 3 implies that the radius of curvature is

$$\rho = \frac{1}{\kappa}, \tag{15}$$

and the fact that $|\mathbf{N}| = 1$ implies that the position vector of the center of curvature is

$$\boldsymbol{\gamma} = \mathbf{r} + \rho\mathbf{N} \quad (\mathbf{r} = \overrightarrow{OP}). \tag{16}$$

EXAMPLE 4 Determine the vectors **T** and **N**, the curvature κ, and the center of curvature of the parabola $y = x^2$ at the point $(1, 1)$.

Solution If the parabola is parametrized by $x = t$, $y = t^2$, then its position vector is $\mathbf{r}(t) = t\mathbf{i} + t^2\mathbf{j}$, so $\mathbf{v}(t) = \mathbf{i} + 2t\mathbf{j}$. The speed is $v(t) = \sqrt{1 + 4t^2}$, so Eq. (7) yields

$$\mathbf{T}(t) = \frac{\mathbf{v}(t)}{v(t)} = \frac{\mathbf{i} + 2t\mathbf{j}}{\sqrt{1 + 4t^2}}.$$

By substituting $t = 1$, we find that the unit tangent vector at $(1, 1)$ is

$$\mathbf{T} = \frac{1}{\sqrt{5}}\mathbf{i} + \frac{2}{\sqrt{5}}\mathbf{j}.$$

Because the parabola is concave upward at $(1, 1)$, the principal unit normal vector is the upward-pointing unit vector

$$\mathbf{N} = -\frac{2}{\sqrt{5}}\mathbf{i} + \frac{1}{\sqrt{5}}\mathbf{j}$$

that is perpendicular to **T**. (Note that $\mathbf{T} \cdot \mathbf{N} = 0$.) If $y = x^2$, then $dy/dx = 2x$ and $d^2y/dx^2 = 2$, so Eq. (13) yields

$$\kappa = \frac{|y''|}{[1 + (y')^2]^{3/2}} = \frac{2}{(1 + 4x^2)^{3/2}}.$$

So at the point $(1, 1)$ we find the curvature and radius of curvature to be

$$\kappa = \frac{2}{5\sqrt{5}} \quad \text{and} \quad \rho = \frac{5\sqrt{5}}{2},$$

respectively.

Next, Eq. (16) gives the center of curvature as

$$\boldsymbol{\gamma} = \langle 1, 1 \rangle + \frac{5\sqrt{5}}{2}\left\langle -\frac{2}{\sqrt{5}}, \frac{1}{\sqrt{5}} \right\rangle = \left\langle -4, \frac{7}{2} \right\rangle.$$

The equation of the osculating circle to the parabola at $(1, 1)$ is therefore

$$(x + 4)^2 + (y - \tfrac{7}{2})^2 = \rho^2 = \tfrac{125}{4}.$$

Figure 12.6.7 shows this large osculating circle at the point $(1, 1)$, as well as the smaller osculating circles that are tangent to the parabola at the points $(0, 0)$, $(\frac{1}{3}, \frac{1}{9})$, and $(\frac{2}{3}, \frac{4}{9})$. Is it clear to you which of these osculating circles is which? ■

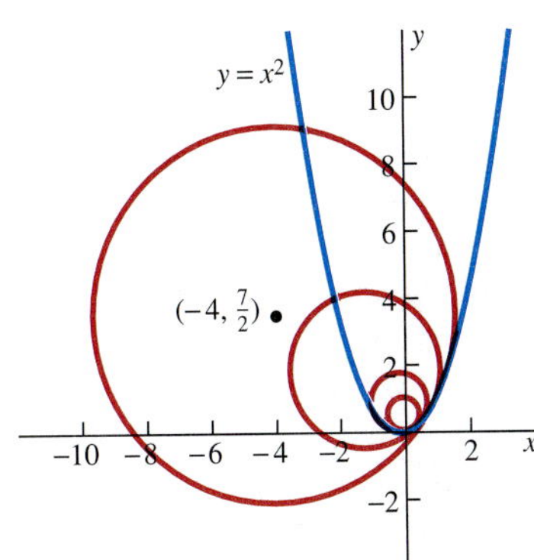

Fig. 12.6.7 Osculating circles for the parabola of Example 4

Curvature of Space Curves

Consider now a moving particle in space with twice-differentiable position vector $\mathbf{r}(t)$. Suppose also that the velocity vector $\mathbf{v}(t)$ is never zero. The **unit tangent vector** at time t is defined, as before, to be

$$\mathbf{T}(t) = \frac{\mathbf{v}(t)}{|\mathbf{v}(t)|} = \frac{\mathbf{v}(t)}{v(t)}, \tag{17}$$

so

$$\mathbf{v} = v\mathbf{T}. \tag{18}$$

We defined the curvature of a plane curve to be $\kappa = |d\phi/ds|$, where ϕ is the angle of inclination of $\mathbf{T}$ from the positive x-axis. For a space curve, there is no single angle that determines the direction of $\mathbf{T}$, so we adopt the following approach (which leads to the same value for curvature when applied to a space curve that happens to lie in the xy-plane). Differentiation of the identity $\mathbf{T}\cdot\mathbf{T} = 1$ with respect to arc length s gives

$$\mathbf{T}\cdot\frac{d\mathbf{T}}{ds} = 0.$$

It follows that the vectors $\mathbf{T}$ and $d\mathbf{T}/ds$ are always perpendicular.

Then we define the **curvature** κ of the curve at the point $\mathbf{r}(t)$ to be

$$\kappa = \left|\frac{d\mathbf{T}}{ds}\right| = \left|\frac{d\mathbf{T}}{dt}\frac{dt}{ds}\right| = \frac{1}{v}\left|\frac{d\mathbf{T}}{dt}\right|. \tag{19}$$

At a point where $\kappa \neq 0$, we define the **principal unit normal vector N** to be

$$\mathbf{N} = \frac{d\mathbf{T}/ds}{|d\mathbf{T}/ds|} = \frac{1}{\kappa}\frac{d\mathbf{T}}{ds}, \tag{20}$$

so

$$\frac{d\mathbf{T}}{ds} = \kappa\mathbf{N}. \tag{21}$$

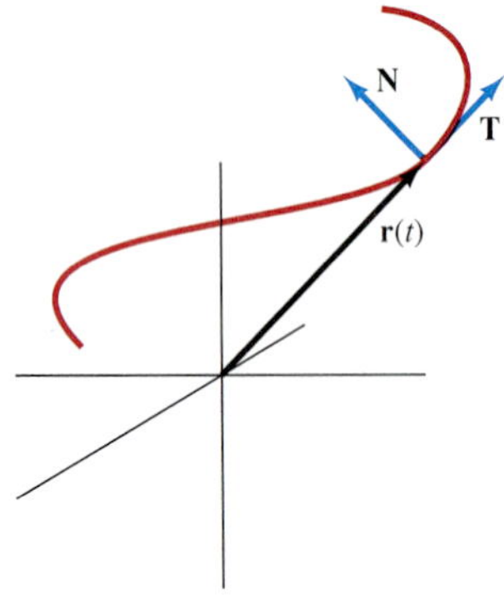

Fig. 12.6.8 The principal unit normal vector **N** points in the direction in which the curve is turning.

Equation (21) shows that $\mathbf{N}$ has the same direction as $d\mathbf{T}/ds$ (Fig. 12.6.8), and Eq. (20) shows that $\mathbf{N}$ is a unit vector. Because Eq. (21) is the same as Eq. (14), we see that the present definitions of κ and $\mathbf{N}$ agree with those given earlier in the two-dimensional case.

EXAMPLE 5 Compute the curvature κ of the helix of Example 1, the helix with parametric equations

$$x(t) = a\cos\omega t, \quad y(t) = a\sin\omega t, \quad z(t) = bt.$$

Solution In Example 7 of Section 12.5, we computed the velocity vector

$$\mathbf{v} = \mathbf{i}(-a\omega\sin\omega t) + \mathbf{j}(a\omega\cos\omega t) + b\mathbf{k}$$

and speed

$$v = |\mathbf{v}| = \sqrt{a^2\omega^2 + b^2}.$$

Hence Eq. (17) gives the unit tangent vector

$$\mathbf{T} = \frac{\mathbf{v}}{v} = \frac{\mathbf{i}(-a\omega \sin \omega t) + \mathbf{j}(a\omega \cos \omega t) + b\mathbf{k}}{\sqrt{a^2\omega^2 + b^2}}.$$

Then

$$\frac{d\mathbf{T}}{dt} = \frac{\mathbf{i}(-a\omega^2 \cos \omega t) + \mathbf{j}(-a\omega^2 \sin \omega t)}{\sqrt{a^2\omega^2 + b^2}},$$

so Eq. (19) gives

$$\kappa = \frac{1}{v}\left|\frac{d\mathbf{T}}{dt}\right| = \frac{a\omega^2}{a^2\omega^2 + b^2}$$

for the curvature of the helix of Example 7 (Section 12.5). Note that the helix has constant curvature. Also note that, if $b = 0$ (so that the helix reduces to a circle of radius a in the xy-plane), our result reduces to $\kappa = 1/a$, in agreement with our computation of the curvature of a circle in Example 3.

Normal and Tangential Components of Acceleration

We may apply Eq. (21) to analyze the meaning of the acceleration vector of a moving particle with velocity vector $\mathbf{v}$ and speed v. Then Eq. (17) gives $\mathbf{v} = v\mathbf{T}$, so the acceleration vector of the particle is

$$\mathbf{a} = \frac{d\mathbf{v}}{dt} = \frac{dv}{dt}\mathbf{T} + v\frac{d\mathbf{T}}{dt} = \frac{dv}{dt}\mathbf{T} + v\frac{d\mathbf{T}}{ds}\frac{ds}{dt}.$$

But $ds/dt = v$, so Eq. (21) gives

$$\mathbf{a} = \frac{dv}{dt}\mathbf{T} + \kappa v^2\mathbf{N}. \tag{22}$$

Because $\mathbf{T}$ and $\mathbf{N}$ are unit vectors tangent and normal to the curve, respectively, Eq. (22) provides a *decomposition of the acceleration vector* into its components tangent to and normal to the trajectory. The **tangential component**

$$a_T = \frac{dv}{dt} \tag{23}$$

is the rate of change of speed of the particle, whereas the **normal component**

$$a_N = \kappa v^2 = \frac{v^2}{\rho} \tag{24}$$

measures the rate of change of its direction of motion. The decomposition

$$\mathbf{a} = a_T\mathbf{T} + a_N\mathbf{N} \tag{25}$$

is illustrated in Fig. 12.6.9.

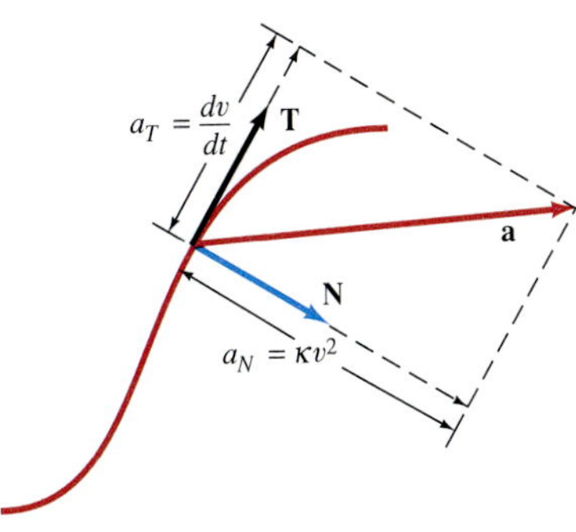

Fig. 12.6.9 Resolution of the acceleration vector **a** into its tangential and normal components

As an application of Eq. (22), think of a train moving along a straight track with constant speed v, so that $a_T = 0 = a_N$ (the latter because $\kappa = 0$ for a straight line). Suppose that at time $t = 0$, the train enters a circular curve of radius ρ. At that instant, it will *suddenly* be subjected to a normal acceleration of magnitude v^2/ρ, proportional to the *square* of the speed of the train. A passenger in the train will experience a sudden jerk to the side. If v is large, the stresses may be great enough to damage the track or derail the train. It is for exactly this reason that railroads are built not with curves shaped like arcs of circles but with *approach curves* in which the curvature, and hence the normal acceleration, build up smoothly.

EXAMPLE 6 A particle moves in the xy-plane with parametric equations

$$x(t) = \tfrac{3}{2}t^2, \quad y(t) = \tfrac{4}{3}t^3.$$

Find the tangential and normal components of its acceleration vector when $t = 1$.

Solution The trajectory and the vectors **N** and **T** appear in Fig. 12.6.10. There **N** and **T** are shown attached at the point of evaluation, at which $t = 1$. The particle has position vector

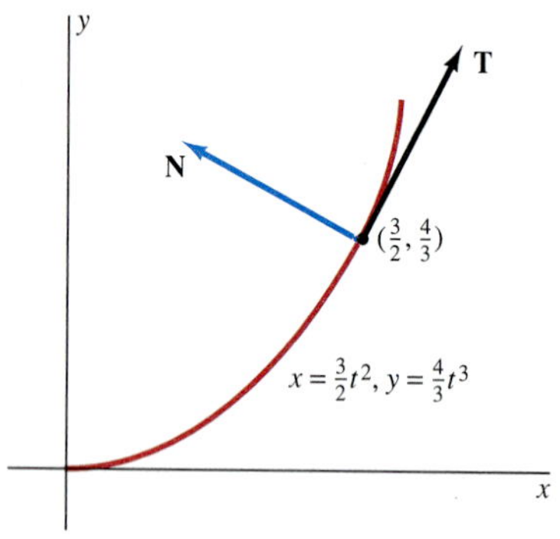

Fig. 12.6.10 The moving particle of Example 6

$$\mathbf{r}(t) = \tfrac{3}{2}t^2\mathbf{i} + \tfrac{4}{3}t^3\mathbf{j}$$

and thus velocity

$$\mathbf{v}(t) = 3t\mathbf{i} + 4t^2\mathbf{j}.$$

Hence its speed is

$$v(t) = \sqrt{9t^2 + 16t^4},$$

from which we calculate

$$a_T = \frac{dv}{dt} = \frac{9t + 32t^3}{\sqrt{9t^2 + 16t^4}}.$$

Thus $v = 5$ and $a_T = \frac{41}{5}$ when $t = 1$.

To use Eq. (12) to compute the curvature at $t = 1$, we compute $dx/dt = 3t$, $dy/dt = 4t^2$, $d^2x/dt^2 = 3$, and $d^2y/dt^2 = 8t$. Thus at $t = 1$ we have

$$\kappa = \frac{|x'y'' - x''y'|}{v^3} = \frac{|3 \cdot 8 - 3 \cdot 4|}{5^3} = \frac{12}{125}.$$

Hence

$$a_N = \kappa v^2 = \tfrac{12}{125} \cdot 5^2 = \tfrac{12}{5}$$

when $t = 1$. As a check (Problem 28), you might compute **T** and **N** when $t = 1$ and verify that

$$\tfrac{41}{5}\mathbf{T} + \tfrac{12}{5}\mathbf{N} = \mathbf{a} = 3\mathbf{i} + 8\mathbf{j}.$$ ■

It remains for us to see how to compute a_T, a_N, and **N** effectively in the case of a space curve. We would prefer to have formulas that explicitly contain only the vectors **r**, **v**, and **a**.

If we compute the dot product of $\mathbf{v} = v\mathbf{T}$ with the acceleration **a** as given in Eq. (22) and use the facts that $\mathbf{T} \cdot \mathbf{T} = 1$ and $\mathbf{T} \cdot \mathbf{N} = 0$, we get

$$\mathbf{v} \cdot \mathbf{a} = v\mathbf{T} \cdot \left(\frac{dv}{dt}\mathbf{T}\right) + (v\mathbf{T}) \cdot (\kappa v^2 \mathbf{N}) = v\frac{dv}{dt}.$$

It follows that

$$a_T = \frac{dv}{dt} = \frac{\mathbf{v} \cdot \mathbf{a}}{v} = \frac{\mathbf{r}'(t) \cdot \mathbf{r}''(t)}{|\mathbf{r}'(t)|}. \tag{26}$$

Similarly, when we compute the cross product of $\mathbf{v} = v\mathbf{T}$ with each side of Eq. (22), we find that

$$\mathbf{v} \times \mathbf{a} = \left(v\mathbf{T} \times \frac{dv}{dt}\mathbf{T}\right) + (v\mathbf{T} \times \kappa v^2 \mathbf{N}) = \kappa v^3 (\mathbf{T} \times \mathbf{N}).$$

Because κ and v are nonnegative and because $\mathbf{T} \times \mathbf{N}$ is a unit vector, we may conclude that

$$\kappa = \frac{|\mathbf{v} \times \mathbf{a}|}{v^3} = \frac{|\mathbf{r}'(t) \times \mathbf{r}''(t)|}{|\mathbf{r}'(t)|^3}. \tag{27}$$

It now follows from Eq. (24) that

$$a_N = \frac{|\mathbf{r}'(t) \times \mathbf{r}''(t)|}{|\mathbf{r}'(t)|}. \tag{28}$$

The curvature of a space curve often is not as easy to compute directly from the definition as we found in the case of the helix of Example 5. It is generally more convenient to use Eq. (27). Once $\mathbf{a}$, $\mathbf{T}$, a_T, and a_N have been computed, we can rewrite Eq. (25) as

$$\mathbf{N} = \frac{\mathbf{a} - a_T\mathbf{T}}{a_N} \tag{29}$$

to find the principal unit normal vector.

EXAMPLE 7 Compute $\mathbf{T}$, $\mathbf{N}$, κ, a_T, and a_N at the point $(1, \frac{1}{2}, \frac{1}{3})$ of the twisted cubic with parametric equations

$$x(t) = t, \quad y(t) = \tfrac{1}{2}t^2, \quad z(t) = \tfrac{1}{3}t^3.$$

Solution Differentiation of the position vector

$$\mathbf{r}(t) = \langle t, \tfrac{1}{2}t^2, \tfrac{1}{3}t^3 \rangle$$

gives

$$\mathbf{r}'(t) = \langle 1, t, t^2 \rangle \quad \text{and} \quad \mathbf{r}''(t) = \langle 0, 1, 2t \rangle.$$

When we substitute $t = 1$, we obtain

$$\mathbf{v}(1) = \langle 1, 1, 1 \rangle \qquad \text{(velocity)},$$

$$v(1) = |\mathbf{v}(1)| = \sqrt{3} \qquad \text{(speed), and}$$

$$\mathbf{a}(1) = \langle 0, 1, 2 \rangle \qquad \text{(acceleration)}$$

at the point $(1, \frac{1}{2}, \frac{1}{3})$. Then Eq. (26) gives the tangential component of acceleration:

$$a_T = \frac{\mathbf{v} \cdot \mathbf{a}}{v} = \frac{3}{\sqrt{3}} = \sqrt{3}.$$

Because

$$\mathbf{v} \times \mathbf{a} = \begin{vmatrix} \mathbf{i} & \mathbf{j} & \mathbf{k} \\ 1 & 1 & 1 \\ 0 & 1 & 2 \end{vmatrix} = \langle 1, -2, 1 \rangle,$$

Eq. (27) gives the curvature:

$$\kappa = \frac{|\mathbf{v} \times \mathbf{a}|}{v^3} = \frac{\sqrt{6}}{(\sqrt{3})^3} = \frac{\sqrt{2}}{3}.$$

The normal component of acceleration is $a_N = \kappa v^2 = \sqrt{2}$. The unit tangent vector is

$$\mathbf{T} = \frac{\mathbf{v}}{v} = \frac{1}{\sqrt{3}}\langle 1, 1, 1 \rangle = \frac{\mathbf{i} + \mathbf{j} + \mathbf{k}}{\sqrt{3}}.$$

Finally, Eq. (29) gives

$$\mathbf{N} = \frac{\mathbf{a} - a_T\mathbf{T}}{a_N} = \frac{1}{\sqrt{2}}(\langle 0, 1, 2 \rangle - \langle 1, 1, 1 \rangle) = \frac{1}{\sqrt{2}}\langle -1, 0\, 1 \rangle = \frac{-\mathbf{i} + \mathbf{k}}{\sqrt{2}}.$$

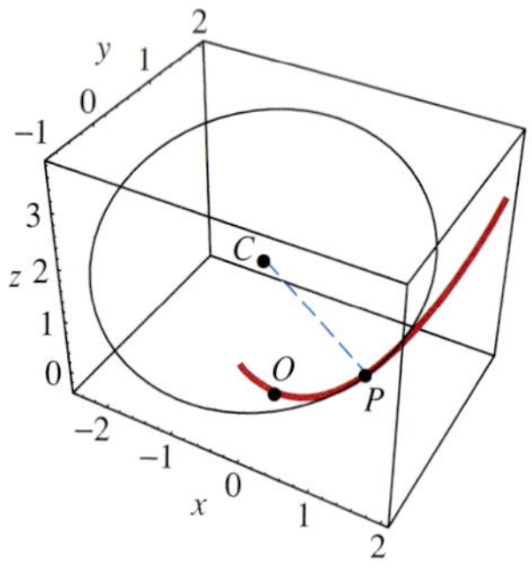

Fig. 12.6.11 Osculating circle for the twisted cubic of Example 7. It is plotted as the parametric curve with position vector $\mathbf{r}(t) = \overrightarrow{OC} - (a \cos t)\mathbf{N} + (a \sin t)\,\mathbf{T}$.

Figure 12.6.11 shows the twisted cubic and its osculating circle at the point $P\left(1, \frac{1}{2}, \frac{1}{3}\right)$. This osculating circle has radius $a = 1/\kappa = \frac{3}{2}\sqrt{2}$, and its center C has position vector $\overrightarrow{OC} = \overrightarrow{OP} + a\mathbf{N} = \left\langle -\frac{1}{2}, \frac{1}{2}, \frac{11}{6} \right\rangle$. ■

Newton, Kepler, and the Solar System

Ancient Greek mathematicians and astronomers developed an elaborate mathematical model to account for the complicated motions of the sun, moon, and six planets then known as viewed from the earth. A combination of uniform circular motions was used to describe the motion of each body around the earth—if the earth is placed at the origin, then each body *does* orbit the earth.

In this system, it was typical for a planet P to travel uniformly around a small circle (the *epicycle*) with center C, which in turn traveled uniformly around a circle centered at the earth, labeled E in Fig. 12.6.12. The radii of the circles and the angular speeds of P and C were chosen to match the observed motion of the planet as closely as possible. For greater accuracy, the ancient Greeks could use secondary circles. In fact, several circles were required for each body in the solar system. The theory of epicycles reached its definitive form in Ptolemy's *Almagest* of the second century A.D.

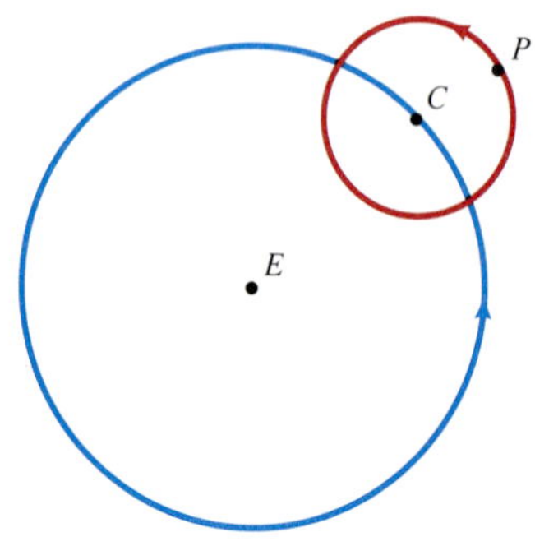

Fig. 12.6.12 The small circle is the epicycle.

In 1543, Copernicus altered Ptolemy's approach by placing the center of each primary circle at the sun rather than at the earth. This change was of much greater philosophical than mathematical importance. For, contrary to popular belief, this *heliocentric system* was *not* simpler than Ptolemy's geocentric system. Indeed, Copernicus's system actually required more circles.

It was Johannes Kepler (1571–1630) who finally got rid of all these circles. On the basis of a detailed analysis of a lifetime of planetary observations by the Danish astronomer Tycho Brahe, Kepler stated the following three propositions, now known as **Kepler's laws of planetary motion.**

1. The orbit of each planet is an ellipse with the sun at one focus.
2. The radius vector from the sun to a planet sweeps out area at a constant rate.
3. The *square* of the period of revolution of a planet is proportional to the *cube* of the major semiaxis of its elliptical orbit.

In his *Principia Mathematica* (1687), Newton showed that Kepler's laws follow from the basic principles of mechanics ($F = ma$, and so on) and the inverse-square law of gravitational attraction. His success in using mathematics to explain natural phenomena ("I now demonstrate the frame of the System of the World") inspired confidence that the universe could be understood and perhaps even mastered. This new confidence permanently altered humanity's perception of itself and of its place in the scheme of things.

Newton employed a powerful but now antiquated form of geometrical calculus in the *Principia.* In the remainder of this section we apply the modern calculus of vector functions to outline the relation between Newton's laws and Kepler's laws.

Radial and Transverse Components of Acceleration

To begin, we set up a coordinate system in which the sun is located at the origin in the plane of motion of a planet. Let $r = r(t)$ and $\theta = \theta(t)$ be the polar coordinates at time t of the planet as it orbits the sun. We want first to split the planet's position, velocity, and acceleration vectors $\mathbf{r}$, $\mathbf{v}$, and $\mathbf{a}$ into *radial* and *transverse* components. To do so, we introduce at each point (r, θ) of the plane (the origin excepted) the *unit* vectors

$$\mathbf{u}_r = \mathbf{i}\cos\theta + \mathbf{j}\sin\theta, \qquad \mathbf{u}_\theta = -\mathbf{i}\sin\theta + \mathbf{j}\cos\theta. \tag{30}$$

If we substitute $\theta = \theta(t)$, then $\mathbf{u}_r$ and $\mathbf{u}_\theta$ become functions of t. The **radial** unit vector $\mathbf{u}_r$ always points directly away from the origin; the **transverse** unit vector $\mathbf{u}_\theta$ is obtained from $\mathbf{u}_r$ by a 90° counterclockwise rotation (Fig. 12.6.13).

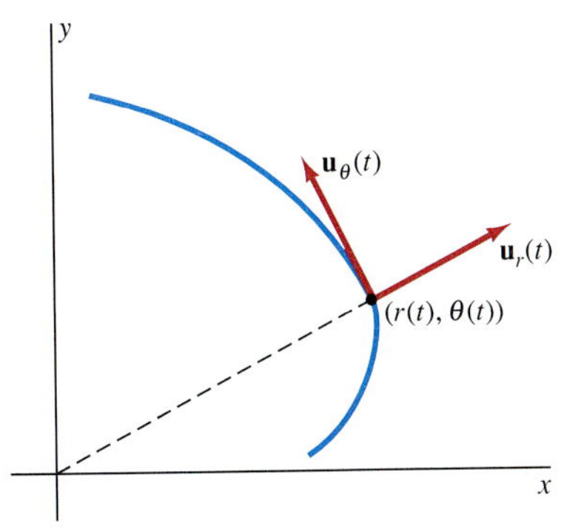

Fig. 12.6.13 The radial and transverse unit vector $\mathbf{u}_r$ and $\mathbf{u}_\theta$

In Problem 68 we ask you to verify, by componentwise differentiation of the equations in (30), that

$$\frac{d\mathbf{u}_r}{dt} = \mathbf{u}_\theta\frac{d\theta}{dt} \quad \text{and} \quad \frac{d\mathbf{u}_\theta}{dt} = -\mathbf{u}_r\frac{d\theta}{dt}. \tag{31}$$

The position vector $\mathbf{r}$ points directly away from the origin and has length $|\mathbf{r}| = r$, so

$$\mathbf{r} = r\mathbf{u}_r. \tag{32}$$

Differentiation of both sides of Eq. (32) with respect to t gives

$$\mathbf{v} = \frac{d\mathbf{r}}{dt} = \mathbf{u}_r\frac{dr}{dt} + r\frac{d\mathbf{u}_r}{dt}.$$

We use the first equation in (31) and find that the planet's velocity vector is

$$\mathbf{v} = \mathbf{u}_r\frac{dr}{dt} + r\frac{d\theta}{dt}\mathbf{u}_\theta. \tag{33}$$

Thus we have expressed the velocity $\mathbf{v}$ in terms of the radial vector $\mathbf{u}_r$ and the transverse vector $\mathbf{u}_\theta$.

We differentiate both sides of Eq. (33) and thereby find that

$$\mathbf{a} = \frac{d\mathbf{v}}{dt} = \left(\mathbf{u}_r \frac{d^2r}{dt^2} + \frac{dr}{dt}\frac{d\mathbf{u}_r}{dt}\right) + \left(\frac{dr}{dt}\frac{d\theta}{dt}\mathbf{u}_\theta + r\frac{d^2\theta}{dt^2}\mathbf{u}_\theta + r\frac{d\theta}{dt}\frac{d\mathbf{u}_\theta}{dt}\right).$$

Then, by using the equations in (31) and collecting the coefficients of $\mathbf{u}_r$ and $\mathbf{u}_\theta$ (Problem 69), we obtain the decomposition

$$\mathbf{a} = \left[\frac{d^2r}{dt^2} - r\left(\frac{d\theta}{dt}\right)^2\right]\mathbf{u}_r + \left[\frac{1}{r}\frac{d}{dt}\left(r^2\frac{d\theta}{dt}\right)\right]\mathbf{u}_\theta \tag{34}$$

of the acceleration vector into its radial and transverse components.

Planets and Satellites

The key to Newton's analysis was the connection between his law of gravitational attraction and Kepler's *second* law of planetary motion. Suppose that we begin with the inverse-square law of gravitation in its vector form

$$\mathbf{F} = m\mathbf{a} = -\frac{GMm}{r^2}\mathbf{u}_r, \tag{35}$$

where M denotes the mass of the sun and m the mass of the orbiting planet. Then the acceleration of the planet is given *also* by

$$\mathbf{a} = -\frac{\mu}{r^2}\mathbf{u}_r, \tag{36}$$

where $\mu = GM$. We equate the transverse components in Eqs. (34) and (36) and thus obtain

$$\frac{1}{r}\cdot\frac{d}{dt}\left(r^2\frac{d\theta}{dt}\right) = 0.$$

We drop the factor $1/r$, then antidifferentiate both sides. We find that

$$r^2\frac{d\theta}{dt} = h \qquad (h \text{ a constant}). \tag{37}$$

We know from Section 10.3 that if $A(t)$ denotes the area swept out by the planet's radius vector from time 0 to time t (Fig. 12.6.14), then

$$A(t) = \int_{\star}^{\star\star}\frac{1}{2}r^2\,d\theta = \int_0^t \frac{1}{2}r^2\frac{d\theta}{dt}\,dt.$$

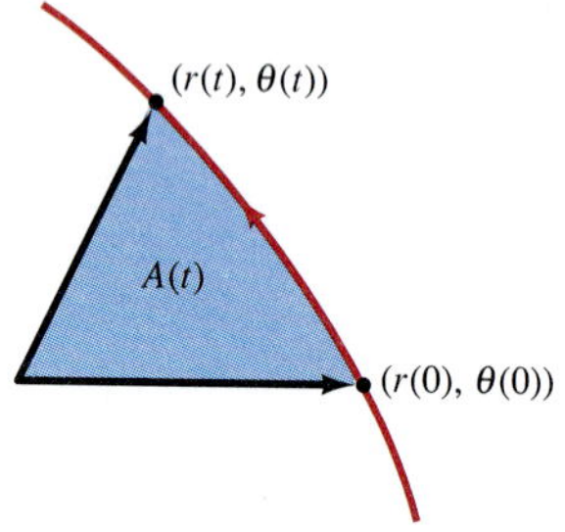

Fig. 12.6.14 Area swept out by the radius vector

Now we apply the fundamental theorem of calculus, which yields

$$\frac{dA}{dt} = \frac{1}{2}r^2\frac{d\theta}{dt}. \tag{38}$$

When we compare Eqs. (37) and (38), we see that

$$\frac{dA}{dt} = \frac{h}{2}. \tag{39}$$

Because $h/2$ is a constant, we have derived Kepler's second law: The radius vector from sun to planet sweeps out area at a constant rate.

Next we outline the derivation of Newton's law of gravitation from Kepler's first and second laws of planetary motion. According to Problem 65, the polar-coordinate equation of an ellipse with eccentricity $e < 1$ and directrix $x = p$ (Fig. 12.6.15) is

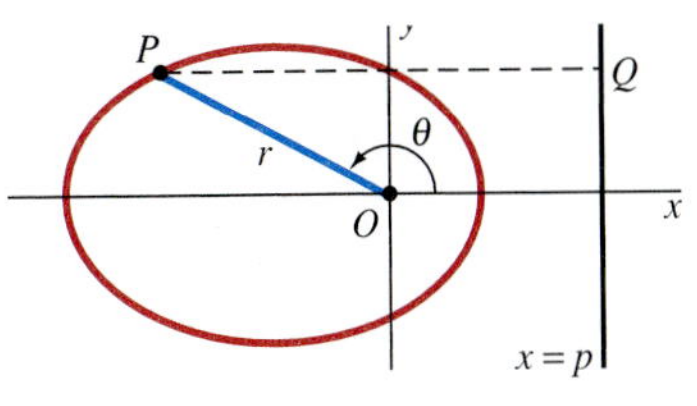

Fig. 12.6.15 A polar coordinate ellipse with eccentricity $e = |OP|/|PQ|$

$$r = \frac{pe}{1 + e\cos\theta}. \tag{40}$$

In Problem 66 we ask you to show by differentiating twice, using the chain rule and Kepler's second law in the form in Eq. (37), that Eq. (40) implies that

$$\frac{d^2r}{dt^2} = \frac{h^2}{r^2}\left(\frac{1}{r} - \frac{1}{pe}\right). \tag{41}$$

Now if Kepler's second law in the form in Eq. (37) holds, then Eq. (34) gives

$$\mathbf{a} = \left[\frac{d^2r}{dt^2} - r\left(\frac{d\theta}{dt}\right)^2\right]\mathbf{u}_r \tag{42}$$

for the planet's acceleration vector. Finally, upon substituting $d\theta/dt = h/r^2$ from Eq. (37) and the expression in Eq. (41) for d^2r/dt^2, we find (Problem 67) that Eq. (42) can be simplified to the form

$$\mathbf{a} = -\frac{h^2}{per^2}\mathbf{u}_r. \tag{43}$$

This is the inverse-square law of gravitation in the form of Eq. (36) with $\mu = h^2/pe$.

Now suppose that the elliptical orbit of a planet around the sun has major semiaxis a and minor semiaxis b. Then the constant

$$pe = \frac{h^2}{\mu}$$

that appears in Eq. (42) satisfies the equations

$$pe = a(1 - e^2) = a\left(1 - \frac{a^2 - b^2}{a^2}\right) = \frac{b^2}{a}.$$

[This follows from Eq. (40) as in Problem 64.] We equate these two expressions for pe and find that $h^2 = \mu b^2/a$.

Now let T denote the period of revolution of the planet—the time required for it to complete one full revolution in its elliptical orbit around the sun. Then we see from Eq. (38) that the area of the ellipse bounded by this orbit is $A = \frac{1}{2}hT = \pi ab$ and thus that

$$T^2 = \frac{4\pi^2a^2b^2}{h^2} = \frac{4\pi^2a^2b^2}{\mu b^2/a}.$$

Therefore,

$$T^2 = \gamma a^3, \tag{44}$$

where the proportionality constant $\gamma = 4\pi^2/\mu = 4\pi^2/GM$ [compare Eqs. (35) and (36)] depends on the gravitational constant G and the sun's mass M. Thus we have derived Kepler's third law of planetary motion from his first two laws and Newton's law of gravitational attraction.

EXAMPLE 8 The period of revolution of Mercury in its elliptical orbit around the sun is $T = 87.97$ days, whereas that of Earth is 365.26 days. Compute the major semiaxis (in astronomical units) of the orbit of Mercury. See Fig. 12.6.16.

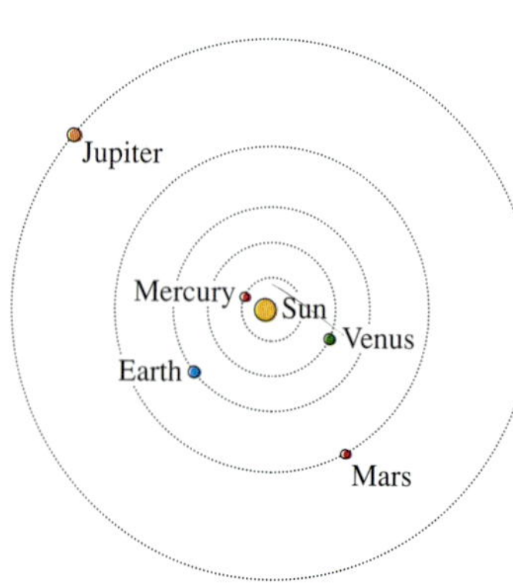

Fig. 12.6.16 The inner planets of the solar system (Example 8)

Solution The major semiaxis of the orbit of the earth is, by definition, 1 AU. So Eq. (44) gives the value of the constant $\gamma = (365.26)^2$ (in day^2/AU^3). Hence the major semiaxis of the orbit of Mercury is

$$a = \left(\frac{T^2}{\gamma}\right)^{1/3} = \left(\frac{(87.97)^2}{(365.26)^2}\right)^{1/3} \approx 0.387 \quad \text{(AU)}.$$

As yet we have considered only planets in orbits around the sun. But Kepler's laws and the equations of this section apply to bodies in orbit around any common central mass, so long as they move solely under the influence of *its* gravitational attraction. Examples include satellites (artificial or natural) orbiting the earth or the moons of Jupiter.

EXAMPLE 9 A communications relay satellite is to be placed in a circular orbit around Earth and is to have a period of revolution of 24 h. This is a *geosynchronous* orbit in which the satellite appears to be stationary in the sky. Assume that the earth's natural moon has a period of 27.32 days in a circular orbit of radius 238,850 mi. What should be the radius of the satellite's orbit? See Fig. 12.6.17.

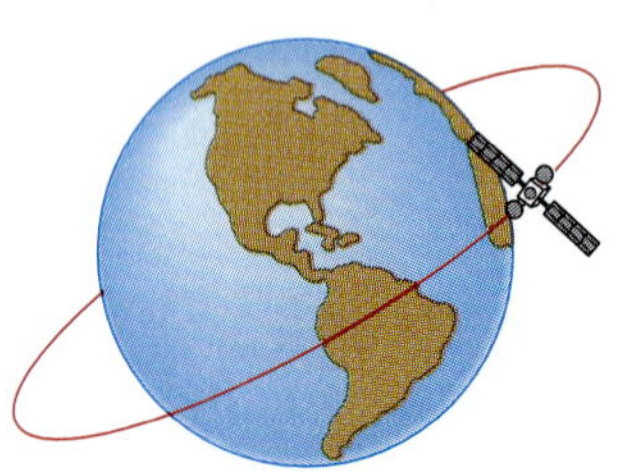

Fig. 12.6.17 A communications satellite in orbit around the earth (Example 9)

Solution Equation (44), when applied to the moon, yields

$$(27.32)^2 = \gamma(238{,}850)^3.$$

For the stationary satellite that has period $T = 1$ (day), it yields $1^2 = \gamma r^3$, where r is the radius of the geosynchronous orbit. To eliminate γ, we divide the second of these equations by the first and find that

$$r^3 = \frac{(238{,}850)^3}{(27.32)^2}.$$

Thus r is approximately 26330 mi. The radius of the earth is about 3960 mi, so the satellite will be 22370 mi above the surface.

12.6 PROBLEMS

Find the arc length of the curves described in Problems 1 through 6.

1. $x = 3\sin 2t,\ y = 3\cos 2t,\ z = 8t;$ from $t = 0$ to $t = \pi$

2. $x = t,\ y = t^2/\sqrt{2},\ z = t^3/3;$ from $t = 0$ to $t = 1$

3. $x = 6e^t\cos t,\ y = 6e^t\sin t,\ z = 17e^t;$ from $t = 0$ to $t = 1$

4. $x = t^2/2,\ y = \ln t,\ z = t\sqrt{2};$ from $t = 1$ to $t = 2$

5. $x = 3t\sin t,\ y = 3t\cos t,\ z = 2t^2;$ from $t = 0$ to $t = 4/5$

6. $x = 2e^t,\ y = e^{-t},\ z = 2t;$ from $t = 0$ to $t = 1$

In Problems 7 through 12, find the curvature of the given plane curve at the indicated point.

7. $y = x^3$ at $(0, 0)$

8. $y = x^3$ at $(-1, -1)$

9. $y = \cos x$ at $(0, 1)$

10. $x = t - 1,\ y = t^2 + 3t + 2$, where $t = 2$

11. $x = 5\cos t,\ y = 4\sin t$, where $t = \pi/4$

12. $x = 5\cosh t,\ y = 3\sinh t$, where $t = 0$

In Problems 13 through 16, find the point or points on the given curve at which the curvature is a maximum.

13. $y = e^x$

14. $y = \ln x$

15. $x = 5\cos t,\ y = 3\sin t$

16. $xy = 1$

For the plane curves in Problems 17 through 21, find the unit tangent and normal vectors at the indicated point.

17. $y = x^3$ at $(-1, -1)$

18. $x = t^3,\ y = t^2$ at $(-1, 1)$

19. $x = 3\sin 2t,\ y = 4\cos 2t$, where $t = \pi/6$

20. $x = t - \sin t,\ y = 1 - \cos t$, where $t = \pi/2$

21. $x = \cos^3 t,\ y = \sin^3 t$, where $t = 3\pi/4$

The position vector of a particle moving in the plane is given in Problems 22 through 26. Find the tangential and normal components of the acceleration vector.

22. $\mathbf{r}(t) = 3\mathbf{i}\sin \pi t + 3\mathbf{j}\cos \pi t$

23. $\mathbf{r}(t) = (2t + 1)\mathbf{i} + (3t^2 - 1)\mathbf{j}$

24. $\mathbf{r}(t) = \mathbf{i}\cosh 3t + \mathbf{j}\sinh 3t$

25. $\mathbf{r}(t) = \mathbf{i}t\cos t + \mathbf{j}t\sin t$ **26.** $\mathbf{r}(t) = \langle e^t \sin t, e^t \cos t\rangle$

27. Use Eq. (13) to compute the curvature of the circle with equation $x^2 + y^2 = a^2$.

28. Verify the equation $\frac{41}{5}\mathbf{T} + \frac{12}{5}\mathbf{N} = 3\mathbf{i} + 8\mathbf{j}$ given at the end of Example 6.

In Problems 29 through 31, find the equation of the osculating circle for the given plane curve at the indicated point.

29. $y = 1 - x^2$ at $(0, 1)$ **30.** $y = e^x$ at $(0, 1)$

31. $xy = 1$ at $(1, 1)$

Find the curvature κ of the space curves with position vectors given in Problems 32 through 36.

32. $\mathbf{r}(t) = t\mathbf{i} + (2t - 1)\mathbf{j} + (3t + 5)\mathbf{k}$

33. $\mathbf{r}(t) = t\mathbf{i} + \mathbf{j}\sin t + \mathbf{k}\cos t$

34. $\mathbf{r}(t) = \langle t, t^2, t^3\rangle$

35. $\mathbf{r}(t) = \langle e^t \cos t, e^t \sin t, e^t\rangle$

36. $\mathbf{r}(t) = \mathbf{i}t\sin t + \mathbf{j}t\cos t + \mathbf{k}t$

37 through **41.** Find the tangential and normal components of acceleration a_T and a_N for the curves of Problems 32 through 36, respectively.

In Problems 42 through 45, find the unit vectors **T** and **N** for the given curve at the indicated point.

42. The curve of Problem 34 at $(1, 1, 1)$

43. The curve of Problem 33 at $(0, 0, 1)$

44. The curve of Problem 3 at $(6, 0, 17)$

45. The curve of Problem 35 at $(1, 0, 1)$

46. Find **T**, **N**, a_T, and a_N as functions of t for the helix of Example 1.

47. Find the arc-length parametrization of the line

$$x(t) = 2 + 4t, \quad y(t) = 1 - 12t, \quad z(t) = 3 + 3t$$

in terms of the arc length s measured from the initial point $(2, 1, 3)$.

48. Find the arc-length parametrization of the circle

$$x(t) = 2\cos t, \quad y(t) = 2\sin t, \quad z = 0.$$

49. Find the arc-length parametrization of the helix

$$x(t) = 3\cos t, \quad y(t) = 3\sin t, \quad z(t) = 4t$$

in terms of the arc length s measured from the initial point $(3, 0, 0)$.

50. Substitute $x = t$, $y = f(t)$, and $z = 0$ into Eq. (27) to verify that the curvature of the plane curve $y = f(x)$ is

$$\kappa(x) = \frac{|f''(x)|}{[1 + (f'(x))^2]^{3/2}}.$$

51. A particle moves under the influence of a force that is always perpendicular to its direction of motion. Show that the speed of the particle must be constant.

52. Deduce from Eq. (20) that

$$\kappa = \frac{\sqrt{a^2 - (a_T)^2}}{v^2} = \frac{\sqrt{(x''(t))^2 + (y''(t))^2 - (v'(t))^2}}{(x'(t))^2 + (y'(t))^2}.$$

53. Apply the formula of Problem 52 to calculate the curvature of the curve

$$x(t) = \cos t + t\sin t, \quad y(t) = \sin t - t\cos t.$$

54. The folium of Descartes with equation $x^3 + y^3 = 3xy$ is shown in Fig. 12.6.18. Find the curvature and center of curvature of this folium at the point $(\frac{3}{2}, \frac{3}{2})$. Begin by calculating dy/dx and d^2y/dx^2 by implicit differentiation.

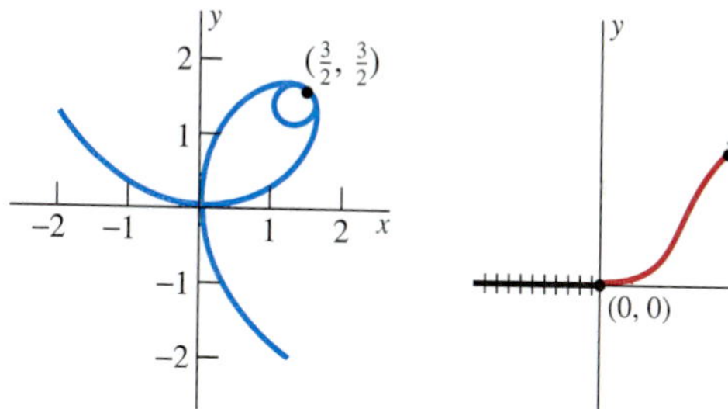

Fig. 12.6.18 The folium of Descartes (Problem 54)

Fig. 12.6.19 Connecting railroad tracks (Problem 55)

55. Determine the constants A, B, C, D, E, and F so that the curve

$$y = Ax^5 + Bx^4 + Cx^3 + Dx^2 + Ex + F$$

does, simultaneously, all of the following:

- Joins the two points $(0, 0)$and $(1, 1)$;
- Has slope 0 at $(0, 0)$and slope 1 at $(1, 1)$;
- Has curvature 0 at both $(0, 0)$and $(1, 1)$.

The curve in question is shown in color in Fig. 12.6.19. Why would this be a good curve to join the railroad tracks, shown in black in the figure?

56. Consider a body in an elliptical orbit with major and minor semiaxes a and b and period of revolution T. (a) Deduce from Eq. (33) that $v = r(d\theta/dt)$ when the body is nearest to and farthest from its foci. (b) Then apply Kepler's second law to conclude that $v = 2\pi ab/(rT)$ at the body's nearest and farthest points.

In Problems 57 through 60, apply the equation of part (b) of Problem 56 to compute the speed (in miles per second) of the given body at the nearest and farthest points of its orbit. Convert 1 AU, the major semiaxis of Earth's orbit, into 92,956,000 mi.

57. Mercury: $a = 0.387$ AU, $e = 0.206$, $T = 87.97$ days

58. Earth: $e = 0.0167$, $T = 365.26$ days

59. Earth's moon: $a = 238{,}900$ mi, $e = 0.055$, $T = 27.32$ days

60. An artificial Earth satellite: $a = 10000$ mi, $e = 0.5$

61. Assuming Earth to be a sphere with radius 3960 mi, find the altitude above Earth's surface of a satellite in a circular orbit that has a period of revolution of 1 h.

62. Given the fact that Jupiter's period of (almost) circular revolution around the Sun is 11.86 yr, calculate the distance of Jupiter from the Sun.

63. Suppose that an Earth satellite in elliptical orbit varies in altitude from 100 to 1000 mi above Earth's surface (assumed spherical). Find this satellite's period of revolution.

64. Substitute $\theta = 0$ and $\theta = \pi$ into Eq. (40) to deduce that $pe = a(1 - e^2)$.

65. Figure 12.6.15 shows an ellipse with eccentricity e and focus at the origin. Derive Eq. (40) from the defining relation $|OP| = e|PQ|$ of the ellipse.

66. (a) Beginning with the polar-coordinates equation of an ellipse in Eq. (40), apply the chain rule and Kepler's second law in the form $d\theta/dt = h/r^2$ to differentiate r with respect to t and thereby show that $dr/dt = (h \sin\theta)/p$. (b) Differentiate again to show that $d^2r/dt^2 = (h^2 \cos\theta)/(pr^2)$. (c) Derive Eq. (41) by solving Eq. (40) for $\cos\theta$ and substituting the result in the formula in part (b).

67. Derive Eq. (43) by substituting the expressions for $d\theta/dt$ and d^2r/dt^2 given by Eqs. (37) and (41), respectively, into Eq. (42).

68. Derive both equations in (31) by differentiation of the equations in (30).

69. Derive Eq. (34) by differentiating Eq. (33).

12.7 CYLINDERS AND QUADRIC SURFACES

Just as the graph of an equation $f(x, y) = 0$ is generally a curve in the xy-plane, the graph of an equation in three variables is generally a surface in space. A function F of three variables associates a real number $F(x, y, z)$ with each ordered triple (x, y, z) of real numbers. The **graph** of the equation

$$F(x, y, z) = 0 \tag{1}$$

is the set of all points whose coordinates (x, y, z) satisfy this equation. We refer to the graph of such an equation as a **surface**. For instance, the graph of the equation

$$x^2 + y^2 + z^2 - 1 = 0$$

is a familiar surface, the unit sphere centered at the origin. But note that the graph of Eq. (1) does not always agree with our intuitive notion of a surface. For example, the graph of the equation

$$(x^2 + y^2)(y^2 + z^2)(z^2 + x^2) = 0$$

consists of the points lying on the three coordinate axes in space, because

- $x^2 + y^2 = 0$ implies that $x = y = 0$ (the z-axis)
- $y^2 + z^2 = 0$ implies that $y = z = 0$ (the x-axis)
- $z^2 + x^2 = 0$ implies that $z = x = 0$ (the y-axis).

We leave for advanced calculus the precise definition of *surface* as well as the study of conditions sufficient to imply that the graph of Eq. (1) actually is a surface.

Planes and Traces

The simplest example of a surface is a plane with linear equation $Ax + By + Cz + D = 0$. In this section we discuss examples of other simple surfaces that frequently appear in multivariable calculus.

When sketching a surface S, it is often helpful to examine its intersections with various planes. The **trace** of the surface S in the plane $\mathcal{P}$ is the intersection of $\mathcal{P}$ and S.

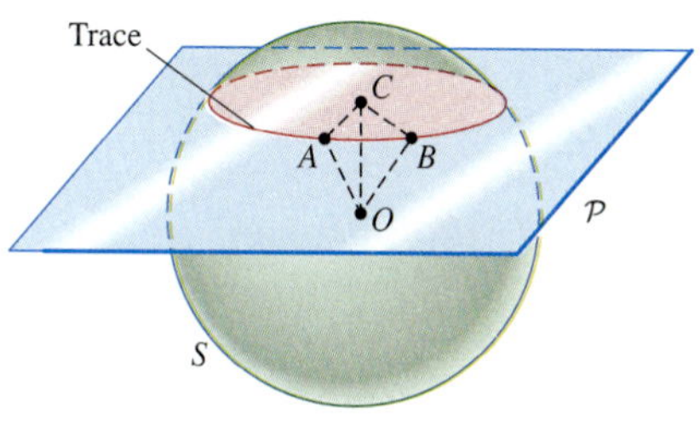

Fig. 12.7.1 The intersection of the sphere S and the plane $\mathcal{P}$ is a circle.

For example, if S is a sphere, then we can verify by the methods of elementary geometry that the trace of S in the plane $\mathcal{P}$ is a circle (Fig. 12.7.1), provided that $\mathcal{P}$ intersects the sphere but is not merely tangent to it (Problem 49). Figure 12.7.2 illustrates the horizontal trace circles that (together with two "polar points") make up the sphere $x^2 + y^2 + z^2 = 25$.

When we want to visualize a specific surface in space, it often suffices to examine its traces in the coordinate planes and possibly a few planes parallel to them, as in Example 1.

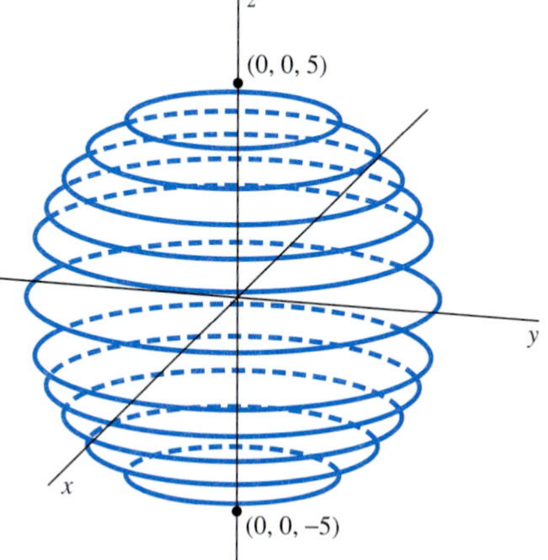

Fig. 12.7.2 A sphere as a union of circles (and two points)

EXAMPLE 1 Consider the plane with equation $3x + 2y + 2z = 6$. We find its trace in the xy-plane by setting $z = 0$. The equation then reduces to the equation $3x + 2y = 6$ of a straight line in the xy-plane. Similarly, when we set $y = 0$, we get the line $3x + 2z = 6$ as the trace of the given plane in the xz-plane. To find its trace in the yz-plane, we set $x = 0$, and this yields the line $y + z = 3$. Figure 12.7.3 shows the portions of these three trace lines that lie in the first octant. Together they give us a good picture of how the plane $3x + 2y + 2z = 6$ is situated in space. ■

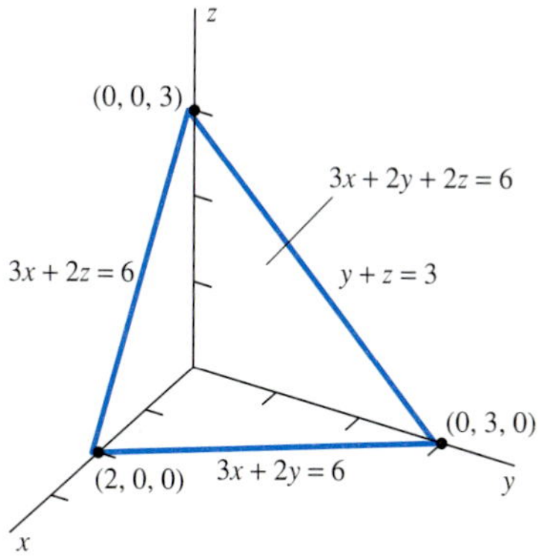

Fig. 12.7.3 Traces of the plane $3x + 2y + 2z = 6$ in the coordinate planes (Example 1)

Cylinders and Rulings

Let C be a curve in a plane and let L be a line not parallel to that plane. Then the set of points on lines parallel to L that intersect C is called a **cylinder**. These straight lines that make up the cylinder are called **rulings** of the cylinder.

EXAMPLE 2 Figure 12.7.4 shows a vertical cylinder for which C is the circle $x^2 + y^2 = a^2$ in the xy-plane. The trace of this cylinder in any horizontal plane $z = c$ is a circle with radius a and center $(0, 0, c)$ on the z-axis. Thus the point (x, y, z) lies on this cylinder if and only if $x^2 + y^2 = a^2$. Hence this cylinder is the graph of the equation $x^2 + y^2 = a^2$, an equation in **three** variables—even though the variable z is technically missing.

The fact that the variable z does not appear explicitly in the equation $x^2 + y^2 = a^2$ means that given any point $(x_0, y_0, 0)$ on the *circle* $x^2 + y^2 = a^2$ in the xy-plane, the point (x_0, y_0, z) lies on the cylinder for any and all values of z. The set of all such points is the vertical line through the point $(x_0, y_0, 0)$. Thus this vertical line is a ruling of the *cylinder* $x^2 + y^2 = a^2$. Figure 12.7.5 exhibits the cylinder as the union of its rulings. ■

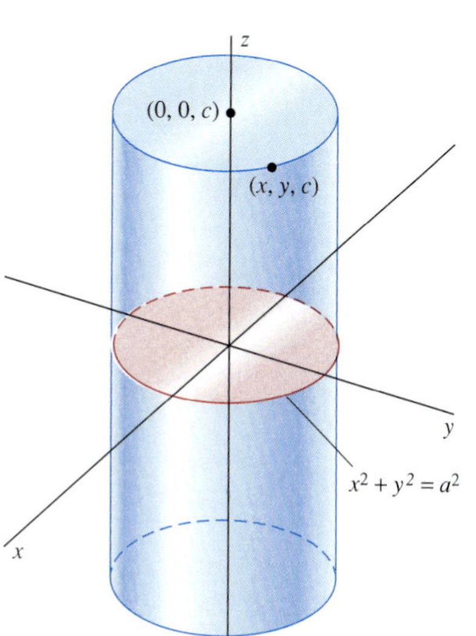

Fig. 12.7.4 A right circular cylinder

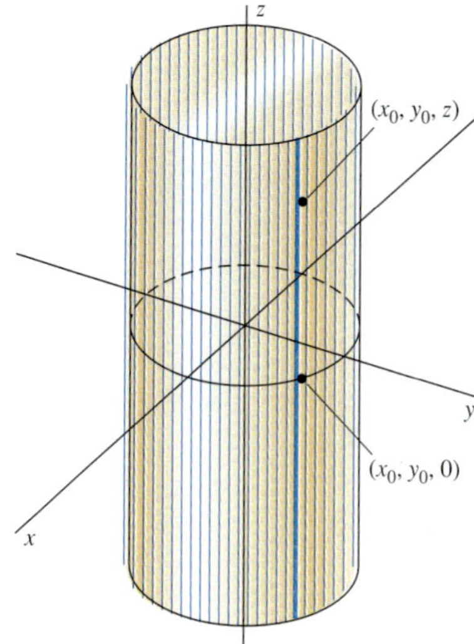

Fig. 12.7.5 The cylinder $x^2 + y^2 = a^2$; its rulings are parallel to the z-axis.

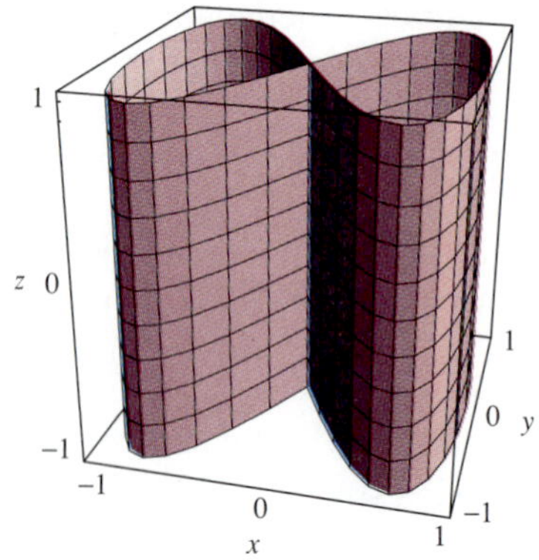

Fig. **12.7.6** The vertical cylinder through the figure-eight curve $x = \sin t$, $y = \sin 2t$

A cylinder need not be circular—that is, the curve C can be a ellipse, a rectangle, or a quite arbitrary curve.

EXAMPLE 3 Figure 12.7.6 shows both horizontal traces and vertical rulings on a vertical cylinder through a figure-eight curve C in the xy-plane (C has the parametric equations $x = \sin t$, $y = \sin 2t$, $0 \leqq t \leqq 2\pi$). ■

If the curve C in the xy-plane has equation

$$f(x, y) = 0, \tag{2}$$

then the cylinder through C with vertical rulings has the same equation in space. This is so because the point $P(x, y, z)$ lies on the cylinder if and only if the point $(x, y, 0)$ lies on the curve C. Similarly, the graph of an equation $g(x, z) = 0$ is a cylinder with rulings parallel to the y-axis, and the graph of an equation $h(y, z) = 0$ is a cylinder with rulings parallel to the x-axis. Thus the graph in space of an equation that includes only two of the three coordinate variables is always a cylinder; its rulings are parallel to the axis corresponding to the *missing* variable.

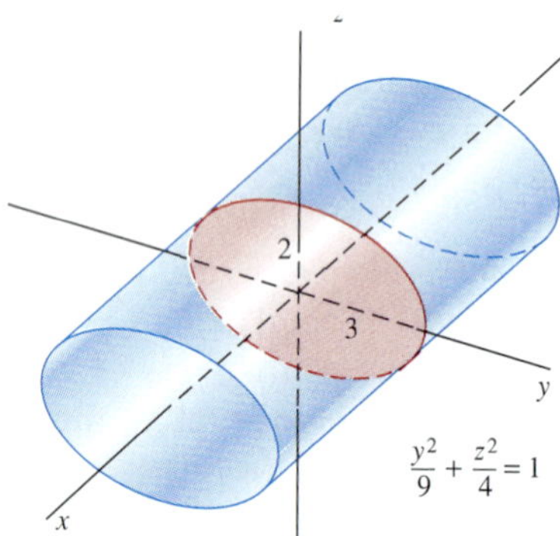

Fig. 12.7.7 An elliptical cylinder (Example 4)

EXAMPLE 4 The graph of the equation $4y^2 + 9z^2 = 36$ is the **elliptic cylinder** shown in Fig. 12.7.7. Its rulings are parallel to the x-axis, and its trace in every plane perpendicular to the x-axis is an ellipse with semiaxes of lengths 3 and 2 (just like the pictured ellipse $y^2/9 + z^2/4 = 1$ in the yz-plane). ■

EXAMPLE 5 The graph of the equation $z = 4 - x^2$ is the **parabolic cylinder** shown in Fig. 12.7.8. Its rulings are parallel to the y-axis, and its trace in every plane perpendicular to the y-axis is a parabola that is a parallel translate of the parabola $z = 4 - x^2$ in the xz-plane. ■

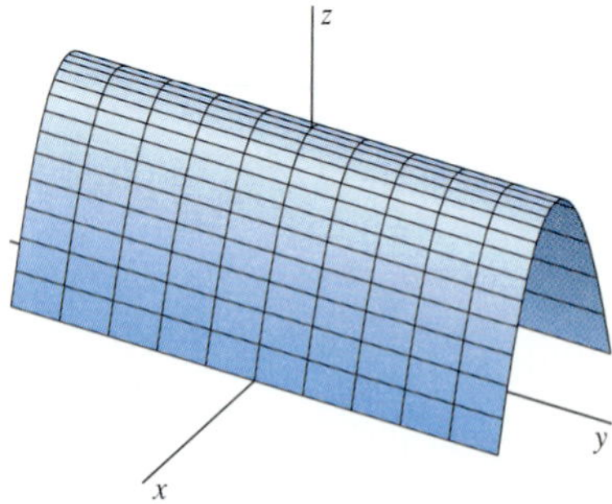

Fig. 12.7.8 The parabolic cylinder $z = 4 - x^2$ (Example 5)

Surfaces of Revolution

Another way to use a plane curve C to generate a surface is to revolve the curve in space around a line L in its plane. This gives a **surface of revolution** with **axis** L. For example, Fig. 12.7.9 shows the surface generated by revolving the curve $f(x, y) = 0$ in the first quadrant of the xy-plane around the x-axis. The typical point $P(x, y, z)$ lies on this surface of revolution provided that it lies on the vertical circle (parallel

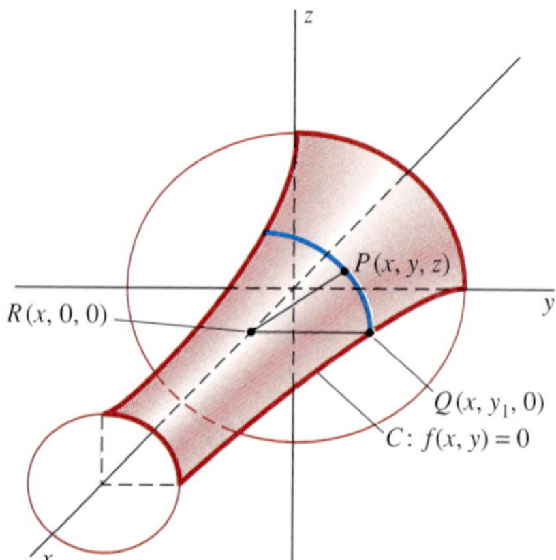

Fig. 12.7.9 The surface generated by rotating C around the x-axis. (For clarity, only a quarter of the surface is shown.)

to the yz-plane) with center $R(x, 0, 0)$ and radius r such that the point $Q(x, r, 0)$ lies on the given curve C, in which case $f(x, r) = 0$. Because

$$r = |RQ| = |RP| = \sqrt{y^2 + z^2},$$

it is therefore necessary that

$$f(x, \sqrt{y^2 + z^2}) = 0. \tag{3}$$

This, then, is the equation of a **surface of revolution around the x-axis**.

The equations of surfaces of revolution around other coordinate axes are obtained similarly. If the first-quadrant curve $f(x, y) = 0$ is revolved instead around the y-axis, then we replace x with $\sqrt{x^2 + z^2}$ to get the equation $f\left(\sqrt{x^2 + z^2}, y\right) = 0$ of the resulting surface of revolution. If the curve $g(y, z) = 0$ in the first quadrant of the yz-plane is revolved around the z-axis, we replace y with $\sqrt{x^2 + y^2}$. Thus the equation of the resulting surface of revolution around the z-axis is $g\left(\sqrt{x^2 + y^2}, z\right) = 0$. These assertions are easily verified with the aid of diagrams similar to Fig 12.7.9.

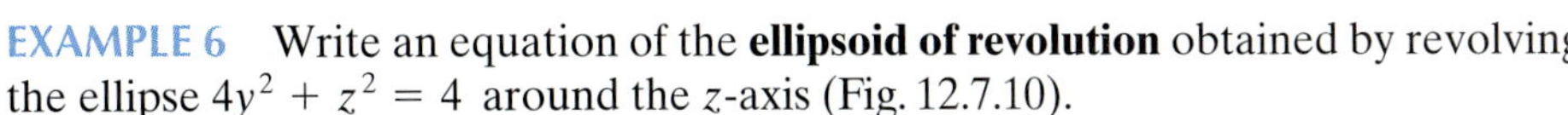

EXAMPLE 6 Write an equation of the **ellipsoid of revolution** obtained by revolving the ellipse $4y^2 + z^2 = 4$ around the z-axis (Fig. 12.7.10).

Solution We replace y with $\sqrt{x^2 + y^2}$ in the given equation. This yields $4x^2 + 4y^2 + z^2 = 4$ as an equation of the ellipsoid. ■

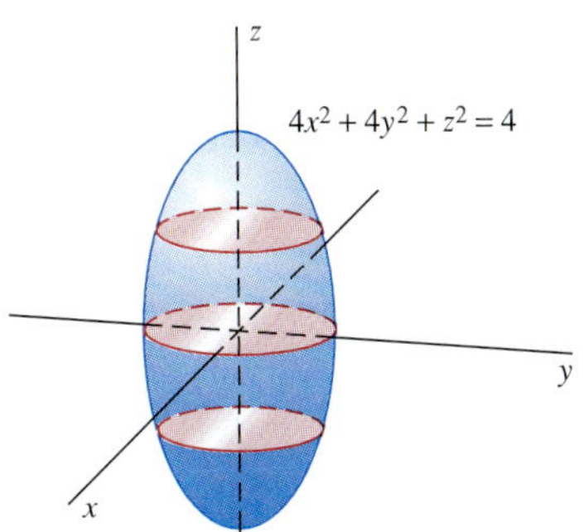

Fig. 12.7.10 The ellipsoid of revolution of Example 6

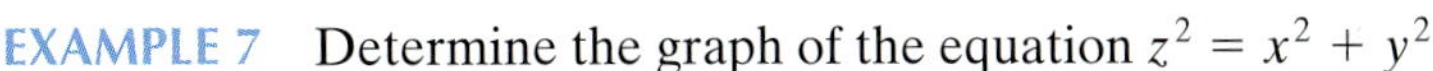

EXAMPLE 7 Determine the graph of the equation $z^2 = x^2 + y^2$.

Solution First we rewrite the given equation in the form $z = \pm\sqrt{x^2 + y^2}$. Thus the surface is symmetric around the xy-plane, and the upper half has equation $z = \sqrt{x^2 + y^2}$. We can obtain this last equation from the simple equation $z = y$ by replacing y with $\sqrt{x^2 + y^2}$. Thus we obtain the upper half of the surface by revolving the line $z = y$ (for $y \geqq 0$) around the z-axis. The graph is the **cone** shown in Fig. 12.7.11. Its upper half has equation $z = \sqrt{x^2 + y^2}$, and its lower half has equation $z = -\sqrt{x^2 + y^2}$. The entire cone $z^2 = x^2 + y^2$ is obtained by revolving the entire line $z = y$ around the z-axis. ■

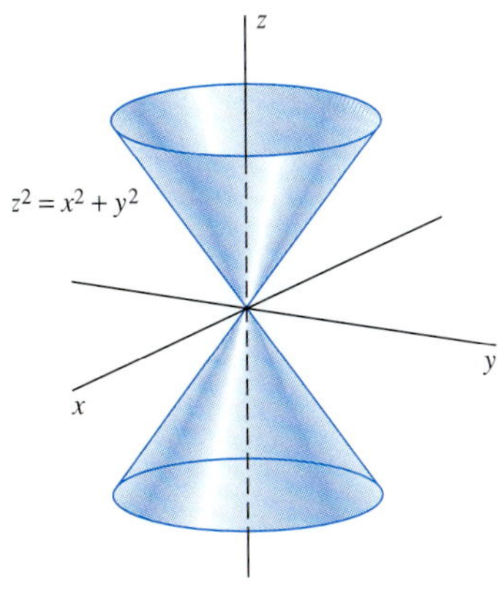

Fig. 12.7.11 The cone of Example 7

Quadric Surfaces

Cones, spheres, circular and parabolic cylinders, and ellipsoids of revolution are all surfaces that are graphs of second-degree equations in x, y, and z. The graph of a second-degree equation in three variables is called a **quadric surface**. We discuss here some important special cases of the equation

$$Ax^2 + By^2 + Cz^2 + Dx + Ey + Fz + H = 0. \tag{4}$$

This is a special second-degree equation in that it contains no terms involving the products xy, xz, or yz.

EXAMPLE 8 The **ellipsoid**

$$\frac{x^2}{a^2} + \frac{y^2}{b^2} + \frac{z^2}{c^2} = 1 \tag{5}$$

is symmetric around each of the three coordinate planes and has intercepts $(\pm a, 0, 0)$, $(0, \pm b, 0)$, and $(0, 0, \pm c)$ on the three coordinate axes. (There is no loss of generality in assuming that a, b, and c are positive.) Each trace of this ellipsoid in

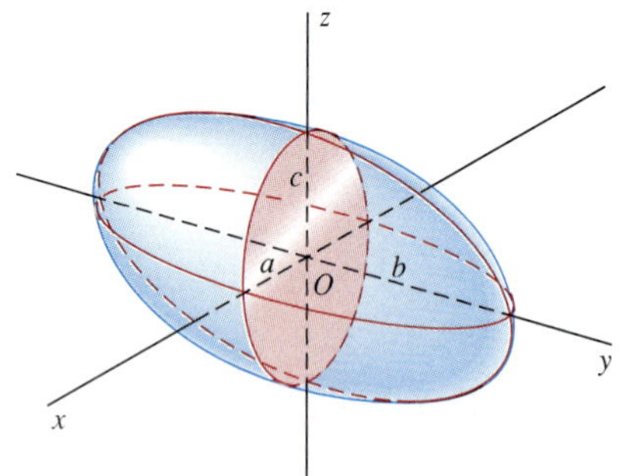

Fig. 12.7.12 The ellipsoid of Example 8

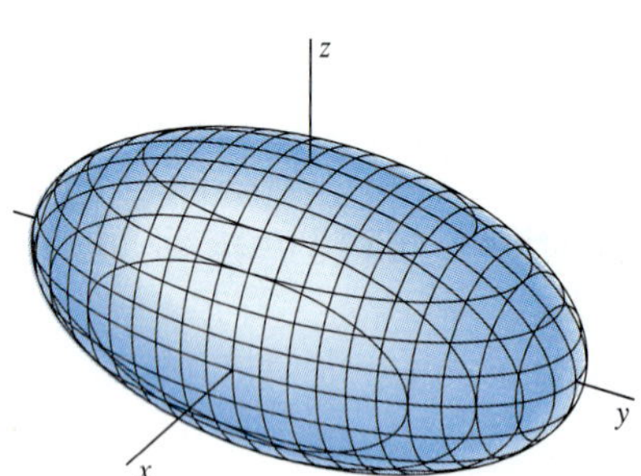

Fig. 12.7.13 The traces of the ellipsoid $\dfrac{x^2}{a^2} + \dfrac{y^2}{b^2} + \dfrac{z^2}{c^2} = 1$ (Example 8)

a plane parallel to one of the coordinate planes is either a single point or an ellipse. For example, if $-c < z_0 < c$, then the trace of the ellipsoid of Eq. (5) in the plane $z = z_0$ has equation

$$\frac{x^2}{a^2} + \frac{y^2}{b^2} = 1 - \frac{z_0^2}{c^2} > 0,$$

which is the equation of an ellipse with semiaxes $(a/c)\sqrt{c^2 - z_0^2}$ and $(b/c)\sqrt{c^2 - z_0^2}$. Figure 12.7.12 shows this ellipsoid with semiaxes a, b, and c labeled. Figure 12.7.13 shows its trace ellipses in planes parallel to the three coordinate planes. ■

EXAMPLE 9 The **elliptic paraboloid**

$$\frac{x^2}{a^2} + \frac{y^2}{b^2} = \frac{z}{c} \tag{6}$$

is shown in Fig. 12.7.14. Its trace in the horizontal plane $z = z_0 > 0$ is the ellipse $x^2/a^2 + y^2/b^2 = z_0/c$ with semiaxes $a\sqrt{z_0/c}$ and $b\sqrt{z_0/c}$. Its trace in any vertical plane is a parabola. For instance, its trace in the plane $y = y_0$ has equation $x^2/a^2 + y_0^2/b^2 = z/c$, which can be written in the form $z - z_1 = k(x - x_1)^2$ by taking $z_1 = cy_0^2/b^2$ and $x_1 = 0$. The paraboloid opens upward if $c > 0$ and downward if $c < 0$. If $a = b$, then the paraboloid is said to be **circular**. Figure 12.7.15 shows the traces of a circular paraboloid in planes parallel to the xz- and yz-planes. ■

EXAMPLE 10 The **elliptical cone**

$$\frac{x^2}{a^2} + \frac{y^2}{b^2} = \frac{z^2}{c^2} \tag{7}$$

is shown in Fig. 12.7.16. Its trace in the horizontal plane $z = z_0 \neq 0$ is an ellipse with semiaxes $a|z_0|/c$ and $b|z_0|/c$. ■

EXAMPLE 11 The **hyperboloid of one sheet** with equation

$$\frac{x^2}{a^2} + \frac{y^2}{b^2} - \frac{z^2}{c^2} = 1 \tag{8}$$

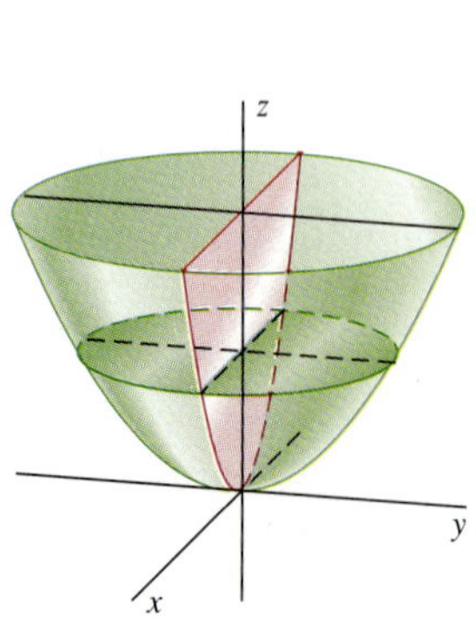

Fig. 12.7.14 An elliptic paraboloid (Example 9)

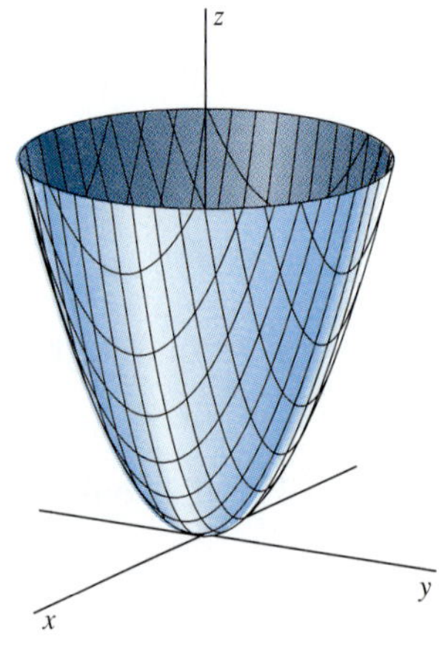

Fig. 12.7.15 Trace parabolas of a circular paraboloid (Example 9)

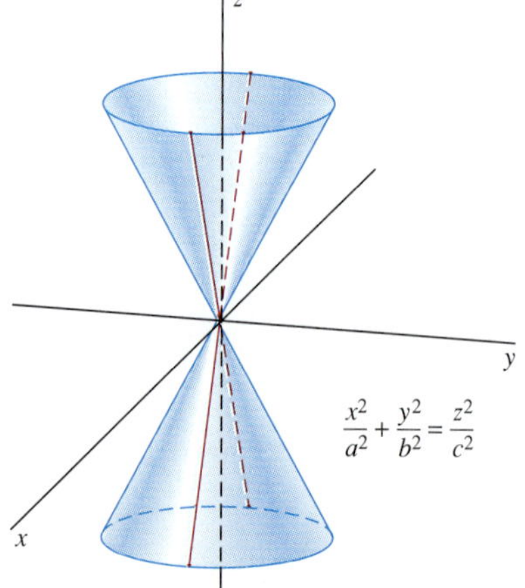

Fig. 12.7.16 An elliptic cone (Example 10)

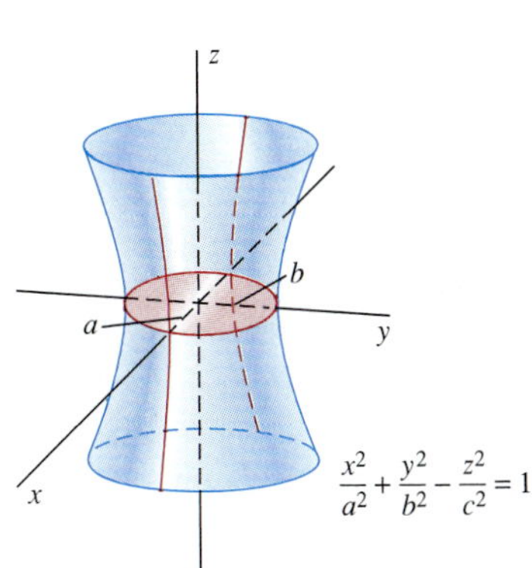

Fig. 12.7.17 A hyperboloid of one sheet (Example 11)

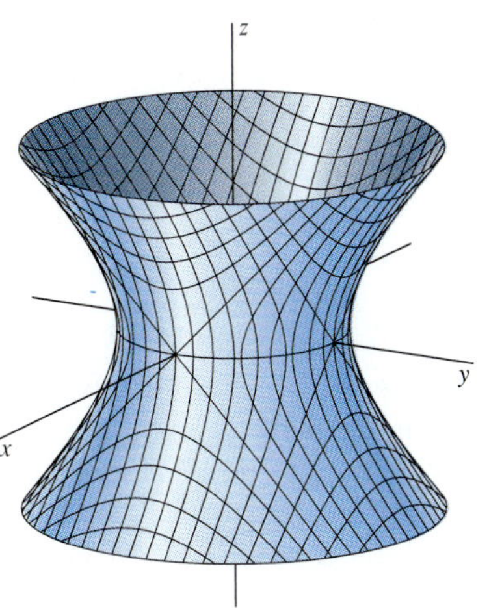

Fig. 12.7.18 A circular hyperboloid of one sheet (Example 11). Its traces in horizontal planes are circles; its traces in vertical planes are hyperbolas.

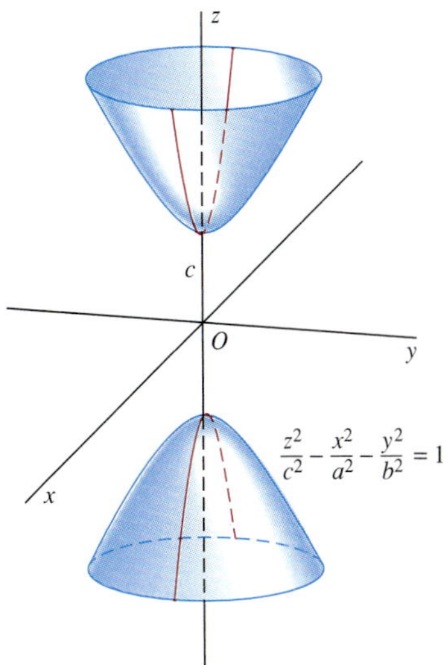

Fig. 12.7.19 A hyperboloid of two sheets (Example 12)

is shown in Fig. 12.7.17. Its trace in the horizontal plane $z = z_0$ is the ellipse $x^2/a^2 + y^2/b^2 = 1 + z_0^2/c^2 > 0$. Its trace in a vertical plane is a hyperbola except when the vertical plane intersects the xy-plane in a line tangent to the ellipse $x^2/a^2 + y^2/b^2 = 1$. In this special case, the trace is a degenerate hyperbola consisting of two intersecting lines. Figure 12.7.18 shows the traces (in planes parallel to the coordinate planes) of a circular ($a = b$) hyperboloid of one sheet.

The graph of the equations

$$\frac{y^2}{b^2}+\frac{z^2}{c^2}-\frac{x^2}{a^2}=1 \quad\text{and}\quad \frac{x^2}{a^2}+\frac{z^2}{c^2}-\frac{y^2}{c^2}=1$$

are also hyperboloids of one sheet, opening along the x- and y-axes, respectively. ■

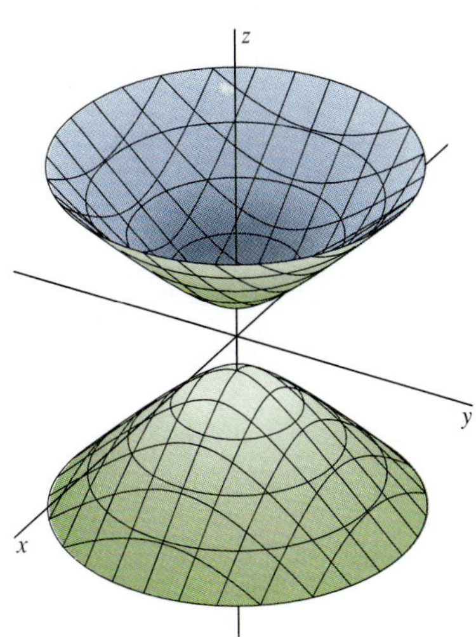

Fig. 12.7.20 A circular hyperboloid of two sheets (Example 12). Its (nondegenerate) traces in horizontal planes are circles; its traces in vertical planes are hyperbolas.

EXAMPLE 12 The **hyperboloid of two sheets** with equation

$$\frac{z^2}{c^2}-\frac{x^2}{a^2}-\frac{y^2}{b^2}=1 \tag{9}$$

consists of two separate pieces, or *sheets* (Fig. 12.7.19). The two sheets open along the positive and negative z-axis and intersect it at the points $(0, 0, \pm c)$. The trace of this hyperboloid in a horizontal plane $z = z_0$ with $|z_0| > c$ is the ellipse

$$\frac{x^2}{a^2}+\frac{y^2}{b^2}=\frac{z_0^2}{c^2}-1>0.$$

Its trace in any vertical plane is a nondegenerate hyperbola. Figure 12.7.20 shows traces of a circular hyperboloid of two sheets.

The graphs of the equations

$$\frac{x^2}{a^2}-\frac{y^2}{b^2}-\frac{z^2}{c^2}=1 \quad\text{and}\quad \frac{y^2}{b^2}-\frac{x^2}{a^2}-\frac{z^2}{c^2}=1$$

are also hyperboloids of two sheets, opening along the x-axis and y-axis, respectively. When the equation of a hyperboloid is written in standard form with $+1$ on the right-hand side [as in Eqs. (8) and (9)], then the number of sheets is equal to the number of negative terms on the left-hand side. ■

EXAMPLE 13 The **hyperbolic paraboloid**

$$\frac{y^2}{b^2} - \frac{x^2}{a^2} = \frac{z}{c} \qquad (c > 0) \tag{10}$$

is saddle-shaped, as indicated in Fig. 12.7.21. Its trace in the horizontal plane $z = z_0$ is a hyperbola (or two intersecting lines if $z_0 = 0$). Its trace in a vertical plane parallel to the xz-plane is a parabola that opens downward, whereas its trace in a vertical plane parallel to the yz-plane is a parabola that opens upward. In particular, the trace of the hyperbolic paraboloid in the xz-plane is a parabola opening downward from the origin, whereas its trace in the yz-plane is a parabola opening upward from the origin. Thus the origin looks like a local maximum from one direction but like a local minimum from another. Such a point on a surface is called a **saddle point**.

Figure 12.7.22 shows the parabolic traces in vertical planes of the hyperbolic paraboloid $z = y^2 - x^2$. Figure 12.7.23 shows its hyperbolic traces in horizontal planes. ■

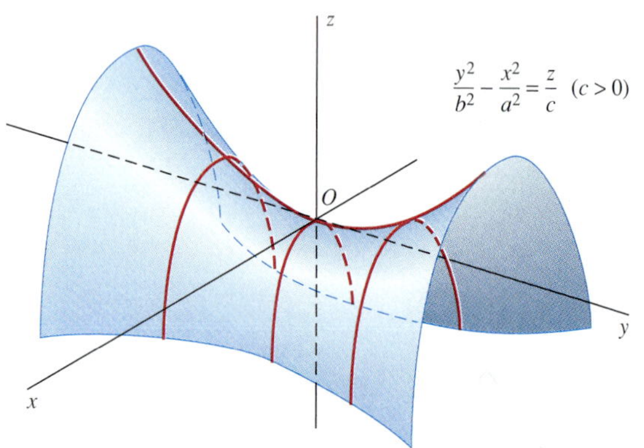

Fig. 12.7.21 A hyperbolic paraboloid is a saddle-shaped surface (Example 13).

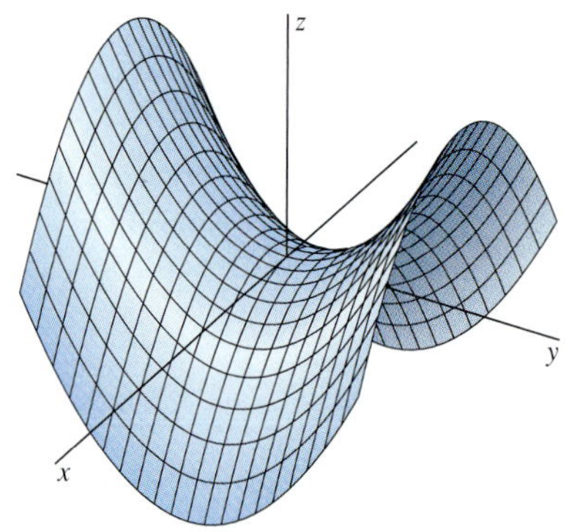

Fig. 12.7.22 The vertical traces of the hyperbolic paraboloid $z = y^2 - x^2$ (Example 13)

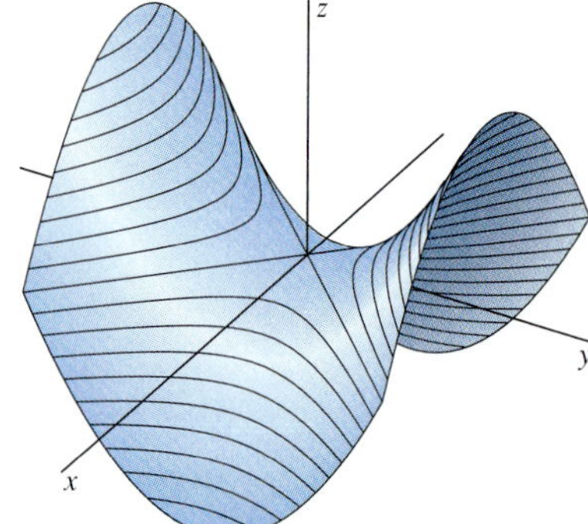

Fig. 12.7.23 The horizontal traces of the hyperbolic paraboloid $z = y^2 - x^2$ (Example 13)

Rotated Conics and Quadrics

In Sections 10.6 through 10.8 we (in effect) studied the second-degree equation

$$ax^2 + cy^2 + dx + ey = f, \tag{11}$$

which is special in that it includes no xy-term. We found that its graph in the xy-plane is always a **conic section**—a parabola, ellipse, or hyperbola—apart from "degenerate cases" of the following five types (where you should verify that the graph of each given equation is of the type claimed):

1. A single point; for example, $x^2 + y^2 = 0$,
2. A straight line; for example, $(x - 1)^2 = 0$,
3. Two parallel lines; for example, $(x - 1)^2 = 1$,
4. Two intersecting lines; for example, $x^2 - y^2 = 0$,
5. The empty set; for example, $x^2 + y^2 = -1$.

We may therefore say that the graph of the special second-degree equation in (11) is always a conic section, possibly **degenerate** (any of the exceptional cases just listed).

The same is true of the *general* second-degree equation

$$ax^2 + bxy + cy^2 + dx + ey = f \tag{12}$$

containing a "cross-product" xy-term. But now the graph of this conic section may be rotated in the xy-plane. That is, it takes a standard form such as $u^2/a^2 \pm v^2/b^2 = 1$ only in some *rotated* (and possibly translated) coordinate system. Figure 12.7.24 illustrates what is meant by a "standard form" conic in a rotated coordinate system. It turns out that the computer-plotted ellipse $73x^2 - 72xy + 52y^2 = 100$ shown there has the standard form $\frac{1}{4}u^2 + v^2 = 1$ in the uv-coordinate system obtained by rotating the xy-coordinate system through a counterclockwise angle of $\alpha = \sin^{-1}(\frac{4}{3}) \approx 53°$.

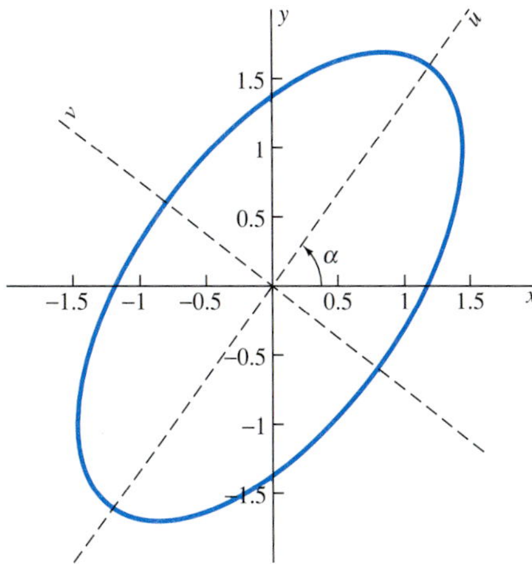

Fig. 12.7.24 The rotated ellipse $73x^2 - 72xy + 52y^2 = 100$

Here we will not discuss the determination of such rotation angles and standard rotated forms. Instead, let us take the view that we can use an appropriate computer system to plot the graph of a given equation of the form

$$ax^2 + 2bxy + cy^2 = f \tag{13}$$

(with only second-degree terms appearing on the left), but we would like to know what determines the type of conic section we should expect to see. For those familiar with matrix notation and multiplication, the reason for writing $2b$ rather than b in Eq. (13) is that this equation then takes the matrix form

$$\begin{bmatrix} x & y \end{bmatrix} \begin{bmatrix} a & b \\ b & c \end{bmatrix} \begin{bmatrix} x \\ y \end{bmatrix} = 1$$

with *coefficient matrix*

$$A = \begin{bmatrix} a & b \\ b & c \end{bmatrix}.$$

The nature of the graph of Eq. (13) is then determined by the two roots of the quadratic equation

$$\begin{vmatrix} a-\lambda & b \\ b & c-\lambda \end{vmatrix} = (a-\lambda)(c-\lambda) - b^2 = \lambda^2 - (a+c)\lambda + (ac - b^2) = 0. \quad \textbf{(14)}$$

The following result is established in elementary linear algebra textbooks.

Theorem 1 ***Classification of Conic Sections***

In an appropriate rotated uv-coordinate system Eq. (13) takes the form $\lambda_1 u^2 + \lambda_2 v^2 = f$, where λ_1 and λ_2 are the two roots of the quadratic equation in (14). Assuming that $f > 0$, it then follows that:

- If both λ_1 and λ_2 are positive, then the graph of Eq. (13) is an ellipse.
- If either λ_1 or λ_2 is positive and the other is negative, then the graph of Eq. (13) is a hyperbola.

EXAMPLE 14 For the equation $73x^2 - 72xy + 52y^2 = 100$ whose graph is shown in Fig. 12.7.24, we have $a = 72$, $b = -36$, and $c = 52$, so Eq. (14) takes the form

$$\begin{vmatrix} 72-\lambda & -36 \\ -36 & 52-\lambda \end{vmatrix} = (72-\lambda)(52-\lambda) - 36^2 = \lambda^2 - 125\lambda + 2500 = 0$$

with roots $\lambda_1 = 100$ and $\lambda_2 = 25$. The fact that both roots are positive is consistent with the (rotated) ellipse we see in the figure. ■

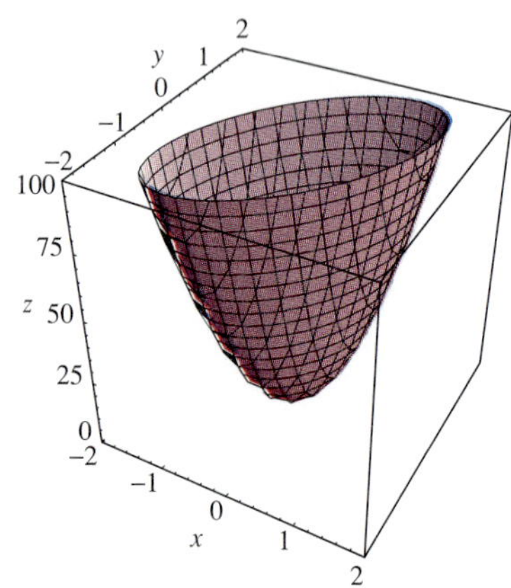

Fig. 12.7.25
$z = 73x^2 - 72xy + 52y^2$

The nature of the graph of Eq. (13) determines the nature of the graph of the three-dimensional equation

$$z = ax^2 + 2bxy + cy^2, \quad \textbf{(15)}$$

a "quadratic surface" in space. An ellipse in Eq. (13) corresponds to elliptical horizontal trace curves on this surface—which is then an elliptic paraboloid, and a hyperbola in Eq. (13) corresponds to hyperbolic horizontal trace curves on a hyperbolic paraboloid. Of course, this paraboloid is also rotated in space in the same fashion that the ellipse or hyperbola is rotated in the plane. Here again, we take the view that we'll ordinarily use a computer to plot the graph of an equation like (15), but we'd like to know what sort of picture to expect to see.

Theorem 2 ***Classification of Quadratic Surfaces***

Let λ_1 and λ_2 be the two roots of the quadratic equation in (14).

- If both λ_1 and λ_2 are nonzero and have the same sign, then the graph of Eq. (15) is an elliptic paraboloid.
- If either λ_1 or λ_2 is positive and the other is negative, then the graph of Eq. (15) is a hyperbolic paraboloid.

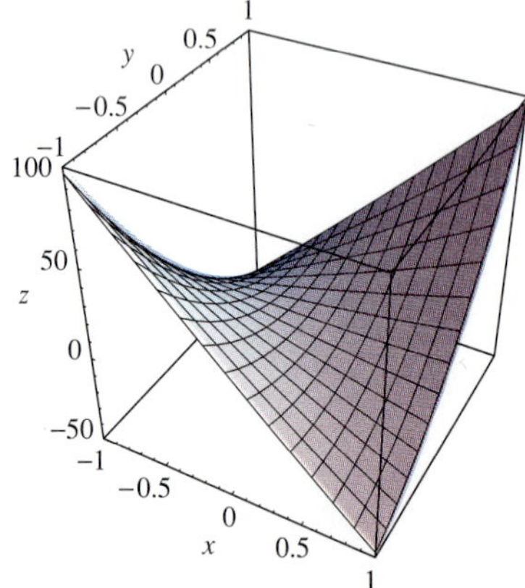

Fig. 12.7.26
$z = 2x^2 + 72xy + 23y^2$

EXAMPLE 15 Figure 12.7.25 shows the elliptic paraboloid $z = 73x^2 - 72xy + 52y^2$ for which $\lambda_1 = 100$ and $\lambda_2 = 25$, as in Example 14. Figure 12.7.26 shows the graph of the equation $z = 2x^2 + 72xy + 23y^2$, for which Eq. (14) takes the form

$$\begin{vmatrix} 2-\lambda & 36 \\ 36 & 23-\lambda \end{vmatrix} = (2-\lambda)(23-\lambda) - 36^2 = \lambda^2 - 25\lambda - 1250 = 0.$$

This equation has roots $\lambda_1 = 50$ and $\lambda_2 = -25$. Theorem 2 then implies that the graph is a hyperbolic paraboloid. Do you see the saddle point in Fig. 12.7.26? ■

Apart from certain degenerate cases, the graph in space of a second-degree equation of the general form

$$ax^2 + by^2 + cz^2 + 2dxy + 2exz + 2fyz + hx + ky + lz = g \tag{16}$$

is a rotated and/or translated quadric surface. For brevity, we will assume that $h = k = l = 0$, so there are no linear terms. Thus the equation takes the form

$$ax^2 + by^2 + cz^2 + 2dxy + 2exz + 2fyz = g. \tag{17}$$

The nature of the graph of Eq. (17) is determined by the roots of the equation

$$\begin{vmatrix} a-\lambda & d & e \\ d & b-\lambda & f \\ e & f & c-\lambda \end{vmatrix} = 0. \tag{18}$$

When the determinant is expanded and coefficients are collected, the result is a cubic equation in λ of the form

$$\lambda^3 + p\lambda^2 + q\lambda + r = 0. \tag{19}$$

Suppose that Eq. (17) is rewritten (if necessary) so that $g > 0$, and let λ_1, λ_2, and λ_3 denote the three solutions of Eq. (19). Then a three-dimensional analogue of Theorem 1 implies that:

- If λ_1, λ_2, and λ_3 are all positive, then the graph of Eq. (17) is an ellipsoid.
- If λ_1, λ_2, and λ_3 are all nonzero but not all positive, then the graph of Eq. (17) is a hyperboloid.

In the hyperbolic case, the number of sheets of the hyperboloid is equal to the number of negative roots.

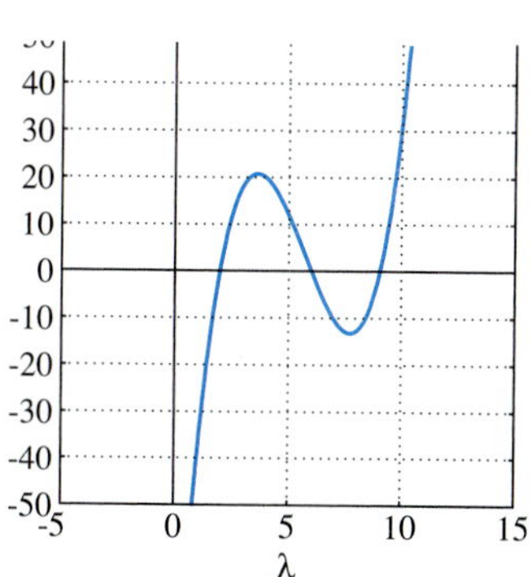

Fig. 12.7.27 The graph $w = \lambda^3 - 17\lambda^2 + 84\lambda - 108$

EXAMPLE 16 Given the second-degree equation

$$5x^2 + 7y^2 + 5z^2 + 2xy + 6xz + 2yz = 30, \tag{20}$$

the equation

$$\begin{vmatrix} 5-\lambda & 1 & 3 \\ 1 & 7-\lambda & 1 \\ 3 & 1 & 5-\lambda \end{vmatrix}$$

$$= (5-\lambda)\begin{vmatrix} 7-\lambda & 1 \\ 1 & 5-\lambda \end{vmatrix} - (1)\begin{vmatrix} 1 & 1 \\ 1 & 5-\lambda \end{vmatrix} + (3)\begin{vmatrix} 1 & 7-\lambda \\ 3 & 1 \end{vmatrix} = 0$$

in (18) simplifies to the cubic equation

$$\lambda^3 - 17\lambda^2 + 84\lambda - 108 = 0. \tag{21}$$

We need only graph such an equation to determine the signs of its solutions. The graph in Fig. 12.7.27 shows that all three solutions ($\lambda_1 = 9$, $\lambda_2 = 6$, and $\lambda_3 = 2$) of Eq. (21) are positive, so the graph of Eq. (20) is an ellipsoid. Figure 12.7.28 shows a computer plot of this quadric. ■

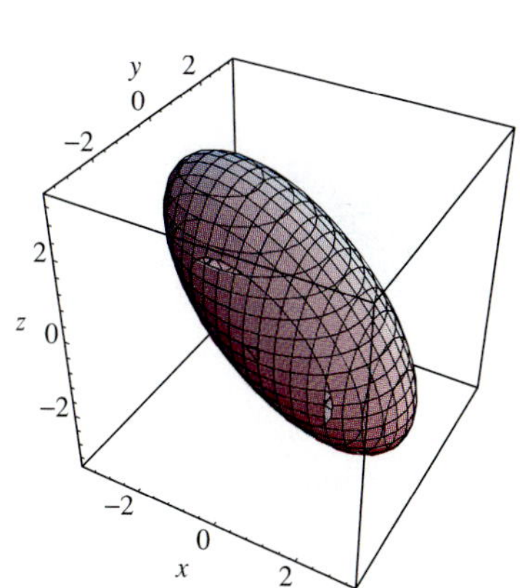

Fig. 12.7.28 $5x^2 + 7y^2 + 5z^2 + 2xy + 6xz + 2yz = 30$

In Section 13.9 we will see how to find the major and minor semiaxes of a rotated ellipse (as in Fig. 12.7.24) or ellipsoid (as in Fig. 12.7.28).

12.7 PROBLEMS

Describe and sketch the graphs of the equations given in Problems 1 through 30.

1. $3x + 2y + 10z = 20$
2. $3x + 2y = 30$
3. $x^2 + y^2 = 9$
4. $y^2 = x^2 - 9$
5. $xy = 4$
6. $z = 4x^2 + 4y^2$
7. $z = 4x^2 + y^2$
8. $4x^2 + 9y^2 = 36$
9. $z = 4 - x^2 - y^2$
10. $y^2 + z^2 = 1$
11. $2z = x^2 + y^2$
12. $x = 1 + y^2 + z^2$
13. $z^2 = 4(x^2 + y^2)$
14. $y^2 = 4x$
15. $x^2 = 4z + 8$
16. $x = 9 - z^2$
17. $4x^2 + y^2 = 4$
18. $x^2 + z^2 = 4$
19. $x^2 = 4y^2 + 9z^2$
20. $x^2 - 4y^2 = z$
21. $x^2 + y^2 + 4z = 0$
22. $x = \sin y$
23. $x = 2y^2 - z^2$
24. $x^2 + 4y^2 + 2z^2 = 4$
25. $x^2 + y^2 - 9z^2 = 9$
26. $x^2 - y^2 - 9z^2 = 9$
27. $y = 4x^2 + 9z^2$
28. $y^2 + 4x^2 - 9z^2 = 36$
29. $y^2 - 9x^2 - 4z^2 = 36$
30. $x^2 + 9y^2 + 4z^2 = 36$

Problems 31 through 40 give the equation of a curve in one of the coordinate planes. Write an equation for the surface generated by revolving this curve around the indicated axis. Then sketch the surface.

31. $x = 2z^2$; the x-axis
32. $4x^2 + 9y^2 = 36$; the y-axis
33. $y^2 - z^2 = 1$; the z-axis
34. $z = 4 - x^2$; the z-axis
35. $y^2 = 4x$; the x-axis
36. $yz = 1$; the z-axis
37. $z = \exp(-x^2)$; the z-axis
38. $(y - z)^2 + z^2 = 1$; the z-axis
39. The line $z = 2x$; the z-axis
40. The line $z = 2x$; the x-axis

In Problems 41 through 48, describe the traces of the given surfaces in planes of the indicated type.

41. $x^2 + 4y^2 = 4$; in horizontal planes (those parallel to the xy-plane)
42. $x^2 + 4y^2 + 4z^2 = 4$; in horizontal planes
43. $x^2 + 4y^2 + 4z^2 = 4$; in planes parallel to the yz-plane
44. $z = 4x^2 + 9y^2$; in horizontal planes
45. $z = 4x^2 + 9y^2$; in planes parallel to the yz-plane
46. $z = xy$; in horizontal planes
47. $z = xy$; in vertical planes through the z-axis
48. $x^2 - y^2 + z^2 = 1$; in both horizontal and vertical planes parallel to the coordinate axes

49. Prove that the triangles OAC and OBC in Fig. 12.7.1 are congruent, and thereby conclude that the trace of a sphere in an intersecting plane is a circle.

50. Prove that the projection into the yz-plane of the curve of intersection of the surfaces $x = 1 - y^2$ and $x = y^2 + z^2$ is an ellipse (Fig. 12.7.29).

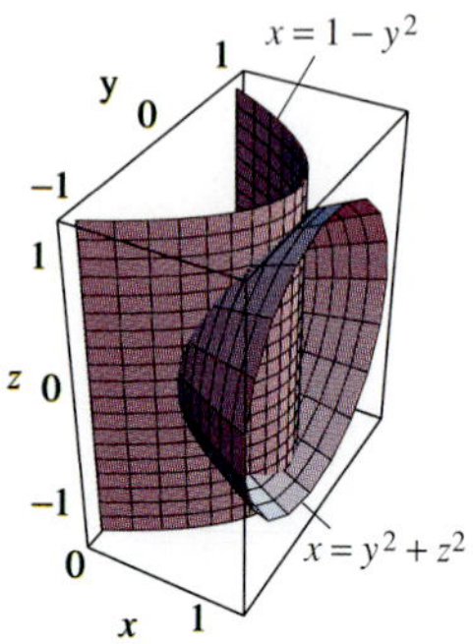

Fig. 12.7.29 The paraboloid and parabolic cylinder of Problem 50

51. Show that the projection into the xy-plane of the intersection of the plane $z = y$ and the paraboloid $z = x^2 + y^2$ is a circle (Fig. 12.7.30).

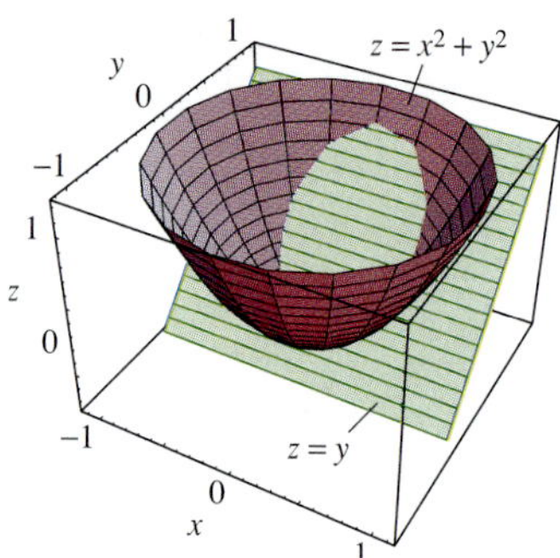

Fig. 12.7.30 The plane and paraboloid of Problem 51

52. Prove that the projection into the xz-plane of the intersection of the paraboloids $y = 2x^2 + 3z^2$ and $y = 5 - 3x^2 - 2z^2$ is a circle (Fig. 12.7.31).

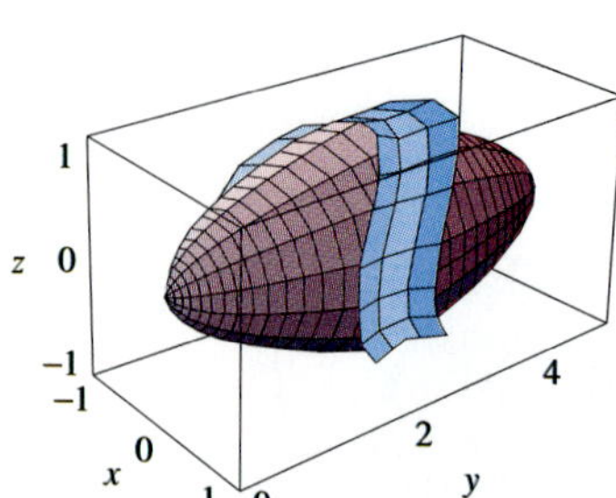

Fig. 12.7.31 The two paraboloids of Problem 52

53. Prove that the projection into the xy-plane of the intersection of the plane $x + y + z = 1$ and the ellipsoid $x^2 + 4y^2 + 4z^2 = 4$ is an ellipse.

54. Show that the curve of intersection of the plane $z = ky$ and the cylinder $x^2 + y^2 = 1$ is an ellipse. (*Suggestion:* Introduce uv-coordinates into the plane $z = ky$ as follows: Let the u-axis be the original x-axis and let the v-axis be the line $z = ky$, $x = 0$.)

In Problems 55 through 60, determine whether the graph (in the xy-plane) of the given equation is an ellipse or a hyperbola. Check your answer graphically if you have access to a computer algebra system with a "contour plotting" facility.

55. $3x^2 + 2xy + 3y^2 = 8$

56. $5x^2 + 2xy + 5y^2 = 12$

57. $9x^2 + 4xy + 6y^2 = 19$

58. $7x^2 + 12xy - 2y^2 = 21$

59. $34x^2 - 24xy + 41y^2 = 99$

60. $40x^2 + 36xy + 25y^2 = 101$

In Problems 61 through 66, determine whether the graph of the given equation is an elliptic or a hyperbolic paraboloid. Check your answer graphically by plotting the surface.

61. $z = 3x^2 - 2xy + 3y^2$

62. $z = x^2 + 6xy + y^2$

63. $z = 4x^2 + 6xy - 4y^2$

64. $z = 9x^2 - 4xy + 6y^2$

65. $z = 33x^2 + 8xy + 18y^2$

66. $z = 9x^2 + 24xy + 16y^2$

In Problems 67 through 72, determine whether the graph of the given equation is a paraboloid or a hyperboloid. Check your answer graphically if you have access to a computer algebra system with a "contour plotting" facility.

67. $x^2 + y^2 + 2z^2 + 6xy = 10$

68. $2x^2 + 5y^2 + 2z^2 + 2xz = 11$

69. $6xy + 8yz = 14$

70. $3x^2 + 2y^2 + 5z^2 - 2xy - 4xz - 2yz = 20$

71. $3x^2 + 2y^2 + 2z^2 + 4xy + 2xz + 6yz = 19$

72. $x^2 + 4y^2 + z^2 + 2xy + 8xz + 2yz = 18$

12.8 CYLINDRICAL AND SPHERICAL COORDINATES

Rectangular coordinates provide only one of several useful ways of describing points, curves, and surfaces in space. Here we discuss two additional coordinate systems in three-dimensional space. Each is a generalization of polar coordinates in the coordinate plane.

Recall from Section 10.2 that the relationship between the rectangular coordinates (x, y) and the polar coordinates (r, θ) of a point in space is given by

$$x = r\cos\theta, \qquad y = r\sin\theta \tag{1}$$

and

$$r^2 = x^2 + y^2, \qquad \tan\theta = \frac{y}{x} \text{ if } x \neq 0. \tag{2}$$

Read these relationships directly from the right triangle in Fig. 12.8.1.

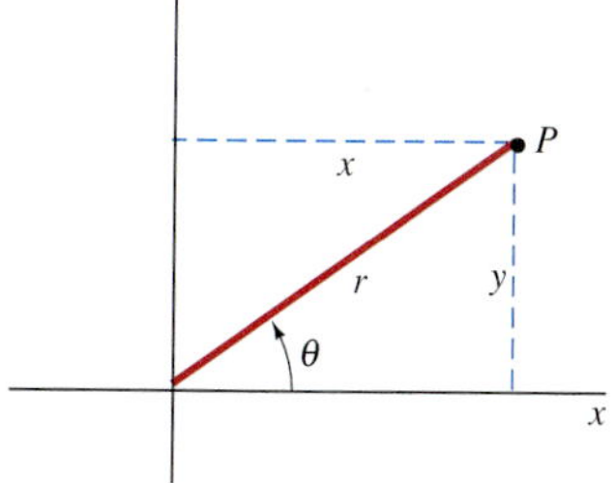

Fig. 12.8.1 The relation between rectangular and polar coordinates in the xy-plane

Cylindrical Coordinates

The **cylindrical coordinates** (r, θ, z) of a point P in space are natural hybrids of its polar and rectangular coordinates. We use the polar coordinates (r, θ) of the point in the plane with rectangular coordinates (x, y) and use the same z-coordinate as in rectangular coordinates. (The cylindrical coordinates of a point P in space are illustrated in Fig. 12.8.2.) This means that we can obtain the relations between the rectangular coordinates (x, y, z) of the point P and its cylindrical coordinates (r, θ, z) by simply adjoining the identity $z = z$ to the equations in (1) and (2):

$$x = r\cos\theta, \qquad y = r\sin\theta, \qquad z = z \tag{3}$$

and

$$r^2 = x^2 + y^2, \qquad \tan\theta = \frac{y}{x}, \qquad z = z. \tag{4}$$

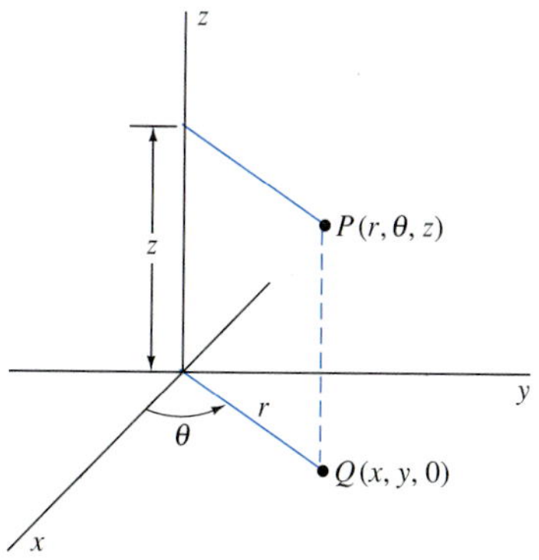

Fig. 12.8.2 Finding the cylindrical coordinates of the point P

We can use these equations to convert from rectangular to cylindrical coordinates and vice versa.

EXAMPLE 1 (a) Find the rectangular coordinates of the point P having cylindrical coordinates $(4, \frac{5}{3}\pi, 7)$. (b) Find the cylindrical coordinates of the point Q having rectangular coordinates $(-2, 2, 5)$.

Solution (a) We apply the equations in (3) to write

$$x = 4\cos\left(\frac{5}{3}\pi\right) = 4\cdot\frac{1}{2} = 2,$$

$$y = 4\sin\left(\frac{5}{3}\pi\right) = 4\cdot\left(-\frac{1}{2}\sqrt{3}\right) = -2\sqrt{3},$$

$$z = 7.$$

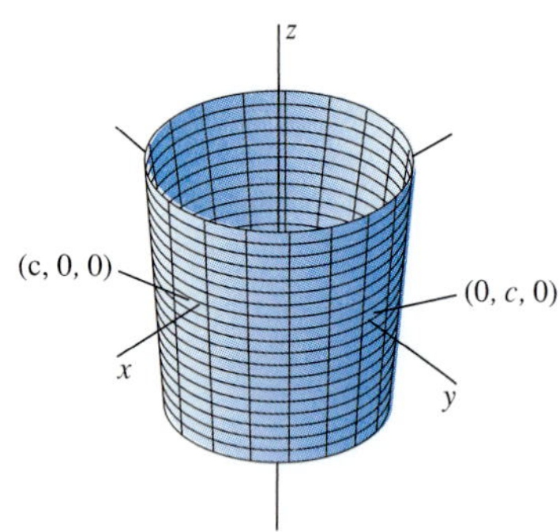

Fig. 12.8.3 The cylinder $r = c$

Thus the point P has rectangular coordinates $(2, -2\sqrt{3}, 7)$.

(b) Noting first that the point Q is in the second quadrant of the xy-plane, we apply the equations in (4) and write

$$r = \sqrt{(-2)^2 + 2^2} = 2\sqrt{2},$$

$$\tan\theta = \frac{-2}{2} = -1, \quad \text{so} \quad \theta = \frac{3\pi}{4},$$

$$z = 5.$$

Thus the point Q has cylindrical coordinates $(2\sqrt{2}, \frac{3}{4}\pi, 5)$. We can add any even integral multiple of π to θ, so other cylindrical coordinates for Q are $(2\sqrt{2}, \frac{11}{4}\pi, 5)$ and $(2\sqrt{2}, -\frac{5}{4}\pi, 5)$. ■

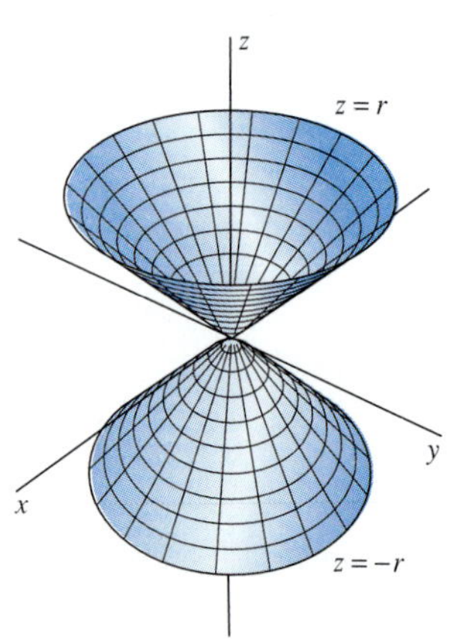

Fig. 12.8.4 The cone $z^2 = r^2$

The **graph** of an equation involving r, θ, and z is the set of all points in space having cylindrical coordinates that satisfy the equation. The name *cylindrical coordinates* arises from the fact that the graph in space of the equation $r = c$ (a constant) is a cylinder of radius c symmetric around the z-axis (Fig. 12.8.3). Cylindrical coordinates are useful in describing other surfaces that are symmetric around the z-axis. The rectangular-coordinates equation of such a surface typically involves x and y only in the combination $x^2 + y^2$, for which we can then substitute r^2 to get the cylindrical-coordinates equation.

EXAMPLE 2 (a) The sphere $x^2 + y^2 + z^2 = a^2$ has cylindrical-coordinates equation $r^2 + z^2 = a^2$.

(b) The cone $z^2 = x^2 + y^2$ has cylindrical-coordinates equation $z^2 = r^2$. Taking square roots, we get $z = \pm r$, and the two signs give (for $r \geqq 0$) the two nappes of the cone (Fig. 12.8.4).

(c) The paraboloid $z = x^2 + y^2$ has cylindrical-coordinates equation $z = r^2$ (Fig. 12.8.5).

(d) The ellipsoid $(x/3)^2 + (y/3)^2 + (z/2)^2 = 1$ has cylindrical-coordinates equation $(r/3)^2 + (z/2)^2 = 1$ (Fig. 12.8.6). ■

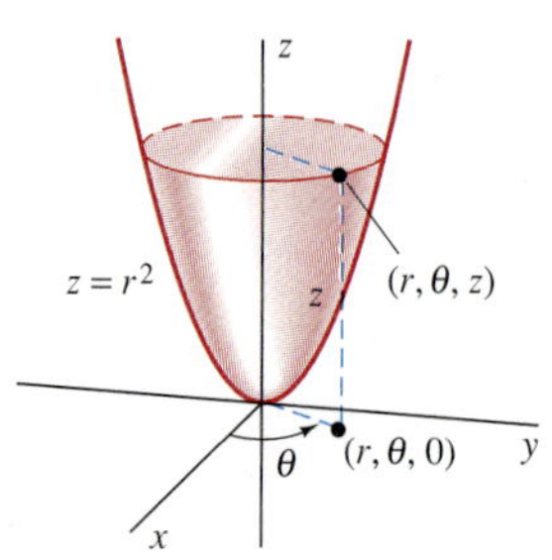

Fig. 12.8.5 The paraboloid $z = r^2$

EXAMPLE 3 Sketch the region that is bounded by the surfaces with cylindrical equations $z = r^2$ and $z = 8 - r^2$.

Solution If we substitute $r^2 = x^2 + y^2$ in the given equations, we get the familiar rectangular equations

$$z = x^2 + y^2 \quad \text{and} \quad z = 8 - x^2 - y^2$$

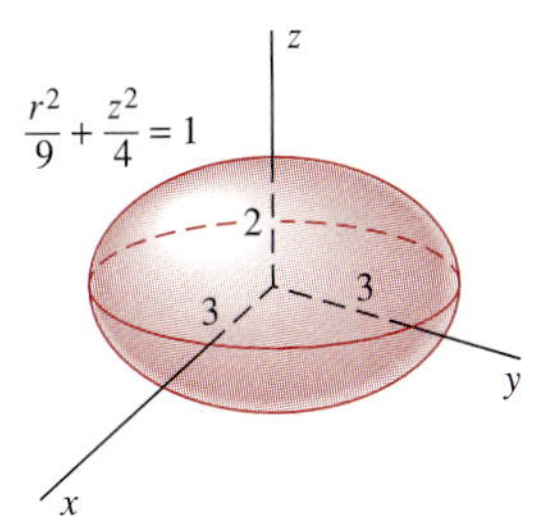

Fig. 12.8.6 The ellipsoid $\dfrac{r^2}{9} + \dfrac{z^2}{4} = 1$

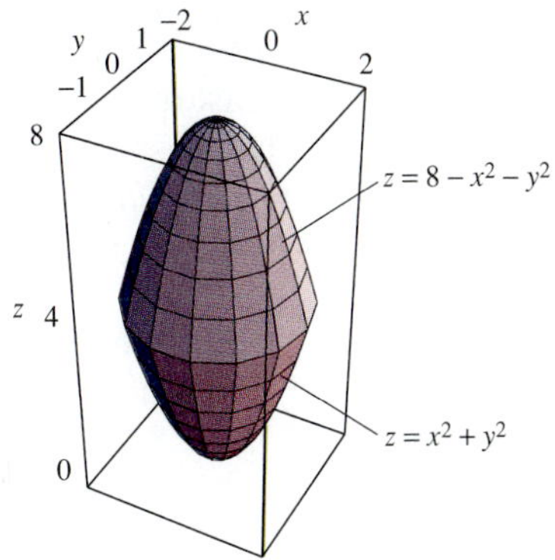

Fig. 12.8.7 The solid of Example 3

that describe paraboloids opening upward from $(0, 0, 0)$ and downward from $(0, 0, 8)$, respectively. Figure 12.8.7 shows a computer plot of the region in space that is bounded below by the paraboloid $z = x^2 + y^2$ and above by the paraboloid $z = 8 - x^2 - y^2$. ■

Remark The relations $x = r\cos\theta$ and $y = r\sin\theta$ play an important role in the computer plotting of figures symmetric around the z-axis. For instance, the paraboloid $z = 8 - r^2$ of Example 3 can be plotted using computer algebra system syntax like the *Maple* command

```
plot3d( [r*cos(t), r*sin(t), 8 - r∧2],
        r=0..2, t=0..2*Pi );
```

or the *Mathematica* command

```
ParametricPlot3D[ {r*Cos[t], r*Sin[t], 8 - r∧2},
                  {r,0,2}, {t,0,2*Pi} ];
```

In either command the paraboloid is described parametrically by giving x, y, and z in terms of r and t (for θ).

Spherical Coordinates

Figure 12.8.8 shows the **spherical coordinates** (ρ, ϕ, θ) of the point P in space. The first spherical coordinate ρ is simply the distance $\rho = |OP|$ from the origin O to P. The second spherical coordinate ϕ is the angle between OP and the positive z-axis. Thus we may always choose ϕ in the interval $[0, \pi]$, although it is not restricted to that domain. Finally, θ is the familiar angle θ of cylindrical coordinates. That is, θ is the angular coordinate of the vertical projection Q of P into the xy-plane. Thus we may always choose θ in the interval $[0, 2\pi]$, although it is not restricted to that domain. Both angles ϕ and θ are always measured in radians.

The name *spherical coordinates* is used because the graph of the equation $\rho = c$ (c is a constant) is a sphere—more precisely, a spherical surface—of radius c centered at the origin. The equation $\phi = c$ (a constant) describes (one nappe of) a cone if $0 < c < \pi/2$ or if $\pi/2 < c < \pi$ (Fig. 12.8.9). The spherical equation of the xy-plane is $\phi = \pi/2$.

From the right triangle OPQ of Fig. 12.8.8, we see that

$$r = \rho\sin\phi \quad \text{and} \quad z = \rho\cos\phi. \tag{5}$$

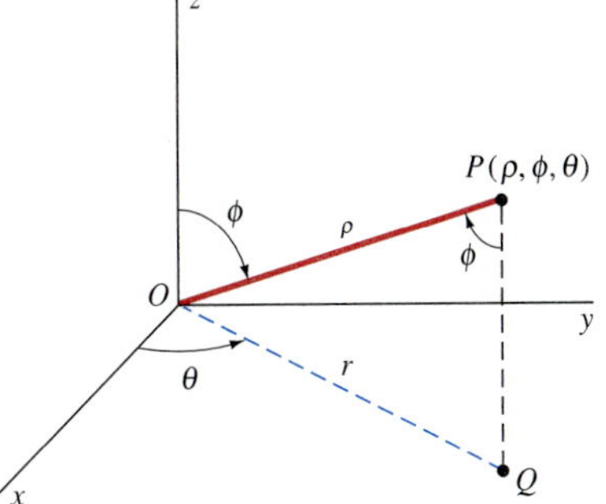

Fig. 12.8.8 Finding the spherical coordinates of the point P

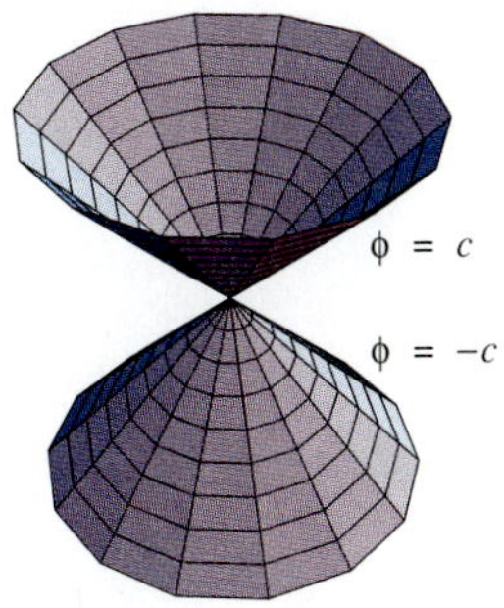

Fig. 12.8.9 The two nappes of a 45° cone; $\phi = \pi/2$ is the spherical equation of the xy-plane.

Indeed, these equations are most easily remembered by visualizing this triangle. Substitution of the equations in (5) into those in (3) yields

$$x = \rho \sin\phi \cos\theta, \quad y = \rho \sin\phi \sin\theta, \quad z = \rho \cos\phi. \tag{6}$$

These three equations give the relationship between rectangular and spherical coordinates. Also useful is the formula

$$\rho^2 = x^2 + y^2 + z^2, \tag{7}$$

a consequence of the distance formula.

It is important to note the order in which the spherical coordinates (ρ, ϕ, θ) of a point P are written—first the distance ρ of P from the origin, then the angle ϕ down from the positive z-axis, and last the counterclockwise angle θ measured from the positive x-axis. You may find this mnemonic device to be helpful: The consonants in the word "raft" remind us, in order, of *r*ho, *f*ee (for phi), and *t*heta. *Warning*: In some other physics and mathematics books, a different order, or even different symbols, may be used.

Given the rectangular coordinates (x, y, z) of the point P, one systematic method for finding the spherical coordinates (ρ, ϕ, θ) of P is this. First we find the cylindrical coordinates r and θ of P with the aid of the triangle in Fig. 12.8.10(a). Then we find ρ and ϕ from the triangle in Fig. 12.8.10(b).

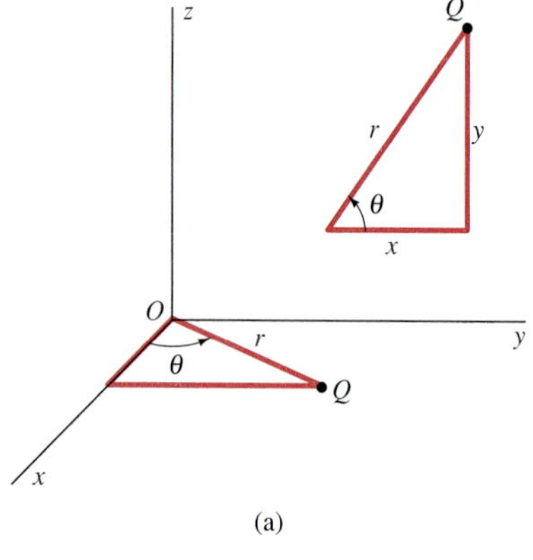

(a)

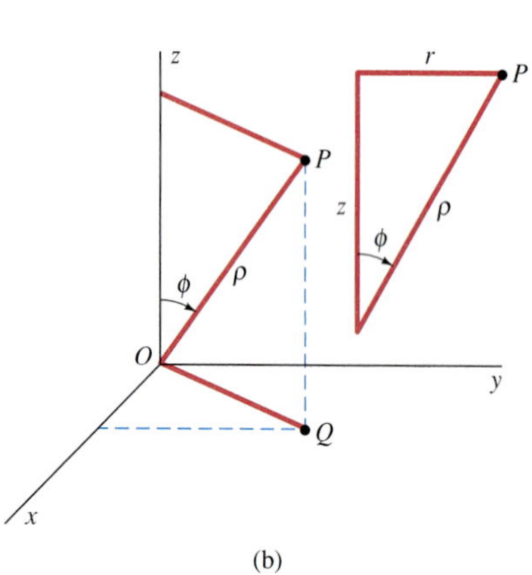

(b)

Fig. 12.8.10 Triangles used in finding spherical coordinates

EXAMPLE 4 (a) Find the rectangular coordinates of the point P having the spherical coordinates $(8, \frac{5}{6}\pi, \frac{1}{3}\pi)$. (b) Approximate the spherical coordinates of the point Q having rectangular coordinates $(-3, -4, -12)$.

Solution We apply the equations in (6) to write

$$x = 8 \sin\left(\tfrac{5}{6}\pi\right) \cos\left(\tfrac{1}{3}\pi\right) = 8 \cdot \tfrac{1}{2} \cdot \tfrac{1}{2} = 2,$$

$$y = 8 \sin\left(\tfrac{5}{6}\pi\right) \sin\left(\tfrac{1}{3}\pi\right) = 8 \cdot \tfrac{1}{2} \cdot \left(\tfrac{1}{2}\sqrt{3}\right) = 2\sqrt{3},$$

$$z = 8 \cos\left(\tfrac{5}{6}\pi\right) = 8 \cdot \left(-\tfrac{1}{2}\sqrt{3}\right) = -4\sqrt{3}.$$

Thus the point P has rectangular coordinates $(2, 2\sqrt{3}, -4\sqrt{3})$.
(b) First we note that $r = \sqrt{(-3)^2 + (-4)^2} = \sqrt{25} = 5$ and that

$$\rho = \sqrt{(-3)^2 + (-4)^2 + (-12)^2} = \sqrt{169} = 13.$$

Next,

$$\phi = \cos^{-1}\left(\frac{z}{\rho}\right) = \cos^{-1}\left(-\frac{12}{13}\right) \approx 2.7468 \quad \text{(rad)}.$$

Finally, the point $(-3, -4)$ lies in the third quadrant of the xy-plane, so

$$\theta = \pi + \tan^{-1}\left(\frac{3}{4}\right) \approx 3.7851 \quad \text{(rad)}.$$

Thus the approximate spherical coordinates of the point Q are $(13, 2.7468, 3.7851)$. ■

EXAMPLE 5 Find a spherical equation of the paraboloid with rectangular equation $z = x^2 + y^2$.

Solution We substitute $z = \rho\cos\phi$ from Eq. (5) and $x^2 + y^2 = r^2 = \rho^2\sin^2\phi$ from Eq. (6). This gives $\rho\cos\phi = \rho^2\sin^2\phi$. Cancellation of ρ gives $\cos\phi = \rho\sin^2\phi$; that is,

$$\rho = \csc\phi\cot\phi$$

is the spherical equation of the paraboloid. We get the whole paraboloid by using ϕ in the range $0 < \phi \leqq \pi/2$. Note that $\phi = \pi/2$ gives the point $\rho = 0$ that might otherwise have been lost by canceling ρ. ■

EXAMPLE 6 Determine the graph of the spherical equation $\rho = 2\cos\phi$.

Solution Multiplication by ρ gives

$$\rho^2 = 2\rho\cos\phi;$$

then substitution of $\rho^2 = x^2 + y^2 + z^2$ and $z = \rho\cos\phi$ yields

$$x^2 + y^2 + z^2 = 2z$$

as the rectangular equation of the graph. Completion of the square in z now gives

$$x^2 + y^2 + (z-1)^2 = 1,$$

so the graph is a sphere with center $(0, 0, 1)$ and radius 1. It is tangent to the xy-plane at the origin (Fig. 12.8.11). ■

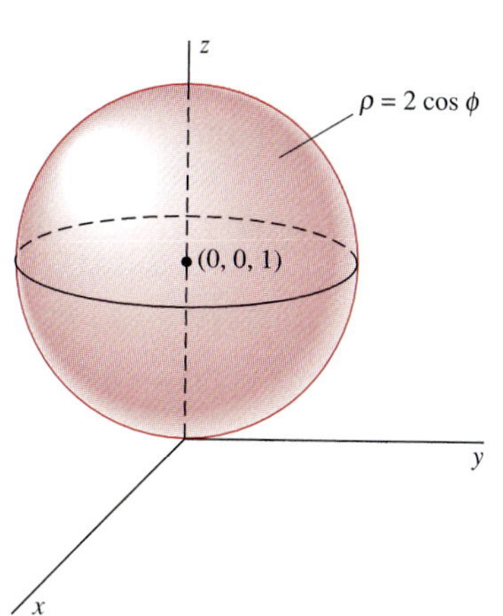

Fig. 12.8.11 The sphere of Example 6

EXAMPLE 7 Determine the graph of the spherical equation $\rho = \sin\phi\sin\theta$.

Solution We first multiply each side by ρ and get $\rho^2 = \rho\sin\phi\sin\theta$. We then use Eqs. (6) and (7) and find that $x^2 + y^2 + z^2 = y$. This is a rectangular equation of a sphere with center $(0, \frac{1}{2}, 0)$ and radius $\frac{1}{2}$. ■

Remark The relations in (6) are used in computer plotting of spherical-coordinate surfaces. For instance, the sphere $\rho = 2\cos\phi$ of Example 6 can be plotted using computer algebra system syntax such as the *Maple* commands

```
p := 2*cos(f);
plot3d( [p*sin(f)*cos(t), p*sin(f)*sin(t), p*cos(f)],
        f = 0..Pi/2, t = 0..2*Pi );
```

or the *Mathematica* commands

```
p = 2 Cos[f];
ParametricPlot3D[
        {p*Sin[f]*Cos[t], p*Sin[f]*Sin[t], p*Cos[f]},
        {f, 0, Pi/2}, {t, 0, 2*Pi} ];
```

In each case the sphere is described parametrically by writing p for ρ and giving x, y, and z in terms of f (for ϕ) and t (for θ).

*Latitude and Longitude

A **great circle** of a spherical surface is a circle formed by the intersection of the surface with a plane through the center of the sphere. Thus a great circle of a spherical surface is a circle (in the surface) that has the same radius of the sphere. Therefore a great circle is a circle of maximum possible circumference that lies on the sphere. It's easy to see that any two points on a spherical surface lie on a great circle (uniquely determined unless the two points lie on the ends of a diameter of the sphere). In the calculus of variations, it is shown that the shortest distance between two such points—measured along the curved surface—is the shorter of the two arcs of the great circle that contains them. The surprise is that the *shortest* distance is found by using the *largest* circle.

The spherical coordinates ϕ and θ are closely related to the latitude and longitude of points on the surface of the earth. Assume that the earth is a sphere with radius $\rho = 3960$ mi. We begin with the **prime meridian** (a **meridian** is a great semicircle connecting the North and South Poles) through Greenwich, England, just outside London. This is the point marked G in Fig. 12.8.12.

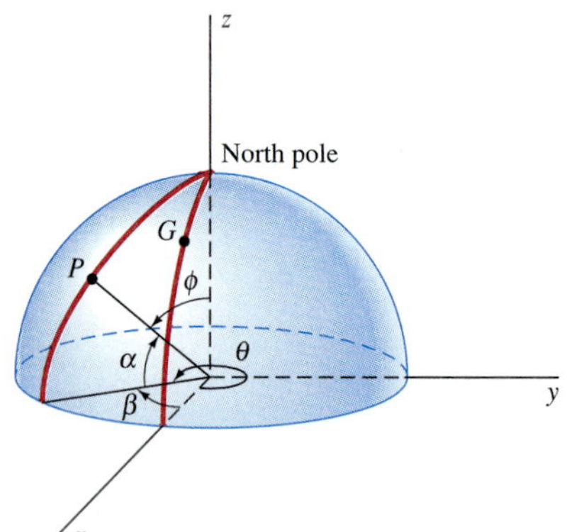

Fig. 12.8.12 The relations among latitude, longitude, and spherical coordinates

We take the z-axis through the North Pole and the x-axis through the point where the prime meridian intersects the equator. The **latitude** α and (west) **longitude** β of a point P in the Northern Hemisphere are given by the equations

$$\alpha = 90^\circ - \phi^\circ \quad \text{and} \quad \beta = 360^\circ - \theta^\circ, \tag{8}$$

where ϕ° and θ° are the angular spherical coordinates, measured in *degrees,* of P. (That is, ϕ° and θ° denote the degree equivalents of the angles ϕ and θ, respectively, which are measured in radians unless otherwise specified.) Thus the latitude α is measured northward from the equator and the longitude β is measured westward from the prime meridian.

EXAMPLE 8 Find the great-circle distance between New York (latitude 40.75° north, longitude 74° west) and London (latitude 51.5° north, longitude 0°). See Fig. 12.8.13.

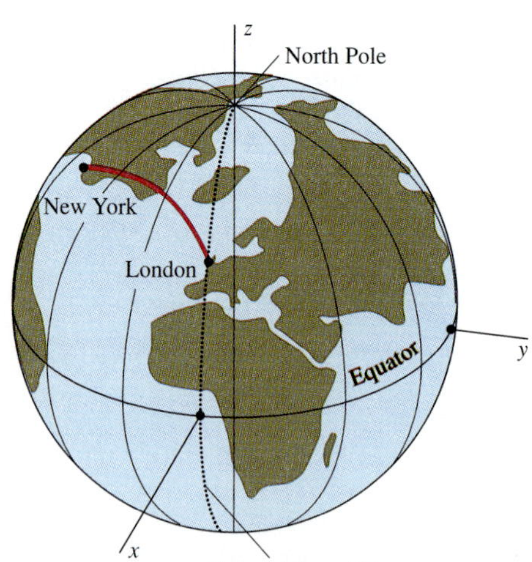

Fig. 12.8.13 Finding the great-circle distance d from New York to London (Example 8)

Solution From the equations in (8), we find that $\phi^\circ = 49.25^\circ$, $\theta^\circ = 286^\circ$ for New York, and $\phi^\circ = 38.5^\circ$, $\theta^\circ = 360^\circ$ (or 0°) for London. Hence the angular spherical coordinates of New York are $\phi = (49.25/180)\pi$, $\theta = (286/180)\pi$, and those of London are $\phi = (38.5/180)\pi$, $\theta = 0$. With these values of ϕ and θ and with $\rho = 3960$ (mi), the equations in (6) give the rectangular coordinates

$$\text{New York:} \quad P_1(826.90, -2883.74, 2584.93)$$

and

$$\text{London:} \quad P_2(2465.16, 0.0, 3099.13).$$

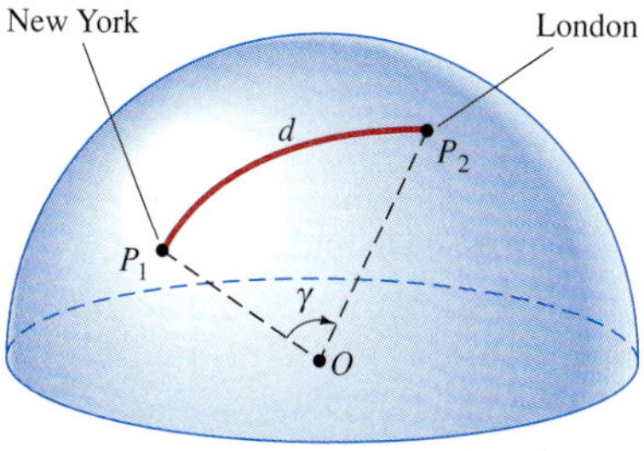

Fig. 12.8.14 The great-circle arc between New York and London (Example 8)

The angle γ between the radius vectors $\mathbf{u} = \overrightarrow{OP_1}$ and $\mathbf{v} = \overrightarrow{OP_2}$ in Fig. 12.8.14 satisfies the equation

$$\cos\gamma = \frac{\mathbf{u}\cdot\mathbf{v}}{|\mathbf{u}||\mathbf{v}|} = \frac{826.90\cdot 2465.16 - 2883.74\cdot 0 + 2584.93\cdot 3099.13}{(3960)^2} \approx 0.641.$$

Thus γ is approximately 0.875 (rad). Hence the great-circle distance between New York and London is close to

$$d = 3960\cdot 0.875 \approx 3465 \quad \text{(mi)},$$

about 5576 km. ■

12.8 PROBLEMS

In Problems 1 through 6, find the rectangular coordinates of the point with the given cylindrical coordinates.

1. $(1, \frac{1}{2}\pi, 2)$ **2.** $(3, \frac{3}{2}\pi, -1)$

3. $(2, \frac{3}{4}\pi, 3)$ **4.** $(3, \frac{7}{6}\pi, -1)$

5. $(2, \frac{1}{3}\pi, -5)$ **6.** $(4, \frac{5}{3}\pi, 6)$

In Problems 7 through 12, find the rectangular coordinates of the points with the given spherical coordinates (ρ, ϕ, θ).

7. $(2, 0, \pi)$ **8.** $(3, \pi, 0)$

9. $(3, \frac{1}{2}\pi, \pi)$ **10.** $(4, \frac{1}{6}\pi, \frac{2}{3}\pi)$

11. $(2, \frac{1}{3}\pi, \frac{3}{2}\pi)$ **12.** $(6, \frac{3}{4}\pi, \frac{4}{3}\pi)$

In Problems 13 through 22, find both the cylindrical coordinates and the spherical coordinates of the point P with the given rectangular coordinates.

13. $P(0, 0, 5)$ **14.** $P(0, 0, -3)$

15. $P(1, 1, 0)$ **16.** $P(2, -2, 0)$

17. $P(1, 1, 1)$ **18.** $P(-1, 1, -1)$

19. $P(2, 1, -2)$ **20.** $P(-2, -1, -2)$

21. $P(3, 4, 12)$ **22.** $P(-2, 4, -12)$

In Problems 23 through 38, describe the graph of the given equation. (It is understood that equations including r are in cylindrical coordinates and those including ρ or ϕ are in spherical coordinates.)

23. $r = 5$ **24.** $\theta = 3\pi/4$

25. $\theta = \pi/4$ **26.** $\rho = 5$

27. $\phi = \pi/6$ **28.** $\phi = 5\pi/6$

29. $\phi = \pi/2$ **30.** $\phi = \pi$

31. $z^2 + 2r^2 = 4$ **32.** $z^2 - 2r^2 = 4$

33. $r = 4\cos\theta$ **34.** $\rho = 4\cos\phi$

35. $r^2 - 4r + 3 = 0$ **36.** $\rho^2 - 4\rho + 3 = 0$

37. $z^2 = r^4$ **38.** $\rho^3 + 4\rho = 0$

In Problems 39 through 44, convert the given equation both to cylindrical and to spherical coordinates.

39. $x^2 + y^2 + z^2 = 25$ **40.** $x^2 + y^2 = 2x$

41. $x + y + z = 1$ **42.** $x + y = 4$

43. $x^2 + y^2 + z^2 = x + y + z$

44. $z = x^2 - y^2$

In Problems 45 through 52, describe and sketch the surface or solid described by the given equations and/or inequalities.

45. $r = 3,\ -1 \leqq z \leqq 1$ **46.** $\rho = 2,\ 0 \leqq \phi \leqq \pi/2$

47. $\rho = 2,\ \pi/3 \leqq \phi \leqq 2\pi/3$

48. $0 \leqq r \leqq 3,\ -2 \leqq z \leqq 2$

49. $1 \leqq r \leqq 3,\ -2 \leqq z \leqq 2$

50. $0 \leqq \rho \leqq 2,\ 0 \leqq \phi \leqq \pi/2$

51. $3 \leqq \rho \leqq 5$

52. $0 \leqq \phi \leqq \pi/6,\ 0 \leqq \rho \leqq 10$

53. The parabola $z = x^2$, $y = 0$ is rotated around the z-axis. Write a cylindrical equation for the surface thereby generated.

54. The hyperbola $y^2 - z^2 = 1$, $x = 0$ is rotated around the z-axis. Write a cylindrical equation for the surface thereby generated.

55. A sphere of radius 2 is centered at the origin. A hole of radius 1 is drilled through the sphere, with the axis of the

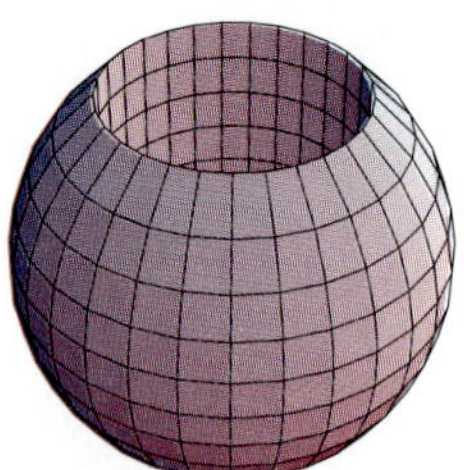

Fig. 12.8.15 The sphere-with-hole of Problem 55

hole lying on the z-axis. Describe the solid region that remains (Fig. 12.8.15) in (a) cylindrical coordinates; (b) spherical coordinates.

56. Find the great-circle distance in miles and in kilometers from Atlanta (latitude 33.75° north, longitude 84.40° west) to San Francisco (latitude 37.78° north, longitude 122.42° west).

57. Find the great-circle distance in miles and in kilometers from Fairbanks (latitude 64.80° north, longitude 147.85° west) to St. Petersburg, Russia (latitude 59.91° north, longitude 30.43° *east* of Greenwich—alternatively, longitude 329.57° west).

58. Because Fairbanks and St. Petersburg, Russia (see Problem 57) are at almost the same latitude, a plane could fly from one to the other roughly along the 62nd parallel of latitude. Accurately estimate the length of such a trip both in kilometers and in miles.

59. In flying the great-circle route from Fairbanks to St. Petersburg, Russia (see Problem 57), how close in kilometers and in miles to the North Pole would a plane fly?

60. The vertex of a right circular cone of radius R and height H is located at the origin and its axis lies on the nonnegative z-axis. Describe the solid cone in cylindrical coordinates.

61. Describe the cone of Problem 60 in spherical coordinates.

62. In flying the great-circle route from New York to London (Example 8), an airplane initially flies generally east-northeast. Does the plane ever fly at a latitude *higher* than that of London? (*Suggestion:* Express the z-coordinate of the plane's route as a function of x, then maximize z.)

12.8 PROJECT: PERSONAL CYLINDRICAL AND SPHERICAL PLOTS

As remarked in this section, the parametric plotting facilities of computing systems such as *Maple* and *Mathematica* can be used to plot cylindrical and spherical coordinate surfaces. Your task in this project is to use a computer to produce some personal plots that are worth keeping. You might warm up by plotting some cylinders and spheres of different sizes, then some specified parts like upper and lower hemispheres.

For practice in combining two or more surfaces in a single figure, try to reproduce the sphere-with-hole of Fig. 12.8.15. You'll need to plan the figure carefully, deciding in advance what the radii of the sphere and cylinder should be, and what should be the z-range of the cylindrical part and the ϕ-range of the spherical part. Once you have constructed two (or more) surfaces with commands of the general form

```
surface1 := plot( ... ); surface2 := plot( ... );
```

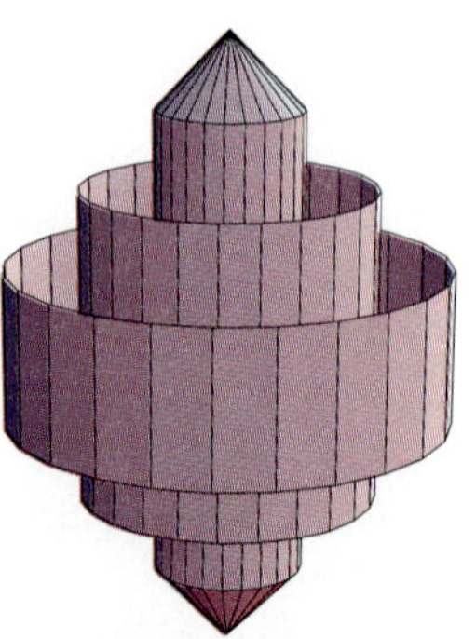

Fig. 12.8.16 Nested cylinders

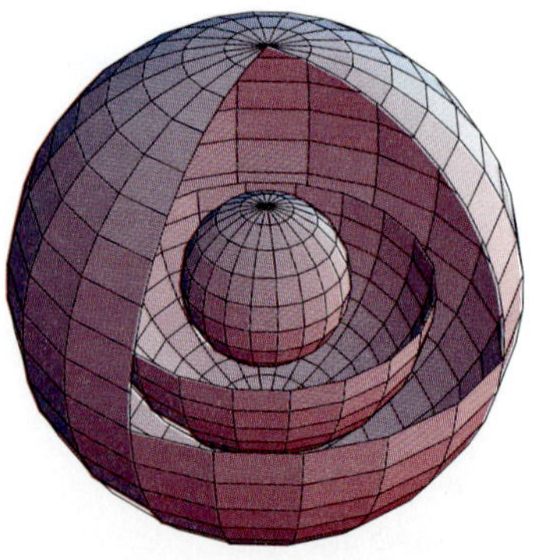

Fig. 12.8.17 Nested spheres

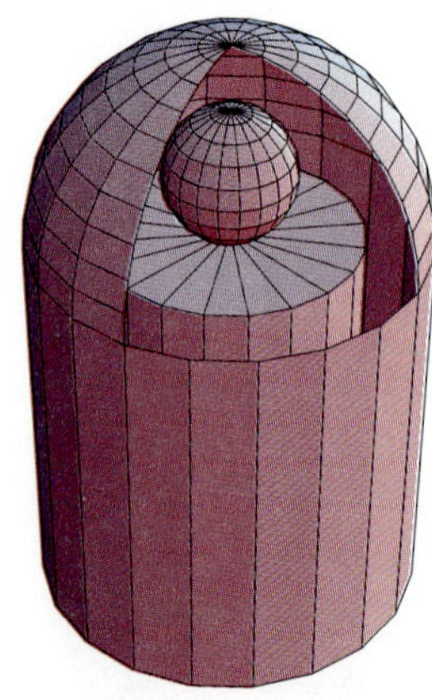

Fig. 12.8.18 An observatory?

you can use syntax such as the *Maple* command `display({surface1,surface2})` or the *Mathematica* command `Show[ surface1, surface2 ]` to exhibit them simultaneously.

Many years ago every engineering student was required to go into a machine shop and construct personally a simple object like a lamp with several parts to be machined and assembled. This project is a computer-graphic version of that experience. You are to produce one or more multi-surface "mechanisms" of your own design, using both cylindrical and spherical coordinates for various parts. The nested cylinders of Fig. 12.8.16, the nested spheres of Fig. 12.8.17, and the observatory of Fig. 12.8.18 may provide some ideas, but incorporate some innovations of your own.

CHAPTER 12 REVIEW: *Concepts, Definitions, Results*

Use the following list as a guide to concepts that you may need to review.

1. Vectors: their definition, length, equality, addition, multiplication by scalars, and dot product
2. The dot (scalar) product of vectors—definition and geometric interpretation
3. Use of the dot product to test perpendicularity of vectors and, more generally, finding the angle between two vectors
4. The cross (vector) product of two vectors—definition and geometric interpretation
5. The scalar triple product of three vectors—definition and geometric interpretation
6. The parametric and symmetric equations of the straight line that passes through a given point and is parallel to a given vector
7. The equation of the plane through a given point normal to a given vector
8. Vector-valued functions, velocity vectors, and acceleration vectors
9. Componentwise differentiation and integration of vector-valued functions
10. The equations of motion of a projectile
11. The velocity and acceleration vectors of a particle moving along a parametric space curve
12. Arc length of a parametric space curve
13. The curvature, unit tangent vector, and principal unit normal vector of a parametric curve in the plane or in space
14. Tangential and normal components of the acceleration vector of a parametric curve
15. Kepler's three laws of planetary motion
16. The radial and transverse unit vectors
17. Polar decomposition of velocity and acceleration vectors
18. Outline of the derivation of Kepler's laws from Newton's law of gravitation
19. Equations of cylinders and of surfaces of revolution
20. The standard examples of quadric surfaces
21. Definition of the cylindrical-coordinate and spherical-coordinate systems, and the equations relating cylindrical and spherical coordinates to rectangular coordinates

CHAPTER 12 *Miscellaneous Problems*

1. Suppose that M is the midpoint of the segment PQ in space and that A is another point. Show that

$$\overrightarrow{AM} = \tfrac{1}{2}(\overrightarrow{AP} + \overrightarrow{AQ}).$$

2. Let $\mathbf{a}$ and $\mathbf{b}$ be nonzero vectors. Define

$$\mathbf{a}_{\parallel} = (\text{comp}_{\mathbf{b}}\mathbf{a})\frac{\mathbf{b}}{|\mathbf{b}|} \quad \text{and} \quad \mathbf{a}_{\perp} = \mathbf{a} - \mathbf{a}_{\parallel}.$$

Prove that $\mathbf{a}_{\perp}$ is perpendicular to $\mathbf{b}$.

3. Let P and Q be different points in space. Show that the point R lies on the line through P and Q *if and only if* there exist numbers a and b such that $a + b = 1$ and $\overrightarrow{OR} = a\overrightarrow{OP} + b\overrightarrow{OQ}$. Conclude that

$$\mathbf{r}(t) = t\overrightarrow{OP} + (1 - t)\overrightarrow{OQ}$$

is a parametric equation of this line.

4. Conclude from the result of Problem 3 that the points P, Q, and R are collinear if and only if there exist numbers a, b, and c, not all zero, such that $a + b + c = 0$ and $a\overrightarrow{OP} + b\overrightarrow{OQ} + c\overrightarrow{OR} = \mathbf{0}$.

5. Let $P(x_0, y_0)$, $Q(x_1, y_1)$, and $R(x_2, y_2)$ be points in the xy-plane. Use the cross product to show that the area of the triangle PQR is

$$A = \tfrac{1}{2}\left|(x_1 - x_0)(y_2 - y_0) - (x_2 - x_0)(y_1 - y_0)\right|.$$

6. Write both symmetric and parametric equations of the line that passes through $P_1(1, -1, 0)$ and is parallel to $\mathbf{v} = \langle 2, -1, 3\rangle$.

7. Write both symmetric and parametric equations of the line that passes through $P_1(1, -1, 2)$ and $P_2(3, 2, -1)$.

8. Write an equation of the plane through $P(3, -5, 1)$ with normal vector $\mathbf{n} = \mathbf{i} + \mathbf{j}$.

9. Show that the lines with symmetric equations

$$x - 1 = 2(y + 1) = 3(z - 2)$$

and

$$x - 3 = 2(y - 1) = 3(z + 1)$$

are parallel. Then write an equation of the plane containing these two lines.

10. Let the lines L_1 and L_2 have symmetric equations

$$\frac{x - x_i}{a_i} = \frac{y - y_i}{b_i} = \frac{z - z_i}{c_i}$$

for $i = 1, 2$. Show that L_1 and L_2 are skew lines if and only if

$$\begin{vmatrix} x_1 - x_2 & y_1 - y_2 & z_1 - z_2 \\ a_1 & b_1 & c_1 \\ a_2 & b_2 & c_2 \end{vmatrix} \neq 0.$$

11. Given the four points $A(2, 3, 2)$, $B(4, 1, 0)$, $C(-1, 2, 0)$, and $D(5, 4, -2)$, find an equation of the plane that passes through A and B and is parallel to the line through C and D.

12. Given the points A, B, C, and D of Problem 11, find points P on the line AB and Q on the line CD such that the line PQ is perpendicular to both AB and CD. What is the perpendicular distance d between the lines AB and CD?

13. Let $P_0(x_0, y_0, z_0)$ be a point of the plane with equation

$$ax + by + cz + d = 0.$$

By projecting $\overrightarrow{OP_0}$ onto the normal vector $\mathbf{n} = \langle a, b, c\rangle$, show that the distance D from the origin to this plane is

$$D = \frac{|d|}{\sqrt{a^2 + b^2 + c^2}}.$$

14. Show that the distance D from the point $P_1(x_1, y_1, z_1)$ to the plane $ax + by + cz + d = 0$ is equal to the distance from the origin to the plane with equation

$$a(x + x_1) + b(y + y_1) + c(z + z_1) + d = 0.$$

Hence conclude from the result of Problem 13 that

$$D = \frac{|ax_1 + by_1 + cz_1 + d|}{\sqrt{a^2 + b^2 + c^2}}.$$

15. Find the perpendicular distance between the parallel planes $2x - y + 2z = 4$ and $2x - y + 2z = 13$.

16. Write an equation of the plane through the point $(1, 1, 1)$ that is normal to the twisted cubic $x = t$, $y = t^2$, $z = t^3$ at this point.

17. Let ABC be an isosceles triangle with $|AB| = |AC|$. Let M be the midpoint of BC. Use the dot product to show that AM and BC are perpendicular.

18. Use the dot product to show that the diagonals of a rhombus (a parallelogram with all four sides of equal length) are perpendicular to each other.

19. The acceleration of a certain particle is

$$\mathbf{a} = \mathbf{i}\sin t - \mathbf{j}\cos t.$$

Assume that the particle begins at time $t = 0$ at the point $(0, 1)$ and has initial velocity $\mathbf{v}_0 = -\mathbf{i}$. Show that its path is a circle.

20. A particle moves in an attracting central force field with force proportional to the distance from the origin. This implies that the particle's acceleration vector is $\mathbf{a} = -\omega^2\mathbf{r}$, where $\mathbf{r}$ is the position vector of the particle. Assume that the particle's initial position is $\mathbf{r}_0 = p\mathbf{i}$ and that its initial velocity is $\mathbf{v}_0 = q\omega\mathbf{j}$. Show that the trajectory of the particle is the ellipse with equation $x^2/p^2 + y^2/q^2 = 1$. (*Suggestion:* If $x''(t) = -k^2x(t)$ (where k is constant), then $x(t) = A\cos kt + B\sin kt$ for some constants A and B.)

21. At time $t = 0$, a ground target is 160 ft from a gun and is moving directly away from it with a constant speed of 80 ft/s. If the muzzle velocity of the gun is 320 ft/s, at what angle of elevation α should it be fired in order to strike the moving target?

22. Suppose that a gun with muzzle velocity v_0 is located at the foot of a hill with a 30° slope. At what angle of elevation (from the horizontal) should the gun be fired in order to maximize its range, as measured up the hill?

23. A particle moves in space with parametric equations $x = t$, $y = t^2$, $z = \frac{4}{3}t^{3/2}$. Find the curvature of its trajectory and the tangential and normal components of its acceleration when $t = 1$.

24. The **osculating plane** to a space curve at a point P of that curve is the plane through P that is parallel to the curve's unit tangent and principal unit normal vectors at P. Write an equation of the osculating plane to the curve of Problem 23 at the point $(1, 1, \frac{4}{3})$.

25. Show that the equation of the plane that passes through the point $P_0(x_0, y_0, z_0)$ and is parallel to the vectors $\mathbf{v}_1 = \langle a_1, b_1, c_1\rangle$ and $\mathbf{v}_2 = \langle a_2, b_2, c_2\rangle$ can be written in the form

$$\begin{vmatrix} x - x_0 & y - y_0 & z - z_0 \\ a_1 & b_1 & c_1 \\ a_2 & b_2 & c_2 \end{vmatrix} = 0.$$

26. Deduce from Problem 25 that the equation of the osculating plane (Problem 24) to the parametric curve $\mathbf{r}(t)$ at the point $\mathbf{r}(t_0)$ can be written in the form

$$[\mathbf{R} - \mathbf{r}(t_0)] \cdot [\mathbf{r}'(t_0) \times \mathbf{r}''(t_0)] = 0,$$

where $\mathbf{R} = \langle x, y, z \rangle$. Note first that the vectors $\mathbf{T}$ and $\mathbf{N}$ are coplanar with $\mathbf{r}'(t)$ and $\mathbf{r}''(t)$.

27. Use the result of Problem 26 to write an equation of the osculating plane to the twisted cubic $x = t$, $y = t^2$, $z = t^3$ at the point $(1, 1, 1)$.

28. Let a parametric curve in space be described by equations $r = r(t)$, $\theta = \theta(t)$, $z = z(t)$ that give the cylindrical coordinates of a moving point on the curve for $a \leqq t \leqq b$. Use the equations relating rectangular and cylindrical coordinates to show that the arc length of the curve is

$$s = \int_a^b \left[\left(\frac{dr}{dt} \right)^2 + \left(r \frac{d\theta}{dt} \right)^2 + \left(\frac{dz}{dt} \right)^2 \right]^{1/2} dt.$$

29. A point moves on the *unit* sphere $\rho = 1$ with its spherical angular coordinates at time t given by $\phi = \phi(t)$, $\theta = \theta(t)$, $a \leqq t \leqq b$. Use the equations relating rectangular and spherical coordinates to show that the arc length of its path is

$$s = \int_a^b \left[\left(\frac{d\phi}{dt} \right)^2 + (\sin^2 \phi) \left(\frac{d\theta}{dt} \right)^2 \right]^{1/2} dt.$$

30. The vector product $\mathbf{B} = \mathbf{T} \times \mathbf{N}$ of the unit tangent vector and the principal unit normal vector is the **unit binormal vector B** of a curve. (a) Differentiate $\mathbf{B} \cdot \mathbf{T} = 0$ to show that $\mathbf{T}$ is perpendicular to $d\mathbf{B}/ds$. (b) Differentiate $\mathbf{B} \cdot \mathbf{B} = 1$ to show that $\mathbf{B}$ is perpendicular to $d\mathbf{B}/ds$. (c) Conclude from parts (a) and (b) that $d\mathbf{B}/ds = -\tau \mathbf{N}$ for some number τ. Called the **torsion** of the curve, τ measures the amount that the curve twists at each point in space.

31. Show that the torsion of the helix of Example 7 of Section 12.5 is constant by showing that its value is

$$\tau = \frac{b\omega}{a^2\omega^2 + b^2}.$$

32. Deduce from the definition of torsion (Problem 30) that $\tau \equiv 0$ for any curve such that $\mathbf{r}(t)$ lies in a fixed plane.

33. Write an equation in spherical coordinates for the spherical surface with radius 1 and center $x = 0 = y$, $z = 1$.

34. Let C be the circle in the yz-plane with radius 1 and center $y = 1$, $z = 0$. Write equations in both rectangular and cylindrical coordinates of the surface obtained by revolving C around the z-axis.

35. Let C be the curve in the yz-plane with equation $(y^2 + z^2)^2 = 2(z^2 - y^2)$. Write an equation in spherical coordinates of the surface obtained by revolving this curve around the z-axis. Then sketch this surface. (*Suggestion:* Remember that $r^2 = 2\cos 2\theta$ is the polar equation of a figure eight curve.)

36. Let A be the area of the parallelogram in space determined by the vectors $\mathbf{a} = \overrightarrow{PQ}$ and $\mathbf{b} = \overrightarrow{RS}$. Let A' be the area of the perpendicular projection of $PQRS$ into a plane that makes an acute angle γ with the plane of $PQRS$. Assuming that $A' = A\cos\gamma$ in such a situation (this is true), prove that the areas of the perpendicular projections of the parallelogram $PQRS$ into the three coordinate planes are

$$|\mathbf{i} \cdot (\mathbf{a} \times \mathbf{b})|, \quad |\mathbf{j} \cdot (\mathbf{a} \times \mathbf{b})|, \quad \text{and} \quad |\mathbf{k} \cdot (\mathbf{a} \times \mathbf{b})|.$$

Conclude that the square of the area of a parallelogram in space is equal to the sum of the squares of the areas of its perpendicular projections into the three coordinate planes.

37. Take $\mathbf{a} = \langle a_1, a_2, a_3 \rangle$ and $\mathbf{b} = \langle b_1, b_2, b_3 \rangle$ in Problem 36. Show that

$$A^2 = \begin{vmatrix} a_2 & a_3 \\ b_2 & b_3 \end{vmatrix}^2 + \begin{vmatrix} a_3 & a_1 \\ b_3 & b_1 \end{vmatrix}^2 + \begin{vmatrix} a_1 & a_2 \\ b_1 & b_2 \end{vmatrix}^2.$$

38. Let C be a curve in a plane $\mathcal{P}$ that is not parallel to the z-axis. Suppose that the projection of C into the xy-plane is an ellipse. Introduce uv-coordinates into the plane $\mathcal{P}$ to prove that the curve C is itself an ellipse.

39. Conclude from Problem 38 that the intersection of a nonvertical plane and an elliptic cylinder with vertical sides is an ellipse.

40. Use the result of Problem 38 to prove that the intersection of the plane $z = Ax + By$ and the paraboloid $z = a^2x^2 + b^2y^2$ is either empty, a single point, or an ellipse.

41. Use the result of Problem 38 to prove that the intersection of the plane $z = Ax + By$ and the ellipsoid $x^2/a^2 + y^2/b^2 + z^2/c^2 = 1$ is either empty, a single point, or an ellipse.

42. Suppose that $y = f(x)$ is the graph of a function for which f'' is continuous, and suppose also that the graph has an inflection point at $(a, f(a))$. Prove that the curvature of the graph at $x = a$ is zero.

43. Find the points on the curve $y = \sin x$ where the curvature is maximal and those where it is minimal.

44. The right branch of the hyperbola $x^2 - y^2 = 1$ may be parametrized by $x(t) = \cosh t$, $y(t) = \sinh t$. Find the point where its curvature is minimal.

45. Find the vectors $\mathbf{N}$ and $\mathbf{T}$ at the point of the curve $x(t) = t\cos t$, $y(t) = t\sin t$ that corresponds to $t = \pi/2$.

46. Find the points on the ellipse $x^2/a^2 + y^2/b^2 = 1$ (with $a > b > 0$) where the curvature is maximal and those where it is minimal.

47. Suppose that the plane curve $r = f(\theta)$ is given in polar coordinates. Write r' for $f'(\theta)$ and r'' for $f''(\theta)$. Show that its curvature is given by

$$\kappa = \frac{|r^2 + 2(r')^2 - rr'|}{[r^2 + (r')^2]^{3/2}}.$$

48. Use the formula in Problem 47 to calculate the curvature $\kappa(\theta)$ at the point (r, θ) of the spiral of Archimedes with equation $r = \theta$. Then show that $\kappa(\theta) \to 0$ as $\theta \to +\infty$.

49. A railway curve must join two straight tracks, one extending due west from $(-1, -1)$ and the other extending due east from $(1, 1)$. Determine A, B, and C so that the curve $y = Ax + Bx^3 + Cx^5$ joins $(-1, -1)$ and $(1, 1)$ and so that the slope and curvature of this connecting curve are zero at both its endpoints.

50. A plane passing through the origin and not parallel to any coordinate plane has an equation of the form $Ax + By + Cz = 0$ and intersects the spherical surface $x^2 + y^2 + z^2 = R^2$ in a great circle. Find the highest point on this great circle; that is, find the coordinates of the point with the largest z-coordinate.

51. Suppose that a tetrahedron in space has a solid right angle at one vertex (like a corner of a cube). Suppose that A is the area of the side opposite the solid right angle and that B, C, and D are the areas of the other three sides. (a) Prove that

$$A^2 = B^2 + C^2 + D^2.$$

(b) Of what famous theorem is this a three-dimensional version?

CHAPTER 13

PARTIAL DIFFERENTIATION

Joseph Louis Lagrange (1736–1813)

Joseph Louis Lagrange is remembered for his great treatises on analytical mechanics and on the theory of functions that summarized much of eighteenth-century pure and applied mathematics. These treatises—*Mécanique analytique* (1788), *Théorie des fonctions analytiques* (1797), and *Leçons sur le calcul des fonctions* (1806)—systematically developed and applied widely the differential and integral calculus of multivariable functions expressed in terms of the rectangular coordinates x, y, z in three-dimensional space. They were written and published in Paris during the last quarter-century of Lagrange's career. But he grew up and spent his first 30 years in Turin, Italy. His father pointed Lagrange toward the law, but by age 17 Lagrange had decided on a career in science and mathematics. Based on his early work in celestial mechanics (the mathematical analysis of the motions of the planets and satellites in our solar system), Lagrange in 1766 succeeded Leonhard Euler as director of the Berlin Academy in Germany.

Lagrange regarded his far-reaching work on maximum-minimum problems as his best work in mathematics. This work, which continued throughout his long career, dated back to a letter to Euler that Lagrange wrote from Turin when he was only 19. This letter outlined a new approach to a certain class of optimization problems that comprise the calculus of variations. A typical example is the *isoperimetric problem,* which asks what curve of a given arc length encloses a plane region with the greatest area. (The answer: a circle.) In the *Mécanique analytique,* Lagrange applied his "method of multipliers" to investigate the motion of a particle in space that is constrained to move on a surface defined by an equation of the form $g(x, y, z) = 0$. Section 13.9 applies the Lagrange multiplier method to the problem of maximizing or minimizing a function $f(x, y, z)$ subject to a "constraint" of the form

$$g(x, y, z) = 0.$$

Today this method has applications that range from minimizing the fuel required for a spacecraft to achieve its desired trajectory to maximizing the productivity of a commercial enterprise limited by the availability of financial, natural, and personnel resources.

Modern scientific visualization often employs computer graphic techniques to present different interpretations of the same data simultaneously in a single figure. This *MATLAB* color graphic shows both a graph of a surface

$$z = f(x, y)$$

and a contour map showing "level curves" of the surface. In Section 13.5 we learn how to locate multivariable maximum-minimum points like those visible on this surface.

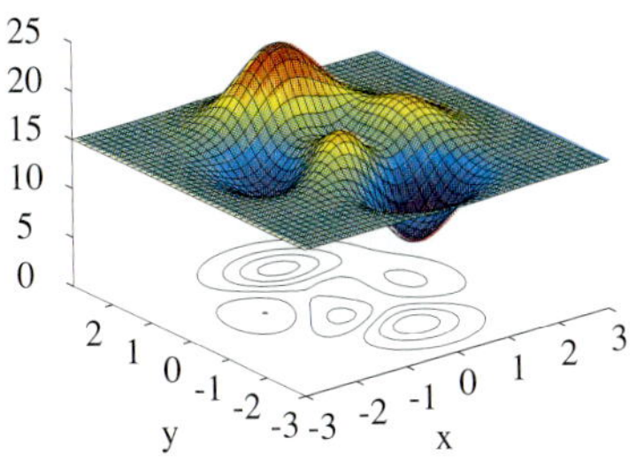

13.1 INTRODUCTION

We turn our attention here and in Chapters 14 and 15 to the calculus of functions of more than one variable. Many real-world functions depend on two or more variables. For example:

- In physical chemistry the ideal gas law $pV = nRT$ (where n and R are constants) is used to express any one of the variables p (pressure), V (volume), and T (temperature) as a function of the other two.
- The altitude above sea level at a particular location on the earth's surface depends on the latitude and longitude of the location.
- A manufacturer's profit depends on sales, overhead costs, the cost of each raw material used, and in many cases, additional variables.
- The amount of usable energy a solar panel can gather depends on its efficiency, its angle of inclination to the sun's rays, the angle of elevation of the sun above the horizon, and other factors.

A typical application may call for us to find an extreme value of a function of several variables. For example, suppose that we want to minimize the cost of making a rectangular box with a volume of 48 ft^3, given that its front and back cost $\$1/\text{ft}^2$, its top and bottom cost $\$2/\text{ft}^2$, and its two ends cost $\$3/\text{ft}^2$. Figure 13.1.1 shows such a box of length x, width y, and height z. Under the conditions given, its total cost will be

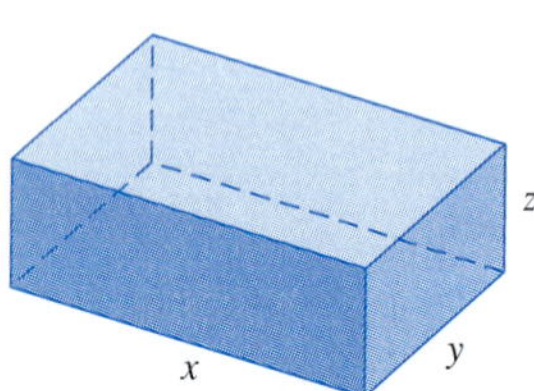

Fig. 13.1.1 A box whose total cost we want to minimize

$$C = 2xz + 4xy + 6yz \qquad \text{(dollars)}.$$

But x, y, and z are not independent variables, because the box has fixed volume

$$V = xyz = 48.$$

We eliminate z, for instance, from the first formula by using the second; because $z = 48/(xy)$, the cost we want to minimize is given by

$$C = 4xy + \frac{288}{x} + \frac{96}{y}.$$

Because neither of the variables x and y can be expressed in terms of the other, the single-variable maximum-minimum techniques of Chapter 3 cannot be applied here. We need new optimization techniques applicable to functions of two or more independent variables. In Section 13.5 we shall return to this problem.

The problem of optimization is merely one example. We shall see in this chapter that all the main ingredients of single-variable differential calculus—limits, derivatives and rates of change, chain rule computations, and maximum-minimum techniques—can be generalized to functions of two or more variables.

13.2 FUNCTIONS OF SEVERAL VARIABLES

Recall from Section 1.1 that a real-valued *function* is a rule or correspondence f that associates a unique real number with each element of a set D. The domain D has always been a subset of the real line for the functions of a single variable that we have studied up to this point. If D is a subset of the plane, then f is a function of *two* variables—for, given a point P of D, we naturally associate with P its rectangular coordinates (x, y).

> **Definition** ***Functions of Two or Three Variables***
>
> A **function of two variables,** defined on the **domain** D in the plane, is a rule f that associates with each point (x, y) in D a unique real number, denoted by $f(x, y)$. A **function of three variables,** defined on the **domain** D in space, is a rule f that associates with each point (x, y, z) in D a unique real number $f(x, y, z)$.

We can typically define a function f of two (or three) variables by giving a formula that specifies $f(x, y)$ in terms of x and y (or $f(x, y, z)$ in terms of x, y, and z). In case the domain D of f is not explicitly specified, we take D to consist of all points for which the given formula is meaningful.

EXAMPLE 1 The domain of the function f with formula

$$f(x, y) = \sqrt{25 - x^2 - y^2}$$

is the set of all (x, y) such that $25 - x^2 - y^2 \geqq 0$—that is, the circular disk $x^2 + y^2 \leqq 25$ of radius 5 centered at the origin. Similarly, the function g defined as

$$g(x, y, z) = \frac{x + y + z}{\sqrt{x^2 + y^2 + z^2}}$$

is defined at all points in space where $x^2 + y^2 + z^2 > 0$. Thus its domain consists of all points in three-dimensional space $\boldsymbol{R}^3$ other than the origin $(0, 0, 0)$. ■

EXAMPLE 2 Find the domain of definition of the function with formula

$$f(x, y) = \frac{y}{\sqrt{x - y^2}}. \tag{1}$$

Find also the points (x, y) at which $f(x, y) = \pm 1$.

Solution For $f(x, y)$ to be defined, the *radicand* $x - y^2$ must be positive—that is, $y^2 < x$. Hence the domain of f is the set of points lying strictly to the right of the parabola $x = y^2$. This domain is shaded in Fig. 13.2.1. The parabola in the figure is dotted to indicate that it is not included in the domain of f; any point for which $x = y^2$ would entail division by zero in Eq. (1).

The function $f(x, y)$ has the value ± 1 whenever

$$\frac{y}{\sqrt{x - y^2}} = \pm 1;$$

that is, when $y^2 = x - y^2$, so $x = 2y^2$. Thus $f(x, y) = \pm 1$ at each point of the parabola $x = 2y^2$ [other than its vertex $(0, 0)$, which is not included in the domain of f]. This parabola is shown as a solid curve in Fig. 13.2.1. ■

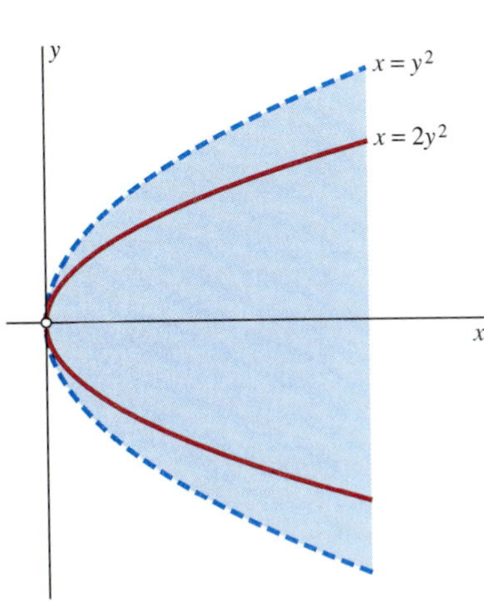

Fig. 13.2.1 The domain of $f(x, y) = \dfrac{y}{\sqrt{x - y^2}}$ (Example 2)

In a geometric, physical, or economic situation, a function typically results from expressing one descriptive variable in terms of others. As we saw in Section 13.1, the cost C of the box discussed there is given by the formula

$$C = 4xy + \frac{288}{x} + \frac{96}{y}$$

in terms of the length x and width y of the box. The value C of this function is a variable that depends on the values of x and y. Hence we call C a **dependent variable,** whereas x and y are **independent variables.** And if the temperature T at the point (x, y, z) in space is given by some formula $T = h(x, y, z)$, then the dependent variable T is a function of the three independent variables x, y, and z.

We can define a function of four or more variables by giving a formula that includes the appropriate number of independent variables. For example, if an amount A of heat is released at the origin in space at time $t = 0$ in a medium with thermal diffusivity k, then—under appropriate conditions—the temperature T at the point (x, y, z) at time $t > 0$ is given by

$$T(x, y, z, t) = \frac{A}{(4\pi kt)^{3/2}} \exp\left(-\frac{x^2 + y^2 + z^2}{4kt}\right).$$

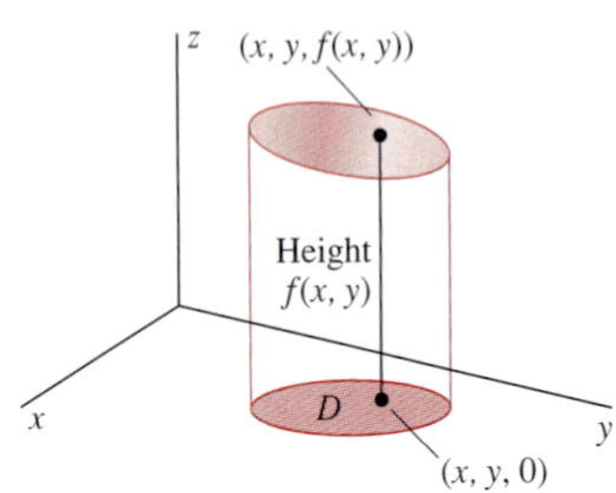

Fig. 13.2.2 The graph of a function of two variables is typically a surface "over" the domain of the function.

This formula gives the temperature T as a function of the four independent variables x, y, z, and t.

We shall see that the main differences between single-variable and multivariable calculus show up when only two independent variables are involved. Hence most of our results will be stated in terms of functions of two variables. Many of these results readily generalize by analogy to the case of three or more independent variables.

Graphs and Level Curves

We can visualize how a function f of two variables "works" in terms of its graph. The **graph** of f is the graph of the equation $z = f(x, y)$. Thus the graph of f is the set of all points in space with coordinates (x, y, z) that satisfy the equation $z = f(x, y)$ (Fig. 13.2.2). The planes and quadric surfaces of Section 12.4 and 12.7 provide some simple examples of graphs of functions of two variables.

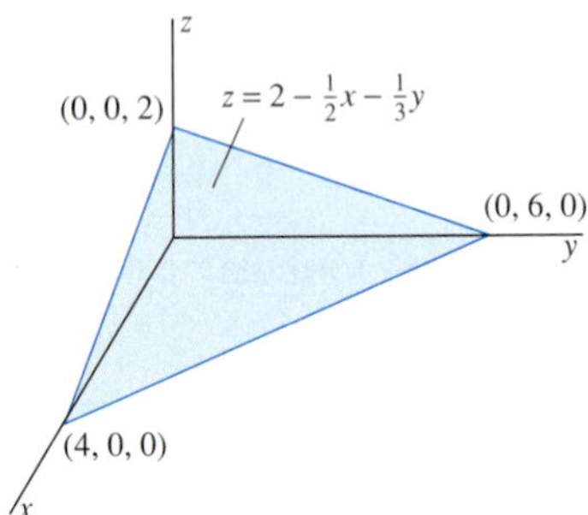

Fig. 13.2.3 The planar graph of Example 3

EXAMPLE 3 Sketch the graph of the function $f(x, y) = 2 - \frac{1}{2}x - \frac{1}{3}y$.

Solution We know from Section 12.4 that the graph of the equation $z = 2 - \frac{1}{2}x - \frac{1}{3}y$ is a plane, and we can visualize it by using its intercepts with the coordinate axes to plot the portion in the first octant of space. Clearly $z = 2$ if $x = y = 0$. Also the equation gives $y = 6$ if $x = z = 0$ and $x = 4$ if $y = z = 0$. Hence the graph looks as pictured in Fig. 13.2.3. ■

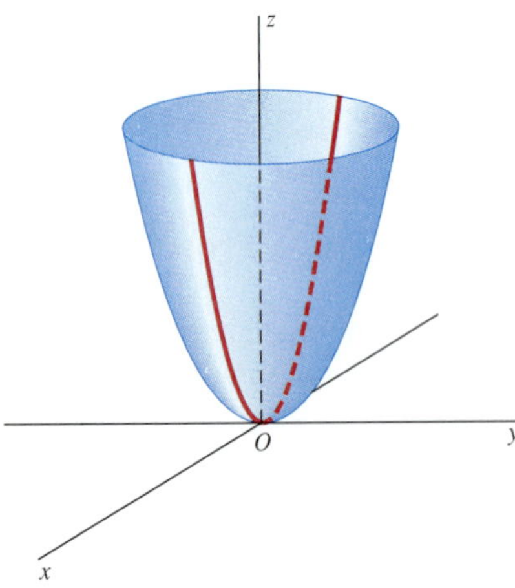

Fig. 13.2.4 The paraboloid is the graph of the function $f(x, y) = x^2 + y^2$.

EXAMPLE 4 The graph of the function $f(x, y) = x^2 + y^2$ is the familiar circular paraboloid $z = x^2 + y^2$ (Section 12.7) shown in Fig. 13.2.4. ■

EXAMPLE 5 Find the domain of definition of the function

$$g(x, y) = \tfrac{1}{2}\sqrt{4 - 4x^2 - y^2} \tag{2}$$

and sketch its graph.

Solution The function g is defined wherever $4 - 4x^2 - y^2 \geqq 0$—that is, $x^2 + \frac{1}{4}y^2 \leqq 1$—so that Eq. (2) does not involve the square root of a negative number. Thus the domain of g is the set of points in the xy-plane that lie on and within the ellipse $x^2 + \frac{1}{4}y^2 = 1$ (Fig. 13.2.5). If we square both sides of the equation $z = \frac{1}{2}\sqrt{4 - 4x^2 - y^2}$ and simplify the result, we get the equation

$$x^2 + \tfrac{1}{4}y^2 + z^2 = 1$$

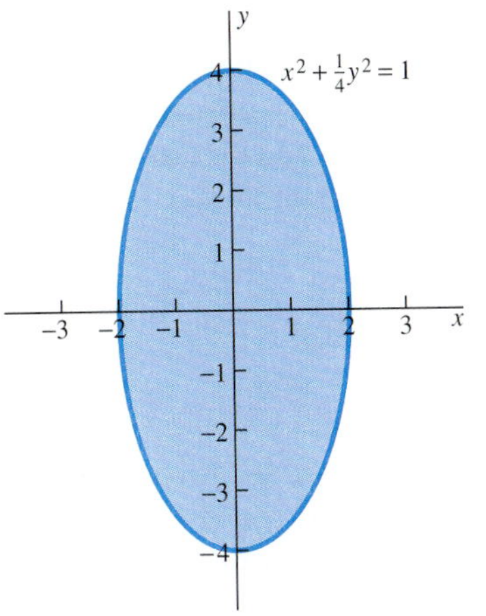

Fig. 13.2.5 The domain of the function $g(x, y) = \frac{1}{2}\sqrt{4 - 4x^2 - y^2}$

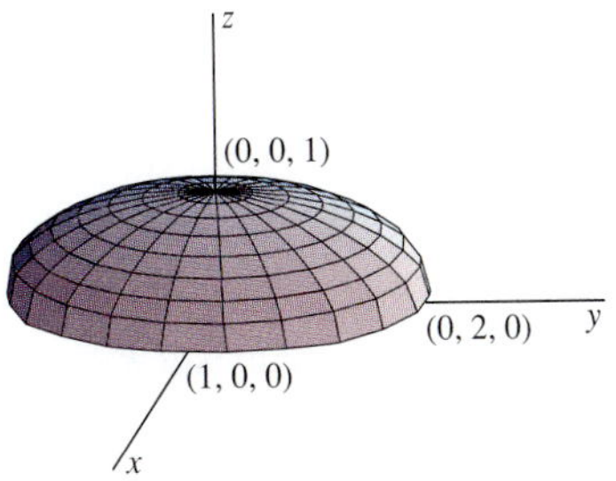

Fig. 13.2.6 The graph of the function g is the upper half of an ellipsoid.

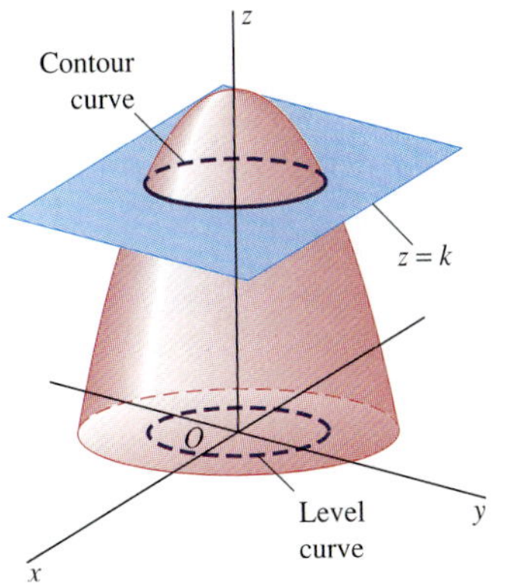

Fig. 13.2.7 A contour curve and the corresponding level curve

of an ellipsoid with semiaxes $a = 1$, $b = 2$, and $c = 1$ (Section 12.7). But $g(x, y)$ as defined in Eq. (2) is nonnegative wherever it is defined, so the graph of g is the upper half of the ellipsoid (Fig. 13.2.6). ■

The intersection of the horizontal plane $z = k$ with the surface $z = f(x, y)$ is called the **contour curve** of **height** k on the surface (Fig. 13.2.7). The vertical projection of this contour curve into the xy-plane is the **level curve** $f(x, y) = k$ of the function f. Thus a level curve of f is simply a set in the xy-plane on which the value $f(x, y)$ is *constant*. On a topographic map, such as the one in Fig. 13.2.8, the level curves are curves of constant height above sea level.

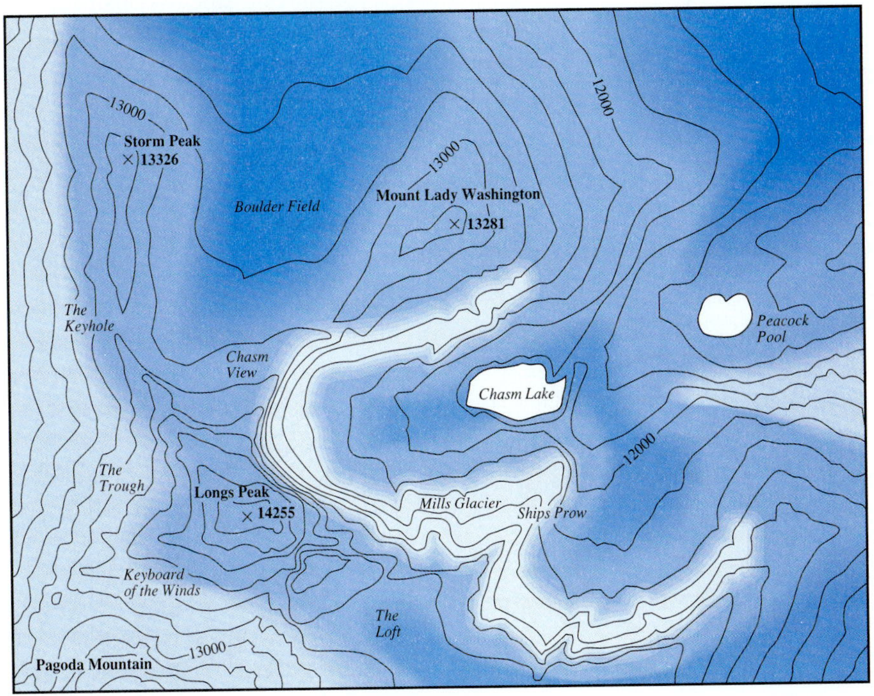

Fig. 13.2.8 The region near Longs Peak, Rocky Mountain National Park, Colorado, showing contour lines at intervals of 200 feet

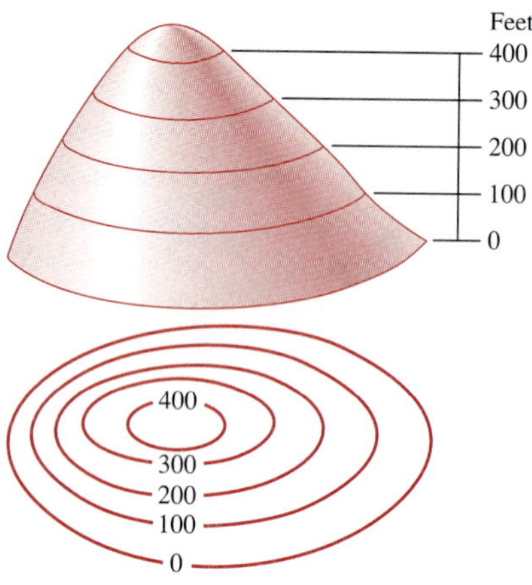

Fig. 13.2.9 Contour curves and level curves for a hill

Level curves give a two-dimensional way of representing a three-dimensional surface $z = f(x, y)$, just as the two-dimensional map in Fig. 13.2.8 represents a three-dimensional mountain range. We do this by drawing typical level curves of $z = f(x, y)$ in the xy-plane, labeling each with the corresponding (constant) value of z. Figure 13.2.9 illustrates this process for a simple hill.

EXAMPLE 6 Figure 13.2.10 shows some typical contour curves on the paraboloid $z = 25 - x^2 - y^2$. Figure 13.2.11 shows the corresponding level curves. ■

EXAMPLE 7 Sketch some typical level curves for the function $f(x, y) = y^2 - x^2$.

Solution If $k \neq 0$, then the curve $y^2 - x^2 = k$ is a hyperbola (Section 10.8). It opens along the y-axis if $k > 0$, along the x-axis if $k < 0$. If $k = 0$, then we have the equation $y^2 - x^2 = 0$, whose graph consists of the two straight lines $y = +x$ and $y = -x$. Figure 13.2.12 shows some of these level curves, each labeled with the corresponding constant value of z. Figure 13.2.13 shows contour curves on the hyperbolic paraboloid $z = y^2 - x^2$ (Section 12.7). Note that the saddle point at the origin on the paraboloid corresponds to the intersection point of the two level curves $y = +x$ and $y = -x$ in Fig. 13.2.12. ■

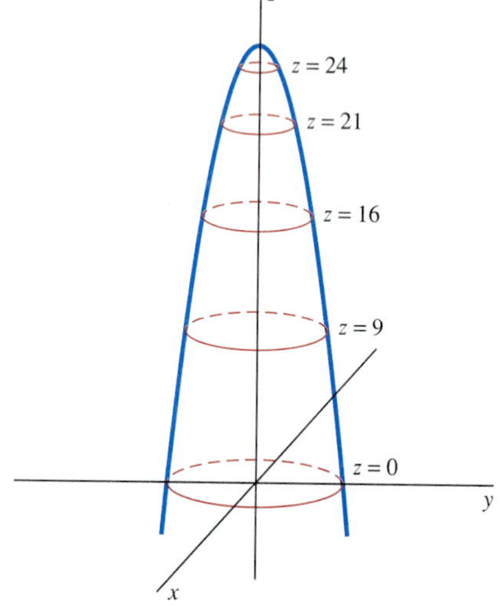

Fig. 13.2.10 Contour curves on $z = 25 - x^2 - y^2$ (Example 6)

The graph of a function $f(x, y, z)$ of three variables cannot be drawn in three dimensions, but we can readily visualize its **level surfaces** of the form $f(x, y, z) = k$. For example, the level surfaces of the function $f(x, y, z) = x^2 + y^2 + z^2$ are spheres (spherical surfaces) centered at the origin. Thus the level surfaces of f are the sets in space on which the value $f(x, y, z)$ is constant.

If the function f gives the temperature at the location (x, y) or (x, y, z), then its level curves or surfaces are called **isotherms.** A weather map typically includes level curves of the ground-level atmospheric pressure; these are called **isobars.** Even though you may be able to construct the graph of a function of two variables, that graph might be so complicated that information about the function (or the situation it describes) is obscure. Frequently the level curves themselves give more information, as in weather maps. For example, Fig. 13.2.14 shows level curves for the annual numbers of days of *high* air pollution forecast at different localities in the United States. The scale of this figure does not show local variations caused by individual cities. But a glance indicates that western Colorado, south Georgia, and central Illinois all expect the same number (10, in this case) of high-pollution days each year.

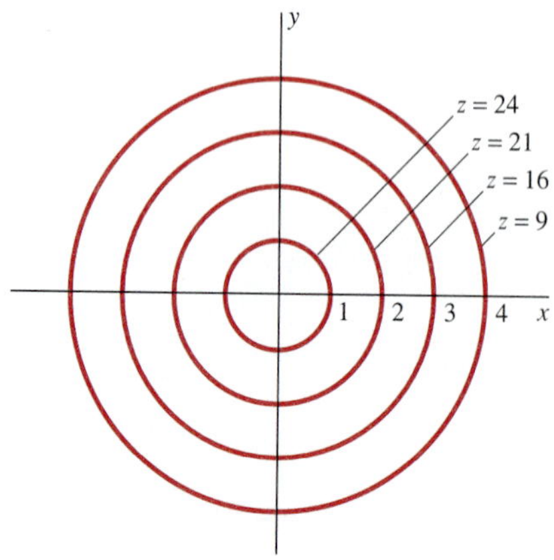

Fig. 13.2.11 Level curves of $f(x, y) = 25 - x^2 - y^2$ (Example 6)

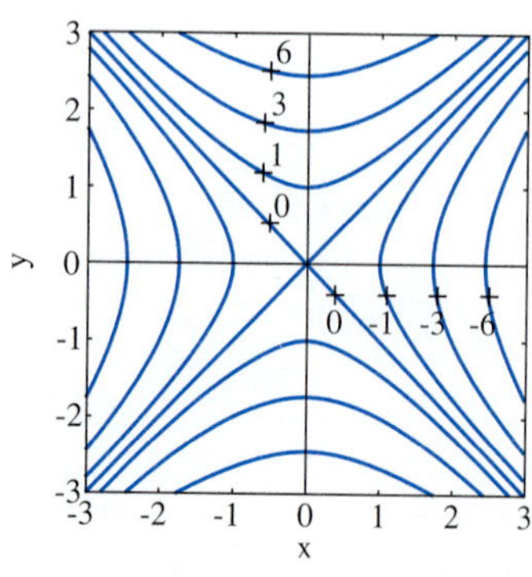

Fig. 13.2.12 Contour curves for the function $f(x, y) = y^2 - x^2$

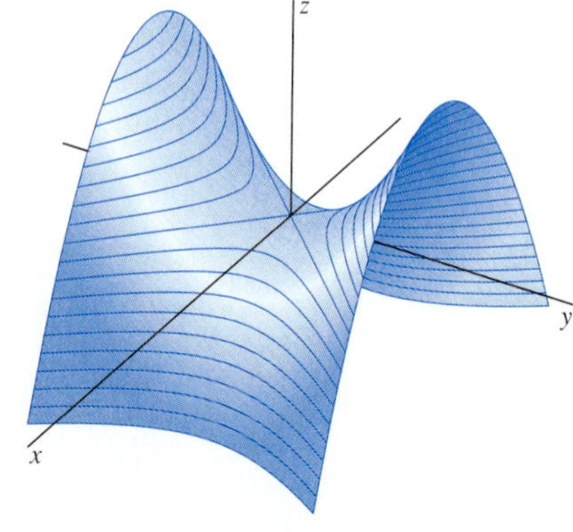

Fig. 13.2.13 Contour curves on $z = y^2 - x^2$ (Example 7)

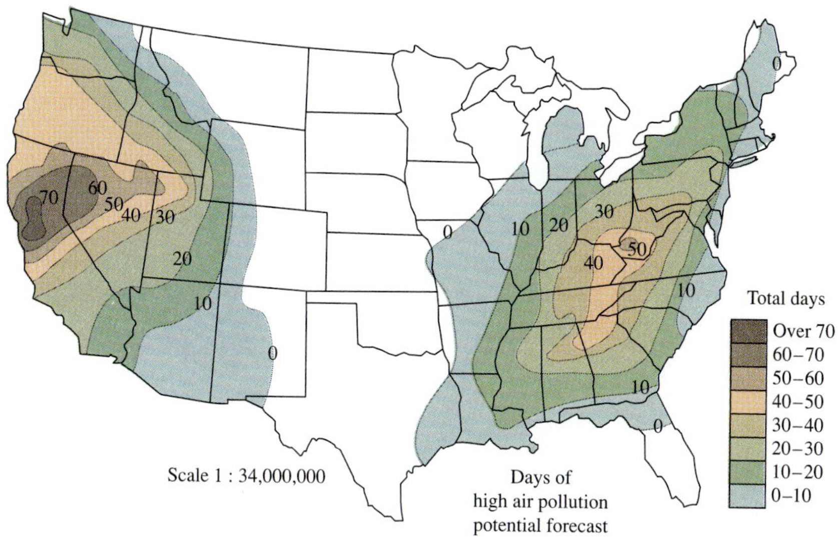

Fig. 13.2.14 Days of high air pollution forecast in the United States (from National Atlas of the United States, U.S. Department of the Interior, 1970)

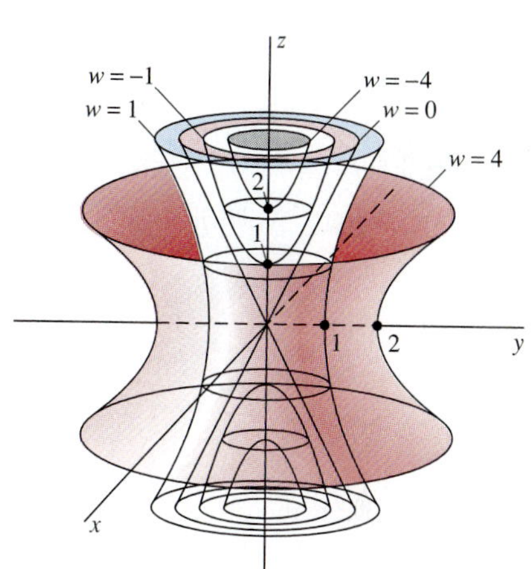

Fig. 13.2.15 Some level surfaces of the function $w = f(x, y, z) = x^2 + y^2 - z^2$ (Example 8)

EXAMPLE 8 Figure 13.2.15 shows some level surfaces of the function

$$f(x, y, z) = x^2 + y^2 - z^2.$$

If $k > 0$, then the graph of $x^2 + y^2 - z^2 = k$ is a hyperboloid of one sheet, whereas if $k < 0$, it is a hyperboloid of two sheets. The cone $x^2 + y^2 - z^2 = 0$ lies between these two types of hyperboloids. ■

Computer Plots

Many computer systems have surface and contour plotting routines like the *Maple* commands

```
plot3d(y∧2 - x∧2, x = -3..3, y = -3..3 );
with(plots): contourplot(y∧2 - x∧2, x = -3..3, y = -3..3 );
```

and the *Mathematica* commands

```
Plot3d[ y∧2 - x∧2, {x,-3,3}, {y,-3,3} ]
ContourPlot[ y∧2 - x∧2, {x,-3,3}, {y,-3,3} ]
```

for the function $f(x, y) = y^2 - x^2$ of Example 7.

EXAMPLE 9 Figure 13.2.16 shows both the graph and some projected contour curves of the function $f(x, y) = (x^2 - y^2)\exp(-x^2 - y^2)$. Observe the characteristic pattern of "nested level curves" that indicate a local maximum or local minimum. It appears that the function $f(x, y)$ attains local minimum values at two points on or near the y-axis and local maximum values at two points on or near the x-axis. The two-dimensional level curve plot in Fig. 13.2.17 suggests that these four extreme value points may well be the points $(0, \pm 1)$ and $(\pm 1, 0)$. ■

Remark In Section 13.5 we will study analytic methods for locating maximum and minimum points of function of two variables *exactly*. But Example 9 indicates that level curve plots provide a valuable tool for locating them *approximately*.

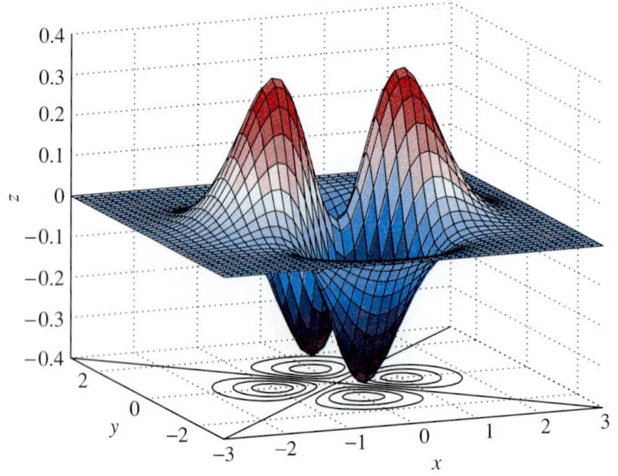

Fig. 13.2.16 The graph and projected contour curves of the function $f(x, y) = (x^2 - y^2)e^{-x^2-y^2}$

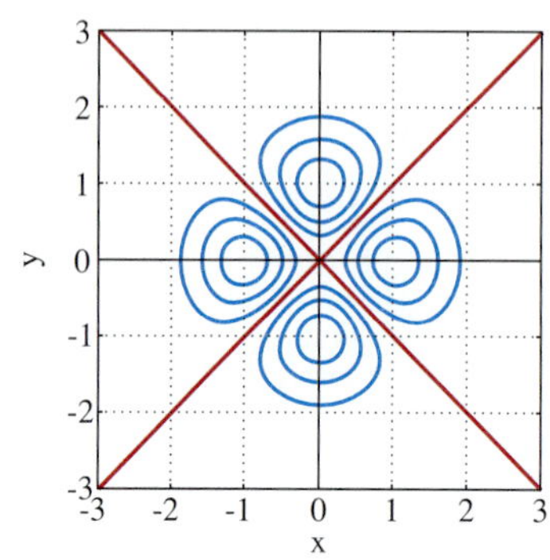

Fig. 13.2.17 Level curves for the function $f(x, y) = (x^2 - y^2)e^{-x^2-y^2}$

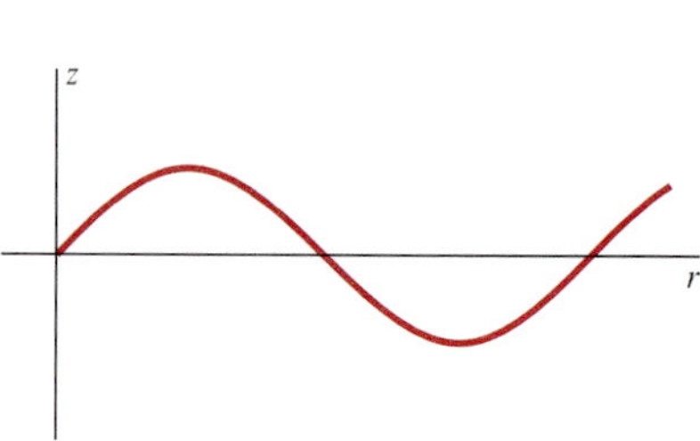

Fig. 13.2.18 The curve $z = \sin r$ (Example 10)

EXAMPLE 10 The surface

$$z = \sin\sqrt{x^2 + y^2} \tag{3}$$

is symmetrical with respect to the z-axis, because Eq. (3) reduces to the equation $z = \sin r$ (Fig. 13.2.18) in terms of the radial coordinate $r = \sqrt{x^2 + y^2}$ that measures perpendicular distance from the z-axis. The *surface* $z = \sin r$ is generated by revolving the curve $z = \sin x$ around the z-axis. Hence its level curves are circles centered at the origin in the xy-plane. For instance, $z = 0$ if r is an integral multiple of π, whereas $z = \pm 1$ if r is any odd integral multiple of $\pi/2$. Figure 13.2.19 shows traces of this surface in planes parallel to the yz-plane. The "hat effect" was achieved by plotting (x, y, z) for those points (x, y) that lie within a certain ellipse in the xy-plane. ■

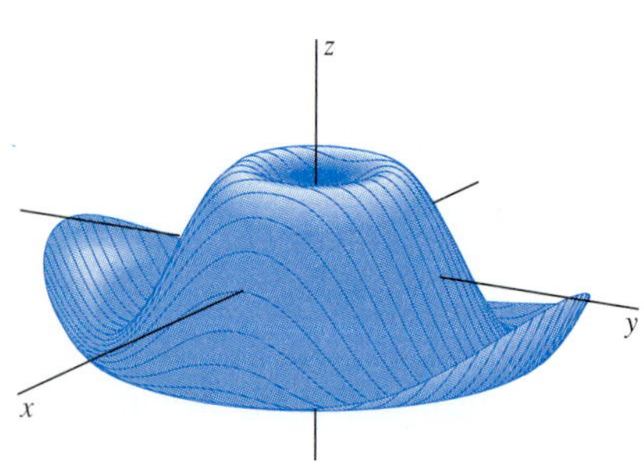

Fig. 13.2.19 The hat surface $z = \sin\sqrt{x^2 + y^2}$ (Example 10)

Given an arbitrary function $f(x, y)$, it can be quite a challenge to construct by hand a picture of the surface $z = f(x, y)$. Example 11 illustrates some special techniques that may be useful. Additional surface-sketching techniques will appear in the remainder of this chapter.

EXAMPLE 11 Investigate the graph of the function

$$f(x, y) = \tfrac{3}{4}y^2 + \tfrac{1}{24}y^3 - \tfrac{1}{32}y^4 - x^2. \tag{4}$$

Solution The key feature in Eq. (4) is that the right-hand side is the *sum* of a function of x and a function of y. If we set $x = 0$, we get the curve

$$z = \tfrac{3}{4}y^2 + \tfrac{1}{24}y^3 - \tfrac{1}{32}y^4 \tag{5}$$

in which the surface $z = f(x, y)$ intersects the yz-plane. But if we set $y = y_0$ in Eq. (4), we get

$$z = \left(\tfrac{3}{4}y_0^2 + \tfrac{1}{24}y_0^3 - \tfrac{1}{32}y_0^4\right) - x^2;$$

that is,

$$z = k - x^2, \tag{6}$$

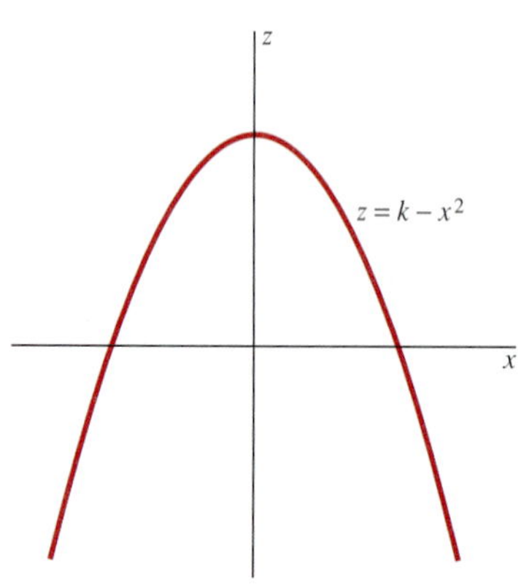

Fig. 13.2.20 The intersection of $z = f(x, y)$ and the plane $y = y_0$ (Example 11)

which is the equation of a parabola in the xz-plane. Hence the trace of $z = f(x, y)$ in each plane $y = y_0$ is a parabola of the form in Eq. (6) (Fig. 13.2.20).

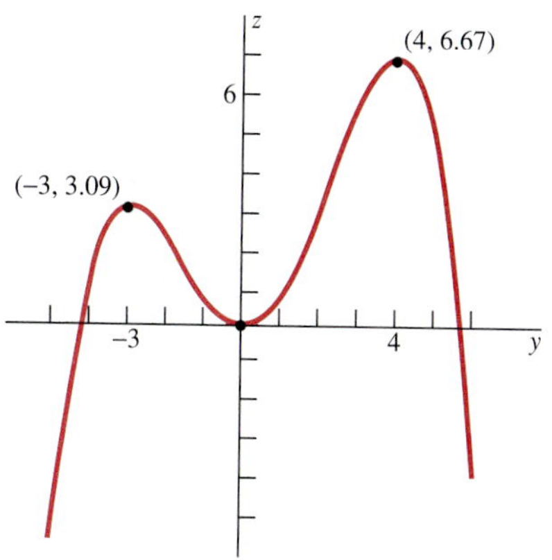

Fig. 13.2.21 The curve $z = \frac{3}{4}y^2 + \frac{1}{24}y^3 - \frac{1}{32}y^4$ (Example 11)

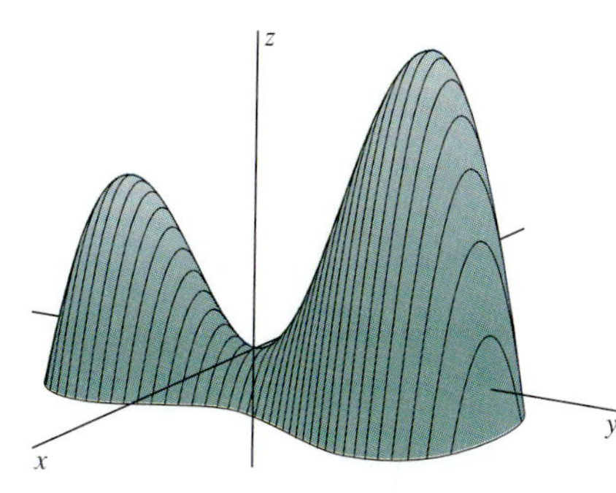

Fig. 13.2.22 Trace parabolas of $z = f(x, y)$ (Example 11)

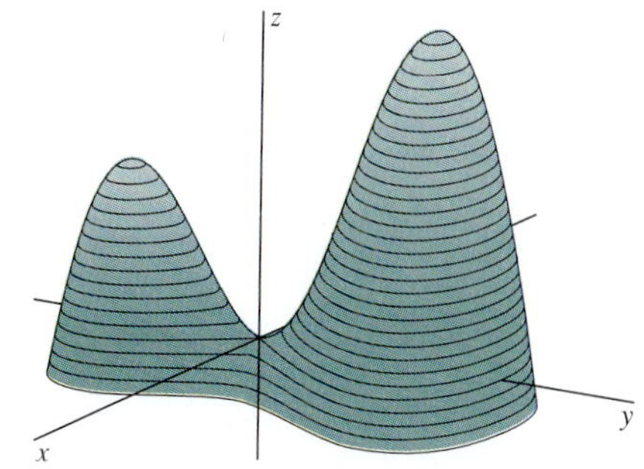

Fig. 13.2.23 Contour curves on $z = f(x, y)$ (Example 11)

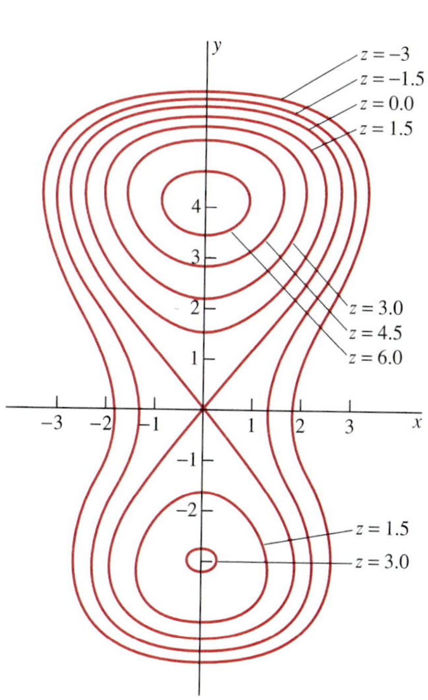

Fig. 13.2.24 Level curves of the function $f(x, y) = \frac{3}{4}y^2 + \frac{1}{24}y^3 - \frac{1}{32}y^4 - x^2$ (Example 11)

We can use the techniques of Section 4.5 to sketch the curve in Eq. (5). Calculating the derivative of z with respect to y, we get

$$\frac{dz}{dy} = \frac{3}{2}y + \frac{1}{8}y^2 - \frac{1}{8}y^3 = -\frac{1}{8}y(y^2 - y - 12) = -\frac{1}{8}y(y + 3)(y - 4).$$

Hence the critical points are $y = -3$, $y = 0$, and $y = 4$. The corresponding values of z are $f(0, -3) \approx 3.09$, $f(0, 0) = 0$, and $f(0, 4) \approx 6.67$. Because $z \to -\infty$ as $y \to \pm\infty$, it follows readily that the graph of Eq. (5) looks like that in Fig. 13.2.21.

Now we can see what the surface $z = f(x, y)$ looks like. Each vertical plane $y = y_0$ intersects the curve in Eq. (5) at a single point, and this point is the vertex of a parabola that opens downward like that in Eq. (6); this parabola is the intersection of the plane and the surface. Thus the surface $z = f(x, y)$ is generated by translating the vertex of such a parabola along the curve

$$z = \tfrac{3}{4}y^2 + \tfrac{1}{24}y^3 - \tfrac{1}{32}y^4,$$

as indicated in Fig. 13.2.22.

Figure 13.2.23 shows some typical contour curves on this surface. They indicate that the surface resembles two peaks separated by a mountain pass. To check this figure, we programmed a microcomputer to plot typical level curves of the function $f(x, y)$. The result is shown in Fig. 13.2.24. The nested level curves around the point $(0, -3)$ and $(0, 4)$ indicate the local maxima of $z = f(x, y)$. The level figure eight curve through $(0, 0)$ marks the *saddle point* we see in Figs. 13.2.22 and 13.2.23. Local extrema and saddle points of functions of two variables are discussed in Sections 13.5 and 13.10. ■

13.2 PROBLEMS

In Problems 1 through 20, state the largest possible domain of definition of the given function f.

1. $f(x, y) = 4 - 3x - 2y$

2. $f(x, y) = \sqrt{x^2 + 2y^2}$

3. $f(x, y) = \dfrac{1}{x^2 + y^2}$

4. $f(x, y) = \dfrac{1}{x - y}$

5. $f(x, y) = \sqrt[3]{y - x^2}$

6. $f(x, y) = \sqrt{2x} + \sqrt[3]{3y}$

7. $f(x, y) = \sin^{-1}(x^2 + y^2)$

8. $f(x, y) = \tan^{-1}\left(\dfrac{y}{x}\right)$

9. $f(x, y) = \exp(-x^2 - y^2)$ (Fig. 13.2.25)

10. $f(x, y) = \ln(x^2 - y^2 - 1)$

11. $f(x, y) = \ln(y - x)$

12. $f(x, y) = \sqrt{4 - x^2 - y^2}$

13. $f(x, y) = \dfrac{1 + \sin xy}{xy}$

14. $f(x, y) = \dfrac{1 + \sin xy}{x^2 + y^2}$ (Fig. 13.2.26)

15. $f(x, y) = \dfrac{xy}{x^2 - y^2}$

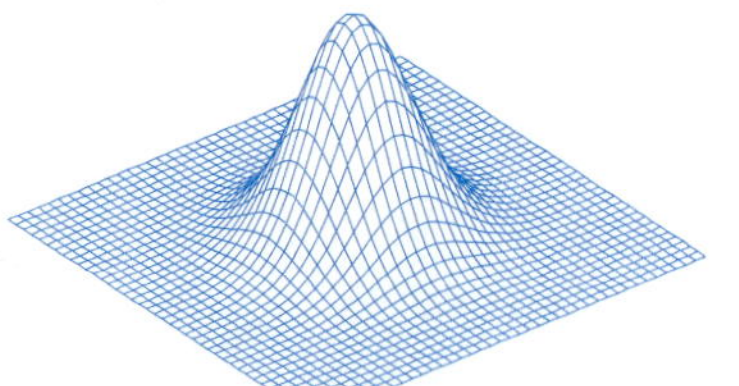

Fig. 13.2.25 The graph of the function of Problem 9

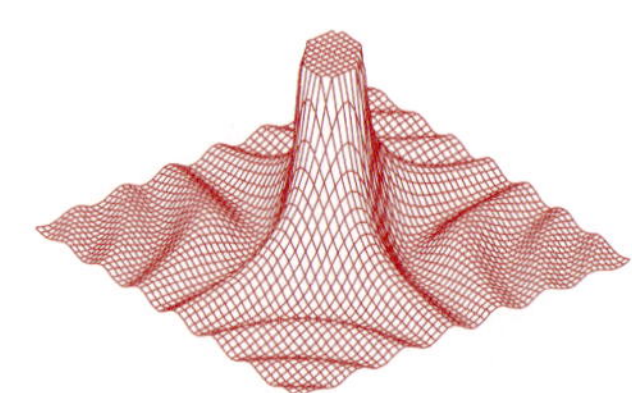

Fig. 13.2.26 The graph of the function of Problem 14

16. $f(x, y, z) = \dfrac{1}{\sqrt{z - x^2 - y^2}}$

17. $f(x, y, z) = \exp\left(\dfrac{1}{x^2 + y^2 + z^2}\right)$

18. $f(x, y, z) = \ln(xyz)$

19. $f(x, y, z) = \ln(z - x^2 - y^2)$

20. $f(x, y, z) = \sin^{-1}(3 - x^2 - y^2 - z^2)$

In Problems 21 through 30, describe the graph of the function f.

21. $f(x, y) = 10$

22. $f(x, y) = x$

23. $f(x, y) = x + y$

24. $f(x, y) = \sqrt{x^2 + y^2}$

25. $f(x, y) = x^2 + y^2$

26. $f(x, y) = 4 - x^2 - y^2$

27. $f(x, y) = \sqrt{4 - x^2 - y^2}$

28. $f(x, y) = 16 - y^2$

29. $f(x, y) = 10 - \sqrt{x^2 + y^2}$

30. $f(x, y) = -\sqrt{36 - 4x^2 - 9y^2}$

In Problems 31 through 40, sketch some typical level curves of the function f.

31. $f(x, y) = x - y$

32. $f(x, y) = x^2 - y^2$

33. $f(x, y) = x^2 + 4y^2$

34. $f(x, y) = y - x^2$

35. $f(x, y) = y - x^3$

36. $f(x, y) = y - \cos x$

37. $f(x, y) = x^2 + y^2 - 4x$

38. $f(x, y) = x^2 + y^2 - 6x + 4y + 7$

39. $f(x, y) = \exp(-x^2 - y^2)$

40. $f(x, y) = \dfrac{1}{1 + x^2 + y^2}$

In Problems 41 through 46, describe the level surfaces of the function f.

41. $f(x, y, z) = x^2 + y^2 - z$

42. $f(x, y, z) = z + \sqrt{x^2 + y^2}$

43. $f(x, y, z) = x^2 + y^2 + z^2 - 4x - 2y - 6z$

44. $f(x, y, z) = z^2 - x^2 - y^2$

45. $f(x, y, z) = x^2 + 4y^2 - 4x - 8y + 17$

46. $f(x, y, z) = x^2 + y^2 + 25$

In Problems 47 through 52, the function $f(x, y)$ is the sum of a function of x and a function of y. Hence you can use the method of Example 11 to construct a sketch of the surface $z = f(x, y)$. Match each function with its graph among Figs. 13.2.27 through 13.2.32.

47. $f(x, y) = x^2 + 2y$

48. $f(x, y) = y - x^2$

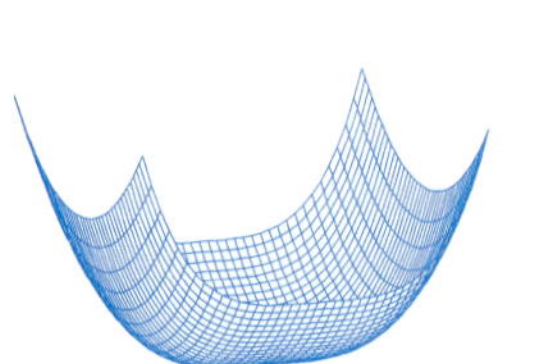

Fig. 13.2.27

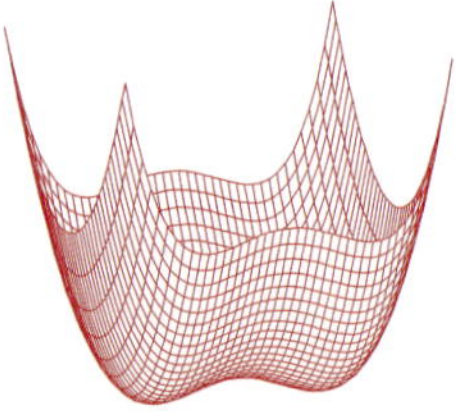

Fig. 13.2.28

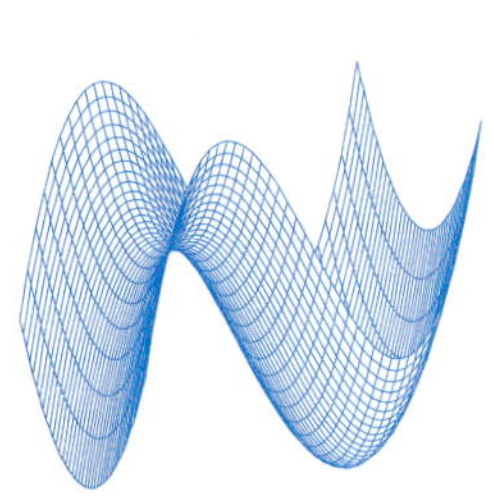

Fig. 13.2.29

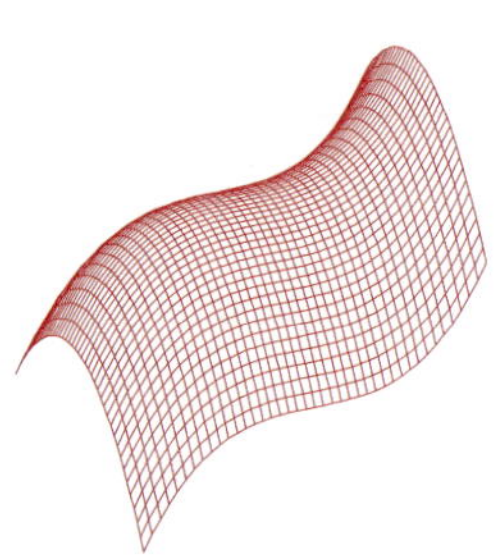

Fig. 13.2.30

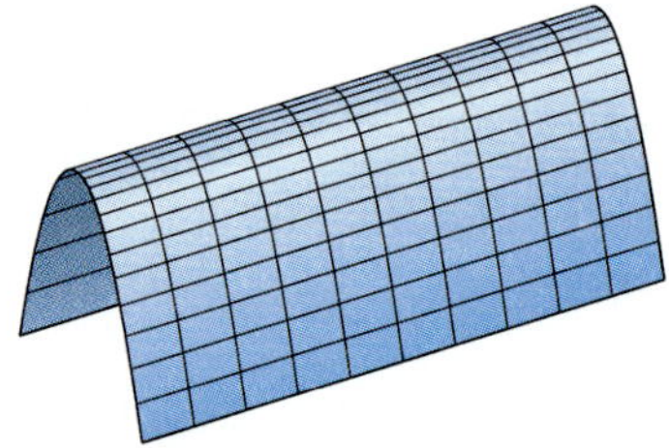

Fig. 13.2.31

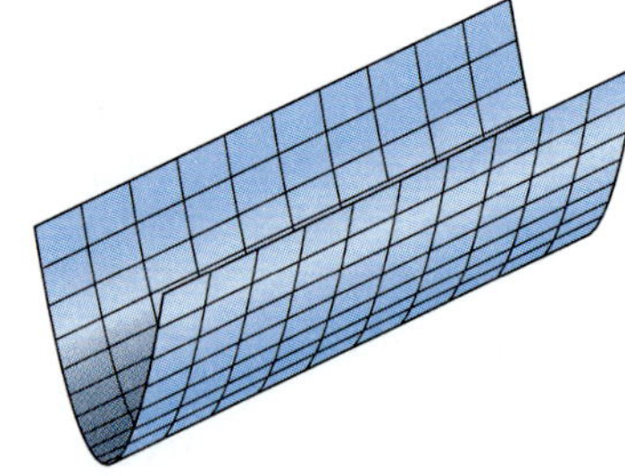

Fig. 13.2.32

49. $f(x, y) = y^3 - x^2$

50. $f(x, y) = y^4 + x^2$

51. $f(x, y) = y^4 - 2y^2 + x^2$

52. $f(x, y) = 2y^3 - 3y^2 - 12y + x^2$

Problems 53 through 58 show the graphs of six functions $z = f(x, y)$ (Figs. 13.2.33–13.2.38). Figures 13.2.39 through 13.2.44 show level curve plots for the same functions but in another order; the level curves in each figure correspond to contours at equally spaced heights on the surface $z = f(x, y)$. Match each surface with its level curves.

53. Fig. 13.2.33

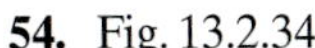

54. Fig. 13.2.34

55. Fig. 13.2.35

56. Fig. 13.2.36

57. Fig. 13.2.37

58. Fig. 13.2.38

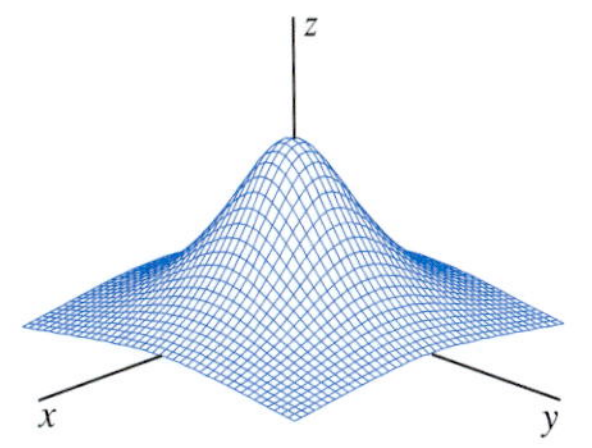

Fig. 13.2.33
$z = \dfrac{1}{1 + x^2 + y^2}$, $|x| \leqq 2$, $|y| \leqq 2$

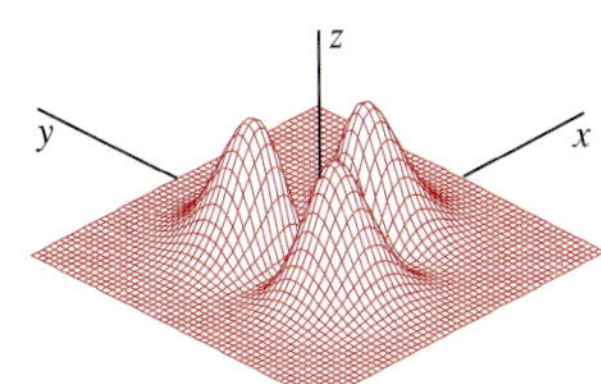

Fig. 13.2.34
$z = r^2 \exp(-r^2)\cos^2\left(\frac{3}{2}\theta\right)$, $|x| \leqq 3$, $|y| \leqq 3$

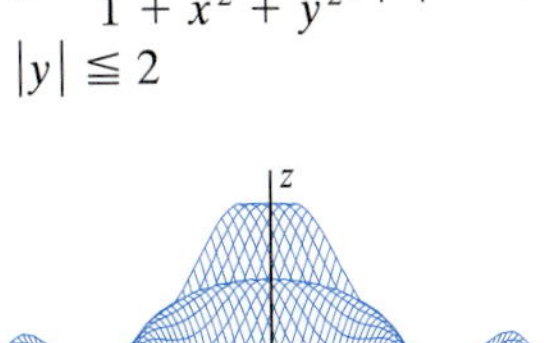
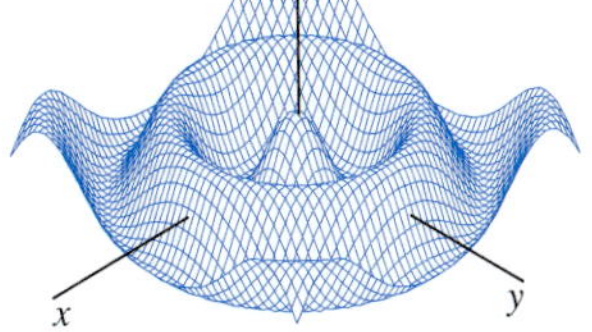

Fig. 13.2.35
$z = \cos\sqrt{x^2 + y^2}$, $|x| \leqq 10$, $|y| \leqq 10$

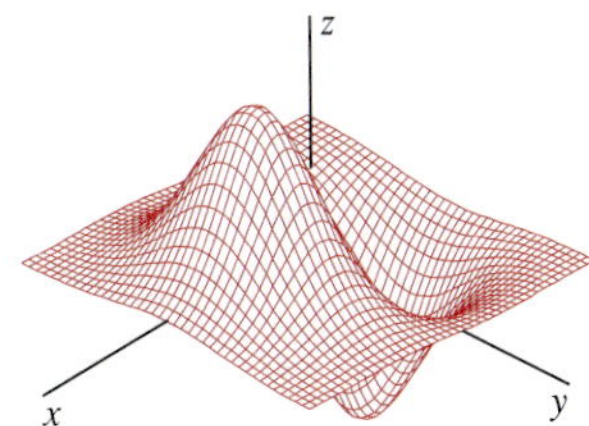

Fig. 13.2.36
$z = x\exp(-x^2 - y^2)$, $|x| \leqq 2$, $|y| \leqq 2$

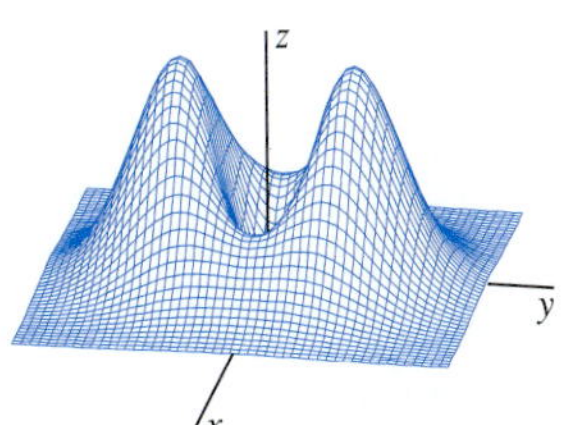

Fig. 13.2.37
$z = 3(x^2 + 3y^2)\exp(-x^2 - y^2)$, $|x| \leqq 2.5$, $|y| \leqq 2.5$

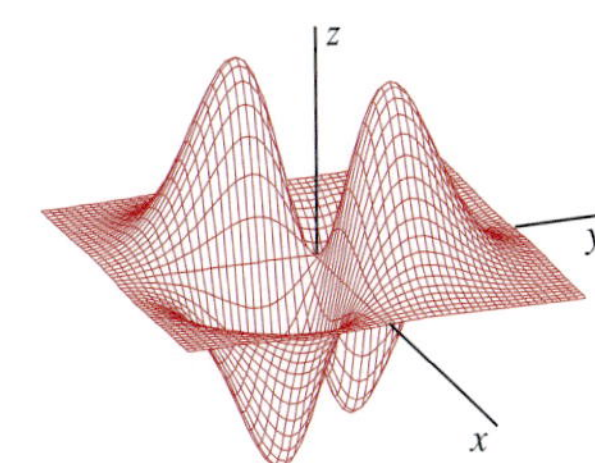

Fig. 13.2.38
$z = xy\exp\left(-\frac{1}{2}(x^2 + y^2)\right)$, $|x| \leqq 3.5$, $|y| \leqq 3.5$

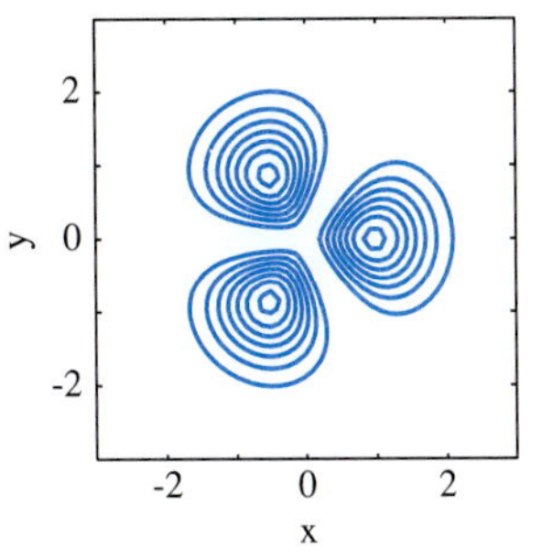

Fig. 13.2.39

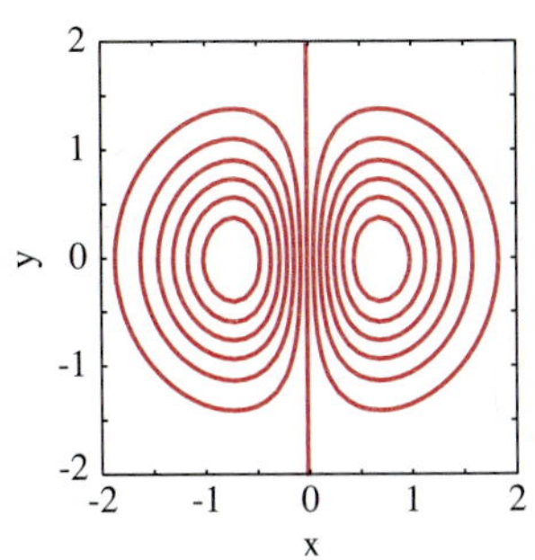

Fig. 13.2.40

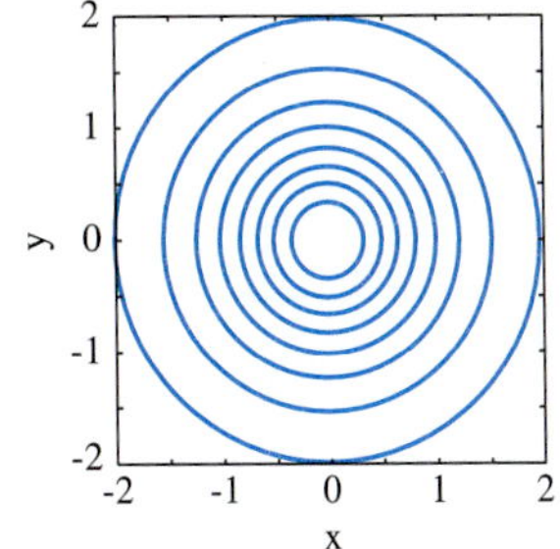

Fig. 13.2.41

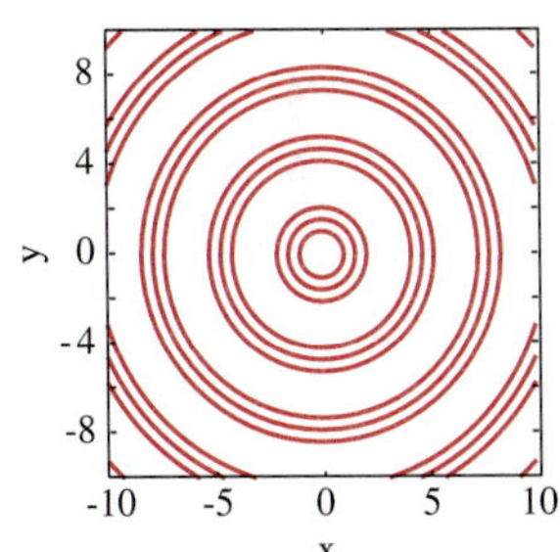

Fig. 13.2.42

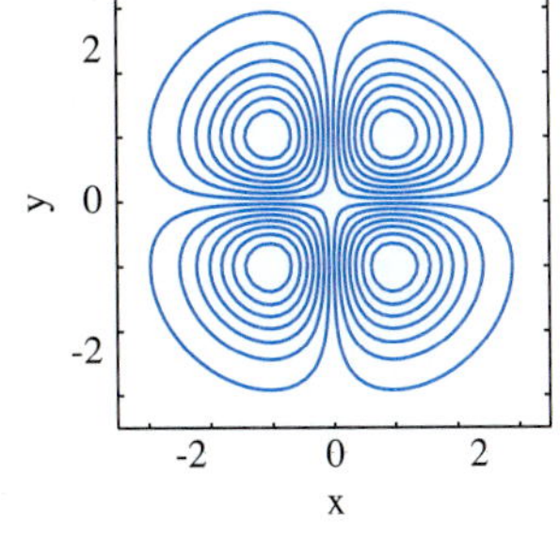

Fig. 13.2.43

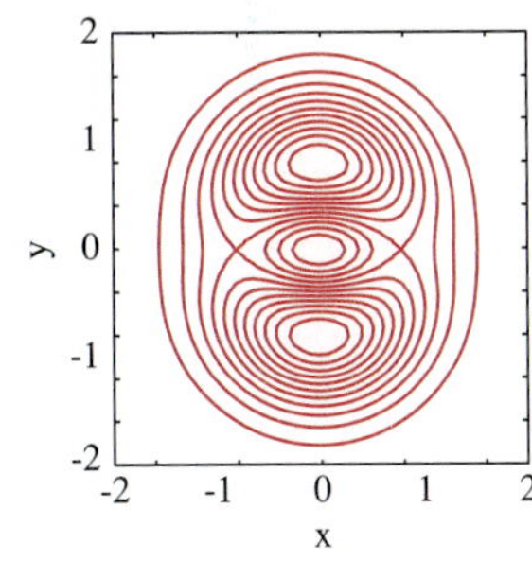

Fig. 13.2.44

13.2 PROJECT: YOUR PERSONAL PORTFOLIO OF SURFACES

Plotting a variety of surfaces with a computer graphing program can help you develop your intuition about graphs of functions of two variables. To get started, graph some of the following functions over rectangular domains $a \leqq x \leqq b$, $c \leqq y \leqq d$ of various sizes to see how the scale affects the picture.

$$f(x, y) = p\cos qx$$

$$f(x, y) = p\cos qy$$

$$f(x, y) = \sin px \sin qy$$

$$\left.\begin{aligned} f(x, y) &= p + qx^2 \\ f(x, y) &= p + qy^2 \\ f(x, y) &= px^2 + qy^2 \end{aligned}\right\}$$ (Use negative *and* positive values of p and q in these three examples.)

$$f(x, y) = px^2 + qxy + ry^2$$

$$f(x, y) = \exp\,(-px^2 - qy^2)$$

$$f(x, y) = (px^2 + qxy + ry^2)\exp\,(-x^2 - y^2)$$

Similarly, vary the numerical parameters p, q, and r and note the resulting changes in the graph. Then make up some functions of your own for experimentation. If you have a computer connected to a printer, assemble a portfolio of your most interesting examples.

13.3 LIMITS AND CONTINUITY

We need limits of functions of several variables for the same reasons that we needed limits of functions of a single variable—so that we can discuss continuity, slopes, and rate of change. Both the definition and the basic properties of limits of functions of several variables are essentially the same as those that we stated in Section 2.2 for functions of a single variable. For simplicity, we shall state them here only for functions of two variables x and y; for a function of three variables, the pair (x, y) should be replaced with the triple (x, y, z).

For a function f of two variables, we ask what number (if any) the values $f(x, y)$ approach as (x, y) approaches the fixed point (a, b) in the coordinate plane. For a function f of three variables, we ask what number (if any) the values $f(x, y, z)$ approach as (x, y, z) approaches the fixed point (a, b, c) in space.

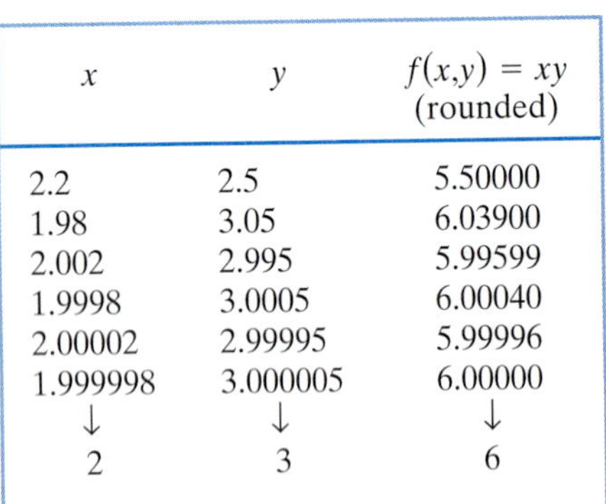

x	y	$f(x,y) = xy$ (rounded)
2.2	2.5	5.50000
1.98	3.05	6.03900
2.002	2.995	5.99599
1.9998	3.0005	6.00040
2.00002	2.99995	5.99996
1.999998	3.000005	6.00000
↓	↓	↓
2	3	6

Fig. 13.3.1 The numerical data of Example 1

EXAMPLE 1 The numerical data in the table of Fig. 13.3.1 suggest that the values of the function $f(x, y) = xy$ approach 6 as $x \to 2$ and $y \to 3$ simultaneously—that is, as (x, y) approaches the point $(2, 3)$. It therefore is natural to write

$$\lim_{(x,\,y)\to(2,\,3)} xy = 6.$$ ■

Our intuitive idea of the limit of a function of two variables is this. We say that the number L is the *limit* of the function $f(x, y)$ as (x, y) approaches the point (a, b), and we write

$$\lim_{(x,\,y)\to(a,\,b)} f(x, y) = L, \tag{1}$$

provided that the number $f(x, y)$ can be made as close as we please to L merely by choosing the point (x, y) sufficiently close to—but not equal to—the point (a, b).

To make this intuitive idea precise, we must specify how close to L—within the distance $\epsilon > 0$, say—we want $f(x, y)$ to be, and then how close to (a, b) the point (x, y) must be to accomplish this. We think of the point (x, y) as being close to (a, b) provided that it lies within a small circular disk (Fig. 13.3.2) with center (a, b) and radius δ, where δ is a small positive number. The point (x, y) lies within this disk if and only if

$$\sqrt{(x - a)^2 + (y - b)^2} < \delta. \tag{2}$$

This observation serves as motivation for the formal definition, with two additional conditions. First, we define the limit of $f(x, y)$ as $(x, y) \to (a, b)$ *only* under the condition that the domain of definition of f contains points $(x, y) \neq (a, b)$ that lie arbitrarily

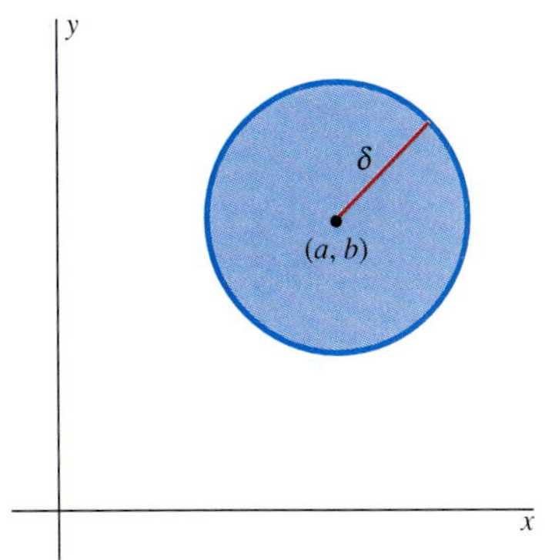

Fig. 13.3.2 The circular disk with center (a, b) and radius δ

close to (a, b)—that is, within *every* disk of the sort shown in Fig. 13.3.2 and thus within any and every preassigned positive distance of (a, b). Hence we do not speak of the limit of f at an isolated point of its domain D. Finally, we do *not* require that f be defined at the point (a, b) itself. Thus we deliberately exclude the possibility that $(x, y) = (a, b)$.

Definition *The Limit of f(x,y)*

We say that the **limit of** $f(x, y)$ **as** (x, y) **approaches** (a, b) **is** L provided that, for every number $\epsilon > 0$, there exists a number $\delta > 0$ with the following property: If (x, y) is a point of the domain of f such that

$$0 < \sqrt{(x-a)^2 + (y-b)^2} < \delta, \tag{2'}$$

then it follows that

$$|f(x, y) - L| < \epsilon. \tag{3}$$

Remark The "extra" inequality $0 < \sqrt{(x-a)^2 + (y-b)^2}$ in Eq. (2′) serves to ensure that $(x, y) \neq (a, b)$.

We ordinarily shall rely on continuity rather than the formal definition of the limit to evaluate limits of functions of several variables. We say that f is **continuous at the point** (a, b) provided that $f(a, b)$ exists and $f(x, y)$ approaches $f(a, b)$ as (x, y) approaches (a, b). That is,

$$\lim_{(x,y)\to(a,b)} f(x, y) = f(a, b).$$

Thus f is continuous at (a, b) if it is defined there and its limit there is equal to its value there, precisely as in the case of a function of a single variable. The function f is said to be **continuous on the set** D if it is continuous at each point of D, again exactly as in the single-variable case.

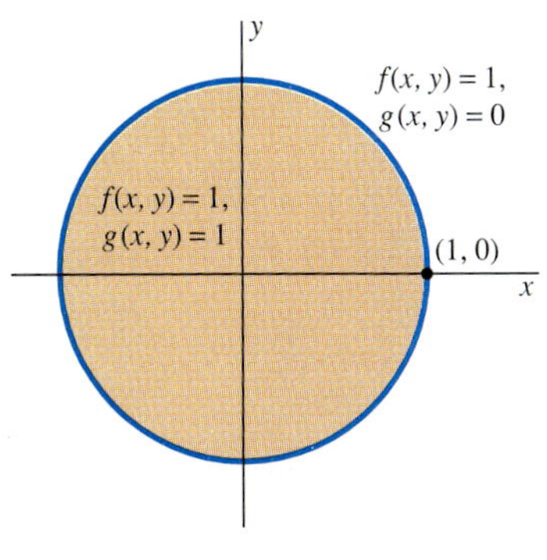

Fig. 13.3.3 The circular disk of Example 2

EXAMPLE 2 Let D be the circular disk consisting of the points (x, y) such that $x^2 + y^2 \leqq 1$, and let $f(x, y) = 1$ at each point of D (Fig. 13.3.3). Then the limit of $f(x, y)$ at each point of D is 1, so f is continuous on D. But let the new function $g(x, y)$ be defined on the entire plane $\mathbf{R}^2$ as follows:

$$g(x, y) = \begin{cases} f(x, y) & \text{if } (x, y) \text{ is in } D, \\ 0 & \text{otherwise.} \end{cases}$$

Then g is *not* continuous on $\mathbf{R}^2$. For instance, the limit of $g(x, y)$ as $(x, y) \to (1, 0)$ does not exist, because there exist both points within D arbitrarily close to $(1, 0)$ at which g has the value 1 and points outside of D arbitrarily close to $(1, 0)$ at which g has the value 0. Thus $g(x, y)$ cannot approach any single value as $(x, y) \to (1, 0)$. Because g has no limit at $(1, 0)$, it cannot be continuous there. ■

The limit laws of Section 2.2 have natural analogues for functions of several variables. If

$$\lim_{(x,y)\to(a,b)} f(x, y) = L \quad \text{and} \quad \lim_{(x,y)\to(a,b)} g(x, y) = M, \tag{4}$$

then the sum, product, and quotient laws for limits are these:

$$\lim_{(x,y)\to(a,b)} [f(x, y) + g(x, y)] = L + M, \tag{5}$$

$$\lim_{(x,y)\to(a,b)} [f(x,y)\cdot g(x,y)] = L\cdot M, \quad \text{and} \tag{6}$$

$$\lim_{(x,y)\to(a,b)} \frac{f(x,y)}{g(x,y)} = \frac{L}{M} \quad \text{if } M \neq 0. \tag{7}$$

EXAMPLE 3 Show that $\lim_{(x,y)\to(a,b)} xy = ab$.

Solution We take $f(x,y) = x$ and $g(x,y) = y$. Then it follows from the definition of limit that

$$\lim_{(x,y)\to(a,b)} f(x,y) = a \quad \text{and} \quad \lim_{(x,y)\to(a,b)} g(x,y) = b.$$

Hence the product law gives

$$\begin{aligned}\lim_{(x,y)\to(a,b)} xy &= \lim_{(x,y)\to(a,b)} f(x,y)g(x,y)\\ &= \left[\lim_{(x,y)\to(a,b)} f(x,y)\right]\left[\lim_{(x,y)\to(a,b)} g(x,y)\right] = ab.\end{aligned}$$

■

More generally, suppose that $P(x,y)$ is a polynomial in the two variables x and y. That is, $P(x,y)$ is a sum of constant multiples of the form $x^i y^j$ where the exponents i and j are nonnegative integers. Thus $P(x,y)$ can be written in the form

$$P(x,y) = \sum c_{ij} x^i y^j.$$

The sum and product laws for limits then imply that

$$\begin{aligned}\lim_{(x,y)\to(a,b)} P(x,y) &= \lim_{(x,y)\to(a,b)} \sum c_{ij} x^i y^j\\ &= \sum \left(\lim_{(x,y)\to(a,b)} c_{ij} x^i y^j\right)\\ &= \sum c_{ij}\left(\lim_{x\to a} x^i\right)\left(\lim_{y\to b} y^j\right)\\ &= \sum c_{ij} a^i b^j = P(a,b).\end{aligned}$$

It follows that *every polynomial in two (or more) variables is an everywhere continuous function.*

EXAMPLE 4 The function $f(x,y) = 2x^4y^2 - 7xy + 4x^2y^3 - 5$ is a polynomial, so we can find its limit at any point (a,b) simply by evaluating $f(a,b)$. For instance,

$$\lim_{(x,y)\to(-1,2)} f(x,y) = f(-1,2) = 2\cdot(-1)^4(2)^2 - 7\cdot(-1)(2) + 4\cdot(-1)^2(2)^3 - 5 = 49.$$

■

Just as in the single-variable case, any composition of continuous multivariable functions is also a continuous function. For example, suppose that the functions f and g are both continuous at (a,b) and that h is continuous at the point $(f(a,b), g(a,b))$. Then the composite function

$$H(x,y) = h(f(x,y), g(x,y))$$

is also continuous at (a,b). As a consequence, any finite combination involving sums, products, quotients, and compositions of the familiar elementary functions is continuous, except possibly at points where a denominator is zero or where the formula for the function is otherwise meaningless. This general rule suffices for the evaluation of most limits that we shall encounter.

EXAMPLE 5 If

$$f(x, y) = e^{xy} \sin \frac{\pi y}{4} + xy \ln \sqrt{y - x},$$

then e^{xy} is the composition of continuous functions, thus continuous; $\sin \frac{1}{4}\pi y$ is continuous for the same reason; their product is continuous because each is continuous. Also $y - x$, a polynomial, is continuous everywhere; $\sqrt{y - x}$ is therefore continuous if $y \geqq x$; $\ln \sqrt{y - x}$ is continuous provided that $y > x$; $xy \ln \sqrt{y - x}$ is the product of functions continuous if $y > x$. And thus the sum

$$f(x, y) = e^{xy} \sin \frac{\pi y}{4} + xy \ln \sqrt{y - x}$$

of functions continuous if $y > x$ is itself continuous if $y > x$. Because $f(x, y)$ is continuous if $y > x$, it follows that

$$\lim_{(x, y) \to (1, 2)} \left[e^{xy} \sin \frac{\pi y}{4} + xy \ln \sqrt{y - x} \right] = f(1, 2) = e^2 \cdot 1 + 2 \ln 1 = e^2.$$ ■

Examples 6 and 7 illustrate techniques that sometimes are successful in handling cases with denominators that approach zero; in such cases the techniques of Examples 4 and 5 cannot be applied.

EXAMPLE 6 Show that $\displaystyle\lim_{(x, y) \to (0, 0)} \frac{xy}{\sqrt{x^2 + y^2}} = 0.$

Solution Let (r, θ) be the polar coordinates of the point (x, y). Then $x = r \cos \theta$ and $y = r \sin \theta$, so

$$\frac{xy}{\sqrt{x^2 + y^2}} = \frac{(r \cos \theta)(r \sin \theta)}{\sqrt{r^2(\cos^2 \theta + \sin^2 \theta)}} = r \cos \theta \sin \theta \qquad \text{for } r > 0.$$

Because $r = \sqrt{x^2 + y^2}$, it is clear that $r \to 0$ as both x and y approach zero. It therefore follows that

$$\lim_{(x, y) \to (0, 0)} \frac{xy}{\sqrt{x^2 + y^2}} = \lim_{r \to 0} r \cos \theta \sin \theta = 0,$$

because $|\cos \theta \sin \theta| \leqq |\cos \theta| |\sin \theta| \leqq 1$ for all θ. So if the function f is defined as

$$f(x, y) = \begin{cases} \dfrac{xy}{\sqrt{x^2 + y^2}} & \text{if } (x, y) \neq (0, 0), \\ 0 & \text{if } x = y = 0, \end{cases}$$

then it follows that f is continuous at the origin $(0, 0)$. Figure 13.3.4 shows the graph of $z = f(x, y)$. It corroborates the zero limit that we found at $(0, 0)$. Near the origin the graph appears to resemble the saddle point on a hyperbolic paraboloid (Fig. 13.2.13), but this doesn't look like a smooth and comfortable saddle. ■

Fig. 13.3.4 The graph $z = \dfrac{xy}{\sqrt{x^2 + y^2}}$ (Example 6)

EXAMPLE 7 Show that

$$\lim_{(x, y) \to (0, 0)} \frac{xy}{x^2 + y^2}$$

does not exist.

Solution Our plan is to show that $f(x, y) = xy/(x^2 + y^2)$ approaches different values as (x, y) approaches $(0, 0)$ from different directions. Suppose that (x, y) approaches $(0, 0)$ along the straight line of slope m through the origin. On this line we have $y = mx$. So, on this line,

$$f(x, y) = f(x, mx) = \frac{x \cdot mx}{x^2 + m^2x^2} = \frac{m}{1 + m^2}$$

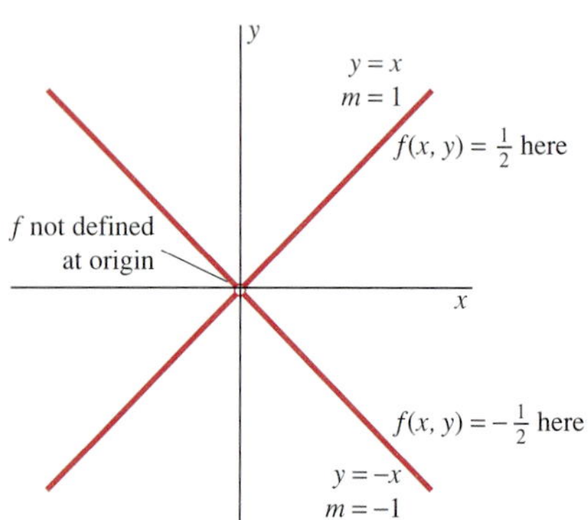

Fig. 13.3.5 The function f of Example 7 takes on both values $+\frac{1}{2}$ and $-\frac{1}{2}$ at points arbitrarily close to the origin.

if $x \neq 0$. If we take $m = 1$, we see that $f(x, y) = \frac{1}{2}$ at every point of the line $y = x$ other than $(0, 0)$. If we take $m = -1$, then $f(x, y) = -\frac{1}{2}$ at every point of the line $y = -x$ other than $(0, 0)$. Thus $f(x, y)$ approaches two different values as (x, y) approaches $(0, 0)$ along these two lines (Fig. 13.3.5). Hence $f(x, y)$ cannot approach any *single* value as (x, y) approaches $(0, 0)$, and this implies that the limit in question cannot exist.

Figure 13.3.6 shows a computer-generated graph of the function $f(x, y) = xy/(x^2 + y^2)$. It consists of linear rays along each of which the polar angular coordinate θ is constant. For each number z between $-\frac{1}{2}$ and $\frac{1}{2}$ (inclusive), there are rays along which $f(x, y)$ has the constant value z. Hence we can make $f(x, y)$ approach any number we please in $\left[-\frac{1}{2}, \frac{1}{2}\right]$ by letting (x, y) approach $(0, 0)$ from the appropriate direction. There are also paths along which (x, y) approaches $(0, 0)$ but the limit of $f(x, y)$ does not exist (Problem 53). ■

REMARK For

$$L = \lim_{(x, y) \to (a, b)} f(x, y)$$

to exist, $f(x, y)$ must approach L for *any and every* mode of approach of (x, y) to (a, b). In Problem 51 we give an example of a function f such that $f(x, y) \to 0$ as $(x, y) \to (0, 0)$ along any straight line through the origin, but $f(x, y) \to 1$ as $(x, y) \to (0, 0)$ along the parabola $y = x^2$. Thus the method of Example 7 cannot be used to show that a limit exists, only that it does not. Fortunately, many important applications, including those we discuss in the remainder of this chapter, involve only functions that exhibit no such exotic behavior as the functions of Problems 51 and 53.

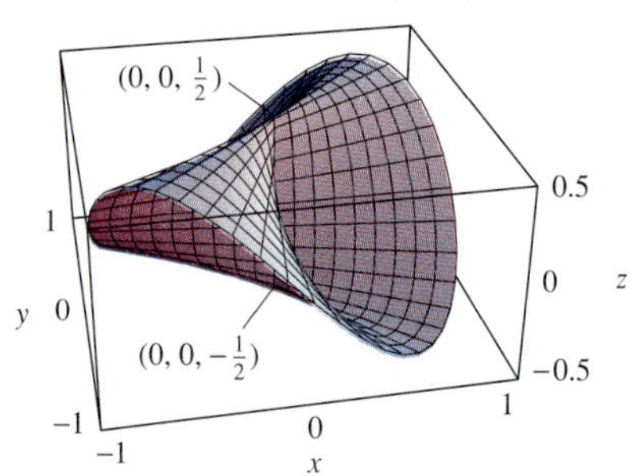

Fig. 13.3.6 The graph of $f(x, y) = \dfrac{xy}{x^2 + y^2}$ (Example 7)

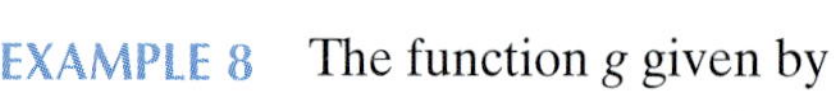
EXAMPLE 8 The function g given by

$$g(x, y) = \begin{cases} \dfrac{xy^2}{x^2 + y^2} & \text{if } (x, y) \neq (0, 0), \\ 0 & \text{if } x = y = 0 \end{cases}$$

differs from the function $f(x, y) = xy/(x^2 + y^2)$ of Example 7 only in the presence of additional factor y in the numerator, so $g(x, y) = y \cdot f(x, y)$. Because $g(x, y)$ is a quotient of polynomials, the function g is continuous except possibly at the origin $(0, 0)$—the only point where its denominator is zero.

To investigate the limit of $g(x, y)$ at $(0, 0)$, we recall from the solution of Example 7 that $f(x, y) = m/(1 + m^2)$ on every straight line $y = mx$ of slope m through the origin in the xy-plane. (The only points of the plane not on such a line are on the y-axis, where $f(0, y) = 0/y^2 \equiv 0$ if $y \neq 0$.) We therefore conclude from the graph of $h(m) = m/(1 + m^2)$ in Fig. 13.3.7 that $|f(x, y)| < 1$ for every point $(x, y) \neq (0, 0)$. It follows that

$$|g(x, y)| = |y \cdot f(x, y)| = |y| \cdot |f(x, y)| \leqq |y|,$$

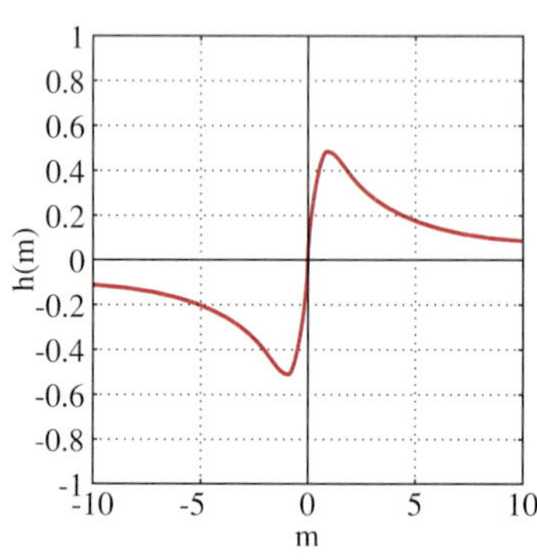

Fig. 13.3.7 $\left|\dfrac{m}{1 + m^2}\right| < 1$ for all m

and hence that

$$\lim_{(x,y)\to(0,0)} g(x, y) = 0 = g(0, 0),$$

because certainly $y \to 0$ as $(x, y) \to (0, 0)$. Thus we see that the function g is continuous at the origin as well. Figure 13.3.8 shows a computer-generated graph $z = g(x, y)$. It appears to exhibit some kind of saddle point behavior near the origin. ■

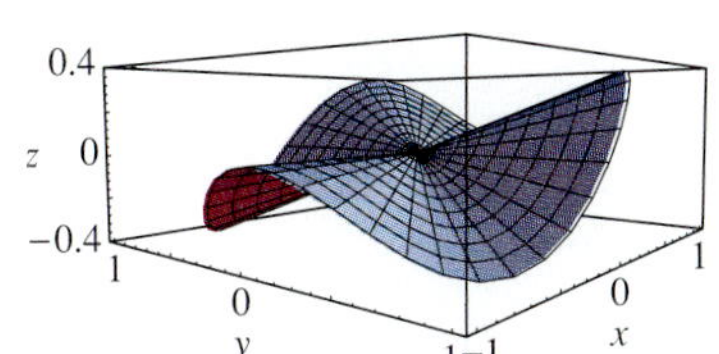

Fig. 13.3.8 The graph of $g(x, y) = \dfrac{xy^2}{x^2 + y^2}$

In this section we have discussed explicitly only functions of two variables, but the concepts of limits and continuity generalize in a straightforward manner to functions of three or more variables. Suppose that $f: \mathbf{R}^n \to \mathbf{R}$ is a function of n variables and we use vector notation $\mathbf{x} = (x_1, x_2, \ldots, x_n)$ for points in n-dimensional space $\mathbf{R}^n$. Then the limit concept takes the precise form of the statement that

$$\lim_{\mathbf{x}\to\mathbf{a}} f(\mathbf{x}) = L \tag{8}$$

provided that, for every number $\epsilon > 0$, there exists a corresponding number $\delta > 0$ such that

$$|f(\mathbf{x}) - L| < \epsilon \quad \text{whenever} \quad 0 < |\mathbf{x} - \mathbf{a}| < \delta. \tag{9}$$

Then the function f is continuous at the point $\mathbf{a} = (a_1, a_2, \ldots, a_n)$ provided that

$$\lim_{\mathbf{x}\to\mathbf{a}} f(\mathbf{x}) = f(\mathbf{a}). \tag{10}$$

An attractive feature of vector notation is that the multidimensional statements in (8) and (9) take precisely the same forms as in the case of functions of a single variable.

13.3 PROBLEMS

Use the limit laws and consequences of continuity to evaluate the limits in Problems 1 through 16.

1. $\displaystyle\lim_{(x,y)\to(0,0)} (7 - x^2 + 5xy)$

2. $\displaystyle\lim_{(x,y)\to(1,-2)} (3x^2 - 4xy + 5y^2)$

3. $\displaystyle\lim_{(x,y)\to(1,-1)} e^{-xy}$

4. $\displaystyle\lim_{(x,y)\to(0,0)} \frac{x + y}{1 + xy}$

5. $\displaystyle\lim_{(x,y)\to(0,0)} \frac{5 - x^2}{3 + x + y}$

6. $\displaystyle\lim_{(x,y)\to(2,3)} \frac{9 - x^2}{1 + xy}$

7. $\displaystyle\lim_{(x,y)\to(0,0)} \ln\sqrt{1 - x^2 - y^2}$

8. $\displaystyle\lim_{(x,y)\to(2,-1)} \ln\frac{1 + x + 2y}{3y^2 - x}$

9. $\displaystyle\lim_{(x,y)\to(0,0)} \frac{e^{xy}\sin xy}{xy}$

10. $\displaystyle\lim_{(x,y)\to(0,0)} \exp\left(-\frac{1}{x^2 + y^2}\right)$

11. $\displaystyle\lim_{(x,y,z)\to(1,1,1)} \frac{x^2 + y^2 + z^2}{1 - x - y - z}$

12. $\displaystyle\lim_{(x,y,z)\to(1,1,1)} (x + y + z)\ln xyz$

13. $\displaystyle\lim_{(x,y,z)\to(1,1,0)} \frac{xy - z}{\cos xyz}$

14. $\displaystyle\lim_{(x,y,z)\to(2,-1,3)} \frac{x + y + z}{x^2 + y^2 + z^2}$

15. $\displaystyle\lim_{(x,y,z)\to(2,8,1)} \sqrt{xy}\tan\frac{3\pi z}{4}$

16. $\displaystyle\lim_{(x,y)\to(1,-1)} \arcsin\frac{xy}{\sqrt{x^2 + y^2}}$

In Problems 17 through 20, evaluate the limits

$$\lim_{h\to 0} \frac{f(x + h, y) - f(x, y)}{h} \quad \text{and} \quad \lim_{k\to 0} \frac{f(x, y + k) - f(x, y)}{k}.$$

17. $f(x, y) = xy$

18. $f(x, y) = x^2 + y^2$

19. $f(x, y) = xy^2 - 2$

20. $f(x, y) = x^2y^3 - 10$

In Problems 21 through 30, find the limit or show that it does not exist.

21. $\displaystyle\lim_{(x,y)\to(1,1)} \frac{1 - xy}{1 + xy}$

22. $\displaystyle\lim_{(x,y)\to(2,-2)} \frac{4 - xy}{4 + xy}$

23. $\displaystyle\lim_{(x,y,z)\to(1,1,1)} \frac{xyz}{yz + xz + xy}$

24. $\displaystyle\lim_{(x,y,z)\to(1,-1,1)} \frac{yz + xz + xy}{1 + xyz}$

25. $\displaystyle\lim_{(x,y)\to(0,0)} \ln(1 + x^2 + y^2)$

26. $\displaystyle\lim_{(x,y)\to(1,1)} \ln(2 - x^2 - y^2)$

27. $\displaystyle\lim_{(x,y)\to(0,0)} \frac{\sin xy}{xy}$

28. $\displaystyle\lim_{(x,y)\to(0,0)} \sin(\ln(1 + x + y))$

29. $\displaystyle\lim_{(x,y)\to(0,0)} \exp\left(-\frac{1}{x^2 + y^2}\right)$

30. $\displaystyle\lim_{(x,y)\to(0,0)} \arctan\left(-\frac{1}{x^2 + y^2}\right)$

In Problems 31 through 36, determine the largest set of points in the xy-plane on which the given formula defines a continuous function.

31. $f(x, y) = \sqrt{x + y}$

32. $f(x, y) = \sin^{-1}(x^2 + y^2)$

33. $f(x, y) = \ln(x^2 + y^2 - 1)$

34. $f(x, y) = \ln(2x - y)$

35. $f(x, y) = \tan^{-1}\left(\dfrac{1}{x^2 + y^2}\right)$

36. $f(x, y) = \tan^{-1}\left(\dfrac{1}{x + y}\right)$

In Problems 37 through 40, evaluate the limit by making the polar coordinates substitution $(x, y) = (r\cos\theta, r\sin\theta)$ and using the fact that $r \to 0$ as $(x, y) \to (0, 0)$.

37. $\displaystyle\lim_{(x,y)\to(0,0)} \frac{x^2 - y^2}{\sqrt{x^2 + y^2}} = 0$

38. $\displaystyle\lim_{(x,y)\to(0,0)} \frac{x^3 - y^3}{x^2 + y^2} = 0$

39. $\displaystyle\lim_{(x,y)\to(0,0)} \frac{x^4 + y^4}{(x^2 + y^2)^{3/2}} = 0$

40. $\displaystyle\lim_{(x,y)\to(0,0)} \frac{\sin\sqrt{x^2 + y^2}}{\sqrt{x^2 + y^2}}$

41. Determine whether or not

$$\lim_{(x,y,z)\to(0,0,0)} \frac{xyz}{x^2 + y^2 + z^2}$$

exists; evaluate it in the former case. (*Suggestion:* Substitute spherical coordinates $x = \rho\sin\phi\cos\theta$, $y = \rho\sin\phi\sin\theta$, $z = \rho\cos\phi$.)

42. Determine whether or not

$$\lim_{(x,y,z)\to(0,0,0)} \arctan\frac{1}{x^2 + y^2 + z^2}$$

exists; evaluate it in the former case. (See the *Suggestion* for Problem 41.)

In Problems 43 and 44, investigate the existence of the given limit by making the substitution $y = mx$.

43. $\displaystyle\lim_{(x,y)\to(0,0)} \frac{x^2 - y^2}{x^2 + y^2}$

44. $\displaystyle\lim_{(x,y)\to(0,0)} \frac{x^3 + y^3}{x^2 + y^2}$

In Problems 45 and 46, show that the given limit does not exist by considering points of the form $(x, 0, 0)$ or $(0, y, 0)$ or $(0, 0, z)$ that approach the origin along one of the coordinate axes.

45. $\displaystyle\lim_{(x,y,z)\to(0,0,0)} \frac{x + y + z}{x^2 + y^2 + z^2}$

46. $\displaystyle\lim_{(x,y,z)\to(0,0,0)} \frac{x^2 + y^2 - z^2}{x^2 + y^2 + z^2}$

In Problems 47 through 50, use a computer-plotted graph to explain why the given limit does not exist.

47. $\displaystyle\lim_{(x,y)\to(0,0)} \frac{x^2 - 2y^2}{x^2 + y^2}$

48. $\displaystyle\lim_{(x,y)\to(0,0)} \frac{x^2y^2}{x^4 + y^4}$

49. $\displaystyle\lim_{(x,y)\to(0,0)} \frac{xy}{2x^2 + 3y^2}$

50. $\displaystyle\lim_{(x,y)\to(0,0)} \frac{x^2 + 4xy + y^2}{x^2 + xy + y^2}$

51. Let

$$f(x, y) = \frac{2x^2y}{x^4 + y^2}.$$

(a) Show that $f(x, y) \to 0$ as $(x, y) \to (0, 0)$ along any and every straight line through the origin. (b) Show that $f(x, y) \to 1$ as $(x, y) \to (0, 0)$ along the parabola $y = x^2$. Conclude that the limit of $f(x, y)$ as $(x, y) \to (0, 0)$ does not exist. The graph of f is shown in Fig. 13.3.9.

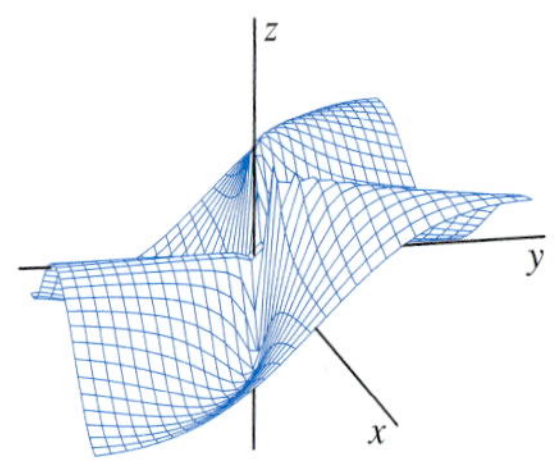

Fig. 13.3.9 The graph of the function f of Problem 51

52. Suppose that $f(x, y) = (x - y)/(x^3 - y)$ except at points of the curve $y = x^3$, where we *define* $f(x, y)$ to be 1. Show that f is not continuous at the point $(1, 1)$. Evaluate the limits of $f(x, y)$ as $(x, y) \to (1, 1)$ along the vertical line $x = 1$ and along the horizontal line $y = 1$. (*Suggestion:* Recall that $a^3 - b^3 = (a - b)(a^2 + ab + b^2)$.)

53. Let

$$\lim_{(x,y)\to(0,0)} \frac{xy}{x^2 + y^2}$$

be the limit of Example 7. Show that as $(x, y) \to (0, 0)$ along the hyperbolic spiral $r\theta = 1$, the limit of $f(x, y)$ does not exist.

13.4 PARTIAL DERIVATIVES

Recall that the derivative of a single-variable function $u = g(x)$ is defined as

$$\frac{du}{dx} = \lim_{\Delta x\to 0} \frac{\Delta u}{\Delta x},$$

where $\Delta u = g(x + h) - g(x)$ is the change in u resulting from the change $h = \Delta x$ in x. This derivative is interpreted as the instantaneous rate of change of u with respect to x. For a function $z = f(x, y)$ of two variables, we need a similar understanding of the rate at which z changes as x and y vary (either singly or simultaneously).

We take a divide-and-conquer approach to this concept. If x is changed by $h = \Delta x$ but y is not changed, then the resulting change in z is

$$\Delta z = f(x + h, y) - f(x, y),$$

and the corresponding instantaneous rate of change of z is

$$\frac{dz}{dx} = \lim_{\Delta x \to 0} \frac{\Delta z}{\Delta x}. \tag{1}$$

On the other hand, if x is not changed but y is changed by the amount $k = \Delta y$, then the resulting change in z is

$$\Delta z = f(x, y + k) - f(x, y),$$

and the corresponding instantaneous rate of change of z is

$$\frac{dz}{dy} = \lim_{\Delta y \to 0} \frac{\Delta z}{\Delta y}. \tag{2}$$

The limits in Eqs. (1) and (2) are the *two* **partial derivatives** of the function $f(x, y)$ with respect to its two independent variables x and y, respectively.

Definition *Partial Derivatives*

The **partial derivatives (with respect to** x **and with respect to** y) of the function $f(x, y)$ are the two functions defined by

$$f_x(x, y) = \lim_{h \to 0} \frac{f(x + h, y) - f(x, y)}{h}, \tag{3}$$

$$f_y(x, y) = \lim_{k \to 0} \frac{f(x, y + k) - f(x, y)}{k} \tag{4}$$

wherever these limits exist.

Note that Eqs. (3) and (4) are simply restatements of Eqs. (1) and (2). Just as with single-variable derivatives, there are several alternative ways of writing partial derivatives.

Notation for Partial Derivatives

If $z = f(x, y)$, then we may express its partial derivatives with respect to x and y, respectively, in these forms:

$$\frac{\partial z}{\partial x} = \frac{\partial f}{\partial x} = f_x(x, y) = \frac{\partial}{\partial x} f(x, y) = D_x[f(x, y)] = D_1[f(x, y)], \tag{5}$$

$$\frac{\partial z}{\partial y} = \frac{\partial f}{\partial y} = f_y(x, y) = \frac{\partial}{\partial y} f(x, y) = D_y[f(x, y)] = D_2[f(x, y)]. \tag{6}$$

Computer algebra systems generally employ variants of the "operator notation" for partial derivatives, such as

`diff(f(x,y), x)` and `D[f[x,y], x]`

in *Maple* and *Mathematica,* respectively.

Note that if we delete the symbol y throughout Eq. (3), the result is the limit that defines the single-variable derivative $f'(x)$. This means that we can calculate $\partial z/\partial x$ as an "ordinary" derivative with respect to x simply by regarding y as a constant during the process of differentiation. Similarly, we can compute $\partial z/\partial y$ as an ordinary derivative by thinking of y as the *only* variable and treating x as a constant during the computation.

Consequently, we seldom need to evaluate directly the limits in Eqs. (3) and (4) in order to calculate partial derivatives. Ordinarily we simply apply familiar differentiation results to differentiate $f(x, y)$ with respect to either independent variable (x or y) while holding the other variable constant. In short,

- To calculate $\partial f/\partial x$, regard y as a constant and differentiate with respect to x.
- To calculate $\partial f/\partial y$ regard x as a constant and differentiate with respect to y.

EXAMPLE 1 Compute the partial derivatives $\partial f/\partial x$ and $\partial f/\partial y$ of the function $f(x, y) = x^2 + 2xy^2 - y^3$.

Solution To compute the partial of f with respect to x, we regard y as a constant. Then we differentiate normally and find that

$$\frac{\partial f}{\partial x} = 2x + 2y^2.$$

When we regard x as a constant and differentiate with respect to y, we find that

$$\frac{\partial f}{\partial y} = 4xy - 3y^2.$$

■

EXAMPLE 2 Find $\partial z/\partial x$ and $\partial z/\partial y$ if $z = (x^2 + y^2)e^{-xy}$.

Solution Because $\partial z/\partial x$ is calculated as if it were an ordinary derivative with respect to x, with y held constant, we use the product rule. This gives

$$\frac{\partial z}{\partial x} = (2x)(e^{-xy}) + (x^2 + y^2)(-ye^{-xy}) = (2x - x^2y - y^3)e^{-xy}.$$

Because x and y appear symmetrically in the expression for z, we get $\partial z/\partial y$ when we interchange x and y in the expression for $\partial z/\partial x$:

$$\frac{\partial z}{\partial y} = (2y - xy^2 - x^3)e^{-xy}.$$

You should check this result by differentiating with respect to y directly in order to find $\partial z/\partial y$. ■

Instantaneous Rates of Change

To get an intuitive feel for the meaning of partial derivatives, we can think of $f(x, y)$ as the temperature at the point (x, y) of the plane. Then $f_x(x, y)$ is the instantaneous rate of change of temperature at (x, y) per unit increase in x (with y held constant). Similarly, $f_y(x, y)$ is the instantaneous rate of change of temperature per unit increase in y (with x held constant).

EXAMPLE 3 Suppose that the xy-plane is somehow heated and that its temperature at the point (x, y) is given by the function $f(x, y) = x^2 + 2xy^2 - y^3$, whose partial derivatives $f_x(x, y) = 2x + 2y^2$ and $f_y(x, y) = 4xy - 3y^2$ were calculated in Example 1. Suppose also that distance is measured in miles and temperature in degrees Celsius (°C). Then at the point $(1, -1)$, one mile east and one mile south of the origin, the rate of change of temperature (in degrees per mile) in the (eastward) positive x-direction is

$$f_x(1, -1) = 2 \cdot (1) + 2 \cdot (-1)^2 = 4 \quad (\text{deg/mi}),$$

and the rate of change in the (northward) positive y-direction is

$$f_y(1, -1) = 4 \cdot 1 \cdot (-1) - 3 \cdot (-1)^3 = -7 \quad (\text{deg/mi}).$$

Thus, if we start at the point $(1, -1)$ and walk $\frac{1}{10}$ mi east, we expect to experience a temperature increase of about $4 \cdot (0.1) = 0.4$°C. If instead we started at $(1, -1)$ and walked 0.2 mi north, we would expect to experience a temperature change of about $(-7) \cdot (0.2) = -1.4$°C; that is, a temperature decrease of about 1.4°C. ■

EXAMPLE 4 The volume V (in cubic centimeters) of 1 mole (mol) of an ideal gas is given by

$$V = \frac{(82.06)T}{p},$$

where p is the pressure (in atmospheres) and T is the absolute temperature (in kelvins (K), where K = °C + 273). Find the rates of change of the volume of 1 mol of an ideal gas with respect to pressure and with respect to temperature when $T = 300$ K and $p = 5$ atm.

Solution The partial derivatives of V with respect to its two variables are

$$\frac{\partial V}{\partial p} = -\frac{(82.06)T}{p^2} \quad \text{and} \quad \frac{\partial V}{\partial T} = \frac{82.06}{p}.$$

With $T = 300$ and $p = 5$, we have the values $\partial V/\partial p = -984.72$ (cm^3/atm) and $\partial V/\partial T = 16.41$ (cm^3/K). These partial derivatives allow us to estimate the effect of a small change in temperature or in pressure on the volume V of the gas, as follows. We are given $T = 300$ and $p = 5$, so the volume of gas with which we are dealing is

$$V = \frac{(82.06)(300)}{5} = 4923.60 \quad (\text{cm}^3).$$

We would expect an increase in pressure of 1 atm (with T held constant) to decrease the volume of gas by approximately 1 L (1000 cm^3), because $-984.72 \approx -1000$. An increase in temperature of 1 K (or 1°C) would, with p held constant, increase the volume by about 16 cm^3, because $16.41 \approx 16$. ■

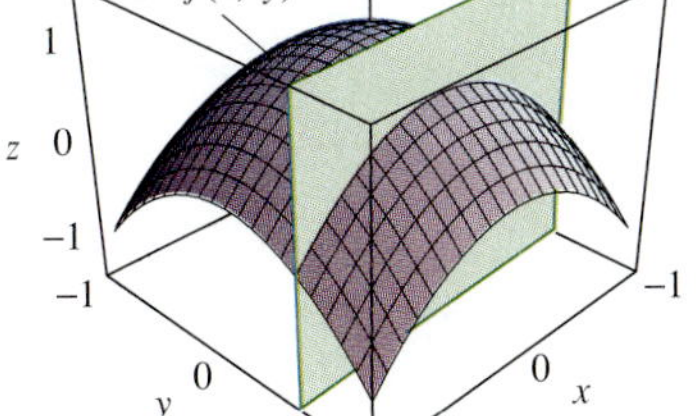

Fig. 13.4.1 A vertical plane parallel to the xz-plane intersects the surface $z = f(x, y)$ in an x-curve.

Geometric Interpretation of Partial Derivatives

The partial derivatives f_x and f_y are the slopes of lines tangent to certain curves on the surface $z = f(x, y)$. Figure 13.4.1 illustrates the intersection of this surface with a vertical plane $y = b$ that is parallel to the xz-coordinate plane. Along the intersection curve, the x-coordinate varies but the y-coordinate is constant: $y = b$ at each point, because the curve lies in the vertical plane $y = b$. A curve of intersection of $z = f(x, y)$ with a vertical plane parallel to the xz-plane is therefore called an **x-curve** on the surface.

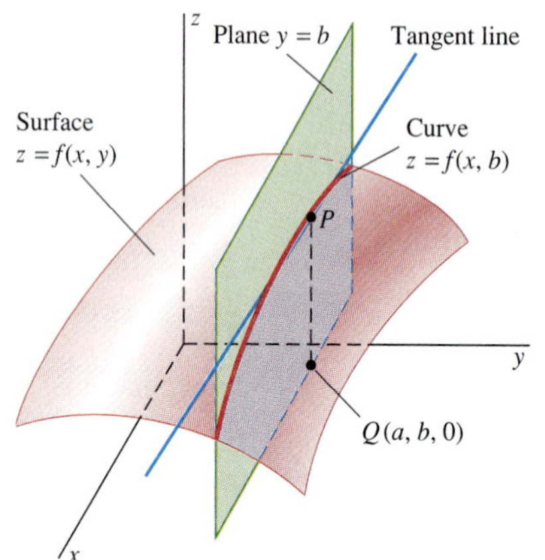

Fig. 13.4.2 An x-curve and its tangent line

Figure 13.4.2 shows a point $P(a, b, c)$ in the surface $z = f(x, y)$, the x-curve through P, and the line tangent to this x-curve at P. Figure 13.4.3 shows the parallel projection of the vertical plane $y = b$ onto the xz-plane itself. We can now "ignore" the presence of $y = b$ and regard $z = f(x, b)$ as a function of the *single* variable x. The slope of the line tangent to the original x-curve through P (see Fig. 13.4.2) is equal to the slope of the tangent line in Fig. 13.4.3. But by familiar single-variable calculus, this latter slope is given by

$$\frac{\partial z}{\partial x} = \lim_{\Delta x \to 0} \frac{\Delta z}{\Delta x} = \lim_{h \to 0} \frac{f(a + h, b) - f(a, b)}{h} = f_x(a, b).$$

Thus we see that the geometric meaning of f_x is this:

The value $\partial z/\partial x = f_x(a, b)$ is the slope of the line tangent at $P(a, b, c)$ to the x-curve through P on the surface $z = f(x, y)$.

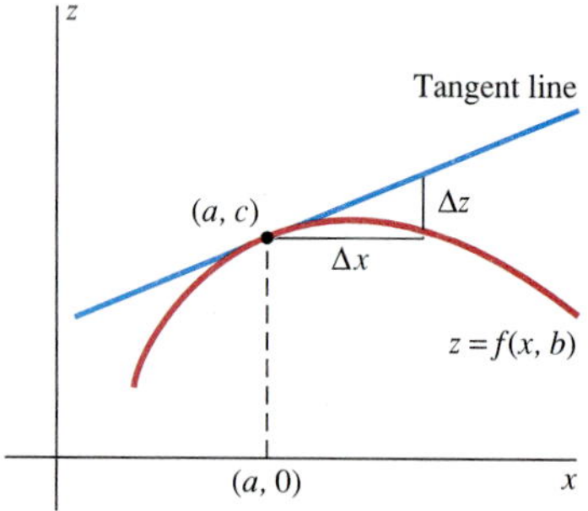

Fig. 13.4.3 Projection into the xz-plane of the x-curve through $P(a, b, c)$ and its tangent line

We proceed in much the same way to investigate the geometric meaning of partial differentiation with respect to y. Figure 13.4.4 illustrates the intersection with the surface $z = f(x, y)$ of a vertical plane $x = a$ that is parallel to the yz-coordinate plane. Now the curve of intersection is a **y-curve** along which y varies but $x = a$ is constant. Figure 13.4.5 shows this y-curve $z = f(a, y)$ and its tangent line at P. The projection of the tangent line in the yz-plane (in Fig. 13.4.6) has slope $\partial z/\partial y = f_y(a, b)$. Thus we see that the geometric meaning of f_y is this:

The value $\partial z/\partial y = f_y(a, b)$ is the slope of the line tangent at $P(a, b, c)$ to the y-curve through P on the surface $z = f(x, y)$.

EXAMPLE 5 Suppose that the graph $z = 5xy\exp(-x^2 - 2y^2)$ in Fig. 13.4.7 represents a terrain featuring two peaks (hills, actually) and two pits. With all distances measured in miles, z is the altitude above the point (x, y) at sea level in the xy-plane. For instance, the height of the pictured point P is $z(-1, -1) = 5e^{-3} \approx 0.2489$ (mi),

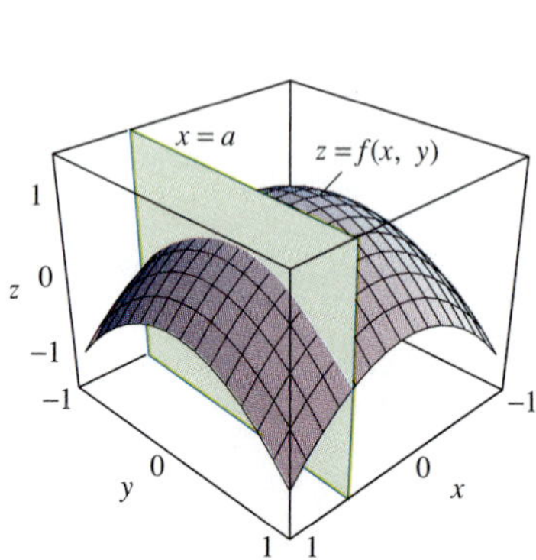

Fig. 13.4.4 A vertical plane parallel to the yz-plane intersects the surface $z = f(x, y)$ in a y-curve.

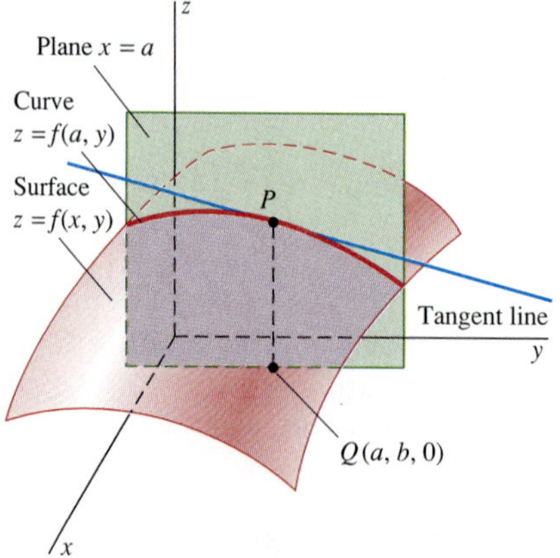

Fig. 13.4.5 A y-curve and its tangent line

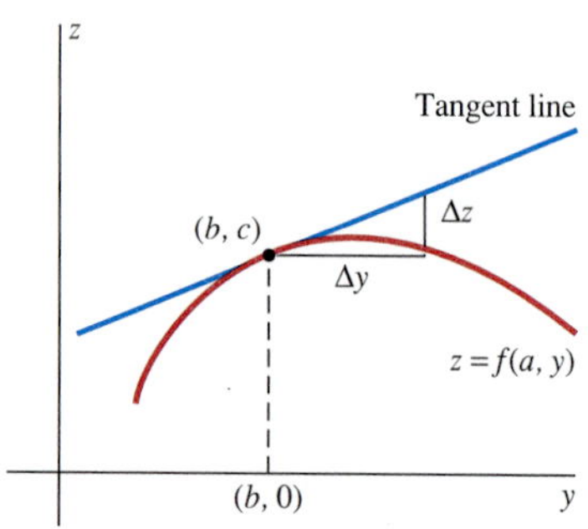

Fig. 13.4.6 Projection into the yz-plane of the y-curve through $P(a, b, c)$ and its tangent line

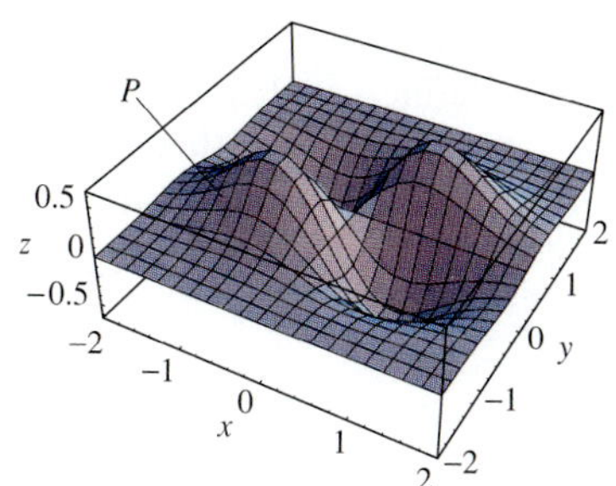

Fig. 13.4.7 The graph $z = 5xy\exp(-x^2 - 2y^2)$

about 1314 ft above sea level. We ask at what rate we climb if, starting at the point $P(-1, -1, 0.2489)$, we head either due east (the positive x-direction) or due north (the positive y-direction). If we calculate the two partial derivatives of $z(x, y)$, we get

$$\frac{\partial z}{\partial x} = 5y(1 - 2x^2)\exp(-x^2 - 2y^2) \quad \text{and} \quad \frac{\partial z}{\partial y} = 5x(1 - 4y^2)\exp(-x^2 - 2y^2).$$

(You should check this!) Substitution of $x = y = -1$ now gives

$$\left.\frac{\partial z}{\partial x}\right|_{(-1,-1)} = 5e^{-3} \approx 0.2489 \quad \text{and} \quad \left.\frac{\partial z}{\partial y}\right|_{(-1,-1)} = 15e^{-3} \approx 0.7468.$$

The units here are in miles per mile—that is, the ratio of rise to run in vertical miles per horizontal mile. So if we head east, we start climbing at an angle of

$$\alpha = \tan^{-1}(0.2489) \approx 0.2439 \quad \text{(rad)},$$

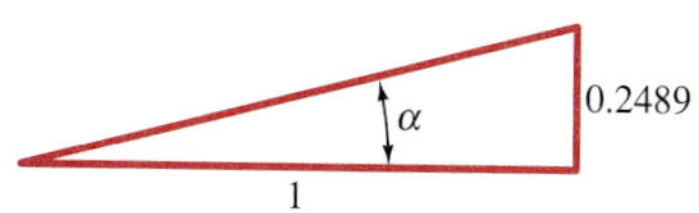

Fig. 13.4.8 The angle of climb in the x-direction

about 13.97° (see Fig. 13.4.8). But if we head north, then we start climbing at an angle of

$$\beta = \tan^{-1}(0.7468) \approx 0.6414 \quad \text{(rad)},$$

approximately 36.75° (see Fig. 13.4.9). Do these result appear consistent with Fig. 13.4.7?.

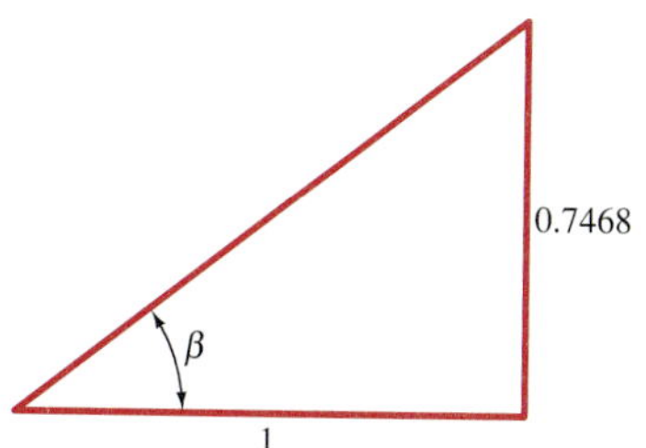

Fig. 13.4.9 The angle of climb in the y-direction

Planes Tangent to Surfaces

The two tangent lines illustrated in Figs. 13.4.2 and 13.4.5 determine a unique plane through the point $P(a, b, f(a, b))$. We will see in Section 13.7 that if the partial derivatives f_x and f_y are continuous functions of x and y, then this plane contains the line tangent at P to *every* smooth curve on the surface $z = f(x, y)$ that passes through P. This plane is therefore (by definition) the plane tangent to the surface at P.

Definition ***Plane Tangent to*** $z = f(x, y)$

Suppose that the function $f(x, y)$ has continuous partial derivatives on a rectangle in the xy-plane containing (a, b) in its interior. Then the **plane tangent** to the surface $z = f(x, y)$ at the point $P(a, b, f(a, b))$ is the plane through P that contains the lines tangent to the two curves

$$z = f(x, b), \qquad y = b \qquad (x\text{-curve}) \tag{7}$$

and

$$z = f(a, y), \qquad x = a \qquad (y\text{-curve}). \tag{8}$$

To find an equation of this tangent plane at the point $P(a, b, c)$ where $c = f(a, b)$, recall from Section 12.4 that a typical nonvertical plane in space that passes through the point P has an equation of the form

$$A(x - a) + B(y - b) + C(z - c) = 0 \tag{9}$$

where $C \neq 0$. If we solve for $z - c$, we get the equation

$$z - c = p(x - a) + q(y - b) \tag{10}$$

where $p = -A/C$ and $q = -B/C$. This plane will be tangent to the surface $z = f(x, y)$ at the point $P(a, b, c)$ provided that the line defined in Eq. (10) with $y = b$ is tangent to the x-curve in Eq. (7), and the line defined in (10) with $x = a$ is tangent to the y-curve in Eq. (8). But the substitution $y = b$ reduces Eq. (10) to

$$z - c = p(x - a), \quad \text{so} \quad \frac{\partial z}{\partial x} = p,$$

and the substitution $x = a$ reduces Eq. (10) to

$$z - c = q(y - b), \quad \text{so} \quad \frac{\partial z}{\partial y} = q.$$

But our discussion of the geometric interpretation of partial derivatives gave

$$\left.\frac{\partial z}{\partial x}\right|_{(a,b)} = f_x(a, b) \quad \text{and} \quad \left.\frac{\partial z}{\partial y}\right|_{(a,b)} = f_y(a, b)$$

for the slopes of the lines through P that are tangent there to the x-curve and y-curve, respectively. Hence we must have $p = f_x(a, b)$ and $q = f_y(a, b)$ in order for the plane in Eq. (10) to be tangent to the surface $z = f(x, y)$ at the point P. Substitution of these values in Eq. (10) yields the following result.

The Plane Tangent to a Surface

The plane tangent to the surface $z = f(x, y)$ at the point $P(a, b, f(a, b))$ has equation

$$z - f(a, b) = f_x(a, b)(x - a) + f_y(a, b)(y - b). \tag{11}$$

If for variety we write (x_0, y_0, z_0) for the coordinates of P, we can rewrite Eq. (11) in the form

$$f_x(x_0, y_0)(x - x_0) + f_y(x_0, y_0)(y - y_0) + (-1)(z - z_0) = 0, \tag{12}$$

from which we see (by consulting Eq. (8) in Section 12.4) that the plane tangent to the surface $z = f(x, y)$ at the point $P(x_0, y_0, z_0)$ has **normal vector**

$$\mathbf{n} = f_x(x_0, y_0)\mathbf{i} + f_y(x_0, y_0)\mathbf{j} - \mathbf{k} = \left\langle \frac{\partial z}{\partial x}, \frac{\partial z}{\partial y}, -1 \right\rangle. \tag{13}$$

Note that $\mathbf{n}$ is a downward-pointing vector (Why?); its negative $-\mathbf{n}$ is the upward-pointing vector shown in Fig. 13.4.10.

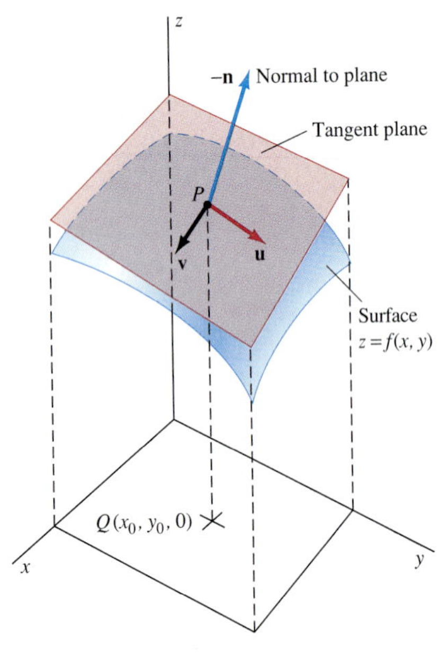

Fig. 13.4.10 The surface $z = f(x, y)$, its tangent plane at $P(x_0, y_0, z_0)$, and the vector $-\mathbf{n}$ normal to both at P

EXAMPLE 6 Write an equation of the plane tangent to the paraboloid $z = 5 - 2x^2 - y^2$ at the point $P(1, 1, 2)$.

Solution If $f(x, y) = 5 - 2x^2 - y^2$, then

$$f_x(x, y) = -4x, \quad f_y(x, y) = -2y;$$
$$f_x(1, 1) = -4, \quad f_y(1, 1) = -2.$$

Hence Eq. (11) gives

$$z - 2 = -4(x - 1) - 2(y - 1)$$

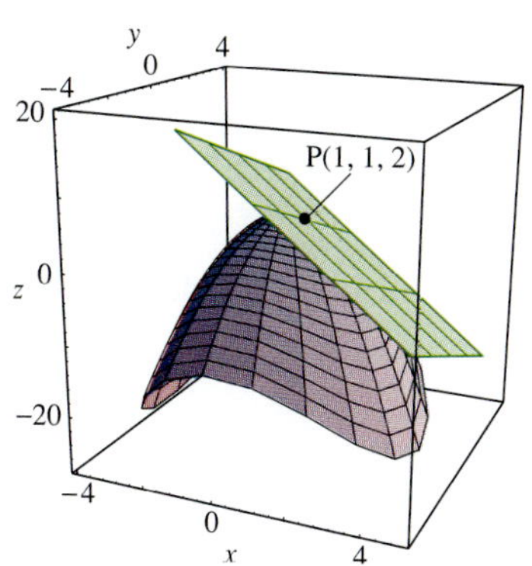

Fig. 13.4.11 The paraboloid and tangent plane of Example 6

(when simplified, $z = 8 - 4x - 2y$) as an equation of the plane tangent to the paraboloid at P. The computer plot in Fig. 13.4.11 corroborates this result.

Higher-Order Partial Derivatives

The **first-order partial derivatives** f_x and f_y are themselves functions of x and y, so they may be differentiated with respect to x or to y. The partial derivatives of $f_x(x, y)$ and $f_y(x, y)$ are called the **second-order partial derivatives** of f. There are four of them because there are four possibilities in the order of differentiation:

$$(f_x)_x = f_{xx} = \frac{\partial f_x}{\partial x} = \frac{\partial}{\partial x}\left(\frac{\partial f}{\partial x}\right) = \frac{\partial^2 f}{\partial x^2},$$

$$(f_x)_y = f_{xy} = \frac{\partial f_x}{\partial y} = \frac{\partial}{\partial y}\left(\frac{\partial f}{\partial x}\right) = \frac{\partial^2 f}{\partial y \partial x},$$

$$(f_y)_x = f_{yx} = \frac{\partial f_y}{\partial x} = \frac{\partial}{\partial x}\left(\frac{\partial f}{\partial y}\right) = \frac{\partial^2 f}{\partial x \partial y},$$

$$(f_y)_y = f_{yy} = \frac{\partial f_y}{\partial y} = \frac{\partial}{\partial y}\left(\frac{\partial f}{\partial y}\right) = \frac{\partial^2 f}{\partial y^2}.$$

If we write $z = f(x, y)$, then we can replace each occurrence of the symbol f here with z.

Note The function f_{xy} is the second-order partial derivative of f with respect to x first and then to y; f_{yx} is the result of differentiating with respect to y first and x second. Although f_{xy} and f_{yx} are not necessarily equal, it is proved in advanced calculus that these two "mixed" second-order partial derivatives are equal if both are continuous. More precisely, if f_{xy} and f_{yx} are continuous on a circular disk centered at the point (a, b), then

$$f_{xy}(a, b) = f_{yx}(a, b). \tag{14}$$

[If both f_{xy} and f_{yx} are defined merely at (a, b), they may well be unequal there.] Because most functions of interest to us have second-order partial derivatives that are continuous everywhere they are defined, we will ordinarily need to deal with only three distinct second-order partial derivatives rather than with four. Similarly, if $f(x, y, z)$ is a function of three variables with continuous second-order partial derivatives, then

$$\frac{\partial^2 f}{\partial x \partial y} = \frac{\partial^2 f}{\partial y \partial x}, \quad \frac{\partial^2 f}{\partial x \partial z} = \frac{\partial^2 f}{\partial z \partial x}, \quad \text{and} \quad \frac{\partial^2 f}{\partial y \partial z} = \frac{\partial^2 f}{\partial z \partial y}.$$

Third-order and higher-order partial derivatives are defined similarly, and the order in which the differentiations are performed is unimportant as long as all derivatives involved are continuous. In such a case, for example, the distinct third-order partial derivatives of the function $z = f(x, y)$ are

$$f_{xxx} = \frac{\partial}{\partial x}\left(\frac{\partial^2 f}{\partial x^2}\right) = \frac{\partial^3 f}{\partial x^3},$$

$$f_{xxy} = \frac{\partial}{\partial y}\left(\frac{\partial^2 f}{\partial x^2}\right) = \frac{\partial^3 f}{\partial y \partial x^2},$$

$$f_{xyy} = \frac{\partial}{\partial y}\left(\frac{\partial^2 f}{\partial y \partial x}\right) = \frac{\partial^3 f}{\partial y^2 \partial x}, \quad \text{and}$$

$$f_{yyy} = \frac{\partial}{\partial y}\left(\frac{\partial^2 f}{\partial y^2}\right) = \frac{\partial^3 f}{\partial y^3}.$$

EXAMPLE 7 Show that the partial derivatives of third and higher orders of the function $f(x, y) = x^2 + 2xy^2 - y^3$ are constant.

Solution We find that

$$f_x(x, y) = 2x + 2y^2 \quad \text{and} \quad f_y(x, y) = 4xy - 3y^2.$$

So

$$f_{xx}(x, y) = 2, \quad f_{xy}(x, y) = 4y, \quad \text{and} \quad f_{yy}(x, y) = 4x - 6y.$$

Finally,

$$f_{xxx}(x, y) = 0, \quad f_{xxy}(x, y) = 0, \quad f_{xyy}(x, y) = 4, \quad \text{and} \quad f_{yyy}(x, y) = -6.$$

The function f is a polynomial, so all its partial derivative are polynomials and are therefore continuous everywhere. Hence we need not compute any other third-order partial derivatives; each is equal to one of these four. Moreover, because all the third-order partial derivatives are constant, all higher-order partial derivatives of f are zero. ■

13.4 PROBLEMS

In Problems 1 through 20, compute the first-order partial derivatives of each function.

1. $f(x, y) = x^4 - x^3y + x^2y^2 - xy^3 + y^4$

2. $f(x, y) = x \sin y$

3. $f(x, y) = e^x(\cos y - \sin y)$

4. $f(x, y) = x^2e^{xy}$

5. $f(x, y) = \dfrac{x + y}{x - y}$

6. $f(x, y) = \dfrac{xy}{x^2 + y^2}$

7. $f(x, y) = \ln(x^2 + y^2)$

8. $f(x, y) = (x - y)^{14}$

9. $f(x, y) = x^y$

10. $f(x, y) = \tan^{-1} xy$

11. $f(x, y, z) = x^2y^3z^4$

12. $f(x, y, z) = x^2 + y^3 + z^4$

13. $f(x, y, z) = e^{xyz}$

14. $f(x, y, z) = x^4 - 16yz$

15. $f(x, y, z) = x^2e^y \ln z$

16. $f(u, v) = (2u^2 + 3v^2)\exp(-u^2 - v^2)$

17. $f(r, s) = \dfrac{r^2 - s^2}{r^2 + s^2}$

18. $f(u, v) = e^{uv}(\cos uv + \sin uv)$

19. $f(u, v, w) = ue^v + ve^w + we^u$

20. $f(r, s, t) = (1 - r^2 - s^2 - t^2)e^{-rst}$

In Problems 21 through 30, verify that $z_{xy} = z_{yx}$.

21. $z = x^2 - 4xy + 3y^2$

22. $z = 2x^3 + 5x^2y - 6y^2 + xy^4$

23. $z = x^2\exp(-y^2)$

24. $z = xye^{-xy}$

25. $z = \ln(x + y)$

26. $z = (x^3 + y^3)^{10}$

27. $z = e^{-3x} \cos y$

28. $z = (x + y) \sec xy$

29. $z = x^2 \cosh(1/y^2)$

30. $z = \sin xy + \tan^{-1} xy$

In Problems 31 through 40, find an equation of the plane tangent to the given surface $z = f(x, y)$ at the indicated point P.

31. $z = x^2 + y^2$; $P = (3, 4, 25)$

32. $z = \sqrt{50 - x^2 - y^2}$; $P = (4, -3, 5)$

33. $z = \sin \dfrac{\pi xy}{2}$; $P = (3, 5, -1)$

34. $z = \dfrac{4}{\pi} \tan^{-1} xy$; $P = (1, 1, 1)$

35. $z = x^3 - y^3$; $P = (3, 2, 19)$

36. $z = 3x + 4y$; $P = (1, 1, 7)$

37. $z = xy$; $P = (1, -1, -1)$

38. $z = \exp(-x^2 - y^2)$; $P = (0, 0, 1)$

39. $z = x^2 - 4y^2$; $P = (5, 2, 9)$

40. $z = \sqrt{x^2 + y^2}$; $P = (3, -4, 5)$

Recall that $f_{xy} = f_{yx}$ for a function $f(x, y)$ with continuous second-order partial derivatives. In Problems 41 through 44, apply this criterion to determine whether there exists a function $f(x, y)$ having the given first-order partial derivatives. If so, try to determine a formula for such a function $f(x, y)$.

41. $f_x(x, y) = 2xy^3$, $f_y(x, y) = 3x^2y^2$

42. $f_x(x, y) = 5xy + y^2,\ f_y(x, y) = 3x^2 + 2xy$

43. $f_x(x, y) = \cos^2(xy),\ f_y(x, y) = \sin^2(xy)$

44. $f_x(x, y) = \cos x \sin y,\ f_y(x, y) = \sin x \cos y$

Figures 13.4.12 through 13.4.17 show the graphs of a certain function $f(x, y)$ and its first- and second-order partial derivatives. In Problems 45 through 50, match that function or partial derivative with its graph.

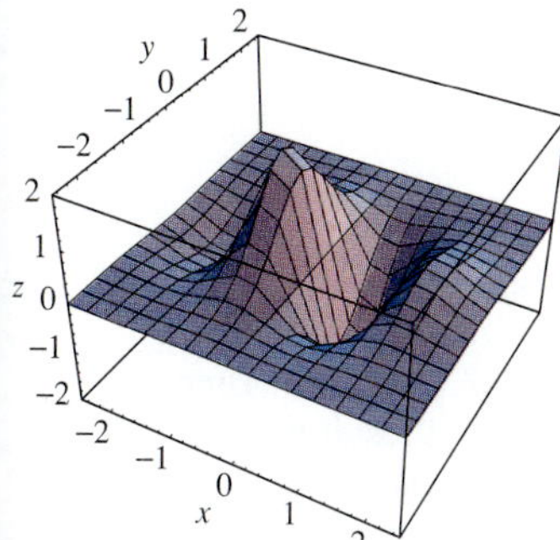

Fig. 13.4.12

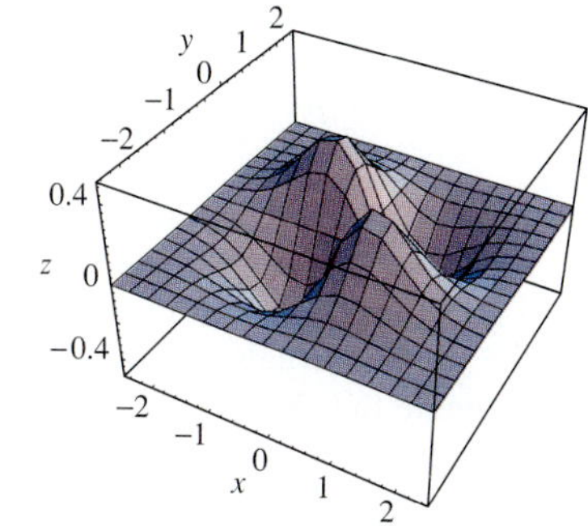

Fig. 13.4.13

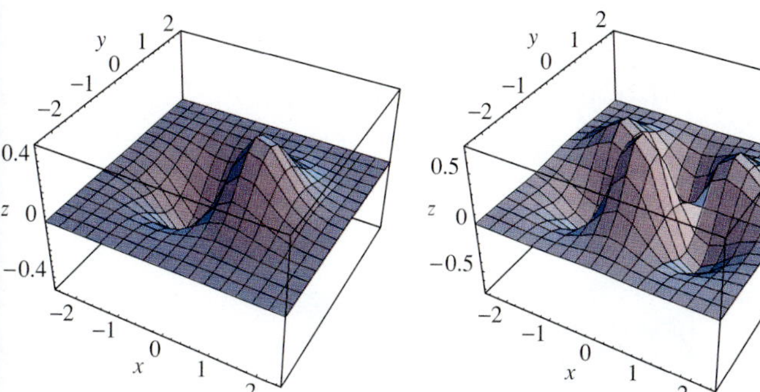

Fig. 13.4.14

Fig. 13.4.15

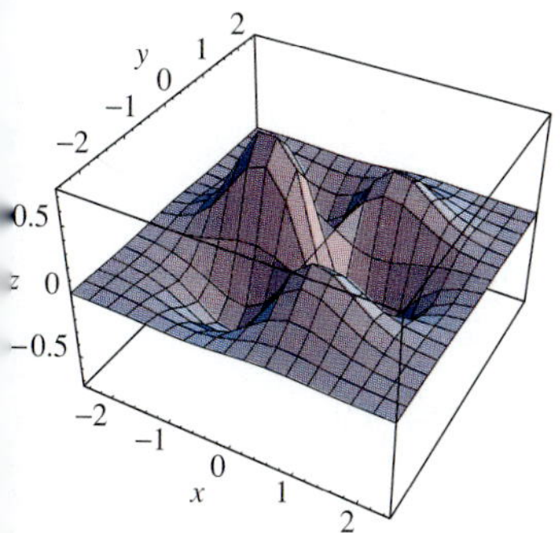

Fig. 13.4.16

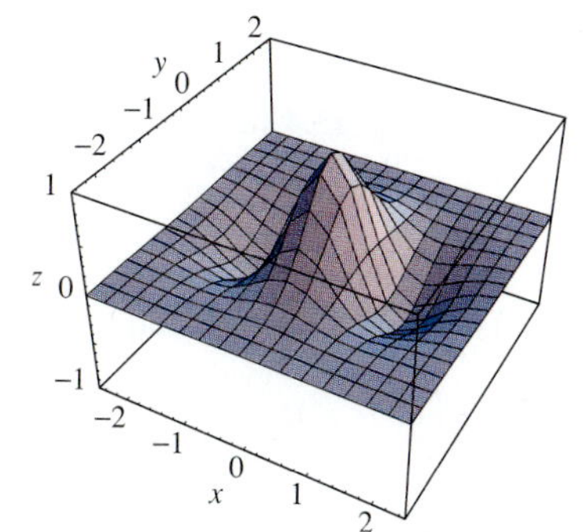

Fig. 13.4.17

45. $f(x, y)$

46. $f_x(x, y)$

47. $f_y(x, y)$

48. $f_{xx}(x, y)$

49. $f_{xy}(x, y)$

50. $f_{yy}(x, y)$

51. Verify that the mixed second-order partial derivatives f_{xy} and f_{yx} are equal if $f(x, y) = x^m y^n$, where m and n are positive integers.

52. Suppose that $z = e^{x+y}$. Show that e^{x+y} is the result of differentiating z first m times with respect to x, then n times with respect to y.

53. Let $f(x, y, z) = e^{xyz}$. Calculate the distinct second-order partial derivatives of f and the third-order partial derivative f_{xyz}.

54. Suppose that $g(x, y) = \sin xy$. Verify that $g_{xy} = g_{yx}$ and that $g_{xxy} = g_{xyx} = g_{yxx}$.

55. It is shown in physics that the temperature $u(x, t)$ at time t at the point x of a long, insulated rod that lies along the x-axis satisfies the *one-dimensional heat equation*

$$\frac{\partial u}{\partial t} = k\frac{\partial^2 u}{\partial x^2} \qquad (k \text{ is a constant}).$$

Show that the function

$$u = u(x, t) = \exp(-n^2 kt) \sin nx$$

satisfies the one-dimensional heat equation for any choice of the constant n.

56. The *two-dimensional heat equation* for an insulated plane is

$$\frac{\partial u}{\partial t} = k\left(\frac{\partial^2 u}{\partial x^2} + \frac{\partial^2 u}{\partial y^2}\right).$$

Show that the function

$$u = u(x, y, t) = \exp(-[m^2 + n^2]kt) \sin mx \cos ny$$

satisfies this equation for any choice of the constants m and n.

57. A string is stretched along the x-axis, fixed at each end, and then set into vibration. It is shown in physics that the displacement $y = y(x, t)$ of the point of the string at location x at time t satisfies the *one-dimensional wave equation*

$$\frac{\partial^2 y}{\partial t^2} = a^2 \frac{\partial^2 y}{\partial x^2},$$

where the constant a depends on the density and tension of the string. Show that the following functions satisfy the one-dimensional wave equation: (a) $y = \sin(x + at)$; (b) $y = \cosh(3[x - at])$; (c) $y = \sin kx \cos kat$ (k is a constant).

58. A steady-state temperature function $u = u(x, y)$ for a thin, flat plate satisfies *Laplace's equation*

$$\frac{\partial^2 u}{\partial x^2} + \frac{\partial^2 u}{\partial y^2} = 0.$$

Determine which of the following functions satisfy Laplace's equation:

(a) $u = \ln\left(\sqrt{x^2 + y^2}\right)$;

(b) $u = \sqrt{x^2 + y^2}$;

(c) $u = \arctan(y/x)$:

(d) $u = e^{-x} \sin y$.

59. Suppose that f and g are twice-differentiable functions of a single variable. Show that $y(x, t) = f(x + at) + g(x - at)$ satisfies the one-dimensional wave equation of Problem 57.

60. The electric potential field of a point charge q is defined (in appropriate units) by $\phi(x, y, z) = q/r$ where $r = \sqrt{x^2 + y^2 + z^2}$. Show that ϕ satisfies the *three-dimensional Laplace equation*

$$\frac{\partial^2 \phi}{\partial x^2} + \frac{\partial^2 \phi}{\partial y^2} + \frac{\partial^2 \phi}{\partial z^2} = 0.$$

61. Let $u(x, t)$ denote the underground temperature at depth x and time t at a location where the seasonal variation of surface $(x = 0)$ temperature is described by

$$u(0, t) = T_0 + a_0 \cos \omega t,$$

where T_0 is the annual average surface temperature and the constant ω is so chosen that the period of $u(0, t)$ is one year. Show that the function

$$u(x, t) = T_0 + a_0 \exp\left(-x\sqrt{\omega/2k}\right) \cos\left(\omega t - x\sqrt{\omega/2k}\right)$$

satisfies both the "surface condition" and the one-dimensional heat equation of Problem 55.

62. The aggregate electrical resistance R of three resistances R_1, R_2, and R_3 connected in parallel satisfies the equation

$$\frac{1}{R} = \frac{1}{R_1} + \frac{1}{R_2} + \frac{1}{R_3}.$$

Show that

$$\frac{\partial R}{\partial R_1} + \frac{\partial R}{\partial R_2} + \frac{\partial R}{\partial R_3} = \left(\frac{1}{R_1^2} + \frac{1}{R_2^2} + \frac{1}{R_3^2}\right) \div \left(\frac{1}{R_1} + \frac{1}{R_2} + \frac{1}{R_3}\right)^2.$$

63. The **ideal gas law** $pV = nRT$ (n is the number of moles of the gas, R is a constant) determines each of the three variables p (pressure), V (volume), and T (temperature) as functions of the other two. Show that

$$\frac{\partial p}{\partial V} \cdot \frac{\partial V}{\partial T} \cdot \frac{\partial T}{\partial p} = -1.$$

64. It is geometrically evident that every plane tangent to the cone $z^2 = x^2 + y^2$ passes through the origin. Show this by methods of calculus.

65. There is only one point at which the plane tangent to the surface

$$z = x^2 + 2xy + 2y^2 - 6x + 8y$$

is horizontal. Find it.

66. Show that the plane tangent to the paraboloid with equation $z = x^2 + y^2$ at the point (a, b, c) intersects the xy-plane in the line with equation $2ax + 2by = a^2 + b^2$. Then show that this line is tangent to the circle with equation $4x^2 + 4y^2 = a^2 + b^2$.

67. According to van der Waals' equation, 1 mol of a gas satisfies the equation

$$\left(p + \frac{a}{V^2}\right)(V - b) = (82.06)T$$

where p, V, and T are as in Example 4. For carbon dioxide, $a = 3.59 \times 10^6$ and $b = 42.7$, and V is 25600 cm^3 when p is 1 atm and $T = 313$ K. (a) Compute $\partial V/\partial p$ by differentiating van der Waals' equation with T held constant. Then estimate the change in volume that would result from an increase of 0.1 atm of pressure with T held at 313 K. (b) Compute $\partial V/\partial T$ by differentiating van der Waals' equation with p held constant. Then estimate the change in volume that would result from an increase of 1 K in temperature with p held at 1 atm.

68. A *minimal surface* has the least surface area of all surfaces with the same boundary. Figure 13.4.18 shows *Scherk's minimal surface.* It has the equation

$$z = \ln(\cos x) - \ln(\cos y).$$

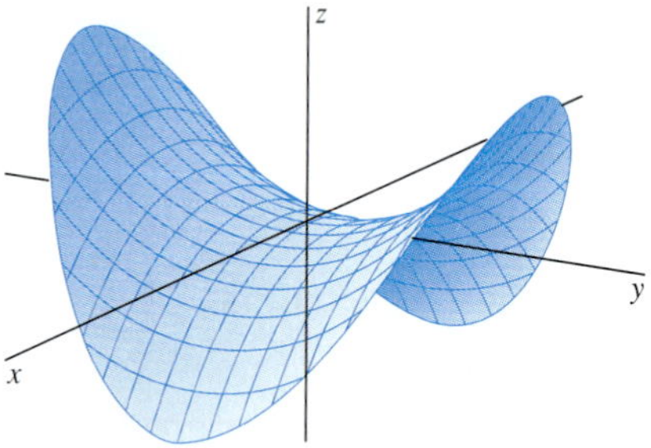

Fig. 13.4.18 Scherk's minimal surface (Problem 68)

A minimal surface $z = f(x, y)$ is known to satisfy the partial differential equation

$$(1 + z_y^2)z_{xx} - z z_x z_y z_{xy} + (1 + z_x^2)z_{yy} = 0.$$

Verify this in the case of Scherk's minimal surface.

69. We say that the function $z = f(x, y)$ is **harmonic** if it satisfies Laplace's equation $z_{xx} + z_{yy} = 0$ (see Problem 58). Show that each of these four functions is harmonic:

(a) $f_1(x, y) = \sin x \sinh(\pi - y)$;

(b) $f_2(x, y) = \sinh 2x \sin 2y$;

(c) $f_3(x, y) = \sin 3x \sinh 3y$;

(d) $f_4(x, y) = \sinh 4(\pi - x) \sin 4y$.

70. Figure 13.4.19 shows the graph of the sum

$$z(x, y) = \sum_{i=1}^{4} f_i(x, y)$$

of the four functions defined in Problem 69. Explain why $z(x, y)$ is a harmonic function.

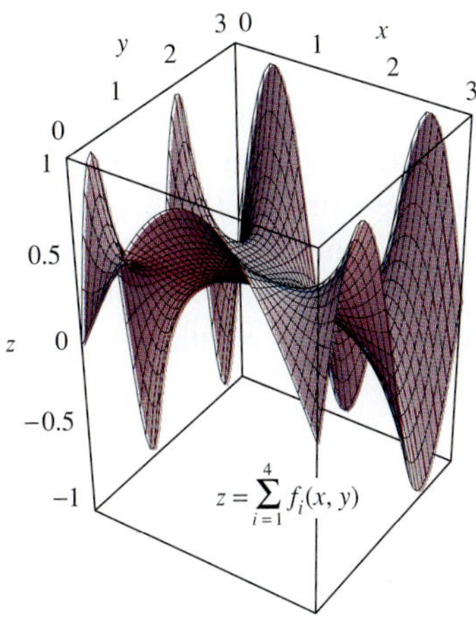

Fig. 13.4.19 The surface $z = f(x, y)$ of Problem 70

71. You are standing at the point where $x = y = 100$ (ft) on a hillside whose height (in feet above sea level) is given by

$$z = 100 + \frac{1}{100}(x^2 - 3xy + 2y^2),$$

with the positive x-axis to the east and the positive y-axis to the north. (a) If you head due east, will you initially be ascending or descending? At what angle (in degrees) from the horizontal? (b) If you head due north, will you initially be ascending or descending? At what angle (in degrees) from the horizontal?

72. Answer questions (a) and (b) in Problem 71, except that now you are standing at the point where $x = 150$ and $y = 250$ (ft) on a hillside whose height (in feet above sea level) is given by

$$z = 1000 + \frac{1}{1000}(3x^2 - 5xy + y^2).$$

73. Figure 13.3.4 shows the graph of the function f defined by

$$f(x, y) = \begin{cases} \dfrac{xy}{\sqrt{x^2 + y^2}} & \text{except at } (0, 0), \\ 0 & \text{if } x = y = 0. \end{cases}$$

In Example 6 of Section 13.3 we saw that f is continuous at $(0, 0)$. Show that the two partial derivatives f_x and f_y exist everywhere and are defined by

$$f_x(x, y) = \frac{y^3}{(x^2 + y^2)^{3/2}} \quad \text{and} \quad f_y(x, y) = \frac{x^3}{(x^2 + y^2)^{3/2}}$$

except at $(0, 0)$, where $f_x(0, 0) = f_y(0, 0) = 0$. Finally, substitute $y = mx$ in these two derivative formulas and conclude that the partial derivatives f_x and f_y are *not* continuous at $(0, 0)$.

13.5 MAXIMA AND MINIMA OF FUNCTIONS OF SEVERAL VARIABLES

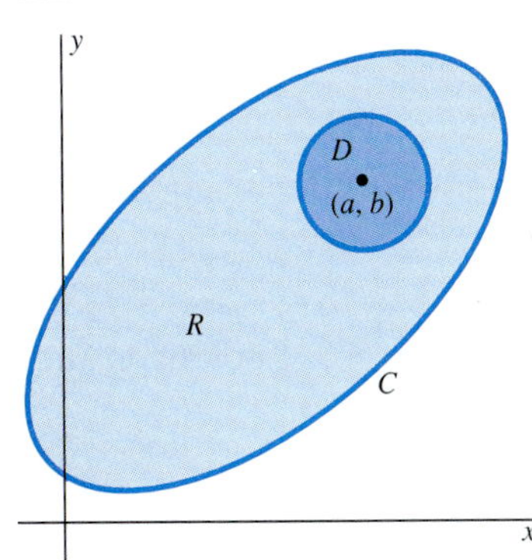

Fig. 13.5.1 A plane region R bounded by the simple closed curve C and a disk D in R centered at the interior point (a, b) of R

The single-variable maximum-minimum techniques of Section 3.5 generalize readily to functions of several variables. We consider first a function f of two variables. Suppose that we are interested in the extreme values attained by $f(x, y)$ on a plane region R that consists of the points on and within a simple closed curve C (Fig. 13.5.1). We say that the function f attains its **absolute, or global, maximum value** M on R at the point (a, b) of R provided that

$$f(x, y) \leqq M = f(a, b)$$

for all points (x, y) of R. Similarly, f attains its **absolute, or global, minimum value** m at the point (c, d) of R provided that $f(x, y) \geqq m = f(c, d)$ for all points (x, y) of R. In plain words, the absolute maximum M and the absolute minimum m are the largest and smallest values (respectively) attained by $f(x, y)$ at points of the domain R of f.

Theorem 1, proved in advanced calculus courses, guarantees the existence of absolute maximum minimum values in many situations of practical interest.

Theorem 1 ***Existence of Extreme Values***

Suppose that the function f is continuous on the region R that consists of the points on and within a simple closed curve C in the plane. Then f attains an absolute maximum value at some point (a, b) of R and attains an absolute minimum value at some point (c, d) of R.

We are interested mainly in the case in which the function f attains its absolute maximum (or minimum) value at an interior point of R. The point (a, b) of R is called an **interior point** of R provided that some circular disk centered at (a, b) lies wholly within R (Fig. 13.5.1). The interior points of a region R of the sort described in Theorem 1 are precisely those that do *not* lie on the boundary curve C.

An absolute extreme value attained by the function at an *interior* point of R is necessarily a local extreme value. We say that $f(a, b)$ is a **local maximum value** of $f(x, y)$ provided that it is the absolute maximum value of f on some disk D that is centered at (a, b) and lies wholly within the domain R. Similarly, a **local minimum value** is an absolute minimum value on some such disk. Thus a local maximum (or minimum) value $f(a, b)$ is not necessarily an absolute maximum (or minimum) value, but is the largest (or smallest) value attained by $f(x, y)$ at points near (a, b).

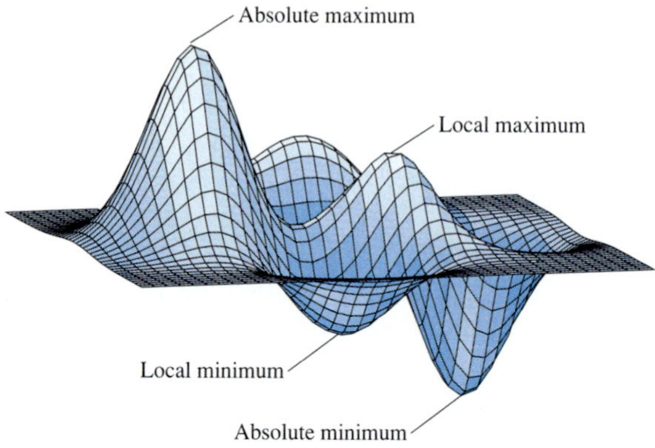

Fig. 13.5.2 Local extrema contrasted with global extrema

EXAMPLE 1 Figure 13.5.2 shows a computer-generated graph of the function

$$f(x, y) = 3(x - 1)^2 e^{-x^2-(y+1)^2} - 10\left(\tfrac{1}{5}x - x^3 - y^5\right)e^{-x^2-y^2} - \tfrac{1}{3}e^{-(x+1)^2-y^2}$$

plotted on the rectangle R for which $-3 \leqq x \leqq 3$ and $-3 \leqq y \leqq 3$. Looking at the labeled extreme values of $f(x, y)$, we see

- A local maximum that is not an absolute maximum,
- A local maximum that is also an absolute maximum,
- A local minimum that is not an absolute minimum, and
- A local minimum that is also an absolute minimum.

We can think of the local maxima on the graph as mountaintops or "peaks" and the local minima as valley bottoms or "pits." ■

Finding Local Extrema

We need a criterion that will provide a practical way to find local extrema of functions of two (or more) variables. The desired result—stated in Theorem 2—is analogous to the single-variable criterion of Section 3.5: If $f(c)$ is a local extreme value of the differentiable single-variable function f, then $x = c$ must be a *critical point* where $f'(c) = 0$.

Suppose, for instance, that $f(a, b)$ is a local maximum value of $f(x, y)$ attained at a point (a, b) where both partial derivatives f_x and f_y exist. We consider vertical plane cross-sectional curves on the graph $z = f(x, y)$, just as when we explored the geometrical interpretation of partial derivatives in Section 13.4. The cross-sectional curves parallel to the xz- and yz-planes are the graphs (in these planes) of the single-variable functions

$$G(x) = f(x, b) \quad \text{and} \quad H(y) = f(a, y)$$

whose derivatives are the partial derivatives of f:

$$f_x(a, b) = G'(a) \quad \text{and} \quad f_y(a, b) = H'(b). \tag{1}$$

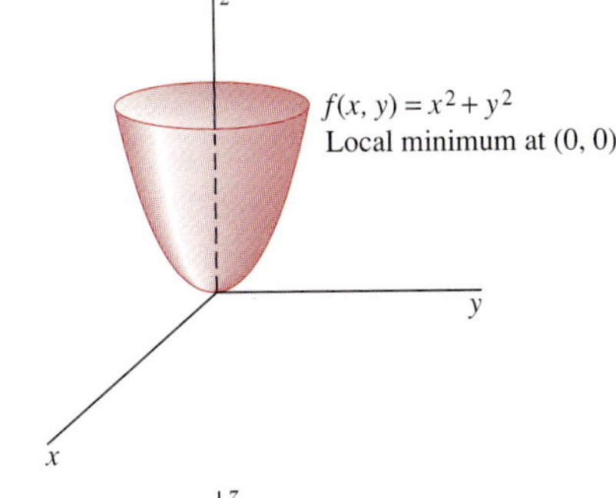

Because $f(a, b)$ is a local maximum value of $f(x, y)$, it follows readily that $G(a)$ and $H(b)$ are local maximum values of $G(x)$ and $H(y)$, respectively. Therefore the single-variable maximum-minimum criterion of Section 3.5 implies that

$$G'(a) = 0 \quad \text{and} \quad H'(b) = 0. \tag{2}$$

Combining (1) and (2), we conclude that

$$f_x(a, b) = 0 \quad \text{and} \quad f_y(a, b) = 0. \tag{3}$$

Essentially the same argument yields the same conclusion if $f(a, b)$ is a local minimum value of $f(x, y)$. This discussion establishes Theorem 2.

z
$g(x, y) = -x^2 - y^2$
Local maximum at (0, 0)
y
x

Theorem 2 *Necessary Conditions for Local Extrema*

Suppose that $f(x, y)$ attains a local maximum value or a local minimum value at the point (a, b) and that both the partial derivatives $f_x(a, b)$ and $f_y(a, b)$ exist. Then

$$f_x(a, b) = 0 = f_y(a, b). \tag{3}$$

The equations in (3) imply that the plane tangent to the surface $z = f(x, y)$ must be horizontal at any local maximum or local minimum point $(a, b, f(a, b))$, in perfect analogy to the single-variable case (in which the tangent line is horizontal at any local maximum or minimum point on the graph of a differentiable function).

EXAMPLE 2 Consider the three familiar surfaces

$$z = f(x, y) = x^2 + y^2,$$

$$z = g(x, y) = -x^2 - y^2, \qquad \text{and}$$

$$z = h(x, y) = y^2 - x^2$$

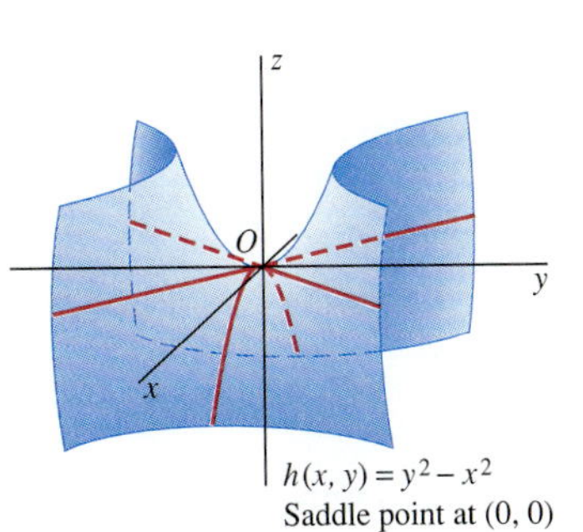

Fig. 13.5.3 When both partial derivatives are zero, there may be (a) a minimum, (b) a maximum, or (c) neither (Example 2)

shown in Fig. 13.5.3. In each case $\partial z/\partial x = \pm 2x$ and $\partial z/\partial y = \pm 2y$. Thus both partial derivatives are zero at the origin $(0, 0)$ (and only there). It is clear from the figure that

$f(x, y) = x^2 + y^2$ has a local minimum at $(0, 0)$. In fact, because a square cannot be negative, $z = x^2 + y^2$ has the global minimum value 0 at $(0, 0)$. Similarly, $g(x, y)$ has a local (indeed, global) maximum value at $(0, 0)$, whereas $h(x, y)$ has neither a local minimum nor a local maximum there—the origin is a *saddle point* of h. This example shows that a point (a, b) where

$$\frac{\partial z}{\partial x} = 0 = \frac{\partial z}{\partial y}$$

may correspond to either a local minimum, a local maximum, or neither. Thus the necessary condition in Eq. (3) is *not* a sufficient condition for a local extremum. ■

EXAMPLE 3 Find all points on the surface

$$z = \tfrac{3}{4}y^2 + \tfrac{1}{24}y^3 - \tfrac{1}{32}y^4 - x^2$$

at which the tangent plane is horizontal.

Solution We first calculate the partial derivatives $\partial z/\partial x$ and $\partial z/\partial y$:

$$\frac{\partial z}{\partial x} = -2x,$$

$$\frac{\partial z}{\partial y} = \tfrac{3}{2}y + \tfrac{1}{8}y^2 - \tfrac{1}{8}y^3 = -\tfrac{1}{8}y(y^2 - y - 12) = -\tfrac{1}{8}y(y + 3)(y - 4).$$

We next equate both $\partial z/\partial x$ and $\partial z/\partial y$ to zero. This yields

$$-2x = 0 \quad \text{and} \quad -\tfrac{1}{8}y(y + 3)(y - 4) = 0.$$

Simultaneous solution of these equations yields exactly three points where both partial derivatives are zero: $(0, -3)$, $(0, 0)$, and $(0, 4)$. The three corresponding points on the surface where the tangent plane is horizontal are $\left(0, -3, \frac{99}{32}\right)$, $(0, 0, 0)$, and $\left(0, 4, \frac{20}{3}\right)$. These three points are indicated on the graph in Fig. 13.5.4 of the surface. (Recall that we constructed this surface in Example 11 of Section 13.2.) ■

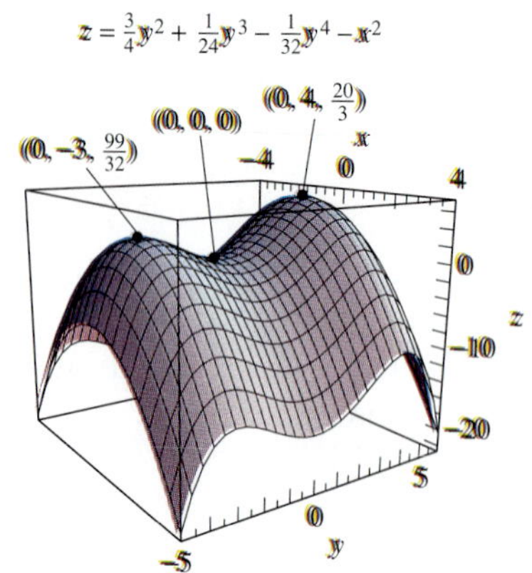

Fig. 13.5.4 The surface of Example 3

Finding Global Extrema

Theorem 2 is a very useful tool for finding the absolute maximum and absolute minimum values attained by a continuous function f on a region R of the type described in Theorem 1. If $f(a, b)$ is the absolute maximum value, for example, then (a, b) is either an interior point of R or a point of the boundary curve C. If (a, b) is an interior point and both the partial derivatives $f_x(a, b)$ and $f_y(a, b)$ exist, then Theorem 2 implies that both these partial derivatives must be zero. Thus we have the following result.

Theorem 3 ***Types of Absolute Extrema***

Suppose that f is continuous on the plane region R consisting of the points on and within a simple closed curve C. If $f(a, b)$ is either the absolute maximum or the absolute minimum value of $f(x, y)$ on R, then (a, b) is either

1. An interior point of R at which

$$\frac{\partial f}{\partial x} = \frac{\partial f}{\partial y} = 0,$$

2. An interior point R where not both partial derivatives exist, or
3. A point of the boundary curve C of R.

A point (a, b) where either condition (1) or condition (2) holds is called a **critical point** of the function f. Thus Theorem 3 says that *any extreme value of the continuous function f on the plane region R must occur at an interior critical point or at a boundary point.* Note the analogy with Theorem 3 of Section 3.5, which implies that an extreme value of a single-variable function $f(x)$ on a closed and bounded interval I must occur either at an interior critical point of I or at an endpoint (boundary point) of I.

As a consequence of Theorem 3, we can find the absolute maximum and minimum values of $f(x, y)$ on R as follows:

1. First, locate the interior critical points.
2. Next, find the possible extreme values of f on the boundary curve C.
3. Finally, compare the values of f at the points found in steps 1 and 2.

The technique to be used in the second step will depend on the nature of the boundary curve C, as illustrated in Examples 4 and 5.

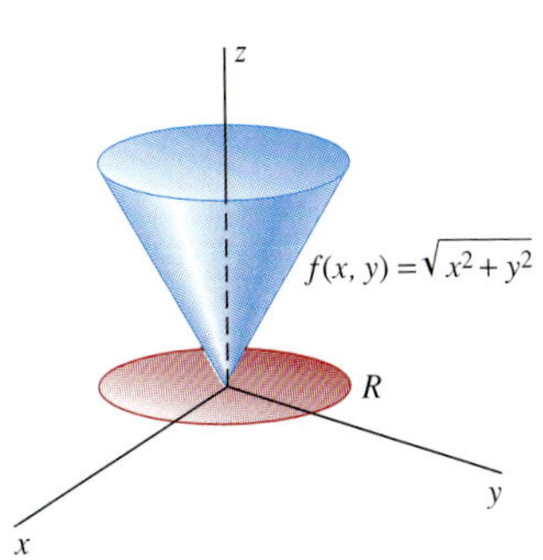

Fig. 13.5.5 The graph of the function of Example 4

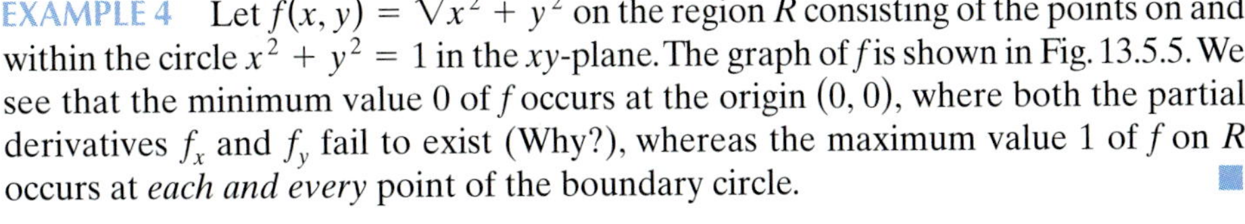

EXAMPLE 4 Let $f(x, y) = \sqrt{x^2 + y^2}$ on the region R consisting of the points on and within the circle $x^2 + y^2 = 1$ in the xy-plane. The graph of f is shown in Fig. 13.5.5. We see that the minimum value 0 of f occurs at the origin $(0, 0)$, where both the partial derivatives f_x and f_y fail to exist (Why?), whereas the maximum value 1 of f on R occurs at *each and every* point of the boundary circle. ■

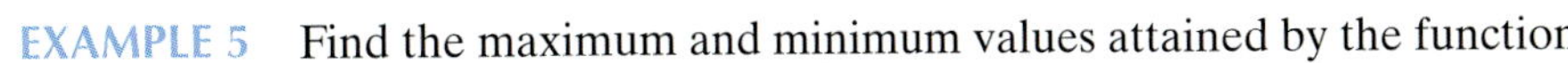

EXAMPLE 5 Find the maximum and minimum values attained by the function

$$f(x, y) = xy - x - y + 3$$

at points of the triangular region R in the xy-plane with vertices at $(0, 0)$, $(2, 0)$, and $(0, 4)$.

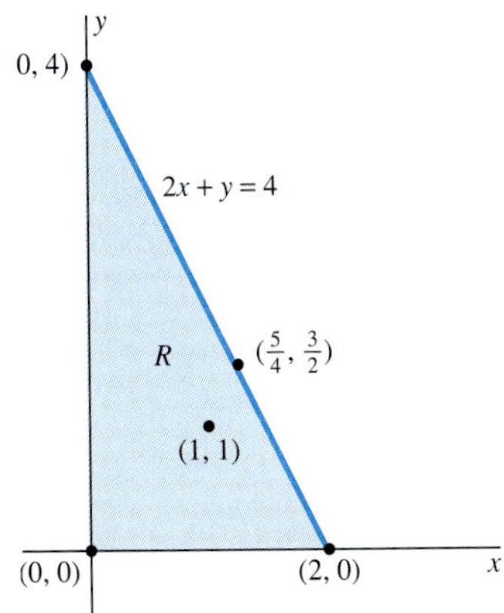

Fig. 13.5.6 The triangular region of Example 5

Solution The region R is shown in Fig. 13.5.6. Its boundary "curve" C consists of the segment $0 \leqq x \leqq 2$ on the x-axis, the segment $0 \leqq y \leqq 4$ on the y-axis, and the part of the line $2x + y = 4$ that lies in the first quadrant. Any interior extremum must occur at a point where both

$$\frac{\partial f}{\partial x} = y - 1 \quad \text{and} \quad \frac{\partial f}{\partial y} = x - 1$$

are zero. Hence the only interior critical point is $(1, 1)$.

Along the edge where $y = 0$: The function $f(x, y)$ takes the form

$$\alpha(x) = f(x, 0) = 3 - x, \qquad 0 \leqq x \leqq 2.$$

Because $\alpha(x)$ is a decreasing function, its extrema for $0 \leqq x \leqq 2$ occur at the endpoints $x = 0$ and $x = 2$. This gives the two possibilities $(0, 0)$ and $(2, 0)$ for locations of extrema of $f(x, y)$.

Along the edge where $x = 0$: The function $f(x, y)$ takes the form

$$\beta(y) = f(0, y) = 3 - y, \qquad 0 \leqq y \leqq 4.$$

The endpoints of this interval yield the points $(0, 0)$ and $(0, 4)$ as possibilities for locations of extrema of $f(x, y)$.

On the edge of R where $y = 4 - 2x$: We may substitute $4 - 2x$ for y into the formula for $f(x, y)$, and thus express f as a function of a single variable:

$$\begin{aligned}\gamma(x) &= x(4 - 2x) - x - (4 - 2x) + 3 \\ &= -2x^2 + 5x - 1, \qquad 0 \leqq x \leqq 2.\end{aligned}$$

To find the extreme values of $\gamma(x)$, we first calculate

$$\gamma'(x) = -4x + 5;$$

$\gamma'(x) = 0$ where $x = \frac{5}{4}$. Thus each extreme value of $\gamma(x)$ on $[0, 2]$ must occur either at the interior point $x = \frac{5}{4}$ of the interval $[0, 2]$ or at one of the endpoints $x = 0$ and $x = 2$. This gives the possibilities $(0, 4)$, $\left(\frac{5}{4}, \frac{3}{2}\right)$, and $(2, 0)$ for locations of extrema of $f(x, y)$.

We conclude by evaluating f at each of the points we have found:

$$\begin{aligned} f(0, 0) &= 3, && \leftarrow \text{maximum} \\ f\left(\tfrac{5}{4}, \tfrac{3}{2}\right) &= 2.125 \\ f(1, 1) &= 2, \\ f(2, 0) &= 1, \\ f(0, 4) &= -1. && \leftarrow \text{minimum}\end{aligned}$$

Thus the maximum value of $f(x, y)$ on the region R is $f(0, 0) = 3$ and the minimum value is $f(0, 4) = -1$. ■

Note the terminology used in this section. In Example 5, the maximum *value* of f is 3, the maximum *occurs at* the point $(0, 0)$ in the domain of f, and the *highest point* on the graph of f is $(0, 0, 3)$.

Highest and Lowest Points of Surfaces

In applied problems we frequently know in advance that the absolute maximum (or minimum) value of $f(x, y)$ on R occurs at an *interior* point of R where both partial derivatives of f exist. In this important case, Theorem 3 tells us that we can locate every possible point at which the maximum (or minimum) might occur by simultaneously solving the two equations

$$f_x(x, y) = 0 \quad \text{and} \quad f_y(x, y) = 0. \tag{4}$$

If we are lucky, these equations will have only one simultaneous solution (x, y) interior to R. If so, then *that* solution must be the location of the desired maximum (or minimum). If we find that the equations in (4) have several simultaneous solutions interior to R, then we simply evaluate f at each solution to determine the largest (or smallest) value of $f(x, y)$.

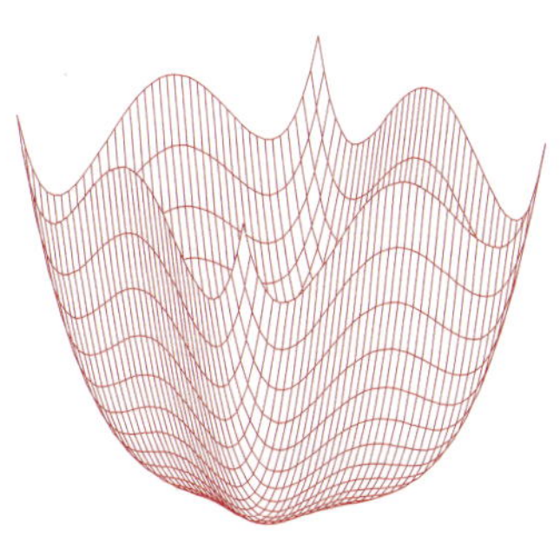

Fig. 13.5.7 The surface $z = x^4 + y^4 - x^2y^2$ opens upward.

We can use this method to find the lowest point on a surface $z = f(x, y)$ that opens upward, as in Fig. 13.5.7. If R is a sufficiently large rectangle, then $f(x, y)$ attains large positive values everywhere on the boundary of R but smaller values at interior points. It follows that the minimum value of $f(x, y)$ must be attained at an interior point of R.

The question of a highest or lowest point is not pertinent for a surface that opens both upward and downward, as in Fig. 13.5.8.

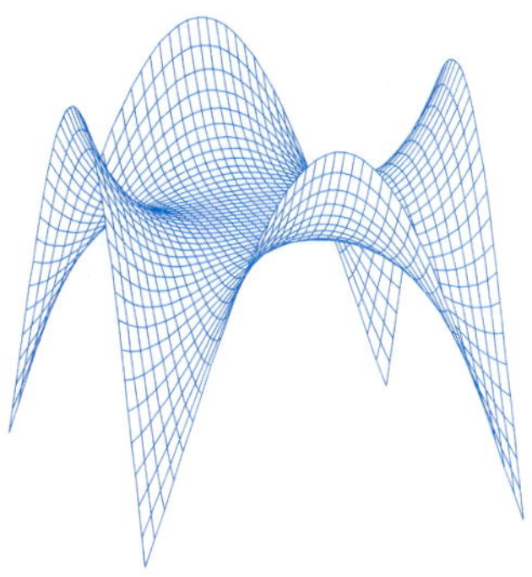

Fig. 13.5.8 The surface $z = x^4 + y^4 - 4x^2y^2$ opens both upward and downward.

EXAMPLE 6 Find the highest point on the surface

$$z = \tfrac{8}{3}x^3 + 4y^3 - x^4 - y^4. \tag{5}$$

Solution Because of the negative fourth-degree terms in Eq. (5) that predominate when $|x|$ and/or $|y|$ is large, this surface opens downward (Fig. 13.5.9). We can verify this observation by writing

$$z = (x^4 + y^4)\left(-1 + \frac{\frac{8}{3}x^3 + 4y^3}{x^4 + y^4}\right)$$

and then substituting $x = r\cos\theta$, $y = r\sin\theta$:

$$z = (x^4 + y^4)\left(-1 + \frac{\frac{8}{3}\cos^3\theta + 4\sin^3\theta}{r(\cos^4\theta + \sin^4\theta)}\right).$$

It is now clear that the fraction approaches zero as $r \to \infty$ and hence that $z < 0$ if either $|x|$ or $|y|$ is large.

But $z = z(x, y)$ does attain positive values, such as $z(1, 1) = \frac{14}{3}$. So let us find the maximum value of z.

Because the partial derivatives of z with respect to x and y exist everywhere, Theorem 3 implies that we need only solve the equations $\partial z/\partial x = 0$ and $\partial z/\partial y = 0$ in Eq. (2)—that is,

$$\frac{\partial z}{\partial x} = 8x^2 - 4x^3 = 4x^2(2 - x) = 0,$$

$$\frac{\partial z}{\partial y} = 12y^2 - 4y^3 = 4y^2(3 - y) = 0.$$

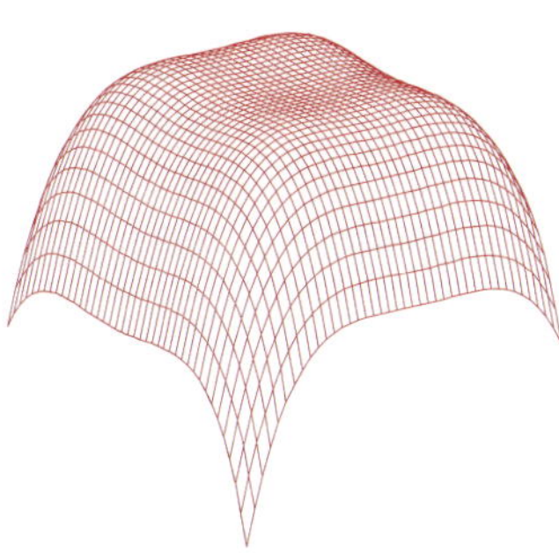

Fig. 13.5.9 The surface $z = \frac{8}{3}x^3 + 4y^3 - x^4 - y^4$ opens downward (Example 6).

If these two equations are satisfied, then

Either $x = 0$ or $x = 2$	and	either $y = 0$ or $y = 3$.

It follows that either

$x = 0$ and $y = 0$	or	$x = 0$ and $y = 3$	or	$x = 2$ and $y = 0$	or	$x = 2$ and $y = 3$.

Consequently, we need only inspect the values

$$z(0, 0) = 0,$$

$$z(2, 0) = \tfrac{16}{3} = 5.333333333\ldots,$$

$$z(0, 3) = 27,$$

$$z(2, 3) = \tfrac{97}{3} = 32.333333333\ldots \quad \leftarrow \text{maximum}$$

Thus the highest point on the surface is the point $(2, 3, \frac{97}{3})$. The four critical points on the surface are indicated in Fig. 13.5.10.

Fig. 13.5.10 The critical points of Example 6

Applied Maximum-Minimum Problems

The analysis of a multivariable applied maximum-minimum problem involves the same general steps that we listed at the beginning of Section 3.6. Here, however, we will express the dependent variable—the quantity to be maximized or minimized—as a function $f(x, y)$ of *two* independent variables. Once we have identified the appropriate region in the xy-plane as the domain of f, the methods of this section are applicable. We often find that a preliminary step is required: If the meaningful domain of definition of f is an unbounded region, then we first restrict f to a *bounded* plane region R on which we know the desired extreme value occurs. This procedure is similar to the one we used with open-interval maximum-minimum problems in Section 4.4.

EXAMPLE 7 Find the minimum cost of a rectangular box with volume 48 ft^3 if the front and back cost $\$1/\text{ft}^2$, the top and bottom cost $\$2/\text{ft}^2$, and the two ends cost $\$3/\text{ft}^2$. (We first discussed such a box in Section 13.1.) This box is shown in Fig. 13.5.11.

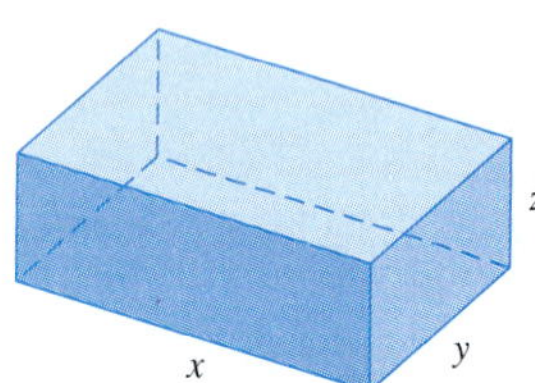

Fig. 13.5.11 A box whose total cost we want to minimize (Example 7)

Solution We found in Section 13.1 that the cost C (in dollars) of this box is given by

$$C(x, y) = 4xy + \frac{288}{x} + \frac{96}{y}$$

in terms of its length x and width y. Let R be a square such as the one shown in Fig. 13.5.12. Two sides of R are so close to the coordinate axes that $288/x > 1000$ on the side nearest the y-axis and $96/y > 1000$ on the side nearest the x-axis. Also, the square is so large that $4xy > 1000$ on both of the other two sides. This means that $C(x, y) > 1000$ at every point (x, y) of the first quadrant that lies on or outside the boundary of the square R. Because $C(x, y)$ attains reasonably small values within R (for instance, $C(1, 1) = 388$), it is clear that the absolute minimum of C must occur at an interior point of R. Thus, although the natural domain of the cost function $C(x, y)$ is the entire first quadrant, we have succeeded in restricting its domain to a region R of the sort to which Theorem 3 applies.

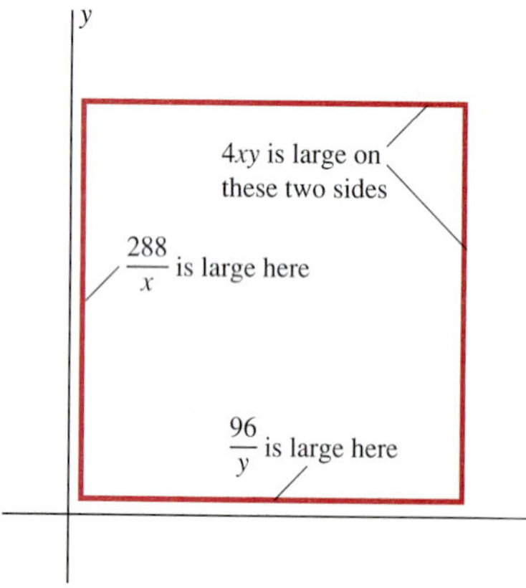

Fig. 13.5.12 The cost function $C(x, y)$ of Example 7 takes on large positive values on the boundary of the square.

We therefore solve the equations

$$\frac{\partial C}{\partial x} = 4y - \frac{288}{x^2} = 0,$$

$$\frac{\partial C}{\partial y} = 4x - \frac{96}{y^2} = 0.$$

We multiply the first equation by x and the second by y. (Ad hoc methods are frequently required in the solution of simultaneous nonlinear equations.) This procedure gives

$$\frac{288}{x} = 4xy = \frac{96}{y},$$

so that $x = 288y/96 = 3y$. We substitute $x = 3y$ into the equation $\partial C/\partial y = 0$ and find that

$$12y - \frac{96}{y^2} = 0, \quad \text{so} \quad 12y^3 = 96.$$

Hence $y = \sqrt[3]{8} = 2$, so $x = 6$. Therefore the minimum cost of this box is $C(6, 2) = 144$ (dollars). Because the volume of the box is $V = xyz = 48$, its height is $z = 48/(6 \cdot 2) = 4$ when $x = 6$ and $y = 2$. Thus the optimal box is 6 ft wide, 2 ft deep, and 4 ft high. ■

REMARK As a check, note that the cheapest surfaces (front and back) are the largest, whereas the most expensive surfaces (the ends) are the smallest.

We have seen that if $f_x(a, b) = 0 = f_y(a, b)$, then $f(a, b)$ may be either a maximum value, a minimum value, or neither. In Section 13.10 we will discuss sufficient conditions for $f(a, b)$ to be either a local maximum or a local minimum. These conditions involve the second-order partial derivatives of f at (a, b).

The methods of this section generalize readily to functions of three or more variables. For example, if the function $f(x, y, z)$ has a local extremum at the point (a, b, c) where its three first-order partial derivatives exist, then all three must be zero there. That is,

$$f_x(a, b, c) = f_y(a, b, c) = f_z(a, b, c) = 0. \tag{6}$$

Example 8 illustrates a "line-through-the-point" method that we can sometimes use to show that a point (a, b, c) where the conditions in (6) hold is neither a local maximum nor a local minimum point. (The method is also applicable to functions of two or of more than three variables.)

EXAMPLE 8 Determine whether the function $f(x, y, z) = xy + yz - xz$ has any local extrema.

Solution The necessary conditions in Eq. (6) give the equations

$$f_x(x, y, z) = y - z = 0,$$

$$f_y(x, y, z) = x + z = 0,$$

$$f_z(x, y, z) = y - x = 0.$$

We easily find that the simultaneous solution of these equations is $x = y = z = 0$. On the line $x = y = z$ through $(0, 0, 0)$, the function $f(x, y, z)$ reduces to x^2, which is minimal at $x = 0$. But on the line $x = -y = z$, it reduces to $-3x^2$, which is maximal when $x = 0$. Hence f can have neither a local maximum nor a local minimum at $(0, 0, 0)$. Therefore it has no extrema, local *or* global.

13.5 PROBLEMS

In Problems 1 through 12, find every point on the given surface $z = f(x, y)$ at which the tangent plane is horizontal.

1. $z = x - 3y + 5$

2. $z = 4 - x^2 - y^2$

3. $z = xy + 5$

4. $z = x^2 + y^2 + 2x$

5. $z = x^2 + y^2 - 6x + 2y + 5$

6. $z = 10 + 8x - 6y - x^2 - y^2$

7. $z = x^2 + 4x + y^3$

8. $z = x^4 + y^3 - 3y$

9. $z = 3x^2 + 12x + 4y^3 - 6y^2 + 5$ (Fig. 13.5.13)

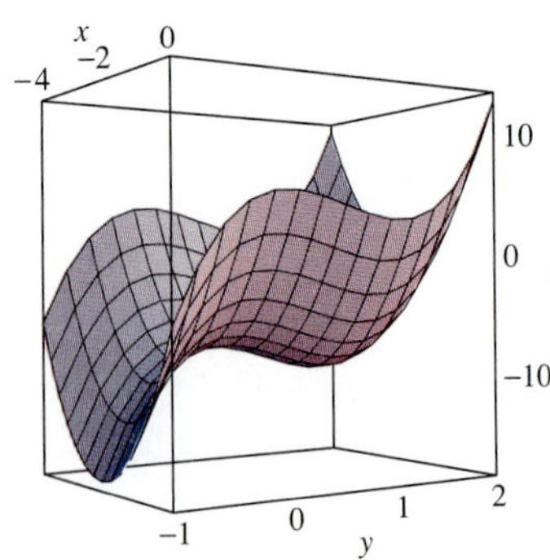

Fig. 13.5.13 The surface of Problem 9

10. $z = \dfrac{1}{1 - 2x + 2y + x^2 + y^2}$

11. $z = (2x^2 + 3y^2)\exp(-x^2 - y^2)$ (Fig. 13.5.14)

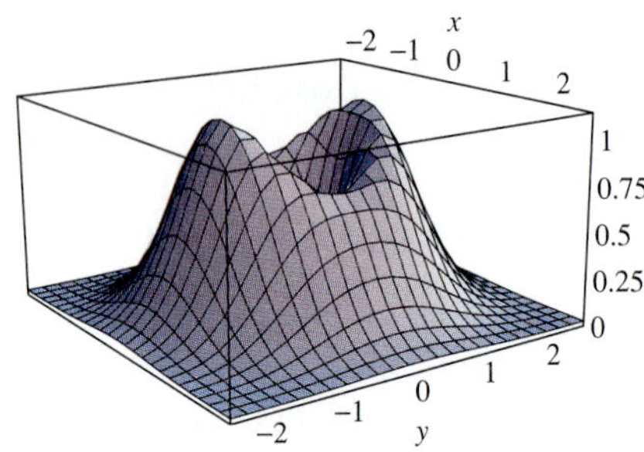

Fig. 13.5.14 The surface of Problem 11

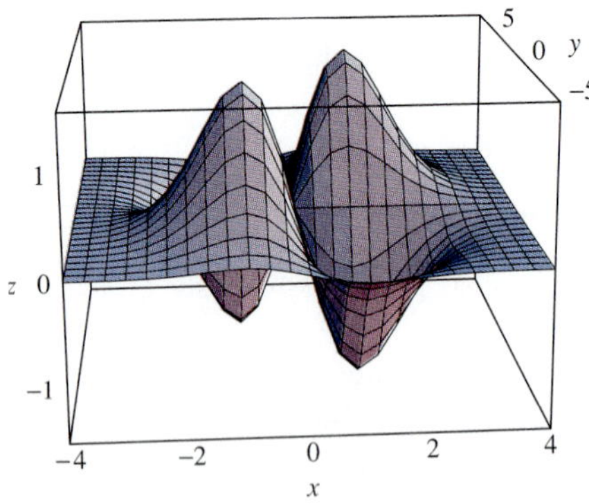

Fig. 13.5.15 The surface of Problem 12

12. $z = 2xy\exp\left(-\frac{1}{8}(4x^2 + y^2)\right)$ (Fig. 13.5.15)

Each of the surfaces defined in Problems 13 through 22 either opens downward and has a highest point or opens upward and has a lowest point. Find this highest or lowest point on the surface $z = f(x, y)$.

13. $z = x^2 - 2x + y^2 - 2y + 3$

14. $z = 6x - 8y - x^2 - y^2$ **15.** $z = 2x - x^2 + 2y^2 - y^4$

16. $z = 4xy - x^4 - y^4$

17. $z = 3x^4 - 4x^3 - 12x^2 + 2y^2 - 12y$

18. $z = 3x^4 + 4x^3 + 6y^4 - 16y^3 + 12y^2$

19. $z = 2x^2 + 8xy + y^4$

20. $z = \dfrac{1}{10 - 2x - 4y + x^2 + y^4}$

21. $z = \exp(2x - 4y - x^2 - y^2)$

22. $z = (1 + x^2)\exp(-x^2 - y^2)$

In Problems 23 through 28, find the maximum and minimum values attained by the given function $f(x, y)$ on the given plane region R.

23. $f(x, y) = x + 2y$; R is the square with vertices at $(\pm 1, \pm 1)$.

24. $f(x, y) = x^2 + y^2 - x$; R is the square of Problem 23.

25. $f(x, y) = x^2 + y^2 - 2x$; R is the triangular region with vertices at $(0, 0)$, $(2, 0)$, and $(0, 2)$.

26. $f(x, y) = x^2 + y^2 - x - y$; R is the region of Problem 25.

27. $f(x, y) = 2xy$; R is the circular disk $x^2 + y^2 \leqq 1$.

28. $f(x, y) = xy^2$; R is the circular disk $x^2 + y^2 \leqq 3$.

In Problems 29 through 34, the equation of a plane or surface is given. Find the first-octant point $P(x, y, z)$ on the surface closest to the given fixed point $Q(x_0, y_0, z_0)$. (*Suggestion:* Minimize the squared distance $|PQ|^2$ as a function of x and y.)

29. The plane $12x + 4y + 3z = 169$ and the fixed point $Q(0, 0, 0)$

30. The plane $2x + 2y + z = 27$ and the fixed point $Q(9, 9, 9)$

31. The plane $2x + 3y + z = 49$ and the fixed point $Q(7, -7, 0)$

32. The surface $xyz = 8$ and the fixed point $Q(0, 0, 0)$

33. The surface $x^2y^2z = 4$ and the fixed point $Q(0, 0, 0)$

34. The surface $x^4y^8z^2 = 8$ and the fixed point $Q(0, 0, 0)$

35. Find the maximum possible product of three positive numbers whose sum is 120.

36. Find the maximum possible volume of a rectangular box if the sum of the lengths of its 12 edges is 6 meters.

37. Find the dimensions of the box with volume 1000 in.^{3} that has minimal total surface area.

38. Find the dimensions of the open-topped box with volume 4000 cm^3 whose bottom and four sides have minimal total surface area.

In Problems 39 through 42, you are to find the dimensions that minimize the total cost of the material needed to construct the rectangular box that is described. It is either *closed* (top, bottom, and four sides) or *open-topped* (four sides and a bottom).

39. The box is to be open-topped with a volume of 600 in.^{3} The material for its bottom costs 6¢/in.^{2}, and the material for its four sides costs 5¢/in.^{2}

40. The box is to be closed with a volume of 48 ft^3. The material for its top and bottom costs \$3/ft^2, and the material for its four sides costs \$4/ft^2.

41. The box is to be closed with a volume of 750 in.^{3} The material for its top and bottom costs 3¢/in.^{2}, the material for its front and back costs 6¢/in.^{2}, and the material for its two ends costs 9¢/in.^{2}

42. The box is to be a closed shipping crate with a volume of 12 m^3. The material for its bottom costs *twice* as much (per square meter) as the material for its top and four sides.

43. A rectangular building is to have a volume of 8000 ft^3. Annual heating and cooling costs will amount to \$2/ft^2 for its top, front, and back, and \$4/ft^2 for the two end walls. What dimensions of the building would minimize these annual costs?

44. You want to build a rectangular aquarium with a bottom made of slate costing 28¢/in.^{2} Its sides will be glass, which costs 5¢/in.^{2}, and its top will be stainless steel, which costs 2¢/in.^{2} The volume of this aquarium is to be 24000 in.^{3} What are the dimensions of the least expensive such aquarium?

45. A rectangular box is inscribed in the first octant with three of its sides in the coordinate planes, their common vertex at the origin, and the opposite vertex on the plane with equation $x + 3y + 7z = 11$. What is the maximum possible volume of such a box?

46. Three sides of a rectangular box lie in the coordinate planes, their common vertex at the origin; the opposite vertex is on the plane with equation

$$\frac{x}{a} + \frac{y}{b} + \frac{z}{c} = 1$$

(a, b, and c are positive constants). In terms of a, b, and c, what is the maximum possible volume of such a box?

47. Find the maximum volume of a rectangular box that a post office will accept for delivery if the sum of its *length* and *girth* cannot exceed 108 in.

48. Repeat Problem 47 for the case of a cylindrical box—one shaped like a hatbox or a fat mailing tube.

49. A rectangular box with its base in the xy-plane is inscribed under the graph of the paraboloid $z = 1 - x^2 - y^2$, $z \geqq 0$. Find the maximum possible volume of the box. (*Suggestion:* You may assume that the sides of the box are parallel to the vertical coordinate planes, and it follows that the box is symmetrically placed around these planes.)

50. What is the maximum possible volume of a rectangular box inscribed in a hemisphere of radius R? Assume that one face of the box lies in the planar base of the hemisphere.

51. A buoy is to have the shape of a right circular cylinder capped at each end by identical right circular cones with the same radius as the cylinder. Find the minimum possible surface area of the buoy, given that it has fixed volume V.

52. A pentagonal window is to have the shape of a rectangle surmounted by an isosceles triangle (with horizontal base, so the window is symmetric around its vertical axis), and the perimeter of the window is to be 24 ft. What are the dimensions of such a window that will admit the most light (because its area is the greatest)?

53. Find the point (x, y) in the plane for which the sum of the squares of its distances from $(0, 1)$, $(0, 0)$, and $(2, 0)$ is a minimum.

54. Find the point (x, y) in the plane for which the sum of the squares of its distances from (a_1, b_1), (a_2, b_2), and (a_3, b_3) is a minimum.

55. An A-frame house is to have fixed volume V. Its front and rear walls are in the shape of equal, parallel isosceles triangles with horizontal bases. The roof consists of two rectangles that connect pairs of upper sides of the triangles. To minimize heating and cooling costs, the total area of the A-frame (excluding the floor) is to be minimized. Describe the shape of the A-frame of minimal area.

56. What is the maximum possible volume of a rectangular box whose longest diagonal has fixed length L?

57. A wire 120 cm long is cut into three *or fewer* pieces, and each piece is bent into the shape of a square. How should this be done to minimize the total area of these squares? to maximize it?

58. You must divide a lump of putty of fixed volume V into three or fewer pieces and form the pieces into cubes. How should you do this to maximize the total surface area of the cubes? to minimize it?

59. A very long rectangle of sheet metal has width L and is to be folded to make a rain gutter (Fig. 13.5.16). Maximize its volume by maximizing the cross-sectional area shown in the figure.

Fig. 13.5.16 Cross section of the rain gutter of Problem 59

60. Consider the function $f(x, y) = (y - x^2)(y - 3x^2)$. (a) Show that $f_x(0, 0) = 0 = f_y(0, 0)$. (b) Show that for every straight line $y = mx$ through $(0, 0)$, the function $f(x, mx)$ has a local minimum at $x = 0$. (c) Examine the values of f at points of the parabola $y = 2x^2$ to show that f does *not* have a local minimum at $(0, 0)$. This tells us that we cannot use the line-through-the-point method of Example 8 to show that a point *is* a local extremum.

61. Suppose that Alpha, Inc. and Beta, Ltd. manufacture competitive (but not identical) products, with the weekly sales of each product determined by the selling price of that product *and* the price of its competition. Suppose that Alpha sets a sales price of x dollars per unit for its product, while Beta sets a sales price of y dollars per unit for its product. Market research shows that the weekly profit made by Alpha is then

$$P(x) = -2x^2 + 12x + xy - y - 10$$

and that the weekly profit made by Beta is

$$Q(y) = -3y^2 + 18y + 2xy - 2x - 15$$

(both in thousands of dollars). The peculiar notation arises from the fact that x is the only variable under the control of Alpha and y is the only variable under the control of Beta. (If this disturbs you, feel free to write $P(x, y)$ in place of $P(x)$ and $Q(x, y)$ in place of $Q(y)$.) (a) Assume that both company managers know calculus, and that each knows that the *other* knows calculus and has some common sense. What price will each manager set to maximize his company's weekly profit? (b) Now suppose that the two managers enter into an agreement (legal or otherwise), by which they plan to maximize their *total* weekly profit. Now what should be the selling price of each product? (We suppose that they will divide the resulting profit in an equitable way, but the details of this intriguing problem are not the issue.)

62. Three firms—Ajax Products (AP), Behemoth Quicksilver (BQ), and Conglomerate Resources (CR)—produce products in quantities A, B, and C, respectively. The weekly profits that accrue to each, in thousands of dollars, obey the following equations:

$$\text{AP: } P = 1000A - A^2 - 2AB,$$

$$\text{BQ: } Q = 2000B - 2B^2 - 4BC,$$

$$\text{CR: } R = 1500C - 3C^2 - 6AC.$$

(a) If each firm acts independently to maximize its weekly profit, what will those profits be? (b) If firms AP and CR join to maximize their total profit while BQ continues to act alone, what effects will this have? Give a *complete* answer to this problem. Assume that the fact of the merger of AP and CR is known to the management of BQ.

63. A farmer can raise sheep, hogs, and cattle. She has space for 80 sheep or 120 hogs or 60 cattle or any combination using the same amount of space; that is, 8 sheep use as much space as 12 hogs or 6 cattle. The anticipated profits per animal are \$10 per sheep, \$8 per hog, and \$20 for each head of cattle. State law requires that a farmer raise at least as many hogs as sheep and cattle combined. How does the farmer maximize her profit?

13.5 PROJECT: EXOTIC CRITICAL POINTS

In Section 13.10 we discuss a systematic "second derivative test" for local extrema of functions of two variables. This project explores a more direct approach that works well for certain functions. A **homogeneous** polynomial is one such as

$$f(x, y) = x^2 + 4xy + y^2 \quad \text{or} \quad F(x, y) = 2x^4 - 7x^2y^2 + y^4 \tag{1}$$

in which each term has the same (total) degree. We see readily that any such polynomial of degree at least 2 has the origin $(0, 0)$ as a critical point. The question is whether this critical point is a local maximum, a local minimum, or neither.

To answer this question, you can substitute polar coordinates $x = r\cos\theta$, $y = r\sin\theta$. For example, with the fourth-degree polynomial $F(x, y)$ in (1) for which $F(0, 0) = 0$, show first that this substitution yields

$$F(x, y) = r^4 g(\theta) \tag{2}$$

where

$$g(\theta) = 2\cos^4\theta - 7\cos^2\theta\sin^2\theta + \sin^4\theta. \tag{3}$$

Then explain carefully why it follows from Eq. (2) that $F(x, y)$ has

- A local maximum at $(0, 0)$ if $g(\theta) > 0$ for $0 \leqq \theta \leqq 2\pi$,
- A local minimum at $(0, 0)$ if $g(\theta) < 0$ for $0 \leqq \theta \leqq 2\pi$,
- Neither a maximum nor a minimum at $(0, 0)$ if $g(\theta)$ attains both positive and negative values for $0 \leqq \theta \leqq 2\pi$.

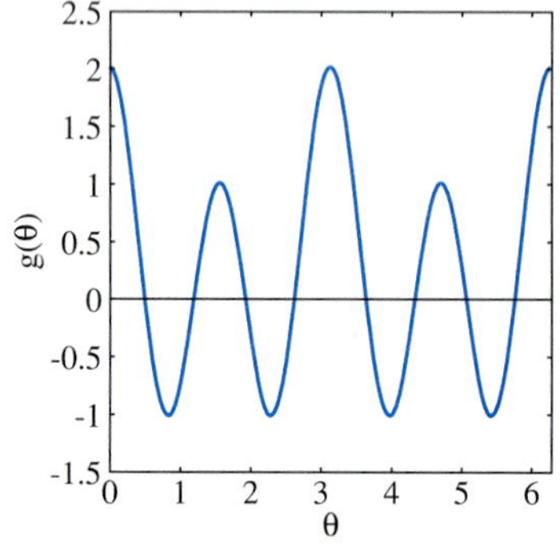

Fig. 13.5.17 The graph of $g(\theta) = 2\cos^4\theta - 7\cos^2\theta\sin^2\theta + \sin^4\theta$ exhibits both positive and negative values.

But the pertinent behavior of $g(\theta)$ is immediately visible in a graph. For instance, Fig. 13.5.17 shows the graph of $g(\theta)$ as defined in Eq. (3), and we see that $g(\theta)$ attains both positive and negative values. Therefore the function $F(x, y)$

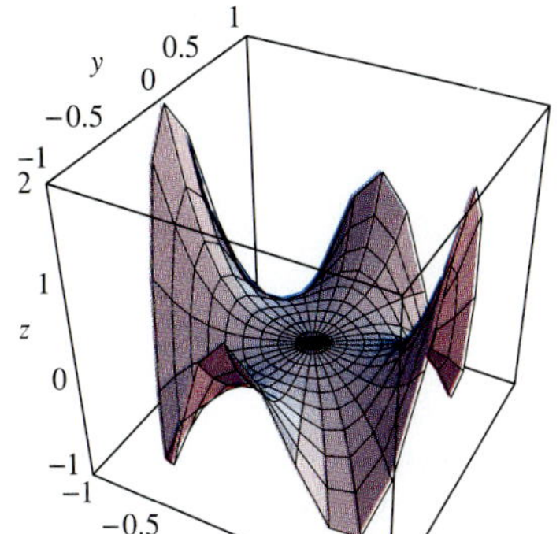

Fig. 13.5.18 A dog-saddle critical point

defined in (1) has neither a local maximum nor a local minimum at the critical point $(0, 0)$. The graph $z = F(x, y)$ shown in Fig. 13.5.18 was plotted by writing

$$x = r\cos t, \qquad y = r\sin t, \qquad z = 2x^4 - 7x^2y^2 + y^4$$

and using a parametric plot command for $0 \leqq r \leqq 1$, $0 \leqq t \leqq 2\pi$. What we see might well be termed a "dog saddle," because four rather than only two legs can be accommodated.

For your own exotic critical points to investigate, try homogeneous polynomials such as

$$f(x, y) = px^2 + qxy + ry^2 \quad \text{or} \quad F(x, y) = px^4 - qx^2y^2 + ry^4$$

or a similar sixth-degree polynomial. Let p, q, and r be selected integers, such as the last three nonzero digits of your student I.D. number. Assemble a portfolio of pictures illustrating local maxima and minima as well as different varieties of saddle points.

13.6 INCREMENTS AND DIFFERENTIALS

In Section 4.2 we used the *differential*

$$df = f'(x)\,\Delta x \tag{1}$$

to approximate the *increment,* or actual change,

$$\Delta f = f(x + \Delta x) - f(x) \tag{2}$$

in the value of a single-variable function that results from the change Δx in the independent variable. Thus

$$\Delta f = f(x + \Delta x) - f(x) \approx f'(x)\,\Delta x = df. \tag{3}$$

We now describe the use of the partial derivatives $\partial f/\partial x$ and $\partial f/\partial y$ to approximate the **increment**

$$\Delta f = f(x + \Delta x, y + \Delta y) - f(x, y) \tag{4}$$

in the value of a two-variable function that results when its independent variables are changed simultaneously. If only x were changed and y were held constant, we could temporarily regard $f(x, y)$ as a function of x alone. Then, with $f_x(x, y)$ playing the role of $f'(x)$, the linear approximation in Eq. (3) would give

$$f(x + \Delta x, y) - f(x, y) \approx f_x(x, y)\,\Delta x \tag{5}$$

for the change in f corresponding to the change Δx in x. Similarly, if only y were changed and x were held constant, then—temporarily regarding $f(x, y)$ as a function of y alone—we would get

$$f(x, y + \Delta y) - f(x, y) \approx f_y(x, y)\,\Delta y \tag{6}$$

for the change in f corresponding to the change Δy in y.

If both x and y are changed simultaneously, we expect the *sum* of the approximations in (5) and (6) to be a good estimate of the resulting increment in the value of f. On this basis we define the **differential**

$$df = f_x(x, y)\,\Delta x + f_y(x, y)\,\Delta y \tag{7}$$

of a function $f(x, y)$ of two independent variables.

According to Theorem 1 of this section, the approximation $\Delta f \approx df$ of the actual increment in Eq. (4) by the differential in Eq. (7) is accurate under appropriate conditions. We therefore write

$$f(x + \Delta x, y + \Delta y) = f(x, y) + \Delta f \quad \text{(exact)}, \tag{8}$$

$$f(x + \Delta x, y + \Delta y) \approx f(x, y) + df \quad \text{(approximation)}. \tag{9}$$

EXAMPLE 1 Find the differential df of the function $f(x, y) = x^2 + 3xy - 2y^2$. Then compare df and the actual increment Δf when (x, y) changes from $P(3, 5)$ to $Q(3.2, 4.9)$.

Solution The differential of f, as given in Eq. (7), is

$$df = \frac{\partial f}{\partial x}\,\Delta x + \frac{\partial f}{\partial y}\,\Delta y = (2x + 3y)\,\Delta x + (3x - 4y)\,\Delta y.$$

At the point $P(3, 5)$ this differential is

$$df = (2\cdot 3 + 3\cdot 5)\,\Delta x + (3\cdot 3 - 4\cdot 5)\,\Delta y = 21\,\Delta x - 11\,\Delta y.$$

With $\Delta x = 0.2$ and $\Delta y = -0.1$, corresponding to change from $P(3, 5)$ to $Q(3.2, 4.9)$, we get

$$df = 21\cdot(0.2) - 11\cdot(-0.1) = 5.3\,.$$

The actual change in the value of f from P to Q is the increment

$$\Delta f = f(3.2, 4.9) - f(3, 5) = 9.26 - 4 = 5.26,$$

so in this example the differential seems to be a good approximation to the increment. ■

At the fixed point $P(a, b)$, the differential

$$df = f_x(a, b)\,\Delta x + f_y(a, b)\,\Delta y \tag{10}$$

is a *linear* function of Δx and Δy; the coefficients $f_x(a, b)$ and $f_y(a, b)$ in this linear function depend on a and b. Thus the differential df is a **linear approximation** to the actual increment Δf. Theorem 1, stated later in this section, implies that if the function f has continuous partial derivatives, then df is a *very good approximation* to Δf when the changes Δx and Δy in x and y are sufficiently small. The approximation

$$f(a + \Delta x, b + \Delta y) \approx f(a, b) + f_x(a, b)\,\Delta x + f_y(a, b)\,\Delta y \tag{11}$$

may then be used to estimate the value of $f(a + \Delta x, b + \Delta y)$ when Δx and Δy are small and the values $f(a, b)$, $f_x(a, b)$, and $f_y(a, b)$ are all known.

EXAMPLE 2 Use the differential to estimate $\sqrt{2\cdot(2.03)^3 + (2.97)^2}$.

Solution We begin by letting $f(x, y) = \sqrt{2x^3 + y^2}$, $a = 2$, and $b = 3$. It is then easy to compute the exact value $f(2, 3) = \sqrt{2\cdot 8 + 9} = \sqrt{25} = 5$. Next,

$$\frac{\partial f}{\partial x} = \frac{3x^2}{\sqrt{2x^3 + y^2}} \quad \text{and} \quad \frac{\partial f}{\partial y} = \frac{y}{\sqrt{2x^3 + y^2}},$$

so

$$f_x(2, 3) = \tfrac{12}{5} \quad \text{and} \quad f_y(2, 3) = \tfrac{3}{5}.$$

Hence Eq. (11) gives

$$\begin{aligned}\sqrt{2 \cdot (2.02)^3 + (2.97)^2} &= f(2.02, 2.97)\\ &\approx f(2, 3) + f_x(2, 3) \cdot (0.02) + f_y(2, 3) \cdot (-0.03)\\ &= 5 + \tfrac{12}{5}(0.02) + \tfrac{3}{5}(-0.03) = 5.03.\end{aligned}$$

The actual value to four decimal places is 5.0305. ■

If $z = f(x, y)$, we often write dz in place of df. So the differential of the dependent variable z at the point (a, b) is $dz = f_x(a, b)\,\Delta x + f_y(a, b)\,\Delta y$. At the arbitrary point (x, y) the differential of z takes the form

$$dz = f_x(x, y)\,\Delta x + f_y(x, y)\,\Delta y.$$

More simply, we can write

$$dz = \frac{\partial z}{\partial x}\,\Delta x + \frac{\partial z}{\partial y}\,\Delta y. \tag{12}$$

It is customary to write dx for Δx and dy for Δy in this formula. When this is done, Eq. (12) takes the form

$$dz = \frac{\partial z}{\partial x}\,dx + \frac{\partial z}{\partial y}\,dy. \tag{13}$$

When we use this notation, we must realize that dx and dy have *no* connotation of being "infinitesimal" or even small. The differential dz is still simply a linear function of the ordinary real variables dx and dy, a function that gives a linear approximation to the change in z when x and y are changed by the amounts dx and dy, respectively.

EXAMPLE 3 In Example 4 of Section 13.4, we considered 1 mol of an ideal gas—its volume V in cubic centimeters given in terms of its pressure p in atmospheres and temperature T in kelvins by the formula $V = (82.06)T/p$. Approximate the change in V when p is increased from 5 atm to 5.2 atm and T is increased from 300 K to 310 K.

Solution The differential of $V = V(p, T)$ is

$$dV = \frac{\partial V}{\partial p}\,dp + \frac{\partial V}{\partial T}\,dT = -\frac{82.06 \cdot T}{p^2}\,dp + \frac{82.06}{p}\,dT.$$

With $p = 5$, $T = 300$, $dp = 0.2$, and $dT = 10$, we compute

$$dV = -\frac{82.06 \cdot 300}{5^2} \cdot 0.2 + \frac{82.06}{5} \cdot 10 = -32.8 \quad (\text{cm}^3).$$

This indicates that the gas will decrease in volume by about 33 cm^3. The actual change is the increment

$$\Delta V = \frac{82.06 \cdot 310}{5.2} - \frac{82.06 \cdot 300}{5} = 4892.0 - 4923.6 = -31.6 \quad (\text{cm}^3).$$ ■

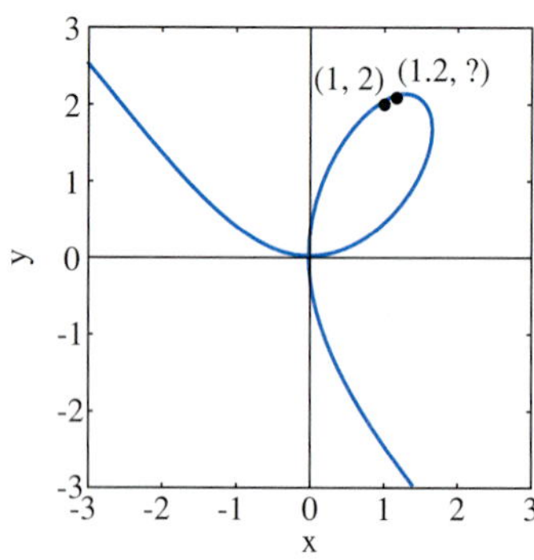

Fig. 13.6.1 The curve of Example 4

EXAMPLE 4 The point $(1, 2)$ lies on the curve (Fig. 13.6.1) with equation

$$f(x, y) = 2x^3 + y^3 - 5xy = 0. \tag{14}$$

Approximate the y-coordinate of the nearby point $(1.2, y)$.

Solution When we compute the differentials in Eq. (14), we get

$$df = \frac{\partial f}{\partial x}\,dx + \frac{\partial f}{\partial y}\,dy = (6x^2 - 5y)\,dx + (3y^2 - 5x)\,dy = 0.$$

Now when we substitute $x = 1$, $y = 2$, and $dx = 0.2$, we obtain the equation $(-4)(0.2) + (7)\,dy = 0$. It then follows that $dy = (0.8)/7 \approx 0.114 \approx 0.1$. This yields $(1.2, 2.1)$ for the approximate coordinates of the nearby point. As a check on the accuracy of this approximation, we can substitute $x = 1.2$ into Eq. (14). This gives the equation

$$2\cdot(1.2)^3 + y^3 - 5\cdot(1.2)y = y^3 - 6y + 3.456 = 0.$$

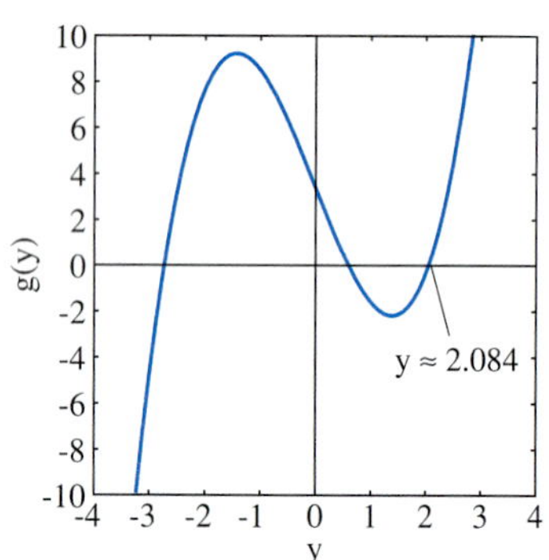

Fig. 13.6.2 The graph of $g(y) = y^3 - 6y + 3.456$

The roots of this equation are the x-intercepts of the curve in Fig. 13.6.2. A calculator or computer with an equation solver (or Newton's method) then yields $y \approx 2.084 \approx 2.1$ for the solution near $y = 2$. ■

Increments and differentials of functions of more than two variables are defined similarly. A function $w = f(x, y, z)$ has **increment**

$$\Delta w = \Delta f = f(x + \Delta x, y + \Delta y, z + \Delta z) - f(x, y, z)$$

and **differential**

$$dw = df = \frac{\partial f}{\partial x}\Delta x + \frac{\partial f}{\partial y}\Delta y + \frac{\partial f}{\partial z}\Delta z;$$

that is,

$$dw = \frac{\partial w}{\partial x}\,dx + \frac{\partial w}{\partial y}\,dy + \frac{\partial w}{\partial z}\,dz,$$

if, as in Eq. (12), we write dx for Δx, dy for Δy, and dz for Δz.

EXAMPLE 5 You have constructed a metal cube that is supposed to have edge length 100 mm, but each of its three measured dimensions x, y, and z may be in error by as much as a millimeter. Use differentials to estimate the maximum resulting error in its calculated volume $V = xyz$.

Solution We need to approximate the increment

$$\Delta V = V(100 + dx, 100 + dy, 100 + dz) - V(100, 100, 100)$$

when the errors dx, dy, and dz in x, y, and z are maximal. The differential of $V = xyz$ is

$$dV = yz\,dx + xz\,dy + xy\,dz.$$

When we substitute $x = y = z = 100$ and $dx = \pm 1$, $dy = \pm 1$, and $dz = \pm 1$, we get

$$dV = 100\cdot 100\cdot(\pm 1) + 100\cdot 100\cdot(\pm 1) + 100\cdot 100\cdot(\pm 1) = \pm 30000.$$

It may surprise you to find that an error of only a millimeter in each dimension of a cube can result in an error of 30000 mm^3 in its volume. (For a cube made of precious metal, an error of 30 cm^3 in its volume could correspond to a difference of hundreds or thousands of dollars in its cost.) ■

The Linear Approximation Theorem

The differential $df = f_x\,dx + f_y\,dy$ is defined provided that both partial derivatives f_x and f_y exist. Theorem 1 gives sufficient conditions for df to be a good approximation to the increment Δf when Δx and Δy are small.

Theorem 1 *Linear Approximation*

Suppose that $f(x, y)$ has continuous first-order partial derivatives in a rectangular region that has horizontal and vertical sides and contains the points

$$P(a, b) \quad \text{and} \quad Q(a + \Delta x, b + \Delta y)$$

in its interior. Let

$$\Delta f = f(a + \Delta x, b + \Delta y) - f(a, b)$$

be the corresponding increment in the value of f. Then

$$\Delta f = f_x(a, b)\,\Delta x + f_y(a, b)\,\Delta y + \epsilon_1\,\Delta x + \epsilon_2\,\Delta y, \tag{15}$$

where ϵ_1 and ϵ_2 are functions of Δx and Δy that approach zero as $\Delta x \to 0$ and $\Delta y \to 0$.

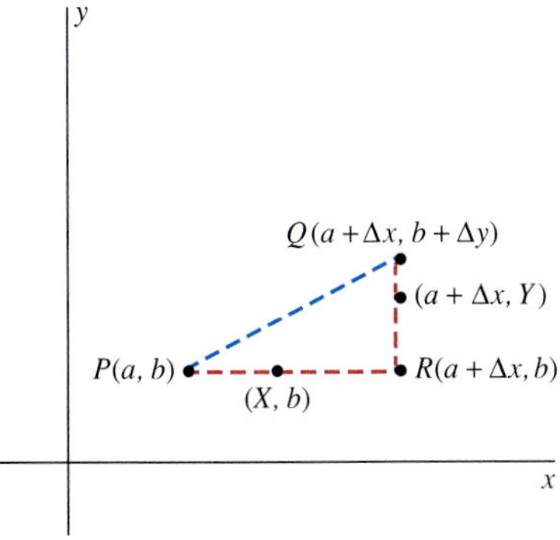

Fig. 13.6.3 Illustration of the proof of the linear approximation theorem

Proof: If R is the point $(a + \Delta x, b)$ indicated in Fig. 13.6.3, then

$$\begin{aligned} \Delta f &= f(Q) - f(P) = [f(R) - f(P)] + [f(Q) - f(R)] \\ &= [f(a + \Delta x, b) - f(a, b)] + [f(a + \Delta x, b + \Delta y) - f(a + \Delta x, b)]. \end{aligned} \tag{16}$$

We consider separately the last two terms in Eq. (16). For the first such term, we define the single-variable function

$$g(x) = f(x, b) \qquad \text{for } x \text{ in } [a, a + \Delta x].$$

Then the mean value theorem gives

$$f(a + \Delta x, b) - f(a, b) = g(a + \Delta x) - g(a) = g'(X)\,\Delta x = f_x(X, b)\,\Delta x$$

for some number X in the open interval $(a, a + \Delta x)$.

For the last term in Eq. (16), we define the single-variable function

$$h(y) = f(a + \Delta x, y) \qquad \text{for } y \text{ in } [b, b + \Delta y].$$

The mean value theorem now yields

$$\begin{aligned} f(a + \Delta x, b + \Delta y) - f(a + \Delta x, b) &= h(b + \Delta y) - h(b) \\ &= h'(Y)\,\Delta y = f_y(a + \Delta x, Y)\,\Delta y \end{aligned}$$

for some number Y in the open interval $(b, b + \Delta y)$.

When we substitute these two results into Eq. (16), we find that

$$\begin{aligned}\Delta f &= f_x(X, b)\,\Delta x + f_y(a + \Delta x, Y)\,\Delta y \\ &= [f_x(a, b) + f_x(X, b) - f_x(a, b)]\,\Delta x + [f_y(a, b) + f_y(a + \Delta x, Y) - f_y(a, b)]\,\Delta y.\end{aligned}$$

So

$$\Delta f = f_x(a, b)\,\Delta x + f_y(a, b)\,\Delta y + \epsilon_1\,\Delta x + \epsilon_2\,\Delta y,$$

where

$$\epsilon_1 = f_x(X, b) - f_x(a, b) \quad \text{and} \quad \epsilon_2 = f_y(a + \Delta x, Y) - f_y(a, b).$$

Finally, because both the points (X, b) and $(a + \Delta x, Y)$ approach (a, b) as $\Delta x \to 0$ and $\Delta y \to 0$, it follows from the continuity of f_x and f_y that both ϵ_1 and ϵ_2 approach zero as Δx and Δy approach zero. This completes the proof. ■

Differentiability of Multivariable Functions

Under the hypotheses of Theorem 1 it follows from Eq. (15) (see Problem 44) that

$$\lim_{(\Delta x, \Delta y)\to(0,0)} \frac{\Delta f - L(\Delta x, \Delta y)}{|(\Delta x, \Delta y)|} = 0 \tag{17}$$

where $|(\Delta x, \Delta y)| = \sqrt{(\Delta x)^2 + (\Delta y)^2}$ and L is the *linear* function of $\Delta x = x - a$ and $\Delta y = y - b$ defined by

$$L(\Delta x, \Delta y) = f_x(a, b)\,\Delta x + f_y(a, b)\,\Delta y. \tag{18}$$

Because

$$\Delta f - L(\Delta x, \Delta y) = f(x, y) - f(a, b) - f_x(a, b)\,\Delta x - f_y(a, b)\,\Delta y,$$

Eq. (17) implies that the linear approximation

$$f(x, y) \approx f(a, b) + f_x(a, b)(x - a) + f_y(a, b)(y - b) \tag{19}$$

is so accurate near (a, b) that the difference between the height of the surface $z = f(x, y)$ and its tangent plane

$$z = f(a, b) + f_x(a, b)(x - a) + f_y(a, b)(y - b) \tag{20}$$

is small even in comparison with $\Delta x = x - a$ and $\Delta y = y - b$.

We know from Theorem 2 in Section 3.4 that a single-variable function is continuous wherever it is differentiable. In contrast, the simple function defined in Problem 43 illustrates the fact that a function $f(x, y)$ of two variables can have partial derivatives at a point without even being continuous there! Thus the mere existence of partial derivatives appears to imply much less for a function of two (or more) variables than it does for a single-variable function.

For this reason, a different definition of differentiability itself is used for multivariable functions. The function $f(x, y)$ is said to be **differentiable** at the point (a, b) provided that there exists a *linear* function L of $\Delta x = x - a$ and $\Delta y = y - b$ such that Eq. (17) holds. Thus the function f is differentiable at (a, b) if it can be approximated sufficiently closely near (a, b) by a *linear* function [as in (19)]. In this case it follows (see Problem 45) that the partial derivatives $f_x(a, b)$ and $f_y(a, b)$ must exist and, moreover, are the coefficients of the linear function L (as in Eq. (18)).

The linear approximation theorem (Theorem 1 of this section) implies that the function $f(x, y)$ is differentiable at the point (a, b) if it is **continuously differentiable** near (a, b)—meaning that the partial derivatives f_x and f_y exist and are continuous near (a, b). On the other hand, the function $f(x, y)$ is continuous wherever it is differentiable—see Problem 46. In brief, it may be said that the linear approximation concept of differentiability lies "between" the concepts of continuity and continuous differentiability.

Polynomials and rational functions of several variables are differentiable wherever defined because their partial derivatives are continuous. Indeed, we will have little explicit need for the linear approximation concept of differentiability, because we deal mainly with functions having continuous partial derivatives wherever they are defined.

Remark The linear approximation theorem generalizes in a natural way to functions of three or more variables. For example, if $w = f(x, y, z)$, then the analogue of Eq. (15) is

$$\begin{aligned}\Delta f = f_x(a, b, c)\,\Delta x + f_y(a, b, c)\,\Delta y + f_z(a, b, c)\,\Delta z \\ + \epsilon_1\,\Delta x + \epsilon_2\,\Delta y + \epsilon_3\,\Delta z,\end{aligned}$$

where ϵ_1, ϵ_2, and ϵ_3 all approach zero as Δx, Δy, and Δz approach zero. The proof for the three-variable case is like the one given here for two variables.

13.6 PROBLEMS

Find the differential dw in Problems 1 through 16.

1. $w = 3x^2 + 4xy - 2y^3$

2. $w = \exp(-x^2 - y^2)$

3. $w = \sqrt{1 + x^2 + y^2}$

4. $w = xye^{x+y}$

5. $w = \arctan\left(\dfrac{x}{y}\right)$

6. $w = xz^2 - yx^2 + zy^2$

7. $w = \ln(x^2 + y^2 + z^2)$

8. $w = \sin xyz$

9. $w = x \tan yz$

10. $w = xye^{uv}$

11. $w = e^{-xyz}$

12. $w = \ln(1 + rs)$

13. $w = u^2\exp(-v^2)$

14. $w = \dfrac{s + t}{s - t}$

15. $w = \sqrt{x^2 + y^2 + z^2}$

16. $w = pqr\exp(-p^2 - q^2 - r^2)$

In Problems 17 through 24, use the exact value $f(P)$ and the differential df to approximate the value $f(Q)$.

17. $f(x, y) = \sqrt{x^2 + y^2}$; $P(3, 4)$, $Q(2.97, 4.04)$

18. $f(x, y) = \sqrt{x^2 - y^2}$; $P(13, 5)$, $Q(13.2, 4.9)$

19. $f(x, y) = \dfrac{1}{1 + x + y}$; $P(3, 6)$, $Q(3.02, 6.05)$

20. $f(x, y, z) = \sqrt{xyz}$; $P(1, 3, 3)$, $Q(0.9, 2.9, 3.1)$

21. $f(x, y, z) = \sqrt{x^2 + y^2 + z^2}$; $P(3, 4, 12)$, $Q(3.03, 3.96, 12.05)$

22. $f(x, y, z) = \dfrac{xyz}{x + y + z}$; $P(2, 3, 5)$, $Q(1.98, 3.03, 4.97)$

23. $f(x, y, z) = e^{-xyz}$; $P(1, 0, -2)$, $Q(1.02, 0.03, -2.02)$

24. $f(x, y) = (x - y)\cos 2\pi xy$; $P(1, \frac{1}{2})$, $Q(1.1, 0.4)$

In Problems 25 through 32, use differentials to approximate the indicated number.

25. $(\sqrt{15} + \sqrt{99})^2$

26. $(\sqrt{26})(\sqrt[3]{28})(\sqrt[4]{17})$

27. $e^{0.4} = \exp(1.1^2 - 0.9^2)$

28. $\dfrac{\sqrt[3]{25}}{\sqrt[5]{30}}$

29. $\sqrt{(3.1)^2 + (4.2)^2 + (11.7)^2}$

30. $\sqrt[3]{(5.1)^2 + 2\cdot(5.2)^2 + 2\cdot(5.3)^2}$

31. The y-coordinate of the point P near $(1, 2)$ on the curve $2x^3 + 2y^3 = 9xy$, if the x-coordinate of P is 1.1

32. The x-coordinate of the point P near $(2, 4)$ on the curve $4x^4 + 4y^4 = 17x^2y^2$, if the y-coordinate of P is 3.9

33. The base and height of a rectangle are measured as 10 cm and 15 cm, respectively, with a possible error of as much as 0.1 cm in each measurement. Use differentials to estimate the maximum resulting error in computing the area of the rectangle.

34. The base radius r and the height h of a right circular cylinder are measured as 3 cm and 9 cm, respectively. There is a possible error of 1 mm in each measurement. Use differentials to estimate the maximum possible error in computing: (a) the volume of the cylinder; (b) the total surface area of the cylinder.

35. The base radius r and height h of a right circular cone are measured as 5 in. and 10 in., respectively. There is a possible error of as much as $\frac{1}{16}$ in. in each measurement. Use differentials to estimate the maximum resulting error that might occur in computing the volume of the cone.

36. The dimensions of a closed rectangular box are found by measurement to be 10 cm by 15 cm by 20 cm, but there is a possible error of 0.1 cm in each. Use differentials to estimate the maximum resulting error in computing the total surface area of the box.

37. A surveyor want to find the area in acres of a certain field (1 acre is 43560 ft^2). She measures two different sides, finding them to be $a = 500$ ft and $b = 700$ ft, with a possible error of as much as 1 ft in each measurement. She finds the angle between these two sides to be $\theta = 30°$, with a possible error of as much as 0.25°. The field is triangular, so its area is given by $A = \frac{1}{2}ab\sin\theta$. Use differentials to estimate the maximum resulting error, in acres, in computing the area of the field by this formula.

38. Use differentials to estimate the change in the volume of the gas of Example 3 if its pressure is decreased from 5 atm to 4.9 atm and its temperature is decreased from 300 K to 280 K.

39. The period of oscillation of a simple pendulum of length L is given (approximately) by the formula $T = 2\pi\sqrt{L/g}$. Estimate the change in the period of a pendulum if its length is increased from 2 ft to 2 ft 1 in. and it is simultaneously moved from a location where g is exactly 32 ft/s^2 to one where $g = 32.2$ ft/s^2.

40. Given the pendulum of Problem 39, show that the relative error in the determination of T is half the difference of the relative errors in measuring L and g—that is, that

$$\frac{dT}{T} = \frac{1}{2}\left(\frac{dL}{L} - \frac{dg}{g}\right).$$

41. The range of a projectile fired (in a vacuum) with initial velocity v_0 and inclination angle α from the horizontal is $R = \frac{1}{32}v_0^2 \sin 2\alpha$. Use differentials to approximate the change in range if v_0 is increased from 400 to 410 ft/s and α is increased from 30° to 31°.

42. A horizontal beam is supported at both ends and supports a uniform load. The deflection, or sag, at its midpoint is given by

$$S = \frac{k}{wh^3}, \tag{20}$$

where w and h are the width and height, respectively, of the beam and k is a constant that depends on the length and composition of the beam and the amount of the load. Show that

$$dS = -S\left(\frac{1}{w}\,dw + \frac{3}{h}\,dh\right).$$

If $S = 1$ in. when $w = 2$ in. and $h = 4$ in., approximate the sag when $w = 2.1$ in. and $h = 4.1$ in. Compare your approximation with the actual value you compute from Eq. (20).

43. Let the function f be defined on the whole xy-plane by $f(x, y) = 1$ if $x = y \neq 0$, whereas $f(x, y) = 0$ otherwise. (a) Show that f is not continuous at $(0, 0)$. (b) Show that both partial derivatives f_x and f_y exist at $(0, 0)$.

44. Show that the linear approximation theorem implies that the function $f(x, y)$ is differentiable at the point (a, b) if it is continuously differentiable near (a, b). (*Suggestion:* Show that Eq. (17) in the text follows from Eq. (15).)

45. Show that if the function $f(x, y)$ is differentiable at (a, b), then the partial derivatives $f_x(a, b)$ and $f_y(a, b)$ both exist and, moreover, are the coefficients of the linear function L in Eq. (17). (*Suggestion:* Let one of the differences Δx and Δy approach zero while the other *is* zero.)

46. Show that a function $f(x, y)$ of two variables is continuous wherever it is differentiable.

47. Show that the function $f(x, y) = \left(\sqrt[3]{x} + \sqrt[3]{y}\right)^3$ is continuous and has partial derivatives at the origin $(0, 0)$ but is not differentiable there.

48. Show that the function f defined by $f(x, y) = y^2 + x^2 \sin(1/x)$ for $x \neq 0$, and $f(0, y) = y^2$, is differentiable at $(0, 0)$, but is not continuously differentiable there because $f_x(x, y)$ is not continuous at $(0, 0)$.

13.7 THE CHAIN RULE

The single-variable chain rule expresses the derivative of a composite function $f(g(t))$ in terms of the derivatives of f and g:

$$D_t f(g(t)) = f'(g(t))\cdot g'(t). \tag{1}$$

With $w = f(x)$ and $x = g(t)$, the chain rule says that

$$\frac{dw}{dt} = \frac{dw}{dx}\,\frac{dx}{dt}. \tag{2}$$

The simplest multivariable chain rule situation involves a function $w = f(x, y)$ where both x and y are functions of the same single variable t: $x = g(t)$ and $y = h(t)$. The composite function $f(g(t), h(t))$ is then a single-variable function of t, and Theorem 1 expresses its derivative in terms of the partial derivatives of f and the ordinary derivatives of g and h. We assume that the stated hypotheses hold on suitable domains such that the composite function is defined.

Theorem 1 ***The Chain Rule***

Suppose that $w = f(x, y)$ has continuous first-order partial derivatives and that $x = g(t)$ and $y = h(t)$ are differentiable functions. Then w is a differentiable function of t, and

$$\frac{dw}{dt} = \frac{\partial w}{\partial x} \cdot \frac{dx}{dt} + \frac{\partial w}{\partial y} \cdot \frac{dy}{dt}. \tag{3}$$

The variable notation of Eq. (3) ordinarily will be more useful than function notation. Remember, in any case, that the partial derivatives in Eq. (3) are to be evaluated at the point $\big(g(t), h(t)\big)$, so in function notation Eq. (3) is

$$D_t\big[f\big(g(t), h(t)\big)\big] = f_x\big(g(t), h(t)\big) \cdot g'(t) + f_y\big(g(t), h(t)\big) \cdot h'(t). \tag{4}$$

A proof of the chain rule is included at the end of this section. In outline, it consists of beginning with the linear approximation

$$\Delta w \approx \frac{\partial w}{\partial x} \Delta x + \frac{\partial w}{\partial y} \Delta y$$

of Section 13.6 and dividing by Δt:

$$\frac{\Delta w}{\Delta t} \approx \frac{\partial w}{\partial x} \frac{\Delta x}{\Delta t} + \frac{\partial w}{\partial y} \frac{\Delta y}{\Delta t}.$$

Then we take the limit as $\Delta t \to 0$ to obtain

$$\frac{dw}{dt} = \frac{\partial w}{\partial x} \cdot \frac{dx}{dt} + \frac{\partial w}{\partial y} \cdot \frac{dy}{dt}.$$

EXAMPLE 1 Suppose that $w = e^{xy}$, $x = t^2$, and $y = t^3$. Then

$$\frac{\partial w}{\partial x} = ye^{xy}, \quad \frac{\partial w}{\partial y} = xe^{yx}, \quad \frac{dx}{dt} = 2t, \quad \text{and} \quad \frac{dy}{dt} = 3t^2.$$

So Eq. (3) yields

$$\begin{aligned} \frac{dw}{dt} &= \frac{\partial w}{\partial x} \cdot \frac{dx}{dt} + \frac{\partial w}{\partial y} \cdot \frac{dy}{dt} = (ye^{xy})(2t) + (xe^{xy})(3t^2) \\ &= \big(t^3 e^{t^5}\big)(2t) + \big(t^2 e^{t^5}\big)(3t^2) = 5t^4 e^{t^5}. \end{aligned}$$

■

Remark Had our purpose not been to illustrate the multivariable chain rule, we could have obtained the same result $dw/dt = 5t^4 \exp{(t^5)}$ more simply by writing

$$w = e^{xy} = e^{(t^2)(t^3)} = e^{t^5}$$

and then differentiating w as a single-variable function of t. But this single-variable approach is available only if the functions $x(t)$ and $y(t)$ are known explicitly. Sometimes, however, we know only the *numerical values* of x and y and/or their rates of change at a given instant. In such cases the multivariable chain rule in (3) can then be used to find the numerical rate of change of w at that instant.

Fig. 13.7.1 Warm sun melting a cylindrical block of ice

EXAMPLE 2 Figure 13.7.1 shows a melting cylindrical block of ice. Because of the sun's heat beating down from above, its height h is decreasing more rapidly than its radius r. If its height is decreasing at 3 cm/h and its radius is decreasing at 1 cm/h when $r = 15$ cm and $h = 40$ cm, what is the rate of change of the volume V of the block at that instant?

Solution With $V = \pi r^2 h$, the chain rule gives

$$\frac{dV}{dt} = \frac{\partial V}{\partial r}\frac{dr}{dt} + \frac{\partial V}{\partial h}\frac{dh}{dt} = 2\pi rh\frac{dr}{dt} + \pi r^2\frac{dh}{dt}.$$

Substituting the given numerical values $r = 15$, $h = 40$, $dr/dt = -1$, and $dh/dt = -3$, we find that

$$\frac{dV}{dt} = 2\pi(15)(40)(-1) + \pi(15)^2(-3) = -1875\pi \approx -5890.49 \quad (\text{cm}^3/\text{h}).$$

Thus the volume of the cylindrical block is decreasing at slightly less than 6 liters per hour at the given instant. ■

In the context of Theorem 1, we may refer to w as the **dependent variable**, x and y as **intermediate variables**, and t as the **independent variable**. Then note that the right-hand side of Eq. (3) has two terms, one for each intermediate variable, both terms like the right-hand side of the single-variable chain rule in Eq. (2). If there are more than two intermediate variables, then there is still one term on the right-hand side for each intermediate variable. For example, if $w = f(x, y, z)$ with x, y, and z each a function of t, then the chain rule takes the form

$$\frac{dw}{dt} = \frac{\partial w}{\partial x}\cdot\frac{dx}{dt} + \frac{\partial w}{\partial y}\cdot\frac{dy}{dt} + \frac{\partial w}{\partial z}\cdot\frac{dz}{dt}. \tag{5}$$

The derivation of Eq. (5) is essentially the same as the derivation of Eq. (3); it requires the linear approximation theorem for three variables rather than for two variables.

You may find it useful to envision the three types of variables—dependent, intermediate, and independent—as though they were lying at three different levels, as in Fig. 13.7.2, with the dependent variable at the top and the independent variable at the bottom. Each variable then depends (either directly or indirectly) on those that lie below it.

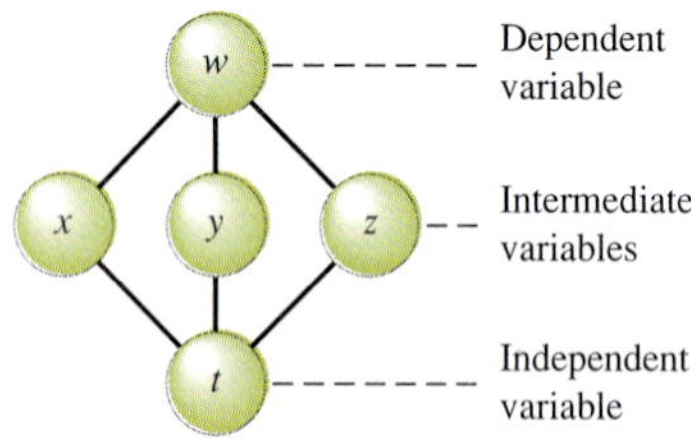

Fig. 13.7.2 Levels of chain rule variables

EXAMPLE 3 Find dw/dt if $w = x^2 + ze^y + \sin xz$ and $x = t$, $y = t^2$, $z = t^3$.

Solution Equation (5) gives

$$\begin{aligned}\frac{dw}{dt} &= \frac{\partial w}{\partial x}\cdot\frac{dx}{dt} + \frac{\partial w}{\partial y}\cdot\frac{dy}{dt} + \frac{\partial w}{\partial z}\cdot\frac{dz}{dt}\\ &= (2x + z\cos xz)(1) + (ze^y)(2t) + (e^y + x\cos xz)(3t^2)\\ &= 2t + (3t^2 + 2t^4)e^{t^2} + 4t^3\cos t^4.\end{aligned}$$ ■

In Example 3 we could check the result given by the chain rule by first writing w as an explicit function of t and then computing the ordinary single-variable derivative of w with respect to t.

Several Independent Variables

There may be several independent variables as well as several intermediate variables. For example, if $w = f(x, y, z)$ where $x = g(u, v)$, $y = h(u, v)$, and $z = k(u, v)$, so that

$$w = f(x, y, z) = f\big(g(u, v), h(u, v), k(u, v)\big),$$

then we have the three intermediate variables x, y, and z and the two independent variables u and v. In this case we would need to compute the *partial* derivatives $\partial w/\partial u$ and $\partial w/\partial v$ of the composite function. The general chain rule in Theorem 2 implies that each partial derivative of the dependent variable w is given by a chain rule formula such as Eq. (3) or (5). The only difference is that the derivatives with respect to the independent variables are partial derivatives. For instance,

$$\frac{\partial w}{\partial u} = \frac{\partial w}{\partial x}\cdot\frac{\partial x}{\partial u} + \frac{\partial w}{\partial y}\cdot\frac{\partial y}{\partial u} + \frac{\partial w}{\partial z}\cdot\frac{\partial z}{\partial u}.$$

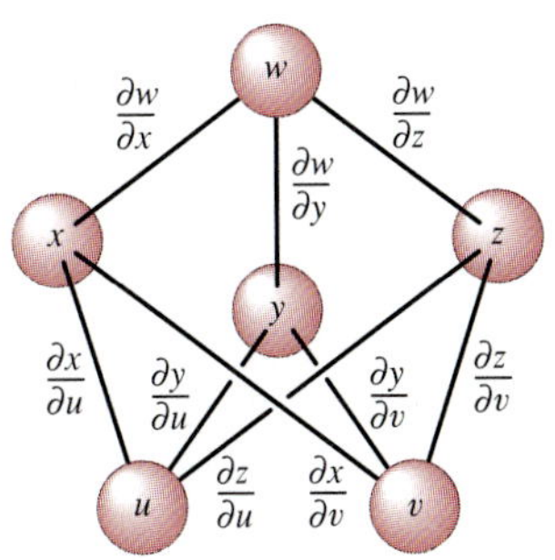

Fig. 13.7.3 Diagram for $w = w(x, y, z)$, where $x = x(u, v)$, $y = y(u, v)$, and $z = z(u, v)$

The "molecular model" in Fig. 13.7.3 illustrates this formula. The "atom" at the top represents the dependent variable w. The atoms at the next level represent the intermediate variables x, y, and z. The atoms at the bottom represent the independent variables u and v. Each "bond" in the model represents a partial derivative involving the two variables (the atoms joined by that bond). Finally, note that the preceding formula expresses $\partial w/\partial u$ as the sum of the products of the partial derivatives taken along all paths from w to u. Similarly, the sum of the products of the partial derivatives along all paths from w to v yields the correct formula

$$\frac{\partial w}{\partial v} = \frac{\partial w}{\partial x}\cdot\frac{\partial x}{\partial v} + \frac{\partial w}{\partial y}\cdot\frac{\partial y}{\partial v} + \frac{\partial w}{\partial z}\cdot\frac{\partial z}{\partial v}.$$

Theorem 2 describes the most general such situation.

Theorem 2 *The General Chain Rule*

Suppose that w is a function of the variables $x_1, x_2, \dots, x_m$ and that each of these variables is a function of the variables $t_1, t_2, \dots, t_n$. If all these functions have continuous first-order partial derivatives, then

$$\frac{\partial w}{\partial t_i} = \frac{\partial w}{\partial x_1}\cdot\frac{\partial x_1}{\partial t_i} + \frac{\partial w}{\partial x_2}\cdot\frac{\partial x_2}{\partial t_i} + \cdots + \frac{\partial w}{\partial x_m}\cdot\frac{\partial x_m}{\partial t_i} \tag{6}$$

for each i, $1 \leqq i \leqq n$.

Thus there is a formula in Eq. (6) for *each* of the independent variables $t_1, t_2, \dots, t_n$, and the right-hand side of each such formula contains one typical chain rule term for each of the intermediate variables $x_1, x_2, \dots, x_m$.

EXAMPLE 4 Suppose that

$$z = f(u, v), \qquad u = 2x + y, \qquad v = 3x - 2y.$$

Given the values $\partial z/\partial u = 3$ and $\partial z/\partial v = -2$ at the point $(u, v) = (3, 1)$, find the values $\partial z/\partial x$ and $\partial z/\partial y$ at the corresponding point $(x, y) = (1, 1)$.

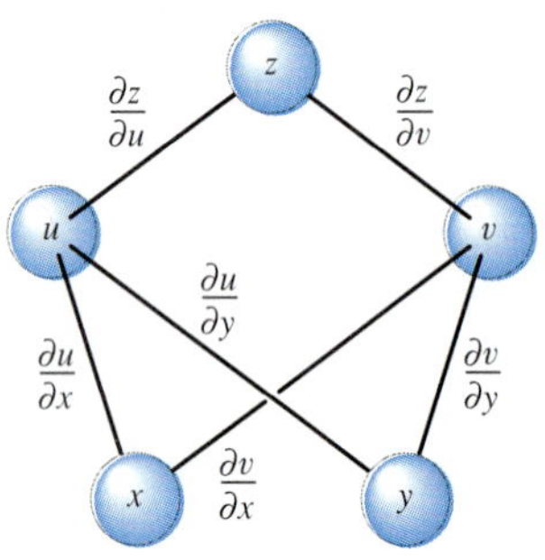

Fig. 13.7.4 Diagram for $z = z(u, v)$, where $u = u(x, y)$ and $v = v(x, y)$ (Example 4)

Solution The relationships among the variables are shown in Fig. 13.7.4. The chain rule gives

$$\frac{\partial z}{\partial x} = \frac{\partial z}{\partial u}\cdot\frac{\partial u}{\partial x} + \frac{\partial z}{\partial v}\cdot\frac{\partial v}{\partial x} = 3\cdot 2 + (-2)\cdot 3 = 0$$

and

$$\frac{\partial z}{\partial y} = \frac{\partial z}{\partial u}\cdot\frac{\partial u}{\partial y} + \frac{\partial z}{\partial v}\cdot\frac{\partial v}{\partial y} = 3\cdot 1 + (-2)\cdot(-2) = 7$$

at the indicated point$(x, y) = (1, 1)$. ■

EXAMPLE 5 Let $w = f(x, y)$ where x and y are given in polar coordinates by the equations $x = r\cos\theta$ and $y = r\sin\theta$. Calculate

$$\frac{\partial w}{\partial r}, \quad \frac{\partial w}{\partial \theta}, \quad \text{and} \quad \frac{\partial^2 w}{\partial r^2}$$

in terms of r, θ, and the partial derivatives of w with respect to x and y (Fig. 13.7.5).

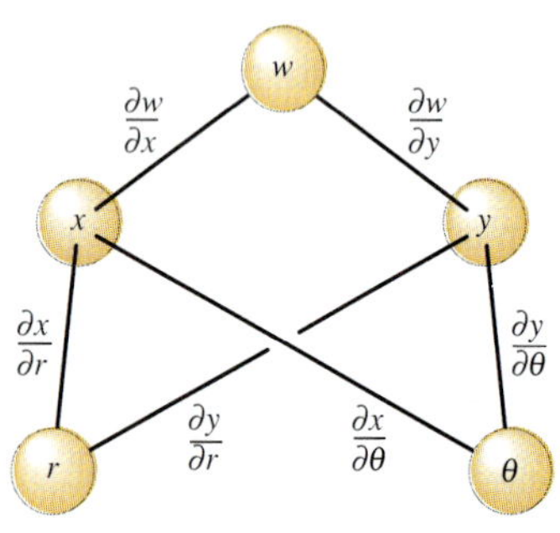

Fig. 13.7.5 Diagram for $w = w(x, y)$, where $x = x(r, \theta)$ and $y = y(r, \theta)$ (Example 5)

Solution Here x and y are intermediate variables; the independent variables are r and θ. First note that

$$\frac{\partial x}{\partial r} = \cos\theta, \quad \frac{\partial y}{\partial r} = \sin\theta, \quad \frac{\partial x}{\partial \theta} = -r\sin\theta, \quad \text{and} \quad \frac{\partial y}{\partial \theta} = r\cos\theta.$$

Then

$$\frac{\partial w}{\partial r} = \frac{\partial w}{\partial x}\cdot\frac{\partial x}{\partial r} + \frac{\partial w}{\partial y}\cdot\frac{\partial y}{\partial r} = \frac{\partial w}{\partial x}\cos\theta + \frac{\partial w}{\partial y}\sin\theta \tag{7a}$$

and

$$\frac{\partial w}{\partial \theta} = \frac{\partial w}{\partial x}\cdot\frac{\partial x}{\partial \theta} + \frac{\partial w}{\partial y}\cdot\frac{\partial y}{\partial \theta} = -r\frac{\partial w}{\partial x}\sin\theta + r\frac{\partial w}{\partial y}\cos\theta. \tag{7b}$$

Next,

$$\begin{aligned}\frac{\partial^2 w}{\partial r^2} &= \frac{\partial}{\partial r}\left(\frac{\partial w}{\partial r}\right) = \frac{\partial}{\partial r}\left(\frac{\partial w}{\partial x}\cos\theta + \frac{\partial w}{\partial y}\sin\theta\right) \\ &= \frac{\partial w_x}{\partial r}\cos\theta + \frac{\partial w_y}{\partial r}\sin\theta,\end{aligned}$$

where $w_x = \partial w/\partial x$ and $w_y = \partial w/\partial y$. We apply Eq. (7a) to calculate $\partial w_x/\partial r$ and $\partial w_y/\partial r$, and we obtain

$$\begin{aligned}\frac{\partial^2 w}{\partial r^2} &= \left(\frac{\partial w_x}{\partial x}\cdot\frac{\partial x}{\partial r} + \frac{\partial w_x}{\partial y}\cdot\frac{\partial y}{\partial r}\right)\cos\theta + \left(\frac{\partial w_y}{\partial x}\cdot\frac{\partial x}{\partial r} + \frac{\partial w_y}{\partial y}\cdot\frac{\partial y}{\partial r}\right)\sin\theta \\ &= \left(\frac{\partial^2 w}{\partial x^2}\cos\theta + \frac{\partial^2 w}{\partial y\,\partial x}\sin\theta\right)\cos\theta + \left(\frac{\partial^2 w}{\partial x\,\partial y}\cos\theta + \frac{\partial^2 w}{\partial y^2}\sin\theta\right)\sin\theta.\end{aligned}$$

Finally, because $w_{yx} = w_{xy}$, we get

$$\frac{\partial^2 w}{\partial r^2} = \frac{\partial^2 w}{\partial x^2}\cos^2\theta + 2\frac{\partial^2 w}{\partial x\,\partial y}\cos\theta\sin\theta + \frac{\partial^2 w}{\partial y^2}\sin^2\theta. \tag{8}$$

■

EXAMPLE 6 Suppose that $w = f(u, v, x, y)$, where u and v are functions of x and y. Here x and y play dual roles as intermediate and independent variables. The chain rule yields

$$\begin{aligned}\frac{\partial w}{\partial x} &= \frac{\partial f}{\partial u}\cdot\frac{\partial u}{\partial x} + \frac{\partial f}{\partial v}\cdot\frac{\partial v}{\partial x} + \frac{\partial f}{\partial x}\cdot\frac{\partial x}{\partial x} + \frac{\partial f}{\partial y}\cdot\frac{\partial y}{\partial x}\\ &= \frac{\partial f}{\partial u}\cdot\frac{\partial u}{\partial x} + \frac{\partial f}{\partial v}\cdot\frac{\partial v}{\partial x} + \frac{\partial f}{\partial x},\end{aligned}$$

because $\partial x/\partial x = 1$ and $\partial y/\partial x = 0$. Similarly,

$$\frac{\partial w}{\partial y} = \frac{\partial f}{\partial u}\cdot\frac{\partial u}{\partial y} + \frac{\partial f}{\partial v}\cdot\frac{\partial v}{\partial y} + \frac{\partial f}{\partial y}.$$

These results are consistent with the paths from w to x and from w to y in the molecular model shown in Fig. 13.7.6. ■

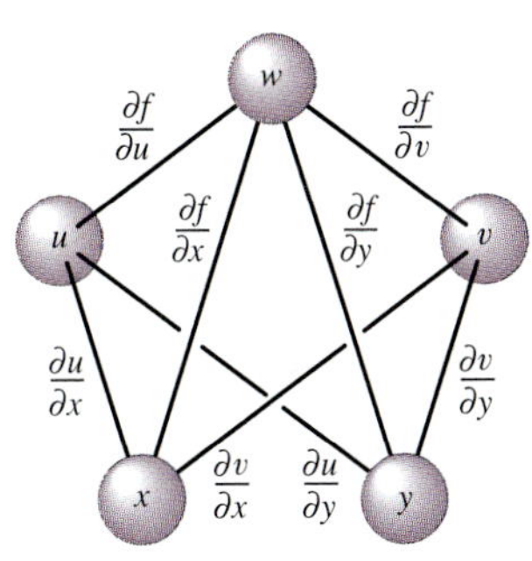

Fig. 13.7.6 Diagram for $w = f(u, v, x, y)$, where $u = u(x, y)$ and $v = v(x, y)$ (Example 6)

EXAMPLE 7 Consider a parametric curve $x = x(t)$, $y = y(t)$, $z = z(t)$ that lies on the surface $z = f(x, y)$ in space. Recall that if

$$\mathbf{T} = \left\langle \frac{dx}{dt}, \frac{dy}{dt}, \frac{dz}{dt}\right\rangle \quad\text{and}\quad \mathbf{N} = \left\langle \frac{\partial z}{\partial x}, \frac{\partial z}{\partial y}, -1\right\rangle,$$

then $\mathbf{T}$ is tangent to the curve and $\mathbf{N}$ is normal to the surface. Show that $\mathbf{T}$ and $\mathbf{N}$ are everywhere perpendicular.

Solution The chain rule in Eq. (3) tells us that

$$\frac{dz}{dt} = \frac{\partial z}{\partial x}\cdot\frac{dx}{dt} + \frac{\partial z}{\partial y}\cdot\frac{dy}{dt}.$$

But this equation is equivalent to the vector equation

$$\left\langle \frac{\partial z}{\partial x}, \frac{\partial z}{\partial y}, -1\right\rangle\cdot\left\langle \frac{dx}{dt}, \frac{dy}{dt}, \frac{dz}{dt}\right\rangle = 0.$$

Thus $\mathbf{N}\cdot\mathbf{T} = 0$, so $\mathbf{N}$ and $\mathbf{T}$ are perpendicular. ■

Theorem 3 ***Implicit Partial Differentiation***

Suppose that the function $F(x, y, z)$ has continuous first-order partial derivatives and that the equation $F(x, y, z) = 0$ implicitly defines a function $z = f(x, y)$ that has continuous first-order partial derivatives. Then

$$\frac{\partial z}{\partial x} = -\frac{F_x}{F_z} \quad\text{and}\quad \frac{\partial z}{\partial y} = -\frac{F_y}{F_z} \tag{9}$$

wherever $F_z = \partial F/\partial z \neq 0$.

Proof Because $w = F\big(x, y, f(x, y)\big)$ is identically zero, differentiation with respect to x yields

$$0 = \frac{\partial w}{\partial x} = \frac{\partial F}{\partial x}\cdot\frac{\partial x}{\partial x} + \frac{\partial F}{\partial y}\cdot\frac{\partial y}{\partial x} + \frac{\partial F}{\partial z}\cdot\frac{\partial z}{\partial x}$$

$$= 1\cdot F_x + 0\cdot F_y + \frac{\partial z}{\partial x}\cdot F_z,$$

so

$$F_x + \frac{\partial z}{\partial x}\cdot F_z = 0.$$

The first formula in (9) now follows. The second is obtained similarly by differentiating w with respect to y. ■

Remark In a specific example it is usually simpler to differentiate the equation

$$F\big(x, y, f(x, y)\big) = 0$$

implicitly than to apply the formulas in (9).

EXAMPLE 8 Find the plane tangent at the point $(1, 3, 2)$ to the surface with equation

$$z^3 + xz - y^2 = 1.$$

Solution Implicit partial differentiation of the given equation with respect to x and with respect to y yields the equations

$$3z^2\frac{\partial z}{\partial x} + z + x\frac{\partial z}{\partial x} = 0 \quad\text{and}\quad 3z^2\frac{\partial z}{\partial y} + x\frac{\partial z}{\partial y} - 2y = 0.$$

When we substitute $x = 1$, $y = 3$, and $z = 2$, we find that $\partial z/\partial x = -\frac{2}{13}$ and $\partial z/\partial y = \frac{6}{13}$. Hence an equation of the tangent plane in question is

$$z - 2 = -\tfrac{2}{13}(x - 1) + \tfrac{6}{13}(y - 3);\ \text{that is,}\ 2x - 6y + 13z = 10.$$ ■

Matrix Form of the Chain Rule

The case $m = n = 2$ of the chain rule corresponds to the case of two intermediate variables (x and y, say) that are functions of two independent variables (u and v, say),

$$x = f(u, v), \qquad y = g(u, v). \tag{10}$$

These functions describe a **transformation** $T{:}R^2_{uv} \to R^2_{xy}$ from the coordinate plane R^2_{uv} of (u, v)-pairs to the coordinate plane R^2_{xy} of (x, y)-pairs. The **image** of the point (u, v) of R^2_{uv} is the point $T(u, v) = \big(f(u, v), g(u, v)\big) = (x, y)$ of R^2_{xy}. The **derivative matrix** of the transformation T at the point (u, v) is then the 2×2 array

$$T'(u, v) = \begin{bmatrix} \dfrac{\partial x}{\partial u} & \dfrac{\partial x}{\partial v} \\[2ex] \dfrac{\partial y}{\partial u} & \dfrac{\partial y}{\partial v} \end{bmatrix} \tag{11}$$

of partial derivatives of the component functions in (10) of the transformation T (all evaluated at the point (u, v)).

EXAMPLE 9 The polar coordinate transformation $T : R^2_{r\theta} \to R^2_{xy}$ is defined by the familiar equations

$$x = r\cos\theta, \qquad y = r\sin\theta. \tag{12}$$

Its derivative matrix is given by

$$T'(r,\theta) = \begin{bmatrix} \dfrac{\partial x}{\partial r} & \dfrac{\partial x}{\partial \theta} \\ \dfrac{\partial y}{\partial r} & \dfrac{\partial y}{\partial \theta} \end{bmatrix} = \begin{bmatrix} \cos\theta & -r\sin\theta \\ \sin\theta & r\cos\theta \end{bmatrix}. \tag{13}$$

Now suppose that the dependent variable w is a function $F(x, y)$ of the intermediate variables x and y, and thereby is the composite function

$$G(u,v) = F(T(u,v)) = F(x(u,v), y(u,v)) \tag{14}$$

of the independent variables u and v (Fig. 13.7.7). The derivative matrices

$$F'(x,y) = \begin{bmatrix} \dfrac{\partial w}{\partial x} & \dfrac{\partial w}{\partial y} \end{bmatrix} \quad \text{and} \quad G'(u,v) = \begin{bmatrix} \dfrac{\partial w}{\partial u} & \dfrac{\partial w}{\partial v} \end{bmatrix} \tag{15}$$

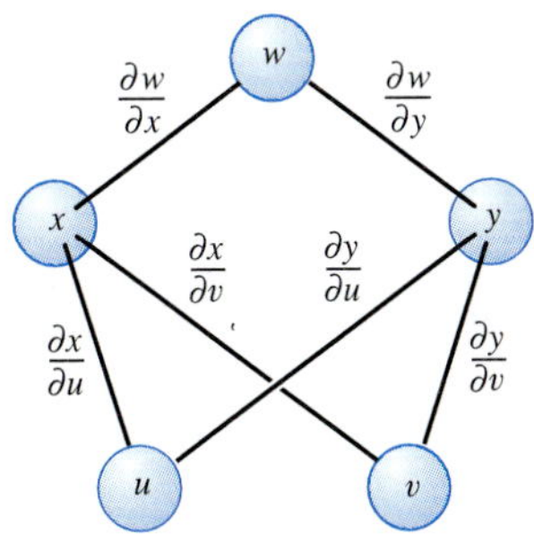

Fig. 13.7.7 Diagram for $w = F(x, y)$ where $(x, y) = \big(x(u, v), y(u, v)\big) = T(u, v)$

of F and G are defined in analogy with (11)—there being a single row in each matrix, corresponding to the single dependent variable w. Those who are familiar with matrix multiplication will recognize that the two chain rule formulas

$$\frac{\partial w}{\partial u} = \frac{\partial w}{\partial x}\frac{\partial x}{\partial u} + \frac{\partial w}{\partial y}\frac{\partial y}{\partial u}, \qquad \frac{\partial w}{\partial v} = \frac{\partial w}{\partial x}\frac{\partial x}{\partial v} + \frac{\partial w}{\partial y}\frac{\partial y}{\partial v}$$

are the "components" of the single matrix equation

$$G'(u,v) = F'(x,y)\,T'(u,v); \tag{16a}$$

that is,

$$\begin{bmatrix} \dfrac{\partial w}{\partial u} & \dfrac{\partial w}{\partial v} \end{bmatrix} = \begin{bmatrix} \dfrac{\partial w}{\partial x} & \dfrac{\partial w}{\partial y} \end{bmatrix} \begin{bmatrix} \dfrac{\partial x}{\partial u} & \dfrac{\partial x}{\partial v} \\ \dfrac{\partial y}{\partial u} & \dfrac{\partial y}{\partial v} \end{bmatrix}. \tag{16b}$$

Thus the chain rule for the situation indicated in Fig. 13.7.7 implies that *the derivative matrix of the composite function $G = F \circ T$ is the matrix product $G' = F'T'$.*

EXAMPLE 10 With the polar-coordinate derivative matrix $T'(r, \theta)$ in (13), the matrix multiplication in Eq. (16b) yields

$$\begin{bmatrix} \dfrac{\partial w}{\partial r} & \dfrac{\partial w}{\partial \theta} \end{bmatrix} = \begin{bmatrix} \dfrac{\partial w}{\partial x} & \dfrac{\partial w}{\partial y} \end{bmatrix} \begin{bmatrix} \cos\theta & -r\sin\theta \\ \sin\theta & r\cos\theta \end{bmatrix}$$

$$= \begin{bmatrix} \dfrac{\partial w}{\partial x}\cos\theta + \dfrac{\partial w}{\partial y}\sin\theta & \quad -r\dfrac{\partial w}{\partial x}\sin\theta + r\dfrac{\partial w}{\partial y}\cos\theta \end{bmatrix}.$$

The components of this matrix equation are the scalar chain rule formulas

$$\frac{\partial w}{\partial r} = \frac{\partial w}{\partial x}\cos\theta + \frac{\partial w}{\partial y}\sin\theta, \qquad \frac{\partial w}{\partial \theta} = -r\frac{\partial w}{\partial x}\sin\theta + r\frac{\partial w}{\partial y}\cos\theta$$

that we saw previously in Example 5.

We have discussed here the 2×2 case of a general $m \times n$ matrix formulation of the multivariable chain rule. The 3×3 case and its application to spherical coordinates are discussed in Problems 58 through 61.

Proof of the Chain Rule

Given that $w = f(x, y)$ satisfies the hypotheses of Theorem 1, we choose a point t_0 at which we wish to compute dw/dt and write

$$a = g(t_0), \qquad b = h(t_0).$$

Let

$$\Delta x = g(t_0 + \Delta t) - g(t_0), \qquad \Delta y = h(t_0 + \Delta t) - h(t_0).$$

Then

$$g(t_0 + \Delta t) = a + \Delta x \quad \text{and} \quad h(t_0 + \Delta t) = b + \Delta y.$$

If

$$\begin{aligned}\Delta w &= f\big(g(t_0 + \Delta t), h(t_0 + \Delta t)\big) - f\big(g(t_0), h(t_0)\big)\\ &= f(a + \Delta x, b + \Delta y) - f(a, b),\end{aligned}$$

then what we need to compute is

$$\frac{dw}{dt} = \lim_{\Delta t \to 0} \frac{\Delta w}{\Delta t}.$$

The linear approximation theorem of Section 13.6 gives

$$\Delta w = f_x(a, b)\,\Delta x + f_y(a, b)\,\Delta y + \epsilon_1\,\Delta x + \epsilon_2\,\Delta y,$$

where ϵ_1 and ϵ_2 approach zero as $\Delta x \to 0$ and $\Delta y \to 0$. We note that both Δx and Δy approach zero as $\Delta t \to 0$, because both the derivatives

$$\frac{dx}{dt} = \lim_{\Delta t \to 0} \frac{\Delta x}{\Delta t} \quad \text{and} \quad \frac{dy}{dt} = \lim_{\Delta t \to 0} \frac{\Delta y}{\Delta t}$$

exist. Therefore,

$$\begin{aligned}\frac{dw}{dt} = \lim_{\Delta t \to 0} \frac{\Delta w}{\Delta t} &= \lim_{\Delta t \to 0} \left[f_x(a, b)\frac{\Delta x}{\Delta t} + f_y(a, b)\frac{\Delta y}{\Delta t} + \epsilon_1 \frac{\Delta x}{\Delta t} + \epsilon_2 \frac{\Delta y}{\Delta t} \right]\\ &= f_x(a, b)\frac{dx}{dt} + f_y(a, b)\frac{dy}{dt} + 0 \cdot \frac{dx}{dt} + 0 \cdot \frac{dy}{dt}.\end{aligned}$$

Hence

$$\frac{dw}{dt} = \frac{\partial w}{\partial x} \cdot \frac{dx}{dt} + \frac{\partial w}{\partial y} \cdot \frac{dy}{dt}.$$

Thus we have established Eq. (3), writing $\partial w/\partial x$ and $\partial w/\partial y$ for the partial derivatives $f_x(a, b)$ and $f_y(a, b)$ in the final step.

13.7 PROBLEMS

In Problems 1 through 4, find dw/dt both by using the chain rule *and* by expressing w explicitly as a function of t before differentiating.

1. $w = \exp(-x^2 - y^2);\quad x = t, y = \sqrt{t}$

2. $w = \dfrac{1}{u^2 + v^2};\quad u = \cos 2t, v = \sin 2t$

3. $w = \sin xyz;\quad x = t, y = t^2, z = t^3$

4. $w = \ln(u + v + z);\quad u = \cos^2 t, v = \sin^2 t, z = t^2$

In Problems 5 through 8, find $\partial w/\partial s$ and $\partial w/\partial t$.

5. $w = \ln(x^2 + y^2 + z^2);\quad x = s - t, y = s + t, z = 2\sqrt{st}$

6. $w = pq \sin r;\quad p = 2s + t, q = s - t, r = st$

7. $w = \sqrt{u^2 + v^2 + z^2};\quad u = 3e^t \sin s, v = 3e^t \cos s, z = 4e^t$

8. $w = yz + zx + xy;\quad x = s^2 - t^2, y = s^2 + t^2, z = s^2t^2$

In Problems 9 through 12, find $\partial r/\partial x$, $\partial r/\partial y$, and $\partial r/\partial z$.

9. $r = e^{u+v+w};\quad u = yz,\ v = xz,\ w = xy$

10. $r = uvw - u^2 - v^2 - w^2;\quad u = y + z, v = x + z, w = x + y$

11. $r = \sin(p/q);\quad p = \sqrt{xy^2z^3}, q = \sqrt{x + 2y + 3z}$

12. $r = \dfrac{p}{q} + \dfrac{q}{s} + \dfrac{s}{p};\quad p = e^{yz},\ q = e^{xz},\ s = e^{xy}$

In Problems 13 through 18, write chain rule formulas giving the partial derivative of the dependent variable p with respect to each independent variable.

13. $p = f(x, y);\quad x = x(u, v, w), y = y(u, v, w)$

14. $p = f(x, y, z);\quad x = x(u, v), y = y(u, v), z = z(u, v)$

15. $p = f(u, v, w);\quad u = u(x, y, z), v = v(x, y, z), w = w(x, y, z)$

16. $p = f(v, w);\quad v = v(x, y, z, t), w = w(x, y, z, t)$

17. $p = f(w);\quad w = w(x, y, z, u, v)$

18. $p = f(x, y, u, v);\quad x = x(s, t), y = y(s, t), u = u(s, t), v = v(s, t)$

In Problems 19 through 24, find $\partial z/\partial x$ and $\partial z/\partial y$ as functions of x, y, and z, assuming that $z = f(x, y)$ satisfies the given equation.

19. $x^{2/3} + y^{2/3} + z^{2/3} = 1$

20. $x^3 + y^3 + z^3 = xyz$

21. $xe^{xy} + ye^{zx} + ze^{xy} = 3$

22. $x^5 + xy^2 + yz = 5$

23. $\dfrac{x^2}{a^2} + \dfrac{y^2}{b^2} + \dfrac{z^2}{c^2} = 1$

24. $xyz = \sin(x + y + z)$

In Problems 25 through 28, use the method of Example 6 to find $\partial w/\partial x$ and $\partial w/\partial y$ as functions of x and y.

25. $w = u^2 + v^2 + x^2 + y^2;\quad u = x - y, v = x + y$

26. $w = \sqrt{uvxy};\quad u = \sqrt{x - y}, v = \sqrt{x + y}$

27. $w = xy \ln(u + v);\quad u = (x^2 + y^2)^{1/3}, v = (x^3 + y^3)^{1/2}$

28. $w = uv - xy;\quad u = \dfrac{x}{x^2 + y^2}, v = \dfrac{y}{x^2 + y^2}$

In Problems 29 through 32, write an equation for the plane tangent at the point P to the surface with the given equation.

29. $x^2 + y^2 + z^2 = 9;\quad P(1, 2, 2)$

30. $x^2 + 2y^2 + 2z^2 = 14;\quad P(2, 1, -2)$

31. $x^3 + y^3 + z^3 = 5xyz;\quad P(2, 1, 1)$

32. $z^3 + (x + y)z^2 + x^2 + y^2 = 13;\quad P(2, 2, 1)$

33. The sun is melting a rectangular block of ice. When the block's height is 1 ft and the edge of its square base is 2 ft, its height is decreasing at 2 in./h and its base edge is decreasing at 3 in./h. What is the block's rate of change of volume V at that instant?

34. A rectangular box has a square base. Find the rate at which its volume and surface area are changing if its base edge is increasing at 2 cm/min and its height is decreasing at 3 cm/min at the instant when each dimension is 1 meter.

35. Falling sand forms a conical sandpile. When the sandpile has a height of 5 ft and its base radius is 2 ft, its height is increasing at 0.4 ft/min and its base radius is increasing at 0.7 ft/min. At what rate is the volume of the sandpile increasing at that moment?

36. A rectangular block has dimensions $x = 3$ m, $y = 2$ m, and $z = 1$ m. If x and y are increasing at 1 cm/min and 2 cm/min, respectively, while z is decreasing at 2 cm/min, are the block's volume and total surface area increasing or are they decreasing? At what rates?

37. The volume V (in cubic centimeters) and pressure p (in atmospheres) of n moles of an ideal gas satisfy the equation $pV = nRT$, where T is its temperature (in degrees Kelvin) and $R = 82.06$. Suppose that a sample of the gas has a volume of 10 L when the pressure is 2 atm and the temperature is 300°K. If the pressure is increasing at 1 atm/min and the temperature is increasing at 10°K/min, is the volume of the gas sample increasing or is it decreasing? At what rate?

38. The aggregate resistance R of three variable resistances R_1, R_2, and R_3 connected in parallel satisfies the *harmonic equation*

$$\frac{1}{R} = \frac{1}{R_1} + \frac{1}{R_2} + \frac{1}{R_3}.$$

Suppose that R_1 and R_2 are 100 Ω and are increasing at 1 Ω/s, while R_3 is 200 Ω and is decreasing at 2 Ω/s. Is R increasing or decreasing at that instant? At what rate?

39. Suppose that $x = h(y, z)$ satisfies the equation $F(x, y, z) = 0$ and that $F_x \neq 0$. Show that

$$\frac{\partial x}{\partial y} = -\frac{\partial F/\partial y}{\partial F/\partial x}.$$

40. Suppose that $w = f(x, y)$, $x = r\cos\theta$, and $y = r\sin\theta$. Show that

$$\left(\frac{\partial w}{\partial x}\right)^2 + \left(\frac{\partial w}{\partial y}\right)^2 = \left(\frac{\partial w}{\partial r}\right)^2 + \frac{1}{r^2}\left(\frac{\partial w}{\partial \theta}\right)^2.$$

41. Suppose that $w = f(u)$ and that $u = x + y$. Show that $\partial w/\partial x = \partial w/\partial y$.

42. Suppose that $w = f(u)$ and that $u = x - y$. Show that $\partial w/\partial x = -\partial w/\partial y$ and that

$$\frac{\partial^2 w}{\partial x^2} = \frac{\partial^2 w}{\partial y^2} = -\frac{\partial^2 w}{\partial x\,\partial y}.$$

43. Suppose that $w = f(x, y)$ where $x = u + v$ and $y = u - v$. Show that

$$\frac{\partial^2 w}{\partial x^2} - \frac{\partial^2 w}{\partial y^2} = \frac{\partial^2 w}{\partial u\,\partial v}.$$

44. Assume that $w = f(x, y)$ where $x = 2u + v$ and $y = u - v$. Show that

$$5\frac{\partial^2 w}{\partial x^2} + 2\frac{\partial^2 w}{\partial x\,\partial y} + 2\frac{\partial^2 w}{\partial y^2} = \frac{\partial^2 w}{\partial u^2} + \frac{\partial^2 w}{\partial v^2}.$$

45. Suppose that $w = f(x, y)$, $x = r\cos\theta$, and $y = r\sin\theta$. Show that

$$\frac{\partial^2 w}{\partial x^2} + \frac{\partial^2 w}{\partial y^2} = \frac{\partial^2 w}{\partial r^2} + \frac{1}{r}\frac{\partial w}{\partial r} + \frac{1}{r^2}\frac{\partial^2 w}{\partial \theta^2}.$$

(*Suggestion:* First find $\partial^2 w/\partial\theta^2$ by the method of Example 5. Then combine the result with Eqs. (7) and (8).)

46. Suppose that

$$w = \frac{1}{r}f\left(t - \frac{r}{a}\right)$$

and that $r = \sqrt{x^2 + y^2 + z^2}$. Show that

$$\frac{\partial^2 w}{\partial x^2} + \frac{\partial^2 w}{\partial y^2} + \frac{\partial^2 w}{\partial z^2} = \frac{1}{a^2}\frac{\partial^2 w}{\partial t^2}.$$

47. Suppose that $w = f(r)$ and that $r = \sqrt{x^2 + y^2 + z^2}$. Show that

$$\frac{\partial^2 w}{\partial x^2} + \frac{\partial^2 w}{\partial y^2} + \frac{\partial^2 w}{\partial z^2} = \frac{d^2 w}{dr^2} + \frac{2}{r}\frac{dw}{dr}.$$

48. Suppose that $w = f(u) + g(v)$, that $u = x - at$, and that $v = x + at$. Show that

$$\frac{\partial^2 w}{\partial t^2} = a^2\frac{\partial^2 w}{\partial x^2}.$$

49. Assume that $w = f(u, v)$ where $u = x + y$ and $v = x - y$. Show that

$$\frac{\partial w}{\partial x}\frac{\partial w}{\partial y} = \left(\frac{\partial w}{\partial u}\right)^2 - \left(\frac{\partial w}{\partial v}\right)^2.$$

50. Given: $w = f(x, y)$, $x = e^u\cos v$, and $y = e^u\sin v$. Show that

$$\left(\frac{\partial w}{\partial x}\right)^2 + \left(\frac{\partial w}{\partial y}\right)^2 = e^{-2u}\left[\left(\frac{\partial w}{\partial u}\right)^2 + \left(\frac{\partial w}{\partial v}\right)^2\right].$$

51. Assume that $w = f(x, y)$ and that there is a constant α such that

$$x = u\cos\alpha - v\sin\alpha \quad \text{and} \quad y = u\sin\alpha + v\cos\alpha.$$

Show that

$$\left(\frac{\partial w}{\partial u}\right)^2 + \left(\frac{\partial w}{\partial v}\right)^2 = \left(\frac{\partial w}{\partial x}\right)^2 + \left(\frac{\partial w}{\partial y}\right)^2.$$

52. Suppose that $w = f(u)$, where

$$u = \frac{x^2 - y^2}{x^2 + y^2}.$$

Show that $xw_x + yw_y = 0$.

Suppose that the equation $F(x, y, z) = 0$ defines implicitly the three functions $z = f(x, y)$, $y = g(x, z)$, and $x = h(y, z)$. To keep track of the various partial derivatives, we use the notation

$$\left(\frac{\partial z}{\partial x}\right)_y = \frac{\partial f}{\partial x}, \qquad \left(\frac{\partial z}{\partial y}\right)_x = \frac{\partial f}{\partial y}, \qquad \textbf{(17a)}$$

$$\left(\frac{\partial y}{\partial x}\right)_z = \frac{\partial g}{\partial x}, \qquad \left(\frac{\partial y}{\partial z}\right)_x = \frac{\partial g}{\partial z}, \qquad \textbf{(17b)}$$

$$\left(\frac{\partial x}{\partial y}\right)_z = \frac{\partial h}{\partial y}, \qquad \left(\frac{\partial x}{\partial z}\right)_y = \frac{\partial h}{\partial z}, \qquad \textbf{(17c)}$$

In short, the general symbol $(\partial w/\partial u)_v$ denotes the derivative of w with respect to u, where w is regarded as a function of the independent variables u and v.

53. Using the notation in the equations in (17), show that

$$\left(\frac{\partial x}{\partial y}\right)_z\left(\frac{\partial y}{\partial z}\right)_x\left(\frac{\partial z}{\partial x}\right)_y = -1$$

[*Suggestion:* Find the three partial derivatives on the right-hand side in (17) in terms of F_x, F_y, and F_z.]

54. Verify the result of Problem 53 for the equation

$$F(x, y, z) = x^2 + y^2 + z^2 - 1 = 0.$$

55. Verify the result of Problem 53 (with p, V, and T in place of x, y, and z) for the equation

$$F(p, V, T) = pV - nRT = 0$$

(n and R are constants), which expresses the ideal gas law.

56. Consider a given quantity of liquid whose pressure p, volume V, and temperature T satisfy a given "state equation" of the form $F(p, V, T) = 0$. The **thermal expansivity**

α and **isothermal compressivity** β of the liquid are defined by

$$\alpha = \frac{1}{V}\frac{\partial V}{\partial T} \quad \text{and} \quad \beta = -\frac{1}{V}\frac{\partial V}{\partial p}.$$

Apply Theorem 3 first to calculate $\partial V/\partial p$ and $\partial V/\partial T$, and then to calculate $\partial p/\partial V$ and $\partial p/\partial T$. Deduce from the results that $\partial p/\partial T = \alpha/\beta$.

57. The thermal expansivity and isothermal compressivity of liquid mercury are $\alpha = 1.8 \times 10^{-4}$ and $\beta = 3.9 \times 10^{-6}$, respectively, in L-atm-°C units. Suppose that a thermometer bulb is exactly filled with mercury at 50°C. If the bulb can withstand an internal pressure of no more than 200 atm, can it be heated to 55°C without breaking? (*Suggestion:* Apply the result of Problem 56 to calculate the increase in pressure with each increase of one degree in temperature.)

58. Suppose that the transformation $T: R^3_{uvw} \to R^3_{xyz}$ is defined by the functions $x = x(u, v, w)$, $y = y(u, v, w)$, $z = z(u, v, w)$. Then its derivative matrix is defined by

$$T'(u, v, w) = \begin{bmatrix} x_u & x_v & x_w \\ y_u & y_v & y_w \\ z_u & z_v & z_w \end{bmatrix}.$$

Calculate the derivative matrix of the linear transformation defined by $x = a_1u + b_1v + c_1w$, $y = a_2u + b_2v + c_2w$, $z = a_3u + b_3v + c_3w$.

59. Calculate the derivative matrix of the spherical coordinate transformation T defined by $x = \rho \sin\phi \cos\theta$, $y = \rho \sin\phi \sin\theta$, $z = \rho \cos\phi$.

60. Suppose that $q = F(x, y, z)$ with 1×3 derivative matrix $F' = [F_x \;\; F_y \;\; F_z]$ and that $(x, y, z) = T(u, v, w)$ as in Prob-lem 58. If $G = F \circ T$, deduce from the chain rule in Theorem 2 that $G' = F'T'$ (matrix product).

61. If $q = F(x, y, z)$, apply the results of Problems 59 and 60 to calculate the partial derivatives of q with respect to the spherical coordinates ρ, ϕ, θ by matrix multiplication.

13.8 DIRECTIONAL DERIVATIVES AND THE GRADIENT VECTOR

The change in the value of the function $w = f(x, y, z)$ from the point $P(x, y, z)$ to the nearby point $Q(x + \Delta x, y + \Delta y, z + \Delta z)$ is given by the increment

$$\Delta w = f(Q) - f(P). \tag{1}$$

The linear approximation theorem of Section 13.6 yields

$$\Delta w \approx \frac{\partial f}{\partial x}\Delta x + \frac{\partial f}{\partial y}\Delta y + \frac{\partial f}{\partial z}\Delta z. \tag{2}$$

We can express this approximation concisely in terms of the **gradient vector** ∇f (read as "del f") of the function f, which is defined to be

$$\nabla f(x, y, z) = \mathbf{i}f_x(x, y, z) + \mathbf{j}f_y(x, y, z) + \mathbf{k}f_z(x, y, z). \tag{3}$$

We also write

$$\nabla f = \left\langle \frac{\partial f}{\partial x}, \frac{\partial f}{\partial y}, \frac{\partial f}{\partial z} \right\rangle = \frac{\partial f}{\partial x}\mathbf{i} + \frac{\partial f}{\partial y}\mathbf{j} + \frac{\partial f}{\partial z}\mathbf{k}.$$

Then Eq. (2) implies that the increment $\Delta w = f(Q) - f(P)$ is given approximately by

$$\Delta w \approx \nabla f(P) \cdot \mathbf{v}, \tag{4}$$

where $\mathbf{v} = \overrightarrow{PQ} = \langle \Delta x, \Delta y, \Delta z \rangle$ is the *displacement vector* from P to Q.

EXAMPLE 1 If $f(x, y, z) = x^2 + yz - 2xy - z^2$, then the definition of the gradient vector in Eq. (3) yields

$$\nabla f(x, y, z) = \frac{\partial f}{\partial x}\mathbf{i} + \frac{\partial f}{\partial y}\mathbf{j} + \frac{\partial f}{\partial z}\mathbf{k} = (2x - 2y)\mathbf{i} + (z - 2x)\mathbf{j} + (y - 2z)\mathbf{k}.$$

For instance, the value of ∇f at the point $P(2, 1, 3)$ is

$$\nabla f(P) = \nabla f(2, 1, 3) = 2\mathbf{i} - \mathbf{j} - 5\mathbf{k}.$$

To apply Eq. (4), we first calculate

$$f(P) = f(2, 1, 3) = 2^2 + 1 \cdot 3 - 2 \cdot 2 \cdot 1 - 3^2 = -6.$$

If Q is the nearby point $Q(1.9, 1.2, 3.1)$, then $\overrightarrow{PQ} = \mathbf{v} = \langle -0.1, 0.2, 0.1 \rangle$, so the approximation in (4) gives

$$f(Q) - f(P) \approx \nabla f(P) \cdot \mathbf{v} = \langle 2, -1, -5 \rangle \cdot \langle -0.1, 0.2, 0.1 \rangle = -0.9.$$

Hence $f(Q) \approx -6 + (-0.9) = -6.9$. In this case we can also readily calculate, for comparison, the exact value $f(Q) = -6.84$. ■

Directional Derivatives

We know that the partial derivatives $f_x(x, y, z)$, $f_y(x, y, z)$, and $f_z(x, y, z)$ give the rates of change of $w = f(x, y, z)$ at the point $P(x, y, z)$ in the x-, y-, and z-directions, respectively. We can now use the gradient vector ∇f to calculate the rate of change of w at P in an *arbitrary* direction. Recall that a "direction" is prescribed by a *unit* vector $\mathbf{u}$.

Let Q be a point on the ray in the direction of $\mathbf{u}$ from the point P (Fig. 13.8.1). The **average rate of change of w with respect to distance between P and Q** is

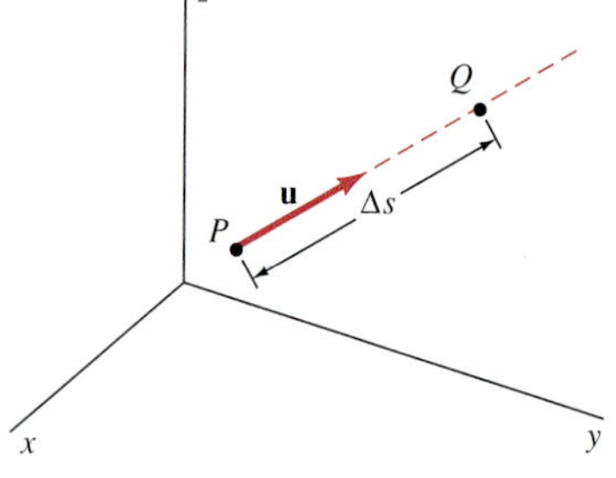

Fig. 13.8.1 The first step in computing the rate of change of $f(x, y, z)$ in the direction of the unit vector $\mathbf{u}$

$$\frac{f(Q) - f(P)}{|\overrightarrow{PQ}|} = \frac{\Delta w}{\Delta s},$$

where $\Delta s = |\overrightarrow{PQ}| = |\mathbf{v}|$ is the distance from P to Q. Then the approximation in (4) yields

$$\frac{\Delta w}{\Delta s} \approx \frac{\nabla f(P) \cdot \mathbf{v}}{|\mathbf{v}|} = \nabla f(P) \cdot \mathbf{u}, \tag{5}$$

where $\mathbf{u} = \mathbf{v}/|\mathbf{v}|$ is the *unit* vector in the direction from P to Q. When we take the limit of the average rate of change $\Delta w/\Delta s$ as $\Delta s \to 0$, we get the *instantaneous* rate of change

$$\frac{dw}{ds} = \lim_{\Delta s \to 0} \frac{\Delta w}{\Delta s} = \nabla f(P) \cdot \mathbf{u}. \tag{6}$$

This computation motivates the *definition*

$$D_{\mathbf{u}} f(P) = \nabla f(P) \cdot \mathbf{u} \tag{7}$$

of the **directional derivative of f at $P(x, y, z)$ in the direction $\mathbf{u}$**. Physics and engineering texts may use the notation

$$\left.\frac{df}{ds}\right|_P = D_{\mathbf{u}} f(P),$$

or simply dw/ds as in Eq. (6), for the rate of change of the function $w = f(x, y, z)$ with *respect to distance s* in the direction of the unit vector $\mathbf{u}$.

REMARK Remember that the vector $\mathbf{u}$ in Eq. (7) is a *unit* vector: $|\mathbf{u}| = 1$. If $\mathbf{u} = \langle a, b, c \rangle$, then Eq. (7) implies simply that

$$D_{\mathbf{u}}f = a\frac{\partial f}{\partial x} + b\frac{\partial f}{\partial y} + c\frac{\partial f}{\partial z}. \tag{8}$$

EXAMPLE 2 Suppose that the temperature at the point (x, y, z), with distance measured in kilometers, is given by

$$w = f(x, y, z) = 10 + xy + xz + yz$$

(in degrees Celsius). Find the rate of change (in degrees per kilometer) of temperature at the point $P(1, 2, 3)$ in the direction of the vector $\mathbf{v} = \mathbf{i} + 2\mathbf{j} - 2\mathbf{k}$.

Solution Because $\mathbf{v}$ is not a unit vector, we must replace it with a unit vector with the same direction before we can use the formulas of this section. So we take

$$\mathbf{u} = \frac{\mathbf{v}}{|\mathbf{v}|} = \left\langle \frac{1}{3}, \frac{2}{3}, -\frac{2}{3} \right\rangle.$$

The gradient vector of f is

$$\nabla f = (y + z)\mathbf{i} + (x + z)\mathbf{j} + (x + y)\mathbf{k},$$

so $\nabla f(1, 2, 3) = 5\mathbf{i} + 4\mathbf{j} + 3\mathbf{k}$. Hence Eq. (7) gives

$$D_{\mathbf{u}}f(P) = \langle 5, 4, 3 \rangle \cdot \langle \tfrac{1}{3}, \tfrac{2}{3}, -\tfrac{2}{3} \rangle = \tfrac{7}{3}$$

(degrees per kilometer) for the desired range of change of temperature with respect to distance. ■

The Vector Chain Rule

The directional derivative $D_{\mathbf{u}}f$ is closely related to a version of the multivariable chain rule. Suppose that the first-order partial derivatives of f are continuous and that

$$\mathbf{r}(t) = x(t)\mathbf{i} + y(t)\mathbf{j} + z(t)\mathbf{k}$$

is a differentiable vector-valued function. Then

$$f(\mathbf{r}(t)) = f(x(t), y(t), z(t))$$

is a differentiable function of t, and its (ordinary) derivative with respect to t is

$$D_t f(\mathbf{r}(t)) = D_t[f(x(t), y(t), z(t))] = \frac{\partial f}{\partial x}\cdot\frac{dx}{dt} + \frac{\partial f}{\partial y}\cdot\frac{dy}{dt} + \frac{\partial f}{\partial z}\cdot\frac{dz}{dt}. \tag{9}$$

Hence

$$D_t f(\mathbf{r}(t)) = \nabla f(\mathbf{r}(t)) \cdot \mathbf{r}'(t), \tag{10}$$

where $\mathbf{r}'(t) = \langle x'(t), y'(t), z'(t) \rangle$ is the velocity vector of the parametric curve $\mathbf{r}(t)$. Equation (10) is the **vector chain rule**. The operation on the right-hand side of Eq. (10) is the *dot* product, because both the gradient of f and the derivative of $\mathbf{r}$ are *vector-valued* functions.

If the velocity vector $\mathbf{v}(t) = \mathbf{r}'(t) \neq \mathbf{0}$, then $\mathbf{v} = v\mathbf{u}$, where $v = |\mathbf{v}|$ is the speed and $\mathbf{u} = \mathbf{v}/v$ is the unit vector tangent to the curve. Then Eq. (10) implies that

$$D_t f(\mathbf{r}(t)) = v D_{\mathbf{u}} f(\mathbf{r}(t)). \tag{11}$$

With $w = f(\mathbf{r}(t))$, $D_{\mathbf{u}} f = dw/ds$, and $v = ds/dt$, Eq. (11) takes the simple chain rule form

$$\frac{dw}{dt} = \frac{dw}{ds} \cdot \frac{ds}{dt}. \tag{12}$$

EXAMPLE 3 If the function

$$w = f(x, y, z) = 10 + xy + xz + yz$$

of Example 2 gives the temperature at the point (x, y, z) of space, what time rate of change (degrees per minute) will a hawk observe as it flies through $P(1, 2, 3)$ at a speed of 2 km/min, heading directly toward the point $Q(3, 4, 4)$?

Solution In Example 2 we calculated $\nabla f(P) = \langle 5, 4, 3 \rangle$, and the unit vector in the direction from P to Q is

$$\mathbf{u} = \frac{\overrightarrow{PQ}}{|\overrightarrow{PQ}|} = \left\langle \frac{2}{3}, \frac{2}{3}, \frac{1}{3} \right\rangle.$$

Then

$$D_{\mathbf{u}} f(P) = \nabla f(P) \cdot \mathbf{u} = \langle 5, 4, 3 \rangle \cdot \langle \tfrac{2}{3}, \tfrac{2}{3}, \tfrac{1}{3} \rangle = 7$$

(degrees per kilometer). Hence Eq. (12) yields

$$\frac{dw}{dt} = \frac{dw}{ds} \cdot \frac{ds}{dt} = \left(7 \frac{\text{deg}}{\text{km}}\right)\left(2 \frac{\text{km}}{\text{min}}\right) = 14 \frac{\text{deg}}{\text{min}}$$

as the hawk's rate of change of temperature at P. ■

REMARK In Section 13.7 we interpreted the chain rule in terms of products of derivative matrices. Recall that a derivative matrix has one row for each dependent variable and one column for each independent variable. Hence the derivative matrix of the real-valued function $w = f(x, y, z)$ is the 1×3 *row matrix*

$$f'(x, y, z) = \begin{bmatrix} f_x(x, y, z) & f_y(x, y, z) & f_z(x, y, z) \end{bmatrix} = \begin{bmatrix} \dfrac{\partial f}{\partial x} & \dfrac{\partial f}{\partial y} & \dfrac{\partial f}{\partial z} \end{bmatrix} \tag{13}$$

whose entries are the partial derivatives of f (and thus are the components of the gradient vector ∇f). The derivative matrix of the vector-valued function $\alpha(t) = \langle x(t), y(t), z(t) \rangle$ is the 3×1 *column matrix*

$$\alpha'(t) = \begin{bmatrix} x'(t) \\ y'(t) \\ z'(t) \end{bmatrix} = \begin{bmatrix} dx/dt \\ dy/dt \\ dz/dt \end{bmatrix} \tag{14}$$

whose entries are the derivatives of the component functions of α (and thus are the components of the velocity vector $\mathbf{v} = \langle x'(t), y'(t), z'(t) \rangle$). In terms of these derivative matrices, Eq. (9) means that the derivative of the composition $g(t) = f(\alpha(t))$ is given by

$$g'(t) = \frac{\partial f}{\partial x}\frac{dx}{dt} + \frac{\partial f}{\partial y}\frac{dy}{dt} + \frac{\partial f}{\partial z}\frac{dz}{dt}$$

$$= \begin{bmatrix} \dfrac{\partial f}{\partial x} & \dfrac{\partial f}{\partial y} & \dfrac{\partial f}{\partial z} \end{bmatrix} \begin{bmatrix} dx/dt \\ dy/dt \\ dz/dt \end{bmatrix} = f'\big(\alpha(t)\big)\alpha'(t). \tag{15}$$

So here again the derivative matrix of the composition $g = f \circ \alpha$ is equal to the matrix product $g' = f'\alpha'$.

Interpretation of the Gradient Vector

As yet we have discussed directional derivatives only for functions of three variables. The formulas for a function of two (or more than three) variables are analogous:

$$\nabla f(x, y) = \left\langle \frac{\partial f}{\partial x}, \frac{\partial f}{\partial y} \right\rangle = \frac{\partial f}{\partial x}\mathbf{i} + \frac{\partial f}{\partial y}\mathbf{j} \tag{16}$$

and

$$D_{\mathbf{u}}f(x, y) = \nabla f(x, y) \cdot \mathbf{u} = a\frac{\partial f}{\partial x} + b\frac{\partial f}{\partial y} \tag{17}$$

if $\mathbf{u} = \langle a, b \rangle$ is a unit vector. If α is the angle of inclination of $\mathbf{u}$ (measured counterclockwise from the positive x-axis, as in Fig. 13.8.2), then $a = \cos\alpha$ and $b = \sin\alpha$, so Eq. (17) takes the form

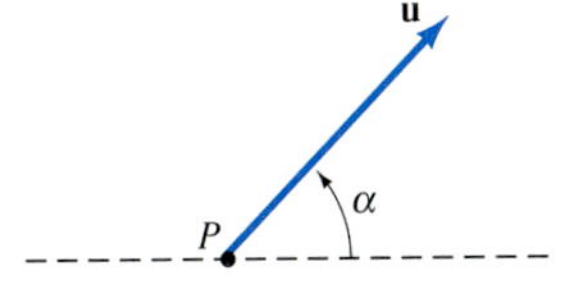

Fig. 13.8.2 The unit vector **u** of Eq. (17)

$$D_{\mathbf{u}}f(x, y) = \frac{\partial f}{\partial x}\cos\alpha + \frac{\partial f}{\partial y}\sin\alpha. \tag{18}$$

The gradient vector ∇f has an important interpretation that involves the *maximal* directional derivative of f. If ϕ is the angle between ∇f at the point P and the unit vector **u** (Fig. 13.8.3), then the formula in Eq. (7) gives

$$D_{\mathbf{u}}f(P) = \nabla f(P) \cdot \mathbf{u} = |\nabla f(P)| \cos\phi,$$

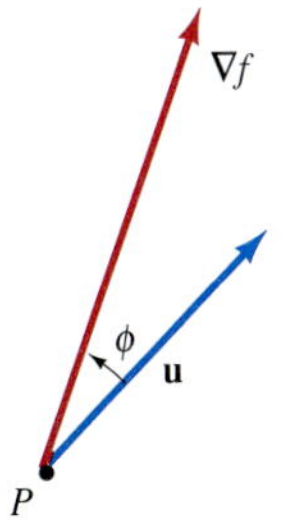

because $|\mathbf{u}| = 1$. The maximum value of $\cos\phi$ is 1, and this occurs when $\phi = 0$. This is so when **u** is the particular unit vector $\nabla f(P)/|\nabla f(P)|$ that points in the direction of the gradient vector itself. In this case the previous formula yields

$$D_{\mathbf{u}}f(P) = |\nabla f(P)|, \tag{19}$$

Fig. 13.8.3 The angle ϕ between ∇f and the unit vector **u**

so the value of the directional derivative is the length of the gradient vector. We have therefore proved Theorem 1.

Theorem 1 ***Significance of the Gradient Vector***

The maximum value of the directional derivative $D_{\mathbf{u}}f(P)$ is obtained when **u** is the vector in the direction of the gradient vector $\nabla f(P)$; that is, when $\mathbf{u} = \nabla f(P)/|\nabla f(P)|$. The value of the maximum directional derivative is $|\nabla f(P)|$, the length of the gradient vector.

Thus *the gradient vector* ∇f *points in the direction in which the function* f *increases the most rapidly, and its length is the rate of increase of* f *(with respect to distance) in that direction.* For instance, if the function f gives the temperature in space, then the gradient vector $\nabla f(P)$ points in the direction in which a bumblebee at P should initially fly to get warmer the fastest.

EXAMPLE 4 Suppose that the temperature w (in degrees Celsius) at the point (x, y) is given by

$$w = f(x, y) = 10 + (0.003)x^2 - (0.004)y^2.$$

In what direction $\mathbf{u}$ should a bumblebee at the point $(40, 30)$ initially fly in order to get warmer fastest? Find the directional derivative $D_{\mathbf{u}} f(40, 30)$ in this optimal direction $\mathbf{u}$.

Solution The gradient vector is

$$\nabla f = \frac{\partial f}{\partial x}\mathbf{i} + \frac{\partial f}{\partial y}\mathbf{j} = (0.006)x\,\mathbf{i} - (0.008)y\,\mathbf{j},$$

so

$$\nabla f(40, 30) = (0.24)\mathbf{i} - (0.24)\mathbf{j} = \left(0.24\sqrt{2}\right)\mathbf{u}.$$

The unit vector

$$\mathbf{u} = \frac{\nabla f(40, 30)}{|\nabla f(40, 30)|} = \frac{\mathbf{i} - \mathbf{j}}{\sqrt{2}}$$

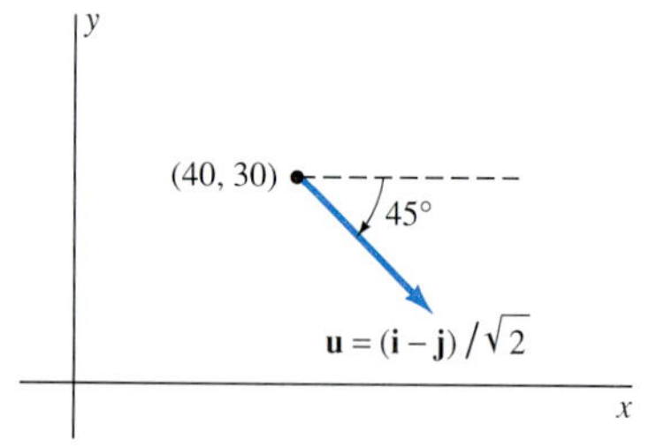

Fig. 13.8.4 The unit vector $\mathbf{u} = \dfrac{\nabla f}{|\nabla f|}$ of Example 4

points southeast (Fig. 13.8.4); this is the direction in which the bumblebee should initially fly. And, according to Theorem 1, the directional derivative of f in this optimal direction is

$$D_{\mathbf{u}} f(40, 30) = |\nabla f(40, 30)| = (0.24)\sqrt{2} \approx 0.34$$

degrees per unit of distance. ■

The Gradient Vector as a Normal Vector

Consider the graph of the equation

$$F(x, y, z) = 0, \tag{20}$$

where F is a function with continuous first-order partial derivatives. According to the **implicit function theorem** of advanced calculus, near every point where $\nabla F \neq \mathbf{0}$—that is, at least one of the partial derivatives of F is nonzero—the graph of Eq. (20) coincides with the graph of an equation of one of the forms

$$z = f(x, y), \quad y = g(x, z), \quad x = h(y, z).$$

Because of this, we are justified in general in referring to the graph of Eq. (20) as a "surface." The gradient vector ∇F is normal to this surface, in the sense of Theorem 2.

Theorem 2 ***Gradient Vector as Normal Vector***

Suppose that $F(x, y, z)$ has continuous first-order partial derivatives, and let $P_0(x_0, y_0, z_0)$ be a point of the graph of the equation $F(x, y, z) = 0$ at which $\nabla F(P_0) \neq \mathbf{0}$. If $\mathbf{r}(t)$ is a differentiable curve on this surface with $\mathbf{r}(t_0) = \langle x_0, y_0, z_0 \rangle$, then

$$\nabla F(P_0) \cdot \mathbf{r}'(t_0) = 0. \tag{21}$$

Thus $\nabla F(P_0)$ is perpendicular to the tangent vector $\mathbf{r}'(t_0)$, as indicated in Fig. 13.8.5.

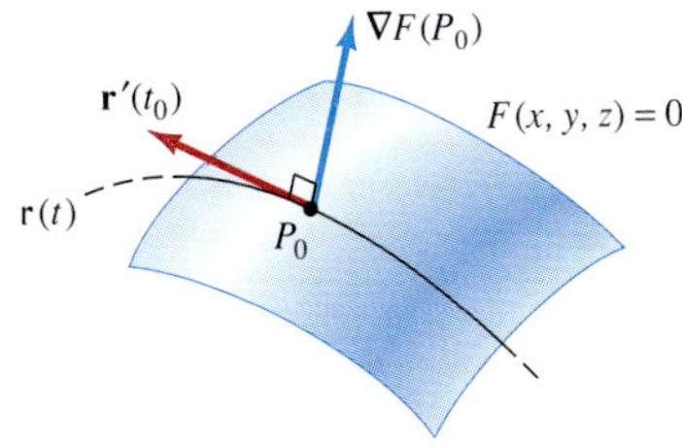

Fig. 13.8.5 The gradient vector ∇F is normal to every curve in the surface $F(x, y, z) = 0$.

PROOF The statement that $\mathbf{r}(t)$ lies on the surface $F(x, y, z) = 0$ means that $F(\mathbf{r}(t)) = 0$ for all t. Hence

$$0 = D_t F(\mathbf{r}(t_0)) = \nabla F(\mathbf{r}(t_0)) \cdot \mathbf{r}'(t_0) = \nabla F(P_0) \cdot \mathbf{r}'(t_0)$$

by the chain rule in the form of Eq. (10). Therefore the vectors $\nabla F(P_0)$ and $\mathbf{r}'(t_0)$ are perpendicular. ■

Because $\nabla F(P_0)$ is perpendicular to every curve on the surface $F(x, y, z)$ through P_0, it is a *normal vector* to the surface at P_0,

$$\mathbf{n} = \frac{\partial F}{\partial x}\mathbf{i} + \frac{\partial F}{\partial y}\mathbf{j} + \frac{\partial F}{\partial z}\mathbf{k}. \tag{22}$$

If we rewrite the equation $z = f(x, y)$ in the form $F(x, y, z) = f(x, y) - z = 0$, then

$$\left\langle \frac{\partial F}{\partial x}, \frac{\partial F}{\partial y}, \frac{\partial F}{\partial z} \right\rangle = \left\langle \frac{\partial f}{\partial x}, \frac{\partial f}{\partial y}, -1 \right\rangle.$$

Thus Eq. (22) agrees with the definition of normal vector that we gave in Section 13.4 (Eq. (13) there).

The **tangent plane** to the surface $F(x, y, z) = 0$ at the point $P_0(x_0, y_0, z_0)$ is the plane through P_0 that is perpendicular to the normal vector $\mathbf{n}$ of Eq. (22). Its equation is

$$F_x(x_0, y_0, z_0)(x - x_0) + F_y(x_0, y_0, z_0)(y - y_0) + F_z(x_0, y_0, z_0)(z - z_0) = 0. \tag{23}$$

EXAMPLE 5 Write an equation of the plane tangent to the ellipsoid $2x^2 + 4y^2 + z^2 = 45$ at the point $(2, -3, -1)$.

Solution If we write $F(x, y, z) = 2x^2 + 4y^2 + z^2 - 45$, then $F(x, y, z) = 0$ is the equation of the ellipsoid. Hence a normal vector is $\nabla F(x, y, z) = \langle 4x, 8y, 2z \rangle$, so

$$\nabla F(2, -3, -1) = 8\mathbf{i} - 24\mathbf{j} - 2\mathbf{k}$$

is normal to the ellipsoid at $(2, -3, -1)$. Equation (18) then gives the answer in the form

$$8(x - 2) - 24(y + 3) - 2(z + 1) = 0;$$

that is,

$$4x - 12y - z = 45.$$ ■

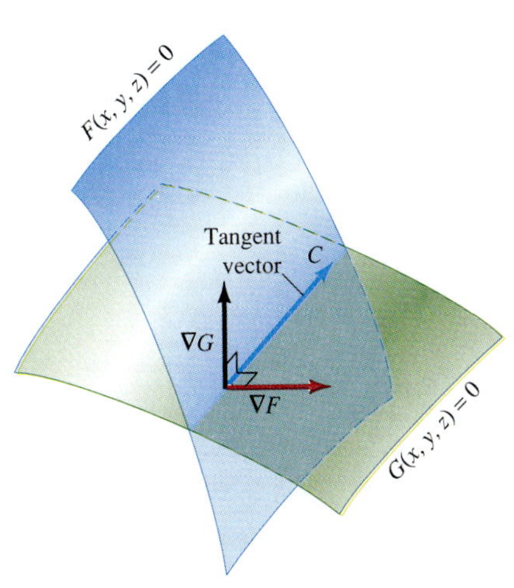

Fig. 13.8.6 $\nabla F \times \nabla G$ is tangent to the curve C of intersection.

The intersection of the two surfaces $F(x, y, z) = 0$ and $G(x, y, z) = 0$ will generally be some sort of curve in space. By the implicit function theorem, we can represent this curve in parametric fashion near every point where the gradient vectors ∇F and ∇G are *not* parallel. This curve C is perpendicular to both normal vectors ∇F and ∇G. That is, if P is a point of C, then the vector tangent to C at P is perpendicular to both vectors $\nabla F(P)$ and $\nabla G(P)$ (Fig. 13.8.6). It follows that the vector

$$\mathbf{T} = \nabla F \times \nabla G \tag{24}$$

is tangent to the curve of intersection of the surfaces $F(x, y, z) = 0$ and $G(x, y, z) = 0$.

EXAMPLE 6 The point $P(1, -1, 2)$ lies on both the paraboloid

$$F(x, y, z) = x^2 + y^2 - z = 0$$

and the ellipsoid

$$G(x, y, z) = 2x^2 + 3y^2 + z^2 - 9 = 0.$$

Write an equation of the plane through P that is normal to the curve of intersection of these two surfaces (Fig. 13.8.7).

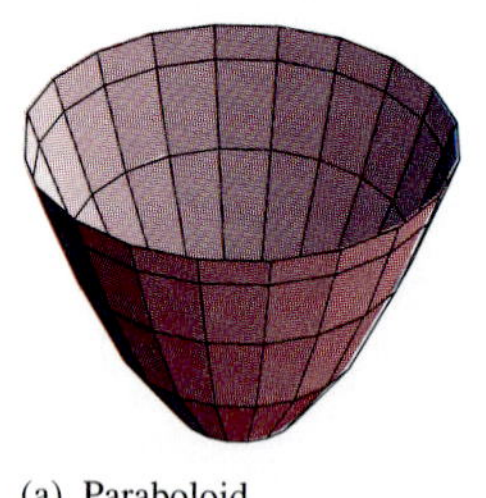

(a) Paraboloid

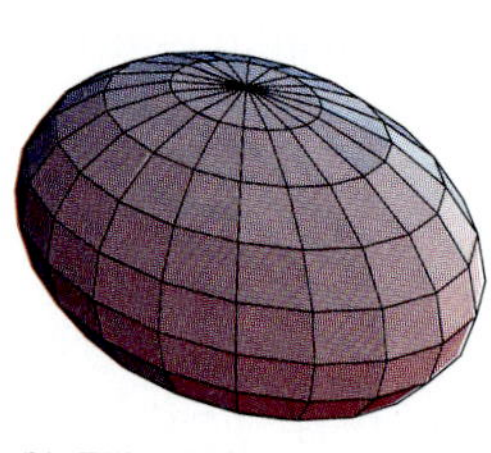

(b) Ellipsoid

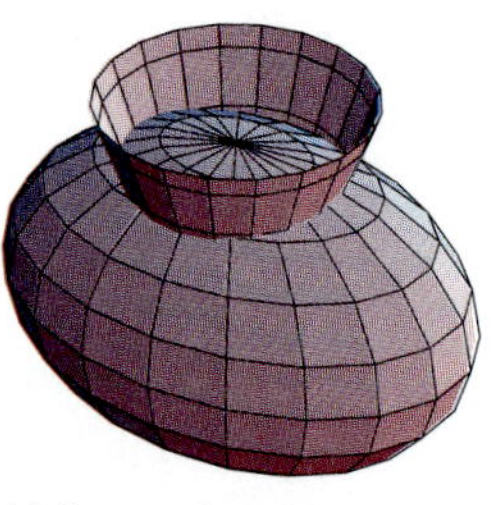

(c) Intersection of paraboloid and ellipsoid

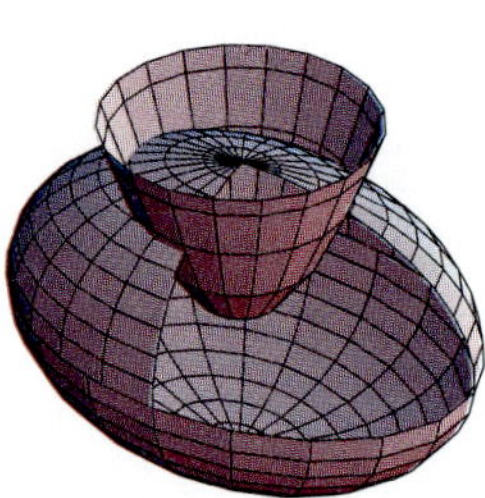

(d) Cutaway view

Fig. 13.8.7 Example 6

Solution First we compute

$$\nabla F = \langle 2x, 2y, -1 \rangle \quad \text{and} \quad \nabla G = \langle 4x, 6y, 2z \rangle.$$

At $P(1, -1, 2)$ these two vectors are

$$\nabla F(1, -1, 2) = \langle 2, -2, -1 \rangle \quad \text{and} \quad \nabla G(1, -1, 2) = \langle 4, -6, 4 \rangle.$$

Hence a vector tangent to the curve of intersection of the paraboloid and the ellipsoid is

$$\mathbf{T} = \nabla F \times \nabla G = \begin{vmatrix} \mathbf{i} & \mathbf{j} & \mathbf{k} \\ 2 & -2 & -1 \\ 4 & -6 & 4 \end{vmatrix} = \langle -14, -12, -4 \rangle.$$

A slightly simpler vector parallel to $\mathbf{T}$ is $\mathbf{n} = \langle 7, 6, 2 \rangle$, and $\mathbf{n}$ is also normal to the desired plane through $(1, -1, 2)$. Therefore, an equation of the plane is

$$7(x - 1) + 6(y + 1) + 2(z - 2) = 0;$$

that is, $7x + 6y + 2z = 5$. ■

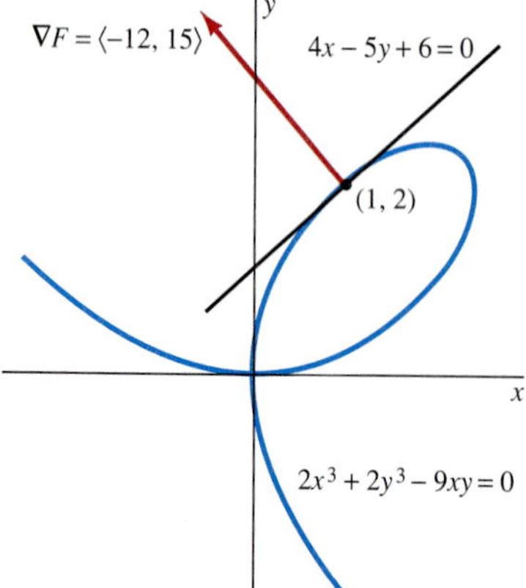

Fig. 13.8.8 The folium and its tangent (Example 7)

A result analogous to Theorem 2 holds in two dimensions (and in higher dimensions). The graph of the equation $F(x, y) = 0$ looks like a *curve* near each point at which $\nabla F \neq \mathbf{0}$, and ∇F is normal to the curve in such cases.

EXAMPLE 7 Write an equation of the line tangent at the point $(1, 2)$ to the folium of Descartes with equation $F(x, y) = 2x^3 + 2y^3 - 9xy = 0$ (Fig. 13.8.8).

Solution The gradient of F is

$$\nabla F(x, y) = (6x^2 - 9y)\mathbf{i} + (6y^2 - 9x)\mathbf{j}.$$

So a vector normal to the folium at $(1, 2)$ is $\nabla F(1, 2) = -12\mathbf{i} + 15\mathbf{j}$. Hence the tangent line has equation $-12(x - 1) + 15(y - 2) = 0$. Simplified, this is $4x - 5y + 6 = 0$.

13.8 PROBLEMS

In Problems 1 through 10, find the gradient vector ∇f at the indicated point P.

1. $f(x, y) = 3x - 7y;\quad P(17, 39)$
2. $f(x, y) = 3x^2 - 5y^2;\quad P(2, -3)$
3. $f(x, y) = \exp(-x^2 - y^2);\quad P(0, 0)$
4. $f(x, y) = \sin \frac{1}{4}\pi xy;\quad P(3, -1)$
5. $f(x, y, z) = y^2 - z^2;\quad P(17, 3, 2)$
6. $f(x, y, z) = \sqrt{x^2 + y^2 + z^2};\quad P(12, 3, 4)$
7. $f(x, y, z) = e^x \sin y + e^y \sin z + e^z \sin x;\quad P(0, 0, 0)$
8. $f(x, y, z) = x^2 - 3yz + z^3;\quad P(2, 1, 0)$
9. $f(x, y, z) = 2\sqrt{xyz};\quad P(3, -4, -3)$
10. $f(x, y, z) = (2x - 3y + 5z)^5;\quad P(-5, 1, 3)$

In Problems 11 through 20, find the directional derivative of f at P in the direction of $\mathbf{v}$; that is, find

$$D_{\mathbf{u}}f(P), \quad \text{where} \quad \mathbf{u} = \frac{\mathbf{v}}{|\mathbf{v}|}.$$

11. $f(x, y) = x^2 + 2xy + 3y^2;\quad P(2, 1), \mathbf{v} = \langle 1, 1 \rangle$
12. $f(x, y) = e^x \sin y;\quad P(0, \pi/4), \mathbf{v} = \langle 1, -1 \rangle$
13. $f(x, y) = x^3 - x^2y + xy^2 + y^3;\quad P(1, -1), \mathbf{v} = 2\mathbf{i} + 3\mathbf{j}$
14. $f(x, y) = \tan^{-1}\left(\dfrac{y}{x}\right);\quad P(-3, 3), \mathbf{v} = 3\mathbf{i} + 4\mathbf{j}$
15. $f(x, y) = \sin x \cos y;\quad P(\pi/3, -2\pi/3), \mathbf{v} = \langle 4, -3 \rangle$
16. $f(x, y, z) = xy + yz + zx;\quad P(1, -1, 2), \mathbf{v} = \langle 1, 1, 1 \rangle$
17. $f(x, y, z) = \sqrt{xyz};\quad P(2, -1, -2), \mathbf{v} = \mathbf{i} + 2\mathbf{j} - 2\mathbf{k}$
18. $f(x, y, z) = \ln(1 + x^2 + y^2 - z^2);\quad P(1, -1, 1)$, $\mathbf{v} = 2\mathbf{i} - 2\mathbf{j} - 3\mathbf{k}$
19. $f(x, y, z) = e^{xyz};\quad P(4, 0, -3), \mathbf{v} = \mathbf{j} - \mathbf{k}$
20. $f(x, y, z) = \sqrt{10 - x^2 - y^2 - z^2};\quad P(1, 1, -2)$, $\mathbf{v} = \langle 3, 4, -12 \rangle$

In Problems 21 through 28, find the maximum directional derivative of f at P and the direction in which it occurs.

21. $f(x, y) = 2x^2 + 3xy + 4y^2;\quad P(1, 1)$
22. $f(x, y) = \arctan\left(\dfrac{y}{x}\right);\quad P(1, -2)$
23. $f(x, y) = \ln(x^2 + y^2);\quad P(3, 4)$
24. $f(x, y) = \sin(3x - 4y);\quad P(\pi/3, \pi/4)$
25. $f(x, y, z) = 3x^2 + y^2 + 4z^2;\quad P(1, 5, -2)$
26. $f(x, y, z) = \exp(x - y - z);\quad P(5, 2, 3)$
27. $f(x, y, z) = \sqrt{xy^2z^3};\quad P(2, 2, 2)$
28. $f(x, y, z) = \sqrt{2x + 4y + 6z};\quad P(7, 5, 5)$

In Problems 29 through 34, use the normal gradient vector to write an equation of the line (or plane) tangent to the given curve (or surface) at the given point P.

29. $\exp(25 - x^2 - y^2) = 1;\quad P(3, 4)$
30. $2x^2 + 3y^2 = 35;\quad P(2, 3)$
31. $x^4 + xy + y^2 = 19;\quad P(2, -3)$
32. $3x^2 + 4y^2 + 5z^2 = 73;\quad P(2, 2, 3)$
33. $x^{1/3} + y^{1/3} + z^{1/3} = 1;\quad P(1, -1, 1)$
34. $xyz + x^2 - 2y^2 + z^3 = 14;\quad P(5, -2, 3)$

The properties of gradient vectors listed in Problems 35 through 38 exhibit the close analogy between the gradient operator ∇ and the single-variable derivative operator D. Verify each, assuming that a and b are constants and that u and v are differentiable functions of x and y.

35. $\nabla(au + bv) = a\nabla u + b\nabla v$.
36. $\nabla(uv) = u\nabla v + v\nabla u$.
37. $\nabla\left(\dfrac{u}{v}\right) = \dfrac{v\nabla u - u\nabla v}{v^2}$ if $v \neq 0$.
38. If n is a positive integer, then $\nabla u^n = nu^{n-1}\nabla u$.
39. Show that the value of a differentiable function f decreases the most rapidly at P in the direction of the vector $-\nabla f(P)$, directly opposite to the gradient vector.
40. Suppose that f is a function of three independent variables x, y, and z. Show that $D_{\mathbf{i}}f = f_x$, $D_{\mathbf{j}}f = f_y$, and $D_{\mathbf{k}}f = f_z$.
41. Show that the equation of the line tangent to the conic section $Ax^2 + Bxy + Cy^2 = D$ at the point (x_0, y_0) is

$$(Ax_0)x + \tfrac{1}{2}B(y_0x + x_0y) + (Cy_0)y = D.$$

42. Show that the equation of the plane tangent to the quadric surface $Ax^2 + By^2 + Cz^2 = D$ at the point (x_0, y_0, z_0) is

$$(Ax_0)x + (By_0)y + (Cz_0)z = D.$$

43. Suppose that the temperature W (in degrees Celsius) at the point (x, y, z) in space is given by

$$W = 50 + xyz.$$

(a) Find the rate of change of temperature with respect to distance at the point $P(3, 4, 1)$ in the direction of the vector $\mathbf{v} = \langle 1, 2, 2 \rangle$. (The units of distance in space are feet.)
(b) Find the maximal directional derivative $D_{\mathbf{u}}W$ at the

point $P(3, 4, 1)$ and the direction $\mathbf{u}$ in which that maximum occurs.

44. Suppose that the temperature at the point (x, y, z) in space (in degrees Celsius) is given by the formula

$$W = 100 - x^2 - y^2 - z^2.$$

The units in space are meters. (a) Find the rate of change of temperature at the point $P(3, -4, 5)$ in the direction of the vector $\mathbf{v} = 3\mathbf{i} - 4\mathbf{j} + 12\mathbf{k}$. (b) In what direction does W increase most rapidly at P? What is the value of the maximal directional derivative at P?

45. Suppose that the altitude z (in miles above sea level) of a certain hill is described by the equation $z = f(x, y)$, where

$$f(x, y) = \tfrac{1}{10}(x^2 - xy + 2y^2).$$

(a) Write an equation (in the form $z = ax + by + c$) of the plane tangent to the hillside at the point $P(2, 1, 0.4)$. (b) Use $\nabla f(2, 1)$ to approximate the altitude of the hill above the point $(2.2, 0.9)$ in the xy-plane. Compare your result with the actual altitude at this point.

46. Find an equation for the plane tangent to the paraboloid $z = 2x^2 + 3y^2$ and, simultaneously, parallel to the plane $4x - 3y - z = 10$.

47. The cone with equation $z^2 = x^2 + y^2$ and the plane with equation $2x + 3y + 4z + 2 = 0$ intersect in an ellipse. Write an equation of the plane normal to this ellipse at the point $P(3, 4, -5)$ (Fig. 13.8.9).

48. It is apparent from geometry that the highest and lowest points of the ellipse of Problem 47 are those points where its tangent line is horizontal. Find those points.

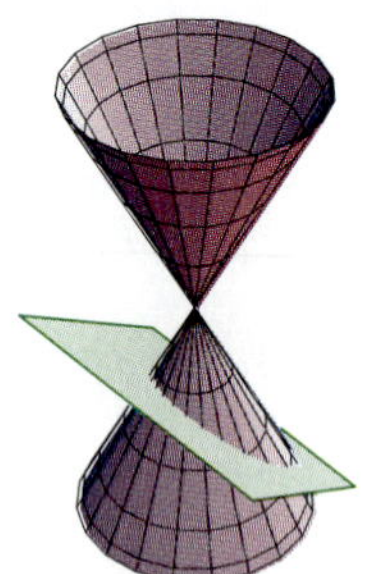

Fig. 13.8.9 The cone and plane of Problems 47 and 48

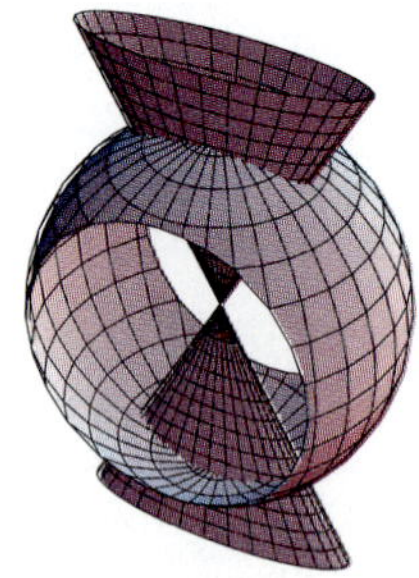

Fig. 13.8.10 A cutaway view of the cone and sphere of Problem 49

49. Show that the sphere $x^2 + y^2 + z^2 = r^2$ and the cone $z^2 = a^2x^2 + b^2y^2$ are orthogonal (that is, have perpendicular tangent planes) at every point of their intersection (Fig. 13.8.10).

In Problems 50 through 55, the function $z = f(x, y)$ describes the shape of a hill; $f(P)$ is the altitude of the hill above the point $P(x, y)$ in the xy-plane. If you start at the point $(P, f(P))$ of this hill, then $D_{\mathbf{u}}f(P)$ is your rate of climb (rise per unit of horizontal distance) as you proceed in the *horizontal* direction $\mathbf{u} = a\mathbf{i} + b\mathbf{j}$. And the angle at which you climb while you walk in this direction is $\gamma = \tan^{-1}(D_{\mathbf{u}}f(P))$, as shown in Fig. 13.8.11.

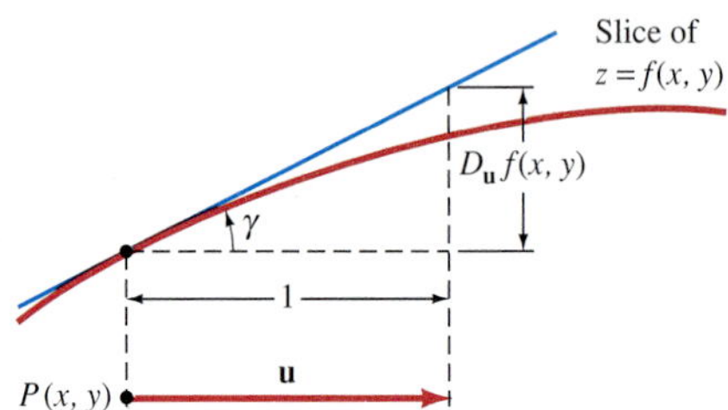

Fig. 13.8.11 The cross section of the part of the graph above $\mathbf{u}$ (Problems 50 through 55)

50. You are standing at the point $(-100, -100, 430)$ on a hill that has the shape of the graph of

$$z = 500 - (0.003)x^2 - (0.004)y^2,$$

with x, y, and z given in feet. (a) What will be your rate of climb (*rise* over *run*) if you head northwest? At what angle from the horizontal will you be climbing? (b) Repeat part (a), except now you head northeast.

51. You are standing at the point $(-100, -100, 430)$ on the hill of Problem 50. In what direction (that is, with what compass heading) should you proceed in order to climb the most steeply? At what angle from the horizontal will you initially be climbing?

52. Repeat Problem 50, but now you are standing at the point $P(100, 100, 500)$ on the hill described by

$$z = \frac{1000}{1 + (0.00003)x^2 + (0.00007)y^2}.$$

53. Repeat Problem 51, except begin at the point $P(100, 100, 500)$ of the hill of Problem 52.

54. You are standing at the point $(30, 20, 5)$ on a hill with the shape of the surface

$$z = 100\exp\left(-\frac{x^2 + 3y^2}{701}\right).$$

(a) In what direction (with what compass heading) should you proceed in order to climb the most steeply? At what angle from the horizontal will you initially be climbing? (b) If, instead of climbing as in part (a), you head directly west (the negative x-direction), then at what angle will you be climbing initially?

55. (a) You are standing at the point where $x = y = 100$ (ft) on the side of a mountain whose height (in feet above sea level) is given by

$$z = \frac{1}{1000}(3x^2 - 5xy + y^2),$$

with the x-axis pointing east and the y-axis pointing north. If you head northeast, will you be ascending or descending? At what angle (in degrees) from the horizontal? (b) If you head 30° north of east, will you be ascending or descending? At what angle (in degrees) from the horizontal?

56. Suppose that the two surfaces $f(x, y, z) = 0$ and $g(x, y, z) = 0$ both pass through the point P where both gradient vectors $\nabla f(P)$ and $\nabla g(P)$ exist. (a) Show that the two surfaces are tangent at P if and only if $\nabla f(P) \times \nabla g(P) = \mathbf{0}$. (b) Show that the two surfaces are orthogonal at P if and only if $\nabla f(P) \cdot \nabla g(P) = 0$.

57. Let $T: R^2_{uv} \to R^2_{xy}$ be a transformation from the uv-plane to the xy-plane with 2×2 derivative matrix $T'(u, v)$ as in Eq. (11) of Section 13.7. If $\alpha(t) = \langle u(t), v(t) \rangle$ is a parametric curve in R^2_{uv} and the parametric curve $\beta(t) = \langle x(t), y(t) \rangle$ is defined by $\beta(t) = T(\alpha(t))$, apply the chain rule to show that $\beta'(t) = T'(\alpha(t))\,\alpha'(t)$. That is, the derivative matrix of the composition $\beta = T \circ \alpha$ is the matrix product $\beta' = T'\alpha'$.

58. Suppose that a parametric curve in the plane R^2_{xy} is described by giving the polar coordinates $r(t)$ and $\theta(t)$ of the moving point (x, y) as functions of time t. Use the matrix chain rule of Problem 57 and the polar coordinate derivative matrix $T'(r, \theta)$ of Example 9 in Section 13.7 to write the components of the velocity vector $\mathbf{v}(t) = \langle x'(t), y'(t) \rangle$ in terms of $r(t)$, $\theta(t)$, and their derivatives.

59. Suppose that a parametric curve in 3-space R^3_{xyz} is described by giving the spherical coordinates $\rho(t)$, $\phi(t)$, and $\theta(t)$ of the moving point (x, y, z) as functions of time t. Use the three-dimensional analogue of the matrix chain rule of Problem 57 and the spherical coordinate derivative matrix $T'(\rho, \phi, \theta)$ of Problem 59 in Section 13.7 to write the components of the velocity vector $\mathbf{v}(t) = \langle x'(t), y'(t), z'(t) \rangle$ in terms of $\rho(t)$, $\phi(t)$, $\theta(t)$, and their derivatives.

60. Suppose that the function $f(x, y)$ has continuous partial derivatives near the point $\mathbf{c} = (a, b)$. Apply the linear approximation theorem of Section 13.6 to show that

$$\lim_{\mathbf{x} \to \mathbf{c}} \frac{f(\mathbf{x}) - f(\mathbf{c}) - \nabla f(\mathbf{c}) \cdot (\mathbf{x} - \mathbf{c})}{|\mathbf{x} - \mathbf{c}|} = 0.$$

13.9 LAGRANGE MULTIPLIERS AND CONSTRAINED MAXIMUM-MINIMUM PROBLEMS

In Section 13.5 we discussed the problem of finding the maximum and minimum values attained by a function $f(x, y)$ at points of the plane region R, in the simple case in which R consists of the points on and within the simple closed curve C. We saw that any local maximum or minimum in the *interior* of R occurs at a point where $f_x(x, y) = 0 = f_y(x, y)$ or at a point where f is not differentiable (the latter usually signaled by the failure of f_x or f_y to exist). Here we discuss the very different matter of finding the maximum and minimum values attained by f at points of the *boundary* curve C.

If the curve C is the graph of the equation $g(x, y) = 0$, then our task is to maximize or minimize the function $f(x, y)$ subject to the **constraint,** or **side condition,**

$$g(x, y) = 0. \qquad \textbf{(1)}$$

We could in principle try to solve this constraint equation for $y = \phi(x)$ and then maximize or minimize the single-variable function $f(x, \phi(x))$ by the standard method of finding its critical points. But what if it is impractical or impossible to solve Eq. (1) explicitly for y in terms of x? An alternative approach that does not require that we first solve this equation is the **method of Lagrange multipliers.** It is named for its discoverer, the Italian-born French mathematician Joseph Louis Lagrange (1736–1813). The method is based on Theorem 1.

Theorem 1 ***Lagrange Multipliers (one constraint)***

Let $f(x, y)$ and $g(x, y)$ be functions with continuous first-order partial derivatives. If the maximum (or minimum) value of f subject to the condition

$$g(x, y) = 0 \qquad \textbf{(1)}$$

occurs at a point P where $\nabla g(P) \neq \mathbf{0}$, then

$$\nabla f(P) = \lambda \nabla g(P) \tag{2}$$

for some constant λ.

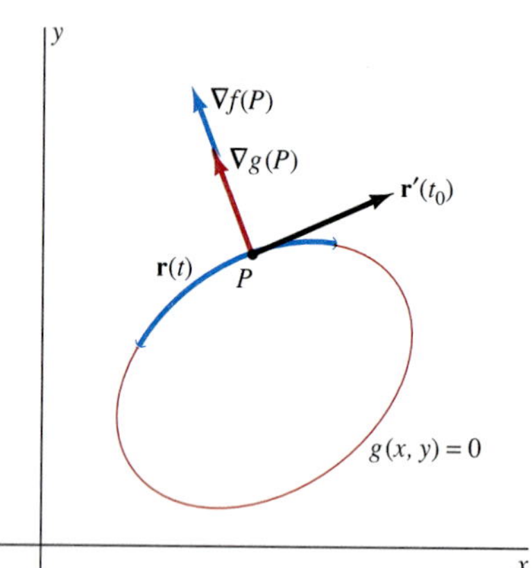

Fig. 13.9.1 The conclusion of Theorem 1 illustrated

Proof By the implicit function theorem mentioned in Section 13.8, the fact that $\nabla g(P) \neq \mathbf{0}$ allows us to represent the curve $g(x, y) = 0$ near P by a parametric curve $\mathbf{r}(t)$, and in such fashion that $\mathbf{r}$ has a nonzero tangent vector near P. Thus $\mathbf{r}'(t) \neq \mathbf{0}$ (Fig. 13.9.1). Let t_0 be the value of t such that $\mathbf{r}(t_0) = \overrightarrow{OP}$. If $f(x, y)$ attains its maximum value at P, then the composite function $f(\mathbf{r}(t))$ attains its maximum value at $t = t_0$, so

$$D_t f(\mathbf{r}(t))\Big|_{t=t_0} = \nabla f(\mathbf{r}(t_0)) \cdot \mathbf{r}'(t_0) = \nabla f(P) \cdot \mathbf{r}'(t_0) = 0. \tag{3}$$

Here we have used the vector chain rule, Eq. (10) of Section 13.8.

Because $\mathbf{r}(t)$ lies on the curve $g(x, y) = 0$, the composite function $g(\mathbf{r}(t))$ is a constant function. Therefore,

$$D_t g(\mathbf{r}(t))\Big|_{t=t_0} = \nabla g(\mathbf{r}(t_0)) \cdot \mathbf{r}'(t_0) = \nabla g(P) \cdot \mathbf{r}'(t_0) = 0. \tag{4}$$

Equations (3) and (4) together imply that both of the vectors $\nabla f(P)$ and $\nabla g(P)$ are perpendicular to the (nonzero) tangent vector $\mathbf{r}'(t_0)$. Hence $\nabla f(P)$ must be a scalar multiple of $\nabla g(P)$, and this is exactly the meaning of Eq. (2). This concludes the proof of the theorem in the case that $f(x, y)$ attains its maximum at P. The proof in the case of a minimum value is almost exactly the same. ∎

The Method

Let's see what steps we should follow to solve a problem by using Theorem 1—the method of Lagrange multipliers. First we need to identify a quantity $z = f(x, y)$ to be maximized or minimized, subject to the constraint $g(x, y) = 0$. Then Eq. (1) and the two scalar components of Eq. (2) yield three equations:

$$g(x, y) = 0 \tag{1}$$

$$f_x(x, y) = \lambda g_x(x, y), \quad \text{and} \tag{2a}$$

$$f_y(x, y) = \lambda g_y(x, y). \tag{2b}$$

Thus we have three equations that we can attempt to solve for the three unknowns x, y, and λ. The points (x, y) that we find (assuming that our efforts are successful) are the only possible locations for the extrema of f subject to the constraint $g(x, y) = 0$. The associated values of λ, called **Lagrange multipliers,** may be revealed as well but often are not of much interest. Finally, we calculate the value $f(x, y)$ at each of the solution points (x, y) in order to spot its maximum and minimum values.

We must bear in mind the additional possibility that the maximum or minimum (or both) of f may occur at a point where $g_x(x, y) = 0 = g_y(x, y)$. The Lagrange multiplier method may fail to locate these exceptional points, but they can usually be recognized as points where the graph $g(x, y) = 0$ fails to be a smooth curve.

EXAMPLE 1 Find the points of the rectangular hyperbola $xy = 1$ that are closest to the origin $(0, 0)$.

Solution We need to minimize the distance $d = \sqrt{x^2 + y^2}$ from the origin of a point $P(x, y)$ on the curve $xy = 1$. But the algebra is simpler if instead we minimize the square

$$f(x, y) = x^2 + y^2$$

of this distance subject to the constraint

$$g(x, y) = xy - 1 = 0$$

that the point P lies on the hyperbola. Because

$$\frac{\partial f}{\partial x} = 2x, \quad \frac{\partial f}{\partial y} = 2y, \quad \text{and} \quad \frac{\partial g}{\partial x} = y, \quad \frac{\partial g}{\partial y} = x,$$

the Lagrange multiplier equations (2a) and (2b) take the form

$$2x = \lambda y, \quad 2y = \lambda x.$$

If we multiply the first of these equations by y and the second by x, we can conclude that

$$2x^2 = \lambda xy = 2y^2$$

at $P(x, y)$. But the fact that $xy = 1 > 0$ implies that x and y have the same sign. Hence the fact that $x^2 = y^2$ implies that $x = y$. Substitution in $xy = 1$ then gives $x^2 = 1$, so it follows finally that either $x = y = +1$ or $x = y = -1$. The two resulting possibilities $(1, 1)$ and $(-1, -1)$ are indicated in Fig. 13.9.2. ■

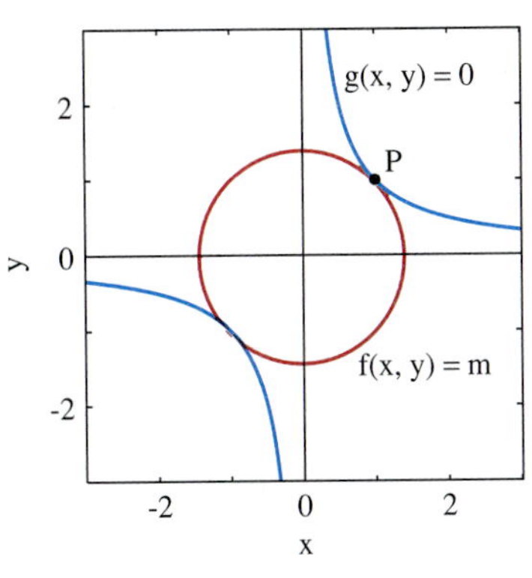

Fig. 13.9.2 The curves $f(x, y) = m$ and $g(x, y) = 0$ are tangent at a constrained maximum or minimum point P of $f(x, y)$.

REMARK Example 1 illustrates an interesting geometric interpretation of Theorem 1. Let $f(P) = m$ be a maximum or minimum value of $f(x, y)$ subject to the constraint $g(x, y) = 0$. Then Theorem 2 of Section 13.8 implies that the gradient vectors $\nabla f(P)$ and $\nabla g(P)$ are normal to the curves $f(x, y) = m$ and $g(x, y) = 0$. The fact that the two normal vectors are collinear then means that *the two curves are tangent at P*. Thus we see in Fig. 13.9.2 that the circle $x^2 + y^2 = 2$ and the hyperbola $xy = 1$ are, indeed, tangent at the two points $(1, 1)$ and $(-1, -1)$ where the squared distance $f(x, y) = x^2 + y^2$ is minimal subject to the constraint $g(x, y) = xy - 1 = 0$.

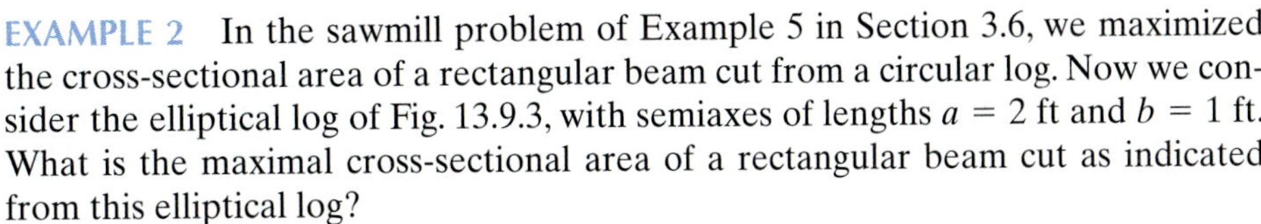

EXAMPLE 2 In the sawmill problem of Example 5 in Section 3.6, we maximized the cross-sectional area of a rectangular beam cut from a circular log. Now we consider the elliptical log of Fig. 13.9.3, with semiaxes of lengths $a = 2$ ft and $b = 1$ ft. What is the maximal cross-sectional area of a rectangular beam cut as indicated from this elliptical log?

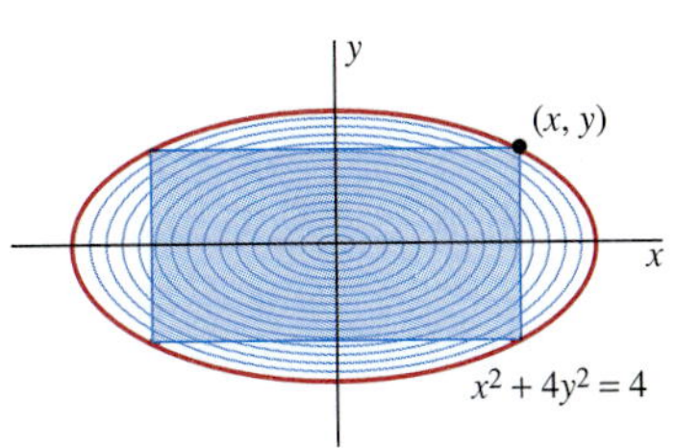

Fig. 13.9.3 Cutting a rectangular beam from an elliptical log (Example 2)

Solution The log is bounded by the ellipse $(x/2)^2 + y^2 = 1$; that is, $x^2 + 4y^2 = 4$. So with the coordinate system indicated in Fig. 13.9.3, we want to maximize the cross-sectional area

$$A = f(x, y) = 4xy \tag{5}$$

of the beam subject to the constraint

$$g(x, y) = x^2 + 4y^2 - 4 = 0. \tag{6}$$

Because

$$\frac{\partial f}{\partial x} = 4y, \quad \frac{\partial f}{\partial y} = 4x \quad \text{and} \quad \frac{\partial g}{\partial x} = 2x, \quad \frac{\partial g}{\partial y} = 8y,$$

Eqs. (2a) and (2b) give

$$4y = 2\lambda x, \qquad 4x = 8\lambda y.$$

It's clear that neither $x = 0$ nor $y = 0$ gives the maximum area, so we can solve these two multiplier equations for

$$\frac{2y}{x} = \lambda = \frac{x}{2y}.$$

Thus $x^2 = 4y^2$ at the desired maximum. Because $x^2 + 4y^2 = 4$, it follows that $x^2 = 4y^2 = 2$. Because we seek (as in Fig. 13.9.3) a first-quadrant solution point (x, y), we conclude that $x = \sqrt{2}$, $y = 1/\sqrt{2}$ gives the maximum possible cross-sectional area $A_{\max} = 4(\sqrt{2})(1/\sqrt{2}) = 4\text{ ft}^2$ of a rectangular beam cut from the elliptical log. Note that this maximum area of 4 ft^2 is about 64% of the total cross-sectional area $A = \pi ab = 2\pi\text{ ft}^2$ of the original log. ■

Remark If we consider all four quadrants, then the condition $x^2 = 4y^2 = 2$ yields the *four* points $(\sqrt{2}, 1/\sqrt{2})$, $(-\sqrt{2}, 1/\sqrt{2})$, $(-\sqrt{2}, -1/\sqrt{2})$, and $(\sqrt{2}, -1/\sqrt{2})$. The function $f(x, y) = 4xy$ in Eq. (5) attains its maximum value $+4$ on the ellipse $x^2 + 4y^2 = 4$ at the first and third of these points and its minimum value -4 at the second and fourth points. The Lagrange multiplier methods thus locates all of the global extrema of $f(x, y)$ on the ellipse.

Remark In the applied maximum-minimum problems of Section 3.6, we typically started with a *formula* such as Eq. (5) of this section, expressing the quantity to be maximized in terms of *two* variables x and y, for example. We then used some available *relation* such as Eq. (6) between the variables x and y to eliminate one of them, such as y. Thus we finally obtained a single-variable *function* by substituting for y in terms of x in the original formula. As in Example 2, the Lagrange multiplier method frees us from the necessity of formulating the problem in terms of a single-variable function and generally leads to a solution process that is algebraically simpler and easier.

Lagrange Multipliers in Three Dimensions

Now suppose that $f(x, y, z)$ and $g(x, y, z)$ have continuous first-order partial derivatives and that we want to find the points of the *surface*

$$g(x, y, z) = 0 \tag{7}$$

at which the function $f(x, y, z)$ attains its maximum and minimum values. With functions of three rather than two variables, Theorem 1 holds precisely as we stated it, with the z-direction taken into account. We leave the details to Problem 45, but an argument similar to the proof of Theorem 1 shows that at a maximum or minimum point P of $f(x, y, z)$ on the surface in Eq. (7), both gradient vectors $\nabla f(P)$ and $\nabla g(P)$ are normal to the surface (Fig. 13.9.4). It follows that

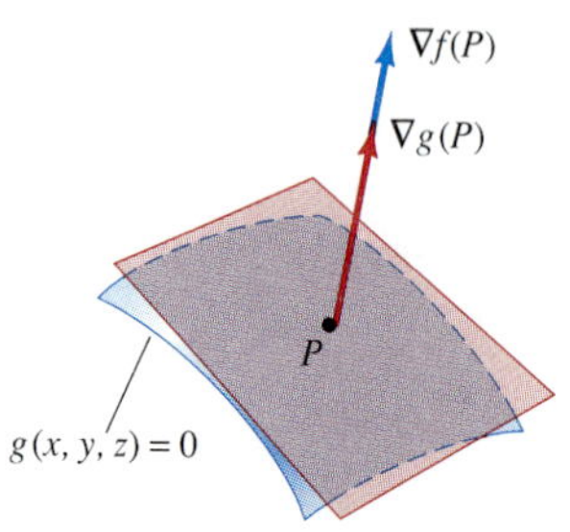

Fig. 13.9.4 The natural generalization of Theorem 1 holds for functions of three variables.

$$\nabla f(P) = \lambda \nabla g(P) \tag{8}$$

for some scalar λ. This vector equation corresponds to three scalar equations. To find the possible locations of the extrema of f subject to the constraint g, we can attempt to solve simultaneously the four equations

$$g(x, y, z) = 0, \tag{7}$$

$$f_x(x, y, z) = \lambda g_x(x, y, z), \tag{8a}$$

$$f_y(x, y, z) = \lambda g_y(x, y, z), \tag{8b}$$

$$f_z(x, y, z) = \lambda g_z(x, y, z) \tag{8c}$$

for the four unknowns x, y, z, and λ. If successful, we then evaluate $f(x, y, z)$ at each of the solution points (x, y, z) to see at which it attains its maximum and minimum values. In analogy to the two-dimensional case, we also check points at which the surface $g(x, y, z) = 0$ fails to be smooth. Thus the Lagrange multiplier method with one constraint is essentially the same in dimension three as in dimension two.

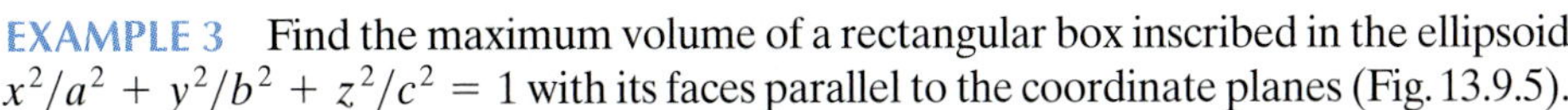

EXAMPLE 3 Find the maximum volume of a rectangular box inscribed in the ellipsoid $x^2/a^2 + y^2/b^2 + z^2/c^2 = 1$ with its faces parallel to the coordinate planes (Fig. 13.9.5).

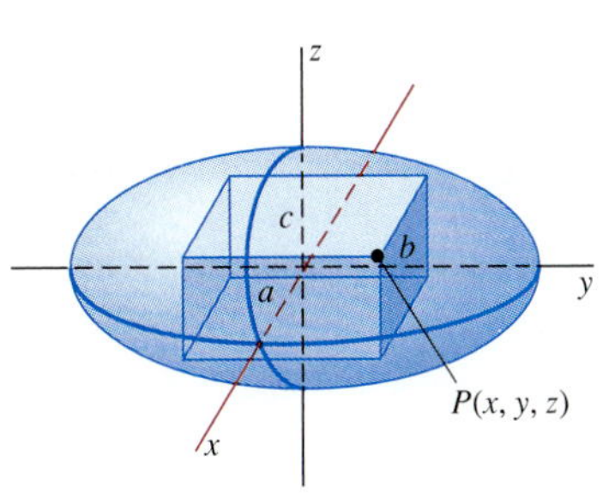

Fig. 13.9.5 A rectangular $2x \times 2y \times 2z$ box inscribed in an ellipsoid with semiaxes a, b, and c. The whole box is determined by its first-octant vertex $P(x, y, z)$.

Solution Let $P(x, y, z)$ be the vertex of the box that lies in the first octant (where x, y, and z are all positive). We want to maximize the volume $V(x, y, z) = 8xyz$ subject to the constraint

$$g(x, y, z) = \frac{x^2}{a^2} + \frac{y^2}{b^2} + \frac{z^2}{c^2} - 1 = 0.$$

Equations (8a), (8b), and (8c) give

$$8yz = \frac{2\lambda x}{a^2}, \qquad 8xz = \frac{2\lambda y}{b^2}, \qquad 8xy = \frac{2\lambda z}{c^2}.$$

Part of the art of mathematics lies in pausing for a moment to find an elegant way to solve a problem rather than rushing in headlong with brute force methods. Here, if we multiply the first equation by x, the second by y, and the third by z, we find that

$$2\lambda\frac{x^2}{a^2} = 2\lambda\frac{y^2}{b^2} = 2\lambda\frac{z^2}{c^2} = 8xyz.$$

Now $\lambda \neq 0$ because (at maximum volume) x, y, and z are nonzero. We conclude that

$$\frac{x^2}{a^2} = \frac{y^2}{b^2} = \frac{z^2}{c^2}.$$

The sum of the last three expressions is 1, because that is precisely the constraint condition in this problem. Thus each of these three expressions is equal to $\frac{1}{3}$. All three of x, y, and z are positive, and therefore

$$x = \frac{a}{\sqrt{3}}, \qquad y = \frac{b}{\sqrt{3}}, \qquad \text{and} \qquad z = \frac{c}{\sqrt{3}}.$$

Therefore the box of maximum volume has volume

$$V = V_{\max} = \frac{8}{3\sqrt{3}}abc.$$

Note that this answer is dimensionally correct—the product of the three *lengths a, b,* and *c* yields a *volume.* But because the volume of the ellipsoid is $V = \frac{4}{3}\pi abc$, and $\left[8/(3\sqrt{3})\right]/(4\pi/3) = 2/(\pi\sqrt{3}) \approx 0.37$, it follows that the maximal box occupies only about 37% of the volume of the circumscribed ellipsoid. Considering the 64% result in Example 2, would you consider this result plausible, or surprising? ■

Problems That Have Two Constraints

Suppose that we want to find the maximum and minimum values of the function $f(x, y, z)$ at points of the curve of intersection of the two surfaces

$$g(x, y, z) = 0 \quad \text{and} \quad h(x, y, z) = 0. \tag{9}$$

This is a maximum-minimum problem with *two* constraints. The Lagrange multiplier method for such situations is based on Theorem 2.

Theorem 2 ***Lagrange Multipliers (two constraints)***

Let $f(x, y, z)$, $g(x, y, z)$, and $h(x, y, z)$ be functions with continuous first-order partial derivatives. If the maximum (or minimum) value of f subject to the two conditions

$$g(x, y, z) = 0 \quad \text{and} \quad h(x, y, z) = 0 \tag{9}$$

occurs at a point P where the vectors $\nabla g(P)$ and $\nabla h(P)$ are nonzero and non-parallel, then

$$\nabla f(P) = \lambda_1 \nabla g(P) + \lambda_2 \nabla h(P) \tag{10}$$

for some two constants λ_1 and λ_2.

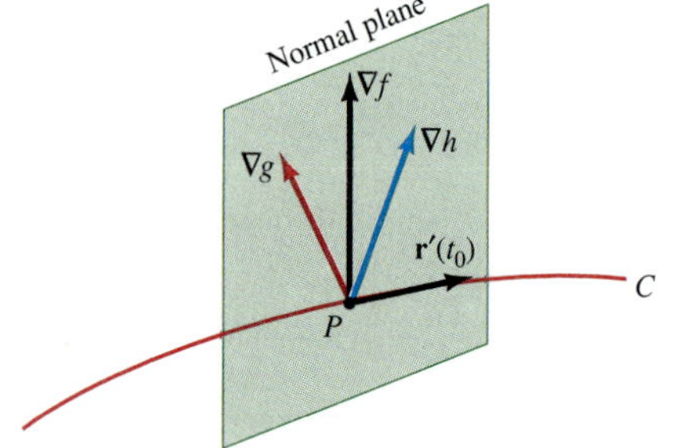

Fig. 13.9.6 The relation between the gradient vectors in the proof of Theorem 2

Outline of Proof: By an appropriate version of the implicit function theorem, the curve C of intersection of the two surfaces (Fig. 13.9.6) may be represented near P by a parametric curve $\mathbf{r}(t)$ with nonzero tangent vector $\mathbf{r}'(t)$. Let t_0 be the value of t such that such that $\mathbf{r}(t_0) = \overrightarrow{OP}$. We compute the derivatives at t_0 of the composite functions $f(\mathbf{r}(t))$, $g(\mathbf{r}(t))$, and $h(\mathbf{r}(t))$. We find —exactly as in the proof of Theorem 1—that

$$\nabla f(P)\cdot\mathbf{r}'(t_0) = 0, \quad \nabla g(P)\cdot\mathbf{r}'(t_0) = 0, \quad \text{and} \quad \nabla h(P)\cdot\mathbf{r}'(t_0) = 0.$$

These three equations imply that all three gradient vectors are perpendicular to the curve C at P and thus that they all lie in a single plane, the plane normal to the curve C at the point P.

Now $\nabla g(P)$ and $\nabla h(P)$ are nonzero and nonparallel, so $\nabla f(P)$ is the sum of its projections onto $\nabla g(P)$ and $\nabla h(P)$ (see Problem 65 of Section 12.2). As illustrated in Fig. 13.9.7, this fact implies Eq. (8). ■

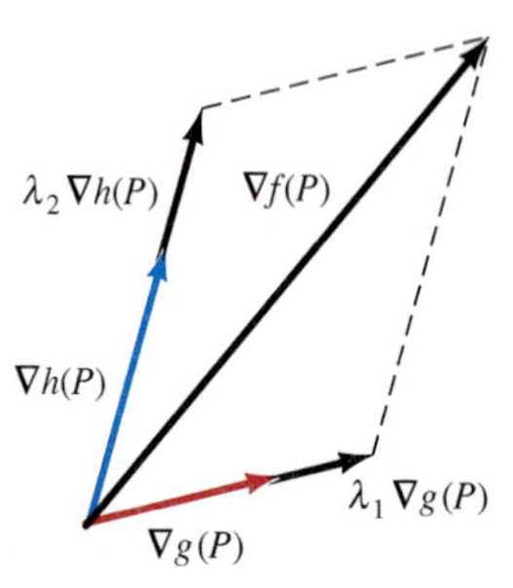

Fig. 13.9.7 Geometry of the equation $\nabla f(P) = \lambda_1 \nabla g(P) + \lambda_2 \nabla h(P)$

In examples we prefer to avoid subscripts by writing λ and μ for the Lagrange multipliers λ_1 and λ_2 in the statement of Theorem 2. The equations in (9) and the three scalar components of the vector equation in (10) then give rise to the five simultaneous equations

$$g(x, y, z) = 0, \tag{9a}$$

$$h(x, y, z) = 0, \tag{9b}$$

$$f_x(x, y, z) = \lambda g_x(x, y, z) + \mu h_x(x, y, z), \tag{10a}$$

$$f_y(x, y, z) = \lambda g_y(x, y, z) + \mu h_y(x, y, z), \tag{10b}$$

$$f_z(x, y, z) = \lambda g_z(x, y, z) + \mu h_z(x, y, z) \tag{10c}$$

in the five unknowns x, y, z, λ, and μ.

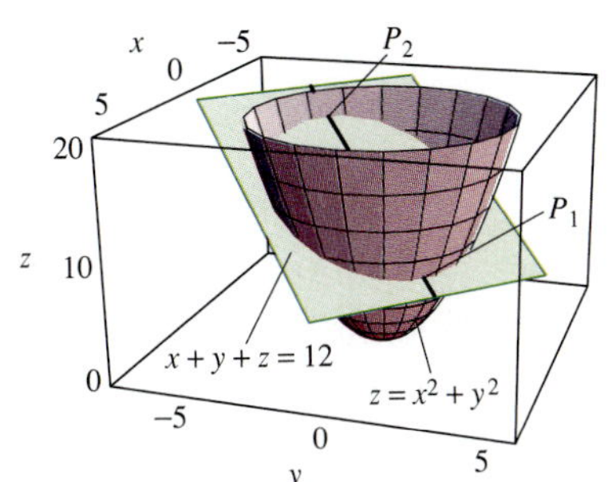

Fig. 13.9.8 The plane and paraboloid intersecting in the ellipse of Example 4

EXAMPLE 4 The plane $x + y + z = 12$ intersects the paraboloid $z = x^2 + y^2$ in an ellipse (Fig. 13.9.8). Find the highest and lowest points on this ellipse.

Solution The height of the point (x, y, z) is z, so we want to find the maximum and minimum values of

$$f(x, y, z) = z \tag{11}$$

subject to the two conditions

$$g(x, y, z) = x + y + z - 12 = 0 \tag{12}$$

and

$$h(x, y, z) = x^2 + y^2 - z = 0. \tag{13}$$

The conditions in (10a) through (10c) yield

$$0 = \lambda + 2\mu x, \tag{14a}$$

$$0 = \lambda + 2\mu y, \quad \text{and} \tag{14b}$$

$$1 = \lambda - \mu. \tag{14c}$$

If μ were zero, then Eq. (14a) would imply that $\lambda = 0$, which contradicts Eq. (14c). Hence $\mu \neq 0$, and therefore the equations

$$2\mu x = -\lambda = 2\mu y$$

imply that $x = y$. Substitution of $x = y$ into Eq. (13) gives $z = 2x^2$, and then Eq. (12) yields

$$2x^2 + 2x - 12 = 0;$$

$$x^2 + x - 6 = 0;$$

$$(x + 3)(x - 2) = 0.$$

Thus we obtain the two solutions $x = -3$ and $x = 2$. Because $y = x$ and $z = 2x^2$, the corresponding points of the ellipse are $P_1(2, 2, 8)$ and $P_2(-3, -3, 18)$. It's clear which is the lowest and which is the highest. ■

REMARK The Lagrange multiplier problems in Examples 1 through 4 are somewhat unusual in that the equations in these examples can be solved exactly with little difficulty. Frequently a Lagrange multiplier problem leads to a system of equations that can be solved only numerically and approximately—for instance, using a computer algebra system. Thus the *Mathematica* commands

```
equations =
{ x + y + z == 12, z == x² + y², λ+ 2 x μ== 0, λ+ 2 y μ== 0, λ- μ== 1 }
```

```
unknowns = {x,y,z,λ,μ}
NSolve[ equations, unknowns ]
{{λ→0.8, μ→-0.2, x→2., y→2., z→8.},
{λ→1.2, μ→0.2, x→-3., y→-3., z→18.}}
```

define and solve (read **NSolve** as "numerically solve") the system in (12)–(14) of equations in Example 4. After entering similarly the equations and unknowns, the *Maple* command

```
fsolve( equations, unknowns )
```

is analogous (read **fsolve** as "floating point solve").

13.9 PROBLEMS

In Problems 1 through 18, find the maximum and minimum values—if any—of the given function f subject to the given constraint or constraints.

1. $f(x, y) = 2x + y;\quad x^2 + y^2 = 1$

2. $f(x, y) = x + y;\quad x^2 + 4y^2 = 1$

3. $f(x, y) = x^2 - y^2;\quad x^2 + y^2 = 4$

4. $f(x, y) = x^2 + y^2;\quad 2x + 3y = 6$

5. $f(x, y) = xy;\quad 4x^2 + 9y^2 = 36$

6. $f(x, y) = 4x^2 + 9y^2;\quad x^2 + y^2 = 1$

7. $f(x, y, z) = x^2 + y^2 + z^2;\quad 3x + 2y + z = 6$

8. $f(x, y, z) = 3x + 2y + z;\quad x^2 + y^2 + z^2 = 1$

9. $f(x, y, z) = x + y + z;\quad x^2 + 4y^2 + 9z^2 = 36$

10. $f(x, y, z) = xyz;\quad x^2 + y^2 + z^2 = 1$

11. $f(x, y, z) = xy + 2z;\quad x^2 + y^2 + z^2 = 36$

12. $f(x, y, z) = x - y + z;\quad z = x^2 - 6xy + y^2$

13. $f(x, y, z) = x^2y^2z^2;\quad x^2 + 4y^2 + 9z^2 = 27$

14. $f(x, y, z) = x^2 + y^2 + z^2;\quad x^4 + y^4 + z^4 = 3$

15. $f(x, y, z) = x^2 + y^2 + z^2;\quad x + y + z = 1$ and $x + 2y + 3z = 6$

16. $f(x, y, z) = z;\quad x^2 + y^2 = 1$ and $2x + 2y + z = 5$

17. $f(x, y, z) = z;\quad x + y + z = 1$ and $x^2 + y^2 = 1$

18. $f(x, y, z) = x;\quad x + y + z = 12$ and $4y^2 + 9z^2 = 36$

19. Find the point on the line $3x + 4y = 100$ that is closest to the origin. Use Lagrange multipliers to minimize the *square* of the distance.

20. A rectangular open-topped box is to have volume 700 in.^{3} The material for its bottom costs 7¢/in.^{2}, and the material for its four vertical sides costs 5¢/in.^{2} Use the method of Lagrange multipliers to find what dimensions will minimize the cost of the material used in constructing this box.

In Problems 21 through 34, use the method of Lagrange multipliers to solve the indicated problem from Section 13.5.

21. Problem 29 **22.** Problem 30

23. Problem 31 **24.** Problem 32

25. Problem 33 **26.** Problem 34

27. Problem 35 **28.** Problem 36

29. Problem 37 **30.** Problem 38

31. Problem 39 **32.** Problem 40

33. Problem 41 **34.** Problem 42

35. Find the point or points of the surface $z = xy + 5$ closest to the origin. (*Suggestion:* Minimize the *square* of the distance.)

36. A triangle with sides x, y, and z has fixed perimeter $2s = x + y + z$. Its area A is given by *Heron's formula:*

$$A = \sqrt{s(s - x)(s - y)(s - z)}.$$

Use the method of Lagrange multipliers to show that, among all triangles with the given perimeter, the one of largest area is equilateral. (*Suggestion:* Consider maximizing A^2 rather than A.)

37. Use the method of Lagrange multipliers to show that, of all triangles inscribed in the unit circle, the one of greatest area is equilateral. (*Suggestion:* Use Fig. 13.9.9 and the fact that the area of a triangle with sides a and b and included angle θ is given by the formula $A = \frac{1}{2}ab\sin\theta$.)

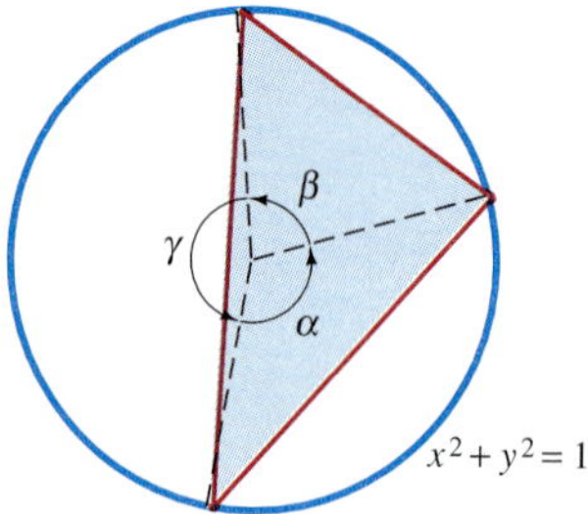

Fig. 13.9.9 A triangle inscribed in a circle (Problem 37)

38. Find the points on the rotated ellipse $x^2 + xy + y^2 = 3$ that are closest to and farthest from the origin. (*Suggestion:* Write the Lagrange multiplier equations in the form

$$ax + by = 0,$$

$$cx + dy = 0.$$

These equations have a nontrivial solution *only if* $ad - bc = 0$. Use this fact to solve first for λ.)

39. Use the method of Problem 38 to find the points of the rotated hyperbola $x^2 + 12xy + 6y^2 = 130$ that are closest to the origin.

40. Find the points of the ellipse $4x^2 + 9y^2 = 36$ that are closest to the point $(1, 1)$ as well as the point or points farthest from it.

41. Find the highest and lowest points on the ellipse formed by the intersection of the cylinder $x^2 + y^2 = 1$ and the plane $2x + y - z = 4$.

42. Apply the method of Example 4 to find the highest and lowest points on the ellipse formed by the intersection of the cone $z^2 = x^2 + y^2$ and the plane $x + 2y + 3z = 3$.

43. Find the points on the ellipse of Problem 42 that are nearest the origin and those that are farthest from it.

44. The ice tray shown in Fig. 13.9.10 is to be made from material that costs 1¢/in.^{2} Minimize the cost function $f(x, y, z) = xy + 3xz + 7yz$ subject to the constraints that each of the 12 compartments is to have a square horizontal cross section and that the total volume (ignoring the partitions) is to be 12 in.^{3}

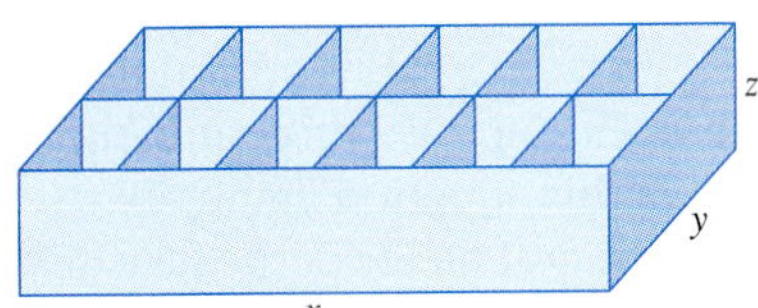

Fig. 13.9.10 The ice tray of Problem 44

45. Prove Theorem 1 for functions of three variables by showing that both of the vectors $\nabla f(P)$ and $\nabla g(P)$ are perpendicular at P to every curve on the surface $g(x, y, z) = 0$.

46. Find the lengths of the semiaxes of the ellipse of Example 4.

47. Figure 13.9.11 shows a right triangle with sides x, y, and z and fixed perimeter P. Maximize its area $A = \frac{1}{2}xy$ subject to the constraints $x + y + z = P$ and $x^2 + y^2 = z^2$. In particular, show that the optimal such triangle is isosceles (by showing that $x = y$).

48. Figure 13.9.12 shows a general triangle with sides x, y, and z and fixed perimeter P. Maximize its area

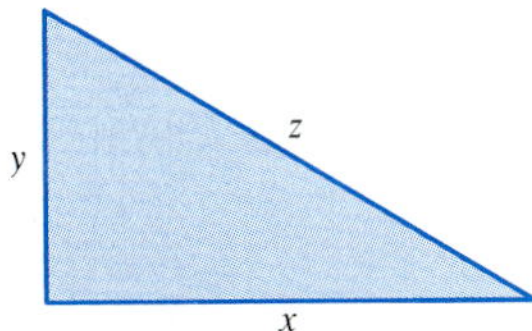

Fig. 13.9.11 A right triangle with fixed perimeter P (Problem 47)

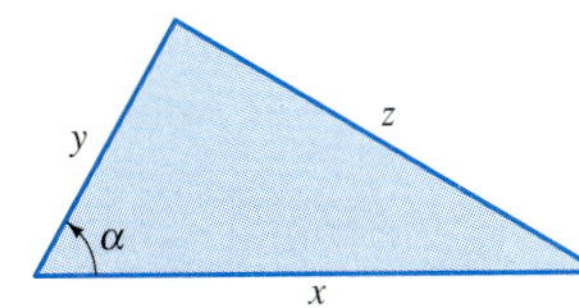

Fig. 13.9.12 A general triangle with fixed perimeter P (Problem 48)

$$A = f(x, y, z, \alpha) = \tfrac{1}{2}xy \sin \alpha$$

subject to the constraints $x + y + z = P$ and

$$z^2 = x^2 + y^2 - 2xy \cos \alpha$$

(the law of cosines). In particular, show that the optimal such triangle is equilateral (by showing that $x = y = z$). [*Note:* The Lagrange multiplier equations for optimizing $f(x, y, z, w)$ subject to the constraint $g(x, y, z, w) = 0$ take the form

$$f_x = \lambda g_x, \quad f_y = \lambda g_y, \quad f_z = \lambda g_z, \quad f_w = \lambda g_w;$$

that is, $\nabla f = \lambda \nabla g$ in terms of the gradient vectors with four components.]

49. Figure 13.9.13 shows a hexagon with vertices $(0, \pm 1)$ and $(\pm x, \pm y)$ inscribed in the unit circle $x^2 + y^2 = 1$. Show that its area is maximal when it is a *regular* hexagon with equal sides and angles.

50. When the hexagon of Fig. 13.9.13 is rotated around the y-axis, it generates a solid of revolution consisting of a cylinder and two cones (Fig. 13.9.14). What radius and cylinder height maximize the volume of this solid?

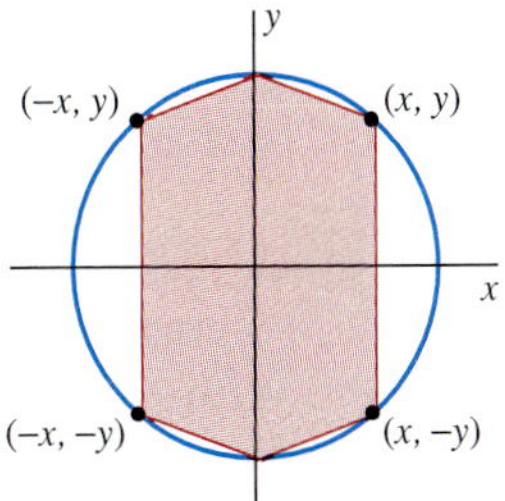

Fig. 13.9.13 The inscribed hexagon of Problem 49

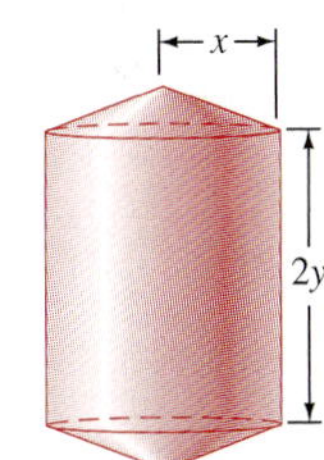

Fig. 13.9.14 The solid of Problem 50

In Problems 51 through 58, consider the *square* of the distance to be maximized or minimized. Use the numerical solution command in a computer algebra system as needed to solve the appropriate Lagrange multiplier equations.

51. Find the points of the parabola $y = (x - 1)^2$ that are closest to the origin.

52. Find the points of the ellipse $4x^2 + 9y^2 = 36$ that are closest to and farthest from the points $(3, 2)$.

53. Find the first-quadrant point of the curve $xy = 24$ that is closest to the point $(1, 4)$.

54. Find the point of the surface $xyz = 1$ that is closest to the point $(1, 2, 3)$.

55. Find the points on the sphere with center $(1, 2, 3)$ and radius 6 that are closest to and farthest from the origin.

56. Find the points of the ellipsoid $4x^2 + 9y^2 + z^2 = 36$ that are closest to and farthest from the origin.

57. Find the points of the ellipse $4x^2 + 9y^2 = 36$ that are closest to and farthest from the straight line $x + y = 10$.

58. Find the points on the ellipsoid $4x^2 + 9y^2 + z^2 = 36$ that are closest to and farthest from the plane $2x + 3y + z = 10$.

59. Find the maximum possible volume of a rectangular box that has its base in the xy-plane and its upper vertices on the elliptic paraboloid $z = 9 - x^2 - 2y^2$.

60. The plane $4x + 9y + z = 0$ intersects the elliptic paraboloid $z = 2x^2 + 3y^2$ in an ellipse. Find the highest and lowest points on this ellipse.

13.9 PROJECT: NUMERICAL INVESTIGATION OF LAGRANGE MULTIPLIER PROBLEMS

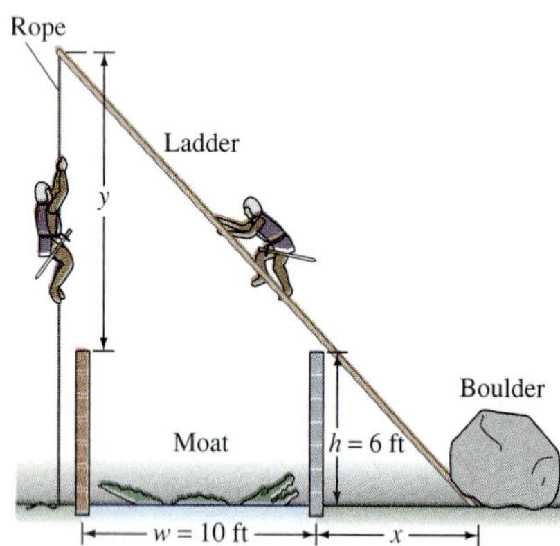

Fig. 13.9.15 The alligator-filled moat of Investigation A

INVESTIGATION A Figure 13.9.15 shows an alligator-filled moat of width $w = 10$ ft. bounded on each side by a wall of height $h = 6$ ft. Soldiers plan to bridge this moat by scaling a ladder that is placed across the wall as indicated and anchored at the ground by a handy boulder, with the upper end directly above the wall on the opposite side of the moat. What is the minimal length L of a ladder that will suffice for this purpose? We outline two approaches.

WITH A SINGLE CONSTRAINT Apply the Pythagorean theorem and the proportionality theorem for similar triangles to show that you need to minimize the (ladder-length-squared) function $f(x, y) = (x + 10)^2 + (y + 6)^2$ subject to the constraint $g(x, y) = xy - 60 = 0$. Then apply the Lagrange multiplier method to derive the fourth-degree equation

$$x^4 + 10x^3 - 360x - 3600 = 0. \qquad \textbf{(1)}$$

You can approximate the pertinent solution of this equation graphically (Fig. 13.9.16). You may even be able to solve this equation manually—if you can first spot an integer solution (which must be an integral factor of the constant term 3600).

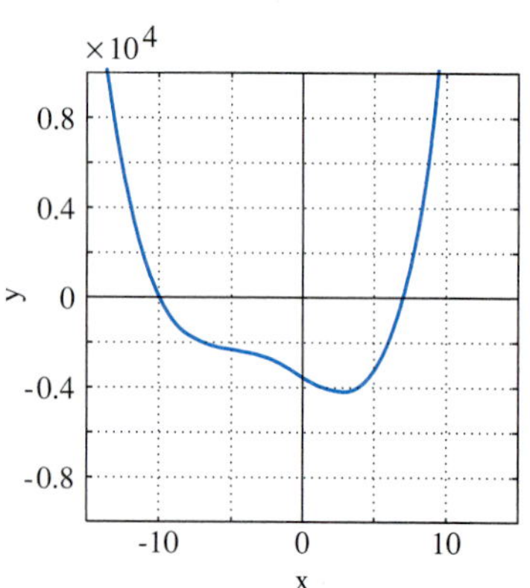

Fig. 13.9.16 $y = x^4 + 10x^3 - 360x - 3600$ (Investigation A)

WITH TWO CONSTRAINTS You can avoid the manual algebra involved in deriving and solving the quartic equation in (1) if a computer algebra system is available to you. With $z = L$ for the length of the ladder, observe directly from Fig. 13.9.15 that you need to minimize the function $f(x, y, z) = z$ subject to the two constraints

$$g(x, y, z) = xy - 60 = 0,$$

$$h(x, y, z) = (x + 10)^2 + (y + 6)^2 - z^2 = 0.$$

This leads to a system of five equations in five unknowns (x, y, z, and the two Lagrange multipliers).

For your own personal moat problem, you might choose w and $h < w$ as the two largest distinct digits in your student I.D. number.

INVESTIGATION B Figure 13.9.17 shows a 14-sided polygon that is almost inscribed in the unit circle. It has vertices $(0, \pm 1)$, $(\pm x, \pm y)$, $(\pm u, \pm v)$, and $(\pm u, \pm y)$. When this polygon is revolved around the y-axis, it generates the "spindle solid" of Fig. 13.9.18, which consists of a solid cylinder of radius x, two solid cylinders of radius u, and two cones. The problem is to determine x, y, u, and v to maximize the volume of this spindle.

First express the volume V of the spindle as a function $V = f(x, y, u, v)$ of four variables. The problem then is to maximize $f(x, y, u, v)$ subject to the two constraints

$$g(x, y, u, v) = x^2 + y^2 - 1 = 0,$$

$$h(x, y, u, v) = u^2 + v^2 - 1 = 0.$$

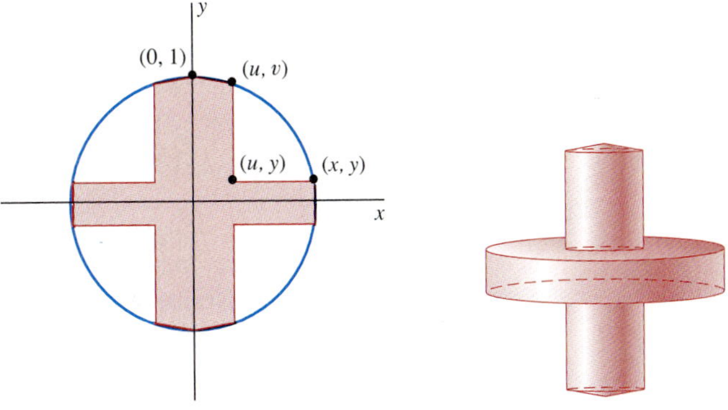

Fig. 13.9.17 The polygon of Investigation B

Fig. 13.9.18 The spindle of Investigation B

The corresponding Lagrange multiplier condition takes the form

$$\nabla f = \lambda \nabla g + \mu \nabla h,$$

where $\nabla f = \langle f_x, f_y, f_u, f_v \rangle$ and ∇g and ∇h are similar 4-vectors of partial derivatives.

All this results in a system of six equations in the six unknowns x, y, u, v, λ, and μ. Set up this system, but you probably should attempt to solve it only if a computer algebra system is available to you.

13.10 THE SECOND DERIVATIVE TEST FOR FUNCTIONS OF TWO VARIABLES

We saw in Section 13.5 that in order for the differentiable function $f(x, y)$ to have either a local minimum or a local maximum at an interior critical point $P(a, b)$ of its domain, it is a *necessary* condition that P be a *critical point* of f—that is, that

$$f_x(a, b) = 0 = f_y(a, b).$$

Here we give conditions *sufficient* to ensure that f has a local extremum at a critical point. The criterion stated in Theorem 1 involves the second-order partial derivatives of f at (a, b) and plays the role of the single-variable second derivative test (Section 4.6) for functions of two variables. To simplify the statement of this result, we use the following abbreviations:

$$A = f_{xx}(a, b), \quad B = f_{xy}(a, b), \quad C = f_{yy}(a, b), \tag{1}$$

and

$$\Delta = AC - B^2 = f_{xx}(a, b)f_{yy}(a, b) - [f_{xy}(a, b)]^2. \tag{2}$$

We outline a proof of Theorem 1 at the end of this section.

Theorem 1 ***Sufficient Conditions for Local Extrema***
Let (a, b) be a critical point of the function $f(x, y)$, and suppose that f has continuous first- and second-order partial derivatives in some circular disk centered at (a, b).

1. If $\Delta > 0$ and $A > 0$, then f has a local minimum at (a, b).
2. If $\Delta > 0$ and $A < 0$, then f has a local maximum at (a, b).
3. If $\Delta < 0$, then f has neither a local minimum nor a local maximum at (a, b). Instead, it has a saddle point there.

Thus f has *either* a local maximum *or* a local minimum at the critical point (a, b) provided that the **discriminant** $\Delta = AC - B^2$ is *positive*. In this case, $A = f_{xx}(a, b)$ plays the role of the second derivative of a single-variable function: There is a local minimum at (a, b) if $A > 0$ and a local maximum if $A < 0$.

If $\Delta < 0$, then f has *neither* a local maximum *nor* a local minimum at (a, b). In this case we call (a, b) a **saddle point** of f, thinking of the appearance of the hyperbolic paraboloid $f(x, y) = x^2 - y^2$ (Fig. 13.10.1), a typical example of this case.

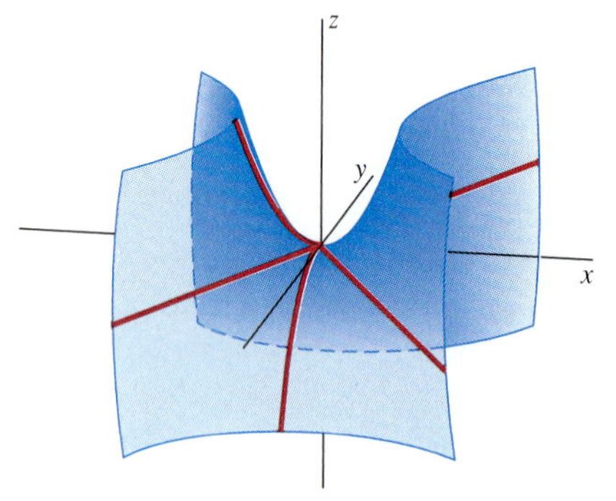

Fig. 13.10.1 The origin is a saddle point of the surface with equation $z = x^2 - y^2$.

Theorem 1 does not answer the question of what happens when $\Delta = 0$. In this case, the two-variable second derivative test fails—it gives no information. Moreover, at such a point (a, b), *anything* can happen, ranging from the local (indeed global) minimum of $f(x, y) = x^4 + y^4$ at $(0, 0)$ to the "monkey saddle" of Example 2.

In the case of a function $f(x, y)$ with several critical points, we must compute the quantities A, B, C, and Δ separately at each critical point in order to apply the test.

EXAMPLE 1 Locate and classify the critical points of

$$f(x, y) = 3x - x^3 - 3xy^2.$$

Solution This function is a polynomial, so all its partial derivatives exist and are continuous everywhere. When we equate its first partial derivatives to zero (to locate the critical points of f), we get

$$f_x(x, y) = 3 - 3x^2 - 3y^2 = 0, \qquad f_y(x, y) = -6xy = 0.$$

The second of these equations implies that x or y must be zero; then the first implies that the other must be ± 1. Thus there are four critical points: $(1, 0)$, $(-1, 0)$, $(0, 1)$, and $(0, -1)$.

The second-order partial derivatives of f are

$$A = f_{xx}(x, y) = -6x, \qquad B = f_{xy}(x, y) = -6y, \qquad C = f_{yy}(x, y) = -6x.$$

Hence $\Delta = 36(x^2 - y^2)$ at each of the critical points. The table in Fig. 13.10.2 summarizes the situation at each of the four critical points, which are labeled in the con-

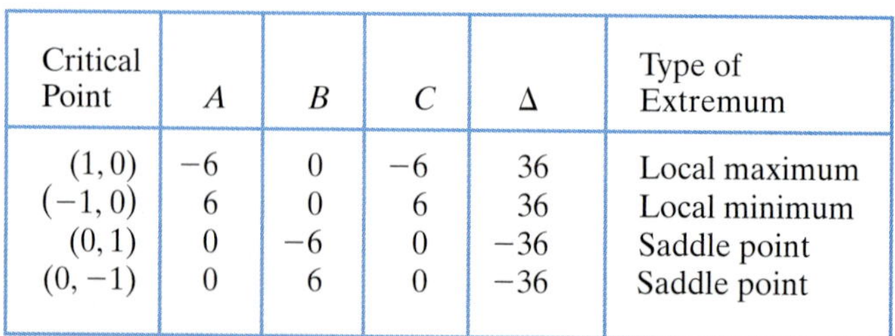

Critical Point	A	B	C	Δ	Type of Extremum
$(1, 0)$	-6	0	-6	36	Local maximum
$(-1, 0)$	6	0	6	36	Local minimum
$(0, 1)$	0	-6	0	-36	Saddle point
$(0, -1)$	0	6	0	-36	Saddle point

Fig. 13.10.2 Critical-point analysis for the function of Example 1

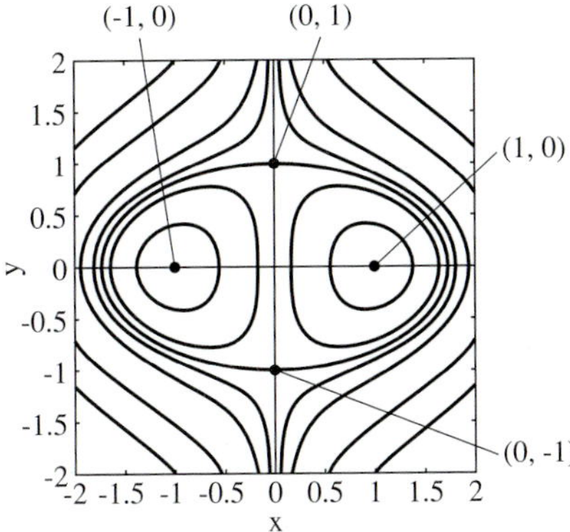

Fig. 13.10.3 Contour curves for the function of Example 1

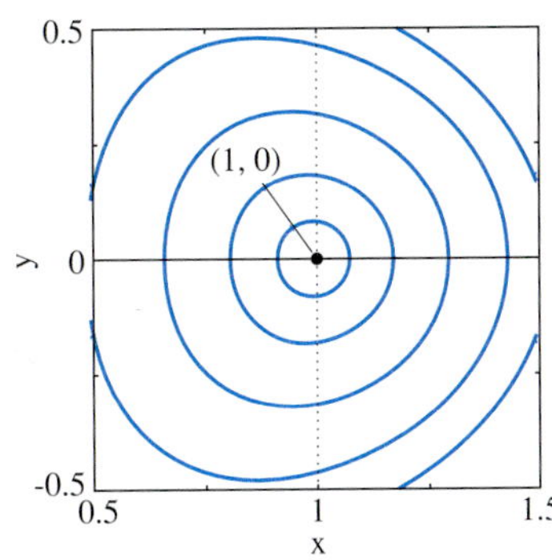

Fig. 13.10.4 Contour curves near the critical point $(1, 0)$

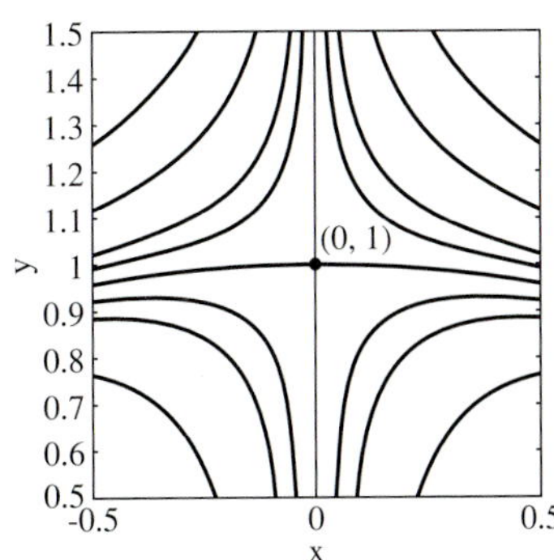

Fig. 13.10.5 Contour curves near the critical point $(0, 1)$

tour plot in Fig. 13.10.3. Near the points $(\pm 1, 0)$ we see the nested "ellipse-like" contours that signal local extrema (Fig. 13.10.4), and near the points $(0, \pm 1)$ we see "hyperbola-like" contours that signal saddle points (Fig. 13.10.5). Figure 13.10.6 shows the critical points on the two-dimensional graph $z = f(x, y)$. ■

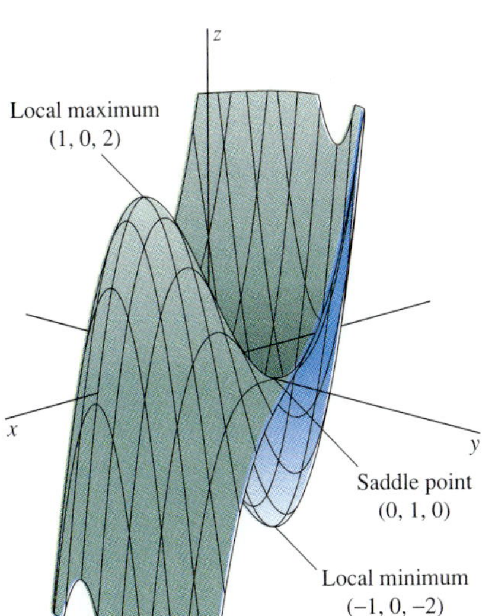

Fig. 13.10.6 Graph of the function of Example 1

EXAMPLE 2 Find and classify the critical points of the function

$$f(x, y) = 6xy^2 - 2x^3 - 3y^4.$$

Solution When we equate the first-order partial derivatives to zero, we get the equations

$$f_x(x, y) = 6y^2 - 6x^2 = 0 \quad \text{and} \quad f_y(x, y) = 12xy - 12y^3 = 0.$$

It follows that

$$x^2 = y^2 \quad \text{and} \quad y(x - y^2) = 0.$$

The first of these equations gives $x = \pm y$. If $x = y$, the second equation implies that $y = 0$ or $y = 1$. If $x = -y$, the second equation implies that $y = 0$ or $y = -1$. Hence there are three critical points: $(0, 0)$, $(1, 1)$, and $(1, -1)$.

The second-order partial derivatives of f are

$$A = f_{xx}(x, y) = -12x, \quad B = f_{xy}(x, y) = 12y, \quad C = f_{yy}(x, y) = 12x - 36y^2.$$

These expressions give the data shown in the table in Fig. 13.10.7. The critical point test fails at $(0, 0)$, so we must find another way to test this point.

Critical point	A	B	C	Δ	Type of Extremum
$(0, 0)$	0	0	0	0	Test fails
$(1, 1)$	-12	12	-24	144	Local maximum
$(1, -1)$	-12	-12	-24	144	Local maximum

Fig. 13.10.7 Critical-point analysis for the function of Example 2

We observe that $f(x, 0) = -2x^3$ and that $f(0, y) = -3y^4$. Hence, as we move away from the origin in the

Positive x-direction: f decreases;

Negative x-direction: f increases;

Positive y-direction: f decreases;

Negative y-direction: f decreases.

Consequently, f has neither a local maximum nor a local minimum at the origin. The graph of f is shown in Fig. 13.10.8. If a monkey were to sit with its rump at the origin and face the negative x-direction, then the directions in which $f(x, y)$ decreases would provide places for both its tail and its two legs to hang. That's why this particular surface is called a *monkey saddle* (Fig. 13.10.9). ■

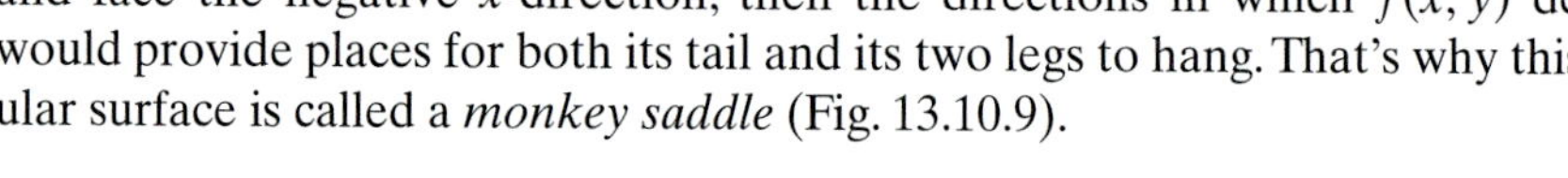

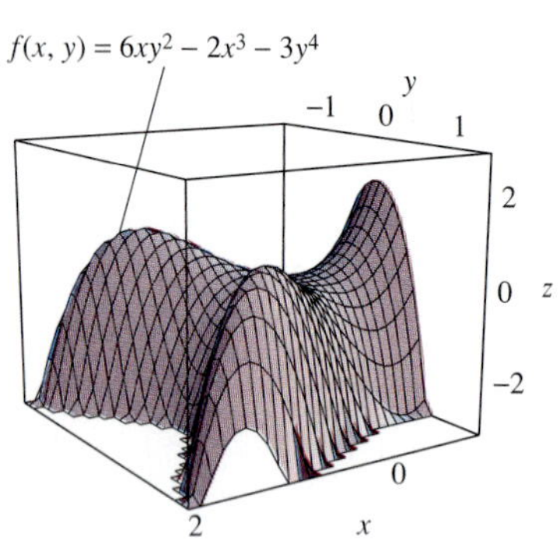

Fig. 13.10.8 The monkey saddle of Example 2

Fig. 13.10.9 The monkey in its saddle (Example 2)

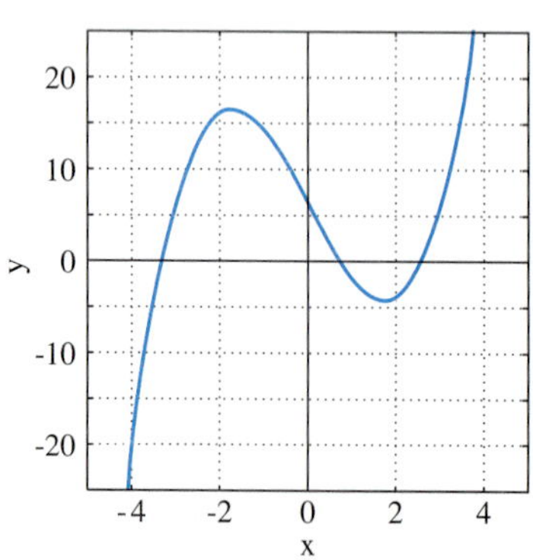

Fig. 13.10.10 The graph of $\phi(x) = x^3 - 9x + 6$ (Example 3)

EXAMPLE 3 Find and classify the critical points of the function

$$f(x, y) = \tfrac{1}{3}x^4 + \tfrac{1}{2}y^4 - 4xy^2 + 2x^2 + 2y^2 + 3.$$

Solution When we equate to zero the first-order partial derivatives of f, we obtain the equations

$$f_x(x, y) = \tfrac{4}{3}x^3 - 4y^2 + 4x = 0, \tag{3}$$

$$f_y(x, y) = 2y^3 - 8xy + 4y = 0, \tag{4}$$

which are not as easy to solve as the corresponding equations in Example 1 and 2. But if we write Eq. (4) in the form

$$2y(y^2 - 4x + 2) = 0,$$

we see that either $y = 0$ or

$$y^2 = 4x - 2. \tag{5}$$

If $y = 0$, then Eq. (3) reduces to the equation

$$\tfrac{4}{3}x^3 + 4x = \tfrac{4}{3}x(x^2 + 3) = 0,$$

whose only solution is $x = 0$. Thus one critical point of f is $(0, 0)$.

If $y \neq 0$, we substitute $y^2 = 4x - 2$ into Eq. (3) to obtain

$$\tfrac{4}{3}x^3 - 4(4x - 2) + 4x = 0;$$

that is,

$$\tfrac{4}{3}x^3 - 12x + 8 = 0.$$

Thus we need to solve the cubic equation

$$\phi(x) = x^3 - 9x + 6 = 0. \tag{6}$$

The graph of $\phi(x)$ in Fig. 13.10.10 shows that this equation has three real solutions with approximate values $x \approx -3$, $x \approx 1$, and $x \approx 3$. Using either graphical techniques or Newton's method (Section 3.9), you can obtain the values

$$x \approx -3.2899, \qquad x \approx 0.7057, \qquad x \approx 2.5842, \tag{7}$$

accurate to four decimal places. The corresponding values of y are given from Eq. (5) by

$$y = \pm\sqrt{4x - 2}, \tag{8}$$

but the first value of x in (7) yields *no* real value at all for y. Thus the two positive values of x in (7) add *four* critical points of $f(x, y)$ to the one critical point $(0, 0)$ already found.

These five critical points are listed in the table in Fig. 13.10.11, together with the corresponding values of

$$A = f_{xx}(x, y) = 4x^2 + 4, \qquad B = f_{xy}(x, y) = -8y,$$

$$C = f_{yy}(x, y) = 6y^2 - 8x + 4, \quad \Delta = AC - B^2$$

Critical point	1	2	3	4	5
x	0.0000	0.7057	0.7057	2.5842	2.5842
y	0.0000	0.9071	−0.9071	2.8874	2.8874
z	3.0000	3.7402	3.7402	−3.5293	−3.5293
A	4.00	5.99	5.99	30.71	30.71
B	0.00	−7.26	7.26	−23.10	23.10
C	4.00	3.29	3.29	33.35	33.35
Δ	16.00	−32.94	−32.94	490.64	490.64
Type	Local minimum	Saddle point	Saddle point	Local minimum	Local minimum

Fig. 13.10.11 Classification of the critical points in Example 3

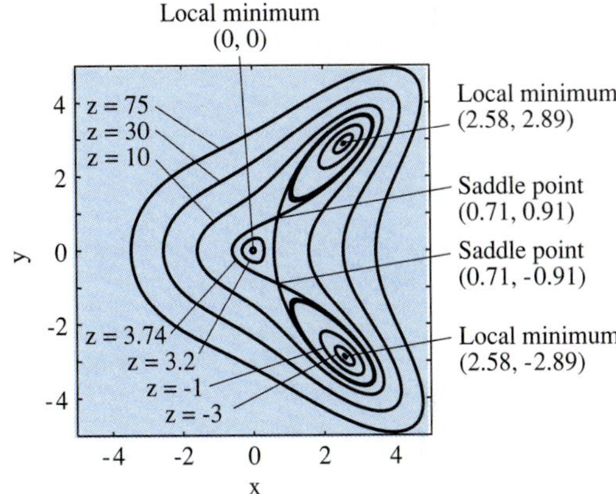

Fig. 13.10.12 Level curves for the function of Example 3

(rounded to two decimal places) at each of these critical points. We see that $\Delta > 0$ and $A > 0$ at $(0, 0)$ and at $(2.5482, \pm 2.8874)$, so these points are local minimum points. But $\Delta < 0$ at $(0.7057, \pm 0.9071)$, so these two are saddle points. The level curve diagram in Fig. 13.10.12 shows how these five critical points fit together.

Finally, we observe that the behavior of $f(x, y)$ is approximately that of $\frac{1}{3}x^4 + \frac{1}{2}y^4$ when $|x|$ or $|y|$ is large, so the surface $z = f(x, y)$ must open upward and therefore have a global low point (but no global high point). Examining the values

$$f(0, 0) = 3 \quad \text{and} \quad f(2.5842, \pm 2.8874) \approx -3.5293,$$

we see that the global minimum value of $f(x, y)$ is approximately -3.5293. ■

Discussion of Theorem 1

A complete proof of Theorem 1 (providing sufficient conditions for local extrema of functions of two variables) is best left for advanced calculus. Here, however, we provide an outline of the main ideas. Given a function $f(x, y)$ with critical point (a, b) that we wish to investigate, the function $f(x - a, y - b)$ would have a critical point of the same type at $(0, 0)$, so let's assume that $a = b = 0$.

To analyze the behavior of $f(x, y)$ near $(0, 0)$, we fix x and y and introduce the single-variable function

$$g(t) = f(tx, ty), \tag{9}$$

whose values agree with those of f on the straight line through $(0, 0)$ and (x, y) in the xy-plane. Its second-degree Taylor formula at $t = 0$ is

$$g(t) = g(0) + g'(0) \cdot t + \tfrac{1}{2}g''(0) \cdot t^2 + R, \tag{10}$$

where the remainder term is of the form $R = g^{(3)}(\tau) \cdot t^3/3!$ for some τ between 0 and t. With $t = 1$ we get

$$g(1) = g(0) + g'(0) + \tfrac{1}{2}g''(0) + R. \tag{11}$$

But

$$g(0) = f(0, 0) \quad \text{and} \quad g(1) = f(x, y) \tag{12}$$

by Eq. (9), and the chain rule gives

$$g'(0) = \frac{\partial f}{\partial x}\frac{d(tx)}{dt} + \frac{\partial f}{\partial y}\frac{d(ty)}{dt} = xf_x + yf_y \tag{13}$$

and

$$\begin{aligned} g''(0) &= \frac{\partial}{\partial x}(xf_x + yf_y)\,\frac{d(tx)}{dt} + \frac{\partial}{\partial y}(xf_x + yf_y)\,\frac{d(ty)}{dt} \\ &= x^2 f_{xx} + 2xyf_{xy} + y^2 f_{yy}, \end{aligned} \tag{14}$$

where the partial derivatives of f are to be evaluated at the point $(0, 0)$.

Because $f_x(0, 0) = f_y(0, 0) = 0$, substitution of Eqs. (12), (13), and (14) into Eq. (11) yields the two-variable Taylor expansion

$$f(x, y) = f(0, 0) + \tfrac{1}{2}(Ax^2 + 2Bxy + Cy^2) + R, \tag{15}$$

where

$$A = f_{xx}(0, 0), \qquad B = f_{xy}(0, 0), \qquad C = f_{yy}(0, 0). \tag{16}$$

If $|x|$ and $|y|$ are sufficiently small, then the remainder term R in (15) is negligible, so the shape of the surface $z = f(x, y)$ resembles that of the quadric surface

$$z = q(x, y) = Ax^2 + 2Bxy + Cy^2. \tag{17}$$

But, according to Theorem 2 in Section 12.7, the shape of this quadric surface is determined by the *signs* of the roots λ_1 and λ_2 of the quadratic equation

$$\begin{aligned} \begin{vmatrix} A - \lambda & B \\ B & C - \lambda \end{vmatrix} &= (A - \lambda)(C - \lambda) - B^2 \\ &= \lambda^2 - (A + C)\lambda + (AC - B^2) = 0. \end{aligned} \tag{18}$$

The quadratic formula gives

$$\begin{aligned} \lambda_1, \lambda_2 &= \tfrac{1}{2}\left(A + C \pm \sqrt{(A + C)^2 - 4(AC - B^2)}\right) \\ &= \tfrac{1}{2}\left(A + C \pm \sqrt{(A + C)^2 - 4\Delta}\right), \end{aligned} \tag{19}$$

where $\Delta = AC - B^2$ as in Theorem 1. Simplification of the first radical here yields

$$\lambda_1, \lambda_2 = \tfrac{1}{2}\left(A + C \pm \sqrt{(A - C)^2 + 4B^2}\right), \tag{20}$$

so it follows (Why?) that both roots λ_1 and λ_2 are real. There are three cases to consider, corresponding to the three parts of Theorem 1.

1. If $\Delta > 0$, then A and C have the same sign, and the radical in Eq. (19) is *less than* $|A + C|$. If $A > 0$ it therefore follows that the roots λ_1 and λ_2 are both *positive.* Hence Theorem 2 in Section 12.7 implies that the surface $z = q(x, y)$ is an elliptic paraboloid that opens *upward,* so the critical point $(0, 0)$ is a *local minimum point* for $q(x, y)$ and also for $f(x, y)$.
2. If $\Delta > 0$ and $A < 0$, it follows similarly that the roots λ_1 and λ_2 are both *negative.* Hence Theorem 2 in Section 12.7 implies that the surface $z = q(x, y)$ is an elliptic paraboloid that opens *downward,* so the critical point $(0, 0)$ is a *local maximum point* for $q(x, y)$ and also for $f(x, y)$.
3. If $\Delta < 0$, then the radical in Eq. (19) is *greater than* $|A + C|$, so it follows that the roots λ_1 and λ_2 have different signs. Hence Theorem 2 in Section 12.7 implies that the surface $z = q(x, y)$ is a (saddle-shaped) hyperbolic paraboloid. Consequently the critical point $(0, 0)$ is a *saddle point* for $q(x, y)$ and also for $f(x, y)$.

13.10 PROBLEMS

Find and classify the critical points of the functions in Problems 1 through 22. If a computer algebra system is available, check your results by means of contour plots like those in Figs. 13.10.3–13.10.5.

1. $f(x, y) = 2x^2 + y^2 + 4x - 4y + 5$
2. $f(x, y) = 10 + 12x - 12y - 3x^2 - 2y^2$
3. $f(x, y) = 2x^2 - 3y^2 + 2x - 3y + 7$ (Fig. 13.10.13)
4. $f(x, y) = xy + 3x - 2y + 4$
5. $f(x, y) = 2x^2 + 2xy + y^2 + 4x - 2y + 1$
6. $f(x, y) = x^2 + 4xy + 2y^2 + 4x - 8y + 3$

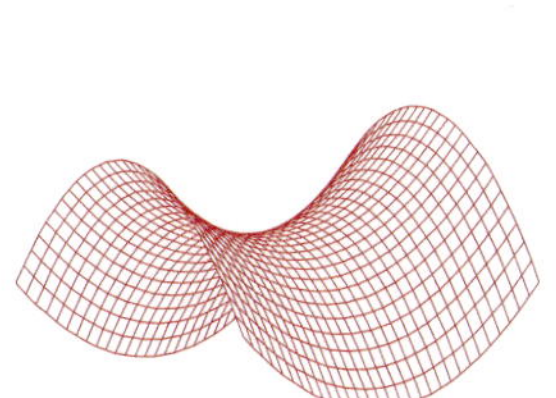

Fig. 13.10.13 Graph for Problem 3

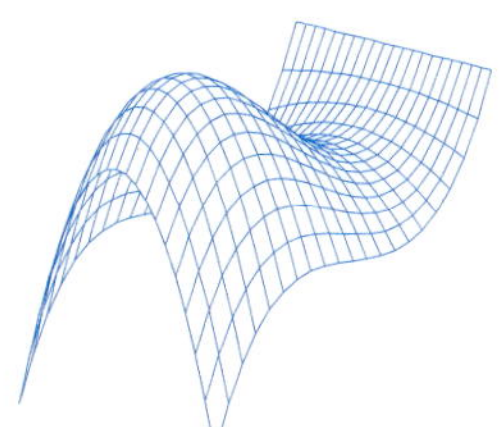

Fig. 13.10.14 Graph for Problem 7

7. $f(x, y) = x^3 + y^3 + 3xy + 3$ (Fig. 13.10.14)
8. $f(x, y) = x^2 - 2xy + y^3 - y$
9. $f(x, y) = 6x - x^3 - y^3$
10. $f(x, y) = 3xy - x^3 - y^3$
11. $f(x, y) = x^4 + y^4 - 4xy$
12. $f(x, y) = x^3 + 6xy + 3y^2$
13. $f(x, y) = x^3 + 6xy + 3y^2 - 9x$
14. $f(x, y) = x^3 + 6xy + 3y^2 + 6x$
15. $f(x, y) = 3x^2 + 6xy + 2y^3 + 12x - 24y$ (Fig. 13.10.15)
16. $f(x, y) = 3x^2 + 12xy + 2y^3 - 6x + 6y$

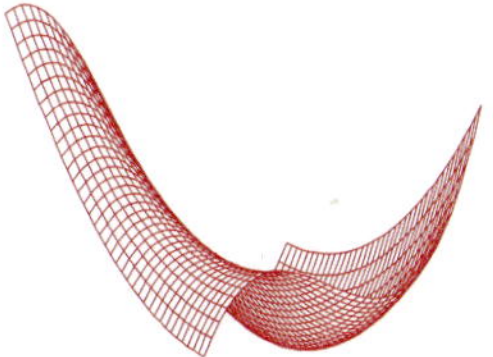

Fig. 13.10.15 Graph for Problem 15

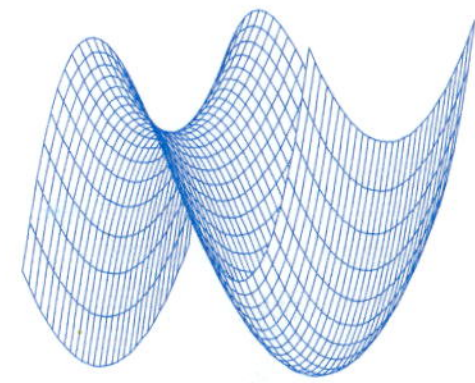

Fig. 13.10.16 Graph for Problem 19

17. $f(x, y) = 4xy - 2x^4 - y^2$
18. $f(x, y) = 8xy - 2x^2 - y^4$
19. $f(x, y) = 2x^3 - 3x^2 + y^2 - 12x + 10$ (Fig. 13.10.16)
20. $f(x, y) = 2x^3 + y^3 - 3x^2 - 12x - 3y$ (Fig. 13.10.17)

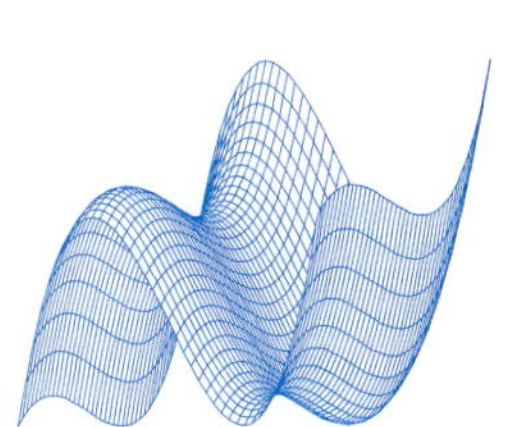

Fig. 13.10.17 Graph for Problem 20

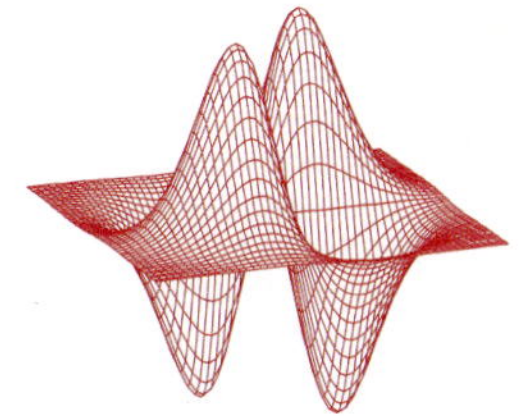

Fig. 13.10.18 Graph for Problem 21

21. $f(x, y) = xy\exp(-x^2 - y^2)$ (Fig. 13.10.18)
22. $f(x, y) = (x^2 + y^2)\exp(x^2 - y^2)$

In Problems 23 through 25, first show that $\Delta = f_{xx}f_{yy} - (f_{xy})^2$ is zero at the origin. Then classify this critical point by imagining what the surface $z = f(x, y)$ looks like.

23. $f(x, y) = x^4 + y^4$
24. $f(x, y) = x^3 + y^3$

25. $f(x, y) = \exp(-x^4 - y^4)$

26. Let $f(x, t)$ denote the *square* of the distance between a typical point of the line $x = t$, $y = t + 1$, $z = 2t$ and a typical point of the line $x = 2s$, $y = s - 1$, $z = s + 1$. Show that the single critical point of f is a local minimum. Hence find the closest points on these two skew lines.

27. Let $f(x, y)$ denote the square of the distance from $(0, 0, 2)$ to a typical point of the surface $z = xy$. Find and classify the critical points of f.

28. Show that the surface

$$z = (x^2 + 2y^2)\exp(1 - x^2 - y^2)$$

looks like two mountain peaks joined by two ridges with a pit between them.

29. A wire 120 cm long is cut into three pieces of lengths x, y, and $120 - x - y$, and each piece is bent into the shape of a square. Let $f(x, y)$ denote the sum of the areas of these squares. Show that the single critical point of f is a local minimum. But surely it is possible to *maximize* the sum of the areas. Explain.

30. Show that the graph of the function

$$f(x, y) = xy\exp\left(\tfrac{1}{8}\left[x^2 + 4y^2\right]\right)$$

has a saddle point but no extrema.

31. Find and classify the critical points of the function

$$f(x, y) = \sin\frac{\pi x}{2}\sin\frac{\pi y}{2}.$$

32. Let $f(x, y) = x^3 - 3xy^2$. (a) Show that its only critical point is $(0, 0)$ and that $\Delta = 0$ there. (b) By examining the behavior of $x^3 - 3xy^2$ on straight lines through the origin, show that the surface $z = x^3 - 3xy^2$ qualifies as a monkey saddle (Fig. 13.10.19).

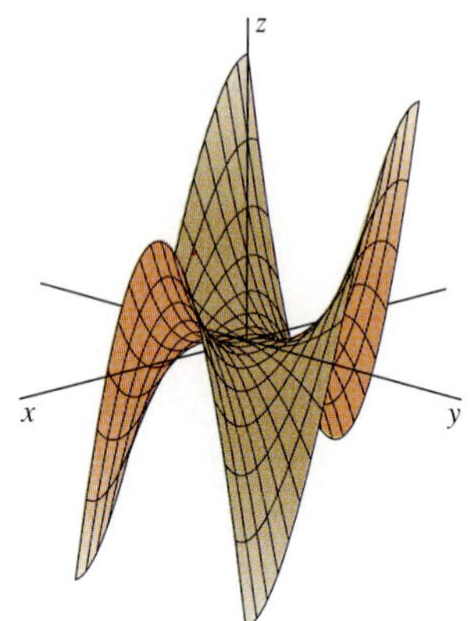

Fig. 13.10.19 The monkey saddle of Problem 32

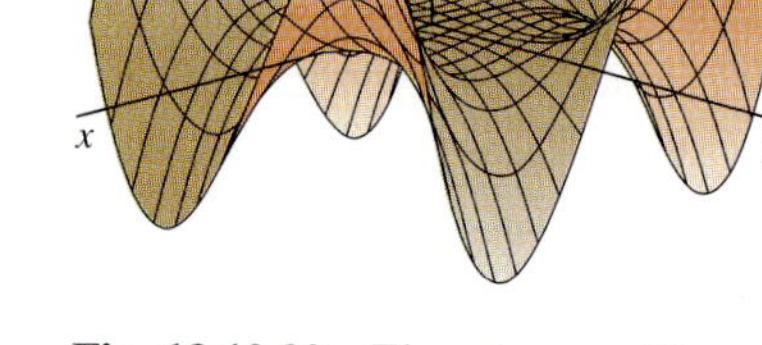

Fig. 13.10.20 The dog saddle of Problem 33

33. Repeat Problem 32 with $f(x, y) = 4xy(x^2 - y^2)$. Show that near the critical point $(0, 0)$ the surface $z = f(x, y)$ qualifies as a "dog saddle" (Fig. 13.10.20).

34. Let

$$f(x, y) = \frac{xy(x^2 - y^2)}{x^2 + y^2}.$$

Classify the behavior of f near the critical point $(0, 0)$.

In Problems 35 through 39, use graphical or numerical methods to find the critical points of f to four-place accuracy. Then classify them.

35. $f(x, y) = 2x^4 - 12x^2 + y^2 + 8x$

36. $f(x, y) = x^4 + 4x^2 - y^2 - 16x$

37. $f(x, y) = x^4 + 12xy + 6y^2 + 4x + 10$

38. $f(x, y) = x^4 + 8xy - 4y^2 - 16x + 10$

39. $f(x, y) = x^4 + 2y^4 - 12xy^2 - 20y^2$

13.10 PROJECT: CRITICAL POINT INVESTIGATIONS

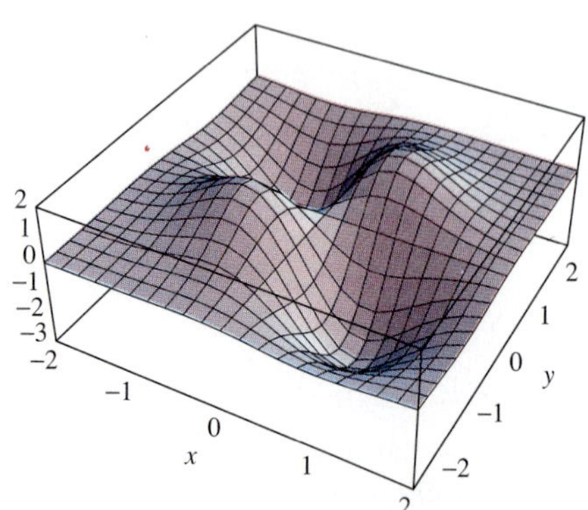

Fig. 13.10.21 The graph $z = f(x, y)$ of the function in Eq. (1)

Investigation A: The graph of the function

$$f(x, y) = 10\exp\left(-x^2 - (0.5)xy - (0.5)y^2\right)\sin x\sin y \qquad \textbf{(1)}$$

is shown in Fig. 13.10.21, and there we see two peaks, two pits, and a saddle point. The contour plot in Fig. 13.10.22 indicates that the approximate locations of the four local extrema are $(\pm 0.5, \pm 0.75)$ and $(\pm 0.75, \mp 1)$. Use these guesses to solve numerically the critical point equations $f_x(x, y) = 0 = f_y(x, y)$ with computer algebra system commands such as

```
FindRoot[ {fx==0, fy==0}, { x, 0.5 }, { y, 0.75 } ]
```

(Mathematica)

or

```
fsolve({fx=0, fy=0}, {x,y}, {x=0.4..0.6, y=0.7..0.8} )
```

(Maple).

Then apply Theorem 1 of this section to classify each critical point found.

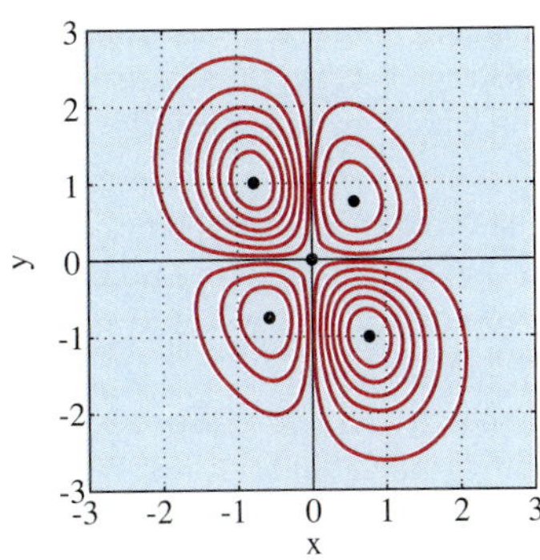

Fig. 13.10.22 Contour plot for the function in Eq. (1)

INVESTIGATION B: Let a and b be the two smallest distinct nonzero digits (in either order) of your student I.D. number. Then find the local maximum, local minimum, and saddle points on the surface

$$z = 5(ax^2 + xy + by^2)\exp(-x^2 - y^2). \tag{2}$$

First plot a graph of the surface to see where the action is. Attempt to choose the range so you can see all the critical points on the surface. Then use a contour plot to estimate the locations of the critical points. Next, use these estimates as initial values to solve numerically the critical point equations $f_x(x, y) = 0 = f_y(x, y)$. Finally, apply the two-variable second derivative test (Theorem 1 of this section) to classify the type of each of these critical points. You might verify that reality agrees with theory by constructing a contour plot of a small neighborhood of one or two selected critical points to verify that the picture looks as it should. Do your numerical results agree with your original surface plot?

INVESTIGATION C: Repeat Investigation B with the surface

$$z = 10\left(x^3 + y^5 + \frac{x}{p}\right)\exp(-x^2 - y^2) + \tfrac{1}{3}\exp(-(x-1)^2 - y^2),$$

where p is the largest digit in your student I.D. number.

CHAPTER 13 REVIEW: *Definitions, Concepts, Results*

Use the following list as a guide to concepts that you may need to review.

1. Graphs and level curves of functions of two variables
2. Limits and continuity of functions of two or three variables
3. Partial derivatives—definition and computation
4. Geometric interpretation of partial derivatives and the plane tangent to the surface $z = f(x, y)$
5. Absolute and local maxima and minima
6. Necessary conditions for a local extremum
7. Increments and differentials of functions of two or three variables
8. The linear approximation theorem
9. The chain rule for functions of several variables
10. Directional derivatives—definition and computation
11. The gradient vector and the chain rule
12. Significance of the length and direction of the gradient vector
13. The gradient vector as a normal vector; tangent plane to a surface $F(x, y, z) = 0$
14. Constrained maximum-minimum problems and the Lagrange multiplier method
15. Sufficient conditions for a local extremum of a function of two variables

CHAPTER 13 *Miscellaneous Problems*

1. Use the method of Example 5 of Section 13.3 to show that

$$\lim_{(x,y)\to(0,0)} \frac{x^2y^2}{x^2+y^2} = 0.$$

2. Use spherical coordinates to show that

$$\lim_{(x,y,z)\to(0,0,0)} \frac{x^3+y^3-z^3}{x^2+y^2+z^2} = 0.$$

3. Suppose that

$$g(x, y) = \frac{xy}{x^2+y^2}$$

if $(x, y) \neq (0, 0)$; we *define* $g(0, 0)$ to be zero. Show that g is not continuous at $(0, 0)$.

4. Compute $g_x(0, 0)$ and $g_y(0, 0)$ for the function g of Problem 3.

5. Find a function $f(x, y)$ such that

$$f_x(x, y) = 2xy^3 + e^x \sin y$$

and

$$f_y(x, y) = 3x^2y^2 + e^x \cos y + 1.$$

6. Prove that there is *no* function f with continuous second-order partial derivatives such that $f_x(x, y) = 6xy^2$ and $f_y(x, y) = 8x^2y$.

7. Find the point or points on the paraboloid $z = x^2 + y^2$ at which the normal line passes through the point $(0, 0, 1)$.

8. Write an equation of the plane tangent to the surface

$$\sin xy + \sin yz + \sin xz = 1$$

at the point $(1, \pi/2, 0)$.

9. Prove that every line normal to the cone with equation $z = \sqrt{x^2 + y^2}$ intersects the z-axis.

10. Show that the function

$$u(x, t) = \frac{1}{\sqrt{4\pi kt}} \exp\left(-\frac{x^2}{4kt}\right)$$

satisfies the one-dimensional heat equation

$$\frac{\partial u}{\partial t} = k\frac{\partial^2 u}{\partial x^2}.$$

11. Show that the function

$$u(x, y, t) = \frac{1}{4\pi kt} \exp\left(-\frac{x^2 + y^2}{4kt}\right)$$

satisfies the two-dimensional heat equation

$$\frac{\partial u}{\partial t} = k\left(\frac{\partial^2 u}{\partial x^2} + \frac{\partial^2 u}{\partial y^2}\right).$$

12. Let

$$f(x, y) = \frac{xy(x^2 - y^2)}{x^2 + y^2}$$

unless $(x, y) = (0, 0)$; we *define* $f(0, 0)$ to be zero. Show that the second-order partial derivatives f_{xx}, f_{xy}, f_{yx}, and f_{yy} all exist at $(0, 0)$ but that $f_{xy}(0, 0) \neq f_{yx}(0, 0)$.

13. Define the partial derivatives $\mathbf{r}_x$ and $\mathbf{r}_y$ of the vector-valued function $\mathbf{r}(x, y) = \mathbf{i}x + \mathbf{j}y + \mathbf{k}f(x, y)$ by componentwise partial differentiation. Then show that the vector $\mathbf{r}_x \times \mathbf{r}_y$ is normal to the surface $z = f(x, y)$.

14. An open-topped rectangular box is to have total surface area 300 cm^2. Find the dimensions that maximize its volume.

15. You must build a rectangular shipping crate with volume 60 ft^3. Its sides cost \$1/ft^2, its top costs \$2/ft^2, and its bottom costs \$3/ft^2. What dimensions would minimize the total cost of the box?

16. A pyramid is bounded by the three coordinate planes and by the plane tangent to the surface $xyz = 1$ at a point in the first octant. Find the volume of this pyramid (it is independent of the point of tangency).

17. Two resistors have resistances R_1 and R_2, respectively. When they are connected in parallel, the total resistance R of the resulting circuit satisfies the equation

$$\frac{1}{R} = \frac{1}{R_1} + \frac{1}{R_2}.$$

Suppose that R_1 and R_2 are measured to be 300 and 600 Ω (ohms), respectively, with a maximum error of 1% in each measurement. Use differentials to estimate the maximum error (in ohms) in the calculated value of R.

18. Consider a gas that satisfies van der Waals' equation (see Problem 67 of Section 13.4). Use differentials to approximate the change in its volume if p is increased from 1 atm to 1.1 atm and T is decreased from 313 K to 303 K.

19. Each of the semiaxes a, b, and c of an ellipsoid with volume $V = \frac{4}{3}\pi abc$ is measured with a maximum percentage error of 1%. Use differentials to estimate the maximum percentage error in the calculated value of V.

20. Two spheres have radii a and b, and the distance between their centers is $c < a + b$. Thus the spheres meet in a common circle. Let P be a point on this circle, and let $\mathcal{P}_1$ and $\mathcal{P}_2$ be the planes tangent at P to the two spheres. Find the angle between $\mathcal{P}_1$ and $\mathcal{P}_2$ in terms of a, b, and c. (*Suggestion:* Recall that the angle between two planes is, by definition, the angle between their normal vectors.)

21. Find every point on the surface of the ellipsoid $x^2 + 4y^2 + 9z^2 = 16$ at which the normal line at the point passes through the center $(0, 0, 0)$ of the ellipsoid.

22. Suppose that

$$F(x) = \int_{g(x)}^{h(x)} f(t)\,dt.$$

Show that

$$F'(x) = f(h(x))h'(x) - f(g(x))g'(x).$$

(*Suggestion:* Write $w = \int_u^v f(t)\,dt$ where $u = g(x)$ and $v = h(x)$.)

23. Suppose that $\mathbf{a}$, $\mathbf{b}$, and $\mathbf{c}$ are mutually perpendicular unit vectors in space and that f is a function of the three independent variables x, y, and z. Show that

$$\nabla f = \mathbf{a}(D_{\mathbf{a}}f) + \mathbf{b}(D_{\mathbf{b}}f) + \mathbf{c}(D_{\mathbf{c}}f).$$

24. Let $\mathbf{R} = \langle \cos\theta, \sin\theta, 0\rangle$ and $\Theta = \langle -\sin\theta, \cos\theta, 0\rangle$ be the polar-coordinates unit vectors. Given $f(x, y, z) = w(r, \theta, z)$, show that

$$D_{\mathbf{R}}f = \frac{\partial w}{\partial r} \quad \text{and} \quad D_{\Theta}f = \frac{1}{r}\frac{\partial w}{\partial \theta}.$$

Then conclude from Problem 23 that the gradient vector is given in cylindrical coordinates by

$$\nabla f = \frac{\partial w}{\partial r}\mathbf{R} + \frac{1}{r}\frac{\partial w}{\partial \theta}\Theta + \frac{\partial w}{\partial z}\mathbf{k}.$$

25. Suppose that you are standing at the point with coordinates $(-100, -100, 430)$ on a hill that has the shape of the graph of

$$z = 500 - (0.003)x^2 - (0.004)y^2$$

(in units of meters). In what (horizontal) direction should you move in order to maintain a constant altitude—that is, to neither climb nor descend the hill?

26. Suppose that the blood concentration in the ocean at the point (x, y) is given by

$$f(x, y) = A\exp(-k[x^2 + 2y^2]),$$

where A and k are positive constants. A shark always swims in the direction of ∇f. Show that its path is a parabola $y = cx^2$. (*Suggestion:* Show that the condition that $\langle dx/dt, dy/dt\rangle$ is a multiple of ∇f implies that

$$\frac{1}{x}\frac{dx}{dt} = \frac{1}{2y}\frac{dy}{dt}.$$

Then antidifferentiate this equation.)

27. Consider a plane tangent to the surface with equation $x^{2/3} + y^{2/3} + z^{2/3} = 1$. Find the sum of the squares of the x-, y-, and z-intercepts of this plane.

28. Find the points on the ellipse $x^2/a^2 + y^2/b^2 = 1$ (with $a \neq b$) where the normal line passes through the origin.

29. (a) Show that the origin is a critical point of the function f of Problem 12. (b) Show that f does not have a local extremum at $(0, 0)$.

30. Find the point of the surface $z = xy + 1$ that is closest to the origin.

31. Use the method of Problem 38 in Section 13.9 to find the semiaxes of the rotated ellipse

$$73x^2 + 72xy + 52y^2 = 100.$$

32. Use the Lagrange multiplier method to show that the longest chord of the sphere $x^2 + y^2 + z^2 = 1$ has length 2. (*Suggestion:* There is no loss of generality in assuming that $(1, 0, 0)$ is one endpoint of the chord.)

33. Use the method of Lagrange multipliers, the law of cosines, and Fig. 13.9.9 to find the triangle of minimum perimeter inscribed in the unit circle.

34. When a current I enters two resistors, with resistances R_1 and R_2, that are connected in parallel, it splits into two currents I_1 and I_2 (with $I = I_1 + I_2$) in such a way as to minimize the total power $R_1I_1^2 + R_2I_2^2$. Express I_1 and I_2 in terms of R_1, R_2, and I. Then derive the formula of Problem 17.

35. Use the method of Lagrange multipliers to find the points of the ellipse $x^2 + 2y^2 = 1$ that are closest to and farthest from the line $x + y = 2$. (*Suggestion:* Let $f(x, y, u, v)$ denote the square of the distance between the point (x, y) of the ellipse and the point (u, v) of the line.)

36. (a) Show that the maximum of

$$f(x, y, z) = x + y + z$$

at points of the sphere $x^2 + y^2 + z^2 = a^2$ is $a\sqrt{3}$.
(b) Conclude from the result of part (a) that

$$(x + y + z)^2 \leqq 3(x^2 + y^2 + z^2)$$

for any three numbers x, y, and z.

37. Generalize the method of Problem 36 to show that

$$\left(\sum_{i=1}^{n} x_i\right)^2 \leqq n\sum_{i=1}^{n} x_i^2$$

for any n real numbers $x_1, x_2, \ldots, x_n$.

38. Find the maximum and minimum values of $f(x, y) = xy - x - y$ at points on and within the triangle with vertices $(0, 0)$, $(0, 1)$, and $(3, 0)$.

39. Find the maximum and minimum values of $f(x, y, z) = x^2 - yz$ at points of the sphere $x^2 + y^2 + z^2 = 1$.

40. Find the maximum and minimum values of $f(x, y) = x^2y^2$ at points of the ellipse $x^2 + 4y^2 = 24$.

Locate and classify the critical points (local maxima, local minima, saddle points, and other points at which the tangent plane is horizontal) of the functions in Problems 41 through 50.

41. $f(x, y) = x^3y - 3xy + y^2$

42. $f(x, y) = x^2 + xy + y^2 - 6x + 2$

43. $f(x, y) = x^3 - 6xy + y^3$

44. $f(x, y) = x^2y + xy^2 + x + y$

45. $f(x, y) = x^3y^2(1 - x - y)$

46. $f(x, y) = x^4 - 2x^2 + y^2 + 4y + 3$

47. $f(x, y) = e^{xy} - 2xy$

48. $f(x, y) = x^3 - y^3 + x^2 + y^2$

49. $f(x, y) = (x - y)(xy - 1)$

50. $f(x, y) = (2x^2 + y^2)\exp(-x^2 - y^2)$

51. Given the data points (x_i, y_i) for $i = 1, 2, \ldots, n$, the **least-squares straight line** $y = mx + b$ is the line that best fits these data in the following sense. Let $d_i = y_i - (mx_i + b)$ be the *deviation* of the predicted value $mx_i + b$ from the true value y_i. Let

$$f(m, b) = d_1^2 + d_2^2 + \cdots + d_n^2 = \sum_{i=1}^{n} [y_i - (mx_i + b)]^2$$

be the sum of the squares of the deviations. The least-squares straight line is the one that minimizes this sum (Fig. 13.MP.1). Show how to choose m and b by minimizing f. (*Note:* The only variables in this computation are m and b.) Apply the two-variable second derivative test of Section 13.10 to verify that the single critical point of f is, indeed, a local minimum point.

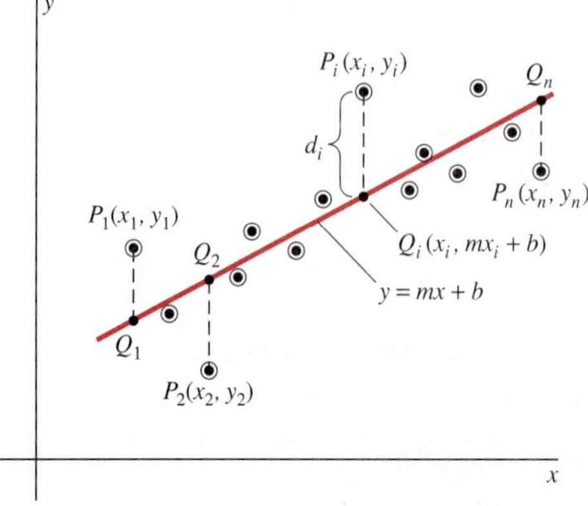

Fig. 13.MP.1 Fitting the best straight line to the data points (x_i, y_i), $1 \leqq i \leqq n$ (Problem 51)

CHAPTER 14

MULTIPLE INTEGRALS

Henri Lebesgue (1875–1941)

Geometric problems of *measure*—dealing with concepts of length, area, and volume—can be traced back 40 centuries to the rise of civilizations in the fertile river valleys of Africa and Asia, when such issues as areas of fields and volume of granaries became important. These problems led ultimately to the *integral,* which is used to calculate (among other things) areas and volumes of curvilinear figures. But only in the early twentieth century were certain longstanding difficulties with measure and integration finally resolved, largely as a consequence of the work of the French mathematician Henri Lebesgue.

In his 1902 thesis presented at the Sorbonne in Paris, Lebesgue presented a new definition of the integral, generalizing Riemann's definition. In essence, to define the integral of the function f from $x = a$ to $x = b$, Lebesgue replaced Riemann's subdivision of the interval $[a, b]$ into nonoverlapping subintervals with a partition of $[a, b]$ into disjoint measurable sets $\{E_i\}$. The Riemann sum $\sum f(x_i^\star)\,\Delta x$ was thereby replaced with a sum of the form $\sum f(x_i^\star)\,m_i$, where m_i is the measure of the ith set E_i and $x_i^\star$ is a number in E_i. To see the advantage of the "Lebesgue integral," consider the fact that there exist differentiable functions whose derivatives are not integrable in the sense of Riemann. For such a function, the fundamental theorem of calculus in the form

$$\int_a^b f'(x)\,dx = f(b) - f(a)$$

fails to hold. But with his new definition of the integral, Lebesgue showed that a derivative function f' is integrable and that the fundamental theorem holds. Similarly, the equality of double and iterated integrals (Section 14.1) holds only under rather drastic restrictions if the Riemann definition of multiple integrals is used, but the Lebesgue integral resolves the difficulty.

For such reasons, the Lebesgue theory of measure and integration predominates in modern mathematical research, both pure and applied. For instance, the Lebesgue integral is basic to such diverse realms as applied probability and mathematical biology, the quantum theory of atoms and nuclei, and the information theory and electric signals processing of modern computer technology.

The Section 14.5 Project illustrates the application of multiple integrals to such concrete problems as the optimal design of race car wheels.

We could use multiple integrals to determine the best design for the wheels of these soapbox derby cars.

14.1 DOUBLE INTEGRALS

This chapter is devoted to integrals of functions of two or three variables. Such integrals are called **multiple integrals**. The applications of multiple integrals include computation of area, volume, mass, and surface area in a wider variety of situations than can be handled with the single integral of Chapters 5 and 6.

The simplest sort of multiple integral is the *double integral*

$$\iint_R f(x, y)\, dA$$

of a continuous function $f(x, y)$ over the *rectangle*

$$R = [a, b] \times [c, d] = \{(x, y) | a \leqq x \leqq b, c \leqq y \leqq d\}$$

in the xy-plane. Just as the definition of the single integral is motivated by the problem of computing areas, the definition of the double integral is motivated by the problem of computing the volume V of the solid of Fig. 14.1.1—a solid bounded above by the graph $z = f(x, y)$ of the nonnegative function f over the rectangle R in the xy-plane.

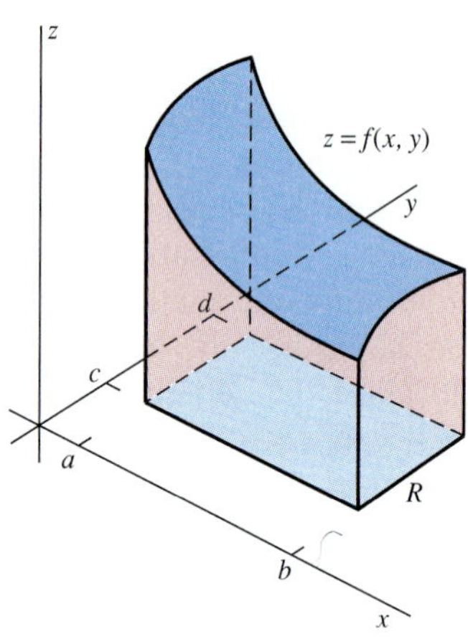

Fig. 14.1.1 We will use a double integral to compute the volume V.

To define the *value*

$$V = \iint_R f(x, y)\, dA$$

of such a double integral, we begin with an approximation to V. To obtain this approximation, the first step is to construct a **partition** $\mathcal{P}$ of R into subrectangles R_1, $R_2, \ldots, R_k$ determined by the points

$$a = x_0 < x_1 < x_2 < \cdots < x_m = b$$

of $[a, b]$ and

$$c = y_0 < y_1 < y_2 < \cdots < y_n = d$$

of $[c, d]$. Such a partition of R into $k = mn$ rectangles is shown in Fig. 14.1.2. The order in which these rectangles are labeled makes no difference.

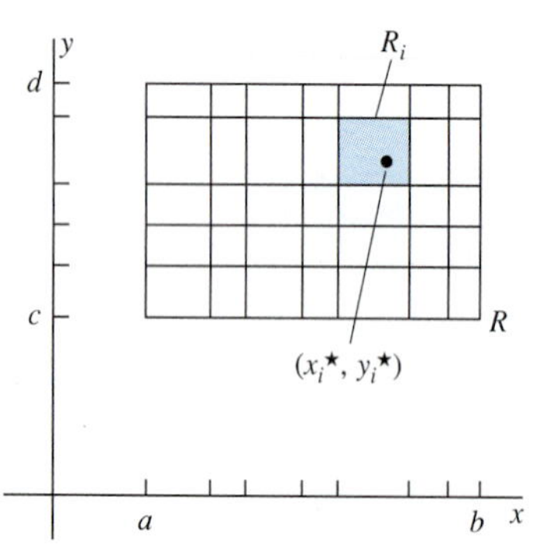

Fig. 14.1.2 A partition $\mathcal{P}$ of the rectangle R

Next we choose an arbitrary point $(x_i^\star, y_i^\star)$ of the ith rectangle R_i for each i (where $1 \leqq i \leqq k$). The collection of points $S = \{(x_i^\star, y_i^\star) | 1 \leqq i \leqq k\}$ is called a **selection** for the partition $\mathcal{P} = \{R_i | 1 \leqq i \leqq k\}$. As a measure of the size of the rectangles of the partition $\mathcal{P}$, we define its **norm** $|\mathcal{P}|$ to be the maximum of the lengths of the diagonals of the rectangles $\{R_i\}$.

Now consider a rectangular column that rises straight up from the xy-plane. Its base is the rectangle R_i and its height is the value $f(x_i^\star, y_i^\star)$ of f at the selected point $(x_i^\star, y_i^\star)$ of R_i. One such column is shown in Fig. 14.1.3. If ΔA_i denotes the area of R_i, then the volume of the ith column is $f(x_i^\star, y_i^\star)\,\Delta A_i$. The sum of the volumes of all such columns (Fig. 14.1.4) is the **Riemann sum**

$$\sum_{i=1}^{k} f(x_i^\star, y_i^\star)\,\Delta A_i, \tag{1}$$

an approximation to the volume V of the solid region that lies above the rectangle R and under the graph $z = f(x, y)$.

We would expect to determine the exact volume V by taking the limit of the Riemann sum in Eq. (1) as the norm $|\mathcal{P}|$ of the partition $\mathcal{P}$ approaches zero. We therefore define the (**double**) **integral** of the function f over the rectangle R to be

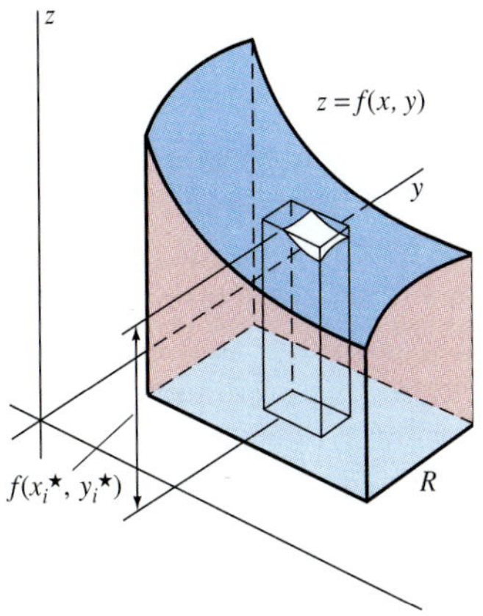

Fig. 14.1.3 Approximating the volume under the surface by summing volumes of towers with rectangular bases

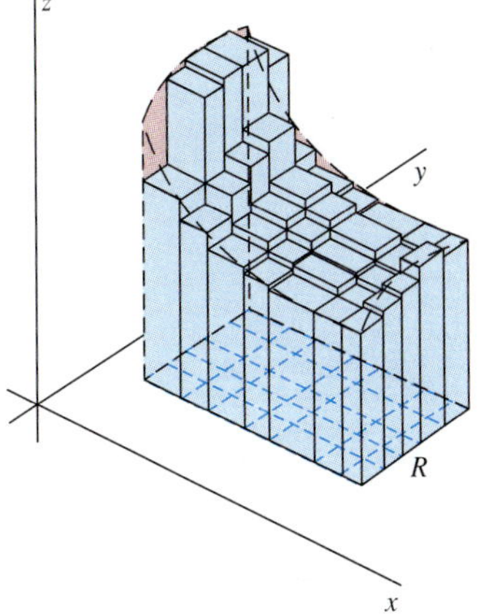

Fig. 14.1.4 Columns corresponding to a partition of the rectangle R

$$\iint_R f(x, y)\, dA = \lim_{|\mathcal{P}| \to 0} \sum_{i=1}^{k} f(x_i^\star, y_i^\star)\, \Delta A_i, \tag{2}$$

provided that this limit exists (we will make the concept of the existence of such a limit more precise in Section 14.2). It is proved in advanced calculus that the limit in Eq. (2) *does* exist if f is continuous on R. To motivate the introduction of the Riemann sum in Eq. (1), we assumed that f was nonnegative on R, but Eq. (2) serves to define the double integral over a rectangle whether or not f is nonnegative.

EXAMPLE 1 Approximate the value of the integral

$$\iint_R (4x^3 + 6xy^2)\, dA$$

over the rectangle $R = [1, 3] \times [-1, 2]$, by calculating the Riemann sum in (1) for the partition illustrated in Fig. 14.1.5, with the ith point $(x_i^\star, y_i^\star)$ selected as the center of the ith rectangle R_i (for each i, $1 \leqq i \leqq 6$).

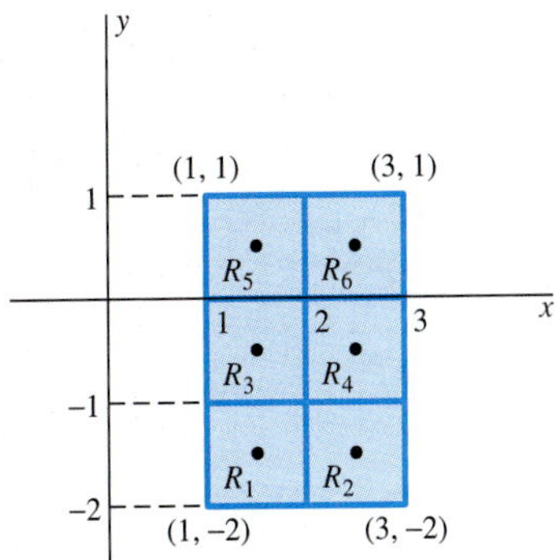

Fig. 14.1.5 The partition in Example 1

Solution Each of the six partition rectangles shown in Fig. 14.1.5 is a unit square with area $\Delta A_i = 1$. With $f(x, y) = 4x^3 + 6xy^2$, the desired Riemann sum is therefore

$$\begin{aligned}
\sum_{i=1}^{6} f(x_i^\star, y_i^\star)\,\Delta A_i &= f(x_1^\star, y_1^\star)\,\Delta A_1 + f(x_2^\star, y_2^\star)\,\Delta A_2 + f(x_3^\star, y_3^\star)\,\Delta A_3 \\
&\quad + f(x_4^\star, y_4^\star)\,\Delta A_4 + f(x_5^\star, y_5^\star)\,\Delta A_5 + f(x_6^\star, y_6^\star)\,\Delta A_6 \\
&= f\left(\tfrac{3}{2}, -\tfrac{3}{2}\right)(1) + f\left(\tfrac{5}{2}, -\tfrac{3}{2}\right)(1) + f\left(\tfrac{3}{2}, -\tfrac{1}{2}\right)(1) \\
&\quad + f\left(\tfrac{5}{2}, -\tfrac{1}{2}\right)(1) + f\left(\tfrac{3}{2}, \tfrac{1}{2}\right)(1) + f\left(\tfrac{5}{2}, \tfrac{1}{2}\right)(1) \\
&= \tfrac{135}{4} \cdot 1 + \tfrac{385}{4} \cdot 1 + \tfrac{63}{4} \cdot 1 + \tfrac{265}{4} \cdot 1 + \tfrac{63}{4} \cdot 1 + \tfrac{265}{4} \cdot 1 = 294.
\end{aligned}$$

Thus we find that

$$\iint_R (4x^3 + 6xy^2)\, dA \approx 294,$$

but our calculation provides no information about the accuracy of this approximation. ■

Remark The single-integral approximation methods of Section 5.9 all have analogues for double integrals. In Example 1 we calculated the *midpoint approximation* to the given double integral using six equal rectangles. If we divide each rectangle in Fig. 14.1.5 into four equal subrectangles, we get a partition of R into 24 rectangles. Suppose that we continue in this way, quadrupling the number of equal rectangles at each step, and use a computer to calculate each time the Riemann sum defined by selecting the center of each rectangle. Then we get the midpoint approximations listed in Fig. 14.1.6. In Example 2 we will see (much more easily) that the exact value of the integral in Example 1 is

$$\iint_R (4x^3 + 6xy^2)\, dA = 312.$$

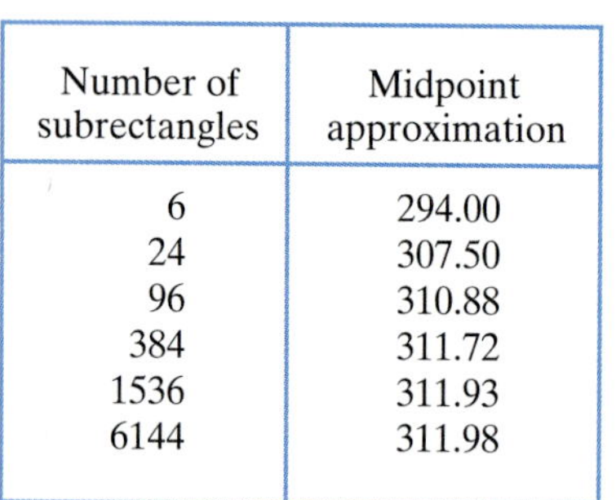

Number of subrectangles	Midpoint approximation
6	294.00
24	307.50
96	310.88
384	311.72
1536	311.93
6144	311.98

Fig. 14.1.6 Midpoint approximations to the integral in Example 1

Iterated Integrals

The direct evaluation of the limit in Eq. (2) is generally even less practical than the direct evaluation of the limit we used in Section 5.4 to define the single-variable integral. In practice, we shall calculate double integrals over rectangles by means of the **iterated integrals** that appear in Theorem 1.

Theorem 1 ***Double Integrals as Iterated Single Integrals***

Suppose that $f(x, y)$ is continuous on the rectangle $R = [a, b] \times [c, d]$. Then

$$\iint_R f(x, y)\, dA = \int_a^b \left(\int_c^d f(x, y)\, dy \right) dx = \int_c^d \left(\int_a^b f(x, y)\, dx \right) dy. \tag{3}$$

Theorem 1 tells us how to compute a double integral by means of two successive (or *iterated*) single-variable integrations, both of which we can compute by using the fundamental theorem of calculus (if the function f is sufficiently well-behaved on R).

Let us explain what we mean by the parentheses in the iterated integral

$$\int_a^b \int_c^d f(x, y)\, dy\, dx = \int_a^b \left(\int_c^d f(x, y)\, dy \right) dx. \tag{4}$$

First we hold x constant and integrate with respect to y, from $y = c$ to $y = d$. The result of this first integration is the **partial integral of f with respect to y**, denoted by

$$\int_c^d f(x, y)\, dy,$$

and it is a function of x alone. Then we integrate this latter function with respect to x, from $x = a$ to $x = b$.

Similarly, we calculate the iterated integral

$$\int_c^d \int_a^b f(x, y)\, dx\, dy = \int_c^d \left(\int_a^b f(x, y) \right) dx\, dy \tag{5}$$

by first integrating from a to b with respect to x (while holding y fixed) and then integrating the result from c to d with respect to y. The order of integration (either first with respect to x and then with respect to y, or the reverse) is determined by the order

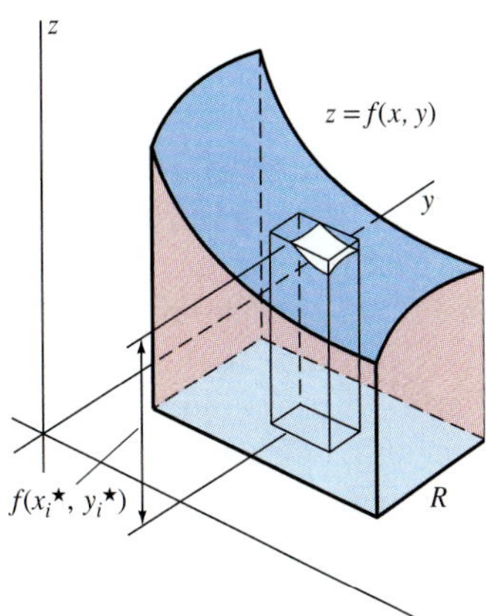

Fig. 14.1.3 Approximating the volume under the surface by summing volumes of towers with rectangular bases

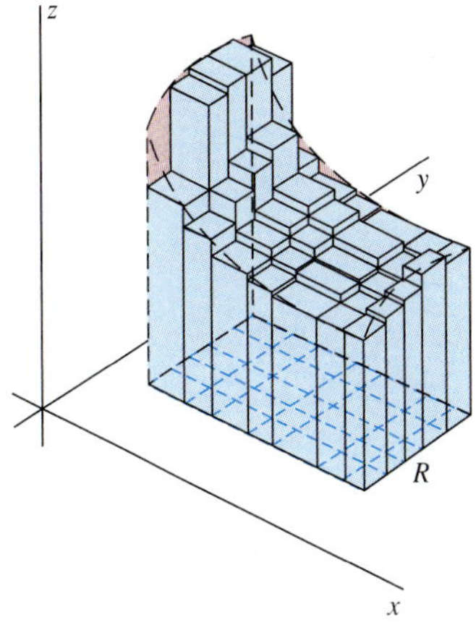

Fig. 14.1.4 Columns corresponding to a partition of the rectangle R

$$\iint_R f(x, y)\, dA = \lim_{|\mathcal{P}|\to 0} \sum_{i=1}^{k} f(x_i^\star, y_i^\star)\,\Delta A_i, \tag{2}$$

provided that this limit exists (we will make the concept of the existence of such a limit more precise in Section 14.2). It is proved in advanced calculus that the limit in Eq. (2) *does* exist if f is continuous on R. To motivate the introduction of the Riemann sum in Eq. (1), we assumed that f was nonnegative on R, but Eq. (2) serves to define the double integral over a rectangle whether or not f is nonnegative.

EXAMPLE 1 Approximate the value of the integral

$$\iint_R (4x^3 + 6xy^2)\, dA$$

over the rectangle $R = [1, 3] \times [-1, 2]$, by calculating the Riemann sum in (1) for the partition illustrated in Fig. 14.1.5, with the ith point $(x_i^\star, y_i^\star)$ selected as the center of the ith rectangle R_i (for each i, $1 \leqq i \leqq 6$).

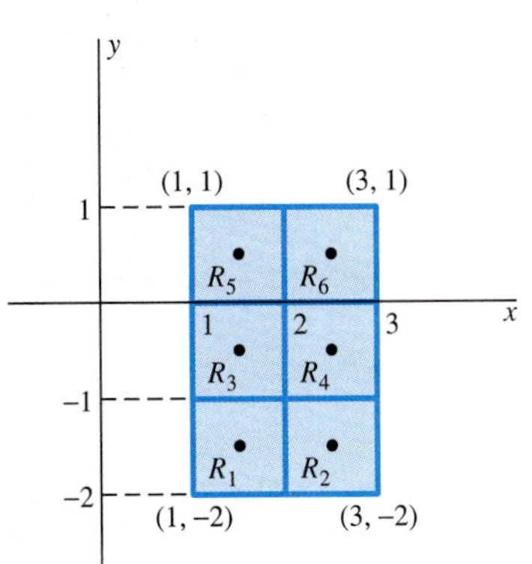

Fig. 14.1.5 The partition in Example 1

Solution Each of the six partition rectangles shown in Fig. 14.1.5 is a unit square with area $\Delta A_i = 1$. With $f(x, y) = 4x^3 + 6xy^2$, the desired Riemann sum is therefore

$$\begin{aligned}\sum_{i=1}^{6} f(x_i^\star, y_i^\star)\,\Delta A_i &= f(x_1^\star, y_1^\star)\,\Delta A_1 + f(x_2^\star, y_2^\star)\,\Delta A_2 + f(x_3^\star, y_3^\star)\,\Delta A_3 \\ &\quad + f(x_4^\star, y_4^\star)\,\Delta A_4 + f(x_5^\star, y_5^\star)\,\Delta A_5 + f(x_6^\star, y_6^\star)\,\Delta A_6 \\ &= f\left(\tfrac{3}{2}, -\tfrac{3}{2}\right)(1) + f\left(\tfrac{5}{2}, -\tfrac{3}{2}\right)(1) + f\left(\tfrac{3}{2}, -\tfrac{1}{2}\right)(1) \\ &\quad + f\left(\tfrac{5}{2}, -\tfrac{1}{2}\right)(1) + f\left(\tfrac{3}{2}, \tfrac{1}{2}\right)(1) + f\left(\tfrac{5}{2}, \tfrac{1}{2}\right)(1) \\ &= \tfrac{135}{4}\cdot 1 + \tfrac{385}{4}\cdot 1 + \tfrac{63}{4}\cdot 1 + \tfrac{265}{4}\cdot 1 + \tfrac{63}{4}\cdot 1 + \tfrac{265}{4}\cdot 1 = 294.\end{aligned}$$

Thus we find that

$$\iint_R (4x^3 + 6xy^2)\, dA \approx 294,$$

but our calculation provides no information about the accuracy of this approximation. ■

REMARK The single-integral approximation methods of Section 5.9 all have analogues for double integrals. In Example 1 we calculated the *midpoint approximation* to the given double integral using six equal rectangles. If we divide each rectangle in Fig. 14.1.5 into four equal subrectangles, we get a partition of R into 24 rectangles. Suppose that we continue in this way, quadrupling the number of equal rectangles at each step, and use a computer to calculate each time the Riemann sum defined by selecting the center of each rectangle. Then we get the midpoint approximations listed in Fig. 14.1.6. In Example 2 we will see (much more easily) that the exact value of the integral in Example 1 is

$$\iint_R (4x^3 + 6xy^2)\, dA = 312.$$

Number of subrectangles	Midpoint approximation
6	294.00
24	307.50
96	310.88
384	311.72
1536	311.93
6144	311.98

Fig. 14.1.6 Midpoint approximations to the integral in Example 1

Iterated Integrals

The direct evaluation of the limit in Eq. (2) is generally even less practical than the direct evaluation of the limit we used in Section 5.4 to define the single-variable integral. In practice, we shall calculate double integrals over rectangles by means of the **iterated integrals** that appear in Theorem 1.

Theorem 1 *Double Integrals as Iterated Single Integrals*

Suppose that $f(x, y)$ is continuous on the rectangle $R = [a, b] \times [c, d]$. Then

$$\iint_R f(x, y)\, dA = \int_a^b \left(\int_c^d f(x, y)\, dy \right) dx = \int_c^d \left(\int_a^b f(x, y)\, dx \right) dy. \tag{3}$$

Theorem 1 tells us how to compute a double integral by means of two successive (or *iterated*) single-variable integrations, both of which we can compute by using the fundamental theorem of calculus (if the function f is sufficiently well-behaved on R).

Let us explain what we mean by the parentheses in the iterated integral

$$\int_a^b \int_c^d f(x, y)\, dy\, dx = \int_a^b \left(\int_c^d f(x, y)\, dy \right) dx. \tag{4}$$

First we hold x constant and integrate with respect to y, from $y = c$ to $y = d$. The result of this first integration is the **partial integral of f with respect to** y, denoted by

$$\int_c^d f(x, y)\, dy,$$

and it is a function of x alone. Then we integrate this latter function with respect to x, from $x = a$ to $x = b$.

Similarly, we calculate the iterated integral

$$\int_c^d \int_a^b f(x, y)\, dx\, dy = \int_c^d \left(\int_a^b f(x, y) \right) dx\, dy \tag{5}$$

by first integrating from a to b with respect to x (while holding y fixed) and then integrating the result from c to d with respect to y. The order of integration (either first with respect to x and then with respect to y, or the reverse) is determined by the order in which the differentials dx and dy appear in the iterated integrals in Eqs. (4) and (5). We almost always work "from the inside out." Theorem 1 guarantees that the value obtained is independent of the order of integration provided that f is continuous on R.

EXAMPLE 2 Compute the iterated integrals in Eqs. (4) and (5) for the function $f(x, y) = 4x^3 + 6xy^2$ on the rectangle $R = [1, 3] \times [-2, 1]$.

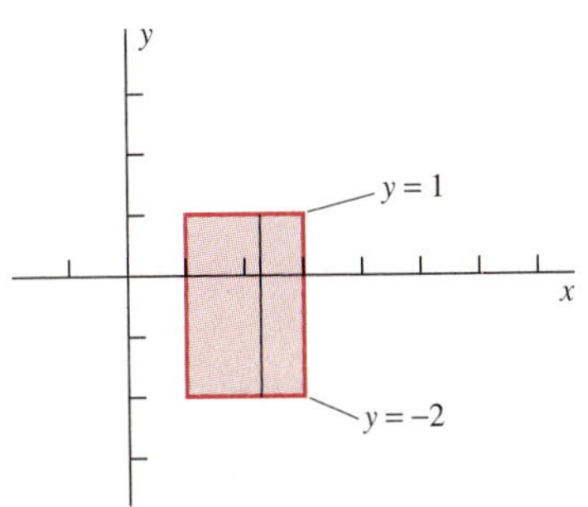

Fig. 14.1.7 The inner limits of the first iterated integral (Example 2)

Solution The rectangle R is shown in Fig. 14.1.7, where the vertical segment (on which x is constant) corresponds to the inner integral in Eq. (4). Its endpoints lie at heights $y = -2$ and $y = 1$, which are therefore the limits on the inner integral. So Eq. (4) yields

$$\begin{aligned}\int_1^3 \left(\int_{-2}^1 (4x^3 + 6xy^2)\, dy \right) dx &= \int_1^3 \left[4x^3 y + 2xy^3 \right]_{y=-2}^1 dx \\ &= \int_1^3 [(4x^3 + 2x) - (-8x^3 - 16x)]\, dx \\ &= \int_1^3 (12x^3 + 18x)\, dx \\ &= \left[3x^4 + 9x^2 \right]_1^3 = 312.\end{aligned}$$

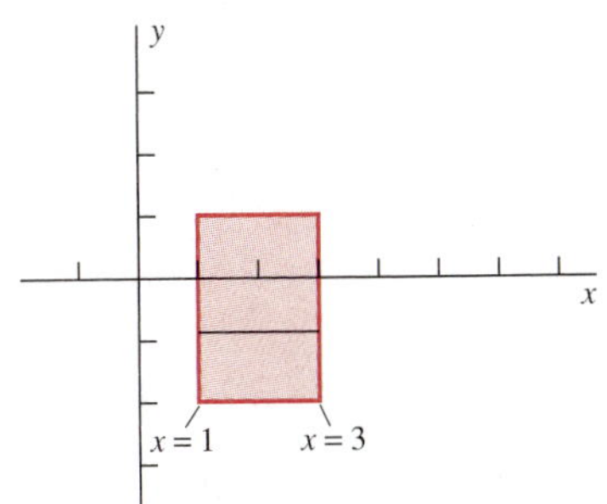

Fig. 14.1.8 The inner limits of the second iterated integral (Example 2)

The horizontal segment (on which y is constant) in Fig. 14.1.8 corresponds to the inner integral in Eq. (5). Its endpoints lie at $x = 1$ and $x = 3$ (the limits of integration for x), so Eq. (5) gives

$$\begin{aligned}\int_{-2}^1 \left(\int_1^3 (4x^3 + 6xy^2)\, dx \right) dy &= \int_{-2}^1 \left[x^4 + 3x^2y^2 \right]_{x=1}^3 dy \\ &= \int_{-2}^1 [(81 + 27y^2) - (1 + 3y^2)]\, dy \\ &= \int_{-2}^1 (80 + 24y^2)\, dy \\ &= \left[80y + 8y^3 \right]_{-2}^1 = 312.\end{aligned}$$

■

When we note that iterated double integrals are almost always evaluated from the inside out, it becomes clear that the parentheses appearing on the right-hand sides in Eqs. (4) and (5) are unnecessary. They are therefore generally omitted, as in Examples 3 and 4. When $dy\, dx$ appears in the integrand, we integrate first with respect to y, whereas the appearance of $dx\, dy$ tells us to integrate first with respect to x.

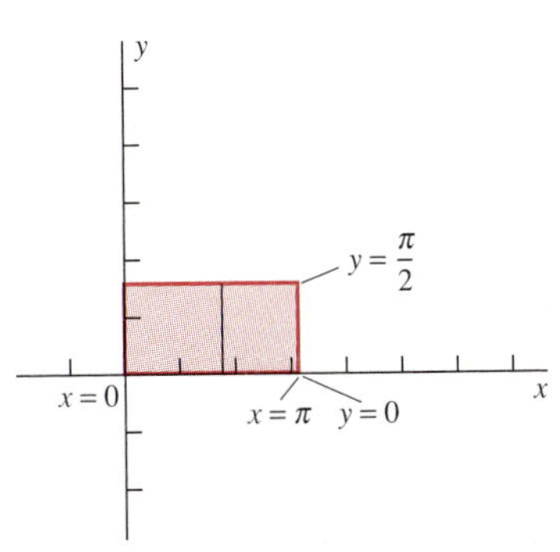

Fig. 14.1.9 Example 3

EXAMPLE 3 See Fig. 14.1.9.

$$\begin{aligned}\int_0^\pi \int_0^{\pi/2} \cos x \cos y\, dy\, dx &= \int_0^\pi \left[\cos x \sin y \right]_{y=0}^{\pi/2} dx \\ &= \int_0^\pi \cos x\, dx = \left[\sin x \right]_0^\pi = 0.\end{aligned}$$

■

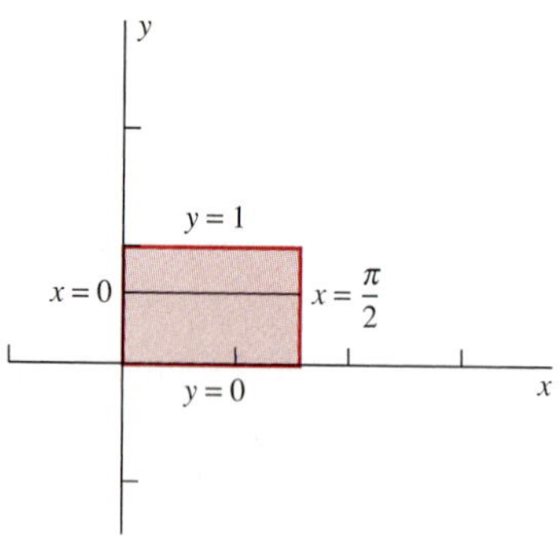

Fig. 14.1.10 Example 4

EXAMPLE 4 See Fig. 14.1.10.

$$\begin{aligned}\int_0^1\int_0^{\pi/2}(e^y+\sin x)\,dx\,dy &= \int_0^1\Big[xe^y-\cos x\Big]_{x=0}^{\pi/2}dy\\ &= \int_0^1\left(\tfrac{1}{2}\pi e^y+1\right)dy\\ &= \Big[\tfrac{1}{2}\pi e^y+y\Big]_0^1=\tfrac{1}{2}\pi(e-1)+1.\end{aligned}$$

Iterated Integrals and Cross Sections

An outline of the proof of Theorem 1 illuminates the relationship between iterated integrals and the method of cross sections (for computing volumes) discussed in Section 6.2. First we partition $[a, b]$ into n equal subintervals, each of length $\Delta x = (b-a)/n$, and we also partition $[c, d]$ into n equal subintervals, each of length $\Delta y = (d-c)/n$. This gives n^2 rectangles, each of which has area $\Delta A = \Delta x\,\Delta y$. Choose a point $x_i^\star$ in $[x_{i-1}, x_i]$ for each i, $1 \leqq i \leqq n$. Then the average value theorem for single integrals (Section 5.6) gives a point $y_{ij}^\star$ in $[y_{j-1}, y_j]$ such that

$$\int_{y_{j-1}}^{y_j} f(x_i^\star, y)\,dy = f(x_i^\star, y_{ij}^\star)\,\Delta y.$$

This gives us the selected point $(x_i^\star, y_{ij}^\star)$ in the rectangle $[x_{i-1}, x_i] \times [y_{j-1}, y_j]$. Then

$$\begin{aligned}\iint_R f(x,y)\,dA \approx \sum_{i,j=1}^n f(x_i^\star, y_{ij}^\star)\,\Delta A &= \sum_{i=1}^n\sum_{j=1}^n f(x_i^\star, y_{ij}^\star)\,\Delta y\,\Delta x\\ &= \sum_{i=1}^n\left(\sum_{j=1}^n\int_{y_{j-1}}^{y_j} f(x_i^\star, y)\,dy\right)\Delta x\\ &= \sum_{i=1}^n\left(\int_c^d f(x_i^\star, y)\,dy\right)\Delta x\\ &= \sum_{i=1}^n A(x_i^\star)\,\Delta x,\end{aligned}$$

where

$$A(x) = \int_c^d f(x,y)\,dy.$$

Moreover, the last sum is a Riemann sum for the integral

$$\int_a^b A(x)\,dx,$$

so the result of our computation is

$$\iint_R f(x,y)\,dA \approx \sum_{i=1}^n A(x_i^\star)\,\Delta x \approx \int_a^b A(x)\,dx = \int_a^b\left(\int_c^d f(x,y)\,dy\right)dx.$$

We can convert this outline into a complete proof of Theorem 1 by showing that the preceding approximations become equalities when we take limits as $n \to +\infty$.

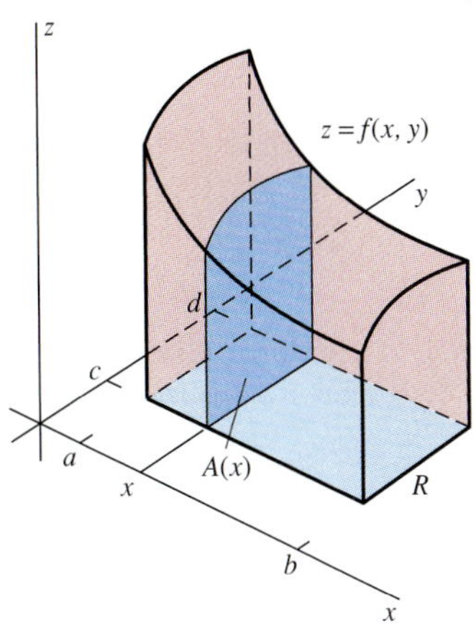

Fig. 14.1.11 The area of the cross section at x is

$$A(x) = \int_c^d f(x, y)\, dy.$$

In case the function f is nonnegative on R, the function $A(x)$ introduced here gives the area of the vertical cross section of R perpendicular to the x-axis (Fig. 14.1.11). Thus the iterated integral in Eq. (4) expresses the volume V as the integral from $x = a$ to $x = b$ of the cross-sectional area function $A(x)$. Similarly, the iterated integral in Eq. (5) expresses V as the integral from $y = c$ to $y = d$ of the function

$$A(y) = \int_a^b f(x, y)\, dx,$$

which gives the area of a vertical cross section in a plane perpendicular to the y-axis. [Although it seems appropriate to use the notation $A(y)$ here, note that $A(x)$ and $A(y)$ are by no means the same function!]

14.1 PROBLEMS

1. Approximate the integral

$$\iint_R (4x^3 + 6xy^2)\, dA$$

of Example 1 using the partition shown in Fig. 14.1.5, but selecting each $(x_i^\star, y_i^\star)$ as (a) the lower left corner of the rectangle R_i; (b) the upper right corner of the rectangle R_i.

2. Approximate the integral

$$\iint_R (4x^3 + 6xy^2)\, dA$$

as in Problem 1, but selecting each $(x_i^\star, y_i^\star)$ as (a) the upper left corner of the rectangle R_i; (b) the lower right corner of the rectangle R_i.

In Problems 3 through 8, calculate the Riemann sum for

$$\iint_R f(x, y)\, dA$$

using the given partition and selection of points $(x_i^\star, y_i^\star)$ for the rectangle R.

3. $f(x, y) = x + y$; $R = [0, 2] \times [0, 2]$; the partition $\mathcal{P}$ consists of four unit squares; each $(x_i^\star, y_i^\star)$ is the center point of the ith rectangle R_i.

4. $f(x, y) = xy$; $R = [0, 2] \times [0, 2]$; the partition $\mathcal{P}$ consists of four unit squares; each $(x_i^\star, y_i^\star)$ is the center point of the ith rectangle R_i.

5. $f(x, y) = x^2 - 2y$; $R = [2, 6] \times [-1, 1]$; the partition $\mathcal{P}$ consists of four equal rectangles of width $\Delta x = 2$ and height $\Delta y = 1$; each $(x_i^\star, y_i^\star)$ is the lower left corner of the ith rectangle R_i.

6. $f(x, y) = x^2 + y^2$; $R = [0, 2] \times [0, 3]$; the partition $\mathcal{P}$ consists of six unit squares; each $(x_i^\star, y_i^\star)$ is the upper right corner of the ith rectangle R_i.

7. $f(x, y) = \sin x \sin y$; $R = [0, \pi] \times [0, \pi]$; the partition $\mathcal{P}$ consists of four equal squares; each $(x_i^\star, y_i^\star)$ is the center point of the ith rectangle R_i.

8. $f(x, y) = \sin 4xy$; $R = [0, 1] \times [0, \pi]$; the partition $\mathcal{P}$ consists of six equal rectangles of width $\Delta x = \frac{1}{2}$ and height $\Delta y = \frac{1}{3}\pi$; each $(x_i^\star, y_i^\star)$ is the center point of the ith rectangle R_i.

In Problems 9 and 10, let L, M, and U denote the Riemann sums calculated for the given function f and the indicated partition $\mathcal{P}$ by selecting the lower left corners, midpoints, and upper right corners (respectively) of the rectangles in $\mathcal{P}$. Without actually calculating any of these Riemann sums, arrange them in increasing order of size.

9. $f(x, y) = x^2y^2$; $R = [1, 3] \times [2, 5]$; the partition $\mathcal{P}$ consists of six unit squares.

10. $f(x, y) = \sqrt{100 - x^2 - y^2}$; $R = [1, 4] \times [2, 5]$; the partition $\mathcal{P}$ consists of nine unit squares.

Evaluate the iterated integrals in Problems 11 through 30.

11. $\displaystyle\int_0^2\int_0^4 (3x + 4y)\, dx\, dy$

12. $\displaystyle\int_0^3\int_0^2 x^2y\, dx\, dy$

13. $\displaystyle\int_{-1}^2\int_1^3 (2x - 7y)\, dy\, dx$

14. $\displaystyle\int_{-2}^1\int_2^4 x^2y^3\, dy\, dx$

15. $\displaystyle\int_0^3\int_0^3 (xy + 7x + y)\, dx\, dy$

16. $\displaystyle\int_0^2\int_2^4 (x^2y^2 - 17)\, dx\, dy$

17. $\displaystyle\int_{-1}^2\int_{-1}^2 (2xy^2 - 3x^2y)\, dy\, dx$

18. $\displaystyle\int_1^3 \int_{-3}^{-1} (x^3y - xy^3)\, dy\, dx$

19. $\displaystyle\int_0^{\pi/2} \int_0^{\pi/2} \sin x \cos y\, dx\, dy$

20. $\displaystyle\int_0^{\pi/2} \int_0^{\pi/2} \cos x \sin y\, dy\, dx$

21. $\displaystyle\int_0^1 \int_0^1 xe^y\, dy\, dx$

22. $\displaystyle\int_0^1 \int_{-2}^2 x^2e^y\, dx\, dy$

23. $\displaystyle\int_0^1 \int_0^\pi e^x \sin y\, dy\, dx$

24. $\displaystyle\int_0^1 \int_0^1 e^{x+y}\, dx\, dy$

25. $\displaystyle\int_0^\pi \int_0^\pi (xy + \sin x)\, dx\, dy$

26. $\displaystyle\int_0^{\pi/2} \int_0^{\pi/2} (y - 1) \cos x\, dx\, dy$

27. $\displaystyle\int_0^{\pi/2} \int_1^e \frac{\sin y}{x}\, dx\, dy$

28. $\displaystyle\int_1^e \int_1^e \frac{1}{xy}\, dy\, dx$

29. $\displaystyle\int_0^1 \int_0^1 \left(\frac{1}{x+1} + \frac{1}{y+1}\right) dx\, dy$

30. $\displaystyle\int_1^2 \int_1^3 \left(\frac{x}{y} + \frac{y}{x}\right) dy\, dx$

In Problems 31 through 34, verify that the values of

$$\iint_R f(x, y)\, dA$$

given by the iterated integrals in Eqs. (4) and (5) are indeed equal.

31. $f(x, y) = 2xy - 3y^2$; $R = [-1, 1] \times [-2, 2]$

32. $f(x, y) = \sin x \cos y$; $R = [0, \pi] \times [-\pi/2, \pi/2]$

33. $f(x, y) = \sqrt{x + y}$; $R = [0, 1] \times [1, 2]$

34. $f(x, y) = e^{x+y}$; $R = [0, \ln 2] \times [0, \ln 3]$

35. Prove that

$$\lim_{n\to\infty} \int_0^1 \int_0^1 x^n y^n\, dx\, dy = 0.$$

36. Suppose that $f(x, y) = k$ is a constant-valued function and $R = [a, b] \times [c, d]$. Use Riemann sums to prove that

$$\iint_R k\, dA = k(b - a)(d - c).$$

37. Use Riemann sums to show, without calculating the value of the integral, that

$$0 \leqq \int_0^\pi \int_0^\pi \sin\sqrt{xy}\, dx\, dy \leqq \pi^2.$$

Problems 38 through 40 list properties of double integrals that are analogous to familiar properties of single integrals. In each case state the corresponding relation between Riemann sums associated with a given partition and selection for the rectangle R.

38. $\displaystyle\iint_R cf(x, y)\, dA = c\iint_R f(x, y)\, dA$ (c is a constant).

39. $\displaystyle\iint_R [f(x, y) + g(x, y)]\, dA$

$$= \iint_R f(x, y)\, dA + \iint_R g(x, y)\, dA.$$

40. If $f(x, y) \leqq g(x, y)$ at each point of R, then

$$\iint_R f(x, y)\, dA \leqq \iint_R g(x, y)\, dA.$$

14.1 PROJECT: MIDPOINT APPROXIMATIONS OF DOUBLE INTEGRALS

This project explores the *midpoint approximation* to the double integral

$$I = \iint_R f(x, y)\, dA \tag{1}$$

of the function $f(x, y)$ over the plane rectangle $R = [a, b] \times [c, d]$. To define the midpoint approximation, let $[a, b]$ be partitioned into m subintervals, all with the same length $h = \Delta x = (b - a)/m$, and let $[c, d]$ be partitioned into n subintervals, all with the same length $k = \Delta y = (d - c)/n$. For each i and j ($1 \leqq i \leqq m$ and $1 \leqq j \leqq n$), let u_i and v_j denote the *midpoints* of the ith subinterval $[x_{i-1}, x_i]$ and the jth subinterval $[y_{j-1}, y_j]$, respectively. Then the corresponding **midpoint approximation** to the double integral I is the sum

$$S_{mn} = \sum_{i=1}^{m} \sum_{j=1}^{n} f(u_i, v_j)\, hk. \tag{2}$$

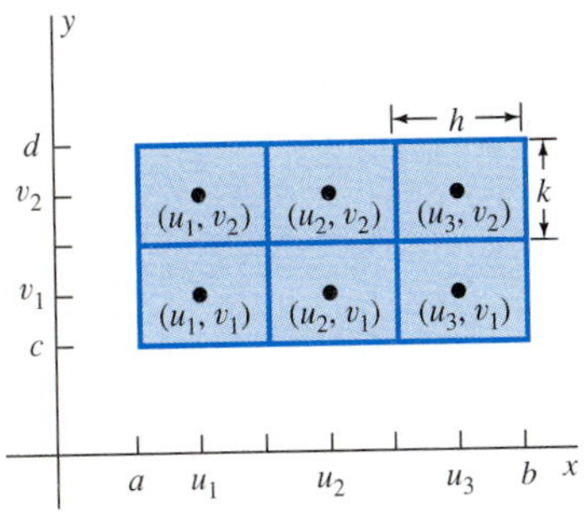

Fig. 14.1.12 The points used in the midpoint approximation

Figure 14.1.12 illustrates the case $m = 3$, $n = 2$, in which $h = (b - a)/3$, $k = (d - c)/2$, and

$$S_{32} = hk[f(u_1, v_1) + f(u_2, v_1) + f(u_3, v_1) + f(u_1, v_2) + f(u_2, v_2) + f(u_3, v_2)].$$

The midpoint sum in Eq. (2) can be calculated painlessly using computer algebra system commands such as

```
sum( sum( f(a-h/2+i*h, c-k/2+j*k), i=1..m), j=1..n)*h*k
```

in *Maple* or

```
Sum[ f[x,y], {x, a+h/2, b-h/2, h}, {y, c+k/2, d-k/2, k} ]*h*k
```

in *Mathematica*. For each of the double integrals in Problems 1 through 6, first calculate the midpoint approximation S_{mn} with the indicated values of m and n. Then try larger values. Compare each numerical approximation with the exact value of the integral.

1. $\displaystyle\int_0^1\int_0^1 (x + y)\,dy\,dx,\quad m = n = 2$

2. $\displaystyle\int_0^3\int_0^2 (2x + 3y)\,dy\,dx,\quad m = 3,\quad n = 2$

3. $\displaystyle\int_0^2\int_0^2 xy\,dy\,dx,\quad m = n = 2$

4. $\displaystyle\int_0^1\int_0^1 x^2y\,dy\,dx,\quad m = n = 3$

5. $\displaystyle\int_0^{\pi/2}\int_0^{\pi/2} \sin x \sin y\,dy\,dx,\quad m = n = 2$

6. $\displaystyle\int_0^{\pi/2}\int_0^1 \frac{\cos x}{1 + y^2}\,dy\,dx,\quad m = n = 2$

14.2 DOUBLE INTEGRALS OVER MORE GENERAL REGIONS

Now we want to define and compute double integrals over regions more general than rectangles. Let the function f be defined on the plane region R, and suppose that R is **bounded**—that is, that R lies within some rectangle S. To define the (double) integral of f over R, we begin with a partition $\mathcal{Q}$ of the rectangle S into subrectangles. Some of the rectangles of $\mathcal{Q}$ will lie wholly within R, some will be outside R, and some will lie partly within and partly outside R. We consider the collection $\mathcal{P} = \{R_1, R_2, \dots, R_k\}$ of all those rectangles in $\mathcal{Q}$ that lie *completely within* the region R. This collection $\mathcal{P}$ is called the **inner partition** of the region R determined by the partition $\mathcal{Q}$ of the rectangle S (Fig. 14.2.1). By the **norm** $|\mathcal{P}|$ of the inner partition $\mathcal{P}$ we mean the norm of the partition $\mathcal{Q}$ that determines $\mathcal{P}$. Note that $|\mathcal{P}|$ depends not only on $\mathcal{P}$ but on $\mathcal{Q}$ as well.

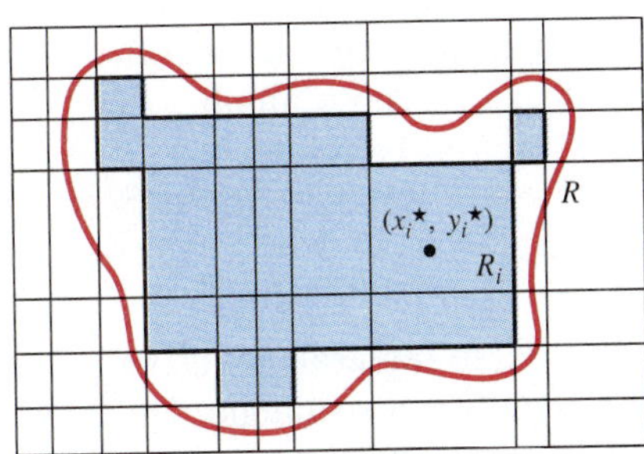

Fig. 14.2.1 The rectangular partition of S produces an associated inner partition (shown shaded) of the region R.

Using the inner partition $\mathcal{P}$ of the region R, we can proceed in much the same way as in Section 14.1. By choosing an arbitrary point $(x_i^\star, y_i^\star)$ in the ith rectangle R_i of $\mathcal{P}$ for $i = 1, 2, 3, \dots, k$, we obtain a **selection** for the inner partition $\mathcal{P}$. Let us denote by ΔA_i the area of R_i. Then this selection gives the **Riemann sum**

$$\sum_{i=1}^{k} f(x_i^\star, y_i^\star)\,\Delta A_i$$

associated with the inner partition $\mathcal{P}$. In case f is nonnegative on R, this Riemann sum approximates the volume of the three-dimensional region that lies under the surface $z = f(x, y)$ and above the region R in the xy-plane. We therefore define the double integral of f over the region R by taking the limit of this Riemann sum as the norm $|\mathcal{P}|$ approaches zero. Thus

$$\iint_R f(x, y)\, dA = \lim_{|\mathcal{P}|\to 0} \sum_{i=1}^{k} f(x_i^\star, y_i^\star)\,\Delta A_i, \tag{1}$$

provided that this limit exists in the sense of the following definition.

Definition ***The Double Integral***

The **double integral** of the bounded function f over the plane region R is the number

$$I = \iint_R f(x, y)\, dA$$

provided that, for every $\epsilon > 0$, there exists a number $\delta > 0$ such that

$$\left| \sum_{i=1}^{k} f(x_i^\star, y_i^\star)\,\Delta A_i - I \right| < \epsilon$$

for every inner partition $\mathcal{P} = \{R_1, R_2, \ldots, R_k\}$ of R that has norm $|\mathcal{P}| < \delta$ and every selection of points $(x_i^\star, y_i^\star)$ in R_i $(i = 1, 2, \ldots, k)$.

Thus the meaning of the limit in Eq. (1) is that the Riemann sum can be made arbitrarily close to the number

$$I = \iint_R f(x, y)\, dA$$

merely by choosing the norm of the inner partition $\mathcal{P}$ sufficiently small. In this case we say that the function f is **integrable** on the region R.

Note If R is a rectangle and we choose $S = R$ (so that an inner partition of R is simply a partition of R), then the preceding definition reduces to our earlier definition of a double integral over a rectangle. In advanced calculus the double integral of the function f over the bounded plane region R is shown to exist provided that f is continuous on R and the *boundary* of R is reasonably well behaved. In particular, it suffices for the boundary of R to consist of a finite number of piecewise smooth simple closed curves (that is, each boundary curve consists of a finite number of smooth arcs).

Evaluation of Double Integrals

For certain common types of regions, we can evaluate double integrals by using iterated integrals in much the same way as when the region is a rectangle. The plane region R is called **vertically simple** if it can be described by means of the inequalities

$$a \leqq x \leqq b, \quad y_1(x) \leqq y \leqq y_2(x), \tag{2}$$

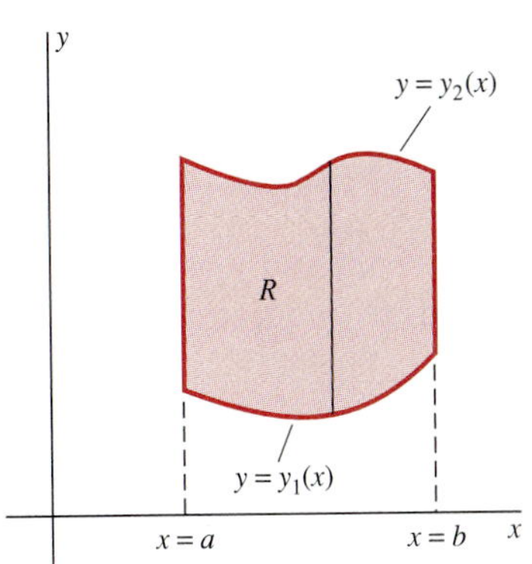

Fig. 14.2.2 A vertically simple region R

where $y_1(x)$ and $y_2(x)$ are continuous functions of x on $[a, b]$. Such a region appears in Fig. 14.2.2. The region R is called **horizontally simple** if it can be described by the inequalities

$$c \leqq y \leqq d, \quad x_1(y) \leqq x \leqq x_2(y), \tag{3}$$

where $x_1(y)$ and $x_2(y)$ are continuous functions of y on $[c, d]$. The region in Fig. 14.2.3 is horizontally simple.

Theorem 1 tells us how to compute by iterated integration a double integral over a region R that is either vertically simple or horizontally simple.

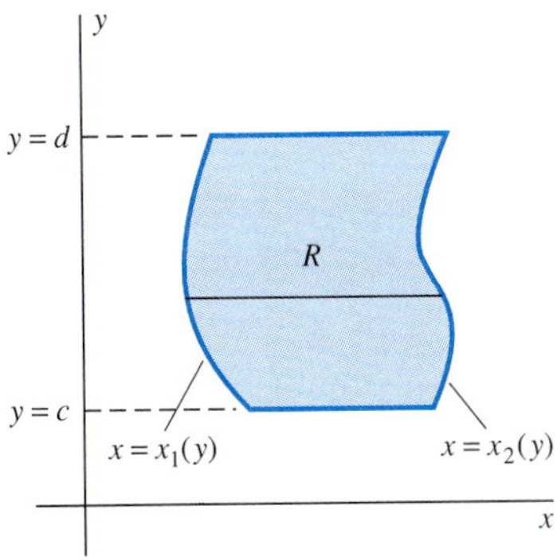

Fig. 14.2.3 A horizontally simple region R

Theorem 1 *Evaluation of Double Integrals*

Suppose that $f(x, y)$ is continuous on the region R. If R is the vertically simple region given in (2), then

$$\iint_R f(x, y)\, dA = \int_a^b \int_{y_1(x)}^{y_2(x)} f(x, y)\, dy\, dx. \tag{4}$$

If R is the horizontally simple region given in (3), then

$$\iint_R f(x, y)\, dA = \int_c^d \int_{x_1(y)}^{x_2(y)} f(x, y)\, dx\, dy. \tag{5}$$

Theorem 1 here includes Theorem 1 of Section 14.1 as a special case (when R is a rectangle), and it can be proved by a generalization of the argument we outlined there.

EXAMPLE 1 Compute in two different ways the integral

$$\iint_R xy^2\, dA,$$

where R is the first-quadrant region bounded by the two curves $y = \sqrt{x}$ and $y = x^3$.

Solution *Always sketch the region R of integration before attempting to evaluate a double integral.* As indicated in Figs. 14.2.4 and 14.2.5, the given region R is both vertically and horizontally simple. The vertical segment in Fig. 14.2.4 with endpoints on the curves $y = x^3$ and $y = \sqrt{x}$ corresponds to integrating first with respect to y:

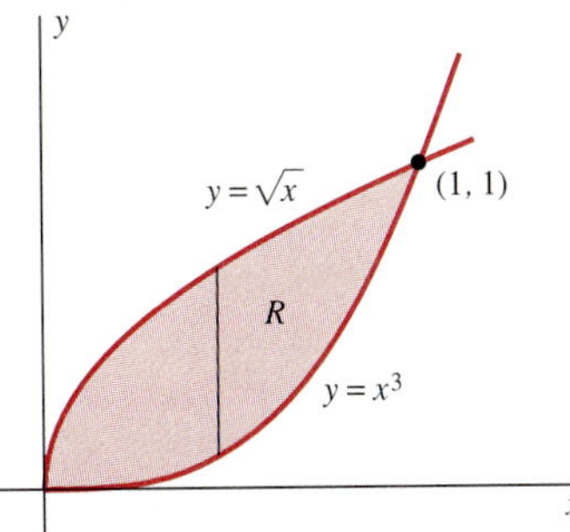

Fig. 14.2.4 The vertically simple region of Example 1

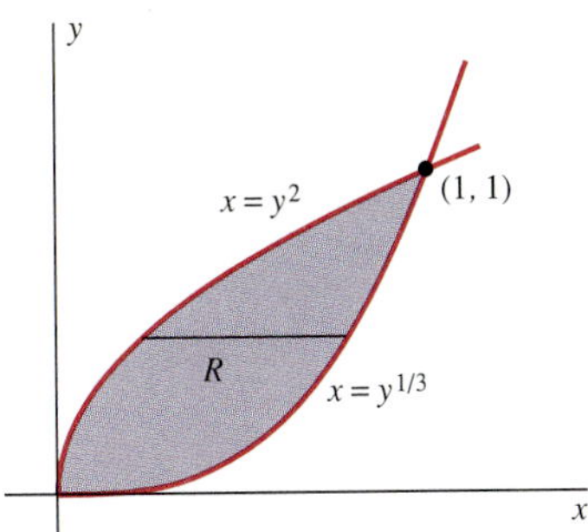

Fig. 14.2.5 The horizontally simple region of Example 1

$$\iint_R xy^2\,dA = \int_0^1\int_{x^3}^{\sqrt{x}} xy^2\,dy\,dx = \int_0^1\left[\frac{1}{3}xy^3\right]_{y=x^3}^{\sqrt{x}}dx$$
$$= \int_0^1\left(\frac{1}{3}x^{5/2} - \frac{1}{3}x^{10}\right)dx = \frac{2}{21} - \frac{1}{33} = \frac{5}{77}.$$

We get $x = y^2$ and $x = y^{1/3}$ when we solve the equations $y = \sqrt{x}$ and $y = x^3$ for x in terms of y. The horizontal segment in Fig. 14.2.5 corresponds to integrating first with respect to x:

$$\iint_R xy^2\,dA = \int_0^1\int_{y^2}^{y^{1/3}} xy^2\,dx\,dy = \int_0^1\left[\frac{1}{2}x^2y^2\right]_{x=y^2}^{y^{1/3}}$$
$$= \int_0^1\left(\frac{1}{2}y^{8/3} - \frac{1}{2}y^6\right)dy = \frac{3}{22} - \frac{1}{14} = \frac{5}{77}.$$ ■

EXAMPLE 2 Evaluate

$$\iint_R (6x + 2y^2)\,dA,$$

where R is the region bounded by the parabola $x = y^2$ and the straight line $x + y = 2$.

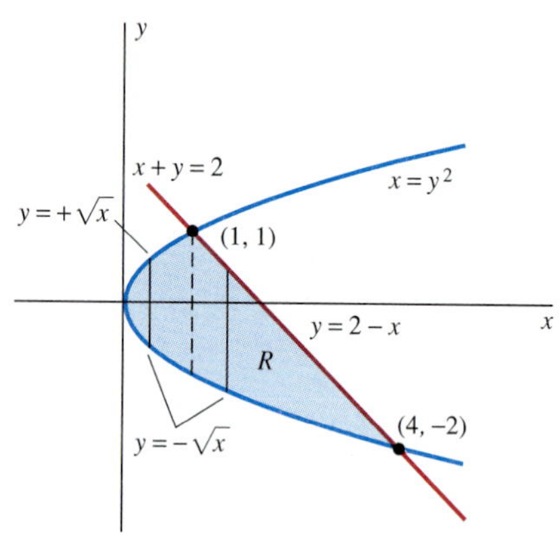

Fig. 14.2.6 The vertically simple region of Example 2

Solution The region R appears in Fig. 14.2.6. It is both horizontally and vertically simple. If we wished to integrate first with respect to y and then with respect to x, we would need to evaluate two integrals:

$$\iint_R f(x, y)\,dA = \int_0^1\int_{-\sqrt{x}}^{\sqrt{x}} (6x + 2y^2)\,dy\,dx + \int_1^4\int_{-\sqrt{x}}^{2-x} (6x + 2y^2)\,dy\,dx.$$

The reason is that the formula of the function $y = y_2(x)$ describing the "top boundary curve" of R changes at the point $(1, 1)$, from $y = \sqrt{x}$ on the left to $y = 2 - x$ on the right. But as we see in Fig. 14.2.7, every *horizontal* segment in R extends from $x = y^2$ on the left to $x = 2 - y$ on the right. Therefore integration first with respect to x requires us to evaluate only *one* iterated integral:

$$\iint_R f(x, y)\,dA = \int_{-2}^1\int_{y^2}^{2-y} (6x + 2y^2)\,dx\,dy$$
$$= \int_{-2}^1\left[3x^2 + 2xy^2\right]_{x=y^2}^{2-y}dy$$
$$= \int_{-2}^1 [3(2-y)^2 + 2(2-y)y^2 - 3(y^2)^2 - 2y^4]\,dy$$
$$= \int_{-2}^1 (12 - 12y + 7y^2 - 2y^3 - 5y^4)\,dy$$
$$= \left[12y - 6y^2 + \frac{7}{3}y^3 - \frac{1}{2}y^4 - y^5\right]_{-2}^1 = \frac{99}{2}.$$ ■

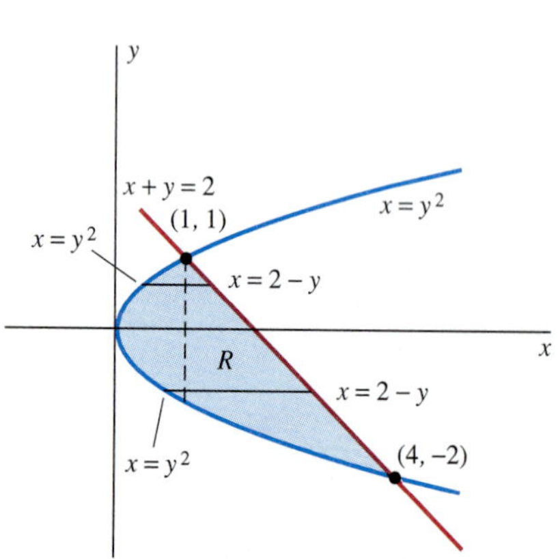

Fig. 14.2.7 The horizontally simple region of Example 2

Example 2 shows that even when the region R is both vertically and horizontally simple, it may be easier to integrate in one order rather than the other because of the shape of R. We naturally prefer the easier route. The choice of the preferable order of integration may be influenced also by the nature of the function $f(x, y)$. It may be difficult—or even impossible—to compute a given iterated integral but easy to do so *after we reverse the order of integration*. Example 3 shows that the key to reversing the order of integration is this:

> *Find and sketch the region R over which the integration is to be performed.*

EXAMPLE 3 Evaluate

$$\int_0^2 \int_{y/2}^1 ye^{x^3}\,dx\,dy.$$

Solution We cannot integrate first with respect to x, as indicated, because $\exp(x^3)$ is known to have no elementary antiderivative. So we try to evaluate the integral by first reversing the order of integration. To do so, we sketch the region of integration specified by the limits in the given iterated integral.

The region R is determined by the inequalities $\frac{1}{2}y \leqq x \leqq 1$ and $0 \leqq y \leqq 2$. Thus all points (x, y) of R lie between the horizontal lines $y = 0$ and $y = 2$ and between the two lines $x = y/2$ and $x = 1$. We draw the four lines $y = 0$, $y = 2$, $x = y/2$, and $x = 1$ and find that the region of integration is the shaded triangle that appears in Fig. 14.2.8.

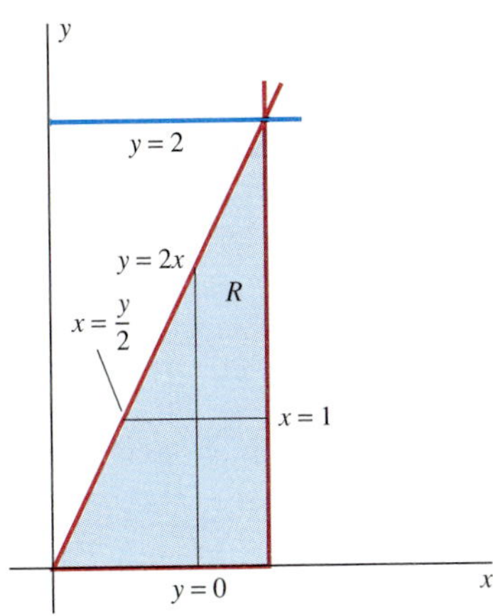

Fig. 14.2.8 The region of Example 3

Integrating first with respect to y, from $y_1(x) \equiv 0$ to $y_2(x) = 2x$, we obtain

$$\int_0^2 \int_{y/2}^1 ye^{x^3}\,dx\,dy = \int_0^1 \int_0^{2x} ye^{x^3}\,dy\,dx = \int_0^1 \left[\tfrac{1}{2}y^2\right]_{y=0}^{2x} e^{x^3}\,dx$$

$$= \int_0^1 2x^2 e^{x^3}\,dx = \left[\frac{2}{3}e^{x^3}\right]_{x=0}^{1} = \frac{2}{3}(e - 1). \quad \blacksquare$$

Properties of Double Integrals

We conclude this section by listing some formal properties of double integrals. Let c be a constant and f and g be continuous functions on a region R on which $f(x, y)$ attains a minimum value m and a maximum value M. Let $a(R)$ denote the area of the region R. If all the indicated integrals exist, then

$$\iint_R cf(x, y)\,dA = c\iint_R f(x, y)\,dA, \tag{6}$$

$$\iint_R [f(x, y) + g(x, y)]\,dA = \iint_R f(x, y)\,dA + \iint_R g(x, y)\,dA, \tag{7}$$

$$m \cdot a(R) \leqq \iint_R f(x, y)\,dA \leqq M \cdot a(R), \tag{8}$$

$$\iint_R f(x, y)\,dA = \iint_{R_1} f(x, y)\,dA + \iint_{R_2} f(x, y)\,dA. \tag{9}$$

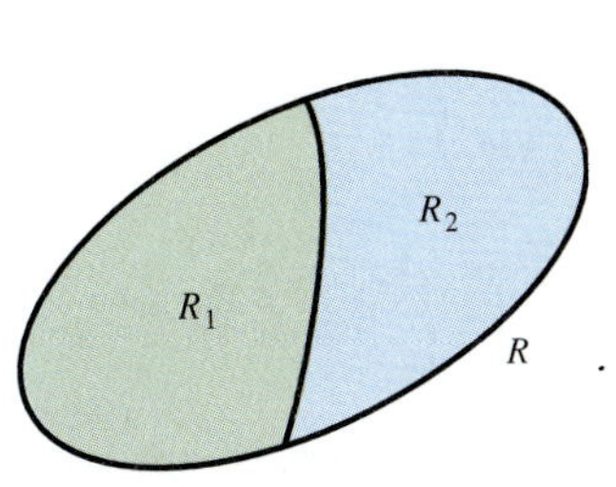

Fig. 14.2.9 The regions of Eq. (9)

In Eq. (9), R_1 and R_2 are simply two nonoverlapping regions (with disjoint interiors) with union R (Fig. 14.2.9). We indicate in Problems 45 through 48 proofs of the properties in (6) through (9) for the special case in which R is a rectangle.

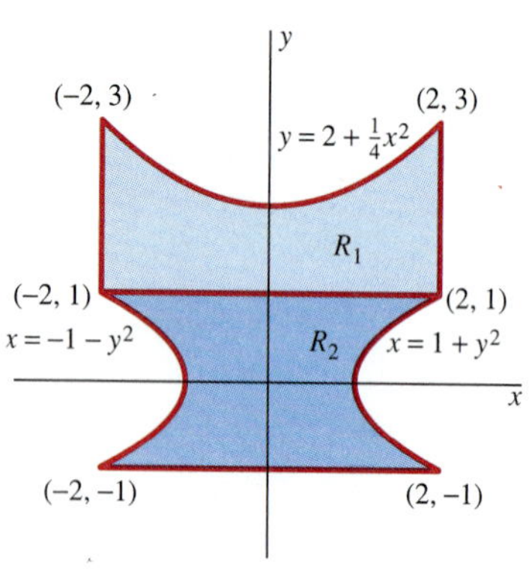

Fig. 14.2.10 The nonsimple region R is the union of the nonoverlapping simple regions R_1 and R_2.

The property in Eq. (9) enables us to evaluate double integrals over a region R that is neither vertically nor horizontally simple. All that is necessary is to divide R into a finite number of simple regions $R_1, R_2, \ldots, R_n$. Then we integrate over each (converting each double integral into an iterated integral, as in the examples of this section) and add the results.

EXAMPLE 4 Let f be a function that is integrable on the region R of Fig. 14.2.10. Note that R is not simple, but is the union of the vertically simple region R_1 and the horizontally simple region R_2. Using the boundary curves labeled in the figure and the appropriate order of integration for each region, we see that

$$\iint_R f(x, y)\,dA = \iint_{R_1} f(x, y)\,dA + \iint_{R_2} f(x, y)\,dA$$

$$= \int_{-2}^{2}\int_{1}^{2+x^2/4} f(x, y)\,dy\,dx + \int_{-1}^{1}\int_{-1-y^2}^{1+y^2} f(x, y)\,dx\,dy.$$ ■

14.2 PROBLEMS

Evaluate the iterated integrals in Problems 1 through 14.

1. $\displaystyle\int_0^1\int_0^x (1 + x)\,dy\,dx$

2. $\displaystyle\int_0^2\int_0^{2x} (1 + y)\,dy\,dx$

3. $\displaystyle\int_0^1\int_y^1 (x + y)\,dx\,dy$ (Fig. 14.2.11)

4. $\displaystyle\int_0^2\int_{y/2}^1 (x + y)\,dx\,dy$ (Fig. 14.2.12)

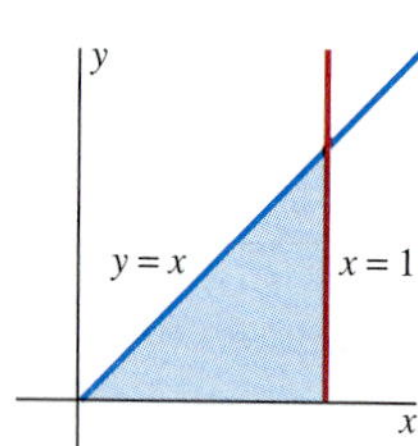

Fig. 14.2.11 Problem 3

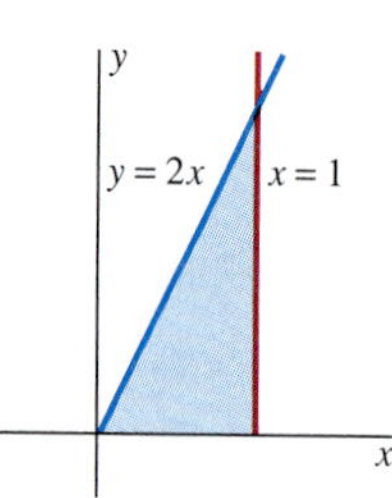

Fig. 14.2.12 Problem 4

5. $\displaystyle\int_0^1\int_0^{x^2} xy\,dy\,dx$

6. $\displaystyle\int_0^1\int_y^{\sqrt{y}} (x + y)\,dx\,dy$

7. $\displaystyle\int_0^1\int_x^{\sqrt{x}} (2x - y)\,dy\,dx$ (Fig. 14.2.13)

8. $\displaystyle\int_0^2\int_{-\sqrt{2y}}^{\sqrt{2y}} (3x + 2y)\,dx\,dy$ (Fig. 14.2.14)

9. $\displaystyle\int_0^1\int_{x^4}^{x} (y - x)\,dy\,dx$

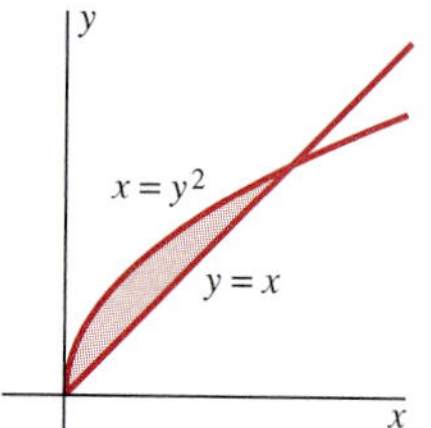

Fig. 14.2.13 Problem 7

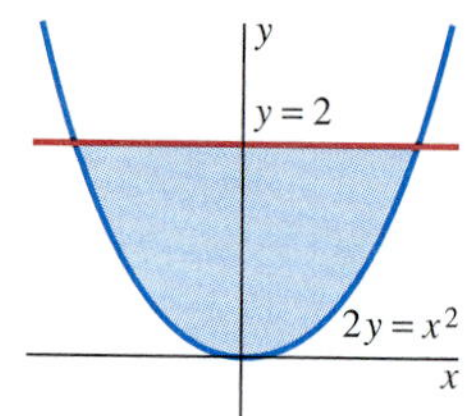

Fig. 14.2.14 Problem 8

10. $\displaystyle\int_{-1}^2\int_{-y}^{y+2} (x + 2y^2)\,dx\,dy$ (Fig. 14.2.15)

11. $\displaystyle\int_0^1\int_0^{x^3} e^{y/x}\,dy\,dx$

12. $\displaystyle\int_0^\pi\int_0^{\sin x} y\,dy\,dx$ (Fig. 14.2.16)

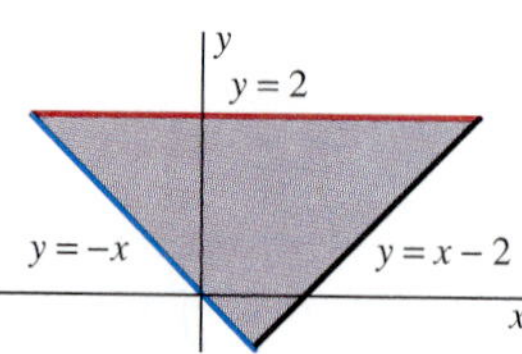

Fig. 14.2.15 Problem 10

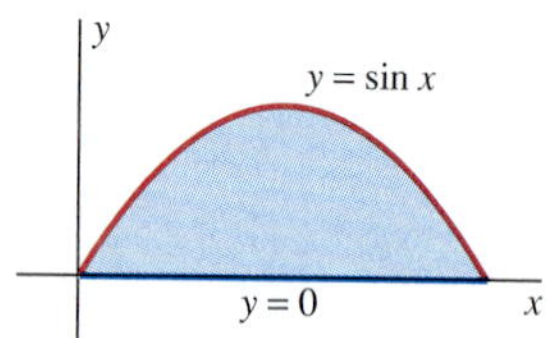

Fig. 14.2.16 Problem 12

13. $\displaystyle\int_0^3\int_0^y \sqrt{y^2 + 16}\,dx\,dy$

14. $\displaystyle\int_1^{e^2}\int_0^{1/y} e^{xy}\,dx\,dy$

In Problems 15 through 24, evaluate the integral of the given function $f(x, y)$ over the plane region R that is described.

15. $f(x, y) = xy$; R is bounded by the parabola $y = x^2$ and the line $y = 4$.

16. $f(x, y) = x^2$; R is bounded by the parabola $y = 2 - x^2$ and the line $y = -4$.

17. $f(x, y) = x$; R is bounded by the parabolas $y = x^2$ and $y = 8 - x^2$.

18. $f(x, y) = y$; R is bounded by the parabolas $x = 1 - y^2$ and $x = y^2 - 1$.

19. $f(x, y) = x$; R is bounded by the x-axis and the curve $y = \sin x$, $0 \leqq x \leqq \pi$.

20. $f(x, y) = \sin x$; R is bounded by the x-axis and the curve $y = \cos x$, $-\pi/2 \leqq x \leqq \pi/2$.

21. $f(x, y) = 1/y$; R is the triangle bounded by the lines $y = 1$, $x = e$, and $y = x$.

22. $f(x, y) = xy$; R is the first-quadrant quarter circle bounded by $x^2 + y^2 = 1$ and the coordinate axes.

23. $f(x, y) = 1 - x$; R is the triangle with vertices $(0, 0)$, $(1, 1)$, and $(-2, 1)$.

24. $f(x, y) = 9 - y$; R is the triangle with vertices $(0, 0)$, $(0, 9)$, and $(3, 6)$.

In Problems 25 through 34, first sketch the region of integration, reverse the order of integration as in Examples 2 and 3, and finally evaluate the resulting integral.

25. $\displaystyle\int_{-2}^{2}\int_{x^2}^{4} x^2 y\,dy\,dx$

26. $\displaystyle\int_{0}^{1}\int_{x^4}^{x} (x - 1)\,dy\,dx$

27. $\displaystyle\int_{-1}^{3}\int_{x^2}^{2x+3} x\,dy\,dx$

28. $\displaystyle\int_{-2}^{2}\int_{y^2-4}^{4-y^2} y\,dx\,dy$

29. $\displaystyle\int_{0}^{2}\int_{2x}^{4x-x^2} 1\,dy\,dx$

30. $\displaystyle\int_{0}^{1}\int_{y}^{1} e^{-x^2}\,dx\,dy$

31. $\displaystyle\int_{0}^{\pi}\int_{x}^{\pi} \frac{\sin y}{y}\,dy\,dx$

32. $\displaystyle\int_{0}^{\sqrt{\pi}}\int_{y}^{\sqrt{\pi}} \sin x^2\,dx\,dy$

33. $\displaystyle\int_{0}^{1}\int_{y}^{1} \frac{1}{1 + x^4}\,dx\,dy$

34. $\displaystyle\int_{0}^{1}\int_{\tan^{-1} y}^{\pi/4} \sec x\,dx\,dy$

In Problems 35 through 40, find the approximate value of

$$\iint_R x\,dA,$$

where R is the region bounded by the two given curves. Before integrating, use a calculator or computer to approximate (graphically or otherwise) the coordinates of the points of intersection of the given curves.

35. $y = x^3 + 1$, $y = 3x^2$

36. $y = x^4$, $y = x + 4$

37. $y = x^2 - 1$, $y = \dfrac{1}{1 + x^2}$

38. $y = x^4 - 16$, $y = 2x - x^2$

39. $y = x^2$, $y = \cos x$

40. $y = x^2 - 2x$, $y = \sin x$

In Problems 41 through 44, the region R is the square with vertices $(\pm 1, 0)$ and $(0, \pm 1)$. Use the symmetry of this region around the coordinate axes to reduce the labor of evaluating the given integrals.

41. $\displaystyle\iint_R x\,dA$

42. $\displaystyle\iint_R x^2\,dA$

43. $\displaystyle\iint_R xy\,dA$

44. $\displaystyle\iint_R (x^2 + y^2)\,dA$

45. Use Riemann sums to prove Eq. (6) for the case in which R is a rectangle with sides parallel to the coordinate axes.

46. Use iterated integrals and familiar properties of single integrals to prove Eq. (7) for the case in which R is a rectangle with sides parallel to the coordinate axes.

47. Use Riemann sums to prove the inequalities in (8) for the case in which R is a rectangle with sides parallel to the coordinate axes.

48. Use iterated integrals and familiar properties of single integrals to prove Eq. (9) if R_1 and R_2 are rectangles with sides parallel to the coordinate axes and the right-hand edge of R_1 is the left-hand edge of R_2.

49. Use Riemann sums to prove that

$$\iint_R f(x, y)\,dA \leqq \iint_R g(x, y)\,dA$$

if $f(x, y) \leqq g(x, y)$ at each point of the region R, a rectangle with sides parallel to the coordinate axes.

50. Suppose that the continuous function f is integrable on the plane region R and that f attains a minimum value m and a maximum value M on R. Assume that R is *connected* in the following sense: For any two points (x_0, y_0) and (x_1, y_1) of R, there is a continuous parametric curve $\mathbf{r}(t)$ in R for which $\mathbf{r}(0) = \langle x_0, y_0\rangle$ and $\mathbf{r}(1) = \langle x_1, y_1\rangle$. Let $a(R)$ denote the area of R. Then deduce from (8) the *average value property* of double integrals:

$$\iint_R f(x, y)\,dA = f(\overline{x}, \overline{y}) \cdot a(R)$$

for some point $(\overline{x}, \overline{y})$ of R. (*Suggestion:* If $m = f(x_0, y_0)$ and $M = f(x_1, y_1)$, then you may apply the intermediate value property of the continuous function $f(\mathbf{r}(t))$.)

14.3 AREA AND VOLUME BY DOUBLE INTEGRATION

Our definition of the double integral $\iint_R f(x, y)\,dA$ in Section 14.2 was *motivated* by the problem of computing the volume of the solid

$$T = \{(x, y, z)\,|\,(x, y) \in R \quad\text{and}\quad 0 \leqq z \leqq f(x, y)\}$$

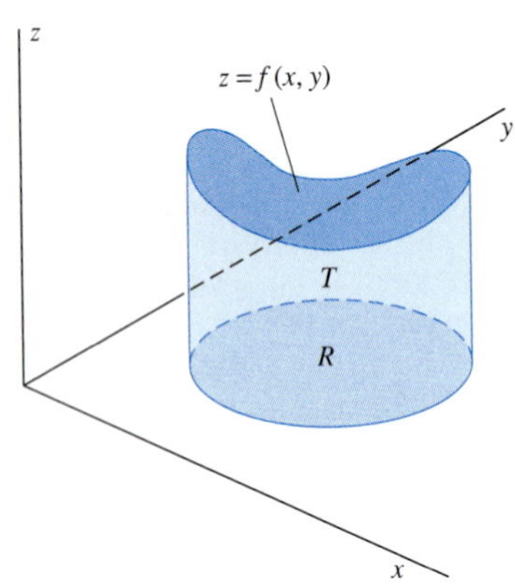

Fig. 14.3.1 A solid region T with vertical sides and base R in the xy-plane

that lies below the surface $z = f(x, y)$ and above the region R in the xy-plane. Such a solid appears in Fig. 14.3.1. Despite this geometric motivation, the actual definition of the double integral as a limit of Riemann sums does not depend on the concept of volume. We may therefore turn matters around and use the double integral to *define* volume.

Definition ***Volume below $z = f(x, y)$***

Suppose that the function f is continuous and nonnegative on the bounded plane region R. Then the **volume** V of the solid that lies below the surface $z = f(x, y)$ and above the region R is defined to be

$$V = \iint_R f(x, y)\, dA, \tag{1}$$

provided that this integral exists.

It is of interest to note the connection between this definition and the cross-sectional approach to volume that we discussed in Section 6.2. If, for example, the region R is vertically simple, then the volume integral in Eq. (1) takes the form

$$V = \iint_R z\, dA = \int_a^b \int_{y_1(x)}^{y_2(x)} f(x, y)\, dy\, dx$$

in terms of iterated integrals. The inner integral

$$A(x) = \int_{y_1(x)}^{y_2(x)} f(x, y)\, dy$$

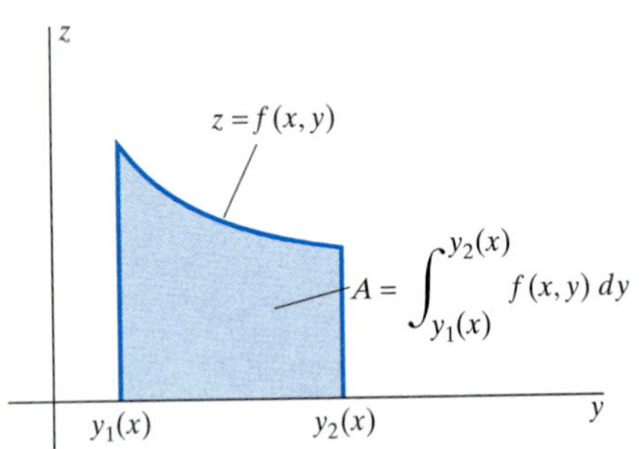

Fig. 14.3.2 The inner integral in Eq. (1) as the area of a region in the yz-plane

is equal to the area of the region in the yz-plane that lies below the curve $z = f(x, y)$ (x fixed) and above the interval $y_1(x) \leqq y \leqq y_2(x)$ (Fig. 14.3.2). But this is the projection of the cross section shown in Fig. 14.3.3. Hence the value of the inner integral is simply the area of the cross section of the solid region T in a plane perpendicular to the x-axis. Thus

$$V = \int_a^b A(x)\, dx,$$

and so in this case Eq. (1) reduces to "volume is the integral of cross-sectional area."

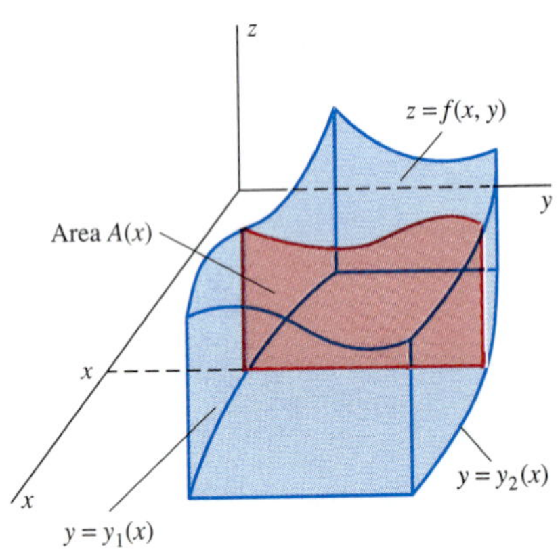

Fig. 14.3.3 The cross-sectional area is $A = \displaystyle\int_{y_1(x)}^{y_2(x)} f(x, y)\, dy.$

Volume by Iterated Integrals

A three-dimensional region T is typically described in terms of the surfaces that bound it. The first step in applying Eq. (1) to compute the volume V of such a region is to determine the region R in the xy-plane over which T lies. The second step is to determine the appropriate order of integration. This may be done in the following way:

If each vertical line in the xy-plane meets R in a *single* line segment, then R is vertically simple, and you may integrate first with respect to y. The limits on y will be the y-coordinates $y_1(x)$ and $y_2(x)$ of the endpoints of this line segment (Fig. 14.3.4). The limits on x will be the endpoints a and b of the interval on the x-axis onto which R projects. Theorem 1 of Section 14.2 then gives

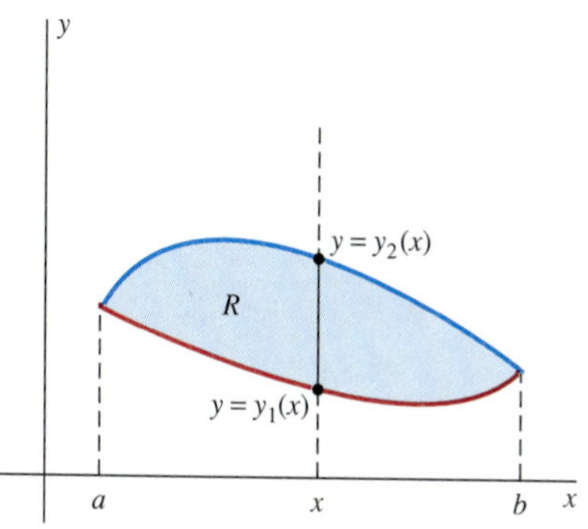

Fig. 14.3.4 A vertically simple region

$$V = \iint_R f(x, y)\, dA = \int_a^b \int_{y_1(x)}^{y_2(x)} f(x, y)\, dy\, dx. \tag{2}$$

Alternatively,

> If each horizontal line in the xy-plane meets R in a *single* line segment, then R is horizontally simple, and you may integrate with respect to x first. In this case,
>
> $$V = \iint_R f(x, y)\, dA = \int_c^d \int_{x_1(y)}^{x_2(y)} f(x, y)\, dx\, dy. \tag{3}$$
>
> As indicated in Fig. 14.3.5, $x_1(y)$ and $x_2(y)$ are the x-coordinates of the endpoints of this horizontal line segment, and c and d are the endpoints of the corresponding interval on the y-axis.

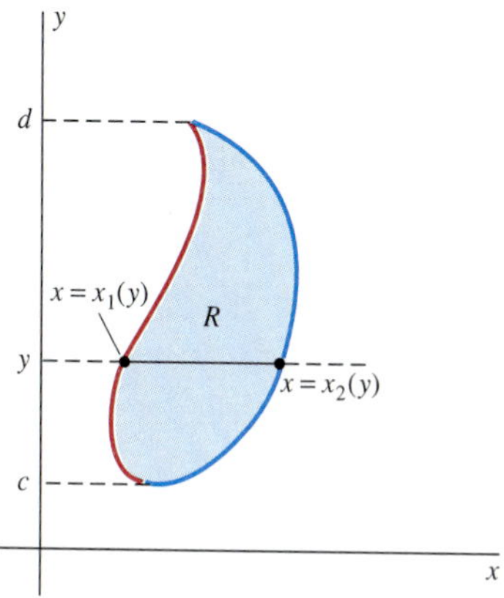

Fig. 14.3.5 A horizontally simple region

If the region R is both vertically simple and horizontally simple, then you have the pleasant option of choosing the order of integration that will lead to the simpler subsequent computations. If R is neither vertically simple nor horizontally simple, then you must first subdivide R into simple regions before you proceed with iterated integration.

EXAMPLE 1 The rectangle R in the xy-plane consists of those points (x, y) for which $0 \leqq x \leqq 2$ and $0 \leqq y \leqq 1$. Find the volume V of the solid that lies below the surface $z = 1 + xy$ and above R (Fig. 14.3.6).

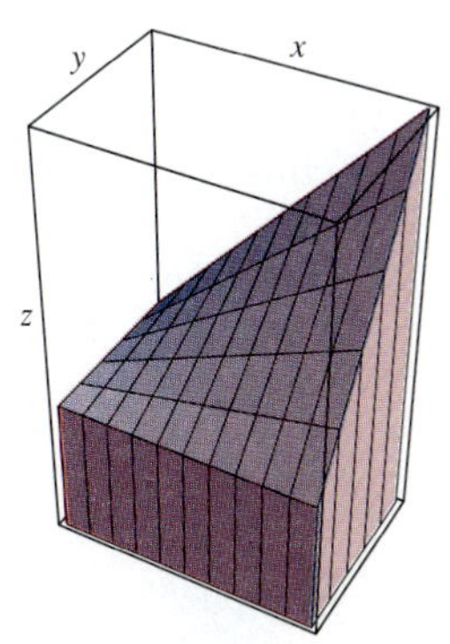

Fig. 14.3.6 The solid of Example 1

Solution Here $f(x, y) = 1 + xy$, so Eq. (1) yields

$$\begin{aligned} V &= \iint_R z\, dA = \int_0^2 \int_0^1 (1 + xy)\, dy\, dx \\ &= \int_0^2 \left[y + \frac{1}{2}xy^2 \right]_{y=0}^1 dx = \int_0^2 \left(1 + \frac{1}{2}x \right) dx = \left[x + \frac{1}{4}x^2 \right]_0^2 = 3. \end{aligned}$$

The special case $f(x, y) \equiv 1$ in Eq. (1) gives the area

$$A = a(R) = \iint_R 1\, dA = \iint_R dA \tag{4}$$

of the plane region R. In this case the solid region T resembles a desert mesa (Fig. 14.3.7)—a solid cylinder with base R of area A and height 1. The volume of any such cylinder—not necessarily circular—is the product of its height and the area of its base. In this case, the iterated integrals in Eqs. (2) and (3) reduce to

$$A = \int_a^b \int_{y_{\text{bot}}}^{y_{\text{top}}} 1\, dy\, dx \quad \text{and} \quad A = \int_c^d \int_{x_{\text{left}}}^{x_{\text{right}}} 1\, dx\, dy,$$

respectively.

EXAMPLE 2 Compute by double integration the area A of the region R in the xy-plane that is bounded by the parabola $y = x^2 - 2x$ and the line $y = x$.

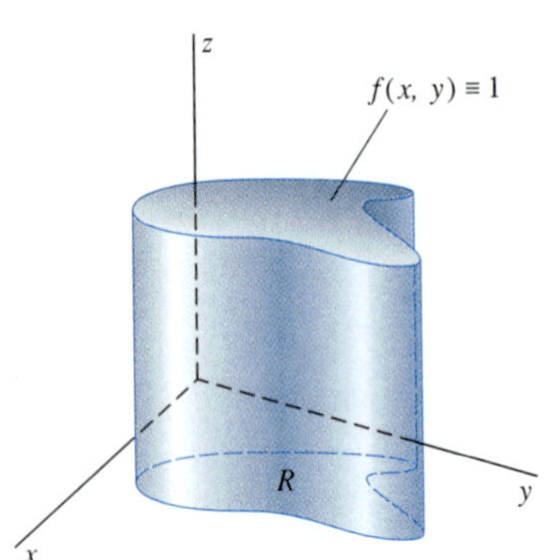

Fig. 14.3.7 The mesa

Solution As indicated in Fig. 14.3.8, the line $y_{\text{top}} = x$ and the parabola $y_{\text{bot}} = x^2 - 2x$ intersect at the points (0, 0) and (3, 3). (These coordinates are easy to find by solving the equation $y_{\text{top}} = y_{\text{bot}}$.) Therefore,

$$A = \int_a^b \int_{y_{\text{bot}}}^{y_{\text{top}}} 1\, dy\, dx = \int_0^3 \int_{x^2-2x}^{x} 1\, dy\, dx$$

$$= \int_0^3 \Big[y \Big]_{y=x^2-2x}^{x} dx = \int_0^3 (3x - x^2)\, dx = \left[\frac{3}{2}x^2 - \frac{1}{3}x^3 \right]_0^3 = \frac{9}{2}.$$

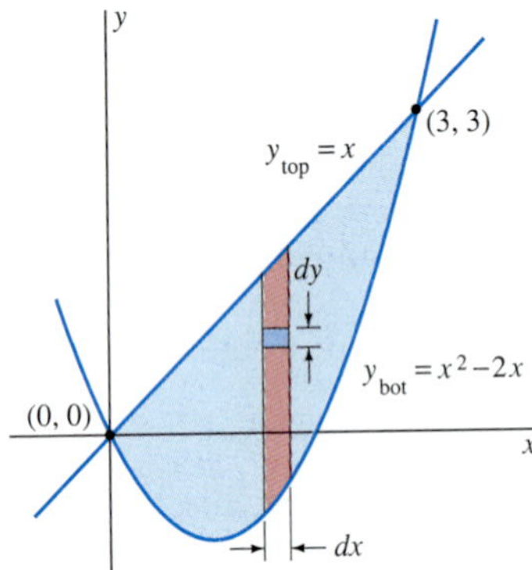

Fig. 14.3.8 The region R of Example 2

EXAMPLE 3 Find the volume of the wedge-shaped solid T that lies above the xy-plane, below the plane $z = x$, and within the cylinder $x^2 + y^2 = 4$. This wedge is shown in Fig. 14.3.9.

Solution The base region R is a semicircle of radius 2, but by symmetry we may integrate over the first-quadrant quarter circle S alone and then double the result. A sketch of the quarter circle (Fig. 14.3.10) helps establish the limits of integration. We could integrate in either order, but integration with respect to x first gives a slightly simpler computation of the volume V:

$$V = \iint_S z\, dA = 2\int_0^2 \int_0^{\sqrt{4-y^2}} x\, dx\, dy = 2\int_0^2 \left[\frac{1}{2}x^2 \right]_{x=0}^{\sqrt{4-y^2}} dy$$

$$= \int_0^2 (4 - y^2)\, dy = \left[4y - \frac{1}{3}y^3 \right]_0^2 = \frac{16}{3}.$$

As an exercise, you should integrate in the other order and verify that the result is the same.

Volume Between Two Surfaces

Suppose now that the solid region T lies above the plane region R, as before, but *between* the surfaces $z = z_1(x, y)$ and $z = z_2(x, y)$, where $z_1(x, y) \leqq z_2(x, y)$ for all (x, y) in R (Fig. 14.3.11). Then we get the volume V of T by subtracting the volume below $z = z_1(x, y)$ from the volume below $z = z_2(x, y)$, so

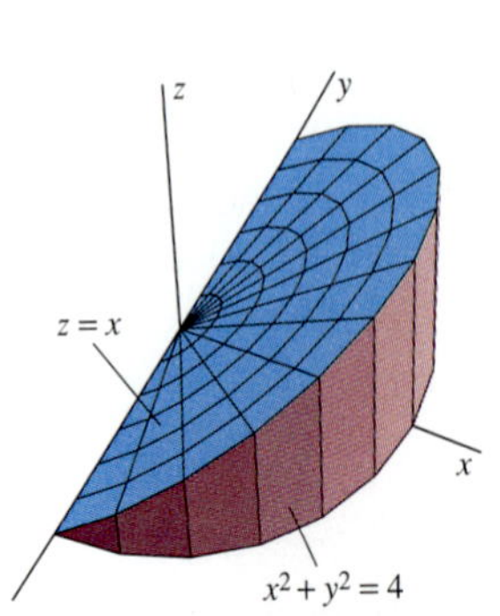

Fig. 14.3.9 The wedge of Example 3

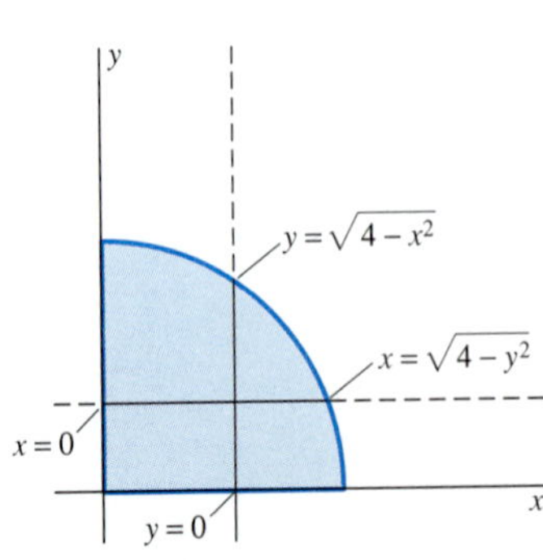

Fig. 14.3.10 *Half* of the base R of the wedge (Example 3)

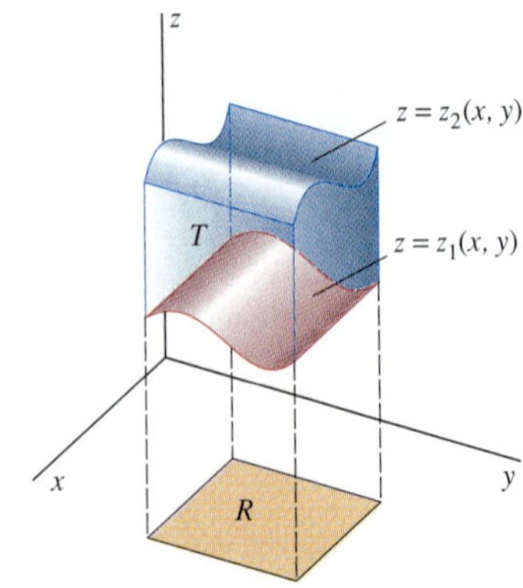

Fig. 14.3.11 The solid T has vertical sides and is bounded above and below by surfaces.

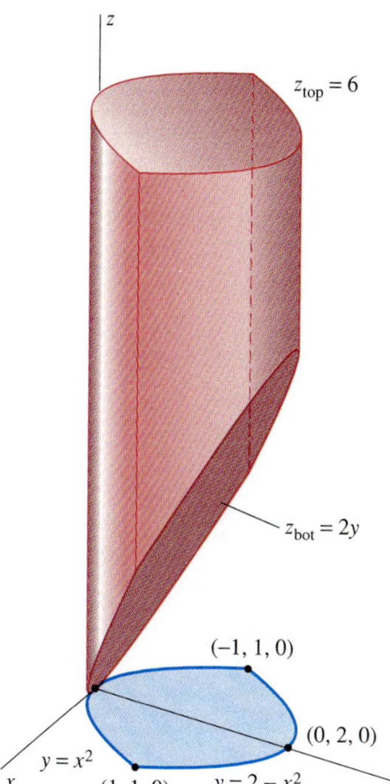

Fig. 14.3.12 The solid T of Example 4

$$V = \iint_R [z_2(x, y) - z_1(x, y)]\, dA. \tag{5}$$

More briefly,

$$V = \iint_R (z_{\text{top}} - z_{\text{bot}})\, dA$$

where $z_{\text{top}} = z_2(x, y)$ describes the top surface and $z_{\text{bot}} = z_1(x, y)$ the bottom surface of T. This is a natural generalization of the formula for the area of the plane region between the curves $y = f_1(x)$ and $y = f_2(x)$ over the interval $[a, b]$. Moreover, like that formula, Eq. (5) is valid even if $f_1(x, y)$, or both $f_1(x, y)$ and $f_2(x, y)$, are negative over part or all of the region R.

EXAMPLE 4 Find the volume V of the solid T bounded by the planes $x = 6$ and $z = 2y$ and by the parabolic cylinders $y = x^2$ and $y = 2 - x^2$. This solid is sketched in Fig. 14.3.12.

Solution Because the given parabolic cylinders are perpendicular to the xy-plane, the solid T has vertical sides. Thus we may think of T as lying between the planes $z_{\text{top}} = 6$ and $z_{\text{bot}} = 2y$ and above the xy-plane region R that is bounded by the parabolas $y = x^2$ and $y = 2 - x^2$. As indicated in Fig. 14.3.13, these parabolas intersect at the points $(-1, 1)$ and $(1, 1)$.

Integrating first with respect to y (for otherwise we would need two integrals), we get

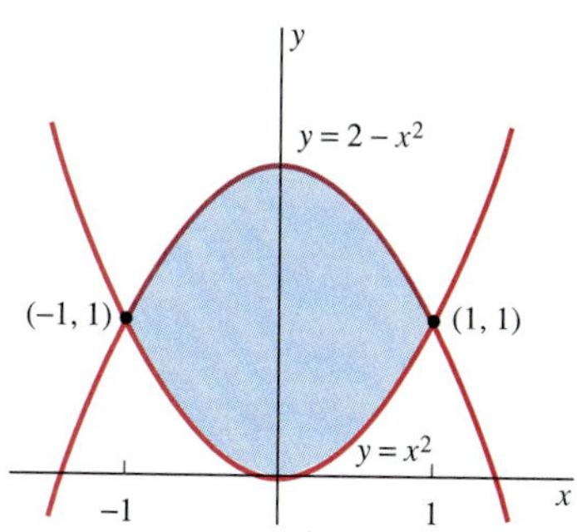

Fig. 14.3.13 The region R of Example 4

$$\begin{aligned} V &= \iint_R (z_{\text{top}} - z_{\text{bot}})\, dA = \int_{-1}^{1} \int_{x^2}^{2-x^2} (6 - 2y)\, dy\, dx \\ &= 2\int_0^1 \Big[6y - y^2 \Big]_{y=x^2}^{2-x^2} dx \quad \text{(by symmetry)} \\ &= 2\int_0^1 \left([6 \cdot (2 - x^2) - (2 - x^2)^2] - [6x^2 - x^4] \right) dx \\ &= 2\int_0^1 (8 - 8x^2)\, dx = 16\left[x - \frac{1}{3}x^3 \right]_0^1 = \frac{32}{3}. \end{aligned}$$

■

14.3 PROBLEMS

In Problems 1 through 10, use double integration to find the area of the region in the xy-plane bounded by the given curves.

1. $y = x$, $y^2 = x$

2. $y = x$, $y = x^4$

3. $y = x^2$, $y = 2x + 3$ (Fig. 14.3.14)

4. $y = 2x + 3$, $y = 6x - x^2$ (Fig. 14.3.15)

5. $y = x^2$, $x + y = 2$, $y = 0$

6. $y = (x - 1)^2$, $y = (x + 1)^2$, $y = 0$

7. $y = x^2 + 1$, $y = 2x^2 - 3$ (Fig. 14.3.16)

8. $y = x^2 + 1$, $y = 9 - x^2$ (Fig. 14.3.17)

9. $y = x$, $y = 2x$, $xy = 2$

10. $y = x^2$, $y = \dfrac{2}{1 + x^2}$

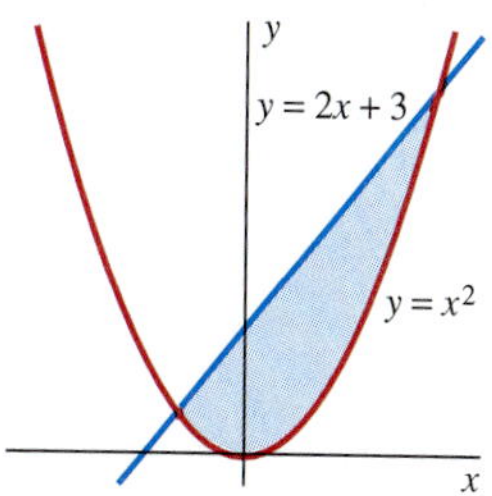

Fig. 14.3.14 Problem 3

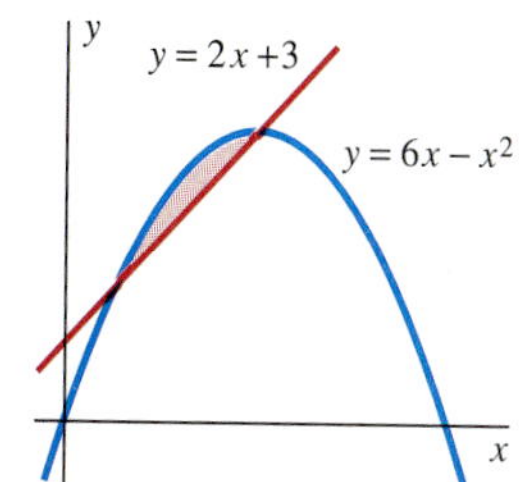

Fig. 14.3.15 Problem 4

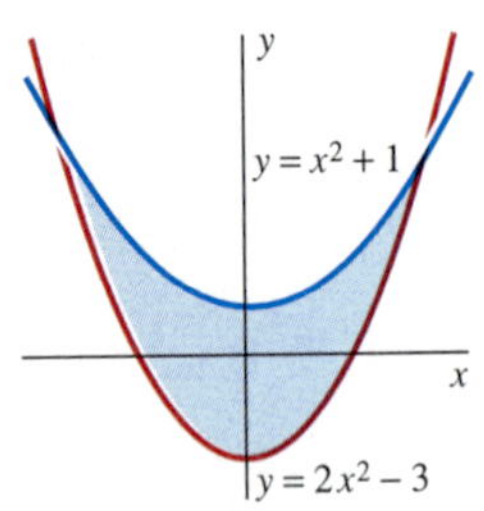

Fig. 14.3.16 Problem 7

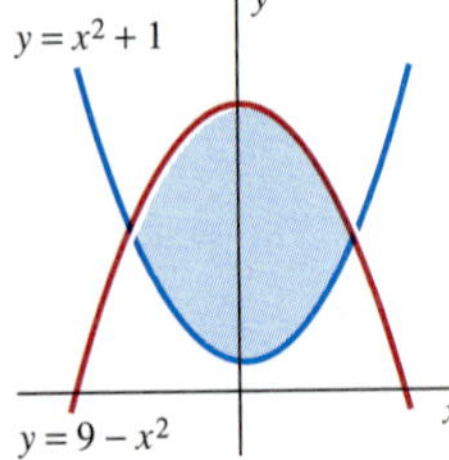

Fig. 14.3.17 Problem 8

In Problems 11 through 26, find the volume of the solid that lies below the surface $z = f(x, y)$ and above the region in the xy-plane bounded by the given curves.

11. $z = 1 + x + y$; $x = 0, x = 1, y = 0, y = 1$

12. $z = 2x + 3y$; $x = 0, x = 3, y = 0, y = 2$

13. $z = y + e^x$; $x = 0, x = 1, y = 0, y = 2$

14. $z = 3 + \cos x + \cos y$; $x = 0, x = \pi, y = 0, y = \pi$ (Fig. 14.3.18)

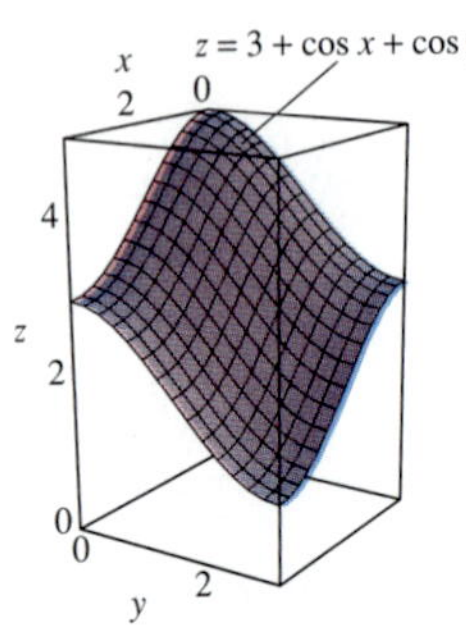

Fig. 14.3.18 The surface of Problem 14

15. $z = x + y$; $x = 0, y = 0, x + y = 1$

16. $z = 3x + 2y$; $x = 0, y = 0, x + 2y = 4$

17. $z = 1 + x + y$; $x = 1, y = 0, y = x^2$

18. $z = 2x + y$; $x = 0, y = 1, x = \sqrt{y}$

19. $z = x^2$; $y = x^2, y = 1$

20. $z = y^2$; $x = y^2, x = 4$

21. $z = x^2 + y^2$; $x = 0, x = 1, y = 0, y = 2$

22. $z = 1 + x^2 + y^2$; $y = x, y = 2 - x^2$

23. $z = 9 - x - y$; $y = 0, x = 3, y = \frac{2}{3}x$

24. $z = 10 + y - x^2$; $y = x^2, x = y^2$

25. $z = 4x^2 + y^2$; $x = 0, y = 0, 2x + y = 2$

26. $z = 2x + 3y$; $y = x^2, y = x^3$

In Problems 27 through 30, find the volume of the given solid.

27. The solid is bounded by the planes $x = 0$, $y = 0$, $z = 0$, and $3x + 2y + z = 6$.

28. The solid is bounded by the planes $y = 0$, $z = 0$, $y = 2x$, and $4x + 2y + z = 8$.

29. The solid lies under the hyperboloid $z = xy$ and above the triangle in the xy-plane with vertices $(1, 2)$, $(1, 4)$, and $(5, 2)$.

30. The solid lies under the paraboloid $z = 25 - x^2 - y^2$ and above the triangle in the xy-plane with vertices $(-3, -4)$, $(-3, 4)$, and $(5, 0)$.

In Problems 31 through 34, first set up an iterated integral that gives the volume of the given solid. Then use a computer algebra system (if available) to evaluate this integral.

31. The solid lies inside the cylinder $x^2 + y^2 = 1$, above the xy-plane, and below the plane $z = x + 1$ (Fig. 14.3.19).

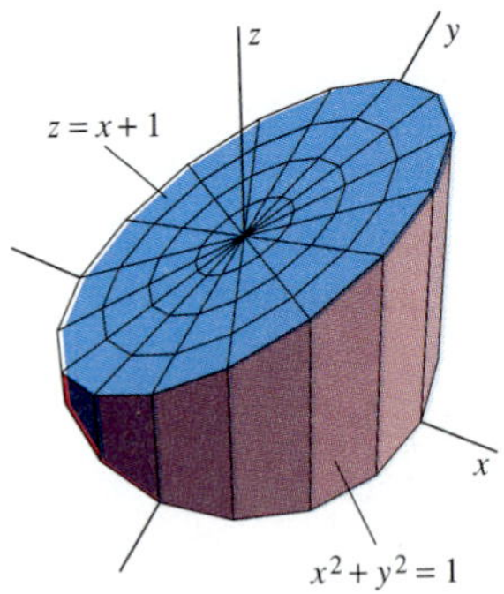

Fig. 14.3.19 The solid of Problem 31

32. The solid lies above the xy-plane and below the paraboloid $z = 9 - x^2 - y^2$.

33. The solid lies inside both the cylinder $x^2 + y^2 = 1$ and the sphere $x^2 + y^2 + z^2 = 4$.

34. The solid lies inside the sphere $x^2 + y^2 + z^2 = 2$ and above the paraboloid $z = x^2 + y^2$.

35. Use double integration to find the volume of the tetrahedron in the first octant that is bounded by the coordinate planes and the plane with equation

$$\frac{x}{a} + \frac{y}{b} + \frac{z}{c} = 1$$

(Fig. 14.3.20). The numbers a, b, and c are positive constants.

36. Suppose that $h > a > 0$. Show that the volume of the solid bounded by the cylinder $x^2 + y^2 = a^2$, the plane $z = 0$, and the plane $z = x + h$ is $\pi a^2 h$.

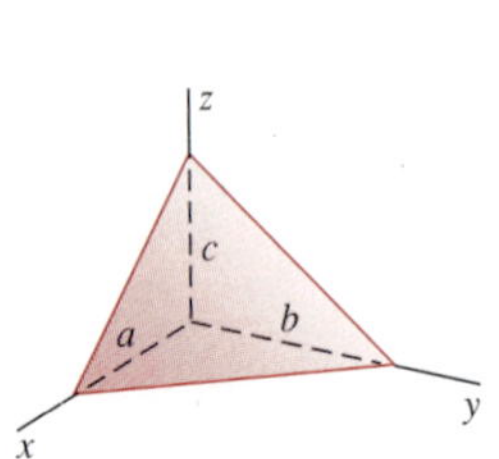

Fig. 14.3.20 The tetrahedron of Problem 35

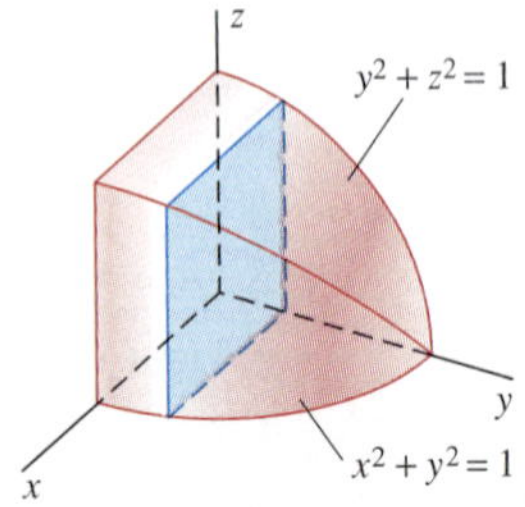

Fig. 14.3.21 The solid of Problem 37

37. Find the volume of the first octant part of the solid bounded by the cylinders $x^2 + y^2 = 1$ and $y^2 + z^2 = 1$ (Fig. 14.3.21). (*Suggestion:* One order of integration is considerably easier than the other.)

38. Find by double integration the volume of the solid bounded by the surfaces $y = \sin x$, $y = -\sin x$, $z = \sin x$, and $z = -\sin x$ for $0 \leqq x \leqq \pi$.

For Problems 39 through 46, you may consult Chapter 9 or the integral table on the inside covers of this book to find antiderivatives of such expressions as $(a^2 - x^2)^{3/2}$.

39. Find the volume of a sphere of radius a by double integration.

40. Use double integration to find the formula $V = V(a, b, c)$ for the volume of an ellipsoid with semiaxes of lengths a, b, and c.

41. Find the volume of the solid bounded below by the xy-plane and above by the paraboloid $z = 25 - x^2 - y^2$ by evaluating a double integral (Fig. 14.3.22).

42. Find the volume of the solid bounded by the two paraboloids $z = x^2 + 2y^2$ and $z = 12 - 2x^2 - y^2$ (Fig. 14.3.23).

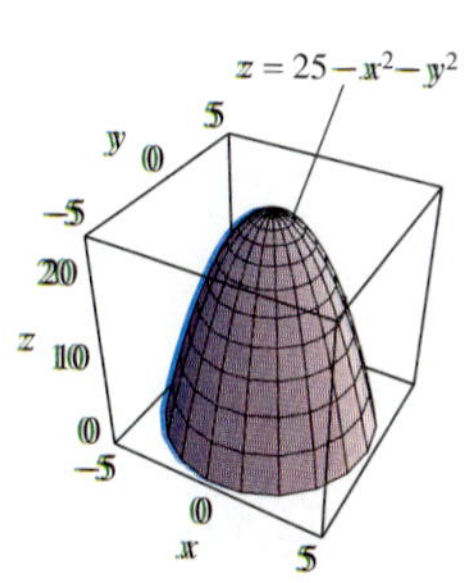

Fig. 14.3.22 The solid paraboloid of Problem 41

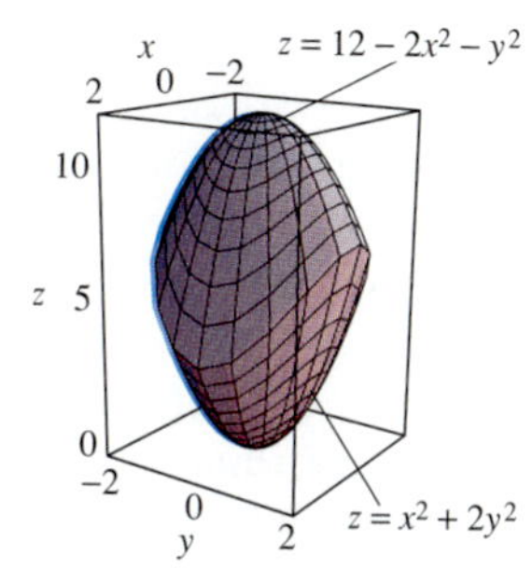

Fig. 14.3.23 The solid of Problem 42

43. Find the volume removed when a vertical square hole of edge length R is cut directly through the center of a long horizontal solid cylinder of radius R.

44. Find the volume of the solid bounded by the two surfaces $z = x^2 + 3y^2$ and $z = 4 - y^2$ (Fig. 14.3.24).

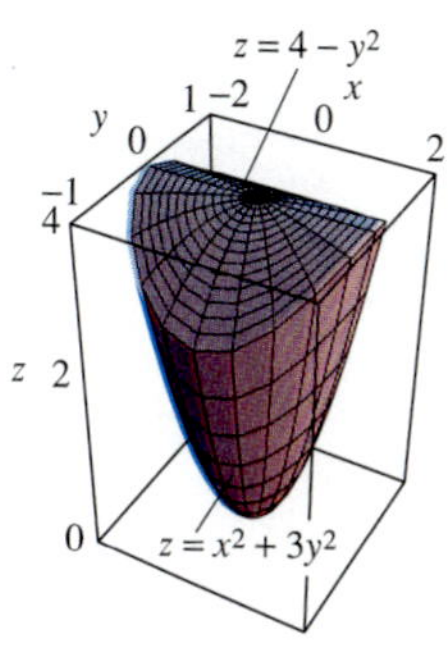

Fig. 14.3.24 The solid of Problem 44

45. Find the volume V of the solid T bounded by the parabolic cylinders $z = x^2$, $z = 2x^2$, $y = x^2$, and $y = 8 - x^2$.

46. Suppose that a square hole with sides of length 2 is cut symmetrically through the center of a sphere of radius 2. Show that the volume removed is given by

$$V = \int_0^1 F(x)\,dx,$$

where

$$F(x) = 4\sqrt{3 - x^2} + 4(4 - x^2)\arcsin\frac{1}{\sqrt{4 - x^2}}.$$

Next, use a computer algebra system (or the **INTEGRATE** key on a calculator) to approximate the volume numerically. Finally use a computer algebra system to determine the exact value

$$V = \frac{2}{3}\left(19\pi + 2\sqrt{2} - 54\arctan\sqrt{2}\right),$$

and verify that your numerical value is consistent with this exact value.

For Problems 47 and 48, use a computer algebra system to find (either approximately or exactly) the volume of the solid that lies under the surface $z = f(x, y)$ and above the region in the xy-plane that is bounded by $y = \cos x$ and $y = -\cos x$ for $-\pi/2 \leqq x \leqq \pi/2$.

47. $f(x, y) = 4 - x^2 - y^2$ **48.** $f(x, y) = \cos y$

14.4 DOUBLE INTEGRALS IN POLAR COORDINATES

A double integral may be easier to evaluate after it has been transformed from rectangular xy-coordinates into polar $r\theta$-coordinates. This is likely to be the case when the region R of integration is a *polar rectangle.* A **polar rectangle** is a region described in polar coordinates by the inequalities

$$a \leqq r \leqq b, \quad \alpha \leqq \theta \leqq \beta. \tag{1}$$

This polar rectangle is shown in Fig. 14.4.1. If $a = 0$, it is a sector of a circular disk of radius b. If $0 < a < b$, $\alpha = 0$, and $\beta = 2\pi$, it is an annular ring of inner radius a and

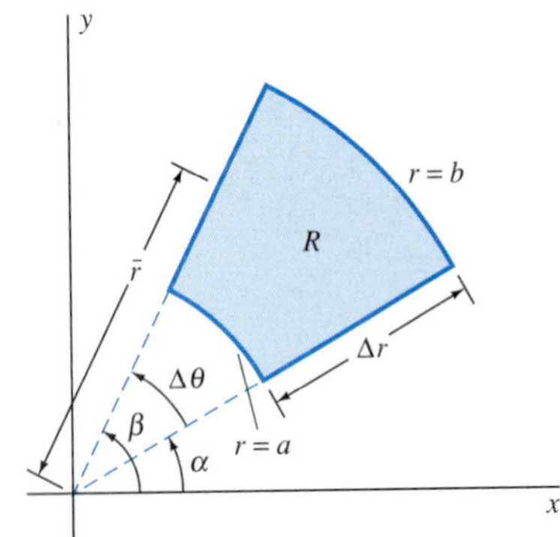

Fig. 14.4.1 A polar rectangle

outer radius b. Because the area of a circular sector with radius r and central angle θ is $\frac{1}{2}r^2\theta$, the area of the polar rectangle in (1) is

$$A = \tfrac{1}{2}b^2(\beta - \alpha) - \tfrac{1}{2}a^2(\beta - \alpha) = \tfrac{1}{2}(a + b)(a - b)(\beta - \alpha) = \bar{r}\,\Delta r\,\Delta\theta, \tag{2}$$

where $\Delta r = b - a$, $\Delta\theta = \beta - \alpha$, and $\bar{r} = \frac{1}{2}(a + b)$ is the *average radius* of the polar rectangle.

Suppose that we want to compute the value of the double integral

$$\iint_R f(x, y)\, dA,$$

where R is the polar rectangle in (1). Thus we want the volume of the solid with base R that lies below the surface $z = f(x, y)$ (Fig. 14.4.2). We defined in Section 14.1 the double integral as a limit of Riemann sums associated with partitions consisting of ordinary rectangles. We can define the double integral in terms of *polar partitions* as well, made up of polar rectangles. We begin with a partition

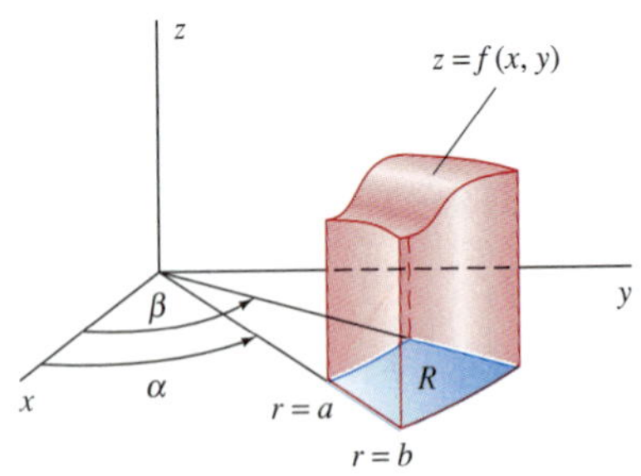

Fig. 14.4.2 A solid region whose base is the polar rectangle R

$$a = r_0 < r_1 < r_2 < \cdots < r_m = b$$

of $[a, b]$ into m subintervals all having the same length $\Delta r = (b - a)/m$ and a partition

$$\alpha = \theta_0 < \theta_1 < \theta_2 < \cdots < \theta_n = \beta$$

of $[\alpha, \beta]$ into n subintervals all having the same length $\Delta\theta = (\beta - \alpha)/n$. This gives the **polar partition** $\mathcal{P}$ of R into the $k = mn$ polar rectangles $R_1, R_2, \dots, R_k$ indicated in Fig. 14.4.3. The **norm** $|\mathcal{P}|$ of this polar partition is the maximum of the lengths of the diagonals of its polar subrectangles.

Let the center point of R_i have polar coordinates $(r_i^\star, \theta_i^\star)$, where $r_i^\star$ is the average radius of R_i. Then the rectangular coordinates of this point are $x_i^\star = r_i^\star \cos\theta_i^\star$ and $y_i^\star = r_i^\star \sin\theta_i^\star$. Therefore the Riemann sum for the function $f(x, y)$ associated with the polar partition $\mathcal{P}$ is

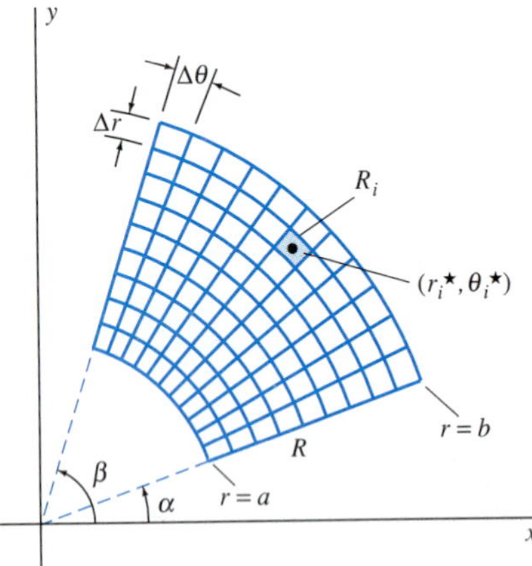

Fig. 14.4.3 A polar partition of the polar rectangle R

$$\sum_{i=1}^{k} f(x_i^\star, y_i^\star)\,\Delta A_i,$$

where $\Delta A_i = r_i^\star \Delta r\, \Delta\theta$ is the area of the polar rectangle R_i [in part a consequence of Eq. (2)]. When we express this Riemann sum in polar coordinates, we obtain

$$\sum_{i=1}^{k} f(x_i^\star, y_i^\star)\,\Delta A_i = \sum_{i=1}^{k} f(r_i^\star \cos\theta_i^\star, r_i^\star \sin\theta_i^\star) r_i^\star\,\Delta r\,\Delta\theta = \sum_{i=1}^{k} g(r_i^\star, \theta_i^\star)\,\Delta r\,\Delta\theta,$$

where $g(r, \theta) = r \cdot f(r\cos\theta, r\sin\theta)$. This last sum is simply a Riemann sum for the double integral

$$\int_\alpha^\beta \int_a^b g(r, \theta)\, dr\, d\theta = \int_\alpha^\beta \int_a^b f(r\cos\theta, r\sin\theta) r\, dr\, d\theta,$$

so it finally follows that

$$\iint_R f(x, y)\, dA = \lim_{|\mathcal{P}| \to 0} \sum_{i=1}^{k} f(x_i^\star, y_i^\star)\,\Delta A_i$$

$$= \lim_{\Delta r, \Delta\theta \to 0} \sum_{i=1}^{k} g(r_i^\star, \theta_i^\star)\,\Delta r\,\Delta\theta = \int_\alpha^\beta \int_a^b g(r, \theta)\,dr\,d\theta.$$

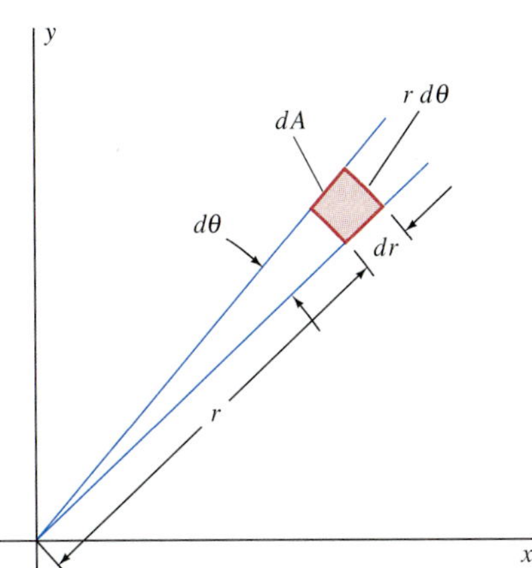

Fig. 14.4.4 The dimensions of the small polar rectangle suggest that its area is $dA = r\,dr\,d\theta$.

That is,

$$\iint_R f(x, y)\,dA = \int_\alpha^\beta \int_a^b f(r\cos\theta, r\sin\theta)r\,dr\,d\theta. \tag{3}$$

Thus we formally transform into polar coordinates a double integral over a polar rectangle of the form in (1) by substituting

$$x = r\cos\theta, \quad y = r\sin\theta, \quad dA = r\,dr\,d\theta \tag{4}$$

and inserting the appropriate limits of integration on r and θ. In particular, *note the "extra" r on the right-hand side* of Eq. (3). You may remember it by visualizing the "infinitesimal polar rectangle" of Fig. 14.4.4, with "area" $dA = r\,dr\,d\theta$ (formally).

EXAMPLE 1 Find the volume V of the solid shown in Fig. 14.4.5. This is the figure bounded below by the xy-plane and above by the paraboloid $z = 25 - x^2 - y^2$.

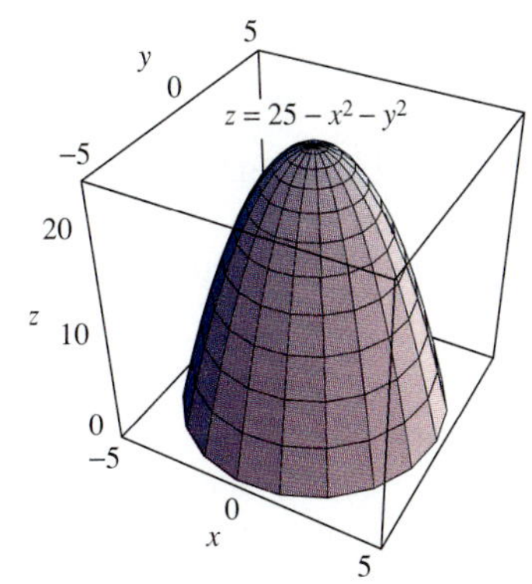

Fig. 14.4.5 The paraboloid of Example 1

Solution The paraboloid intersects the xy-plane in the circle $x^2 + y^2 = 25$. We can compute the volume of the solid by integrating over the quarter of that circle that lies in the first quadrant (Fig. 14.4.6) and then multiplying the result by 4. Thus

$$V = 4\int_0^5 \int_0^{\sqrt{25-x^2}} (25 - x^2 - y^2)\,dy\,dx.$$

There is no difficulty in performing the integration with respect to y, but then we are confronted with the integrals

$$\int \sqrt{25 - x^2}\,dx, \quad \int x^2\sqrt{25 - x^2}\,dx, \quad \text{and} \quad \int (25 - x^2)^{3/2}\,dx.$$

Let us instead transform the original integral into polar coordinates. Because $25 - x^2 - y^2 = 25 - r^2$ and because the quarter of the circular disk in the first quadrant is described by

$$0 \leqq r \leqq 5, \quad 0 \leqq \theta \leqq \pi/2,$$

Eq. (3) yields the volume

$$V = 4\int_0^{\pi/2} \int_0^5 (25 - r^2)r\,dr\,d\theta$$

$$= 4\int_0^{\pi/2} \left[\frac{25}{2}r^2 - \frac{1}{4}r^4\right]_{r=0}^{5} d\theta = 4 \cdot \frac{625}{4} \cdot \frac{\pi}{2} = \frac{625\pi}{2}.$$ ■

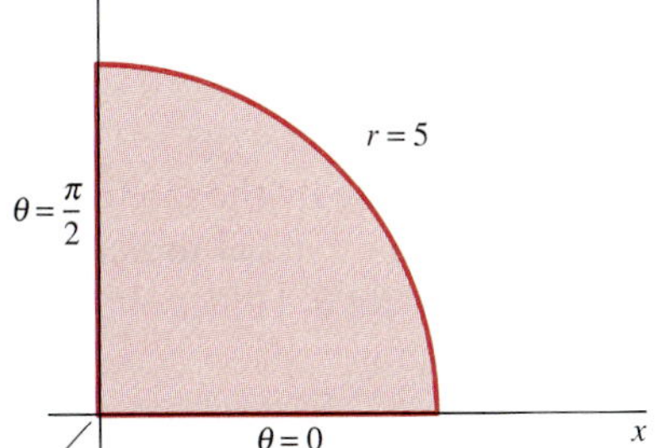

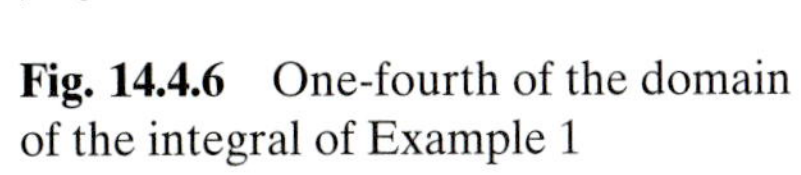

Fig. 14.4.6 One-fourth of the domain of the integral of Example 1

More General Polar-Coordinate Regions

If R is a more general region, then we can transform into polar coordinates the double integral

$$\iint_R f(x, y)\,dA$$

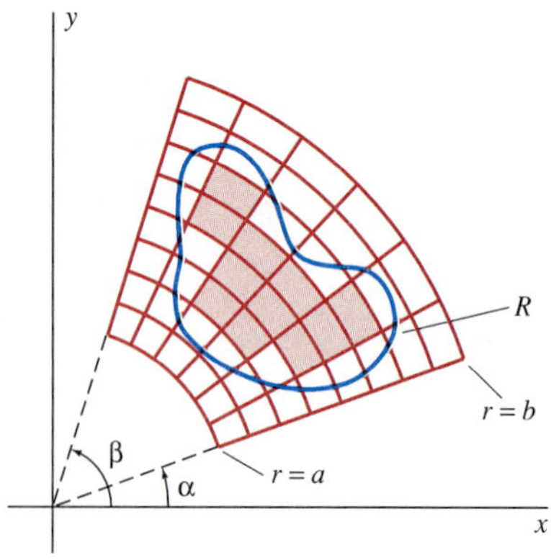

Fig. 14.4.7 A polar inner partition of the region R

by expressing it as a limit of Riemann sums associated with "polar inner partitions" of the sort indicated in Fig. 14.4.7. Instead of giving the detailed derivation—a generalization of the preceding derivation of Eq. (3)—we shall simply give the results in one special case of practical importance.

Figure 14.4.8 shows a *radially simple* region R consisting of those points with polar coordinates that satisfy the inequalities

$$\alpha \leqq \theta \leqq \beta, \qquad r_1(\theta) \leqq r \leqq r_2(\theta).$$

In this case, the formula

$$\iint_R f(x, y)\, dA = \int_\alpha^\beta \int_{r_1(\theta)}^{r_2(\theta)} f(r\cos\theta, r\sin\theta) r\, dr\, d\theta \tag{5}$$

gives the evaluation in polar coordinates of a double integral over R (under the usual assumption that the indicated integrals exist). Note that we integrate first with respect to r, with the limits $r_1(\theta)$ and $r_2(\theta)$ being the r-coordinates of a typical radial segment in R (Fig. 14.4.8).

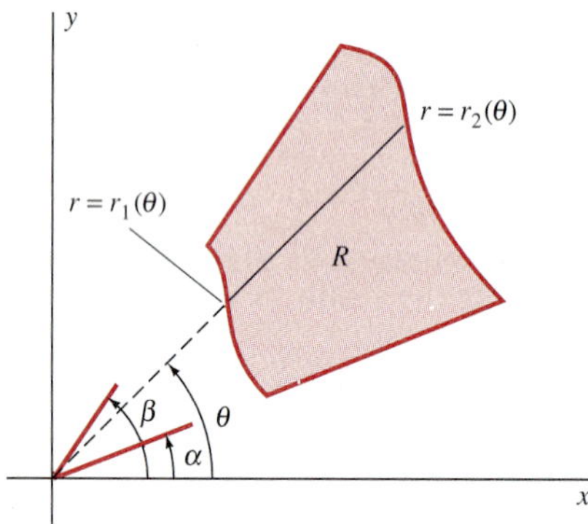

Fig. 14.4.8 A radially simple region R

Figure 14.4.9 shows how we can set up the iterated integral on the right-hand side of Eq. (5) in a formal way. First, a typical area element $dA = r\, dr\, d\theta$ is swept radially from $r = r_1(\theta)$ to $r = r_2(\theta)$. Second, the resulting strip is rotated from $\theta = \alpha$ to $\theta = \beta$ to sweep out the region R. Equation (5) yields the volume formula

$$V = \int_\alpha^\beta \int_{r_{\text{inner}}}^{r_{\text{outer}}} zr\, dr\, d\theta \tag{6}$$

for the volume V of the solid that lies above the region R of Fig. 14.4.8 and below the surface $z = f(x, y) = f(r\cos\theta, r\sin\theta)$.

Observe that Eqs. (3) and (5) for the evaluation of a double integral in polar coordinates take the form

$$\iint_R f(x, y)\, dA = \iint_S f(r\cos\theta, r\sin\theta) r\, dr\, d\theta. \tag{7}$$

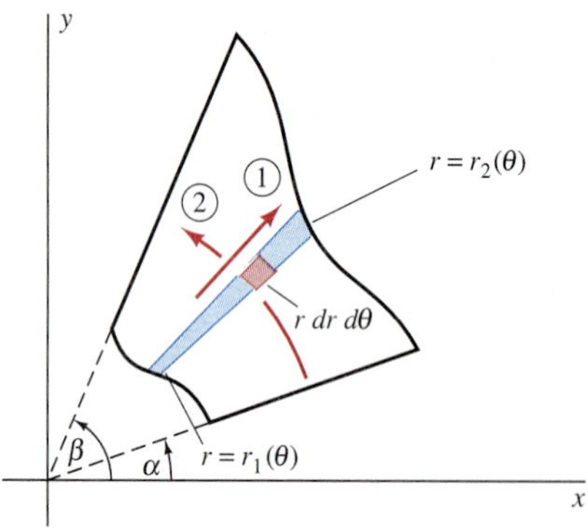

Fig. 14.4.9 Integrating first with respect to r and then with respect to θ

The symbol S on the right-hand side represents the appropriate limits on r and θ such that the region R is swept out in the manner indicated in Fig. 14.4.9.

With $f(x, y) \equiv 1$, Eq. (7) reduces to the formula

$$A = a(R) = \iint_S r\, dr\, d\theta \tag{8}$$

for computing the area $a(R)$ of R by double integration in polar coordinates. Note again that the symbol S refers not to a new region in the xy-plane, but to a new description—in terms of polar coordinates—of the original region R.

EXAMPLE 2 Figure 14.4.10 shows the region R bounded on the inside by the circle $r = 1$ and on the outside by the limaçon $r = 2 + \cos\theta$. By following a typical radial line outward from the origin, we see that $r_{\text{inner}} = 1$ and $r_{\text{outer}} = 2 + \cos\theta$. Hence the area of R is

$$\begin{aligned} A &= \int_\alpha^\beta \int_{r_{\text{inner}}}^{r_{\text{outer}}} r\, dr\, d\theta \\ &= 2\int_0^\pi \int_1^{2+\cos\theta} r\, dr\, d\theta \qquad \text{(symmetry)} \end{aligned}$$

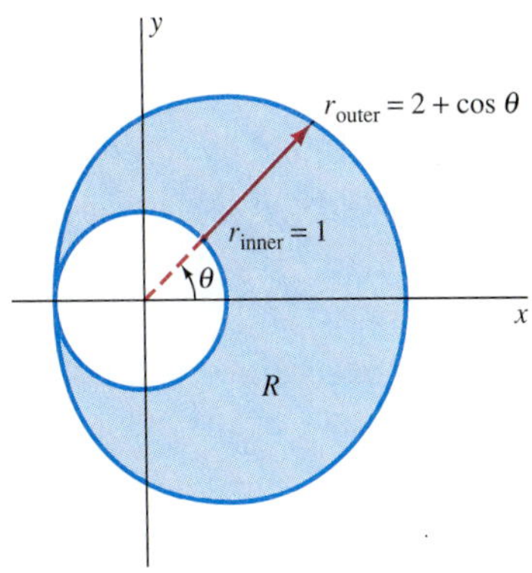

Fig. 14.4.10 The region R of Example 2

$$
\begin{aligned}
&= 2\int_0^{\pi} \frac{1}{2}[(2+\cos\theta)^2 - 1^2]\,d\theta \\
&= \int_0^{\pi} (3 + 4\cos\theta + \cos^2\theta)\,d\theta \\
&= \int_0^{\pi} \left(3 + 4\cos\theta + \frac{1}{2} + \frac{1}{2}\cos 2\theta\right) d\theta \\
&= \int_0^{\pi} \left(3 + \frac{1}{2}\right) d\theta = \frac{7}{2}\pi.
\end{aligned}
$$

The cosine terms in the next-to-last integral contribute nothing, because upon integration they yield sine terms that are zero at both limits of integration. ■

EXAMPLE 3 Find the volume of the solid region that is interior to both the sphere $x^2 + y^2 + z^2 = 4$ of radius 2 and the cylinder $(x-1)^2 + y^2 = 1$. This is the volume of material removed when an off-center hole of radius 1 is bored just tangent to a diameter all the way through a sphere of radius 2 (Fig. 14.4.11).

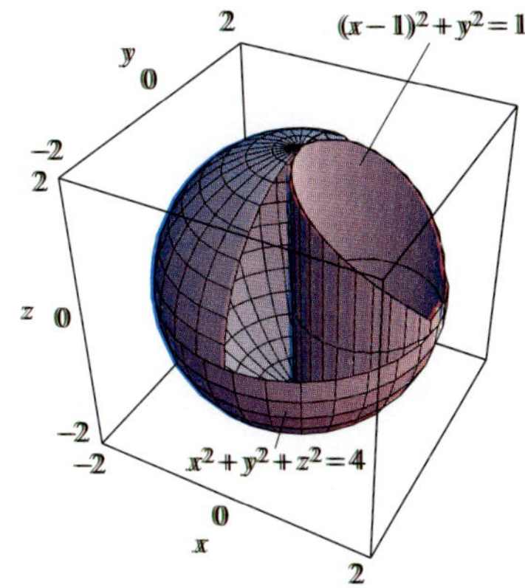

Fig. 14.4.11 The sphere with off-center hole (Example 3)

Solution We need to integrate the function $f(x, y) = \sqrt{4 - x^2 - y^2}$ over the disk R that is bounded by the circle with center $(1, 0)$ and radius 1 (Fig. 14.4.12). The desired volume V is twice that of the part above the xy-plane, so

$$V = 2\iint_R \sqrt{4 - x^2 - y^2}\,dA.$$

But this integral would be awkward to evaluate in rectangular coordinates, so we change to polar coordinates.

The circle of radius 1 in Fig. 14.4.12 is familiar from Chapter 10; its polar equation is $r = 2\cos\theta$. Therefore the region R is described by the inequalities

$$0 \leqq r \leqq 2\cos\theta, \quad -\pi/2 \leqq \theta \leqq \pi/2.$$

We shall integrate only over the upper half of R, taking advantage of the symmetry of the sphere-with-hole. This involves doubling, for a second time, the integral we write. So—using Eq. (5)—we find that

$$
\begin{aligned}
V &= 4\int_0^{\pi/2}\int_0^{2\cos\theta} \sqrt{4 - r^2}\,r\,dr\,d\theta \\
&= 4\int_0^{\pi/2} \left[-\frac{1}{3}(4 - r^2)^{3/2}\right]_{r=0}^{2\cos\theta} d\theta = \frac{32}{3}\int_0^{\pi/2} (1 - \sin^3\theta)\,d\theta.
\end{aligned}
$$

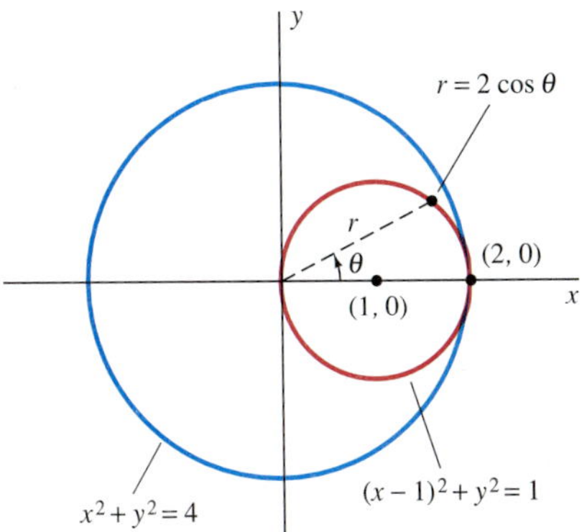

Fig. 14.4.12 The small circle is the domain of the integral of Example 3.

Now we see from Formula (113) on the inside back cover that

$$\int_0^{\pi/2} \sin^3\theta\,d\theta = \frac{2}{3},$$

and therefore

$$V = \tfrac{16}{3}\pi - \tfrac{64}{9} \approx 9.64405. \quad ■$$

In Example 4 we use a polar-coordinates version of the familiar volume formula

$$V = \iint_R (z_{\text{top}} - z_{\text{bot}})\,dA.$$

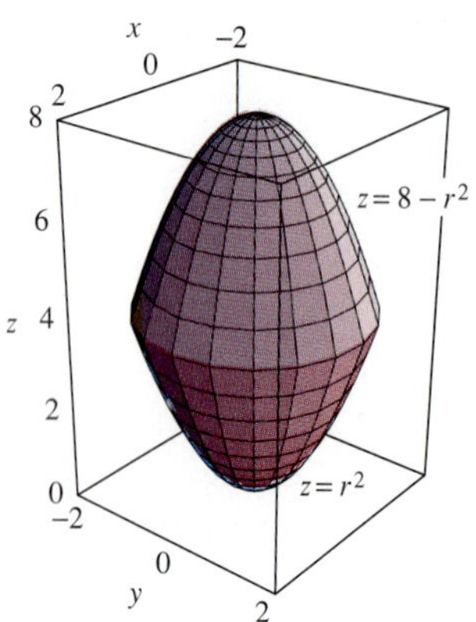

Fig. 14.4.13 The solid of Example 4

EXAMPLE 4 Find the volume of the solid that is bounded above by the paraboloid $z = 8 - r^2$ and below by the paraboloid $z = r^2$ (Fig. 14.4.13).

Solution The curve of intersection of the two paraboloids is found by simultaneous solution of the equations of the two surfaces. We eliminate z to obtain

$$r^2 = 8 - r^2; \quad \text{that is,} \quad r^2 = 4.$$

Hence the solid lies above the plane circular disk D with polar description $0 \leqq r \leqq 2$, and so the volume of the solid is

$$\begin{aligned} V &= \iint_D (z_{\text{top}} - z_{\text{bot}})\, dA = \int_0^{2\pi}\int_0^2 [(8 - r^2) - r^2]r\, dr\, d\theta \\ &= \int_0^{2\pi}\int_0^2 (8r - 2r^3)\, dr\, d\theta = 2\pi\left[4r^2 - \frac{1}{2}r^4\right]_0^2 = 16\pi. \end{aligned}$$

EXAMPLE 5 Here we apply a standard polar-coordinates technique to show that

$$I = \int_0^\infty e^{-x^2}\, dx = \frac{\sqrt{\pi}}{2}. \tag{9}$$

REMARK This important improper integral converges because

$$\int_1^b e^{-x^2}\, dx \leqq \int_1^b e^{-x}\, dx \leqq \int_1^\infty e^{-x}\, dx = \frac{1}{e}.$$

(The first inequality is valid because $e^{-x^2} \leqq e^x$ for $x \geqq 1$.) It follows that

$$\int_1^b e^{-x^2}\, dx$$

is a bounded and increasing function of b.

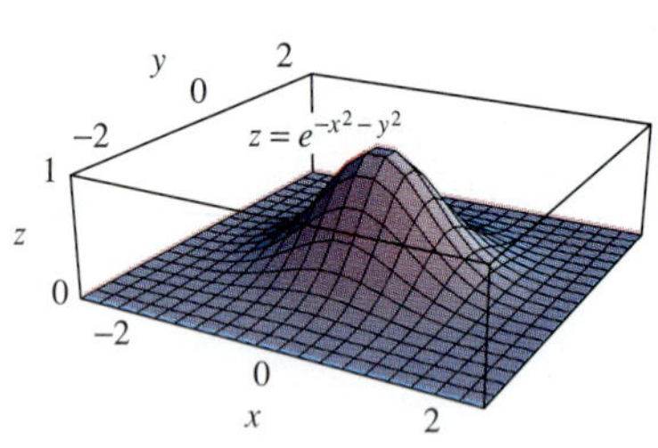

Fig. 14.4.14 The surface $z = e^{-x^2-y^2}$ (Example 5)

Solution Let V_b denote the volume of the region that lies below the surface $z = e^{-x^2-y^2}$ and above the square with vertices $(\pm b, \pm b)$ in the xy-plane (Fig. 14.4.14). Then

$$\begin{aligned} V_b &= \int_{-b}^b\int_{-b}^b e^{-x^2-y^2}\, dx\, dy = \int_{-b}^b e^{-y^2}\left(\int_{-b}^b e^{-x^2}\, dx\right) dy \\ &= \left(\int_{-b}^b e^{-x^2}\, dx\right)\left(\int_{-b}^b e^{-y^2}\, dy\right) = \left(\int_{-b}^b e^{-x^2}\, dx\right)^2 = 4\left(\int_0^b e^{-x^2}\, dx\right)^2. \end{aligned}$$

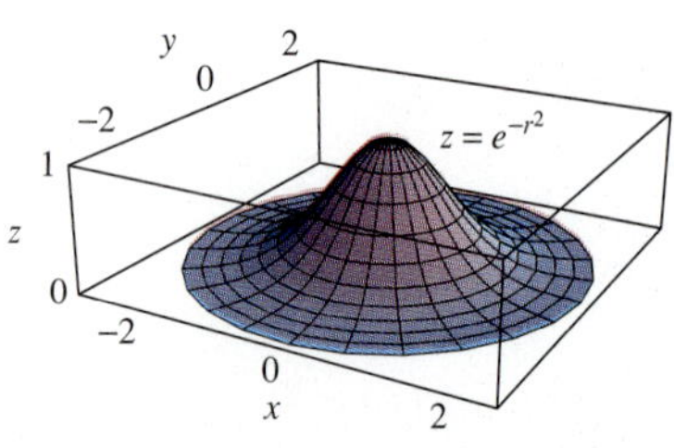

Fig. 14.4.15 The surface $z = e^{-r^2}$ (Example 5)

It follows that the volume below $z = e^{-x^2-y^2}$ and above the entire xy-plane is

$$V = \lim_{b\to\infty} V_b = \lim_{b\to\infty} 4\left(\int_0^b e^{-x^2}\, dx\right)^2 = 4\left(\int_0^\infty e^{-x^2}\, dx\right)^2 = 4I^2.$$

Now we compute V by another method—by using polar coordinates. We take the limit, as $b \to \infty$, of the volume below $z = e^{-x^2-y^2} = e^{-r^2}$ and above the circular disk with center $(0, 0)$ and radius b (Fig. 14.4.15). This disk is described by $0 \leqq r \leqq b$, $0 \leqq \theta \leqq 2\pi$, so we obtain

$$V = \lim_{b\to\infty} \int_0^{2\pi} \int_0^b re^{-r^2}\,dr\,d\theta = \lim_{b\to\infty} \int_0^{2\pi} \left[-\frac{1}{2}e^{-r^2} \right]_{r=0}^b d\theta$$

$$= \lim_{b\to\infty} \int_0^{2\pi} \frac{1}{2}\left[1 - e^{-b^2} \right] d\theta = \lim_{b\to\infty} \pi(1 - e^{-b^2}) = \pi.$$

We equate these two values of V and it follows that $4I^2 = \pi$. Therefore $I = \frac{1}{2}\sqrt{\pi}$, as desired. ■

14.4 PROBLEMS

In Problems 1 through 7, find the indicated area by double integration in polar coordinates.

1. The area bounded by the circle $r = 1$
2. The area bounded by the circle $r = 3\sin\theta$
3. The area bounded by the cardioid $r = 1 + \cos\theta$ (Fig. 14.4.16)
4. The area bounded by one loop of $r = 2\cos 2\theta$ (Fig. 14.4.17)

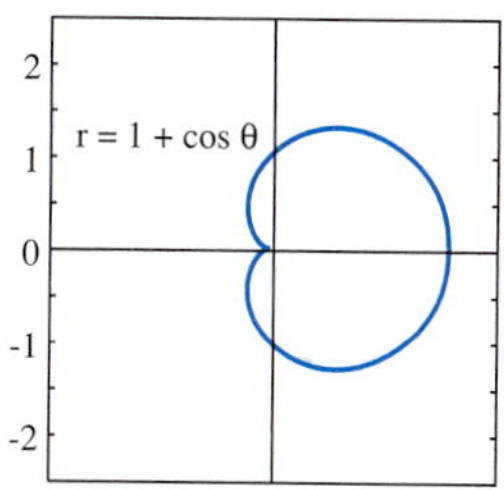

Fig. 14.4.16 The cardioid of Problem 3

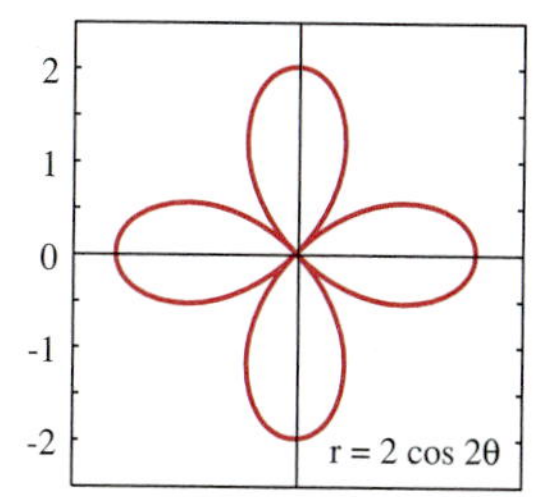

Fig. 14.4.17 The rose of Problem 4

5. The area inside both the circles $r = 1$ and $r = 2\sin\theta$
6. The area inside $r = 2 + \cos\theta$ and outside the circle $r = 2$
7. The area inside the smaller loop of $r = 1 - 2\sin\theta$ (Fig. 14.4.18)

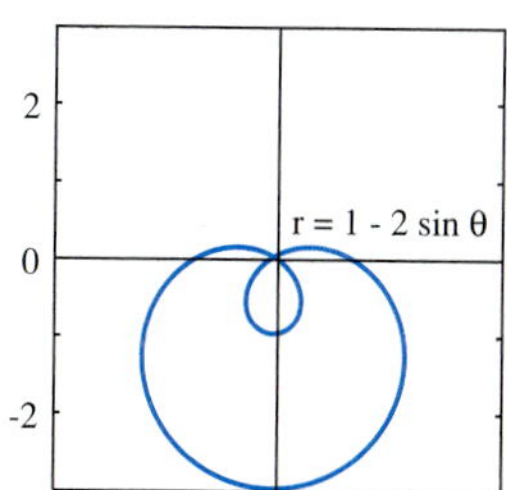

Fig. 14.4.18 The limaçon of Problem 7

In Problems 8 through 12, use double integration in polar coordinates to find the volume of the solid that lies below the given surface and above the plane region R bounded by the given curve.

8. $z = x^2 + y^2$; $r = 3$
9. $z = \sqrt{x^2 + y^2}$; $r = 2$
10. $z = x^2 + y^2$; $r = 2\cos\theta$
11. $z = 10 + 2x + 3y$; $r = \sin\theta$
12. $z = a^2 - x^2 - y^2$; $r = a$

In Problems 13 through 18, evaluate the given integral by first converting to polar coordinates.

13. $\displaystyle\int_0^1 \int_0^{\sqrt{1-y^2}} \frac{1}{1 + x^2 + y^2}\,dx\,dy$ (Fig. 14.4.19)

14. $\displaystyle\int_0^1 \int_0^{\sqrt{1-x^2}} \frac{1}{\sqrt{4 - x^2 - y^2}}\,dy\,dx$ (Fig. 14.4.19)

15. $\displaystyle\int_0^2 \int_0^{\sqrt{4-x^2}} (x^2 + y^2)^{3/2}\,dy\,dx$

16. $\displaystyle\int_0^1 \int_x^1 x^2\,dy\,dx$

17. $\displaystyle\int_0^1 \int_0^{\sqrt{1-y^2}} \sin(x^2 + y^2)\,dx\,dy$

18. $\displaystyle\int_1^2 \int_0^{\sqrt{2x-x^2}} \frac{1}{\sqrt{x^2 + y^2}}\,dy\,dx$ (Fig. 14.4.20)

In Problems 19 through 22, find the volume of the solid that is bounded above and below by the given surfaces $z = z_1(x, y)$

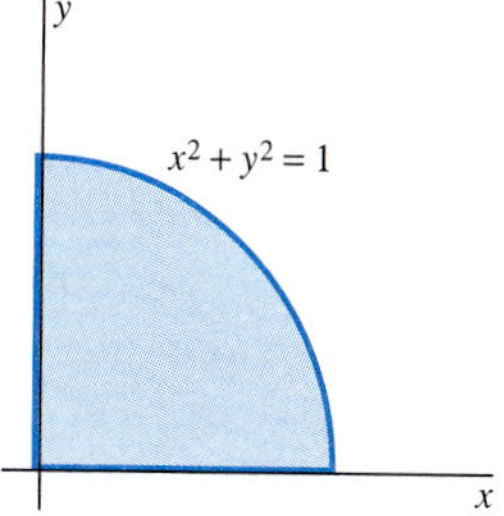

Fig. 14.4.19 The quarter-circle of Problems 13 and 14

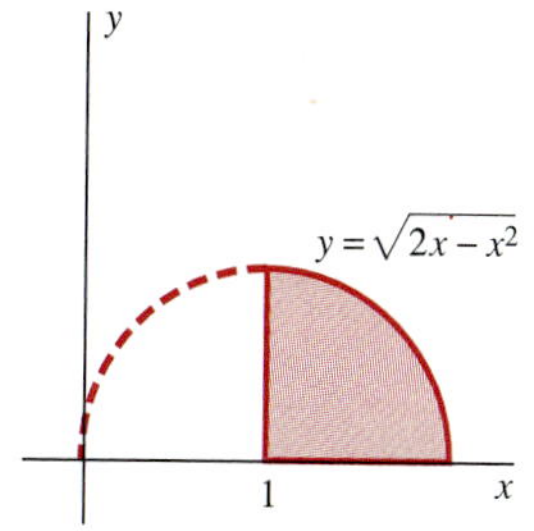

Fig. 14.4.20 The quarter-circle of Problem 18

and $z = z_2(x, y)$ and lies above the plane region R bounded by the given curve $r = g(\theta)$.

19. $z = 1, z = 3 + x + y; \quad r = 1$

20. $z = 2 + x, z = 4 + 2x; \quad r = 2$

21. $z = 0, z = 3 + x + y; \quad r = 2\sin\theta$

22. $z = 0, z = 1 + x; \quad r = 1 + \cos\theta$

Solve Problems 23 through 32 by double integration in polar coordinates.

23. Find the volume of a sphere of radius a by double integration.

24. Find the volume of the solid bounded by the paraboloids $z = 12 - 2x^2 - y^2$ and $z = x^2 + 2y^2$.

25. Suppose that $h > a > 0$. Show that the volume of the solid bounded by the cylinder $x^2 + y^2 = a^2$, the plane $z = 0$, and the plane $z = x + h$ is $\pi a^2 h$.

26. Find the volume of the wedge-shaped solid described in Example 3 of Section 14.3 (Fig. 14.4.21).

27. Find the volume bounded by the paraboloids $z = x^2 + y^2$ and $z = 4 - 3x^2 - 3y^2$.

28. Find the volume bounded by the paraboloids $z = x^2 + y^2$ and $z = 2x^2 + 2y^2 - 1$.

29. Find the volume of the "ice-cream cone" bounded by the sphere $x^2 + y^2 + z^2 = a^2$ and the cone $z = \sqrt{x^2 + y^2}$. When $a = 1$ this solid is the one shown in Fig. 14.4.22.

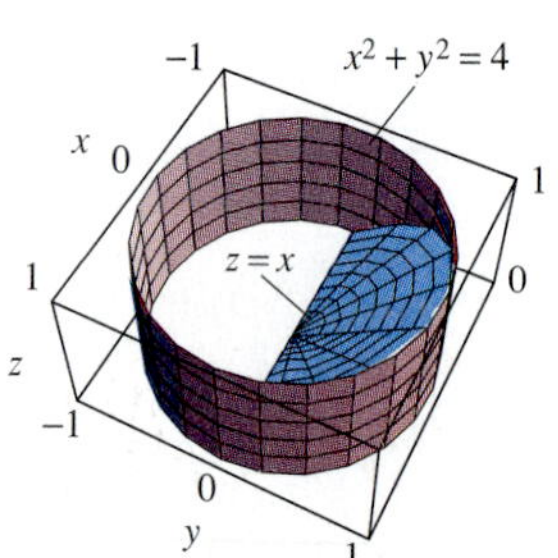

Fig. 14.4.21 The wedge of Problem 26

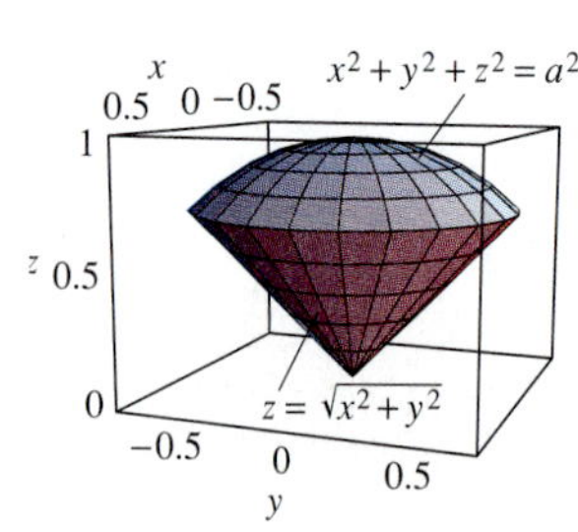

Fig. 14.4.22 The fat ice-cream cone of Problem 29

30. Find the volume bounded by the paraboloid $z = r^2$, the cylinder $r = 2a\sin\theta$, and the plane $z = 0$.

31. Find the volume that lies below the paraboloid $z = r^2$ and above one loop of the lemniscate with equation $r^2 = 2\sin\theta$.

32. Find the volume that lies inside both the cylinder $x^2 + y^2 = 4$ and the ellipsoid $2x^2 + 2y^2 + z^2 = 18$.

33. If $0 < h < a$, then the plane $z = a - h$ cuts off a spherical segment of height h and radius b from the sphere $x^2 + y^2 + z^2 = a^2$ (Fig. 14.4.23). (a) Show that $b^2 = 2ah - h^2$. (b) Show that the volume of the spherical segment is $V = \frac{1}{6}\pi h(3b^2 + h^2)$.

34. Show by the method of Example 5 that

$$\int_0^\infty \int_0^\infty \frac{1}{(1 + x^2 + y^2)^2}\, dx\, dy = \frac{\pi}{4}.$$

35. Find the volume of the solid torus obtained by revolving the disk $r \leqq a$ around the line $x = b > a$ (Fig. 14.4.24). (*Suggestion:* If the area element $dA = r\,dr\,d\theta$ is revolved around the line, the volume generated is $dV = 2\pi(b - x)\,dA$. Express everything in polar coordinates.)

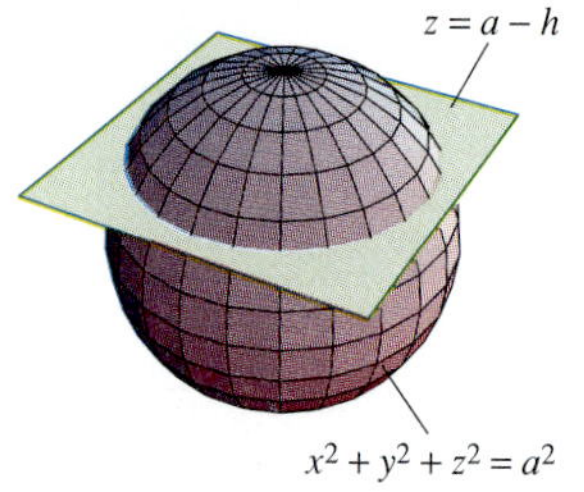

Fig. 14.4.23 The spherical segment of Problem 33

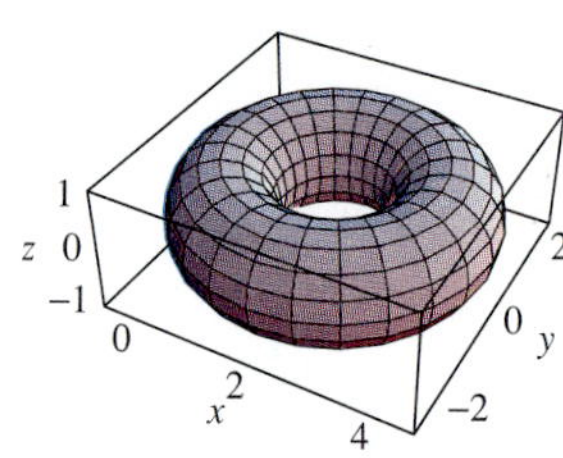

Fig. 14.4.24 The torus of Problem 35 (the case $a = 1, b = 2$ is shown)

In Problems 36 through 40, use double integrals in polar coordinates to find the volumes of the indicated solids.

36. The solid lies above the plane $z = -3$ and below the paraboloid $z = 15 - 2x^2 - 2y^2$.

37. The solid is bounded above by the plane $z = y + 4$ and below by the paraboloid $z = x^2 + y^2 + y$.

38. The solid lies inside the cylinder $x^2 + y^2 = 4$, above the xy-plane, and below the plane $z = x + y + 3$.

39. The solid is bounded by the elliptical paraboloids $z = x^2 + 2y^2$ and $z = 12 - 2x^2 - y^2$.

40. The solid lies inside the ellipsoid $4x^2 + 4y^2 + z^2 = 80$ and above the paraboloid $z = 2x^2 + 2y^2$.

14.5 APPLICATIONS OF DOUBLE INTEGRALS

We can use the double integral to find the mass m and the *centroid* $(\bar{x}, \bar{y})$ of a plane *lamina,* or thin plate, that occupies a bounded region R in the xy-plane. We suppose that the density of the lamina (in units of mass per unit *area*) at the point (x, y) is given by the continuous function $\delta(x, y)$.

Let $\mathcal{P} = \{R_1, R_2, \dots, R_n\}$ be an inner partition of R, and choose a point $(x_i^\star, y_i^\star)$ in each subrectangle R_i (Fig. 14.5.1). Then the mass of the part of the lamina occupying

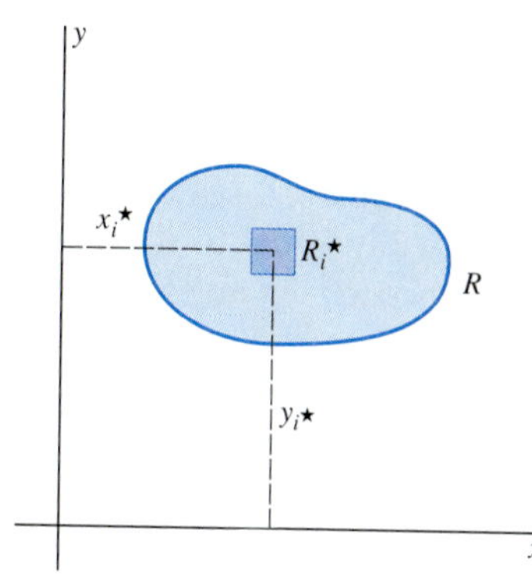

Fig. 14.5.1 The area element $\Delta A_i = a(R_i)$

R_i is approximately $\delta(x_i^\star, y_i^\star)\,\Delta A_i$, where ΔA_i denotes the area $a(R_i)$ of R_i. Hence the mass of the entire lamina is given approximately by

$$m \approx \sum_{i=1}^{n} \delta(x_i^\star, y_i^\star)\,\Delta A_i.$$

As the norm $|\mathcal{P}|$ of the inner partition $\mathcal{P}$ approaches zero, this Riemann sum approaches the corresponding double integral over R. We therefore *define* the **mass** m of the lamina by means of the formula

$$m = \iint_R \delta(x, y)\,dA. \tag{1}$$

In brief,

$$m = \iint_R \delta\,dA = \iint_R dm$$

in terms of the density δ and the mass element $dm = \delta\,dA$.

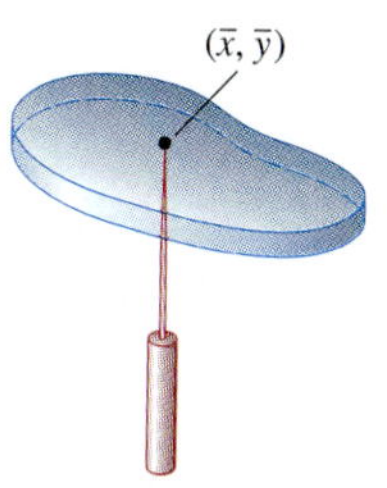

Fig. 14.5.2 A lamina balanced on its centroid

The coordinates $(\bar{x}, \bar{y})$ of the **centroid,** or *center of mass,* of the lamina are defined to be

$$\bar{x} = \frac{1}{m}\iint_R x\,\delta(x, y)\,dA, \tag{2}$$

$$\bar{y} = \frac{1}{m}\iint_R y\,\delta(x, y)\,dA. \tag{3}$$

You may remember these formulas in the form

$$\bar{x} = \frac{1}{m}\iint_R x\,dm, \quad \bar{y} = \frac{1}{m}\iint_R y\,dm.$$

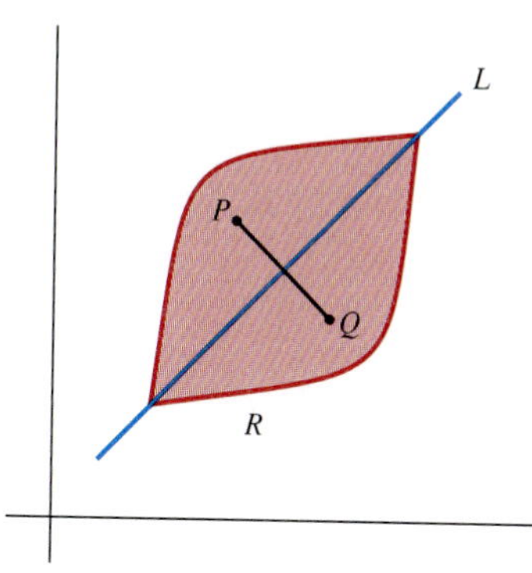

Fig. 14.5.3 A line of symmetry

Thus $\bar{x}$ and $\bar{y}$ are the *average values* of x and y *with respect to mass* in the region R. The centroid $(\bar{x}, \bar{y})$ is the point of the lamina where it would balance horizontally if placed on the point of an ice pick (Fig. 14.5.2).

If the density function δ has the *constant* value $k > 0$, then the coordinates of $\bar{x}$ and $\bar{y}$ are independent of the specific value of k. (Why?) In such a case we will generally take $\delta \equiv 1$ in our computations. Moreover, in this case m will have the same numerical value as the area A of R, and $(\bar{x}, \bar{y})$ is then called the **centroid of the plane region** R.

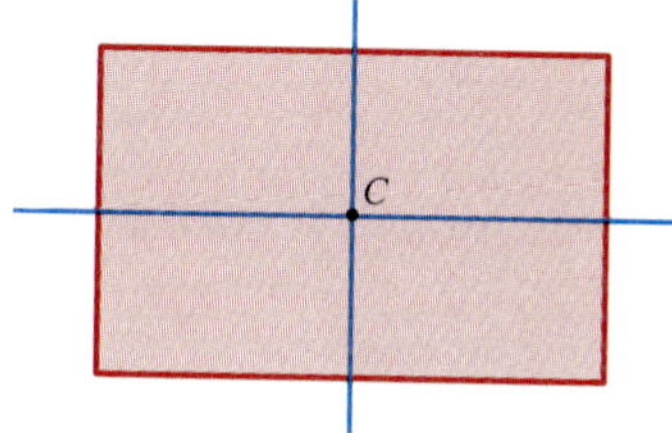

Fig. 14.5.4 The centroid of a rectangle

Generally, we must calculate all three integrals in Eqs. (1) through (3) in order to find the centroid of a lamina. But sometimes we can take advantage of the following *symmetry principle:* If the plane region R (considered to be a lamina of constant density) is symmetric with respect to the line L—that is, if R is carried onto itself when the plane is rotated through an angle of 180° around the line L—then the centroid of R lies on L (Fig. 14.5.3). For example, the centroid of a rectangle (Fig. 14.5.4) is the point where the perpendicular bisectors of its sides meet, because these bisectors are also lines of symmetry.

In the case of a nonconstant density function δ, we require (for symmetry) that δ—as well as the region itself—be symmetric about the geometric line L of symmetry. That is, $\delta(P) = \delta(Q)$ if, as in Fig. 14.5.3, the points P and Q are symmetrically located with respect to L. Then the centroid of the lamina R will lie on the line L of symmetry.

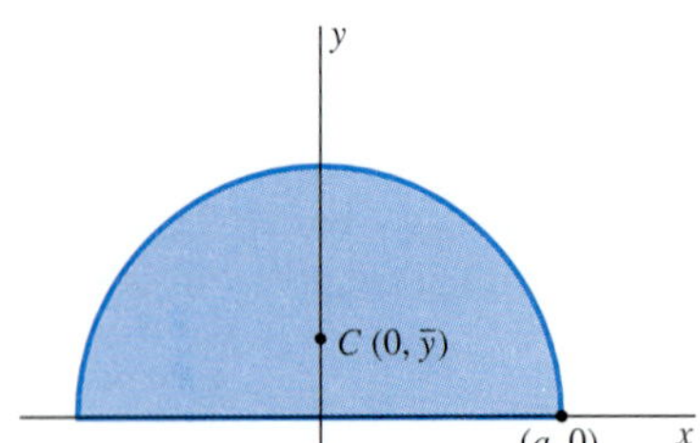

Fig. 14.5.5 The centroid of a semicircular disk (Example 1)

EXAMPLE 1 Consider the semicircular disk of radius a shown in Fig. 14.5.5. If it has constant density $\delta \equiv 1$, then its mass is $m = \frac{1}{2}\pi a^2$ (numerically equal to its area), and by symmetry its centroid $C(\overline{x}, \overline{y})$ lies on the y-axis. Hence $\overline{x} = 0$, and we need only compute

$$\overline{y} = \frac{1}{m}\iint_R y\,dm = \frac{2}{\pi a^2}\int_0^{\pi}\int_0^{a}(r\sin\theta)\,r\,dr\,d\theta \qquad \text{(polar coordinates)}$$

$$= \frac{2}{\pi a^2}\Big[-\cos\theta\Big]_0^{\pi}\left[\frac{1}{3}r^3\right]_0^{a} = \frac{2}{\pi a^2}\cdot 2\cdot\frac{a^3}{3} = \frac{4a}{3\pi}.$$

Thus the centroid of the semicircular lamina is located at the point $(0, 4a/3\pi)$. Note that the computed value for $\overline{y}$ has the dimensions of length (because a is a length), as it should. Any answer that has other dimensions would be suspect.

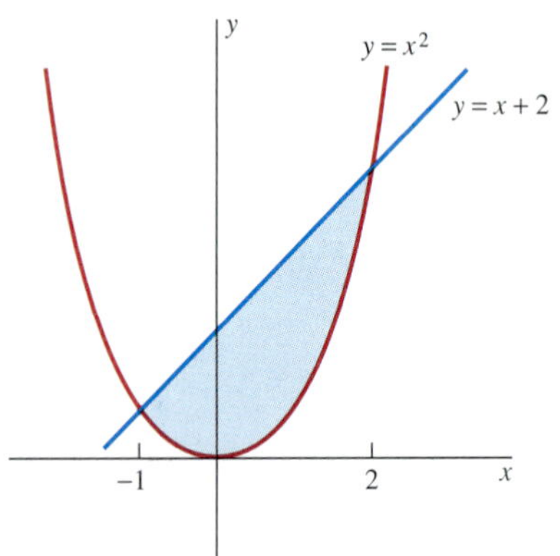

Fig. 14.5.6 The lamina of Example 2

EXAMPLE 2 A lamina occupies the region bounded by the line $y = x + 2$ and the parabola $y = x^2$ (Fig. 14.5.6). The density of the lamina at the point $P(x, y)$ is proportional to the square of the distance of P from the y-axis—thus $\delta(x, y) = kx^2$ (where k is a positive constant). Find the mass and centroid of the lamina.

Solution The line and the parabola intersect in the two points $(-1, 1)$ and $(2, 4)$, so Eq. (1) gives mass

$$m = \int_{-1}^{2}\int_{x^2}^{x+2} kx^2\,dy\,dx = k\int_{-1}^{2}\Big[x^2y\Big]_{y=x^2}^{x+2}\,dx$$

$$= k\int_{-1}^{2}(x^3 + 2x^2 - x^4)\,dx = \frac{63}{20}k.$$

Then Eqs. (2) and (3) give

$$\overline{x} = \frac{20}{63k}\int_{-1}^{2}\int_{x^2}^{x+2} kx^3\,dy\,dx = \frac{20}{63}\int_{-1}^{2}\Big[x^3y\Big]_{y=x^2}^{x+2}\,dx$$

$$= \frac{20}{63}\int_{-1}^{2}(x^4 + 2x^3 - x^5)\,dx = \frac{20}{63}\cdot\frac{18}{5} = \frac{8}{7};$$

$$\overline{y} = \frac{20}{63k}\int_{-1}^{2}\int_{x^2}^{x+2} kx^2y\,dy\,dx = \frac{20}{63}\int_{-1}^{2}\left[\frac{1}{2}x^2y^2\right]_{y=x^2}^{x+2}$$

$$= \frac{10}{63}\int_{-1}^{2}(x^4 + 4x^3 + 4x^2 - x^6)\,dx = \frac{10}{63}\cdot\frac{531}{35} = \frac{118}{49}.$$

Thus the lamina of this example has mass $63k/20$, and its centroid is located at the point $\left(\frac{8}{7}, \frac{118}{49}\right)$.

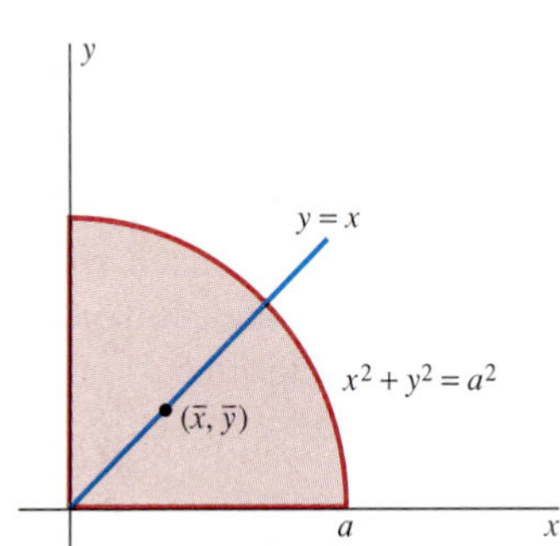

Fig. 14.5.7 Finding mass and centroid (Example 3)

EXAMPLE 3 A lamina is shaped like the first-quadrant quarter-circle of radius a shown in Fig. 14.5.7. Its density is proportional to distance from the origin—that is, its density at (x, y) is $\delta(x, y) = k\sqrt{x^2 + y^2} = kr$ (where k is a positive constant). Find its mass and centroid.

Solution First we change to polar coordinates, because both the shape of the boundary of the lamina and the formula for its density suggest that this will make the computations much simpler. Equation (1) then yields the mass to be

$$m = \iint_R \delta\, dA = \int_0^{\pi/2} \int_0^a kr^2\, dr\, d\theta$$

$$= k\int_0^{\pi/2} \left[\frac{1}{3}r^3\right]_{r=0}^a d\theta = k\int_0^{\pi/2} \frac{1}{3}a^3\, d\theta = \frac{k\pi a^3}{6}.$$

By symmetry of the lamina and its density function, the centroid lies on the line $y = x$. So Eq. (3) gives

$$\bar{x} = \bar{y} = \frac{1}{m}\iint_R y\delta\, dA = \frac{6}{k\pi a^3}\int_0^{\pi/2}\int_0^a kr^3 \sin\theta\, dr\, d\theta$$

$$= \frac{6}{\pi a^3}\int_0^{\pi/2}\left[\frac{1}{4}r^4 \sin\theta\right]_{r=0}^a d\theta = \frac{6}{\pi a^3}\cdot\frac{a^4}{4}\int_0^{\pi/2} \sin\theta\, d\theta = \frac{3a}{2\pi}.$$

Thus the given lamina has mass $\frac{1}{6}k\pi a^3$; its centroid is located at the point $(3a/2\pi, 3a/2\pi)$.

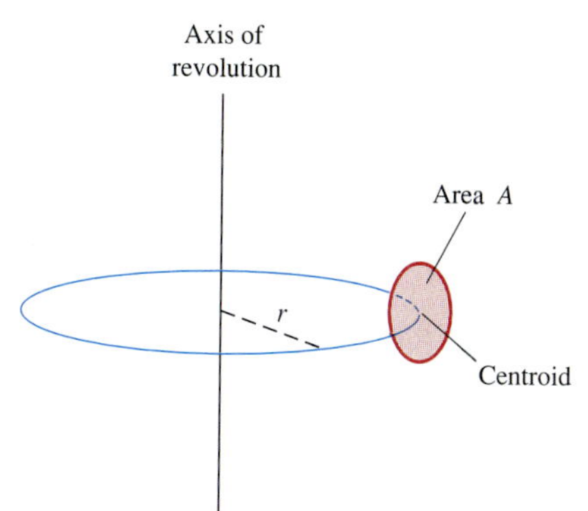

Fig. 14.5.8 A solid of volume $V = A \cdot d$ is generated by the area A as its centroid travels the distance $d = 2\pi r$ around a circle of radius r.

Volume and the First Theorem of Pappus

An important theorem relating centroids and volumes of revolution is named for the Greek mathematician who stated it during the third century A.D.

First Theorem of Pappus *Volume of Revolution*

Suppose that a plane region R is revolved around an axis in its plane (Fig. 14.5.8), generating a solid of revolution with volume V. Assume that the axis does not intersect the interior of R. Then the volume

$$V = A \cdot d$$

of the solid is the product of the area A of R and the distance d traveled by the centroid of R.

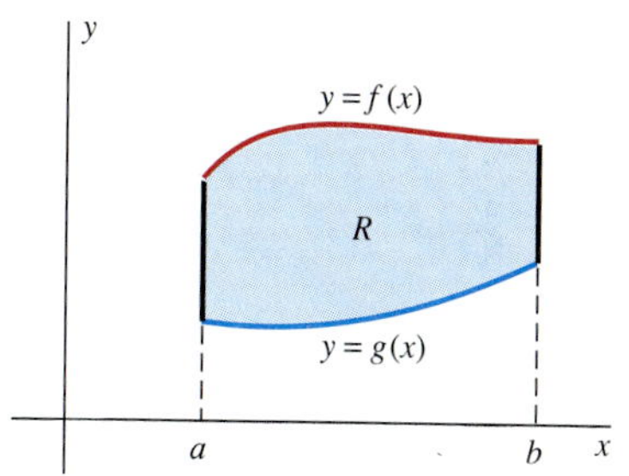

Fig. 14.5.9 A region R between the graphs of two functions

Proof We treat only the special case of a region like that in Fig. 14.5.9. This is the region between the two graphs $y = f(x)$ and $y = g(x)$ for $a \leqq x \leqq b$, with $f(x) \geqq g(x)$ for such x and with the axis of revolution being the y-axis. Then, in a revolution around the y-axis, the distance traveled by the centroid of R is $d = 2\pi\bar{x}$. By the method of cylindrical shells [see Eq. (4) of Section 6.3 and Fig. 14.5.10], the volume of the solid generated is

$$V = \int_a^b 2\pi x[f(x) - g(x)]\, dx = \int_a^b \int_{g(x)}^{f(x)} 2\pi x\, dy\, dx$$

$$= 2\pi \iint_R x\, dA = 2\pi\bar{x} \cdot A$$

[by Eq. (2), with $\delta \equiv 1$.] Thus $V = d \cdot A$, as desired.

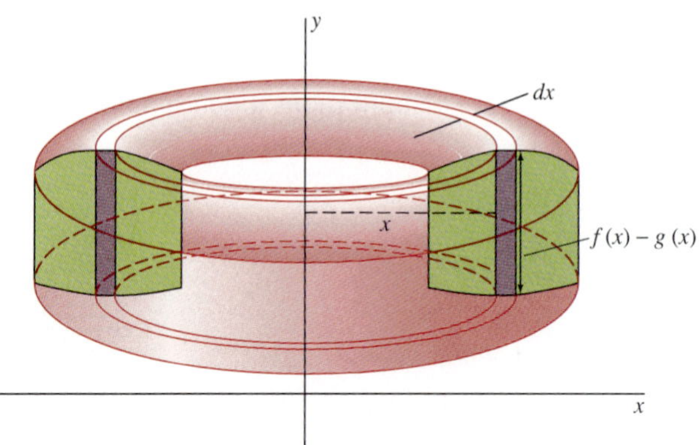

Fig. 14.5.10 A solid of revolution consisting of cylindrical shells

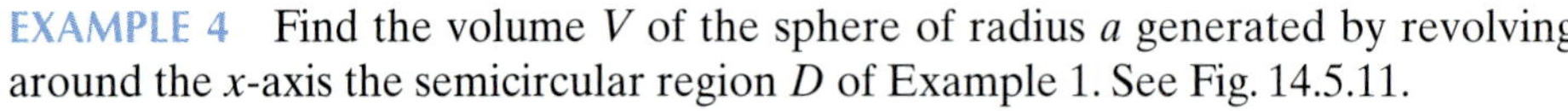

EXAMPLE 4 Find the volume V of the sphere of radius a generated by revolving around the x-axis the semicircular region D of Example 1. See Fig. 14.5.11.

Solution The area of D is $A = \frac{1}{2}\pi a^2$, and we found in Example 1 that $\bar{y} = 4a/3\pi$. Hence Pappus's theorem gives

$$V = 2\pi\bar{y}A = 2\pi \cdot \frac{4a}{3\pi} \cdot \frac{\pi a^2}{2} = \frac{4}{3}\pi a^3.$$

■

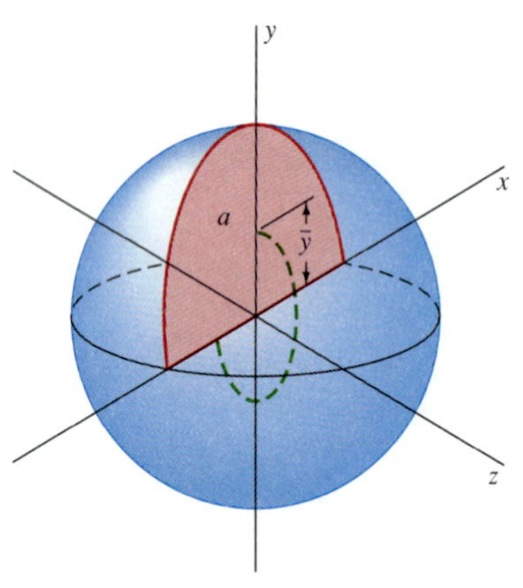

Fig. 14.5.11 A sphere of radius a generated by revolving a semicircular region of area $A = \frac{1}{2}\pi a^2$ around its diameter on the x-axis (Example 4). The centroid of the semicircle travels along a circle of circumference $d = 2\pi\bar{y}$.

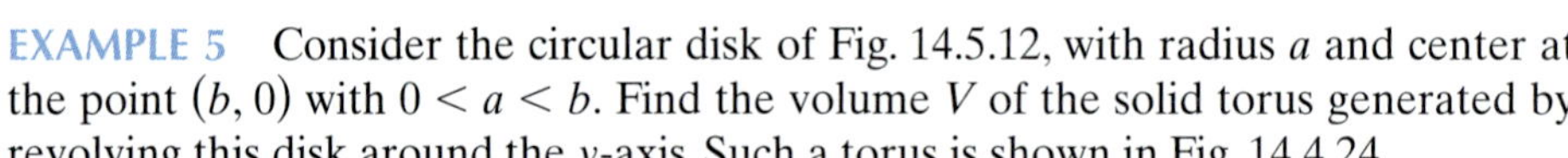

EXAMPLE 5 Consider the circular disk of Fig. 14.5.12, with radius a and center at the point $(b, 0)$ with $0 < a < b$. Find the volume V of the solid torus generated by revolving this disk around the y-axis. Such a torus is shown in Fig. 14.4.24.

Solution The centroid of the circle is at its center $(b, 0)$, so $\bar{x} = b$. Hence the centroid is revolved through the distance $d = 2\pi b$. Consequently,

$$V = d \cdot A = 2\pi b \cdot \pi a^2 = 2\pi^2 a^2 b.$$

Note that this result is dimensionally correct. ■

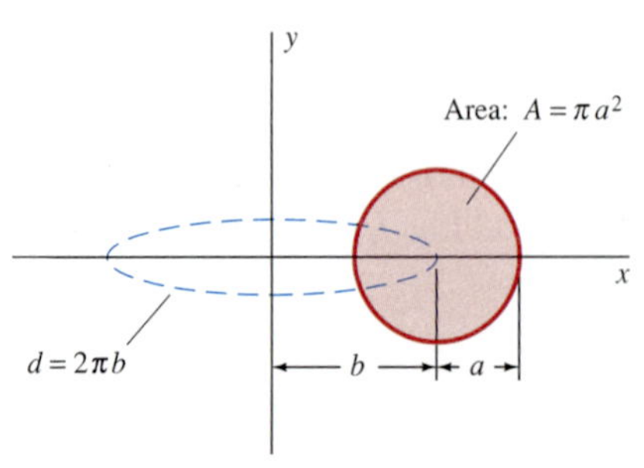

Fig. 14.5.12 Rotating the circular disk around the y-axis to generated a torus (Example 5)

Surface Area and the Second Theorem of Pappus

Centroids of plane *curves* are defined in analogy with the method for plane regions, so we shall present this topic in less detail. It will suffice for us to treat only the case of constant density $\delta \equiv 1$ (like a wire with unit mass per unit length). Then the centroid $(\bar{x}, \bar{y})$ of the plane curve C is defined by the formulas

$$\bar{x} = \frac{1}{s}\int_C x\,ds, \qquad \bar{y} = \frac{1}{s}\int_C y\,ds \tag{4}$$

where s is the arc length of C.

The meaning of the integrals in Eq. (4) is that of the notation of Section 6.4. That is, ds is a symbol to be replaced (before the integral is evaluated) with either

$$ds = \sqrt{1 + \left(\frac{dy}{dx}\right)^2}\,dx \quad \text{or} \quad ds = \sqrt{1 + \left(\frac{dx}{dy}\right)^2}\,dy,$$

depending on whether C is a smooth arc of the form $y = f(x)$ or one of the form $x = g(y)$. Alternatively, we may have

$$ds = \sqrt{(dx)^2 + (dy)^2} = \sqrt{\left(\frac{dx}{dt}\right)^2 + \left(\frac{dy}{dt}\right)^2}\, dt$$

if C is presented in parametric form, as in Section 10.5.

EXAMPLE 6 Let J denote the upper half of the *circle* (not the disk) of radius a and center $(0, 0)$, represented parametrically by

$$x = a\cos t, \qquad y = a\sin t, \qquad 0 \leqq t \leqq \pi.$$

The arc J is shown in Fig. 14.5.13. Find its centroid.

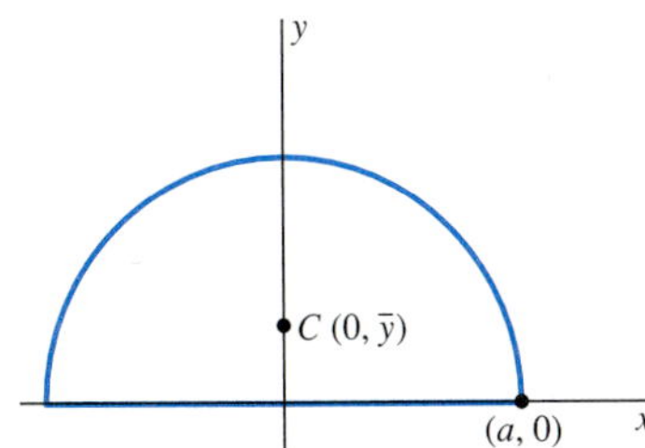

Fig. 14.5.13 The semicircular arc of Example 6

Solution Note first that $\bar{x} = 0$ by symmetry. The arc length of J is $s = \pi a$; the arc-length element is

$$ds = \sqrt{(-a\sin t)^2 + (a\cos t)^2}\, dt = a\, dt.$$

Hence the second formula in (4) yields

$$\bar{y} = \frac{1}{\pi a}\int_0^{\pi} (a\sin t)\, a\, dt = \frac{a}{\pi}\Big[-\cos t\Big]_0^{\pi} = \frac{2a}{\pi}.$$

Thus the centroid of the semicircular arc is located at the point $(0, 2a/\pi)$ on the y-axis. Note that the answer is both plausible and dimensionally correct. ■

The first theorem of Pappus has an analogue for surface area of revolution.

Second Theorem of Pappus ***Surface Area of Revolution***

Let the plane curve C be revolved around an axis in its plane that does not intersect the curve. Then the area

$$A = s \cdot d$$

of the surface of revolution generated is equal to the product of the length s of C and the distance d traveled by the centroid of C.

PROOF We treat only the special case in which C is a smooth arc described by $y = f(x)$, $a \leqq x \leqq b$, and the axis of revolution is the y-axis. The distance traveled by the centroid of C is $d = 2\pi\bar{x}$. By Eq. (11) of Section 6.4, the area of the surface of revolution is

$$A = \int_{\star}^{\star\star} 2\pi x\, ds = \int_a^b 2\pi x\sqrt{1 + [f'(x)]^2}\, dx = 2\pi s \cdot \frac{1}{s}\int_C x\, ds = 2\pi s\bar{x}$$

by Eq. (4). Therefore $A = d \cdot s$, as desired. ■

EXAMPLE 7 Find the surface area A of the sphere of radius a generated by revolving around the x-axis the semicircular arc of Example 6.

Solution Because we found that $\bar{y} = 2a/\pi$ and we know that $s = \pi a$, the second theorem of Pappus gives

$$A = 2\pi \bar{y} s = 2\pi \cdot \frac{2a}{\pi} \cdot \pi a = 4\pi a^2.$$ ■

EXAMPLE 8 Find the surface area A of the torus of Example 5.

Solution Now we think of revolving around the y-axis the circle (*not* the disk) of radius a centered at the point $(b, 0)$. Of course, the centroid of the circle is located at its center $(b, 0)$; this follows from the symmetry principle or can be verified by using computations such as those in Example 6. Hence the distance traveled by the centroid is $d = 2\pi b$. Because the circumference of the circle is $s = 2\pi a$, the second theorem of Pappus gives

$$A = 2\pi b \cdot 2\pi a = 4\pi^2 ab.$$ ■

Moments of Inertia

Let R be a plane lamina and L a straight line that may or may not lie in the xy-plane. Then the **moment of inertia** I of R around the axis L is defined to be

$$I = \iint_R w^2 \, dm, \tag{5}$$

where $w = w(x, y)$ denotes the perpendicular distance to L from the point (x, y) of R.

The most important case is that in which the axis of revolution is the z-axis, so $w = r = \sqrt{x^2 + y^2}$ (Fig. 14.5.14). In this case we call $I = I_0$ the **polar moment of inertia** of the lamina R. Thus the polar moment of inertia of R is defined to be

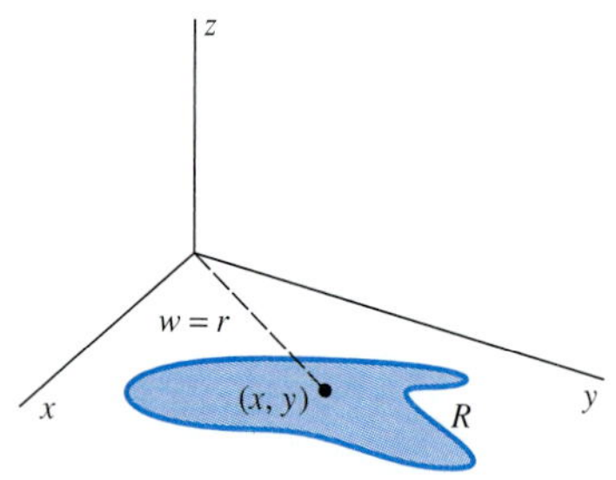

Fig. 14.5.14 A lamina in the xy-plane in space

$$I_0 = \iint_R r^2 \delta(x, y) \, dA = \iint_R (x^2 + y^2) \, dm. \tag{6}$$

It follows that $I_0 = I_x + I_y$, where

$$I_x = \iint_R y^2 \, dm = \iint_R y^2 \delta \, dA \tag{7}$$

and

$$I_y = \iint_R x^2 \, dm = \iint_R x^2 \delta \, dA. \tag{8}$$

Here I_x is the moment of inertia of the lamina around the x-axis and I_y is its moment of inertia around the y-axis.

An important application of moments of inertia involves *kinetic energy of rotation.* Consider a circular disk that is revolving around its center (the origin) with angular speed ω radians per second. A mass element dm at distance r from the origin is moving with (linear) velocity $v = r\omega$ (Fig. 14.5.15). Thus the kinetic energy of the mass element is

$$\tfrac{1}{2}(dm)v^2 = \tfrac{1}{2}\omega^2 r^2 \, dm.$$

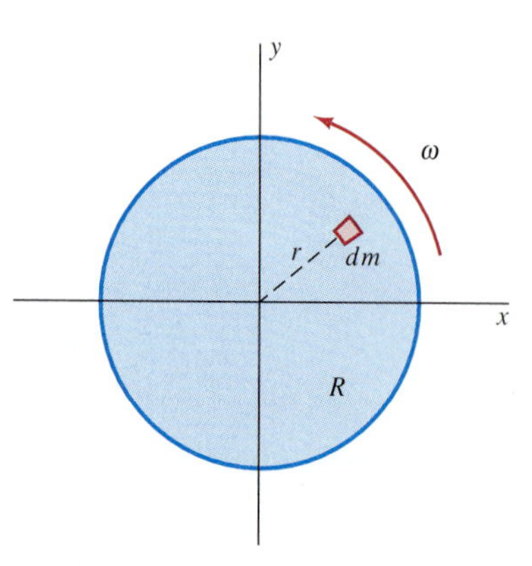

Fig. 14.5.15 The rotating disk

Summing by integration over the whole disk, we find that its kinetic energy due to rotation at angular speed ω is

$$\text{KE}_{\text{rot}} = \iint_R \frac{1}{2} \omega^2 r^2 \, dm = \tfrac{1}{2}\omega^2 \iint_R r^2 \, dm;$$

that is,

$$\mathrm{KE}_{\mathrm{rot}} = \tfrac{1}{2} I_0 \omega^2. \tag{9}$$

Because linear kinetic energy has the formula $\mathrm{KE} = \frac{1}{2}mv^2$, Eq. (9) suggests (correctly) that moment of inertia is the rotational analogue of mass.

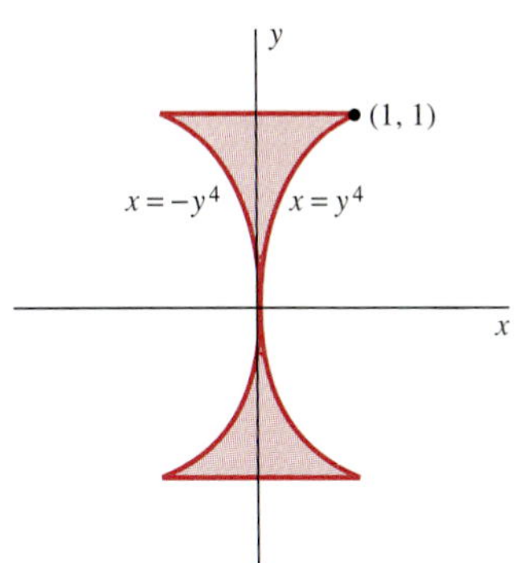

Fig. 14.5.16 The lamina of Example 9

EXAMPLE 9 Compute I_x for a lamina of constant density $\delta \equiv 1$ that occupies the region bounded by the curves $x = \pm y^4$, $-1 \leqq y \leqq 1$ (Fig. 14.5.16).

Solution Equation (7) gives

$$I_x = \int_{-1}^{1}\int_{-y^4}^{y^4} y^2\,dx\,dy = \int_{-1}^{1}\Big[xy^2\Big]_{x=-y^4}^{y^4}\,dy = \int_{-1}^{1} 2y^6\,dy = \frac{4}{7}.$$

The region of Example 9 resembles the cross section of an I beam. It is known that the stiffness, or resistance to bending, of a horizontal beam is proportional to the moment of inertia of its cross section with respect to a horizontal axis through the centroid of the cross section of the beam. Let us compare our I beam with a rectangular beam of equal height 2 and equal area

$$A = \int_{-1}^{1}\int_{-y^4}^{y^4} 1\,dx\,dy = \frac{4}{5}.$$

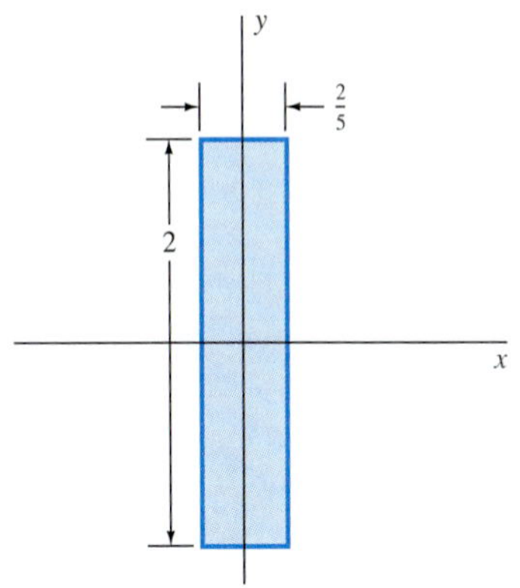

Fig. 14.5.17 A rectangular beam for comparison with the I beam of Example 9

The cross section of such a rectangular beam is shown in Fig. 14.5.17. Its width is $\frac{2}{5}$, and the moment of inertia of its cross section is

$$I_x = \int_{-1}^{1}\int_{-1/5}^{1/5} y^2\,dx\,dy = \frac{4}{15}.$$

Because the ratio of $\frac{4}{7}$ to $\frac{4}{15}$ is $\frac{15}{7}$, we see that the I beam is more than twice as strong as a rectangular beam of the same cross-sectional area. This strength is why I beams are commonly used in construction.

EXAMPLE 10 Find the polar moment of inertia of a circular lamina R of radius a and constant density δ centered at the origin.

Solution In Cartesian coordinates, the lamina R occupies the plane region $x^2 + y^2 \leqq a^2$; in polar coordinates, this region has the much simpler description $0 \leqq r \leqq a$, $0 \leqq \theta \leqq 2\pi$. Equation (6) then gives

$$I_0 = \iint_R r^2\delta\,dA = \int_0^{2\pi}\int_0^a r^3\delta\,dr\,d\theta = \frac{\delta\pi a^4}{2} = \frac{1}{2}ma^2,$$

where $m = \delta\pi a^2$ is the mass of the circular lamina.

Finally, the **radius of gyration** $\hat{r}$ of a lamina of mass m around an axis is defined to be

$$\hat{r} = \sqrt{\frac{I}{m}}, \tag{10}$$

where I is the moment of inertia of the lamina around that axis. For example, the radii of gyration $\hat{x}$ and $\hat{y}$ around the y-axis and x-axis, respectively, are given by

$$\hat{x} = \sqrt{\frac{I_y}{m}} \quad \text{and} \quad \hat{y} = \sqrt{\frac{I_x}{m}}. \tag{11}$$

Now suppose that this lamina lies in the right half-plane $x > 0$ and is symmetric around the x-axis. If it represents the face of a tennis racquet whose handle (considered of negligible weight) extends along the x-axis from the origin to the face, then the point $(\hat{x}, 0)$ is a plausible candidate for the racquet's "sweet spot" that delivers the maximum impact and control (see Problem 56).

The definition in Eq. (10) is motivated by consideration of a plane lamina R rotating with angular speed ω around the z-axis (Fig. 14.5.18). Then Eq. (10) yields

$$I_0 = m\hat{r}^2,$$

so it follows from Eq. (9) that the kinetic energy of the lamina is

$$\text{KE} = \tfrac{1}{2}\, m(\hat{r}\,\omega)^2.$$

Thus the kinetic energy of the rotating lamina equals that of a single particle of mass m revolving with velocity $v = \hat{r}\omega$ at the distance $\hat{r}$ from the axis of revolution.

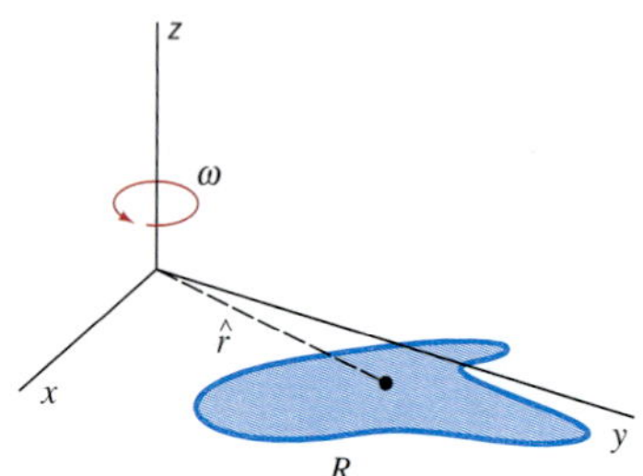

Fig. 14.5.18 A plane lamina rotating around the z-axis

14.5 PROBLEMS

In Problems 1 through 10, find the centroid of the plane region bounded by the given curves. Assume that the density is $\delta \equiv 1$ for each region.

1. $x = 0,\ x = 4,\ y = 0,\ y = 6$

2. $x = 1,\ x = 3,\ y = 2,\ y = 4$

3. $x = -1,\ x = 3,\ y = -2,\ y = 4$

4. $x = 0,\ y = 0,\ x + y = 3$

5. $x = 0,\ y = 0,\ x + 2y = 4$

6. $y = 0,\ y = x,\ x + y = 2$

7. $y = 0,\ y = x^2,\ x = 2$

8. $y = x^2,\ y = 9$

9. $y = 0,\ y = x^2 - 4$

10. $x = -2,\ x = 2,\ y = 0,\ y = x^2 + 1$

In Problems 11 through 30, find the mass and centroid of the plane lamina with the indicated shape and density.

11. The triangular region bounded by $x = 0$, $y = 0$, and $x + y = 1$, with $\delta(x, y) = xy$

12. The triangular region of Problem 11, with $\delta(x, y) = x^2$

13. The region bounded by $y = 0$ and $y = 4 - x^2$, with $\delta(x, y) = y$

14. The region bounded by $x = 0$ and $x = 9 - y^2$, with $\delta(x, y) = x^2$

15. The region bounded by the parabolas $y = x^2$ and $x = y^2$, with $\delta(x, y) = xy$

16. The region of Problem 15, with $\delta(x, y) = x^2 + y^2$

17. The region bounded by the parabolas $y = x^2$ and $y = 2 - x^2$, with $\delta(x, y) = y$

18. The region bounded by $x = 0$, $x = e$, $y = 0$, and $y = \ln x$ for $1 \leqq x \leqq e$, with $\delta(x, y) \equiv 1$

19. The region bounded by $y = 0$ and $y = \sin x$ for $0 \leqq x \leqq \pi$, with $\delta(x, y) \equiv 1$

20. The region bounded by $y = 0$, $x = -1$, $x = 1$, and $y = \exp(-x^2)$, with $\delta(x, y) = |xy|$

21. The square with vertices $(0, 0)$, $(0, a)$, (a, a), and $(a, 0)$, with $\delta(x, y) = x + y$

22. The triangular region bounded by the coordinate axes and the line $x + y = a$ $(a > 0)$, with $\delta(x, y) = x^2 + y^2$

23. The region bounded by $y = x^2$ and $y = 4$; $\delta(x, y) = y$

24. The region bounded by $y = x^2$ and $y = 2x + 3$; $\delta(x, y) = x^2$

25. The region of Problem 19; $\delta(x, y) = x$

26. The semicircular region $x^2 + y^2 \leqq a^2$, $y \geqq 0$; $\delta(x, y) = y$

27. The region of Problem 26; $\delta(x, y) = r$ (the radial polar coordinate)

28. The region bounded by the cardioid with polar equation $r = 1 + \cos\theta$; $\delta(r, \theta) = r$ (Fig. 14.5.19)

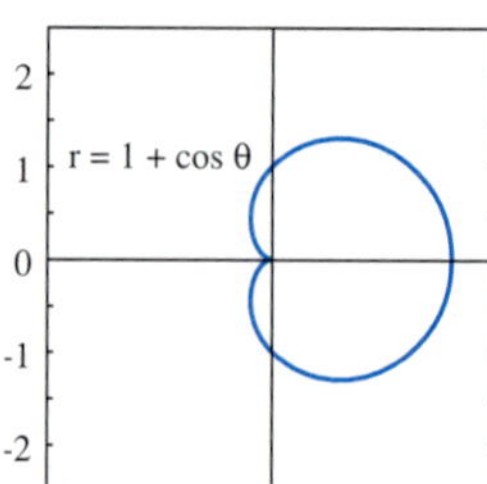

Fig. 14.5.19 The cardiod of Problem 28

29. The region inside the circle $r = 2\sin\theta$ and outside the circle $r = 1$; $\delta(x, y) = y$

30. The region inside the limaçon $r = 1 + 2\cos\theta$ and outside the circle $r = 2$; $\delta(r, \theta) = r$ (Fig. 14.5.20)

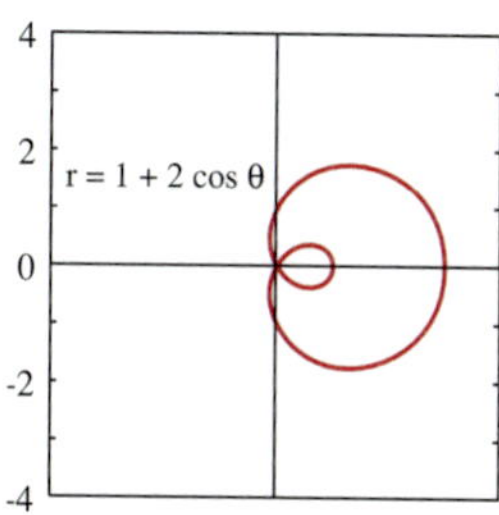

Fig. 14.5.20 The limaçon of Problem 30

In Problems 31 through 35, find the polar moment of inertia I_0 of the indicated lamina.

31. The region bounded by the circle $r = a$; $\delta(x, y) = r^n$, where n is a fixed positive integer

32. The lamina of Problem 26

33. The disk bounded by $r = 2\cos\theta$; $\delta(x, y) = k$ (a positive constant)

34. The lamina of Problem 29

35. The region bounded by the right-hand loop of the lemniscate $r^2 = \cos 2\theta$; $\delta(x, y) = r^2$ (Fig. 14.5.21)

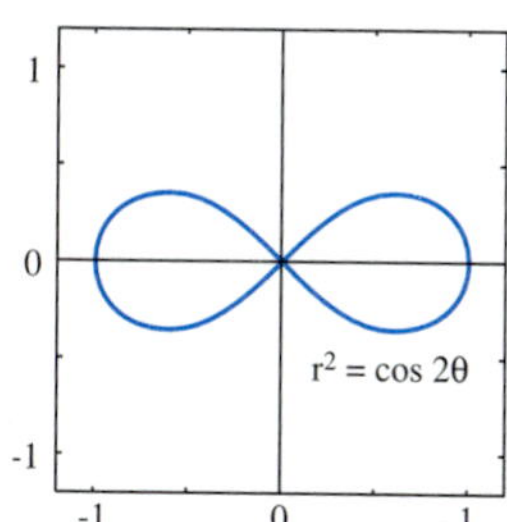

Fig. 14.5.21 The leminscate of Problem 35

In Problems 36 through 40, find the radii of gyration $\hat{x}$ and $\hat{y}$ of the indicated lamina around the coordinate axes.

36. The lamina of Problem 21

37. The lamina of Problem 23

38. The lamina of Problem 24

39. The lamina of Problem 27

40. The lamina of Problem 33

41. Find the centroid of the first quadrant of the circular disk $x^2 + y^2 \leqq r^2$ by direct computation, as in Example 1.

42. Apply the first theorem of Pappus to find the centroid of the first quadrant of the circular disk $x^2 + y^2 \leqq r^2$. Use the facts that $\overline{x} = \overline{y}$ (by symmetry) and that revolution of this quarter-disk around either coordinate axis gives a solid hemisphere with volume $V = \frac{2}{3}\pi r^3$.

43. Find the centroid of the arc that consists of the first-quadrant portion of the circle $x^2 + y^2 = r^2$ by direct computation, as in Example 6.

44. Apply the second theorem of Pappus to find the centroid of the quarter-circular arc of Problem 43. Note that $\overline{x} = \overline{y}$ (by symmetry) and that rotation of this arc around either coordinate axis gives a hemisphere with surface area $A = 2\pi r^2$.

45. Show by direct computation that the centroid of the triangle with vertices $(0, 0)$, $(r, 0)$, and $(0, h)$ is the point $(r/3, h/3)$. Verify that this point lies on the line from the vertex $(0, 0)$ to the midpoint of the opposite side of the triangle and two-thirds of the way from the vertex to the midpoint.

46. Apply the first theorem of Pappus and the result of Problem 45 to verify the formula $V = \frac{1}{3}\pi r^2 h$ for the volume of the cone obtained by revolving the triangle around the y-axis.

47. Apply the second theorem of Pappus to show that the lateral surface area of the cone of Problem 46 is $A = \pi r L$, where $L = \sqrt{r^2 + h^2}$ is the slant height of the cone.

48. (a) Find the centroid of the trapezoid shown in Fig. 14.5.22. (b) Apply the first theorem of Pappus and the result of part (a) to show that the volume of the conical frustum generated by revolving the trapezoid around the y-axis is

$$V = \frac{\pi h}{3}(r_1^2 + r_1 r_2 + r_2^2).$$

Fig. 14.5.22 The trapezoid of Problem 48

49. Apply the second theorem of Pappus to show that the lateral surface area of the conical frustum of Problem 48 is $a = \pi(r_1 + r_2)L$, where

$$L = \sqrt{(r_1 - r_2)^2 + h^2}$$

is its slant height.

50. (a) Apply the second theorem of Pappus to verify that the curved surface area of a right circular cylinder of height h and base radius r is $A = 2\pi rh$. (b) Explain how this follows also from the result of Problem 49.

51. (a) Find the centroid of the plane region shown in Fig. 14.5.23, which consists of a semicircular region of radius a sitting atop a rectangular region of width $2a$ and height b whose base is on the x-axis. (b) Then apply the

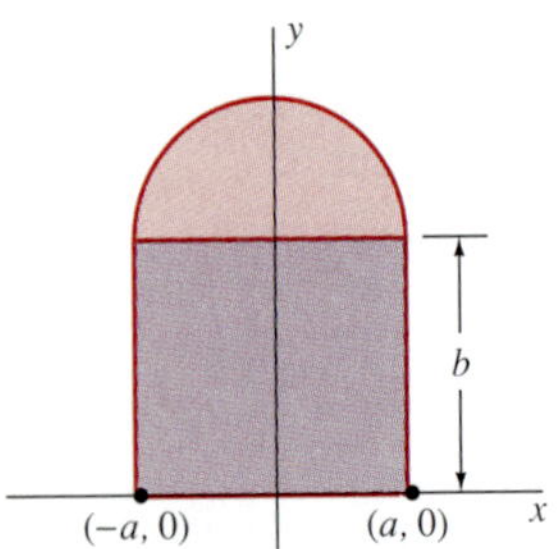

Fig. 14.5.23 The plane region of Problem 51(a)

first theorem of Pappus to find the volume generated by rotating this region around the x-axis.

52. (a) Consider the plane region of Fig. 14.5.24, bounded by $x^2 = 2py$, $x = 0$, and $y = h = r^2/2p$ $(p > 0)$. Show that its area is $A = \frac{2}{3}rh$ and that the x-coordinate of its centroid is $\bar{x} = \frac{3}{8}r$. (b) Use Pappus's theorem and the result of part (a) to show that the volume of a paraboloid of revolution with radius r and height h is $V = \frac{1}{2}\pi r^2 h$.

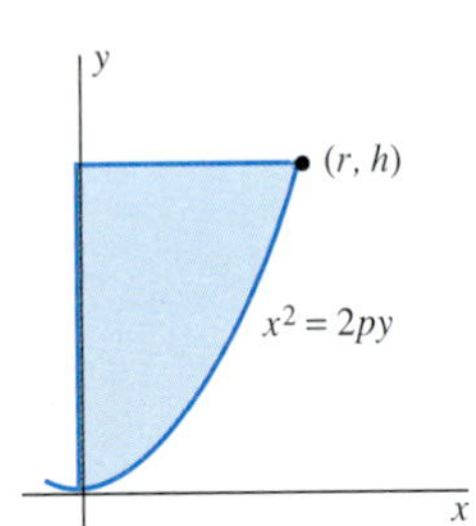

Fig. 14.5.24 The region of Problem 52

53. A uniform rectangular plate with base length a, height b, and mass m is centered at the origin. Show that its polar moment of inertia is $I_0 = \frac{1}{2}m(a^2 + b^2)$.

54. The centroid of a uniform plane region is at $(0, 0)$, and the region has total mass m. Show that its moment of inertia about an axis perpendicular to the xy-plane at the point (x_0, y_0) is

$$I = I_0 + m(x_0^2 + y_0^2).$$

55. Suppose that a plane lamina consists of two nonoverlapping laminae. Show that its polar moment of inertia is the sum of theirs. Use this fact together with the results of Problems 53 and 54 to find the polar moment of inertia of the T-shaped lamina of constant density $\delta = k > 0$ shown in Fig. 14.5.25.

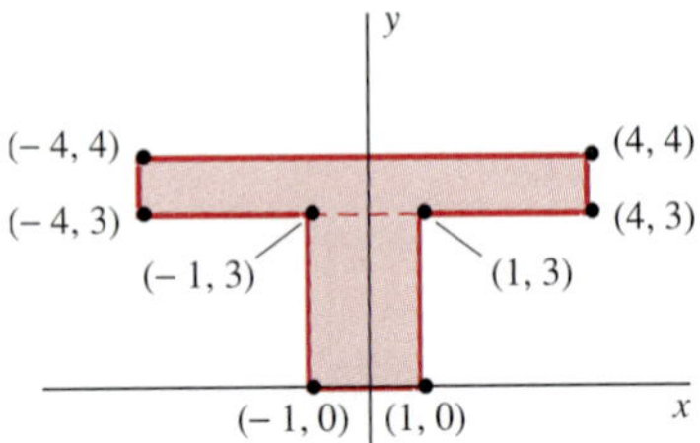

Fig. 14.5.25 One lamina made of two simpler ones (Problem 55)

56. A racquet consists of a uniform lamina that occupies the region inside the right-hand loop of $r^2 = \cos 2\theta$ on the end of a handle (assumed to be of negligible mass) corresponding to the interval $-1 \leqq x \leqq 0$ (Fig. 14.5.26). Find the radius of gyration of the racquet around the line $x = -1$. Where is its sweet spot?

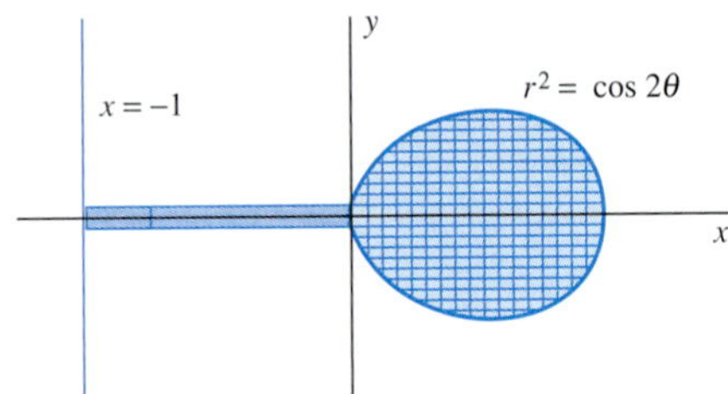

Fig. 14.5.26 The racquet of Problem 56

In Problems 57 through 60, find the mass m and centroid $(\bar{x}, \bar{y})$ of the indicated plane lamina R. You may use either a computer algebra system or the sine-cosine integrals of Formula (113) on the inside back cover.

57. R is bounded by the circle with polar equation $r = 2\sin\theta$ and has density function $\delta(x, y) = y$.

58. R is bounded by the circle with polar equation $r = 2\sin\theta$ and has density function $\delta(x, y) = y\sqrt{x^2 + y^2}$.

59. R is the semicircular disk bounded by the x-axis and the upper half of the circle with polar equation $r = 2\cos\theta$ and has density function $\delta(x, y) = x$.

60. R is the semicircular disk bounded by the x-axis and the upper half of the circle with polar equation $r = 2\cos\theta$ and has density function $\delta(x, y) = x^2y^2$.

14.5 PROJECT: OPTIMAL DESIGN OF DOWNHILL RACE CAR WHEELS

To see moments of inertia in action, suppose that your club is designing an unpowered race car for the annual downhill derby. You have a choice of solid wheels, bicycle wheels with thin spokes, or even solid spherical wheels (like giant ball bearings). Which wheels will make the race car go the fastest?

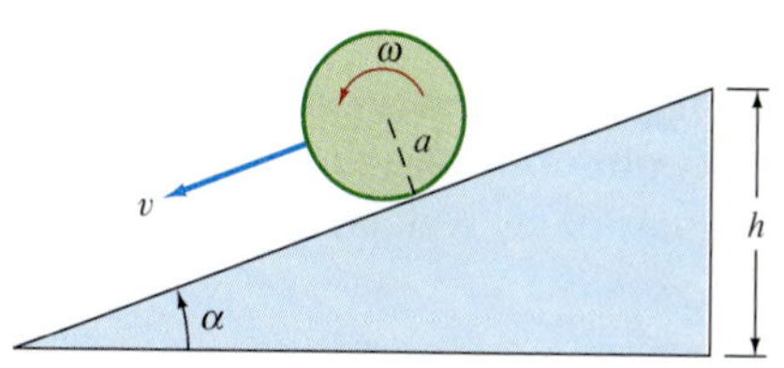

Fig. 14.5.27 A circular object rolling down an incline

Imagine an experiment in which you roll various types of wheels down an incline to see which reaches the bottom the fastest (Fig. 14.5.27). Suppose that a wheel of radius a and mass M starts from rest at the top with potential energy PE $= Mgh$ and reaches the bottom with angular speed ω and (linear) velocity $v = a\omega$. Then (by conservation of energy) the wheel's initial potential energy has been transformed into a sum $\text{KE}_{\text{tr}} + \text{KE}_{\text{rot}}$ of translational kinetic energy $\text{KE}_{\text{tr}} = \frac{1}{2}Mv^2$ and rotational kinetic energy

$$\text{KE}_{\text{rot}} = \frac{1}{2}I_0\omega^2 = \frac{I_0v^2}{2a^2}, \tag{1}$$

a consequence of Eq. (9) of this section. Thus

$$Mgh = \frac{1}{2}Mv^2 + \frac{I_0v^2}{2a^2}. \tag{2}$$

Problems 1 through 8 explore the implications of this formula.

1. Suppose that the wheel's (polar) moment of inertia is given by

$$I_0 = kMa^2 \tag{3}$$

for some constant k. (For instance, Example 10 gives $k = \frac{1}{2}$ for a wheel in the shape of a uniform solid disk.) Then deduce from Eq. (2) that

$$v = \sqrt{\frac{2gh}{1+k}}. \tag{4}$$

Thus the smaller k is (and hence the smaller the wheel's moment of inertia), the faster the wheel will roll down the incline.

In Problems 2 through 8, take $g = 32$ ft/s^2 and assume that the vertical height of the incline is $h = 100$ ft.

2. Why does it follow from Eq. (4) that, whatever the wheel's design, the maximum velocity a circular wheel can attain on this incline is 80 ft/s (just under 55 mi/h)?

3. If the wheel is a uniform solid disk (like a medieval wooden wagon wheel) with $I_0 = \frac{1}{2}Ma^2$, what is its speed v at the bottom of the incline?

4. Answer Problem 3 if the wheel is shaped like a narrow bicycle tire, with its entire mass, in effect, concentrated at the distance a from its center. In this case, $I_0 = Ma^2$. (Why?)

5. Answer Problem 3 if the wheel is shaped like an annular ring (or washer) with outer radius a and inner radius b.

Do not attempt Problems 6 through 8 until you have studied Example 3 of Section 14.7. In Problems 6 through 8, what is the velocity of the wheel when it reaches the bottom of the incline?

6. The wheel is a uniform solid sphere of radius a.

7. The wheel is a very thin, spherical shell whose entire mass is, in effect, concentrated at the distance a from its center.

8. The wheel is a spherical shell with outer radius a and inner radius $b = \frac{1}{2}a$.

Finally, what is your conclusion? What is the shape of the wheels that will yield the fastest downhill race car?

14.6 TRIPLE INTEGRALS

The definition of the triple integral is the three-dimensional version of the definition of the double integral of Section 14.2. Let $f(x, y, z)$ be continuous on the bounded space region T, and suppose that T lies inside the rectangular block R determined by the inequalities

$$a \leqq x \leqq b, \quad c \leqq y \leqq d, \quad \text{and} \quad p \leqq z \leqq q.$$

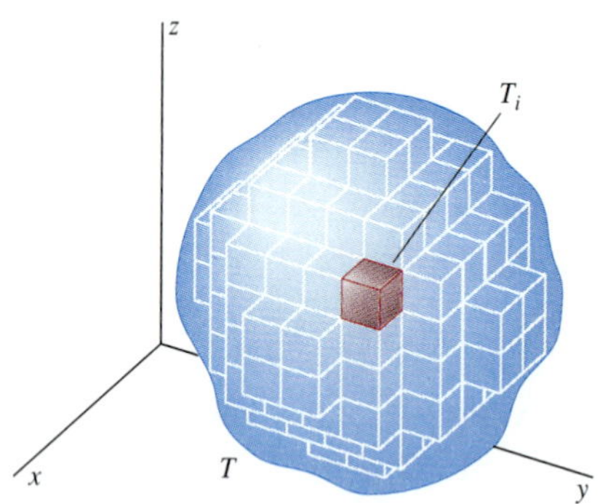

Fig. 14.6.1 One small block in an inner partition of the bounded space region T

We divide $[a, b]$ into subintervals of equal length Δx, $[c, d]$ into subintervals of equal length Δy, and $[p, q]$ into subintervals of equal length Δz. This generates a partition of R into smaller rectangular blocks (as in Fig. 14.6.1), each of volume $\Delta V = \Delta x\, \Delta y\, \Delta z$. Let $\mathcal{P} = \{T_1, T_2, \dots, T_n\}$ be the collection of these smaller blocks that lie wholly within T. Then $\mathcal{P}$ is called an **inner partition** of the region T. The **norm** $|\mathcal{P}|$ of $\mathcal{P}$ is the length of a longest diagonal of any of the blocks T_i. If $(x_i^\star, y_i^\star, z_i^\star)$ is an arbitrarily selected point of T_i (for each $i = 1, 2, \dots, n$), then the **Riemann sum**

$$\sum_{i=1}^{n} f(x_i^\star, y_i^\star, z_i^\star)\, \Delta V$$

is an approximation to the triple integral of f over the region T.

For example, if T is a solid body with density function f, then such a Riemann sum approximates its total mass. We define the **triple integral of f over T** by means of the equation

$$\iiint_T f(x, y, z)\, dV = \lim_{|\mathcal{P}| \to 0} \sum_{i=1}^{n} f(x_i^\star, y_i^\star, z_i^\star)\, \Delta V. \qquad \textbf{(1)}$$

It is proved in advanced calculus that this limit of Riemann sums exists as the norm $|\mathcal{P}|$ approaches zero provided that f is continuous on T and that the boundary of the region T is reasonably well behaved. (For instance, it suffices for the boundary of T to be *piecewise smooth*, consisting of a finite number of smooth surfaces.)

Just as with double integrals, we ordinarily compute triple integrals by means of iterated integrals. If the region of integration is a rectangular block, as in Example 1, then we can integrate in any order we wish.

EXAMPLE 1 If $f(x, y, z) = xy + yz$ and T consists of those points (x, y, z) in space that satisfy the inequalities

$$-1 \leqq x \leqq 1, \quad 2 \leqq y \leqq 3, \quad \text{and} \quad 0 \leqq z \leqq 1$$

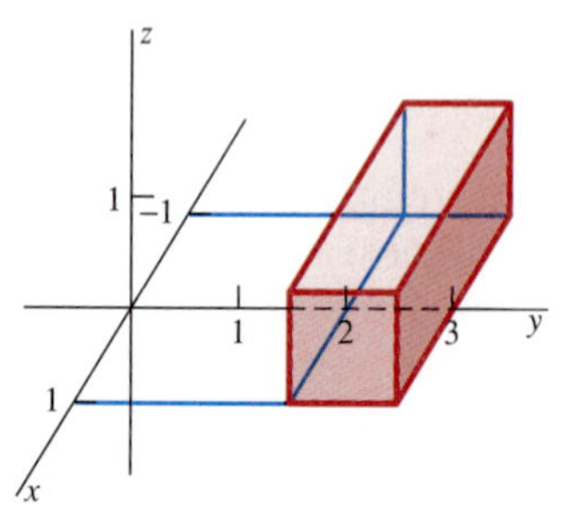

Fig. 14.6.2 The rectangular block T of Example 1, for which $-1 \leqq x \leqq 1$, $2 \leqq y \leqq 3$, and $0 \leqq z \leqq 1$

(Fig. 14.6.2), then

$$\begin{aligned}
\iiint_T f(x, y, z)\, dV &= \int_{-1}^{1}\int_{2}^{3}\int_{0}^{1} (xy + xz)\, dz\, dy\, dx \\
&= \int_{-1}^{1}\int_{2}^{3} \left[xyz + \frac{1}{2}yz^2 \right]_{z=0}^{1} dy\, dx \\
&= \int_{-1}^{1}\int_{2}^{3} \left(xy + \frac{1}{2}y \right) dy\, dx \\
&= \int_{-1}^{1} \left[\frac{1}{2}xy^2 + \frac{1}{4}y^2 \right]_{y=2}^{3} dx
\end{aligned}$$

$$= \int_{-1}^{1} \left(\frac{5}{2}x + \frac{5}{4} \right) dx = \left[\frac{5}{4}x^2 + \frac{5}{4}x \right]_{-1}^{1} = \frac{5}{2}.$$ ■

The applications of double integrals that we saw in earlier sections generalize immediately to triple integrals. If T is a solid body with the density function $\delta(x, y, z)$, then its **mass** m is given by

$$m = \iiint_T \delta \, dV. \tag{2}$$

The case $\delta \equiv 1$ gives the **volume**

$$V = \iiint_T dV \tag{3}$$

of T. The coordinates of its **centroid** are

$$\bar{x} = \frac{1}{m} \iiint_T x\delta \, dV, \tag{4a}$$

$$\bar{y} = \frac{1}{m} \iiint_T y\delta \, dV, \quad \text{and} \tag{4b}$$

$$\bar{z} = \frac{1}{m} \iiint_T z\delta \, dV. \tag{4c}$$

The **moments of inertia** of T around the three coordinate axes are

$$I_x = \iiint_T (y^2 + z^2)\delta \, dV, \tag{5a}$$

$$I_y = \iiint_T (x^2 + z^2)\delta \, dV, \quad \text{and} \tag{5b}$$

$$I_z = \iiint_T (x^2 + y^2)\delta \, dV. \tag{5c}$$

Iterated Triple Integrals

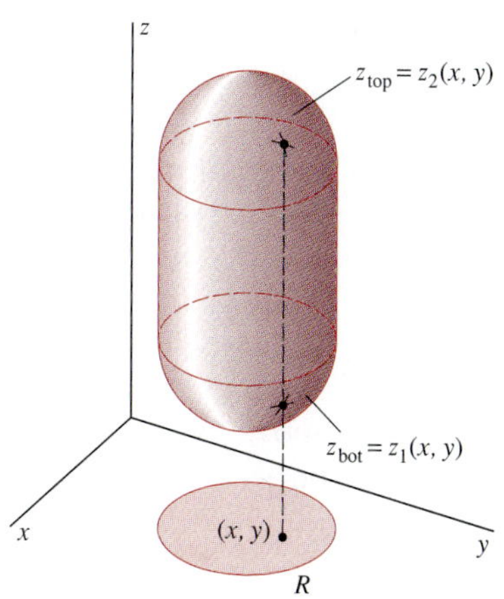

Fig. 14.6.3 Obtaining the limits of integration for z

As indicated previously, we almost always evaluate triple integrals by iterated single integration. Suppose that the region T with piecewise smooth boundary is **z-simple:** Each line parallel to the z-axis intersects T (if at all) in a single line segment. In effect, this means that T can be described by the inequalities

$$z_1(x, y) \leqq z \leqq z_2(x, y), \qquad (x, y) \text{ in } R,$$

where R is the vertical projection of T into the xy-plane. Then

$$\iiint_T f(x, y, z) \, dV = \iint_R \left(\int_{z_1(x, y)}^{z_2(x, y)} f(x, y, z) \, dz \right) dA. \tag{6}$$

In Eq. (6), we take $dA = dx\,dy$ or $dA = dy\,dx$, depending on the preferred order of integration over the set R. The limits $z_1(x, y)$ and $z_2(x, y)$ are the z-coordinates of the endpoints of the line segment in which the vertical line at (x, y) meets T (Fig. 14.6.3).

If the region R has the description

$$y_1(x) \leqq y \leqq y_2(x), \qquad a \leqq x \leqq b,$$

then (integrating last with respect to x),

$$\iiint_T f(x, y, z)\, dV = \int_a^b \int_{y_1(x)}^{y_2(x)} \int_{z_1(x,y)}^{z_2(x,y)} f(x, y, z)\, dz\, dy\, dx.$$

Thus the triple integral reduces in this case to three iterated single integrals. These can (in principle) be evaluated by using the fundamental theorem of calculus.

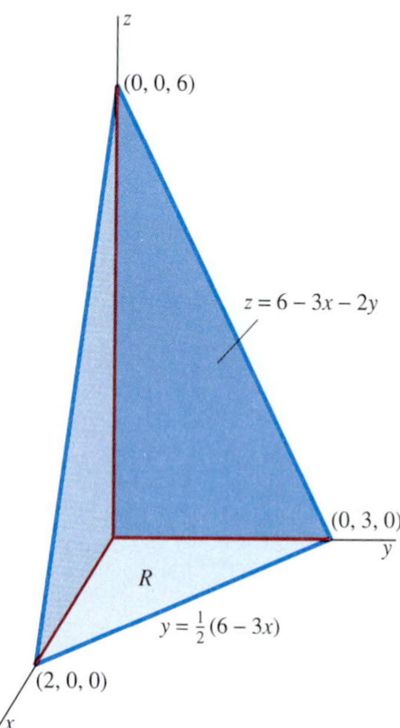

Fig. 14.6.4 The pyramid T of Example 2; its base is the triangle R in the xy-plane.

EXAMPLE 2 Find the mass m of the pyramid T of Fig. 14.6.4 if its density function is given by $\delta(x, y, z) = z$.

Solution The region T is bounded below by the xy-plane $z = 0$ and above by the plane $z = 6 - 3x - 2y$. Its base is the plane region R bounded by the x- and y-axes and the line $y = \frac{1}{2}(6 - 3x)$. Hence Eqs. (2) and (6) yield

$$\begin{aligned}
m &= \int_0^2 \int_0^{(6-3x)/2} \int_0^{6-3x-2y} z\, dz\, dy\, dx = \int_0^2 \int_0^{(6-3x)/2} \left[\frac{1}{2}z^2\right]_{z=0}^{6-3x-2y} dy\, dx \\
&= \frac{1}{2}\int_0^2 \int_0^{(6-3x)/2} (6 - 3x - 2y)^2\, dy\, dx = \frac{1}{2}\int_0^2 \left[-\frac{1}{6}(6 - 3x - 2y)^3\right]_{y=0}^{(6-3x)/2} dx \\
&= \frac{1}{12}\int_0^2 (6 - 3x)^3\, dx = \frac{1}{12}\left[-\frac{1}{12}(6 - 3x)^4\right]_{x=0}^{2} = \frac{6^4}{12^2} = 9.
\end{aligned}$$

We leave as an exercise (Problem 45) to show that the coordinates of the centroid $(\bar{x}, \bar{y}, \bar{z})$ of the pyramid are given by

$$\bar{x} = \frac{1}{9}\int_0^2 \int_0^{(6-3x)/2} \int_0^{6-3x-2y} xz\, dz\, dy\, dx = \frac{2}{5},$$

$$\bar{y} = \frac{1}{9}\int_0^2 \int_0^{(6-3x)/2} \int_0^{6-3x-2y} yz\, dz\, dy\, dx = \frac{3}{5},$$

$$\bar{z} = \frac{1}{9}\int_0^2 \int_0^{(6-3x)/2} \int_0^{6-3x-2y} z^2\, dz\, dy\, dx = \frac{12}{5}.$$

■

If the solid T is bounded by the *two* surfaces $z = z_1(x, y)$ and $z = z_2(x, y)$ (as in Fig. 14.6.5), then we can find the "base region" R in Eq. (6) as follows. Note that the equation $z_1(x, y) = z_2(x, y)$ determines a vertical cylinder (not necessarily circular) that passes through the curve of intersection of the two surfaces. (Why?) This cylinder intersects the xy-plane in the boundary curve C of the plane region R. In essence, we obtain the equation of the curve C by equating the height functions of the surfaces that form the top and bottom of the space region T.

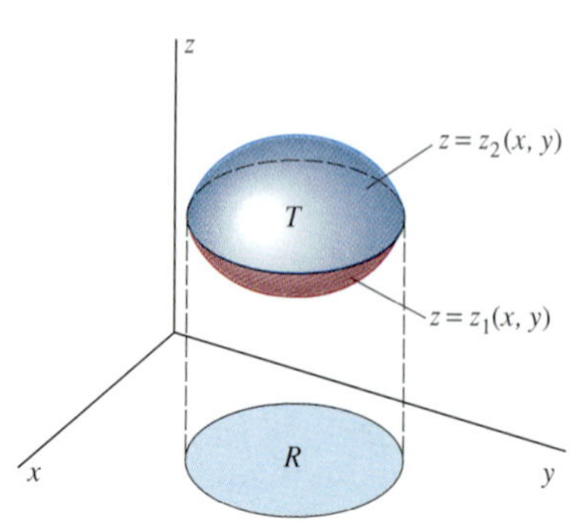

Fig. 14.6.5 To find the boundary of R, solve the equation $z_1(x, y) = z_2(x, y)$.

EXAMPLE 3 Figure 14.6.6 shows the solid T that is bounded above by the plane $z = y + 2$ and below by the paraboloid $z = x^2 + y^2$. The equation

$$x^2 + y^2 = y + 2; \quad \text{that is,} \quad x^2 + \left(y - \tfrac{1}{2}\right)^2 = \tfrac{9}{4}$$

describes the boundary circle of the disk R of radius $\frac{3}{2}$ and with center $\left(0, \frac{1}{2}\right)$ in the xy-plane (Fig. 14.6.7). Because this disk is not centered at the origin, the volume integral

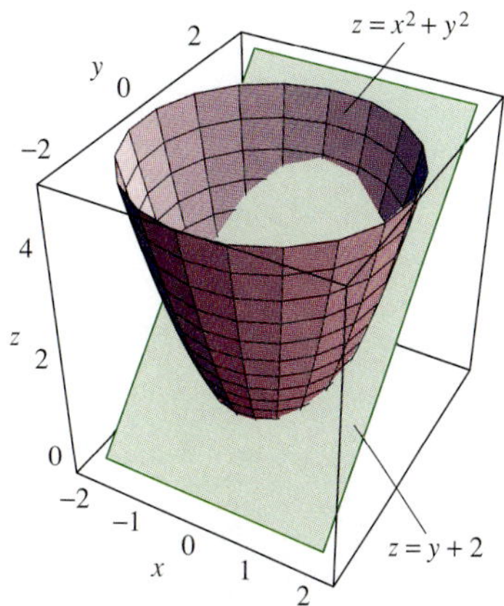

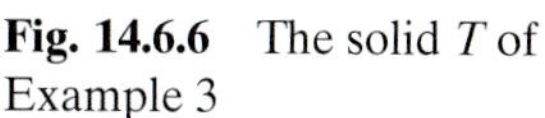

Fig. 14.6.6 The solid T of Example 3

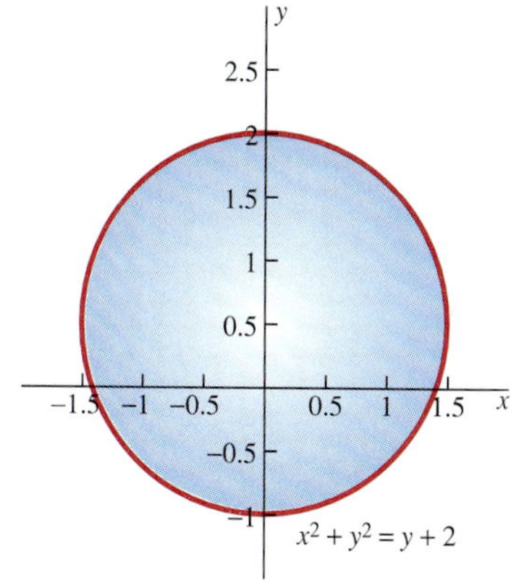

Fig. 14.6.7 The circular disk R of Example 3

$$V = \iint_R \left(\int_{z=x^2+y^2}^{y+2} dz \right) dA$$

is awkward to evaluate directly. In Example 5 we calculate V by integrating in a different order. ■

We may integrate first with respect to either x or y if the space region T is either **x-simple** or **y-simple**. Such situations, as well as a z-simple solid, appear in Fig. 14.6.8. For example, suppose that T is y-simple, so that it has a description of the form

$$y_1(x, z) \leqq y \leqq y_2(x, z), \qquad (x, z) \text{ in } R,$$

where R is the projection of T into the xz-plane. Then

$$\iiint_T f(x, y, z)\, dV = \iint_R \left(\int_{y_1(x, z)}^{y_2(x, z)} f(x, y, z)\, dy \right) dA, \tag{7}$$

where $dA = dx\, dz$ or $dA = dz\, dx$ and the limits $y_1(x, z)$ and $y_2(x, z)$ are the y-coordinates of the endpoints of the line segment in which a typical line parallel to the y-axis intersects T. If T is x-simple, we have

$$\iiint_T f(x, y, z)\, dA = \iint_R \left(\int_{x_1(y, z)}^{x_2(y, z)} f(x, y, z)\, dx \right) dA, \tag{8}$$

where $dA = dy\, dz$ or $dA = dz\, dy$ and R is the projection of T into the yz-plane.

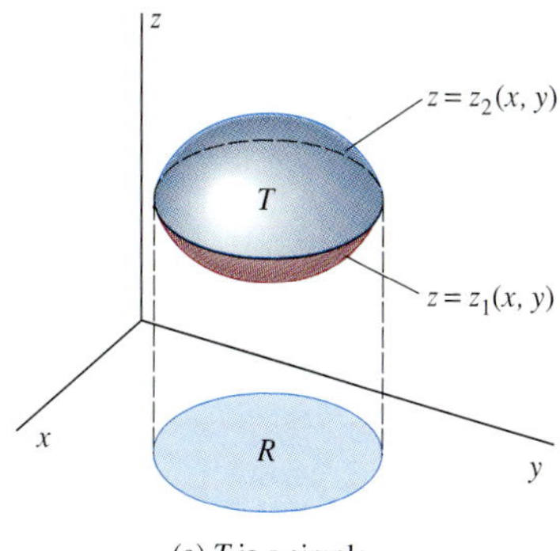

(a) T is z-simple

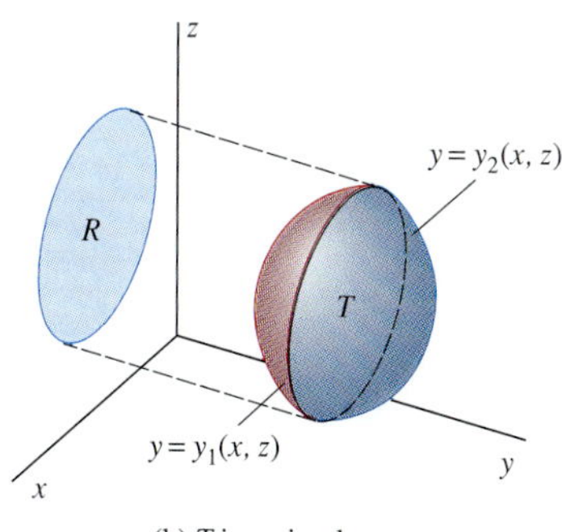

(b) T is y-simple

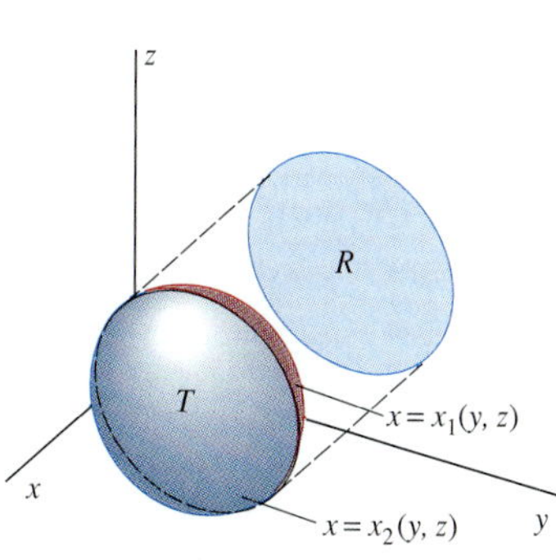

(c) T is x-simple

Fig. 14.6.8 Solids that are (a) z-simple, (b) y-simple, and (c) x-simple

EXAMPLE 4 Compute by triple integration the volume of the region T that is bounded by the parabolic cylinder $x = y^2$ and the planes $z = 0$ and $x + z = 1$. Also find the centroid of T given that it has constant density $\delta \equiv 1$.

Comment The three segments in Fig. 14.6.9 parallel to the coordinate axes indicate that the region T is simultaneously x-simple, y-simple, and z-simple. We may therefore integrate in any order we choose, so there are six ways to evaluate the integral. Here are three computations of the volume V of T.

Solution 1 The projection of T into the xy-plane is the region shown in Fig. 14.6.10, bounded by $x = y^2$ and $x = 1$. So Eq. (6) gives

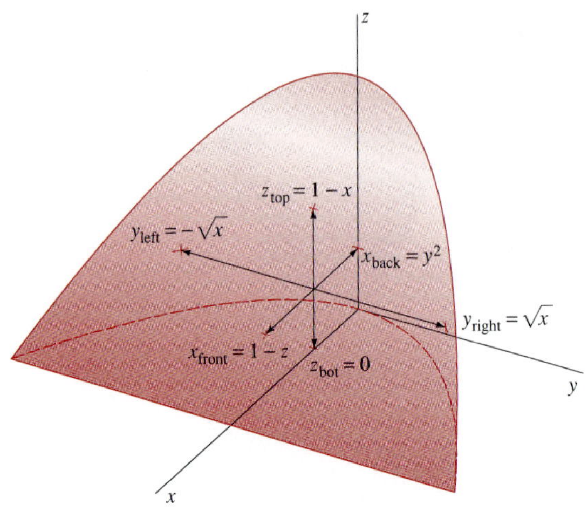

Fig. 14.6.9 The region T of Example 4 is x-simple, y-simple, and z-simple.

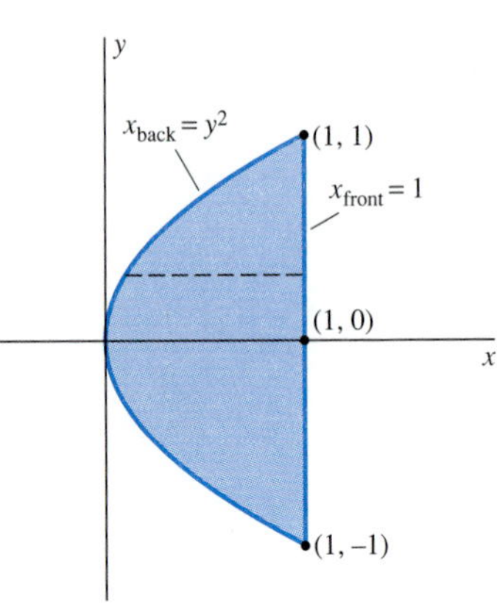

Fig. 14.6.10 The vertical projection of the solid region T into the xy-plane (Example 4, Solution 1)

$$V = \int_{-1}^{1}\int_{y^2}^{1}\int_{0}^{1-x} dz\,dx\,dy = 2\int_{0}^{1}\int_{y^2}^{1}(1-x)\,dx\,dy$$

$$= 2\int_{0}^{1}\left[x - \frac{1}{2}x^2\right]_{x=y^2}^{1} dy = 2\int_{0}^{1}\left(\frac{1}{2} - y^2 + \frac{1}{2}y^4\right)dy = \frac{8}{15}.$$

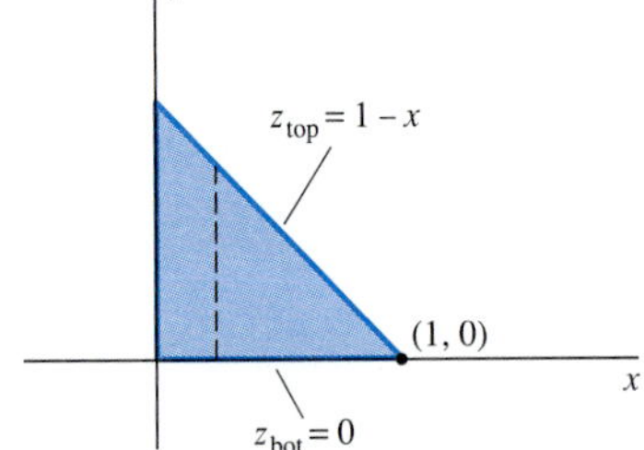

Fig. 14.6.11 The vertical projection of the solid region T into the xz-plane (Example 4, Solution 2)

Solution 2 The projection of T into the xz-plane is the triangle bounded by the coordinate axes and the line $x + z = 1$ (Fig. 14.6.11), so Eq. (7) gives

$$V = \int_{0}^{1}\int_{0}^{1-x}\int_{-\sqrt{x}}^{\sqrt{x}} dy\,dz\,dx = 2\int_{0}^{1}\int_{0}^{1-x}\sqrt{x}\,dz\,dx$$

$$= 2\int_{0}^{1}(x^{1/2} - x^{3/2})\,dx = \frac{8}{15}.$$

Solution 3 The projection of T into the yz-plane is the region bounded by the y-axis and the parabola $z = 1 - y^2$ (Fig. 14.6.12), so Eq. (8) yields

$$V = \int_{-1}^{1}\int_{0}^{1-y^2}\int_{y^2}^{1-z} dx\,dz\,dy,$$

and evaluation of this integral again gives $V = \frac{8}{15}$.

Now for the centroid of T. Because the region T is symmetric with respect to the xz-plane, its centroid lies in this plane, and so $\bar{y} = 0$. We compute $\bar{x}$ and $\bar{z}$ by integrating first with respect to y:

$$\bar{x} = \frac{1}{V}\iiint_T x\,dV = \frac{15}{8}\int_{0}^{1}\int_{0}^{1-x}\int_{-\sqrt{x}}^{\sqrt{x}} x\,dy\,dz\,dx$$

$$= \frac{15}{4}\int_{0}^{1}\int_{0}^{1-x} x^{3/2}\,dz\,dx = \frac{15}{4}\int_{0}^{1}(x^{3/2} - x^{5/2})\,dx = \frac{3}{7};$$

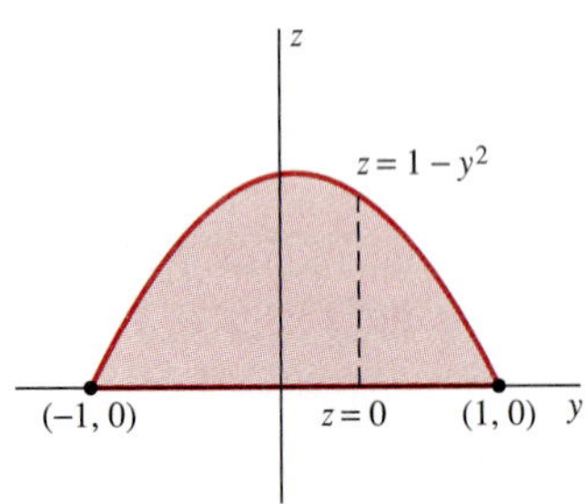

Fig. 14.6.12 The vertical projection of the solid region T into the yz-plane (Example 4, Solution 3)

similarly,

$$\bar{z} = \frac{1}{V}\iiint_R z\,dV = \frac{15}{8}\int_0^1\int_0^{1-x}\int_{-\sqrt{x}}^{\sqrt{x}} z\,dy\,dz\,dx = \frac{2}{7}.$$

Thus the centroid of T is located at the point $\left(\frac{3}{7}, 0, \frac{2}{7}\right)$. ■

EXAMPLE 5 Find the volume of the *oblique segment of a paraboloid* bounded by the paraboloid $z = x^2 + y^2$ and the plane $z = y + 2$ (Fig. 14.6.13).

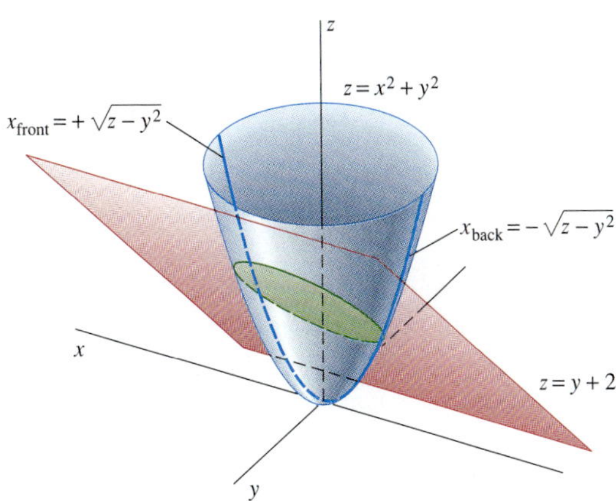

Fig. 14.6.13 An oblique segment of a paraboloid (Example 5)

Solution The given region T is z-simple, but its projection into the xy-plane is bounded by the graph of the equation $x^2 + y^2 = y + 2$, which is a translated circle. It would be possible to integrate first with respect to z, but perhaps another choice will yield a simpler integral.

The region T is also x-simple, so we may integrate first with respect to x. The projection of T into the yz-plane is bounded by the line $z = y + 2$ and the parabola $z = y^2$, which intersect at the points $(-1, 1)$ and $(2, 4)$ (Fig. 14.6.14). The endpoints of a line segment in T parallel to the x-axis have x-coordinates $x = \pm\sqrt{z - y^2}$. Because T is symmetric with respect to the yz-plane, we can integrate from $x = 0$ to $x = \sqrt{z - y^2}$ and double the result. Hence T has volume

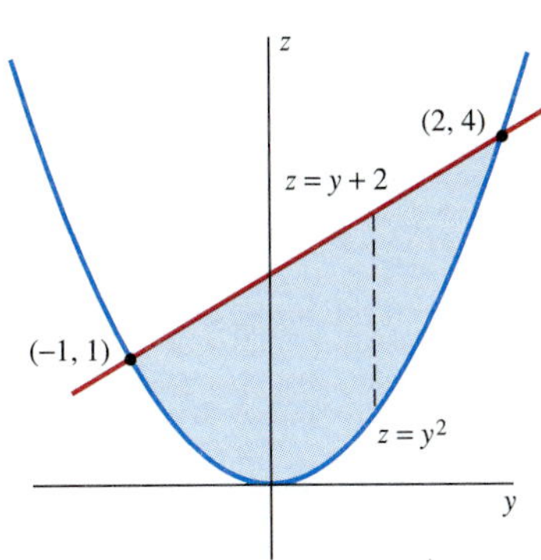

Fig. 14.6.14 Projection of the segment of the paraboloid into the yz-plane (Example 5)

$$\begin{aligned}
V &= 2\int_{-1}^{2}\int_{y^2}^{y+2}\int_0^{\sqrt{z-y^2}} dx\,dz\,dy = 2\int_{-1}^{2}\int_{y^2}^{y+2}\sqrt{z - y^2}\,dz\,dy \\
&= 2\int_{-1}^{2}\left[\frac{2}{3}\left(z - y^2\right)^{3/2}\right]_{z=y^2}^{y+2} dy = \frac{4}{3}\int_{-1}^{2}(2 + y - y^2)^{3/2}\,dy \\
&= \frac{4}{3}\int_{-3/2}^{3/2}\left(\frac{9}{4} - u^2\right)^{3/2} du \qquad \left(\text{completing the square; } u = y - \frac{1}{2}\right) \\
&= \frac{27}{4}\int_{-\pi/2}^{\pi/2}\cos^4\theta\,d\theta \qquad \left(u = \frac{3}{2}\sin\theta\right) \\
&= \frac{27}{4}\cdot 2\cdot\frac{1}{2}\cdot\frac{3}{4}\cdot\frac{\pi}{2} = \frac{81\pi}{32}.
\end{aligned}$$

In the final evaluation, we used Formula (113) (on the inside back cover). ■

14.6 PROBLEMS

In Problems 1 through 10, compute the value of the triple integral

$$\iiint_T f(x, y, z)\, dV.$$

1. $f(x, y, z) = x + y + z$; T is the rectangular box $0 \leqq x \leqq 2, 0 \leqq y \leqq 3, 0 \leqq z \leqq 1$.
2. $f(x, y, z) = xy \sin z$; T is the cube $0 \leqq x \leqq \pi$, $0 \leqq y \leqq \pi, 0 \leqq z \leqq \pi$.
3. $f(x, y, z) = xyz$; T is the rectangular block $-1 \leqq x \leqq 3$, $0 \leqq y \leqq 2, -2 \leqq z \leqq 6$.
4. $f(x, y, z) = x + y + z$; T is the rectangular block of Problem 3.
5. $f(x, y, z) = x^2$; T is the tetrahedron bounded by the coordinate planes and the first octant part of the plane with equation $x + y + z = 1$.
6. $f(x, y, z) = 2x + 3y$; T is a first-octant tetrahedron as in Problem 5, except that the plane has equation $2x + 3y + z = 6$.
7. $f(x, y, z) = xyz$; T lies below the surface $z = 1 - x^2$ and above the rectangle $-1 \leqq x \leqq 1$, $0 \leqq y \leqq 2$ in the xy-plane.
8. $f(x, y, z) = 2y + z$; T lies below the surface with equation $z = 4 - y^2$ and above the rectangle $-1 \leqq x \leqq 1$, $-2 \leqq y \leqq 2$ in the xy-plane.
9. $f(x, y, z) = x + y$; T is the region between the surfaces $z = 2 - x^2$ and $z = x^2$ for $0 \leqq y \leqq 3$ (Fig. 14.6.15).

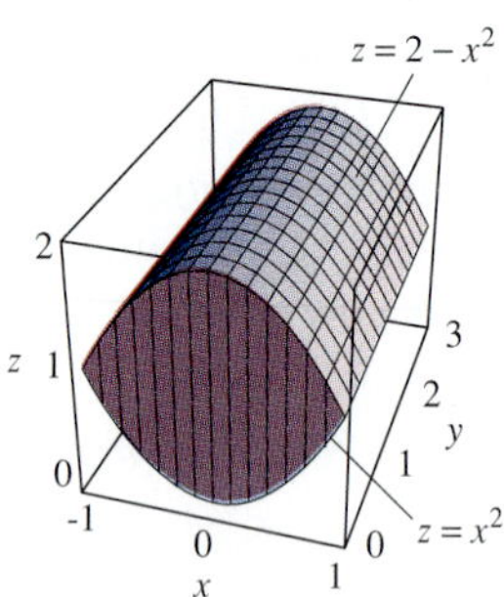

Fig. 14.6.15 The solid of Problem 9

10. $f(x, y, z) = z$; T is the region between the surfaces $z = y^2$ and $z = 8 - y^2$ for $-1 \leqq x \leqq 1$.

In Problems 11 through 20, sketch the solid bounded by the graphs of the given equations. Then find its volume by triple integration.

11. $2x + 3y + z = 6, \quad x = 0, \quad y = 0, \quad z = 0$
12. $z = y, \quad y = x^2, \quad y = 4, \quad z = 0$ (Fig. 14.6.16)
13. $y + z = 4, \quad y = 4 - x^2, \quad y = 0, \quad z = 0$

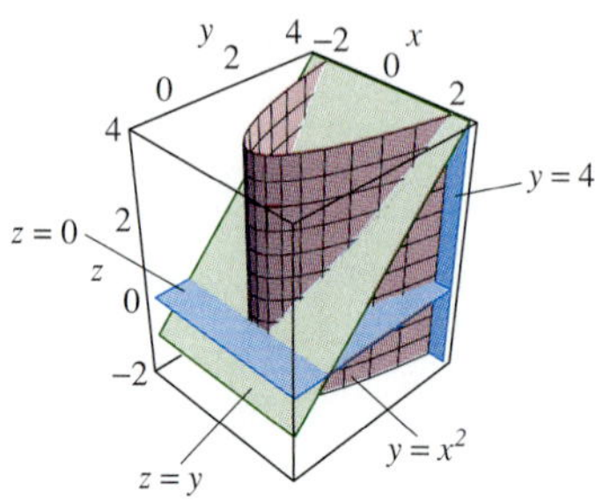

Fig. 14.6.16 The surfaces of Problem 12

14. $z = x^2 + y^2, \quad z = 0, \quad x = 0, \quad y = 0, \quad x + y = 1$
15. $z = 10 - x^2 - y^2, \quad y = x^2, \quad x = y^2, \quad z = 0$
16. $x = z^2, \quad x = 8 - z^2, \quad y = -1, \quad y = -3$
17. $z = x^2, \quad y + z = 4, \quad y = 0, \quad z = 0$
18. $z = 1 - y^2, \quad z = y^2 - 1, \quad x + z = 1, \quad x = 0$ (Fig. 14.6.17)

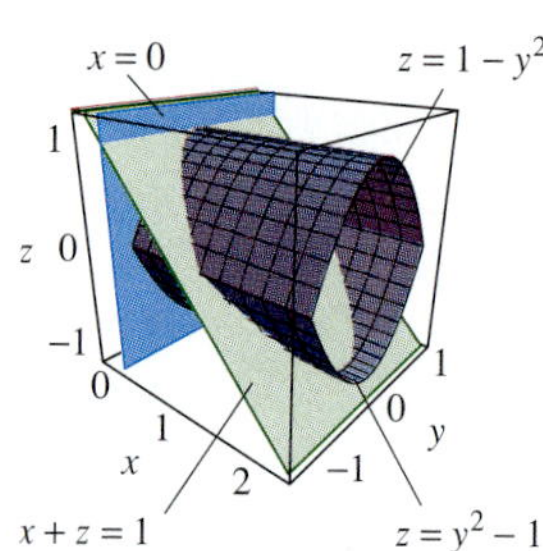

Fig. 14.6.17 The surfaces of Problem 18

19. $y = z^2, \quad z = y^2, \quad x + y + z = 2, \quad x = 0$
20. $y = 4 - x^2 - z^2, \quad x = 0, \quad y = 0, \quad z = 0, \quad x + z = 2$

In Problems 21 through 32, assume that the indicated solid has constant density $\delta \equiv 1$.

21. Find the centroid of the solid of Problem 12.
22. Find the centroid of the hemisphere

$$x^2 + y^2 + z^2 \leqq R^2, \qquad z \geqq 0.$$

23. Find the centroid of the solid of Problem 17.
24. Find the centroid of the solid bounded by $z = 1 - x^2$, $z = 0, y = -1$, and $y = 1$.
25. Find the centroid of the solid bounded by $z = \cos x$, $x = -\pi/2, x = \pi/2, y = 0, z = 0$, and $y + z = 1$.
26. Find the moment of inertia around the z-axis of the solid of Problem 12.
27. Find the moment of inertia around the y-axis of the solid of Problem 24.

28. Find the moment of inertia around the z-axis of the solid cylinder $x^2 + y^2 \leqq R^2, 0 \leqq z \leqq H$.

29. Find the moment of inertia around the z-axis of the solid bounded by $x + y + z = 1, x = 0, y = 0$, and $z = 0$.

30. Find the moment of inertia around the z-axis of the cube with vertices $\left(\pm\frac{1}{2}, 3, \pm\frac{1}{2}\right)$ and $\left(\pm\frac{1}{2}, 4, \pm\frac{1}{2}\right)$.

31. Consider the solid paraboloid bounded by $z = x^2 + y^2$ and the plane $z = h > 0$. Show that its centroid lies on its axis of symmetry, two-thirds of the way from its "vertex" $(0, 0, 0)$ to its base.

32. Show that the centroid of a right circular cone lies on the axis of the cone and three-fourths of the way from the vertex to the base.

In Problems 33 through 40, the indicated solid has uniform density $\delta \equiv 1$ unless otherwise indicated.

33. For a cube with edge length a, find the moment of inertia around one of its edges.

34. The density at $P(x, y, z)$ of the first-octant cube with edge length a, faces parallel to the coordinate planes, and opposite vertices $(0, 0, 0)$ and (a, a, a) is proportional to the square of the distance from P to the origin. Find the coordinates of the centroid of this cube.

35. Find the moment of inertia around the z-axis of the cube of Problem 34.

36. The cube bounded by the coordinate planes and the planes $x = 1, y = 1$, and $z = 1$ has density $\delta = kz$ at the point $P(x, y, z)$ (k is a positive constant). Find its centroid.

37. Find the moment of inertia around the z-axis of the cube of Problem 36.

38. Find the moment of inertia around a diameter of a solid sphere of radius a.

39. Find the centroid of the first-octant region that is interior to the two cylinders $x^2 + z^2 = 1$ and $y^2 + z^2 = 1$ (Figs. 14.6.18 and 14.6.19).

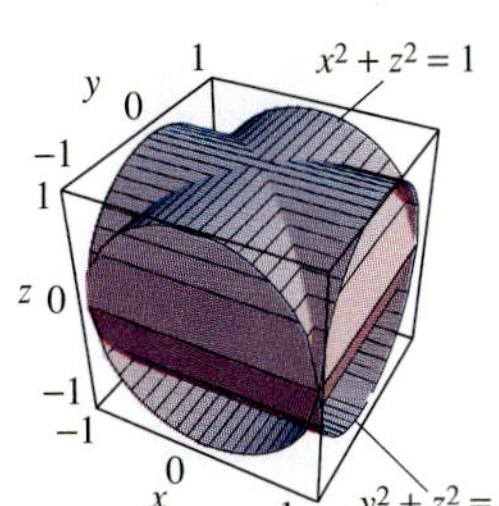

Fig. 14.6.18 The intersecting cylinders of Problem 39

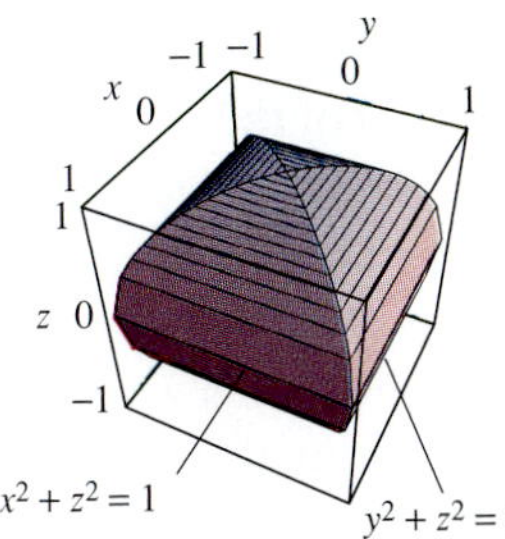

Fig. 14.6.19 The solid of intersection in Problem 39

40. Find the moment of inertia around the z-axis of the solid of Problem 39.

41. Find the volume bounded by the elliptic paraboloids $z = 2x^2 + y^2$ and $z = 12 - x^2 - 2y^2$. Note that this solid projects onto a circular disk in the xy-plane.

42. Find the volume bounded by the elliptic paraboloid $y = x^2 + 4z^2$ and the plane $y = 2x + 3$.

43. Find the volume of the elliptical cone bounded by $z = \sqrt{x^2 + 4y^2}$ and the plane $z = 1$. (*Suggestion:* Integrate first with respect to x.)

44. Find the volume of the region bounded by the paraboloid $x = y^2 + 2z^2$ and the parabolic cylinder $x = 2 - y^2$ (Fig. 14.6.20).

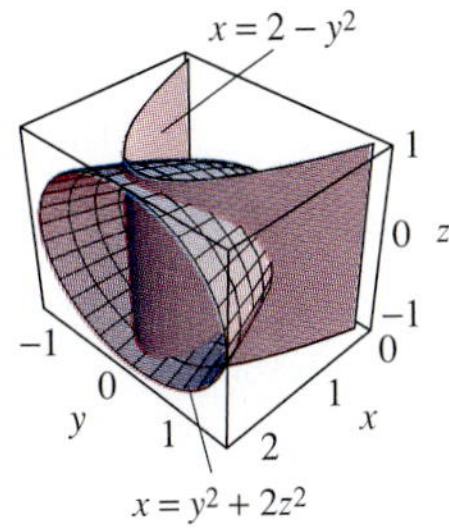

Fig. 14.6.20 The surfaces of Problem 44

45. Find the centroid of the pyramid in Example 2 with density $\delta(x, y, z) = z$.

46. Find the centroid of the parabolic segment (with density $\delta \equiv 1$) in Example 5.

For Problems 47 through 53, the **average value** $\bar{f}$ of the function $f(x, y, z)$ at points of the space region T is defined to be

$$\bar{f} = \frac{1}{V}\iiint_T f(x, y, z)\, dV$$

where V is the volume of T. For instance, if T is a solid with density $\delta \equiv 1$, then the coordinates $\bar{x}, \bar{y}$, and $\bar{z}$ of its centroid are the average values of the "coordinate functions" x, y, and z at points of T.

47. Find the average value of the density function $\delta(x, y, z) = z$ at points of the pyramid T of Example 2.

48. Suppose that T is the unit cube in the first octant with diagonally opposite vertices $(0, 0, 0)$ and $(1, 1, 1)$. Find the average of the "squared distance" $f(x, y, z) = x^2 + y^2 + z^2$ of points of T from the origin.

49. Let T be the cube of Problem 48. Find the average squared distance of points of T from its centroid.

50. Let T be the cube of Problem 48, but with density function $\delta(x, y, z) = x + y + z$ that varies linearly from 0 at the origin to 3 at the opposite vertex of T. Find the average value $\bar{\delta}$ of the density of T. Can you guess the value of $\bar{\delta}$ before evaluating the triple integral?

51. Find the average squared distance from the origin of points of the pyramid of Example 2.

52. Suppose that T is the pyramid of Example 2, but with density function $\delta \equiv 1$. Find the average squared distance of points of T from its centroid.

53. Find the average distance of points of the cube T of Problem 48 from the origin. *Warning:* Do not attempt this problem by hand; try it only if you have a computer algebra system. *Answer:*

$$\tfrac{1}{24}\left[6\sqrt{3} - \pi + 8\ln\left(\sqrt{3} + \tfrac{1}{2}\sqrt{2}\right) - 8\ln 2 + 16\ln\left(1 + \sqrt{3}\right)\right] \approx 0.960591956455.$$

14.6 PROJECT: ARCHIMEDES' FLOATING PARABOLOID

Archimedes was interested in floating bodies, and he studied the possible positions (see Fig. 14.6.21) of a floating right circular paraboloid of uniform density. For a paraboloid that floats in an "inclined position," he discovered how to determine its angle of inclination in terms of the volume and centroid of the "oblique segment" of the paraboloid that lies beneath the water line. The principles he introduced for this investigation (over 22 centuries ago) are still important in modern naval architecture.

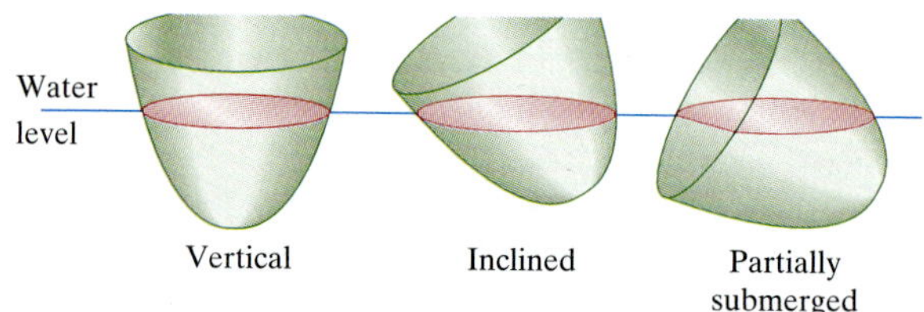

Fig. 14.6.21 How a uniform solid paraboloid might float

For your own personal paraboloid to investigate, let T be the three-dimensional solid region that is bounded below by the paraboloid $z = x^2 + y^2$ and above by the plane $z = (b - a)y + ab$, where a and b are the smallest and largest nonzero digits (respectively) of your student I.D. number. (If $a = 1$ and $b = 2$, then T is the solid of Example 5.) In the following problems you can evaluate the triple integrals either by hand—consulting an integral table if you wish—or by using a computer algebra system.

1. Find the volume V of the solid oblique paraboloid T. Sketch a picture of T similar to Fig. 14.6.13. Can you see that T is symmetric with respect to the yz-plane? Describe the region R in the yz-plane that is the horizontal projection of T. This plane region will determine the z-limits and the y-limits of your triple integral (as in Example 5).
2. Find the coordinates $(\bar{x}, \bar{y}, \bar{z})$ of the centroid C of T (assume that T has density $\delta \equiv 1$).
3. Find the coordinates of the point P at which a plane parallel to the original top plane $z = (b - a)y + ab$ is tangent to the paraboloid. Also find the coordinates of the point Q in which a vertical line through P intersects the top plane. According to Archimedes, the centroid C of Problem 2 should lie on the line PQ two-thirds of the way from P to Q. Is this so, according to your computations? (Compare with Problem 31 of this section.)

14.7 INTEGRATION IN CYLINDRICAL AND SPHERICAL COORDINATES

Suppose that $f(x, y, z)$ is a continuous function defined on the z-simple region T, which—because it is z-simple—can be described by

$$z_1(x, y) \leqq z \leqq z_2(x, y) \qquad \text{for} \quad (x, y) \text{ in } R$$

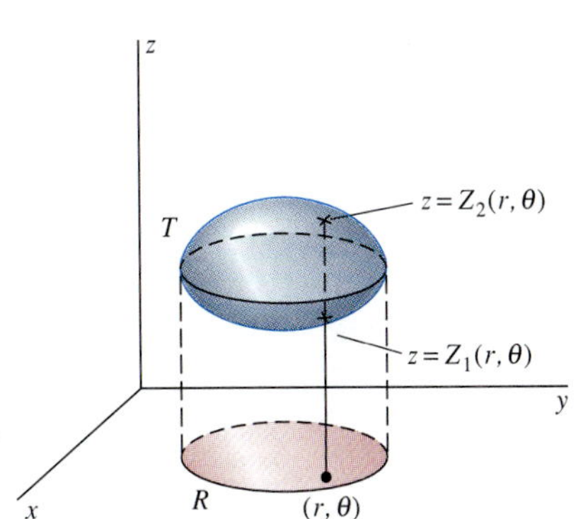

Fig. 14.7.1 The limits on z in a triple integral in cylindrical coordinates are determined by the lower and upper surfaces.

(where R is the projection of T into the xy-plane, as usual). We saw in Section 14.6 that

$$\iiint_T f(x, y, z)\,dV = \iint_R \left(\int_{z_1(x,y)}^{z_2(x,y)} f(x, y, z)\,dz \right) dA. \tag{1}$$

If we can describe the region R more naturally in polar coordinates than in rectangular coordinates, then it is likely that the integration over the plane region R will be simpler if it is carried out in polar coordinates.

We first express the inner partial integral of Eq. (1) in terms of r and θ by writing

$$\int_{z_1(x,y)}^{z_2(x,y)} f(x, y, z)\,dz = \int_{Z_1(r,\theta)}^{Z_2(r,\theta)} F(r, \theta, z)\,dz, \tag{2}$$

where

$$F(r, \theta, z) = f(r\cos\theta, r\sin\theta, z) \tag{3a}$$

and

$$Z_i(r, \theta) = z_i(r\cos\theta, r\sin\theta) \tag{3b}$$

for $i = 1, 2$. Substitution of Eq. (2) into Eq. (1) with (**important**) $dA = r\,dr\,d\theta$ gives

$$\iiint_T f(x, y, z)\,dV = \iint_S \left(\int_{Z_1(r,\theta)}^{Z_2(r,\theta)} F(r, \theta, z)\,dz \right) r\,dr\,d\theta, \tag{4}$$

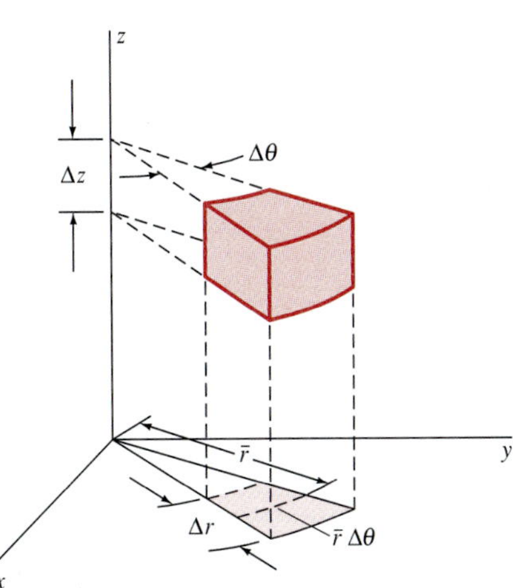

Fig. 14.7.2 The volume of the cylindrical block is $\Delta V = \bar{r}\,\Delta z\,\Delta r\,\Delta\theta$.

where F, Z_1, and Z_2 are the functions given in (3) and S represents the appropriate limits on r and θ needed to describe the plane region R in polar coordinates (as discussed in Section 14.4). The limits on z are simply the z-coordinates (in terms of r and θ) of a typical line segment joining the lower and upper boundary surfaces of T, as indicated in Fig. 14.7.1.

Thus the general formula for **triple integration in cylindrical coordinates** is

$$\iiint_T f(x, y, z)\,dV = \iiint_U f(r\cos\theta, r\sin\theta, z) r\,dz\,dr\,d\theta, \tag{5}$$

where U is not a region in space, but—as in Section 14.4—a representation of limits on z, r, and θ appropriate to describe the space region T in cylindrical coordinates. Before we integrate, we must replace the variables x and y with $r\cos\theta$ and $r\sin\theta$, respectively, but z is left unchanged. The cylindrical-coordinates volume element

$$dV = r\,dz\,dr\,d\theta$$

may be regarded formally as the product of dz and the polar-coordinates area element $dA = r\,dr\,d\theta$. It is a consequence of the formula $\Delta V = \bar{r}\,\Delta z\,\Delta r\,\Delta\theta$ for the volume of the *cylindrical block* shown in Fig. 14.7.2.

Integration in cylindrical coordinates is particularly useful for computations associated with solids of revolution. So that the limits of integration will be the simplest, the solid should usually be placed so that the axis of revolution is the z-axis.

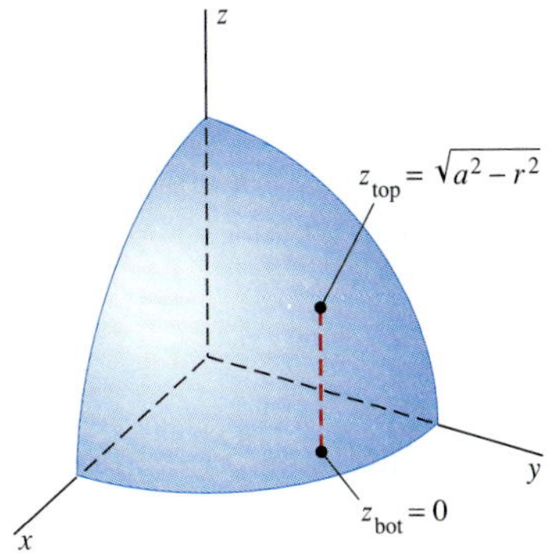

Fig. 14.7.3 The first octant of the sphere (Example 1)

EXAMPLE 1 Find the centroid of the first-octant portion T of the solid ball bounded by the sphere $r^2 + z^2 = a^2$. The solid T appears in Fig. 14.7.3.

Solution The volume of the first octant of the solid ball is $V = \frac{1}{8} \cdot \frac{4}{3}\pi a^3 = \frac{1}{6}\pi a^3$. Because $\bar{x} = \bar{y} = \bar{z}$ by symmetry, we need calculate only

$$\begin{aligned}
\bar{z} = \frac{1}{V}\iiint_T z\,dV &= \frac{6}{\pi a^3}\int_0^{\pi/2}\int_0^a\int_0^{\sqrt{a^2-r^2}} zr\,dz\,dr\,d\theta \\
&= \frac{6}{\pi a^3}\int_0^{\pi/2}\int_0^a \frac{1}{2}r(a^2 - r^2)\,dr\,d\theta \\
&= \frac{3}{\pi a^3}\int_0^{\pi/2}\left[\frac{1}{2}a^2r^2 - \frac{1}{4}r^4\right]_{r=0}^{a} d\theta = \frac{3}{\pi a^3}\cdot\frac{\pi}{2}\cdot\frac{a^4}{4} = \frac{3a}{8}.
\end{aligned}$$

Thus the centroid is located at the point $\left(\frac{3}{8}a, \frac{3}{8}a, \frac{3}{8}a\right)$. Observe that the answer is both plausible and dimensionally correct. ■

EXAMPLE 2 Find the volume and centroid of the solid T that is bounded by the paraboloid $z = b(x^2 + y^2)$ $(b > 0)$ and the plane $z = h$ $(h > 0)$.

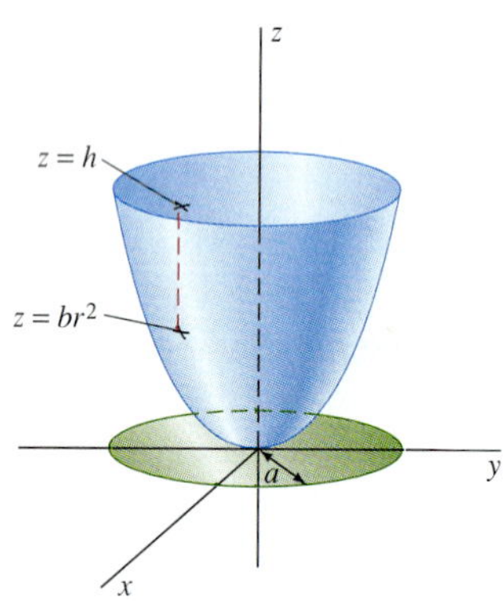

Fig. 14.7.4 The paraboloid of Example 2

Solution Figure 14.7.4 makes it clear that we get the radius of the circular top of T by equating $z = b(x^2 + y^2) = br^2$ and $z = h$. This gives $a = \sqrt{h/b}$ for the radius of the circle over which the solid lies. Hence Eq. (4), with $f(x, y, z) \equiv 1$, gives the volume,

$$\begin{aligned}
V = \iiint_T dV &= \int_0^{2\pi}\int_0^a\int_{br^2}^{h} r\,dz\,dr\,d\theta = \int_0^{2\pi}\int_0^a (hr - br^3)\,dr\,d\theta \\
&= 2\pi\left(\frac{1}{2}ha^2 - \frac{1}{4}ba^4\right) = \frac{\pi h^2}{2b} = \frac{1}{2}\pi a^2 h
\end{aligned}$$

(because $a^2 = h/b$).

By symmetry, the centroid of T lies on the z-axis, so all that remains is to compute $\bar{z}$,

$$\begin{aligned}
\bar{z} = \frac{1}{V}\iiint_T z\,dV &= \frac{2}{\pi a^2 h}\int_0^{2\pi}\int_0^a\int_{br^2}^{h} rz\,dz\,dr\,d\theta \\
&= \frac{2}{\pi a^2 h}\int_0^{2\pi}\int_0^a\left(\frac{1}{2}h^2 r - \frac{1}{2}b^2 r^5\right)dr\,d\theta \\
&= \frac{4}{a^2 h}\left(\frac{1}{4}h^2 a^2 - \frac{1}{12}b^2 a^6\right) = \frac{2}{3}h,
\end{aligned}$$

again using the fact that $a^2 = h/b$. Therefore the centroid of T is located at the point $(0, 0, \frac{2}{3}h)$. Again, this answer is both plausible and dimensionally correct. ■

We can summarize the results of Example 2 as follows: The volume of a right circular paraboloid is *half* that of the circumscribed cylinder (Fig. 14.7.5), and its centroid lies on its axis of symmetry *two-thirds* of the way from the "vertex" at $(0, 0, 0)$ to its circular "base" at the top.

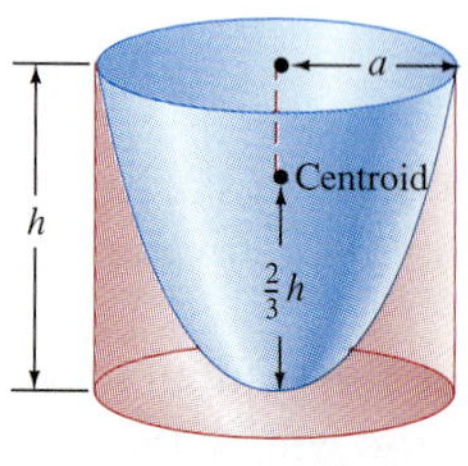

Fig. 14.7.5 Volume and centroid of a right circular paraboloid in terms of the circumscribed cylinder

Spherical Coordinate Integrals

When the boundary surfaces of the region T of integration are spheres, cones, or other surfaces with simple descriptions in spherical coordinates, it is generally advantageous to transform a triple integral over T into spherical coordinates. Recall from Section 12.8 that the relationship between spherical coordinates (ρ, ϕ, θ) (shown in Fig. 14.7.6) and rectangular coordinates (x, y, z) is given by

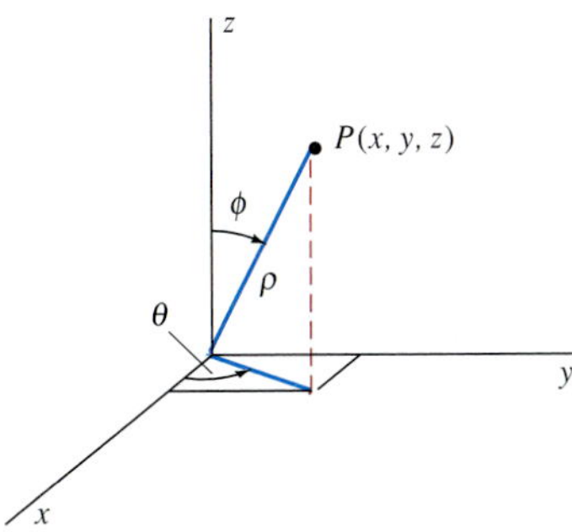

Fig. 14.7.6 The spherical coordinates (ρ, ϕ, θ) of the point P

$$x = \rho \sin \phi \cos \theta, \qquad y = \rho \sin \phi \sin \theta, \qquad z = \rho \cos \phi. \tag{6}$$

Suppose, for example, that T is the **spherical block** determined by the simple inequalities

$$\begin{aligned} \rho_1 &\leqq \rho \leqq \rho_2 = \rho_1 + \Delta\rho, \\ \phi_1 &\leqq \phi \leqq \phi_2 = \phi_1 + \Delta\phi, \\ \theta_1 &\leqq \theta \leqq \theta_2 = \theta_1 + \Delta\theta. \end{aligned} \tag{7}$$

As indicated by the dimensions labeled in Fig. 14.7.7, this spherical block is (if $\Delta\rho$, $\Delta\phi$, and $\Delta\theta$ are small) *approximately* a rectangular block with dimensions $\Delta\rho$, $\rho_1\,\Delta\phi$, and $\rho_1 \sin\phi_2\,\Delta\theta$. Thus its volume is approximately $\rho_1^2 \sin\phi_2\,\Delta\rho\,\Delta\phi\,\Delta\theta$. It can be shown (see Problem 19 of Section 14.8) that the *exact* volume of the spherical block described in (7) is

$$\Delta V = \hat{\rho}^2 \sin\hat{\phi}\,\Delta\rho\,\Delta\phi\,\Delta\theta \tag{8}$$

for certain numbers $\hat{\rho}$ and $\hat{\phi}$ such that $\rho_1 < \hat{\rho} < \rho_2$ and $\phi_1 < \hat{\phi} < \phi_2$.

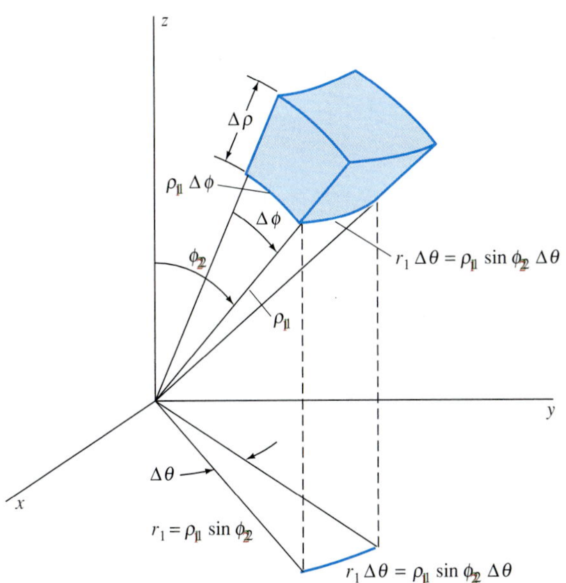

Fig. 14.7.7 The volume of the spherical block is approximately $\rho_1^2 \sin\phi_2\,\Delta\rho\,\Delta\phi\,\Delta\theta$.

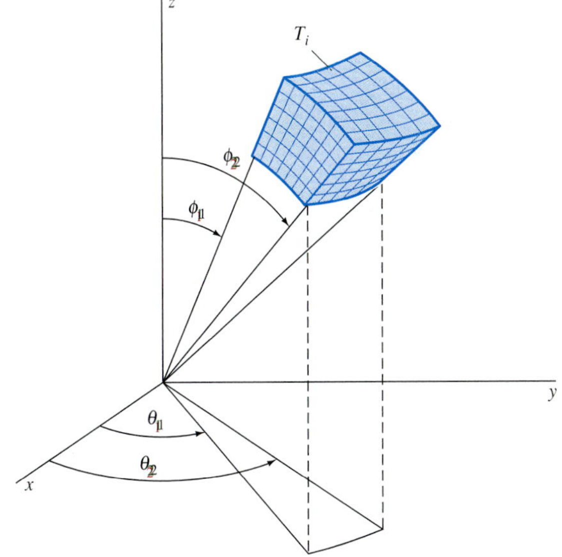

Fig. 14.7.8 The spherical block T divided into k smaller spherical blocks

Now suppose that we partition each of the intervals $[\rho_1, \rho_2]$, $[\phi_1, \phi_2]$, and $[\theta_1, \theta_2]$ into n subintervals of lengths

$$\Delta\rho = \frac{\rho_2 - \rho_1}{n}, \qquad \Delta\phi = \frac{\phi_2 - \phi_1}{n}, \qquad \text{and} \qquad \Delta\theta = \frac{\theta_2 - \theta_1}{n},$$

respectively. This produces a **spherical partition** $\mathcal{P}$ of the spherical block T into $k = n^3$ smaller spherical blocks $T_1, T_2, \ldots, T_k$; see Fig. 14.7.8. By Eq. (8), there exists a point $(\hat{\rho}_i, \hat{\phi}_i, \hat{\theta}_i)$, of the spherical block T_i such that its volume is $\Delta V_i = \hat{\rho}_i^2 \sin\hat{\phi}_i\,\Delta\rho\,\Delta\phi\,\Delta\theta$. The **norm** $|\mathcal{P}|$ of $\mathcal{P}$ is the length of the longest diagonal of any of the small spherical blocks $T_1, T_2, \ldots, T_k$.

If $(x_i^\star, y_i^\star, z_i^\star)$ are the rectangular coordinates of the point with spherical coordinates $(\hat{\rho}_i, \hat{\phi}_i, \hat{\theta}_i)$, then the definition of the triple integral as a limit of Riemann sums as the norm $|\mathcal{P}|$ approaches zero gives

$$\begin{aligned}\iiint_T f(x, y, z)\, dV &= \lim_{|\mathcal{P}|\to 0} \sum_{i=1}^{k} f(x_i^\star, y_i^\star, z_i^\star)\, \Delta V_i \\ &= \lim_{|\mathcal{P}|\to 0} \sum_{i=1}^{k} F(\hat{\rho}_i, \hat{\phi}_i, \hat{\theta}_i)\, \hat{\rho}_i^2 \sin \hat{\phi}_i\, \Delta\rho\, \Delta\phi\, \Delta\theta, \end{aligned} \tag{9}$$

where

$$F(\rho, \phi, \theta) = f(\rho \sin\phi \cos\theta, \rho \sin\phi \sin\theta, \rho \cos\phi) \tag{10}$$

is the result of substituting Eq. (6) into $f(x, y, z)$. But the right-hand sum in Eq. (9) is simply a Riemann sum for the triple integral

$$\int_{\theta_1}^{\theta_2} \int_{\phi_1}^{\phi_2} \int_{\rho_1}^{\rho_2} F(\rho, \phi, \theta) \rho^2 \sin\phi\, d\rho\, d\phi\, d\theta.$$

It therefore follows that

$$\iiint_T f(x, y, z)\, dV = \int_{\theta_1}^{\theta_2} \int_{\phi_1}^{\phi_2} \int_{\rho_1}^{\rho_2} F(\rho, \phi, \theta)\, \rho^2 \sin\phi\, d\rho\, d\phi\, d\theta. \tag{11}$$

Thus we transform the integral

$$\iiint_T f(x, y, z)\, dV$$

into spherical coordinates by replacing the rectangular-coordinate variables x, y, and z with their expressions in Eq. (6) in terms of the spherical-coordinate variables ρ, ϕ, and θ. In addition, we write

$$dV = \rho^2 \sin\phi\, d\rho\, d\phi\, d\theta$$

for the volume element in spherical coordinates.

More generally, we can transform the triple integral

$$\iiint_T f(x, y, z)\, dV$$

into spherical coordinates whenever the region T is **centrally simple**—that is, whenever it has a spherical-coordinates description of the form

$$\rho_1(\phi, \theta) \leqq \phi \leqq \rho_2(\phi, \theta), \quad \phi_1 \leqq \phi \leqq \phi_2, \quad \theta_1 \leqq \theta \leqq \theta_2. \tag{12}$$

If so, then

$$\iiint_T f(x, y, z)\, dV = \int_{\theta_1}^{\theta_2} \int_{\phi_1}^{\phi_2} \int_{\rho_1(\phi,\theta)}^{\rho_2(\phi,\theta)} F(\rho, \phi, \theta)\, \rho^2 \sin\phi\, d\rho\, d\phi\, d\theta. \tag{13}$$

The limits on ρ are simply the ρ-coordinates (in terms of ϕ and θ) of the endpoints of a typical radial segment that joins the "inner" and "outer" parts of the boundary of T (Fig. 14.7.9). Thus the general formula for **triple integration in spherical coordinates** is

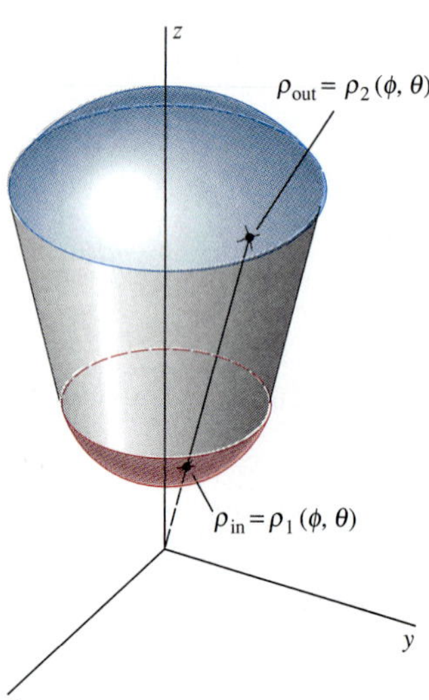

Fig. 14.7.9 A centrally simple region

$$\iiint_T f(x, y, z)\,dV = \iiint_U f(\rho\sin\phi\cos\theta,\ \rho\sin\phi\sin\theta,\ \rho\cos\phi)\,\rho^2\sin\phi\,d\rho\,d\phi\,d\theta, \tag{14}$$

where, as before, U does not denote a region in space but rather indicates limits on ρ, ϕ, and θ appropriate to describe the region T in spherical coordinates.

EXAMPLE 3 A solid ball T with constant density δ is bounded by the spherical surface with equation $\rho = a$. Use spherical coordinates to compute its volume V and its moment of inertia I_z around the z-axis.

Solution The points of the ball T are described by the inequalities

$$0 \leqq \rho \leqq a, \qquad 0 \leqq \phi \leqq \pi, \qquad 0 \leqq \theta \leqq 2\pi.$$

We take $f = F \equiv 1$ in Eq. (11) and thereby obtain

$$\begin{aligned} V &= \iiint_T dV = \int_0^{2\pi}\int_0^{\pi}\int_0^{a} \rho^2\sin\phi\,d\rho\,d\phi\,d\theta \\ &= \frac{1}{3}a^3\int_0^{2\pi}\int_0^{\pi}\sin\phi\,d\phi\,d\theta \\ &= \frac{1}{3}a^3\int_0^{2\pi}\Big[-\cos\phi\Big]_{\phi=0}^{\pi}d\theta = \frac{2}{3}a^3\int_0^{2\pi}d\theta = \frac{4}{3}\pi a^3. \end{aligned}$$

The distance from the typical point (ρ, ϕ, θ) of the sphere to the z-axis is $r = \rho\sin\phi$, so the moment of inertia of the sphere around that axis is

$$\begin{aligned} I_z &= \iiint_T r^2\delta\,dV = \int_0^{2\pi}\int_0^{\pi}\int_0^{a}\delta\rho^4\sin^3\phi\,d\rho\,d\phi\,d\theta \\ &= \frac{1}{5}\delta a^5\int_0^{2\pi}\int_0^{\pi}\sin^3\phi\,d\phi\,d\theta \\ &= \frac{2}{5}\pi\delta a^5\int_0^{\pi}\sin^3\phi\,d\phi = \frac{2}{5}\pi\delta a^5\cdot 2\cdot\frac{2}{3} = \frac{2}{5}ma^2, \end{aligned}$$

where $m = \frac{4}{3}\pi a^3\delta$ is the mass of the ball. The answer is dimensionally correct because it is the product of mass and the square of a distance. The answer is plausible because it implies that, for purposes of rotational inertia, the sphere acts as if its mass were concentrated about 63% of the way from the axis to the equator. ■

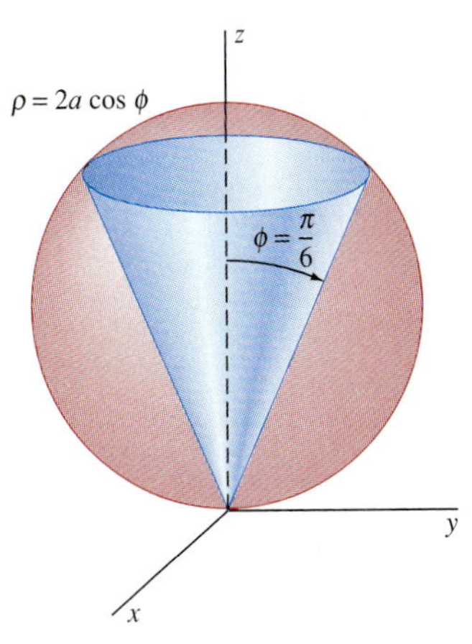

Fig. 14.7.10 The ice-cream cone of Example 4 is the part of the cone that lies within the sphere.

EXAMPLE 4 Find the volume and centroid of the uniform "ice-cream cone" C that is bounded by the cone $\phi = \pi/6$ and the sphere $\rho = 2a\cos\phi$ of radius a. The sphere and the part of the cone within it are shown in Fig. 14.7.10.

Solution The ice-cream cone is described by the inequalities

$$0 \leqq \theta \leqq 2\pi, \qquad 0 \leqq \phi \leqq \frac{\pi}{6}, \qquad 0 \leqq \rho \leqq 2a\cos\phi.$$

Using Eq. (13) to compute its volume, we get

$$V = \int_0^{2\pi}\int_0^{\pi/6}\int_0^{2a\cos\phi}\rho^2\sin\phi\,d\rho\,d\phi\,d\theta$$

$$= \frac{8}{3}a^3 \int_0^{2\pi} \int_0^{\pi/6} \cos^3\phi \sin\phi \, d\phi \, d\theta$$

$$= \frac{16}{3}\pi a^3 \left[-\frac{1}{4}\cos^4\phi \right]_0^{\pi/6} = \frac{7}{12}\pi a^3.$$

Now for the centroid. It is clear by symmetry that $\bar{x} = \bar{y} = 0$. We may also assume that C has density $\delta \equiv 1$, so that the mass of C is numerically the same as its volume. Because $z = \rho\cos\phi$, the z-coordinate of the centroid of C is

$$\bar{z} = \frac{1}{V}\iiint_C z \, dV = \frac{12}{7\pi a^3}\int_0^{2\pi}\int_0^{\pi/6}\int_0^{2a\cos\phi} \rho^3 \cos\phi \sin\phi \, d\rho \, d\phi \, d\theta$$

$$= \frac{48a}{7\pi}\int_0^{2\pi}\int_0^{\pi/6} \cos^5\phi \sin\phi \, d\phi \, d\theta = \frac{96a}{7}\left[-\frac{1}{6}\cos^6\phi \right]_0^{\pi/6} = \frac{37a}{28}.$$

Hence the centroid of the ice-cream cone is located at the point $\left(0, 0, \frac{37}{28}a\right)$. ■

14.7 PROBLEMS

Solve Problems 1 through 20 by triple integration in cylindrical coordinates. Assume throughout that each solid has unit density unless another density function is specified.

1. Find the volume of the solid bounded above by the plane $z = 4$ and below by the paraboloid $z = r^2$.
2. Find the centroid of the solid of Problem 1.
3. Derive the formula for the volume of a sphere of radius a.
4. Find the moment of inertia around the z-axis of the solid sphere of Problem 3 given that the z-axis passes through its center.
5. Find the volume of the region that lies inside both the sphere $x^2 + y^2 + z^2 = 4$ and the cylinder $x^2 + y^2 = 1$.
6. Find the centroid of the half of the region of Problem 5 that lies on or above the xy-plane.
7. Find the mass of the cylinder $0 \leqq r \leqq a$, $0 \leqq z \leqq h$ if its density at (x, y, z) is z.
8. Find the centroid of the cylinder of Problem 7.
9. Find the moment of inertia around the z-axis of the cylinder of Problem 7.
10. Find the volume of the region that lies inside both the sphere $x^2 + y^2 + z^2 = 4$ and the cylinder $x^2 + y^2 - 2x = 0$ (Fig. 14.7.11).
11. Find the volume and centroid of the region bounded by the plane $z = 0$ and the paraboloid $z = 9 - x^2 - y^2$.
12. Find the volume and centroid of the region bounded by the paraboloids $z = x^2 + y^2$ and $z = 12 - 2x^2 - 2y^2$.
13. Find the volume of the region bounded by the paraboloids $z = 2x^2 + y^2$ and $z = 12 - x^2 - 2y^2$.
14. Find the volume of the region bounded below by the paraboloid $z = x^2 + y^2$ and above by the plane $z = 2x$ (Fig. 14.7.12).
15. Find the volume of the region bounded above by the spherical surface $x^2 + y^2 + z^2 = 2$ and below by the paraboloid $z = x^2 + y^2$ (Fig. 14.7.13).

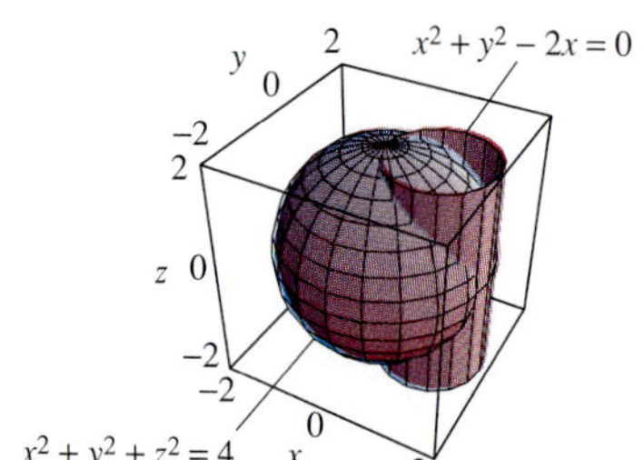

Fig. 14.7.11 The sphere and cylinder of Problem 10

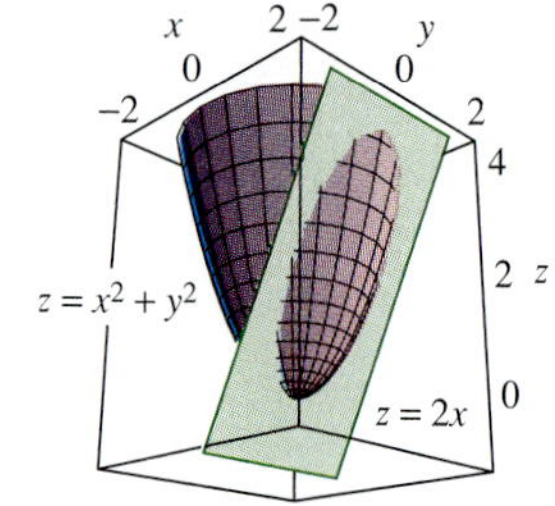

Fig. 14.7.12 The plane and paraboloid of Problem 14

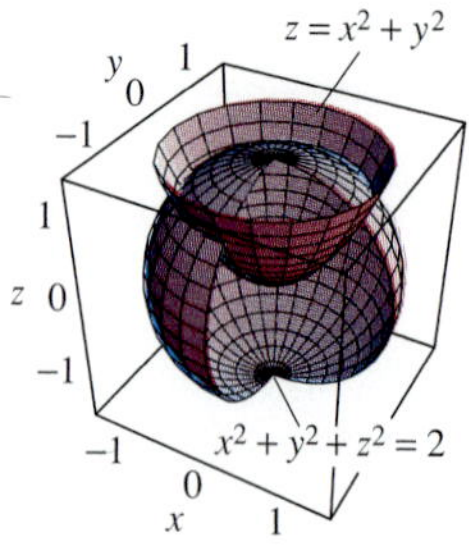

Fig. 14.7.13 The sphere and paraboloid of Problem 15

16. A homogeneous solid cylinder has mass m and radius a. Show that its moment of inertia around its axis of symmetry is $\frac{1}{2}ma^2$.
17. Find the moment of inertia I of a homogeneous solid right circular cylinder around a diameter of its base. Express I in terms of the radius a, the height h, and the (constant) density δ of the cylinder.

18. Find the centroid of a homogeneous solid right circular cylinder of radius a and height h.

19. Find the volume of the region bounded by the plane $z = 1$ and the cone $z = r$.

20. Show that the centroid of a homogeneous solid right circular cone lies on its axis three-quarters of the way from its vertex to its base.

Solve Problems 21 through 30 by triple integration in spherical coordinates.

21. Find the centroid of a homogeneous solid hemisphere of radius a.

22. Find the mass and centroid of the solid hemisphere $x^2 + y^2 + z^2 \leqq a^2$, $z \geqq 0$ if its density δ is proportional to distance z from its base—so $\delta = kz$ (where k is a positive constant).

23. Solve Problem 19 by triple integration in spherical coordinates.

24. Solve Problem 20 by triple integration in spherical coordinates.

25. Find the volume and centroid of the uniform solid that lies inside the sphere $\rho = a$ and above the cone $r = z$.

26. Find the moment of inertia I_z of the solid of Problem 25.

27. Find the moment of inertia around a tangent line of a solid homogeneous sphere of radius a and total mass m.

28. A spherical shell of mass m is bounded by the spheres $\rho = a$ and $\rho = 2a$, and its density function is $\delta = \rho^2$. Find its moment of inertia around a diameter.

29. Describe the surface $\rho = 2a \sin\phi$ and compute the volume of the region it bounds.

30. Describe the surface $\rho = 1 + \cos\phi$ and compute the volume of the region it bounds. Figure 14.7.14 may be useful.

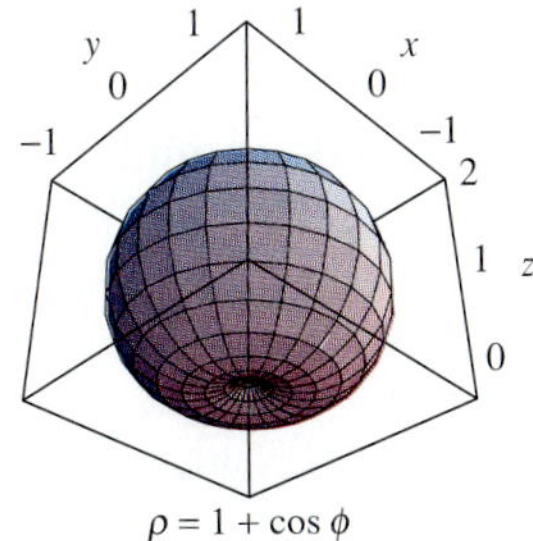

Fig. 14.7.14 The surface of Problem 30

31. Find the moment of inertia around the x-axis of the region that lies inside both the cylinder $r = a$ and the sphere $\rho = 2a$.

32. Find the moment of inertia around the z-axis of the ice-cream cone of Example 4.

33. Find the mass and centroid of the ice-cream cone of Example 4 if its density at (x, y, z) is $\delta(x, y, z) = z$.

34. Find the moment of inertia of the ice-cream cone of Problem 33 around the z-axis.

35. Suppose that a gaseous spherical star of radius a has density function $\delta = k(1 - \rho^2/a^2)$, so its density varies from $\delta = k$ at its center to $\delta = 0$ at its boundary $\rho = a$. Show that its mass is $\frac{2}{5}$ that of a similar star with uniform density k.

36. Find the moment of inertia around a diameter of the gaseous spherical star of Problem 35.

37. (a) Use spherical coordinates to evaluate the integral

$$\iiint_B \exp(-\rho^3)\, dV$$

where B is the solid ball of radius a centered at the origin.
(b) Let a fi ∞ in the result of part (a) to show that

$$\int_{-\infty}^{\infty}\int_{-\infty}^{\infty}\int_{-\infty}^{\infty} \exp\left(-(x^2 + y^2 + z^2)^{3/2}\right) dx\, dy\, dz = \frac{4}{3}\pi.$$

38. Use the method of Problem 37 to show that

$$\int_{-\infty}^{\infty}\int_{-\infty}^{\infty}\int_{-\infty}^{\infty} (x^2 + y^2 + z^2)^{1/2} \exp(-x^2 - y^2 - z^2)\, dx\, dy\, dz = 2\pi.$$

39. Find the average distance of points of a solid ball of radius a from the center of the ball. (The definition of the average value of a function precedes Problem 47 in Section 14.6.)

40. Find the average distance of the points of a solid ball of radius a from a fixed boundary point of the ball.

41. Consider a homogeneous spherical ball of radius a centered at the origin, with density δ and mass $M = \frac{4}{3}\pi a^3 \delta$. Show that the gravitational force $\mathbf{F}$ exerted by this ball on a point mass m located at the point $(0, 0, c)$, where $c > a$ (Fig. 14.7.15), is the same as though all the mass of the ball were concentrated at its center $(0, 0, 0)$. That is, show that $|\mathbf{F}| = GMm/c^2$. (*Suggestion:* By symmetry you may assume that the force is vertical, so that $\mathbf{F} = F_z\mathbf{k}$. Set up the integral

$$F_z = -\int_0^{2\pi}\int_0^{a}\int_0^{\pi} \frac{Gm\delta \cos\alpha}{w^2} \rho^2 \sin\phi\, d\phi\, d\rho\, d\theta.$$

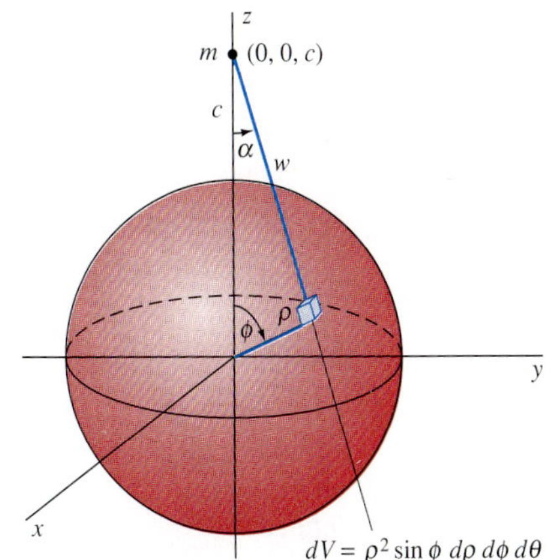

Fig. 14.7.15 The system of Problem 41

Change the first variable of integration from ϕ to w by using the law of cosines:

$$w^2 = \rho^2 + c^2 - 2\rho c \cos\phi.$$

Then $2w\,dw = 2\rho c \sin\phi\, d\phi$ and $w\cos\alpha + \rho\cos\phi = c$. (Why?))

42. Consider now the spherical shell $a \leqq r \leqq b$ with uniform density δ. Show that this shell exerts *no* net force on a point mass m located at the point $(0, 0, c)$ *inside* it—that is, with $|c| < a$. The computation will be the same as in Problem 41 except for the limits of integration on ρ and w.

14.7 PROJECT: THE EARTH'S MANTLE

If the earth were a perfect sphere with radius $R = 6370$ km, *uniform* density δ, and mass $M = \frac{4}{3}\pi\delta R^3$, then (according to Example 3) its moment of inertia around its polar axis would be $I = \frac{2}{5}MR^2$. In actuality, however, it turns out that

$$I = kMR^2, \tag{1}$$

where $k < 0.4 = \frac{2}{5}$. The reason is that, instead of having a uniform interior, the earth has a dense core covered with a lighter mantle a few thousand kilometers thick (Fig. 14.7.16). The density of the core is $\delta_1 \approx 11 \times 10^3$ (kg/m^2) and that of the mantle is $\delta_2 \approx 5 \times 10^3$ (kg/m^2).

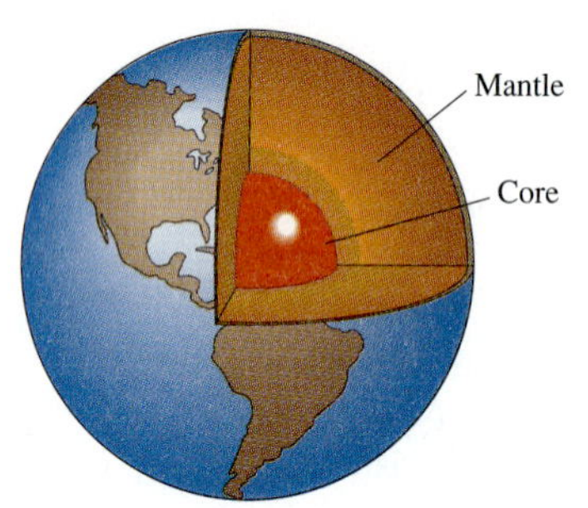

Fig. 14.7.16 The core and mantle of the earth

The numerical value of k in Eq. (1) can be determined from certain earth satellite observations. If the earth's polar moment of inertia I and mass M (for the core-mantle model) are expressed in terms of the unknown radius x of the spherical core, then substitution of these expressions into Eq. (1) yields an equation that can be solved for x.

Show that this equation can be written in the form

$$2(\delta_1 - \delta_2)x^5 - 5k(\delta_1 - \delta_2)R^2x^3 + (2 - 5k)\,\delta_2 R^5 = 0. \tag{2}$$

Given the measured numerical value $k = 0.371$, solve this equation (graphically or numerically) to find x and from this solution determine the thickness of the earth's mantle.

14.8 SURFACE AREA

Until now our concept of a surface has been the graph $z = f(x, y)$ of a function of two variables. Occasionally we have seen such a surface defined implicitly by an equation of the form $F(x, y, z) = 0$. Now we want to introduce the more precise concept of a *parametric surface*—the two-dimensional analogue of a parametric curve.

A **parametric surface** S is the *image* of a function or transformation $\mathbf{r}$ that is defined on a region R in the uv-plane (Fig. 14.8.1) and has values in xyz-space (Fig. 14.8.2). The **image** under $\mathbf{r}$ of each point (u, v) in R is the point in xyz-space with position vector

$$\mathbf{r}(u, v) = \langle x(u, v), y(u, v), z(u, v)\rangle. \tag{1}$$

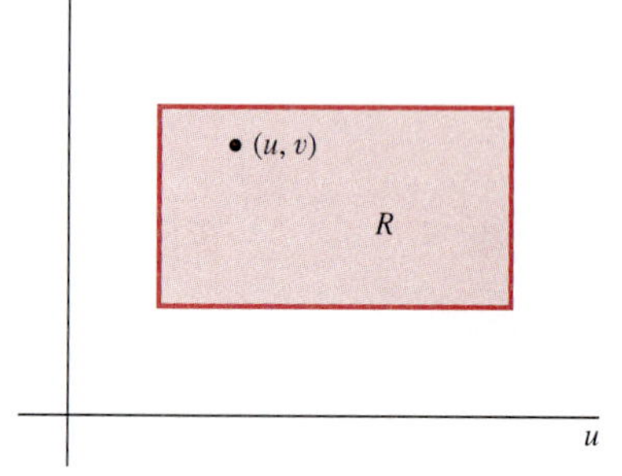

Fig. 14.8.1 The uv-region R on which the transformation $\mathbf{r}$ is defined

We shall assume throughout this section that the component functions of $\mathbf{r}$ have continuous partial derivatives with respect to u and v and also that the vectors

$$\mathbf{r}_u = \frac{\partial \mathbf{r}}{\partial u} = \langle x_u, y_u, z_u\rangle = \frac{\partial x}{\partial u}\mathbf{i} + \frac{\partial y}{\partial u}\mathbf{j} + \frac{\partial z}{\partial u}\mathbf{k} \tag{2}$$

and

$$\mathbf{r}_v = \frac{\partial \mathbf{r}}{\partial v} = \langle x_v, y_v, z_v\rangle = \frac{\partial x}{\partial v}\mathbf{i} + \frac{\partial y}{\partial v}\mathbf{j} + \frac{\partial z}{\partial v}\mathbf{k} \tag{3}$$

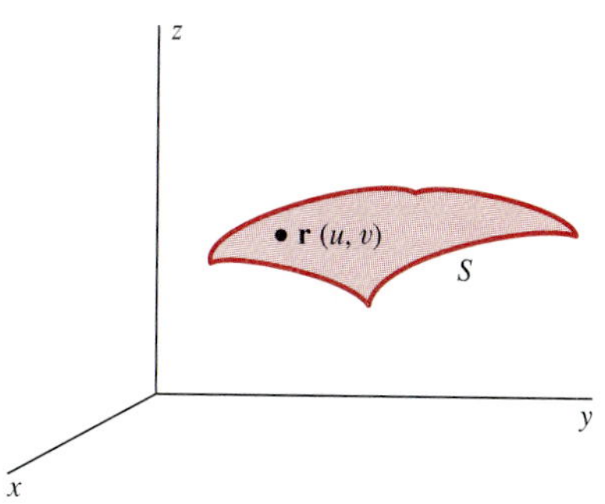

Fig. 14.8.2 The parametric surface S in xyz-space

are nonzero and nonparallel at each interior point of R. (Compare this with the definition of *smooth* parametric curve $\mathbf{r}(t)$ in Section 10.4.) We call the variables u and v the *parameters* for the surface S, in analogy with the single parameter t for a parametric curve.

EXAMPLE 1 (a) We may regard the graph $z = f(x, y)$ of a function as a parametric surface with parameters x and y. In this case the transformation $\mathbf{r}$ from the xy-plane to xyz-space has the component functions

$$x = x, \quad y = y, \quad z = f(x, y). \tag{4}$$

(b) Similarly, we may regard a surface given in cylindrical coordinates by the graph of $z = g(r, \theta)$ as a parametric surface with parameters r and θ. The transformation $\mathbf{r}$ from the $r\theta$-plane (Fig. 14.8.3) to xyz-space (Fig. 14.8.4) is then given by

$$x = r\cos\theta, \quad y = r\sin\theta, \quad z = g(r, \theta). \tag{5}$$

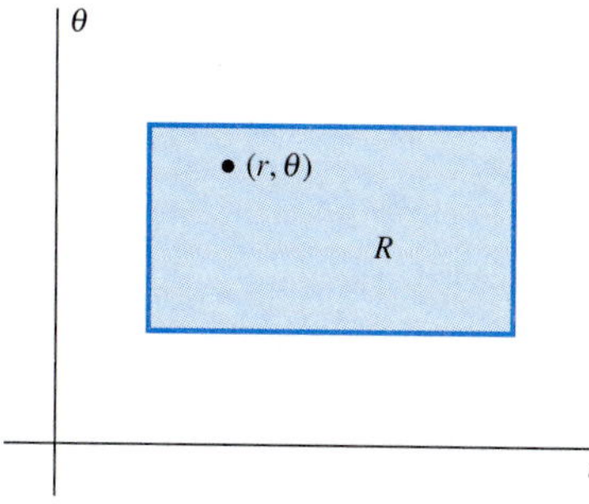

Fig. 14.8.3 A rectangle in the $r\theta$-plane; the domain of the function $z = g(r, \theta)$ of Example 1

(c) We may regard a surface given in spherical coordinates by $\rho = h(\phi, \theta)$ as a parametric surface with parameters ϕ and θ, and the corresponding transformation from the $\phi\theta$-plane to xyz-space is then given by

$$x = h(\phi, \theta)\sin\phi\cos\theta, \quad y = h(\phi, \theta)\sin\phi\sin\theta, \quad z = h(\phi, \theta)\cos\phi. \tag{6}$$

The concept of a parametric surface lets us treat all these special cases, and many others, with the same techniques. ■

Now we want to define the *surface area* of the general parametric surface given in Eq. (1). We begin with an inner partition of the region R—the domain of $\mathbf{r}$ in the uv-plane—into rectangles $R_1, R_2, \ldots, R_n$, each with dimensions Δu and Δv. Let (u_i, v_i) be the lower left-hand corner of R_i (as in Fig. 14.8.5). The image S_i of R_i under $\mathbf{r}$ will not generally be a rectangle in xyz-space; it will look more like a *curvilinear figure* on the image surface S, with $\mathbf{r}(u_i, v_i)$ as one "vertex" (Fig. 14.8.6). Let ΔS_i denote the area of this curvilinear figure S_i.

The parametric curves $\mathbf{r}(u, v_i)$ and $\mathbf{r}(u_i, v)$—with parameters u and v, respectively—lie on the surface S and meet at the point $\mathbf{r}(u_i, v_i)$. At this point of intersection, these two curves have the tangent vectors $\mathbf{r}_u(u_i, v_i)$ and $\mathbf{r}_v(u_i, v_i)$ shown in Fig. 14.8.7. Hence their vector product

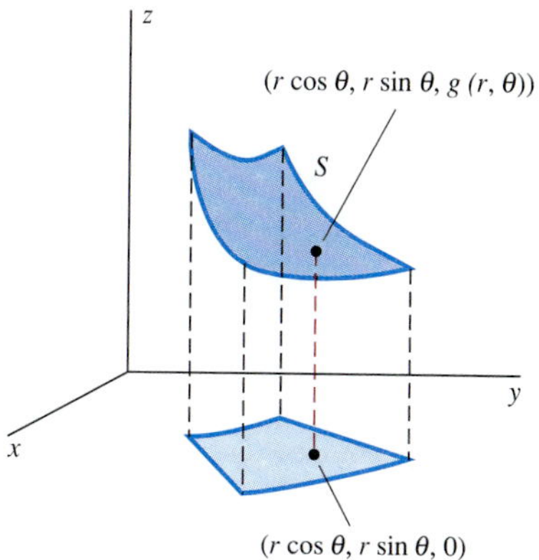

Fig. 14.8.4 A cylindrical-coordinates surface in xyz-space (Example 1)

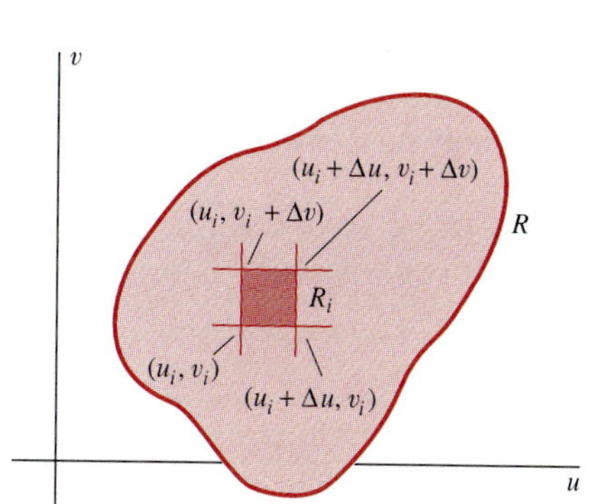

Fig. 14.8.5 The rectangle R_i in the uv-plane

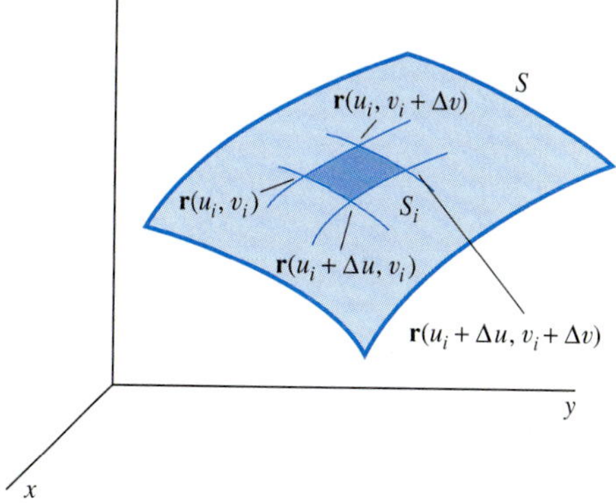

Fig. 14.8.6 The image of R_i is a curvilinear figure.

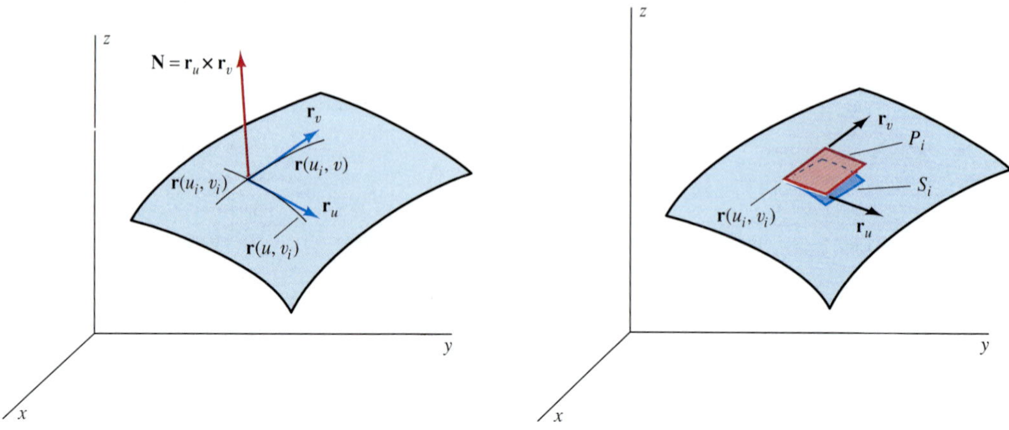

Fig. 14.8.7 The vector **N** normal to the surface at $\mathbf{r}(u_i, v_i)$

Fig. 14.8.8 The area of the parallelogram P_i is an approximation to the area of the curvilinear figure S_i.

$$\mathbf{N}(u_i, v_i) = \mathbf{r}_u(u_i, v_i) \times \mathbf{r}_v(u_i, v_i) \tag{7}$$

is a vector normal to S at the point $\mathbf{r}(u_i, v_i)$.

Now suppose that both Δu and Δv are small. Then the area ΔS_i of the curvilinear figure S_i will be approximately equal to the area ΔP_i of the parallelogram P_i with adjacent sides $\mathbf{r}_u(u_i, v_i)\ \Delta u$ and $\mathbf{r}_v(u_i, v_i)\ \Delta v$ (Fig. 14.8.8). But the area of this parallelogram is

$$\Delta P_i = |\mathbf{r}_u(u_i, v_i)\ \Delta u \times \mathbf{r}_v(u_i, v_i)\ \Delta v| = |\mathbf{N}(u_i, v_i)|\ \Delta u\ \Delta v.$$

This means that the area $a(S)$ of the surface S is given approximately by

$$a(S) = \sum_{i=1}^{n} \Delta S_i \approx \sum_{i=1}^{n} \Delta P_i,$$

so

$$a(S) \approx \sum_{i=1}^{n} |\mathbf{N}(u_i, v_i)|\ \Delta u\ \Delta v.$$

But this last sum is a Riemann sum for the double integral

$$\iint_R |\mathbf{N}(u, v)|\ du\ dv.$$

We are therefore motivated to *define* the **surface area** A of the parametric surface S by

$$A = a(S) = \iint_R |\mathbf{N}(u, v)|\ du\ dv = \iint_R \left|\frac{\partial \mathbf{r}}{\partial u} \times \frac{\partial \mathbf{r}}{\partial v}\right|\ du\ dv. \tag{8}$$

Surface Area in Rectangular Coordinates

In the case of the surface $z = f(x, y)$ for (x, y) in the region R in the xy-plane, the component functions of $\mathbf{r}$ are given by the equations in (4) with parameters x and y (in place of u and v). Then

$$\mathbf{N} = \frac{\partial \mathbf{r}}{\partial x} \times \frac{\partial \mathbf{r}}{\partial y} = \begin{vmatrix} \mathbf{i} & \mathbf{j} & \mathbf{k} \\ 1 & 0 & \dfrac{\partial f}{\partial x} \\ 0 & 1 & \dfrac{\partial f}{\partial y} \end{vmatrix} = -\frac{\partial f}{\partial x}\mathbf{i} - \frac{\partial f}{\partial y}\mathbf{j} + \mathbf{k},$$

so Eq. (8) takes the special form

$$A = a(S) = \iint_R \sqrt{1 + \left(\frac{\partial f}{\partial x}\right)^2 + \left(\frac{\partial f}{\partial y}\right)^2}\,dx\,dy$$
$$= \iint_R \sqrt{1 + z_x^2 + z_y^2}\,dx\,dy. \tag{9}$$

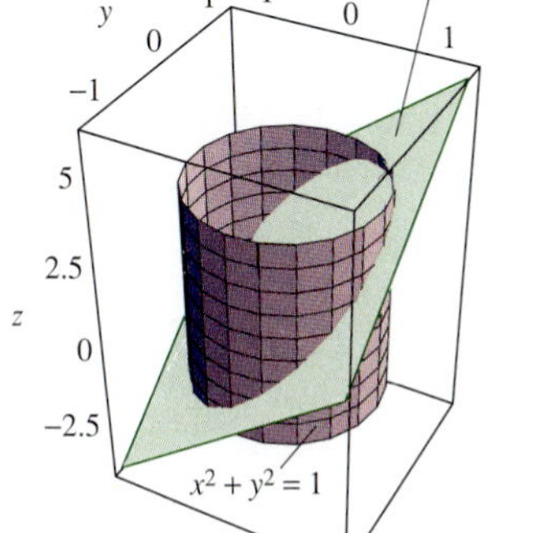

Fig. 14.8.9 The cylinder and plane of Example 2

EXAMPLE 2 Find the area of the ellipse cut from the plane $z = 2x + 2x + 1$ by the cylinder $x^2 + y^2 = 1$ (Fig. 14.8.9).

Solution Here, R is the unit circle in the xy-plane with area

$$\iint_R 1\,dx\,dy = \pi,$$

so Eq. (9) gives the area of the ellipse to be

$$A = \iint_R \sqrt{1 + z_x^2 + z_y^2}\,dx\,dy$$
$$= \iint_R \sqrt{1 + 2^2 + 2^2}\,dx\,dy = \iint_R 3\,dx\,dy = 3\pi.$$

■

REMARK Computer-generated figures such as Fig. 14.8.9 could not be constructed without using parametric surfaces. For example, the vertical cylinder in Fig. 14.8.9 was generated by instructing the computer to plot the parametric surface defined on the $z\theta$-rectangle $-5 \leqq z \leqq 5, 0 \leqq \theta \leqq 2\pi$ by $\mathbf{r}(z, \theta) = \langle \cos\theta, \sin\theta, z\rangle$. Is it clear that the image of this transformation is the cylinder $x^2 + y^2 = 1, -5 \leqq z \leqq 5$?

Surface Area in Cylindrical Coordinates

Now consider a cylindrical-coordinates surface $z = g(r, \theta)$ parametrized by the equations in (5) for (r, θ) in a region R of the $r\theta$-plane. Then the normal vector is

$$\mathbf{N} = \frac{\partial \mathbf{r}}{\partial r} \times \frac{\partial \mathbf{r}}{\partial \theta} = \begin{vmatrix} \mathbf{i} & \mathbf{j} & \mathbf{k} \\ \cos\theta & \sin\theta & \dfrac{\partial z}{\partial r} \\ -r\sin\theta & r\cos\theta & \dfrac{\partial z}{\partial \theta} \end{vmatrix}$$

$$= \mathbf{i}\left(\frac{\partial z}{\partial \theta}\sin\theta - r\frac{\partial z}{\partial r}\cos\theta\right) - \mathbf{j}\left(\frac{\partial z}{\partial \theta}\cos\theta + r\frac{\partial z}{\partial r}\sin\theta\right) + r\mathbf{k}.$$

After some simplifications, we find that

$$|\mathbf{N}| = \sqrt{r^2 + r^2\left(\frac{\partial z}{\partial r}\right)^2 + \left(\frac{\partial z}{\partial \theta}\right)^2}.$$

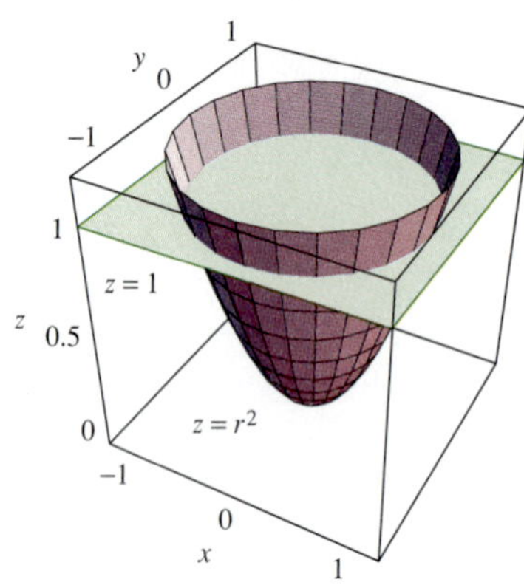

Fig. 14.8.10 The part of the paraboloid $z = r^2$ inside the cylinder $r = 1$ (Example 3) is the same as the part beneath the plane $z = 1$. (Why?)

Then Eq. (8) yields the formula

$$A = \iint_R \sqrt{r^2 + (rz_r)^2 + (z_\theta)^2}\, dr\, d\theta \tag{10}$$

for surface area in cylindrical coordinates.

EXAMPLE 3 Find the surface area cut from the paraboloid $z = r^2$ by the cylinder $r = 1$ (Fig. 14.8.10).

Solution Equation (10) gives surface area

$$A = \int_0^{2\pi}\int_0^1 \sqrt{r^2 + r^2\cdot(2r)^2}\, dr\, d\theta = 2\pi\int_0^1 r\sqrt{1 + 4r^2}\, dr$$

$$= 2\pi\left[\frac{2}{3}\cdot\frac{1}{8}(1 + 4r^2)^{3/2}\right]_0^1 = \frac{\pi}{6}\left(5\sqrt{5} - 1\right) \approx 5.3304.$$

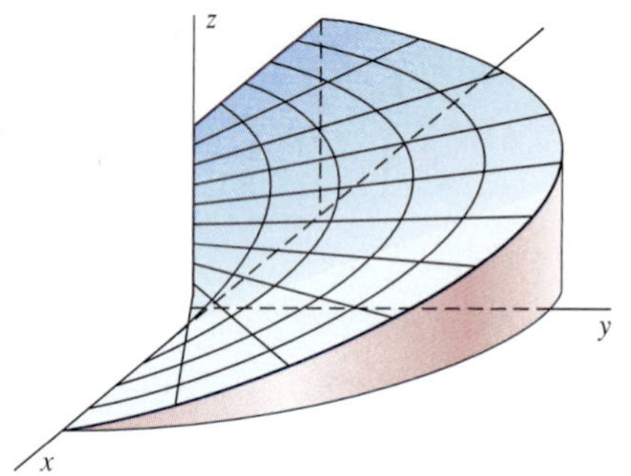

Fig. 14.8.11 The spiral ramp of Example 4

In Example 3, you would get the same result if you first wrote $z = x^2 + y^2$, used Eq. (9), which gives

$$A = \iint_R \sqrt{1 + 4x^2 + 4y^2}\, dx\, dy,$$

and then changed to polar coordinates. In Example 4 it would be less convenient to begin with rectangular coordinates.

EXAMPLE 4 Find the surface area of the *spiral ramp* $z = \theta$, $0 \leqq r \leqq 1$, $0 \leqq \theta \leqq \pi$. This is the upper surface of the solid shown in Fig. 14.8.11.

Solution Equation (10) gives surface area

$$A = \int_0^{\pi}\int_0^1 \sqrt{r^2 + 1}\, dr\, d\theta = \frac{\pi}{2}\left[\sqrt{2} + \ln\left(1 + \sqrt{2}\right)\right] \approx 3.6059.$$

We avoided a trigonometric substitution by using the table of integrals on the inside back cover.

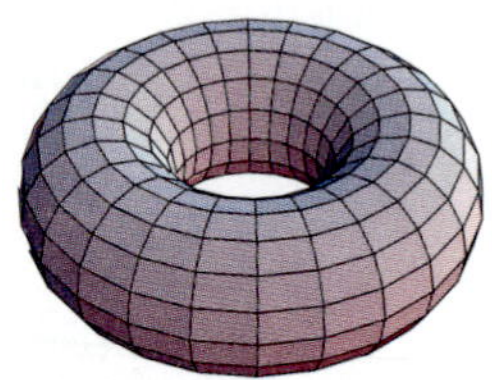

Fig. 14.8.12 The torus of Example 5

EXAMPLE 5 Find the surface area of the torus generated by revolving the circle

$$(x - b)^2 + z^2 = a^2 \quad (0 < a < b)$$

in the xz-plane around the z-axis (Fig. 14.8.12).

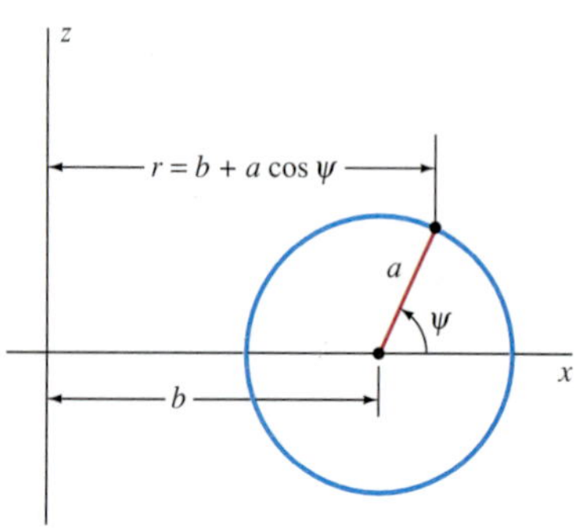

Fig. 14.8.13 The circle that generates the torus of Example 5

Solution With the ordinary polar coordinate θ and the angle ψ of Fig. 14.8.13, the torus is described for $0 \leqq \theta \leqq 2\pi$ and $0 \leqq \psi \leqq 2\pi$ by the parametric equations

$$x = r\cos\theta = (b + a\cos\psi)\cos\theta,$$

$$y = r\sin\theta = (b + a\cos\psi)\sin\theta,$$

$$z = a\sin\psi.$$

When we compute $\mathbf{N} = \mathbf{r}_\theta \times \mathbf{r}_\psi$ and simplify, we find that $|\mathbf{N}| = a(b + a\cos\psi)$. Hence the general surface-area formula, Eq. (8), gives area

$$A = \int_0^{2\pi}\int_0^{2\pi} a(b + a\cos\psi)\,d\theta\,d\psi = 2\pi a\Big[b\psi + a\sin\psi\Big]_0^{2\pi} = 4\pi^2 ab.$$

We obtained the same result in Section 14.5 with the aid of Pappus's theorem. ■

14.8 PROBLEMS

1. Find the area of the portion of the plane $z = x + 3y$ that lies inside the elliptical cylinder with equation $x^2/4 + y^2/9 = 1$.

2. Find the area of the region in the plane $z = 1 + 2x + 2y$ that lies directly above the region in the xy-plane bounded by the parabolas $y = x^2$ and $x = y^2$.

3. Find the area of the part of the paraboloid $z = 9 - x^2 - y^2$ that lies above the plane $z = 5$.

4. Find the area of the part of the surface $2z = x^2$ that lies directly above the triangle in the xy-plane with vertices at $(0, 0)$, $(1, 0)$, and $(1, 1)$.

5. Find the area of the surface that is the graph of $z = x + y^2$ for $0 \leqq x \leqq 1$, $0 \leqq y \leqq 2$.

6. Find the area of that part of the surface of Problem 5 that lies above the triangle in the xy-plane with vertices at $(0, 0)$, $(0, 1)$, and $(1, 1)$.

7. Find by integration the area of the part of the plane $2x + 3y + z = 6$ that lies in the first octant.

8. Find the area of the ellipse that is cut from the plane $2x + 3y + z = 6$ by the cylinder $x^2 + y^2 = 2$.

9. Find the area that is cut from the saddle-shaped surface $z = xy$ by the cylinder $x^2 + y^2 = 1$.

10. Find the area that is cut from the surface $z = x^2 - y^2$ by the cylinder $x^2 + y^2 = 4$.

11. Find the surface area of the part of the paraboloid $z = 16 - x^2 - y^2$ that lies above the xy-plane.

12. Show by integration that the surface area of the conical surface $z = br$ between the planes $z = 0$ and $z = h = ab$ is given by $A = \pi aL$, where L is the slant height $\sqrt{a^2 + h^2}$ and a is the radius of the base of the cone.

13. Let the part of the cylinder $x^2 + y^2 = a^2$ between the planes $z = 0$ and $z = h$ be parametrized by $x = a\cos\theta$, $y = a\sin\theta$, $z = z$. Apply Eq. (8) to show that the area of this zone is $A = 2\pi ah$.

14. Consider the meridional zone of height $h = c - b$ that lies on the sphere $r^2 + z^2 = a^2$ between the planes $z = b$ and $z = c$, where $0 \leqq b < c \leqq a$. Apply Eq. (10) to show that the area of this zone is $A = 2\pi ah$.

15. Find the area of the part of the cylinder $x^2 + z^2 = a^2$ that lies within the cylinder $r^2 = x^2 + y^2 = a^2$.

16. Find the area of the part of the sphere $r^2 + z^2 = a^2$ that lies within the cylinder $r = a\sin\theta$.

17. (a) Apply Eq. (8) to show that the surface area of the surface $y = f(x, z)$, for (x, z) in the region R of the xz-plane, is given by

$$A = \iint_R \sqrt{1 + \left(\frac{\partial f}{\partial x}\right)^2 + \left(\frac{\partial f}{\partial z}\right)^2}\,dx\,dz.$$

(b) State and derive a similar formula for the area of the surface $x = f(y, z)$ for (y, z) in the region R of the yz-plane.

18. Suppose that R is a region in the $\phi\theta$-plane. Consider the part of the sphere $\rho = a$ that corresponds to (ϕ, θ) in R, parametrized by the equations in (6) with $h(\phi, \theta) = a$. Apply Eq. (8) to show that the surface area of this part of the sphere is

$$A = \iint_R a^2 \sin\phi\,d\phi\,d\theta.$$

19. (a) Consider the "spherical rectangle" defined by

$$\rho = a, \quad \phi_1 \leqq \phi \leqq \phi_2 = \phi_1 + \Delta\phi, \quad \theta_1 \leqq \theta \leqq \theta_2 = \theta_1 + \Delta\theta.$$

Apply the formula of Problem 18 and the average value property (see Problem 50 in Section 14.2) to show that the area of this spherical rectangle is $A = a^2 \sin\hat{\phi}\,\Delta\phi\,\Delta\theta$ for some $\hat{\phi}$ in (ϕ_1, ϕ_2). (b) Conclude from the result of part (a) that the volume of the spherical block defined by

$$\rho_1 \leqq \rho \leqq \rho_2 = \rho_1 + \Delta\rho, \quad \phi_1 \leqq \phi \leqq \phi_2, \quad \theta_1 \leqq \theta \leqq \theta_2$$

is

$$\Delta V = \tfrac{1}{3}(\rho_2^3 - \rho_1^3)\sin\hat{\phi}\,\Delta\phi\,\Delta\theta.$$

Finally, derive Eq. (8) of Section 14.7 by applying the mean value theorem to the function $f(\rho) = \rho^3$ on the interval $[\rho_1, \rho_2]$.

20. Describe the surface $\rho = 2a\sin\phi$. Why is it called a *pinched torus*? It is parametrized as in Eq. (6) with $h(\phi, \theta) = 2a\sin\theta$. Show that its surface area is $A = 4\pi^2 a^2$. Figure 14.8.14 may be helpful.

21. The surface of revolution obtained when we revolve the curve $x = f(z)$, $a \leqq z \leqq b$, around the z-axis is parametrized in terms of θ $(0 \leqq \theta \leqq 2\pi)$ and z $(a \leqq z \leqq b)$ by $x = f(z)\cos\theta$, $y = f(z)\sin\theta$, $z = z$. From Eq. (8) derive the surface-area formula

$$A = \int_0^{2\pi}\int_a^b f(z)\sqrt{1 + [f'(z)]^2}\,dz\,d\theta.$$

This formula agrees with the area of a surface of revolution as defined in Section 6.4.

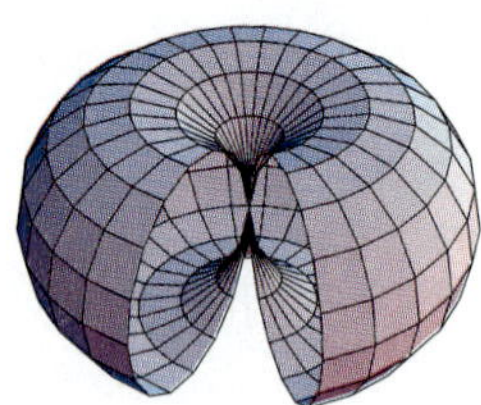

Fig. 14.8.14 Cutaway view of the pinched torus of Problem 20

22. Apply the formula of Problem 18 in both parts of this problem. (a) Verify the formula $A = 4\pi a^2$ for the surface area of a sphere of radius a. (b) Find the area of that part of a sphere of radius a and center $(0, 0, 0)$ that lies inside the cone $\phi = \pi/6$.

23. Apply the result of Problem 21 to verify the formula $A = 2\pi rh$ for the lateral surface area of a right circular cylinder of radius r and height h.

24. Apply Eq. (9) to verify the formula $A = 2\pi rh$ for the lateral surface area of the cylinder $x^2 + z^2 = r^2, 0 \leqq y \leqq h$ of radius r and height h.

In Problems 25 through 28, use a computer algebra system first to plot and then to approximate (with four-place accuracy) the area of the part of the given surface S that lies above the square in the xy-plane defined by: (a) $-1 \leqq x \leqq 1,\ -1 \leqq y \leqq 1$; (b) $|x| + |y| \leqq 1$.

25. S is the paraboloid $z = x^2 + y^2$.

26. S is the cone $z = \sqrt{x^2 + y^2}$.

27. S is the hyperboloid $z = 1 + xy$.

28. S is the sphere $x^2 + y^2 + z^2 = 4$.

29. An ellipsoid with semiaxes a, b, and c is defined by the parametrization

$$x = a \sin\phi \cos\theta, \quad y = b \sin\phi \sin\theta, \quad z = c \cos\phi$$

$(0 \leqq \phi \leqq \pi, 0 \leqq \theta \leqq 2\pi)$ in terms of the angular spherical coordinates ϕ and θ. Use a computer algebra system to approximate (to four-place accuracy) the area of the ellipsoid with $a = 4, b = 3$, and $c = 2$.

30. (a) Generalize Example 5 to derive the parametric equations

$$x = (b + a\cos\psi)\cos\theta, y = (b + a\cos\psi)\sin\theta, z = c\sin\psi$$

$(0 \leqq \psi \leqq 2\pi,\ 0 \leqq \theta \leqq 2\pi)$ of the "elliptical torus" obtained by revolving around the z-axis the ellipse $(x - b)^2/a^2 + z^2/c^2 = 1$ (where $0 < a < b$) in the xz-plane. (b) Use a computer algebra system to approximate (to four-place accuracy) the area of the elliptical torus obtained as in part (a) with $a = 2, b = 3$, and $c = 1$. (c) Also approximate the perimeter of the ellipse of part (a). Are your results consistent with Pappus's theorem for the area of a surface of revolution?

14.8 PROJECT: COMPUTER-GENERATED PARAMETRIC SURFACES

Parametric surfaces are used in most serious computer graphics work. Common computer algebra systems provide parametric plotting instructions such as the *Maple* command

```
plot3d( [f(u,v), g(u,v), h(u,v)], u = a..b, v = c..d )
```

and the *Mathematica* command

```
ParametricPlot3D[ {f[u,v],g[u,v],h[u,v]},
                  {u, a, b}, {v, c, d} ]
```

to plot the parametric surface defined by

$$x = f(u, v), \quad y = g(u, v), \quad z = h(u, v)$$

for $a \leqq u \leqq b, c \leqq v \leqq d$. In each of the following problems, first verify the given parametrization for the indicated surface. Then plot the surface, or a typical part of it, with selected numerical values of any constants that appear. Finally, calculate the area of your plotted surface.

1. The cylindrical-coordinates parametrization

$$x = r\cos\theta, \quad y = r\sin\theta, \quad z = f(r, \theta)$$

of a polar-coordinates surface such as a cone $z = kr$ or a circular paraboloid $z = kr^2$. Then try something more exotic, such as $z = (\sin r)/r$.

2. The parametrization

$$x = a\cos\theta, \quad y = b\sin\theta, \quad z = z$$

$(0 \leqq \theta \leqq 2\pi)$ of the elliptic cylinder $(x/a)^2 + (y/b)^2 = 1$.

3. The parametrization

$$x = au\cos\theta, \quad y = bu\sin\theta, \quad z = u^2$$

of the elliptic paraboloid $z = (x/a)^2 + (y/b)^2$, or the analogous elliptic cone.

4. The parametrization

$$x = a\sin u\cos v, \quad y = b\sin u\sin v, \quad z = c\cos u$$

of the ellipsoid $(x/a)^2 + (y/b)^2 + (z/c)^2 = 1$.

5. The parametrization

$$x = a\cosh u\cos v, \quad y = b\cosh u\sin v, \quad z = c\sinh u$$

of the hyperboloid $(x/a)^2 + (y/b)^2 - (z/c)^2 = 1$ of one sheet.

6. The parametrization

$$x = a\sinh u\cos v, \quad y = b\sinh u\sin v, \quad z = c\cosh u$$

of the hyperboloid $(z/c)^2 - (x/a)^2 - (y/b)^2 = 1$ of two sheets.

*14.9 CHANGE OF VARIABLES IN MULTIPLE INTEGRALS

We have seen in preceding sections that we can evaluate certain multiple integrals by transforming them from rectangular coordinates into polar or spherical coordinates. The technique of changing coordinate systems to evaluate a multiple integral is the multivariable analogue of substitution in a single integral. Recall from Section 5.7 that if $x = g(u)$, then

$$\int_a^b f(x)\,dx = \int_c^d f(g(u))\,g'(u)\,du, \tag{1}$$

where $a = g(c)$ and $b = g(d)$. The method of substitution involves a "change of variables" that is tailored to the evaluation of a given integral.

Suppose that we want to evaluate the double integral

$$\iint_R F(x, y)\,dx\,dy.$$

A change of variables for this integral is determined by a **transformation** T from the uv-plane to the xy-plane—that is, a function T that associates with the point (u, v) a point $(x, y) = T(u, v)$ given by equations of the form

$$x = f(u, v), \quad y = g(u, v). \tag{2}$$

The point (x, y) is called the **image** of the point (u, v) under the transformation T. If no two different points in the uv-plane have the same image point in the xy-plane, then the transformation T is said to be **one-to-one**. In this case it may be possible to solve the equations in (2) for u and v in terms of x and y and thus obtain the equations

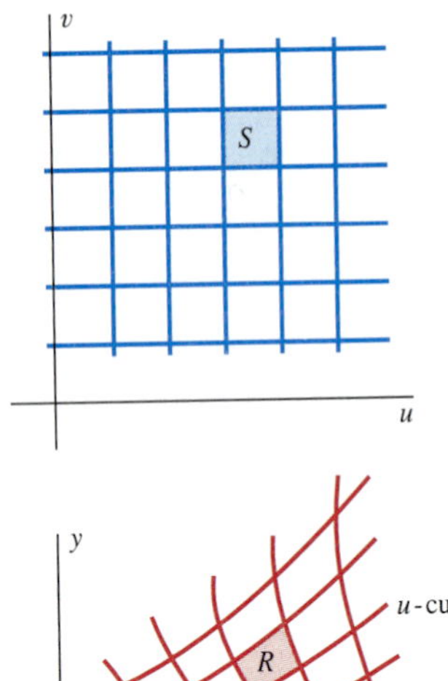

Fig. 14.9.1 The transformation T turns the rectangle S into the curvilinear figure R.

$$u = h(x, y), \qquad v = k(x, y) \tag{3}$$

of the **inverse transformation** T^{-1} from the xy-plane to the uv-plane.

Often it is convenient to visualize the transformation T geometrically in terms of its *u-curves* and *v-curves.* The **u-curves** of T are the images in the xy-plane of the *horizontal* lines in the uv-plane—on each such curve the value of u varies but v is constant. The **v-curves** of T are the images of the *vertical* lines in the uv-plane—on each of these, the value of v varies but u is constant. Note that the image under T of a rectangle bounded by horizontal and vertical lines in the uv-plane is a *curvilinear figure* bounded by u-curves and v-curves in the xy-plane (Fig. 14.9.1). If we know the equations in (3) of the inverse transformation, then we can find the u-curves and the v-curves quite simply by writing the equations

$$k(x, y) = C_1 \qquad (u\text{-curve on which } v = C_1 \text{ is constant}),$$

$$h(x, y) = C_2 \qquad (v\text{-curve on which } u = C_2 \text{ is constant}).$$

EXAMPLE 1 Determine the u-curves and the v-curves of the transformation T whose inverse T^{-1} is specified by the equations $u = xy$, $v = x^2 - y^2$.

Solution The u-curves are the hyperbolas

$$x^2 - y^2 = v = C_1 \qquad \text{(constant)},$$

and the v-curves are the rectangular hyperbolas

$$xy = u = C_2 \qquad \text{(constant)}.$$

These two familiar families of hyperbolas are shown in Fig. 14.9.2. ■

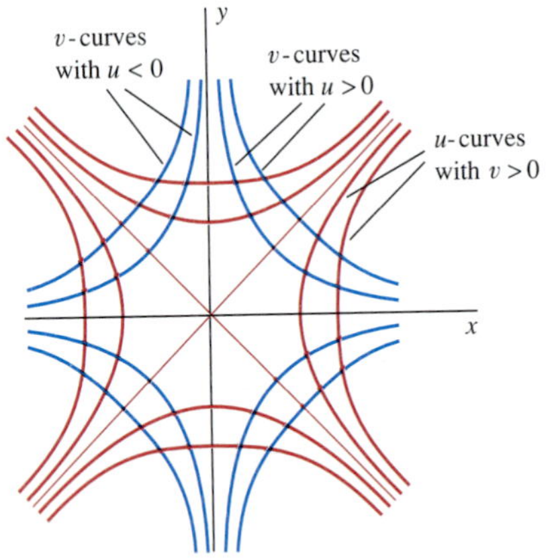

Fig. 14.9.2 The u-curves and v-curves of Example 1

Change of Variables in Double Integrals

Now we shall describe the change of variables in a double integral that corresponds to the transformation T specified by the equations in (2). Let the region R in the xy-plane be the image under T of the region S in the uv-plane. Let $F(x, y)$ be continuous on R and let $\{S_1, S_2, \dots, S_n\}$ be an inner partition of S into rectangles each with dimensions Δu by Δv. Each rectangle S_i is transformed by T into a curvilinear figure R_i in the xy-plane (Fig. 14.9.3). The images $\{R_1, R_2, \dots, R_n\}$ under T of the rectangles S_i then constitute an inner partition of the region R (though into curvilinear figures rather than rectangles).

Let $(u_i^\star, v_i^\star)$ be the lower left-hand corner point of S_i, and write

$$(x_i^\star, y_i^\star) = \big(f(u_i^\star, v_i^\star), g(u_i^\star, v_i^\star)\big)$$

for its image under T. The u-curve through $(x_i^\star, y_i^\star)$ has velocity vector

$$\mathbf{t}_u = \mathbf{i}f_u(u_i^\star, v_i^\star) + \mathbf{j}g_u(u_i^\star, v_i^\star) = \frac{\partial x}{\partial u}\mathbf{i} + \frac{\partial y}{\partial u}\mathbf{j},$$

and the v-curve through $(x_i^\star, y_i^\star)$ has velocity vector

$$\mathbf{t}_v = \mathbf{i}f_v(u_i^\star, v_i^\star) + \mathbf{j}g_v(u_i^\star, v_i^\star) = \frac{\partial x}{\partial v}\mathbf{i} + \frac{\partial y}{\partial v}\mathbf{j}.$$

Thus we can approximate the curvilinear figure R_i by a parallelogram P_i with edges that are "copies" of the vectors $\mathbf{t}_u\,\Delta u$ and $\mathbf{t}_v\,\Delta v$. These edges and the approximating parallelogram appear in Fig. 14.9.3.

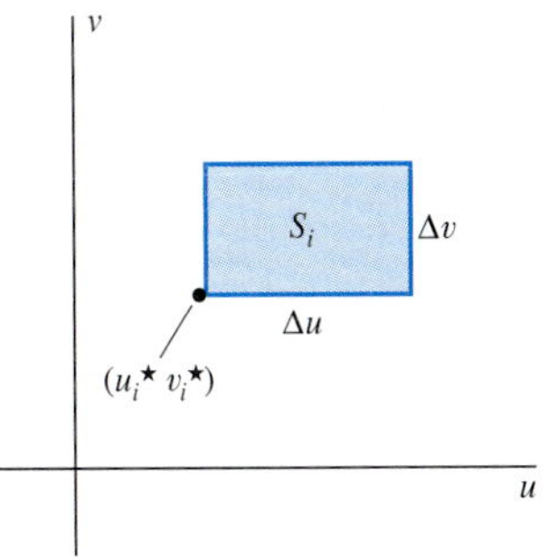

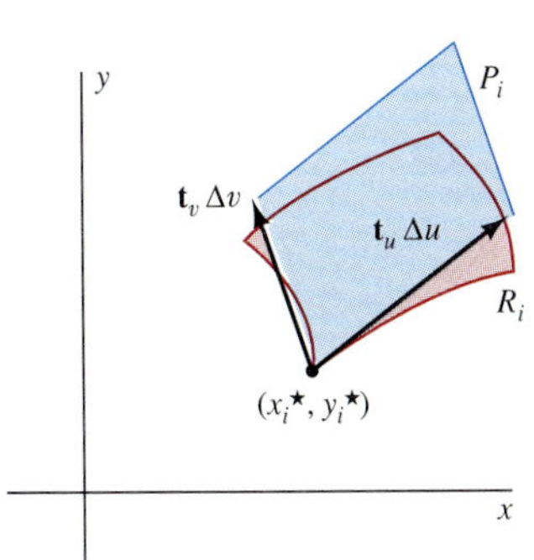

Fig. 14.9.3 The effect of the transformation T. We estimate the area of $R_i = T(S_i)$ by computing the area of the parallelogram P_i.

Now the area ΔA_i of R_i is also approximated by the area $a(P_i)$ of the parallelogram P_i, and we can compute the latter. Indeed,

$$\Delta A_i \approx a(P_i) = |(\mathbf{t}_u\,\Delta u) \times (\mathbf{t}_v\,\Delta v)| = |\mathbf{t}_u \times \mathbf{t}_v|\,\Delta u\,\Delta v.$$

But

$$\mathbf{t}_u \times \mathbf{t}_v = \begin{vmatrix} \mathbf{i} & \mathbf{j} & \mathbf{k} \\ \dfrac{\partial x}{\partial u} & \dfrac{\partial y}{\partial u} & 0 \\ \dfrac{\partial x}{\partial v} & \dfrac{\partial y}{\partial v} & 0 \end{vmatrix} = \begin{vmatrix} \dfrac{\partial x}{\partial u} & \dfrac{\partial x}{\partial v} \\ \dfrac{\partial y}{\partial u} & \dfrac{\partial y}{\partial v} \end{vmatrix} \mathbf{k}.$$

The two-by-two determinant on the right is called the **Jacobian** of the transformation T, after the German mathematician Carl Jacobi (1804–1851), who first investigated general changes of variables in multiple integrals. The Jacobian of the transformation T is a function of u and v, and we denote it by $J_T = J_T(u, v)$. Thus

$$J_T(u, v) = \begin{vmatrix} f_u(u, v) & f_v(u, v) \\ g_u(u, v) & g_v(u, v) \end{vmatrix}. \tag{4}$$

A common and particularly suggestive notation for the Jacobian is

$$J_T = \frac{\partial(x, y)}{\partial(u, v)}.$$

The computation preceding Eq. (4) shows that the area ΔA_i of R_i is given approximately by $\Delta A_i \approx |J_T(u_i^\star, v_i^\star)|\,\Delta u\,\Delta v$. Therefore, when we set up Riemann sums for approximating double integrals, we find that

$$\begin{aligned} \iint_R F(x, y)\,dx\,dy &\approx \sum_{i=1}^{n} F(x_i^\star, y_i^\star)\,\Delta A_i \\ &\approx \sum_{i=1}^{m} F\big(f(u_i^\star, v_i^\star), g(u_i^\star, v_i^\star)\big)\,|J_T(u_i^\star, v_i^\star)|\,\Delta u\,\Delta v \\ &\approx \iint_S F\big(f(u, v), g(u, v)\big)\,|J_T(u, v)|\,du\,dv. \end{aligned}$$

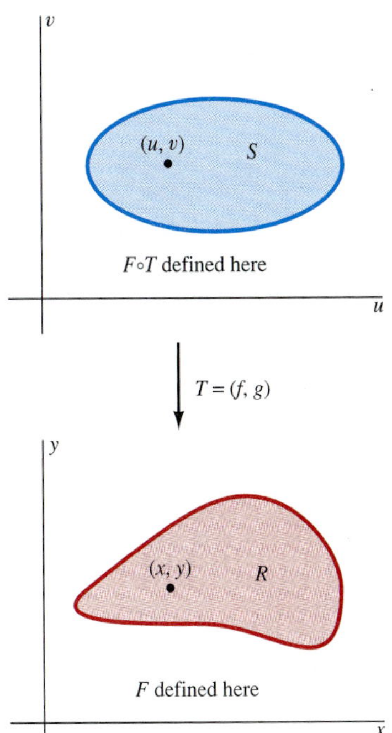

Fig. 14.9.4 The domains of $F(x, y)$ and $F(T(u, v)) = F(f(x, y), g(x, y))$

This discussion is, in fact, an outline of a proof of the following general **change-of-variables** theorem. We assume that T transforms the bounded region S in the uv-plane into the bounded region R in the xy-plane (Fig. 14.9.4) and that T is one-to-one from the interior of S to the interior of R. Suppose that the function $F(x, y)$ and the first-order partial derivatives of the component functions of T are continuous functions. Finally, to ensure the existence of the indicated double integrals, we assume that the boundaries of both regions R and S consist of a finite number of piecewise smooth curves.

Theorem 1 *Change of Variables*

If the transformation T with component functions $x = f(u, v)$ and $y = g(u, v)$ satisfies the conditions in the preceding paragraph, then

$$\iint_R F(x, y)\,dx\,dy = \iint_S F\big(f(u, v), g(u, v)\big)\,|J_T(u, v)|\,du\,dv. \tag{5}$$

If we write $G(u, v) = F(f(u, v), g(u, v))$, then the change-of-variables formula, Eq. (5), becomes

$$\iint_R F(x, y)\,dx\,dy = \iint_S G(u, v)\left|\frac{\partial(x, y)}{\partial(u, v)}\right| du\,dv. \tag{5a}$$

Thus we formally transform $\iint_R F(x, y)\,dA$ by replacing the variables x and y with $f(u, v)$ and $g(u, v)$, respectively, and writing

$$dA = \left|\frac{\partial(x, y)}{\partial(u, v)}\right| du\,dv$$

for the area element in terms of u and v. Note the analogy between Eq. (5a) and the single-variable formula in Eq. (1). In fact, if $g'(x) \neq 0$ on $[c, d]$ and we denote by α the smaller, and by β the larger, of the two limits c and d in Eq. (1), then Eq. (1) takes the form

$$\int_a^b f(x)\,dx = \int_\alpha^\beta f(g(u))\,|g'(u)|\,du. \tag{1a}$$

Thus the Jacobian in Eq. (5a) plays the role of the derivative $g'(u)$ in Eq. (1).

EXAMPLE 2 Suppose that the transformation T from the $r\theta$-plane to the xy-plane is determined by the polar equations

$$x = f(r, \theta) = r\cos\theta, \qquad y = g(r, \theta) = r\sin\theta.$$

The Jacobian of T is

$$\frac{\partial(x, y)}{\partial(r, \theta)} = \begin{vmatrix} \cos\theta & -r\sin\theta \\ \sin\theta & r\cos\theta \end{vmatrix} = r > 0,$$

so Eq. (5) or (5a) reduces to the familiar formula

$$\iint_R F(x, y)\,dx\,dy = \iint_S F(r\cos\theta, r\sin\theta)\,r\,dr\,d\theta.$$

■

Given a particular double integral $\iint_R f(x, y)\,dx\,dy$, how do we find a *productive* change of variables? One standard approach is to choose a transformation T such that the boundary of R consists of u-curves and v-curves. In case it is more convenient to express u and v in terms of x and y, we can first compute $\partial(u, v)/\partial(x, y)$ explicitly and then find the needed Jacobian $\partial(x, y)/\partial(u, v)$ from the formula

$$\frac{\partial(x, y)}{\partial(u, v)} \cdot \frac{\partial(u, v)}{\partial(x, y)} = 1. \tag{6}$$

Equation (6) is a consequence of the chain rule (see Problem 18).

EXAMPLE 3 Suppose that R is the plane region of unit density that is bounded by the hyperbolas

$$xy = 1, \quad xy = 3 \quad \text{and} \quad x^2 - y^2 = 1, \quad x^2 - y^2 = 4.$$

Find the polar moment of inertia

$$I_0 = \iint_R (x^2 + y^2)\,dx\,dy$$

of this region.

Solution The hyperbolas bounding R are u-curves and v-curves if $u = xy$ and $v = x^2 - y^2$, as in Example 1. We can most easily write the integrand $x^2 + y^2$ in terms of u and v by first noting that

$$4u^2 + v^2 = 4x^2y^2 - (x^2 - y^2)^2 = (x^2 + y^2)^2,$$

so $x^2 + y^2 = \sqrt{4u^2 + v^2}$. Now

$$\frac{\partial(u, v)}{\partial(x, y)} = \begin{vmatrix} y & x \\ 2x & -2y \end{vmatrix} = -2(x^2 + y^2).$$

Hence Eq. (6) gives

$$\frac{\partial(x, y)}{\partial(u, v)} = -\frac{1}{2(x^2 + y^2)} = -\frac{1}{2\sqrt{4u^2 + v^2}}.$$

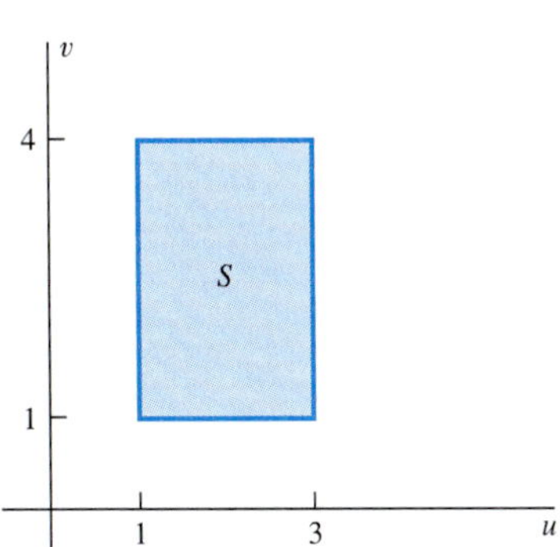

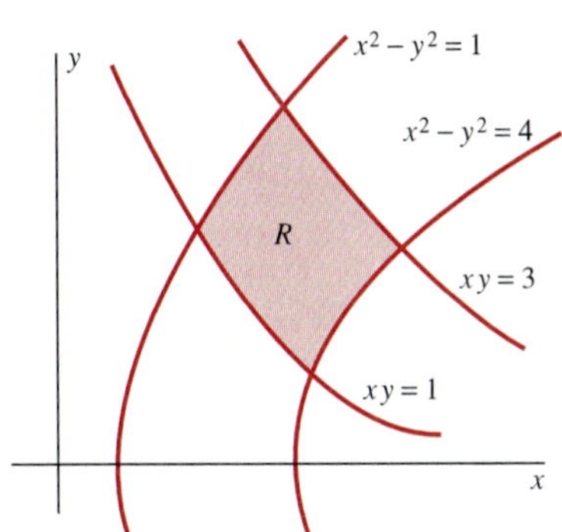

Fig. 14.9.5 The transformation T and the new region S constructed in Example 3

We are now ready to apply the change-of-variables theorem, with the regions S and R as shown in Fig. 14.9.5. With $f(x, y) = x^2 + y^2$, Eq. (5a) gives

$$I_0 = \iint_R (x^2 + y^2)\,dx\,dy = \int_1^4 \int_1^3 \sqrt{4u^2 + v^2}\,\frac{1}{2\sqrt{4u^2 + v^2}}\,du\,dv$$

$$= \int_1^4 \int_1^3 \frac{1}{2}\,du\,dv = 3.$$ ■

Example 4 is motivated by an important application. Consider an engine with an operating cycle that consists of alternate expansion and compression of gas in a piston. During one cycle the point (p, V), which gives the pressure and volume of this gas, traces a closed curve in the pV-plane. The work done by the engine—ignoring friction and related losses—is then equal (in appropriate units) to the area *enclosed by this curve,* called the *indicator diagram* of the engine. The indicator diagram for an ideal *Carnot engine* consists of two *isotherms* $xy = a$, $xy = b$ and two *adiabatics* $xy^\gamma = c$, $xy^\gamma = d$, where γ is the heat capacity ratio of the working gas in the piston. A typical value is $\gamma = 1.4$.

EXAMPLE 4 Find the area of the region R bounded by the curves $xy = 1$, $xy = 3$ and $xy^{1.4} = 1$, $xy^{1.4} = 2$ (Fig. 14.9.6).

Solution To force the given curves to be u-curves and v-curves, we define our change of variables transformation by $u = xy$ and $v = xy^{1.4}$. Then

$$\frac{\partial(u, v)}{\partial(x, y)} = \begin{vmatrix} y & x \\ y^{1.4} & (1.4)xy^{0.4} \end{vmatrix} = (0.4)xy^{1.4} = (0.4)v.$$

So

$$\frac{\partial(x, y)}{\partial(u, v)} = \frac{1}{\partial(u, v)/\partial(x, y)} = \frac{2.5}{v}.$$

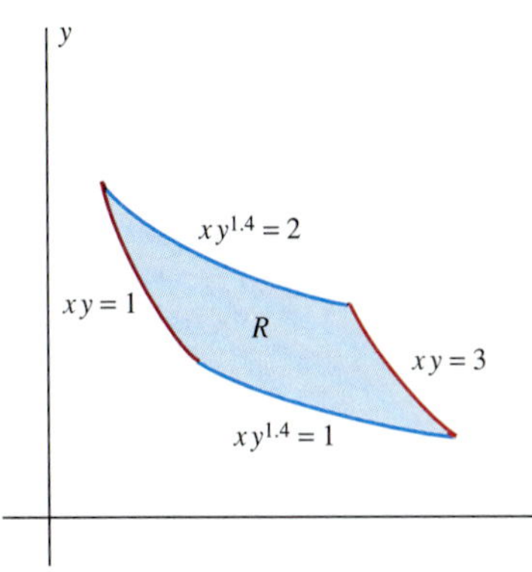

Fig. 14.9.6 Finding the area of the region R (Example 4)

Consequently, the change-of-variables theorem gives the formula

$$A = \iint_R 1\,dx\,dy = \int_1^2 \int_1^3 \frac{2.5}{v}\,du\,dv = 5\ln 2.$$ ■

Change of Variables in Triple Integrals

The change-of-variables formula for triple integrals is similar to Eq. (5). Let S and R be regions that correspond under the one-to-one transformation T from uvw-space to xyz-space, where the coordinate functions that comprise T are

$$x = f(u, v, w), \qquad y = g(u, v, w), \qquad z = h(u, v, w). \tag{7}$$

The Jacobian of T is

$$J_T(u, v, w) = \frac{\partial(x, y, z)}{\partial(u, v, w)} = \begin{vmatrix} \dfrac{\partial x}{\partial u} & \dfrac{\partial x}{\partial v} & \dfrac{\partial x}{\partial w} \\ \dfrac{\partial y}{\partial u} & \dfrac{\partial y}{\partial v} & \dfrac{\partial y}{\partial w} \\ \dfrac{\partial z}{\partial u} & \dfrac{\partial z}{\partial v} & \dfrac{\partial z}{\partial w} \end{vmatrix}. \tag{8}$$

Then the change-of-variables formula for triple integrals is

$$\iiint_R F(x, y, z)\,dx\,dy\,dz = \iiint_S G(u, v, w) \left| \frac{\partial(x, y, z)}{\partial(u, v, w)} \right| du\,dv\,dw, \tag{9}$$

where $G(u, v, w) = F\big(f(u, v, w), g(u, v, w), h(u, v, w)\big)$ is the function obtained from $F(x, y, z)$ by expressing the variables x, y, and z in terms of u, v, and w.

EXAMPLE 5 If T is the spherical-coordinates transformation given by

$$x = \rho \sin\phi \cos\theta, \qquad y = \rho \sin\phi \sin\theta, \qquad z = \rho \cos\phi,$$

then the Jacobian of T is

$$\frac{\partial(x, y, z)}{\partial(u, v, w)} = \begin{vmatrix} \sin\phi\cos\theta & \rho\cos\phi\cos\theta & -\rho\sin\phi\sin\theta \\ \sin\phi\sin\theta & \rho\cos\phi\sin\theta & \rho\sin\phi\cos\theta \\ \cos\phi & -\rho\sin\phi & 0 \end{vmatrix} = \rho^2 \sin\phi.$$

Thus Eq. (9) reduces to the familiar formula

$$\iiint_R F(x, y, z)\,dx\,dy\,dz = \iiint_S G(\rho, \phi, \theta)\,\rho^2 \sin\phi\,d\rho\,d\phi\,d\theta.$$

The sign is correct because $\rho^2 \sin\phi \geqq 0$ for ϕ in $[0, \pi]$. ■

EXAMPLE 6 Find the volume of the solid torus R obtained by revolving around the z-axis the circular disk

$$(x - b)^2 + z^2 \leqq a^2, \qquad 0 < a < b \tag{10}$$

in the xz-plane.

Solution This is the torus of Example 5 of Section 14.8. Let us write u for the ordinary polar coordinate angle θ, v for the angle ψ of Fig. 14.8.12, and w for the distance from the center of the circular disk described by the inequality in (10). We then define the transformation T by means of the equations

$$x = (b + w\cos v)\cos u, \qquad y = (b + w\cos v)\sin u, \qquad z = w\sin v.$$

Then the solid torus R is the image under T of the region in uvw-space described by the inequalities

$$0 \leqq u \leqq 2\pi, \qquad 0 \leqq v \leqq 2\pi, \qquad 0 \leqq w \leqq a.$$

By a routine computation, we find that the Jacobian of T is

$$\frac{\partial(x, y, z)}{\partial(u, v, w)} = w(b + w\cos v).$$

Hence Eq. (9) with $F(x, y, z) \equiv 1$ yields volume

$$V = \iiint_T dx\,dy\,dz = \int_0^{2\pi}\int_0^{2\pi}\int_0^a (bw + w^2\cos v)\,dw\,du\,dv$$
$$= 2\pi\int_0^{2\pi}\left(\frac{1}{2}a^2b + \frac{1}{3}a^3\cos v\right)dv = 2\pi^2a^2b,$$

which agrees with the value $V = 2\pi b \cdot \pi a^2$ given by Pappus's first theorem (Section 14.5). ■

14.9 PROBLEMS

In Problems 1 through 6, solve for x and y in terms of u and v. Then compute the Jacobian $\partial(x, y)/\partial(u, v)$.

1. $u = x + y, \quad v = x - y$

2. $u = x - 2y, \quad v = 3x + y$

3. $u = xy, \quad v = y/x$

4. $u = 2(x^2 + y^2), \quad v = 2(x^2 - y^2)$

5. $u = x + 2y^2, \quad v = x - 2y^2$

6. $u = \dfrac{2x}{x^2 + y^2}, \quad v = -\dfrac{2y}{x^2 + y^2}$

7. Let R be the parallelogram bounded by the lines $x + y = 1$, $x + y = 2$ and $2x - 3y = 2$, $2x - 3y = 5$. Substitute $u = x + y$, $v = 2x - 3y$ to find its area

$$A = \iint_R dx\,dy.$$

8. Substitute $u = xy$, $v = y/x$ to find the area of the first-quadrant region bounded by the lines $y = x$, $y = 2x$ and the hyperbolas $xy = 1$, $xy = 2$ (Fig. 14.9.7).

9. Substitute $u = xy$, $v = xy^3$ to find the area of the first-quadrant region bounded by the curves $xy = 2$, $xy = 4$ and $xy^3 = 3$, $xy^3 = 6$ (Fig. 14.9.8).

10. Find the area of the first-quadrant region bounded by the curves $y = x^2$, $y = 2x^2$ and $x = y^2$, $x = 4y^2$ (Fig. 14.9.9). (*Suggestion:* Let $y = ux^2$ and $x = vy^2$.)

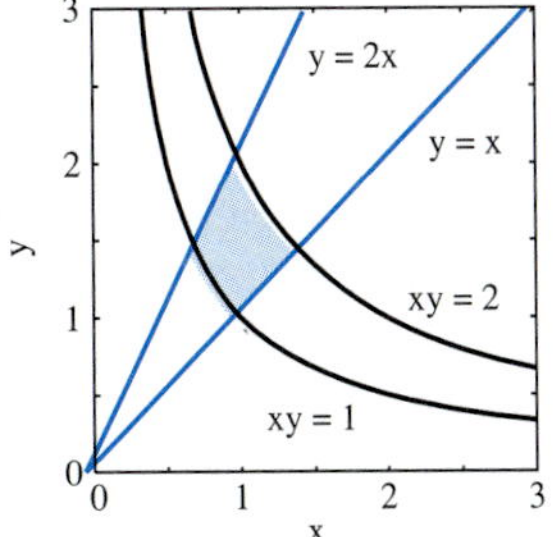

Fig. 14.9.7 The region of Problem 8

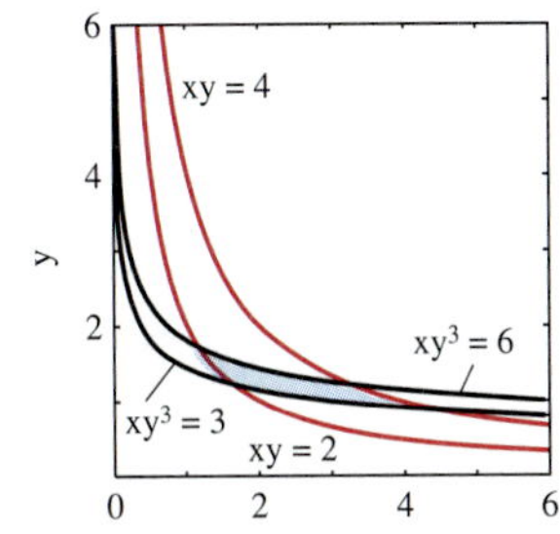

Fig. 14.9.8 The region of Problem 9

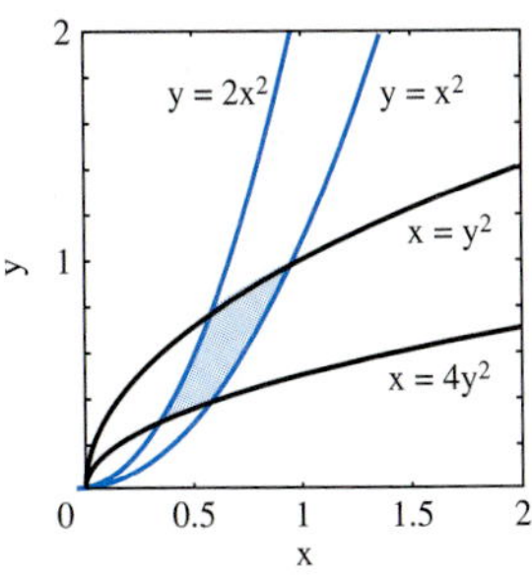

Fig. 14.9.9 The region of Problem 10

11. Use the method of Problem 10 to find the area of the first-quadrant region bounded by the curves $y = x^3$, $y = 2x^3$ and $x = y^3$, $x = 4y^3$.

12. Let R be the first-quadrant region bounded by the circles $x^2 + y^2 = 2x$, $x^2 + y^2 = 6x$ and the circles $x^2 + y^2 = 2y$, $x^2 + y^2 = 8y$. Use the transformation

$$u = \frac{2x}{x^2 + y^2}, \qquad v = \frac{2y}{x^2 + y^2}$$

to evaluate the integral

$$\iint_R \frac{1}{(x^2 + y^2)^2}\, dx\, dy.$$

13. Use elliptical coordinates $x = 3r\cos\theta$, $y = 2r\sin\theta$ to find the volume of the region bounded by the xy-plane, the paraboloid $z = x^2 + y^2$, and the elliptic cylinder

$$\frac{x^2}{9} + \frac{y^2}{4} = 1.$$

14. Let R be the solid ellipsoid with outer boundary surface

$$\frac{x^2}{a^2} + \frac{y^2}{b^2} + \frac{z^2}{c^2} = 1.$$

Use the transformation $x = au$, $y = bv$, $z = cw$ to show that the volume of this ellipsoid is

$$V = \iiint_R 1\, dx\, dy\, dz = \frac{4}{3}\pi abc.$$

15. Find the volume of the region in the first octant that is bounded by the hyperbolic cylinders $xy = 1, xy = 4$; $xz = 1, xz = 9$; and $yz = 4, yz = 9$. (*Suggestion:* Let $u = xy$, $v = xz$, $w = yz$, and note that $uvw = x^2y^2z^2$.)

16. Use the transformation

$$x = \frac{r}{t}\cos\theta, \qquad y = \frac{r}{t}\sin\theta, \qquad z = r^2$$

to find the volume of the region R that lies between the paraboloids $z = x^2 + y^2$, $z = 4(x^2 + y^2)$ and the planes $z = 1$, $z = 4$.

17. Let R be the rotated elliptical region bounded by the graph of $x^2 + xy + y^2 = 3 = 3u^2 + v^2$, where $x = u + v$ and $y = u - v$. Show that

$$\iint_R \exp(-x^2 - xy - y^2)\, dx\, dy = 2\iint_S \exp(-3u^2 - v^2)\, du\, dv.$$

Then substitute $u = r\cos\theta$, $v = \sqrt{3}(r\sin\theta)$ to evaluate the latter integral.

18. From the chain rule and from the following property of determinants, derive the relation in Eq. (6) between the Jacobians of a transformation and its inverse.

$$\begin{vmatrix} a_1 & b_1 \\ c_1 & d_1 \end{vmatrix} \cdot \begin{vmatrix} a_2 & b_2 \\ c_2 & d_2 \end{vmatrix} = \begin{vmatrix} a_1a_2 + b_1c_2 & a_1b_2 + b_1d_2 \\ a_2c_1 + c_2d_1 & b_2c_1 + d_1d_2 \end{vmatrix}.$$

19. Change to spherical coordinates to show that, for $k > 0$,

$$\int_{-\infty}^{+\infty}\int_{-\infty}^{+\infty}\int_{-\infty}^{+\infty} \sqrt{x^2 + y^2 + z^2}\, \exp\left(-k(x^2 + y^2 + z^2)\right) dx\, dy\, dz = \frac{2\pi}{k^2}.$$

20. Let R be the solid ellipsoid with constant density δ and boundary surface

$$\frac{x^2}{a^2} + \frac{y^2}{b^2} + \frac{z^2}{c^2} = 1.$$

Use ellipsoidal coordinates $x = a\rho\sin\phi\cos\theta$, $y = b\rho\sin\phi\sin\theta$, $z = c\rho\cos\phi$ to show that the mass of R is $M = \frac{4}{3}\pi\delta abc$.

21. Show that the moment of inertia of the ellipsoid of Problem 20 around the z-axis is $I_z = \frac{1}{5}M(a^2 + b^2)$.

In Problems 22 through 26, use a computer algebra system (if necessary) to find the indicated centroids and moments of inertia.

22. The centroid of the plane region of Problem 8 (Fig. 14.9.7)

23. The centroid of the plane region of Problem 9 (Fig. 14.9.8)

24. The centroid of the plane region of Problem 10 (Fig. 14.9.9)

25. The moment of inertia around each coordinate axis of the solid ellipsoid of Problem 20

26. The centroid of the solid of Problem 16 and its moments of inertia around the coordinate axes

27. Write the triple integral that gives the average distance of points of the solid ellipsoid of Problem 20 from the origin. Then approximate that integral in the case $a = 4$, $b = 3$, and $c = 2$.

CHAPTER 14 REVIEW: *Definitions, Concepts, Results*

Use the following list as a guide to concepts that you may need to review.

1. Definition of the double integral as a limit of Riemann sums
2. Evaluation of double integrals by iterated single integration
3. Use of the double integral to find the volume between two surfaces above a given plane region
4. Transformation of the double integral $\iint_R f(x, y)\, dA$ into polar coordinates

5. Application of double integrals to find mass, centroids, and moments of inertia of plane laminae
6. The two theorems of Pappus
7. Definition of the triple integral as a limit of Riemann sums
8. Evaluation of triple integrals by iterated single integration
9. Applications of triple integrals to find volume, mass, centroids, and moments of inertia
10. Transformation of the triple integral $\iiint_T f(x, y, z)\, dV$ into cylindrical and spherical coordinates
11. The surface area of a parametric surface
12. The area of a surface $z = f(x, y)$ for (x, y) in the plane region R
13. The Jacobian of a transformation of coordinates
14. The transformation of a double or triple integral corresponding to a given change of variables

CHAPTER 14 *Miscellaneous Problems*

In Problems 1 through 5, evaluate the given integral by first reversing the order of integration.

1. $\displaystyle\int_0^1\int_{y^{1/3}}^1 \frac{1}{\sqrt{1+x^2}}\,dx\,dy$ **2.** $\displaystyle\int_0^1\int_y^1 \frac{\sin x}{x}\,dx\,dy$

3. $\displaystyle\int_0^1\int_x^1 \exp(-y^2)\,dy\,dx$ **4.** $\displaystyle\int_0^8\int_{x^{2/3}}^4 x\cos y^4\,dy\,dx$

5. $\displaystyle\int_0^4\int_{\sqrt{y}}^2 \frac{y\exp(x^2)}{x^3}\,dx\,dy$

6. The double integral

$$\int_0^\infty\int_x^\infty \frac{e^{-y}}{y}\,dy\,dx$$

is an improper integral over the unbounded region in the first quadrant between the lines $y = x$ and $x = 0$. Assuming that it is valid to reverse the order of integration, evaluate this integral by integrating first with respect to x.

7. Find the volume of the solid T that lies below the paraboloid $z = x^2 + y^2$ and above the triangle R in the xy-plane that has vertices at $(0, 0, 0)$, $(1, 1, 0)$, and $(2, 0, 0)$.

8. Find by integration in cylindrical coordinates the volume bounded by the paraboloids $z = 2x^2 + 2y^2$ and $z = 48 - x^2 - y^2$.

9. Use integration in spherical coordinates to find the volume and centroid of the solid region that is inside the sphere $\rho = 3$, below the cone $\phi = \pi/3$, and above the xy-plane $\phi = \pi/2$.

10. Find the volume of the solid bounded by the elliptic paraboloids $z = x^2 + 3y^2$ and $z = 8 - x^2 - 5y^2$.

11. Find the volume bounded by the paraboloid $y = x^2 + 3z^2$ and the parabolic cylinder $y = 4 - z^2$.

12. Find the volume of the region bounded by the parabolic cylinders $z = x^2$, $z = 2 - x^2$ and the planes $y = 0$, $y + z = 4$.

13. Find the volume of the region bounded by the elliptical cylinder $y^2 + 4z^2 = 4$ and the planes $x = 0$, $x = y + 2$.

14. Show that the volume of the solid bounded by the elliptical cylinder

$$\frac{x^2}{a^2} + \frac{y^2}{b^2} = 1$$

and the planes $z = 0$, $z = h + x$ (where $h > a > 0$) is $V = \pi abh$.

15. Let R be the first-quadrant region bounded by the curve $x^4 + x^2y^2 = y^2$ and the line $y = x$. Use polar coordinates to evaluate

$$\iint_R \frac{1}{(1 + x^2 + y^2)^2}\,dA.$$

In Problems 16 through 20, find the mass and centroid of a plane lamina with the given shape and density δ.

16. The region bounded by $y = x^2$ and $x = y^2$; $\delta(x, y) = x^2 + y^2$

17. The region bounded by $x = 2y^2$ and $y^2 = x - 4$; $\delta(x, y) = y^2$

18. The region between $y = \ln x$ and the x-axis over the interval $1 \leqq x \leqq 2$; $\delta(x, y) = 1/x$

19. The circle bounded by $r = 2\cos\theta$; $\delta(r, \theta) = k$ (a constant)

20. The region of Problem 19; $\delta(r, \theta) = r$

21. Use the first theorem of Pappus to find the y-coordinate of the centroid of the upper half of the ellipse

$$\frac{x^2}{a^2} + \frac{y^2}{b^2} = 1.$$

Employ the facts that the area of this semiellipse is $A = \pi ab/2$ and the volume of the ellipsoid it generates when rotated around the x-axis is $V = \frac{4}{3}\pi ab^2$.

22. (a) Use the first theorem of Pappus to find the centroid of the first-quadrant portion of the annular ring with boundary circles $x^2 + y^2 = a^2$ and $x^2 + y^2 = b^2$ (where $0 < a < b$). (b) Show that the limiting position of this centroid as b fi a is the centroid of a quarter-circular arc, as we found in Problem 44 of Section 14.5.

23. Find the centroid of the region in the xy-plane bounded by the x-axis and the parabola $y = 4 - x^2$.

24. Find the volume of the solid that lies below the parabolic cylinder $z = x^2$ and above the triangle in the xy-plane bounded by the coordinate axes and the line $x + y = 1$.

25. Use cylindrical coordinates to find the volume of the ice-cream cone bounded above by the sphere $x^2 + y^2 + z^2 = 5$ and below by the cone $z = 2\sqrt{x^2 + y^2}$.

26. Find the volume and centroid of the ice-cream cone bounded above by the sphere $\rho = a$ and below by the cone $\phi = \pi/3$.

27. A homogeneous solid circular cone has mass M and base radius a. Find its moment of inertia around its axis of symmetry.

28. Find the mass of the first octant of the ball $\rho \leqq a$ if its density at (x, y, z) is $\delta(x, y, z) = xyz$.

29. Find the moment of inertia around the x-axis of the homogeneous solid ellipsoid with unit density and boundary surface

$$\frac{x^2}{a^2} + \frac{y^2}{b^2} + \frac{z^2}{c^2} = 1.$$

30. Find the volume of the region in the first octant that is bounded by the sphere $\rho = a$, the cylinder $r = a$, the plane $z = a$, the xz-plane, and the yz-plane.

31. Find the moment of inertia around the z-axis of the homogeneous region of unit density that lies inside both the sphere $\rho = 2$ and the cylinder $r = 2\cos\theta$.

In Problems 32 through 34, a volume is generated by revolving a plane region R around an axis. To find the volume, set up a double integral over R by revolving an area element dA around the indicated axis to generate a volume element dV.

32. Find the volume of the solid obtained by revolving around the y-axis the region inside the circle $r = 2a\cos\theta$.

33. Find the volume of the solid obtained by revolving around the x-axis the region enclosed by the cardioid $r = 1 + \cos\theta$.

34. Find the volume of the solid torus obtained by revolving the disk $0 \leqq r \leqq a$ around the line $x = -b$, $|b| \geqq a$.

35. Assume that the torus of Problem 34 has uniform density δ. Find its moment of inertia around its natural axis of symmetry.

Problems 36 through 42 deal with *average distance.* The **average distance** $\bar{d}$ of the point (x_0, y_0) from the points of the plane region R with area A is defined to be

$$\bar{d} = \frac{1}{A}\iint_R \sqrt{(x - x_0)^2 + (y - y_0)^2}\, dA.$$

The average distance of a point (x_0, y_0, z_0) from the points of a space region is defined analogously.

36. Show that the average distance of the points of a disk of radius a from its center is $2a/3$.

37. Show that the average distance of the points of a disk of radius a from a fixed point on its boundary is $32a/9\pi$.

38. A circle of radius 1 is interior to and tangent to a circle of radius 2. Find the average distance of the point of tangency from the points that lie between the two circles.

39. Show that the average distance of the points of a spherical ball of radius a from its center is $3a/4$.

40. Show that the average distance of the points of a spherical ball of radius a from a fixed point on its surface is $6a/5$.

41. A sphere of radius 1 is interior to and tangent to a sphere of radius 2. Find the average distance of the point of tangency from the set of all points between the two spheres.

42. A right circular cone has radius R and height H. Find the average distance of points of the cone from its vertex.

43. Find the surface area of the part of the paraboloid $z = 10 - r^2$ that lies between the two planes $z = 1$ and $z = 6$.

44. Find the surface area of the part of the surface $z = y^2 - x^2$ that is inside the cylinder $x^2 + y^2 = 4$.

45. Let A be the surface area of the zone on the sphere $\rho = a$ between the planes $z = z_1$ and $z = z_2$ (where $-a \leqq z_1 < z_2 \leqq a$). Use the formula of Problem 18 in Section 14.8 to show that $A = 2\pi ah$, where $h = z_2 - z_1$.

46. Find the surface area of the part of the sphere $\rho = 2$ that is inside the cylinder $x^2 + y^2 = 2x$.

47. A square hole with side length 2 is cut through a cone of height 2 and base radius 2; the centerline of the hole is the axis of symmetry of the cone. Find the area of the surface removed from the cone.

48. Numerically approximate the surface area of the part of the parabolic cylinder $2z = x^2$ that lies inside the cylinder $x^2 + y^2 = 1$.

49. A "fence" of variable height $h(t)$ stands above the plane curve $\big(x(t), y(t)\big)$. Thus the fence has the parametrization $x = x(t)$, $y = y(t)$, $z = z$ for $a \leqq t \leqq b$, $0 \leqq z \leqq h(t)$. Apply Eq. (8) of Section 14.8 to show that the area of the fence is

$$A = \int_a^b \int_0^{h(t)} \left[\left(\frac{dx}{dt}\right)^2 + \left(\frac{dy}{dt}\right)^2\right]^{1/2} dz\, dt.$$

50. Apply the formula of Problem 49 to compute the area of the part of the cylinder $r = a\sin\theta$ that lies inside the sphere $r^2 + z^2 = a^2$.

51. Find the polar moment of inertia of the first-quadrant region of constant density δ that is bounded by the hyperbolas $xy = 1$, $xy = 3$ and $x^2 - y^2 = 1$, $x^2 - y^2 = 4$.

52. Substitute $u = x - y$ and $v = x + y$ to evaluate

$$\iint_R \exp\left(\frac{x - y}{x + y}\right) dx\, dy,$$

where R is bounded by the coordinate axes and the line $x + y = 1$.

53. Use ellipsoidal coordinates $x = a\rho\sin\phi\cos\theta$, $y = b\rho\sin\phi\sin\theta$, $z = c\rho\cos\phi$ to find the mass of the solid ellipsoid

$$\frac{x^2}{a^2}+\frac{y^2}{b^2}+\frac{z^2}{c^2}\leqq 1$$

if its density at the point (x, y, z) is given by

$$\delta(x,y,z)=1-\frac{x^2}{a^2}-\frac{y^2}{b^2}-\frac{z^2}{c^2}.$$

54. Let R be the first-quadrant region bounded by the lemniscates $r^2 = \cos 2\theta$, $r^2 = 4\cos 2\theta$ and $r^2 = \sin 2\theta$, $r^2 = 4\sin 2\theta$ (Fig. 14.MP.1). Show that its area is $A = \frac{1}{4}(2\sqrt{17} - 5\sqrt{2})$. [*Suggestion:* Define the transformation T from the uv-plane to the $r\theta$-plane by $r^2 = u^{1/2}\cos 2\theta$, $r^2 = v^{1/2}\sin 2\theta$. Show first that

$$r^4=\frac{uv}{u+v},\qquad \theta=\frac{1}{2}\arctan\frac{u^{1/2}}{v^{1/2}}.$$

Then show that

$$\frac{\partial(r,\theta)}{\partial(u,v)}=-\frac{1}{16r(u+v)^{3/2}}.\Bigg]$$

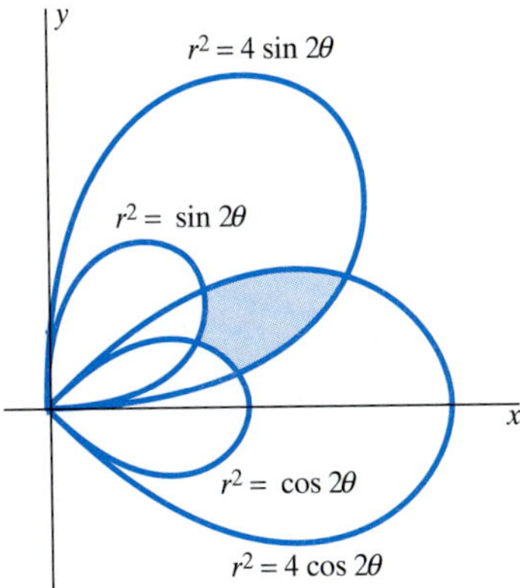

Fig. 14.MP.1 The region R of Problem 54

55. A 2-by-2 square hole is cut symmetrically through a sphere of radius $\sqrt{3}$ (see Fig. 14.MP.2). (a) Show that the total surface area of the two pieces cut from the sphere is

$$A=\int_0^1 8\sqrt{3}\arcsin\left(\frac{1}{\sqrt{3-x^2}}\right)dx.$$

Then use Simpson's rule to approximate this integral. (b) (Difficult!) Show that the exact value of the integral in part (a) is $A = 4\pi(\sqrt{3}-1)$. (*Suggestion:* First integrate by parts, then substitute $x = \sqrt{2}\sin\theta$.)

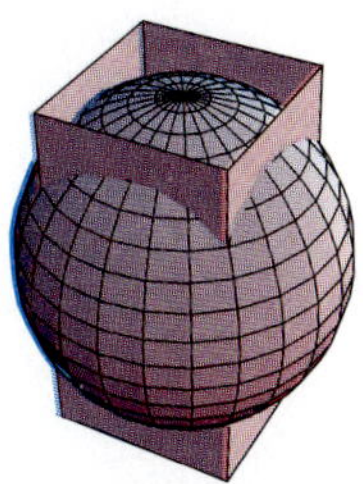

Fig. 14.MP.2
Cutting a square hole through the sphere of Problem 55

56. Show that the volume enclosed by the surface

$$x^{2/3}+y^{2/3}+z^{2/3}=a^{2/3}$$

is $V = \frac{4}{35}\pi a^3$. (*Suggestion:* Substitute $y = b\sin^3\theta$.)

57. Show that the volume enclosed by the surface

$$x^{1/3}+y^{1/3}+z^{1/3}=a^{1/3}$$

is $V = \frac{1}{210}a^3$. (*Suggestion:* Substitute $y = b\sin^6\theta$.)

58. Find the average of the *square* of the distance of points of the solid ellipsoid $(x/a)^2 + (y/b)^2 + (z/c)^2 \leqq 1$ from the origin.

59. A cube C of edge length 1 is rotated around a line passing through two opposite vertices, thereby sweeping out a solid S of revolution. Find the volume of S. (*Answer:* $\pi/\sqrt{3} \approx 1.8138$.)

CHAPTER 15

VECTOR CALCULUS

C. F. Gauss (1777–1855)

It is customary to list Archimedes, Newton, and Carl Friedrich Gauss as history's three preeminent mathematicians. Gauss was a precocious infant in a poor and uneducated family. He learned to calculate before he could talk and taught himself to read before beginning school in his native Brunswick, Germany. At age 14 he was already familiar with elementary geometry, algebra, and analysis. By age 18, when he entered the University of Göttingen, he had discovered empirically the "prime number theorem," which implies that the number of primes p between 1 and n is about $n/(\ln n)$. This theorem was not proved rigorously until a century later.

During his first year at university, Gauss investigated ruler-and-compass constructions of regular polygons and demonstrated the constructability of the regular 17-gon (the first advance in this area since the similar construction of the regular pentagon in Euclid's *Elements* 2000 years earlier). In 1801 Gauss published his great treatise *Disquisitiones arithmeticae,* which set the pattern for nineteenth-century research in number theory. This book established Gauss as a mathematician of uncommon stature, but another event thrust him into the public eye. On January 1, 1801, the new asteroid Ceres was observed, but it disappeared behind the sun a month later. In the following weeks, astronomers searched the skies in vain for Ceres' reappearance. It was Gauss who developed the method of least-squares approximations to predict the asteroid's future orbit on the basis of a handful of observations. When Gauss's three-month long computation was finished, Ceres was soon spotted in the precise location he had predicted.

In 1807 Gauss become director of the astronomical observatory in Göttingen, where he remained until his death. His published work thereafter dealt mainly with physical science, although his unpublished papers show that he continued to work on theoretical mathematics ranging from infinite series and special functions to non-Euclidean geometry. His work on the shape of the earth's surface established the new subject of differential geometry, and his studies of the earth's magnetic and gravitational fields involved results such as the divergence theorem of Section 15.6.

The concept of curved space-time in Albert Einstein's general relativity theory traces back to the discovery of non-Euclidean geometry and Gauss's early investigations of differential geometry. A current application of relativity theory is the study of black holes. Space is itself thought to be severely warped in the vicinity of a black hole, with its immense gravitational attraction, and the mathematics required begins with the vector calculus of Chapter 15.

Artists conception of a black hole (lower left)

15.1 VECTOR FIELDS

This chapter is devoted to topics in the calculus of vector fields of importance in science and engineering. A **vector field** defined on a region T in space is a vector-valued function $\mathbf{F}$ that associates with each point (x, y, z) of T a vector

$$\mathbf{F}(x, y, z) = \mathbf{i}P(x, y, z) + \mathbf{j}Q(x, y, z) + \mathbf{k}R(x, y, z). \tag{1}$$

We may more briefly describe the vector field $\mathbf{F}$ in terms of its *component functions* P, Q, and R by writing

$$\mathbf{F} = P\mathbf{i} + Q\mathbf{j} + R\mathbf{k} \quad \text{or} \quad \mathbf{F} = \langle P, Q, R \rangle.$$

Note that the components P, Q, and R of a vector function are *scalar* (real-valued) functions.

A **vector field** in the plane is similar except that neither z-components nor z-coordinates are involved. Thus a vector field on the plane region R is a vector-valued function $\mathbf{F}$ that associates with each point (x, y) of R a vector

$$\mathbf{F}(x, y) = \mathbf{i}P(x, y) + \mathbf{j}Q(x, y) \tag{2}$$

or, briefly, $\mathbf{F} = P\mathbf{i} + Q\mathbf{j}$ or $\mathbf{F} = \langle P, Q \rangle$.

It is useful to be able to visualize a given vector field $\mathbf{F}$. One common way is to sketch a collection of typical vectors $\mathbf{F}(x, y)$, each represented by an arrow of length $|\mathbf{F}(x, y)|$ and located with (x, y) as its initial point. This procedure is illustrated in Example 1.

EXAMPLE 1 Describe the vector field $\mathbf{F}(x, y) = x\mathbf{i} + y\mathbf{j}$.

Solution For each point (x, y) in the coordinate plane, $\mathbf{F}(x, y)$ is simply its position vector. It points directly away from the origin and has length

$$|\mathbf{F}(x, y)| = |x\mathbf{i} + y\mathbf{j}| = \sqrt{x^2 + y^2} = r,$$

equal to the distance from the origin to (x, y). Figure 15.1.1 shows some typical vectors representing this vector field. ■

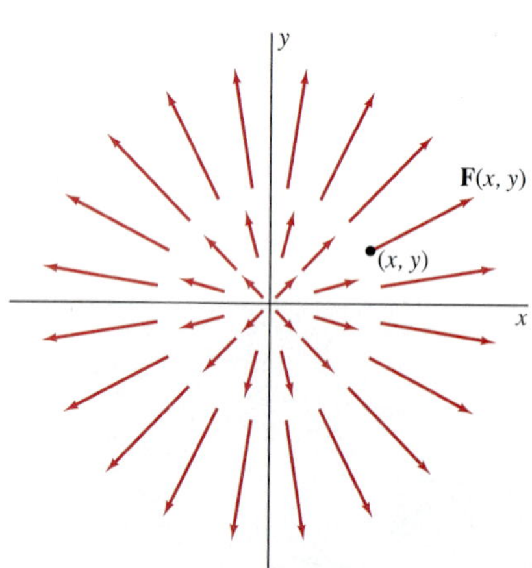

Fig. 15.1.1 The vector field $\mathbf{F}(x, y) = x\mathbf{i} + y\mathbf{j}$

Among the most important vector fields in applications are velocity vector fields. Imagine the steady flow of a fluid, such as the water in a river or the solar wind. By a *steady flow* we mean that the velocity vector $\mathbf{v}(x, y, z)$ of the fluid flowing through each point (x, y, z) is independent of time (although not necessarily independent of x, y, and z), so the pattern of the flow remains constant. Then $\mathbf{v}(x, y, z)$ is the **velocity vector field** of the fluid flow.

EXAMPLE 2 Suppose that the horizontal xy-plane is covered with a thin sheet of water that is revolving (rather like a whirlpool) around the origin with constant angular speed ω radians per second in the counterclockwise direction. Describe the associated velocity vector field.

Solution In this case we have a two-dimensional vector field $\mathbf{v}(x, y)$. At each point (x, y) the water is moving with speed $v = r\omega$ and tangential to the circle of radius $r = \sqrt{x^2 + y^2}$. The vector field

$$\mathbf{v}(x, y) = \omega(-y\mathbf{i} + x\mathbf{j}) \tag{3}$$

has length $r\omega$ and points in a generally counterclockwise direction, and

$$\mathbf{v} \cdot \mathbf{r} = \omega(-y\mathbf{i} + x\mathbf{j}) \cdot (x\mathbf{i} + y\mathbf{j}) = 0,$$

so $\mathbf{v}$ is tangent to the circle just mentioned. The velocity field determined by Eq. (3) is illustrated in Fig. 15.1.2. ■

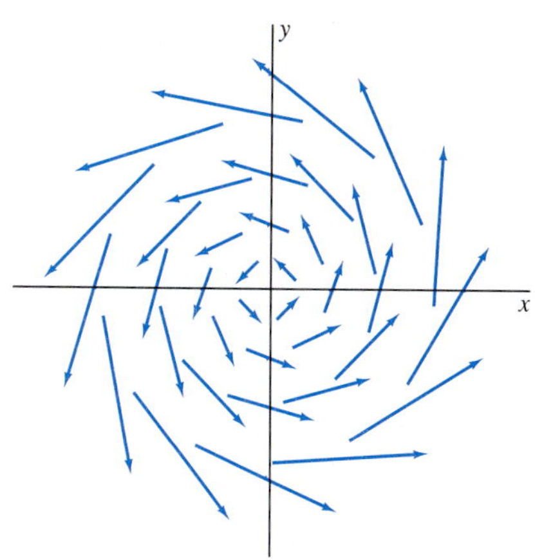

Fig. 15.1.2 The velocity vector field $\mathbf{v}(x, y) = \omega(-y\mathbf{i} + x\mathbf{j})$, drawn for $\omega = 1$ (Example 2)

Remark Most computer algebra systems have the facility to plot vector fields. For instance, either the *Maple* command `fieldplot( [x,y], x=-2..2, y=-2..2 )` or the *Mathematica* command `PlotVectorField[ {x,y}, {x,-2,2}, {y,-2,2} ]` generates a computer plot like Fig. 15.1.3 of the vector field $\mathbf{F} = x\mathbf{i} + y\mathbf{j}$ of Example 1. The computer program has scaled the vectors to a fixed maximum length so that the length of each vector as plotted is proportional to its actual length. Figure 15.1.4 shows a similar computer plot of the vector field $\mathbf{F} = -y\mathbf{i} + x\mathbf{j}$ of Example 2 (with $\omega = 1$).

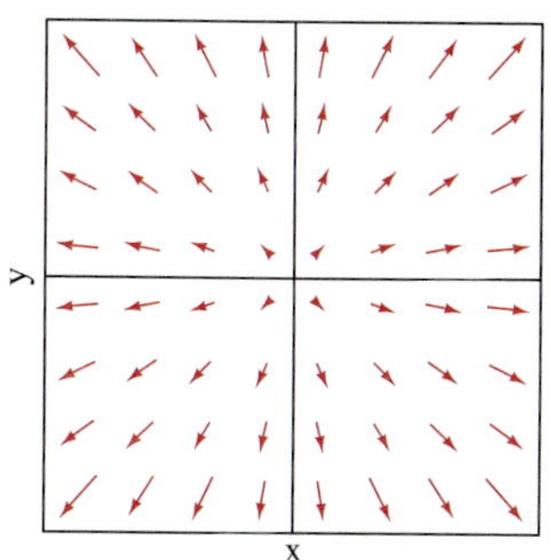

Fig. 15.1.3 The vector field $\mathbf{F} = x\mathbf{i} + y\mathbf{j}$

Equally important in physical applications are *force fields.* Suppose that some circumstance (perhaps gravitational or electrical in character) causes a force $\mathbf{F}(x, y, z)$ to act on a particle when it is placed at the point (x, y, z). Then we have a force field $\mathbf{F}$. Example 3 deals with what is perhaps the most common force field perceived by human beings.

EXAMPLE 3 Suppose that a mass M is fixed at the origin in space. When a particle of unit mass is placed at the point (x, y, z) other than the origin, it is subjected to a force $\mathbf{F}(x, y, z)$ of gravitational attraction directed toward the mass M at the origin. By Newton's inverse-square law of gravitation, the magnitude of $\mathbf{F}$ is $F = GM/r^2$, where $r = \sqrt{x^2 + y^2 + z^2}$ is the length of the position vector $\mathbf{r} = x\mathbf{i} + y\mathbf{j} + z\mathbf{k}$. It follows immediately that

$$\mathbf{F}(x, y, z) = -\frac{k\mathbf{r}}{r^3}, \tag{4}$$

where $k = GM$, because this vector has both the correct magnitude and the correct direction (toward the origin, for $\mathbf{F}$ is a multiple of $-\mathbf{r}$). A force field of the form in Eq. (4) is called an *inverse-square* force field. Note that $\mathbf{F}(x, y, z)$ is not defined at the origin and that $|\mathbf{F}| \to +\infty$ as $r \to 0^+$. Figure 15.1.5 illustrates an inverse-square force field. ■

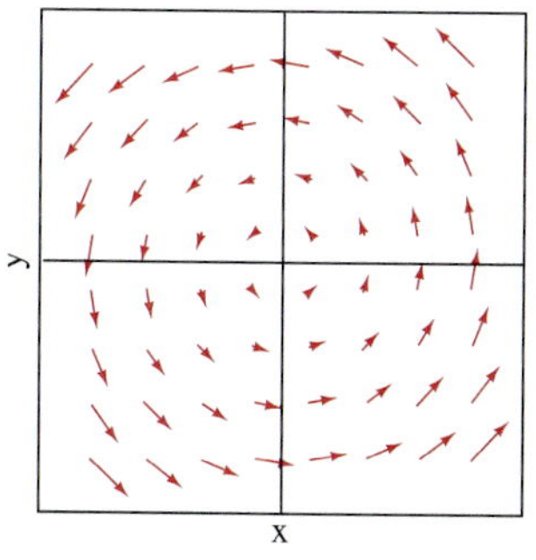

Fig. 15.1.4 The vector field $\mathbf{F} = -y\mathbf{i} + x\mathbf{j}$

The Gradient Vector Field

In Section 13.8 we introduced the gradient vector of the real-valued function $f(x, y, z)$. It is the vector ∇f defined as follows:

$$\nabla f = \mathbf{i}\frac{\partial f}{\partial x} + \mathbf{j}\frac{\partial f}{\partial y} + \mathbf{k}\frac{\partial f}{\partial z}. \tag{5}$$

The partial derivatives on the right-hand side of Eq. (5) are evaluated at the point (x, y, z). Thus $\nabla f(x, y, z)$ is a vector field: It is the **gradient vector field** of the function f. According to Theorem 1 of Section 13.8, the vector $\nabla f(x, y, z)$ points in the direction in which the maximal directional derivative of f at (x, y, z) is obtained. For example, if $f(x, y, z)$ is the temperature at the point (x, y, z) in space, then you should move in the direction $\nabla f(x, y, z)$ in order to warm up the most quickly.

In the case of a two-variable scalar function $f(x, y)$, we suppress the third component in Eq. (5), so $\nabla f = \langle f_x, f_y \rangle = f_x\mathbf{i} + f_y\mathbf{j}$ defines a plane vector field.

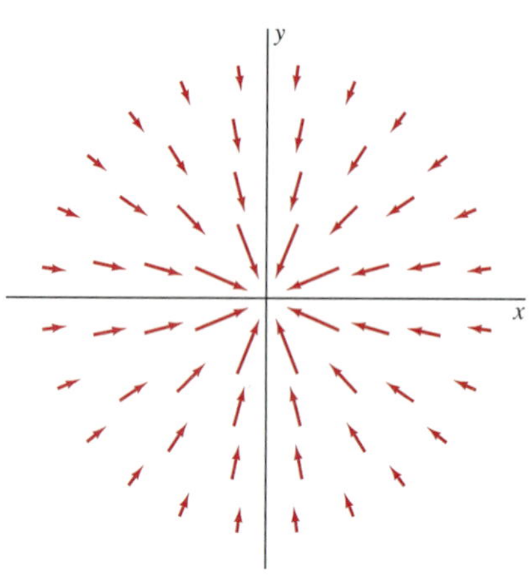

Fig. 15.1.5 An inverse-square force field (Example 3)

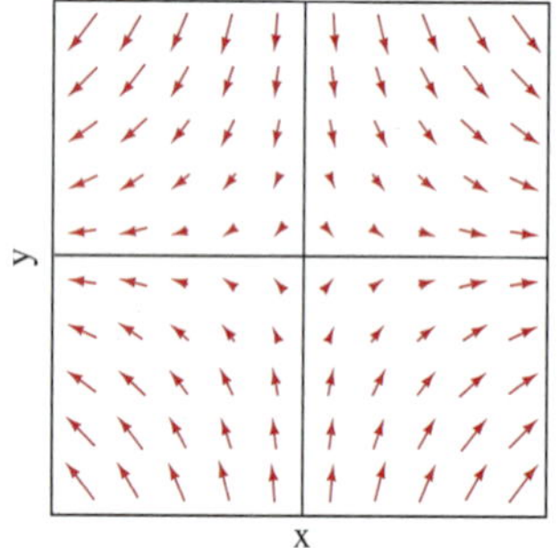

Fig. 15.1.6 The gradient vector field $\nabla f = 2x\mathbf{i} - 2y\mathbf{j}$ of Example 4

EXAMPLE 4 With $f(x, y) = x^2 - y^2$, the gradient vector field $\nabla f = 2x\mathbf{i} - 2y\mathbf{j}$ plotted in Fig. 15.1.6 should remind you of a contour plot near a saddle point. ■

The notation in Eq. (5) suggests the formal expression

$$\nabla = \mathbf{i}\frac{\partial}{\partial x} + \mathbf{j}\frac{\partial}{\partial y} + \mathbf{k}\frac{\partial}{\partial z}. \tag{6}$$

It is fruitful to think of ∇ as a *vector differential operator.* That is, ∇ is the operation that, when applied to the scalar function f, yields its gradient vector field ∇f. This operation behaves in several familiar and important ways like the operation D_x of single-variable differentiation. For a familiar example of this phenomenon, recall that in Chapter 13 we found the critical points of a function f of several variables to be those points at which $\nabla f(x, y, z) = \mathbf{0}$ and those at which $\nabla f(x, y, z)$ does not exist. As a computationally useful instance, suppose that f and g are functions and that a and b are constants. It then follows readily from Eq. (5) and from the linearity of partial differentiation that

$$\nabla(af + bg) = a\nabla f + b\nabla g. \tag{7}$$

Thus the gradient operator is *linear.* It also satisfies the product rule, as demonstrated in Example 5.

EXAMPLE 5 Given the differentiable functions $f(x, y, z)$ and $g(x, y, z)$, show that

$$\nabla(fg) = f\,\nabla g + g\nabla f. \tag{8}$$

Solution We apply the definition in Eq. (5) and the product rule for partial differentiation. Thus

$$\begin{aligned}\nabla(fg) &= \mathbf{i}\frac{\partial(fg)}{\partial x} + \mathbf{j}\frac{\partial(fg)}{\partial y} + \mathbf{k}\frac{\partial(fg)}{\partial z}\\ &= \mathbf{i}(fg_x + gf_x) + \mathbf{j}(fg_y + gf_y) + \mathbf{k}(fg_z + gf_z)\\ &= f\cdot(\mathbf{i}g_x + \mathbf{j}g_y + \mathbf{k}g_z) + g\cdot(\mathbf{i}f_x + \mathbf{j}f_y + \mathbf{k}f_z) = f\,\nabla g + g\nabla f.\end{aligned}$$ ■

The Divergence of a Vector Field

Suppose that we are given the vector-valued function

$$\mathbf{F}(x, y, z) = \mathbf{i}P(x, y, z) + \mathbf{j}Q(x, y, z) + \mathbf{k}R(x, y, z)$$

with differentiable component functions P, Q, and R. Then the **divergence** of $\mathbf{F}$ is the scalar function div $\mathbf{F}$ defined as follows:

$$\text{div}\,\mathbf{F} = \nabla\cdot\mathbf{F} = \frac{\partial P}{\partial x} + \frac{\partial Q}{\partial y} + \frac{\partial R}{\partial z}. \tag{9}$$

Here *div* is an abbreviation for "divergence," and the alternative notation $\nabla\cdot\mathbf{F}$ is consistent with the formal expression for ∇ in Eq. (6). That is,

$$\nabla\cdot\mathbf{F} = \left\langle\frac{\partial}{\partial x}, \frac{\partial}{\partial y}, \frac{\partial}{\partial z}\right\rangle\cdot\langle P, Q, R\rangle = \frac{\partial P}{\partial x} + \frac{\partial Q}{\partial y} + \frac{\partial R}{\partial z}.$$

We will see in Section 15.6 that if $\mathbf{v}$ is the velocity vector field of a steady fluid flow, then the value of div $\mathbf{v}$ at the point (x, y, z) is essentially the net rate per unit volume at which fluid mass is flowing away (or "diverging") from the point (x, y, z).

EXAMPLE 6 If the vector field $\mathbf{F}$ is given by

$$\mathbf{F}(x, y, z) = (xe^y)\mathbf{i} + (z \sin y)\mathbf{j} + (xy \ln z)\mathbf{k},$$

then $P(x, y, z) = xe^y$, $Q(x, y, z) = z \sin y$, and $R(x, y, z) = xy \ln z$. Hence Eq. (9) yields

$$\operatorname{div} \mathbf{F} = \frac{\partial}{\partial x}(xe^y) + \frac{\partial}{\partial y}(z \sin y) + \frac{\partial}{\partial z}(xy \ln z) = e^y + z \cos y + \frac{xy}{z}.$$

For instance, the value of div $\mathbf{F}$ at the point $(-3, 0, 2)$ is

$$\nabla \cdot \mathbf{F}(-3, 0, 2) = e^0 + 2 \cos 0 + 0 = 3.$$ ■

The analogues of Eqs. (7) and (8) for divergence are the formulas

$$\nabla \cdot (a\mathbf{F} + b\mathbf{G}) = a\,\nabla \cdot \mathbf{F} + b\,\nabla \cdot \mathbf{G}, \tag{10}$$

$$\nabla \cdot (f\mathbf{G}) = (f)(\nabla \cdot \mathbf{G}) + (\nabla f) \cdot \mathbf{G}. \tag{11}$$

We ask you to verify these formulas in the problems. Note that Eq. (11)—in which f is a scalar function and $\mathbf{G}$ is a vector field—is consistent in that f and $\nabla \cdot \mathbf{G}$ are scalar functions, whereas ∇f and $\mathbf{G}$ are vector fields, so the sum on the right-hand side makes sense (and is a scalar function).

The Curl of a Vector Field

The **curl** of the vector field $\mathbf{F} = P\mathbf{i} + Q\mathbf{j} + R\mathbf{k}$ is the following vector field, abbreviated as curl $\mathbf{F}$:

$$\operatorname{curl} \mathbf{F} = \nabla \times \mathbf{F} = \begin{vmatrix} \mathbf{i} & \mathbf{j} & \mathbf{k} \\ \dfrac{\partial}{\partial x} & \dfrac{\partial}{\partial y} & \dfrac{\partial}{\partial z} \\ P & Q & R \end{vmatrix}. \tag{12}$$

When we evaluate the formal determinant in Eq. (12), we obtain

$$\operatorname{curl} \mathbf{F} = \mathbf{i}\left(\frac{\partial R}{\partial y} - \frac{\partial Q}{\partial z}\right) + \mathbf{j}\left(\frac{\partial P}{\partial z} - \frac{\partial R}{\partial x}\right) + \mathbf{k}\left(\frac{\partial Q}{\partial x} - \frac{\partial P}{\partial y}\right). \tag{13}$$

Although you could try to memorize this formula, we recommend—because you will generally find it simpler—that in practice you set up and evaluate directly the formal determinant in Eq. (12). Example 7 shows how easy this is.

EXAMPLE 7 For the vector field $\mathbf{F}$ of Example 6, Eq. (12) yields

$$\operatorname{curl} \mathbf{F} = \begin{vmatrix} \mathbf{i} & \mathbf{j} & \mathbf{k} \\ \dfrac{\partial}{\partial x} & \dfrac{\partial}{\partial y} & \dfrac{\partial}{\partial z} \\ xe^y & z \sin y & xy \ln z \end{vmatrix}$$

$$= \mathbf{i}(x \ln z - \sin y) + \mathbf{j}(-y \ln z) + \mathbf{k}(-xe^y).$$

In particular, the value of curl **F** at the point $(3, \pi/2, e)$ is

$$\{\nabla \times \mathbf{F}\}(3, \pi/2, e) = 2\mathbf{i} - \tfrac{1}{2}\pi\mathbf{j} - 3e^{\pi/2}\mathbf{k}.$$

We will see in Section 15.7 that if $\mathbf{v}$ is the velocity vector of a fluid flow, then the value of the vector curl $\mathbf{v}$ at the point (x, y, z) (where that vector is nonzero) determines the axis through (x, y, z) about which the fluid is rotating (or whirling or "curling") as well as the angular velocity of the rotation.

The analogues of Eqs. (10) and (11) for curl are the formulas

$$\nabla \times (a\mathbf{F} + b\,\mathbf{G}) = a(\nabla \times \mathbf{F}) + b\,(\nabla \times \mathbf{G}), \tag{14}$$

$$\nabla \times (f\mathbf{G}) = (f)(\nabla \times \mathbf{G}) + (\nabla f) \times \mathbf{G} \tag{15}$$

that we ask you to verify in the problems.

EXAMPLE 8 If the function $f(x, y, z)$ has continuous second-order partial derivatives, show that $\text{curl}(\text{grad} f) = \mathbf{0}$.

Solution Direct computation yields

$$\nabla \times \nabla f = \begin{vmatrix} \mathbf{i} & \mathbf{j} & \mathbf{k} \\ \dfrac{\partial}{\partial x} & \dfrac{\partial}{\partial y} & \dfrac{\partial}{\partial z} \\ \dfrac{\partial f}{\partial x} & \dfrac{\partial f}{\partial y} & \dfrac{\partial f}{\partial z} \end{vmatrix}$$

$$= \mathbf{i}\left(\frac{\partial^2 f}{\partial y\,\partial z} - \frac{\partial^2 f}{\partial z\,\partial y}\right) + \mathbf{j}\left(\frac{\partial^2 f}{\partial z\,\partial x} - \frac{\partial^2 f}{\partial x\,\partial z}\right) + \mathbf{k}\left(\frac{\partial^2 f}{\partial x\,\partial y} - \frac{\partial^2 f}{\partial y\,\partial x}\right).$$

Therefore $\nabla \times \nabla f = \mathbf{0}$ because of the equality of continuous mixed second-order partial derivatives.

15.1 PROBLEMS

In Problems 1 through 10, illustrate the given vector field **F** by sketching several typical vectors in the field.

1. $\mathbf{F}(x, y) = \mathbf{i} + \mathbf{j}$ **2.** $\mathbf{F}(x, y) = 3\mathbf{i} - 2\mathbf{j}$

3. $\mathbf{F}(x, y) = x\mathbf{i} - y\mathbf{j}$ **4.** $\mathbf{F}(x, y) = 2\mathbf{i} + x\mathbf{j}$

5. $\mathbf{F}(x, y) = (x^2 + y^2)^{1/2}\,(x\mathbf{i} + y\mathbf{j})$

6. $\mathbf{F}(x, y) = (x^2 + y^2)^{-1/2}\,(x\mathbf{i} + y\mathbf{j})$

7. $\mathbf{F}(x, y, z) = \mathbf{j} + \mathbf{k}$ **8.** $\mathbf{F}(x, y, z) = \mathbf{i} + \mathbf{j} - \mathbf{k}$

9. $\mathbf{F}(x, y, z) = -x\mathbf{i} - y\mathbf{j}$ **10.** $\mathbf{F}(x, y, z) = x\mathbf{i} + y\mathbf{j} + z\mathbf{k}$

Match the gradient vector fields of the functions in Problems 11 through 14 with the computer-generated plots in Figs. 15.1.7 through 15.1.10.

11. $f(x, y) = xy$ **12.** $f(x, y) = 2x^2 + y^2$

13. $f(x, y) = \sin \tfrac{1}{2}(x^2 + y^2)$

14. $f(x, y) = \sin \tfrac{1}{2}(y^2 - x^2)$

In Problems 15 through 24, calculate the divergence and curl of the given vector field **F**.

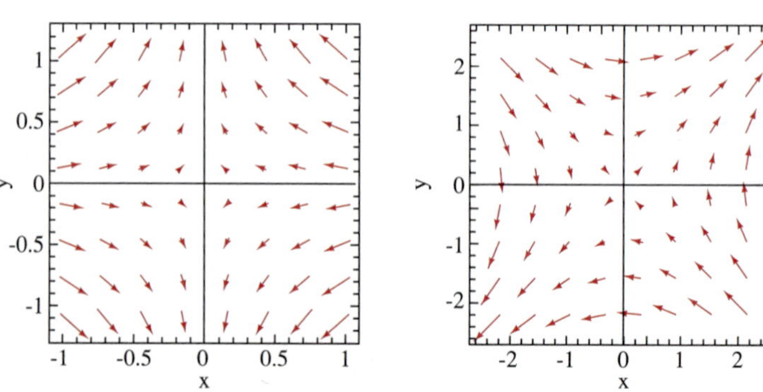

Fig. 15.1.7 **Fig. 15.1.8**

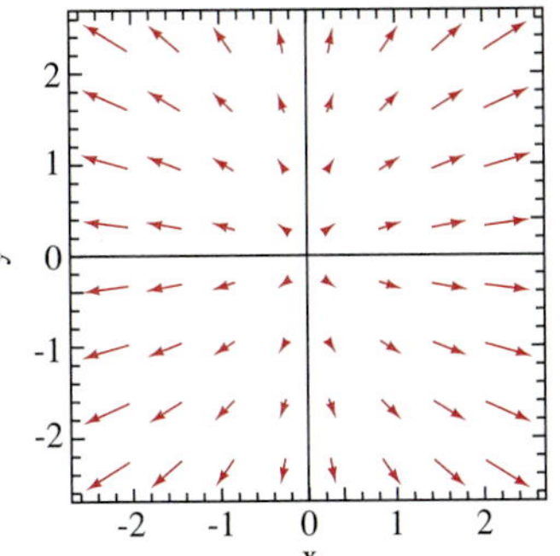

Fig. 15.1.9

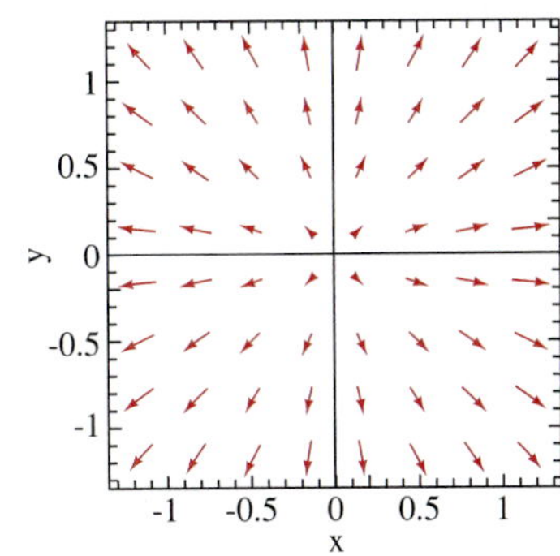

Fig. 15.1.10

15. $\mathbf{F}(x, y, z) = x\mathbf{i} + y\mathbf{j} + z\mathbf{k}$
16. $\mathbf{F}(x, y, z) = 3x\mathbf{i} - 2y\mathbf{j} - 4z\mathbf{k}$
17. $\mathbf{F}(x, y, z) = yz\mathbf{i} + xz\mathbf{j} + xy\mathbf{k}$
18. $\mathbf{F}(x, y, z) = x^2\mathbf{i} + y^2\mathbf{j} + z^2\mathbf{k}$
19. $\mathbf{F}(x, y, z) = xy^2\mathbf{i} + yz^2\mathbf{j} + zx^2\mathbf{k}$
20. $\mathbf{F}(x, y, z) = (2x - y)\mathbf{i} + (3y - 2z)\mathbf{j} + (7z - 3x)\mathbf{k}$
21. $\mathbf{F}(x, y, z) = (y^2 + z^2)\mathbf{i} + (x^2 + z^2)\mathbf{j} + (x^2 + y^2)\mathbf{k}$
22. $\mathbf{F}(x, y, z) = (e^{xz} \sin y)\mathbf{j} + (e^{xy} \cos z)\mathbf{k}$
23. $\mathbf{F}(x, y, z) = (x + \sin yz)\mathbf{i} + (y + \sin xz)\mathbf{j} + (z + \sin xy)\mathbf{k}$
24. $\mathbf{F}(x, y, z) = (x^2e^{-z})\mathbf{i} + (y^3 \ln x)\mathbf{j} + (z \cosh y)\mathbf{k}$

Apply the definitions of gradient, divergence, and curl to establish the identities in Problems 25 through 31, in which a and b denote constants, f and g denote differentiable scalar functions, and $\mathbf{F}$ and $\mathbf{G}$ denote differentiable vector fields.

25. $\nabla(af + bg) = a\,\nabla f + b\,\nabla g$
26. $\nabla \cdot (a\mathbf{F} + b\mathbf{G}) = a\,\nabla \cdot \mathbf{F} + b\,\nabla \cdot \mathbf{G}$
27. $\nabla \times (a\mathbf{F} + b\mathbf{G}) = a(\nabla \times \mathbf{F}) + b(\nabla \times \mathbf{G})$
28. $\nabla \cdot (f\mathbf{G}) = (f)(\nabla \cdot \mathbf{G}) + (\nabla f) \cdot \mathbf{G}$
29. $\nabla \times (f\mathbf{G}) = (f)(\nabla \times \mathbf{G}) + (\nabla f) \times \mathbf{G}$
30. $\nabla\left(\dfrac{f}{g}\right) = \dfrac{g\,\nabla f - f\,\nabla g}{g^2}$
31. $\nabla \cdot (\mathbf{F} \times \mathbf{G}) = \mathbf{G} \cdot (\nabla \times \mathbf{F}) - \mathbf{F} \cdot (\nabla \times \mathbf{G})$

Establish the identities in Problems 32 through 34 under the assumption that the scalar functions f and g and the vector field $\mathbf{F}$ are twice differentiable.

32. $\operatorname{div}(\operatorname{curl} \mathbf{F}) = 0$
33. $\operatorname{div}(\nabla fg) = f \operatorname{div}(\nabla g) + g \operatorname{div}(\nabla f) + 2(\nabla f) \cdot (\nabla g)$
34. $\operatorname{div}(\nabla f \times \nabla g) = 0$

Verify the identities in Problems 35 through 44, in which $\mathbf{a}$ is a constant vector, $\mathbf{r} = x\mathbf{i} + y\mathbf{j} + z\mathbf{k}$, and $r = |\mathbf{r}|$. Problems 37 and 38 imply that both the divergence and the curl of an inverse-square vector field vanish identically.

35. $\nabla \cdot \mathbf{r} = 3$ and $\nabla \times \mathbf{r} = \mathbf{0}$
36. $\nabla \cdot (\mathbf{a} \times \mathbf{r}) = 0$ and $\nabla \times (\mathbf{a} \times \mathbf{r}) = 2\mathbf{a}$
37. $\nabla \cdot \dfrac{\mathbf{r}}{r^3} = 0$
38. $\nabla \times \dfrac{\mathbf{r}}{r^3} = \mathbf{0}$
39. $\nabla r = \dfrac{\mathbf{r}}{r}$
40. $\nabla\left(\dfrac{1}{r}\right) = -\dfrac{\mathbf{r}}{r^3}$
41. $\nabla \cdot (r\mathbf{r}) = 4r$
42. $\nabla \cdot (\nabla r) = 0$
43. $\nabla(\ln r) = \dfrac{\mathbf{r}}{r^2}$
44. $\nabla(r^{10}) = 10r^8\mathbf{r}$

15.2 LINE INTEGRALS

The single integral $\int_a^b f(x)\,dx$ might be described as an integral along the x-axis. We now define integrals along curves in space (or in the plane). Such integrals are called *line integrals* (although the phrase *curve integrals* might be more appropriate).

To motivate the definition of the line integral of the function f along the smooth space curve C with parametrization

$$x = x(t), \quad y = y(t), \quad z = z(t) \qquad (a \leqq t \leqq b), \tag{1}$$

we imagine a thin wire shaped like C (Fig. 15.2.1). Suppose that $f(x, y, z)$ denotes the density of the wire at the point (x, y, z), measured in units of mass per unit length—for example, grams per centimeter. Then we expect to compute the total mass m of the curved wire as some kind of integral of the function f. To approximate m, we begin with a partition

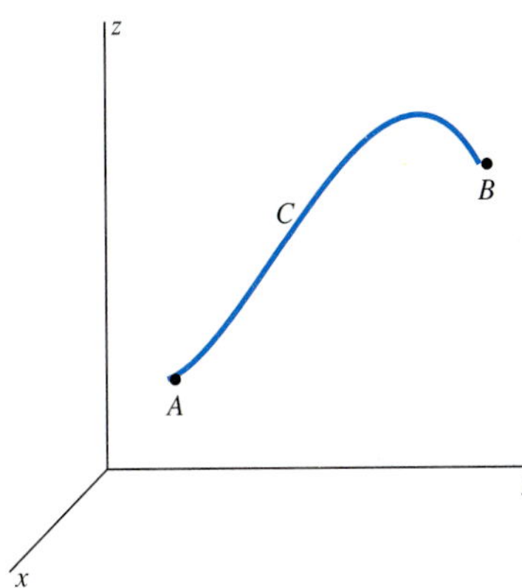

Fig. 15.2.1 A wire of variable density in the shape of the smooth curve C from A (where $t = a$) to B (where $t = b$)

$$a = t_0 < t_1 < t_2 < \cdots < t_{n-1} < t_n = b$$

of $[a, b]$ into n subintervals, all with the same length $\Delta t = (b - a)/n$. These subdivision points of $[a, b]$ produce, via our parametrization, a physical subdivision of the wire into short curve segments (Fig. 15.2.2). We let P_i denote the point $(x(t_i), y(t_i), z(t_i))$ for $i = 0, 1, 2, \ldots, n$. Then the points $P_0, P_1, \ldots, P_n$ are the subdivision points of C.

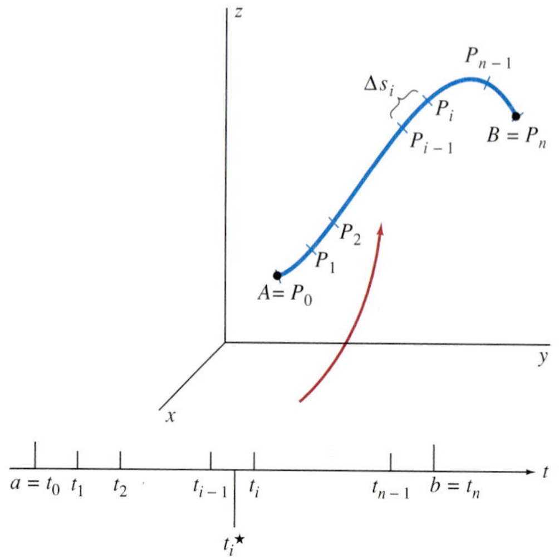

Fig. 15.2.2 The partition of the interval $[a, b]$ determines a related partition of the curve C into short arcs.

From our study of arc length in Sections 10.4 and 12.6, we know that the arc length Δs_i of the segment of C from P_{i-1} to P_i is

$$\begin{aligned}\Delta s_i &= \int_{t_{i-1}}^{t_i} \sqrt{[x'(t)]^2 + [y'(t)]^2 + [z'(t)]^2}\, dt \\ &= \sqrt{[x'(t_i^\star)]^2 + [y'(t_i^\star)]^2 + [z'(t_i^\star)]^2}\, \Delta t \end{aligned} \tag{2}$$

for some number $t_i^\star$ in the interval $[t_{i-1}, t_i]$. This is a consequence of the average value theorem for integrals of Section 5.5.

Denote $x(t_i^\star)$ by $x_i^\star$ and similarly define $y_i^\star$ and $z_i^\star$. If we multiply the density of the wire at the point $(x_i^\star, y_i^\star, z_i^\star)$ by the length Δs_i of the segment of C containing that point, we obtain an estimate of the mass of that segment of C. So, after we sum over all the segments, we have an estimate of the total mass m of the wire:

$$m \approx \sum_{i=1}^{n} f(x(t_i^\star), y(t_i^\star), z(t_i^\star))\, \Delta s_i .$$

The limit of this sum as $\Delta t \to 0$ should be the actual mass m. This is our motivation for the definition of the line integral of the function f along the curve C, denoted by

$$\int_C f(x, y, z)\, ds .$$

Definition ***Line Integral of a Function Along a Curve***

Suppose that the function $f(x, y, z)$ is defined at each point of the smooth curve C parametrized as in (1). Then the **line integral of f along C** is defined by

$$\int_C f(x, y, z)\, ds = \lim_{\Delta t \to 0} \sum_{i=1}^{n} f(x(t_i^\star), y(t_i^\star), z(t_i^\star))\, \Delta s_i , \tag{3}$$

provided that this limit exists.

Remark 1 It can be shown that the limit in (3) always exists if the function f is continuous at each point of C. Recall from Section 10.4 that the curve C is *smooth* provided that the component functions in its parametrization have continuous derivatives that are never simultaneously zero. When we substitute Eq. (2) into Eq. (3), we recognize the result as the limit of a Riemann sum. Therefore

$$\int_C f(x, y, z)\, ds = \int_a^b f(x(t), y(t), z(t))\sqrt{[x'(t)]^2 + [y'(t)]^2 + [z'(t)]^2}\, dt. \tag{4}$$

Thus we may evaluate the line integral $\int_C f(x, y, z)\, ds$ by expressing everything in terms of the parameter t, including the symbolic arc-length element

$$ds = \sqrt{\left(\frac{dx}{dt}\right)^2 + \left(\frac{dy}{dt}\right)^2 + \left(\frac{dz}{dt}\right)^2}\, dt.$$

As a consequence, the right-hand side in Eq. (4) is *evaluated as an ordinary single integral with respect to the real variable t.* Because of the appearance of the arc-length element ds, the line integral $\int_C f(x, y, z)\, ds$ is sometimes called the line integral of the function f **with respect to arc length** along the curve C.

Remark 2 A curve C that lies in the xy-plane may be regarded as a space curve for which z [and $z'(t)$] are zero. In this case we simply suppress the variable z in Eq. (4) and write

$$\int_C f(x, y)\, ds = \int_a^b f(x(t), y(t))\sqrt{[x'(t)]^2 + [y'(t)]^2}\, dt. \tag{5}$$

In the case that f is positive-valued, Fig. 15.2.3 illustrates an interpretation of the line integral in Eq. (5) as the area of a "fence" whose base is the curve C in the xy-plane, with the height of the fence above the point (x, y) given by $f(x, y)$.

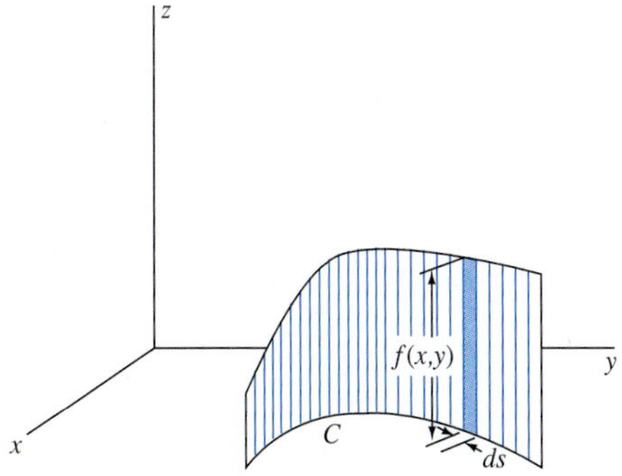

Fig. 15.2.3 The vertical strip with base ds and height $f(x, y)$ has area $dA = f(x, y)\, ds$, so the whole fence with base curve C has area

$$A = \int dA = \int_C f(x, y)\, ds.$$

EXAMPLE 1 Evaluate the line integral

$$\int_C xy\, ds$$

where C is the first-quadrant quarter-circle of radius 1 parametrized by $x = \cos t$, $y = \sin t$, $0 \leqq t \leqq \pi/2$ (Fig. 15.2.4).

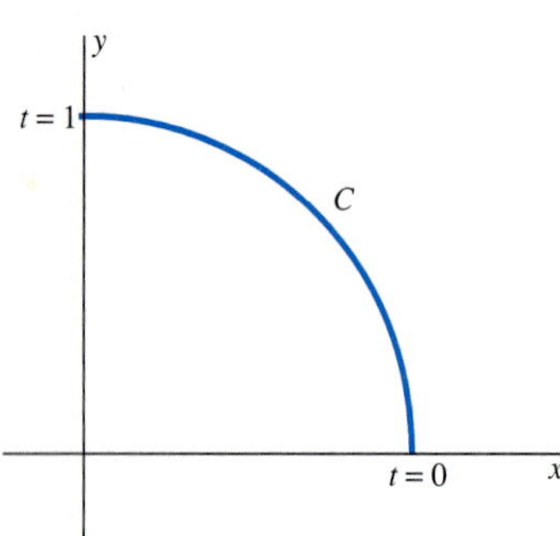

Fig. 15.2.4 The quarter-circle of Example 1

Solution Here $ds = \sqrt{(-\sin t)^2 + (\cos t)^2}\, dt = dt$, so Eq. (5) yields

$$\int_C xy\, ds = \int_{t=0}^{\pi/2} \cos t \sin t\, dt = \left[\frac{1}{2}\sin^2 t\right]_0^{\pi/2} = \frac{1}{2}.$$

Let us now return to the physical wire and denote its density function by $\delta(x, y, z)$. The mass of a small piece of length Δs is $\Delta m = \delta\, \Delta s$, so we write

$$dm = \delta(x, y, z)\, ds$$

for its (symbolic) element of mass. Then the **mass** m of the wire and its **centroid** $(\overline{x}, \overline{y}, \overline{z})$ are defined as follows:

$$m = \int_C dm = \int_C \delta\, ds, \qquad \overline{x} = \frac{1}{m}\int_C x\, dm,$$
$$\overline{y} = \frac{1}{m}\int_C y\, dm, \qquad \overline{z} = \frac{1}{m}\int_C z\, dm. \tag{6}$$

Note the analogy with Eqs. (2) and (4) of Section 14.6. The **moment of inertia** of the wire around a given axis is

$$I = \int_C w^2\, dm, \tag{7}$$

where $w = w(x, y, z)$ denotes the perpendicular distance from the point (x, y, z) of the wire to the axis in question.

EXAMPLE 2 Find the centroid of a wire that has density $\delta = kz$ and the shape of the helix C (Fig. 15.2.5) with parametrization

$$x = 3\cos t, \quad y = 3\sin t, \quad z = 4t, \quad 0 \leqq t \leqq \pi.$$

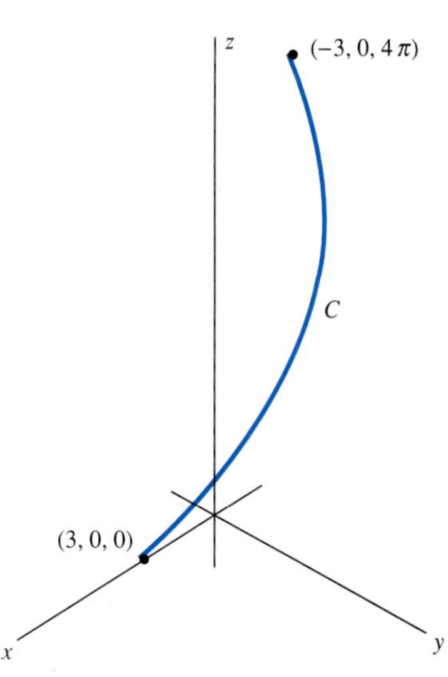

Fig. 15.2.5 The helical wire of Example 2. Does the centroid $(-1.22, 1.91, 8.38)$ lie on the wire?

Solution The mass element of the wire is

$$dm = \delta\, ds = kz\, ds = 4kt\sqrt{(-3\sin t)^2 + (3\cos t)^2 + 4^2}\, dt = 20kt\, dt.$$

Hence the formulas in (6) yield

$$m = \int_C \delta\, ds = \int_0^{\pi} 20kt\, dt = 10k\pi^2;$$

$$\overline{x} = \frac{1}{m}\int_C \delta x\, ds = \frac{1}{10k\pi^2}\int_0^{\pi} 60kt\cos t\, dt$$
$$= \frac{6}{\pi^2}\Big[\cos t + t\sin t\Big]_0^{\pi} = -\frac{12}{\pi^2} \approx -1.22;$$

$$\overline{y} = \frac{1}{m}\int_C \delta y\, ds = \frac{1}{10k\pi^2}\int_0^{\pi} 60kt\sin t\, dt$$

$$= \frac{6}{\pi^2}\left[\sin t - t\cos t\right]_0^{\pi} = \frac{6}{\pi} \approx 1.91;$$

$$\overline{z} = \frac{1}{m}\int_C \delta z\, ds = \frac{1}{10k\pi^2}\int_0^{\pi} 80kt^2\, dt$$

$$= \frac{8}{\pi^2}\left[\frac{1}{3}t^3\right]_0^{\pi} = \frac{8\pi}{3} \approx 8.38.$$

So the centroid of the wire is located close to the point $(-1.22, 1.91, 8.38)$. ■

Line Integrals with Respect to Coordinate Variables

We obtain a different kind of line integral by replacing Δs_i in Eq. (3) with

$$\Delta x_i = x(t_i) - x(t_{i-1}) = x'(t_i^\star)\,\Delta t.$$

The **line integral of f along C with respect to** x is defined to be

$$\int_C f(x, y, z)\, dx = \lim_{\Delta t \to 0} \sum_{i=1}^{n} f(x(t_i^\star), y(t_i^\star), z(t_i^\star))\,\Delta x_i.$$

Thus

$$\int_C f(x, y, z)\, dx = \int_a^b f(x(t), y(t), z(t))\, x'(t)\, dt. \tag{8a}$$

Similarly, the **line integrals of f along C with respect to** y and **with respect to** z are given by

$$\int_C f(x, y, z)\, dy = \int_a^b f(x(t), y(t), z(t))\, y'(t)\, dt, \tag{8b}$$

$$\int_C f(x, y, z)\, dz = \int_a^b f(x(t), y(t), z(t))\, z'(t)\, dt. \tag{8c}$$

The three integrals in (8) typically occur together. If P, Q, and R are continuous functions of the variables x, y, and z, then we write (indeed, *define*)

$$\int_C P\, dx + Q\, dy + R\, dz = \int_C P\, dx + \int_C Q\, dy + \int_C R\, dz. \tag{9}$$

The line integrals in Eqs. (8) and (9) are evaluated by expressing x, y, z, dx, dy, and dz in terms of t as determined by a suitable parametrization of the curve C. The result is an ordinary single-variable integral. For instance, if C is a parametric plane curve parametrized over the interval $[a, b]$ by $\mathbf{r}(t) = \langle x(t), y(t)\rangle$, then

$$\int_C P dx + Q\, dy = \int_a^b [P(x(t), y(t))\cdot x'(t) + Q(x(t), y(t))\cdot y'(t)]\, dt.$$

EXAMPLE 3 Evaluate the line integral

$$\int_C y\, dx + z\, dy + x\, dz,$$

where C is the parametric curve $x = t, y = t^2, z = t^3, 0 \leqq t \leqq 1$.

Solution Because $dx = dt, dy = 2t\,dt$, and $dz = 3t^2\,dt$, substitution in terms of t yields

$$\int_C y\,dx + z\,dy + x\,dz = \int_0^1 t^2\,dt + t^3(2t\,dt) + t(3t^2\,dt)$$

$$= \int_0^1 (t^2 + 3t^3 + 2t^4)\,dt = \left[\frac{1}{3}t^2 + \frac{3}{4}t^4 + \frac{2}{5}t^5\right]_0^1 = \frac{89}{60}.$$ ■

The given parametrization of a smooth curve C determines an **orientation** or "positive direction" along the curve. As the parameter t increases from $t = a$ to $t = b$, the point $(x(t), y(t))$ moves along the curve from its initial point A to its terminal point B. Now think of a curve $-C$ with the *opposite orientation.* This new curve consists of the same points as C, but the parametrization of $-C$ traces these points in the opposite direction, from initial point B to terminal point A (Fig. 15.2.6). Because the arc-length differential $ds = \sqrt{[x'(t)]^2 + [y'(t)]^2 + [z'(t)]^2}\,dt$ is always positive (the square root is positive), the value of the line integral with respect to arc length is not affected by the reversal of orientation. That is,

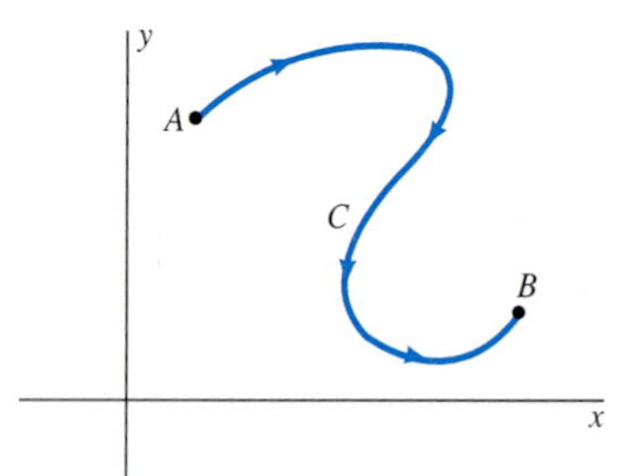

$$\int_{-C} f(x, y, z)\,ds = \int_C f(x, y, z)\,ds. \tag{10}$$

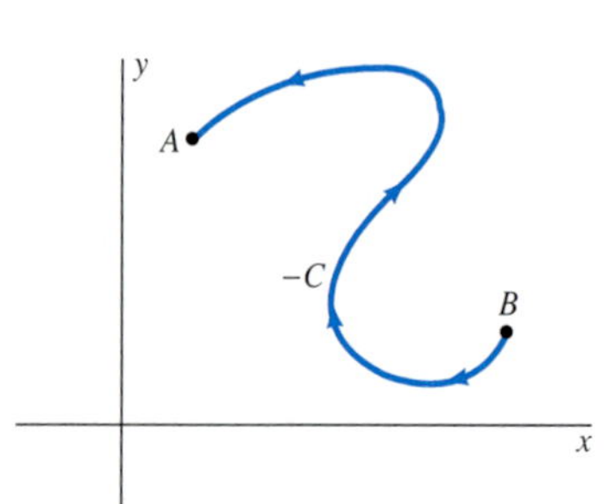

Fig. 15.2.6 $\int_{-C} f\,ds = \int_C f\,ds$ but $\int_{-C} P\,dx + Q\,dy = -\int_C P\,dx + Q\,dy.$

In contrast, the signs of the derivatives $x'(t), y'(t)$, and $z'(t)$ in Eqs. (8a), (8b), and (8c) are changed when the direction of the parametrization is reversed, so it follows that

$$\int_{-C} P\,dx + Q\,dy + R\,dz = -\int_C P\,dx + Q\,dy + R\,dz. \tag{11}$$

Thus changing the orientation of the curve changes the *sign* of a line integral with respect to coordinate variables but does not affect the value of a line integral with respect to arc length. It is proved in advanced calculus that, for either type of line integral, two one-to-one parametrizations of the same smooth curve give the same value if they agree in orientation.

EXAMPLE 4 The parametrization $x = 1 + 8t,\ y = 2 + 6t\ (0 \leqq t \leqq 1)$ of the line segment C from $A(1, 2)$ to $B(9, 8)$ in Fig. 15.2.7 gives $dx = 8\,dt,\ dy = 6\,dt$, and $ds = 10\,dt$. Hence we easily verify that

$$\int_C xy\,ds = \int_0^1 (1 + 8t)(2 + 6t) \cdot 10\,dt = 290$$

and

$$\int_C y\,ds + x\,dy = \int_0^1 [(2 + 6t) \cdot 8 + (1 + 8t) \cdot 6]\,dt = 70.$$

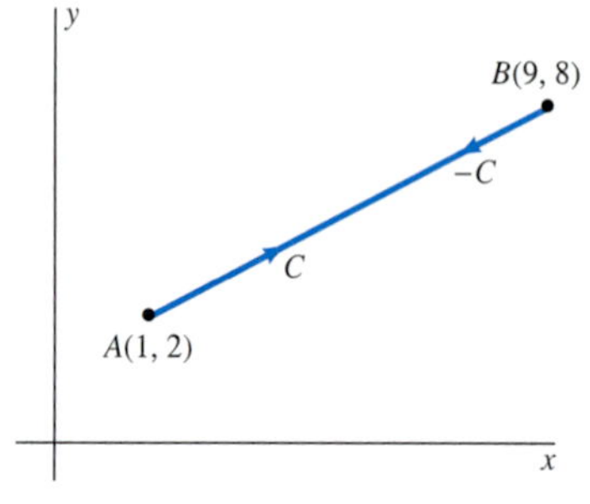

Fig. 15.2.7 The line segment of Example 4

The parametrization $x = 9 - 4t,\ y = 8 - 3t\ (0 \leqq t \leqq 2)$ of the oppositely oriented segment $-C$ from $B(9, 8)$ to $A(1, 2)$ gives $dx = -4\,dt, dy = -3\,dt$, and $ds = 5\,dt$, and we easily verify that

$$\int_{-C} xy\,ds = \int_0^2 (9 - 4t)(8 - 3t) \cdot 5\,dt = 290,$$

whereas

$$\int_{-C} y\,dx + x\,dy = \int_0^2 [(8 - 3t)\cdot(-4) + (9 - 4t)\cdot(-3)]\,dt = -70.$$

If the curve C consists of a finite number of smooth curves joined at consecutive corner points, then we say that C is **piecewise smooth.** In such a case the value of a line integral along C is defined to be the sum of its values along the smooth segments of C. For instance, with the piecewise smooth curve $C = C_1 + C_2$ of Fig. 15.2.8, we have

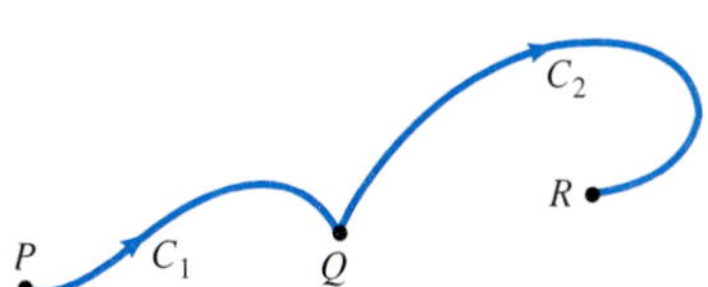

Fig. 15.2.8 The curve $C = C_1 + C_2$ from P to R

$$\int_C f(x, y, z)\,ds = \int_{C_1 + C_2} f(x, y, z)\,ds = \int_{C_1} f(x, y, z)\,ds + \int_{C_2} f(x, y, z)\,ds.$$

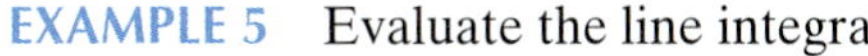

EXAMPLE 5 Evaluate the line integral

$$\int_C y\,dx + 2x\,dy$$

for each of these three curves C (Fig. 15.2.9):

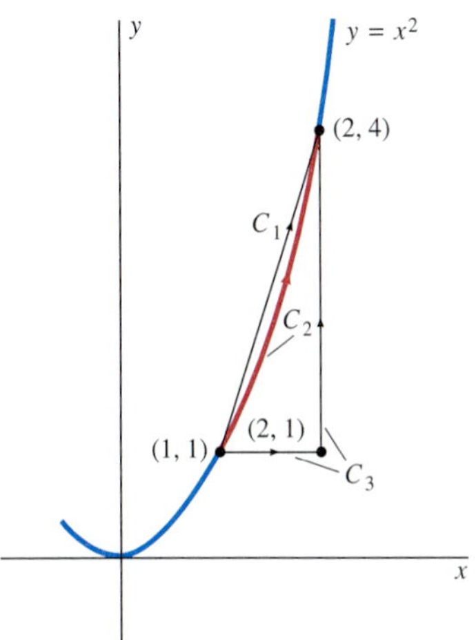

Fig. 15.2.9 The three arcs of Example 5

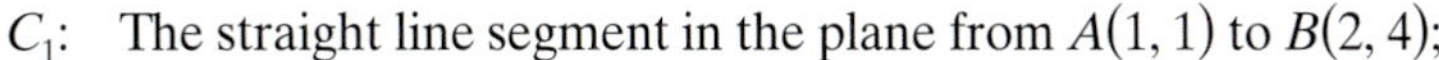

C_1: The straight line segment in the plane from $A(1, 1)$ to $B(2, 4)$;

C_2: The plane path from $A(1, 1)$ to $B(2, 4)$ along the graph of the parabola $y = x^2$; and

C_3: The straight line in the plane from $A(1, 1)$ to $Q(2, 1)$ followed by the straight line from $Q(2, 1)$ to $B(2, 4)$.

Solution The straight line segment C_1 from A to B can be parametrized by $x = 1 + t$, $y = 1 + 3t$, $0 \leqq t \leqq 1$. Hence

$$\int_{C_1} y\,dx + 2x\,dy = \int_0^1 (1 + 3t)\,dt + 2(1 + t)(3\,dt)$$
$$= \int_0^1 (7 + 9t)\,dt = \frac{23}{2}.$$

Next, the arc C_2 of the parabola $y = x^2$ from A to B is "self-parametrizing": It has the parametrization $x = x$, $y = x^2$, $1 \leqq x \leqq 2$. So

$$\int_{C_2} y\,dx + 2x\,dy = \int_1^2 (x^2)(dx) + 2(x)(2x\,dx) = \int_1^2 5x^2\,dx = \frac{35}{3}.$$

Finally, along the straight line segment from $(1, 1)$ to $(2, 1)$, we have $y \equiv 1$ and (because y is a constant) $dy = 0$. Along the vertical segment from $(2, 1)$ to $(2, 4)$, we have $x \equiv 2$ and $dx = 0$. Therefore

$$\int_{C_3} y\,dx + 2x\,dy = \int_{x=1}^2 [(1)(dx) + (2x)(0)] + \int_{y=1}^4 [(y)(0) + (4)(dy)]$$
$$= \int_{x=1}^2 1\,dx + \int_{y=1}^4 4\,dy = 13.$$

Example 5 shows that we may well obtain different values for the line integral from A to B if we evaluate it along different curves from A to B. Thus this line integral is *path-dependent.* We shall give in Section 15.3 a sufficient condition for the line integral

$$\int_C P\,ds + Q\,dy + R\,dz$$

to have the same value for *all* smooth or piecewise smooth curves C from A to B, and thus for the integral to be *independent of path.*

Line Integrals and Vector Fields

Suppose now that $\mathbf{F} = P\mathbf{i} + Q\mathbf{j} + R\mathbf{k}$ is a force field defined on a region that contains the curve C from the point A to the point B. Suppose also that C has a parametrization

$$\mathbf{r}(t) = \mathbf{i}x(t) + \mathbf{j}y(t) + \mathbf{k}z(t), \qquad t \text{ in } [a, b],$$

with a *nonzero* velocity vector

$$\mathbf{v} = \mathbf{i}\frac{dx}{dt} + \mathbf{j}\frac{dy}{dt} + \mathbf{k}\frac{dz}{dt}.$$

The speed associated with this velocity vector is

$$v = |\mathbf{v}| = \sqrt{\left(\frac{dx}{dt}\right)^2 + \left(\frac{dy}{dt}\right)^2 + \left(\frac{dz}{dt}\right)^2}.$$

Recall from Section 12.6 that the *unit tangent vector* to the curve C is

$$\mathbf{T} = \frac{\mathbf{v}}{v} = \frac{1}{v}\left(\frac{dx}{dt}\mathbf{i} + \frac{dy}{dt}\mathbf{j} + \frac{dz}{dt}\mathbf{k}\right).$$

We want to approximate the work W done by the force field $\mathbf{F}$ in moving a particle along the curve C from A to B. Subdivide C as indicated in Fig. 15.2.10. Think of $\mathbf{F}$ moving the particle from P_{i-1} to P_i, two consecutive division points of C. The work ΔW_i done is approximately the product of the distance Δs_i from P_{i-1} to P_i (measured along C) and the tangential component $\mathbf{F}\cdot\mathbf{T}$ of the force $\mathbf{F}$ at a typical point $(x(t_i^\star), y(t_i^\star), z(t_i^\star))$ between P_{i-1} and P_i. Thus

$$\Delta W_i \approx \mathbf{F}(x(t_i^\star), y(t_i^\star), z(t_i^\star)) \cdot \mathbf{T}(t_i^\star)\,\Delta s_i,$$

so the total work W is given approximately by

$$W \approx \sum_{i=1}^{n} \mathbf{F}(x(t_i^\star), y(t_i^\star), z(t_i^\star)) \cdot \mathbf{T}(t_i^\star)\,\Delta s_i.$$

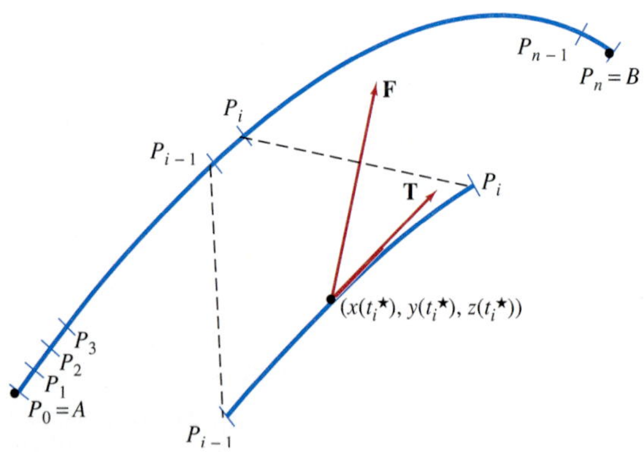

Fig. 15.2.10 The component of $\mathbf{F}$ along C from P_{i-1} to P_i is $\mathbf{F}\cdot\mathbf{T}$.

This approximation suggests that we *define* the **work** W as

$$W = \int_C \mathbf{F} \cdot \mathbf{T}\, ds. \tag{12}$$

Thus *work is the integral with respect to arc length of the tangential component of the force.* Intuitively, we may regard $dW = \mathbf{F} \cdot \mathbf{T}\, ds$ as the infinitesimal element of work done by the tangential component $\mathbf{F} \cdot \mathbf{T}$ of the force in moving the particle along the arc-length element ds. The line integral in Eq. (12) is then the "sum" of all these infinitesimal elements of work.

It is customary to write formally

$$\mathbf{r} = x\mathbf{i} + y\mathbf{j} + z\mathbf{k}, \qquad d\mathbf{r} = \mathbf{i}\, dx + \mathbf{j}\, dy + \mathbf{k}\, dz,$$

and

$$\mathbf{T}\, ds = \left(\frac{dx}{dt}\mathbf{i} + \frac{dy}{dt}\mathbf{j} + \frac{dz}{dt}\mathbf{k} \right) ds = d\mathbf{r}.$$

With this notation, Eq. (12) takes the form

$$W = \int_C \mathbf{F} \cdot d\mathbf{r} \tag{13}$$

that is common in engineering and physics texts.

To evaluate the line integral in Eqs. (12) or (13), we express its integrand and limits of integration in terms of the parameter t, as usual. Thus

$$\begin{aligned} W &= \int_C \mathbf{F} \cdot \mathbf{T}\, ds \\ &= \int_a^b (P\mathbf{i} + Q\mathbf{j} + R\mathbf{k}) \cdot \frac{1}{v} \left(\frac{dx}{dt}\mathbf{i} + \frac{dy}{dt}\mathbf{j} + \frac{dz}{dt}\mathbf{k} \right) v\, dt \\ &= \int_a^b \left(P\frac{dx}{dt} + Q\frac{dy}{dt} + R\frac{dz}{dt} \right) dt. \end{aligned}$$

Therefore

$$W = \int_C P\, dx + Q\, dy + R\, dz. \tag{14}$$

This computation reveals an important relation between the two types of line integrals we have defined here.

Theorem 1 ***Equivalent Line Integrals***

Suppose that the vector field $\mathbf{F} = P\mathbf{i} + Q\mathbf{j} + R\mathbf{k}$ has continuous component functions and that $\mathbf{T}$ is the unit tangent vector to the smooth curve C. Then

$$\int_C \mathbf{F} \cdot \mathbf{T}\, ds = \int_C P\, dx + Q\, dy + R\, dz. \tag{15}$$

Remark If the orientation of the curve C is reversed, then the sign of the right-hand integral in Eq. (15) is changed according to Eq. (11), whereas the sign of the left-hand integral is changed because $\mathbf{T}$ is replaced with $-\mathbf{T}$.

EXAMPLE 6 The work done by the force field $\mathbf{F} = y\mathbf{i} + z\mathbf{j} + x\mathbf{k}$ in moving a particle from $(0, 0, 0)$ to $(1, 1, 1)$ along the twisted cubic $x = t$, $y = t^2$, $z = t^3$ is given by the line integral

$$W = \int_C \mathbf{F}\cdot d\mathbf{r} = \int_C \mathbf{F}\cdot\mathbf{T}\,ds = \int_C y\,dx + z\,dy + x\,dz,$$

and we computed the value of this integral in Example 3. Hence $W = \frac{89}{60}$. ■

EXAMPLE 7 Find the work done by the inverse-square force field

$$\mathbf{F}(x, y, z) = \frac{k\mathbf{r}}{r^3} = \frac{k(x\mathbf{i} + y\mathbf{j} + z\mathbf{k})}{(x^2 + y^2 + z^2)^{3/2}}$$

in moving a particle along the straight line segment C from $(0, 4, 0)$ to $(0, 4, 3)$.

Solution Along C we have $x = 0$, $y = 4$, and z varying from 0 to 3. Thus we choose z as the parameter:

$$x \equiv 0, \quad y \equiv 4, \quad \text{and} \quad z = z, \quad 0 \leqq z \leqq 3.$$

Because $dx = 0 = dy$, Eq. (14) gives

$$W = \int_C \frac{k(x\,dx + y\,dy + z\,dz)}{(x^2 + y^2 + z^2)^{3/2}}$$

$$= \int_0^3 \frac{kz}{(16 + z^2)^{3/2}}\,dz = \left[\frac{-k}{\sqrt{16 + z^2}}\right]_0^3 = \frac{k}{20}.$$ ■

15.2 PROBLEMS

In Problems 1 through 5, evaluate the line integrals

$$\int_C f(x, y)\,ds, \quad \int_C f(x, y)\,dx, \quad \text{and} \quad \int_C f(x, y)\,dy$$

along the indicated parametric curve.

1. $f(x, y) = x^2 + y^2$; $x = 4t - 1, y = 3t + 1, -1 \leqq t \leqq 1$

2. $f(x, y) = x$; $x = t, y = t^2, 0 \leqq t \leqq 1$

3. $f(x, y) = x + y$; $x = e^t + 1, y = e^t - 1, 0 \leqq t \leqq \ln 2$

4. $f(x, y) = 2x - y$; $x = \sin t, y = \cos t, 0 \leqq t \leqq \pi/2$

5. $f(x, y) = xy$; $x = 3t, y = t^4, 0 \leqq t \leqq 1$

In Problems 6 through 10, evaluate

$$\int_C P(x, y)\,dx + Q(x, y)\,dy.$$

6. $P(x, y) = xy$, $Q(x, y) = x + y$; C is the part of the graph of $y = x^2$ from $(-1, 1)$ to $(2, 4)$.

7. $P(x, y) = y^2$, $Q(x, y) = x$; C is the part of the graph of $x = y^3$ from $(-1, -1)$ to $(1, 1)$.

8. $P(x, y) = y\sqrt{x}$, $Q(x, y) = x\sqrt{x}$; C is the part of the graph of $y^2 = x^3$ from $(1, 1)$ to $(4, 8)$.

9. $P(x, y) = x^2y$, $Q(x, y) = xy^3$; C consists of the line segments from $(-1, 1)$ to $(2, 1)$ and from $(2, 1)$ to $(2, 5)$.

10. $P(x, y) = x + 2y$, $Q(x, y) = 2x - y$; C consists of the line segments from $(3, 2)$ to $(3, -1)$ and from $(3, -1)$ to $(-2, -1)$.

In Problems 11 through 15, evaluate the line integral

$$\int_C \mathbf{F}\cdot d\mathbf{r} = \int_C \mathbf{F}\cdot\mathbf{T}\,ds$$

along the indicated path C.

11. $\mathbf{F}(x, y, z) = z\mathbf{i} + x\mathbf{j} - y\mathbf{k}$; C is parametrized by $x = t$, $y = t^2$, $z = t^3$, $0 \leqq t \leqq 1$.

12. $\mathbf{F}(x, y, z) = yz\mathbf{i} + xz\mathbf{j} + xy\mathbf{k}$; C is the straight line segment from $(2, -1, 3)$ to $(4, 2, -1)$.

13. $\mathbf{F}(x, y, z) = y\mathbf{i} - x\mathbf{j} + z\mathbf{k}$; $x = \sin t,\ y = \cos t,\ z = 2t$, $0 \leqq y \leqq \pi$.

14. $\mathbf{F}(x, y, z) = (2x + 3y)\mathbf{i} + (3x + 2y)\mathbf{j} + 3z^2\mathbf{k}$; C is the path from $(0, 0, 0)$ to $(4, 2, 3)$ that consists of three line segments parallel to the x-axis, the y-axis, and the z-axis, in that order.

15. $\mathbf{F}(x, y, z) = yz^2\mathbf{i} + xz^2\mathbf{j} + 2xyz\mathbf{k}$; C is the path from $(-1, 2, -2)$ to $(1, 5, 2)$ that consists of three line segments parallel to the z-axis, the x-axis, and the y-axis, in that order.

In Problems 16 through 18, evaluate

$$\int_C f(x, y, z)\, ds$$

for the given function $f(x, y, z)$ and the given path C.

16. $f(x, y, z) = xyz$; C is the straight line segment from $(1, -1, 2)$ to $(3, 2, 5)$.

17. $f(x, y, z) = 2x + 9xy$; C is the curve $x = t,\ y = t^2$, $z = t^3, 0 \leqq t \leqq 1$.

18. $f(x, y, z) = xy$; C is the elliptical helix $x = 4\cos t$, $y = 9\sin t, z = 7t, 0 \leqq t \leqq 5\pi/2$.

19. Find the centroid of a uniform thin wire shaped like the semicircle $x^2 + y^2 = a^2, a \geqq 0, y \geqq 0$.

20. Find the moments of inertia around the x- and y-axes of the wire of Problem 19.

21. Find the mass and centroid of a wire that has constant density $\delta = k$ and is shaped like the helix $x = 3\cos t$, $y = 3\sin t, z = 4t, 0 \leqq t \leqq 2\pi$.

22. Find the moment of inertia $I_z = \int_C (x^2 + y^2)\, dm$ around the z-axis of the helical wire of Problem 21.

23. A wire shaped like the first-quadrant portion of the circle $x^2 + y^2 = a^2$ has density $\delta = kxy$ at the point (x, y). Find its mass, centroid, and moment of inertia around each coordinate axis.

24. A wire is shaped like the arch $x = t - \sin t, y = 1 - \cos t$ $(0 \leqq t \leqq 2\pi)$ of a cycloid C and has constant density $\delta(x, y) \equiv k$. Find its mass, centroid, and moment of inertia $I_x = \int_C y^2\, dm$ around the x-axis.

25. A wire is shaped like the astroid $x = \cos^3 t,\ y = \sin^3 t$ $(0 \leqq t \leqq 2\pi)$ and has constant density $\delta(x, y) \equiv k$. Find its moment of inertia $I_0 = \int_C (x^2 + y^2)\, dm$ around the origin.

The **average distance** $\overline{D}$ from the fixed point P to points of the parametrized curve C is defined by

$$\overline{D} = \frac{1}{s}\int_C D(x, y)\, ds$$

where s is the length of C and $D(x, y)$ denotes the distance from P to the variable point (x, y) of C. In Problems 26 through 31, compute the average distance, exactly if possible, or use a computer algebra system to find it (either symbolically or, if necessary, numerically).

26. Use the standard trigonometric parametrization of a circle C of radius a to verify that the average distance of points of C from its center is $\overline{D} = a$.

27. Find (exactly) the average distance from the point $(a, 0)$ to points of the circle of radius a centered at the origin. (*Suggestion:* Use the law of cosines to find $D(x, y)$.)

28. Find the average distance from the origin to points of the cycloidal arch of Problem 24.

29. Find the average distance from the origin to points of the astroid of Problem 25.

30. Find the average distance from the origin to points of the helix of Problem 21.

31. The spiral parametrized by $x = e^{-t}\cos t,\ y = e^{-t}\sin t$ starts at $(1, 0)$ when $t = 0$ and closes in on the origin as $t \to \infty$. Use improper integrals to calculate the average distance from the origin to points of this spiral.

32. Find the work done by the inverse-square force field of Example 7 in moving a particle from $(1, 0, 0)$ to $(0, 3, 4)$. Integrate first along the line segment from $(1, 0, 0)$ to $(5, 0, 0)$ and then along a path on the sphere with equation $x^2 + y^2 + z^2 = 25$. The second integral is automatically zero. (Why?)

33. Imagine an infinitely long and uniformly charged wire that coincides with the z-axis. The electric force that it exerts on a unit charge at the point $(x, y) \neq (0, 0)$ in the xy-plane is

$$\mathbf{F}(x, y) = \frac{k(x\mathbf{i} + y\mathbf{j})}{x^2 + y^2}.$$

Find the work done by $\mathbf{F}$ in moving a unit charge along the straight line segment from (a) $(1, 0)$ to $(1, 1)$; (b) $(1, 1)$ to $(0, 1)$.

34. Show that if $\mathbf{F}$ is a *constant* force field, then it does zero work on a particle that moves once uniformly counterclockwise around the unit circle in the xy-plane.

35. Show that if $\mathbf{F} = k\mathbf{r} = k(x\mathbf{i} + y\mathbf{j})$, then $\mathbf{F}$ does zero work on a particle that moves once uniformly counterclockwise around the unit circle in the xy-plane.

36. Find the work done by the force field $\mathbf{F} = -y\mathbf{i} + x\mathbf{j}$ in moving a particle counterclockwise once around the unit circle in the xy-plane.

37. Let C be a curve on the unit sphere $x^2 + y^2 + z^2 = 1$. Explain why the inverse-square force field of Example 7 does zero work in moving a particle along C.

In Problems 38 through 40, the given curve C joins the points P and Q in the xy-plane. The point P represents the top of a ten-story building, and Q is a point on the ground 100 ft from the base of the building. A 150-lb person slides down a frictionless slide shaped like the curve C from P to Q under the influence of the gravitational force $\mathbf{F} = -150\mathbf{j}$. In each problem show that $\mathbf{F}$ does the same amount of work on the person,

$W = 15000$ ft · lb, as if he or she had dropped straight down to the ground.

38. C is the straight line segment $y = x$ from $P(100, 100)$ to $Q(0, 0)$.

39. C is the circular arc $x = 100 \sin t,\ y = 100 \cos t$ from $P(0, 100)$ to $Q(100, 0)$.

40. C is the parabolic arc $y = x^2/100$ from $P(100, 100)$ to $Q(0, 0)$.

41. Now suppose that the 100-ft ten-story building of Problems 38 through 40 is a circular tower with a radius of 25 ft, and the fire-escape slide is a spiral (helical) ramp that encircles the tower once every two floors. Use a line integral to compute the work done by the gravitational force field $\mathbf{F} = -200\mathbf{k}$ on a 200-lb person who slides down this (frictionless) ramp from the top of the building to the ground.

42. An electric current I in a long straight wire generates a magnetic field $\mathbf{B}$ in the space surrounding the wire. The vector $\mathbf{B}$ is tangent to any circle C that is centered on the wire and lies in a plane perpendicular to the wire. *Ampere's law* implies that

$$\int_C \mathbf{B} \cdot d\mathbf{r} = \mu I,$$

where μ is a certain electromagnetic constant. Deduce from this fact that the magnitude $B = |\mathbf{B}|$ of the magnetic field is proportional to the current I and inversely proportional to the distance r from the wire.

15.3 THE FUNDAMENTAL THEOREM AND INDEPENDENCE OF PATH

The fundamental theorem of calculus says, in effect, that differentiation and integration are inverse processes for single-variable functions. Specifically, part 2 of the fundamental theorem in Section 5.6 implies that

$$\int_a^b G'(t)\,dt = G(b) - G(a) \tag{1}$$

if the derivative G' is continuous on $[a, b]$. Theorem 1 here can be interpreted as saying that "gradient vector differentiation" and "line integration" are, similarly, inverse processes for multivariable functions.

Theorem 1 ***The Fundamental Theorem for Line Integrals***

Let f be a function of two or three variables and let C be a smooth curve parametrized by the vector-valued function $\mathbf{r}(t)$ for $a \leqq t \leqq b$. If f has continuous partial derivatives at each point of C, then

$$\int_C \nabla f \cdot d\mathbf{r} = f(\mathbf{r}(b)) - f(\mathbf{r}(a)). \tag{2}$$

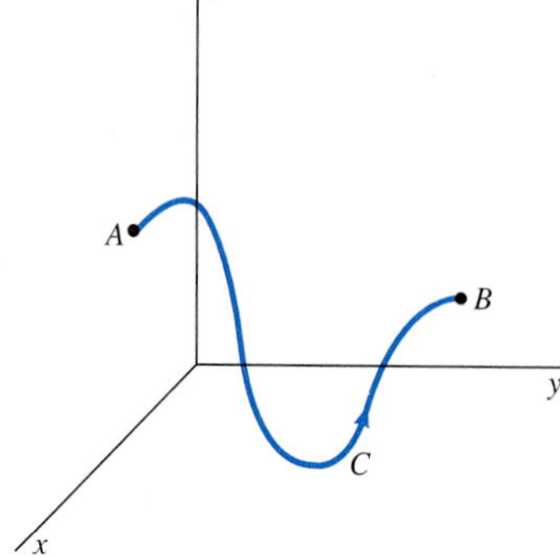

Fig. 15.3.1 The path C of Theorem 1

PROOF: We consider the three-dimensional case $f(x, y, z)$ illustrated in Fig. 15.3.1. Then $\nabla f = \langle \partial f/\partial x, \partial f/\partial y, \partial f/\partial z \rangle$, so Theorem 1 in Section 15.2 yields

$$\int_C \nabla f \cdot d\mathbf{r} = \int_C \frac{\partial f}{\partial x}\,dx + \frac{\partial f}{\partial y}\,dy + \frac{\partial f}{\partial z}\,dz$$

$$= \int_a^b \left(\frac{\partial f}{\partial x}\frac{dx}{dt} + \frac{\partial f}{\partial y}\frac{dy}{dt} + \frac{\partial f}{\partial z}\frac{dz}{dt} \right) dt.$$

By the multivariable chain rule (Section 13.7), the integrand here is the derivative $G'(t)$ of the composite function $G(t) = f(\mathbf{r}(t)) = f(x(t), y(t), z(t))$. Therefore it follows that

$$\int_C \nabla f \cdot d\mathbf{r} = \int_a^b G'(t)\,dt = G(b) - G(a) \qquad \text{(by Eq. (1))}$$
$$= f(\mathbf{r}(b)) - f(\mathbf{r}(a)),$$

and so we have established Eq. (2), as desired. ■

Remark 1 If we write A and B for the endpoints $\mathbf{r}(a)$ and $\mathbf{r}(b)$ of C, then Eq. (2) takes the form

$$\int_C \nabla f \cdot d\mathbf{r} = f(B) - f(A), \tag{3}$$

which is quite similar to Eq. (1).

Remark 2 If

$$f(x, y, z) = -\frac{k}{r} = -\frac{k}{\sqrt{x^2 + y^2 + z^2}},$$

then a brief computation shows that $\nabla f = \mathbf{F}$ is the inverse-square force field

$$\mathbf{F}(x, y, z) = \frac{k(x\mathbf{i} + y\mathbf{j} + z\mathbf{k})}{(x^2 + y^2 + z^2)^{3/2}}$$

of Example 7 in Section 15.2, where we calculated directly the work $W = k/20$ done by the force field $\mathbf{F}$ in moving a particle along a straight line segment from the point $A(0, 4, 0)$ to the point $B(0, 4, 3)$. Indeed, using Theorem 1, we find that the work done by $\mathbf{F}$ in moving a particle along *any* smooth path from A to B (that does not pass through the origin) is given by

$$W = \int_C \mathbf{F} \cdot d\mathbf{r} = \int_C \nabla f \cdot d\mathbf{r}$$
$$= f(0, 4, 3) - f(0, 4, 0) \qquad \text{(by Eq. (3))}$$
$$= \left(-\frac{k}{5}\right) - \left(-\frac{k}{4}\right) = \frac{k}{20}.$$

Independence of Path

We next apply the fundamental theorem for line integrals to discuss the question whether the integral

$$\int_C \mathbf{F} \cdot \mathbf{T}\,ds = \int_C \mathbf{F} \cdot d\mathbf{r} = \int_C P\,dx + Q\,dt + R\,dz \tag{4}$$

(where $\mathbf{F} = \langle P, Q, R \rangle$) has the *same value* for *any* two curves with the same initial and terminal points.

Definition *Independence of Path*

The line integral in Eq. (4) is said to be **independent of the path in the region** D provided that, given any two points A and B of D, the integral has the same value along every piecewise smooth curve, or **path,** in D from A to B. In this case we may write

$$\int_C \mathbf{F}\cdot\mathbf{T}\,ds = \int_A^B \mathbf{F}\cdot\mathbf{T}\,ds \qquad \mathbf{(5)}$$

because the value of the integral depends only on the points A and B, not on the particular choice of the path C joining them.

For a tangible interpretation of independence of path, let us think of walking along the curve C from point A to point B in the plane where a wind with velocity vector $\mathbf{w}(x, y)$ is blowing. Suppose that when we are at (x, y), the wind exerts a force $\mathbf{F} = k\mathbf{w}(x, y)$ on us, k being a constant that depends on our size and shape (and perhaps other factors as well). Then, by Eq. (12) of Section 15.2, the amount of work the wind does on us as we walk along C is given by

$$W = \int_C \mathbf{F}\cdot\mathbf{T}\,ds = k\int_C \mathbf{w}\cdot\mathbf{T}\,ds. \qquad \mathbf{(6)}$$

This is the wind's contribution to our trip from A to B. In this context, the question of independence of path is whether the wind's work W depends on *which* path we choose from point A to point B.

EXAMPLE 1 Suppose that a steady wind blows toward the northeast with velocity vector $\mathbf{w} = 10\mathbf{i} + 10\mathbf{j}$ in fps units; its speed is $|\mathbf{w}| = 10\sqrt{2} \approx 14$ ft/s—about 10 mi/h. Assume that $k = 0.5$, so the wind exerts 0.5 lb of force for each foot per second of its velocity. Then $\mathbf{F} = 5\mathbf{i} + 5\mathbf{j}$, so Eq. (6) yields

$$W = \int_C \langle 5, 5\rangle\cdot\mathbf{T}\,ds = \int_C 5\,dx + 5\,dy \qquad \mathbf{(7)}$$

for the work done on us by the wind as we walk along C.

For instance, if C is the straight path $x = 10t$, $y = 10t$ $(0 \leqq x \leqq 1)$ from $(0, 0)$ to $(10, 10)$, then Eq. (7) gives

$$W = \int_0^1 5\cdot 10\,dt + 5\cdot 10\,dt = 100\int_0^1 1\,dt = 100$$

ft·lb of work. Or, if C is the parabolic path $y = \frac{1}{10}x^2$, $0 \leqq x \leqq 10$ from the same initial point $(0, 0)$ to the same terminal point $(10, 10)$, then Eq. (7) yields

$$W = \int_0^{10} 5\,dx + 5\cdot\frac{1}{5}x\,dx = \int_0^{10} (5 + x)\,dx$$

$$= \left[5x + \frac{1}{2}x^2\right]_0^{10} = 100$$

ft·lb of work, the same as before. We shall see that it follows from Theorem 2 of this section that the line integral in Eq. (7) is independent of path, so the wind does 100 ft·lb of work along any path from $(0, 0)$ to $(10, 10)$. ■

EXAMPLE 2 Suppose that $\mathbf{w} = -2y\mathbf{i} + 2x\mathbf{j}$. This wind is blowing counterclockwise around the origin, as in a hurricane with its eye at $(0, 0)$. With $k = 0.5$ as before, $\mathbf{F} = -y\mathbf{i} + x\mathbf{j}$, so the work integral is

$$W = \int_C \mathbf{F} \cdot \mathbf{T}\, ds = \int_C -y\, dx + x\, dy. \tag{8}$$

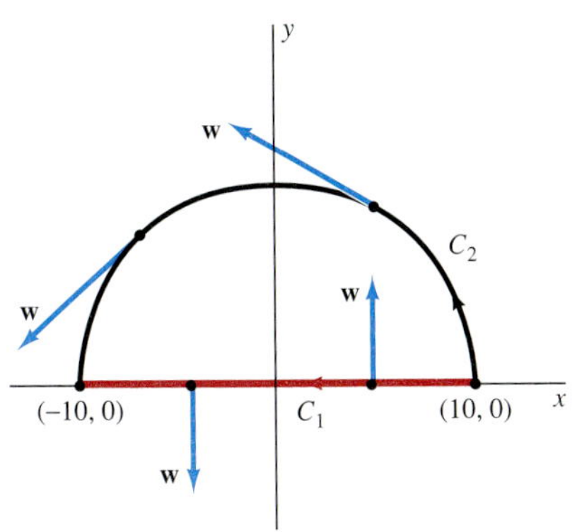

Fig. 15.3.2 Around and through the eye of the hurricane (Example 2)

If we walk from $(10, 0)$ to $(-10, 0)$ along the straight path C_1 through the eye of the hurricane, then the wind is always perpendicular to our unit tangent vector $\mathbf{T}$ (Fig. 15.3.2). Hence $\mathbf{F} \cdot \mathbf{T} = 0$, and therefore

$$W = \int_{C_1} \mathbf{F} \cdot \mathbf{T}\, ds = \int_{C_1} -y\, dx + x\, dy = 0.$$

But if we walk along the semicircular path C_2 shown in Fig. 15.3.2, then $\mathbf{w}$ remains tangent to our path, so $\mathbf{F} \cdot \mathbf{T} = |\mathbf{F}| = 10$ at each point. In this case,

$$W = \int_{C_2} -y\, dx + x\, dy = \int_{C_2} \mathbf{F} \cdot \mathbf{T}\, ds = 10 \cdot 10\pi = 100\pi.$$

The fact that we get different values along different paths from $(10, 0)$ to $(-10, 0)$ shows that the line integral in Eq. (8) is *not* independent of path. ■

Theorem 2 tells us when a given line integral is independent of path and when it is not.

Theorem 2 ***Independence of Path***

The line integral $\int_C \mathbf{F} \cdot \mathbf{T}\, ds$ is independent of path in the region D if and only if $\mathbf{F} = \nabla f$ for some function f defined on D.

PROOF: Suppose that $\mathbf{F} = \nabla f = \langle \partial f/\partial x, \partial f/\partial y, \partial f/\partial z \rangle$ and that C is a piecewise smooth curve from A to B in D parametrized as usual with parameter t in $[a, b]$. Then the fundamental theorem in the form in Eq. (3) gives

$$\int_C \mathbf{F} \cdot \mathbf{T}\, ds = \int_C \nabla f \cdot \mathbf{T}\, ds = \int_C \nabla f \cdot d\mathbf{r} = f(B) - f(A).$$

This result shows that the value of the line integral depends only on the points A and B and is therefore independent of the choice of the particular path C. This proves the *if* part of Theorem 1.

To prove the *only if* part of Theorem 2, we suppose that the line integral is independent of path in D. Choose a *fixed* point $A_0 = A_0(x_0, y_0, z_0)$ in D, and let $B = B(x, y, z)$ be an arbitrary point in D. Given any path C from A_0 to B in D, we *define* the function f by means of the equation

$$f(x, y, z) = \int_C \mathbf{F} \cdot \mathbf{T}\, ds = \int_{(x_0, y_0, z_0)}^{(x, y, z)} \mathbf{F} \cdot \mathbf{T}\, ds. \tag{9}$$

Because of the hypothesis of independence of path, the resulting value of $f(x, y, z)$ depends only on (x, y, z), not on the particular path C used. We shall omit the verification that $\nabla f = \mathbf{F}$ (see Problem 21 of Section 15.7). ■

As an application of Theorem 2, we see that the vector field $\mathbf{F} = -y\mathbf{i} + x\mathbf{j}$ of Example 2 is not the gradient of any scalar function f because $\int \mathbf{F} \cdot \mathbf{T}\, ds$ is not independent of path. More precisely, $\int \mathbf{F} \cdot \mathbf{T}\, ds$ is not independent of path in any region that either includes or encloses the origin.

Definition ***Conservative Fields and Potential Functions***

The vector field $\mathbf{F}$ defined on a region D is **conservative** provided that there exists a scalar function f defined on D such that

$$\mathbf{F} = \nabla f \tag{10}$$

at each point of D. In this case f is called a **potential function** for the vector field $\mathbf{F}$.

COMMENT In some physical applications the scalar function f is called a *potential function* for the vector field $\mathbf{F}$ provided that $\mathbf{F} = -\nabla f$.

If the line integral $\int_C \mathbf{F} \cdot \mathbf{T}\, ds$ is known to be independent of path, then Theorem 2 guarantees that the vector field $\mathbf{F}$ is conservative and that Eq. (9) yields a potential function f for $\mathbf{F}$. In this case—because the value of the integral does not depend on the specific curve C from A to B—we may well write Eq. (3) in the form

$$\int_A^B \mathbf{F} \cdot \mathbf{T}\, ds = \int_A^B \nabla f \cdot d\mathbf{r} = f(B) - f(A) \tag{11}$$

that is still more reminiscent of the ordinary fundamental theorem.

EXAMPLE 3 Find a potential function for the conservative vector field

$$\mathbf{F}(x, y) = (6xy - y^3)\mathbf{i} + (4y + 3x^2 - 3xy^2)\mathbf{j}. \tag{12}$$

Solution Because we are given the information that $\mathbf{F}$ is a conservative field, the line integral $\int \mathbf{F} \cdot \mathbf{T}\, ds$ is independent of path by Theorem 2. Therefore we may apply Eq. (9) to find a scalar potential function $\mathbf{F}$. Let C be the straight-line path from $A(0, 0)$ to $B(x_1, y_1)$ parametrized by $x = x_1 t$, $y = y_1 t$, $0 \leqq t \leqq 1$. Then Eq. (9) yields

$$\begin{aligned}
f(x_1, y_1) &= \int_A^B \mathbf{F} \cdot \mathbf{T}\, ds \\
&= \int_A^B (6xy - y^3)\, dx + (4y + 3x^2 - 3xy^2)\, dy \\
&= \int_0^1 (6x_1 y_1 t^2 - y_1^3 t^3)(x_1\, dt) + (4y_1 t + 3x_1^2 t^2 - 3x_1 y_1^2 t^3)(y_1\, dt) \\
&= \int_0^1 (4y_1^2 t + 9x_1^2 y_1 t^2 - 4x_1 y_1^3 t^3)\, dt \\
&= \Big[2y_1^2 t^2 + 3x_1^2 y_1 t^3 - x_1 y_1^3 t^4 \Big]_0^1 = 2y_1^2 + 3x_1^2 y_1 - x_1 y_1^3.
\end{aligned}$$

At this point we delete the subscripts, because (x_1, y_1) is an arbitrary point of the plane. Thus we obtain the potential function

$$f(x, y) = 2y^2 + 3x^2 y - xy^3$$

for the vector field $\mathbf{F}$ in Eq. (12). As a check, we can differentiate f to obtain

$$\frac{\partial f}{\partial x} = 6xy - y^3, \quad \frac{\partial f}{\partial y} = 4y + 3x^2 - 3xy^2.$$

■

But how did we know in advance that the vector field $\mathbf{F}$ was conservative? The answer is provided by Theorem 3.*

Theorem 3 ***Conservative Fields and Potential Functions***

Suppose that the functions $P(x, y)$ and $Q(x, y)$ are continuous and have continuous first-order partial derivatives in the open rectangle $R = \{(x, y) \mid a < x < b, c < y < d\}$. Then the vector field $\mathbf{F} = P\mathbf{i} + Q\mathbf{j}$ is conservative in R—and hence has a potential function $f(x, y)$ defined on R—if and only if, at each point of R,

$$\frac{\partial P}{\partial y} = \frac{\partial Q}{\partial x}. \tag{13}$$

Observe that the vector field $\mathbf{F}$ in Eq. (12), where $P(x, y) = 6xy - y^3$ and $Q(x, y) = 4y + 3x^2 - 3xy^2$, satisfies the criterion in Eq. (13) because

$$\frac{\partial P}{\partial y} = 6x - 3y^2 = \frac{\partial Q}{\partial x}.$$

When this sufficient condition for the existence of a potential function is satisfied, the method illustrated in Example 4 is usually an easier way to find a potential function than the evaluation of the line integral in Eq. (9)—the method used in Example 3.

EXAMPLE 4 Given $P(x, y) = 6xy - y^3$ and $Q(x, y) = 4y + 3x^2 - 3xy^2$, note that P and Q satisfy the condition $\partial P/\partial y = \partial Q/\partial x$. Find a potential function $f(x, y)$ such that

$$\frac{\partial f}{\partial x} = 6xy - y^3 \quad \text{and} \quad \frac{\partial f}{\partial y} = 4y + 3x^2 - 3xy^2. \tag{14}$$

Solution Upon integrating the first of these equations with respect to x, we get

$$f(x, y) = 3x^2y - xy^3 + \xi(y), \tag{15}$$

where $\xi(y)$ is an "arbitrary function" of y alone; it acts as a "constant of integration" with respect to x, because its derivative with respect to x is zero. We next determine $\xi(y)$ by imposing the second condition in (14):

$$\frac{\partial f}{\partial y} = 3x^2 - 3xy^2 + \xi'(y) = 4y + 3x^2 - 3xy^2.$$

It follows that $\xi'(y) = 4y$, so $\xi(y) = 2y^2 + C$. When we set $C = 0$ and substitute the result into Eq. (15), we get the same potential function

$$f(x, y) = 3x^2y - xy^3 + 2y^2$$

that we found by entirely different methods in Example 3. ■

*A proof of Theorem 3 can be found in Section 1.6 of C. H. Edwards, Jr. and David E. Penney, *Differential Equations: Computing and Modeling* (Englewood Cliffs, N.J.: Prentice-Hall, 1996), pp. 60–61.

Conservative Force Fields and Conservation of Energy

Given a conservative force field $\mathbf{F}$, it is customary in physics to introduce a minus sign and write $\mathbf{F} = -\nabla V$. Then $V(x, y, z)$ is called the **potential energy** at the point (x, y, z). With $f = -V$ in Eq. (11), we have

$$W = \int_A^B \mathbf{F} \cdot d\mathbf{r} = V(A) - V(B), \tag{16}$$

and this means that the work W done by $\mathbf{F}$ in moving a particle from A to B is equal to the *decrease* in potential energy.

Here is the reason the expression *conservative field* is used. Suppose that a particle of mass m moves from A to B under the influence of the conservative force $\mathbf{F}$, with position vector $\mathbf{r}(t)$, $a \leqq t \leqq b$. Then Newton's law $\mathbf{F}(\mathbf{r}(t)) = m\mathbf{r}''(t) = m\mathbf{v}'(t)$ with $d\mathbf{r} = \mathbf{r}'(t)\,dt = \mathbf{v}(t)\,dt$ gives

$$\begin{aligned} \int_A^B \mathbf{F} \cdot d\mathbf{r} &= \int_a^b m\mathbf{v}'(t) \cdot \mathbf{v}(t)\,dt \\ &= \int_a^b mD_t\left[\frac{1}{2}\mathbf{v}(t) \cdot \mathbf{v}(t)\right] dt = \left[\frac{1}{2}m[v(t)]^2\right]_a^b. \end{aligned}$$

Thus with the abbreviations v_A for $v(a)$ and v_B for $v(b)$, we see that

$$\int_A^B \mathbf{F} \cdot d\mathbf{r} = \frac{1}{2}m(v_B)^2 - \frac{1}{2}m(v_A)^2. \tag{17}$$

By equating the right-hand sides of Eqs. (16) and (17), we get the formula

$$\tfrac{1}{2}m(v_A)^2 + V(A) = \tfrac{1}{2}m(v_B)^2 + V(B). \tag{18}$$

This is the law of **conservation of mechanical energy** for a particle moving under the influence of a *conservative* force field: Its **total energy**—the sum of its kinetic energy $\frac{1}{2}mv^2$ and its potential energy V—remains *constant.*

15.3 PROBLEMS

Determine whether the vector fields in Problems 1 through 16 are conservative. Find potential functions for those that are conservative (either by inspection or by using the method of Example 4).

1. $\mathbf{F}(x, y) = (2x + 3y)\mathbf{i} + (3x + 2y)\mathbf{j}$

2. $\mathbf{F}(x, y) = (4x - y)\mathbf{i} + (6y - x)\mathbf{j}$

3. $\mathbf{F}(x, y) = (3x^2 + 2y^2)\mathbf{i} + (4xy + 6y^2)\mathbf{j}$

4. $\mathbf{F}(x, y) = (2xy^2 + 3x^2)\mathbf{i} + (2x^2y + 4y^3)\mathbf{j}$

5. $\mathbf{F}(x, y) = (2y + \sin 2x)\mathbf{i} + (3x + \cos 3y)\mathbf{j}$

6. $\mathbf{F}(x, y) = (4x^2y - 5y^4)\mathbf{i} + (x^3 - 20xy^3)\mathbf{j}$

7. $\mathbf{F}(x, y) = \left(x^3 + \dfrac{y}{x}\right)\mathbf{i} + (y^2 + \ln x)\mathbf{j}$

8. $\mathbf{F}(x, y) = (1 + ye^{xy})\mathbf{i} + (2y + xe^{xy})\mathbf{j}$

9. $\mathbf{F}(x, y) = (\cos x + \ln y)\mathbf{i} + \left(\dfrac{x}{y} + e^y\right)\mathbf{j}$

10. $\mathbf{F}(x, y) = (x + \arctan y)\mathbf{i} + \dfrac{x + y}{1 + y^2}\mathbf{j}$

11. $\mathbf{F}(x, y) = (x \cos y + \sin y)\mathbf{i} + (y \cos x + \sin x)\mathbf{j}$

12. $\mathbf{F}(x, y) = e^{x-y}[(xy + y)\mathbf{i} + (xy + x)\mathbf{j}]$

13. $\mathbf{F}(x, y) = (3x^2y^3 + y^4)\mathbf{i} + (3x^3y^2 + y^4 + 4xy^3)\mathbf{j}$

14. $\mathbf{F}(x, y) = (e^x \sin y + \tan y)\mathbf{i} + (e^x \cos y + x \sec^2 y)\mathbf{j}$

15. $\mathbf{F}(x, y) = \left(\dfrac{2x}{y} - \dfrac{3y^2}{x^4}\right)\mathbf{i} + \left(\dfrac{2y}{x^3} - \dfrac{x^2}{y^2} + \dfrac{1}{\sqrt{y}}\right)\mathbf{j}$

16. $\mathbf{F}(x, y) = \dfrac{2x^{5/2} - 3y^{5/3}}{2x^{5/2}y^{2/3}}\mathbf{i} + \dfrac{3y^{5/3} - 2x^{5/2}}{3x^{3/2}y^{5/3}}\mathbf{j}$

In Problems 17 through 20, apply the method of Example 3 to find a potential function for the indicated vector field.

17. The vector field of Problem 3

18. The vector field of Problem 4

19. The vector field of Problem 13
20. The vector field of Problem 8

In Problems 21 through 26, show that the given line integral is independent of path in the entire xy-plane, and then calculate the value of the line integral.

21. $\displaystyle\int_{(0,0)}^{(1,2)} (y^2 + 2xy)\,dx + (x^2 + 2xy)\,dy$

22. $\displaystyle\int_{(0,0)}^{(1,1)} (2x - 3y)\,dx + (2y - 3x)\,dy$

23. $\displaystyle\int_{(0,0)}^{(1,-1)} 2xe^y\,dx + x^2e^y\,dy$

24. $\displaystyle\int_{(0,0)}^{(2,\pi)} \cos y\,dx - x\sin y\,dy$

25. $\displaystyle\int_{(\pi/2,\pi/2)}^{(\pi,\pi)} (\sin y + y\cos x)\,dx + (\sin x + x\cos y)\,dy$

26. $\displaystyle\int_{(0,0)}^{(1,-1)} (e^y + ye^x)\,dx + (e^x + xe^y)\,dy$

Find a potential function for each of the conservative vector fields in Problems 27 through 29.

27. $\mathbf{F}(x, y, z) = yz\,\mathbf{i} + xz\,\mathbf{j} + xy\,\mathbf{k}$
28. $\mathbf{F}(x, y, z) = (2x - y - z)\mathbf{i} + (2y - x)\mathbf{j} + (2z - x)\mathbf{k}$
29. $\mathbf{F}(x, y, z) = (y\cos z - yze^x)\mathbf{i} + (x\cos z - ze^x)\mathbf{j} - (xy\sin z + ye^x)\mathbf{k}$
30. Let $\mathbf{F}(x, y) = (-y\mathbf{i} + x\mathbf{j})/(x^2 + y^2)$ for x and y not both zero. Calculate the values of

$$\int_C \mathbf{F}\cdot\mathbf{T}\,ds$$

along both the upper and the lower halves of the circle $x^2 + y^2 = 1$ from $(1, 0)$ to $(-1, 0)$. Is there a function $f = f(x, y)$ defined for x and y not both zero such that $\nabla f = \mathbf{F}$? Why?

31. Show that if the force field $\mathbf{F} = P\mathbf{i} + Q\mathbf{j}$ is conservative, then $\partial P/\partial y = \partial Q/\partial x$. Show that the force field of Problem 30 satisfies the condition $\partial P/\partial y = \partial Q/\partial x$ but nevertheless is *not* conservative.

32. Suppose that the force field $\mathbf{F} = P\mathbf{i} + Q\mathbf{j} + R\mathbf{k}$ is conservative. Show that

$$\frac{\partial P}{\partial y} = \frac{\partial Q}{\partial x}, \quad \frac{\partial P}{\partial z} = \frac{\partial R}{\partial x}, \quad \text{and} \quad \frac{\partial Q}{\partial z} = \frac{\partial R}{\partial y}.$$

33. Apply Theorem 2 and the result of Problem 32 to show that

$$\int_C 2xy\,dx + x^2\,dy + y^2\,dz$$

is not independent of path.

34. Let $\mathbf{F}(x, y, z) = yz\,\mathbf{i} + (xz + y)\mathbf{j} + (xy + 1)\mathbf{k}$. Define the function f by

$$f(x, y, z) = \int_C \mathbf{F}\cdot\mathbf{T}\,ds,$$

where C is the straight line segment from $(0, 0, 0)$ to (x, y, z). Determine f by evaluating this line integral, and then show that $\nabla f = \mathbf{F}$.

35. Let $f(x, y) = \tan^{-1}(y/x)$, which if $x > 0$ equals the polar angle θ for the point (x, y). (a) Show that

$$\mathbf{F} = \nabla f = \frac{-y\mathbf{i} + x\mathbf{j}}{x^2 + y^2}.$$

(b) Suppose that $A(x_1, y_1) = (r_1, \theta_1)$ and $B(x_2, y_2) = (r_2, \theta_2)$ are two points in the right half-plane $x > 0$ and that C is a smooth curve from A to B. Explain why it follows from the fundamental theorem for line integrals that $\int_C \mathbf{F}\cdot\mathbf{T}\,ds = \theta_2 - \theta_1$. (c) Suppose that C_1 is the upper half of the unit circle from $(1, 0)$ to $(-1, 0)$ and that C_2 is the lower half, oriented also from $(1, 0)$ to $(-1, 0)$. Show that

$$\int_{C_1} \mathbf{F}\cdot\mathbf{T}\,ds = \pi \quad \text{whereas} \quad \int_{C_2} \mathbf{F}\cdot\mathbf{T}\,ds = -\pi.$$

Why does this not contradict the fundamental theorem?

36. Let $\mathbf{F} = k\mathbf{r}/r^3$ be the inverse-square force field of Example 7 in Section 15.2. Show that the work done by $\mathbf{F}$ in moving a particle from a point at distance r_1 from the origin to a point at distance r_2 from the origin is given by

$$W = k\left(\frac{1}{r_1} - \frac{1}{r_2}\right).$$

37. Suppose that an Earth satellite with mass $m = 10000$ kg travels in an elliptical orbit whose apogee (farthest point) and perigee (closest point) are, respectively, 11000 km and 9000 km from the center of the Earth. Calculate the work done against Earth's gravitational force field $\mathbf{F} = -GMm\mathbf{r}/r^3$ in lifting the satellite from perigee to apogee. Use the values $M = 5.97 \times 10^{24}$ kg for the mass of Earth and $G = 6.67 \times 10^{-11}$ N·m²/kg² for the universal gravitational constant.

38. Calculate the work that must be done against the sun's gravitational force field in transporting the satellite of Problem 37 from Earth to Mars. Use the values $M = 1.99 \times 10^{30}$ kg for the mass of the sun, $r_E = 1.50 \times 10^8$ km for the distance from the sun to Earth, and $r_M = 2.29 \times 10^8$ km for the distance from the sun to Mars.

15.4 GREEN'S THEOREM

Green's theorem relates a line integral around a simple closed plane curve C to an ordinary double integral over the plane region R bounded by C. Suppose that the curve C is piecewise smooth—it consists of a finite number of parametric arcs with continuous nonzero velocity vectors. Then C has a unit tangent vector $\mathbf{T}$ everywhere except possibly at a finite number of *corner points.* The **positive,** or **counterclockwise,** direction along C is the direction determined by a parametrization $\mathbf{r}(t)$ of C such that the region R remains on the *left* as the point $\mathbf{r}(t)$ traces the boundary curve C. That is, the vector obtained from the unit tangent vector $\mathbf{T}$ by a counterclockwise rotation through $90°$ always points *into* the region R (Fig. 15.4.1). The symbol

$$\oint_C P\,dx + Q\,dy$$

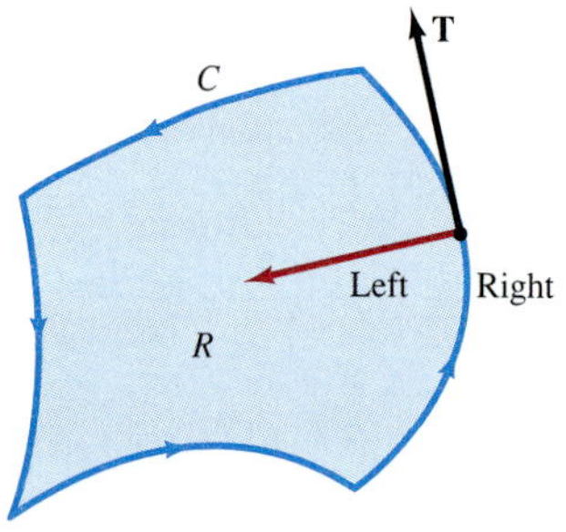

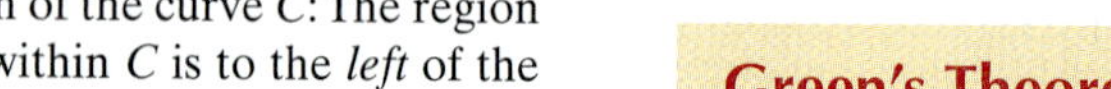

Fig. 15.4.1 Positive orientation of the curve C: The region R within C is to the *left* of the unit tangent vector $\mathbf{T}$.

denotes a line integral along or around C in this positive direction. A reversed arrow on the circle through the integral sign indicates a line integral around C in the opposite direction, which we call the **negative,** or the **clockwise,** direction.

The following result first appeared (in an equivalent form) in a booklet on the applications of mathematics to electricity and magnetism, published privately in 1828 by the self-taught English mathematical physicist George Green (1793–1841).

Green's Theorem

Let C be a piecewise smooth simple closed curve that bounds the region R in the plane. Suppose that the functions $P(x, y)$ and $Q(x, y)$ have continuous first-order partial derivatives on R. Then

$$\oint_C P\,dx + Q\,dy = \iint_R \left(\frac{\partial Q}{\partial x} - \frac{\partial P}{\partial y}\right) dA. \qquad \textbf{(1)}$$

PROOF: First we give a proof for the case in which the region R is both horizontally simple and vertically simple. Then we indicate how to extend the result to more general regions.

Recall from Section 14.2 that if R is vertically simple, then it has a description of the form $g_1(x) \leqq y \leqq g_2(y), a \leqq x \leqq b$. The boundary curve C is then the union of the four arcs C_1, C_2, C_3, and C_4 of Fig. 15.4.2, positively oriented as indicated there. Hence

$$\oint_C P\,dx = \int_{C_1} P\,dx + \int_{C_2} P\,dx + \int_{C_3} P\,dx + \int_{C_4} P\,dx.$$

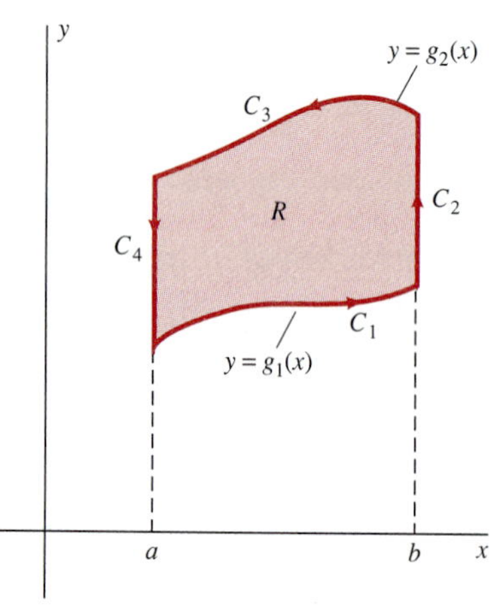

Fig. 15.4.2 The boundary curve C is the union of the four arcs C_1, C_2, C_3, and C_4,

The integrals along both C_2 and C_4 are zero, because on those two curves $x(t)$ is constant, so that $dx = x'(t)\,dt = 0$. Thus we need compute only the integrals along C_1 and C_3.

The point $(x, g_1(x))$ traces C_1 as x increases from a to b, whereas the point $(x, g_2(x))$ traces C_3 as x decreases from b to a. Hence

$$\oint_C P\,dx = \int_a^b P(x, g_1(x))\,dx + \int_b^a P(x, g_2(x))\,dx$$

$$= -\int_a^b [P(x, g_2(x)) - P(x, g_1(x))]\,dx = -\int_a^b \int_{g_1(x)}^{g_2(x)} \frac{\partial P}{\partial y}\,dy\,dx$$

by the fundamental theorem of calculus. Thus

$$\oint_C P\,dx = -\iint_R \frac{\partial P}{\partial y}\,dA. \tag{2}$$

In Problem 36 we ask you to show in a similar way that

$$\oint_C Q\,dy = +\iint_R \frac{\partial Q}{\partial x}\,dA \tag{3}$$

if the region R is horizontally simple. We then obtain Eq. (1), the conclusion of Green's theorem, simply by adding Eqs. (2) and (3). ∎

The complete proof of Green's theorem for more general regions is beyond the scope of an elementary text. But the typical region R that appears in practice can be divided into smaller regions $R_1, R_2, \ldots, R_k$ that are both vertically and horizontally simple. Green's theorem for the region R then follows from the fact that it holds for each of the regions $R_1, R_2, \ldots, R_k$ (see Problem 29).

For example, we can divide the horseshoe-shaped region R of Fig. 15.4.3 into the two regions R_1 and R_2, both of which are horizontally simple and vertically simple. We also subdivide the boundary C of R accordingly and write $C_1 \cup D_1$ for the boundary of R_1 and $C_2 \cup D_2$ for the boundary of R_2 (Fig. 15.4.3). Applying Green's theorem separately to the regions R_1 and R_2, we get

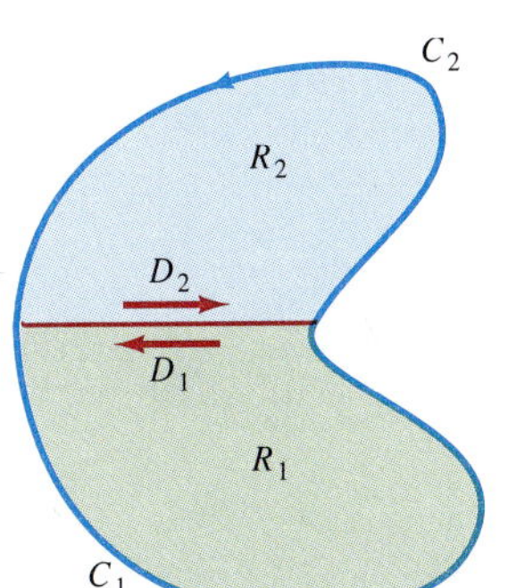

Fig. 15.4.3 Decomposing the region R into two horizontally and vertically simple regions by using a crosscut

$$\oint_{C_1 \cup D_1} P\,dx + Q\,dy = \iint_{R_1} \left(\frac{\partial Q}{\partial x} - \frac{\partial P}{\partial y}\right) dA$$

and

$$\oint_{C_2 \cup D_2} P\,dx + Q\,dy = \iint_{R_2} \left(\frac{\partial Q}{\partial x} - \frac{\partial P}{\partial y}\right) dA.$$

When we add these two equations, the result is Eq. (1), Green's theorem for the region R, because the two line integrals along D_1 and D_2 cancel. This occurs because D_1 and D_2 represent the same curve with opposite orientations, so

$$\int_{D_2} P\,dx + Q\,dy = -\int_{D_1} P\,dx + Q\,dy$$

by Eq. (11) of Section 15.2. It therefore follows that

$$\int_{C_1 \cup D_1 \cup C_2 \cup D_2} P\,dx + Q\,dy = \oint_{C_1 \cup C_2} P\,dx + Q\,dy = \oint_C P\,dx + Q\,dy.$$

Similarly, we could establish Green's theorem for the region shown in Fig. 15.4.4 by dividing it into the four simple regions indicated there.

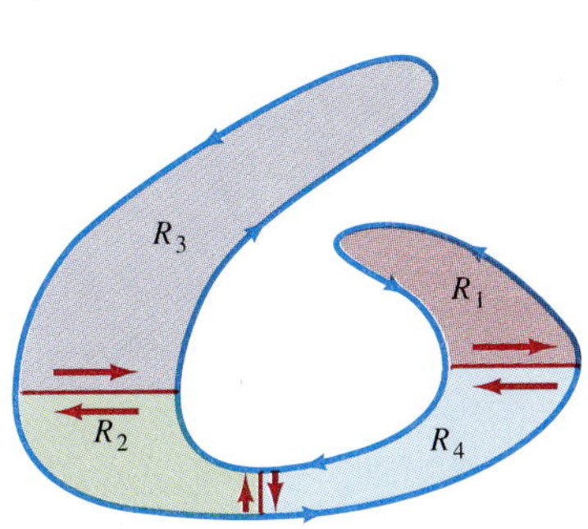

Fig. 15.4.4 Many important regions can be decomposed into simple regions by using one or more crosscuts.

EXAMPLE 1 Use Green's theorem to evaluate the line integral

$$\oint_C \left(2y + \sqrt{9 + x^3}\right) dx + \left(5x + e^{\arctan y}\right) dy,$$

where C is the circle $x^2 + y^2 = 4$.

Solution With $P(x, y) = 2y + \sqrt{9 + x^3}$ and $Q(x, y) = 5x + e^{\arctan y}$, we see that

$$\frac{\partial Q}{\partial x} - \frac{\partial P}{\partial y} = 5 - 2 = 3.$$

Because C bounds R, a circular disk with area 4π, Green's theorem therefore implies that the given line integral is equal to

$$\iint_R 3\,dA = 3 \cdot 4\pi = 12\pi.$$

REMARK Suppose that the force field $\mathbf{F}$ is defined by

$$\mathbf{F}(x, y) = \left(2y + \sqrt{9 + x^2}\right)\mathbf{i} + (5x + e^{\arctan y})\mathbf{j} = P(x, y)\mathbf{i} + Q(x, y)\mathbf{j},$$

using the notation in Example 1. Then (as in Section 15.2) the work W done by the force field $\mathbf{F}$ in moving a particle counterclockwise once around the circle C of radius 2 is given by

$$W = \oint_C \mathbf{F} \cdot \mathbf{T}\,ds = \oint_C P\,dx + Q\,dy = \iint_R \left(\frac{\partial Q}{\partial x} - \frac{\partial P}{\partial y}\right) dA = \iint_R 3\,dA = 12\pi$$

as in Example 1.

EXAMPLE 2 Evaluate the line integral

$$\oint_C 3xy\,dx + 2x^2\,dy,$$

where C is the boundary of the region R shown in Fig. 15.4.5. It is bounded above by the line $y = x$ and below by the parabola $y = x^2 - 2x$.

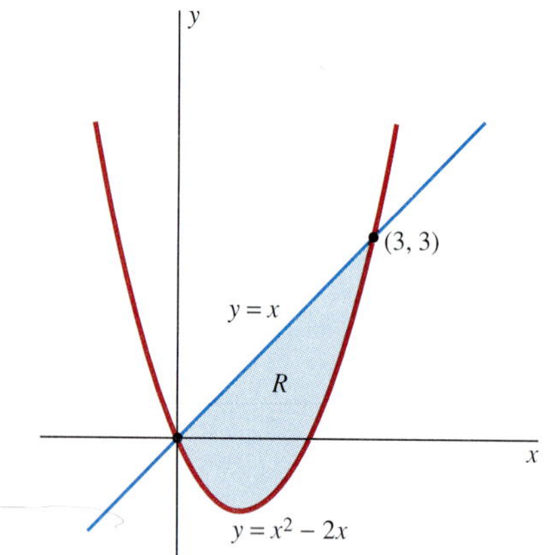

Fig. 15.4.5 The region of Example 2

Solution To evaluate the line integral directly, we would need to parametrize separately the line and the parabola. Instead, we apply Green's theorem with $P(x, y) = 3xy$ and $Q(x, y) = 2x^2$, so

$$\frac{\partial Q}{\partial x} - \frac{\partial P}{\partial y} = 4x - 3x = x.$$

Then

$$\begin{aligned}\oint_C 3xy\,dx + 2x^2\,dy &= \iint_R x\,dA \\ &= \int_0^3 \int_{x^2-2x}^{x} x\,dy\,dx = \int_0^3 \Big[xy\Big]_{y=x^2-2x}^{x} dx \\ &= \int_0^3 (3x^2 - x^3)\,dx = \left[x^3 - \frac{1}{4}x^4\right]_0^3 = \frac{27}{4}.\end{aligned}$$

In Examples 1 and 2 we found the double integral easier to evaluate directly than the line integral. Sometimes the situation is the reverse. The following consequence of Green's theorem illustrates the technique of evaluating a double integral $\iint_R f(x, y)\,dA$ by converting it into a line integral

$$\oint_C P\,dx + Q\,dy.$$

To do this, we must be able to find functions $P(x, y)$ and $Q(x, y)$ such that

$$\frac{\partial Q}{\partial x} - \frac{\partial P}{\partial y} = f(x, y).$$

As in the proof of the following result, this is sometimes easy.

Corollary to Green's Theorem

The area A of the region R bounded by the piecewise smooth simple closed curve C is given by

$$A = \frac{1}{2}\oint_C -y\,dx + x\,dy = -\oint_C y\,dx = \oint_C x\,dy. \tag{4}$$

PROOF: With $P(x, y) = -y$ and $Q(x, y) \equiv 0$, Green's theorem gives

$$-\oint_C y\,dx = \iint_R 1\,dA = A.$$

Similarly, with $P(x, y) \equiv 0$ and $Q(x, y) = x$, we obtain

$$\oint_C x\,dy = \iint_R 1\,dA = A.$$

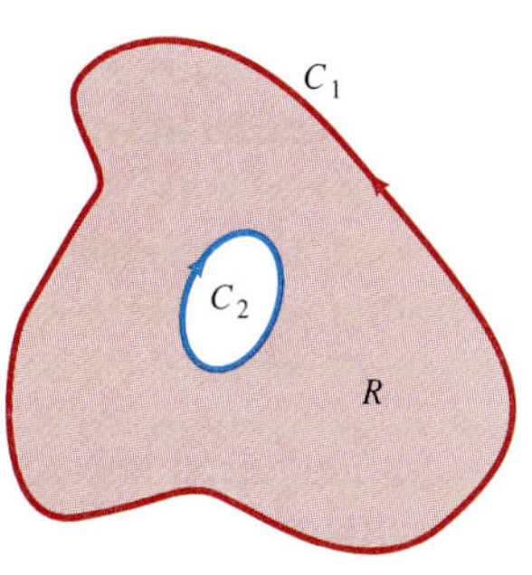

Fig. 15.4.6 An annular region—the boundary consists of two simple closed curves, one within the other.

The third result may be obtained by averaging the left- and right-hand sides in the last two equations. Alternatively, with $P(x, y) = -y/2$ and $Q(x, y) = x/2$, Green's theorem gives

$$\frac{1}{2}\oint_C -y\,dx + x\,dy = \iint_R \left(\tfrac{1}{2} + \tfrac{1}{2}\right) dA = A.$$ ■

EXAMPLE 3 Apply the corollary to Green's theorem to find the area A of the region R bounded by the ellipse $x^2/a^2 + y^2/b^2 = 1$.

Solution With the parametrization $x = a\cos t$, $y = b\sin t$, $0 \leqq t \leqq 2\pi$, Eq. (4) gives

$$A = \oint_R x\,dy = \int_0^{2\pi} (a\cos t)(b\cos t\,dt)$$

$$= \tfrac{1}{2}ab\int_0^{2\pi} (1 + \cos 2t)\,dt = \pi ab.$$ ■

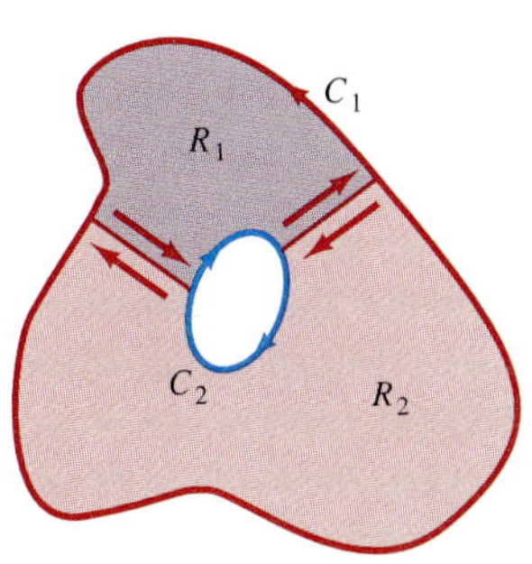

Fig. 15.4.7 Two crosscuts convert the annular region into the union of two ordinary regions R_1 and R_2, each bounded by a single closed curve.

By using the technique of subdividing a region into simpler ones, we can extend Green's theorem to regions with boundaries that consist of two or more simple closed curves. For example, consider the annular region R of Fig. 15.4.6, with boundary C consisting of the two simple closed curves C_1 and C_2. The positive direction along C—the direction for which the region R always lies on the left—is counterclockwise on the outer curve C_1 but clockwise on the inner curve C_2.

We divide R into two regions R_1 and R_2 by using two crosscuts, as shown in Fig. 15.4.7. Applying Green's theorem to each of these subregions, we get

$$\iint_R (Q_x - P_y)\,dA = \iint_{R_1} (Q_x - P_y)\,dA + \iint_{R_2} (Q_x - P_y)\,dA$$

$$= \oint_{C_1} (P\,dx + Q\,dy) + \oint_{C_2} (P\,dx + Q\,dy)$$

$$= \oint_C P\,dx + Q\,dy.$$

Thus we obtain Green's theorem for the given region R. What makes this proof work is that the opposite line integrals along the two crosscuts cancel each other. You may, of course, use any finite number of crosscuts.

EXAMPLE 4 Suppose that C is a smooth simple closed curve that encloses the origin $(0, 0)$. Show that

$$\oint_C \frac{-y\,dx + x\,dy}{x^2 + y^2} = 2\pi,$$

and also show that this integral is zero if C does *not* enclose the origin.

Solution With $P(x, y) = -y/(x^2 + y^2)$ and $Q(x, y) = x/(x^2 + y^2)$, a brief computation gives $\partial Q/\partial x - \partial P/\partial y \equiv 0$ when x and y are not both zero. If the region R bounded by C does not contain the origin, then P and Q and their derivatives are continuous on R. Hence Green's theorem implies that the integral in question is zero.

If C does enclose the origin, then we enclose the origin in a small circle C_a of radius a so small that C_a lies wholly within C (Fig. 15.4.8). We parametrize this circle by $x = a\cos t$, $y = a\sin t$, $0 \leqq t \leqq 2\pi$. Then Green's theorem, applied to the region R between C and C_a, gives

$$\oint_C \frac{-y\,dx + x\,dy}{x^2 + y^2} + \oint_{C_a} \frac{-y\,dx + x\,dy}{x^2 + y^2} = \iint_R 0\,dA = 0.$$

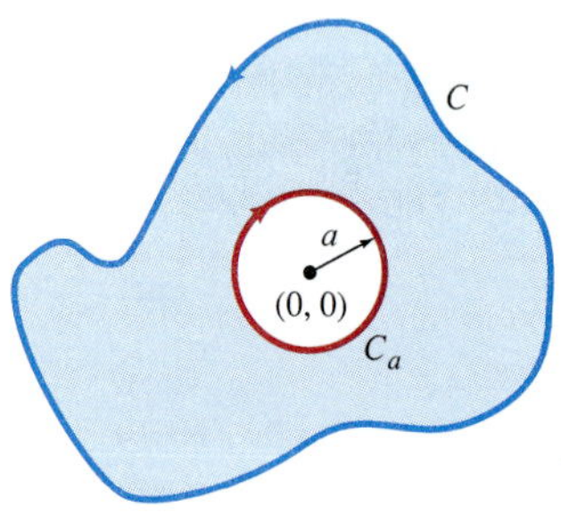

Fig. 15.4.8 Use the small circle C_a if C encloses the origin (Example 4).

IMPORTANT Note the *reversed* arrow in the second line integral, required because the parametrization we chose is the clockwise (*negative*) orientation for C_a. Therefore

$$\oint_C \frac{-y\,dx + x\,dy}{x^2 + y^2} = \oint_{C_a} \frac{-y\,dx + x\,dy}{x^2 + y^2}$$

$$= \int_0^{2\pi} \frac{(-a\sin t)(-a\sin t\,dt) + (a\cos t)(a\cos t\,dt)}{(a\cos t)^2 + (a\sin t)^2}$$

$$= \int_0^{2\pi} 1\,dt = 2\pi.$$

REMARK The result of Example 4 can be interpreted in terms of the polar-coordinate angle $\theta = \arctan(y/x)$. Because

$$d\theta = \frac{-y\,dx + x\,dy}{x^2 + y^2},$$

the line integral of Example 4 measures the net change in θ as we go around the curve C once in a counterclockwise direction. This net change is 2π if C encloses the origin and is zero otherwise.

The Divergence and Flux of a Vector Field

Now let us consider the steady flow of a thin layer of fluid in the plane (perhaps like a sheet of water spreading across a floor). Let $\mathbf{v}(x, y)$ be its velocity vector field and $\delta(x, y)$ be the density of the fluid at the point (x, y). The term *steady flow* means that $\mathbf{v}$ and δ depend only on x and y, *not* on time t. We want to compute the rate at which the fluid flows out of the region R bounded by a simple closed curve C (Fig. 15.4.9). We seek the net rate of outflow—the actual outflow minus the inflow (Fig. 15.4.10).

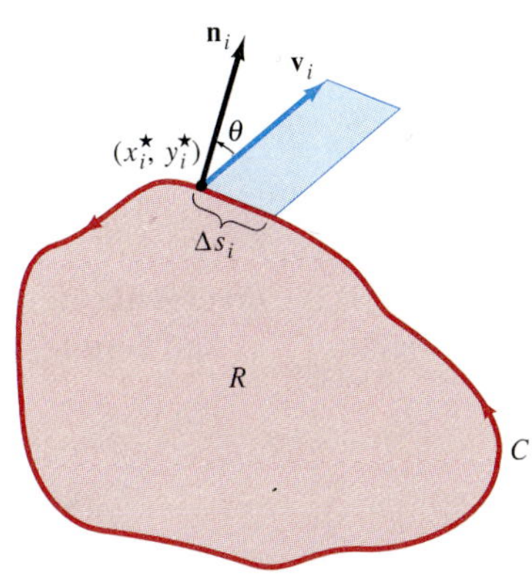

Fig. 15.4.9 The area of the parallelogram approximates the fluid flow across Δs_i in unit time.

Let Δs_i be a short segment of the curve C, and let $(x_i^\star, y_i^\star)$ be an endpoint of Δs_i. Then the area of the portion of the fluid that flows out of R across Δs_i per unit time is approximately the area of the parallelogram in Fig. 15.4.9. This is the parallelogram spanned by the segment Δs_i and the vector $\mathbf{v}_i = \mathbf{v}(x_i^\star, y_i^\star)$. Suppose that $\mathbf{n}_i$ is the unit normal vector to C at the point $(x_i^\star, y_i^\star)$, the normal that points *out* of R. Then the area of this parallelogram is

$$(|\mathbf{v}_i| \cos\theta)\ \Delta s_i = \mathbf{v}_i \cdot \mathbf{n}_i\ \Delta s_i,$$

where θ is the angle between $\mathbf{n}_i$ and $\mathbf{v}_i$.

We multiply this area by the density $\delta_i = \delta(x_i^\star, y_i^\star)$ and then add these terms over those values of i that correspond to a subdivision of the entire curve C. This gives the (net) total mass of fluid leaving R per unit of time; it is approximately

$$\sum_{i=1}^{n} \delta_i \mathbf{v}_i \cdot \mathbf{n}_i\ \Delta s_i = \sum_{i=1}^{n} \mathbf{F}_i \cdot \mathbf{n}_i\ \Delta s_i,$$

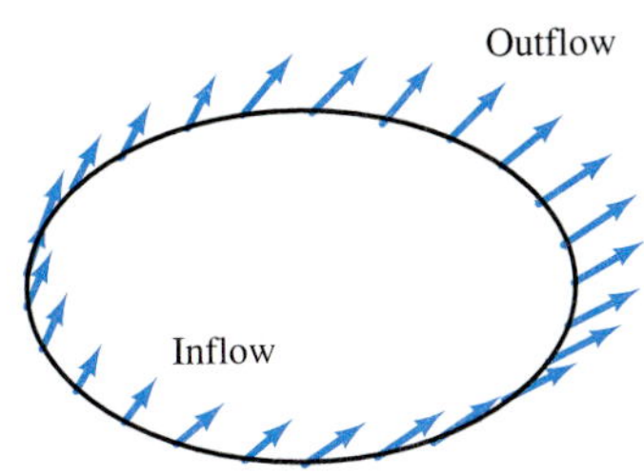

Fig. 15.4.10 The flux ϕ of the vector field $\mathbf{F}$ across the curve C is the net outflow minus the net inflow.

where $\mathbf{F} = \delta\mathbf{v}$. The line integral around C that this sum approximates is called the **flux of the vector field F across the curve** C. Thus the flux ϕ of $\mathbf{F}$ across C is given by

$$\oint_C \mathbf{F} \cdot \mathbf{n}\, ds, \tag{5}$$

where $\mathbf{n}$ is the *outer* unit normal vector to C (Fig. 15.4.11).

In the present case of fluid flow with velocity vector $\mathbf{v}$, the flux ϕ of $\mathbf{F} = \delta\mathbf{v}$ is the rate at which the fluid is flowing out of R across the boundary curve C, in units of mass per unit of time. But the same terminology is used for an arbitrary vector field $\mathbf{F} = M\mathbf{i} + N\mathbf{j}$. For example, we may speak of the flux of an electric or gravitational field across a curve C.

From Fig. 15.4.11 we see that the outer unit normal vector $\mathbf{n}$ is equal to $\mathbf{T} \times \mathbf{k}$. The unit tangent vector $\mathbf{T}$ to the curve C is

$$\mathbf{T} = \frac{1}{v}\left(\mathbf{i}\frac{dx}{dt} + \mathbf{j}\frac{dy}{dt}\right) = \mathbf{i}\frac{dx}{ds} + \mathbf{j}\frac{dy}{ds}$$

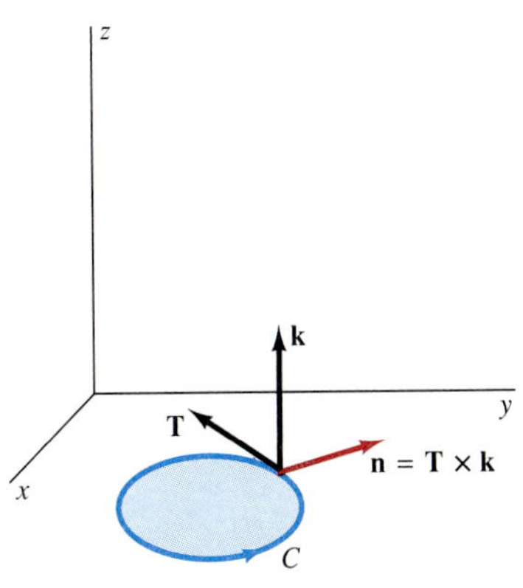

Fig. 15.4.11 Computing the outer unit normal vector $\mathbf{n}$ from the unit tangent vector $\mathbf{T}$

because $v = ds/dt$. Hence

$$\mathbf{n} = \mathbf{T} \times \mathbf{k} = \left(\mathbf{i}\frac{dx}{ds} + \mathbf{j}\frac{dy}{ds}\right) \times \mathbf{k}.$$

But $\mathbf{i} \times \mathbf{k} = -\mathbf{j}$ and $\mathbf{j} \times \mathbf{k} = \mathbf{i}$. Thus we find that

$$\mathbf{n} = \mathbf{i}\frac{dy}{ds} - \mathbf{j}\frac{dx}{ds}. \tag{6}$$

Substitution of the expression in Eq. (6) into the flux integral of Eq. (5) gives

$$\oint_C \mathbf{F}\cdot\mathbf{n}\,ds = \oint_C (M\mathbf{i}+N\mathbf{j})\cdot\left(\mathbf{i}\frac{dy}{ds}-\mathbf{j}\frac{dx}{ds}\right)ds = \oint_C -N\,dx + M\,dy.$$

Applying Green's theorem to the last line integral with $P = -N$ and $Q = M$, we get

$$\oint_C \mathbf{F}\cdot\mathbf{n}\,ds = \iint_R \left(\frac{\partial M}{\partial x}+\frac{\partial N}{\partial y}\right)dA \tag{7}$$

for the flux of $\mathbf{F} = M\mathbf{i} + N\mathbf{j}$ across C.

The scalar function $\partial M/\partial x + \partial N/\partial y$ that appears in Eq. (7) is the **divergence** of the two-dimensional vector field $\mathbf{F} = M\mathbf{i} + N\mathbf{j}$ as defined in Section 15.1 and denoted by

$$\operatorname{div}\mathbf{F} = \nabla\cdot\mathbf{F} = \frac{\partial M}{\partial x}+\frac{\partial N}{\partial y}. \tag{8}$$

When we substitute Eq. (8) into Eq. (7), we obtain a **vector form of Green's theorem**:

$$\oint_C \mathbf{F}\cdot\mathbf{n}\,ds = \iint_R \nabla\cdot\mathbf{F}\,dA, \tag{9}$$

with the understanding that $\mathbf{n}$ is the *outer* unit normal to C. Thus the flux of a vector field across a simple closed curve C is equal to the double integral of its divergence over the region R bounded by C.

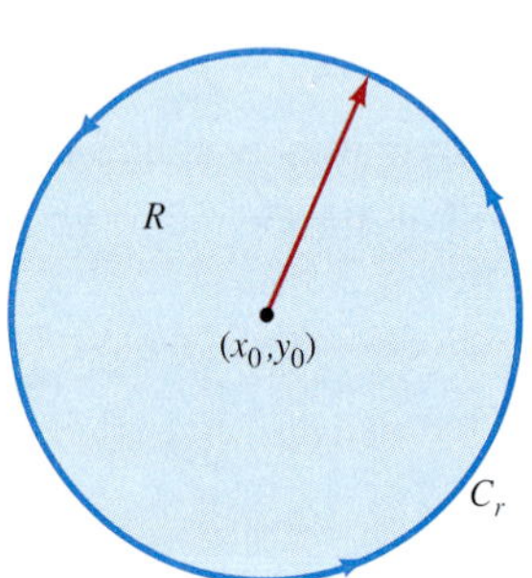

Fig. 15.4.12 The circular disk R of radius r centered at (x_0, y_0)

If the region R is bounded by a circle C_r of radius r centered at the point (x_0, y_0), then

$$\oint_{C_r} \mathbf{F}\cdot\mathbf{n}\,ds = \iint_R \nabla\cdot\mathbf{F}\,dA = \pi r^2\cdot\{\nabla\cdot\mathbf{F}\}(\overline{x},\overline{y})$$

for some point $(\overline{x},\overline{y})$ in R (Fig. 15.4.12); this is a consequence of the average value property of double integrals (see Problem 50 of Section 14.2). We divide by πr^2 and then let r approach zero. Thus we find that

$$\{\nabla\cdot\mathbf{F}\}(x_0, y_0) = \lim_{r\to 0}\frac{1}{\pi r^2}\oint_{C_r}\mathbf{F}\cdot\mathbf{n}\,ds \tag{10}$$

because $(\overline{x},\overline{y}) \to (x_0, y_0)$ as $r \to 0$.

In the case of our original fluid flow, with $\mathbf{F} = \delta\mathbf{v}$, Eq. (10) implies that the value of $\nabla\cdot\mathbf{F}$ at (x_0, y_0) is a measure of the rate at which the fluid is "diverging away" from the point (x_0, y_0).

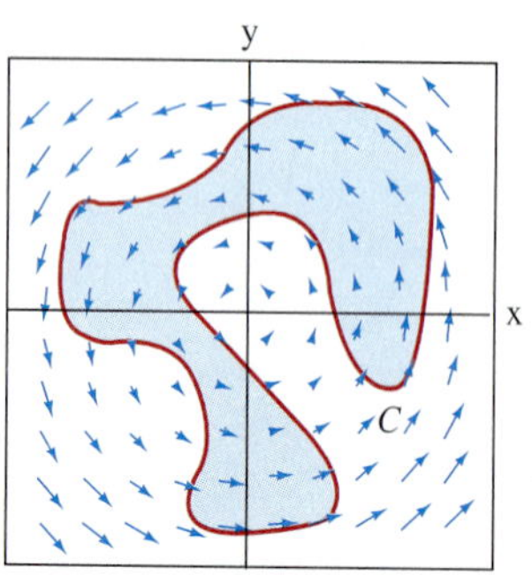

Fig. 15.4.13 The flux $\oint \mathbf{F}\cdot\mathbf{n}\,ds$ of the vector field $\mathbf{F} = -y\mathbf{i} + x\mathbf{j}$ across the curve C is zero.

EXAMPLE 5 The vector field $\mathbf{F} = -y\mathbf{i} + x\mathbf{j}$ is the velocity field of a steady-state counterclockwise rotation around the origin. Show that the flux of $\mathbf{F}$ across any simple closed curve C is zero (Fig. 15.4.13).

Solution This follows immediately from Eq. (9) because

$$\nabla\cdot\mathbf{F} = \frac{\partial}{\partial x}(-y) + \frac{\partial}{\partial y}(x) = 0.$$

■

15.4 PROBLEMS

In Problems 1 through 12, apply Green's theorem to evaluate the integral

$$\oint_C P\,dx + Q\,dy$$

around the specified closed curve C.

1. $P(x, y) = x + y^2$, $Q(x, y) = y + x^2$; C is the square with vertices $(\pm 1, \pm 1)$.
2. $P(x, y) = x^2 + y^2$, $Q(x, y) = -2xy$; C is the boundary of the triangle bounded by the lines $x = 0$, $y = 0$, and $x + y = 1$.
3. $P(x, y) = y + e^x$, $Q(x, y) = 2x^2 + \cos y$; C is the boundary of the triangle with vertices $(0, 0)$, $(1, 1)$, and $(2, 0)$.
4. $P(x, y) = x^2 - y^2$, $Q(x, y) = xy$; C is the boundary of the region bounded by the line $y = x$ and the parabola $y = x^2$.
5. $P(x, y) = -y^2 + \exp(e^x)$, $Q(x, y) = \arctan y$; C is the boundary of the region between the parabolas $y = x^2$ and $x = y^2$.
6. $P(x, y) = y^2$, $Q(x, y) = 2x - 3y$; C is the circle $x^2 + y^2 = 9$.
7. $P(x, y) = x - y$, $Q(x, y) = y$; C is the boundary of the region between the x-axis and the graph of $y = \sin x$ for $0 \leqq x \leqq \pi$.
8. $P(x, y) = e^x \sin y$, $Q(x, y) = e^x \cos y$; C is the right-hand loop of the graph of the polar equation $r^2 = 4\cos\theta$.
9. $P(x, y) = y^2$, $Q(x, y) = xy$; C is the ellipse with equation $x^2/9 + y^2/4 = 1$.
10. $P(x, y) = y/(1 + x^2)$, $Q(x, y) = \arctan x$; C is the oval with equation $x^4 + y^4 = 1$.
11. $P(x, y) = xy$, $Q(x, y) = x^2$; C is the first-quadrant loop of the graph of the polar equation $r = \sin 2\theta$.
12. $P(x, y) = x^2$, $Q(x, y) = -y^2$; C is the cardioid with polar equation $r = 1 + \cos\theta$.

In Problems 13 through 16, use the corollary to Green's theorem to find the area of the indicated region.

13. The circle bounded by $x = a\cos t$, $y = a\sin t$, $0 \leqq t \leqq 2\pi$
14. The region between the x-axis and one arch of the cycloid with parametric equations $x = a(t - \sin t)$, $y = a(1 - \cos t)$
15. The region bounded by the astroid with parametric equations $x = \cos^3 t$, $y = \sin^3 t$, $0 \leqq t \leqq 2\pi$
16. The region between the graphs of $y = x^2$ and $y = x^3$

In Problems 17 through 20, use Green's theorem to calculate the work

$$W = \oint_C \mathbf{F} \cdot \mathbf{T}\,ds$$

done by the given force field $\mathbf{F}$ in moving a particle counterclockwise once around the indicated curve C.

17. $\mathbf{F} = -2y\mathbf{i} + 3x\mathbf{j}$ and C is the ellipse $x^2/9 + y^2/4 = 1$.
18. $\mathbf{F} = (y^2 - x^2)\mathbf{i} + 2xy\mathbf{j}$ and C is the circle $x^2 + y^2 = 9$.
19. $\mathbf{F} = 5x^2y^3\mathbf{i} + 7x^3y^2\mathbf{j}$ and C is the triangle with vertices $(0, 0)$, $(3, 0)$, and $(0, 6)$.
20. $\mathbf{F} = xy^2\mathbf{i} + 3x^2y\mathbf{j}$ and C is the boundary of the semicircular disk bounded by the x-axis and the circular arc $y = \sqrt{4 - x^2} \geqq 0$.

In Problems 21 through 24, use Green's theorem in the vector form of Eq. (9) to calculate the outward flux

$$\phi = \oint_C \mathbf{F} \cdot \mathbf{n}\,ds$$

of the given vector field across the indicated closed curve C.

21. $\mathbf{F} = 2x\mathbf{i} + 3y\mathbf{j}$ and C is the ellipse of Problem 17.
22. $\mathbf{F} = x^3\mathbf{i} + y^3\mathbf{j}$ and C is the circle of Problem 18.
23. $\mathbf{F} = \left(3x + \sqrt{1 + y^2}\right)\mathbf{i} + \left(2y - \sqrt[3]{1 + x^4}\right)\mathbf{j}$ and C is the triangle of Problem 19.
24. $\mathbf{F} = (3xy^2 + 4x)\mathbf{i} + (3x^2y - 4y)\mathbf{j}$ and C is the closed curve of Problem 20.
25. Suppose that f is a twice-differentiable scalar function of x and y. Show that

$$\nabla^2 f = \text{div}(\nabla f) = \frac{\partial^2 f}{\partial x^2} + \frac{\partial^2 f}{\partial y^2}.$$

26. Show that $f(x, y) = \ln(x^2 + y^2)$ satisfies **Laplace's equation** $\nabla^2 f = 0$ except at the point $(0, 0)$.
27. Suppose that f and g are twice-differentiable functions. Show that

$$\nabla^2(fg) = f\,\nabla^2 g + g\,\nabla^2 f + 2\,\nabla f \cdot \nabla g.$$

28. Suppose that the function $f(x, y)$ is twice continuously differentiable in the region R bounded by the piecewise smooth curve C. Prove that

$$\oint_C \frac{\partial f}{\partial x}\,dy - \frac{\partial f}{\partial y}\,dx = \iint_R \nabla^2 f\,dx\,dy.$$

29. Let R be the plane region with area A enclosed by the piecewise smooth simple closed curve C. Use Green's theorem to show that the coordinates of the centroid of R are

$$\overline{x} = \frac{1}{2A}\oint_C x^2\,dy, \quad \overline{y} = -\frac{1}{2A}\oint_C y^2\,dx.$$

30. Use the result of Problem 29 to find the centroid of: (a) a semicircular region of radius a; (b) a quarter-circular region of radius a.

31. Suppose that a lamina shaped like the region of Problem 29 has constant density δ. Show that its moments of inertia around the coordinate axes are

$$I_x = -\frac{\delta}{3}\oint_C y^3\,dx, \quad I_y = \frac{\delta}{3}\oint_C x^3\,dy.$$

32. Use the result of Problem 31 to show that the polar moment of inertia $I_0 = I_x + I_y$ of a circular lamina of radius a, centered at the origin and of constant density δ, is $\frac{1}{2}Ma^2$, where M is the mass of the lamina.

33. The loop of the folium of Descartes (with equation $x^3 + y^3 = 3xy$) appears in Fig. 15.4.14. Apply the corollary to Green's theorem to find the area of this loop. (*Suggestion:* Set $y = tx$ to discover a parametrization of the loop. To obtain the area of the loop, use values of t that lie in the interval $[0, 1]$. This gives the half of the loop that lies below the line $y = x$.)

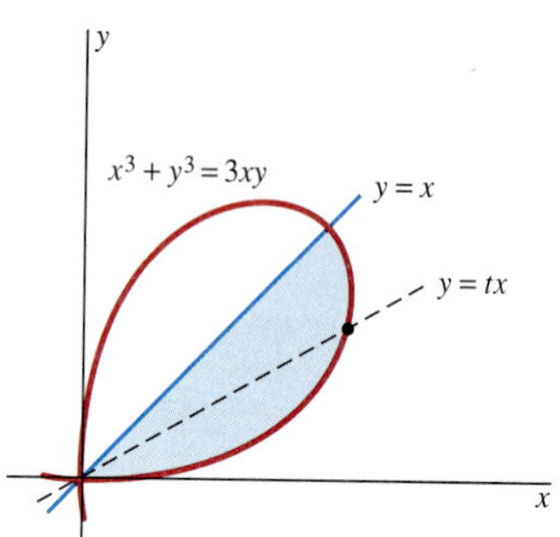

Fig. 15.4.14 The loop of Problem 33

34. Find the area bounded by one loop of the curve $x = \sin 2t$, $y = \sin t$.

35. Let f and g be functions with continuous second-order partial derivatives in the region R bounded by the piecewise smooth simple closed curve C. Apply Green's theorem in vector form to show that

$$\oint_C f\,\nabla g\cdot\mathbf{n}\,ds = \iint_R [f\nabla\cdot\nabla g + \nabla f\cdot\nabla g]\,dA.$$

It was this formula rather than Green's theorem itself that appeared in Green's book of 1828.

36. Complete the proof of the simple case of Green's theorem by showing directly that

$$\oint_C Q\,dy = \iint_R \frac{\partial Q}{\partial x}\,dA$$

if the region R is horizontally simple.

37. Suppose that the bounded plane region R is divided into the nonoverlapping subregions $R_1, R_2, \dots, R_k$. If Green's theorem, Eq. (1), holds for each of these subregions, explain why it follows that Green's theorem holds for R. State carefully any assumptions that you need to make.

38. (a) If C is the line segment from (x_1, y_1) to (x_2, y_2), show by direct evaluation of the line integral that

$$\int_C x\,dy - y\,dx = x_1y_2 - x_2y_1.$$

(b) Let $(0, 0)$, (x_1, y_1), and (x_2, y_2) be the vertices of a triangle taken in counterclockwise order. Deduce from part (a) and Green's theorem that the area of this triangle is $A = \frac{1}{2}(x_1y_2 - x_2y_1)$.

39. Use the result of Problem 38 to find the area of (a) the equilateral triangle with vertices $(1, 0)$, $(\cos\frac{2}{3}\pi, \sin\frac{2}{3}\pi)$, and $(\cos\frac{4}{3}\pi, \sin\frac{4}{3}\pi)$; (b) the regular pentagon with vertices $(1, 0)$, $(\cos\frac{2}{5}\pi, \sin\frac{2}{5}\pi)$, $(\cos\frac{4}{5}\pi, \sin\frac{4}{5}\pi)$, $(\cos\frac{6}{5}\pi, \sin\frac{6}{5}\pi)$, and $(\cos\frac{8}{5}\pi, \sin\frac{8}{5}\pi)$.

40. Let T be a one-to-one transformation from the region S (with boundary curve J) in the uv-plane to the region R (with boundary curve C) in the xy-plane. Then the change-of-variables formula in Section 14.9 implies that the area A of the region R is given by

$$\iint_R dx\,dy = \iint_S \left|\frac{\partial(x, y)}{\partial(u, v)}\right| du\,dv. \qquad \textbf{(11)}$$

Establish this formula by carrying out the following steps. (a) Use Eq. (4) to convert the left-hand side in Eq. (11) to a line integral around C. (b) Use the coordinate functions $x(u, v)$ and $y(u, v)$ of the transformation T to convert the line integral in part (a) to a line integral around J. (c) Apply Green's theorem to the line integral in part (b).

15.4 PROJECT: GREEN'S THEOREM AND LOOP AREAS

Figure 15.4.15 shows the first-quadrant loop of the generalized folium of Descartes that is defined implicitly by the equation

$$x^{2n+1} + y^{2n+1} = (2n + 1)x^n y^n \qquad \textbf{(1)}$$

(where n is a positive integer). Your task here is to calculate the area A_n of the region bounded by this loop. Begin by substituting $y = tx$ to discover the parametrization

$$x = \frac{(2n + 1)t^n}{t^{2n+1} + 1}, \quad y = \frac{(2n + 1)t^{n+1}}{t^{2n+1} + 1} \quad (0 \leqq t < \infty) \qquad \textbf{(2)}$$

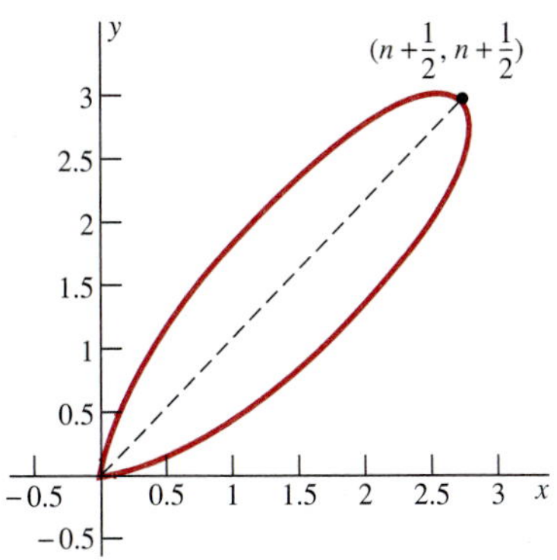

Fig. 15.4.15 The first-quadrant loop of the generalized folium

of the loop. A computer algebra system may be useful in showing that with this parametrization the area formula in Eq. (4) of this section yields

$$A_n = (2n+1)^2 \int_0^{\infty} \left[\frac{(2n+1)t^{2n}}{(t^{2n+1}+1)^3} - \frac{nt^{2n}}{(t^{2n+1}+1)^2} \right] dt. \tag{3}$$

You can now calculate A_n for $n = 1, 2, 3, \ldots$; you should find that $A_n = n + \frac{1}{2}$. (Do you need a computer algebra system for this?) But the improper integral in Eq. (3) should give you pause. Check your result by calculating (and then doubling) the area of the lower half of the loop indicated in Fig. 15.4.15—this involves only the integral from $t = 0$ to $t = 1$ (Why?).

15.5 SURFACE INTEGRALS

A *surface integral* is to surfaces in space what a line (or "curve") integral is to curves in the plane. Consider a curved, thin metal sheet shaped like the surface S. Suppose that this sheet has variable density, given at the point (x, y, z) by the known continuous function $f(x, y, z)$ in units such as grams per square centimeter of surface. We want to define the surface integral

$$\iint_S f(x, y, z)\, dS$$

in such a way that—upon evaluation—it gives the total mass of the thin metal sheet. In case $f(x, y, z) \equiv 1$, the numerical value of the integral should also equal the surface area of S.

As in Section 14.8, we assume that S is a parametric surface described by the function or transformation

$$\mathbf{r}(u, v) = \langle x(u, v), y(u, v), z(u, v) \rangle = x\mathbf{i} + y\mathbf{j} + z\mathbf{k}$$

for (u, v) in a region D in the uv-plane. We suppose throughout that the component functions of $\mathbf{r}$ have continuous partial derivatives and also that the vectors $\mathbf{r}_u = \partial\mathbf{r}/\partial u$ and $\mathbf{r}_v = \partial\mathbf{r}/\partial v$ are nonzero and nonparallel at each interior point of D.

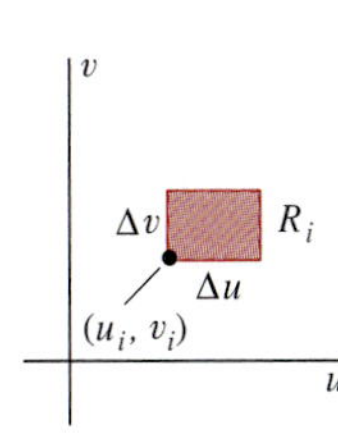

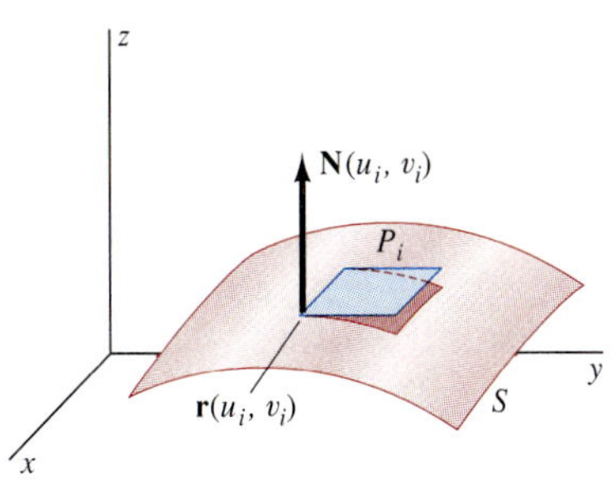

Fig. 15.5.1 Approximating surface area with parallelograms

Recall how we computed the surface area A of S in Section 14.8. We began with an inner partition of D consisting of n rectangles $R_1, R_2, \ldots, R_n$, each Δu by Δv in size. The images under $\mathbf{r}$ of the rectangles are curvilinear figures filling most of the surface S, and these pieces of S are themselves approximated by parallelograms P_i of the sort shown in Fig. 15.5.1. This gave us the approximation

$$A \approx \sum_{i=1}^{n} \Delta S_i = \sum_{i=1}^{n} |\mathbf{N}(u_i, v_i)|\, \Delta u\, \Delta v, \tag{1}$$

where the vector

$$\mathbf{N} = \frac{\partial \mathbf{r}}{\partial u} \times \frac{\partial \mathbf{r}}{\partial v} = \begin{vmatrix} \mathbf{i} & \mathbf{j} & \mathbf{k} \\ \dfrac{\partial x}{\partial u} & \dfrac{\partial y}{\partial u} & \dfrac{\partial z}{\partial u} \\ \dfrac{\partial x}{\partial v} & \dfrac{\partial y}{\partial v} & \dfrac{\partial z}{\partial v} \end{vmatrix} \tag{2}$$

is normal to S at the point $\mathbf{r}(u, v)$ and $\Delta S_i = |\mathbf{N}(u_i, v_i)|\,\Delta u\,\Delta v$ is the area of the parallelogram P_i that is tangent to the surface S at the point $\mathbf{r}(u_i, v_i)$.

If the surface S also has a density function $f(x, y, z)$, then we can approximate the total mass m of the surface by first multiplying each parallelogram area ΔS_i in Eq. (1) by the density $f(\mathbf{r}(u_i, v_i))$ at $\mathbf{r}(u_i, v_i)$ and then summing these estimates over all such parallelograms. Thus we obtain the approximation

$$m \approx \sum_{i=1}^{n} f(\mathbf{r}(u_i, v_i))\,\Delta S_i = \sum_{i=1}^{n} f(\mathbf{r}(u_i, v_i))\,|\mathbf{N}(u_i, v_i)|\,\Delta u\,\Delta v. \tag{3}$$

This approximation is a Riemann sum for the **surface integral of the function** f **over the surface** S, denoted by

$$\iint_S f(x, y, z)\,dS = \iint_D f(\mathbf{r}(u, v))\,|\mathbf{N}(u, v)|\,du\,dv$$
$$= \iint_D f(\mathbf{r}(u, v)) \left| \frac{\partial \mathbf{r}}{\partial u} \times \frac{\partial \mathbf{r}}{\partial v} \right| du\,dv. \tag{4}$$

To evaluate the surface integral $\iint_S f(x, y, z)\,dS$, we simply use the parametrization $\mathbf{r}$ to express the variables x, y, and z in terms of u and v and formally replace the **surface area element** dS with

$$dS = |\mathbf{N}(u, v)|\,du\,dv = \left| \frac{\partial \mathbf{r}}{\partial u} \times \frac{\partial \mathbf{r}}{\partial v} \right| du\,dv. \tag{5}$$

Expansion of the cross product determinant in Eq. (2) gives

$$\mathbf{N} = \frac{\partial \mathbf{r}}{\partial u} \times \frac{\partial \mathbf{r}}{\partial v} = \frac{\partial(y, z)}{\partial(u, v)}\mathbf{i} + \frac{\partial(z, x)}{\partial(u, v)}\mathbf{j} + \frac{\partial(x, y)}{\partial(u, v)}\mathbf{k} \tag{6}$$

in the Jacobian notation of Section 14.9, so the surface integral in Eq. (4) takes the form

$$\iint_S f(x, y, z)\,dS$$
$$= \iint_D f(x(u, v), y(u, v), z(u, v)) \sqrt{\left[\frac{\partial(y, z)}{\partial(u, v)}\right]^2 + \left[\frac{\partial(z, x)}{\partial(u, v)}\right]^2 + \left[\frac{\partial(x, y)}{\partial(u, v)}\right]^2}\,du\,dv. \tag{7}$$

This formula converts the surface integral into an *ordinary double integral* over the region D in the uv-plane and is analogous to the formula [Eq. (4) of Section 15.2]

$$\int_C f(x, y, z)\,ds = \int_a^b f(x(t), y(t), z(t)) \sqrt{\left(\frac{dx}{dt}\right)^2 + \left(\frac{dy}{dt}\right)^2 + \left(\frac{dz}{dt}\right)^2}\,dt$$

that converts a line integral into an ordinary single integral.

In the important special case of a surface S described as a graph $z = h(x, y)$ of a function h defined on a region D in the xy-plane, we may use x and y (rather than u and v) as the parameters. The surface area element then takes the form

$$dS = \sqrt{1 + \left(\frac{\partial h}{\partial x}\right)^2 + \left(\frac{\partial h}{\partial y}\right)^2}\,dx\,dy \tag{8}$$

(as in Eq. (9) of Section 14.8). The surface integral of f over S is then given by

$$\iint_S f(x, y, z)\, dS = \iint_D f(x, y, h(x, y))\sqrt{1 + \left(\frac{\partial h}{\partial x}\right)^2 + \left(\frac{\partial h}{\partial y}\right)^2}\, dx\, dy. \tag{9}$$

Centroids and moments of inertia for surfaces are computed in much the same way as for curves (see Section 15.2, using surface integrals in place of line integrals). For example, suppose that the surface S has density $\delta(x, y, z)$ at the point (x, y, z) and total mass m. Then the z-component $\overline{z}$ of its centroid and its moment of inertia I_z around the z-axis are given by

$$\overline{z} = \frac{1}{m}\iint_S z\delta(x, y, z)\, dS \quad \text{and} \quad I_z = \iint_S (x^2 + y^2)\delta(x, y, z)\, dS.$$

EXAMPLE 1 Find the centroid of the unit-density hemispherical surface

$$z = \sqrt{a^2 - x^2 - y^2}, \qquad x^2 + y^2 \leqq a^2.$$

Solution By symmetry, $\overline{x} = 0 = \overline{y}$. A simple computation gives $\partial z/\partial x = -x/z$ and $\partial z/\partial y = -y/z$, so Eq. (8) takes the form

$$dS = \sqrt{1 + \left(\frac{\partial z}{\partial x}\right)^2 + \left(\frac{\partial z}{\partial y}\right)^2}\, dx\, dy = \sqrt{1 + \left(\frac{x}{z}\right)^2 + \left(\frac{y}{z}\right)^2}\, dx\, dy$$

$$= \frac{1}{z}\sqrt{x^2 + y^2 + z^2}\, dx\, dy = \frac{a}{z}\, dx\, dy.$$

Hence

$$\overline{z} = \frac{1}{2\pi a^2}\iint_D z \cdot \frac{a}{z}\, dx\, dy = \frac{1}{2\pi a}\iint_D 1\, dx\, dy = \frac{a}{2}.$$

Note in the final step that D is a circular disk of radius a in the xy-plane. This simplifies the computation of the last integral. ■

EXAMPLE 2 Find the moment of inertia around the z-axis of the spherical surface $x^2 + y^2 + z^2 = a^2$, assuming that it has constant density $\delta = k$.

Solution The spherical surface of radius a is most easily parametrized in spherical coordinates:

$$x = a \sin\phi \cos\theta, \qquad y = a \sin\phi \sin\theta, \qquad z = a \cos\phi$$

for $0 \leqq \phi \leqq \pi$ and $0 \leqq \theta \leqq 2\pi$. Hence the sphere S is defined parametrically by

$$\mathbf{r}(\phi, \theta) = \mathbf{i}a \sin\phi \cos\theta + \mathbf{j}a \sin\phi \sin\theta + \mathbf{k}a \cos\phi.$$

As in Problem 18 of Section 14.8, the surface area element is then

$$dS = \left|\frac{\partial \mathbf{r}}{\partial \phi} \times \frac{\partial \mathbf{r}}{\partial \theta}\right| = a^2 \sin\phi\, d\phi\, d\theta.$$

Because

$$x^2 + y^2 = a^2 \sin^2\phi \cos^2\theta + a^2 \sin^2\phi \sin^2\theta = a^2 \sin^2\phi,$$

it follows that

$$I_z = \iint_S (x^2 + y^2)\delta \, dS = \int_0^{2\pi}\int_0^{\pi} k(a^2 \sin^2 \phi)a^2 \sin\phi \, d\phi \, d\theta$$

$$= 2\pi \cdot ka^4 \cdot 2\int_0^{\pi/2} \sin^3\phi \, d\phi = 4\pi ka^4 \cdot \frac{2}{3} \quad \text{(by integral formula 113)}$$

$$= \tfrac{2}{3} \cdot 4\pi ka^2 \cdot a^2 = \tfrac{2}{3} ma^2,$$

using in the final step the fact that the mass of the spherical surface with density k is $m = 4\pi ka^2$. Is this result both plausible and dimensionally correct? ■

Surface Integrals with Respect to Coordinate Elements

The surface integral $\iint_S f(x, y, z)\, dS$ is an integral **with respect to surface area** and thus is analogous to the line integral $\int_C f(x, y)ds$ with respect to arc length. A second type of surface integral of the form

$$\iint_S P\, dy\, dz + Q\, dz\, dx + R\, dx\, dy$$

is analogous to the line integral $\int_C P\, dx + Q\, dy$ with respect to coordinate variables.

The definition of the integral $\iint_S R\, dx\, dy$, for instance—with $R(x, y, z)$ a scalar function (instead of f) and $dx\, dy$ an area element in the xy-plane (instead of the area element dS on the surface S)—is motivated by replacing the area element $\Delta S_i = |\mathbf{N}(u_i, v_i)|\, \Delta u\, \Delta v$ in the Riemann sum in Eq. (3) with the area $\Delta S_i \cos\gamma$ of its projection into the xy-plane (Fig. 15.5.2). The result is the Riemann sum

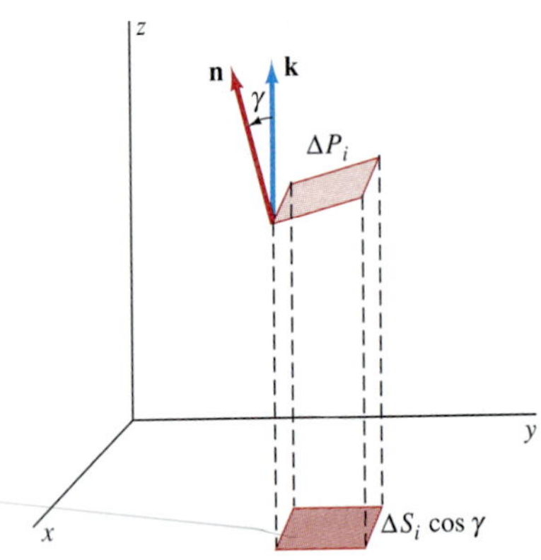

Fig. 15.5.2 Finding the area of the projected parallelogram

$$\sum_{i=1}^{n} R(\mathbf{r}(u_i, v_i)) \cos\gamma |\mathbf{N}(u_i, v_i)|\, \Delta u\, \Delta v \approx \iint_D R(\mathbf{r}(u, v)) \cos\gamma |\mathbf{N}(u, v)|\, du\, dv. \quad \textbf{(10)}$$

To calculate the factor $\cos\gamma$ in Eq. (10), we consider the unit normal vector

$$\mathbf{n} = \frac{\mathbf{N}}{|\mathbf{N}|} = \mathbf{i}\cos\alpha + \mathbf{j}\cos\beta + \mathbf{k}\cos\gamma \quad \textbf{(11)}$$

with "direction cosines" $\cos\alpha$, $\cos\beta$, and $\cos\gamma$. Using Eq. (6) we find that

$$\cos\alpha = \mathbf{n} \cdot \mathbf{i} = \frac{\mathbf{N} \cdot \mathbf{i}}{|\mathbf{N}|} = \frac{1}{|\mathbf{N}|}\frac{\partial(y, z)}{\partial(u, v)} \quad \text{and, similarly,}$$

$$\cos\beta = \frac{1}{|\mathbf{N}|}\frac{\partial(z, x)}{\partial(u, v)}, \qquad \cos\gamma = \frac{1}{|\mathbf{N}|}\frac{\partial(x, y)}{\partial(u, v)}. \quad \textbf{(12)}$$

Substitution for $\cos\gamma$ in (10) now yields the *definition*

$$\iint_S R(x, y, z)\, dx\, dy = \iint_S R(x, y, z) \cos\gamma\, dS$$

$$= \iint_D R(\mathbf{r}(u, v)) \frac{\partial(x, y)}{\partial(u, v)}\, du\, dv. \quad \textbf{(13)}$$

Similarly, we *define*

$$\iint_S P(x, y, z)\, dy\, dz = \iint_S P(x, y, z) \cos\alpha \, dS$$

$$= \iint_D P(\mathbf{r}(u, v)) \frac{\partial(y, z)}{\partial(u, v)}\, du\, dv \qquad \textbf{(14)}$$

and

$$\iint_S Q(x, y, z)\, dz\, dx = \iint_S Q(x, y, z) \cos\beta \, dS$$

$$= \iint_D Q(\mathbf{r}(u, v)) \frac{\partial(z, x)}{\partial(u, v)}\, du\, dv. \qquad \textbf{(15)}$$

Note The symbols z and x appear in reverse of alphabetical order in Eq. (15). It is important to write them in the correct order because

$$\frac{\partial(x, z)}{\partial(u, v)} = \begin{vmatrix} x_u & x_v \\ z_u & z_v \end{vmatrix} = -\begin{vmatrix} z_u & z_v \\ x_u & x_v \end{vmatrix} = -\frac{\partial(z, x)}{\partial(u, v)}.$$

This implies that

$$\iint_S f(x, y, z)\, dx\, dz = -\iint_S f(x, y, z)\, dz\, dx.$$

In an ordinary *double integral,* the order in which the differentials are written simply indicates the order of integration. But in a *surface integral,* it instead indicates the order of appearance of the corresponding variables in the Jacobians in Eqs. (13) through (15).

The three integrals in Eqs. (13) through (15) typically occur together, and the general **surface integral with respect to coordinate area elements** is the sum

$$\iint_S P\, dy\, dz + Q\, dz\, dx + R\, dx\, dy$$

$$= \iint_S (P\cos\alpha + Q\cos\beta + R\cos\gamma)\, dS \qquad \textbf{(16)}$$

$$= \iint_D \left(P\frac{\partial(y, z)}{\partial(u, v)} + Q\frac{\partial(z, x)}{\partial(u, v)} + R\frac{\partial(x, y)}{\partial(u, v)} \right) du\, dv. \qquad \textbf{(17)}$$

Equation (17) gives the evaluation procedure for the surface integral in Eq. (16): Substitute for x, y, z, and their derivatives in terms of u and v, then integrate over the appropriate region D in the uv-plane.

The relation between surface integrals with respect to surface area and with respect to coordinate areas is somewhat analogous to the formula

$$\int_C \mathbf{F}\cdot\mathbf{T}\, ds = \int_C P\, dx + Q\, dy + R\, dz$$

relating line integrals with respect to arc length and with respect to coordinates. Given the vector field $\mathbf{F} = P\mathbf{i} + Q\mathbf{j} + R\mathbf{k}$, Eq. (11) implies that

$$\mathbf{F}\cdot\mathbf{n} = P\cos\alpha + Q\cos\beta + R\cos\gamma, \qquad \textbf{(18)}$$

so the equations in (12) yield

$$\iint_S \mathbf{F}\cdot\mathbf{n}\,dS = \iint_S P\,dy\,dz + Q\,dz\,dx + R\,dx\,dy. \tag{19}$$

Only the sign of the right-hand surface integral in Eq. (19) depends on the parametrization of S. The unit normal vector on the left-hand side is the vector provided by the parametrization of S via the equations in (12). In the case of a surface given by $z = h(x, y)$, with x and y used as the parameters u and v, this will be the *upper* normal, as you will see in Example 3.

EXAMPLE 3 Suppose that S is the surface $z = h(x, y)$, (x, y) in D. Then show that

$$\iint_S P\,dy\,dz + Q\,dz\,dx + R\,dx\,dy = \iint_D \left(-P\frac{\partial z}{\partial x} - Q\frac{\partial z}{\partial y} + R\right) dx\,dy, \tag{20}$$

where P, Q, and R in the second integral are evaluated at $(x, y, h(x, y))$.

Solution This is simply a matter of computing the three Jacobians in Eq. (17) with the parameters x and y. We note first that $\partial x/\partial x = 1 = \partial y/\partial y$ and that $\partial x/\partial y = 0 = \partial y/\partial x$. Hence

$$\frac{\partial(y, z)}{\partial(x, y)} = \begin{vmatrix} y_x & y_y \\ z_x & z_y \end{vmatrix} = -\frac{\partial z}{\partial x}, \quad \frac{\partial(z, x)}{\partial(x, y)} = \begin{vmatrix} z_x & z_y \\ x_x & x_y \end{vmatrix} = -\frac{\partial z}{\partial y},$$

and

$$\frac{\partial(x, y)}{\partial(x, y)} = \begin{vmatrix} x_x & x_y \\ y_x & y_y \end{vmatrix} = 1.$$

Equation (20) is an immediate consequence. ■

The Flux of a Vector Field

One of the most important applications of surface integrals involves the computation of the flux of a vector field. To define the flux of the vector field $\mathbf{F}$ across the surface S, we assume that S has a unit normal vector field $\mathbf{n}$ that varies *continuously* from point to point of S. This condition excludes from our consideration one-sided (*nonorientable*) surfaces, such as the Möbius strip of Fig. 15.5.3. If S is a two-sided (*orientable*) surface, then there are two possible choices for $\mathbf{n}$. For example, if S is a closed surface (such as a torus or sphere) that separates space, then we may choose for $\mathbf{n}$ either the outer normal vector (at each point of S) or the inner normal vector (Fig. 15.5.4). The unit normal vector defined in Eq. (11) may be either the outer normal or the inner normal; which of the two it is depends on how S has been parametrized.

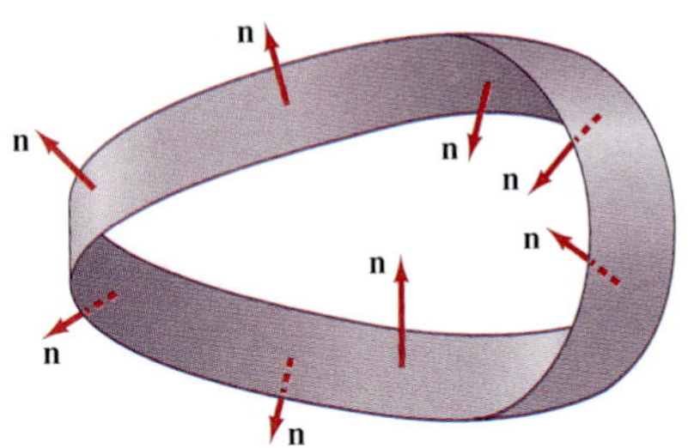

Fig. 15.5.3 The Möbius strip is an example of a one-sided surface.

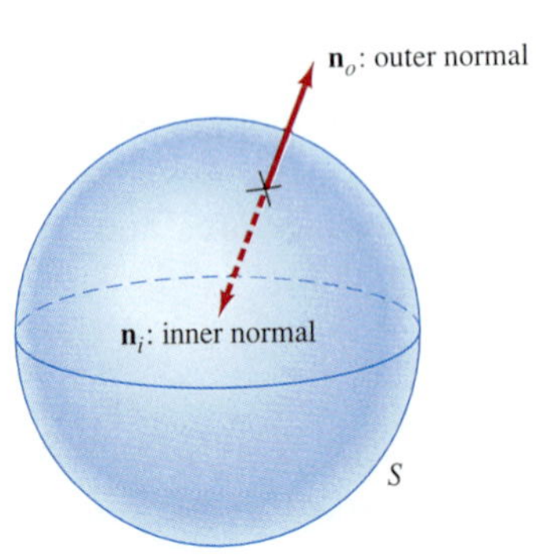

Fig. 15.5.4 Inner and outer normal vectors to a two-sided closed surface

To define the concept of flux, suppose that we are given the vector field $\mathbf{F}$, the orientable surface S, and a continuous unit normal vector field $\mathbf{n}$ on S. Then, in analogy with Eq. (5) in Section 15.4, we define the **flux ϕ across S in the direction of $\mathbf{n}$** by

$$\phi = \iint_S \mathbf{F}\cdot\mathbf{n}\,dS. \tag{21}$$

For example, if $\mathbf{F} = \delta\mathbf{v}$, where $\mathbf{v}$ is the velocity vector field corresponding to the steady flow in space of a fluid of density δ and $\mathbf{n}$ is the *outer* unit normal vector for a closed surface S that bounds the space region T, then the flux determined by Eq. (21) is the net rate of flow of the fluid *out* of T across its boundary surface S in units such as grams per second.

EXAMPLE 4 Calculate the flux $\iint_S \mathbf{F}\cdot\mathbf{n}\,dS$, where $\mathbf{F} = v_0\mathbf{k}$ and S is the hemispherical surface of radius a with equation $z = \sqrt{a^2 - x^2 - y^2}$ and with outer unit normal vector $\mathbf{n}$ (see Fig. 15.5.5).

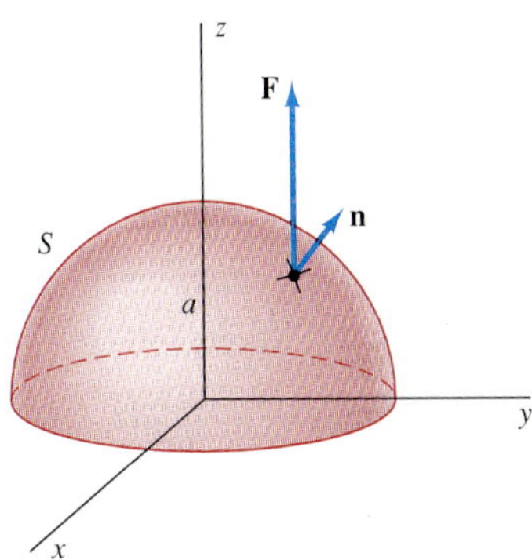

Fig. 15.5.5 The hemisphere S of Example 4

Solution If we think of $\mathbf{F} = v_0\mathbf{k}$ as the velocity vector field of a fluid that is flowing upward with constant speed v_0, then we can interpret the flux in question as the rate of flow (in cubic centimeters per second, for example) of the fluid across S. To calculate this flux, we note that

$$\mathbf{n} = \frac{x\mathbf{i} + y\mathbf{j} + z\mathbf{k}}{\sqrt{x^2 + y^2 + z^2}} = \frac{1}{a}(x\mathbf{i} + y\mathbf{j} + z\mathbf{k}).$$

Hence

$$\mathbf{F}\cdot\mathbf{n} = v_0\mathbf{k}\cdot\frac{1}{a}(x\mathbf{i} + y\mathbf{j} + z\mathbf{k}) = \frac{v_0}{a}z,$$

so

$$\iint_S \mathbf{F}\cdot\mathbf{n}\,dS = \iint_S \frac{v_0}{a}z\,dS.$$

If we introduce spherical coordinates $z = a\cos\phi$, $dS = a^2\sin\phi\,d\phi\,d\theta$ on the hemispherical surface, we get

$$\begin{aligned}\iint_S \mathbf{F}\cdot\mathbf{n}\,dS &= \frac{v_0}{a}\int_0^{2\pi}\int_0^{\pi/2}(a\cos\phi)(a^2\sin\phi)\,d\phi\,d\theta\\ &= 2\pi a^2 v_0\int_0^{\pi/2}\cos\phi\sin\phi\,d\phi = 2\pi a^2 v_0\left[\frac{1}{2}\sin^2\phi\right]_0^{\pi/2};\end{aligned}$$

thus

$$\iint_S \mathbf{F}\cdot\mathbf{n}\,dS = \pi a^2 v_0.$$

This last quantity is equal to the flux of $\mathbf{F} = v_0\mathbf{k}$ across the disk $x^2 + y^2 \leqq a^2$ of area πa^2. If we think of the hemispherical region T bounded by the hemisphere S and the circular disk D that forms its base, it should be no surprise that the rate of inflow of an incompressible fluid across the disk D is equal to its rate of outflow across the hemisphere S. ■

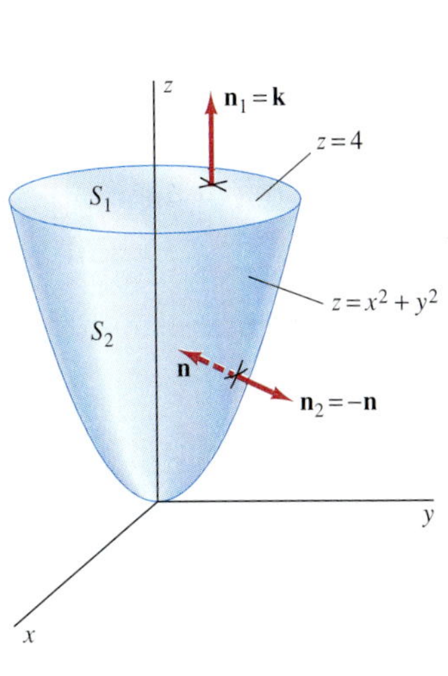

Fig. 15.5.6 The surface of Example 5

EXAMPLE 5 Find the flux of the vector field $\mathbf{F} = x\mathbf{i} + y\mathbf{j} + 3\mathbf{k}$ out of the region T bounded by the paraboloid $z = x^2 + y^2$ and the plane $z = 4$ (Fig. 15.5.6).

Solution Let S_1 denote the circular top, which has outer unit normal vector $\mathbf{n}_1 = \mathbf{k}$. Let S_2 be the parabolic part of this surface, with outer unit normal vector $\mathbf{n}_2$. The flux across S_1 is

$$\iint_{S_1} \mathbf{F}\cdot\mathbf{n}_1\,dS = \iint_{S_1} 3\,dS = 12\pi$$

because S_1 is a circular disk of radius 2.

Next, the computation in Example 3 gives

$$\mathbf{N} = \left\langle -\frac{\partial z}{\partial x}, -\frac{\partial z}{\partial y}, 1 \right\rangle = \langle -2x, -2y, 1\rangle$$

for a vector normal to the paraboloid $z = x^2 + y^2$. Then $\mathbf{n} = \mathbf{N}/|\mathbf{N}|$ is an upper—and thus an *inner*—unit normal vector to the surface S_2. The unit *outer* normal vector is therefore $\mathbf{n}_2 = -\mathbf{n}$, opposite to the direction of $\mathbf{N} = \langle -2x, -2y, 1\rangle$. With parameters (x, y) in the circular disk $x^2 + y^2 \leqq 4$ in the xy-plane, the surface-area element is $dS = |\mathbf{N}|\,dx\,dy$. Therefore the outward flux across S_2 is

$$\iint_{S_2} \mathbf{F}\cdot\mathbf{n}_2\,dS = -\iint_{S_2} \mathbf{F}\cdot\mathbf{n}\,dS = -\iint_D \mathbf{F}\cdot\frac{\mathbf{N}}{|\mathbf{N}|}\,|\mathbf{N}|\,dx\,dy$$

$$= -\iint_D [(x)(-2x) + (y)(-2y) + (3)(1)]\,dx\,dy.$$

We change to polar coordinates in the disk D of radius 2—so that $3 - 2x^2 - 2y^2 = 3 - 2r^2$ and $dx\,dy = r\,dr\,d\theta$—and find that

$$\iint_{S_2} \mathbf{F}\cdot\mathbf{n}_2\,dS = \int_0^{2\pi}\int_0^2 (2r^2 - 3)r\,dr\,d\theta = 2\pi\left[\frac{1}{2}r^4 - \frac{3}{2}r^2\right]_0^2 = 4\pi.$$

Hence the total flux of $\mathbf{F}$ out of T is $12\pi + 4\pi = 16\pi \approx 50.27$. ■

Another physical application of flux is to the flow of heat, which is mathematically quite similar to the flow of a fluid. Suppose that a body has temperature $u = u(x, y, z)$ at the point (x, y, z). Experiments indicate that the flow of heat in the body is described by the heat-flow vector

$$\mathbf{q} = -K\,\nabla u. \tag{22}$$

The number K—normally, but not always, a constant—is the *heat conductivity* of the body. The vector $\mathbf{q}$ points in the direction of heat flow, and its length is the rate of flow of heat across a unit area normal to $\mathbf{q}$. This flow rate is measured in units such as calories per second per square centimeter. If S is a closed surface within the body bounding the solid region T and $\mathbf{n}$ denotes the outer unit normal vector for S, then

$$\iint_S \mathbf{q}\cdot\mathbf{n}\,dS = -\iint_S K\,\nabla u\cdot\mathbf{n}\,dS \tag{23}$$

is the net rate of heat flow (in calories per second, for example) out of the region T across its boundary surface S.

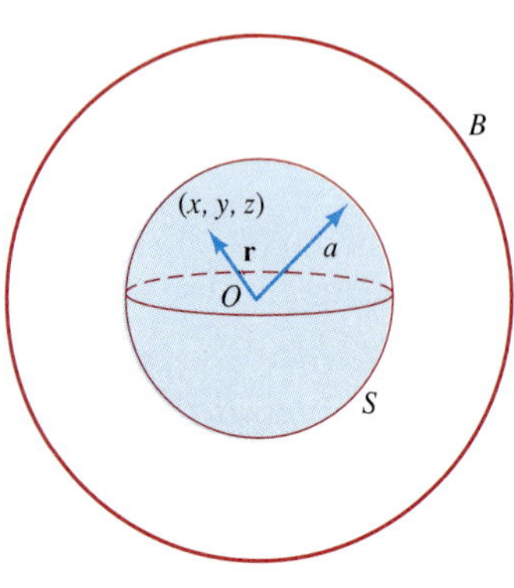

Fig. 15.5.7 The solid ball B of Example 6

EXAMPLE 6 Suppose that a uniform solid ball B of radius R is centered at the origin (Fig. 15.5.7) and that the temperature u within it is given by

$$u(x, y, z) = c(R^2 - x^2 - y^2 - z^2).$$

Thus the temperature of B is maximal at its center and is zero on its boundary. Find the rate of flow of heat across a sphere S of radius $a < R$ centered at the origin.

Solution Writing $\mathbf{r} = x\mathbf{i} + y\mathbf{j} + z\mathbf{k}$ for the position vector of a point (x, y, z) of B, we find that the heat flow vector $\mathbf{q}$ in Eq. (22) is

$$\mathbf{q} = -K\nabla u = -K\cdot c(-2x\mathbf{i} - 2y\mathbf{j} - 2z\mathbf{k}) = 2Kc\mathbf{r}.$$

Obviously the outer unit normal vector $\mathbf{n}$ at a point (x, y, z) of the sphere S is $\mathbf{n} = \mathbf{r}/a$, so

$$\mathbf{q} \cdot \mathbf{n} = 2Kc\mathbf{r} \cdot \frac{\mathbf{r}}{a} = 2Kca$$

because $\mathbf{r} \cdot \mathbf{r} = a^2$ at points of S. Therefore the heat flow across the sphere S (with area $A(S) = 4\pi a^2$) is

$$\iint_S \mathbf{q} \cdot \mathbf{n}\, dS = \iint_S 2Kca\, dS = 2Kca \cdot 4\pi a^2 = 8Kc\pi a^3.$$

Still other applications of flux involve force fields rather than flow fields. For example, suppose that $\mathbf{F}$ is the gravitational field of a collection of fixed masses in space, so $\mathbf{F}(\mathbf{r})$ is the net force exerted on a unit mass located at $\mathbf{r}$. Then **Gauss's law** for inverse-square gravitational fields says that the (outward) flux of $\mathbf{F}$ across the closed surface S is

$$\phi = \iint_S \mathbf{F} \cdot \mathbf{n}\, dS = -4\pi GM \tag{24}$$

where M is the total mass enclosed by S and G is the universal gravitational constant.

Gauss's law also applies to inverse-square electric fields. The electric field at $\mathbf{r}$ of a charge q located at the origin is $\mathbf{E} = q\mathbf{r}/(4\pi\epsilon_0|\mathbf{r}|^3)$, where $\epsilon_0 \approx 8.85 \times 10^{-12}$ in mks units (charge in coulombs). Then **Gauss's law** for electric fields says that the (outward) flux of $\mathbf{E}$ across the closed surface S is

$$\phi = \iint_S \mathbf{E} \cdot \mathbf{n}\, dS = \frac{Q}{\epsilon_0} \tag{25}$$

where Q is the net (positive minus negative) charge enclosed by S.

15.5 PROBLEMS

In Problems 1 through 6, evaluate the surface integral $\iint_S f(x, y, z)\, dS$.

1. $f(x, y, z) = x + y$; S is the first-octant part of the plane $x + y + z = 1$.

2. $f(x, y, z) = xyz$; S is the triangle with vertices $(3, 0, 0)$, $(0, 2, 0)$, and $(0, 0, 6)$.

3. $f(x, y, z) = y + z + 3$; S is the part of the plane $z = 2x + 3y$ that lies inside the cylinder $x^2 + y^2 = 9$.

4. $f(x, y, z) = z^2$; S is the part of the cone $z = \sqrt{x^2 + y^2}$ that lies inside the cylinder $x^2 + y^2 = 4$.

5. $f(x, y, z) = xy + 1$; S is the part of the paraboloid $z = x^2 + y^2$ that lies inside the cylinder $x^2 + y^2 = 4$.

6. $f(x, y, z) = (x^2 + y^2)z$; S is the hemisphere $z = \sqrt{1 - x^2 - y^2}$.

In Problems 7 through 12, find the moment of inertia $\iint_S (x^2 + y^2)\, dS$ of the given surface S. Assume that S has constant density $\delta \equiv 1$.

7. S is the part of the plane $z = x + y$ that lies inside the cylinder $x^2 + y^2 = 9$.

8. S is the part of the surface $z = xy$ that lies inside the cylinder $x^2 + y^2 = 25$.

9. S is the part of the cylinder $x^2 + z^2 = 1$ that lies between the planes $y = -1$ and $y = 1$. As parameters on the cylinder use y and the polar angular coordinate in the xz-plane.

10. S is the part of the cone $z = \sqrt{x^2 + y^2}$ that lies between the planes $z = 2$ and $z = 5$.

11. S is the part of the sphere $x^2 + y^2 + z^2 = 25$ that lies above the plane $z = 3$.

12. S is the part of the sphere $x^2 + y^2 + z^2 = 25$ that lies outside the cylinder $x^2 + y^2 = 9$.

In Problems 13 through 18, evaluate the surface integral $\iint_S \mathbf{F} \cdot \mathbf{n}\, dS$, where $\mathbf{n}$ is the upward-pointing unit normal vector to the given surface S.

13. $\mathbf{F} = x\mathbf{i} + y\mathbf{j}$; S is the hemisphere $z = \sqrt{9 - x^2 - y^2}$.

14. $\mathbf{F} = x\mathbf{i} + y\mathbf{j} + z\mathbf{k}$; S is the first-octant part of the plane $2x + 2y + z = 3$.

15. $\mathbf{F} = 2y\mathbf{i} + 3z\mathbf{k}$; S is the part of the plane $z = 3x + 2$ that lies within the cylinder $x^2 + y^2 = 4$.

16. $\mathbf{F} = z\mathbf{k}$; S is the upper half of the spherical surface $\rho = 2$. (*Suggestion:* Use spherical coordinates.)
17. $\mathbf{F} = y\mathbf{i} - x\mathbf{j}$; S is the part of the cone $z = r$ that lies within the cylinder $r = 3$.
18. $\mathbf{F} = 2x\mathbf{i} + 2y\mathbf{j} + 3\mathbf{k}$; S is the part of the paraboloid $z = 4 - x^2 - y^2$ that lies above the xy-plane.

In Problems 19 through 24, calculate the outward flux of the vector field **F** across the given closed surface S.

19. $\mathbf{F} = x\mathbf{i} + 2y\mathbf{j} + 3z\mathbf{k}$; S is the boundary of the first-octant unit cube with opposite vertices $(0, 0, 0)$ and $(1, 1, 1)$.
20. $\mathbf{F} = 2x\mathbf{i} - 3y\mathbf{j} + z\mathbf{k}$; S is the boundary of the solid hemisphere $0 \leqq z \leqq \sqrt{4 - x^2 - y^2}$.
21. $\mathbf{F} = x\mathbf{i} - y\mathbf{j}$; S is the boundary of the solid first-octant pyramid bounded by the coordinate planes and the plane $3x + 4y + z = 12$.
22. $\mathbf{F} = 2x\mathbf{i} + 2y\mathbf{j} + 3\mathbf{k}$; S is the boundary of the solid paraboloid bounded by the xy-plane and $z = 4 - x^2 - y^2$.
23. $\mathbf{F} = z^2\mathbf{k}$; S is the boundary of the solid bounded by the paraboloids $z = x^2 + y^2$ and $z = 18 - x^2 - y^2$.
24. $\mathbf{F} = x^2\mathbf{i} + 2y^2\mathbf{j} + 3z^2\mathbf{k}$; S is the boundary of the solid bounded by the cone $z = \sqrt{x^2 + y^2}$ and the plane $z = 3$.
25. The first-octant part of the spherical surface $\rho = a$ has unit density. Find its centroid.
26. The conical surface $z = r$, $r \leqq a$, has constant density $\delta = k$. Find its centroid and its moment of inertia around the z-axis.
27. The paraboloid $z = r^2$, $r \leqq a$, has constant density δ. Find its centroid and moment of inertia around the z-axis.
28. Find the centroid of the part of the spherical surface $\rho = a$ that lies within the cone $r = z$.
29. Find the centroid of the part of the spherical surface $x^2 + y^2 + z^2 = 4$ that lies both inside the cylinder $x^2 + y^2 = 2x$ and above the xy-plane.
30. Suppose that the toroidal surface of Example 5 of Section 14.8 has uniform density and total mass M. Show that its moment of inertia around the z-axis is $\frac{1}{2}M(3a^2 + 2b^2)$.

In Problems 31 and 32, use a table of integrals or a computer algebra system (if necessary) to find the moment of inertia around the z-axis of the given surface S. Assume that S has constant density $\delta \equiv 1$.

31. S is the part of the parabolic cylinder $z = 4 - y^2$ that lies inside the rectangular cylinder $-1 \leqq x \leqq 1$, $-2 \leqq y \leqq 2$.
32. S is the part of the paraboloid $z = 4 - x^2 - y^2$ that lies inside the square cylinder $-1 \leqq x \leqq 1$, $-1 \leqq y \leqq 1$.
33. Let S denote the surface $z = h(x, y)$ for (x, y) in the region D in the xy-plane, and let γ be the angle between **k** and the upper normal vector **N** to S. Prove that

$$\iint_S f(x, y, z)\, dS = \iint_S f(x, y, h(x, y)) \sec\gamma\, dx\, dy.$$

34. Find a formula for

$$\iint_S P\, dy\, dz + Q\, dz\, dx + R\, dx\, dy$$

analogous to Eq. (20), but for the case of a surface S described explicitly by $x = h(y, z)$.

35. A uniform solid ball has radius 5, and its temperature u is proportional to the square of the distance from its center, with $u = 100$ at the boundary of the ball. If the heat conductivity of the ball is $K = 2$, find the rate of flow of heat across a concentric sphere of radius 3.
36. A uniform solid cylinder has radius 5 and height 10, and its temperature u is proportional to the square of the distance from its vertical axis, with $u = 100$ at the outer curved boundary of the cylinder. If the heat conductivity of the cylinder is $K = 2$, find the rate of flow of heat across a concentric cylinder of radius 3 and height 10.

In Problems 37 through 39, set up integrals giving the area and moment of inertia around the z-axis of the given surface S (assuming that S has constant density $\delta \equiv 1$). Use a computer algebra system to evaluate these integrals, symbolically if possible, numerically if necessary (with the numerical values $a = 4$, $b = 3$, and $c = 2$ of the given parameters).

37. S is the elliptic paraboloid $z = (x/a)^2 + (y/b)^2$ with parametrization $x = au\cos v$, $y = bu\sin v$, $z = u^2$, $0 \leqq u \leqq c$, $0 \leqq v \leqq 2\pi$.
38. S is the ellipsoid $(x/a)^2 + (y/b)^2 + (z/c)^2 = 1$ with parametrization $x = a\sin u\cos v$, $y = b\sin u\sin v$, $z = c\cos u$, $0 \leqq u \leqq \pi$, $0 \leqq v \leqq 2\pi$.
39. S is the hyperboloid $(x/a)^2 + (y/b)^2 - z^2 = 1$ with parametrization $x = a\cosh u\cos v$, $y = b\cosh u\sin v$, $z = \sinh u$, $-c \leqq u \leqq c$, $0 \leqq v \leqq 2\pi$. See Fig. 15.5.8, where the u-curves are hyperbolas and the v-curves are ellipses.
40. The Möbius strip in Fig. 15.5.9 was generated by plotting the points

$$x = (4 + t\cos\tfrac{1}{2}\theta)\cos\theta, \quad y = (4 + t\cos\tfrac{1}{2}\theta)\sin\theta,$$

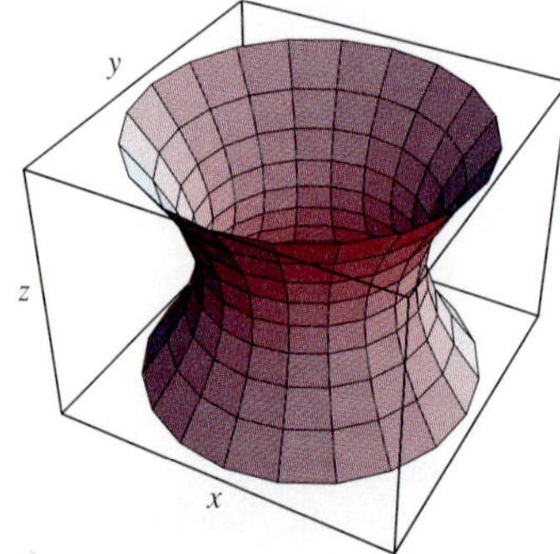

Fig. 15.5.8 The hyperboloid of Problem 39

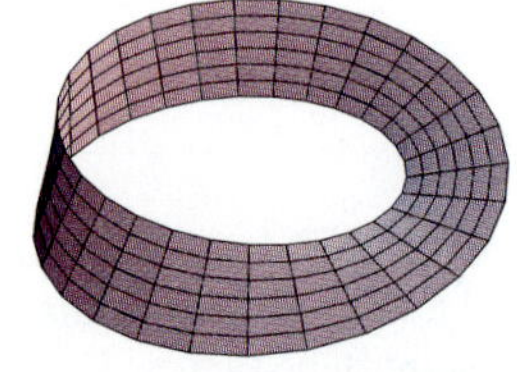

Fig. 15.5.9 The Möbius strip of Problem 40

$$z = t \sin \tfrac{1}{2}\theta$$

for $-1 \leqq t \leqq 1, 0 \leqq \theta \leqq 2\pi$. This Möbius strip has width 2 and a circular centerline of radius 4. Set up integrals giving its area and moment of inertia (assume constant density $\delta \equiv 1$) around the z-axis, and use a computer algebra system to evaluate them numerically.

41. Consider a homogeneous thin spherical shell S of radius a centered at the origin, with density δ and total mass $M = 4\pi a^2 \delta$. A particle of mass m is located at the point $(0, 0, c)$ with $c > a$. Use the method and notation of Problem 41 of Section 14.7 to show that the gravitational force of attraction between the particle and the spherical shell is

$$F = \iint_S \frac{Gm\delta \cos\alpha}{w^2}\, dS = \frac{GMm}{c^2}.$$

15.5 PROJECT: SURFACE INTEGRALS AND ROCKET NOSE CONES

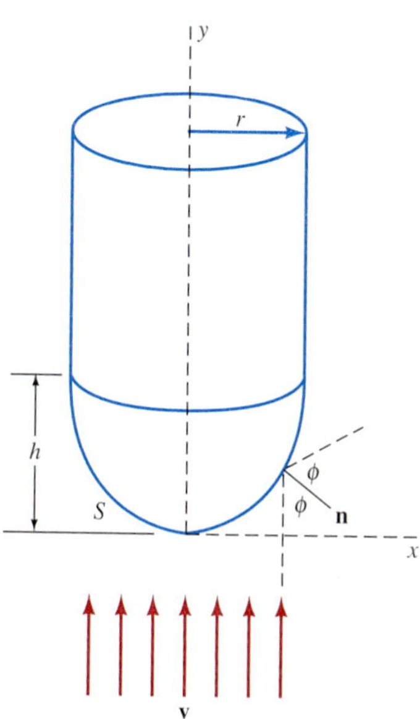

Fig. 15.5.10 The nose-cone S of height h and radius r

Figure 15.5.10 shows a (curved) nose cone S of height $h = 1$ attached to a cylindrical rocket of radius $r = 1$ that is moving downward with velocity v through air of density δ (or, equivalently, the rocket is stationary and the air is streaming upward). In the *Principia Mathematica,* Newton showed that (under plausible assumptions) the force of air resistance the rocket experiences is given by $F = 2\pi R \delta v^2$ and thus is proportional both to the density of the air and to the square of its velocity. The *drag coefficient* R is given by the surface integral

$$R = \frac{1}{\pi} \iint_S \cos^3 \phi \, dS,$$

where ϕ is the angle between the unit normal **n** and the direction of motion.

1. If the surface S of the nose cone is obtained by revolving the curve $y = y(x)$, with $y(0) = 0$ and $y(1) = 1$, around the y-axis, use the fact that $\cos\phi = dx/ds$ to show that

$$R = \int_0^1 \frac{2x}{1 + [y'(x)]^2}\, dx.$$

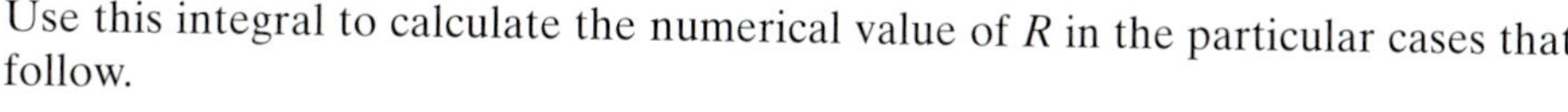

Use this integral to calculate the numerical value of R in the particular cases that follow.

2. $y = x$, so that S is an actual cone with 90° vertex angle

3. $y = 1 - \sqrt{1 - x^2}$, so that S is a hemisphere

4. $y = x^2$, so that S is a paraboloid

5. For the flat-tipped conical frustum nose cone illustrated in Fig. 15.5.11 (still with $r = h = 1$), show that

$$R = \cos^2\alpha - 2\cos\alpha \sin\alpha + 2\sin^2\alpha$$

where α is the indicated angle. Then show that this drag coefficient is minimal when $\tan 2\alpha = 2$.

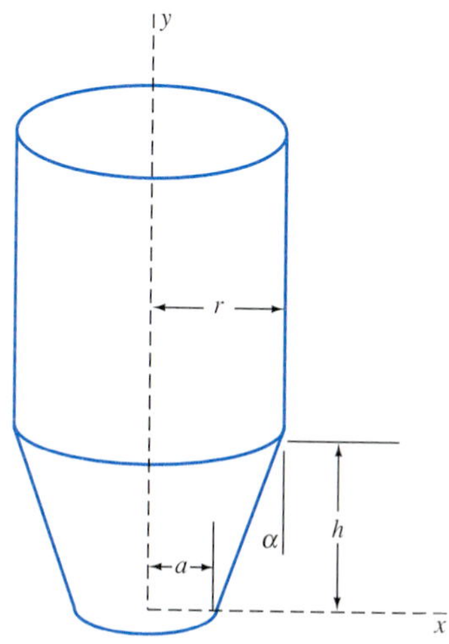

Fig. 15.5.11 The flat-tipped nose cone

If you compare your numerical results, you should find that

- the cone and hemisphere offer the same resistance;
- the paraboloid offers less resistance than either, and
- the optimal flat-tipped conical frustum offers still less!

In an extraordinary tour de force, Newton determined the nose cone with minimum possible air resistance, allowing both a circular flat tip and a curved surface of revolution connecting the tip to the cylindrical body of the rocket—see C. Henry Edwards, "Newton's Nose-Cone Problem," *The Mathematica Journal 7* (Winter 1997), pp. 75–82.

15.6 THE DIVERGENCE THEOREM

The *divergence theorem* is to surface integrals what Green's theorem is to line integrals. It lets us convert a surface integral over a closed surface into a triple integral over the enclosed region, or vice versa. The divergence theorem is known also as *Gauss's theorem,* and in some eastern European countries it is called *Ostrogradski's theorem.* The German "prince of mathematics" Carl Friedrich Gauss (1777–1855) used it to study inverse-square force fields; the Russian Michel Ostrogradski (1801–1861) used it to study heat flow. Both did their work in the 1830s.

The surface S is called **piecewise smooth** if it consists of a finite number of smooth parametric surfaces. It is called **closed** if it is the boundary of a bounded region in space. For example, the boundary of a cube is a closed piecewise smooth surface, as are the boundary of a pyramid and the boundary of a solid cylinder.

The Divergence Theorem

Suppose that S is a closed piecewise smooth surface that bounds the space region T. Let $\mathbf{F} = P\mathbf{i} + Q\mathbf{j} + R\mathbf{k}$ be a vector field with component functions that have continuous first-order partial derivatives on T. Let $\mathbf{n}$ be the *outer* unit normal vector to S. Then

$$\iint_S \mathbf{F} \cdot \mathbf{n}\, dS = \iiint_T \nabla \cdot \mathbf{F}\, dV. \tag{1}$$

Equation (1) is a three-dimensional analogue of the vector form of Green's theorem that we saw in Eq. (9) of Section 15.4:

$$\oint_C \mathbf{F} \cdot \mathbf{n}\, ds = \iint_R \nabla \cdot \mathbf{F}\, dA,$$

where $\mathbf{F}$ is a vector field in the plane, C is a piecewise smooth curve that bounds the plane region R, and $\mathbf{n}$ is the outer unit normal vector to C. The left-hand side of Eq. (1) is the flux of $\mathbf{F}$ across S in the direction of the outer unit normal vector $\mathbf{n}$.

Recall from Section 15.1 that the *divergence* $\nabla \cdot \mathbf{F}$ of the vector field $\mathbf{F} = \langle P, Q, R \rangle$ is given in the three-dimensional case by

$$\operatorname{div} \mathbf{F} = \nabla \cdot \mathbf{F} = \frac{\partial P}{\partial x} + \frac{\partial Q}{\partial y} + \frac{\partial R}{\partial z}. \tag{2}$$

If $\mathbf{n}$ is given in terms of its direction cosines, as $\mathbf{n} = \langle \cos\alpha, \cos\beta, \cos\gamma \rangle$, then we can write the divergence theorem in scalar form:

$$\iint_S (P\cos\alpha + Q\cos\beta + R\cos\gamma)\, dS = \iiint_T \left(\frac{\partial P}{\partial x} + \frac{\partial Q}{\partial y} + \frac{\partial R}{\partial z} \right) dV. \tag{3}$$

It is best to parametrize S so that the normal vector given by the parametrization is the outer normal. Then we can write Eq. (3) entirely in Cartesian form:

$$\iint_S P\, dy\, dz + Q\, dz\, dx + R\, dx\, dy = \iiint_T \left(\frac{\partial P}{\partial x} + \frac{\partial Q}{\partial y} + \frac{\partial R}{\partial z} \right) dV. \tag{4}$$

PROOF OF THE DIVERGENCE THEOREM We shall prove the divergence theorem only for the case in which the region T is simultaneously x-simple, y-simple, and z-simple. This guarantees that every straight line parallel to a coordinate axis intersects T, if at all, in a single point or a single line segment. It suffices for us to derive separately the equations

$$\iint_S P\,dy\,dz = \iiint_T \frac{\partial P}{\partial x}\,dV,$$
$$\iint_S Q\,dz\,dx = \iiint_T \frac{\partial Q}{\partial y}\,dV, \quad \text{and} \tag{5}$$
$$\iint_S R\,dx\,dy = \iiint_T \frac{\partial R}{\partial z}\,dV.$$

Then the sum of the equations in (5) is Eq. (4).

Because T is z-simple, it has the description

$$z_1(x, y) \leqq z \leqq z_2(x, y)$$

for (x, y) in D, the projection of T into the xy-plane. As in Fig. 15.6.1, we denote the lower surface $z = z_1(x, y)$ of T by S_1, the upper surface $z = z_2(x, y)$ by S_2, and the lateral surface between S_1 and S_2 by S_3. In the case of some simple surfaces, such as a spherical surface, there may be no S_3 to consider. But even if there is,

$$\iint_{S_3} R\,dx\,dy = \iint_{S_3} R\cos\gamma\,dS = 0, \tag{6}$$

because $\gamma = 90°$ at each point of the vertical cylinder S_3.

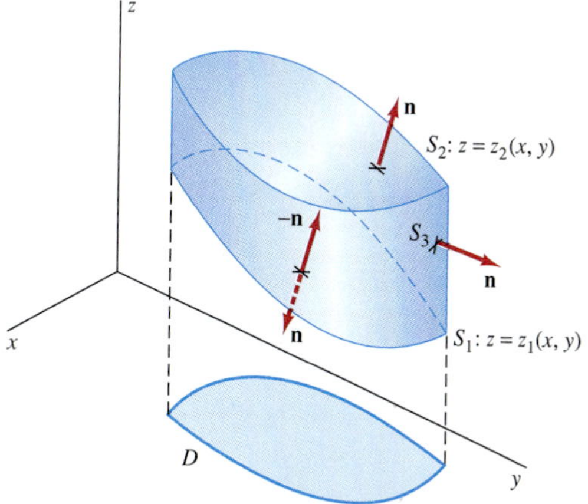

Fig. 15.6.1 A z-simple space region bounded by the surfaces S_1, S_2, and S_3

On the upper surface S_2, the unit upper normal vector corresponding to the parametrization $z = z_2(x, y)$ is the given outer unit normal vector **n**, so Eq. (20) of Section 15.5 yields

$$\iint_{S_2} R\,dx\,dy = \iint_D R(x, y, z_2(x, y))\,dx\,dy. \tag{7}$$

But on the lower surface S_1, the unit upper normal vector corresponding to the parametrization $z = z_1(x, y)$ is the inner normal vector $-\mathbf{n}$, so we must reverse the sign. Thus

$$\iint_{S_1} R\,dx\,dy = -\iint_D R(x, y, z_1(x, y))\,dx\,dy. \tag{8}$$

We add Eqs. (6), (7), and (8). The result is that

$$\begin{aligned}\iint_S R\,dx\,dy &= \iint_D \left[R(x, y, z_2(x, y)) - R(x, y, z_1(x, y))\right] dx\,dy \\ &= \iint_D \left(\int_{z_1(x,y)}^{z_2(x,y)} \frac{\partial R}{\partial z}\,dz\right) dx\,dy.\end{aligned}$$

Therefore

$$\iint_S R\,dx\,dy = \iiint_T \frac{\partial R}{\partial z}\,dV.$$

This is the third equation in (5), and we can derive the other two in the same way. ■

EXAMPLE 1 Let S be the surface (with outer unit normal vector $\mathbf{n}$) of the region T bounded by the planes $z = 0$, $y = 0$, $y = 2$, and the paraboloid $z = 1 - x^2$ (Fig. 15.6.2). Apply the divergence theorem to compute

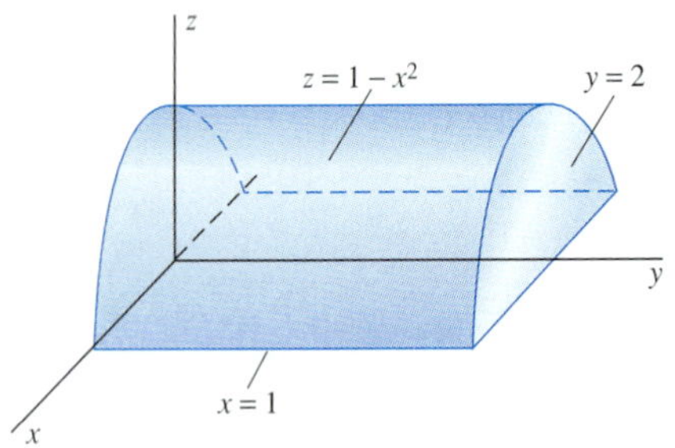

Fig. 15.6.2 The region of Example 1

$$\iint_S \mathbf{F}\cdot\mathbf{n}\,dS$$

given $\mathbf{F} = (x + \cos y)\mathbf{i} + (y + \sin z)\mathbf{j} + (z + e^x)\mathbf{k}$.

Solution Evaluating the surface integral directly would be a lengthy project. But $\operatorname{div}\mathbf{F} = 1 + 1 + 1 = 3$, so we can apply the divergence theorem easily:

$$\iint_S \mathbf{F}\cdot\mathbf{n}\,dS = \iiint_T \operatorname{div}\mathbf{F}\,dV = \iiint_T 3\,dV.$$

We examine Fig. 15.6.2 to find the limits for the volume integral and thus obtain

$$\iint_S \mathbf{F}\cdot\mathbf{n}\,dS = \int_{-1}^{1}\int_0^2\int_0^{1-x^2} 3\,dz\,dy\,dx = 12\int_0^1 (1 - x^2)\,dx = 8.$$ ■

EXAMPLE 2 Let S be the surface of the solid cylinder T bounded by the planes $z = 0$ and $z = 3$ and the cylinder $x^2 + y^2 = 4$. Calculate the outward flux

$$\iint_S \mathbf{F}\cdot\mathbf{n}\,dS$$

given $\mathbf{F} = (x^2 + y^2 + z^2)(x\mathbf{i} + y\mathbf{j} + z\mathbf{k})$.

Solution If we denote by P, Q, and R the component functions of the vector field $\mathbf{F}$, we find that

$$\frac{\partial P}{\partial x} = 2x\cdot x + (x^2 + y^2 + z^2)\cdot 1 = 3x^2 + y^2 + z^2.$$

Similarly,

$$\frac{\partial Q}{\partial y} = 3y^2 + z^2 + x^2 \quad \text{and} \quad \frac{\partial R}{\partial z} = 3z^2 + x^2 + y^2,$$

so $\text{div}\,\mathbf{F} = 5(x^2 + y^2 + z^2)$. Therefore the divergence theorem yields

$$\iint_S \mathbf{F} \cdot \mathbf{n}\, dS = \iiint_T 5(x^2 + y^2 + z^2)\, dV.$$

Using cylindrical coordinates to evaluate the triple integral, we get

$$\begin{aligned}\iint_S \mathbf{F} \cdot \mathbf{n}\, dS &= \int_0^{2\pi}\int_0^2\int_0^3 5(r^2 + z^2)r\, dz\, dr\, d\theta \\ &= 10\pi\int_0^2 \left[r^3 z + \tfrac{1}{3}rz^3\right]_{z=0}^3 dr \\ &= 10\pi\int_0^2 (3r^3 + 9r)\, dr = 10\pi\left[\tfrac{3}{4}r^4 + \tfrac{9}{2}r^2\right]_0^2 = 300\pi.\end{aligned}$$

■

EXAMPLE 3 Suppose that the region T is bounded by the closed surface S with a parametrization that gives the outer unit normal vector. Show that the volume V of T is given by

$$V = \frac{1}{3}\iint_S x\, dy\, dz + y\, dz\, dx + z\, dx\, dy. \tag{9}$$

Solution Equation (9) follows immediately from Eq. (4) if we take $P(x, y, z) = x$, $Q(x, y, z) = y$, and $R(x, y, z) = z$. For example, if S is the spherical surface $x^2 + y^2 + z^2 = a^2$ with volume V, surface area A, and outer unit normal vector

$$\mathbf{n} = \langle \cos\alpha, \cos\beta, \cos\gamma \rangle = \left\langle \frac{x}{a}, \frac{y}{b}, \frac{z}{c} \right\rangle,$$

then Eq. (9) yields

$$\begin{aligned}V &= \tfrac{1}{3}\iint_S x\, dy\, dz + y\, dz\, dx + z\, dx\, dy \\ &= \tfrac{1}{3}\iint_S (x\cos\alpha + y\cos\beta + z\cos\gamma)\, dS \\ &= \tfrac{1}{3}\iint_S \frac{x^2 + y^2 + z^2}{a}\, dS = \tfrac{1}{3}a\iint_S 1\, dS = \tfrac{1}{3}aA.\end{aligned}$$

You should confirm that this result is consistent with the familiar formulas $V = \frac{4}{3}\pi a^3$ and $A = 4\pi a^2$.

■

EXAMPLE 4 Show that the divergence of the vector field $\mathbf{F}$ at the point P is given by

$$\{\text{div}\,\mathbf{F}\}(P) = \lim_{r \to 0} \frac{1}{V_r}\iint_{S_r} \mathbf{F} \cdot \mathbf{n}\, dS, \tag{10}$$

where S_r is the sphere of radius r centered at P and $V_r = \frac{4}{3}\pi r^3$ is the volume of the ball B_r that the sphere bounds.

Solution The divergence theorem gives

$$\iint_{S_r} \mathbf{F} \cdot \mathbf{n}\, dS = \iiint_{B_r} \operatorname{div} \mathbf{F}\, dV.$$

Then we apply the average value property of triple integrals, a result analogous to the double integral result of Problem 50 in Section 14.2. This yields

$$\iiint_{B_r} \operatorname{div} \mathbf{F}\, dV = V_r \cdot \{\operatorname{div} \mathbf{F}\}(P^\star)$$

for some point $P^\star$ of B_r; here, we write $\{\operatorname{div} \mathbf{F}\}(P^\star)$ for the value of div **F** at the point $P^\star$. We assume that the component functions of **F** have continuous first-order partial derivatives at P, so it follows that

$$\{\operatorname{div} \mathbf{F}\}(P^\star) \to \{\operatorname{div} \mathbf{F}\}(P) \quad \text{as} \quad P^\star \to P.$$

Equation (10) follows after we divide both sides by V_r and then take the limit as $r \to 0$. ■

For instance, suppose that $\mathbf{F} = \delta \mathbf{v}$ is a fluid flow vector field. We can interpret Eq. (10) as saying that $\{\operatorname{div} \mathbf{F}\}(P)$ is the net rate per unit volume that fluid mass is flowing away (or "diverging") from the point P. For this reason the point P is called a **source** if $\{\operatorname{div} \mathbf{F}\}(P) > 0$ but a **sink** if $\{\operatorname{div} \mathbf{F}\}(P) < 0$.

Heat in a conducting body can be treated mathematically as though it were a fluid flowing through the body. Miscellaneous Problems 25 through 27 at the end of this chapter ask you to apply the divergence theorem to show that if $u = u(x, y, z, t)$ is the temperature at the point (x, y, z) at the time t in a body through which heat is flowing, then the function u must satisfy the equation

$$\frac{\partial^2 u}{\partial x^2} + \frac{\partial^2 u}{\partial y^2} + \frac{\partial^2 u}{\partial z^2} = \frac{1}{k} \cdot \frac{\partial u}{\partial t}, \tag{11}$$

where k is a constant (the *thermal diffusivity* of the body). This is a *partial differential equation* called the **heat equation**. If both the initial temperature $u(x, y, z, 0)$ and the temperature on the boundary of the body are given, then its interior temperatures at future times are determined by the heat equation. A large part of advanced applied mathematics consists of techniques for solving such partial differential equations.

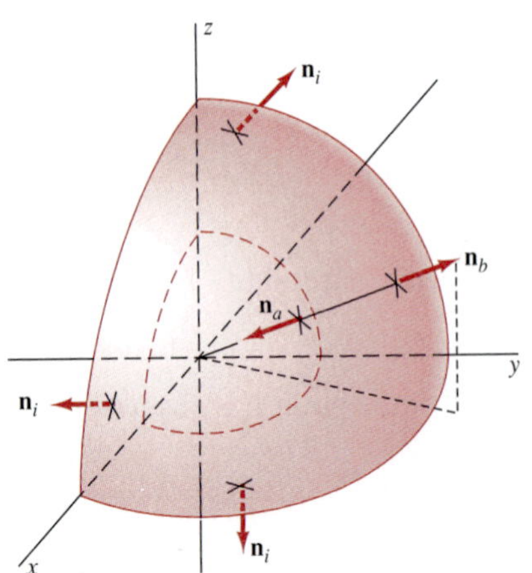

Fig. 15.6.3 One octant of the shell between S_a and S_b

More General Regions and Gauss's Law

We can establish the divergence theorem for more general regions by the device of subdividing T into simpler regions, regions for which the preceding proof holds. For example, suppose that T is the shell between the concentric spherical surfaces S_a and S_b of radii a and b, with $0 < a < b$. The coordinate planes separate T into eight regions $T_1, T_2, \dots, T_8$, each shaped as in Fig. 15.6.3. Let Σ_i denote the boundary of T_i and let $\mathbf{n}_i$ be the outer unit normal vector to Σ_i. We apply the divergence theorem to each of these eight regions and obtain

$$\begin{aligned}\iiint_T \nabla \cdot \mathbf{F}\, dV &= \sum_{i=1}^{8} \iiint_{T_i} \nabla \cdot \mathbf{F}\, dV \\ &= \sum_{i=1}^{8} \iint_{\Sigma_i} \mathbf{F} \cdot \mathbf{n}_i\, dS \quad \text{(divergence theorem)}\end{aligned}$$

$$= \iint_{S_a} \mathbf{F} \cdot \mathbf{n}_a \, dS + \iint_{S_b} \mathbf{F} \cdot \mathbf{n}_b \, dS.$$

Here we write $\mathbf{n}_a$ for the inner normal vector on S_a and $\mathbf{n}_b$ for the outer normal vector on S_b. The last equality holds because the surface integrals over the internal boundary surfaces (the surfaces in the coordinate planes) cancel in pairs—the normals are oppositely oriented there. As the boundary S of T is the union of the spherical surfaces S_a and S_b, it now follows that

$$\iiint_T \nabla \cdot \mathbf{F} \, dV = \iint_S \mathbf{F} \cdot \mathbf{n} \, dS.$$

This is the divergence theorem for the spherical shell T.

In a similar manner, the divergence theorem can be established for a region T that is bounded by two smooth closed surfaces S_1 and S_2 with S_1 interior to S_2, as in Fig. 15.6.4, where $\mathbf{n}_1$ and $\mathbf{n}_2$ denote the outward-pointing unit normal vectors to the two surfaces. Then the boundary S of T is the union of S_1 and S_2, and the outer unit normal vector field $\mathbf{n}$ on S consists of $-\mathbf{n}_1$ on the inner surface S_1 and $\mathbf{n}_2$ on the outer surface S_2 (both pointing out of T). Hence the divergence theorem takes the form

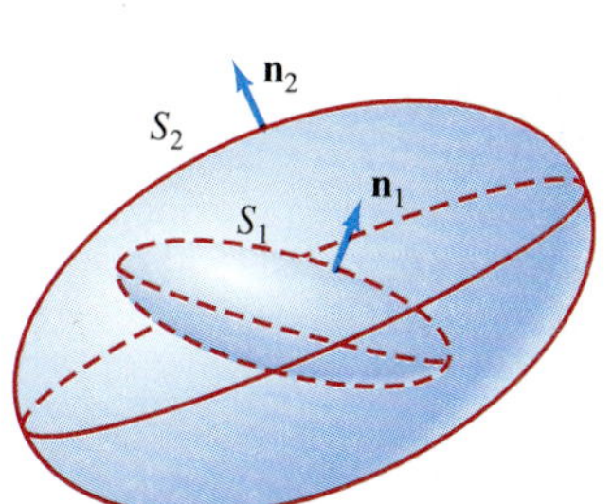

Fig. 15.6.4 Nested closed surfaces S_1 and S_2

$$\iiint_T \nabla \cdot \mathbf{F} \, dV = \iint_S \mathbf{F} \cdot \mathbf{n} \, dS = \iint_{S_2} \mathbf{F} \cdot \mathbf{n}_2 \, dS - \iint_{S_1} \mathbf{F} \cdot \mathbf{n}_1 \, dS. \tag{12}$$

For example, suppose that $\mathbf{F} = -GM\mathbf{r}/|\mathbf{r}|^3$ is the inverse-square gravitational force field of a mass M located at the origin. According to Problem 22, $\nabla \cdot \mathbf{F} = 0$ except at the origin. If S is a smooth surface enclosing M and S_a is a small sphere of radius a within S enclosing M, it then follows from Eq. (12) that (with $\mathbf{n}$ denoting the outer unit normal on each surface)

$$\begin{aligned} \iint_S \mathbf{F} \cdot \mathbf{n} \, dS &= \iint_{S_a} \mathbf{F} \cdot \mathbf{n} \, dS \\ &= \iint_{S_a} -\frac{GM\mathbf{r}}{|\mathbf{r}|^3} \cdot \frac{\mathbf{r}}{|\mathbf{r}|} \, dS = -\frac{GM}{a^2} \iint_{S_a} 1 \, dS = -4\pi GM. \end{aligned} \tag{13}$$

Thus we have established Gauss's law [Eq. (24) in Section 15.5] for the special case of a single point mass. The more general case of a collection of point masses within S can be established by enclosing each in its own small sphere. If we replace the constant GM in Eq. (13) with $Q/4\pi\epsilon_0$, we obtain similarly Gauss's law

$$\iint_S \mathbf{E} \cdot \mathbf{n} \, dS = \frac{Q}{\epsilon_0} \tag{14}$$

for the inverse-square electric field $\mathbf{E} = -Q\mathbf{r}/(4\pi\epsilon_0|\mathbf{r}|^3)$ of a charge Q lying within S.

Another impressive consequence of the divergence theorem is Archimedes' law of buoyancy; see Problem 21 here and Problem 22 of Section 15.7.

15.6 PROBLEMS

In Problems 1 through 5, verify the divergence theorem by direct computation of both the surface integral and the triple integral of Eq. (1).

1. $\mathbf{F} = x\mathbf{i} + y\mathbf{j} + z\mathbf{k}$; S is the spherical surface with equation $x^2 + y^2 + z^2 = 1$.

2. $\mathbf{F} = |\mathbf{r}|\mathbf{r}$, where $\mathbf{r} = x\mathbf{i} + y\mathbf{j} + z\mathbf{k}$; S is the spherical surface with equation $x^2 + y^2 + z^2 = 9$.

3. $\mathbf{F} = x\mathbf{i} + y\mathbf{j} + z\mathbf{k}$; S is the surface of the cube bounded by the three coordinate planes and the three planes $x = 2$, $y = 2$, and $z = 2$.

4. $\mathbf{F} = xy\mathbf{i} + yz\mathbf{j} + xz\mathbf{k}$; S is the surface of Problem 3.

5. $\mathbf{F} = (x + y)\mathbf{i} + (y + z)\mathbf{j} + (z + x)\mathbf{k}$; S is the surface of the tetrahedron bounded by the three coordinate planes and the plane $x + y + z = 1$.

In Problems 6 through 14, use the divergence theorem to evaluate $\iint_S \mathbf{F} \cdot \mathbf{n}\, dS$, where $\mathbf{n}$ is the outer unit normal vector to the surface S.

6. $\mathbf{F} = x^2\mathbf{i} + y^2\mathbf{j} + z^2\mathbf{k}$; S is the surface of Problem 3.

7. $\mathbf{F} = x^3\mathbf{i} + y^3\mathbf{j} + z^3\mathbf{k}$; S is the surface of the cylinder bounded by $x^2 + y^2 = 9$, $z = -1$, and $z = 4$.

8. $\mathbf{F} = (x^2 + y^2)(x\mathbf{i} + y\mathbf{j})$; S is the surface of the region bounded by the plane $z = 0$ and the paraboloid $z = 25 - x^2 - y^2$.

9. $\mathbf{F} = (x^2 + e^{-yz})\mathbf{i} + (y + \sin xz)\mathbf{j} + (\cos xy)\mathbf{k}$; S is the surface of Problem 5.

10. $\mathbf{F} = (xy^2 + e^{-y}\sin z)\mathbf{i} + (x^2y + e^{-x}\cos z)\mathbf{j} + (\tan^{-1}xy)\mathbf{k}$; S is the surface of the region bounded by the paraboloid $z = x^2 + y^2$ and the plane $z = 9$.

11. $\mathbf{F} = (x^2 + y^2 + z^2)(x\mathbf{i} + y\mathbf{j} + z\mathbf{k})$; S is the surface of Problem 8.

12. $\mathbf{F} = \mathbf{r}/|\mathbf{r}|$, where $\mathbf{r} = x\mathbf{i} + y\mathbf{j} + z\mathbf{k}$; S is the sphere $\rho = 2$ of radius 2 centered at the origin.

13. $\mathbf{F} = x\mathbf{i} + y\mathbf{j} + 3\mathbf{k}$; S is the boundary of the region bounded by the paraboloid $z = x^2 + y^2$ and the plane $z = 4$.

14. $\mathbf{F} = (x^3 + e^z)\mathbf{i} + x^2y\mathbf{j} + (\sin xy)\mathbf{k}$; S is the boundary of the region bounded by the paraboloid $z = 4 - x^2$ and the planes $y = 0$, $z = 0$, and $y + z = 5$.

15. The **Laplacian** of the twice-differentiable scalar function f is defined to be $\nabla^2 f = \text{div}\,(\text{grad}\, f) = \nabla \cdot \nabla f$. Show that

$$\nabla^2 f = \frac{\partial^2 f}{\partial x^2} + \frac{\partial^2 f}{\partial y^2} + \frac{\partial^2 f}{\partial z^2}.$$

16. Let $\partial f/\partial n = (\nabla f) \cdot \mathbf{n}$ denote the directional derivative of the scalar function f in the direction of the outer unit normal vector $\mathbf{n}$ to the surface S that bounds the region T. Show that

$$\iint_S \frac{\partial f}{\partial n}\, dS = \iiint_T \nabla^2 f\, dV.$$

17. Suppose that $\nabla^2 f \equiv 0$ in the region T with boundary surface S. Show that

$$\iint_S f\frac{\partial f}{\partial n}\, dS = \iiint_T |\nabla f|^2\, dV.$$

(See Problems 15 and 16 for the notation.)

18. Apply the divergence theorem to $\mathbf{F} = f\nabla g$ to establish **Green's first identity,**

$$\iint_S f\frac{\partial g}{\partial n}\, dS = \iiint_T (f\,\nabla^2 g + \nabla f \cdot \nabla g)\, dV.$$

19. Interchange f and g in Green's first identity (Problem 18) to establish **Green's second identity,**

$$\iint_S \left(f\frac{\partial g}{\partial n} - g\frac{\partial f}{\partial n}\right) dS = \iiint_T (f\,\nabla^2 g - g\,\nabla^2 f)\, dV.$$

20. Suppose that f is a differentiable scalar function defined on the region T of space and that S is the boundary of T. Prove that

$$\iint_S f\mathbf{n}\, dS = \iiint_T \nabla f\, dV.$$

(*Suggestion:* Apply the divergence theorem to $\mathbf{F} = f\mathbf{a}$, where $\mathbf{a}$ is an arbitrary constant vector. *Note:* Integrals of vector-valued functions are defined by componentwise integration.)

21. *Archimedes' Law of Buoyancy* Let S be the surface of a body T submerged in a fluid of constant density δ. Set up coordinates so that positive values of z are measured *downward* from the surface. Then the pressure at depth z is $p = \delta g z$. The buoyant force exerted on the body by the fluid is

$$\mathbf{B} = -\iint_S p\mathbf{n}\, dS.$$

(Why?) Apply the result of Problem 20 to show that $\mathbf{B} = -W\mathbf{k}$, where W is the weight of the fluid displaced by the body. Because z is measured downward, the vector $\mathbf{B}$ is directed upward.

22. Let $\mathbf{r} = \langle x, y, z\rangle$, let $\mathbf{r}_0 = \langle x_0, y_0, z_0\rangle$ be a fixed point, and suppose that

$$\mathbf{F}(x, y, z) = \frac{\mathbf{r} - \mathbf{r}_0}{|\mathbf{r} - \mathbf{r}_0|^3}.$$

Show that $\text{div}\,\mathbf{F} = 0$ except at the point $\mathbf{r}_0$.

23. Apply the divergence theorem to compute the outward flux

$$\iint_S \mathbf{F} \cdot \mathbf{n}\, dS,$$

where $\mathbf{F} = |\mathbf{r}|\mathbf{r}$, $\mathbf{r} = x\mathbf{i} + y\mathbf{j} + z\mathbf{k}$, and S is the surface of Problem 8. (*Suggestion:* Integrate in cylindrical coordinates, first with respect to r and then with respect to z. For the latter integration, make a trigonometric substitution and then consult Eq. (9) of Section 9.3 for the antiderivative of $\sec^5\theta$.)

24. Assume that Gauss's law in (13) holds for a uniform solid ball of mass M centered at the origin. Also assume by symmetry that the force $\mathbf{F}$ exerted by this mass on an exterior particle of unit mass is directed toward the origin. Apply Gauss's law with S being a sphere of radius r to show that $|\mathbf{F}| = GM/r^2$ at each point of S. Thus it follows that the solid ball acts (gravitationally) like a single point-mass M concentrated at its center.

25. Let **F** be the gravitational force field due to a uniform distribution of mass in the shell bounded by the concentric spherical surfaces $\rho = a$ and $\rho = b > a$. Apply Gauss's law in (13), with S being the sphere $\rho = r < a$, to show that **F** is zero at all points interior to this spherical shell.

26. Consider a solid spherical ball of radius a and constant density δ. Apply Gauss's law to show that the gravitational force on a particle of unit mass located at a distance $r < a$ from the center of the ball is given by $F = GM_r/r^2$, where $M_r = \frac{4}{3}\pi\delta r^3$ is the mass enclosed by a sphere of radius r. Thus the mass at a greater distance from the center of the ball exerts no net gravitational force on the particle.

27. Consider an infinitely long vertical straight wire with a uniform positive charge of q coulombs per meter. Assume by symmetry that the electric field vector **E** is at each point a horizontal radial vector directed away from the wire. Apply Gauss's law in (14), with S being a cylinder with the wire as its axis, to show that $|\mathbf{E}| = q/(2\pi\epsilon_0 r)$. Thus the electric field intensity is inversely proportional to distance r from the wire.

15.7 STOKES' THEOREM

In Section 15.4 we gave Green's theorem,

$$\oint_C P\,dx + Q\,dy = \iint_R \left(\frac{\partial Q}{\partial x} - \frac{\partial P}{\partial y}\right) dA, \tag{1}$$

in a vector form that is equivalent to a two-dimensional version of the divergence theorem. Another vector form of Green's theorem involves the curl of a vector field. Recall from Section 15.1 that if $\mathbf{F} = P\mathbf{i} + Q\mathbf{j} + R\mathbf{k}$ is a vector field, then curl **F** is the vector field given by

$$\text{curl}\,\mathbf{F} = \nabla \times \mathbf{F} = \begin{vmatrix} \mathbf{i} & \mathbf{j} & \mathbf{k} \\ \dfrac{\partial}{\partial x} & \dfrac{\partial}{\partial y} & \dfrac{\partial}{\partial z} \\ P & Q & R \end{vmatrix}$$

$$= \left(\frac{\partial R}{\partial y} - \frac{\partial Q}{\partial z}\right)\mathbf{i} + \left(\frac{\partial P}{\partial z} - \frac{\partial R}{\partial x}\right)\mathbf{j} + \left(\frac{\partial Q}{\partial x} - \frac{\partial P}{\partial y}\right)\mathbf{k}. \tag{2}$$

The **k**-component of $\nabla \times \mathbf{F}$ is the integrand of the double integral in Eq. (1). We know from Section 15.2 that we can write the line integral in Eq. (1) as

$$\oint_C \mathbf{F}\cdot\mathbf{T}\,ds,$$

where **T** is the positive-directed unit tangent vector to C. Consequently, we can rewrite Green's theorem in the form

$$\oint_C \mathbf{F}\cdot\mathbf{T}\,ds = \iint_R (\text{curl}\,\mathbf{F})\cdot\mathbf{k}\,dA. \tag{3}$$

Stokes' theorem is the generalization of Eq. (3) that we get by replacing the plane region R with a floppy two-dimensional version: an oriented bounded surface S in three-dimensional space with boundary C that consists of one or more simple closed curves in space.

An *oriented* surface is a surface together with a chosen continuous unit normal vector field **n**. The positive orientation of the boundary C of an oriented surface S corresponds to the unit tangent vector **T** such that $\mathbf{n} \times \mathbf{T}$ always points *into* S (Fig. 15.7.1). Check that for a plane region with unit normal vector **k**, the positive orientation of its outer boundary is counterclockwise.

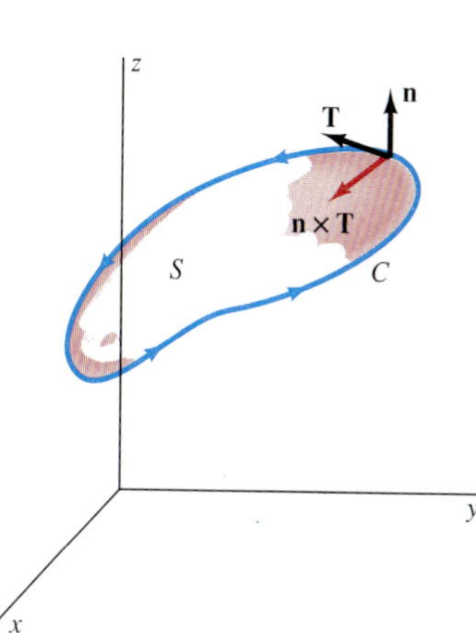

Fig. 15.7.1 Vectors, surface, and boundary curve mentioned in the statement of Stokes' theorem

Stokes' Theorem

Let S be an oriented, bounded, and piecewise smooth surface in space with positively oriented boundary C. Suppose that the components of the vector field $\mathbf{F}$ have continuous first-order partial derivatives in a space region that contains S. Then

$$\oint_C \mathbf{F}\cdot\mathbf{T}\,ds = \iint_S (\operatorname{curl}\mathbf{F})\cdot\mathbf{n}\,dS. \tag{4}$$

Thus Stokes' theorem says that *the line integral around the boundary curve of the tangential component of* $\mathbf{F}$ *equals the surface integral of the normal component of* curl $\mathbf{F}$. Compare Eqs. (3) and (4).

This result first appeared publicly as a problem posed by George Stokes (1819–1903) on a prize examination for Cambridge University students in 1854. It had been stated in an 1850 letter to Stokes from the physicist William Thomson (Lord Kelvin, 1824–1907).

In terms of the components of $\mathbf{F} = P\mathbf{i} + Q\mathbf{j} + R\mathbf{k}$ and those of curl $\mathbf{F}$, we can recast Stokes' theorem—with the aid of Eq. (19) of Section 15.5—in its scalar form

$$\oint_C P\,dx + Q\,dy + R\,dz = \iint_S \left(\frac{\partial R}{\partial y} - \frac{\partial Q}{\partial z}\right) dy\,dz + \left(\frac{\partial P}{\partial z} - \frac{\partial R}{\partial x}\right) dz\,dx + \left(\frac{\partial Q}{\partial x} - \frac{\partial P}{\partial y}\right) dx\,dy. \tag{5}$$

Here, as usual, the parametrization of S must correspond to the given unit normal vector $\mathbf{n}$.

To prove Stokes' theorem, we need only establish the equation

$$\oint_C P\,dx = \iint_S \left(\frac{\partial P}{\partial z}\,dz\,dx - \frac{\partial P}{\partial y}\,dx\,dy\right) \tag{6}$$

and the corresponding two equations that are the Q and R "components" of Eq. (5). Equation (5) itself then follows by adding the three results.

Partial Proof Suppose first that S is the graph of a function $z = f(x, y)$, (x, y) in D, where S has an upper unit normal vector and D is a region in the xy-plane bounded by the simple closed curve J (Fig. 15.7.2). Then

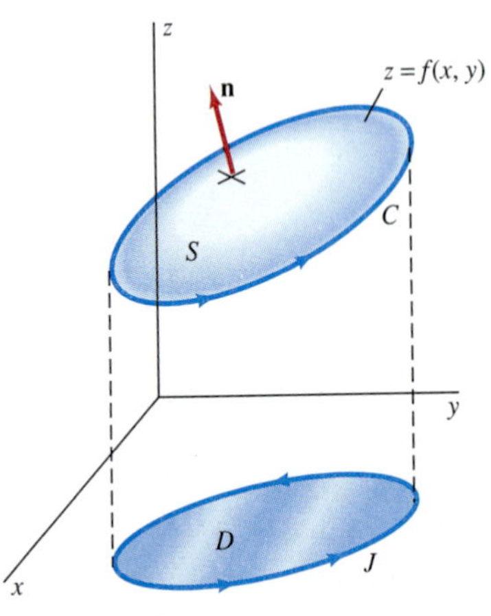

Fig. 15.7.2 The surface S

$$\begin{aligned}\oint_C P\,dx &= \oint_J P\big(x, y, f(x, y)\big)\,dx \\ &= \oint_J p(x, y)\,dx \qquad \big[\text{where } p(x, y) \equiv P\big(x, y, f(x, y)\big)\big] \\ &= -\iint_D \frac{\partial p}{\partial y}\,dx\,dy \qquad \text{(by Green's theorem).}\end{aligned}$$

We now use the chain rule to compute $\partial p/\partial y$ and find that

$$\oint_C P\,dx = -\iint_D \left(\frac{\partial P}{\partial y} + \frac{\partial P}{\partial z}\frac{\partial z}{\partial y}\right) dx\,dy. \tag{7}$$

Next, we use Eq. (20) of Section 15.5:

$$\iint_S P\,dy\,dz + Q\,dz\,dx + R\,dx\,dy = \iint_D \left(-P\frac{\partial z}{\partial x} - Q\frac{\partial z}{\partial y} + R \right) dx\,dy.$$

In this equation we replace P with 0, Q with $\partial P/\partial z$, and R with $-\partial P/\partial y$. This gives

$$\iint_S \left(\frac{\partial P}{\partial z}\,dz\,dx - \frac{\partial P}{\partial y}\,dx\,dy \right) = \iint_D \left(-\frac{\partial P}{\partial z}\frac{\partial z}{\partial y} - \frac{\partial P}{\partial y} \right) dx\,dy. \tag{8}$$

Finally, we compare Eqs. (7) and (8) and see that we have established Eq. (6). If we can write the surface S in the forms $y = g(x, z)$ and $x = h(y, z)$, then we can derive the Q and R "components" of Eq. (5) in much the same way. This proves Stokes' theorem for the special case of a surface S that can be represented as a graph in all three coordinate directions. Stokes' theorem may then be extended to a more general oriented surface by the now-familiar method of subdividing it into simpler surfaces, to each of which the preceding proof is applicable. ■

EXAMPLE 1 Apply Stokes' theorem to evaluate

$$\oint_C \mathbf{F}\cdot\mathbf{T}\,ds,$$

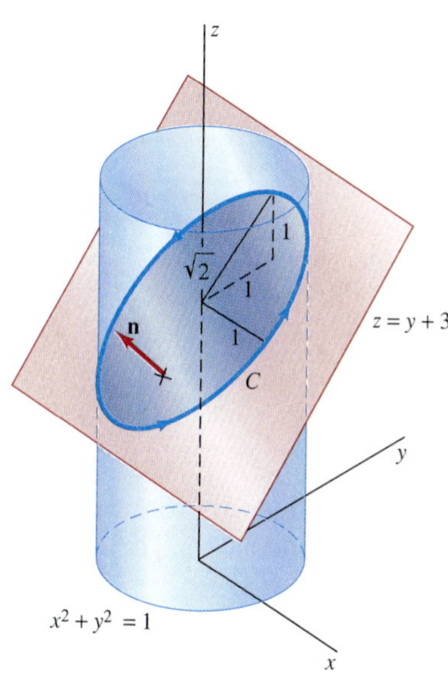

Fig. 15.7.3 The ellipse of Example 1

where C is the ellipse in which the plane $z = y + 3$ intersects the cylinder $x^2 + y^2 = 1$. Orient the ellipse counterclockwise as viewed from above and take $\mathbf{F}(x, y, z) = 3z\,\mathbf{i} + 5x\,\mathbf{j} - 2y\,\mathbf{k}$.

Solution The plane, cylinder, and ellipse appear in Fig. 15.7.3. The given orientation of C corresponds to the upward unit normal vector $\mathbf{n} = (-\mathbf{j} + \mathbf{k})/\sqrt{2}$ to the elliptical region S in the plane $z = y + 3$ bounded by C. Now

$$\text{curl}\,\mathbf{F} = \begin{vmatrix} \mathbf{i} & \mathbf{j} & \mathbf{k} \\ \dfrac{\partial}{\partial x} & \dfrac{\partial}{\partial y} & \dfrac{\partial}{\partial z} \\ 3z & 5x & -2y \end{vmatrix} = -2\mathbf{i} + 3\mathbf{j} + 5\mathbf{k},$$

so

$$(\text{curl}\,\mathbf{F})\cdot\mathbf{n} = (-2\mathbf{i} + 3\mathbf{j} + 5\mathbf{k})\cdot\frac{1}{\sqrt{2}}(-\mathbf{j} + \mathbf{k}) = \frac{-3 + 5}{\sqrt{2}} = \sqrt{2}.$$

Hence by Stokes' theorem,

$$\oint_C \mathbf{F}\cdot\mathbf{T}\,ds = \iint_S (\text{curl}\,\mathbf{F})\cdot\mathbf{n}\,dS = \iint_S \sqrt{2}\,dS = \sqrt{2}\cdot\text{area}(S) = 2\pi,$$

because we can see from Fig. 15.7.3 that S is an ellipse with semiaxes 1 and $\sqrt{2}$. Thus its area is $\pi\sqrt{2}$. ■

EXAMPLE 2 Apply Stokes' theorem to evaluate

$$\iint_S (\nabla \times \mathbf{F})\cdot\mathbf{n}\,dS,$$

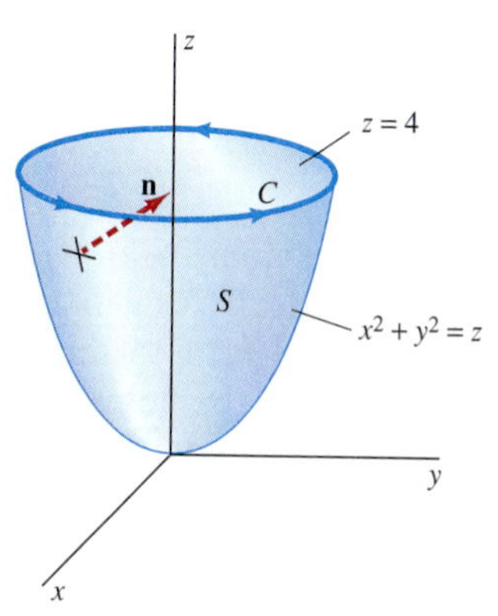
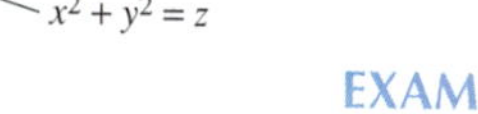

Fig. 15.7.4 The parabolic surface of Example 2

where $\mathbf{F} = 3z\,\mathbf{i} + 5x\,\mathbf{j} - 2y\,\mathbf{k}$ and S is the part of the surface $z = x^2 + y^2$ that lies below the plane $z = 4$ and whose orientation is given by the upper unit normal vector (Fig. 15.7.4).

Solution We parametrize the boundary circle C of S by $x = 2\cos t$, $y = 2\sin t$, $z = 4$ for $0 \leqq t \leqq 2\pi$. Then $dx = -2\sin t\,dt$, $dy = 2\cos t\,dt$, and $dz = 0$. So Stokes' theorem yields

$$\iint_S (\nabla \times \mathbf{F}) \cdot \mathbf{n}\,dS = \oint_C \mathbf{F} \cdot \mathbf{T}\,ds = \oint_C 3z\,dx + 5x\,dy - 2y\,dz$$

$$= \int_0^{2\pi} 3 \cdot 4 \cdot (-2\sin t\,dt) + 5 \cdot (2\cos t)(2\cos t\,dt) + 2 \cdot (2\sin t) \cdot 0$$

$$= \int_0^{2\pi} (-24\sin t + 20\cos^2 t)\,dt = \int_0^{2\pi} (-24\sin t + 10 + 10\cos 2t)\,dt$$

$$= \Big[24\cos t + 10t + 5\sin 2t \Big]_0^{2\pi} = 20\pi.$$

Just as the divergence theorem yields a physical interpretation of div $\mathbf{F}$ [Eq. (10) of Section 15.6], Stokes' theorem yields a physical interpretation of curl $\mathbf{F}$. Let S_r be a circular disk of radius r, centered at the point P in space and perpendicular to the unit vector $\mathbf{n}$. Let C_r be the boundary circle of S_r (Fig. 15.7.5). Then Stokes' theorem and the average value property of double integrals together give

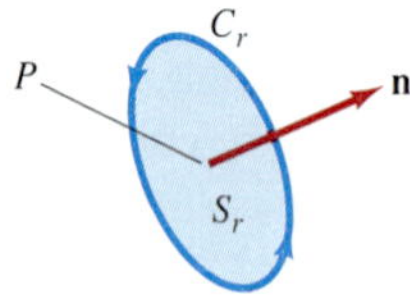

Fig. 15.7.5 A physical interpretation of the curl of a vector field

$$\oint_{C_r} \mathbf{F} \cdot \mathbf{T}\,ds = \iint_{S_r} (\text{curl}\,\mathbf{F}) \cdot \mathbf{n}\,dS = \pi r^2 \cdot \{(\text{curl}\,\mathbf{F}) \cdot \mathbf{n}\}(P^\star)$$

for some point $P^\star$ of S_r, where $\{(\text{curl}\,\mathbf{F}) \cdot \mathbf{n}\}(P^\star)$ denotes the value of $(\text{curl}\,\mathbf{F}) \cdot \mathbf{n}$ at the point $P^\star$. We divide this equality by πr^2 and then take the limit as $r \to 0$. This gives

$$\big\{(\text{curl}\,\mathbf{F}) \cdot \mathbf{n}\big\}(P) = \lim_{r \to 0} \frac{1}{\pi r^2} \oint_{C_r} \mathbf{F} \cdot \mathbf{T}\,ds. \tag{9}$$

Equation (9) has a natural physical meaning. Suppose that $\mathbf{F} = \delta\mathbf{v}$, where $\mathbf{v}$ is the velocity vector field of a steady-state fluid flow with constant density δ. Then the value of the integral

$$\Gamma(C) = \oint_C \mathbf{F} \cdot \mathbf{T}\,ds \tag{10}$$

measures the rate of flow of fluid mass *around* the curve C and is therefore called the **circulation** of $\mathbf{F}$ around C. We see from Eq. (9) that

$$\big\{(\text{curl}\,\mathbf{F}) \cdot \mathbf{n}\big\}(P^\star) \approx \frac{\Gamma(C_r)}{\pi r^2}$$

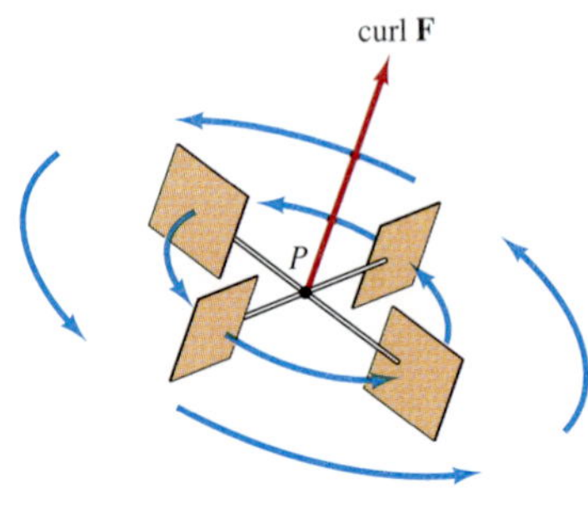

Fig. 15.7.6 The paddle-wheel interpretation of curl $\mathbf{F}$

if C_r is a circle of very small radius r centered at P and perpendicular to $\mathbf{n}$. If it should happen that $\{\text{curl}\,\mathbf{F}\}(P) \neq \mathbf{0}$, it follows that $\Gamma(C_r)$ is greatest (for r fixed and small) when the unit vector $\mathbf{n}$ points in the direction of $\{\text{curl}\,\mathbf{F}\}(P)$. Hence the line through P determined by $\{\text{curl}\,\mathbf{F}\}(P)$ is the axis about which the fluid near P is revolving the most rapidly. A tiny paddle wheel placed in the fluid at P (see Fig. 15.7.6) would rotate the fastest if its axis lay along this line. It follows from Miscellaneous Problem 32 at the end of this chapter that $|\text{curl}\,\mathbf{F}| = 2\delta\omega$ in the case of a fluid revolving steadily around a fixed axis with constant angular speed ω (in radians per second). Thus $\{\text{curl}\,\mathbf{F}\}(P)$ indicates both the direction *and* rate of rotation of the fluid near P. Because of this interpretation, some older books use the notation "rot **F**" for the curl, an abbreviation that we are happy has disappeared from general use.

If curl $\mathbf{F} = \mathbf{0}$ everywhere, then the fluid flow and the vector field **F** are said to be **irrotational**. An infinitesimal straw placed in an irrotational fluid flow would be translated parallel to itself without rotating. A vector field **F** defined on a simply connected region D is irrotational if and only if it is conservative, which in turn is true if and only if the line integral

$$\int_C \mathbf{F} \cdot \mathbf{T}\, ds$$

is independent of the path in D. (The region D is said to be **simply connected** if every simple closed curve in D can be continuously shrunk to a point while staying inside D. The interior of a torus is an example of a space region that is *not* simply connected. It is true, though not obvious, that any piecewise smooth simple closed curve in a simply connected region D is the boundary of a piecewise smooth oriented surface in D.)

Theorem 1 ***Conservative and Irrotational Fields***

Let **F** be a vector field with continuous first-order partial derivatives in a simply connected region D in space. Then the vector field **F** is irrotational if and only if it is conservative; that is, $\nabla \times \mathbf{F} = \mathbf{0}$ if and only if $\mathbf{F} = \nabla\phi$ for some scalar function ϕ defined on D.

PARTIAL PROOF A complete proof of the *if* part of Theorem 1 is easy; by Example 8 of Section 15.1, $\nabla \times (\nabla\phi) = \mathbf{0}$ for any twice-differentiable scalar function ϕ.

Here is a description of how we might show the *only if* part of the proof of Theorem 1. Assume that **F** is irrotational. Let $P_0(x_0, y_0, z_0)$ be a fixed point of D. Given an arbitrary point $P(x, y, z)$ of D, we would like to define

$$\phi(x, y, z) = \int_{C_1} \mathbf{F} \cdot \mathbf{T}\, ds, \tag{11}$$

where C_1 is a path in D from P_0 to P. But we must show that any *other* path C_2 from P_0 to P would give the same value for $\phi(x, y, z)$.

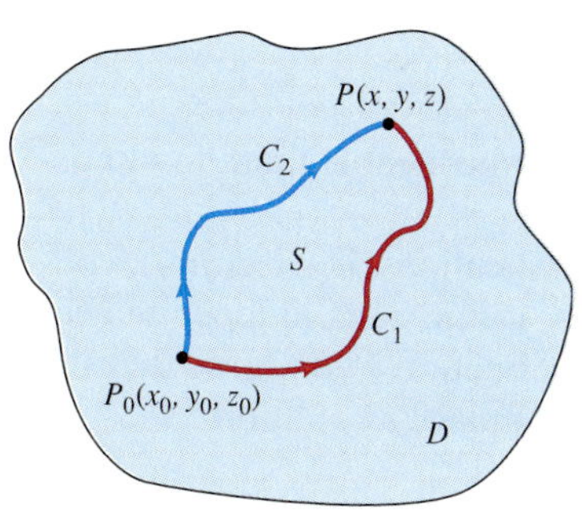

Fig. 15.7.7 Two paths from P_0 to P in the simply connected space region D

We may assume, as suggested by Fig. 15.7.7, that C_1 and C_2 intersect only at their endpoints. Then the simple closed curve $C = C_1 \cup (-C_2)$ bounds an oriented surface S in D, and Stokes' theorem gives

$$\int_{C_1} \mathbf{F} \cdot \mathbf{T}\, ds - \int_{C_2} \mathbf{F} \cdot \mathbf{T}\, ds = \oint_C \mathbf{F} \cdot \mathbf{T}\, ds = \iint_S (\nabla \times \mathbf{F}) \cdot \mathbf{n}\, dS = 0$$

because of the hypothesis that $\nabla \times \mathbf{F} \equiv \mathbf{0}$. This shows that the line integral $\int_C \mathbf{F} \cdot \mathbf{T}\, ds$ is *independent of path,* just as desired. In Problem 21 we ask you to complete this proof by showing that the function ϕ of Eq. (11) is the one whose existence is claimed in Theorem 1. That is, $\mathbf{F} = \nabla\phi$. ■

EXAMPLE 3 Show that the vector field $\mathbf{F} = 3x^2\mathbf{i} + 5z^2\mathbf{j} + 10yz\mathbf{k}$ is irrotational. Then find a potential function $\phi(x, y, z)$ such that $\nabla\phi = \mathbf{F}$.

Solution To show that **F** is irrotational, we calculate

$$\nabla \times \mathbf{F} = \begin{vmatrix} \mathbf{i} & \mathbf{j} & \mathbf{k} \\ \dfrac{\partial}{\partial x} & \dfrac{\partial}{\partial y} & \dfrac{\partial}{\partial z} \\ 3x^2 & 5z^2 & 10yz \end{vmatrix} = (10z - 10z)\mathbf{i} = \mathbf{0}.$$

He ce Theorem 1 implies that $\mathbf{F}$ has a potential function ϕ. We can apply Eq. (11) to find ϕ explicitly. If C_1 is the straight line segment from $(0, 0, 0)$ to (u, v, w) that is parametrized by $x = ut$, $y = vt$, $z = wt$ for $0 \leqq t \leqq 1$, then Eq. (11) yields

$$\begin{aligned} \phi(u, v, w) &= \int_{C_1} \mathbf{F} \cdot \mathbf{T}\, ds = \int_{(0,0,0)}^{(u,v,w)} 3x^2\, dx + 5z^2\, dy + 10yz\, dz \\ &= \int_{t=0}^{1} (3u^2t^2)(u\, dt) + (5w^2t^2)(v\, dt) + (10vtwt)(w\, dt) \\ &= \int_{t=0}^{1} (3u^3t^2 + 15vw^2t^2)\, dt = \Big[u^3t^3 + 5vwt^3 \Big]_{t=0}^{1}, \end{aligned}$$

and thus $\phi(u, v, w) = u^3 + 5vw^2$. But because (u, v, w) is an arbitrary point of space, we have found that $\phi(x, y, z) = x^3 + 5yz^2$ is a scalar potential for $\mathbf{F}$. As a check, we note that $\phi_x = 3x^2$, $\phi_y = 5z^2$, and $\phi_z = 10yz$, so $\nabla\phi = \mathbf{F}$, as desired. ■

Application Suppose that $\mathbf{v}$ is the velocity field of a steady fluid flow that is both irrotational and incompressible—the density δ of the fluid is constant. Suppose that S is any closed surface that bounds a region T. Then, because of conservation of mass, the flux of $\mathbf{v}$ across S must be zero; the mass of fluid within S remains constant. Hence the divergence theorem gives

$$\iiint_T \operatorname{div} \mathbf{v}\, dV = \iint_S \mathbf{v} \cdot \mathbf{n}\, dS = 0.$$

Because this holds for *any* region T, it follows from the usual average value property argument that $\operatorname{div} \mathbf{v} = 0$ everywhere. The scalar function ϕ provided by Theorem 1, for which $\mathbf{v} = \nabla\phi$, is called the **velocity potential** of the fluid flow. We substitute $\mathbf{v} = \nabla\phi$ into the equation $\operatorname{div} \mathbf{v} = 0$ and thereby obtain

$$\operatorname{div}(\nabla\phi) = \frac{\partial^2\phi}{\partial x^2} + \frac{\partial^2\phi}{\partial y^2} + \frac{\partial^2\phi}{\partial z^2} = 0. \tag{12}$$

Thus the velocity potential ϕ of an irrotational and incompressible fluid flow satisfies *Laplace's equation.*

Laplace's equation appears in numerous other applications. For example, consider a heated body whose temperature function $u = u(x, y, z)$ is independent of time t. Then $\partial u/\partial t \equiv 0$ in the heat equation, Eq. (11) of Section 15.6, shows that the "steady-state temperature" function $u(x, y, z)$ satisfies Laplace's equation

$$\frac{\partial^2 u}{\partial x^2} + \frac{\partial^2 u}{\partial y^2} + \frac{\partial^2 u}{\partial z^2} = 0. \tag{13}$$

These brief remarks should indicate how the mathematics of this chapter forms the starting point for investigations in a number of areas, including acoustics, aerodynamics, electromagnetism, meteorology, and oceanography. Indeed, the entire subject of vector calculus stems historically from its mathematical applications rather than from abstract mathematical considerations. The modern form of the subject is due primarily to J. Willard Gibbs (1839–1903), the first great American physicist, and the English electrical engineer Oliver Heaviside (1850–1925).

15.7 PROBLEMS

In Problems 1 through 5, use Stokes' theorem for the evaluation of

$$\iint_S (\text{curl } \mathbf{F}) \cdot \mathbf{n} \, dS.$$

1. $\mathbf{F} = 3y\mathbf{i} - 2x\mathbf{j} + xyz\mathbf{k}$; S is the hemispherical surface $z = \sqrt{4 - x^2 - y^2}$ with upper unit normal vector.

2. $\mathbf{F} = 2y\mathbf{i} + 3x\mathbf{j} + e^z\mathbf{k}$; S is the part of the paraboloid $z = x^2 + y^2$ below the plane $z = 4$ and with upper unit normal vector.

3. $\mathbf{F} = \langle xy, -2, \arctan x^2 \rangle$; S is the part of the paraboloid $z = 9 - x^2 - y^2$ above the xy-plane and with upper unit normal vector.

4. $\mathbf{F} = yz\mathbf{i} + xz\mathbf{j} + xy\mathbf{k}$; S is the part of the cylinder $x^2 + y^2 = 1$ between the two planes $z = 1$ and $z = 3$ and with outer unit normal vector.

5. $\mathbf{F} = \langle yz, -xz, z^3 \rangle$; S is the part of the cone $z = \sqrt{x^2 + y^2}$ between the two planes $z = 1$ and $z = 3$ and with upper unit normal vector.

In Problems 6 through 10, use Stokes' theorem to evaluate

$$\oint_C \mathbf{F} \cdot \mathbf{T} \, ds.$$

6. $\mathbf{F} = 3y\mathbf{i} - 2x\mathbf{j} + 4x\mathbf{k}$; C is the circle $x^2 + y^2 = 9$, $z = 4$, oriented counterclockwise as viewed from above.

7. $\mathbf{F} = 2z\mathbf{i} + x\mathbf{j} + 3y\mathbf{k}$; C is the ellipse in which the plane $z = x$ meets the cylinder $x^2 + y^2 = 4$, oriented counterclockwise as viewed from above.

8. $\mathbf{F} = y\mathbf{i} + z\mathbf{j} + x\mathbf{k}$; C is the boundary of the triangle with vertices $(0, 0, 0)$, $(2, 0, 0)$, and $(0, 2, 2)$, oriented counterclockwise as viewed from above.

9. $\mathbf{F} = \langle y - x, x - z, x - y \rangle$; C is the boundary of the part of the plane $x + 2y + z = 2$ that lies in the first octant, oriented counterclockwise as viewed from above.

10. $\mathbf{F} = y^2\mathbf{i} + z^2\mathbf{j} + x^2\mathbf{k}$; C is the intersection of the plane $z = y$ and the cylinder $x^2 + y^2 = 2y$, oriented counterclockwise as viewed from above.

In Problems 11 through 14, first show that the given vector field $\mathbf{F}$ is irrotational; then apply the method of Example 3 to find a potential function $\phi = \phi(x, y, z)$ for $\mathbf{F}$.

11. $\mathbf{F} = (3y - 2z)\mathbf{i} + (3x + z)\mathbf{j} + (y - 2x)\mathbf{k}$

12. $\mathbf{F} = (3y^3 - 10xz^2)\mathbf{i} + 9xy^2\mathbf{j} - 10x^2z\mathbf{k}$

13. $\mathbf{F} = (3e^z - 5y \sin x)\mathbf{i} + (5 \cos x)\mathbf{j} + (17 + 3xe^z)\mathbf{k}$

14. $\mathbf{F} = r^3\mathbf{r}$, where $\mathbf{r} = x\mathbf{i} + y\mathbf{j} + z\mathbf{k}$ and $r = |\mathbf{r}|$

15. Suppose that $\mathbf{r} = x\mathbf{i} + y\mathbf{j} + z\mathbf{k}$ and that $\mathbf{a}$ is a constant vector. Show that

(a) $\nabla \cdot (\mathbf{a} \times \mathbf{r}) = 0$;

(b) $\nabla \times (\mathbf{a} \times \mathbf{r}) = 2\mathbf{a}$;

(c) $\nabla \cdot [(\mathbf{r} \cdot \mathbf{r})\mathbf{a}] = 2\mathbf{r} \cdot \mathbf{a}$;

(d) $\nabla \times [(\mathbf{r} \cdot \mathbf{r})\mathbf{a}] = 2(\mathbf{r} \times \mathbf{a})$.

16. Prove that

$$\iint_S (\text{curl } \mathbf{F}) \cdot \mathbf{n} \, dS$$

has the same value for all oriented surfaces S that have the same oriented boundary curve C.

17. Suppose that S is a closed surface. Prove in two different ways that

$$\iint_S (\text{curl } \mathbf{F}) \cdot \mathbf{n} \, dS = 0:$$

(a) by using the divergence theorem, with T the region bounded outside by S, and (b) by using Stokes' theorem, with the aid of a simple closed curve C on S.

Line integrals, surface integrals, and triple integrals of vector-valued functions are defined by componentwise integration. Such integrals appear in Problems 18 through 20.

18. Suppose that C and S are as described in the statement of Stokes' theorem and that is a scalar function. Prove that

$$\oint_C \phi\mathbf{T} \, ds = \iint_S \mathbf{n} \times \nabla\phi \, dS.$$

(*Suggestion:* Apply Stokes' theorem with where $\mathbf{a}$ is an arbitrary constant vector.)

19. Suppose that $\mathbf{a}$ and $\mathbf{r}$ are as in Problem 15. Prove that

$$\oint_C (\mathbf{a} \times \mathbf{r}) \cdot \mathbf{T} \, ds = 2\mathbf{a} \cdot \iint_S \mathbf{n} \, dS.$$

20. Suppose that S is a closed surface that bounds the region T. Prove that

$$\iint_S \mathbf{n} \times \mathbf{F} \, dS = \iiint_T \nabla \times \mathbf{F} \, dV.$$

(*Suggestion:* Apply the divergence theorem to where $\mathbf{a}$ is an arbitrary constant vector.)

REMARK The formulas of Problem 20, the divergence theorem, and Problem 20 of Section 15.6 all fit the pattern

$$\iint_S \mathbf{n} * (\ ) \, dS = \iiint_T \nabla * (\ ) \, dV,$$

where $*$ denotes either ordinary multiplication, the dot product, or the vector product, and either a scalar function or a vector-valued function is placed within the parentheses, as appropriate.

21. Suppose that the line integral $\int_C \mathbf{F}\cdot\mathbf{T}\,ds$ is independent of path. If

$$\phi(x, y, z) = \int_{P_0}^{P_1} \mathbf{F}\cdot\mathbf{T}\,ds$$

as in Eq. (11), show that $\nabla\phi = \mathbf{F}$. (*Suggestion:* If L is the line segment from (x, y, z) to $(x + \Delta x, y, z)$, then

$$\phi(x + \Delta x, y, z) - \phi(x, y, z) = \int_L \mathbf{F}\cdot\mathbf{T}\,ds = \int_x^{x+\Delta x} P\,dx.)$$

22. Let T be the submerged body of Problem 21 in Section 15.6, with centroid

$$\mathbf{r}_0 = \frac{1}{V}\iiint_T \mathbf{r}\,dV.$$

The torque about $\mathbf{r}_0$ of Archimedes' buoyant force $\mathbf{B} = -W\mathbf{k}$ is given by

$$\mathbf{L} = \iint_S (\mathbf{r} - \mathbf{r}_0) \times (-\delta g z\,\mathbf{n})\,dS.$$

(Why?) Apply the result of Problem 20 of this section to prove that $\mathbf{L} = \mathbf{0}$. It follows that $\mathbf{B}$ acts along the vertical line through the centroid $\mathbf{r}_0$ of the submerged body. (Why?)

CHAPTER 15 REVIEW: *Definitions, Concepts, Results*

Use the following list as a guide to concepts that you may need to review.

1. Definition and evaluation of the line integral
$$\int_C f(x, y, z)\,ds$$
2. Definition and evaluation of the line integral
$$\int_C P\,dx + Q\,dy + R\,dz$$
3. Relationship between the two types of line integrals; the line integral of the tangential component of a vector field
4. Line integrals and independence of path
5. Green's theorem
6. Flux and the vector form of Green's theorem
7. The divergence of a vector field
8. Definition and evaluation of the surface integral
$$\iint_S f(x, y, z)\,dS$$
9. Definition and evaluation of the surface integral
$$\iint_S P\,dy\,dz + Q\,dz\,dx + R\,dx\,dy$$
10. Relationship between the two types of surface integrals; the flux of a vector field across a surface
11. The divergence theorem in vector and in scalar notation
12. The curl of a vector field
13. Stokes' theorem in vector and in scalar notation
14. The circulation of a vector field around a simple closed curve
15. Physical interpretation of the divergence and the curl of a vector field

CHAPTER 15 *Miscellaneous Problems*

1. Evaluate the line integral

$$\int_C (x^2 + y^2)\,ds,$$

where C is the straight line segment from $(0, 0)$ to $(3, 4)$.

2. Evaluate the line integral

$$\int_C y^2\,dx + x^2\,dy,$$

where C is the part of the graph of $y = x^2$ from $(-1, 1)$ to $(1, 1)$.

3. Evaluate the line integral

$$\int_C \mathbf{F}\cdot\mathbf{T}\,ds,$$

where $\mathbf{F} = x\mathbf{i} + y\mathbf{j} + z\mathbf{k}$ and C is the curve $x = e^{2t}$, $y = e^t$, $z = e^{-t}$, $0 \leqq t \leqq \ln 2$.

4. Evaluate the line integral

$$\int_C xyz\,ds,$$

where C is the path from $(1, 1, 2)$ to $(2, 3, 6)$ consisting of three straight line segments, the first parallel to the x-axis, the second parallel to the y-axis, and the third parallel to the z-axis.

5. Evaluate the line integral

$$\int_C \sqrt{z}\,dx + \sqrt{x}\,dy + y^2\,dz,$$

where C is the curve $x = t$, $y = t^{3/2}$, $z = t^2$, $0 \leqq t \leqq 4$.

6. Apply Theorem 2 of Section 15.3 to show that the line integral

$$\int_C y^2\,dx + 2xy\,dy + z\,dz$$

is independent of the path C from A to B.

7. Apply Theorem 2 of Section 15.3 to show that the line integral

$$\int_C x^2y\,dx + xy^2\,dy$$

is not independent of the path C from $(0, 0)$ to $(1, 1)$.

8. A wire shaped like the circle $x^2 + y^2 = a^2$, $z = 0$ has constant density and total mass M. Find its moment of inertia around (a) the z-axis; (b) the x-axis.

9. A wire shaped like the parabola $y = \frac{1}{2}x^2$, $0 \leqq x \leqq 2$, has density function $\delta = x$. Find its mass and its moment of inertia around the y-axis.

10. Find the work done by the force field $\mathbf{F} = z\mathbf{i} - x\mathbf{j} + y\mathbf{k}$ in moving a particle from $(1, 1, 1)$ to $(2, 4, 8)$ along the curve $y = x^2$, $z = x^3$.

11. Apply Green's theorem to evaluate the line integral

$$\oint_C x^2y\,dx + xy^2\,dy,$$

where C is the boundary of the region between the two curves $y = x^2$ and $y = 8 - x^2$.

12. Evaluate the line integral

$$\oint_C x^2\,dy,$$

where C is the cardioid with polar equation $r = 1 + \cos\theta$, by first applying Green's theorem and then changing to polar coordinates.

13. Let C_1 be the circle $x^2 + y^2 = 1$ and C_2 the circle $(x - 1)^2 + y^2 = 9$. Show that if $\mathbf{F} = x^2y\mathbf{i} - xy^2\mathbf{j}$, then

$$\oint_{C_1} \mathbf{F}\cdot\mathbf{n}\,ds = \oint_{C_2} \mathbf{F}\cdot\mathbf{n}\,ds.$$

14. (a) Let C be the straight line segment from (x_1, y_1) to (x_2, y_2). Show that

$$\int_C -y\,dx + x\,dy = x_1y_2 - x_2y_1.$$

(b) Suppose that the vertices of a polygon are (x_1, y_1), (x_2, y_2), . . . , (x_n, y_n), named in counterclockwise order around the polygon. Apply the result in part (a) to show that the area of the polygon is

$$A = \frac{1}{2}\sum_{i=1}^{n} (x_iy_{i+1} - x_{i+1}y_i),$$

where $x_{n+1} = x_1$ and $y_{n+1} = y_1$.

15. Suppose that the line integral $\int_C P\,dx + Q\,dy$ is independent of the path in the plane region D. Prove that

$$\oint_C P\,dx + Q\,dy = 0$$

for every piecewise smooth simple closed curve C in D.

16. Use Green's theorem to prove that

$$\oint_C P\,dx + Q\,dy = 0$$

for every piecewise smooth simple closed curve C in the plane region D if and only if $\partial P/\partial y = \partial Q/\partial x$ at each point of D.

17. Evaluate the surface integral

$$\iint_S (x^2 + y^2 + 2z)\,dS,$$

where S is the part of the paraboloid $z = 2 - x^2 - y^2$ that lies above the xy-plane.

18. Suppose that $\mathbf{F} = (x^2 + y^2 + z^2)(x\mathbf{i} + y\mathbf{j} + z\mathbf{k})$ and that S is the spherical surface $x^2 + y^2 + z^2 = a^2$. Evaluate

$$\iint_S \mathbf{F}\cdot\mathbf{n}\,dS$$

without actually performing an antidifferentiation.

19. Let T be the solid bounded by the paraboloids

$$z = x^2 + 2y^2 \quad \text{and} \quad z = 12 - 2x^2 - y^2,$$

and suppose that $\mathbf{F} = x\mathbf{i} + y\mathbf{j} + z\mathbf{k}$. Find by evaluation of surface integrals the outward flux of $\mathbf{F}$ across the boundary of T.

20. Give a reasonable definition—in terms of a surface integral—of the average distance of the point P from points of the surface S. Then show that the average distance of a fixed point of a spherical surface of radius a from all points of the surface is $\frac{4}{3}a$.

21. Suppose that the surface S is the graph of the equation $x = g(y, z)$ for (y, z) in the region D of the yz-plane. Prove that

$$\iint_S P\,dy\,dz + Q\,dz\,dx + R\,dx\,dy$$

$$= \iint_D \left(P - Q\frac{\partial x}{\partial y} - R\frac{\partial x}{\partial z} \right) dy\, dz.$$

22. Suppose that the surface S is the graph of the equation $y = g(x, z)$ for (x, z) in the region D of the xz-plane. Prove that

$$\iint_S f(x, y, z)\, dS = \iint_D f(x, g(x, z), z) \sec\beta\, dx\, dz,$$

where $\sec\beta = \sqrt{1 + (\partial y/\partial x)^2 + (\partial y/\partial z)^2}$.

23. Let T be a region in space with volume V, boundary surface S, and centroid $(\bar{x}, \bar{y}, \bar{z})$. Use the divergence theorem to show that

$$\bar{z} = \frac{1}{2V} \iint_S z^2\, dx\, dy.$$

24. Apply the result of Problem 23 to find the centroid of the solid hemisphere

$$x^2 + y^2 + z^2 \leqq a^2, \qquad z \geqq 0.$$

Problems 25 through 27 outline the derivation of the heat equation for a body with temperature $u = u(x, y, z, t)$ at the point (x, y, z) at time t. Denote by K its heat conductivity and by c its heat capacity, both assumed constant, and let $k = K/c$. Let B be a small solid ball within the body, and let S denote the boundary sphere of B.

25. Deduce from the divergence theorem and Eq. (23) of Section 15.5 that the rate of heat flow across S into B is

$$R = \iiint_B k\nabla^2 u\, dV.$$

26. The meaning of heat capacity is that, if Δu is small, then $(c\,\Delta u)\,\Delta V$ calories of heat are required to raise the temperature of the volume ΔV by Δu degrees. It follows that the rate at which the volume ΔV is absorbing heat is $c(\partial u/\partial t)\,\Delta V$. (Why?) Conclude that the rate of heat flow into B is

$$R = \iiint_B c\frac{\partial u}{\partial t}\, dV.$$

27. Equate the results of Problem 25 and 26, apply the average value property of triple integrals, and then take the limit as the radius of the ball B approaches zero. You should thereby obtain the heat equation

$$\frac{\partial u}{\partial t} = k\nabla^2 u.$$

28. For a *steady-state* temperature function (one that is independent of time t), the heat equation reduces to Laplace's equation,

$$\nabla^2 u = \frac{\partial^2 u}{\partial x^2} + \frac{\partial^2 u}{\partial y^2} + \frac{\partial^2 u}{\partial z^2} = 0.$$

(a) Suppose that u_1 and u_2 are two solutions of Laplace's equation in the region T and that u_1 and u_2 agree on its boundary surface S. Apply Problem 17 of Section 15.6 to the function $f = u_1 - u_2$ to conclude that $\nabla f = \mathbf{0}$ at each point of T. (b) From the facts that $\nabla f = \mathbf{0}$ in T and $f \equiv 0$ on S, conclude that $f \equiv 0$, so $u_1 \equiv u_2$. Thus the steady-state temperatures within a region are *determined* by the boundary-value temperatures.

29. Suppose that $\mathbf{r} = x\mathbf{i} + y\mathbf{j} + z\mathbf{k}$ and that $\phi(r)$ is a scalar function of $r = |\mathbf{r}|$. Compute:

(a) $\nabla\phi(r)$; (b) div $[\phi(r)\mathbf{r}]$; (c) curl $[\phi(r)\mathbf{r}]$.

30. Let S be the upper half of the torus obtained by revolving around the z-axis the circle $(y - a)^2 + z^2 = b^2$ in the yz-plane, with upper unit normal vector. Describe how to subdivide S to establish Stokes' theorem for it. How are the two boundary circles oriented?

31. Explain why the method of subdivision is not sufficient to establish Stokes' theorem for the Möbius strip of Fig. 15.5.3.

32. (a) Suppose that a fluid or a rigid body is rotating with angular speed ω radians per second around the line through the origin determined by the unit vector $\mathbf{u}$. Show that the velocity of the point with position vector $\mathbf{r}$ is $\mathbf{v} = \boldsymbol{\omega} \times \mathbf{r}$, where $\boldsymbol{\omega} = \omega\mathbf{u}$ is the angular velocity vector. Note that $|\mathbf{v}| = \omega\,|\mathbf{r}|\sin\theta$, where θ is the angle between $\mathbf{r}$ and $\boldsymbol{\omega}$. (b) Use the fact that $\mathbf{v} = \boldsymbol{\omega} \times \mathbf{r}$, established in part(a), to show that curl $\mathbf{v} = 2\boldsymbol{\omega}$.

33. Consider an incompressible fluid flowing in space (no sources or sinks) with variable density $\delta(x, y, z, t)$ and velocity field $\mathbf{v}(x, y, z, t)$. Let B be a small ball with radius r and spherical surface S centered at the point (x_0, y_0, z_0). Then the amount of fluid within S at time t is

$$Q(t) = \iiint_B \delta\, dV,$$

and differentiation under the integral sign yields

$$Q'(t) = \iiint_B \frac{\partial\delta}{\partial t}\, dV.$$

(a) Consider fluid flow across S to get

$$Q'(t) = -\iint_S \delta\mathbf{v}\cdot\mathbf{n}\, dS,$$

where $\mathbf{n}$ is the outer unit normal vector to S. Now apply the divergence theorem to convert this into a volume integral. (b) Equate your two volume integrals for $Q'(t)$, apply the mean value theorem for integrals, and finally take limits as $r \to 0$ to obtain the **continuity equation**

$$\frac{\partial\delta}{\partial t} + \nabla\cdot(\delta\mathbf{v}) = 0.$$

APPENDICES

APPENDIX A: REAL NUMBERS AND INEQUALITIES

The **real numbers** are already familiar to you. They are just those numbers ordinarily used in most measurements. The mass, velocity, temperature, and charge of a body are measured with real numbers. Real numbers can be represented by **terminating** or **nonterminating** decimal expansions; in fact, every real number has a nonterminating decimal expansion because a terminating expansion can be padded with infinitely many zeros:

$$\tfrac{3}{8} = 0.375 = 0.375000000\ldots.$$

Any **repeating** decimal, such as

$$\tfrac{7}{22} = 0.31818181818\ldots$$

represents a **rational** number, one that is the ratio of two integers. Conversely, every rational number is represented by a repeating decimal like the two displayed above. But the decimal expansion of an **irrational** number (a real number that is not rational), such as

$$\sqrt{2} = 1.414213562\ldots \quad \text{or} \quad \pi = 3.14159265358979\ldots$$

is both nonterminating and nonrepeating.

The geometric interpretation of real numbers as points on the **real line** (or *real number line*) $\boldsymbol{R}$ should also be familiar to you. Each real number is represented by precisely one point of $\boldsymbol{R}$, and each point of $\boldsymbol{R}$ represents precisely one real number. By convention, the positive numbers lie to the right of zero and the negative numbers to the left, as in Fig. A.1.

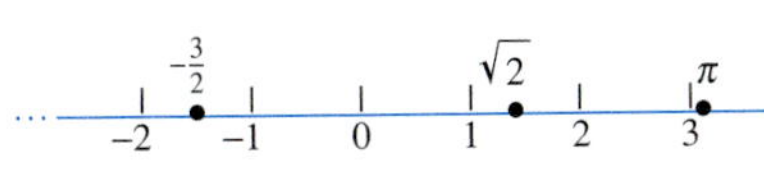

Fig. A.1 The real line $\boldsymbol{R}$

The following properties of inequalities of real numbers are fundamental and often used:

$$
\begin{aligned}
&\text{If } a < b \text{ and } b < c, \text{ then } a < c.\\
&\text{If } a < b, \text{ then } a + c < b + c.\\
&\text{If } a < b \text{ and } c > 0, \text{ then } ac < bc.\\
&\text{If } a < b \text{ and } c < 0, \text{ then } ac > bc.
\end{aligned}
\qquad \textbf{(1)}
$$

The last two statements mean that an inequality is preserved when its members are multiplied by a *positive* number but is *reversed* when they are multiplied by a *negative* number.

Absolute Value

The (nonnegative) distance along the real line between zero and the real number a is the **absolute value** of a, written $|a|$. Equivalently,

$$|a| = \begin{cases} a & \text{if } a \geqq 0, \\ -a & \text{if } a < 0. \end{cases} \tag{2}$$

The notation $a \geqq 0$ means that a is *either* greater than zero *or* equal to zero. Equation (2) implies that $|a| \geqq 0$ for every real number a and that $|a| = 0$ if and only if $a = 0$.

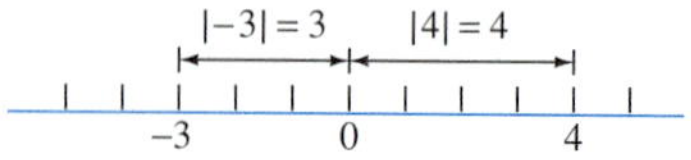

Fig. A.2 The absolute value of a real number is simply its distance from zero (Example 1).

EXAMPLE 1 As Fig. A.2 shows,

$$|4| = 4 \quad \text{and} \quad |-3| = 3.$$

Moreover, $|0| = 0$ and $|\sqrt{2} - 2| = 2 - \sqrt{2}$, the latter being true because $2 > \sqrt{2}$. Thus $\sqrt{2} - 2 < 0$, and hence

$$|\sqrt{2} - 2| = -(\sqrt{2} - 2) = 2 - \sqrt{2}.$$

The following properties of absolute values are frequently used:

$$\begin{aligned} |a| &= |-a| = \sqrt{a^2} \geqq 0, \\ |ab| &= |a|\,|b|, \\ -|a| &\leqq a \leqq |a|, \quad \text{and} \\ |a| &< b \quad \text{if and only if} \quad -b < a < b. \end{aligned} \tag{3}$$

The **distance** between the real numbers a and b is defined to be $|a - b|$ (or $|b - a|$; there's no difference). This distance is simply the length of the line segment of the real line $\boldsymbol{R}$ with endpoints a and b (Fig. A.3).

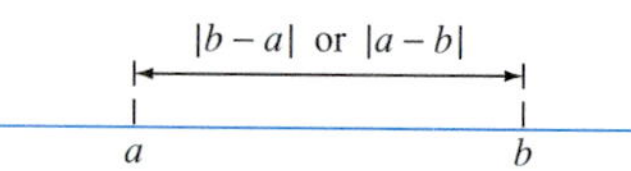

Fig. A.3 The distance between a and b

The properties of inequalities and of absolute values in Eqs. (1) through (3) imply the following important theorem.

Theorem 1 ***Triangle Inequality***

For all real numbers a and b,

$$|a + b| \leqq |a| + |b|. \tag{4}$$

PROOF There are several cases to consider, depending on whether the two numbers a and b are positive or negative and which has the larger absolute value. If both are positive, then so is $a + b$; in this case,

$$|a + b| = a + b = |a| + |b|. \tag{5}$$

If $a > 0$ but $b < 0$ and $|b| < |a|$, then

$$0 < a + b < a,$$

so

$$|a + b| = a + b < a = |a| < |a| + |b|, \tag{6}$$

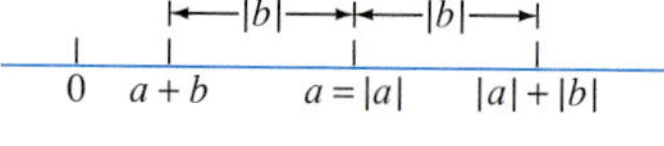

Fig. A.4 The triangle inequality with $a > 0$, $b < 0$, and $|b| < |a|$

as illustrated in Fig. A.4. The other cases are similar. In particular, we see that the triangle inequality is actually an equality [as in Eq. (5)] unless a and b have different signs, in which case it is a strict inequality [as in Eq. (6)]. ■

Intervals

Suppose that S is a set (collection) of real numbers. It is common to describe S by the notation

$$S = \{x : \text{condition}\},$$

where the "condition" is true for those numbers x in S and false for those numbers x not in S. The most important sets of real numbers in calculus are **intervals**. If $a < b$, then the **open interval** (a, b) is defined to be the set

$$(a, b) = \{x : a < x < b\}$$

of real numbers, and the **closed interval** $[a, b]$ is

$$[a, b] = \{x : a \leqq x \leqq b\}.$$

Thus a closed interval contains its endpoints, whereas an open interval does not. We also use the **half-open intervals**

$$[a, b) = \{x : a \leqq x < b\} \quad \text{and} \quad (a, b] = \{x : a < x \leqq b\}.$$

Thus the open interval $(1, 3)$ is the set of those real numbers x such that $1 < x < 3$, the closed interval $[-1, 2]$ is the set of those real numbers x such that $-1 \leqq x \leqq 2$, and the half-open interval $(-1, 2]$ is the set of those real numbers x such that $-1 < x \leqq 2$. In Fig. A.5 we show examples of such intervals as well as some **unbounded** intervals, which have forms such as

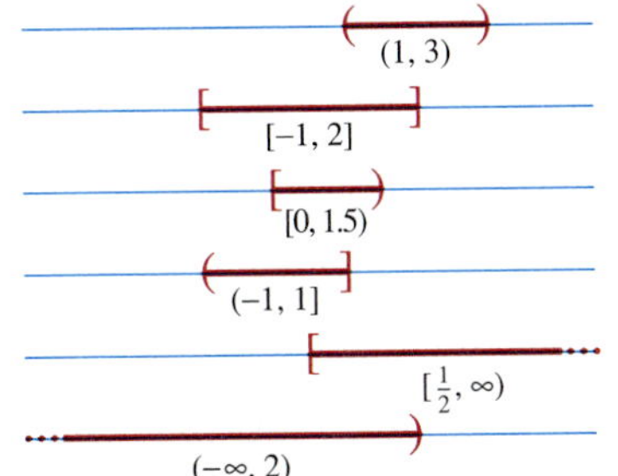

Fig. A.5 Some examples of intervals of real numbers

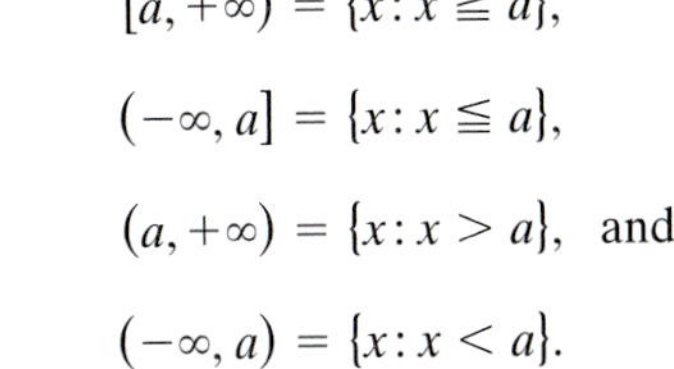

$$[a, +\infty) = \{x : x \geqq a\},$$

$$(-\infty, a] = \{x : x \leqq a\},$$

$$(a, +\infty) = \{x : x > a\}, \quad \text{and}$$

$$(-\infty, a) = \{x : x < a\}.$$

The symbols $+\infty$ and $-\infty$, denoting "plus infinity" and "minus infinity," are merely notational conveniences and do *not* represent real numbers—the real line $\boldsymbol{R}$ does *not* have "endpoints at infinity." The use of these symbols is motivated by the brief and natural descriptions $[\pi, +\infty)$ and $(-\infty, 2)$ for the sets

$$\{x : x \geqq \pi\} \quad \text{and} \quad \{x : x < 2\}$$

of all real numbers x such that $x \geqq \pi$ and $x < 2$, respectively.

Inequalities

The set of solutions of an inequality involving a variable x is often an interval or a union of intervals, as in the next examples. The **solution set** of such an inequality is simply the set of all those real numbers x that satisfy the inequality.

EXAMPLE 2 Solve the inequality $2x - 1 < 4x + 5$.

Solution Using the properties of inequalities listed in (1), we proceed much as if we were solving an equation for x: We isolate x on one side of the inequality. Here we begin with

$$2x - 1 < 4x + 5$$

and it follows that

$$-1 < 2x + 5;$$

$$-6 < 2x;$$

$$-3 < x.$$

Hence the solution set is the unbounded interval $(-3, \infty)$. ■

EXAMPLE 3 Solve the inequality $-13 < 1 - 4x \leqq 7$.

Solution We simplify the given inequality as follows:

$$-13 < 1 - 4x \leqq 7;$$

$$-7 \leqq 4x - 1 < 13;$$

$$-6 \leqq 4x < 14;$$

$$-\tfrac{3}{2} \leqq x < \tfrac{7}{2}.$$

Thus the solution set of the given inequality is the half-open interval $[-\frac{3}{2}, \frac{7}{2})$. ■

EXAMPLE 4 Solve the inequality $|3 - 5x| < 2$.

Solution From the fourth property of absolute values in (3), we see that the given inequality is equivalent to

$$-2 < 3 - 5x < 2.$$

We now simplify as in the previous two examples:

$$-5 < -5x < -1;$$

$$\tfrac{1}{5} < x < 1.$$

Thus the solution set is the open interval $\left(\frac{1}{5}, 1\right)$. ■

EXAMPLE 5 Solve the inequality

$$\frac{5}{|2x - 3|} < 1.$$

Solution It is usually best to begin by eliminating a denominator containing the unknown. Here we multiply each term by the *positive* quantity $|2x - 3|$ to obtain the equivalent inequality

$$|2x - 3| > 5.$$

It follows from the last property in (3) that this is so if and only if either

$$2x - 3 < -5 \quad \text{or} \quad 2x - 3 > 5.$$

The solutions of these *two* inequalities are the open intervals $(-\infty, -1)$ and $(4, \infty)$, respectively. Hence the solution set of the original inequality consists of all those numbers x that lie in *either* of these two open intervals. ■

The **union** of the two sets S and T is the set $S \cup T$ given by

$$S \cup T = \{x : \text{either } \; x \in S \; \text{ or } \; x \in T \; \text{ or both}\}.$$

Thus the solution set in Example 5 can be written in the form $(-\infty, -1) \cup (4, \infty)$.

EXAMPLE 6 In accord with Boyle's law, the pressure p (in pounds per square inch) and volume V (in cubic inches) of a certain gas satisfy the condition $pV = 100$. Suppose that $50 \leqq V \leqq 150$. What is the range of possible values of the pressure p?

Solution If we substitute $V = 100/p$ in the given inequality $50 \leqq V \leqq 150$, we get

$$50 \leqq \frac{100}{p} \leqq 150.$$

It follows that *both*

$$50 \leqq \frac{100}{p} \quad \text{and} \quad \frac{100}{p} \leqq 150;$$

that is, that both

$$p \leqq 2 \quad \text{and} \quad p \geqq \tfrac{2}{3}.$$

Thus the pressure p must lie in the closed interval $[\frac{2}{3}, 2]$. ■

The **intersection** of the two sets S and T is the set $S \cap T$ defined as follows:

$$S \cap T = \{x : \text{both } \; x \in S \; \text{ and } \; x \in T\}.$$

Thus the solution set in Example 6 is the set $(-\infty, 2] \cap [\frac{2}{3}, \infty) = [\frac{2}{3}, 2]$.

APPENDIX A PROBLEMS

Simplify the expressions in Problems 1 through 12 by writing each without using absolute value symbols.

1. $|3 - 17|$

2. $|-3| + |17|$

3. $|-0.25 - \frac{1}{4}|$

4. $|5| - |-7|$

5. $|(-5)(4 - 9)|$

6. $\dfrac{|-6|}{|4| + |-2|}$

7. $|(-3)^3|$

8. $|3 - \sqrt{3}|$

9. $|\pi - \frac{22}{7}|$

10. $-|7 - 4|$

11. $|x - 3|$, given $x < 3$

12. $|x - 5| + |x - 10|$, given $|x - 7| < 1$

Solve the inequalities in Problems 13 through 31. Write each solution set in interval notation.

13. $2x - 7 < -3$

14. $1 - 4x > 2$

15. $3x - 4 \geqq 17$

16. $2x + 5 \leqq 9$

17. $2 - 3x < 7$

18. $6 - 5x > -9$

19. $-3 < 2x + 5 < 7$

20. $4 \leqq 3x - 5 \leqq 10$

21. $-6 \leqq 5 - 2x < 2$

22. $3 < 1 - 5x < 7$

23. $|3 - 2x| < 5$

24. $|5x + 3| \leqq 4$

25. $|1 - 3x| > 2$

26. $1 < |7x - 1| < 3$

27. $2 \leqq |4 - 5x| \leqq 4$

28. $\dfrac{1}{2x + 1} > 3$

29. $\dfrac{2}{7 - 3x} \leqq -5$

30. $\dfrac{2}{|3x - 4|} < 1$

31. $\dfrac{1}{|1 - 5x|} \geqq -\dfrac{1}{3}$

32. Solve the inequality $x^2 - x - 6 > 0$. (*Suggestion:* Conclude from the factorization $x^2 - x - 6 = (x - 3)(x + 2)$ that the quantities $x - 3$ and $x + 2$ are either both positive or both negative. Consider the two cases separately to deduce that the solution set is $(-\infty, -2) \cup (3, \infty)$.)

Use the method of Problem 32 to solve the inequalities in Problems 33 through 36.

33. $x^2 - 2x - 8 > 0$

34. $x^2 - 3x + 2 < 0$

35. $4x^2 - 8x + 3 \geqq 0$

36. $2x \geqq 15 - x^2$

37. In accord with Boyle's law, the pressure p in pounds (per square inch) and volume V (in cubic inches) of a certain gas satisfy the condition $pV = 800$. What is the range of possible values of the pressure, given $100 \leqq V \leqq 200$?

38. The relationship between the Fahrenheit temperature F and the Celsius temperature C is given by $F = 32 + \frac{9}{5}C$. If the temperature on a certain day ranged from a low of 70° F to a high of 90° F, what was the range of the temperature in degrees Celsius?

39. An electrical circuit contains a battery supplying E volts in series with a resistance of R ohms, as shown in Fig. A.6. Then the current of I amperes that flows in the circuit satisfies Ohm's law, $E = IR$. If $E = 100$ and $25 < R < 50$, what is the range of possible values of I?

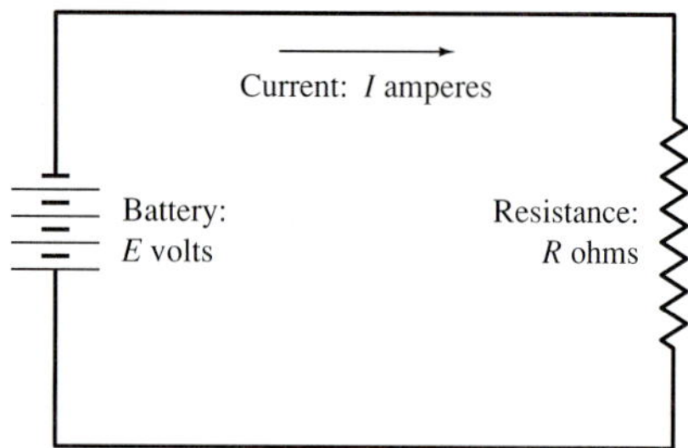

Fig. A.6 A simple electric circuit

40. The period T (in seconds) of a simple pendulum of length L (in feet) is given by $T = 2\pi\sqrt{L/32}$. If $3 < L < 4$, when is the range of possible values of T?

41. Use the properties of inequalities in (1) to show that the sum of two positive numbers is positive.

42. Use the properties of inequalities in (1) to show that the product of two positive numbers is positive.

43. Prove that the product of two negative numbers is positive and that the product of a positive number and a negative number is negative.

44. Suppose that $a < b$ and that a and b are either both positive or both negative. Prove that $1/a > 1/b$.

45. Apply the triangle inequality twice to show that

$$|a + b + c| \leqq |a| + |b| + |c|$$

for arbitrary real numbers a, b, and c.

46. Write $a = (a - b) + b$ to deduce from the triangle inequality that

$$|a| - |b| \leqq |a - b|$$

for arbitrary real numbers a and b.

47. Deduce from the definition in (2) that $|a| < b$ if and only if $-b < a < b$.

APPENDIX B: THE COORDINATE PLANE AND STRAIGHT LINES

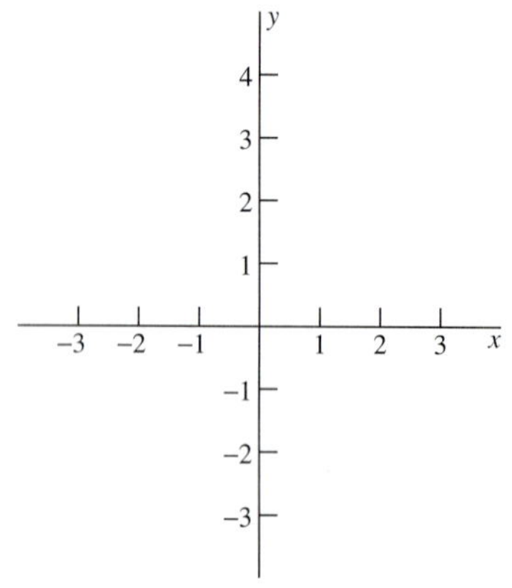

Fig. B.1 The coordinate plane

Imagine the flat, featureless, two-dimensional plane of Euclid's geometry. Install a copy of the real number line $\boldsymbol{R}$, with the line horizontal and the positive numbers to the right. Add another copy of $\boldsymbol{R}$ perpendicular to the first, with the two lines crossing where the number zero is located on each. The vertical line should have the positive numbers above the horizontal line, as in Fig. B.1; the negative numbers thus will be below it. The horizontal line is called the x-**axis** and the vertical line is called the y-**axis**.

With these added features, we call the plane the **coordinate plane,** because it's now possible to locate any point there by a pair of numbers, called the *coordinates of the point*. Here's how: If P is a point in the plane, draw perpendiculars from P to the coordinate axes, as shown in Fig. B.2. One perpendicular meets the x-axis at the x-**coordinate** (or **abscissa**) of P, labeled x_1 in Fig. B.2. The other meets the y-axis in the y-**coordinate** (or **ordinate**) y_1 of P. The pair of numbers (x_1, y_1), in that order, is called the **coordinate pair** for P, or simply the **coordinates** of P. To be concise, we speak of "the point $P(x_1, y_1)$."

This coordinate system is called the **rectangular coordinate system,** or the **Cartesian coordinate system** (because its use was popularized, beginning in the 1630s, by the French mathematician and philosopher René Descartes [1596–1650]). The plane, thus coordinatized, is denoted by $\boldsymbol{R}^2$ because two copies of $\boldsymbol{R}$ are used; it is known also as the **Cartesian plane.**

Rectangular coordinates are easy to use, because $P(x_1, y_1)$ and $Q(x_2, y_2)$ denote the same point if and only if $x_1 = x_2$ and $y_1 = y_2$. Thus when you know that P and Q are two different points, you may conclude that P and Q have different abscissas, different ordinates, or both.

The point of symmetry $(0, 0)$ where the coordinate axes meet is called the **origin.** All points on the x-axis have coordinates of the form $(x, 0)$. Although the *real number* x is not the same as the geometric point $(x, 0)$, there are situations in which it is useful to think of the two as the same. Similar remarks apply to points $(0, y)$ on the y-axis.

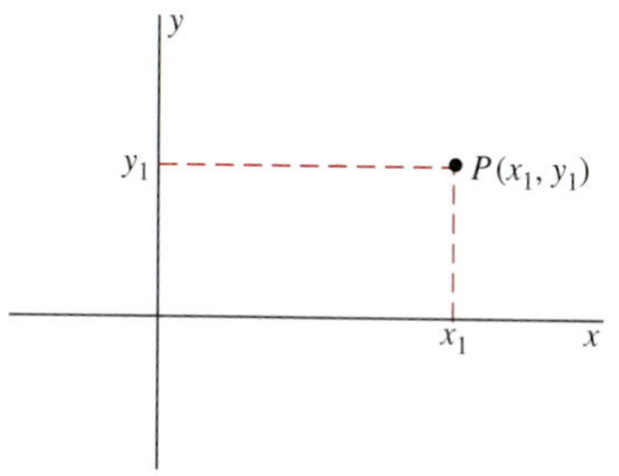

Fig. B.2 The point P has rectangular coordinates (x_1, y_1).

The concept of distance in the coordinate plane is based on the **Pythagorean theorem**: If ABC is a right triangle with its right angle at the point C, with hypotenuse of length c and the other two sides of lengths a and b (as in Fig. B.3), then

$$c^2 = a^2 + b^2. \tag{1}$$

The converse of the Pythagorean theorem is also true: If the three sides of a given triangle satisfy the Pythagorean relation in Eq. (1), then the angle opposite side c must be a right angle.

The *distance* $d(P_1, P_2)$ between the points P_1 and P_2 is, by definition, the length of the straight-line segment joining P_1 and P_2. The following formula gives $d(P_1, P_2)$ in terms of the coordinates of the two points.

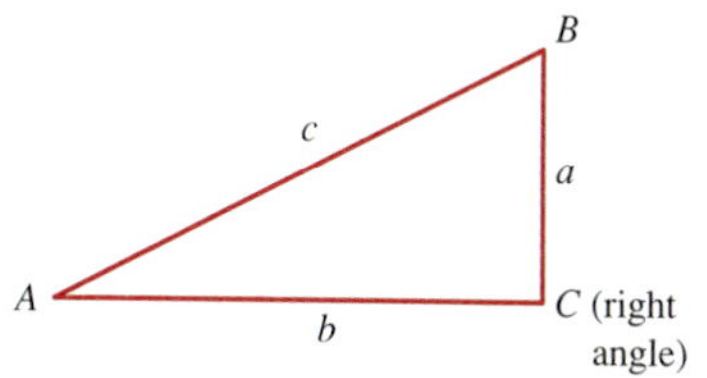

Fig. B.3 The Pythagorean theorem

Distance Formula

The **distance** between the two points $P_1(x_1, y_1)$ and $P_2(x_2, y_2)$ is

$$d(P_1, P_2) = \sqrt{(x_2 - x_1)^2 + (y_2 - y_1)^2}. \tag{2}$$

PROOF If $x_1 \neq x_2$ and $y_1 \neq y_2$, then Eq. (2) follows from the Pythagorean theorem. Use the right triangle with vertices P_1, P_2, and $P_3(x_2, y_1)$ shown in Fig. B.4.

If $x_1 = x_2$, then P_1 and P_2 lie in a vertical line. In this case

$$d(P_1, P_2) = |y_1 - y_2| = \sqrt{(y_1 - y_2)^2}.$$

This agrees with Eq. (2) because $x_1 = x_2$. The remaining case $(y_1 = y_2)$ is similar. ■

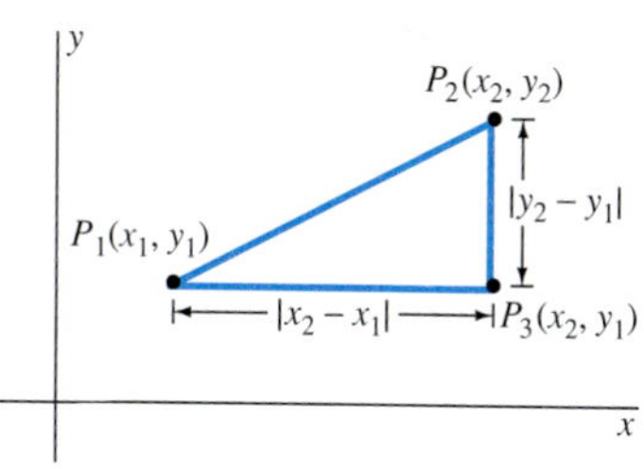

Fig. B.4 Use this triangle to deduce the distance formula.

EXAMPLE 1 Show that the triangle PQR with vertices $P(1, 0)$, $Q(5, 4)$, and $R(-2, 3)$ is a right triangle (Fig. B.5).

Solution The distance formula gives

$$a^2 = [d(P, Q)]^2 = (5 - 1)^2 + (4 - 0)^2 = 32,$$

$$b^2 = [d(P, R)]^2 = (-2 - 1)^2 + (3 - 0)^2 = 18, \quad \text{and}$$

$$c^2 = [d(Q, R)]^2 = (-2 - 5)^2 + (3 - 4)^2 = 50.$$

Because $a^2 + b^2 = c^2$, the *converse* of the Pythagorean theorem implies that RPQ is a right angle. (The right angle is at P because P is the vertex opposite the longest side, QR.) ■

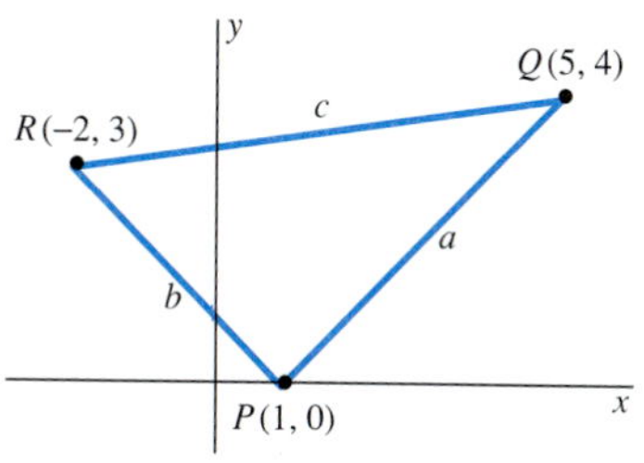

Fig. B.5 Is this a right triangle (Example 1)?

Another application of the distance formula is an expression for the coordinates of the midpoint M of the line segment P_1P_2 with endpoint P_1 and P_2 (Fig. B.6). Recall from geometry that M is the one (and only) point of the line segment P_1P_2 that is equally distant from P_1 and P_2. The following formula tells us that the coordinates of M are the *averages* of the corresponding coordinates of P_1 and P_2.

Midpoint Formula

The **midpoint** of the line segment with endpoints $P_1(x_1, y_1)$ and $P_2(x_2, y_2)$ is the point $M(\bar{x}, \bar{y})$ with coordinates

$$\bar{x} = \tfrac{1}{2}(x_1 + x_2) \quad \text{and} \quad \bar{y} = \tfrac{1}{2}(y_1 + y_2). \tag{3}$$

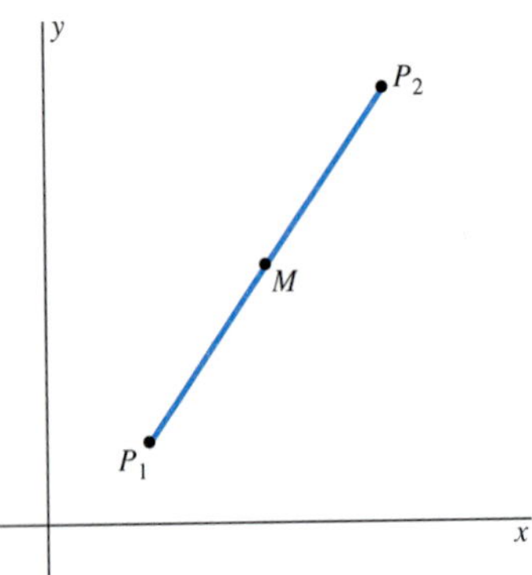

Fig. B.6 The midpoint M

PROOF If you substitute the coordinates of P_1, M, and P_2 in the distance formula, you find that $d(P_1, M) = d(P_2, M)$. All that remains is to show that M lies on the line segment P_1P_2. We ask you to do this, and thus complete the proof, in Problem 31. ■

Straight Lines and Slope

We want to define the *slope* of a straight line, a measure of its rate of rise or fall from left to right. Given a nonvertical straight line L in the coordinate plane, choose two points $P_1(x_1, y_1)$ and $P_2(x_2, y_2)$ on L. Consider the **increments** Δx and Δy (read "delta x" and "delta y") in the x- and y-coordinates from P_1 to P_2. These are defined as follows:

$$\Delta x = x_2 - x_1 \quad \text{and} \quad \Delta y = y_2 - y_1. \tag{4}$$

Engineers (and others) call Δx the **run** from P_1 to P_2 and Δy the **rise** from P_1 to P_2, as in Fig. B.7. The **slope** m of the nonvertical line L is then defined to be the ratio of the rise to the run:

$$m = \frac{\Delta y}{\Delta x} = \frac{y_2 - y_1}{x_2 - x_1}. \tag{5}$$

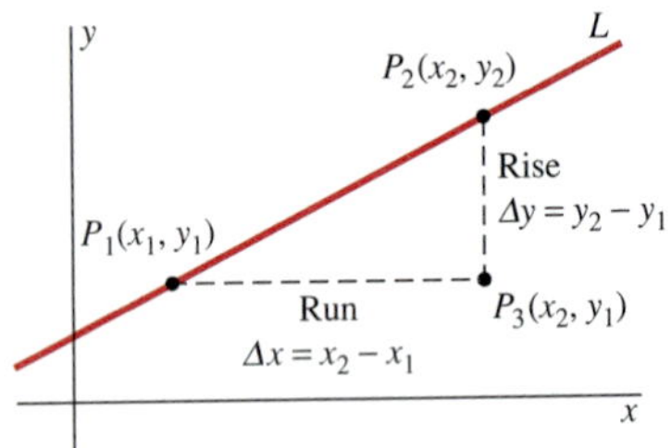

Fig. B.7 The slope of a straight line

This is also the definition of a line's slope in civil engineering (and elsewhere). In a surveying text you are likely to find the memory aid

$$\text{“slope} = \frac{\text{rise}}{\text{run}}.\text{”}$$

Recall that corresponding sides of similar (that is, equal-angled) triangles have equal ratios. Hence, if $P_3(x_3, y_3)$ and $P_4(x_4, y_4)$ are two other points of L, then the similarity of the triangles in Fig. B.8 implies that

$$\frac{y_4 - y_3}{x_4 - x_3} = \frac{y_2 - y_1}{x_2 - x_1}.$$

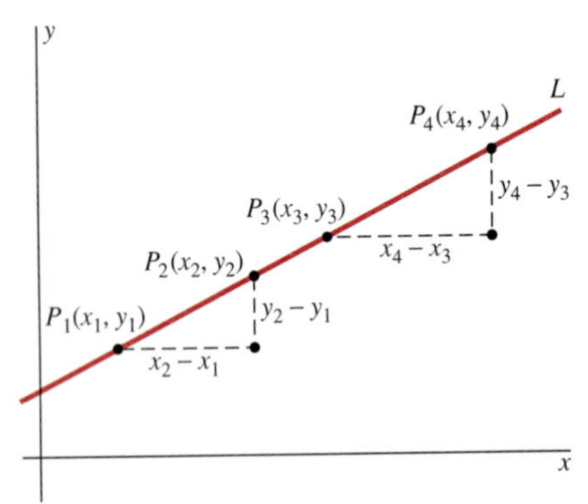

Fig. B.8 The result of the slope computation does not depend on which two points of L are used.

Therefore, the slope m as defined in Eq. (5) does *not* depend on the particular choice of P_1 and P_2.

If the line L is horizontal, then $\Delta y = 0$. In this case Eq. (5) gives $m = 0$. If L is vertical, then $\Delta x = 0$, so the slope of L is *not defined*. Thus we have the following statements:

- Horizontal lines have slope zero.
- Vertical lines have no defined slope.

EXAMPLE 2 (a) The slope of the line through the points $(3, -2)$ and $(-1, 4)$ is

$$m = \frac{4 - (-2)}{(-1) - 3} = \frac{6}{-4} = -\frac{3}{2}.$$

(b) The points $(3, -2)$ and $(7, -2)$ have the same y-coordinate. Therefore, the line through them is horizontal and thus has slope zero.
(c) The points $(3, -2)$ and $(3, 4)$ have the same x-coordinate. Thus the line through them is vertical, and so its slope is undefined. ■

Equations of Straight Lines

Our immediate goal is to be able to write equations of given straight lines. That is, if L is a straight line in the coordinate plane, we wish to construct a mathematical sentence—an equation—about points (x, y) in the plane. We want this equation to be *true* when (x, y) is a point on L and *false* when (x, y) is not a point on L. Clearly this equation will involve x and y and some numerical constants determined by L itself. If we are to write this equation, the concept of the slope of L is essential.

Suppose, then, that $P(x_0, y_0)$ is a fixed point on the nonvertical line L of slope m. Let $P(x, y)$ be any *other* point on L. We apply Eq. (5) with P and P_0 in place of P_1 and P_2 to find that

$$m = \frac{y - y_0}{x - x_0};$$

that is,

$$y - y_0 = m(x - x_0). \tag{6}$$

Because the point (x_0, y_0) satisfies Eq. (6), as does every other point of L, and because no other points of the plane can do so, Eq. (6) is indeed an equation for the given line L. In summary, we have the following result.

The Point-Slope Equation

The point $P(x, y)$ lies on the line with slope m through the fixed point (x_0, y_0) if and only if its coordinates satisfy the equation

$$y - y_0 = m(x - x_0). \tag{6}$$

Equation (6) is called the **point-slope** equation of L, partly because the coordinates of the point (x_0, y_0) and the slope m of L may be read directly from this equation.

EXAMPLE 3 Write an equation for the straight line L through the points $P_1(1, -1)$ and $P_2(3, 5)$.

Solution The slope m of L may be obtained from the two given points:

$$m = \frac{5 - (-1)}{3 - 1} = 3.$$

Either P_1 or P_2 will do for the fixed point. We use $P_1(1, -1)$. Then, with the aid of Eq. (6), the point-slope equation of L is

$$y + 1 = 3(x - 1).$$

If simplification is appropriate, we may write $3x - y = 4$ or $y = 3x - 4$. ■

Equation (6) can be written in the form

$$y = mx + b \tag{7}$$

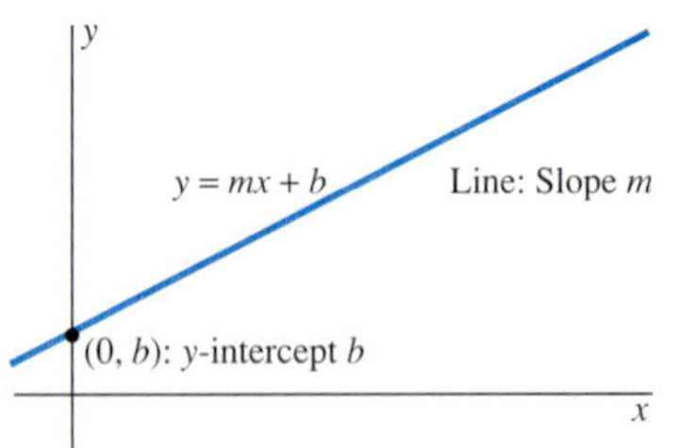

Fig. B.9 The straight line with equation $y = mx + b$ has a slope m and y-intercept b.

where $b = y_0 - mx_0$ is a constant. Because $y = b$ when $x = 0$, the y-**intercept** of L is the point $(0, b)$ shown in Fig. B.9. Equations (6) and (7) are different forms of the equation of a straight line.

The Slope-Intercept Equation

The point $P(x, y)$ lies on the line with slope m and y-intercept b if and only if the coordinates of P satisfy the equation

$$y = mx + b. \tag{7}$$

Perhaps you noticed that both Eq. (6) and Eq. (7) can be written in the form of the general linear equation

$$Ax + By = C, \tag{8}$$

where A, B, and C are constants. Conversely, if $B \neq 0$, then Eq. (8) can be written in the form of Eq. (7) if we divide each term by B. Therefore, Eq. (8) represents a straight line with its slope being the coefficient of x *after* solution of the equation for y. If $B = 0$, then Eq. (8) reduces to the equation of a vertical line: $x = K$ (where K is a constant). If $A = 0$ and $B \neq 0$, then Eq. (8) reduces to the equation of a horizontal line: $y = H$ (where H is a constant). Thus we see that Eq. (8) is always an equation of a straight line unless $A = B = 0$. Conversely, every straight line in the coordinate plane—even a vertical one—has an equation of the form in (8).

Parallel Lines and Perpendicular Lines

If the line L is not horizontal, then it must cross the x-axis. Then its **angle of inclination** is the angle ϕ measured counterclockwise from the positive x-axis to L. It follows that $0° < \phi < 180°$ if ϕ is measured in degrees. Figure B.10 makes it clear that this angle ϕ and the slope m of a nonvertical line are related by the equation

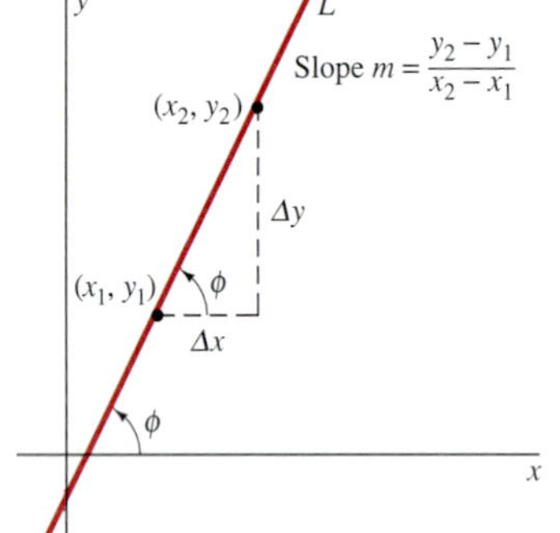

Fig. B.10 How is the angle of inclination ϕ related to the slope m?

$$m = \frac{\Delta y}{\Delta x} = \tan\phi. \tag{9}$$

This is true because if ϕ is an acute angle in a right triangle, then $\tan\phi$ is the ratio of the leg opposite ϕ to the leg adjacent to ϕ.

Your intuition correctly assures you that two lines are parallel if and only if they have the same angle of inclination. So it follows from Eq. (9) that two parallel nonvertical lines have the same slope and that two lines with the same slope must be parallel. This completes the proof of Theorem 1.

Theorem 1 *Slopes of Parallel Lines*

Two nonvertical lines are parallel if and only if they have the same slope.

Theorem 1 can also be proved without the use of the tangent function. The two lines shown in Fig. B.11 are parallel if and only if the two right triangles are similar, which is equivalent to the slopes of the lines being equal.

EXAMPLE 4 Write an equation of the line L that passes through the point $P(3, -2)$ and is parallel to the line L' with the equation $x + 2y = 6$.

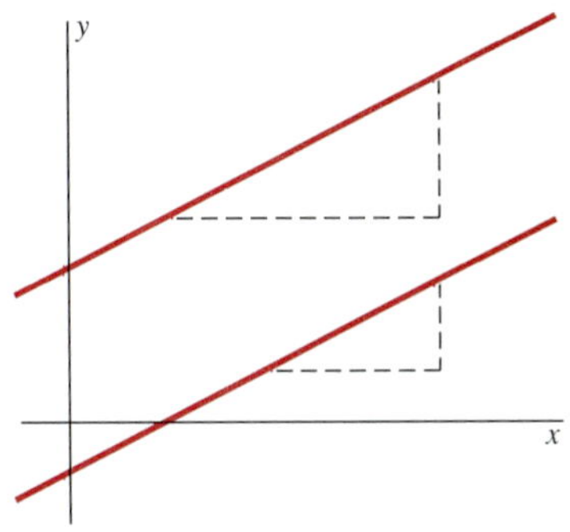

Fig. B.11 Two parallel lines

Solution When we solve the equation of L' for y, we get $y = -\frac{1}{2}x + 3$. So L' has slope $m = -\frac{1}{2}$. Because L has the same slope, its point-slope equation is then

$$y + 2 = -\tfrac{1}{2}(x - 3);$$

if you prefer, $x + 2y = -1$. ■

Theorem 2 ***Slopes of Perpendicular Lines***

Two lines L_1 and L_2 with slopes m_1 and m_2, respectively, are perpendicular if and only if

$$m_1 m_2 = -1. \tag{10}$$

That is, the slope of each is the *negative reciprocal* of the slope of the other.

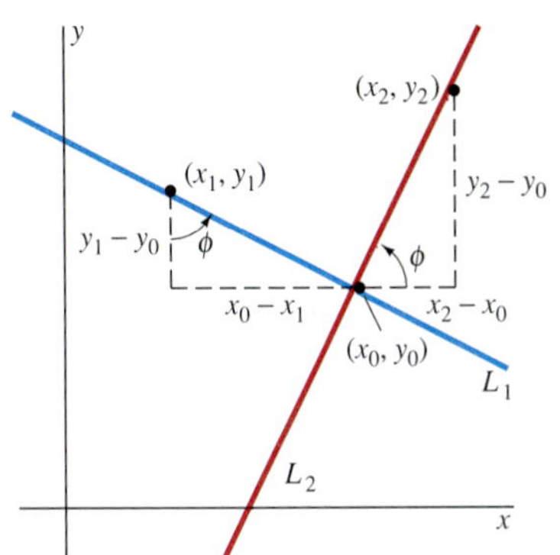

Fig. B.12 Illustration of the proof of Theorem 2

Proof If the two lines L_1 and L_2 are perpendicular and the slope of each exists, then neither is horizontal or vertical. Thus the situation resembles the one shown in Fig. B.12, in which the two lines meet at the point (x_0, y_0). It is easy to see that the two right triangles of the figure are similar, so equality of ratios of corresponding sides yields

$$m_2 = \frac{y_2 - y_0}{x_2 - x_0} = \frac{x_0 - x_1}{y_1 - y_0} = -\frac{x_1 - x_0}{y_1 - y_0} = -\frac{1}{m_1}.$$

Thus Eq. (10) holds if the two lines are perpendicular. This argument can be reversed to prove the converse—that the lines are perpendicular if $m_1 m_2 = -1$. ■

EXAMPLE 5 Write an equation of the line L through the point $P(3, -2)$ that is perpendicular to the line L' with equation $x + 2y = 6$.

Solution As we saw in Example 4, the slope of L' is $m' = -\frac{1}{2}$. By Theorem 2, the slope of L is $m = -1/m' = 2$. Thus L has the point-slope equation

$$y + 2 = 2(x - 3);$$

equivalently, $2x - y = 8$. ■

You will find it helpful to remember that the sign of the slope m of the line L indicates whether L runs upward or downward as your eyes move from left to right. If $m > 0$, then the angle of inclination ϕ of L must be an acute angle, because $m = \tan\phi$. In this case, L "runs upward" to the right. If $m < 0$, then ϕ is obtuse, so L "runs downward." Figure B.13 shows the geometry behind these observations.

Graphical Investigation

Many mathematical problems require the simultaneous solution of a pair of linear equations of the form

$$\begin{aligned} a_1 x + b_1 y &= c_1 \\ a_2 x + b_2 y &= c_2. \end{aligned} \tag{11}$$

The graph of these two equations consists of a pair of straight lines in the xy-plane. If these two lines are not parallel, then they must intersect at a single point whose

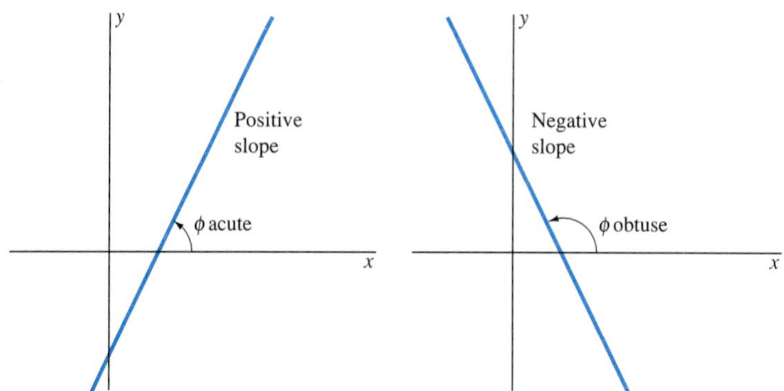

Fig. B.13 Positive and negative slope; effect on ϕ

coordinates (x_0, y_0) constitute the solution of (11). That is, $x = x_0$ and $y = y_0$ are the (only) values of x and y for which both equations in (11) are true.

In elementary algebra you studied various elimination and substitution methods for solving linear systems such as the one in (11). Example 6 illustrates an alternative *graphical method* that is sometimes useful when a graphing utility—a graphics calculator or a computer with a graphing program—is available.

EXAMPLE 6 We want to investigate the simultaneous solution of the linear equations

$$\begin{aligned} 10x - 8y &= 17 \\ 15x + 18y &= 67. \end{aligned} \tag{12}$$

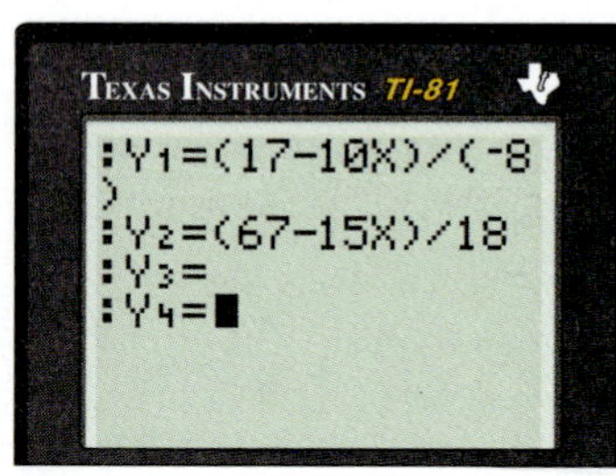

Fig. B.14 A calculator prepared to graph the lines in Eq. (12) (Example 6)

With many graphics calculators, it is necessary first to solve each equation for y:

$$\begin{aligned} y &= (17 - 10x)/(-8) \\ y &= (67 - 15x)/18. \end{aligned} \tag{13}$$

Figure B.14 shows a calculator prepared to graph the two lines represented by the equations in (12), and Fig. B.15 shows the result in the viewing window $-5 \leqq x \leqq 5$, $-5 \leqq y \leqq 5$.

Before proceeding, note that in Fig. B.15 the two lines *appear* to be perpendicular. But their slopes, $(-10)/(-8) = \frac{5}{4}$ and $(-15)/18 = -\frac{5}{6}$, are *not* negative reciprocals of each other. It follows from Theorem 2 that the two lines are *not* perpendicular.

Figures B.16, B.17, and B.18 show successive magnifications produced by "zooming in" on the point of intersection of the two lines. The dashed-line box in each figure is the viewing window for the next figure. Looking at Fig. B.18, we see that the intersection point is given by the approximations

$$x \approx 2.807, \qquad y \approx 1.383, \tag{14}$$

rounded to three decimal places.

The result in (14) can be checked by equating the right-hand sides in (13) and solving for x. This gives $x = 421/150 \approx 2.8067$. Substitution of the exact value of x into either equation in (13) then yields $y = 83/60 \approx 1.3833$. ■

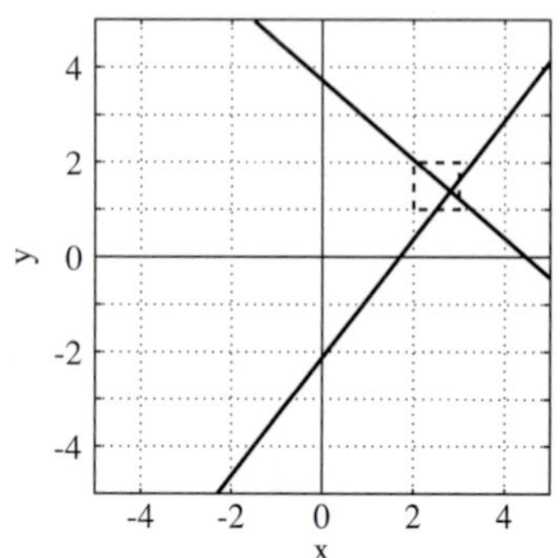

Fig. B.15 $-5 \leqq x \leqq 5$, $-5 \leqq y \leqq 5$ (Example 6)

The graphical method illustrated by Example 6 typically produces approximate solutions that are sufficiently accurate for practical purposes. But the method is especially useful for *nonlinear* equations, for which exact algebraic techniques of solution may not be available.

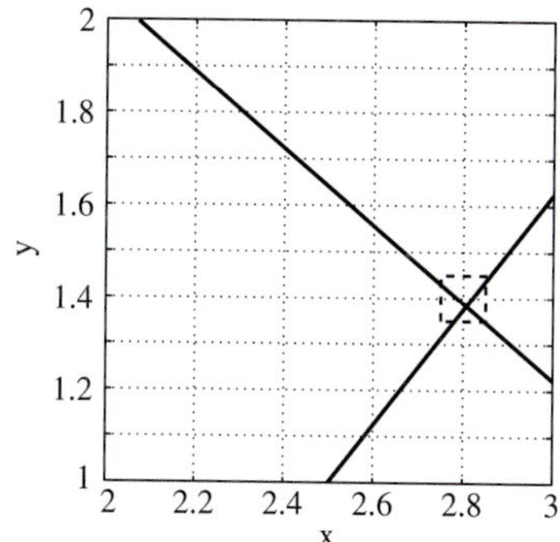

Fig. B.16 $2 \leqq x \leqq 3, 1 \leqq y \leqq 2$ (Example 6)

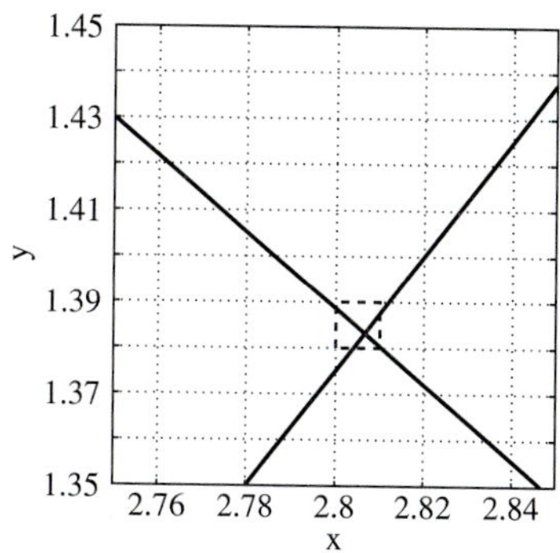

Fig. B.17 $2.75 \leqq x \leqq 2.85$, $1.35 \leqq y \leqq 1.45$ (Example 6)

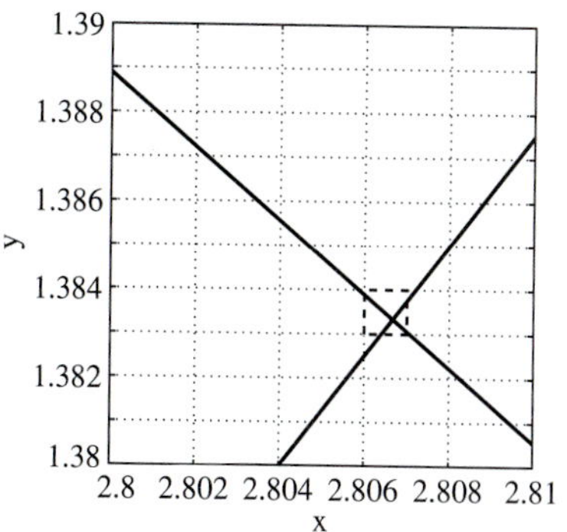

Fig. B.18 $2.80 \leqq x \leqq 2.81$, $1.38 \leqq y \leqq 1.39$ (Example 6)

APPENDIX B PROBLEMS

Three points A, B, and C lie on a single straight line if and only if the slope of AB is equal to the slope of BC. In Problems 1 through 4, plot the three given points and then determine whether they lie on a single line.

1. $A(-1, -2)$, $B(2, 1)$, $C(4, 3)$

2. $A(-2, 5)$, $B(2, 3)$, $C(8, 0)$

3. $A(-1, 6)$, $B(1, 2)$, $C(4, -2)$

4. $A(-3, 2)$, $B(1, 6)$, $C(8, 14)$

In Problems 5 and 6, use the concept of slope to show that the four points given are the vertices of a parallelogram.

5. $A(-1, 3)$, $B(5, 0)$, $C(7, 4)$, $D(1, 7)$

6. $A(7, -1)$, $B(-2, 2)$, $C(1, 4)$, $D(10, 1)$

In Problems 7 and 8, show that the three given points are the vertices of a right triangle.

7. $A(-2, -1)$, $B(2, 7)$, $C(4, -4)$

8. $A(6, -1)$, $B(2, 3)$, $C(-3, -2)$

In Problems 9 through 13, find the slope m and y-intercept b of the line with the given equation. Then sketch the line.

9. $2x = 3y$

10. $x + y = 1$

11. $2x - y + 3 = 0$

12. $3x + 4y = 6$

13. $2x = 3 - 5y$

In Problems 14 through 23, write an equation of the straight line L described.

14. L is vertical and has x-intercept 7.

15. L is horizontal and passes through $(3, -5)$.

16. L has x-intercept 2 and y-intercept -3.

17. L passes through $(2, -3)$ and $(5, 3)$.

18. L passes through $(-1, -4)$ and has slope $\frac{1}{2}$.

19. L passes through $(4, 2)$ and has angle of inclination $135°$.

20. L has slope 6 and y-intercept 7.

21. L passes through $(1, 5)$ and is parallel to the line with equation $2x + y = 10$.

22. L passes through $(-2, 4)$ and is perpendicular to the line with equation $x + 2y = 17$.

23. L is the perpendicular bisector of the line segment that has endpoints $(-1, 2)$ and $(3, 10)$.

24. Find the perpendicular distance from the point $(2, 1)$ to the line with equation $y = x + 1$.

25. Find the perpendicular distance between the parallel lines $y = 5x + 1$ and $y = 5x + 9$.

26. The points $A(-1, 6)$, $B(0, 0)$, and $C(3, 1)$ are three consecutive vertices of a parallelogram. What are the coordinates of the fourth vertex? (What happens if the word *consecutive* is omitted?)

27. Prove that the diagonals of the parallelogram of Problem 26 bisect each other.

28. Show that the points $A(-1, 2)$, $B(3, -1)$, $C(6, 3)$, and $D(2, 6)$ are the vertices of a **rhombus**—a parallelogram with all four sides having the same length. Then prove that the diagonals of this rhombus are perpendicular to each other.

29. The points $A(2, 1)$, $B(3, 5)$, and $C(7, 3)$ are the vertices of a triangle. Prove that the line joining the midpoints of AB and BC is parallel to AC.

30. A **median** of a triangle is a line joining a vertex to the midpoint of the opposite side. Prove that the medians of the triangle of Problem 29 intersect in a single point.

31. Complete the proof of the midpoint formula in Eq. (3). It is necessary to show that the point M lies on the segment P_1P_2. One way to do this is to show that the slope of MP_1 is equal to the slope of MP_2.

32. Let $P(x_0, y_0)$ be a point of the circle with center $C(0, 0)$ and radius r. Recall that the line tangent to the circle at the point P is perpendicular to the radius CP. Prove that the equation of this tangent line is $x_0x + y_0y = r^2$.

33. The Fahrenheit temperature F and the absolute temperature K satisfy a linear equation. Moreover, $K = 273.16$ when $F = 32$, and $K = 373.16$ when $F = 212$. Express K in terms of F. What is the value of F when $K = 0$?

34. The length L (in centimeters) of a copper rod is a linear function of its Celsius temperature C. If $L = 124.942$ when $C = 20$ and $L = 125.134$ when $C = 110$, express L in terms of C.

35. The owner of a grocery store finds that she can sell 980 gal of milk each week at \$1.69/gal and 1220 gal of milk each week at \$1.49/gal. Assume a linear relationship between price and sales. How many gallons would she then expect to sell each week at \$1.56/gal?

36. Figure B.19 shows the graphs of the equations

$$17x - 10y = 57$$

$$25x - 15y = 17.$$

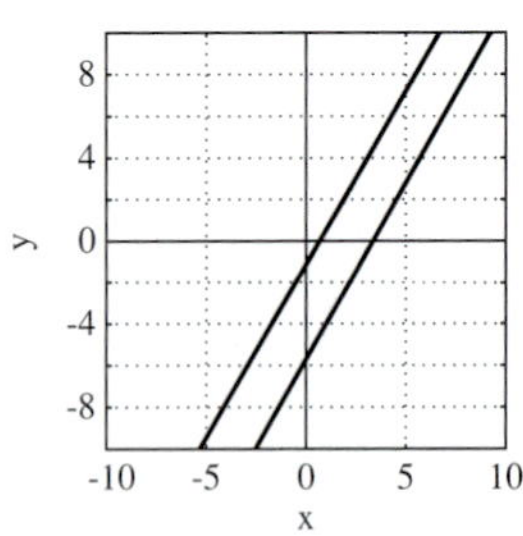

Fig. B.19 The lines of Problem 36

Are these two lines parallel? If not, find their point of intersection. If you have a graphing utility, find the solution by graphical approximation as well as by exact algebraic methods.

In Problems 37 through 46, use a graphics calculator or computer to approximate graphically (with three digits to the right of the decimal correct or correctly rounded) the solution of the given linear equation. Then check your approximate solution by solving the system by an exact algebraic method.

37. $2x + 3y = 5$, $2x + 5y = 12$

38. $6x + 4y = 5$, $8x - 6y = 13$

39. $3x + 3y = 17$, $3x + 5y = 16$

40. $2x + 3y = 17$, $2x + 5y = 20$

41. $4x + 3y = 17$, $5x + 5y = 21$

42. $4x + 3y = 15$, $5x + 5y = 29$

43. $5x + 6y = 16$, $7x + 10y = 29$

44. $5x + 11y = 21$, $4x + 10y = 19$

45. $6x + 6y = 31$, $9x + 11y = 37$

46. $7x + 6y = 31$, $11x + 11y = 47$

47. Justify the phrase "no other point of the plane can do so" that follows the first appearance of Eq. (6).

48. The discussion of the linear equation $Ax + By = C$ in Eq. (8) does not include a description of the graph of this equation if $A = B = 0$. What is the graph in this case?

APPENDIX C: REVIEW OF TRIGONOMETRY

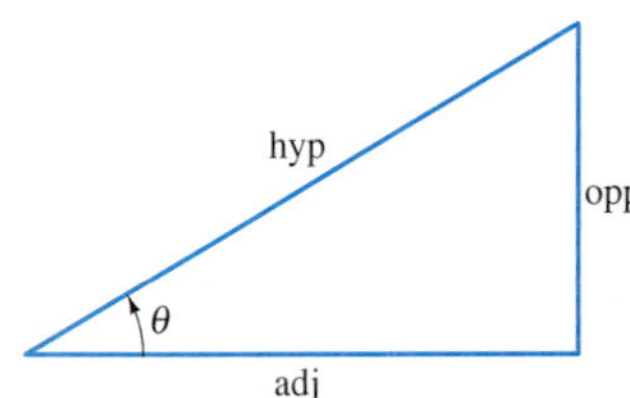

Fig. C.1 The sides and angle θ of a right triangle

In elementary trigonometry, the six basic trigonometric functions of an acute angle θ in a right triangle are defined as ratios between pairs of sides of the triangle. As in Fig. C.1, where "adj" stands for "adjacent," "opp" for "opposite," and "hyp" for "hypotenuse,"

$$\cos\theta = \frac{\text{adj}}{\text{hyp}}, \quad \sin\theta = \frac{\text{opp}}{\text{hyp}}, \quad \tan\theta = \frac{\text{opp}}{\text{adj}},$$
$$\sec\theta = \frac{\text{hyp}}{\text{adj}}, \quad \csc\theta = \frac{\text{hyp}}{\text{opp}}, \quad \cot\theta = \frac{\text{adj}}{\text{opp}}. \tag{1}$$

We generalize these definitions to *directed* angles of arbitrary size in the following way. Suppose that the initial side of the angle θ is the positive x-axis, so its vertex is at the origin. The angle is **directed** if a direction of rotation from its initial side to its terminal side is specified. We call θ a **positive angle** if this rotation is counterclockwise and a **negative angle** if it is clockwise.

Let $P(x, y)$ be the point at which the terminal side of θ intersects the unit circle $x^2 + y^2 = 1$. Then we define

$$\cos\theta = x, \quad \sin\theta = y, \quad \tan\theta = \frac{y}{x},$$
$$\sec\theta = \frac{1}{x}, \quad \csc\theta = \frac{1}{y}, \quad \cot\theta = \frac{x}{y}. \tag{2}$$

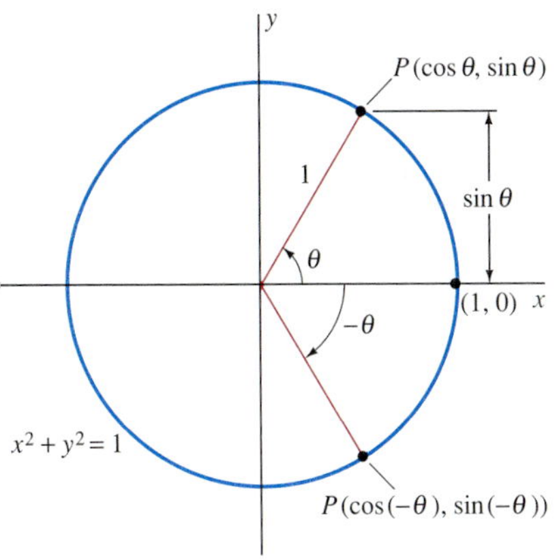

Fig. C.2 Using the unit circle to define the trigonometric functions

Sine
Cosecant
All
Tangent
Cotangent
Cosine
Secant

Positive in quadrants shown

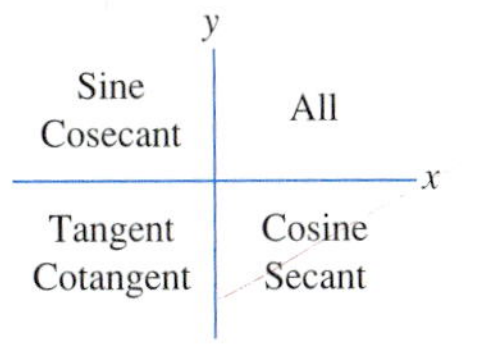

Fig. C.3 The signs of the trigonometric functions

We assume that $x \neq 0$ in the case of $\tan\theta$ and $\sec\theta$ and that $y \neq 0$ in the case of $\cot\theta$ and $\csc\theta$. If the angle θ is positive and acute, then it is clear from Fig. C.2 that the definitions in (2) agree with the right triangle definitions in (1) in terms of the coordinates of P. A glance at the figure also shows which of the functions are positive for angles in each of the four quadrants. Figure C.3 summarizes this information.

Here we discuss primarily the two most basic trigonometric functions, the sine and the cosine. From (2) we see immediately that the other four trigonometric functions are defined in terms of $\sin\theta$ and $\cos\theta$ by

$$\tan\theta = \frac{\sin\theta}{\cos\theta}, \quad \sec\theta = \frac{1}{\cos\theta}, \quad \cot\theta = \frac{\cos\theta}{\sin\theta}, \quad \csc\theta = \frac{1}{\sin\theta}. \tag{3}$$

Next, we compare the angles θ and $-\theta$ in Fig. C.4. We see that

$$\cos(-\theta) = \cos\theta \quad \text{and} \quad \sin(-\theta) = -\sin\theta. \tag{4}$$

Because $x = \cos\theta$ and $y = \sin\theta$ in (2), the equation $x^2 + y^2 = 1$ of the unit circle translates immediately into the **fundamental identity of trigonometry,**

$$\cos^2\theta + \sin^2\theta = 1. \tag{5}$$

Dividing each term of this fundamental identity by $\cos^2\theta$ gives the identity

$$1 + \tan^2\theta = \sec^2\theta. \tag{5$'$}$$

Similarly, dividing each term in Eq. (5) by $\sin^2\theta$ yields the identity

$$1 + \cot^2\theta = \csc^2\theta. \tag{5$''$}$$

(See Problem 9 of this appendix.)

In Problems 15 and 16 we outline derivations of the **addition formulas**

$$\sin(\alpha + \beta) = \sin\alpha\cos\beta + \cos\alpha\sin\beta, \tag{6}$$

$$\cos(\alpha + \beta) = \cos\alpha\cos\beta - \sin\alpha\sin\beta. \tag{7}$$

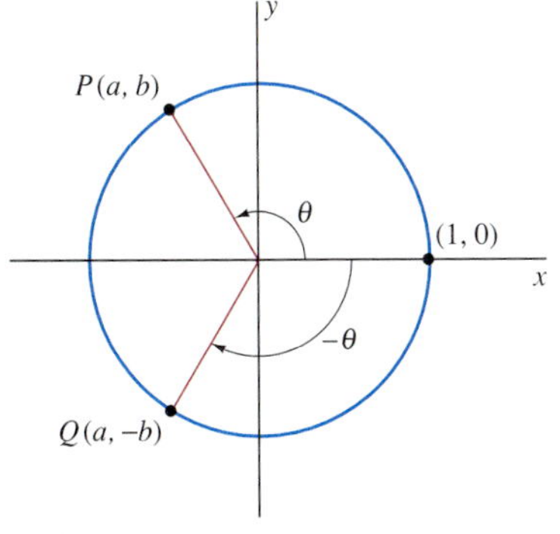

Fig. C.4 The effect of replacing θ with $-\theta$ in the sine and cosine functions

With $\alpha = \theta = \beta$ in Eqs. (6) and (7), we get the **double-angle formulas**

$$\sin 2\theta = 2\sin\theta\cos\theta, \tag{8}$$

$$\cos 2\theta = \cos^2\theta - \sin^2\theta \tag{9}$$

$$= 2\cos^2\theta - 1 \tag{9a}$$

$$= 1 - 2\sin^2\theta, \tag{9b}$$

where Eqs. (9a) and (9b) are obtained from Eq. (9) by use of the fundamental identity in Eq. (5).

If we solve Eq. (9a) for $\cos^2\theta$ and Eq. (9b) for $\sin^2\theta$, we get the **half-angle formulas**

$$\cos^2\theta = \tfrac{1}{2}(1 + \cos 2\theta), \tag{10}$$

$$\sin^2\theta = \tfrac{1}{2}(1 - \cos 2\theta). \tag{11}$$

Equations (10) and (11) are especially important in integral calculus.

Radian Measure

In elementary mathematics, angles frequently are measured in *degrees,* with 360° in one complete revolution. In calculus it is more convenient—and often essential—to measure angles in radians. The **radian measure** of an angle is the length of the arc it subtends in (that is, the arc it cuts out of) the unit circle when the vertex of the angle is at the center of the circle (Fig. C.5).

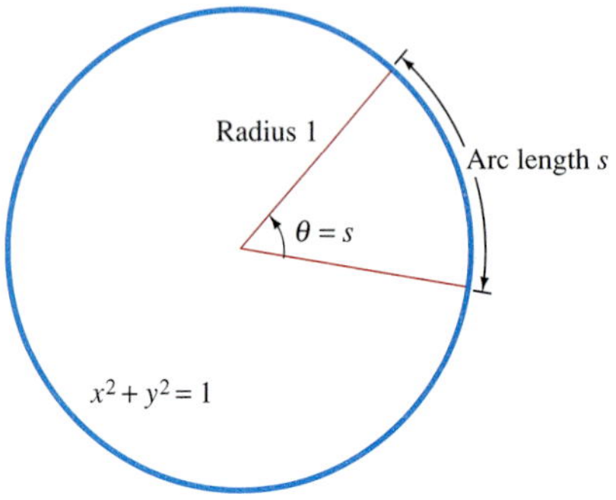

Fig. C.5 The radian measure of an angle

Recall that the area A and circumference C of a circle of radius r are given by the formulas

$$A = \pi r^2 \quad \text{and} \quad C = 2\pi r,$$

where the irrational number π is approximately 3.14159. Because the circumference of the unit circle is 2π and its central angle is 360°, it follows that

$$2\pi \text{ rad} = 360^\circ; \qquad 180^\circ = \pi \text{ rad} \approx 3.14159 \text{ rad}. \tag{12}$$

Using Eq. (12), we can easily convert back and forth between radians and degrees:

$$1 \text{ rad} = \frac{180^\circ}{\pi} \approx 57^\circ 17' 44.8'', \tag{12a}$$

$$1^\circ = \frac{\pi}{180} \text{ rad} \approx 0.01745 \text{ rad}. \tag{12b}$$

Radians	Degrees
0	0
$\pi/6$	30
$\pi/4$	45
$\pi/3$	60
$\pi/2$	90
$2\pi/3$	120
$3\pi/4$	135
$5\pi/6$	150
π	180
$3\pi/2$	270
2π	360
4π	720

Fig. C.6 Some radian-degree conversions

Figure C.6 shows radian-degree conversions for some common angles.

Now consider an angle of θ radians at the center of a circle of radius r (Fig. C.7). Denote by s the length of the arc subtended by θ; denote by A the area of the sector of the circle bounded by this angle. Then the proportions

$$\frac{s}{2\pi r} = \frac{A}{\pi r^2} = \frac{\theta}{2\pi}$$

give the formulas

$$s = r\theta \qquad (\theta \text{ in radians}) \tag{13}$$

and

$$A = \tfrac{1}{2}r^2\theta \qquad (\theta \text{ in radians}). \tag{14}$$

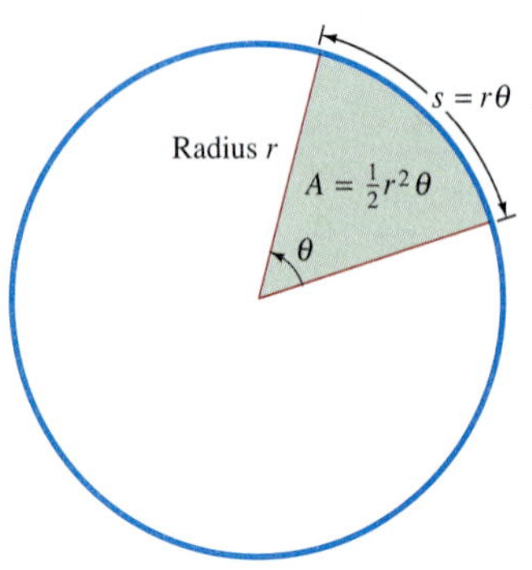

Fig. C.7 The area of a sector and arc length of a circular arc

The definitions in (2) refer to trigonometric functions of angles rather than trigonometric functions of numbers. Suppose that t is a real number. Then the number $\sin t$ is, *by definition,* the sine of an angle of t radians—recall that a positive angle is directed counterclockwise from the positive x-axis, whereas a negative angle is directed clockwise. Briefly, $\sin t$ is the sine of an angle of t radians. The other trigonometric functions of the number t have similar definitions. Hence, when we write $\sin t$, $\cos t$, and so on, with t a real number, it is always in reference to an angle of t radians.

When we need to refer to the sine of an angle of t degrees, we will henceforth write $\sin t^\circ$. The point is that $\sin t$ and $\sin t^\circ$ are quite different functions of the variable t. For example, you would get

$$\sin 1^\circ \approx 0.0175 \quad \text{and} \quad \sin 30^\circ = 0.5$$

on a calculator set in degree mode. But in radian mode, a calculator would give

$$\sin 1 \approx 0.8415 \quad \text{and} \quad \sin 30 \approx -0.9880.$$

The relationship between the functions $\sin t$ and $\sin t^\circ$ is

$$\sin t^\circ = \sin\left(\frac{\pi t}{180}\right). \tag{15}$$

The distinction extends even to programming languages. In FORTRAN, the function `SIN` is the radian sine function, and you must write $\sin t^\circ$ in the form `SIND(T)`. In BASIC you must write `SIN(Pi*T/180)` to get the correct value of the sine of an angle of t degrees.

An angle of 2π rad corresponds to one revolution around the unit circle. This implies that the sine and cosine functions have **period** 2π, meaning that

$$\begin{aligned} \sin(t + 2\pi) &= \sin t, \\ \cos(t + 2\pi) &= \cos t. \end{aligned} \tag{16}$$

It follows from the equations in (16) that

$$\sin(t + 2n\pi) = \sin t \quad \text{and} \quad \cos(t + 2n\pi) = \cos t \tag{17}$$

for every integer n. This periodicity of the sine and cosine functions is evident in their graphs (Fig. C.8). From the equations in (3), the other four trigonometric functions also must be periodic, as their graphs in Figs. C.9 and C.10 show.

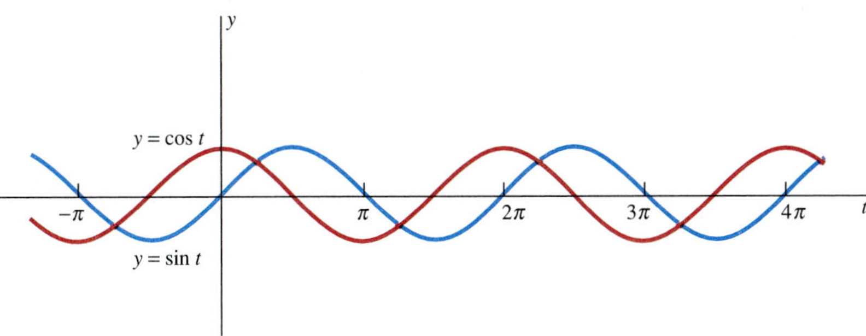

Fig. C.8 Periodicity of the sine and cosine functions

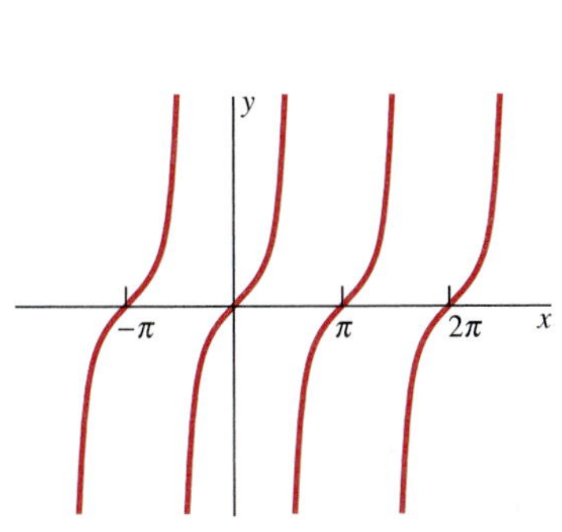

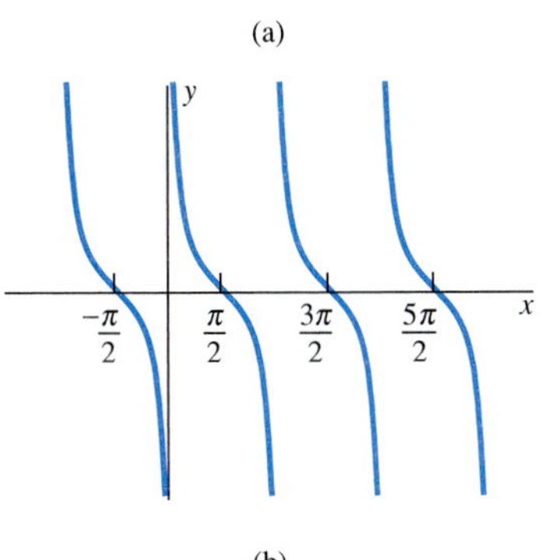

Fig. C.9 The graphs of (a) the tangent function and (b) the cotangent function

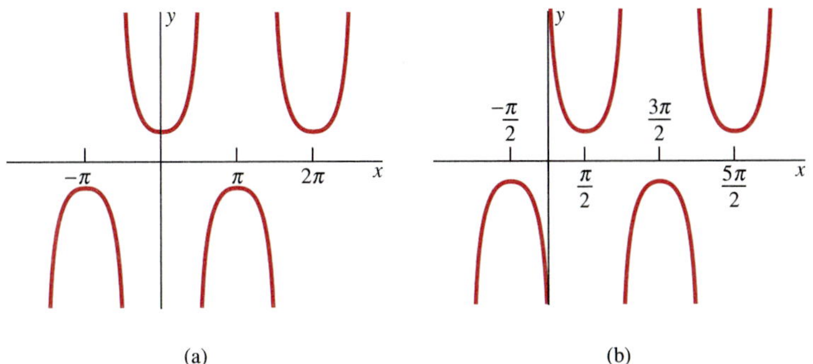

Fig. C.10 The graphs of (a) the secant function and (b) the cosecant function

We see from the equations in (2) that

$$\sin 0 = 0, \quad \sin\frac{\pi}{2} = 1, \quad \sin\pi = 0,$$
$$\cos 0 = 1, \quad \cos\frac{\pi}{2} = 0, \quad \cos\pi = -1. \tag{18}$$

The trigonometric functions of $\pi/6$, $\pi/4$, and $\pi/3$ (the radian equivalents of 30°, 45°, and 60°, respectively) are easy to read from the well-known triangles of Fig. C.11. For instance,

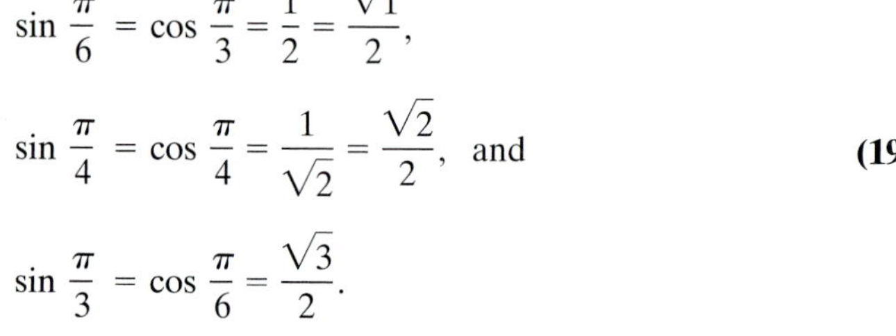

$$\sin\frac{\pi}{6} = \cos\frac{\pi}{3} = \frac{1}{2} = \frac{\sqrt{1}}{2},$$
$$\sin\frac{\pi}{4} = \cos\frac{\pi}{4} = \frac{1}{\sqrt{2}} = \frac{\sqrt{2}}{2}, \quad \text{and} \tag{19}$$
$$\sin\frac{\pi}{3} = \cos\frac{\pi}{6} = \frac{\sqrt{3}}{2}.$$

Fig. C.11 Familiar right triangles

To find the values of trigonometric functions of angles larger than $\pi/2$, we can use their periodicity and the identities

$$\sin(\pi \pm \theta) = \mp\sin\theta,$$
$$\cos(\pi \pm \theta) = -\cos\theta, \quad \text{and} \tag{20}$$
$$\tan(\pi \pm \theta) = \pm\tan\theta$$

(Problem 14) as well as similar identities for the cosecant, secant, and cotangent functions.

EXAMPLE 1

$$\sin\frac{5\pi}{4} = \sin\left(\pi + \frac{\pi}{4}\right) = -\sin\frac{\pi}{4} = -\frac{\sqrt{2}}{2};$$
$$\cos\frac{2\pi}{3} = \cos\left(\pi - \frac{\pi}{3}\right) = -\cos\frac{\pi}{3} = -\frac{1}{2};$$
$$\tan\frac{3\pi}{4} = \tan\left(\pi - \frac{\pi}{4}\right) = -\tan\frac{\pi}{4} = -1;$$
$$\sin\frac{7\pi}{6} = \sin\left(\pi + \frac{\pi}{6}\right) = -\sin\frac{\pi}{6} = -\frac{1}{2};$$
$$\cos\frac{5\pi}{3} = \cos\left(2\pi - \frac{\pi}{3}\right) = \cos\left(-\frac{\pi}{3}\right) = \cos\frac{\pi}{3} = \frac{1}{2};$$
$$\sin\frac{17\pi}{6} = \sin\left(2\pi + \frac{5\pi}{6}\right) = \sin\frac{5\pi}{6}$$
$$= \sin\left(\pi - \frac{\pi}{6}\right) = \sin\frac{\pi}{6} = \frac{1}{2}.$$ ■

EXAMPLE 2 Find the solutions (if any) of the equation

$$\sin^2 x - 3\cos^2 x + 2 = 0$$

that lie in the interval $[0, \pi]$.

Solution Using the fundamental identity in Eq. (5), we substitute $\cos^2 x = 1 - \sin^2 x$ into the given equation to obtain

$$\sin^2 x - 3(1 - \sin^2 x) + 2 = 0;$$

$$4\sin^2 x - 1 = 0;$$

$$\sin x = \pm\tfrac{1}{2}.$$

Because $\sin x \geqq 0$ for x in $[0, \pi]$, $\sin x = -\frac{1}{2}$ is impossible. But $\sin x = \frac{1}{2}$ for $x = \pi/6$ and for $x = \pi - \pi/6 = 5\pi/6$. These are the solutions of the given equation that lie in $[0, \pi]$. ■

APPENDIX C PROBLEMS

Express in radian measure the angles in Problems 1 through 5.

1. $40°$ **2.** $-270°$

3. $315°$ **4.** $210°$

5. $-150°$

In Problems 6 through 10, express in degrees the angles given in radian measure.

6. $\dfrac{\pi}{10}$ **7.** $\dfrac{2\pi}{5}$

8. 3π **9.** $\dfrac{15\pi}{4}$

10. $\dfrac{23\pi}{60}$

In Problems 11 through 14, evaluate the six trigonometric functions of x at the given values.

11. $x = -\dfrac{\pi}{3}$ **12.** $x = \dfrac{3\pi}{4}$

13. $x = \dfrac{7\pi}{6}$ **14.** $x = \dfrac{5\pi}{3}$

Find all solutions x of each equation in Problems 15 through 23.

15. $\sin x = 0$ **16.** $\sin x = 1$

17. $\sin x = -1$ **18.** $\cos x = 0$

19. $\cos x = 1$ **20.** $\cos x = -1$

21. $\tan x = 0$ **22.** $\tan x = 1$

23. $\tan x = -1$

24. Suppose that $\tan x = \frac{3}{4}$ and that $\sin x < 0$. Find the values of the other five trigonometric functions of x.

25. Suppose that $\csc x = -\frac{5}{3}$ and that $\cos x > 0$. Find the values of the other five trigonometric functions of x.

Deduce the identities in Problems 26 and 27 from the fundamental identity

$$\cos^2\theta + \sin^2\theta = 1$$

and from the definitions of the other four trigonometric functions.

26. $1 + \tan^2\theta = \sec^2\theta$

27. $1 + \cot^2\theta = \csc^2\theta$

28. Deduce from the addition formulas for the sine and cosine the addition formula for the tangent:

$$\tan(x + y) = \frac{\tan x + \tan y}{1 - \tan x \tan y}.$$

In Problems 29 through 36, use the method of Example 1 to find the indicated values.

29. $\sin\dfrac{5\pi}{6}$ **30.** $\cos\dfrac{7\pi}{6}$

31. $\sin\dfrac{11\pi}{6}$ **32.** $\cos\dfrac{19\pi}{6}$

33. $\sin\dfrac{2\pi}{3}$ **34.** $\cos\dfrac{4\pi}{3}$

35. $\sin\dfrac{5\pi}{3}$ **36.** $\cos\dfrac{10\pi}{3}$

37. Apply the addition formula for the sine, cosine, and tangent functions (the latter from Problem 28) to show that if $0 < \theta < \pi/2$, then

(a) $\cos\left(\dfrac{\pi}{2} - \theta\right) = \sin\theta$; (b) $\sin\left(\dfrac{\pi}{2} - \theta\right) = \cos\theta$;

(c) $\cot\left(\dfrac{\pi}{2} - \theta\right) = \tan\theta$.

The prefix *co-* is an abbreviation for the adjective *complementary,* which describes two angles whose sum is $\pi/2$. For example, $\pi/6$ and $\pi/3$ are complementary angles, so (a) implies that $\cos \pi/6 = \sin \pi/3$.

Suppose that $0 < \theta < \pi/2$. Derive the identities in Problems 38 through 40.

38. $\sin(\pi \pm \theta) = \mp\sin\theta$ **39.** $\cos(\pi \pm \theta) = -\cos\theta$

40. $\tan(\pi \pm \theta) = \pm\tan\theta$

41. The points $A(\cos\theta, -\sin\theta)$, $B(1, 0)$, $C(\cos\phi, \sin\phi)$, and $D(\cos(\theta + \phi), \sin(\theta + \phi))$ are shown in Fig. C.12; all are points on the unit circle. Deduce from the fact that the line segments AC and BD have the same length (because they are subtended by the same central angle $\theta + \phi$) that

$$\cos(\theta + \phi) = \cos\theta\cos\phi - \sin\theta\sin\phi.$$

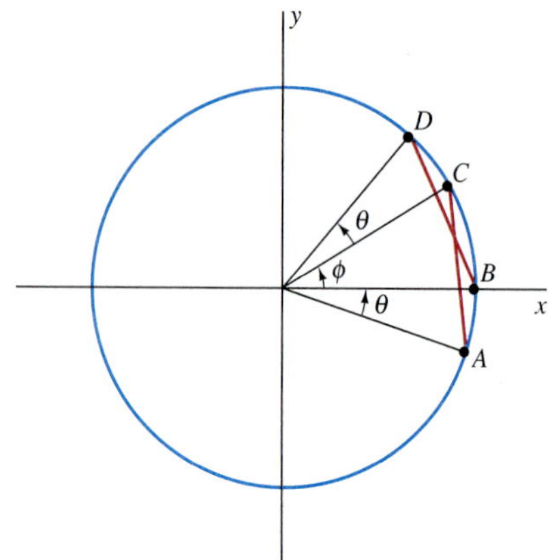

Fig. C.12 Deriving the cosine addition formula (Problem 41)

42. (a) Use the triangles shown in Fig. C.13 to deduce that

$$\sin\left(\theta + \frac{\pi}{2}\right) = \cos\theta \quad \text{and} \quad \cos\left(\theta + \frac{\pi}{2}\right) = -\sin\theta.$$

(b) Use the results of Problem 41 and part (a) to derive the addition formula for the sine function.

In Problems 43 through 48, find all solutions of the given equation that lie in the interval $[0, \pi]$.

43. $3\sin^2 x - \cos^2 x = 2$ **44.** $\sin^2 x = \cos^2 x$

45. $2\cos^2 x + 3\sin^2 x = 3$ **46.** $2\sin^2 x + \cos x = 2$

47. $8\sin^2 x\cos^2 x = 1$ **48.** $\cos 2\theta - 3\cos\theta = -2$

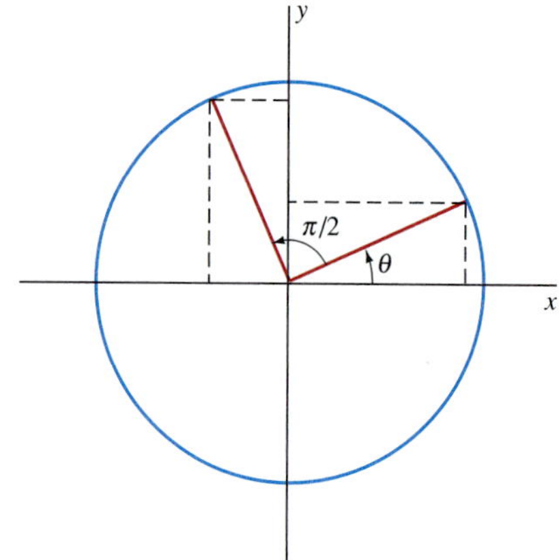

Fig. C.13 Deriving the identities of Problem 42

APPENDIX D: PROOFS OF THE LIMIT LAWS

Recall the definition of the limit:

$$\lim_{x \to a} F(x) = L$$

provided that, given $\epsilon > 0$, there exists a number $\delta > 0$ such that

$$0 < |x - a| < \delta \quad \text{implies that} \quad |F(x) - L| < \epsilon. \qquad \textbf{(1)}$$

Note that the number ϵ comes *first.* Then a value of $\delta > 0$ must be found so that the implication in (1) holds. To prove that $F(x) \to L$ as $x \to a$, you must, in effect, be able to stop the next person you see and ask him or her to pick a positive number ϵ at random. Then you must *always* be ready to respond with a positive number δ. This number δ must have the property that the implication in (1) holds for your number δ and the given number ϵ. The **only** restriction on x is that

$$0 < |x - a| < \delta,$$

as given in (1).

To do all this, you will ordinarily need to give an explicit method—a recipe or formula—for producing a value of δ that works for each value of ϵ. As Examples 1 through 3 show, the method will depend on the particular function F under study as well as the values of a and L.

EXAMPLE 1 Prove that $\lim_{x\to 3}(2x-1)=5$.

Solution Given $\epsilon > 0$, we must find $\delta > 0$ such that

$$|(2x-1)-5)| < \epsilon \quad \text{if} \quad 0 < |x-3| < \delta.$$

Now

$$|(2x-1)-5| = |2x-6| = 2|x-3|,$$

so

$$0 < |x-3| < \frac{\epsilon}{2} \quad \text{implies that} \quad |(2x-1)-5| < 2\cdot\frac{\epsilon}{2} = \epsilon.$$

Hence, given $\epsilon > 0$, it suffices to choose $\delta = \epsilon/2$. This illustrates the observation that the required number δ is generally a function of the given number ϵ. ■

EXAMPLE 2 Prove that $\lim_{x\to 2}(3x^2+5)=17$.

Solution Given $\epsilon > 0$, we must find $\delta > 0$ such that

$$0 < |x-2| < \delta \quad \text{implies that} \quad |(3x^2+5)-17| < \epsilon.$$

Now

$$|(3x^2+5)-17| = |3x^2-12| = 3\cdot|x+2|\cdot|x-2|.$$

Our problem, therefore, is to show that $|x+2|\cdot|x-2|$ can be made as small as we please by choosing $x-2$ sufficiently small. The idea is that $|x+2|$ cannot be too large if $|x-2|$ is fairly small. For example, if $|x-2| < 1$, then

$$|x+2| = |(x-2)+4| \leqq |x-2| + 4 < 5.$$

Therefore,

$$0 < |x-2| < 1 \quad \text{implies that} \quad |(3x^2+5)-17| < 15\cdot|x-2|.$$

Consequently, let us choose δ to be the minimum of the two numbers 1 and $\epsilon/15$. Then

$$0 < |x-2| < \delta \quad \text{implies that} \quad |(3x^2+5)-17| < 15\cdot\frac{\epsilon}{15} = \epsilon,$$

as desired. ■

EXAMPLE 3 Prove that

$$\lim_{x\to a}\frac{1}{x} = \frac{1}{a} \quad \text{if} \quad a \neq 0.$$

Solution For simplicity, we will consider only the case in which $a > 0$ (the case $a < 0$ is similar).

Suppose that $\epsilon > 0$ is given. We must find a number δ such that

$$0 < |x - a| < \delta \quad \text{implies that} \quad \left|\frac{1}{x} - \frac{1}{a}\right| < \epsilon.$$

Now

$$\left|\frac{1}{x} - \frac{1}{a}\right| = \left|\frac{a - x}{ax}\right| = \frac{|x - a|}{a|x|}.$$

The idea is that $1/|x|$ cannot be too large if $|x - a|$ is fairly small. For example, if $|x - a| < a/2$, then $a/2 < x < 3a/2$. Therefore,

$$|x| > \frac{a}{2}, \quad \text{so} \quad \frac{1}{|x|} < \frac{2}{a}.$$

In this case it would follow that

$$\left|\frac{1}{x} - \frac{1}{a}\right| < \frac{2}{a^2} \cdot |x - a|$$

if $|x - a| < a/2$. Thus, if we choose δ to be the minimum of the two numbers $a/2$ and $a^2\epsilon/2$, then

$$0 < |x - a| < \delta \quad \text{implies that} \quad \left|\frac{1}{x} - \frac{1}{a}\right| < \frac{2}{a^2} \cdot \frac{a^2\epsilon}{2} = \epsilon.$$

Therefore,

$$\lim_{x \to a} \frac{1}{x} = \frac{1}{a} \quad \text{if} \quad a \neq 0,$$

as desired. ■

We are now ready to give proofs of the limit laws stated in Section 2.2.

Constant Law

If $f(x) \equiv C$, a constant, then

$$\lim_{x \to a} f(x) = \lim_{x \to a} C = C.$$

Proof Because $|C - C| = 0$, we merely choose $\delta = 1$, regardless of the previously given value of $\epsilon > 0$. Then, if $0 < |x - a| < \delta$, it is automatic that $|C - C| < \epsilon$. ■

Addition Law

If $\lim_{x \to a} F(x) = L$ and $\lim_{x \to a} G(x) = M$, then

$$\lim_{x \to a} [F(x) + G(x)] = L + M.$$

Proof Let $\epsilon > 0$ be given. Because L is the limit of $F(x)$ as $x \to a$, there exists a number $\delta_1 > 0$ such that

$$0 < |x - a| < \delta_1 \quad \text{implies that} \quad |F(x) - L| < \frac{\epsilon}{2}.$$

Because M is the limit of $G(x)$ as $x \to a$, there exists a number $\delta_2 > 0$ such that

$$0 < |x - a| < \delta_2 \quad \text{implies that} \quad |G(x) - M| < \frac{\epsilon}{2}.$$

Let $\delta = \min\{\delta_1, \delta_2\}$. Then $0 < |x - a| < \delta$ implies that

$$|(F(x) + G(x)) - (L + M)| \leqq |F(x) - L| + |G(x) - M| < \frac{\epsilon}{2} + \frac{\epsilon}{2} = \epsilon.$$

Therefore,

$$\lim_{x \to a} [F(x) + G(x)] = L + M,$$

as desired. ■

Product Law

If $\lim_{x \to a} F(x) = L$ and $\lim_{x \to a} G(x) = M$, then

$$\lim_{x \to a} [F(x) \cdot G(x)] = L \cdot M.$$

Proof Given $\epsilon > 0$, we must find a number $\delta > 0$ such that

$$0 < |x - a| < \delta \quad \text{implies that} \quad |F(x) \cdot G(x) - L \cdot M| < \epsilon.$$

But first, the triangle inequality gives the result

$$\begin{aligned} |F(x) \cdot G(x) - L \cdot M| &= |F(x) \cdot G(x) - L \cdot G(x) + L \cdot G(x) - L \cdot M| \\ &\leqq |G(x)| \cdot |F(x) - L| + |L| \cdot |G(x) - M|. \end{aligned} \tag{2}$$

Because $\lim_{x \to a} F(x) = L$, there exists $\delta_1 > 0$ such that

$$0 < |x - a| < \delta_1 \quad \text{implies that} \quad |F(x) - L| < \frac{\epsilon}{2(|M| + 1)}. \tag{3}$$

And because $\lim_{x \to a} G(x) = M$, there exists $\delta_2 > 0$ such that

$$0 < |x - a| < \delta_2 \quad \text{implies that} \quad |G(x) - M| < \frac{\epsilon}{2(|L| + 1)}. \tag{4}$$

Moreover, there is a *third* number $\delta_3 > 0$ such that

$$0 < |x - a| < \delta_3 \quad \text{implies that} \quad |G(x) - M| < 1,$$

which in turn implies that

$$|G(x)| < |M| + 1, \tag{5}$$

We now choose $\delta = \min\{\delta_1, \delta_2, \delta_3\}$. Then we substitute (3), (4), and (5) into (2) and, finally, see that $0 < |x - a| < \delta$ implies that

$$|F(x)\cdot G(x) - L\cdot M| < (|M| + 1)\cdot\frac{\epsilon}{2(|M| + 1)} + |L|\cdot\frac{\epsilon}{2(|L| + 1)}$$

$$< \frac{\epsilon}{2} + \frac{\epsilon}{2} = \epsilon,$$

as desired. The use of $|M| + 1$ and $|L| + 1$ in the denominators avoids the technical difficulty that arises should either L or M be zero. ■

Substitution Law

If $\lim_{x\to a} g(x) = L$ and $\lim_{x\to L} f(x) = f(L)$, then

$$\lim_{x\to a} f(g(x)) = f(L).$$

PROOF Let $\epsilon > 0$ be given. We must find a number $\delta > 0$ such that

$$0 < |x - a| < \delta \quad \text{implies that} \quad |f(g(x)) - f(L)| < \epsilon.$$

Because $f(y) \to f(L)$ as $y \to L$, there exists $\delta_1 > 0$ such that

$$0 < |y - L| < \delta_1 \quad \text{implies that} \quad |f(y) - f(L)| < \epsilon. \qquad \textbf{(6)}$$

Also, because $g(x) \to L$ as $x \to a$, we can find $\delta > 0$ such that

$$0 < |x - a| < \delta \quad \text{implies that} \quad |g(x) - L| < \delta_1;$$

that is, such that

$$|y - L| < \delta_1,$$

where $y = g(x)$. From (6) we see that $0 < |x - a| < \delta$ implies that

$$|f(g(x)) - f(L)| = |f(y) - f(L)| < \epsilon,$$

as desired. ■

Reciprocal Law

If $\lim_{x\to a} g(x) = L$ and $L \neq 0$, then

$$\lim_{x\to a}\frac{1}{g(x)} = \frac{1}{L}.$$

PROOF Let $f(x) = 1/x$. Then, as we saw in Example 3,

$$\lim_{x\to a} f(x) = \lim_{x\to a}\frac{1}{x} = \frac{1}{L} = f(L).$$

Hence the substitution law gives the result

$$\lim_{x\to a}\frac{1}{g(x)} = \lim_{x\to a} f(g(x)) = f(L) = \frac{1}{L},$$

as desired. ■

Quotient Law

If $\lim_{x \to a} F(x) = L$ and $\lim_{x \to a} G(x) = M \neq 0$, then

$$\lim_{x \to a} \frac{F(x)}{G(x)} = \frac{L}{M}.$$

Proof It follows immediately from the product and reciprocal laws that

$$\lim_{x \to a} \frac{F(x)}{G(x)} = \lim_{x \to a} F(x) \cdot \frac{1}{G(x)} = \left(\lim_{x \to a} F(x)\right)\left(\lim_{x \to a} \frac{1}{G(x)}\right) = L \cdot \frac{1}{M} = \frac{L}{M},$$

as desired. ■

Squeeze Law

Suppose that $f(x) \leqq g(x) \leqq h(x)$ in some deleted neighborhood of a and that

$$\lim_{x \to a} f(x) = L = \lim_{x \to a} h(x).$$

Then

$$\lim_{x \to a} g(x) = L.$$

Proof Given $\epsilon > 0$, we choose $\delta_1 > 0$ and $\delta_2 > 0$ such that

$$0 < |x - a| < \delta_1 \quad \text{implies that} \quad |f(x) - L| < \epsilon$$

and

$$0 < |x - a| < \delta_2 \quad \text{implies that} \quad |h(x) - L| < \epsilon.$$

Let $\delta = \min\{\delta_1, \delta_2\}$. Then $\delta > 0$. Moreover, if $0 < |x - a| < \delta$, then both $f(x)$ and $h(x)$ are points of the open interval $(L - \epsilon, L + \epsilon)$. So

$$L - \epsilon < f(x) \leqq g(x) \leqq h(x) < L + \epsilon.$$

Thus

$$0 < |x - a| < \delta \quad \text{implies that} \quad |g(x) - L| < \epsilon,$$

as desired. ■

APPENDIX D PROBLEMS

In Problems 1 through 10, apply the definition of the limit to establish the given equality.

1. $\lim_{x \to a} x = a$

2. $\lim_{x \to 2} 3x = 6$

3. $\lim_{x \to 2} (x + 3) = 5$

4. $\lim_{x \to -3} (2x + 1) = -5$

5. $\lim_{x \to 1} x^2 = 1$

6. $\lim_{x \to a} x^2 = a^2$

7. $\lim_{x \to -1} (2x^2 - 1) = 1$

8. $\lim_{x \to a} \frac{1}{x^2} = \frac{1}{a^2}$

9. $\lim_{x \to a} \frac{1}{x^2 + 1} = \frac{1}{a^2 + 1}$

10. $\lim_{x \to a} \frac{1}{\sqrt{x}} = \frac{1}{\sqrt{a}}$ if $a > 0$

11. Suppose that

$$\lim_{x \to a} f(x) = L \quad \text{and} \quad \lim_{x \to a} f(x) = M.$$

Apply the definition of the limit to prove that $L = M$. Thus the limit of the function f at $x = a$ is unique if it exists.

12. Suppose that C is a constant and that $f(x) \to L$ as $x \to a$. Apply the definition of the limit to prove that

$$\lim_{x \to a} C \cdot f(x) = C \cdot L.$$

13. Suppose that $L \neq 0$ and that $f(x) \to L$ as $x \to a$. Use the method of Example 3 and the definition of the limit to show directly that

$$\lim_{x \to a} \frac{1}{f(x)} = \frac{1}{L}.$$

14. Use the algebraic identity

$$x^n - a^n = (x - a)(x^{n-1} + x^{n-2}a + x^{n-3}a^2 + \cdots + xa^{n-2} + a^{n-1})$$

to show directly from the definition of the limit that $\lim_{x \to a} x^n = a^n$ if n is a positive integer.

15. Apply the identity

$$\left|\sqrt{x} - \sqrt{a}\right| = \frac{|x - a|}{\sqrt{x} + \sqrt{a}}$$

to show directly from the definition of the limit that $\lim_{x \to a} \sqrt{x} = \sqrt{a}$ if $a > 0$.

16. Suppose that $f(x) \to f(a) > 0$ as $x \to a$. Prove that there exists a neighborhood of a on which $f(x) > 0$; that is, prove that there exists $\delta > 0$ such that

$$|x - a| < \delta \quad \text{implies that} \quad f(x) > 0.$$

APPENDIX E: THE COMPLETENESS OF THE REAL NUMBER SYSTEM

Here we present a self-contained treatment of those consequences of the completeness of the real number system that are relevant to this text. Our principal objective is to prove the intermediate value theorem and the maximum value theorem. We begin with the least upper bound property of the real numbers, which we take to be an axiom.

Definition *Upper Bound and Lower Bound*

The set S of real numbers is said to be **bounded above** if there is a number b such that $x \leqq b$ for every number x in S, and the number b is then called an **upper bound** for S. Similarly, if there is a number a such that $x \geqq a$ for every number x in S, then S is said to be **bounded below,** and a is called a **lower bound** for S.

Definition *Least Upper Bound and Greatest Lower Bound*

The number λ is said to be a **least upper bound** for the set S of real numbers provided that

1. λ is an upper bound for S, and
2. If b is an upper bound for S, then $\lambda \leqq b$.

Similarly, the number γ is said to be a **greatest lower bound** for S if γ is a lower bound for S and $\gamma \geqq a$ for every lower bound a of S.

Exercise Prove that if the set S has a least upper bound λ, then it is unique. That is, prove that if λ and μ are least upper bounds for S, then $\lambda = \mu$.

It is easy to show that the greatest lower bound γ of a set S, if any, is also unique. At this point you should construct examples to illustrate that a set with a least upper bound λ may or may not contain λ and that a similar statement is true of the set's greatest lower bound.

We now state the *completeness axiom* of the real number system.

Least Upper Bound Axiom

If the nonempty set S of real numbers has an upper bound, then it has a least upper bound.

By working with the set T consisting of the numbers $-x$, where x is in S, it is not difficult to show the following consequence of the least upper bound axiom: If the nonempty set S of real numbers is bounded below, then S has a greatest lower bound. Because of this symmetry, we need only one axiom, not two; results for least upper bounds also hold for greatest lower bounds, provided that some attention is paid to the directions of the inequalities.

The restriction that S be nonempty is annoying but necessary. If S is the "empty" set of real numbers, then 15 is an upper bound for S, but S has no least upper bound because $14, 13, 12, \dots, 0, -1, -2, \dots$ are also upper bounds for S.

Definition *Increasing, Decreasing, and Monotonic Sequences*

The infinite sequence $x_1, x_2, x_3, \dots x_k, \dots$ is said to be **nondecreasing** if $x_n \leqq x_{n+1}$ for every $n \geqq 1$. This sequence is said to be **nonincreasing** if $x_n \geqq x_{n+1}$ for every $n \geqq 1$. If the sequence $\{x_n\}$ is either nonincreasing or nondecreasing, then it is said to be **monotonic**.

Theorem 1 gives the **bounded monotonic sequence property** of the set of real numbers. (Recall that a set S of real numbers is said to be **bounded** if it is contained in an interval of the form $[a, b]$.)

Theorem 1 *Bounded Monotonic Sequences*

Every bounded monotonic sequence of real numbers converges.

PROOF Suppose that the sequence

$$S = \{x_n\} = \{x_1, x_2, x_3, \dots, x_k, \dots\}$$

is bounded and nondecreasing. By the least upper bound axiom, S has a least upper bound λ. We claim that λ is the limit of the sequence $\{x_n\}$. Consider an open interval centered at λ—that is, an interval of the form $I = (\lambda - \epsilon, \lambda + \epsilon)$, where $\epsilon > 0$. Some terms of the sequence must lie within I, or else $\lambda - \epsilon$ would be an upper bound for S that is less than its least upper bound λ. But if x_N is in I, then—because we are dealing with a nondecreasing sequence—$x_N \leqq x_k \leqq \lambda$ for all $k \geqq N$. That is, x_k is in I for all $k \geqq N$. Because ϵ is an arbitrary positive number, λ is by definition (Section 11.2) the limit of the sequence $\{x_n\}$. Thus we have shown that a bounded nondecreasing sequence converges. A similar proof can be constructed for nonincreasing sequences by working with the greatest lower bound. ■

Therefore, the least upper bound axiom implies the bounded monotonic sequence property of the real numbers. With just a little effort, you can prove that the two are logically equivalent. That is, if you take the bounded monotonic sequence property as an axiom, then the least upper bound property follows as a theorem. The *nested interval property* of Theorem 2 is also equivalent to the least upper bound property, but we shall prove only that it follows from the least upper bound property, because we have chosen the latter as the fundamental completeness axiom for the real number system.

Theorem 2 *Nested Interval Property of the Real Numbers*

Suppose that $I_1, I_2, I_3, \ldots, I_n, \ldots$ is a sequence of closed intervals (so I_n is of the form $[a_n, b_n]$ for each positive integer n) such that

1. I_n contains I_{n+1} for each $n \geqq 1$, and
2. $\lim_{n\to\infty} (b_n - a_n) = 0$.

Then there exists exactly one real number c such that c belongs to I_n for all n. Thus

$$\{c\} = I_1 \cap I_2 \cap I_3 \cap \ldots .$$

PROOF It is clear from hypothesis (2) of Theorem 2 that there is at most one such number c. The sequence $\{a_n\}$ of the left-hand endpoints of the intervals is a bounded (by b_1) nondecreasing sequence and thus has a limit a by the bounded monotonic sequence property. Similarly, the sequence $\{b_n\}$ has a limit b. Because $a_n \leqq b_n$ for all n, it follows easily that $a \leqq b$. It is clear that $a_n \leqq a \leqq b \leqq b_n$ for all $n \geqq 1$, so a and b belong to every interval I_n. But then hypothesis (2) of Theorem 2 implies that $a = b$, and clearly this common value—call it c—is the number satisfying the conclusion of Theorem 2. ■

We can now use these results to prove several important theorems used in the text.

Theorem 3 *Intermediate Value Property of Continuous Functions*

If the function f is continuous on the interval $[a, b]$ and $f(a) < K < f(b)$, then $K = f(c)$ for some number c in (a, b).

PROOF Let $I_1 = [a, b]$. Suppose that I_n has been defined for $n \geqq 1$. We describe (inductively) how to define I_{n+1}, and this shows in particular how to define I_2, I_3, and so forth. Let a_n be the left-hand endpoint of I_n, b_n be its right-hand endpoint, and m_n be its midpoint. There are now three cases to consider: $f(m_n) > K$, $f(m_n) < K$, and $f(m_n) = K$.

If $f(m_n) > K$, then $f(a_n) < K < f(m_n)$; in this case, let $a_{n+1} = a_n$, $b_{n+1} = m_n$, and $I_{n+1} = [a_{n+1}, b_{n+1}]$.

If $f(m_n) < K$, then let $a_{n+1} = m_n$, $b_{n+1} = b_n$, and $I_{n+1} = [a_{n+1}, b_{n+1}]$.

If $f(m_n) = K$, then we simply let $c = m_n$ and the proof is complete. Otherwise, at each stage we bisect I_n and let I_{n+1} be the half of I_n on which f takes on values both above and below K.

It is easy to show that the sequence $\{I_n\}$ of intervals satisfies the hypotheses of Theorem 2. Let c be the (unique) real number common to all the intervals I_n. We will show that $f(c) = K$, and this will conclude the proof.

The sequence $\{b_n\}$ has limit c, so by the continuity of f, the sequence $\{f(b_n)\}$ has limit $f(c)$. But $f(b_n) > K$ for all n, so the limit of $\{f(b_n)\}$ can be no less than K; that is, $f(c) \geqq K$. By considering the sequence $\{a_n\}$, it follows that $f(c) \leqq K$ as well. Therefore, $f(c) = K$. ■

Lemma 1

If f is continuous on the closed interval $[a, b]$, then f is bounded there.

PROOF Suppose by way of contradiction that f is not bounded on $I_1 = [a, b]$. Bisect I_1 and let I_2 be the half of I_1 on which f is unbounded. (If f is unbounded on both halves, let I_2 be the left half of I_1.) In general, let I_{n+1} be a half of I_n on which f is unbounded.

Again it is easy to show that the sequence $\{I_n\}$ of closed intervals satisfies the hypotheses of Theorem 2. Let c be the number common to them all. Because f is continuous, there exists a number $\epsilon > 0$ such that f is bounded on the interval $(c - \epsilon, c + \epsilon)$. But for sufficiently large values of n, I_n is a subset of $(c - \epsilon, c + \epsilon)$. This contradiction shows that f must be bounded on $[a, b]$. ■

Theorem 4 ***Maximum Value Property of Continuous Functions***

If the function f is continuous on the closed and bounded interval $[a, b]$, then there exists a number c in $[a, b]$ such that $f(x) \leqq f(c)$ for all x in $[a, b]$.

PROOF Consider the set $S = \{f(x) | a \leqq x \leqq b\}$. By Lemma 1, this set is bounded, and it is certainly nonempty. Let λ be the least upper bound of S. Our goal is to show that λ is a value $f(x)$ of f.

With $I_1 = [a, b]$, bisect I_1 as before. Note that λ is the least upper bound of the values of f on at least one of the two halves of I_1; let I_2 be that half. Having defined I_n, let I_{n+1} be the half of I_n on which λ is the least upper bound of the values of f. Let c be the number common to all these intervals. It then follows from the continuity of f, much as in the proof of Theorem 3, that $f(c) = \lambda$. And it is clear that $f(x) \leqq \lambda$ for all x in $[a, b]$. ■

The technique we are using in these proofs is called the *method of bisection*. We now use it once again to establish the *Bolzano-Weierstrass property* of the real number system.

Definition ***Limit Point***

Let S be a set of real numbers. The number p is said to be a **limit point** of S if every open interval containing p also contains points of S other than p.

Theorem 5 ***Bolzano-Weierstrass Theorem***

Every bounded infinite set of real numbers has a limit point.

PROOF Let I_0 be a closed interval containing the bounded infinite set S of real numbers. Bisect I_0. Let I_1 be one of the resulting closed half-intervals of I_0 that contains infinitely many points of S. If I_n has been chosen, let I_{n+1} be one of the closed half-intervals of I_n containing infinitely many points of S. An application of Theorem 2 yields a number p common to all the intervals I_n. If J is an open interval containing p, then J contains I_n for some sufficiently large value of n and thus contains infinitely many points of S. Therefore p is a limit point of S. ■

Our final goal is in sight: We can now prove that a sequence of real numbers converges if and only if it is a *Cauchy sequence*.

Definition ***Cauchy Sequence***

The sequence $\{a_n\}_1^\infty$ is said to be a **Cauchy sequence** if, for every $\epsilon > 0$, there exists an integer N such that

$$|a_m - a_n| < \epsilon$$

for all $m, n \geqq N$.

Lemma 2 ***Convergent Subsequences***

Every bounded sequence of real numbers has a convergent subsequence.

Proof If $\{a_n\}$ has only a finite number of values, then the conclusion of Lemma 2 follows easily. We therefore focus our attention on the case in which $\{a_n\}$ is an infinite set. It is easy to show that this set is also bounded, and thus we may apply the Bolzano-Weierstrass theorem to obtain a limit point p of $\{a_n\}$.

For each integer $k \geqq 1$, let $a_{n(k)}$ be a term of the sequence $\{a_n\}$ such that

1. $n(k+1) > n(k)$ for all $k \geqq 1$, and

2. $|a_{n(k)} - p| < \dfrac{1}{k}$.

It is then easy to show that $\{a_{n(k)}\}$ is a convergent (to p) subsequence of $\{a_n\}$. ■

Theorem 6 ***Convergence of Cauchy Sequences***

A sequence of real numbers converges if and only if it is a Cauchy sequence.

Proof It follows immediately from the triangle inequality that every convergent sequence is a Cauchy sequence. Thus suppose that the sequence $\{a_n\}$ is a Cauchy sequence.

Choose N such that

$$|a_m - a_n| < 1$$

if $m, n \geqq N$. It follows that if $n \geqq N$, then a_n lies in the closed interval $[a_N - 1, a_N + 1]$. This implies that the sequence $\{a_n\}$ is bounded, and thus by Lemma 2 it has a convergent subsequence $\{a_{n(k)}\}$. Let p be the limit of this subsequence.

We claim that $\{a_n\}$ itself converges to p. Given $\epsilon > 0$, choose M such that

$$|a_m - a_n| < \frac{\epsilon}{2}$$

if $m, n \geqq M$. Next choose K such that $n(K) \geqq M$ and

$$|a_{n(K)} - p| < \frac{\epsilon}{2}.$$

Then if $n \geqq M$,

$$|a_n - p| \leqq |a_n - a_{n(K)}| + |a_{n(K)} - p| < \epsilon.$$

Therefore $\{a_n\}$ converges to p by definition. ■

APPENDIX F: PROOF OF THE CHAIN RULE

To prove the chain rule, we need to show that if f is differentiable at a and g is differentiable at $f(a)$, then

$$\lim_{h\to 0} \frac{g(f(a+h)) - g(f(a))}{h} = g'(f(a))\cdot f'(a). \tag{1}$$

If the quantities h and

$$k(h) = f(a+h) - f(a) \tag{2}$$

are nonzero, then we can write the difference quotient on the left-hand side of Eq. (1) as

$$\frac{g(f(a+h)) - g(f(a))}{h} = \frac{g(f(a)+k(h)) - g(f(a))}{k(h)} \cdot \frac{k(h)}{h}. \tag{3}$$

To investigate the first factor on the right-hand side of Eq. (3), we define a new function ϕ as follows:

$$\phi(k) = \begin{cases} \dfrac{g(f(a)+k) - g(f(a))}{k} & \text{if } k \neq 0; \\ g'(f(a)) & \text{if } k = 0. \end{cases} \tag{4}$$

By the definition of the derivative of g, we see from Eq. (4) that ϕ is continuous at $k = 0$; that is,

$$\lim_{k\to 0} \phi(k) = g'(f(a)). \tag{5}$$

Next,

$$\lim_{h\to 0} k(h) = \lim_{h\to 0} [f(a+h) - f(a)] = 0 \tag{6}$$

because f is continuous at $x = a$, and $\phi(0) = g'(f(a))$. It therefore follows from Eq. (5) that

$$\lim_{h\to 0} \phi(k(h)) = g'(f(a)). \tag{7}$$

We are now ready to assemble all this information. By Eq. (3), if $h \neq 0$, then

$$\frac{g(f(a+h)) - g(f(a))}{h} = \phi(k(h)) \cdot \frac{f(a+h) - f(a)}{h} \tag{8}$$

even if $k(h) = 0$, because in this case both sides of Eq. (8) are zero. Hence the product rule for limits yields

$$\begin{aligned} \lim_{h\to 0} \frac{g(f(a+h)) - g(f(a))}{h} &= \lim_{h\to 0} \phi(k(h)) \cdot \frac{f(a+h) - f(a)}{h} \\ &= g'(f(a))\cdot f'(a), \end{aligned}$$

a consequence of Eq. (7) and the definition of the derivative of the function f. We have therefore established the chain rule in the form of Eq. (1). ■

APPENDIX G: EXISTENCE OF THE INTEGRAL

When the basic computational algorithms of the calculus were discovered by Newton and Leibniz in the latter half of the seventeenth century, the logical rigor that had been a feature of the Greek method of exhaustion was largely abandoned. When computing the area A under the curve $y = f(x)$, for example, Newton took it as intuitively obvious that the area function existed, and he proceeded to compute it as the antiderivative of the height function $f(x)$. Leibniz regarded A as an infinite sum of infinitesimal area elements, each of the form $dA = f(x)\,dx$, but in practice computed the area

$$A = \int_a^b f(x)\,dx$$

by antidifferentiation just as Newton did—that is, by computing

$$A = \Big[D^{-1} f(x) \Big]_a^b .$$

The question of the existence of the area function—of the conditions that a function f must satisfy in order for its integral to exist—did not at first seem to be of much importance. Eighteenth-century mathematicians were mainly occupied (and satisfied) with the impressive applications of calculus to the solution of real-world problems and did not concentrate on the logical foundations of the subject.

The first attempt at a precise definition of the integral and a proof of its existence for continuous functions was that of the French mathematician Augustin Louis Cauchy (1789–1857). Curiously enough, Cauchy was trained as an engineer, and much of his research in mathematics was in fields that we today regard as applications-oriented: hydrodynamics, waves in elastic media, vibrations of elastic membranes, polarization of light, and the like. But he was a prolific researcher, and his writings cover the entire spectrum of mathematics, with occasional essays into almost unrelated fields.

Around 1824, Cauchy defined the integral of a continuous function in a way that is familiar to us, as a limit of left-endpoint approximations:

$$\int_a^b f(x)\,dx = \lim_{\Delta x \to 0} \sum_{i=1}^{n} f(x_{i-1})\,\Delta x.$$

This is a much more complicated sort of limit than the ones we discussed in Chapter 2. Cauchy was not entirely clear about the nature of the limit process involved in this equation, nor was he clear about the precise role that the hypothesis of the continuity of f played in proving that the limit exists.

A complete definition of the integral, as we gave in Section 5.4, was finally produced in the 1850s by the German mathematician Georg Bernhard Riemann. Riemann was a student of Gauss; he met Gauss upon his arrival at Göttingen, Germany, for the purpose of studying theology, when he was about 20 years old and Gauss was about 70. Riemann soon decided to study mathematics and became known as one of the truly great mathematicians of the nineteenth century. Like Cauchy, he was particularly interested in applications of mathematics to the real world; his research emphasized electricity, heat, light, acoustics, fluid dynamics, and—as you might infer from the fact that Wilhelm Weber was a major influence of Riemann's education—magnetism. Riemann also made significant contributions to mathematics itself, particularly in the field of complex analysis. A major conjecture of his, involving the zeta function

$$\zeta(s) = \sum_{n=1}^{\infty} \frac{1}{n^s}, \tag{1}$$

remains unsolved to this day. This conjecture has important consequences in the theory of the distribution of prime numbers because

$$\zeta(k) = \prod \left(1 - \frac{1}{p^k}\right)^{-1},$$

where the product Π is taken over all primes p. [The zeta function is defined in Eq. (1) for complex numbers s to the right of the vertical line at $x = 1$ and is extended to other complex numbers by the requirement that it be differentiable.] Riemann died of tuberculosis shortly before his fortieth birthday.

Here we give a proof of the existence of the integral of a continuous function. We will follow Riemann's approach. Specifically, suppose that the function f is continuous on the closed and bounded interval $[a, b]$. We will prove that the definite integral

$$\int_a^b f(x)\,dx$$

exists. That is, we will demonstrate the existence of a number I that satisfies the following condition: For every $\epsilon > 0$ there exists $\delta > 0$ such that, for every Riemann sum R associated with any partition P with $|P| < \delta$,

$$|I - R| < \epsilon.$$

(Recall that the norm $|P|$ of the partition P is the length of the longest subinterval in the partition.) In other words, every Riemann sum associated with every sufficiently "fine" partition is close to the number I. If this happens, then the definite integral

$$\int_a^b f(x)\,dx$$

is said to **exist,** and I is its **value**.

Now we begin the proof. Suppose throughout that f is a function continuous on the closed interval $[a, b]$. Given $\epsilon > 0$, we need to show the existence of a number $\delta > 0$ such that

$$\left| I - \sum_{i=1}^{n} f(x_i^\star)\,\Delta x_i \right| < \epsilon \tag{2}$$

for every Riemann sum associated with any partition P of $[a, b]$ with $|P| < \delta$.

Given a partition P of $[a, b]$ into n subintervals that are *not necessarily of equal length,* let p_i be a point in the subinterval $[x_{i-1}, x_i]$ at which f attains its minimum value $f(p_i)$. Similarly, let $f(q_i)$ be its maximum value there. These numbers exist for $i = 1, 2, 3, \ldots, n$ because of the maximum value property of continuous functions (Theorem 4 of Appendix E).

In what follows we will denote the resulting lower and upper Riemann sums associated with P by

$$L(P) = \sum_{i=1}^{n} f(p_i)\,\Delta x_i \tag{3a}$$

and

$$U(P) = \sum_{i=1}^{n} f(q_i)\,\Delta x_i, \tag{3b}$$

respectively. Then Lemma 1 is obvious.

Lemma 1

For any partition P of $[a, b]$, $L(P) \leqq U(P)$.

Now we need a definition. The partition P' is called a **refinement** of the partition P if each subinterval of P' is contained in some subinterval of P. That is, P' is obtained from P by adding more points of subdivision to P.

Lemma 2

Suppose that P' is a refinement of P. Then

$$L(P) \leqq L(P') \leqq U(P') \leqq U(P). \tag{4}$$

Proof The inequality $L(P') \leqq U(P')$ is a consequence of Lemma 1. We will show that $L(P) \leqq L(P')$; the proof that $U(P') \leqq U(P)$ is similar.

The refinement P' is obtained from P by adding one or more points of subdivision to P. So all we need show is that the Riemann sum $L(P)$ cannot be decreased by adding a single point of subdivision. Thus we will suppose that the partition P' is obtained from P by dividing the kth subinterval $[x_{k-1}, x_k]$ of P into two subintervals $[x_{k-1}, z]$ and $[z, x_k]$ by means of the new subdivision point z.

The only resulting effect on the corresponding Riemann sum is to replace the term

$$f(p_k)\cdot(x_k - x_{k-1})$$

in $L(P)$ with the two-term sum

$$f(u)\cdot(z - x_{k-1}) + f(v)\cdot(x_k - z),$$

where $f(u)$ is the minimum of f on $[x_{k-1}, z]$ and $f(v)$ is the minimum of f on $[z, x_k]$. But

$$f(p_k) \leqq f(u) \quad \text{and} \quad f(p_k) \leqq f(v).$$

Hence

$$\begin{aligned} f(u)\cdot(z - x_{k-1}) + f(v)\cdot(x_k - z) &\geqq f(p_k)\cdot(z - x_{k-1}) + f(p_k)\cdot(x_k - z) \\ &= f(p_k)\cdot(z - x_{k-1} + x_k - z) \\ &= f(p_k)\cdot(x_k - x_{k-1}). \end{aligned}$$

So the replacement of $f(p_k)\cdot(x_k - x_{k-1})$ cannot decrease the sum $L(P)$ in question, and therefore $L(P) \leqq L(P')$. Because this is all we needed to show, we have completed the proof of Lemma 2. ■

To prove that all the Riemann sums for sufficiently fine partitions are close to some number I, we must first give a construction of I. This is accomplished through Lemma 3.

Lemma 3

Let P_n denote the regular partition of $[a, b]$ into 2^n subintervals of equal length. Then the (sequential) limit

$$I = \lim_{n\to\infty} L(P_n) \tag{5}$$

exists.

Proof We begin with the observation that each partition P_{n+1} is a refinement of P_n, so (by Lemma 2)

$$L(P_1) \leqq L(P_2) \leqq \cdots \leqq L(P_n) \leqq \cdots.$$

Therefore, $\{L(P_n)\}$ is a nondecreasing sequence of real numbers. Moreover,

$$L(P_n) = \sum_{i=1}^{2^n} f(p_i)\,\Delta x_i \leqq M \sum_{i=1}^{2^n} \Delta x_i = M(b - a),$$

where M is the maximum value of f on $[a, b]$.

Theorem 1 of Appendix E guarantees that a bounded monotonic sequence of real numbers must converge. Thus the number

$$I = \lim_{n\to\infty} L(P_n)$$

exists. This establishes Eq. (5), and the proof of Lemma 3 is complete. ■

It is proved in advanced calculus that if f is continuous on $[a, b]$, then—for every number $\epsilon > 0$—there exists a number $\delta > 0$ such that

$$|f(u) - f(v)| < \epsilon$$

for every two points u and v of $[a, b]$ such that

$$|u - v| < \delta.$$

This property of a function is called **uniform continuity** of f on the interval $[a, b]$. Thus the theorem from advanced calculus that we need to use states that every continuous function on a closed and bounded interval is uniformly continuous there.

Note The fact that f is continuous on $[a, b]$ means that for each number u in the interval and each $\epsilon > 0$, there exists $\delta > 0$ such that if v is a number in the interval with $|u - v| < \delta$, then $|f(u) - f(v)| < \epsilon$. But *uniform* continuity is a more stringent condition. It means that given $\epsilon > 0$, you can find not only a value δ_1 that "works" for u_1, a value δ_2 that works for u_2, and so on, but more: You can find a universal value of δ that works for *all* values of u in the interval. This should not be obvious when you notice the possibility that $\delta_1 = 1$, $\delta_2 = \frac{1}{2}$, $\delta_3 = \frac{1}{3}$, and so on. In any case, it is clear that uniform continuity of f on an interval implies its continuity there.

Remember that throughout we have a continuous function f defined on the closed interval $[a, b]$.

Lemma 4

Suppose that $\epsilon > 0$ is given. Then there exists a number $\delta > 0$ such that if P is a partition of $[a, b]$ with $|P| < \delta$ and P' is a refinement of P, then

$$|R(P) - R(P')| < \frac{\epsilon}{3} \qquad \textbf{(6)}$$

for any two Riemann sums $R(P)$ associated with P and $R(P')$ associated with P'.

Proof Because f must be uniformly continuous on $[a, b]$, there exists a number $\delta > 0$ such that if

$$|u - v| < \delta, \quad \text{then} \quad |f(u) - f(v)| < \frac{\epsilon}{3(b - a)}.$$

Suppose now that P is a partition of $[a, b]$ with $|P| < \delta$. Then

$$|U(P) - L(P)| = \sum_{i=1}^{n} |f(q_i) - f(p_i)|\,\Delta x_i < \frac{\epsilon}{3(b - a)} \sum_{i=1}^{n} \Delta x_i = \frac{\epsilon}{3}.$$

This is valid because $|p_i - q_i| < \delta$, for both p_i and q_i belong to the same subinterval $[x_{i-1}, x_i]$ of P, and $|P| < \delta$.

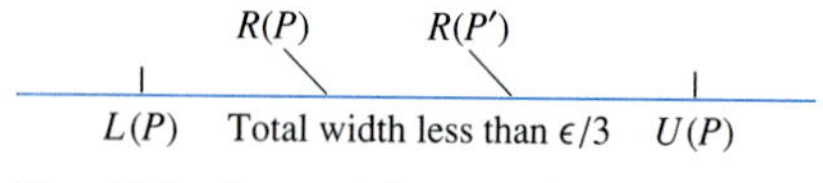

Fig. G.1 Part of the proof of Lemma 4

Now, as shown in Fig. G.1, we know that $L(P)$ and $U(P)$ differ by less than $\epsilon/3$. We know also that

$$L(P) \leqq R(P) \leqq U(P)$$

for every Riemann sum $R(P)$ associated with P. But

$$L(P) \leqq L(P') \leqq U(P') \leqq U(P)$$

by Lemma 2, because P' is a refinement of P; moreover,

$$L(P') \leqq R(P') \leqq U(P')$$

for every Riemann sum $R(P')$ associated with P'.

As Fig. G.1 shows, both the numbers $R(P)$ and $R(P')$ lie in the interval $[L(P), U(P)]$ of length less than $\epsilon/3$, so Eq. (6) follows. This concludes the proof of Lemma 4. ■

Theorem 1 ***Existence of the Integral***
If f is continuous on the closed and bounded interval $[a, b]$, then the integral

$$\int_a^b f(x)\,dx$$

exists.

Proof Suppose that $\epsilon > 0$ is given. We must show the existence of a number $\delta > 0$ such that, for every partition P of $[a, b]$ with $|P| < \delta$, we have

$$|I - R(P)| < \epsilon,$$

where I is the number given in Lemma 3 and $R(P)$ is an arbitrary Riemann sum for f associated with P.

We choose the number δ provided by Lemma 4 such that

$$|R(P) - R(P')| < \frac{\epsilon}{3}$$

if $|P| < \delta$ and P' is a refinement of P.

By Lemma 3, we can choose an integer N so large that

$$|P_N| < \delta \quad \text{and} \quad |L(P_N) - I| < \frac{\epsilon}{3}. \tag{7}$$

Given an arbitrary partition P such that $|P| < \delta$, let P' be a common refinement of both P and P_N. You can obtain such a partition P', for example, by using all the points of subdivision of both P and P_N to form the subintervals of $[a, b]$ that constitute P'.

Because P' is a refinement of both P and P_N and both the latter partitions have norm less than δ, Lemma 4 implies that

$$|R(P) - R(P')| < \frac{\epsilon}{3} \quad \text{and} \quad |L(P_N) - R(P')| < \frac{\epsilon}{3}. \tag{8}$$

Here $R(P)$ and $R(P')$ are (arbitrary) Riemann sums associated with P and P', respectively.

Given an arbitrary Riemann sum $R(P)$ associated with the partition P with norm less than δ, we see that

$$\begin{aligned} |I - R(P)| &= |I - L(P_N) + L(P_N) - R(P') + R(P') - R(P)| \\ &\leqq |I - L(P_N)| + |L(P_N) - R(P')| + |R(P') - R(P)|. \end{aligned}$$

In the last sum, both of the last two terms are less than $\epsilon/3$ by virtue of the inequalities in (8). We also know, by (7), that the first term is less than $\epsilon/3$. Consequently,

$$|I - R(P)| < \epsilon.$$

This establishes Theorem 1. ■

We close with an example that shows that some hypothesis of continuity is required for integrability.

Example 1: Suppose that f is defined for $0 \leqq x \leqq 1$ as follows:

$$f(x) = \begin{cases} 1 & \text{if } x \text{ is irrational,} \\ 0 & \text{if } x \text{ is rational.} \end{cases}$$

Then f is not continuous anywhere. (Why?) Give a partition P of $[0, 1]$, let p_i be a rational point and q_i an irrational point of the ith subinterval of P for each i, $1 \leqq i \leqq n$. As before, f attains its minimum value 0 at each p_i and its maximum value 1 at each q_i. Also

$$L(P) = \sum_{i=1}^{n} f(p_i)\,\Delta x_i = 0, \quad \text{whereas} \quad U(P) = \sum_{i=1}^{n} f(q_i)\,\Delta x_i = 1.$$

Thus if we choose $\epsilon = \frac{1}{2}$, then there is no number I that can lie within ϵ of both $L(P)$ and $U(P)$, no matter how small the norm of P. It follows that f is *not* integrable on $[0, 1]$. ■

REMARK This is not the end of the story of the integral. Integrals of highly discontinuous functions are important in many applications of physics, and near the beginning of the twentieth century a number of mathematicians, most notably Henri Lebesgue (1875–1941), developed more powerful integrals. The Lebesgue integral, in particular, always exists when the Riemann integral does and gives the same value; but the Lebesgue integral is sufficiently powerful to integrate even functions that are continuous nowhere. It reports that

$$\int_0^1 f(x)\,dx = 1$$

for the function f of Example 1. Other mathematicians have developed integrals with domains far more general than sets of real numbers or subsets of the plane or space.

APPENDIX H: APPROXIMATIONS AND RIEMANN SUMS

Several times in Chapter 6 our attempt to compute some quantity Q led to the following situation. Beginning with a regular partition of an appropriate interval $[a, b]$ into n subintervals, each of length Δx, we found an approximation A_n to Q of the form

$$A_n = \sum_{i=1}^{n} g(u_i)h(v_i)\,\Delta x, \tag{1}$$

where u_i and v_i are two (generally different) points of the ith subinterval $[x_{i-1}, x_i]$. For example, in our discussion of surface area of revolution that precedes Eq. (8) of Section 6.4, we found the approximation

$$\sum_{i=1}^{n} 2\pi f(u_i)\sqrt{1 + [f'(v_i)]^2}\,\Delta x \tag{2}$$

to the area of the surface generated by revolving the curve $y = f(x)$, $a \leqq x \leqq b$, around the x-axis. (In Section 6.4 we wrote $x_i^{\star\star}$ for u_i and $x_i^{\star}$ for v_i.) Note that the expression in (2) is the same as the right-hand side in Eq. (1); take $g(x) = 2\pi f(x)$ and $h(x) = \sqrt{1 + [f'(x)]^2}$.

In such a situation we observe that if u_i and v_i were the same point $x_i^{\star}$ of $[x_{i-1}, x_i]$ for each i $(i = 1, 2, 3, \dots, n)$, then the approximation in Eq. (1) would be a Riemann sum for the function $g(x)h(x)$ on $[a, b]$. This leads us to suspect that

$$\lim_{\Delta x \to 0} \sum_{i=1}^{n} g(u_i)h(v_i)\,\Delta x = \int_a^b g(x)h(x)\,dx. \tag{3}$$

In Section 6.4, we assumed the validity of Eq. (3) and concluded from the approximation in (2) that the surface area of revolution ought to be defined to be

$$A = \lim_{\Delta x \to 0} \sum_{i=1}^{n} 2\pi f(u_i)\sqrt{1 + [f'(v_i)]^2}\,\Delta x = \int_a^b 2\pi f(x)\sqrt{1 + [f'(x)]^2}\,dx.$$

Theorem 1 guarantees that Eq. (3) holds under mild restrictions on the functions g and h.

Theorem 1 ***A Generalization of Riemann Sums***
Suppose that h and g' are continuous on $[a, b]$. Then

$$\lim_{\Delta x \to 0} \sum_{i=1}^{n} g(u_i)h(v_i)\,\Delta x = \int_a^b g(x)h(x)\,dx, \tag{3}$$

where u_i and v_i are arbitrary points of the ith subinterval of a regular partition of $[a, b]$ into n subintervals, each of length Δx.

Proof Let M_1 and M_2 denote the maximum values on $[a, b]$ of $|g'(x)|$ and $|h(x)|$, respectively. Note that

$$\sum_{i=1}^{n} g(u_i)h(v_i)\,\Delta x = R_n + S_n, \quad \text{where} \quad R_n = \sum_{i=1}^{n} g(v_i)h(v_i)\,\Delta x$$

is a Riemann sum approaching $\int_a^b g(x)h(x)\,dx$ as $\Delta x \to 0$, and

$$S_n = \sum_{i=1}^{n} [g(u_i) - g(v_i)]h(v_i)\,\Delta x.$$

To prove Eq. (3), it is sufficient to show that $S_n \to 0$ as $\Delta x \to 0$. The mean value theorem gives

$$\begin{aligned} |g(u_i) - g(v_i)| &= |g'(\bar{x}_i)| \cdot |u_i - v_i| \qquad [\bar{x}_i \text{ in } (u_i, v_i)] \\ &\leqq M_1\,\Delta x, \end{aligned}$$

because both u_i and v_i are points of the interval $[x_{i-1}, x_i]$ of length Δx. Then

$$\begin{aligned} |S_n| &\leqq \sum_{i=1}^{n} |g(u_i) - g(v_i)| \cdot |h(v_i)|\,\Delta x \leqq \sum_{i=1}^{n} (M_1\,\Delta x) \cdot (M_2\,\Delta x) \\ &= (M_1 M_2\,\Delta x) \sum_{i=1}^{n} \Delta x = M_1 M_2 (b - a)\,\Delta x, \end{aligned}$$

from which it follows that $S_n \to 0$ as $\Delta x \to 0$, as desired. ■

As an application of Theorem 1, let us give a rigorous derivation of Eq. (2) of Section 6.3,

$$V = \int_a^b 2\pi x f(x)\,dx, \tag{4}$$

for the volume of the solid generated by revolving around the y-axis the region between the graph of $y = f(x)$ and the x-axis for $a \leqq x \leqq b$. Beginning with the usual regular partition of $[a, b]$, let $f(x_i^\flat)$ and $f(x_i^\sharp)$ denote the minimum and maximum values of f on the ith subinterval $[x_{i-1}, x_i]$. Denote by $x_i^\star$ the midpoint of this subinterval. From Fig. H.1, we see that the part of the solid generated by revolving the region below $y = f(x)$, $x_{i-1} \leqq x \leqq x_i$, contains a cylindrical shell with average radius $x_i^\star$, thickness Δx, and height $f(x_i^\flat)$ and is contained in another cylindrical shell

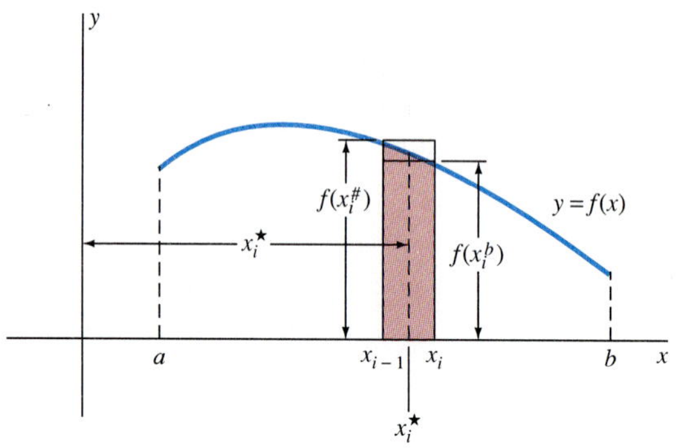

Fig. H.1 A careful estimate of the volume of a solid of revolution around the y-axis

with the same average radius and thickness but with height $f(x_i^\sharp)$. Hence the volume ΔV_i of this part of the solid satisfies the inequalities

$$2\pi x_i^\star f(x_i^\flat)\,\Delta x \leqq \Delta V_i \leqq 2\pi x_i^\star f(x_i^\sharp)\,\Delta x.$$

We add these inequalities for $i = 1, 2, 3, \dots, n$ and find that

$$\sum_{i=1}^{n} 2\pi x_i^\star f(x_i^\flat)\,\Delta x \leqq V \leqq \sum_{i=1}^{n} 2\pi x_i^\star f(x_i^\sharp)\,\Delta x.$$

Because Theorem 1 implies that both of the last two sums approach $\int_a^b 2\pi x f(x)\,dx$, the squeeze law of limits now implies Eq. (4).

We will occasionally need a generalization of Theorem 1 that involves the notion of a continuous function $F(x, y)$ of two variables. We say that F is *continuous* at the point (x_0, y_0) provided that the value $F(x, y)$ can be made arbitrarily close to $F(x_0, y_0)$ merely by choosing the point (x, y) sufficiently close to (x_0, y_0). We discussed continuity of functions of two variables in Chapter 13. Here it will suffice to accept the following facts: If $g(x)$ and $h(y)$ are continuous functions of the single variables x and y, respectively, then simple combinations such as

$$g(x) \pm h(y), \qquad g(x)h(y), \qquad \text{and} \qquad \sqrt{[g(x)]^2 + [h(y)]^2}$$

are continuous functions of the two variables x and y.

Now consider a regular partition of $[a, b]$ into n subintervals, each of length Δx, and let u_i and v_i denote arbitrary points of the ith subinterval $[x_{i-1}, x_i]$. Theorem 2—we omit the proof—tells us how to find the limit as $\Delta x \to 0$ of a sum such as

$$\sum_{i=1}^{n} F(u_i, v_i)\,\Delta x.$$

Theorem 2 ***A Further Generalization***

Let $F(x, y)$ be continuous for x and y both in the interval $[a, b]$. Then, in the notation of the preceding paragraph,

$$\lim_{\Delta x \to 0} \sum_{i=1}^{n} f(u_i, v_i)\,\Delta x = \int_a^b F(x, x)\,dx. \tag{5}$$

Theorem 1 is the special case $f(x, y) = g(x)h(y)$ of Theorem 2. Moreover, the integrand $F(x, x)$ on the right in Eq. (5) is merely an ordinary function of the single variable x. As a formal matter, the integral corresponding to the sum in Eq. (5) is obtained by replacing the summation symbol with an integral sign, changing both u_i and v_i to x, replacing Δx with dx, and inserting the correct limits of integration. For example, if the interval $[a, b]$ is $[0, 4]$, then

$$\begin{aligned}\lim_{\Delta x \to 0} \sum_{i=1}^{n} \sqrt{9u_i^2 + v_i^4}\ \Delta x &= \int_0^4 \sqrt{9x^2 + x^4}\ dx \\ &= \int_0^4 x(9 + x^2)^{1/2}\ dx = \left[\tfrac{1}{3}(9 + x^2)^{3/2}\right]_0^4 \\ &= \tfrac{1}{3}\left[(25)^{3/2} - (9)^{3/2}\right] = \tfrac{98}{3}.\end{aligned}$$

APPENDIX H PROBLEMS

In Problems 1 through 7, u_i and v_i are arbitrary points of the ith subinterval of a regular partition of $[a, b]$ into n subintervals, each of length Δx. Express the given limit as an integral from a to b, then compute the value of this integral.

1. $\displaystyle \lim_{\Delta x \to 0} \sum_{i=1}^{n} u_i v_i\ \Delta x;\quad a = 0, b = 1$

2. $\displaystyle \lim_{\Delta x \to 0} \sum_{j=1}^{n} (3u_j + 5v_j)\ \Delta x;\quad a = -1, b = 3$

3. $\displaystyle \lim_{\Delta x \to 0} \sum_{i=1}^{n} u_i\sqrt{4 - v_i^2}\ \Delta x;\quad a = 0, b = 2$

4. $\displaystyle \lim_{\Delta x \to 0} \sum_{i=1}^{n} \frac{u_i}{\sqrt{16 + v_i^2}}\ \Delta x;\quad a = 0, b = 3$

5. $\displaystyle \lim_{\Delta x \to 0} \sum_{i=1}^{n} \sin u_i \cos v_i\ \Delta x;\quad a = 0, b = \pi/2$

6. $\displaystyle \lim_{\Delta x \to 0} \sum_{i=1}^{n} \sqrt{\sin^2 u_i + \cos^2 v_i}\ \Delta x;\quad a = 0, b = \pi$

7. $\displaystyle \lim_{\Delta x \to 0} \sum_{k=1}^{n} \sqrt{u_k^4 + v_k^7}\ \Delta x;\quad a = 0, b = 2$

8. Explain how Theorem 1 applies to show that Eq. (8) of Section 6.4 follows from the discussion that precedes it in that section.

9. Use Theorem 1 to derive Eq. (10) of Section 6.4.

APPENDIX I: L'HÔPITAL'S RULE AND CAUCHY'S MEAN VALUE THEOREM

Here we give a proof of l'Hôpital's rule,

$$\lim_{x \to a} \frac{f(x)}{g(x)} = \lim_{x \to a} \frac{f'(x)}{g'(x)}, \tag{1}$$

under the hypotheses of Theorem 1 in Section 8.3. The proof is based on a generalization of the mean value theorem due to the French mathematician Augustin Louis Cauchy. Cauchy used this generalization in the early nineteenth century to give rigorous proofs of several calculus results not previously established firmly.

Cauchy's Mean Value Theorem

Suppose that the functions f and g are continuous on the closed and bounded interval $[a, b]$ and differentiable on (a, b). Then there exists a number c in (a, b) such that

$$[f(b) - f(a)]g'(c) = [g(b) - g(a)]f'(c). \tag{2}$$

REMARK 1 To see that this theorem is indeed a generalization of the (ordinary) mean value theorem, we take $g(x) = x$. Then $g'(x) \equiv 1$, and the conclusion in Eq. (2) reduces to the fact that

$$f(b) - f(a) = (b - a)f'(c)$$

for some number c in (a, b).

REMARK 2 Equation (2) has a geometric interpretation like that of the ordinary mean value theorem. Let us think of the equations $x = g(t)$, $y = f(t)$ as describing the motion of a point $P(x, y)$ moving along a curve C in the xy-plane as t increases from a to b (Fig. I.1). That is, $P(x, y) = P(g(t), f(t))$ is the location of the point P at time t. Under the assumption that $g(b) \neq g(a)$, the slope of the line L connecting the endpoints of the curve C is

$$m = \frac{f(b) - f(a)}{g(b) - g(a)}. \tag{3}$$

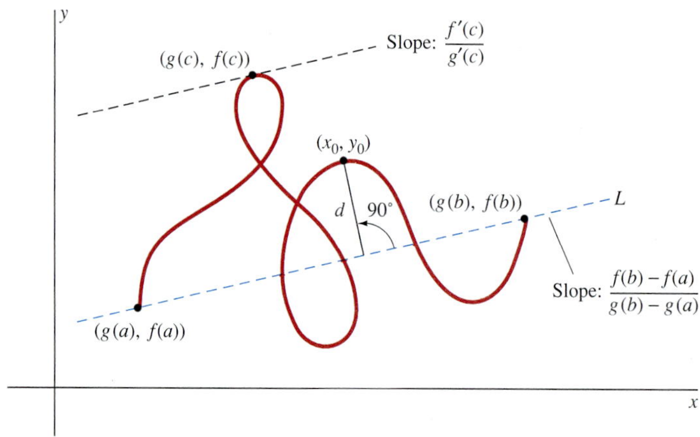

Fig. I.1 The idea of Cauchy's mean value theorem

But if $g'(c) \neq 0$, then the chain rule gives

$$\frac{dy}{dx} = \frac{dy/dt}{dx/dt} = \frac{f'(c)}{g'(c)} \tag{4}$$

for the slope of the line tangent to the curve C at the point $(g(c), f(c))$. So if $g(b) \neq g(a)$ and $g'(c) \neq 0$, then Eq. (2) may be written in the form

$$\frac{f(b) - f(a)}{g(b) - g(a)} = \frac{f'(c)}{g'(c)}, \tag{5}$$

so the two slopes in Eqs. (3) and (4) are equal. Thus Cauchy's mean value theorem implies that (under our assumptions) there is a point on the curve C where the tangent line is *parallel* to the line joining the endpoints of C. This is exactly what the (ordinary) mean value theorem says for an explicitly defined curve $y = f(x)$. This geometric interpretation motivates the following proof of Cauchy's mean value theorem.

Proof The line L through the endpoints in Fig. I.1 has point-slope equation

$$y - f(a) = \frac{f(b) - f(a)}{g(b) - g(a)}\,[x - g(a)],$$

which can be rewritten in the form $Ax + By + C = 0$ with

$$A = f(b) - f(a), \quad B = -[g(b) - g(a)], \quad \text{and}$$

$$C = f(a)[g(b) - g(a)] - g(a)[f(b) - f(a)]. \tag{6}$$

According to Miscellaneous Problem 71 at the end of Chapter 3, the (perpendicular) distance from the point (x_0, y_0) to the line L is

$$d = \frac{|Ax_0 + By_0 + C|}{\sqrt{A^2 + B^2}}.$$

Figure I.1 suggests that the point $(g(c), f(c))$ will maximize this distance d for points on the curve C.

We are motivated, therefore, to define the auxiliary function

$$\phi(t) = Ag(t) + Bf(t) + C, \tag{7}$$

with the constants A, B, and C as defined in (6). Thus $\phi(t)$ is essentially a constant multiple of the distance from $(g(t), f(t))$ to the line L in Fig. I.1.

Now $\phi(a) = 0 = \phi(b)$ (Why?), so Rolle's theorem (Section 4.3) implies the existence of a number c in (a, b) such that

$$\phi'(c) = Ag'(c) + Bf'(c) = 0. \tag{8}$$

We substitute the values of A and B from Eq. (6) into (8) and obtain the equation

$$[f(b) - f(a)]g'(c) - [g(b) - g(a)]f'(c) = 0.$$

This is the same as Eq. (2) in the conclusion of Cauchy's mean value theorem, and the proof is complete. ■

Note Although the assumptions that $g(b) \neq g(a)$ and $g'(c) \neq 0$ were needed for our geometric interpretation of the theorem, they were not used in its proof—only in the motivation for the method of proof.

Proof of l'Hôpital's Rule

Suppose that $f(x)/g(x)$ has the indeterminate form $0/0$ at $x = a$. We may invoke continuity of f and g to allow the assumption that $f(a) = 0 = f(b)$. That is, we simply define $f(a)$ and $g(a)$ to be zero in case their values at $x = a$ are not originally given.

Now we restrict our attention to values of x in a fixed deleted neighborhood of a on which both f and g are differentiable. Choose one such value of x and hold it temporarily constant. Then apply Cauchy's mean value theorem on the interval $[a, x]$. (If $x < a$, use the interval $[x, a]$.) We find that there is a number z between a and x that behaves as c does in Eq. (2). Hence, by virtue of Eq. (2), we obtain the equation

$$\frac{f(x)}{g(x)} = \frac{f(x) - f(a)}{g(x) - g(a)} = \frac{f'(z)}{g'(z)}.$$

Now z depends on x, but z is trapped between x and a, so z is forced to approach a as $x \to a$. We conclude that

$$\lim_{x \to a} \frac{f(x)}{g(x)} = \lim_{z \to a} \frac{f'(z)}{g'(z)} = \lim_{x \to a} \frac{f'(x)}{g'(x)},$$

under the assumption that the right-hand limit exists. Thus we have verified l'Hôpital's rule in the form of Eq. (1). ■

APPENDIX J: PROOF OF TAYLOR'S FORMULA

Several different proofs of Taylor's formula (Theorem 2 of Section 11.4) are known, but none of them seems very well motivated—each requires some "trick" to begin the proof. The trick we employ here (suggested by C. R. MacCluer) is to begin by introducing an auxiliary function $F(x)$, defined as follows:

$$F(x) = f(b) - f(x) - f'(x)(b - x) - \frac{f''(x)}{2!}(b - x)^2 - \cdots - \frac{f^{(n)}(x)}{n!}(b - x)^n - K(b - x)^{n+1}, \qquad \textbf{(1)}$$

where the *constant* K is chosen so that $F(a) = 0$. To see that there *is* such a value of K, we could substitute $x = a$ on the right and $F(x) = F(a) = 0$ on the left in Eq. (1) and then solve routinely for K, but we have no need to do this explicitly.

Equation (1) makes it quite obvious that $F(b) = 0$ as well. Therefore, Rolle's theorem (Section 3.2) implies that

$$F'(z) = 0 \qquad \textbf{(2)}$$

for some point z of the open interval (a, b) (under the assumption that $a < b$). To see what Eq. (2) means, we differentiate both sides of Eq. (1) and find that

$$\begin{aligned} F'(x) = {} & -f'(x) + [f'(x) - f''(x)(b - x)] \\ & + \left[f''(x)(b - x) - \frac{1}{2!}f^{(3)}(x)(b - x)^2\right] \\ & + \left[\frac{1}{2!}f^{(3)}(x)(b - x)^2 - \frac{1}{3!}f^{(4)}(x)(b - x)^3\right] \\ & + \cdots + \left[\frac{1}{(n-1)!}f^{(n)}(x)(b - x)^{n-1} - \frac{1}{n!}f^{(n+1)}(x)(b - x)^n\right] \\ & + (n + 1)K(b - x)^n. \end{aligned}$$

Upon careful inspection of this result, we see that all terms except the final two cancel in pairs. Thus the sum "telescopes" to give

$$F'(x) = (n + 1)K(b - x)^n - \frac{f^{(n+1)}(x)}{n!}(b - x)^n. \qquad \textbf{(3)}$$

Hence Eq. (2) means that

$$(n + 1)K(b - z)^n - \frac{f^{(n+1)}(z)}{n!}(b - z)^n = 0.$$

Consequently we can cancel $(b - z)^n$ and solve for

$$K = \frac{f^{(n+1)}(z)}{(n+1)!}. \tag{4}$$

Finally, we return to Eq. (1) and substitute $x = a$, $f(x) = 0$, and the value of K given in Eq. (4). The result is the equation

$$0 = f(b) - f(a) - f'(a)(b-a) - \frac{f''(a)}{2!}(b-a)^2 - \cdots - \frac{f^{(n)}(a)}{n!}(b-a)^n - \frac{f^{(n+1)}(z)}{(n+1)!}(b-a)^{n+1},$$

which is equivalent to the desired Taylor's formula, Eq. (11) of Section 11.4. ■

APPENDIX K: CONIC SECTIONS AS SECTIONS OF A CONE

The parabola, hyperbola, and ellipse that we studied in Chapter 10 were originally introduced by the ancient Greek mathematicians as plane sections (traces) of a right circular cone. Here we show that the intersection of a plane and a cone is indeed one of the three conic sections as defined in Chapter 10.

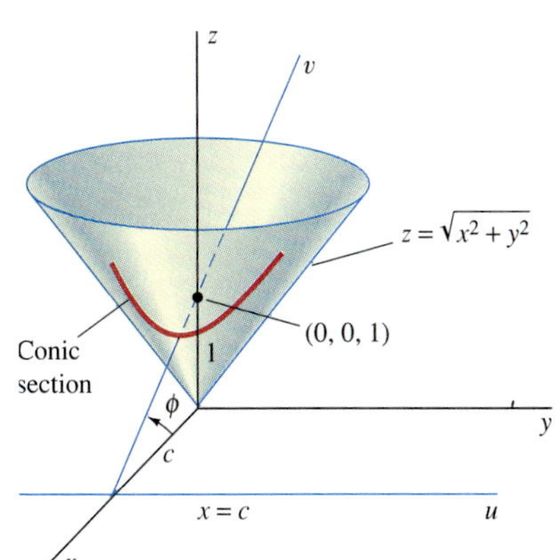

Fig. K.1 Finding an equation for a conic section

Figure K.1 shows the cone with equation $z = \sqrt{x^2 + y^2}$ and its intersection with a plane $\mathcal{P}$ that passes through the point $(0, 0, 1)$ and the line $x = c > 0$ in the xy-plane. The equation of $\mathcal{P}$ is

$$z = 1 - \frac{x}{c}. \tag{1}$$

The angle between $\mathcal{P}$ and the xy-plane is $\phi = \tan^{-1}(1/c)$. We want to show that the conic section obtained by intersecting the cone and the plane is

A parabola if $\phi = 45°$ $(c = 1)$,

An ellipse if $\phi < 45°$ $(c > 1)$,

A hyperbola if $\phi > 45°$ $(c < 1)$.

We begin by introducing uv-coordinates in the plane $\mathcal{P}$ as follows. The u-coordinate of the point (x, y, z) of $\mathcal{P}$ is $u = y$. The v-coordinate of the same point is its perpendicular distance from the line $x = c$. This explains the u- and v-axes indicated in Fig. K.1. Figure K.2 shows the cross section in the plane $y = 0$ exhibiting the relation between v, x, and z. We see that

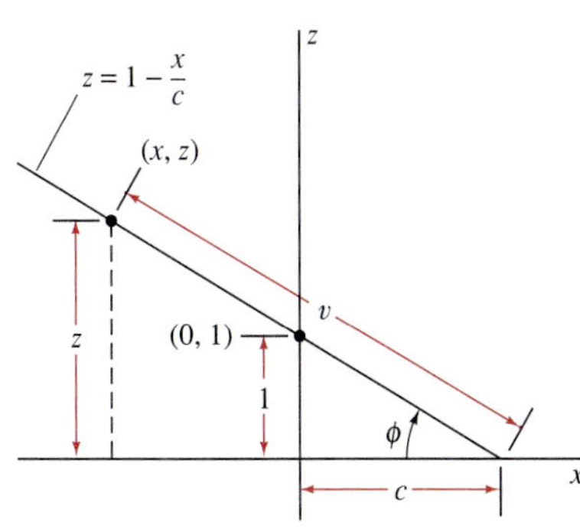

Fig. K.2 Computing coordinates in the uv-plane

$$z = v \sin \phi = \frac{v}{\sqrt{1 + c^2}}. \tag{2}$$

Equations (1) and (2) give

$$x = c(1 - z) = c\left(1 - \frac{v}{\sqrt{1 + c^2}}\right). \tag{3}$$

We had $z^2 = x^2 + y^2$ for the equation of the cone. We make the following substitution in this equation: Replace y with u, and replace z and x with the expressions on the right-hand sides of Eqs. (2) and (3), respectively. These replacements yield

$$\frac{v^2}{1+c^2} = c^2\left(1 - \frac{v}{\sqrt{1+c^2}}\right)^2 + u^2.$$

After we simplify, this last equation takes the form

$$u^2 + \frac{c^2-1}{c^2+1}v^2 - \frac{2c^2}{\sqrt{1+c^2}}v + c^2 = 0. \qquad \textbf{(4)}$$

This is the equation of the curve in the uv-plane. We examine the three cases for the angle ϕ.

Suppose first that $\phi = 45°$. Then $c = 1$, so Eq. (4) contains a term that includes u^2, another term that includes v, and a constant term. So the curve is a parabola; see Eq. (6) of Section 10.6.

Suppose next that $\phi < 45°$. Then $c > 1$, and both the coefficients of u^2 and v^2 in Eq. (4) are positive. Thus the curve is an ellipse; see Eq. (7) of Section 10.7.

Finally, if $\phi > 45°$, then $c < 1$, and the coefficients of u^2 and v^2 in Eq. (4) have opposite signs. So the curve is a hyperbola; see Eq. (8) of Section 10.8.

APPENDIX L: UNITS OF MEASUREMENT AND CONVERSION FACTORS

MKS Scientific Units

- *Length* in meters (m); *mass* in kilograms (kg), *time* in seconds (s)
- *Force* in newtons (N); a force of 1 N imparts an acceleration of 1 m/s^2 to a mass of 1 kg.
- *Work* in joules (J); 1 J is the work done by a force of 1 N acting through a distance of 1 m.
- *Power* in watts (W); 1 W is 1 J/s.

British Engineering Units (fps)

- *Length* in feet (ft), *force* in pounds (lb), *time* in seconds (s)
- *Mass* in slugs; 1 lb of force imparts an acceleration of 1 ft/s^2 to a mass of 1 slug. A mass of m slugs at the surface of the earth has a *weight* of $w = mg$ pounds (lb), where $g \approx 32.17\ ft/s^2$.
- *Work* in ft · lb, *power* in ft · lb/s.

Conversion Factors

$$1\text{ in.} = 2.54\text{ cm} = 0.0254\text{ m},\ 1\text{ m} \approx 3.2808\text{ ft}$$

$$1\text{ mi} = 5280\text{ ft};\ 60\text{ mi/h} = 88\text{ ft/s}$$

$$1\text{ lb} \approx 4.4482\text{ N};\ 1\text{ slug} \approx 14.594\text{ kg}$$

$$1\text{ hp} = 550\text{ ft}\cdot\text{lb/s} \approx 745.7\text{ W}$$

- *Gravitational acceleration:* $g \approx 32.17\ ft/s^2 = 9.807\ m/s^2$.
- *Atmospheric pressure:* 1 atm is the pressure exerted by a column of mercury 76 cm high; 1 atm $\approx$ 14.70 lb/in.^{2} = $1.013 \times 10^5\ N/m^2$.
- *Heat energy:* 1 Btu ≈ 778 ft · lb ≈ 252 cal, 1 cal ≈ 4.184 J.

APPENDIX M: FORMULAS FROM ALGEBRA, GEOMETRY, AND TRIGONOMETRY

Laws of Exponents

$$a^m a^n = a^{m+n}, \qquad (a^m)^n = a^{mn}, \qquad (ab)^n = a^n b^n, \qquad a^{m/n} = \sqrt[n]{a^m};$$

in particular,

$$a^{1/2} = \sqrt{a}.$$

If $a \neq 0$, then

$$a^{m-n} = \frac{a^m}{a^n}, \qquad a^{-n} = \frac{1}{a^n}, \qquad \text{and} \qquad a^0 = 1.$$

Quadratic Formula

The quadratic equation

$$ax^2 + bx + c = 0 \qquad (a \neq 0)$$

has solutions

$$x = \frac{-b \pm \sqrt{b^2 - 4ac}}{2a}.$$

Factoring

$$\begin{aligned}
a^2 - b^2 &= (a - b)(a + b) \\
a^3 - b^3 &= (a - b)(a^2 + ab + b^2) \\
a^4 - b^4 &= (a - b)(a^3 + a^2b + ab^2 + b^3) \\
&= (a - b)(a + b)(a^2 + b^2) \\
a^5 - b^5 &= (a - b)(a^4 + a^3b + a^2b^2 + ab^3 + b^4)
\end{aligned}$$

(The pattern continues.)

$$\begin{aligned}
a^3 + b^3 &= (a + b)(a^2 - ab + b^2) \\
a^5 + b^5 &= (a + b)(a^4 - a^3b + a^2b^2 - ab^3 + b^4) \\
a^7 + b^7 &= (a + b)(a^6 - a^5b + a^4b^2 - a^3b^3 + a^2b^4 - ab^5 + b^6)
\end{aligned}$$

(The pattern continues for odd exponents.)

Binomial Formula

$$(a + b)^n = a^n + na^{n-1}b + \frac{n(n-1)}{1 \cdot 2} a^{n-2}b^2 + \frac{n(n-1)(n-2)}{1 \cdot 2 \cdot 3} a^{n-3}b^3 + \cdots + nab^{n-1} + b^n$$

if n is a positive integer.

Area and Volume

In Fig. M.1, the symbols have the following meanings.

A: area	b: length of base	r: radius
B: area of base	C: circumference	V: volume
h: height	l: length	w: width

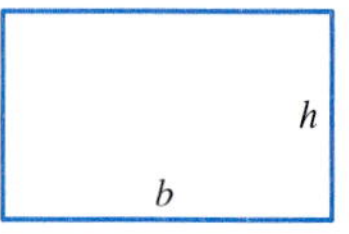

Rectangle: $A = bh$

Parallelogram: $A = bh$

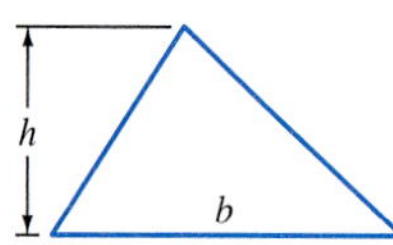

Triangle: $A = \frac{1}{2}bh$

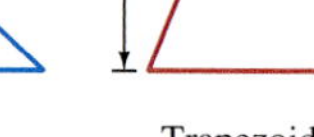

Trapezoid: $A = \frac{1}{2}(b_1 + b_2)h$

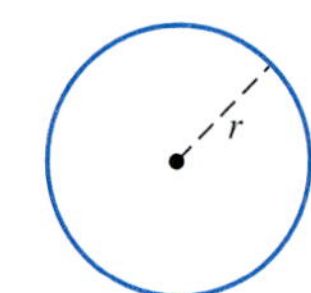

Circle: $C = 2\pi r$ and $A = \pi r^2$

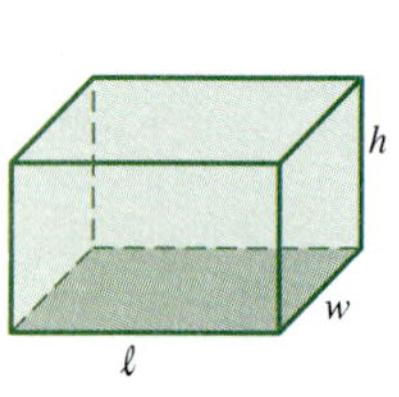

Rectangular parallelepiped: $V = \ell wh$

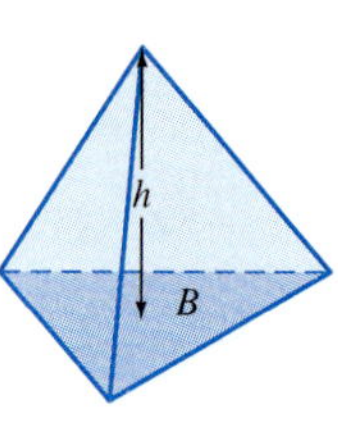

Pyramid: $V = \frac{1}{3}Bh$

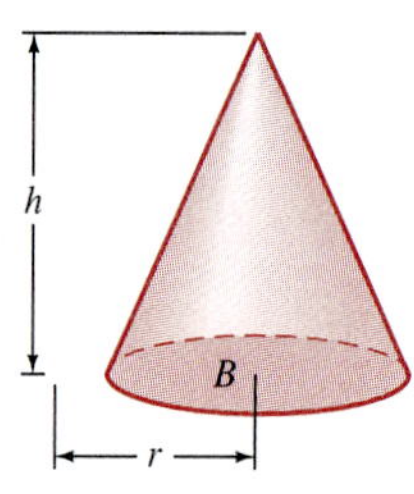

Right circular cone: $V = \frac{1}{3}\pi r^2 h = \frac{1}{3}Bh$

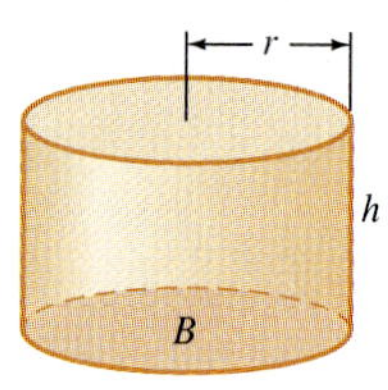

Right circular cylinder: $V = \pi r^2 h = Bh$

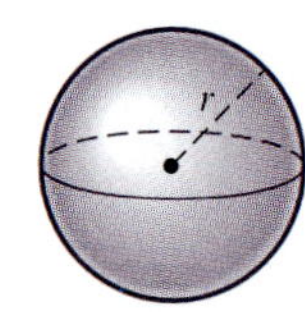

Sphere: $V = \frac{4}{3}\pi r^3$ and $A = 4\pi r^2$

Fig. M.1 The basic geometric shapes

Pythagorean Theorem

In a right triangle with legs a and b and hypotenuse c,

$$a^2 + b^2 = c^2.$$

Formulas from Trigonometry

$$\sin(-\theta) = -\sin\theta$$

$$\cos(-\theta) = \cos\theta$$

$$\sin^2\theta + \cos^2\theta = 1$$

$$\sin 2\theta = 2\sin\theta\cos\theta$$

$$\cos 2\theta = \cos^2\theta - \sin^2\theta$$

$$\sin(\alpha + \beta) = \sin\alpha\cos\beta + \cos\alpha\sin\beta$$

$$\cos(\alpha + \beta) = \cos\alpha\cos\beta - \sin\alpha\sin\beta$$

$$\tan(\alpha + \beta) = \frac{\tan\alpha + \tan\beta}{1 - \tan\alpha\tan\beta}$$

$$\sin^2\frac{\theta}{2} = \frac{1-\cos\theta}{2}$$

$$\cos^2\frac{\theta}{2} = \frac{1+\cos\theta}{2}$$

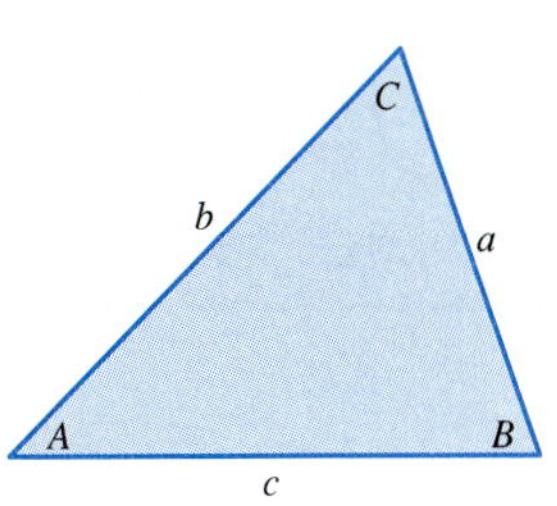

Fig. M.2 An arbitrary triangle

For an arbitrary triangle (Fig. M.2):

Law of cosines: $c^2 = a^2 + b^2 - 2ab\cos C.$

Law of sines: $\dfrac{\sin A}{a} = \dfrac{\sin B}{b} = \dfrac{\sin C}{c}.$

APPENDIX N: THE GREEK ALPHABET

Α	α	alpha	Ι	ι	iota	Ρ	ρ	rho
Β	β	beta	Κ	κ	kappa	Σ	σ	sigma
Γ	γ	gamma	Λ	λ	lambda	Τ	τ	tau
Δ	δ	delta	Μ	μ	mu	Υ	υ	upsilon
Ε	ϵ	epsilon	Ν	ν	nu	Φ	ϕ	phi
Ζ	ζ	zeta	Ξ	ξ	xi	Χ	χ	chi
Η	η	eta	Ο	o	omicron	Ψ	ψ	psi
Θ	θ	theta	Π	π	pi	Ω	ω	omega

ANSWERS TO ODD-NUMBERED PROBLEMS

Section 1.1 (page 9)

1. (a) $-\dfrac{1}{a}$; (b) a; (c) $\dfrac{1}{\sqrt{a}}$; (d) $\dfrac{1}{a^2}$

3. (a) $\dfrac{1}{a^2+5}$; (b) $\dfrac{a^2}{1+5a^2}$; (c) $\dfrac{1}{a+5}$; (d) $\dfrac{1}{a^4+5}$

5. $\frac{1}{3}$ 7. ± 3 9. 100

11. $3h$ 13. $2ah + h^2$ 15. $-\dfrac{h}{a(a+h)}$

17. $\{-1, 0, 1\}$ 19. $\{-1, 1\}$ 21. $\boldsymbol{R}$

23. $\boldsymbol{R}$ 25. $\left[\frac{5}{3}, \infty\right)$ 27. $t \leqq \frac{1}{2}$

29. All real numbers other than 3

31. $\boldsymbol{R}$ 33. $[0, 16]$

35. All real numbers other than zero

37. $C(A) = 2\sqrt{\pi A}$, $A \geqq 0$ (or $A > 0$)

39. $C(F) = \frac{5}{9}(F - 32)$, $F > -459.67$

41. $A(x) = x\sqrt{16 - x^2}$, $0 \leqq x \leqq 4$ (or $0 < x < 4$)

43. $C(x) = 3x^2 + \dfrac{1296}{x}$, $x > 0$

45. $A(r) = 2\pi r^2 + \dfrac{2000}{r}$, $r > 0$

47. $V(x) = x(50 - 2x)^2$, $0 \leqq x \leqq 25$ (or $0 < x < 25$)

49. Drill 10 new wells.

51. For each x, choose that integer n such that $n < x \leqq n + 1$.

53. The integral multiples of $\frac{1}{10}$.

55. ROUND4$(x) = \frac{1}{10000}$ROUND$(10000x)$

57. 0.38 59. 1.24 61. 0.72 63. 3.21 65. 1.62

Section 1.2 (page 21)

1. $2y = 3x$ 3. $y = -5$

5. $y - 3 = 2(x - 5)$ 7. $y - 2 = 4 - x$

9. $y - 5 = -2(x - 1)$ 11. Center $(2, 0)$, radius 2

13. Center $(-1, -1)$, radius 2

15. Center $\left(-\frac{1}{2}, \frac{1}{2}\right)$, radius 1

17. Opens upward, vertex at $(3, 0)$

19. Opens upward, vertex at $(-1, 3)$

21. Opens upward, vertex at $(-2, 3)$

23. Circle, center $(3, -4)$, radius 5

25. There are no points on the graph.

27. The graph is the straight line segment joining and including the two points $(-1, 7)$ and $(1, -3)$.

29. Parabola, opening downward, vertex at $(0, 10)$

31.

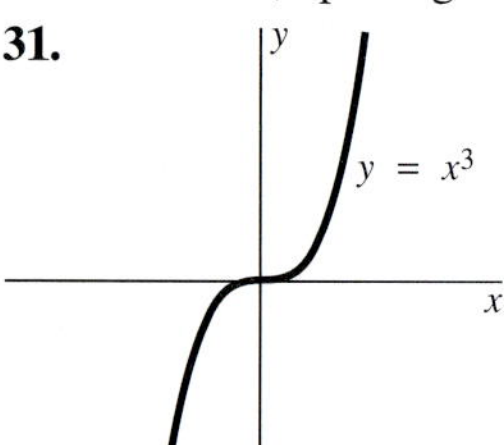

33.

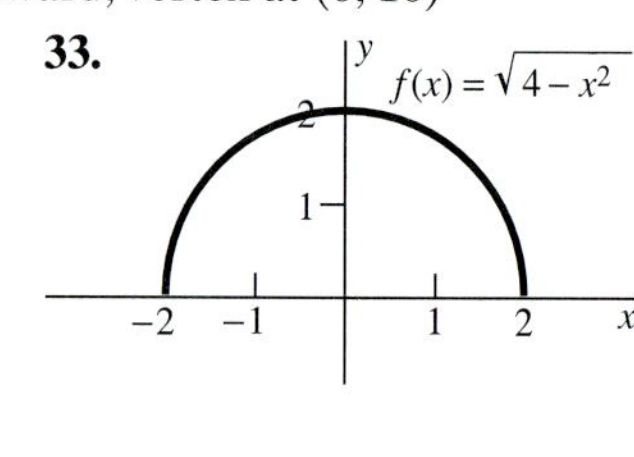

35.

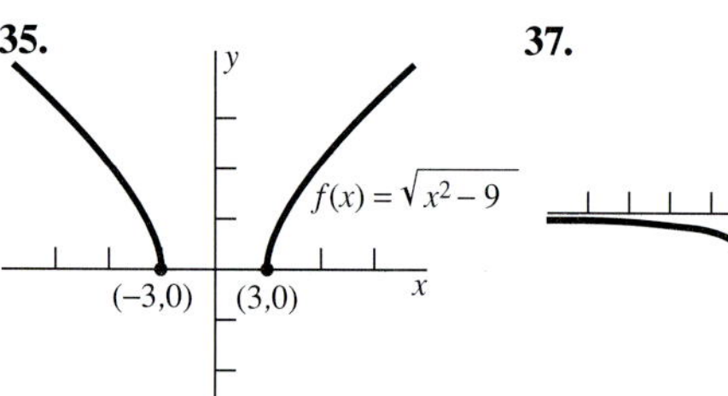

37.

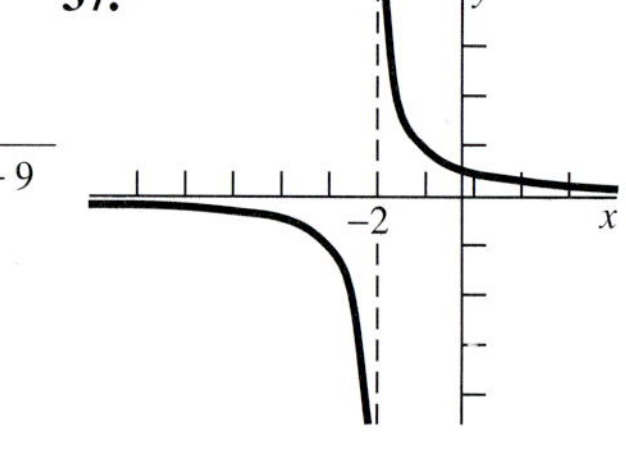

39.

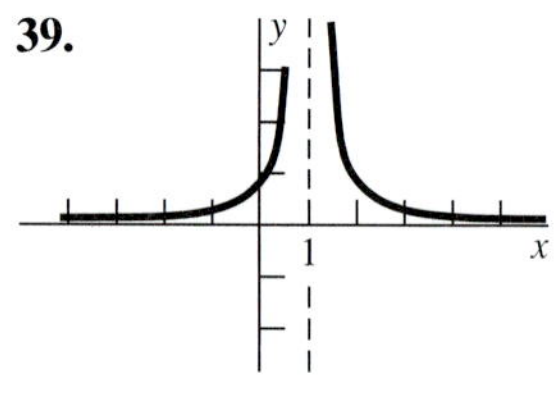

41.

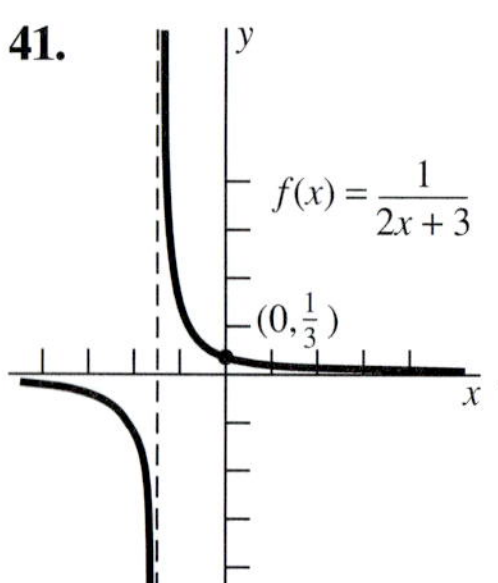

43.

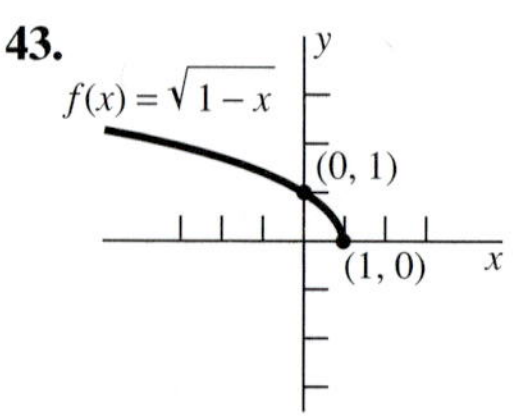

45.

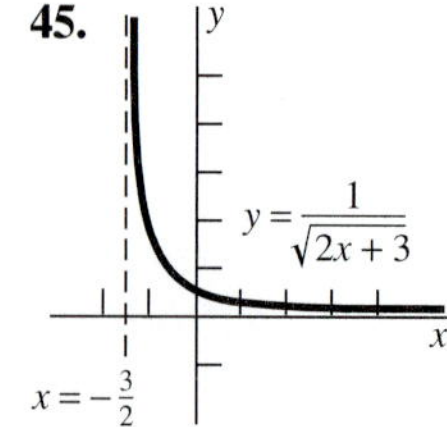

47.

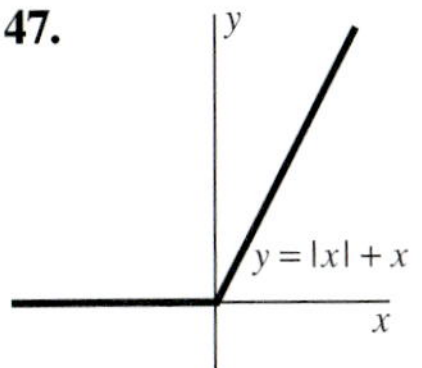

49.

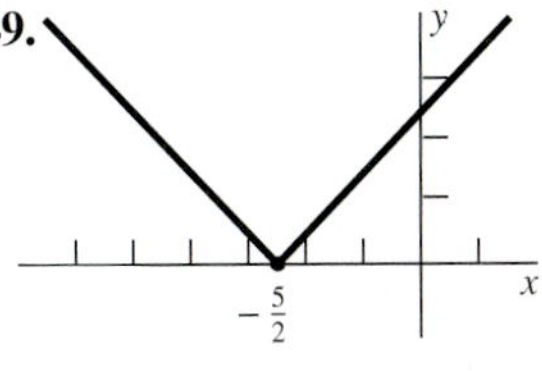

51.

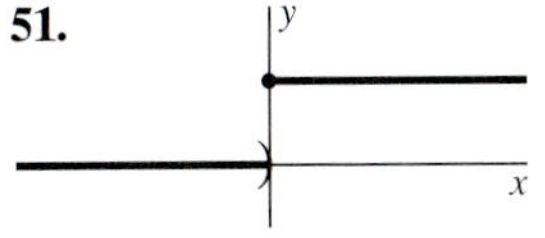

53.

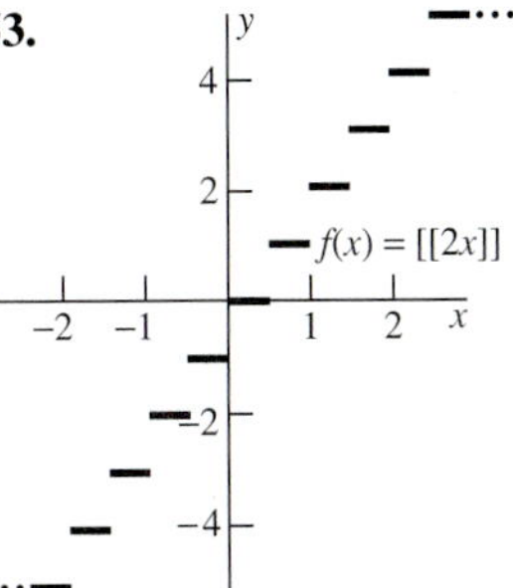

55.

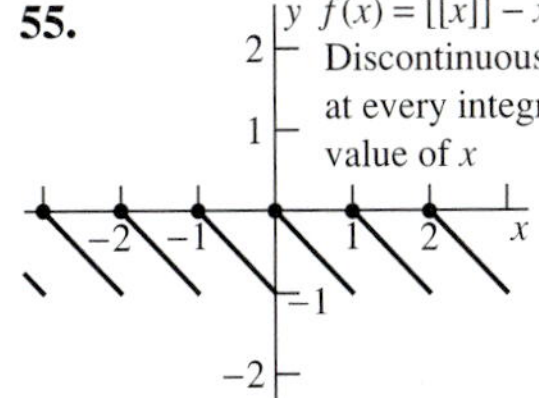

57. $(1.5, 2.5)$

59. $(2.25, 1.75)$

61. $(2.25, 8.5)$

63. $\left(-\frac{4}{3}, \frac{25}{3}\right)$

65. 144 ft

67. 625

69. $f(x) = |x + 1|, \ -2 \leqq x \leqq 2$

71. $f(x) = [\![2x]\!], \ -1 \leqq x < 2$

73. $x(t) = \begin{cases} 45t & \text{if } 0 \leqq t \leqq 1, \\ 75t - 30 & \text{if } 1 < t \leqq 2 \end{cases}$

75. $x(t) = \begin{cases} 60t & \text{if } 0 \leqq t \leqq 1, \\ -30t + 90 & \text{if } 1 < t \leqq 3 \end{cases}$

77. (a) $C(p) = (3p + 68)/100, \ 0 < p \leqq 100$; when $p = 50$, $C = \$2.18$. (b) Fixed cost: \$0.68; marginal cost: 3¢/page

79. $C(x) = \begin{cases} 8 & \text{if } 0 < x \leqq 8, \\ 8 - \frac{4}{5}[\![8 - x]\!] & \text{if } 8 < x \leqq 16 \end{cases}$

81. $V(0.5) \approx 3.35$; $V(5) \approx 0.33$

Section 1.3 (page 31)

1. $(f + g)(x) = x^2 + 3x - 2$, domain $\boldsymbol{R}$;
$(f \cdot g)(x) = x^3 + 3x^2 - x - 3$, domain $\boldsymbol{R}$;
$$\left(\frac{f}{g}\right)(x) = \frac{x + 1}{x^2 + 2x - 3}, \quad \text{domain } x \neq 1, -3$$

3. $(f + g)(x) = \sqrt{x} + \sqrt{x - 2}$, domain $x \geqq 2$;
$(f \cdot g)(x) = \sqrt{x^2 - 2x}$, domain $x \geqq 2$;
$$\left(\frac{f}{g}\right)(x) = \sqrt{\frac{x}{x - 2}}, \quad \text{domain } x > 2$$

5. $(f + g)(x) = \sqrt{x^2 + 1} + \dfrac{1}{\sqrt{4 - x^2}}$, $(f \cdot g)(x) = \dfrac{\sqrt{x^2 + 1}}{\sqrt{4 - x^2}}$,
$(f/g)(x) = \sqrt{4 + 3x^2 - x^4}$, each with domain $(-2, 2)$

7.

9.

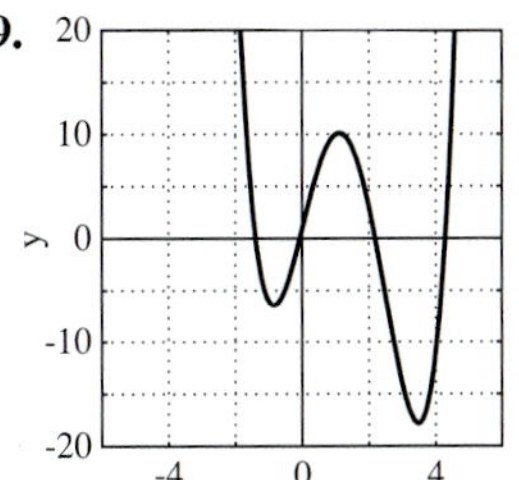

11.

13.

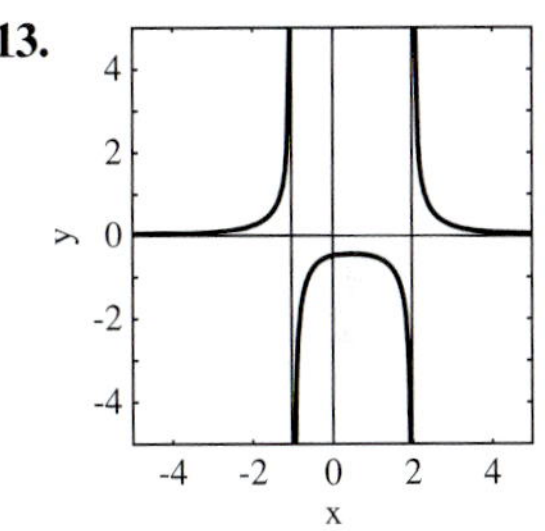

15.

17.

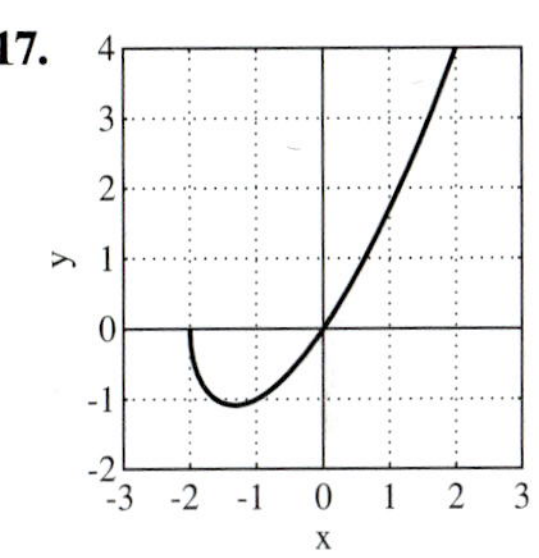

19.

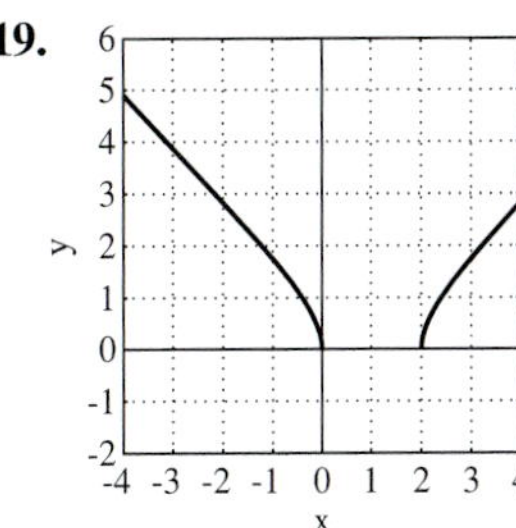

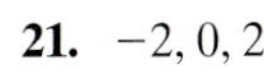

21. $-2, 0, 2$

23. -2

25. $-1, 1, 2$

27. $-6, -2, 8$

29. $-10, -7, 2, 3, 7$

31.

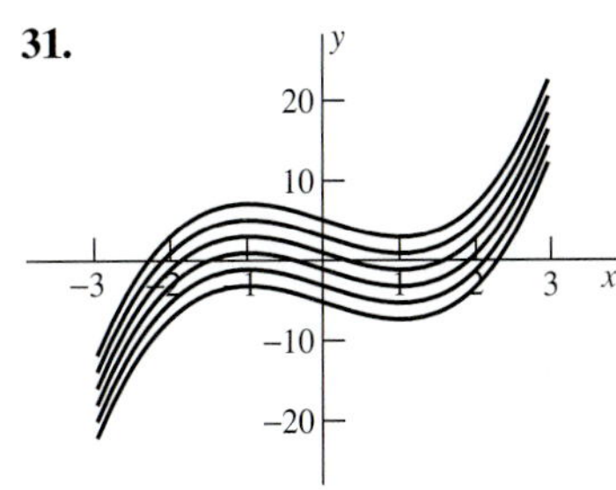

33.

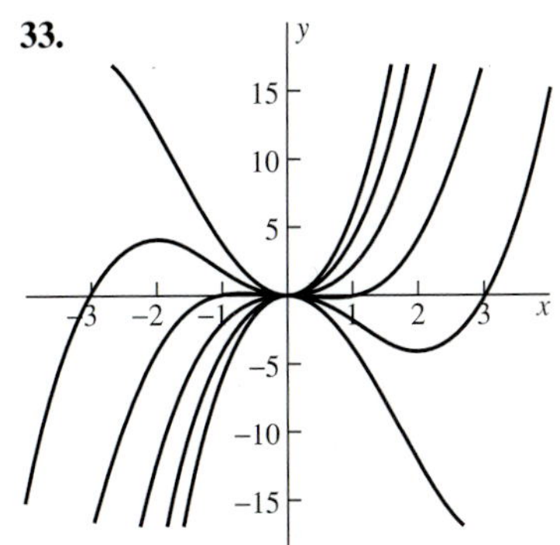

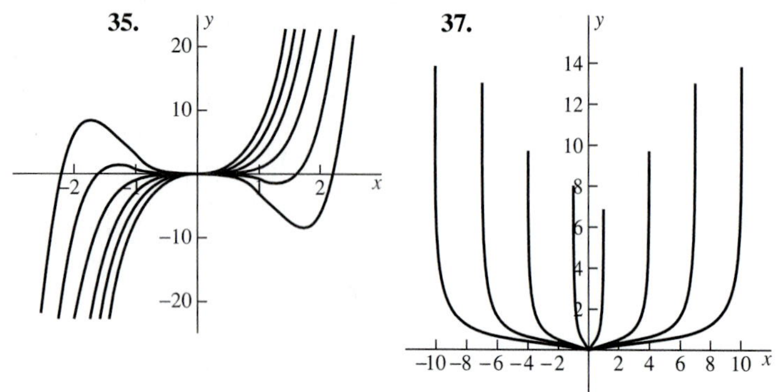

41. The graph resembles that of $y = 1/x$, but with vertical asymptote $x = -5$ instead.

43. The graph resembles that of $y = |x|$, but with the "vertex" at $(3, 0)$.

47. $(-\infty, -2) \cup (3, \infty)$ **49.** $(-\infty, -2) \cup (4, \infty)$

51. $-1.140, 6.140$ **53.** $1.191, 2.309$

55. $-5.021, 0.896$ **57.** $\left(\frac{5}{2}, \frac{3}{4}\right)$

59. $\left(\frac{7}{4}, -\frac{5}{4}\right)$ **61.** $\left(-\frac{33}{16}, \frac{31}{32}\right)$

63. $x \approx 0.4505$ **65.** 3

67. 3 **69.** 3

Section 1.4 (page 42)

1. 1.4.29 **3.** 1.4.27 **5.** 1.4.35 **7.** 1.4.31 **9.** 1.4.34

11. $f(g(x)) = -4x^2 - 12x - 8, g(f(x)) = -2x^2 + 5$

13. $f(g(x)) = \sqrt{x^4 + 6x^2 + 6}, g(f(x)) = x^2$

15. $f(g(x)) = g(f(x)) = x$

17. $f(g(x)) = \sin x^3, g(f(x)) = \sin^3 x$

19. $f(g(x)) = 1 + \tan^2 x, g(f(x)) = \tan(1 + x^2)$

Problems 21 through 30 have many correct answers. We give only the most natural.

21. $k = 2, g(x) = 2 + 3x$ **23.** $k = \frac{1}{2}, g(x) = 2x - x^2$

25. $k = \frac{3}{2}, g(x) = 5 - x^2$ **27.** $k = -1, g(x) = x + 1$

29. $k = -\frac{1}{2}, g(x) = x + 10$

31. 1 **33.** 1 **35.** 5

37. 3 **39.** 6

41. Approximately 3.32 months

43. Approximately 27 years

45. Approximately 98 years

47. Approximately 4.849

Chapter 1 Miscellaneous Problems (page 49)

1. $x \geqq 4$ **3.** $x \neq \pm 3$ **5.** $x \geqq 0$

7. $x \leqq \frac{2}{3}$ **9.** $\mathbf{R}$ **11.** $4 \leqq p \leqq 8$

13. $2 < I < 4$

15. $V(S) = (S/6)^{3/2}, 0 < S < \infty$

17. $A = A(P) = \dfrac{\sqrt{3}}{36}P^2, 0 < P < \infty$

19. $y - 5 = 2(x + 3)$ **21.** $2y = x - 10$

23. $x + 2y = 11$ **25.** 1.MP.6 **27.** 1.MP.4

29. 1.MP.3 **31.** 1.MP.7 **33.** 1.MP.8

35. The straight line with intercepts $\left(0, -\frac{7}{5}\right)$ and $\left(\frac{7}{2}, 0\right)$

37. The circle of radius 1 with center $(1, 0)$

39. The parabola that opens upward with vertex at $(1, -3)$ and y-intercept $(0, -1)$

Section 2.1 (page 62)

1. $m(a) \equiv 0; y \equiv 5$ **3.** $m(a) = 2a; y = 4x - 4$

5. $m(a) \equiv 4; y = 4x - 5$

7. $m(a) = 4a - 3; y = 5x - 4$

9. $m(a) = 4a + 6; y = 14x - 8$

11. $m(a) = 2 - \frac{1}{50}a; y = \frac{1}{25}(49x + 1)$

13. $m(a) = 8a; y = 16x - 15$

15. $(0, 10)$ **17.** $(1, 0)$ **19.** $(50, 25)$

21. $(1, -16)$ **23.** $(35, 1225)$

25. $m(x) = 2x; y - 4 = -4(x + 2), y - 4 = \frac{1}{4}(x + 2)$

27. $m(x) = 4x + 3; y - 9 = 11(x - 2), y - 9 = -\frac{1}{11}(x - 2)$

29. $y(3) = 144$ (ft)

31. 625 **33.** $y = 12(x - 3)$ **35.** $(1, 1)$

37. 12 **39.** $\frac{1}{2}$ **41.** -1

Section 2.2 (page 74)

1. 36 **3.** 0 **5.** $\frac{2}{3}$ **7.** 125 **9.** 3

11. $16\sqrt{2}$ **13.** 1 **15.** 4 **17.** 0 **19.** $-\frac{1}{3}$

21. $-\frac{3}{2}$ **23.** 0 **25.** 0 **27.** $\frac{3}{4}$ **29.** $-\frac{1}{9}$

31. 4 **33.** $\frac{1}{4}$ **35.** -32

37. $m(x) = 3x^2; y - 8 = 12(x - 2)$

39. $m(x) = -\dfrac{2}{x^3}; y - \frac{1}{4} = -\frac{1}{4}(x - 2)$

41. $m(x) = -\dfrac{2}{(x - 1)^2}; y - 2 = -2(x - 2)$

43. $m(x) = -\dfrac{1}{2(x + 2)^{3/2}}; y - \frac{1}{2} = -\frac{1}{16}(x - 2)$

45. $m(x) = \dfrac{1}{\sqrt{2x + 5}}; y - 3 = \frac{1}{3}(x - 2)$

47. 2 **49.** $\frac{1}{6}$

51. $-\frac{3}{8}$ **53.** 1

55. $\frac{1}{6}$ **57.** $e \approx 2.71828$

59. -0.3333

61. This limit does not exist.

Section 2.3 (page 85)

1. 0 **3.** $\frac{1}{2}$
5. $-\infty$ (or "does not exist") **7.** 5
9. Does not exist **11.** $\frac{1}{3}$
13. 0 **15.** 1 **17.** $\frac{1}{2}$
19. $\frac{1}{2}$ **21.** $\frac{1}{3}$ **23.** $\frac{1}{4}$
25. 0 **27.** 0 **29.** 3
31. Does not exist **33.** 0 **35.** 0
37. $+\infty$ (or "does not exist") **39.** -1
41. 1 **43.** -1
45. 2 **47.** -1
49. $f(x) \to +\infty$ as $x \to 1^+$, $f(x) \to -\infty$ as $x \to 1^-$.
51. $f(x) \to +\infty$ as $x \to -1^-$, $f(x) \to -\infty$ as $x \to -1^+$.
53. $f(x) \to +\infty$ as $x \to -2^-$, $f(x) \to -\infty$ as $x \to -2^+$.
55. $f(x) \to +\infty$ as $x \to 1$.
57. $f(x) \to -\infty$ as $x \to -2^+$, $f(x) \to +\infty$ as $x \to -2^-$, and $f(x) \to -\frac{1}{4}$ as $x \to 2$.
59. $f(x) \to 4$ as $x \to 2^+$, $f(x) \to -4$ as $x \to 2^-$.
61. If n is an integer, then $f(x) \to 2$ as $x \to n$.
63. If n is an integer, then $f(x) \to 10n - 1$ as $x \to n^-$, $f(x) \to 10n$ as $x \to n^+$. The two-sided limit exists at each real number a that is not an integral multiple of $\frac{1}{10}$.
65. If n is an integer, then $f(x) \to \frac{1}{2}$ as $x \to n^-$, whereas $f(x) \to -\frac{1}{2}$ as $x \to n^+$.
67. If n is an integer, then $f(x) \to -1$ as $x \to n$.
69. The limit of $g(x)$ exists at each real number that is not an integral multiple of $\frac{1}{10}$.
71. Note that $-x^2 \leqq f(x) \leqq x^2$ for all x.
73. $f(x) \to 1$ as $x \to 0$.

Section 2.4 (page 94)

15. $\boldsymbol{R}$ **17.** $(-\infty, -3) \cup (-3, +\infty)$
19. $\boldsymbol{R}$ **21.** $(-\infty, 5) \cup (5, +\infty)$
23. $(-\infty, 2) \cup (2, +\infty)$ **25.** $(-\infty, 1) \cup (1, +\infty)$
27. $(-\infty, 0) \cup (0, 1) \cup (1, +\infty)$
29. $(-2, 2)$ **31.** $(-\infty, 0) \cup (0, +\infty)$
33. At every real number other than the integral multiples of $\pi/2$
35. $\boldsymbol{R}$
37. Nonremovable discontinuity at $x = -3$
39. Removable discontinuity at $x = 2$; let $f(2) = \frac{1}{4}$. Nonremovable discontinuity at $x = -2$
41. Nonremovable discontinuity at $x = \pm 1$
43. Nonremovable discontinuity at $x = 17$
45. Removable discontinuity at $x = 0$; let $f(0) = 0$
47. Removable discontinuity at $x = 0$; let $f(0) = 1$
49. $c = 4$ **51.** $c = 0$
59. $f(-3) = -14$, $f(-2) = 1$, $f(-1) = 4$, $f(0) = 1$, $f(1) = -2$, $f(2) = 1$, $f(3) = 16$
61. The function f is discontinuous at each integral value of t.
63. It is easy to see that $h(a)$ and $h(b)$ have opposite signs and that h is continuous on $[a, b]$. Therefore $h(c) = 0$ for some number c in (a, b), and it follows that $f(c) = g(c)$.
65. Begin as follows: Given $a > 0$, let $f(x) = x^2 - a$. It is easy to see that $f(0) < 0$. To show that $f(a + 1) > 0$, it suffices to show that $(a + 1)^2 > a$ if $a > 0$.
69. Show that f has no limit at each real number a. Do this by using the fact that every deleted neighborhood of a contains both rational numbers and irrational numbers.
71. To six places, the solution of the equation is 0.739085, so the answer is 0.74.

Chapter 2 Miscellaneous Problems (page 96)

1. 4 **3.** 0 **5.** $-\frac{5}{3}$
7. -2 **9.** 0 **11.** 4
13. 8 **15.** $\frac{1}{6}$ **17.** $-\frac{1}{54}$
19. -1 **21.** 1 **23.** Does not exist
25. $+\infty$ (or "does not exist") **27.** $+\infty$
29. $-\infty$ **31.** 3 **33.** $\frac{3}{2}$
35. 0 **37.** $\frac{9}{4}$ **39.** 2
41. $m(x) = 4x$; $y - 5 = 4(x - 1)$
43. $m(x) = 6x + 4$; $y - 2 = 10(x - 1)$
45. $m(x) = 4x - 3$; $y = x - 1$
47. $m(x) = 4x + 3$
49. $m(x) = \dfrac{1}{(3 - x)^2}$ **51.** $m(x) = 1 + \dfrac{1}{x^2}$
53. $m(x) = -\dfrac{2}{(x - 1)^2}$ **55.** $a = 3 \pm \sqrt{5}$
57. Nonremovable discontinuity at $x = -1$; removable discontinuity at $x = 1$; let $f(1) = \frac{1}{2}$.
59. Nonremovable discontinuity at $x = -3$; removable discontinuity at $x = 1$; let $f(1) = \frac{3}{4}$.
61. Proof: Let $f(x) = x^5 + x - 1$. Then $f(0) = -1$ and $f(1) = 1$. Because f is continuous on $[0, 1]$ ($f(x)$ is a polynomial), the intermediate value theorem now implies that there exists a number c in $(0, 1)$ such that $f(c) = 0$. That is, $c^5 + c = 1$. Therefore the equation $x^5 + x = 1$ has a [real] solution. (There is only one; its value is approximately 0.754878.)
65. Three; their slopes are $\frac{1}{2}$, $\frac{1}{4}$, and $-\frac{1}{6}$.

Section 3.1 (page 111)

1. $f'(x) = 4$ **3.** $h'(z) = 25 - 2z$
5. $\dfrac{dy}{dx} = 4x + 3$ **7.** $\dfrac{dz}{du} = 10u - 3$

9. $\dfrac{dx}{dy} = -10y + 17$

11. $f'(x) = 2$

13. $f'(x) = 2x$

15. $f'(x) = -\dfrac{2}{(2x+1)^2}$

17. $f'(x) = \dfrac{1}{\sqrt{2x+1}}$

19. $f'(x) = \dfrac{1}{(1-2x)^2}$

21. $x(0) = 100$

23. $x(2.5) = 99$

25. $x(-2) = 120$

27. $y(2) = 64$ (ft)

29. $y(3) = 194$ (ft)

31. (e)

33. (f)

35. (d)

37. $\dfrac{dA}{dC} = \dfrac{C}{2\pi}$

39. 500 ft; 10 s

41. (a) 2.5 months; (b) 50 chipmunks per month

43. *Very* roughly: $v(20) = 40$ mi/h; $v(40) = 75$ mi/h

47. $V'(30) = -\frac{25}{12}\pi$ in.^{3}/s; that is, air is leaking out at about 6.545 in.^{3}/s then.

49. (a) $V'(6) = -144\pi$ cm^3/h; (b) -156π cm^3/h

51. At $t = 2$ s, $v = 0$ m/s.

Section 3.2 (page 122)

1. $f'(x) = 6x - 1$

3. $f'(x) = 2(3x-2) + 3(2x+3)$

5. $h'(x) = 3x^2 + 6x + 3$

7. $f'(y) = 12y^2 - 1$

9. $g'(x) = \dfrac{1}{(x-1)^2} - \dfrac{1}{(x+1)^2}$

11. $h'(x) = -\dfrac{3(2x+1)}{(x^2+x+1)^2}$

13. $g'(t) = (t^2+1)(3t^2+2t) + (t^3+t^2+1)(2t)$

15. $g'(z) = -\dfrac{1}{2z^2} + \dfrac{2}{3z^3}$

17. $g'(y) = 30y^4 + 48y^3 + 48y^2 - 8y - 6$

19. $g'(t) = \dfrac{3-t}{(t+1)^3}$

21. $v'(t) = -\dfrac{3}{(t-1)^4}$

23. $g'(x) = -\dfrac{6x^3+15}{(x^3+7x-5)^2}$

25. $g'(x) = \dfrac{4x^3 - 13x^2 + 12x}{(2x-3)^2}$

27. $\dfrac{dy}{dx} = 3x^2 - 30x^4 - 6x^{-5}$

29. $\dfrac{dy}{dx} = \dfrac{2x^5 + 4x^2 - 15}{x^4}$

31. $\dfrac{dy}{dx} = 3 + \dfrac{1}{2x^3}$

33. $\dfrac{dy}{dx} = \dfrac{2x - 1 - 4x^2}{3x^2(x-1)^2}$

35. $\dfrac{dy}{dx} = \dfrac{x^4 + 31x^2 - 10x - 36}{(x^2+9)^2}$

37. $\dfrac{dy}{dx} = \dfrac{30x^5(5x^5-8)}{(15x^2-4)^2}$

39. $\dfrac{dy}{dx} = \dfrac{x^2+2x}{(x+1)^2}$

41. $12x - y = 16$

43. $x + y = 3$

45. $5x - y = 10$

47. $18x - y = -25$

49. $3x + y = 0$

51. (a) It contracts; (b) -0.06427 cm^3/°C

53. $14{,}400\pi \approx 45{,}239$ cm^3/cm

55. $y = 3x + 2$

57. Suppose that some line is tangent at both (a, a^2) and (b, b^2). Use the derivative to show that $a = b$.

59. $x = \dfrac{n-1}{n}x_0$

65. $g'(x) = 17(x^3 - 17x + 35)^{16}(3x^2 - 17)$

71. $0, \pm\sqrt{3}$

Section 3.3 (page 132)

1. $\dfrac{dy}{dx} = 15(3x+4)^4$

3. $\dfrac{dy}{dx} = -3(3x-2)^{-2}$

5. $\dfrac{dy}{dx} = 3(x^2+3x+4)^2(2x+3)$

7. $\dfrac{dy}{dx} = -4(2-x)^3(3+x)^7 + 7(2-x)^4(3+x)^6$

9. $\dfrac{dy}{dx} = -\dfrac{6x+22}{(3x-4)^4}$

11. $\dfrac{dy}{dx} = 12[1+(1+x)^3]^3(1+x)^2$

13. $\dfrac{dy}{dx} = -\dfrac{6}{x^3}\left(\dfrac{1}{x^2}+1\right)^2$

15. $\dfrac{dy}{dx} = 48[1+(4x-1)^4]^2(4x-1)^3$

17. $\dfrac{dy}{dx} = \dfrac{12(1-x^{-4})^2}{x^9} - \dfrac{4(1-x^{-4})^3}{x^5} = -\dfrac{4(x^{12} - 6x^8 + 9x^4 - 4)}{x^{17}}$

19. $\dfrac{dy}{dx} = -4x^{-5}(x^{-2} - x^{-8})^3 + 3x^{-4}(8x^{-9} - 2x^{-3})(x^{-2} - x^{-8})^2 = -\dfrac{2(x^3-1)^2(x^3+1)^2(5x^6-14)}{x^{29}}$

21. $u(x) = 2x - x^2$, $n = 3$; $f'(x) = 3(2x - x^2)^2(2-2x)$

23. $u(x) = 1 - x^2$, $n = -4$; $f'(x) = 8x(1-x^2)^{-5}$

25. $u(x) = \dfrac{x+1}{x-1}$, $n = 7$; $f'(x) = -\dfrac{14(x+1)^6}{(x-1)^8}$

27. $g'(y) = 1 + 10(2y-3)^4$

29. $F'(s) = 3(s - s^{-2})^2(1 + 2s^{-3})$

31. $f'(u) = 8u(u+1)^3(u^2+1)^3 + 3(u+1)^2(u^2+1)^4$

33. $h'(v) = 2(v-1)(v^2-2v+2)(v^{-3})(2-v)^{-3}$

35. $F'(z) = 10(4 - 25z^4)(3 - 4z + 5z^5)^{-11}$

37. $\dfrac{dy}{dx} = 4(x^3)^3 \cdot 3x^2 = 12x^{11}$

39. $\dfrac{dy}{dx} = 2(x^2-1)(2x) = 4x^3 - 4x$

41. $\dfrac{dy}{dx} = 4(x+1)^3 = 4x^3 + 12x^2 + 12x + 4$

43. $\dfrac{dy}{dx} = -\dfrac{2x}{(x^2+1)^2}$ **45.** $f'(x) = 3x^2 \cos x^3$

47. $g'(z) = 6(\sin 2z)^2 \cos 2z$

49. 40π in.^{2}/s **51.** 40 in.^{2}/s

53. 600 in.^{3}/h **55.** -18

57. $400\pi \approx 1256.64$ cm^3/s **59.** 5 cm

61. Total melting time: $2/(2 - 4^{1/3}) \approx 4.85$ h; all melted by about 2:50:50 P.M. that day.

Section 3.4 (page 138)

1. $f'(x) = 10x^{3/2} - x^{-3/2}$ **3.** $f'(x) = (2x+1)^{-1/2}$

5. $f'(x) = -3x^{-3/2} - \frac{3}{2}x^{1/2}$

7. $f'(x) = 3(2x+3)^{1/2}$ **9.** $f'(x) = 6x(3 - 2x^2)^{-5/2}$

11. $f'(x) = \dfrac{3x^2}{2\sqrt{x^3+1}}$ **13.** $f'(x) = 2x(2x^2+1)^{-1/2}$

15. $f'(t) = 3t^2(2t^3)^{-1/2} = \frac{3}{2}\sqrt{2t}$

17. $f'(x) = \frac{3}{2}(2x^2 - x + 7)^{1/2}(4x - 1)$

19. $g'(x) = -\frac{4}{3}(x - 2x^3)^{-7/3}(1 - 6x^2)$

21. $f'(x) = (1 - x^2)^{1/2} - x^2(1 - x^2)^{-1/2}$
$= (1 - 2x^2)(1 - x^2)^{-1/2}$

23. $f'(t) = \frac{1}{2}\left(\dfrac{t^2+1}{t^2-1}\right)^{-1/2} \cdot \dfrac{(t^2-1)(2t) - (2t)(t^2+1)}{(t^2-1)^2}$
$= -2t(t^2+1)^{-1/2}(t^2-1)^{-3/2}$

25. $f'(x) = 3\left(x - \dfrac{1}{x}\right)^2\left(1 + \dfrac{1}{x^2}\right)$

27. $f'(v) = -\dfrac{v+2}{2v^2\sqrt{v+1}}$

29. $f'(x) = \frac{1}{3}(1 - x^2)^{-2/3}(-2x)$

31. $f'(x) = (3 - 4x)^{1/2} - 2x(3 - 4x)^{-1/2}$

33. $f'(x) = (-2x)(2x+4)^{1/3} + \frac{8}{3}(1 - x^2)(2x+4)^{-2/3}$

35. $g'(t) = -2t^{-2}(1 + t^{-1})(3t^2+1)^{1/2}$
$+ 3t(1 + t^{-1})^2(3t^2+1)^{-1/2}$
$= \dfrac{3t^4 - 3t^2 - 2t - 2}{t^3\sqrt{3t^2+1}}$

37. $f'(x) = \dfrac{2(3x+4)^5 - 15(3x+4)^4(2x-1)}{(3x+4)^{10}} = \dfrac{23 - 24x}{(3x+4)^6}$

39. $f'(x) = \dfrac{(3x+4)^{1/3}(2x+1)^{-1/2} - (3x+4)^{-2/3}(2x+1)^{1/2}}{(3x+4)^{2/3}}$
$= \dfrac{x+3}{(3x+4)^{4/3}(2x+1)^{1/2}}$

41. $h'(y) = \dfrac{(1+y)^{-1/2} - (1-y)^{-1/2}}{2y^{5/3}}$
$- \dfrac{5[(1+y)^{1/2} + (1-y)^{1/2}]}{3y^{8/3}}$
$= \dfrac{(7y - 10)\sqrt{1+y} - (7y+10)\sqrt{1-y}}{6y^{8/3}\sqrt{1-y^2}}$

43. $g'(t) = \frac{1}{2}\left[t + (t + t^{1/2})^{1/2}\right]^{-1/2} \times$
$\left[1 + \frac{1}{2}(t + t^{1/2})^{-1/2}\left(1 + \frac{1}{2}t^{-1/2}\right)\right]$
$= \dfrac{1 + \dfrac{1 + \dfrac{1}{2\sqrt{t}}}{2\sqrt{t + \sqrt{t}}}}{2\sqrt{t + \sqrt{t + \sqrt{t}}}}$

45. No horizontal tangents; vertical tangent line at $(0, 0)$

47. Horizontal tangent line at $\left(\frac{1}{3}, \frac{2}{9}\sqrt{3}\right)$; vertical tangent line at $(0, 0)$

49. No horizontal or vertical tangents

51. $y - 4 = \frac{1}{2}(x - 4)$ **53.** $y - 3 = -2(x + 1)$

55. $y = 2x$ **57.** Matches Fig. 3.4.13(d)

59. Matches Fig. 3.4.13(b) **61.** Matches Fig. 3.4.13(e)

63. $\frac{1}{32}\pi^2 \approx 0.3084$ (s/ft)

65. $\left(\frac{2}{5}\sqrt{5}, \frac{1}{5}\sqrt{5}\right)$ and $\left(-\frac{2}{5}\sqrt{5}, -\frac{1}{5}\sqrt{5}\right)$

67. $x + 4y = 18$

69. $3x + 2y = 5$ and $3x - 2y = -5$

71. Equation (3) is an *identity,* and if two functions have identical graphs on an interval, then their derivatives are also identically equal on that interval.

Section 3.5 (page 147)

In the answers to Problems 1 through 33, we give first the maximum (if any), then the minimum (if any).

1. 2; none **3.** None; 0 **5.** 2; 0

7. 2; 0 **9.** $-\frac{1}{6}$; $-\frac{1}{2}$ **11.** 7; -8

13. 3; -5 **15.** 9; 0 **17.** 52; -2

19. 5; 4 **21.** 5; 1 **23.** 9; -16

25. 10; -22 **27.** 56; -56 **29.** 13; 5

31. 17; 0 **33.** $\frac{3}{4}$; 0

35. Maximum: $\frac{1}{2}$ (at $x = -1$); minimum: $-\frac{1}{6}$ (at $x = 3$)

37. Maximum: $\frac{1}{2}$ (at $x = 1/\sqrt{2}$); minimum: $-\frac{1}{2}$ (at $x = -1/\sqrt{2}$)

39. Maximum: $3 \cdot 2^{-4/3}$ (at $x = \frac{3}{2}$); minimum: -3 (at $x = 3$)

41. Consider the cases $A = 0$ and $A \neq 0$.

47. (c) **49.** (d) **51.** (a)

Section 3.6 (page 158)

1. 25 and 25 **3.** 1250 **5.** 500 in.^{3}

7. 1152 **9.** 250 **11.** 11,250 yd^2

13. 128 **15.** Approximately 3.9665°C

17. 1000 cm^3

19. 0.25 m^3 (all cubes, no open-top boxes)

21. Two equal pieces yield minimum total area 200 in.^{2}; no cut yields one square of maximum area 400 in.^{2}

23. 30000 m^2

25. Approximately 9259.26 in.^{3}

27. Five presses

29. The minimizing value of x is $-2 + \frac{10}{3}\sqrt{6}$ in. To the nearest integer, use $x = 6$ in. of insulation for an annual saving of \$285.

31. Either \$1.10 or \$1.15

33. Radius $\frac{2}{3}R$, height $\frac{1}{3}H$

35. Let R denote the radius of the circle, and remember that R is constant.

37. $\dfrac{2000\pi\sqrt{3}}{27}$

39. Max.: 4; min.: $\sqrt[3]{16}$

41. $\frac{1}{2}\sqrt{3}$

43. Each plank has width $\dfrac{-3\sqrt{2} + \sqrt{34}}{8} \approx 0.198539$, height $\dfrac{\sqrt{7 - \sqrt{17}}}{2} \approx 0.848071$, and area $\dfrac{\sqrt{7 - \sqrt{17}}\left(-3\sqrt{2} + \sqrt{34}\right)}{16} \approx 0.673500$.

45. $\frac{2}{3}\sqrt{3} \approx 1.1547$ km from the point nearest the island.

47. $\frac{1}{3}\sqrt{3}$

49. Actual $x \approx 3.45246$.

51. To minimize the sum, choose the radius of the sphere to be $5\sqrt{10/(\pi + 6)}$ and the edge length of the cube to be twice that amount. To maximize the sum, choose the edge length of the cube to be zero.

55. In Problem 53, $x = 4$ maximizes $V(x)$, and $V(4) = \frac{128}{3}\sqrt{5} \approx 95.4056$. In Problem 54, $x = 8$ maximizes $V(x)$, and $V(8) = \frac{256}{3}\sqrt{10} \approx 269.8477$.

Section 3.7 (page 170)

1. $f'(x) = 6\sin x\cos x$

3. $f'(x) = \cos x - x\sin x$

5. $f'(x) = \dfrac{x\cos x - \sin x}{x^2}$

7. $f'(x) = \cos^3 x - 2\sin^2 x\cos x$

9. $g'(t) = 4(1 + \sin t)^3\cos t$

11. $g'(t) = \dfrac{\sin t - \cos t}{(\sin t + \cos t)^2}$

13. $f'(x) = 2\sin x + 2x\cos x - 6x\cos x + 3x^2\sin x$

15. $f'(x) = 3\cos 2x\cos 3x - 2\sin 2x\sin 3x$

17. $g'(t) = 3t^2\sin^2 2t + 4t^3\sin 2t\cos 2t$

19. $g'(t) = -\frac{5}{2}(\cos 3t + \cos 5t)^{3/2}(3\sin 3t + 5\sin 5t)$

21. $\dfrac{dy}{dx} = \dfrac{1}{\sqrt{x}}\sin\sqrt{x}\cos\sqrt{x}$

23. $\dfrac{dy}{dx} = 2x\cos(3x^2 - 1) - 6x^3\sin(3x^2 - 1)$

25. $\dfrac{dy}{dx} = 2\cos 2x\cos 3x - 3\sin 2x\sin 3x$

27. $\dfrac{dy}{dx} = -\dfrac{3\sin 5x\sin 3x + 5\cos 5x\cos 3x}{\sin^2 5x}$

29. $\dfrac{dy}{dx} = 4x\sin x^2\cos x^2$

31. $\dfrac{dy}{dx} = \dfrac{\cos 2\sqrt{x}}{\sqrt{x}}$

33. $\dfrac{dy}{dx} = \sin x^2 + 2x^2\cos x^2$

35. $\dfrac{dy}{dx} = \frac{1}{2}x^{-1/2}\sin x^{1/2} + \frac{1}{2}\cos x^{1/2}$

37. $\dfrac{dy}{dx} = \frac{1}{2}x^{-1/2}(x - \cos x)^3 + 3x^{1/2}(x - \cos x)^2(1 + \sin x)$

39. $\dfrac{dy}{dx} = -2x[\sin(\sin x^2)]\cos x^2$

41. $\dfrac{dx}{dt} = 7t^6\sec^2 t^7$

43. $\dfrac{dx}{dt} = 7\sec^2 t\tan^6 t$

45. $\dfrac{dx}{dt} = 5t^7\sec^2 5t + 7t^6\tan 5t$

47. $\dfrac{dx}{dt} = \dfrac{\sec\sqrt{t} + \sqrt{t}\sec\sqrt{t}\tan\sqrt{t}}{2\sqrt{t}}$

49. $\dfrac{dx}{dt} = \dfrac{2\cot\dfrac{1}{t^2}\csc\dfrac{1}{t^2}}{t^3}$

51. $\dfrac{dx}{dt} = \dfrac{5\tan 3t\sec 5t\tan 5t - 3\sec 5t\sec^2 3t}{\tan^2 3t}$
$= 5\cot 3t\sec 5t\tan 5t - 3\csc^2 3t\sec 5t$

53. $\dfrac{dx}{dt} = \sec t\csc t + t\sec t\tan t\csc t - t\sec t\csc t\cot t$
$= t\sec^2 t + \sec t\csc t - t\csc^2 t$

55. $\dfrac{dx}{dt} = [\sec(\sin t)\tan(\sin t)]\cos t$

57. $\dfrac{dx}{dt} = \dfrac{\sec t\cos t - \sin t\sec t\tan t}{\sec^2 t} = \cos^2 t - \sin^2 t$

59. $\dfrac{dx}{dt} = -\dfrac{5\csc^2 5t}{2\sqrt{1 + \cot 5t}}$

61. $y = -x$

63. $y - \dfrac{4}{\pi} = 2(x - 1)$

65. At every integral multiple of $\pi/2$

67. At every odd integral multiple of $\pi/4$

69. $y = x \pm 2$

73. $\pi/4$

75. $\dfrac{\pi}{18}\sec^2\dfrac{5\pi}{18} \approx 0.4224$ mi/s (about 1521 mi/h)

77. $\dfrac{2000\pi}{27}$ ft/s (about 158.67 mi/h)

79. $\pi/3$

81. $\frac{8}{3}\pi R^3$, twice the volume of the sphere!

83. $\frac{3}{4}\sqrt{3}$

85. *Suggestion*: $A(\theta) = \dfrac{s^2(\theta - \sin\theta)}{2\theta^2}$

Section 3.8 (page 179)

1. $\dfrac{dy}{dx} = \dfrac{x}{y} = \pm\dfrac{x}{\sqrt{x^2-1}}$

3. $\dfrac{dy}{dx} = -\dfrac{16x}{25y} = \mp\dfrac{16x}{5\sqrt{400-16x^2}}$

5. $\dfrac{dy}{dx} = -\sqrt{\dfrac{y}{x}}$ **7.** $\dfrac{dy}{dx} = -\left(\dfrac{y}{x}\right)^{1/3}$

9. $\dfrac{dy}{dx} = \dfrac{3x^2 - 2xy - y^2}{x^2 + 2xy + 3y^2}$ **11.** $\dfrac{dy}{dx} = -\dfrac{\sin y + y\cos x}{\sin x + x\cos y}$

13. $\dfrac{dy}{dx} = -\dfrac{\cos(x+y) + 3\sin x\cos^2 x}{\cos(x+y) + 3\sin y\cos^2 y}$

15. $\dfrac{dy}{dx} = -\dfrac{x}{y}$; $3x - 4y = 25$

17. $\dfrac{dy}{dx} = \dfrac{1-2xy}{x^2}$; $3x + 4y = 10$

19. $\dfrac{dy}{dx} = -\dfrac{2xy + y^2}{2xy + x^2}$; $y \equiv -2$

21. $\dfrac{dy}{dx} = \dfrac{25y - 24x}{24y - 25x}$; $4x = 3y$

23. $\dfrac{dy}{dx} = -\dfrac{y^4}{x^4}$; $x + y = 2$

25. $\dfrac{dy}{dx} = -\dfrac{y^{1/3}}{x^{1/3}}$; $x + 2y = 10$

27. $\dfrac{dy}{dx} = \dfrac{2x^3 - 25y + 2xy^2}{25x - 2x^2y - 2y^3}$; $11y - 2x = 40$

29. $\dfrac{dy}{dx} = \dfrac{x^2 - 3y}{3x - y^2}$; $5y - 4x = 12$; $x + y = 9$

31. $(2, 2 \pm 2\sqrt{2})$

33. $y = 2(x-3)$, $y = 2(x+3)$

35. Horizontal tangents at all four points $\left(\pm\frac{1}{4}\sqrt{6}, \pm\frac{1}{4}\sqrt{2}\right)$; vertical tangents at the two points $(\pm 1, 0)$

37. $\dfrac{4}{5\pi} \approx 0.25645$ ft/s **39.** $\dfrac{32\pi}{125} \approx 0.80425$ m/h

41. 20 cm²/s **43.** 0.25 cm/s

45. 6 ft/s **47.** 384 mi/h

49. (a) $-\dfrac{191}{1300\pi} \approx 0.047$ ft/min; (b) $-\dfrac{1337}{5100\pi} \approx 0.083$ ft/min

51. $\frac{160}{9} \approx 17.78$ ft/s

53. Increasing at 16π cm³/s

55. 6000 mi/h

57. (a) $\frac{11}{15}\sqrt{21} \approx 3.36$ ft/s downward; (b) approximately 176 ft/s downward; (c) about 4470 ft/s (over 3000 mi/h!)

59. At $t = 12$ min; $32\sqrt{13} \approx 115.38$ mi

61. $\dfrac{50}{81\pi} \approx 0.1965$ ft/s **63.** $-\dfrac{10}{81\pi} \approx -0.0383$ in./min

65. $300\sqrt{2} \approx 424.26$ mi/h **67.** 2 ft/s

Section 3.9 (page 191)

1. 2.2361 **3.** 2.5119 **5.** 0.3028

7. −0.7402 **9.** 0.7391 **11.** 1.2361

13. 2.3393 **15.** 2.0288 **17.** 2.1544

19. 1.8022 **21.** (b) 1.25992 **23.** 0.45018

25. The first formula yields $x = 2.8794$, the *wrong root*. The second yields the same wrong root, and convergence is *extremely slow*.

27. Use the intermediate value theorem to show that there is at *least* one real root. Use the fact that $f(x) = x^5 + x - 1$ is increasing on $\boldsymbol{R}$ to show that there is at *most* one real root. To four places, it is 0.7549.

29. Three solutions: 0 and ±1.8955.

31. Use the intermediate value theorem to show that the equation has at least one real solution. There are three real solutions, with approximate values 1.257, 0.715, and −1.358.

33. $x \approx 0.8654740331$ **35.** 3.4525

37. 0.2261 **39.** $\alpha_1 \approx 2.029$, $\alpha_2 \approx 4.913$

41. $t \approx 2.0369$ (−3.7454 is not possible); $w \approx 13.0164$.

Chapter 3 Miscellaneous Problems (page 195)

Some of these answers have been simplified considerably, including use of trigonometric identities.

1. $\dfrac{dy}{dx} = 2x - 6x^{-3}$ **3.** $\dfrac{dy}{dx} = \frac{1}{2}x^{-1/2} - \frac{1}{3}x^{-4/3}$

5. $\dfrac{dy}{dx} = (x-1)^6(3x+2)^8(48x - 13)$

7. $\dfrac{dy}{dx} = 4(3x - \frac{1}{2}x^{-2})^3(3 + x^{-3})$

9. $\dfrac{dy}{dx} = -\dfrac{y}{x} = -\dfrac{9}{x^2}$ **11.** $\dfrac{dy}{dx} = \dfrac{3(1-3x^2)}{2(x^3-x)^{5/2}}$

13. $\dfrac{dy}{dx} = \dfrac{4x(x^2+1)}{(x^4+2x^2+2)^2}$

15. $\dfrac{dy}{dx} = \dfrac{7}{3}(\sqrt{x} + (2x)^{1/3})^{4/3}\left(\dfrac{1}{2\sqrt{x}} + \dfrac{(2x)^{1/3}}{3x}\right)$

17. $\dfrac{dy}{dx} = -\dfrac{1}{(-1+\sqrt{x+1})^2\sqrt{x+1}}$

19. $\dfrac{dy}{dx} = -\dfrac{1-2xy^2}{1-2x^2y}$ **21.** $\dfrac{dy}{dx} = \dfrac{1 + \dfrac{2 + \dfrac{\sqrt{3}}{2\sqrt{x}}}{2\sqrt{2x+\sqrt{3x}}}}{2\sqrt{x+\sqrt{2x+\sqrt{3x}}}}$

23. $\dfrac{dy}{dx} = -\dfrac{y^{2/3}}{x^{2/3}}$

25. $\dfrac{dy}{dx} = -\dfrac{18(x^3 + 3x^2 + 3x + 3)^2}{(x + 1)^{10}}$

27. $\dfrac{dy}{dx} = \dfrac{\sin x}{2\sqrt{1 - \cos x}}$

29. $\dfrac{dy}{dx} = -\dfrac{3\cos 2x \cos 3x + 4 \sin 2x \sin 3x}{2(\sin 3x)^{3/2}}$

31. $\dfrac{dy}{dx} = 6 \sin^2 2x \cos 3x \cos 5x$

33. $\dfrac{dy}{dx} = 5\left(1 - \dfrac{1}{x^2}\right) \sin^4\left(x + \dfrac{1}{x}\right) \cos\left(x + \dfrac{1}{x}\right)$

35. $\dfrac{dy}{dx} = -\dfrac{4x^3 \cos^2((x^4 + 1)^{1/3}) \sin((x^4 + 1)^{1/3})}{(x^4 + 1)^{2/3}}$

37. $x = 1$ **39.** $x = 0$ **41.** 0.5 ft/min

43. $\dfrac{1}{3}$ **45.** $\dfrac{1}{4}$ **47.** 0

49. $h'(x) = -\dfrac{x}{(x^2 + 25)^{3/2}}$ **51.** $h'(x) = \dfrac{5}{3}(x - 1)^{2/3}$

53. $h'(x) = -2x \sin(x^2 + 1)$

55. $V'(A) = \dfrac{1}{4}\sqrt{\dfrac{A}{\pi}}$

57. $\dfrac{\pi}{10} \sec^2 \dfrac{\pi}{3} = \dfrac{2\pi}{5} \approx 1.2566$ (mi/s)

59. R^2

61. For maximum surface area, make two equal spheres. For minimum area, make only one sphere.

63. $\dfrac{32}{81}\pi R^3$ **65.** $\dfrac{M}{2}$

67. 36 (ft^3) **69.** $3\sqrt{3}$

73. Two miles from the point on the shore nearest the "one-mile" town

75. (a) $\dfrac{m^2 v^2}{64(m^2 + 1)}$; (b) $m = 1$; $\alpha = \pi/4$

77. 2.6458 **79.** 2.3714 **81.** −0.3473

83. 0.7402 **85.** −0.7391 **87.** −1.2361

89. About 1.5478525724 ft

91. −2.7225, 0.8013, 2.3100

93. The derivative of the sum of a *constant* number of functions is equal to the sum of their derivatives.

97. 4 in.^2/s **99.** $-\dfrac{50}{9\pi}$ ft/min **101.** 1 in./min

Section 4.2 (page 207)

1. $(6x + 8x^{-3})\,dx$

3. $[1 + \frac{3}{2}x^2(4 - x^3)^{-1/2}]\,dx$

5. $[6x(x - 3)^{3/2} + \frac{9}{2}x^2(x - 3)^{1/2}]\,dx$

7. $[(x^2 + 25)^{1/4} + \frac{1}{2}x^2(x^2 + 25)^{-3/4}]\,dx$

9. $-\frac{1}{2}x^{-1/2} \sin x^{1/2}\,dx$

11. $(2\cos^2 2x - 2\sin^2 2x)\,dx$

13. $\left(\frac{2}{3}x^{-1} \cos 2x - \frac{1}{3}x^{-2} \sin 2x\right) dx$

15. $(\sin x + x \cos x)(1 - x \sin x)^{-2}\,dx$

17. $f(x) \approx 1 + x$ **19.** $f(x) \approx 1 + 2x$

21. $f(x) \approx 1 - 3x$ **23.** $f(x) \approx x$

25. $3 - \frac{2}{27} \approx 2.926$ **27.** $2 - \frac{1}{32} \approx 1.969$

29. $\frac{95}{1536} \approx 0.06185$ **31.** $\dfrac{1 + \frac{1}{90}\pi}{\sqrt{2}} \approx 0.7318$

33. $\sin \dfrac{\pi}{2} - \dfrac{\pi}{90} \cos \dfrac{\pi}{2} = 1.000$

35. $\dfrac{dy}{dx} = -\dfrac{x}{y}$ **37.** $\dfrac{dy}{dx} = \dfrac{y - x^2}{y^2 - x}$

41. $-4\ \text{in.}^2$ **43.** $-405\pi/2\ \text{cm}^3$

45. 10 ft **47.** 6 W

49. $25\pi \approx 78.54\ \text{in.}^3$ **51.** $4\pi \approx 12.57\ \text{m}^2$

53. (0.56, 1.44) **55.** (1.74, 2.30)

57. (−0.67, 0.67) **59.** (0.54, 1.01)

Section 4.3 (page 216)

1. Increasing for $x < 0$, decreasing for $x > 0$; (c)

3. Decreasing for $x < -2$, increasing for $x > -2$; (f)

5. Increasing for $x < -1$ and for $x > 2$; decreasing on $(-1, 2)$; (d)

7. $f(x) = 2x^2 + 5$ **9.** $f(x) = 2 - \dfrac{1}{x}$

11. Increasing on $\boldsymbol{R}$

13. Increasing for $x < 0$, decreasing for $x > 0$,

15. Increasing for $x < \frac{3}{2}$, decreasing for $x > \frac{3}{2}$

17. Increasing on $(-1, 0) \cup (1, +\infty)$, decreasing on $(-\infty, -1) \cup (0, 1)$

19. Increasing on $(-2, 0) \cup (1, +\infty)$, decreasing on $(-\infty, -2) \cup (0, 1)$

21. Increasing for $x < 2$, decreasing for $x > 2$

23. Increasing on $(-\infty, -\sqrt{3}) \cup (-\sqrt{3}, 1) \cup (3, +\infty)$, decreasing on $(1, \sqrt{3}) \cup (\sqrt{3}, 3)$

25. $f(0) = 0 = f(2)$, $f'(x) = 2x - 2$; $c = 1$

27. $f(-1) = 0 = f(1)$, $f'(x) = -\dfrac{4x}{(1 + x^2)^2}$; $c = 0$

29. $f'(0)$ does not exist. **31.** $f(0) \neq f(1)$

33. $c = -\frac{1}{2}$ **35.** $c = \frac{35}{27}$

37. The average slope is $\frac{1}{3}$, but $|f'(x)| = 1$ where $f'(x)$ exists.

39. The average slope is 1, but $f'(x) = 0$ wherever it exists.

41. If $g(x) = x^5 + 2x - 3$, then $g'(x) > 0$ for all x in $[0, 1]$ and $g(1) = 0$. So $x = 1$ is the only solution of the equation in the given interval.

43. If $g(x) = x^4 - 3x - 20$, then $g(2) = -10$ and $g(3) = 52$. If x is in $[2, 3]$, then

$$g'(x) = 4x^3 - 3 \geqq 4 \cdot 2^3 - 3 = 29 > 0,$$

so g is an increasing function on $[2, 3]$. Hence $g(x)$ can have at most one zero in $[2, 3]$. It has at least one zero because $g(2) < 0 < g(3)$ and g is continuous.

49. Note that $f'(x) = \frac{3}{2}\left(-1 + \sqrt{x + 1}\right)$.

51. Assume that $f'(x)$ has the form

$$a_0 + a_1 x + \cdots + a_{n-1}x^{n-1}.$$

Construct a polynomial $p(x)$ such that $p'(x) = f'(x)$. Conclude that $f(x) = p(x) + C$ on $[a, b]$.

Section 4.4 (page 226)

1. Global minimum at $x = 2$

3. Local maximum at $x = 0$, local minimum at $x = 2$

5. No extremum at the critical point $x = 1$

7. Local minimum at $x = -2$, local maximum at $x = 5$

9. Global minima at $x = \pm 1$, local maximum at $x = 0$

11. Local maximum at $x = -1$, local minimum at $x = 1$

13. Local minimum at $x = 1$

15. Vertical tangent and global maximum at $(0, 3)$

17. Global maximum at $x = \pi/2$

19. Global minimum at $(-\pi/2, -1)$, global maximum at $(\pi/2, 1)$

21. Global minimum at $(0, 0)$

23. Global minimum at $(-\pi, -\pi)$, global maximum at (π, π), no extremum at $(0, 0)$

25. Global maximum at $(\pi/4, 1)$

27. -10 and 10

29. $(1, 1)$

31. Base 9 by 18, height 6

33. Radius $5\pi^{-1/3}$, height $10\pi^{-1/3}$

37. Base 5 by 5 in., height 2.5 in.

39. Radius $(25/\pi)^{1/3}$, height $20(5\pi)^{-1/3}$

41. $\left(\pm\frac{1}{2}\sqrt{6}, \frac{3}{2}\right)$

43. 8 cm

45. $L = \sqrt{20 + 12 \cdot 4^{1/3} + 24 \cdot 2^{1/3}} \approx 8.324$ m

49. Height $(6V)^{1/3}$, base edge $\left(\frac{9}{2}V^2\right)^{1/6}$

Section 4.5 (page 235)

1. (c)

3. (d)

5. Critical point: $x_1 = \frac{5}{2}$; decreasing on $(-\infty, x_1)$, increasing on $(x_1, +\infty)$.

7. Critical points: $x_1 = -\frac{5}{2}$, $x_2 = 3$; decreasing on (x_1, x_2), increasing on $(-\infty, x_1)$ and on $(x_2, +\infty)$

9. Critical points: $x_1 = -3$, $x_2 = 0$, $x_3 = 2$; decreasing on $(-\infty, x_1)$ and on (x_2, x_3), increasing on (x_1, x_2) and on $(x_3, +\infty)$

11. Critical points: $x_1 = -4$, $x_2 = -2$, $x_3 = 2$, $x_4 = 4$; decreasing on (x_1, x_2) and on (x_3, x_4), increasing on $(-\infty, x_1)$, on (x_2, x_3), and on $(x_4, +\infty)$

13. Critical points: $x_1 = -4$, $x_2 = -2$, $x_3 = 0$, $x_4 = 2$, $x_5 = 4$; decreasing on (x_1, x_2) and on (x_4, x_5), increasing on $(-\infty, x_1)$, on (x_2, x_4), and on $(x_5, +\infty)$

15. Critical point: $x_1 = 1$. Decreasing on $(-\infty, x_1)$, increasing on $(1, +\infty)$; local (global) minimum at $(1, 2)$

17. Increasing if $|x| > 2$, decreasing on $(-2, 2)$; local maximum at $(-2, 16)$, local minimum at $(2, -16)$

19. Increasing for $x < 1$ and for $x > 3$, decreasing on $(1, 3)$; local maximum at $(1, 4)$, local minimum at $(3, 0)$

21. Increasing for all x, no extrema

23. Decreasing on $(-\infty, -2)$ and on $\left(-\frac{1}{2}, 1\right)$, increasing on $\left(-2, -\frac{1}{2}\right)$ and on $(1, +\infty)$; global minima at $(-2, 0)$ and $(1, 0)$, local maximum at $\left(-\frac{1}{2}, 5.0625\right)$

25. Increasing on $(0, 1)$, decreasing on $(1, +\infty)$; global maximum at $(1, 2)$, local minimum at $(0, 0)$, no graph for $x < 0$

27. Increasing on $(-\infty, -1)$ and on $(1, +\infty)$, decreasing on $(-1, 1)$ (but with a horizontal tangent at $(0, 0)$); local maximum at $(-1, 2)$, local minimum at $(1, -2)$

29. Decreasing on $(-\infty, -2)$ and on $(0, 2)$, increasing on $(-2, 0)$ and on $(2, +\infty)$; global minima at $(-2, -9)$ and $(2, -9)$, local maximum at $(0, 7)$

31. Decreasing on $\left(-\infty, \frac{3}{4}\right)$, increasing on $\left(\frac{3}{4}, +\infty\right)$; global minimum at $\left(\frac{3}{4}, -\frac{81}{8}\right)$

33. Decreasing on $(-2, 1)$, increasing on $(-\infty, -2)$ and on $(1, +\infty)$; local maximum at $(-2, 20)$, local minimum at $(1, -7)$

35. Increasing on $\left(-\infty, \frac{3}{5}\right)$ and on $\left(\frac{4}{5}, +\infty\right)$, decreasing on $\left(\frac{3}{5}, \frac{4}{5}\right)$; local maximum at $\left(\frac{3}{5}, \frac{81}{5}\right)$, local minimum at $\left(\frac{4}{5}, 16\right)$

37. Increasing on $(-1, 0)$ and on $(2, +\infty)$, decreasing on $(-\infty, -1)$ and on $(0, 2)$; local maximum at $(0, 8)$, local minimum at $(-1, 3)$, global minimum at $(2, -24)$

39. Increasing on $(-\infty, -2)$ and on $(2, +\infty)$, decreasing on $(-2, 2)$; local maximum at $(-2, 64)$, local minimum at $(2, -64)$

41. Increasing everywhere; no extrema. The graph passes through the origin; its minimum slope is $\frac{9}{2}$, which occurs where $x = -\frac{1}{2}$

43. Increasing on $\left(-\infty, -\sqrt{2}\right)$ and on $\left(0, \sqrt{2}\right)$, decreasing on $\left(-\sqrt{2}, 0\right)$ and on $\left(\sqrt{2}, +\infty\right)$; global maxima at $\left(-\sqrt{2}, 16\right)$ and $\left(\sqrt{2}, 16\right)$, local minimum at $(0, 0)$

45. Increasing on $(-\infty, 1)$, decreasing on $(1, +\infty)$; global maximum at $(1, 3)$, vertical tangent at $(0, 0)$ (not an extremum)

47. Increasing on $\left(-\infty, \frac{3}{5}\right)$ and on $(1, +\infty)$, decreasing on $\left(\frac{3}{5}, 1\right)$; local minimum and a cusp at $(1, 0)$, local maximum at $\left(\frac{3}{5}, 0.3527\right)$ (ordinate approximate)

49. Plot $y = 2x^3 + 3x^2 - 36x - 3$ to see the graph.

51. Plot $y = -2x^3 - 3x^2 + 36x + 15$ to see the graph.

53. Plot $y = 3x^4 - 8x^3 - 30x^2 + 72x + 45$ to see the graph.

55. (b) $x^3 - 3x + 3 \approx (x + 2.1038)(x^2 - 2.1038x + 1.42599)$; (c) the complex conjugate roots are approximately $1.051901701 \pm 0.5652358517i$.

57. $f'(x) = 2x(2x - 1)(x - 1)(6x^2 - 6x + 1)$

Section 4.6 (page 248)

1. $8x^3 - 9x^2 + 6,\ 24x^2 - 18x,\ 48x - 18$

3. $-\dfrac{8}{(2x-1)^3},\ \dfrac{48}{(2x-1)^4},\ -\dfrac{384}{(2x-1)^5}$

5. $4(3t-2)^{1/3},\ \dfrac{4}{(3t-2)^{2/3}},\ -\dfrac{8}{(3t-2)^{5/3}}$

7. $\dfrac{1}{(y+1)^2},\ -\dfrac{2}{(y+1)^3},\ \dfrac{6}{(y+1)^4}$

9. $-\frac{1}{4}t^{-3/2} - (1-t)^{-4/3},\ \frac{3}{8}t^{-5/2} - \frac{4}{3}(1-t)^{-7/3},$
$-\frac{15}{16}t^{-7/2} - \frac{28}{9}(1-t)^{-10/3}$

11. $3\cos 3x,\ -9\sin 3x,\ -27\cos 3x$

13. $\cos^2 x - \sin^2 x,\ -4\sin x\cos x,\ 4\sin^2 x - 4\cos^2 x$

15. $\dfrac{x\cos x - \sin x}{x^2},\ \dfrac{2\sin x - 2x\cos x - x^2\sin x}{x^3},$
$\dfrac{3x^2\sin x - x^3\cos x + 6x\cos x - 6\sin x}{x^4}$

17. $-\dfrac{2x+y}{x+2y},\ -\dfrac{18}{(x+2y)^3}$

19. $-\dfrac{1+2x}{3y^2},\ \dfrac{2y^3-42}{9y^5}$

21. $\dfrac{y}{\cos y - x},\ \dfrac{2y\cos y - 2xy + y^2\sin y}{(\cos y - x)^3}$

23. $(-3, 81), (5, -175), (1, -47)$

25. $\left(\frac{9}{2}, -\frac{941}{2}\right), \left(-\frac{7}{2}, \frac{1107}{2}\right), \left(\frac{1}{2}, \frac{83}{2}\right)$

27. $(3\sqrt{3}, -492),\ (-3\sqrt{3}, -492),\ (0, 237),\ (3, -168),$ $(-3, -168)$

29. $\left(\frac{16}{3}, -\frac{181144}{81}\right), (0, 1000), (4, -1048)$

31. Local minimum at $(2, -1)$; no inflection points

33. Local maximum at $(-1, 3)$, local minimum at $(1, -1)$; inflection point at $(0, 1)$

35. No local extrema; inflection point at $(0, 0)$

37. No local extrema; inflection point at $(0, 0)$

39. Local maximum at $\left(\frac{1}{2}, \frac{1}{16}\right)$, local minima at $(0, 0)$ and $(1, 0)$; the abscissas of the inflection points are the roots of $6x^2 - 6x + 1 = 0$, and the inflection points are at (approximately) $(0.79, 0.03)$ and $(1.21, 0.03)$.

41. Global maximum at $(\pi/2, 1)$, global minimum at $(3\pi/2, -1)$; inflection point at $(\pi, 0)$

43. Inflection point at $(0, 0)$

45. Global maxima at $(0, 1)$ and $(\pi, 1)$, global minimum at $(\pi/2, 0)$; inflection points at $(-\pi/4, 1/2)$, $(\pi/4, 1/2)$, $(3\pi/4, 1/2)$, and $(5\pi/4, 1/2)$

47. Global maximum at $(\pi/4, \sqrt{2})$, global minimum at $(5\pi/4, -\sqrt{2})$; inflection points at $(3\pi/4, 0)$ and $(5\pi/4, 0)$

49. Global maximum at

$$\left(\arctan \tfrac{1}{2},\ \sqrt{5}\right) \approx (0.4636, 2.2351),$$

global minimum at

$$\left(\pi + \arctan \tfrac{1}{2}, -\sqrt{5}\right) \approx (3.6052, -2.2361);$$

inflection points at $(\pi - \arctan 2, 0)$ and $(2\pi - \arctan 2, 0)$

63. Increasing for $x < -1$ and for $x > 2$, decreasing on $(-1, 2)$, local maximum at $(-1, 10)$, local minimum at $(2, -17)$, inflection point at $\left(\frac{1}{2}, -\frac{7}{2}\right)$

65. Increasing for $x < -2$ and on $(0, 2)$, decreasing for $x > 2$ and on $(-2, 0)$, global maximum at $(\pm 2, 22)$, local minimum at $(0, 6)$, inflection points at $\left(\pm\frac{2}{3}\sqrt{3}, \frac{134}{9}\right)$

67. Decreasing for $x < -1$ and on $(0, 2)$, increasing for $x > 2$ and on $(-1, 0)$, local maximum at $(0, -1)$, global minimum at $(2, -33)$, inflection points with approximate coordinates $(1.22, -19.36)$ and $(-0.55, -3.68)$

69. Local maximum at $\left(\frac{3}{7}, \frac{6912}{823543}\right) \approx (0.43, 0.0084)$, local minimum at $(1, 0)$, inflection points at $(0, 0)$ and at the two solutions of $7x^2 - 6x + 1 = 0$: approximately $(0.22, 0.0042)$ and $(0.63, 0.0047)$

71. The graph is increasing everywhere, with a vertical tangent and inflection point at $(0, 1)$; the other intercept is $(-1, 0)$ and there are no extrema

73. Global minimum at $(0, 0)$, increasing for all $x > 0$, inflection point at $(1, 4)$, concave upward for $x > 1$, vertical tangent at the origin

75. Increasing for $x < 1$, decreasing for $x > 1$, vertical tangent and inflection point at $(0, 0)$, another inflection point at $(-2, -7.56)$ (ordinate approximate), global maximum at $(1, 3)$

77. (c) **79.** (b) **81.** (d)

89. $a = 3pV^2 \approx 3{,}583{,}858.8$, $b = \frac{1}{3}V = 42.7$,

$$R = \frac{8pV}{3T} \approx 81.81421$$

Section 4.7 (page 261)

1. 1 **3.** 3 **5.** 2
7. 1 **9.** 4 **11.** 0
13. 2 **15.** $+\infty$ **17.** (g)
19. (a) **21.** (f) **23.** (j)
25. (l) **27.** (k)

29. No critical points or inflection points, vertical asymptote $x = 3$, horizontal asymptote $y = 0$, sole intercept at $\left(0, -\frac{2}{3}\right)$

31. No critical points or inflection points, vertical asymptote $x = -2$, horizontal asymptote $y = 0$, sole intercept at $\left(0, \frac{3}{4}\right)$

33. No critical points or inflection points, vertical asymptote $x = \frac{3}{2}$, horizontal asymptote $y = 0$, sole intercept. $\left(0, -\frac{1}{27}\right)$

35. Global minimum at $(0, 0)$, inflection points where $3x^2 = 1$ $\left(\text{and } y = \frac{1}{4}\right)$, horizontal asymptote $y = 1$

37. Local maximum at $\left(0, -\frac{1}{9}\right)$, no inflection points, vertical asymptotes $x = \pm 3$, horizontal asymptote $y = 0$

39. Local maximum at $\left(-\frac{1}{2}, -\frac{4}{25}\right)$, horizontal asymptote $y = 0$, vertical asymptotes $x = -3$ and $x = 2$, no inflection points

41. Local minimum at $(1, 2)$, local maximum at $(-1, -2)$, no inflection points, vertical asymptote $x = 0$, slant asymptote $y = x$

43. Local minimum at $(2, 4)$, local maximum at $(0, 0)$, no inflection points, vertical asymptote $x = 1$, slant asymptote $y = x + 1$

45. No extrema or inflection points, vertical asymptote $x = 1$, horizontal asymptote $y = 0$, sole intercept $(0, 1)$

47. No extrema or inflection points, vertical asymptote $x = -1$, horizontal asymptote $y = 1$, sole intercept $(0, 0)$

49. Local maximum at $\left(\frac{1}{2}, -\frac{4}{9}\right)$. vertical asymptotes $x = -1$ and $x = 2$, horizontal asymptote $y = 0$, sole intercept $\left(0, -\frac{1}{2}\right)$

51. No extrema or inflection points, intercepts $(-2, 0)$ and $(2, 0)$, vertical asymptote $x = 0$, slant asymptote $y = x$

53. Local maximum at $(-2, -3)$, no inflection points, sole intercept $\left(\sqrt[3]{4}, 0\right)$, vertical asymptote $x = 0$, slant asymptote $y = x$

In the answers to Problems 55 through 67, most decimals have been rounded to five places.

55. Local minima at $(-1.90952, -0.31324)$ and $(1.39070, 3.26486)$, local maximum at $(4.51882, 0.16297)$, inflection points at $(-2.81191, -0.27678)$ and $(6.06299, 0.14485)$, vertical asymptotes $x = 0$ and $x = 2$, horizontal asymptote $y = 0$

57. Local minima at $(-2.81730, -0.17815)$ and $(1.46950, 5.54443)$, local maxima at $(-1, 0)$ and $(4.34780, 0.19976)$, inflection points at $(-4.34109, -0.15761)$, $(-1.25693, -0.04345)$, and $(5.70076, 0.17687)$, intercepts at $(-1, 0)$ and $(3, 0)$, vertical asymptotes $x = 0$ and $x = 2$, horizontal asymptote $y = 0$

59. Local minima at $(-2.66433, -0.21599)$, $(1.24713, 14.1117)$, and $(3, 0)$, local maxima at $(-1, 0)$ and $(5.41720, 0.12961)$, inflection points at

$$(-4.05618, -0.19000), \quad (-1.24685, -0.05378),$$

$$(3.32642, 0.03080), \quad \text{and} \quad (7.49694, 0.11467),$$

vertical asymptotes $x = 0$ and $x = 2$, horizontal asymptote $y = 0$, intercepts $(-1, 0)$ and $(3, 0)$

61. Local minimum at $(0, 0)$, local maximum at $(1.25992, -0.90081)$, no inflection points, vertical asymptotes $x = -0.53209$, $x = 0.65270$, and $x = 2.87939$, horizontal asymptote $y = 0$, intercept $(0, 0)$

63. Local minima at $(0.82120, 0.61460)$ and $(5.08273, 11.0886)$, local maxima at

$$(-2.35622, -1.92916) \quad \text{and} \quad (2.37606, 18.5247),$$

inflection points at $(1.94325, 11.3790)$ and $(2.70396, 16.8013)$, slant asymptote $y = x + 3$, vertical asymptote $x = -1.10380$, no intercepts

65. Local maximum at $(0.22010, 0.60008)$, local minimum at $(0.82216, -2.96905)$, inflection points at $(-2.24169, -1.27822)$, $(-0.59463, -0.12114)$, $(0.67009, -1.68203)$, and $(0.96490, -2.25012)$, slant asymptote $y = \frac{1}{2}x$, intercepts

$$(-0.49269, 0), \quad (0.50842, 0), \quad (1.52864, 0), \quad \text{and} \quad \left(0, \tfrac{1}{2}\right)$$

67. Local minima at $(-0.53789, -0.35913)$ and $(1.87859, 1.43883)$, local maxima at $(-3.15941, -2.36646)$ and $(1.33805, 1.77918)$, inflection points at $(0, 0)$, $(0.53241, 0.48053)$, $(1.16066, 1.42940)$, and $(1.46265, 1.67271)$, vertical asymptote $x = -1.72766$, slant asymptote $y = \frac{1}{2}x$, intercepts $(-1, 0)$ and $(0, 0)$

69. Local minimum at $(1, 3)$, inflection point at $\left(-\sqrt[3]{2}, 0\right)$, vertical asymptote $x = 0$

Chapter 4 Miscellaneous Problems (page 264)

1. $dy = 3(4x - x^2)^{1/2}(2 - x)\,dx$

3. $dy = -2(x - 1)^{-2}\,dx$

5. $dy = \left(2x \cos x^{1/2} - \frac{1}{2}x^{3/2} \sin x^{1/2}\right) dx$

7. $\frac{12801}{160} = 80.00625$ (true value: about 80.00624975588)

9. $\frac{128192}{125} = 1025.536$ (true value: about 1025.537037)

11. $\frac{601}{60} \approx 10.016667$ (true value: about 10.016639)

13. 132.5 (true value: about 132.574507)

15. 2.03125 (true value: about 2.030543185)

17. $7.5\ (\text{in.}^3)$

19. $10\pi \approx 31.416\ (\text{cm}^3)$

21. $\pi/96 \approx 0.0327$ s

23. $c = \sqrt{3}$

25. $c = 1$

27. $c = \sqrt[4]{2.2}$

29. Decreasing for $x < 3$, increasing for $x > 3$, global minimum at $(3, -5)$, concave upward everywhere

31. Increasing everywhere, no extrema; inflection points where $x = 0, x = \pm 1/\sqrt{2}$

33. Increasing for $x < \frac{1}{4}$, decreasing for $x > \frac{1}{4}$, vertical tangent at $(0, 0)$, global maximum at $x = \frac{1}{4}$; inflection points where $x = 0, x = -\frac{1}{2}$

35. $3x^2 - 2, 6x, 6$

37. $-\dfrac{1}{t^2} + \dfrac{2}{(2t+1)^2}, \ \dfrac{2}{t^3} - \dfrac{8}{(2t+1)^3}, \ -\dfrac{6}{t^4} + \dfrac{48}{(2t+1)^4}$

39. $3t^{1/2} - 4t^{1/3}, \frac{3}{2}t^{-1/2} - \frac{4}{3}t^{-2/3}, -\frac{3}{4}t^{-3/2} + \frac{8}{9}t^{-5/3}$

41. $-\dfrac{4}{(t-2)^2}, \ \dfrac{8}{(t-2)^3}, \ -\dfrac{24}{(t-2)^4}$

43. $-\frac{4}{3}(5 - 4x)^{-2/3}, -\frac{32}{9}(5 - 4x)^{-5/3}, -\frac{640}{27}(5 - 4x)^{-8/3}$

45. $\dfrac{dy}{dx} = -\left(\dfrac{y}{x}\right)^{2/3}, \ \dfrac{d^2y}{dx^2} = \dfrac{2y^{1/3}}{3x^{5/3}}$

47. $\dfrac{dy}{dx} = \dfrac{1}{2\sqrt{x}(5y^4 - 4)}$, $\dfrac{d^2y}{dx^2} = -\dfrac{20y^3\sqrt{x} + (5y^4 - 4)^2}{4x\sqrt{x}(5y^4 - 4)^3}$

49. $\dfrac{dy}{dx} = \dfrac{5y - 2x}{2y - 5x}$ **51.** $\dfrac{dy}{dx} = \dfrac{2xy}{3y^2 - x^2 - 1}$

53. Global minimum at $(2, -48)$, x-intercepts 0 and (approximately) 3.1748, no inflection points or asymptotes, concave upward everywhere

55. Local maximum at $(0, 0)$, global minimum where $x^2 = \frac{4}{3}$ $\left(\text{and } y = -\frac{32}{27}\right)$, inflection points where $x^2 = \frac{4}{5}$ $\left(\text{and } y = -\frac{96}{125}\right)$, no asymptotes

57. Global maximum at $(3, 3)$, inflection points where $x = 6$ and $(4, 0)$ (and a vertical tangent at the latter point), no asymptotes

59. Local maximum at $\left(0, -\frac{1}{4}\right)$, no inflection points, horizontal asymptote $y = 1$, vertical asymptotes $x = \pm 2$

61. The inflection point has abscissa the only real solution of $x^3 + 6x^2 + 4 = 0$, approximately -6.10724.

63.

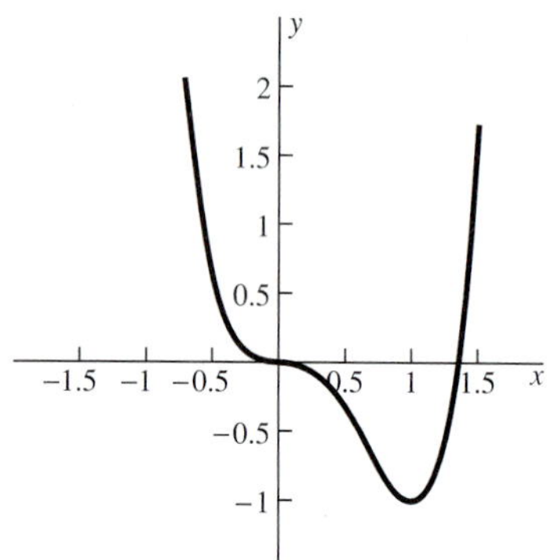

Inflection points at $(0, 0)$ and $\left(\frac{2}{3}, -\frac{16}{27}\right)$, global minimum at $(1, -1)$

65.

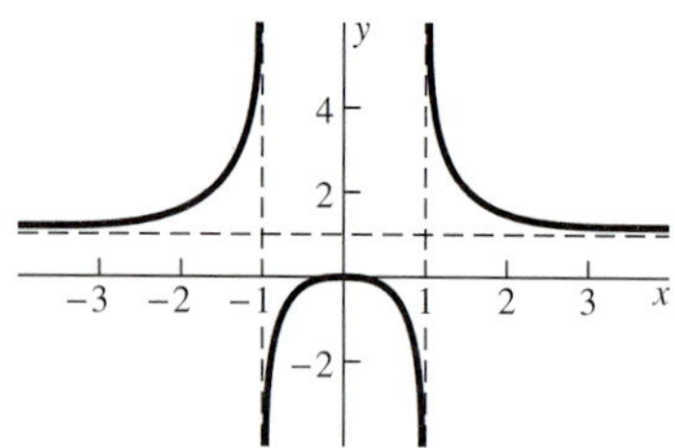

Local maximum at $(0, 0)$, vertical asymptotes $x = \pm 1$, horizontal asymptote $y = 1$

67.

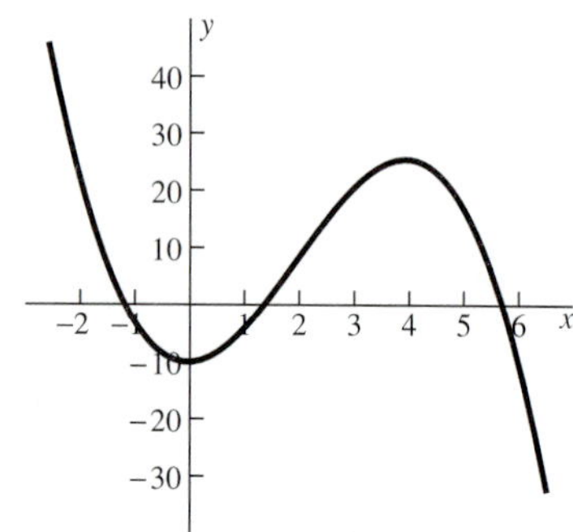

Local minimum at $(0, -10)$, local maximum at $(4, 22)$, inflection point at $(2, 6)$

69.

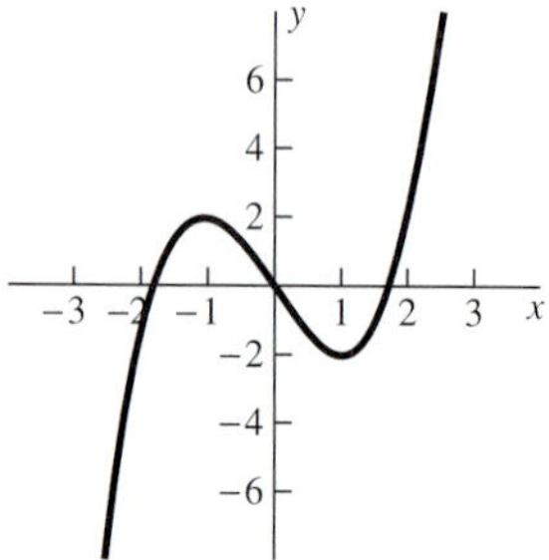

Local maximum at $(-1, 2)$, local minimum at $(1, -2)$, inflection point at $(0, 0)$

71.

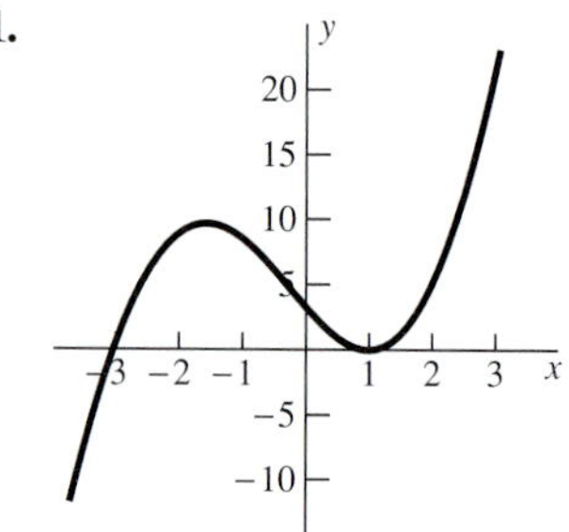

Local maximum at $\left(-\frac{5}{3}, \frac{256}{27}\right)$, local minimum at $(1, 0)$, inflection point at $\left(-\frac{1}{3}, \frac{128}{27}\right)$

73. Maximum value 1 at $x = -1$

75. Base 15×30, height 10

77. Base 5×10, height 8

79. $100 \cdot \left(\frac{2}{9}\right)^{2/5} \approx 54.79$ mi/h

81. Two horizontal tangents, where $x \approx 0.42$ and $y \approx \pm 0.62$; intercepts at $(0, 0)$, $(1, 0)$, and $(2, 0)$; vertical tangent at each intercept; inflection points corresponding to the only positive solution of $3x^2(x - 2)^2 = 4$—that is, $x \approx 2.46789$, $y \approx \pm 1.30191$; no asymptotes

83. 240 ft **85.** $2\sqrt{2A(n + 2)}$ ft

87. There is neither a maximum nor a minimum

89. 288 in.^{2} **91.** 270 cm^2

93. In both cases, $m = 1$ and $b = -\frac{2}{3}$

Section 5.2 (page 280)

1. $x^3 + x^2 + x + C$ **3.** $x - \frac{2}{3}x^3 + \frac{3}{4}x^4 + C$

5. $-\frac{3}{2}x^{-2} + \frac{4}{5}x^{5/2} - x + C$

7. $t^{3/2} + 7t + C$ **9.** $\frac{3}{5}x^{5/3} - 16x^{-1/4} + C$

11. $x^4 - 2x^2 + 6x + C$ **13.** $7x + C$

15. $\frac{1}{5}(x + 1)^5 + C$ **17.** $-\frac{1}{6}(x - 10)^{-6} + C$

19. $\frac{2}{3}x^{3/2} - \frac{4}{5}x^{5/2} + \frac{2}{7}x^{7/2} + C$

21. $\frac{2}{21}x^3 - \frac{3}{14}x^2 - \frac{5}{7}x^{-1} + C$

23. $\frac{1}{54}(9t + 11)^6 + C$ **25.** $-7(x + 77)^{-1} + C$

27. $\frac{1}{2}\sin 10x + 2\cos 5x + C$

29. $\dfrac{3\sin \pi t}{\pi} + \dfrac{\sin 3\pi t}{3\pi} + C$

33. $\frac{1}{2}(x - \sin x\cos x) + C$; $\frac{1}{2}(x + \sin x\cos x) + C$

35. $y = f(x) = x^2 + x + 3$

37. $y = f(x) = \frac{2}{3}(x^{3/2} - 8)$

39. $y = f(x) = 2\sqrt{x+2} - 5$

41. $y = f(x) = \frac{3}{4}x^4 - 2x^{-1} + \frac{9}{4}$

43. $y = f(x) = \frac{1}{4}(x-1)^4 + \frac{7}{4}$

45. $y = f(x) = 2\sqrt{x - 13} - 2$

47. $x(t) = 2t^3 - 2t^2 - 10t$

49. $x(t) = \frac{1}{6}t^4 + 3t - 7$ **51.** $x(t) = t - \sin t$

53.

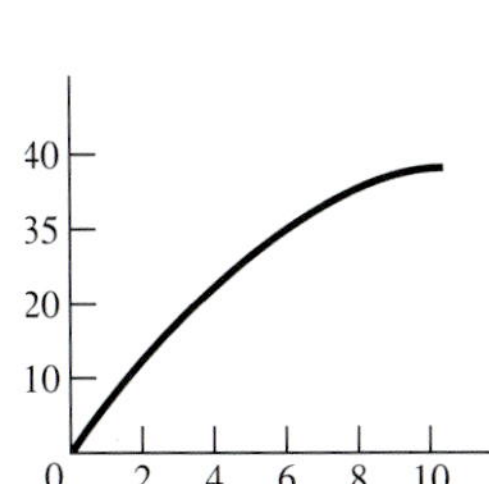

55.

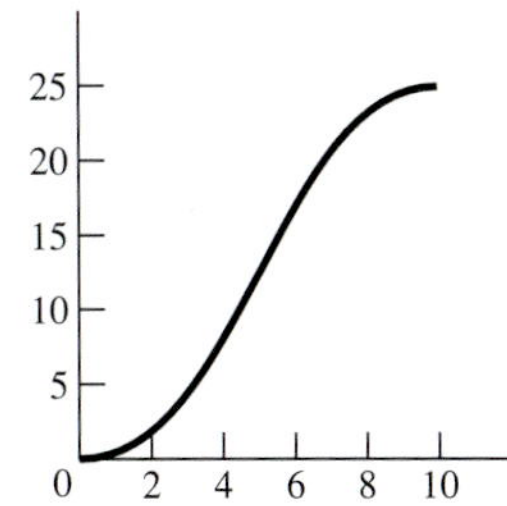

57. 144 ft; 6 s **59.** 144 ft

61. 5 s; 112 ft/s

63. $\sqrt{60} \approx 7.75$ s; $32\sqrt{60} \approx 247.87$ ft/s

65. 120 ft/s **67.** 5 s; -160 ft/s

69. 400 ft

71. $\frac{1}{4}\left(-5 + 2\sqrt{145}\right) \approx 4.77$ s; $16\sqrt{145} \approx 192.6655$ ft/s

73. $\frac{544}{3}$ ft/s **75.** 22 ft/s^2

77. Approximately 886.154 ft

79. $\frac{25}{3}\sqrt{42} \approx 54$ (mi/h)

Section 5.3 (page 292)

1. $3 + 9 + 27 + 81 + 243$

3. $\dfrac{1}{2} + \dfrac{1}{3} + \dfrac{1}{4} + \dfrac{1}{5} + \dfrac{1}{6}$

5. $1 + \dfrac{1}{4} + \dfrac{1}{9} + \dfrac{1}{16} + \dfrac{1}{25} + \dfrac{1}{36}$

7. $x + x^2 + x^3 + x^4 + x^5$

9. $\sum_{i=1}^{5} i^2$ **11.** $\sum_{n=1}^{5} \dfrac{1}{n}$ **13.** $\sum_{j=1}^{6} \dfrac{1}{2^j}$

15. $\sum_{k=1}^{5} \left(\dfrac{2}{3}\right)^k$ **17.** $\sum_{n=1}^{10} \dfrac{x^n}{n}$ **19.** 190

21. 1165 **23.** 224 **25.** 350

27. 338,350 **29.** $\frac{1}{3}$ **31.** n^2

33. $\frac{2}{5}, \frac{3}{5}$ **35.** $\frac{33}{2}, \frac{39}{2}$ **37.** $\frac{6}{25}, \frac{11}{25}$

39. $\frac{378}{25}, \frac{513}{25}$ **41.** $\frac{81}{400}, \frac{121}{400}$

45. $\dfrac{n(n+1)}{2n^2} \to \dfrac{1}{2}$ as $n \to \infty$

47. $\dfrac{81n^2(n+1)^2}{4n^4} \to \dfrac{81}{4}$ as $n \to \infty$

49. $5n \cdot \dfrac{1}{n} - \dfrac{3n(n+1)}{2n^2} \to \dfrac{7}{2}$ as $n \to \infty$

Section 5.4 (page 301)

1. $\displaystyle\int_1^3 (2x - 1)\,dx$ **3.** $\displaystyle\int_0^{10} (x^2 + 4)\,dx$

5. $\displaystyle\int_4^9 \sqrt{x}\,dx$ **7.** $\displaystyle\int_3^8 \frac{1}{\sqrt{1+x}}\,dx$

9. $\displaystyle\int_0^{1/2} \sin 2\pi x\,dx$ **11.** 0.44

13. 1.45 **15.** 19.5 **17.** 58.8

19. $-\frac{1}{6}\pi$ **21.** 0.24 **23.** $\frac{137}{60} \approx 2.28333$

25. 16.5 **27.** 26.4 **29.** $\frac{1}{6}\pi$

31. 0.33 **33.** $\frac{6086}{3465} \approx 1.75462$ **35.** 18

37. 40.575 **39.** 0 **41.** 1.83753

43. $\frac{8}{3}$ **45.** 12 **47.** 30

Section 5.5 (page 312)

1. $\frac{55}{12}$ **3.** $\frac{49}{60}$ **5.** $-\frac{1}{20}$ **7.** $\frac{1}{4}$

9. $\frac{16}{3}$ **11.** 24 **13.** 0 **15.** $\frac{32}{3}$

17. 0 **19.** $\frac{93}{5}$ **21.** $\frac{17}{6}$ **23.** $\frac{28}{3}$

25. $\frac{52}{5}$ **27.** $\frac{122}{9}$ **29.** $\frac{1}{4}$ **31.** $\frac{2}{5}$

33. $-\frac{1}{3}$ **35.** $4/\pi$ **37.** 0 **39.** $\frac{1}{2}$

41. $\frac{2}{3}$ **43.** 5 **45.** $-\frac{5}{2}$ **47.** $\frac{25}{4}\pi$

55. $\dfrac{1}{2} \leqq \displaystyle\int_0^1 \frac{1}{x+1}\,dx \leqq 1.$ **57.** $\dfrac{\pi}{8} \leqq \displaystyle\int_0^{\pi/6} \cos^2 x\,dx \leqq \frac{\pi}{6}.$

63. $1000 + \displaystyle\int_0^{30} V'(t)\,dt = 160$ (gal)

65. $\frac{2}{3} \leqq \ln 2 \leqq \frac{3}{4}$

Section 5.6 (page 321)

1. $\frac{16}{5}$ **3.** $\frac{26}{3}$ **5.** 0 **7.** $\frac{125}{4}$

9. $\frac{14}{9}$ **11.** 0 **13.** 4 **15.** $\frac{1}{3}$

17. $-\frac{22}{81} \approx -0.271605$ **19.** 0

21. $\frac{35}{24} \approx 1.45833$ **23.** 0 **25.** 4

27. $\frac{38}{3}$ **29.** $\frac{31}{20}$ **31.** $\frac{81}{4} + \frac{81}{4} = \frac{81}{2}$

33. Average height: $\frac{800}{3} \approx 266.67$ ft; average velocity: -80 ft/s

35. $\frac{5000}{3} \approx 1666.67$ (L) **37.** $\frac{200}{3}$ **39.** $\pi/3$

41. (a) $A(x) = 27 - 3x^2$, $-3 \leqq x \leqq 3$; (b) 18; (c) two

43. (a) $A(x) = 2x\sqrt{16 - x^2}$, $0 \leqq x \leqq 4$; (b) $\frac{32}{3}$; (c) two

45. $f'(x) = (x^2 + 1)^{17}$ **47.** $h'(z) = (z - 1)^{1/3}$

49. $f'(x) = -x - x^{-1}$ **51.** $G'(x) = (x + 4)^{1/2}$
53. $G'(x) = (x^3 + 1)^{1/2}$ **55.** $g'(x) = 3\sin 9x^2$
57. $f'(x) = 2x \sin x^2$ **59.** $f'(x) = 2x/(x^2 + 1)$
61. $y(x) = \int_1^x \frac{1}{t}\,dt$
63. $y(x) = 10 + \int_5^x \sqrt{1 + t^2}\,dt$
65. The integral does not exist.
67. (a) $g(0) = 0,\ g(2) = 4,\ g(4) = 8,\ g(6) = 4,\ g(8) = -4,\ g(10) = -8$. (b) Increasing on $(0, 4)$, decreasing on $(4, 10)$. (c) Maximum: 8; minimum: -8.
69. (a) $x = 0,\ \pi,\ 2\pi,\ 3\pi$, and 4π. (b) Global maximum at $x = 3\pi$, global minimum at $x = 4\pi$. (c) Inflection points occur at points whose x-coordinates are the solutions of the equation $x + \tan x = 0$; Newton's method yields the approximate values 2.028757838, 4.913180439, 7.978665712, and 11.08553841.

Section 5.7 (page 329)

1. $\frac{1}{54}(3x - 5)^{18} + C$ **3.** $\frac{1}{3}(x^2 + 9)^{3/2} + C$
5. $-\frac{1}{5}\cos 5x + C$ **7.** $-\frac{1}{4}\cos 2x^2 + C$
9. $\frac{1}{6}(1 - \cos x)^6 + C$ **11.** $\frac{1}{7}(x + 1)^7 + C$
13. $-\frac{1}{24}(4 - 3x)^8 + C$ **15.** $\frac{2}{7}\sqrt{7x + 5} + C$
17. $-\frac{1}{\pi}\cos(\pi x + 1) + C$ **19.** $\frac{1}{2}\sec 2\theta + C$
21. $\frac{1}{3}(x^2 - 1)^{3/2} + C$ **23.** $-\frac{1}{9}(2 - 3x^2)^{3/2} + C$
25. $\frac{1}{6}(x^4 + 1)^{3/2} + C$ **27.** $\frac{1}{6}\sin 2x^3 + C$
29. $-\frac{1}{9}(x^3 + 5)^{-3} + C$ **31.** $-\frac{1}{4}\cos^4 x + C$
33. $\frac{1}{4}\tan^4\theta + C$ **35.** $2\sin\sqrt{x} + C$
37. $\frac{1}{10}(x + 1)^{10} + C = \frac{1}{10}(x^2 + 2x + 1)^5 + C$
39. $\frac{1}{4}(6t + t^3)^{4/3} + C$ **41.** $\frac{5}{72}$
43. $\frac{98}{3}$ **45.** $\frac{1192}{15}$ **47.** $\frac{15}{128}$
49. $\frac{62}{15}$ **51.** $\frac{128}{15}$ **53.** $\frac{1}{2\pi}$
55. $\frac{1}{2}x - \frac{1}{4}\sin 2x + C$ **57.** $\frac{\pi}{2}$
59. $-x + \tan x + C$

Section 5.8 (page 339)

1. $\frac{256}{3}$ **3.** $\frac{9}{2}$ **5.** 32 **7.** $\frac{128}{3}$
9. $\frac{11}{2}\sqrt{33}$ **11.** $\frac{1}{4}$ **13.** $\frac{1}{20}$ **15.** $\frac{4}{9}$
17. $\frac{32}{3}$ **19.** $\frac{128}{3}\sqrt{2}$ **21.** $\frac{4}{3}$ **23.** $\frac{500}{3}$
25. $\frac{32}{3}$ **27.** $\frac{125}{6}$ **29.** $\frac{1}{3}$ **31.** $\frac{16}{3}$
33. $\frac{16}{3}$ **35.** $\frac{27}{2}$ **37.** $\frac{320}{3}$ **39.** $\frac{8}{15}$
41. $\frac{4}{3}$ **43.** $\frac{5}{12} + \frac{8}{3} = \frac{37}{12}$ **45.** $\frac{45}{2}\pi$
51. $\frac{253}{6}$ **53.** $\frac{40}{3}\sqrt{5}$ **55.** 1.09475 **57.** 3.00044
59. $k = 18$

Section 5.9 (page 354)

1. $T_4 = 8$; true value: 8 **3.** $T_5 \approx 0.65$; true value: $\frac{2}{3}$
5. $T_3 = 0.98$; true value: 1 **7.** $M_4 = 8$; true value: 8
9. $M_5 \approx 0.67$; true value: $\frac{2}{3}$ **11.** $M_3 \approx 1.01$; true value: 1
13. $T_4 = 8.75$; $S_4 \approx 8.6667$; true value: $\frac{26}{3}$
15. $T_4 \approx 0.0973$; $S_4 \approx 0.0940$; true value: 0.09375
17. $T_6 \approx 3.2599$; $S_6 \approx 3.2411$; true value: approximately 3.2413
19. $T_8 \approx 8.5499$; $S_8 \approx 8.5509$; true value: approximately 8.5507
21. (a) 3.0200; (b) 3.0717 **23.** (a) 2519; (b) 2521
25. (a) 2.09; (b) 2.15 **27.** 19

Chapter 5 Miscellaneous Problems (page 358)

1. $\frac{1}{3}x^3 + 2x^{-1} - \frac{5}{2}x^{-2} + C$
3. $-\frac{1}{30}(1 - 3x)^{10} + C$ **5.** $\frac{3}{16}(4x + 9)^{4/3} + C$
7. $\frac{1}{24}(x^4 + 1)^6 + C$ **9.** $-\frac{3}{8}(1 - x^2)^{4/3} + C$
11. $\frac{7}{5}\sin 5x + \frac{5}{7}\cos 7x + C$ **13.** $\frac{1}{6}(1 + x^4)^{3/2} + C$
15. $-\frac{2}{1 + \sqrt{x}} + C$ **17.** $\frac{1}{2}\sin 4x^3 + C$
19. $\frac{1}{30}(x^2 + 1)^{15} + C$
21. $\frac{2}{5}(4 - x)^{5/2} - \frac{8}{3}(4 - x)^{3/2} + C$
23. $\sqrt{1 + x^4} + C$ **25.** $y(x) = x^3 + x^2 + 5$
27. $y(x) = \frac{1}{12}(2x + 1)^6 + \frac{23}{12}$
29. $y(x) = \frac{3}{2}x^{2/3} - \frac{1}{2}$ **31.** 6 s; 396 ft
33. 120 ft/s
35. Impact time: $t = 2\sqrt{10} \approx 6.32$ s; impact speed: $20\sqrt{10} \approx 63.25$ ft/s
37. 176 ft **39.** 1700
41. 2845 **43.** $2\left(\sqrt{2} - 1\right) \approx 0.82843$
45. $\frac{2}{3}\pi\left(2\sqrt{2} - 1\right) \approx 3.82945$
47. Show that every Riemann sum is equal to $c(b - a)$.
51. $\frac{2x\sqrt{2x}}{3} + \frac{2}{\sqrt{3x}} + C$ **53.** $-\frac{2}{x} - \frac{1}{4}x^2 + C$
55. $\frac{2}{3}\sin x^{4/3} + C$ **57.** $\cos\left(\frac{1}{t}\right) + C$
59. $-\frac{3}{8}(1 + u^{4/3})^{-2} + C$ **61.** $\frac{38}{3}$
63. $\frac{1}{3}x^{-3}(4x^2 - 1)^{3/2} + C$ (use $u = 1/x$)
65. $\frac{1}{30}$ **67.** $\frac{44}{15}$
69. $\frac{125}{6}$
71. Semicircle, center $(1, 0)$, radius 1: area $\pi/2$
73. $f(x) = \sqrt{4x^2 - 1}$
75. Use $n \geqq 9$. $L_{10} \approx 1.12767$, $R_{10} \approx 1.16909$. The integrand is increasing on $[0, 1]$, so the average of these approximations yields 1.1483 ± 0.05.
77. $M_5 \approx 0.28667$, $T_5 \approx 0.28971$. They bound the true value of the integral because the second derivative of the integrand is positive on $[1, 2]$.

Section 6.1 (page 368)

1. 1 **3.** $2/\pi$ **5.** $\frac{98}{3}$ **7.** 4 **9.** -1

11. $\int_1^4 2\pi x f(x)\,dx$ **13.** $\int_0^{10} \sqrt{1+[f(x)]^2}\,dx$

15. 1000 **17.** $\frac{500}{3}$ **19.** -320; 320

21. -50; 106.25 **23.** 65; 97 **25.** 1; 1

27. 0; $4/\pi$ **29.** $\frac{70}{3}$; 65 **31.** 0.75; 7.7333

33. π; 4.2638 **37.** $625\pi/2$ **39.** 550 (gal)

41. 695,000 **43.** 109.5 (in.) **45.** $\frac{3}{4}$

47. $f(x) = 400\pi x^2(1+x)$; $700\pi/3$ (lb)

Section 6.2 (page 377)

1. $\pi/5$ **3.** 8π **5.** $\pi^2/2$

7. $3\pi/10$ **9.** $512\pi/3$ **11.** $16\pi/15$

13. $\pi/2$ **15.** 8π **17.** $121\pi/210$

19. 8π **21.** $49\pi/30$ **23.** $17\pi/10$

25. $\pi^2/2$ **27.** $\pi/2$ **29.** $\pi - (\pi/2)^2$

31. 2.99832; 267.442 **33.** 3.67743

35. 9π **37.** $\frac{4}{3}\pi a^2 b$ **39.** $\frac{16}{3}a^3$

41. $\frac{4}{3}a^3\sqrt{3}$ **47.** $\frac{16}{3}a^3$

53. (a) \$3000; (b) \$3000

Section 6.3 (page 386)

1. 8π **3.** $625\pi/2$ **5.** 16π

7. π **9.** $6\pi/5$ **11.** $256\pi/15$

13. $4\pi/15$ **15.** $11\pi/15$ **17.** $56\pi/5$

19. $8\pi/3$ **21.** $2\pi/15$ **23.** $\pi/2$

25. $16\pi/3$ **27.** 64π **29.** 23.2991

31. 1.06027 **33.** 8.13345 **37.** $\frac{4}{3}\pi a^2 b$

39. $V = 2\pi^2 a^2 b$ **41.** $V = 2\pi^2 a^3$ **43.** (a) $V = \frac{1}{6}\pi h^3$

45. (a) $625\pi/12$; (b) $\frac{400}{7}\pi\sqrt{5}$; (c) $\frac{1600}{21}\pi\sqrt{5}$

Section 6.4 (page 396)

1. $\int_0^1 \sqrt{1+4x^2}\,dx$ **3.** $\int_0^2 \sqrt{1+36x^2(x-1)^2}\,dx$

5. $\int_0^{100} \sqrt{1+4x^2}\,dx$ **7.** $\int_{-1}^2 \sqrt{1+16y^6}\,dy$

9. $\int_1^2 \frac{\sqrt{x^4+1}}{x^2}\,dx$ **11.** $\int_0^4 2\pi x^2\sqrt{1+4x^2}\,dx$

13. $\int_0^1 2\pi(x-x^2)\sqrt{2-4x+4x^2}\,dx$

15. $\int_0^1 2\pi(2-x)\sqrt{1+4x^2}\,dx$

17. $\int_1^4 \pi\sqrt{4x+1}\,dx$ **19.** $\int_1^4 \pi(x+1)\sqrt{4+9x}\,dx$

21. $\frac{22}{3}$ **23.** $\frac{14}{3}$

25. $\frac{123}{32} = 3.84375$

27. $\frac{1}{27}\left(104\sqrt{13}-125\right) \approx 9.25842$

29. $\frac{1}{6}\pi\left(5\sqrt{5}-1\right) \approx 5.3304$

31. $\frac{339}{16}\pi \approx 66.5625$

33. $\frac{1}{9}\pi\left(82\sqrt{82}-1\right) \approx 258.8468$

35. 4π

37. 3.8194 (true value: approximately 3.820197789)

41. Avoid the problem when $x = 0$ as follows:

$$L = 8\int_{1/(2\sqrt{2})}^{1} \frac{1}{\sqrt[3]{x}}\,dx = 6.$$

Section 6.5 (page 404)

1. $y(x) = \left(\frac{1}{2}x^2 + C\right)^2$ **3.** $y(x) = 3/(C - x^3)$

5. $y(x) = 1 + \left(\frac{1}{2}x^2 + C\right)^2$

7. $3y + 2y^{3/2} = 3x + 2x^{3/2} + C$

9. $y^3 + y = x - x^{-1} + C$ **11.** $y(x) = (1-x)^{-1}$

13. $y(x) = (x+1)^{1/4}$ **15.** $y(x) = \left(\frac{1}{2} - \frac{1}{3}x^{3/2}\right)^{-2}$

17. $x^2 + y^2 = 169$ **19.** $y(x) = (1 + x - x^3)^{-1}$

21. 20 weeks **23.** 169,000; early in 2011

25. $P(t) \to +\infty$ as $t \to 6^-$ **27.** 3 h

29. 1 h 18 min 40 s after the plug is pulled

31. 1:20 P.M. **33.** About 6 min 2.9 s

35. $f(x) = (\pi/86{,}400)^2 x^4$; the radius of the hole should be $\sqrt{3}/720$ ft, about 0.02887 in.

Section 6.6 (page 413)

1. 30 **3.** 9 **5.** 0 **7.** 15 ft·lb

9. 2.816×10^9 ft·lb (with $R = 4000$ mi, $g = 32$ ft/s^2)

11. $13000\pi \approx 40841$ ft·lb

13. $125{,}000\pi/3 \approx 130{,}900$ ft·lb

15. $156{,}000\pi \approx 490{,}088$ ft·lb

17. $4{,}160{,}000\pi \approx 13{,}069{,}025$ ft·lb

19. 8750 ft·lb **21.** 11250 ft·lb

23. $25000 \cdot [1 - 10^{-0.4}]$ in.·lb ≈ 1254 ft·lb

25. 16π ft·lb **27.** $1{,}382{,}400\pi$ ft·lb

29. About 690.53 ft·lb **31.** 249.6 lb

33. 748.8 lb **35.** 19500 lb

37. $700\rho/3 \approx 14560$ lb **39.** About 32574 tons

Chapter 6 Miscellaneous Problems (page 417)

1. $-\frac{3}{2}; \frac{31}{6}$ **3.** 1; 3 **5.** $\frac{14}{3}$
7. $\frac{2}{15}\pi$ **9.** 12 in. **11.** $\frac{41}{105}\pi$
13. $10.625\pi \approx 33.379$ g **19.** $f(x) = \sqrt{1+3x}$
21. $\dfrac{24 - 2\pi^2}{3\pi} \approx 0.4521$ **23.** $\frac{10}{3}$
25. $\frac{63}{8}$ **27.** $\frac{52}{5}\pi$
31. $y(x) = x^2 + \sin x$ **33.** $y(x) = (C - x)^{-1} - 1$
35. $y(x) = (1 - x^3)^{-1}$ **37.** $y(x) = (C - 3x^{-1})^{1/3}$
39. $y(x) = (1 - \sin x)^{-1}$
41. $(2y^{1/2} - 1)x = (Cx + 2x^{1/2} - 1)y$
43. 1 ft **45.** $W = 4\pi R^4 \rho$
47. 10,454,400 ft·lb **49.** 36400 tons
51. There is no maximum volume; $c = \frac{1}{3}\sqrt{5}$ minimizes the volume V.

Section 7.1 (page 430)

1. (a) 128; (b) 243; (c) 64; (d) 65536; 1
3. (a) 4; (b) 3; (c) 3
5. (a) $3 \ln 2$; (b) $2 \ln 3$; (c) $(\ln 2) + (\ln 3)$
7. (a) $(3 \ln 2) - (3 \ln 3)$; (b) $(2 \ln 2) + (\ln 3) - 2 \ln 5$
9. $2^{(3^4)} = 2^{81} > 2^{12} = (2^3)^4$.
11. $x = 2, x = 4$
13. $2^x < e^x < 3^x$ if $x > 0$; the inequalities are reversed if $x < 0$.
15. The graphs are symmetric around the y-axis.
17. $x = 6$ **19.** $x = -2$
21. $x = 1, x = 2$ **23.** $x = 3^4 = 81$
25. $x = 0$ **27.** $dy/dx = (x+1)e^x$
29. $dy/dx = \left(x^{1/2} + \frac{1}{2}x^{-1/2}\right)e^x$
31. $dy/dx = x^{-3}(x-2)e^x$ **33.** $dy/dx = 1 + \ln x$
35. $dy/dx = x^{-1/2}\left(1 + \frac{1}{2}\ln x\right)$
37. $dy/dx = (1-x)e^{-x}$ **39.** $dy/dx = 3/x$
41. $dy/dx = \frac{1}{2}(e^x + e^{-x})(e^x - e^{-x})^{-1/2}$
43. $dy/dx = e^{3x}(3\cos 4x - 4\sin 4x)$
45. $y = 3e^2x - 2e^2$ **47.** $y = x - 1$
49. $f^{(n)}(x) = 2^n e^{2x}$ **51.** e^{-2}
53. $-e^{-1/2}, 9e^{-3}$
55. Approximately 0.780380 and -0.462285
57. (a) $(\ln 4)/(\ln 3)$; (b) $162 \ln 3$
59. (a) About 9.00647 years; (b) about \$113.08 dollars per year

Section 7.2 (page 440)

1. $f'(x) = \dfrac{3}{3x-1}$ **3.** $f'(x) = \dfrac{1}{1+2x}$
5. $f'(x) = \dfrac{3x^2-1}{3x^3-3x}$ **7.** $f'(x) = -\dfrac{1}{x}\sin(\ln x)$
9. $f'(x) = -\dfrac{1}{x(\ln x)^2}$ **11.** $f'(x) = \dfrac{1}{x} + \dfrac{x}{x^2+1}$
13. $f'(x) = -\tan x$
15. $f'(t) = 2t\ln(\cos t) - t^2 \tan t$
17. $g'(t) = (\ln t)^2 + 2\ln t$ **19.** $f'(x) = \dfrac{6}{2x+1} + \dfrac{8x}{x^2-4}$
21. $f'(x) = -\dfrac{x}{4-x^2} - \dfrac{x}{9+x^2}$
23. $f'(x) = \dfrac{1}{x+1} - \dfrac{1}{x-1}$ **25.** $g'(t) = \dfrac{2}{t} - \dfrac{2t}{t^2+1}$
27. $f'(x) = -\dfrac{1}{x} + \cot x$ **29.** $\dfrac{dy}{dx} = \dfrac{y\ln y}{y-x}$
31. $\dfrac{dy}{dx} = \dfrac{y}{-x + \cot y}$ **33.** $\frac{1}{2}\ln|2x-1| + C$
35. $\frac{1}{6}\ln(1+3x^2) + C$ **37.** $\frac{1}{4}\ln|2x^2+4x+1| + C$
39. $\frac{1}{3}(\ln x)^3 + C$ **41.** $\ln|x+1| + C$
43. $\ln(x^2+x+1) + C$ **45.** $\frac{1}{2}(\ln x)^2 + C$
47. $\frac{1}{2}\ln(1 - \cos 2x) + C$ **49.** $\frac{1}{3}\ln|x^3 - 3x^2 + 1| + C$
51. 0 **53.** 0
55. 0
59. $m \approx -0.2479$, $k \approx 291.7617$
61. 3.693 **63.** 0.464; 5.357
65. 1.446; 3.665; 4.100 **67.** 3.0597
69. 8.895
75. $y \to 0$ as $x \to 0^+$; $dy/dx \to 0$ as $x \to 0^+$. The point $(0, 0)$ is not on the graph. Intercept at $(1, 0)$; global minimum where $x = e^{-1/2}$ (the coordinates are approximately $(0.61, -0.18)$); inflection point where $x = e^{-3/2}$ (the coordinates are approximately $(0.22, -0.07)$).
77. $y \to -\infty$ as $x \to 0^+$; $y \to 0$ as $x \to +\infty$. Global maximum at $(e^2, 2/e)$; sole intercept $(1, 0)$; inflection point where $x = e^{8/3}$. The x-axis is a horizontal asymptote and the y-axis is a vertical asymptote.
81. Midpoint estimate: approximately 872.47; trapezoidal estimate: approximately 872.60; true value of integral: approximately 872.5174.

Section 7.3 (page 448)

1. $f'(x) = 2e^{2x}$
3. $f'(x) = 2xe^{(x^2)} = 2xe^{x^2} = 2x\exp(x^2)$
5. $f'(x) = -\dfrac{2}{x^3}\exp\left(\dfrac{1}{x^2}\right)$ **7.** $g'(t) = \left(1 + \frac{1}{2}\sqrt{t}\right)e^{\sqrt{t}}$
9. $g'(t) = (1 + 2t - t^2)e^{-t}$
11. $g'(t) = -e^{\cos t}\sin t$
13. $f'(x) = -e^{-x}\sin(1 - e^{-x})$

15. $f'(x) = \dfrac{1 - e^{-x}}{x + e^{-x}}$

17. $f'(x) = e^{-2x}(3\cos 3x - 2\sin 3x)$

19. $g'(t) = 15(e^t - t^{-1})(e^t - \ln t)^4$

21. $f'(x) = -(5 + 12x)e^{-4x}$

23. $g'(t) = (e^{-t} + te^{-t} - 1)t^{-2}$

25. $f'(x) = (x - 2)e^{-x}$ **27.** $f'(x) = e^x \exp(e^x)$

29. $f'(x) = 2e^x \cos 2e^x$ **31.** $\dfrac{dy}{dx} = \dfrac{e^y}{1 - y}$

33. $\dfrac{dy}{dx} = \dfrac{e^x - ye^{xy}}{xe^{xy} - e^y}$ **35.** $\dfrac{dy}{dx} = \dfrac{xy - y}{xy + x}$

37. $-\frac{1}{2}e^{1-2x} + C$ **39.** $\frac{1}{9}\exp(3x^3 - 1) + C$

41. $\frac{1}{2}\ln(1 + e^{2x}) + C$ **43.** $\frac{1}{2}\exp(1 - \cos 2x) + C$

45. $\frac{1}{2}\ln(x^2 + e^{2x}) + C$ **47.** $-\exp\left(-\frac{1}{2}t^2\right) + C$

49. $2e^{\sqrt{x}} + C$ **51.** $\ln(1 + e^x) + C$

53. $-\frac{2}{3}\exp(-x^{3/2}) + C$ **55.** e^2

57. e **59.** $+\infty$ **61.** $+\infty$

63. Global minimum and intercept at $(0, 0)$, local maximum at $(2, 4e^{-2})$; inflection points where $x = 2 \pm \sqrt{2}$. The x-axis is an asymptote.

65. Global maximum at $(0, 1)$, the only intercept; inflection points where $x = \pm\frac{1}{2}\sqrt{2}$ The x-axis is the only asymptote.

67. $\frac{1}{2}\pi(e^2 - 1) \approx 10.0359$ **69.** $\frac{1}{2}(e - e^{-1}) \approx 1.1752$

71. 1.278 **73.** 0.884, 2.194 **75.** 0.212, 1.002

77. 1.118 **79.** 174.002

81. $f'(x) = 0$ when $x = 0$ and when $x = n$; f is increasing on $(0, n)$ and decreasing for $x > n$. Thus $x = n$ yields the global maximum value of $f(x)$ for $x \geqq 0$. The x-axis is a horizontal asymptote, and there are inflection points where $x = n \pm \sqrt{n}$.

85. $y(x) = e^{-2x} + 4e^x$

Section 7.4 (page 455)

1. $f'(x) = 10^x \ln 10$

3. $f'(x) = 3^x 4^{-x} \ln 3 - 3^x 4^{-x} \ln 4 = \left(\frac{3}{4}\right)^x \ln\left(\frac{3}{4}\right)$

5. $f'(x) = -(7^{\cos x})(\ln 7)(\sin x)$

7. $f'(x) = 2^{x\sqrt{x}}\left(\frac{3}{2}\ln 2\right)\sqrt{x}$

9. $f'(x) = x^{-1}2^{\ln x}\ln 2$ **11.** $f'(x) = 17^x \ln 17$

13. $f'(x) = -x^{-2}10^{1/x}\ln 10$

15. $f'(x) = (2^{2^x}\ln 2)(2^x \ln 2)$

17. $f'(x) = \dfrac{1}{\ln 3} \cdot \dfrac{x}{x^2 + 4}$ **19.** $f'(x) = \dfrac{\ln 2}{\ln 3} = \log_3 2$

21. $f'(x) = \dfrac{1}{x(\ln 2)(\ln x)}$ **23.** $f'(x) = \dfrac{\exp(\log_{10} x)}{x \ln 10}$

25. $\dfrac{3^{2x}}{2\ln 3} + C$ **27.** $\dfrac{2 \cdot 2^{\sqrt{x}}}{\ln 2} + C$

29. $\dfrac{7^{x^3+1}}{3\ln 7} + C$ **31.** $\dfrac{(\ln x)^2}{2\ln 2} + C$

33. $\dfrac{dy}{dx} = \left(\dfrac{x}{x^2 - 4} + \dfrac{1}{4x + 2}\right)(x^2 - 4)^{1/2}(2x + 1)^{1/4}$

35. $\dfrac{dy}{dx} = 2^x \ln 2$ **37.** $\dfrac{dy}{dx} = \dfrac{(x^{\ln x})(2\ln x)}{x}$

39. $\dfrac{dy}{dx} = \dfrac{y}{3}\left(\dfrac{1}{x + 1} + \dfrac{1}{x + 2} - \dfrac{2x}{x^2 + 1} - \dfrac{2x}{x^2 + 2}\right)$

41. $\dfrac{dy}{dx} = (\ln x)^{\sqrt{x}}\left[\frac{1}{2}x^{-1/2}\ln(\ln x) + x^{-1/2}(\ln x)^{-1}\right]$

43. $\dfrac{dy}{dx} = \left(\dfrac{3x}{1 + x^2} - \dfrac{4x^2}{1 + x^3}\right)(1 + x^2)^{3/2}(1 + x^3)^{-4/3}$

45. $\dfrac{dy}{dx} = \left[\dfrac{2x^3}{x^2 + 1} + 2x\ln(x^2 + 1)\right](x^2 + 1)^{(x^2)}$

47. $\dfrac{dy}{dx} = \frac{1}{4}x^{-1/2}(2 + \ln x)(\sqrt{x})^{\sqrt{x}}$

49. $\dfrac{dy}{dx} = x^{\exp x}e^x(x^{-1} + \ln x)$

51. The high point occurs where $x = (\ln 2)^{-1}$; the ordinate is

$$\frac{1}{2^{1/\ln 2}\ln 2} \approx 0.5307378454.$$

53. 1.343088

61. Note that $\ln \dfrac{x^x}{e^x} = x \ln \dfrac{x}{e}$.

63. $\dfrac{dy}{dx} = -\dfrac{\ln 2}{x(\ln x)^2}$

Section 7.5 (page 464)

1. \$119.35; \$396.24 **3.** About 3.8685 h

5. About 686 years old

7. (a) 9.308%; (b) 9.381%; (c) 9.409%; (d) 9.416%; (e) 9.417%

9. \$44.52

11. After an additional 32.26 days

13. About 35 years

15. About 4.2521×10^9 years old

17. 2.40942 minutes

19. (a) 20.486 in.; 9.604 in. (b) 3.4524 mi, about 18230 ft

21. About 325.07 days

Section 7.6 (page 471)

1. $y(x) = -1 + 2e^x$ **3.** $y(x) = \frac{1}{2}(e^{2x} + 3)$

5. $x(t) = 1 - e^{2t}$ **7.** $x(t) = 27e^{5t} - 2$

9. $v(t) = 10(1 - e^{-10t})$ **11.** 4,870,238

15. About 46 days after the rumor starts

19. $400/\ln 2 \approx 577$ ft

23. (b) \$1,308,283

25. (a) $x(t) = 100{,}000 - 80000e^{-kt}$, where $k = \frac{1}{14}\ln 2$; (b) on March 29; (c) everyone gets the flu

63. (b) $v(10) = 176(1 - e^{-1}) \approx 111.2532$ ft/s, about 75.85 mi/h. The limiting velocity is $a/\rho = 176$ ft/s, exactly 120 mi/h.

65. (a) The minimum of $f(x) - g(x)$ occurs when $x = 4$ and is $2(1 - \ln 2) > 0$. Hence $f(x) > g(x)$ for all $x > 0$. (b) The (large) solution of $g(x) = h(x)$ is (by Newton's method) approximately 93.354460835. (c) $p = e$

Chapter 7 Miscellaneous Problems (page 472)

1. $f'(x) = \dfrac{1}{2x}$ **3.** $f'(x) = \dfrac{1 - e^x}{x - e^x}$

5. $f'(x) \equiv \ln 2$ **7.** $f'(x) = (2 + 3x^2)e^{-1/x^2}$

9. $f'(x) = x^{-1}(1 + \ln \ln x)$

11. $f'(x) = x^{-1}2^{\ln x}\ln 2$

13. $f'(x) = -\dfrac{2}{(x-1)^2}\exp\left(\dfrac{x+1}{x-1}\right)$

15. $f'(x) = \dfrac{3}{2}\left(\dfrac{1}{x-1} + \dfrac{8x}{3 - 4x^2}\right)$

17. $f'(x) = \dfrac{(\sin x\cos x)\exp\left(\sqrt{1 + \sin^2 x}\right)}{\sqrt{1 + \sin^2 x}}$

19. $f'(x) = \cot x + \ln 3$ **21.** $f'(x) = \dfrac{x^{1/x}(1 - \ln x)}{x^2}$

23. $f'(x) = \left(\dfrac{1 + \ln \ln x}{x}\right)(\ln x)^{\ln x}$

25. $-\frac{1}{2}\ln|1 - 2x| + C$ **27.** $\frac{1}{2}\ln|1 + 6x - x^2| + C$

29. $-\ln(2 + \cos x) + C$ **31.** $\dfrac{2\cdot 10^{\sqrt{x}}}{\ln 10} + C$

33. $\frac{2}{3}(1 + e^x)^{3/2} + C$ **35.** $\dfrac{6^x}{\ln 6} + C$

37. $x(t) = t^2 + 17$ **39.** $x(t) = 1 + e^t$

41. $x(t) = \frac{1}{3}(2 + 7e^{3t})$ **43.** $x(t) = \sqrt{2}e^{\sin t}$

45. Horizontal asymptote: the x-axis; global maximum where $x = \frac{1}{2}$; inflection points where $x = \frac{1}{2}(1 + \sqrt{2})$—approximately $(1.21, 0.33)$; global minimum and intercept at $(0, 0)$, with a vertical tangent there as well.

47. Global minimum at $(4, 2 - \ln 4)$, inflection point at $(16, 1.23)$ (ordinate approximate). The y-axis is a vertical asymptote; there is no horizontal asymptote (the graph continues to rise for large increasing x).

49. Inflection point at $(0.5, e^{-2})$. The horizontal line $y = 1$ and the y-axis are asymptotes. The point $(0, 0)$ is *not* on the graph. As $x \to 0^+$, $y \to 0$; as $x \to 0^-$, $y \to +\infty$. As $|x| \to +\infty$, $y \to 1$.

51. Sell immediately!

53. (b) The minimizing value is about 10.516. But because the batch size must be an integer, 11 (rather than 10) minimizes $f(x)$. (c) \$977.85

57. 20 weeks

59. (a) \$925.20; (b) \$1262.88

61. About 22.567 h after the power failure; that is, at about 9:34 P.M. the following evening

Section 8.2 (page 484)

1. (a) $\pi/6$, (b) $-\pi/6$, (c) $\pi/4$, (d) $-\pi/3$

3. (a) 0, (b) $\pi/4$, (c) $-\pi/4$, (d) $\pi/3$

5. $f'(x) = 100x^{99}(1 - x^{200})^{-1/2}$

7. $f'(x) = \dfrac{1}{x|\ln x|\sqrt{(\ln x)^2 - 1}}$

9. $f'(x) = \dfrac{\sec^2 x}{\sqrt{1 - \tan^2 x}}$ **11.** $f'(x) = \dfrac{e^x}{\sqrt{1 - e^{2x}}}$

13. $f'(x) = -\dfrac{2}{\sqrt{1 - x^2}}$ **15.** $f(x) = -\dfrac{2}{x\sqrt{x^4 - 1}}$

17. $f'(x) = -\dfrac{1}{(1 + x^2)(\arctan x)^2}$

19. $f'(x) = \dfrac{1}{x[1 + (\ln x)^2]}$ **21.** $f'(x) = \dfrac{2e^x}{1 + e^{2x}}$

23. $f'(x) = \dfrac{\cos(\arctan x)}{1 + x^2}$ **25.** $f'(x) = \dfrac{1 - 4x\arctan x}{(1 + x^2)^3}$

27. $\dfrac{dy}{dx} = -\dfrac{1 + y^2}{1 + x^2}$; $x + y = 2$

29. $\dfrac{dy}{dx} = -\dfrac{\sqrt{1 - y^2}\arcsin y}{\sqrt{1 - x^2}\arcsin x}$; $x + y = \sqrt{2}$

31. $\pi/4$ **33.** $\pi/12$ **35.** $\pi/12$

37. $\frac{1}{2}\arcsin 2x + C$ **39.** $\frac{1}{5}\operatorname{arcsec}|x/5| + C$

41. $\arctan(e^x) + C$ **43.** $\frac{1}{15}\operatorname{arcsec}|x^3/5| + C$

45. Both $\arcsin(2x - 1) + C$ and $2\arcsin\sqrt{x} + C$ are correct.

47. $\frac{1}{50}\arctan(x^{50}) + C$ **49.** $\arctan(\ln x) + C$

51. $\pi/4$ **53.** $\pi/2$ **55.** $\pi/12$

65. 8 m **69.** $\pi/2$

75. Approximately (2.68922, 0.92834)

Section 8.3 (page 491)

1. $\frac{1}{2}$ **3.** $\frac{2}{5}$ **5.** 0 **7.** 0 **9.** $\frac{1}{2}$

11. 2 **13.** 0 **15.** 1 **17.** 1 **19.** $\frac{3}{5}$

21. $\frac{3}{2}$ **23.** $\frac{1}{3}$ **25.** $\dfrac{\ln 2}{\ln 3}$ **27.** $\frac{1}{2}$ **29.** 1

31. $\frac{1}{3}$ **33.** $-\frac{1}{2}$ **35.** 1 **37.** $\frac{1}{4}$ **39.** $\frac{2}{3}$

41. 6 **43.** $\frac{4}{3}$ **45.** $\frac{2}{3}$ **47.** 0

49. $y \to 0$ as $x \to 0$ **51.** $y \to -1$ as $x \to \pi$

53. $y \to \frac{1}{2}$ as $x \to 0$ **55.** $y \to 0$ as $x \to +\infty$

57. $y \to 0$ as $x \to +\infty$ **59.** $y \to 0$ as $x \to +\infty$

63. There is a local maximum at $x = n$ and inflection points at $x = n \pm \sqrt{n}$.

65. 1 **69.** 1

Section 8.4 (page 496)

1. 1 **3.** $\frac{3}{8}$ **5.** 1 **7.** 1 **9.** 0

11. -1 **13.** $-\infty$ **15.** $-\infty$ **17.** $-\frac{1}{2}$ **19.** 0

21. 1 **23.** 1 **25.** $e^{-1/6}$ **27.** $e^{-1/2}$ **29.** 1

31. e^{-1} **33.** $-\infty$

35. $f(x) \to 0$ as $x \to 0^+$; $f(x) \to 1$ as $x \to +\infty$; the high point on the graph is $(e, e^{1/e})$.

37. $f(x) \to 0$ as $x \to 0^+$; $f(x) \to 1$ as $x \to +\infty$; the high point on the graph is $(e, e^{2/e})$.

39. $f(x) \to 0$ as $x \to 0^+$; $f(x) \to 1$ as $x \to +\infty$; the high point on the graph is about (1.98029, 2.23612).

41. $f(x) \to 0$ as $x \to 0^+$; $f(x) \to 1$ as $x \to +\infty$; the high point on the graph is about (1.209599, 1.879360).

45. Approximately (0.409978, 0.678741)

49. $f(x) \to +\infty$ as $x \to 0^+$; $f(x) \to 1$ as $x \to +\infty$. There is a global minimum at (1, 0), a local maximum near (5.831200, 1.102147), and inflection points near (1.116391, 0.138777) and (8.928008, 1.091727).

51. $f(x) \to +\infty$ as $x \to 0^+$ and as $x \to +\infty$; $f(x) \to 1$ as $x \to 1$; f is not differentiable at $x = 1$.

Section 8.5 (page 505)

1. $f'(x) = 3\sinh(3x - 2)$

3. $f'(x) = 2x\tanh(1/x) - \operatorname{sech}^2(1/x)$

5. $f'(x) = -12\coth^2 4x \operatorname{csch}^2 4x$

7. $f'(x) = -e^{\operatorname{csch} x}\operatorname{csch} x \coth x$

9. $f'(x) = (\cosh x)\cos(\sinh x)$

11. $f'(x) = 4x^3\cosh x^4$

13. $f'(x) = -\dfrac{1 + \operatorname{sech}^2 x}{(x + \tanh x)^2}$

15. $\frac{1}{2}\cosh x^2 + C$ **17.** $x - \frac{1}{3}\tanh 3x + C$

19. $\frac{1}{6}\sinh^3 2x + C$ **21.** $-\frac{1}{2}\operatorname{sech}^2 x + C$

23. $-\frac{1}{2}\operatorname{csch}^2 x + C$ **25.** $\ln(1 + \cosh x) + C$

27. $\frac{1}{4}\tanh x + C$ **29.** $f'(x) = \dfrac{2}{\sqrt{4x^2 + 1}}$

31. $f'(x) = \frac{1}{2}x^{-1/2}(1 - x)^{-1}$

33. $f'(x) = (x^2 - 1)^{-1/2}$

35. $f'(x) = \frac{3}{2}(\sinh^{-1} x)^{1/2}(x^2 + 1)^{-1/2}$

37. $f'(x) = (1 - x^2)^{-1}(\tanh^{-1} x)^{-1}$

39. $\arcsin(x/3) + C$ **41.** $\frac{1}{4}\ln\frac{9}{5} \approx 0.14695$

43. $-\frac{1}{3}\operatorname{sech}^{-1}|3x/2| + C$ **45.** $\sinh^{-1}(e^x) + C$

47. $-\operatorname{sech}^{-1}(e^x) + C$ **53.** $\sinh a$

57. $\ln(1 + \sqrt{2}) \approx 0.881373587$

67. Approximately (0.440687, 0.171753)

Chapter 8 Miscellaneous Problems (page 508)

1. $f'(x) = \dfrac{3}{\sqrt{1 - 9x^2}}$ **3.** $g'(t) = \dfrac{2}{t\sqrt{t^4 - 1}}$

5. $f'(x) = -\dfrac{\sin x}{\sqrt{1 - \cos^2 x}} = -\dfrac{\sin x}{|\sin x|}$

7. $g'(t) = \dfrac{10}{\sqrt{100t^2 - 1}}$ **9.** $f'(x) = -\dfrac{2}{x\sqrt{x^4 - 1}}$

11. $f'(x) = \dfrac{1}{2\sqrt{x}\sqrt{1 - x}}$ **13.** $f'(x) = \dfrac{2x}{x^4 + 2x^2 + 2}$

15. $f'(x) = e^x \sinh e^x + e^{2x}\cosh e^x$

17. $f'(x) \equiv 0$ **19.** $f'(x) = \dfrac{x}{|x|\sqrt{x^2 + 1}}$

21. $\frac{1}{2}\arcsin 2x + C$ **23.** $\arcsin\left(\frac{1}{2}x\right) + C$

25. $\sin^{-1}(e^x) + C$ **27.** $\frac{1}{2}\arcsin\left(\frac{2}{3}x\right) + C$

29. $\frac{1}{3}\arctan(x^3) + C$ **31.** $\sec^{-1}|2x| + C$

33. $\operatorname{arcsec}(e^x) + C$ **35.** $2\cosh\sqrt{x} + C$

37. $\frac{1}{2}(\arctan x)^2 + C$ **39.** $\frac{1}{2}\sinh^{-1}\left(\frac{2}{3}x\right) + C$

41. $\frac{1}{4}$ **43.** $\frac{1}{2}$ **45.** $-\frac{1}{6}$ **47.** 1 **49.** $-\infty$

51. $+\infty$ **53.** e^2 **55.** $-\frac{1}{2}e$ **57.** $\frac{1}{6}\pi^2$

61. $x \approx 4.730041$

Section 9.2 (page 515)

1. $-\frac{1}{15}(2 - 3x)^5 + C$ **3.** $\frac{1}{9}(2x^3 - 4)^{3/2} + C$

5. $\frac{3}{4}(2x^2 + 3)^{2/3} + C$ **7.** $-2\csc\sqrt{y} + C$

9. $\frac{1}{6}(1 + \sin\theta)^6 + C$ **11.** $e^{-\cot x} + C$

13. $\frac{1}{11}(\ln t)^{11} + C$ **15.** $\frac{1}{3}\arcsin 3t + C$

17. $\frac{1}{2}\arctan(e^{2x}) + C$ **19.** $\frac{3}{2}\arcsin(x^2) + C$

21. $\frac{1}{15}\tan^5 3x + C$ **23.** $\tan^{-1}(\sin\theta) + C$

25. $\frac{2}{5}\left(1 + \sqrt{x}\right)^5 + C$ **27.** $\ln|\arctan t| + C$

29. $\sec^{-1}(e^x) + C$

31. $\frac{2}{7}(x - 2)^{7/2} + \frac{8}{5}(x - 2)^{5/2} + \frac{8}{3}(x - 2)^{3/2} + C$

33. $\frac{1}{3}(2x + 3)^{1/2}(x - 3) + C$

35. $\frac{3}{10}(x + 1)^{2/3}(2x - 3) + C$

37. $\frac{1}{60}\ln\left|\dfrac{3x + 10}{3x - 10}\right| + C$

39. $\frac{1}{2}x(4 + 9x^2)^{1/2} + \frac{2}{3}\ln|3x + (4 + 9x^2)^{1/2}| + C$

41. $\frac{1}{32}x(16x^2 + 9)^{1/2} - \frac{9}{128}\ln|4x + (16x^2 + 9)^{1/2}| + C$

43. $\frac{1}{128}x(32x^2 - 25)(25 - 16x^2)^{1/2} + \frac{625}{512}\arcsin\left(\frac{4}{5}x\right) + C$

45. The substitution $u = e^x$ leads to an integral in the form of formula (44) in the Table of Integrals inside the back cover. The answer is

$\frac{1}{2}e^x\sqrt{9+e^{2x}}+\frac{9}{2}\ln\left[e^x+\sqrt{9+e^{2x}}\right]+C.$

47. With $u=x^2$ and formula (47) in the Table of Integrals, we obtain

$$\tfrac{1}{2}\left[\sqrt{x^4-1}-\operatorname{arcsec}(x^2)\right]+C.$$

49. With $u=\ln x$ and formula (48) in the Table of Integrals, we obtain

$$\tfrac{1}{8}\left\{(\ln x)[2(\ln x)^2+1]\sqrt{(\ln x)^2+1}-\ln\left|(\ln x)+\sqrt{(\ln x)^2+1}\right|\right\}+C.$$

53. $\sin^{-1}(x-1)+C$

55. $\frac{1}{2}\tan^{-1}x^2=\frac{1}{4}\pi-\frac{1}{2}\tan^{-1}x^{-2}$ for all $x\neq 0$. Both *are* antiderivatives of $x/(1+x^4)$.

Section 9.3 (page 522)

1. $\frac{1}{2}xe^{2x}-\frac{1}{4}e^{2x}+C$ **3.** $-t\cos t+\sin t+C$

5. $\frac{1}{3}x\sin 3x+\frac{1}{9}\cos 3x+C$ **7.** $\frac{1}{4}x^4\ln x-\frac{1}{16}x^4+C$

9. $x\arctan x-\frac{1}{2}\ln(1+x^2)+C$

11. $\frac{2}{3}y^{3/2}\ln y-\frac{4}{9}y^{3/2}+C$ **13.** $t(\ln t)^2-2t\ln t+2t+C$

15. $\frac{2}{3}x(x+3)^{3/2}-\frac{4}{15}(x+3)^{5/2}+C$

17. $\frac{2}{9}x^3(x^3+1)^{3/2}-\frac{4}{45}(x^3+1)^{5/2}+C$

19. $-\frac{1}{2}(\csc\theta\cot\theta+\ln|\csc\theta+\cot\theta|)+C$

21. $\frac{1}{3}x^3\arctan x-\frac{1}{6}x^2+\frac{1}{6}\ln(1+x^2)+C$

23. $x\operatorname{arcsec}x^{1/2}-(x-1)^{1/2}+C$

25. $(x+1)\arctan x^{1/2}-x^{1/2}+C$

27. $-x\cot x+\ln|\sin x|+C$

29. $\frac{1}{2}x^2\sin x^2+\frac{1}{2}\cos x^2+C$

31. $-2x^{-1/2}(2+\ln x)+C$ **33.** $x\sinh x-\cosh x+C$

35. $\frac{1}{2}(\sin x^2-x^2\cos x^2)+C$

37. $-(2+2\sqrt{x})\exp(-\sqrt{x})+C$

39. $\pi(\pi-2)$ **41.** $\frac{1}{2}\pi(e^2+1)$

43. 1.06027 **45.** 22.78941

47. $\frac{1}{2}(x-1)e^x\sin x+\frac{1}{2}xe^x\cos x+C$

55. $6-2e\approx 0.563436$ **57.** $6-2e$

65. $\frac{1}{80}(2\pi^6-10\pi^4+15\pi^2)\approx 13.709144$

Section 9.4 (page 529)

1. $\frac{1}{4}(2x-\sin 2x\cos 2x)+C$

3. $2\tan\dfrac{x}{2}+C$ **5.** $\frac{1}{3}\ln|\sec 3x|+C$

7. $\frac{1}{3}\ln|\sec 3x+\tan 3x|+C$

9. $\frac{1}{2}(x-\sin x\cos x)+C$ **11.** $\frac{1}{3}\cos^3 x-\cos x+C$

13. $\frac{1}{3}\sin^3\theta-\frac{1}{5}\sin^5\theta+C$

15. $\frac{1}{5}\sin^5 x-\frac{2}{3}\sin^3 x+\sin x+C$

17. $\frac{2}{5}(\cos x)^{5/2}-2(\cos x)^{1/2}+C$

19. $-\frac{1}{14}\cos^7 2z+\frac{1}{5}\cos^5 2z-\frac{1}{6}\cos^3 2z+C$

21. $\frac{1}{4}(\sec 4x+\cos 4x)+C$

23. $\frac{1}{3}\tan^3 t+\tan t+C$

25. $-\frac{1}{4}\csc^2 2x-\frac{1}{2}\ln|\sin 2x|+C$

27. $\frac{1}{12}\tan^6 2x+C$

29. $-\frac{1}{10}\cot^5 2t-\frac{1}{3}\cot^3 2t-\frac{1}{2}\cot 2t+C$

31. $\frac{1}{4}\cos^4\theta-\frac{1}{2}\cos^2\theta+C_1=\frac{1}{4}\sin^4\theta+C_2$

33. $\frac{2}{3}(\sec t)^{3/2}+2(\sec t)^{-1/2}+C$

35. $\frac{1}{3}\sin^3\theta+C$ **37.** $\frac{1}{5}\sin 5t-\frac{1}{15}\sin^3 5t+C$

39. $t+\frac{1}{3}\cot 3t-\frac{1}{9}\cot^3 3t+C$

41. $-\frac{1}{5}(\cos 2t)^{5/2}+\frac{2}{9}(\cos 2t)^{9/2}-\frac{1}{13}(\cos 2t)^{13/2}+C$

43. $\frac{1}{2}\sin^2 x-\cos x+C_1=-\frac{1}{2}\cos^2 x-\cos x+C_2$

45. $\frac{4}{3}$ **47.** $\frac{1}{8}(4+3\pi)$ **49.** 0

51. $\frac{3}{8}\pi^2$ **53.** $\frac{2}{3}\pi(4\pi-3\sqrt{3})$

55. (a) $\pi/4$; (b) $\frac{1}{4}\pi(8-\pi)$ **59.** $\frac{1}{4}\cos 2x-\frac{1}{16}\cos 8x+C$

61. $\frac{1}{6}\sin 3x+\frac{1}{10}\sin 5x+C$

67. *Mathematica's* answer is correct. Another version of *Mathematica* gives

$$\int\tan^4 x\,dx=x-\tfrac{4}{3}\tan x+\tfrac{1}{3}\sec^2 x\tan x.$$

Section 9.5 (page 538)

1. $\frac{1}{2}x^2-x+\ln|x+1|+C$

3. $\frac{1}{3}\ln\left|\dfrac{x-3}{x}\right|+C$ **5.** $\frac{1}{5}\ln\left|\dfrac{x-2}{x+3}\right|+C$

7. $\frac{1}{4}\ln|x|-\frac{1}{8}\ln(x^2+4)+C$

9. $\frac{1}{3}x^3-4x+8\arctan\left(\frac{1}{2}x\right)+C$

11. $x-2\ln|x+1|+C$ **13.** $x+\dfrac{1}{x+1}+C$

15. $\frac{1}{4}\ln\left|\dfrac{x-2}{x+2}\right|+C$

17. $\frac{3}{2}\ln|2x-1|-\ln|x+3|+C$

19. $\ln|x|+\dfrac{2}{x+1}+C$

21. $\frac{3}{2}\ln|x^2-4|+\frac{1}{2}\ln|x^2-1|+C$

23. $\ln|x+2|+\dfrac{4}{x+2}-\dfrac{2}{(x+2)^2}+C$

25. $\frac{1}{2}\ln\left(\dfrac{x^2}{x^2+1}\right)+C$

27. $\frac{1}{2}\ln\left(\dfrac{x^2}{x^2+4}\right)+\frac{1}{2}\arctan\left(\dfrac{x}{2}\right)+C$

29. $-\frac{1}{2}\ln|x+1|+\frac{1}{4}\ln(x^2+1)+\frac{1}{2}\arctan x+C$

31. $\arctan\left(\dfrac{x}{2}\right)-\frac{3}{2}\sqrt{2}\arctan(x\sqrt{2})+C$

33. $\dfrac{1}{\sqrt{2}}\arctan\dfrac{x}{\sqrt{2}}+\frac{1}{2}\ln(x^2+3)+C$

35. $x+\frac{1}{2}\ln|x-1|-\dfrac{5}{2x-2}+\frac{3}{4}\ln(x^2+1)+2\arctan x+C$

37. $\dfrac{1-2e^{2t}}{4(e^{2t}-1)^2}+C$

39. $\frac{1}{4}\ln|3+2\ln t|+\dfrac{1}{4(3+2\ln t)}+C$

41. $5\ln 2$ **43.** $\frac{1}{3}(23\ln 2-7\ln 5)$

45. $2\pi(1+6\ln 2)$ **47.** $2\pi(7\ln 5-10\ln 2-2)$

49. $\frac{1}{2}\pi(13+16\ln 2)$ **51.** $\frac{2}{3}\pi(3\ln 2-2)$

53. $49\ln|x-5|+93\ln|x-7|-44\ln x-\dfrac{280}{x}+C$

55. $\dfrac{39}{2(x+5)}-\dfrac{48}{x-4}+\frac{567}{16}\ln|x-3|$
$-\frac{104}{3}\ln|x-4|-\frac{37}{48}\ln|x+5|+C$

57. $x(t)=\dfrac{2e^t}{2e^t-1}$ **59.** $x(t)=\dfrac{2e^{2t}+1}{2e^{2t}-1}$

61. $x(t)=\dfrac{21e^t-16}{8-7e^t}$ **63.** About 153,700,000

65. (a) 1.37 s; (b) 200 g

67. $P(t)=\dfrac{200}{2-e^{t/100}}$;
(a) $t=100\ln(1.8)\approx 58.8$ (days);
(b) $t=100\ln 2\approx 69.3$ (days)

Section 9.6 (page 545)

1. $\arcsin\left(\dfrac{x}{4}\right)+C$ **3.** $-\dfrac{\sqrt{4-x^2}}{4x}+C$

5. $8\arcsin\left(\dfrac{x}{4}\right)-\dfrac{x\sqrt{16-x^2}}{2}+C$

7. $\dfrac{x}{9\sqrt{9-16x^2}}+C$

9. $\ln\left|x+\sqrt{x^2-1}\right|-\dfrac{\sqrt{x^2-1}}{x}+C$

11. $\frac{1}{80}[(9+4x^2)^{5/2}-15(9+4x^2)^{3/2}]+C$

13. $\sqrt{1-4x^2}-\ln\left|\dfrac{1+\sqrt{1-4x^2}}{2x}\right|+C_1$
$=\sqrt{1-4x^2}+\ln\left|\dfrac{x}{1+\sqrt{1-4x^2}}\right|+C_2$

15. $\frac{1}{2}\ln\left|2x+\sqrt{9+4x^2}\right|+C$

17. $\frac{25}{2}\arcsin\left(\dfrac{x}{5}\right)-\frac{1}{2}x\sqrt{25-x^2}+C$

19. $\frac{1}{2}x\sqrt{x^2+1}-\frac{1}{2}\ln\left|x+\sqrt{1+x^2}\right|+C$

21. $\frac{1}{18}x\sqrt{4+9x^2}-\frac{2}{27}\ln\left|3x+\sqrt{4+9x^2}\right|+C$

23. $\dfrac{x}{\sqrt{1+x^2}}+C$

25. $\frac{1}{512}\left[3\ln\left|\dfrac{x+2}{x-2}\right|-\dfrac{12x}{x^2-4}+\dfrac{32x}{(x^2-4)^2}\right]+C$

27. $\frac{1}{2}x\sqrt{9+16x^2}+\frac{9}{8}\ln\left|4x+\sqrt{9+16x^2}\right|+C$

29. $\sqrt{x^2-25}-5\,\mathrm{arcsec}\left(\dfrac{x}{5}\right)+C$

31. $\frac{1}{8}x(2x^2-1)\sqrt{x^2-1}-\frac{1}{8}\ln\left|x+\sqrt{x^2-1}\right|+C$

33. $-\dfrac{x}{\sqrt{4x^2-1}}+C$

35. $-\dfrac{\sqrt{x^2-5}}{x}+\ln\left|x+\sqrt{x^2-5}\right|+C$

37. $\mathrm{arcsinh}\left(\dfrac{x}{5}\right)+C$

39. $\mathrm{arccosh}\left(\dfrac{x}{2}\right)-\dfrac{\sqrt{x^2-4}}{x}+C$

41. $\frac{1}{8}x(1+2x^2)\sqrt{1+x^2}-\frac{1}{8}\,\mathrm{arcsinh}\,x+C$

45. $\frac{1}{32}\pi[18\sqrt{5}-\ln(2+\sqrt{5})]\approx 3.8097$

47. $\sqrt{5}-\sqrt{2}+\ln\left(\dfrac{2+2\sqrt{2}}{1+\sqrt{5}}\right)\approx 1.222016$

51. $2\pi[\sqrt{2}+\ln(1+\sqrt{2})]\approx 14.4236$

55. \$$6\frac{2}{3}$ million

Section 9.7 (page 550)

1. $\arctan(x+2)+C$

3. $11\arctan(x+2)-\frac{3}{2}\ln(x^2+4x+5)+C$

5. $\arcsin\left(\dfrac{x+1}{2}\right)+C$

7. $-2\arcsin\left(\dfrac{x+1}{2}\right)-\frac{1}{2}(x+1)\sqrt{3-2x-x^2}$
$-\frac{1}{3}(3-2x-x^2)^{3/2}+C$

9. $\frac{5}{16}\ln|2x+3|+\frac{7}{16}\ln|2x-1|+C$

11. $\frac{1}{3}\arctan\left(\dfrac{x+2}{3}\right)+C$ **13.** $\frac{1}{4}\ln\left|\dfrac{1+x}{3-x}\right|+C$

15. $\ln(x^2+2x+2)-7\arctan(x+1)+C$

17. $\frac{2}{9}\arcsin\left(x-\frac{2}{3}\right)-\frac{1}{9}\sqrt{5+12x-9x^2}+C$

19. $\frac{75}{4}\arcsin\left(\dfrac{2x-4}{5}\right)+\frac{3}{2}(x-2)\sqrt{9+16x-4x^2}$
$+\frac{1}{6}(9+16x-4x^2)^{3/2}+C$

21. $\dfrac{7x-12}{9\sqrt{6x-x^2}}+C$ **23.** $-\dfrac{1}{16x^2+48x+52}+C$

25. $\frac{3}{2}\ln(x^2+x+1)-\frac{5}{3}\sqrt{3}\arctan\left(\dfrac{(2x+1)\sqrt{3}}{3}\right)+C$

27. $\frac{1}{32}\ln\left|\dfrac{x+2}{x-2}\right|-\dfrac{x}{8(x^2-4)}+C$

29. $\ln|x|-\frac{2}{3}\sqrt{3}\arctan\left(\dfrac{(2x+1)\sqrt{3}}{3}\right)+C$

31. $\frac{1}{4}\ln|x+1|-\frac{1}{4}\ln|x-1|-\dfrac{5}{4(x-1)}-\dfrac{5}{4(x+1)}+C$

33. $-\dfrac{x+7}{4(x^2+2x+5)} - \frac{1}{8}\arctan\left(\dfrac{x+1}{2}\right) + C$

37. $\left[\frac{1}{2}\arctan\left(\dfrac{x-1}{2}\right)\right]_0^5 = \frac{1}{2}\left[\arctan 2 + \arctan\left(\frac{1}{2}\right)\right] = \frac{1}{4}\pi$

39. $\dfrac{\pi}{16}\cdot\left[\dfrac{2x-2}{x^2-2x+5} + \arctan\left(\dfrac{x-1}{2}\right)\right]_0^5$

$= \dfrac{\pi}{80}\left[4 + 5\arctan\left(\frac{1}{2}\right) + 5\arctan 2\right] = \dfrac{\pi}{160}(5\pi + 8)$

41. $\dfrac{\pi}{4}\cdot\left[5\arctan\left(\dfrac{2x-5}{2}\right) + \ln(4x^2 - 20x + 29)\right]_1^4$

$= \frac{5}{2}\pi\arctan\left(\frac{3}{2}\right)$

43. About 3.694048722 mi

45. $\ln|x-1| - \frac{1}{2}\ln(x^2+2x+2) + \arctan(x+1) + C$

47. $\frac{1}{2}x^2 + \ln|x-1| + \frac{1}{2}\ln(x^2+x+1)$

$+ \frac{1}{3}\sqrt{3}\arctan\left(\dfrac{(2x+1)\sqrt{3}}{3}\right) + C$

49. $\frac{1}{2}\ln(x^4+x^2+1) - \frac{2}{3}\sqrt{3}\arctan\left(\dfrac{\sqrt{3}}{2x^2+1}\right) + C$

51. The integrand splits into

$$\frac{7}{x-1} - \frac{6x-5}{(x^2+2x+2)^2},$$

and the antiderivative is

$$7\ln|x-1| + \frac{11}{2}\arctan(x+1) + \frac{11x+17}{2(x^2+2x+2)} + C.$$

53. The integrand splits into

$$\frac{7}{8(x-5)} + \frac{9}{8(x+3)} + \frac{3x-4}{(4x^2+4x+5)^2},$$

and the antiderivative is

$$\tfrac{7}{8}\ln|x-5| + \tfrac{9}{8}\ln|x+3| + \tfrac{11}{64}\arctan\left(\frac{2}{2x+1}\right)$$

$$- \frac{22x+23}{32(4x^2+4x+5)} + C.$$

Section 9.8 (page 558)

1. $\sqrt{2}$ **3.** $+\infty$ **5.** $+\infty$ **7.** 1 **9.** $+\infty$
11. $-\frac{1}{2}$ **13.** $\frac{9}{2}$ **15.** $+\infty$ **17.** Does not exist
19. $2(e-1)$ **21.** $\frac{1}{9}$ **23.** $\frac{1}{2}$ **25.** $\frac{1}{2}\pi$
27. Does not exist **29.** $+\infty$ **31.** $1/\ln 2$
33. 2 **35.** -1 **37.** $-\infty$

39. The first integral does not exist; the second (from 1 to ∞) converges to $\ln 2$.

41. Both integrals converge to $\frac{1}{2}\pi$.

43. This integral converges to $1/(1-k)$ if $k < 1$ and diverges otherwise.

45. This integral converges to $-1/(k+1)^2$ if $k > -1$ and diverges otherwise.

47. Let $u = u(x) = x^{t-1}$ and $dv = e^{-x}\,dx$ to show first that

$$\Gamma(t) = (t-1)\Gamma(t-1) \quad \text{if } t > 1,$$

then replace t with $x + 1$.

59. Integrate $2\pi x e^{-x^2}$ from $x = 0$ to $+\infty$ to obtain volume π.

Chapter 9 Miscellaneous Problems (page 563)

1. $2\arctan\sqrt{x} + C$ **3.** $\ln|\sec x| + C$

5. $\frac{1}{2}\sec^2\theta + C$

7. $x\tan x - \frac{1}{2}x^2 + \ln|\cos x| + C$

9. $\frac{2}{15}(2-x^3)^{5/2} - \frac{4}{9}(2-x^3)^{3/2} + C$

11. $\frac{1}{2}x\sqrt{25+x^2} - \frac{25}{2}\ln\left|x + \sqrt{25+x^2}\right| + C$

13. $\frac{2}{3}\sqrt{3}\arctan\left(\dfrac{(2x-1)\sqrt{3}}{3}\right) + C$

15. $\frac{103}{87}\sqrt{29}\arctan\left(\dfrac{(3x-2)\sqrt{29}}{29}\right) +$

$\frac{5}{6}\ln(9x^2 - 12x + 33) + C$

17. $\frac{2}{9}(1+x^3)^{3/2} + C$

19. $\arcsin\left(\frac{1}{2}\sin x\right) + C$

21. $-\ln|\ln\cos x| + C$

23. $(x+1)\ln(x+1) - x + C$

25. $\frac{1}{2}x\sqrt{x^2+9} + \frac{9}{2}\ln\left|x + \sqrt{x^2+9}\right| + C$

27. $\frac{1}{2}(x-1)\sqrt{2x-x^2} + \frac{1}{2}\arcsin(x-1) + C$

29. $\frac{1}{3}x^2 + 2x - \sqrt{2}\ln\left|\dfrac{x+\sqrt{2}}{x-\sqrt{2}}\right| + C$

31. $\frac{1}{2}(x^2+x)(x^2+2x+2)^{-1} - \frac{1}{2}\arctan(x+1) + C$

33. $\dfrac{\sin 2\theta}{2(1+\cos 2\theta)} + C = \frac{1}{2}\tan\theta + C$

35. $\frac{1}{5}\sec^5 x - \frac{1}{3}\sec^3 x + C$

37. $\frac{1}{8}x^2[4(\ln x)^3 - 6(\ln x)^2 + 6\ln x - 3] + C$

39. $\frac{1}{2}e^x\sqrt{1+e^{2x}} + \frac{1}{2}\ln\left(e^x + \sqrt{1+e^{2x}}\right) + C$

41. $\frac{1}{18}x^{-2}\sqrt{x^2-9} - \frac{1}{54}\arctan\left(\dfrac{3}{\sqrt{x^2-9}}\right) + C$

43. $\ln|x| + \frac{1}{2}\arctan 2x + C$

45. $\frac{1}{2}(\sec x\tan x - \ln|\sec x + \tan x|) + C$

47. $\ln|x+1| - \frac{2}{3}x^{-3} + C$

49. $\ln|x-1| + \ln(x^2+x+1) + (x-1)^{-1}$

$- 2(x^2+x+1)^{-1} + C$

51. $x[(\ln x)^6 - 6(\ln x)^5 + 30(\ln x)^4 - 120(\ln x)^3$

$+ 360(\ln x)^2 - 720\ln x + 720] + C$

53. $\frac{1}{3}(\arcsin x)^3 + C$

55. $\frac{1}{2}\sec^2 z + \ln|\cos z| + C$

57. $\frac{1}{2}\arctan(\exp(x^2)) + C$

59. $-\frac{1}{2}(x^2+1)\exp(-x^2) + C$

61. $-\dfrac{1}{x}\arcsin x - \ln\left|\dfrac{1+\sqrt{1-x^2}}{x}\right| + C$

63. $\frac{1}{8}\arcsin x + \frac{1}{8}x(2x^2-1)\sqrt{1-x^2} + C$

65. $\frac{1}{4}\ln|2x+1| + \frac{5}{4}(2x+1)^{-1} + C$
67. $\frac{1}{2}\ln|e^{2x}-1| + C$
69. $2\ln|x+1| + 3(x+1)^{-1} - \frac{5}{3}(x+1)^{-3} + C$
71. $\frac{1}{2}\ln(x^2+1) + \arctan x - \frac{1}{2}(x+1)^{-1} + C$
73. $\frac{1}{45}(x^3-1)^{1/2}(6x^6-2x^3-4) + C$
75. $\frac{2}{3}(1+\sin x)^{3/2} + C$
77. $\frac{1}{2}\ln|\sec x + \tan x| + C$
79. $-2\sqrt{1-\sin t} + C$
81. $-2x + \sqrt{3}\arctan\left(\frac{(2x+1)\sqrt{3}}{3}\right) + \frac{1}{2}(2x+1)\ln(x^2+x+1) + C$
83. $-\frac{1}{x}\arctan x + \ln\left|\frac{x}{\sqrt{1+x^2}}\right| + C$
85. $\frac{1}{2}\ln(x^2+1) + \frac{1}{2}(x^2+1)^{-1} + C$
87. $\frac{x-6}{2\sqrt{x^2+4}} + C$
89. $\frac{1}{3}(1+\sin^2 x)^{3/2} + C$
91. $\frac{1}{2}e^x(x\sin x - x\cos x + \cos x) + C$
93. $-\frac{\arctan x}{2(x-1)^2} + \frac{1}{8}\ln(x^2+1) - \frac{1}{4}\ln|x-1| - \frac{1}{4(x-1)} + C$
95. $\frac{11}{9}\arcsin\left(\frac{3x-1}{2}\right) - \frac{2}{9}\sqrt{3+6x-9x^2} + C$
97. $\frac{1}{3}x^3 + x^2 + 3x + 4\ln|x-1| - (x-1)^{-1} + C$
99. $x\,\mathrm{arcsec}\sqrt{x} - \sqrt{x-1} + C$
101. $\frac{1}{4}\pi(e^2 - e^{-2} + 4)$
103. (a) $A_t = \pi\left(\sqrt{2} - e^{-t}\sqrt{1+e^{-2t}} + \ln\left[\frac{1+\sqrt{2}}{e^{-t}+\sqrt{1+e^{-2t}}}\right]\right)$; (b) $\pi\left[\sqrt{2} + \ln\left(1+\sqrt{2}\right)\right] \approx 7.2118$
105. $\frac{\pi\sqrt{2}}{2}\left[2\sqrt{14} - \sqrt{2} + \ln\left(\frac{1+\sqrt{2}}{2\sqrt{2}+\sqrt{7}}\right)\right] \approx 11.66353$
109. $\frac{5}{4}\pi \approx 3.92699$
111. The value of the integral is $\frac{1}{630}$.
113. $\frac{5\sqrt{6}-3\sqrt{2}}{2} + \frac{1}{2}\ln\left(\frac{1+\sqrt{2}}{\sqrt{3}+\sqrt{2}}\right) \approx 3.869983$
115. (a) The substitution is $u = e^x$; the value of the integral is $\frac{2\sqrt{3}}{3}\arctan\left(\frac{\sqrt{3}}{3}(1+2e^x)\right) + C.$
119. $\frac{1}{4}\sqrt{2}\ln\left|\frac{1+\tan\theta-\sqrt{2}\tan\theta}{1+\tan\theta+\sqrt{2}\tan\theta}\right| + \frac{1}{2}\sqrt{2}\left[\arctan\left(-1+\sqrt{2}\tan\theta\right) + \arctan\left(1+\sqrt{2}\tan\theta\right)\right] + C$
121. $\frac{2}{25515}(3x-2)^{1/2}(2835x^4 - 270x^3 - 216x^2 - 192x - 256) + C$
123. $\frac{3}{4}(x^2-1)^{-1/3}(x^2-3) + C$
125. $\frac{2}{9}(x^3+1)^{1/2}(x^3-2) + C$
127. $2\arctan\sqrt{\frac{1+x}{1-x}} - (1-x)\sqrt{\frac{1+x}{1-x}} + C$
129. $3(x+1)^{1/3} - \sqrt{3}\arctan\left(\frac{1+2(x+1)^{1/3}}{\sqrt{3}}\right) + \ln|(x+1)^{1/3}-1| - \frac{1}{2}\ln|(x+1)^{2/3} + (x+1)^{1/3} + 1| + C$
131. $\sqrt{1+e^{2x}} + \frac{1}{2}\ln\left(\frac{\sqrt{1+e^{2x}}-1}{\sqrt{1+e^{2x}}+1}\right) + C$
133. $\frac{8}{15}$
135. $\tan\frac{\theta}{2} + C = \frac{1-\cos\theta}{\sin\theta} + C = \frac{\sin\theta}{1+\cos\theta} + C$
137. $-\frac{2\sin\theta}{1+\sin\theta-\cos\theta} + C$
139. $\frac{\sqrt{2}}{2}\ln\left|\frac{1-\cos\theta+(\sqrt{2}-1)\sin\theta}{1-\cos\theta-(\sqrt{2}+1)\sin\theta}\right| + C$
141. $\ln\frac{1-\cos\theta}{1-\cos\theta+\sin^2\theta} + C$

Section 10.1 (page 572)

1. $x + 2y + 3 = 0$ **3.** $4y + 25 = 3x$
5. $x + y = 1$
7. Center $(-1, 0)$, radius $\sqrt{5}$
9. Center $(2, -3)$, radius 4
11. Center $\left(\frac{1}{2}, 0\right)$, radius 1
13. Center $\left(\frac{1}{2}, -\frac{3}{2}\right)$, radius 3
15. Center $\left(-\frac{1}{3}, \frac{4}{3}\right)$, radius 2
17. The point $(3, 2)$ **19.** No points
21. $(x+1)^2 + (y+2)^2 = 34$
23. $(x-6)^2 + (y-6)^2 = \frac{4}{5}$
25. $2x + y = 13$
27. $(x-6)^2 + (y-11)^2 = 18$
29. $\left(\frac{x}{5}\right)^2 + \left(\frac{y}{3}\right)^2 = 1$
31. $y - 7 + 4\sqrt{3} = \left(4 - 2\sqrt{3}\right)\left(x - 2 + \sqrt{3}\right)$, $y - 7 - 4\sqrt{3} = (4 + 2\sqrt{3})(x - 2 - \sqrt{3})$
33. $y - 1 = 4(x-4)$ and $y + 1 = 4(x+4)$
35. $a = \sqrt{h^2 - p^2}$, $b = a\sqrt{e^2-1}$, $h = \frac{p(e^2+1)}{1-e^2}$

Section 10.2 (page 578)

1. (a) $\left(\frac{1}{2}\sqrt{2}, \frac{1}{2}\sqrt{2}\right)$ (b) $\left(1, -\sqrt{3}\right)$
(c) $\left(\frac{1}{2}, -\frac{1}{2}\sqrt{3}\right)$ (d) $(0, -3)$
(e) $\left(\sqrt{2}, \sqrt{2}\right)$ (f) $\left(\sqrt{3}, -1\right)$ (g) $\left(-\sqrt{3}, 1\right)$

3. $r\cos\theta = 4$

5. $\theta = \arctan\left(\frac{1}{3}\right)$ **7.** $r^2\cos\theta\sin\theta = 1$

9. $r = \tan\theta\sec\theta$ **11.** $x^2 + y^2 = 9$

13. $x^2 + 5x + y^2 = 0$ **15.** $(x^2 + y^2)^3 = 4y^4$

17. $x = 3$ **19.** $x = 2;\ r = 2\sec\theta$

21. $x + y = 1;\ r = \dfrac{1}{\cos\theta + \sin\theta}$

23. $y = x + 2;\ r = \dfrac{2}{\sin\theta - \cos\theta}$

25. $x^2 + y^2 + 8y = 0;\ r = -8\sin\theta$

27. $x^2 + y^2 = 2x + 2y;\ r = 2(\cos\theta + \sin\theta)$

29. Fig. 10.2.17 **31.** Fig. 10.2.18

33. Fig. 10.2.20 **35.** Fig. 10.2.19

37. Center $\left(\frac{1}{2}a, \frac{1}{2}b\right)$, radius $\frac{1}{2}\sqrt{a^2 + b^2}$—unless $a = b = 0$, in which case the graph consists of the single point $(0, 0)$.

39. Symmetric around the x-axis

41. Symmetric around the x-axis

43. Symmetric around the x-axis

45. Symmetric around the origin

47. Symmetric around both axes and the origin

49. Symmetric around the x-axis

51. Symmetric around the y-axis

53. $(1, 0)$

55. $(0, 0)$, $\left(\frac{1}{2}, \pi/6\right)$, $\left(\frac{1}{2}, 5\pi/6\right)$, $(1, \pi/2)$

57. The pole, the point $(r, \theta) = (2, \pi)$, and the two points $r = 2\left(\sqrt{2} - 1\right)$, $|\theta| = \arccos\left(3 - 2\sqrt{2}\right)$

59. (a) $r\cos(\theta - \alpha) = p$

61. The polar equation is $r = \pm a + b\sin\theta$: a limaçon.

Section 10.3 (page 584)

1.

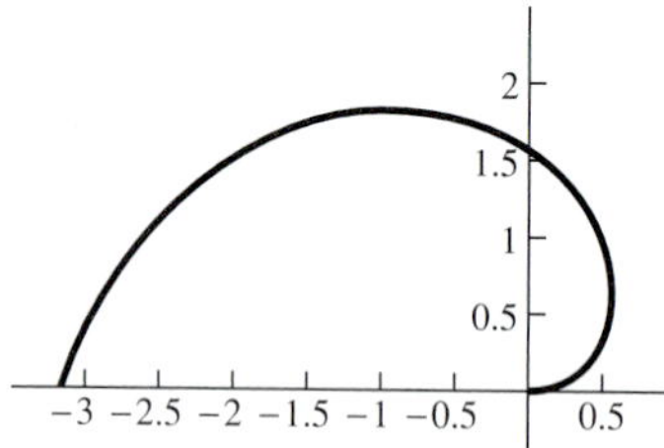

3.

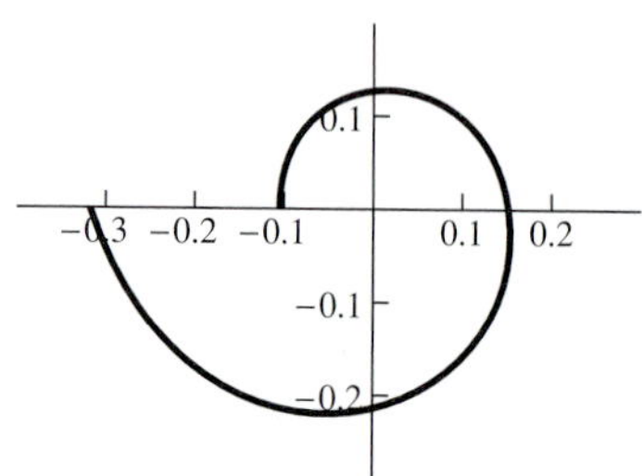

5.

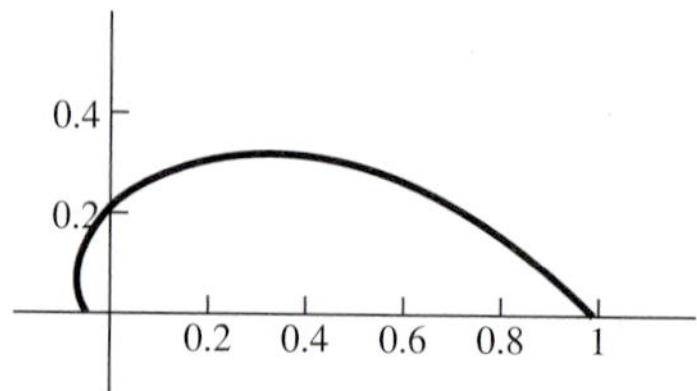

7. π **9.** $3\pi/2$

11. $9\pi/2$ **13.** 4π

15. $19\pi/2$ **17.** $\pi/2$ (one of *four* loops)

19. $\pi/4$ (one of 8 loops) **21.** 2 (one of *two* loops)

23. 4 (one of *two* loops) **25.** $\frac{1}{6}\left(2\pi + 3\sqrt{3}\right)$

27. $\frac{1}{24}\left(5\pi - 6\sqrt{3}\right)$ **29.** $\frac{1}{6}\left(39\sqrt{3} - 10\pi\right)$

31. $\frac{1}{2}\left(2 - \sqrt{2}\right)$ **33.** $\frac{1}{6}\left(20\pi + 21\sqrt{3}\right)$

35. $\frac{1}{2}(2 + \pi)$ **37.** $\pi/2$

41. (a) $\frac{5}{2}(1 - e^{-2\pi/5})^2 \approx 1.279458764$;
(b) $\frac{5}{2}e^{-2n\pi/5}(e^{2\pi/5} - 1)^2$

43. The point of intersection in the second quadrant is located where $\theta = \alpha \approx 2.326839$. Using symmetry, the total area of the shaded region R is approximately

$$2\int_0^{\alpha} \frac{1}{2}(e^{-t/5})^2\,dt + 2\int_{\alpha}^{\pi} \frac{1}{2}[2(1 + \cos t)]^2\,dt \approx 1.58069.$$

Section 10.4 (page 592)

1. $y = 2x - 3$ **3.** $y^2 = x^3$

5. $y = 2x^2 - 5x + 2$ **7.** $y = 4x^2, x > 0$

9. $\left(\dfrac{x}{5}\right)^2 + \left(\dfrac{y}{3}\right)^2 = 1$ **11.** $\left(\dfrac{x}{2}\right)^2 - \left(\dfrac{y}{3}\right)^2 = 1, x > 0$

13. $x^2 + y^2 = 1$ **15.** $x + y = 1, 0 \leqq x \leqq 1$

17. (a) $y - 5 = \frac{9}{4}(x - 3)$; (b) $\dfrac{d^2y}{dx^2} = \frac{9}{16}t^{-1}$

19. (a) $y = -\frac{1}{2}\pi\left(x - \frac{1}{2}\pi\right)$; (b) concave downward

21. $\psi = \frac{1}{6}\pi$ (constant)

23. $\psi = \frac{1}{2}\pi$

25. There are horizontal tangents at $(1, 2)$ and $(1, -2)$. There is a vertical tangent line at $(0, 0)$. There is no tangent line at the other x-intercept $(3, 0)$ because the curve crosses itself with two different slopes there, namely the slopes $\pm\sqrt{3}$.

27. There are horizontal tangents at the points corresponding to $\theta = \pm\pi/3$. The corresponding value of r is $\frac{3}{2}$, and the rectangular coordinates of these two points are $\left(\frac{3}{4}, \pm\frac{3}{4}\sqrt{3}\right)$. There is a vertical tangent at $(2, 0)$.If a line tangent to the curve C at the point P is simply a line through P that approximates the curve's shape very very well at and near P, then there is a horizontal tangent line at $(0, 0)$.

31. $x = p/m^2$, $y = 2p/m$

41. Given: $x^5 + y^5 = 5x^2y^2$. Substitute $y = tx$ to obtain $x^5 + t^5x^5 = 5x^4t^2$, then solve for

$$x = \frac{5t^2}{1+t^5}, \quad y = \frac{5t^3}{1+t^5}, \quad 0 \leqq t < +\infty.$$

Section 10.5 (page 600)

1. $\frac{22}{5}$
3. $\frac{4}{3}$
5. $\frac{1}{2}(1 + e^{\pi}) \approx 12.0703$
7. $\frac{358}{35}\pi \approx 32.13400$
9. $\frac{16}{15}\pi \approx 3.35103$
11. $\frac{74}{3}$
13. $\frac{1}{4}\pi\sqrt{2} \approx 1.11072$
15. $(e^{2\pi} - 1)\sqrt{5} \approx 1195.1597$
17. $\frac{8}{3}\pi\left(5\sqrt{5} - 2\sqrt{2}\right) \approx 69.96882$
19. $\frac{2}{27}\pi\left(13\sqrt{13} - 8\right) \approx 9.04596$
21. $16\pi^2 \approx 157.91367$
23. $5\pi^2a^3$
25. (a) $A = \pi ab$; (b) $V = \frac{4}{3}\pi ab^2$
27. $\pi\sqrt{1 + 4\pi^2} + \frac{1}{2}\ln\left(2\pi + \sqrt{1 + 4\pi^2}\right) \approx 21.25629$
29. $\frac{3}{8}\pi a^2$
31. $\frac{12}{5}\pi a^2$
33. $\frac{216}{5}\sqrt{3} \approx 78.8246$
35. $\frac{243}{4}\pi\sqrt{3} \approx 330.5649$
37. 4.91749
39. $6\pi^3a^3$
40. $\frac{1}{2}\pi^2a$
41. $\frac{5}{6}\pi^3a^2$
43. 20.04734
45. 19.3769
47. 61.0036
49. 16.3428
51. $S_x \approx 16.0570$, $V_x = \frac{16}{15}\pi$
53. 24.603
55. 39.4036

Section 10.6 (page 607)

1. $y^2 = 12x$
3. $(x-2)^2 = -8(y-3)$
5. $(y-3)^2 = -8(x-2)$
7. $x^2 = -6\left(y + \frac{3}{2}\right)$
9. $x^2 = 4(y+1)$
11. $y^2 = 12x$; vertex $(0, 0)$, axis the x-axis
13. $y^2 = -6x$; vertex $(0, 0)$, axis the x-axis
15. $x^2 - 4x - 4y = 0$; vertex $(2, -1)$, axis the line $x = 2$
17. $4y = -12 - (2x+1)^2$; vertex $\left(-\frac{1}{2}, -3\right)$, axis the line $x = -\frac{1}{2}$
23. About 0.693 days; that is, about 16 h 38 min
27. $\alpha = \frac{1}{2}\arcsin(0.49) \approx 0.256045$ (14° 40′ 13″), $\alpha = \frac{1}{2}[\pi - \arcsin(0.49)] \approx 1.314751$ (75° 19′ 47″)
29. *Suggestion:* $x^2 - 2xy + y^2 - 2ax - 2ay + a^2 = 0$

Section 10.7 (page 608)

1. $\left(\frac{x}{4}\right)^2 + \left(\frac{y}{5}\right)^2 = 1$
3. $\left(\frac{x}{15}\right)^2 + \left(\frac{y}{17}\right)^2 = 1$
5. $\frac{x^2}{16} + \frac{y^2}{7} = 1$
7. $\frac{x^2}{100} + \frac{y^2}{75} = 1$
9. $\frac{x^2}{16} + \frac{y^2}{12} = 1$
11. $\frac{(x-2)^2}{16} + \frac{(y-3)^2}{4} = 1$
13. $\frac{(x-1)^2}{25} + \frac{(y-1)^2}{16} = 1$
15. $\frac{(x-1)^2}{81} + \frac{(y-2)^2}{72} = 1$
17. Center $(0, 0)$, foci $(\pm 2\sqrt{5}, 0)$, major axis 12, minor axis 8
19. Center $(0, 4)$, foci $(0, 4 \pm \sqrt{5})$, major axis 6, minor axis 4
21. (a) About 3466.36 AU—that is, about 3.22×10^{11} mi, or about 20 light-days
27. $\frac{(x-1)^2}{4} + \frac{y^2}{16/3} = 1$

Section 10.8 (page 618)

1. $\frac{x^2}{1} - \frac{y^2}{15} = 1$
3. $\frac{x^2}{16} - \frac{y^2}{9} = 1$
5. $\frac{y^2}{25} - \frac{x^2}{25} = 1$
7. $\frac{y^2}{9} - \frac{x^2}{27} = 1$
9. $\frac{x^2}{4} - \frac{y^2}{12} = 1$
11. $\frac{(x-2)^2}{9} - \frac{(y-2)^2}{27} = 1$
13. $\frac{(y+2)^2}{9} - \frac{(x-1)^2}{4} = 1$
15. Center $(1, 2)$, foci $\left(1 \pm \sqrt{2}, 2\right)$, asymptotes $y - 2 = \pm(x-1)$
17. Center $(0, 3)$, foci $\left(0, 3 \pm 2\sqrt{3}\right)$, asymptotes $y = 3 \pm x\sqrt{3}$
19. Center $(-1, 1)$, foci $\left(-1 \pm \sqrt{13}, 1\right)$, asymptotes $y = \frac{1}{2}(3x+5)$, $y = -\frac{1}{2}(3x+1)$
21. There are no points on the graph if $c > 15$.
25. $16x^2 + 50xy + 16y^2 = 369$
27. About 16.42 mi north of B and 8.66 mi west of B; that is, about 18.56 mi from B at a bearing of 27°48′ west of north

Chapter 10 Miscellaneous Problems (page 619)

1. Circle, center $(1, 1)$, radius 2
3. Circle, center $(3, -1)$, radius 1
5. Parabola, vertex $(4, -2)$, opening downward
7. Ellipse, center $(2, 0)$, major axis 6, minor axis 4
9. Hyperbola, center $(-1, 1)$, vertical axis, foci at $\left(-1, 1 \pm \sqrt{3}\right)$
11. There are no points on the graph.
13. Hyperbola, center $(1, 0)$, horizontal axis
15. Circle, center $(4, 1)$, radius 1

17. The graph consists of the line $y = -x$ together with the isolated point $(2, 2)$.
19. Circle, center $(1, 0)$, radius 1
21. The straight line $y = x + 1$
23. The horizontal line $y = 3$
25. Two ovals tangent to each other and to the y-axis at $(0, 0)$
27. Apple-shaped curve, symmetric around the y-axis
29. Ellipse, one focus at $(0, 0)$, directrix $x = 4$, eccentricity $e = \frac{1}{2}$
31. $\frac{1}{2}(\pi - 2)$
33. $\frac{1}{6}(39\sqrt{3} - 10\pi) \approx 6.02234$
35. 2
37. $\frac{5}{4}\pi$
39. The straight line $y = x + 2$
41. The circle $(x - 2)^2 + (y - 1)^2 = 1$
43. Equation: $y^2 = (x - 1)^3$
45. $y - 2\sqrt{2} = -\frac{4}{3}\left(x - \frac{3}{2}\sqrt{2}\right)$
47. $2\pi y + 4x = \pi^2$
49. 24
51. 3π
53. $\frac{1}{27}\left(13\sqrt{13} - 8\right) \approx 1.4397$
55. $\frac{43}{6}$
57. $\frac{1}{8}(4\pi - 3\sqrt{3}) \approx 0.92128$
59. $\frac{471,295}{1024}\pi \approx 1445.915$
61. $\frac{1}{2}\pi(e^{\pi} + 1)\sqrt{5} \approx 84.7919$
63. $x = a\theta - b\sin\theta,\ y = a - b\cos\theta$
65. *Suggestion:* Compute $r^2 = x^2 + y^2$.
67. $6\pi^3 a^3$
69. $r = 2p\cos(\theta - \alpha)$
71. If $a > b$, then the maximum is $2a$ and the minimum is $2b$.
73. $b^2 y = 4hx(b - x)$; alternatively,

$$r = b\sec\theta - \frac{b^2}{4h}\sec\theta\tan\theta.$$

75. *Suggestion:* Let θ be the angle that QR makes with the x-axis.
79. The curve is a hyperbola with one focus at the origin, directrix $x = -\frac{3}{2}$, and eccentricity $e = 2$.
81. $\frac{3}{2}$

Section 11.2 (page 632)

1. $a_n = n^2$ **3.** $a_n = \dfrac{1}{3^n}$
5. $a_n = \dfrac{1}{3n - 1}$ **7.** $a_n = 1 + (-1)^n$
9. $\frac{2}{5}$ **11.** 0 **13.** 1
15. Does not converge **17.** 0
19. 0 **21.** 0 **23.** 1
25. 0 **27.** 0 **29.** 0
31. 0 **33.** e **35.** e^{-2}
37. 2 **39.** 1
41. Does not converge **43.** 1
45. 2 **47.** 1 **49.** π
57. (b) 4

Section 11.3 (page 642)

1. $\frac{3}{2}$ **3.** Diverges **5.** Diverges
7. 6 **9.** Diverges **11.** Diverges
13. Diverges **15.** $2 + \sqrt{2}$ **17.** Diverges
19. $\frac{1}{12}$ **21.** $\dfrac{e}{\pi - e}$ **23.** Diverges
25. $\frac{65}{12}$ **27.** $\frac{247}{8}$ **29.** $\frac{1}{4}$
31. Diverges **33.** Diverges
35. $\dfrac{\arctan 1}{1 - \arctan 1} \approx 3.659792$
37. Diverges **39.** $\frac{47}{99}$
41. $\frac{41}{333}$ **43.** $\frac{314,156}{99999}$
45. $-3 < x < 3;\ \dfrac{x}{3 - x}$ **47.** $-1 < x < 5;\ \dfrac{x - 2}{5 - x}$
49. $-2 < x < 2;\ \dfrac{5x^2}{16 - 4x^2}$ **51.** $S_n = \dfrac{1}{6} - \dfrac{1}{9n + 6};\ \dfrac{1}{6}$
53. $S_n = \dfrac{1}{4} - \dfrac{1}{16n + 4};\ \dfrac{1}{4}$
55. $S_n = \dfrac{1}{2} + \dfrac{1}{4} - \dfrac{1}{2n + 2} - \dfrac{1}{2n + 4};\ \dfrac{3}{4}$
57. $S_n = 2 - \dfrac{1}{n + 1} - \dfrac{1}{2n + 1};\ 2$
59. $S_n = \dfrac{1}{3} - \dfrac{1}{n + 1} + \dfrac{2}{n + 2} - \dfrac{1}{n + 3};\ \dfrac{1}{3}$
65. 4.5 s **67.** (a) $M_n = (0.95)^n M_0$; (b) 0
69. Peter $\frac{4}{7}$, Paul $\frac{2}{7}$, Mary $\frac{1}{7}$ **71.** $\frac{1}{12}$

Section 11.4 (page 657)

1. $e^{-x} = 1 - \dfrac{x}{1!} + \dfrac{x^2}{2!} - \dfrac{x^3}{3!} + \dfrac{x^4}{4!} - \dfrac{x^5}{5!} + \dfrac{x^6}{6!}e^{-z}$ for some z between 0 and x.

3. $\cos x = 1 - \dfrac{x^2}{2!} + \dfrac{x^4}{4!} - \dfrac{x^5}{5!}\sin z$ for some z between 0 and x.

5. $\sqrt{1 + x} = 1 + \dfrac{x}{1!2} - \dfrac{x^2}{2!4} + \dfrac{3x^3}{3!8} - \dfrac{5x^4}{128}(1 + z)^{-7/2}$ for some z between 0 and x.

7. $\tan x = \dfrac{x}{1!} + \dfrac{2x^3}{3!} + \dfrac{x^4}{4!}(16\sec^4 z \tan z + 8\sec^2 z \tan^3 z)$ for some z between 0 and x.

9. $\sin^{-1} x = \dfrac{x}{1!} + \dfrac{x^3}{3!} \cdot \dfrac{1 + 2z^2}{(1 - z^2)^{5/2}}$ for some z between 0 and x.

11. $e^x = e + \dfrac{e}{1!}(x-1) + \dfrac{e}{2!}(x-1)^2 + \dfrac{e}{3!}(x-1)^3 + \dfrac{e}{4!}(x-1)^4 + \dfrac{e^z}{5!}(x-1)^5$ for some z between 1 and x.

13. $\sin x = \dfrac{1}{2} + \dfrac{\sqrt{3}}{1!2}(x - \tfrac{1}{6}\pi) - \dfrac{1}{2!2}(x - \tfrac{1}{6}\pi)^2 - \dfrac{\sqrt{3}}{3!2}(x - \tfrac{1}{6}\pi)^3 + \dfrac{\sin z}{4!}(x - \tfrac{1}{6}\pi)^4$ for some z between $\tfrac{1}{6}\pi$ and x.

15. $\dfrac{1}{(x-4)^2} = 1 - 2(x-5) + 3(x-5)^2 - 4(x-5)^3 + 5(x-5)^4 - 6(x-5)^5 + \dfrac{7}{(z-4)^8}(x-5)^6$ for some z between 5 and x.

17. $\cos x = -1 + \dfrac{(x-\pi)^2}{2!} - \dfrac{(x-\pi)^4}{4!} - \dfrac{(x-\pi)^5}{5!}\sin z$ for some z between π and x.

19. $x^{3/2} = 1 + \dfrac{3(x-1)}{2} + \dfrac{3(x-1)^2}{8} - \dfrac{(x-1)^3}{16} + \dfrac{3(x-1)^4}{128} - \dfrac{3(x-1)^5}{256z^{7/2}}$ for some z between 1 and x.

21. $e^{-x} = 1 - x + \dfrac{x^2}{2!} - \dfrac{x^3}{3!} + \dfrac{x^4}{4!} - \cdots$

23. $e^{-3x} = 1 - 3x + \dfrac{9x^2}{2!} - \dfrac{27x^3}{3!} + \dfrac{81x^4}{4!} - \cdots$

25. $\sin 2x = 2x - \dfrac{(2x)^3}{3!} + \dfrac{(2x)^5}{5!} - \dfrac{(2x)^7}{7!} + \dfrac{(2x)^9}{9!} - \cdots$

$$= \sum_{n=0}^{\infty} \frac{(-1)^n 2^{2n+1} x^{2n+1}}{(2n+1)!} = 2x - \frac{4x^3}{3} + \frac{4x^5}{15} - \frac{8x^7}{315} + \frac{4x^9}{2835} - \cdots$$

27. $\sin x^2 = x^2 - \dfrac{x^6}{3!} + \dfrac{x^{10}}{5!} - \dfrac{x^{14}}{7!} + \dfrac{x^{18}}{9!} - \cdots$

29. $\ln(1+x) = x - \dfrac{x^2}{2} + \dfrac{x^3}{3} - \dfrac{x^4}{4} + \dfrac{x^5}{5} - \cdots$

31. $e^{-x} = 1 - x + \dfrac{x^2}{2!} - \dfrac{x^3}{3!} + \dfrac{x^4}{4!} - \cdots$

33. $\ln x = (x-1) - \dfrac{(x-1)^2}{2} + \dfrac{(x-1)^3}{3} - \dfrac{(x-1)^4}{4} + \cdots$

35. $\cos x = \dfrac{\sqrt{2}}{2} - \dfrac{\sqrt{2}}{1!2}(x - \tfrac{1}{4}\pi) - \dfrac{\sqrt{2}}{2!2}(x - \tfrac{1}{4}\pi)^2 + \dfrac{\sqrt{2}}{3!2}(x - \tfrac{1}{4}\pi)^3 + \dfrac{\sqrt{2}}{4!2}(x - \tfrac{1}{4}\pi)^4 - \dfrac{\sqrt{2}}{5!2}(x - \tfrac{1}{4}\pi)^5 - \dfrac{\sqrt{2}}{6!2}(x - \tfrac{1}{4}\pi)^6 + \dfrac{\sqrt{2}}{7!2}(x - \tfrac{1}{4}\pi)^7 + \cdots$

37. $\dfrac{1}{x} = 1 - (x-1) + (x-1)^2 - (x-1)^3 + (x-1)^4 - \cdots$

39. $\sin x = \dfrac{\sqrt{2}}{2}\left[1 + (x - \tfrac{1}{4}\pi) - \dfrac{1}{2!}(x - \tfrac{1}{4}\pi)^2 - \dfrac{1}{3!}(x - \tfrac{1}{4}\pi)^3 + \dfrac{1}{4!}(x - \tfrac{1}{4}\pi)^4 + \cdots\right]$

45. Here is a plot of e^{-x} and its Taylor polynomials, center zero, of degrees 3, 5, and 7.

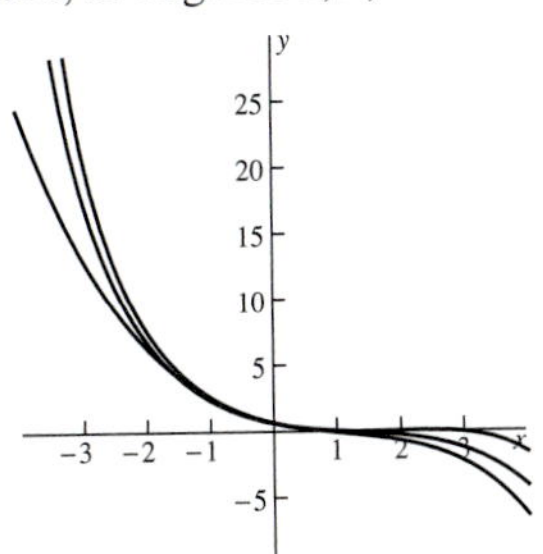

47. Here is a plot of $\cos x$ and its Taylor polynomials, center zero, of degrees 3, 4, and 5.

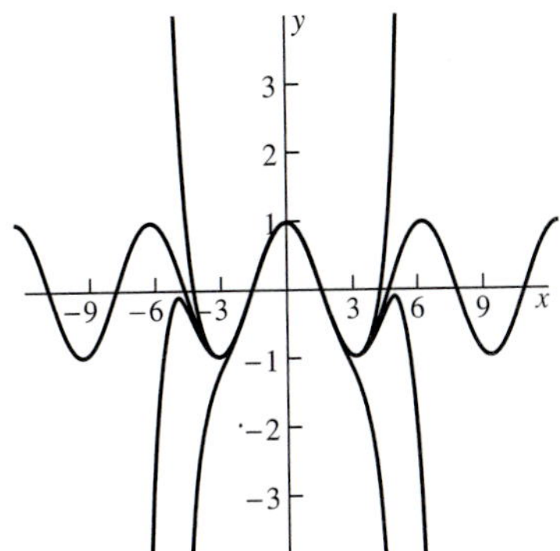

49. Here is a plot of $1/(1+x)$ and its Taylor polynomials, center zero, of degrees 2, 3, and 4.

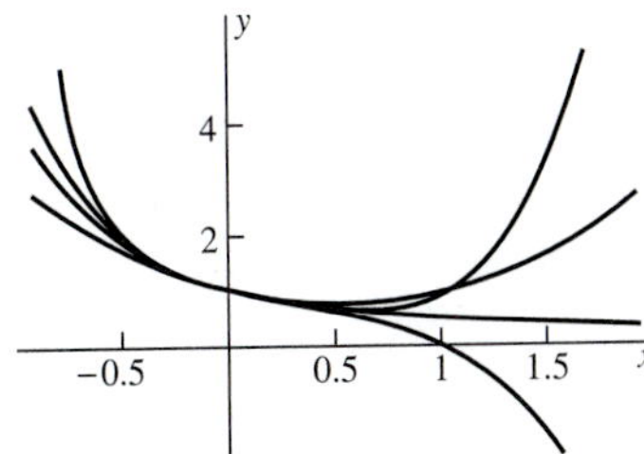

51. Here is a plot of $g(x)$ and the partial sums of the series for $f(x)$ of degrees 2, 3, and 4.

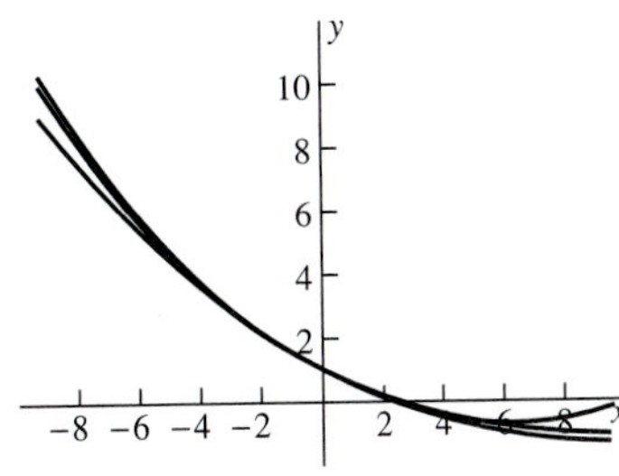

59. The first column of the table gives the degree of the partial sum, the second column its value using the Maclaurin series of Problem 56, and the third column its value using the Maclaurin series of the second series of Problem 58. It is clear that the latter series converges much more rapidly to ln 2.

5	0.783333333	0.693146047
10	0.645634921	0.693147181
15	0.725371850	0.693147181
20	0.668771403	0.693147181
25	0.712747500	0.693147181
30	0.676758138	0.693147181

Section 11.5 (page 665)

1. Diverges **3.** Diverges **5.** Converges
7. Diverges **9.** Converges **11.** Converges
13. Converges **15.** Converges **17.** Diverges
19. Converges **21.** Diverges **23.** Diverges
25. Converges **27.** Converges **29.** Converges
31. The terms are not nonnegative.
33. The terms are not decreasing.
35. $|p| > 1$ **37.** $p > 1$ **39.** $n > 10000$
41. $n > 100$ **43.** 2.62 **45.** 1.03693
47. $p > 1$ **49.** About a million centuries

Section 11.6 (page 672)

1. Converges **3.** Diverges **5.** Converges
7. Diverges **9.** Converges **11.** Converges
13. Diverges **15.** Converges **17.** Converges
19. Converges **21.** Converges **23.** Converges
25. Converges **27.** Diverges **29.** Diverges
31. Converges **33.** Converges **35.** Diverges
37. 0.98; 0.10 **39.** 0.53; 0.05 **41.** 10; 0.68
43. 4; 0.10

Section 11.7 (page 680)

1. Converges **3.** Diverges **5.** Diverges
7. Diverges **9.** Converges **11.** Converges
13. Converges **15.** Converges **17.** Diverges
19. Diverges **21.** Converges absolutely
23. Converges conditionally
25. Converges absolutely **27.** Converges absolutely
29. Diverges **31.** Converges absolutely
33. Converges conditionally
35. Diverges **37.** Diverges
39. Converges absolutely **41.** Converges absolutely
43. 0.90 **45.** 0.632
47. 0.7 **49.** $n = 6$; 0.947
51. $n = 5$ (six terms); 0.6065
53. $n = 4$ (five terms); 0.86603
63. $1 + \frac{1}{3} - \frac{1}{2} + \frac{1}{5} - \frac{1}{4} + \frac{1}{7} + \frac{1}{9} - \frac{1}{6} + \frac{1}{11} + \frac{1}{13} - \frac{1}{8} + \frac{1}{15} + \cdots$
65. The sum of the first 100,000 terms of the series is approximately -0.0000031249414, which (correctly) suggests that the sum of the series is zero.

Section 11.8 (page 691)

1. $(-1, 1)$ **3.** $(-2, 2)$ **5.** $\{0\}$
7. $\left[-\frac{1}{3}, \frac{1}{3}\right]$ **9.** $\left(-\frac{1}{2}, \frac{1}{2}\right)$ **11.** $[-2, 2]$
13. $(-3, 3)$ **15.** $\left(\frac{2}{5}, \frac{4}{5}\right)$ **17.** $\left[\frac{5}{2}, \frac{7}{2}\right]$
19. Converges only for $x = 0$ **21.** $(-4, 2)$
23. $[2, 4]$ **25.** Converges only for $x = 5$
27. $(-1, 1)$ **29.** $(-\infty, +\infty)$
31. $x + x^2 + x^3 + x^4 + \cdots$; $(-1, 1)$
33. $x^2 - \dfrac{3x^3}{1!} + \dfrac{3^2x^4}{2!} - \dfrac{3^3x^5}{3!} + \cdots$; $(\infty, +\infty)$
35. $x^2 - \dfrac{x^6}{3!} + \dfrac{x^{10}}{5!} - \dfrac{x^{14}}{7!} + \cdots$; $(-\infty, +\infty)$
37. $(1 - x)^{1/3} = 1 - \dfrac{x}{3} - \dfrac{2x^2}{2!3^2} - \dfrac{2 \cdot 5x^3}{3!3^3} - \dfrac{2 \cdot 5 \cdot 8x^4}{4!3^4} - \cdots$; $R = 1$
39. $f(x) = 1 - 3x + 6x^2 - 10x^3 + 15x^4 - \cdots$; $R = 1$
41. $f(x) = 1 - \dfrac{x}{2} + \dfrac{x^2}{3} - \dfrac{x^3}{4} + \dfrac{x^4}{5} - \cdots$; $R = 1$
43.
$$f(x) = \frac{x^4}{4} - \frac{x^{10}}{3!10} + \frac{x^{16}}{5!16} - \frac{x^{22}}{7!22} + \cdots = \sum_{n=0}^{\infty} \frac{(-1)^n}{(2n+1)!(6n+4)} x^{6n+4}$$
45.
$$f(x) = x - \frac{x^4}{4} + \frac{x^7}{2!7} - \frac{x^{10}}{3!10} + \frac{x^{13}}{4!13} - \cdots = \sum_{n=0}^{\infty} \frac{(-1)^n}{n!(3n+1)} x^{3n+1}$$
47.
$$f(x) = x - \frac{x^3}{2!3} + \frac{x^5}{3!5} - \frac{x^7}{4!7} + \frac{x^9}{5!9} - \cdots = \sum_{n=1}^{\infty} \frac{(-1)^{n+1}}{n!(2n-1)} x^{2n-1}$$
49. $\dfrac{x}{(1 - x)^2}$

51. $\dfrac{x(1+x)}{(1-x)^3}$

63. Using six terms: $3.14130878 < \pi < 3.1416744$

Section 11.9 (page 700)

1. $65^{1/3} = 4 \cdot (1 + \frac{1}{64})^{1/3}$. The first four terms of the binomial series give $65^{1/3} \approx 4.020726$; answer: 4.021.

3. Three terms of the usual sine series give 0.479427 with error less than 0.000002; answer: 0.479.

5. Five terms of the usual arctangent series give 0.463684 with error less than 0.000045; answer: 0.464.

7. 0.309 **9.** 0.174 **11.** 0.9461

13. 0.4872 **15.** 0.09761 **17.** 0.4438

19. 0.7468 **21.** 0.5133 **23.** $-\frac{1}{2}$

25. $\frac{1}{2}$ **27.** 0 **29.** 0.9848

31. 0.681998 **33.** Six places **35.** Five places

37. 1.3956 **41.** 8.9105 **43.** 15.3162

49. The first five coefficients are $1, 0, \frac{1}{2}, 0$, and $\frac{5}{24}$.

55. *Mathematica* Version 2.2.2 gives

$$\int_0^{1/2} \frac{1}{1+x^2+x^4}\,dx$$

$$= \left[\frac{1}{2\sqrt{3}}\arctan\left(\frac{2x-1}{\sqrt{3}}\right) + \frac{1}{2\sqrt{3}}\arctan\left(\frac{2x+1}{\sqrt{3}}\right) + \frac{1}{4}\ln\left(\frac{x^2+x+1}{x^2-x+1}\right)\right]_0^{1/2}$$

$$= \frac{1}{2\sqrt{3}}\arctan\left(\frac{2}{\sqrt{3}}\right) + \frac{1}{4}\ln\left(\frac{7}{3}\right) \approx 0.45924.$$

A series approximation is

$$\int_0^{1/2} \frac{1}{1+x^2+x^4}\,dx$$

$$= \int_0^{1/2} (1 - x^2 + x^6 - x^8 + x^{12} - x^{14} + \cdots + x^{60} - x^{62})\,dx$$

$\approx 0.459239825.$

57. The polynomial approximations are the graphs approaching $+\infty$ as $x \to \pm\infty$; they are the Taylor polynomials of degree 2, 4, and 6.

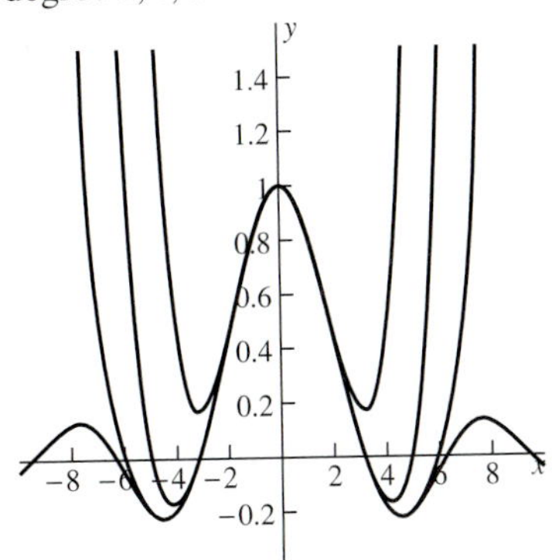

59. We used

$$\sin x - \tan x \approx -\tfrac{1}{2}x^3 - \tfrac{1}{8}x^5 - \tfrac{13}{240}x^7$$

and

$$\arcsin x - \arctan x \approx \tfrac{1}{2}x^3 - \tfrac{1}{8}x^5 + \tfrac{3}{16}x^7$$

to obtain

$$\frac{\sin x - \tan x}{\arcsin x - \arctan x} \approx \frac{120 + 30x^2 + 13x^4}{-120 + 30x^2 - 45x^4} \to -1$$

as $x \to 0$.

Chapter 11 Miscellaneous Problems (page 703)

1. 1 **3.** 10 **5.** 0

7. 0 **9.** No limit **11.** 0

13. $+\infty$ (or "no limit") **15.** 1

17. Converges **19.** Converges **21.** Converges

23. Diverges **25.** Converges **27.** Converges

29. Diverges **31.** $(-\infty, +\infty)$ **33.** $[-2, 4)$

35. $[-1, 1]$ **37.** Converges only for $x = 0$

39. $(-\infty, +\infty)$ **41.** Converges for *no* x

43. Converges for all x

51. Seven terms of the binomial series give 1.084.

53. 0.461 **55.** 0.797

65. $$\sqrt{5} = 2 + \cfrac{1}{4 + \cfrac{1}{4 + \cfrac{1}{4 + \cfrac{1}{4 + \cfrac{1}{4 + \cdots}}}}}$$

Section 12.1 (page 713)

1. $\mathbf{v} = \langle 2, 3 \rangle$ **3.** $\mathbf{v} = \langle -10, -20 \rangle$

5. $\mathbf{u} + \mathbf{v} = \langle 4, 2 \rangle$ **7.** $\mathbf{u} + \mathbf{v} = 5\mathbf{i} - 2\mathbf{j}$

9. $\sqrt{5}, 2\sqrt{13}, 4\sqrt{2}, \langle -2, 0 \rangle, \langle 9, -10 \rangle$

11. $2\sqrt{2}, 10, \sqrt{5}, \langle -5, -6 \rangle, \langle 0, 2 \rangle$

13. $\sqrt{10}, 2\sqrt{29}, \sqrt{65}; 3\mathbf{i} - 2\mathbf{j}; -\mathbf{i} + 19\mathbf{j}$

15. $4, 14, \sqrt{65}; 4\mathbf{i} - 7\mathbf{j}, 12\mathbf{i} + 14\mathbf{j}$

17. $\mathbf{u} = \langle -\frac{3}{5}, -\frac{4}{5} \rangle; \mathbf{v} = \langle \frac{3}{5}, \frac{4}{5} \rangle$

19. $\mathbf{u} = \frac{8}{17}\mathbf{i} + \frac{15}{17}\mathbf{j}; \mathbf{v} = -\frac{8}{17}\mathbf{i} - \frac{15}{17}\mathbf{j}$

21. $-4\mathbf{j}$ **23.** $8\mathbf{i} - 14\mathbf{j}$ **25.** Yes

27. No **29.** $\mathbf{i} = 3\mathbf{b} - 4\mathbf{a}, \mathbf{j} = 3\mathbf{a} - 2\mathbf{b}$

31. $\mathbf{c} = -\frac{1}{2}\mathbf{a} + \frac{5}{2}\mathbf{b}$ **33.** (a) $15\mathbf{i} - 21\mathbf{j}$; (b) $\frac{5}{3}\mathbf{i} - \frac{7}{3}\mathbf{j}$

35. (a) $\frac{35}{58}\mathbf{i}\sqrt{58} - \frac{15}{58}\mathbf{j}\sqrt{58}$; (b) $-\frac{40}{89}\mathbf{i}\sqrt{89} - \frac{25}{89}\mathbf{j}\sqrt{89}$

37. $c = 0$

43. The tension in each cable is 100 lb.

45. The tension in the right-hand cable is approximately 71.970925645 lb, and the tension in the left-hand cable is approximately 96.121326055 lb.

47. $\mathbf{v}_a = (500 + 25\sqrt{2})\mathbf{i} + 25\mathbf{j}\sqrt{2}$

49. $\mathbf{v}_a = -225\mathbf{i}\sqrt{2} + 275\mathbf{j}\sqrt{2}$

55. To rotate the vector **a** counterclockwise through an angle of 90°, add $\pi/2$ to θ.

Section 12.2 (page 723)

1. (a) $\langle 5, 8, -11\rangle$; (b) $\langle 2, 23, 0\rangle$; (c) 4; (d) $\sqrt{51}$; (e) $(\frac{1}{15}\sqrt{5})\langle 2, 5, -4\rangle$

3. (a) $2\mathbf{i} + 3\mathbf{j} + \mathbf{k}$; (b) $3\mathbf{i} - \mathbf{j} + 7\mathbf{k}$; (c) 0; (d) $\sqrt{5}$; (e) $(\frac{1}{3}\sqrt{3})(\mathbf{i} + \mathbf{j} + \mathbf{k})$

5. (a) $4\mathbf{i} - \mathbf{j} - 3\mathbf{k}$; (b) $6\mathbf{i} - 7\mathbf{j} + 12\mathbf{k}$; (c) -1; (d) $\sqrt{17}$; (e) $(\frac{1}{5}\sqrt{5})(2\mathbf{i} - \mathbf{j})$

7. 81° **9.** 90° **11.** 98°

13. $\text{comp}_\mathbf{b}\,\mathbf{a} = \frac{2}{7}\sqrt{14}$, $\text{comp}_\mathbf{a}\,\mathbf{b} = \frac{4}{15}\sqrt{5}$

15. $\text{comp}_\mathbf{b}\,\mathbf{a} = 0$, $\text{comp}_\mathbf{a}\,\mathbf{b} = 0$

17. $\text{comp}_\mathbf{b}\,\mathbf{a} = -\frac{1}{10}\sqrt{10}$, $\text{comp}_\mathbf{a}\,\mathbf{b} = -\frac{1}{5}\sqrt{5}$

19. $x^2 + y^2 + z^2 - 6x - 2y - 4z = 11$

21. $x^2 + y^2 + z^2 - 10x - 8y + 2z + 33 = 0$

23. $x^2 + y^2 + z^2 - 4z = 0$

25. Center $(-2, 3, 0)$, radius $\sqrt{13}$

27. Center $(0, 0, 3)$, radius 5

29. The xy-plane

31. The horizontal plane (parallel to the xy-plane) meeting the z-axis at the point $(0, 0, 10)$

33. All points on any of the three coordinate planes

35. The origin $(0, 0, 0)$

37. The single point $(3, -4, 0)$

39. They are parallel: $\mathbf{b} = \frac{3}{2}\mathbf{a}$.

41. They are parallel: $\mathbf{b} = -\frac{3}{4}\mathbf{a}$.

43. Yes, on one line, because $\overrightarrow{QR} = 3\overrightarrow{PQ}$.

45. All angles 60° exactly

47. 79°, 64°, 37°

49. [Approximately] 74.2068°, 47.1240°, 47.1240°

51. [Approximately] 64.8959°, 55.5501°; [exactly] 45°

53. 3

55. Approximately 7323.385 calories

57. If there's no friction, the work done is mgh.

61. $\mathbf{w} = \langle w_1, -\frac{7}{2}w_1, -2w_1\rangle$ where w_1 is arbitrary but nonzero

67. $2x + 9y - 5z = 23$; the plane through the midpoint of the segment AB perpendicular to AB

69. 60°

Section 12.3 (page 732)

1. $\langle 0, -14, 7\rangle$ **3.** $\langle -10, -7, 1\rangle$

5. $\langle 0, 0, 22\rangle$ **7.** $\pm\frac{1}{13}\langle 12, -3, 4\rangle$

11. $(\mathbf{a} \times \mathbf{b}) \times \mathbf{c} = \langle -1, 1, 0\rangle$, $\mathbf{a} \times (\mathbf{b} \times \mathbf{c}) = \langle 0, 0, -1\rangle$

15. $A = \frac{1}{2}\sqrt{2546} \approx 25.229$ **17.** (a) 55; (b) $\frac{55}{6}$

19. Coplanar **21.** Not coplanar; 1

23. 4395.657 (m^2) **25.** 31271.643 (ft^2)

29. (b) $\frac{1}{38}\sqrt{9842} \approx 2.6107$

Section 12.4 (page 740)

1. $x = t,\ y = 2t,\ z = 3t$

3. $x = 4 + 2t,\ y = 13,\ z = -3 - 3t$

5. $x = -6t,\ y = 3t,\ z = 5t$

7. $x = 3 + 3t,\ y = 5,\ z = 7 - 3t$

9. $x = 2 + t,\ y = 3 - t,\ z = -4 - 2t$; $x - 2 = -y + 3 = \dfrac{-z - 4}{2}$

11. $x = 1,\ y = 1,\ z = 1 + t$; $x - 1 = 0 = y - 1$, z arbitrary

13. $x = 2 + 2t,\ y = -3 - t,\ z = 4 + 3t$; $\dfrac{x - 2}{2} = -y - 3 = \dfrac{z - 4}{3}$

15. The lines meet at the single point $(2, -1, 3)$.

17. These are skew lines.

19. These lines are parallel (and do not intersect).

21. $x + 2y + 3z = 0$ **23.** $x - z + 8 = 0$

25. $y = 7$ **27.** $7x + 11y = 114$

29. $3x + 4y - z = 0$ **31.** $2x - y - z = 0$

33. $2x - 7y + 17z = 78$

35. The line and the plane are parallel (and do not intersect).

37. They meet at the single point $(\frac{9}{2}, \frac{9}{4}, \frac{17}{4})$.

39. $\theta = \cos^{-1}\left(\dfrac{1}{\sqrt{3}}\right) \approx 54.736°$

41. The planes are parallel: $\theta = 10$.

43. Technically, the line doesn't have symmetric equations; elimination of the parameter yields the Cartesian equations $x = 10$, $y + z = 0$.

45. The two planes are parallel and do not coincide, so there is no line of intersection.

47. $\dfrac{x - 3}{2} = y - 3 = \dfrac{-z + 1}{5}$

49. $3x + 2y + z = 6$ **51.** $7x - 5y - 2z = 9$

53. $x - 2y + 4z = 3$ **55.** $\frac{10}{3}\sqrt{3}$

59. (b) $\dfrac{133}{\sqrt{501}} \approx 5.942$

61. L and $\mathcal{P}$ are parallel (and have no points in common).

63. L and $\mathcal{P}$ meet at the single point $(7, 1, 1)$. The angle between them is approximately 40.367°.

65. The planes coincide.

67. The planes meet at an angle of about 33.41°. Parametric equations of their line of intersection are $x = -39t$, $y = 235t$, $z = 64t$.

Section 12.5 (page 753)

1. Matches Fig. 12.5.17. **3.** Matches Fig. 12.5.16.

5. $\mathbf{0}, \mathbf{0}$ **7.** $2\mathbf{i} - \mathbf{j}, 4\mathbf{i} + \mathbf{j}$ **9.** $6\pi\mathbf{i}, 12\pi^2\mathbf{j}$

11. $\mathbf{v} = \mathbf{i} + 2t\mathbf{j} + 3t^2\mathbf{k}$, $\mathbf{a} = 2\mathbf{j} + 6t\mathbf{k}$, $v = \sqrt{1 + 4t^2 + 9t^4}$

13. $\mathbf{v} = \mathbf{i} + 3e^t\mathbf{j} + 4e^t\mathbf{k}$, $\mathbf{a} = 3e^t\mathbf{j} + 4e^t\mathbf{k}$, $v = \sqrt{1 + 25e^{2t}}$

15. $\mathbf{v} = (-3 \sin t)\mathbf{i} + (3 \cos t)\mathbf{j} - 4\mathbf{k}$, $\mathbf{a} = (-3 \cos t)\mathbf{i} - (3 \sin t)\mathbf{j}$, $v = 5$

17. $\frac{1}{2}(2 - \sqrt{2})\mathbf{i} + \mathbf{j}\sqrt{2}$ **19.** $\frac{484}{15}\mathbf{i}$

21. 11 **23.** 0

25. $\mathbf{r}(t) = \langle 1, 0, t \rangle$ **27.** $\mathbf{r}(t) = t^2\mathbf{i} + 10t\mathbf{j} - 2t^2\mathbf{k}$

29. $\mathbf{r}(t) = 2\mathbf{i} + t^2\mathbf{j} + (5t - t^3)\mathbf{k}$

31. $\mathbf{r}(t) = (10 + \frac{1}{6}t^3)\mathbf{i} + (10t + \frac{1}{12}t^4)\mathbf{j} + \frac{1}{20}t^5\mathbf{k}$

33. $\mathbf{r}(t) = (1 - t - \cos t)\mathbf{i} + (1 + t - \sin t)\mathbf{j} + 5t\mathbf{k}$

35. $\mathbf{v} = (3\sqrt{2})(\mathbf{i} + \mathbf{j}) + 8\mathbf{k}$, $v = 10$, $\mathbf{a} = (6\sqrt{2})(-\mathbf{i} + \mathbf{j})$

39. *Suggestion*: Compute $\dfrac{d}{dt}(\mathbf{r} \cdot \mathbf{r})$.

41. 100 (ft), $\sqrt{6400 + 25} \approx 80.15609$ (ft/s)

43. $v_0 = 32\sqrt{165} \approx 411.047$ (ft/s)

47. (a) 100 ft, $400\sqrt{3}$ ft; (b) 200, 800; (c) 300, $400\sqrt{3}$

49. $70\sqrt{10} \approx 221.36$ (m/s)

51. Inclination angle

$$\alpha = \arctan\left(\frac{8049 - 280\sqrt{20}}{8000}\right) \approx 0.730293 \text{ (about}$$

41° 50′ 34″); initial velocity

$$v_0 = \frac{5600}{(20\sqrt{10} - 7)\cos\alpha} \approx 133.64595 \text{ (m/s)}$$

55. Begin with $\dfrac{d}{dt}(\mathbf{v} \cdot \mathbf{v}) = 0$.

57. A repulsive force acting directly away from the origin, with magnitude proportional to distance from the origin.

63. 5 ft

65. (a) $\mathbf{v}(t) = \langle 2t, 200, 160 - 32t \rangle$; $\mathbf{r}(t) = \langle t^2, 200t, 384 + 160t - 16t^2 \rangle$. (b) 12 s. (c) 2400 ft north, 144 ft east. (d) 784 ft.

Section 12.6 (page 770)

1. 10π **3.** $19(e - 1) \approx 32.647$

5. $2 + \frac{9}{10}\ln 3 \approx 2.9888$ **7.** 0 **9.** 1

11. $\dfrac{40\sqrt{2}}{41\sqrt{41}} \approx 0.2155$ **13.** At $(-\frac{1}{2}\ln 2, \frac{1}{2}\sqrt{2})$

15. Maximum curvature $\frac{5}{9}$ at $(\pm 5, 0)$, minimum curvature $\frac{3}{25}$ at $(0, \pm 3)$

17. $\mathbf{T} = (\frac{1}{10}\sqrt{10})(\mathbf{i} + 3\mathbf{j})$, $\mathbf{N} = (\frac{1}{10}\sqrt{10})(3\mathbf{i} - \mathbf{j})$

19. $\mathbf{T} = (\frac{1}{57}\sqrt{57})(3\mathbf{i} - 4\mathbf{j}\sqrt{3})$, $\mathbf{N} = (\frac{1}{57}\sqrt{57})(-4\mathbf{i}\sqrt{3} - 3\mathbf{j})$

21. $\mathbf{T} = (-\frac{1}{2}\sqrt{2})(\mathbf{i} + \mathbf{j})$, $\mathbf{N} = (\frac{1}{2}\sqrt{2})(-\mathbf{i} + \mathbf{j})$

23. $a_T = \dfrac{18t}{\sqrt{9t^2 + 1}}$, $a_N = \dfrac{6}{\sqrt{9t^2 + 1}}$

25. $a_T = \dfrac{t}{\sqrt{1 + t^2}}$, $a_N = \dfrac{2 + t^2}{\sqrt{1 + t^2}}$

27. $\dfrac{1}{a}$ **29.** $x^2 + (y - \frac{1}{2})^2 = \frac{1}{4}$

31. $(x - 2)^2 + (y - 2)^2 = 2$

33. $\frac{1}{2}$ **35.** $\frac{1}{3}e^{-t}\sqrt{2}$

37. $a_T = 0 = a_N$

39. $a_T = \dfrac{4t + 18t^3}{\sqrt{1 + 4t^2 + 9t^4}}$, $a_N = \dfrac{2\sqrt{1 + 9t^2 + 9t^4}}{\sqrt{1 + 4t^2 + 9t^4}}$

41. $a_T = \dfrac{t}{\sqrt{t^2 + 2}}$, $a_N = \dfrac{\sqrt{t^4 + 5t^2 + 8}}{\sqrt{t^2 + 2}}$

43. $\mathbf{T} = \left(\frac{1}{2}\sqrt{2}\right)\langle 1, \cos t, -\sin t \rangle$, $\mathbf{N} = \langle 0, -\sin t, -\cos t \rangle$; at $(0, 0, 1)$, $\mathbf{T} = (\frac{1}{2}\sqrt{2})\langle 1, 1, 0 \rangle$, $\mathbf{N} = \langle 0, 0, -1 \rangle$

45. $\mathbf{T} = \left(\frac{1}{3}\sqrt{3}\right)(\mathbf{i} + \mathbf{j} + \mathbf{k})$, $\mathbf{N} = \left(\frac{1}{2}\sqrt{2}\right)(-\mathbf{i} + \mathbf{j})$

47. $x = 2 + \frac{4}{13}s$, $y = 1 - \frac{12}{13}s$, $z = 3 + \frac{3}{13}s$

49. $x(s) = 3 \cos \frac{1}{5}s$, $y(s) = 3 \sin \frac{1}{5}s$, $z(s) = \frac{4}{5}s$

51. Begin with $\dfrac{d}{dt}(\mathbf{v} \cdot \mathbf{v})$. **53.** $\dfrac{1}{|t|}$

55. $A = 3$, $B = -8$, $C = 6$, $D = E = F = 0$

57. 36.65 mi/s, 24.13 mi/s **59.** 0.672 mi/s, 0.602 mi/s

61. Approximately −795 mi, and thus it's not possible

63. About 1.962 h

Section 12.7 (page 782)

1. A plane with intercepts $\left(\frac{20}{3}, 0, 0\right)$, $(0, 10, 0)$, and $(0, 0, 2)$

3. A vertical circular cylinder with radius 3

5. A vertical cylinder intersecting the xy-plane in the rectangular hyperbola $xy = 4$

7. An elliptical paraboloid opening upward from its vertex at the origin

9. A circular paraboloid opening downward from its vertex at $(0, 0, 4)$

11. A paraboloid opening upward, vertex at the origin, axis the z-axis

13. A cone, vertex the origin, axis the z-axis (both nappes)

15. A parabolic cylinder perpendicular to the xz-plane, its trace there the parabola opening upward with axis the z-axis and vertex at $(x, z) = (0, -2)$

17. An elliptical cylinder perpendicular to the xy-plane, its trace there the ellipse with center $(0, 0)$ and intercepts $(\pm 1, 0)$ and $(0, \pm 2)$

19. An elliptical cone, vertex $(0, 0, 0)$, axis the x-axis

21. A paraboloid opening downward, vertex at the origin, axis the z-axis

23. A hyperbolic paraboloid, saddle point at the origin, meeting the xz-plane in a parabola with vertex the origin and opening downward, meeting the xy-plane in a parabola with vertex the origin and opening upward, meeting each plane parallel to the yz-plane in a hyperbola with directrices parallel to the y-axis

25. A hyperboloid of one sheet, axis the z-axis, trace in the xy-plane the circle with center $(0, 0)$ and radius 3, traces in parallel planes larger circles, and traces in planes parallel to the z-axis hyperbolas

27. An elliptical paraboloid, axis the y-axis, vertex at the origin

29. A hyperboloid of two sheets, axis the y-axis

31. A paraboloid, axis the x-axis, vertex at the origin, equation $x = 2(y^2 + z^2)$

33. Hyperboloid of one sheet, equation $x^2 + y^2 - z^2 = 1$

35. A paraboloid, vertex at the origin, axis the x-axis, equation $y^2 + z^2 = 4x$

37. The surface resembles a rug covering a turtle: highest point $(0, 0, 1)$; $z \to 0$ from above as $|x|$ or $|y|$ (or both) increase without bound; equation $z = \exp(-x^2 - y^2)$.

39. A circular cone with axis of symmetry the z-axis

41. Ellipses with semiaxes 2 and 1

43. Circles

45. Parabolas opening upward

47. Parabolas opening upward if $k > 0$, downward if $k < 0$

51. The projection of the intersection has equation $x^2 + y^2 = y$; it is the circle with center $(0, \frac{1}{2})$ and radius $\frac{1}{2}$.

53. Equation: $5x^2 + 8xy + 8y^2 - 8x - 8y = 0$. The roots $\lambda_1 \approx 2.23$ and $\lambda_2 \approx 10.77$ are both positive, so by Theorem 1 the curve is a rotated ellipse.

55. $\lambda_1 = 2, \lambda_2 = 4$; ellipse **57.** $\lambda_1 = 5, \lambda_2 = 10$; ellipse

59. $\lambda_1 = 25, \lambda_2 = 50$; ellipse

61. $\lambda_1 = 2, \lambda_2 = 4$; elliptic paraboloid

63. $\lambda_1 = -5, \lambda_2 = 5$; hyperbolic paraboloid

65. $\lambda_1 = 17, \lambda_2 = 34$; elliptic paraboloid

67. $\lambda_1 = -2, \lambda_2 = 2, \lambda_3 = 4$; hyperboloid

69. $\lambda_1 = -5, \lambda_2 = 0, \lambda_3 = 5$; hyperbolic cylinder

71. $\lambda_1 \approx -1.14644, \lambda_2 \approx 1.78156, \lambda_3 \approx 6.36488$; hyperboloid

Section 12.8 (page 789)

1. $(0, 1, 2)$ **3.** $(-\sqrt{2}, \sqrt{2}, 3)$ **5.** $(1, \sqrt{3}, -5)$

7. $(0, 0, 2)$ **9.** $(-3, 0, 0)$ **11.** $(0, -\sqrt{3}, 1)$

13. $(0, 0, 5)_{\text{cyl}}$, $(5, 0, 0)_{\text{sph}}$

15. $\left(\sqrt{2}, \frac{\pi}{4}, 0\right)_{\text{cyl}}$, $\left(\sqrt{2}, \frac{\pi}{2}, \frac{\pi}{4}\right)_{\text{sph}}$

17. $\left(\sqrt{2}, \frac{\pi}{4}, 1\right)_{\text{cyl}}$, $\left(\sqrt{3}, \tan^{-1}\sqrt{2}, \frac{\pi}{4}\right)_{\text{sph}}$

19. $(\sqrt{5}, \tan^{-1}\frac{1}{2}, -2)_{\text{cyl}}$,
$\left(3, \frac{\pi}{2} + \tan^{-1}(\frac{2}{5}\sqrt{5}), \tan^{-1}(\frac{1}{2})\right)_{\text{sph}}$

21. $(5, \tan^{-1}\frac{4}{3}, 12)_{\text{cyl}}$, $(13, \tan^{-1}\frac{5}{12}, \tan^{-1}\frac{4}{3})_{\text{sph}}$

23. A cylinder, radius 5, axis the z-axis

25. The *plane* $y = x$

27. The upper nappe of the cone $x^2 + y^2 = 3z^2$

29. The xy-plane

31. An ellipsoidal surface centered at $(0, 0, 0)$

33. The vertical circular cylinder $(x - 2)^2 + y^2 = 2^2$

35. Two cylinders concentric around the z-axis, one of radius 1, the other of radius 3

37. Because $z = \pm r^2$, we have two paraboloids symmetric around the z-axis, each with vertex at $(0, 0, 0)$, one opening upward, the other downward.

39. $r^2 + z^2 = 25$; $\rho = 5$

41. $r(\cos\theta + \sin\theta) + z = 1$;
$\rho(\sin\phi\cos\theta + \sin\phi\sin\theta + \cos\phi) = 1$

43. $r^2 + z^2 = r(\cos\theta + \sin\theta) + z$;
$\rho = \sin\phi\cos\theta + \sin\phi\sin\theta + \cos\phi$

45. The segment of the vertical cylinder $r = 3$ between the horizontal planes $z = -1$ and $z = 1$. One way to see this surface: Execute the *Mathematica* command

```
ParametricPlot3D[ {3*Cos[t], 3*Sin[t], z},
   {t, 0, 2*Pi}, {z, -1, 1} ];
```

47. The segment of the spherical surface with radius 2 and center $(0, 0, 0)$ that lies between the horizontal planes $z = -1$ and $z = 1$. One way to see this surface: Execute the *Mathematica* command

```
ParametricPlot3D[ {2*Sin[phi]*Cos[theta],
2*Sin[phi]*Sin[theta],
2*Cos[phi]}, {phi, Pi/3, 2*Pi/3}, {theta,
0, 2*Pi}];
```

49. The region between the concentric cylinders $r = 1$ and $r = 3$ and between the horizontal planes $z = -2$ and $z = 2$. One way to see this solid: Execute the *Mathematica* commands

```
p1 = ParametricPlot3D[ {3*Cos[t], 3*Sin[t], z},
   {t, 0, 2*Pi}, {z, -2, 2}];
p2 = ParametricPlot3D[ {r*Cos[t], r*Sin[t], 2},
   {t, 0, 2*Pi}, {r, 1, 3} ];
p3 = ParametricPlot3D[ {Cos[t], Sin[t], z},
   {t, 0, 2*Pi}, {z, -2, 2} ];
Show[ p1, p2, p3]
```

51. This is the solid region between two spherical surfaces centered at the origin, the inner surface of radius 3, the outer of radius 5. To see the bottom half of this solid, execute the *Mathematica* commands

```
p1 = ParametricPlot3D[ {3*Sin[phi]*Cos[theta],
   3*Sin[phi]*Sin[theta], 3*Cos[phi]}, {phi,
   Pi/2, Pi}, {theta, 0, 2*Pi} ];
p2 = ParametricPlot3D[ {5*Sin[phi]*Cos[theta],
   5*Sin[phi]*Sin[theta],5*Cos[phi]}, {phi,
   Pi/2, Pi},{theta, 0, 2*Pi} ];
p3 = ParametricPlot3D[ {r*Cos[theta],
   r*Sin[theta], 0},{r, 3, 5}, {t, 0, 2*Pi} ];
Show[ p1, p2, p3 ]
```

53. $z = r^2$

55. (a) $1 \leqq r^2 \leqq 4 - z^2$; (b) $\csc \phi \leqq \rho \leqq 2$ (and, as a consequence, $\frac{1}{6}\pi \leqq \phi \leqq \frac{5}{6}\pi$)

57. About 3821 mi (about 6149 km)

59. Just under 50 km (about 31 mi)

61. $0 \leqq \rho \leqq H \sec \phi$, $0 \leqq \phi \leqq \arctan(R/H)$, θ arbitrary

Chapter 12 Miscellaneous Problems (page 791)

7. $x = 1 + 2t$, $y = -1 + 3t$, $z = 2 - 3t$; $\dfrac{x-1}{2} = \dfrac{y+1}{3} = \dfrac{-z+2}{3}$

9. $-13x + 22y + 6z = -23$

11. $x - y + 2z = 3$

15. 3

21. Two solutions: $\alpha \approx 0.033364$ rad (about $1°54'53''$) and $\alpha \approx 1.29116$ rad (about $73°58'40''$)

23. $\frac{1}{9}$; $a_T = 2$, $a_N = 1$

27. $3x - 3y + z = 1$

33. $\rho = 2\cos\phi$

35. $\rho^2 = 2\cos 2\phi$; shaped like an hourglass with rounded ends

43. The curvature is zero when x is an integral multiple of π and reaches the maximum value 1 when x is an odd integral multiple of $\pi/2$.

45. $\mathbf{N} = -\dfrac{2}{\sqrt{\pi^2+4}}\mathbf{i} - \dfrac{\pi}{\sqrt{\pi^2+4}}\mathbf{j}$,

$\mathbf{T} = -\dfrac{\pi}{\sqrt{\pi^2+4}}\mathbf{i} + \dfrac{2}{\sqrt{\pi^2+4}}\mathbf{j}$

49. $A = \frac{15}{8}$, $B = -\frac{5}{4}$, $C = \frac{3}{8}$

51. This is a variation of Problem 36.

Section 13.2 (page 803)

1. The xy-plane

3. All points of the xy-plane other than $(0, 0)$

5. All points of the xy-plane

7. All points (x, y) such that $x^2 + y^2 \leqq 1$; that is, all points on and within the unit circle centered at the origin

9. All (x, y)

11. All points of the xy-plane for which $y > x$; that is, all points above the line with equation $y = x$

13. Except on the coordinate axes $x = 0$ and $y = 0$

15. Except on the lines $y = \pm x$

17. Except at the origin $(0, 0, 0)$

19. All points (x, y, z) of space for which $x^2 + y^2 < z$; that is, all points below the paraboloid with equation $z = x^2 + y^2$

21. The horizontal plane with equation $z = 10$

23. A plane that makes a $45°$ angle with the xy-plane, intersecting it in the line $x + y = 0$

25. A circular paraboloid opening upward from its vertex at the origin

27. The upper hemispherical surface of radius 2 centered at the origin

29. A circular cone opening downward from its vertex at $(0, 0, 10)$

31. Straight lines of slope 1

33. Ellipses centered at $(0, 0)$, each with major axis twice the minor axis and lying on the x-axis

35. Vertical (y-direction) translates of the curve $y = x^3$

37. Circles centered at the point $(2, 0)$

39. Circles centered at the origin

41. Circular paraboloids opening upward, each with its vertex on the z-axis

43. Spheres centered at $(2, 1, 3)$

45. Elliptical cylinders, each with axis the vertical line through $(2, 1, 0)$ and with the length of the x-semiaxis twice that of the y-semiaxis

47.

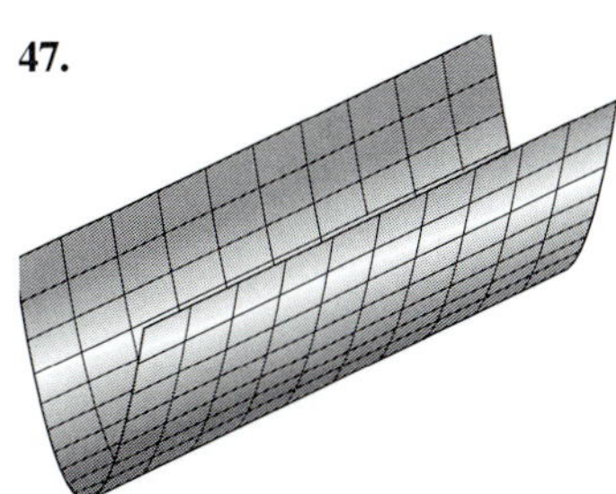

49.

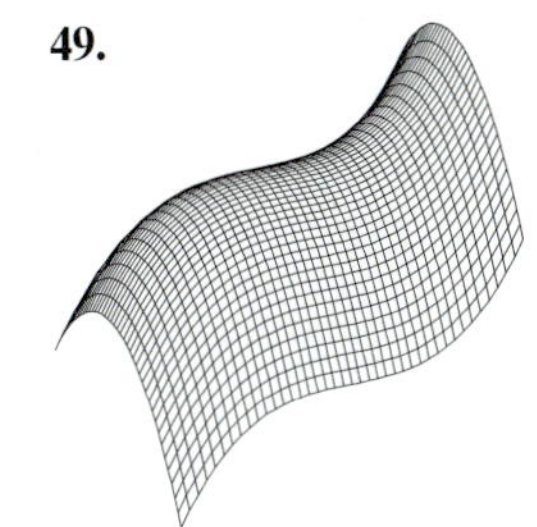

51.

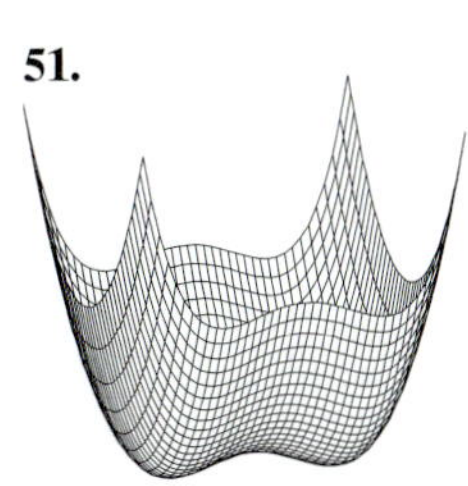

53.

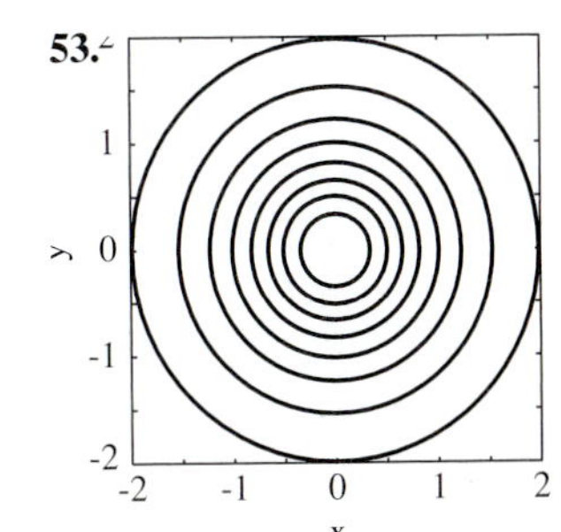

55.

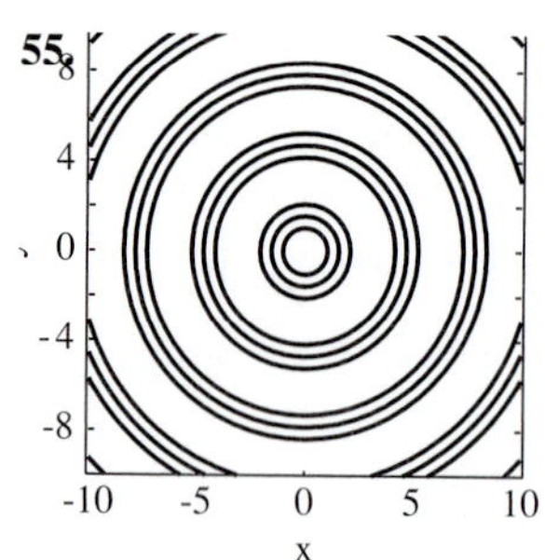

57.

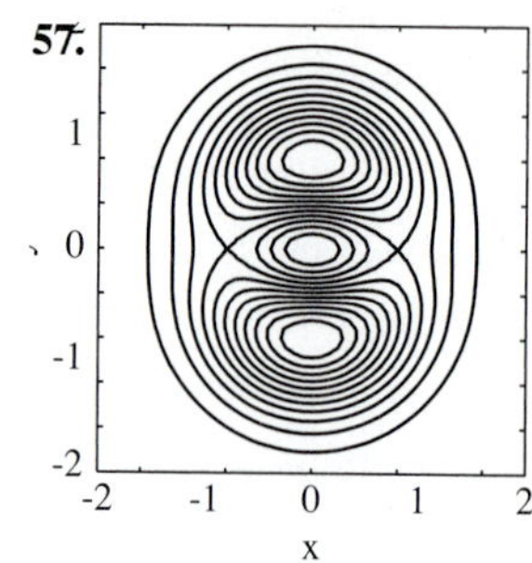

Section 13.3 (pages 811)

1. 7 **3.** e **5.** $\frac{5}{3}$ **7.** 0

9. 1 **11.** $-\frac{3}{2}$ **13.** 1 **15.** -4

17. y, x **19.** $y^2, 2xy$ **21.** 0 **23.** $\frac{1}{3}$

25. 0 **27.** 1 **29.** 0

31. All points above (not "on") the graph of $y = -x$

33. All points *outside* the circle with equation $x^2 + y^2 = 1$

35. All points in the xy-plane other than the origin $(0, 0)$

41. This limit does not exist.

43. This limit does not exist.

Section 13.4 (page 820)

1. $\frac{\partial f}{\partial x} = 4x^3 - 3x^2y + 2xy^2 - y^3$,

$\frac{\partial f}{\partial y} = -x^3 + 2x^2y - 3xy^2 + 4y^3$

3. $\frac{\partial f}{\partial x} = e^x(\cos y - \sin y), \frac{\partial f}{\partial y} = -e^x(\cos y + \sin y)$

5. $\frac{\partial f}{\partial x} = -\frac{2y}{(x-y)^2}, \frac{\partial f}{\partial y} = \frac{2x}{(x-y)^2}$

7. $\frac{\partial f}{\partial x} = \frac{2x}{x^2+y^2}, \frac{\partial f}{\partial y} = \frac{2y}{x^2+y^2}$

9. $\frac{\partial f}{\partial x} = yx^{y-1}, \frac{\partial f}{\partial y} = x^y \ln x$

11. $\frac{\partial f}{\partial x} = 2xy^3z^4, \frac{\partial f}{\partial y} = 3x^2y^2z^4, \frac{\partial f}{\partial z} = 4x^2y^3z^3$

13. $\frac{\partial f}{\partial x} = yze^{xyz}, \frac{\partial f}{\partial y} = xze^{xyz}, \frac{\partial f}{\partial z} = xye^{xyz}$

15. $\frac{\partial f}{\partial x} = 2xe^y \ln z, \frac{\partial f}{\partial y} = x^2e^y \ln z, \frac{\partial f}{\partial z} = \frac{x^2e^y}{z}$

17. $\frac{\partial f}{\partial r} = \frac{4rs^2}{(r^2+s^2)^2}, \frac{\partial f}{\partial s} = -\frac{4r^2s}{(r^2+s^2)^2}$

19. $\frac{\partial f}{\partial u} = e^v + we^u, \frac{\partial f}{\partial v} = e^w + ue^v, \frac{\partial f}{\partial w} = e^u + ve^w$

21. $z_{xy} = z_{yx} \equiv -4$

23. $z_{xy} = z_{yx} = -4xy\exp(-y^2)$

25. $z_{xy} = z_{yx} = \frac{1}{(x+y)^2}$

27. $z_{xy} = z_{yx} = 3e^{-3x} \sin y$

29. $z_{xy} = z_{yx} = -4xy^{-3} \sinh(y^{-2})$

31. $6x + 8y - z = 25$ **33.** $z \equiv 1$

35. $27x - 12y - z = 38$ **37.** $x - y + z = 1$

39. $10x - 16y - z = 9$ **41.** $f(x, y) = x^2y^3 + C$

43. There is no such function f.

45.

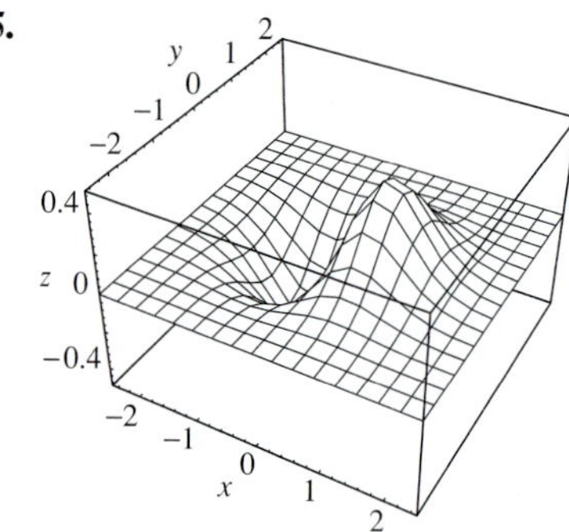

47.

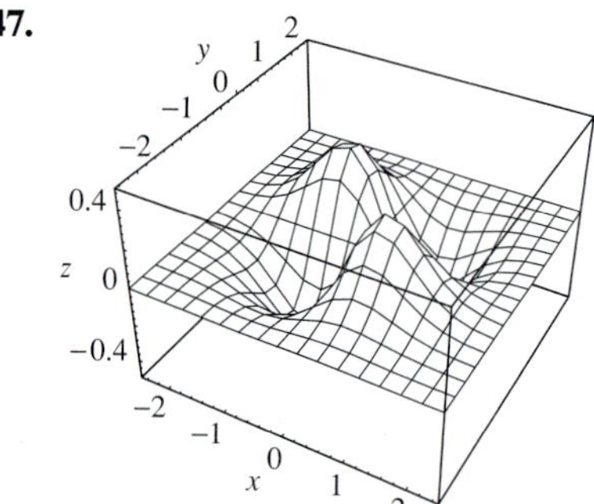

49.

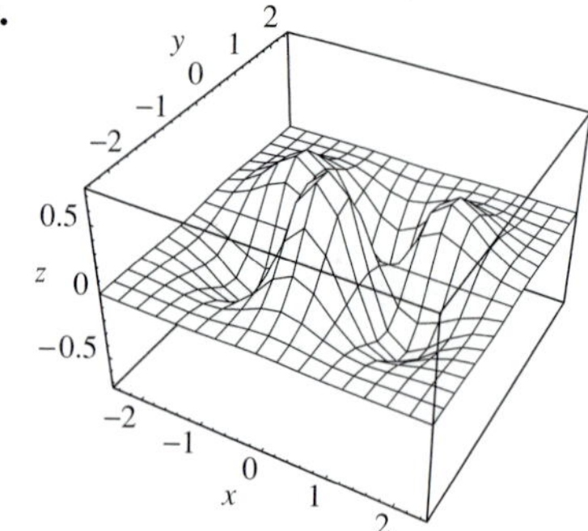

53. $f_{xyz}(x, y, z) = (x^2y^2z^2 + 3xyz + 1)e^{xyz}$

65. $(10, -7, -58)$

67. (a) A decrease of about 2750 cm^3; (b) an increase of about 82.5 cm^3

71. (a) Descending at an angle of 45°; (b) ascending at an angle of 45°

Section 13.5 (page 831)

1. None
3. (0, 0, 5)
5. (3, −1, −5)
7. (−2, 0, −4)
9. (−2, 0, −7) and (−2, 1, −9)
11. (0, 0, 0), (1, 0, 2/e), (−1, 0, 2/e), (0, 1, 3/e), and (0, −1, 3/e)
13. Minimum value: $f(1, 1) = 1$
15. Maximum value: $f(1, \pm 1) = 2$
17. Minimum value: $f(2, 3) = -50$
19. Minimum value: $f(4, -2) = f(-4, 2) = -16$
21. Maximum value: $f(1, -2) = e^5$
23. Maximum value: $f(1, 1) = 3$; minimum value: $f(-1, -1) = -3$
25. Maximum value: $f(0, 2) = 4$; minimum value: $f(1, 0) = -1$
27. Let $t = 1/\sqrt{2}$. Maximum value: $f(t, t) = f(-t, -t) = 1$; minimum value: $f(t, -t) = f(-t, t) = -1$
29. (12, 4, 3)
31. (15, 5, 4)
33. $(\sqrt{2}, \sqrt{2}, 1)$
35. 64000
37. 10 in. along each edge
39. Base 10 by 10 in., height 6 in.
41. 15 in. wide, 10 in. deep, and 5 in. high
43. Height 10 ft, front and back 40 ft wide, sides 20 ft deep
45. $\frac{1331}{567} \approx 2.347443$
47. 11664 in.^{3}
49. $\frac{1}{2}$
51. $(18\pi\sqrt{5}V^2)^{1/3}$
53. $(\frac{2}{3}, \frac{1}{3})$
55. Each triangular end should have height $h = 2^{-1/6}V^{1/3}$ and base $b = 2h$; the depth of the A-frame should be $h\sqrt{2}$.
59. Maximum when
$$x = \frac{7 - \sqrt{5}}{22}L \quad \text{and} \quad \theta = \frac{2}{5}\pi;$$
the maximum value of the area is approximately $(0.130534)L^2$.
61. (a) $x = \frac{45}{11}$, about \$4.09, and $y = \frac{48}{11}$, about \$4.36, for weekly profits of \$19107 and \$33942, respectively. (b) $x = \frac{37}{5}$, exactly \$7.40, and $y = \frac{98}{15}$, about \$6.53, for total weekly profit of \$67533, much greater than the sum of the profits in part (a).
63. Raise 40 hogs and 40 head of cattle per unit of land, and no sheep.

Section 13.6 (page 841)

1. $dw = (6x + 4y)\,dx + (4x - 6y^2)\,dy$
3. $dw = \dfrac{x\,dx + y\,dy}{\sqrt{1 + x^2 + y^2}}$
5. $dw = \dfrac{y\,dx - x\,dy}{x^2 + y^2}$
7. $dw = \dfrac{2x\,dx + 2y\,dy + 2z\,dz}{x^2 + y^2 + z^2}$
9. $dw = (\tan yz)\,dx + (xz\sec^2 yz)\,dy + (xy\sec^2 yz)\,dz$
11. $dw = -e^{-xyz}(yz\,dx + xz\,dy + xy\,dz)$
13. $dw = \exp(-v^2)(2u\,du - 2u^2v\,dv)$
15. $dw = \dfrac{x\,dx + y\,dy + z\,dz}{\sqrt{x^2 + y^2 + z^2}}$
17. $\Delta f \approx 0.014$ (true value: about 0.01422975)
19. $\Delta f \approx -0.0007$
21. $\Delta f \approx \frac{53}{1300} \approx 0.04077$
23. $\Delta f \approx 0.06$
25. 191.1
27. 1.4
29. 12.8077
31. 2.08
33. 2.5 cm^2
35. 8.18 in.^{3}
37. 0.022 acres
39. The period increases by about 0.0278 s.
41. About 303.8 ft
47. Note that $f_x(0, 0) = \lim_{h \to 0} \dfrac{f(0 + h, 0) - f(0, 0)}{h} = 1.$

Section 13.7 (page 851)

1. $-(2t + 1)\exp(-t^2 - t)$
3. $6t^5 \cos t^6$
5. $\dfrac{\partial w}{\partial s} = \dfrac{\partial w}{\partial t} = \dfrac{2}{s + t}$
7. $\dfrac{\partial w}{\partial s} \equiv 0, \dfrac{\partial w}{\partial t} = 5e^t$
9. $\dfrac{\partial r}{\partial x} = (y + z)\exp(yz + xy + xz),$
$\dfrac{\partial r}{\partial y} = (x + z)\exp(yz + xy + xz),$
$\dfrac{\partial r}{\partial z} = (x + y)\exp(yz + xy + xz)$
11. $\dfrac{\partial r}{\partial x} = \dfrac{(2y + 3z)\sqrt{xy^2z^3}}{2x(x + 2y + 3z)^{3/2}} \cos\dfrac{\sqrt{xy^2z^3}}{\sqrt{x + 2y + 3z}},$
$\dfrac{\partial r}{\partial y} = \dfrac{(x + y + 3z)\sqrt{xy^2z^3}}{y(x + 2y + 3z)^{3/2}} \cos\dfrac{\sqrt{xy^2z^3}}{\sqrt{x + 2y + 3z}},$
$\dfrac{\partial r}{\partial z} = \dfrac{(3x + 6y + 6z)\sqrt{xy^2z^3}}{2z(x + 2y + 3z)^{3/2}} \cos\dfrac{\sqrt{xy^2z^3}}{\sqrt{x + 2y + 3z}}.$
13. $\dfrac{\partial p}{\partial u} = \dfrac{\partial p}{\partial x}\dfrac{\partial x}{\partial u} + \dfrac{\partial p}{\partial y}\dfrac{\partial y}{\partial u},$
$\dfrac{\partial p}{\partial v} = \dfrac{\partial p}{\partial x}\dfrac{\partial x}{\partial v} + \dfrac{\partial p}{\partial y}\dfrac{\partial y}{\partial v},$
$\dfrac{\partial p}{\partial w} = \dfrac{\partial p}{\partial x}\dfrac{\partial x}{\partial w} + \dfrac{\partial p}{\partial y}\dfrac{\partial y}{\partial w}.$
15. $\dfrac{\partial p}{\partial x} = \dfrac{\partial p}{\partial u}\dfrac{\partial u}{\partial x} + \dfrac{\partial p}{\partial v}\dfrac{\partial v}{\partial x} + \dfrac{\partial p}{\partial w}\dfrac{\partial w}{\partial x},$
$\dfrac{\partial p}{\partial y} = \dfrac{\partial p}{\partial u}\dfrac{\partial u}{\partial y} + \dfrac{\partial p}{\partial v}\dfrac{\partial v}{\partial y} + \dfrac{\partial p}{\partial w}\dfrac{\partial w}{\partial y},$
$\dfrac{\partial p}{\partial z} = \dfrac{\partial p}{\partial u}\dfrac{\partial u}{\partial z} + \dfrac{\partial p}{\partial v}\dfrac{\partial v}{\partial z} + \dfrac{\partial p}{\partial w}\dfrac{\partial w}{\partial z}.$

17. $\dfrac{\partial p}{\partial x} = \dfrac{dp}{dw}\dfrac{\partial w}{\partial x}, \dfrac{\partial p}{\partial y} = \dfrac{dp}{dw}\dfrac{\partial w}{\partial y}, \dfrac{\partial p}{\partial z} = \dfrac{dp}{dw}\dfrac{\partial w}{\partial z}, \dfrac{\partial p}{\partial u} = \dfrac{dp}{dw}\dfrac{\partial w}{\partial u}, \dfrac{\partial p}{\partial v} = \dfrac{dp}{dw}\dfrac{\partial w}{\partial v}.$

19. $\dfrac{\partial z}{\partial x} = -\left(\dfrac{z}{x}\right)^{1/3}, \dfrac{\partial z}{\partial y} = -\left(\dfrac{z}{y}\right)^{1/3}$

21. $\dfrac{\partial z}{\partial x} = -\dfrac{yz(e^{xy} + e^{xz}) + (xy + 1)e^{xy}}{e^{xy} + xye^{yz}},$

$\dfrac{\partial z}{\partial y} = -\dfrac{x(x + z)e^{xy} + e^{xz}}{xye^{xz} + e^{xy}}$

23. $\dfrac{\partial z}{\partial x} = -\dfrac{c^2x}{a^2z}, \dfrac{\partial z}{\partial y} = -\dfrac{c^2y}{b^2z}$

25. $w_x = 6x$ and $w_y = 6y$.

27. $w_x = y\ln\left(\sqrt{x^3 + y^3} + \sqrt[3]{x^2 + y^2}\right) + \dfrac{2x^2y}{3\left(\sqrt{x^3 + y^3} + \sqrt[3]{x^2 + y^2}\right)(x^2 + y^2)^{2/3}} + \dfrac{3x^3y}{2\left(\sqrt{x^3 + y^3} + \sqrt[3]{x^2 + y^2}\right)\sqrt{x^3 + y^3}},$

$w_y = x\ln\left(\sqrt{x^3 + y^3} + \sqrt[3]{x^2 + y^2}\right) + \dfrac{2xy^2}{3\left(\sqrt{x^3 + y^3} + \sqrt[3]{x^2 + y^2}\right)(x^2 + y^2)^{2/3}} + \dfrac{3xy^3}{2\left(\sqrt{x^3 + y^3} + \sqrt[3]{x^2 + y^2}\right)\sqrt{x^3 + y^3}}.$

29. $x + 2y + 2z = 9$

31. $x - y - z = 0$

33. -2880 (in.^{3}/h)

35. Increasing at $\frac{26}{5}\pi$ (ft^3/min)

37. Decreasing at $\frac{14}{3}$ L/min

41. $\dfrac{\partial w}{\partial x} = f'(u)\cdot\dfrac{\partial u}{\partial x} = f'(u)$, and so on.

43. Show that $w_u = w_x + w_y$. Then note that

$$w_{uv} = \frac{\partial}{\partial v}w_u = \frac{\partial w_u}{\partial x}\cdot\frac{\partial x}{\partial v} + \frac{\partial w_u}{\partial y}\cdot\frac{\partial y}{\partial v}.$$

57. A 5°C-increase in temperature multiplies the initial pressure (1 atm) by $\frac{3000}{13} \approx 230.77$, so the bulb will burst.

59. $T'(\rho, \phi, \theta) = \rho^2 \sin\phi$

Section 13.8 (Page 861)

1. $\langle 3, -7\rangle$ **3.** $\langle 0, 0\rangle$ **5.** $\langle 0, 6, -4\rangle$

7. $\langle 1, 1, 1\rangle$ **9.** $\langle 2, -\frac{3}{2}, -2\rangle$ **11.** $8\sqrt{2}$

13. $\frac{12}{13}\sqrt{13}$ **15.** $-\frac{13}{20}$ **17.** $-\frac{1}{6}$

19. $-6\sqrt{2}$

21. Maximum: $\sqrt{170}$; direction: $\langle 7, 11\rangle$

23. Maximum: $\frac{2}{5}$; direction: $\langle 3, 4\rangle$

25. Maximum: $14\sqrt{2}$; direction: $\langle 3, 5, -8\rangle$

27. Maximum: $2\sqrt{14}$; direction: $\langle 1, 2, 3\rangle$

29. $3x + 4y = 25$

31. $29(x - 2) - 4(y + 3) = 0$

43. (a) $\frac{34}{3}$°C/ft; (b) 13 °C/ft and $\langle 4, 3, 12\rangle$

45. (a) $z = \frac{3}{10}x + \frac{1}{5}y - \frac{2}{5}$; (b) 0.44 (true value: 0.448)

47. $x - 2y + z + 10 = 0$

51. Compass heading about 36.87°; climbing at 45°

53. Compass heading about 203.2°; climbing at about 75.29°

55. (a) Descending at about 8.049467°; (b) Descending at about 3.627552°

59. $\dfrac{dx}{dt} = \dfrac{d\phi}{dt}\rho\cos\phi\cos\theta + \dfrac{d\rho}{dt}\sin\phi\cos\theta - \dfrac{d\theta}{dt}\rho\sin\phi\sin\theta,$

$\dfrac{dy}{dt} = \dfrac{d\phi}{dt}\rho\cos\phi\sin\theta + \dfrac{d\rho}{dt}\sin\phi\sin\theta + \dfrac{d\theta}{dt}\rho\sin\phi\cos\theta,$

$\dfrac{dz}{dt} = \dfrac{d\rho}{dt}\cos\phi - \dfrac{d\phi}{dt}\rho\sin\phi.$

Section 13.9 (pages 870)

1. Maximum: $\sqrt{5}$, at $\left(\frac{2}{5}\sqrt{5}, \frac{1}{5}\sqrt{5}\right)$; minimum: $-\sqrt{5}$, at $\left(-\frac{2}{5}\sqrt{5}, -\frac{1}{5}\sqrt{5}\right)$

3. Maximum: 4, at $(\pm 2, 0)$; minimum: -4, at $(0, \pm 2)$

5. Maximum: 3, at $\left(\frac{3}{2}\sqrt{2}, \sqrt{2}\right)$ and $\left(-\frac{3}{2}\sqrt{2}, -\sqrt{2}\right)$; minimum: -3 at $\left(-\frac{3}{2}\sqrt{2}, \sqrt{2}\right)$ and $\left(\frac{3}{2}\sqrt{2}, -\sqrt{2}\right)$

7. Minimum: $\frac{18}{7}$, at $\left(\frac{9}{7}, \frac{6}{7}, \frac{3}{7}\right)$; no maximum

9. Maximum: 7, at $\left(\frac{36}{7}, \frac{9}{7}, \frac{4}{7}\right)$; minimum: -7, at $\left(-\frac{36}{7}, -\frac{9}{7}, -\frac{4}{7}\right)$

11. Maximum: 20, at $(4, 4, 2)$ and at $(-4, -4, 2)$; minimum: -20, at $(-4, 4, -2)$ and at $(4, -4, -2)$

13. Maximum: $\frac{81}{4}$, at all eight of the critical points $\left(\pm 3, \pm\frac{3}{2}, \pm 1\right)$. Minimum: Zero, at all points on the ellipses in which the ellipsoid intersects the coordinate planes.

15. Minimum: $\frac{25}{3}$, at $\left(-\frac{5}{3}, \frac{1}{3}, \frac{7}{3}\right)$. There is no maximum because, in effect, we seek the extrema of the square of the distance between the origin and a point (x, y, z) on an unbounded straight line.

17. Maximum: $1 + \sqrt{2}$, at $(-1/\sqrt{2}, -1/\sqrt{2}, 1 + \sqrt{2})$; minimum: $1 - \sqrt{2}$, at $(1/\sqrt{2}, 1/\sqrt{2}, 1 - \sqrt{2})$

19. $(12, 16)$ **21.** $(12, 4, 3)$ **23.** $(15, 5, 4)$

25. $\left(\sqrt{2}, \sqrt{2}, 1\right)$ **27.** 64000 **29.** A 10-in. cube

31. Base 10 in. by 10 in., height 6 in.

33. Base 15 in. by 10 in., height 5 in.

35. $(2, -2, 1)$ and $(-2, 2, 1)$ **39.** $(2, 3)$ and $(-2, -3)$

31. Base 10 in. by 10 in., height 6 in.

33. Base 15 in. by 10 in., height 5 in.

35. $(2, -2, 1)$ and $(-2, 2, 1)$ **39.** $(2, 3)$ and $(-2, -3)$

41. Highest: $\left(\frac{2}{5}\sqrt{5}, \frac{1}{5}\sqrt{5}, \sqrt{5} - 4\right)$; lowest: $\left(-\frac{2}{5}\sqrt{5}, -\frac{1}{5}\sqrt{5}, -\sqrt{5} - 4\right)$

43. Farthest: $x = -\frac{1}{20}\left(15 + 9\sqrt{5}\right)$, $y = 2x$, $z = \frac{1}{4}\left(9 + 3\sqrt{5}\right)$; nearest: $x = -\frac{1}{20}\left(15 - 9\sqrt{5}\right)$, $y = 2x$, $z = \frac{1}{4}\left(9 - 3\sqrt{5}\right)$

47. $\frac{1}{4}\left(3 - 2\sqrt{2}\right)P^2$ **51.** $(0.4102, 0.3478)$ **53.** $(4, 6)$

55. The point of the sphere closest to $(0, 0, 0)$ is approximately
$$(-0.60357, -1.20713, -1.81070),$$
at approximate distance 2.25834, and the point farthest from $(0, 0, 0)$ is approximately
$$(2.60357, 5.20713, 7.81070),$$
at approximate distance 9.74166.

57. The point $(x, y) \approx (2.49615, 1.10940)$ is closest, at approximate distance 3.48501 from the point $(u, v) \approx (5.69338, 4.30662)$ of the line. The point $(x, y) \approx (-2.49615, -1.10940)$ is farthest, at approximate distance 6.94268 from the point $(u, v) \approx (4.30662, 5.69338)$ of the line.

59. $\frac{81}{4}\sqrt{2} \approx 28.637825$

Section 13.10 (page 879)

1. Minimum $(-1, 2, -1)$, no other extrema
3. Saddle point $\left(-\frac{1}{2}, -\frac{1}{2}, \frac{29}{4}\right)$. no extrema
5. Minimum $(-3, 4, -9)$, no other extrema
7. Saddle point $(0, 0, 3)$, local maximum $(-1, -1, 4)$
9. No extrema
11. Saddle point $(0, 0, 0)$, local minima $(-1, -1, -2)$ and $(1, 1, -2)$
13. Saddle point $(-1, 1, 5)$, local minimum $(3, -3, -27)$
15. Saddle point $(0, -2, 32)$, local minimum $(-5, 3, -93)$
17. Saddle point $(0, 0, 0)$, local maxima $(1, 2, 2)$ and $(-1, -2, 2)$
19. Saddle point $(-1, 0, 17)$, local minimum $(2, 0, -10)$
21. Saddle point $(0, 0, 0)$, local (actually, global) maxima
$$\left(\frac{\sqrt{2}}{2}, \frac{\sqrt{2}}{2}, \frac{1}{2e}\right) \text{ and } \left(-\frac{\sqrt{2}}{2}, -\frac{\sqrt{2}}{2}, \frac{1}{2e}\right),$$
local (actually, global) minima
$$\left(-\frac{\sqrt{2}}{2}, \frac{\sqrt{2}}{2}, -\frac{1}{2e}\right) \text{ and } \left(\frac{\sqrt{2}}{2}, -\frac{\sqrt{2}}{2}, -\frac{1}{2e}\right)$$
23. Local (actually, global) minimum
25. Local (actually, global) maximum
27. Minimum value 3 at $(1, 1)$ and at $(-1, -1)$
29. See Problem 57 of Section 13.5 and its answer
31. The critical points are those of the form (m, n), where both m and n are even integers or odd integers. The critical point (m, n) is a saddle point if both m and n are even, but a local maximum if both m and n are of the form $4k + 1$ or of the form $4k + 3$. It is a local minimum in the remaining cases.
35. Local minima at $(-1.8794, 0)$ and $(1.5321, 0)$, saddle point at $(0.3473, 0)$
37. Local minima at $(-1.8794, 1.8794)$ and $(1.5321, -1.5321)$ and a saddle point at the point $(0.3473, -0.3473)$
39. Local minima at $(3.6247, 3.9842)$ and $(3.6247, -3.9842)$, saddle point at $(0, 0)$

Chapter 13 Miscellaneous Problems (page 881)

3. On the line $y = x$, $g(x, y) \equiv \frac{1}{2}$, except that $g(0, 0) = 0$
5. $f(x, y) = x^2y^3 + e^x \sin y + y + C$
7. All points of the form $\left(a, b, \frac{1}{2}\right)$ $\left(\text{so } a^2 + b^2 = \frac{1}{2}\right)$ together with $(0, 0, 0)$
9. The normal to the cone at the point $\left(a, b, \sqrt{a^2 + b^2}\right)$ meets the z-axis at the point $(0, 0, 2\sqrt{a^2 + b^2})$.
15. Base $2\sqrt[3]{3} \times 2\sqrt[3]{3}$ ft, height $5\sqrt[3]{3}$ ft
17. $200 \pm 2\ \Omega$ **19.** 3%
21. $(\pm 4, 0, 0)$, $(0, \pm 2, 0)$, and $\left(0, 0, \pm\frac{4}{3}\right)$
25. Parallel to the vector $\langle 4, -3\rangle$; that is, at an approximate bearing of either 126.87° or 306.87°
27. 1 **31.** Semiaxes 1 and 2
33. There is no such triangle of minimum perimeter, unless we consider as a triangle the figure with all sides of length zero—a single point on the circumference of the circle. The triangle of maximum perimeter is equilateral, with perimeter $3\sqrt{3}$.
35. Closest: $\left(\frac{1}{3}\sqrt{6}, \frac{1}{6}\sqrt{6}\right)$; farthest: $\left(-\frac{1}{3}\sqrt{6}, -\frac{1}{6}\sqrt{6}\right)$
39. Maximum: 1; minimum: $-\frac{1}{2}$
41. Local minimum -1 at $(1, 1)$ and at $(-1, -1)$; horizontal tangent plane (but no extrema) at $(0, 0, 0)$, $(\sqrt{3}, 0, 0,)$, and $(-\sqrt{3}, 0, 0,)$
43. Local minimum -8 at $(2, 2)$, horizontal tangent plane at $(0, 0, 0)$
45. Local maximum $\frac{1}{432}$ at $\left(\frac{1}{2}, \frac{1}{3}\right)$. All points on the intervals $(-\infty, 0)$ and $(1, +\infty)$ on the x-axis are local minima (value: 0), and all points on the interval $(0, 1)$ on the x-axis are local maxima (value: 0). There is a saddle point at $(0, 1)$. There is a horizontal tangent plane at the origin, but it isn't really a saddle point.
47. There is a saddle point at $(0, 0)$; each point on the hyperbola $xy = \ln 2$ yields a global minimum.
49. There are no extrema, only saddle points at $(1, 1)$ and $(-1, -1)$.

Section 14.1 (page 891)

1. 198; 480 **3.** 8 **5.** 88
7. $\frac{1}{2}\pi^2$ **9.** $L \leqq M \leqq U$ **11.** 80
13. -78 **15.** $\frac{513}{4}$ **17.** $-\frac{9}{2}$
19. 1 **21.** $\frac{1}{2}(e - 1)$ **23.** $2(e - 1)$
25. $2\pi + \frac{1}{4}\pi^4$ **27.** 1 **29.** $2 \ln 2$
31. -32 **33.** $\frac{4}{15}\left(9\sqrt{3} - 8\sqrt{2} + 1\right)$

Section 14.2 (page 898)

1. $\frac{5}{6}$ **3.** $\frac{1}{2}$ **5.** $\frac{1}{12}$ **7.** $\frac{1}{20}$
9. $-\frac{1}{18}$ **11.** $\frac{1}{2}(e - 2)$ **13.** $\frac{61}{3}$ **15.** 0

17. 0 **19.** π **21.** 1 **23.** 2

25. $\int_0^4 \int_{-\sqrt{y}}^{\sqrt{y}} x^2 y\,dx\,dy = \frac{512}{21}$

27. $\int_0^1 \int_{-\sqrt{y}}^{\sqrt{y}} x\,dx\,dy + \int_1^9 \int_{(y-3)/2}^{\sqrt{y}} x\,dx\,dy = \frac{32}{3}$

29. $\int_0^4 \int_{2-\sqrt{4-y}}^{y/2} 1\,dx\,dy = \frac{4}{3}$

31. $\int_0^{\pi} \int_0^{y} \frac{\sin y}{y}\,dx\,dy = 2$

33. $\int_0^1 \int_0^x \frac{1}{1+x^4}\,dy\,dx = \frac{\pi}{8}$

35. The curves bound two bounded plane regions. The area of the one on the left is approximately 0.02767; the area of the region on the right is approximately 7.92408.

37. 0 **39.** 0 **41.** 0 **43.** 0

Section 14.3 (page 903)

1. $\frac{1}{6}$ **3.** $\frac{32}{3}$ **5.** $\frac{5}{6}$ **7.** $\frac{32}{3}$ **9.** $2\ln 2$
11. 2 **13.** $2e$ **15.** $\frac{1}{3}$ **17.** $\frac{41}{60}$ **19.** $\frac{4}{15}$
21. $\frac{10}{3}$ **23.** 19 **25.** $\frac{4}{3}$ **27.** 6 **29.** 24
31. π **33.** $\left(\frac{32}{3} - 4\sqrt{3}\right)\pi$ **35.** $\frac{1}{6}abc$ **37.** $\frac{2}{3}$
41. $\frac{625}{2}\pi$ **43.** $\frac{1}{6}\left(2\pi + 3\sqrt{3}\right)R^3 \approx (1.913)R^3$ **45.** $\frac{256}{15}$
47. $\frac{208}{9} - \pi^2 \approx 13.241506710022.$

Section 14.4 (page 911)

3. $\frac{3}{2}\pi$ **5.** $\frac{1}{6}\left(4\pi - 3\sqrt{3}\right) \approx 1.22837$
7. $\frac{1}{2}\left(2\pi - 3\sqrt{3}\right) \approx 0.5435$
9. $\frac{16}{3}\pi$ **11.** $\frac{23}{8}\pi$ **13.** $\frac{1}{4}\pi\ln 2$ **15.** $\frac{16}{5}\pi$
17. $\frac{1}{4}\pi(1 - \cos 1) \approx 0.36105$
19. 2π **21.** 4π **27.** 2π
29. $\frac{1}{3}\pi\left(2 - \sqrt{2}\right)a^3 \approx (0.6134)a^3$
31. $\frac{1}{4}\pi$ **35.** $2\pi^2 a^2 b$ **37.** 8π **39.** 24π

Section 14.5 (page 920)

1. $(2, 3)$ **3.** $(1, 1)$ **5.** $\left(\frac{4}{3}, \frac{2}{3}\right)$
7. $\left(\frac{3}{2}, \frac{6}{5}\right)$ **9.** $\left(0, -\frac{8}{5}\right)$ **11.** $\frac{1}{24}; \left(\frac{2}{5}, \frac{2}{5}\right)$
13. $\frac{256}{15}, \left(0, \frac{16}{7}\right)$ **15.** $\frac{1}{12}, \left(\frac{9}{14}, \frac{9}{14}\right)$ **17.** $\frac{8}{3}, \left(0, \frac{43}{25}\right)$
19. $2, \left(\frac{1}{2}\pi, \frac{1}{8}\pi\right)$ **21.** $a^3, \left(\frac{7}{12}a, \frac{7}{12}a\right)$ **23.** $\frac{128}{5}, \left(0, \frac{20}{7}\right)$

25. π; $\bar{x} = \frac{\pi^2 - 4}{\pi} \approx 1.87$, $\bar{y} = \frac{\pi}{8} \approx 0.39$

27. $\frac{1}{3}\pi a^3$; $\bar{x} = 0$, $\bar{y} = \frac{3a}{2\pi}$

29. $\frac{2}{3}\pi + \frac{1}{4}\sqrt{3}$; $\bar{x} = 0$, $\bar{y} = \frac{36\pi + 33\sqrt{3}}{32\pi + 12\sqrt{3}} \approx 1.4034$

31. $\frac{2\pi a^{n+4}}{n+4}$ **33.** $\frac{3}{2}\pi k$ **35.** $\frac{1}{9}$

37. $\hat{x} = \frac{2}{21}\sqrt{105}, \hat{y} = \frac{4}{3}\sqrt{5}$ **39.** $\hat{x} = \hat{y} = \frac{1}{10}a\sqrt{30}$

41. $\left(\frac{4r}{3\pi}, \frac{4r}{3\pi}\right)$ **43.** $\left(\frac{2r}{\pi}, \frac{2r}{\pi}\right)$

51. (a) $\bar{x} = 0, \bar{y} = \frac{4a^2 + 3\pi ab + 6b^2}{3\pi a + 12b}$;
(b) $\frac{1}{3}\pi a(4a^2 + 2\pi ab + 6b^2)$

53. $\left(1, \frac{1}{4}\right)$ **55.** $\frac{484}{3}k$
57. Mass $\pi, \bar{x} = 0, \bar{y} = \frac{5}{4}$
59. Mass $\frac{1}{2}\pi, \bar{x} = \frac{5}{4}, \bar{y} = 4/(3\pi)$

Section 14.6 (page 930)

1. 18 **3.** 128 **5.** $\frac{1}{60}$ **7.** 0 **9.** 12

11. $V = \int_0^3 \int_0^{3-(2x/3)} \int_0^{6-2x-2y} 1\,dz\,dy\,dx = 6$

13. $\frac{128}{5}$ **15.** $\frac{332}{105}$ **17.** $\frac{256}{15}$ **19.** $\frac{11}{30}$
21. $\left(0, \frac{20}{7}, \frac{10}{7}\right)$ **23.** $\left(0, \frac{8}{7}, \frac{12}{7}\right)$

25. $\bar{x} = 0, \bar{y} = \frac{44 - 9\pi}{72 - 9\pi}, \bar{z} = \frac{9\pi - 16}{72 - 9\pi}$

27. $\frac{8}{7}$ **29.** $\frac{1}{30}$ **33.** $\frac{2}{3}a^5$ **35.** $\frac{38}{45}ka^7$
37. $\frac{1}{3}k$ **39.** $\left(\frac{9}{64}\pi, \frac{9}{64}\pi, \frac{3}{8}\right)$ **41.** 24π
43. $\frac{1}{6}\pi$ **47.** $\frac{3}{2}$ **49.** $\frac{1}{4}$ **51.** $\frac{49}{10}$

Section 14.7 (page 938)

1. 8π **5.** $\frac{4}{3}\pi\left(8 - 3\sqrt{3}\right)$ **7.** $\frac{1}{2}\pi a^2 h^2$
9. $\frac{1}{4}\pi a^4 h^2$ **11.** $\frac{81}{2}\pi, (0, 0, 3)$ **13.** 24π
15. $\frac{1}{6}\pi\left(8\sqrt{2} - 7\right)$ **17.** $\frac{1}{12}\pi\delta a^2 h(3a^2 + 4h^2)$
19. $\frac{1}{3}\pi$ **21.** $\left(0, 0, \frac{3}{8}a\right)$
23. $\frac{1}{3}\pi$
25. $\frac{1}{3}\pi a^3(2 - \sqrt{2}); \bar{x} = 0 = \bar{y}, \bar{z} = \frac{3}{16}\left(2 + \sqrt{2}\right)a$
27. $\frac{7}{5}ma^2$
29. The surface obtained by rotating the circle in the xz-plane with center $(a, 0)$ and radius a around the z-axis—a doughnut with an infinitesimal hole; $2\pi^2 a^3$
31. $\frac{2}{15}\left(128 - 51\sqrt{3}\right)\pi a^5$
33. $\frac{37}{48}\pi a^4, \bar{z} = \frac{105}{74}a$
35. The gaseous star has mass $\frac{8}{15}k\pi a^3$.
37. The value of the integral in part (a) is $\frac{4}{3}\pi[1 - \exp(-a^3)]$.
39. $\frac{3}{4}a$

Section 14.8 (page 945)

1. $6\pi\sqrt{11}$ **3.** $\frac{1}{6}\pi\left(17\sqrt{17}-1\right)$
5. $3\sqrt{2}+\frac{1}{2}\ln\left(3+2\sqrt{2}\right)\approx 5.124$
7. $3\sqrt{14}$ **9.** $\frac{2}{3}\pi\left(2\sqrt{2}-1\right)\approx 3.829$
11. $\frac{1}{6}\pi\left(65\sqrt{65}-1\right)$ **15.** $8a^2$
25. (a) 7.4463; (b) 3.0046 **27.** (a) 5.1232; (b) 2.3023
29. 111.5458

Section 14.9 (page 953)

1. $x=\frac{1}{2}(u+v),\ y=\frac{1}{2}(u-v);\ J=-\frac{1}{2}$
3. $x=\sqrt{u/v},\ y=\sqrt{uv};\ J=1/(2v)$
5. $x=\frac{1}{2}(u+v),\ y=\frac{1}{2}\sqrt{u-v};\ J=-\frac{1}{4}(u-v)^{-1/2}$
7. $\frac{3}{5}$ **9.** $\ln 2$ **11.** $\frac{1}{8}\left(2-\sqrt{2}\right)$
13. $\frac{39}{2}\pi$ **15.** 8
17. S is the region $3u^2+v^2\leqq 3$; the value of the integral is $\dfrac{2\pi(e^3-1)\sqrt{3}}{3e^3}$.
23. $\bar{x}\approx 2.5707$
$\bar{y}\approx 1.2126$
25. $I_x=\frac{1}{5}M(b^2+c^2),\ I_y=\frac{1}{5}M(a^2+c^2)$, and $I_z=\frac{1}{5}M(a^2+b^2)$, where M is the mass of the ellipsoid
27. $\displaystyle\frac{1}{V}\int_0^{2\pi}\int_0^{\pi}\int_0^{1}(abc\rho^2\sin\phi)\rho\sqrt{(a\sin\phi\cos\theta)^2+(b\sin\phi\sin\theta)^2+(c\cos\phi)^2}\,d\rho\,d\phi\,d\theta$ where V is the volume of the ellipsoid. If $a=4$, $b=3$, and $c=2$, the value of this expression is approximately 2.30027.

Chapter 14 Miscellaneous Problems (page 955)

1. $\frac{1}{3}\left(2-\sqrt{2}\right)$ **3.** $\dfrac{e-1}{2e}$ **5.** $\frac{1}{4}(e^4-1)$
7. $\frac{4}{3}$ **9.** $9\pi;\ \bar{z}=\frac{9}{16}$ **11.** 4π
13. 4π **15.** $\frac{1}{16}(\pi-2)$ **17.** $\frac{128}{15},\left(\frac{32}{7},0\right)$
19. $k\pi,(1,0)$ **21.** $\bar{y}=\dfrac{4b}{3\pi}$ **23.** $\left(0,\frac{8}{5}\right)$
25. $\frac{10}{3}\pi\left(\sqrt{5}-2\right)\approx 2.4721$
27. $\frac{3}{10}Ma^2$ **29.** $\frac{1}{5}M(b^2+c^2)$
31. $\frac{128}{225}\delta(15\pi-26)\approx(12.017)\delta$, where δ is the (constant) density
33. $\frac{8}{3}\pi$
35. $\frac{1}{4}M(3a^2+4b^2)$, where M is the mass of the torus
41. $\frac{18}{7}$ **43.** $\frac{1}{6}\pi\left(37\sqrt{37}-17\sqrt{17}\right)\approx 81.1418$
47. $4\sqrt{2}$ **48.** Approximately 3.49608
51. 3δ **53.** $\frac{8}{15}\pi abc$

Section 15.1 (page 964)

1.

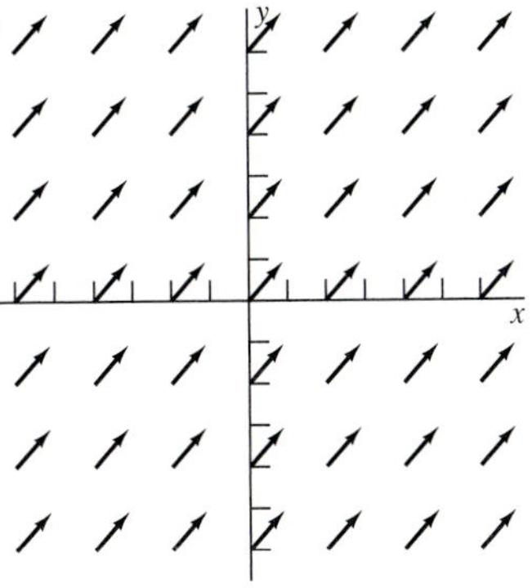

3.

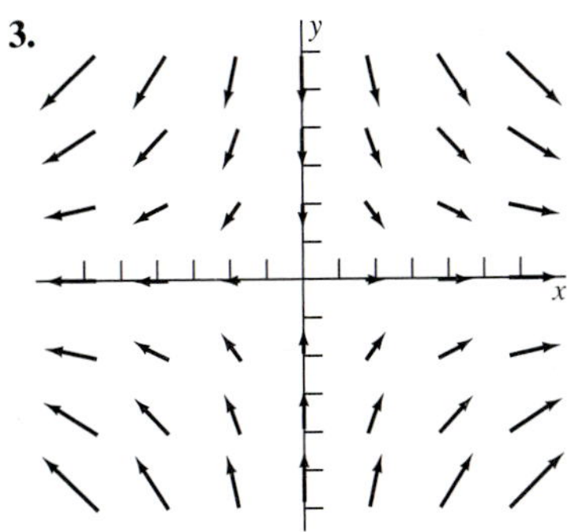

5.

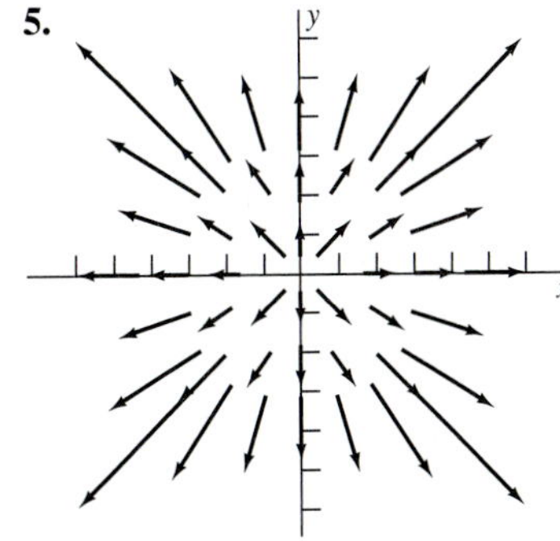

7.

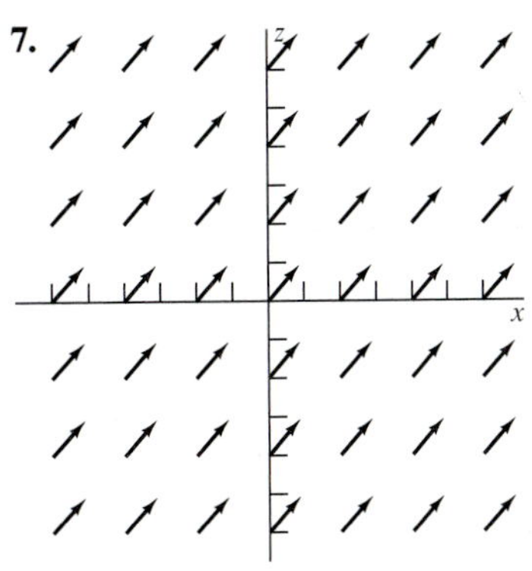

9.

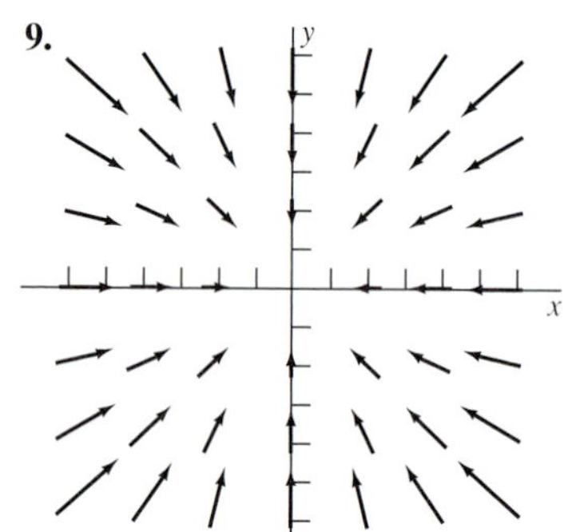

11.

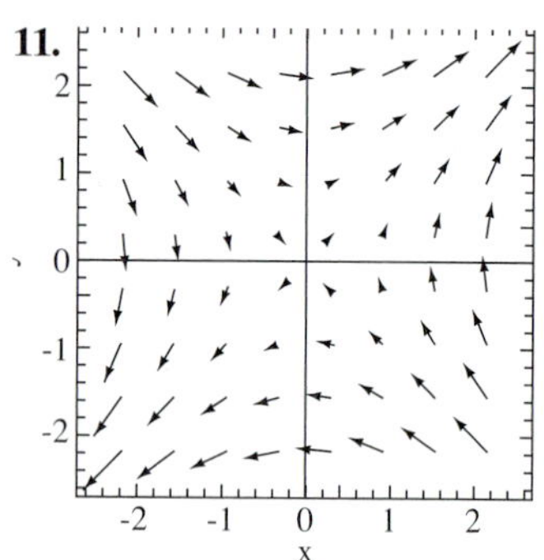

13.

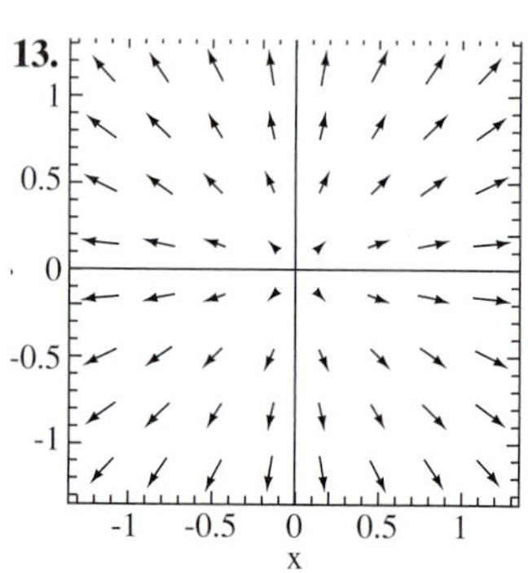

15. 3; 0 **17.** 0; 0

19. $x^2 + y^2 + z^2$; $\langle -2yz, -2xz, -2xy\rangle$

21. 0; $\langle 2y - 2z, 2z - 2x, 2x - 2y\rangle$

23. 3; $\langle x\cos xy - x\cos xz, y\cos yz - y\cos xy, z\cos xz - z\cos yz\rangle$

Section 15.2 (page 974)

1. $\frac{310}{3}, \frac{248}{3}, 62$ **3.** $3\sqrt{2}, 3, 3$ **5.** $\frac{49}{24}, \frac{3}{2}, \frac{4}{3}$

7. $\frac{6}{5}$ **9.** 315 **11.** $\frac{19}{60}$

13. $\pi + 2\pi^2$ **15.** 28

17. $\frac{1}{6}\left(14\sqrt{14} - 1\right) \approx 8.563867$ **19.** $(0, 2a/\pi)$

21. $10k\pi$, $(0, 0, 4\pi)$

23. $\frac{1}{2}ka^3$, $\left(\frac{2}{3}a, \frac{2}{3}a\right)$, $I_x = I_y = \frac{1}{4}ka^5$

25. $I_0 = 3k = \frac{1}{2}m$ where m is the mass of the wire

27. $4a/\pi$ **29.** $\frac{1}{2}$ **31.** $\frac{1}{2}$

33. (a) $\frac{1}{2}k\ln 2$; (b) $-\frac{1}{2}k\ln 2$ **41.** 20000 ft·lb

Section 15.3 (page 982)

1. $f(x, y) = x^2 + 3xy + y^2$ **3.** $f(x, y) = x^3 + 2xy^2 + 2y^3$

5. Not conservative

7. $f(x, y) = \frac{1}{4}x^4 + \frac{1}{3}y^3 + y\ln x$

9. $f(x, y) = \sin x + x\ln y + e^y$

11. Not conservative

13. $f(x, y) = x^3y^3 + xy^4 + \frac{1}{5}y^5$

15. $f(x, y) = \dfrac{x^2}{y} + \dfrac{y^2}{x^3} + 2\sqrt{y}$

21. 6 **23.** $\dfrac{1}{e}$ **25.** $-\pi$ **27.** $f(x, y, z) = xyz$

29. $f(x, y, z) = xy\cos z - yze^x$

31. **F** is not conservative on any region containing $(0, 0)$.

33. $Q_z = 0 \neq 2y = R_y$

37. $W \approx 8.04 \times 10^{10}$ N·m

Section 15.4 (page 991)

1. 0 **3.** 3 **5.** $\frac{3}{10}$ **7.** 2 **9.** 0

11. $\frac{16}{105}$ **13.** πa^2 **15.** $\frac{3}{8}\pi$ **33.** $\frac{3}{2}$

Section 15.5 (page 1001)

1. $\frac{1}{3}\sqrt{3}$ **3.** $27\pi\sqrt{14}$

5. $\frac{1}{6}\pi\left(17\sqrt{17} - 1\right)$ **7.** $\frac{81}{2}\pi\sqrt{3}$

9. $\frac{10}{3}\pi$

11. $\dfrac{520}{3}\pi \approx 544.54272662$

13. 36π **15.** 24π **17.** 0 **19.** $\phi = 6$

21. 0 **23.** Net flux: 1458π **25.** $\left(\frac{1}{2}a, \frac{1}{2}a, \frac{1}{2}a\right)$

27. $\bar{z} = \dfrac{1 + (24a^4 + 2a^2 - 1)\sqrt{1 + 4a^2}}{10[(1 + 4a^2)^{3/2} - 1]}$,

$I_z = \frac{1}{60}\pi\delta\left[1 + (24a^4 + 2a^2 - 1)\sqrt{1 + 4a^2}\right]$

29. $\bar{x} = \dfrac{4}{3\pi - 6} \approx 1.16796$, $\bar{y} = 0$,

$\bar{z} = \dfrac{\pi}{2\pi - 4} \approx 1.13797$

31. $I_z = \frac{115}{12}\sqrt{17} + \frac{13}{48}\sinh^{-1}4$

35. -1728π

37. Area approximately 194.703, $I_z \approx 5157.17$

39. Area approximately 1057.35, $I_z \approx 98546.9$

Section 15.6 (page 1009)

1. Both values: 4π **3.** Both values: 24

5. Both values: $\frac{1}{2}$ **7.** $\frac{2385}{2}\pi$ **9.** $\frac{1}{4}$

11. $\frac{703{,}125}{4}\pi$ **13.** 16π
23. $\frac{1}{48}\pi(482{,}620 + 29403\ln 11) \approx 36201.967$

Section 15.7 (page 1017)

1. -20π **3.** 0 **5.** -52π **7.** -8π **9.** -2
11. $\phi(x, y, z) = yz - 2xz + 3xy$
13. $\phi(x, y, z) = 3xe^z + 5y\cos x + 17z$

Chapter 15 Miscellaneous Problems (page 1018)

1. $\frac{125}{3}$
3. $\frac{69}{8}$ (Use the fact that $\int \mathbf{F}\cdot\mathbf{T}\,ds$ is independent of the path.)
5. $\frac{2148}{5}$
9. $\frac{1}{3}\left(5\sqrt{5} - 1\right) \approx 3.3934$; $I_y = \frac{1}{15}\left(2 + 50\sqrt{5}\right) \approx 7.5869$
11. $\frac{2816}{7}$ **17.** $\frac{371}{30}\pi$ **19.** 72π
29. (a) $\phi'(r)\frac{\mathbf{r}}{r}$; (b) $3\phi(r) + r\phi'(r)$; (c) $\mathbf{0}$

Appendix A (page A–5)

1. 14 **3.** $\frac{1}{2}$ **5.** 25
7. 27 **9.** $\frac{22}{7} - \pi \approx 0.001264489$
11. $3 - x$ **13.** $(-\infty, 2)$ **15.** $[7, \infty)$
17. $\left(-\frac{5}{3}, \infty\right)$ **19.** $(-4, 1)$ **21.** $\left(\frac{3}{2}, \frac{11}{2}\right)$
23. $(-1, 4)$ **25.** $\left(-\infty, -\frac{1}{3}\right) \cup (1, \infty)$
27. $\left[0, \frac{2}{5}\right] \cup \left[\frac{6}{5}, \frac{8}{5}\right]$ **29.** $\left(\frac{7}{3}, \frac{37}{15}\right]$
31. $\left(-\infty, \frac{1}{5}\right) \cup \left(\frac{1}{5}, \infty\right)$ **33.** $(-\infty, -2) \cup (4, \infty)$
35. $\left(-\infty, \frac{1}{2}\right] \cup \left[\frac{3}{2}, \infty\right)$ **37.** $4 \leqq p \leqq 8$
39. $2 < I < 4$

Appendix B (page A–13)

1. AB and BC have slope 1.
3. AB has slope -2, but BC has slope $-\frac{4}{3}$.
5. AB and CD have slope $-\frac{1}{2}$; BC and DA have slope 2.
7. AB has slope 2, and AC has slope $-\frac{1}{2}$.
9. $m = \frac{2}{3}, b = 0$ **11.** $m = 2, b = 3$
13. $m = -\frac{2}{5}, b = \frac{3}{5}$ **15.** $y = -5$
17. $y - 3 = 2(x - 5)$ **19.** $y - 2 = 4 - x$
21. $y - 5 = -2(x - 1)$ **23.** $x + 2y = 13$
25. $\frac{4}{13}\sqrt{26} \approx 1.568929$
33. $K = \dfrac{125F + 57461}{225}$; $F = -459.688$ when $K = 0$.
35. 1136 gal/week **37.** $x = -2.75, y = 3.5$
39. $x = \frac{37}{6}, y = -\frac{1}{2}$ **41.** $x = \frac{22}{5}, y = -\frac{1}{5}$
43. $x = -\frac{7}{4}, y = \frac{33}{8}$ **45.** $x = \frac{119}{12}, y = -\frac{19}{4}$

Appendix C (page A–19)

1. $2\pi/9$ **3.** $7\pi/4$ **5.** $-5\pi/6$ **7.** 72° **9.** 675°
11. $\sin x = -\frac{1}{2}\sqrt{3}$, $\cos x = \frac{1}{2}$, $\tan x = -\sqrt{3}$, $\csc x = -2/\sqrt{3}$, $\sec x = 2$, $\cot x = -1/\sqrt{3}$
13. $\sin x = -\frac{1}{2}$, $\cos x = -\frac{1}{2}\sqrt{3}$, $\cot x = \sqrt{3}$, $\csc x = -2$, $\sec x = -2/\sqrt{3}$, $\tan x = 1/\sqrt{3}$
15. $x = n\pi$ (n any integer)
17. $x = \frac{3}{2}\pi + 2n\pi$ (n any integer)
19. $x = 2n\pi$ (n any integer)
21. $x = n\pi$ (n any integer)
23. $x = \frac{3}{4}\pi + n\pi$ (n any integer)
25. $\sin x = -\frac{3}{5}$, $\cos x = \frac{4}{5}$, $\tan x = -\frac{3}{4}$, $\sec x = \frac{5}{4}$, $\cot x = -\frac{4}{3}$
29. $\frac{1}{2}$ **31.** $-\frac{1}{2}$ **33.** $\frac{1}{2}\sqrt{3}$ **35.** $-\frac{1}{2}\sqrt{3}$
43. $\pi/3, 2\pi/3$ **45.** $\pi/2$
47. $\pi/8, 3\pi/8, 5\pi/8, 7\pi/8$

Appendix H (page A–41)

1. $\displaystyle\int_0^1 x^2\,dx = \frac{1}{3}$

3. $\displaystyle\int_0^2 x\sqrt{4 - x^2}\,dx = \frac{8}{3}$

5. $\displaystyle\int_0^{\pi/2} \sin x\cos x\,dx = \frac{1}{2}$

7. $\displaystyle\int_0^2 \sqrt{x^4 + x^7}\,dx = \frac{52}{9}$

References for Further Study

References 2, 3, 7, and 10 may be consulted for historical topics pertinent to calculus. Reference 14 provides a more theoretical treatment of single-variable calculus topics than ours. References 4, 5, 8, and 15 include advanced topics in multivariable calculus. Reference 11 is a standard work on infinite series. References 1, 9, and 13 are differential equations textbooks. Reference 6 discusses topics in calculus together with computing and programming in BASIC. Those who would like to pursue the topic of fractals should look at Reference 12.

1. Boyce, William E. and Richard C. DiPrima, *Elementary Differential Equations* (5th ed.). New York: John Wiley, 1991.
2. Boyer, Carl B., *A History of Mathematics* (2nd ed.). New York: John Wiley, 1991.
3. Boyer, Carl B., *The History of the Calculus and Its Conceptual Development*. New York: Dover Publications, 1959.
4. Buck, R. Creighton, *Advanced Calculus* (3rd ed.). New York: McGraw-Hill, 1977.
5. Courant, Richard and Fritz John, *Introduction to Calculus and Analysis.* Vols. I and II. New York: Springer-Verlag, 1982.
6. Edwards, C. H., Jr., *Calculus and the Personal Computer.* Englewood Cliffs, N.J.: Prentice-Hall, 1986.
7. Edwards, C. H., Jr., *The Historical Development of the Calculus.* New York: Springer-Verlag, 1979.
8. Edwards, C. H., Jr., *Advanced Calculus of Several Variables.* New York: Academic Press, 1973.
9. Edwards, C. H., Jr. and David E. Penney, *Differential Equations with Boundary Value Problems: Computing and Modeling.* Englewood Cliffs, N.J.: Prentice Hall, 1996.
10. Kline, Morris, *Mathematical Thought from Ancient to Modern Times*. Vols. I, II, and III. New York: Oxford University Press, 1972.
11. Knopp, Konrad, *Theory and Application of Infinite Series* (2nd ed.). New York: Hafner Press, 1990.
12. Peitgen, H.-O. and P. H. Richter, *The Beauty of Fractals.* New York: Springer-Verlag, 1986.
13. Simmons, George F., *Differential Equations with Applications and Historical Notes.* New York: McGraw-Hill, 1972.
14. Spivak, Michael E., *Calculus* (2nd ed.). Berkeley: Publish or Perish, 1980.
15. Taylor, Angus E. and W. Robert Mann, *Advanced Calculus* (3rd ed.). New York: John Wiley, 1983.

INDEX

D

E

Q

R

S

T

U

V

W

X

Y

Z

TABLE OF INTEGRALS

ELEMENTARY FORMS

1 $\int u\,dv = uv - \int v\,du$

2 $\int u^n\,du = \frac{1}{n+1}u^{n+1} + C \quad \text{if } n \neq -1$

3 $\int \frac{du}{u} = \ln|u| + C$

4 $\int e^u\,du = e^u + C$

5 $\int a^u\,du = \frac{a^u}{\ln a} + C$

6 $\int \sin u\,du = -\cos u + C$

7 $\int \cos u\,du = \sin u + C$

8 $\int \sec^2 u\,du = \tan u + C$

9 $\int \csc^2 u\,du = -\cot u + C$

10 $\int \sec u \tan u\,du = \sec u + C$

11 $\int \csc u \cot u\,du = -\csc u + C$

12 $\int \tan u\,du = \ln|\sec u| + C$

13 $\int \cot u\,du = \ln|\sin u| + C$

14 $\int \sec u\,du = \ln|\sec u + \tan u| + C$

15 $\int \csc u\,du = \ln|\csc u - \cot u| + C$

16 $\int \frac{du}{\sqrt{a^2 - u^2}} = \sin^{-1}\frac{u}{a} + C$

17 $\int \frac{du}{a^2 + u^2} = \frac{1}{a}\tan^{-1}\frac{u}{a} + C$

18 $\int \frac{du}{a^2 - u^2} = \frac{1}{2a}\ln\left|\frac{u+a}{u-a}\right| + C$

19 $\int \frac{du}{u\sqrt{u^2 - a^2}} = \frac{1}{a}\sec^{-1}\left|\frac{u}{a}\right| + C$

TRIGONOMETRIC FORMS

20 $\int \sin^2 u\,du = \frac{1}{2}u - \frac{1}{4}\sin 2u + C$

21 $\int \cos^2 u\,du = \frac{1}{2}u + \frac{1}{4}\sin 2u + C$

22 $\int \tan^2 u\,du = \tan u - u + C$

23 $\int \cot^2 u\,du = -\cot u - u + C$

24 $\int \sin^3 u\,du = -\frac{1}{3}(2 + \sin^2 u)\cos u + C$

25 $\int \cos^3 u\,du = \frac{1}{3}(2 + \cos^2 u)\sin u + C$

26 $\int \tan^3 u\,du = \frac{1}{2}\tan^2 u + \ln|\cos u| + C$

27 $\int \cot^3 u\,du = -\frac{1}{2}\cot^2 u - \ln|\sin u| + C$

28 $\int \sec^3 u\,du = \frac{1}{2}\sec u \tan u + \frac{1}{2}\ln|\sec u + \tan u| + C$

29 $\int \csc^3 u\,du = -\frac{1}{2}\csc u \cot u + \frac{1}{2}\ln|\csc u - \cot u| + C$

30 $\int \sin au \sin bu\,du = \frac{\sin(a-b)u}{2(a-b)} - \frac{\sin(a+b)u}{2(a+b)} + C \quad \text{if } a^2 \neq b^2$

31 $\int \cos au \cos bu\,du = \frac{\sin(a-b)u}{2(a-b)} + \frac{\sin(a+b)u}{2(a+b)} + C \quad \text{if } a^2 \neq b^2$

32 $\int \sin au \cos bu\,du = -\frac{\cos(a-b)u}{2(a-b)} - \frac{\cos(a+b)u}{2(a+b)} + C \quad \text{if } a^2 \neq b^2$

33 $\int \sin^n u\,du = -\frac{1}{n}\sin^{n-1} u \cos u + \frac{n-1}{n}\int \sin^{n-2} u\,du$

34 $\int \cos^n u\,du = \frac{1}{n}\cos^{n-1} u \sin u + \frac{n-1}{n}\int \cos^{n-2} u\,du$

35 $\int \tan^n u\,du = \frac{1}{n-1}\tan^{n-1} u - \int \tan^{n-2} u\,du \quad \text{if } n \neq 1$

36 $\int \cot^n u\,du = -\frac{1}{n-1}\cot^{n-1} u - \int \cot^{n-2} u\,du \quad \text{if } n \neq 1$

37 $\int \sec^n u\,du = \frac{1}{n-1}\sec^{n-2} u \tan u + \frac{n-2}{n-1}\int \sec^{n-2} u\,du \quad \text{if } n \neq 1$

38 $\int \csc^n u\,du = -\frac{1}{n-1}\csc^{n-2} u \cot u + \frac{n-2}{n-1}\int \csc^{n-2} u\,du \quad \text{if } n \neq 1$

39a $\int \sin^n u \cos^m u\,du = -\frac{\sin^{n-1} u \cos^{m+1} u}{n+m} + \frac{n-1}{n+m}\int \sin^{n-2} u \cos^m u\,du \quad \text{if } n \neq -m$

39b $\int \sin^n u \cos^m u\,du = -\frac{\sin^{n+1} u \cos^{m-1} u}{n+m} + \frac{m-1}{n+m}\int \sin^n u \cos^{m-2} u\,du \quad \text{if } m \neq -n$

40 $\int u \sin u\,du = \sin u - u \cos u + C$

41 $\int u \cos u\,du = \cos u + u \sin u + C$

42 $\int u^n \sin u\,du = -u^n \cos u + n\int u^{n-1} \cos u\,du$

43 $\int u^n \cos u\,du = u^n \sin u - n\int u^{n-1} \sin u\,du$

(Table of Integrals continues from previous page)

FORMS INVOLVING $\sqrt{u^2 \pm a^2}$

44 $\displaystyle\int \sqrt{u^2 \pm a^2}\, du = \frac{u}{2}\sqrt{u^2 \pm a^2} \pm \frac{a^2}{2}\ln\left|u + \sqrt{u^2 \pm a^2}\right| + C$

45 $\displaystyle\int \frac{du}{\sqrt{u^2 \pm a^2}} = \ln\left|u + \sqrt{u^2 \pm a^2}\right| + C$

46 $\displaystyle\int \frac{\sqrt{u^2 + a^2}}{u}\, du = \sqrt{u^2 + a^2} - a\ln\left(\frac{a + \sqrt{u^2 + a^2}}{u}\right) + C$

47 $\displaystyle\int \frac{\sqrt{u^2 - a^2}}{u}\, du = \sqrt{u^2 - a^2} - a\sec^{-1}\frac{u}{a} + C$

48 $\displaystyle\int u^2\sqrt{u^2 \pm a^2}\, du = \frac{u}{8}(2u^2 \pm a^2)\sqrt{u^2 \pm a^2} - \frac{a^4}{8}\ln\left|u + \sqrt{u^2 \pm a^2}\right| + C$

49 $\displaystyle\int \frac{u^2\, du}{\sqrt{u^2 \pm a^2}} = \frac{u}{2}\sqrt{u^2 \pm a^2} \mp \frac{a^2}{2}\ln\left|u + \sqrt{u^2 \pm a^2}\right| + C$

50 $\displaystyle\int \frac{du}{u^2\sqrt{u^2 \pm a^2}} = \mp\frac{\sqrt{u^2 \pm a^2}}{a^2 u} + C$

51 $\displaystyle\int \frac{\sqrt{u^2 \pm a^2}}{u^2}\, du = -\frac{\sqrt{u^2 \pm a^2}}{u} + \ln\left|u + \sqrt{u^2 \pm a^2}\right| + C$

52 $\displaystyle\int \frac{du}{(u^2 \pm a^2)^{3/2}} = \pm\frac{u}{a^2\sqrt{u^2 \pm a^2}} + C$

53 $\displaystyle\int (u^2 \pm a^2)^{3/2}\, du = \frac{u}{8}(2u^2 \pm 5a^2)\sqrt{u^2 \pm a^2} + \frac{3a^4}{8}\ln\left|u + \sqrt{u^2 \pm a^2}\right| + C$

FORMS INVOLVING $\sqrt{a^2 - u^2}$

54 $\displaystyle\int \sqrt{a^2 - u^2}\, du = \frac{u}{2}\sqrt{a^2 - u^2} + \frac{a^2}{2}\sin^{-1}\frac{u}{a} + C$

55 $\displaystyle\int \frac{\sqrt{a^2 - u^2}}{u}\, du = \sqrt{a^2 - u^2} - a\ln\left|\frac{a + \sqrt{a^2 - u^2}}{u}\right| + C$

56 $\displaystyle\int \frac{u^2\, du}{\sqrt{a^2 - u^2}} = -\frac{u}{2}\sqrt{a^2 - u^2} + \frac{a^2}{2}\sin^{-1}\frac{u}{a} + C$

57 $\displaystyle\int u^2\sqrt{a^2 - u^2}\, du = \frac{u}{8}(2u^2 - a^2)\sqrt{a^2 - u^2} + \frac{a^4}{8}\sin^{-1}\frac{u}{a} + C$

58 $\displaystyle\int \frac{du}{u^2\sqrt{a^2 - u^2}} = -\frac{\sqrt{a^2 - u^2}}{a^2 u} + C$

59 $\displaystyle\int \frac{\sqrt{a^2 - u^2}}{u^2}\, du = -\frac{\sqrt{a^2 - u^2}}{u} - \sin^{-1}\frac{u}{a} + C$

60 $\displaystyle\int \frac{du}{u\sqrt{a^2 - u^2}} = -\frac{1}{a}\ln\left|\frac{a + \sqrt{a^2 - u^2}}{u}\right| + C$

61 $\displaystyle\int \frac{du}{(a^2 - u^2)^{3/2}} = \frac{u}{a^2\sqrt{a^2 - u^2}} + C$

62 $\displaystyle\int (a^2 - u^2)^{3/2}\, du = \frac{u}{8}(5a^2 - 2u^2)\sqrt{a^2 - u^2} + \frac{3a^4}{8}\sin^{-1}\frac{u}{a} + C$

EXPONENTIAL AND LOGARITHMIC FORMS

63 $\displaystyle\int ue^u\, du = (u - 1)e^u + C$

64 $\displaystyle\int u^n e^u\, du = u^n e^u - n\int u^{n-1}e^u\, du$

65 $\displaystyle\int \ln u\, du = u\ln u - u + C$

66 $\displaystyle\int u^n \ln u\, du = \frac{u^{n+1}}{n+1}\ln u - \frac{u^{n+1}}{(n+1)^2} + C$

67 $\displaystyle\int e^{au}\sin bu\, du = \frac{e^{au}}{a^2 + b^2}(a\sin bu - b\cos bu) + C$

68 $\displaystyle\int e^{au}\cos bu\, du = \frac{e^{au}}{a^2 + b^2}(a\cos bu + b\sin bu) + C$

INVERSE TRIGONOMETRIC FORMS

69 $\displaystyle\int \sin^{-1} u\, du = u\sin^{-1} u + \sqrt{1 - u^2} + C$

70 $\displaystyle\int \tan^{-1} u\, du = u\tan^{-1} u - \frac{1}{2}\ln(1 + u^2) + C$

71 $\displaystyle\int \sec^{-1} u\, du = u\sec^{-1} u - \ln\left|u + \sqrt{u^2 - 1}\right| + C$

72 $\displaystyle\int u\sin^{-1} u\, du = \frac{1}{4}(2u^2 - 1)\sin^{-1} u + \frac{u}{4}\sqrt{1 - u^2} + C$

73 $\displaystyle\int u\tan^{-1} u\, du = \frac{1}{2}(u^2 + 1)\tan^{-1} u - \frac{u}{2} + C$

74 $\displaystyle\int u\sec^{-1} u\, du = \frac{u^2}{2}\sec^{-1} u - \frac{1}{2}\sqrt{u^2 - 1} + C$

75 $\displaystyle\int u^n\sin^{-1} u\, du = \frac{u^{n+1}}{n+1}\sin^{-1} u - \frac{1}{n+1}\int \frac{u^{n+1}}{\sqrt{1 - u^2}}\, du \quad \text{if } n \neq -1$

76 $\displaystyle\int u^n\tan^{-1} u\, du = \frac{u^{n+1}}{n+1}\tan^{-1} u - \frac{1}{n+1}\int \frac{u^{n+1}}{1 + u^2}\, du \quad \text{if } n \neq -1$

77 $\displaystyle\int u^n\sec^{-1} u\, du = \frac{u^{n+1}}{n+1}\sec^{-1} u - \frac{1}{n+1}\int \frac{u^n}{\sqrt{u^2 - 1}}\, du \quad \text{if } n \neq -1$